Anatomy of the Human Body

Henry Gray appears in the foreground, third from the left.
(Reproduced by permission of the Governors of St. George's Hospital.)

Anatomy of the Human Body

by **Henry Gray, F.R.S.**

Late Fellow of the Royal College of Surgeons; Lecturer on Anatomy at St. George's Hospital Medical School, London

Twenty-Ninth American Edition, edited by

Charles Mayo Goss, A.B., M.D.

Visiting Professor of Anatomy, George Washington University School of Medicine, Member of the International Anatomical Nomenclature Committee, Former President of the American Association of Anatomists, Former Editor of the Anatomical Record

With new drawings by Don M. Alvarado, Professor of Medical Illustration

1178 Illustrations, mostly in Color

 **LEA & FEBIGER**

PHILADELPHIA

ISBN 0-8121-0377-7

Library of Congress Catalog Card Number 73-170735

Printed in the United States of America

Print number: 7

to

Henry Gray and **H. Van Dyke Carter**

in admiration of their imagination and skill in
originating this book

and to

The Many Students and Teachers

Who Have Used It

this one hundred and fourteenth year edition is dedicated
with deep appreciation

Henry Gray, *F.R.S., F.R.C.S.*

THE ORIGINATOR of this now famous book, which has reached its one hundred and fourteenth year of continuous publication in America, was born in Windsor, England, in 1825. A letter written to Lea & Febiger, by Henry Gray's great nephew, F. Lawrence Gray, has supplied this date as well as some interesting biographical information.

Henry's father, William Gray, was born in 1787. He is reported to have been an only child. Whether for protection or from special interest he was placed in the royal household where he was brought up. In 1811, before the Regency, and while he was in his twenties, he was Deputy Treasurer to the Household of the Prince of Wales. On the accession of George IV as Regent in 1820, William Gray became the King's private messenger and continued in this capacity, with William IV, until his death at the age of forty-seven. He married Ann Walker in 1817 and during George IV's reign lived at Windsor Castle in accommodations provided for him and his wife. When William IV acceded to the throne in 1830, the Grays moved to London in order to be near Buckingham Palace, and either immediately or soon after that No. 8 Wilton Street became the family residence.

Henry Gray had two brothers and one sister. His younger brother, Robert, trained to become a naval surgeon. He died at sea in his twenty-second year. Henry's sister, younger than he, apparently never was married. His older brother, Thomas William, adopted the legal profession, and was Attorney to the Queen's Bench in 1841, and, in 1846, Solicitor of the High Court of Chancery. He had a large family—seven sons and three daughters. His third son, Charles, the nephew whom Henry treated for smallpox in 1861, lived to the ripe age of fifty-three. His fifth son, Frederick, moved to South Africa, married in 1881 and had six children, one of whom was the F. Lawrence Gray mentioned above.

After his father's death in 1834, Henry Gray continued living at 8 Wilton Street with his mother, brothers and sister. This is undoubtedly the address from which he later carried on his short practice and the place where he wrote his famous book. He and his brothers were educated at one of the large London schools, possibly Westminster, Charterhouse, or St. Paul's. He was engaged to be married at the time of his death.

Henry Gray's signature appears on the pupil's book at St. George's Hospital, London, as a "perpetual student" entering on the 6th of May, 1845. Four years later, in 1849, his name appears in the Proceedings of the Royal Society with the M.R.C.S. after it, the approximate equivalent of our M.D. While still a student in 1848, he won the triennial prize of the Royal College of Surgeons for an essay entitled "The Origin, Connections and Distribution of the Nerves to the Human Eye and Its Appendages, Illustrated by Comparative Dissections of the Eye in Other Vertebrate Animals." He was appointed for the customary year as house surgeon to St. George's Hospital in 1850. Successively thereafter he held the posts of demonstrator of anatomy, curator of the museum, and lecturer on anatomy at St. George's Hospital. In 1861 he was surgeon to St. James Infirmary and was a candidate for the post of assistant surgeon to St. George's Hospital and would certainly have been elected had he not died from confluent smallpox which he contracted from his nephew, Charles, whom he was treating. He was buried at Highgate Cemetery in June 1861. Sir Benjamin Brodie, president of the Royal Society, wrote, "I am much grieved about poor Gray. His death, just as he was on the point of obtaining the reward of his labors, is a sad event indeed

. . . Gray is a great loss to the hospital and school. Who is there to take his place?"

During his lifetime Henry Gray received outstanding recognition for his original investigations. That they have received so little mention since his death is as surprising as the lack of information about his life. The study of the eye which won him the Royal College of Surgeons prize was expanded into an embryological work, "On the development of the retina and optic nerve, and of the membranous labyrinth and auditory nerve," published in the Philosophical Transactions of the Royal Society in 1850. It contains the earliest description of the histogenesis of the retina. Two years later the Transactions contained another article, "On the development of the ductless glands in the chick." This must have stimulated Gray's interest in the spleen, which he classed as a ductless gland, because he obtained an allotment of funds for further study from the annual grant placed at the disposal of the Royal Society by Parliament for the promotion of science. The result was a monograph of 380 pages, "On the Structure and Use of the Spleen," which won him the triennial Astley Cooper Prize of £300 (about $1,500.00) in 1853. It was published by J. W. Parker and Son in 1854, but appears now to be excessively rare. Numerous "first observations" recorded in this book have escaped notice by all subsequent authors writing about the spleen. Gray described, among other things, the origin of the spleen from the dorsal mesogastrium ten years before Müller who is usually given credit for the discovery.

As a result of his ability and accomplishment, Henry Gray was made a Fellow of the Royal Society at the very young age of twenty-five. Besides anatomy, his interests also included pathological and clinical investigation. In 1853 he had a paper in the Medico-Chirurgical Proceedings entitled, "An Account of a Dissection of an Ovarian Cyst Which Contained Brain," and in 1856 a more extensive treatise entitled, "On Myeloid and Myelo-Cystic Tumours of Bone; Their Structure, Pathology, and Mode of Diagnosis."

His crowning achievement, however, and the one which is the source of his lasting fame is the publication, *Anatomy, Descriptive and Surgical,* now widely known as *Gray's Anatomy.*

114 YEARS OF GRAY'S ANATOMY

The first edition of *Gray's Anatomy* was published in London by J. W. Parker and Son in 1858 and in June of the following year in Philadelphia by Blanchard and Lea who had purchased the American rights for the book. The American edition was identical with the English, except that many typographical errors had been corrected, the index considerably improved, and the binding made more rugged. It contained xxxii + 754 pages and 363 figures. The drawings were the work of Dr. H. Van Dyke Carter of whom Gray writes in his preface, "The Author gratefully acknowledges the great services he has derived, in the execution of this work, from the assistance of his friend, Dr. H. V. Carter, late Demonstrator of Anatomy at St. George's Hospital. All the drawings, from which the engravings were made, were executed by him. In the majority of cases, they have been copied from, or corrected by, recent dissections, made jointly by the Author and Dr. Carter."

Blanchard and Lea obtained the services of Dr. R. J. Dunglison to edit the first and the next four American editions. Dunglison corrected the typographical and other small errors and improved the index but made very few alterations or additions in the text. Almost no adaptation was required because medical education in this country and in England were still much alike.

Henry Gray had just finished preparing the second edition before his untimely death. This was published in 1860 in England and in 1862 in this country under the editorship of Dunglison. The next edition published in America was a "New American from the 5th English Edition." This appeared in 1870 and was bound in either cloth or sheep. Another pause, and there appeared the "New American from the 8th English" in 1878, and "The New American from the 10th English" in 1883. Dunglison was again the editor, but a new editor, W. W. Keen, revised extensively the section on topographical anatomy at the end of the book. The following edition, a

"New American from the Eleventh English," published in 1887, was edited and thoroughly revised by W. W. Keen. Color was used for the first time in the chapters on blood vessels and nerves. The "New American from the Thirteenth English" appeared in 1893, apparently edited by Keen and others.

In 1896 the reference to the English edition was dropped and we have the fourteenth edition with the editors Bern B. Gallaudet, F. J. Brockway, and J. P. McMurrich. J. C. DaCosta, editor of the sixteenth edition in 1905, expanded the book, introducing much new material. E. A. Spitzka assisted DaCosta with the seventeenth edition in 1908 and edited the next two alone in 1910 and 1913. The publishers also tried a "New American from the Eighteenth English Edition" in 1913, but its sale was much smaller than that of the American edition of that year. In 1918, W. H. Lewis began his editorship, giving the book a scholarly treatment, reducing its length, improving the sections on Embryology, and generally giving a more straightforward treatment of the various chapters. For his last edition, the twenty-fourth, published in 1942, he had the assistance of six associate editors.

Beginning with the twentieth edition in 1918, new editions appeared at regular intervals of six years; their dates and the names of the editors are listed on page x. The interval was changed to five years for the twenty-seventh edition, published in 1959, in order to have its publication fall on the 100th year.

During the last sixty-seven years the American and English editions have tended to drift apart. Even from an early date, new American editions were less frequent than the English, and if the American editions had been numbered independently from the beginning the present edition would be the twenty-fourth, whereas the current English edition is the thirty-fifth. Somewhat more of the imprint of Henry Gray has been preserved in the American edition, in the illustrations especially, as it still has over 200 drawings based on the original illustrations by Carter but only some twenty-five have been retained in the latest English edition. Nevertheless, both books are *Gray's Anatomy*, both have the unmistakable stamp of their originator, and both uphold his fine tradition.

AMERICAN EDITIONS OF GRAY'S ANATOMY

Date	Edition	Editor
June 1859	First American Edition	Dr. R. J. Dunglison
February 1862	Second American Edition	Dr. R. J. Dunglison
May 1870	New Third American from Fifth English Edition	Dr. R. J. Dunglison
July 1878	New American from the Eighth English Edition	Dr. R. J. Dunglison
August 1883	New American from the Tenth English Edition	Dr. R. J. Dunglison
September 1887	New American from the Eleventh English Edition	Dr. W. W. Keen
September 1893	New American from the Thirteenth English Edition	Dr. W. W. Keen
September 1896	Fourteenth Edition	Drs. Gallaudet, Brockway and McMurrich
October 1901	Fifteenth Edition	Drs. Gallaudet, Brockway and McMurrich
October 1905	Sixteenth Edition	Dr. J. C. DaCosta
September 1908	Seventeenth Edition	Drs. DaCosta and Spitzka
October 1910	Eighteenth Edition	Dr. E. A. Spitzka
July 1913	Nineteenth Edition	Dr. E. A. Spitzka
October 1913	New American from Eighteenth English Edition	Dr. R. Howden
September 1918	Twentieth Edition	Dr. W. H. Lewis
August 1924	Twenty-first Edition	Dr. W. H. Lewis
August 1930	Twenty-second Edition	Dr. W. H. Lewis
July 1936	Twenty-third Edition	Dr. W. H. Lewis
May 1942	Twenty-fourth Edition	Dr. W. H. Lewis
August 1948	Twenty-fifth Edition	Dr. C. M. Goss
July 1954	Twenty-sixth Edition	Dr. C. M. Goss
August 1959	Twenty-seventh Edition	Dr. C. M. Goss
August 1966	Twenty-eighth Edition	Dr. C. M. Goss
January 1973	Twenty-ninth Edition	Dr. C. M. Goss

Preface

When Henry Gray undertook the writing of his new Anatomy book for students and practitioners of medicine he was fortunate in having a fellow anatomist with skill and originality to make his illustrations. These drawings from original dissections by Van Dyke Carter, with their diagrammatic clarity, have left an imprint almost as characteristic as the author's design for the text.

In keeping with the constant purpose of preserving the familiar atmosphere of Gray's Anatomy while bringing it up to date I have been desirous of continuing Carter's style in the illustrations. During the last two or three editions, however, a number of borrowed illustrations have been used to supplement the original figures. These have been colored halftones, for the most part, and although they have served their purpose admirably they have left something to be desired in clarity and register. In this edition I have again had the collaboration of Don Alvarado, Professor of Medical Illustration at Louisiana State University in replacing these illustrations. His pen has captured much of the style of the engravings on wood by Carter.

The trend in medical education toward earlier and earlier introduction of clinical subjects into the curriculum continues to place a burden on both faculty and students in the basic sciences by reducing the number of hours allotted to them. Each teacher of Anatomy will doubtless provide his class with his own outline, therefore, but the students must have a more complete, detailed, and accurate source of information as a supplement. This edition is well adapted for this purpose by its logical arrangement and clear illustrations. It is not commonly realized that the book may be used as an atlas as well as a text. Almost every structure in the body is illustrated somewhere in the book. In order to avoid increasing the size of the book, however, illustrations are seldom repeated. Many references between chapters are given in the text and as a student becomes familiar with the book, he may supplement these with his own index of illustrations. The new drawings in the chapter on the Peripheral Nerves especially may be used as if they were plates in a regional atlas for the visualization of blood vessels and other structures as well as nerves. Also in this edition the number of x-ray pictures has been somewhat increased and they are appropriately placed throughout the book instead of being included in a single chapter.

The nomenclature in this edition again follows the Nomina Anatomica adopted by the International Anatomical Nomenclature Committee and the International Congress of Anatomists. Following the recommendation of the I.A.N.C., however, the English equivalent of the Latin term is frequently used. A fundamental change in the terms of position and direction in the N.A. is inevitable because the newly adopted Nomina Embryologica and Nomina Histologica have been required to use such terms as dorsal and ventral or cranial and caudal in place of anterior and posterior or superior and inferior. These newer terms are occasionally used in appropriate places in anticipation of these changes. The older names still in common use by the teachers in clinical departments are included in parentheses for the benefit of students reviewing Anatomy for later examinations.

Grateful acknowledgement is made to Professor Frank Allan and an artist of his Visual Aids Department, Robert Edwards, for histological drawings. Acknowledgement is also made to Professor Don Fawcett for an illustration of the spermatozoon and to Professor Barry Anson for drawings of the liver from his new edition of *Morris' Human Anatomy.*

Gratitude is expressed to my colleagues

in the Department of Anatomy at George Washington University, Professors Ira Telford, Paul Calabrisi, John Christensen, Tom Johnson, Ernest Albert, and Marilyn Koering as well as to the students for their help and encouragement. Thanks are extended to the publishers for the loyal confidence, patience, and encouragement they have shown.

CHARLES MAYO GOSS

Bethesda, Maryland

Contents

12. THE PERIPHERAL NERVOUS SYSTEM

13. THE ORGANS OF THE SENSES

14. THE INTEGUMENT

1 | *Introduction*

N ancient Greece, at the time of Hippocrates (460 B.C.), the word anatomy ἀνατομή meant a dissection, from τομή, a cutting and the prefix ἀνά meaning up. Today anatomy is still closely associated in our minds with the dissection of a human cadaver, but the term was extended very early to include the whole field of knowledge dealing with the structure of living things. Even before human dissection was practiced by Herophilos and Erasistratos (300–250 B.C.), Aristotle (384–323 B.C.) dissected animals, wrote a treatise on their anatomy, and laid the groundwork for a scientific study of their form and structure. Anatomy, therefore, became the branch of knowledge concerned with **structure** or **morphology**, and it now applies to plants as well as animals. Our fund of anatomical information has been increased greatly during the last three or four centuries by a study of minute structure through the microscope, by following the development of the embryo, and by the addition of new and refined technical methods. The whole field of anatomy has become very large and as a result a number of subdivisions of the subject have been recognized and named, usually to correspond with a specialized interest or avenue of approach.

Gross Anatomy or **Macroscopic Anatomy** is the study of morphology by means of dissection with the unaided eye or with low magnification such as that provided by a hand lens. The study of the structure of organs is **Organology**. The study of the tissues is **Histology**.

Microscopic Anatomy is the study of structure with the aid of a microscope and since it deals largely with the tissues, is often rather inaccurately called **Histology**. Microscopic anatomy has another branch, the study of the cell, or **Cytology**. An enormous increase in research into the finer structure of the cell has taken place in recent years, due to the adaptation of the **Electron Microscope** to the observation of ultrathin sections of the tissues.

Embryology or **Developmental Anatomy** deals with the growth and differentiation of the organism from the single-celled ovum to birth. When applied to the life history of an individual it is called *ontogeny*. It is impossible to understand clearly all the structures found in the adult body without some knowledge of embryology. Since human embryos, particularly the younger stages, are difficult to obtain for study, a large part of our knowledge has been gained from animals, that is, from **comparative embryology**. The second chapter in this book is devoted to a brief summary of **human embryology**. The development of each system is outlined in the appropriate chapters and sections.

Comparative Anatomy deals with the structure of all living creatures, in contradistinction to human anatomy which deals only with man. A comparison of all the known animal forms, both living and fossil, indicates that they can be arranged in a scale which begins with the simplest forms and progresses through various gradations of complexity and specialization to the highest forms. The unfolding of a particular race or species is called *phylogeny*. Many of the earlier stages of development in man and other higher animals resemble the adult stages of animals lower in the scale, hence it has been said that ontogeny repeats phylogeny. Although this is not strictly true, the study of comparative anatomy has contributed vitally to the understanding of human anatomy and physiology.

Terms of Position and Direction

The Anatomical Position.—The traditional anatomical position, which has long been agreed upon, places the body in the erect posture with the feet together, the arms hanging at the side, and the thumbs pointing away from the body. This position is used in giving topographic relationships, especially those which are medial and lateral when referred to the limbs. The muscle actions and motions at the joints are given with reference to this position unless it is stated otherwise. It is particularly important that this position be remembered in descriptions using anterior and posterior, or the more general and ambiguous terms such as up, down, over, under, below, above, etc.

The **median plane** is a vertical plane which divides the body into right and left halves; it passes approximately through the sagittal suture of the skull, and therefore any plane parallel to it is called a **sagittal plane. Frontal planes** are vertical planes, also passing from head to feet, but they are at right angles to the sagittal, and since one of them passes approximately through the coronal suture of the skull, they are also called **coronal planes. Transverse** or **horizontal planes** cut across the body at right angles to both sagittal and frontal planes.

Ventral and **dorsal** refer to the front or belly and the back of the body, and are synonymous with **anterior** and **posterior.** Because of man's erect posture, this is not the same usage as that of comparative anatomy. In the hands and forearms, **palmar** or **volar** are substituted for ventral or anterior, and the sole of the foot is called **plantar** while **dorsal** is retained for the top or opposite surface of the foot.

Cranial and **caudal** indicate the head and tail respectively, and with human subjects in the anatomical position, superior and inferior are synonymous with them, but in comparative anatomy the latter terms might be confused with dorsal and ventral when referring to four-footed beasts.

Median refers to structures in the middle line or median sagittal plane of the body or middle of a limb.

Medial and **lateral** refer to structures nearer or farther away from the median plane. It should be noticed that the Latin word *"medius"* means middle, not medial. The Latin word for medial is *"medialis."*

Superficial and **deep** indicate the relative depth from the surface, and are preferable to over and under, and above or below, because the latter may be confused with superior and inferior.

Proximal and **distal** indicate a direction toward or away from the attached end of a limb, the origin of a structure, or the center of the body.

External and **internal** are most commonly used for describing the body wall, or the walls of cavities and hollow viscera, but they may be synonymous with superficial and deep, or with lateral and medial.

Nomenclature

Although the majority of the common anatomical names have been taken from those used by the ancient Greeks, a considerable amount of disagreement and confusion has arisen because of conflicting loyalties to teachers, schools, or national traditions. Shortly before the end of the last century, however, the need for a comprehensive system of nomenclature was realized, and a Commission of eminent authorities from the various countries of Europe and the United States was organized for this purpose. The system devised by this Commission was adopted by the German Anatomical Society in 1895 at their meeting in Basel, Switzerland, and it has since been called the **Basle Nomina Anatomica** or the **BNA** (His, 1895).

The **BNA** gives the names in Latin, which is our closest approach to an international language, and many nations, including the United States, adopted it, translating the terms into their own language whenever desirable. The French anatomists, however, did not entertain it with much enthusiasm, preferring their own quite ancient tradition. The British anatomists gradually found certain inaccuracies and inconsistencies with their time-honored tradition which favored human anatomy related to the arbitrary "anatomical position." They broke away from the BNA and adopted their own revision in 1933 which they called the **Birmingham Revision** or **BR.** The German anatomists became restless under the restric-

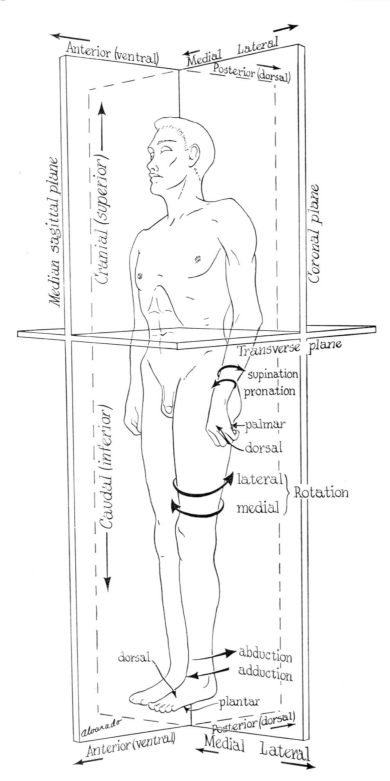

Fig. 1–1.—Diagram illustrating the planes of the body and the terms of position and direction.

tions of the BNA adoption of the "anatomical position" and preferred an attempt to establish a scientific language applicable to the comparative anatomy of vertebrates as well as to human anatomy. In 1937 the Anatomische Gesellschaft met in Jena and adopted a revision which is known as the **Jena Nomina Anatomica** or **INA.** The anatomists of the United States remained faithful to the BNA.

At the Fifth International Congress of Anatomists held at Oxford in 1950 a committee was appointed to attempt a new revision of anatomical names. This International Nomenclature Committee met at Paris in July 1955, and approved the "**Nomina Anatomica**" which was then submitted to the Sixth International Congress of Anatomists at Paris, also in July 1955, now known as the **PNA.**

The International Nomenclature Committee met again in 1960 and submitted a minor revision to the International Congress in New York which made a few additional changes and restored a number of the **BNA** terms that had been discarded in 1955. This new nomenclature, although an improvement, is still a compromise between the anatomical position and a scientific terminology, with some confusing inconsistencies, cumbersome phrases, and pseudofunctional terms. In this edition, the entire book has been rewritten incorporating the new **NA** or **PNA** except in a few instances where it has been abandoned deliberately in the interests of clarity and consistency. Since Latin is becoming unfamiliar to students, names are usually given in English with the official Latin name italicized in parentheses. Other recognized names are also added in parentheses.

Anatomical Eponyms, that is, designations by the names of persons, are popular among clinicians and specialists. Many of them are added to the other names in parentheses in order that students may not be totally ignorant of the language of clinical medicine.

Human Anatomy is presented from three principal points of view: (*a*) **Descriptive** or **Systemic Anatomy,** (*b*) **Regional** or **Topographical Anatomy,** and (*c*) **Applied** or **Practical Anatomy.** This book is primarily concerned with Systemic Anatomy, but Regional and Applied Anatomy are brought in at appropriate places.

Systemic Anatomy.—The body as a whole is composed of a number of systems whose parts are related to each other by physiological as well as anatomical considerations. Each system is composed of similar parts or tissues and assists in the performance of particular functions. Although the study of anatomy is concerned primarily with morphology, the knowledge of structure becomes understandable and of practical value only if the close association between structure and function is kept continually in mind.

The systems of the body are as follows:

The **Skeletal System,** a study of which is called *Osteology,* is composed of bones and cartilage, and its function is to support and protect the soft parts of the body.

The **System of Joints** or **Articulations** (*Arthrology*) makes the rigid segments of bone moveable, but holds them together with strong fibrous bands, the ligaments.

The **Muscles** (*Myology*) form the fleshy parts of the body and put the bones and joints into useful motion.

The **Vascular System** (*Angiology*) includes the **circulatory system** or the heart and blood vessels and the **lymphatic system** which transports the lymph and tissue fluids.

The **Nervous System** (*Neurology*) includes the **Central Nervous System,** which is composed of the brain and spinal cord, the **Peripheral Nervous System,** which is composed of nerves and ganglia, and the **Sense Organs,** such as the eye and ear. Its function is to control and coordinate all the other organs and structures, and to relate the individual to his environment.

The **Integumentary System** is composed of the skin, hair and nails.

The **Alimentary System** is composed of the food passages and the associated digestive glands.

The **Respiratory System** is composed of the air passages and lungs.

The **Urogenital System** includes the kidneys and urinary passages and the reproductive organs in both sexes.

The **Endocrine System** includes the thyroid, suprarenal, pituitary, and other ductless glands which control certain functions of the whole body or of specific remote target organs by secreting hormones into the blood stream.

Splanchnology is the name given to the study of all the internal organs, especially those in the thorax and abdomen.

An account of the anatomy of any system would be incomplete without consideration of its microscopic anatomy and embryology, and sections dealing with them will be found in their proper places.

Topographical or Regional Anatomy.— Regional Anatomy is more strictly morphological than Systemic Anatomy because it deals with the structural relationships within the various parts of the body. Students in the laboratory must approach the subject of anatomy regionally because they dissect by regions rather than systems. Although the acquisition and organization of anatomical knowledge are easier for beginners if followed by systems, students in the medical fields must continually be on the alert to learn the relationship of the various parts to each other and to the surface of the body because the final purpose of their study is to visualize them in living subjects. In addition to the dissection of the body, the study of topographical anatomy is carried out by the study of **Surface Anatomy, Cross Sectional Anatomy,** and **Radiographic** or **X-ray Anatomy.**

Applied Anatomy has a number of subdivisions, and is concerned with the practical application of anatomical knowledge in some field or specialty. **Surgical Anatomy, Pathological Anatomy,** and **Radiological Anatomy** are probably the largest fields.

Physical Anthropology is closely related to Human Anatomy because it relies heavily upon anatomical studies and measurements of man and other primates, but it is a much broader science of human biology with particular emphasis on racial development, evolution, genetics, and paleontology.

REFERENCES

BARCLAY, JOHN. 1803. *A New Anatomical Nomenclature Related to the Terms which are Expressive of Position and Aspect in the Animal System.* viii + 182 pages, 5 plates. Ross and Blackwood, Edinburgh.

DONATH, T. 1969. Anatomical Dictionary with Nomenclature and Explanatory Notes. English Translation edited by G. N. C. Crawford. 634 pages. Pergamon Press, Oxford.

HIS, WILHELM. 1895. *Die anatomische Nomenclature. Nomina Anatomica.* Verzeichniss der von der anatomischen Gesellschaft auf ihrer IX. Versammlung in Basel angenommenen Namen. Archiv fuer Anatomie und Physiologie, anatomische Abteilung, Supplement-band. 1895. iv + 180 pages, 30 figures, 2 plates. Viet & Comp., Leipzig.

KNESE, K. H. 1957. *Nomina Anatomica,* 5th edition. xi + 155 pages. Georg Thieme Verlag, Stuttgart.

Nomina Anatomica. 1965. Third edition with index. Revised by the International Anatomical Nomenclature Committee and adopted at the eighth International Congress of Anatomists. Wiesbaden. v + 164 pages. Excerpta Medical Foundation, Amsterdam.

Nomina Embryologica. 1970. Compiled by the Embryology Subcommittee of the I.A.N.C. and adopted at the ninth International Congress of Anatomists, Leningrad 1970. vi + 70 pages, FASEB, 9650 Rockville Pike, Bethesda, Md. 20014.

Nomina Histologica. 1970. Compiled by the Histology Subcommittee of the I.A.N.C. and adopted at the ninth International Congress of Anatomists, Leningrad 1970. 111 pages. Leningrad.

Rufus of Ephesus. Second Century. *On Names of the Parts of the Human Body.* In: Oeuvres de Rufus d'Ephese by Daremberg and Ruelle, 1879. Paris. 133-185.

TRIEPEL, HERMAN. 1957. *Die Anatomischen Namen.* Ihre Ableitung und Aussprache. 25th edition by Robert Herrlinger. 82 pages. J. F. Bergmann, München.

SKINNER, H. A. 1963. *The Origin of Medical Terms.* 2nd edition. x + 437 pages, illustrated. The Williams & Wilkins Company, Baltimore.

2 | *Embryology*

THE CELL

The human body is composed of structural units of protoplasm, called **cells,** together with large amounts of **intercellular material** especially evident in the bones, cartilages, tendons, and ligaments. The individual cells are microscopic in size and occur in many gradations of shape and composition corresponding to their diversified functions. Every human being begins his or her existence as a single cell, the **ovum,** the development of which entails an increase in the number of cells through cell division, followed by growth and differentiation of the cells, and accompanied by the functional activity broadly known as metabolism. All cells are composed of two parts, the outer soft, jelly-like **cytoplasm** and an inner **nucleus.** During the interval between periodic divisions of the cell, the nucleus is called a resting nucleus; during periods of division, the nucleus goes through a series of changes called mitosis.

The **cytoplasm** of a living cell is composed of an optically homogeneous matrix with a thin outer layer, the ectoplasm, and an internal portion, the endoplasm, in which various intracellular bodies and organelles are embedded (Fig. 2–1). Although most of these structures are recognizable with the light microscope, their finer details have only been seen through the higher magnifications with the electron microscope. Living cells, however, cannot be viewed with the electron microscope. Thus the outer surface of the cell, scarcely distinguishable with the light microscope, becomes the **cell membrane** or **plasma membrane,** frequently studded with microvilli (Fig. 2–1). The membrane is a most important functional organ because it controls the exchange of materials between the cell and its environment by osmosis, phagocytosis, pinocytosis, secretion and other phenomena. The **central body** or **centrosome** with its **centrioles,** not always visible with the light microscope, lies at one side of the nucleus near the central part of the cell. It serves as a dynamic center with other intracellular structures such as the Golgi complex arranged around it, and it is especially important during cell division as described below. The **mitochondria** are visible in living cells in the form of threads, rodlets, and granules, but they are likely to be destroyed by the action of lipid solvents during the preparation of routine, preserved and stained microscopic slides. Their internal organization revealed by the electron microscope includes membranes and internal folds named **cristae** (Palade '52). They are important centers of respiratory and other metabolic processes. The **Golgi bodies** or **complex** can be demonstrated in nearly all cells by special methods of preparation but are seldom identifiable in living cells. With the electron microscope they have been shown to include large and small vacuoles, granules, and membranes associate with vesicles or clefts (Dalton and Felix '54). The cytoplasm, which appears to be more or less homogeneous with the light microscope, has been shown by electron micrographic studies to contain a membranous complex named the **endoplasmic reticulum.** In certain cells the cytoplasm is colored by the same basic dyes which stain the chromatin of the nucleus, due to the presence of **chromophilic substance** or the ergastoplasm of light microscopy. Electron micrographic studies have demonstrated that the coloration is due to minute granules associated with ribonucleic acid and protein and are named RNA granules or **ribosomes.** These granules, about 150

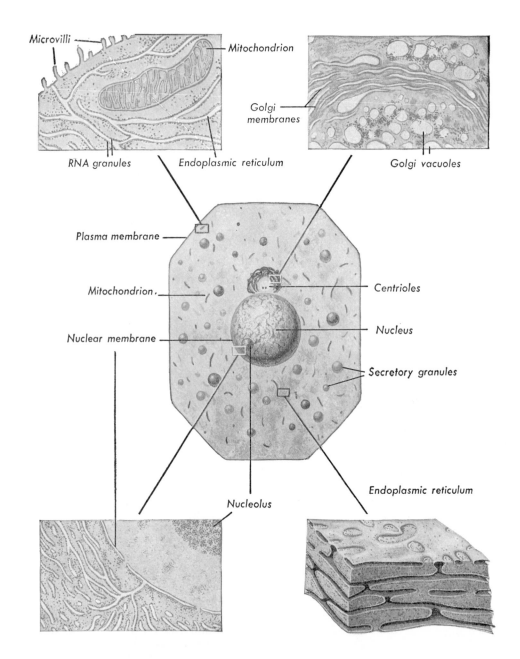

Fɪɢ. 2–1.—Schematic representation of a cell as seen with the light microscope and enlargements showing the fine structure of some of the cell constituents as revealed by electron microscopy. The endoplasmic reticulum is shown in three dimensions, other structures are in section. (From *Bailey's Histology* by W. M. Copenhaver, The Williams & Wilkins Co., 1965.)

angstrom units in diameter, are commonly associated with the endoplasmic reticulum, but may be scattered throughout the cell (Palade '55). Many cells contain granules of metabolic products such as glycogen, secretion granules, vacuoles, lipid inclusions, and inert or ingested material.

The **nucleus** of a living cell in the intermitotic or resting stage is a spherical or elongated body with a thin nuclear membrane and a clear, translucent internal nucleoplasm containing one to several semisolid bodies called **nucleoli.** In nuclei which have been preserved by chemical fixation, dark masses of material called **chromatin** may be brought out by staining the nuclei with various dyes, the one most commonly used being hematoxylin. The chromatin material is invisible in living resting nuclei except possibly with phase microscopy, but during cell division it is aggregated first into threads and then into refractive bodies called chromosomes. The process is called **mitosis** from the Greek word mitos ($\mu\iota\tau\sigma\varsigma$) meaning thread. The chromosomes of the germ cells carry the factors for the genetic or hereditary constitution of the individual.

CELL DIVISION

Although the division of a cell by mitosis is a continuous process, it is the custom to divide it into four stages; prophase, metaphase, anaphase, and telophase. In preparation for the prophase, the **centrosome** divides into two parts which become oriented at opposite poles of the nucleus. During the **prophase** the chromatin which has been scattered throughout the nucleus becomes more condensed into **chromomeres** scattered along a thread or **chromonema.** The threads are individual entities composed especially of deoxyribonucleic acid (DNA); they become condensed into more compact bodies, the **chromosomes.** During the **metaphase** the chromosomes become oriented in a plane midway between the two centrosomes, the **equatorial plate,** with a clear area, the **spindle,** leading toward each centrosome. At this time each chromosome has become double. The two parts are named **chromatids** and are attached to each other at a particular part, the **centromere.** In late metaphase and early

anaphase, each centromere divides. During **anaphase** the two chromatids of each chromosome separate, pulled apart by a spindle fiber leading to one or the other of the two centrosomes. The two sets of chromatids, now called daughter chromosomes, are grouped around the two centrosomes, which move to opposite ends of the cell. In **telophase** the chromosomes form a compact mass, lose their individuality and disperse into the chromatin of the intermitotic nucleus.

The division of living cells cannot be observed with high magnifications of the microscope in the intact body. It was discovered by Harrison ('10), however, that embryonic cells would survive and divide on glass slides if they were placed in an appropriate medium with aseptic precautions. Since that time a great many observations have been made with this method, known as *tissue culture*, on a wide range of different animal forms, including tissues from human fetuses, normal adults, and tumor growth. The following description of cell division is based upon observations of connective tissue cells or fibroblasts in cultures of tissues from chick embryos by Warren H. Lewis.

Mitosis in Tissue Culture.—An hour or two before division begins, the centrosome material is divided into two parts, placed at opposite poles of the nucleus (Fig. 2–2, a–e).

During **prophase** (Fig. 2–2, g, h, i), the chromatin is formed into threads and then chromosomes. The cell retracts its processes and tends to become spherical with the nucleus centrally placed and mitochondria or other intracellular bodies arranged around it. Near the end of the prophase, which takes about an hour, the nuclear membrane disappears and the chromosomes are released and move into the equatorial plane.

During **metaphase,** about six minutes, the chromosomes oscillate back and forth in short paths in the equatorial region as if they were pulled first toward one pole and then toward the other by strings or fibers. These fibers are not visible in living fibroblasts (Fig. 2–2, k) but are visible in many kinds of cells, especially after fixation, and together with the chromosomes make up

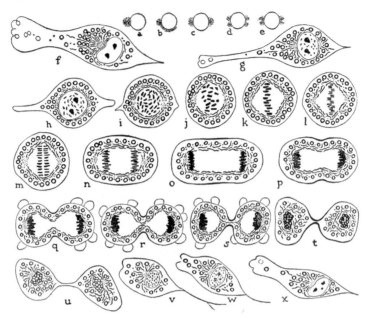

FIG. 2–2.—Cell division. **a–e.** From motion picture of a living monkey egg. **f–x.** From a living fibroblast dividing in a tissue culture. **f.** Resting stage. **g.** Beginning prophase. Chromosomes are beginning to appear as granules. **h.** Mid-prophase. Chromosomes are more evident in the otherwise homogeneous nucleoplasm. **i.** Late prophase. Chromosomes nearly fill the nucleus. **j.** Transition from prophase to metaphase. The chromosomes are being pulled into the metaphase plate. **k.** Metaphase. The chromosomes are in the equatorial plane. **l.** Late metaphase. Each chromosome has split longitudinally into two equal parts. **m.** Early anaphase. Each group of chromosomes has begun to move toward the poles of the spindle. **n.** Late anaphase. The chromosomes have nearly reached the poles. **o.** End of anaphase. Chromosomes form small compact daughter nuclei. **p.** Beginning telophase. The cleavage furrow has begun to appear. **q.** Early telophase. The cleavage furrow has deepened. A few blebs have appeared on the surface of the cell. **r.** Mid-telophase. **s.** Late telophase. **t.** End of telophase. The daughter cells are connected by the ectoplasmic stalk. The compact daughter nuclei have begun to show clear areas and to enlarge. **u.** The daughter cells have moved in opposite directions. The nuclei have larger clear areas and less visible chromosome material. **v.** A nuclear membrane has developed. **w.** Some of the chromosomal granules have begun to agglutinate to form the nucleoli; the others have begun to disappear. The connecting stalk is broken. **x.** All the chromosomal granules except those which have agglutinated to form the nucleoli have disappeared into homogeneous nucleoplasm. (W. H. Lewis.)

the *mitotic spindle* (Fig. 2–3). Each chromosome splits longitudinally into two equal and similar daughter chromosomes.

During **anaphase,** about three minutes, one set of daughter chromosomes moves to each pole of the spindle, or is pulled there by contraction of the spindle fibers, and is clumped into a compact mass. As the chromosomes approach the poles, the cell becomes elongated and a groove encircles it at the equator.

During **telophase,** about three minutes, the groove sinks farther into the cell, gradually pinching it into two daughter cells which remain connected for a short time by a stalk. Each daughter cell receives approxi-

mately one-half the mitochondria, but the division of fat droplets and other intracellular bodies is more haphazard. The chromosomes lose their identity in the compact mass of the daughter nuclei. The daughter nuclei slowly increase in size, as clear areas split the compact mass into granules, and a nuclear membrane forms. As the clear areas increase, the chromosomal granules become invisible except for those which become the nucleoli. It takes several hours for the nucleus to attain the final resting or intermitotic condition.

Amitosis or direct division of the cell has been described but probably does not occur under normal conditions. Degenerating cells

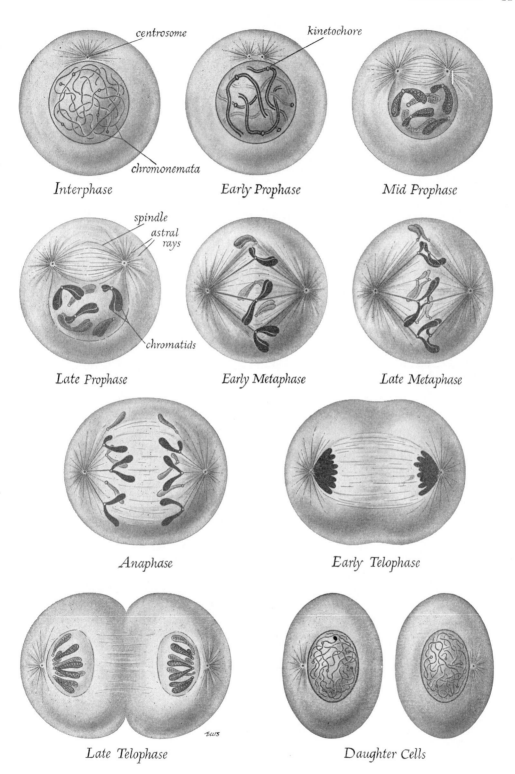

Fig. 2–3.—Diagrammatic representation of various stages in mitosis. (Winchester, A. M.: *Genetics*, ed. 2, courtesy of Houghton Mifflin.)

may show **nuclear fragmentation** without division of the cytoplasm or the centrosome.

Binucleate cells are not uncommon, especially in certain organs, and are formed either by mitotic division of the nucleus without cleavage of the cytoplasm, or by some process not clearly understood such as fragmentation.

Chromosomes and **Heredity.**—In each animal species there is a characteristic number of chromosomes. In man there are 23 pairs or 46 chromosomes. Each pair consists of a maternal and a paternal chromosome. The random distribution of maternal and paternal chromosomes is such that in the germ cells during maturation over 16,000,000 possible combinations will result in the daughter cells. Of the 23 pairs in man, 22 are called **autosomes** and the remaining pair are the **sex chromosomes** which can be identified under the microscope by their size and shape. The sex pair in all the cells of a male organism consists of a large X and a small Y chromosome. The sex pair of a female contains two X but no Y chromosomes. Al-

though a chromosome usually shows but little visible internal structure, experiment has shown that it consists of a large number of submicroscopic structures, the **genes,** spaced along a thread in lineal order. The genes are the centers for control of particular hereditary traits, and the similar genes of a pair of chromosomes react with each other and with their environment to establish the characteristics of a particular individual or **genotype.**

Genetics.—In the study of human genetics, living cells from the blood or other tissues are prepared in a tissue culture which allows them to proliferate and divide by mitosis. When individual cells are found in the appropriate stage of division, that is, the metaphase, the cell is flattened out on a glass slide in order to separate the chromosomes (Figs. 2–4, 2–5). When examined under the microscope, the individual chromosomes can be identified and mapped or arranged in a diagram called an **idiogram** (Figs. 2–6, 2–7). Thus they may be compared with normal chromosomes and if abnormalities are iden-

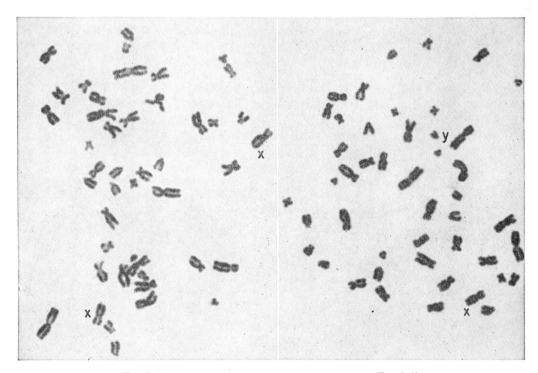

Fig. 2–4

Fig. 2–5

Fig. 2–4, female and Figure 2–5, male, photographs of colchicine metaphase of human cells grown in vitro, showing chromosomes. (From Tijo and Puck, Somatic Chromosomes of Man, Proc. Nat. Acad. Sci., 1958.)

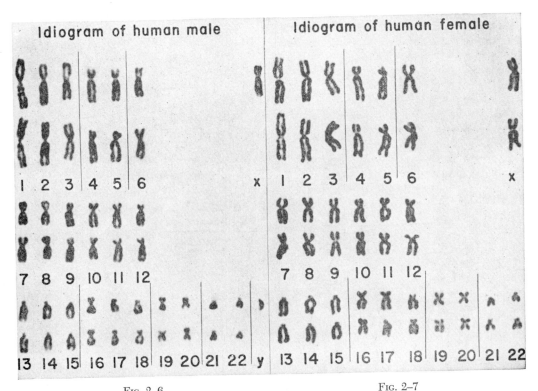

FIG. 2–6   FIG. 2–7

FIGS. 2–6 and 2–7.—Idiograms of human chromosomes. (From Tijo and Puck, Proc. Nat. Acad. Sci., 1958.)

tified they may point to the diagnosis of some hereditary aberration in the individual.

Germ Cells.—For an individual to begin life and develop into an embryo it is necessary for specialized cells, the germ cells from each sex, to be combined at fertilization. The female germ cell is the ovum, the male germ cell is the spermatozoön.

THE SPERMATOZOÖN

The spermatozoa or male germ cells are produced in the testis and are present in the seminal fluid in enormous numbers. Each mature human spermatozoön is a small but highly specialized cell resembling a free-swimming protozoan and possessing a prominent head and a long tail (Figs. 2–8, 2–9). The tail is subdivided into a middle piece, principal piece, and end piece. The **head** is a somewhat flattened oval body 4 or 5 μ long and 2 or 3 μ thick. The bulk of it is composed of a condensed nucleus the anterior two-thirds of which is covered by a **head cap** with an **acrosome** at the tip. Running the entire length of the tail is an **axial core** composed of filaments identical with those of a cilium as revealed by the electron microscope. These filaments consist of a central core of two single filaments surrounded by nine uniformly spaced double filaments. The cylindrical **middle piece** surrounds the axial structures from the head to the ring-shaped **centriole** or annulus. It is 5 to 7 μ long, 1 μ thick, and contains a single layer of spirally arranged thread-like mitochondria. The **principal piece** is 45 μ long and ½ μ thick composed of circularly arranged dense fibers. In the **end piece** the axial filaments are covered only by the **plasma membrane** covering the entire spermatozoön. The length of the entire spermatozoön is 52 to 62 μ. The axial filaments are assumed to be the contractile elements which provide the undulating motion of the tail. The development of the spermatozoa within the seminiferous tubules can be considered as

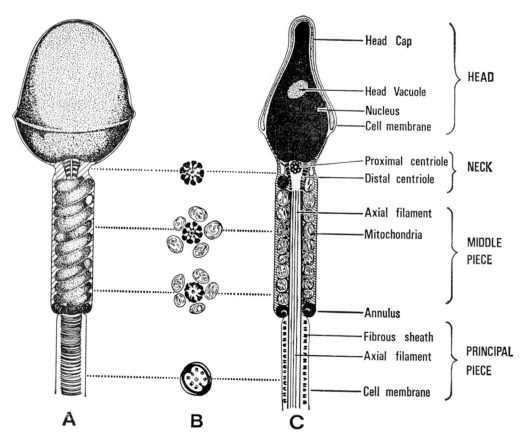

FIG. 2–8.—Diagram of the structure of the human spermatozoan based on studies with the electron microscope. Most of the principal piece and end piece have been omitted. *A.* Depicted as though the cell membrane and cytoplasmic matrix had been removed. *B.* Diagrammatic cross sections showing the arrangement of the axial fibrils and their relation to adjacent structures. *C.* Diagram of a longitudinal section passing through the center of *A* in a plane perpendicular to the page. (Fawcett, 1958. International Review of Cytology, 7, 206. Academic Press.)

a twofold process, one involving the nucleus, spermatogenesis, the other involving the cytoplasm, spermiogenesis or sperm differentiation (see Testis in Chapter 17).

Spermatogenesis.—When the primary germ cells divide by mitosis they provide a number of cells called **spermatogonia** which form a prominent layer next to the limiting membrane of the seminiferous tubule. When they proliferate, some of the daughter cells form more spermatogonia, others migrate toward the lumen and enlarge into the **primary spermatocytes.** The primary spermatocytes begin the process of **maturation** by dividing into two secondary spermatocytes, and each **secondary spermatocyte** in the second maturation division divides into two spermatids. Thus each primary sperma-

tocyte provides four spermatids all of which develop into motile spermatozoa (Fig. 2–9).

Meiosis.—The most important part of maturation of the germ cells is the distribution of chromosomes and their constituent carriers of heredity, the genes. This takes place by a special type of division called meiosis. During the prophase of a usual somatic mitosis each threadlike chromosome initiates a longitudinal splitting. In the **first maturation division** of meiosis in the primary spermatocyte, however, the individual chromosomes do not carry the splitting to completion and the two chromosomes of each pair except the sex chromosomes come into close union by a process known as **synapsis.** The resulting chromosomal configuration is known as a **tetrad** from its quadrupli-

SPERMATOGENESIS — OOGENESIS

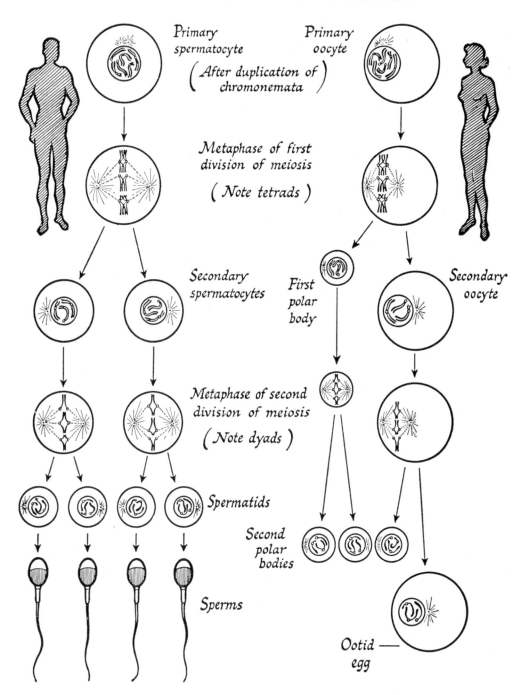

Primary spermatocyte — Primary oocyte

(*After duplication of chromonemata*)

Metaphase of first division of meiosis

(*Note tetrads*)

Secondary spermatocytes — First polar body — Secondary oocyte

Metaphase of second division of meiosis

(*Note dyads*)

Spermatids

Second polar bodies

Sperms

Ootid — egg

Fig. 2–9.—Diagrammatic representation of gametogenesis. Compare the mechanism involved in oögenesis and spermatogenesis, particularly with reference to their differences. Simplification of the mechanism is effected through the use of 6 of the 46 chromosomes in man. (Winchester, A. M.: *Genetics*, ed. 2, courtesy of Houghton Mifflin.)

cate appearance (Painter '23). The individual chromosomes of the synaptic pair, without splitting, separate from each other and one chromosome of each pair goes to each daughter cell. As a result the number of chromosomes is reduced by one-half or to 23 in the secondary spermatocytes, one of which contains an X chromosome and the other a Y chromosome. In the **secondary maturation division** the chromosomes of each secondary spermatocyte now complete their splitting and each divides into two spermatids. The result is four viable spermatids, two with X chromosomes and two with Y chromosomes.

Spermiogenesis.—The differentiation of the spermatid into the mature sperm begins with the formation at one pole of the nucleus of a fluid-filled **acrosomal vesicle** from the Golgi complex. The centrioles migrate toward the opposite pole of the nucleus where a slender **flagellum** is forming. The acrosomal vesicle loses its fluid and fits over the nucleus at the anterior end as a head cap with the acrosome at the tip. One of the centrioles forms the base of the flagellum which then elongates and develops its axial filaments. The other centriole forms a ring or annulus which migrates away from the nucleus to the caudal end of the middle piece and the spiral mitochondrial sheath

occupies the rest of the middle piece. The residual cytoplasm is finally cast off, leaving the plasma membrane covering the entire mature spermatozoön (Fawcett).

THE OVUM

The **human ovum** (Fig. 2–10) is a large cell, about 0.14 mm. in diameter, and contains the visible internal structures found in most cells. The nucleus with nucleolus and chromatin is known as the **germinal vesicle.** The cytoplasm contains a centrosome, mitochondria, granules, fat droplets, Golgi complex, and yolk material. The ovum or **vitellus** is enclosed in a tough transparent membrane about 12 μ in thickness, named the **zona pellucida.** Within the ovary, the mature ovum is contained in a spherical vesicle about 1 cm. in diameter called an **ovarian** or **Graafian follicle** (see Index). The ovum is held at one side of the follicle by a mass of follicular cells named the **cumulus oöphorus;** the follicular cells immediately around the zona pellucida are radially arranged and make up the **corona radiata** (Fig. 2–10). While in the ovary and at the time of ovulation, the vitellus completely fills the cavity of the zona, but it shrinks shortly after ovulation and a perivitelline space filled with clear fluid is developed.

COMPARATIVE SIZE OF LIVING ONE-CELL TUBAL EGGS AFTER THE FORMATION OF THE PERIVITELLINE SPACE (LEWIS AND WRIGHT)

Volume and surface area of vitellus in cubic and square microns estimated by authors. ODZ: outside diameter of zona. IDZ: inside diameter of zona. DV: Diameter of vitellus in microns.

Animal	Eggs	Author	ODZ	IDZ	DV	Volume*	Surface*
Mouse	26	Lewis and Wright	113.0	87.8	71.6	192000	16100
Guinea-pig	1	Squier	121.3	96.8	84.3	314000	22300
Macaque I	1	Corner		109.0*	86.0	333000	23400
Macaque two-cell	1	Lewis and Hartman	150.0	125.0	103.0*	562000	
Macaque IV	1	Allen	178.5	138.5*	104.0	589000	34000
Human	1	Allen *et al.*	176.0*	139.0*	104.0*	589000	34000
Human (abnormal)	1	Lewis	148.0	136.0			
Rabbit	2	Gregory	174.0	126.0*			
Rabbit	9	Lewis	174.2	126.5	111.0	718000	38700
Pig	4	Heuser and Streeter	160.0	130.0	111.0	718000	38700
Cow (non-fertile)	1	Hartman and Lewis	170.0	143.0	120.0	907000	45300
Cow two-cell	1	Miller and Swett	162.5	135.0		740000	
Dog	3	Hartman	172.0	141.0	120.0	907000	45300
Sheep	6	Clark	178.0*	150.0*	133.0*	1232000	55700

*Determined or estimated from illustrations and data in articles quoted.

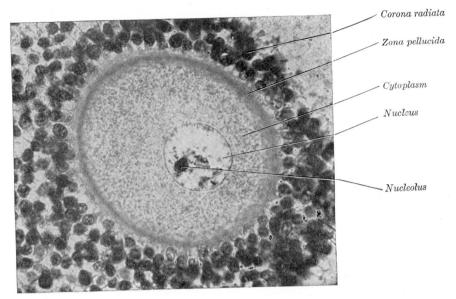

Corona radiata

Zona pellucida

Cytoplasm

Nucleus

Nucleolus

Fig. 2–10.—Mature human ovum. (Carnegie collection.)

Ovulation occurs when the mature Graafian follicle ruptures through the outer wall of the ovary and the ovum is discharged into the peritoneal cavity. Normally it is not allowed to go free in the peritoneal cavity because the fimbriated end of the uterine tube covers the point of rupture and the ovum is immediately swept into the lumen of the tube, probably by ciliary action (Rock and Hertig, '44).

Maturation of the Ovum.—Before an ovum can be fertilized it must have reached a particular stage in its maturation which begins just previous to ovulation. During the **first maturation division,** the ovum, now known as the **primary oöcyte,** divides into one large cell, the **secondary oöcyte,** and one small cell which is known as the first **polar body** or **polocyte** (N.E. *polocytum*) (Fig. 2–8). The secondary oöcyte quickly initiates the **second maturation division** by forming a mitotic spindle but it is arrested in the metaphase stage until after it escapes from the follicle and is penetrated by a spermatozoön. During the second maturation division, which can be completed normally only if the ovum is fertilized, a large **mature oötid** and a small **second polocyte** are formed (Fig. 2–11). The first polocyte may or may not divide into two smaller polocytes.

Meiosis.—When the ovum or primary oöcyte undergoes the first maturation division there is a reduction in the number of chromosomes in the same manner as in spermatogenesis. The secondary oöcyte and the first polar body both contain 22 autosomes and one sex chromosome. Since it is a female cell, the primary oöcyte has two X chromosomes and the secondary oöcyte also must have an X chromosome. When it undergoes its second maturation division therefore, there must be only an X chromosome in the oötid and the second polocyte. It will be seen from this that every ovum from a ruptured follicle can provide only one fertilizable oötid and that it will have an X chromosome. Since there are equal numbers of spermatozoa with X chromosomes and with Y chromosomes in the ejaculate, there is an equal opportunity for the fertilized ovum to contain two X or an X and a Y chromosome and the genetic equality of the sexes is assured.

FERTILIZATION OF THE OVUM

Ovulation takes place in a woman with regular menstrual cycles approximately ten to fourteen days after the onset of the preceding menstrual flow and the released ovum immediately begins its course through the uterine tube. It must be met by sperm in the ampulla of the tube within approximately six to twelve hours or it will begin

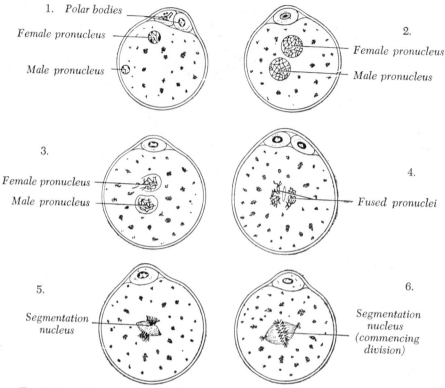

1. *Polar bodies*

Female pronucleus

Male pronucleus

2.

Female pronucleus

Male pronucleus

3.

Female pronucleus

Male pronucleus

4.

Fused pronuclei

5.

Segmentation nucleus

6.

Segmentation nucleus (commencing division)

FIG. 2–11.—The process of fertilization in the ovum of a mouse. (After Sobotta.)

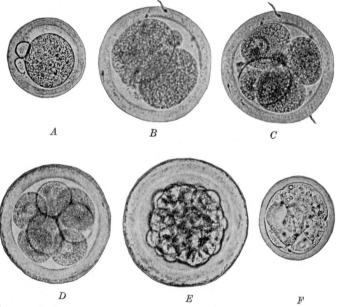

A   B   C

D   E   F

FIG. 2–12.—Photographs of living eggs, × 200. (A) Fertile one-cell mouse egg showing the zona pellucida, perivitelline space, vitellus, first and second polar bodies (Lewis and Wright). (B) Two-cell monkey egg. (C) Four-cell stage of same egg, extra sperm in zona (Lewis and Hartman). (D) Eight-cell rabbit egg. (E) Rabbit morula (Gregory). (F) Early blastocyst, seventy-six-hour mouse egg with a small amount of fluid.

to show signs of degeneration and no longer be fertilizable.

Fertilization takes place when a spermatozoön enters the ovum (Fig. 2–11). Normally only one sperm takes part in the process; its entrance causes the peripheral layer of the ovum to change into the vitelline membrane which prevents the entrance of additional sperm. Once the spermatozoön has penetrated the ovum, the tail portion separates, and the head and middle piece expand into the **male pronucleus** and the **centrosome.** The nucleus of the ovum, known as the **female pronucleus,** and the male pronucleus migrate toward each other, and, breaking down into their constituent chromosomes near the center of the cell, are combined into the new **segmentation nucleus.** The mitotic division of this nucleus and the cleavage of the cell produce the **two-celled stage** which is the beginning of the embryological development of the individual.

DEVELOPMENT OF THE FERTILIZED OVUM

These first two cells, known as **blastomeres** (Fig. 2–12B), soon undergo division, and the process of **cleavage** or **segmentation** continues until a grape-like cluster of daughter cells, the **morula,** has been produced. During segmentation the ovum does not enlarge; the morula is about the same size as the single-celled ovum. The exact timing is not known but probably two or three days are required for these five or six cleavage divisions, and while they are taking place the ovum makes its passage through the uterine tube.

Morula.—Although there is no visible difference between the early daughter cells, a segregation of materials and potencies takes place during cleavage which first becomes evident in the morula. Two parts become recognizable, an outer layer of cells known as the **trophoblast,** and an inner cluster of cells, known as the **inner cell mass** or **embryoblast.**

The **blastocyst** (Fig. 2–13) is formed by the secretion of fluid into the interior of the morula by the trophoblast, and as it enlarges into a vesicle, the trophoblast cells multiply, spread out, and become flattened against the zona pellucida which, in turn, is stretched into a thin membrane and disappears. The inner cell mass remains as a relatively small cluster of cells attached to the inner surface of the trophoblast at a region which is known as the **animal pole.** The blastocyst stage is reached soon after the ovum enters the uterus and it remains free in the uterine cavity for an estimated three to five days.

IMPLANTATION

On the seventh or eighth day after fertilization, the zona pellucida has disappeared and the blastocyst comes into direct contact with the uterine mucosa or endometrium and adheres to it. The trophoblast works its way into the endometrium by digesting the uterine tissue and on the eighth to tenth day has become a thickened invasive mass buried in the mucosa with a thin wall of

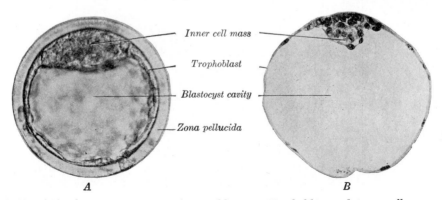

Inner cell mass

Trophoblast

Blastocyst cavity

Zona pellucida

A   *B*

Fig. 2–13.—(*A*) Blastocyst, ninety-two-hour rabbit egg. Trophoblast and inner cell mass. × 200 (Gregory). (*B*) Blastocyst, nine-day monkey egg. Stained section. The zona has disappeared, the trophoblast is thin, and the inner cell mass small. × 200. (Carnegie collection.)

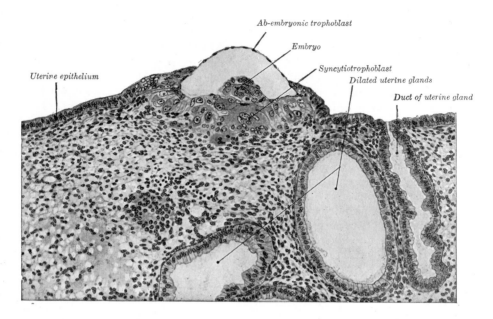

FIG. 2–14.—A human ovum (Carnegie 8020), fertilization age seven to seven and one-half days, in process of embedding in the uterine mucosa. In the actual specimen the abembryonic trophoblast had collapsed on the inner cell mass, but for the purposes of clarity it is shown projecting into the uterine cavity. Drawing from a photomicrograph. × 150. (Hertig and Rock, Am. J. Obst. & Gynec.)

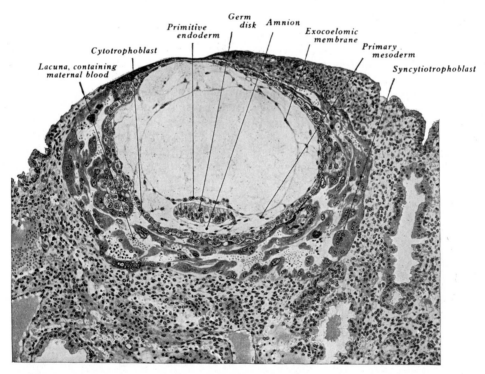

FIG. 2–15.—A human ovum (Carnegie 7700), fertilization age twelve to twelve and one-half days, embedded in the stratum compactum of the endometrium. Drawing from a photomicrograph. × 105. (Hertig and Rock.)

blastocyst still protruding into the uterine cavity (Fig. 2–14).

By the twelfth day the embryo has become completely buried and the defect in the endometrium has been closed over by the uterine epithelium (Fig. 2–15). The trophoblast has formed a spongy mass which has broken into the walls of some of the maternal vessels and the strands of cells are bathed in maternal blood. The trophoblast continues to grow rapidly, later combining with mesoderm to form the **chorion,** the extraembryonic membrane which protects the embryo and makes contact with the maternal blood for the absorption of oxygen and food substances, and for the elimination of waste.

Germ Layers.—While the blastocyst is becoming embedded in the uterine mucosa, the inner cell mass proliferates rapidly and within its cluster two groups of cells become distinguishable. The group nearest to the animal pole is the outer germ layer or **ectoderm.** The group protruding into the blastocyst cavity is the inner germ layer or **endoderm.** In the interior of each group, the cells pull away from each other and the area of separation quickly expands into a cavity.

The cells surrounding the cavity within the ectoderm stretch out into a thin-roofed vesicle known as the **amnion,** the cavity becoming the **amniotic cavity.** The cells of the endoderm stretch out into the vesicle which protrudes into the blastocyst cavity, and is known as the **yolk sac,** its cavity being the **yolk sac cavity.**

The parts of the ectoderm and endoderm which are adjacent to each other remain in close contact, sharing in the formation of a thickened plate named the **embryonic** or **germ disk** (Fig. 2–15), from which the embryo proper will be developed. At about this time, scattered cells migrate out from the border of the germ disk, forming a loose network of cellular processes which stretch across the blastocyst cavity and occupy the space between the trophoblast and the yolk sac, helping to hold the latter in place. This

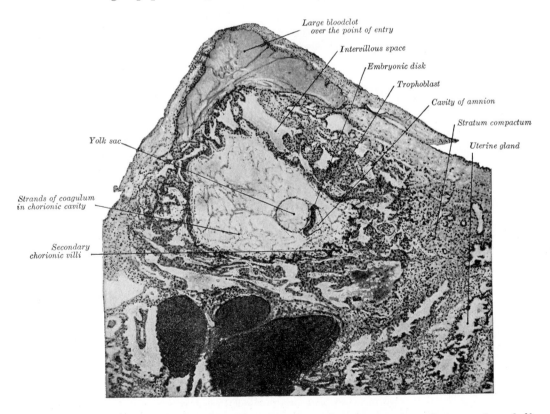

Fig. 2–16.—A human ovum embedded in the stratum compactum. Estimated age thirteen and one-half days. × 35. (Heuser, Hertig and Rock.)

is the first representation of the **middle germ layer** or **mesoderm,** and since it occupies the blastocyst cavity rather than the germ disk it is called the **extraembryonic mesoderm** (Fig. 2–16).

As the blastocyst cavity expands, the entire embryonic mass, including the yolk sac and amnion, is carried away from its close contact with the trophoblast and becomes suspended in the chorionic cavity by a concentration of mesoderm, the **primitive body stalk** (Fig. 2–17).

The **Extraembryonic Coelom.**—The first representation of the body cavity or coelom

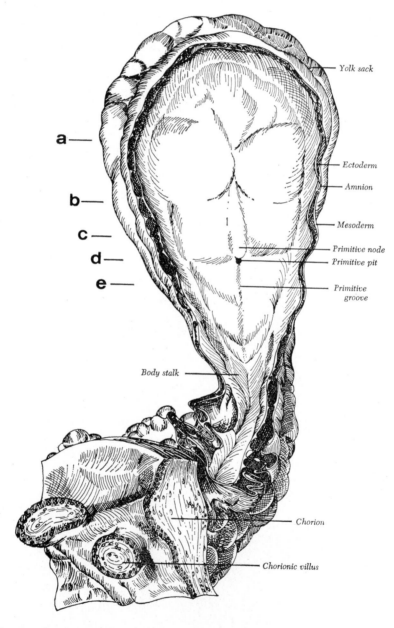

Fig. 2–17.—Dorsal or ectodermal view of a presomite human embryo with a definite chorda canal redrawn from Heuser '32. The letters mark the level of the sections illustrated in Fig. 2–18. × 100.

in the human embryo occurs very early, at a time when the amniotic cavity and yolk sac are still very small and the germ disk has just been established (Fig. 2–15). The extra-embryonic mesoderm, just beginning to form, spreads out in a layer over the amnion and yolk sac, and around the inner surface of the trophoblast, as well as filling in the cavity of the blastocyst with a loose mesh-work (Fig. 2–16, labeled chorionic cavity). Between the layers of mesoderm covering the embryo and lining the blastocyst cavity, the loosely arranged mesodermal cells separate from each other and leave a space, the

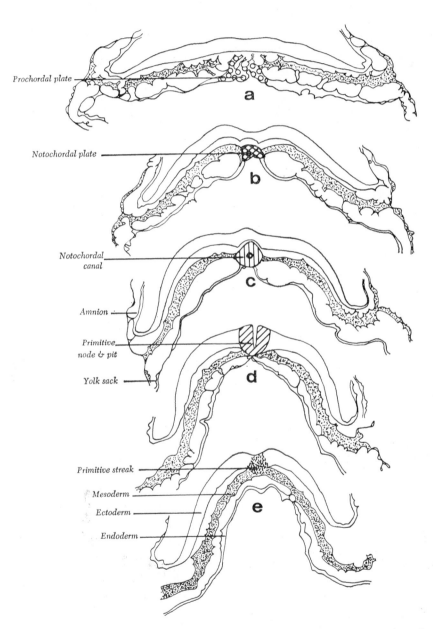

Fig. 2–18.—Outline drawings of sections as marked in Fig. 2–17 to show position of axial structures, ectoderm, endoderm, and mesoderm. × 100. (Heuser '32.)

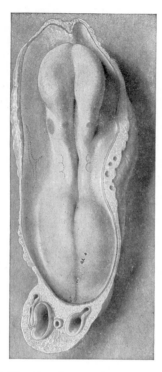

Fig. 2–19.—Dorsal view human embryo 1.38 mm. in length, medullary groove open, about eighteen days old. × 52.5 (Ingalls.)

extraembryonic coelom (Fig. 2–42). The body cavity of the embryo proper is formed somewhat later and is at first independent of this extraembryonic coelom, but, during the period when the yolk sac is being constricted at the body stalk, the two cavities become confluent for a short time (Fig. 2–21).

THE DEVELOPMENT OF THE EMBRYO

Embryonic Disk.—As the thickened layers of the germ disk (Fig. 2–16) grow out into a flat oval plate, they begin the formation of the embryo proper. The narrow end of the disk (Fig. 2–17) is attached at the **body stalk** and represents the caudal end of the embryo. The median line or axis, in the caudal part of the embryonic disk, is marked by the **primitive groove.** Lying along the bottom of the primitive groove is an elongated mass of rapidly proliferating cells known as the **primitive streak.** The cells formed by this rapid proliferation spread out laterally between the ectoderm and en-

doderm as the **definitive mesoderm** of the embryo, part of which migrates forward as far as the cranial portion of the embryo (Fig. 2–18). Between the cranial end of the primitive streak and this anterior growth of mesoderm the endoderm is thickened into a plate known as the **prechordal plate.** The endoderm and ectoderm remain in close contact at this point without the intervention of mesoderm and the **bilaminar area** later becomes the buccopharyngeal membrane of the future stomodeum.

The **Notochord.**—At the cranial end of the primitive streak there appears a knot of ectodermal cells known as the **primitive node** (*Henson's node*) (Fig. 2–17) from which a strand of cells grows cranialward between the ectoderm and endoderm in the midline until it is blocked at the prechordal plate. This is the **notochordal** or **head process.** Since the primitive streak must grow by a caudalward migration when the neural groove lengthens, the growth of the notochordal process must also grow by caudalward migration of the primitive node. During this migration the notochordal cells form a flattened strand or plate which displaces the endoderm in the midline for a short time. The plate soon becomes a groove and then a tube, the **notochordal canal.** The tube separates from the endoderm which then closes the defect, leaving the caudal end of the tube with an opening into the yolk sac cavity. The other end of the canal retains an opening through the ectoderm at the primitive node and is known as the **primitive pit.** The primitive pit is the **blasto-**

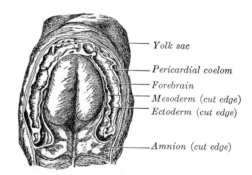

Fig. 2–20.—Human embryo with first somite forming, 1.5 mm in length. Dorsal view with amnion removed and ectoderm and mesoderm cut away to show pericardial cavity. Carnegie collection No. 5080. (Redrawn from Davis '27, courtesy Carnegie Institution.)

Labels on Fig. 2–20:
- Yolk sac
- Pericardial coelom
- Forebrain
- Mesoderm (cut edge)
- Ectoderm (cut edge)
- Amnion (cut edge)

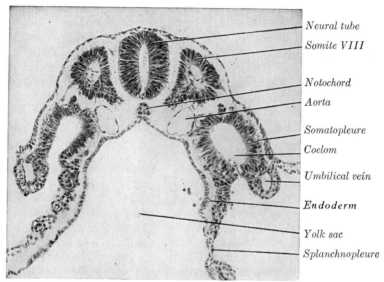

Neural tube
Somite VIII

Notochord
Aorta

Somatopleure
Coelom

Umbilical vein

Endoderm

Yolk sac
Splanchnopleure

FIG. 2–21.—Transverse section through somites VIII (primitive segments) of the 14-somite human embryo shown in Fig. 2–22. × 150. (Heuser.)

pore and corresponds to the structure of that name in the **gastrulation** of lower forms. The notochordal canal with its openings through both ectoderm and endoderm becomes the **neurenteric canal.** These early notochordal structures are transient, lasting only a short time before the somites are formed but the notochord remains in the embryo as a solid rod of cells lying between the neural groove and the endoderm in the midline (Fig. 2–21) as a structural guiding axis for the growing vertebral column. Only remnants of it persist into the adult.

EARLY GROWTH OF THE EMBRYO

Neural Tube.—The broad region of ectoderm in front of the primitive streak in the embryonic disk (Fig. 2–17) becomes thickened and elevated into ridges on either side of the midline forming the **neural folds,** the first evidence of the nervous system. As the primitive streak migrates caudalward, the neural folds and the intervening **neural groove** become elongated (Fig. 2–19). The crests of the folds rise into ridges which gradually curve over the groove until they join, closing off the **neural tube** (Fig. 2–22). The closing of the neural tube occurs first along the early somites in the region of the future hindbrain (Fig. 2–22) and progresses both cranialward and caudalward. Toward

the end of the third week the cranial opening (anterior neuropore) of the neural tube finally closes at the future forebrain. The caudal part of the neural groove remains open for some time and, having a rhomboidal outline, is called the *sinus rhomboidalis.*

As the neural folds close, a narrow angu-

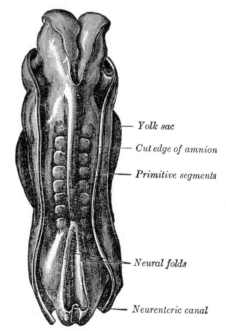

Yolk sac
Cut edge of amnion
Primitive segments

Neural folds

Neurenteric canal

FIG. 2–22.—Dorsum of human embryo, 2.11 mm in length. (After Eternod.)

lar interval or groove is left along the neural tube between it and the primitive body ectoderm (Fig. 2–21). Cells at the top of the neural tube in this junction push out into the groove forming what is known as the **neural crest** or **ganglionic ridge** (Fig. 11–4). These cells migrate out lateralward and ventralward beside the neural tube, eventually forming certain ganglionic nerve and sheath cell components of the cranial, spinal, and autonomic nerves (see Chap. 11).

The cranial end of the neural tube is expanded into a large vesicle with three subdivisions which correspond to the future **forebrain** (*prosencephalon*), **midbrain** (*mesencephalon*), and **hindbrain** (*rhombencephalon*) (Fig. 2–27). The walls develop into the nervous tissue and neuroglia of the brain, and the cavity is modified into the cerebral ventricles. The more caudal portion of the neural tube develops into the spinal cord or medulla spinalis.

Primitive Gut.—The embryo increases rapidly in size once the neural tube is established, but it does so by the overgrowth of the central region of the embryonic disk, while the margin of the disk, or line of junction between the embryo and amnion, stops growing and later even gradually becomes narrower. The constriction thus formed, corresponding in position to the future umbilicus, gradually pinches off the part of the yolk sac included in the disk so that it becomes enclosed in the embryo proper as the **primitive gut** (Fig. 2–24).

The embryo grows more rapidly in length than in width, with the result that the cranial and caudal ends push out beyond the margin of the embryonic disk, and are bent ventrally into **head** and **tail folds** (Figs. 2–23, 2–24). The diverticulum of the primitive gut which occupies the head is known as the **foregut,** and its most cranial extremity, the **buccopharyngeal membrane,** marks the location of the future opening into the mouth, the **stomodeum.** The caudal end of

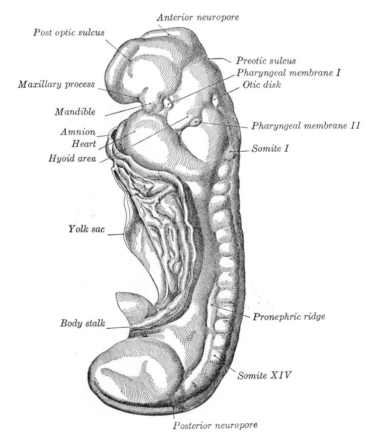

Fig. 2–23.—Lateral view of a 14-somite human embryo. × 50. (Heuser.)

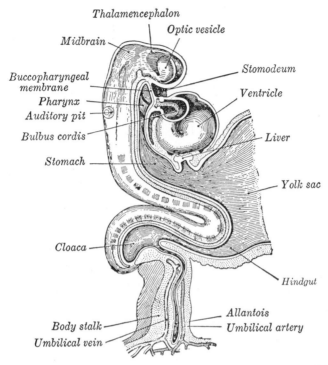

Thalamencephalon

Optic vesicle

Midbrain

Stomodeum

Buccopharyngeal membrane

Ventricle

Pharynx

Auditory pit

Bulbus cordis

Liver

Stomach

Yolk sac

Cloaca

Hindgut

Body stalk

Allantois

Umbilical artery

Umbilical vein

Fig. 2–24.—Human embryo about fifteen days old. Brain and heart represented from right side. Digestive tube and yolk sac in median section. (After His.)

the embryo is at first connected to the chorion by a band of mesoderm called the body stalk, but with the formation of the tail fold, the body stalk is moved into the position of the **umbilicus,** and a diverticulum of the yolk sac called the **hindgut** is formed. For a time the opening into the yolk sac between the fore- and hindgut remains wide, but the communication is gradually reduced to a slender tube known as the **vitelline duct** or **yolk stalk** and the yolk sac itself, being rudimentary, remains a small pear-shaped sac (Figs. 2–24, 16–6B).

MESODERM AND SOMITES

The cells of the mesoderm which migrate forward from the primitive streak on either side of the primitive node and prechordal plate join across the midline in the most cranial portion of the embryo. They form a horseshoe-shaped plate around the cranial ends of the neural folds. At about the time of the formation of the first somite (Fig. 2–20) the cells in the interior of the plate separate from each other leaving small ves-

icles which soon coalesce to form a U-shaped cavity. This is the earliest representative of the **embryonic body cavity** or **coelom** and is destined to be the **pericardial cavity.** The mesoderm of the outer coelomic wall becomes the **parietal pericardium** and combined with the overlying ectoderm is known as the **somatopleure.** The mesodermal cells of the inner wall of the coelom will form the **myocardium** and **epicardium** of the **heart** and combined with the endoderm are called the **splanchnopleure** (Fig. 2–21).

The median portion of the pericardial coelom expands greatly to accommodate the very rapidly enlarging heart. As the rim of the embryonic shield is contracted in pinching off the foregut, the heart and pericardial cavity protrude caudalward from the cranial ends of the neural folds, gradually becoming ventral to them with the foregut in between (Figs. 7–1 to 7–4). Since the heart is the most rapidly growing part of the embryo at this time it forms a prominent **cardiac swelling** between the yolk sac and the head fold of the embryo (Fig. 2–23).

Vascular Mesenchyme.—During the time when the primitive pericardial cavity begins to expand, scattered cells separate from the mesoderm of the pericardial splanchnopleure and gather in the interval between the mesoderm and endoderm (Fig. 2–21). These cells are called **mesenchyme** (*mesenchyma*) or **angioblasts** and represent the primordium of the endothelial lining of the vascular system. They proliferate rapidly and in this cardiac region form a tubular sack, the **primitive endocardium** of the heart. A strand of cells grows out from the endocardium on either side of the cranial end of the foregut to become the primordium of the **first pair** of **aortic arches.** The latter join similar strands of cells lying along the notochord which become the **dorsal aortae** (Figs. 7–1 to 7–4).

At about this time also mesenchymal cells appear between the endoderm of the yolk sac and its covering of mesoderm. They proliferate very rapidly, some gathering into groups to form the solid knots which are called **blood islands;** others form strands which acquire a lumen and become the **yolk sac capillaries.** The capillaries join into larger vessels which become the **vitelline** or **omphalomesenteric veins** leading to the heart (Fig. 2–26).

Blood.—The cells forming the blood islands of the vascular mesenchyme of the yolk sac are at first in compact groups. The fluid within the lumen of the capillaries, the **primitive blood plasma,** gradually separates the individual cells of the group into free cells which are swept into the circulation when the final circuit is established. These are the **primitive hemocytoblasts** and most of them rapidly acquire hemoglobin in their cytoplasm to become the **primitive erythroblasts** and **primitive erythrocytes** or red blood cells.

It is commonly theorized that the vascular endothelium or perhaps even connective tissue cells of the later embryo, fetus, or adult in various parts of the body may revert into hemocytoblasts and produce hemoglobin-containing blood cells. Another theoretical point of view, however, is that the tissue as highly specialized as the hemoglobin-containing cells must logically come from a specific primordium in the embryo. The mechanism of this would be the seeding of undifferentiated descendants of the primitive blood island hemocytoblasts in various tissues such as the liver, spleen, and bone marrow. In favor of this view are the experiments with Amphibian embryos in which the entire blood island was removed surgically. The resulting embryos grew into free-swimming larvae with complete vascular systems including hearts and livers but no erythrocytes. The larvae were kept alive with an increased oxygen supply. They fed normally and the circulation of clear plasma pumped by the heart could be detected by occasional passage of particles and non-erythrocytic cells in the capillaries of the gills (Goss, '28).

The **beginning of circulation** occurs at about the eight-somite stage, during the third to fourth week interval in the human embryo. By this time the vascular lumen has become continuous from the endocardium, through the first aortic arches, aortae, yolk sac capillaries, and vitelline veins. The myocardium has become a muscular tube contracting with its own regular intrinsic rhythm. The cells within the blood islands have become free-swimming hemocytoblasts and are swept into the circulation by the primitive blood plasma, propelled by the pumping of the heart.

During this change, the endocardium retains its close association with the endoderm, occupying the foregut portal, and thus reversing its relation from ventral to the early splanchnic mesoderm in the presomite stage into a position dorsal to the expanded myocardium in the three or four somite stage (Figs. 7–1 to 7–4). This development of the heart differs essentially from that in the much-studied chick in that the heart begins as a median rather than a lateral structure and progresses much more rapidly, both morphologically and physiologically. The precocious development of the heart and circulation is necessary for the survival of the embryo because of the deficiency in stores of nutriment within the ovum (see Chap. 7).

Somites.—As the neural folds increase in size the mesoderm along either side of the notochord expands into a column of cells, the **paraxial mesoderm,** between the neural tissue and ectoderm (Fig. 2–21). Early in the third week the cells of these columns become organized into blocks called **primitive body segments** or **somites** which lie just under the ectoderm, where they are visible in the intact embryo and can be counted (Fig. 2–22). They are responsible for the segmentation of the future muscular, skeletal, and nervous systems. The first somite to differentiate is in the future occipital region, and the separation of new somites progresses in a caudal direction until eventually there are 36 to 38 somites. The number of somites in each part of the human embryo are 4 occipital, 8 cervical, 12 thoracic, 5 lumbar, 5 sacral and 6 or 8 coccygeal. The first occipital and last few coccygeal somites regress and disappear. The other occipital somites are incorporated into structures of the head. Each of the remaining somites differentiates into three parts, a sclerotome, a myotome, and a dermatome.

At first the somite contains a central cavity, called the **myocele,** surrounded by cells in a columnar epithelium-like arrangement (Fig. 2–21). The medial wall soon breaks down into irregular-shaped cells which migrate toward the notochord giving rise to the **sclerotome.** These cells multiply rapidly, surround the notochord, and extend up beside the neural tube to become the primordium of the vertebral column. The cells remaining in the dorsal and ventral parts of the somite next migrate into the region adjacent to the sclerotome, become elongated into myoblasts, and constitute the **myotome** from which the segmentally arranged musculature is developed. The remaining lateral part of the somite also breaks down, its cells migrate close under the ectoderm as the

dermatome which provides the future dermis.

As the somites become blocked out of the medial portion of the paraxial mesoderm, the lateral portion spreads out into two thin sheets, one lying against the ectoderm, the other against the endoderm (Fig. 2–21). The space left between the two sheets is part of the embryonic coelom. The ectoderm with its layer of mesoderm is known as the **somatopleure** which forms the skin, muscles, bones, and fascias of the body wall. The endoderm with its layer of mesoderm is the **splanchnopleure** which forms the serous membranes of the future body cavities. At the junction between the somites and the lateral mesoderm a special strand of cells appears which is known as the **intermediate cell mass.** It is the source of the future genitourinary system.

DEVELOPMENT OF THE BODY CAVITIES

The **embryonic coelom** originates in the mesoderm which is the earliest product of the primitive streak and occupies the most cranial area of the embryonic disk as described above. It begins with the formation of scattered vesicles in the layer of mesoderm at the periphery of the embryonic disk encircling the cranial end of the neural folds (Fig. 2–20). The vesicles quickly coalesce into a horseshoe-shaped cavity which is continuous across the midline in front of the

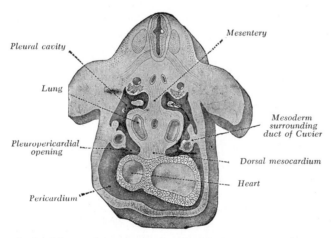

Fig. 2–25.—Figure obtained by combining several successive sections of a human embryo of about the fourth week. (From Kollmann.) The upper arrow is in the pleuroperitoneal opening, the lower in the pleuropericardial.

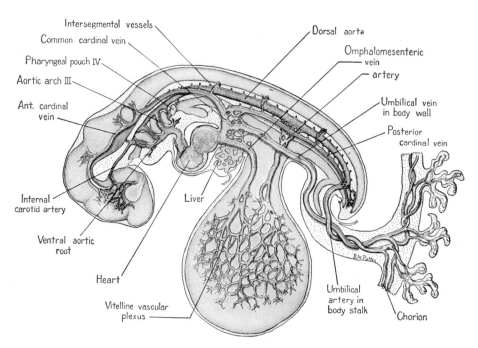

Fig. 2–26.—Semischematic diagram to show basic vascular plan of human embryo at end of first month. For the sake of simplicity the paired vessels are shown only on side toward observer. (Patten's *Human Embryology,* courtesy of the Blakiston Company.)

forebrain but is at first closed caudally in the region of the early somites. This cavity is the primitive pericardial portion of the primitive coelomic cavity. The two lateral ends of the U-shaped cavity progress caudally, and in a short time become confluent with the extraembryonic coelom at each side of the body of the embryo (Fig. 2–21).

As the pericardial cavity is moved into its position ventral to the foregut by the formation of the head fold, a thick plate of splanchnic mesoderm comes to lie between the greatly enlarged heart and the constricted portion of the yolk sac. This plate is known as the **septum transversum.** Through it the vitelline veins from the yolk sac reach the heart and into it grows the diverticulum from the gut which forms the liver (Fig. 2–26). The more caudal parts of the coelom remain as narrow extensions on each side of the foregut in the region of the somites. This narrowed portion of the coelom dorsal to the septum transversum is called the **pleural canal** because a little later the diverticula from the foregut which develop into the lungs push out into this canal

and expand it into the pleural cavities (Fig. 2–25).

At this early stage, the coelomic cavity is continuous from the pericardial cavity around the heart, through the pleural canals around the lung buds, into the peritoneal cavity in the abdomen. Later, it is broken up into the pericardial, pleural, and peritoneal cavities by the formation of two septa. One septum begins as a fold of tissue from the cranial border of the septum transversum and the lateral body wall; it protrudes into the pleural canal cranial to the lung bud and is known as the **pleuropericardial fold.** It continues to grow out until it completely closes the canal as a septum, separating the pericardial cavity from the pleural cavities (Fig. 2–25).

The second fold appears at the lower part of the septum transversum in the region where the common cardinal veins (*ducts of Cuvier*) open into the sinus venosus at the base of the heart. This fold, known as the **pleuroperitoneal fold,** grows dorsally from the septum transversum, and, as the growing lung buds expand the thoracic cavities at both sides of the heart, it contributes an important part of the tissue which forms the diaphragm. Eventually the diaphragm closes off the coelom and separates the pleural cavities from the peritoneal cavity.

THE BRANCHIAL REGION

The Branchial or Visceral Arches and Pharyngeal Pouches.—In the lateral walls of the anterior part of the foregut five *pharyngeal pouches* appear (Fig. 2–26); each of the upper four pouches is prolonged into a dorsal and a ventral diverticulum. Over these pouches corresponding indentations of the ectoderm occur, forming what are known as the **branchial** or **outer pharyngeal grooves** (Fig. 2–27). The intervening mesoderm is pressed aside and the ectoderm comes for a time into contact with the endodermal lining of the foregut, and the two layers unite along the floors of the grooves to form thin **closing membranes** between the foregut and the exterior. Later the mesoderm again penetrates between the endoderm and the ectoderm. In gill-bearing animals the closing membranes disappear, and the grooves become complete clefts, the **gill clefts**, opening from the pharynx out to the exterior; perforation, however, does not occur in birds or mam-

mals. The grooves separate a series of rounded bars or arches, the **branchial** or **visceral arches,** in which thickening of the mesoderm takes place (Figs. 2–26, 2–27), in order to contain the aortic arches (Figs. 8–2 and 8–4). The dorsal ends of these arches are attached to the sides of the head, while the ventral extremities ultimately meet in the middle line of the neck. In all, six arches make their appearance, but of these only the first four are visible externally. The first arch is named the mandibular, and the

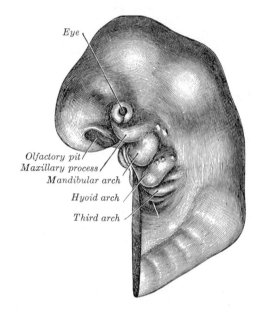

FIG. 2–28.—Head end of human embryo, about the end of the fourth week. (From model by Peter.)

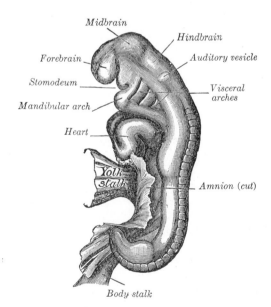

FIG. 2–27.—Embryo of the fourth week showing pharyngeal grooves between visceral arches. (His.)

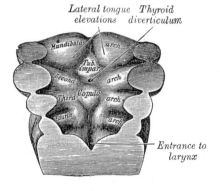

FIG. 2–29.—Floor of pharynx of the embryo in Fig. 2–27 showing the pharyngeal pouches.

second the hyoid; the others have no distinctive names. The mandibular and hyoid arches are the first ones to appear and are recognizable in the 14-somite stage (Fig. 2–23). In each arch a cartilaginous bar, consisting of right and left halves, is developed, and with each of these there is one of the primitive aortic arches.

The **mandibular arch** lies between the first branchial groove and the stomodeum; from it are developed the lower lip, the mandible, the muscles of mastication, and the anterior part of the tongue. Its cartilaginous bar is formed by what are known as **Meckel's cartilages** (right and left) (Figs. 2–30, 4–6); above this the incus is developed. The dorsal end of each cartilage is connected with the ear capsule and is ossified to form the malleus; the ventral ends meet each other in the region of the symphysis menti. Most of the cartilage disappears; the portion immediately adjacent to the malleus is replaced by fibrous membrane, which constitutes the sphenomandibular ligament, while from the connective tissue covering the remainder of the cartilage the greater part of the mandible is ossified. From the dorsal ends of the mandibular arch a triangular process, the **maxillary process**, grows forward on either side and forms the cheek and lateral part of the upper lip. The **second** or **hyoid arch** assists in forming the side and front of the neck. From its cartilage are developed the styloid process, stylohyoid ligament, and lesser cornu of the hyoid bone. The stapes probably arises in the upper part of this arch. The cartilage of the **third arch** gives origin to the greater cornu of the hyoid bone. The ventral ends of the second and third arches unite with those of the opposite side, and form a transverse band, from which the body of the hyoid bone and the posterior part of the tongue are developed. The ventral portions of the cartilages of the **fourth arch** unite to form the thyroid cartilage; from the cartilages of the **fifth arch** the cricoid and arytenoid cartilages are developed. The mandibular and hyoid arches grow more rapidly than those behind them, with the result that the latter become, to a certain extent, telescoped within the former, and a deep depression, the **sinus cervicalis**, is formed on either side of the neck. This sinus is bounded in front by the hyoid arch, and behind by the thoracic wall; it is ultimately obliterated by the fusion of its walls.

From the first **branchial groove** the concha auriculae and external acoustic meatus are developed, while around the groove there appear, on the mandibular and hyoid arches, a number of swellings from which the auricula or pinna is formed (Fig. 2–39). The first pharyngeal pouch is prolonged dorsally to form the auditory tube and the tympanic cavity; the closing membrane between the

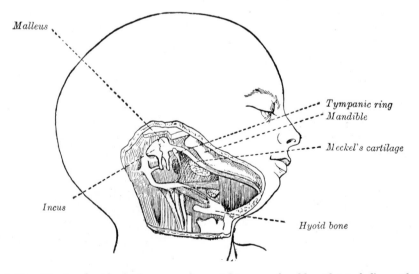

Fig. 2–30.—Head and neck of a human embryo eighteen weeks old, with Meckel's cartilage and hyoid bar exposed. (After Kölliker.)

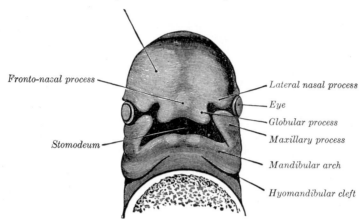

Membranous capsule over cerebral hemisphere

Fronto-nasal process

Lateral nasal process

Eye

Globular process

Maxillary process

Stomodeum

Mandibular arch

Hyomandibular cleft

FIG. 2–31.—Undersurface of the head of a human embryo about twenty-nine days old. (After His.)

mandibular and hyoid arches is invaded by mesoderm, and forms the tympanic membrane. No traces of the second, third, and fourth branchial grooves persist. The inner part of the second pharyngeal pouch is named the **sinus tonsillaris;** in it the tonsil is developed, above which a trace of the sinus persists as the supratonsillar fossa. The fossa of Rosenmüller or lateral recess of the pharynx is regarded by some as a persistent part of the second pharyngeal pouch, but it is probably developed as a secondary formation. From the third pharyngeal pouch the thymus arises as an endodermal diverticulum on either side, and from the fourth pouches small diverticula project and become incorporated with the thymus, but in man these diverticula probably never form true thymus tissue. The parathyroids also arise as diverticula from the third and fourth pouches. From the fifth pouches the ultimobranchial bodies originate and are enveloped by the lateral prolongations of the median thyroid rudiment; they do not, however, form true thyroid tissue, nor are any traces of them found in the human adult (see: Development of Thyroid Gland, Chap. 18).

The Nose and Face.—During the third week two areas of thickened ectoderm, the **olfactory areas,** appear immediately under the forebrain in the anterior wall of the stomodeum, one on either side of a region termed the **frontonasal process** (Fig. 2–31). By the upgrowth of the surrounding parts these areas are converted into pits, the **olfactory pits,** which indent the frontonasal process and divide it into a **medial** and two **lateral nasal processes** (Fig. 2–32). The rounded lateral angles of the medial process constitute the **globular processes** (*His*). The olfactory pits form the rudiments of the nasal cavities, and from their ectodermal lining the epithelium of the nasal cavities, with the exception of that of the inferior meatuses, is derived. The globular processes are prolonged

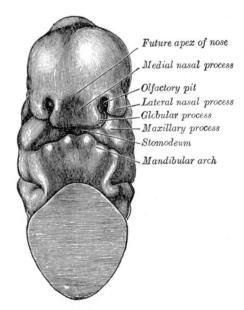

Future apex of nose

Medial nasal process

Olfactory pit

Lateral nasal process

Globular process

Maxillary process

Stomodeum

Mandibular arch

FIG. 2–32.—Head end of human embryo of about thirty to thirty-one days. (From model by Peters.)

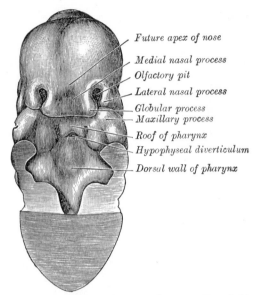

Future apex of nose
Medial nasal process
Olfactory pit
Lateral nasal process
Globular process
Maxillary process
Roof of pharynx
Hypophyseal diverticulum
Dorsal wall of pharynx

Fig. 2–33.—Same embryo as shown in Fig. 2–32, with front wall of pharynx removed.

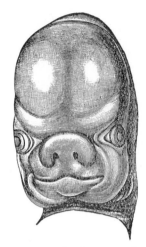

Fig. 2–34.—Head of a human embryo of about eight weeks, in which the nose and mouth are formed. (His.)

backward as plates, termed the **nasal laminae:** these laminae are at first some distance apart, but, gradually approaching, they ultimately fuse and form the nasal septum; the processes themselves meet in the middle line, and form the premaxillae and the philtrum or central part of the upper lip (Fig. 2–35). The depressed part of the medial nasal process between the globular processes forms the lower part of the nasal septum or **columella,** while above this is seen a prominent angle, which becomes the future apex (Figs. 2–32, 2–33), and still higher a flat area, the future bridge, of the nose. The lateral nasal processes form the alae of the nose.

Continuous with the dorsal end of the mandibular arch, and growing forward from its cranial border, is a triangular process, the **maxillary process,** the ventral extremity of which is separated from the mandibular arch by a > shaped notch (Fig. 2–31). The maxillary process forms the lateral wall and floor of the orbit, and in it are ossified the zygomatic bone and the greater part of the maxilla; it meets with the lateral nasal process, from which, however, it is separated for a time by a groove, the **naso-optic furrow,** that extends from the furrow encircling the eyeball to the olfactory pit. The maxillary processes ultimately fuse

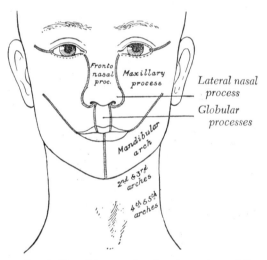

Fronto nasal proc.
Maxillary process
Lateral nasal process
Globular processes
Mandibular arch
2nd & 3rd arches
4th & 5th arches

Fig. 2–35.—Diagram showing the regions of the adult face and neck related to the fronto-nasal process and the branchial arches.

with the lateral nasal and globular processes, and form the lateral parts of the upper lip and the posterior boundaries of the nares (Figs. 2–34, 2–35). From the third to the fifth month the nares are filled by masses of epithelium, on the breaking down and disappearance of which the permanent openings are produced. The maxillary process also gives rise to the lower portion of the lateral wall of the nasal cavity. The roof of the nose and the remaining parts of the lateral wall, viz., the ethmoidal labyrinth, the inferior nasal concha, the lateral cartilage, and the lateral crus of the alar cartilage, are developed in the lateral nasal process. By the fusion of the maxillary and nasal processes in the roof of the stomodeum the **primitive palate** (Fig. 2–36) is formed, and the olfactory pits extend backward above it. The posterior end of each pit is closed by an epithelial membrane, the **bucconasal membrane**, formed by the apposition of the nasal and stomodeal epithelium. By the rupture of these membranes the **primitive choanae** or openings between

the olfactory pits and the stomodeum are established.

The floor of the nasal cavity is completed by the development of a pair of shelf-like **palatine processes** which extend medialward from the maxillary processes (Figs. 2–36, 2–37); these coalesce with each other in the middle line, and constitute the entire palate, except a small part in front which is formed by the premaxillary bones. Two apertures persist for a time between the palatine processes and the premaxillae and

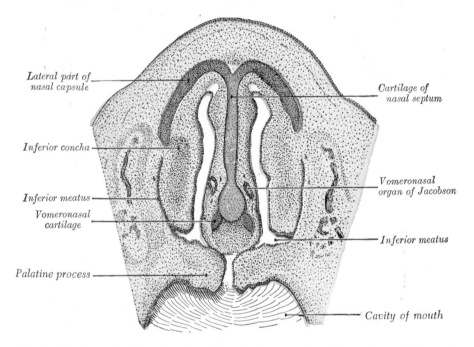

Fig. 2–36.—Frontal section of nasal cavities of a human embryo 28 mm long. (Kollmann.)

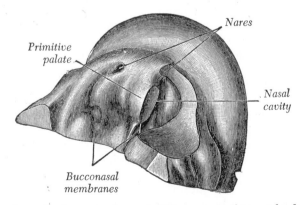

Fig. 2–37.—Primitive palate of a human embryo of thirty-seven to thirty-eight days. (From model by Peters.) On the left side the lateral wall of the nasal cavity has been removed.

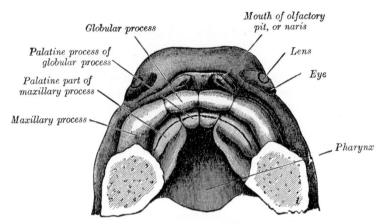

FIG. 2–38.—The roof of the mouth of a human embryo, aged about two and a half months, showing the mode of formation of the palate. (His.)

represent the permanent channels which in the lower animals connect the nose and mouth. The union of the parts which form the palate commences in front, the pre-maxillary and palatine processes joining in the eighth week, while the region of the future hard palate is completed by the ninth, and that of the soft palate by the eleventh week. By the completion of the palate the **permanent choanae** are formed and are situated a considerable distance behind the primitive choanae. The deformity known as cleft palate results from a non-union of the palatine processes, and that of hare lip through a non-union of the maxillary and globular processes (see above).

The nasal cavity becomes divided by a vertical septum, which extends downward and backward from the medial nasal process and nasal laminae, and unites below with the palatine processes. Into this septum a plate of cartilage extends from the under aspect of the ethmoid plate of the chondro-cranium. The anterior part of this cartilaginous plate persists as the septal cartilage of the nose and the medial crus of the alar cartilage, but the posterior and upper parts are replaced by the vomer and perpendicular plate of the ethmoid. On either side of the nasal septum, at its lower and anterior part, the ectoderm is invaginated to form a blind pouch or diverticulum, which extends backward and upward into the nasal septum and is supported by a curved plate of cartilage. These pouches form the

rudiments of the **vomero-nasal organs** of Jacobson, which open below, close to the junction of the premaxillary and maxillary bones (Fig. 2–36).

THE LIMBS

The limbs begin to make their appearance in the fourth week as small elevations or buds at the side of the trunk (Figs. 2–27, 2–50). Unsegmented somatic meso-derm pushes into the limb buds and multiplies by division of its cells into closely packed cellular masses. The intrinsic mus-cles of the limbs differentiate *in situ* from the peripheral portions of this unsegmented mesoderm. The upper limb begins its differ-entiation in the neck region and receives its nerve supply from the fourth cervical to the second thoracic before it migrates caudally. The lower limb arises in the region from the twelfth thoracic to the fourth sacral inclusive receiving nerves from these segments before its caudal migration. The axial part of the mesoderm of the limb bud becomes condensed and converted into its cartilaginous skeleton, and by the ossifi-cation of this the bones of the limbs are formed. By the sixth week the three chief divisions of the limbs are marked off by furrows—the upper into arm, forearm, and hand; the lower into thigh, leg, and foot (Fig. 2–39). The limbs are at first directed backward nearly parallel to the long axis of the trunk, and each presents two surfaces and two borders. Of the surfaces, one—the

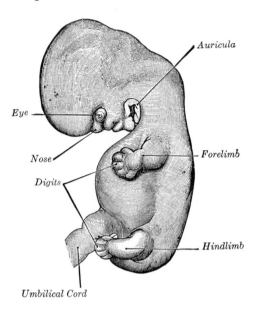

FIG. 2–39.—Embryo of about six weeks. (His.)

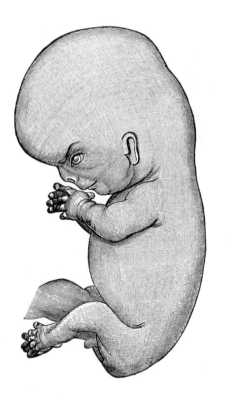

FIG. 2–40.—Human embryo about eight and a half weeks old. (His.)

future *flexor* surface of the limb—is directed ventrally; the other, the *extensor* surface, dorsally; one border, the *preaxial*, looks forward toward the cranial end of the embryo, and the other, the *postaxial*, backward toward the caudal end. The lateral epicondyle of the humerus, the radius, and the thumb lie along the preaxial border of the upper limb; and the medial epicondyle of the femur, the tibia, and the great toe along the corresponding border of the lower limb. The preaxial part is derived from the anterior segments, the postaxial from the posterior segments of the limb bud; and this explains, to a large extent, the innervation of the adult limb, the nerves of the more anterior segments being distributed along the preaxial (radial or tibial), and those of the more posterior along the postaxial (ulnar or fibular) border of the limb. The limbs next undergo a rotation through an angle of 90° around their long axes the rotation being effected almost entirely at the limb girdles. In the upper limb the rotation is outward and forward; in the lower limb, inward and backward. As a consequence of this rotation the preaxial (radial) border of the forelimb is directed lateralward, and the preaxial (tibial) border of the hindlimb is directed medialward; thus the flexor surface of the forelimb is turned forward, and that of the hindlimb backward.

FETAL MEMBRANES

In the early blastocyst stage, the trophoblast consists of a single layer of cells. As the blastocyst becomes embedded in the endometrium, two layers are formed in the trophoblast; the outer layer, rich in nuclei but with no evident cell boundaries, is called the **syncytiotrophoblast**; the inner layer is cellular and is called the **cytotrophoblast** (layer of Langhans) (Fig. 2–51). The trophoblast, as it invades the uterine wall, becomes converted into a thick sponge-like mass of strands and sheets with cores of cytotrophoblast and outer coverings of syncytiotrophoblast. The walls of the uterine vessels which are in the path become eroded and maternal blood seeps into the spaces between the strands of trophoblast, now called the **intervillous space** (Fig. 2–52).

Chorion.—As some of the strands of trophoblast become anchored to the uterine tissue and others protrude into the intervillous space as primary villi, the resulting structure is known as the chorion (Figs. 2–42, 2–51). At first its internal cavity, developed from the blastocyst cavity, contains only fluid and loose strands of extraembryonic mesoderm, but later the amnion expands so rapidly that it encroaches upon the space until it comes into contact with the inner wall of the chorion and obliterates its cavity. The chorion continues to expand throughout pregnancy, however, to accommodate the fetus and serve as the outer barrier between it and the uterus.

grow and ramify, the ingrowth of mesoderm and umbilical vessels keeps pace and the structural pattern is established for the exchange of substances between the fetal and maternal circulation (Fig. 2–52).

Until about the end of the second month, the villi sprout from the entire outer surface of the chorion and are more or less uniform in size. After this time the villi nearest the body stalk of the embryo (Fig. 2–43) grow more exuberantly than the rest, becoming the fetal portion of the placenta. This is called the **chorion frondosum** and the portion of the endometrium to which it is attached is known as the **decidua basalis** (see below). The villi in the rest or non-placental

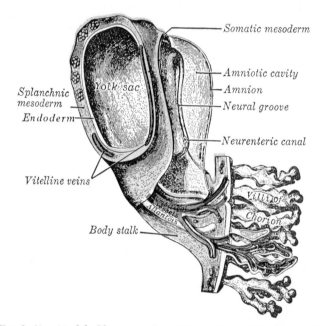

Fig. 2–41.—Model of human embryo 1.3 mm long. (After Eternod.)

The Chorionic Villi.—The **primary villi,** as mentioned above, develop from the solid strands of trophoblast which extend into the uterine wall and consist of an inner cluster of cytotrophoblast covered with an irregular layer of syncytiotrophoblast (Fig. 2–51). The primary villi are converted into the **secondary** or **true chorionic villi** by the ingrowth of a central core of mesoderm accompanied by sprouts from the umbilical vessels which have grown out into the extraembryonic mesoderm by way of the body stalk. As the chorionic villi continue to

chorion gradually atrophy and by the end of the fourth month scarcely a trace of them is left. The resulting smooth surface has given this part of the chorion the name **chorion laeve.**

Maternal Tissues.—By the time the blastocyst comes into contact with the uterus, the endometrium has undergone the necessary changes in preparation for its arrival. These changes are also called premenstrual because the tissue is lost in the event that a fertilized ovum does not become implanted. After the implantation of an ovum

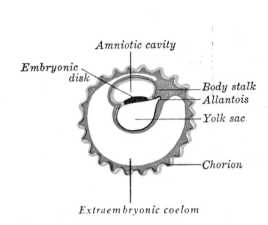

FIG. 2–42.—Diagram illustrating early formation of allantois and differentiation of body stalk.

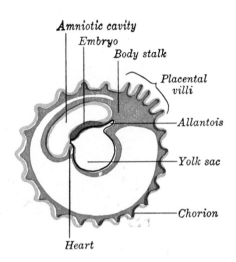

FIG. 2–43.—Diagram showing later stage of allantoic development with commencing constriction of the yolk sac.

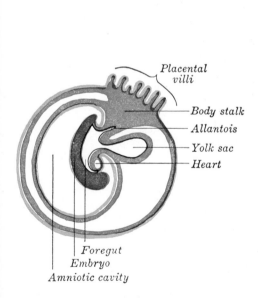

FIG. 2–44.—Diagram showing the expansion of amnion and constriction of the yolk sac.

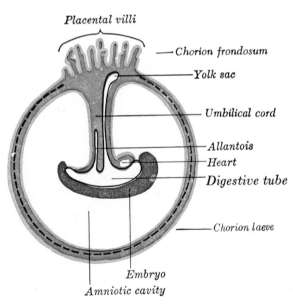

FIG. 2–45.—Diagram illustrating a later stage in the development of the umbilical cord.

Chorionic cavity

Decidua capsularis

Villi

Embryo

Uterine gland

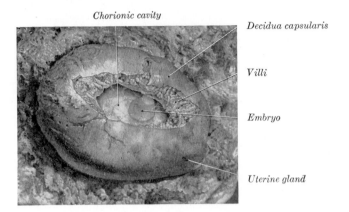

Fig. 2–46.—Human ovum, about fourteen days old, embedded in the wall of the uterus. A window has been cut through the decidua capsularis (reflexa) and the chorion to show the embryo with its large yolk sac and amnion. × 4.5 diameter. (Carnegie collection.)

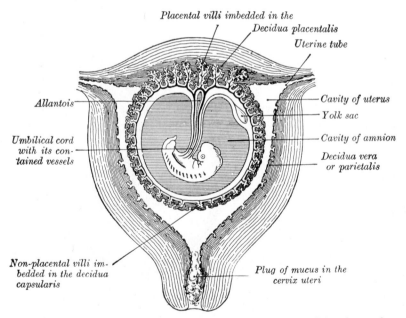

Placental villi imbedded in the
Decidua placentalis
Uterine tube

Allantois

Cavity of uterus

Yolk sac

Umbilical cord
with its con-
tained vessels

Cavity of amnion

Decidua vera
or parietalis

Non-placental villi im-
bedded in the decidua
capsularis

Plug of mucus in the
cervix uteri

Fig. 2–47.—Sectional plan of the gravid uterus in the third and fourth month.
(Modified from Wagner.)

takes place, however, the thickness and vascularity of the endometrium are greatly increased; its glands are elongated with funnel-shaped orifices at the surface, and their deeper portions are dilated into irregular, tortuous spaces. These changes are well advanced by the second month of pregnancy and the tissue between the glands has become crowded with enlarged connective tissue cells known as **decidual cells.** The **endometrium** consists of the following strata (Fig. 2–48): (a) the **stratum compactum,** next to the free surface, is traversed by the necks of the uterine glands and contains a rather compact layer of interglandular tissue; (b) the **stratum spongiosum** contains the tortuous and dilated glands and only a small amount of interglandular tissue; (c) the **stratum basale,** or boundary layer next to the uterine muscular wall, contains the deepest parts of the uterine glands. At each menstrual flow and at the termination of pregnancy the stratum compactum and spongiosum are cast off, and together are known as the **pars functionalis** of the endometrium. The stratum basale remains and is the source of epithelium for the regeneration of the mucous membrane after the functional layer is lost.

The **decidua** is the name given to the pars functionalis of the endometrium during pregnancy, from the Latin word meaning to fall away. The various parts of the decidua are given special names according to their relation to the chorion of the embryo and fetus. The portion which closes over the embryo after it has burrowed into the uterine tissue is the **decidua capsularis** (Fig. 2–46); the deeper part of the endometrium between the embryo and the muscular wall of the uterus is the **decidua basalis,** or, since it is the part involved in the placenta, the **decidua placentalis;** and the part which lines the rest of the uterus is the **decidua parietalis** or the **decidua vera.**

During the growth of the embryo, the decidua capsularis is stretched out but is not broken, and it gradually fills in the cavity of the uterus until it comes in contact with the decidua vera and, by the third month, has obliterated the cavity of the uterus. By the fifth month the remnant of the capsularis has practically disappeared and during the succeeding months the decidua vera also undergoes atrophy from the increased pressure. The glands of the stratum compactum are obliterated, and in the spongiosum they are compressed into slit-like fissures with degenerated epithelium. In the stratum basale, the glandular epithelium is retained.

The Amnion.—The amnion in man arises in the ectodermal part of the inner cell mass which lies next to the trophoblast. Fluid is secreted within the cluster of cells, causing the formation of the **amniotic cavity** (Fig.

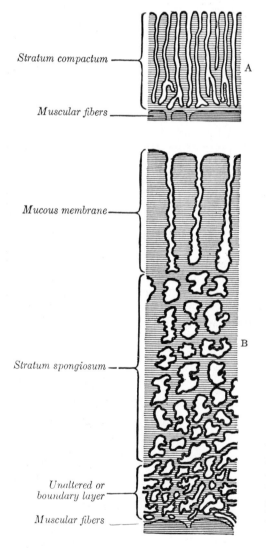

Stratum compactum

Muscular fibers

A

Mucous membrane

B

Stratum spongiosum

Unaltered or boundary layer

Muscular fibers

Fig. 2–48.—Diagrammatic sections of the uterine mucous membrane: A, The non-pregnant uterus. B, The pregnant uterus, showing the thickened mucous membrane and the altered condition of the uterine glands. (Kundrat and Engelmann.)

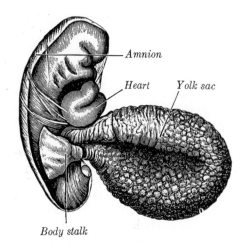

Fig. 2–49.—Human embryo of 2.6 mm. (His.)

2–16). The ectodermal cells in contact with the endoderm are part of the embryonic disk, the other ectodermal cells enclosing the cavity are the amnion. As the amnion increases in size its cells become flat, and the outer surface is separated from the cytotrophoblast (Fig. 2–16) by a layer of extraembryonic mesoderm. About the fourth week, fluid (*liquor amnii*) accumulates and expands the amnion until it eventually comes into contact with the inner wall of the chorionic sac by about the end of the second month and obliterates the primitive blastocyst cavity. It also grows around the body stalk and yolk sac and thus forms the covering of the umbilical cord (Figs. 2–42 to 2–45).

The liquor amnii increases in quantity up to the sixth or seventh month of pregnancy, after which it diminishes somewhat; at the end of pregnancy it amounts to about 1 liter. It allows the fetus free movement and protects it from mechanical injury during the later stages of pregnancy.

The Yolk Sac.—The yolk sac develops within the cluster of endoderm cells in the inner cell mass (Fig. 2–16). As it expands into a vesicle, two parts become distinguishable. The thicker part lying against the ectoderm in the embryonic disk is the future primitive gut of the embryo; the thinner expansion growing out into the cavity of the chorion is the yolk sac. The extraembryonic mesoderm on the outer surface of the endodermal yolk sac becomes a compact layer

which differentiates quite early. During the presomite and early somite stages, the cells of this mesodermal layer proliferate rapidly, producing clusters which are known as **blood islands** and which give the yolk sac a lumpy appearance (Fig. 2–49). The outermost cells of the blood islands gradually become flattened into primitive endothelium; the inner cells become separated from each other by a small amount of primitive blood plasma, and most of them elaborate hemoglobin in their cytoplasm. The blood vessels of the yolk sac remain as the **vitelline plexus**, fed by the vitelline or **omphalomesenteric artery** and drained by the **vitelline** or **omphalomesenteric vein**; but in the human embryo, after it has supplied the blood cells for the beginning of circulation, the yolk sac appears to have no further function and undergoes regression (Fig. 2–50).

At the end of the fourth week the yolk sac presents the appearance of a small pear-shaped vesicle (umbilical vesicle) opening into the digestive tube by the long narrow yolk stalk. As the amnion spreads around the body stalk and over the inner surface of the chorion, the proximal part of the yolk stalk becomes enclosed in the umbilical cord; the distal part extends to the placenta and ends in the yolk sac vesicle which lies between amnion and chorion either on the placenta or a short distance from it (Fig. 2–47). The vesicle can be seen in the **afterbirth** as a small oval body whose diameter

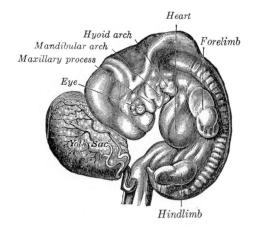

Fig. 2–50.—Human embryo from thirty-one to thirty-four days. (His.)

varies from 1 to 5 mm. As a rule the yolk stalk undergoes complete obliteration during the seventh week, but occasionally it persists within the embryo, and in about 2 per cent of adult bodies is found as a diverticulum from the small intestine, *Meckel's diverticulum,* which is situated about 1 meter proximal to the ileocolic junction and may be attached by a fibrous cord to the abdominal wall at the umbilicus.

The **allantois** (Figs. 2–42 to 2–45) arises as a tubular diverticulum from the posterior part of the yolk sac endoderm. When the hindgut is developed the allantois retains its opening into the terminal part of the hindgut or the cloaca. It grows out into the body stalk as a diverticulum lined by endoderm and covered by mesoderm. With it are carried the **allantoic vessels** which extend into the chorionic mesoderm, eventually forming the **umbilical vessels** and their branches in the chorionic villi of the placenta. In reptiles, birds, and many mammals the allantois expands into a large sac, acquires an elaborate blood supply, and plays an important role in the early nutrition of the embryo. In man and other primates, however, the allantois itself remains rudimentary, but its blood vessels become functionally significant as the umbilical vessels.

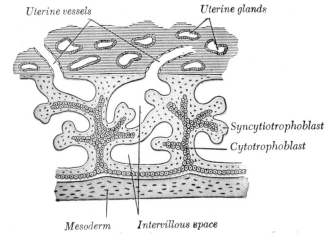

FIG. 2–51.—Primary chorionic villi. Diagrammatic. (Modified from Bryce.)

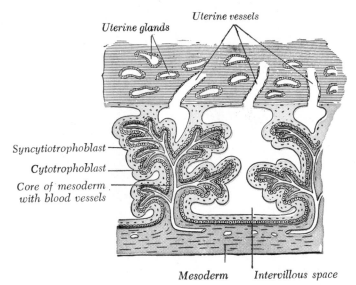

FIG. 2–52.—Secondary or true chorionic villi. Diagrammatic. (Modified from Bryce.)

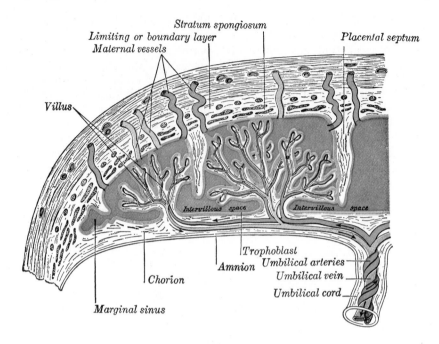

FIG. 2–53.—Scheme of placental circulation.

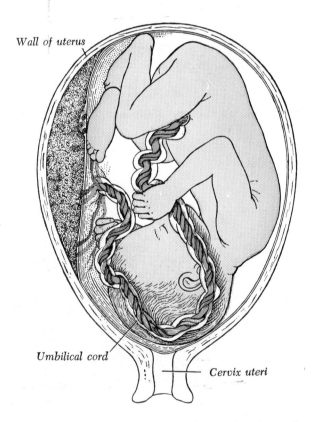

FIG. 2–54.—Fetus in utero, between fifth and sixth months.

The Body Stalk and Umbilical Cord.—
The inner cell mass is at first directly in
contact with the trophoblast. After the
development of the embryonic disk, chorionic mesoderm anchors the embryo to the
inner wall of the chorion. As the embryo
and the chorion grow, this mesodermal
attachment becomes elongated into the
body stalk (Fig. 2–42), which extends from
the posterior end of the embryo to the
chorion, and when the tail fold is formed,
the attachment is moved to the midventral
region of the embryo (Figs. 2–43 to 2–45),
becoming the umbilical cord. The umbilical
blood vessels grow beside the allantois into
the body stalk and ramify as the chorionic
vessels which provide the vascular pathways
to and from the embryo and the chorion.
With the further development of the embryo
and the expansion of the amnion, the body

stalk and yolk sac are enclosed in the umbilical cord. The mesoderm of the body stalk
blends with that of the yolk sac and its
elongated yolk stalk, being converted into
a gelatinous tissue known as *Wharton's jelly*.
The vitelline vessels, together with the right
umbilical vein, undergo atrophy and disappear; and thus the cord, at birth, contains
a pair of umbilical arteries and one (the
left) umbilical vein. It attains a length of
about 50 cm.

THE PLACENTA

The **placenta** is the highly specialized
organ by means of which the fetus makes
its functional contact with the uterine wall.
It has a fetal and a maternal portion.

The **fetal portion** of the placenta consists
of the villi of the chorion frondosum. The

MENSTRUAL AGE WITH MEAN SITTING HEIGHT AND WEIGHT OF FETUS. (STREETER.)

Menstrual age, weeks	Sitting height at end of week, mm	Increment in height		Formalin weight, grams	Increment in weight	
		mm	per cent		grams	per cent
8	23	. . .	. . .	1.1		
9	31	8	26.0	2.7	1.6	59.3
10	40	9	22.5	4.6	1.9	41.3
11	50	10	20.0	7.9	3.3	41.8
12	61	11	18.0	14.2	6.3	44.4
13	74	13	17.6	26.0	11.8	45.4
14	87	13	15.0	45.0	19.0	42.2
15	101	14	14.0	72.0	27.0	37.5
16	116	15	13.0	108.0	36.0	33.3
17	130	14	10.8	150.0	42.0	28.0
18	142	12	8.4	198.0	48.0	24.2
19	153	11	7.2	253.0	55.0	21.7
20	164	11	6.7	316.0	63.0	20.0
21	175	11	6.3	385.0	69.0	18.0
22	186	11	6.0	460.0	75.0	16.3
23	197	11	5.6	542.0	82.0	15.0
24	208	11	5.3	630.0	88.0	14.0
25	218	10	4.6	723.0	93.0	13.0
26	228	10	4.4	823.0	100.0	12.2
27	238	10	4.2	830.0	107.0	11.5
28	247	9	3.6	1045.0	115.0	11.0
29	256	9	3.5	1174.0	129.0	11.0
30	265	9	3.4	1323.0	149.0	11.3
31	274	9	3.3	1492.0	169.0	11.3
32	283	9	3.1	1680.0	188.0	11.2
33	293	10	3.4	1876.0	196.0	10.4
34	302	9	3.0	2074.0	198.0	9.5
35	311	9	3.0	2274.0	200.0	8.8
36	321	10	3.1	2478.0	204.0	8.2
37	331	10	3.0	2690.0	212.0	8.0
38	341	10	3.0	2914.0	224.0	7.7
39	352	11	3.1	3150.0	236.0	7.5
40	362	10	2.8	3405.0	255.0	7.5

greatly ramified villi are suspended in the intervillous space, and are bathed in maternal blood. The branches of the umbilical arteries enter each of the villi and end in capillary plexuses, and these in turn are drained by tributaries of the umbilical vein. Within a villus, the endothelium of the blood vessels is surrounded by a thin layer of gelatinous mesodermal connective tissue and the trophoblast. During the first half of pregnancy the trophoblast consists of two layers, the deeper stratum next to the connective tissue is the cellular cytotrophoblast or Langhans layer, and the superficial layer in contact with the maternal blood, the

space, the latter being lined by the syncytiotrophoblast. The portions of the stratum compactum which persist in the form of septa extend from the basal plate through the thickness of the placenta and subdivide it into the **lobules** or **cotyledons** which are characteristic markings on the detached surface of the placenta seen at parturition.

The fetal blood currents pass through the blood vessels of the placental villi, the maternal blood through the intervillous space (Fig. 2–51). The two currents do not intermingle, being separated from each other by the delicate walls of the villi. Nevertheless, the fetal blood is able to

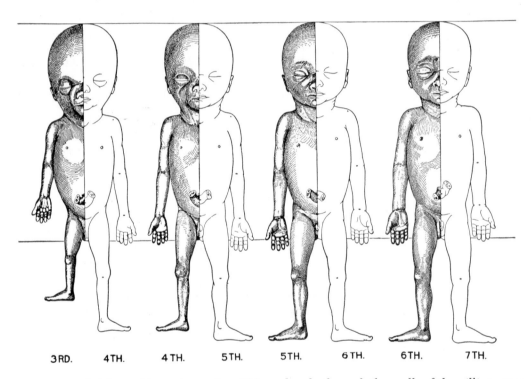

3RD.	4TH.	4TH.	5TH.	5TH.	6TH.	6TH.	7TH.

syncytiotrophoblast. After about the fifth month only a single stratum, that of the syncytiotrophoblast, is present.

The **maternal portion** of the placenta is formed by the pars functionalis of the decidua placentalis. The changes involve the conversion of the greater portion of the stratum compactum and the stratum spongiosum into a **basal plate** and **placental septa** (Fig. 2–51), through which the uterine arteries and veins pass to and from the intervillous space. The endothelial lining of the uterine vessels ceases at the intervillous

absorb, through the walls of the villi, oxygen and nutritive materials from the maternal blood, and give up to the latter its waste products. The blood, so purified, is carried back to the fetus by the umbilical vein.

The placenta is usually attached near the fundus of the uterus, and more frequently on the posterior than on the anterior wall. It may occupy a lower position, however, and, in rare cases, its site is close to the internal os or opening into the cervix, which it may occlude, giving rise to the condition known as placenta previa.

Separation of the Placenta.—After the child is born, the placenta and membranes are expelled from the uterus as the **after-birth.** The separation of the placenta and membranes from the uterine wall takes place through the stratum spongiosum, and necessarily causes rupture of the uterine vessels. The vessels are tightly compressed and closed, however, by the firm contraction of the uterine muscular fibers, and thus postpartum hemorrhage is controlled. During the postpartum period, the epithelial lining of the uterus is restored by the proliferation and extension of the epithelium which lines the persistent portions of the uterine glands in the stratum basale of the decidua.

The expelled placenta is a discoid mass which weighs about 450 gm and has a diameter of from 15 to 20 cm. Its average thickness is about 3 cm, but this diminishes rapidly toward the circumference of the disk, which is continuous with the membranes. Its uterine surface is divided by a series of fissures into lobules or cotyledons, the fissures containing the remains of the septa which extended between the maternal and fetal portions. Most of these septa end in irregular or pointed processes; others, especially those near the edge of the pla-

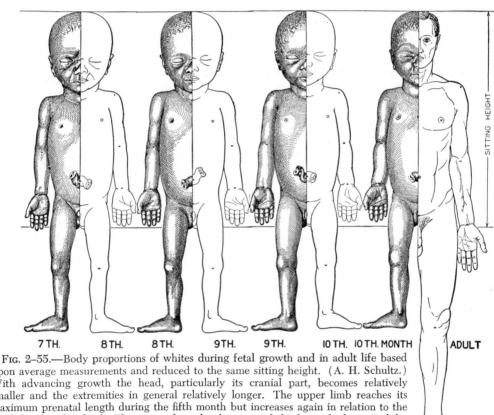

7 TH.   8 TH.   8 TH.   9 TH.   9 TH.   10 TH.   10 TH. MONTH   ADULT

Fig. 2–55.—Body proportions of whites during fetal growth and in adult life based upon average measurements and reduced to the same sitting height. (A. H. Schultz.) With advancing growth the head, particularly its cranial part, becomes relatively smaller and the extremities in general relatively longer. The upper limb reaches its maximum prenatal length during the fifth month but increases again in relation to the trunk height after birth. The tremendous development of the lower limb, typical for man, does not become evident until postnatal growth. The forearm grows faster than the upper arm and the leg faster than the thigh. The width between the hips increases at a more rapid rate than the width between the shoulders. In fetuses the shoulders are relatively higher above the suprasternal notch, the nipples relatively higher on the chest, and the umbilicus relatively lower on the abdominal wall than in adults. With advance in growth the breadth of the head decreases in relation to the head length, the ears become relatively larger, the eyes move closer together, the nose increases in height in relation to the face height, and decreases in width in relation to the face width.

Racial differences in body proportions develop as early as the human form can be recognized. Individual differences *i.e.,* variability, are fully as pronounced in fetal as in adult life. Asymmetries become evident long before birth.

3

centa, pass through its thickness and are attached to the chorion. The fetal surface of the placenta is smooth, being closely invested by the amnion. Seen through the latter, the chorion presents a mottled appearance, consisting of gray, purple, or yellowish areas. The **umbilical cord** is usually attached near the center of the placenta, but may be inserted anywhere between the center and the margin; in some cases it is inserted into the membranes, *i.e.*, the velamentous insertion. From the attachment of the cord the larger branches of the umbilical vessels radiate under the amnion, the veins being deeper and larger than the arteries. The remains of the yolk stalk and yolk sac may be observed beneath the amnion, close to the cord, the former as an attenuated thread, the latter as a minute sac.

THE GROWTH OF THE EMBRYO AND FETUS

First Week.—No fertile human ova of the first week have been examined. From what we know of the monkey and other mammals it seems probable that the egg is fertilized in the upper end of the uterine tube and segments into about eight cells before it passes into the uterus at the end of the third day. In the uterus it continues to segment and develop into a blastocyst with a trophoblast and an inner cell mass.

Second Week.—The blastocyst enlarges, loses its zona pellucida and becomes implanted in the uterine mucosa. The trophoblast, enclosing the blastocyst cavity, develops into an actively invading outer syncytiotrophoblast and inner cytotrophoblast and forms primitive chorionic villi into which first mesoderm and then blood vessels grow. The inner cell mass becomes the embryonic disk, amnion, and yolk sac. The primitive streak differentiates, and mesoderm and notochord are formed.

Third Week.—During the first part of the third week the neural folds appear (Figs. 2–18, 2–40), the allantoic duct begins to develop, the yolk sac enlarges and blood vessels begin to form. Before the end of the week the neural folds begin to unite (Fig. 2–21). The neurenteric canal opens. The primitive segments begin to form. The changes during this week occur with great rapidity.

Fourth Week.—During the fourth week (Figs 2–26, 2–27, 2–53) the neural folds close, the primitive segments increase in number, the branchial arches appear and the connection of the yolk sac with the embryo becomes considerably narrowed so that the embryo assumes a more definite form. The limb buds begin to show and the heart increases greatly in size, producing a prominent bulge in the branchial region.

Fifth Week.—The embryo becomes markedly curved, the head increases greatly in size and the limb buds show segments (Fig. 2–53). The branchial arches undergo profound changes and partly disappear. The superficial nose, eye and ear rudiments become prominent.

Sixth Week.—The curvature of the embryo is further diminished. The branchial grooves—except the first—have disappeared, and the rudiments of the fingers and toes can be recognized (Fig. 2–38).

Seventh and Eighth Weeks.—The flexure of the head is gradually reduced and the neck is somewhat lengthened. The upper lip is completed and the nose is more prominent. The nostrils are directed forward and the palate is not completely developed. The eyelids are present in the shape of folds above and below the eye, and the different parts of the auricula are distinguishable. By the end of the second month the fetus measures from 28 to 30 mm in length (Fig. 2–39).

Third Month.—The head is extended and the neck is lengthened. The eyelids meet and fuse, remaining closed until the end of the sixth month. The limbs are well developed and nails appear on the digits. The external generative organs are so far differentiated that it is possible to distinguish the sex. By the end of this month the length of the fetus is about 7 cm, but if the legs are included it is from 9 to 10 cm.

Fourth Month.—The loop of gut which projected into the umbilical cord is withdrawn within the fetus. The hairs begin to make their appearance. There is a general increase in size so that by the end of the fourth month the fetus is from 12 to 13 cm in length, but if the legs are included it is from 16 to 20 cm.

Fifth Month.—It is during this month that the first movements of the fetus are usually observed. The eruption of hair on the head commences, and the *vernix caseosa* begins to be deposited. By the end of this month the total length of the fetus, including the legs, is from 25 to 27 cm.

Sixth Month.—The body is covered by fine

hairs (*lanugo*) and the deposit of vernix caseosa is considerable. The papillae of the skin are developed and the free border of the nail projects from the corium of the dermis. Measured from vertex to heels, the total length of the fetus at the end of this month is from 30 to 32 cm.

Seventh Month.—The pupillary membrane atrophies and the eyelids are open. The testis descends with the vaginal sac of the peritoneum. From vertex to heels the total length at the end of the seventh month is from 35 to 36 cm. The weight is a little over three pounds.

Eighth Month.—The skin assumes a pink

color and is now entirely coated with vernix caseosa, and the lanugo begins to disappear. Subcutaneous fat has been developed to a considerable extent, and the fetus presents a plump appearance. The total length, *i.e.*, from head to heels at the end of the eighth month is about 40 cm, and the weight varies between four and one-half and five and one-half pounds.

Ninth Month.—The lanugo has largely disappeared from the trunk. The umbilicus is almost in the middle of the body and the testes are in the scrotum. At full time the fetus weighs from six and one-half to eight pounds, and measures from head to heels about 50 cm.

REFERENCES

THE CELL

DALTON, A. J. and M. D. FELIX. 1954. Cytologic and cytochemical characteristics of the Golgi substance of epithelial cells of the epididymis—in situ, in homogenates and after isolation. Amer. J. Anat., 94, 171–207.

FAWCETT, D. W. 1958. The structure of the mammalian spermatozoon. Internat. Rev. Cytol., 7, 195–234.

FAWCETT, D. 1961. Cilia and Flagella. *The Cell: Biochemistry, Physiology, Morphology* (Brachet, J. and A. E. Mirsky, editors). Vol. 2, 217–297. Academic Press, New York.

LEWIS, WARREN H. 1947. Interphase (resting) nuclei, chromosomal vesicles and amitosis. Anat. Rec., 97, 433–445.

LEWIS, W. H. and M. R. LEWIS. 1917. The duration of the various phases of mitosis in the mesenchyme cells of tissue cultures. Anat. Rec., 13, 359–367.

PAINTER, T. S. 1923. Studies in mammalian spermatogenesis. II. The spermatogenesis of man. J. Exp. Zool., 37, 291–336.

PALADE, G. E. 1952. The fine structure of mitochondria. Anat. Rec., 114, 427–452.

PALAY, S. L. 1958. *Frontiers in Cytology*, xiv + 529 pages. Yale University Press, New Haven.

PORTER, K. R. 1953. Observations on a submicroscopic basophilic component of cytoplasm. J. Exp. Med., 97, 727–750.

WHITE, EMIL H. 1964. *Cell Physiology and Biochemistry*. viii + 120 pages, illustrated. Prentice-Hall, Inc., Englewood Cliffs, N.J.

WINCHESTER, A. M. 1958. *Genetics*. 2nd edition. Houghton Mifflin Co., Boston.

ELECTRON MICROSCOPY AND EXPERIMENTAL CYTOLOGY

BARRNETT, R. J. and G. E. PALADE. 1957. Histochemical demonstration of the sites of activity of dehydrogenase systems with the electron microscope. J. Biophys. Biochem. Cytol., 3, 577–588.

FREEMAN, JAMES A. 1964. *Cellular Fine Structure, An Introductory Student Text and Atlas.* ix +198 pages. McGraw Hill Book Company, New York.

HARRISON, R. G. 1910. The development of peripheral nerve fibers in altered surroundings. Roux's Arch. f. Entwicklungsmechanik d. Org., *30*, 15–31.

MATTHEWS, J. L. and J. H. MARTIN. 1971. *Atlas of Human Histology and Ultrastructure.* xiii + 382 pages, 150 plates. Lea & Febiger, Philadelphia.

MORELAND, JAMES EDMOND. 1962. Electron microscopic studies of mitochondria in cardiac and skeletal muscle from hibernated ground squirrels. Anat. Rec., *142*, 155–167.

PORTER, K. R. and M. A. BONNEVILLE. 1968. *Fine Structure of Cells and Tissues.* 3rd edition, vii + 196 pages, illustrated. Lea & Febiger, Philadelphia.

CELL DIVISION, CHROMOSOMES, GENETICS AND EVOLUTION

KINDRED, JAMES E. 1963. The chromosomes of the ovary of the human fetus. Anat. Rec., *147*, 295–311.

LEWIS, W. H. and C. G. HARTMAN, 1933. Early cleavage stages of the egg of the monkey (Macacus rhesus). Carneg. Inst., Contr. Embryol., *24*, 187–201.

McKUSICK, VICTOR A. 1964. *Human Genetics.* xii + 148 pages, illustrated. Prentice-Hall, Inc., Englewood Cliffs, N.J.

OSBORNE, RICHARD H. and FRANCIS V. DE GEORGE. 1959. *Genetic Basis of Morphological Variation, an Evaluation and Application of the Twin Study Method.* xxii + 204 pages. Harvard University Press, Cambridge.

TIJO, J. H. and T. T. PUCK. 1958. The somatic chromosomes of man. Proc. Nat. Acad. Sci., *44*, 1229–1236.

HISTOLOGY

BARGMAN, W. 1964. *Histologie und Mikroscopische Anatomie des Menschen.* xv + 856 pages. Georg Thieme Verlag, Stuttgart.

BLOOM W. and D. W. FAWCETT. 1968. *A Textbook of Histology.* xvi + 858 pages, illustrated. W. B. Saunders Company, Philadelphia.

COPENHAVER, WILFRED M. 1964. *Bailey's Textbook of Histology.* xiii + 679 pages, illustrated. The Williams & Wilkins Company, Baltimore.

FINERTY, J. C. and E. V. COWDRY. 1960. *A Textbook of Histology.* 573 pages, illustrated. Lea & Febiger, Philadelphia.

GREEP, R. O. 1966. *Histology.* 2nd edition, x + 914 pages, 1073 illustrations. The Blakiston Division, McGraw-Hill Book Company, New York.

HAM, ARTHUR WORTH. 1957. *Histology.* xv + 894 pages, illustrated. J. B. Lippincott Company, Philadelphia.

KÖLLIKER, A. 1854. *Manual of Human Histology.* xiv + 498 pages. Sydenham Soc., London.

STÖHR, PHILIPP, WILHELM VON MOLLENDORF, and K. GOERTTLER. 1959. *Lehrbuch der Histologie und der mikroskopischen Anatomie des Menschen.* xvi + 560 pages, illustrated. Gustav Fischer Verlag, Jena.

WINDLE, WILLIAM F. 1960. *Textbook of Histology.* xii + 573 pages, illustrated. McGraw-Hill Book Company, Inc., New York.

EMBRYOLOGY, GROWTH, AND AGEING

ABERCROMBIE, M. and JEAN BRACHET. 1961. *Advances in Morphogenesis.* Vol. *1*, xiii + 445 pages, illustrated. Academic Press, Inc., New York.

ALTMAN, PHILIP L. and DOROTHY S. DITTMER. 1962. *Growth, Including Reproduction and Morphological Development.* xii + 608 pages. Federation of American Societies for Experimental Biology.

BLECHSCHMIDT, E. 1963. *The Human Embryo. Documentations on Kinetic Anatomy.* xiv + 105 pages, 47 plates. Friedrich-Kart Schattauer-Verlag, Stuttgart.

CLARK, R. B. 1964. *Dynamics in Metazoan Evolution—The origin of the coelom and segments.* x + 313 pages. Claredon Press, Oxford.

COMFORT, ALEX. 1964. *Ageing: The Biology of Senescence.* xvi + 365 pages. Holt, Rinehart, & Winston, Inc., New York.

CRELIN, E. S. 1969. *Anatomy of the Newborn: an Atlas.* xiii + 256 pages, 351 illustrations. Lea & Febiger, Philadelphia.

LANGMAN, JAN. 1963. *Medical Embryology.* ix + 335 pages, illustrated. The Williams & Wilkins Company, Baltimore.

PATTEN, B. M. 1968. *Human Embryology.* 3rd edition, xix + 651 pages, illustrated. The Blakiston Division, McGraw-Hill Book Company, New York.

RUDNICK, DOROTHEA. 1958. *Embryonic Nutrition.* xi + 113 pages, illustrated. The University of Chicago Press, Chicago.

TURNER, R. S. 1963. Some limitations of the theoretical approach to problems in morphology. Anat. Rec., *146*, 293–298.

IMPLANTATION, FETAL MEMBRANES, AND PLACENTA

DALLENBACH-HELLWEG, G. and G. NETTE. 1964. Morphological and histochemical observations on trophoblast and decidua of the basal plate on the human placenta at term. Am. J. Anat., *115*, 309–326.

ECKSTEIN, P. 1959. *Implantation of Ova.* vii + 97 pages, illustrated. Cambridge University Press, New York.

FREDA V. J. 1962. Placental transfer of antibodies in man. Amer. J. Obstet. Gynec., *84*, 1756–1777.

HAMILTON, W. J. and J. D. BOYD. 1960. Development of the human placenta in the first three months of gestation. J. Anat. (Lond.), *94*, 297–328.

LARSEN, JORGEN F. and JACK DAVIES. 1962. The paraplacental chorion and accessory fetal membranes of the rabbit. Histology and electron microscopy. Anat. Rec., *143*, 27–45.

STERNBERG, J. 1962. Placental transfers: modern methods of study. Amer. J. Obstet. Gynec., *84*, 1731–1748.

VILLEE, CLAUDE A. 1960. *The Placenta and Fetal Membranes.* xi + 404 pages. The Williams and Wilkins Company, Baltimore.

OOGENESIS, OVULATION, AND FERTILIZATION

BEDFORD, J. M. 1972. An electron microscopic study of sperm penetration into the rabbit egg after natural mating. Am. J. Anat. *133*, 213–253.

BLANDAU, R. J. and D. L. ODOR. 1949. The total number of spermatozoa reaching various segments of the reproductive tract in the female albino rat at intervals after insemination. Anat. Rec., *103*, 93–109.

CHANG, M. C. 1955. The maturation of rabbit oocytes in culture and their maturation, activation, fertilization, and subsequent development in the fallopian tubes. J. Exp. Zool., *128*, 379–406.

CHIQUOINE, A. D. 1960. The development of the zona pellucida of the mammalian ovum. Amer. J. Anat., *106*, 149–169.

EVERETT, J. W. 1956. The time of release of ovulating hormone from the rat hypophysis. Endocrinology, *59*, 580–585.

EVERETT, J. W. and C. H. SAWYER. 1949. A neural timing factor in the mechanism by which progesterone advances ovulation in the cyclic rat. Endocrinology, *45*, 581–595.

HARTMAN, C. G. and G. W. CORNER. 1941. The first maturation division of the macaque ovum. Carneg. Instn., Contr. Embryol., *29*, 1–7.

LEWIS, W. H. and C. G. HARTMAN. 1941. Tubal ova of the rhesus monkey. Carneg. Instn., Contr. Embryol., *29*, 7–15.

POZHIDAEV, E. A. 1964. Morphogenetic processes during cleavage of rat ovum in connection with cytoplasmic differentiation of oocyte in oogenesis. Fed. Proc., *23*, T404–T409.

PURSHOTTAM, N. and G. PINCUS. 1961. In vitro cultivation of mammalian eggs. Anat. Rec., *140*, 51–55.

RAVEN, C. P. 1961. *Oogenesis: The Storage of Developmental Information.* viii + 274 pages. Pergamon Press, New York.

ROCK, J. and A. T. HERTIG. 1944. Information regarding the time of human ovulation derived from a study of 3 unfertilized and 11 fertilized ova. Amer. J. Obstet. Gynec., *47*, 343–356.

ANOMALIES, MULTIPLE BIRTHS, AND TERATOLOGY

BRIDGES, J. B. and W. R. M. MORTON. 1964. Multiple anomalies in a human fetus associated with absence of one umbilical artery. Anat. Rec., *148*, 103–109.

SMITH, W. N. ADAMS. 1963. The site of action of trypan blue in cardiac teratogenesis. Anat. Rec., *147*, 507–523.

WILSON, JAMES G. and J. WARKANY. 1965. *Teratology Principles and Techniques.* viii + 279 pages, illustrated. The University of Chicago Press, Chicago.

EARLY HUMAN EMBRYOS

CORNER, G. W. 1929. A well-preserved human embryo of 10 somites. Carneg. Instn., Contr. Embryol., *20*, 81–102.

GEORGE, W. C. 1942. A presomite human embryo with chorda canal and prochordal plate. Carneg. Instn., Contr. Embryol., *30*, 1–8.

GITLIN, G. 1968. Mode of union of right and left coelomic channels during development of the peritoneal cavity in the human embryo. Acta Anat., 71, 45–52.

HERTIG, A. T. and J. ROCK. 1941. Two human ova of the pre-villous stage, having an ovulation age of about eleven and twelve days respectively. Carneg. Instn., Contr. Embryol., *29*, 127–156.

HERTIG, A. T. and J. ROCK. 1945. Two human ova of the pre-villous stage, having a developmental age of about seven and nine days respectively. Carneg. Instn., Contr. Embryol., *31*, 65–84.

HEUSER, C. H. 1930. A human embryo with fourteen pairs of somites. Carneg. Instn., Contr. Embryol., *22*, 135–154.

HEUSER, C. H. 1932. A presomite human embryo with a definite chorda canal. Carneg. Instn., Contr. Embryol., *23*, 251–267.

HEUSER, C. H., J. ROCK, and A. T. HERTIG. 1945. Two human embryos showing early stages of the definitive yolk sac. Carneg. Instn., Contr. Embryol., *31*, 85–99.

INGALLS, N. W. 1920. A human embryo at the beginning of segmentation, with special reference to the vascular system. Carneg. Instn., Contr. Embryol., *11*, 61–90.

PAYNE, F. 1925. General description of a seven somite human embryo. Carneg. Instn., Contr. Embryol., *16*, 115–124.

SHANER, R. F. 1945. A human embryo of two to three pairs of somites. Canad. J. Res. E., *23*, 235–243.

WEST, C. M. 1930. Description of a human embryo of eight somites. Carneg. Instn., Contr. Embryol., *21*, 25–36.

Surface and Topographical Anatomy

ranium.—The covering of the cranial part of the head is the scalp, a structure composed of the following layers: (1) skin; (2) subcutaneous tissue; (3) Occipitofrontalis muscle, including the galea aponeurotica; (4) subaponeurotic fascial cleft; and (5) pericranium or periosteum of the bones (Fig. 3–1). These layers are of rather uniform thickness, so that the conformation of the head is largely that of the underlying frontal, parietal, and parts of the occipital and temporal bones.

Bones.—Clearly visible landmarks are the **parietal** and **frontal eminences,** and the opening of the **external auditory meatus of** the ear. In addition, certain prominences can be recognized by palpation. In the occipital region or back of the head, the **external occipital protuberance** or **inion** is a bony prominence in the midline, at the junction of the head and neck. The **superior nuchal line** is a slight, upward curving ridge which extends laterally from the protuberance to the mastoid process of the temporal bone. The muscles of the neck stop at the superior nuchal line; above it, the cranium is covered only by scalp. The **mastoid process** projects downward and forward from behind the ear; its anterior border lies immediately behind the concha, and the apex is on a level with the lobe of the auricula.

Face.—The main contours of the face are governed by the bony landmarks, most of which are evident to the eye but which require palpation for establishment of their details (Fig. 3–2).

The **superciliary ridge** is the part of the frontal bone above the eye. It is marked by a depression, the supraorbital notch or foramen.

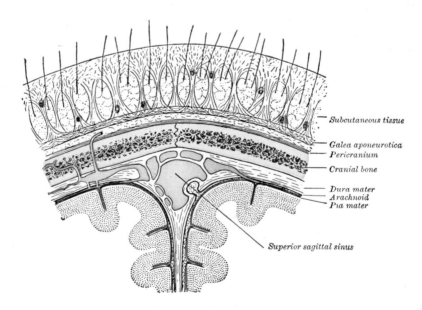

Subcutaneous tissue

Galea aponeurotica
Pericranium

Cranial bone

Dura mater
Arachnoid
Pia mater

Superior sagittal sinus

FIG. 3–1.—Diagrammatic section of scalp.

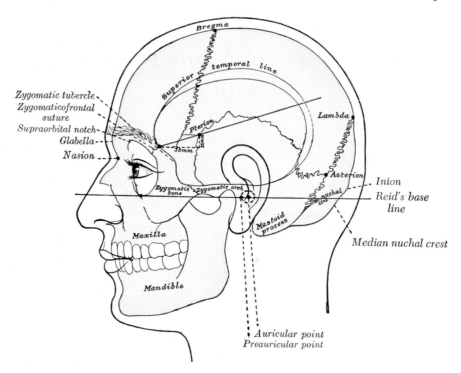

Fɪɢ. 3–2.—Side view of head, showing surface relations of bones.

The rim of the hollow of the orbit is the **orbital margin,** formed by the frontal, zygomatic, and maxillary bones.

The **zygomatic arch** is formed by part of the frontal, zygomatic, and temporal bones.

The **superior temporal line** arches high on the parietal bone between the zygomatic process of the frontal bone and the region behind the ear.

The **mandible** can be recognized throughout most of its extent, including the prominence of the chin or mental protuberance, the angle of the jaw, the alveolar portion containing the teeth, and the condyle which articulates with the temporal bone.

The **temporomandibular joint** can be made easily recognizable by opening the jaws widely; this draws the mandible forward and a depression can be felt between the condyle of the mandible and fossa in the temporal bone, just in front of the tragus of the ear.

Some of these recognizable bony points are used in surgery and in anthropological measurements and therefore are given special names; other bony landmarks, not recognizable by observation or palpation, can be located by using the identifiable structures as points of reference (Fig. 3–2).

Auricular Point.—The center of the opening of the external acoustic meatus.

Preauricular Point.—At the root of the zygomatic arch immediately in front of the external acoustic meatus.

Asterion.—A point 4 cm behind and 12 mm above the auricular point. It marks the meeting of the lambdoidal, occipitomastoid, and parietomastoid sutures.

Zygomatic Tubercle.—A prominence on the posterior margin of the zygomatic bone at the level of the lateral palpebral commissure of the eye.

Frontozygomatic Suture.—A slight depression on the posterior margin of the zygomatic bone, 1 cm above the zygomatic tubercle.

Pterion.—A point 35 mm behind and 12 mm above the level of the frontozygomatic suture. It marks the point where the great wing of the sphenoid meets the sphenoidal angle of the parietal bone.

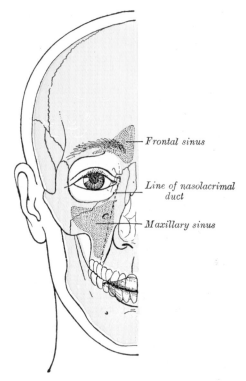

FIG. 3–3.—Outline of bones of face, showing position of air sinuses.

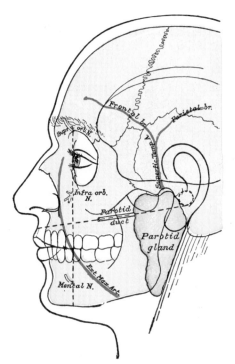

FIG. 3–4.—Outline of side of face, showing chief surface markings.

Inion.—The external occipital protuberance.

Lambda.—The point of meeting of the lambdoidal and sagittal sutures, about 6.5 cm above the inion.

Bregma.—The meeting point of the coronal and sagittal sutures, at the intersection of a line drawn vertically upward from the preauricular point and the midline at the top of the head.

Lambdoidal Suture.—The upper two-thirds of a line from the mastoid process to the lambda.

Sagittal Suture.—The midline between the lambda and the bregma.

Obelion.—The point on the sagittal suture between the parietal emissary foramina, a flattened area at the top of the head.

Coronal Suture.—Is approximated by a line from the bregma to the middle of the zygomatic arch.

Glabella.—A flattened triangular area between the two superciliary ridges.

Nasion.—A slight depression at the root of the nose marking the frontonasal suture.

Reid's base line passes through the inferior margin of the orbit and the auricular point.

The **fontanelles** or soft spots in the skull of a newborn infant (page 150) correspond to the above-mentioned points and sutures as follows: (*a*) the anterior fontanelle, the largest, is at the bregma; (*b*) the posterior is at the lambda; (*c*) the lateral or sphenoidal fontanelle at the pterion; and (*d*) the mastoid fontanelle at the asterion.

Muscles and Soft Parts.—The *Masseter* produces the fullness of the posterior part of the cheek, between the angle of the mandible and the zygomatic arch. Its posterior border is masked by the substance of the *parotid gland,* lying over the muscle and between it and the ear (Fig. 3–4). The anterior border may be palpated readily when the muscle is contracted by clenching the teeth; in front of it the fullness of the cheek is produced by the *buccopharyngeal fat pad* and the *facial muscles.*

The *Temporalis* occupies the temporal fossa, extending from the hollow behind the zygomatic arch to the superior temporal line. The thick temporal fascia, covering the muscle, makes the superior border of the zygomatic arch difficult to palpate.

Arteries (Fig. 3–4).—The pulsations of the *superficial temporal artery* may be felt just above the zygomatic arch in front of the ear, and its frontal branch frequently is visible making its serpentine course across the temple, especially in older individuals. The **facial** may be felt as it crosses the margin of the mandible at the anterior border of the Masseter; it has a tortuous course from this point upward across the face to the angle between the eye and the root of the nose.

Veins.—The facial vein crosses the margin of the mandible with the facial artery, and takes a relatively straight course to the angle between the eye and nose.

Nerves.—The *supraorbital nerve* crosses the superciliary ridge through the supraorbital notch (or foramen). The *infraorbital nerve* becomes subcutaneous at the infraorbital foramen, just below the orbit. The *mental nerve* emerges from the mandible above and lateral to the mental protuberance. The *facial nerve* emerges from the

substance of the anterior part of the parotid gland and spreads over the face like a fan.

Lymph nodes usually are not palpable in the head, but if enlarged, the following nodes may be felt: posterior auricular, parotid, occipital, buccal, and submandibular.

The Eye.—The palpebral fissure, between the two lids, has as its extremities, the medial and lateral commissures. At the medial commissure are the caruncula lacrimalis, the plica semilunaris, and the puncta lacrimalia. The nasolacrimal duct runs from the puncta to an opening in the inferior meatus of the nose (*see* Sense Organ Chapter). The lacrimal sac, at the top of the duct, lies behind the medial palpebral ligament which may be felt at the medial commissure if the eyelids are drawn laterally to tighten the skin. The palpebral conjunctiva lines the lids and at the fornix conjunctivae, it doubles back over the eyeball as the bulbar conjunctiva. The colored part of the eye is covered by the transparent cornea,

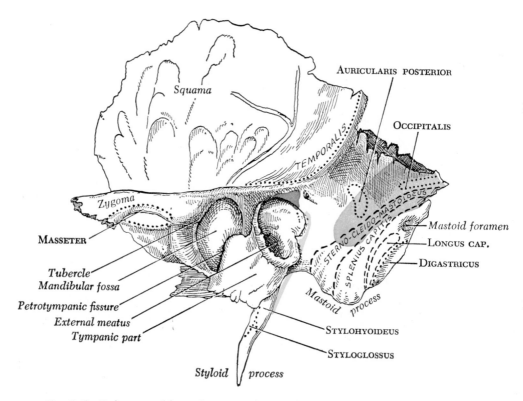

Fig. 3–5.—Left temporal bone showing surface markings for the tympanic antrum (red), transverse sinus (blue), and facial nerve (yellow).

and through its central aperture, the pupil, the lens, the vessels of the retina, and the optic disk can be seen with an ophthalmoscope.

The Ear.—The auricula is marked by various prominences and fossae (*see* Sense Organs). The opening of the external acoustic meatus is more fully exposed by drawing the tragus forward; the orifice is guarded by crisp hairs and contains a coating of wax, the secretion of the ceruminous glands. The interior of the meatus can be examined through a speculum or otoscope more easily if the auricula is drawn upward, backward, and slightly outward in order to straighten the slight curvature at the junction of the cartilaginous and bony portions of the wall. At the interior end of the meatus is the *tympanic membrane* which shows certain structures and markings when viewed with an otoscope. The *suprameatal triangle* (Fig. 3–5) is an important landmark for internal structures such as the tympanic antrum, facial nerve, and transverse sinus.

The **brain** can be placed approximately by certain external landmarks and measurements (Figs. 3–6, 3–7). The *lateral sulcus* (Sylvian) can be located by a point, termed the Sylvian point, which practically corresponds to the pterion, 3.5 cm behind and 12 mm above the level of the frontozygomatic suture. The position of the *lateral ventricle* may be shown by an x-ray ventriculogram after air is injected into it (Fig. 11–83). The branches of the *internal carotid artery* may be visualized in an x-ray after injection of the opaque medium diotrast (Fig. 8–28).

The **middle meningeal artery** lies at the level of the middle of the superior border of the zygomatic arch; its anterior branch may be reached through a trephined opening slightly anterior and below the pterion (Fig 3–6). The **transverse dural sinus** may be approximated by a line through the asterion curving downward toward the mastoid process (Fig. 3–6).

Air sinuses or **paranasal sinuses** (Fig. 3–3) vary greatly in size, shape, and position. The

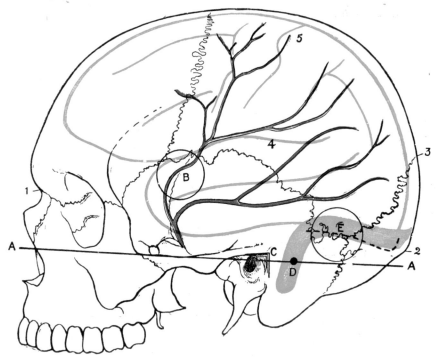

Fig. 3–6.—Relations of the brain and middle meningeal artery to the surface of the skull. 1. Nasion. 2. Inion. 3. Lambda. 4. Lateral cerebral fissure. 5. Central sulcus. *AA.* Reid's base line. *B.* Point for trephining the anterior branch of the middle meningeal artery. *C.* Suprameatal triangle. *D.* Sigmoid bend of the transverse sinus. *E.* Point for trephining over the straight portion of the transverse sinus, exposing dura mater of both cerebrum and cerebellum. Outline of cerebral hemisphere indicated in blue; course of middle meningeal artery in red.

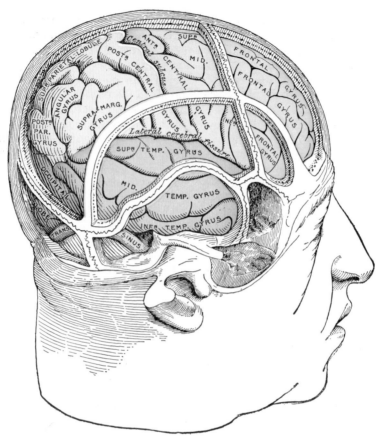

Fig. 3–7.—Drawing of a cast by Cunningham to illustrate the relations of the brain to the skull.

frontal sinus occupies the area in the bone deep to the medial part of the superciliary ridge. The *maxillary sinus* occupies the body of the maxilla, the area between the orbit, nasal cavity, and upper teeth.

THE NECK

Bones.—The **vertebral column** is deeply placed in the neck but certain processes can be identified by palpation. The tip of the **transverse process of the atlas** can be felt about 1 cm below and in front of the mastoid process. The **anterior tubercle** on the **transverse process of the sixth cervical vertebra** may be felt deep in the neck at the anterior border of the Sternocleidomastoideus; it has been named the **carotid** or **Chassaignac's tubercle,** and is the point of preference for compressing the common carotid artery to stop bleeding. The **hyoid**

bone may be felt in the receding angle between the chin and anterior part of the neck. It is at the level of the fourth cervical vertebra and its greater cornu extends back on a level with the angle of the mandible (Fig. 3–8).

The **Larynx and Trachea.**—The laryngeal prominence of the **thyroid cartilage** is visible in the midline, 1 or 2 cm below the hyoid bone. The upper margin of the thyroid cartilage is connected with the hyoid bone by the thyrohyoid membrane, and its lower margin to the cricoid cartilage by the cricothyroid ligament. The level of the vocal folds corresponds to the middle of the anterior border of the thyroid cartilage. The **cricoid cartilage** is at the level of the sixth cervical vertebra. Below it the **trachea** can be felt, but only in thin subjects can the individual rings be distinguished. As a rule there are seven or eight rings above the jugular notch of the sternum, and the

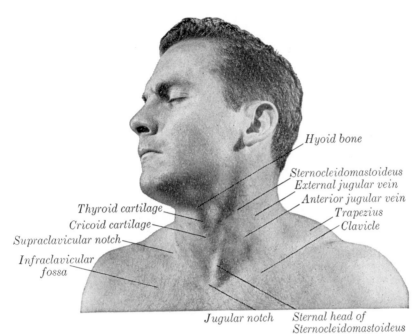

Hyoid bone

Sternocleidomastoideus
External jugular vein
Anterior jugular vein
Trapezius
Clavicle

Thyroid cartilage
Cricoid cartilage
Supraclavicular notch
Infraclavicular fossa

Jugular notch   *Sternal head of Sternocleidomastoideus*

Fig. 3–8.—Anterior and lateral view of head and neck showing surface markings. (Courtesy Dr. C. John Gourgott. Photograph by Roy Trahan.)

isthmus of the **thyroid gland** covers the second, third, and fourth rings.

Muscles.—The **Sternocleidomastoideus** is the most prominent muscle in the neck. Its entire extent, from the mastoid process to the sternal and clavicular heads (Fig. 3–8) is clearly visible in most persons. The **Trapezius** forms the upper border of the shoulder, sloping upward from the point of the shoulder toward the back of the head. The **Platysma** is a very thin superficial sheet of muscle overlying the anterior neck; it can be detected only if thrown into prominence by being contracted, as it is when one attempts to relieve the pressure of a tight collar.

Triangles of the Neck.—The lack of prominent topographical features and the large number of important nerves and blood vessels contained within the neck have made it customary to subdivide this region into smaller areas. These areas, triangular in shape and related to some of the more superficial muscles, are known as the triangles of the neck (Fig. 3–9).

ANTERIOR TRIANGLE.—The anterior triangle is bounded anteriorly by the **midline,** and posteriorly by the anterior border of

the **Sternocleidomastoideus;** its apex is below at the sternum, and its base is formed by the lower margin of the **body of the mandible** and an extension of this line to the mastoid process. This larger triangle is subdivided by the Digastricus above, and the superior belly of the Omohyoideus below into four smaller triangles: the inferior carotid, superior carotid, submandibular, and suprahyoid triangles.

The **Inferior Carotid** or **Muscular Triangle** is bounded in front by the midline of the neck; behind, by the superior belly of the Omohyoideus above, and the anterior margin of the lower part of the Sternocleidomastoideus below. In this triangle, under cover of the skin, subcutaneous fascia, Platysma, and deep fascia, are the Sternohyoideus, Sternothyroideus, the isthmus of the thyroid gland, the larynx, and the trachea. Although not within the triangle as bounded above, the following structures can be approached surgically through the triangle by displacing the Sternocleidomastoideus laterally: the lower part of the common carotid artery in the carotid sheath with the internal jugular vein and vagus nerve, the

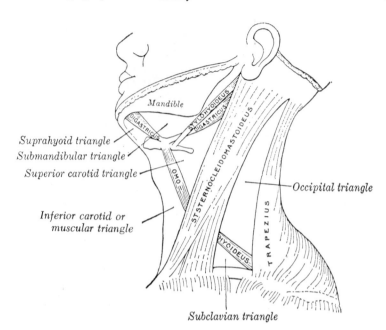

Mandible

Suprahyoid triangle
Submandibular triangle
Superior carotid triangle

Occipital triangle

Inferior carotid or muscular triangle

Subclavian triangle

Fig. 3–9.—The triangles of the neck. Anterior and posterior triangles, not labeled, are separated by the Sternocleidomastoideus.

ansa hypoglossi, the sympathetic trunk, recurrent nerve, and the esophagus.

The **Superior Carotid** or **Carotid Triangle** is bounded behind, by the Sternocleidomastoideus; below, by the superior belly of the Omohyoideus; and above, by the Stylohyoideus and posterior belly of the Digastricus. It is covered by the skin, subcutaneous fascia, Platysma, and deep fascia, layers which contain the ramifications of the cutaneous cervical nerves and the cervical branch of the facial nerve. The floor or deepest part of the triangle is formed by the Thyrohyoideus, Hyoglossus and the Constrictores pharyngis medius and inferior. In the triangle, especially if it is enlarged by displacing the Sternocleidomastoideus backward, is the upper part of the common carotid artery which bifurcates at the level of the upper border of the thyroid cartilage into the internal and external carotids. The internal carotid is here posterior and somewhat lateral to the external and has no branches. The branches of the external carotid at the triangle are the superior thyroid, the lingual, the facial, the occipital, and the ascending pharyngeal. Enclosed with the arteries in a fascial membrane, the carotid sheath, are the internal

jugular vein and the vagus nerve. The tributaries of the vein correspond with the branches of the external carotid: superior thyroid, lingual, facial, ascending pharyngeal, and sometimes the occipital. Superficial to the carotid sheath is the ansa hypoglossi (NA *ansa cervicalis*); and deep to it or imbedded in its substance is the sympathetic trunk. The hypoglossal nerve crosses both the internal and external carotids in the upper part of the triangle, curving around the origin of the occipital artery. Between the external carotid and the pharynx, just below the hyoid bone, is the internal branch of the superior laryngeal nerve; the external branch is slightly lower. The accessory nerve may cross the uppermost corner of the triangle. The carotid sinus and carotid body lie at the level below the hyoid bone, between it and the superior border of the thyroid cartilage (Fig. 3–10).

The **Submandibular** or **Digastric Triangle** corresponds to the region of the neck immediately below the body of the mandible. It is bounded, above, by the lower border of the body of the mandible and an extension of this line to the mastoid process; below, by the posterior belly of the Digastricus and the Stylohyoideus; and in front, by the anterior belly of the Digastricus. It is covered by the skin, subcutaneous fascia, Platysma, and deep fascia, ramifying in which are branches of the facial nerve and ascending filaments from the cervical cutaneous nerves. Its floor is formed by the Mylohyoideus and Hyoglossus. It is divided into an anterior and a posterior part by the stylomandibular ligament at the angle of the mandible. The anterior part contains the submandibular gland, which is crossed by the anterior facial vein and facial artery. Deep to the gland are the submental artery and the mylohyoid artery and nerve. The posterior part of this triangle is largely occupied by the parotid gland. If the triangle is enlarged by displacing the muscles slightly and by turning the head and raising the chin, certain deeper structures may be included. The external carotid is deep in the substance of the parotid gland, where it is superficial to the internal carotid, is crossed by the facial nerve, and gives off its posterior auricular, superficial temporal, and maxillary branches. The internal jugular vein and vagus nerve are still deeper, separated from the external carotid by the Styloglossus, Stylopharyngeus, and the glossopharyngeal nerve.

The **Suprahyoid Triangle** is bounded by the anterior belly of the Digastricus, the midline, and the body of the hyoid bone. Its floor is the Mylohyoideus. It contains one or two lymph nodes and small tributaries of the anterior jugular vein.

POSTERIOR TRIANGLE.—The posterior triangle is bounded, in front, by the **Sternocleidomastoideus**; behind, by the anterior margin of the **Trapezius**; its base is the middle third of the **clavicle**, and its apex is at the occipital bone. About 2.5 cm above the clavicle, it is crossed obliquely by the inferior belly of the Omohyoideus, dividing it into an upper or occipital and a lower or subclavian triangle.

The **Occipital Triangle** is bounded, in front, by the Sternocleidomastoideus; behind, by the Trapezius; and below, by the Omohyoideus. Its floor is formed by the Splenius capitis, Levator scapulae, and Scalenus medius and posterior, and it is covered by the skin, subcutaneous fascia, and deep fascia. About one-half or two-thirds of the way up, the triangle is crossed

by the accessory nerve; somewhat lower the cervical cutaneous and supraclavicular nerves emerge from under the posterior border of the Sternocleidomastoideus; and the upper part of the brachial plexus crosses the lowest part of the triangle.

The **Subclavian Triangle** is bounded, above, by the inferior belly of the Omohyoideus; below, by the clavicle; and in front by the Sternocleidomastoideus. Its floor is formed by the first rib and the first digitation of the Serratus anterior. Its size depends upon the extent of attachment of the Sternocleidomastoideus and Trapezius to the clavicle and the position of the Omohyoideus; some of the structures listed below might have to be reached by displacement of the muscles, and the triangle is increased in size by drawing the shoulder downward to lower the clavicle. It is covered by the skin, subcutaneous fascia, Platysma, and deep fascia containing the supraclavicular nerves. In the medial part of the triangle, the third part of the subclavian artery emerges from behind the Scalenus anterior and curves down behind the clavicle. Sometimes the arch of the artery rises 4 cm above the clavicle, and this is the most commonly chosen place for ligature. The subclavian vein usually remains down behind the clavicle, but may be found partly in the triangle. The transverse scapular vessels cross the lower part, the superficial cervical vessels the upper part of the triangle. The external jugular vein enters the anterior part of the triangle from its position on the surface of the Sternocleidomastoideus and empties into the subclavian vein; its tributaries, the transverse scapular and transverse cervical veins, make a plexus superficial to the subclavian artery. The brachial plexus of nerves crosses the lateral part of the triangle coming into close relationship with the subclavian artery as it passes behind the clavicle.

Arteries.—The position of the *common* and *external carotid arteries* is indicated by a line drawn from the upper part of the sternal end of the clavicle to a point midway between the tip of the mastoid process and the angle of the mandible. Above the bifurcation at the upper border of the thyroid cartilage, this line overlies

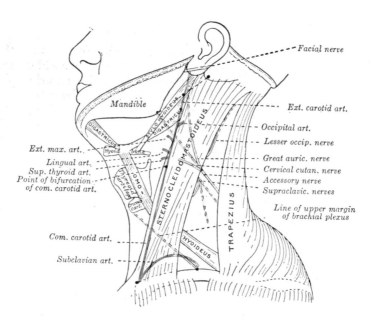

Fig. 3–10.—Side of neck, showing chief surface markings.

the internal carotid and the external carotid until the latter arches backward slightly toward the external acoustic meatus. The *main branches* of the *external carotid* originate near the tip of the greater cornu of the hyoid bone. The *subclavian artery* underlies a line arching upward from the sternoclavicular joint to the middle of the clavicle (Fig. 3–10).

Veins.—The *internal jugular vein* is parallel and slightly lateral to the internal carotid artery. The *external jugular* runs from the angle of the mandible to the middle of the clavicle (Fig. 3–10).

Nerves.—The exit of the *facial nerve* from the stylomastoid foramen is about 2.5 cm below the surface of the skin, opposite the anterior border of the mastoid process. The *accessory nerve* is quite superficial as it crosses the posterior triangle, running downward from under the upper middle part of the posterior border of the Sternocleidomastoideus to pass under the anterior border of the Trapezius. Also emerging from under the posterior border of the Sternocleidomastoideus in this same region are the *cutaneous nerves:* the lesser occipital, the greater auricular, the anterior cervical cutaneous, and the supraclavicular nerves. The *phrenic nerve* begins at the level of the

middle of the thyroid cartilage and passes deep to and about half way between the two borders of the Sternocleidomastoideus. The upper part of the *brachial plexus* is indicated by a line from the cricoid cartilage to the middle of the clavicle. The *vagus nerve* and *sympathetic trunk* run parallel and deep to the internal carotid artery.

Lymph Nodes.—Lymph nodes may be felt between the ramus of the mandible and the Sternocleidomastoideus, along the course of the internal jugular vein at the anterior border of this muscle, and along the posterior border of the same muscle in the upper and lower portions of the posterior triangle.

THE BACK

Bones.—The furrow down the middle of the back lies over the tips of the *spinous processes of the vertebrae.* The upper cervicals cannot be felt, but the seventh can be distinguished easily and is next above the more prominent *first thoracic.* Other thoracic spines can be identified by counting from these. The root of the *spine of the scapula* is on a level with the third thoracic spine, the inferior angle with the seventh. The highest point of the *crest of the ilium*

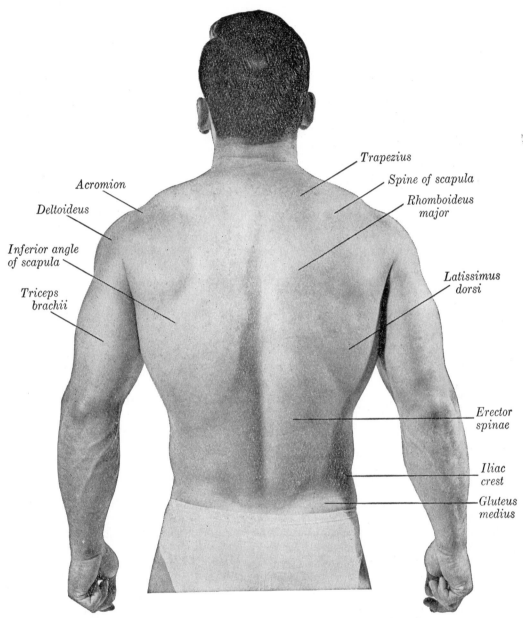

Fig. 3–11.—Surface anatomy of back. (Gourgott.)

is on a level with the fourth lumbar spinous process and the *posterior superior iliac spine* with that of the second sacral (Fig. 3–11).

The **spinal cord** extends down to the level of the spinous process of the second lumbar vertebra (Fig. 3–12) in the adult body, but as far as the fourth in an infant. The **subarachnoid space**, containing the spinal fluid, extends down to the third sacral vertebra.

The **muscles** of the back are large and usually can be identified with ease in a living subject (Fig. 3–31).

THE THORAX

The surface of the thorax is covered by several large muscles belonging to the musculature of the upper extremity (Fig. 3–18). At each side, the **axilla** or hollow of

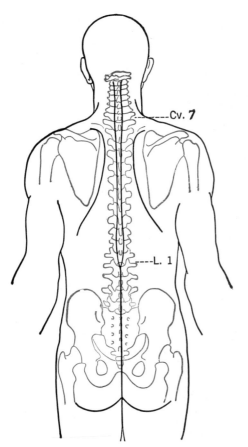

Fig. 3–12.—Diagram showing the relation of the spinal cord to the dorsal surface of the trunk. The bones are outlined in red.

the armpit (Fig. 3–35) is limited by two fleshy folds; the **anterior axillary fold** is the prominence caused by the Pectoralis major; the **posterior axillary fold** is the prominence of the Latissimus dorsi. The size of the **breast** or **mamma** is subject to great variation. In most adult nulliparous females it extends vertically from the second to the sixth ribs, and transversely from the sternum to the midaxillary line. In males and nulliparous females, the **mammary papilla** or **nipple** is situated in the fourth intercostal space (Fig. 3–14).

Bony Landmarks.—The *sternum, ribs, scapula* and *clavicle* can be seen in many individuals and can be felt in all but the very muscular or obese subjects. The upper border of the sternum is marked by the **jugular notch,** between the sternal heads of the two Sternocleidomastoidei (Fig. 3–8).

In the midline the sternum is subcutaneous, leaving the sternal furrow between the origins of the two Pectorales majores. The junction between the two parts of the sternum, the manubrium above and the body or gladiolus below, is marked by a well defined transverse ridge about 5 cm below the jugular notch. This ridge, called the **sternal angle** or **angle of Louis,** is opposite the sternochondral junction of the second rib (Fig. 3–18). At the lower end of the body of the sternum is the infrasternal notch between the sternal attachments of the seventh costal cartilages. In the triangular depression below the notch, the epigastric fossa, the **xiphoid process** of the sternum can be felt.

The **ribs** can usually be felt at the sternum in front, at the sides, and in back as far as their angles, although they are mostly covered by muscles. The first rib is difficult to palpate because it is deep and partly hidden by the clavicle. The **second rib** is the one most reliably identified because of its attachment at the angle of Louis. The lower boundary of the thorax is formed by the xiphoid process, the cartilages of the seventh, eighth, ninth, and tenth ribs and the ends of the eleventh and twelfth cartilages (Figs. 3–17, 3–19).

Lines for Orientation (Fig. 3–21).—The **midsternal line** is the midline of the body over the sternum. The **midclavicular** or **mammary line** is a vertical line, parallel with the midsternal, through a point midway between the center of the jugular notch and the tip of the acromion or point of the shoulder. The **lateral sternal line** is a vertical line along the sternal margin. The **anterior** and **posterior axillary lines** are vertical lines drawn on the corresponding folds; the **midaxillary line,** half way between them, passes through the apex of the axilla. On the back, the **scapular line** is drawn vertically through the inferior angle of the Scapula (Fig. 3–11).

Vertebral Level.—The **jugular notch** is at the same horizontal level as the lower border of the body of the second thoracic vertebra. The **sternal angle** is at the level of the fifth thoracic vertebra, and the **xiphisternal junction** at the level of the disk between the ninth and tenth thoracic vertebrae.

The Lungs (Figs. 3–14, 3–15).—The **apex of the lung** lies behind the medial third of the clavicle and extends up into the neck from 1 to 5 cm, usually about 2.5 cm. The **anterior border** of the right lung approaches the midsternal line, that of the left does likewise as far as the fourth costal cartilage where it deviates laterally because of the cardiac notch. The **lower border** at expiration follows a curving line downward from the sixth sternocostal junction to the spinous process of the tenth thoracic vertebra. This line crosses the midclavicular line at the sixth and the midaxillary line at the eighth rib. The **posterior border** is parallel with the midline, 2 to 3 cm from it, and extends from the spinous process of the seventh cervical vertebra to that of the tenth thoracic vertebra.

The **pleura** corresponds in general with the lungs, but is more extensive. The **parietal pleurae** of the two sides of the body come

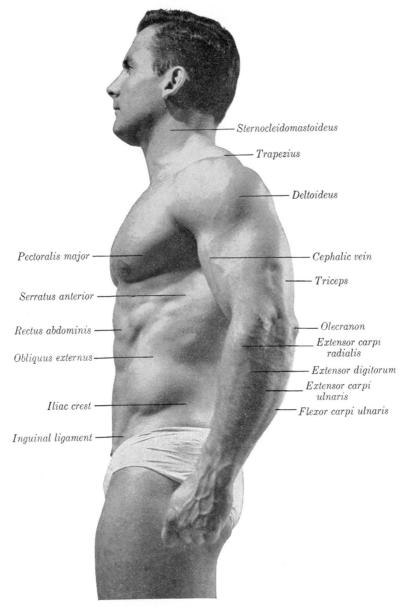

Sternocleidomastoideus

Trapezius

Deltoideus

Pectoralis major

Cephalic vein

Triceps

Serratus anterior

Rectus abdominis

Olecranon

Extensor carpi radialis

Obliquus externus

Extensor digitorum

Extensor carpi ulnaris

Iliac crest

Flexor carpi ulnaris

Inguinal ligament

FIG. 3–13.—Surface anatomy of left side. (Gourgott.)

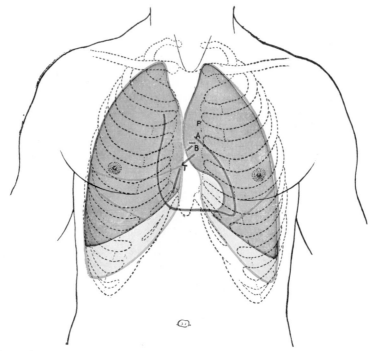

Fɪɢ. 3–14.—Front of thorax, showing surface relations of bones, lungs (purple), pleura (blue) and heart (red outline). *P.* Pulmonary valve. *A.* Aortic valve. *B.* Bicuspid valve. *T.* Tricuspid valve.

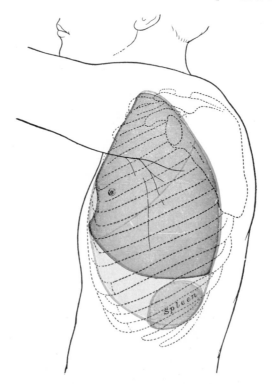

Fɪɢ. 3–15.—Side of thorax, showing surface markings for bones, lungs (purple), pleura (blue), and spleen (green).

close to each other in the midline anteriorly, at expiration, opposite the second to fourth costal cartilages. The medial 1 or 2 cm of the pleural cavity is unoccupied by lung, the potential space being known as the **costomediastinal sinus.** The inferior limit of the pleura is 2 to 5 cm below that of the lung at expiration, leaving the **phrenicocostal sinus** unoccupied by lung. This is a favorite place for the introduction of a needle into the pleural cavity to drain fluid. The **attachment of the diaphragm** along the costal margin is 2 or 3 cm below the inferior limit of the pleura.

The Heart.—The **apex of the heart** usually may be felt by its pulsation in the fifth intercostal space just below the nipple or about 9 cm to the left of the midsternal line. In an approximate projection of the heart on the anterior chest wall, the **superior border** is marked by a horizontal line at the third sternochondral attachment; the **right margin** corresponds to a vertical line drawn 2.5 cm lateral to the sternal margin; the **inferior or diaphragmatic margin** is marked by a line sloping slightly downward and to the left at the xiphisternal junction, and the **left border** angles upward from the apex to the second intercostal space 2.5 cm from the sternal margin. The superior margin is also the **base of the heart** and marks the beginning of the great vessels. The position of the chambers, sulci, and valves are shown in Figure 3–16.

The **internal thoracic vessels** run vertically 1 cm lateral to the sternal margin as far as the sixth cartilage. The **intercostal vessels** and **nerves** generally lie along the

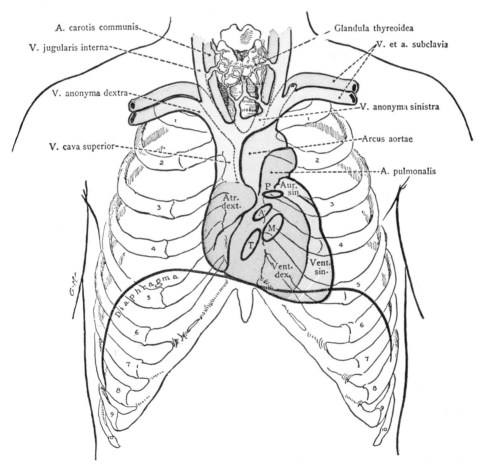

Fig. 3–16.—The heart and cardiac valves projected on the anterior chest wall, showing their relation to the ribs, sternum and diaphragm. (Eycleshymer and Jones.)

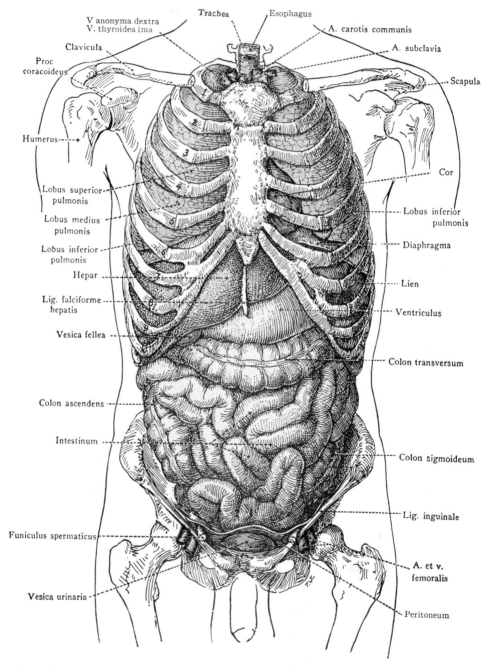

Fig. 3–17.—Thoracic and abdominal viscera shown in their normal relations to the skeleton. Anterior view. (Eycleshymer and Jones.)

inferior border of the rib in their intercostal spaces anteriorly and tend to lie deep to the rib posteriorly.

THE ABDOMEN

The contours of the abdomen are established largely by the muscles, with modifi-cations brought about by the accumulation of adipose tissue in the subcutaneous layers. A groove in the midline separates the medial margin of the **Rectus abdominis** of one side from that of the other side and lies over an aponeurotic junction known as the **linea alba** (Figs. 3–18, 3–28). The **umbilicus** interrupts the linea alba about half way

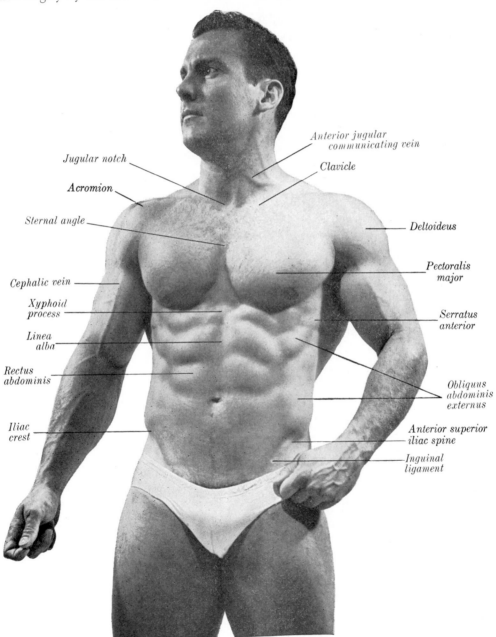

Jugular notch

Acromion

Sternal angle

Cephalic vein

Xyphoid process

Linea alba

Rectus abdominis

Iliac crest

Anterior jugular communicating vein

Clavicle

Deltoideus

Pectoralis major

Serratus anterior

Obliquus abdominis externus

Anterior superior iliac spine

Inguinal ligament

Fig. 3–18.—Anterior view showing surface markings. (Courtesy of Dr. Gourgott. Photograph by Roy Trahan.)

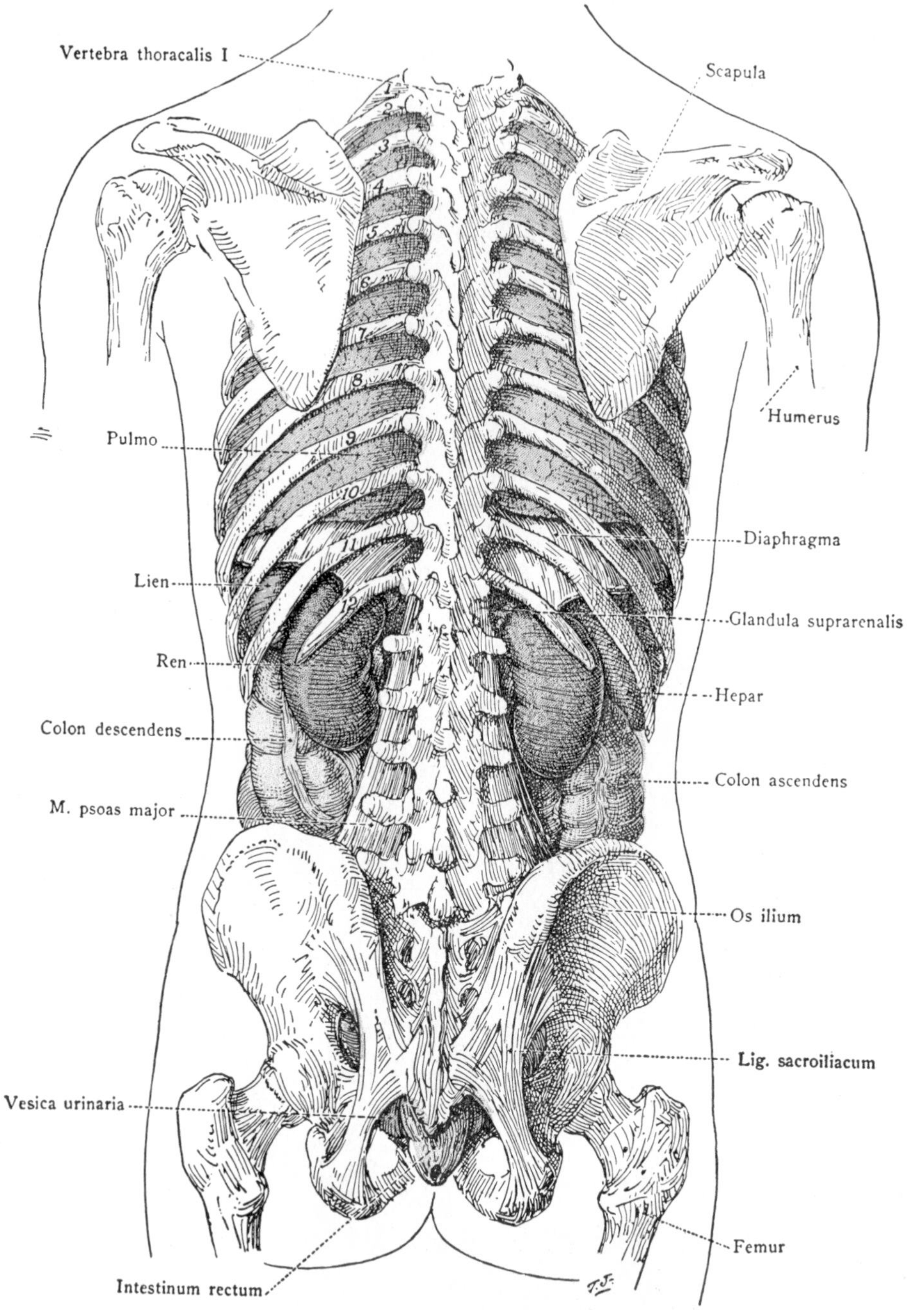
Vertebra thoracalis I
Scapula
Humerus
Pulmo
Diaphragma
Lien
Glandula suprarenalis
Ren
Hepar
Colon descendens
Colon ascendens
M. psoas major
Os ilium
Lig. sacroiliacum
Vesica urinaria
Femur
Intestinum rectum
1
2
3
4
5
6
7
8
9
10
11
12
T.J.

FIG. 3–19.—Thoracic and abdominal viscera shown in their normal relations to the skeleton. Posterior view. (Eycleshymer and Jones.)

between the infrasternal notch and the pubic symphysis. The lateral margin of the Rectus abdominis corresponds to another groove, the **linea semilunaris** (Fig. 3–28), which is not as clearly marked as the linea alba except in muscular subjects. At the level of the umbilicus, the linea semilunaris is about 7 cm from the midline and it curves medially as it nears the pubis. Lateral to the Rectus, the **Obliquus externus abdominis** (Fig. 3–18) is the superficial muscle; it interdigitates with the **Serratus anterior** above, and posteriorly it is partly overlapped by the **Latissimus dorsi.** A separation between the Externus and Latissimus at their attachment to the crest of the ilium is known as the *lumbar triangle* of Petit. At the lower border of the Obliquus externus, the **inguinal ligament** lies deep to the groove of the groin. The surface of the Rectus abdominis, especially in a muscular subject, has three transverse furrows caused by the tendinous inscriptions; one is a little below the xiphoid process, one at the umbilicus, and the third in between these two.

The **umbilicus** is at the level of the intervertebral disk between the third and fourth lumbar vertebrae.

The **superficial inguinal ring** (Fig. 3–28) is situated 1 cm above and lateral to the pubic tubercle; the **deep inguinal ring** lies 1 to 2 cm above the middle of the inguinal ligament. The position of the **inguinal canal** is indicated by a line joining these two points.

The **Bony Landmarks** of the abdomen are: above, the lower border of the thorax which has already been described, and below, the bony pelvis. The **crest of the ilium** may be identified by the prominent **tubercle** on the outer lip which can be seen or palpated. The **anterior superior spine** of the **ilium** may be felt in the groin at the lateral end of the inguinal ligament (Fig. 3–18), and the **pubic tubercle** may be felt at the medial end of the ligament, a point which may also be identified as the inferior attachment of the Rectus abdominis in the region known as the **mons pubis.**

Surface Lines.—For convenience of description and reference, the abdomen is divided into nine regions by imaginary planes, two horizontal and two sagittal, indicated by the following lines drawn on

the surface of the body (Fig. 3–21). (1) An upper transverse, the **transpyloric,** half way between the jugular notch and the upper border of the symphysis pubis; this cuts through the pylorus, the tips of the ninth costal cartilages and the lower border of the first lumbar vertebra; (2) a lower transverse line termed the **transtubercular,** corresponds to the iliac tubercles and cuts the body of the fifth lumbar vertebra.

By means of these horizontal planes the abdomen is divided into three zones named from above, the **subcostal, umbilical,** and **hypogastric zones.** Each of these is further subdivided into three regions by the two sagittal planes which are indicated on the surface by a right and a left lateral line drawn vertically through points half way between the anterior superior iliac spines and the midline. The middle region of the upper zone is called the **epigastric,** and the two lateral regions the **right** and **left hypochondriac.** The central region of the middle zone is the **umbilical,** and the two lateral regions the **right** and **left lumbar.** The middle region of the lower zone is the **hypogastric** or **pubic,** and the lateral are the **right** and **left iliac** or **inguinal.**

Another more simplified but quite useful subdivision of the abdomen is by two planes at right angles to each other, one corresponding to the midsagittal plane of the body, the other a transverse plane through the umbilicus. The resulting four portions are known as quadrants, the upper and lower, right and left.

Viscera.—Under normal conditions the various portions of the digestive tube cannot be identified by simple palpation. The greater part of the liver lies under cover of the ribs and cartilages, especially in the supine position, but during a deep inspiration it may be pushed out below the costal margin on the right side and be felt. Other viscera can only be palpated in emaciated subjects with lax abdominal walls, or if they are the seat of disease or tumors (Fig. 3–24).

Stomach (Fig. 3–22).—The shape of the stomach is constantly undergoing alteration; it is affected by the particular phase of digestion, by the state of the surrounding viscera, and by the amount and character of its contents. Its position also varies with that

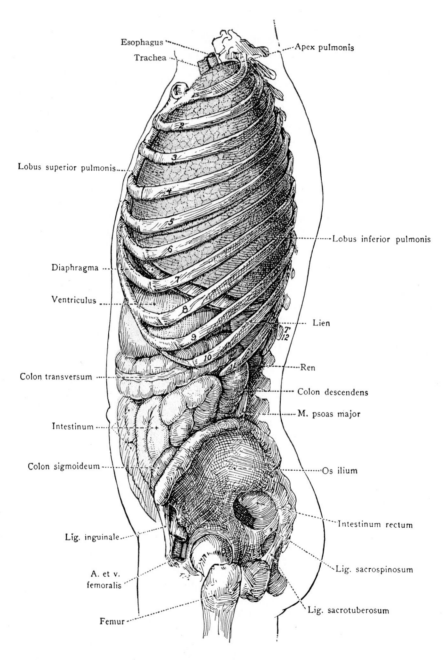

Esophagus

Trachea

Apex pulmonis

Lobus superior pulmonis

Lobus inferior pulmonis

Diaphragma

Ventriculus

Lien

Colon transversum

Ren

Colon descendens

M. psoas major

Intestinum

Colon sigmoideum

Os ilium

Intestinum rectum

Lig. inguinale

A. et v. femoralis

Lig. sacrospinosum

Lig. sacrotuberosum

Femur

FIG. 3–20.—Thoracic and abdominal viscera shown in their normal relations to the skeleton, from the left side. (Eycleshymer and Jones.)

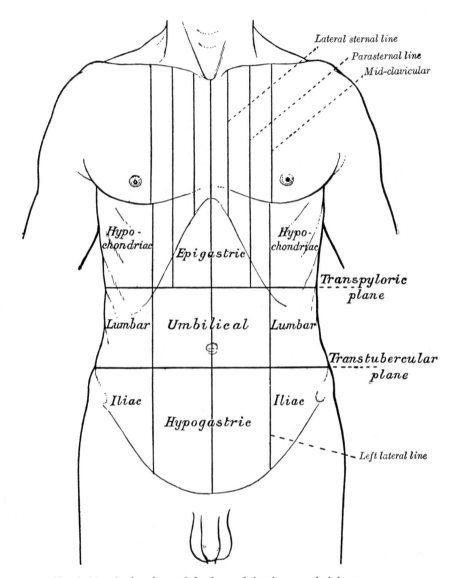

Fig. 3–21.—Surface lines of the front of the thorax and abdomen.

of the body, so that it is impossible to indicate it on the surface with any degree of accuracy. The measurements given refer to a moderately filled stomach with the body in the supine position (see Digestive Apparatus Chapter for X-rays).

The cardiac orifice is opposite the seventh left costal cartilage about 2.5 cm from the side of the sternum; it corresponds to the level of the tenth thoracic vertebra. The pyloric orifice is on the transpyloric line about 1 cm to the right of the midline, or alternately 5 cm below the seventh right

sternocostal articulation; it is at the level of the first lumbar vertebra.

Duodenum (Fig. 3–24).—The superior part is horizontal and extends from the pylorus to the right lateral line; the descending part is situated medial to the right lateral line, from the transpyloric line to a point midway between the transpyloric and transtubercular lines. The horizontal part runs with a slight upward slope from the end of the descending part to the left of the midline; the ascending part is vertical, and reaches the transpyloric line, where it ends

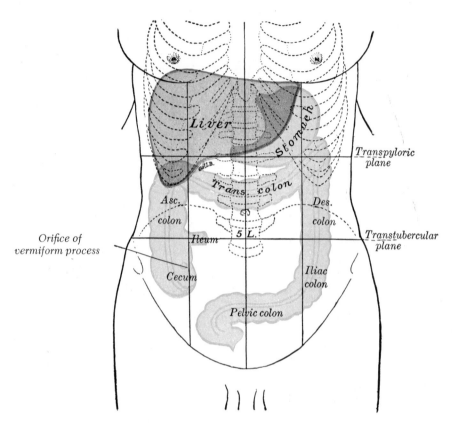

Fɪɢ. 3–22.—Front of abdomen, showing surface markings for liver, stomach, and large intestine.

in the duodenojejunal flexure, about 2.5 cm to the left of the midline.

Small Intestine.—The coils of small intestine occupy most of the abdomen. Frequently the coils of the jejunum are situated on the left side, the coils of the ileum toward the right and partly within the pelvis. The end of the ileum, *i.e.*, the **ileocolic junction,** is slightly below and medial to the intersection of the right lateral and transtubercular lines.

Cecum and Vermiform Appendix.—The cecum is in the right iliac and hypogastric regions; its position varies with its degree of distention, but a line drawn from the right anterior superior iliac spine to the upper margin of the symphysis pubis will mark approximately the middle of its lower border. The position of the base of the **appendix** is indicated by a point on the lateral line on a level with the anterior superior iliac spine (Fig. 3–22) (see Chapter 16 for X-rays).

Ascending Colon.—The ascending colon passes upward through the right lumbar region, lateral to the right lateral line. The **right colic flexure** is situated in the upper and right angle of intersection of the subcostal and right lateral lines.

Transverse Colon.—The transverse colon crosses the abdomen in the umbilical and epigastric regions, its lower border being on a level slightly above the umbilicus, its upper border just below the greater curvature of the stomach (Fig. 3–17).

Descending Colon.—The **left colic flexure** is situated in the upper left angle of the intersection between the left lateral and transpyloric lines. The descending colon courses down through the left lumbar region, lateral to the left lateral line, as far as the iliac crest.

Iliac Colon.—The line of the iliac colon is from the end of the descending colon to the left lateral line at the level of the anterior superior iliac spine.

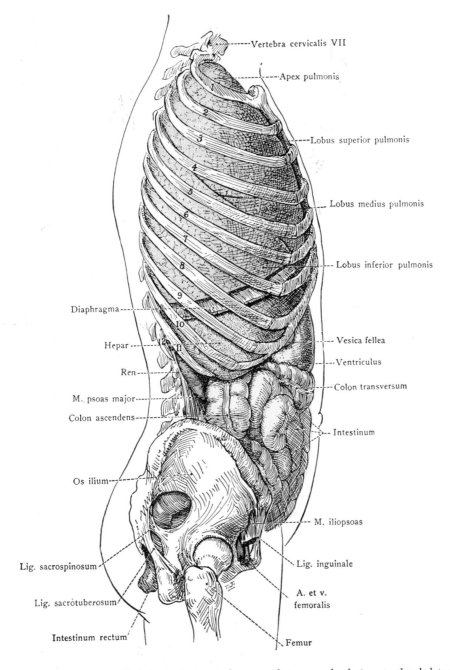

Vertebra cervicalis VII

Apex pulmonis

Lobus superior pulmonis

Lobus medius pulmonis

Lobus inferior pulmonis

Diaphragma

Hepar

Ren

M. psoas major

Colon ascendens

Os ilium

Lig. sacrospinosum

Lig. sacrotuberosum

Intestinum rectum

Vesica fellea

Ventriculus

Colon transversum

Intestinum

M. iliopsoas

Lig. inguinale

A. et v. femoralis

Femur

FIG. 3–23.—Thoracic and abdominal viscera shown in their normal relations to the skeleton, from the right side. (Eycleshymer and Jones.)

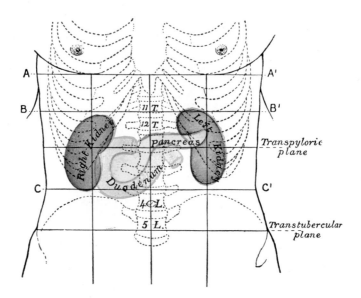

Fig. 3–24.—Front of abdomen, showing surface markings for duodenum, pancreas, and kidneys. *A A'*, Plane through joint between body and xiphoid process of sternum. *B B'*, Plane midway between *A A'* and transpyloric plane. *C C'*, Plane midway between transpyloric and transtubercular planes.

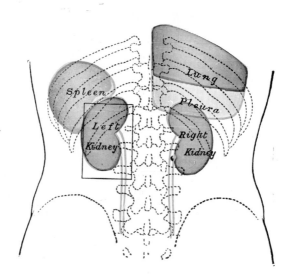

Fig. 3–25.—Back of lumbar region, showing surface markings for kidneys, ureters, and spleen. The lower portions of the lung and pleura are shown on the right side.

Liver (Fig. 3–22).—The upper limit of the right lobe of the liver, in the midline, is at the level of the junction between the body of the sternum and the xiphoid process; on the right side the line must be carried upward as far as the fifth costal cartilage in the mammary line, and then downward to reach the seventh rib at the side of the thorax. The upper limit of the left lobe can be defined by continuing this line downward and to the left to the sixth costal cartilage, 5 cm from the midline. The lower limit can be indicated by a line drawn 1 cm below the lower margin of the thorax on the right side as far as the ninth costal cartilage, thence obliquely upward to the eighth left costal cartilage, crossing the midline just above the transpyloric plane and finally, with a slight left convexity, to the end of the line indicating the upper limit.

The fundus of the **gallbladder** approaches the surface behind the anterior end of the ninth right costal cartilage close to the lateral margin of the Rectus abdominis.

Pancreas (Fig. 3–24).—The pancreas lies in front of the second lumbar vertebra. Its head occupies the curve of the duodenum and is therefore indicated by the same lines as that viscus; its neck corresponds to the pylorus. Its body extends along the transpyloric line, the bulk of it lying above this line; the tail is in the left hypochondriac region slightly to the left of the lateral line and above the transpyloric line.

Spleen (Figs. 3–17, 3–19, 3–20).—The long axis of the spleen corresponds to that of the tenth rib, and it is situated between

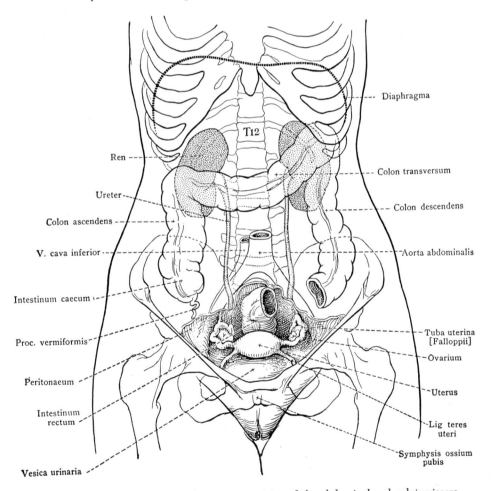

FIG. 3–26.—Projection showing the average position of the abdominal and pelvic viscera in the female. Anterior view. (Eycleshymer and Jones.)

the upper border of the ninth and the lower border of the eleventh ribs. Its medial end is about 4 cm from the midline in back, its lateral end in the midaxillary line at the ninth intercostal space; the highest point is at the ninth rib in the scapular line and the lowest point is at the level of the spine of the first lumbar vertebra in the posterior axillary line. It is posterior to the stomach more than lateral to it.

Kidneys (Figs. 3–24, 3–25).—The right kidney usually lies about 1 cm lower than the left, but for practical purposes similar surface markings are taken for each.

On the front of the abdomen the upper pole lies midway between the plane of the lower end of the body of the sternum and the transpyloric plane, 5 cm from the midline. The lower pole is situated midway between the transpyloric and intertubercular planes, 7 cm from the midline. The hilum is on the transpyloric plane, 5 cm from the midline. Around these three points a kidney-shaped figure 4 to 5 cm broad is drawn, two-thirds of which lies medial to the lateral line. To indicate the position of the kidney from the back, the parallelogram of Morris is used; two vertical lines are drawn, the first 2.5 cm, the second 9.5 cm from the midline; the parallelogram is completed by two hori-

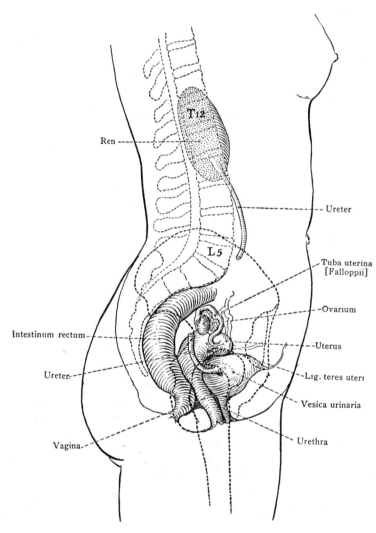

Fig. 3–27.—Projection showing the average position of the female pelvic organs.
Lateral view. (Eycleshymer and Jones.)

zontal lines drawn at the level of the tip of the spinous process of the eleventh thoracic and the lower border of the spinous process of the third lumbar vertebra. The hilum is 5 cm from the midline at the level of the spinous process of the first lumbar vertebra (see Urogenital System Chapter for X-rays).

Ureters.—On the front of the abdomen, the line of the ureter runs from the hilum of the kidney to the pubic tubercle; on the back, from the hilum vertically downward, passing practically across the posterior superior iliac spine (Fig. 3-26).

Vessels (Fig. 3–28).—The **inferior epigastric artery** can be marked out by a line from a point midway between the anterior superior iliac spine and the pubic symphysis to the umbilicus. This line also indicates the lateral boundary of **Hesselbach's triangle** (important in inguinal hernia); the other boundaries being the lateral edge of Rectus abdominis, and the medial half of the inguinal ligament. The **abdominal aorta** begins in the midline about 4 cm above the

transpyloric line and extends to a point 2 cm below and to the left of the umbilicus (AA', Fig. 3–28). The point of termination of the abdominal aorta corresponds to the level of the fourth lumbar vertebra; a line drawn from it to a point midway between the anterior superior iliac spine and the symphysis pubis indicates the common and external iliac arteries.

Of the larger branches of the abdominal aorta, the **celiac artery** is 4 cm, the **superior mesenteric** 2 cm above the transpyloric line; the **renal arteries** are 2 cm below the same line. The **inferior mesenteric** artery is 4 cm above the bifurcation of the abdominal aorta.

Nerves.—The thoracic nerves on the anterior abdominal wall are represented by lines continuing those of the bony ribs. The termination of the seventh nerve is at the level of the xiphoid process, the tenth reaches the vicinity of the umbilicus, the twelfth ends about midway between the umbilicus and the upper border of the symphysis pubis. The first lumbar is parallel

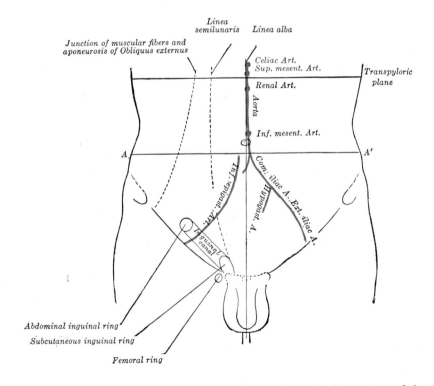

FIG. 3–28.—The front of the abdomen, showing the surface markings for the arteries and the inguinal canal. The line A A' is drawn at the level of the highest points of the iliac crests.

to the thoracic nerves; its iliohypogastric branch becomes cutaneous above the subcutaneous inguinal ring; its ilioinguinal branch at the ring.

THE PERINEUM

A line drawn transversely across in front of the ischial tuberosities divides the perineum into a posterior or rectal and an anterior or urogenital triangle (see Fig. 6–29). This line passes through the central point of the perineum, which is situated about 2.5 cm in front of the center of the anal aperture, or, in the male, midway between the anus and the reflection of the skin on to the scrotum.

Rectal Examination.—A finger inserted through an anal orifice is compressed by the Sphincter ani externus, passes into the region of the Sphincter ani internus, and higher up encounters the resistance of the Pubo-rectalis; beyond this it may reach the lowest of the transverse rectal folds. In front, the urethral bulb and membranous part of the urethra are first identified, and then about 4 cm above the anal orifice the prostate is felt; beyond this the seminal vesicles, if enlarged, and the fundus of the bladder, when distended, can be recognized. On either side is the ischiorectal fossa. Behind are the anococcygeal body, the pelvic surfaces of the coccyx and lower end of the sacrum, and the sacrospinous ligaments.

In the female the posterior wall and fornix of the vagina, and the cervix and body of the uterus can be felt in front, while somewhat laterally the ovaries can just be reached.

Male Urogenital Organs (see Urogenital System).—The **corpora cavernosa penis** can be followed backward to the crura which are attached to the sides of the pubic arch. The **glans penis,** covered by the prepuce, and the external urethral orifice can be examined, and the course of the urethra traced along the under surface of the penis to the bulb which is situated immediately in front of the central point of the perineum. Through the wall of the **scrotum** on either side the **testis** can be palpated; it lies toward the back of the scrotum, and along its posterior border the **epididymis** can be felt; passing upward along the medial side of the epididymis is the **spermatic cord,** which can be traced upward to the superficial inguinal ring.

Female Urogenital Organs (Figs. 3–26, 3–27).—In the **pudendal cleft** (see Index) between the labia minora are the openings of the **vagina** and **urethra.** In the virgin the vaginal opening is partly closed by the **hymen**—after coitus the remains of the hymen are represented by the carunculae hymenales. Between the hymen and the frenulum of the labia is the **fossa navicularis,** while in the groove between the hymen and the labium minus, on either side, the small opening of the **greater vestibular** (*Bartholin's*) **gland** can be seen. These glands when enlarged can be felt on either side of the posterior part of the vaginal orifice.

Vaginal Examination.—With the examining finger inserted into the vagina the following structures can be palpated through its wall (Fig. 3–27). Behind from below upward, are the **anal canal,** the **rectum,** and the **rectouterine excavation.** Projecting into the roof of the vagina is the vaginal portion of the cervix uteri with the external uterine orifice; in front of and behind the cervix the anterior and posterior **vaginal fornices** respectively can be examined. With the finger in the vagina and the other hand exerting pressure on the abdominal wall the whole of the **cervix** and **body of the uterus,** the **uterine tubes,** and the **ovaries** can be palpated. If a speculum be introduced into the vagina, the walls of the passage, the vaginal portion of the cervix, and the external uterine orifice can all be exposed for visual examination.

The external urethral orifice lies in front of the vaginal opening; the angular gap in which it is situated between the two converging labia minora is termed the **vestibule.** The urethral canal in the female is very dilatable and can be explored with the finger. About 2.5 cm in front of the external orifice of the urethra are the **glans** and **prepuce of the clitoris,** and still farther forward is the **mons pubis.**

THE UPPER LIMB

Bones.—The **clavicle** can be felt throughout its entire length (Fig. 3–18).

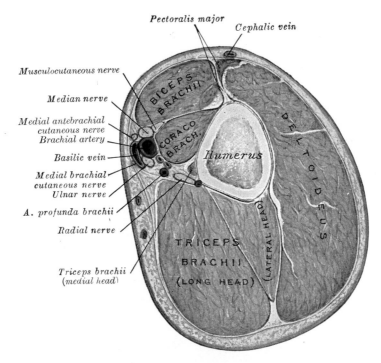

Fig. 3–29.—Cross section through the arm at the junction of the proximal with the intermediate one-third of the humerus.

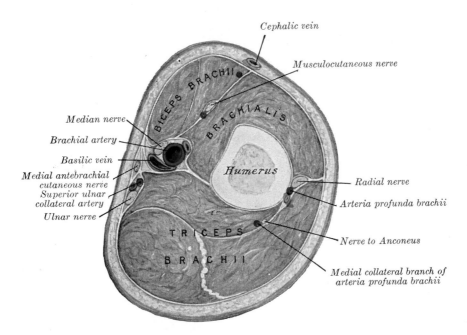

Fig. 3–30.—Cross section through the arm, a little below the middle of the body of the humerus.

Extensor digitorum

Ext. carpi ulnaris

Ext. carpi radialis

Vertebra prominens

Lateral epicondyle

Biceps brachii

Trapezius

Acromion

Triceps, lat. head

Spine of scapula

Deltoideus

Infraspinatus

Inferior angle of scapula

Latissimus dorsi

Erector spinae

Fig. 3–31.—Back of muscular subject demonstrating muscles. (Gourgott.)

The only parts of the **scapula** that are truly subcutaneous are the spine and acromion, but the coracoid process, the vertebral border, the inferior angle, and to a lesser extent the axillary border can also be readily defined. The acromion and spine are easily recognizable throughout their entire extent, forming with the clavicle the arch of the shoulder. The acromion forms the point of the shoulder; it joins the clavicle at an acute angle—the acromial angle—slightly medial to, and behind the tip of the acromion. The spine can be felt as a distinct ridge, marked on the surface as an oblique depression which becomes less distinct and ends in a slight dimple a little **lateral** to the spinous processes of the

vertebrae (Figs. 3–11, 3–31). The coracoid process is situated about 2 cm below the junction of the intermediate and lateral thirds of the clavicle; it is covered by the anterior border of Deltoideus, and thus lies a little lateral to the infraclavicular fossa or depression which marks the interval between the Pectoralis major and Deltoideus.

The **humerus** is almost entirely surrounded by muscles, and the only parts which are strictly subcutaneous are small portions of the medial and lateral epicondyles; in addition to these, however, the tubercles and a part of the head of the bone can be felt under the skin and muscles by which they are covered. The greater tubercle forms the most prominent bony point of the shoulder, extending beyond the acromion. On either side of the elbow joint and just above it are the medial and lateral epicondyles. Of these, the former is the more prominent (Figs. 3–31, 3–35) but the medial supracondylar ridge passing upward from it is much less marked than the lateral, and as a rule is not palpable.

The most prominent part of the ulna, the olecranon, can always be identified at the back of the elbow joint. The prominent dorsal border can be felt in its whole length and the styloid process forms a prominent tubercle continuous above with the dorsal border and ending below in a blunt apex at the level of the wrist joint (Figs. 3–33, 3–34).

Below the lateral epicondyle of the humerus a portion of the head of the radius is palpable; its position is indicated on the surface by a little dimple, which is best seen when the arm is extended. The upper half of the body of the bone is obscured by muscles; the lower half, though not subcutaneous, can be traced downward to a lozenge-shaped convex surface on the lateral side of the base of the styloid process.

On the front of the wrist are two subcutaneous eminences, one, on the radial side, the larger and flatter, produced by the tuberosity of the scaphoid and the ridge on the trapezoid bone; the other, on the ulnar side, by the pisiform. The rest of the palmar surface of the bony carpus is covered by tendons and the transverse carpal ligament, and is entirely concealed, with the exception of the hamulus of the hamate bone, which, however, is difficult to define. On the dorsal surface of the carpus only the triangular bone can be clearly made out (Figs. 5–43, 5–44).

Distal to the carpus the dorsal surfaces of the metacarpal bones, covered by the Extensor tendons are visible only in very thin hands; the dorsal surface of the fifth is, however, subcutaneous throughout almost its whole length. The heads of the metacarpal bones can be plainly seen and felt, rounded in contour and standing out in bold relief under the skin when the fist is clenched; the head of the third is the most prominent.

The enlarged ends of the phalanges can be easily felt. When the digits are bent the proximal phalanges form prominences, which in the joints between the first and

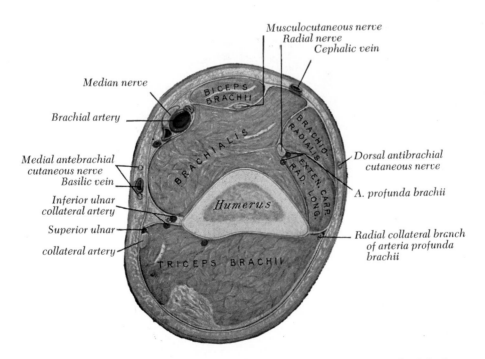

Fig. 3–32.—Cross section through the arm, 2 cm proximal to the medial epicondyle of the humerus.

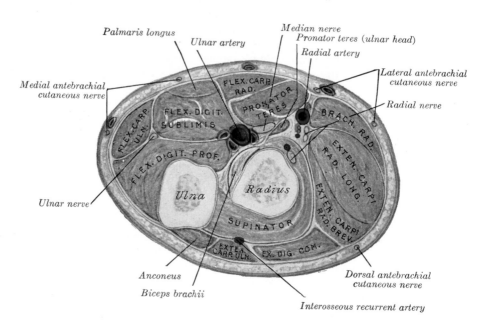

Fig. 3–33.—Cross section through the forearm at the level of the radial (bicipital) tuberosity.

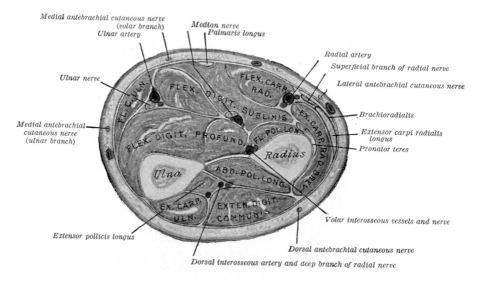

Fig. 3–34.—Cross section through the middle of the forearm.

second phalanges are slightly hollow, but flattened and square-shaped in those between the second and third.

Articulations.—The **sternoclavicular joint** is subcutaneous, and its position is indicated by the enlarged sternal extremity of the clavicle, lateral to the long cord-like sternal head of the Sternocleidomastoideus. If this muscle be relaxed a depression between the end of the clavicle and the sternum can be felt, defining the exact position of the joint.

The position of the **acromioclavicular joint** can generally be ascertained by determining the slightly enlarged acromial end of the clavicle which projects above the level of the acromion; sometimes this enlargement is so considerable as to form a rounded eminence.

The **shoulder joint** (Fig. 5–31) is deeply seated and cannot be palpated. If the forearm be slightly flexed a curved crease or fold with its convexity downward is seen in front of the elbow, extending from one epicondyle to the other; the **elbow joint** is

slightly distal to the center of the fold. The position of the **radiohumeral joint** can be ascertained by feeling for a slight groove or depression between the head of the radius and the capitulum of the humerus, at the back of the elbow joint (Figs. 5–34, 5–35).

The position of the **proximal radioulnar joint** is marked on the surface at the back of the elbow by the dimple which indicates the position of the head of the radius. The site of the **distal radioulnar joint** can be defined by feeling for the slight groove at the back of the wrist between the prominent head of the ulna and the lower end of the radius, when the forearm is in a state of almost complete pronation.

Of the three transverse skin furrows on the front of the wrist, the middle corresponds fairly accurately with the **wrist joints**, while the most distal indicates the position of the **midcarpal articulation.**

The **metacarpophalangeal** and **interphalangeal joints** are readily available for surface examination; the former are situated

Flexor digitorum
Palmaris longus
Biceps
Sternocleidomastoideus
Triceps, medial head
Triceps, lateral head
Pectoralis major
Teres major
Subscapularis
Serratus anterior
Obliquus externus

FIG. 3–35.—Anterior of head and chest of muscular subject demonstrating muscles. (Gourgott.)

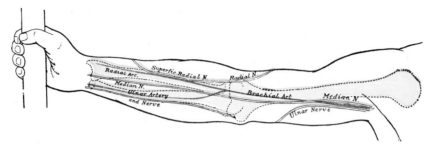

FIG. 3–36.—Front of right upper extremity, showing surface markings for bones, arteries, and nerves.

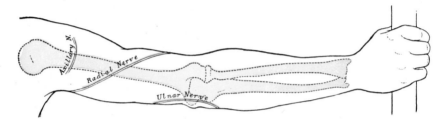

FIG. 3–37.—Back of right upper extremity, showing surface markings for bones and nerves.

just distal to the prominences of the knuckles; the latter are sufficiently indicated by the furrows on the palmar and the wrinkles on the dorsal surfaces.

Muscles.—The large muscles which arise on the trunk and insert on the shoulder girdle form the outer layers of muscle of the thorax; the **Pectoralis major** (Figs. 3–18, 3–35), **Latissimus dorsi** (Fig. 3–31), and **Trapezius** (Fig. 3–11) are easily visible in most subjects, but the **Serratus anterior** (Figs. 3–13, 3–35) is mostly concealed by the scapula and the **Pectoralis minor** is completely hidden by the Pectoralis major.

The **Deltoideus** (Fig. 3–13) and the **Teres major** (Fig. 3–35) are prominent on the shoulder. The **Biceps, Coracobrachialis,** and **Brachialis** (Fig. 3–13) are prominent on the anterior, and the **Triceps** (Fig. 3–35) on the posterior aspect of the arm. The muscles of the forearm form two groups, the flexor and pronator group arising from the medial epicondyle and making the medial group, the extensor and supinator group arising from the lateral epicondyle and making the lateral group. The tendons of insertion of many of the individual muscles can be identified at the wrist (Figs. 3–13, 3–35).

The **antecubital fossa** (Fig. 3–18), in front of the elbow joint, is triangular in

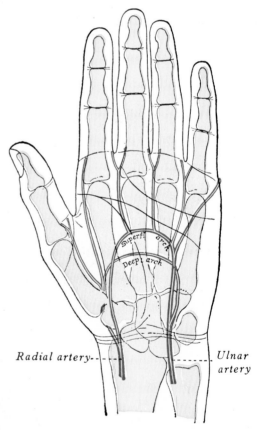

FIG. 3–38.—Palm of left hand, showing position of skin creases and bones, and surface markings for the palmar arterial arches.

shape, and its boundaries are the Brachioradialis, a member of the lateral group of antebrachial muscles, the Pronator teres, a member of the medial group, and the tendon of the Biceps and Brachialis above. In the fossa the brachial artery and median nerve may be palpated, and through it the median cubital vein communicates with the deeper veins.

The **Synovial Tendon Sheaths** of the palm and dorsum of the hand cannot be identified except as they follow the various tendons across the wrist and out in the fingers (see Chapter 6).

Arteries.—Above the middle of the clavicle the pulsation of the **subclavian artery** can be detected by pressing downward, backward, and medialward against the first rib. The pulsation of the **axillary artery** as it crosses the second rib can be felt below the middle of the clavicle just medial to the coracoid process; along the lateral wall of the axilla the course of the artery can be easily followed close to the medial border of Coracobrachialis. The **brachial artery** can be recognized in practically the whole of its extent, along the medial margin of the Biceps. Over the lower end of the radius, between the styloid process and Flexor carpi radialis, a portion of the **radial artery** is superficial and is used clinically for observations on the pulse (Fig. 3–36).

The **superficial palmar arch** (Fig. 3–38) can be indicated by a line starting from the radial side of the pisiform bone and curving distalward and lateralward as far as the base of the thumb, with its convexity toward the fingers. The summit of the arch is usually on a level with the ulnar border of the outstretched thumb. The **deep palmar arch** is practically transverse, and is situated about 1 cm nearer to the carpus than the superficial palmar arch.

Veins.—The superficial veins of the upper limb are easily rendered visible by compressing the proximal trunks; their arrangement is described in Chapter 9.

Nerves (Figs. 3–36, 3–37).—The uppermost trunks of the **brachial plexus** are palpable for a short distance above the clavicle as they emerge from under the lateral border of Sternocleidomastoideus; the larger nerves derived from the plexus

can be rolled under the finger against the lateral axillary wall but cannot be identified. The **ulnar nerve** can be detected in the groove behind the medial epicondyle of the humerus.

THE LOWER LIMB

Bones.—The **hip bones** are largely covered with muscles, so that only at a few points do they approach the surface. In front the anterior superior iliac spine is easily recognized, and the iliac crest can be traced to the posterior superior iliac spine, the site of which is indicated by a slight depression; on the outer lip of the crest, about 5 cm behind the anterior superior spine, is the prominent iliac tubercle. In thin subjects the pubic tubercle is very apparent, but in the obese it is obscured by the pubic fat; it can, however, be detected by following up the tendon of origin of the Adductor longus. Another part of the bony pelvis which is accessible to touch is the ischial tuberosity, situated beneath the Gluteus maximus, and, when the hip is flexed, easily felt, as it is then not covered by muscle.

The **femur** is enveloped by muscles, so that the only accessible parts are the lateral surface of the greater trochanter and the lower expanded end of the bone. The greater trochanter is generally indicated by a depression, owing to the thickness of the Glutei medius and minimus which project above it. The lateral condyle is more easily felt than the medial; both epicondyles can be readily identified, and at the upper part of the medial condyle the sharp adductor tubercle can be recognized without difficulty.

The anterior surface of the **patella** is subcutaneous.

A considerable portion of the **tibia** is subcutaneous. At the upper end the condyles can be felt just below the knee. In front of the upper end of the bone, between the condyles, is an oval eminence, the tuberosity, which is continuous below with the anterior crest of the bone. The medial malleolus forms a broad prominence, situated at a higher level and somewhat farther forward than the lateral malleolus.

The only subcutaneous parts of the **fibula** are the head, the lower part of the body, and the lateral malleolus. The lateral malleolus is a narrow elongated prominence, from which the lower third or half of the lateral surface of the body of the bone can be traced upward.

On the dorsum of the tarsus the individual bones cannot be distinguished, with the exception of the head of the **talus**, which forms a rounded projection in front of the ankle joint when the foot is forcibly extended. The whole dorsal surface of the foot has a smooth convex outline, the summit of which is the ridge formed by the head of the talus, the navicular, the intermediate cuneiform, and the second metatarsal bone. On the medial side of the foot the medial process of the tuberosity of the **calcaneus** is in front of this, and below the medial malleolus is the sustentaculum tali. The

tuberosity of the **navicular** is palpable about 2.5 to 3 cm in front of the medial malleolus.

Farther forward, the ridge formed by the base of the **first metatarsal bone** can be obscurely felt; beneath the base of the first phalanx is the medial sesamoid bone. On the lateral side of the foot the most posterior bony point is the lateral process of the tuberosity of the calcaneus; in front of this the greater part of the lateral surface is subcutaneous. Farther forward the base of the **fifth metatarsal bone** is prominent.

The dorsal surfaces of the **metatarsal bones** are easily defined; the plantar surfaces are obscured by muscles. The **phalanges** in their whole extent are readily palpable.

Articulations.—The **hip joint** is deeply seated and cannot be palpated (Fig. 5–45).

The interval between the tibia and femur

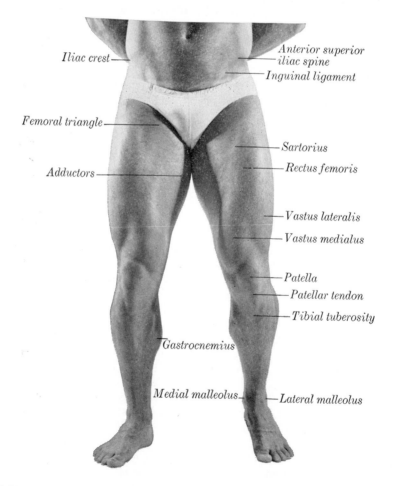

Fig. 3–39.—Anterior of lower limbs and abdomen to show surface markings. (Gourgott.)

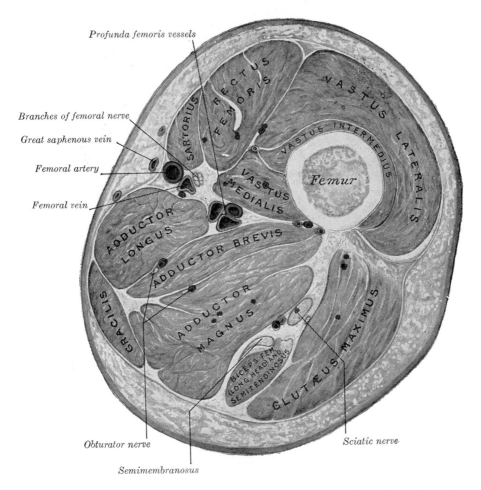

Profunda femoris vessels

Branches of femoral nerve

Great saphenous vein

Femoral artery

Femoral vein

Obturator nerve

Semimembranosus

Sciatic nerve

Fig. 3–40.—Cross section through the thigh at the level of the apex of the femoral triangle.
Four-fifths of natural size.

can always be felt; if the **knee joint** be extended this interval is on a higher level than the apex of the patella, but if the joint be slightly flexed it is directly behind the apex.

The **ankle joint** can be felt on either side of the Extensor tendons, and during extension of the joint the superior articular surface of the talus presents below the anterior border of the lower end of the tibia (Fig. 5–65).

Muscles.—The prominent muscles on the anterior aspect of the thigh are the **Quadriceps** femoris, Sartorius and **Tensor fasciae latae** (Fig. 3–43). At the medial side of the thigh are the **Adductors** (Fig. 3–39), and at the back of the thigh the hamstrings, medially the **Semimembranosus**

and **Semitendinosus** and laterally the **Biceps femoris** (Fig. 3–43). Above, the prominence of the buttock is caused mostly by the **Gluteus maximus.** At the back of the leg are the **Gastrocnemius** and **Soleus,** terminating at the heel in the tendon of Achilles or tendo calcaneus (Fig. 3–46). On the lateral aspect of the leg, the **Peronei** and **Tibialis anterior** can usually be recognized (Fig. 3–46).

The **femoral triangle** is bounded above by the inguinal ligament, laterally by the medial border of Sartorius, and medially by the medial border of Adductor longus. In the triangle is the hiatus saphenus, through which the great saphenous vein dips to join the femoral; the center of this hiatus is about 4 cm below and lateral to the **pubic**

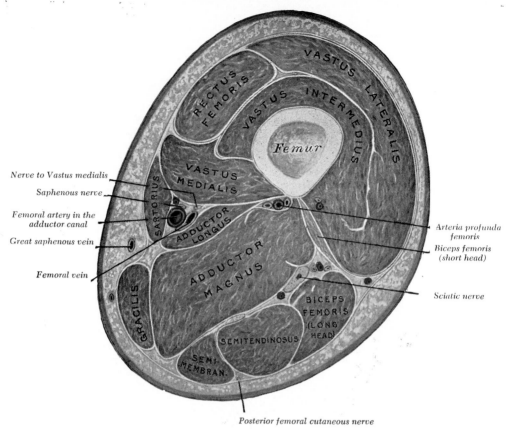

FIG. 3–41.—Cross section through the middle of the thigh. Four-fifths of natural size.

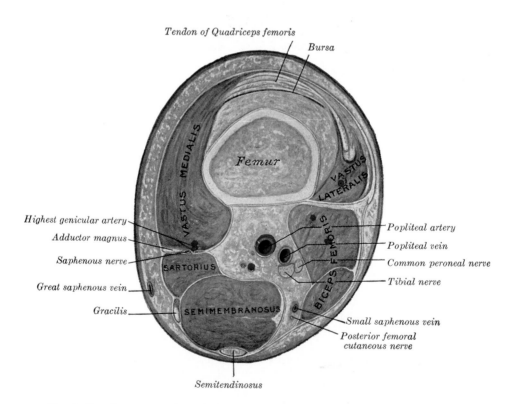

FIG. 3–42.—Cross section through the thigh, 4 cm proximal to the adductor tubercle of the femur. Four-fifths of natural size.

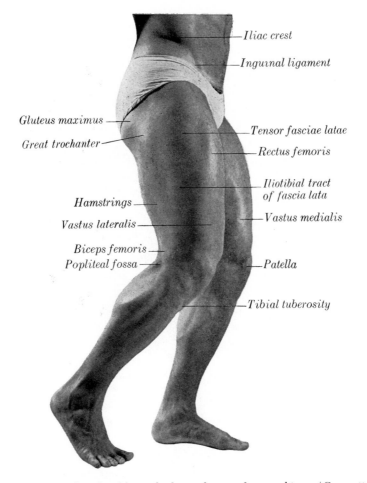

Iliac crest

Inguinal ligament

Gluteus maximus

Great trochanter

Tensor fasciae latae

Rectus femoris

Iliotibial tract
of fascia lata

Hamstrings

Vastus lateralis

Vastus medialis

Biceps femoris

Popliteal fossa

Patella

Tibial tuberosity

FIG. 3–43.—Right side of lower limbs to show surface markings. (Gourgott.)

tubercle; its vertical diameter measures about 4 cm and its transverse about 1.5 cm. The femoral ring is about 1.25 cm lateral to the pubic tubercle.

The **adductor canal** occupies the medial part of the middle third of the thigh; it begins at the apex of the femoral triangle and lies deep to the vertical part of Sartorius (Fig. 3–41).

The popliteal fossa, at the back of the knee, is bounded above by the medial and lateral hamstrings, and below by the medial and lateral heads of the Gastrocnemius. In this fossa the popliteal artery and sciatic nerve are covered only by subcutaneous tissue and fat, and through it the small saphenous vein joins the popliteal vein.

Synovial Sheaths.—The positions of the synovial sheaths around the tendons about

the ankle joints are described and illustrated in the Muscle Chapter.

Arteries.—The **femoral artery** as it crosses the brim of the pelvis is readily felt; in its course down the thigh its pulsation becomes gradually more difficult of recognition. When the knee is flexed the pulsation of the **popliteal artery** can easily be detected in the popliteal fossa.

On the lower part of the front of the tibia the **anterior tibial artery** becomes superficial and can be traced over the ankle into the **dorsalis pedis**; the latter can be followed to the proximal end of the first intermetatarsal space. The pulsation of the **posterior tibial artery** becomes evident near the lower end of the back of the tibia, and is easily detected behind the medial malleolus.

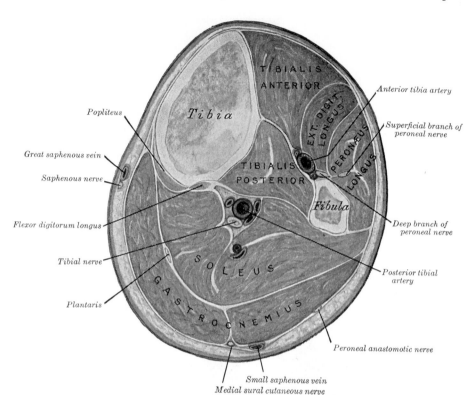

FIG. 3–44.—Cross section through the leg, 9 cm distal to the knee joint.

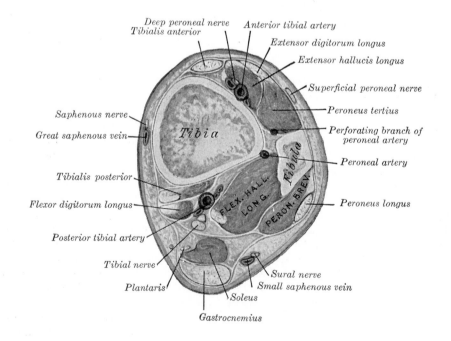

FIG. 3–45.—Cross section through the leg, 6 cm proximal to the tip of the medial malleolus.

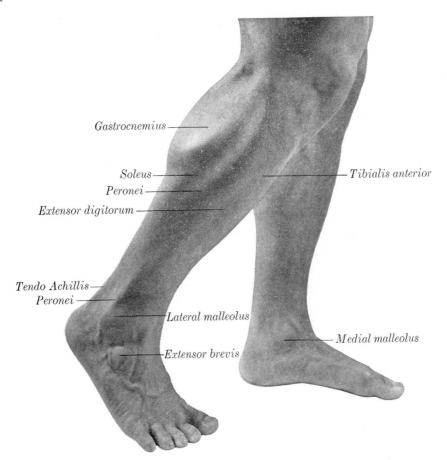

Gastrocnemius

Soleus
Peronei
Extensor digitorum

Tibialis anterior

Tendo Achillis
Peronei

Lateral malleolus

Medial malleolus

Extensor brevis

FIG. 3–46.—Right side of leg to show surface markings. (Gourgott.)

Veins.—By compressing the proximal trunks, the venous arch on the dorsum of the foot, together with the great and small saphenous veins leading from it are rendered visible.

Nerves.—The only nerve of the lower limb which can be located by palpation is the **common peroneal** as it winds around the lateral side of the neck of the fibula.

REFERENCES

ANSON, B. J. 1966. *Morris' Human Anatomy. A Complete Systemic Treatise*. 12th edition. xi + 1623 pages. The Blakiston Division, McGraw-Hill Book Co., New York.

BARGMANN, LEONHARDT, and TÖNDURY. 1968. *Rauber-Kopsch Lehrbuch und Atlas der Anatomie des Menschen*. 20th edition. 3 volumes. Georg Thieme Verlag, Stuttgart.

BASSETT, D. L. 1952–63. *A Stereoscopic Atlas of Human Anatomy*. 8 sections, 23 volumes, 1,351 views. Sawyer's, Inc., Portland, and C. V. Mosby, St. Louis.

CROUCH, JAMES E. 1972. *Functional Human Anatomy*. 2nd edition, xviii + 649 pages, 393 illustrations. Lea & Febiger, Philadelphia.

FIGGE, H. J. FRANK, 1963. *Sobotta's Atlas of Human Anatomy*. Vol. 3, part 1. xx + 175 figures. Part 2, xx + figures 176–481. Hafner Publishing Co., New York.

GOERTTLER, K. 1968. *Benninghoff's Lehrbuch der Anatomie des Menschen*. 10th edition. 3 volumes. Urben & Schwartzenberg, München.

HOLLINSHEAD, W. H. 1971. *Anatomy for Surgeons*. 2nd edition, 3 volumes. Harper-Hoeber, New York.

HUBER, G. C. 1930. *Piersol's Human Anatomy, Including Structure, Development, and Practical Considerations*. 9th edition. xx + 2104 pages. J. B. Lippincott Company, Philadelphia.

KAPLAN, EMANUEL B. 2nd edition. 1965. *Functional and Surgical Anatomy of the Hand.* xiv + 337 pages, 182 figures. J. B. Lippincott Co., Philadelphia.

KISS, F. and J. SZENTAGOTHAI. 1964. *Atlas of Human Anatomy.* 3 volumes, illustrated. Pergamon Press, New York.

LACHMAN, ERNST. 1965. *Case Studies in Anatomy.* xii + 238 pages, illustrated. Oxford University Press, New York.

LAST, R. J. 1959. *Anatomy Regional and Applied.* 2nd edition. xvi + 741 pages, 372 figures. Little, Brown and Company, Boston.

LOPEZ-ANTUNEZ, L. 1971. *Atlas of Human Anatomy.* Translated by H. Monsen, illustrated by L. A. Gasparo. xxxi + 366 pages, 163 plates. W. B. Saunders Company, Philadelphia.

PANSKY, B. and E. L. HOUSE. 1969. *Review of Gross Anatomy, a Dynamic Approach.* 2nd edition. xiii + 494 pages, illustrated. The Macmillan Company, New York.

PERNKOPF, EDWARD. 1963. *Atlas of Topographical and Applied Anatomy.* Edited by Helmut Ferner, translated by Harry Monssen. Vol. 1, Head and Neck, xv + 356 pages, 332 figures. Vol. 2, Thorax, Abdomen and Extremities, xvi + 421 pages, 378 figures. W. B. Saunders Company, Philadelphia.

RASCH, P. J. and R. K. BURKE. 1967. *Kinesiology and Applied Anatomy.* 3rd edition. xv + 488 pages, illustrated. Lea & Febiger, Philadelphia.

REHMAN, IRVING and N. HIATT. 1965. *Descriptive Atlas of Surgical Anatomy.* x + 180 pages + 106 plates. Blakiston Division, McGraw-Hill Book Co., New York.

ROUVIERE, H. 1954. *Anatomie Humaine–Descriptive et topographique.* 7th edition, 3 volumes. Masson et Cie, Editeurs, Paris.

TÖNDURY, GIAN. 1965. *Angewandte und topographische Anatomie.* 3rd edition. xv + 657 pages, 499 figures. Georg Thieme Verlag, Stuttgart.

WARFEL, J. H. 1967. *Quiring's The Head, Neck and Trunk.* 3rd edition, 128 pages, 111 illustrations. Lea & Febiger Philadelphia.

WARFEL, J. H. 1967. *Quiring's The Extremitics.* 3rd edition. 124 pages, 110 illustrations. Lea & Febiger, Philadelphia.

WOLF-HEIDEGGER, G. 1962. *Atlas of Systematic Human Anatomy.* 3 vols. in one, viii + 223 + 245 + 167 pages, illustrated. Hafner Publishing Company, New York.

4 | *Osteology*

study of bones is called osteology. The bones provide a framework for the body and when they are assembled in their proper position they make up the skeleton. In the living body they are held in position by strong fibrous bands, the ligaments, and they are moved about by the muscles which are attached to them. In many places they provide a protective covering for more delicate vital organs. Because of their large mineral content, they survive for many years after death, becoming the symbols of grave yards and the specimens of museums collected for the study of comparative anatomy, physical anthropology, archeology, and paleontology.

In the skeleton of the adult there are 206 distinct bones, as follows:—

Axial Skeleton	Vertebral column .	26	
	Skull	22	
	Hyoid bone . .	1	
	Ribs and sternum .	25	
		—	74
Appendicular Skeleton	Upper extremities .	64	
	Lower extremities .	62	
		—	126
Auditory ossicles			6
			—
Total			206

The patellae are included in this enumeration, but the smaller sesamoid bones are not.

Development of the Skeleton

The skeleton is derived from the mesoderm which grows out from the primitive streak. Its cells multiply, undergo certain changes, migrate out into the appropriate regions, and form aggregates or sheets which can be recognized as the cellular or **membranous skeleton.** The cells of these primordia, except in certain parts of the cranium, soon elaborate an intercellular substance which converts them into cartilage. After the **cartilaginous skeleton** is well established, centers of ossification appear which gradually spread to form the **bony skeleton.** In some of the cranial bones ossification centers arise in the membranous skeleton without the intervention of cartilage (page 98).

The Vertebral Column.—The central axis of the embryo around which the vertebrae develop is the notochord (page 24). This rod of cells lies in the median line between the neural tube and the primitive gut, extending from the level of the midbrain to the end of the tail (Fig. 4–4). The somites in the paraxial mesoderm on either side of the neural tube give rise to three primordia, the sclerotome, myotome, and dermotome (see page 27). The cells of the sclerotome multiply rapidly and migrate toward the notochord forming the **sclerotogenous layer** on either side. The **segmental primordia** of the individual vertebrae are recognizable within each segment containing a densely arranged caudal portion and loosely arranged cranial portion. The densely arranged cells soon migrate laterally and cranially until they reach the level of the center of the original somite where they spread back to the midline around the notochord. Thus the original loosely arranged portion is separated into **cranial** and **caudal halves.** The **dense portion** becomes the future **intervertebral disk** and the **loosely** arranged **portions** of two adjacent somites left in close juxtaposition combine into the future **vertebral body** or **centrum.** In a similar way, the extension of the sclerotome dorsalward around the neural tube forms the future **vertebral arch** from its loosely arranged cells and the ligaments from the dense portion. The ventral extension of the sclerotome forms the future **costal process** from its looser portion

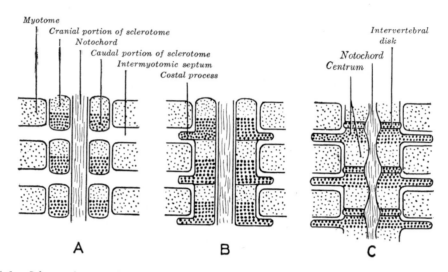

FIG. 4-1.—Scheme showing the manner in which each vertebral centrum is developed from portions of two adjacent somites or body segments.

and the ligaments from the dense portion (Fig. 4-1).

This stage is succeeded by that of the **cartilaginous vertebral column.** In the fourth week two cartilaginous centers make their appearance, one on either side of the notochord; these extend around the notochord and form the body of the cartilaginous vertebra. A second pair of cartilaginous foci appear in the lateral parts of the vertebral bow, and grow dorsally on either side of the neural tube to form the cartilaginous vertebral arch, and a separate cartilaginous center appears for each costal process. By the eighth week the cartilaginous arch has fused with the body, and in the fourth month the two halves of the arch are joined on the dorsal aspect of the neural tube. The spinous process is developed from the junction of the two halves of the neural or vertebral arch. The transverse process grows out from the vertebral arch dorsal to the costal process.

In the more cranial cervical vertebrae a band of mesodermal tissue connects the ends of the vertebral arches across the ventral surfaces of the intervertebral disks. This is termed the **hypochordal bar** or **brace;** in all except the first it is transitory and disappears by fusing with the disks. In the atlas, however, the entire bow persists and undergoes chondrification; it

develops into the ventral arch of the bone, while the cartilage representing the body of the atlas forms the dens or odontoid process which fuses with the body of the second cervical vertebra.

The portions of the notochord which are surrounded by the bodies of the vertebrae atrophy, and ultimately disappear, while those which lie in the centers of the intervertebral disks undergo enlargement, and persist throughout life as part of the central **nucleus pulposus** of the disks.

The Ribs.—The ribs are formed from the ventral or costal processes of the primitive vertebral bows, the processes extending between the muscle plates. In the *thoracic region* of the vertebral column the costal processes grow lateralward to form a series of arches, the **primitive costal arches.** As already described, the transverse process grows out dorsal to the vertebral end of each arch. It is at first connected to the costal process by continuous mesoderm, but this becomes differentiated later to form the costotransverse ligament; between the costal process and the tip of the transverse process the costotransverse joint is formed by absorption. The costal process becomes separated from the vertebral bow by the development of the costocentral joint. In the *cervical vertebrae* (Fig. 4-2) the transverse process forms the posterior boundary

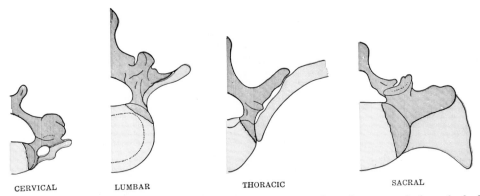

CERVICAL   LUMBAR   THORACIC   SACRAL

Fig. 4–2.—Diagrams showing the portions of the adult vertebrae derived respectively from the bodies, vertebral arches and costal processes of the embryonic vertebrae. The bodies are represented in yellow, the vertebral arches in red, and the costal processes in blue.

of the foramen transversarium, while the costal process corresponding to the head and neck of the rib fuses with the body of the vertebra, and forms the anterolateral boundary of the foramen. The distal portions of the primitive costal arches remain undeveloped; occasionally the arch of the seventh cervical vertebra undergoes greater development, and by the formation of costovertebral joints is separated off as a rib. In the *lumbar region* the distal portions of the primitive costal arches fail; the proximal portions fuse with the transverse processes to form the transverse processes of descriptive anatomy. Occasionally a movable rib is developed in connection with the first lumbar vertebra. In the *sacral region* costal processes are developed only in connection with the upper three, or it may be four, vertebrae; the processes of adjacent segments fuse with one another to form the lateral parts of the sacrum. The *coccygeal vertebrae* are devoid of costal processes.

The Sternum.—The ventral ends of the ribs become joined by a unilateral longitudinal bar termed the **sternal plate.** Opposite the first seven pairs of ribs the sternal plates of the two sides fuse in the middle line to form the manubrium and body of the sternum. The xiphoid process is formed by a caudal extension of the sternal plates.

The Skull.—The first indications of the membranous skull are found in the basioccipital and basisphenoid and about the auditory vesicles. The condensation of the mesoderm gradually extends from these areas around the brain until the latter is enclosed by the **membranous cranium.** This is incomplete in the region where the large nerves and vessels pass into or out of the cranium. Before the membranous cranium is complete, chondrification begins to show in the basioccipital. Two centers appear, one on either side of the notochord near where it enters the occipital blastema or condensed mesoderm. Chondrification gradually spreads from these centers, medially around the notochord, laterally about the roots of the hypoglossal nerve, and cranialward to unite with the spreading cartilaginous center of the basisphenoid. This forms an elongated basal plate of cartilage extending from the foramen magnum to the cranial end of the sphenoid. It continues into the blastema of the ethmoid region which later becomes chondrified. When the auditory capsules begin to chondrify they are quite widely separated from the basal plate. By the time the embryo is 20 mm in length the cochlear portion of the auditory or otic capsule is fused to the widened basal plate and the jugular foramen has become separated from the foramen lacerum (Fig. 4–3). From the lateral region of the occipital cartilage a broad thin plate of cartilage (tectum posterius or nuchal plate) extends around the caudal region of the brain in a complete ring (Fig. 4–5) forming the primitive foramen magnum. The complete **chondrocranium** is shown in Figures 4–5 and 4–7. There are other minor cartilaginous centers which unite with the main continuous mass. The chondrocranium forms only a small part of the future ossified skull.

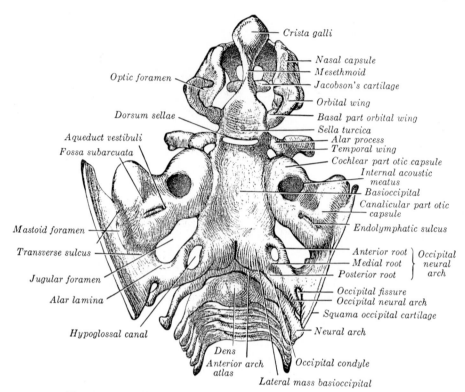

Fig. 4–3.—Cartilaginous skull of 21-mm human embryo. (Lewis.)

Various centers of ossification develop in the cartilage and give rise to all of the occipital bone, except the upper part of the squama, to the petrous and mastoid portions of the temporal, to the sphenoid, except its medial pterygoid plates and part of the temporal wings, and to the ethmoid.

The **membrane bones** of the cranial vault and face ossify directly in the mesoderm of the membranous cranium. They comprise the upper part of the occipital squama (interparietal), the squamae and tympanic parts of the temporals, the parietals, the frontal, the vomer, the medial pterygoid plates, and the bones of the face. Some of them remain distinct throughout life, e.g., parietal and frontal, while others join with the bones of the chondrocranium, e.g., interparietal, squamae of temporals, and medial pterygoid plates.

The ventral and dorsal thirds of the cranial notochord become surrounded by the cartilage of the basal plate and its middle part lies between the middle part of the basal plate and the wall of the pharynx

(Fig. 4–4). The cranial end is embedded in the basisphenoid. There are very distinct indications of an occipital vertebra at the caudal end of the occipital cartilage in embryos about 20 mm in length.

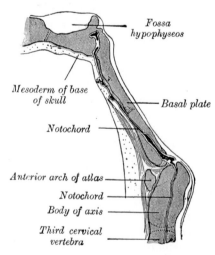

Fig. 4–4.—Sagittal section of cephalic end of notochord. (Kiebel.)

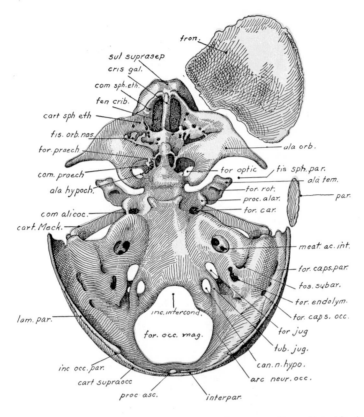

Fig. 4–5.—Cartilaginous skull of a 43-mm human embryo. (Macklin.)

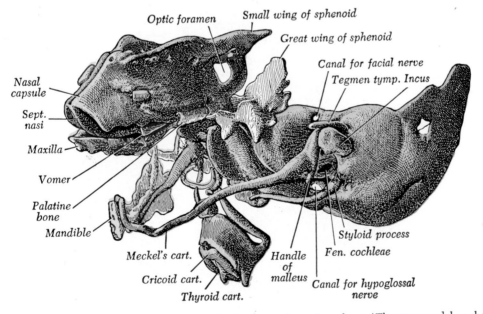

Fig. 4–6.—Model of the chondrocranium of a human embryo, 8 cm long. (The same model as shown in Fig. 4–7 from the left side.) Certain of the membrane bones of the right side are represented in yellow. (Hertwig.)

THE AXIAL SKELETON

The Vertebral Column

The **Vertebral Column** or **Backbone** (*columna vertebralis; spinal column*) (Fig. 4–9) is formed of a series of bones called vertebrae. The thirty-three vertebrae are grouped under the names cervical, thoracic, lumbar, sacral, and coccygeal, according to the regions they occupy. There are seven in the cervical region, twelve in the thoracic, five in the lumbar, five in the sacral, and four in the coccygeal.

The vertebrae in the three more cranial regions of the column remain distinct throughout life, and are known as **true** or **movable vertebrae**; those of the sacral and coccygeal regions, on the other hand, are termed **false** or **fixed vertebrae**, because they are united with one another in the adult to form two bones, five forming the sacrum, and four the terminal bone or coccyx.

A typical vertebra consists of two essential parts—viz., a ventral segment, the **body**, and a dorsal part, the **vertebral arch** (*arcus vertebralis; neural arch*), which encloses the vertebral foramen. When the bodies of the vertebrae are joined or articulated by means of fibrocartilaginous intervertebral disks, they form a strong pillar for the support of the head and trunk, and the vertebral foramina form a tube, the vertebral canal, which encloses the spinal cord.

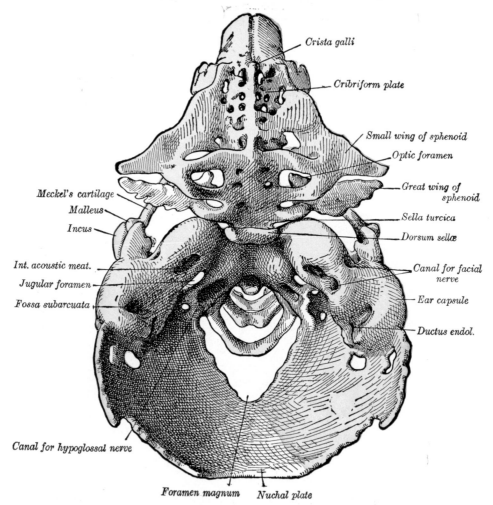

FIG. 4–7.—Model of the chondrocranium of a human embryo, 8 cm long. (Hertwig.) The membrane bones are not represented.

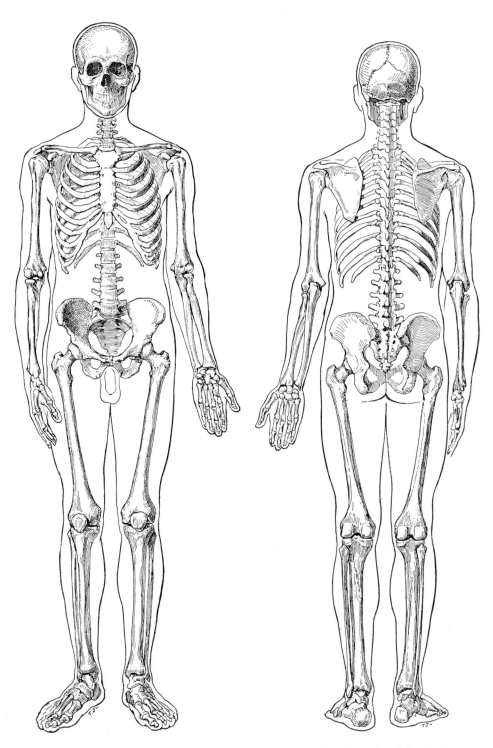

Fig. 4–8.—The skeleton as projected on the surface of the body. Ventral and dorsal views. (Eycleshymer and Jones.)

The Vertebral Column as a Whole

The average length of the vertebral column in the male is about 71 cm. Of this length the cervical part measures 12.5 cm, the thoracic about 28 cm, the lumbar 18 cm, and the sacrum and coccyx 12.5 cm. The female column is about 61 cm in length.

Curvatures.—A lateral view (Fig. 4–9) of the vertebral column presents several curves, which correspond with the different regions of the column, and are called cervical, thoracic, lumbar, and pelvic. The **cervical curve,** convex ventrally, from the apex of the dens to the middle of the second thoracic vertebra is the least marked of all the curves. The **thoracic curve,** concave ventrally, begins at the middle of the second and ends at the middle of the twelfth thoracic vertebra. Its most prominent point dorsally corresponds to the spinous process of the seventh thoracic vertebra. The **lumbar curve** is more marked in the female than in the male; it begins at the middle of the last thoracic vertebra, and ends at the sacrovertebral angle. It is convex ventrally, the convexity of the more caudal three vertebrae being much greater than that of the more cranial two. The **pelvic curve** begins at the sacrovertebral articulation, and ends at the point of the coccyx; its concavity is directed caudally and ventrally. The thoracic and pelvic curves are termed **primary curves,** because they alone are present during fetal life. The cervical and lumbar are **compensatory** or **secondary curves,** and are developed after birth, the former when the child is able to hold up its head (at three or four months), and to sit upright (at nine months), the latter at twelve or eighteen months, when the child begins to walk.

The vertebral column also has a slight **lateral curvature,** the convexity of which is directed toward the right side. This may be produced by muscular action, most persons using the right arm in preference to the left, especially in making long-continued efforts, when the body is curved to the right side. In support of this explanation it has been found that in individuals who were left-handed, the convexity was to the left side. By others this curvature is regarded as being produced by the aortic arch and upper part of the descending thoracic aorta —a view which is supported by the fact that in cases where the viscera are transposed and the aorta is on the right side, the convexity of the curve is directed to the left side.

Surfaces. Ventral Surface.—The width of the bodies of the vertebrae increases from the second cervical to the first thoracic, diminishes slightly in the next three, again progressively increases caudal to this as far as the sacrovertebral angle. From this point there is a rapid diminution, to the apex of the coccyx.

Dorsal Surface (Fig. 4–8).—The median line of the dorsal surface of the vertebral column is occupied by the spinous processes. In the cervical region (with the exception of the second and seventh vertebrae) these are short and horizontal, with bifid extremities. In the cranial part of the thoracic region they are directed obliquely caudalward; in the middle they are almost vertical, and in the more caudal part they are nearly horizontal. In the lumbar region they are nearly horizontal. The spinous processes are separated by considerable intervals in the lumbar region, by narrower intervals in the neck, and are closely approximated in the middle of the thoracic region. Occasionally one of these processes deviates a little from the median line—a fact to be remembered in practice, as irregularities of this sort are attendant also on fractures or displacements of the vertebral column. On either side of the spinous processes is the **vertebral groove** formed by the laminae in the cervical and lumbar regions, where it is shallow, and by the laminae and transverse processes in the thoracic region, where it is deep and broad; these grooves lodge the deep muscles of the back. Lateral to the vertebral grooves are the articular processes, and still more laterally the transverse processes. In the thoracic region, the transverse processes are on a plane considerably dorsal to that of the same processes in the cervical and lumbar regions. In the cervical region, the transverse processes are placed ventral to the articular processes, lateral to the pedicles and between the intervertebral foramina. In the thoracic region they are dorsal to the pedicles, intervertebral foramina, and articular processes. In the lumbar region they are ventral to the articular processes, but dorsal to the intervertebral foramina.

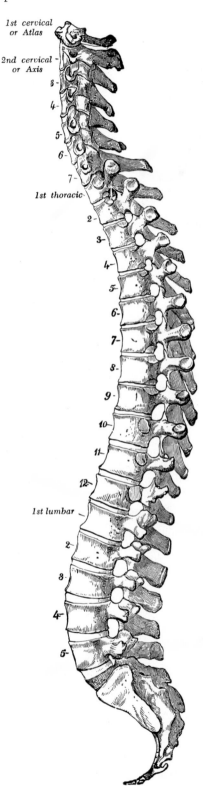

1st cervical
or Atlas

2nd cervical
or Axis

3 -

4-

5-

6-

7 -

1st thoracic

2 -

3-

4-

5-

6-

7-

8-

9-

10—

11-

12-

1st lumbar

2-

3-

4-

5-

Fig. 4–9.—Lateral view of the vertebral column.

Lateral Surfaces (Fig. 4–9).—The lateral surfaces are separated from the dorsal surface by the articular processes in the cervical and lumbar regions, and by the transverse processes in the thoracic region. They present, ventrally, the sides of the bodies of the vertebrae, marked in the thoracic region by the facets for articulation with the heads of the ribs. More dorsally the **intervertebral foramina** are formed by the juxtaposition of the vertebral notches, oval in shape, smallest in the cervical and cranial part of the thoracic regions, and gradually increasing in size to the last lumbar. They transmit the spinal nerves and are situated between the transverse processes in the cervical region, and ventral to them in the thoracic and lumbar regions.

Vertebral Canal.—The vertebral canal follows the different curves of the column; it is large and triangular in those parts of the column which enjoy the greatest freedom of movement, viz., the cervical and lumbar regions; and is small and rounded in the thoracic region, where motion is more limited.

General Characteristics of a Vertebra

With the exception of the first and second cervical, the true or movable vertebrae present certain characteristics in common which may be studied in a more or less typical example from the middle of the thoracic region (Fig. 4–10).

The **body** (*corpus vertebrae; centrum*) is the largest part of a vertebra, and is more or less cylindrical in shape. Its cranial and caudal surfaces are roughened for the attachment of the intervertebral disks, and there is a slightly elevated rim around the circumference. The middle portion of the body is somewhat constricted and the compact bone of the ventral surface presents a few small apertures for the passage of nutrient blood vessels. The dorsal surface, which faces the vertebral canal, has one or more large, irregular apertures (*foramen vasculare*) for the passage of the basivertebral veins (Fig. 4–11).

The **vertebral arch** (*arcus vertebralis*) consists of a pair of pedicles and a pair of laminae which together support seven proc-

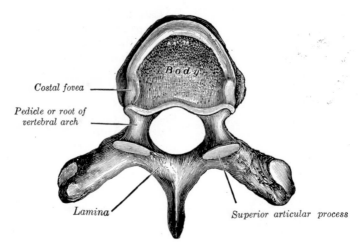

Costal fovea

Pedicle or root of
vertebral arch

Body

Lamina   Superior articular process

FIG. 4–10.—A typical thoracic vertebra, cranial aspect.

esses—viz., four articular, two transverse, and one spinous.

The **pedicles** (*pediculi arci vertebrae; radices arci vertebrae*) are two short, thick processes which project dorsally, one on either side, from the cranial part of the body, at the junction of its dorsal and lateral surfaces. They connect the transverse processes to the body and are constricted in the middle, leaving the **superior** and **inferior vertebral notches** which are combined to form the **intervertebral foramina** when the vertebrae are articulated.

The **laminae** (*laminae arci vertebrae*) are two broad plates directed dorsally and medially from the pedicles. They fuse in the middle line to complete the dorsal part of the arch and provide a base for the spinous process. Their cranial borders and the caudal parts of their ventral surfaces are rough for the attachment of the ligamenta flava.

Processes.—The **transverse processes** (*processus transversi*) project laterally at either side from the point where the lamina joins the pedicle, between the superior and inferior articular processes. They serve for the attachment of muscles and ligaments.

The **spinous process** (*processus spinosus*) is directed dorsally from the junction of the laminae, and serves for the attachment of muscles and ligaments.

The **articular processes** (*processus articulares*), two superior and two inferior, spring from the junctions of the pedicles and laminae. The superior project cranially, and their articular surfaces are directed more or less dorsally; the inferior project caudally, and their surfaces look more or less ventrally. The articular surfaces are coated with hyaline cartilage.

Structure of a Vertebra (Fig. 4–11).—The body is composed of cancellous tissue, covered by a thin coating of compact bone, whereas the arches and processes have a thick covering of compact bone. The interior of the body is traversed by one or two large canals, for the basivertebral veins which converge toward a single large, irregular aperture (*foramen vasculare*) at the dorsal part of the body. The thin bony lamellae of the cancellous tissue are more pronounced in lines perpendicular to the upper and lower surfaces and are developed in response to greater pressure in this direction. The arch and processes projecting from it have thick coverings of compact tissue.

The Cervical Vertebrae

The **Cervical Vertebrae** (*vertebrae cervicales*) (Figs. 4–12; 4–13) are the smallest of the true vertebrae, and can readily be distinguished from those of the thoracic or lumbar regions by the presence of a foramen in each transverse process. The first, second, and seventh present exceptional features and must be described separately; the following characteristics are common to the remaining four.

The **body** is small, oval, and broader in

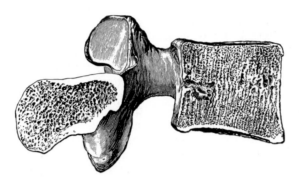

FIG. 4–11.—Sagittal section of a lumbar vertebra showing the foramen vasculare for the basivertebral vein.

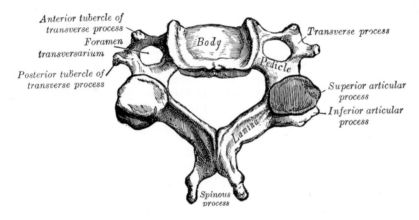

FIG. 4–12.—A cervical vertebra.

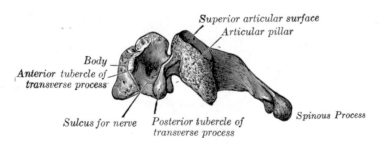

FIG. 4–13.—Side view of a typical cervical vertebra.

its transverse diameter. The ventral and dorsal surfaces are flattened and of equal depth; the ventral projects more caudally than the dorsal and its inferior border is prolonged so as to overlap the subjacent vertebra in the articulated column (Fig. 4–9). The cranial surface is concave transversely, and presents a projecting lip on either side; the caudal surface is convex from side to side, and presents laterally shallow concavities which receive the corresponding projecting lips of the subjacent vertebra. The **pedicles** are attached to the body midway between its cranial and caudal borders, so that the superior **vertebral notch** is as deep as the inferior, but it is, at the same time, narrower. The **laminae** are narrow, and the superior border thinner than the inferior. The **vertebral foramen** is large, and triangular in shape. The **spinous process** is short and bifid, the two divisions often being of unequal size. The superior and inferior **articular processes** are fused to form an articular pillar, which projects laterally from the junction of the pedicle and lamina. The articular facets are flat and oval: the superior look dorsally, cranially, and slightly medially: the inferior ventrally, caudally, and slightly laterally.

The **transverse processes** are each pierced by the **foramen transversarium**, which, in the first six vertebrae, transmits the vertebral artery and vein, accompanied by a plexus of sympathetic nerves. Each process consists of a ventral and a dorsal part. The portion ventral to the foramen is the homologue of the rib, and is therefore named the *costal process* or *costal element*: it arises from the side of the body, and ends in the

anterior tubercle. The dorsal part, the true transverse process, springs from the vertebral arch dorsal to the foramen and is directed ventrally and laterally; it ends in a flattened vertical prominence, the **posterior tubercle.** The bar of bone joining the two parts lateral to the foramen exhibits a deep sulcus on its cranial surface for the passage of the corresponding spinal nerve. The costal element of a cervical vertebra includes not only the portion which springs from the side of the body, but also the anterior and posterior tubercles and the bar of bone which connects them (Fig. 4–2).

The **First Cervical Vertebra** (Fig. 4–14) is named the **atlas** after a Titan in Greek mythology because it supports the globe of the head. Its chief peculiarity is that it has no body. This is due to the fact that during development the primordium of the body of the atlas was fused with that of the second vertebra. Its other peculiarities are that it has no spinous process, is ring-like, and consists of an anterior and a posterior arch and two lateral masses.

The **anterior arch** forms about one-fifth of the ring; the center of its ventral surface presents the **anterior tubercle** for attachment of the Longus colli muscles. Its dorsal or internal surface is marked by a smooth, oval or circular facet (*fovea dentis*), for articulation with the dens (odontoid process) of the axis. The cranial border gives attachment to the anterior atlantoöccipital membrane which connects it with the occipital bone; the caudal border to the anterior atlantoaxial ligament which connects it with the axis.

The **posterior arch** forms about two-fifths

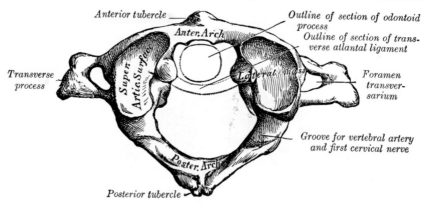

FIG. 4–14.—First cervical vertebra, or atlas.

of the circumference of the ring: it ends dorsally in the **posterior tubercle,** which is the rudiment of a spinous process and gives origin to the Recti capitis posteriores minores. The diminutive size of this process obviates interference with the movements between the atlas and the skull. The cranial surface of the dorsal part of the arch presents a rounded edge for the attachment of the posterior atlantoöccipital membrane, but immediately dorsal to each superior articular process is a groove (*sulcus arteriae vertebralis*), sometimes converted into a foramen by a delicate bony spicule growing from the posterior end of the superior articular process. It corresponds to a superior vertebral notch and transmits the suboccipital (first spinal) nerve. It also serves for the transmission of the vertebral artery, which, after ascending through the foramen in the transverse process, winds around the lateral mass in a dorsal and medial direction. On the caudal surface of the posterior arch, dorsal to the articular facets, are two shallow grooves, the **inferior vertebral notches.** The inferior border gives attachment to the posterior atlantoaxial ligament.

The **lateral masses** are the most bulky and solid parts of the atlas and support the weight of the head. Each has a superior and an inferior articular facet. The **superior facets** are of large size, oval, concave, and converge ventrally; they face cranialward,

medialward, and dorsalward, each forming a cup for the corresponding condyle of the occipital bone. Not infrequently they are partially subdivided by indentations which encroach upon their margins. The **inferior articular facets** are circular in form, are flattened or slightly convex, and are directed caudalward and medialward, articulating with the axis. Just inferior to the medial margin of each superior facet is a small tubercle for the attachment of the transverse atlantal ligament which stretches across the ring of the atlas and divides the vertebral foramen into two unequal parts. The ventral or smaller part receives the dens of the axis, the dorsal transmits the spinal cord and its membranes. This part of the vertebral canal is of considerable size, much greater than is required for the accommodation of the spinal cord, and hence lateral displacement of the atlas may occur as a result of trauma without compression of the cord. The **transverse processes** project laterally from the lateral masses, and serve for the attachment of muscles which assist in rotating the head. They are long, their anterior and posterior tubercles are fused into one mass, and the foramen transversarium is slanted dorsalward.

The **Second Cervical Vertebra** (Figs. 4–15 and 4–16) is named the **axis** or **epistropheus** because it forms the pivot upon which the first vertebra, carrying the head, rotates. Its most distinctive characteristic

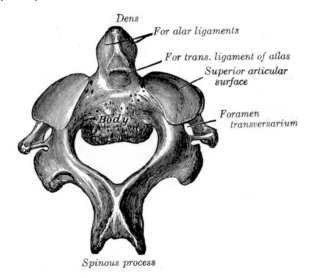

Dens

For alar ligaments

For trans. ligament of atlas

Superior articular surface

Body

Foramen transversarium

Spinous process

Fɪɢ. 4–15.—Second cervical vertebra or axis, cranial aspect.

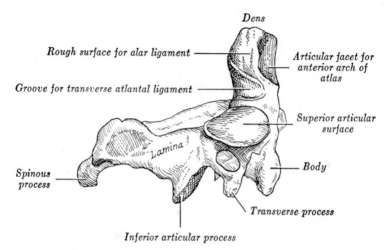

Dens

Rough surface for alar ligament ——————

Articular facet for anterior arch of atlas

Groove for transverse atlantal ligament ——————

Superior articular surface

Lamina

Spinous process

Body

Transverse process

Inferior articular process

Fɪɢ. 4–16.—Second cervical vertebra, axis or epistropheus, from the side.

is the cranial extension of the body into a strong bony process, the **dens** (*odontoid process*). The ventral surface of the body is marked by a median ridge for the attachment of the Longus colli and is prolonged into a lip which overlaps the third vertebra when articulated.

The **dens** exhibits a slight constriction or *neck* where it joins the body. On its ventral surface is an oval or nearly circular facet for articulation with the facet on the anterior arch of the atlas. A shallow groove on the dorsal surface of the neck and frequently extending on to its lateral surfaces, is for the transverse atlantal ligament which keeps the process in position. The *apex* is pointed, and gives attachment to the apical dental ligament; near the body, the process is somewhat enlarged and its sides have rough impressions for the attachment of the alar ligaments which connect it to the occipital bone. The internal structure of the dens is more compact than that of the body.

The **pedicles** are broad and strong, especially ventrally, where they coalesce with the sides of the body and the root of the dens. They are covered cranially by the superior articular surfaces. The **laminae** are thick and strong, and the vertebral foramen is large, but smaller than that of the atlas. The small **transverse processes** end in a single tubercle, perforated by the foramen transversarium, which has an obliquely lateral direction. The superior **articular surfaces** are round, slightly convex, directed

cranially and laterally and are supported by the body, pedicles, and transverse processes. The inferior articular surfaces have the same direction as those of the other cervical vertebrae. The superior **vertebral notches** are very shallow, and lie dorsal to the articular processes; the inferior lie ventral to the articular processes, as in the other cervical vertebrae. The **spinous process** is large, very strong, deeply channeled on its caudal surface, and presents a bifid, tuberculated extremity.

The **Seventh Cervical Vertebra** (Fig. 4–17).—The most distinctive characteristic

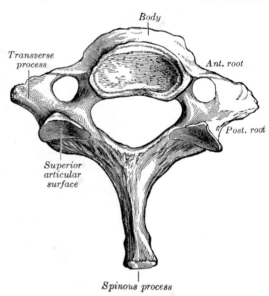

Body

Transverse process

Ant. root

Post. root

Superior articular surface

Spinous process

Fɪɢ. 4–17.—Seventh cervical vertebra.

of this vertebra is its long and prominent spinous process. This protrudes beyond the other cervical spines in the articulated skeleton, and since it is the most cranial spinous process that is palpable, it is a useful landmark in identifying and counting the other spines. It is named, therefore, the **vertebra prominens**. This process is thick, resembling that of a thoracic vertebra, nearly horizontal in direction, not bifurcated, and terminating in a tubercle to which the caudal end of the ligamentum nuchae is attached. The transverse processes are of considerable size, their dorsal roots are large and prominent, their ventral, small and faintly marked. The foramen transversarium may be as large as that in the other cervical vertebrae, but is generally smaller on one or both sides; occasionally it is double, sometimes it is absent. On the left side it occasionally gives passage to the vertebral artery; more frequently the vertebral vein traverses it on both sides; but the usual arrangement is for both artery and vein to pass ventral to the transverse process, and not through the foramen.

The Thoracic Vertebrae

The **Thoracic Vertebrae** (*vertebrae thoracicae*) (Fig. 4–18) are intermediate in size between the cervical and lumbar, providing a gradual transition from the small cervical cranially to the large lumbar vertebrae caudally. They are distinguished by the presence of facets on the sides of the bodies for articulation with the heads of the ribs, and facets on the transverse processes of all except the eleventh and twelfth, for articulation with the tubercles of the ribs.

The **bodies** in the middle of the thoracic region are heart-shaped, and as broad in the anteroposterior as in the transverse direction. They are flat and slightly thickened dorsally. They present two **costal demi-facets** on either side, one superior, near the root of the pedicle, the other inferior, ventral to the inferior vertebral notch. They are covered with cartilage in the intact body, and, when the vertebrae are articulated with one another, form oval surfaces for the reception of the heads of the ribs. The **pedicles** are directed slightly cranialward, and the inferior vertebral notches are of large size and deeper than in any other region of the vertebral column. The **laminae** are broad, thick, and imbricated—that is to say, they overlap those of subjacent vertebrae like tiles on a roof. The **vertebral foramen** is small, and circular in shape. The **spinous process** is long, triangular in coronal section, directed obliquely caudalward, and ends in a tuberculated extremity. These processes are most oblique and overlap from the fifth to the eighth, but the cranial and caudal ones are less oblique. The superior

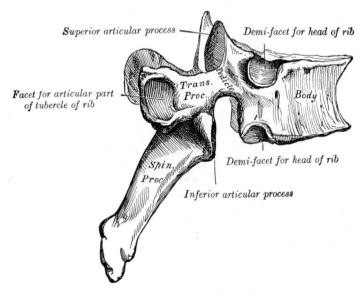

Superior articular process

Demi-facet for head of rib

Facet for articular part
of tubercle of rib

*Trans.
Proc.*

Pedicle

Body

*Spin.
Proc.*

Demi-facet for head of rib

Inferior articular process

Fig. 4–18.—A thoracic vertebra.

articular processes are thin plates of bone projecting up from the junctions of the pedicles and laminae; their articular facets are practically flat, and are directed dorsally and a little laterally. The inferior articular processes are fused to a considerable extent with the laminae, and project but slightly beyond their inferior borders; their facets are directed ventralward and a little medialward and caudalward. The **transverse processes** arise from the arch dorsal to the superior articular processes and pedicles; they are long, thick, strong, and with an

obliquely dorsal and lateral direction. Each ends in a clubbed extremity, on the ventral aspect of which is a small, concave surface, for articulation with the tubercle of a rib.

The first, ninth, tenth, eleventh, and twelfth thoracic vertebrae present certain peculiarities, and must be considered separately (Fig. 4–19).

The **First Thoracic Vertebra** has an **entire articular facet** for the head of the first rib and a **demifacet** for the cranial half of the head of the second rib, on either side of the body. The body is like that of a cervical

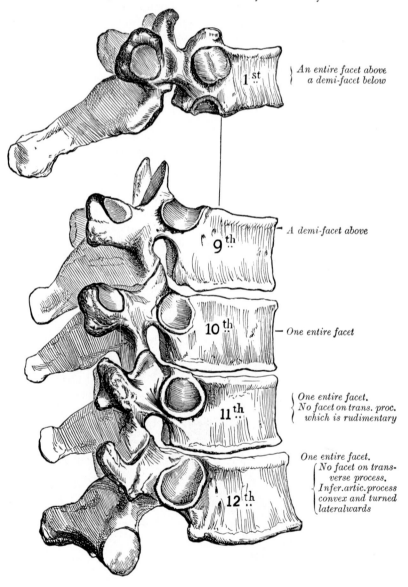

An entire facet above
a demi-facet below

A demi-facet above

One entire facet

One entire facet.
No facet on trans. proc.
which is rudimentary

One entire facet.
No facet on transverse process.
Infer. artic. process convex and turned lateralwards

Fɪɢ. 4–19.—Peculiar thoracic vertebrae.

vertebra, being broad transversely. The superior articular surfaces are directed cranially and dorsally; the spinous process is thick, long, and almost horizontal. The transverse processes are long, and the superior vertebral notches are deeper than those of the other thoracic vertebrae.

The **Ninth Thoracic Vertebra** may have no demifacets caudally. In some subjects however, it has two demifacets on either side; when this occurs the tenth has only demifacets at the cranial part.

The **Tenth Thoracic Vertebra** has (except as just mentioned) an entire articular facet on either side, which is placed partly on the lateral surface of the pedicle.

In the **Eleventh Thoracic Vertebra** the body approaches the lumbar vertebrae in form and size. The articular facets for the heads of the ribs are of large size, and placed chiefly on the pedicles, which are thicker and stronger in this and the next vertebra than in any other part of the thoracic region. The spinous process is short, and nearly horizontal in direction. The transverse processes are very short, tuberculated at their extremities, and have no articular facets.

The **Twelfth Thoracic Vertebra** has the same general characteristics as the eleventh, but may be distinguished from it: by its inferior articular surfaces being convex and directed laterally, like those of the lumbar vertebrae; by the general form of the body, laminae, and spinous process, in which it resembles the lumbar vertebrae; and by each transverse process being subdivided into three elevations, the superior, infe-

rior, and lateral **tubercles.** The superior and inferior tubercles correspond to the mammillary and accessory processes of the lumbar vertebrae. Traces of similar elevations are found on the transverse processes of the tenth and eleventh thoracic vertebrae.

The Lumbar Vertebrae

The **Lumbar Vertebrae** (*vertebrae lumbales*) (Figs. 4–20 and 4–21) are the largest segments of the movable part of the vertebral column, and can be distinguished by the absence of a foramen in the transverse process, and by the absence of facets on the sides of the body. The foramen vasculare for the basivertebral vein is larger than in the other vertebrae.

The **body** is large, wider transversely, and a little thicker ventrally. It is flattened or slightly concave superiorly and inferiorly and deeply constricted ventrally at the sides. The strong **pedicles** unite with the cranial part of the body, leaving deep inferior **vertebral notches.** The **laminae** are broad, short, and strong. The **vertebral foramen** is triangular, larger than in the thoracic, but smaller than in the cervical region. The **spinous process** is thick, broad, and somewhat quadrilateral ending in a rough, uneven border, thickest caudally where it is occasionally notched. The superior and inferior **articular processes** are well defined, projecting from the junctions of pedicles and laminae. The facets on the superior processes are concave, and look dorsally and medially; those on the inferior are convex, and are directed ventrally and laterally. The

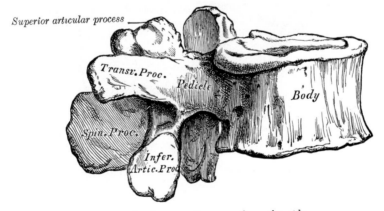

Superior articular process

Transv. Proc.

Pedicle

Body

Spin. Proc.

Infer. Artic. Proc.

Fig. 4–20.—A lumbar vertebra seen from the side.

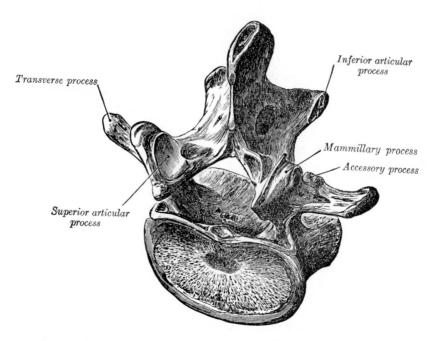

Transverse process

Inferior articular process

Mammillary process

Accessory process

Superior articular process

FIG. 4–21.—A lumbar vertebra, craniodorsal aspect.

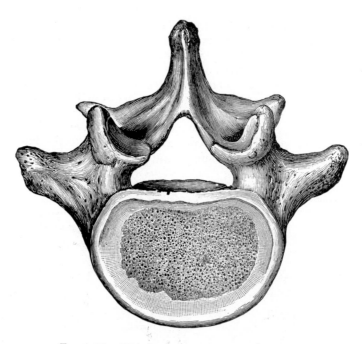

FIG. 4–22.—Fifth lumbar vertebra, cranial aspect.

former are wider apart than the latter, since in the articulated column the inferior articular processes are embraced by the superior processes of the subjacent vertebra. The **transverse processes** are long, slender, and horizontal in the cranial three lumbar vertebrae, but incline a little in the caudal two. In the upper three vertebrae they arise from the junctions of the pedicles and laminae, but in the lower two they are set farther ventrally and spring from the pedicles and dorsal parts of the bodies. They are situated ventral to the articular processes instead of dorsal to them as in the thoracic vertebrae, and are homologous with the ribs. Of the three tubercles noticed in connection with the transverse processes of the more caudal thoracic vertebrae, the cranial one is connected in the lumbar region with the dorsal part of the superior articular process, and is named the **mammillary process**; the inferior is situated at the dorsal part of the base of the transverse process, and is called the **accessory process** (Fig. 4–21).

The **Fifth Lumbar Vertebra** (Fig. 4–22) is characterized by its body being much deeper ventrally than dorsally, which accords with the prominence of the lumbosacral articulation; by the smaller size of its spinous process; by the wide interval between the inferior articular processes; and by the thickness of its transverse processes, which spring from the body as well as from the pedicles.

Variations.—The last lumbar vertebra is subject to certain defects described as *bifid* and *separate neural arches*, the latter occurring three times as frequently as the former. Both defects result in weakness of the column; the bifid arch by impairing ligamentous attachments; the separate arch through loss of bony anchorage of the column to its base.

The Sacral and Coccygeal Vertebrae

The sacral and coccygeal vertebrae consist at an early period of life of nine separate segments which, in the adult, are united to form two bones, five entering into

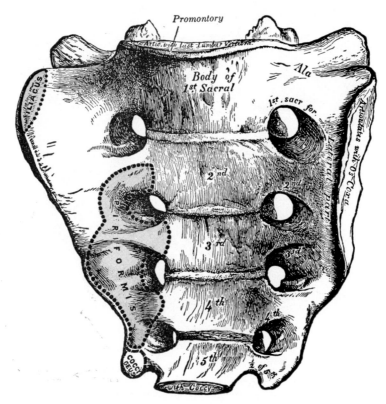

FIG. 4–23.—Sacrum, pelvic surface.

the formation of the sacrum, four into that of the coccyx.

The Sacrum (*os sacrum*).—The sacrum is a large, triangular bone, situated at the dorsal part of the pelvis (Fig. 4–175). Its base articulates with the last lumbar vertebra, its apex with the coccyx. It is inserted like a wedge between the two hip bones, its base projects ventrally and forms the prominent **sacrovertebral angle** when articulated with the last lumbar vertebra. Its central part is projected dorsally (Fig. 4–174).

The **pelvic surface** (*facies pelvina*) (Fig. 4–23) is concave, giving increased capacity to the pelvic cavity. Its median part is crossed by four transverse ridges, the positions of which correspond with the original planes of separation between the five segments of the fetal bone (Fig. 4–38). The portions of bone intervening between the ridges are the **bodies** of the sacral vertebrae. The body of the first segment is large, and

in form resembles that of a lumbar vertebra; the succeeding ones diminish in size, are flattened, and are curved with the form of the whole sacrum. At the ends of the ridges the four rounded **ventral sacral foramina** (*foramina sacralia pelvina*) also diminishing in size caudally. From them the ventral divisions of the sacral nerves make their exit and the lateral sacral arteries enter. Lateral to these foramina are the **lateral parts of the sacrum** which consist of five separate segments at an early period of life, but which are blended with the bodies and with each other in the adult. Four broad, shallow grooves which lodge the ventral divisions of the sacral nerves extend laterally from the foramina and are separated by prominent ridges of bone which give origin to the Piriformis muscle.

The **dorsal surface** (*facies dorsalis*) (Fig. 4–24) is convex and narrower than the pelvic surface. In the middle line, the

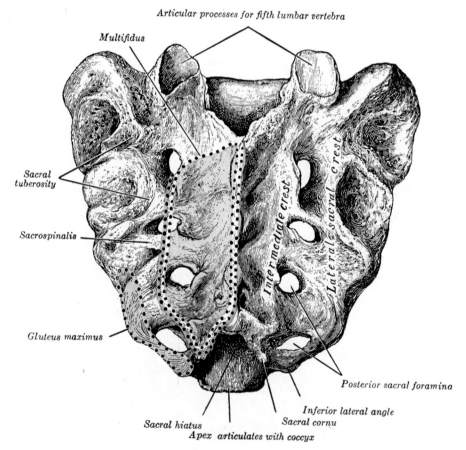

Articular processes for fifth lumbar vertebra

Multifidus

Intermediate crest

Lateral sacral crest

Sacral tuberosity

Sacrospinalis

Gluteus maximus

Posterior sacral foramina

Inferior lateral angle
Sacral cornu

Sacral hiatus

Apex articulates with coccyx

Fig. 4–24.—The sacrum, dorsal aspect.

middle sacral crest (*crista sacralis media*) is surmounted by three or four tubercles, the rudimentary **spinous processes** of the first three or four sacral segments. On either side of the middle sacral crest, a shallow **sacral groove** is formed by the united **laminae** of the corresponding vertebrae and gives origin to the Multifidus. The Erector spinae and the Latissimus dorsi have part of their origins from the middle crest. The laminae of the fifth sacral vertebra fail to meet in the midline dorsally, leaving an enlarged aperture into the sacral canal, the **sacral hiatus.** On the lateral aspect of the sacral groove is a linear series of tubercles produced by the fusion of the **articular processes** which together form the indistinct **intermediate crests** (*crista sacralis intermedia; articular crests*). The articular

processes of the first sacral vertebra are large and oval in shape; their facets are concave, look dorsalward and medialward, and articulate with the facets on the inferior processes of the fifth lumbar vertebra. The tubercles which represent the inferior articular processes of the fifth sacral vertebra are prolonged downward as rounded processes, are named the **sacral cornua,** and are connected to the cornua of the coccyx. Lateral to the intermediate crest are the four **dorsal sacral foramina;** they are smaller in size and less regular in form than the ventral. They transmit the dorsal divisions of the sacral nerves. Lateral to the dorsal foramina, a series of tubercles represent the **transverse processes** of the sacral vertebrae, and form the **lateral crests** of the sacrum (*crista sacralis lateralis*). The tubercles of the first

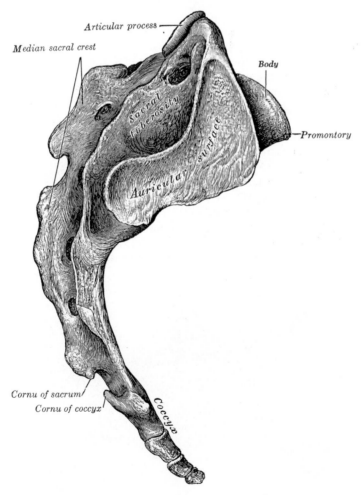

FIG. 4-25.—Lateral surfaces of sacrum and coccyx.

and second segments receive the attachment of the horizontal parts of the posterior sacroiliac ligaments; those of the third vertebra give attachment to the oblique fasciculi of the dorsal sacroiliac ligaments; and those of the fourth and fifth to the sacrotuberous ligaments.

The cranial half of the **lateral surface** (Fig. 4–25) is broad and irregular. It presents an ear-shaped surface, the **auricular surface** (*facies auricularis; articular surface*), covered with cartilage in the fresh state for articulation with the ilium. Dorsal to it is a rough surface, the **sacral tuberosity,** on which are three deep and uneven impressions for the attachment of the posterior sacroiliac ligament. The caudal half of the lateral surface is thin and ends in a projection called the **inferior lateral angle;** medial to this angle is a notch which is converted by the transverse process of the first piece of the coccyx into a foramen for transmission of the ventral division of the fifth sacral nerve. The thin caudal half of the lateral surface gives attachment to the sacrotuberous and sacrospinous ligaments, to some fibers of the Gluteus maximus, and to the Coccygeus.

The **base of the sacrum** (*basis oss. sacri*) (Fig. 4–26), is broad and expanded laterally. In the middle is the upper surface of the body of the first sacral vertebra, its ventral border projecting into the pelvis as the **promontory.** It is connected with the caudal surface of the body of the last lumbar vertebra by an intervertebral disk. Dorsal to this the orifice of the sacral canal is completed

by the laminae and spinous process. The superior articular processes on either side are directed dorsalward and medialward, like the superior articular processes of a lumbar vertebra. They are attached to the body of the first sacral vertebra and to the alae by short thick pedicles; on the cranial surfaces of each pedicle is a vertebral notch which forms the caudal part of the foramen between the last lumbar and first sacral vertebrae. On either side of the body, a large triangular surface called the ala, supports the Psoas major and the lumbosacral trunk, and receives the attachment of a few fibers of the Iliacus. The dorsal fourth of the ala represents the transverse process, and its ventral three-fourths the costal process of the first sacral segment (Fig. 4–26).

The **apex** (*apex ossis sacri*) is the caudal extremity of the bone and presents an oval facet for articulation with the coccyx.

The **sacral canal** (*canalis sacralis*) (Fig. 4–27), the vertebral canal in the sacrum, is incomplete due to the non-development of the laminae and spinous processes of the last one or two segments. The resultant widened aperture into the caudal end of the sacral canal, the **sacral hiatus,** is used by anaesthesiologists for the insertion of a flexible needle in producing caudal analgesia. The canal lodges the sacral nerves, and its walls are perforated by the dorsal and ventral sacral foramina through which these nerves pass.

Structure.—The sacrum consists of cancellous tissue enveloped by a thin layer of compact bone. In a sagittal section through the center

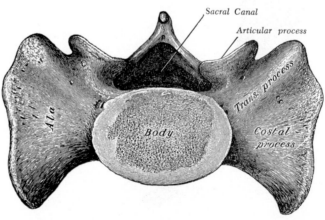

Fɪɢ. 4–26.—Base of sacrum.

of the sacrum (Fig. 4–27), the bodies appear united at their circumferences by bone, but there are wide intervals centrally which, in the intact body, are filled by the intervertebral disks. This union may be more complete between the caudal than the cranial segments in some bones.

Articulations.—The sacrum articulates with four bones; the last lumbar vertebra cranially, the coccyx caudally, and the hip bone on either side.

Differences in the Sacrum of the Male and Female.—In the female the sacrum is shorter and wider than in the male; the caudal half forms a greater angle with the cranial, which is nearly straight, and the caudal half presents the greatest amount of curvature. The bone is also more oblique and dorsally directed which increases the size of the pelvic cavity and renders the sacrovertebral angle more prominent. In the male the curvature is greater and more evenly distributed over the whole length of the bone.

Variation.—The sacrum may consist of the remnants of six segments; occasionally the number is reduced to four. The bodies of the

first and second may fail to unite. Sometimes the uppermost transverse tubercles are not joined to the rest of the ala on one or both sides, or the sacral canal may be open throughout a considerable part of its length, in consequence of the imperfect development of the laminae and spinous processes. The sacrum also varies considerably with respect to its degree of curvature.

The **Coccyx** (*os coccygis*) (Figs. 4–28 and 4–29) is usually formed of four rudimentary vertebrae; the number may, however, be increased to five or diminished to three. Rudimentary bodies and articular and transverse processes can be identified in the first three segments, but they are destitute of pedicles, laminae, and spinous processes. The last piece is a mere nodule of bone. The first is the largest; it resembles the last sacral vertebra, and often exists as a separate piece; the last three diminish in size, and are usually fused with one another.

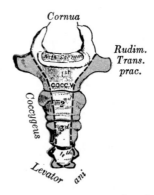

Fig. 4–28.—Coccyx, ventral surface.

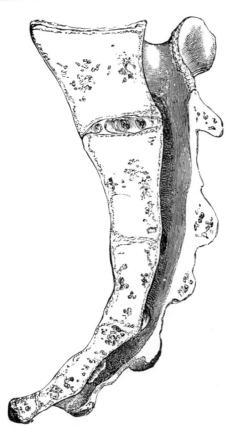

Fig. 4–27.—Median sagittal section of the sacrum.

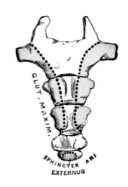

Fig. 4–29.—Coccyx, dorsal surface.

The **pelvic surface** is marked with three transverse grooves which indicate the junctions of the different segments. It gives attachment to the ventral sacrococcygeal ligament and the Levatores ani, and supports part of the rectum. The **dorsal surfaces** on either side has a row of tubercles, the rudimentary articular processes; the large first pair are called the **coccygeal cornua;** they articulate with the cornua of the sacrum, and on either side complete the foramen for the transmission of the dorsal division of the fifth sacral nerve.

On the **lateral borders** a series of small eminences represent the transverse processes. The first is the largest, and often joins the thin lateral edge of the sacrum, thus completing the foramen for the transmission of the ventral division of the fifth sacral nerve. The narrow borders of the coccyx receive the attachment of the sacrotuberous and sacrospinous ligaments laterally, the Coccygeus ventrally, and the Gluteus maximus dorsally. The oval surface of the base articulates with the sacrum. The rounded apex has the tendon of the Sphincter ani externus attached to it and may be bifid, or be deflected to one side or the other.

Variations and Anomalies of the Vertebral Column.—Number of Vertebrae.—Variation in the total number of vertebrae cranial to the coccyx is less common than variations within the different regions. The number of segments in the coccyx is the most variable, the sacral next, then the thorax, and finally that in the cervical region the least. The number of vertebrae in one region may be increased at the expense of the neighboring group, however. Changes in the lumbosacral and cervicothoracic region are rather frequent and usually can be traced to variations in the development of the costal processes (Fig. 4–2).

The **costal processes** of the 12th thoracic vertebra may be wanting, in which case there will be eleven thoracic and six lumbar vertebrae. Conversely, the costal element of the 1st lumbar may develop into a rib, resulting in thirteen thoracic and four lumbar vertebrae. The costal element of the last lumbar may unite with the first sacral, a condition known as *sacralization of the lumbar vertebra,* resulting in four lumbar and six sacral segments. True sacralization does not take place, however, unless the 5th lumbar enters into the articulation with the ilium. Again, the first sacral may fail to unite with the other sacrals, resulting in six lumbar and four sacral segments. There may be a secondary sacral promontory.

Cervical rib is the condition in which the costal element of the 7th cervical is elongated into a separate rib-like structure which in rare instances may reach an articulation with the sternum. The unusual bony structure may cause serious symptoms by interference with blood flow in the subclavian artery or by paralysis due to pressure on the brachial plexus.

Asymmetry.—The development of one side of a vertebra may be defective to a greater or lesser degree. One side of the centrum or body may be defective causing *congenital scoliosis* or lateral curvature. The overgrowth or failure of the costal process may be on one side, especially in the lumbosacral region, or there may be one-sided defects in the pedicles and laminae.

Spina bifida.—The laminae fail to unite and form a spinous process, not uncommonly, and when it involves only one or two vertebrae, causing few symptoms, it is called *spina bifida occulta.* Depending on the size of the defect, there may be a lipoma in its place, but larger defects may contain a *meningocele* or a *myelomeningocele,* produced by protrusion of corresponding tissues from the spinal canal. The commonest site of spina bifida is the lumbar region, less common is the cervical region. The absence of spinous processes may extend distally through the entire sacrum, leaving the sacral canal exposed. Other parts of the neural arch may have similar defects. The union of the pedicles with the body may fail on one or both sides; the resulting weakness of the arch may cause displacement of one vertebra over the other, the condition named *spondylolisthesis.*

Ossification of the Vertebral Column.—Each cartilaginous vertebra is ossified from **three primary centers** (Fig. 4–30), two for the vertebral arch and one for the body. Ossification of the vertebral arches begins in the upper cervical vertebrae about the seventh or eighth week of fetal life, and gradually extends caudalward. The ossific granules first appear in the situations where the transverse processes afterward project, and spread dorsally to the spinous process, ventrally into the pedicles, and laterally into the transverse and articular processes. Ossification of the bodies begins about the eighth week in the lower thoracic region, and subsequently extends cranially and caudally along the column. The center for the body does not give rise to the entire body of the adult vertebra, the dorsolateral portions being ossified by extensions from the vertebral arch centers.

During the first few years of life the body of the vertebra, therefore, shows two neurocentral synchondroses, traversing it along the planes of junction of the three centers (Fig. 4–31). In the thoracic region, the facets for the heads of the ribs lie dorsal to the neurocentral synchondroses and are ossified from the centers for the vertebral arch.

At birth the vertebra consists of three pieces, the body and the halves of the vertebral arch. During the first year the halves of the arch unite dorsally, union taking place first in the lumbar region and then extending caudalward through the thoracic and cervical regions. About the third year the bodies of the upper cervical vertebrae are joined to the arches on either side; in the caudal lumbar vertebrae the union is not completed until the sixth year. Before

By 3 primary centers

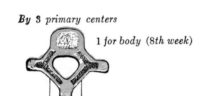

1 for body (8th week)

1 for each vertebral arch (7th or 8th week)

Fig. 4–30.

By 3 secondary centers

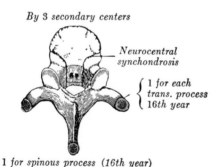

Neurocentral synchondrosis

{ 1 for each trans. process 16th year

1 for spinous process (16th year)

Fig. 4–31.

By 2 additional plates

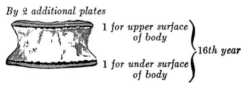

1 for upper surface of body
1 for under surface of body }16th year

Fig. 4–32.

Figs. 4–30—4–32.—Ossification of a vertebra.

puberty, a gradual increase of these primary centers occurs while the cranial and caudal surfaces of the bodies and the ends of the transverse and spinous processes remain cartilaginous. About the sixteenth year (Fig. 4–31), **five secondary centers** appear, one for the tip of each transverse process, one for the extremity of the spinous process, one for the cranial and one for the caudal surface of the body (Fig. 4–32). These fuse with the rest of the bone about the age of twenty-five.

Exceptions to this mode of development occur in the first, second and seventh cervical vertebrae, and in the lumbar vertebrae.

By 3 centers

1 for anter. arch (end of 1st year)
1 for each lateral mass } 7th week

Fig. 4–33.—Atlas.

Atlas.—The atlas is usually ossified from three centers (Fig. 4–33). Two of these appear in the lateral masses about the seventh week of fetal life. At birth, the dorsal portions of bone are separated from one another by a narrow interval filled with cartilage. Between the third and fourth years they unite either directly or through the medium of a separate center developed in the cartilage. At birth, the ventral arch consists of cartilage; a separate center appears about the end of the first year after birth and joins the lateral masses between the sixth and eighth years—the lines of union extending across the ventral portions of the superior articular facets. Occasionally there is no separate center for the ventral arch, or it may be ossified from two centers, one on either side of the middle line.

Axis or Epistropheus.—The axis is ossified from five primary and two secondary centers (Fig. 4–34). The body and arch are ossified in the same manner as the corresponding parts in the other vertebrae, viz., one center for the body, and two for the vertebral arch. The centers for the arch appear about the seventh or eighth week of fetal life, that for the body about the fourth or fifth month. The **dens** consists originally of a cranial continuation of the cartilaginous mass in which the caudal part of the body is formed. About the sixth month of fetal life, two centers make their appearance in the base of this process: they are placed laterally, and join before birth to form a conical bilobed mass deeply cleft cranially; the interval between the sides of the cleft and the summit of the process is formed by a wedge-shaped piece of cartilage. The base of the process is separated from the body by a cartilaginous disk, which gradually becomes ossified at its circumference, but remains cartilaginous in its center until advanced age. In this cartilage, rudiments of the caudal epiphyseal lamella of the atlas and the cranial epiphyseal lamella of the axis may sometimes be found. The apex of the dens has a separate center which appears in the second and joins about the twelfth year; this is the cranial epiphyseal lamella of the atlas. In addition to these there is a secondary center for a thin epiphyseal plate on the caudal surface of the body of the bone.

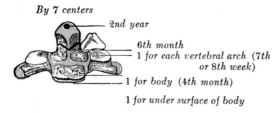

By 7 centers

— *2nd year*

6th month
— *1 for each vertebral arch (7th or 8th week)*

— *1 for body (4th month)*

1 for under surface of body

Fig. 4–34.—Axis.

2 additional centers for mammillary processes

Fig. 4–35.—Lumbar vertebra.

The Seventh Cervical Vertebra.—The ventral or costal part (Fig. 4–2) of the transverse process of this vertebra is sometimes ossified from a separate center which appears about the sixth month of fetal life, and joins the body and dorsal part of the transverse process between the fifth and sixth years.

Lumbar Vertebrae.—Each lumbar vertebra (Fig. 4–35) has two additional centers for the mammillary processes. The transverse process of the first lumbar is sometimes developed as a separate piece which may remain permanently

ununited with the rest of the bone, thus forming the rare peculiarity, a lumbar rib.

Sacrum (Figs. 4–36–4–39).—The body of each sacral vertebra is ossified from a primary center and two epiphyseal plates, while each vertebral arch is ossified from two centers. The ventral portions of the lateral parts have six additional centers, two for each of the three vertebrae; these represent the costal elements, and make their appearance cranial and lateral to the pelvic sacral foramina (Figs. 4–36, 4–37). On each lateral surface two epiphyseal plates are developed (Figs. 4–38, 4–39); one for the auricular surface, and another for the remaining part of the thin lateral edge of the bone.

Periods of Ossification.—About the eighth or ninth week of fetal life, ossification of the central part of the body of the first sacral vertebra commences, and is rapidly followed by deposit of ossific matter in the second and third: ossification does not commence in the bodies of the last two segments until between the fifth and eighth months of fetal life. Between the sixth and eighth months ossification of the vertebral arches takes place, and about the same time the costal centers for the lateral parts make their appearance. The junctions of the vertebral arches with the bodies take place in the caudal vertebrae as early as the second year, but are not effected in the more cranial until the fifth or sixth year. About the sixteenth year the epiphyseal plates for the surfaces of the bodies are formed; and between the eighteenth and twentieth years, those for the lateral parts make their appearance. The bodies of the sacral vertebrae are, during early life, separated from each other by intervertebral disks, but about the eighteenth year the last two segments become united by bone, and the process of bony union gradually extends cranially, with the result that between the twenty-fifth and thirtieth years of life all the segments are united. In a sagittal section of the sacrum, the situations of the intervertebral fibrocartilages are indicated by a series of oval cavities (Fig. 4–27).

Coccyx.—The coccyx is ossified from four centers, one for each segment. The ossific nuclei make their appearance in the following order: in the first segment between the first and fourth years; in the second between the fifth and tenth years; in the third between the tenth and fifteenth years; in the fourth between the fourteenth and twentieth years. As age advances, the segments unite, the union between the first and second segments being frequently delayed until after the age of twenty-five or thirty. At a late period of life, especially in females, the coccyx often fuses with the sacrum.

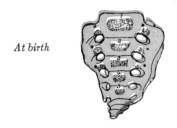

At birth

Fig. 4–36.—(°) Additional centers for costal elements.

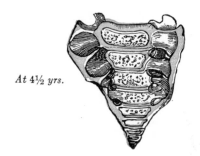

At 4½ yrs.

Fig. 4–37.

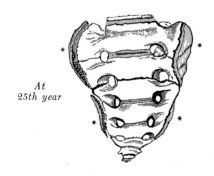

At 25th year

Fig. 4–38.—(°) Two epiphyseal plates for each lateral surface.

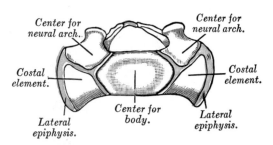

Center for neural arch.

Center for neural arch.

Costal element.

Costal element.

Lateral epiphysis.

Center for body.

Lateral epiphysis.

Fig. 4–39.—Base of young sacrum.

Figs. 4–36—4–39.—Ossification of the sacrum.

THE THORAX

The skeleton of the **thorax** or **chest** (Figs. 4–40, 4–41, 4–42) is an osseocartilaginous cage, containing the principal organs of respiration and circulation and covering part of the abdominal organs. It is conical, and is somewhat kidney-shaped on transverse section because of the projection of the vertebral bodies into the cavity.

Boundaries.—The **dorsal surface** is formed by the twelve thoracic vertebrae and the dorsal parts of the twelve ribs. The **ventral surface** is formed by the sternum and costal cartilages. The **lateral surfaces** are formed by the ribs, separated from each other by the eleven intercostal spaces which are occupied by the Intercostal muscles and membranes.

The **superior opening** of the thorax is broader from side to side than dorsoventrally. It is formed by the first thoracic vertebra, the cranial margin of the sternum, and the first rib on either side. Its dorsoventral diameter is about 5 cm, and its transverse diameter about 10 cm. The plane of the aperture slopes caudally from the vertebrae, causing the opening to face ventralward as well as cranialward.

The **inferior opening** is formed by the twelfth thoracic vertebra dorsally, by the eleventh and twelfth ribs at the sides, and ventrally by the cartilages of the tenth, ninth, eighth, and seventh ribs, which slope on either side to form the **subcostal angle** at the xiphoid process. The caudal opening is wider transversely than dorsoventrally, and inclines obliquely caudalward and dorsalward. It is closed by the diaphragm which forms the floor of the thorax.

The thorax of the female differs from that of the male as follows: 1. Its capacity is less. 2. The sternum is shorter. 3. The cranial margin of the sternum is on a level with the caudal part of the body of the third thoracic vertebra, whereas in the male it is on a level with the caudal part of the body of the second. 4. The upper ribs are more movable, and so allow a greater enlargement of the cranial part of the thorax.

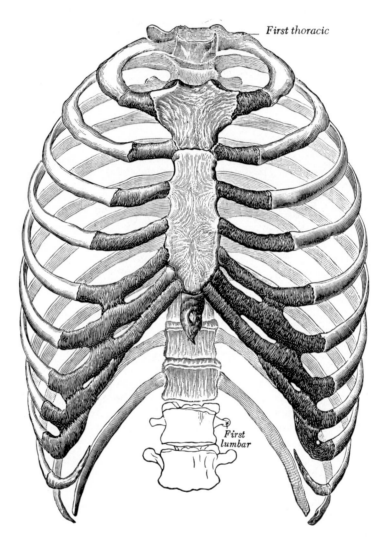

FIG. 4–40.—The thorax, ventral aspect. (Spalteholz.)

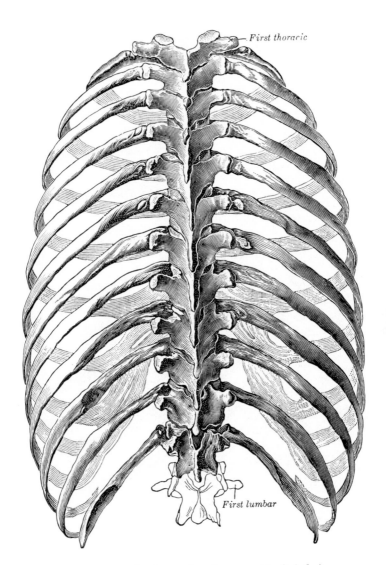

FIG. 4–41.—The thorax, dorsal aspect. (Spalteholz.)

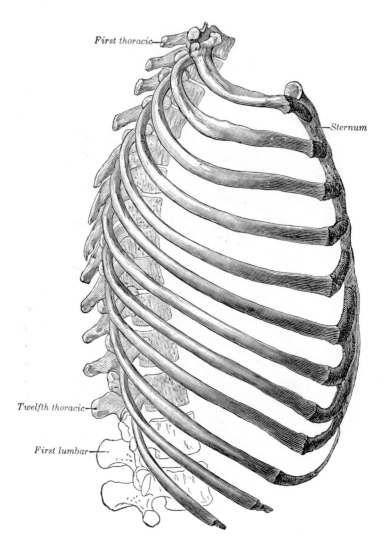

First thoracic

Sternum

Twelfth thoracic

First lumbar

Fig. 4–42.—The thorax from the right. (Spalteholz.)

The Sternum

The **Sternum** (*breast bone*) (Figs. 4–43 to 4–44) is an elongated, flattened bone, forming the middle portion of the ventral wall of the thorax. Its cranial end supports the clavicles, and its margins articulate with the cartilages of the first seven pairs of ribs. It consists of three parts, the manubrium, the body or gladiolus, and the xiphoid process; in early life the body consists of four segments or **sternebrae.** In its natural position the caudal end of the bone is inclined away from the vertebral column. Its average length in the adult is about 17 cm., and is rather greater in the male than in the female.

The **Manubrium** (*manubrium sterni*) is somewhat quadrangular in shape, broad and thick cranially, and narrow caudally at its junction with the body.

Its **ventral surface** is smooth, and affords attachment on either side to the sternal origins of the Pectoralis major and Sterno-

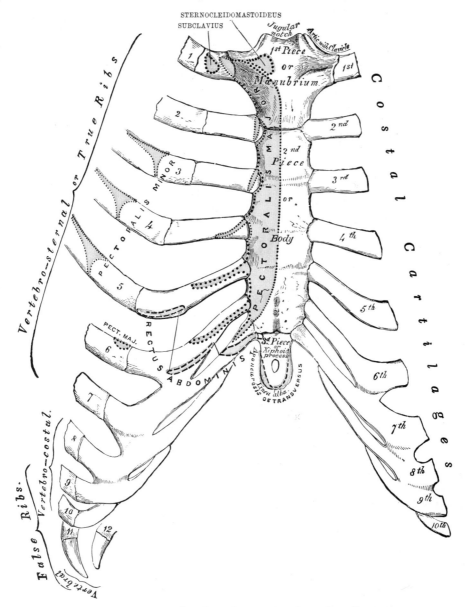

Fig. 4–43.—Ventral surface of sternum and costal cartilages.

cleidomastoideus. Sometimes the ridges limiting the attachments of these muscles are very distinct. Its **dorsal surface,** concave and smooth, affords attachment on either side to the Sternohyoideus and Sternothyroideus (Fig. 4–44).

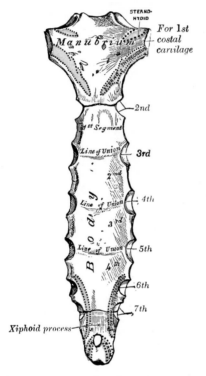

FIG. 4–44.—Dorsal surface of sternum.

The **cranial border** is the thickest and presents at its center the **jugular** (*presternal*) **notch;** on either side of which is an oval surface for articulation with the sternal end of the clavicle. The **caudal border,** oval and rough, is covered in the intact body with a thin layer of cartilage for articulation with the body. The **lateral borders** are each marked cranially by a depression for the first costal cartilage, and caudally by a small facet which forms part of the notch for the reception of the costal cartilage of the second rib.

The **Body** (*corpus sterni; gladiolus*) is considerably longer, narrower, and thinner than the manubrium.

Its **ventral surface** is nearly flat, and marked by three transverse ridges which cross the bone opposite the third, fourth,

and fifth articular depressions. It affords attachment on either side to the sternal origin of the Pectoralis major. At the junction of the third and fourth pieces of the body a hole, the sternal foramen, of varying size and form is occasionally seen. The **dorsal** or **internal surface,** slightly concave, is also marked by three transverse lines, less distinct, however, than the ventral; from its caudal part, on either side, the Transversus thoracis takes origin.

The **cranial border** is oval and articulates with the manubrium, the junction of the two forming the **sternal angle** (*angle of Louis*). The **caudal border** is narrow, and articulates with the xiphoid process. Each **lateral border** (Fig. 4–45), at its cranial angle, has a small facet, which with a similar facet on the manubrium, forms a cavity for the cartilage of the second rib; caudal to this are four angular depressions which receive the cartilages of the third, fourth, fifth, and sixth ribs, while the caudal angle has a small facet, which, with a corresponding one on the xiphoid process, forms a notch for the cartilage of the seventh rib. These articular depressions are separated by a series of

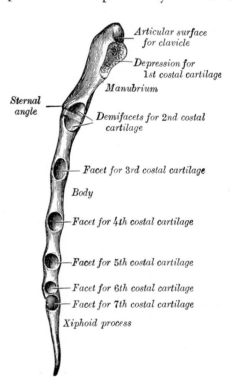

FIG. 4–45.—Lateral border of sternum.

curved interarticular intervals, which diminish in length caudally, and correspond to the intercostal spaces. Most of the cartilages belonging to the true ribs, as will be seen from the foregoing description, articulate with the sternum at the lines of junction of its primitive component segments. This is well seen in many of the lower animals, in which the sternebrae remain distinct longer than in man.

The **Xiphoid Process** (*processus xiphoideus; ensiform* or *xiphoid appendix*) is the smallest of the three pieces; it is thin and elongated, cartilaginous in youth, but more or less ossified at its proximal part in the adult.

Its **ventral surface** affords attachment on either side to the ventral costoxiphoid ligament and a small part of the Rectus abdominis; its **dorsal surface,** to the dorsal costoxiphoid ligament and to some of the fibers of the Diaphragm and Transversus thoracis; its **lateral borders,** to the aponeuroses of the abdominal muscles. It articulates with the body, and at each cranial angle presents a facet for part of the cartilage of the seventh rib; its pointed caudal extremity gives attachment to the linea alba. The xiphoid process varies much in form; it may be broad and thin, pointed, bifid, perforated, curved, or deflected considerably to one side or the other.

Structure.—The sternum is composed of highly vascular cancellous tissue, covered by a thin layer of compact bone which is thickest in the manubrium between the articular facets for the clavicles. The accessibility of the sternum is utilized by hematologists for *sternal puncture* in which a needle of large bore is thrust through the thin subcutaneous cortical bone and a specimen of marrow aspirated.

Ossification.—The sternum originally consists of two cartilaginous bars, situated one on either side of the median plane and connected with the cartilages of the upper nine ribs of its own side. These two bars fuse with each other along the middle line to form the cartilaginous sternum which is ossified from six centers: one for the manubrium, four for the body, and one for the xiphoid process (Fig. 4–46). The ossific centers appear in the intervals between the articular depressions for the costal cartilages, in the following order: in the manubrium and first piece of the body during the sixth month; in the second and third pieces of the body dur-

ing the seventh month of fetal life; in its fourth piece during the first year after birth; and in the xiphoid process between the fifth and eighteenth years. The centers make their appearance at the cranial parts of the segments and proceed caudalward. To these may be added the occasional existence of two small episternal centers which make their appearance on either side of the jugular notch; they are probably vestiges of the episternal bone of the monotremes and lizards.

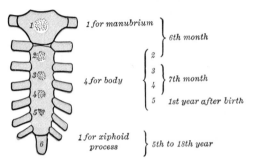

FIG. 4–46.—Time of appearance. Ossification of the sternum.

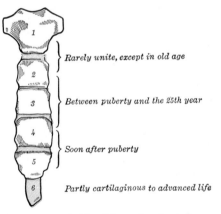

FIG. 4–47.—Time of union. Ossification of the sternum.

Occasionally some of the segments are formed from more than one center, the number and position of which vary. Thus, the first piece may have two, three, or even six centers. When two are present, they are generally situated one above the other, the cranial being the larger; the second piece seldom has more than one; the third, fourth, and fifth pieces are often formed from two centers placed laterally, the irregular union of which explains the rare occurrence of the sternal foramen, or of the vertical fissure which occasionally intersects this part of the bone constituting the malformation

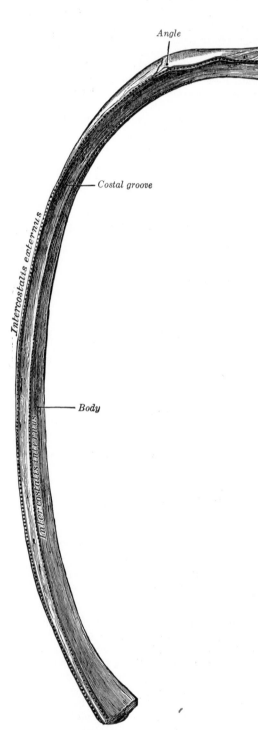

Non-articular part of tubercle

Articular part of tubercle

Angle

Neck

Head

Costal groove

Intercostalis externus

Body

Intercostalis internus

FIG. 4–48.—A central rib of the left side.
Caudal aspect.

known as **fissura sterni;** these conditions are further explained by the manner in which the cartilaginous sternum is formed. More rarely still the upper end of the sternum may be divided by a fissure.

Union of the various centers of the body begins about puberty, and proceeds cranially (Fig. 4–47); by the age of twenty-five they are all united. The xiphoid process may become joined to the body before the age of thirty, but this occurs more frequently after forty; on the other hand, it sometimes remains ununited in old age. In advanced life the manubrium is occasionally joined to the body by bone. When this takes place, however, the bony tissue is generally only superficial, the central portion of the intervening cartilage remaining unossified.

Articulations.—The sternum articulates on either side with the clavicle and first seven costal cartilages.

The Ribs

The **Ribs** (*costae*) are elastic arches of bone, which form a large part of the thoracic skeleton. There are twelve on either side, the first seven connected dorsally with the vertebral column, and ventrally, through the intervention of the costal cartilages, with the sternum (Fig. 4–41); they are called true or **vertebrosternal** ribs. The remaining five are false ribs; of these, the first three have their cartilages attached to the cartilage of the rib above (**vertebro-chondral**): the last two are free at their ventral extremities and are termed **floating** or **vertebral ribs.** The ribs vary in their direction, the upper ones being less oblique than the lower; the obliquity reaches its maximum at the ninth rib, and gradually decreases from that rib to the twelfth. The ribs are separated by the intercostal spaces. The length of each space corresponds to that of the adjacent ribs and their cartilages; the breadth is greater ventrally and between the more cranial ribs. The ribs increase in length from the first to the seventh, then diminish to the twelfth.

Common Characteristics of the Ribs (Figs. 4–48, 4–49).—A rib from the middle of the series should be used to study the common characteristics of these bones. Each rib has two extremities, a dorsal or vertebral, and a ventral or sternal, and an intervening portion, the body or shaft.

The **dorsal** or **vertebral extremity** presents for examination a head, neck, and tubercle.

The **head** is marked by a kidney-shaped articular surface, divided by a horizontal crest into two facets for articulation with the depression formed on the bodies of two adjacent thoracic vertebrae; the cranial facet is the smaller; to the crest is attached the interarticular ligament.

The **neck** is the flattened portion which extends lateralward from the head; it is about 2.5 cm long, and is placed ventral to the transverse process of the more caudal of the two vertebrae with which the head articulates. Its ventral surface is flat and smooth, its dorsal rough for the attachment of the ligament of the neck, and perforated by numerous foramina. Of its two borders the cranial presents a rough **crest** (*crista colli costae*) for the attachment of the ante-

rior costotransverse ligament; its caudal border is rounded. The **tubercle** is an eminence on the posterior surface at the junction of the neck and body; it consists of an articular and a non-articular portion. The articular portion, the more caudal and medial of the two, presents a small, oval surface for articulation with the end of the transverse process of the more caudal of the two vertebrae to which the head is connected. The non-articular portion is a rough elevation, and affords attachment to the ligament of the tubercle. The tubercle is much more prominent in the more cranial ribs.

The **body** or **shaft** is thin and flat, with two surfaces, an external and an internal; and two borders, a superior and an inferior. The **external surface** is marked, a little beyond the tubercle, by a prominent line which gives attachment to a tendon of the Iliocostalis, and is called the **angle**. The rib is bent in two directions, and at the same time twisted on its long axis. If it is placed upon its inferior border, the portion dorsal to the angle is bent medialward and at the same time tilted cranialward; as the result of the twisting, the external surface, dorsal to the angle, looks caudalward, and ventral to the angle, slightly cranialward. The distance between the angle and the tubercle is progressively greater from the second to the tenth ribs. The portion between the angle and the tubercle is rounded, rough, and irregular, and serves for the attachment

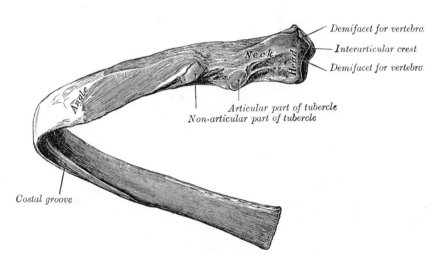

FIG. 4–49.—A central rib of the left side, dorsal aspect.

of the Longissimus thoracis. The **internal surface** is concave, smooth, and marked by a ridge which commences at the head; and is strongly marked as far as the angle, but gradually becomes lost at the junction of the ventral and middle thirds of the bone. Between it and the **inferior border** is the **costal groove,** for the intercostal vessels and nerve. This groove is on the inferior border dorsally but just ventral to the angle, where it is deepest and broadest, it is on the internal surface. The superior edge of the groove is rounded and serves for the attachment of the Intercostalis internus; the inferior edge corresponds to the margin of the rib, and gives attachment to the Intercostalis externus. Within the groove are seen the orifices of numerous small foramina for nutrient vessels. The **superior border,** thick and rounded, is marked by an external and an internal lip, more distinct dorsally, which serve for the attachment of Intercostales externi and interni. The inferior border is thin, and has attached to it an Intercostalis externus.

The **ventral** or **sternal extremity** is flattened, and presents a porous, oval depression, into which the costal cartilage is received.

Peculiar Ribs.—The first, second, tenth, eleventh, and twelfth ribs present certain variations from the common characteristics described above, and require special consideration.

The **First Rib** (Fig. 4–50) is the most curved and usually the shortest of all the ribs; it is broad and flat, its surface looking cranially and caudally, and its borders internally and externally. The head is small, rounded, and possesses only a single articular facet for articulation with the body of the first thoracic vertebra. The neck is narrow and rounded. The tubercle, thick and prominent, is placed on the outer border. There is no angle, but at the tubercle the rib is slightly bent, with cranial convexity. The cranial surface of the body is marked by two shallow grooves, separated from each other by a slight ridge prolonged internally into a tubercle, the **scalene tubercle,** for the attachment of the Scalenus anterior; the **ventral groove** transmits the subclavian vein, the **dorsal groove,** the subclavian artery and the inferior trunk of the brachial plexus. Beyond the dorsal groove is a rough area for the attachment of the Scalenus medius. The caudal surface is smooth, and destitute of a costal groove. The outer border is convex, thick, and rounded, and at its dorsal part gives attachment to the

Fig. 4–50.

Fig. 4–51.

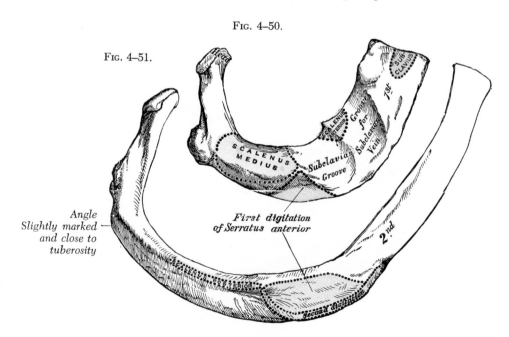

Figs. 4–50 and 4–51.—Peculiar ribs.

first digitation of the Serratus anterior. The inner border is concave, thin, sharp, and marked about its center by the scalene tubercle. The ventral extremity is larger and thicker than that of any of the other ribs and gives attachment to the Subclavius muscle.

The **Second Rib** (Fig. 4–51) is much longer than the first, but has a very similar curvature. The non-articular portion of the tubercle is occasionally only feebly marked. The angle is slight, and situated close to the tubercle. The body is not twisted, so that both ends touch any plane surface upon which it may be laid; but there is a bend with cranial convexity similar to although smaller than that found in the first rib. The body is not flattened horizontally like that of the first rib. Near the middle of its external surface is a rough eminence for the origin of part of the first and the whole of the second digitation of the Serratus anterior, and more dorsally the Scalenus posterior is attached. The internal surface is smooth, and concave, and on its posterior part there is a short costal groove.

The **Tenth Rib** (Fig. 4–52) has only a single articular facet on its head.

The **Eleventh** and **Twelfth Ribs** (Figs. 4–53 and 4–54) have each a single articular facet which is of rather large size on the head; they have no necks or tubercles, and are narrow or pointed at their anterior ends. The eleventh has a slight angle and a shal-

low costal groove. The twelfth has neither; it is much shorter than the eleventh, and its head is inclined slightly. Sometimes the twelfth rib is even shorter than the first.

The Costal Cartilages

The **Costal Cartilages** (*cartilagines costales*) (Fig. 4–43) are bars of hyaline cartilage which serve to prolong the ribs ventrally. The first seven pairs are connected with the sternum; the next three are each articulated with the lower border of the cartilage of the preceding rib; the last two have pointed extremities which end in the wall of the abdomen. They increase in length from the first to the seventh, then gradually decrease to the twelfth. Their breadth, as well as that of the intervals between them, diminishes from the first to the last. They are broad at their attachments to the ribs, and taper toward their sternal extremities, excepting the first two, which are of the same breadth throughout, and the sixth, seventh, and eighth, which are enlarged where their margins are in contact. They also vary in direction; the first three are horizontal, the others angular. Each costal cartilage presents two surfaces, two borders, and two extremities.

The **ventral surface** of the first gives attachment to the costoclavicular ligament and the Subclavius muscle; those of the first six or seven at their sternal ends, to

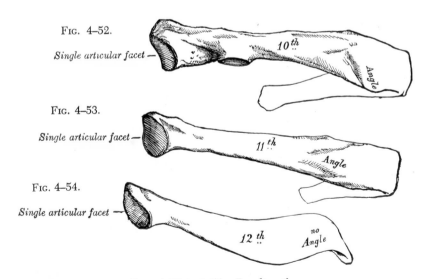

FIG. 4–52.

Single articular facet —

10th   Angle

FIG. 4–53.

Single articular facet —

11th   Angle

FIG. 4–54.

Single articular facet —

12th   no Angle

FIGS. 4–52 to 4–54.—Peculiar ribs.

the Pectoralis major. The others are covered by, and give partial attachment to, some of the flat muscles of the abdomen. The **dorsal surface** of the first gives attachment to the Sternothyroideus, those of the third to the sixth inclusive to the Transversus thoracis, and the last six or seven to the Transversus abdominis and the Diaphragm.

Of the two **borders** the superior is concave, the inferior convex; they afford attachment to the Intercostales interni. The superior border of the sixth gives attachment also to the Pectoralis major. The inferior borders of the sixth, seventh, eighth, and ninth cartilages present heel-like projections at the points of greatest convexity. These projections carry smooth oblong facets which articulate respectively with facets on slight projections from the superior borders of the seventh, eighth, ninth, and tenth cartilages.

Extremities.—The lateral end of each cartilage is continuous with the osseous tissue of the rib to which it belongs. The medial end of the first is continuous with the sternum; the medial ends of the six succeeding ones are rounded and are received into shallow concavities on the lateral margins of the sternum. The medial ends of the eighth, ninth, and tenth costal cartilages are pointed, and each is connected with the cartilage immediately above. Those of the eleventh and twelfth are pointed and free. In old age the costal cartilages are prone to undergo superficial ossification.

Structure.—The ribs consist of highly vascular cancellous tissue, enclosed in a thin layer of compact bone.

Ossification.—Each rib, with the exception of the last two, is ossified from four centers; a primary center for the body, and three epiphyseal centers, one for the head and one each for the articular and non-articular parts of the tubercle. The eleventh and twelfth ribs have each only two centers, those for the tubercles being wanting. Ossification begins near the angle toward the end of the second month of fetal life, and is seen first in the sixth and seventh ribs. The epiphyses for the head and tubercle make their appearance between the sixteenth and twentieth years, and are united to the body about the twenty-fifth year.

Variations and Clinical Interest.—**Cervical ribs** derived from the costal process of the **seventh** cervical vertebra (Fig. 4–2) are of not infrequent occurrence, and are important clinically because they may give rise to obscure nervous or vascular symptoms. The cervical rib may be a mere epiphysis articulating only with the transverse process of the vertebra, but more commonly it consists of a defined head, neck, and tubercle, with or without a body. It extends lateralward into the posterior triangle of the neck where it may terminate in a free end or may join the first thoracic rib, the first costal cartilage, or the sternum. It varies much in shape, size, direction, and mobility. If it reaches far enough ventrally, part of the brachial plexus and the subclavian artery and vein cross over it and are apt to suffer compression in so doing. Pressure on the artery may obstruct the circulation so much that arterial thrombosis results, causing gangrene of the finger tips. Pressure on the nerves is common, and affects the eighth cervical and first thoracic nerves, causing paralysis of the muscles they supply, and neuralgic pains and paresthesia in the area of skin to which they are distributed; no oculopupillary changes are to be found.

The thorax is frequently altered in shape in certain diseases.

In **rickets,** the ends of the ribs where they join the costal cartilages become enlarged, giving rise to the so-called "rickety rosary," which in mild cases is only found on the internal surface of the thorax. Lateral to these enlargements the softened ribs sink in, leaving a groove along either side of the sternum, which in turn is forced outward by the bending of the ribs, and the dorsoventral diameter of the chest is increased. The ribs affected are the second to the eighth, the lower ones being prevented from falling in by the presence of the liver, stomach, and spleen; and when the abdomen is distended, as it often is in rickets, the lower ribs may be pushed outward, causing a transverse groove (**Harrison's sulcus**) just above the costal arch. This deformity or forward projection of the sternum, often asymmetrical, is known as **pigeon breast,** and may be taken as evidence of active or old rickets except in cases of primary spinal curvature. In many instances it is associated with obstruction in the upper air passages due to enlarged tonsils or adenoid growths. In some rickety children or adults, and also in others who give no history or further evidence of having had rickets, an opposite condition obtains. The lower part of the sternum and often the xiphoid process as well are deeply depressed backward, producing an oval hollow in the lower sternal and upper epigastric regions. This is known as **funnel breast;** it never appears to produce the least disturbance of any of the vital functions.

The *phthisical chest* is often long and narrow, and with great obliquity of the ribs and projection of the scapulae. In *pulmonary emphysema* the chest is enlarged in all its diameters, and presents on section an almost circular outline. It has received the name of **barrel chest.** In severe cases of *scoliosis* or lateral curvature of the vertebral column the thorax becomes much distorted. Due to rotation of the bodies of the vertebrae, the ribs opposite the convexity of the dorsal curve become extremely convex behind, being thrown out and bulging, and at the same time flattened in front, so that the two ends of the same rib are almost parallel. Coincidentally with this the ribs on the opposite side, on the concavity of the curve, are sunk and depressed behind, and bulging and convex in front.

THE SKULL

The **Skull** or bony structure of the head may be divided into two parts: (1) the **cranium,** which lodges and protects the brain, and consists of eight bones, and (2) the **skeleton of the face,** composed of fourteen bones. These bones are flat or irregular and except for the mandible are joined to each other by immovable articulations called sutures. Since these individual bones can be studied separately only in special preparations, the skull as a whole is described first, the separate bones later. The exterior of the skull viewed from above is the Norma Verticalis; from the side, the Norma Lateralis; toward the face, the Norma Frontalis; from the back, the Norma Occipitalis; and from below, the Norma Basalis.

The Exterior of the Skull

Norma Verticalis.—The superior aspect is also called the skull cap or **calvaria.** It varies greatly in different skulls; in some it is more or less oval, in others more nearly circular. It is traversed by three sutures: (1) the **coronal suture,** nearly transverse in direction, between the frontal and parietal bones; (2) the **sagittal suture,** placed in the midline, between the two parietal, and (3) the upper part of the **lambdoidal suture,** between the parietals and the occipital. The point of junction of the sagittal and coronal sutures is named the **bregma** (Fig. 3–2), that of the sagittal and lambdoid sutures, the **lambda;** they indicate respectively the positions of the anterior and posterior fontanelles, or the so-called soft spots in the skull of an infant. On either side of the sagittal suture are the **parietal eminences** where the skull is often somewhat flattened, and close to the suture and not far from the lambda there may be a parietal foramen.

The term **obelion** is applied to the point on the sagittal suture which is at this level. The frontal eminences form the two prominences of the forehead and below them two superciliary arches are on either side of the **glabella.** Immediately above the glabella in a small percentage of skulls the frontal suture persists and extends along the middle line to the bregma. The temporal lines and zygomatic arches may or may not be seen.

Norma Lateralis (Fig. 4–55).—The lateral aspect of the skull consists of both cranial and facial bones. Entering into its formation are the frontal, the parietal, the occipital, the temporal, the zygomatic, and the great wing of the sphenoid. The mandible also forms part, but its description here is omitted. These bones are joined to one another and to the zygomatic by the following sutures: the **zygomaticotemporal, zygomaticofrontal,** and the sutures surrounding the great wing of the sphenoid, viz., the **sphenozygomatic,** the **sphenofrontal, sphenoparietal,** and **sphenosquamosal sutures.** The sphenoparietal suture varies in length in different skulls, and is absent when the frontal articulates with the temporal squama. The point corresponding with the posterior end of the sphenoparietal suture is named the **pterion;** it is situated about 3 cm posterior to, and a little superior to the level of the zygomatic process of the frontal bone.

The **squamosal suture** arches posteriorly from the pterion and lies between the temporal squama and the inferior border of the parietal; this suture is continuous posteriorly with the short, nearly horizontal **parietomastoid suture** which unites the mastoid process of the temporal with the region of the mastoid angle of the parietal. Extending across the cranium are the

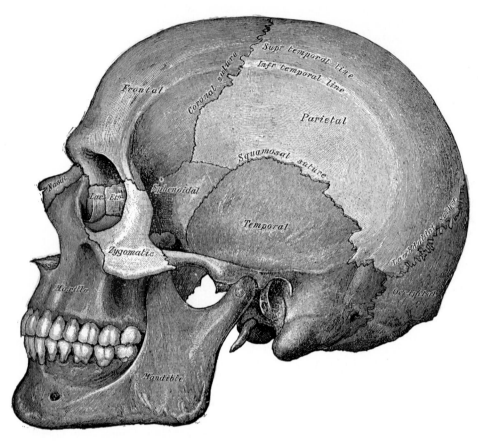

Fig. 4–55.—Side view of the skull.

coronal and lambdoidal sutures. The **lamb-doidal suture** is continuous with the occipi-tomastoid suture between the occipital and the mastoid portion of the temporal. In or near the last suture the mastoid foramen transmits an emissary vein. The point of meeting of the parietomastoid, occipitomas-toid, and lambdoidal sutures is known as the **asterion**. Arching across the side of the cranium are the temporal lines, which mark the superior limit of the temporal fossa.

The **Temporal Fossa** (*fossa temporalis*) (Fig. 4–55) is surrounded by the **temporal lines**, which extend from the zygomatic process of the frontal bone across the frontal and parietal bones, and then curve anterior-ly to become continuous with the supramas-toid crest and the posterior root of the zygomatic arch. The point where the **superior temporal line** cuts the coronal suture is named the **stephanion**. The tem-poral fossa is bounded anteriorly by the

frontal and zygomatic bones. It is separated from the infratemporal fossa by the **infra-temporal crest** (Fig. 4–56) on the great wing of the sphenoid, and by a ridge, continuous with this crest, which is carried posteriorly across the temporal squama to the anterior root of the zygomatic process. Laterally the fossa is limited by the zygo-matic arch. The fossa communicates with the orbital cavity through the inferior orbital or sphenomaxillary fissure. The floor of the fossa is formed by the zygomatic, frontal, parietal, sphenoid, and temporal bones. The temporal fossa contains the Temporalis muscle and its vessels and nerves, together with the zygomatico-temporal nerve.

The **zygomatic arch** is formed by the zygomatic process of the temporal and the temporal process of the zygomatic, the two being united by an oblique suture. The tendon of the Temporalis passes under, *i.e.,* medial to the arch to gain insertion into

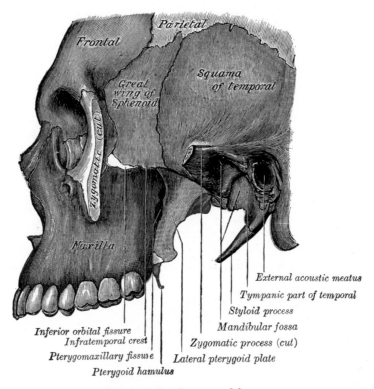

Fig. 4–56.—Left infratemporal fossa.

the coronoid process of the mandible. The zygomatic process of the temporal arises by two **roots**, an **anterior**, directed inward anterior to the mandibular fossa, where it expands to form the articular tubercle, and a **posterior**, which runs posteriorly above the external acoustic meatus and is continuous with the supramastoid crest. The superior border of the arch gives attachment to the temporal fascia; the inferior border and medial surface give origin to the Masseter.

Between the posterior root of the zygomatic arch and the mastoid process is the elliptical orifice of the **external acoustic meatus**, bounded by the tympanic part of the temporal bone, to whose outer margin the cartilaginous segment of the external acoustic meatus is attached. The small triangular area between the posterior root of the zygomatic arch and the posterosuperior part of the orifice is termed the **suprameatal triangle**, on the anterior border of which a small spinous process, the **suprameatal spine**, is sometimes seen. Between the

tympanic part and the articular tubercle is the **mandibular fossa**, divided into two parts by the **petrotympanic fissure.** The anterior and larger part of the fossa is articulated with the condyle of the mandible; the posterior part is non-articular and sometimes lodges a portion of the parotid gland. The **styloid process** extends downward and forward for a variable distance from the tympanic part, and gives attachment to the Styloglossus, Stylohyoideus, and Stylopharyngeus, and to the stylohyoid and stylomandibular ligaments. Posterior to the external acoustic meatus is the **mastoid process**, to the outer surface of which the Sternocleidomastoideus, Splenius capitis, and Longissimus capitis are attached.

The **Infratemporal Fossa** (*fossa infratemporalis; zygomatic fossa*) (Fig. 4–56) is an irregularly shaped cavity medial or deep to the zygomatic arch. It is bounded laterally by the infratemporal surface of the maxilla and the ridge from its zygomatic process; by the articular tubercle of the temporal and the spine of the sphenoid; by the great wing of the sphenoid and the inferior surface of the temporal squama; by the alveolar border of the maxilla; and by the lateral pterygoid plate. It contains part of the Temporalis, the Pterygoidei medialis and lateralis, the maxillary vessels, the mandibular and maxillary nerves and the pterygoid plexus of veins. The foramen ovale, foramen spinosum and the alveolar canals open into it. At its superior and medial part are two fissures which meet at right angles, the horizontal limb being named the inferior orbital fissure, and the vertical limb the pterygomaxillary fissure.

The **inferior orbital fissure** (*fissura orbitalis inferior; sphenomaxillary fissure*), opens into the posterolateral part of the orbit. It is bounded by the inferior border of the orbital surface of the great wing of the sphenoid; by the lateral border of the orbital surface of the maxilla and the orbital process of the palatine bone; by a small part of the zygomatic bone, and it joins the pterygomaxillary fissure at right angles. Through the inferior orbital fissure, the orbit communicates with the temporal, infratemporal, and pterygopalatine fossae; it transmits the maxillary nerve, its zygomatic branch, and the ascending branches from its sphenopalatine branch, the infraorbital vessels and the clinically important vein which connects the inferior ophthalmic vein with the pterygoid venous plexus.

The **pterygomaxillary fissure**, at right angles with the medial end of the preceding, is a triangular interval, formed by the diver-

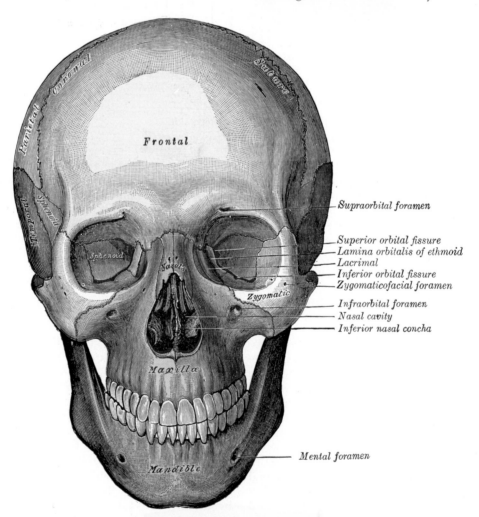

FIG. 4–57.—The skull from the front.

gence of the maxilla from the pterygoid process of the sphenoid. It connects the infratemporal with the pterygopalatine fossa, and transmits the terminal part of the maxillary artery and veins.

The **pterygopalatine fossa** (*fossa pterygopalatina; sphenomaxillary fossa*) (Figs. 4–60, 4–61) is the space at the junction of the inferior orbital and pterygomaxillary fissures in the inferior part of the apex of the orbit. It is bounded by the body, the base of the pterygoid process, and the anterior surface of the great wing of the sphenoid; by the infratemporal surface of the maxilla; and by the vertical part of the palatine bone with its orbital and sphenoidal processes. This fossa communicates with the orbit by the inferior orbital fissure, with the nasal cavity by the sphenopalatine foramen, and with the infratemporal fossa by the pterygomaxillary fissure. Of the five foramina which open into it, three are on the posterior wall, viz., the foramen rotundum, the pterygoid canal, and the pharyngeal canal. On the medial wall is the sphenopalatine foramen and the pterygopalatine canal. The fossa contains the maxillary nerve, the pterygopalatine ganglion, and the terminal part of the maxillary artery.

Norma Frontalis (Fig. 4–57).—The facial aspect of the skull is outlined by the frontal bone superiorly, the body of the mandible inferiorly, and the zygomatic bones and rami of the mandible laterally. The **frontal eminences** stand out more or less prominently above the **superciliary arches,** which merge into one another at the glabella (Fig. 3–2). Below the **glabella** is the **frontonasal suture,** the mid-point of which is termed the **nasion.** Lateral to the frontonasal suture the frontal bone articulates with the frontal process of the maxilla and with the lacrimal. Below the superciliary arches, the superior part of the margin of the orbit is thin and prominent in its lateral two-thirds, rounded in its medial third. At the junction of these two portions is the **supraorbital notch** or foramen for the supraorbital nerve and vessels. The **supraorbital margin** ends laterally in the zygomatic process which articulates with the zygomatic bone. The **bridge of the nose** is formed by the two nasal bones supported in the middle line by the perpendicular plate of the ethmoid, and laterally

by the frontal processes of the maxillae which form the lower and medial part of the margin of each orbit. The anterior aperture of the nose is bounded by sharp margins laterally to which the lateral and alar cartilages of the nose are attached; the inferior margins are thicker and end in the anterior **nasal spine** medially. In the anterior part of the nasal cavity, the bony septum has a large triangular deficiency anteriorly, which is filled by septal cartilage in the intact body. On the lateral wall of each nasal cavity the anterior part of the inferior nasal concha is visible. The anterior surface of the maxilla is perforated near the inferior margin of the orbit by the **infraorbital foramen** for the passage of the infraorbital nerve and vessels. The zygomatic bone forms the prominence of the cheek, the inferior and lateral portion of the orbital cavity, and the anterior part of the zygomatic arch. It articulates medially with the maxilla and the zygomatic process of the frontal; it is perforated by the zygomaticofacial foramen for the passage of the zygomaticofacial nerve.

On the **body of the mandible** are the **mental protuberance,** the **mental tubercles,** and a median ridge indicating the position of the fetal **symphysis menti.** Below the incisor teeth is the incisive fossa, and inferior to the second premolar tooth is the **mental foramen** which transmits the mental nerve and vessels. The **oblique line** runs posteriorly from the mental tubercle and is continuous with the anterior border of the ramus. The posterior border of the ramus runs from the condyle to the angle. The latter frequently is more or less everted. The upper teeth in the alveolar processes of the maxillae overlap the teeth of the mandible anteriorly.

The **Orbits** (*orbitae*) (Fig. 4–58).—The conical cavities of the orbits are so placed that their medial walls are approximately parallel with each other and with the middle line, but their lateral walls are widely divergent. As a result, the axes of the two orbits diverge widely in an anterior direction but if prolonged in a posterior direction they come together near the center of the skull. It is customary to describe the orbits in terms of a roof, a floor, a medial wall, a lateral wall, a base, and an apex.

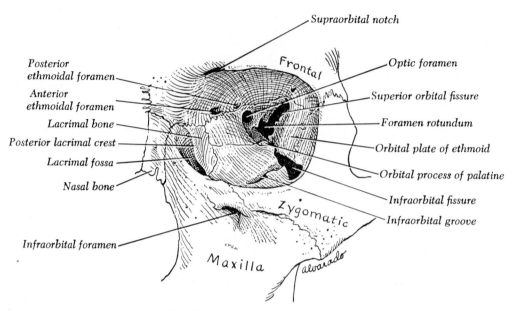

FIG. 4–58.—The left orbital cavity, anterior aspect.

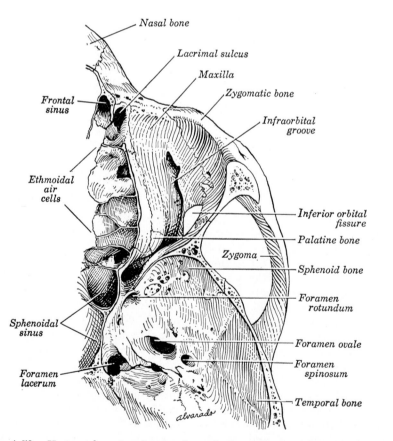

FIG. 4–59.—Horizontal section showing floor of orbit and surrounding structures.

The **roof** (Fig. 4–85) is formed by the orbital plate of the frontal and the small wing of the sphenoid bones. It presents medially the **trochlear fovea** for the attachment of the cartilaginous pulley of the Obliquus superior oculi; laterally, the **lacrimal fossa** for the lacrimal gland; and posteriorly, the suture between the frontal bone and the small wing of the sphenoid.

The **floor** (Fig. 4–59) is of less extent than the roof and is formed by the orbital surface of the maxilla, the orbital process of the zygomatic bone, and to a small extent by the orbital process of the palatine. Near the orbital margin medially is the lacrimal sulcus for the **nasolacrimal canal**, and immediately lateral to this a depression for the origin of the Obliquus inferior oculi. On its lateral part is the suture between the maxilla and zygomatic bone, and at its posterior part that between the maxilla and the orbital process of the palatine. Near the middle of the floor is the **infraorbital groove**, which leads anteriorly into the **infraorbital**
canal and transmits the infraorbital nerve and vessels.

The **medial wall** (Fig. 4–60) is nearly vertical, and is formed by the frontal process of the maxilla, the lacrimal, the lamina orbitalis of the ethmoid, and a small part of the body of the sphenoid. It exhibits three vertical **sutures:** the **lacrimomaxillary, lacrimoethmoidal,** and **sphenoethmoidal.** Anteriorly the **lacrimal groove** lodges the lacrimal sac, and posterior to the groove is the **posterior lacrimal crest,** from which the lacrimal part of the Orbicularis oculi arises. At the junction of the medial wall and the roof are the **frontomaxillary, frontolacrimal, frontoethmoidal,** and **sphenofrontal sutures.** The point of junction of the anterior border of the lacrimal with the frontal is named the **dacryon.** In the frontoethmoidal suture are the **anterior** and **posterior ethmoidal foramina,** the former transmitting the nasociliary nerve and anterior ethmoidal vessels, the latter the posterior ethmoidal nerve and vessels.

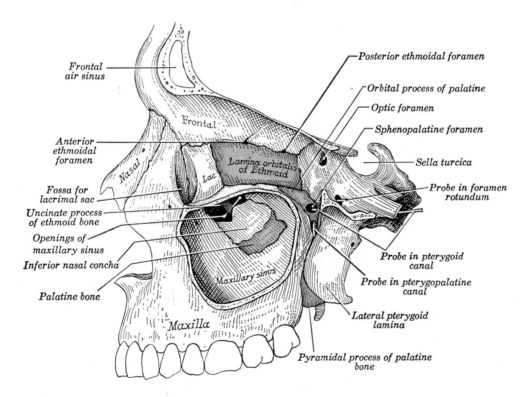

Fig. 4–60.—Medial wall of left orbit and medial wall of maxillary sinus exposed.

The **lateral wall** (Fig. 4–61) is formed by the orbital process of the zygomatic and the orbital surface of the great wing of the sphenoid; the sphenozygomatic suture terminates at the anterior end of the inferior orbital fissure. Between the roof and the lateral wall, near the apex of the orbit, is the **superior orbital fissure.** Through this fissure the oculomotor, the trochlear, the ophthalmic division of the trigeminal, and the abducent nerves enter the orbital cavity,

ethmoidal foramina, zygomatic foramen, and the canal for the nasolacrimal duct.

Norma Occipitalis.—In the middle line of the posterior aspect of the cranium is the sagittal suture between the posterior parts of the parietal bones. From the posterior end of the sagittal suture the deeply serrated **lambdoidal suture** lies between the parietals and the occipital and continues into the **parietomastoid** and **occipitomastoid** sutures. It frequently contains one or more **sutural**

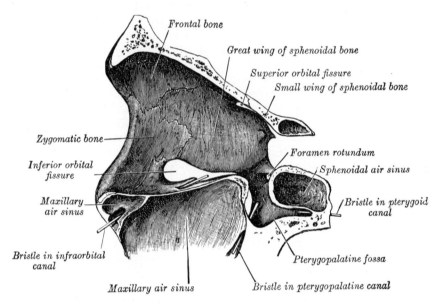

Frontal bone

Great wing of sphenoidal bone

Superior orbital fissure

Small wing of sphenoidal bone

Zygomatic bone

Inferior orbital fissure

Maxillary air sinus

Bristle in infraorbital canal

Maxillary air sinus

Foramen rotundum

Sphenoidal air sinus

Bristle in pterygoid canal

Pterygopalatine fossa

Bristle in pterygopalatine canal

Fig. 4–61.—The lateral wall of the right orbit.

also some filaments from the cavernous plexus of the sympathetic and the orbital branches of the middle meningeal artery. Leaving through the fissure are the superior ophthalmic vein and the recurrent branch from the lacrimal artery to the dura mater. The lateral wall and the floor are separated posteriorly by the inferior orbital fissure.

The **base** of the orbit is formed by the orbital margin, already described, and the **apex** by the optic foramen which continues as a short, cylindrical canal for the optic nerve and ophthalmic artery.

Nine openings communicate with each orbit: the optic foramen, superior and inferior orbital fissures, supraorbital foramen, infraorbital canal, anterior and posterior

bones. Near the middle of the occipital squama is the **external occipital protuber-ance** or **inion,** and extending laterally from it on either side is the **superior nuchal** and the faintly marked **highest nuchal lines.** The part of the squama above the inion and highest lines is named the **planum occipitale,** and is covered by the Occipitalis muscle; the part below is termed the **planum nuchale,** and is divided by a ridge, the **median nuchal line,** which runs from the inion to the foramen magnum; and gives attachment to the ligamentum nuchae. Anterior to the planum nuchale are the mastoid processes, and near the occipitomastoid suture is the **mastoid foramen** for the passage of the mastoid emissary vein.

Norma Basalis (Fig. 4–62).—The inferior surface of the base of the skull with the mandible removed is formed by the palatine processes of the maxillae and palatine bones, the vomer, the pterygoid processes, the inferior surfaces of the great wings, spinous processes, and part of the body of the sphenoid, the inferior surfaces of the squamae and petrous portions of the temporals, and the inferior surface of the occipital bone. The anterior part or **hard palate** projects beyond the level of the rest of the surface and is bounded anteriorly and laterally by the alveolar arch containing the sixteen teeth of the maxillae. The vault of the hard palate is marked by depressions for the palatine glands, and traversed by a cruciate suture formed by the junction of the four bones of which it is composed. At either posterior angle of the hard palate is the **greater palatine foramen,** for the transmission of the greater palatine vessels and nerve. Posterior to the foramen is the pyramidal process of the palatine bone, perforated by one or more **lesser palatine foramina,** and marked by the commencement of a transverse ridge, for the attachment of the tendinous expansion of the Tensor veli palatini. Projecting backward from the center of the posterior border of the hard palate is the **posterior nasal spine,** for the attachment of the Musculus uvulae.

Behind and above the hard palate are the **choanae** or posterior openings into the nasal cavities. They are separated by the vomer, and each is bounded by the body of the sphenoid, by the horizontal part of the palatine bone, and laterally by the medial pterygoid plate of the sphenoid. At the superior border of the vomer, the expanded alae of this bone may be seen, receiving between them the rostrum of the sphenoid. The medial pterygoid plate is long and narrow; on the lateral side of its base is the scaphoid fossa, for the origin of the Tensor veli palatini, and at its lower extremity the **pterygoid hamulus,** around which the tendon of this muscle turns. The lateral pterygoid plate is broad; its lateral surface forms the medial boundary of the infratemporal fossa, and affords attachment to the Pterygoideus lateralis.

Posterior to the vomer is the basilar portion of the occipital bone, presenting near its center the **pharyngeal tubercle** for the attachment of the fibrous raphe of the pharynx, with depressions on either side for the insertions of the Rectus capitis anterior and Longus capitis. At the base of the lateral pterygoid plate is the **foramen ovale,** and lateral to this is the prominent **sphenoidal spine** (*spina angularis*) which gives attachment to the sphenomandibular ligament and the Tensor veli palatini. It is pierced by the **foramen spinosum** which transmits the middle meningeal vessels. Lateral to the spine is the **mandibular fossa,** divided into two parts by the petrotympanic fissure; the anterior portion, concave, smooth, and bounded in front by the articular tubercle, articulates with the condyle of the mandible; the posterior portion is rough and bounded behind by the tympanic part of the temporal. Emerging from between the laminae of the vaginal process of the tympanic part is the **styloid process;** and at the base of this process is the **stylomastoid foramen.** Upon the medial side of the mastoid process is the **mastoid notch** for the posterior belly of the Digastricus, and medial to the notch, the **occipital groove** for the occipital artery.

Appearing in a dried skull at the base of the medial pterygoid plate is an aperture, irregular in shape and variable in size, named the **foramen lacerum.** It is not a complete foramen in the intact body, because its inferior portion is closed over by the fibrocartilaginous plate containing the auditory tube. It is bounded by the great wing of the sphenoid, by the apex of the petrous portion of the temporal bone, by the body of the sphenoid, and by the basilar portion of the occipital bone. Concealed within its anterior aspect is the aperture of the **carotid canal.** The internal carotid artery emerging from the carotid canal traverses only the upper part of the foramen lacerum. Lateral to this aperture is a groove, the **sulcus tubae auditivae,** between the petrous part of the temporal and the great wing of the sphenoid. This sulcus lodges the cartilaginous part of the auditory tube which is continuous with the bony part within the temporal bone. At the bottom of this sulcus is a narrow cleft, the **petrosphenoidal fissure.** Near the apex of the petrous portion of the temporal bone, the quadrilateral rough surface affords attachment to the Levator veli palatini; lateral to this surface is the orifice or entrance of the **carotid canal.**

Posterior to the carotid canal is the **jugular foramen,** a large aperture, formed by the petrous portion of the temporal and the occipital; it is generally larger on the right than on the left side, and may be subdivided into three compartments. The anterior compartment transmits the inferior petrosal sinus; the intermediate, the glossopharyngeal, vagus, and accessory nerves; the posterior, the transverse sinus and some meningeal branches from the occipital and ascending pharyngeal arteries. Extending anteriorly from the jugular foramen to the foramen lacerum is the **petroöccipital fissure** occupied, in the intact body, by a plate of cartilage.

Posterior to the basilar portion of the occipital bone is the **foramen magnum,** bounded laterally by the occipital condyles. Lateral to each condyle is the jugular process which gives attachment to the Rectus capitis lateralis muscle and the lateral atlantoöccipital ligament. The foramen magnum transmits the medulla oblongata and its membranes, the accessory nerves, the vertebral arteries, the anterior and posterior spinal arteries, and the ligaments connecting the occipital bone with the axis. The midpoints on the anterior and posterior margins of the foramen are respectively termed the **basion** and the **opisthion.** Anterior to each condyle is the **canal** for the passage of the **hypoglossal** nerve and a meningeal artery. Posterior to each condyle is the condyloid fossa, perforated on one or both sides by the **condyloid canal,** for the transmission of a vein from the transverse sinus. Posterior to the foramen magnum is the median nuchal line ending superiorly at the external occipital protuberance, while on either side are the superior and inferior **nuchal lines;** these, as well as the surfaces of bone between them, are roughened for the attachment of the muscles which are enumerated on Figure 4–79.

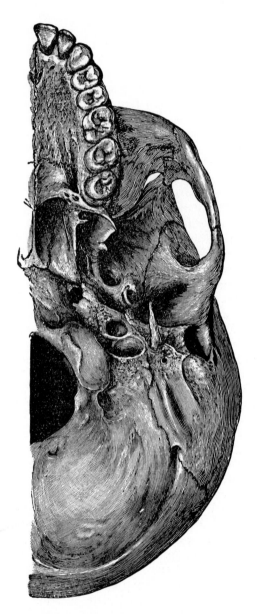

Fig. 4–62.—The external surface of the left half of the base of the skull. (Norma basalis.)

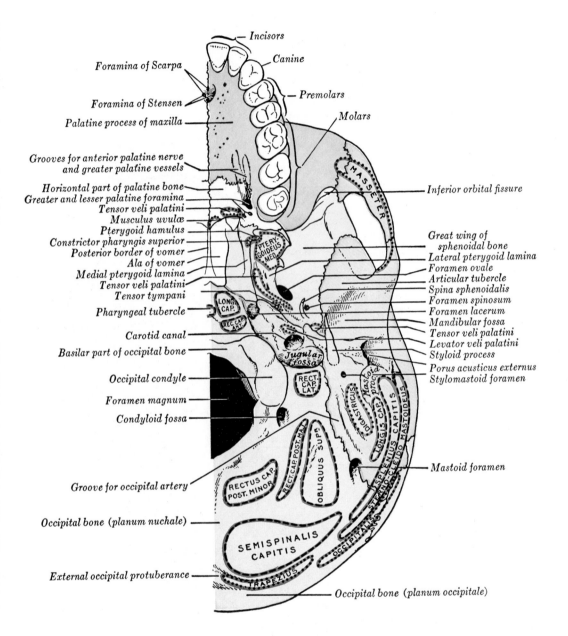

Fig. 4–63.—Key to Figure 4–62.

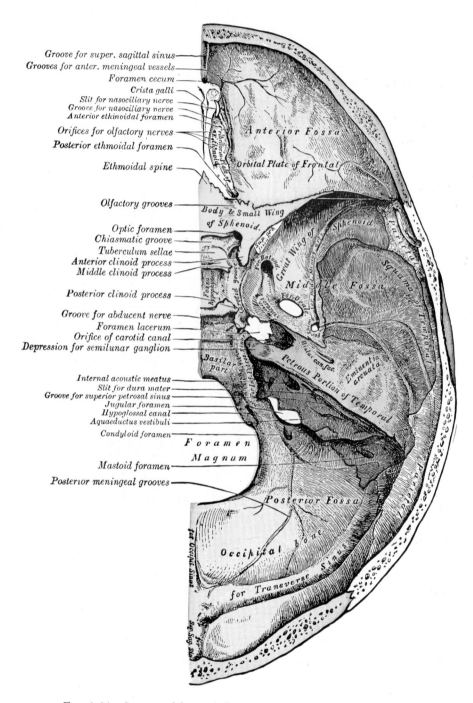

Groove for super. sagittal sinus
Grooves for anter. meningeal vessels
Foramen cecum
Crista galli
Slit for nasociliary nerve
Groove for nasociliary nerve
Anterior ethinoidal foramen
Orifices for olfactory nerves
Posterior ethmoidal foramen
Ethmoidal spine
Olfactory grooves
Optic foramen
Chiasmatic groove
Tuberculum sellae
Anterior clinoid process
Middle clinoid process
Posterior clinoid process
Groove for abducent nerve
Foramen lacerum
Orifice of carotid canal
Depression for semilunar ganglion
Internal acoustic meatus
Slit for dura mater
Groove for superior petrosal sinus
Jugular foramen
Hypoglossal canal
Aquaeductus vestibuli
Condyloid foramen
Mastoid foramen
Posterior meningeal grooves

Anterior Fossa
Orbital Plate of Frontal
Body & Small Wing of Sphenoid
Great Wing of Sphenoid
Middle Fossa
Squama of Temporal
Petrous Portion of Temporal
Eminentia arcuata
Basilar part
Foramen Magnum
Posterior Fossa
Occipital bone
for Transverse Sinus

FIG. 4–64.—Interior of base of the skull, showing the cranial fossae.

The Interior of the Skull

Inner Surface of the Skullcap.—The inner surface of the **calvaria** is marked by depressions for the convolutions of the cerebrum, and by numerous furrows for the branches of the meningeal vessels. Along the middle line is a longitudinal groove, narrow where it commences at the frontal crest, but broader posteriorly. It lodges the superior sagittal sinus, and its margins afford attachment to the falx cerebri. On either side of it are several **foveolae granulares** for the arachnoid granulations, and posteriorly, the openings of the **parietal foramina** when these are present. It is crossed by the coronal and lambdoidal sutures; the sagittal suture lies in the median plane between the parietal bones.

Internal or Superior Surface of the Base of the Skull (Fig. 4–64).—The superior surface of the base of the skull forms the **floor of the cranial cavity** and is divided into three fossae, called the anterior, middle, and posterior cranial fossae.

Anterior Cranial Fossa (*fossa cranii anterior*).—The floor of the anterior fossa is formed by the orbital plates of the frontal, the cribriform plate of the ethmoid, and the small wings and anterior part of the body of the sphenoid; it is limited by the posterior borders of the small wings of the sphenoid and by the anterior margin of the chiasmatic groove. It is traversed by the frontoethmoidal, and sphenofrontal sutures. Its lateral portions or **orbital plates** support the frontal lobes of the cerebrum. The median portion corresponds with the roof of the nasal cavity, on either side of the **crista galli.** The **frontal crest** for the attachment of the falx cerebri ends at the **foramen cecum,** between the frontal bone and the crista galli which usually transmits a small vein from the nasal cavity to the superior sagittal sinus. On either side of the crista galli, the olfactory groove of the **cribriform plate** supports the olfactory bulb and is perforated by foramina for the transmission of the olfactory nerves, and rostrally, by a slit for the nasociliary nerve (Fig. 4–64). In the lateral wall of the olfactory groove are the internal openings of the anterior and posterior ethmoidal foramina. The anterior transmits the anterior ethmoidal vessels and the nasociliary nerve; the posterior ethmoidal foramen opens at the back part of the margin under cover of the projecting lamina of the sphenoid, and transmits the posterior ethmoidal vessels and nerve. More posteriorly are the **ethmoidal spine** of the sphenoid and the anterior margin of the **chiasmatic groove** which runs lateralward on either side to the upper margin of the optic foramen.

The **Middle Cranial Fossa** (*fossa cranii media*).—The middle fossa on either side is bounded anteriorly by the posterior margins of the small wings of the sphenoid, the anterior clinoid processes, and the ridge forming the anterior margin of the chiasmatic groove; posteriorly, by the superior angles of the petrous portion of the temporals and the dorsum sellae; laterally by the temporal squamae, sphenoidal angles of the parietals, and great wings of the sphenoid.

In the median region are the **chiasmatic groove** and **tuberculum sellae.** The chiasmatic groove ends on either side at the optic foramen. Posterior to the optic foramen the **anterior clinoid process** gives attachment to the tentorium cerebelli. Posterior to the tuberculum sellae, the **fossa hypophyseos** forms a deep depression in the **sella turcica** which lodges the hypophysis, and presents on its anterior wall the **middle clinoid processes.** The sella turcica is bounded posteriorly by a quadrilateral plate of bone, the **dorsum sellae,** the free angles of which are surmounted by the **posterior clinoid processes:** these afford attachment to the tentorium cerebelli, and below each is a notch for the abducent nerve. On either side of the sella turcica is the carotid groove, which is broad, shallow, and curved somewhat like the italic letter *f.* It extends from the foramen lacerum to the medial side of the anterior clinoid process, where it is sometimes converted into a *foramen* (*caротico-clinoid*) by the union of the anterior with the middle clinoid process; it is bounded laterally by the **lingula.**

The lateral parts of the middle fossa are of considerable depth, and support the temporal lobes of the brain. They are traversed by furrows for the **anterior** and posterior branches of the **middle meningeal vessels.** From near the **foramen spinosum,** the anterior runs to the sphenoidal angle of the parietal, where it is sometimes converted into a bony canal; the posterior runs lateralward across the temporal squama and passes on to the parietal near the middle of its inferior border. The following apertures are also to be seen. Rostrally the **superior orbital fissure** is bounded by the small wing, by the great wing, and medially, by the body of the sphenoid; it is usually completed laterally by the orbital plate of the frontal bone. Posterior to the medial end of the superior orbital fissure are the **foramen rotundum** and the **foramen ovale.** Between them the **foramen of Vesalius,** often absent, opens at the lateral side of the scaphoid fossa inferiorly, and transmits a small vein. Lateral to the foramen ovale is the foramen spinosum. Medial to the foramen ovale the **foramen lacerum** may be seen in a dried skull but in the fresh state the inferior part of this aperture is occupied by a layer of fibrocartilage. The internal carotid artery, surrounded by a plexus of sympathetic nerves, passes across the superior part of the foramen but does not traverse its whole extent. The small nerve of the pterygoid canal and a small meningeal branch from the ascending pharyngeal artery alone pierce the layer of fibrocartilage and are thus the only structures to pass through the foramen lacerum. On the anterior surface of the petrous portion of the temporal bone, the **arcuate eminence** is caused by the projection of the superior semicircular canal; anterior to and a little lateral to this is a depression corresponding to the roof of the tympanic cavity; and, near the apex of the bone, is the orifice of the **carotid canal.**

The **Posterior Cranial Fossa** (*fossa cranii posterior*) is the largest and deepest of the three. It is formed by the dorsum sellae and clivus of the sphenoid, the occipital, the petrous and mastoid portions of the temporals, and the mastoid angles of the parietal bones. It is crossed by the occipitomastoid and the parietomastoid sutures, and lodges the cerebellum, pons, and medulla oblongata. It is separated from the middle fossa in and near the median line by the dorsum sellae of the sphenoid and on either side by the superior margin of the petrous portion of the temporal bone, which gives attachment to the tentorium cerebelli, and is grooved for the superior petrosal sinus. In its center is the **foramen magnum,** on either side of which in a rough tubercle is the **hypoglossal canal.** Anterior to the foramen magnum the basilar portion of the occipital and the posterior part of the body of the sphenoid form a grooved surface which supports the medulla oblongata and pons. In the young skull these bones are joined by a synchondrosis. On either side the **petroöccipital fissure** is occupied in the intact body by a plate of cartilage and is continuous with the jugular foramen, and its margins are grooved for the inferior petrosal sinus.

The **jugular foramen** is situated between the lateral part of the occipital and the petrous part of the temporal. The anterior portion of this foramen transmits the inferior petrosal sinus; the posterior portion, the transverse sinus and some meningeal branches from the occipital and ascending pharyngeal arteries; and the intermediate portion, the glossopharyngeal, vagus and accessory nerves. Superior to the jugular foramen is the **internal acoustic meatus,** for the facial and acoustic (*otic, vestibulocochlear, stato-acoustic*) nerves and internal auditory artery. The inferior occipital fossae, which support the hemispheres of the cerebellum, are separated from one another by the **internal occipital crest,** which serves for the attachment of the falx cerebelli, and lodges the occipital sinus. The posterior fossae are bounded posteriorly by deep **grooves for the transverse sinuses.** Each of these channels, in its passage to the jugular foramen, grooves the occipital, the mastoid angle of the parietal, the mastoid portion of the temporal, and the jugular process of the occipital, and ends at the jugular foramen. Where this sinus grooves the mastoid portion of the temporal, the orifice of the **mastoid foramen** may be seen; and, just previous to its termination, the **condyloid canal** opens into it; neither opening is constant.

The **Nasal Cavity** (*cavum nasi; nasal fossa*).—The nasal cavities open on the face through the pear-shaped **anterior nasal aperture,** and the posterior openings, the **choanae,** communicate, in the intact body, with the nasal part of the pharynx. They are much narrower superiorly than inferiorly, and each cavity is bounded by a roof, a floor, a medial, and a lateral wall.

The **roof** (Figs. 4–65, 4–67) is formed anteriorly by the nasal bone and the spine of the frontal; in the middle, by the cribriform plate of the ethmoid; and posteriorly, by the body of the sphenoid, the sphenoidal concha, the ala of the vomer and the sphenoidal process of the palatine bone.

nasal concha; posteriorly, by the vertical plate of the palatine bone, and the medial pterygoid plate of the sphenoid. On this wall are three irregular anteroposterior passages, termed the superior, middle, and inferior meatuses of the nose. The **superior meatus**, occupies the middle third of the lateral wall between the superior and middle nasal conchae; the posterior ethmoidal cells

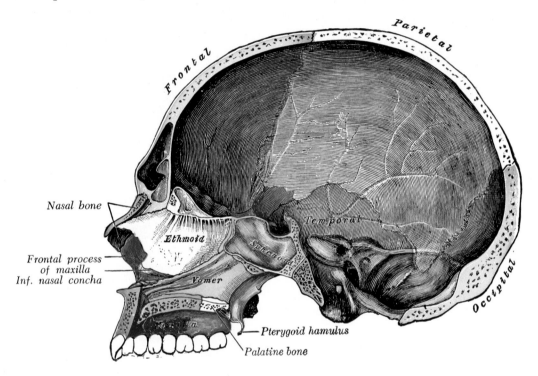

Nasal bone

Frontal process
of maxilla
Inf. nasal concha

Pterygoid hamulus
Palatine bone

FIG. 4–65.—Sagittal section of skull.

The **floor** is formed by the palatine process of the maxilla and the horizontal part of the palatine bone; near its anterior end is the opening of the **incisive canal.**

The **medial wall** (*septum nasi*) (Fig. 4–65), is frequently deflected to one side or the other, more often to the left than to the right. It is formed anteriorly by the crest of the ethmoid; posteriorly by the vomer and the rostrum of the sphenoid; inferiorly, by the crest of the maxillae and the palatine bones.

The **lateral wall** (Fig. 4–66) is formed, anteriorly, by the frontal process of the maxilla and by the lacrimal bone; in the middle, by the ethmoid, maxilla, and inferior

open into it. The sphenoidal sinus opens into the **sphenoethmoidal recess,** which is the most superior and posterior part of the nasal cavity, formed by the angle of junction of the sphenoid and ethmoid bones. The **middle meatus** is situated between the middle and inferior conchae (Fig. 4–66), and extends from the anterior to the posterior end of the latter. The lateral wall of this meatus can be satisfactorily studied only after the removal of the middle concha (Fig. 4–67). A curved fissure, the **hiatus semilunaris,** lies between the **uncinate process** and the **bulla ethmoidalis;** the middle ethmoidal cells are contained within the bulla and open on or near it. The middle

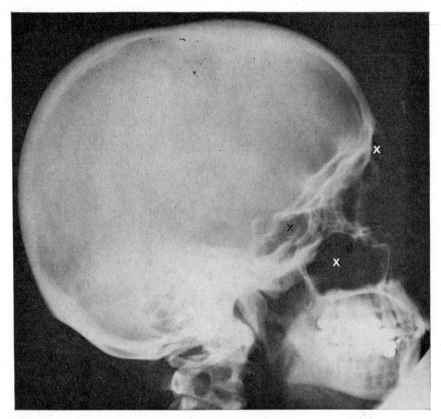

Fig. 4–66.—Adult skull. Lateral view. Crosses are placed on the frontal, the maxillary and the sphe-noidal sinuses. Behind the last-named, the hypophyseal fossa can be identified. The dense white area below and behind the fossa is due to the petrous part of the temporal bone (Courtesy of Doctors Moreland and Arndt).

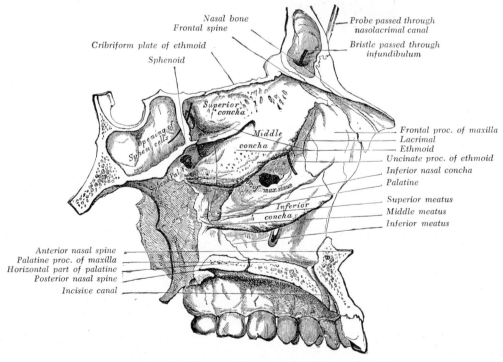

Fig. 4–67.—Roof, floor, and lateral wall of left nasal cavity.

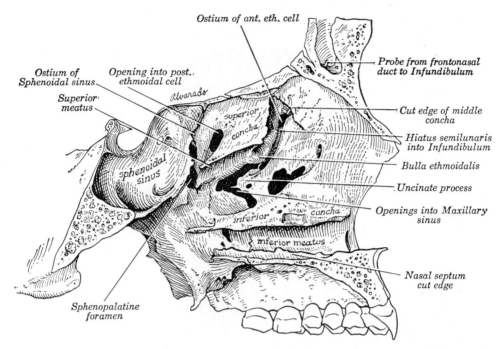

Fig. 4–68.—Lateral wall of nasal cavity with middle concha removed to show middle meatus.

meatus communicates through the hiatus semilunaris with a curved passage termed the **infundibulum,** into which the anterior ethmoidal cells open. The **fronto-nasal duct** leading into the frontal sinus opens directly into the anterior part of the meatus or into the anterior end of the infundibulum. Below the bulla ethmoidalis and hidden by the uncinate process of the ethmoid is the **opening** of the **maxillary sinus** (*ostium maxillare*) and an **accessory opening** frequently is present above the posterior part of the inferior nasal concha. The **inferior meatus** is the space between the inferior concha and the floor of the nasal cavity. It extends almost the entire length of the lateral wall of the nose, and presents anteriorly the **inferior orifice** of the **nasolacrimal canal.**

The **anterior nasal aperture** (Fig. 4–57) is a heart-shaped or pyriform opening, bounded superiorly by the inferior borders of the nasal bones; laterally by the thin, sharp margins which separate the anterior from the nasal surfaces of the maxillae; and inferiorly by the same borders, where they curve medialward to join each other at the anterior nasal spine.

The two **choanae,** or posterior nasal apertures, are bounded by the body of the sphenoid and ala of the vomer; by the posterior border of the horizontal part of the palatine bone; by the medial pterygoid plate; and they are separated from each other in the middle line by the posterior border of the vomer.

DIFFERENCES IN THE SKULL DUE TO AGE

At birth the skull is large in proportion to the other parts of the skeleton, but its facial portion is small, and equals only about one-eighth of the bulk of the cranium as compared with one-half in the adult. The frontal and parietal eminences are prominent, and the greatest width of the skull is at the level of the latter; on the other hand, the glabella, superciliary arches, and mastoid processes are not developed. Ossification of the skull bones is not completed, and many of them, *e.g.,* the occipital, temporals, sphenoid, frontal, and mandible, consist of more than one piece. Unossified membranous intervals, termed *fontanelles,* are seen at the angles of the parietal bones; these fontanelles are six in number; two, an anterior and a posterior, are situated in the

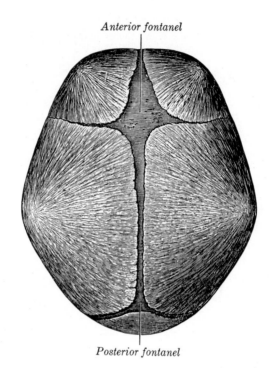

Anterior fontanel

Posterior fontanel

Fig. 4–69.—Skull at birth, showing frontal and occipital fontanelles.

middle line, and two, an antero-lateral and a postero-lateral, on either side.

The **anterior** or **bregmatic fontanelle** (*fonticulus anterior*) (Fig. 4–69) is the largest, and is placed at the junction of the sagittal, coronal, and frontal sutures; it is lozenge-shaped, and measures about 4 cm in its anteroposterior and 2.5 cm in its transverse diameter. The **posterior fontanelle** is triangular in form and is situated at the junction of the sagittal and lambdoidal sutures. The **sphenoidal and mastoid fontanelles** (Fig. 4–70) are small, irregular in shape, and correspond respectively to the sphenoidal and mastoid angles of the parietal bones. The posterior and lateral fontanelles are obliterated within a month or two after birth, but the anterior is not completely closed until about the middle of the second year.

The smallness of the **face at birth** is mainly accounted for by the rudimentary condition of the maxillae and mandible, the non-eruption of the teeth, and the small size of the maxillary air sinuses and nasal cavities. At birth the nasal cavities lie almost entirely between the orbits, and the lower border of the anterior nasal aperture is only a little below the level of the orbital floor. With the eruption of the deciduous teeth there is an enlargement of the face and jaws, and these changes are still more marked after the second dentition.

The skull grows rapidly from birth to the seventh year, by which time the foramen magnum and petrous parts of the temporals have reached their full size and the orbital cavities are only a little smaller than those of the adult. Growth is slow from the seventh year until the approach of puberty, when a second period of activity occurs: this results in an increase in all directions, but it is especially marked in the frontal and facial regions, where it is associated with the development of the air sinuses.

Suture closure begins at twenty-two years in the sagittal and sphenofrontal, at twenty-four years in the coronal and at twenty-six years in the lambdoid and occipitomastoid. The process is most rapid from twenty-six to thirty years then slows down and may not be complete until old age. The sphenoparietal, sphenotemporal, parietomastoid and squamous begin to close at twenty-nine, thirty, and thirty-seven years. Closure progresses very slowly with a final burst of activity in old age. There is considerable individual variation.

The most striking feature of the old skull is the diminution in the size of the maxillae and mandible consequent on the loss of the teeth and the absorption of the alveolar processes. This is associated with a marked reduction in the (occlusal) vertical measurement of the face

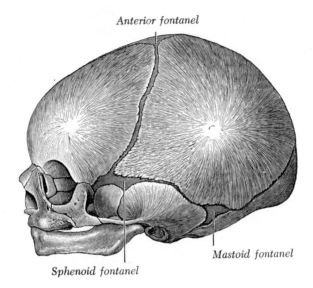

FIG. 4–70.—Skull at birth, showing sphenoidal and mastoid fontanelles.

and with an alteration in the condylar angles of the mandible (Fig. 4–77).

SEX DIFFERENCES IN THE SKULL

Until the age of puberty there is little difference between the skull of the female and that of the male. The skull of an adult female is, as a rule, lighter and smaller, and its cranial capacity about 10 per cent less than that of the male. Its walls are thinner and its muscular ridges less strongly marked; the glabella, superciliary arches, and mastoid processes are less prominent, and the corresponding air sinuses are smaller or may be rudimentary. The upper margin of the orbit is sharp, the forehead vertical, the frontal and parietal eminences prominent, and the vault somewhat flattened. The contour of the face is more rounded, the facial bones are smoother, and the maxillae and mandible and their contained teeth smaller. From what has been said it will be seen that more of the infantile characteristics are retained in the skull of the adult female than in that of the adult male. A well-marked male or female skull can easily be recognized as such, but in some cases the respective characteristics are so indistinct that the determination of the sex may be difficult or impossible.

FRACTURES OF THE SKULL

Those portions of the skull which are most exposed to external violence are thicker than those which are shielded from injury by overlying muscles. The skull-cap is thick and dense, whereas the temporal squamae, protected by the temporal muscles, and the inferior occipital fossae, shielded by the muscles at the back of the neck, are thin and fragile. Fracture of the skull is further prevented by its elasticity, its rounded shape, and its construction of a number of secondary elastic arches, each made up of a single bone. The manner in which vibrations are transmitted through the bones of the skull is of importance as a protective mechanism, especially as far as the base is concerned. In the vault, the bones being of a fairly equal thickness and density, vibrations are transmitted in a uniform manner in all directions, but in the base, owing to the varying thickness and density of the bones, this is not so; in this situation there are special buttresses which serve to carry the vibrations in certain definite directions. At the front of the skull, on either side, is the ridge which separates the anterior from the middle fossa of the base; and behind, the ridge or buttress which separates the middle from the posterior fossa; and if any violence is applied to the vault, the vibrations would be carried along these buttresses to the sella turcica, where they meet. In like manner, when violence is applied to the base of the skull, the vibrations are carried backward through the occipital crest, and forward through the basilar part of the occipital and body of the sphenoid to the vault of the skull.

MALFORMATIONS OF THE SKULL

In connection with the bones of the face a common malformation is **cleft palate**. The cleft usually starts posteriorly, and its most elementary form is a bifid uvula; or the cleft may extend through the soft palate; or the posterior part or the whole of the hard palate may be involved, the cleft extending as far forward as the incisive foramen. In the severest forms, the cleft extends through the alveolus and passes between the incisive or premaxillary bone and the rest of the maxilla; that is to say, between the lateral incisor and canine teeth.

In some instances, the cleft runs between the central and lateral incisor teeth; and this has induced some anatomists to believe that the premaxillary bone is developed from two centers and not from one. The cleft may affect one or both sides; if the latter, the central part is frequently displaced forward and remains united to the septum of the nose, the deficiency in the alveolus being complicated with a cleft in the lip (*hare-lip*).

INDIVIDUAL BONES OF THE SKULL

The Mandible

The **Mandible** (*mandibula; inferior maxillary bone; lower jaw*) (Figs. 4–71 and 4–72) contains the lower teeth and consists of a horizontal portion, the body, and two perpendicular portions, the rami, which join the body nearly at right angles.

The **body** (*corpus mandibulae*) is curved somewhat like a horseshoe, and has two surfaces and two borders.

The **external surface** (Figs. 4–57 and 4–71) is marked in the median line by a faint ridge, indicating the **symphysis menti** or line of junction of the two pieces of which the bone is composed in the fetus. This ridge splits inferiorly to enclose a triangular eminence, the **mental protuberance,** the base of which is depressed in the center but raised on either side to form the **mental tubercle.** On either side of the symphysis is a depression, the incisive fossa, which gives origin to the Mentalis and a small portion of the Orbicularis oris. Inferior to the second premolar tooth on either side is the **mental foramen,** for the passage of the mental vessels and nerve. Running posteriorly from each mental tubercle is a faint ridge, the **oblique line,** which is continuous with the anterior border of the ramus; it affords attachment to the Depressor labii inferioris and Depressor anguli oris. The Platysma is attached near the inferior border.

The **internal surface** (Fig. 4–72).—Near the inferior part of the symphysis is a pair of laterally placed spines, termed the **mental spines,** which give origin to the Genioglossi. Immediately inferior to these is a second pair of spines, or more frequently a median ridge or impression, for the origin of the Geniohyoidei. On either side of the middle line is an oval depression for the attachment of the anterior belly of the Digastricus. Extending posteriorly on either side of the symphysis is the **mylohyoid line,** which gives origin to the Mylohyoideus. The posterior part of this line, near the alveolar margin, gives attachment to a small part of the Constrictor pharyngis superior, and to the pterygomandibular raphe. Superior to the anterior part of this line is a smooth triangular area against which the sublingual gland rests, and inferior to the posterior part, an oval fossa for the submandibular gland.

The **superior** or **alveolar border** is hollowed into sockets for the reception of the sixteen teeth. The Buccinator is attached to the outer lip of the superior border as far forward as the first molar tooth. The **inferior border** is rounded, thicker anteriorly, and at the point where it joins the lower border of the ramus there is a shallow groove for the facial artery.

The **ramus** (*ramus mandibulae; perpendicular portion*) has two surfaces, four borders, and two processes.

The **lateral surface** of the ramus (Fig. 4–71) is flat and marked by oblique ridges where it gives attachment to the Masseter. The **medial surface** (Fig. 4–72) presents about its center the oblique **mandibular foramen,** for the entrance of the inferior alveolar vessels and nerve. The margin of this opening presents a prominent ridge, surmounted by a sharp spine anteriorly, the **lingula mandibulae,** which gives attachment to the sphenomandibular ligament. The **mylohyoid groove** runs obliquely downward and forward from the foramen and lodges the mylohyoid vessels and nerve. Posterior to this groove is a rough surface, for the insertion of the Pterygoideus medialis.

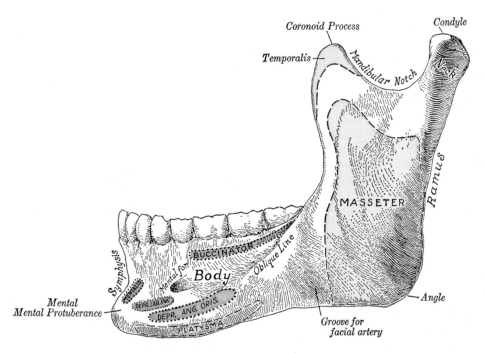

FIG. 4–71.—The left half of the mandible. Lateral aspect.

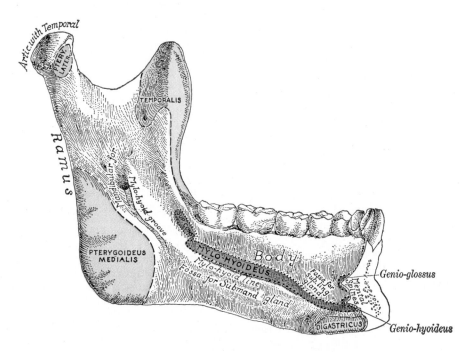

FIG. 4–72.—The left half of the mandible. Medial aspect.

Within the substance of the ramus, the **mandibular canal** runs obliquely downward and forward and then horizontally forward in the body, where it is placed under the alveoli and communicates with them by small openings. At the bicuspid teeth, it has an opening to the exterior, the **mental foramen,** but continues toward the symphysis, giving off two small canals to the incisor teeth. In the posterior two-thirds of the bone the canal is situated nearer the internal surface of the mandible; and in the anterior third, nearer its external surface. It contains the inferior alveolar vessels and nerve, from which branches are distributed to the teeth.

The **inferior border** of the ramus is thick, straight, and at its junction with the posterior border is the **angle of the mandible,** which has the stylomandibular ligament attached to it. The **posterior border** is thick, smooth, rounded, and covered by the parotid gland. The **anterior border** is continuous with the oblique line. The **superior border** is thin, and is surmounted by two processes, the coronoid anteriorly and the condylar posteriorly, separated by the mandibular notch.

The **coronoid process** (*processus coronoideus*) is thin, triangular, and varies in shape and size. Its anterior border is convex and is continuous with the anterior border of the ramus; its posterior border is concave and forms the anterior boundary of the mandibular notch. Both medial and lateral surfaces receive the insertion of the Temporalis. Its medial presents a ridge which begins near the apex of the process and runs downward and forward to the inner side of the last molar tooth. Between this ridge and the anterior border is a grooved, triangular area, to which the Temporalis, and some fibers of the Buccinator are attached.

The **condylar process** (*processus condylaris*) is thicker than the coronoid, and consists of two portions: the articular condyle, and the constricted portion which supports it, the **neck.** The **condyle** presents an oval articular surface for articulation with the articular disk of the temporomandibular joint. Its long axis is approximately transverse, and if prolonged would meet the axis of the opposite condyle near the anterior margin of the foramen magnum. At the lateral extremity of the condyle is a small tubercle for the attachment of the lateral (temporomandibular) ligament. The **neck** is flattened and strengthened by ridges which descend from the anterior and lateral parts of the condyle. Its posterior surface is convex; its anterior presents a depression for the attachment of the Pterygoideus lateralis. On the medial surface of the neck just posterior to the insertion of the Pterygoideus lateralis is a blunt crest. It is the most important trajectory of the mandible, called the dental trajectory or the **mandibular crest.**

The **mandibular notch,** separating the two processes, is a deep semilunar depression, and is crossed by the masseteric vessels and nerve.

Ossification.—The mandible is ossified in the fibrous membrane covering the outer surface of *Meckel's cartilages* (see page 97). Their proximal or cranial ends are connected with the ear capsules, and their distal extremities are joined at the symphysis by mesodermal tissue. From the proximal end of each cartilage the malleus and incus, two of the bones of the middle ear, are developed; the next succeeding portion, as far as the lingula, is replaced by fibrous tissue, which persists to form the sphenomandibular ligament. Between the lingula and the canine tooth the cartilage disappears; the remaining portion becomes ossified and incorporated in this part of the mandible.

Ossification takes place in the membrane covering the outer surface of the ventral end of Meckel's cartilage (Figs. 4–73 to 4–76), and each half of the bone is formed from a single center which appears near the mental foramen at about the sixth week of fetal life. By the tenth week the portion of Meckel's cartilage which lies below and behind the incisor teeth is surrounded and invaded by the membrane bone. Somewhat later, accessory nuclei of cartilage make their appearance, viz., a wedge-shaped nucleus in the condylar process and extending downward through the ramus; a small strip along the anterior border of the coronoid process; and smaller nuclei in the front part of both alveolar walls and along the front of the lower border of the bone. These accessory nuclei possess no separate ossific centers, but are invaded by the surrounding membrane bone and undergo absorption. The inner alveolar border, usually described as arising from a separate ossific center (splenial center), is formed in the human mandible by an ingrowth from the main mass of the bone.

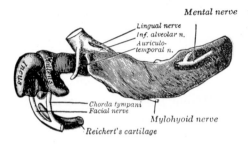

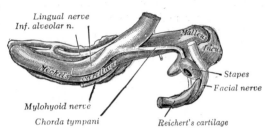

FIG. 4–73.—Mandible of human embryo, 24 mm long. Outer aspect. (From model by Low.)

FIG. 4–74.—Mandible of human embryo 24 mm long. Inner aspect. (From model by Low.)

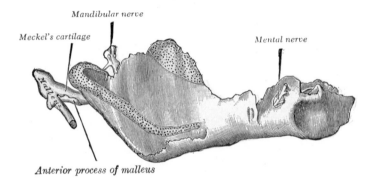

FIG. 4–75.—Mandible of human embryo 95 mm long. Outer aspect. Nuclei of cartilage stippled. (From model by Low.)

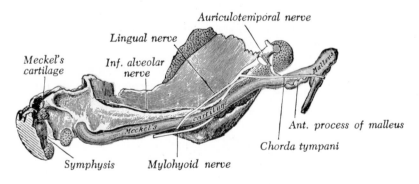

FIG. 4–76.—Mandible of human embryo 95 mm long showing symphysis. Inner aspect. Nuclei of cartilage stippled. (From model by Low.)

At birth the bone consists of two parts, united by a fibrous symphysis in which ossification takes place during the first year.

Articulations.—The mandible articulates with the two temporal bones.

CHANGES PRODUCED IN THE MANDIBLE BY AGE

At **birth** (Fig. 4–77 A and B) the body of the bone is a mere shell, containing the sockets of the two incisors, the canine, and the two deciduous molar teeth, imperfectly partitioned off from one another. The mandibular canal is of large size, and runs near the lower border of the bone; the mental foramen opens beneath the socket of the first deciduous molar tooth. The angle is obtuse (175°), and the condylar portion is nearly in line with the body. The coronoid process is of comparatively large size, and projects above the level of the condyle.

After birth (Fig. 4–77 C).—The two segments of the bone become joined at the symphysis in the first year (Fig. 4–77 A and B), but a trace of separation may be visible in the beginning of the second year near the alveolar margin. The body becomes elongated in its whole length, but more especially posterior to the mental foramen to provide space for the three additional teeth developed in this part. The depth of the body increases owing to increased growth of the alveolar part, to afford room for the roots of the teeth, and by thickening of the subdental portion which enables the jaw to withstand the powerful action of the masticatory muscles; but the alveolar portion is the deeper of the two, and consequently, the chief part of the body lies above the oblique line. The mandibular canal, after the second *dentition* (Fig. 4–77 D), is situated just above the level of the mylohyoid line and the mental foramen occupies the definitive position. The angle becomes more acute, owing to the separation of the jaws by the teeth and by the fourth year it is 140°.

In the **adult** (Fig. 4–77 E) the alveolar and subdental portions of the body are usually of equal depth. The mental foramen opens midway between the upper and lower borders of the bone, and the mandibular canal runs nearly parallel with the mylohyoid line. The ramus is almost vertical in direction, the angle measuring from 110° to 120°.

In **old age** (Fig. 4–77 F) the bone becomes greatly reduced in size, because, with the loss of the teeth, the alveolar process is absorbed, leaving the chief bulk of the bone below the oblique line. The mandibular canal with the mental foramen opening from it is close to the alveolar border. The angle becomes more obtuse and measures about 140°, and the neck of the condyle is bent slightly backward.

The Hyoid Bone

The **Hyoid Bone** (*os hyoideum; lingual bone*) (Fig. 4–78) is named from its resemblance to the Greek letter "U". It is suspended from the tips of the styloid processes of the temporal bones by the stylohyoid ligaments. It is composed of five portions: a body, two greater cornua, and two lesser cornua.

The **body** or **basihyal** (*corpus ossis hyoidei*) or central part is rather square and flat. Its **ventral surface** (Fig. 4–78) is convex and faces somewhat cranially. It is crossed by a well-marked transverse ridge, and in many cases a vertical median ridge divides it into two lateral halves. The Geniohyoideus is inserted into the greater part of the ventral surface; a portion of the origin of the Hyoglossus encroaches upon the lateral margin of the Geniohyoideus attachment. Inferior to the transverse ridge the Mylohyoideus, Sternohyoideus, and Omohyoideus are inserted. The **dorsal surface** is smooth, concave, faces somewhat caudally, and is separated from the epiglottis by the thyrohyoid membrane, with an intervening bursa and a quantity of loose areolar tissue. The **cranial border** is rounded, and gives attachment to the thyrohyoid membrane and some aponeurotic fibers of the Genioglossus. The **caudal border** receives the insertion of the Sternohyoideus medially, the Omohyoideus laterally, and occasionally a portion of the Thyrohyoideus. It also gives attachment to the Levator glandulae thyroideae, when this muscle is present. In early life the lateral borders are connected to the greater cornua by synchondroses; after middle life usually by bony union.

The **greater cornu** or **thyrohyal** (*cornu majus*) projects dorsally from the lateral border of the body; it is flattened and diminishes in size, ending in a tubercle to which is fixed the lateral thyrohyoid ligament. The cranial surface is roughened near its lateral border for muscular attachments. The largest of these are the origins of the Hyoglossus and Constrictor pharyngis

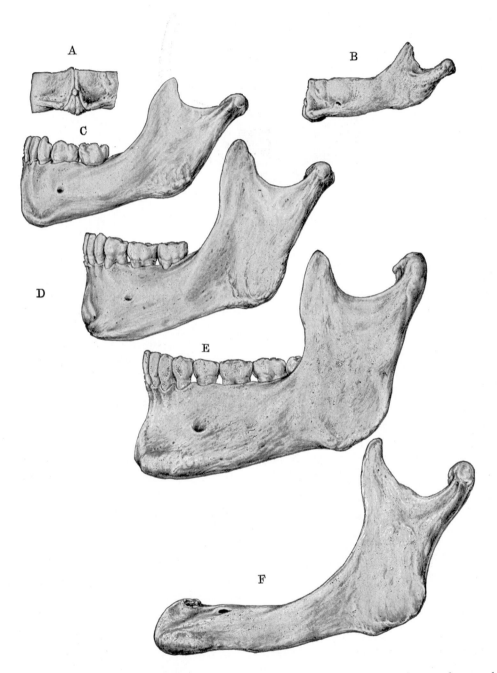

Fɪɢ. 4–77.—The mandible at different periods of life. A, at birth. Anterior aspect, showing the ossicula mentalia. B, at birth. Left lateral aspect. C, at four years. Full deciduous dentition. D, at eight years. The permanent incisor and first molar teeth have erupted; the deciduous molars are in process of being shed. E, adult. F, old age.

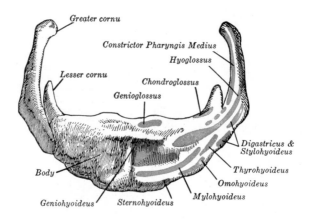

Greater cornu

Constrictor Pharyngis Medius

Hyoglossus

Lesser cornu

Chondroglossus

Genioglossus

*Digastricus &
Stylohyoideus*

Body

Thyrohyoideus

Omohyoideus

Mylohyoideus

Geniohyoideus   *Sternohyoideus*

Fig. 4–78.—Hyoid bone. Ventral surface. Enlarged.

medius which extend along the whole length of the cornu. The Digastricus and Stylohyoideus have small insertions near the junction of the cornu with the body. The thyrohyoid membrane is attached to the medial border and the Thyrohyoideus to the ventral half of the lateral border.

The **lesser cornu** or **ceratohyal** (*cornu minus*) is a small, conical eminence, attached by its base to the angle of junction between the body and greater cornu. It is connected to the body of the bone by fibrous tissue, and occasionally to the greater cornu by a distinct synovial joint, which may either persist throughout life, or become ankylosed.

The lesser cornu is situated in the line of the transverse ridge on the body and appears to be a morphological continuation of it. The stylohyoid ligament is attached to the apex of the cornu and the Chondroglossus arises from the medial side of the base.

Ossification.—The hyoid is ossified from six centers: two for the body, and one for each cornu. Ossification commences in the greater cornua toward the end of fetal life, in the body shortly afterward, and in the lesser cornua during the first or second year after birth.

THE CRANIAL BONES
(OSSA CRANII)

The Occipital Bone

The **Occipital Bone** (*os occipitale*) has a trapezoid outline and is rather cup-like in

shape. It is pierced by a large oval aperture, the foramen magnum, through which the cranial cavity communicates with the vertebral canal. The curved, expanded plate posterior to the foramen magnum is named the squama; the thick, somewhat quadrilateral piece anterior to the foramen is called the basilar part, and on either side of the foramen is the lateral portion.

The **squama** (*squama occipitalis*).—The **external surface** is convex and midway between the summit of the bone and the foramen magnum presents the **external occipital protuberance**. Extending lateralward from this on either side are two curved lines, one a little above the other. The superior, often faintly marked, is named the **highest nuchal line,** and to it the galea aponeurotica is attached. The lower is termed the **superior nuchal line.** That part of the squama which lies superior to the highest nuchal lines is named the **planum occipitale,** and is covered by the Occipitalis muscle; that below, termed the **planum nuchale,** is rough and irregular for the attachment of several muscles. From the external occipital protuberance a ridge or crest, the **median nuchal line,** often faintly marked, runs to the foramen magnum and affords attachment to the ligamentum nuchae; running from the middle of this line across either half of the nuchal plane is the **inferior nuchal line.** The muscles attached to the external surface of the squama are as follows (Fig. 4–79): the origin of the Occipitalis and Trapezius; the

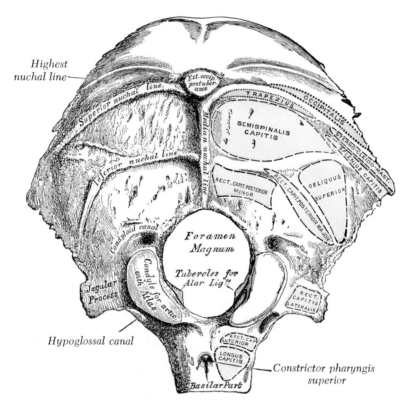

Fig. 4–79.—Occipital bone. Outer surface.

insertion of the Sternocleidomastoideus, Splenius capitis, Semispinalis capitis, Obliquus capitis superior, and the Recti capitis posteriores major and minor. The posterior atlantoöccipital membrane is attached around the posterolateral part of the foramen magnum, just outside its margin.

The **internal surface** is divided into four fossae by a **cruciate eminence.** The two superior fossae are triangular and lodge the occipital lobes of the cerebrum; the inferior two are quadrilateral and accommodate the hemispheres of the cerebellum. At the point of intersection of the four divisions of the cruciate eminence is the **internal occipital protuberance.** From this protuberance the superior division of the cruciate eminence runs to the superior angle of the bone, and on one side of it (generally the right) is a deep groove, the **sagittal sulcus,** which lodges the posterior part of the superior sagittal sinus; to the margins of this sulcus the falx cerebri is attached. The inferior

division of the cruciate eminence is named the **internal occipital crest;** it bifurcates near the foramen magnum and gives attachment to the falx cerebelli; in the attached margin of this falx is the occipital sinus, which is sometimes duplicated. In the superior part of the internal occipital crest, a small depression is sometimes distinguishable; it is termed the vermian fossa since it is occupied by part of the vermis of the cerebellum. Transverse grooves extend on either side from the internal occipital protuberance to the lateral angles of the bone; they accommodate the transverse sinuses, and their prominent margins give attachment to the tentorium cerebelli. The groove on the right side is usually larger than that on the left, and is continuous with that for the superior sagittal sinus. Exceptions to this condition are, however, not infrequent; the left may be larger than the right or the two may be almost equal in size. The point of junction between the superior sagittal and transverse sinuses is

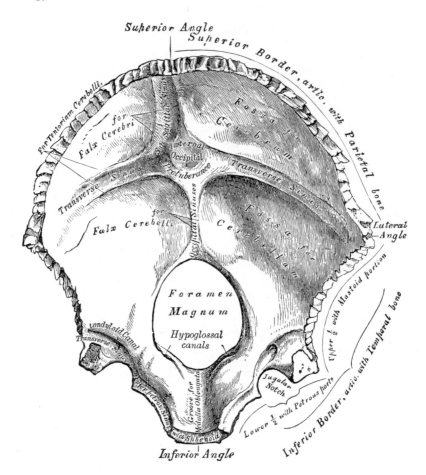

Fig. 4–80.—Occipital bone. Inner surface.

named the confluence of the sinuses (*torcular Herophili*), and its position is indicated by a depression situated on one side or the other of the protuberance.

The **lateral parts** (*pars lateralis*) are situated at the sides of the foramen magnum; on their **inferior surfaces** are the condyles for articulation with the superior facets of the atlas. The **condyles** are oval in shape, with their anterior extremities encroaching on the basilar portion of the bone, and their posterior extremities diverging toward the middle of the foramen magnum. The articular surface of the condyle is convex, facing laterally and caudally. To its margin is attached the capsule of the atlantoöccipital articulation, and on the medial side of each is a rough impression or tubercle for the alar ligament. At the base of the condyle the bone is

tunnelled by a short canal, the **hypoglossal canal** (*anterior condyloid foramen*). It gives exit to the hypoglossal or twelfth cranial nerve, and entrance to a meningeal branch of the ascending pharyngeal artery and may be partially or completely divided into two by a spicule of bone. Posterior to either condyle is a depression, the **condyloid fossa**, which receives the posterior margin of the superior facet of the atlas when the head is bent backward; the floor of this fossa is sometimes perforated by the **condyloid canal**, through which an emissary vein passes. Extending laterally from the posterior half of the condyle is a quadrilateral plate of bone, the **jugular process**, excavated anteriorly by the **jugular notch**, which, in the articulated skull, forms the posterior part of the jugular foramen. The jugular notch may be divided into two by a bony spicule, the intrajugular process, which projects laterally above the hypoglossal canal. The inferior surface of the jugular process is rough, and gives attachment to the Rectus capitis lateralis muscle and the lateral atlantoöccipital ligament. From this surface an eminence, the *paramastoid process,* sometimes projects downward, and may be of sufficient length to reach and articulate with the transverse process of the atlas. Laterally the jugular process presents a rough quadrilateral or triangular area which is joined to the jugular surface of the temporal bone by a plate of cartilage; after the age of twenty-five this plate tends to ossify.

The **superior surface** of the lateral part presents an oval eminence, the **jugular tubercle,** which overlies the hypoglossal canal and is sometimes crossed by an oblique groove for the glossopharyngeal, vagus, and accessory nerves. On the superior surface of the jugular process is a deep groove which curves medially and anteriorly and is continuous with the jugular notch. This groove lodges the terminal part of the transverse sinus, and opening into it, close to its medial margin, is the orifice of the condyloid canal.

The **basilar part** (*pars basilaris*) extends anteriorly from the foramen magnum, and in the young skull, is joined to the body of the sphenoid by a plate of cartilage (Fig. 4–81). By the twenty-fifth year this cartilaginous plate is ossified, and the occipital and sphenoid form a continuous bone.

On its **inferior surface,** about 1 cm anterior to the foramen magnum, is the **pharyngeal tubercle** which gives attachment to the fibrous raphe of the pharynx. On either side of the middle line, the Longus capitis and Rectus capitis anterior are inserted, and immediately anterior to the foramen magnum the anterior atlantoöccipital membrane is attached.

The **superior surface** presents a broad, shallow groove which supports the medulla oblongata, and near the margin of the foramen magnum gives attachment to the membrana tectoria. On the lateral margins of this surface are faint grooves for the inferior petrosal sinuses.

The **foramen magnum** is a large oval aperture with its long diameter anteroposterior. It transmits the medulla oblongata

and its membranes, the accessory nerves, the vertebral arteries, the anterior and posterior spinal arteries, and the membrana tectoria and alar ligaments.

The **superior angle** of the occipital bone articulates with the occipital angles of the parietal bones and, in the fetal skull, corresponds in position with the posterior fontanelle (Fig. 4–69). The **inferior angle** is fused with the body of the sphenoid. The **lateral angles** are received into the interval between the mastoid angle of the parietal and the mastoid part of the temporal.

The **superior borders** (*margo lambdoideus*) extend from the superior to the lateral angles: they are deeply serrated for articulation with the occipital borders of the parietals, and form by this union the lambdoidal suture. The **inferior borders** (*margo mastoideus*) extend from the lateral angles to the inferior angle; the posterior half of each articulates with the mastoid portion of the corresponding temporal, the anterior half with the petrous part of the same bone. These two portions of the inferior border are separated from one another by the jugular process, the notch on the anterior surface of which forms the posterior part of the jugular foramen.

Structure.—The occipital, like the other cranial bones, consists of two compact lamellae, called the outer and inner tables, between which is the cancellous tissue or diploë; the bone is especially thick at the ridges, protuberances, condyles, and anterior part of the basilar part; in the inferior fossae it is thin, semitransparent, and destitute of diploë.

Ossification (Fig. 4–81).—The planum occipitale of the squama is developed in membrane, and may remain separate throughout life in which case it constitutes the *interparietal bone;* the rest of the bone is developed in cartilage. The number of nuclei for the planum occipitale is usually given as four, two appearing near the middle line about the second month, and two some little distance from the middle line about the third month of fetal life. The planum nuchale of the squama is ossified from two centers, which appear about the seventh week of fetal life and soon unite to form a single piece. Union of the upper and lower portions of the squama takes place in the third month of fetal life. Each of the lateral parts begins to ossify from a single center during the eighth week of fetal life. The basilar portion is ossified from one or two centers, which appear about

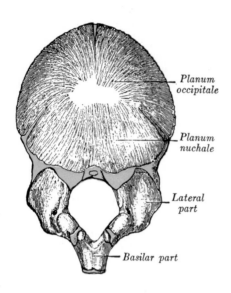

FIG. 4–81.—Occipital bone at birth.

the sixth week of fetal life. About the fourth year the squama and the two lateral portions unite, and about the sixth year the bone consists of a single piece. Between the eighteenth and twenty-fifth years the occipital and sphenoid become united, forming a single bone.

Articulations.—The occipital articulates with six bones; the two parietals, the two temporals, the sphenoid, and the atlas.

The Parietal Bone

The **Parietal Bones** (*os parietale*) form the sides and roof of the cranium. Each has two surfaces, four borders, and four angles.

The **external surface** (Fig. 4–82) is convex, smooth, and marked near the center by the **parietal eminence** (*tuber parietale*), which indicates the point where ossification commenced. Crossing the middle of the bone are two curved lines, the superior and inferior **temporal lines**; the former gives attachment to the temporal fascia, and the

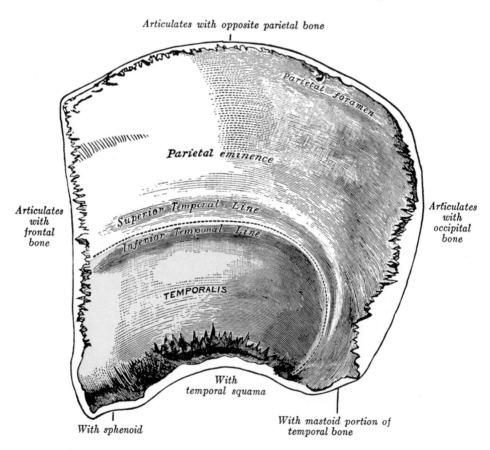

FIG. 4–82.—Left parietal bone. External surface.

latter indicates the upper limit of the muscular origin of the Temporalis. Above these lines the bone is covered by the galea aponeurotica. Close to the upper or sagittal border is the **parietal foramen,** which transmits a vein to the superior sagittal sinus, and sometimes a small branch of the occipital artery; it is not constantly present, and its size varies considerably.

The **internal surface** (Fig. 4–83) is concave and presents depressions corresponding to the cerebral convolutions, and numerous furrows for the ramifications of the middle meningeal vessels. Along the superior margin is a shallow groove, which combines with that on the opposite parietal to form the sagittal sulcus for the superior sagittal sinus; the edges afford attachment to the falx cerebri. Near the groove are several depressions (*foveolae granulares*), best marked in the skulls of old persons, for the **arachnoid granulations** (*Pacchionian bodies*). In the groove is the internal opening of the parietal foramen when that aperture exists.

The **sagittal border** articulates with its fellow of the opposite side, forming the sagittal suture. The **squamous border** is divided into three parts: of these, the anterior is thin and pointed, and is overlapped by the tip of the great wing of the sphenoid; the middle portion is arched, and is overlapped by the squama of the temporal; the posterior part is thick and serrated for articulation with the mastoid portion of the temporal. The **frontal border** is deeply serrated and articulates with the frontal bone, forming one-half of the coronal suture. The **occipital border,** deeply denticulated, articulates with the occipital, forming one-half of the lambdoidal suture.

The **frontal angle** is practically a right angle, and corresponds with the point of meeting of the sagittal and coronal sutures; this point is named the **bregma;** in the fetal skull and for about a year and a half after birth this region is membranous, and is called the **anterior fontanelle** (*fonticulus anterior*) (Fig. 4–70). The **sphenoidal angle,** thin and acute, is received into the interval between the frontal bone and the great wing of the sphenoid. Its inner surface is

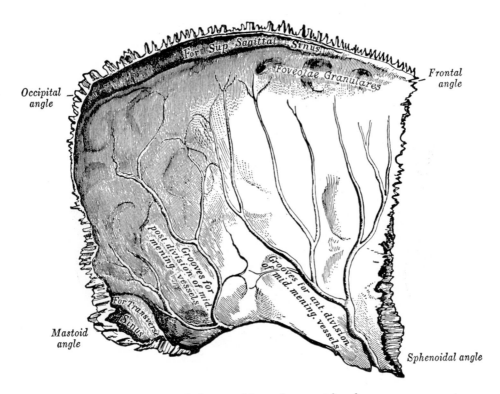

FIG. 4–83.—Left parietal bone. Intracranial surface.

marked by a deep groove, sometimes a canal, for the anterior divisions of the middle meningeal artery. The **occipital angle** is rounded and corresponds with the point of meeting of the sagittal and lambdoidal sutures—a point which is termed the **lambda**; in the fetus this part of the skull is membranous, and is called the **posterior fontanelle** (*fonticulus posterior*). The **mastoid angle** is truncated; it articulates with the occipital bone and with the mastoid portion of the temporal, and presents on its inner surface a broad, shallow groove which lodges part of the transverse sinus. The point of meeting of this angle with the occipital and the mastoid part of the temporal is named the **asterion.**

Ossification.—The parietal bone is ossified in membrane from a single center, which appears at the parietal eminence about the eighth week of fetal life. Ossification gradually extends in a radial manner from the center toward the margins of the bone; the angles are consequently the parts last formed, and it is here that the fontanelles exist. Occasionally the parietal bone is divided into two parts by an anteroposterior suture.

Articulations.—The parietal articulates with five bones: the opposite parietal, the occipital, frontal, temporal, and sphenoid.

The Frontal Bone

The **Frontal Bone** (*os frontale*) consists of two portions, a vertical portion, the squama, corresponding with the forehead, and an orbital or horizontal portion, which enters into the formation of the roofs of the orbital and nasal cavities.

The **squama** (*squama frontalis*).—The **external surface** (Fig. 4–84) of this portion is convex and usually exhibits, in part of the middle line, the remains of the **frontal** or **metopic suture**; in infancy this suture divides the bone into two, a condition which may persist throughout life. On either side of this suture, about 3 cm superior to the supraorbital margin, is a rounded elevation, the **frontal eminence** (*tuber frontale*). Inferior to the frontal eminences, and separated from them by a shallow groove, are the **superciliary arches**; prominent medially, and joined to one another by a smooth elevation named the **glabella.** They are

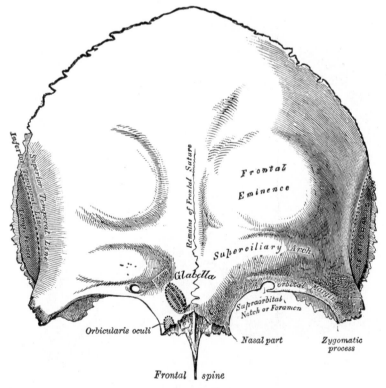

Fig. 4–84.—Frontal bone. Outer surface.

larger in the male than in the female, and their degree of prominence depends to some extent on the size of the frontal air sinuses. Inferior to each superciliary arch is a curved and prominent margin, the supraorbital margin, which forms the boundary of the orbit. The lateral part of this margin is sharp and prominent, affording protection to the eye; its medial part is rounded and at the junction of the medial and intermediate thirds is the **supraorbital notch** or **foramen,** which transmits the supraorbital vessels and nerve. A small aperture in the notch transmits a vein from the diploë to join the supraorbital vein. The supraorbital margin ends laterally in the zygomatic process, which is strong and prominent, and articulates with the zygomatic bone. Running from this process is the temporal line, which divides into upper and lower lines, continuous in the articulated skull with the corresponding lines on the parietal bone. Between the supraorbital margins, the squama projects inferiorly as the nasal part

of the bone and presents a rough, uneven interval, the **nasal notch,** which articulates with the nasal bone, the frontal process of the maxilla, and the lacrimal. The term **nasion** is applied to the middle of the frontonasal suture. From the center of the notch the nasal process projects inferiorly toward the bridge of the nose, ending in the sharp **nasal spine,** on either side of which is a small grooved surface that enters into the formation of the roof of the corresponding nasal cavity. The spine forms part of the septum of the nose, articulating with the crest of the nasal bones and with the perpendicular plate of the ethmoid.

The **internal surface** (Fig. 4–85) of the squama is concave and presents in the middle line a vertical groove, the sagittal sulcus, which ends in the **frontal crest;** the sulcus lodges the superior sagittal sinus; its margins and the crest afford attachment to the falx cerebri. The crest ends in a small notch which is converted into the **foramen cecum,** by articulation with the ethmoid.

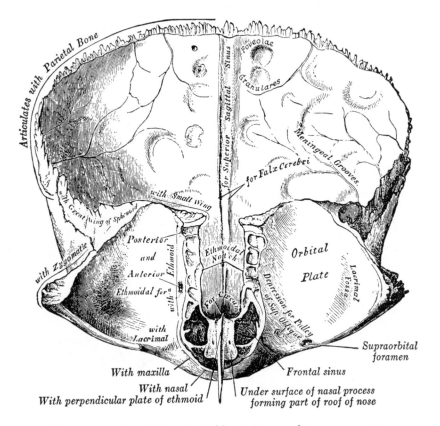

FIG. 4–85.—Frontal bone. Inner surface.

This foramen varies in size in different subjects, and is frequently impervious; when open, it transmits a vein from the nose to the superior sagittal sinus. On either side of the middle line the bone presents depressions for the convolutions of the brain, and numerous small furrows for the arachnoid granulations.

The **orbital** or **horizontal part** (*pars orbitalis*) consists of the orbital plates, which form the vaults of the orbits, and are separated from one another by the ethmoidal notch. The inferior surface (Fig. 4–85) is smooth and concave, and presents a shallow depression laterally, the **lacrimal fossa,** for the lacrimal gland; near the nasal part is a depression, the **fovea trochlearis,** or occasionally a small trochlear spine, for the attachment of the cartilaginous pulley of the Obliquus superior oculi. The superior or intracranial surface is convex, and marked by depressions for the convolutions of the frontal lobes of the brain, and faint grooves for the meningeal branches of the ethmoidal vessels.

The **ethmoidal notch** separates the two orbital plates; it is filled, in the articulated skull, by the cribriform plate of the ethmoid. The margins of the notch present several half cells which, when united with corresponding half cells on the superior surface of the ethmoid, complete the ethmoidal air cells. Two grooves crossing these edges transversely are converted by articulation with the ethmoid into the anterior and posterior ethmoidal canals which open on the medial wall of the orbit.

The **frontal air sinuses** are anterior to the ethmoidal notch, on either side of the frontal spine. The hollows of the sinuses extend for a variable distance between the two tables of the skull and may penetrate into the orbital plates. They are separated from one another by a thin bony septum, which often deviates to one side or the other, with the result that the sinuses are rarely symmetrical. They vary in size in different persons, and are larger in men than in women. They are lined by mucous membrane, and communicate with the nasal cavity by means of a passage called the **frontonasal duct** which opens into the middle meatus or the infundibulum.

The **border** of the squama is thick, strongly serrated, beveled at the expense of the inner table, where it rests upon the parietal bones, and at the expense of the outer table on either side, where it receives the lateral pressure of those bones. This border is continued into a triangular, rough surface, which articulates with the great wing of the sphenoid. The posterior borders of the orbital plates are thin and serrated, and articulate with the small wings of the sphenoid.

Structure.—The squama and the zygomatic processes are very thick, consisting of diploic tissue contained between two compact laminae; the diploic tissue is absent in the regions occupied by the frontal air sinuses. The orbital portion is thin, translucent, and composed entirely of compact bone; hence the facility with which instruments can penetrate the cranium through this part of the orbit.

Ossification (Fig. 4–86).—The frontal bone is ossified in membrane from two primary centers, one for each half, which appear toward the end of the second month of fetal life, one above each supraorbital margin. From each of these centers ossification extends upward to form the corresponding half of the squama, and backward to form the orbital plate. The spine is ossified from a pair of secondary centers, on either side of the middle line; similar centers appear in the nasal part and zygomatic processes. At birth the bone consists of two pieces, separated by the frontal suture, which is usually obliterated by the eighth year except at its lower part, but occasionally persists throughout life. It is generally maintained that the development of the frontal sinuses begins at the end of the first or beginning of the second year.

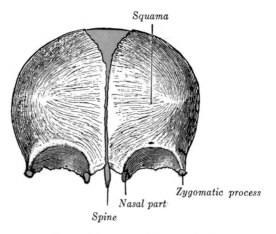

Fig. 4–86.—Frontal bone at birth.

The sinuses are of considerable size by the seventh or eighth year, but do not attain their full proportions until after puberty.

Articulations.—The frontal articulates with twelve bones: the sphenoid, the ethmoid, the two parietals, the two nasals, the two maxillae, the two lacrimals, and the two zygomatics.

The Temporal Bone

The **Temporal Bone** (*os temporale*) (Figs. 4–87, 4–88) consists of three parts, the squama, the petrous, and the tympanic parts.

The **squama** (*pars squamosa*) is scale-like, thin, and translucent, and forms the anterior and superior part of the bone. Its **external surface** (Fig. 4–86) is smooth and convex, forms part of the temporal fossa, and affords attachment to the Temporalis muscle. A curved line, the **temporal line,** limits the origin of the Temporalis muscle. The boundary between the squama and the mastoid process of the bone, as indicated by traces of the original suture, lies about 1 cm below this line.

The **zygomatic process** is a long arch projecting from the inferior part of the squama. Its superior border is long, thin, and sharp, and serves for the attachment of the temporal fascia; its inferior border, short, thick and arched, has some fibers of the Masseter attached to it. The lateral surface is convex and subcutaneous; the medial is concave, and affords attachment to the Masseter. The anterior end is deeply serrated and articulates with the zygomatic bone. The posterior end arises from the squama by two roots. The **posterior root,** a prolongation of the upper border, runs above the external acoustic meatus (*porus acusticus*) and is continuous with the temporal line. The **anterior root,** continuous with the inferior border, is short but broad and strong; it is directed medialward and ends in a rounded eminence, the **articular tubercle** (*tuberculum articulare*). This tubercle forms the anterior boundary of the mandibular fossa, and in the intact body is covered with cartilage. Anterior to the articular tubercle is a small triangular area which assists in forming the infratemporal fossa; this area is separated from the outer surface of the squama by a ridge which is

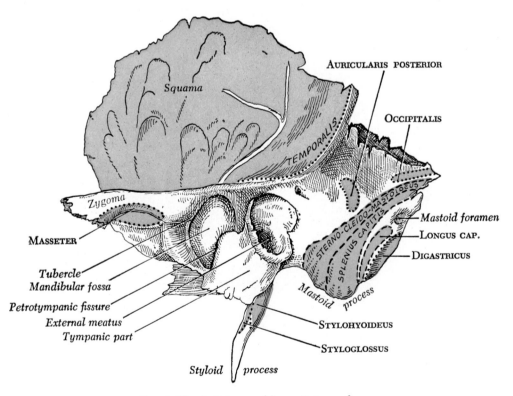

Fig. 4–87.—Left temporal bone. Outer surface.

continuous posteriorly with the anterior root of the zygomatic process, and anteriorly, in the articulated skull, with the infratemporal crest on the great wing of the sphenoid. Between the posterior wall of the external acoustic meatus and the posterior root of the zygomatic process is the area called **suprameatal triangle**, or **mastoid fossa**, through which an instrument may be pushed into the tympanic antrum. At the junction of the anterior root with the zygomatic process is a projection for the attachment of the lateral temporomandibular ligament; and posterior to the anterior root is an oval depression, forming part of the mandibular fossa, for the reception of the condyle of the mandible.

The **mandibular fossa** (*glenoid fossa*) is bounded by the articular tubercle and by the tympanic part of the bone, which separates it from the external acoustic meatus. It is divided into two parts by a narrow slit, the **petrotympanic fissure** (*Glaserian fissure*). The anterior part, formed by the squama, is smooth, covered

in the intact body with cartilage, and articulates with the condyle of the mandible. Posterior to this part of the fossa is a small conical eminence; this is the representative of a prominent tubercle which, in some mammals, projects behind the condyle of the mandible, and prevents its backward displacement. The posterior part of the mandibular fossa is formed by the tympanic part of the bone and is non-articular. The petrotympanic fissure leads into the middle ear or tympanic cavity where it lodges the anterior process of the malleus, and transmits the tympanic branch of the maxillary artery. The chorda tympani nerve passes through a canal (*of Huguier*), separated from the anterior edge of the petrotympanic fissure by a thin scale of bone and situated on the lateral side of the auditory tube in the retiring angle between the squama and the petrous portion of the temporal.

The **internal surface** of the squama (Fig. 4–88) presents depressions corresponding to the convolutions of the temporal lobe of the brain, and grooves for the branches of the

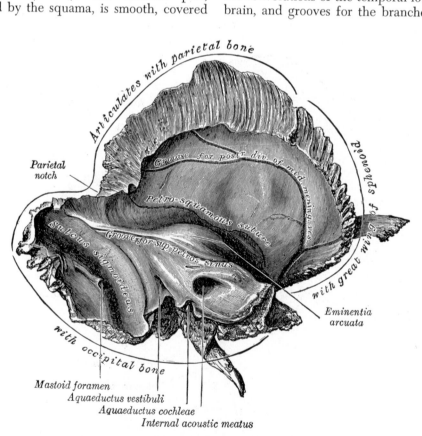

Fig. 4–88.—Left temporal bone. Inner surface.

middle meningeal vessels. The **superior border** is thin, and bevelled at the expense of the internal table, so as to overlap the squamous border of the parietal bone, forming with it the squamosal suture. Posteriorly, the superior border forms an angle, the parietal notch, with the mastoid portion of the bone. The **anteroinferior** border is thick, serrated, and bevelled at the expense of the inner table above and of the outer below, for articulation with the great wing of the sphenoid.

The **mastoid portion** (NA–*Processus mastoideus*) forms the posterior part of the bone. The **external surface** (Fig. 4–87) is rough, and gives attachment to the Occipitalis and Auricularis posterior. It is perforated by numerous foramina; one of these, of large size, situated near the posterior border, is termed the mastoid foramen; it transmits a vein to the transverse sinus and a small branch of the occipital artery to the dura mater. The position and size of this foramen are variable; it is not always present, and may be situated in the occipital bone, or in the suture between the temporal and the occipital. The **mastoid process** is a conical projection, the size and form of

which vary; it serves for the attachment of the Sternocleidomastoideus, Splenius capitis, and Longissimus capitis. On the medial side of the process is a deep groove, the **mastoid notch** (*digastric fossa*), for the attachment of the Digastricus; medial to this is a shallow furrow, the **occipital groove,** which lodges the occipital artery.

The **internal surface** of the mastoid portion presents a deep, curved groove, the sigmoid sulcus, which lodges part of the transverse sinus; in it may be seen the opening of the mastoid foramen. The groove for the transverse sinus is separated from the innermost of the mastoid air cells by a very thin lamina of bone, and even this may be partly deficient. The **superior border** of the mastoid portion is broad and serrated, for articulation with the mastoid angle of the parietal. The **posterior border,** also serrated, articulates with the inferior border of the occipital between the lateral angle and jugular process. Anteriorly, the mastoid portion is fused with the descending process of the squama and it enters into the formation of the external acoustic meatus and the tympanic cavity.

A section of the mastoid process (Fig.

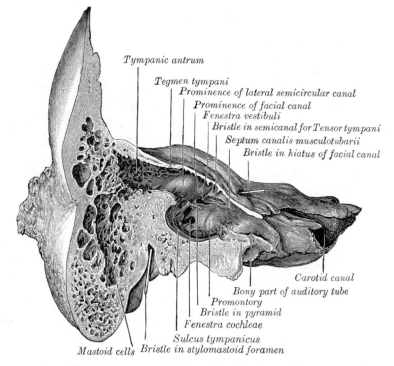

Tympanic antrum
Tegmen tympani
Prominence of lateral semicircular canal
Prominence of facial canal
Fenestra vestibuli
Bristle in semicanal for Tensor tympani
Septum canalis musculotubarii
Bristle in hiatus of facial canal

Carotid canal
Bony part of auditory tube
Promontory
Bristle in pyramid
Fenestra cochleae
Sulcus tympanicus
Mastoid cells *Bristle in stylomastoid foramen*

Fig. 4–89.—Coronal section of right temporal bone.

4–89) shows it to be hollowed out into a number of spaces, the **mastoid cells,** which exhibit the greatest possible variety as to their size and number. At the superior part of the process they are large and irregular and contain air, but toward the inferior part they diminish in size, while those at the apex of the process are frequently quite small and contain marrow; occasionally they are entirely absent, and the mastoid is then solid throughout. In addition, a large irregular cavity at the superior and anterior part of the process is called the **tympanic antrum,** and must be distinguished from the mastoid cells, although it communicates with them. Like the mastoid cells it contains air and is lined by a prolongation of the mucous membrane of the tympanic cavity with which it communicates. A thin plate of bone, the **tegmen tympani,** separates it from the middle fossa of the cranial cavity; and the lateral semicircular canal of the internal ear projects into its cavity. It opens anteriorly into that portion of the tympanic cavity which is known as the **attic** or **epitympanic recess.** The tympanic antrum is a cavity of some considerable size at birth; the mastoid air cells may be regarded as diverticula from the antrum, and begin to appear at or before birth; by the fifth year they are well marked, but their development is not completed until toward puberty.

The **petrous portion** (*pars petrosa; pyramis*) or **pyramid** is wedged in between the sphenoid and occipital bones at the base of the skull. It presents for examination a base, an apex, three surfaces, and two margins, and contains, in its interior, the essential parts of the organ of hearing and equilibration.

The **base** is fused with the internal surfaces of the squama and mastoid portion.

The **apex,** rough and uneven, is received into the interval between the posterior border of the great wing of the sphenoid and the basilar part of the occipital and sphenoid; it presents the anterior or internal orifice of the carotid canal and forms the posterolateral boundary of the foramen lacerum.

The **anterior surface** forms the posterior part of the middle cranial fossa and is continuous with the inner surface of the squamous portion, to which it is united by

the petrosquamous suture, the remains of which are distinct even at a late period of life. It is marked by depressions for the convolutions of the brain, and presents six points for examination: (1) near the center, the **eminentia arcuata** which indicates the situation of the superior semicircular canal; (2) anterior to and a little lateral to this eminence, a depression indicating the position of the tympanic cavity, the **tegmen tympani;** (3) a shallow groove, sometimes double, leading into the **hiatus** of the **facial canal,** for the passage of the greater petrosal nerve and the petrosal branch of the middle meningeal artery; (4) lateral to the hiatus, a smaller opening, occasionally seen, for the passage of the lesser petrosal nerve; (5) near the apex of the bone, the termination of the **carotid canal,** the wall of which in this situation is deficient (6) and at the apex the shallow **trigeminal impression** for the trigeminal (*semilunar*) ganglion.

The **posterior surface** of the petrous portion (Fig. 4–64) forms the anterior part of the posterior cranial fossa. Near the center is the large orifice of the **internal acoustic meatus** leading into the internal acoustic canal (*porus acusticus internus*); its margins are smooth and rounded and form a short canal about 1 cm in length. It

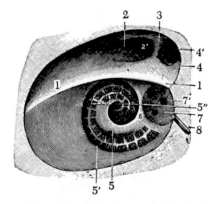

FIG. 4–90.—Diagrammatic view of the fundus of the right internal acoustic meatus. (Testut.) 1. Crista falciformis. 2. Area facialis, with (2′) internal opening of the facial canal. 3. Ridge separating the area facialis from the area cribrosa superior. 4. Area cribrosa superior, with (4′) openings for nerve filaments. 5. Anterior inferior cribriform area, with (5′) the tractus spiralis foraminosus, and (5″) the canalis centralis of the cochlea. 6. Ridge separating the tractus spiralis foraminosus from the area cribrosa media. 7. Area cribrosa media, with (7′) orifices for nerves to saccule. 8. Foramen singulare.

transmits the facial and acoustic nerves, the nervus intermedius, and the internal auditory branch of the basilar artery. The meatus is closed by a vertical plate which is divided by a horizontal crest, the **crista falciformis** (NA—*crista transversa*), into two unequal portions (Fig. 4–90). Each portion is further subdivided by a vertical ridge into an anterior and a posterior part. Inferior to the crista falciformis are three sets of foramina; one group, situated in the *area cribrosa media* just below the posterior part of the crest, consists of several small openings for the nerves to the saccule; posterior to this area is the *foramen singulare,* or opening for the nerve to the posterior semicircular duct; anterior to the first is the *tractus spiralis foraminosus,* consisting of a number of small spirally arranged openings which encircle the canalis centralis cochlea. Above the crista falciformis, the *area cribrosa superior* (NA —*area vestibularis superior*) is pierced by a series of small openings for the passage of the nerves to the utricle and the superior and lateral semicircular ducts; anteriorly the

area facialis contains one large opening, the commencement of the **canal** for the **facial nerve** (*Canalis facialis; aquaeductus Fallopii*). Posterior to the internal acoustic meatus, a small slit almost hidden by a thin plate of bone leads to a canal, the aquaeductus vestibuli, which transmits the ductus endolymphaticus together with a small artery and vein. Above and between these two openings is an irregular depression which lodges a process of the dura mater and transmits a small vein; in the infant this depression is represented by a large fossa, the *subarcuate fossa,* which extends backward as a blind tunnel under the superior semicircular canal.

The **inferior surface** of the petrous portion (Fig. 4–91) is rough and irregular, and forms part of the exterior of the base of the skull. It presents eleven points of interest for examination: (1) near the apex is a rough quadrilateral surface partly for the attachment of the Levator veli palatini and the cartilaginous portion of the auditory tube, and partly for connection with the basilar part of the occipital bone through

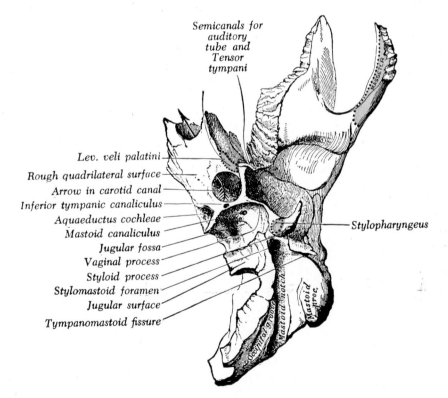

Semicanals for
auditory
tube and
Tensor
tympani

Lev. veli palatini
Rough quadrilateral surface
Arrow in carotid canal
Inferior tympanic canaliculus
Aquaeductus cochleae
Mastoid canaliculus
Jugular fossa
Vaginal process
Styloid process
Stylomastoid foramen
Jugular surface
Tympanomastoid fissure

Stylopharyngeus

FIG. 4–91.—Left temporal bone. Inferior surface.

the intervention of some dense fibrous tissue; (2) posterior to this is the large circular aperture of the **carotid canal,** which transmits the internal carotid artery and the carotid plexus of nerves into the cranium; (3) medial and posterior to the carotid opening and anterior to the jugular fossa is a triangular depression at the apex of which is a small opening, the **aquaeductus cochleae.** It lodges a tubular prolongation of the dura mater establishing a communication between the perilymphatic space and the subarachnoid space and it transmits a vein from the cochlea to the internal jugular; (4) posterior to these openings is a deep depression, the **jugular fossa,** which lodges the bulb of the internal jugular vein; (5) in the bony ridge between the carotid canal and the jugular fossa is the small **inferior tympanic canaliculus** for the passage of the tympanic branch of the glossopharyngeal nerve; (6) in the lateral part of the jugular fossa is the **mastoid canaliculus** for the passage of the auricular branch of the vagus nerve; (7) posterior to the jugular fossa is a quadrilateral area, the jugular surface, covered with cartilage in the intact body, and articulating with the jugular process of the occipital bone; (8) extending posteriorly from the carotid canal is the vaginal process, a sheath-like plate of bone, which divides behind into two laminae; the lateral lamina is continuous with the tympanic part of the bone, the medial with the lateral margin of the jugular surface; (9) between these laminae is a sharp spine, the **styloid process;** (10) between the styloid and mastoid processes is the **stylomastoid foramen;** it is the termination of the facial canal, and transmits the facial nerve and stylomastoid artery; (11) situated between the tympanic portion and the mastoid process is the tympanomastoid fissure, through which the auricular branch of the vagus nerve makes exit.

The **superior margin** (*superior angle*), is grooved for the superior petrosal sinus, and gives attachment to the tentorium cerebelli; at its medial extremity is a notch, in which the trigeminal nerve lies. The **posterior margin** (*posterior angle*) is intermediate in length between the superior and the anterior. Its medial half is marked by a sulcus which forms, with a corresponding sulcus on the occipital bone, the channel for the

inferior petrosal sinus. Its lateral half presents an excavation, the jugular fossa, which, with the jugular notch on the occipital, forms the **jugular foramen;** an eminence occasionally projects from the center of the fossa, and divides the foramen into two. The **anterior angle** is joined to the squama by the petrosquamous suture, the remains of which are more or less distinct; it also articulates with the sphenoid. At the angle of junction of the petrous part and the squama are two canals, one above the other, and separated by a thin plate of bone, the septum canalis musculotubarii (*processus cochleariformis*); both canals lead into the tympanic cavity. The upper one (*semicanalis m. tensoris tympani*) transmits the Tensor tympani, the lower one (*semicanalis tubae auditivae*) forms the bony part of the auditory tube.

The tympanic cavity, auditory ossicles, and internal ear are described with the organ of hearing in the chapter on the Organs of the Senses.

The **tympanic portion** (*pars tympanica*) is a curved plate of bone lying inferior to the squama and anterior to the mastoid process (Fig. 4–87). Its **posterior superior surface** is concave, and forms the anterior wall, the floor, and part of the posterior wall of the bony external acoustic meatus. Medially, it presents a narrow furrow, the **tympanic sulcus,** for the attachment of tympanic membrane. Its **anterior inferior surface** is quadrilateral and slightly concave; it constitutes the posterior boundary of the mandibular fossa, and is in contact with the retromandibular part of the parotid gland. Its **lateral border** is free and rough, and gives attachment to the cartilaginous part of the external acoustic meatus. Internally, the tympanic part is fused with the petrous portion, and appears in the retreating angle between it and the squama, where it lies inferior and lateral to the orifice of the auditory tube. Posteriorly, it blends with the squama and mastoid part, and forms the anterior boundary of the tympanomastoid fissure. Its **superior border** fuses laterally with the postglenoid process, while medially it bounds the petrotympanic fissure. The medial part of the **inferior border** is thin and sharp; its lateral part splits to enclose the root of the styloid process, and is therefore named the **vaginal process.** The central portion of the tympanic part is thin, and in a considerable percentage of skulls is perforated by a small hole, the *foramen of Huschke.*

The **external acoustic meatus** is nearly 2 cm long and is directed slightly forward; at the same time it forms a slight curve, so that the floor of the canal is convex superiorly. The anterior wall, the floor, and the inferior part of the posterior wall are formed by the tympanic portion; the roof and superior part of the posterior wall by the squama. Its inner end is closed in the intact body by the tympanic membrane; the superior limit of its outer orifice is formed by the posterior root of the zygomatic process, immediately below which there is sometimes seen a small spine, the supra-meatal spine, situated at the upper and posterior part of the orifice.

The **styloid process** (*processus styloideus*) (Fig. 4–91) is slender, pointed, and of variable length; it projects downward and forward, from the inferior surface of the temporal bone. Its proximal part (*tympanohyal*) is ensheathed by the vaginal process of the tympanic portion; its distal part (*stylohyal*) gives attachment to the stylohyoid and stylomandibular ligaments, and to the Styloglossus, Stylohyoideus, and Stylopharyngeus muscles. The stylohyoid ligament extends from the apex of the process to the lesser cornu of the hyoid bone, and in some instances is partially, in others completely, ossified.

Structure.—The structure of the squama is like that of the other cranial bones: the mastoid portion is spongy, and the petrous portion dense and hard.

Ossification.—The temporal bone is ossified from eight centers, exclusive of those for the internal ear and the tympanic ossicles, viz., one for the squama including the zygomatic process, one for the tympanic part, four for the petrous and mastoid parts, and two for the styloid process. Just before the close of fetal life (Fig. 4–92) the temporal bone consists of three principal parts: 1. The *squama* is ossified in membrane from a single nucleus, which appears near the root of the zygomatic process about the second month. 2. The *petromastoid* part is developed from four centers, which make their appearance in the cartilaginous ear capsule about the fifth or sixth month. One (proötic)

appears in the neighborhood of the eminentia arcuata, spreads anterior and superior to the internal acoustic meatus and extends to the apex of the bone; it forms part of the cochlea, vestibule, superior semicircular canal, and medial wall of the tympanic cavity. A second (*opisthotic*) appears at the promontory on the medial wall of the tympanic cavity and surrounds the fenestra cochleae; it forms the floor of the tympanic cavity and vestibule, surrounds the carotid canal, invests the lateral and lower part of the cochlea, and spreads medially below the internal acoustic meatus. A third (*pterotic*) roofs in the tympanic cavity and antrum; while the fourth (*epiotic*) appears near the posterior semicircular canal and extends to form the mastoid process. 3. The *tympanic ring* is an incomplete circle, in the concavity of which is a groove, the tympanic sulcus, for the attachment of the circumference of the tympanic membrane. This ring expands to form the tympanic part, and is ossified in membrane from a single center which appears about the third month. The *styloid process* is developed from the proximal part of the cartilage of the second branchial or hyoid arch by two centers: one for the proximal part, the *tympanohyal*, appears before birth; the other, comprising the rest of the process, is named the *stylohyal*, and does not appear until after birth. The tympanic ring unites with the squama shortly before birth; the petromastoid part and squama join during the first year, and the tympanohyal portion of the styloid process about the same time (Figs. 4–93, 4–94). The stylohyal does not unite with the rest of the bone until after puberty, and in some skulls never at all.

The chief subsequent changes in the temporal bone apart from increase in size are: (1) The tympanic ring extends outward and backward to form the tympanic part. This extension does not take place at an equal rate all around the circumference of the ring, but occurs most rapidly on its anterior and posterior portions. These outgrowths meet and blend, and thus, for a time, there exists in the floor of the meatus a foramen, the *foramen of Huschke;* this foramen is usually closed about the fifth year, but may persist throughout life. (2) The mandibular fossa is at first extremely shallow, and faces laterally as well as inferiorly; it becomes deeper and is ultimately directed inferiorly. Its change in direction is accounted for as follows: the part of the squama which forms the fossa lies at first inferior to the level of the zygomatic process. As the base of the skull increases in width, however, this lower part of the squama is directed horizontally inward to contribute to the middle fossa of the skull, and its surfaces therefore come to face

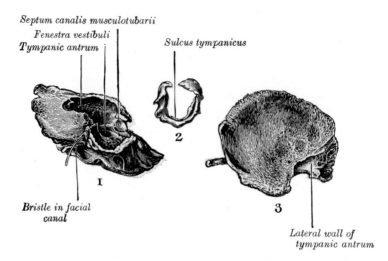

Fig. 4–92.—The three principal parts of the temporal bone at birth. 1. Outer surface of petromastoid part. 2. Outer surface of tympanic ring. 3. Inner surface of squama.

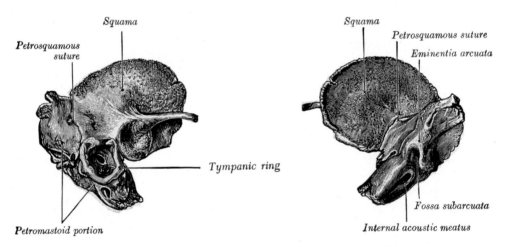

Fig. 4–93.—Temporal bone at birth. Outer aspect.   Fig. 4–94.—Temporal bone at birth. Inner aspect.

superiorly and inferiorly; the attached portion of the zygomatic process also becomes everted, and projects like a shelf at right angles to the squama. (3) The mastoid portion is at first quite flat, and the stylomastoid foramen and rudimentary styloid process lie immediately behind the tympanic ring. With the development of the air cells the outer part of the mastoid portion grows inferiorly and anteriorly to form the mastoid process, and the styloid process and stylomastoid foramen now come to lie on the inferior surface. The descent of the foramen is necessarily accompanied by a corresponding lengthening of the facial canal. (4) The growth of the mastoid process also pushes forward the tympanic part, so that the

portion of it which formed the original floor of the meatus and contained the foramen of Huschke is ultimately found in the anterior wall. (5) The fossa subarcuata becomes filled up and almost obliterated.

Articulations.—The temporal articulates with five bones: occipital, parietal, sphenoid, mandible, and zygomatic.

The Sphenoid Bone

The **Sphenoid Bone** (*os sphenoidale*) (Figs. 4–95, 4–96) is situated at the base of the skull anterior to the temporals and basilar part of the occipital. It somewhat resembles a bat with its wings extended, and is divided into a median portion or body, two great and two small wings extending outward from the sides of the body, and two pterygoid processes which project from its inferior surface.

The **body** (*corpus sphenoidale*), more or less cubical in shape, is hollowed out to form two large cavities, the sphenoidal air sinuses, which are separated from each other by a thin septum.

The **superior** or **intracranial surface** of the body (Fig. 4–95) presents the **ethmoidal spine** rostrally, for articulation with the cribriform plate of the ethmoid; posterior to this is a smooth surface slightly raised in the middle line, and grooved on either side for the olfactory lobes of the brain. This surface is bounded posteriorly by a ridge which forms the anterior border of the **chiasmatic groove** (*optic groove*), on which the optic chiasma lies; the groove ends on

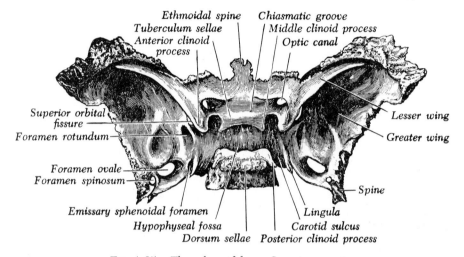

Fig. 4–95.—The sphenoid bone. Superior aspect.

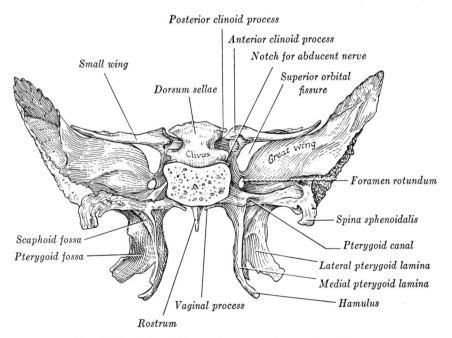

Fig. 4–96.—Sphenoid bone. Superior and posterior surfaces.

either side in the **optic canal** (*foramen*), which transmits the optic nerve and ophthalmic artery into the orbital cavity. Posterior to the chiasmatic groove is the **tuberculum sellae**; and still more posteriorly, a deep depression, the **sella turcica**, the deepest part of which lodges the hypophysis cerebri and is known as the **fossa hypophyseos**. The anterior boundary of the sella turcica is completed by two small eminences called the **middle clinoid processes**. The posterior boundary is formed by a square plate of bone, the **dorsum sellae**, ending at its superior angles in two tubercles, the **posterior clinoid processes**, the size and form of which vary considerably in different skulls. The posterior clinoid processes deepen the sella turcica, and give attachment to the tentorium cerebelli. On either side of the dorsum sellae is a notch for the passage of the abducent nerve, and inferior to the notch a sharp process, the **petrosal process**, which articulates with the apex of the petrous portion of the temporal bone, and forms the medial boundary of the foramen lacerum. Posterior to the dorsum sellae is a shallow depression, the **clivus**, which slopes posteriorly and is continuous with the groove on the basilar portion of the occipital bone; it supports the upper part of the pons.

The **lateral surfaces** of the body are united with the great wings and the medial pterygoid plates. Above the attachment of each great wing is a broad groove, curved something like the italic letter *f* named the **carotid sulcus**. Along the posterior part of the lateral margin of this groove, in the angle between the body and great wing, is a ridge of bone, called the **lingula**.

The **posterior surface** (Fig. 4–96) is joined, during infancy and adolescence, to the basilar part of the occipital bone by a plate of cartilage which becomes ossified between the eighteenth and twenty-fifth years.

The **anterior surface** of the body (Fig. 4–97) forms the posterior wall of the nasal cavity and presents, in the middle line, the **sphenoidal crest**, which articulates with the perpendicular plate of the ethmoid, and forms part of the septum of the nose. On either side of the crest is an irregular opening leading into the corresponding **sphenoidal air sinus**. These sinuses vary considerably in form and size, are seldom symmetrical, are often partially subdivided by irregular bony laminae, and occasionally extend into the basilar part of the occipital nearly as far as the foramen magnum. They are partially closed by two thin, curved plates of bone, the **sphenoidal conchae**. The

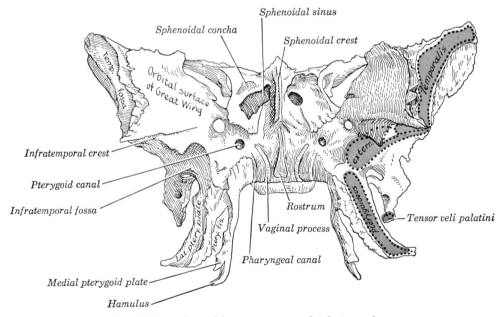

FIG. 4–97.—Sphenoid bone. Anterior and inferior surfaces.

lateral margin of the anterior surface is serrated, and articulates with the lamina orbitalis of the ethmoid, completing the posterior ethmoidal cells, the inferior margin articulates with the orbital process of the palatine bone, and the superior with the orbital plate of the frontal bone.

The **inferior surface** also forms part of the posterior wall of the nasal cavity and presents, in the middle line, a triangular spine, the **sphenoidal rostrum,** which is continuous with the sphenoidal crest on the anterior surface, and is received into a deep fissure between the alae of the vomer. On either side of the rostrum is a projecting lamina, the **vaginal process,** directed medially from the base of the medial pterygoid plate.

The **great wings** (*alae majores*), or alisphenoids, are two strong processes of bone, which arise from the sides of the body. The posterior part of each projects as a triangular process which fits into the angle between the squama and the petrous portion of the temporal and presents at its apex an inferiorly directed process, the **sphenoidal spine** (*spina angularis*).

The **superior** or **cerebral surface** of each great wing (Figs. 4–64, 4–95) forms part of the middle cranial fossa; it is deeply concave, and presents depressions for the convolutions of the temporal lobe of the brain. At its anterior and medial part is a circular aperture, the **foramen rotundum,** for the transmission of the maxillary nerve. Posterior and lateral to this is the **foramen ovale,** for the transmission of the mandibular nerve, the accessory meningeal artery, and sometimes the lesser petrosal nerve. Medial to the foramen ovale, a small aperture, the **foramen Vesalii,** may occasionally be seen. It opens inferiorly near the scaphoid fossa, and transmits a small vein from the cavernous sinus to the pterygoid plexus. In the posterior angle, near the spine, is the **foramen spinosum,** sometimes double, which transmits the middle meningeal vessels and a recurrent branch from the mandibular nerve. The lesser petrosal nerve sometimes passes through a special canal (*canaliculus innominatus of Arnold*) situated medial to the foramen spinosum.

The **temporal surface** (*lateral surface*) (Fig. 4–56) is convex, and divided by a transverse ridge, the **infratemporal crest,** into two portions. The superior or temporal portion gives attachment to the Temporalis; the inferior or infratemporal, smaller in size, enters into the formation of the infratemporal fossa and together with the infratemporal crest, affords attachment to the Pterygoideus lateralis. It is pierced by the foramen ovale and foramen spinosum, and at its posterior part is the **sphenoidal spine,** which frequently is grooved on its medial surface for the chorda tympani nerve. To the sphenoidal spine are attached the sphenomandibular ligament and the Tensor veli palatini. Medial to the anterior extremity of the infratemporal crest is a triangular process which serves to increase the attachment of the Pterygoideus lateralis; extending downward and medialward from this process on the anterior part of the lateral pterygoid plate is a ridge which forms the anterior limit of the infratemporal surface, and in the articulated skull, the posterior boundary of the pterygomaxillary fissure.

The **orbital surface** of the great wing (Fig. 4–97) forms the posterior part of the lateral wall of the orbit. Its superior serrated edge articulates with the orbital plate of the frontal. Its inferior rounded border forms the posterior lateral boundary of the inferior orbital fissure. Its medial sharp margin forms the inferior boundary of the superior orbital fissure and has projecting from about its center a little tubercle which gives attachment to the inferior head of the Rectus lateralis oculi; at the upper part of this margin is a notch for the transmission of a recurrent branch of the lacrimal artery. Its lateral margin is serrated and articulates with the zygomatic bone. Inferior to the medial end of the superior orbital fissure is a grooved surface, which forms the posterior wall of the pterygopalatine fossa, and is pierced by the foramen rotundum.

The **margin of great wing** (Fig. 4–95).— The *posterior* portion of the *margin* or circumference of the great wing near the body forms the anterior wall of the foramen lacerum, in which is the posterior aperture of the **pterygoid canal** (Fig. 4–96). Its lateral half articulates, by means of a synchondrosis, with the petrous portion of the temporal, and between the two bones on the inferior surface of the skull, is a furrow, the **sulcus tubae auditivae,** for the lodgment of the cartilaginous part of the auditory tube. Anterior to the spine, the **margo squamosus,** bevelled at the expense of the inner table inferiorly, and of the outer table superiorly, articulates with the temporal squama. At the tip of the great wing is a triangular portion, the *margo parietalis* bevelled at the expense of the internal surface, for articulation with the sphenoidal angle of the parietal bone. This region is named the **pterion.** Medial to this is a triangular, serrated surface, the *margo frontalis,* for articulation with the frontal bone; this surface is continuous medially with the sharp edge which forms the inferior boundary of the superior orbital fissure, and laterally with the *margo zygomaticus* for articulation with the zygomatic bone.

The **small wings** (*alae minores*) or **orbitosphenoids** are two thin triangular plates, which arise from the superior and anterior parts of the body, and, projecting laterally, end in sharp points (Fig. 4–95).

The **superior surface** is flat and supports part of the frontal lobe of the brain. The **inferior surface** forms the posterior part of the roof of the orbit, and the superior boundary of the superior orbital fissure (page 140). The **anterior border** is serrated for articulation with the frontal bone. The **posterior border,** smooth and rounded, is received into the lateral fissure of the brain; the medial end of this border forms the **anterior clinoid process,** and gives attachment to the tentorium cerebelli. The small wing is connected to the body by two roots, superior and inferior, between which is the **optic canal** for the transmission of the optic nerve and ophthalmic artery.

The **pterygoid process** (*processus pterygoideus*), on either side, projects perpendicularly from the inferior region where the body and great wing unite. It consists of a medial and a lateral plate, the superior parts of which are fused anteriorly; the **pterygopalatine groove** lies in the line of fusion. The plates are separated inferiorly by an angular cleft, the **pterygoid fissure,** the margins of which are rough for articulation with the pyramidal process of the palatine bone. The two plates diverge posteriorly and enclose between them a V-shaped fossa,

the **pterygoid fossa,** which contains the Pterygoideus medialis and Tensor veli palatini. Superior to this fossa is a small, oval depression, the **scaphoid fossa,** which gives origin to the Tensor veli palatini. The anterior surface of the pterygoid process is broad and triangular near its root, where it forms the posterior wall of the pterygopalatine fossa.

The **lateral pterygoid plate** (*lamina lateralis*) is broad, thin, and everted; its lateral surface forms part of the medial wall of the infratemporal fossa, and gives attachment to the Pterygoideus lateralis; its medial surface forms part of the pterygoid fossa, and gives attachment to the Pterygoideus medialis.

The **medial pterygoid plate** (*lamina medialis*) is narrower and longer than the lateral; it curves laterally at its inferior extremity into a hook-like process, the **pterygoid hamulus,** around which the tendon of the Tensor veli palatini glides. The lateral surface of this plate forms part of the pterygoid fossa, the medial surface constitutes the lateral boundary of the choana or posterior aperture of the corresponding nasal cavity. Superiorly the medial plate is prolonged on to the inferior surface of the body as a thin lamina, named the **vaginal process,** which articulates anteriorly with the sphenoidal process of the palatine and posteriorly with the ala of the vomer. The angular prominence between the posterior margin of the vaginal process and the medial border of the scaphoid fossa is named the pterygoid tubercle, and immediately superior to this is the posterior opening of the **pterygoid canal.** On the inferior surface of the vaginal process is a furrow which is converted into a canal by the sphenoidal process of the palatine bone, for the transmission of the pharyngeal branch of the maxillary artery and the pharyngeal nerve from the pterygopalatine nerves. The pharyngeal aponeurosis is attached to the entire length of the posterior edge of the medial plate, and the Constrictor pharyngis superior takes origin from its inferior third. Projecting dorsally from near the middle of the posterior edge of this plate is an angular process, the **processus tubarius,** which supports the pharyngeal end of the auditory tube. The anterior margin of the plate articulates with the posterior border of the vertical part of the palatine bone.

The **pterygoid canal** (*Vidian canal*), 1 or 2 mm in diameter, runs anteroposteriorly in the part of the body of the sphenoid to which each pterygoid process is attached. When the sphenoidal sinus is extensive, the canal is covered by a ridge of bone in the floor of the sinus. The posterior orifice opens into the foramen lacerum (Fig. 4–96); the anterior into the pterygopalatine fossa (Fig. 4–97). It transmits the nerve of the pterygoid canal (*Vidian nerve*) and a small branch of the pterygopalatine portion of the maxillary artery of the same name.

The **sphenoidal conchae** (*conchae sphenoidales; sphenoidal turbinated processes*) (Fig. 4–97) are two thin, curved plates, situated at the anterior part of the body of the sphenoid. An aperture of variable size exists in the anterior wall of each, and through this the **sphenoidal sinus** opens into the sphenoethmoidal recess of the nasal cavity. Their inferior surface is convex, and forms part of the roof of the nasal cavity.

The sphenoid bone articulates anteriorly with the ethmoid, laterally with the palatine; its pointed posterior extremity is placed about the vomer, and is received between the root of the pterygoid process laterally and the rostrum of the sphenoid medially. A small portion of the sphenoidal concha sometimes enters into the formation of the medial wall of the orbit between the lamina orbitalis of the ethmoid, the orbital plate of the palatine, and the frontal bone.

Ossification.—The greater part of the sphenoid bone is first formed in cartilage. After ossification has set in, the bone may be considered in two parts which persist until the seventh or eighth month of fetal life. One part, anterior to the tuberculum sellae named the presphenoid, is continuous with the small wings; the other, comprising the sella turcica and dorsum sellae, is the postsphenoid and is associated with the great wings and pterygoid processes. There are fourteen centers in all, six for the presphenoid and eight for the postsphenoid.

Presphenoid.—About the ninth week of fetal life an ossific center appears for each of the small wings (orbitosphenoids) just lateral to the optic foramen; shortly afterward two nuclei appear in the presphenoid part of the body. The sphenoidal conchae are each developed from four centers which make their appearance

about the fifth month; at birth they consist of small triangular laminae, and it is not until the third year that they become hollowed out and cone-shaped; about the fourth year they join the labyrinths of the ethmoid, and between the ninth and twelfth years they unite with the sphenoid.

Postsphenoid.—The first ossific nuclei are those for the great wings (ali-sphenoids). One makes its appearance in each wing between the foramen rotundum and foramen ovale about the eighth week. The orbital plate and that part of the sphenoid which is found in the temporal fossa, as well as the lateral pterygoid plate, are ossified in membrane. The centers for the postsphenoid part of the body appear soon after, one on either side of the sella turcica, and become blended together about the middle of fetal life. Each medial pterygoid plate (with the exception of its hamulus) is ossified in membrane, and its center probably appears about the ninth or tenth week. The hamulus becomes chondrified during the third month, and almost at once undergoes ossification. The medial joins the lateral pterygoid plate about the sixth month. About the fourth month a center appears for each lingula and speedily joins the rest of the bone.

The presphenoid is united to the postsphenoid portion of the body about the eighth month, and at birth the bone is in three pieces (Fig. 4–98): a central, consisting of the body and small wings, and two lateral, each comprising a great wing and pterygoid process. In the first year after birth the great wings and body unite, and the small wings extend inward above the anterior part of the body, and, meeting with each other in the middle line, form an elevated smooth surface, termed the *jugum sphenoidale*. By the twenty-fifth year the sphenoid and occipital are completely fused.

The *craniopharyngeal canal*, through which the hypophyseal diverticulum is connected with buccal ectoderm in fetal life, occasionally persists between the presphenoid and postsphenoids.

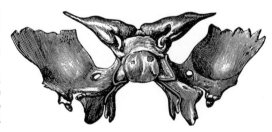

Fig. 4–98.—Sphenoid bone at birth. Posterior aspect.

The *sphenoidal sinuses* are present as minute cavities at the time of birth, but do not attain their full size until after puberty.

Intrinsic Ligaments of the Sphenoid.—The more important of these are: The pterygospinous, stretching between the sphenoidal spine and the lateral pterygoid plate (see cervical fascia); the interclinoid, a fibrous process joining the anterior to the posterior clinoid process; and the caroticoclinoid, connecting the anterior to the middle clinoid process. These ligaments occasionally ossify.

Articulations.—The sphenoid articulates with twelve bones: four single, the vomer, ethmoid, frontal, and occipital; and four paired, the parietal, temporal, zygomatic, and palatine. It also sometimes articulates with the tuberosity of the maxilla.

The Ethmoid Bone

The **Ethmoid Bone** (*os ethmoidale*) (Figs. 4–99, 4–102) is exceedingly light and spongy; it is situated at the anterior part of the base of the cranium, between the two orbits, and forms most of the walls of the superior part of the nasal cavity. It consists of four parts: a horizontal or cribriform plate, forming part of the base of the cranium; a perpendicular plate, constituting part of the nasal septum; and two lateral masses or labyrinths.

The **cribriform plate** (*lamina cribrosa; horizontal lamina*) (Fig. 4–99) is received into the ethmoidal notch of the frontal bone and roofs in the nasal cavities. Projecting superiorly into the cranial fossa from the

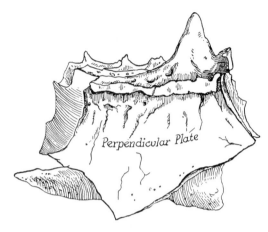

Fig. 4–100.—Perpendicular plate of ethmoid. Shown by removing the right labyrinth.

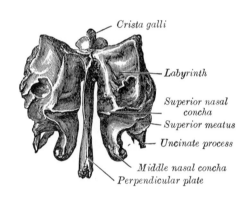

Fig. 4–101.—Ethmoid bone from behind.

middle line of this plate is a triangular process, the **crista galli,** so called from its resemblance to a cock's comb. The long thin posterior border of the crista galli serves for the attachment of the falx cerebri; its anterior border, short and thick, articulates with the frontal bone, and presents two small projecting alae, which are received into corresponding depressions in the frontal bone and complete the foramen cecum; its sides are smooth, and sometimes bulge from the presence of a small air sinus in the interior. On either side of the crista galli, the cribriform plate is narrow and deeply grooved; it supports the olfactory bulb and is perforated by the **olfactory foramina** for the passage of the olfactory nerves. At the anterior part of the cribriform plate, on either side of the crista galli, is a small fissure which is occupied by a process of dura

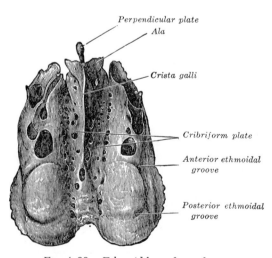

Fig. 4–99.—Ethmoid bone from above.

mater. Lateral to this fissure is a notch or foramen which transmits the nasociliary nerve; from this notch a groove extends backward to the anterior ethmoid foramen.

The **perpendicular plate** (*lamina perpendicularis; vertical plate*) (Figs. 4–100, 4–101) is a flattened lamina, which assists in forming the septum of the nose. The anterior border articulates with the spine of the frontal bone and the crest of the nasal bones. The posterior border articulates by its superior half with the sphenoidal crest, by its inferior half with the vomer. The inferior border is thicker than the posterior, and serves for the attachment of the septal cartilage of the nose. The surfaces of the plate are smooth except where numerous grooves and canals lead from the medial foramina in the cribriform plate and lodge filaments of the olfactory nerves.

The **labyrinth** or **lateral mass** (*labyrinthus ethmoidals*) consists of a number of thin-walled cellular cavities, the **ethmoidal cells**, interposed between two vertical plates of bone and arranged in three groups, anterior, middle and posterior. The lateral plate forms part of the orbit, the medial, part of the corresponding nasal cavity. In the disarticulated bone many of the cells are incomplete, but in the articulated skull, they are closed in at every part except where their ostia open into the nasal cavity.

The **superior surface** of the labyrinth (Fig. 4–99) presents a number of half-broken cells, the walls of which are completed, in the articulated skull, by the edges of the ethmoidal notch of the frontal bone. Crossing this surface on each side are two grooves, converted into canals by articulation with the frontal; they are the **anterior** and **poste-**rior ethmoidal canals, and open on the inner wall of the orbit. The posterior surface presents large irregular cellular cavities, which are closed by articulation with the sphenoidal conchae and orbital processes of the palatines. The lateral surface (Fig. 4–102) is formed of a thin, smooth, oblong plate, the **lamina orbitalis** (*lamina papyracea*), which covers in the middle and posterior ethmoidal cells and forms a large part of the medial wall of the orbit. It articulates with the orbital plate of the frontal bone, with the maxilla, with the orbital process of the palatine, with the lacrimal, and with the sphenoid.

Anterior to the lamina orbitalis are some broken air cells which are overlapped and completed by the lacrimal bone and the frontal process of the maxilla. A curved lamina, the **uncinate process**, projects inferiorly and posteriorly from this part of the labyrinth; it forms part of the medial wall of the maxillary sinus and articulates with the ethmoidal process of the inferior nasal concha (Fig. 4–102).

The **medial surface of the labyrinth** (Fig. 4–103) forms part of the lateral wall of the corresponding nasal cavity. It consists of a thin, rough lamella, which is marked by numerous grooves for the branches of the olfactory nerves. It ends inferiorly in a free, convoluted margin, the **middle nasal concha**. The posterior part of the medial surface of the labyrinth is subdivided by a narrow anteroposterior fissure, the **superior meatus** of the nose, which separates the **superior nasal concha** from the middle concha. The middle concha extends along the whole anteroposterior extent of the medial surface of the labyrinth, and its

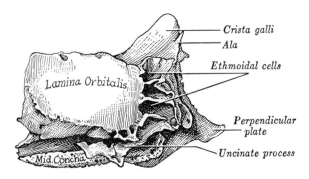

FIG. 4–102.—Ethmoid bone from the right side.

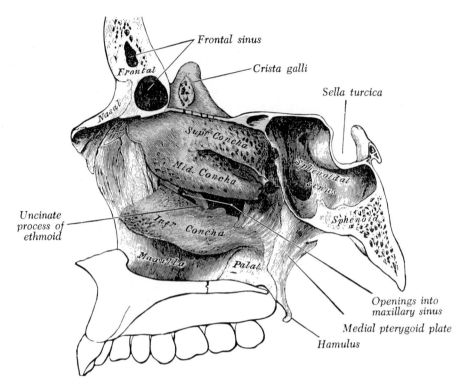

Frontal sinus

Crista galli

Sella turcica

Frontal

Nasal

Supr. Concha

Mid. Concha

Sphenoidal sinus

Sphenoid

Uncinate
process of
ethmoid

Inf. Concha

Maxilla

Palat.

Openings into
maxillary sinus

Medial pterygoid plate

Hamulus

Fig. 4–103.—Lateral wall of nasal cavity, showing ethmoid bone in position.

margin is free and thick; the superior concha extends about half as far anteriorly and its free margin is less thickened. The superior meatus extends superiorly as a deep groove under cover of the superior concha; the posterior ethmoidal air cells open into it. The **middle meatus** of the nose is an extensive narrow space extending superiorly under cover of the middle concha. The lateral wall of the middle meatus contains a deep curved passage, the **infundibulum,** between the **uncinate process** inferiorly and the **bulla ethmoidalis** superiorly. The crescentic opening by which this passage communicates with the meatus is named the **hiatus semilunaris** (*hiatus ethmoidalis*) (Fig. 4–68). The **ostium** of the **maxillary sinus** opens into the bottom of the middle part of the infundibulum. The portion of the wall of the meatus posterior to the infundibulum is lacking in bone and may fail to be closed by mucous membrane in the intact body thus providing an **accessory opening** into the **maxillary sinus.**

The middle ethmoidal air cells lie within

the bulla and open above it; whereas, the anterior ethmoidal cells open into the infundibulum. The **ostia** of the **frontal sinus** and the **anterior ethmoidal cells** may open into the anterior end of the infundibulum or into the anterior superior terminal recess of the middle meatus.

Ossification.—The ethmoid is ossified in the cartilage of the nasal capsule by three centers: one for the perpendicular plate, and one for each labyrinth.

The labyrinths are developed first, ossific granules making their appearance between the fourth and fifth months of fetal life in the region of the lamina orbitalis and extending into the conchae. At birth, the bone consists of the two labyrinths, which are small and undeveloped. During the first year after birth, the perpendicular plate and crista galli begin to ossify from a single center, and are joined to the labyrinths about the beginning of the second year. The cribriform plate is ossified partly from the perpendicular plate and partly from the labyrinths. The development of the ethmoidal cells begins after birth.

Articulations.—The ethmoid articulates with

thirteen bones: two of the cranium—the frontal, and the sphenoid; and eleven of the face—the two nasals, two maxillae, two lacrimals, two palatines, two inferior nasal conchae and the vomer.

Sutural Bones.—In addition to the usual centers of ossification of the cranium, others may occur in the course of the sutures, giving rise to irregular, isolated bones, termed **sutural** or **Wormian bones.** They occur most frequently in the course of the lambdoidal suture, but are occasionally seen at the fontanelles, especially the posterior. They have a tendency to be more or less symmetrical on the two sides of the skull, and vary much in size. Their number is generally limited to two or three; but more than a hundred have been found in the skull of an adult hydrocephalic subject. A pterion ossicle sometimes exists between the sphenoidal angle of the parietal and the great wing of the sphenoid.

THE FACIAL BONES (OSSA FACIEI)

The Nasal Bones

The **Nasal Bones** (*ossa nasalia*), two small oblong bones, are placed side by side to form, by their junction, "the bridge" of the nose (Fig. 4–57).

The **external surface** (Fig. 4–105) is convex from side to side and is perforated about its center by a foramen for the transmission

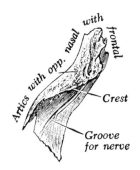

FIG. 4–104.—Right nasal bone. Inner surface.

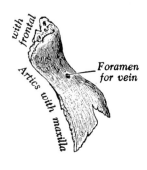

FIG. 4–105.—Right nasal bone. Outer surface.

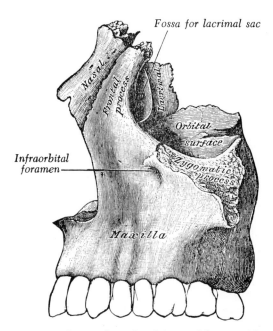

FIG. 4–106.—Articulation of nasal and lacrimal bones with maxilla.

of a small vein. It is covered by the Procerus and Nasalis. The **internal surface** (Fig. 4–104) is concave and is traversed by a groove for the passage of a branch of the naso-ciliary nerve.

The **superior border** is narrow, thick, and serrated for articulation with the nasal notch of the frontal bone. The **inferior border** is thin, and gives attachment to the lateral cartilage of the nose; near its middle is a notch which marks the end of the naso-ciliary groove. The **lateral border** is serrated, bevelled at the expense of the inner surface above, and of the outer below, to articulate with the frontal process of the maxilla. The **medial border** articulates with its fellow of the opposite side and is prolonged posteriorly into a vertical crest which forms part of the nasal septum: this crest articulates with the spine of the frontal, the perpendicular plate of the ethmoid, and the septal cartilage of the nose.

Ossification.—Each bone is ossified from one center which appears at the beginning of the third month of fetal life in the membrane overlying the anterior part of the cartilaginous nasal capsule.

Articulations.—The nasal articulates with four bones: two of the cranium, the frontal and ethmoid, and two of the face, the opposite nasal and the maxilla.

The Maxillary Bone

The **Maxillae** (*upper jaw*) are the largest bones of the face, excepting the mandible, and form, by their union, the whole of the upper jaw. Each assists in forming the boundaries of four cavities, viz., the roof of the mouth, the floor and lateral wall of the nose, the floor of the orbit, and the maxillary sinus; it also enters into the formation of two fossae, the infratemporal and pterygopalatine, and two fissures, the inferior orbital and pterygomaxillary. Each bone consists of a body and four processes—zygomatic, frontal, alveolar, and palatine.

The **body** (*corpus maxillae*) is somewhat pyramidal in shape, and contains a large cavity, the maxillary sinus (antrum of Highmore). It has four surfaces—an anterior, a posterior or infratemporal, a superior or orbital, and a medial or nasal.

The **anterior** or **facial surface** (Fig. 4–57) presents at its inferior part a series of eminences corresponding to the positions of the roots of the teeth. Superior to those of the incisor teeth is a depression, the incisive fossa, which gives origin to the Depressor septi. To the alveolar border below the fossa a slip of the Orbicularis oris is attached; above and a little lateral to it, the Nasalis arises. Lateral to the incisive fossa is another depression, the canine fossa; it is larger and deeper than the incisive fossa, and is separated from it by a vertical ridge, the canine eminence, corresponding to the socket of the canine tooth; the canine fossa gives origin to the Levator anguli oris. Above the fossa is the **infraorbital foramen** for the infraorbital vessels and nerve. Above the foramen is the margin of the orbit which affords attachment to part of the Levator labii superioris. Medially, the anterior surface is limited by a deep concavity, the nasal notch, the margin of which gives attachment to the Dilatator naris and ends below in a pointed process which, with its fellow of the opposite side, forms the **anterior nasal spine.**

The **infratemporal surface** (Fig. 4–107) is separated from the anterior surface by the zygomatic process and by a strong ridge, extending upward from the socket of the first molar tooth. It is pierced about its center by the apertures of the alveolar canals which transmit the posterior superior alveolar vessels and nerves. At the inferior part of this surface is a rounded eminence, the **maxillary tuberosity,** especially prominent after the growth of the wisdom tooth; it is rough on its medial side for articulation with the pyramidal process of the palatine bone and in some cases articulates with the lateral pterygoid plate of the sphenoid. It gives origin to a few fibers of the Pterygoideus medialis. Immediately above this is a smooth surface which forms the anterior boundary of the pterygopalatine fossa, and presents a groove for the maxillary nerve.

The **orbital surface** (Fig. 4–59) forms the greater part of the floor of the orbit. It is bounded medially by an irregular margin, the **lacrimal notch;** and posterior to this notch the margin articulates with the lacrimal, the lamina orbitalis of the ethmoid, and the orbital process of the palatine. It is bounded posteriorly by a smooth rounded

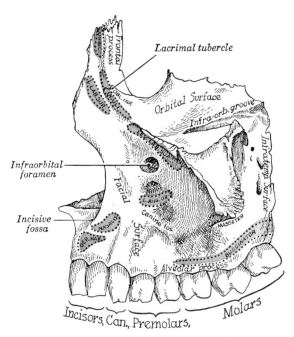

FIG. 4–107.—Left maxilla. Outer surface.

edge which forms the anterior margin of the inferior orbital fissure, and sometimes articulates at its lateral extremity with the orbital surface of the great wing of the sphenoid. It is limited anteriorly by the part of the circumference of the orbit which is continuous medially with the frontal process, and laterally with the zygomatic process. Near the middle of the posterior part of the orbital surface is the **infraorbital groove,** for the passage of the infraorbital vessels and nerve. The groove begins at the middle of the posterior border, where it is continuous with that for the maxillary nerve, and, passing rostrally, ends in a canal which subdivides into two branches. One of the canals, the **infraorbital canal,** opens just below the margin of the orbit; the other, which is smaller, runs downward in the substance of the anterior wall of the maxillary sinus, and transmits the anterior superior alveolar vessels and nerve to the front teeth. From the posterior part of the infraorbital canal, a second small canal is sometimes given off; it runs in the lateral wall of the sinus, and conveys the middle alveolar nerve to the premolar teeth. At the medial and anterior part of the orbital surface, just lateral to the lacrimal groove, is a depression which gives origin to the Obliquus inferior oculi.

The **nasal surface** (Fig. 4–108) presents a large, irregular opening leading into the **maxillary sinus.** At the superior border of this aperture are some broken air cells which, in the articulated skull, are closed in by the ethmoid and lacrimal bones. Below the aperture is a smooth concavity which forms part of the inferior meatus of the nasal cavity, and posterior to it is a rough surface for articulation with the perpendicular part of the palatine bone. A groove commencing near the middle of the posterior border and running obliquely downward and forward is converted into the pterygopalatine canal by articulation with the palatine bone. Rostral to the opening of the sinus is the **lacrimal groove,** which is converted into the nasolacrimal canal by the lacrimal bone and inferior nasal concha; this canal opens into the inferior meatus and transmits the nasolacrimal duct. More anteriorly is an oblique ridge, the conchal crest, for articulation with the inferior nasal concha. The shallow concavity superior to this ridge forms part of the atrium of the middle meatus, and that inferior to it, part of the inferior meatus.

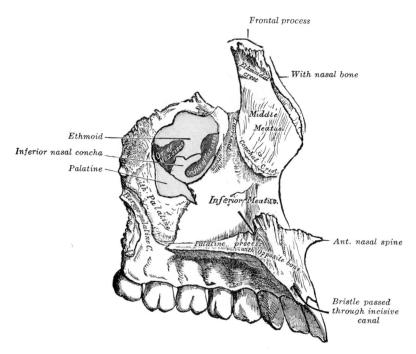

Fig. 4–108.—Left maxilla. Nasal surface.

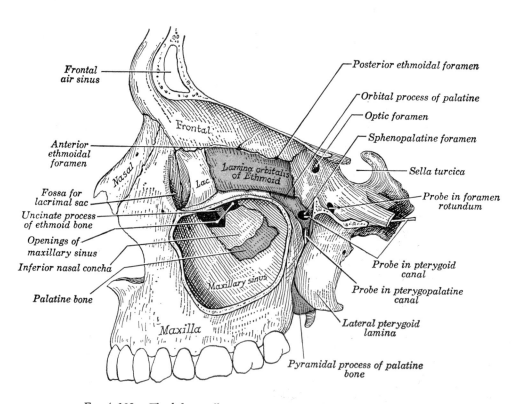

Fig. 4–109.—The left maxillary air sinus. Opened from the lateral side.

The **maxillary sinus** (*sinus maxillaris, antrum of Highmore*) (Fig. 4–109) is a large pyramidal cavity within the body of the maxilla. Its walls are thin and correspond to the nasal, orbital, anterior, and infratemporal surfaces of the body of the bone. Its nasal wall presents, in the disarticulated bone, a large, irregular aperture, communicating with the nasal cavity. In the articulated skull this aperture is partly closed by the following bones: the uncinate process of the ethmoid, the ethmoidal process of the inferior nasal concha, the vertical part of the palatine, and a small part of the lacrimal (Figs. 4–67, 4–68). The sinus communicates with the middle meatus, generally by two small apertures left between the above-mentioned bones. In the intact body usually only one small opening exists near the upper part of the cavity; the other is closed by mucous membrane. On the posterior wall are the alveolar canals, transmitting the posterior superior alveolar vessels and nerves to the molar teeth. The floor is formed by the alveolar process of the maxilla, and if the sinus is of an average size, is on a level with the floor of the nose; if the sinus is large it reaches below this level.

Projecting into the floor of the sinus are several conical processes corresponding to the roots of the teeth; in some cases the floor is perforated by the fangs of the teeth. The infraorbital canal usually projects into the cavity as a well-marked ridge extending from the roof to the anterior wall; additional ridges are sometimes seen in the posterior wall of the cavity, and are caused by the alveolar canals. The size of the cavity varies in different skulls, and even on the two sides of the same skull.

The **zygomatic process** (*processus zygomaticus; malar process*) is a rough triangular eminence, situated at the angle of separation of the anterior, infratemporal, and orbital surfaces. It articulates with the zygomatic bone and forms part of the anterior facial surface and the infratemporal fossa, the prominent arched border marking the division between them.

The **frontal process** (*processus frontalis; nasal process*) (Fig. 4–107) is a strong plate which forms part of the lateral boundary of the nose. Its lateral surface is smooth, and gives attachment to the Levator labii superi-

oris alaeque nasi, the Orbicularis oculi, and the medial palpebral ligament. Its medial surface forms part of the lateral wall of the nasal cavity, closing in the anterior ethmoidal cells. Inferior to this is the ethmoidal crest, the posterior end of which articulates with the middle nasal concha, and forms the upper limit of the atrium of the middle meatus while the anterior part is termed the **agger nasi.** The superior border articulates with the frontal bone and the anterior with the nasal; the posterior border is thick, and hollowed into a groove which is continuous below with the lacrimal groove on the nasal surface of the body and which articulates with the lacrimal to form the lacrimal fossa for the lodgment of the lacrimal sac. The lateral margin of the groove is named the *anterior lacrimal crest,* and at its junction with the orbital surface is a small tubercle, the **lacrimal tubercle,** which serves as a guide to the position of the lacrimal sac.

The **alveolar process** (*processus alveolaris*) is excavated into deep cavities for the reception of the teeth. These eight cavities vary in size and depth according to the teeth they contain. The Buccinator arises from the outer surface of this process, as far anteriorly as the first molar tooth. When the maxillae are articulated with each other, their alveolar processes together form the **alveolar arch;** the center of the anterior margin of this arch is named the alveolar point. The teeth are described in the chapter on the Digestive System.

The **palatine process** (*processus palatinus; palatal process*) (Fig. 4–108) is horizontal and projects medially from the nasal surface of the bone. It forms a considerable part of the floor of the nose and the roof of the mouth. Its inferior surface (Fig. 4–110) is concave, rough, and uneven, and forms, with the palatine process of the opposite bone, the anterior three-fourths of the **hard palate.** It is perforated by numerous foramina for the passage of nutrient vessels, and presents little depressions for the lodgment of the palatine glands. At the posterior part near the last molar is a groove, sometimes a **canal,** for the transmission of the **greater palatine** vessels and nerve. When the two maxillae are articulated, a funnel-shaped opening, the **incisive foramen,** is seen in the

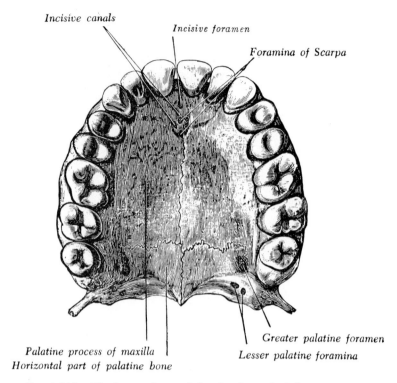

Incisive canals

Incisive foramen

Foramina of Scarpa

Greater palatine foramen
Lesser palatine foramina

Palatine process of maxilla
Horizontal part of palatine bone

FIG. 4–110.—The bony palate and the alveolar arch. Inferior aspect.

middle line, immediately behind the incisor teeth. In this opening the orifices of two lateral canals are visible; they are named the incisive canals or *foramina of Stenson;* through each of them passes the terminal branch of the descending septal artery and the nasopalatine nerve. Occasionally two additional canals are present in the middle line; they are termed the *foramina of Scarpa,* and when present transmit the nasopalatine nerves, the left passing through the anterior, and the right through the posterior canal. On the inferior surface of the palatine process, a delicate linear suture, well seen in young skulls, may sometimes be noticed extending laterally and rostrally on either

FIG. 4–112.—Inferior surface of maxilla at birth.

side from the incisive foramen to the interval between the lateral incisor and the canine tooth. The small part anterior to this suture constitutes the **premaxilla** (*os incisivum*), which in most vertebrates forms an independent bone. It includes the whole thickness of the alveolus, the corresponding part of the floor of the nose, and the anterior nasal spine, and contains the sockets of the incisor teeth. The medial border is thicker anteriorly, and is raised into a ridge, the nasal crest, which, with the corresponding ridge of the opposite bone, forms a groove for the reception of the vomer. The anterior part of this ridge is named the incisor crest

FIG. 4–111.—Anterior surface of maxilla at birth.

which is prolonged into a sharp process, the **anterior nasal spine**. The posterior border is serrated for articulation with the horizontal part of the palatine bone.

Ossification.—The maxilla is ossified in membrane from two centers, one for the maxilla proper and one for the premaxilla. These centers appear during the sixth week of fetal life and unite in the beginning of the third month, but the suture between the two portions persists on the palate until nearly middle life.

Articulations.—The maxilla articulates with nine bones: two of the cranium, the frontal and ethmoid, and seven of the face, viz., the nasal, zygomatic, lacrimal, inferior nasal concha, palatine, vomer, and its fellow of the opposite side. Sometimes it articulates with the orbital surface, and sometimes with the lateral pterygoid plate of the sphenoid.

CHANGES PRODUCED IN THE MAXILLA BY AGE

At birth the transverse and anteroposterior diameters of the bone are each greater than the vertical. The frontal process is well marked and the body of the bone consists of little more than the alveolar process, the teeth sockets reaching almost to the floor of the orbit. The maxillary sinus is represented by a furrow on the lateral wall of the nasal cavity. In the adult the vertical diameter is the greatest, owing to the development of the alveolar process and the increase in size of the sinus. In old age the bone reverts in some measure to the infantile condition; its height is diminished, and after loss of the teeth the alveolar process is absorbed, and the lower part of the bone is contracted and reduced in thickness.

The Palatine Bone

The **Palatine Bone** (*os palatinum; palate bone*) forms the posterior part of the hard palate, part of the floor and lateral wall of the nasal cavity, and the floor of the orbit. It enters into the formation of three fossae, the pterygopalatine, pterygoid, and infratemporal fossae; and one fissure, the inferior orbital fissure. The palatine bone somewhat resembles the letter L, and consists of a horizontal and a vertical part and three outstanding processes—viz., the pyramidal, the orbital, and sphenoidal processes.

The **horizontal part** (*lamina horizontalis; horizontal plate*) (Figs. 4–113, 4–114) is quadrilateral, and has two surfaces and three borders.

The **nasal** or **superior surface** forms the back part of the floor of the nasal cavity. The **palatine** or **inferior surface** forms, with the corresponding surface of the opposite bone, the posterior fourth of the **hard palate**. Near its posterior margin may be seen a more or less marked transverse ridge for the attachment of part of the aponeurosis of the Tensor veli palatini.

The **anterior border** is serrated, and articulates with the palatine process of the maxilla. The **posterior border** is concave, free, and serves for the attachment of the soft palate. Its medial end is sharp and pointed, and, when united with that of the opposite bone, forms a projecting process, the posterior nasal spine for the attachment of the Musculus uvulae. The **medial border** is serrated for articulation with its fellow of the opposite side; its superior edge is raised into a ridge, which, united with the ridge of the opposite bone, forms the nasal crest for articulation with the posterior part of the edge of the vomer.

The **vertical part** (*pars perpendicularis; perpendicular plate*) (Figs. 4–114, 4–115) is thin, oblong, and presents two surfaces and four borders.

The inferior part of the **nasal surface** exhibits a broad, shallow depression which forms part of the inferior meatus of the nose. Superior to this is the conchal crest, for articulation with the inferior nasal concha; still higher is a second broad, shallow depression, which forms part of the middle meatus, and is limited by the ethmoidal crest, for articulation with the middle nasal concha. Above the ethmoidal crest is a narrow, horizontal groove, which forms part of the superior meatus.

The **maxillary surface** is rough and irregular throughout the greater part of its extent, for articulation with the nasal surface of the maxilla; its posterosuperior part is smooth where it enters into the formation of the pterygopalatine fossa; it is also smooth anteriorly, where it forms the posterior part of the medial wall of the maxillary sinus. On the posterior part of this surface is a deep vertical groove, converted into the pterygopalatine canal, by articulation with the maxilla; this canal transmits the greater palatine vessels and nerve.

The **anterior border** is thin and irregular;

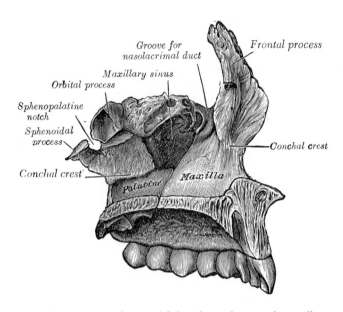

Fig. 4–113.—Articulation of left palatine bone with maxilla.

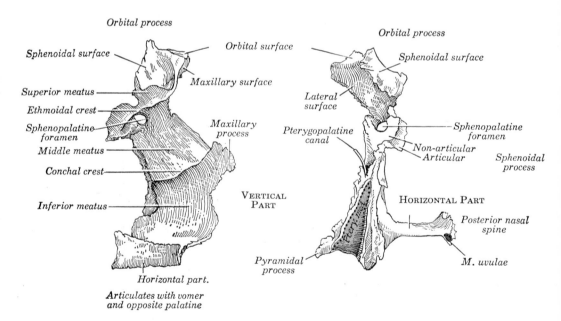

Fig. 4–114.—Left palatine bone. Nasal aspect. Enlarged.

Fig. 4–115.—Left palatine bone. Posterior aspect. Enlarged.

opposite the conchal crest is a pointed, projecting lamina, the maxillary process which closes in the lower and back part of the defect in the medial wall of the maxillary sinus (Fig. 4–109). The **posterior border** presents a deep groove, the edges of which are serrated for articulation with the medial pterygoid plate of the sphenoid. This border is continuous with the sphenoidal process and expands into the pyramidal process. Between the orbital process and the sphenoidal process on the superior border is the **sphenopalatine notch,** which is converted into the **sphenopalatine foramen** by articulation with the body of the sphenoid. In the articulated skull this foramen leads from the pterygopalatine fossa into the posterior part of the nasal cavity and transmits the pterygopalatine vessels and nerves. The inferior border is fused with the lateral edge of the horizontal part, and immediately anterior to the pyramidal process is grooved by the lower end of the pterygopalatine canal.

The **pyramidal process** or **tuberosity** (*processus pyramidalis*) projects laterally from the junction of the horizontal and vertical parts, and is received into the angular interval between the inferior extremities of the pterygoid plates. On its posterior surface, rough articular furrows articulate with the pterygoid plates, while the grooved intermediate area completes the pterygoid fossa and gives origin to a few fibers of the Pterygoideus medialis. The anterior part of the lateral surface is rough, for articulation with the tuberosity of the maxilla; its posterior part consists of a smooth triangular area which appears, in the articulated skull, between the tuberosity of the maxilla and the inferior part of the lateral pterygoid plate, and completes the infratemporal fossa. On the base of the pyramidal process, close to its union with the horizontal part, are the **lesser palatine foramina** for the transmission of the lesser palatine nerves.

The **orbital process** (*processus orbitalis*) is directed superiorly and laterally from the vertical part, to which it is connected by a constricted neck. It presents five surfaces, which enclose an air cell. Of these surfaces, three are articular and two non-articular. The articular surfaces are: (1) the anterior or maxillary, of an oblong form, and rough for articulation with the maxilla; (2) the posterior or sphenoidal presents the opening of the air cell which usually communicates with the sphenoidal sinus; the margins of the opening are serrated for articulation with the sphenoidal concha; (3) the medial or ethmoidal articulates with the labyrinth of the ethmoid. In some cases the air cell opens on this surface of the bone and then communicates with the posterior ethmoidal cells. More rarely it opens on both surfaces, and then communicates with the posterior ethmoidal cells and the sphenoidal sinus. The non-articular surfaces are: (1) the superior or orbital which is triangular in shape, and forms the posterior part of the floor of the orbit; and (2) the lateral, of an oblong form and separated from the orbital surface by a rounded border which enters into the formation of the inferior orbital fissure.

The **sphenoidal process** (*processus sphenoidalis*) is a thin, compressed plate, much smaller than the orbital, and directed medially. It presents three surfaces and two borders. The superior surface articulates with the root of the pterygoid process and the sphenoidal concha, its medial border reaching as far as the ala of the vomer; it presents a groove which contributes to the formation of the pharyngeal canal. The medial surface is concave, and forms part of the lateral wall of the nasal cavity. The lateral surface is divided into an articular and a non-articular portion: the former is rough, for articulation with the medial pterygoid plate; the latter is smooth, and forms part of the pterygopalatine fossa. The anterior border forms the posterior boundary of the sphenopalatine notch. The posterior border articulates with the medial pterygoid plate.

The **orbital** and **sphenoidal processes** are separated from one another by the **sphenopalatine notch.** Sometimes the two processes are united above, and form between them a complete foramen (Fig. 4–114), or the notch may be crossed by one or more spicules of bone, giving rise to two or more foramina.

Ossification.—The palatine bone is ossified in membrane from a single center, which makes its appearance about the sixth or eighth week of fetal life at the angle of junction of the two parts of the bone. From this point ossification spreads medialward to the horizontal part, downward into the pyramidal process, and upward into the vertical part. Some authorities describe the bone as ossifying from four centers: one for the pyramidal process and portion of the vertical part behind the pterygopalatine groove; a second for the rest of the vertical and the horizontal parts; a third for the orbital, and a fourth for the sphenoidal process. At the time of birth the height of the vertical part is about equal to the transverse width of the horizontal part, whereas in the adult the former measures about twice as much as the latter.

Articulations.—The palatine articulates with six bones: the sphenoid, ethmoid, maxilla, inferior nasal concha, vomer, and opposite palatine.

The Inferior Nasal Concha

The **Inferior Nasal Concha** (*concha nasalis inferior; inferior turbinated bone*) extends horizontally along the lateral wall of the nasal cavity (Fig. 4–68) and consists of a lamina of spongy bone, curled upon itself like a scroll. It has two surfaces, two borders, and two extremities.

The **medial surface** (Fig. 4–116) is convex, perforated by numerous apertures, and traversed by longitudinal grooves for the lodgment of vessels. The **lateral surface** is concave (Fig. 4–117), and forms part of the inferior meatus. Its **superior border** is thin, irregular, and connected to various bones along the lateral wall of the nasal cavity. It may be divided into three portions: of these, the anterior articulates with the conchal crest of the maxilla; the posterior with the conchal crest of the palatine; the middle portion presents three well-marked processes, which vary much in their size and form. Of these, the anterior or **lacrimal process** is small and pointed and is situated

at the junction of the anterior fourth with the posterior three-fourths of the bone: it articulates, by its apex, with the descending process of the lacrimal bone and by its margins, with the groove on the back of the frontal process of the maxilla, and thus assists in forming the canal for the nasolacrimal duct. Posterior to this process a broad, thin plate, the **ethmoidal process**, joins the uncinate process of the ethmoid; from its inferior border a thin lamina, the **maxillary process**, curves downward and lateralward; it articulates with the maxilla and forms a part of the medial wall of the maxillary sinus. The **inferior border** is free, thick, and cellular in structure, more especially in the middle of the bone. Both extremities are more or less pointed, the posterior being the more tapering.

Ossification.—The inferior nasal concha is ossified from a single center which appears about the fifth month of fetal life in the lateral wall of the cartilaginous nasal capsule.

Articulations.—The inferior nasal concha articulates with four bones: the ethmoid, maxilla, lacrimal, and palatine.

The Vomer

The **Vomer** is situated in the median plane, but its anterior portion is frequently deviated to one side or the other. It forms the posterior and inferior part of the **nasal septum** (Fig. 4–118); it has two surfaces and four borders. The surfaces (Fig. 4–119) are marked by small furrows for blood vessels, and on each is the nasopalatine groove, which lodges the nasopalatine nerve and vessels. The superior border, the thickest, presents a deep furrow, bounded on either side by a horizontal projecting **ala** of bone; the furrow receives the rostrum of the sphenoid, while the margins of the alae

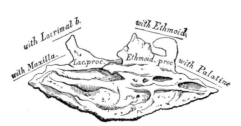

Fig. 4–116.—Right inferior nasal concha. Medial surface.

Fig. 4–117.—Right inferior nasal concha. Lateral surface.

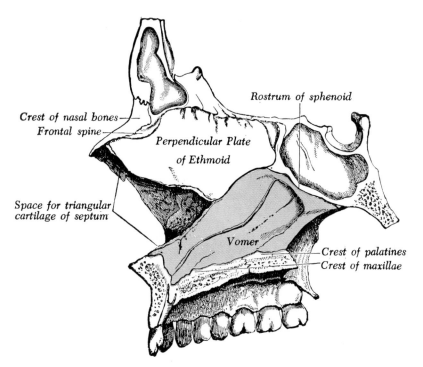

Crest of nasal bones
Frontal spine
Perpendicular Plate of Ethmoid
Rostrum of sphenoid
Space for triangular cartilage of septum
Vomer
Crest of palatines
Crest of maxillae

FIG. 4–118.—Median wall of left nasal cavity showing vomer *in situ*.

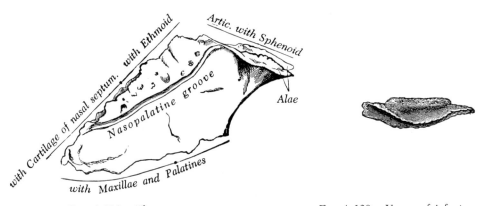

Artic. with Sphenoid
with Cartilage of nasal septum. with Ethmoid
Nasopalatine groove
Alae
with Maxillae and Palatines

FIG. 4–119.—The vomer.

FIG. 4–120.—Vomer of infant.

articulate with the vaginal processes of the medial pterygoid plates of the sphenoid, and with the sphenoidal processes of the palatine bones. The inferior border articulates with the crest formed by the maxillae and palatine bones. The anterior border is the longest; its superior half is fused with the perpendicular plate of the ethmoid; its inferior half is grooved for the inferior margin of the septal cartilage of the nose. The posterior border is free, concave, and separates the choanae.

Ossification.—At an early period the septum of the nose consists of a plate of cartilage, the ethmovomerine cartilage. The posterior superior part of this cartilage is ossified to form the perpendicular plate of the ethmoid; its anterior-inferior portion persists as the septal cartilage, while the vomer is ossified in the membrane covering its posterior inferior part. Two ossific centers, one on either side of the middle line, appear about the eighth week of fetal life in this part of the membrane, and hence the vomer consists primarily of two lamellae. About the third month these unite below, and thus a deep groove is formed in which the cartilage is lodged. As growth proceeds, the union of the lamellae extends anteriorly and superiorly, and at the same time the intervening plate of cartilage undergoes absorption. By puberty the lamellae are almost completely united to form a median plate, but evidence of the bilaminar origin of the bone is seen in the everted alae of its upper border and the groove on its anterior margin.

Articulations.—The vomer articulates with six bones: two of the cranium, the sphenoid and ethmoid; and four of the face, the two maxillae and the two palatine bones; it also articulates with the septal cartilage of the nose.

The Lacrimal Bone

The **Lacrimal Bone** (*os lacrimale*), the smallest and most fragile bone of the face, is situated at the anterior part of the medial wall of the orbit (Fig. 4–106). It has two surfaces and four borders.

The lateral or orbital surface (Fig. 4–121) is divided by a vertical ridge, the **posterior lacrimal crest,** into two parts. The anterior part is the **lacrimal sulcus** (*sulcus lacrimalis*), which unites with the frontal process of the maxilla to form the **lacrimal fossa.** The upper part of this fossa lodges the lacrimal sac; the lower part, the nasolacrimal duct. The posterior portion forms part of the medial wall of the orbit. The crest gives origin to the lacrimal part of the Orbicularis oculi and ends in a small, hook-like projection, the **lacrimal hamulus,** which articulates with the lacrimal tubercle of the maxilla, and completes the orifice of the lacrimal canal; it sometimes exists as a

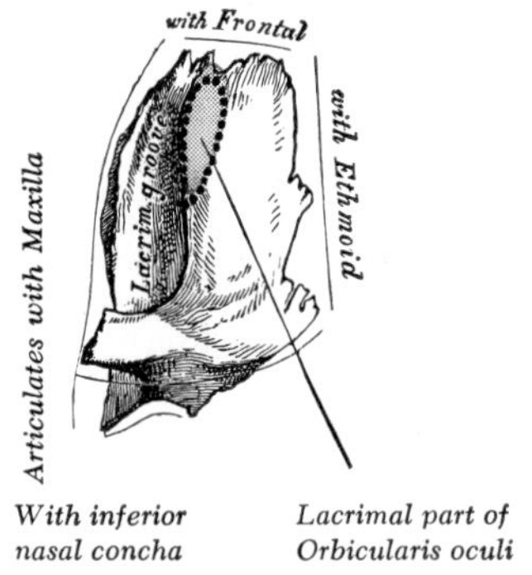

FIG. 4–121.—The left lacrimal bone. Lateral aspect. (Enlarged.)

separate piece, and is then called the *lesser lacrimal bone.*

The medial or nasal surface presents a longitudinal furrow, corresponding to the crest on the lateral surface. The area anterior to this furrow forms part of the middle meatus of the nose; that posterior to it articulates with the ethmoid, and completes some of the anterior ethmoidal cells.

Of the four borders the anterior articulates with the frontal process of the maxilla; the posterior with the lamina orbitalis of the ethmoid; the superior with the frontal bone. The inferior is divided by the lower edge of the posterior lacrimal crest into two parts: the posterior part articulates with the orbital plate of the maxilla; the anterior is prolonged downward as the descending process, which articulates with the lacrimal process of the inferior nasal concha, and assists in forming the canal for the nasolacrimal duct.

Ossification.—The lacrimal is ossified from a single center, which appears about the twelfth week in the membrane covering the cartilaginous nasal capsule.

Articulations.—The lacrimal articulates with four bones: two of the cranium, the frontal and ethmoid, and two of the face, the maxilla and the inferior nasal concha.

The Zygomatic Bone

The **Zygomatic Bone** (*os zygomaticum; malar bone*) forms the prominence of the cheek, part of the lateral wall and floor of the orbit, and parts of the temporal and infratemporal fossae (Fig. 4–122). It presents a malar, orbital, and temporal surface; frontal, maxillary, and temporal processes, and four borders.

The **malar surface** (Fig. 4–123) is convex and perforated near its center by a small aperture, the **zygomaticofacial foramen,** for the passage of the zygomaticofacial nerve and vessels; below this foramen is a slight elevation, which gives origin to the Zygomaticus major and the Levator labii superioris.

The **temporal surface** (Fig. 4–124) is concave, presenting medially a rough, triangular area, for articulation with the maxilla, and laterally a smooth, concave surface

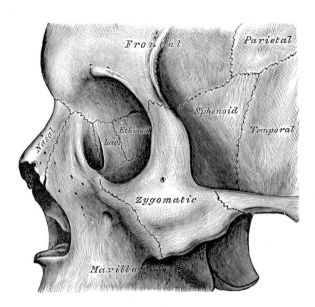

FIG. 4–122.—Left zygomatic bone *in situ.*

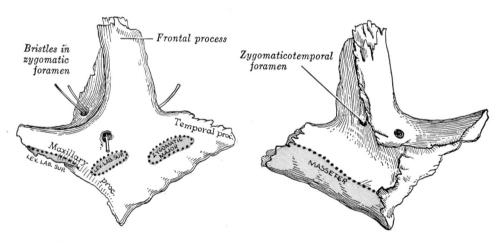

FIG. 4–123.—Left zygomatic bone.
Malar surface.

FIG. 4–124.—Left zygomatic bone.
Temporal surface.

which forms the anterior boundary of the temporal fossa and part of the infratemporal fossa. Near the center of this surface is the **zygomaticotemporal foramen** for the transmission of the zygomaticotemporal nerve.

The **orbital surface** forms, by its junction with the orbital surface of the maxilla and with the great wing of the sphenoid, part of the floor and lateral wall of the orbit. Usually it presents the orifices of two canals, the **zygomaticoörbital foramina,** for the zygomaticotemporal and zygomaticofacial nerves.

The **frontal process** is thick and serrated, and articulates with the zygomatic process of the frontal bone. On its orbital surface, just within the orbital margin and about 11 mm below the zygomaticofrontal suture is the tuberculum marginale. Its anterior margin, smooth and rounded, is part of the circumference of the orbit. Its superior margin articulates with the frontal bone. Its posterior margin is serrated for articulation with the great wing of the sphenoid and the orbital surface of the maxilla. At the angle of junction of the sphenoidal and maxillary portions, a short, concave, nonarticular part is generally seen; this forms the anterior boundary of the inferior orbital fissure: occasionally, this non-articular part is absent, the fissure then being completed by the junction of the maxilla and sphenoid, or by the interposition of a small sutural bone in the angular interval between them.

The **maxillary process** presents a rough, triangular surface which articulates with the maxilla.

The **temporal process,** long, narrow, and serrated, articulates with the zygomatic process of the temporal.

The posterior superior or temporal **border** of the zygomatic bone, curved like an italic letter *f,* is continuous with the commencement of the temporal line, and with the border of the zygomatic arch; the temporal fascia is attached to it. The posterior inferior or zygomatic border affords attachment by its rough edge to the Masseter.

Ossification.—The zygomatic bone is generally described as ossifying from three centers, one for the malar and two for the orbital portion; these appear about the eight week and fuse about the fifth month of fetal life. After birth the bone is sometimes divided by a horizontal suture into an upper large, and a lower smaller division. In some quadrumana the zygomatic bone consists of two parts, an orbital and a malar.

Articulations.—The zygomatic articulates with four bones: the frontal, sphenoidal, temporal, and maxilla.

THE APPENDICULAR SKELETON

The bones by which the upper and lower limbs are attached to the trunk constitute respectively the shoulder and pelvic girdles. The shoulder girdle or girdle of the superior limb is formed by the scapulae and clavicles. Ventrally it is completed by the cranial end of the sternum, with which the medial ends of the clavicles articulate. Dorsally, it is widely deficient, the scapulae being connected to the trunk by muscles only. The pelvic girdle or girdle of the inferior limb is formed by the hip bones, which articulate with each other, at the symphysis pubis. It is imperfect posteriorly, but the gap is filled in by the sacrum. The pelvic girdle, with the sacrum, is a complete ring, massive and comparatively rigid, in marked contrast to the light and mobile shoulder girdle.

THE BONES OF THE UPPER LIMB (OSSA MEMBRI SUPERIORIS)

The Shoulder Girdle (Cingulum Membri Superioris; Pectoral Girdle)

The Clavicle

The **Clavicle** (*clavicula; collar bone*) (Figs. 4–125, 4–126) forms the ventral portion of the shoulder girdle. It is a long bone, curved somewhat like the italic letter *f,* and is placed nearly horizontally, immediately above the first rib. It articulates medially with the manubrium sterni, and laterally with the acromion of the scapula. It presents a double curvature; at the sternal end the

Sternal extremity   *Acromial extremity*

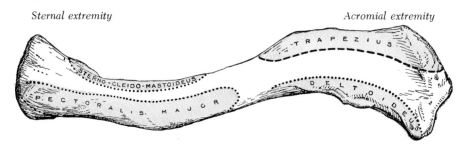

Fig. 4–125.—Left clavicle. Superior surface.

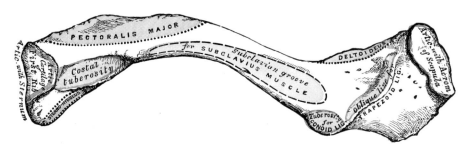

Fig. 4–126.—Left clavicle. Inferior surface. Attachment of articular capsules indicated by solid blue lines.

convexity is directed ventrally, at the acromial end the convexity is dorsal. It has, for descriptive purposes, two extremities, two surfaces, and two borders.

Acromial Extremity (*extremitas acromialis*).—The cranial or superior surface of the lateral third is flat, rough, and marked toward the borders by impressions for the attachment of the Trapezius and Deltoideus; between these impressions a strip of bone is smooth and subcutaneous. The caudal or inferior surface at the dorsal border near the flat end presents a rough eminence, the **conoid tubercle** (*tuberculum conoideum; coracoid tuberosity*), for the attachment of the conoid ligament. An oblique ridge, the **trapezoid line** (*linea trapezoidea; trapezoid ridge*) runs ventrally and laterally from this tubercle and affords attachment to the trapezoid ligament. The dorsal border of the lateral third is convex, rough, thicker than the ventral, and receives the insertion of the Trapezius. The ventral border is concave, thin, rough, and gives origin to the Deltoideus. The distal or acromial end of the clavicle, more on the inferior surface, presents a small, oval surface for articulation with the acromion of the scapula. The circumference of the **acromioclavicular articu-** lar facet is rough for the attachment of the ligaments.

The **medial two-thirds** of the bone are rounded or prismatic and curved with a ventral convexity. The medial third of the **ventral** or anterior **border** presents an elliptical surface for the attachment of the clavicular portion of the Pectoralis major. Between the attachments of the Pectoralis major and Deltoideus the surface is smooth. The **superior border** is smooth and rounded laterally, but becomes rough toward the medial third for the attachment of the Sternocleidomastoideus. The **dorsal** or subclavian **border,** from the coracoid tuberosity to the costal tuberosity forms the posterior boundary of the groove for the Subclavius, and gives attachment to a layer of cervical fascia which envelops the Omohyoideus. The **ventral surface** of the middle portion is smooth, convex, and nearly subcutaneous, being covered only by the Platysma. The medial portion is divided by a narrow subcutaneous area into two parts: an inferior elliptical in form, for the attachment of the Pectoralis major; and a superior for the attachment of the Sternocleidomastoideus. The **dorsal** or **cervical surface** is smooth, and looks toward the root of the neck. It is concave mediolaterally, and is in relation with the transverse scapular vessels. This surface, at the junction of the curves of the bone, is also in relation with the brachial plexus of nerves and the subclavian vessels. It gives attachment, near the sternal extremity, to part of the Sternohyoideus; and presents, near the middle, an oblique foramen directed lateralward, which transmits the chief nutrient artery of the bone. The **inferior** or subclavian **surface** is narrowed medially, but gradually increases in width laterally, and is continuous with the inferior surface of the flat portion. On its medial part is a broad, rough surface, the **costal tuberosity** (*rhomboid impression*), rather more than 2 cm in length, for the attachment of the costoclavicular ligament. The rest of this surface is occupied by a groove, which gives attachment to the Subclavius; the clavipectoral fascia, which splits to enclose the muscle, is attached to the margins of the groove.

The **sternal extremity** (*extremitas sternalis; internal extremity*) of the clavicle is directed medialward, and presents an **articular facet** which articulates with the manubrium sterni through the intervention of an **articular disk.** The inferior part of the facet is continued on to the inferior surface of the bone as a small semioval area for articulation with the cartilage of the first rib. The circumference of the articular surface is rough, for the attachment of numerous ligaments and the superior angle to the articular disk.

Structure.—The clavicle consists of cancellous tissue, enveloped by a compact layer, which is much thicker in the intermediate part than at the extremities of the bone.

In the female, the clavicle is generally shorter, thinner, less curved, and smoother than in the male. In those persons who perform considerable manual labor it becomes thicker and more curved, and its ridges for muscular attachment are prominently marked.

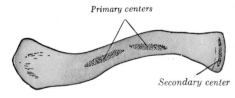

Primary centers

Secondary center

Fig. 4–127.—Diagram showing the three centers of ossification of the clavicle.

Ossification.—The clavicle begins to ossify before any other bone in the body; it is ossified from three centers—viz., two primary centers, a medial and a lateral, for the body, which appear during the fifth or sixth week of fetal life; and a secondary center for the sternal end, which appears about the eighteenth or twentieth year and unites with the rest of the bone about the twenty-fifth year.

The Scapula

The **Scapula** (*shoulder blade*) forms the dorsal part of the shoulder girdle. It is a large, flat, triangular bone, with two surfaces, three borders, and three angles.

Surfaces.—The **costal surface** (*facies costalis; ventral surface*) (Fig. 4–128) presents a broad concavity, the **subscapular fossa.** The medial two-thirds of the fossa are marked by several oblique ridges for the tendinous attachments between the surfaces for the origin of the fleshy fibers of the Subscapularis. The lateral third of the fossa is smooth and also covered by the fibers of this muscle. The fossa is separated from the two extremities of the vertebral border by smooth triangular areas at both the medial and inferior angles, and in the interval between these by a narrow ridge which is often deficient. These triangular areas and the intervening ridge receive the insertion of the Serratus anterior.

The **dorsal surface** (*facies dorsalis*) (Fig. 4–129) is subdivided into two unequal parts by the spine; the smaller portion above the spine is called the supraspinatous fossa, and that below it the infraspinatous fossa.

The **supraspinatous fossa** is a smooth concavity, which is broader at its vertebral than at its humeral end; its medial two-thirds give origin to the Supraspinatus.

The **infraspinatous fossa** has a shallow concavity toward the cranial part of its vertebral margin, a prominent convexity at its center, and a deep groove near the axillary border. The medial two-thirds of the fossa give origin to the Infraspinatus; the lateral third is covered by this muscle. An elevated ridge runs along the surface near the axillary border, down to the inferior angle, and around on the vertebral border for about 2.5 cm. The ridge serves for the attachment of a fibrous septum which sepa-

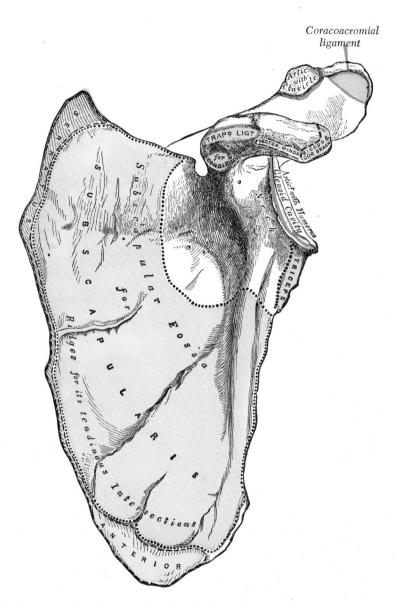

FIG. 4–128.—Left scapula. Costal surface. Attachment of articular capsules indicated by solid blue lines.

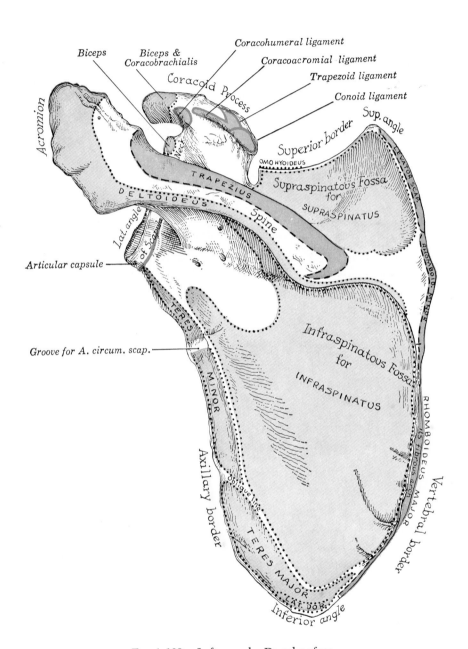

Fig. 4–129.—Left scapula. Dorsal surface.

rates the Infraspinatus from the Teres major and Teres minor. The surface between the ridge and the axillary border is narrow in the cranial two-thirds of its extent, and is crossed near its center by a groove for the passage of the scapular circumflex vessels; it affords attachment to the Teres minor. Its caudal third presents a broader, somewhat triangular surface, which gives origin to the Teres major, and over which the Latissimus dorsi glides; frequently the latter muscle takes origin by a few fibers from this part. An oblique line from the axillary border to the elevated ridge has attached to it a fibrous septum which separates the Teres muscles from each other.

The **spine** (*spina scapulae*) is a prominent plate of bone, which projects from the dorsal surface of the scapula and separates the supra- and infraspinatous fossae. It begins at the vertebral border by a smooth, triangular area over which the tendon of insertion of the Trapezius glides, and gradually becoming more elevated, ends in the acromion, which overhangs the shoulder joint. Its cranial surface is concave; it assists in forming the supraspinatous fossa, and gives origin to part of the Supraspinatus. Its caudal surface forms part of the infraspinatous fossa, gives origin to a portion of the Infraspinatus, and presents near its center the orifice of a nutrient canal. The **dorsal border,** or **crest of the spine,** is broad, and presents two lips with an intervening rough interval. The Trapezius is attached to the cranial lip, and a rough tubercle is generally seen on that portion of the spine which receives the tendon of insertion of the caudal part of this muscle. The Deltoideus is attached to the whole length of the caudal lip. The interval between the lips is subcutaneous and partly covered by the tendinous fibers of these muscles. The **lateral border** or **base** is slightly concave; its edge, thick and round, is continuous with the under surface of the acromion and the neck of the scapula. It forms the medial boundary of the **great scapular notch,** which connects the supra- and infraspinatous fossae.

The **acromion** forms the summit of the shoulder, and overhangs the glenoid cavity. Its cranial surface is convex, rough, and gives attachment to some fibers of the Deltoideus, and in the rest of its extent is sub-

cutaneous. Its caudal surface is smooth and concave. Its lateral border is thick and irregular, and presents three or four tubercles for the tendinous origins of the Deltoideus. Its medial border, shorter than the lateral, is concave, gives attachment to a portion of the Trapezius, and presents about its center a small, oval **surface for articulation** with the acromial end of the clavicle. Its apex, which corresponds to the point of meeting of these two borders is thin, and has attached to it the coracoacromial ligament.

Borders.—The **superior border** (*margo superior*) of the scapula is the shortest and thinnest; it is concave, and extends from the medial angle to the base of the coracoid process. At its lateral part is a semicircular notch, the **scapular notch,** formed partly by the base of the coracoid process. This notch is converted into a foramen by the superior transverse ligament, sometimes ossified, and serves for the passage of the suprascapular nerve. The adjacent part of the superior border affords attachment to the Omohyoideus. The **axillary border** (NA *margo lateralis*) is the thickest of the three. It begins at the margin of the glenoid cavity, and inclines obliquely to the inferior angle. Immediately below the glenoid cavity is a rough impression, the **infraglenoid tubercle,** about 2.5 cm in length, which gives origin to the long head of the Triceps brachii. The caudal third is thin and sharp, and serves for the attachment of a few fibers of the Teres major and Subscapularis. The **vertebral border** (PNA *margo medialis*) is the longest of the three, and extends from the superior to the inferior angle. It is arched and the portion of it cranial to the spine forms an obtuse angle with the caudal part. This border presents a ventral and a dorsal lip, and an intermediate narrow area. The **ventral lip** affords attachment to the Serratus anterior; the **dorsal lip,** to the Supraspinatus above the spine, the Infraspinatus below; the area between the two lips, to the Levator scapulae cranial to the triangular surface at the commencement of the spine, to the Rhomboideus minor on the edge of that surface, and to the Rhomboideus major caudal to it; this last is attached by means of a fibrous arch, connected to the triangular surface at the base of the spine, and to the caudal part of the border.

Angles.—The **superior angle** (*medial angle*) is thin, smooth, rounded, and gives attachment to a few fibers of the Levator scapulae. The **inferior angle,** thick and rough, affords attachment to the Teres major and frequently to a few fibers of the Latissimus dorsi. The **lateral angle** is broadened into a thick process, sometimes called the **head of the scapula,** which is connected with the rest of the bone by a slightly constricted **neck.** This is the part of the scapula that enters into the shoulder joint and is hollowed out to form the **glenoid cavity.** The surface of this shallow cavity faces laterally and in the intact body is covered with articular cartilage for reception of the head of the humerus. Its margins or lips are covered by a fibrocartilaginous ring, the glenoid labrum, which broadens and deepens the cavity. At its apex, near the base of the coracoid process, is a slight elevation, the **supraglenoid tubercle,** to which the long head of the Biceps brachii is attached.

The **coracoid process** (*processus coracoideus*) (Fig. 4-128), shaped like a raven's beak, is a thick, curved process attached by a broad base at the superior border near the neck of the scapula. Its concave surface faces laterally and is smooth to accommodate the passage of the tendon of the Subscapularis. The distal portion is horizontal and its convex outer surface is rough and irregular for the attachment of the Pectoralis minor. Its medial and lateral borders are rough; the former gives attachment to the Pectoralis minor and the latter to the coracoacromial ligament; the **apex** is embraced by the conjoined tendon of origin of the Coracobrachialis and short head of the Biceps brachii and gives attachment to the clavipectoral fascia. On the medial part of the root of the coracoid process is a rough impression for the attachment of the conoid ligament; and running from it obliquely to the convex surface of the horizontal portion, is an elevated ridge for the attachment of the trapezoid ligament.

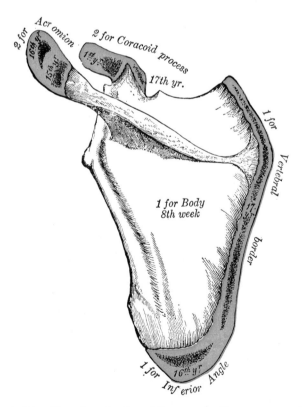

Fɪɢ. 4–130.—Plan of ossification of the scapula. From seven centers.

Structure.—The head, processes, and the thickened parts of the bone, contain cancellous tissue; the rest consists of a thin layer of compact tissue. The central part of the supraspinatous and infraspinatous fossae, but especially the former, are usually so thin as to be semitransparent; occasionally the bone is deficient in this situation, and the adjacent muscles are separated only by fibrous tissue.

Ossification (Fig. 4–130).—The scapula is ossified from seven or more centers: one for the body, two for the coracoid process, two for the acromion, one for the vertebral border, and one for the inferior angle.

Ossification of the body begins about the second month of fetal life, by the formation of an irregular quadrilateral plate of bone adjacent to the glenoid cavity. This plate extends so as to form the chief part of the bone, the spine growing up from its dorsal surface about the third month. At birth, a large part of the scapula is osseous, but the glenoid cavity, the coracoid process, the acromion, the vertebral border, and the inferior angle are cartilaginous. From the fifteenth to the eighteenth month after birth, ossification takes place in the middle of the coracoid process, which as a rule becomes joined with the rest of the bone about the fifteenth year. Between the fourteenth and twentieth years, ossification of the remaining parts takes place in quick succession, and usually in the following order: first, in the root of the coracoid process, in the form of a broad scale; secondly, near the base of the acromion; thirdly, in the inferior angle and contiguous part of the vertebral border; fourthly, near the extremity of the acromion; fifthly, in the vertebral border. The base of the acromion is formed by an extension from the spine; the two sepa-

rate nuclei of the acromion unite, and then join with the extension from the spine. The cranial third of the glenoid cavity is ossified from a separate center (subcoracoid), which makes its appearance between the tenth and eleventh years and joins between the sixteenth and eighteenth. Further, an epiphyseal plate appears for the caudal part of the glenoid cavity, while the tip of the coracoid process frequently presents a separate nucleus. These various epiphyses are joined to the bone by the twenty-fifth year. Failure of bony union between the acromion and spine sometimes occurs, the junction being effected by fibrous tissue, or by an imperfect articulation; in some cases of supposed fracture of the acromion with ligamentous union, it is probable that the detached segment was never united to the rest of the bone.

The Humerus

The **Humerus** (*arm bone*) (Figs. 4–132, 4–133) is the longest and largest bone of the upper limb; it is divisible into a body, a head, and a condyle.

The **head** (*caput humeri*) (Fig. 4–131), nearly hemispherical in form, articulates with the glenoid cavity of the scapula. The circumference of its articular surface is slightly constricted and is termed the anatomical neck, in contradistinction to a constriction below the tubercles called the surgical neck which is frequently the seat of fracture. Fracture of the anatomical neck rarely occurs.

The **anatomical neck** (*collum anatomi-*

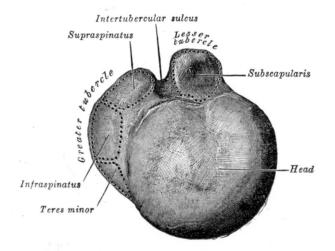

FIG. 4–131.—The upper end of the left humerus. Superior aspect.

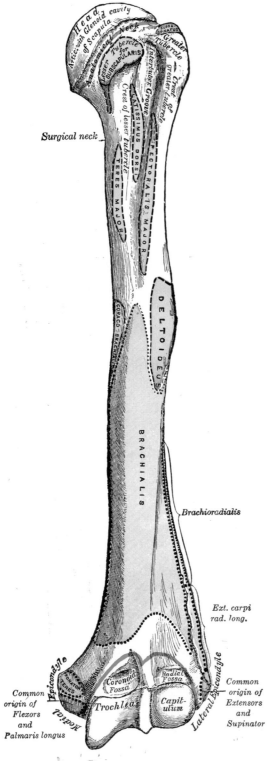

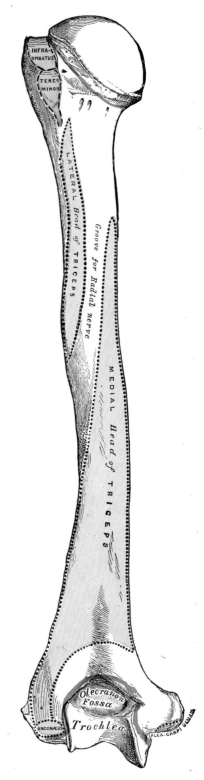

Fig. 4–132.—Left humerus. Ventral view.
Articular capsule in blue line.

Fig. 4–133.—Left humerus. Dorsal view.
Articular capsule in blue line.

cum) is obliquely directed, forming an obtuse angle with the body. It is best marked in the distal half of its circumference; in the proximal half it is represented by a narrow groove separating the head from the tubercles. It affords attachment to the articular capsule of the shoulder joint, and is perforated by numerous vascular foramina.

The **greater tubercle** (*tuberculum majus; greater tuberosity*) is situated lateral to the head and lesser tubercle. Its proximal surface is rounded and marked by three flat impressions: the highest of these gives insertion to the Supraspinatus, the middle to the Infraspinatus, the lowest to the Teres minor. The lateral surface of the greater tubercle is convex, rough, and merges distally into the lateral surface of the body.

The **lesser tubercle** (*tuberculum minus; lesser tuberosity*), although smaller, is as prominent as the greater, projecting medially just distal to the neck. Its ventral surface serves for the insertion of the tendon of the Subscapularis.

The tubercles are separated from each other by a deep groove, the **intertubercular groove** (*bicipital groove*), which lodges the tendon of the long head of the Biceps brachii and transmits a branch of the anterior humeral circumflex artery to the shoulder joint. In the intact body its proximal part is coated with a thin layer of cartilage and enclosed by a prolongation of the synovial membrane of the shoulder joint. Its distal portion becomes shallow and receives the insertion of the tendon of the Latissimus dorsi. Its lips are called, respectively, the **crests** of the **greater** and **lesser tubercles** (*bicipital ridges*).

The **body** or **shaft** (*corpus humeri*) is almost cylindrical proximally, prismatic and flattened distally, and has two borders and three surfaces (Figs. 4–132, 4–133).

The **lateral border** (*margo lateralis*) runs from the dorsal part of the greater tubercle to the lateral epicondyle, and separates the ventrolateral (NA *facies anterior lateralis*) from the dorsal surface (NA *facies posterior*). Its proximal half is rounded and indistinctly marked, serving for the attachment of part of the insertion of the Teres minor, and the origin of the lateral head of the Triceps brachii; its center is traversed by a broad but shallow oblique depression,

the **radial sulcus** (*sulcus nervi radialis; musculospiral groove*). Its distal part forms a prominent, rough margin, the **lateral supracondylar ridge,** which presents an anterior lip for the origin of the Brachioradialis and Extensor carpi radialis longus, a posterior lip for the Triceps brachii, and an intermediate ridge for the attachment of the lateral intermuscular septum.

The **medial border** extends from the lesser tubercle to the medial epicondyle. Its proximal third consists of the **crest of the lesser tubercle,** which receives the insertion of the tendon of the Teres major. About its center is a rough impression for the insertion of the Coracobrachialis, and just distal to this is the entrance of the nutrient canal. Sometimes there is a second nutrient canal at the commencement of the radial sulcus. The distal third of this border is raised into a slight ridge, the **medial supracondylar ridge,** which becomes very prominent distally; it presents an anterior lip for the origin of the Brachialis, a posterior lip for the medial head of the Triceps brachii, and an intermediate ridge for the attachment of the medial intermuscular septum.

Surfaces.—The **ventral surface** is divided longitudinally into two parts by an oblique ridge beginning laterally at the greater tubercle and ending near the medial epicondyle. The **ventrolateral surface** (NA *facies anterior lateralis*) receives the insertion of the pectoralis major along the distal part of the crest of the greater tubercle. Lateral to this it is smooth, rounded, and covered by the Deltoideus. Distally there is a broad area, slightly concave, which gives origin to part of the Brachialis. About the middle of this surface is a rough, triangular elevation, the **deltoid tuberosity** for the insertion of the Deltoideus; beyond this is the radial sulcus, directed obliquely distalward marking the path of the radial nerve and profunda artery.

The **ventromedial surface** proximally receives the tendon of insertion of the Latissimus dorsi in the intertubercular groove. Somewhat distal and medial to this, at the medial border, is the insertion of the Teres major. The distal part of the surface gives origin to the Brachialis.

The **dorsal surface** appears somewhat twisted and is almost completely covered

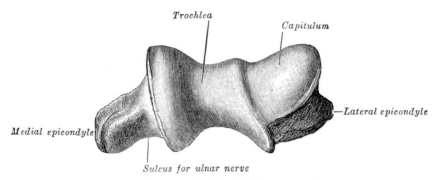

Trochlea

Capitulum

Lateral epicondyle

Medial epicondyle

Sulcus for ulnar nerve

Fig. 4–134.—The lower end of the left humerus. Inferior aspect.

by the lateral and medial heads of the Triceps brachii, the former arising proximal, the latter distal to the radial sulcus.

The **condyle** (*condylus humeri; the lower extremity*) (Fig. 4–134).—The distal extremity is flattened, and ends in a broad, articular surface which is divided into two parts by a slight ridge. The lateral portion of this surface consists of a smooth, rounded eminence, named the **capitulum** of the humerus; it articulates with the fovea on the head of the radius. On the medial side of this eminence is a shallow groove, in which is received the medial margin of the head of the radius, when the forearm is flexed. The slight depression, the **radial fossa**, which receives the anterior border of the head of the radius, when the forearm is flexed. The medial portion of the articular surface is named the **trochlea**, and presents a deep groove between two well-marked borders. The lateral border separates it from the groove which articulates with the margin of the head of the radius. The medial border is thicker and projects farther distally than the lateral. The grooved portion fits accurately within the trochlear notch of the ulna; it is broader and deeper on the dorsal than on the ventral aspect of the bone. Ventral and proximal to the trochlea is a small depression, the **coronoid fossa** which receives the coronoid process of the ulna during flexion of the forearm. Dorsal to the trochlea is a deep triangular depression, the **olecranon fossa**, into which the summit of the olecranon is received in extension of the forearm. These fossae are separated from one another by a thin, transparent lamina of bone which is sometimes perforated to produce a **supratrochlear foramen**; they are

lined in the intact body by the synovial membrane of the elbow joint, and their margins afford attachment to the ventral and dorsal ligaments of this articulation.

The **lateral epicondyle** is a small, tuberculated eminence giving attachment to the radial collateral ligament of the elbow joint, and to the common tendon of origin of the Supinator and Extensor muscles of the forearm. The **medial epicondyle**, larger and more prominent than the lateral, gives attachment to the ulnar collateral ligament of the elbow joint, to the Pronator teres, and to the common tendon of origin of the Flexor muscles of the forearm; the ulnar nerve runs in a groove on the dorsum of this epicondyle.

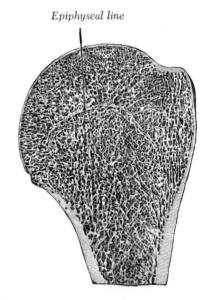

Epiphyseal line

Fig. 4–135.—Longitudinal section of head of left humerus.

Structure.—The extremities consist of cancellous tissue, covered with a thin, compact layer (Fig. 4–135); the body is composed of a cylinder of compact tissue, thicker at the center than toward the extremities, and contains a large medullary canal which extends along its whole length.

Ossification (Figs. 4–137, 4–138).—The humerus is ossified from eight centers, one for each of the following parts: The body, the head, the greater tubercle, the capitulum, the trochlea, and one for each epicondyle. The center for the body appears near the middle of the bone in the eighth week of fetal life, and soon extends toward the extremities. At birth the humerus is ossified in nearly its whole length, only the extremities remaining cartilaginous. During the first year, sometimes before birth, ossification commences in the head of the bone, and during the third year the center for the greater tubercle, and during the fifth that for the lesser tubercle, make their appearance. By the sixth year the centers for the head and tubercles have joined, so as to form a single

large epiphysis, which fuses with the body about the twentieth year. The conical shape of the proximal end of the diaphysis, where the epiphyseal cap fits over it as shown by the epiphyseal line in Figure 4–135, is an adult condition. In the fetus and newborn, the end of the diaphysis is flat; the conical shape is established during the first year and gradually reaches the height of the adult by the twelfth year. The distal end of the humerus is ossified as follows: At the end of the second year ossification begins in the capitulum, and extends medialward to form the chief part of the articular end of the bone; the center for the medial part of the trochlea appears about the age of twelve. Ossification begins in the medial epicondyle about the fifth year, and in the lateral about the thirteenth or fourteenth year. About the sixteenth or seventeenth year, the lateral epicondyle and both portions of the articulating surface, having already joined, unite with the body, and at the eighteenth year the medial epicondyle becomes joined to it.

Torsion of the Humerus is the term used to

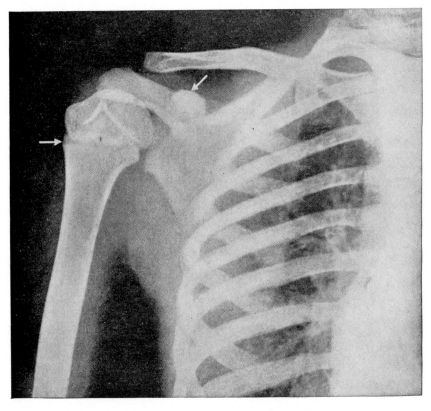

Fig. 4–136.—Shoulder of a child aged six years. The upper arrow indicates the coracoid process; the lower arrow indicates the epiphyseal line. Note that the upper end of the diaphysis is conical and projects into the center of the epiphysis. The centers for the head of the humerus and the tuberosities have fused to form a single epiphysis.

refer to the twisting of the bone about its longitudinal axis, that is, the change in the relation of the transverse, articular axes of the two ends of the bone to each other. This is entirely distinct and independent of the rotation of the whole limb referred to in the section on Embryology. The angle formed by the axes of the two ends of the bone is called the torsion angle. According to Evans and Krahl (1945), the angle measures 74° in the adult bone, the distal end having been twisted medially in relation to the proximal end. Two factors have been identified; a primary torsion which is hereditary and evolutionary, and a secondary which is ontogenetic. The primary torsion is found in all tetrapods, and in the evolutionary scale of mammals, the angle increases from 27° to 74°. The secondary torsion takes place at the proximal epiphyseal junction before the age of twenty, while cartilage is still present, and apparently is due to muscular pull since the lateral rotators are attached proximal and the medial rotators distal to the junction. There is no torsion in the diaphysis; the apparent twisting of this part of the bone is due to the surface markings associated with the spiral course of the radial nerve.

Variations.—A small, hook-shaped process of bone, the **supracondylar process,** varying from

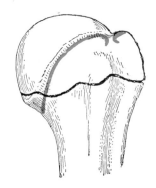

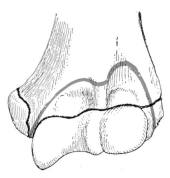

Fig. 4–138.—Epiphyseal lines of humerus in a young adult. Ventral aspect. The lines of attachment of the articular capsules are in blue.

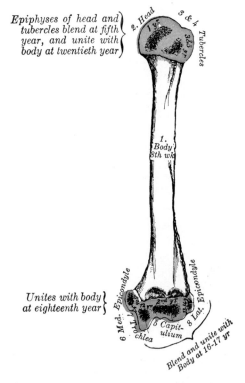

Epiphyses of head and tubercles blend at fifth year, and unite with body at twentieth year

1. Body 8th wk

Unites with body at eighteenth year

Fig. 4–137.—Plan of ossification of the humerus.

2 to 20 mm in length, is not infrequently found projecting from the ventromedial surface of the body of the humerus 5 cm cranial to the medial epicondyle. It is curved distally and ventrally, and its pointed end is connected to the medial border, just cranial to the medial epicondyle, by a fibrous band which gives origin to a portion of the Pronator teres; through the arch completed by this fibrous band the median nerve and brachial artery pass, when these structures deviate from their usual course. Sometimes the nerve alone is transmitted through it, or the nerve may be accompanied by the ulnar artery, in cases of high division of the brachial. A well-marked groove is usually found dorsal to the process, in which the nerve and artery are lodged. This arch is the homologue of the **supracondyloid foramen** found in many animals, and probably serves in them to protect the nerve and artery from compression during the contraction of the muscles in this region.

The Ulna

The **Ulna** (*elbow bone*) (Figs. 4–140, 4–146) occupies the medial or little finger side of the forearm, and lies parallel with the radius. It is divisible into a body and two extremities. Its proximal extremity, of great thickness and strength, forms a large part of the elbow joint; the bone diminishes in size distally.

The **proximal extremity** (Fig. 4–139) presents two curved processes, the olecranon and coronoid processes; and two concave, articular cavities, the trochlear and radial notches.

The **olecranon** (*olecranon process*) is a large, thick, curved eminence, forming the point of the elbow. At the summit is a prominent lip which is received into the olecranon fossa of the humerus when the forearm is extended. Its base is narrow

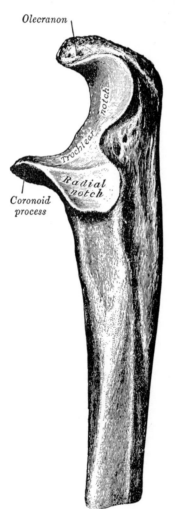

Olecranon

Trochlear notch

Radial notch

Coronoid process

Fig. 4–139.—Proximal extremity of left ulna. Lateral aspect.

where it joins the body. Its posterior surface is triangular, smooth, subcutaneous, and covered by a bursa. Its proximal surface is marked by a rough impression for the insertion of the Triceps brachii, and near the margin by a slight transverse groove for the attachment of part of the posterior ligament of the elbow joint. Its anterior surface is smooth, concave, and forms part of the trochlear notch. Its borders serve for the attachment of the ulnar collateral ligament medially, and the posterior ligament laterally. Part of the Flexor carpi ulnaris arises from the medial border; the Anconeus is attached to the lateral border.

The **coronoid process** (*processus coronoideus*) projects from the proximal and anterior part of the body of the bone. Its apex is pointed, slightly curved, and in flexion of the forearm is received into the coronoid fossa of the humerus. Its proximal surface is smooth, concave, and forms the lower part of the trochlear notch. Its distal surface is concave, rough, and at its junction with the front of the body is a rough eminence, the **tuberosity of the ulna**, which gives insertion to the Brachialis and the oblique cord. Its lateral surface presents a narrow, oblong, articular depression, the **radial notch**. Its medial surface, by its prominent, free margin, serves for the attachment of part of the ulnar collateral ligament. At the front part of this surface is a small rounded eminence for the origin of one head of the Flexor digitorum superficialis; and descending from it is a ridge which gives origin to one head of the Pronator teres. Frequently, the Flexor pollicis longus arises from the distal part of the coronoid process by a rounded bundle of muscular fibers.

The **trochlear notch** (*incisura trochlearis; semilunar notch; greater sigmoid cavity*).— The trochlear notch is a large depression, formed by the olecranon and the coronoid processes, for the articulation with the trochlea of the humerus.

The **radial notch** (*incisura radialis; lesser sigmoid cavity*) is a narrow, oblong, articular depression on the lateral side of the coronoid process; it receives the circumferential articular surface of the head of the radius. Its prominent extremities serve for the attachment of the annular ligament.

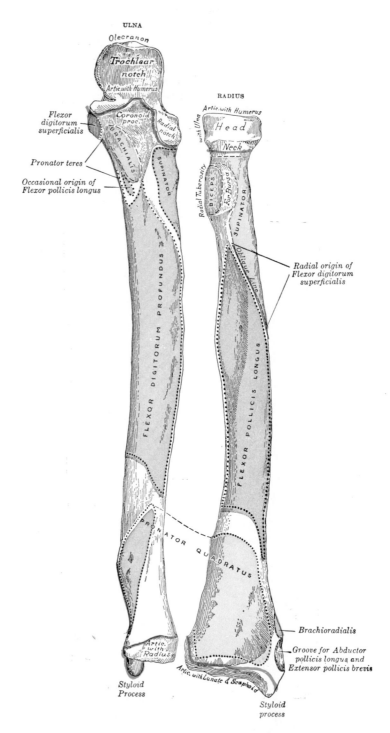

Fig. 4–140.—The bones of the left forearm. Anterior aspect. Attachment of articular capsules outlined in blue.

The **body or shaft** (*corpus ulnae*) is curved so as to be convex dorsally and laterally; its central part is straight; its distal part is rounded, smooth, and a little concave lateralward. It tapers gradually and has three borders and three surfaces.

The **anterior border** (*margo anterior; volar border*) begins at the prominent medial angle of the coronoid process, and ends at the styloid process. Its proximal part, well defined, and its middle portion smooth, and rounded, give origin to the Flexor digitorum profundus; its distal fourth, called the **pronator ridge**, serves for the origin of the Pronator quadratus. This border separates the anterior from the medial surface.

The **dorsal border** (NA *margo posterior; posterior border*) begins at the apex of the triangular subcutaneous surface of the olecranon, and ends at the styloid process; it is well marked in the proximal three-fourths, and gives attachment to an aponeurosis which affords a common origin to the Flexor carpi ulnaris, the Extensor carpi ulnaris, and the Flexor digitorum profundus; its distal fourth is smooth and rounded. This border separates the medial from the posterior surface.

The **interosseous crest** (*margo interossea; external or interosseous border*) begins proximally by the union of two lines which converge from the extremities of the radial notch and enclose between them a triangular space for the origin of part of the Supinator; it ends at the head of the ulna. Its proximal part is sharp, its distal fourth smooth and rounded. This crest gives attachment to the interosseous membrane, and separates the anterior from the posterior surface.

The **anterior surface** (*facies anterior; volar surface*) is much broader proximally where it gives origin to the Flexor digitorum profundus; its distal fourth, covered by the Pronator quadratus, is separated from the remaining portion by an oblique ridge, which marks the extent of origin of the Pronator quadratus. At the junction of the proximal with the middle third of the bone is the nutrient canal.

The **dorsal surface** (NA *facies posterior; posterior surface*) is broad and concave above; convex and somewhat narrower in the middle; narrow, smooth, and rounded

distally. On its proximal part is an oblique ridge, which runs distally from the dorsal end of the radial notch to the dorsal border; the triangular surface proximal to this ridge receives the insertion of the Anconeus, while the proximal part of the ridge affords attachment to the Supinator. Distally the surface is subdivided by a longitudinal ridge, sometimes called the perpendicular line, into two parts: the medial part is smooth, and covered by the Extensor carpi ulnaris; the lateral portion, wider and rougher, gives origin to the Supinator, the Abductor pollicis longus, the Extensor pollicis longus, and the Extensor indicis.

The **medial surface** (*facies medialis; internal surface*) is broad and concave proximally, narrow and convex distally. Its proximal three-fourths give origin to the Flexor digitorum profundus; its distal fourth is subcutaneous.

The **distal extremity** (Fig. 4–142) of the ulna is small, and presents two eminences; the lateral and larger is a rounded, articular eminence, termed the **head of the ulna;** the medial, narrower and more projecting, is a non-articular eminence, the **styloid process.** The head presents an articular surface, part of which articulates with the proximal surface of the triangular articular disk which separates it from the wrist joint; the remaining portion is received into the ulnar notch of the radius. The styloid process projects from the medial and posterior part of the bone; its rounded end affords attachment to the ulnar collateral ligament of the wrist joint. The head is separated from the styloid process by a depression for the attachment of the apex of the triangular articular disk, and posteriorly, by a shallow groove for the tendon of the Extensor carpi ulnaris.

Structure.—The long, narrow medullary cavity is enclosed in a strong wall of compact tissue which is thickest along the interosseous border and dorsal surface. At the extremities the compact layer becomes thinner. The compact layer is continued onto the back of the olecranon as a plate of close spongy bone with parallel lamellae. From the inner surface of this plate and the compact layer distal to it trabeculae arch toward the olecranon and coronoid, and cross other trabeculae passing posteriorly over the medullary cavity from the proximal part of the shaft distal to the coronoid.

Distal to the coronoid process there is a small area of compact bone from which trabeculae curve proximally to end obliquely to the surface of the semilunar notch which is coated with a thin layer of compact bone. The trabeculae at the distal end have more longitudinal direction.

Ossification (Figs. 4–141, 4–143).—The ulna is ossified from three centers: one each for the

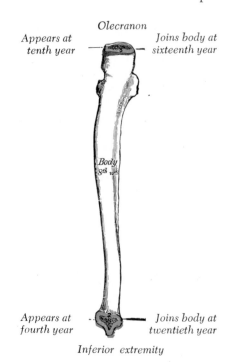

Olecranon

Appears at tenth year — *Joins body at sixteenth year*

Body 8th wk

Appears at fourth year — *Joins body at twentieth year*

Inferior extremity

Fig. 4–143.—Plan of ossification of the ulna. From three centers.

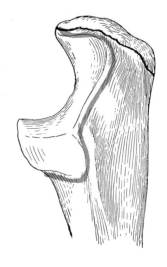

Fig. 4–141.—Epiphyseal lines of ulna in a young adult. Lateral aspect. The lines of attachment of the articular capsules are in blue.

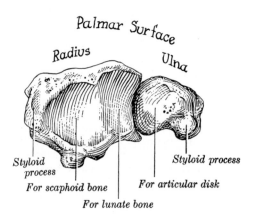

Palmar Surface

Radius *Ulna*

Styloid process *Styloid process*

For scaphoid bone | *For articular disk*

For lunate bone

Fig. 4–142.—The lower ends of the right radius and ulna. Inferior aspect.

body, the distal extremity, and the top of the olecranon. Ossification begins near the middle of the body about the eighth week of fetal life, and soon extends through the greater part of the bone. At birth the ends are cartilaginous. About the fourth year, a center appears in the middle of the head and soon extends into the styloid process. About the tenth year, a center appears in the olecranon near its extremity, the chief part of this process being formed by a proximal extension of the body. The proximal epiphysis joins the body about the sixteenth, the distal about the twentieth year.

Articulations.—The ulna articulates with the humerus and radius. Although it is customary to exclude the ulna from the wrist joint, because of the intervention of the articular disk, it should be noticed that it enters into this joint in much the same way as the clavicle into the sternoclavicular and the mandible into the temporomandibular joints.

The Radius

The **Radius** (Figs. 4–140, 4–146) lies parallel with the ulna and is on the lateral, that is, the thumb side of the forearm. Its proximal end is small, and forms only a small part of the elbow joint, but its distal end is large, and forms the chief part of the wrist joint. It has a body and two extremities.

The **proximal extremity** presents a head, neck, and tuberosity. The **head** (*caput radii*) is cylindrical, and on its proximal surface is a shallow cup or **fovea** for articulation with the capitulum of the humerus. The circumference of the head is smooth; it is broad medially where it articulates with the radial notch of the ulna, narrow in the rest of its extent where it is embraced by the annular ligament. The head is supported on a round, smooth, and constricted portion called the **neck,** on the back of which is a slight ridge for the insertion of part of the Supinator. Distal to the neck, on the medial side, is an eminence, the **radial tuberosity;** its surface is divided into a posterior, rough portion, for the insertion of the tendon of the Biceps brachii, and an anterior, smooth portion, on which a bursa is interposed between the tendon and the bone.

The **body** or **shaft** (*corpus radii*) is narrower proximally, and slightly curved, convex laterally. It presents three borders and three surfaces.

The **anterior border** (*margo anterior; volar border*) extends from the lower part of the tuberosity to the anterior part of the base of the styloid process, and separates the anterior from the lateral surface. Its proximal third is elevated to form the **oblique line** of the radius, which gives origin to the Flexor digitorum superficialis and Flexor pollicis longus. The surface proximal to the line receives the insertion of part of the Supinator. The distal fourth of this border receives the insertion of the Pronator quadratus, and attachment of the dorsal carpal ligament. It ends in a small tubercle, into which the tendon of the Brachioradialis is inserted.

The **dorsal border** (PNA *margo posterior; posterior border*) begins proximally at the neck, and ends distally at the posterior part of the base of the styloid process; it separates the posterior from the lateral surface. It is indistinct at the ends, but well marked in the middle third of the bone.

The **interosseous crest** (*margo interossea; internal or interosseous border*) where it begins at the tuberosity is rounded and indistinct; it becomes sharp and prominent, and at its distal part divides into two ridges which are continued to the anterior and posterior margins of the ulnar notch. The triangular surface between the ridges receives the insertion of part of the Pronator quadratus. This crest separates the anterior from the posterior surface, and gives attachment to the interosseous membrane.

The **anterior surface** (*facies anterior; volar surface*) is concave in its proximal three-fourths, and gives origin to the Flexor pollicis longus; it is broad and flat in its distal fourth, and receives the insertion of the Pronator quadratus. A prominent ridge limits the insertion of the Pronator quadratus distally, and between this and the inferior border is a triangular rough surface for the attachment of the anterior radiocarpal ligament. At the middle of the anterior surface is the nutrient foramen.

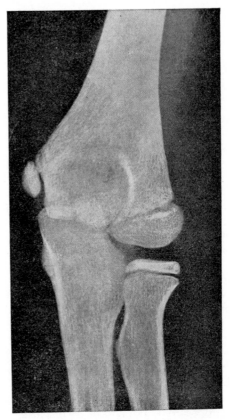

Fig. 4–144.—Elbow of a child aged eleven years. Frontal view. The upper epiphysis of the radius, the epiphysis for the medial epicondyle, and the center for the capitulum and lateral part of the trochlea can be recognized without difficulty.

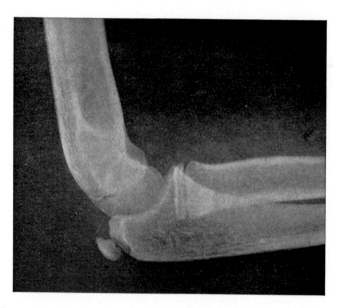

Fig. 4–145.—Elbow of a child aged ten years. Lateral view. The upper epiphysis of the radius, the olecranon epiphysis and the center for the capitulum and the lateral part of the trochlea can be recognized without difficulty.

The **dorsal surface** (NA *facies posterior; posterior surface*) is convex, smooth in the proximal third of its extent, and covered by the Supinator. Its middle third is broad, slightly concave, and gives origin to the Abductor pollicis longus and Extensor pollicis brevis. Its distal third is broad, convex, and covered by the tendons of the muscles which subsequently run in the grooves on the distal end of the bone.

The **lateral surface** (*facies lateralis; external surface*) is convex throughout its entire extent. Its upper third receives the insertion of the Supinator. About its center is a rough ridge, for the insertion of the Pronator teres. Its lower part is narrow, and covered by the tendons of the Abductor pollicis longus and Extensor pollicis brevis.

The **distal extremity** is large, club-shaped, and provided with two articular surfaces—one on the distal surface for the carpus, and another at the medial side for the ulna. The **carpal articular surface** is concave (Fig. 4–142), smooth, and divided by a slight anteroposterior ridge into two parts. Of these, the lateral, articulates with the scaphoid bone; the medial, with the lunate bone. The articular surface for the ulna is called the **ulnar notch** (*sigmoid cavity*) of the radius. These two articular surfaces are separated by a prominent ridge, to which is attached the base of the triangular articular disk which separates the wrist joint from the distal radioulnar articulation. This end of the bone has three non-articular surfaces: anterior, dorsal, and lateral. The palmar or **anterior surface**, rough and irregular, affords attachment to the radiocarpal ligament. The **dorsal** or posterior **surface** is convex, affords attachment to the dorsal radiocarpal ligament, and is marked by three grooves. Beginning at the lateral or thumb side, the first groove is broad, but shallow, and subdivided into two by a slight ridge; the lateral of these two transmits the tendon of the Extensor carpi radialis longus, the medial the tendon of the Extensor carpi radialis brevis. The second is deep but narrow and bounded laterally by a sharply defined oblique ridge; it transmits the tendon of the Extensor pollicis longus. The third is broad, for the passage of the tendons of the Extensor indicis and Extensor digitorum. The **lateral surface** is prolonged distally into a strong, conical projection, the **styloid process**, which gives attachment by its base to the tendon of the Brachioradialis, and by its apex to the radial collateral liga-

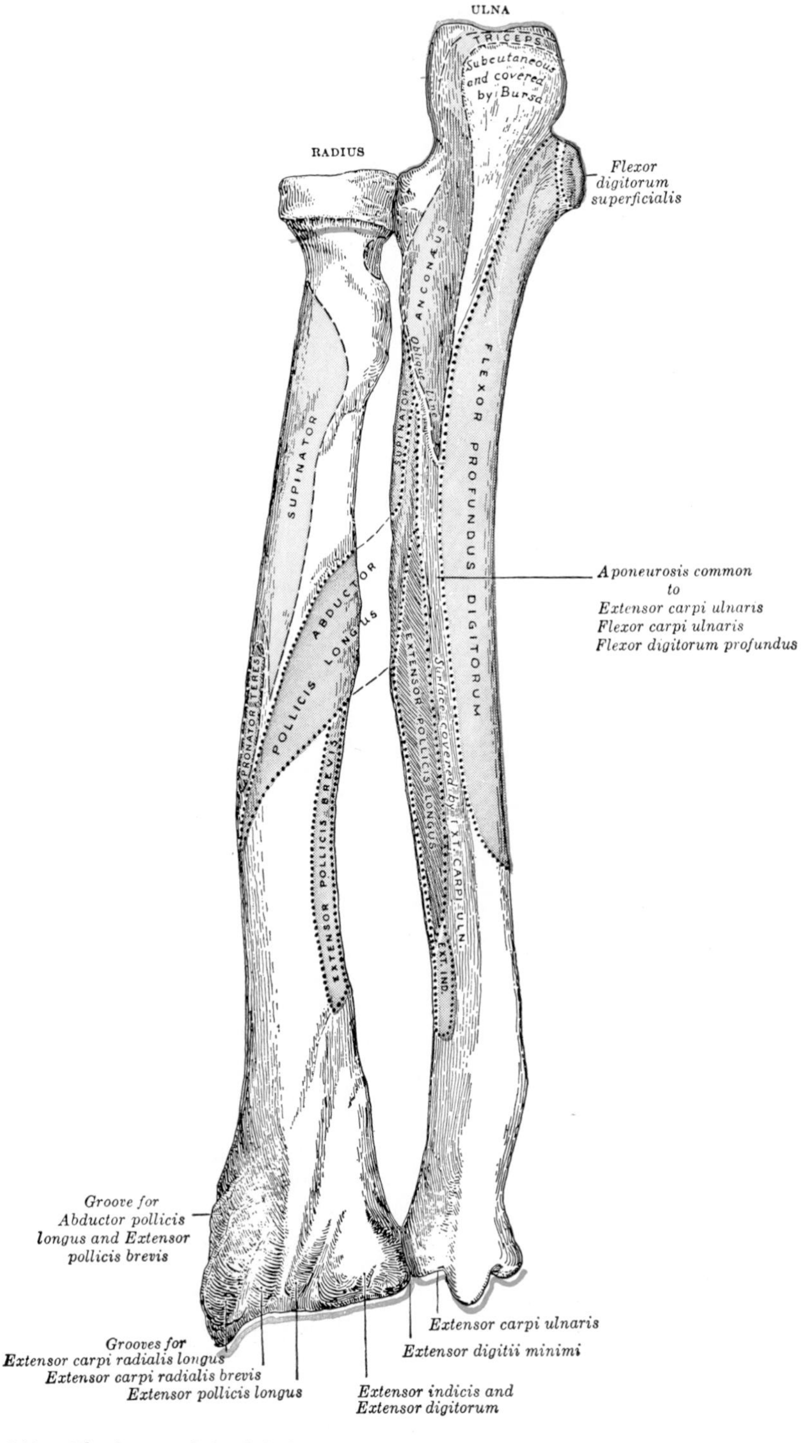

FIG. 4–146.—The bones of the left forearm. Dorsal aspect. Attachment of articular
capsules outlined in blue.

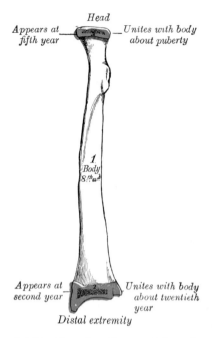

Head

Appears at fifth year — *Unites with body about puberty*

1 Body 8th w^k

Appears at second year — *Unites with body about twentieth year*

Distal extremity

Fig. 4–147.—Plan of ossification of the radius. From three centers.

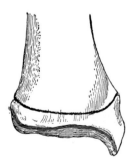

Fig. 4–148.—Epiphyseal lines of radius in a young adult. Anterior aspect. The line of attachment of the articular capsule of the wrist joint is in blue.

ment of the wrist joint. The lateral surface of this process is marked by a flat groove, for the tendons of the Abductor pollicis longus and Extensor pollicis brevis.

Structure.—The long narrow medullary cavity is enclosed in a strong wall of compact tissue which is thickest along the interosseous border and thinnest at the extremities except over the cup-shaped articular surface (fovea) of the head where it is thickened. The trabeculae of the spongy tissue are somewhat arched at the proximal end and pass proximally from the compact layer of shaft to the fovea capituli; they are crossed by others parallel with the surface of the fovea. The arrangement at the distal end is somewhat similar.

Ossification (Figs. 4–147, 4–148).—The radius is ossified from three centers: one for the body, and one for either extremity. That for the body makes its appearance near the center of the bone, during the eighth week of fetal life. About the end of the second year, ossification commences in the distal end; and at the fifth year, in the proximal end. The proximal epiphysis fuses with the body at the age of seventeen or eighteen years, the distal about the age of twenty. An additional center sometimes found in the radial tuberosity, appears about the fourteenth or fifteenth year.

Articulations.—The radius articulates with the humerus, ulna, scaphoid, and lunate bones.

WRIST

The skeleton of the hand (Figs. 4–157, 4–158) is subdivided into three segments: the carpus or wrist bones; the metacarpus or bones of the palm; and the phalanges or bones of the digits.

The **Carpal Bones** (*ossa carpi*), eight in number, are arranged in two rows. Those of the proximal row, from the radial to the ulnar side, are named the scaphoid, lunate, triangular, and pisiform; those of the distal row, in the same order, are named the trapezium, trapezoideum, capitate, and hamate.

Common Characteristics of the Carpal Bones.—Each bone (except the pisiform) presents six surfaces. Of these the palmar or volar and the dorsal surfaces are rough, for ligamentous attachment; the dorsal surfaces being the broader, except in the scaphoid and lunate. The proximal and distal surfaces are articular, the proximal generally convex, the distal concave; the

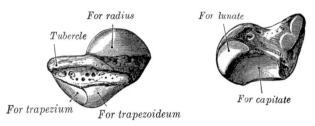

FIG. 4–149.—The left scaphoid bone.

medial or ulnar, and the lateral or radial surfaces are also articular where they are in contact with contiguous bones, otherwise they are rough and tuberculated. The structure in all is similar, viz. cancellous tissue enclosed in a layer of compact bone.

Bones of the Proximal Row.—The **Scaphoid Bone** (*os scaphoideum; navicular bone*) (Fig. 4–149) is the largest bone of the proximal row, and has received its name from its fancied resemblance to a boat. It is situated at the radial side of the carpus. The **proximal surface** is convex, smooth, of triangular shape, and articulates with the radius. The **distal surface** is also smooth, convex, and triangular, and is divided by a slight ridge into two parts, the lateral articulating with the trapezium, the medial with the trapezoideum. On the **dorsal surface** is a narrow, rough groove, which runs the entire length of the bone, and serves for the attachment of ligaments. The **palmar surface** is concave, and elevated at its lateral part into a rounded projection, the **tubercle,** to which is attached the flexor retinaculum and sometimes the origin of a few fibers of the Abductor pollicis brevis. The **lateral** or **radial surface** is rough and narrow, and gives attachment to the radial collateral ligament of the wrist. The **medial** or **ulnar surface** presents two articular facets; of these, the proximal or smaller is flattened, of semilunar form, and articulates with the lunate bone; the distal or larger is concave, forming with the lunate a concavity for the head of the capitate bone.

Articulations.—The scaphoid articulates with five bones: the radius proximally, trapezium and trapezoideum distally, and capitate and lunate medially.

The **Lunate Bone** (*os lunatum; semilunar bone*) (Fig. 4–150) may be distinguished by

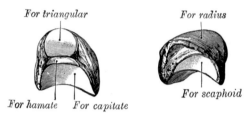

FIG. 4–150.—The left lunate bone.

its deep concavity and crescentic outline. It is situated in the center of the proximal row of the carpus, between the scaphoid and triangular. The **proximal surface,** convex and smooth, articulates with the radius. The **distal surface** is deeply concave, and articulates with the head of the capitate, and by a long, narrow facet (separated by a ridge from the general surface), with the hamate. The **dorsal** and **palmar surfaces** are rough, for the attachment of ligaments. The **lateral surface** presents a narrow, flattened, semilunar facet for articulation with the scaphoid. The medial surface is marked by a smooth, quadrilateral facet, for articulation with the triangular.

Articulations.—The lunate articulates with five bones: the radius proximally, capitate and hamate distally, scaphoid laterally, and triangular medially.

The **Triangular Bone** (*os triquetrum; cuneiform bone*) (Fig. 4–151) may be distinguished by its pyramidal shape, and by an oval isolated facet for articulation with the pisiform bone. It is situated at the proximal and ulnar side of the carpus. The **proximal surface** presents a medial, rough, nonarticular portion, and a lateral convex articular portion which articulates with the triangular articular disk of the wrist. The **distal surface,** directed lateralward, is concave, sinuously curved and smooth for articulation with the hamate. The **dorsal surface** is rough for the attachment of ligaments. The **palmar surface** presents, on its medial part, an oval facet for articulation with the pisiform; its lateral part is rough for ligamentous attachment. The **lateral surface**, the base of the pyramid, is marked by a flat, quadrilateral facet, for articulation with the lunate. The **medial surface**, the summit of the pyramid, is pointed and roughened for the attachment of the ulnar collateral ligament of the wrist.

Articulations.—The triangular articulates with three bones: the lunate laterally, the pisiform anteriorly, the hamate distally, and with the triangular articular disk which separates it from the lower end of the ulna proximaliy.

The **Pisiform Bone** (*os pisiforme*) (Fig. 4–152) may be recognized by its small size, and by its presenting a single articular facet. It is situated on a plane palmar or anterior to the other carpal bones and is spheroidal in form. Its **dorsal surface** presents a smooth, **oval facet,** for articulation with the triangular; this facet approaches the proximal, but

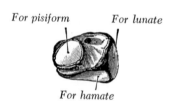

For pisiform   *For lunate*

For hamate

Fig. 4–151.—The left triangular bone.

For triangular

Fig. 4–152.—The left pisiform bone.

not the distal border of the bone. The palmar surface is rounded and rough, and gives attachment to the flexor retinaculum, and to the Flexor carpi ulnaris and Abductor digiti minimi. The lateral and medial surfaces are also rough and usually convex.

Articulation.—The pisiform articulates with one bone, the triangular.

Bones of the Distal Row.—The **Trapezium** (*os trapezium; greater multangular*) (Fig. 4–153) may be distinguished by a deep groove on its palmar surface. It is situated at the radial side of the carpus, between the scaphoid and the first metacarpal bone. The **proximal surface** is smooth medially, and articulates with the scaphoid; laterally it is rough and continuous with the lateral surface. The **distal surface** is oval, saddle-shaped, and articulates with the base of the first metacarpal bone. The **dorsal surface** is rough. The **palmar surface** is narrow and rough. At its proximal part is a deep groove which transmits the tendon of the Flexor carpi radialis, and is bounded laterally by an oblique ridge. This surface gives origin to the Opponens pollicis and to the Abductor and Flexor brevis; it also affords attachment to the flexor retinaculum. The **lateral surface** is broad and rough, for the attachment of ligaments. The **medial surface** presents two facets; the proximal, large and concave, articulates with the trapezoideum; the distal, small and oval, with the base of the second metacarpal.

Articulations.—The trapezium articulates with four bones: the scaphoid proximally, the first metacarpal distally, and the trapezoideum and second metacarpal medially.

The **Trapezoid Bone** (*os trapezoideum; lesser multangular bone*) (Fig. 4–154) is the smallest bone in the distal row. It may be

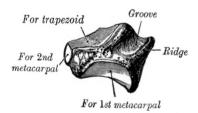

For trapezoid   *Groove*

For 2nd metacarpal   — *Ridge*

For 1st metacarpal

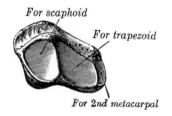

For scaphoid

For trapezoid

For 2nd metacarpal

Fig. 4–153.—The left trapezium.

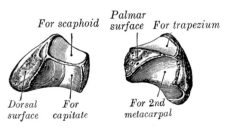

FIG. 4–154.—The left trapezoideum.

known by its wedge shape, the broad end of the wedge constituting the dorsal, the narrow end the palmar surface; and by its having four articular facets touching each other, separated by sharp edges. The **proximal surface** articulates with the scaphoid. The **distal surface** subdivided by an elevated ridge into two unequal facets, articulates with the proximal end of the second metacarpal bone. The **dorsal** and **palmar surfaces** are rough for the attachment of ligaments, the former being the larger of the two. The **lateral surface**, convex and smooth, articulates with the trapezium. The **medial surface** is concave and smooth anteriorly for articulation with the capitate but rough dorsally for the attachment of an interosseous ligament.

Articulations.—The trapezoideum articulates with four bones: the scaphoid proximally, second metacarpal distally, trapezium laterally, and capitate medially.

The **Capitate Bone** (*os capitatum; os magnum*) (Fig. 4–155) is the largest of the carpal bones, and occupies the center of the wrist. It presents a rounded portion or **head,** which is received into the concavity formed by the scaphoid and lunate; a constricted portion or **neck,** and a **body.** The **proximal surface** is round, smooth, and articulates

with the lunate. The **distal surface** is divided by two ridges into three facets, for articulation with the second, third, and fourth metacarpal bones, that for the third being the largest. The **dorsal surface** is broad and rough. The **palmar surface** is narrow, rounded, and rough, for the attachment of ligaments and a part of the Adductor pollicis obliquus. The **lateral surface** articulates with the trapezoid by a small facet and a rough depression for the attachment of an interosseous ligament. Proximal to this is a deep, rough groove, forming part of the neck, and serving for the attachment of ligaments; it is bounded proximally by a smooth, convex surface for articulation with the scaphoid. The **medial surface** articulates with the hamate by a smooth, concave, oblong facet, which occupies its proximal and dorsal parts; its palmar part is rough for the attachment of an interosseous ligament.

Articulations.—The capitate articulates with seven bones: the scaphoid and lunate, proximally; the second, third, and fourth metacarpals distally; the trapezoid on the radial side, and the hamate on the ulnar side.

The **Hamate Bone** (*os hamatum; unciform bone*) (Fig. 4–156) may readily be distinguished by the hook-like process which projects from its palmar surface. It is situated at the medial and distal angle of the carpus, resting on the fourth and fifth metacarpal bones. The **proximal surface** is narrow, convex, smooth, and articulates with the lunate. The **distal surface** articulates with the fourth and fifth metacarpal bones by concave facets which are separated by a ridge. The **dorsal surface** is triangular

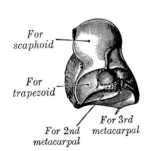

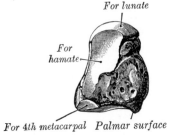

FIG. 4–155.—The left capitate bone.

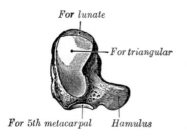

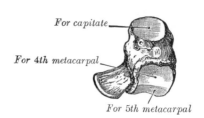

FIG. 4–156.—The left hamate bone.

and rough for ligamentous attachment. The **palmar surface** presents, at its distal and ulnar side, a curved, hook-like process, the **hamulus**. This process gives attachment, by its apex, to the flexor retinaculum and the Flexor carpi ulnaris and by its medial surface, to the Flexor brevis and Oppenens digiti minimi. Its lateral side is grooved for the passage of the Flexor tendons into the palm of the hand. It is one of the four eminences on the front of the carpus to which the flexor retinaculum of the wrist is attached; the others being the pisiform medially, the oblique ridge of the trapezium and the tubercle of the navicular laterally. The **ulnar surface** of the hamate articulates with the triangular bone by an oblong facet. The **radial surface** articulates with the capitate by its proximal part, the remaining portion being rough, for the attachment of ligaments.

Articulations.—The hamate articulates with five bones: the lunate proximally, the fourth and fifth metacarpals distally, the triangular medially, the capitate laterally.

HAND
The Metacarpus

The **Metacarpus** (Figs. 4–157, 4–158) consists of five slender bones which are numbered from the radial, lateral, or thumb side (**ossa metacarpalia I-V**); each consists of a body and two extremities.

Common Characteristics of the Metacarpal Bones.—The **body** is curved longitudinally, and presents three surfaces: medial, lateral, and dorsal. The **medial** and **lateral surfaces** are concave, for the attachment of the Interossei, and separated from one another by a prominent palmar ridge.

The **dorsal surface** presents in its distal two-thirds a smooth, triangular, flattened area which is covered in the intact body by the tendons of the Extensor muscles. This surface is bounded by two lines which commence in small tubercles on either side of the distal extremity and converge some distance proximal to the center of the bone to form a ridge which runs along the rest of the dorsal surface to the proximal extremity. This ridge separates two sloping surfaces for the attachment of the Interossei dorsales. To the tubercles on the digital extremities are attached the collateral ligaments of the metacarpophalangeal joints.

The **base** (*basis*), **proximal** or **carpal extremity** is of a cuboidal form and broader dorsally; it articulates with the carpus, and with the adjoining metacarpal bones; its dorsal and palmar surfaces are rough, for the attachment of ligaments.

The **head** (*caput*), **distal** or **digital extremity** presents an oblong convex surface flattened from side to side; and extending farther proximally on the palmar than on the dorsal aspect. It articulates with the proximal phalanx. On either side of the head is a tubercle for the attachment of the collateral ligament of the metacarpophalangeal joint. The dorsal surface, broad and flat, supports the Extensor tendons; the palmar surface is grooved in the middle line for the passage of the Flexor tendons and marked on either side by an articular eminence continuous with the terminal articular surface.

Characteristics of the Individual Metacarpal Bones.—The **First Metacarpal Bone** (*os metacarpale I; metacarpal bone of the thumb*) (Fig. 4–159) is shorter and stouter than the others, diverges to a greater degree from the carpus, and is rotated about its

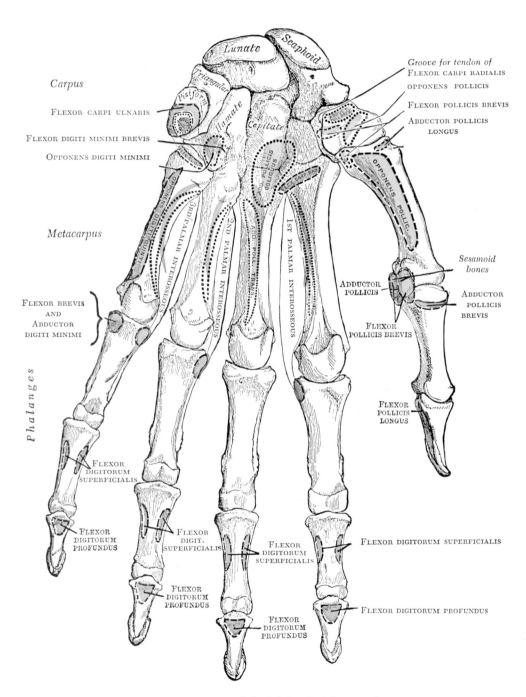

Fig. 4–157.—Bones of the left hand. Palmar surface.

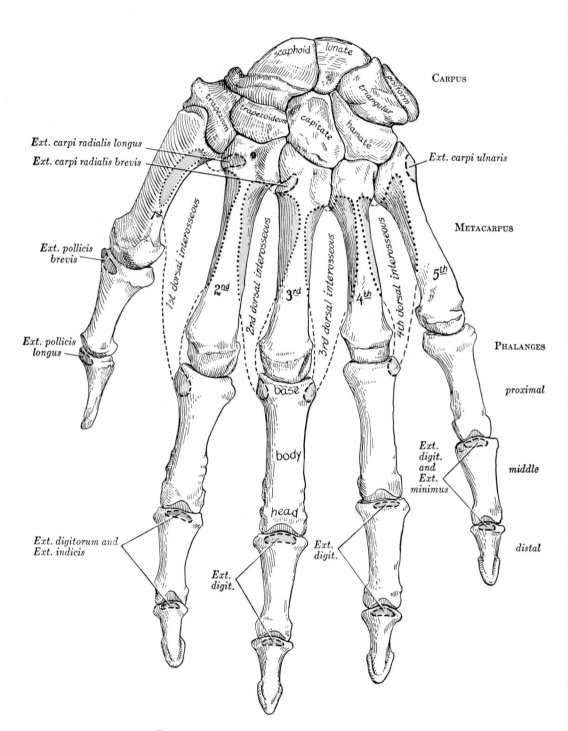

FIG. 4–158.—Bones of the left hand. Dorsal surface.

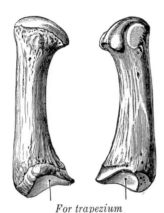

For trapezium

FIG. 4-159.—The first metacarpal. (Left)

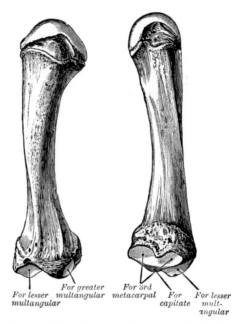

For lesser multangular   For greater multangular   For 3rd metacarpal   For capitate   For lesser multangular

FIG. 4-160.—The second metacarpal. (Left.)

long axis so that its flexor surface is directed toward the side of the second metacarpal. The **body** is flattened and broad and does not present the ridge which is found on the other metacarpal bones; its palmar surface is concave. On its radial border is inserted the Opponens pollicis; its ulnar border gives origin to the lateral head of the first Interosseus dorsalis. The **base** presents a concavo-convex surface, for articulation with the trapezium; it has no facets on its sides, but on its radial side is a tubercle for the insertion of the Abductor pollicis longus. The head is less convex than those of the other metacarpal bones, and is broader from side to side. On its palmar surface are two articular eminences, of which the lateral is larger, for the two sesamoid bones in the tendons of the two heads of the Flexor pollicis brevis.

The **Second Metacarpal Bone** (*os metacarpale II; metacarpal bone of the index finger*) (Fig. 4-160) is the longest, and its base the largest, of the four remaining bones. Its **base** is prolonged proximally and medially, forming a prominent ridge. It presents four articular facets: three on the proximal surface and one on the ulnar side. Of the facets on the proximal surface the intermediate is the largest and is for articulation with the trapezoid; the lateral is small, flat and oval for articulation with the trapezium; the medial, on the summit of the ridge, is long and narrow for articulation with the capitate. The facet on the ulnar side articulates with the third meta-

carpal. The Extensor carpi radialis longus is inserted on the dorsal surface and the Flexor carpi radialis on the palmar surface of the base.

The **Third Metacarpal Bone** (*os metacarpale III; metacarpal bone of the middle finger*) (Fig. 4-161) is a little smaller than the second. The dorsal aspect of its **base** presents on its radial side a pyramidal eminence, the **styloid process**, which extends proximally dorsal to the capitate; immediately distal to this is a rough surface for the attachment of the Extensor carpi radialis brevis. The carpal articular facet is concave and articulates with the capitate. On the radial side is a smooth, concave facet for articulation with the second metacarpal, and on the ulnar side two small oval facets for the fourth metacarpal.

The **Fourth Metacarpal Bone** (*os metacarpale IV; metacarpal bone of the ring finger*) (Fig. 4-162) is shorter and smaller than the third. The base is small and quadrilateral; its proximal surface presents two facets, a large one medially for articulation with the hamate, and a small one laterally for the capitate. On the radial side are two oval facets, for articulation with the third metacarpal; and on the ulnar side a single concave facet, for the fifth metacarpal.

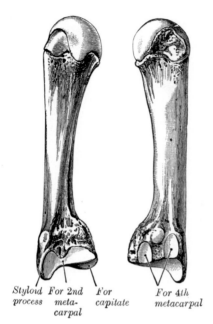

Styloid For 2nd For   For 4th
process meta-   capitate metacarpal
         carpal

FIG. 4–161.—The third metacarpal. (Left.)

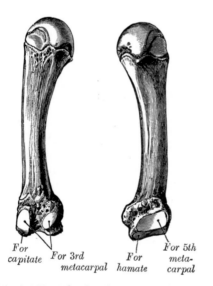

For   For   For 5th
capitate For 3rd   For   meta-
          metacarpal hamate carpal

FIG. 4–162.—The fourth metacarpal. (Left.)

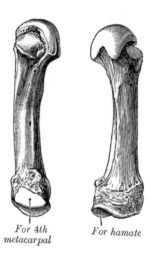

For 4th   For hamate
metacarpal

FIG. 4–163.—The fifth metacarpal. (Left.)

insertion of the tendon of the Extensor carpi ulnaris. The dorsal surface of the body is divided by an oblique ridge, which extends from near the ulnar side of the base to the radial side of the head. The lateral part of this surface serves for the attachment of the fourth Interosseus dorsalis; the medial part is smooth, triangular, and covered by the Extensor tendons of the little finger.

Articulations.—Besides their phalangeal articulations, the metacarpal bones articulate as follows: the first with the trapezium; the second with the trapezium, trapezoid, capitate, and third metacarpal; the third with the capitate, and second and fourth metacarpals; the fourth with the capitate, hamate, and third and fifth metacarpals; and the fifth with the hamate and fourth metacarpal.

The Phalanges of the Hand

The **Phalanges** (*phalanges digitorum manus*) (Figs. 4–157, 4–158) are fourteen in number, three for each finger, and two for the thumb. Each consists of a body and two extremities. The **body** tapers distally, is convex dorsally and flat from side to side; its sides are marked by rough ridges which give attachment to the fibrous sheaths of the Flexor tendons. The **proximal extremities** of the bones of the first or **proximal row** present oval, concave articular surfaces, broader from side to side. The **proximal extremities** of each of the bones of the

The **Fifth Metacarpal Bone** (*os metacarpale V; metacarpal bone of the little finger*) (Fig. 4–163) presents on its base one facet on its proximal surface, which is concavo-convex and articulates with the hamate, and one on its radial side, which articulates with the fourth metacarpal. On its ulnar side is a prominent **tubercle** for the

second and **third rows** present double concavities separated by a median ridge. The **distal extremities** are smaller than the proximal, and each ends in two condyles separated by a shallow groove; the articular surface extends farther on the palmar than on the dorsal surface, a condition best marked in the bones of the first row.

The **distal** or **ungual phalanges** are convex on their dorsal and flat on their palmar surfaces; they are recognized by their small size, and by a roughened, horseshoe-shaped elevation on the palmar surface of the distal extremity of each which supports the finger nail and the sensitive pulp.

Articulations.—In the four fingers the phalanges of the first row articulate with those of the second row and with the metacarpals; the phalanges of the second row with those of the first and third rows, and the ungual phalanges with those of the second row. In the thumb, which has only two phalanges, the first phalanx

articulates by its proximal extremity with the metacarpal bone and by its distal with the ungual phalanx.

Ossification of the Bones of the Hand.—The **carpal bones** are each ossified from a single center, and ossification proceeds in the following order (Fig. 4–164): in the capitate and hamate, during the first year, the former preceding the latter; in the triangular, during the third year, in the lunate and trapezium, during the fifth year, the former preceding the latter; in the scaphoid, during the sixth year; in the trapezoid, during the eighth year; and in the pisiform, about the twelfth year. Ossification is usually bilaterally symmetrical.

The metacarpal bones are each ossified from two centers: one for the body and one for the distal extremity of each of the second, third, fourth, and fifth bones; one for the body and one for the base of the first metacarpal bone. The first metacarpal bone is therefore ossified in the same manner as the phalanges and this has led some anatomists to regard the thumb as being made up of three phalanges, and not of a metacarpal bone and two phalanges. Ossifi-

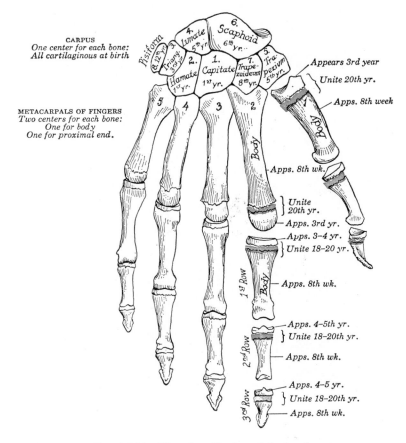

FIG. 4–164.—Plan of ossification of the hand.

cation commences in the middle of the body about the eighth or ninth week of fetal life, the centers for the second and third metacarpals being the first, and that for the first metacarpal, the last to appear; about the third year the distal extremities of the metacarpals of the fingers, and the base of the metacarpal of the thumb, begin to ossify; they unite with the bodies about the twentieth year.

The phalanges are each ossified from two centers: one for the body, and one for the proximal extremity. Ossification begins in the body, about the eighth week of fetal life. Ossification of the proximal extremity commences in the bones of the first row between the third and fourth years, and a year later in those of the second and third rows. The two centers become united in each row between the eighteenth and twentieth years.

In the ungual phalanges the centers for the bodies appear at the distal extremities of the phalanges, instead of at the middle of the bodies, as in other phalanges. Moreover, of all the bones of the hand, the ungual phalanges are the first to ossify.

Variations.—Occasionally an additional bone, the *os centrale*, is found on the dorsum of the carpus, lying between the scaphoid, trapezoid, and capitate. During the second month of fetal life it is represented by a small cartilaginous nodule, which usually fuses with the cartilaginous scaphoid. Sometimes the **styloid process** of the third metacarpal is detached and forms an additional bone.

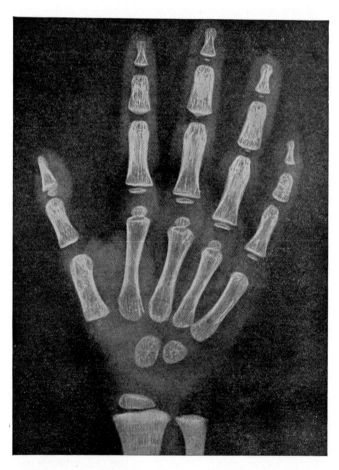

FIG. 4–165.—Hand and wrist of a child aged two and a half years. The capitate and hamate bones are in process of ossification, but the other carpal bones are still cartilaginous. The center for the head of the ulna has not yet appeared, but the center for the lower epiphysis of the radius is present. Note the condition of the metacarpal bones and phalanges.

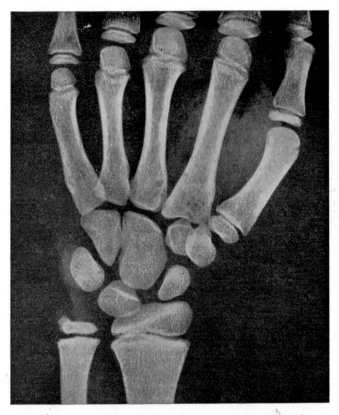

FIG. 4–166.—Hand and wrist of a child aged eleven years. All the centers of ossification are present except that for the pisiform bone. Note how the first metacarpal differs from the other metacarpal bones.

THE BONES OF THE LOWER LIMB (OSSA MEMBRI INFERIORIS)

The Hip Bone

The **Hip Bone** (*os coxae; innominate bone*) (Figs. 4–167, 4–168) is a large flattened, irregularly shaped bone, which forms the greater part of the pelvis. It consists of three parts, the ilium, ischium, and pubis, which are distinct from each other in the child (Fig. 4–171), but are fused in the adult. The union of the three parts takes place in and around a large cup-shaped articular cavity, the **acetabulum,** which is situated near the middle of the outer surface of the bone. The ilium, so called because it supports the flank, is the superior broad and expanded portion which extends cranialward from the acetabulum. The ischium is the most inferior and strongest portion; it proceeds downward from the acetabulum, expands into a large tuberosity, and then, curving ventrally,

forms, with the pubis, a large aperture, the obturator foramen. The pubis extends medialward from the acetabulum and articulates in the middle line with the bone of the opposite side where it forms the front of the pelvis and supports the external organs of generation.

The **Ilium** (*os ilii*) (Figs. 4–167, 4–168) is divisible into two parts, the body and the ala; the separation is indicated on the internal surface by the **arcuate line,** and on the external surface by the margin of the acetabulum.

The **body** (*corpus oss. ilii*) enters into the formation of the acetabulum, of which it forms rather less than two-fifths. Its external surface is partly articular, partly non-articular; the articular segment forms part of the lunate surface of the acetabulum, the non-articular portion contributes to the

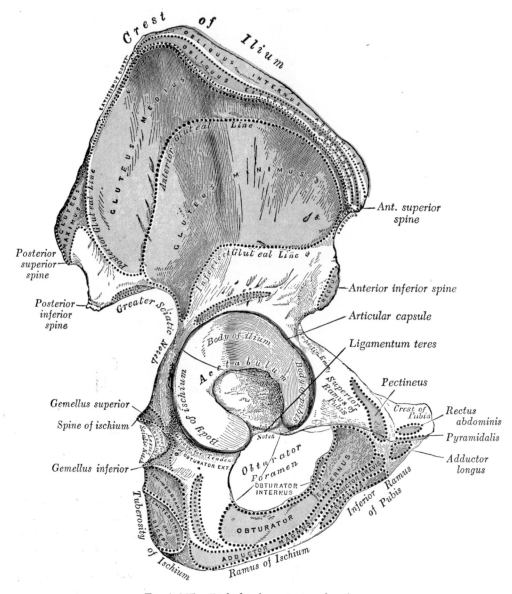

FIG. 4–167.—Right hip bone. External surface.

acetabular fossa. The internal surface of the body is part of the wall of the lesser pelvis and gives origin to some fibers of the Obturator internus. It is continuous with the pelvic surfaces of the ischium and pubis, only a faint line indicating the place of union.

The **ala** (*ala oss. ilii*) is the large expanded portion which bounds the greater pelvis laterally. It presents for examination two surfaces, a crest, and two borders.

The **external** or **gluteal surface** (*facies glutea; dorsum ilii*) (Fig. 4–167) is smooth, convex ventrally, deeply concave dorsally; bounded by the crest, the superior border of the acetabulum, and the ventral and dorsal borders. This surface is crossed in an arched direction by three lines—the posterior, anterior, and inferior gluteal lines. The **posterior gluteal line** (*superior curved line*), the shortest of the three, begins at the crest, about 5 cm distally from its dorsal extremity; it is at first distinctly marked, but as it passes distally toward its termina-

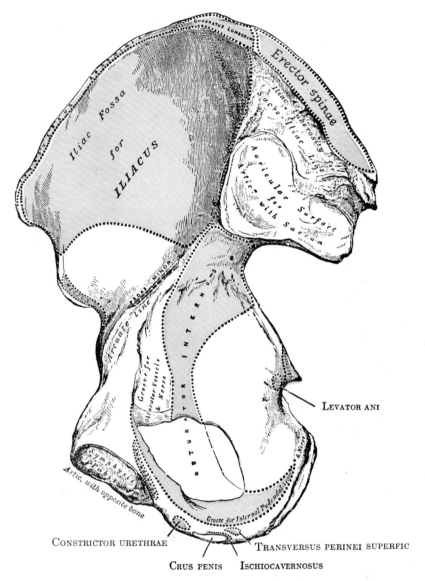

Fig. 4–168.—Right hip bone. Internal surface.

tion at the greater sciatic notch, it becomes less distinct and is often altogether lost. Dorsal to this line is a narrow semilunar surface which is rough and gives origin to a portion of the Gluteus maximus. The **anterior gluteal line** (*middle curved line*), the longest of the three, begins at the crest, about 4 cm from its ventral extremity, and, taking a curved direction, ends at the greater sciatic notch. Near the middle of this line a nutrient foramen is often seen. The space between the anterior and poste-

rior gluteal lines, and the crest, is concave and gives origin to the Gluteus medius. The **inferior gluteal line** (*inferior curved line*), the least distinct of the three, begins ventrally at the notch on the anterior border, and, curving posteriorly and distally, end near the middle of the greater sciatic notch. The surface of bone included between the anterior and inferior gluteal lines gives origin to the Gluteus minimus. Between the inferior gluteal line and the upper part of the acetabulum is a rough, shallow groove, from which the reflected tendon of the Rectus femoris arises.

The **internal** or **pelvic surface** (*facies sacropelvina*) (Fig. 4–168) of the **ala** is bounded by the crest, the arcuate line (iliopectineal line. Fig. 4–168), and the ventral and dorsal borders. It presents a large, smooth, concave surface, called the **iliac fossa**, which gives origin to the Iliacus and is perforated at its inner part by a nutrient canal. Dorsal to the iliac fossa is a rough surface, divided into two portions, a ventral and dorsal. The ventral part or **auricular surface** (*facies auricularis*) so called from its fancied resemblance in shape to the ear, is coated with cartilage, and articulates with a similar surface on the side of the sacrum. The dorsal portion, known as the **iliac tuberosity**, is elevated and rough, for the attachment of the dorsal sacroiliac ligament and for the origins of the Sacrospinalis and Multifidus. Ventral to the auricular surface is the **preauricular sulcus**, more commonly present and better marked in the female than in the male; to it is attached the pelvic portion of the ventral sacroiliac ligament.

The **crest of the ilium** is like an arch in its general outline but is sinuously curved. It is thinner at the center than at the extremities, and ends in the **anterior** and **posterior superior iliac spines.** The surface of the crest is broad, and divided into **external** and **internal lips,** and an intermediate line. About 5 cm dorsal to the anterior superior iliac spine there is a prominent **tubercle** on the outer lip. To the external lip are attached the Tensor fasciae lata; to the internal lip, the transversus abdominis, Quadratus lumborum, Sacrospinalis, and Iliacus.

The **ventral border** of the ala is concave. It presents two projections, separated by a notch. Of these, the more cranial, situated at the junction of the crest and ventral border, is called the **anterior superior iliac spine;** its outer border gives attachment to the fascia lata and the Tensor fasciae latae, its inner border to the Iliacus, and its extremity affords attachment to the inguinal ligament and the origin of the Sartorius. Inferior to this eminence is a notch from which the Sartorius takes origin and across which the lateral femoral cutaneous nerve passes. Inferior to the notch is the **anterior inferior iliac spine,** which ends in the ventral lip of the acetabulum; it gives attachment to the straight tendon of the Rectus femoris and to the iliofemoral ligament of the hip joint. Medial to the anterior inferior spine is a broad, shallow groove, over which the Iliacus and Psoas major pass. This groove is bounded medially by an eminence, the **iliopectineal eminence,** which marks the point of union of the ilium and pubis.

The **dorsal border** of the ala also presents two projections separated by a notch, the **posterior superior iliac spine** and the **posterior inferior iliac spine.** The former serves for the attachment of the oblique portion of the dorsal sacroiliac ligaments and the Multifidus; the latter corresponds with the dorsal extremity of the auricular surface. Inferior to the posterior inferior spine is a deep notch, the **greater sciatic notch.**

The **Ischium** (*os ischii*) (Figs. 4–167, 4–168) forms the inferior and dorsal part of the hip bone. It is divisible into a body and a ramus.

The **body** (*corpus oss. ischii*) enters into and constitutes a little more than two-fifths of the acetabulum. Its external surface forms part of the lunate surface of the acetabulum and a portion of the acetabular fossa. Its internal surface is part of the wall of the lesser pelvis; it gives origin to some fibers of the Obturator internus. Its ventral border projects as the **obturator tubercle;** from its dorsal border there extends a thin and pointed triangular eminence, the **ischial spine,** more or less elongated in different subjects. The external surface of the spine gives attachment to the Gemellus superior, its internal surface to the Coccygeus, Levator ani, and the pelvic fascia. The sacrospinous ligament is attached to the pointed extremity. Superior to the spine is a large notch, the **greater sciatic notch,** converted into a foramen by the sacrospinous ligament; it transmits the Piriformis, the superior and inferior gluteal vessels and nerves, the sciatic and posterior femoral cutaneous nerves, the internal pudendal vessels and pudevidal nerve, and the nerves to the Obturator internus and Quadratus femoris. Of these, the superior gluteal vessels and nerve pass out superior to the Piriformis, the other structures inferior to it. Also inferior to the spine is a smaller notch, the **lesser sciatic notch;** it is smooth, coated with cartilage, and presents two or three ridges corresponding to the subdivisions of the tendon of the Obturator internus, which glides over it. It is converted into a foramen by the sacrotuberous and sacrospinous ligaments and transmits the tendon of the Obturator internus, the nerve which supplies that muscle, and the internal pudendal vessels and nerve.

The portion of the body of the ischium which projects caudally and dorsally from the acetabulum, formerly called the **superior** or **descending ramus,** has three surfaces: external, internal, and dorsal. The **external surface** is crossed by a groove which lodges the tendon of the Obturator externus; it is limited by the dorsal margin of the obturator foramen, and a prominent margin separates it from the dorsal surface. Ventral to this margin the surface gives origin to the Quadratus femoris and to some of the fibers of origin of the Obturator externus; the distal part of the surface gives origin to part of the Adductor magnus. The **internal surface** forms part of the bony wall of the lesser pelvis; ventrally it gives origin to the Transversus perinei and Ischiocavernosus. Dorsally a large swelling, the **tuberosity of the ischium,** is divided into two portions: a distal, rough, somewhat triangular part, and a proximal, smooth, quadrilateral portion. The distal portion is subdivided by a prominent longitudinal ridge into two parts; the outer gives attachment to the Adductor magnus, the inner to the sacrotuberous ligament. An oblique ridge subdivides the proximal portion into a proximal area from which the Semimembranosus arises and a distal area for the head of the Biceps femoris and the Semitendinosus.

The **ramus** (*ramus inferior oss. ischii; ascending ramus*) (Fig. 4–167) is the thin, flattened part of the ischium which joins the inferior ramus of the pubis, the junction being indicated in the adult by a raised line. The combined rami are sometimes called the **ischiopubic ramus.** The **external surface** is uneven for the origin of the Obturator externus and some of the fibers of the Adductor magnus; its **internal surface** forms part of the anterior wall of the pelvis. Its **medial border** is thick, rough, slightly everted, forms part of the outlet of the pelvis, and presents two ridges and an inter-

vening space. The ridges are continuous with similar ones on the inferior ramus of the pubis: to the outer is attached the deep layer of the superficial perineal fascia (fascia of Colles), and to the inner the inferior fascia of the urogenital diaphragm. These two ridges join with each other just dorsal to the origin of the Transversus perinei profundus. The transversus perinei is attached to the intervening space and ventral to this a portion of the crus penis (vel clitoridis) and the Ischiocavernosus are attached. Its lateral border is thin and sharp, and forms part of the medial margin of the obturator foramen.

The **Pubis** (*os pubis*) (Figs. 4–167, 4–168). –The **pubis** is divisible into a body, a superior and an inferior ramus.

The **body** (*corpus oss. pubis*) forms one-fifth of the acetabulum, contributing by its external surface both to the lunate surface and the acetabular fossa. Its internal surface enters into the formation of the wall of the lesser pelvis and gives origin to a portion of the Obturator internus.

The **superior ramus** (*ramus superior oss. pubis; ascending ramus*) extends from the body to the median plane where it articulates with its fellow of the opposite side. It is conveniently described in two portions, a medial flattened part and a narrow lateral prismoid portion.

The **medial portion** of the superior ramus (formerly described as the body of the pubis) presents for examination two surfaces and three borders. The **external surface** is rough and serves for the origin of certain muscles. The Adductor longus arises from the proximal and medial angle immediately distal to the crest; more distally, the Obturator externus, the Adductor brevis, and the proximal part of the Gracilis take origin. The **internal surface** is smooth, and forms part of the anterior wall of the pelvis. It gives origin to the Levator ani and Obturator internus, attachment to the puboprostatic ligaments, and to a few muscular fibers prolonged from the bladder. The **cranial border** presents a prominent elevation, the **pubic tubercle** (*pubic spine*), which projects ventrally; the inferior crus of the superficial inguinal ring (*external or subcutaneous ring*), and the inguinal ligament (*Poupart's ligament*) are attached to it. Passing laterally from the pubic tubercle is a well-defined ridge forming a part of the **pecten pubis** which marks the brim of the lesser pelvis: to it are attached a portion of the inguinal falx (*conjoined tendon of Obliquus internus and Transversus*), the lacunar ligament (*Gimbernat's ligament*), and the reflected inguinal ligament (*triangular fascia*). Medial to the pubic tubercle is the **pubic crest**, which extends to the medial end of the bone. It affords attachment to the inguinal falx, and to the Rectus abdominis and Pyramidalis. The point of junction of the crest with the medial border of the bone is called the **angle**; to it, as well as to the symphysis, the superior crus of the superficial inguinal ring is attached. The **medial border** is articular; it is oval, and is marked by eight or nine transverse ridges, or a series of nipple-like processes arranged in rows, separated by grooves; they serve for the attachment of a thin layer of cartilage, which intervenes between it and the interpubic fibrocartilaginous lamina or disk. The **lateral border** presents a sharp margin, the **obturator crest**, which forms part of the circumference of the obturator foramen and affords attachment to the obturator membrane.

The **lateral portion** of the superior ramus

has three surfaces: superior, inferior, and dorsal. The **superior surface** presents the **iliopectineal line,** a continuation of the pecten pubis, already mentioned as commencing at the pubic tubercle. Ventral to this line, the surface of the bone is triangular in form, wider laterally than medially, and is covered by the Pectineus. The surface is bounded, laterally, by the iliopectineal eminence, which indicates the point of junction of the ilium and pubis, and by a prominent ridge which extends from the acetabular notch to the pubic tubercle. The **inferior surface** forms the superior boundary of the obturator foramen, and presents, laterally, an oblique groove for the passage of the obturator vessels and nerve. Medially, a sharp margin forms part of the obturator crest. The dorsal or **intrapelvic surface** constitutes part of the ventral boundary of the lesser pelvis. It is smooth and gives origin to some fibers of the Obturator internus.

The **inferior ramus** (*ramus inferior oss. pubis; descending ramus*) (Fig. 4–167) passes caudally from the medial end of the superior ramus; it becomes narrower as it descends and joins with the ramus of the ischium along the obturator foramen. Its **external surface** is rough, for the origin of

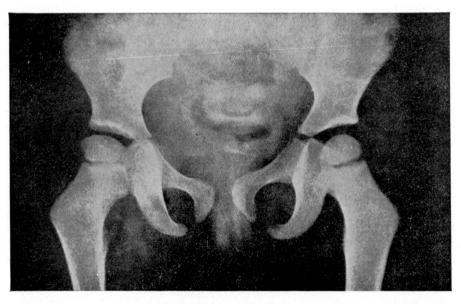

Fig. 4–169.—Pelvis of a child aged three and a half years. The epiphysis for the head of the femur is well formed, but the center for the greater trochanter has not yet appeared. The rami of the pubis and ischium are still connected by cartilage and the triradiate cartilage in the acetabulum is wide.

muscles—the Gracilis along its medial border, a portion of the Obturator externus where it enters into the formation of the obturator foramen, and between these two, the Adductores brevis and magnus, the former being the more medial. The **internal surface** is smooth, and gives origin to the Obturator internus, and, close to the medial margin, to the Constrictor urethrae. The **medial border** is thick, rough, and everted, especially in females. It presents two ridges, separated by an intervening space. The ridges are continuous with similar ridges on the ramus of the ischium; to the external is attached the fascia of Colles, and to the internal the inferior fascia of the urogenital diaphragm. The **lateral border** is thin and sharp, forms part of the circumference of the obturator foramen, and gives attachment to the obturator membrane.

The **acetabulum** (*cotyloid cavity*) (Fig. 4–167) is a deep, cup-shaped, hemispherical depression formed by the pubis medially, by the ilium superiorly, and by the ischium laterally and inferiorly; a little less than two-fifths is contributed by the ilium, a little more than two-fifths by the ischium, and the remaining fifth by the pubis. It is bounded by a prominent uneven rim, which is thick and strong superiorly, and serves for the attachment of the acetabular labrum (*cotyloid ligament*), which deepens the surface for articulation. It presents a deep notch inferiorly, the **acetabular notch,** which is continuous with a circular non-articular depression, the **acetabular fossa,** at the bottom of the cavity; this depression is perforated by numerous apertures and lodges a mass of fat. The notch is converted into a foramen by the transverse acetabular ligament; through the foramen nutrient vessels and nerves enter the joint; the margins of the notch serve for the attachment of the ligamentum capitis. The rest of the acetabulum is formed by a curved articular surface, the **lunate surface,** for articulation with the head of the femur.

The **obturator foramen** (*foramen obturatum; thyroid foramen*) (Fig. 4–167) is a large rounded aperture, situated between the ischium and pubis. It is bounded by a thin, uneven margin, to which a strong membrane is attached. Superiorly, the **obturator groove** runs from the pelvis obliquely medialward along the pubis and is converted into a canal by a ligamentous band, a specialized part of the obturator membrane, attached to two tubercles: one,

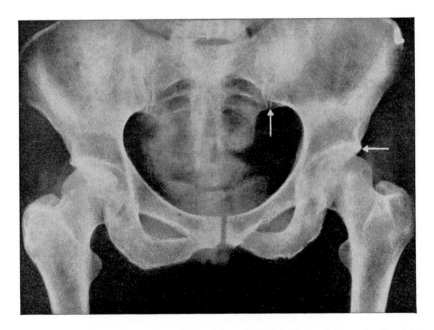

FIG. 4–170.—Adult pelvis. The upper arrow indicates the line of the sacro-iliac joint; the lower arrow points to the anterior inferior iliac spine.

the **dorsal obturator tubercle,** on the medial border of the ischium, just ventral to the acetabular notch; the other, the **ventral obturator tubercle,** on the obturator crest of the superior ramus of the pubis. Through the canal the obturator vessels and nerve pass out of the pelvis.

Structure.—The thicker parts of the bone consist of cancellous tissue, enclosed between two layers of compact tissue; the thinner parts, as at the bottom of the acetabulum and center of the iliac fossa, are usually semitransparent, and composed entirely of compact tissue.

Ossification (Fig. 4–171).—The hip bone is ossified from eight centers: three primary—one each for the ilium, ischium, and pubis; and five secondary—one each for the crest of the ilium, the anterior inferior spine (said to occur more frequently in the male than in the female), the tuberosity of the ischium, the pubic symphysis (more frequent in the female than in the male), and one or more for the Y-shaped piece at the bottom of the acetabulum. The centers appear in the following order: in the lower part of the ilium, immediately above the greater sciatic notch, about the eighth or ninth week of fetal life; in the superior ramus of the ischium, about the third month; in the superior ramus of the

pubis, between the fourth and fifth months. At birth, the three primary centers are quite separate, the crest, the bottom of the acetabulum, the ischial tuberosity and the inferior rami of the ischium and pubis being still cartilaginous. By the seventh or eighth year, the inferior rami of the pubis and ischium are almost completely united by bone. About the thirteenth or fourteenth year, the three primary centers have extended their growth into the bottom of the acetabulum, and are there separated from each other by a Y-shaped portion of cartilage, which now presents traces of ossification, often by two or more centers. One of these, the os acetabuli, appears about the age of twelve, between the ilium and pubis, and fuses with them about the age of eighteen; it forms the pubic part of the acetabulum. The ilium and ischium then become joined, and lastly the pubis and ischium, through the intervention of this Y-shaped portion. At about the age of puberty, ossification takes place in each of the remaining portions, and they join with the rest of the bone between the twentieth and twenty-fifth years. Separate centers are frequently found for the pubic tubercle and the ischial spine, and for the crest and angle of the pubis.

Articulations.—The hip bone articulates with its fellow of the opposite side, and with the sacrum and femur.

By eight centers $\begin{cases} \textit{Three primary (Ilium, Ischium, and Pubis)} \\ \textit{Five secondary} \end{cases}$

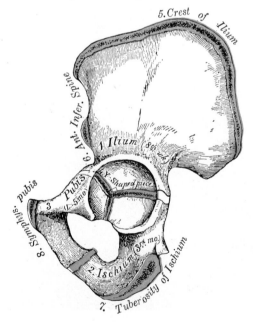

Fig. 4–171.—Plan of ossification of the hip bone. The three primary centers unite through a Y-shaped piece about puberty. Epiphyses appear about puberty, and unite about twenty-fifth year.

The Pelvis as a Whole

The **Pelvis**, so called from its resemblance to a basin, is a bony ring, interposed between the movable part of the vertebral column which it supports, and the lower limbs upon which it rests. It is stronger and more massively constructed than the wall of the cranial or thoracic cavities, and is composed of four bones: the two hip bones laterally and ventrally and the sacrum and coccyx dorsally.

The pelvis is divided into the greater and the lesser pelvis by an oblique plane passing through the prominence of the sacrum, the arcuate and pectineal lines, and the superior margin of the symphysis pubis. The circumference of this plane is termed the **linea terminalis** or **pelvic brim**.

The **greater** or **false pelvis** (*pelvis major*) is the expanded portion of the cavity situated cranial and ventral to the pelvic brim. It is bounded on either side by the ilium; ventrally it is incomplete, presenting between the ventral borders of the ilia a wide interval, which is occupied in the intact body by the anterior wall of the abdomen; dorsally there is a deep notch on either side between the ilium and the base of the sacrum.

The **lesser or true pelvis** (*pelvis minor*) is that part of the pelvic cavity which is situated distal to the pelvic brim. Its bony walls are more complete than those of the greater pelvis. For convenience of description, it is divided into a cavity, an inlet bounded by the superior circumference, and an outlet bounded by the inferior circumference.

The **superior circumference** forms the brim of the pelvis, the included space being called the **superior aperture** or **inlet** (*apertura pelvis minoris superior*) (Fig. 4–172). It is formed laterally by the pectineal and arcuate lines, ventrally by the crests of the pubes, and dorsally by the anterior margin of the base of the sacrum and sacrovertebral angle. The superior aperture is somewhat heart-shaped, obtusely pointed ventrally, diverging on either side, and encroached upon dorsally by the projection of the promontory of the sacrum. It has three principal diameters: anteroposterior, transverse, and oblique. The **anteroposterior** or **conjugate diameter** extends from the sacrovertebral angle to the symphysis pubis; its average measurement is about 11 cm in the female. The **transverse diameter** extends across the greatest width of the superior aperture, from the middle of the brim on one side to the same point on the opposite; its average measurement is about 13.5 cm in the female. The **oblique diameter** extends from the iliopectineal eminence of one side to the sacroiliac articulation of the opposite side; its average measurement is about 12.5 cm in the female.

The **cavity** of the **lesser pelvis** (Fig. 4–172) is bounded by the pubic symphysis and the superior rami of the pubes; by the pelvic surfaces of the sacrum and coccyx, by a

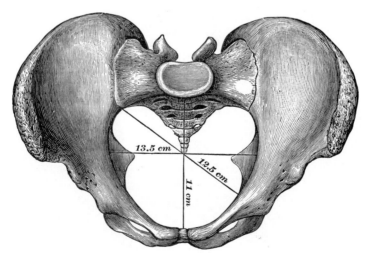

F𝗂ɢ. 4–172.—Diameters of superior aperture of lesser pelvis (female).

broad, smooth, quadrangular area of bone, corresponding to the inner surfaces of the body and superior ramus of the ischium and that part of the ilium which is below the arcuate line. The cavity of the lesser pelvis is a short, curved canal, considerably deeper on its dorsal than on its ventral wall. In the intact body, it contains the pelvic colon, rectum, bladder and some of the organs of generation. The rectum is placed at the dorsum of the pelvis in the curve of the sacrum and coccyx; the bladder is against the dorsal surface of the pubic symphysis. In the female the uterus and vagina occupy the interval between these viscera.

The **inferior circumference** of the **pelvis** is very irregular; the space enclosed by it is named the **inferior aperture** or **outlet** (*apertura pelvis minoris inferior*) (Fig. 4–173), and is bounded dorsally by the point of the coccyx, and laterally by the ischial tuberosities. These eminences are separated by three notches: one ventral to the tuberosities, the pubic arch is formed by the convergence of the inferior rami of the pubes. The dorsal notches, one on either side, extend deeply between the sacrum and the ischium; they are called the sciatic notches (*incisurae ischiadicae*). In the intact body they are converted into foramina by the sacrotuberous and sacrospinous ligaments. When the ligaments are *in situ*, the inferior aperture of the pelvis is lozenge-shaped, bounded ventrally by the pubic arcuate ligament and the inferior rami of the pubes; laterally, by the ischial tuberosities; and dorsally, by the sacrotuberous ligaments and the tip of the coccyx (see Figs. 5–25, 5–26).

The two **diameters** of the **outlet** of the **pelvis** are anteroposterior and transverse. The **anteroposterior diameter** extends from the tip of the coccyx to the interior part of the pubic symphysis; its measurement is from 9 to 11.5 cm in the female. It varies with the length of the coccyx and may be increased or diminished by the mobility of that bone. The **transverse diameter**, measured between the dorsal parts of the ischial tuberosities, is about 11 cm in the female.

Axis of the Pelvis (Fig. 4–174).—A line at right angles to the plane of the superior aperture at its center would, if prolonged, pass through the umbilicus and the middle of the coccyx and is therefore directed dorsally and distally. The axis of the inferior aperture, if prolonged, would touch the promontory of the sacrum. The axis of the cavity between the superior and inferior apertures is curved like the cavity itself, corresponding to the concavity of the sacrum and coccyx. A knowledge of the direction of these axes is especially important for an understanding of the course of the fetus in its passage through the pelvis during parturition.

Position of the Pelvis (Fig. 4–174).—In the erect posture the plane of the superior aperture forms an angle of from 50° to 60° with the horizontal plane, and that of the inferior aperture one of about 15°. The pelvic sur-

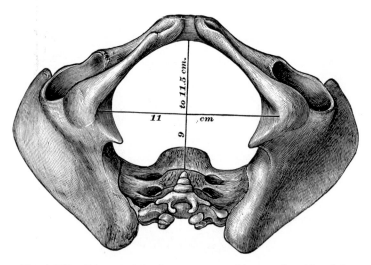

Fig. 4–173.—Diameters of inferior aperture of lesser pelvis (female).

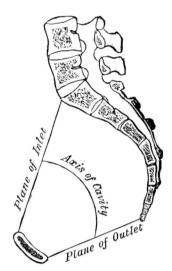

Fig. 4–174.—Median sagittal section of pelvis.

the two sexes, but also in different members of the same sex, and does not appear to be influenced by the height of the individual, although women of short stature, as a rule, have broad pelves. Occasionally the pelvis is equally contracted in all its dimensions, even in otherwise well-formed women of average height. The principal divergences, however, are found in the shape of the superior aperture, and dorsoventral and transverse diameters. Thus the superior aperture may be elliptical either in a transverse or dorsoventral direction; the transverse diameter, in the former and the dorsoventral in the latter, greatly exceeding the other diameters; in other instances it is almost circular.

In the fetus, and for several years after birth, the pelvis is smaller in proportion than in the adult, and the projection of the sacrovertebral angle less marked. The characteristic differences between the male and female pelvis are distinctly indicated as early as the fourth month of fetal life.

The Femur

The **Femur** (*thigh bone*) (Figs. 4–178, 4–179), is the longest and strongest bone in the skeleton, and is almost perfectly cylindrical in the greater part of its extent. In the erect posture it is not vertical, being separated proximally from its fellow by a considerable interval which corresponds to the breadth of the pelvis, but inclining gradually toward its fellow distally, bringing the knee joint near the line of gravity of the body. The degree of this inclination varies in different persons, and is greater in the female than in the male on account of the greater breadth of the pelvis. The femur, like other long bones, has a body or shaft and two extremities.

The **proximal extremity** (*upper extremity*) (Fig. 4–177) presents for examination a head, a neck, a greater and a lesser trochanter.

The **head** (*caput femoris*) is globular, forming rather more than a hemisphere. Its surface is smooth, and coated with cartilage in the intact body except over an ovoid depression, the **fovea capitis femoris**, near the center of the head, where the ligamentum acetabulare is attached.

face of the symphysis pubis looks cranially and dorsally, the concavity of the sacrum and coccyx caudally and ventrally. The position of the pelvis in the erect posture may be indicated by holding it so that the anterior superior iliac spines and the top of the symphysis pubis are in the same vertical plane.

Differences between the Male and Female Pelvis.—The female pelvis (Fig. 4–176) is contrasted with that of the male (Fig. 4–175) by its bones being more delicate. The whole female pelvis is less massive, and its muscular impressions are less marked. The ilia flare more laterally and the anterior iliac spines are more widely separated, causing the greater lateral prominence of the hips. The superior aperture of the lesser pelvis is larger than in the male; it is more nearly circular, and its obliquity is greater. The cavity is shallower and wider; the sacrum is shorter, wider and less curved; the obturator foramina are more triangular in shape and smaller in size than in the male. The inferior aperture is larger and the coccyx more movable. The sciatic notches are wider and shallower, and the spines of the ischia project less. The acetabula are smaller and face more ventrally. The ischial tuberosities and the acetabula are wider apart, and the former are more everted. The pubic symphysis is less deep, and the pubic arch is wider and more rounded than in the male.

The **size of the pelvis** varies not only in

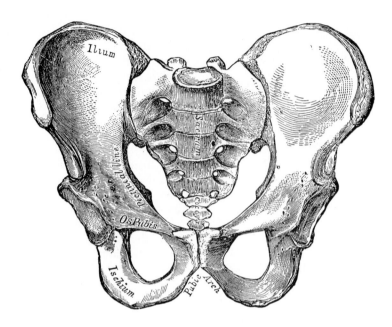

FIG. 4–175.—Male pelvis.

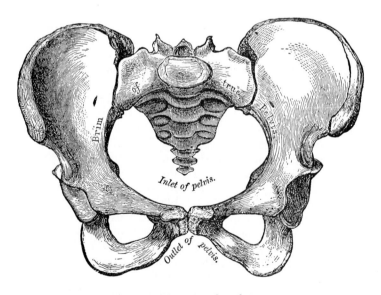

FIG. 4–176.—Female pelvis.

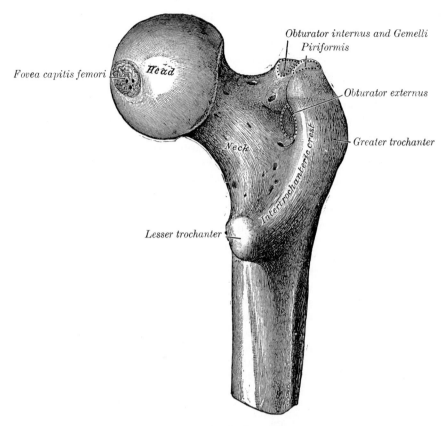

Obturator internus and Gemelli
Piriformis

Fovea capitis femori

Head

Obturator externus

Neck

Intertrochanteric crest

Greater trochanter

Lesser trochanter

Fig. 4–177.—Upper extremity of right femur. Posterior superior aspect.

The **Neck** (*collum femoris*) is a pyramidal process of bone connecting the head with the body, set at almost a right angle with the shaft. The angle is widest in infancy, and becomes lessened during growth, so that at puberty it forms a gentle curve from the axis of the body of the bone. In the adult, the neck forms an angle of about 125° with the body, but this varies in inverse proportion to the development of the pelvis and the stature. In the female, in consequence of the increased width of the pelvis, the neck of the femur forms more nearly a right angle with the body than it does in the male. The angle decreases during the period of growth, but after full growth has been attained it does not usually undergo any change even in old age; it varies considerably in different persons of the same age, and is smaller in short than in long bones, or when the pelvis is wide. In the articulated skeleton, the neck projects somewhat ventrally, as well as **cranially**

and medially from the body of the femur. The amount of this projection is extremely variable, but on an average is from 12° to 14°.

The **neck** is flattened anteroposteriorly, is constricted in the middle, and is broader laterally than medially. The vertical diameter of the lateral half is increased by the obliquity of the lower edge which slopes to join the body at the level of the lesser trochanter, so that it measures one-third more than the anteroposterior diameter. The medial half is smaller and of a more cylindrical shape. The **anterior surface** of the neck is perforated by numerous vascular foramina. Along the line of junction of the anterior surface with the head is a shallow groove, best marked in elderly subjects, which lodges the orbicular fibers of the capsule of the hip joint. The **posterior surface** is smooth, and is broader and more concave than the anterior: the posterior part of the capsule of the hip joint is attached to it about 1 cm above the intertrochanteric crest. The **superior border** is short and thick, and ends laterally at the greater trochanter; its surface is perforated by large foramina. The **inferior border,** long and narrow, ends at the lesser trochanter.

The **greater trochanter** (*trochanter major; great trochanter*) is a large, irregular, quadrilateral eminence, projecting from the angle of the junction between the neck and the body. It has two surfaces and four borders. The **lateral surface** is broad, rough, convex, and marked by a diagonal impression for the insertion of the tendon of the Gluteus medius. Above the impression is a triangular surface, sometimes rough for part of the tendon of the same muscle, sometimes smooth for the interposition of a bursa between the tendon and the bone. Distal to the diagonal impression is a smooth triangular surface, over which the Gluteus maximus plays, a bursa being interposed. The **medial surface,** of much less extent than the lateral, presents at its base a deep depression, the **trochanteric fossa** (*digital fossa*), where the Obturator externus inserts, and anterior to this an impression for the insertion of the Obturator internus and Gemelli. The **superior border** is thick and irregular, and marked near the center by an impression for the insertion of the Piriformis. The **inferior border** corresponds to the line of junction of the base of the trochanter with the lateral surface of the body; it is marked by a rough, prominent, slightly curved ridge, which gives origin to the proximal part of the Vastus lateralis. The **anterior border** is prominent and somewhat irregular; it affords insertion at its lateral part to the Gluteus minimus. The **posterior border** is very prominent and appears as a free, rounded edge which bounds the trochanteric fossa.

The **lesser trochanter** (*trochanter minor; small trochanter*) is a conical eminence which varies in size in different subjects; projecting posteriorly from the base of the neck. Three well-marked **borders** extend from its **apex.** The medial is continuous with the lower border of the neck, the lateral with the intertrochanteric crest, the inferior with the middle division of the linea aspera. The summit of the trochanter is **roughened**

for the insertion of the tendon of the Psoas major.

The **intertrochanteric crest** (Fig. 4–177) is a prominent ridge on the **posterior surface,** running in an oblique curve from the summit of the greater to the lesser trochanter. Its proximal half forms the posterior border of the greater trochanter. A slight ridge just distal to its middle portion, called the linea quadrata, receives the insertion of the Quadratus femoris and a few fibers of origin of the Adductor magnus.

The **intertrochanteric line** (*spiral line of the femur*) runs obliquely from the greater to the lesser trochanter on the **anterior surface.** It begins at the **tubercle of the femur,** a prominence of variable size at the junction of the greater trochanter and the distal part of the neck of the femur. The intertrochanteric line winds around the medial surface of the body distal to the lesser trochanter and ends in the linea aspera. Its proximal half is rough where the iliofemoral ligament is attached; its distal half gives origin to the proximal part of the Vastus medialis.

The **body** or **shaft** (*corpus femoris*) is almost cylindrical but a little broader proximally than in the center and broadest and somewhat flattened distally. It is slightly arched, convex anteriorly, and concave posteriorly where it is strengthened by a prominent longitudinal ridge, the linea aspera. It has three borders and three surfaces.

The **linea aspera** (Fig. 4–179), the posterior border of the shaft, is a prominent longitudinal ridge or crest on the surface of the middle third of the bone, having a medial and a lateral lip, and a narrow, rough, intervening band. The linea aspera is prolonged **proximally** by three ridges. The lateral ridge is very rough and runs almost vertically upward to the base of the greater trochanter. It is termed the **gluteal tuberosity,** and gives attachment to the Gluteus maximus: its proximal part is often elongated into a roughened crest, on which a more or less well-marked, rounded tubercle, the *third trochanter*, is occasionally developed. The intermediate ridge or **pectineal line** is continued to the base of the lesser trochanter and gives attachment to the Pectineus. The **medial ridge** winds ante-

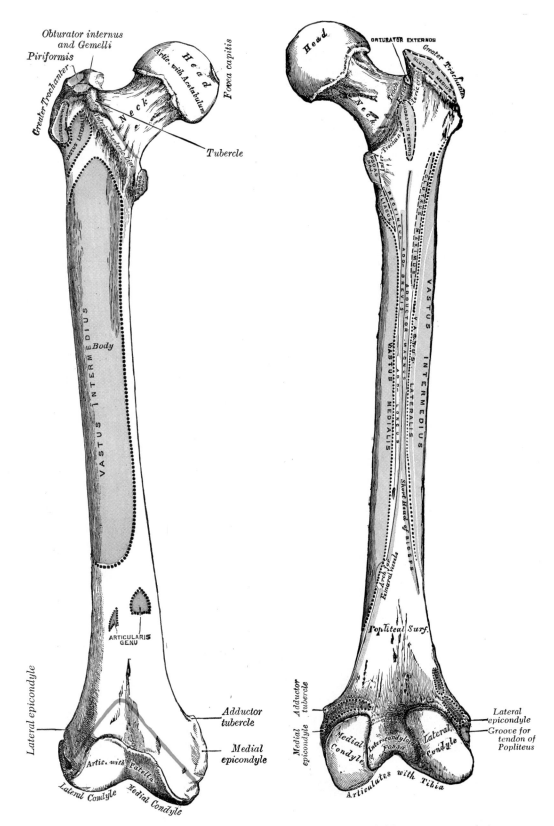

Fig. 4–178.—Right femur. Anterior surface.
Articular capsule outlined in blue.

Fig. 179.–Right femur. Posterior surface.

(241)

riorly to the intertrochanteric line distal to the area where the Iliacus is inserted. The linea aspera is prolonged **distally** into two ridges, enclosing between them a triangular area, the **popliteal surface.** Of these two ridges, the lateral is the more prominent, and descends to the summit of the lateral condyle. The medial is less marked, especially at its proximal part, where it is crossed by the femoral artery. It ends distally at the summit of the medial condyle in the **adductor tubercle,** which receives the insertion of the tendon of the Adductor magnus.

The **medial lip** of the **linea aspera** and its prolongations proximally and distally give origin to the Vastus medialis; the **lateral lip** and its proximal prolongation to the Vastus lateralis. The Adductor magnus is inserted into the linea aspera, and to its lateral prolongation proximally, and its medial prolongation distally. Between the Vastus lateralis and the Adductor magnus two muscles are attached, the gluteus maximus, and the short head of the Biceps femoris. Between the Adductor magnus and the Vastus medialis four muscles are inserted: the Iliacus and Pectineus proximally; the Adductor brevis and Adductor longus distally. The linea aspera is perforated a little below its center by the **nutrient canal.**

Two other borders of the femur are only slightly marked: the **lateral border** extends from the anterior inferior angle of the greater trochanter to the anterior extremity of the lateral condyle; the **medial border** from the intertrochanteric line, at a point opposite the lesser trochanter, to the anterior extremity of the medial condyle.

The **anterior surface** is between the lateral and medial borders. The Vastus intermedius arises from the upper three-fourths of this surface; the distal fourth is separated from the muscle by the synovial membrane of the knee joint, a bursa, and the origin of the Articularis genu. The **lateral surface** between the lateral border and the linea aspera is continuous with the corresponding surface of the greater trochanter proximally, and with that of the lateral condyle distally. The vastus intermedius takes origin from its proximal three-fourths. The **medial surface** between the medial border and the linea aspera, is continuous with the distal border of the neck proximally, and with the medial side of the medial condyle distally. It is covered by the Vastus medialis.

The **distal extremity** (*lower extremity*) (Fig. 4–180) is greatly expanded in all directions to form two prominences, the medial and lateral condyles, which are joined together anteriorly but are separated posteriorly. The distal surface is coated with cartilage for articulation with the tibia. The articular surface of the lateral condyle is longer and narrower than that of the medial condyle and its axis is more nearly anteroposterior.

The **medial condyle** diverges more from the anteroposterior plane and projects somewhat farther distally than the **lateral.** At the junction of the articular surfaces of the two condyles anteriorly, there is a shallow depression, the patellar surface. Between

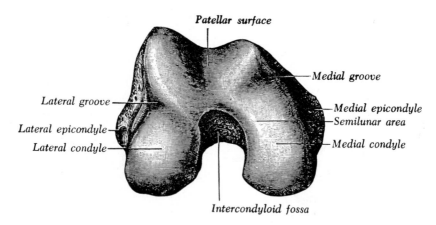

Fig. 4–180.—Lower extremity of right femur viewed from below.

the condyles posteriorly is a deep notch, the **intercondylar fossa.** The walls of the fossa are rough and concave. It is limited proximally by the intercondylar line and distally by the margin of the patellar surface. The posterior cruciate ligament of the knee joint is attached to the distal and anterior part of the medial wall of the fossa and the anterior cruciate ligament to an impression on the proximal and posterior part of its lateral wall.

The **epicondyles** are two roughened prominences proximal to the condyles. The **medial epicondyle** receives the attachment of the medial collateral ligament of the knee joint. At its proximal part is the **adductor tubercle** for the insertion of the Adductor magnus and posterior to this, the medial head of the Gastrocnemius takes origin. The **lateral epicondyle,** smaller and less prominent than the medial, gives attachment to the fibular collateral ligament of the knee joint. Directly distal to it is a small depression from which a smooth well-marked groove curves proximally to the posterior extremity of the condyle. This groove is separated from the articular surface of the condyle by a prominent lip across which a second, shallower groove runs vertically from the depression. In the intact body these grooves are covered with cartilage. The Popliteus arises from the depression; its tendon lies in the oblique groove when the knee is flexed and in the vertical groove when the knee is extended. Proximal and posterior to the lateral epicondyle is an area for the origin of the lateral head of the Gastrocnemius, and proximal to this for the Plantaris.

The **articular surface** of the distal end of the femur occupies the anterior, distal and posterior surfaces of the condyles. Its anterior part, the **patellar surface,** articulates with the patella; it presents a median groove which extends toward the intercondylar fossa, and two convexities, the lateral of which is broader, more prominent, and extends farther proximally than the medial. The distal posterior parts of the articular surface constitute the **tibial surfaces** for articulation with the corresponding condyles of the tibia and menisci. These surfaces are separated from the patellar surface by faint grooves which extend obliquely across the condyles. The lateral groove is the better marked; it runs laterally and anteriorly from the intercondylar fossa, and expands to form a triangular depression. When the knee joint is fully extended, the triangular depression rests upon the anterior portion of the lateral meniscus, and the medial part of the groove comes into contact with the medial margin of the lateral articular surface of the tibia anterior to the lateral tubercle of the tibial intercondylar eminence. The medial groove is less distinct than the lateral. It does not reach as far as the intercondylar fossa and therefore exists only on the medial part of the condyle; it receives the anterior edge of the medial meniscus when the knee joint is extended. Where the groove ceases laterally, the patellar surface is continued posteriorly as a semilunar area close to the anterior part of the intercondylar fossa; this semilunar area articulates with the medial vertical facet of the patella in forced flexion of the knee joint. The tibial surfaces of the condyles are convex from side to side and anteroposteriorly. Each presents a double curve, its posterior segment being an arc of a circle, its anterior, part of a cycloid.

The Architecture of the Femur.—The femur obeys the mechanical laws that govern other elastic bodies under stress; the relation between the computed internal stresses due to the load on the femur head, and the internal structure of the different portions of the femur is in very close agreement with the theoretical relations that should exist between stress and structure for maximum economy and efficiency (Figs. 4–181, 4–182).

The Inner Architecture of the Femur.—The spongy bone of the proximal end of the femur as far as the distal limit of the lesser trochanter is composed of two distinct systems of trabeculae arranged in curved paths: one, which has its origin in the medial (inner) side of the shaft and curving upward in a fan-like radiation to the opposite side of the bone; the other, having origin in the lateral (outer) portion of the shaft and arching upward and medially to end in the upper surface of the greater trochanter, neck, and head. These two systems intersect each other at right angles.

The trabeculae, as shown in frontal sections, are arranged in two general systems, compressive and tensile, which correspond in position with the lines of maximum and minimum stresses determined by the mathematical analy-

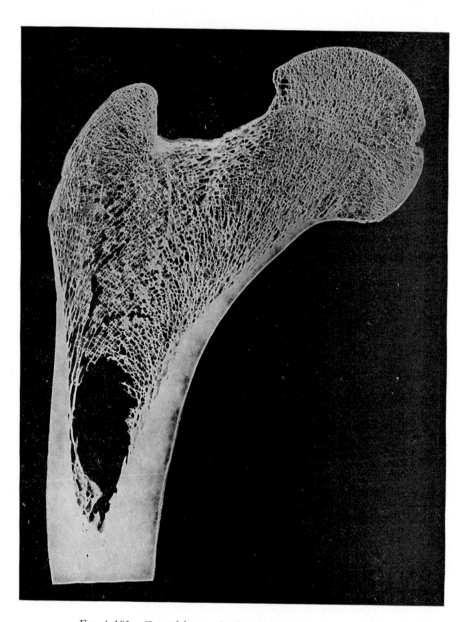

Fig. 4–181.—Frontal longitudinal midsection of upper femur.

sis of the femur as a mechanical structure. The thickness and spacing of the trabeculae vary with the intensity of the maximum stresses at various points, being thickest and most closely spaced in the regions where the greatest stresses occur. The amount of bony material in the spongy bone varies in proportion to the intensity of the shearing force at the various sections. The arrangement of the trabeculae in the positions of maximum stresses is such that the greatest strength is secured with a minimum of material.

Significance of the Inner Architecture of the Shaft.—The shearing stresses are at a minimum in the shaft. Very little if any material is in the central space, practically the only material near the neutral plane being in the compact bone.

The hollow shaft of the femur is an efficient structure for resisting **bending stresses**, all of the material in the shaft being relatively at a considerable distance from the neutral axis. It provides for resisting bending movement, not only due to the load on the femur head, but also from any other loads tending to produce bending in other planes. The structure of the shaft is such as to secure great strength with a relatively small amount of material.

The Distal Portion of the Femur.—In the distal 6 inches of the femur there are two main

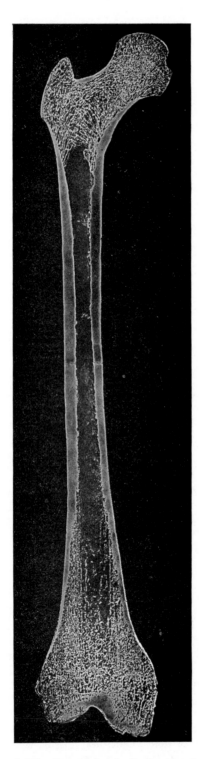

FIG. 4–182.—Frontal longitudinal midsection of right femur, 4/9 of natural size. (After Koch.)

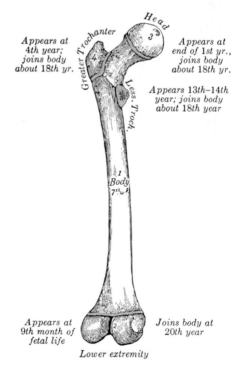

Appears at 4th year; joins body about 18th yr.

Appears at end of 1st yr., joins body about 18th yr.

Appears 13th–14th year; joins body about 18th year

Appears at 9th month of fetal life

Joins body at 20th year

Lower extremity

FIG. 4–183.—Plan of ossification of the femur. From five centers.

systems of trabeculae, a longitudinal and a
transverse system. The large expansion of the
bone is produced by the gradual transition of
the hollow shaft of compact bone to cancellated
bone, resulting in the production of a much
larger volume. The trabeculae are given off
from the shaft in lines parallel with the longitudinal axis, and are braced transversely by
two series of trabeculae at right angles to each
other, in the same manner as required theoretically for economy.

Ossification (Figs. 4–183, 4–184, 4–185).—
The femur is ossified from five centers: one for
the body, one for the head, one for each trochanter, and one for the distal extremity. Of
all the long bones, except the clavicle, it is the
first to show traces of ossification; this commences in the middle of the body, at about the
seventh week of fetal life, and rapidly extends
proximally and distally. The centers in the *epiphyses* appear in the following order: in the
distal end of the bone, at the ninth month of
fetal life (from this center the condyles and epicondyles are formed); in the head, at the
end of the first year after birth; in the greater
trochanter, during the fourth year; and in the

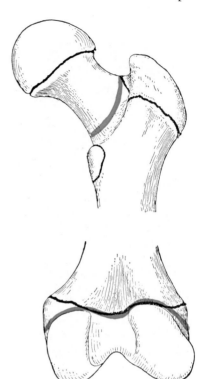

Fig. 4–185.—Epiphyseal lines of femur in a
young adult. Posterior aspect. The lines of attachment of the articular capsules are in blue.

lesser trochanter, between the thirteenth and
fourteenth years. The order in which the
epiphyses are joined to the body is the reverse
of that of their appearance; they are not united
until after puberty, the lesser trochanter being
first joined, then the greater, then the head,
and, lastly, the distal extremity, which is not
united until the twentieth year.

The Patella

The **Patella** (*knee cap*) (Figs. 4–186, 4–
187) is a flat, rounded, triangular bone,
situated on the front of the knee joint. It is
usually regarded as a sesamoid bone, developed in the tendon of the Quadriceps
femoris, and resembles these bones (1) in
being developed in a tendon; (2) in its
center of ossification presenting a knotty or
tuberculated outline; (3) in being composed
mainly of dense cancellous tissue. It serves
to protect the front of the joint, and
increases the leverage of the Quadriceps
femoris by making it act at a greater angle.

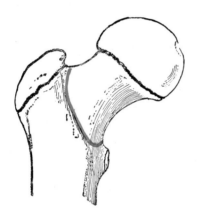

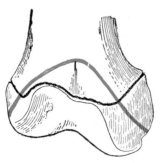

Fig. 4–184.—Epiphyseal lines of femur in a
young adult. Anterior aspect. The lines of attachment of the articular capsules are in blue.

FIG. 4–186.—Right patella. Anterior surface.

FIG. 4–187.—Right patella. Posterior surface.

It has an anterior and a posterior surface, three borders, and an apex.

The **anterior surface** is convex, perforated by small apertures for the passage of nutrient vessels, and marked by numerous rough, longitudinal striae. This surface is covered, in the intact body, by an expansion from the tendon of the Quadriceps femoris which is continuous distally with the superficial fibers of the ligamentum patellae. It is separated from the integument by a bursa. The **posterior surface** presents a smooth, oval, articular area divided into two facets by a vertical ridge; the ridge corresponds to the groove on the patellar surface of the femur, and the facets to the medial and lateral parts of the same surface; the lateral facet is the broader and deeper. Distal to the articular surface is a rough, convex, non-articular area, the distal half of which gives attachment to the ligamentum patellae; the proximal half is separated from the head of the tibia by adipose tissue.

The **base** or **proximal border** is thick, and gives attachment to that portion of the Quadriceps femoris which is derived from the Rectus femoris and Vastus intermedius. The **medial** and **lateral borders** are thinner and converge distally: they give attachment to those portions of the Quadriceps femoris

which are derived from the Vasti lateralis and medialis.

Apex.—The apex is pointed, and gives attachment to the ligamentum patellae.

Structure.—The patella consists of a nearly uniform dense cancellous tissue, covered by a thin compact lamina. The cancelli immediately beneath the anterior surface are arranged parallel with it. In the rest of the bone they radiate from the articular surface toward the other parts of the bone.

Ossification.—The patella is ossified from a single center, which usually makes its appearance in the second or third year, but may be delayed until the sixth year. More rarely, the bone is developed by two centers, placed side by side. Ossification is completed about the age of puberty.

Articulation.—The patella articulates with the femur.

The Tibia

The **tibia** (*shin bone*) (Figs. 4–190, 4–191), situated at the medial side of the leg, is the second longest bone of the skeleton. It is expanded proximally, where it enters into the knee joint, and again enlarged but to a lesser extent distally. It has a body and two extremities.

The **proximal extremity** (*upper extremity*) is expanded into two eminences, the **medial** and **lateral condyles.** The superior articular surface presents two smooth articular facets (Fig. 4–188). The medial facet is oval in shape and slightly concave. The lateral, nearly circular, is concave from side to side, but slightly convex anteroposteriorly, especially where it is prolonged over to the posterior surface for a short distance. The central portions of these facets articulate with the condyles of the femur, while their peripheral portions support the menisci of the knee joint, which here intervene between the two bones. Between the articular facets, but nearer the posterior than the anterior aspect of the bone, is the **intercondylar eminence** (*spine of tibia*), surmounted on either side by a prominent **tubercle,** on to the sides of which the articular facets are prolonged; anterior and posterior to the intercondylar eminence are rough depressions for the attachment of the anterior and posterior cruciate ligaments and the menisci. The anterior surface of the

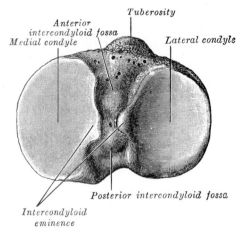

FIG. 4–188.—Proximal surface of right tibia.

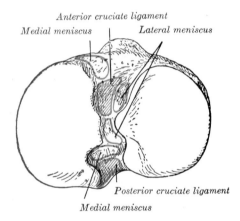

FIG. 4–189.—An outline of Figure 4–180 showing the attachment of the menisci and cruciate ligaments.

proximal extremity of the bone is a flattened triangular area; it is perforated by large vascular foramina and ends in a large oblong elevation, the **tuberosity of the tibia,** which gives attachment to the ligamentum patellae. A bursa intervenes between the deep surface of the ligament and the part of the bone immediately proximal to the tuberosity. Posteriorly, the condyles are separated from each other by a shallow depression, the **posterior intercondylar fossa,** which gives attachment to part of the posterior cruciate ligament of the knee joint. The **medial condyle** presents a deep transverse groove posteriorly for the insertion of the tendon of the Semimembranosus. Its medial surface is rough, and gives attachment to the tibial collateral ligament. The **lateral condyle**

presents a flat articular facet posteriorly, for articulation with the head of the fibula. A rough eminence is situated on a level with the proximal border of the tuberosity at the junction of its anterior and lateral surfaces for the attachment of the iliotibial band. Just distal to this, a part of the Extensor digitorum longus takes origin and a slip from the tendon of the Biceps femoris is inserted.

The **body** or **shaft** (*corpus tibiae*) has three borders and three surfaces.

The **anterior crest** or **border** (*margo anterior*), the most prominent of the three, commences proximally at the tuberosity, and ends distally at the anterior margin of the medial malleolus. It is sinuous and prominent in the proximal two-thirds of its extent, but smooth and rounded distally; it gives attachment to the deep fascia of the leg. The **medial border** is smooth and rounded but prominent in the center. It begins at the posterior part of the medial condyle, and ends at the posterior border of the medial malleolus; its proximal part gives attachment to the tibial collateral ligament of the knee joint to the extent of about 5 cm and insertion to some fibers of the Popliteus; from its middle third some fibers of the Soleus and Flexor digitorum longus take origin.

The **interosseous crest** or **lateral border** (*margo interossea*) is thin and prominent, especially its central part, and gives attachment to the interosseous membrane; it commences at the fibular articular facet, and bifurcates distally to form the boundaries of a triangular rough surface for the attachment of the interosseous ligament connecting the tibia and fibula.

The **medial surface** (*facies medialis*) is smooth, convex, and broad; its proximal third, directed anteriorly and medially, is covered by the aponeurosis derived from the tendon of the Sartorius, and by the tendons of the Gracilis and Semitendinosus, all of which are inserted nearly as far anteriorly as the anterior crest; in the rest of its extent it is subcutaneous.

The **lateral surface** (*facies lateralis*) is narrower than the medial; its proximal two-thirds present a shallow groove for the origin of the Tibialis anterior; its distal thirds present a shallow groove for the

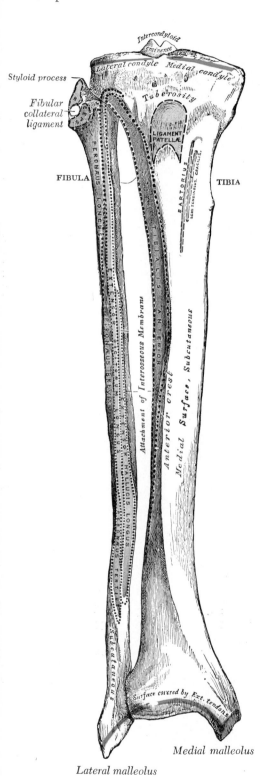

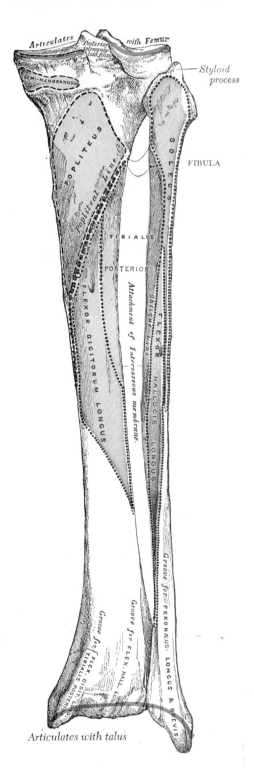

Fig. 4–190.—Bones of the right leg. Anterior surface. Articular capsules outlined in blue.

Fig. 4–191.—Bones of the right leg. Posterior surface.

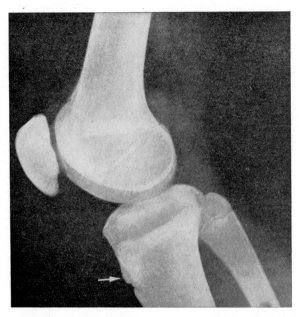

Fig. 4–192.—Knee of a boy aged sixteen years. Lateral view. Note that the upper epiphysis of the tibia includes the tibial tubercle, which is indicated by the arrow.

the anterior aspect of the bone, and is covered by the tendons of the Tibialis anterior, Extensor hallucis longus, and Extensor digitorum longus, arranged in this order from the medial side.

The **posterior surface** (*facies posterior*) (Fig. 4–191) presents, at its proximal part, a prominent ridge, the **linea m. solei** (*popliteal line*), which extends distally from the proximal and middle thirds; it marks the distal limit of the insertion of the Popliteus, serves for the attachment of the fascia covering this muscle, and gives origin to part of the Soleus, Flexor digitorum longus, and Tibialis posterior. The triangular area proximal to this line, receives the insertion of the Popliteus. The middle third of the posterior surface is divided by a vertical ridge into two parts; the ridge begins at the popliteal line and is well marked proximally, but indistinct distally; the medial and broader portion gives origin to the Flexor digitorum longus, the lateral and narrower portion to part of the Tibialis posterior. The remaining part of the posterior surface is smooth and covered by the Tibialis posterior, Flexor digitorum longus, and Flexor hallucis longus. Immediately distal to the popliteal line is the large, obliquely directed **nutrient foramen.**

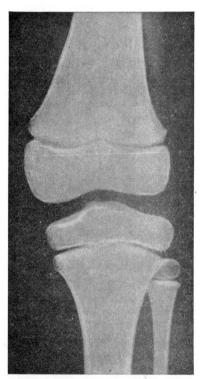

Fig. 4–193.—Knee of a child aged seven and a half years. Note that the styloid process of the head of the fibula and the tubercles of the intercondylar eminence of the tibia are still cartilaginous and therefore cannot be recognized.

The **distal extremity** (*lower extremity*), much smaller than the proximal, presents five surfaces; it is prolonged distally on its medial side as a strong process, the medial malleolus.

The **inferior surface**, continuous with that on the medial malleolus, is smooth for articulation with the talus, and is traversed anteroposteriorly by a slight elevation separating two depressions.

The **anterior surface** of the lower extremity is smooth, rounded proximally, and covered by the tendons of the Extensor muscles. Its distal margin presents a rough transverse depression for the attachment of the articular capsule of the ankle joint.

The **posterior surface** is traversed by a shallow groove which is continuous with a similar groove on the posterior surface of the talus and serves for the passage of the tendon of the Flexor hallucis longus.

The **lateral surface** presents a triangular rough depression for the attachment of the inferior interosseous ligament connecting it with the fibula; the lower part of this depression is smooth, covered with cartilage in the intact body, and articulates with the fibula. The surface is bounded by two prominent borders continuous proximally with the interosseous crest; they afford attachment to the anterior and posterior ligaments of the lateral malleolus.

The **medial surface** is prolonged distally to form a strong pyramidal process, the **medial malleolus.** The medial surface of this process is convex and subcutaneous; its lateral or articular surface is smooth, slightly concave, and articulates with the talus. Its anterior border is rough, for the attachment of the anterior fibers of the deltoid ligament of the ankle joint; its posterior border presents a broad groove, the **malleolar sulcus,** which occasionally is double and lodges the tendons of the Tibialis posterior and Flexor digitorum longus. The summit of the medial malleolus is marked by a rough depression for the attachment of the deltoid ligament.

Structure.—The structure of the tibia is like that of the other long bones. The compact wall of the body is thickest at the junction of the middle and lower thirds of the bone.

Ossification.—The tibia is ossified from three centers (Figs. 4–194, 4–195): one for the body and one for either extremity. Ossification begins

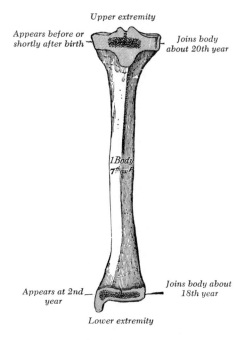

FIG. 4–194.—Plan of ossification of the tibia. From three centers.

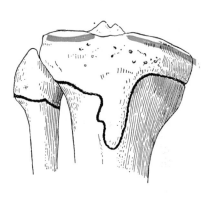

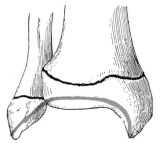

FIG. 4–195.—Epiphyseal lines of tibia and fibula in a young adult. Anterior aspect.

in the center of the body, about the seventh week of fetal life, and gradually extends toward the extremities. The center for the upper epiphysis appears before or shortly after birth; it is flattened in form, and has a thin tongue-shaped process in front which forms the tuberosity (Fig. 4–191); that for the lower epiphysis appears in the second year. The lower epiphysis joins the body at about the eighteenth, and the upper one joins about the twentieth year. Two additional centers occasionally exist, one for the tongue-shaped process of the upper epiphysis, which forms the tuberosity, and one for the medial malleolus.

The Fibula

The **Fibula** (*calf bone*) (Figs. 4–190, 4–191) is lateral to and smaller than the tibia, and, in proportion to its length, is the most slender of all the long bones. Its proximal extremity is placed slightly posterior to the head of the tibia, below the level of the knee joint, and is excluded from the formation of this joint. Its distal extremity inclines a little anteriorly and projects beyond the tibia, forming the lateral part of the ankle joint. The bone has a head and a body.

The **head** (*caput fibulae; proximal or upper extremity*) (Fig. 4–190) is irregular in form and presents a flattened surface for articulation with a corresponding surface on the lateral condyle of the tibia. On the lateral side is a thick and rough prominence continued posteriorly into a pointed eminence, the **apex** (*styloid process*), which gives attachment to the tendon of the Biceps femoris and to the fibular collateral ligament of the knee joint, the ligament dividing the tendon into two parts. The remaining part of the circumference of the head is rough, for the attachment of muscles and ligaments. Anteriorly, it presents a tubercle for the origin of the proximal and anterior fibers of the Peroneus longus, and a surface for the attachment of the anterior ligament of the head; posteriorly, another tubercle gives attachment to the posterior ligament of the head and the origin of the proximal fibers of the Soleus.

The **body** or **shaft** (*corpus fibulae*) presents three borders and three surfaces (Figs. 4–190, 4–191).

The **anterior border** (*crista anterior*) begins at the head, runs vertically and distally to a little below the middle of the bone, then curving somewhat laterally, it bifurcates so as to embrace a triangular subcutaneous surface immediately above the lateral malleolus. This border gives attachment to an intermuscular septum which separates the Extensor muscles on the anterior surface of the leg from the Peronei longus and brevis on the lateral surface.

The **interosseous border** or **crest** (*crista interossea*) is situated close to the medial side of the preceding, and runs nearly parallel with it in the proximal third of its extent, but diverges from it in the distal two-thirds. It begins at the head of the bone (sometimes it is quite indistinct for about 2.5 cm), and ends at the apex of a rough triangular surface immediately proximal to the articular facet of the lateral malleolus (Fig. 4–197). It serves for the attachment of the interosseous membrane which separates the extensor muscles anteriorly from the flexor muscles posteriorly.

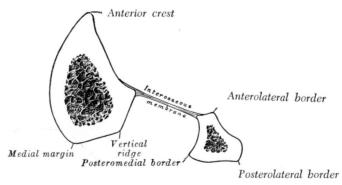

FIG. 4–196.—A transverse section through the right tibia and fibula, showing the attachment of the crural interosseous membrane.

The **posterior border** (*margo posterior; crista lateralis*) begins at the apex, and ends in the posterior border of the lateral malleolus. It gives attachment to an aponeurosis which separates the Peronei on the lateral surface from the Flexors on the posterior surface.

The **medial surface** (*facies medialis*) is the interval between the anterior and interosseous borders. It is extremely narrow and flat in the proximal third of its extent; broader and grooved longitudinally in its distal third; it serves for the origin of three muscles: the Extensor digitorum longus, Extensor hallucis longus, and Peroneus tertius.

The **crista medialis** (*postero-medial border, oblique line*) begins proximally at the medial side of the head, and ends by becoming continuous with the interosseous crest at the distal fourth of the bone. It gives attachment to an aponeurosis which separates the Tibialis posterior from the Soleus and Flexor hallucis longus.

The **posterior surface** (*facies posterior*) is the space between the posterior and interosseous borders. It is divided into

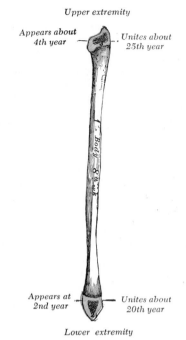

Upper extremity

Appears about 4th year — Unites about 25th year

Body

Appears at 2nd year — Unites about 20th year

Lower extremity

Fig. 4–198A.—Plan of ossification of the fibula. From three centers.

posteromedial and posterolateral portions by the crista medialis. Thus the **postero-lateral portion** is included between the posterior border and the crista medialis. It is continuous distally with the triangular area proximal to the articular surface of the lateral malleolus. Its proximal third is rough, for the origin of the Soleus; its distal part presents a triangular surface connected to the tibia by a strong interosseous ligament; the intervening part of the surface is covered by the fibers of origin of the Flexor hallucis longus. Near the middle of this surface is the **nutrient foramen**. The **posteromedial portion** is the interval included between the interosseous border and the crista medialis. It is grooved for the origin of the Tibialis posterior.

The **lateral surface** (*facies lateralis*) is the space between the anterior and posterior borders. It is broad, often deeply grooved and gives origin to the Peronei longus and brevis.

The **lateral malleolus** (*malleolus lateralis; distal extremity; external malleolus*) (Fig. 4–197).—The lateral malleolus is the expanded distal portion of the fibula, an enlargement not unlike the head. Its

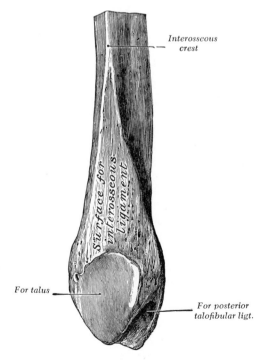

Interosseous crest

Surface for interosseous ligament

For talus

For posterior talofibular ligt.

Fig. 4–197.—Distal extremity of right fibula. Medial aspect.

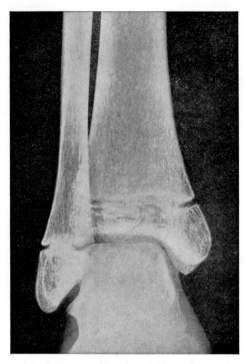

FIG. 4–198B.—Ankle of a child aged ten years. Note that the inferior epiphyseal line of the fibula is opposite the ankle joint.

extremity is less narrow and pointed than the apex, however, and its articular surface broader and more oval in shape with a deep longitudinally running groove beside it. The **lateral surface** is convex, subcutaneous, and continuous with the triangular, subcutaneous surface on the lateral side of the body. The **medial surface** presents a smooth oval surface anteriorly which articulates with a corresponding surface on the lateral side of the talus. Posterior to the **articular surface** is a rough depression, which gives attachment to the posterior talofibular ligament. The **anterior border** is thick and rough, and marked by a depression for the attachment of the anterior talofibular ligament. The **posterior border** is broad and presents the shallow malleolar sulcus for the passage of the tendons of the Peronei longus and brevis. The **summit** is rounded, and gives attachment to the calcaneofibular ligament.

Ossification (Fig. 4–198).—The fibula is ossified from three centers: one for the body, and

one for either end. Ossification begins in the body about the eighth week of fetal life, and extends toward the extremities. At birth the ends are cartilaginous. Ossification commences in the distal end in the second year, and in the proximal about the fourth year. The distal epiphysis, the first to ossify, unites with the body about the twentieth year; the proximal epiphysis, about the twenty-fifth year (Fig. 4–198).

THE FOOT

The skeleton of the foot (Figs. 4–199 and 4–200) consists of three parts: tarsus, metatarsus, and phalanges.

The Tarsus

There are seven **tarsal bones** (*ossa tarsi*): the talus, calcaneus, cuboid, navicular, and the three cuneiforms.

The **Talus** (*astragalus; ankle bone*) (Figs. 4–201 to 4–204) is the second largest of the tarsal bones. It occupies the proximal part of the tarsus, supporting the tibia, resting upon the calcaneus, articulating on either side with the malleoli and with the navicular anteriorly. It consists of a body, a neck, and a head.

The **body** (*corpus tali*).—The **superior** or **proximal portion** of the body, the **trochlea**, is covered by a smooth articular surface. The superior surface of the trochlea is convex, forming a large arc for articulation with the tibia; its lateral surface articulates with the fibula and its medial with medial malleolus of the tibia.

The **inferior** or **distal surface** of the body presents two articular areas, the posterior and middle calcaneal surfaces, separated by a deep groove, the **sulcus tali.** In the articulated foot it lies above a similar groove upon the proximal surface of the calcaneus, and forms the sinus tarsi occupied by the interosseous talocalcaneal ligament. The **posterior calcaneal articular surface** (Fig. 4–202) is large and articulates with the corresponding facet on the surface of the calcaneus. The **middle calcaneal articular surface** is small, and articulates with the facet on the sustentaculum tali of the calcaneus.

The **medial surface** is continuous superiorly with the trochlea and presents a

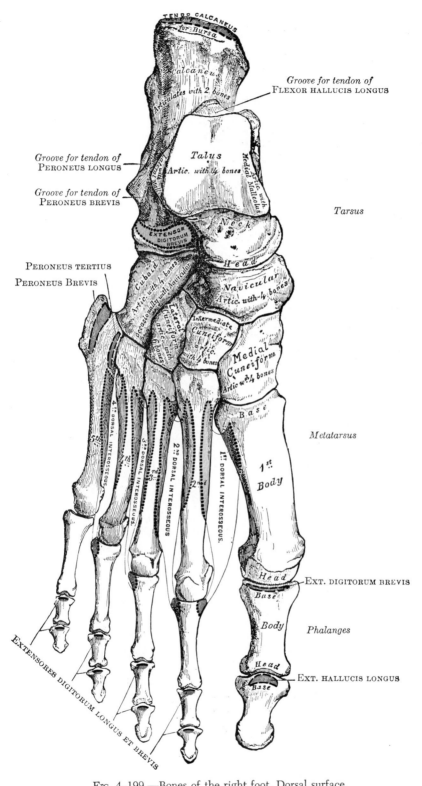

FIG. 4–199.—Bones of the right foot. Dorsal surface.

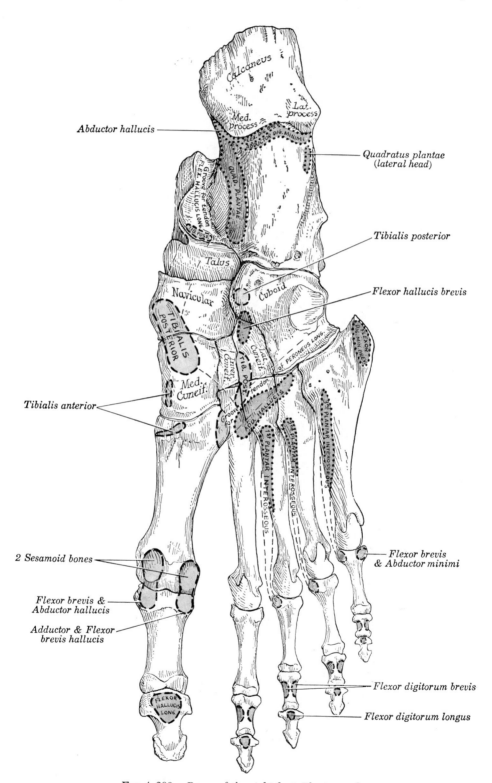

Fig. 4–200.—Bones of the right foot. Plantar surface.

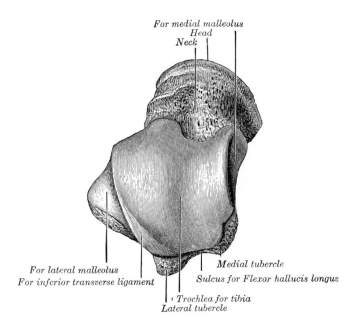

For medial malleolus
Head
Neck

For lateral malleolus
For inferior transverse ligament

Medial tubercle
Sulcus for Flexor hallucis longus

Trochlea for tibia
Lateral tubercle

FIG. 4–201.—Left talus, proximal aspect.

pear-shaped articular facet for the medial malleolus. Below the articular surface is a rough depression for the attachment of the deep portion of the deltoid ligament of the ankle joint.

The **lateral surface** which also is continuous with the trochlea carries a large triangular facet for articulation with the lateral malleolus. Anteriorly there is a rough depression for the attachment of the anterior talofibular ligament. Between the posterior half of the lateral border of the trochlea and the posterior part of the base of the fibular articular surface is a triangular facet which comes into contact with the transverse inferior tibiofibular ligament during flexion of the ankle joint; below the base of this facet is a groove which affords attachment to the posterior talofibular ligament.

The **posterior surface** is narrow, and is traversed by a groove for the tendon of the Flexor hallucis longus. Lateral to the groove is a prominent tubercle, the **posterior process,** to which the posterior talofibular ligament is attached; this process is sometimes separated from the rest of the talus, and is then known as the **os trigonum.** Medial to the groove is a second smaller tubercle.

The **neck** (*collum tali*) is the constricted

portion of the bone between the body and the head. Its superior and medial surfaces are rough for the attachment of ligaments; its lateral surface is concave and is continuous with the deep groove for the interosseous talocalcaneal ligament.

The **head** (*caput tali*) (Fig. 4–201).—The anterior surface of the head is covered by a large, oval, convex facet for articulation with the navicular. Its inferior surface has two facets, which are best seen in the fresh condition. The medial, anterior to the middle calcaneal facet, is semi-oval in shape, and rests on the plantar calcaneonavicular ligament; the lateral, named the anterior calcaneal articular surface, articulates with the facet on the superior surface of the anterior part of the calcaneus.

Articulations.—The talus articulates with four bones: tibia, fibula, calcaneus, and navicular.

The **Calcaneus** (*os calcis*) (Figs. 4–205 to 4–208) is the largest of the tarsal bones. It is situated at the posterior part of the foot, forming the heel, serving to transmit the weight of the body to the ground, and providing a strong lever for the muscles of the calf. It is somewhat cuboidal and presents six surfaces for examination.

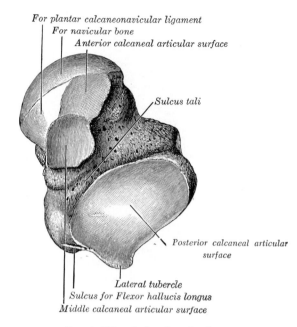

For plantar calcaneonavicular ligament
For navicular bone
Anterior calcaneal articular surface

Sulcus tali

Posterior calcaneal articular
surface

Lateral tubercle
Sulcus for Flexor hallucis longus
Middle calcaneal articular surface

Fig. 4–202.—Left talus, distal aspect.

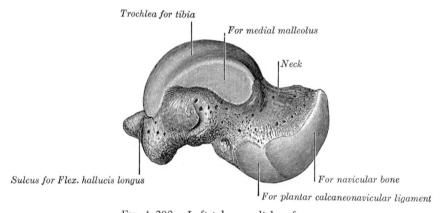

Trochlea for tibia

For medial malleolus

Neck

Sulcus for Flex. hallucis longus

For navicular bone

For plantar calcaneonavicular ligament

Fig. 4–203.—Left talus, medial surface.

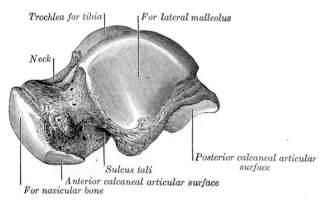

Trochlea for tibia   For lateral malleolus

Neck

Posterior calcaneal articular
surface

Sulcus tali
Anterior calcaneal articular surface
For navicular bone

Fig. 4–204.—Left talus, lateral surface.

The **superior** or **proximal surface** is divided into two parts which vary in relative size in different individuals. The anterior part contains three facets which face proximally for articulation with the talus. The **posterior articular surface** is the largest and faces somewhat anteriorly. Medial to this facet is the **middle articular surface**, supported by the sustentaculum tali. The **anterior articular facet** is anterior and lateral to the middle facet and is in close proximity or joined with it. All three facets articulate with those of similar name on the distal surface of the talus. The **calcaneal sulcus** (*sulcus calcanei*) is a deep depression between the posterior and middle articular facets. In the intact foot it combines with a similar groove on the talus to form the **sinus tarsi** which contains the interosseous talocalcaneal ligament. Anterior and lateral to the sulcus and facets, the rough surface provides attachment for ligaments and for the origin of the Extensor digitorum brevis. The posterior part of the superior surface is a rather smooth convex area extending from the articular facets to the tuberosity.

The **distal** or **plantar surface** is bounded posteriorly by a transverse elevation, the **calcaneal tuberosity** (*tuber calcanei*), which is depressed in the middle and prolonged at either side. The **lateral process,** small, prominent, and rounded, gives origin to part of the Abductor digiti minimi; the **medial process,** broader and larger, gives attachment by its prominent medial margin to the Abductor hallucis, the Flexor digitorum brevis, and the plantar aponeurosis; the depression between the processes gives origin to the Abductor digiti minimi. The rough surface anterior to the processes gives attachment to the long plantar ligament, and to the lateral head of the Quadratus plantae; the plantar calcaneocuboid ligament is attached to a prominent tubercle nearer the anterior part of this surface, as well as to a transverse groove anterior to the tubercle.

The **lateral surface** is broad, flat, and almost subcutaneous; near its center is a tubercle for the attachment of the calcaneofibular ligament. Anterior to the tubercle is a narrow surface marked by two oblique grooves, separated by an elevated ridge, the

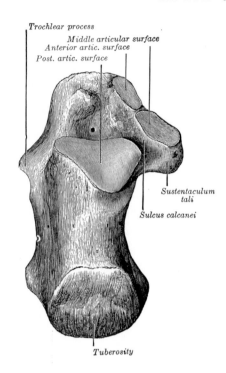

Trochlear process
Middle articular surface
Anterior artic. surface
Post. artic. surface
Sustentaculum tali
Sulcus calcanei
Tuberosity

Fig. 4–205.—Left calcaneus, proximal surface.

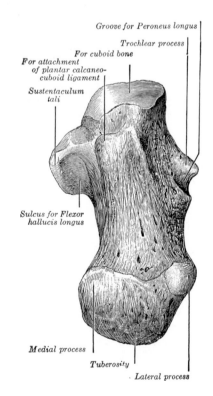

Groove for Peroneus longus
Trochlear process
For cuboid bone
For attachment of plantar calcaneo- cuboid ligament
Sustentaculum tali
Sulcus for Flexor hallucis longus
Medial process
Tuberosity
Lateral process

Fig. 4–206.—Left calcaneus, plantar surface.

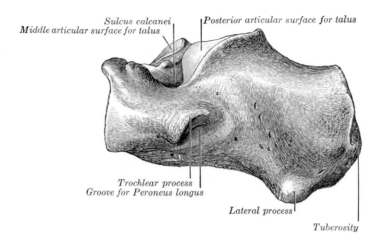

Sulcus calcanei
Middle articular surface for talus   *Posterior articular surface for talus*

Trochlear process
Groove for Peroneus longus

Lateral process

Tuberosity

Fig. 4–207.—Left calcaneus, lateral surface.

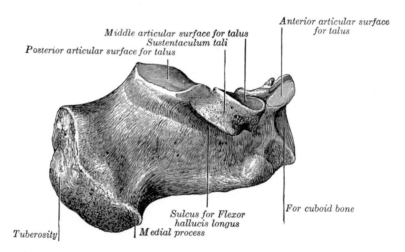

Anterior articular surface for talus

Middle articular surface for talus
Sustentaculum tali
Posterior articular surface for talus

Sulcus for Flexor hallucis longus

For cuboid bone

Tuberosity   *Medial process*

Fig. 4–208.—Left calcaneus, medial surface.

trochlea peronealis (*trochlear process; peroneal tubercle*), which varies much in size in different bones. The superior groove transmits the tendon of the Peroneus brevis; the inferior groove, that of the Peroneus longus.

The **medial surface** is deeply concave, and serves for the transmission of the plantar vessels and nerves into the sole of the foot; it affords origin to part of the Quadratus plantae. At its anterior superior part is a horizontal eminence, the **sustentaculum tali,** which gives attachment to a slip of the tendon of the Tibialis posterior. The superior surface of this eminence contains the facet for articulation with the middle calcaneal articular surface of the talus. The

inferior surface is grooved for the tendon of the Flexor hallucis longus; its anterior margin gives attachment to the plantar calcaneonavicular ligament, and its medial, to a part of the deltoid ligament of the ankle joint.

The **anterior surface** is taken up by a facet for articulation with the cuboid bone. The plantar calcaneonavicular ligament is attached to its medial border.

The **posterior surface** protrudes as the prominence of the heel. It is divisible into three areas. The inferior is rough and covered by the fatty, fibrous tissue of the heel. The middle, also rough, is the **tuberosity** and receives the insertion of the tendo calcaneus and Plantaris. The superior portion is smooth, and is covered by a bursa which intervenes between it and the tendo calcaneus.

Articulations.—The calcaneus articulates with two bones: the talus and cuboid.

The **Cuboid Bone** (*os cuboideum*) (Figs. 4–209, 4–210) is placed on the lateral side of the foot, proximal to the fourth and fifth metatarsal bones. The **dorsal surface** is rough for the attachment of ligaments. The **plantar surface** presents a deep oblique groove, the **peroneal sulcus,** which lodges the tendon of the Peroneus longus. It is bounded posteriorly by a prominent ridge, to which the long plantar ligament is attached. The ridge ends laterally in the **tuberosity,** the surface of which presents an oval facet over which the sesamoid bone or cartilage in the tendon of the Peroneus longus glides. The surface of bone posterior to the groove is rough, for the attachment of the plantar calcaneocuboid ligament, a few fibers of the Flexor hallucis brevis, and a fasciculus from the tendon of the Tibialis posterior. The **lateral surface** presents a deep notch formed by the commencement of the peroneal sulcus. The **posterior surface** is smooth, for articulation with the anterior

surface of the calcaneus; its medial inferior angle projects posteriorly as a process which underlies and supports the anterior end of the calcaneus. The **anterior surface** is divided by a vertical ridge into two facets: the medial articulates with the fourth metatarsal; the lateral articulates with the fifth. The medial surface is broad and presents at its middle and superior part a smooth oval facet for articulation with the lateral cuneiform; posterior to this (occasionally) is a smaller facet for articulation with the navicular; it is rough in the rest of its extent for the attachment of strong interosseous ligaments.

Articulations.—The cuboid articulates with four bones: the calcaneus, lateral cuneiform, and fourth and fifth metatarsals; occasionally with a fifth, the navicular.

The **Navicular Bone** (*os naviculare pedis; scaphoid bone*) (Figs. 4–211, 4–212) is situated at the medial side of the tarsus, between the talus and the cuneiform bones. The **anterior surface** is subdivided by two ridges into three facets, for articulation with the three cuneiform bones. The **poste-**

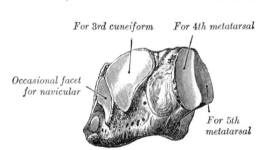

For 3rd cuneiform   For 4th metatarsal

Occasional facet for navicular

For 5th metatarsal

FIG. 4–209.—The left cuboid. Anteromedial view.

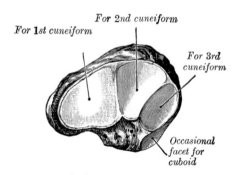

For 2nd cuneiform

For 1st cuneiform

For 3rd cuneiform

Occasional facet for cuboid

FIG. 4–211.—The left navicular. Anterolateral view

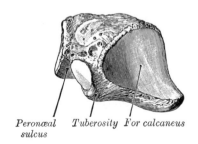

Peroneal sulcus   Tuberosity For calcaneus

FIG. 4–210.—The left cuboid. Posterolateral view.

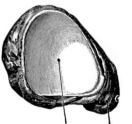

For talus   Tuberosity

FIG. 4–212.—The left navicular. Posterior view.

rior surface is oval and articulates with the rounded head of the talus. The **dorsal surface** is convex and rough for the attachment of ligaments. The **plantar surface** is irregular, and also rough for the attachment of ligaments. The **medial surface** presents a rounded tuberosity, the lower part of which gives attachment to part of the tendon of the Tibialis posterior. The **lateral surface** is rough and irregular for the attachment of ligaments, and occasionally presents a small facet for articulation with the cuboid bone.

Articulations.—The navicular articulates with four bones: the talus and the three cuneiforms; occasionally with a fifth, the cuboid.

The **Medial Cuneiform Bone** (*os cuneiforme mediale; first or internal cuneiform*) (Figs. 4–213, 4–214).—The medial cuneiform (wedge-shaped) bone is the largest of the three cuneiforms. It is situated at the medial side of the foot, between the navicular and the base of the first metatarsal. The **medial surface** is subcutaneous, broad, and at its anterior plantar angle is a smooth oval impression, into which part of the tendon of the Tibialis anterior is inserted; in the rest of its extent it is rough for the attachment of ligaments. The **lateral surface** is concave, presenting along its superior and posterior borders a narrow L-shaped surface which articulates with the intermediate cuneiform, and at the anterior part with the second metatarsal bone; the rest of this surface is rough for the attachment of ligaments and

part of the tendon of the Peroneus longus. The **anterior surface**, kidney-shaped, articulates with the first metatarsal bone. The posterior surface articulates with the most medial and largest of the three facets on the anterior surface of the navicular. The **plantar surface** is rough, and forms the base of the wedge; at its posterior part is a tuberosity for the insertion of part of the tendon of the Tibialis anterior. The **dorsal surface** is the narrow end of the wedge, and is rough for the attachment of ligaments.

Articulations.—The first cuneiform articulates with four bones: the navicular, intermediate cuneiform, and first and second metatarsals.

The **Intermediate Cuneiform Bone** (*os cuneiforme intermedium, second or middle cuneiform*) (Figs. 4–215, 4–216), the smallest of the three, is of a regular wedge shape, the thin end being the plantar. The **anterior surface** articulates with the base of the second metatarsal bone. The **posterior surface** articulates with the intermediate facet on the anterior surface of the navicular. The **medial surface** carries an L-shaped articular facet, running along the superior and posterior borders, for articulation with the medial cuneiform, and is rough in the rest of its extent for the attachment of ligaments. The **lateral surface** presents posteriorly a smooth facet for articulation with the lateral cuneiform bone. The **dorsal surface** forms the base of the wedge; it is quadrilateral and rough for the attachment of ligaments. The **plantar surface**, sharp and tuberculated, is also rough for the attachment of ligaments, and for the insertion of a slip from the tendon of the Tibialis posterior.

For 1st metatarsal

For 2nd metatarsal   For 2nd cuneiform

For tendon of
Tibialis anterior

For navicular

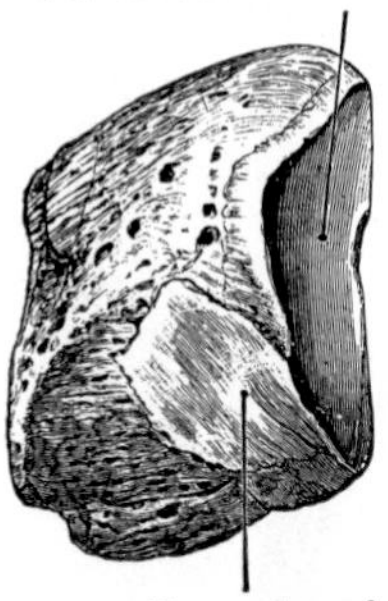

Fig. 4–213.—The left medial cuneiform. Anteromedial view.

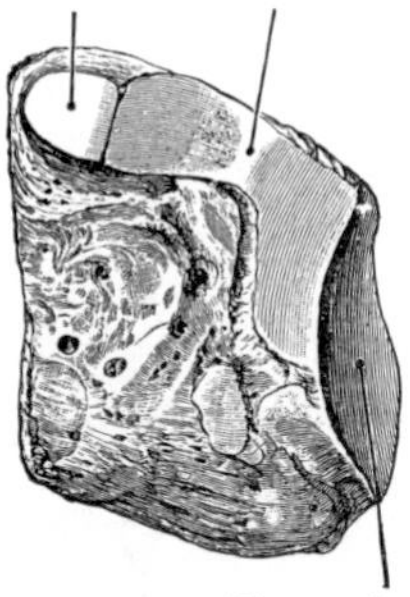

Fig. 4–214.—The left medial cuneiform. Posterolateral view.

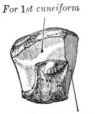

For 1st cuneiform

For 2nd metatarsal

Fig. 4–215.—The left intermediate cuneiform. Anteromedial view.

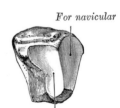

For navicular

For 3rd cuneiform

Fig. 4–216.—The left intermediate cuneiform. Posterolateral view.

Articulations.—The intermediate cuneiform articulates with four bones: the navicular, medial and lateral cuneiforms, and second metatarsal.

The **Lateral Cuneiform bone** (*os cuneiforme laterale; third or external cuneiform*) (Figs. 4–217, 4–218) is intermediate in size between the two preceding. It occupies the center of the front row of the tarsal bones, between the intermediate cuneiform medially, the cuboid laterally, the navicular posteriorly, and the third metatarsal anteriorly. The **anterior surface** articulates with the third metatarsal bone. The **posterior surface** articulates with the lateral facet on the anterior surface of the navicular, and is rough inferiorly for the attachment of ligamentous fibers. The **medial surface** presents an anterior and a posterior articular facet, separated by a rough depression: the anterior, sometimes divided, articulates with the lateral side of the base of the second metatarsal bone; the posterior skirts the posterior border and articulates with the intermediate cuneiform; the rough depression gives attachment to an interosseous ligament. The **lateral surface** also presents

two articular facets, separated by a rough non-articular area; the anterior facet, situated at the superior angle of the bone, is small, semi-oval in shape, and articulates with the medial side of the base of the fourth metatarsal bone; the posterior and larger one is triangular or oval, and articulates with the cuboid; the rough, non-articular area serves for the attachment of an interosseous ligament. The three facets for articulation with the three metatarsal bones are continuous with one another; those for articulation with the intermediate cuneiform and navicular are also continuous, but that for articulation with the cuboid is usually separate. The **dorsal surface** is oblong, its posterior lateral angle being prolonged posteriorly. The **plantar surface** is a rounded margin, and serves for the attachment of part of the tendon of the Tibialis posterior, part of the Flexor hallucis brevis, and for ligaments.

Articulations.—The lateral cuneiform articulates with six bones: the navicular, intermediate cuneiform, cuboid, and second, third, and fourth metatarsals.

The Metatarsus

The **Metatarsus** (Figs. 4–199, 4–200) consists of five bones which are numbered from the medial side (ossa metatarsalia I-V); each presents for examination a base, body, and head.

Common Characteristics of the Metatarsal Bones.—The **body** is long and slender, tapers gradually from the tarsal to the phalangeal extremity, and is curved so as to be convex dorsally. The **base** or **proximal extremity** is wedge-shaped, articulating with the tarsal bones, and by its sides with the contiguous metatarsal bones: its **dorsal** and **plantar surfaces** are rough for the attachment of ligaments. The **head** or **distal extremity** presents a convex articular surface the plantar portion extending farther proximally than the dorsal. Its sides are flattened, and on each is a depression, surmounted by a tubercle, for ligamentous attachment. Its plantar surface is grooved anteroposteriorly for the passage of the Flexor tendons, and marked on either side by an articular eminence continuous with the terminal articular surface.

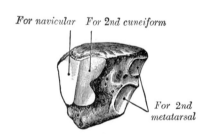

For navicular   For 2nd cuneiform

For 2nd metatarsal

Fɪɢ. 4–217.—The left lateral cuneiform. Posteromedial view.

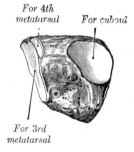

For 4th metatarsal   For cuboid

For 3rd metatarsal

Fɪɢ. 4–218.—The left lateral cuneiform. Anterolateral view.

Characteristics of the Individual Metatarsal Bones.–The **First Metatarsal Bone** (*os metatarsale I; metatarsal bone of the great toe*) (Fig. 4–219) is remarkable for its great thickness, and is the shortest of the metatarsal bones. The **body** is strong, and of well-marked prismoid form. The **base** presents, as a rule, no articular facets on its sides, but occasionally on the lateral side there is an oval facet, by which it articulates with the second metatarsal. Its proximal articular surface is of large size and kidney-shaped; its circumference is grooved, for the tarsometatarsal ligaments; medially it receives the insertion of part of the tendon of the Tibialis anterior and laterally it presents a rough oval prominence for the insertion of the tendon of the Peroneus longus. The **head** is large; on its plantar surface are two grooved facets, on which glide sesamoid bones; the facets are separated by a smooth elevation.

The **Second Metatarsal Bone** (*os metatarsale II*) (Fig. 4–220) is the longest of the metatarsal bones, being prolonged proximally to fit into the recess formed by the three cuneiform bones. Its **base** is broad dorsally, narrow and rough below. It presents four articular surfaces: a posterior

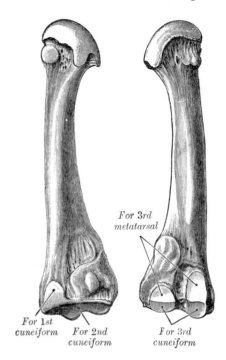

For 3rd
metatarsal

For 1st
cuneiform   For 2nd   For 3rd
cuneiform   cuneiform

FIG. 4–220.—The second metatarsal. (Left.)

for articulation with the intermediate cuneiform; one at its medial surface for articulation with the medial cuneiform; and two on its lateral surface separated by a rough non-articular interval. Each of these lateral articular surfaces is divided into two by a vertical ridge; the two anterior facets articulate with the third metatarsal; the two posterior (sometimes continous) with the lateral cuneiform. A fifth facet is occasionally present for articulation with the first metatarsal; it is oval in shape, and is situated on the medial side of the body near the base.

The **Third Metatarsal Bone** (*os metatarsale III*) (Fig. 4–221) articulates proximally by means of a triangular smooth surface with the lateral cuneiform; medially, by two facets, with the second metatarsal; and laterally, by a single facet, with the fourth metatarsal. This last facet is situated at the dorsal angle of the base.

The **Fourth Metatarsal Bone** (*os metatarsale IV*) (Fig. 4–222) is smaller in size than the preceding; its base presents an oblique quadrilateral surface for articulation with the cuboid; a smooth facet on the medial side is divided by a ridge into an

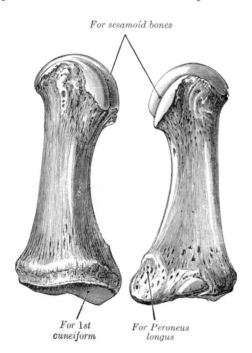

For sesamoid bones

For 1st
cuneiform   For Peroneus
longus

FIG. 4–219.—The first metatarsal. (Left.)

anterior portion for articulation with the third metatarsal, and a posterior portion for articulation with the lateral cuneiform; on the lateral side a single facet is for articulation with the fifth metatarsal.

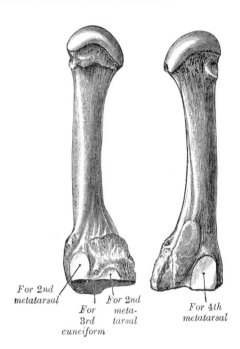

For 2nd
metatarsal

For
3rd
cuneiform

For 2nd
meta-
tarsal

For 4th
metatarsal

Fig. 4–221.—The third metatarsal. (Left.)

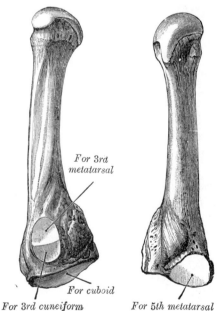

For 3rd
metatarsal

For cuboid

For 3rd cuneiform

For 5th metatarsal

Fig. 4–222.—The fourth metatarsal. (Left.)

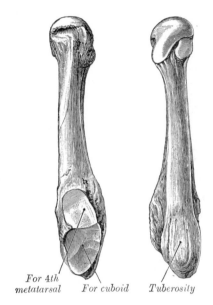

For 4th
metatarsal   For cuboid   Tuberosity

Fig. 4–223.—The fifth metatarsal. (Left.)

The **Fifth Metatarsal Bone** (*os metatarsale* V) (Fig. 4–223) is recognized by a rough eminence, the **tuberosity**, on the lateral side of its base. The base articulates posteriorly with the cuboid; and medially, with the fourth metatarsal. On the medial part of its dorsal surface is inserted the tendon of the Peroneus tertius and on the dorsal surface of the tuberosity that of the Peroneus brevis. A strong band of the plantar aponeurosis connects the projecting part of the tuberosity with the lateral process of the tuberosity of the calcaneus. The plantar surface of the base is grooved for the tendon of the Abductor digiti minimi, and gives origin to the Flexor digiti minimi brevis.

Articulations.—The base of each metatarsal bone articulates with one or more of the tarsal bones, and the head with one of the first row of phalanges. The first metatarsal articulates with the medial cuneiform, the second with all three cuneiforms, the third with the lateral cuneiform, the fourth with the lateral cuneiform and the cuboid, and the fifth with the cuboid.

The Phalanges of the Foot

The **Phalanges** of the foot (*phalanges digitorum pedis*) (Figs. 4–199, 4–200) correspond in number and general arrangement with those of the hand; there are two in the great toe and three in each of the other toes. They differ from them, however, in their size, the bodies being much reduced in length, and especially in the proximal row, laterally compressed.

Proximal Row (*first row*).—The body of each is compressed from side to side, convex dorsally. The base is concave; and the head presents a trochlear surface for articulation with the middle phalanx.

Middle Row (*second row*).—The phalanges of the second row are remarkably small and short, but rather broader than those of the proximal row.

Distal Row.—The **ungual phalanges,** resemble those of the fingers, but are smaller and more flattened; each presents a broad base for articulation with the corresponding bone of the middle row, and an expanded distal extremity for the support of the nail and end of the toe.

Articulations.—In the second, third, fourth, and fifth toes the phalanges of the proximal row articulate with the metatarsal bones, and with the middle phalanges, which in their turn articulate with the proximal and distal. The ungual phalanges articulate with the middle.

Ossification of the bones of the foot (Fig. 4–224).—The **tarsal bones** are each ossified from a single center, excepting the calcaneus, which has an epiphysis for its posterior extremity. The centers make their appearance in the following order: calcaneus at the sixth month of fetal life: talus, about the seventh month; cuboid, at the ninth month; lateral cuneiform, during the first year; medial cuneiform in the third year; intermediate cuneiform and navicular, in the fourth year. The epiphysis for the posterior extremity of the calcaneus appears at the tenth year, and unites with the rest of the bone soon after puberty. The posterior process of the talus is sometimes ossified from a separate center, and may remain distinct from the main mass of

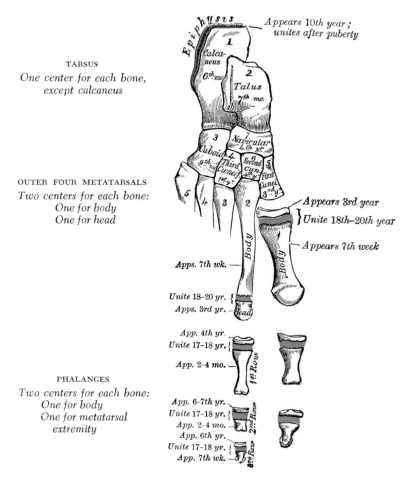

FIG. 4–224.—Plan of ossification of the foot.

the bone, in which event it is named the *os trigonum*.

The **metatarsal bones** are each ossified from two centers: one for the body, and one for the head, of the second, third, fourth and fifth metatarsals; one for the body, and one for the base, of the first metatarsal. Ossification commences in the center of the body about the ninth week, and extends toward either extremity. The center for the base of the first metatarsal appears about the third year; the centers for the heads of the other bones between the fifth and eighth years; they join the bodies between the eighteenth and twentieth years.

The **phalanges** are each ossified from two centers: one for the body, and one for the base. The center for the body appears about the tenth week, that for the base between the fourth and tenth years; it joins the body about the eighteenth year.

The Sesamoid Bones

Sesamoid bones (*ossa sesamoidea*) are small rounded masses of bone embedded in certain tendons where the latter are subjected to compression as well as to their usual tensile stresses. They are genuine parts of the skeleton, however, and are found in the fetus as well as in phylogenetically distantly related animal forms, often in greater abundance than in the adult human body.

In the pectoral limb, the sesamoid bones occur normally only in relation to the joints of the palmar surface of the hand. The two most constant are in the tendons of the two heads of the Flexor pollicis brevis at the metacarpophalangeal joint, the medial or deep head having the larger bone. A corresponding one in the Flexor minimi brevis is frequently present at the metacarpophalangeal joint and one (or two) at the same joint of the index finger. Sesamoids are also found occasionally at the metacarpophalangeal joint of the middle and ring fingers, at the interphalangeal joint of the thumb, and at the distal interphalangeal joint of the index finger.

In the pelvic limb the largest sesamoid bone is the patella, in the tendon of the Quadriceps femoris at the knee. Two constant sesamoids occur in the tendons of the Flexor hallucis brevis at the metatarsophalangeal joint, the medial being the larger. One is occasionally found at the same joint of the second and fifth toes, or even of the third and fourth toes, and one at the interphalangeal joint of the great toe.

Sesamoid bones apart from joints are seldom found in the tendons of the upper limb; one is sometimes seen in the tendon of the Biceps brachii opposite the radial tuberosity. They are, however, present in several of the tendons of the lower limb, viz., one in the tendon of the Peroneus longus, where it glides on the cuboid; one, appearing late in life, in the tendon of the Tibialis anterior, opposite the smooth facet of the medial cuneiform bone; one in the tendon of the Tibialis posterior, opposite the medial side of the head of the talus; one in the lateral head of the Gastrocnemius, behind the lateral condyle of the femur; and one in the tendon of the Psoas major, where it glides over the pubis. Sesamoid bones are found occasionally in the tendons which wind around the medial and lateral malleoli, and in the tendon of the Gluteus maximus, as it passes over the greater trochanter.

Comparison of the Bones of the Hand and Foot

The hand and foot are constructed on somewhat similar principles, each consisting of a proximal part, the carpus or the tarsus, a middle portion, the metacarpus, or the metatarsus, and a terminal portion, the phalanges. The proximal part consists of a series of more or less cubical bones which allow a slight amount of gliding on one another and are chiefly concerned in distributing forces transmitted to or from the bones of the arm or leg. The middle part is made up of slightly movable long bones which assist the carpus or tarsus in distributing forces and also give greater breadth for the reception of such forces. The separation of the individual bones from one another allows the attachments of the Interossei and protects the dorsal to palmar and dorsal to plantar vascular anastomoses. The terminal portion is the most movable, and its separate elements enjoy a varied range of movements, the chief of which are flexion and extension.

The functions of the hand and foot are very different, however, and the general similarity between them is greatly modified to meet these requirements. Thus the foot forms a firm basis of support for the body

Fig. 4–225.—Skeleton of foot. Medial aspect.

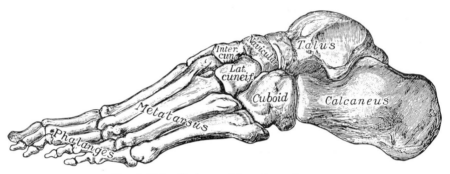

Fig. 4–226.—Skeleton of foot. Lateral aspect.

in the erect posture, and is therefore more solid and its component parts are less movable on each other than those of the hand. In the case of the phalanges the difference is readily noticeable; those of the foot are smaller and their movements are more limited than those of the hand. Very much more marked is the difference between the metacarpal bone of the thumb and the metatarsal bone of the great toe. The metacarpal bone of the thumb is constructed to permit great mobility, is directed at an acute angle from that of the index finger, and is capable of a considerable range of movements at its articulation with the carpus. The metatarsal bone of the great toe assists in supporting the weight of the body, is constructed with great solidity, lies parallel with the other metatarsals, and has a very limited degree of mobility. The carpus is small in proportion to the rest of the hand, is placed in line with the forearm, and forms a transverse arch, the concavity of which constitutes a bed for the Flexor tendons and the palmar vessels and nerves. The tarsus forms a considerable part of the foot, and is placed at right angles to the leg, a position which is almost peculiar to man, and has relation to his erect posture. In order to allow of their supporting the weight of the body with the least expenditure of material the tarsus and a part of the metatarsus are constructed in a series of arches (Figs. 4–225, 4–226), the disposition of which will be considered after the articulations of the foot have been described.

HISTOLOGIC STRUCTURE OF CONNECTIVE TISSUES

The connective tissues in their broadest sense, include the supporting tissues such as bone and cartilage. They are derived from the mesoderm of the embryo and contain large amounts of intercellular material as well as cells. The intercellular material is composed of fibers in a matrix or ground substance which may be liquid, as in tissue fluid, a gelatinous fluid, or solid as in bone and cartilage. Both the fibers and the ground substance vary according to functional requirements and the different types of connective tissue are named according to their content of intercellular material.

The **specific cells** which are resident in the connective tissues and preside over the formation and maintenance of the intercellular material are called **fibroblasts.** In a mature individual they may be called fibrocytes, but the dynamic, ever-changing conditions within the tissue in an adult as well as in a fetus have made it customary to call them fibroblasts at all ages. They are polymorphic cells containing a nucleus and relatively large amount of clear cytoplasm and the usual organelles such as mitochondria. Although they have little or no visible specializations when found in different types of connective tissue, they may receive specific names such as osteoblasts and osteocytes, or chondroblasts and chondrocytes in bone and cartilage.

Many other types of cells are found in the connective tissues. Probably the most common is the ubiquitous **histiocyte** or **macrophage.** The latter is similar to other phagocytic cells occurring in certain organs and because of this likeness, it is said to belong to the great **phagocytic system** of the body, the **reticuloendothelial system.** Another resident of the connective tissue, particularly in certain regions of the body, is the **mast cell.** It is readily identified by its large amount of cytoplasm filled with granules which stain with certain basic dyes. **Plasma cells** are found in the connective tissue of certain organs or areas. All types of white blood cells are frequent visitors in the connective tissue, having escaped from their capillaries. The polymorphonuclear **leukocytes** accumulate in an area where there is acute inflammation, and **lymphocytes** where the disease is more chronic. Red blood corpuscles are usually not found in the connective tissue unless there has been an injury to the blood vessels.

The **fibers of connective tissue** are of two primary types, **collagenous** and **elastic.** The unit fibrils of collagen are very minute, made visible only by the electron microscope. When they are combined into fine fibers, just visible with a light microscope, they are called **reticular fibrils.** When combined into larger bundles they form the white, glistening, **collagenous fibers.** These collagenous fibers form the bulk of tendons, ligaments, and fascia, where they are pliable but tough and practically not extensible. The **elastic fibers** are stretchable and rubbery, as their name implies. They vary in thickness and length but seem to be homogeneous rather than composed of unit fibrils such as collagen.

The **matrix** or **ground substance** surrounding both cells and fibers may be fluid, semifluid, or solid, and although not composed of living material such as protoplasm, it is not inert. It may vary with the general condition of the body and has its own special diseases as well as carrying out important functions in transport of materials for metabolism, nutrition and waste throughout the body.

The three principal types of connective tissue based on the intercellular material are: fibrous connective tissue, cartilage, and bone. Fibrous tissue and cartilage are again subdivided according to their content of fibers. Fibrous tissue is as follows: (1) dense fibrous, (a) organized or (b) unorganized; (2) loose fibrous (a) fibroelastic (b) areolar (c) reticular (d) adipose.

Dense Fibrous Tissue.—The *organized* dense fibrous tissues are the tendons, aponeuroses, and ligaments. The collagenous fibers are arranged in compact parallel bundles and impart a glistening, white color which is characteristic of the tissue. They are pliable and inelastic, and withstand great tensional stress particularly in one direction, but endure only small compressional stress. The *unorganized* dense fibrous tissues have their collagenous fibers interwoven rather than arranged into parallel bundles. They are the fascial membranes, corium or dermis of the skin, periosteum, and capsules of organs. They are able to resist strong tensional stress in many directions and in this respect differ from the unidirectional resistance of tendons. As might be expected, the distinction between organized and unorganized tissue is rather arbitrary; some membranes have the collagenous bundles in layers rather than interwoven and thus may have a white, glistening appearance similar to that of aponeuroses, as for example in the dura mater and certain of the named fascias.

Tendons are white, glistening fibrous bands which attach the muscles to the bones. They vary in length and thickness, sometimes round, sometimes flattened, of great tensile strength, flexible, and practically inelastic. They consist of collagenous bundles which are firmly united together and whose fibrils have a parallel course. Except where they are attached, the tendons have a sheath of delicate fibroelastic connective tissue, and the larger ones have a stroma of thin internal septa as well. They are very sparingly supplied with blood vessels, the smaller tendons presenting no trace of them in their interior. They are supplied with sensory nerves whose fibers have specialized terminations called the organs of Golgi and which mediate a special stereognostic sensibility.

Aponeuroses are fibrous membranes, of a pearly white color, iridescent, and glistening, which represent very much flattened tendons. They consist of closely packed, parallel, collagenous bundles, and by this characteristic may be differentiated from the fibrous membranes of fascia which have their collagenous bundles more irregularly interwoven. They are only sparingly supplied with blood vessels.

Ligaments resist tensions predominantly in one direction and their collagenous bundles accordingly are mainly parallel; they lack the glistening whiteness of tendons, however, because there is a greater admixture of elastic and fine collagenous fibers woven among the parallel bundles. They are as strong as tendons and have in addition a slight amount of elasticity. They are of different shapes, such as cords, bands or sheets, and are attached to bones at both ends. Their deep surface may form part of the synovial membrane of a joint, but their superficial surface is covered with a fibroelastic tissue which blends with the surrounding connective tissue.

Fascia is the term used in gross anatomy for all the fibrous connective structures not otherwise specifically named. It varies in thickness and density according to functional demands, and is usually in the form of membranous sheets. In places where two fibrous sheets of fascia are easily separated, a loose areolar connective tissue intervenes and in subsequent pages this is called a fascial cleft (see introductory paragraphs of the Muscle Chapter). Some of the membranous sheets which are named fascia attach muscles to bone and should be named aponeuroses except for established tradition. French Anatomists call most of the fascial membranes aponévroses.

Loose fibrous connective tissue is the most pervasive of all tissues, with the exception of the blood. It has a strong binding power but is very pliable and somewhat stretchable since it contains elastic fibers and its collagenous bundles are loosely interwoven and easily displaced. Between the fibers there may be comparatively wide interstices or areolae which are filled with liquid ground substance or tissue fluid. When this tissue is closely woven, as in the capsules of organs, it is called fibroelastic or fibrous tissue. When it is loosely woven and weak, as in the fascial clefts, it is called fibroareolar or **areolar tissue.**

Reticular tissue is composed of delicate reticular fibrils woven into a loose meshwork which surrounds individual or small groups of cells not belonging to the connective tissue. Its most typical form is found in such organs as the lymph nodes, the spleen, and the bone marrow, where the individual lymphocytes are retained in its meshes. The fibrils are difficult to see in the usually prepared histological slides stained with hematoxylin and eosin because they are masked by the cells. They can be made visible with silver impregnation, however, and in these preparations the delicate fibrils merge with the collagenous bundles of the capsule of the organ. The same delicate meshwork of reticular fibrils surrounds the individual muscle fibers, glandular acini, nerve fibers, capillary blood vessels, etc., in other parts of the body but usually will not be given the specific name of reticular tissue. Within the central nervous system the connective tissue is confined to the blood vessels and meninges.

Adipose tissue is the name given to tissue in which there is an accumulation of connective tissue cells, each containing a large vesicle of fat. In an obese individual these accumulations may occur wherever there is loose fibrous connective tissue. In all but emaciated individuals, however, they are found in certain areas where they act as packing or padding tissue. Examples of the latter are the orbit, perirenal fat pad, and various areas of the subcutaneous tissue.

CARTILAGE

Cartilage is a non-vascular structure which is found in various parts of the body —in adult life chiefly in the joints, in the parietes of the thorax, and in various tubes, which must be kept permanently open, such as the larynx, trachea and bronchi, nose, and ears. In the fetus, at an early period, the greater part of the skeleton is cartilaginous; as this cartilage is afterward replaced by bone, it is called **temporary,** in contradistinction to that which remains unossified during the whole of life, and is called **permanent.**

Cartilage is divided, according to the composition of its matrix, into **hyaline cartilage, white fibrocartilage,** and **yellow** or **elastic fibrocartilage.**

Hyaline cartilage consists of a gristly mass of a firm consistence, but of considerable elasticity and pearly bluish color. Except where it coats the articular ends of bones, it is covered externally by a fibrous membrane, the **perichondrium,** from the vessels of which it imbibes its nutritive fluids, being itself destitute of blood vessels.

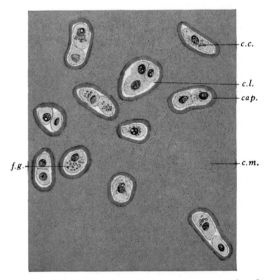

Fig. 4–227.—Hyaline cartilage. *cap.,* capsule of lacuna; *c.c.,* cartilage cell; *c.l.,* cartilage lacuna; *c.m.* cartilage matrix; *f.g.,* fat globules in cartilage cell. × 1000. (Redrawn from Sobotta.)

It contains no nerves. Its intimate structure is very simple. If a thin slice be examined under the microscope, it will be found to consist of cells of a rounded or bluntly angular form, lying in groups of two or more in a granular or almost homogeneous matrix (Fig. 4–227). The cells, when arranged in groups of two or more, have generally straight outlines where they are in contact with each other, and in the rest of their circumference are rounded. They consist of clear translucent protoplasm in which fine interlacing filaments and minute granules are sometimes present; imbedded in this are one or two round nuclei, having the usual intranuclear network. The cavities in the matrix which contain the cells are called **cartilage lacunae;** around these the matrix is arranged in concentric lines, as if it had been formed in successive portions around the cartilage cells. This constitutes the so-called **capsule of the space.** Each lacuna is generally occupied by a single cell, but during the division of the cells it may contain two, four, or eight cells.

The matrix is translucent and apparently without grossly visible structure, but resembling ground glass. It has been shown that the matrix of hyaline cartilage, and especially of the articular variety, after prolonged maceration, can be broken up into fine fibrils. These fibrils are of the same nature, chemically, as the white fibers of connective tissue.

In **articular cartilage** (Fig. 4–227), which shows no tendency to ossification, the matrix is finely granular; the cells and nuclei are small, and are disposed parallel to the surface in the superficial part, while nearer to the bone they are arranged in vertical rows. The free surface of articular cartilage, where it is exposed to friction, is not covered by perichondrium, although a layer of connective tissue continuous with that of the synovial membrane can be traced over a small part of its circumference, and here the cartilage cells are more or less branched and pass insensibly into the branched connective tissue corpuscles of the synovial membrane. Articular cartilage forms a thin incrustation upon the joint surfaces of the bones, and its elasticity enables it to break the force of concussions, while its smoothness affords ease and freedom of movement.

It varies in thickness according to the shape of the articular surface on which it lies; where this is convex the cartilage is thickest at the center, the reverse being the case on concave articular surfaces. It appears to derive its nutriment partly from the vessels of the neighboring synovial membrane and partly from those of the bone upon which it is implanted. The minute vessels of the cancellous tissue as they approach the articular lamella dilate and form arches, and then return into the substance of the bone.

In **costal cartilage** the cells and nuclei are large, and the matrix has a tendency to fibrous striation, especially in old age (Fig. 4–228). In the thickest parts of the costal cartilages a few large vascular channels may be detected. This appears, at first sight, to be an exception to the statement that cartilage is a non-vascular tissue, but is not so really, for the vessels give no branches to the cartilage substance itself, and the channels may rather be looked upon as involutions of the perichondrium. The

xiphoid process and the cartilages of the nose, larynx, and trachea (except the epiglottis and corniculate cartilages of the larynx, which are composed of elastic fibrocartilage) resemble the costal cartilages in microscopic characteristics. The arytenoid cartilage of the larynx shows a transition from hyaline cartilage at its base to elastic cartilage at the apex.

The hyaline cartilages, especially in adult and advanced life, are prone to calcify— that is to say, to have their matrix permeated by calcium salts without any appearance of true bone. The process of calcification occurs frequently, in such cartilages as those of the trachea and in the costal cartilages, where it may be succeeded by conversion into true bone.

Elastic cartilage (Fig. 4–229) has a yellowish color, is more opaque than hyalin cartilage, and is more flexible or rubbery. These characteristics are imparted by the elastic fibers embedded in the cartilage matrix. The lacunae containing the cartilage

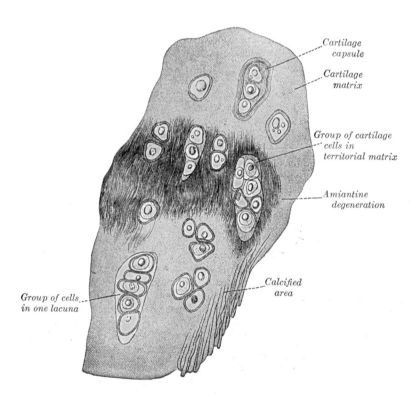

Cartilage capsule

Cartilage matrix

Group of cartilage cells in territorial matrix

Amiantine degeneration

Calcified area

Group of cells in one lacuna

Fɪɢ. 4–228.—Cross section of costal cartilage from an old man. Approx. 400 ×. (From Rauber-Kopsch, *Lehrbuch u. Atlas d. Anatomie d. Menschen,* 19th Edition, Vol. I, courtesy of Georg Thieme Verlag, Stuttgart, 1955.)

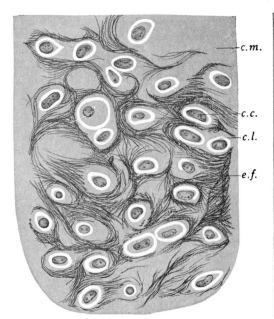

FIG. 4–229.—Elastic cartilage from human ear. *c.c.*, cartilage cell; *c.l.*, cartilage lacuna; *c.m.*, cartilage matrix; *e.f.*, elastic fibers imbedded in matrix. × 1000. (Redrawn from Sobotta.)

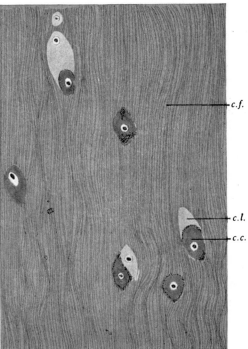

FIG. 4–230.—Fibrocartilage from human intervertebral disk. *c.c.*, cartilage cell; *c.f.*, collagenous fibrils imbedded in cartilaginous matrix; *c.l.*, cartilage lacuna. × 1000. (Redrawn from Sobotta.)

cells are rounded, similar to those of hyaline cartilage, and have a capsule of clear matrix. The elastic fibers may be sparsely scattered or densely arranged and they run in all directions. The fibers at the periphery are continuous with the elastic fibers of the perichondrium. Elastic cartilage is found in the external ear, the cartilage of the auditory (*Eustachian*) tube, the epiglottis and certain of the small laryngeal cartilages.

White fibrocartilage consists of a mixture of white fibrous tissue and cartilaginous tissue in various proportions; to the former of these constituents it owes its flexibility and toughness, and to the latter its elasticity. It is made up of fibrous connective tissue arranged in bundles, with cartilage cells between the bundles; the cells to a certain extent resemble tendon cells, but may be distinguished from them by being surrounded by a concentrically striated area of cartilage matrix and by being less flattened (Fig. 4–230). The white fibrocartilages admit of arrangement into four groups—**interarticular, connecting, circumferential,** and **stratiform.**

1. The **interarticular fibrocartilages** (*menisci*) are flattened fibrocartilaginous

plates, of a round, oval, triangular, or sickle-like form, interposed between the articular cartilages of certain joints. They are free on both surfaces, usually thinner toward the center than at the circumference, and held in position by the attachment of their margins and extremities to the surrounding ligaments. The synovial surfaces of the joints are prolonged over them. They are found in the temporomandibular, sternoclavicular, acromioclavicular, wrist, and knee joints—*i.e.*, in those joints which are most exposed to violent concussion and subject to frequent movement. Their uses are to obliterate the intervals between opposed surfaces in their various motions; to increase the depths of the articular surfaces and give ease to the gliding movements; to moderate the effects of great pressure and deaden the intensity of the shocks to which the parts may be subjected. It should be pointed out that these interarticular fibrocartilages serve an important purpose in increasing the varieties of movement in a joint. Thus in the knee joint there are two kinds of motion, viz., angular movement and rotation, although it is a hinge joint, in which, as a rule, only one variety of motion is permitted; the former movement takes place between the condyles of the femur and the interarticular cartilages, the latter between the cartilages and the head of the tibia.

2. The **connecting fibrocartilages** are interposed between the bony surfaces of those joints which admit of only slight mobility, as between the bodies of the vertebrae. They form disks which are closely adherent to the opposed surfaces. Each disk is composed of concentric rings of fibrous tissue, with cartilaginous laminae interposed, the former tissue predominating toward the circumference, the latter toward the center.

3. The **circumferential fibrocartilages** consist of rims of fibrocartilage, which surround the margins of some of the articular cavities, e.g., the glenoidal labrum of the hip, and of the shoulder; they serve to deepen the articular cavities and to protect their edges.

4. The **stratiform fibrocartilages** are those which form a thin coating to osseous grooves through which the tendons of certain muscles glide. Small masses of fibrocartilage are also developed in the tendons of some muscles, where they glide over bones, as in the tendons of the Peroneus longus and Tibialis posterior.

BONE

Structure and Physical Properties.—Bone is the hardest structure of the animal body with the exception of enamel and dentin. It is tough and slightly elastic, and will withstand both tension and compression to a remarkable degree.

If one examines a section cut through a dried bone, one sees that it is composed of two kinds of osseous tissue. One is dense in texture, like ivory, and is termed compact bone; the other consists of slender spicules, trabeculae, and lamellae, joined into a spongy structure which is called, from its resemblance to latticework, cancellous bone. The compact bone is always placed on the exterior of the bone, the cancellous in the interior. The relative quantity of these two kinds of tissue varies in different bones, and in different parts of the same bone, according to functional requirements. Close examination of the compact bone shows it to be porous, so that the difference in structure between it and the cancellous bone depends merely upon the different amount of solid matter, and the size and number of spaces in each; the cavities are small in the compact bone and the solid matter between them abundant, while in the cancellous bone the spaces are large and the solid matter is in smaller quantity (Fig. 4–231).

Bone during life is permeated by vessels, and is enclosed, except where it is coated with articular cartilage, in a fibrous membrane, the **periosteum,** by means of which many of these vessels reach the hard tissue. If the periosteum be stripped from the surface of living bone, small bleeding points mark the entrance of the periosteal vessels; and on section during life every part of the bone exudes blood from the minute vessels which ramify in it. The interior of each

THE STRENGTH OF BONE COMPARED WITH OTHER MATERIALS

Substance	Weight in pounds per cubic foot	Ultimate strength Pounds per square inch		
		Tension	Compression	Shear
Medium steel	490	65,000	60,000	40,000
Granite	170	1,500	15,000	2,000
Oak, white	46	12,500[1]	7,000[1]	4,000[2]
Compact bone (low)	119	12,200[1]	18,000[1]	11,800[2]
Compact bone (high)	. . . . . .	17,700[1]	24,000[1]	7,150[1]

[1] Indicates stresses with the grain, i.e., when the load is parallel to the long axis of the material, or parallel to the direction of the fibers of the material.

[2] Indicates unit stresses across the grain, i.e., at right angles to the direction of the fibers of the material.

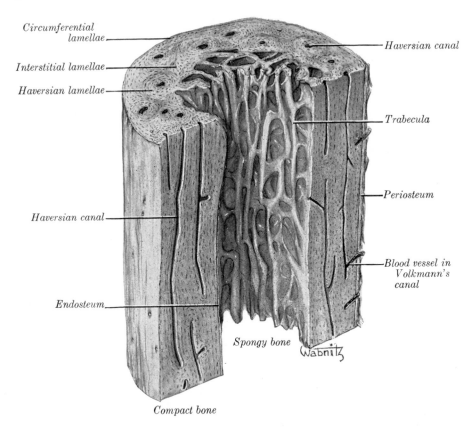

Circumferential lamellae

Interstitial lamellae

Haversian lamellae

Haversian canal

Endosteum

Haversian canal

Trabecula

Periosteum

Blood vessel in Volkmann's canal

Spongy bone

Wabnitz

Compact bone

FIG. 4–231.—Portion of finger bone from which marrow has been removed to show spongy bone. 10 ×.

of the long bones of the limbs presents a cylindrical cavity filled with **bone marrow** and lined by a vascular areolar membrane, called the **endosteum.**

Periosteum.—The periosteum adheres to the surface of the bone in every part except the cartilaginous extremities. It consists of two layers closely united together, the outer one formed chiefly of collagenous tissue, containing occasionally a few fat cells; the inner one, of elastic fibers of the finer kind, forming dense membranous networks, which again can be separated into layers. When strong tendons or ligaments are attached to a bone, the periosteum is incorporated with them. In young bones the periosteum is thick and very vascular, and is intimately connected at either end of the bone with the epiphyseal cartilage, but less closely with the body of the bone, from which it is separated by a layer of soft tissue, containing a number of **osteoblasts,** cells which preside over the process of ossification on the exterior of the young bone. Later in life the periosteum is thinner and less vascular, and the osteoblasts are converted into an epithelioid layer on the deep surface of the periosteum. The smaller vessels ramify in the periosteum previous to their distribution in the bone; hence the liability of bone to exfoliate or necrose when denuded of this membrane by injury or disease. Fine nerves and lymphatics, which generally accompany the arteries, may also be demonstrated in the periosteum.

Bone Marrow.—Bone marrow fills the cavities of the bones. **Yellow marrow is** found in the large cavities of the long bones. It consists for the most part of fat cells and a few primitive blood cells. It may be replaced by red marrow in anemia. **Red marrow** is the site for the production of the granular leukocytes (neutrophil, eosinophil, and basophil leukocytes) and the red blood cells (erythrocytes). It is found in the flat and short bones, the articular ends of the long bones, the bodies of the vertebrae, the cranial diploë, and the sternum and ribs. It consists, for the most part, of myeloid cells, namely, primitive blood cells in immature stages. Macrophages, fat cells and megakaryocytes (giant cells) are always present. Both types of marrow have a supporting connective tissue and numerous blood vessels. Thin-walled sinusoids are supposed to connect the terminal arterioles with the veins in the red bone marrow. Great numbers of recently matured blood cells (both red and white) pass into the blood stream. The mechanism whereby the mature cells escape into the blood stream and immature ones are held back is not known.

Vessels and Nerves of Bone.—The **blood vessels** of bone are very numerous. Those of the compact tissue are derived from a close and dense network of vessels ramifying in the periosteum. From this membrane vessels pass into the minute orifices in the compact tissue, and run through the canals which traverse its substance. The cancellous tissue is supplied in a similar way, but by less numerous and larger vessels, which, perforating the outer compact tissue, are distributed to the cavities of the spongy portion of the bone. In the long bones, numerous apertures may be seen at the ends near the articular surfaces; some of these give passage to the arteries of the larger set of vessels referred to; but the most numerous and largest apertures are for some of the veins of the cancellous tissue, which emerge apart from the arteries. The marrow in the body of a long bone is supplied by one large artery (or sometimes more), which enters the bone at the nutrient foramen situated in most cases near the center of the body. The *medullary* or *nutrient* artery, usually accompanied by one or two veins, sends branches proximally and distally, which ramify in the marrow and give twigs to the adjoining canals. The ramifications of this vessel anastomose with the arteries of the cancellous and compact tissues. In most of the flat, and in many of the short spongy bones, one or more large apertures are observed, which transmit to the central parts of the bone vessels corresponding to the nutrient arteries and veins. The **veins** emerge from the long bones in three places

(Kölliker): (1) one or two large veins accompany the artery; (2) numerous large and small veins emerge at the articular extremities; (3) many small veins pass out of the compact substance. In the flat cranial bones the veins are large, very numerous, and run in tortuous canals in the diploic tissue, the sides of the canals being formed by thin lamellae of bone, perforated here and there for the passage of branches from the adjacent cancelli. The same condition is also found in all cancellous tissue, the veins being enclosed and supported by osseous material, and having exceedingly thin coats. When a bone is divided, the vessels remain patulous, and do not contract in the canals in which they are contained. **Lymphatic vessels,** in addition to those found in the periosteum, have been traced by Cruikshank into the substance of bone, and Klein describes them as running in the Haversian canals. **Nerves** are distributed freely to the periosteum, and accompany the nutrient arteries into the interior of the bone. They are said by Kölliker to be most numerous in the articular extremities of the long bones, in the vertebrae and in the larger flat bones.

Minute Anatomy.—A transverse slice of compact bone cut with a saw may be ground down until it is sufficiently thin to be examined under the microscope with transmitted light. If this be examined with a rather low power the bone will be seen to be mapped out into a number of circular districts each consisting of a central hole surrounded by a number of concentric rings (Fig. 4–232). These districts are termed

Fig. 4–232.—Transverse section of compact bone tissue. Magnified. (Sharpey.)

Haversian systems; the central hole is an **Haversian canal,** and the rings are layers of bony tissue arranged concentrically around the central canal, and termed **lamellae.** Moreover, on closer examination it will be found that between these lamellae, and therefore also arranged concentrically around the central canal, are a number of little dark spots, the **lacunae,** and that these lacunae are connected with each other and with the central Haversian canal by a number of fine dark lines, which radiate like the spokes of a wheel and are called **canaliculi.** Filling in the irregular intervals which are left between these circular systems are other lamellae, with their lacunae and canaliculi running in various directions, but more or less curved (Fig. 4–231); they are termed **interstitial lamellae.** Again, other lamellae, found on the surface of the bone, are arranged parallel to its circumference; they are termed **circumferential,** or by some authors **primary** or **fundamental lamellae,** to distinguish them from those laid down around the axes of the Haversian canals, which are then termed **secondary** or **special lamellae** (Fig. 4–231).

The **Haversian canals,** seen in a transverse section of bone as round holes at the center of each Haversian system, appear as true canals in a longitudinal section (Fig. 4–233). The canals run parallel with the longitudinal axis of the bone for a short distance and then branch and communicate through cross channels (*Volkmann's canals*). They vary considerably in size, some being as much as 0.12 mm in diameter; the average size is, however, about 0.05 mm. Near the medullary cavity the canals are larger than those near the surface of the bone. Each canal contains one or two blood vessels, with a small quantity of delicate connective tissue and some nerve filaments. In the larger ones there are also lymphatic vessels, and cells with branching processes which communicate, through the canaliculi, with the branched processes of certain bone cells in the substance of the bone. Those canals near the surface of the bone open upon it by minute orifices, and those near the medullary cavity open in the same way into this space, so that the whole of the bone is permeated by a system of blood vessels running through the bony canals in the centers of the Haversian systems.

The **lamellae** are thin plates of bony tissue, and for the sake of illustration may be compared to sheets of paper pasted one over another around a central hollow cylinder. They are composed of collagenous bundles running in a parallel direction within each lamella. The bundles of adjacent lamellae, however, are more or less perpendicular to each other, with the result that in a cross section of stained decalcified bone, lamellae with bundles cut longitudi-

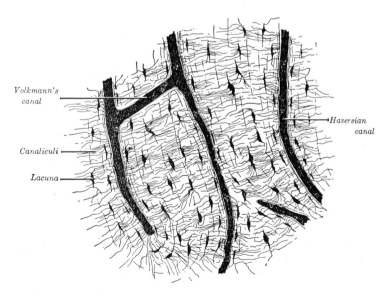

Fig. 4–233.—Longitudinal section of compact bone tissue. Magnified.

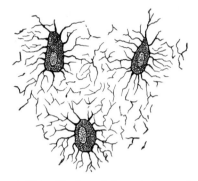

Fig. 4–234.—Nucleated bone cells and their processes, contained in the bone lacunae and their canaliculi respectively. From a section through the vertebra of an adult mouse. (Klein and Noble Smith.)

nally (*striated lamellae*) alternate with bundles cut across (*punctate lamellae*). Between the bundles and lamellae is a darker staining cement material containing deposits of calcareous salts. In many places the lamellae appear to be held together by **perforating fibers** which run obliquely through them, pinning them together. Similar fibers (*Sharpey's fibers*) run from the surface of the bone into the substance of the periosteum, anchoring the latter to the bone. These perforating fibers may be demonstrated in the dissecting room by pulling the entire scalp away from the calvaria which then is covered with whiskery projections representing the Sharpey's fibers as well as larger bundles containing minute blood vessels.

The **lacunae** are situated between the lamellae, and consist of a number of oblong spaces. In a microscopic section of dried bone, viewed by transmitted light, they appear as fusiform opaque spots. Each lacuna is occupied during life by a branched cell, termed a **bone cell or osteocyte,** the processes from which extend into the canaliculi (Fig. 4–234).

The **canaliculi** are exceedingly minute channels, crossing the lamellae and connecting the lacunae with neighboring lacunae and also with the Haversian canal. From the Haversian canal a number of canaliculi are given off, which radiate from it, and open into the first set of lacunae between the first and second lamellae. From these lacunae a second set of canaliculi is given off; these run outward to the next series of lacunae,

and so on until the periphery of the Haversian system is reached; here the canaliculi given off from the last series of lacunae do not as a rule communicate with the lacunae of neighboring Haversian systems, but after passing outward for a short distance form loops and return to their own lacunae. Thus every part of an Haversian system is supplied with nutrient fluids derived from the vessels in the Haversian canal and distributed through the canaliculi and lacunae.

The **bone cells** are contained in the lacunae, which, however, they do not completely fill. They are flattened nucleated branched cells, homologous with those of connective tissue; the branches, especially in young bones, pass into the canaliculi from the lacunae.

In thin plates of bone (as in the walls of the spaces of cancellous tissue) the Haversian canals are absent, and the canaliculi open into the spaces of the cancellous tissue (medullary spaces), which thus have the same function as the Haversian canals.

Chemical Composition.—Bone consists of an animal and an earthy part intimately combined together.

The animal part may be obtained by immersing a bone for a considerable time in dilute mineral acid, after which process the bone comes out exactly the same shape as before, but perfectly flexible, so that a long bone (one of the ribs, for example) can easily be tied in a knot. If now a transverse section is made the same general arrangement of the Haversian canals, lamellae, lacunae, and canaliculi is seen.

The earthy part may be separately obtained by calcination, by which the animal matter is completely burnt out. The bone will still retain its original form, but it will be white and brittle, will have lost about one-third of its original weight, and will crumble down with the slightest force. The earthy matter is composed chiefly of calcium phosphate, about 58 per cent of the weight of the bone, calcium carbonate about 7 per cent, calcium fluoride and magnesium phosphate from 1 to 2 per cent each and sodium chloride less than 1 per cent; they confer on bone its hardness and rigidity, while the animal matter (*ossein*) determines its tenacity.

Ossification

Some bones are preceded by membrane, such as those forming the roof and sides of the skull; others, such as the bones of the limbs, are preceded by rods of cartilage. Hence two kinds of ossification are described: the **intramembranous** and the **intracartilaginous**.

INTRAMEMBRANOUS OSSIFICATION.—In the case of bones which are developed in membrane, no cartilaginous mold precedes the appearance of the bony tissue. The membrane which occupies the place of the future bone is of the nature of connective tissue, and ultimately forms the periosteum; it is composed of fibers and granular cells in a matrix. The peripheral portion is more fibrous, while, in the interior the cells or *osteoblasts* predominate; the whole tissue is richly supplied with blood vessels. At the outset of the process of bone formation a little network of spicules is noticed radiating from the point or center of ossification. These rays consist at their growing points of a network of fine clear fibers and granular cells with an intervening ground substance. The fibers are termed **osteogenetic fibers**, and are made up of fine fibrils differing little from those of white fibrous tissue. The membrane soon assumes a dark and granular appearance from the deposition of calcareous granules in the fibers and in the intervening matrix, and in the calcified material some of the granular cells or osteoblasts are enclosed. By the fusion of the calcareous granules the tissue again assumes a more transparent appearance, but the fibers are no longer so distinctly seen. The involved osteoblasts form the corpuscles of the future bone, the spaces in which they are enclosed constituting the lacunae. As the osteogenetic fibers grow out to the periphery they continue to ossify, and give rise to fresh bone spicules. Thus a network of bone is formed, the meshes of which contain the blood vessels and a delicate connective tissue crowded with osteoblasts. The bony trabeculae thicken by the addition of fresh layers of bone formed by the osteoblasts on their surface, and the meshes are correspondingly encroached upon. Subsequently successive layers of bony tissue are deposited under the periosteum and around the larger vascular channels which become the Haversian canals, as the bone increases in thickness.

INTRACARTILAGINOUS OSSIFICATION.—Just before ossification begins the mass is entirely cartilaginous, and in a long bone, which may be taken as an example, the process commences in the center and proceeds toward the extremities, which for some time remain cartilaginous. Subsequently a similar process commences in one or more places in those extremities and gradually extends through them. The extremities do not, however, become joined to the body of the bone by bony tissue until growth has ceased; between the body and either extremity a layer of cartilaginous tissue termed the **epiphyseal cartilage** persists for a definite period.

The first step in the ossification of the cartilage is that the cartilage cells, at the point where ossification is commencing and which is termed a **center of ossification**, enlarge and arrange themselves in rows (Fig. 4–235). The matrix in which they are imbedded increases in quantity, so that the cells become separated from each other. A deposit of calcareous material now takes place in this matrix, between the rows of cells, so that they become separated from each other by longitudinal columns of calcified matrix, presenting a granular and opaque appearance. Here and there the matrix between two cells of the same row also becomes calcified, and transverse bars of calcified substance stretch across from one calcareous column to another. Thus there are longitudinal groups of the cartilage cells enclosed in oblong cavities, the walls of which are formed of calcified matrix which cuts off all nutrition from the cells; the cells, in consequence, atrophy, leaving spaces called the primary areolae.

SUBPERIOSTEAL OSSIFICATION.—At the same time that this process is going on in the center of the solid bar of cartilage, certain changes are taking place on its surface. This is covered by a very vascular membrane, the **perichondrium**, entirely similar to the embryonic connective tissue already described as constituting the basis of membrane bone; on the inner surface of this—that is to say, on the surface in contact with the cartilage—are gathered the formative

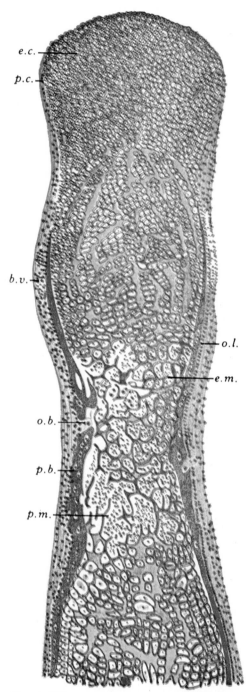

e.c.—

p.c.—

b.v.—

—o.l.

—e.m.

o.b.—

p.b.—

p.m.—

Fig. 4–235.—Osteogenesis. Subperiosteal or inframembranous stage of ossification in an embryonic long bone. *b.v.*, blood vessel in periosteum; *e.c.*, epiphyseal cartilage; *e.m.*, embryonic marrow space; *o.b.*, site of entrance of osteogenic bud from periosteal blood vessel; *o.l.*, osteoblastic layer; *p.b.*, subperiosteal bone; *p.c.*, perichondrium; *p.m.*, primitive marrow cavity. × 20 (Redrawn from Sobotta.)

cells, the **osteoblasts.** By the agency of these cells a thin layer of bony tissue is formed between the perichondrium and the cartilage, by the *intramembranous* mode of ossification just described. There are then, in this first stage of ossification, two processes going on simultaneously: in the center of the cartilage the formation of a number of oblong spaces, formed of calcified matrix and containing the withered cartilage cells, and on the surface of the cartilage the formation of a layer of true membrane bone.

The second stage of the intracartilaginous ossification (Fig. 4–235) consists in the perforation of the membrane bone by a process of the deeper or osteogenetic layer of the perichondrium, which has now become periosteum, consisting of blood vessels and bone-forming cells—**osteoblasts,** and **osteoclasts,** or **bone destroyers.** The latter are similar to the giant cells found in marrow and are associated with the absorption of the matrix which excavates passages through newly formed areas. Wherever these passages come in contact with the calcified walls of the primary areolae they absorb them, and thus cause a fusion of the original cavities and the formation of larger spaces, which are termed the secondary areolae or **medullary spaces.** These secondary spaces become filled with embryonic marrow, consisting of osteoblasts and vessels, derived, in the manner described above, from the osteogenetic layer of the periosteum (Fig. 4–236).

Thus far there has been traced the formation of enlarged medullary spaces, the perforated walls of which are still formed by calcified cartilage matrix, containing an **embryonic marrow** derived from the processes sent in from the osteogenetic layer of the periosteum, and consisting of blood vessels and osteoblasts. The walls of these secondary areolae are at this time of only inconsiderable thickness, but they become thickened by the deposition of layers of true bone on their surface. This process takes place in the following manner: Some of the osteoblasts of the embryonic marrow, after undergoing rapid division, arrange themselves as an epithelioid layer on the surface of the wall of the space. This layer of osteoblasts forms a bony stratum, and thus the wall of the space becomes gradually

covered with a layer of true osseous substance in which some of the bone-forming
cells are included as bone corpuscles. The
next stage in the process consists in the
removal of these primary bone spicules by
the osteoclasts. One of these giant cells may
be found lying in a Howship's foveola at
the free end of each spicule. The removal
of the primary spicules goes on *pari passu*
with the formation of permanent bone by

the periosteum, and in this way the medullary cavity of the body of the bone is formed.

This series of changes has been gradually
proceeding toward the end of the body of
the bone, so that in the ossifying bone all
the changes described above may be seen
in different parts, from the true bone at
the center of the body to the hyaline
cartilage at the extremities.

While the ossification of the cartilaginous

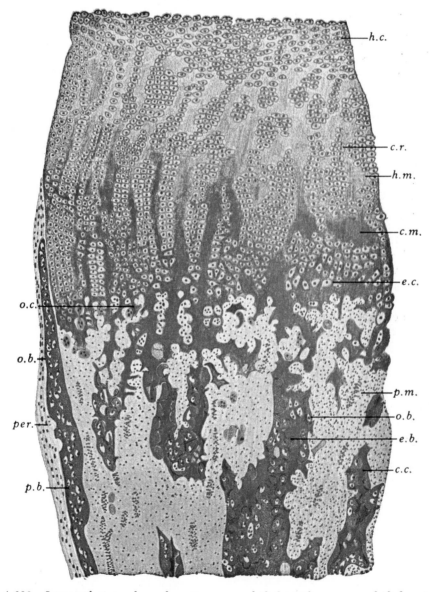

FIG. 4–236.—Intracartilaginous bone formation. *c.c.* calcified cartilage; *c.m.* calcified matrix; *c.r.*,
cartilage cells in rows; *e.b.*, endochondral bone; *e.c.*, enlarged cartilage cell and lacuna; *h.c.* hyaline cartilage; *h.m.*, hyaline matrix; *o.b.*, osteoblast; *o.c.*, osteoclast in Howship's lacuna; *p.b.*, subperiosteal bone;
per., periosteum; *p.m.* primitive marrow with blood vessel. ✕ 40. (Redrawn from Sobotta.)

body is extending toward the articular ends, the cartilage immediately in advance of the osseous tissue continues to grow until the length of the adult bone is reached.

During the period of growth the articular end, or epiphysis, remains for some time entirely cartilaginous, then a bony center appears within it, and initiates the process of intracartilaginous ossification. This is a secondary center of ossification. The epiphysis remains separated from the body by a narrow cartilaginous layer for a definite length of time, but ultimately ossifies. The distinction between body and epiphysis is obliterated, and the bone assumes its completed form and shape. The same remarks also apply to such processes of bone as are separately ossified, e.g., the trochanters of the femur. The bones therefore continue to grow until the body has acquired its full stature. They increase in length by ossification continuing to extend behind the epiphyseal cartilage, which goes on growing in advance of the ossifying process. They increase in circumference by deposition of new bone from the deeper layer of the periosteum, and at the same time an absorption takes place from within, by which the medullary cavities are increased.

The permanent bone formed by the periosteum when first laid down is cancellous in structure. Later the osteoblasts contained in its spaces become arranged in the concentric layers characteristic of the Haversian systems, and are included as bone corpuscles.

The number of ossific centers varies in different bones. In most of the short bones ossification commences at a single point near the center, and proceeds toward the surface. In the long bones there is a central point of ossification for the body or diaphysis: and one or more for each extremity, the epiphysis. That for the body is the first to appear. The times of union of the epiphyses with the body vary inversely with the dates at which their ossifications began (with the exception of the fibula) and regulate the direction of the nutrient arteries of the bones. Thus, the nutrient arteries of the bones of the arm and forearm are directed toward the elbow, since the epiphyses at this joint become united to the bodies before those at the opposite extremities. In the lower limb, on the other hand, the nutrient arteries are directed away from the knee: that is, upward in the femur, downward in the tibia and fibula; and in them it is observed that the upper epiphysis of the femur, and the lower epiphyses of the tibia and fibula, unite first with the bodies. Where there is only one epiphysis, the nutrient artery is directed toward the other end of the bone; as toward the acromial end of the clavicle, toward the distal ends of the metacarpal bone of the thumb and the metatarsal bone of the great toe, and toward the proximal ends of the other metacarpal and metatarsal bones.

REFERENCES

EMBRYOLOGY, OSSIFICATION, AND GROWTH

BRODIE, A. G. 1941. On the growth pattern of the human head from the third month to the eighth year of life. Amer. J. Anat., 68, 209–262.

FAWCETT, E. 1905. Ossification of the lower jaw in man. J. Amer. Med. Ass., 45, 696–705.

FRANCIS, C. C., P. P. WERLE and A. BEHM. 1939. The appearance of centers of ossification from birth to 5 years. Amer. J. Phys. Anthrop., 24, 273–299.

GRAY, D. J., E. GARDNER and R. O'RAHILLY. 1957. The prenatal development of the skeleton and joints of the human hand. Amer. J. Anat., 101, 169–224.

GRAY, D. L. 1969. The prenatal development of the human humerus. Amer. J. Anat., 124, 431–445.

GREULICH, W. W. and S. I. PYLE. 1950. Radiographic Atlas of Skeletal Development of the Hand and Wrist. xiii + 190 pages. Stanford University Press, Stanford, California.

HILL, A. H. 1939. Fetal age assessment by centers of ossification. Amer. J. Phys. Anthrop., 24, 251–272.

KRAUS, B. S. 1961. Sequence of appearance of primary centers of ossification in the human foot. Amer. J. Anat., 109, 103–115.

LAURENSON, R. D. 1964. The primary ossification of the human ilium. Anat. Rec., 148, 209–218.

LOWRANCE, E. W. 1955. Roentgenographic record of growth of the femur of the rabbit. Growth, 19, 247–256.

McKERN, T. W. and T. D. STEWART. 1957. Skeletal Age Changes in Young American Males. Technical Report. viii + 179 pages. Quartermaster Research & Development Center, Natick, Mass.

NOBACK, C. R. 1954. The appearance of ossification centers and the fusion of bones. Amer. J. Phys. Anthrop., 12, 63–70.

ORTIZ, MANUEL HIGINIO and A. G. BRODIE. 1949. On the growth of the human head from birth to the third month of life. Anat. Rec., *103*, 311–333.

PYLE, S. I. and N. L. HOERR. 1955. *Radiographic Atlas of Skeletal Development of the Knee. A Standard of Reference.* viii + 82 pages. Charles C Thomas, Springfield, Illinois.

ROBERTS, W. H. 1962. Femoral torsion in normal human development as related to dysplasia. Anat. Rec., *143*, 369–375.

ROCHE, ALEXANDER F. 1964. Epiphyseal ossification and shaft elongation in human metatarsal bones. Anat. Rec., *149*, 449–452.

WARWICK, R. 1950. The relation of the direction of the mental foramen to the growth of the human mandible. J. Anat. (Lond.), *84*, 116–120.

WOO, JU-KANG. 1949. Ossification and growth of the human maxilla, premaxilla and palate bone. Anat. Rec., *105*, 737–762.

HISTOLOGY AND OSTEOGENESIS

BARBOUR, E. P. 1950. A study of the structure of fresh and fossil human bone by means of the electron microscope. Amer. J. Phys. Anthrop., *8*, 315–330.

BEVELANDER, G. and P. L. JOHNSON. 1950. A histochemical study of the development of membrane bone. Anat. Rec., *108*, 1–22.

BOHATIRCHUK, F. 1963. Microradiology of mammalian bone. J. Canad. Ass. Radiol., *14*, 29–38.

COHEN, J. and W. H. HARRIS. 1958. The three-dimensional anatomy of Haversian systems. J. Bone Jt. Surg., *40-A*, 419–434.

DUDLEY, H. R. and D. SPIRO. 1961. The fine structure of bone cells. J. biophys. biochem. Cytol., *11*, 627–649.

ENLOW, DONALD H. 1962. A study of the post-natal growth and remodeling of bone. Amer. J. Anat., *110*, 79–101. Functions of the Haversian system. Ibid., *110*, 269–305.

EVANS, F. GAYNOR. 1958. Relations between the microscopic structure and tensile strength of human bone. Acta Anat. (Basel), *35*, 285–301.

FROST, H. M. 1964. *Mathematical Elements of Lamellar Bone Remodeling.* ix + 127 pages. Charles C Thomas, Springfield, Illinois.

FROST, H. M. 1964. *The Laws of Bone Structure.* xiii + 167 pages. Charles C Thomas, Springfield, Illinois.

HAVERS, CLOPTON. 1691. Osteolgia nova, or some new observations on the bones, and the parts belonging to them. London.

KNESE, KARL-HEINRICH. 1958. *Knochenstruktur als Verbundau. Versuch einer technischen Deutung der Materialstruktur des Knochens.* v + 56 pages. Georg Thieme Verlag, Stuttgart.

LACROIX, P. 1951. *The Organization of Bones.* Trans. by S. Gilder. viii + 235 pages. The Blakiston Co., Philadelphia.

STILLWELL, D. L., JR. and D. J. GRAY. 1954. The microscopic structure of periosteum in areas of tendinous contact. Anat. Rec., *120*, 663–678.

THORPE, E. J., B. B. BELLONY and R. F. SELLERS. 1963. Ultrasonic decalcification of bone. J. Bone Jt. Surg., *45-A*, 1257–1259.

OSTEOLOGY, VARIATIONS, AND PHYSICAL ANTHROPOLOGY

ANGEL, J. L. 1946. Skeletal change in ancient Greece. Amer. J. Phys. Anthrop., *4*, 69–98.

BREATHNACH, A. S. 1960. 5th Edition. *Frazer's Anatomy of the Human Skeleton.* viii + 247 pages, 197 figures. J. & A. Churchill, Ltd., London, and Little, Brown & Company, Boston.

CLARK, W. E. LeGros. 1964. *The Fossil Evidence for Human Evolution.* xii + 201 pages. University of Chicago Press, Chicago.

GOLDSTEIN, M. S. 1957. Skeletal pathology of early Indians in Texas. Amer. J. Phys. Anthrop., *15*, 299–312.

LOWRANCE, E. W. and H. B. LATIMER. 1957. Weights and linear measurements of 105 human skeletons from Asia. Amer. J. Anat., *101*, 445–460.

MERZ, A. L., M. TROTTER and R. R. PETERSON. 1956. Estimation of skeleton weight in the living. Amer. J. Phys. Anthrop., *14*, 589–610.

TROTTER, M. and G. C. GLESER. 1958. A re-evaluation of estimation of stature based on measurements of stature taken during life and of long bones after death. Amer. J. Phys. Anthrop., *16*, 79–124.

VERTEBRAL COLUMN AND RIBS

BRANNON, E. W. 1963. Cervical rib syndrome. J. Bone Jt. Surg., *45-A*, 977–998.

DAVIS, P. R. 1961. Human lower lumbar vertebrae: some mechanical and osteological considerations. J. Anat. (Lond.), *95*, 337–344.

EVANS, F. GAYNOR and H. R. LISSNER. 1959. Biomechanical studies on the lumbar spine and pelvis. J. Bone Jt. Surg., *41A*, 278–290.

FAWCETT, E. 1931. A note on the identification of the lumbar vertebrae of man. J. Anat. (Lond.), *66*, 384–386.

FRANCIS, C. C. 1955. Dimensions of the cervical vertebrae. Anat. Rec., *122*, 603–610.

HADLEY, LEE A. 1964. *Anatomico-Roentgenographic Studies of the Spine.* ix + 545 pages, illustrated. Charles C Thomas, Springfield, Illinois.

HAINES, R. WHEELER. 1946. Movements of the first rib. J. Anat. (Lond.), *80*, 94–100.

KEEGAN, J. JAY. 1953. Alterations of the lumbar curve related to posture and seating. J. Bone Jt. Surg., *35A*, 589–603.

KENDRICK, GEORGE S. and NORMAN L. BIGGS. 1963. Incidence of the ponticulus posticus of the first cervical vertebra between ages six to seventeen. Anat. Rec., *145*, 449–453.

LANIER, R. R. 1954. Some factors to be considered in the study of lumbosacral fusion. Amer. J. Phys. Anthrop., *12*, 363–372.

ROWE, G. G. and M. B. ROCHE. 1953. The etiology of separate neural arch. J. Bone Jt. Surg., *35A*, 102–110.

RUSSELL, H. E. and G. T. AITKEN. 1963. Congenital absence of the sacrum and lumbar vertebrae with prosthetic management. J. Bone Jt. Surg., *45-A*, 501–508.

STEWART, T. D. 1954. Metamorphosis of the joints of the sternum in relation to age changes in other bones. Amer. J. Phys. Anthrop., *12*, 519–536.

TROTTER, M. 1947. Variations of the sacral canal: Their significance in the administration of caudal analgesia. Anesth. Analgesia, *26*, 192–202.

WELLS, L. H. 1963. Congenital deficiency of the vertebral pedicle. Anat. Rec., *145*, 193–196.

SKULL

COBB, W. M. 1943. The cranio-facial union and the maxillary tuber in mammals. Amer. J. Anat., *72*, 39–111.

DI DIO, LIBERATO J. A. 1962. The presence of the eminentia orbitalis in the os zygomaticum of Hindu Skulls. Anat. Rec., *142*, 31–39.

ETTER, LEWIS E. 1955. *Atlas of Roentgen Anatomy of the Skull*. xv + 215 pages. Charles C Thomas, Springfield, Illinois.

MILLER, WILLIAM A. 1965. *The Keys to Orthopedic Anatomy*. vii + 155 pages. Charles C Thomas, Springfield, Illinois.

OSCHINSKY, LAWRENCE. 1960. Two recently discovered human mandibles from Cape Dorset Sites on Sugluk and Mansel Islands. Anthropologica N.S., *11*, 1–16.

PARSONS, F. G. 1909. The topography and morphology of the human hyoid bone. J. Anat. Phys., *43*, 279–291.

SHILLER, W. R. and O. B. WISWELL. 1954. Lingual foramina of the mandible. Anat. Rec., *119*, 387–390.

SINCLAIR, J. B. and JACK MCKAY. 1945. Median hare lip, cleft palate and glossal agenesis. Anat. Rec., *91*, 155–160.

VAN DER KLAAUW, C. J. 1963. *Projections, Deepenings and Undulations of the Skull in Relation to the Attachment of Muscle*. 247 pages, illustrated. North Holland Publishing Company, Amsterdam, Holland.

WALENSKY, NORMAN AARON. 1964. A re-evaluation of the mastoid region of contemporary and fossil man. Anat. Rec., *149*, 67–72.

UPPER LIMB

CARROLL, S. E. 1963. A study of the nutrient foramina of the humeral diaphysis. J. Bone Jt. Surg., *45-B*, 176–181.

EVANS, F. G., A. ALFARO and S. ALFARO. 1950. An unusual anomaly of the superior extremities in a Tarascan Indian girl. Anat. Rec., *106*, 37–48.

FRANTZ, C. H. and R. O'RAHILLY. 1961. Congenital skeletal limb deficiencies. J. Bone Jt. Surg., *43-A*, 1202–1224.

GRAY, D. J. and ERNEST GARDNER. 1969. The prenatal development of the human humerus. Am. J. Anat., *124*, 431–445.

HANDFORTH, J. R. 1950. Polydactylism of the hand in southern Chinese. Anat. Rec., *106*, 119–125.

LACHMAN, ERNEST. 1953. Pseudo-epiphyses in hand and foot. Amer. J. Roentgenol., *70*, 149–151.

LOWER LIMB

BARNETT, C. H. 1962. The normal orientation of the human hallux and the effect of footwear. J. Anat. (Lond.), *96*, 489–494.

DAVIS, G. G. 1927. Os Vesalianum pedis. Amer. J. Roentgenol., *17*, 551–553.

EVANS, F. GAYNOR and H. R. LISSNER. 1955. Studies on pelvic deformations and fractures. Anat. Rec., *121*, 141–166.

HAXTON, H. 1945. The function of the patella and the effects of its excision. Surg. Gynec. Obstet., *80*, 389.

HENDERSON, R. S. 1963. Os intermetatarseum and a possible relationship to hallux valgas. J. Bone Jt. Surg., *45-B*, 117–121.

IRANI, ROSHEN and M. SHERMAN. 1963. The pathological anatomy of club foot. J. Bone Jt. Surg., *45-A*, 45–52.

MORTON, D. J. 1935. *The Human Foot—Its Evolution, Physiology, and Functional Disorders*. xiii + 244 pages. Columbia University Press, New York.

NELSON, EDWARD M. 1963. A report of a 7-toed foot. Anat. Rec., *147*, 1–3.

ROGERS, L. 1928. The styloid epiphysis of the fifth metatarsal bone. J. Bone Jt. Surg., *10*, 197.

SHANDS, ALFRED R. and IRL W. WENTZ. 1953. Congenital anomalies, accessor bones, and osteochondritis in the feet of 850 children. Surg. Clin. N.A., 1643–1666.

SINGH, INDERBIR. 1959. Squatting facets on the talus and tibia in Indians. J. Anat. (Lond.), *93*, 540–550.

SINGH, I. 1960. Variations in the metatarsal bones. J. Anat. (Lond.), *94*, 345–350.

WRAY, JAMES B. and C. N. HERNDON. 1963. Hereditary transmission of congenital coalition of the calcaneus to the navicular. J. Bone Jt. Surg., *45-A*, 365–372.

BLOOD VESSELS AND NERVES

CROCK, H. V. 1965. A revision of the anatomy of the arteries supplying the upper end of the human femur. J. Anat. (Lond.), *99*, 77–88.

LAING, P. G. 1956. The arterial supply of the adult humerus. J. Bone Jt. Surg., *38-A*, 1105–1116.

MILLER, MALCOLM R. and MICHIKO KASAHARA. 1963. Observations on the innervation of human long bones. Anat. Rec., *145*, 13–23.

SHERMAN, MARY S. 1963. The nerves of bone. J. Bone Jt. Surg., *45-A*, 522–528.

TUCKER, F. R. 1949. Arterial supply to the femoral head and its clinical importance. J. Bone Jt. Surg., *31-A*, 82–93.

EXPERIMENTAL

CRELIN, E. S. 1960. The development of bony pelvic sexual dimorphism in mice. Ann. N. Y. Acad. Sci., *84*, 479–512.

HAMRE, CHRISTOPHER J. and VERNON L. YEAGER. 1957. Influence of muscle section on exostoses of lathyric rats. A.M.A. Arch. Path., *64*, 426–433.

HOLTZER, HOWARD and S. R. DETWILER. 1953. An experimental analysis of the development of the spinal column. J. exp. Zool., *123*, 335–370.

HOROWITZ, S. L. and H. H. SHAPIRO. 1955. Modification of skull and jaw architecture following removal of the masseter muscle in the rat. Amer. J. Phys. Anthrop., *13*, 301–308.

MEDNICK, LOIS and S. L. WASHBURN. 1956. The role of the sutures in the growth of the braincase of the infant pig. Amer. J. Phys. Anthrop., *14*, 175–192.

PFEIFFER, C. A. 1948. Development of bone from transplanted marrow in mice. Anat. Rec., *102*, 225–244.

RUTH, ELBERT B. 1961. Basophilic islands in osseous tissue and their relation to resorption. Anat. Rec., *140*, 307–320.

WILLIAMS, ROY G. 1962. Comparison of living autogenous and homogenous grafts of cancellous bone heterotopically placed in rabbits. Anat. Rec., *143*, 93–105.

WILLIAMS, ROY G. 1963. Studies of cartilage and osteoid arising spontaneously and experimental attempts to induce their formation in ear chambers. Anat. Rec., *146*, 93–108.

YOUNG, RICHARD W. 1962. Autoradiographic studies on postnatal growth of the skull in young rats injected with tritiated glycine. Anat. Rec., *143*, 1–14.

PATHOLOGY

BENDA, CLEMENS E. 1940. Growth disorder of the skull in mongolism. Amer. J. Path., *16*, 71–86.

GOLDENBERG, R. R. 1951. The skull in Paget's disease. J. Bone Jt. Surg., *33-A*, 911–922.

JIT, I. 1960. Lead shots amongst fragments of burnt bones. J. anat. Soc. India, Calcutta, *9*, 196.

MAITLAND, G. D. 1964. *Vertebral Manipulation.* viii + 146 pages, 53 figures. Butterworth, Inc., London and Washington.

STECHER, ROBERT M. 1958. Osteoarthritis in the gorilla. Description of a skeleton with involvement of the knee and the spine. Lab. Invest., *7*, 445–457.

SWANSON, ALFRED B. 1963. Restoration of hand function by the use of partial or total prosthetic replacement. J. Bone Jt. Surg., *45-A*, 276-283.

THOMSON, JAMES E. M. 1942. Fracture of the patella treated by removal of the loose fragments and plastic repair of the tendon. Surg. Gynec. Obstet., *74*, 860–866.

CONNECTIVE TISSUE

BARRNETT, R. J. and E. G. BALL. 1960. Metabolic and ultrastructural changes induced in adipose tissue by insulin. J. Biophys. Biochem. Cytol., *8*, 83-101.

BILLINGHAM, R. E. and W. K. SILVERS. 1961. *Transplantation of Tissues and Cells.* x + 149 pages, illustrated. Wistar Institute Press, Philadelphia.

CLARK, E. R. and E. L. CLARK. 1940. Microscopic studies of the new formation of fat in living adult rabbits. Amer. J. Anat., *67*, 255-285.

CONN, H. J. 1961. *Biological Stains.* x + 355 pages, illustrated. Williams & Wilkins Company, Baltimore.

DOWNEY, H. 1955. The development of histiocytes and macrophages from lymphocytes. J. Lab. Clin. Med., *45*, 499-507.

FAWCETT, D. W. and I. C. JONES. 1949. The effects of hypophysectomy, adrenalectomy and of thiouracil feeding on the cytology of brown adipose tissue. Endocrinology, *45*, 609-621.

GLICK, D. 1962. *Quantitative Chemical Techniques of Histo- and Cytochemistry.* Vol. 1, xxi + 470 pages, 192 figures. Interscience Publishers, John Wiley & Sons, New York.

HALL, D. 1963. *International Review of Connective Tissue Research.* Vol. 1, xiii + 401 pages, illustrated. Academic Press, New York.

HODGE, A. J. and F. O. SCHMITT. 1961. M. V. Edds, Jr., Editor. The Tropocollagen Macromolecule and Its Properties of Ordered Interaction, *in Macromolecular Complexes.* Ronald Press, New York. 6th Ann. Symp. Publication, 19.

KELSALL, M. A. and E. B. CRABB. 1959. *Lymphocytes and Mast Cells.* xvi + 399 pages, 31 figures. The Williams and Wilkins Company, Baltimore.

LILLIE, R. D. 1965. *Histopathologic Technic and Practical Histochemistry.* 3rd Edition. xii + 715 pages, illustrated. McGraw-Hill Book Company, New York.

ODOR, D. L. 1954. Observations of the rat mesothelium with the electron and phase microscopes. Amer. J. Anat., *95*, 433-465.

PAFF, G. H. and D. D. MERGENTHALER. 1955. Vacuolation in normal mast cells and in mast cells treated with protamine sulfate. Anat. Rec., *121*, 579-591.

PORTER, K. R. and G. D. PAPPAS. 1959. Collagen formation by fibroblasts of the chick embryo dermis. J. Biophys. Biochem. Cytol., *5*, 153-166.

RILEY, J. F. 1959. *The Mast Cells.* x + 182 pages, 65 figures. E. and S. Livingstone Ltd., Edinburgh, and The Williams and Wilkins Company, Baltimore.

SCHMITT, F. O., J. GROSS, and J. H. HIGHBERGER. 1955. Tropocollagen and the properties of fibrous collagen. Exp. Cell. Res., Suppl., *3*, 326-334.

TÖRÖ, I. 1961. *Makrophagen und Phagozytose.* 153 pages, illustrated. Adadémiai Kiádo, Budapest.

5 | *Joints and Ligaments*

rticulations or joints are the functional connections between the different bones of the skeleton. In some of the articulations the bones are held together by immovable attachments, as in the sutures or articulations between practically all the bones of the skull. The adjacent margins of the bones are almost in contact, being separated merely by a thin layer of fibrous membrane, named the **sutural ligament.** In certain regions at the base of the skull this fibrous membrane is replaced by a layer of cartilage. Where *slight movement* combined with great strength is required, the osseous surfaces are united by tough and elastic **fibrocartilages,** as in the joints between the vertebral bodies, and in the interpubic articulation. In the *freely movable* joints the surfaces are completely separated; the bones forming the articulation are expanded for greater convenience of mutual connection, covered by **cartilage** and enveloped by **capsules** of fibrous tissue. The cells lining the interior of the fibrous capsule form an imperfect membrane—the **synovial membrane**—which secretes a lubricating fluid. The joints are strengthened by strong fibrous bands called **ligaments,** which extend between the bones forming the joint.

DEVELOPMENT OF THE JOINTS

The mesoderm from which the different parts of the skeleton are formed shows at first no differentiation into masses corresponding with the individual bones. Thus continuous cores of mesoderm form the axes of the limb buds and a continuous column of mesoderm the future vertebral column. The first indications of the bones and joints are circumscribed condensations of the mesoderm; these condensed parts become chondrified and finally ossified to form the bones of the skeleton. The intervening non-condensed portions consist at first of undifferentiated mesoderm, which may develop in one of three directions. It may be converted into fibrous tissue as in the case of the skull bones, a fibrous joint being the result, or it may become partly cartilaginous, in which case a slightly movable joint is formed. Again, it may become looser in texture and a cavity ultimately appear in its midst; the cells lining the sides of this cavity form a synovial membrane and thus a freely movable joint is developed.

The tissue surrounding the original mesodermal core forms fibrous sheaths for the developing bones, *i.e.,* periosteum and perichondrium, which are continued between the ends of the bones over the synovial membrane as the capsules of the joints. These capsules are not of uniform thickness, so that in them may be recognized especially strengthened bands which are described as ligaments. This, however, is not the only method of formation of ligaments. In some cases by modification of, or derivations from, the tendons surrounding the joint, additional ligamentous bands are provided to further strengthen the articulations.

In several of the movable joints the mesoderm which originally existed between the ends of the bones does not become completely absorbed—a portion of it persists and forms an articular disk. These disks may be intimately associated in their development with the muscles surrounding the joint, *e.g.,* the menisci of the knee joint, or with cartilaginous elements, representatives of skeletal structures, which are vestigial in human anatomy, *e.g.,* the articular disk of the sternoclavicular joint.

CLASSIFICATION OF JOINTS

The **articulations** or **joints** *(juncturae ossium)* are divided into the three following classes according to their structural composition and movability: *juncturae fibrosae* (synarthroses) or immovable, *juncturae cartilagineae* (amphiarthroses) or slightly movable, and *juncturae synoviales* (diarthroses) or freely movable joints.

Juncturae fibrosae *(synarthoses; immovable* articulations) include all those articulations in which the surfaces of the bones are in almost direct contact, fastened together by intervening connective tissue or hyaline cartilage, and in which there is no appreciable motion, as in the joints between the bones of the skull, except those of the mandible. There are three varieties of juncturae fibrosae: syndesmosis, sutura, and gomphosis.

Syndesmosis is that form of articulation in which two bones are united by interosseous ligaments, as in the inferior tibiofibular articulation.

Sutura is that form of articulation where the contiguous margins of the bones are united by a thin layer of fibrous tissue; it is met with only in the skull (Fig. 5–1). When the margins of the bones are connected by a series of processes, and indentations interlocked together, the articulation is termed a **true suture** *(sutura vera)*; and of

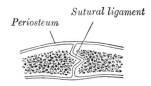

Periosteum   Sutural ligament

FIG. 5–1.—Section across the sagittal suture.

this there are three varieties: sutura dentata, serrata, and limbosa. The margins of the bones are not in direct contact, being separated by a thin layer of fibrous tissue, continuous externally with the pericranium, internally with the dura mater. The **sutura dentata** is so called from the tooth-like form of the projecting processes, as in the suture between the parietal bones. In the **sutura serrata** the edges of the bones are serrated like

the teeth of a fine saw, as between the two portions of the frontal bone. In the **sutura limbosa**, there is besides the interlocking, a certain degree of bevelling of the articular surfaces, so that the bones overlap one another, as in the suture between the parietal and frontal bones. When the articulation is formed by roughened surfaces placed in apposition with one another, it is termed a **false suture** *(sutura notha)*, of which there are two kinds: the **sutura squamosa**, formed by the overlapping of contiguous bones by broad bevelled margins, as in the squamosal suture between the temporal and parietal, and the **sutura plana** *(harmonia)*, where there is simple apposition of contiguous rough surfaces, as in the articulation between the maxillae, or between the horizontal parts of the palatine bones. **Schindylesis** is that form of articulation in which a thin plate of bone is received into a cleft or fissure formed by the separation of two laminae in another bone, as in the articulation of the rostrum of the sphenoid and perpendicular plate of the ethmoid with the vomer, or in the reception of the latter in the fissure between the maxillae and between the palatine bones.

Gomphosis is articulation by the insertion of a conical process into a socket; this is not illustrated by any articulation between bones, properly so called, but is seen in the articulations of the roots of the teeth with the alveoli of the mandible and maxilla.

Junctura cartilaginea *(amphiarthrosis; slightly movable articulation)* is that form of articulation in which the contiguous bony surfaces are united by cartilage; there are two varieties: synchondrosis and symphysis.

Synchondrosis (Fig. 5–2) is a temporary form of joint, for the cartilage is converted into bone before adult life. Such joints are

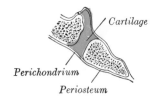

Cartilage

Perichondrium

Periosteum

FIG. 5–2.—Section through occipitosphenoid synchondrosis of an infant.

found between the epiphyses and bodies of long bones, between the occipital and the sphenoid at birth and for some years after, and between the petrous portion of the temporal and the jugular process of the occipital.

Symphysis is the articulation in which the contiguous bony surfaces are connected by broad, flattened disks of fibrocartilage of a more or less complex structure, as in the articulations between the bodies of the vertebrae or the two pubic bones (Fig. 5–3).

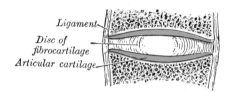

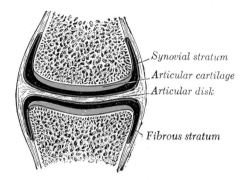

FIG. 5–5.—Diagrammatic section of a synovial joint with an articular disk.

FIG. 5–3.—Diagrammatic section of a symphysis.

Juncturae synoviales (*diarthroses; freely movable articulations*) include most of the joints in the body. In a junctura synovialis the contiguous bony surfaces are covered with articular cartilage, and connected by ligaments lined by synovial membrane (*articulatio simplex*) (Fig. 5–4). The joint

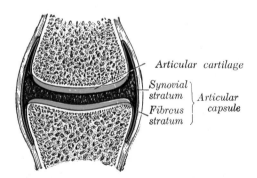

FIG. 5–4.—Diagrammatic section of a synovial (diarthrodial) joint.

may be divided, completely or incompletely, by an articular disk or meniscus, the periphery of which is continuous with the fibrous capsule while its free surfaces are covered by synovial membrane (*articulatio composita*) (Fig. 5–5).

The varieties of joints in this class have been determined by the kind of motion

permitted in each. There are two varieties in which the movement is **uniaxial**, that is to say, all movements take place around one axis. In one form, the **ginglymus** or hinge joint, this axis is, practically speaking, transverse; in the other, the **trochoid** or pivot joint, it is longitudinal. There are two varieties where the movement is *biaxial,* or around two horizontal axes at right angles to each other, or at any intervening axis between the two. These are the **condyloid** and the **saddle joint.** There is one form where the movement is **polyaxial,** the **enarthrosis** or ball-and-socket joint; and finally there are the **arthrodia** or gliding joints.

Ginglymus or Hinge Joint.—In this form the articular surfaces are molded to each other in such a manner as to permit motion only in one plane, the extent of motion at the same time being considerable. The direction which the distal bone takes in this motion is seldom in the same plane as that of the axis of the proximal bone; there is usually a certain amount of deviation from the straight line during flexion. The articular surfaces are connected together by strong collateral ligaments, which form their chief bond of union. The best examples of ginglymus are the interphalangeal joints and the joint between the humerus and ulna; the knee and ankle joints are less typical, as they allow a slight degree of rotation or of side-to-side movement in certain positions of the limb.

Trochoid or Pivot Joint (*articulatio trochoidea; rotary joint*).—Where the movement is limited to rotation, the joint is formed by a pivot-like process turning within a ring, or a ring on a pivot, the ring being formed partly of bone, partly of ligament. In the proximal radioulnar articulation, the ring is formed by the radial notch of the ulna and the annular ligament; here, the head of the radius rotates within the ring. In the articulation of the dens of the axis with the atlas the ring is formed in front by the anterior arch, and behind by the transverse ligament of the atlas; here, the ring rotates around the dens.

Condyloid Articulation (*articulatio ellipsoidea*).—In this form of joint, an ovoid articular surface, or condyle, is received into an elliptical cavity in such a manner as to permit of flexion, extension, adduction, abduction, and circumduction, but no axial rotation. The wrist joint is an example of this form of articulation.

Saddle Joint (*articulatio sellaris*).—In this variety the opposing surfaces are reciprocally concavo-convex. The movements are the same as in the preceding form; that is to say, flexion, extension, adduction, abduction, and circumduction are allowed; but no axial rotation. The best example of this form is the carpometacarpal joint of the thumb.

Spheroidea (*enarthrosis; ball-and-socket joint*).—A ball-and-socket joint is one in which the distal bone is capable of motion around an indefinite number of axes, which have one common center. It is formed by the reception of a globular head into a cup-like cavity, hence the name "ball-and-socket." Examples of this form of articulation are found in the hip and shoulder.

Articulatio plana (*arthrodia; gliding joints*) is a joint which admits of only gliding movement; it is formed by the apposition of plane surfaces, or one slightly concave, the other slightly convex, the amount of motion between them being limited by the ligaments or osseous processes surrounding the articulation. It is the form present in the joints between the articular processes of the vertebrae, the carpal joints, except that of the capitate with the navicular and lunate, and the tarsal joints with the exception of that between the talus and the navicular.

The Kinds of Movement Admitted in Joints

The movements admissible in joints may be divided into four kinds: gliding and angular movements, circumduction, and rotation. These movements are often, however, more or less combined in the various joints, so as to produce an infinite variety, and it is seldom that only one kind of motion is found in any particular joint.

Gliding movement is the simplest kind of motion that can take place in a joint, one surface gliding or moving over another without any angular or rotatory movement. It is common to all movable joints; but in some, as in most of the articulations of the carpus and tarsus, it is the only motion permitted. This movement is not confined to plane surfaces, but may exist between any two contiguous surfaces, of whatever form.

Angular movements increase or decrease the angle between two adjoining bones. Flexion and extension, abduction and adduction are the common examples.

Flexion occurs when the angle between two bones is decreased; a typical example is the bending of the arm at the elbow. In some instances rather arbitrary definitions must be given. For example, bringing the femur forward or ventrally at the hip is flexion, but this motion includes the entire arc beginning in a position with the limb held back and then brought forward, as in kicking a football, and includes bringing the thigh against the trunk as in the high kick of a chorus dancer.

Extension occurs when the angle between the bones is increased, as when the arm or leg is straightened. Hyperextension is the term used for moving through an arc in the same direction, but beyond the straight position, as in bringing the arm back before launching a bowling ball.

Abduction occurs when an arm or leg is moved away from the midsagittal plane, or when the fingers and toes are moved away from the median longitudinal axis of the hand or foot.

Adduction occurs when an arm or leg is moved toward or beyond the midsagittal plane or when the fingers or toes are moved toward the median longitudinal axis of the hand or foot.

Circumduction.—Circumduction is that form of motion which takes place between the head of a bone and its articular cavity when the bone is made to circumscribe a conical space; the base of the cone is described by the distal end of the bone; the apex is in the articular cavity. It is a combination of flexion, abduction, extension, and adduction acting in sequence. This kind of motion is best seen in the shoulder and hip joints.

Rotation.—Rotation is a form of movement in which a bone moves around a central axis without undergoing any displacement from this axis; this axis of rotation may lie in a separate bone, as in the case of the pivot formed by the dens of the axis around which the atlas turns; or a bone may rotate around its own longitudinal axis, as in the rotation of the humerus at the shoulder joint; or the axis of rotation may not be quite parallel to the long axis of the bone, as in the movement of the radius on the ulna during pronation and supination of the hand, where it is represented by a line connecting the center of the head of the radius proximally with the center of the head of the ulna distally.

In **supination** the radius and ulna are parallel, and the palm faces ventralward or cranialward.

In **pronation** the radius is rotated diagonally across the ulna and the palm faces dorsalward or caudalward.

Structures Composing Movable Joints

Ligaments.—Ligaments are composed mainly of bundles of **collagenous fibers** placed parallel with, or closely interlaced with one another, and present a **white,** shining, silvery appearance. They are pliant and flexible, so as to allow perfect freedom of movement, but strong, tough, and inextensible, so as not to yield readily to applied force. Some ligaments consist entirely of **yellow elastic tissue,** as the ligamenta flava which connect together the laminae of adjacent vertebrae, and the ligamentum nuchae in the lower animals. In these instances the elasticity of the ligament acts as a substitute for muscular power.

The **Articular Capsules** form complete envelopes for the freely movable joints. Each capsule consists of two strata—an **external** (*stratum fibrosum*) composed of white fibrous tissue, and an **internal** (*stratum synoviale*) which is a specialized layer, and is usually described separately as the synovial membrane.

The **fibrous capsule** is attached to the whole circumference of the articular end of each bone entering into the joint, and thus entirely surrounds the articulation.

The **synovial membrane** covers the inner surface of the fibrous capsule, forming a closed sac called the **synovial cavity.** It is composed of loose connective tissue, cellular in some places, fibrous in others, and it has a free surface which elaborates a thick, viscous, glairy fluid, similar to the white of an egg and termed, therefore, **synovia** or **synovial fluid.** The synovial membrane covers tendons which pass through the joint, such as the tendon of the Popliteus in the knee, and the long head of the Biceps in the shoulder. The membrane is not closely applied to the inner surface of the fibrous capsule, but is thrown into folds, fringes, or projections which are composed of connective tissue, fat, and blood vessels. These folds commonly surround the margin of the articular cartilage, filling in clefts and crevices, and in some joints, such as the knee, forming large pads of fat. The synovial cavity of a normal joint contains only enough synovial fluid to moisten and lubricate the synovial surfaces, but in an injured or inflamed joint, the fluid may accumulate in painful amounts. Part of the lining of the synovial cavity is provided by the surface of the articular cartilage which is moistened by the synovia but is not covered by the synovial membrane.

Similar to the synovial cavities of true joints are the synovial tendon sheaths and synovial bursae which have an inner lining equivalent to the synovial membrane of a joint and are lubricated by a fluid very similar to synovial fluid.

The **Synovial Tendon Sheaths** (*vaginae mucosae*) facilitate the gliding of tendons which pass through fibrous and bony tunnels such as those under the flexor retinaculum of the wrist. These sheaths are closed sacs, one layer of the synovial membrane lining the tunnel, the other reflected over the surface of the tendon.

Synovial Bursae (*bursae mucosae*) are clefts in the connective tissue between muscles, tendons, ligaments, and bones. They are made into closed sacs by a synovial lining, similar to that of a true joint, which may in some cases be continuous through an opening in the wall with the lining of a joint cavity. They facilitate the gliding of muscles or tendons over bony or ligamentous prominences, and are named according to their location, subcutaneous, submuscular, and subtendinous.

Ligamentous Action of Muscles.—The movements of the different joints of a limb are combined by means of the long muscles passing over more than one joint. These, when relaxed and stretched to their greatest extent, act as elastic ligaments in restraining certain movements of one joint, except when combined with corresponding movements of the other—the latter movements being usually in the opposite direction. Thus the shortness of the hamstring muscles prevents complete flexion of the hip, unless the knee joint is also flexed so as to bring their attachments nearer together. The uses of this arrangement are threefold: (1) it coördinates the kinds of movements which are the most habitual and necessary and enables them to be performed with the least expenditure of power. (2) It enables the short muscles which pass over only one joint to act upon more than one. (3) It provides the joints with ligaments which, while they are of very great power in resisting movements to an extent incompatible with the mechanism of the joint, at the same time spontaneously yield when necessary.

ARTICULATIONS OF THE AXIAL SKELETON

These may be divided into the following groups:

  I. Of the Mandible.
 II. Of the Vertebral Column with the Cranium.
III. Of the Atlas with the Axis.
 IV. Of the Vertebral Column.
  V. Of the Ribs with the Vertebrae.

 VI. Of the Cartilages of the Ribs with the Sternum, and with Each Other.
VII. Of the Sternum.
VIII. Of the Vertebral Column with the Pelvis.
 IX. Of the Pelvis.

I. Articulation of the Mandible

The **Temporomandibular Joint** (*articulatio temporomandibularis*) (Figs. 5–6, 5–7, 5–8) is a combined ginglymus and gliding joint. The parts entering into its formation on either side are: the anterior part of the mandibular fossa of the temporal bone, the articular tubercle, and the condyle of the mandible. The ligaments of the joint are the following:

> Articular Capsule
> Lateral
> Sphenomandibular
> Articular Disk
> Stylomandibular

The **Articular Capsule** (*capsula articularis; capsular ligament*) is a thin, loose envelope, attached to the circumference of the mandibular fossa and the articular tubercle, and to the neck of the condyle of the mandible.

The **Lateral Ligament** (*ligamentum laterale; external lateral ligament; temporomandibular ligament*) (Fig. 5–6) consists of two short, narrow fasciculi, one anterior to the other, attached to the lateral surface of the zygomatic arch and to the tubercle on its inferior border superiorly, and to the lateral surface and posterior border of the neck of the mandible inferiorly. It is broader superiorly, and its fibers are directed obliquely inferiorly and posteriorly. It is covered by the parotid gland, and by the integument.

The **Sphenomandibular Ligament** (*ligamentum sphenomandibulare; internal lateral ligament*) (Fig. 5–7) is a flat, thin band which is attached to the spine of the sphenoid bone, and, becoming broader as it descends, is fixed to the lingula of the mandibular foramen. Its lateral surface is in relation with the Pterygoideus lateralis; inferiorly, it is separated from the neck of the condyle by the maxillary vessels. The

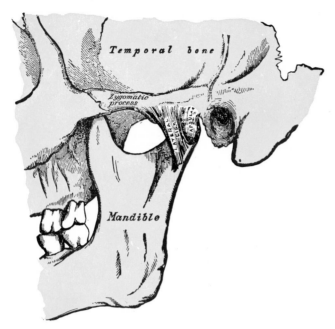

FIG. 5–6.—Articulation of the mandible. Lateral aspect.

inferior alveolar vessels and nerve and a lobule of the parotid gland lie between it and the ramus of the mandible. Its medial surface is in relation with the Pterygoideus medialis.

The **Articular Disk** (*discus articularis; interarticular fibrocartilage; articular menis-*

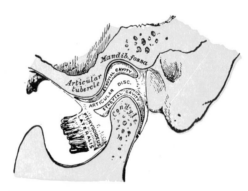

FIG. 5–8.—Sagittal section of the articulation of the mandible.

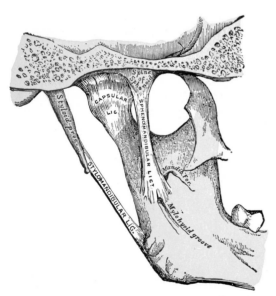

FIG. 5–7.—Articulation of the mandible. Medial aspect.

cus) (Fig. 5–8) is a thin, oval plate, placed between the condyle of the mandible and the mandibular fossa. Its superior surface is concavo-convex, to accommodate itself to the form of the mandibular fossa and the articular tubercle. Its inferior surface, in contact with the condyle, is concave. Its circumference is connected to the articular capsule and to the tendon of the Pterygoideus lateralis. It is thicker at its periphery than at its center. The fibers of which it is composed have a concentric arrangement, more apparent at the circumference than at the center. It divides the joint into two cavities, each of which is furnished with a synovial membrane.

The **Synovial Membranes.**—The synovial cavity above the articular disk, the larger and looser of the two, is continued from the margin of the cartilage covering the mandibular fossa and articular tubercle on to the superior surface of the disk. The separate cavity below the disk passes from the under surface of the disk to the neck of the condyle, being prolonged a little farther posteriorly. The articular disk is sometimes perforated in its center, and the two cavities then communicate with each other.

The **Stylomandibular Ligament** (*ligamentum stylomandibulare; stylomaxillary ligament*) (Fig. 5–7) is a specialized band of the cervical fascia which extends from near the apex of the styloid process of the temporal bone to the angle and posterior border of the ramus of the mandible, between the Masseter and Pterygoideus medialis. This ligament separates the parotid from the submandibular gland, and from its deep surface some fibers of the Styloglossus take origin. Although classed among the ligaments of the temporomandibular joint, it can only be considered as accessory to it.

The **nerves** of the temporomandibular joint are derived from the auriculotemporal and masseteric branches of the mandibular nerve, the **arteries** from the superficial temporal branch of the external carotid.

Movements.—The movements in this articulation are the opening and closing of the jaws, protrusion of the mandible, and lateral displacement of the mandible. It must be borne in mind that there are two parts to this articulation—one between the condyle and the articular disk, and the other between the disk and the mandibular fossa. When the jaws are opened and closed, motion takes place in both parts; the disk glides anteriorly on the articular tubercle, and the condyle moves on the disk like a hinge, causing the mandible to rotate about a center of suspension near the center of the ramus of the mandible. This somewhat movable center is provided by the attachment of the sphenomandibular ligament to the lingula, and the sling formed by the Masseter and the Pterygoideus medialis. When the jaws are opened, the angle of the mandible moves posteriorly while the condyle glides forward as the short arm of a lever, and the chin, as the

long arm of the lever, describes a wide arc. The motion between the condyle and the articular disk is largely one of accommodation to the change in position. When the jaws are closed, some of the force is applied to the condyle as a fulcrum, especially in biting with the incisors, but in chewing with the molars, the pressure comes more directly between the teeth, the condyle acting as a guide more than as a fulcrum. In protruding the mandible, both disks glide forward in the mandibular fossa, the usual rotation of opening the jaws being prevented by the synergetic action of the closing muscles. In lateral displacement of the mandible, one disk glides forward while the other remains in place. In grinding or chewing movements there is first a lateral displacement of the mandible by a forward movement of one condyle, and then the mandible is brought back into place by the action of the closing muscles and the meshing of the teeth. The condyles may be displaced alternately, or the same one may be displaced repeatedly as in chewing with the teeth of one side.

The jaws are opened, that is, the mandible is depressed, by the Pterygoideus lateralis, assisted by the Digastricus, Mylohyoideus, and Geniohyoideus. The jaws are closed, that is, the mandible elevated, by the Masseter, Pterygoideus medialis, and Temporalis. It is protruded by the simultaneous action of the lateral Pterygoidei of both sides and the synergetic action of the closing muscles. It is drawn backward by the posterior fibers of the Temporalis, and displaced laterally by the action of the Pterygoideus lateralis of the opposite side.

II. Articulations of the Vertebral Column with the Cranium

Articulation of the Atlas with the Occipital Bone (*articulatio atlantoöccipitalis*). —The **atlantoöccipital joint** of each side lies between the superior articular facet of the lateral mass of the atlas and the condyle of the occipital bone; it is condyloid in type. The articular surfaces are reciprocally curved. The ligaments connecting the bones are:

> Two Articular Capsules
> Anterior Atlantoöccipital Membrane
> Posterior Atlantoöccipital Membrane
> Two Lateral Atlantoöccipital

The **Articular Capsules** (*capsulae articulares; capsular ligaments*) surround the condyles of the occipital bone, and connect them with the articular processes of the atlas: they are thin and loose (Fig. 5–9).

The **Anterior Atlantoöccipital Membrane** (*membrana atlantoöccipitalis anterior; anterior atlantoöccipital ligament*) (Fig. 5–9) is broad and composed of densely woven fibers, which pass between the anterior margin of the foramen magnum and the cranial border of the anterior arch of the

defective over the groove for the vertebral artery, and forms with this groove an opening for the entrance of the artery and the exit of the suboccipital nerve. The free border of the membrane, arching over the artery and nerve, is sometimes ossified. The membrane is in relation with the Recti capitis posteriores minores and Obliqui capitis superiores dorsally, and with the

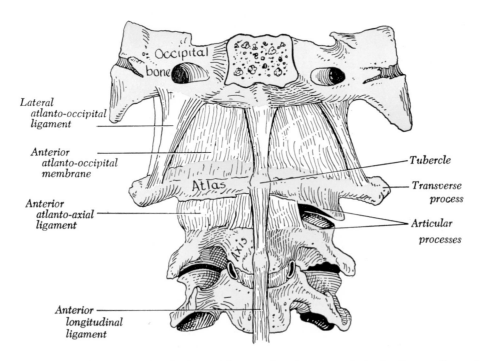

FIG. 5–9.—The atlantoöccipital and atlantoaxial joints. Anterior aspect. On each side a small occasional synovial joint is shown between the lateral part of the upper surface of the body of the third cervical vertebra and the bevelled, inferior surface of the axis. The joint cavities have been opened.

atlas; laterally, it is continuous with the articular capsules; ventrally, it is strengthened in the middle line by a strong, rounded cord which connects the basilar part of the occipital bone to the tubercle on the anterior arch of the atlas. This membrane is in relation with the Recti capitis anteriores and with the alar ligaments.

The **Posterior Atlantoöccipital Membrane** (*membrana atlantoöccipitalis posterior; posterior atlantoöccipital ligament*) (Fig. 5–10), broad but thin, is connected to the posterior margin of the foramen magnum and to the superior border of the posterior arch of the atlas. On either side this membrane is

dura mater of the vertebral canal, to which it is intimately adherent ventrally.

The **Lateral Ligaments** are thickened portions of the articular capsules reinforced by bundles of fibrous tissue, and are directed obliquely upward and medialward; they are attached to the jugular processes of the occipital bone, and to the bases of the transverse processes of the atlas.

Synovial Membranes.—There are two synovial membranes: one lining each of the articular capsules. The joints frequently communicate with that between the dorsal surface of the dens and the transverse ligament of the atlas.

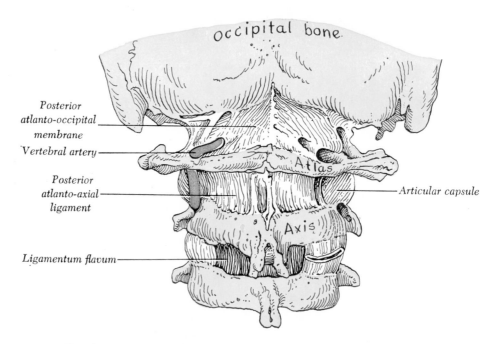

FIG. 5–10.—Posterior atlantoöccipital membrane and atlantoaxial ligament.

Movements.—The movements permitted in this joint are (a) flexion and extension, which give rise to the ordinary forward and backward nodding of the head, and (b) slight lateral motion to one or the other side. *Flexion* is produced mainly by the action of the Longi capitis and Recti capitis anteriores; *extension* by the Recti capitis posteriores major and minor, the Obliquus superior, the Semispinalis capitis, Splenius capitis, Sternocleidomastoideus, and upper fibers of the Trapezius. The Recti laterales are concerned in the *lateral movement*, assisted by the Trapezius, Splenius capitis, Semispinalis capitis, and the Sternocleidomastoideus of the same side, all acting together.

Ligaments Connecting the Axis with the Occipital Bone.—

    Membrana Tectoria
    Two Alar
    Apical Dental

The **Membrana Tectoria** (*occipitoaxial ligament*) (Figs. 5–12, 5–13) is a broad, strong band which covers the dens and its ligaments within the vertebral canal, and appears to be a cranial prolongation of the posterior longitudinal ligament of the vertebral column. It is fixed to the dorsal surface of the body of the axis, and expanding as it ascends is attached to the basilar groove of the occipital bone ventral to the foramen magnum, where it blends with the cranial dura mater. Its ventral surface is in relation with the transverse ligament of the atlas, and its dorsal surface with the dura mater.

The **Alar Ligaments** (*ligamenta alaria; odontoid ligaments*) (Fig. 5–12) are strong, rounded cords, which arise one on either side of the cranial part of the dens and, passing obliquely upward and lateralward, are inserted into the rough depressions on the medial sides of the condyles of the occipital bone. In the triangular interval between these ligaments is another fibrous cord, the **apical dental ligament** (Fig. 5–13), which extends from the tip of the process to the anterior margin of the foramen magnum, being intimately blended with the deep portion of the anterior atlantoöccipital membrane and superior crus of the transverse ligament of the atlas. It is regarded as a rudimentary intervertebral disk, and in it traces of the notochord may persist. The alar ligaments limit rotation of the cranium and therefore receive the name of **check ligaments.**

In addition to the ligaments which unite the atlas and axis to the skull, the ligamentum nuchae (page 302) must be regarded as one of the ligaments connecting the vertebral column with the cranium.

III. Articulation of the Atlas with the Axis (Articulatio Atlantoaxial Mediana et Lateralis)

The articulation of the atlas with the axis comprises three joints. There is a pivot articulation, the **median atlantoaxial joint,** between the dens of the axis and the ring formed by the anterior arch and the transverse ligament of the atlas (see Fig. 5–12); here there are two synovial cavities: one between the posterior surface of the anterior

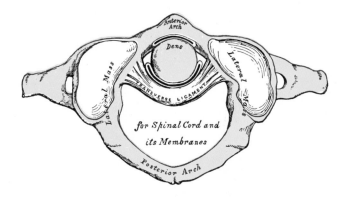

Fig. 5–11.—Articulation between the dens and atlas.

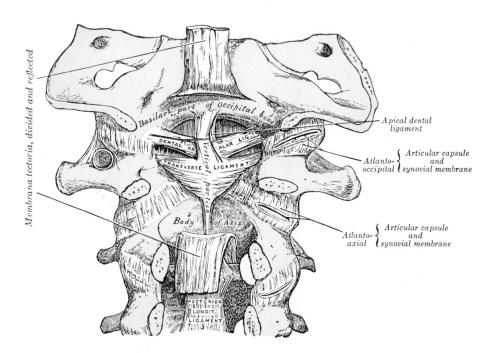

Fig. 5–12.—Membrana tectoria, transverse, and alar ligaments.

arch of the atlas and the dens; the other between the anterior surface of the ligament and the dens. Between the articular processes of the two bones there is a gliding joint, the **lateral atlantoaxial joint.** The ligaments are:

Two Articular Capsules
Anterior Atlantoaxial
Posterior Atlantoaxial
Transverse

The **Articular Capsules** (*capsulae articulares; capsular ligaments*) are thin and loose, and connect the margins of the lateral masses of the atlas with those of the posterior articular surfaces of the axis. Each is strengthened at its posterior and medial part by an **accessory ligament,** which is attached to the body of the axis near the base of the dens, and to the lateral mass of the atlas near the transverse ligament.

The **Anterior Atlantoaxial Ligament** (Figs. 5–9, 5–13) is a strong membrane, fixed to the inferior border of the anterior arch of the atlas, and to the ventral surface of the body of the axis. It is strengthened in the middle line by a rounded cord, which connects the tubercle on the anterior arch of the atlas to the body of the axis, and is a continuation cranially of the anterior longitudinal ligament. The ligament is in relation with the Longi capitis ventrally.

The **Posterior Atlantoaxial Ligament** (Figs. 5–10, 5–13) is a broad, thin membrane attached to the inferior border of the posterior arch of the atlas and to the superior edges of the laminae of the axis. It supplies the place of the ligamenta flava, and is in relation with the Obliqui capitis inferiores dorsally.

The **Transverse Ligament of the Atlas** (*ligamentum transversum atlantis*) (Figs. 5–11, 5–12, 5–13) is a thick, strong band, which arches across the ring of the atlas, and retains the dens in contact with the anterior arch. It is broader and thicker in the middle

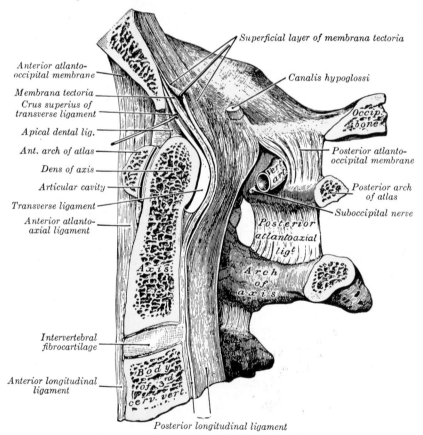

FIG. 5–13.—Median sagittal section through the occipital bone and first three cervical vertebrae. (Spalteholz.)

than at the ends, and is firmly attached on either side to a small tubercle on the medial surface of the lateral mass of the atlas. As it crosses the dens, a small fasciculus (*crus superius*) is prolonged cranially, and another (*crus inferius*) caudally, from the superficial or posterior fibers of the ligament. The former is attached to the basilar part of the occipital bone, in close relation with the membrana tectoria; the latter is fixed to the posterior surface of the body of the axis, hence forming a cross named the **cruciate ligament of the atlas.** The transverse ligament divides the circular opening in the atlas into two unequal parts: of these, the posterior and larger transmits the spinal cord and its membranes, and the accessory nerves; the anterior and smaller contains the dens. The neck of the dens is constricted where it is embraced posteriorly by the transverse ligament, so that this ligament suffices to hold it in position after all the other ligaments have been divided.

Synovial Membranes.—There is a synovial membrane for each joint; the joint cavity between the dens and the transverse ligament is often continuous with those of the atlanto-occipital articulations.

Movements.—The opposed articular surfaces of the atlas and axis are not reciprocally curved; both surfaces are convex in their long axes. When the superior facet glides ventrally on the inferior, it also descends; the fibers of the articular capsule are relaxed in a vertical direction, and will then permit movement in an anteroposterior direction. By this means a shorter capsule suffices and the strength of the joint is materially increased.

This joint allows the rotation of the atlas (and, with it, the skull) upon the axis, the extent of rotation being limited by the alar ligaments.

The principal muscles by which these movements are produced are the Sternocleidomastoideus and Semispinalis capitis of one side, acting with the Longus capitis, Splenius, Longissimus capitis, Rectus capitis posterior major, and Obliqui capitis superior and inferior of the other side.

IV. Articulations of the Vertebral Column

The articulations of the vertebral column consist of (1) a series of symphyses between the vertebral bodies, and (2) a series of joints between the vertebral arches.

1. **Articulations of Vertebral Bodies** (*intercentral ligaments*).—The articulations between the bodies of the vertebrae are symphyses, and the individual vertebrae move only slightly on each other. When this slight degree of movement between the pairs of bones takes place in all the joints of the vertebral column, however, the total range of movement is very considerable. The ligaments of these articulations are the following:

Anterior Longitudinal
Posterior Longitudinal
Intervertebral Disks

The **Anterior Longitudinal Ligament** (*ligamentum longitudinale anterius; anterior common ligament*) (Figs. 5–14, 5–17).—The anterior longitudinal ligament is a broad and strong band of fibers, which extends along the surfaces of the bodies of the vertebrae, from the axis to the sacrum. It is thicker in the thoracic than in the cervical and lumbar regions, and somewhat thicker opposite the bodies of the vertebrae than opposite the intervertebral disks. It is attached, cranially, to the body of the axis, where it is continuous with the anterior atlantoaxial ligament, and extends down as far as the sacrum. It consists of dense longitudinal fibers which are intimately adherent to the intervertebral disks and the prominent margins of the vertebrae, but not to the middle parts of the bodies. In the latter situation the ligament is thick and serves to fill up the concavities on the ventral surfaces, and to make the front of the vertebral column more even. It is composed of several layers of fibers which vary in length, but are closely interlaced with each other. The most superficial fibers are the longest and extend between four or five vertebrae. A second, subjacent set extends between two or three vertebrae; while a third set, the shortest and deepest, reaches from one vertebra to the next. At the sides of the bodies the ligament consists of a few short fibers which pass from one vertebra to the next, separated from the concavities of the vertebral bodies by oval apertures for the passage of vessels.

The **Posterior Longitudinal Ligament** (*ligamentum longitudinale posterius; posterior common ligament*) (Figs. 5–14, 5–15).

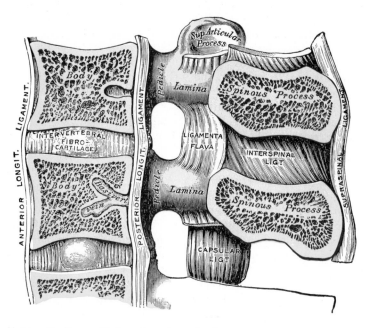

Fig. 5–14.—Median sagittal section of two lumbar vertebrae and their ligaments.

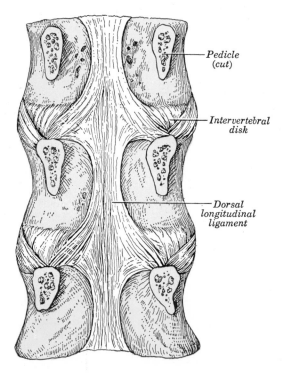

Fig. 5–15.—Posterior longitudinal ligament of the
vertebrae in the lumbar region.

—The posterior longitudinal ligament extends along the dorsal surfaces of the bodies of the vertebrae, within the vertebral canal, from the axis, where it is continuous with the membrana tectoria, to the sacrum. It is broader cranially, and thicker in the thoracic than in the cervical and lumbar regions. Over the intervertebral disks and contiguous margins of the vertebrae, the ligament is more intimately adherent, broad, and in the thoracic and lumbar regions presents a series of dentations with intervening concave margins. It is narrow and thick over the centers of the bodies, from which it is separated by the basivertebral veins. This ligament is composed of longitudinal fibers, denser and more compact than those of the anterior ligament, and consists of superficial layers occupying the interval between three or four vertebrae, and deeper layers which extend between adjacent vertebrae.

The **Intervertebral Disks** (*disci interverte-brales; intervertebral fibrocartilages*) (Figs. 5–14, 5–18) are interposed between the adjacent surfaces of the bodies of the vertebrae, from the axis to the sacrum, and form the chief bonds of connection between the vertebrae. They vary in shape, size, and thickness, in different parts of the vertebral column. In shape and size they correspond with the surfaces of the bodies between which they are placed, except in the cervical region, where they are slightly smaller from side to side than the corresponding bodies. In thickness they vary not only in the different regions of the column, but in different parts of the same disk; they are thicker ventrally in the cervical and lumbar regions, and thus contribute to the anterior convexities of these parts of the column; while they are of nearly uniform thickness in the thoracic region, the anterior concavity of this part of the column being almost entirely owing to the shape of the vertebral bodies. The intervertebral disks constitute about one-fourth of the length of the vertebral column, exclusive of the first two vertebrae; but this amount is not equally distributed between the various bones, the cervical and lumbar portions having, in proportion to their length, a much greater amount than the thoracic region, corresponding to the greater pliancy and freedom of movement of these parts. The intervertebral disks are

adherent to thin layers of hyaline cartilage which cover the superior and inferior surfaces of the bodies of the vertebrae; in the lower cervical vertebrae, however, small joints lined by synovial membrane are occasionally present between the superior surfaces of the bodies and the margins of the disks on either side. By their circumferences the intervertebral disks are closely connected to the ventral and dorsal longitudinal ligaments. In the thoracic region they are joined laterally, by means of the interarticular ligaments to the heads of those ribs which articulate with two vertebrae.

Structure of the Intervertebral Disks.—Each is composed, at its circumference, of laminae of fibrous tissue and fibrocartilage, forming the **anulus fibrosus**; and, at its center, of a soft, pulpy, highly elastic substance, of a yellowish color, which projects considerably above the surrounding level when the disk is divided horizontally. This pulpy substance (**nucleus pulposus**), especially well developed in the lumbar region, contains the remains of the notochord. The laminae are arranged concentrically; the outermost consist of ordinary fibrous tissue, the others of white fibrocartilage. The laminae are not quite vertical in their direction, those near the circumference being curved outward and closely approximated; while those nearest the center curve in the opposite direction, and are somewhat more widely separated. The fibers in each lamina are directed obliquely for the most part, those of adjacent laminae passing in opposite directions, crossing one another, like the limbs of the letter X. This laminar arrangement belongs to about the outer half of each fibrocartilage. The pulpy substance, however, consists of a fine fibrous matrix containing angular cells united to form a reticular structure.

The intervertebral fibrocartilages are important shock absorbers. Under pressure the highly elastic nucleus pulposus becomes flatter and broader and pushes the more resistant fibrous laminae outward in all directions.

2. Articulations of Vertebral Arches.— The joints between the articular processes of the vertebrae belong to the gliding variety and are enveloped by capsules lined by synovial membrane.

The **Articular Capsules** (*capsulae articulares; capsular ligaments*) (Fig. 5–14) are thin and loose, and are attached to the margins of the articular processes of adjacent vertebrae. They are longer and looser in the cervical than in the thoracic and lumbar regions.

The laminae, spinous, and transverse processes are connected by the following ligaments:

Ligamenta Flava
Supraspinal
Ligamentum Nuchae
Interspinal
Intertransverse

The **Ligamenta Flava** (*ligamenta subflava*) (Fig. 5–16) connect the laminae of adjacent vertebrae, from the axis to the first segment of the sacrum. They are best seen from the interior of the vertebral canal; when looked at from the outer surface they appear short, being overlapped by the laminae. Each ligament consists of two lateral portions which commence on either side of the roots of the articular processes, and extend dorsally to the point where the laminae meet to form the spinous process; the dorsal margins of the two portions are in contact except for slight intervals being left for the

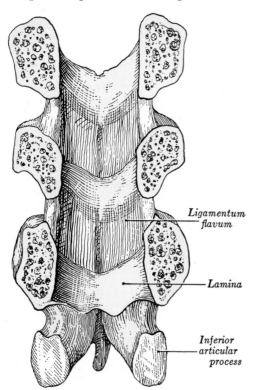

passage of small vessels. Each consists of yellow elastic tissue, the fibers of which, almost perpendicular in direction, are attached to the ventral surface of the superior lamina and to the dorsal surface and superior margin of the lamina below. In the cervical region the ligaments are thin, but broad and long; they are thicker in the thoracic region and thickest in the lumbar region. Their marked elasticity serves to preserve the upright posture.

The **Supraspinal Ligament** (*ligamentum supraspinale; supraspinous ligament*) (Fig. 5–14) is a strong fibrous cord, which connects together the apices of the spinous processes from the seventh cervical vertebra to the sacrum; at the points of attachment to the tips of the spinous processes fibrocartilage is developed in the ligament. It is thicker and broader in the lumbar than in the thoracic region, and intimately blended, in both situations, with the neighboring fascia. The most superficial fibers of this ligament extend over three or four vertebrae; those more deeply seated pass between two or three vertebrae, while the deepest connect the spinous processes of neighboring vertebrae. Between the spinous processes it is continuous with the interspinal ligaments. It is continued upward to the external occipital protuberance and median nuchal line as the ligamentum nuchae.

The **Ligamentum Nuchae** is a fibrous membrane, which, in the neck, represents the supraspinal ligaments of the lower vertebrae. It reaches from the external occipital protuberance and median nuchal line to the spinous process of the seventh cervical vertebra. Penetrating deeply from it is a fibrous lamina, attached to the posterior tubercle of the atlas, and to the spinous processes of the cervical vertebrae, forming a septum between the muscles on either side of the neck. In man it is merely the rudiment of an important elastic ligament, which, in some of the grazing animals, serves to sustain the weight of the head.

The **Interspinal Ligaments** (*ligamenta interspinalia; interspinous ligaments*) (Fig. 5–14).—The thin and membranous interspinal ligaments connect adjoining spinous processes and extend from the root to the apex of each process. They meet the liga-

FIG. 5–16.—The ligamenta flava of the lumbar region. Anterior aspect.

Ligamentum flavum

Lamina

Inferior articular process

menta flava ventrally and the supraspinal ligament dorsally. They are narrow and elongated in the thoracic region; broader, thicker, and quadrilateral in the lumbar region; and only slightly developed in the neck.

The **Intertransverse Ligaments** (*ligamenta intertransversaria*) are interposed between the transverse processes. In the cervical region they consist of a few irregular, scattered fibers; in the thoracic region they are rounded cords intimately connected with the deep muscles of the back; in the lumbar region they are thin and membranous.

Movements.—The movements permitted in the vertebral column are: *flexion, extension, lateral flexion, circumduction*, and *rotation.*

In **flexion,** or bending ventrally, the anterior longitudinal ligament is relaxed, and the intervertebral fibrocartilages are compressed ventrally; the posterior longitudinal ligament, the ligamenta flava, and the inter- and supraspinal ligaments are stretched, as well as the dorsal fibers of the intervertebral disks. The interspaces between the laminae are widened, and the inferior articular processes glide upward upon the superior articular processes of the subjacent vertebrae. Tension of the extensor muscles of the back is the most important factor in limiting the movement. Flexion is the most extensive of all the movements of the vertebral column, and is freest in the cervical region.

In **extension,** or bending dorsally, an exactly opposite disposition of the parts takes place. This movement is limited by the anterior longitudinal ligament, and by the approximation of the spinous processes. It is freest in the cervical and lumbar regions.

In **lateral flexion,** the sides of the intervertebral disks are compressed, the extent of motion being limited by the resistance offered by the surrounding ligaments. This movement may take place in any part of the column, but is freest in the cervical and lumbar regions.

Circumduction is very limited, and is merely a succession of the preceding movements.

Rotation is produced by the twisting of the intervertebral disks; this, although only slight between any two vertebrae, allows a considerable extent of movement when it takes place in the whole length of the column. This movement occurs in the cervical, thoracic and lumbar regions.

The extent and variety of the movements are influenced by the shape and direction of the articular surfaces. In the *cervical* region the inclination of the superior articular surfaces

allows free flexion and extension. *Extension* can be carried farther than flexion; at the cranial end of the region it is checked by the locking of the posterior edges of the superior atlantal facets in the condyloid fossae of the occipital bone; at the caudal end it is limited by a mechanism whereby the inferior articular processes of the seventh cervical vertebra slip into grooves behind and below the superior articular processes of the first thoracic. *Flexion* is arrested just beyond the point where the cervical convexity is straightened; the movement is checked by the apposition of the projecting lower lips of the bodies of the vertebrae with the shelving surfaces on the bodies of the subjacent vertebrae. *Lateral flexion* and rotation are free in the cervical region; they are, however, always combined. The inclinations of the superior articular surfaces impart a rotatory movement during lateral flexion, while pure rotation is prevented by the slight medial slope of these surfaces.

In the **thoracic region,** notably in its cephalic part, all the movements are limited in order to reduce interference with respiration to a minimum. The almost complete absence of an upward inclination of the superior articular surfaces prohibits any marked flexion, while extension is checked by the contact of the inferior articular margins with the laminae, and the contact of the spinous processes with one another. The mechanism between the seventh cervical and the first thoracic vertebrae, which limits extension of the cervical region, will also serve to limit flexion of the thoracic region when the neck is extended. Rotation is free in the thoracic region: the superior articular processes are segments of a cylinder whose axis is in the mid-ventral line of the vertebral bodies. The direction of the articular facets would allow of free lateral flexion, but this movement is considerably limited by the resistance of the ribs and sternum.

In the **lumbar region** extension is free and wider in range than flexion. The inferior articular facets are not in close apposition with the superior facets of the subjacent vertebrae, and on this account a considerable amount of lateral flexion is permitted. For the same reason a slight amount of rotation can be carried out, but this is so soon checked by the interlocking of the articular surfaces that it is negligible.

The *principal muscles* which produce *flexion* are the Sternocleidomastoideus, Longus capitis, and Longus colli, the Scaleni, the abdominal muscles and the Psoas major. *Extension* is produced by the intrinsic muscles of the back, assisted in the neck by the Splenius, Semispinales, and the Multifidus. *Lateral* motion is produced by the intrinsic muscles of the back by the Splenius, the Scaleni, the Quadratus lumborum, and the Psoas major, the muscles of one side only acting; and *rotation* by the action of the following muscles of one side only, viz., the Sternocleidomastoideus, the Longus capitis, the Scaleni, the Multifidus, the Semispinalis capitis, and the abdominal muscles.

V. Costovertebral Articulations (Articulationes Costovertebrales)

The articulations of the ribs with the vertebral column may be divided into two sets, one connecting the heads of the ribs with the bodies of the vertebrae, another uniting the necks and tubercles of the ribs with the transverse processes.

1. **Articulations of the Heads of the Ribs** (*articulationes capitis costae; costocentral articulations*) (Fig. 5–17).—These constitute a series of gliding joints, and are formed by the articulation of the heads of the typical ribs with the facets on the contiguous margins of the bodies of the thoracic vertebrae and with the intervertebral disks between them; the first, tenth, eleventh, and twelfth ribs each articulate with a single vertebra. The ligaments of the joints are:

> Articular Capsule
> Radiate
> Intraarticular

The **Articular Capsule** (*capsula articularis; capsular ligament*) surrounds the joint, being composed of short, strong fibers, connecting the head of the rib with the circumference of the articular cavity formed by the intervertebral disk and the adjacent vertebrae. It is most distinct at the superior and inferior parts of the articulation; some of its upper fibers pass through the intervertebral foramen to the back of the intervertebral disk, while its posterior fibers are continuous with the ligament of the neck of the rib.

The **Radiate Ligament** (*ligamentum capituli costae radiatum; anterior costovertebral* or *stellate ligament*) connects the anterior part of the head of each rib with the side of the bodies of two vertebrae, and the intervertebral fibrocartilage between them. It consists of three flat fasciculi, which are attached to the anterior part of the head of the rib, just beyond the articular surface. The superior fasciculus is connected with the body of the vertebra above; the inferior

one to the body of the vertebra below; the middle one, the smallest and least distinct, is horizontal and is attached to the intervertebral disk. The radiate ligament is in relation, anteriorly, with the thoracic ganglia of the sympathetic trunk, the pleura, and, on the right side, with the azygos vein; dorsally, with the interarticular ligament and synovial membranes (Fig. 5–17).

In the case of the first rib, this ligament is not divided into three fasciculi, but its fibers are attached to the body of the last cervical vertebra, as well as to that of the first thoracic. In the articulations of the heads of the tenth, eleventh, and twelfth ribs, each of which articulates with a single vertebra, the triradiate arrangement does not exist; but the fibers of the ligament in each case are connected to the vertebra above, as well as to that with which the rib articulates.

The **Intraarticular Ligament** (*ligamentum capitis costae intraarticulare*) is situated in the interior of the joint. It consists of a short flattened band of fibers, attached by one extremity to the crest separating the two articular facets on the head of the rib, and by the other to the intervertebral disk; it divides the joint into two cavities. Each cavity has a synovial membrane. In the joints of the first, tenth, eleventh, and twelfth ribs, the intraarticular ligament does not exist; consequently, there is but one cavity and one synovial membrane in each of these articulations.

2. The **Costotransverse Articulations** (*articulationes costotransversariae*) (Fig. 5–18).—The articular portion of the tubercle of the ribs forms a gliding joint with the articular surface on the adjacent transverse process.

In the eleventh and twelfth ribs this articulation is wanting.

The ligaments of the joint are:

> Articular Capsule
> Superior Costotransverse
> Posterior Costotransverse
> Ligament of the Neck of the Rib
> Ligament of the Tubercle of the Rib

The **Articular Capsule** (*capsula articularis; capsular ligament*).—The articular capsule is a thin membrane attached to the circumferences of the articular surfaces, and lined by a synovial membrane.

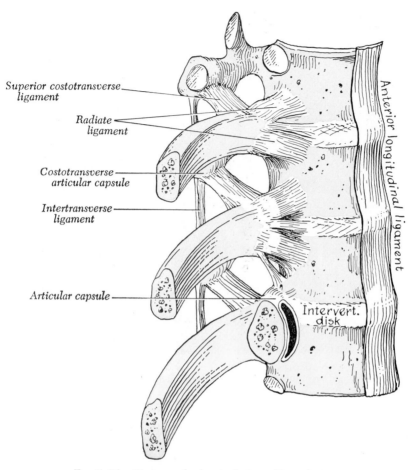

FIG. 5–17.—Costovertebral articulations. Ventral view.

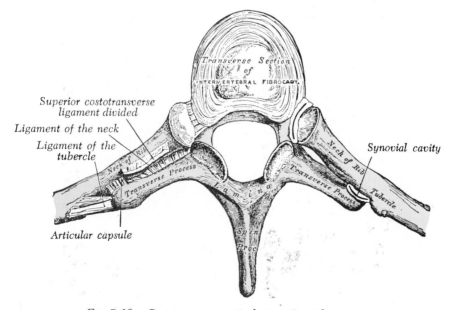

FIG. 5–18.—Costotransverse articulation. Cranial aspect.

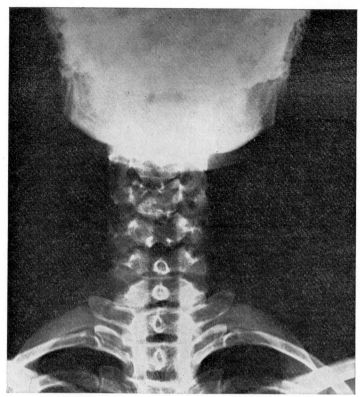

Fig. 5–19.—Antero-posterior view of lower cervical and upper thoracic vertebrae of adult. The skull obscures the upper cervicals. (Courtesy of Doctors Moreland and Arndt.)

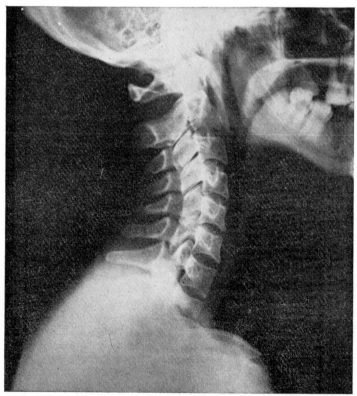

Fig. 5–20.—Lateral view of adult neck. Notice spacing and slope between vertebral bodies, articular process, and spinous processes. (Courtesy of Doctors Moreland and Arndt.)

The **Superior Costotransverse Ligament** (*ligamentum costotransversarium superius; anterior ligament*) is attached to the sharp crest on the superior border of the neck of the rib, and passes obliquely lateralward to the lower border of the transverse process immediately above. It is in relation with the intercostal vessels and nerves; its medial border is thickened and free, and bounds an aperture which transmits the posterior branches of the intercostal vessels and nerves; its lateral border is continuous with a thin aponeurosis, which covers the Intercostalis externus.

The first rib has no superior costotransverse ligament. The neck of the twelfth rib is connected to the base of the transverse process of the first lumbar vertebra by a band of fibers, the **lumbocostal ligament,** in series with the superior costotransverse ligaments. It is merely a thickened portion of the lumbocostal aponeurosis or anterior layer of the thoracolumbar fascia.

The **Posterior Costotransverse Ligament** (*ligamentum costotransversarium posterius*) is a feeble band which is attached to the neck of the rib and passes upward and medialward to the base of the transverse process and lateral border of the inferior articular process of the vertebra above.

The **Ligament of the Neck of the Rib** (PNA — *ligamentum costotransversarium; ligamentum colli costae; middle costotransverse or interosseous ligament*).—The ligament of the neck of the rib consists of short but strong fibers, connecting the rough surface on the back of the neck of the rib with the anterior surface of the adjacent transverse process. A rudimentary ligament may be present in the case of the eleventh and twelfth ribs.

The **Ligament of the Tubercle of the Rib** (PNA—*ligamentum costotransversarium laterale; ligamentum tuberculi costae; posterior costotransverse ligament*) is a short but thick and strong fasciculus, which passes obliquely from the apex of the transverse process to the rough non-articular portion of the tubercle of the rib. The ligaments attached to the superior ribs ascend from the transverse processes; they are shorter and more oblique than those attached to the inferior ribs, which descend slightly.

Movements.—The heads of the ribs are so closely connected to the bodies of the vertebrae by the radiate and interarticular ligaments that only slight gliding movements of the articular surfaces on one another can take place. Similarly, the strong ligaments binding the necks and tubercles of the ribs to the transverse processes limit the movements of the costotransverse joints to slight gliding, the nature of which is determined by the shape and direction of the articular surfaces. In the upper six ribs the articular surfaces on the tubercles are oval in shape and convex from above downward; they fit into corresponding concavities on the *anterior surfaces* of the transverse process, so that upward and downward movements of the tubercles are associated with rotation of the rib neck on its long axis. In the seventh, eighth, ninth, and tenth ribs the articular surfaces on the tubercles are flat, and are directed obliquely downward, medialward, and backward. The surfaces with which they articulate are placed on the *upper margins* of the transverse processes; when, therefore, the tubercles are drawn up they are at the same time carried backward and medialward. The two joints, costocentral and costotransverse, move simultaneously and in the same directions, the total effect being that the neck of the rib moves as if on a single joint, of which the costocentral and costotransverse articulations form the ends. In the upper six ribs the neck of the rib moves but slightly upward and downward; its chief movement is one of rotation around its own long axis, rotation backward being associated with depression, rotation forward with elevation. In the seventh, eighth, ninth, and tenth ribs the neck of the rib moves upward, backward, and medialward, or downward, forward, and lateralward; very slight rotation accompanies these movements.

VI. Sternocostal Articulations (Articulationes Sternocostales; Costosternal Articulations) (Fig. 5–21)

The articulations of the cartilages of the true ribs with the sternum are gliding joints, with the exception of the first, in which the cartilage is directly united with the sternum, and which is, therefore, a synchondrosis articulation. The ligaments connecting them are:

> Articular Capsules
> Radiate Sternocostal
> Intraarticular Sternocostal
> Costoxiphoid

The **Articular Capsules** (*capsulae articulares; capsular ligaments*) surround the joints between the cartilages of the true ribs and the sternum. They are very thin, intimately blended with the radiate sternocostal ligaments, and strengthened at the upper and lower parts of the articulations by a few fibers, which connect the cartilages to the side of the sternum.

The **Radiate Sternocostal Ligaments** (*ligamenta sternocostalia radiata; chondro-*sternal or sternocostal *ligaments*) consist of broad and thin membranous bands that radiate from the ventral and dorsal surfaces of the sternal ends of the cartilages of the true ribs to the anterior and posterior surfaces of the sternum. They are composed of fasciculi which pass in different directions. The **superior fasciculi** ascend obliquely, the **inferior fasciculi** descend obliquely, and the **middle fasciculi** run horizontally. The superficial fibers are the

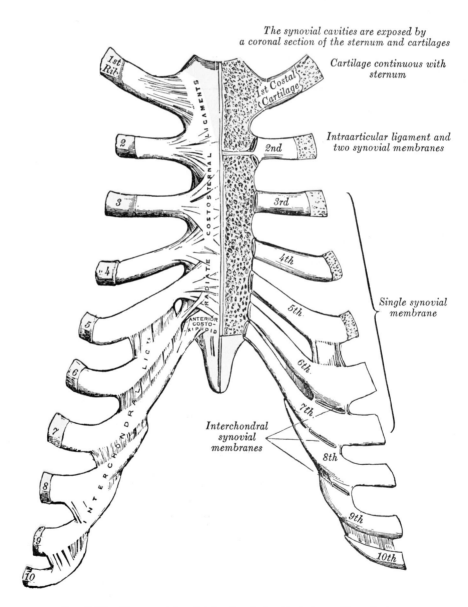

The synovial cavities are exposed by a coronal section of the sternum and cartilages

Cartilage continuous with sternum

Intraarticular ligament and two synovial membranes

Single synovial membrane

Interchondral synovial membranes

Fig. 5–21.—Sternocostal and interchondral articulations. Ventral view.

longest; they intermingle with the fibers of the ligaments above and below them, with those of the opposite side and ventrally with the tendinous fibers of origin of the Pectoralis major, forming a thick fibrous membrane (**membrana sterni**) which envelops the sternum. This is more distinct at the inferior than at the superior part of the bone.

The **Intraarticular Sternocostal Ligament** (*ligamentum sternocostale intraarticulare; interarticular chondrosternal ligament*) is found constantly only between the second costal cartilages and the sternum. The cartilage of the *second rib* is connected with the sternum by means of an intraarticular ligament, attached by one end to the cartilage of the rib, and by the other to the fibrocartilage which unites the manubrium and body of the sternum. This articulation is provided with two synovial membranes. Occasionally the cartilage of the *third rib* is connected with the first and second pieces of the body of the sternum by an interarticular ligament. Still more rarely, similar ligaments are found in the other four joints of the series. In the lower two the ligament sometimes completely obliterates the cavity, so as to convert the articulation into synchondrosis.

The **Costoxiphoid Ligaments** (*ligamenta costoxiphoidea; chondroxiphoid ligaments*) connect the anterior and posterior surfaces of the seventh costal cartilage, and sometimes those of the sixth, to the surfaces of the xiphoid process. They vary in length and breadth in different subjects and the dorsal are less distinct than the ventral.

Synovial Membranes.—There is no synovial membrane between the first costal cartilage and the sternum as this cartilage is directly continuous with the manubrium. There are two in the articulation of the second costal cartilage and generally one in each of the other joints; but those of the sixth and seventh sternocostal joints are sometimes absent; where an intraarticular ligament is present, there are two synovial cavities. After middle life the articular surfaces lose their polish, become roughened, and the synovial membranes apparently disappear. In old age, the cartilages of most of the ribs become continuous with the sternum, and the joint cavities are consequently obliterated.

Movements.—Slight gliding movements are permitted in the sternocostal articulations.

Interchondral Articulations (*articulationes interchondrales; articulations of the cartilages of the ribs with each other*) (Fig. 5–21).—The contiguous borders of the sixth, seventh, and eighth, and sometimes those of the ninth and tenth, costal cartilages articulate with each other by small, smooth, oblong facets. Each articulation is enclosed in a thin **articular capsule**, lined by **synovial membrane** and strengthened laterally and medially by ligamentous fibers (**interchondral ligaments**) which pass from one cartilage to the other. Sometimes the fifth costal cartilages, more rarely the ninth and tenth, articulate by their lower borders with the adjoining cartilages by small oval facets; more frequently the connection is by a few ligamentous fibers.

Costochondral Articulations.—The lateral end of each costal cartilage is received into a depression in the sternal end of the rib, and the two are held together by the periosteum.

VII. Articulations of the Sternum

The **Manubriosternal Articulation.**—The manubrium is united to the body of the sternum by fibrocartilage. It occasionally ossifies in advanced life. About one-third of the fibrocartilages develop a synovial cavity. The two bones are also connected by fibrous tissue.

The **Xiphisternal Articulation** between the xiphoid process and the body of the sternum is cartilaginous. It is usually ossified by the fifteenth year.

Mechanism of the Thorax.—Each rib possesses its own range and variety of movements, but the movements of all are combined in the respiratory excursions of the thorax. Each rib may be regarded as a lever, the fulcrum of which is situated immediately outside the costotransverse articulation, so that when the body of the rib is elevated the neck is depressed and *vice versa;* from the disproportion in length of the arms of the lever a slight movement at the vertebral end of the rib is greatly magnified at the anterior extremity.

The ventral ends of the ribs lie on a plane inferior to the dorsal; when the body of the rib is elevated, therefore, the ventral extremity is also thrust ventrally. Since the middle of the body of the rib lies in a plane inferior to that

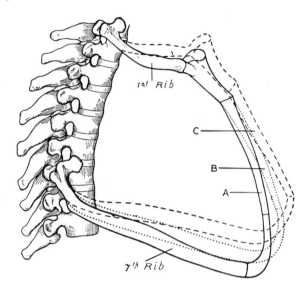

Fig. 5–22.—Lateral view of first and seventh ribs in position, showing the movements of the sternum and ribs in *A*, ordinary expiration; *B*, quiet inspiration; *C*, deep inspiration.

passing through the two extremities, the body is elevated at the same time that it is carried outward from the median plane of the thorax. Further, each rib forms the segment of a curve which is greater than that of the rib immediately above, and therefore the elevation of a rib increases the transverse diameter of the thorax. The modifications of the rib movements at their vertebral ends have already been described (page 307). Further modifications result from the attachments of their ventral extremities, and it is convenient therefore to consider the movements of the ribs of the three groups—vertebrosternal, vertebrochondral, and vertebral separately.

Vertebrosternal Ribs (Figs. 5–22, 5–23).—The first rib differs from the others of this group in that its attachment to the sternum is a rigid one, but its head possesses no interarticular ligament, and is therefore more movable. The first pair of ribs with the manubrium sterni move as a single piece, the ventral portion being elevated by rotatory movements at the vertebral extremities. In normal quiet respiration the movement of this arc is practically *nil;* when it does occur the ventral part is raised and carried ventrally, increasing both diameters of this region of the chest. The movement of the second rib is also slight in normal respiration because its anterior extremity is fixed to the manubrium, and prevented therefore from moving upward. The sternocostal articulation, however, allows the middle of the body of the rib to be drawn up, and in this way the transverse thoracic diameter is increased. Elevation

of the third, fourth, fifth, and sixth ribs raises and thrusts their anterior extremities ventrally, the greater part of the movement being effected by the rotation of the rib neck. The joint between the body and the manubrium allows the relative position of these two parts to be adjusted when the anteroposterior thoracic

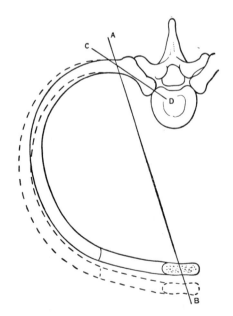

Fig. 5–23.—Diagram showing the axes of movement (*A B* and *C D*) of a vertebrosternal rib. The interrupted lines indicate the position of the rib in inspiration.

diameter is increased. This movement is soon arrested, however, and the force is then expended in raising the middle part of the body of the rib, everting its lower border and increasing the costochondral angle.

Vertebrochondral Ribs (Fig. 5–24).—The seventh rib is included with this group, as it conforms more closely to their type. While the movements of these ribs assist in enlarging the thorax for respiratory purposes, they are also concerned in increasing the upper abdominal space for viscera displaced by the action of the diaphragm. The costal cartilages articulate with

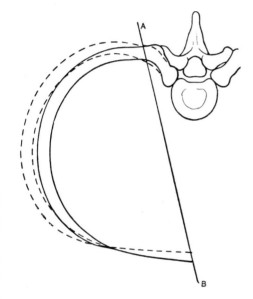

FIG. 5–24.—Diagram showing the axis of movement (*A B*) of a vertebrochondral rib. The interrupted lines indicate the position of the rib in inspiration.

one another, so that each pushes up that above it, the final thrust being directed to pushing forward and upward the lower end of the body of the sternum. The amount of elevation of the anterior extremities is limited on account of the very slight rotation of the rib neck. Elevation of the shaft is accompanied by an outward and backward movement; the outward movement everts the anterior end of the rib and opens up the subcostal angle, while the backward movement pulls back the anterior extremity and counteracts the forward thrust due to its elevation; this latter is most noticeable in the lower ribs, which are the shortest. The total result is a considerable increase in the transverse and a diminution in the median anteroposterior diameter of the upper part of

the abdomen; at the same time, however, the lateral anteroposterior diameters of the abdomen are increased.

Vertebral Ribs.—Since these ribs have free anterior extremities and only costocentral articulations with no interarticular ligaments, they are capable of slight movements in all directions. When the other ribs are elevated these are depressed and fixed to form points of action for the diaphragm.

VIII. Articulation of the Vertebral Column with the Pelvis

The ligaments connecting the fifth lumbar vertebra with the sacrum are similar to those which join the movable segments of the vertebral column with each other—viz.: 1. The continuation caudally of the anterior and posterior longitudinal ligaments. 2. The intervertebral disk connecting the body of the fifth lumbar to that of the first sacral vertebra. 3. Ligamenta flava, uniting the laminae of the fifth lumbar vertebra with those of the first sacral. 4. Capsules connecting the articular processes. 5. Inter- and supraspinal ligaments.

On either side an additional ligament, the **iliolumbar**, connects the pelvis with the vertebral column.

The **Iliolumbar Ligament** (*ligamentum iliolumbale*) (Fig. 5–25) is attached to the transverse process of the fifth lumbar vertebra. It radiates as it passes lateralward and is attached by two main bands to the pelvis. The more caudal bands run to the base of the sacrum, blending with the anterior sacroiliac ligament; the superior is attached to the crest of the ilium immediately ventral to the sacroiliac articulation, and is continuous with the fascia thoracolumbalis. It is in relation with the Psoas major ventrally and with the muscles occupying the vertebral groove dorsally, and with the Quadratus lumborum superiorly.

IX. Articulations of the Pelvis

The ligaments connecting the bones of the pelvis with each other may be divided into four groups: 1. Those connecting the sacrum and ilium. 2. Those passing between the sacrum and ischium. 3. Those uniting the sacrum and coccyx. 4. Those between the two pubic bones.

1. **Sacroiliac Articulation** (*articulatio sacroiliaca*).—The sacroiliac articulation is a synchondrosis, formed between the auricular surfaces of the sacrum and the ilium. The articular surface of each bone is covered with a thin plate of cartilage, thicker on the sacrum than on the ilium. These cartilaginous plates are in close contact with each other and to a certain extent are united together by irregular patches of softer fibrocartilage, and at their proximal and dorsal part by fine interosseous fibers. In a considerable part of their extent, especially in advanced life, they are separated by a space containing a synovial fluid, hence the joint presents the characteristics of a gliding joint. The ligaments of the joint are:

Ventral Sacroiliac
Dorsal Sacroiliac
Interosseous

The **Ventral Sacroiliac Ligament** (*ligamentum sacroiliacum ventrale; anterior sacroiliac ligament*) (Fig. 5–25) consists of numerous thin bands, which connect the ventral surface of the lateral part of the sacrum to the margin of the auricular surface of the ilium and to the preauricular sulcus.

The **Dorsal Sacroiliac Ligament** (*ligamentum sacroiliacum dorsale; posterior sacroiliac ligament*) (Fig. 5–26) is situated in a deep depression between the sacrum and ilium; it is strong and forms the chief bond of union between the bones. It consists of numerous fasciculi which pass between the bones in various directions. The superior part (**short posterior sacroiliac ligament**) is nearly horizontal in direction, and passes from the first and second transverse tubercles on the dorsum of the sacrum to the tuberosity of the ilium. The inferior part (**long posterior sacroiliac ligament**) is oblique in direction; it is attached by one extremity to the third transverse tubercle of the dorsum of the sacrum, and by the other

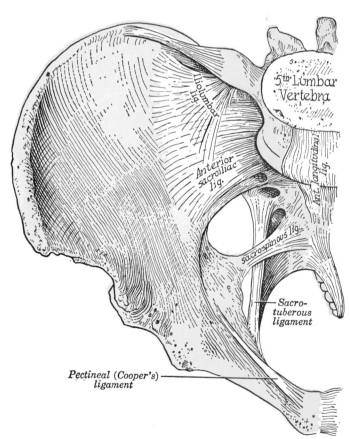

Fig. 5–25.—The joints and ligaments of the right side of the pelvis. Anterior superior aspect.

to the posterior superior spine of the ilium, where it merges with the superior part of the sacrotuberous ligament.

The **Interosseous Sacroiliac Ligament** (*ligamentum sacroiliacum interosseum*) lies deep to the dorsal ligament, and consists of a series of short, strong fibers connecting the tuberosities of the sacrum and ilium.

2. Ligaments Connecting the Sacrum and Ischium (Fig. 5–26)

Sacrotuberous
Sacrospinous

The **Sacrotuberous Ligament** (*ligamentum sacrotuberosum; great* or *posterior sacrosciatic ligament*) (Fig. 5–26) is a broad, flat, fan-shaped complex of fibers stretching from the posterior inferior spine of the ilium, the fourth and fifth transverse tubercles and the caudal part of the lateral margin of the sacrum and the coccyx, to the inner margin of the tuberosity of the ischium. Before the converging fibers reach the tuberosity they form a strong, thick band but again fan out as they attach to the tuberosity, some of them being prolonged ventrally along the ramus as the **falciform process,** to which the obturator fascia is attached. The caudal border of the ligament is directly continuous with the tendon of origin of the Biceps femoris, and is believed by many to be the proximal end of this tendon, interrupted by the projection of the tuberosity of the ischium. Many of its proximal fibers are shared with the long posterior sacroiliac ligament and its deep surface merges with the sacrospinous ligament.

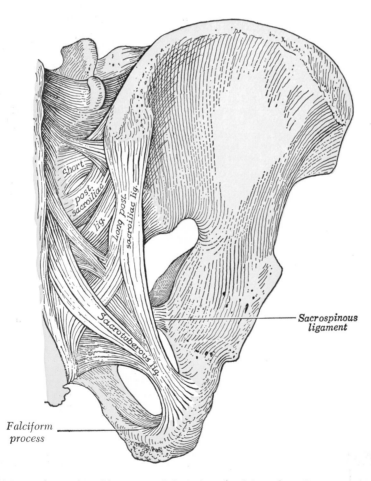

FIG. 5–26.—The joints and ligaments of the right side of the pelvis. Posterior aspect.

Relations.—The *dorsal surface* of this ligament gives origin, by its whole extent, to the Gluteus maximus. Its *ventral surface* is in part united to the sacrospinous ligament. Its *superior border* forms the posterior boundary of the greater sciatic foramen, and the posterior boundary of the lesser sciatic foramen. Its *inferior border* forms part of the boundary of the perineum. It is pierced by the coccygeal nerve and the coccygeal branch of the inferior gluteal artery.

The **Sacrospinous Ligament** (*ligamentum sacrospinale; small* or *anterior sacrosciatic ligament*) (Fig. 5–25) is a thin, triangular sheet attached by its broad base to the lateral margins of the sacrum and coccyx where its fibers are intermingled with those of the intrapelvic surface of the sacrotuberous ligament, and by its apex to the spine of the ischium.

Relations.—It is in relation, *anteriorly*, with the Coccygeus muscle, to which it is closely connected; *posteriorly,* it is covered by the sacrotuberous ligament, and crossed by the internal pudendal vessels and nerve. Its *superior border* forms the lower boundary of the greater sciatic foramen; its *inferior border*, part of the margin of the lesser sciatic foramen.

These two ligaments convert the sciatic notches into foramina. The **greater sciatic foramen** is bounded by the posterior border of the hip bone, by the sacrotuberous ligament, and by the sacrospinous ligament. It is partially occupied in the recent state by the Piriformis which leaves the pelvis through it. Cranial to this muscle, the superior gluteal vessels and nerve emerge from the pelvis; and caudal to it, the inferior gluteal vessels and nerve, the internal pudendal vessels and nerve, the sciatic and posterior femoral cutaneous nerves, and the nerves to the Obturator internus and Quadratus femoris make their exit from the pelvis. The **lesser sciatic foramen** is bounded by the tuberosity of the ischium, by the spine of the ischium and sacrospinous ligament, and by the sacrotuberous ligament. It transmits the tendon of the Obturator internus, its nerve, and the internal pudendal vessels and nerve.

3. **Sacrococcygeal Joint** (*junctura sacrococcygea; articulation of the sacrum and coccyx*).—This articulation is a slightly movable joint between the oval surface at the apex of the sacrum and the base of the coccyx. It is homologous with the joints between the bodies of the vertebrae, and is connected by similar ligaments. They are:

> Ventral Sacrococcygeal
> Dorsal Sacrococcygeal
> Lateral Sacrococcygeal
> Interposed Fibrocartilage
> Interarticular

The **Ventral Sacrococcygeal Ligament** (*ligamentum sacrococcygeum anterius*) consists of a few irregular fibers, which descend from the ventral surface of the sacrum to that of the coccyx, blending with the periosteum.

The **Dorsal Sacrococcygeal Ligament** (*ligamentum sacrococcygeum posterius*) is a flat band, which arises from the margin of the distal orifice of the sacral canal, and descends to be inserted into the dorsal surface of the coccyx. This ligament completes the distal part of the sacral canal, and is divisible into a short **deep portion** and a longer **superficial part**. It is in relation dorsally with the Gluteus maximus.

The **Lateral Sacrococcygeal Ligament** (*ligamentum sacrococcygeum laterale; intertransverse ligament*) exists on either side and connects the transverse process of the coccyx to the lower lateral angle of the sacrum; it completes the foramen for the fifth sacral nerve.

A disk of fibrocartilage is interposed between the contiguous surfaces of the sacrum and coccyx; it differs from those between the bodies of the vertebrae in that it is thinner, and its central part is firmer in texture. It is somewhat thicker ventrally and dorsally than at the sides. Occasionally the coccyx is freely movable on the sacrum, most notably during pregnancy; in such cases a synovial membrane is present.

The **Interarticular Ligaments** are thin bands which unite the cornua of the two bones.

The different segments of the coccyx are connected together by the extension distally of the ventral and dorsal sacrococcygeal ligaments, thin annular disks of fibrocartilage being interposed between the segments. In the adult male, all the pieces become ossified together at a comparatively early period; but in the female, this does not commonly occur until a later period of life.

At more advanced age the joint between the sacrum and coccyx becomes ankylosed.

Movements.—The movements which take place between the sacrum and coccyx, and between the different pieces of the latter bone, are ventral and dorsal; they are very limited. Their extent increases during pregnancy.

4. **The Pubic Symphysis** (*symphysis pubica; articulation of the pubic bones*) (Fig. 5–27).—The articulation between the pubic bones is a slightly movable joint between the two oval articular surfaces of the bones. The ligaments of this articulation are:

> Superior Pubic
> Arcuate Pubic
> Interpubic Disk

The **Superior Pubic Ligament** (*ligamentum pubicum superius*) connects the two pubic bones together superiorly and extends laterally as far as the pubic tubercles.

The **Arcuate Pubic Ligament** (*ligamentum arcuatum pubis; inferior pubic* or *subpubic ligament*) is a thick band of fibers connecting together the two pubic bones and forming the boundary of the pubic arch. It is blended with the interpubic fibrocartilaginous disk; it is attached to the inferior rami of the pubic bones and its free border is separated from the fascia of the urogenital diaphragm by an opening through which the deep dorsal vein of the penis passes into the pelvis.

The **Interpubic Disk** (*discus interpubicus*).—An interpubic fibrocartilaginous lamina or disk connects the opposed surfaces of the pubic bones. Each of these surfaces is covered by a thin layer of hyaline cartilage firmly joined to the bone by a series of nipple-like processes which accurately fit into corresponding depressions on the osseous surfaces. These opposed cartilaginous surfaces are connected together by an intermediate lamina of fibrocartilage which varies in thickness in different subjects. It often contains a cavity in its interior, probably formed by the softening and absorption of the fibrocartilage, since it rarely appears before the tenth year of life and is not lined by synovial membrane (Fig. 5–25). The disk is strengthened by decussating fibers which pass obliquely from one bone to the other and interlace with fibers

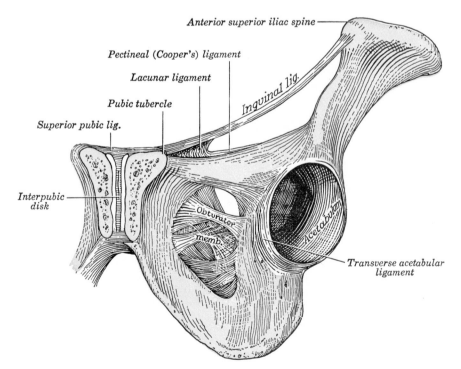

FIG. 5–27.—Symphysis pubis exposed by a coronal section.

of the aponeuroses of the Oblique externi and the medial tendons of origin of the Recti abdominis.

Mechanism of the Pelvis.—The pelvic girdle supports and protects the contained viscera and affords surfaces for the attachments of the trunk and lower limb muscles. In man, however, because of his upright posture, its most important mechanical function is to transmit the weight of the trunk and upper limbs to the lower limbs.

It may be divided into two arches by a vertical plane passing through the acetabular cavities; the dorsal of these arches is the one chiefly concerned in the function of transmitting the weight. Its essential parts are the upper three sacral vertebrae and two strong pillars of bone running from the sacroiliac articulations to the acetabular cavities. For the reception and diffusion of the weight each acetabular cavity is strengthened by two additional bars running toward the pubis and ischium. In order to lessen concussion in rapid changes of distribution of the weight, joints (sacroiliac articulations) are interposed between the sacrum and the iliac bones; an accessory joint (pubic symphysis) exists in the middle of the ventral

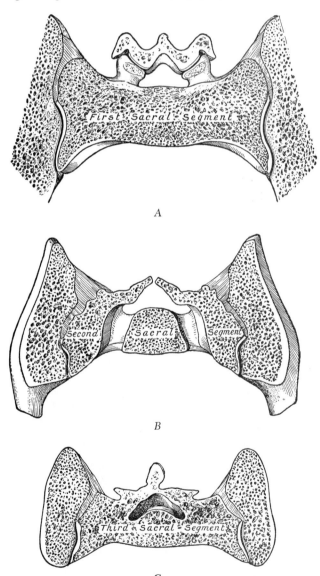

A

B

C

Fɪɢ. 5–28.—Coronal sections through the sacrum: *A*, through first sacral segment; *B*, through second sacral segment; *C*, through third sacral segment.

arch. The sacrum forms the summit of the dorsal arch; the weight transmitted falls on it at the lumbosacral articulation and, theoretically, has a component in each of two directions. One component of the force is expended in driving the sacrum caudally and dorsally between the iliac bones, while the other thrusts the cranial end of the sacrum caudally and ventrally toward the pelvic cavity.

The movements of the sacrum are regulated by its form. Viewed as a whole, it presents the shape of a wedge with its base cranial and ventral. The first component of the force is therefore acting against the resistance of the wedge, and its tendency to separate the iliac bones is resisted by the sacroiliac and iliolumbar ligaments and by the ligaments of the pubic symphysis.

If a series of coronal sections of the sacroiliac joints be made, it will be found possible to divide the articular portion of the sacrum into three segments; ventral, middle, and dorsal. In the **first segment** (Fig. 5–28A), which involves the first sacral vertebra, the articular surfaces show slight sinuosities and are almost parallel to one another; the distance between their dorsal margins is, however, slightly greater than that between their ventral margins. This segment therefore presents a slight wedge shape with the truncated apex caudally. The **second segment** (Fig. 5–28B) is a narrow band across the centers of the articulations. Its dorsal width is distinctly greater than its ventral, so that the segment is more definitely wedge-shaped, the truncated apex being again directed caudally. Each articular surface presents in the center a marked concavity into which a corresponding convexity of the iliac articular surface fits, forming an interlocking mechanism. In the **third segment** (Fig. 5–28C) the ventral width is greater than the dorsal, so that the wedge form is the reverse of those of the other segments— *i.e.*, the truncated apex is directed cranially. The articular surfaces are only slightly concave.

Dislocation caudally and ventrally of the sacrum by the second component of the force applied to it is prevented therefore by the middle segment, which interposes the resistance of its wedge shape and that of the interlocking mechanism on its surfaces; a rotatory movement, however, is produced by which the ventral segment is tilted caudally and the dorsal cranially; the axis of this rotation passes through the dorsal part of the middle segment. The movement of the ventral segment is slightly limited by its wedge form, but chiefly by the dorsal and interosseous sacroiliac ligaments; that of the dorsal segment is checked to a slight extent by its wedge form, but the chief limiting factors are the sacrotuberous and sacrospinous ligaments. In all these movements the effect of the sacroiliac and iliolumbar ligaments and the ligaments of the symphysis pubis in resisting the separation of the iliac bones must be recognized.

During pregnancy the pelvic joints and ligaments are relaxed, and capable therefore of more extensive movements. When the fetus is being expelled the force is applied to the ventral sacrum. Cranial dislocation is again prevented by the interlocking mechanism of the middle segment. As the fetal head passes the ventral segment the latter is carried cranially, enlarging the dorsoventral diameter of the pelvic inlet; when the head reaches the dorsal segment this also is pressed cranially against the resistance of its wedge, the movement being possible only by the laxity of the joints and the stretching of the sacrotuberous and sacrospinous ligaments.

ARTICULATIONS OF THE UPPER LIMB

The articulations of the Upper Limb may be arranged as follows:

I. Sternoclavicular	VII. Intercarpal
II. Acromioclavicular	VIII. Carpometacarpal
III. Shoulder	IX. Intermetacarpal
IV. Elbow	X. Metacarpophalangeal
V. Radioulnar	XI. Articulations of the Digits
VI. Wrist	

I. Sternoclavicular Articulation

The **sternoclavicular articulation** (*articulatio sternoclavicularis*) (Fig. 5–29) is a double gliding joint. The parts entering into its formation are the sternal end of the clavicle, the superior and lateral part of the

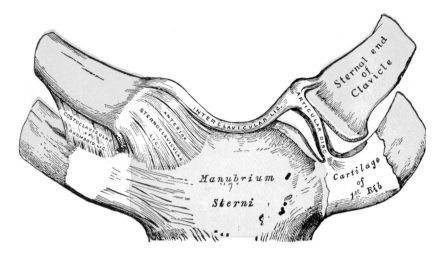

Fig. 5–29.—Sternoclavicular articulation. Ventral aspect.

manubrium sterni, and the cartilage of the first rib. The articular surface of the clavicle is much larger than that of the sternum, and is invested with a layer of fibrocartilage, which is thicker than that on the latter bone. The ligaments of this joint are:

> Articular Capsule
> Anterior Sternoclavicular
> Posterior Sternoclavicular
> Interclavicular
> Costoclavicular
> Articular Disk

The **Articular Capsule** (*capsula articularis; capsular ligament*) surrounds the articulation and varies in thickness and strength. Ventrally and dorsally it is of considerable thickness, and forms the anterior and posterior sternoclavicular ligaments, otherwise it is thin and partakes more of the character of areolar than of true fibrous tissue.

The **Anterior Sternoclavicular Ligament** (*ligamentum sternoclaviculare anterior*) is a broad band of fibers covering the anterior surface of the articulation; it is attached to the superior and ventral part of the sternal end of the clavicle, and passing obliquely downward and medialward, is attached to the ventral part of the manubrium sterni. This ligament is covered by the sternal portion of the Sternocleidomastoideus and the integument; and is in relation with the

capsule, the articular disk, and the two synovial membranes dorsally.

The **Posterior Sternoclavicular Ligament** (*ligamentum sternoclaviculare posterius*) is a similar band of fibers covering the dorsal surface of the articulation; it is attached to the superior part of the sternal end of the clavicle, and, passing obliquely medialward, is fixed below to the dorsal superior part of the manubrium sterni. It is in relation with the articular disk and synovial membranes ventrally and with the Sternohyoideus and Sternothyroideus dorsally.

The **Interclavicular Ligament** (*ligamentum interclaviculare*) is a flattened band which varies considerably in form and size in different individuals. It passes in a curved direction from the superior part of the sternal end of one clavicle to that of the other, and is also attached to the superior margin of the sternum. It is in relation, ventrally, with the integument and Sternocleidomastoidei; dorsally, with the Sternothyroidei.

The **Costoclavicular Ligament** (*ligamentum costoclaviculare; rhomboid ligament*) is short, flat, strong, and rhomboid in form. It is attached to the superior and medial part of the cartilage of the first rib, and to the costal tuberosity on the inferior surface of the clavicle. It is in relation with the tendon of origin of the Subclavius ventrally and with the subclavian vein dorsally.

The **Articular Disk** (*discus articularis*) is flat, nearly circular, and is interposed between the articulating surfaces of the sternum and clavicle. It is attached, *above*, to the dorsal superior border of the articular surface of the clavicle; *below*, to the cartilage of the first rib near its junction with the sternum, and by its circumference to the interclavicular and anterior and posterior sternoclavicular ligaments. It is thicker at the circumference, especially its superior and dorsal part, than at its center. It divides the joint into two cavities, each of which is furnished with a synovial membrane.

Synovial Membranes.—Of the two synovial membranes found in this articulation, the lateral is reflected from the sternal end of the clavicle, over the adjacent surface of the articular disk, and around the margin of the facet on the cartilage of the first rib; the medial is attached to the margin of the articular surface of the sternum and clothes the adjacent surface of the articular disk; the latter is the larger of the two.

Movements.—This articulation admits a limited amount of motion in nearly every direction. When movements take place in the joint, the clavicle carries the scapula with it, the latter gliding on the surface of the chest. This joint therefore forms the center from which all movements of the supporting arch of the shoulder originate, and is the only point of articulation of the shoulder girdle with the trunk. The movements attendant on elevation and depression of the shoulder take place between the clavicle and the articular disk, the bone rotating upon the ligament on an axis drawn through its own articular facet; when the shoulder is moved ventrally and dorsally, the clavicle, with the articular disk, rolls to and fro on the articular surface of the sternum, revolving, with a sliding movement, around an axis drawn nearly vertically through the sternum; in the circumduction of the shoulder, which is a combination of the other movements, the clavicle revolves upon the articular disk and the latter, with the clavicle, rolls upon the sternum. Elevation of the shoulder is limited principally by the costoclavicular ligament; depression, by the interclavicular ligament and articular disk. The muscles which raise the shoulder are the cranial fibers of the Trapezius, the Levator scapulae, and the clavicular head of the Sternocleidomastoideus, assisted to a certain extent by the Rhomboidei. The *depression* of the shoulder is principally effected by gravity assisted by the Subclavius, Pectoralis minor and caudal fibers of the Trapezius. The shoulder is drawn dorsally by the Rhomboidei and the middle and caudal fibers of the Trapezius, and ventrally by the Serratus anterior and Pectoralis minor.

II. Acromioclavicular Articulation

The **acromioclavicular articulation** (*articulatio acromioclavicularis*) (Fig. 5–31) is a gliding joint between the acromial end of the clavicle and the medial margin of the acromion of the scapula. Its ligaments are:

Articular Capsule
Acromioclavicular
Articular Disk
Coracoclavicular
  Trapezoid Ligament
  Conoid Ligament

The **Articular Capsule** (*capsula articularis; capsular ligament*).—The articular capsule completely surrounds the articular margins, and is strengthened above and below by the superior and inferior acromioclavicular ligaments.

The **Acromioclavicular Ligament** (*ligamentum acromioclaviculare*).—The superior part of this ligament is a quadrilateral band covering the superior part of the articulation, and extending between the acromial end of the clavicle and the adjoining part of the acromion. It is composed of parallel fibers, which interlace with the aponeuroses of the Trapezius and Deltoideus; it is in contact with the articular disk when this is present.

The inferior part of this ligament is somewhat thinner than the preceding; it covers the under part of the articulation, and is attached to the adjoining surfaces of the two bones. It is in relation in rare cases with the articular disk or with the tendon of the Supraspinatus.

The **Articular Disk** (*discus articularis*) is frequently absent in this articulation. When present, it generally only partially separates the articular surfaces, and occupies the superior part of the articulation. More rarely, it completely divides the joint into two cavities.

The Synovial Membrane.—There is usually only one synovial membrane in this articulation, but when a complete articular disk is present, there are two.

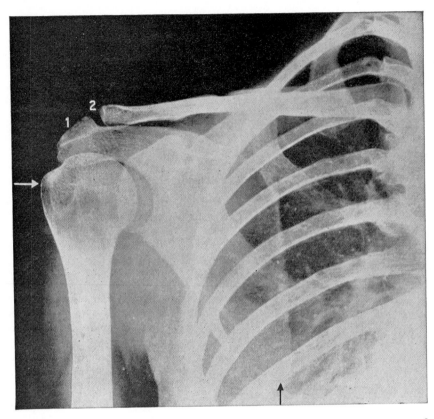

Fig. 5–30.—Adult shoulder, *1*, acromion; *2*, acromioclavicular joint. The lower arrow indicates the inferior angle of the scapula, the upper arrow the greater tuberosity. Note that the shadow of the head of the humerus overlaps the shadow of the acromial angle and a part of the glenoid cavity.

The **Coracoclavicular Ligament** (*ligamentum coracoclaviculare*) (Fig. 5–31) serves to connect the clavicle with the coracoid process of the scapula. Although it is placed at some distance from this articulation, it is usually described with it because it forms a most efficient means of retaining the clavicle in contact with the acromion. It consists of two fasciculi, called the **trapezoid** and **conoid** ligaments.

The **Trapezoid Ligament** (*ligamentum trapezoideum*), the ventral and lateral fasciculus, is broad, thin, and quadrilateral, running obliquely between the coracoid process and the clavicle. It is attached to the superior surface of the coracoid process, and to the oblique ridge on the inferior surface of the clavicle. Its ventral border is free; its dorsal border is joined with the conoid ligament, the two forming an angle by their junction.

The **Conoid Ligament** (*ligamentum co-*

noideum), the dorsal and medial fasciculus, is a dense band of fibers, conical in form, with its base directed superiorly. It is attached by its apex to a rough impression at the base of the coracoid process, medial to the trapezoid ligament, and by its expanded base, to the coracoid tuberosity on the inferior surface of the clavicle, and to a line proceeding medialward from it for 1.25 cm. These ligaments are in relation with the Subclavius and Deltoideus ventrally and with the Trapezius dorsally.

Movements.—The movements of this articulation are of two kinds: (1) a gliding motion of the articular end of the clavicle on the acromion; (2) rotation of the scapula upon the clavicle. The extent of this rotation is limited by the coracoclavicular ligament.

The acromioclavicular and sternoclavicular joints function in cooperation whenever the scapula changes position on the chest wall. The acromion is held at a fixed distance from

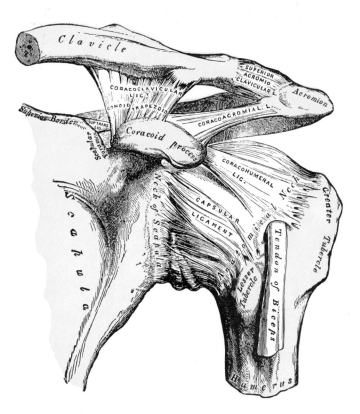

FIG. 5–31.—The left shoulder and acromioclavicular joints, and the proper
ligaments of the scapula.

the sternum by the clavicle which thus serves
as the radius of the circle through which the
acromion must move. The diameter of this
circle being greater than that of the curvature
of the chest wall, the vertebral border of the
scapula would be swung out, away from the
chest when the scapula is moved ventrally.
The acromioclavicular joint, therefore, allows
the scapula to adjust its position so that it
remains in close contact with the chest.

Ligaments of the Scapula.—The following
ligaments complete the structural features
of the scapula and are not related functionally to any joint.

Coracoacromial
Superior Transverse
Inferior Transverse

The **Coracoacromial Ligament** (*ligamentum coracoacromiale*) (Fig. 5–31) is a strong
triangular band extending between the
coracoid process and the acromion. It is
attached to the summit of the acromion

just ventral to the articular surface for the
clavicle, and by a broad base to the whole
length of the lateral border of the coracoid
process. This ligament, together with the
coracoid process and the acromion, completes the arch covering the superior aspect
of the head of the humerus. It is in relation,
superiorly, with the clavicle and under surface of the Deltoideus, and inferiorly, with
the tendon of the Supraspinatus, a bursa
being interposed. Its lateral border is continuous with a dense lamina that passes
deep to the Deltoideus upon the tendons of
the Supraspinatus and Infraspinatus. The
ligament is sometimes described as consisting of two marginal bands and a thinner
intervening portion, the two bands being
attached respectively to the apex and the
base of the coracoid process, and joining
together at the acromion. When the Pectoralis minor is inserted, as occasionally occurs,
into the capsule of the shoulder joint instead
of into the coracoid process, it passes

between these two bands, and the intervening portion of the ligament is then deficient.

The **Superior Transverse Ligament** (*ligamentum transversum scapulae superius; transverse* or *suprascapular ligament*) converts the scapular notch into a foramen. It is a thin, flat fasciculus, attached by one end to the base of the coracoid process, and by the other to the medial end of the scapular notch. The suprascapular nerve runs through the foramen; the suprascapular vessels cross over the ligament. The ligament is sometimes ossified.

The **Inferior Transverse Ligament** (*ligamentum transversum scapulae inferius; spinoglenoid ligament*) is a weak membranous band stretching from the lateral border of the spine to the margin of the glenoid cavity. It forms an arch under which the suprascapular vessels and nerve enter the infraspinatous fossa.

III. Humeral Articulation or Shoulder Joint

The **Shoulder Joint** (*articulatio humeri*) (Fig. 5–31) is a spheroidea or ball-and-socket joint. The bones entering into its formation are the hemispherical head of the humerus and the shallow glenoid cavity of the scapula, an arrangement which permits of very considerable movement, while the joint itself is protected against displacement by the ligaments and tendons which surround it. The ligaments do not maintain the joint surfaces in apposition, because when they alone remain the humerus can be separated to a considerable extent from the glenoid cavity; their use, therefore, is to limit the amount of movement. The joint is protected above by the arch formed by the coracoid process, acromion, and coracoacromial ligament. The articular cartilage on the head of the humerus is thicker at the center than at the circumference, the reverse being the case with the articular cartilage of the glenoid cavity. The ligaments of the shoulder are:

> Articular Capsule
> Coracohumeral
> Glenohumeral
> Glenoidal Labrum
> Transverse Humeral

The **Articular Capsule** (*capsula articularis; capsular ligament*) (Figs. 5–31, 5–32, 5–33) completely encircles the joint, being attached to the circumference of the glenoid

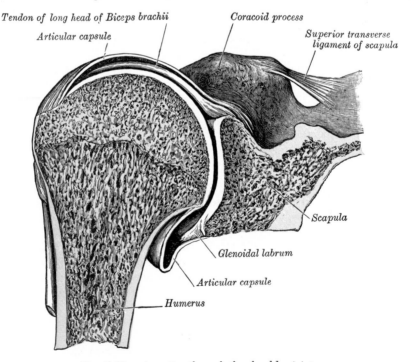

Tendon of long head of Biceps brachii

Articular capsule

Coracoid process

Superior transverse ligament of scapula

Scapula

Glenoidal labrum

Articular capsule

Humerus

Fig. 5–32.—A section through the shoulder joint.

cavity beyond the glenoidal labrum and to the anatomical neck of the humerus, approaching nearer to the articular cartilage superiorly than in the rest of its extent. It is thicker superiorly and inferiorly than elsewhere, and is so remarkably loose and lax, that it has no action in keeping the bones in contact, but allows them to be separated from each other more than 2.5 cm, an evident provision for that extreme freedom of movement which is peculiar to this articulation. It is reinforced superiorly by the Supraspinatus; inferiorly, by the long head of the Triceps brachii; dorsally, by the tendons of the Infraspinatus and Teres minor; and ventrally, by the tendon of the Subscapularis. There are usually three openings in the capsule. One ventrally, below the coracoid process, establishes a communication between the joint and a bursa beneath the tendon of the Subscapularis. The second, which is not constant, is at the posterior part, where an opening sometimes exists between the joint and a bursal sac under the tendon of the Infraspinatus. The third is between the tubercles of the humerus, for the passage of the long tendon of the Biceps brachii.

The **Coracohumeral Ligament** (*ligamentum coracohumerale*) is a broad band which strengthens the upper part of the capsule (Fig. 5–31). From the lateral border of the base of the coracoid process, it takes a slightly spreading and oblique course laterally to the anterior border of the greater tubercle of the humerus, lateral to the tendon of the long head of the Biceps brachii. At its humeral attachment it blends with the tendon of the Supraspinatus superficial to it, and with the capsule deep to it except for its ventral free border.

The **Glenohumeral Ligaments** (*ligg. glenohumeralia*) are robust thickenings of the articular capsule over the ventral part of the joint. They are described as three ligaments, but their individuality is often difficult to demonstrate. They may be seen in dissection, however, if the joint is disarticulated (Fig. 5–33), or they may be palpated through a wide opening cut in the dorsal part of the capsule. The **superior glenohumeral ligament** is the portion attached to the apex of the glenoid cavity close to the base of the coracoid process and is attached to a small depression at the top of the lesser tubercle. It runs along the medial edge of the tendon of the long head of the Biceps; the tendon separates it from the coracohumeral ligament and it lies between the long head tendon and the tendon of the Subscapularis. The **middle glenohumeral ligament** is attached to the medial edge of the

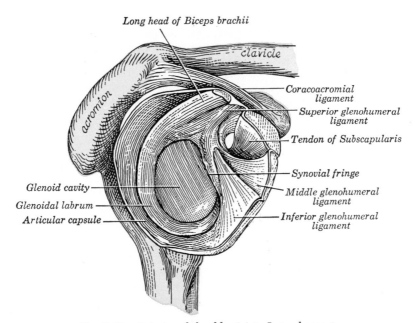

Long head of Biceps brachii

clavicle

acromion

Coracoacromial ligament

Superior glenohumeral ligament

Tendon of Subscapularis

Synovial fringe

Glenoid cavity

Glenoidal labrum

Articular capsule

Middle glenohumeral ligament

Inferior glenohumeral ligament

FIG. 5–33.—Interior of shoulder joint. Lateral aspect.

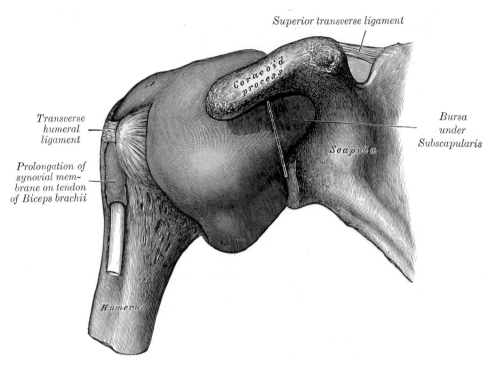

Superior transverse ligament

Transverse humeral ligament

Prolongation of synovial membrane on tendon of Biceps brachii

Bursa under Subscapularis

Fig. 5–34.—Synovial membrane of shoulder joint (distended). Anterior aspect.

glenoid cavity and the inferior part of the lesser tubercle. When there is an opening between the synovial cavity and the subscapular bursa it is between the superior and middle ligaments. The **inferior glenohumeral ligament** extends from the inferior edge of the glenoid cavity to the inferior part of the anatomical neck of the humerus.

The **Transverse Humeral Ligament** (Fig. 5–34) is a narrow sheet of short transverse fibers passing from the lesser to the greater tubercle of the humerus, and always limited to that portion of the bone which lies superior to the epiphyseal line (Fig. 5–32). It converts the intertubercular groove into a tunnel for the tendon of the long head of the Biceps.

The **Glenoidal Labrum** (*labrium glenoidale; glenoid ligament*) is a fibrocartilaginous rim attached around the margin of the glenoid cavity. It is triangular on section, the base being fixed to the circumference of the cavity, while the free edge is thin and sharp. It is continuous with the tendon of the long head of the Biceps brachii, which gives off two fasciculi to blend with the fibrous tissue of the labrum. It deepens

the articular cavity, and protects the edges of the bone (Fig. 5–33).

Synovial Membrane.—The synovial membrane is reflected from the margin of the glenoid cavity over the labrum; it is then reflected over the inner surface of the capsule, and covers inferior part and sides of the anatomical neck of the humerus as far as the articular cartilage on the head of the bone. The tendon of the long head of the Biceps brachii passes through the capsule and is enclosed in a tubular sheath of synovial membrane, which is reflected upon it from the summit of the glenoid cavity and is continued around the tendon into the intertubercular groove as far as the surgical neck of the humerus (Fig. 5–34). The tendon thus traverses the articulation, but it is not contained within the synovial cavity.

Bursae Associated with the Shoulder (Fig. 6–38).—The **Subscapular Bursa** is of constant occurrence situated between the tendon of the Subscapularis and the underlying joint capsule. It usually communicates with the synovial cavity through an opening in the ventral part of the capsule.

The **Subdeltoid Bursa** is a large bursa between the deep surface of the Deltoideus and the joint capsule; it does not communicate with the synovial cavity.

The **Subacromial Bursa** lies between the inferior surface of the acromion and the joint capsule; it usually extends under the coraco-acromial ligament and frequently is continuous with the subdeltoid bursa.

A **Subcoracoid Bursa** may lie between the coracoid and the capsule or it may be an extension from the subacromial bursa.

The **Coracobrachialis Bursa** intervenes between the Coracobrachialis, but is inconstant.

The **Infraspinatus Bursa,** between the tendon of the Infraspinatus and the capsule, may communicate with the synovial cavity.

The **Latissimus dorsi, Teres major,** and **Pectoralis major Bursae** lie between these muscles and the humerus at their insertions.

The **Subcutaneous Acromial Bursa** is of considerable extent over the superficial surface of the acromion.

The **muscles** in relation with the joint are: *superior,* the Supraspinatus; *inferior,* the long head of the Triceps brachii; *anterior,* the Subscapularis; *posterior,* the Infraspinatus and Teres minor; *within,* the tendon of the long head of the Biceps brachii. The Deltoideus covers the articulation anteriorly, posteriorly and laterally.

The **arteries** supplying the joint are articular branches of the anterior and posterior humeral circumflex, and suprascapular.

The **nerves** are derived from the axillary and suprascapular.

Movements.—The shoulder joint is capable of every variety of movement, flexion, extension, abduction, adduction, circumduction, and rotation. The humerus is *flexed* (drawn forward) by the Pectoralis major, anterior fibers of the Deltoideus, Coracobrachialis, and when the forearm is flexed, by the Biceps brachii; *extended* (drawn backward) by the Latissimus dorsi, Teres major, posterior fibers of the Deltoideus, and, when the forearm is extended, by the Triceps brachii; it is *abducted* by the Deltoideus and Supraspinatus; it is *adducted* by the Subscapularis, Pectoralis major, Latissimus dorsi, and Teres major, and by the weight of the limb; it is *rotated outward* by the Infraspinatus and Teres minor; and it is *rotated inward* by the Subscapularis, Latissimus dorsi, Teres major, Pectoralis major, and the anterior fibers of the Deltoideus.

The most striking peculiarities in this joint are: (1) The large size of the head of the humerus in comparison with the depth of the glenoid cavity, even when this latter is supplemented by the glenoidal labrum. (2) The looseness of the capsule of the joint. (3) The intimate connection of the capsule with the muscles attached to the head of the humerus.

(4) The peculiar relation of the tendon of the long head of the Biceps brachii to the joint.

It is in consequence of the relative sizes of the two articular surfaces and the looseness of the articular capsule that the joint enjoys such free movement in all directions. The arm can be carried considerably farther by the movements of the scapula, involving, of course, motion at the acromio- and sternoclavicular joints. These joints are therefore to be regarded as accessory structures to the shoulder joint (see pages 320 and 321). The extent of the scapular movements is very considerable, especially in extreme elevation of the arm, a movement best accomplished when the arm is thrown somewhat forward and outward, because the margin of the head of the humerus is by no means a true circle; its greatest diameter is from the intertubercular groove, downward, medialward, and backward, and the greatest elevation of the arm can be obtained by rolling its articular surface in the direction of this measurement. The great width of the central portion of the humeral head also allows of very free horizontal movement when the arm is raised to a right angle, in which movement the arch formed by the acromion, the coracoid process and the coracoacromial ligament constitutes a sort of supplemental articular cavity for the head of the bone.

The looseness of the capsule is so great that the humeral head will fall away from the scapula about 2.5 cm when the muscles are dissected from around the joint. The articular surfaces of the two bones are held in contact, not so much by the capsule as by the surrounding muscles, an arrangement which allows very easy movement in the joint, especially when the muscles are not under tension. In all ordinary positions of the joint, the capsule is not put on the stretch, but extreme movements are checked by the tension of appropriate portions of the capsule, as well as by other ligaments and certain muscles.

The scapula is capable of being moved superiorly and inferiorly, ventrally and dorsally, or, by a combination of these movements, circumducted on the wall of the chest. The muscles which *raise* the scapula are the superior fibers of the Trapezius, the Levator scapulae, and the Rhomboidei; those which *depress* it are the inferior fibers of the Trapezius, the Pectoralis minor, and, through the clavicle, the Subclavius. The scapula is drawn posteriorly by the Rhomboidei and the middle and inferior fibers of the Trapezius, and anteriorly by the Serratus anterior and Pectoralis minor, assisted, when the arm is fixed, by the Pectoralis major. The mobility of the scapula is very considerable, and greatly assists the movements of the arm at the shoulder joint. This mobility is of special importance in ankylosis of the shoulder joint, the movements of this bone compensating to a very great extent for the immobility of the joint.

Raising of the arm above the head, either by carrying it forward in flexion or to the side in abduction, is brought about by the combined activity of the shoulder joint and rotation of the scapula on the chest wall. Although it has been customary to separate the movement into two parts, one, raising the arm to the horizontal, the other, from horizontal to overhead, such a distinction is artificial. During practically all of both parts of the movement, the ratio of motion in the two articulations is that of two parts glenohumeral to one part scapulothoracic. If motion in the glenohumeral articulation is destroyed by ankylosis, therefore, only one-third of the whole movement of 60° of motion will be retained in the compensatory motion of the scapulothoracic articulation.

The intimate union of the tendons of the Supraspinatus, Infraspinatus, Teres minor and Subscapularis with the capsule converts these muscles into elastic and spontaneously acting ligaments of the joint.

The peculiar relations of the tendon of the long head of the Biceps brachii to the shoulder joint appear to subserve various purposes. In the first place, by its connection with both the shoulder and elbow the muscle harmonizes the action of the two joints, and acts as an elastic ligament in all positions, in the manner previously discussed (see page 292). It strengthens the upper part of the articular cavity and prevents the head of the humerus from being pressed up against the acromion when the Deltoideus contracts; it thus fixes the head of the humerus as the center of motion in the glenoid cavity. By its passage along the intertubercular groove it assists in steadying the head of the humerus in the various movements of the arm. When the arm is raised from the side it assists the Supraspinatus and Infraspinatus in rotating the head of the humerus in the glenoid cavity. It also holds the head of the bone firmly in contact with the glenoid cavity, and prevents its slipping over its lower edge, or being displaced by the action of the Latissimus dorsi and Pectoralis major, as in climbing and many other movements.

IV. Elbow Joint

The **Elbow Joint** (*articulatio cubiti*) (Figs. 5–36, 5–37, 5–38) is a ginglymus or hinge joint. The trochlea of the humerus fits into the trochlear notch of the ulna, and the

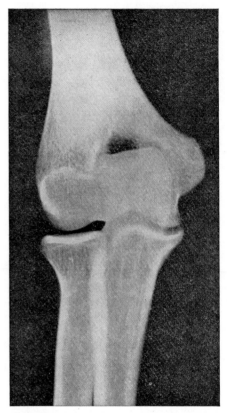

FIG. 5–35.—Adult elbow. Frontal view. The shadow of the olecranon extends upwards to the olecranon fossa and obscures the outline of the trochlea. The gap between the humerus and the bones of the forearm is occupied by the articular cartilage of the bones concerned.

capitulum of the humerus articulates with the fovea on the head of the radius. The articular surfaces are connected together by a **capsule**, which is thickened medially and laterally into the **ulnar collateral** and the **radial collateral** ligaments.

The **Articular Capsule** (Fig. 5–38).—The anterior part is a thin fibrous layer covering the anterior aspect of the joint; it is attached to the medial epicondyle and to the humerus immediately proximal to the coronoid and radial fossae, and to the anterior surface of the coronoid process of the ulna and to the annular ligament (page 330), being continuous on either side with the collateral ligaments. Its superficial fibers pass obliquely from the medial epicondyle of the humerus to the annular ligament. The middle fibers pass from the proximal part

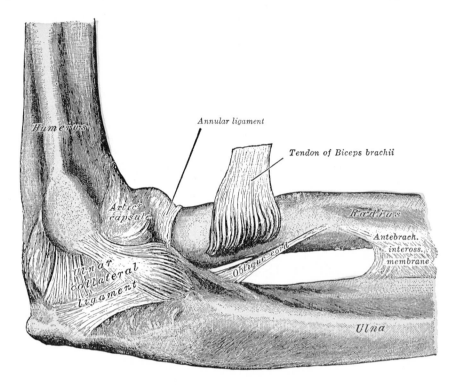

FIG. 5–36.—The left elbow joint. Medial aspect.

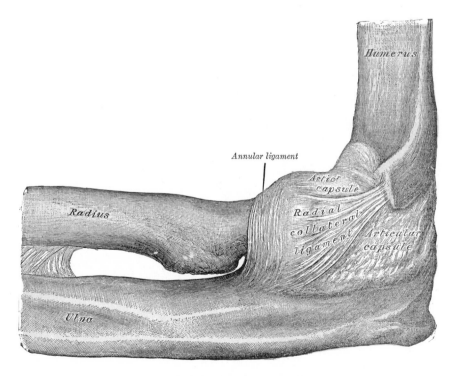

FIG. 5–37.—The left elbow joint. Lateral aspect.

of the coronoid depression and become partly blended with the preceding, but are inserted mainly into the anterior surface of the coronoid process. The deep or transverse set intersects these at right angles. Its anterior relation is with the Brachialis, except at its most lateral part.

The posterior part (Fig. 5–38) is thin and membranous, and consists of transverse and oblique fibers. Proximally it is attached to the humerus immediately behind the capitulum and close to the medial margin of the trochlea, to the margins of the olecranon fossa, and to the back of the lateral epicondyle some little distance from the trochlea. Distally, it is fixed to the proximal and lateral margins of the olecranon, to the posterior part of the annular ligament, and to the ulna posterior to the radial notch. The transverse fibers form a strong band which bridges across the olecranon fossa; under cover of this band a pouch of synovial membrane

and a pad of fat project into the upper part of the fossa when the joint is extended. In the fat are a few scattered fibrous bundles, which pass from the deep surface of the transverse band to the proximal part of the fossa. It is in relation, posteriorly, with the tendon of the Triceps brachii and the Anconeus.

The **Ulnar Collateral Ligament** (*ligamentum collaterale ulnare; internal lateral ligament*) (Fig. 5–36) is a thick triangular band consisting of two portions, an anterior and posterior united by a thinner intermediate portion. The **anterior portion** is directed obliquely from the anterior part of the medial epicondyle of the humerus to the medial margin of the coronoid process. The **posterior portion** is attached to the distal and posterior part of the medial epicondyle, and to the medial margin of the olecranon. Between these two bands a few intermediate fibers descend from the medial epicon-

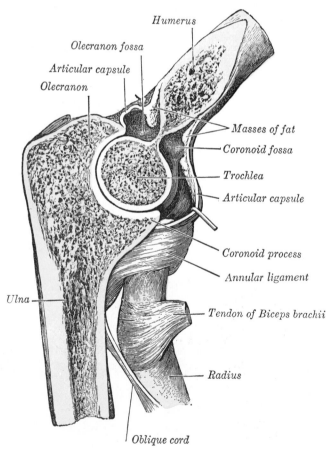

FIG. 5–38.—Sagittal section through the left elbow joint.

dyle to blend with a *transverse band* which bridges across the notch between the olecranon and the coronoid process. This ligament is in relation with the Triceps brachii and Flexor carpi ulnaris and the ulnar nerve, and gives origin to part of the Flexor digitorum superficialis.

The **Radial Collateral Ligament** (*ligamentum collaterale radiale; external lateral ligament*) (Fig. 5–37) is a triangular fibrous band, less distinct than the ulnar collateral, attached to a depression distal to the lateral epicondyle of the humerus proximally, and to the annular ligament distally, some of its most posterior fibers passing over that ligament to be inserted into the lateral margin of the ulna. It is intimately blended with the tendon of origin of the Supinator.

Synovial Membrane (Figs. 5–39, 5–40).—The synovial membrane is very extensive. It extends from the margin of the articular surface of the humerus, and lines the coronoid, radial and olecranon fossae on that bone; it is reflected over the deep surface of the capsule and forms a pouch between the radial notch, the deep surface of the annular ligament, and the circumference of the head of the radius. Projecting between the radius and ulna into the cavity is a crescentic fold of synovial membrane, suggesting the division of the joint into two: one, the humeroradial; the other, the humeroulnar.

Between the capsule and the synovial membrane are three masses of fat: the largest, over the olecranon fossa, is pressed into the fossa by the Triceps brachii during the flexion; the second, over the coronoid fossa, and the third, over the radial fossa, are pressed by the Brachialis into their respective fossae during extension.

The **muscles** in relation with the joint are anteriorly the Brachialis; the Triceps brachii and Anconeus posteriorly; the Supinator, and the common tendon of origin of the Extensor

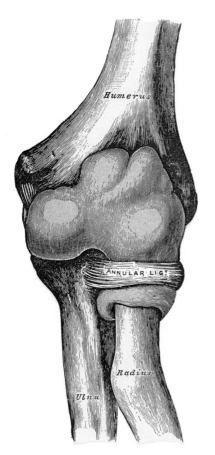

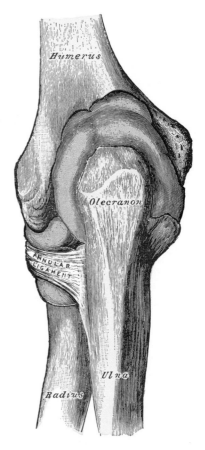

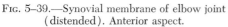

Fig. 5–39.—Synovial membrane of elbow joint (distended). Anterior aspect.

Fig. 5–40.—Synovial membrane of capsule of elbow joint (distended). Dorsal aspect.

muscles laterally; and the common tendon of origin of the Flexor muscles, and the Flexor carpi ulnaris medially.

The arteries supplying the joint are derived from the anastomosis between the profunda and the superior and inferior ulnar collateral branches of the brachial, with the anterior, posterior, and interosseous recurrent branches of the ulnar, and the recurrent branch of the radial. These vessels form a complete anastomotic network around the joint.

The nerves of the joint are a twig from the ulnar, as it passes between the medial condyle and the olecranon; a filament from the musculocutaneous, and two from the median.

Movements.—The elbow joint comprises three different portions—viz., the joint between the ulna and humerus, that between the head of the radius and the humerus, and the proximal radioulnar articulation, described below. All these articular surfaces are enveloped by a common synovial membrane, and the movements of the whole joint should be studied together. The combination of the movements of flexion and extension of the forearm with those of pronation and supination of the hand, which is ensured by the two being performed at the same joint, is essential to the accuracy of the various minute movements of the hand.

The portion of the joint between the ulna and humerus is a simple hinge joint, and allows movements of flexion and extension only. Owing to the obliquity of the trochlea of the humerus, this movement does not take place in the anteroposterior plane of the body of the humerus. When the forearm is extended and supinated, the axes of the arm and forearm are not in the same line; the arm forms an obtuse angle with the forearm, the hand and forearm being directed lateralward. During flexion, however, the forearm and the hand tend to approach the middle line of the body, and thus enable the hand to be easily carried to the face. The accurate adaptation of the trochlea of the humerus, with its prominences and depressions, to the semilunar notch of the ulna, prevents any lateral movement. *Flexion* is produced by the action of the Biceps brachii and Brachialis, assisted by the Brachioradialis and the muscles arising from the medial epicondyle of the humerus; *extension,* by the Triceps brachii and Anconeus, assisted by the Extensors of the wrist, the Extensor digitorum, and the Extensor digiti minimi.

The joint between the head of the radius and the capitulum of the humerus is a gliding joint. The bony surfaces would of themselves constitute an enarthrosis and allow of movement in all directions, were it not for the annular ligament, by which the head of the radius is bound

to the radial notch of the ulna, and which prevents any separation of the two bones laterally. It is to the same ligament that the head of the radius owes its security from dislocation which would otherwise tend to occur, from the shallowness of the cup-like surface on the head of the radius. In fact, but for this ligament, the tendon of the Biceps brachii would be liable to pull the head of the radius out of the joint. The head of the radius is not in complete contact with the capitulum of the humerus in all positions of the joint. The capitulum occupies only the anterior and inferior surfaces of the lower end of the humerus, so that in complete extension a part of the radial head can be plainly felt projecting at the back of the articulation. In full flexion the movement of the radial head is hampered by the compression of the surrounding soft parts, so that the freest rotatory movement of the radius on the humerus (pronation and supination) takes place in semiflexion in which position the two articular surfaces are in most intimate contact.

In any position of flexion or extension, the radius, carrying the hand with it, can be rotated in the proximal radioulnar joint. The hand is directly articulated to the lower surface of the radius only, and the ulnar notch on the lower end of the radius travels around the lower end of the ulna. The latter bone is excluded from the wrist joint by the articular disk. Thus, rotation of the head of the radius around an axis passing through the center of the radial head of the humerus imparts circular movement to the hand through a very considerable arc.

V. Radioulnar Articulation
(Articulatio Radioulnaris)

The articulation of the radius with the ulna is effected by ligaments which connect together the extremities as well as the bodies of these bones. The ligaments may, consequently, be subdivided into three sets: 1, those of the proximal radioulnar articulation; 2, the middle radioulnar ligaments; 3, those of the distal radioulnar articulation.

Proximal Radioulnar Articulation (*articulatio radioulnaris proximalis; superior radioulnar joint*).—This articulation is a trochoid or pivot joint between the circumference of the head of the radius and the ring formed by the radial notch of the ulna and the *annular ligament.*

The **Annular Ligament** (*ligamentum anulare radii; orbicular ligament*) (Fig. 5–41) is a strong band of fibers which

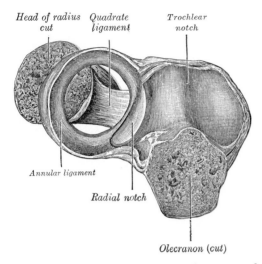

Head of radius cut *Quadrate ligament* *Trochlear notch*

Annular ligament

Radial notch

Olecranon (cut)

Fig. 5–41.—Annular ligament of radius, proximal aspect. The head of the radius has been sawn off and the bone dislodged from the ligament.

encircles the head of the radius and retains it in contact with the radial notch of the ulna. It forms about four-fifths of the osseofibrous ring, and is attached to the anterior and posterior margins of the notch; a few of its fibers are continued distal to the notch and form a complete fibrous ring. Its proximal border blends with the anterior and posterior ligaments of the elbow, while from its distal border a thin loose membrane is attached to the neck of the radius; a thickened band which extends from the inferior border of the annular ligament to the neck of the radius is known as the **quadrate ligament**. The superficial surface of the annular ligament is strengthened by the radial collateral ligament of the elbow, and affords origin to part of the Supinator. Its deep surface is smooth, and lined by synovial membrane which is continuous with that of the elbow joint.

Movements.—The movements allowed in this articulation are limited to rotatory movements of the head of the radius within the ring formed by the annular ligament and the radial notch of the ulna; rotation which moves the thumb from lateral to medial is called *pronation;* rotation in the opposite direction is *supination.* Supination is performed by the Biceps brachii and Supinator. Pronation is performed by the Pronator teres and Pronator quadratus.

Middle Radioulnar Union.—The shafts of the radius and ulna are held together by

the Interosseous Membrane and the Oblique Cord.

The **Interosseous Membrane** (*membrana interossea antebrachii*) is a broad plane of fibrous tissue connecting almost the whole length of the interosseous crest of the radius to that of the ulna; the distal part of the membrane is attached to the posterior of the two lines into which the interosseous crest of the radius divides. It is deficient proximally, commencing about 2.5 cm from the tuberosity of the radius; it is broader in the middle than at either end and presents an oval aperture a little above its distal margin for the passage of the palmar interosseous vessels to the back of the forearm. This membrane serves to increase the extent of surface for the attachment of the deep muscles as well as to connect the bones. Between its proximal border and the oblique cord is a gap, through which the dorsal interosseous vessels pass. The direction of the fibrous bundles in the membrane is oblique, extending from the radius proximally to the ulna distally. Two or three bands are occasionally found on the dorsal surface of this membrane running obliquely in a direction contrary to that of the other fibers. The membrane is in relation, in its proximal three-fourths, with the Flexor pollicis longus on the radial side, and with the Flexor digitorum profundus on the ulnar, lying in the interval between which are the palmar interosseous vessels and nerve; by its distal fourth with the Pronator quadratus; dorsally, with the Supinator, Abductor pollicis longus, Extensor pollicis brevis, Extensor pollicis longus, Extensor indicis; and, near the wrist, with the palmar interosseous artery and dorsal interosseous nerve.

The **Oblique Cord** (*chorda obliqua; oblique ligament*) (Fig. 5–36) is a small, flattened band extending distally and laterally from the lateral side of the tubercle of the ulna at the base of the coronoid process to the radius a little distal to the radial tuberosity. Its fibers run in the opposite direction to those of the interosseous membrane. It is sometimes wanting.

Distal Radioulnar Articulation (*articulatio radioulnaris distalis; inferior radioulnar joint*).—This is a pivot joint formed between the head of the ulna and the ulnar notch on the lower end of the radius. The articular surfaces are connected together by the following ligaments:

Articular Capsule

Articular Disk

The **Articular Capsule** (*capsula articularis*) (Figs. 5–43, 5–44, not labelled) **is** composed of bands of fibers attached to the margins of the ulnar notch and to the head of the ulna. It has two thickenings, named according to their position the **palmar radioulnar ligament** and the **dorsal radioulnar ligament.** The portion of the capsule between these two bands is loose and extends proximally between the radius and the ulna to form the **recessus sacciformis.**

The **Articular Disk** (*discus articularis; triangular fibrocartilage*) (Fig. 5–45) is triangular in shape, and is placed transversely beneath the head of the ulna, binding the lower ends of the ulna and radius firmly together. Its periphery is thicker than its center, which is occasionally perforated. It is attached by its apex to a depression between the styloid process and the head of the ulna; and by its base, which is thin, to the prominent edge of the radius, which separates the ulnar notch from the carpal articular surface. Its margins are united to the ligaments of the wrist joint. Its **proximal surface,** smooth and concave, articulates with the head of the ulna, forming a gliding joint; its **distal surface,** also concave and smooth, forms part of the wrist joint and articulates with the triangular bone and medial part of the lunate. Both surfaces are clothed by synovial membrane; the proximal, by that of the distal radioulnar articulation, the distal, by that of the wrist.

Synovial Membrane (Fig. 5–45).—Both surfaces of the articular disk are covered with a synovial layer. The proximal surface enters into the radioulnar, the distal into the wrist joint. The synovial surface of the radioulnar articulation lines the articular capsule and the recessus sacciformis.

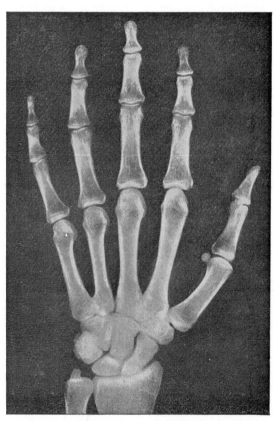

Fɪɢ. 5–42.—Adult hand.

Movements.—The movements in the distal radioulnar articulation consist of rotation of the distal end of the radius around an axis which passes through the center of the head of the ulna. When the radius rotates toward the palm, *pronation* of the forearm and hand is the result; and when toward the dorsum, *supination*. Thus in pronation and supination the radius describes the segment of a cone, the axis of which extends from the center of the head of the radius to the middle of the head of the ulna. In this movement the head of the ulna is stationary, but appears to describe a curve in a direction opposite to that taken by the head of the radius.

VI. Radiocarpal Articulation or Wrist Joint

The **Wrist Joint** (*articulatio radiocarpea*) (Figs. 5–43, 5–44) is a condylar articulation.

The parts forming it are the distal end of the radius and distal surface of the articular disk with the scaphoid, lunate, and triangular bones. The articular surfaces of the radius and the articular disk form together a transversely elliptical concave surface. The superior articular surfaces of the scaphoid, lunate, and triangular form a smooth convex surface, the **condyle**, which is received into the concavity. The joint is surrounded by a capsule and strengthened by the following ligaments:

Palmar Radiocarpal
Dorsal Radiocarpal
Ulnar Collateral
Radial Collateral

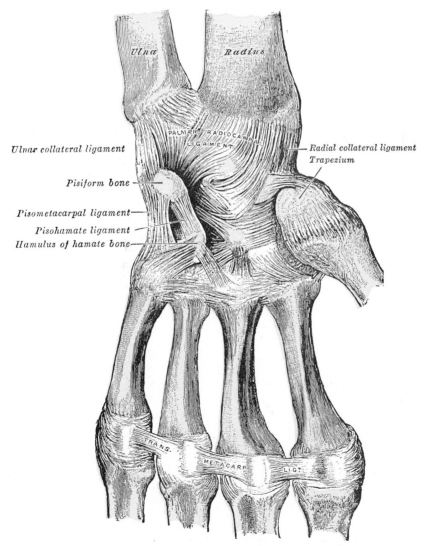

FIG. 5–43.—The ligaments of the left wrist and metacarpus. Palmar aspect.

The **Palmar Radiocarpal Ligament** (*ligamentum radiocarpeum volare; anterior ligament*) (Fig. 5–43) is a broad membranous band attached to the anterior margin of the distal end of the radius, to its styloid process, and to the palmar aspect of the distal end of the ulna; its fibers pass distally and ulnaward to be inserted into the palmar surfaces of the scaphoid, lunate, and triangular bones, some being continued to the capitate. In addition to this broad membrane, there is a rounded fasciculus, superficial to the rest, which reaches from the base of the styloid process of the ulna to the lunate and triangular bones. The ligament is perforated by apertures for the passage of vessels, and its palmar surface is in relation with the tendons of the Flexor digitorum profundus and Flexor pollicis longus, and it is closely adherent to the anterior border of the articular disk of the distal radioulnar articulation.

The **Dorsal Radiocarpal Ligament** (*ligamentum radiocarpeum dorsale; posterior ligament*) (Fig. 5–44) is less thick and strong than the palmar and is attached proximally to the posterior border of the distal end of the radius; its fibers are directed obliquely distally and ulnaward, and are fixed, distally, to the dorsal surfaces of the scaphoid, lunate, and triangular, being continuous with those of the dorsal intercarpal ligaments. Its dorsal surface is in relation with the Extensor tendons of the fingers; its ventral is blended with the articular disk.

The **Ulnar Collateral Ligament** (*ligamentum collaterale carpi ulnare; internal lateral ligament*) (Fig. 5–44) is a rounded cord attached proximally to the end of the styloid process of the ulna, and divides distally into two fasciculi, one of which is attached to the medial side of the triangular bone, the other to the pisiform and transverse carpal ligament.

The **Radial Collateral Ligament** (*ligamentum collaterale carpi radiale; external lateral ligament*) (Fig. 5–44) extends from the tip of the styloid process of the radius to the radial side of the scaphoid, some of its fibers being prolonged to the trapezium and the flexor retinaculum. It is in relation with the radial artery, which separates the ligament from the tendons of the Abductor pollicis longus and Extensor pollicis brevis.

Synovial Membrane (Fig. 5–45).—The synovial layer lines the deep surfaces of the ligaments described above, extending from the margin of the distal end of the radius and articular disk to the margins of the articular

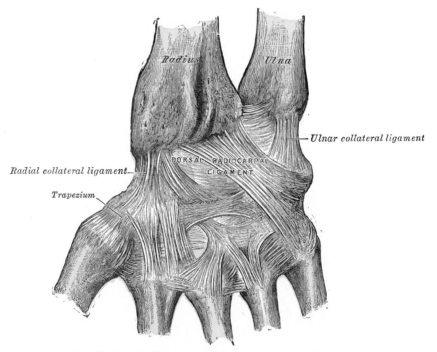

FIG. 5–44.—The ligaments of the left wrist. Dorsal aspect.

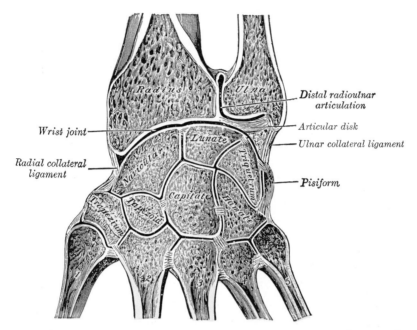

Fig. 5–45.—Vertical section through the articulations at the wrist, showing the synovial cavities.

surfaces of the carpal bones. It is loose and lax, and presents numerous folds, especially dorsally.

The wrist joint is covered by the Flexor and Extensor tendons.

The **arteries** supplying the joint are the palmar and dorsal carpal branches of the radial and ulnar, the palmar and dorsal metacarpals, and some ascending branches from the deep palmar arch.

The **nerves** are derived from the ulnar and dorsal interosseous.

Movements.—The movements permitted in this joint are flexion, extension, abduction, adduction, and circumduction. They will be studied with those of the carpus, with which they are combined.

VII. Intercarpal Articulations
(Articulationes Intercarpeae; Articulations of the Carpus)

These articulations may be subdivided into three sets:

1. The Articulations of the Proximal Row of Carpal Bones.
2. The Articulations of the Distal Row of Carpal Bones.
3. The Articulations of the Two Rows with Each Other.

Articulations of the Proximal Row of Carpal Bones.—These are gliding joints. The scaphoid, lunate, and triangular are connected by dorsal, palmar, and interosseous ligaments.

The **Dorsal Ligaments** (*ligamenta intercarpea dorsalia*).—The dorsal ligaments, two in number, are placed transversely behind the bones of the first row; they connect the scaphoid and lunate, and the lunate and triangular.

The **Palmar Ligaments** (*ligamenta intercarpea palmaria; volar ligaments*).—The palmar ligaments, also two, connect the scaphoid and lunate, and the lunate and triangular; they are less strong than the dorsal, and placed deep to the Flexor tendons and the palmar radiocarpal ligament.

The **Interosseous Ligaments** (*ligamenta intercarpea interossea*) (Fig. 5–45).—The interosseous ligaments are two narrow bundles, one connecting the lunate with the scaphoid, the other joining it to the triangular. They are on a level with the proximal surfaces of these bones; their surfaces are smooth, and form part of the convex articular surface of the wrist joint.

The **ligaments connecting the pisiform bone** are the articular capsule and the two palmar ligaments. The **articular capsule** is a thin membrane which connects the pisiform to the triangular; it is lined by synovial membrane. The two **palmar ligaments** are strong fibrous bands; one, the **pisohamate ligament,** connects the pisiform to the hamate, the other, the **pisometacarpal ligament,** joins the pisiform to the base of the fifth metacarpal bone (Fig. 5–43). These ligaments are in reality prolongations of the tendon of the Flexor carpi ulnaris.

Articulations of the Distal Row of Carpal Bones.—These also are gliding joints; the bones are connected by dorsal, palmar, and interosseous ligaments.

The **Dorsal Ligaments** (*ligamenta intercarpea dorsalia*), three in number, extend transversely from one bone to another on the dorsal surface, connecting the trapezium with the trapezoid, the trapezoid with the capitate, and the capitate with the hamate.

The **Palmar Ligaments** (*ligamenta intercarpea palmaria; volar ligaments*), also three, have a similar arrangement on the palmar surface.

The **Interosseous Ligaments** (*ligamenta intercarpea interossea*).—The three interosseous ligaments are much thicker than those of the first row; one is placed between the capitate and the hamate, a second between the capitate and the trapezoid, and a third between the trapezium and trapezoid. The first is much the strongest, and the third is sometimes wanting.

Articulations of the Two Rows of Carpal Bones with Each Other.—The joint between the scaphoid, lunate, and triangular proximally, and the second row of carpal bones distally, is named the **midcarpal joint,** and is made up of three distinct portions; in the center the head of the capitate and the superior surface of the hamate articulate with the deep cup-shaped cavity formed by the scaphoid and lunate, and constitute a sort of ball-and-socket joint. On the radial side the trapezium and trapezoid articulate with the scaphoid, and on the ulnar side the hamate articulates with the triangular, forming gliding joints.

The ligaments are: palmar, dorsal, ulnar and radial collateral.

The **Palmar Ligaments** (*ligamenta intercarpea palmaria; anterior* or *volar ligaments*).—The palmar ligaments consist of short fibers, which pass, for the most part, from the palmar surfaces of the bones of the first row to the front of the capitate.

The **Dorsal Ligaments** (*ligamenta intercarpea dorsalia; posterior ligaments*).—The dorsal ligaments consist of short, irregular bundles passing between the dorsal surfaces of the bones of the first and second rows.

The **Collateral Ligaments** (*lateral ligaments*).—The collateral ligaments are very short; one is placed on the radial, the other on the ulnar side of the carpus; the former, the stronger and more distinct, connects the scaphoid and trapezium, the latter the triangular and hamate; they are continuous with the collateral ligaments of the wrist joint. In addition to these ligaments, a slender interosseous band sometimes connects the capitate and the scaphoid.

Synovial Membrane.—The synovial layer of the carpus is very extensive (Fig. 5–45), and bounds a synovial cavity of very irregular shape. The proximal portion of the cavity intervenes between the distal surfaces of the scaphoid, lunate, and triangular bones and the proximal surfaces of the bones of the second row. It sends two proximal prolongations—between the scaphoid and lunate, and the lunate and triangular—and three distal prolongations between the four bones of the second row. The prolongation between the trapezium and trapezoid or that between the trapezoid and capitate, is, owing to the absence of the interosseous ligament, often continuous with the cavity of the carpometacarpal joints, sometimes of the second, third, fourth, and fifth metacarpal bones, sometimes of the second and third only. In the latter condition the joint between the hamate and the fourth and fifth metacarpal bones has a separate synovial membrane. The synovial cavities of these joints are prolonged for a short distance between the bases of the metacarpal bones. There is a separate synovial membrane between the pisiform and triangular.

Movements.—The articulation of the hand and wrist considered as a whole involves four articular surfaces: (*a*) the distal surfaces of the radius and articular disk; (*b*) the proximal surfaces of the scaphoid, lunate, and triangular, the pisiform having no essential part in the movement of the hand; (*c*) the S-shaped surface formed by the inferior surfaces of the scaphoid, lunate, and triangular; (*d*) the reciprocal surface formed by the upper surfaces of the bones of the second row. These four surfaces form two joints: (1) a proximal, the wrist joint proper; and (2) a distal, the mid-carpal joint.

1. The wrist joint proper is a true condyloid articulation, and therefore all movements but rotation are permitted. Flexion and extension are the most free, and of these a greater amount of extension than of flexion is permitted, since the articulating surfaces extend farther on the dorsal than on the palmar surfaces of the carpal bones. In this movement the carpal bones swing on a transverse axis drawn between the tips of the styloid processes of the radius and ulna. A certain amount of adduction (or ulnar flexion) and abduction (or radial flexion) is also permitted. The former is considerably greater in extent than the latter on account of the shortness of the styloid process of the ulna, abduction being soon limited by the contact of the styloid process of the radius with the trapezium. In this movement the carpus moves about an anteroposterior axis drawn through the center of the wrist. Finally, circumduction is permitted by the combined and consecutive movements of adduction, extension, abduction, and flexion. No rotation is possible, but the effect of rotation is obtained by the pronation and supination of the radius on the ulna. The movement of *flexion* is performed by the Flexor carpi radialis, the Flexor carpi ulnaris, and the Palmaris longus; *extension* by the Extensores carpi radiales longus and brevis and the Extensor carpi ulnaris; *adduction* (ulnar flexion) by the Flexor carpi ulnaris and the Extensor carpi ulnaris; and *abduction* (radial flexion) by the Extensores carpi radiales longus and brevis and the Flexor carpi radialis. When the fingers are extended, flexion of the wrist is performed by the Flexores carpi radialis and ulnaris and extension is aided by the Extensor digitorum communis. When the fingers are flexed, flexion of the wrist is aided by the Flexores digitorum superficialis and profundus, and extension is performed by the Extensores carpi radiales and ulnaris.

2. The chief movements permitted in the mid-carpal joint are flexion and extension and a slight amount of rotation. In flexion and extension, which are the movements most freely enjoyed, the trapezium and trapezoid on the radial side and the hamate on the ulnar side glide forward and backward on the scaphoid and triangular respectively, while the head of the capitate and the superior surface of the hamate rotate in the cup-shaped cavity of the scaphoid and lunate. Flexion at this point is freer than extension. A very trifling amount of rotation is also permitted, the head of the capitate rotating around a vertical axis drawn through its own center, while at the same time a slight gliding movement takes place in the lateral and medial portions of the joint.

VIII. Carpometacarpal Articulations (Articulationes Metacarpaphalangeae) (Figs. 5–43, 5–44)

Carpometacarpal Articulation of the Thumb (*articulatio carpometacarpea pollicis*).—This is a joint of reciprocal reception between the first metacarpal and the trapezium; it enjoys great freedom of movement on account of the configuration of its articular surfaces, which are saddle-shaped. The joint is surrounded by a capsule, which is thick but loose, and passes from the circumference of the base of the metacarpal bone to the rough edge bounding the articular surface of the trapezium; it is thickest laterally and dorsally, and is lined by synovial membrane.

Movements.—In this articulation the movements permitted are flexion and extension in the plane of the palm of the hand, abduction and adduction in a plane at right angles to the palm, circumduction, and opposition. It is by the movement of opposition that the tip of the thumb is brought into contact with the palmar surfaces of the slightly flexed fingers. This movement is effected through the medium of a small sloping facet on the anterior lip of the saddle-shaped articular surface of the trapezium.

Flexion is produced by the Flexores pollicis longus and brevis, with the Adductor and Opponeus acting in a somewhat synergetic fashion to hold the thumb close to the palm. Extension is effected by the Extensor longus. The Extensor brevis acts to produce a position midway between extension and abduction, which is the very commonly used reciprocal of opposition. Abduction is produced by the Abductores longus and brevis. Adduction is carried out by the Adductor with assistance from the short Flexor. By the movement of opposition the metacarpal bone is rotated to a position in which the palmar surface of the thumb faces the palmar surface of the fingers. This is brought about by the Opponeus, assisted by the Abductors. After the thumb is in position, the strong pressure between thumb and fingers is produced by the long Flexors.

Articulations of the Other Four Metacarpal Bones with the Carpus (*articulationes carpometacarpeae*).—The joints between the carpus and the second, third, fourth, and fifth metacarpal bones are gliding. The bones are united by dorsal, palmar, and interosseous ligaments.

The **Dorsal Ligaments** (*ligamenta carpometacarpea dorsalia*), the strongest and most distinct, connect the carpal and metacarpal bones on their dorsal surfaces. The second metacarpal bone receives two fasciculi, one from the trapezium, the other from the trapezoideum; the third metacarpal receives two, one each from the trapezoideum and capitate; the fourth two, one each from the capitate and hamate; the fifth receives a single fasciculus from the hamate, and this is continuous with a similar ligament on the palmar surface, forming an incomplete capsule.

The **Palmar Ligaments** (*ligamenta carpometacarpea palmaria; volar ligaments*).— The palmar ligaments have a somewhat similar arrangement, with the exception of those of the third metacarpal, which are three in number: a lateral one from the trapezium, situated superficial to the sheath of the tendon of the Flexor carpi radialis; an intermediate one from the capitate; and a medial one from the hamate.

The **Interosseous Ligaments.**—The interosseous ligaments consist of short, thick fibers, and are limited to one part of the carpometacarpal articulation; they connect the contiguous inferior angles of the capitate and hamate with the adjacent surfaces of the third and fourth metacarpal bones.

Synovial Membrane.—The synovial cavity is a continuation of that of the intercarpal joints. Occasionally, the joint between the hamate and the fourth and fifth metacarpal bones has a separate synovial cavity.

The synovial cavities of the wrist and carpus (Fig. 5–45) are thus seen to be five in number. The *first* passes from the distal end of the ulna to the ulnar notch of the radius, and includes the proximal surface of the articular disk. The *second* passes from the articular disk and the distal end of the radius to the bones of the first row. The *third*, the most extensive, passes between the contiguous margins of the two rows of carpal bones, and sometimes, in the event of one of the interosseous ligaments being absent, between the bones of the second row to the carpal extremities of the second, third, fourth, and fifth metacarpal bones. The *fourth* extends from the margin of the trapezium to

the metacarpal bone of the thumb. The *fifth* runs between the adjacent margins of the triangular and pisiform bones. Occasionally the fourth and fifth carpometacarpal joints have a separate synovial cavity.

Movements.—The movements permitted in the carpometacarpal articulations of the fingers are limited to slight gliding of the articular surfaces upon each other, the extent of which varies in the different joints. The metacarpal bone of the little finger is most movable, then that of the ring finger; the metacarpal bones of the index and middle fingers are almost immovable.

IX. Intermetacarpal Articulations (Articulationes Intermetacarpeae)

The bases of the second, third, fourth and fifth metacarpal bones articulate with one another by small surfaces covered with cartilage, and are connected together by dorsal, palmar, and interosseous ligaments.

The **dorsal** (*ligamenta metacarpea dorsalia*) and **palmar ligaments** (*ligamenta metacarpea palmaria; volar ligaments*) pass transversely from one bone to another on the dorsal and palmar surfaces. The **interosseous ligaments** (*ligamenta metacarpea interossea*) connect their contiguous surfaces just distal to their collateral articular facets.

The **synovial membrane** for these joints is continuous with that of the carpometacarpal articulations.

The **Transverse Metacarpal Ligament** (*ligamentum metacarpea transversum profundum*) (Fig. 5–43).—This ligament is a narrow fibrous band, which runs across the palmar surfaces of the heads of the second, third, fourth and fifth metacarpal bones, connecting them together. It is blended with the palmar (glenoid) ligaments of the metacarpophalangeal articulations. Its palmar surface is concave where the Flexor tendons pass over it; behind it the tendons of the Interossei pass to their insertions.

X. Metacarpophalangeal Articulations (Articulationes Metacarpophalangeae) (Figs. 5–46, 5–47)

These articulations are of the condyloid kind, formed by the reception of the rounded heads of the metacarpal bones into

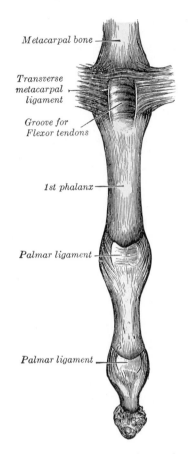

FIG. 5–46.—Metacarpophalangeal articulation and articulations of digit. Palmar aspect.

shallow cavities on the proximal ends of the first phalanges, with the exception of that of the thumb, which presents more of the characters of a ginglymoid joint. Each joint has a palmar and two collateral ligaments.

The **Palmar Ligaments** (*glenoid ligaments of Cruveilhier; volar or vaginal ligaments*).— The palmar ligaments are thick, dense, fibrocartilaginous structures, placed upon the palmar surfaces of the joints in the intervals between the collateral ligaments, to which they are connected; they are loosely united to the metacarpal bones, but are very firmly attached to the bases of the first phalanges. Their palmar surfaces are intimately blended with the transverse metacarpal ligament, and present grooves for the passage of the Flexor tendons, the sheaths surrounding which are connected to the sides of the grooves. Their deep surfaces

form parts of the articular facets for the heads of the metacarpal bones, and are lined by synovial membranes.

The **Collateral Ligaments** (*ligamenta collateralia; lateral ligaments*).—The collateral ligaments are strong, rounded cords placed on the sides of the joints; each is attached by one extremity to the posterior tubercle and adjacent depression on the side of the head of the metacarpal bone, and by the other to the contiguous extremity of the phalanx.

The dorsal surfaces of these joints are covered by the expansions of the Extensor tendons, together with some loose areolar tissue which connects the deep surfaces of the tendons to the bones.

Movements.—The movements which occur in these joints are flexion, extension, adduction, abduction, and circumduction; the movements of abduction and adduction are performed by the Interossei, but cannot be performed when the fingers are fully flexed.

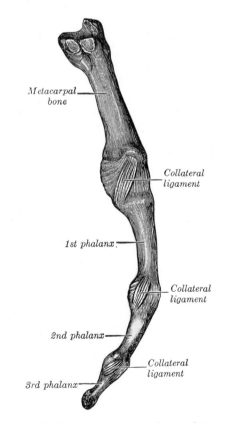

FIG. 5–47.—Metacarpophalangeal articulation and articulations of digit. Ulnar aspect.

XI. Articulations of the Digits (Articulationes Digitorum Manus; Interphalangeal Joints) (Figs. 5–46, 5–47)

The interphalangeal articulations are hinge joints; each has a palmar and two collateral ligaments. The arrangement of these ligaments is similar to those in the metacarpophalangeal articulations. The Extensor tendons supply the place of posterior ligaments.

Movements.—The only movements permitted in the interphalangeal joints are flexion and extension; these movements are more extensive between the proximal and middle phalanges than between the middle and distal. The amount of flexion is very considerable, but extension is limited by the palmar and collateral ligaments.

Muscles Acting on the Joints of the Digits.—Flexion of the metacarpophalangeal joints of the fingers is effected by the Flexores digitorum superficialis and profundus, Lumbricales, and Interossei, assisted in the case of the little finger by the Flexor digiti minimi. Extension is produced by the Extensor digitorum, Extensor indicis, and Extensor digiti minimi.

Flexion of the interphalangeal joints of the fingers is accomplished by the Flexor digitorum profundus acting on the proximal and distal joints and by the Flexor digitorum superficialis acting on the proximal joints. Extension is effected mainly by the Lumbricales and Interossei, the long Extensors having little or no action upon these joints.

Flexion of the metacarpophalangeal joint of the thumb is effected by the Flexores pollicis longus and brevis; extension by the Extensores pollicis longus and brevis. Flexion of the interphalangeal joint is accomplished by the Flexor pollicis longus, and extension by the Extensor pollicis longus.

ARTICULATIONS OF THE LOWER LIMB

The articulations of the Lower Limb are the following:

I. Hip
II. Knee
III. Tibiofibular
IV. Ankle
V. Intertarsal

VI. Tarsometatarsal
VII. Intermetatarsal
VIII. Metatarsophalangeal
IX. Articulations of the Digits

I. Coxal Articulation or Hip Joint (Articulatio Coxae)

This articulation is a ball-and-socket joint (*articulatio spheroidea; enarthrodial joint*), formed by the reception of the head of the femur into the cup-shaped cavity of the acetabulum. The articular cartilage on the head of the femur, thicker at the center than at the circumference, covers the entire surface with the exception of the fovea capitis femoris, to which the ligamentum capitis is attached; that on the acetabulum forms an incomplete marginal ring, the lunate surface. Within the lunate surface there is a circular depression devoid of cartilage, occupied in the intact body by a mass of fat covered by synovial membrane. The ligaments of the joint are:

Articular Capsule
Iliofemoral
Ischiofemoral

Pubofemoral
Ligamentum Capitis Femoris
Acetabular Labrum
Transverse Acetabular

The **Articular Capsule** (*capsula articularis; capsular ligament*) (Figs. 5–48, 5–49). —The articular capsule is strong and dense. It is attached to the margin of the acetabulum 5 to 6 mm beyond the acetabular labrum posteriorly; but anteriorly it is attached to the outer margin of the labrum, and, opposite to the notch where the margin of the cavity is deficient, it is connected to the transverse ligament, and by a few fibers to the edge of the obturator foramen. It surrounds the neck of the femur, and is attached anteriorly to the intertrochanteric line; proximally, to the base of the neck; posteriorly, to the neck, about 1.25 cm above the intertrochanteric crest; and inferiorly, to the lower part of the neck, close to the lesser trochanter. From its femoral attachment some of the fibers are reflected proximally along the neck as longitudinal bands, termed **retinacula.** The capsule is much thicker at the proximal and anterior part of the joint, where the greatest amount of resistance is required; posteriorly and distally, it is thin and loose. It consists of two sets of fibers, circular and longitudinal. The circular fibers, **zona orbicularis,** are most abundant at the distal and posterior part of the capsule (Fig. 5–52), and form a sling or collar around the neck of the femur. Anteriorly they blend with the deep surface of the iliofemoral ligament, and gain an attachment to the anterior inferior iliac spine. The longitudinal fibers are greatest in amount at the proximal and anterior part of the capsule, where they are reinforced by distinct bands, or accessory ligaments, of which the most important is the **iliofemoral ligament.** The other accessory bands are known as the **pubofemoral** and the **ischiofemoral ligaments.** The external surface of the capsule is rough, covered by numerous muscles, and separated anteriorly from the Psoas major and Iliacus by a bursa which not infrequently communicates by a circular aperture with the cavity of the joint.

The **Iliofemoral Ligament** (*ligamentum iliofemorale; Y-ligament; ligament of Bigelow*) (Fig. 5–48) is a band of great strength which lies anterior to the joint; it is intimately connected with the capsule, and serves to strengthen it in this situation. It is attached to the lower part of the anterior inferior iliac spine; distally, it divides into two bands, one of which passes distally and is fixed to the distal part of the intertrochanteric line; the other is directed distally and laterally and is attached to the proximal part of the same line. Between the two bands is a thinner part of the capsule. In some cases there is no division, and the ligament spreads out into a flat triangular band which is attached to the whole length of the intertrochanteric line. This ligament is frequently called the Y-shaped ligament of Bigelow, and its lateral band is sometimes named the **iliotrochanteric ligament.**

The **Pubofemoral Ligament** (*ligamentum pubofemorale; pubocapsular ligament*) is attached to the obturator crest and the

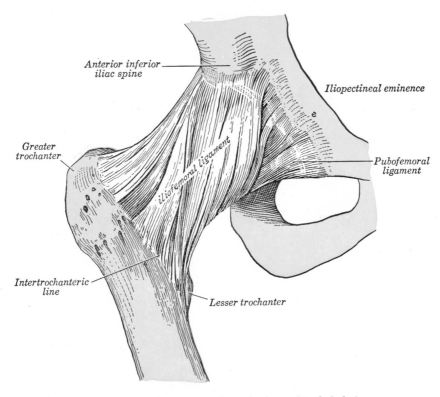

Fig. 5–48.—Right hip joint from the front. (Spalteholz.)

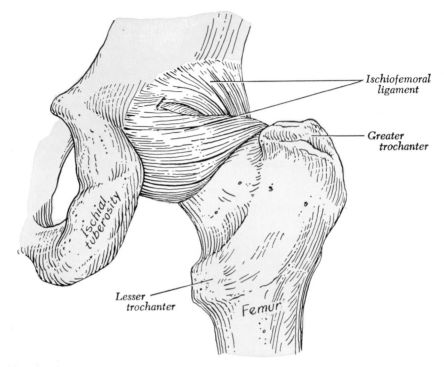

FIG. 5–49.—The hip joint. Posterior aspect.

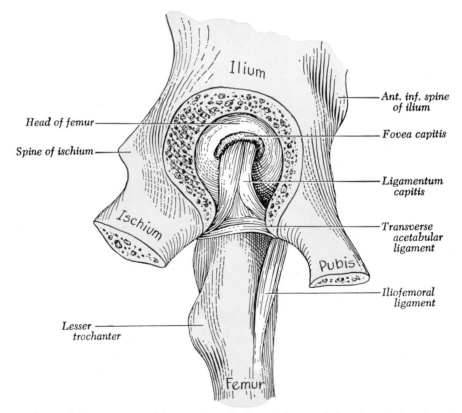

FIG. 5–50.—Left hip joint, opened by removing the floor of the acetabulum from within the pelvis.

superior ramus of the pubis; distally, it blends with the capsule and with the deep surface of the vertical band of the iliofemoral ligament.

The **Ischiofemoral Ligament** (*ligamentum ischiofemorale; ischiofemoral band; ligament of Bertin*) consists of a triangular band of strong fibers, which spring from the ischium distal and posterior to the acetabulum and blend with the circular fibers of the capsule (Fig. 5–49).

The **Ligamentum Capitis Femoris** (*ligamentum teres*) (Fig. 5–50) is a triangular, somewhat flattened band implanted by its apex into the fovea capitis femoris; its base is attached by two bands, one into either side of the acetabular notch, and between these bony attachments it blends with the transverse ligament. It is ensheathed by the synovial membrane, and varies greatly in strength in different subjects; occasionally only the synovial fold exists, and in rare cases even this is absent. The ligament is made tense when the thigh is semiflexed and the limb then adducted or rotated outward; it is, on the other hand, relaxed when the

limb is abducted. It has, however, but little influence as a ligament.

The **Acetabular Labrum** (*labrum acetabulare; glenoid labrum; cotyloid ligament*) is a fibrocartilaginous rim attached to the margin of the acetabulum, the cavity of which it deepens; at the same time it protects the edge of the bone, and fills up the inequalities of its surface. It bridges over the notch as the **transverse ligament,** and thus forms a complete circle, which closely surrounds the head of the femur and assists in holding it in its place. It is triangular on section, its base being attached to the margin of the acetabulum, while its opposite edge is free and sharp. Its two free surfaces are invested by synovial membrane, the external one being in contact with the capsule, the internal one being inclined inward so as to narrow the acetabulum and embrace the cartilaginous surface of the head of the femur (Fig. 5–51).

The **Transverse Acetabular Ligament** (*ligamentum transversum acetabuli; transverse ligament*) is in reality a portion of the acetabular labrum, though differing from it

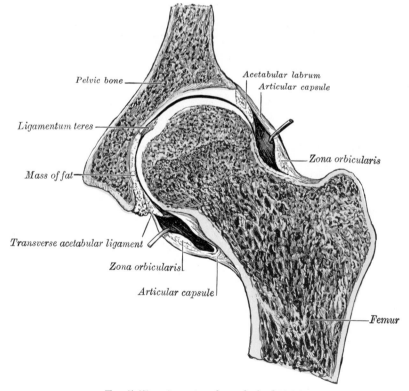

FIG. 5–51.—A section through the hip joint.

in having no cartilage cells among its fibers. It consists of strong, flattened fibers which cross the acetabular notch and convert it into a foramen through which the nutrient vessels enter the joint.

Synovial Membrane (Fig. 5–52).—The synovial surface is very extensive. Commencing at the margin of the cartilaginous surface of the head of the femur, it covers the portion of the neck which is contained within the joint; from the neck it is reflected on the internal surface of the capsule, covers both surfaces of the labrum and the mass of fat contained in the depression at the bottom of the acetabulum, and ensheathes the ligamentum capitis as far as the head of the femur. The joint cavity sometimes communicates through a hole in the capsule between the vertical band of the iliofemoral ligament and the pubofemoral ligament with a bursa situated on the deep surfaces of the Psoas major and Iliacus.

The **muscles** in relation with the joint *anteriorly* are the Psoas major and Iliacus, separated from the capsule by a bursa; *proximally,* the reflected head of the Rectus femoris and Gluteus minimus, the latter being closely adherent to the capsule; *medially,* the Obturator externus and Pectineus; *posteriorly,* the Piriformis, Gemellus superior, Obturator internus, Gemellus inferior, Obturator externus, and Quadratus femoris (Fig. 5–53).

The **arteries** supplying the joint are derived from the obturator, medial femoral circumflex, and superior and inferior gluteals.

The **nerves** are articular branches from the sacral plexus, sciatic, obturator, accessory obturator, and a filament from the branch of the femoral supplying the Rectus femoris.

Movements.—The movements of the hip are very extensive, and consist of flexion, extension, adduction, abduction, circumduction, and rotation.

Although the hip joint is a freely movable ball-and-socket joint, it is essential to observe that the femur cannot rotate around the long axis of the shaft because this would necessitate throwing the head out of its socket. Rotation of the thigh, therefore, takes place around an imaginary axis, that is, a line drawn from the point of attachment of the ligamentum capitis to the medial condyle. Further explanations of this peculiarity rest with the fact that the neck

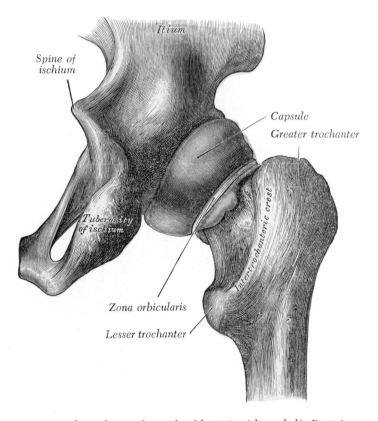

FIG. 5–52.—Synovial membrane of capsule of hip joint (distended). Posterior aspect.

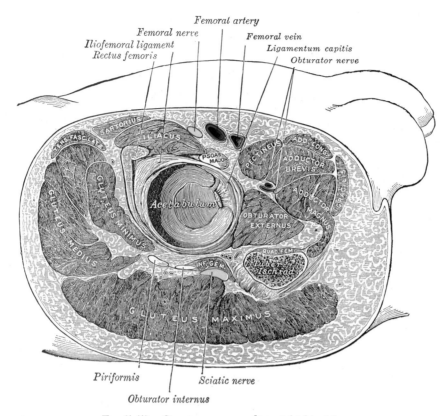

Femoral artery
Femoral nerve
Iliofemoral ligament   Femoral vein
Rectus femoris   Ligamentum capitis
                              Obturator nerve

Piriformis   Sciatic nerve
Obturator internus

FIG. 5–53.—Structures surrounding right hip joint.

of the femur is a segment of bone of appreciable length and that it joins the shaft at an abrupt angle. All movements of the thigh must be translated into movements of the neck segment rather than the shaft. For example, flexion and extension of the thigh as a whole are caused by rotation of the head and neck segment. Consequently rotation of the thigh is the result of flexion and extension of the head-neck segment.

Abduction and adduction of the thigh, on the other hand, move both segments within the same plane, the one established by their angle, and are produced by a bending in the joint with some rotation caused by the normally backward inclination of the neck.

The hip joint presents a very striking contrast to the shoulder joint in the much more complete mechanical arrangements for its security and for the limitation of its movements. In the shoulder, as has been seen, the head of the humerus is not adapted at all in size to the glenoid cavity, and is hardly restrained in any of its ordinary movements by the capsule. In the hip joint, on the contrary, the head of the femur is closely fitted to the acetabulum for an area extending over nearly half a sphere, and at the margin of the bony cup it is still more

closely embraced by the glenoidal labrum, so that the head of the femur is held in its place by that ligament even when the fibers of the capsule have been quite divided. The iliofemoral ligament is the strongest of all the ligaments in the body, and is put on the stretch by any attempt to extend the femur beyond a straight line with the trunk. That is to say, this ligament is the chief agent in maintaining the erect position without muscular fatigue; for a vertical line passing through the center of gravity of the trunk falls behind the centers of rotation in the hip joints, and therefore the pelvis tends to fall backward, but is prevented by the tension of the iliofemoral ligaments. The security of the joint may be provided for also by the two bones being directly united through the ligamentum capitis; but it is doubtful whether this ligament has much influence upon the mechanism of the joint. When the knee is flexed, flexion of the hip joint is arrested by the soft parts of the thigh and abdomen being brought into contact, and when the knee is extended, by the action of the hamstring muscles; extension is checked by the tension of the iliofemoral ligament; adduction by the thighs coming into contact; adduction with flexion by the lateral band of the iliofemoral ligament and the lateral part of the capsule; abduction by the medial band of the iliofemoral ligament and the pubofemoral ligament; rotation outward by the lateral band of the iliofemoral ligament; and rotation inward by the ischiofemoral ligament and the posterior part of the capsule.

In the interpretation of the actions of the muscles on the joint it must be taken into account that the angle between the neck and shaft of the femur establishes a plane. As described above, therefore, any muscle originating ventral to this plane must rotate the thigh medially and any muscle originating dorsal to the plane must rotate laterally. For example, the dorsal insertion of the Iliopsoas might be expected to rotate the thigh laterally but rotation about the long axis being impossible, the ventral origin of the muscle would make it a medial rotator of the thigh. According to electromyographic recordings it is not much used in rotation, however, probably because of its closeness to the joint capsule. The actions given below assume that the motion begins with limb in the usual upright posture. A whole new interpretation would be necessary if the limb began in some other posture.

The muscles which *flex* the femur on the pelvis are the Psoas major, Iliacus, Tensor fasciae latae, Rectus femoris, Sartorius, Pectineus, Adductores longus and brevis, and the anterior fibers of the Glutei medius and minimus. *Extension* is mainly performed by the Gluteus maximus, assisted by the hamstring muscles and the ischial head of the Adductor magnus. The thigh is *adducted* by the Adductores magnus, longus, and brevis, the Pectineus, and the Gracilis, and *abducted* by the Glutei medius and minimus. The muscles which *rotate* the thigh *inward* are the Gluteus minimus and the anterior fibers of the Gluteus medius, the Tensor fasciae latae, the Adductores longus, brevis, and magnus, the Pectineus, and the Iliacus and Psoas major; while those which rotate it *outward* are the posterior fibers of the Gluteus medius, the Piriformis, Obturatores externus and internus, Gemelli superior and inferior, Quadratus femoris, Gluteus maximus, and the Sartorius.

II. The Knee Joint (Articulatio Genus)

The knee joint was formerly described as a ginglymus or hinge joint, but is really of a much more complicated character. It must be regarded as consisting of three articulations in one: two condyloid joints, one between each condyle of the femur and the corresponding meniscus and condyle of the tibia; and a third between the patella and the femur, partly arthrodial, but not completely so, since the articular surfaces are not mutually adapted to each other, so that the movement is not a simple gliding one. This view of the construction of the knee joint receives confirmation from the study of the articulation in some of the lower mammals, where three synovial cavities corresponding to these three subdivisions are sometimes found, either entirely distinct or connected together only by small communications. This view is further rendered probable by the existence in the middle of the joint of the two cruciate ligaments, which must be regarded as the collateral ligaments of the medial and lateral joints. The existence of the patellar fold of synovial membrane would further indicate a tendency to separation of the synovial cavity into two minor sacs, one corresponding to the lateral and the other to the medial joint.

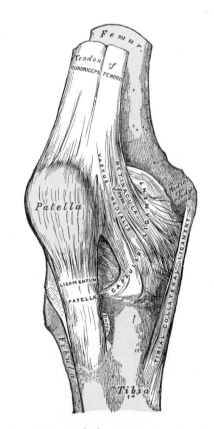

Fig. 5–54.—Right knee joint. Anterior view.

The bones are connected together by the following ligaments:

Articular Capsule
Ligamentum Patellae
Oblique Popliteal
Arcuate Popliteal
Tibial Collateral
Fibular Collateral
Anterior Cruciate
Posterior Cruciate
Medial and Lateral Menisci
Transverse
Coronary

The **Articular Capsule** (*capsula articularis; capsular ligament*) (Fig. 5–54) consists of a thin, but strong, fibrous membrane which is strengthened in almost its entire extent by various bands inseparably connected with it. Beneath the tendon of the Quadriceps femoris, it is represented only by the synovial membrane. Its chief strengthening bands are derived from the fascia lata and from the tendons surrounding the joint. *Anteriorly*, expansions from the Vasti and from the fascia lata and its iliotibial band fill in the intervals between the anterior and collateral ligaments, constituting the **medial** and **lateral patellar retinacula.** *Posteriorly*, the capsule consists of vertical fibers which arise from the condyles and from the sides of the intercondylar fossa of the femur; the posterior part of the capsule is therefore situated on the sides of and anterior to the cruciate ligaments, which are thus excluded from the joint cavity. Posterior to the cruciate ligaments is the oblique popliteal ligament which is augmented by fibers derived from the tendon of the Semimembranosus. *Laterally*, a prolongation from the iliotibial band fills in the interval between the oblique popliteal and the fibular collateral ligaments, and partly covers the latter. *Medially*, expansions from the Sartorius and Semimembranosus pass proximally to the tibial collateral ligament and strengthen the capsule.

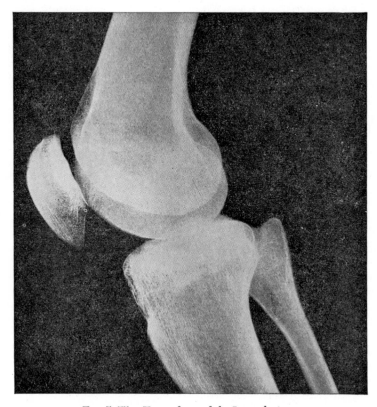

Fig. 5–55.—Knee of an adult. Lateral view.

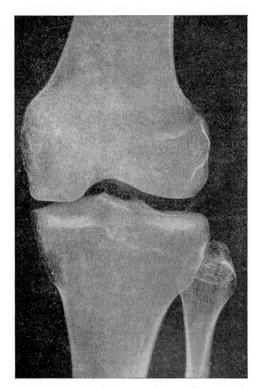

Fig. 5–56.—Adult knee. The gap between the
lateral condyles of the femur and tibia is occupied
by the articular cartilage of the two bones and the
lateral semilunar cartilage.

The **Patellar Ligament** (*ligamentum
patellae; anterior ligament*) (Fig. 5–54) is
the central portion of the common tendon
of the Quadriceps femoris, which is continued from the patella to the tuberosity of
the tibia. It is a strong, flat, ligamentous
band, about 8 cm in length, attached,
proximally, to the apex and adjoining
margins of the patella and the rough
depression on its posterior surface, and
distally to the tuberosity of the tibia; its
superficial fibers are continuous over the
front of the patella with those of the tendon
of the Quadriceps femoris. The medial and
lateral portions of the tendon of the Quadriceps pass down on either side of the
patella, to be inserted into the proximal
extremity of the tibia on either side of the
tuberosity; these portions merge into the
capsule forming the **medial** and **lateral
patellar retinacula.** The posterior surface of
the ligamentum patellae is separated from
the synovial membrane of the joint by a
large infrapatellar pad of fat, and from the
tibia by a bursa.

The **Oblique Popliteal Ligament** (*ligamentum popliteum obliquum; posterior
ligament*) (Fig. 5–57) is a broad, flat, fibrous
band, formed of fasciculi separated from
one another by apertures for the passage of
vessels and nerves on the posterior aspect
of the joint. It is attached proximally to the
margin of the intercondylar fossa and
posterior surface of the femur close to the
articular margins of the condyles, and
distally to the posterior margin of the head
of the tibia. Superficial to the main part of
the ligament is a strong fasciculus, derived
from the tendon of the Semimembranosus
and passing obliquely from the posterior
part of the medial condyle of the tibia
proximally and laterally to the posterior
part of the lateral condyle of the femur.
The oblique popliteal ligament forms part
of the floor of the popliteal fossa, and the
popliteal artery rests upon it.

The **Arcuate Popliteal Ligament** (Fig.
5–57) arches distally from the lateral condyle of the femur to the posterior surface of

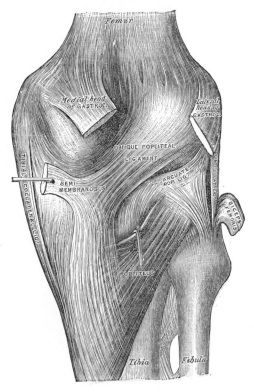

Fig. 5–57.—The right knee joint. Posterior aspect.

the capsular ligament. It is connected to the styloid process of the head of the fibula by two converging bands.

The **Tibial Collateral Ligament** (*ligamentum collaterale tibiale; internal lateral ligament*) (Fig. 5–54) is a broad, flat, membranous band, situated toward the posterior part of the joint. It is attached to the medial condyle of the femur immediately distal to the adductor tubercle proximally, and to the medial condyle and medial surface of the body of the tibia distally. The fibers of the posterior part of the ligament are short and are inserted into the tibia proximal to the groove for the Semimembranosus. The anterior part of the ligament is a flattened band, about 10 cm long, which is inserted into the medial surface of the body of the tibia about 2.5 cm distal to the level of the condyle. It is crossed, at its distal part, by the tendons of the Sartorius, Gracilis, and Semitendinosus, a bursa being interposed. Its deep surface covers the inferior medial genicular vessels and nerve and the anterior portion of the tendon of the Semimembranosus, with which it is connected by a few fibers; it is

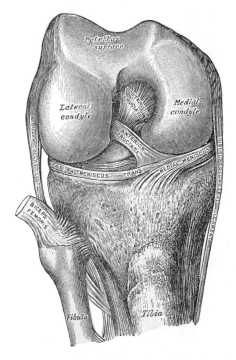

FIG. 5–58.—The right knee joint. Flexed and dissected anteriorly.

intimately adherent to the medial meniscus.

The **Fibular Collateral Ligament** (*ligamentum collaterale fibulare; external lateral or long external lateral ligament*) (Fig. 5–58) is a strong, rounded, fibrous cord, attached to the posterior part of the lateral condyle of the femur, immediately proximal to the groove for the tendon of the Popliteus, and to the lateral side of the head of the fibula anterior to the styloid process. The greater part of its lateral surface is covered by the tendon of the Biceps femoris, although the tendon divides at its insertion into two parts which are separated by the ligament. Deep to the ligament are the tendon of the Popliteus, and the inferior lateral genicular vessels and nerve. The ligament has no attachment to the lateral meniscus.

An inconstant bundle of fibers, the **short fibular collateral ligament**, is placed posterior to and parallel with the preceding, attached to the lateral condyle of the femur, and to the summit of the styloid process of the fibula. Passing deep to it are the tendon of the Popliteus, and the inferior lateral genicular vessels and nerve.

The **Cruciate Ligaments** (*ligamenta cruciata genu; crucial ligaments*) are of considerable strength, situated in the middle of the joint, nearer to its posterior than to its anterior surface. They are called *cruciate* because they cross each other somewhat like the lines of the letter X, and have received the names **anterior** and **posterior**, from the position of their attachments to the tibia.

The **anterior cruciate** ligament (*ligamentum cruciatum anterius; external crucial ligament*) (Fig. 5–58) is attached to the depression anterior to the intercondylar eminence of the tibia, being blended with the anterior extremity of the lateral meniscus; passing posteriorly and laterally, it is fixed into the posterior part of the medial surface of the lateral condyle of the femur. (See Fig. 4–180.)

The **posterior cruciate** ligament (*ligamentum cruciatum posterius; internal crucial ligament*) (Fig. 5–59) is stronger, but shorter and less oblique in its direction, than the anterior. It is attached to the posterior intercondylar fossa of the tibia, and to the posterior extremity of the lateral meniscus;

it passes anteriorly and medially, to be fixed into the anterior part of the medial surface of the condyle of the femur. (See Fig. 5–58.)

The **Menisci** (*semilunar fibrocartilages*) (Fig. 5–60) are two crescentic lamellae which serve to deepen the surfaces of the articular fossae of the head of the tibia for reception of the condyles of the femur. The peripheral border of each meniscus is thick, convex, and attached to the inside of the capsule of the joint; the opposite border

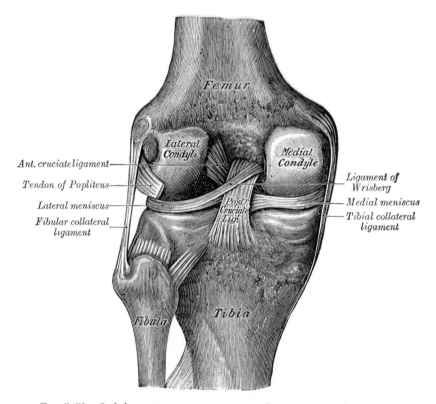

Fig. 5–59.—Left knee joint, posterior aspect, showing interior ligaments.

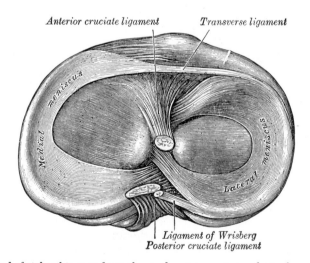

Fig. 5–60.—Head of right tibia seen from above, showing menisci and attachments of ligaments.

tapers to a thin free edge. The proximal surfaces of the menisci are concave, and in contact with the condyles of the femur; their distal surfaces are flat, and rest upon the head of the tibia; both surfaces are smooth, and covered by synovial surface. Each meniscus covers approximately the peripheral two-thirds of the corresponding articular surface of the tibia.

The **medial meniscus** (*meniscus medialis; internal semilunar fibrocartilage*) is nearly semicircular in form, a little elongated, and broader posteriorly; its anterior end, thin and pointed, is attached to the anterior intercondylar fossa of the tibia, in front of the anterior cruciate ligament; its posterior end is fixed to the posterior intercondylar

fossa of the tibia, between the attachments of the lateral meniscus and the posterior cruciate ligament.

The **lateral meniscus** (*meniscus lateralis; external semilunar fibrocartilage*) is nearly circular and covers a larger portion of the articular surface than the medial one. It is grooved laterally for the tendon of the Popliteus, which separates it from the fibular collateral ligament. Its anterior end is attached anterior to the intercondylar eminence of the tibia, lateral and posterior to the anterior cruciate ligament with which it blends; the posterior end is attached posterior to the intercondylar eminence of the tibia and anterior to the posterior end of the medial meniscus. The anterior attach-

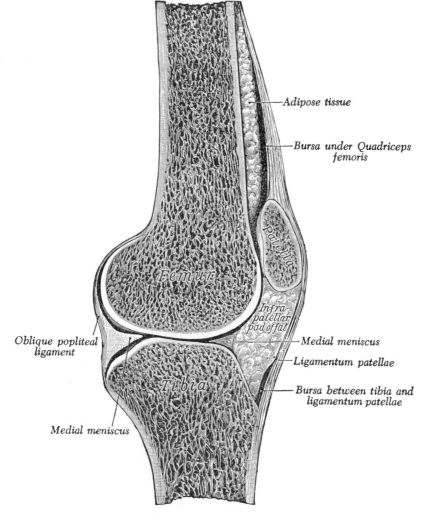

Fig. 5–61.—Sagittal section of right knee joint.

ment of the lateral meniscus is twisted on itself, resting on a sloping shelf of bone on the lateral process of the intercondylar eminence. Close to its posterior attachment it sends off a strong fasciculus, the **ligament of Wrisberg** (*lig. meniscofemorale posterius*) (Figs. 5–59, 5–60), which passes proximally and medially, to be inserted into the medial condyle of the femur, immediately posterior to the attachment of the posterior cruciate ligament. Occasionally a small fasciculus passes forward to be inserted into the lateral part of the anterior cruciate ligament (*lig. meniscofemorale anterius*). The lateral meniscus gives off from its anterior convex margin a fasciculus which forms the transverse ligament.

The **Transverse Ligament** (*ligamentum transversum genus*) connects the anterior convex margin of the lateral meniscus to the anterior end of the medial meniscus; its thickness varies considerably in different subjects, and it is sometimes absent.

The **coronary ligaments** are merely portions of the capsule which connect the periphery of each meniscus with the margin of the head of the tibia.

Synovial Membrane.—The synovial membrane of the knee joint is the largest and most extensive in the body. Commencing at the proximal border of the patella, it forms a large cul-de-sac beneath the Quadriceps femoris (Figs. 5–61, 5–62) on the distal part of the front of the femur, and frequently communicates with a bursa interposed between the tendon and the bone. The pouch of synovial membrane between the Quadriceps and the femur is supported, during the movements of the knee, by a small muscle, the Articularis genu, which is inserted into it. On either side of the patella, the synovial membrane extends beneath the aponeuroses of the Vasti, and more especially beneath that of the Vastus medialis. Distal to the patella it is separated from the ligamentum patellae by a considerable quantity of fat, known as the **infrapatellar pad.** From the medial and lateral borders of the articular

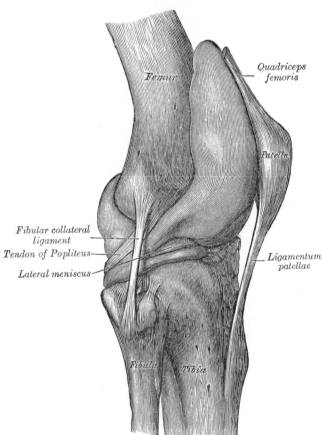

Fig. 5–62.—Synovial membrane of capsule of right knee joint (distended). Lateral aspect.

surface of the patella, reduplications of the synovial membrane project into the interior of the joint. These form two fringe-like folds termed the **alar folds,** which converge distally and are continued as a single band, the **patellar fold** (*plica synovialis infrapatellaris*) to the intercondylar fossa of the femur. On either side of the joint, the synovial membrane passes from the femur, lining the capsule to its point of attachment to the menisci; it may then be traced over the proximal surfaces of these to their free borders, and thence along their distal surfaces to the tibia (Figs. 5–62, 5–63). At the posterior part of the lateral meniscus it forms a cul-de-sac between the groove on its surface and the tendon of the Popliteus; it is reflected anteriorly across the cruciate ligaments, which are therefore situated outside the synovial cavity.

Bursae.—The bursae near the knee joint are the following: Anteriorly there are *four* bursae: a large one is interposed between the patella and the skin, a small one between the upper part of the tibia and the ligamentum patellae, a third between the lower part of the tuberosity of the tibia and the skin, and a fourth between the anterior surface of the lower part of the femur and the deep surface of the Quadriceps femoris, usually communicating with the knee joint. Laterally, there are four bursae: (1) one (which sometimes communicates with the joint) between the lateral head of the Gastrocnemius and the capsule; (2) one between the fibular collateral ligament and the tendon of the Biceps; (3) one between the fibular collateral ligament and the tendon of the Popliteus (this is sometimes only an expansion from the next bursa); (4) one between the tendon of the Popliteus and the lateral condyle of the femur, usually an extension from the synovial membrane of the joint. Medially, there are five bursae: (1) one between the medial head of the Gastrocnemius and the capsule; this sends a prolongation between the tendon of the medial head of the Gastrocnemius and the tendon of the Semimembranosus and often communicates with the joint; (2) one superficial to the tibial collateral ligament, between it and the tendons of the Sartorius, Gracilis, and Semitendinosus; (3) one deep to the tibial collateral ligament, between it and the tendon of the Semimembranosus (this is sometimes only an expansion from the next bursa); (4) one between the tendon of the Semimembrano-

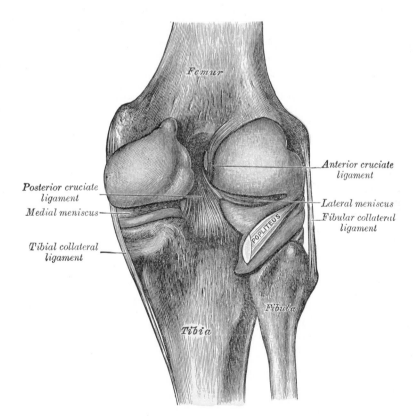

FIG. 5–63.—Synovial membrane of capsule of right knee joint (distended). Posterior aspect.

sus and the head of the tibia; (5) occasionally there is a bursa between the tendons of the Semimembranosus and Semitendinosus.

Structures Around the Joint.—Anteriorly and at the sides is the Quadriceps femoris; laterally, the tendons of the Biceps femoris and Popliteus and the common peroneal nerve; medially, the Sartorius, Gracilis, Semitendinosus, and Semimembranosus; posteriorly, the popliteal vessels and the tibial nerve, Popliteus, Plantaris, and medial and lateral heads of the Gastrocnemius, some lymph nodes, and fat.

The **arteries** supplying the joint are the descending genicular (*anastomotica magna*), a branch of the femoral, the genicular branches of the popliteal, the recurrent branches of the anterior tibial, and the descending branch from the lateral femoral circumflex of the profunda femoris.

Movements.—The movements which take place at the knee joint are flexion and extension, and, in certain positions of the joint, medial and lateral rotation. The movements of flexion and extension at this joint differ from those in a typical hinge joint, such as the elbow, in that (*a*) the axis around which motion takes place is not a fixed one, but shifts anteriorly during extension and posteriorly during flexion; (*b*) the commencement of flexion and the end of extension are accompanied by rotatory movements associated with the fixation of the limb in a position of great stability. The movement from full flexion to full extension may therefore be described in three phases:

1. In the fully flexed condition the posterior parts of the femoral condyles rest on the corresponding portions of the meniscotibial surfaces, and in this position a slight amount of simple rolling movement is allowed.

2. During the passage of the limb from the flexed to the extended position a gliding movement is superposed on the rolling, so that the axis, which at the commencement is represented by a line through the inner and outer condyles of the femur, gradually shifts anteriorly. In this part of the movement, the posterior two-thirds of the tibial articular surfaces of the two femoral condyles are involved, and as these have similar curvatures and are parallel to one another, they move anteriorly equally.

3. The lateral condyle of the femur is brought almost to rest by the tightening of the anterior cruciate ligament; it moves, however, slightly anteriorly and medialward, pushing before it the anterior part of the lateral meniscus. The tibial surface on the medial condyle is prolonged farther anteriorly than that on the lateral, and this prolongation is directed lateralward. When the anterior movement of the condyles is checked by the anterior cruciate ligament,

continued muscular action causes the medial condyle, dragging with it the meniscus, to travel posteriorly and medially, thus producing a medial rotation of the thigh on the leg. When the position of full extension is reached the lateral part of the groove on the lateral condyle is pressed against the anterior part of the corresponding meniscus, while the medial part of the groove rests on the articular margin anterior to the lateral process of the tibial intercondylar eminence. Into the groove on the medial condyle is fitted the anterior part of the medial meniscus, while the anterior cruciate ligament and the articular margin in front of the medial process of the tibial intercondylar eminence are received into the intercondylar fossa of the femur. This third phase by which all these parts are brought into accurate apposition is known as the "screwing home," or locking movement of the joint.

The complete movement of flexion is the converse of that described above, and is therefore preceded by a lateral rotation of the femur which unlocks the extended joint.

The axes around which the movements of flexion and extension take place are not precisely at right angles to either bone; in flexion, the femur and tibia are in the same plane, but in extension the one bone forms an angle, opening lateralward with the other.

In addition to the rotatory movements associated with the completion of extension and the initiation of flexion, rotation medialward or lateralward can be effected when the joint is partially flexed; these movements take place mainly between the tibia and the menisci, and are freest when the leg is bent at right angles with the thigh.

Movements of Patella.—The articular surface of the patella is indistinctly divided into seven facets—proximal, middle, and distal horizontal pairs, and a medial perpendicular facet (Fig. 5–64). When the knee is forcibly flexed, the medial perpendicular facet is in contact with the semilunar surface on the lateral part of

FIG. 5–64.—Posterior surface of the right patella, showing diagrammatically the areas of contact with the femur in different positions of the knee.

the medial condyle; this semilunar surface is a prolongation backward of the medial part of the patellar surface. As the leg is carried from the flexed to the extended position, first the proximal pair, then the middle pair, and lastly the distal pair of horizontal facets is successively brought into contact with the patellar surface of the femur. In the extended position, when the Quadriceps femoris is relaxed, the patella lies loosely on the end of the femur.

During flexion, the ligamentum patellae is put upon the stretch, and in extreme flexion the posterior cruciate and the oblique popliteal ligaments are tense, but the two collateral ligaments and to a slight extent the anterior cruciate ligament are relaxed. Flexion is checked during life by the contact of the leg with the thigh. When the knee joint is fully extended the oblique popliteal and collateral ligaments, the anterior cruciate ligament, and the posterior cruciate ligament, are rendered tense; in the act of extending the knee, the ligamentum patellae is tightened by the Quadriceps femoris, but in full extension with the heel supported it is relaxed. Medial rotation is resisted by the cruciate ligaments, especially the anterior and by the fibular collateral ligament. Lateral rotation is resisted by the tibial collateral ligament but the cruciate ligaments tend to be relaxed. The main function of the cruciate ligaments is to act as a direct bond between the tibia and femur and to prevent the former bone from being carried too far backward or forward. They also assist the collateral ligaments in resisting any bending of the joint to either side. The menisci are intended, as it seems, to adapt the surfaces of the tibia to the shape of the femoral condyles to a certain extent, so as to fill up the intervals which would otherwise be left in the varying positions of the joint, and to obviate the jars which would be so frequently transmitted up the limb in jumping or by falls on the feet; also to permit the two varieties of motion, flexion and extension, and rotation, as explained above. The patella is a great defense to the front of the knee joint, and distributes upon a large and tolerably even surface, during kneeling, the pressure which would otherwise fall upon the prominent ridges of the condyles; it also affords leverage to the Quadriceps femoris.

When one is standing erect in the attitude of "attention," the weight of the body falls in front of a line carried across the centers of the knee joints, and therefore tends to produce overextension of the articulations; this, however, is prevented by the tension of the anterior cruciate, oblique popliteal, and collateral ligaments.

Extension of the leg on the thigh is performed by the Quadriceps femoris; *flexion* by the Biceps femoris, Semitendinosus, and Semimembranosus, assisted by the Gracilis, Sartorius, Gastrocnemius, Popliteus, and Plantaris. *Rotation lateralward* is effected by the Biceps femoris, and *rotation medialward* by the Popliteus, Semitendinosus, and, to a slight extent, the Semimembranosus, the Sartorius, and the Gracilis. The Popliteus comes into action especially at the commencement of the movement of flexion of the knee; by its contraction the leg is rotated medialward, or, if the tibia be fixed, the thigh is rotated lateralward, and the knee joint is unlocked.

III. Articulations between the Tibia and Fibula

The tibia and fibula are connected by: (1) the tibiofibular articulation; (2) the interosseous membrane; (3) the tibiofibular syndesmosis.

Tibiofibular Articulation (*articulatio tibiofibularis; superior tibiofibular articulation*). —This articulation is a gliding joint between the lateral condyle of the tibia and the head of the fibula. The contiguous surfaces of the bones present flat, oval facets covered with cartilage and connected together by an articular capsule and by anterior and posterior ligaments.

The **Articular Capsule** (*capsula articularis; capsular ligament*) surrounds the articulation, being attached around the margins of the articular facets on the tibia and fibula; it is much thicker anteriorly than posteriorly.

The **Anterior Ligament** (*ligamentum capitis fibulae anterius*) of the head of the fibula (Fig. 5–58) consists of two or three broad, flat bands which pass obliquely from the head of the fibula to the anterior part of the lateral condyle of the tibia.

The **Posterior Ligament** (*lig. cap. fibulae posterius*) of the head of the fibula (Fig. 5–59) is a single thick, broad band which passes obliquely from the head of the fibula to the posterior part of the lateral condyle of the tibia. It is covered by the tendon of the Popliteus.

Synovial Membrane.—A synovial membrane lines the capsule; occasionally it is continuous with that of the knee joint.

Interosseous Membrane (*membrana interossea cruris; middle tibiofibular ligament*).— An interosseous membrane extends between the interosseous crests of the tibia and fibula, and separates the muscles on the anterior from those on the posterior aspect of the leg. It consists of a thin, aponeurotic lamina composed of oblique fibers, which for the most part run distally and laterally; some few fibers, however, pass in the opposite direction. Its proximal margin does not quite reach the tibiofibular joint, but presents a free concave border, above which is a large, oval aperture for the passage of the anterior tibial vessels to the front of the leg. In its distal part is an opening for the passage of the anterior peroneal vessels. It is continuous with the interosseous ligament of the tibiofibular syndesmosis, and presents numerous perforations for the passage of small vessels. It is in relation anteriorly with the Tibialis anterior, Extensor digitorum longus, Extensor hallucis longus, Peroneus tertius, and the anterior tibial vessels and deep peroneal nerve, and posteriorly with the Tibialis posterior and Flexor hallucis longus.

The **Tibiofibular Syndesmosis** (*syndesmosis tibiofibularis; inferior tibiofibular articulation*) is formed by the rough, convex surface of the medial side of the distal end of the fibula, and a rough concave surface on the lateral side of the tibia. For about 4 mm distal to the syndesmosis, these surfaces are smooth, and covered with cartilage which is continuous with that of the ankle joint. The ligaments are: anterior, posterior, inferior transverse, and interosseous.

The **Anterior Tibiofibular Ligament** (*ligamentum tibiofibulare anterius*) (Fig. 5–66) is a flat band of fibers which extends obliquely distally and laterally between the adjacent margins of the tibia and fibula, on the anterior aspect of the syndesmosis. It is in relation anteriorly with the Peroneus tertius, the aponeurosis of the leg, and the integument, and posteriorly with the interosseous ligament, and lies in contact with the cartilage covering the talus.

The **Posterior Tibiofibular Ligament** (*ligamentum tibiofibulare posterius*) (Fig. 5–66), smaller than the preceding, is disposed in a similar manner on the posterior surface of the syndesmosis.

The **Inferior Transverse Ligament** lies anterior to the posterior ligament, and is a strong, thick band of yellowish fibers which passes transversely across the posterior of the joint, from the lateral malleolus to the posterior border of the articular surface of the tibia, almost as far as its malleolar process. This ligament projects below the margin of the bones, and forms part of the articulating surface for the talus.

The **Interosseous Ligament.**—The interosseous ligament consists of numerous short, strong, fibrous bands which pass between the contiguous rough surfaces of the tibia and fibula, and constitute the chief bond of union between the bones. It is continuous proximally with the interosseous membrane (Figs. 5–68, 5–69).

Synovial Membrane.—The synovial membrane associated with the small articular part of this joint is continuous with that of the ankle joint.

IV. Talocrural Articulation or Ankle Joint (Articulatio Talocruralis; Tibiotarsal Articulation)

The ankle joint is a ginglymus, or hinge joint. The structures entering into its formation are the distal end of the tibia and its malleolus, the malleolus of the fibula, and the inferior transverse ligament, which together form a mortise for the reception of the proximal convex surface of the talus and its medial and lateral facets. The bones are connected by the following ligaments:

> Articular Capsule
> Deltoid
> Anterior Talofibular
> Posterior Talofibular
> Calcaneofibular

The **Articular Capsule** (*capsula articularis; capsular ligament*) surrounds the joint, and is attached to the borders of the articular surfaces of the tibia and malleoli proximally, and to the talus around its articular surface distally. The anterior part of the capsule (*anterior ligament*) is a broad, thin, membranous layer, attached to the anterior margin of the distal end of the tibia, and to the talus, anterior to its superior

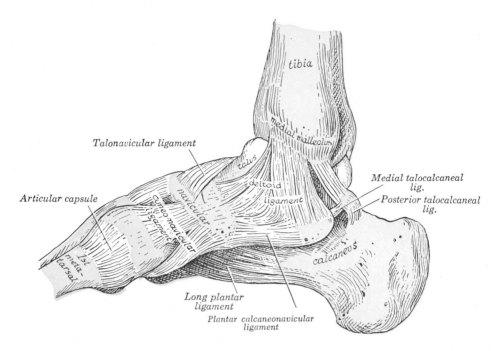

Fig. 5–65.—The ligaments of the right ankle and tarsus. Medial aspect.

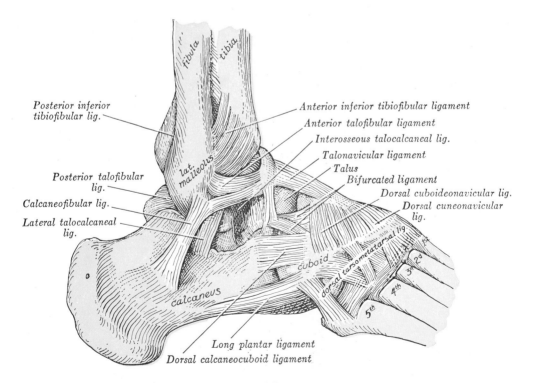

Fig. 5–66.—The ligaments of the right ankle and tarsus. Lateral aspect.

articular surface. It is in relation with the Extensor tendons of the toes, the tendons of the Tibialis anterior and Peroneus tertius, and the anterior tibial vessels and deep peroneal nerve. The posterior part of the capsule (*posterior ligament*) is very thin, and consists principally of transverse fibers. It is attached to the margin of the articular surface of the tibia, blending with the transverse ligament, and to the talus posterior to its superior articular facet. Laterally, it is somewhat thickened, and is attached to the hollow on the medial surface of the lateral malleolus.

The **Deltoid Ligament** (*ligamentum deltoideum; internal lateral ligament*) (Fig. 5–65) is a strong, flat, triangular band, attached to the apex and anterior and posterior borders of the medial malleolus proximally. It consists of two sets of fibers, superficial and deep. Of the superficial fibers the most anterior (*tibionavicular*) are inserted into the tuberosity of the navicular bone, and immediately posterior to this they blend with the medial margin of the plantar calcaneonavicular ligament; the middle (*calcaneotibial*) descend almost perpendic-

ularly to be inserted into the whole length of the sustentaculum tali of the calcaneus; the posterior fibers (*posterior talotibial*) pass lateralward to be attached to the inner side of the talus, and to the prominent tubercle on its posterior surface, medial to the groove for the tendon of the Flexor hallucis longus. The deep fibers (*anterior talotibial*) are attached to the tip of the medial malleolus, and to the medial surface of the talus. The deltoid ligament is covered by the tendons of the Tibialis posterior and Flexor digitorum longus.

The anterior and posterior talofibular and the calcaneofibular ligaments were formerly described as the three fasciculi of the *external lateral ligament* of the ankle joint.

The **Anterior Talofibular Ligament** (*ligamentum talofibulare anterius*) (Fig. 5–66), the shortest of the three, passes anteriorly and medially from the anterior margin of the fibular malleolus to the talus, anterior to its lateral articular facet.

The **Posterior Talofibular Ligament** (*ligamentum talofibulare posterius*) (Fig. 5–66), the strongest and most deeply seated, runs almost horizontally from the depression at

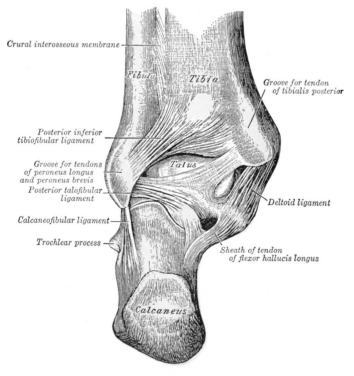

FIG. 5–67.—The left ankle joint. Posterior aspect.

the medial and posterior part of the fibular malleolus to a prominent tubercle on the posterior surface of the talus immediately lateral to the groove for the tendon of the Flexor hallucis longus.

The **Calcaneofibular Ligament** (*ligamentum calcaneofibulare*) (Fig. 5–66), the longest of the three, is a narrow, rounded cord, running from the apex of the fibular malleolus to a tubercle on the lateral surface of the calcaneus. It is covered by the tendons of the Peronei longus and brevis.

Synovial Membrane (Fig. 5–68).—The synovial membrane invests the deep surfaces of the ligaments, and sends a small process proximally between the distal ends of the tibia and fibula.

Relations.—The tendons, vessels, and nerves in connection with the joint are anteriorly the Tibialis anterior, Extensor hallucis longus, anterior tibial vessels, deep peroneal nerve, Extensor digitorum longus, and Peroneus tertius, and posteriorly the Tibialis posterior, Flexor digitorum longus, posterior tibial vessels, tibial nerve, Flexor hallucis longus, and, in the groove behind the fibular malleolus, the tendons of the Peronei longus and brevis.

The **arteries** supplying the joint are derived from the malleolar branches of the anterior tibial and the peroneal.

The **nerves** are derived from the deep peroneal and tibial.

Movements.—When the body is in the erect position, the foot is at right angles to the leg. The movements of the joint are those of dorsiflexion and plantar flexion; dorsiflexion consists in the approximation of the dorsum of the foot to the front of the leg, while in plantar flexion the heel is drawn up and the toes pointed downward. The range of movement varies in different individuals from about 50° to 90°. The transverse axis about which movement takes place is slightly oblique. The malleoli tightly embrace the talus in all positions of the joint, so that any slight degree of side-to-side movement which may exist is simply due to stretching of the ligaments of the talofibular syndesmosis, and slight bending of the body of the fibula. The superior articular surface of the talus is broader in front than behind. In dorsiflexion, therefore, greater space is required between the two malleoli. This is obtained by a slight outward rotatory movement of the lower end of the fibula and a stretching of the ligaments of the syndesmosis; this lateral movement is facilitated by a slight gliding at the tibiofibular articulation, and possibly also by the bending of the body of the fibula. Of the ligaments, the deltoid is of very great

power—so much so that it usually resists a force which fractures the process of bone to which it is attached. Its middle portion, together with the calcaneofibular ligament, binds the bones of the leg firmly to the foot, and resists displacement in every direction. Its anterior and posterior fibers limit plantar flexion and dorsiflexion of the foot and the anterior fibers also limit abduction. The posterior talofibular ligament assists the calcaneofibular in resisting the displacement of the foot backward, and deepens the cavity for the reception of the talus. The anterior talofibular is a security against the displacement of the foot forward, and limits plantar flexion of the joint.

The movements of inversion and eversion of the foot, together with the minute changes in form by which it is applied to the ground or takes hold of an object in climbing, etc., are mainly effected in the tarsal joints; the joint which enjoys the greatest amount of motion being that between the talus and calcaneus behind and the navicular and cuboid in front. This is often called the **transverse tarsal joint**, and it can, with the subordinate joints of the tarsus, replace the ankle joint in a great measure when the latter has become ankylosed.

Plantar flexion of the foot upon the tibia and fibula is produced by the Gastrocnemius, Soleus, Plantaris, Tibialis posterior, Peronei longus and brevis, Flexor digitorum longus, and Flexor hallucis longus; *dorsiflexion*, by the Tibialis anterior, Peroneus tertius, Extensor digitorum longus, and Extensor hallucis longus.

V. Intertarsal Articulations (Articulations Intertarseae; Articulations of the Tarsus)

Subtalar Articulation (*articulatio subtalaris; talocalcaneal articulation; calcaneoastragaloid articulation*).—The articulations between the calcaneus and talus are two in number—anterior and posterior. Of these, the anterior forms part of the talocalcaneonavicular joint, and will be described with that articulation. The posterior or talocalcaneal articulation is formed between the posterior calcaneal facet on the inferior surface of the talus, and the posterior facet on the superior surface of the calcaneus. It is a gliding joint, and the two bones are connected by an articular capsule and by anterior, posterior, lateral, medial, and interosseous talocalcaneal ligaments (Figs. 5–65, 5–69).

The **Articular Capsule** (*capsula articularis*) envelops the joint, and consists for

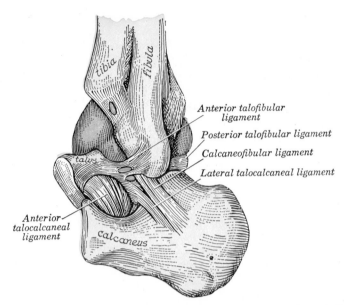

Fig. 5–68.—Synovial membrane of the capsule of left talocrural articulation (distended). Lateral aspect.

the most part of short fibers which are split up into distinct slips; between these there is only a weak fibrous investment.

The **Anterior Talocalcaneal Ligament** (*ligamentum talocalcaneum anterius; anterior calcaneo-astragaloid ligament*) (Figs. 5–68, 5–71) extends from the anterior and lateral surface of the neck of the talus to the superior surface of the calcaneus. It forms the posterior boundary of the talocalcaneonavicular joint, and is sometimes described as the **anterior interosseous ligament**.

The **Posterior Talocalcaneal Ligament** (*ligamentum talocalcaneum posterius; posterior calcaneo-astragaloid ligament*) (Fig. 5–65) connects the lateral tubercle of the talus with the proximal and medial part of the calcaneus; it is a short band, and its fibers radiate from their narrow attachment to the talus.

The **Lateral Talocalcaneal Ligament** (*ligamentum talocalcaneum laterale; external calcaneo-astragaloid ligament*) (Figs. 5–66, 5–68) is a short, strong fasciculus, passing from the lateral surface of the talus immediately beneath its fibular facet to the lateral surface of the calcaneus. It is placed anterior to, but on a deeper plane than, the calcaneofibular ligament, with the fibers of which it is parallel.

The **Medial Talocalcaneal Ligament** (*ligamentum talocalcaneum mediale; internal calcaneo-astragaloid ligament*) connects the medial tubercle of the talus with the posterior of the sustentaculum tali. Its fibers blend with those of the plantar calcaneonavicular ligament (Fig. 5–65).

The **Interosseous Talocalcaneal Ligament** (*ligamentum talocalcaneum interosseum*) (Figs. 5–69, 5–71) forms the chief bond of union between the bones. It is, in fact, a portion of the united capsules of the talocalcaneonavicular and the talocalcaneal joints, and consists of two partially united layers of fibers, one belonging to the former and the other to the latter joint. It is attached to the groove between the articular facets of the distal surface of the talus, and to a corresponding depression on the proximal surface of the calcaneus. It is very thick and strong, being at least 2.5 cm in breadth from side to side, and serves to bind the calcaneus and talus firmly together.

Synovial Membrane (Fig. 5–68).—The synovial membrane lines the capsule of the joint, and is distinct from the other synovial membranes of the tarsus.

Movements.—The movements permitted between the talus and calcaneus are limited to gliding of the one bone on the other anteroposteriorly and from side to side.

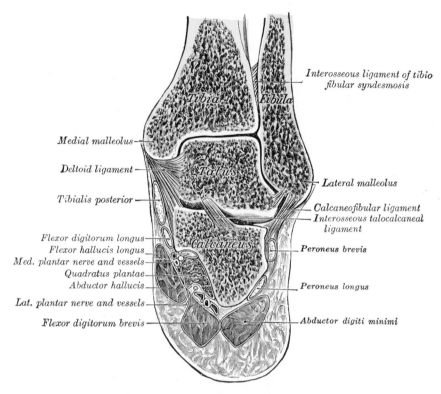

Medial malleolus

Deltoid ligament

Tibialis posterior

Flexor digitorum longus
Flexor hallucis longus
Med. plantar nerve and vessels
Quadratus plantae
Abductor hallucis

Lat. plantar nerve and vessels

Flexor digitorum brevis

Interosseous ligament of tibiofibular syndesmosis

Lateral malleolus

Calcaneofibular ligament
Interosseous talocalcaneal
ligament

Peroneus brevis

Peroneus longus

Abductor digiti minimi

Tibia   Fibula

Talus

Calcaneus

Fig. 5–69.—Coronal section through right talocrural and talocalcaneal joints.

The **Talocalcaneonavicular Articulation** (*articulatio talocalcaneonavicularis*) is a gliding joint: the rounded head of the talus being received into the concavity formed by the posterior surface of the navicular, the anterior articular surface of the calcaneus, and the proximal surface of the plantar calcaneonavicular ligament. There are two ligaments in this joint: the articular capsule and the dorsal talonavicular.

The **Articular Capsule** (*capsula articularis*) is imperfectly developed except posteriorly, where it is considerably thickened and forms, with a part of the capsule of the talocalcaneal joint, the strong interosseous ligament which fills in the canal formed by the opposing grooves on the calcaneus and talus, as mentioned above.

The **Dorsal Talonavicular Ligament** (*ligamentum talonaviculare; superior astragalonavicular ligament*) (Fig. 5–65).—This ligament is a broad, thin band which connects the neck of the talus to the dorsal surface of the navicular bone; it is covered by the Extensor tendons. The plantar

calcaneonavicular supplies the place of a plantar ligament for this joint.

Synovial Membrane.—The synovial membrane lines all parts of the capsule of the joint.

Movements.—This articulation permits a considerable range of gliding movements, and some rotation; its feeble construction allows occasional dislocation of the other bones of the tarsus from the talus.

Calcaneocuboid Articulation (*articulatio calcaneocuboidea; articulation of the calcaneus with the cuboid*).—The ligaments connecting the calcaneus with the cuboid are five in number, viz., the articular capsule, the dorsal calcaneocuboid, part of the bifurcated, the long plantar, and the plantar calcaneocuboid.

The **Articular Capsule** (*capsula articularis*).—The articular capsule is an imperfectly developed investment, containing certain strengthened bands which form the other ligaments of the joint.

The **Dorsal Calcaneocuboid Ligament** (*ligamentum calcaneocuboideum; superior* *calcaneocuboid ligament*) (Fig. 5–66) is a thin but broad fasciculus, which passes between the contiguous surfaces of the calcaneus and cuboid on the dorsal surface of the joint.

The **Bifurcated Ligament** (*ligamentum bifurcatum; internal calcaneocuboid; interosseous ligament*) (Figs. 5–66, 5–71) is a strong band, attached posteriorly to the deep hollow on the proximal surface of the calcaneus and dividing anteriorly into a calcaneocuboid and a calcaneonavicular part. The **calcaneocuboid part** is fixed to the medial side of the cuboid and forms one

of the principal bonds between the first and second rows of the tarsal bones. The **calcaneonavicular part** is attached to the lateral side of the navicular.

The **Long Plantar Ligament** (*ligamentum plantare longum; long calcaneocuboid ligament; superficial long plantar ligament*) (Fig. 5–70) is the longest of all the ligaments of the tarsus; it is attached to the plantar surface of the calcaneus anterior to the tuberosity, and posteriorly to the tuberosity on the plantar surface of the cuboid bone, the more superficial fibers being continued distally to the bases of the

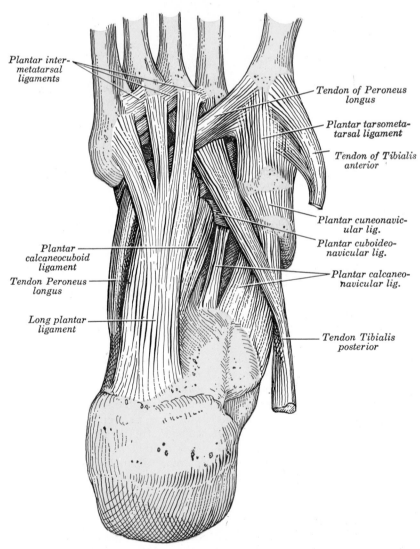

Fig. 5–70.—Ligaments of the sole of the foot, with the tendons of the Peroneus longus, Tibialis posterior and Tibialis anterior muscles.

third, fourth and fifth, and occasionally also the second metatarsal bones. This ligament converts the groove on the plantar surface of the cuboid into a canal for the tendon of the Peroneus longus.

The **Plantar Calcaneocuboid Ligament** (*ligamentum calcaneocuboideum plantare; short calcaneocuboid ligament; short plantar ligament*) (Fig. 5–70) lies nearer to the bones than the preceding, from which it is separated by a little areolar tissue. It is a short but wide band of great strength which extends from the tubercle and the depression anterior to it on the distal part of the plantar surface of the calcaneus to the plantar surface of the cuboid proximal to the peroneal groove.

Synovial Membrane.—The synovial membrane lines the inner surface of the capsule and is distinct from that of the other tarsal articulations (Fig. 5–72).

Movements.—The movements permitted between the calcaneus and cuboid are limited to slight gliding movements of the bones upon each other.

The **transverse tarsal joint** (*articulatio tarsi transversum*) is formed by the articulation of the calcaneus with the cuboid, and the articulation of the talus with the navicular. The movement which takes place in this joint is more extensive than that in the other tarsal joints, and consists of a sort of rotation by means of which the foot may be slightly flexed or extended, the sole being at the same time carried medially (inverted) or laterally (everted).

The **Ligaments Connecting the Calcaneus and Navicular.**—Although the calcaneus and navicular do not articulate directly, they are connected by two ligaments: the calcaneonavicular part of the bifurcated, and the plantar calcaneonavicular.

The **calcaneonavicular part of the bifurcated ligament** is described on page 362.

The **Plantar Calcaneonavicular Ligament** (*ligamentum calcaneonaviculare plantare; inferior* or *internal calcaneonavicular ligament; calcaneonavicular ligament*) (Figs. 5–65, 5–70) is a broad and thick band of fibers which connects the anterior margin of the sustentaculum tali of the calcaneus to the plantar surface of the navicular. This ligament not only serves to connect the calca-

neus and navicular, but supports the head of the talus, forming part of the articular cavity in which it is received. The **dorsal surface** of the ligament presents a fibrocartilaginous facet, lined by the synovial membrane, and upon this a portion of the head of the talus rests. Its **plantar surface** is supported by the tendon of the Tibialis posterior; its **medial border** is blended with the anterior part of the deltoid ligament of the ankle joint.

The plantar calcaneonavicular ligament, by supporting the head of the talus, is principally concerned in maintaining the arch of the foot. When it yields, the head of the talus is pressed medially and distally by the weight of the body, and the foot becomes pronated, everted, and turned lateralward, exhibiting the condition known as *flat-foot*. This ligament contains a considerable amount of elastic fibers, so as to give elasticity to the arch and spring to the foot; hence it is sometimes called the "spring" ligament. It is supported, on its plantar surface, by the tendon of the Tibialis posterior, which spreads out at its insertion into a number of fasciculi, to be attached to most of the tarsal and metatarsal bones.

Cuneonavicular Articulation (*articulatio cuneonavicularis; articulation of the navicular with the cuneiform bones.*)—The navicular is connected to the three cuneiform bones by dorsal and plantar ligaments.

The **Dorsal Ligaments** (*ligamenta navicularicuneiformia dorsalia*) are three small bundles, one attached to each of the cuneiform bones. The bundle connecting the navicular with the medial cuneiform is continuous around the medial side of the articulation with the plantar ligament which unites these two bones (Figs. 5–65, 5–66).

The **Plantar Ligaments** (*ligamenta navicularicuneiformia plantaria*) have an arrangement similar to the dorsal, and are strengthened by slips from the tendon of the Tibialis posterior (Fig. 5–70).

Synovial Membrane.—The synovial membrane of these joints is part of the great tarsal synovial membrane (Fig. 5–72).

Movements.—Mere gliding movements are permitted between the navicular and cuneiform bones.

Cuboideonavicular Articulation.—The navicular bone is connected with the

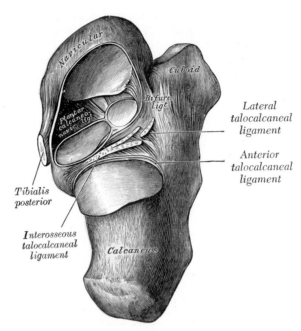

FIG. 5–71.—Talocalcaneal and talocalcaneonavicular articulations exposed from above by removing the talus.

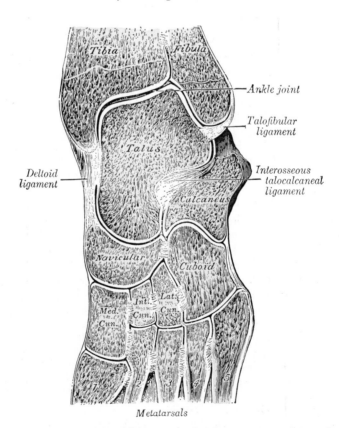

FIG. 5–72.—Oblique section of left intertarsal and tarsometatarsal articulations, showing the synovial cavities.

cuboid by dorsal, plantar, and interosseous ligaments.

The **Dorsal Ligament** (*ligamentum cuboidenaviculare dorsale*) extends obliquely distalward and lateralward from the navicular to the cuboid bone (Fig. 5–66).

The **Plantar Ligament** (*ligamentum cuboidenaviculare plantare*) passes nearly transversely between these two bones (Fig. 5–70).

The **Interosseous Ligament.**—The interosseous ligament consists of strong transverse fibers, and connects the rough non-articular portions of the adjacent surfaces of the two bones (Fig. 5–72).

Synovial Membrane.—The synovial membrane of this joint is part of the great tarsal synovial membrane (Fig. 5–72).

Movements.—The movements permitted between the navicular and cuboid bones are limited to a slight gliding upon each other.

Intercuneiform and Cuneocuboid Articulations.—The three cuneiform bones and the cuboid are connected together by dorsal, plantar, and interosseous ligaments.

The **Dorsal Ligaments** (*ligamenta intercuneiformia dorsalia*) consist of three transverse bands: one connects the medial with the intermediate cuneiform, another the intermediate with the lateral cuneiform, and another the lateral cuneiform with the cuboid.

The **Plantar Ligaments** (*ligamenta intercuneiformia plantaria*) have a similar arrangement to the dorsal, and are strengthened by slips from the tendon of the Tibialis posterior.

The **Interosseous Ligaments** (*ligamenta intercuneiformia interossea*).—The interosseous ligaments consist of strong transverse fibers which pass between the rough non-articular portions of the adjacent surfaces of the bones (Fig. 5–72).

Synovial Membrane.—The synovial membrane of these joints is part of the great tarsal synovial membrane (Fig. 5–72).

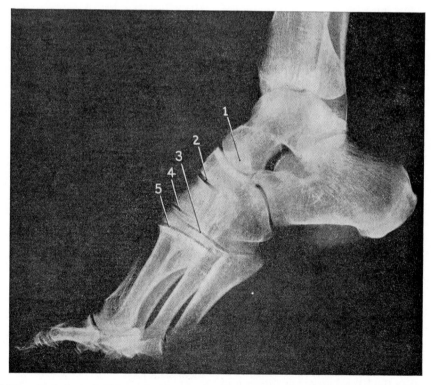

Fig. 5–73.—Adult foot. *1,* Tuberosity of navicular bone, partly obscured by the shadow of the head of the talus; *2,* cuneo-navicular joint; *3,* joint between metatarsal III and the lateral cuneiform bone; *4,* joint between metatarsal II and the intermediate cuneiform bone; *5,* joint between metatarsal I and the medial cuneiform bone.

Movements.—The movements permitted between these bones are limited to a slight gliding upon each other.

VI. Tarsometatarsal Articulations (Articulationes Tarsometatarseae)

These are gliding joints. The bones entering into their formation are the medial, intermediate, and lateral cuneiforms, and the cuboid, which articulate with the bases of the metatarsal bones. The first metatarsal bone articulates with the medial cuneiform; the second is deeply wedged in between the medial and intermediate cuneiforms articulating by its base with the intermediate cuneiform; the third articulates with the lateral cuneiform; the fourth, with the cuboid and lateral cuneiform; and the fifth, with the cuboid. The bones are connected by dorsal, plantar, and interosseous ligaments.

The **Dorsal Ligaments** (*ligamenta tarsometatarsea dorsalia*) are strong, flat bands. The first metatarsal is joined to the medial cuneiform by a broad thin band; the second has three, one from each cuneiform bone; the third has one from the lateral cuneiform; the fourth has one from the lateral cuneiform and one from the cuboid; and the fifth, one from the cuboid (Figs. 5–65, 5–66).

The **Plantar Ligaments** (*ligamenta tarsometatarsea plantaria*) consist of longitudinal and oblique bands disposed with less regularity than the dorsal ligaments. Those for the first and second metatarsals are the strongest; the second and third metatarsals are joined by oblique bands to the medial cuneiform; the fourth and fifth metatarsals are connected by a few fibers to the cuboid (Fig. 5–70).

The **Interosseous Ligaments** (*ligamenta cuneometatarsea interossea*) are two or three in number. The first is the strongest, and passes from the lateral surface of the medial cuneiform to the adjacent angle of the second metatarsal. The second or middle which is the smallest and is less constant than the others passes from the lateral cuneiform to the lateral aspect of the second metatarsal. The third connects the lateral angle of the lateral cuneiform with the adjacent side of the base of the third metatarsal.

Synovial Membrane (Fig. 5–72).—The synovial membrane between the medial cuneiform and the first metatarsal forms a distinct sac. The synovial membrane between the intermediate and lateral cuneiforms, and the second and third metatarsal bones, is part of the great tarsal synovial membrane. Two prolongations are sent forward from it, one between the adjacent sides of the second and third, and another between those of the third and fourth metatarsal bones. The synovial membrane between the cuboid and the fourth and fifth metatarsal bones forms a distinct sac. From it a prolongation is sent forward between the fourth and fifth metatarsal bones.

Movements.—The movements permitted between the tarsal and metatarsal bones are limited to slight gliding of the bones upon each other.

Nerve Supply.—The intertarsal and tarsometatarsal joints are supplied by the deep peroneal nerve.

VII. Intermetatarsal Articulations (Articulationes Intermetatarseae)

The bases of the four lateral metatarsals are connected by the dorsal, plantar, and interosseous ligaments.

The first metatarsal is connected with the second by interosseous fibers only; the fibers are weak and may be largely replaced by a bursa between indistinct facets on the two bones.

The **Dorsal Ligaments** (*ligamenta basium* [*oss. metatars.*] *dorsalia*) pass transversely between the dorsal surfaces of the bases of the adjacent metatarsal bones.

The **Plantar Ligaments** (*ligamenta basium* [*oss. metatars.*] *plantaria*) have a similar arrangement to the dorsal.

The **Interosseous Ligaments** (*ligamenta basium* [*oss. metatars.*] *interossea*) consist of strong transverse fibers which connect the rough non-articular portions of the adjacent surfaces.

Synovial Membranes (Fig. 5–72).—The synovial membranes between the second and third and the third and fourth metatarsal bones are part of the great tarsal synovial membrane; that between the fourth and fifth is a prolongation of the synovial membrane of the cuboideometatarsal joint.

Movements.—The movement permitted between the tarsal ends of the metatarsal bones is limited to a slight gliding of the articular surfaces upon one another.

The heads of all the metatarsal bones are connected together by the transverse metatarsal ligament.

The **Transverse Metatarsal Ligament** is a narrow band which runs across and connects together the heads of all the metatarsal bones; it is blended anteriorly with the plantar (glenoid) ligaments of the metatarsophalangeal articulations. Its plantar surface is concave where the Flexor tendons run below it; above it the tendons of the Interossei pass to their insertions. It differs from the transverse metacarpal ligament in that it connects the first metatarsal to the others.

The Synovial Membranes in the Tarsal and Tarsometatarsal Joints (Fig. 5–72).—The synovial membranes found in the articulations of the tarsus and metatarsus are six in number: one for the talocalcaneal articulation; a second for the talocalcaneonavicular articulation; a third for the calcaneocuboid articulation; and a fourth for the cuneonavicular, intercuneiform, and cuneocuboid articulations, the articulations of the intermediate and lateral cuneiforms with the bases of the second and third metatarsal bones, and the adjacent surfaces of the bases of the second, third, and fourth metatarsal bones; a fifth for the medial cuneiform with the metatarsal bone of the great toe; and a sixth for the articulation of the cuboid with the fourth and fifth metatarsal bones. A small synovial cavity is sometimes found between the contiguous surfaces of the navicular and cuboid bones.

VIII. Metatarsophalangeal Articulations (Articulationes Metatarsophalangeae)

The metatarsophalangeal articulations are of the condyloid kind, formed by the reception of the rounded heads of the metatarsal bones in shallow cavities on the ends of the proximal phalanges.

The ligaments are the plantar and two collateral.

The **Plantar Ligaments** (*ligamenta accessoria plantaria; glenoid ligaments of Cruveilhier*) are thick, dense, fibrous structures. They are placed on the plantar surfaces of the joints in the intervals between the collateral ligaments, to which they are connected; they are loosely united to the metatarsal bones, but very firmly to the bases of the first phalanges. Their plantar surfaces are intimately blended with the

transverse metatarsal ligament, and grooved for the passage of the Flexor tendons, the sheaths surrounding which are connected to the sides of the grooves. Their deep surfaces form part of the articular facets for the heads of the metatarsal bones, and are lined by synovial membrane.

The **Collateral Ligaments** (*ligamenta collateralia; lateral ligaments*) are strong, rounded cords, placed one on either side of each joint, and attached, by one end, to the posterior tubercle on the side of the head of the metatarsal bone, and, by the other, to the contiguous extremity of the phalanx.

The place of **dorsal ligaments** is supplied by the Extensor tendons on the dorsal surfaces of the joints.

Movements.—The movements permitted in the metatarsophalangeal articulations are flexion extension, abduction, and adduction.

IX. Articulations of the Digits (Articulationes Digitorum Pedis; Articulations of the Phalanges)

The interphalangeal articulations are ginglymoid joints, and each has a plantar and two collateral ligaments.

The arrangement of these ligaments is similar to that in the metatarsophalangeal articulations: the Extensor tendons supply the places of dorsal ligaments.

Movements.—The only movements permitted in the joints of the digits are flexion and extension; these movements are more extensive between the proximal and middle phalanges than between the middle and distal. The amount of flexion is very considerable, but extension is limited by the plantar and collateral ligaments.

Arches of the Foot

The **longitudinal arch** is formed by the seven tarsal and the five metatarsal bones (Figs. 4–225, 4–226) and the ligaments which bind them together. The multiplicity of parts gives resiliency. The arch rests posteriorly on the calcaneal tuberosity and anteriorly on the heads of the five metatarsals. In the standing position, 25 per cent of the body weight is distributed to each calcaneus and 25 per cent to the heads of the five metatarsals of each foot, in proportion of about 1 part for metatarsal I to 2.5 parts for metatarsals II to V. The greater part of the tension stress on the longitudinal arch is borne by the plantar ligaments (Fig. 5–70). Only about 15 to 20 per cent of the stress is borne by the Tibialis posterior and the peroneal muscles. When the body is raised on the ball of one foot, the stress on the arch is increased four times.

In addition to the longitudinal arch the foot presents a series of **transverse arches.** At the posterior part of the metatarsus and the anterior part of the tarsus the arches are complete, but in the middle of the tarsus they present more the characters of half-domes, the concavities of which are directed downward and medialward, so that when the medial borders of the feet are placed in apposition a complete tarsal dome is formed. The transverse arches are strengthened by the interosseous, plantar, and dorsal ligaments, and by the Peroneus longus, whose tendon stretches across between the piers of the arches.

REFERENCES

EMBRYOLOGY AND HISTOLOGY

ANDERSEN, HELGE and FREDE BRO-RASSMUSSEN. 1961. Histochemical studies on the histogenesis of the joints in human fetuses with special reference to the development of the joint cavities in the hand and foot. Amer. J. Anat., 108, 111-122.

BAILEY, J. P., J. B. CUBBERLEY, C. A. L. STEPHENS, JR., and A. B. STANFIELD. 1963. Vascular networks from explants of human synovialis in vitro. Anat. Rec., 147, 525-531.

BARLAND, P., A. B. NOVIKOFF, and D. HAMERMAN. 1962. Electron microscopy of the human synovial membrane. J. Cell. Biol., 14, 207-220.

COGGESHALL, H. C., C. F. WARREN, and W. BAUER. 1940. The cytology of normal human synovial fluid. Anat. Rec., 77, 129-144.

CRELIN, E. S., and W. O. SOUTHWICK. 1964. Changes induced by sustained pressure in the knee joint articular cartilage of adult rabbits. Anat. Rec., 149, 113-133.

DAVIES, D. V. 1946. The lymphatics of the synovial membrane. J. Anat. (Lond.), 80, 21-23.

GARDNER, E. and D. J. GRAY. 1950. Prenatal development of the human hip joint. Amer. J. Anat., 87, 163-211.

GARDNER, E. and D. J. GRAY. 1953. Prenatal development of the human shoulder and acromioclavicular joints. Amer. J. Anat., 92, 219-276.

GRAY, D. J. and E. GARDNER. 1950. Prenatal development of the human knee and superior tibiofibular joints. Amer. J. Anat., 86, 235-287.

GRAY, D. J. and E. GARDNER. 1951. Prenatal development of the human elbow joint. Amer. J. Anat., 88, 429-469.

HACKETT, GEORGE STUART. 1965. Joint Ligament Relation Treated by Fibro-Osseous Prolification. xiv + 97 pages, illustrated. Charles C Thomas Publishers, Springfield, Illinois.

HAINES, R. W. 1947. The development of joints. J. Anat. (Lond.), 81, 33-55.

AXIAL AND SKULL

ANGLE, J. L. 1948. Factors in temporomandibular joint form. Amer. J. Anat., 83, 223-246.

ASHLEY, G. T. 1954. The morphological and pathological significance of synostosis at the manubriosternal joint. Thorax, 9, 159-166.

DAVIS, P. R., J. D. G. TROUP, and J. H. BURNARD. 1965. Movements of the thoracic and lumbar spine when lifting: a chrono-cyclophotographic study. J. Anat. (Lond.), 99, 13-26.

DONISCH, E. W., and W. TRAPP. 1971. The cartilage endplates of the human vertebral column. Anat. Rec., 169, 705-715.

FRANCIS, C. C. 1955. Variations in the articular facets of the cervical vertebrae. Anat. Rec., 122, 589-609.

GRAY, D. J. and E. D. GARDNER. 1943. The human sternochondral joints. Anat. Rec., 87, 235-253.

LEONHART, GEORGE P. 1914. A case of stylo-hyoid ossification. Anat. Rec., 8, 325-332.

LEWIN, T. 1968. Anatomical variations in lumbosacral synovial joints with particular reference to subluxation. Acta Anat., 71, 229-248.

MILES, MERYL and WALTER E. SULLIVAN. 1961. Lateral bending at the lumbar and lumbosacral joints. Anat. Rec., 139, 387-398.

POWELL, THOMAS V. and ALLAN G. BRODIE. 1963. Closure of the spheno-occipital synchondrosis. Anat. Rec., 147, 15-23.

ROGERS, LAMBERT C. and E. E. PAYNE. 1961. The dura mater at the craniovertebral-junction. J. Anat. (Lond.), 95, 586-588.

SARNAT, B. G., Editor. 1951. The Temporomandibular Joint. xviii + 148 pages. Charles C Thomas, Springfield, Illinois.

SHAPIRO, H. H. and R. C. TRUEX. 1943. The temporomandibular joint and the auditory function. J. Amer. Dent. Ass., 30, 1147-1168.

SPURLING, R. G. 1956. Lesions of the Cervical Intervertebral Disc. xi + 134 pages. Charles C Thomas, Springfield, Illinois.

STEIN, M. R. 1939. The "mandibular sling." Dental Survey, 15, 883-887.

STEWART, T. D. 1938. Accessory sacro-iliac articulations in the higher primates and their significance. Amer. J. Phys. Anthrop., 24, 43-59.

TROTTER, M. 1940. A common anatomical variation in the sacro-iliac region. J. Bone Jt. Surg., 22, 293-299.

CLAVICULAR AND SHOULDER JOINTS

BEARN, J. G. 1967. Direct observation on the function of the capsule of the sternoclavicular joint in clavicular support. J. Anat. (Lond.), 101, 159-170.

Cave, A. J. E. 1961. The nature and morphology of the costoclavicular ligament. J. Anat. (Lond.), 95, 170-179.

Gardner, E. 1948. The innervation of the shoulder joint. Anat. Rec., 102, 1-18.

Inman, V. T., J. B. deC. M. Saunders, and L. C. Abbott. 1944. Observations on the function of the shoulder joint. J. Bone Jt. Surg., 26, 1-30.

Kaplan, E. B. 1943. The coraco-humeral ligament of the human shoulder. Bull. Hosp. Jt. Dis., 4, 62-65.

Lewis, O. J. 1959. The coraco-clavicular joint. J. Anat. (Lond.), 93, 296-303.

Martin, C. P. 1940. The movements of the shoulder-joint, with special reference to rupture of the supraspinatus tendon. Amer. J. Anat., 66, 213-234.

Moseley, H. F. 1952. Ruptures of the Rotator Cuff. xii + 90 pages. Charles C Thomas, Springfield, Illinois.

ELBOW, WRIST, AND HAND

Gardner, E. 1948. The innervation of the elbow joint. Anat. Rec., 102, 161-174.

Halls, Albert A. and Anthony Travill. 1964. Transmission of pressures across the elbow joint. Anat. Rec., 150, 243-247.

Haxton, Herbert A. 1945. A comparison of the action of extension of the knee and elbow joints in man. Anat. Rec., 93, 279-286.

Kaplan, E. B. 1936-37. Extension deformities of the proximal interphalangeal joints of the fingers. J. Bone Jt. Surg., 18, 781-983. Correction, 19, 1144.

Kilburn, P., J. G. Sweeney, and F. F. Silk. 1962. Three cases of compound posterior dislocation of the elbow with rupture of the brachial artery. J. Bone Jt. Surg., 44-B, 119-121.

Kropp, B. N. 1945. A note on the piso-triquetral joint. Anat. Rec., 92, 91-92.

Landsmeer, J. M. F. 1955. Anatomical and functional investigations on the articulation of the human fingers. Acta Anat., 25, Suppl. 24 = 2, 1-69.

Lewis, O. J. 1970. The development of the human wrist joint during the fetal period. Anat. Rec., 166, 499-515.

Lewis, O. J., R. M. Hamshere, and T. M. Bucknill. 1970. The anatomy of the wrist joint. J. Anat. (Lond.), 106, 539-552.

Robbins, H. 1963. Anatomical study of the median nerve in the carpal tunnel and etiologies of the carpal-tunnel syndrome. J. Bone Jt. Surg., 45-A, 953-966.

Roston, J. B. and R. W. Haines. 1947. Cracking in the metacarpo-phalangeal joint. J. Anat. (Lond.), 81, 165-173.

Smith, R. D. and G. R. Holcomb. 1958. Articular surface interrelationships in finger joints. Acta Anat., 32, 217-229.

Stecher, Robert M. 1958. Ankylosis of the finger joints in rheumatoid arthritis. Ann. Rheum. Dis., 17, 365-375.

HIP AND KNEE JOINTS

Gardner, E. 1948. The innervation of the hip joint. Anat. Rec., 101, 353-371.

Gardner, E., and R. O'Rahilly. 1968. The early development of the knee joint in staged human embryos. J. Anat. (Lond.), 102, 289-299.

Hart, V. L. 1952. Congenital Dysplasia of the Hip Joint and Sequelae in the Newborn and Early Postnatal Life. xv + 187 pages, illustrated. Charles C Thomas, Springfield, Illinois.

Milch, Henry. 1961. The measurement of pelvi-femoral motion. Anat. Rec., 140, 135-145.

Moore, Jack Beals, and John Orren Vaughn. 1928. The range of active motion at the hip joint of men. J. Bone Jt. Surg., 10, 248-257.

Roberts, W. H. 1963. The locking mechanism of the hip joint. Anat. Rec., 147, 321-324.

Scapinelli, R. 1968. Studies on the vasculature of the human knee joint. Acta Anat., 70, 305-331.

Stanisavljevic, Stanko. 1964. Diagnosis and Treatment of Congenital Hip Pathology in the Newborn. xv + 94 pages, illustrated. The Williams & Wilkins Company, Baltimore.

BURSAE

Black, B. M. 1934. The prenatal incidence, structure, and development of some human synovial bursae. Anat. Rec., 60, 333-355.

Black, B. M. 1934. A large bilocular synovial bursa in a plantaris of man. J. Bone Jt. Surg., 16, 919-924.

Codman, E. A. 1937. Rupture of the supraspinatus tendon. J. Bone Jt. Surg., 19, 643, 652.

Horwitz, M. T. and L. M. Tocantins. 1938. An anatomical study of the role of the long thoracic nerve and the related scapular bursae in the pathogenesis of local paralysis of the serratus anterior muscle. Anat. Rec., 71, 375-385.

Schneider, C. L. 1943. Trabeculae traversing human bursae. Anat. Rec., 87, 151-163.

PHYSIOLOGY AND GENERAL

Barnett, C. H., D. V. Davies, and M. A. MacConaill. 1961. Synovial Joints, Their Structure and Mechanics. xi + 304 pages, illustrated. Charles C Thomas, Springfield, Illinois.

Bauer, W., M. W. Ropes, and H. Waine. 1940. The physiology of articular structures. Physiol. Rev., 20, 272-312.

Corbin, K. B. and J. C. Hinsey. 1939. Influence of the nervous system on bone and joints. Anat. Rec., 75, 307-317.

Dintenfass, Leopold. 1963. Lubrication in synovial joints: A theoretical analysis. J. Bone Jt. Surg., 45-A, 1241-1256.

Gardner, E. 1963. Physiology of joints. J. Bone Jt. Surg., 45-A, 1061-1066.

Lewis, Raymond W. 1955. The Joints of the Extremities, a Radiographic Study. vii + 108 pages, illustrated. Charles C Thomas, Springfield, Illinois.

Morton, D. J. 1952. Human Locomotion and Body Form. xii + 285 pages. Williams & Wilkins Company, Baltimore.

6 | *Muscles and Fasciae*

oluntary motion in the various parts of the body is brought about by the contraction of muscles. The energy of their contraction is made mechanically effective by means of the tendons, aponeuroses, and fasciae which secure the ends of the muscles and control the direction of their pull. They form the dark, reddish masses that are popularly known as flesh, and account for approximately 40 per cent of the body weight. They vary greatly in size. The Gastrocnemius forms the bulk of the calf of the leg; the Sartorius is nearly 61 cm in length, and the Stapedius, a tiny muscle of the middle ear, weighs 0.1 gm and is 2 to 3 mm in length. In addition to these muscles, which are properly called **voluntary, skeletal, or striated muscles**, there are other muscular tissues which are not under voluntary control, such as the cardiac muscle of the heart and the smooth muscle of the intestines. They are described in chapters dealing with the viscera.

DEVELOPMENT OF THE MUSCLES

Both the cross striated and smooth muscles, with the exception of a few that are of ectodermal origin, arise from the mesoderm. The intrinsic muscles of the trunk are derived from the myotomes, while the muscles of the head and limbs differentiate directly from the mesoderm.

The Myotomic Muscles.—The intrinsic muscles of the trunk which are derived directly from the myotomes are conveniently treated in two groups, the deep muscles of the back and the thoracoabdominal muscles.

The deep muscles of the back extend from the sacral to the occipital region and vary much in length and size. They act chiefly on the vertebral column. The shorter muscles, such as the Interspinales, Intertransversarii, the deeper layers of the Multifidus, the Rotatores, Levatores costarum, Obliquus capitis inferior, Obliquus capitis superior and Rectus capitis posterior minor which extend between adjoining vertebrae, retain the primitive segmentation of the myotomes. Other muscles, such as the Splenius capitis, Splenius cervicis, Sacrospinalis, Semispinalis, Multifidus, Iliocostalis, Longissimus, Spinales, Semispinales, and Rectus capitis posterior major, which extend over several vertebrae, are formed by the fusion of successive myotomes and the splitting into longitudinal columns.

The fascia lumbodorsalis develops between the true myotomic muscles and the more superficial ones which migrate over the back such as the Trapezius, Rhomboideus, and Latissimus.

The anterior vertebral muscles, the Longus colli, Longus capitis, Rectus capitis anterior and Rectus capitis lateralis are derived from the ventral part of the cervical myotomes as are probably also the Scaleni.

The thoracoabdominal muscles arise through the ventral extension of the thoracic myotomes into the body wall. This process takes place coincidently with the ventral extension of the ribs. In the thoracic region the primitive myotomic segments still persist as the intercostal muscles, but over the abdomen these ventral myotomic processes fuse into a sheet which splits in various ways to form the Rectus, the Obliquus externus and internus, and the Transversus. Such muscles as the Pectoralis major and minor and the Serratus anterior do not belong to the above group.

The Ventrolateral Muscles of the Neck.— The intrinsic muscles of the tongue, the infrahyoid muscles and the diaphragm are derived from a more or less continuous

(371)

premuscle mass which extends on each side from the tongue into the lateral region of the upper half of the neck and into it early extend the hypoglossal and branches of the upper cervical nerves. The two halves which form the infrahyoid muscles and the diaphragm are at first widely separated from each other by the heart. As the latter descends into the thorax the diaphragmatic portion of each lateral mass is carried with its nerve down into the thorax and the laterally placed infrahyoid muscles move toward the midventral line of the neck.

Muscles of the Shoulder Girdle and Arm. —The Trapezius and Sternocleidomastoideus arise from a common premuscle mass in the occipital region just caudal to the last branchial arch; as the mass increases in size it spreads caudalward to the shoulder girdle to which it later becomes attached. It also spreads dorsalward and caudalward to the spinous processes, gaining attachment at a still later period.

The Levator scapulae, Serratus anterior and the Rhomboids arise from premuscle tissue in the lower cervical region and undergo extensive migration.

The Latissimus dorsi and Teres major are associated in their origin from the premuscle sheath of the arm as are also the two Pectoral muscles when the arm bud lies in the lower cervical region.

The intrinsic muscles of the arm develop *in situ* from the mesoderm of the arm bud and probably do not receive cells or buds from the myotomes. The nerves to these muscles enter the arm bud when it still lies in the cervical region and as the arm shifts caudally over the thorax these cervical nerves unite to form the brachial plexus.

The Muscles of the Leg.—The muscles of the leg like those of the arm develop *in situ* from the mesoderm of the leg bud, the myotomes apparently taking no part in their formation.

The Muscles of the Head.—The muscles of the orbit arise from the mesoderm over the dorsal and caudal sides of the optic stalk.

The muscles of mastication arise from mesoderm of the mandibular arch. The mandibular division of the trigeminal nerve enters this premuscle mass before it splits into the Temporal, Masseter and Pterygoidei.

The facial muscles (muscles of expression) arise from the mesoderm of the hyoid arch. The facial nerve enters this mass before it begins to split, and as the muscle mass spreads out over the face and head and neck it splits more or less incompletely into the various muscles.

The early differentiation of the muscular system apparently goes on independently of the nervous system and only later does it appear that muscles are dependent on the functional stimuli of the nerves for their continued existence and growth. Although the nervous system does not influence muscle differentiation, the nerves, owing to their early attachments to the muscle rudiments, are in a general way indicators of the position of origin of many of the muscles and likewise in many instances the nerves indicate the paths along which the developing muscles have migrated during development. The muscle of the diaphragm, for example, has its origin in the region of the fourth and fifth cervical segments. The phrenic nerve enters the muscle mass while the latter is in this region and is drawn out as the diaphragm migrates through the thorax. The Trapezius and Sternocleidomastoideus arise in the lateral occipital region as a common muscle mass, into which the nervus accessorius grows at a very early period and as the muscle mass migrates and expands caudally the nerve is carried with it. The Pectoralis major and minor arise in the cervical region, receive their nerves while in this position and as the muscle mass migrates and expands caudally over the thorax the nerves are carried along. The Latissimus dorsi and Serratus anterior are excellent examples of migrating muscles whose nerve supply indicates their origin in the cervical region. The Rectus abdominis and the other abdominal muscles migrate or shift from a lateral to a ventrolateral or abdominal position, carrying with them the nerves.

The facial nerve, which early enters the common facial muscle mass of the second branchial or hyoid arch, is dragged about with the muscle as it spreads over the head, face and neck, and as the muscle splits into the various muscles of expression, the nerve is correspondingly split. The mandibular division of the trigeminal nerve enters at an early time the muscle mass in the mandibular arch and as this mass splits and migrates apart to form the muscles of mastication the nerve splits into its various branches.

The nerve supply then serves as a key to the common origin of certain groups of muscles. The muscles supplied by the oculomotor nerve arise from a single mass in the eye region; the lingual muscles arise from a common mass supplied by the hypoglossal nerve.

ARRANGEMENT OF MUSCLES AND FASCIAE

A number of non-contractile connective tissue elements are necessary for the organization of the contractile muscle fibers into effective mechanical instruments. Thus the fibers are bound together into fasciculi by the fibroelastic perimysium; the ends of the muscle are attached to the bones by tendons and aponeuroses, and the whole muscle is held in its proper place by connective tissue sheets called fasciae.

FASCIAE

The dissectible, fibrous connective tissues of the body other than the specifically organized structures, tendons, aponeuroses, and ligaments are called fasciae. This same term is used also in a more restricted sense to indicate local connective tissue membranes which enclose a part of the body, or invest muscles or other structures. Although the term will be used most commonly with its restricted meaning of fibrous membranes, it is essential that the concept be borne in mind that the latter are part of the general connective tissues. This allows one to regard all the fascial structures as a part of a functional as well as morphological system in which the connective tissue varies in thickness, in density, in accumulation of fat, and in relative amounts of collagenous fibers, elastic fibers, and tissue fluid according to local requirements.

The entire fascial system is made up of three subdivisions: the **subcutaneous fascia,** the **deep fascia,** and the **subserous fascia.** The deep fascia is the principal somatic

fascia which invests and penetrates between the structures which form the body wall and appendages. It is the most extensive of the three and calls for the major part of our attention. The subcutaneous fascia is the layer which intervenes between the deep fascia and the skin. The subserous fascia lies within the body cavities; it forms the fibrous layer of the serous membranes (pleura, pericardium, and peritoneum), covers and supports the viscera, and attaches the perietal layer of the serous membranes to the deep fascia of the internal surface of the body wall.

The **Subcutaneous Fascia** (*tela subcutanea*) is continuous over the entire body between the skin and the deep fascial investment of the specialized structures of the body, such as muscles. It is composed of two layers. The outer one, often called the *panniculus adiposus*, normally contains an accumulation of fat. The latter may be several centimeters thick, or it may, in emaciated individuals, be almost entirely lacking. The inner layer is a thin membrane which ordinarily has no fat and has a generous amount of elastic tissue. The two layers are quite adherent in most regions but they can be separated by careful dissection, particularly in the lower anterior abdominal wall. Between the two layers lie the superficial arteries, veins, nerves, and lymphatics, the mammary glands, most of the facial muscles, the platysma, and one or two other muscles.

The subcutaneous fascia in many parts of the body glides freely over the deep fascia, producing the characteristic movability of the skin in an area such as the back of the hand. In these areas, the two fasciae can be dissected apart easily by a probing finger or blunt instrument. They are separated, in other words, by a fascial cleft. At certain other points on the body surface, especially over bony prominences, the two fasciae are closely adherent. They retain their individuality even here, however, and do not become continuous with each other.

The **deep fascia** is represented characteristically by the gray felt-like membranes immediately covering the muscles. It comprises a rather intricate series of sheets and bands which hold the muscles and other structures in their proper relative positions, separating them from each other for independent function as well as joining them together into an integrated whole. The intrinsic connective tissue of the capsules or stroma of these structures is not included in this fascia. In the case of a muscle, the epimysium may be fused with the overlying fascia and lose its identity, as in the Triceps, or it may be separated from the fascia by a cleft and retain its individuality, as in the Biceps.

The membranes of the deep fascia are organized into a continuous or never-ending system. The periosteum of the bones (and perichondrium of cartilage), and the ligaments may assist in establishing this continuity. The membranes split, on occasion, in order to surround (invest) muscles or other structures and unite again into single sheets. These phenomena of splitting and fusing are important because it is by their means that any sheet of fascia can be traced to any other sheet and can be shown to make eventual attachment to the skeleton.

The deep fascia, although a continuous system, may be subdivided for the purpose of description into three parts. First, the **outer investing layer** (*Deep Subcutaneous System of Gallaudet '31*) is an extensive sheet which covers the trunk, neck, limbs, and part of the head, and lies just under the subcutaneous superficial fascia. Second, in the trunk there is another extensive sheet, the **internal investing layer** (*Deep Subserous System of Gallaudet*), which covers the internal surface of the body wall, that is, it lines the thoracic and abdominal cavities and, in turn, is covered internally by the Subserous Fascia (see below). The third portion comprises the manifold **intermediate membranes** which are derived from the two investing layers by splitting and attachment, and which lie between the muscles and other structures throughout the body.

The mechanical function of fascia is particularly well developed in the deep fascia and is responsible for its many local variations and specializations. A membrane may be thickened, either for strength or padding; it may be fused with another membrane or split into several sheets; it may be separated from another membrane by a plane of cleavage; or it may combine with other membranes to form compart-

ments for groups of muscles. These specializations will be described in greater detail.

The thickening of a membrane for greater strength, especially if it receives the direct pull of a muscle, is by the addition of parallel bundles of collagenous fibers which impart to it the white, glistening appearance of an aponeurosis. A membrane of this type may lie between the origins of two muscles whose fibers pull in approximately the same direction, as, for example, the forearm muscles originating on the epicondyles of the humerus, in which case it is called an intermuscular septum. Such a strengthened membrane may cover a muscle and be used by it for a surface of attachment as in the case of the outer investing layer of the forearm. An extreme instance is that of the fascia lata of the thigh, whose iliotibial band is in fact the principal tendon of insertion for the Gluteus maximus and the Tensor fasciae latae muscles. This aponeurotic function of fascia has led to some confusion. Certain fibrous membranes retain the name fascia when they are actually aponeuroses and, conversely, some authors, particularly the French, are inclined to call even the unspecialized membranes aponeuroses.

A band of fascia may act as a ligament. This is the case with the greatly strengthened portion of the clavipectoral fascia, the costocoracoid membrane, which, by its attachment to the coracoid process and ribs, serves as a ligament for the articulations of the clavicle. Another important specialization is shown by the annular ligaments and retinacula at the wrist and ankle which provide tunnels for the long tendons of the hand and foot.

Lamination of a fascial membrane is found where there is a thickening without a corresponding increase in strength. The membrane splits into two leaves which are separated by a pad of connective tissue containing fat and an occasional blood vessel or lymph node. An example of this is the lamination of the outer cervical fascia above the sternum which is ambiguously called the suprasternal (Burns') space.

A **fascial compartment** is a portion of the body which is walled off by fascial membranes. Characteristically, it contains a muscle or a group of muscles but in some instances other structures are included. A typical example is offered by the flexor and extensor compartments of the arm, where the substance of the arm, enclosed by brachial fascia, is divided into the two compartments by the medial and lateral intermuscular septa. In many descriptions of fascia, such compartments are ambiguously called spaces or potential spaces. For example, the mediastinum is frequently called a "space" containing the heart, great vessels, esophagus, etc., whereas in reality it is a compartment enclosed by mediastinal fascia. Any "potential space" must be sought in the area of cleavage which separates the parietal pericardium from the sternum except at the pericardiosternal ligaments. Another confusing use of the word space was mentioned in the preceding paragraph, that is, to refer to a lamination and thickening of a fascial membrane.

The **fascial cleft** is an important specialization which is greatly in need of emphasis. It is a place of cleavage which separates two contiguous fascial surfaces. It may be described also as a stratum rich in fluid but poor in traversing fibers which allows two fascial surfaces to move more or less freely over each other and makes them easy to separate in dissection. The degree of separation may vary from an almost complete detachment to a comparatively strong adhesion, depending on the need for motion between the parts. The cleft between the superficial fascia and the deep fascia has already been mentioned. The cleft between a muscle and an overlying, restraining fascial membrane, like the Biceps and brachial fascia, is actually between the epimysium of the muscle and the true fascial sheet. The cleft between two adjacent muscles is likely to be between the simple epimysium of each muscle, but the latter may, in some instances, be thickened into a true fascial sheet.

A **bursa** represents the final step in the development of an efficient device for freedom of motion between contiguous connective tissue surfaces. It is a relatively small, circumscribed area in which all traversing fibers have been lost. The result is a pocket of complete separation, the lining of which provides two opposed, lubricated surfaces similar to the synovial membranes of a joint. Characteristically, a bursa is found where a tendon glides directly over the periosteum of a bone. The **synovial tendon sheaths** of the hand and foot are specialized bursae.

Just as there are adaptations for separation of membranes, there are adaptations for attachment. Fascial membranes may fuse with each other, as in the case of the outer investing and middle cervical fasciae near the hyoid bone. They may attach to bones, as the clavipectoral fascia to the clavicle. The attachment in some instances is very secure; in others it is separable by dissection. The relation between two contiguous fascial membranes, therefore, may vary from the complete separation at a bursa, or the functional separation of a fascial cleft, through progressive degrees of adhesion and attachment up to complete fusion.

The names of particular parts of the fascia are derived most commonly and most appropriately from the regions of the body which they occupy or the structures which they cover. For example, the brachial fascia encloses the arm and the deltoid fascia covers the Deltoideus muscle. Some fasciae have descriptive names, such as the fascia lata from its broad extent on the thigh or the fascia cribrosa from its many holes. A few are named from their attachments, for example, the fascia clavipectoralis or coracoclavicularis.

Eponyms which are taken from the names of persons who first described or emphasized particular portions or concepts of fascia are used very commonly by authors of clinical treatises. In most instances such nomenclature is of doubtful value anatomically, but is worthy of preservation because it emphasizes the importance of a certain fascia in operative procedures or in pathological processes such as the spread of infection. The identification of any specific fascial membrane is intrinsically difficult because the fascia has the same histological structure as the ligament, aponeurosis, and periosteum to which it may be attached, and the continuity of the whole system makes the setting of exact boundaries and limitations a matter of arbitrary definition.

The **Subserous** or **Visceral Fascia** (*tela subserosa; superficial subserous fascia of Gallaudet*) lies between the internal investing layer of deep fascia and the serous membranes lining the body cavities, in much the same way as the subcutaneous fascia lies between the skin and the deep fascia. It is very thin in some areas, as between the pleura and the chest wall. It is thick in other areas and, except in emaciated individuals, forms a pad of adipose tissue like that surrounding the kidney. It is not separable into outer and inner layers, as is the superficial fascia, but it may be irregularly laminated in the adipose accumulations, especially in well-nourished individuals. As a general rule, only the fascia of the parietal serous membrane is given in a description, but it should be remembered that this parietal layer is continuous with the visceral layer carried over to the organs at the reflections of the serous membranes and at the mesenteries.

A cleft of variable prominence separates the subserous fascia from the deep fascia as it does in the case of the subcutaneous and deep fasciae. It allows a considerable amount of sliding motion between the two fasciae and makes it possible to dissect them apart. Where both fasciae are thin and delicate, it is difficult to identify and separate them.

Complications and problems of identification and naming of fasciae arise in the regions where internal structures penetrate the wall of the body cavity. For example, the rectum penetrates the pelvic diaphragm, the spermatic cord penetrates the abdominal wall through the inguinal canal, and the trachea and esophagus leave the thoracic cavity to enter the neck, and, in so doing, they introduce transitions between the subserous, deep, and subcutaneous fasciae.

Careful study of the fascia has been made, for the most part, in restricted areas instead of throughout the body as a complete system. Interest has stemmed from its obvious importance in surgery and pathology. It is logical, however, to weave the fascia into a functional system. Its function is predominantly mechanical in the normal body, except for the activity of the various types of cells which are visitors within its meshes and which are beyond the scope of this discussion. The understanding and learning of the fasciae are much easier on the basis of this mechanical function than on the basis of surgical and pathological impor-

tance, and the latter becomes easily comprehensible only with the knowledge of function as a background. One very important mechanical function must not be overlooked even if it is seldom mentioned, namely, that of supporting and carrying the blood vessels, nerves, and lymphatics.

Although much advantage might be gained by presenting the fasciae in a separate chapter, it has been decided to retain the usual method of describing them with the muscles. Should the reader desire a systematic treatment of the fasciae, he may obtain it by leafing through the pages of the chapter on muscles, reading only the sections on fasciae. A study of the muscles or other structures in a particular region is recommended as a preliminary to the reading of the description of the fascia.

STRUCTURE AND ATTACHMENT OF MUSCLES

The tendons and aponeuroses are parts of the muscles. They are included in the description of a muscle as the ultimate attachment, that is, the origin and insertion of the muscle are the terminal attachments of the tendon or aponeurosis to a bone, not the attachment of the contractile fibers to the tendon.

The **arrangement of the fasciculi,** and the manner in which they approach the tendons has many variations. In some muscles, the fasciculi are parallel with the longitudinal axis and terminate at either end in flat tendons. In others, the fasciculi converge, like the plumes of a feather, to one side of a tendon which runs the entire length of a muscle, forming a **penniform** muscle like the Semimembranosus. If they converge to both sides of a tendon, as in the Rectus femoris, they are called **bipenniform,** or if they converge to several tendons, as in the Deltoideus, they are called **multipenniform.** The fasciculi may converge from a broad surface to a narrow tendinous point, as in the Temporalis, and be called **radiated.**

This arrangement of fasciculi is correlated with the power of the muscles. Those with comparatively few fasciculi, extending the length of the muscle, have a greater range of motion but not as much power. Penniform muscles, with a large number of fasciculi distributed along their tendons, have greater power but smaller range of motion.

The names of the muscles have been derived from: (a) their situation, as the Brachialis, Pectoralis, Supraspinatus; (b) their direction, as the Rectus, Obliquus, and Transversus abdominis; (c) their action, as Flexors, Extensors; (d) their shape, as the Deltoideus, Trapezius, Rhomboideus; (e) the number of divisions, as the Biceps, Triceps, Quadriceps; (f) their points of attachment, as the Sternocleidomastoideus, Omohyoideus.

Origin and Insertion of Muscles.—The attachments of the two ends of a muscle are called the origin and the insertion. In the text it is customary to describe the muscle as *arising* from the origin and *inserting* at the insertion. The origin is the more fixed and proximal end, the insertion the more movable and distal end. For example, the Pectoralis major arises or has its origin from the ribs and clavicle, and its insertion is into the humerus. If the individual were climbing a tree, however, the origin and insertion might seem to be reversed, and the hand be more fixed and the body more movable. The designations of the origins and insertions in the following descriptions are more or less arbitrary, therefore, and a matter of convention among anatomists.

Illustrations of areas of attachment of the muscles are given with the descriptions of the bones and should be consulted constantly. The origins are marked in red, the insertions in blue.

MUSCLE ACTION

When a muscle contracts, it acts upon movable parts to bring about certain movements. These actions of the muscle should be studied from three points of view: (a) *individual action*, (b) *group actions*, and (c) *action correlated with the nerve supply*. The individual action is closely associated with the anatomy of a muscle because mechanically the action is the direct result of the attachment of its two ends. It is not necessarily true, however, that the action in the living body is the same as that deduced from observing its attachments, nor even from pulling on it in a dead subject,

because incidental actions may not be utilized or may even be suppressed in the living body. A knowledge of individual action is of practical value to a surgeon in the diagnosis and treatment of displacements due to fractures. Group actions are related to the functions as well as the anatomy of the muscles. It is seldom possible for a person to make a single muscle contract at will. In other words, the movements, not the muscles, are represented in the central nervous system. A muscle may be associated with one group for one action and a different group for another, possibly even antagonistic action. A correlation of the knowledge of the action with the nerve supply is of practical value in the diagnosis of lesions of the peripheral and central nervous system, and in the treatment of such lesions. Frequently there is a correspondence between groups of muscles arranged according to nerve supply and those arranged according to common actions.

Practically every muscle acting upon a joint is matched by another muscle which has an opposite action. Each muscle of such a pair is the **antagonist** of the other, for example, the Biceps brachii, a flexor, and the Triceps, an extensor, are antagonists at the elbow. The performance of most movements requires the combined action of a number of muscles; those which act directly to bring about the desired movement are called the **prime movers**; those which act to hold the part of the body in an appropriate position are called **fixation muscles**. It happens frequently that the prime movers have actions other than the one desired, in which case the antagonists of the undesired action come into play; these are the **synergists**. For example, in closing the fist, the prime movers are the Flexores digitorum sublimis and profundus, the Flexor pollicis longus, and the small muscles of the thumb; the fixation muscles are the Triceps, Biceps, Brachialis, and the muscles about the shoulder which hold the arm in position; the synergetic muscles are the Extensores carpi radialis and ulnaris, which prevent flexion of the wrist. In some instances, when an act is performed with extreme force, muscles which are not required for a moderate performance come to the assistance of the prime movers, and these are known as *emergency muscles*. For example, the flexors of the fingers may flex the wrist in emergency. A further point which must be borne in mind is that the force of gravity may be the prime mover, in which case, the antagonists of the muscles which might be expected to be the prime movers are the muscles which act, and they do so to retard and control the movement caused by gravity.

Individual muscles cannot always be treated as single mechanical units with regard to their actions. Different parts of the same muscle may have different and even antagonistic actions; for example, the anterior part of the Deltoideus flexes, but the posterior portion extends the arm. Two adjacent muscles like the Infraspinatus and the Teres minor, on the other hand, may have the same action.

No study of muscles is complete without observations of the muscles in their normal positions in the living body. It is recommended that students find an opportunity to make this study. The surface markings associated with the muscles are illustrated in the second chapter of the book.

THE MUSCLES AND FASCIAE OF THE HEAD

The muscles of the head may be arranged in groups, of which the following two will be described in this chapter:

I. The Facial Muscles
II. The Muscles of Mastication

In addition to these two groups, other muscles occupying positions in the head are described in other, more appropriate parts of the book: (1) The Ocular Muscles; (2) The Muscles of the Auditory Ossicles; (3) The Muscles of the Tongue; and (4) The Muscles of the Pharynx.

I. THE FACIAL MUSCLES

The facial muscles (*muscles of expression*) are cutaneous muscles, lying within the layers of subcutaneous fascia. In general, they arise either from fascia or from the bones of the face, and insert into the skin.

The individual muscles seldom remain separate and distinct throughout their length because of a tendency to merge with their neighbors at their terminations or attachments. They may be grouped into: (1) the muscles of the scalp; (2) the extrinsic muscles of the ear; (3) the muscles of the eyelid; (4) the muscles of the nose; (5) the muscles of the mouth. An additional muscle, the Platysma, really belongs to the facial group but will be described with the muscles of the neck.

1. The Muscles and Fasciae of the Scalp

Epicranius
Occipitofrontalis
Temporoparietalis

The **Subcutaneous fascia** (*tela subcutanea*) of the head invests the facial muscles and carries the superficial blood vessels and nerves. It varies considerably in thickness and texture in different areas but everywhere has an abundant blood and nerve supply. Above the superior nuchal and temporal lines, and anterior to the Masseter muscle there is no deep fascia underlying it other than the periosteum of the bones. Under the scalp, the subcutaneous fascia is very thick and tough, and over the cranial vault, the muscular stratum is represented by the broad epicranial aponeurosis or galea aponeurotica. A fascial cleft such as that commonly found under the subcutaneous fascia in the rest of the body is very prominent in this region and separates the galea from the pericranium or cranial periosteum. It accounts for the movability of the scalp and makes possible the sudden accumulation of large amounts of blood in the hematomas following blows upon the head. Over the forehead, the superficial layers of the fascia are much thinner than in the scalp and the skin is closely attached to the underlying Frontalis. Over the eyelids, it is devoid of fat and is composed of a loose areolar tissue which is easily distended and infiltrated with tissue fluid in edema, or blood in ecchymosis or hemorrhage. On the cheeks and lips, it contains a considerable amount of fat and is tougher and more fibrous, especially in men. The subcutaneous fascia is reduced over the cartilages of the nose and external ear, the skin being closely bound to the underlying perichondrium. The subcutaneous fascia of the face is directly continuous over the mandible with that of the neck, and that of the scalp merges posteriorly with the similar fibrous layer of the back of the neck.

The skin of the scalp is thicker than in any other part of the body. The hair follicles are closely set together, have numerous sebaceous glands, and extend deeply into the subcutaneous fascia. The subcutaneous fat is broken up into granular lobules, and is mattressed into a firm layer by the many

fibrous bands which secure the skin to the deeper layers of the subcutaneous fascia.

The **Epicranius** (Fig. 6–1) is a broad muscular and tendinous layer which covers the top and sides of the skull from the occipital bone to the eyebrow. It consists of thin, broad, muscular bellies, connected by an extensive intermediate aponeurosis, the galea aponeurotica. The **occipital belly** (*Venter occipitalis; Occipitalis*), quadrilateral in form, *arises* by short tendinous fibers from the lateral two-thirds of the superior nuchal line of the occipital bone, and from the mastoid part of the temporal. The muscular fasciculi ascend in a parallel

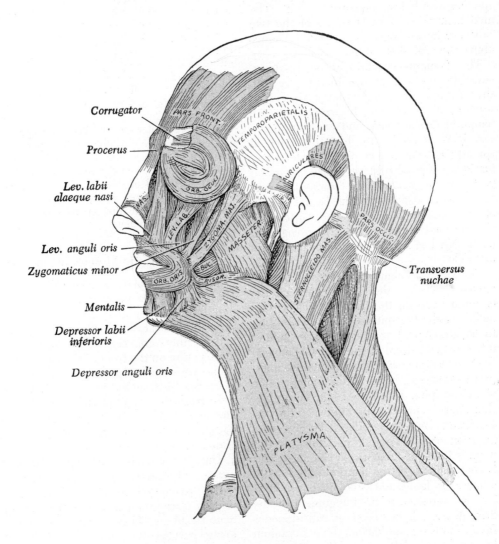

Fig. 6–1.—Muscles of the head, face, and neck.

course toward the vertex and *end* in the galea aponeurotica. Between the muscles of the two sides there is a considerable, though variable, interval which is occupied by a prolongation of the galea. The **frontal belly** (*Venter frontalis; Frontalis*), also quadrilateral in form, is broader than the occipital belly and its fasciculi are longer, finer and paler in color. It has no bony attachments. Its medial fibers are continuous with those of the procerus; its intermediate fibers blend with the Corrugator and Orbicularis oculi; and its lateral fibers are blended with the latter muscle also, over the zygomatic process of the frontal bone. The fibers are directed upward, and join the galea aponeurotica below the coronal suture. The medial margins of the muscles of the two sides are joined together for some distance above the root of the nose.

The **Temporoparietalis** is newly designated in the NA. It is a broad very thin sheet formerly included with the Auricularis superior and anterior. It *arises* from the temporal fascia above and anterior to the ear. It is divided into three parts which spread out like a fan over the temporal fascia, a temporal part anteriorly, a parietal part superiorly, with a triangular part between. It *inserts* into the lateral border of the galea aponeurotica. The Auriculares anterior and superior are still recognized in the NA but refer to small muscle bellies more closely associated with the auricula.

The **galea aponeurotica** (*epicranial aponeurosis*) covers the upper part of the cranium between the frontal and occipital bellies of the Occipitofrontalis. In addition to its attachment to these muscle bellies, it is attached behind, in the interval between the two Occipitales, to the external occipital protuberance and the highest nuchal line of the occipital bone. In front, it forms a short, narrow prolongation between the two Frontales. On either side, it receives the insertion of the Temporoparietalis; at this point it loses its aponeurotic character, and is continuous over the temporal fascia with a layer of laminated areolar tissue. It is closely connected to the integument by the firm, dense, adipose layer of subcutaneous fascia and is separated from the pericranium by the fascial cleft which allows the apo-

neurosis, carrying with it the integument, to move the scalp through a considerable distance.

Action.—The occipital and frontal bellies of the Occipitofrontalis acting together draw the scalp back raising the eyebrows and wrinkling the forehead as in an expression of surprise. The frontal bellies acting alone raise the eyebrows, either on one or both sides. The Temporoparietalis tightens the scalp and draws back the skin of the temples to combine with the Occipitofrontalis in wrinkling the forehead and widening the eyes in an expression of fright or horror. The Temporoparietalis raises the auricula.

Nerves.—The frontal belly and the Temporoparietalis are supplied by the temporal branches, and the occipital belly by the posterior auricular branch of the facial nerve.

Variations.—Both frontal and occipital bellies may vary considerably in size and in extent; either may be absent; the muscles of the two sides may fuse in the middle line; the frontal bellies may interdigitate across the line; the occipital belly may fuse with the Auricularis posterior.

A thin muscular slip, the **Transversus nuchae** or **Occipitalis minor**, is present in 25 per cent of the bodies; it arises from the external occipital protuberance or from the superior nuchal line, either superficial or deep to the Trapezius; it is frequently inserted with the Auricularis posterior, but may join the posterior edge of the Sternocleidomastoideus.

2. The Extrinsic Muscles of the Ear

The **Auricularis anterior** (*Attrahens aurem*) (Fig. 13–44) is thin, pale, delicate, and indistinct. It *arises* from the anterior portion of the fascia in the temporal area, and its fibers converge to be *inserted* into a projection on the front of the helix.

The **Auricularis superior** (*Attollens aurem*) (Fig. 6–1) is thin and fan-shaped. Its fibers *arise* from the fascia of the temporal area, and converge to be *inserted* by a thin flattened tendon into the upper part of the cranial surface of the auricula.

The **Auricularis posterior** (*Retrahens aurem*) (Fig. 6–1) consists of two or three fleshy fasciculi which *arise* from the mastoid portion of the temporal bone by short aponeurotic fibers. They are *inserted* into the lower part of the cranial surface of the concha (see Transversus nuchae above).

Actions.—The Auricularis anterior draws the auricula forward and upward, the Auricularis superior draws it upward, and the posterior draws it backward. In man these muscles seem to act more in conjunction with the Occipitofrontalis to move the scalp than to move the auricula, but in some individuals they can be used to execute voluntary movements of the auricula.

Nerves.—The Auriculares anterior and superior are supplied by the temporal branches, the Auricularis posterior by the posterior auricular branch of the facial nerve.

Variations.—The auricular muscles vary greatly in thickness and extent or rarely may be absent.

3. The Muscles of the Eyelids

Levator palpebrae superioris
Orbicularis oculi
Corrugator

The Levator palpebrae superioris is described with the Anatomy of the Eye.

The **Orbicularis oculi** (*Orbicularis palpebrarum*) (Fig. 6–1) *arises* from the nasal part of the frontal bone, from the frontal process of the maxilla in front of the lacrimal groove, and from the anterior surface and borders of a short fibrous band, the **medial palpebral ligament**. From this origin, the fibers are directed lateralward, forming a broad and thin layer, which occupies the eyelids or palpebrae, surrounds the circumference of the orbit, and spreads over the temple, and downward on the cheek. The **palpebral portion** of the muscle is thin and pale; it *arises* from the bifurcation of the medial palpebral ligament, forms a series of concentric curves, and is *inserted* into the lateral palpebral raphe. The **orbital portion** is thicker and of a reddish color; its fibers form a complete ellipse without interruption at the lateral palpebral commissure; the upper fibers of this portion blend with the Frontalis and Corrugator. The **lacrimal part** (*Tensor tarsi*) is a small, thin muscle, about 6 mm in breadth and 12 mm in length, situated behind the medial palpebral ligament and lacrimal sac (Fig. 6–2). It *arises* from the posterior crest and adjacent part of the orbital surface of the lacrimal bone,

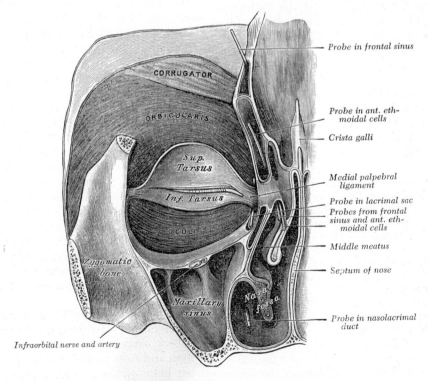

FIG. 6–2.—Left orbicularis oculi, seen from behind.

and passing behind the lacrimal sac, divides into two slips, upper and lower, which are *inserted* into the superior and inferior tarsi medial to the puncta lacrimalia; occasionally it is very indistinct.

The **medial palpebral ligament** (*tendo oculi*), about 4 mm in length and 2 mm in breadth, is attached to the frontal process of the maxilla in front of the lacrimal groove. Crossing the lacrimal sac, it divides into two parts, upper and lower, each attached to the medial end of the corresponding tarsus. As the ligament crosses the lacrimal sac, a strong aponeurotic lamina is given off from its posterior surface; this expands over the sac, and is attached to the posterior lacrimal crest.

The **lateral palpebral raphe** is a much weaker structure than the medial palpebral ligament. It is attached to the margin of the frontosphenoidal process of the zygomatic bone, and passes medialward to the lateral commissure of the eyelids, where it divides into two slips, which are attached to the margins of the respective tarsi.

The **Corrugator supercilii** (*Corrugator*) is a small, narrow, pyramidal muscle, placed at the medial end of the eyebrow, beneath the Frontalis and Orbicularis oculi. It *arises* from the medial end of the superciliary arch, and its fibers pass upward and lateralward, between the palpebral and orbital portions of the Orbicularis oculi, and are *inserted* into the deep surface of the skin, above the middle of the orbital arch. The **depressor supercilii** is the name given to the part of the muscle which draws the eyebrows downward.

Actions.—The Orbicularis oculi is the sphincter muscle of the eyelids. The palpebral portion closes the lids gently, as in blinking or in sleep; the orbital portion is used as well in stronger closing, like winking with one eye. When the entire muscle is brought into action, the skin of the forehead, temple, and cheek is drawn toward the medial angle of the orbit, and the eyelids are firmly closed, as in photophobia. The skin thus drawn upon is thrown into folds, especially radiating from the lateral angle of the eyelids; these folds become permanent in old age, and form the so-called "crow's feet." The Levator palpebrae superioris is the direct antagonist of this muscle; it raises the upper eyelid and exposes the front of the bulb of the

eye. Each time the eyelids are closed through the action of the Orbicularis, the medial palpebral ligament is tightened, the wall of the lacrimal sac is thus drawn lateralward and forward, so that a vacuum is made in it, and the tears are sucked along the lacrimal canals into it. The lacrimal part of the Orbicularis oculi draws the eyelids and the ends of the lacrimal canals medialward and compresses them against the surface of the globe of the eye, thus placing them in the most favorable situation for receiving the tears; it also compresses the lacrimal sac. The Corrugator draws the eyebrow downward and medialward, producing the vertical wrinkles of the forehead. It is the "frowning" muscle, and may be regarded as the principal muscle in the expression of suffering.

Nerves.—Nerves from the temporal and zygomatic branches of the facial nerve.

Variations.—The Orbicularis varies in extent and may be fused with neighboring muscles.

4. The Muscles of the Nose (Fig. 6–1)

> Procerus
> Nasalis
> Depressor septi

The **Procerus** (*Pyramidalis nasi*) is a small pyramidal slip *arising* by tendinous fibers from the fascia covering the lower part of the nasal bone and upper part of the lateral nasal cartilage; it is *inserted* into the skin over the lower part of the forehead between the two eyebrows, its fibers decussating with those of the Frontalis.

The **Nasalis** consists of two parts, transverse and alar. The **transverse part** (*Compressor naris*) *arises* from the maxilla, above and lateral to the incisive fossa; its fibers proceed upward and medialward, expanding into a thin aponeurosis which is continuous on the bridge of the nose with that of the muscle of the opposite side, and with the aponeurosis of the Procerus. The **alar part** (*Dilatator naris*) is attached by one end to the greater alar cartilage, and by the other to the integument at the point of the nose.

The **Depressor septi** (*Depressor alae nasi*) *arises* from the incisive fossa of the maxilla; its fibers ascend to be *inserted* into the septum and back part of the ala of the nose. It lies between the mucous membrane and muscular structure of the lip.

Actions.—The Procerus draws down the medial angle of the eyebrows and produces transverse wrinkles over the bridge of the nose. The alar part of the nasalis (*Dilatator naris*) enlarges the aperture of the nares. Its action in ordinary breathing is to resist the tendency of the nostrils to close from atmospheric pressure, but in difficult breathing, as well as in some emotions, such as anger, it contracts strongly. The Depressor septi is a direct antagonist of the other muscles of the nose, drawing the ala of the nose downward, and thereby constricting the aperture of the nares. The transverse part of the Nasalis depresses the cartilaginous part of the nose and draws the ala toward the septum.

Nerves.—Nerves from the buccal branches of the facial nerve.

Variations.—These muscles vary in size and strength or may be absent.

5. The Muscles of the Mouth (Fig. 6–1)

Levator labii superioris
Levator labii superioris alaeque nasi
Levator anguli oris
Zygomaticus minor
Zygomaticus major
Risorius
Depressor labii inferioris
Depressor anguli oris
Mentalis
Transversus menti
Orbicularis oris
Buccinator

The **Levator labii superioris** (*Quadratus labii superioris, infraorbital head*) has a rather broad *origin* from the lower margin of the orbit immediately above the infraorbital foramen, some of its fibers being attached to the maxilla, others to the zygomatic bone. Its fibers converge to be *inserted* into the muscular substance of the upper lip between the Levator anguli oris and the Levator labii superioris alaeque nasi.

The **Levator labii superioris alaeque nasi** (*Quadratus labii superioris, angular head*) *arises* by a pointed extremity from the upper part of the frontal process of the maxilla and passing obliquely downward and lateralward divides into two slips. One of these is *inserted* into the greater alar cartilage and skin of the nose; the other is prolonged into the upper lip, blending with the Levator labii superioris.

The **Levator anguli oris** (*Caninus*) *arises*

from the canine fossa, immediately below the infraorbital foramen; its fibers are *inserted* into the angle of the mouth, intermingling with those of the Zygomaticus major, Depressor anguli oris, and Orbicularis oris.

The **Zygomaticus minor** (*Quadratus labii superioris, zygomatic head*) *arises* from the malar surface of the zygomatic bone immediately behind the zygomaticomaxillary suture and passes downward and medialward as a narrow slip to be *inserted* into the upper lip between the Levator labii superioris and the Zygomaticus major.

The **Zygomaticus major** (*Zygomaticus*) *arises* from the zygomatic bone, in front of the zygomaticotemporal suture, and descending obliquely with a medial inclination, is *inserted* into the angle of the mouth, where it blends with the fibers of the Levator and Depressor anguli oris and Orbicularis oris.

Actions.—The Levator labii superioris is the proper elevator of the upper lip, carrying it at the same time a little forward. The Levator labii superioris alaeque nasi also dilates the naris, and together with the former and the Zygomaticus minor forms the nasolabial furrow which is deepened in expressions of sadness. When these three muscles act in conjunction with the Levator anguli oris the furrow is deepened into an expression of contempt or disdain. The Zygomaticus major draws the angle of the mouth upward and backward in laughing.

Nerves.—Nerves from the buccal branches of the facial nerve.

Variations.—These muscles, especially the Zygomatic minor, vary in extent and the degree of fusion with each other or with neighboring muscles.

The **Risorius** *arises* in the fascia over the Masseter and, passing horizontally forward, superficial to the Platysma, is *inserted* into the skin at the angle of the mouth (Fig. 6–1).

The **Depressor labii inferioris** (*Quadratus labii inferioris; Quadratus menti*) is a small quadrilateral muscle. It *arises* from the oblique line of the mandible, between the symphysis and the mental foramen, and passes upward and medialward, to be *inserted* into the integument of the lower lip, its fibers blending with the Orbicularis oris, and with those of its fellow of the opposite side. At its origin it is continuous with the fibers of the Platysma. Much yellow fat is intermingled with the fibers of this muscle.

The **Depressor anguli oris** (*Triangularis*) *arises* from the oblique line of the mandible, whence its fibers converge, to be *inserted*, by a narrow fasciculus, into the angle of the mouth. At its origin it is continuous with the Platysma, and at its insertion with the Orbicularis oris and Risorius; some of its fibers are directly continuous with those of the Levator anguli oris.

The **Mentalis** (*Levator menti*) is a small conical fasciculus, situated at the side of the frenulum of the lower lip. It *arises* from the incisive fossa of the mandible, and descends to be *inserted* into the integument of the chin.

The **Transversus menti,** found in more than half the bodies, is a small muscle which crosses the midline just under the chin. It is frequently continuous with the Depressor anguli oris.

Actions.—The Risorius retracts the angle of the mouth. The Depressor labii inferioris draws the lower lip directly downward and a little lateralward, as in the expression of irony. The Depressor anguli oris depresses the angle of the mouth, being the antagonist of the Levator anguli oris and Zygomaticus major; acting with the Levator, it draws the angle of the mouth medialward. The Mentalis raises and protrudes the lower lip, and at the same time wrinkles the skin of the chin, as in pouting or expressing doubt or disdain. The Platysma (page 390) acts with this group, retracting and depressing the angle of the mouth.

Nerves.—Nerves from the mandibular and buccal branches of the facial nerve.

Variations.—The Risorius varies greatly; it may be absent, doubled, greatly enlarged, or blended with the Platysma. The Depressor labii inferioris is continuous with the Platysma to a greater or lesser extent. The Mentalis varies in size and connection with the Platysma. The Depressor anguli oris may be in two or three separate parts; its anterior fibers may cross under the chin to join the Transversus menti.

The **Orbicularis oris** (Fig. 6–3) is not a simple sphincter muscle like the Orbicularis oculi; it consists of numerous strata of muscular fibers surrounding the orifice of the mouth but having different direction. It consists partly of fibers derived from the other facial muscles which are inserted into the lips, and partly of fibers proper to the

lips. Of the former, a considerable number are derived from the Buccinator and form the deeper stratum of the Orbicularis. Some of the Buccinator fibers—namely, those near the middle of the muscle—decussate at the angle of the mouth, those arising from the maxilla passing to the lower lip, and those from the mandible to the upper lip. The uppermost and lowermost fibers of the Buccinator pass across the lips from side to side without decussation. Superficial to this stratum is a second, formed on either side

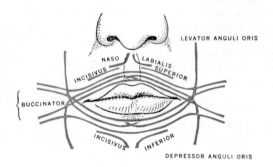

FIG. 6–3.—Scheme showing arrangement of fibers of Orbicularis oris.

by the Levator and Depressor anguli oris which cross each other at the angle of the mouth; those from the Levator passing to the lower lip, and those from the Depressor to the upper lip, along which they run, to be inserted into the skin near the median line. In addition to these there are fibers from the Levator labii superioris, the Zygomaticus major, and the Depressor labii inferioris; these intermingle with the transverse fibers described above, and have principally an oblique direction. The proper fibers of the lips are oblique, and pass from the under surface of the skin to the mucous membrane through the thickness of the lip. Finally there are fibers by which the muscle is connected with the maxillae and the septum of the nose above and with the mandible below. In the upper lip these consist of two bands, lateral and medial, on either side of the middle line; the **lateral band** (*m. incisivus labii superioris*) *arises* from the alveolar border of the maxilla, opposite the lateral incisor tooth, and arching lateralward is continuous with the other muscles at the angle of the mouth; the **medial band** (*m. nasolabialis*) connects the upper lip to the back of the septum of the nose. The interval between the two medial bands corresponds with the depression, called the **philtrum**, seen on the lip beneath the septum of the nose. The additional fibers for the lower lip constitute a slip (*m. incisivus labii inferioris*) on either side of the middle line; this arises from the mandible, lateral to the Mentalis, and intermingles with the other muscles at the angle of the mouth.

Actions.—The Orbicularis oris in its ordinary action effects the direct closure of the lips; by

its deep fibers, assisted by the oblique ones, it closely applies the lips to the alveolar arch. The superficial part, consisting principally of the decussating fibers, brings the lips together and also protrudes them forward.

Nerves.—The buccal branches of the facial nerve.

The **Buccinator** (Fig. 6–4) is the principal muscle of the cheek and forms the lateral wall of the oral cavity. It lies deeper than the other facial muscles, is quadrilateral in form, and occupies the interval between the maxilla and mandible, lateral to the teeth.

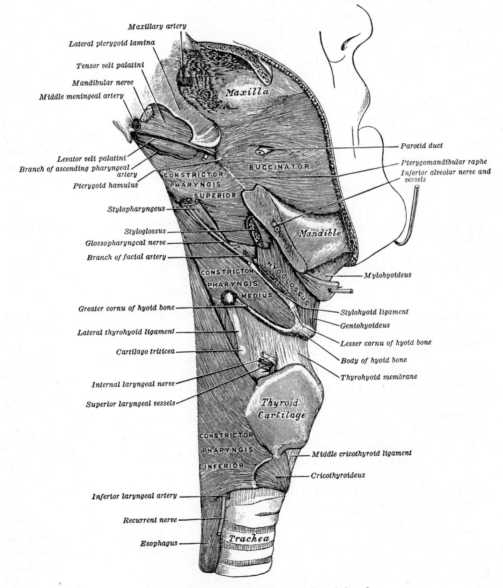

FIG. 6–4.—The Buccinator and the muscles of the pharynx.

It *arises* from the outer surfaces of the alveolar processes of the maxilla above and mandible below, alongside of the three molar teeth, and in between, from the pterygomandibular raphe, a tendinous inscription giving origin to both the Buccinator and the Constrictor pharyngis superior. The fibers of the upper and lower portions follow a slightly converging course forward, and *insert* by blending with the deeper stratum of muscle fibers in the corresponding lips. The fibers of the central portion converge toward the angle of the mouth and decussate, those from above becoming continuous with the Orbicularis oris of the lower lip, those from below with that of the upper lip. The superficial surface of the Buccinator is covered by the **buccopharyngeal fascia** and the buccal fat pad; the deep surface is in relation with the buccal glands and mucous membrane of the mouth. It is pierced by the duct of the parotid gland opposite the upper second molar tooth.

Action.—The Buccinator compresses the cheek and is, therefore, an important accessory muscle of mastication, holding the food under the immediate pressure of the teeth. When the cheeks have been distended with air, the Buccinators compress it and tend to force it out between the lips as in blowing a trumpet (Latin buccinator, a trumpet player). Electromyographic studies show that the Orbicularis oris must act with the Buccinator to compress the cheek (De Sousa '65).

Nerve.—The motor fibers to the Buccinator come from the facial nerve through its buccal branches. The buccal nerve (from the trigeminal and formerly called the buccinator nerve) is sensory only in this area.

The **Pterygomandibular Raphe** (*pterygomandibular ligament*) (Fig. 6–4) is a tendinous inscription between the Buccinator and the Constrictor pharyngis superior which gives origin to the middle portion of both muscles. Except for this tendinous interruption, the Constrictor, Buccinator and Orbicularis oris would form a continuous sphincter-like band of muscle. The raphe is held in place by its attachment superiorly to the pterygoid hamulus and inferiorly to the posterior end of the mylohyoid line of the mandible. Its relations are the same as those of the two muscles, its medial surface being covered by the mucous

membrane of the mouth and its lateral surface being separated from the ramus of the mandible by a quantity of adipose tissue, the buccal fat pad.

The **buccopharyngeal fascia** covers the lateral surface of the Buccinator, the raphe, and the Constrictor muscles, providing a facial surface for the fascial cleft which allows these muscles to move freely.

Buccal Fat Pad (*corpus adiposum buccae, suctorial pad*), a circumscribed or encapsulated mass of fat, lies superficial to the Buccinator at the anterior border of the Masseter. A well-defined fascial cleft separates it from the subcutaneous fascia and facial muscles. From this main mass of adipose tissue, narrow prolongations extend deeply between the Masseter and the Temporalis and upward under the deep temporal fascia. Some of the tissue continues still more deeply into the infratemporal fossa, separating the Pterygoideus lateralis and Temporalis from the maxilla, and filling in between the various structures and the bony fossae. The mass superficial to the Buccinator is particularly prominent in infants, and is called the *suctorial pad* because it is supposed to assist in the act of sucking.

II. THE MUSCLES OF MASTICATION

Temporalis
Masseter
Pterygoideus medialis
Pterygoideus lateralis

The **Temporal Fascia** (*fascia temporalis*) (Fig. 6–7) is a strong, fibrous sheet, aponeurotic in appearance, which covers the Temporalis and is used by it for the attachment of its fibers. It is the most cranial extension of the deep fascia; farther cranially, the deep fascia is represented only by the pericranium. It is covered by the skin and tela subcutanea which includes the Galea aponeurotica and Auricularis superior, the Orbicularis oculi anteriorly, and just anterior to the ear it is crossed by the superficial temporal vessels and auriculotemporal nerve. Its uppermost portion is a thin, single sheet, attached to the entire extent of the superior temporal line. Its lower portion, near the attachment to the zygomatic arch, is thickened and laminated. The inner leaf (*lamina profunda*) ends by attaching to the medial border of the arch; the outer leaf (*lamina superficialis*), after attaching to the lateral border, is continued inferior to the arch as the masseteric fascia. Between the leaves is a small quantity of fat, the orbital branch of the superficial temporal artery, and a filament from the zygomatic branch of the maxillary nerve.

The **parotideomasseteric fascia** (*fascia parotideomasseterica*) (Fig. 6–7) covers the lateral surface of the Masseter and splits to enclose the parotid gland. It is attached to the zygomatic arch cranially, is continuous with the suprahyoid portion of the cervical fascia more caudally, and with the cervical fascia over the Sternocleidomastoideus posteriorly. The sheet which covers the superficial surface of the gland is fused with the dense and tough subcutaneous fascia. It is intimately mingled with its capsule, and sends numerous irregular septa into its substance so that this gland cannot be shelled out, as can the submandibular gland. The layer on the deep surface of the gland follows this surface behind the ramus of the mandible and there fuses with the fascia of the posterior belly of the Digastricus into a strong band, the stylomandibular ligament. The fascia covering the Masseter, the **masseteric fascia**, terminates anteriorly by encircling the ramus of the mandible and becoming continuous with the fascia of the Pterygoideus medialis deep to the bone. The masseteric fascia is attached to the border of the mandible inferiorly and posteriorly, completing a compartment which encloses the muscle except at its upper, deep portion where there is a communication with the tissue spaces about the insertion of the Temporalis.

The **pterygoid fascia** invests the Pterygoideus medialis and lateralis muscles. It is continuous below, at the angle of the mandible, with the masseteric and investing cervical fascias where the latter are attached to the bone. In this region also, it is continuous with the thickened band known as the stylomandibular ligament. It extends upward and forward along the deep surface of the Pterygoideus medialis to be attached with the origin of the muscle to the pterygoid process of the sphenoid bone. This sheet of fascia is attached to the mandible at both the borders of the inferior half of the muscle, but as the muscle angles away from the mandible toward its origin, the fascia wraps around the muscle forming a sheet on its superficial surface. This superficial sheet, continuing upward, splits to invest the Pterygoideus lateralis and is attached to the skull with the origin of this muscle. The fascia between the two Pterygoidei is attached to the skull along a line extending from the lateral pterygoid plate to the spine of the sphenoid bone. The part attached to the spine is thickened into a strong band which is attached below to the lingula of the mandible, forming the sphenomandibular ligament (page 300). Another band, the **pterygospinous ligament,** extends from the spine, between the two Pterygoidei, to the posterior margin of the lateral pterygoid plate. Occasionally this band is ossified, creating, between its upper border and the skull, a **pterygospinous foramen** which transmits the branches of the mandibular division of the trigeminal nerve to the muscles of mastication. Between the sphenomandibular ligament and the neck of the mandible, there is an interval which affords a passage for the maxillary vessels into the infratemporal fossa. The fascia on the surface of the Pterygoideus lateralis is in relation with the pterygoid plexus of veins. Deep to the pterygoid and deep temporal fasciae the layer of soft adipose tissue which is an extension of the buccal fat pad separates these fasciae from the buccopharyngeal fascia and neighboring structures.

The masticator compartment contains the four muscles of mastication and the ramus and posterior part of the body of the mandible. It is enclosed superficially by the masseteric and temporal fasciae and deeply by the pterygoid and deep temporal fasciae.

The **Temporalis** (*Temporal muscle*) (Fig. 6–5) is a broad, radiating muscle, situated at the side of the head. It *arises* from the whole of the temporal fossa and from the deep surface of the temporal fascia. Its fibers converge as they descend, and end in a tendon which passes deep to the zygomatic arch and is *inserted* into the medial surface, apex, and anterior border of the coronoid process, and the anterior border of the ramus of the mandible nearly as far anteriorly as the last molar tooth.

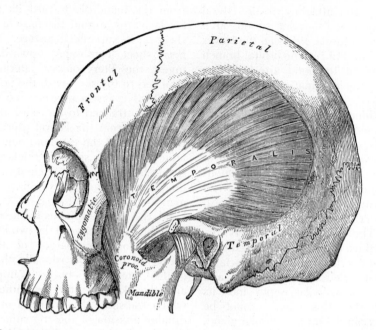

Fig. 6–5.—The Temporalis; the zygomatic arch and Masseter have been removed.

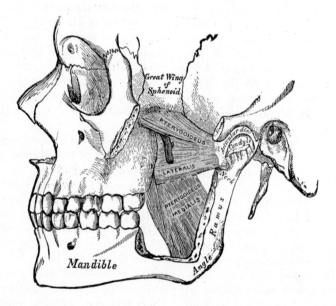

Fig. 6–6.—The Pterygoidei; the zygomatic arch and a portion of the ramus of the mandible have been removed, showing position of the maxillary artery.

Action.—Closes the jaws. The posterior portion retracts the mandible.

Nerves.—Anterior and posterior deep temporal nerves from the mandibular division of the trigeminal nerve.

The **Masseter** (Fig. 6–1) is a thick, somewhat quadrilateral muscle, consisting of two portions, superficial and deep. The *superficial portion,* the larger, *arises* by a thick, tendinous aponeurosis from the zygomatic process of the maxilla, and from the anterior two-thirds of the inferior border of the zygomatic arch: its fibers pass inferiorly and posteriorly, to be *inserted* into the angle and inferior half of the lateral surface of the ramus of the mandible. The *deep portion* is much smaller, and more muscular in texture; it *arises* from the posterior third of the inferior border and from the whole of the medial surface of the zygomatic arch; its fibers pass anteriorly and inferiorly, to be *inserted* into the superior half of the ramus and the lateral surface of the coronoid process of the mandible. The deep portion of the muscle is partly concealed anteriorly by the superficial portion; posteriorly, it is covered by the parotid gland. The fibers of the two portions are continuous at their insertion.

Action.—Closes the jaws.
Nerve.—The masseteric nerve from the mandibular division of the trigeminal nerve.

The **Pterygoideus medialis** (*Internal pterygoid muscle*) (Fig. 6–6) is a thick, quadrilateral muscle occupying a position on the inside of the ramus of the mandible similar to that of the Masseter on the outside. It *arises* from the medial surface of the lateral pterygoid plate and the grooved surface of the pyramidal process of the palatine bone; it has a second slip of origin from the lateral surfaces of the pyramidal process of the palatine and tuberosity of the maxilla. The second slip lies superficial to the Pterygoideus lateralis, while the main mass of the muscle lies deep. Its fibers pass inferiorly, laterally, and posteriorly, and are *inserted,* by a strong tendinous lamina, into the inferior and posterior part of the medial surface of the ramus and angle of the mandibular foramen. The superior portion of

the muscle is separated from the mandible by the sphenomandibular ligament, the maxillary vessels, the inferior alveolar vessels and nerve, and the lingual nerve. The medial surface of the muscle is closely related to the Tensor veli palatini and to the Constrictor pharyngis superior.

Action.—Closes the jaws.
Nerve.—The medial pterygoid nerve from the mandibular division of the trigeminal nerve.

The **Mandibular Sling.**—The Masseter and the Pterygoideus internus are so placed that they suspend the angle of the mandible in a sling. They form a functional articulation between the mandible and the maxilla, with the temporomandibular joint acting as a guide, in a fashion similar to the articulation between the scapula and the thorax, with the clavicle as a guide. When the mouth is opened and closed, the mandible moves about a center of rotation established by the attachment of the sling and the sphenomandibular ligament.

The **Pterygoideus lateralis** (*External pterygoid muscle*) (Fig. 6–6) is a short, thick muscle, somewhat conical in form, which extends almost horizontally between the infratemporal fossa and the condyle of the mandible. It *arises* by two heads; a superior from the inferior part of the lateral surface of the great wing of the sphenoid and from the infratemporal crest; an inferior from the lateral surface of the lateral pterygoid plate. Its fibers pass horizontally backward and lateralward, to be *inserted* into a depression in the anterior part of the neck of the condyle of the mandible, and into the anterior margin of the articular disk of the temporomandibular articulation.

Action.—Opens the jaws; protrudes the mandible; moves mandible from side to side.
Nerve.—The lateral pterygoid nerve from the mandibular division of the trigeminal nerve.
Group Actions.—The Temporalis, Masseter and Pterygoideus medialis close the jaws. Biting with the incisor teeth is performed by the Masseter and Pterygoideus medialis primarily, to some extent by the anterior portion of the Temporalis. Biting or chewing with the molars calls all three into maximal action. Opening of the jaws is performed primarily by the Pterygoideus lateralis pulling forward on the condyle and rotating the mandible about the center of rotation near the angle (see page 294). It is

assisted, at the beginning of the action, by the Mylohyoideus, Digastricus, and Geniohyoideus. When the mouth is opened against great resistance, in addition to the above, the infrahyoid muscles act to fix the hyoid, and other suprahyoid muscles probably come into action. The Platysma is practically without action unless the corners of the mouth are widely drawn back. The Pterygoideus lateralis protrudes the jaw when accompanied by appropriate synergetic action of the closing muscles. The

Pterygoideus medialis assists in this action only as a synergist, along with the other closing muscles, when they prevent the rotation which opens the jaws widely. If the Pterygoideus lateralis of one side acts, the corresponding side of the mandible is drawn forward while the opposite condyle remains comparatively fixed, and side-to-side movements, such as those occurring in the trituration of food, take place. The mandible is retracted by the posterior fibers of the Temporalis.

THE FASCIAE AND MUSCLES OF THE ANTEROLATERAL REGION OF THE NECK

The anterolateral muscles of the neck may be arranged into the following groups:

    I. Superficial Cervical
    II. Lateral Cervical
    III. Suprahyoid
    IV. Infrahyoid
    V. Anterior Vertebral
    VI. Lateral Vertebral

I. THE SUPERFICIAL CERVICAL MUSCLE

The **subcutaneous fascia** (*tela subcutanea; superficial fascia*) in the anterior and lateral regions of the neck is thinner and less dense than the facial portion with which it is continuous over the border of the mandible and the parotid gland. It has imbedded in its deeper layers the fibers of the Platysma muscle, and it is separated from the deep fascia by a distinct fascial cleft which facilitates the action of the muscle and increases the movability of the skin in this region. It is continuous over the clavicle with the tela of the pectoral and deltoid regions. Posteriorly, it is continuous with the tela of the back of the neck which is thick, tough, fibrous, and adherent to the deep fascia.

The **Platysma** (Fig. 6–1) is a broad sheet *arising* from the fascia covering the superior parts of the Pectoralis major and Deltoideus; its fibers cross the clavicle, and proceed obliquely cranialward and medialward along the side of the neck. The anterior fibers interlace, inferior and posterior to the symphysis menti, with the fibers of the muscle of the opposite side; the posterior fibers cross the mandible, some being inserted into the bone below the oblique

line, others into the skin and subcutaneous tissue of the lower part of the face, many of these fibers blending with the muscles about the angle and lower part of the mouth. Sometimes fibers can be traced to the Zygomaticus major, or to the margin of the Orbicularis oculi. Under cover of the Platysma, the external jugular vein descends from the angle of the mandible to the clavicle.

Action.—Draws the lower lip and corner of the mouth laterally and inferiorly, partially opening the mouth as in an expression of horror. When all the fibers act maximally the skin over the clavicle is drawn up toward the mandible increasing the diameter of the neck and relieving the pressure of a tight collar. Electromyography has shown that it is not used in motions of the jaw or the head, nor in laughing (De Sousa '64).

Nerve.—The cervical branch of the facial nerve.

Variations.—The platysma may be composed of delicate, pale, scattered fasciculi, or may form a broad layer of robust, dark fasciculi; it may be deficient or reach well below the clavicle; it may extend into the face for a very short distance or may continue as high as the zygoma or the ear. Decussation of fasciculi in the middle line anteriorly is common. The muscle may be absent.

The **Occipitalis minor** may extend, as a more or less independent band, from the fascia over the Trapezius to the fascia over the insertion of the Sternocleidomastoideus.

CERVICAL FASCIAE

The deep cervical fascia (Fig. 6–7) forms important transitions and connections, as might be expected, because the neck itself is a connecting structure, joining the head with the thorax and making many contribu-

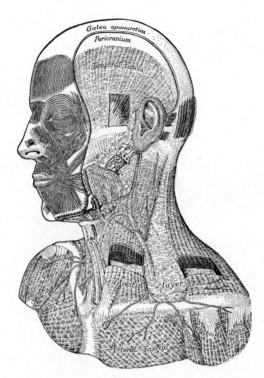

FIG. 6–7.—The external investing layer of deep fascia of the head and neck.

tions to the upper limb. Its components are complex and form various compartments and fascial clefts which are of major surgical interest because of these associations.

Cervical Triangles (Fig. 3–10).—Two triangular areas are formed in the neck by the oblique course of the Sternocleidomastoideus muscle. The **anterior triangle** is bounded by the middle line anteriorly, the Sternocleidomastoideus laterally, and the body of the mandible superiorly. The **posterior triangle** is bounded by the clavicle inferiorly, and by the adjacent borders of the Sternocleidomastoideus and Trapezius superiorly.

The **fascia coli** may be divided first, according to area, into suprahyoid and infrahyoid portions. Both of these, in turn, may be subdivided into smaller portions for the purpose of description. The suprahyoid subdivisions are: (1) the investing fascia, and (2) the deeper portion which is associated with the mandible and the floor of the mouth. The infrahyoid may be subdivided into: (1) the investing fascia; (2) the prevertebral fascia; (3) the middle cervical

fascia; (4) the visceral fascia, and (5) the carotid sheath.

Fascia of the Suprahyoid Region.—Since the suprahyoid region is as much a part of the head as of the neck, it will be necessary to include descriptions of certain head fasciae for the sake of clarity and continuity. It is convenient and logical, moreover, to look upon the fascia of the head as the cranial portion of the cervical fascia, and trace them both to the same superior termination and attachment.

The **investing fascia of the suprahyoid region** (Fig. 6–9) extends cranialward from its attachment to the hyoid bone and is attached to the whole length of the inferior border of the mandible. It covers the anterior belly of the Digastricus, is adherent to its sheath, and is continuous across the middle line. More laterally, it splits to enclose the submandibular gland in a sheath which is separated from the intrinsic capsule of the gland by a fascial cleft. The sheet on the deep surface of the gland lies over the Stylohyoideus and the intermediate tendon

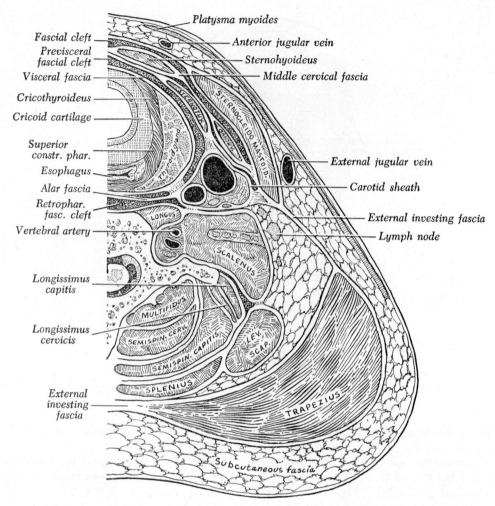

FIG. 6–8.—Section of the neck at about the level of the sixth cervical vertebra. Showing the arrangement of the fascia colli.

of the Digastricus, and, by a fusion with their fascial covering, forms a band which is carried up to the styloid process, prolonging the inferior boundary of the suprahyoid compartment posteriorly from the hyoid bone. The sheets of the superficial and deep surfaces of the submandibular gland come together for a short distance near the angle of the mandible and separate again to ensheathe the parotid gland. The external layer of the parotid portion extends cranialward over the angle of the mandible as the parotideomasseteric fascia and attaches to the zygomatic arch. It is closely adherent to the capsule of the gland which cannot, therefore, be shelled out readily, as is the case with the submandibular gland. The fascia at the posterior border of the parotid

gland is very tough where the superficial layer joins the deeper layer. It splits again to enclose the Sternocleidomastoideus, is attached to the mastoid process of the temporal bone, and is then continuous with the fascia of the back of the neck.

The deeper layers of fascia in the anterior portion of the suprahyoid region form individual sheaths for the muscles and are attached to the hyoid bone, inferiorly, and to the mandible, the styloid process, or the tongue, superiorly. More laterally, the portion of the investing layer between the submandibular and parotid glands extends deeply to fuse with the fascia of the posterior belly of the Digastricus. The result is a strong band which continues upward between the deep surface of the parotid and the posterior belly of the Digastricus; it is attached, superiorly, to the styloid process and inferiorly, to the angle of the mandible, and is known as the stylomandibular ligament.

The **suprahyoid compartment** is closed by the attachment of the investing fascia to the border of the mandible superiorly and to the hyoid bone inferiorly. It is continuous across the middle line anteriorly and reaches up to the floor of the mouth in the region of the sublingual gland and tongue. The fascial clefts between structures in this compartment are continuous posteriorly into the fascial cleft which lies superficial to the buccopharyngeal fascia (the lateral pharyngeal cleft) and into the region of the deep extensions of the buccal fat pad. The fascial clefts in the floor of the mouth and about the sublingual gland communicate with the cleft surrounding the submandibular gland by extending around the posterior border of the Mylohyoideus.

The **fascia of the infrahyoid region** includes most of what is commonly called the fascia coli (deep cervical fascia).

The **investing layer of cervical fascia** (Fig. 6–7) in the infrahyoid region is not sharply marked off from the investing fascia of adjacent regions with which it is continuous, and the fascia of one side is continuous across the middle line with the fascia of the other side. It splits into two sheets to invest the two prominent superficial muscles, the Sternocleidomastoideus and the Trapezius, but it covers the anterior and posterior triangles as a single sheet except just above the sternum. It is continuous, superiorly, with the fascia of the suprahyoid region, and inferiorly, with the pectoral and deltoid fasciae. It has bony attachments superiorly, inferiorly, and posteriorly. The anterior portion of its superior attachment is to the hyoid bone; the lateral and posterior portion is to the mandible, mastoid process, and superior nuchal line through its continuity with the suprahyoid and posterior cervical fasciae. Through its continuity with the posterior cervical fascia also, it is attached posteriorly to the spinous process of the seventh cervical vertebra and the ligamentum nuchae. The inferior attachment is to the acromion, the clavicle, and the manubrium sterni. Extending cranialward from the manubrial attachment

between the sternal origins of the Sternocleidomastoidei for 3 or 4 cm, there is a thickening due to lamination. The outer lamina is attached to the anterior border of the manubrium, the inner lamina to the posterior border and the interclavicular ligament. The shallow interval between the two laminae, mostly filled with fat, is called the **suprasternal space** (*Space of Burns*) (Fig. 6–9). It contains the lower portions of the anterior jugular veins and their transverse connecting branch (arcus venosus), the sternal heads of the Sternocleidomastoidei, and sometimes a lymph node. The anterior jugulars, in order to reach the external jugulars, traverse extensions of the laminated interval which are prolonged laterally deep to the heads of the Sternocleidomastoidei (*cul de sac of Gruber*) (Fig. 6–9). The external and anterior jugular veins, through most of their course in the neck, appear to lie between the superficial and deep fasciae but actually are imbedded in the superficial surface of the investing sheet.

The **prevertebral fascia** (Fig. 6–8) is the anterior portion of a larger complex, the vertebral fascia, which encloses the vertebral column and its muscles. The cervical portion of the prevertebral fascia is part of a larger sheet which goes by that name and which lies on the ventral surface of the vertebral column from the skull to the coccyx. In the neck, it extends laterally across the anterior surface of the Longus colli and capitis, and the Rectus capitis anterior and lateralis muscles, and is then secured to the tips of the transverse processes. From this attachment, it is continuous laterally with the fascia which covers the Levator scapulae and Splenius, and completes the enclosure of the vertebral compartment posteriorly by attaching to the spinous processes of the vertebrae. Inferiorly, it extends over the superficial surface of the Scalenus anterior, medius, and posterior muscles to become continuous with the fascia of the thoracic wall. The fascia on the deep surface of the scalenus group of muscles forms part of a conical, fibrous dome, called **Sibson's fascia**, which arches over the cupula of the lung. It varies considerably in its thickness and composition; it is reinforced frequently by fibrous bands, and, in some cases, by muscle

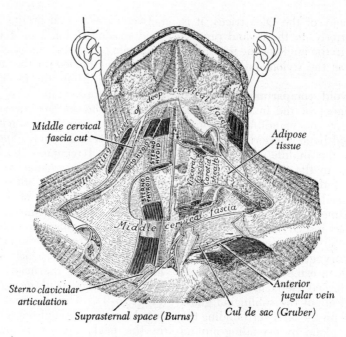

Fig. 6–9.—The middle cervical fascia.

fibers, the latter being called the **Scalenus minimus.** It is attached to the transverse process of the seventh cervical vertebra and to the medial border of the first rib, and merges with the carotid sheath where the latter is pierced by the subclavian artery. Below the first rib it becomes continuous with the endothoracic fascia. As the spinal nerves emerge from between the Scalenus medius and anterior muscles on their way to the brachial plexus, they are covered by a prolongation from the scalenus portion of the prevertebral fascia. This prolongation encloses the nerves and the subclavian artery and vein and extends under the clavicle into the axilla as the axillary sheath.

The prevertebral fascia is separated from the visceral fascia by the **retropharyngeal fascial cleft** (*retropharyngeal space*). In the lateral region, the prevertebral fascia is adherent to the investing fascia of the neck under the Sternocleidomastoideus and superiorly in the Posterior Triangle. In the lower part of the Posterior Triangle a considerable pad of adipose tissue occupies the interval between the prevertebral and investing sheets and surrounds the axillary sheath as it passes under the clavicle.

The **middle cervical fascia** (Fig. 6–9)

invests the two layers of infrahyoid muscles and has, therefore, a superficial, a deep, and a middle sheet. Superiorly, all three sheets are attached to the hyoid bone and the outer sheet is also fused with the external investing fascia for a short distance inferior to the bone. At the lateral border of the Omohyoideus, the superficial and deep sheets come together and are fused to the deep surface of the investing membrane of the neck. No independent representation of this fascia, therefore, is found in the posterior triangle above the Omohyoideus. Inferiorly, all three layers are attached to the posterior surface of the sternum along with the muscles they invest. In the supraclavicular region, the fascia is securely fastened to the clavicle and is looped over the inferior belly of the Omohyoideus like a sleeve with the medial part thickened into a pulley for the intermediate tendon. Under cover of the Sternocleidomastoideus, the lateral border of the fascia is attached to the carotid sheath. A fascial cleft separates the deep surface of the fascia from the underlying visceral fascia, especially inferior to the thyroid cartilage.

The **cervical visceral fascia** (*lamina pretrachialis*) (Fig. 6–8) is a roughly tubular prolongation of the visceral fascia of the mediastinum. It forms a compartment enclosing the esophagus and trachea as they enter the neck from the thorax, and farther superiorly, the pharynx, larynx, and thyroid gland. It extends superiorly into the head and is attached to the base of the skull at the pharyngeal tubercle, with the Constrictor pharyngis superior and the pharyngeal aponeurosis, and to the pterygoid hamulus and the mandible with the pterygomandibular raphe. The portion covering the Constrictor superior continues anteriorly over the Buccinator and is called the **buccopharyngeal fascia**. That covering the Constrictor medius is continuous anteriorly with the fascia over the Hyoglossus and Genioglossus, and with them, is attached to the hyoid bone.

A **perivisceral fascial cleft** (Fig. 6–8) almost completely surrounds the visceral fascia and separates it from the middle cervical fascia anteriorly and laterally, from the carotid sheath laterally, and from the prevertebral fascia posteriorly, leaving the enclosed esophagus and pharynx relatively free for the movement of swallowing. Posteriorly and laterally, however, the visceral fascia has a narrow attachment along its whole length to the tips of the transverse processes, at which point it is fused also with the carotid sheath and prevertebral fascia. This attachment subdivides the entire perivisceral fascial cleft into anterior and posterior portions. The anterior portion, sometimes named the **previsceral cleft**, is in relation to that part of the visceral fascia frequently called the **pretracheal fascia**, that is, the part covering the trachea, larynx, and thyroid gland. The posterior portion of the cleft, between the pharynx and the prevertebral fascia, is called the **retropharyngeal cleft** (see above) and is of surgical importance because of its continuation caudalward behind the esophagus into the thorax.

The **carotid sheath** (*vagina carotica*) (Fig. 6–8) forms a tubular investment for the carotid artery, internal jugular vein and vagus nerve. It is attached medially to the visceral fascia by means of a sheet, the **alar fascia**, which is fused with the latter along the posterior middle line of the pharynx from the skull to the level of the seventh cervical vertebra. The sheath is attached posteriorly to the prevertebral fascia along the line of the tips of the transverse processes. Laterally, it has an attachment to the investing fascia on the deep surface of the Sternocleidomastoideus, and anteriorly, it has an attachment with the middle cervical fascia along the lateral border of the Sternothyroideus. In the superior part of the neck the sheath is fused with the fascia of the Stylohyoideus and posterior belly of the Digastricus as it passes deep to them and finally is fastened to the skull with its enclosed structures. In the root of the neck the sheath is adherent to the sternum and first rib, fuses with the scalenus fascia, and finally becomes continuous with the fibrous pericardium. The cervical sympathetic trunk is embedded in the fascia of the posterior wall of the sheath and is not actually within the sheath.

II. THE LATERAL CERVICAL MUSCLES

Trapezius and Sternocleidomastoideus.— The Trapezius is described on page 447.

The **Sternocleidomastoideus** (*Sternomastoid muscle*) (Fig. 6–10) passes obliquely across the side of the neck. It is thick and narrow at its central part, but broader and thinner at either end. It *arises* from the sternum and clavicle by two heads. The **medial** or **sternal head** is a rounded fasciculus, tendinous ventrally, fleshy dorsally, which *arises* from the cranial part of the ventral surface of the manubrium sterni, and is directed cranialward and dorsalward. The **lateral** or **clavicular head** composed of fleshy and aponeurotic fibers, *arises* from the superior border and anterior surface of the medial third of the clavicle; it is directed almost vertically cranialward. The two heads are separated from each other at their origins by a triangular interval, but gradually blend, below the middle of the neck, into a thick, rounded muscle which is *inserted*, by a strong tendon, into the lateral surface of the mastoid process, from its apex to its superior border, and by a thin aponeurosis into the lateral half of the superior nuchal line of the occipital bone.

Action.—The muscle of one side bends the cervical vertebral column laterally, drawing the head toward the shoulder of the same side, and at the same time rotates it, pointing the chin cranially, and to the opposite side. Both

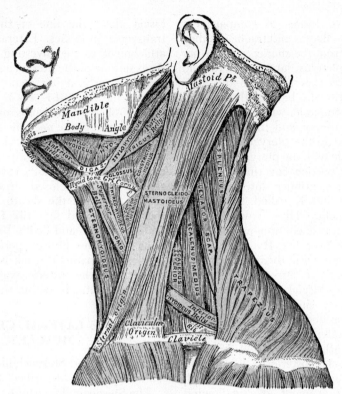

Fig. 6–10.—Muscles of the neck. Lateral view.

muscles acting together flex the vertebral column, bringing the head ventrally and at the same time elevating the chin.

Nerves.—The spinal part of the accessory nerve and branches from the anterior rami of the second and third cervical nerves.

Variations.—The Sternocleidomastoideus varies much in the extent of its origin from the clavicle: in some cases the clavicular head may be as narrow as the sternal; in others it may be as much as 7.5 cm in breadth. When the clavicular origin is broad, it is occasionally subdivided into several slips, separated by narrow intervals. More rarely, the adjoining margins of the Sternocleidomastoideus and Trapezius have been found in contact. The *Supraclavicularis muscle* arises from the manubrium behind the Sternocleidomastoideus and passes behind the Sternocleidomastoideus to the cranial surface of the clavicle.

III. THE SUPRAHYOID MUSCLES
(Figs. 6–10, 6–11)

Digastricus
Stylohyoideus
Mylohyoideus
Geniohyoideus

The **Digastricus** (*Digastric muscle*) consists of two fleshy bellies united by an intermediate rounded tendon. It lies inferior to the body of the mandible, and extends, in a curved form, from the mastoid process to the symphysis menti. The **posterior belly**, longer than the anterior, *arises* from the mastoid notch of the temporal bone and passes anteriorly and inferiorly. The **anterior belly** *arises* from a depression on the inner side of the inferior border of the mandible, close to the symphysis, and passes posteriorly and inferiorly. The two bellies end in an intermediate tendon which perforates the Stylohyoideus muscle, and is held in connection with the side of the body and the greater cornu of the hyoid bone by a fibrous loop, which is sometimes lined by a synovial sheath. A broad aponeurotic layer is given off from the tendon of the Digastricus on either side, to be attached to the body and greater cornu of the hyoid bone; this is termed the **suprahyoid aponeurosis**.

Action.—Raises the hyoid bone; assists in opening the jaws. The anterior belly draws the hyoid forward, the posterior backward.

Nerves.—Anterior belly by the mylohyoid nerve from the inferior alveolar branch of the mandibular division of the trigeminal; posterior belly by a branch of the facial nerve.

Variations are numerous. The posterior belly may arise partly or entirely from the styloid process, or be connected by a slip to the middle or inferior constrictor; the anterior belly may be double or extra slips from this belly may pass to the jaw or Mylohyoideus or decussate with a similar slip on opposite side; anterior belly may be absent and posterior belly inserted into the middle of the jaw or hyoid bone. The tendon may pass in front, more rarely behind the Stylohyoideus. The *Mentohyoideus muscle* passes from the body of hyoid bone to chin.

The Digastricus divides the anterior triangle of the neck into three smaller triangles: (1) the **submandibular triangle,** bounded superiorly by the inferior border of the body of the mandible, and a line drawn from its angle to the Sterno-cleidomastoideus, inferiorly by the posterior belly of the Digastricus and the Stylohyoideus, ventrally by the anterior belly of the Digastricus; (2) the **carotid triangle,** bounded superiorly by the posterior belly of the Digastricus and Stylohyoideus, posteriorly by the Sternocleidomastoideus, inferiorly by the Omohyoideus; (3) the **suprahyoid** or **submental triangle,** bounded laterally by the anterior belly of the Digastricus, medially by the middle line of the neck from the hyoid bone to the symphysis menti, and inferiorly by the body of the hyoid bone (Fig. 3–10).

The **Stylohyoideus** (*Stylohyoid muscle*) is a slender muscle, lying anterior, and superior to, the posterior belly of the Digastricus. It *arises* from the posterior and lateral surface of the styloid process, near the base; and, passing anteriorly and inferiorly, is *inserted* into the body of the hyoid bone, at its junction with the greater cornu, and just above the Omohyoideus. It is perforated, near its insertion, by the tendon of the Digastricus.

Action.—Draws the hyoid bone superiorly and posteriorly.

Nerve.—A branch of the facial nerve.

Variations.—It may be absent or doubled, lie beneath the carotid artery, or be inserted into the Omohyoideus, Thyrohyoideus, or Mylohyoideus.

The **Stylohyoid Ligament** (*ligamentum stylohyoideum*).—In connection with the Stylohyoideus muscle a ligamentous band, the **stylohyoid ligament,** may be described. It is a fibrous cord which is attached to the tip of the styloid process of the temporal and the lesser cornu of the hyoid bone. It frequently contains a little cartilage in its center, is often partially ossified, and in many animals forms a distinct bone, the **epihyal.**

The **Mylohyoideus** (*Mylohyoid muscle*), flat and triangular, is situated immediately superior to the anterior belly of the Digastricus, and forms, with its fellow of the opposite side, a muscular floor for the cavity of the mouth. It *arises* from the whole length of the mylohyoid line of the mandible, extending from the symphysis in front to the last molar tooth. The posterior fibers pass medialward and slightly downward, to be *inserted* into the body of the hyoid bone. The middle and anterior fibers are *inserted* into a median fibrous raphe extending from the symphysis menti to the hyoid bone, where they join at an angle with the fibers of the opposite muscle. This median raphe is sometimes wanting; the fibers of the two muscles are then continuous.

Action.—Raises the hyoid bone and tongue. Electromyography—active in mastication, deglutition, sucking, and blowing. Anterior fibers more active in mandibular, posterior lateral in tongue motions (Lehr *et al.* '71).

Nerve.—The mylohyoid nerve from the inferior alveolar branch of the mandibular division of the trigeminal nerve.

Variations.—It may be united to or replaced by the anterior belly of the Digastricus; accessory slips to other hyoid muscles are frequent.

The **Geniohyoideus** (*Geniohyoid muscle*) is a narrow muscle, situated deep to the medial border of the Mylohyoideus. It *arises* from the inferior mental spine on the inner surface of the symphysis menti, and runs posteriorly and slightly inferiorly, to be *inserted* into the anterior surface of the body of the hyoid bone; it lies in contact with its fellow of the opposite side.

Action.—Draws the hyoid bone and tongue anteriorly.

Nerve.—A branch of the first cervical nerve, through the hypoglossal nerve.

Variations.—It may be blended with the one on the opposite side or double; slips to greater cornu of hyoid bone and Genioglossus occur.

Group Actions.—These muscles perform two very important actions. During deglutition, they raise the hyoid bone, and with it the base of the tongue; when the hyoid bone is fixed by its depressors and those of the larynx, they depress the mandible. During the first act of deglutition, when the mass of food is being driven from the mouth into the pharynx, the hyoid bone and with it the tongue is carried upward and forward by the anterior bellies of the Digastrici, the Mylohyoidei, and Geniohyoidei. In the second act, when the mass is passing through the pharynx, the direct elevation of the hyoid bone takes place by the combined action of all the muscles; and after the food has passed, the hyoid bone is carried upward and backward by the posterior bellies of the Digastrici and the Stylohyoidei, which assist in preventing the return of the food into the mouth.

IV. THE INFRAHYOID MUSCLES
(Figs. 6–10, 6–11)

Sternohyoideus
Sternothyroideus
Thyrohyoideus
Omohyoideus

The **Sternohyoideus** (*Sternohyoid muscle*) is a thin, narrow muscle, which *arises* from the posterior surface of the medial end of the clavicle, the posterior sternoclavicular ligament, and the superior and posterior part of the manubrium sterni. Passing

superiorly and medially, it is *inserted,* by short, tendinous fibers, into the inferior border of the body of the hyoid bone. This muscle is separated inferiorly from its fellow by a considerable interval; but the two muscles come into contact with one another in the middle of their course, and thereafter lie side by side. It sometimes presents a transverse tendinous inscription immediately above its origin.

Action.—Draws the hyoid bone inferiorly.

Nerve.—Branch of the ansa cervicalis (hypoglossi) containing fibers from the first, second, and third cervical nerves.

Variations.—Doubling; accessory slips (Cleidohyoideus); absence.

The **Sternothyroideus** (*Sternothyroid muscle*) is shorter and wider than the preceding muscle, deep to which it is situated. It *arises* from the dorsal surface of the manubrium sterni, caudal to the origin of the Sternohyoideus, and from the edge of the cartilage of the first rib, and sometimes that of the second rib, it is *inserted* into the oblique line on the lamina of the thyroid cartilage. This muscle is in close contact with its fellow at the lower part of the neck, but diverges somewhat as it

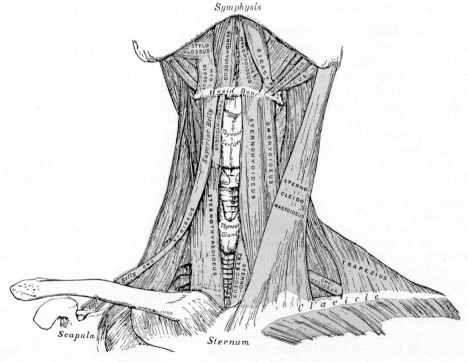

Fɪɢ. 6–11.—Muscles of the neck. Anterior view.

ascends; it is occasionally traversed by a transverse or oblique tendinous inscription.

Action.—Draws the thyroid cartilage caudalward.

Nerve.—Branch of the ansa cervicalis (hypoglossi) containing fibers from the first three cervical nerves.

Variations.—Doubling; absence; accessory slips to Thyrohyoideus, Inferior constrictor, or carotid sheath.

The **Thyrohyoideus** (*Thyrohyoid muscle*) is a small, quadrilateral muscle appearing to be a cranialward continuation of the Sternothyroideus. It *arises* from the oblique line on the lamina of the thyroid cartilage, and is *inserted* into the inferior border of the greater cornu of the hyoid bone.

Action.—Draws the hyoid bone inferiorly, or if the latter is fixed, draws the thyroid cartilage superiorly.

Nerve.—Fibers from the first (and second?) cervical nerves by way of a communication to the hypoglossal nerve and through its descendens hypoglossi branch.

The **Omohyoideus** (*Omohyoid muscle*) consists of two fleshy bellies united by a central tendon. It *arises* from the cranial border of the scapula, and occasionally from the superior transverse ligament which crosses the scapular notch, its extent of attachment to the scapula varying from a few millimeters to 2.5 cm. From this origin, the **inferior belly** forms a flat, narrow fasciculus, which inclines ventrally and slightly cranially across the caudal part of the neck, being bound down to the clavicle by a fibrous expansion; it then passes deep to the Sternocleidomastoideus, becomes tendinous and changes its direction, forming an obtuse angle. It ends in the **superior belly,** which passes almost vertically cranialward, close to the lateral border of the Sternohyoideus, to be inserted into the caudal border of the body of the hyoid bone, lateral to the insertion of the Sternohyoideus. The central tendon of this muscle varies much in length and form, and is held in position by a process of the deep cervical fascia, which sheaths it, and is prolonged caudally to be attached to the clavicle and first rib; it is by this means that the angular form of the muscle is maintained.

Action.—Draws the hyoid bone caudalward.

Nerves.—Branches of the ansa cervicalis (hypoglossi) containing fibers from the first(?), second, and third cervical nerves.

Variations.—Doubling; absence; origin from clavicle; absence or doubling of either belly.

The inferior belly of the Omohyoideus divides the posterior triangle of the neck into a cranial or **occipital triangle** and a caudal or **subclavian triangle,** while its superior belly divides the anterior triangle into a cranial or **carotid triangle** and a caudal or **muscular triangle.**

Group Actions.—These muscles depress the larynx and hyoid, after they have been drawn up with the pharynx in the act of deglutition. The Omohyoidei not only depress the hyoid bone, but carry it dorsalward and to one or the other side. They are concerned especially in prolonged inspiratory efforts; for by rendering the lower part of the cervical fascia tense they lessen the inward suction of the soft parts, which would otherwise compress the great vessels and the apices of the lungs.

V. THE ANTERIOR VERTEBRAL MUSCLES (Fig. 6-12)

Longus colli
Longus capitis
Rectus capitis anterior
Rectus capitis lateralis

The **Longus colli** is situated on the anterior surface of the vertebral column, between the atlas and the third thoracic vertebra. It is broad in the middle, narrow and pointed at either end, and consists of three portions, a superior oblique, an inferior oblique, and a vertical. The **superior oblique portion** *arises* from the anterior tubercles of the transverse processes of the third, fourth, and fifth cervical vertebrae; and, ascending obliquely with a medial inclination, is *inserted* by a narrow tendon into the tubercle on the anterior arch of the atlas. The **inferior oblique portion,** the smallest part of the muscle, *arises* from the anterior surface of the bodies of the first two or three thoracic vertebrae; and, ascending obliquely in a lateral direction, is *inserted* into the anterior tubercles of the transverse processes of the fifth and sixth cervical vertebrae. The **vertical portion** *arises* from the anterior surface of the bodies of the first three thoracic and last three cervical vertebrae, and is *inserted* into the anterior surface of the bodies of the second, third, and fourth cervical vertebrae.

Action.—Flexes the neck and slightly rotates the cervical portion of the vertebral column.

Nerves.—Branches from the second to the seventh cervical nerves.

The **Longus capitis** (*Rectus capitis anticus major*), broad and thick superiorly, narrow inferiorly, *arises* by four tendinous slips from the anterior tubercles of the transverse processes of the third, fourth, fifth, and sixth cervical vertebrae, and ascends, converging toward its fellow of the opposite side, to be *inserted* into the inferior surface of the basilar part of the occipital bone.

Action.—Flexes the head.

Nerves.—Branches from the first, second, and third cervical nerves.

The **Rectus capitis anterior** (*Rectus capitis anticus minor*) is a short, flat muscle, situated immediately deep to the superior part of the Longus capitis. It *arises* from the anterior surface of the lateral mass of the atlas, and from the root of its transverse process, and passing obliquely cranialward and medialward, is *inserted* into the inferior surface of the basilar part of the occipital bone immediately anterior to the foramen magnum.

Action.—Flexes the head.

Nerve.—Branch of the loop between the first and second cervical nerves.

The **Rectus capitis lateralis,** a short, flat muscle, *arises* from the superior surface of the transverse process of the atlas, and is *inserted* into the inferior surface of the jugular process of the occipital bone.

Action.—Bends the head laterally.

Nerve.—Branch of the loop between the first and second cervical nerves.

Group Actions.—The Longus capitis and Rectus anterior are the direct antagonists of the muscles at the back of the neck, serving to restore the head to its natural position after it has been drawn backward. These muscles also flex the head, and from their obliquity, rotate it, so as to turn the face to one or the other side. The Rectus lateralis, acting on one side, bends the head laterally.

VI. THE LATERAL VERTEBRAL MUSCLES (Fig. 6–12)

>Scalenus anterior
>Scalenus medius
>Scalenus posterior

The **Scalenus anterior** (*Scalenus anticus*) lies deeply at the side of the neck, deep to the Sternocleidomastoideus. It *arises* from the anterior tubercles of the transverse processes of the third, fourth, fifth, and sixth cervical vertebrae, and descending almost vertically, is *inserted* by a narrow, flat tendon into the scalene tubercle on the inner border of the first rib, and into the ridge on the cranial surface of the rib ventral to the subclavian groove.

The **Scalenus medius,** the largest and longest of the three Scaleni, *arises* from the posterior tubercles of the transverse processes of the last six cervical vertebrae, and descending along the side of the vertebral column, is *inserted* by a broad attachment into the cranial surface of the first rib, between the tubercle and the subclavian groove.

Action.—Raise the first rib; bend and slightly rotate the neck.

Nerves.—Branches of the lower cervical nerves.

The **Scalenus posterior** (*Scalenus posticus*), the smallest and most deeply seated of the three Scaleni, *arises,* by two or three separate tendons, from the posterior tubercles of the transverse processes of the last two or three cervical vertebrae, and is *inserted* by a thin tendon into the outer surface of the second rib, deep to the attachment of the Serratus anterior. It is occasionally blended with the Scalenus medius.

Action.—Raises the second rib; bends and slightly rotates the neck.

Nerves.—Branches of ventral primary divisions of last three cervical nerves.

Group Actions.—When the Scaleni are fixed superiorly, they elevate the first and second ribs, and are, therefore, inspiratory muscles. Acting from below, they bend the vertebral column to one side or the other; if the muscles of both sides act, the vertebral column is flexed slightly.

Variations.—The Scaleni muscles vary considerably in their attachments and in the arrangement of their fibers. A slip from the Scalenus anterior may pass behind the subclavian artery. The Scalenus posterior may be absent or extend to the third rib. The *Scalenus*

pleuralis muscle or *scalenus minimus* extends from the transverse process of the seventh cervical vertebra to the fascia supporting the dome of the pleura and inner border of first rib.

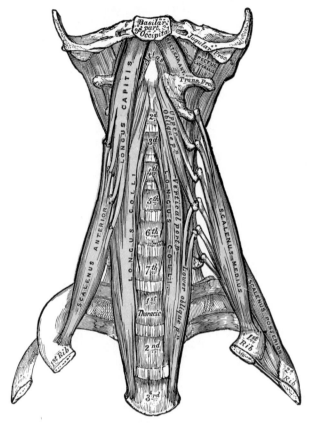

Fig. 6–12.—The ventral vertebral muscles.

THE FASCIAE AND MUSCLES OF THE TRUNK

The muscles of the trunk may be arranged in six groups:

    I. Deep Muscles of the Back
   II. Suboccipital Muscles
 III. Muscles of the Thorax
 IV. Muscles of the Abdomen
  V. Muscles of the Pelvis
 VI. Muscles of the Perineum

I. THE DEEP MUSCLES OF THE BACK (Fig. 6–14)

The deep or intrinsic muscles of the back consist of a complex, serially arranged group of muscles, extending from the pelvis to the skull, which may be looked upon as a single muscle functionally, the extensor of the vertebral column. Two subgroups

may be identified: (*A*) A superficial stratum with fasciculi mainly crossing laterally as they ascend may be called the transversocostal and splenius group:

    Splenius capitis
    Splenius cervicis
    Erector spinae (Iliocostalis,
        Longissimus, Spinalis)

(*B*) The deeper stratum has fasciculi crossing medially as they ascend, and may be called the transversospinal group:

    Semispinalis
    Multifidus
    Rotatores
    Interspinales
    Intertransversarii

The **nuchal fascia** (*fascia nuchae*) (Fig. 6–8) is the cervical portion of the more extensive vertebral fascia and is continuous caudally with the fascia thoracolumbalis. It covers the Splenius capitis and cervicis, and near the skull, the upper portion of the Semispinalis capitis. With these muscles, it is attached to the skull just inferior to the superior nuchal line, the ligamentum nuchae, and the spinous processes of the seventh cervical and first six thoracic vertebrae. In the cranial part of the neck it is more or less adherent to the fascia of the deep surface of the Trapezius. More caudally, a distinct fascial cleft separates it from the fascia of the Serratus posterior superior and Rhomboidei.

The deeper muscles of the neck are enclosed by fascial septa which form compartments for each muscle. A fascial cleft separates the Splenius from the Semispinalis capitis. A considerable layer of adipose tissue and a fascial cleft intervene between the latter and the Semispinalis cervicis. In this adipose layer are found the deep cervical blood vessels. The fascia covering the Semispinalis cervicis continues cranialward from the atlas to form the thick adherent covering of the suboccipital muscles. The fasciae of the several muscles which attach to the transverse processes of the cervical vertebrae are either fused or continuous, the scalenus fascia becoming continuous with the splenius and serratus posterior superior fascia under cover of the Levator scapulae.

Fascia thoracolumbalis (*lumbodorsal fascia*) (Figs. 6–13, 6–21, 6–33).—The name fascia thoracolumbalis or lumbodorsal fascia is given to a rather varied fascial complex. It is, in general terms, the subdivision of the vertebral fascia which forms the sheath of the Sacrospinalis muscle. It should be looked upon primarily as an intermediate stratum, derived from the fascia of the trunk deep to the large limb muscles. It becomes part of the investing fascia of the body, however, in the caudal half of the trunk, because in that region the aponeurosis of the Latissimus dorsi is incorporated in it. Cranially, it is continuous with the fascia nuchae. Medially, it is attached to the spines of the vertebrae, the supraspinal ligaments, and the medial crest

of the sacrum; caudally, to the iliac crests and lateral crests of the sacrum. Laterally, in the thorax, it is attached to the angles of the ribs and intercostal fascia, and in the aponeurosis of origin of the Transversus abdominis muscle. In the cranial part of the lumbar region, it is continuous with the thorax, where it is covered by the Rhomboidei, it is thin, gray and transparent. It retains this consistency also under the fleshy fibers of the cranial part of the Latissimus dorsi. More caudally, it is a thick, white, glistening sheet which serves as the origin of the Latissimus and is called the lumbar aponeurosis. Caudal to the twelfth rib, since it can no longer attach to the ribs, the line of attachment along the lateral border of the Sacrospinalis becomes a line of fusion between the fascia covering the dorsal surface of this muscle (**the posterior layer of the fascia thoracolumbalis**) (*lumbodorsal*) and the fascia on the deep surface of the muscle (**the anterior layer of the fascia thoracolumbalis**). From this fusion, the fascia extends laterally as a single sheet, the aponeurosis of origin of the Transversus abdominis muscle. This portion of the fascia may be described in another way. The aponeurosis of origin of the Transversus abdominis, in seeking to attach to the vertebrae, meets interference at the lateral border of the Sacrospinalis and splits, therefore, to enclose the latter in a superficial and deep sheet. The superficial sheet, the **lumbar aponeurosis** or fascia thoracolumbalis (posterior layer), extends over the dorsal surface of the Sacrospinalis and attaches to the spines. The deep sheet, the **lumbocostal aponeurosis** (anterior layer of the fascia thoracolumbalis), extends over the deep surface of the Sacrospinalis and attaches to the transverse processes. The lumbocostal aponeurosis is a strong sheet reaching from the lower border of the twelfth rib to the crest of the ilium. Its fiber bundles radiate out from attachments to the tips of the transverse processes of the lumbar vertebrae. It lies deep to the Sacrospinalis, and superficial to the Quadratus lumborum and Psoas major muscles. The upper portion, attached to the twelfth rib and the transverse process of the first lumbar vertebra, is more specifically named the **lumbocostal ligament.**

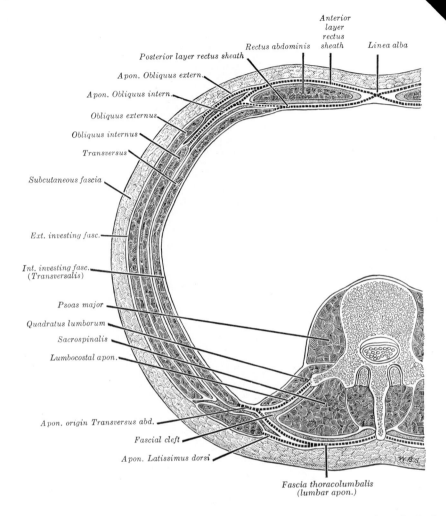

Fig. 6–13.—Fasciae and aponeuroses of abdominal wall in cross section through third lumbar vertebra. Semidiagrammatic.

A. Transversocostal and Splenius Muscles

The **Splenius capitis** (Fig. 6–33) *arises* from the caudal half of the ligamentum nuchae, from the spinous process of the seventh cervical vertebra, and from the spinous processes of the first three or four thoracic vertebrae. The fibers of the muscle are directed cranialward and lateralward and are *inserted* into the rough surface on the occipital bone just inferior to the lateral third of the superior nuchal line, and, under cover of the Sternocleidomastoideus, into the mastoid process of the temporal bone.

The **Splenius cervicis** (*Splenius colli*)

(Fig. 6–33) *arises* by a narrow tendinous band from the spinous processes of the third to the sixth thoracic vertebrae; it is *inserted*, by tendinous fasciculi, into the posterior tubercles of the transverse processes of the upper two or three cervical vertebrae.

Action.—Draw the head and neck dorsally and laterally, and rotate them, turning the face toward the same side. Both sides acting together extend the head and neck.

Nerves.—Lateral branches of the dorsal primary divisions of the middle and lower cervical nerves.

Variations.—The origin is frequently moved up or down one or two vertebrae. Accessory slips are occasionally found.

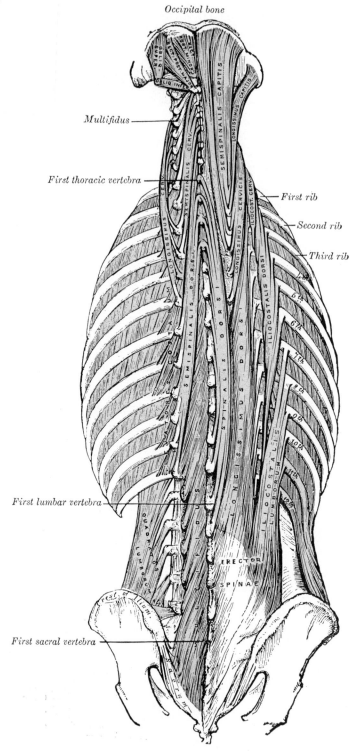

Occipital bone

Multifidus

First thoracic vertebra

First rib

Second rib

Third rib

First lumbar vertebra

ERECTOR

SPINAE

First sacral vertebra

Fig. 6–14.—Deep muscles of the back.

The **Sacrospinalis** (*Erector spinae*) (Fig. 6–14), and its prolongations in the thoracic and cervical regions, lie in the groove on the side of the vertebral column. They are covered in the lumbar and thoracic regions by the thoracolumbar fascia, and in the cervical region by the nuchal fascia. This large muscular and tendinous mass varies in size and structure at different parts of the vertebral column. In the sacral region it is narrow and pointed, and at its origin chiefly tendinous in structure. In the lumbar region it is larger, and forms a thick fleshy mass which, in its cranial course, is subdivided into three columns; these gradually diminish in size as they ascend to be inserted into the vertebrae and ribs.

The Sacrospinalis *arises* from the anterior surface of a broad and thick tendon, which is attached to the middle crest of the sacrum, to the spinous processes of the lumbar and the eleventh and twelfth thoracic vertebrae, and the supraspinal ligament, to the back part of the inner lip of the iliac crests and to the lateral crests of the sacrum, where it blends with the sacrotuberous and posterior sacroiliac ligaments. Some of its fibers are continuous with the fibers of origin of the Gluteus maximus. The muscular fibers form a large fleshy mass which splits, in the upper lumbar region into three columns, viz., a lateral, the **Iliocostalis**, an intermediate, the **Longissimus**, and a medial, the **Spinalis**. Each of these consists of three parts (Fig. 6–14).

Lateral Column

Iliocostalis
    I. lumborum
    I. thoracis
    I. cervicis

The **Iliocostalis lumborum** (*Iliocostalis muscle; Sacrolumbalis muscle*) is *inserted* by six or seven flattened tendons, into the inferior borders of the angles of the last six or seven ribs.

The **Iliocostalis thoracis** (*Musculus accessorius*) *arises* by flattened tendons from the upper borders of the angles of the lower six ribs medial to the tendons of insertion of the Iliocostalis lumborum; these become muscular, and are *inserted* into the cranial borders of the angles of the first six ribs

and into the dorsum of the transverse process of the seventh cervical vertebra.

The **Iliocostalis cervicis** (*Cervicalis ascendens*) *arises* from the angles of the third, fourth, fifth, and sixth ribs, and is *inserted* into the posterior tubercles of the transverse processes of the fourth, fifth, and sixth cervical vertebrae.

Action.—Extend the vertebral column and bend it to one side; lumborum and thoracis draw the ribs caudalward.

Nerves.—Branches of the dorsal primary divisions of the spinal nerves.

Intermediate Column

Longissimus
    L. thoracis
    L. cervicis
    L. capitis

The **Longissimus thoracis** is the intermediate and largest of the continuations of the Sacrospinalis. In the lumbar region, where it is as yet blended with the Iliocostalis lumborum and Spinalis, some of its fibers are attached to the whole length of the posterior surfaces of the transverse processes and the accessory processes of the lumbar vertebrae, and to the anterior layer of the lumbocostal aponeurosis. In the thoracic region it is *inserted*, by rounded tendons, into the tips of the transverse processes of all the thoracic vertebrae, and by fleshy processes into the lower nine or ten ribs between their tubercles and angles.

The **Longissimus cervicis** (*Transversalis cervicis*), situated medial to the Longissimus thoracis, *arises* by long thin tendons from the summits of the transverse processes of the upper four or five thoracic vertebrae, and is *inserted* by similar tendons into the posterior tubercles of the transverse processes of the cervical vertebrae from the second to the sixth inclusive.

Action.—Extend the vertebral column and bend it to one side; draw the ribs caudalward.

Nerves.—Branches of the dorsal primary divisions of the spinal nerves.

The **Longissimus capitis** (*Trachelomastoid muscle*) lies medial to the Longissimus cervicis, between it and the Semispinalis capitis. It *arises* by tendons from the transverse processes of the upper four or five thoracic vertebrae along with the cervicis, and the articular processes of the last three or four cervical vertebrae and is *inserted* into the posterior margin of the mastoid process, deep to the Splenius capitis and Sternocleidomastoideus. In the cranial part of the neck, where the capitis and cervicis diverge toward their insertions, the Longissimus mass is crossed by the Splenius cervicis. The latter's insertion separates that of the Longissimus capitis from that of the Longissimus cervicis, and where it crosses them, their muscular fasciculi are replaced by a tendinous inscription.

Action.—Extends the head; the muscle of one side acting alone bends the head to the same side and rotates the face toward that side.

Nerves.—Branches of the dorsal primary divisions of the middle and lower cervical nerves.

Medial Column

Spinalis
S. thoracis
S. cervicis
S. capitis

The **Spinalis thoracis**, the medial continuation of the Sacrospinalis, is scarcely separable as a distinct muscle. It is situated at the medial side of the Longissimus thoracis, and is intimately blended with it; it *arises* by three or four tendons from the spinous processes of the first two lumbar and the last two thoracic vertebrae: these, uniting, form a small muscle which is *inserted* by separate tendons into the spinous processes of the upper thoracic vertebrae, the number varying from four to eight. It is intimately united with the Semispinalis dorsi, situated deep to it.

The **Spinalis cervicis** (*Spinalis colli*) is an inconstant muscle, which *arises* from the caudal part of the ligamentum nuchae, the spinous process of the seventh cervical, and sometimes from the spinous processes of the first and second thoracic vertebrae, and is *inserted* into the spinous process of the axis, and occasionally into the spinous processes of the two vertebrae caudal to it.

The **Spinalis capitis** (*Biventer cervicis*) is usually inseparably connected with the Semispinalis capitis (see below).

Action.—Extend the vertebral column.

Nerves.—Branches of the dorsal primary divisions of the spinal nerves.

B. Transversospinal Muscles
(*m. transversospinalis*) (Fig. 6–14)

The **Semispinalis thoracis** consists of thin, narrow, fleshy fasciculi, interposed between tendons of considerable length. It *arises* by a series of small tendons from the transverse processes of the sixth to the tenth thoracic vertebrae, and is *inserted*, by tendons, into the spinous processes of the first four thoracic and last two cervical vertebrae.

The **Semispinalis cervicis** (*Semispinalis colli*), thicker than the preceding, *arises* by a series of tendinous and fleshy fibers from the transverse processes of the first five or six thoracic vertebrae, and is inserted into the cervical spinous processes, from the axis to the fifth inclusive. The fasciculus connected with the axis is the largest, and is chiefly muscular in structure.

Action.—Extend the vertebral column and rotate it toward the opposite side.

Nerves.—Branches of the dorsal primary divisions of the spinal nerves.

The **Semispinalis capitis** (*Complexus*) is situated at the cranial and dorsal part of the neck, deep to the Splenius, and medial to the Longissimus cervicis and capitis. It *arises* by a series of tendons from the tips of the transverse processes of the first six or seven thoracic and the seventh cervical vertebrae, and from the articular processes of the three cervical cranial to this. The tendons, uniting, form a broad muscle, which passes cranialward, and is *inserted* between the superior and inferior nuchal lines of the occipital bone. The medial part, usually more or less distinct from the remainder of the muscle, is frequently termed the **Spinalis capitis**; it is also named the **Biventer cervicis** since it is traversed by an imperfect tendinous intersection.

Action.—Extends the head and rotates it toward the opposite side.

Nerves.—Branches of the dorsal primary divisions of the cervical nerves.

The **Multifidus** (*Multifidus spinae*) (Fig. 6–14) consists of a number of fleshy and tendinous fasciculi which fill up the groove on either side of the spinous processes of the vertebrae from the sacrum to the axis. In the sacral region, these fasciculi *arise* from the back of the sacrum, as low as the fourth sacral foramen, from the aponeurosis of origin of the Sacrospinalis, from the medial surface of the posterior superior iliac spine, and from the posterior sacro-iliac ligaments; in the lumbar region, from all the mammillary processes; in the thoracic region, from all the transverse processes; and in the cervical region, from the articular processes of the last four vertebrae. Each fasciculus ascends oblique-ly, crossing over from two to four vertebrae in its course toward the middle line, and is inserted into the spinous process of one of the vertebrae, from the last lumbar to the axis. The fasciculi vary in length and depth of position; the longest and most superficial pass from one vertebra to the fifth above; those somewhat deeper are shorter and cross three vertebrae; the deepest and short-est cross two. The Rotatores longi (see below) are sometimes included in the Multifidus.

Action.—Extends the vertebral column and rotates it toward the opposite side.
Nerves.—Branches of the dorsal primary divisions of the spinal nerves.

The **Rotatores** (*Rotatores spinae*) are a series of small muscles which form the deepest layer in the groove between the spinous and transverse processes. They lie deep to the Multifidus and cannot be distin-guished readily from its deepest fibers. They are found along the entire length of the vertebral column from the sacrum to the axis. They arise from the transverse process of one vertebra and insert at the base of the spinous process of the vertebra above. The **Rotatores longi** cross one vertebra in their oblique course. The **Rotatores breves** insert in the next succeed-ing vertebra and run in an almost horizontal direction.

Action.—Extend the vertebral column and rotate it toward the opposite side.
Nerves.—Branches of the dorsal primary divisions of the spinal nerves.

The **Interspinales** fasciculi, placed in pa spinous processes of the c brae, one on either side of th ligament. In the *cervical region* most distinct, and consist of six pa first being situated between the axis third vertebra, and the last between seventh cervical and the first thoracic. They are small narrow bundles, attached to the apices of the spinous processes. In the *thoracic region,* they are found between the first and second vertebrae, and some-times between the second and third, and between the eleventh and twelfth. In the *lumbar region* there are four pairs in the intervals between the five lumbar vertebrae. There is also occasionally one between the last thoracic and first lumbar, and one between the fifth lumbar and the sacrum.

Action.—Extend the vertebral column.
Nerves.—Branches of the dorsal primary divisions of the spinal nerves.
The **Extensor coccygis** is a slender muscular fasciculus, which is not always present; it extends over the caudal part of the posterior surface of the sacrum and coccyx. It *arises* by tendinous fibers from the last segment of the sacrum, or first piece of the coccyx, and passes caudalward to be *inserted* into the lower part of the coccyx. It is a rudiment of the Extensor muscle of the caudal vertebrae of the lower animals.

The **Intertransversarii** (*Intertransversales*) are small muscles placed between the transverse processes of the vertebrae. In the *cervical region* they are best developed, consisting of rounded muscular and tendi-nous fasciculi, and are placed in pairs, passing between the anterior and the poste-rior tubercles respectively of the transverse processes of two contiguous vertebrae, and separated from one another by a ventral primary division of the cervical nerve which lies in the groove between them. The muscles connecting the anterior tubercles are termed the **Intertransversarii anteriores;** those between the posterior tubercles, the **Intertransversarii posteriores.** There are seven pairs of these muscles, the first pair being between the atlas and axis, and the last pair between the seventh cervical and first thoracic vertebrae. In the *thoracic*

trans-
oracic
sverse
e first
y are
f the
g the
sverse
, the
r set,
from
to the

... are short muscular
... irs between the
... tiguous verte-
... e interspinal
... they are
... irs, the
... and
... he

Action.—Bend the vertebral column laterally.

Nerves.—The anteriores, posteriores, and laterales by branches of the ventral primary divisions of the spinal nerves; the mediales by branches of the dorsal primary divisions.

II. THE SUBOCCIPITAL MUSCLES

(Figs. 6–14, 6–15)

Rectus capitis posterior major
Rectus capitis posterior minor
Obliquus capitis inferior
Obliquus capitis superior

The **Rectus capitis posterior major** (*Rectus capitis posticus major*) *arises* by a pointed tendon from the spinous process of the axis, and, becoming broader as it ascends, is *inserted* into the lateral part of the inferior nuchal line of the occipital bone and the surface of the bone immediately inferior to the line. As the muscles of the two sides pass laterally and cranially, they leave between them a triangular space,

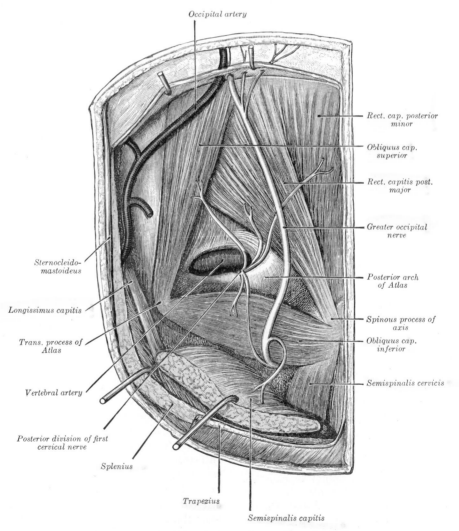

Occipital artery

Rect. cap. posterior minor

Obliquus cap. superior

Rect. capitis post. major

Greater occipital nerve

Posterior arch of Atlas

Spinous process of axis

Obliquus cap. inferior

Semispinalis cervicis

Sternocleido-mastoideus

Longissimus capitis

Trans. process of Atlas

Vertebral artery

Posterior division of first cervical nerve

Splenius

Trapezius

Semispinalis capitis

Fig. 6–15.—The left suboccipital triangle and muscles.

which the Recti capitis posteriores minores occupy.

Action.—Extends the head and rotates it to the same side.

Nerve.—A branch of the dorsal ramus of the suboccipital nerve.

The **Rectus capitis posterior minor** (*Rectus capitis posticus minor*) *arises* by a narrow pointed tendon from the tubercle on the posterior arch of the atlas, and, widening as it ascends, is *inserted* into the medial part of the inferior nuchal line of the occipital bone and the surface between it and the foramen magnum.

Action.—Extends the head.

Nerve.—A branch of the dorsal primary division of the suboccipital nerve.

The **Obliquus capitis inferior** (*Obliquus inferior*), the larger of the two Oblique muscles, *arises* from the apex of the spinous process of the axis, and passes laterally and slightly upward, to be *inserted* into the inferior and dorsal part of the transverse process of the atlas.

Action.—Rotates the atlas, turning the face toward the same side.

Nerve.—A branch of the dorsal primary division of the suboccipital nerve.

The **Obliquus capitis superior** (*Obliquus superior*) *arises* by narrow tendinous fibers from the superior surface of the transverse process of the atlas, joining with the insertion of the preceding. It passes cranialward and medialward, and is *inserted* into the occipital bone, between the superior and inferior nuchal lines, lateral to the Semispinalis capitis.

Action.—Extends the head and bends it laterally.

Nerve.—A branch of the dorsal primary division of the suboccipital nerve.

The **Suboccipital Triangle.**—Between the Obliqui and the Rectus capitis posterior major is the **suboccipital triangle.** It is bounded, *superiorly* and *medially*, by the Rectus capitis posterior major; *superiorly* and *laterally*, by the Obliquus capitis superior; *inferiorly* and *laterally*, by the Obliquus capitis inferior. It is covered by a layer of dense fibro-fatty tissue, situated deep to the Semispinalis capitis. The floor is formed by the posterior occipitoatlantal

membrane, and the posterior arch of the atlas. In the deep groove on the surface of the posterior arch of the atlas are the vertebral artery and the first cervical or suboccipital nerve (Fig. 6–15).

Group Actions.—The Erector spinae and its cranial continuations and the Spinales maintain the vertebral column in the erect posture; they also serve to bend the trunk dorsally when it is required to counterbalance the influence of any weight at the front of the body—as, for instance, when a heavy weight is suspended from the neck, or when there is any great abdominal distension, as in pregnancy or dropsy; the peculiar gait under such circumstances depends upon the vertebral column being drawn dorsally by the counterbalancing action of the Erector spinae. The muscles which form the continuation of the Erector spinae on to the head and neck steady those parts and fix them in the upright position. If the Iliocostalis lumborum and Longissimus thoracis of one side act, they serve to draw down the chest and vertebral column to the corresponding side. The Iliocostales cervicis, taking their fixed points from the cervical vertebrae, elevate those ribs to which they are attached; taking their fixed points from the ribs, both muscles help to extend the neck, while one muscle bends the neck to its own side. The Multifidus acts successively upon the different parts of the column; thus, the sacrum furnishes a fixed point from which the fasciculi of this muscle act upon the lumbar region, which in turn becomes the fixed point for the fasciculi moving the thoracic region, and so on throughout the entire length of the column. The Multifidus also serves to rotate the column, so that the front of the trunk is turned to the side opposite to that from which the muscle acts, this muscle being assisted in its action by the Obliquus externus abdominis.

III. THE MUSCLES OF THE THORAX

Intercostales externi
Intercostales interni
Subcostales
Transversus thoracis
Levatores costarum
Serratus posterior superior
Serratus posterior inferior
Diaphragm

Fascia.—The tela subcutanea (superficial fascia) and the outer layers of deep fascia of the thorax are described with the pectoral region (page 450) and back (page 447). The thoracic cage proper, composed of ribs and intercostal muscles, is covered inside and outside by thin membranes of deep fascia. The outer membrane is the external intercostal fascia, the inner one is the endothoracic fascia.

The **external intercostal fascia** covers the external surface of the Intercostales externi, the anterior intercostal membranes and the intervening surfaces of the ribs, costal cartilages and sternum. Cranially, it is continuous with the scalenus fascia; caudally, with the fascia between the external and internal oblique muscles of the abdomen. Dorsally, it splits at the border of the Erector spinae into an outer membrane, the fascia thoracolumbalis, and an inner membrane which continues on the surface of the Intercostales externi and Levatores costarum and forms an intermuscular septum between these muscles and the Erector spinae.

Between the Intercostales externi and interni is a thin layer of fascia to which both muscles are adherent.

The **endothoracic fascia** (*fascia endothoracica*) is the internal investing fascia, that is, the deep fascia which lines the inside of the thoracic cavity. It covers the internal surface of the Intercostales interni and intervening ribs, the Subcostales, Transversus thoracis, and Diaphragm, and dorsally it includes the thoracic portion of the prevertebral fascia which covers the bodies of the vertebrae and intervertebral disks. It is continuous, cranially, with the cervical prevertebral fascia, with the scalenus fascia (Sibson's) along the inner border of the first rib, and, behind the sternum, with the middle cervical fascia. It covers the entire thoracic surface of the diaphragm and is continuous with the internal investing fascia of the abdominal cavity (transversalis fascia, endoabdominal fascia) dorsal to the diaphragm at the lumbocostal arches and through the aortic hiatus.

The **subserous fascia** (*visceral fascia*) (page 375) intervenes between the endothoracic fascia and the pleura, and provides the connective tissue investment for the mediastinal structures. It is more fully described with the lungs.

The **Intercostales** (*Intercostal muscles*) (Fig. 6–20) are two thin planes of muscular and tendinous fibers occupying each of the intercostal spaces. They are named **external** and **internal** from their surface relations— the external being superficial to the internal.

The **Intercostales externi** (*External intercostals*) are *eleven* in number on either side. They extend from the tubercles of the ribs dorsally to the cartilages of the ribs ventrally, where they end in thin membranes, the **anterior intercostal membranes**, which are continued forward to the sternum. Each *arises* from the caudal border of a rib, and is *inserted* into the cranial border of the rib below. In the last two spaces they extend to the ends of the cartilages, and in the first two or three spaces they do not quite reach the ends of the ribs. They are thicker than the Intercostales interni, and their fibers are directed obliquely caudally and laterally on the dorsum of the thorax, and caudally, ventrally and medially on the ventrum.

Action.—Draw adjacent ribs together. With the first rib fixed by the Scaleni, they lift the ribs, increasing the volume of the thoracic cavity.

Nerves.—Intercostal nerves.

Variations.—Continuation with the Obliquus externus or Serratus anterior: A *Supracostalis muscle*, from the anterior end of the first rib down to the second, third or fourth ribs occasionally occurs.

The **Intercostales interni** (*Internal intercostals*) are also *eleven* in number on either side. They commence ventrally at the sternum, in the interspaces between the cartilages of the true ribs, and at the ventral extremities of the cartilages of the false ribs, and extend dorsalward as far as the angles of the ribs, whence they are continued to the vertebral column by thin aponeuroses, the **posterior intercostal membranes**. Each *arises* from the ridge on the inner surface of a rib, as well as from the corresponding costal cartilage, and is *inserted* into the cranial border of the rib below. Their fibers are also directed obliquely, but pass in a direction perpendicular to those of the Intercostales externi.

The **Subcostales** (*Intracostales*) consist of muscular and aponeurotic fasciculi which are usually well developed only in the lower part of the thorax; each *arises* from the inner surface of one rib near its angle, and is *inserted* into the inner surface of the second or third rib below. Their fibers run in the same direction as those of the Intercostales interni.

Action.—Draw adjacent ribs together. With the last rib fixed by the Quadratus lumborum, they lower the ribs, decreasing the volume of the thoracic cavity.

Nerves.—Intercostal nerves.

The **Transversus thoracis** (*Triangularis sterni*) is a thin plane of muscular and tendinous fibers situated upon the inner surface of the ventral wall of the chest (Fig. 6–16). It *arises* on either side from the caudal third of the inner surface of the body of the sternum, from the dorsal surface of the xiphoid process, and from the sternal ends of the costal cartilages of the last three or four true ribs. Its fibers diverge cranially and laterally, to be *inserted* by slips into the caudal borders and inner surfaces of the costal cartilages of the second, third, fourth, fifth, and sixth ribs. The most caudal fibers of this muscle are horizontal in their direc-

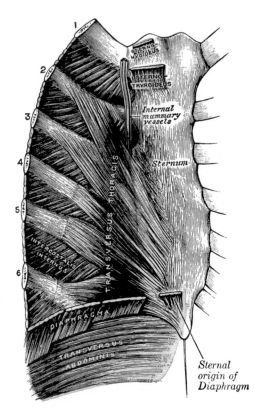

Fig. 6–16.—Dorsal surface of sternum and costal cartilages, showing Transversus thoracis.

tion, and are continuous with those of the Transversus abdominis; the intermediate fibers are oblique, while the most cranial are almost vertical. This muscle varies in its attachments, not only in different subjects, but on opposite sides of the same subject.

Action.—Draws the ventral portion of the ribs caudalward, decreasing the thoracic cavity.

Nerves.—Branches of the intercostal nerves.

The **Levatores costarum** (Fig. 6–14), *twelve* in number on either side, are small tendinous and fleshy bundles which *arise* from the ends of the transverse processes of the seventh cervical and upper eleven thoracic vertebrae; they pass obliquely downward and lateralward, like the fibers of the Intercostales externi, and each is *inserted* into the outer surface of the rib immediately caudal to the vertebra from which it takes origin, between the tubercle and the angle (**Levatores costarum breves**). Each of the four more caudal muscles divides into two fasciculi, one of which is inserted as above described; the other crosses to the second rib below its origin (**Levatores costarum longi**).

Action.—Raise the ribs, increasing the thoracic cavity; extend the vertebral column, bend it laterally and rotate it slightly toward the opposite side.

Nerves.—Branches of the intercostal nerves.

The **Serratus posterior superior** (*Serratus posticus superior*) is a thin, quadrilateral muscle, situated at the dorsal and cranial part of the thorax. It *arises* by a thin and broad aponeurosis from the caudal part of the ligamentum nuchae, from the spinous processes of the seventh cervical and first two or three thoracic vertebrae and from the supraspinal ligament. Inclining caudalward and lateralward it becomes muscular, and is *inserted*, by four fleshy digitations, into the cranial borders of the second, third, fourth, and fifth ribs, a little beyond their angles.

Action.—Raises the ribs to which it is attached, increasing the thoracic cavity.

Nerves.—Branches of the ventral primary divisions of the first four thoracic nerves.

Variations.—Increase or decrease in size and number of slips or entire absence.

The **Serratus posterior inferior** (*Serratus posticus inferior*) (Fig. 6–32) is situated at the junction of the thoracic and lumbar regions: it is of an irregularly quadrilateral form, broader than the preceding, and separated from it by a wide interval. It *arises* by a thin aponeurosis from the spinous processes of the last two thoracic and first two or three lumbar vertebrae, and from the supraspinal ligament. Passing obliquely upward and lateralward, it becomes fleshy, and divides into four flat digitations, which are *inserted* into the inferior borders of the last four ribs, a little beyond their angles. The thin aponeurosis of origin is intimately blended with the fascia thoracolumbalis and aponeurosis of the Latissimus dorsi.

Action.—Draws the ribs to which it is attached outward and downward, counteracting the inward pull of the Diaphragm.

Nerves.—Branches of the ventral primary divisions of the ninth to twelfth thoracic nerves.

Variations.—Increase or decrease in size and number of slips or entire absence.

The **Diaphragm** (Fig. 6–17) is a dome-shaped musculofibrous septum which separates the thoracic from the abdominal cavity, its convex cranial surface forming the floor of the former, and its concave caudal surface the roof of the latter. Its peripheral part consists of muscular fibers which take origin from the circumference of the thoracic outlet and converge to be *inserted* into a central tendon.

The muscular fibers may be grouped according to their origins into three parts—sternal, costal, and lumbar. The **sternal part** *arises* by two fleshy slips from the dorsum of the xiphoid process; the **costal part** from the inner surfaces of the cartilages and adjacent portions of the last six ribs on either side, interdigitating with the Transversus abdominis; and the **lumbar part** from aponeurotic arches, named the lumbo-

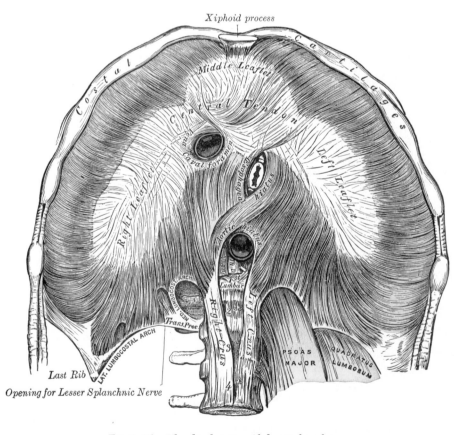

Fig. 6–17.—The diaphragm. Abdominal surface.

costal arches, and from the lumbar vertebrae by two pillars or **crura.** There are two lumbocostal arches, a medial and a lateral, on either side.

The **Medial Lumbocostal Arch** (*arcus lumbocostalis medialis [Halleri]; internal arcuate ligament*) is a tendinous arch in the fascia covering the cranial part of the Psoas major; medially, it is continuous with the lateral tendinous margin of the corresponding crus, and is attached to the side of the body of the first or second lumbar vertebra; laterally, it is fixed to the front of the transverse process of the first and, sometimes also, to that of the second lumbar vertebra.

The **Lateral Lumbocostal Arch** (*arcus lumbocostalis lateralis [Halleri]; external arcuate ligament*) arches across the cranial part of the Quadratus lumborum, and is attached, medially, to the ventral surface of the transverse process of the first lumbar vertebra, and, laterally, to the tip and caudal margin of the twelfth rib.

The Crura.—At their origins the crura are tendinous in structure, and blend with the ventral longitudinal ligament of the vertebral column. The **right crus,** larger and longer than the left, *arises* from the ventral surfaces of the bodies and intervertebral fibrocartilages of the first three lumbar vertebrae, while the **left crus** *arises* from the corresponding parts of the first two only. The medial tendinous margins of the crura pass ventrally and medially, and meet in the middle line to form an arch across the ventral aspect of the aorta; this arch is often poorly defined.

From this series of origins the fibers of the diaphragm converge to be inserted into the central tendon. The fibers arising from the xiphoid process are very short, and occasionally aponeurotic; those from the medial and lateral lumbocostal arches, and more especially those from the ribs and their cartilages, are longer, and describe marked curves as they ascend and converge to their insertion. The fibers of the crura diverge as they ascend, the most lateral being directed cranialward and lateralward to the central tendon. The medial fibers of the right crus ascend on the left side of the esophageal hiatus, and occasionally a fascic-

ulus of the left crus crosses the aorta and runs obliquely through the fibers of the right crus toward the vena caval foramen.

The **Central Tendon** of the diaphragm is a thin but strong aponeurosis situated near the center of the vault formed by the muscle, but somewhat closer to the ventral than to the dorsal part of the thorax, so that the dorsal muscular fibers are the longer. It is situated immediately caudal to the pericardium, with which it is partially blended. It is shaped somewhat like a trefoil leaf, consisting of three divisions or leaflets separated from one another by slight indentations. The right leaflet is the largest, the middle, directed toward the xiphoid process, the next in size, and the left the smallest. In structure the tendon is composed of several planes of fibers, which intersect one another at various angles and unite into straight or curved bundles—an arrangement which gives it additional strength.

Openings in the Diaphragm.—The diaphragm is pierced by a series of apertures to permit of the passage of structures between the thorax and abdomen. Three large openings—the **aortic,** the **esophageal,** and the **vena caval**—and a series of smaller ones are described.

The **aortic hiatus** is the most caudal and dorsal of the large apertures; it lies at the level of the twelfth thoracic vertebra. Strictly speaking, it is not an aperture in the diaphragm but an osseoaponeurotic opening between it and the vertebral column, and therefore behind the diaphragm; occasionally some tendinous fibers are prolonged across the bodies of the vertebrae from the medial parts of the caudal ends of the crura and pass dorsal to the aorta, thus converting the hiatus into a fibrous ring. The hiatus is situated slightly to the left of the middle line, and is bounded ventrally by the crura, and dorsally by the body of the first lumbar vertebra. Through it pass the aorta, the azygos vein, and the thoracic duct; occasionally the azygos vein is transmitted through the right crus.

The **esophageal hiatus** is situated in the muscular part of the diaphragm at the level of the tenth thoracic vertebra, and is elliptical in shape. It is placed cranial, ventral and a little to the left of the aortic hiatus, and transmits the esophagus, the vagus nerves, and some small esophageal blood vessels.

The **vena caval foramen** is the most cranial of the three, and is situated at about the level of the fibrocartilage between the eighth and ninth thoracic vertebrae. It is quadrilateral in form, and is placed at the junction of the right and middle leaflets of the central tendon, so that its margins are tendinous. It transmits the inferior vena cava, the wall of which is adherent to the margins of the opening, and some branches of the right phrenic nerve.

Of the **lesser apertures**, two in the right crus transmit the greater and lesser right splanchnic nerves; three in the left crus give passage to the greater and lesser left splanchnic nerves and the hemiazygos vein. The trunks of the sympathetic usually enter the abdominal cavity dorsal to the diaphragm, under the medial lumbocostal arches.

On either side two small intervals exist at which the muscular fibers of the diaphragm are deficient and are replaced by areolar tissue. One between the sternal and costal parts transmits the superior epigastric branch of the internal mammary artery and some lymphatics from the abdominal wall and convex surface of the liver. The other, between the fibers springing from the medial and lateral lumbocostal arches, is less constant; when this interval exists, the upper and back part of the kidney is separated from the pleura by areolar tissue only.

Action.—Draws the central tendon downward. This action has two effects: (*a*) it tends to increase the volume and decrease the pressure within the thoracic cavity, and (*b*) it tends to decrease the volume and increase the pressure within the abdominal cavity. During inspiration, the lowering of the diaphragm decreases the pressure within the thorax, and air is forced into the lungs through the open larynx and trachea by the pressure of the atmosphere. At the same time, the descending diaphragm presses against the abdominal viscera, forcing them downward against the passive resistance of the abdominal and pelvic muscles, and causing the anterior abdominal wall to protrude slightly. The pressure within the abdominal cavity is greatly increased when

the abdominal muscles and the diaphragm contract actively at the same time, and this increase in pressure tends to make the abdominal viscera discharge their contents as in micturition, defecation, emesis, and parturition. The portion of the diaphragm about the esophageal hiatus is supposed to have a sphincteric action on the esophagus.

Nerve.—The phrenic nerve from the cervical plexus, containing mainly fibers from the fourth, but also some from the third and fifth cervical nerves.

Variations.—The sternal portion of the muscle is sometimes wanting and more rarely defects occur in the lateral part of the central tendon or adjoining muscle fibers.

The **Movements of Respiration** are inspiration, caused by an increase in the thoracic cavity, and expiration, caused by a decrease in the cavity. The increase in the volume of the cavity is the result of muscular action and is brought about in two ways: (*a*) by the descent of the diaphragm from contraction of its muscle, and (*b*) by the expansion of the thoracic wall through the action of certain muscles on the ribs, sternum, and vertebral column. The decrease in volume with expiration may be passive, due to the elastic recoil of the thoracic wall and the tissues of the lungs and bronchi. The decrease may also be the result of muscular action, in which case, (*a*) the abdominal muscles force the diaphragm upward by increasing the abdominal pressure, and (*b*) the thoracic wall is contracted by the action of certain muscles on the ribs and vertebral column.

Quiet Inspiration.—The diaphragm contracts, increasing the vertical diameter of the thoracic cavity. The first and second ribs remain fixed by the inertia and resistance of the cervical structures, and the remaining ribs, except the last two, are brought upward toward them by the contraction of the Intercostales externi. The upward motion of the ribs, due to their position and to the obliquity of their axis of rotation, enlarges the anteroposterior and transverse diameters of the thorax according to the movements described on page 309.

Traditionally, there are two types of quiet inspiration, diaphragmatic and costal, but under normal conditions there is a mixture of both types, although one or the other may show decided predominance in individual cases. Diaphragmatic breathing is also called abdominal because the visible result of contraction of the diaphragm is protrusion of the abdominal wall. To contrast with this, costal breathing is also called thoracic breathing. Costal breathing predominates in recumbency and it is said to be more frequent in women, while diaphragmatic is more frequent in men.

Electromyographic studies indicate that both intercostal muscles are slightly active during quiet inspiration but with no rhythmic increase and decrease, acting more to keep the ribs in a constant position than to expand the thorax. The scalene muscles, however, do show rhythmic activity (Jones and Pauly '57). There is some evidence that only the first three intercostals are thus active and that the fourth to seventh are not (Koepke *et al.* '58). Further analysis of the variations in the electromyographic studies is given by Basmajian ('62).

Quiet Expiration.—The normal resting position of the thorax is that found at the end of a quiet expiration. This position is restored without muscular effort after a quiet inspiration by the recoil of the structures which were displaced by the inspiratory act. The displacement of the anterior abdominal wall is overcome by the tonus of the abdominal muscles. The ribs are restored from their displacement by the elasticity of the ligaments and cartilages which hold them in place. The extensive network of elastic fibers which permeates the pulmonary tissue retracts the lungs wherever possible, and the bronchial tree, which has been elongated by the descent of the diaphragm, helps to draw the latter back up by its elastic recoil.

Deep Inspiration.—All the actions of quiet inspiration are increased in extent. In addition, the first two ribs are raised by the Scaleni and the Sternocleidomastoideus, and the remaining ribs are raised more forcibly by the additional action of the Levatores costarum and the Serratus posterior superior. The ribs are raised still farther by a straightening of the vertebral column through contraction of the Sacrospinalis. After the abdominal viscera have been forced downward by the diaphragm to a considerable extent, the abdominal muscles offer increased resistance, and the viscera may then act as a point of fixation for the diaphragm so that its further contraction raises the ribs.

Forced Inspiration.—In patients with great air hunger, all the muscles of the body seem to combine and coördinate to assist in breathing. The Levator scapulae, the Trapezius, and the Rhomboidei elevate and fix the scapula which is then used as an origin by the Pectoralis minor to draw the ribs upward. If the patient further fixes the shoulder girdle by grasping the back of a chair or end of the bed, the Pectoralis major and the Serratus anterior will also raise the ribs.

Forced Expiration.—In forced expiration, muscles are called into play. The last two ribs are pulled downward and fixed by the Quadratus lumborum, and the other ribs are drawn downward toward them by the Intercostales interni and the Serrati posteriores inferiores. The muscles of the abdominal wall, by pressing on the abdominal viscera, force the diaphragm upward, and the same muscles, by flexing the vertebral column, assist in lowering the ribs.

Position of the Diaphragm.—The height of the diaphragm is constantly varying during respiration; it also varies with the degree of distension of the stomach and intestines and with the size of the liver. After a forced expiration the right cupola is on a level ventrally with the fourth costal cartilage, at the side with the fifth, sixth, and seventh ribs, and dorsally with the eighth rib; the left cupola is a little lower than the right. The absolute range of movement between deep inspiration and deep expiration averages in the male and female 30 mm on the right side and 28 mm on the left; in quiet respiration the average movement is 12.5 mm on the right side and 12 mm on the left.

Radiography shows that the height of the diaphragm in the thorax varies considerably with the position of the body. It stands highest when the body is horizontal and the patient on his back, and in this position it performs the largest respiratory excursions with normal breathing. When the body is erect the dome of the diaphragm falls, and its respiratory movements become smaller. The dome falls still lower when the sitting posture is assumed, and in this position its respiratory excursions are smallest. These facts may, perhaps, explain why it is that patients suffering from severe dyspnea are most comfortable and least short of breath when they sit up. When the body is horizontal and the patient on his side, the two halves of the diaphragm do not behave alike. The uppermost half sinks to a level lower even than when the patient sits, and moves little with respiration; the lower half rises higher in the thorax than it does when the patient is supine, and its respiratory excursions are much increased. In unilateral disease of the pleura or lungs analogous interference with the position or movement of the diaphragm can generally be observed radiographically.

It appears that the position of the diaphragm in the thorax depends upon three main factors, viz.: (1) the elastic retraction of the lung tissue, tending to pull it upward; (2) the pressure exerted on its under surface by the viscera; this naturally tends to be a negative pressure, or downward suction, when the patient sits or stands, and positive, or an upward pressure, when he reclines; (3) the intra-abdominal tension due to the abdominal muscles. These are in a state of contraction in the standing position and not in the sitting; hence the diaphragm, when the patient stands, is pushed up higher than when he sits.

The following figures represent the average changes which occur during deepest possible respiration. The manubrium sterni moves 30 mm in an upward and 14 mm in a forward direction; the width of the subcostal angle, at a level of 30 mm below the articulation between the body of the sternum and the xiphoid process, is increased by 26 mm; the umbilicus is retracted and drawn upward for a distance of 13 mm.

IV. THE FASCIAE AND MUSCLES OF THE ABDOMEN

The muscles of the abdomen may be divided into two groups: (1) the **antero-lateral muscles**; (2) the **posterior muscles**.

1. The Anterolateral Muscles of the Abdomen

Obliquus externus abdominis
Obliquus internus abdominis
Transversus abdominis
Rectus abdominis
Pyramidalis

The **subcutaneous fascia** (*tela subcutanea; superficial fascia*) (Fig. 6–20) of the anterior abdominal wall is soft and movable, and likely to contain fat. It is continuous, cranially, with the subcutaneous fascia of the thorax; caudally, with that of the thigh and external genitalia; and laterally, it gradually becomes tougher and more resistant as it changes into the fascia of the back. In the caudal portion, below the umbilicus, its superficial and deep layers are unusually distinct and can be separated easily by dissection. This unaccustomed divisibility and certain other peculiar features have been emphasized by giving the two layers in this region special names, Camper's fascia and Scarpa's fascia.

The **superficial layer of the subcutaneous fascia** (*Camper's fascia*) is a genuine panniculus adiposus. It may be several centimeters thick in obese individuals, in which case it is likely to be irregularly divisible into laminae. It is continuous over the inguinal ligament with the similar and corresponding layer of the thigh. In the

male, as it continues down on the penis and scrotum, it loses its fat and, fusing with the deep layer, assists in the formation of the special fascia of these organs called the dartos. In the female, it retains some of the adipose tissue as it is continued into the labia majora. In both sexes, it is prolonged dorsalward in the groove between the external genitalia and the thigh and is there continuous with the superficial layer of the subcutaneous fascia of the perineum and medial surface of the thigh.

The **deep layer of the subcutaneous fascia** (*Scarpa's fascia*) is a membranous sheet which usually contains little or no adipose tissue. It is composed, in considerable part, of yellow elastic fibers, and probably corresponds to the tunica abdominalis, an elastic layer which contributes to the support of the viscera and inguinal mammae in some lower mammals. It forms a continuous sheet across the middle line, and is attached to the linea alba as it passes across it. Superiorly and laterally, it loses its identity as a special layer in the subcutaneous fascia of the upper abdomen and back. Inferiorly, it passes over the inguinal ligament and is securely attached either to the ligament itself or to the fascia lata just beyond it. Inferior to the ligament, the corresponding layer is called the fascia cribrosa as it covers and fills in the saphenous opening (*fossa ovalis*). At the medial end of the inguinal ligament, it passes over the external inguinal ring without being attached and continues into the penis and scrotum. It continues along the groove between the scrotum (labium majus) and thigh into the perineum where it is called the fascia of Colles. As the subcutaneous fascia comes to lie under the skin of the scrotum and penis, its two layers are fused into a single tunic called the dartos. Here the superficial layer loses its fat and acquires a layer of scattered smooth muscle cells which attach to the skin and throw it into folds or rugae. In the middle line, over the symphysis pubis, it is thickened by the addition of numerous, closely set, strong bands which extend down to the dorsum and sides of the penis forming the **ligamentum fundiforme penis**.

The fascial cleft which separates the deep layer of the subcutaneous fascia (Scarpa's) from the deep fascia over the caudal portion of the aponeurosis of the Obliquus externus is quite definite and of considerable clinical interest because of its continuity with a similar cleft in the perineum. It is limited superiorly, toward the umbilicus, by an adhesion between the subcutaneous fascia (Scarpa's) to the deep fascia, and laterally, by the former's closer attachment over the muscular portion of the external oblique. It is limited, inferiorly and laterally, by a firm attachment either in the inguinal ligament or to the fascia lata just inferior to it. Over the medial portion of the inguinal ligament and the superficial inguinal ring, however, the two fasciae are not attached, so that the cleft follows along the narrow groove between the scrotum (labium) and thigh, and is continuous with the cleft between the subcutaneous (Colles') fascia and the external perineal deep fascia. From this groove, it is continuous medially with the cleft under the very movable dartos of the scrotum and penis, but it is abruptly limited laterally by an attachment to the deep fascia over the pubic ramus where the adductor muscles of the medial side of the thigh originate. In obese individuals there may be accumulations of adipose tissue between the cleft and the deep surface of Scarpa's fascia. The superficial inferior epigastric and circumflex iliac blood vessels lie between Camper's and Scarpa's fasciae but are attached to the superficial surface of Scarpa's layer.

Deep fascia (*Fascia innominata; Gallaudet's fascia*).—The **outer investing layer** of deep fascia is easily identified in the lateral portion of the anterior abdominal wall where it covers the fleshy fibers of the Obliquus externus abdominis. It is continuous with the fascia of the Latissimus dorsi and Pectoralis major. More medially, over the aponeurosis of the Obliquus externus, it is so firmly adherent that it may escape recognition. Its presence is easily demonstrated in dissection, however, by scraping it back until the glistening fibers of the aponeurosis beneath are revealed. Cranially, it covers the superior end of the rectus sheath and is continuous with the pectoral fascia. Caudally, it is firmly attached to the inguinal ligament and joins the deep fascia emerging from under that ligament to become the fascia lata of the thigh. It covers

the superficial inguinal ring as a distinct and separate layer, and there, reinforced by the fascia of the inner surface of the aponeurosis, gives rise to a tubular prolongation, the external spermatic (intercrural) fascia, which is the coat of the spermatic cord and testis just deep to the dartos. Near the middle line, it is attached to the pubic bone and is then continuous with deep fascia investing the penis. Over the lower end of the linea alba it is thickened into a strong, fibrous triangle, the **suspensory ligament of the penis**, which attaches the dorsum of the penis to the symphysis and arcuate pubic ligament. At the medial end of the inguinal ligament and lowest medial portion of the aponeurosis of the external oblique, it is attached to the pubic ramus and arcuate pubic ligament and is then continuous posteriorly, over the Ischiocavernosus muscle, with the external perineal fascia. Laterally in this region, beyond its attachment to the ramus, it is continuous with the fascia covering the adductor muscles of the medial side of the thigh.

The **Obliquus externus abdominis** (*External or descending oblique muscle*) (Fig. 6–18), situated on the lateral and ventral parts of the abdomen, is the largest and the most superficial of the three flat muscles in this region. It is broad, thin, and irregularly quadrilateral, its muscular portion occupying the lateral, its aponeurosis the ventral wall of the abdomen. It *arises*, by eight fleshy digitations, from the external surfaces and inferior borders of the lower eight ribs; these digitations are arranged in an oblique line which runs caudalward and dorsalward, the cranial ones being attached close to the cartilages of the corresponding ribs, the most caudal to the apex of the cartilage of the last rib, the intermediate ones to the ribs at some distance from their cartilages. The five superior serrations increase in size caudally, and are received between corresponding processes of the Serratus anterior; the three lower ones diminish in size and receive between them corresponding processes from the Latissimus dorsi. The muscular fasciculi from the last two ribs are nearly vertical and are *inserted* into the anterior half of the outer lip of the iliac crest; the rest of the fasciculi, directed caudalward

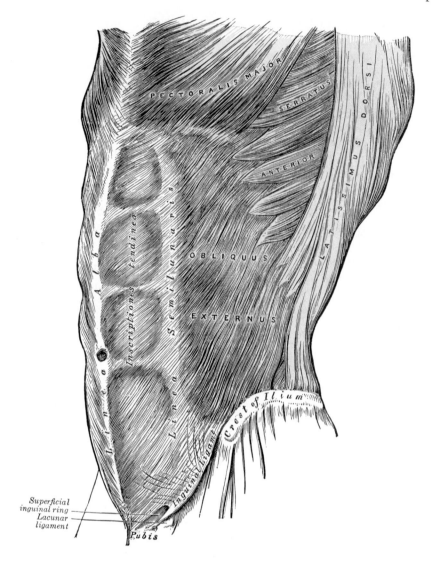

Fig. 6–18.—The Obliquus externus abdominis.

and ventralward, terminate in the broad abdominal aponeurosis by means of which most of the muscle reaches its final *insertion*, the linea alba.

Action.—Compresses the abdominal contents, assisting in micturition, defecation, emesis, parturition, and forced expiration. Both sides acting together flex the vertebral column, drawing the pubis toward the xiphoid process. One side alone bends the vertebral column laterally and rotates it, bringing the shoulder of the same side forward.

Nerves.—Branches of the eighth to twelfth intercostal, and the iliohypogastric and ilio-inguinal nerves.

The **aponeurosis of the Obliquus externus Abdominis** (Figs. 6–18, 6–19) is a strong membrane whose tendinous bundles continue, for the most part, in the direction of the muscular fasciculi caudalward and ventralward. It covers the entire ventral surface of the abdomen, lying superficial to the Rectus abdominis and helping to form its sheath. The fibers of the two sides interlace in the middle line to form the linea alba, the real insertion of the muscle, which extends from the xiphoid process to the symphysis pubis. The cranialmost part of the aponeurosis serves as the origin for the inferior fibers of the Pectoralis major. The caudalmost portion ends in a tendinous border which is termed the **inguinal ligament** although it is aponeurotic rather than ligamentous in function. Two different portions can be identified. The lateral portion, formed by the tendinous fibers of the muscular fasciculi arising from the tenth rib, is attached to the anterior superior iliac spine in passing and is inserted into the iliac fascia as the latter leaves the abdomen to become part of the fascia lata of the thigh (see page 490). The fibers of the Iliopsoas pass deep to this band of insertion. The medial portion of the border of the aponeurosis, formed by the tendinous fibers of fasciculi arising from the ninth rib, arch over the femoral vessels as they emerge from the abdomen into the thigh, and are inserted into the pectineal fascia and adjacent part of the superior ramus of the pubic bone. This portion is named the **superficial inguinal arch** (Boyle '71). The deep arch is formed by the free border of the Transversus as it joins the Obliquus internus in forming the falx inguinalis. The bundles of the arched border near the medial end are curled inward to a more deeply placed attachment to the pectineal fascia thus providing a kind of sling for the spermatic cord (Fig. 6–19). Part of this attachment along the pectineal fascia and pectineal line lateral to the pubic tubercle leaves a crescentic border named the **lacunar ligament.** Other fibers at the deeper attachment double back cranialward toward the linea alba deep to the main aponeurosis in a small triangular sheet called the **reflected inguinal ligament.** The tendinous bundles just above the arch separate from each other at the pubic tubercle leaving a narrow triangular opening, called the **superficial inguinal ring,** through which pass the spermatic cord in a male or the round ligament in a female. The bundles bordering the ring are called the crura. Lateral to the opening the aponeurosis contains, in addition to the bundles running in the usual direction, some scattered, transverse, reinforcing strands which sweep cranialward and toward the midline in curved lines. These are called **intercrural fibers** and are part of the aponeurosis not to be confused with the intercrural fascia (page 421). In the above description the aponeurosis has been treated

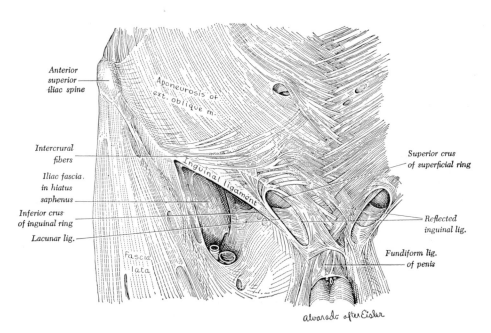

Fig. 6–19.—The aponeurosis of the Obliquus externus abdominis and the ligaments of the inguinal region. (Redrawn from Eisler.)

as if it belonged solely to the Obliquus externus. It also serves, however, as a part of the sheath of the Rectus abdominis and shares its band of attachment to the iliac fascia with the Obliquus internus and Transversus abdominis.

The **Inguinal Ligament** (*ligamentum inguinale; Poupart's ligament*) is the name given to the lower border of the aponeurosis of the Obliquus externus abdominis extending from the anterior superior iliac spine to the pubic tubercle. Its lateral half or third is firmly attached to the band in the iliac fascia where the latter is fused with the fascia innominata and transversalis fascia and continues into the thigh as fascia lata. The medial portion of the border, named the **superficial inguinal arch,** is a free border where the femoral sheath and vessels pass deep to it on their way into the thigh. The fibers of the arch curl under the adjacent aponeurosis to reach their insertion into the pectineal fascia and pubic bone lateral to the tubercle, thereby forming a narrow sling for support of the spermatic cord. The curving of the free border at this attachment is named the **lacunar ligament** (Fig. 6–19). At their attachment some fibers of this curved border continue lateralward along the pectineal line producing a tendinous ridge called the **pectineal ligament** (*Cooper's ligament*) (Fig. 5–25). Other fibers at the deeper attachment spread out medialward in a small triangular sheet called the **reflected inguinal ligament.** The ligamentous appearance of this border of the aponeurosis is produced if the band of attachment to the iliac fascia is dissected free and lifted up producing a dense and sometimes thickened band extending from the anterior superior spine of the ilium to the pubic tubercle. In this condition it appears that fibers of the Obliquus internus and Transversus have an origin from the band, that is, the ligament, although the effective origin is from the iliac fascia.

The **Lacunar Ligament** (*ligamentum lacunare; Gimbernat's ligament*) (Fig. 5–27) is the medial end of the inguinal ligament which is rolled under the spermatic cord and is attached along the pectineal fascia just lateral to the pubic tubercle. When it is viewed through the superficial inguinal ring, after the spermatic cord has

been removed, it appears to be a triangular fibrous membrane, about 1.25 cm long, with a crescentic base, concave laterally, and with an apex medially at the pubic tubercle (Fig. 6–19). It lies almost horizontally, in the erect posture, with the spermatic cord resting on its superior surface. Its concave lateral border is separated from the medial wall of the femoral canal by the femoral sheath.

The **Reflected Inguinal Ligament** (*ligamentum inguinale reflexum [Collesi]; triangular fascia*) (Fig. 6–22) is a triangular, tendinous sheet 2 or 3 cm wide extending from the medial part of the inguinal ring to the linea alba. The fibers arise from the attachment of the inguinal ligament to the pectineal line and course medialward spreading out cranialward deep to the main aponeurosis of the external oblique and interlacing with the fibers of the latter in the linea alba. The ligament may be independent, but more often it is fused either with the aponeurosis of the external oblique or with the falx inguinalis which lies deep to it. Frequently it seems to be entirely lacking.

The **Superficial Inguinal Ring** (*anulus inguinalis superficialis; subcutaneous inguinal ring; external inguinal ring*) (Fig. 6–19) is the opening in the aponeurosis of the Obliquus externus abdominis just cranial and lateral to the pubis, through which the spermatic cord (round ligament of the uterus) passes. It is a narrow triangle, pointing laterally and cranially in the direction of the fibers of the aponeurosis. Its base is at the crest of the pubis; the sides are the margins of the opening in the aponeurosis and are called the crura of the ring. The **lateral (inferior) crus** (*external pillar*) is the stronger and is formed by the portion of the inguinal ligament which is attached to the pubic tubercle; it is curved and turned under into a narrow sling upon which the spermatic cord rests. The **medial (superior) crus** (*internal pillar*), thin and flat, is merely the part of the aponeurosis next to the opening and is not marked off except as it is attached to the front of the symphysis pubis. The triangular opening in the aponeurosis is converted by fascia into an oval ring, 2.5 cm long and 1.25 cm wide. The fascia of the superficial surface of the

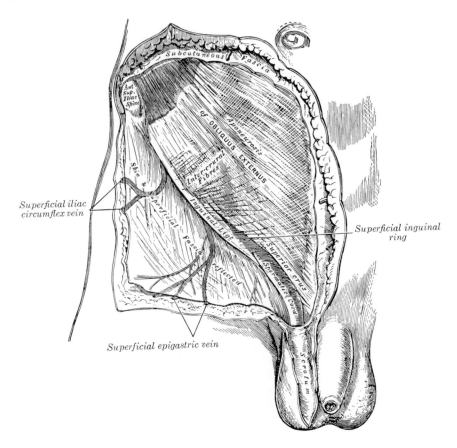

Fig. 6–20.—The superficial inguinal ring.

aponeurosis, called the fascia innominata by Gallaudet, fuses with the fascia of the deep surface and fills in the angular, lateral portion of the opening. A tubular prolongation of this fascia is continued down into the scrotum, enclosing the spermatic cord and testis in a sheath called the external spermatic fascia (*intercrural or intercolumnar fascia*). This intercrural fascia, so named because it occupies the space between the crura of the ring, is not to be confused with the intercrural fibers which are a part of the aponeurosis itself and reinforce the latter above the ring. The superficial inguinal ring gives passage to the ilioinguinal nerve as well as to the spermatic cord or round ligament; it is larger in men than in women.

The **Pectineal Ligament** (*Ligamentum pectinale; Cooper's ligament*) (Fig. 5–27) is a narrow band of strong aponeurotic fibers which continues laterally from the

lacunar ligament along the pectineal line of the pubis. It is firmly attached to the bone along this line and its medial end is continuous with the lacunar ligament. It diminishes in size gradually toward its lateral extremity at the iliopectineal eminence. It is aponeurotic rather than fascial in origin, and to it are attached parts of the iliopectineal, pectineal, and transversalis fasciae. It forms the posterior or deep boundary of the lacuna vasorum through which the femoral vessels pass under the inguinal ligament.

The **Obliquus internus abdominis** (*Internal* or *ascending oblique muscle*) (Fig. 6–21), situated in the lateral and ventral part of the abdominal wall, is of an irregularly quadrilateral form, and is smaller and thinner than the Obliquus externus under which it lies. It *arises* by fleshy fibers from the lateral half of the inguinal ligament and the nearby iliac fascia, from the anterior two-thirds of the middle lip of the iliac

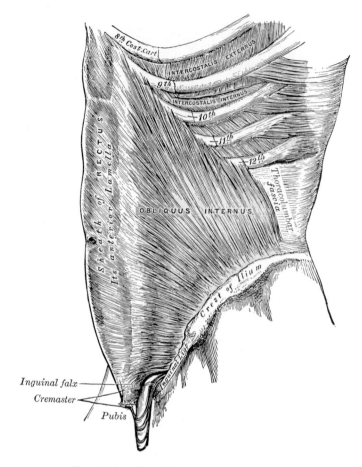

Fɪɢ. 6–21.—The Obliquus internus abdominis.

crest, and from the lower portion of the lumbar aponeurosis (posterior layer of the thoracolumbar fascia) near the crest. The posterior fasciculi are almost vertical and are *inserted* into the inferior borders of the cartilages of the last three or four ribs by fleshy digitations which appear to be continuations of the internal intercostal muscles. The remainder of the fasciculi which arise from the iliac crest diverge as they spread over the side of the abdomen and, in the region of the linea semilunaris, terminate in an aponeurosis which fuses with the aponeuroses of the externus and the transversus at a variable distance from the middle line. By means of the aponeurosis the muscle makes its final *insertion* into the linea alba. The fibers of the aponeurosis assist in the formation of the rectus sheath, some passing ventral and some dorsal to

the latter muscle. The fasciculi arising from the inguinal ligament are less compact and paler than the rest, and descending, form an arch over the spermatic cord (round ligament). They terminate in a tendinous sheet, which they share with the Transversus, and which is not fused with the aponeurosis of the externus but is independently inserted into the pubis and medial part of the pectineal line behind the lacunar ligament, forming what is known as the **conjoined tendon** (*falx inguinalis*).

Action.—Compresses the abdominal contents, assisting in micturition, defecation, emesis, parturition, and forced expiration. Both sides acting together flex the vertebral column, drawing the costal cartilages caudalward toward the pubis. One side acting alone bends the vertebral column laterally and rotates it, bringing the shoulder of the opposite side ventralward.

Nerves.—Branches of the eighth to twelfth intercostal, and the iliohypogastric and ilioinguinal nerves.

Variations.—Occasionally, tendinous inscriptions occur from the tips of the tenth or eleventh cartilages or even from the ninth; an additional slip to the ninth cartilage is sometimes found; separation between iliac and inguinal parts may occur.

The **Cremaster** (Fig. 6–22) is a thin muscular layer whose fasciculi are separate and spread out over the spermatic cord in a series of loops. It *arises* from the middle of the inguinal ligament as a continuation of the Obliquus internus, and is *inserted* by a small pointed tendon into the tubercle and crest of the pubis and into the front of the sheath of the Rectus abdominis. The fasciculi form a compact layer as they lie within the inguinal canal, but after they pass out of the superficial inguinal ring they form a series of loops, the longest of which extend down as far as the testis and are

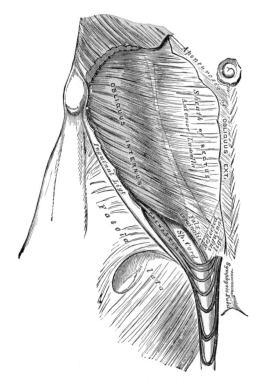

FIG. 6–22.—The Cremaster, falx inguinalis, and reflected inguinal ligament.

attached to the tunica vaginalis. The interval between the loops is occupied by fascia which is a fused continuation of the fasciae of the deep and superficial surfaces of the Obliquus internus and which may be called the cremasteric fascia. The muscular loops and the fascia together make up a single layer which forms the middle tunic of the spermatic cord.

Action.—Draws the testis up toward the superficial inguinal ring.

Nerve.—The genital branch of the genitofemoral nerve.

The **Transversus abdominis** (*Transversalis muscle*) (Fig. 6–23), so called from the direction of its fibers, is the most internal of the flat muscles of the abdomen, being placed immediately deep to the Obliquus internus. It *arises* by fleshy fibers from the lateral third of the inguinal ligament, from the anterior three-fourths of the inner lip of the iliac crest, from the thoracolumbar fascia, and from the inner surface of the cartilages of the last six ribs. The origin from the ribs is by means of digitations which are separated from similar digitations of the Diaphragm (Fig. 6–16) by a narrow fibrous raphe; viewed from the interior of the abdomen, the two muscles appear to be components of a single stratum of muscle. The fasciculi of the Transversus, except the caudalmost, pass horizontally ventralward, and terminate in an aponeurosis which fuses with the aponeurosis of the Obliquus internus, and joins the aponeurosis of the opposite side to form the *insertion* of the muscle into the linea alba. The aponeurosis assists in the formation of the sheath of the Rectus as follows: the more cranial portion passes entirely dorsal to the Rectus; a portion extending for a variable distance below the umbilicus splits and interdigitates with the aponeurosis of the Obliquus internus contributing to both the anterior and posterior layers of the sheath; and the most caudal part beyond a curved fibrous border approximately half way between the umbilicus and the pubis, the arcuate line, passes entirely ventral to the Rectus. The fasciculi of the lowest portion of the muscle pass caudalward as well as ventralward, *terminating* in the falx inguinalis or conjoined tendon along with the lowest fasciculi of the Obliquus internus. The muscle ends inferiorly in a free border which forms an arch extending from the lateral part of the inguinal ligament to the pubis, a short distance above the deep inguinal ring and spermatic cord.

Action.—Constricts the abdomen, compressing the contents and assisting in micturition, defecation, emesis, parturition, and forced expiration.

Nerves.—Branches of the seventh to twelfth intercostal, and the iliohypogastric and ilioinguinal nerves.

Variations.—It may be more or less fused with the Obliquus internus or absent. The spermatic cord may pierce its lower border. Slender muscle slips from the iliopectineal line to transversalis fascia, the aponeurosis of the Transversus abdominis, or the outer end of the linea semicircularis, and other slender slips are occasionally found.

The **Inguinal Falx** (*falx inguinalis; conjoined tendon of Internal oblique* and *Transversalis muscles*) (Figs. 6–22, 6–24, 6–25) is the inferior terminal portion of the common aponeurosis of the Obliquus internus and Transversus abdominis muscles. It is inserted into the crest of the pubis and pectineal line immediately deep to the superficial inguinal ring, giving strength to a potentially weak point in the anterior abdominal wall. There is a wide variation in its width, strength, composition, and degree of union with neighboring aponeurotic and fascial structures. It may be narrow with a high arch, scarcely reaching the lateral part of the superficial inguinal ring, or it may be a broad, strong band, arching close to the inguinal ligament, and greatly reinforcing the abdominal wall in the region of the ring. In many cases it could be called the conjoined muscle instead of the conjoined tendon because the muscular fasciculi continue almost to the pubis. Not infrequently it is intimately fused with the reflected inguinal ligament which lies between it and the aponeurosis of the Obliquus externus. It may be reinforced on its deep surface by a fascial expansion of the Rectus tendon, **Henle's ligament.**

The **Rectus abdominis** (Fig. 6–23) is a long flat muscle, which extends along the whole length of the ventral aspect of the abdomen, and is separated from its fellow of the opposite side by the linea alba. It is much broader, but thinner superiorly. It *arises* by two tendons; the lateral or larger is attached to the crest of the pubis, the medial interlaces with its fellow of the opposite side and is connected with the ligaments covering the ventral surface of the symphysis pubis. The muscle is *inserted* by three portions of unequal size into the cartilages of the fifth, sixth, and seventh ribs. The superior portion, attached principally to the cartilage of the fifth rib, usually has some fibers of insertion into the ventral extremity of the rib itself. Some fibers are occasionally connected with the costoxiphoid ligaments, and the side of the xiphoid process.

The Rectus is crossed by fibrous bands, three in number, which are named the **tendinous intersections** (*inscriptiones tendineae*); one is usually situated opposite the umbilicus, one at the extremity of the xiphoid process, and the third about midway between the xiphoid process and the umbilicus. These inscriptions pass transversely or obliquely across the muscle in a zigzag course; they rarely extend completely through its substance and may pass only halfway across it; they are intimately adherent to the sheath of the muscle ventrally. Sometimes one or two additional inscriptions, generally incomplete, are present below the umbilicus.

Action.—Flexes the vertebral column, particularly the lumbar portion, drawing the sternum toward the pubis; tenses the anterior abdominal wall, and assists in compressing the abdominal contents.

Nerves.—Branches of the seventh to twelfth intercostal nerves; the seventh supplies the portion above the first tendinous inscription, the eighth the portion between the first and second inscriptions, and the ninth the portion between the lower two inscriptions.

Variations.—The Rectus may insert as high as the fourth or third rib or may fail to reach the fifth. Fibers may spring from the lower part of the linea alba. Both the aponeurotic composition and the position of the linea semicircularis vary considerably.

The **sheath of the Rectus abdominis** (*vagina m. recti abdominis*) (Figs. 6–21, 6–23).—The Rectus abdominis is enclosed in a sheath which holds it in position but does not restrict its motion during contraction because it is separated from the muscle by a fascial cleft. The sheath is formed by the aponeuroses of the Obliquus externus, Obliquus internus, and Transversus which preserve their identity in some regions, but fuse or interlace in others. At the lateral border of the Rectus, they form two membranes, the anterior and posterior lamellae of the sheath. The aponeurosis of the externus keeps its position ventral to the Rectus throughout its entire length, but fuses with that of the internus at a variable distance from the middle line. The line of

fusion is close to the linea alba at the pubis and below the umbilicus, and gradually angles laterally as well as superiorly, but remains medial to the lateral border of the Rectus. Thus a surgical incision over the Rectus, below the umbilicus, will go through the aponeuroses of the externus and internus as separate layers before it reaches the Rectus muscle itself.

The aponeurosis of the internus, above the umbilicus, splits into two lamellae, one of which passes ventral to the Rectus and fuses with the externus as just described; the other passes dorsal to the Rectus and

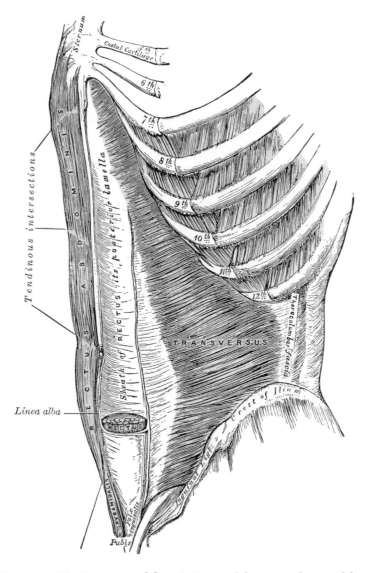

FIG. 6–23.—The Transversus abdominis, Rectus abdominis, and Pyramidalis.

fuses with the aponeurosis of the Transversus to form the posterior lamella of the sheath. Above the costal margin, to which the internus is attached, the costal cartilages and xiphoid process of the sternum take the place of the aponeurotic sheath, dorsally. The posterior lamella of the internus aponeurosis is attached to the cartilages; the anterior lamella ends abruptly in a fibrous band similar to the linea arcuata, the Rectus above this point being covered only by the externus aponeurosis.

The aponeurosis of the Transversus, above the umbilicus, passes entirely dorsal to the Rectus and fuses with the internus as just described. Below the umbilicus, the behavior of the aponeuroses is more complicated and more variable. At an inconstant distance above the pubis, usually about half way, the contribution of the aponeuroses to the posterior lamella of the sheath ceases abruptly in a curved line, the **linea arcuata** (*semicircular line of Douglas*). Caudal to this line all the aponeurotic fibers of all three muscles pass ventral to the Rectus. Between the umbilicus and the semicircular line, the aponeuroses of the internus and Transversus fuse and interdigitate, and tendinous bundles from

both may pass either central or dorsal to the Rectus. Usually a short distance above the linea arcuata, the internus terminates its contribution to the posterior lamella and only the Transversus remains; hence, in the majority of instances, the linea is formed by the fibers of the Transversus alone.

Caudal to the linea, the sheath is formed by a portion of the endoabdominal or transversalis fascia. This portion of the sheath is occasionally reinforced by scattered tendinous bundles from the Transversus and by thickened laminae of the subserous fascia. The sheath contains, besides the Rectus muscle, the Pyramidalis muscle, the superior and inferior epigastric arteries and veins, and branches of the intercostal nerves.

The **Pyramidalis** (Fig. 6–23) is a small triangular muscle, placed at the lower part of the abdomen, ventral to the Rectus, and contained in the sheath of that muscle. It *arises* by tendinous fibers from the ventral surface of the pubis and the anterior pubic ligament; the fleshy portion of the muscle extends superiorly, diminishing in size as it ascends, and ends by a pointed extremity which is *inserted* into the linea alba, midway between the umbilicus and pubis.

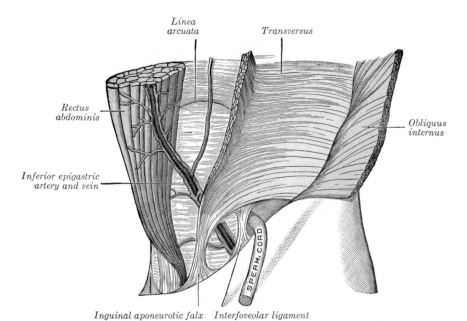

Linea arcuata   *Transversus*

Rectus abdominis

Obliquus internus

Inferior epigastric artery and vein

SPERM. CORD

Inguinal aponeurotic falx   *Interfoveolar ligament*

Fig. 6–24.—The interfoveolar ligament, ventral aspect. The spermatic cord passes through the deep inguinal ring lateral to the ligament. (Modified from Braune.)

Action.—Tenses the linea alba.

Nerve.—Branch of the twelfth thoracic nerve.

Variations.—The Pyramidalis is wanting in 10 per cent; the inferior end of the Rectus then becomes proportionately large. It may vary from 1.5 to 12 cm in length, averaging 6.8 cm; it is occasionally double on one or both sides, and the two sides may be unequal.

The **Linea alba** (Fig. 6–18) is the name given to the portion of the anterior abdominal aponeurosis or Rectus sheath in the middle line. It represents the insertion of the Obliquus externus and internus and the Transversus by the fusion of their aponeuroses with those of the opposite side; the fibers interlace and the three aponeuroses are fused into a single tendinous band which extends from the xiphoid process to the symphysis pubis. It is broader cranialward, where the Recti are separated from each other by a considerable interval; a surgical incision above the umbilicus in the middle line will go through the linea as a single aponeurotic layer. It is narrower more caudally, where the Recti are more closely placed; a surgical incision in the middle line below the umbilicus seldom follows the linea and will, in consequence, go through the anterior and posterior lamellae of the Rectus sheath as individual layers. The inferior end of the linea alba has a double attachment; the superficial one passes ventral to the medial heads of the Recti to the symphysis pubis; the deeper one spreads out into a triangular sheet dorsal to the Recti, attaches to the posterior lip of the crest of the pubis, and is named the **adminiculum lineae albae.** The umbilicus, which is an aperture for the passage of the umbilical vessels in the fetus, is closed in the adult and has the form of a hard fibrous ring or plate of scar tissue within the linea alba.

The **Linea Arcuata** (*semicircular line* or *fold of Douglas*) (Fig. 6–24) is a curved tendinous band, with convexity cranialward, in the posterior lamella of the Rectus sheath below the umbilicus. It marks the inferior limit of the aponeurotic portion of the posterior lamella, the latter being composed of transversalis fascia inferior to the line. Its origin from the lateral border of the sheath may be from 4 to 13 cm, average 8 cm, from the pubis; its arched course may be horizontal or cranialward, but is most frequently caudalward; its insertion is into the linea alba, or occasionally as far down as the pubic crest. The tendinous bundles of the linea arcuata are usually derived from the aponeurosis of the Transversus abdominis, but may be from the Obliquus internus, or from an interlacing of fibers from both the Transversus and the Internus. A secondary linea may be found superior to the primary one, especially in those instances which have the primary linea formed by the Transversus, in which case the secondary is formed by the Obliquus internus.

The **Linea Semilunaris** (Fig. 6–18) is a slightly curved line on the ventral abdominal wall running approximately parallel with the median line, and lying about half way between the latter and the side of the body. It marks the lateral border of the Rectus abdominis, and can be seen as a shallow groove in the living subject when that muscle is tensed (see Fig. 3–26). The abdominal wall is thin along this line, particularly its inferior part, where it is composed only of the aponeuroses of the Obliqui and Transversus, and their fasciae. At its superior end, it is thicker where the muscular fasciculi of the Transversus extend under the Rectus for a variable distance.

Group Actions.—If pelvis and thorax are fixed, the abdominal muscles compress the abdominal viscera by constricting the cavity of the abdomen, in which action they are materially assisted by the descent of the diaphragm. By these means assistance is given in expelling the feces from the rectum, the urine from the bladder, the fetus from the uterus, and the contents of the stomach in vomiting.

If the pelvis and vertebral column be fixed, these muscles compress the inferior part of the thorax, materially assisting expiration. If the pelvis alone be fixed, the thorax is bent directly ventralward when the muscles of both sides act; when the muscles of only one side contract, the trunk is bent toward that side and rotated toward the opposite side.

If the thorax be fixed, the muscles, acting together, draw the pelvis cranialward, as in climbing; or, acting singly, they draw the pelvis cranialward, and bend the vertebral column to one side or the other. The Recti, acting from below, depress the thorax, and consequently flex the vertebral column; when acting from above, they flex the pelvis upon the vertebral column.

The abdominal muscles are the **flexors of the vertebral column,** an important action which is frequently overlooked.

The **Transversalis fascia** (*endoabdominal fascia*) (Fig. 8–57).—The **internal investing layer** of deep fascia which lines the entire wall of the abdomen is now generally called the Transversalis Fascia. It may even include the pelvic portion of this internal layer. Formerly, the name was applied only to the deep fascia covering the internal surface of the Transversus (Transversalis) abdominis muscle, but, because this muscle occupies a large proportion of the surface of the abdominal cavity, the name has gradually become adopted for the entire internal sheet. Various subdivisions of it are still referred to by their more specific designations; for example, the names iliac, psoas, or obturator fascia are used for the portions covering these muscles. This internal fascia is of great surgical interest and is extremely complex; in different areas it covers muscles, aponeurosis, bones, and ligaments; it may be very thin and adherent in one place or thickened and independent in another; it gives rise to specialized structures, such as tubular investments, and from it are derived certain components of important extra-abdominal fasciae. It is a gray, felt-like membrane sometimes transparent, sometimes thickened into strong bands, but seldom aponeurotic in appearance and, except in obese individuals, contains no fat. Between this membrane and the peritoneum there is a layer of subserous fascia which may contain fat.

The definitive part of the transversalis fascia, that covering the internal surface of the muscular portion of the Transversus muscle, is readily identified by dissection. Over the aponeurosis of this muscle, however, it is thin and so closely adherent that only with difficulty can it be dissected free. Cranialward from the muscular portion of the Transversus, the fascia continues onto the diaphragm and covers its entire abdominal surface. It is thin and adherent over the muscular portion as well as over the central tendon. Ventrally, the fascia over the Transversus aponeurosis of one side is continuous across the middle line with that of the other side. Dorsally, as it leaves the muscular fasciculi of the Transversus, it continues for a short distance over the aponeurosis of origin of this muscle and then covers the Quadratus lumborum and Psoas muscles. From the Psoas it covers the crura of the diaphragm, the bodies and disks of the lumbar vertebrae, and is then continuous with the fascia of the Psoas of the other side. As the dorsal origin of the diaphragm crosses the Quadratus and Psoas, the fascia is thickened into the strong, fibrous lumbocostal arches. Caudalward from the muscular portion of the Transversus and the Quadratus lumborum, the fascia is attached to the bone along the crest of the ilium and continues into the greater pelvis on the surface of the Iliacus muscle. It is continuous, from the Iliacus and Psoas muscles, with the pelvic fascia which is described in the section dealing with the muscles of that region. The fascia on the internal surface of the anterior wall below the umbilicus requires an especially detailed description.

In a medial direction from the muscular fasciculi of the Transversus, the transversalis fascia below the umbilicus continues on the aponeurosis toward the middle line. The portion caudal to the linea arcuata, however, splits at the lateral border of the Rectus into two sheets; the thin anterior sheet continues on the aponeurosis and passes with it anterior to the Rectus; a thick posterior sheet forms the posterior lamina of the Rectus sheath and represents the internal investing or endoabdominal fascia in this area. Lateral to the caudal part of the Rectus, the transversalis fascia continues downward on the Transversus aponeurosis to its caudal limit, covering there the falx inguinalis, and passing over the free border of the Transversus aponeurosis as it forms an arch extending from the crest of the pubic bone to the lateral part of the inguinal ligament. The fascial membranes on both the deep and superficial surfaces of the Transversus fuse into a single sheet at this free border, providing a reinforced transversalis to bridge the interval between the **deep inguinal arch** and the inguinal ligament (Fig. 6–25).

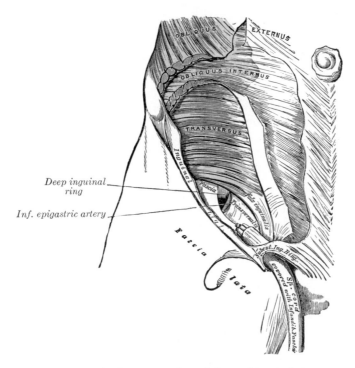

Fɪɢ. 6–25.—The deep, internal, or abdominal inguinal ring.

At a point just above the middle of the inguinal ligament, a tubular prolongation of this reinforced fascia is carried outward on the ductus deferens and testicular vessels as they leave the abdominal cavity. This tubular investment is the inner coat of the spermatic cord and testis and is known as the *internal spermatic* or *infundibuliform fascia*. The ductus and vessels leave the abdominal cavity at a point which is called the deep (abdominal or internal) inguinal ring and they follow an oblique course through the abdominal wall in a tunnel called the inguinal canal. The lateral part of this canal, that is, before it has begun to penetrate the internal oblique muscle, has transversalis fascia for its posterior wall. The fascia is loosely attached to the inguinal ligament and has two thickenings near the ring, one extending cranialward called the interfoveolar ligament, one extending caudalward called the deep crural arch.

The **Interfoveolar Ligament** (*Hesselbach's ligament*) (Fig. 6–24) forms a crescentic medial boundary for the deep inguinal ring. It may be poorly defined or

it may be a strong band whose fibers fan out medially as it follows the upward course of the deep inferior epigastric artery. It forms a slight ridge, not always visible but usually palpable from the interior of the abdominal cavity, which extends cranialward from the middle of the inguinal ligament, dividing the shallow fossa above the ligament into two parts, a **medial** and a **lateral inguinal fovea.** The deep inguinal ring, through which the ductus deferens leaves the abdominal cavity, is in the lateral fovea and it is here that the sac of an indirect inguinal hernia penetrates the abdominal wall. In the medial fovea, a triangular area is marked out by the inguinal ligament, the lateral boundary of the Rectus abdominis, and the inferior deep epigastric artery. This is *Hesselbach's triangle* and is the site of a direct inguinal hernia.

The **Deep Crural Arch** (*iliopubic tract*) is the caudalward extension of the transversalis fascia from the region of the internal inguinal ring. It arches across the external iliac vessels as they pass under the inguinal ligament and marks the transition from transversalis fascia to femoral sheath. Laterally, it is attached to the iliac fascia where the latter gives origin to the inferior fibers of the Transversus muscle. Medially, it follows the downward curve of the lacunar ligament. It may be a strong band or it may be poorly defined and appear to be merely the proximal part of the femoral sheath.

The **Deep Inguinal Ring** (*anulus inguinalis profundus; abdominal* or *internal inguinal ring*) (Figs. 6–25, 8–57) is the name given to the interruption in the transversalis fascia where the spermatic cord (round ligament in the female) penetrates the anterior abdominal wall, carrying with it a sleeve-like investment of the transversalis fascia called the internal spermatic fascia. It is situated midway between the anterior superior iliac spine and the symphysis pubis, and about 1.25 cm above the inguinal ligament (Fig. 6–24). It is of an oval form, the long axis of the oval being vertical. It varies in size, fitting rather closely about the penetrating structures unless it has been distended by the sac of an indirect inguinal hernia. It is bounded, superiorly, by the arched inferior margin of the Transversus abdominis; medially, by the interfoveolar ligament accompanying the inferior deep epigastric vessels, and inferiorly, by the iliopubic tract.

The **Inguinal Canal** (*canalis inguinalis; spermatic canal*) is the tunnel in the lower anterior abdominal wall through which the spermatic cord (round ligament in the female) passes. Its internal (lateral) end is at the deep inguinal ring; its external (medial) end is at the superficial inguinal ring. It is about 4 cm long and takes an oblique course parallel with and a little above the inguinal ligament. It is bounded superficially by the skin, subcutaneous fascia, aponeurosis of the Obliquus externus, and, in its lateral third, by the Obliquus internus; deeply, from medial to lateral ends, by the reflected inguinal ligament, the inguinal falx, the transversalis fascia, subserous fascia and peritoneum; proximally, by the arched fibers of the Obliquus internus and Transversus; distally, by the inguinal ligament, and at its medial end, the lacunar ligament. Through it pass the spermatic cord (round ligament), the ilioinguinal nerve, the testicular vessels, and the Cremaster muscle.

Subserous fascia (*tela subserosa; subperitoneal fascia, superficial subserous fascia [Gallaudet], extraperitoneal connective tissue*).—Intervening between the internal investing layer of deep fascia of the abdominal wall and the peritoneum is the fibroelastic connective tissue which supports the peritoneum. This subserous fascia resembles the subcutaneous superficial fascia in that it supports a free surface epithelial layer; it commonly contains adipose tissue in varying thicknesses, and it is frequently separated from the deep fascia by a fascial cleft. Not only does it have a parietal portion supporting the peritoneum of the abdominal wall, but it has a visceral portion which continues out over the viscera at the peritoneal reflections, and which accompanies the blood vessels into the mesenteries. The subserous fascia has localized thickenings and accumulations of fat which are associated with the particular requirements of the different regions; these are described in the chapters dealing with the viscera.

The subserous fascia in the pelvis is uninterruptedly continuous with that of the abdomen. It is described in the chapter dealing with the pelvic viscera.

2. The Posterior Muscles of the Abdomen

> Psoas major
> Psoas minor
> Iliacus
> Quadratus lumborum

The Psoas major, the Psoas minor, and the Iliacus, with the fasciae covering them, will be described with the muscles of the lower limb.

Fascia covering the quadratus lumborum.—The transversalis (internal investing, endoabdominal) fascia covers the lateral portion of the Quadratus lumborum on its ventral surface. Since the medial portion of the muscle is overlapped by the Psoas, the fascia continues medially between the muscles and is attached to the bases of the transverse processes of the lumbar vertebrae. Its superior portion is thickened into a strong band, called the lateral lumbocostal arch (see Diaphragm), which is attached to the transverse process of the first lumbar vertebra, and the apex and inferior border of the last rib. Inferiorly, the fascia is attached to the crest of the ilium, and is then continuous with the iliac fascia. At the lateral border of the muscle, the fascia fuses with the combined fascia and aponeurosis of origin of the Transversus. The latter, extending medially from this point of fusion, covers the dorsal surface of the Quadratus. The more medial portion of this aponeurotic sheet lies between the Quadratus and the Sacrospinalis and is named the lumbocostal aponeurosis (anterior layer of the thoracolumbar fascia).

The **Quadratus lumborum** (Fig. 6–14, page 404) is irregularly quadrilateral in shape, broader caudally. It *arises* by aponeurotic fibers from the iliolumbar ligament and the adjacent portion of the iliac crest for about 5 cm, and is *inserted* into the inferior border of the last rib for about half its length, and by four small tendons into the apices of the transverse processes of the first four lumbar vertebrae. Occasionally a second portion of this muscle is found ventral to the preceding. It *arises* from the superior borders of the transverse processes of the last three or four lumbar vertebrae, and is *inserted* into the inferior margin of the last rib. Ventral to the Quadratus lumborum are the colon, the kidney, the Psoas major and minor, and the diaphragm; between the fascia and the muscle are the twelfth thoracic, ilioinguinal, and iliohypogastric nerves.

Action.—Draws the last rib toward the pelvis and flexes the lumbar vertebral column laterally toward the side of the muscle acting. Fixes the last two ribs in forced expiration.

Nerves.—Branches of the twelfth thoracic and first lumbar nerves.

Variations.—The number of attachments to the vertebrae and the extent of its attachment to the last rib vary.

V. THE MUSCLES AND FASCIAE OF THE PELVIS

> Levator ani
> Coccygeus
> Obturator internus
> Piriformis

The muscles within the pelvis may be divided into two groups: (1) the true pelvic muscles, the Levator ani and Coccygeus;

(2) the muscles of the lower limb which originate within the pelvis and thus form part of the pelvic wall, the Obturator internus and the Piriformis. The muscles of the second group will be described later with the muscles of the lower limb, but their fasciae will be considered here because they form an important part of the pelvic fascia.

The **Pelvic Diaphragm** (*diaphragma pelvis*) is composed of the Levator ani and Coccygeus muscles together with the fasciae covering their internal and external surfaces. It stretches across the pelvic cavity like a hammock. It is the most caudal portion of the body wall, closing the abdominopelvic cavity, restraining the abdominal contents, and giving support to the pelvic viscera. It is pierced by the anal canal, the urethra, and the vagina, and is reinforced in the perineum by the special muscles and fasciae associated with these structures. The pelvic diaphragm and the structures of the perineum are intimately associated both structurally and functionally and an accurate knowledge of one cannot be obtained without study of the other.

The **Levator ani** (Fig. 6–26) is a broad, thin muscle forming the hammock-like floor of the pelvis. Although the muscles of the two sides are separated from each other ventrally, and are inserted into a raphe posteriorly, they function as a single sheet across the middle line forming the principal part of the pelvic diaphragm. It *arises*, ventrally, from the inner surface of the superior ramus of the pubis lateral to the symphysis; dorsally, from the inner surface of the spine of the ischium; and between these two points, from the *arcus tendineus musculi levatoris ani*. The latter is a thickened band of the obturator fascia attached dorsally to the spine of the ischium and ventrally to the pubic bone at the ventral margin of the obturator membrane. The fasciculi pass dorsalward and medialward in the floor of the pelvis and are *inserted* into the side of the last two segments of the coccyx, the anococcygeal raphe, the Sphincter ani externus, and the central tendinous point of the perineum. The *anococcygeal raphe* is the narrow fibrous band extending from the coccyx to the posterior margin of the anus where the muscles of the two sides join each other in the middle line.

The Levator ani generally shows a separation into two parts, more distinct in lower mammals than in man, the Pubococcygeus and the Iliococcygeus.

The **Pubococcygeus** *arises* from the dorsal surface of the pubis along an oblique line extending from the inferior part of the symphysis to the obturator canal. The muscular fasciculi pass dorsalward, more or less parallel with the middle line, and terminate by joining the fibers from the other side. The lateral margin of the muscle may be separated from the Iliococcygeus by a narrow interval or it may be overlapped by the latter muscle. The medial margin is separated from the muscle of the other side by an interval known as the genital hiatus which allows the passage of the urethra, vagina, and rectum. The most ventral fasciculi, which are also the most medial, pass in close relation to the prostate and insert into the central tendinous point just in front of the anus. This portion has been called the *Levator prostatae*. In the

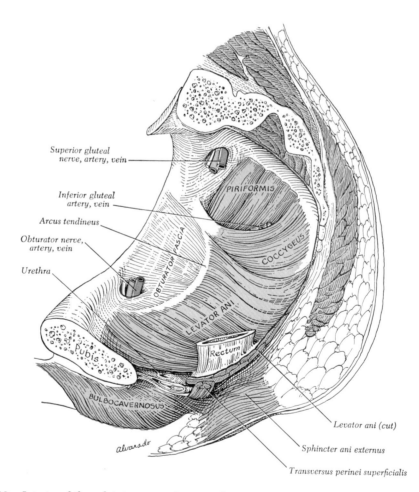

Superior gluteal nerve, artery, vein

Inferior gluteal artery, vein

Arcus tendineus

Obturator nerve, artery, vein

Urethra

PIRIFORMIS

COCCYGEUS

OBTURATOR FASCIA

LEVATOR ANI

Pubis

Rectum

BULBOCAVERNOSUS

alvarado

Levator ani (cut)

Sphincter ani externus

Transversus perinei superficialis

Fig. 6–26.—Interior of the pelvis in a sagittal section showing the Levator ani, obturator fascia, and arcus tendineus of the Levator. (Redrawn from Spalteholz.)

female, this anterior portion has a similar relationship to the vagina. The majority of the fasciculi of the Pubococcygeus pass horizontally dorsalward beside the anal canal, the superficial ones joining the anococcygeal raphe, the deeper ones joining the muscle of the other side to form a loop or sling about the rectum. This portion is called the *Puborectalis muscle.*

The **Iliococcygeus** *arises* from the arcus tendineus m. levatoris ani and the spine of the ischium, and inserts into the last two segments of the coccyx and the anococcygeal raphe. Its fasciculi pass medially as well as caudalward which helps to distinguish them from the fasciculi of the Pubococcygeus.

Action.—Supports and slightly raises the pelvic floor, resisting increased intra-abdominal pressure, as during forced expiration. The Pubococcygeus draws the anus toward the pubis and constricts it.

Nerves.—Branches of the pudendal plexus, containing fibers from the fourth, sometimes also the third or fifth sacral nerves.

Variations.—The degree of distinctness of the Pubococcygeus and Iliococcygeus varies; the latter may be replaced by fibrous tissue.

The **Coccygeus** (Fig. 6–26) is situated dorsal to the preceding. It is a triangular plane of muscular and tendinous fibers, *arising* by its apex from the spine of the ischium and sacrospinous ligament, and *inserted* by its base into the margin of the coccyx and into the side of the last piece of the sacrum. It assists the Levator ani and Piriformis in closing in the back part of the outlet of the pelvis.

Action.—Draws the coccyx ventrally, supporting the pelvic floor against intra-abdominal pressure.

Nerves.—Branches of the pudendal plexus, containing fibers from the fourth and fifth sacral nerves.

Variations.—The iliosacralis is an occasional band of muscle, ventral to the Coccygeus, attached to the iliopectineal line and the lateral border of the sacrum.

The **Sacrococcygeus ventralis** is a muscular or tendinous slip from the lower sacral vertebrae to the coccyx, representing the vestige of the Depressor caudae of lower mammals.

The **Sacrococcygeus dorsalis** is a slip from the dorsal aspect of the sacrum to the coccyx.

Pelvic fascia (*fascia pelvis*).—The **internal investing fascia of the pelvis** covers the Levator ani and Coccygeus and the intrapelvic portions of the Obturator internus and Piriformis muscles. It belongs to the same category of fascia as the endoabdominal or transversalis and is directly continuous with the latter over the brim of the lesser pelvis where it is attached to or fused with the periosteum of the symphysis and superior ramus of the pubis, the ilium along the arcuate or iliopectineal line, and the promontory of the sacrum. From these attachments, it sweeps caudalward and across the middle line, attaching to the anococcygeal raphe dorsally and blending with the fasciae of the anal canal and urogenital structures ventrally. Although it is a continuous sheet, for convenience in description it is divided into (1) piriformis fascia, (2) obturator fascia and (3) supraänal fascia.

The **fascia of the Piriformis** is very thin and is attached to the ventral surface of the sacrum and the sides of the greater sciatic foramen; it is prolonged outward through the greater sciatic foramen, joins the fascia of the external surface of the muscle at its distal border, and, becoming extrapelvic, forms part of the deep gluteal fascia. At its sacral attachment around the margins of the anterior sacral foramina it comes into intimate association with and ensheathes the nerves emerging from these foramina. Hence the sacral nerves are frequently described as lying dorsal to the fascia. The internal iliac vessels and their branches, on the other hand, lie in the subperitoneal tissue ventral to the fascia, and the branches to the gluteal region emerge in special sheaths of this tissue, proximal and distal to the Piriformis muscle.

Obturator fascia (*fascia obturatoria*) (Fig. 6–26).—The fascia covering the Obturator internus muscle is partly intrapelvic and partly extrapelvic. This condition can best be understood if the intrapelvic portion is pictured as having incorporated in it the aponeurosis of origin of the Levator ani. It is as if the Levator had at one time been attached to the pelvic brim cranial to the Obturator (a condition found in lower primates) but had slipped distally for a variable distance, usually about half the length of the muscle. At this point, the origin of the Levator is visible as a thickened band called the **arcus tendineus musculi levatoris ani.** Posteriorly, the arcus always ends by attaching to the spine of the ischium; anteriorly, it varies, but usually attaches to the anterior margin of the obturator membrane or the pubic bone medial to it. Since the Levator closes the aperture of the pelvis, the arcus marks the boundary between the intra- and extra-pelvic portions of the obturator fascia. The intrapelvic portion usually is not aponeu-rotic in appearance and may be quite thin except at the arcus tendineus. Ventrally, the obturator fascia is attached to the superior border of the obturator membrane or the pubic bone just ventral to the obturator canal. It contributes a tubular investment for the obturator nerve and vessels as they leave the pelvis through the obturator canal. Its attachment to the bone gradually angles cranialward from the obturator membrane until it reaches the iliopectineal line near the sacroiliac articulation. Dorsally, it is attached to the margin of the greater sciatic notch down to the spine of the ischium. At the arcus tendineus of the Levator ani, the obturator fascia splits into

three sheets; the outer one continues as the extrapelvic obturator fascia, the other two cover the external and internal surfaces of the pelvic diaphragm. The extrapelvic portion of the obturator fascia follows the muscle distalward into the ischiorectal fossa and will be described with the perineum.

The **supraänal fascia** (*Fascia diaphragmatis pelvis superior*) (Fig. 6–27) covers the internal surface of the Levator ani and Coccygeus muscles. Ventrally, above the Pubococcygeus part of the Levator, it is attached to the pubic bone. Laterally, above the Iliococcygeus, it is continuous with the intrapelvic obturator fascia at the arcus tendineus of the Levator ani. Dorsally, it covers the Coccygeus and becomes continuous with the fascia of the Piriformis. Behind the rectum, the fascia of the two sides is continuous across the middle line over the anococcygeal raphe. In front of the rectum, in the part of the genital hiatus where the medial borders of the Pubococcygei of the two sides do not meet, the supraänal fascia joins the infraänal fascia to assist in the formation of the deep layer of the urogenital diaphragm. (See perineum.)

Subserous fascia (*tela subserosa*) (Fig. 6–28).—The subserous fascia of the pelvis not only covers the parietal wall and the viscera, but also acts as a padding tissue

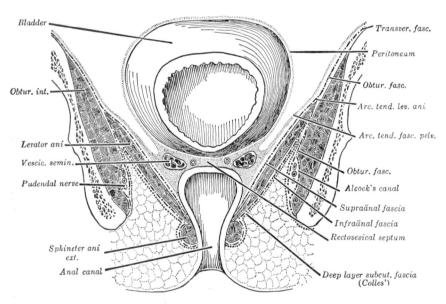

Fig. 6–27.—Fasciae of pelvis and anal region of perineum. Diagram of frontal section.

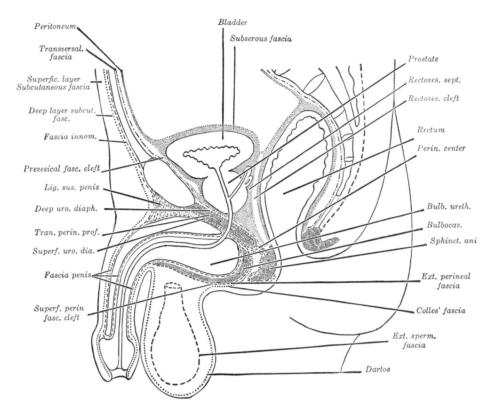

Fig. 6–28.—Fasciae of pelvis and perineum in median sagittal section. Diagram.

for the viscera in the lower part of the pelvis. It forms important ligaments, folds, and bands which are described in the chapter dealing with the pelvic viscera.

VI. THE MUSCLES AND FASCIAE OF THE PERINEUM

The perineum corresponds to the outlet of the pelvis. Its deep boundaries are—*ventrally,* the pubic arch and the arcuate ligament of the pubis; *dorsally,* the tip of the coccyx; and *on either side* the inferior rami of the pubis and ischium, and the sacrotuberous ligament. The space is somewhat lozenge-shaped and is limited on the surface of the body by the scrotum ventrally, by the buttocks dorsally, and laterally by the medial side of the thigh. A line drawn transversely between the ischial tuberosities divides the space into two portions. The dorsal contains the termination of the anal canal and is known as the **anal region;** the ventral, which contains the external uro-genital organs, is termed the **urogenital region.**

The **subcutaneous fascia** (*tela subcutanea, fascia superficialis perinei*) (Fig. 6–29) of the perineum is divisible into two layers, superficial and deep, which are similar to and continuous with the corresponding layers of the anterior abdominal wall. The deep layer is called the fascia of Colles instead of Scarpa. The superficial layer, corresponding to Camper's fascia, usually is not specifically named, although it has been called Cruveilhier's fascia.

The **superficial layer of subcutaneous fascia** in the ventral part of the perineum, that is, over the urogenital region, contains a considerable amount of adipose tissue and is likely to be irregularly laminated. In the groove between the scrotum and the thigh, it is directly continuous with the corresponding layer on the anterior abdominal wall. More medially, it is combined with the deep layer into the dartos tunic of the scrotum in the male, while in the female, it forms the greater part of the labium majus. Laterally, it is continuous with the superficial layer of the thigh. Dorsally, it becomes the superficial layer of the anal region.

The superficial layer in the anal region is greatly thickened into a mass of fat which occupies the ischiorectal fossa. It is continuous posteriorly with the superficial fascia of the gluteal region and posterior thigh. Covering the ischial tuberosities, the fibrous tissue is increased in amount, forming a tough pad.

Ischiorectal Fossa (*fossa ischiorectalis*) (Fig. 6–29).—The fossa is somewhat prismatic, in the shape of a wedge, with its base at the perineum and its apex deep in the pelvis where the Levator ani and Obturator internus come together. It is bounded medially by the infraänal fascia covering the Levator ani and Sphincter ani externus; laterally, by the obturator fascia over the extrapelvic portion of the Obturator internus, and by the tuberosity of the ischium; ventrally, by the posterior borders of the Transversus perinei superficialis and profundus; dorsally, by the fascia over the Gluteus maximus and the sacrotuberous ligament. The superficial boundary is the skin, and the fossa is occupied by a mass of adipose tissue belonging to the superficial layer of the subcutaneous fascia. The superficial and deep layers of the subcutaneous fascia are not separable in the fossa and both are securely attached to the deep fascia covering the entire surface of the fossa. The dorsal portion of the Transversus perinei profundus is separated from the Levator ani for a short distance, so that the ischiorectal mass of fat has an *anterior process* which projects under the posterior border of the Transversus muscles. The adipose tissue is traversed by fibrous bands and incomplete septa. It is crossed transversely by the inferior rectal vessels and nerves; at the dorsal part are the perineal and perforating cutaneous branches of the pudendal plexus; while from the ventral part, the posterior scrotal (or labial) vessels and nerves emerge. The internal pudendal vessels and pudendal nerve lie deep to the obturator fascia on the lateral wall of the fossa in a special reduplication of the latter fascia known as Alcock's canal.

The **deep layer of subcutaneous fascia of the perineum** (*Colles' fascia*) (Figs. 6–28, 6–29, 6–30) in the urogenital region is a distinctive structure. It is a strong membrane but does not have the white glistening appearance of an aponeurosis. It has a slightly yellow color due to its content of elastic fibers and is smooth in texture, its fibrous nature not being detectable with the naked eye. This characteristic texture is of assistance frequently in differentiating it from the deep fascia in the region. Ventrally, it is directly continuous with the deep layer of subcutaneous abdominal fascia (Scarpa's fascia) in the groove between the scrotum (labium) and thigh. More medially, it joins the superficial layer in the formation of the **dartos tunic** of the scrotum. In the middle line, it is attached to the superficial layer along the raphe and continues ventrally into the **septum of the scrotum**. Laterally, it is firmly adherent to the medial surface of the thigh along the ischiopubic ramus at the origin of the adductor muscles. In the ventral part of this area, it is continuous with the fascia cribrosa which covers the saphenous opening. Dorsally, it dips inward toward the ischiorectal fossa around the posterior border of the Transversus perinei superficialis and becomes firmly attached to the deep fascia along the posterior border of the Transversus perinei profundus. It is attached also, with all the other layers, to the central tendinous point of the perineum. There is a distinct fascial cleft between it (Colles' fascia) and the external perineal fascia (deep fascia) over the Bulbocavernosus, Ischiocavernosus, and Transversus perinei superficialis muscles. This **superficial perineal cleft** is continuous with the cleft under Scarpa's fascia on the anterior abdominal wall but is closed off laterally and posteriorly by the attachments described above.

In the anal region, the deep layer of the subcutaneous fascia is adherent both to the superficial layer and to the deep fascia. From its attachment to the posterior border of the Transversus perinei profundus mentioned above, it continues deeply into the ischiorectal fossa in close apposition to the infraänal fascia on the Levator ani muscle. At the origin of this muscle, it folds back sharply over the extrapelvic portion of the

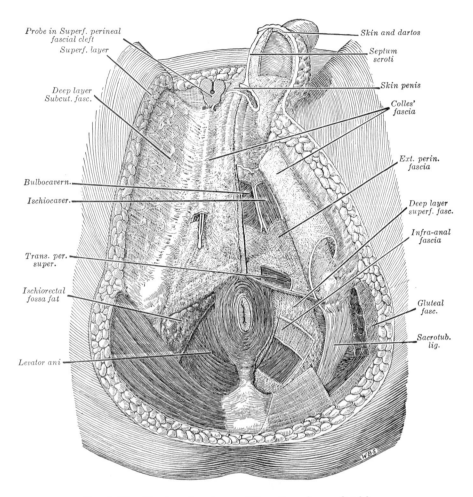

Fig. 6–29.—Fasciae of perineum. Dissection of superficial layers.

obturator fascia. Posteriorly, it leaves the ischiorectal fossa along the border of the gluteal region and posterior thigh.

Deep fascia.—The deep fascia of the perineum consists of the obturator fascia, the infraänal fascia, and the fasciae over two groups of small muscles which are associated with the urogenital organs and occupy a position superficial to the pelvic diaphragm in the urogenital region of the perineum. The phenomena of splitting, fusion, and cleavage are prominent features of the fascia in this region. The clinical interest and importance of these fasciae have made them the subject of numerous treatises in which the various parts are segregated and given special names. In the following description these parts will be considered individually but an attempt will

be made to correlate the diverse terminology into a consistent description. A certain amount of repetition will be unavoidable.

Obturator fascia (*fascia obturatoria*) (Fig. 6–27).—The fascia covering the extrapelvic portion of the Obturator internus forms the lateral wall of the ischiorectal fossa, and is, therefore, a portion of the external investing layer of deep fascia of the body. Superiorly, it meets the infraänal fascia at a sharp angle in the deepest part of the fossa. Its inferior portion, extending from the lesser sciatic foramen to the ischial tuberosity, is thickened and splits to enclose the pudendal nerve and internal pudendal vessels in a fibrous tunnel called **Alcock's canal.** Anteriorly, it is attached to the ischiopubic ramus, and the fibrous sheath of Alcock's canal merges with the external perineal fascia and the superficial fascia of the urogenital diaphragm where branches of the nerve and vessels enter the superficial and deep perineal compartments described below.

The **infraänal fascia** (*fascia diaphragmatis pelvis inferior*) (Fig. 6–28) is the deep fascia of the superficial (inferior, external) surface of the Levator ani and Coccygeus muscles. It is adherent to the muscle throughout. Superiorly, the portion on the Iliococcygeus is continuous with the extrapelvic obturator fascia at the arcus tendineus of the Levator ani; the portion over the Pubococcygeus is attached to the ischiopubic ramus and the pubic bone at the origin of the muscle. Between the medial borders of the two Pubococcygei, it joins the supraänal fascia to form a thick sheet in the genital hiatus. Dorsal to the rectum, it is continuous from side to side over the anococcygeal raphe where it is firmly attached. More laterally, it bridges the slight interval between the Levator ani and Coccygeus, and then binds the latter muscle to the inferior edge of the sacrospinous ligament. From this ligament, it passes outward from the ischiorectal fossa along the overhanging distal border of the Gluteus maximus where it becomes continuous with the gluteal fascia. In the region of the anus, it invests the Sphincter ani externus and just anterior to the anus, it is firmly attached to the other perineal layers at the central tendinous point.

If the infraänal fascia is traced anteriorly from the ischiorectal fossa, it will be seen to split into three sheets at a transverse line connecting the anterior extremities of the ischial tuberosities. The deepest of the three continues on the surface of the Levator ani, and, lying between this muscle and the Transversus perinei profundus, it is called the deep layer of the urogenital diaphragm. This is the sheet which joins the supraänal fascia in the genital hiatus to form the thick membrane which binds the two medial borders of the Pubococcygei together. The middle sheet covers the superficial surface of the Transversus profundus and is called the superficial layer of the urogenital diaphragm. The most superficial of the three sheets curves outward around the posterior

border of the Transversus perinei superficialis and is the external perineal fascia.

The **Urogenital Diaphragm** (*diaphragma urogenitale*).—The Transversus perinei profundus muscle is covered internally and externally by fascial membranes which are called respectively the deep and superficial layers of the urogenital diaphragm (Fig. 6–30). The muscle and the two fasciae taken together constitute the urogenital diaphragm. According to this, it is synonymous with the deep perineal compartment (pouch, interspace) and its contents, which will be described below. The urogenital diaphragm, as it has just been defined, is assisted in its rôle of a supporting structure by the portion of the Levator ani over which it lies and by the superficial perineal muscles and their fasciae.

The **genital hiatus** is the interval in the middle line between the medial borders of the two Pubococcygeus portions of the Levator ani muscle. It is through this hiatus that the urethra passes in both sexes and the vagina in the female. It is filled in by fibrous tissue derived from the fascia of the Levator ani (the supraänal and infraänal fasciae) and the deep layer of the urogenital diaphragm.

The **deep** (*superior; internal*) **layer of the urogenital diaphragm** (*fascia diaphragmatis urogenitalis superior*) is a flat triangular membrane stretching across the interval between the ischiopubic rami. It lies between the Transversus perinei profundus and the Pubococcygeus portions of the Levator ani and represents, therefore, the fused fascial membranes of both these muscles. It represents also, in the genital hiatus between the medial borders of the two Pubococcygei, a fusion with still a third membrane, the supraänal fascia. It is securely attached to the symphysis pubis anteriorly and joins the other perineal layers in the central tendinous point posteriorly. Laterally, it is attached to the medial borders of the ischiopubic rami along with the superficial layer of the diaphragm and there it is continuous with the obturator fascia. The middle portion, which occupies the genital hiatus, is thickened to fill in the gap between the two Pubococcygei and bind their medial borders together. It is pierced by the urethra and the vagina and blends with their walls as they pass through. The prostate gland rests on its pelvic surface and the connective tissue of the inferior portion of the gland's capsule blends with it intimately. The attachment of the fascia to the pubic bone and the blending with the prostatic capsule make the anterior part of the fascia a true **ligament of the prostate.** The tissue attachments may form three strands, a **middle puboprostatic ligament** to the symphysis and two **lateral puboprostatic ligaments** to the pubic bone at the points where the anterior ends of the arcus tendinei of the Levatores ani are attached. At the posterior border of the deep Transversus, it is fused with the superficial layer of the urogenital diaphragm, closing the deep perineal compartment.

The **superficial** (*inferior, external*) **layer of the urogenital diaphragm** (*fascia diaphragmatis urogenitalis inferior*) (Figs. 6–28, 6–30) is a flat triangular membrane which, like the preceding, bridges the angular interval between the ischiopubic rami. It is attached laterally to the medial borders of the rami from the arcuate pubic ligament to the ischial tuberosities. Along these same borders, the deep layer of the diaphragm

is attached more deeply and the crura of the penis more superficially. The middle portion of the fascia is pierced by the urethra and vagina and blends with their walls. It is perforated also by the arteries to the bulb, the ducts of the bulbourethral glands, the deep arteries of the penis, and the dorsal arteries and nerves of the penis. The part posterior to the urethra blends with the fascia of the bulb and, because of its attachment to the rami laterally, is sometimes called the **ligament of the bulb.** At these lateral attachments, the fascia blends with that of the crura also. The part anterior to the urethra arches across the subpubic angle and is sometimes called the **ligamentum transversum pelvis.** The latter is separated from the symphysis and arcuate pubic ligament by an opening for the passage of the dorsal vein of the penis. At the posterior border of the deep Transversus muscle, the superficial and deep layers of the diaphragm fuse into a single layer and blend with the infraänal fascia of the ischiorectal fossa.

The **external perineal fascia** (*inferior perineal fascia of Gallaudet*) (Figs. 6–29, 6–30) is the external investing fascia, that is, the most superficial layer of the deep fascia

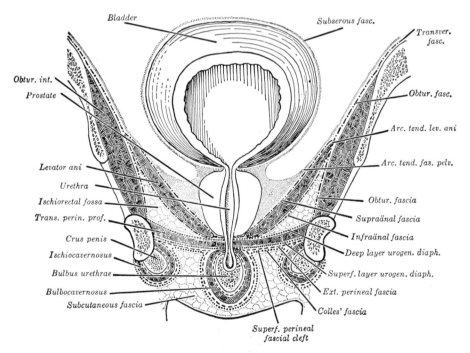

FIG. 6–30.—Fasciae of pelvis and urogenital region of perineum. Diagram of frontal section.

in the urogenital region of the perineum. It covers approximately the same triangular area as the preceding but it is not flat because it must accommodate itself to the contours of the superficial perineal muscles. It is attached laterally to the ischiopubic ramus at the outer border of the Ischiocavernosus. From this attachment it passes anteriorly along the groove between the scrotum and thigh, there becoming continuous with the external oblique fascia (fascia innominata) of the anterior abdominal wall and, more laterally, with the fascia lata of the medial surface of the thigh. It invests the Ischiocavernosus and Bulbocavernosus muscles and, in the interval between them, dips down to the level of the superficial layer of the urogenital diaphragm for a short distance and is there adherent to it. It follows the muscles just mentioned ventrally to the root of the penis and, at their insertion, it joins the fascia of the crura and bulb to become the **deep fascia of the penis** (*Buck's fascia*). Posteriorly, the fascia invests the Transversus perinei superficialis. It passes deeply around the posterior border of this muscle into the ischiorectal fossa until it meets the deep Transversus where the two Transversi are in contact with each other. Here it fuses with the fascia of the urogenital diaphragm, and with it blends into the single layer of infraänal fascia in the ischiorectal fossa. The fusion of this fascia with the fascia of the urogenital diaphragm posteriorly, its attachment to the ischiopubic rami laterally, and its fusion with the fascia of the penis anteriorly make it the superficial layer of a closed compartment, the superficial perineal compartment (pouch, interspace). The deep layer of the compartment is the superficial layer of the urogenital diaphragm. The fascial cleft which is superficial to the external perineal fascia, that is, which lies between it and Colles' fascia, should not be included in the compartment.

The **Perineal Center** is often called the **central tendinous point** (Fig. 6–28) in the male and often simply the **perineum in the female**. It is the mass of tissue in the middle line between the anus and the bulb in the male and between the anus and vagina in the female. It is approximately 2 cm in width and depth in the male and about twice this dimension in the female. It is composed predominantly of fibrous tissue since it represents the fusion of the following: infra-anal fascia (deep layer of the urogenital diaphragm), superficial layer of the urogenital diaphragm, external perineal fascia, and Colles' fascia. It contains a few muscular fibers also, principally from the Pubococcygeus and Sphincter ani externus, and it has attached to it besides these two muscles, the Transversus perinei profundus and superficialis, and the Bulbocavernosus. It is directly continuous anteriorly with the fibrous tissue which fills in the genital hiatus between the two Pubococcygei. It is continuous deeply into the pelvis with the rectoprostatic and rectovesical septum in the male and the rectovaginal septum in the female. It is the time-honored route of approach to the bladder and prostate from the perineum and it is the site of the perineal tears which are frequently a result of child bearing.

The **Triangular Ligament** is a name frequently given to the urogenital diaphragm, that is, the Transversus perinei profundus and its superficial and deep fascial membranes. The name is less commonly used for the superficial layer of the urogenital diaphragm alone, without the muscle or the other layer.

The **Deep Perineal Compartment** (*deep perineal pouch; deep perineal interspace*) is formed by the deep and superficial layers of the urogenital diaphragm. The compartment and its contents, therefore, form the urogenital diaphragm. The compartment is principally occupied by the Transversus perinei profundus but contains also: the Sphincter urethrae and the membranous portion of the urethra; the bulbourethral glands (vestibular glands in the female) and their ducts; the internal pudendal vessels; the deep dorsal vein of the penis, and the dorsal nerve of the penis. The internal pudendal artery enters the compartment posteriorly and its branches to the bulb, to the urethra, the deep and the dorsal arteries of the penis, pierce the superficial fascia of the urogenital diaphragm to reach their destination.

The **Superficial Perineal Compartment** (*pouch; interspace*) is bounded by the superficial layer of the urogenital diaphragm and the external perineal fascia. It contains the Bulbocavernosus, Ischiocavernosus, and Transversus perinei superficialis muscles, and is traversed by the perineal vessels and nerve. The usual description of this compartment, in which Colles' fascia forms the superficial boundary, is erroneous. The superficial boundary is a layer of deep fascia, the external investing perineal fascia (inferior perineal fascia, Gallaudet). There is a distinct **superficial perineal fascial cleft** between the latter and Colles' fascia (Figs. 6–28, 6–30).

Clinical Considerations.—The fascial cleft between the external perineal fascia and Colles' fascia has long been included erroneously in the superficial perineal pouch. When extravasated urine or hemorrhage finds its way into the tissue under Colles' fascia, it is not in the pouch but in the fascial cleft superficial to it. If it were in the pouch, it would infiltrate the superficial muscles and be restricted in its spread to the area covered by these muscles. Clinical experience has shown that it actually does spread along the groove between the scrotum and thigh to the anterior abdominal wall in the fascial cleft under Scarpa's fascia and into the scrotum and penis in the fascial cleft under the dartos. The spread is restricted posteriorly by the termination of the fascial cleft where Colles' fascia is attached to the infraänal fascia at the posterior border of the Transversus perinei profundus muscle, and laterally where it is attached to the fascia lata.

Muscles of the Perineum

The muscles of the perineum may be divided into two groups:

1. Those of the urogenital region: A, In the male; B, In the female.
2. Those of the anal region.

1. A. The Muscles of the Urogenital Region in the Male (Fig. 6–31)

Superficial Group:

Transversus perinei superficialis
Bulbocavernosus
Ischiocavernosus

Deep Group:

Transversus perinei profundus
Sphincter urethrae

The **Transversus perinei superficialis** (*Transversus perinei; Superficial transverse perineal muscle*) is a narrow muscular slip, which passes more or less transversely across the perineal area in front of the anus. It *arises* by tendinous fibers from the inner and anterior part of the tuberosity of the ischium, and, running medialward, is inserted into the central tendinous point of the perineum, joining in this situation with the muscle of the opposite side, with the Sphincter ani externus posteriorly, and with the Bulbocavernosus anteriorly. In some instances, the fibers of the deeper layer of the Sphincter ani externus decussate in front of the anus and are continued into this muscle. Occasionally it gives off fibers which join with the Bulbocavernosus of the same side.

Actions.—The simultaneous contraction of the two muscles serves to fix the central tendinous point of the perineum.

Variations are numerous. It may be absent or double, or insert into Bulbocavernosus or External sphincter.

The **Bulbocavernosus** (*Ejaculator urinae; Accelerator urinae*) is placed in the middle line of the perineum, ventral to the anus. It consists of two symmetrical parts, united along the median line by a tendinous raphe. It *arises* from the central tendinous point of the perineum and its ventral extension into the median raphe. Its fibers diverge like the barbs of a feather; the most posterior form a thin layer, which is lost on the superficial fascia of the urogenital diaphragm; the middle fibers encircle the bulk and adjacent parts of the corpus spongiosum penis, and join with the fibers of the opposite side, on the upper part of the corpus cavernosum penis, in a strong aponeurosis; the anterior fibers spread out over the side of the corpus cavernosum penis to be inserted partly into that body, anterior to the Ischiocavernosus, occasionally extending to the pubis, and partly ending in a tendinous expansion which covers the dorsal vessels of the penis. The latter fibers are seen by dividing the muscle longitudinally,

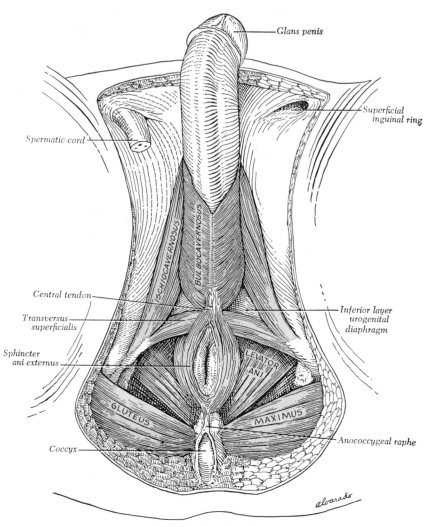

Glans penis

Superficial
inguinal ring

Spermatic cord

ISCHIOCAVERNOSUS

BULBOCAVERNOSUS

Central tendon

Transversus
superficialis

Inferior layer
urogenital
diaphragm

Sphincter
ani externus

LEVATOR
ANI

GLUTEUS

MAXIMUS

Coccyx

Anococcygeal raphe

alvarado

FIG. 6–31.—Muscles of male perineum.

and reflecting it from the surface of the corpus spongiosum penis.

Actions.—This muscle serves to empty the canal of the urethra, after the bladder has expelled its contents; during the greater part of the act of micturition its fibers are relaxed, and it only comes into action at the end of the process. The middle fibers are supposed by Krause to assist in the erection of the corpus cavernosum urethrae, by compressing the erectile tissue of the bulb. The anterior fibers, according to Tyrrel, also contribute to the erection of the penis by compressing the deep dorsal vein of the penis as they are inserted into the fascia of the penis.

The **Ischiocavernosus** (*Erector penis*) covers the crus penis. It is an elongated muscle, broader in the middle than at either end, and situated on the lateral boundary of the perineum. It *arises* by tendinous and fleshy fibers from the inner surface of the tuberosity of the ischium dorsal to the crus penis, and from the rami of the pubis and ischium on either side of the crus. From these points, fleshy fibers run ventrally along the crus, and end in an aponeurosis which is *inserted* into the sides and under surface of the crura as they become the body of the penis.

Action.—The Ischiocavernosus compresses the crus penis, and retards the return of the blood through the veins, and thus serves to maintain the organ erect.

The **Transversus perinei profundus** *arises* from the inferior rami of the ischium and runs to the median line, where it interlaces in a tendinous raphe with its fellow of the opposite side. It lies in the same plane as the Sphincter urethrae; formerly the two muscles were described together as the **Constrictor urethrae.**

The **Sphincter urethrae** surrounds the whole length of the membranous portion of the urethra, and is enclosed in the fasciae of the urogenital diaphragm. Its *external* fibers *arise* from the junction of the inferior rami of the pubis and ischium to the extent of 1.25 to 2 cm, and from the neighboring fasciae. They arch across the front of the urethra and bulbourethral glands, pass around the urethra, and behind it unite with the muscle of the opposite side, by means of a tendinous raphe. Its *innermost* fibers form

a continuous circular investment for the membranous urethra.

Actions.—The muscles of both sides act together as a sphincter, compressing the membranous portion of the urethra. During the transmission of fluids they, like the Bulbocavernosus, are relaxed, and only come into action at the end of the process to eject the last drops of the fluid.

Nerve Supply.—The perineal branch of the pudendal nerve supplies this group of muscles.

1. B. The Muscles of the Urogenital Region in the Female (Fig. 6–32)

Transversus perinei superficialis
Bulbocavernosus
Ischiocavernosus
Transversus perinei profundus
Sphincter urethrae

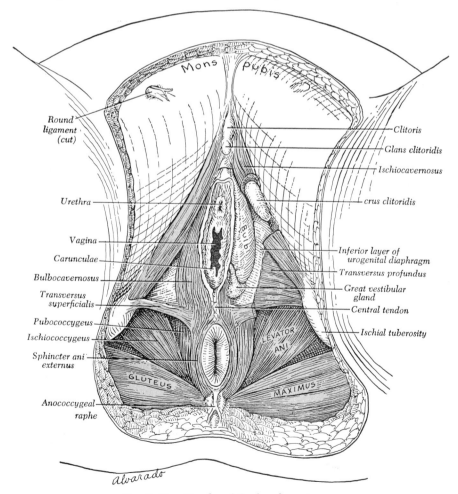

Fig. 6–32.—Muscles of the female perineum.

The **Transversus perinei superficialis** (*Transversus perinei; Superficial transverse perineal muscle*) in the female is a narrow muscular slip, which *arises* by a small tendon from the inner and ventral part of the tuberosity of the ischium, and is *inserted* into the central tendinous point of the perineum, joining in this situation with the muscle of the opposite side, the Sphincter ani externus, and the Bulbocavernosus.

Action.—The simultaneous contraction of the two muscles serves to fix the central tendinous point of the perineum.

The **Bulbocavernosus** (*sphincter vaginae*) surrounds the orifice of the vagina. It covers the lateral parts of the vestibular bulbs, and is attached posteriorly to the central tendinous point of the perineum, where it blends with the Sphincter ani externus. Its fibers pass ventrally on either side of the vagina to be inserted into the corpora cavernosa clitoridis, a fasciculus crossing over the body of the organ so as to compress the deep dorsal vein.

Actions.—The Bulbocavernosus diminishes the orifice of the vagina. The anterior fibers contribute to the erection of the clitoris, as they are inserted into and are continuous with the fascia of the clitoris, compressing the deep dorsal vein during the contraction of the muscle.

The **Ischiocavernosus** (*Erector clitoridis*) is smaller than the corresponding muscle in the male. It covers the unattached surface of the crus clitoridis. It is an elongated muscle, broader at the middle than at either end, and situated on the side of the lateral boundary of the perineum. It *arises* by tendinous and fleshy fibers from the inner surface of the tuberosity of the ischium, dorsal to the crus clitoridis; from the surface of the crus; and from the adjacent portion of the ramus of the ischium. From these points fleshy fibers extend ventrally, and end in an aponeurosis which is *inserted* into the sides and under surface of the crus clitoridis.

Actions.—The Ischiocavernosus compresses the crus clitoridis and retards the return of blood through the veins, and thus serves to maintain the organ erect.

The **fascia of the urogenital diaphragm** in the female is not so strong as in the male. It is attached to the pubic arch, its apex being connected with the arcuate pubic ligament. It is divided in the middle line by the aperture of the vagina, with the external coat of which it becomes blended, and ventral to this is perforated by the urethra. Its posterior border is continuous, as in the male, with the deep layer of the superficial fascia around the Transversus perinei superficialis.

Like the corresponding fascia in the male, it consists of two layers, between which are to be found the following structures: the deep dorsal vein of the clitoris, a portion of the urethra and the Constrictor urethrae muscle, the larger vestibular glands and their ducts; the internal pudendal vessels and the dorsal nerves of the clitoris; the arteries and nerves of the bulbi vestibuli, and a plexus of veins.

The **Transversus perinei profundus** *arises* from the inferior rami of the ischium and runs across to the side of the vagina. The **Sphincter urethrae** (*Constrictor urethrae*), like the corresponding muscle in the male, consists of external and internal fibers. The *external* fibers *arise* on either side from the margin of the inferior ramus of the pubis. They are directed across the pubic arch in front of the urethra, and pass around it to blend with the muscular fibers of the opposite side, between the urethra and vagina. The *innermost* fibers encircle the lower end of the urethra.

Nerve Supply.—The muscles of this group are supplied by the perineal branch of the pudendal.

2. The Muscles of the Anal Region

Corrugator cutis ani
Sphincter ani externus
Sphincter ani internus

The **Corrugator cutis ani.**—Around the anus is a thin stratum of involuntary muscular fibers, which radiates from the orifice. *Medially* the fibers fade off into the submucous tissue, while *laterally* they blend with the true skin. By its contraction it raises the skin into ridges around the margin of the anus.

The **Sphincter ani externus** (*External sphincter ani*) (Fig. 6–32) is a flat plane of muscular fibers, elliptical in shape and intimately adherent to the integument surrounding the margin of the anus. It measures about 8 to 10 cm in length, from its anterior to its posterior extremity, and is about 2.5 cm broad opposite the anus. It consists of two strata, superficial and deep. The *superficial,* constituting the main portion of the muscle, arises from a narrow tendinous band, the **anococcygeal raphe,** which stretches from the tip of the coccyx to the posterior margin of the anus; it forms two flattened planes of muscular tissue, which encircle the anus and meet in front to be inserted into the central tendinous point of the perineum, joining with the Transversus perinei superficialis, the Levator ani, and the Bulbocavernosus. The *deeper portion* forms a complete sphincter to the anal canal. Its fibers surround the canal, closely applied to the Sphincter ani internus, and in front blend with the other muscles at the central point of the perineum. In a considerable proportion of individuals the fibers decussate in front of the anus, and are continuous with the Transversi perinei superficiales. Posteriorly, they are not attached to the coccyx, but are continuous with those of the opposite side behind the anal canal. The upper edge of the muscle is ill defined, since fibers█ off from it to join the Levator ani.█

Actions.—The action of this muscle█ peculiar. (1) It is, like other muscles, alway█ in a state of tonic contraction, and having no antagonistic muscle it keeps the anal canal and orifice closed. (2) It can be put into a condition of greater contraction under the influence of the will, so as more firmly to occlude the anal aperture, in expiratory efforts unconnected with defecation. (3) Taking its fixed point at the coccyx, it helps to fix the central point of the perineum, so that the Bulbocavernosus may act from this fixed point.

Nerve Supply.—A branch from the fourth sacral and twigs from the inferior rectal branch of the pudendal supply the muscle.

The **Sphincter ani internus** (*Internal sphincter ani*) is a muscular ring which surrounds about 2.5 cm of the anal canal; its inferior border is in contact with, but quite separate from, the Sphincter ani externus. It is about 5 mm thick, and is formed by an aggregation of the involuntary circular fibers of the intestine. Its distal border is about 6 mm from the orifice of the anus (Fig. 16–92).

Actions.—Its action is entirely involuntary. It helps the Sphincter ani externus to occlude the anal aperture and aids in the expulsion of the feces.

THE MUSCLES AND FASCIAE OF THE UPPER LIMB

The muscles of the upper limb are divisible into groups, corresponding with the different regions of the limb:

I. Muscles Connecting the Limb to the Vertebral Column
II. Muscles Connecting the Limb to the Anterior and Lateral Thoracic Walls
III. Muscles of the Shoulder
IV. Muscles of the Arm
V. Muscles of the Forearm
VI. Muscles of the Hand

I. THE MUSCLES CONNECTING THE UPPER LIMB TO THE VERTEBRAL COLUMN

Trapezius
Latissimus dorsi
Rhomboideus major
Rhomboideus minor
Levator scapulae

The **subcutaneous fascia** (*tela subcutanea*) of the back is a thick fibrous and fatty layer which extends from the scalp to the gluteal region as a comparatively uniform sheet. At the sides of the neck and trunk, it gradually changes into the thinner or softer fascia of the ventral regions. The dermal fibrous layer is thick, and is bound down to the deeper layers by numerous heavy bands and septa which divide the fat into small granular lobules, and mattress the entire layer into a tough, resilient pad. The fascial cleft, usually present between the subcutaneous and deep fascia, is lacking in the dorsal area, making the subcutaneous fascia firmer and less movable than in the ventral area of the body.

Column 445

are given

is

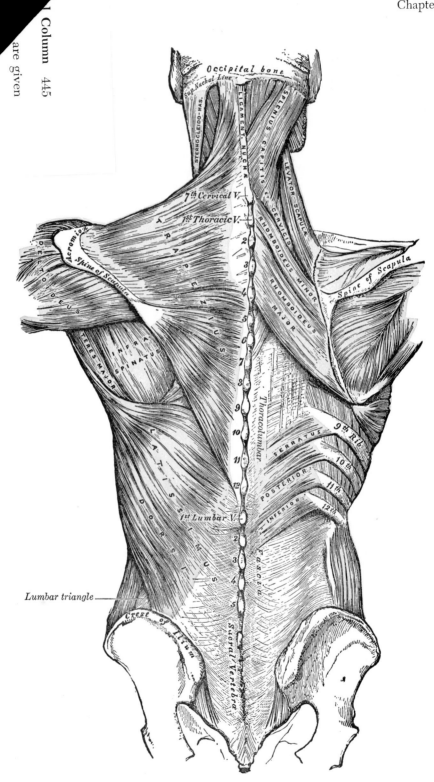

FIG. 6–33.—Muscles connecting the upper limb to the vertebral column.

Deep fascia.—The investing layer of deep fascia is attached in the middle line to the ligamentum nuchae, the supraspinal ligament, and to the spinous processes of all vertebrae caudal to the sixth cervical. It splits to enclose the Trapezius and the fleshy portion of the Latissimus dorsi, but in the neck it covers the posterior triangle as a single layer and there becomes continuous with the anterior cervical fascia. It is attached, over the shoulder, to the acromion and spine of the scapula, and is then continuous with the deltoid fascia. Laterally from the Latissimus, it is continuous with the axillary fascia and the fascia covering the Obliquus abdominis externus. Over the fleshy portions of the Trapezius and Latissimus, it is gray and felt-like, but strong and adherent both to the superficial fascia and to the muscles. In the triangular area between the Trapezius, Deltoideus, and Latissimus it is white and glistening, forming the aponeurosis of the Infraspinatus. In the lumbar region it is greatly strengthened by having incorporated in it the aponeurosis of origin of the Latissimus dorsi. This portion is called the lumbar aponeurosis or the posterior layer of the thoracolumbar fascia (page 402).

The fascia of the deep surface of the Trapezius is thickened by an accumulation of adipose tissue similar to that under the Pectoralis major. It contains the branches of the superficial cervical and transverse cervical arteries, and the accessory nerve. It is separated from the underlying structures by a distinct fascial cleft. The fascia of the deep surface of the fleshy portion of the Latissimus is similar to that of the Trapezius, but over the aponeurosis it loses its identity as a separate layer, and is fused with the thoracolumbar fascia.

The Rhomboidei and the Levator scapulae are enclosed in their own proper fascial sheaths which are attached to the vertebrae and to the vertebral border of the scapula. The fascia of the superficial surface, after attaching to the border of the scapula, continues laterally as the supra- and infraspinatus fasciae. That of the deep surface continues laterally on the deep surface of the Serratus anterior. A distinct fascial cleft separates both superficial and deep surfaces of the Rhomboidei from contiguous structures.

The fascia of the Levator is more adherent to surrounding structures and is continuous with the scalenus fascia in the posterior triangle of the neck. In the interval between the superior and the inferior Serrati posteriores, it is possible to dissect out a thin membrane which lies in the same plane as these muscles, but distinct from the fascia of the underlying Sacrospinalis and the overlying Latissimus and Rhomboidei. This may represent the vestige of a continuous Serratus muscle like that found in some lower animals.

The **Trapezius** (Fig. 6–33) is a flat, triangular muscle, covering the upper and back part of the neck and shoulders. It *arises* from the external occipital protuberance and the medial third of the superior nuchal line of the occipital bone, from the ligamentum nuchae, the spinous process of the seventh cervical, and the spinous processes of all the thoracic vertebrae, and from the corresponding portion of the supraspinal ligament. From this origin, the superior fibers proceed caudally and laterally, the inferior cranially and laterally, and the middle horizontally; the superior fibers are *inserted* into the posterior border of the lateral third of the clavicle; the middle fibers into the medial margin of the acromion, and into the superior lip of the posterior border of the spine of the scapula; the inferior fibers converge near the scapula, and end in an aponeurosis which glides over the smooth triangular surface on the medial end of the scapular spine, to be inserted into a tubercle at the apex of this smooth triangular surface. At its occipital origin, the Trapezius is connected to the bone by a thin fibrous lamina, firmly adherent to the skin. At the middle it is connected to the spinous processes by a broad semi-elliptical aponeurosis, which reaches from the sixth cervical to the third thoracic vertebra, and forms, with that of the opposite muscle, a tendinous ellipse. The rest of the muscle arises by numerous short tendinous fibers. The two Trapezius muscles together resemble a trapezium, or diamond-shaped quadrangle: two angles corresponding to the shoulders; a third to the occipital protuberance; and the fourth to the spinous process of the twelfth thoracic vertebra.

Action.—All parts, acting together, rotate the scapula, raising the point of the shoulder in full abduction and flexion of the arm. They also adduct the scapula, that is, draw it medially toward the vertebral column. The upper part acting alone draws the scapula cranially, bracing the shoulder. The lower part acting alone draws the scapula downward. The upper part of one side draws the head toward the same side, and turns the face to the opposite side; both sides together draw the head directly dorsalward.

Nerve.—The spinal accessory (spinal part of eleventh cranial) nerve and sensory branches from the ventral primary divisions of the third and fourth cervical nerves.

Variations.—The attachments to the thoracic vertebrae are often reduced, the caudal ones being absent. The occipital attachment may be small or wanting. The clavicular attachment may be reduced, but is more often increased, and may cover the posterior triangle. The cervical and thoracic portions may be separate. The muscles of the two sides are seldom symmetrical, and complete absence has been described. Aberrant bundles are not uncommon.

The **Latissimus dorsi** (Fig. 6–33) is a large triangular muscle which covers the lumbar and lower half of the posterior thoracic region. Its *origin* is principally in a broad aponeurosis, the lumbar aponeurosis (posterior layer of the thoracolumbar fascia, see page 402), by means of which it is attached to the spinous processes of the lower six thoracic, the lumbar, and the sacral vertebrae, to the supraspinal ligament, and to the posterior part of the crest of the ilium. It also *arises* by muscular fasciculi from the external lip of the crest of the ilium lateral to the margin of the Sacrospinalis, and from the caudal three or four ribs by fleshy digitations which are interposed between similar processes of the Obliquus externus abdominis (Fig. 6–18). From this extensive origin, the fasciculi converge toward the shoulder; those of the upper part are almost horizontal, and, as they pass over the inferior angle of the scapula, are joined by additional fasciculi arising from this bone. The muscle curves around the lower border of the Teres major, and is twisted upon itself, so that the superior fibers become at first posterior and then inferior, and the vertical fibers at first anterior and then superior. It ends in a quadrilateral tendon, about 7 cm long,

which passes anterior to the tendon of the Teres major, and is *inserted* into the bottom of the intertubercular groove of the humerus; its insertion extends farther proximally on the humerus than that of the tendon of the Pectoralis major. The inferior border of its tendon is united with that of the Teres major, the surfaces of the two being separated near their insertions by a bursa; another bursa is sometimes interposed between the muscle and the inferior angle of the scapula. The tendon of the muscle gives off an expansion to the deep fascia of the arm.

Action.—Extends, adducts, and rotates the arm medially; draws the shoulder downward and backward.

Nerve.—The thoracodorsal (long subscapular) nerve from the brachial plexus, containing fibers from the sixth, seventh and eighth cervical nerves.

Variations.—The number of thoracic vertebrae to which it is attached varies from four to seven or eight; the number of costal attachments varies; muscle fibers may or may not reach the crest of the ilium.

A muscular slip, the **axillary arch,** varying from 7 to 10 cm in length, and from 5 to 15 mm in breadth, occasionally springs from the cranial border of the Latissimus dorsi about the middle of the posterior fold of the axilla, and crosses the axilla ventral to the axillary vessels and nerves, to join the under surface of the tendon of the Pectoralis major, the Coracobrachialis, or the fascia over the Biceps brachii. This axillary arch crosses the axillary artery, just above the spot usually selected for the application of a ligature, and may mislead the surgeon during the operation. It is present in about 7 per cent of subjects and may be easily recognized by the transverse direction of its fibers.

A fibrous slip usually passes from the inferior border of the tendon of the Latissimus dorsi, near its insertion, to the long head of the Triceps brachii. This is occasionally muscular, and is the representative of the *Dorsoepitrochlearis brachii* of apes.

Clinical Considerations.—The **lumbar triangle of Petit** is a small triangular interval which separates the lateral margin of the caudal portion of the Latissimus dorsi from the Obliquus externus abdominis just cranial to the ilium. The base of the triangle is the iliac crest, and its floor is the Obliquus internus abdominis. It is occasionally the site of a hernia.

The **triangle of auscultation** is a triangle

associated with the cranial portion of the Latissimus. It is bounded superiorly by the Trapezius, inferiorly by the Latissimus dorsi, and laterally by the vertebral border of the scapula. The floor is partly formed by the Rhomboideus major. If the scapula is drawn ventrally by folding the arms across the chest, and the trunk bent forward, parts of the sixth and seventh ribs and their interspace become subcutaneous and accessible for auscultation.

The **Rhomboideus major** (Fig. 6–33) *arises* by tendinous fibers from the spinous processes of the second, third, fourth, and fifth thoracic vertebrae and the supraspinal ligament, and is *inserted* into a narrow tendinous arch, attached to the inferior part of the triangular surface at the root of the spine of the scapula cranially and to the inferior angle caudally, the arch being connected to the vertebral border by a thin membrane. When the arch extends, as it occasionally does, only a short distance, the muscular fibers are inserted directly into the scapula.

The **Rhomboideus minor** (Fig. 6–33) *arises* from the inferior part of the ligamentum nuchae and from the spinous processes of the seventh cervical and first thoracic vertebrae. It is *inserted* into the base of the triangular smooth surface at the root of the spine of the scapula, and is usually separated from the Rhomboideus major by a slight interval, but the adjacent margins of the two muscles are occasionally united.

Action.—Adduct the scapula, that is, draw it medially toward the vertebral column, at the same time supporting it and drawing it slightly cranialward. The lower part of the major rotates the scapula to depress the lateral angle, assisting in adduction of the arm.

Nerve.—The dorsal scapular nerve from the brachial plexus, containing fibers from the fifth cervical nerve.

Variations.—The veterbral and scapular attachments of the two muscles vary in extent. A small slip from the scapula to the occipital bone occasionally occurs close to the minor, the *Rhomboideus occipitalis muscle.*

The **Levator scapulae** (*Levator anguli scapulae*) (Fig. 6–33) is situated at the dorsal and lateral part of the neck. It *arises* by tendinous slips from the transverse processes of the atlas and axis and from the posterior tubercles of the transverse proc-

esses of the third and fourth cervical vertebrae. It is *inserted* into the vertebral border of the scapula, between the superior angle and the triangular smooth surface at the root of the spine.

Action.—Raises the scapula, tending to draw it medialward and rotate it to lower the lateral angle. With the scapula fixed, it bends the neck laterally and rotates it slightly toward the same side.

Nerves.—Branches of the third and fourth cervical nerves from the cervical plexus, and frequently the lower portion by a branch of the dorsal scapular nerve containing fibers from the fifth cervical nerve.

Variations.—The number of vertebral attachments varies; a slip may extend to the occipital or mastoid, to the Trapezius, Scalene or Serratus anterior, or to the first or second rib. The muscle may be subdivided into several distinct parts from origin to insertion. A slip from the transverse processes of one or two cranial cervical vertebrae to the outer end of the clavicle corresponds to the *Levator claviculae* muscle of lower animals. More or less union with the Serratus anterior is not uncommon.

II. THE MUSCLES CONNECTING THE UPPER LIMB TO THE ANTERIOR AND LATERAL THORACIC WALLS

> Pectoralis major
> Pectoralis minor
> Subclavius
> Serratus anterior

The **subcutaneous fascia** (*tela subcutanea*) of the pectoral region is continuous with that of the neck and upper limb cranially, the abdomen caudally, and the axilla laterally. The fasciculi of the Platysma muscle extend down from the neck for a variable distance between its superficial and deep layers. In the female, the adipose tissue is increased and molded into a rounded mass which gives the bulk and form to the mamma. The parenchyma of the mammary gland is imbedded in this fat. The connective tissue stroma, distributed between the lobes of the gland, is thickened into fibrous bands, called **ligamenta suspensoria** or **Cooper's ligaments,** which secure the skin to the deep layer of the subcutaneous fascia. A fascial cleft separates the subcutaneous fascia from the deep fascia.

Through these fascial structures, a carcinoma may make its presence known either by pulling on Cooper's ligaments and dimpling the skin like an orange peel, or by interfering with the normal movability of the gland through adhesions between the subcutaneous and deep fasciae.

Pectoral fascia (*fascia pectoralis*) (Fig. 6–34).—The pectoral fascia is a membranous sheet of deep fascia which consists of the external investing fascia superficial to the Pectoralis major, and a deeper layer enclosing its inner surface. It is more or less adherent throughout and attached, with the origin of the muscle, to the clavicle and sternum, and with the insertion, to the humerus. The external layer is continuous medially, across the middle line, with the pectoral fascia of the other side; superiorly and laterally, with the brachial fascia which covers the Coracobrachialis and Biceps; and inferiorly, with the axillary, thoracic, and abdominal investing fasciae. At the more caudal sternocostal and abdominal origins of the muscle, it is aponeurotic and is blended with the sheath of the Rectus abdominis. The deeper layer, covering the

deep surface of the muscle and adherent to it, is thickened by a considerable pad of fat in which the thoracoacromial blood vessels and pectoral nerves are imbedded. A definite fascial cleft separates this layer from the underlying clavipectoral fascia. The external and the deeper layers come together at both borders of the muscle, forming a closed compartment. At the superior border, the fascia separates again to enclose the Deltoideus; at the inferior border, however, a single sheet is formed which becomes immediately the axillary fascia.

The **clavipectoral fascia** (*fascia coracoclavicularis*) (Fig. 6–34) is an intermediate stratum of deep fascia which lies between the Pectoralis major and the thoracic wall. It invests the Pectoralis minor and Subclavius, and extends from the clavicle to the axillary fascia. Its attachment to the clavicle is by means of two membranes which lie superficial and deep to the Subclavius, and which are separated from each other by the insertion of the muscle. The two sheets form a compartment for the muscle by fusing, at its inferior border, into

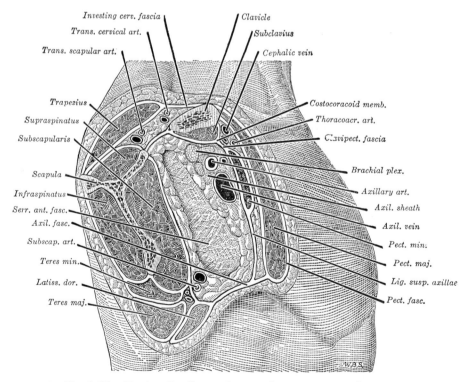

FIG. 6–34.—Fasciae of axillary and pectoral regions in sagittal section.

a single sheet which stretches across the interval between the Subclavius and the Pectoralis minor. The portion superficial to the Subclavius is greatly strengthened by the addition of fibrous bundles which continue laterally beyond the muscle, and attach to the coracoid process, forming the **costocoracoid ligament.** This strong band, attached to the coracoid process, the clavicle, and the first rib, serves as an important ligament for the clavicular articulations. From its attachment to the coracoid process, the fascia passes caudalward as a thin sheet, investing the Pectoralis minor on both its surfaces, and attaching to the ribs with the origin of the latter muscle. Between the ribs it is continuous with the external intercostal fascia. At the superior border of the Pectoralis minor, the two layers investing the muscle fuse into the single sheet which bridges the triangular interval between the Pectoralis minor and Subclavius. This single sheet is thin, caudal to the ligament, and has one or more holes or defects through which pass the thoracoacromial artery and vein, the cephalic vein, and the lateral pectoral nerve. The portion of the fascia between the cranial border of the Pectoralis minor and the clavicle, including the costocoracoid ligament, has been named the **fascia coracoclavicularis.** At the inferior or lateral border of the Pectoralis minor, the two layers combine again into a single sheet which passes into the axilla,

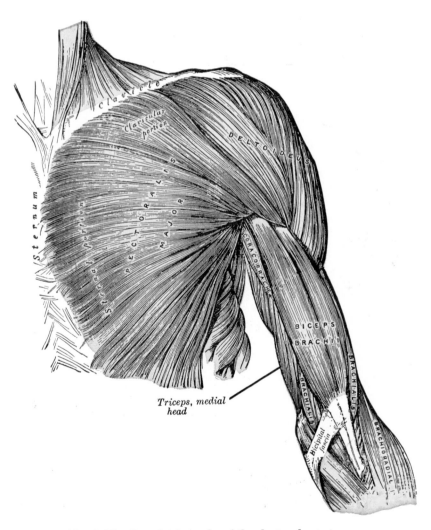

FIG. 6–35.—Superficial muscles of the chest and anterior arm.

and fuses with the deep surface of the axillary fascia a short distance from the lateral border of the Pectoralis major. The axillary sheath, enclosing the axillary vessels and the nerves of the brachial plexus, passes deep to the lateral portion of the clavipectoral fascia, but is partially separated from it by some of the adipose tissue of the deep axillary fossa.

Axillary fascia (*fascia axillaris*) (Fig. 6–34).—The portion of the investing fascia which crosses the interval between the lateral borders of the Pectoralis major and Latissimus dorsi dips inward to form the hollow of the armpit. It is adherent to the subcutaneous fascia, and there are openings through which the adipose tissue of the latter is continuous with that in the deeper axillary fossa. It is continuous with the pectoral, latissimus, serratus, and brachial fasciae. Fused with its deep surface, in the hollow of the armpit, is the termination of

the clavipectoral fascia which has continued laterally from the Pectoralis minor. The latter fascia has been named the **suspensory ligament of the axilla** because it is believed that the hollow, seen when the arm is abducted, is produced mainly by the traction of this fascia on the axillary floor.

The **axilla** (Fig. 6–36, "Axillary Space"), in anatomical usage comprises more than the externally visible armpit. It includes the fossa between the medial side of the arm and the lateral surface of the chest wall, inside or deep to the axillary investing layer described above. It is commonly called a space, but should be recognized as a pyramidal compartment filled with adipose tissue, vessels, nerves, and lymph nodes. The walls of the fossa are formed by the fascial coverings of the following muscles: ventrally, the Pectoralis major and minor; dorsally, the Latissimus dorsi, Teres

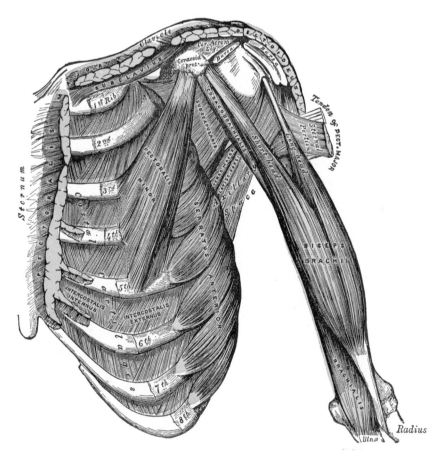

Fɪɢ. 6–36.—Deep muscles of the chest and front of the arm, with the boundaries of the axilla.

major, and Subscapularis; medially, the Serratus anterior; laterally, the Coracobrachialis and Biceps. At the apex of the fossa, between the first rib, clavicle, and base of the coracoid process, the adipose padding tissue is continuous with the mass of similar tissue in the posterior triangle of the neck. At the apex also, the fascia of the first two ribs and first intercostal space is continuous with the scalenus fascia. In this region, the latter fascia gives a tubular investment, called the **axillary sheath,** to the large vessels and nerves of the arm. The sheath is partially adherent to the clavipectoral fascia on the deep surface of the Subclavius and Pectoralis minor as it passes under them, and then, after traversing the lateral wall of the axilla along the Coracobrachialis, it becomes fused with the anterior surface of the medial intermuscular septum of the arm.

The **Pectoralis major** (Fig. 6–35) is a thick, fan-shaped muscle, situated at the ventral and superior part of the chest. It *arises* from the anterior surface of the sternal half of the clavicle; from half the breadth of the ventral surface of the sternum, as far caudalward as the attachment of the cartilage of the sixth or seventh rib; from the cartilages of all the true ribs, with the exception, frequently, of the first or seventh, or both, and from the aponeurosis of the Obliquus externus abdominis. From this extensive origin the fibers converge toward their insertion; those arising from the clavicle pass obliquely caudalward and lateralward, and are usually separated from the rest by a slight interval; those from the inferior part of the sternum, and the cartilages of the lower true ribs, run cranialward and lateralward, while the middle fibers pass horizontally. They all end in a flat tendon, about 5 cm broad, which is *inserted* into the crest of the greater tubercle of the humerus. This tendon consists of two laminae, placed one in front of the other, and usually blended together below. The ventral lamina, the thicker, receives the clavicular and the more cranial sternal fibers; they are inserted in the same order as that in which they arise: that is to say, the most lateral of the clavicular fibers are inserted at the more cranial part of the anterior lamina; the more cranial sternal fibers pass down to the

inferior part of the lamina which extends as far caudally as the tendon of the Deltoideus and joins with it. The dorsal lamina of the tendon receives the attachment of the greater part of the sternal portion and the deep fibers, *i.e.,* those from the costal cartilages. These deep fibers, and particularly those from the lower costal cartilages, ascend more cranially, turning dorsalward successively behind the superficial and superior ones, so that the tendon appears to be twisted. The posterior lamina reaches more proximally on the humerus than the anterior one, and from it an expansion is given off which covers the intertubercular groove and blends with the capsule of the shoulder joint. From the deepest fibers of this lamina at its insertion, an expansion is given off which lines the intertubercular groove, while from the inferior border of the tendon a third expansion passes distalward to the fascia of the arm.

Action.—Flexes, adducts, and rotates the arm medially. The clavicular portion draws the arm or the shoulder, if the arm is at the side, cranialward, ventrally and medially; the sternocostal portion draws the arm or shoulder ventrally, medially, and caudalward.

Nerves.—Medial and lateral pectoral nerves from the brachial plexus, containing fibers from the fifth, sixth, seventh, eighth cervical, and first thoracic nerves.

Variations.—The more frequent variations are greater or less extent of attachment to the ribs and sternum, varying size of the abdominal part or its absence, greater or less extent of separation of sternocostal and clavicular parts, fusion of clavicular part with deltoid, decussation in front of the sternum. Deficiency or absence of the sternocostal part is not uncommon. Absence of the clavicular part is less frequent. Rarely the whole muscle is wanting.

Costocoracoideus is a muscular band occasionally found arising from the ribs or aponeurosis of the External oblique between the Pectoralis major and Latissimus dorsi and inserted into the coracoid process.

Chondro-epitrochlearis is a muscular slip occasionally found arising from the costal cartilages or from the aponeurosis of the External oblique below the Pectoralis major or from the Pectoralis major itself. The insertion is variable on the inner side of the arm to fascia, intermuscular septum or medial epicondyle.

The *Sternalis* is a small superficial muscle at the sternal end of the Pectoralis major at right angles to its fibers and parallel with the margin of the sternum. It is supplied by the pectoral nerves and is probably a misplaced part of the pectoralis.

The **Pectoralis minor** (Fig. 6–36) is a thin, triangular muscle, situated at the cranial part of the thorax, deep to the Pectoralis major. It *arises* from the cranial margins and outer surfaces of the third, fourth, and fifth ribs, near their cartilages, and from the aponeuroses covering the Intercostales; the fibers pass cranialward and lateralward and converge to form a flat tendon which is *inserted* into the medial border and superior surface of the coracoid process of the scapula.

Action.—Draws the scapula ventralward and caudalward, and rotates it to lower the lateral angle, as in adduction of the arm. Raises the third, fourth, and fifth ribs in forced inspiration, the scapula being fixed by the Levator scapulae.

Nerve.—The medial pectoral nerve from the brachial plexus, containing fibers from the eighth cervical and first thoracic nerves.

Variations.—Origin from second, third and fourth or fifth ribs. The tendon of insertion may extend over the coracoid process to the greater tubercle. The muscle may be split into several parts. Absence is rare.

The *Pectoralis minimus* is a slip from the cartilage of first rib to coracoid process. Rare.

The **Subclavius** (Fig. 6–36) is a small cylindrical muscle, placed between the clavicle and the first rib. It *arises* by a short, thick tendon from the first rib and its cartilage at their junction; the fleshy fibers proceed obliquely, cranialward and lateralward, to be *inserted* into the groove on the inferior surface of the clavicle between the costoclavicular and conoid ligaments.

Actions.—Draws the shoulder ventralward and caudalward.

Nerve.—A special nerve from the lateral trunk of the brachial plexus containing fibers from the fifth and sixth cervical nerve.

Variations.—Insertion may be into the coracoid process instead of the clavicle or into both clavicle and coracoid process. A *sternoscapular* fasciculus is to the upper border of the scapula, a *Sternoclavicularis* is from the manubrium to the clavicle between the Pectoralis major and coracoclavicular fascia.

The **Serratus anterior** (*Serratus magnus*) (Fig. 6–36) is a thin muscular sheet, situated between the ribs and the scapula, spreading over the lateral part of the chest. It *arises* by fleshy digitations from the outer surfaces and superior borders of the first eight or nine ribs, and from the aponeuroses covering the intervening Intercostales. Each digitation (except the first) arises from the corresponding rib; the first springs from the first and second ribs and from the fascia covering the first intercostal space. From this extensive attachment the fibers pass dorsalward, closely applied to the chest-wall, to the vertebral border of the scapula, and are inserted into its ventral surface in the following manner. The first digitation is *inserted* into a triangular area on the ventral surface of the superior angle. The next two digitations spread out to form a thin, triangular sheet, the base of which is directed dorsalward and is inserted into nearly the whole length of the ventral surface of the vertebral border. The lower five or six digitations converge to form a fan-shaped mass, the apex of which is inserted, by muscular and tendinous fibers, into a triangular impression on the ventral surface of the inferior angle. The lower four slips interdigitate at their origins with the upper five slips of the Obliquus externus abdominis.

Action.—Rotates the scapula, raising the point of the shoulder as in full flexion and abduction of the arm. Draws the scapula forward as in the act of pushing. The upper digitation may draw the scapula downward and forward; the lower digitations draw the scapula downward.

Nerve.—The long thoracic nerve from the brachial plexus, containing fibers from the fifth, sixth, and seventh cervical nerves.

Variations.—Attachment to tenth rib. Absence of attachments to first rib, to one or more of the lower ribs. Division into three parts; absence or defect of middle part. Union with Levator scapulae, External intercostals or External oblique.

III. THE MUSCLES AND FASCIAE OF THE SHOULDER

Deltoideus
Subscapularis
Supraspinatus

Infraspinatus
Teres minor
Teres major

Deltoid fascia.—The deltoid portion of the investing fascia covers the Deltoideus. Cranially, it is attached to the clavicle, acromion, and spine of the scapula. Caudally, it is continuous with the brachial fascia. Ventrally, it bridges the narrow triangular interval between the adjacent borders of the Deltoideus and the Pectoralis major. In this interval, called *Mohrenheim's triangle,* the fascia is thick but is pierced by the cephalic vein and deltoid branch of the thoracoacromial artery. The deltoid fascia is stronger posteriorly, and continues into the Infraspinatus fascia. Along both borders of the Deltoideus, the investing layer joins the fascia of the deep surface to form a closed compartment. Both layers are adherent to the muscle. The deep layer may contain adipose tissue, and is separated by a distinct fascial cleft from the underlying humerus, subdeltoid bursa, shoulder joint, and associated tendons and ligament.

The **Deltoideus** (*Deltoid muscle*) (Fig. 6–35) is a large, thick, triangular muscle, which covers the shoulder joint ventrally, dorsally, and laterally. It *arises* from the anterior border and superior surface of the lateral third of the clavicle; from the lateral margin and superior surface of the acromion, and from the inferior lip of the posterior border of the spine of the scapula, as far dorsally as the triangular surface at its medial end. From this extensive origin the fibers converge toward their insertion, the middle passing vertically, the anterior obliquely dorsalward and lateralward, the posterior obliquely ventralward and lateralward; they unite in a thick tendon, which is *inserted* into the deltoid prominence on the middle of the lateral side of the body of the humerus. At its insertion the muscle gives off an expansion to the deep fascia of the arm. This muscle is remarkably coarse in texture, and the arrangement of its fibers is somewhat peculiar; the central portion of the muscle—that is to say, the part arising from the acromion—consists of oblique fibers; these arise in a bipenniform manner from the sides of the interspersed tendons, generally four in number, which

are attached above to the acromion and pass distally parallel to one another in the substance of the muscle. The oblique fibers thus formed are inserted into similar interspersed tendons, generally three in number, which pass proximally from the insertion of the muscle and alternate with the descending septa. The portions of the muscle arising from the clavicle and spine of the scapula are not arranged in this manner, but are inserted into the margins of the inferior tendon.

Action.—As a whole abducts the arm. The clavicular and adjacent part of the acromial portions flex the arm; the spinous and adjacent part of the acromial portions extend the arm. The most ventral portion rotates the arm medially, the most dorsal portion laterally.

Nerve.—The axillary nerve from the brachial plexus, containing fibers from the fifth and sixth cervical nerves.

Variations.—Large variations are uncommon. More or less splitting is common. Continuation into the Trapezius; fusion with the Pectoralis major; additional slips from the vertebral border of the scapula, infraspinous fascia and axillary border of scapula are not uncommon. Insertion varies in extent or rarely is prolonged to origin of Brachioradialis.

Subscapular fascia (*fascia subscapularis*).—The subscapular fascia is a thin membrane attached to the entire circumference of the subscapular fossa, and affording attachment by its deep surface to some of the fibers of the Subscapularis.

The **Subscapularis** (Fig. 6–36) is a large triangular muscle which fills the subscapular fossa, and *arises* from its medial two-thirds and from the inferior two-thirds of the groove on the axillary border of the bone. Some fibers *arise* from tendinous laminae which intersect the muscle and are attached to ridges on the bone; others from an aponeurosis, which separates the muscle from the Teres major and the long head of the Triceps brachii. The fibers pass lateralward, and, gradually converging, end in a tendon which is *inserted* into the lesser tubercle of the humerus and the ventral part of the capsule of the shoulder joint. The tendon of the muscle is separated from the neck of the scapula by a large bursa, which usually communicates with the cavity of the shoulder joint through an aperture in the capsule.

Action.—Rotates the arm medially. It assists in both flexion and extension, and abduction and adduction, depending on the position of the arm. Draws the humerus toward the glenoid fossa strengthening the shoulder joint.

Nerves.—The upper and lower subscapular nerves from the brachial plexus, containing fibers from the fifth and sixth cervical nerves.

Supraspinatous fascia (*fascia supraspinata*).—The supraspinatous fascia completes the osseofibrous case in which the Supraspinatus muscle is contained; it affords attachment, by its deep surface, to some of the fibers of the muscle. It is thick medially, but thinner laterally under the coracoacromial ligament.

The **Supraspinatus** (Fig. 6–37) occupies the whole of the supraspinatous fossa, *arising* from its medial two-thirds, and from the strong supraspinatous fascia. The muscular fibers converge to a tendon which

crosses the upper part of the shoulder joint, and is *inserted* into the highest of the three impressions on the greater tubercle of the humerus; the tendon is intimately adherent to the tendon of the Infraspinatus and to the capsule of the shoulder joint.

Action.—Abducts the arm. Draws the humerus toward the glenoid fossa, strengthening the shoulder joint. It is a weak lateral rotator and flexor.

Nerve.—Branches of the suprascapular nerve from the brachial plexus, containing fibers from the fifth cervical nerve.

Infraspinatous fascia (*fascia infraspinata*).—The infraspinatous fascia is a dense fibrous membrane, covering the Infraspinatous muscle and fixed to the circumference of the infraspinatous fossa; it affords attachment, by its deep surface, to some fibers of that muscle. It is intimately attached to the deltoid fascia along the overlapping border of the Deltoideus.

The **Infraspinatus** (Fig. 6–37) is a thick

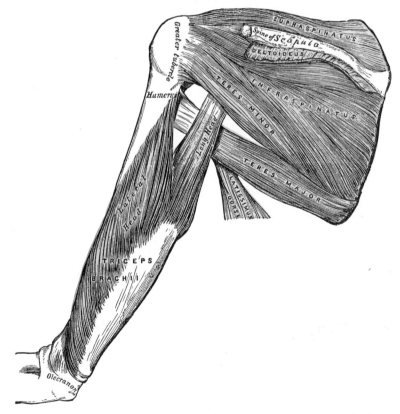

Fig. 6–37.—Muscles on the dorsum of the scapula, and the Triceps brachii, showing the quadrangular and triangular spaces between the two Teres muscles.

triangular muscle, which occupies the chief part of the infraspinatous fossa; it *arises* by fleshy fibers from its medial two-thirds, and by tendinous fibers from the ridges on its surface; it also arises from the infraspinatous fascia which covers it, and separates it from the Teres major and minor. The fibers converge to a tendon which glides over the lateral border of the spine of the scapula, and, passing across the posterior part of the capsule of the shoulder joint, is *inserted* into the middle impression on the greater tubercle of the humerus, where it is fused with its neighbors. The tendon of this muscle is sometimes separated from the capsule of the shoulder joint by a bursa, which may communicate with the joint cavity.

Action.—Rotates the arm laterally. The upper part abducts, the lower part adducts. Draws the humerus toward the glenoid fossa, strengthening the shoulder joint.

Nerve.—The suprascapular nerve from the brachial plexus, containing fibers from the fifth and sixth cervical nerves.

The **Teres minor** (Fig. 6–37) is a cylindrical, elongated muscle, which *arises* from the dorsal surface of the axillary border of the scapula for the cranial two-thirds of its extent, and from two aponeurotic laminae, one of which separates it from the Infraspinatus, the other from the Teres major. Its fibers run obliquely cranialward and lateralward; the upper ones end in a tendon which is *inserted* into the most inferior of the three impressions on the greater tubercle of the humerus; the most inferior fibers are *inserted* directly into the humerus immediately distal to this impression. The tendon of this muscle passes across, and is united with, the posterior part of the capsule of the shoulder joint.

Action.—Rotates the arm laterally and weakly adducts it. Draws the humerus toward the glenoid fossa, strengthening the shoulder joint.

Nerve.—A branch of the axillary nerve, containing fibers from the fifth cervical.

Variations.—It is sometimes inseparable from the Infraspinatus.

The **Teres major** (Fig. 6–37) is a thick but somewhat flattened muscle, which *arises*

from the oval area on the dorsal surface of the inferior angle of the scapula, and from the fibrous septa interposed between the muscle and the Teres minor and Infraspinatus; the fibers are directed cranialward and lateralward, and end in a flat tendon, about 5 cm long, which is *inserted* into the crest of the lesser tubercle of the humerus. The tendon, at its insertion, lies dorsal to that of the Latissimus dorsi, from which it is separated by a bursa, the two tendons being, however, united along their lower borders for a short distance.

Action.—Adducts, extends, and rotates the arm medially.

Nerve.—A branch of the lower subscapular nerve from the brachial plexus, containing fibers from the fifth and sixth cervical nerves.

Group Action of Muscles About the Shoulder.—Flexion of the arm is brought about by the anterior part of the Deltoideus, Coracobrachialis, and short head of the Biceps, acting on the shoulder joint and by the Trapezius and Serratus anterior rotating the scapula on the chest wall to raise the point of the shoulder. The clavicular part of the Pectoralis major also acts until the arm is raised above the shoulder. Extension of the arm is brought about by the Latissimus dorsi, acting on both the shoulder joint and the scapula, and it is strongly assisted by the lower part of the Pectoralis major except in hyperextension. The Teres major, posterior Deltoideus, and long head of the Triceps act on the shoulder joint, and the scapula is rotated downward by the Pectoralis minor, and drawn backward by the Rhomboidei and Trapezius. The arm is abducted by the Deltoideus and Supraspinatus, acting on the shoulder joint and by the Trapezius and Serratus anterior rotating the scapula to raise the shoulder. The arm is adducted by the Pectoralis major and Latissimus dorsi, by the Coracobrachialis and Teres major acting on the shoulder joint, by the Pectoralis minor rotating the scapula downward, and by the lower portion of the Trapezius drawing the scapula downward. Medial rotation is brought about primarily by the Subscapularis and Teres major when it is performed as a voluntary act, but the Pectoralis major and Latissimus dorsi have strong medial rotating power incidental to their contraction. Lateral rotation is brought about by the Infraspinatus and Teres minor. Independent movements of the scapula are elevation by the Levator scapulae and upper part of the Trapezius; depression by the Pectoralis minor, lower Trapezius and lower Serratus anterior; drawing it forward

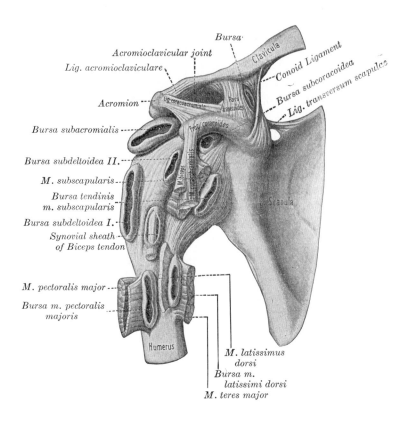

FIG. 6–38.—Muscular insertions and bursae around the shoulder joint. (Rauber-Kopsch, *Lehrbuch u. Atlas d. Anatomie d. Menschen,* 19th Edition, Vol. I, courtesy of Georg Thieme Verlag, Stuttgart.)

(abduction of scapula) by the Serratus anterior, as in pushing; drawing it backward (adduction of scapula) by the Rhomboidei and Trapezius.

IV. THE MUSCLES AND FASCIAE OF THE ARM

Coracobrachialis
Biceps brachii
Brachialis
Triceps brachii

Brachial fascia (*fascia brachii; deep fascia of the arm*) (Fig. 6–39).—The portion of the investing fascia which covers the arm is a strong membrane but is not, for the most part, distinctly aponeurotic. It is continuous proximally with the deltoid, pectoral, and axillary fasciae; it it attached distally to the epicondyles of the humerus and the olecranon, and is then continuous with the antebrachial fascia. Beginning at

the attachment to the epicondyles and prolonged proximally into the arm are two intermuscular septa, medial and lateral, which divide the arm into flexor and extensor compartments. The **lateral intermuscular septum** (*septum intermusculare [humeri] laterale*) is attached along the lateral supracondylar ridge and is fused with the internal surface of the investing brachial fascia. Its distal extremity is the lateral epicondyle, its proximal, the insertion of the Deltoideus where it continues into the deltoid fascia. Its dorsal surface is used by the Triceps for the origin of some of its fibers; the ventral surface by the Brachialis, Brachioradialis, and Extensor carpi radialis longus. Its distal portion is pierced by the radial nerve and the radial collateral branch of the profunda artery. The **medial intermuscular septum** (*septum intermusculare [humeri] mediale*) is attached to the medial supracondylar ridge and extends from the medial epicondyle distally, to the Teres major and Latissimus dorsi insertions, proximally. Some of the fibers of the Triceps originate on its dorsal surface and some of the Brachialis on its ventral surface. It is pierced, near the epicondyle, by the ulnar nerve and superior ulnar collateral artery. The medial septum appears very much thicker than the lateral because the axillary sheath, containing the main vessels and nerves of the arm, blends with its ventral surface, and the nerves and vessels continue this close association down to the elbow. The two intermuscular septa and the investing fascia of the posterior aspect of the arm form the posterior or extensor compartment which contains the Triceps, radial nerve, and profunda artery. The anterior or flexor compartment contains the Biceps, Brachialis, part of the Coracobrachialis, the brachial vessels, and the median and ulnar nerves. The relationship of the investing fascia to the muscles is different on the dorsal and ventral aspects of the arm. That over the Triceps is adherent to the muscle and is used in part for its origin. That over the Biceps is separated from the muscle by a distinct fascial cleft which is continued around the deep surface of the muscle, also separating it from the

Brachialis. The ventral investing fascia, medially, just distal to the middle of the arm, is pierced by the basilic vein.

The **Coracobrachialis** (Fig. 6–36), the smallest of the three muscles in this region, is situated at the upper and medial part of the arm. It *arises* from the apex of the coracoid process, in common with the short head of the Biceps brachii, and from the intermuscular septum between the two muscles; it is *inserted* by means of a flat tendon into an impression at the middle of the medial surface and border of the body of the humerus between the origins of the Triceps brachii and Brachialis. It is perforated by the musculocutaneous nerve.

Action.—Flexes and adducts the arm.

Nerve.—A branch of the musculocutaneous nerve, containing fibers from the sixth and seventh cervical nerves.

Variations.—A bony head may reach the medial epicondyle; a short head more rarely found may insert into the lesser tubercle.

The **Biceps brachii** (*Biceps; Biceps flexor cubiti*) (Fig. 6–36) is a long fusiform muscle, placed on the anterior aspect of the arm, and *arising* by two heads, from which circumstance it has received its name. The **short head** *arises* by a thick flattened

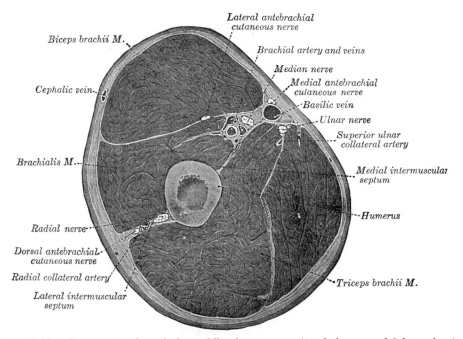

FIG. 6–39.—Cross section through the middle of upper arm. (Eycleshymer and Schoemaker.)

tendon from the apex of the coracoid process, in common with the Coracobrachialis. The **long head** *arises* from the supraglenoid tuberosity at the superior margin of the glenoid cavity, and is continuous with the glenoidal labrum. This tendon, enclosed in a special sheath of the synovial membrane of the shoulder joint, arches over the head of the humerus; it emerges from the capsule through an opening close to the humeral attachment of the ligament, and descends in the intertubercular groove; it is retained in the groove by the transverse humeral ligament and by a fibrous prolongation from the tendon of the Pectoralis major. The tendon of each head is succeeded by an elongated muscular belly, and the two bellies, although closely applied to each other, can readily be separated until within about 7.5 cm of the elbow joint. Here they end in a flattened tendon which is *inserted* into the rough posterior portion of the tuberosity of the radius, a bursa being interposed between the tendon and the anterior part of the tuberosity. As the tendon of the muscle approaches the radius it is twisted upon itself, so that its anterior surface becomes lateral and is applied to the tuberosity of the radius at its insertion. Opposite the bend of the elbow the tendon gives off, from its medial side, a broad aponeurosis, the **aponeurosis m. bicipitis brachii** (*lacertus fibrosus; bicipital fascia*) which passes obliquely distalward and medialward across the brachial artery, and is continuous with the deep fascia covering the origins of the Flexor muscles of the forearm (Fig. 6–35).

Action.—Flexes the arm, flexes the forearm, and supinates the hand. The long head draws the humerus toward the glenoid fossa, strengthening the shoulder joint.

Nerves.—Branches of the musculocutaneous nerve, containing fibers from the fifth and sixth cervical nerves.

Variations.—A third head (10 per cent) to the Biceps brachii is occasionally found, arising at the proximal and medial part of the Brachialis, with the fibers of which it is continuous, and inserted into the bicipital fascia and medial side of the tendon of the muscle. In most cases this additional slip lies deep to the brachial artery in its course down the arm. In some instances the third head consists of two

slips, which pass down, one superficial and the other deep to the artery, concealing the vessel in the lower half of the arm. More rarely a fourth head occurs arising from the outer side of the humerus, from the intertubercular groove, or from the greater tubercle. Other heads are occasionally found. Slips sometimes pass from the inner border of the muscle over the brachial artery to the medial intermuscular septum, or the medial epicondyle; more rarely to the Pronator teres or Brachialis. The long head may be absent or arise from the intertubercular groove.

The **Brachialis** (*Brachialis anticus*) (Fig. 6–36) covers the anterior aspect of the elbow joint and the distal half of the humerus. It *arises* from the distal half of the anterior aspect of the humerus, commencing at the insertion of the Deltoideus, which it embraces by two angular processes. Its origin extends distalward to within 2.5 cm of the margin of the articular surface of the elbow joint. It also arises from the intermuscular septa, but more extensively from the medial than the lateral; it is separated from the lateral distally by the Brachioradialis and Extensor carpi radialis longus. Its fibers converge to a thick tendon, which is *inserted* into the tuberosity of the ulna and the rough depression on the anterior surface of the coronoid process.

Action.—Flexes the forearm.

Nerve.—A branch of the musculocutaneous nerve, containing fibers from the fifth and sixth cervical nerves; usually an additional small branch of the radial and occasionally of the median nerve.

Variations.—It is occasionally doubled; additional slips to the Supinator, Pronator teres, Biceps, bicipital fascia, or radius are more rarely found.

The **Triceps brachii** (*Triceps; Triceps extensor cubiti*) (Fig. 6–37) is situated on the dorsal aspect of the arm, extending the entire length of the dorsal surface of the humerus. It is of large size, and *arises* by three heads (long, lateral, and medial), hence its name.

The **long head** *arises* by a flattened tendon from the infraglenoid tuberosity of the scapula, being blended at its proximal part with the capsule of the shoulder joint; the muscular fibers pass distalward between the two other heads of the muscle, and join with them in the tendon of insertion.

The **lateral head** *arises* from the posterior surface of the body of the humerus, between the insertion of the Teres minor and the proximal part of the groove for the radial nerve, and from the lateral border of the humerus and the lateral intermuscular septum; the fibers from this origin converge toward the tendon of insertion.

The **medial head,** which really should be called the **deep head,** *arises* from the posterior surface of the body of the humerus, distal to the groove for the radial nerve; it is narrow and pointed proximally, and extends from the insertion of the Teres major to within 2.5 cm of the trochlea: it also arises from the medial border of the humerus and from the whole length of the medial intermuscular septum. Some of the fibers are directed downward to the olecranon, while others converge to the tendon of insertion (Figs. 6–35, 3–35).

The **tendon of the Triceps brachii** begins about the middle of the muscle: it consists of two aponeurotic laminae, one of which is subcutaneous and covers the superficial surface of the distal half of the muscle; the other is more deeply seated in the substance of the muscle. After receiving the attachment of the muscular fibers, the two lamellae join together proximal to the elbow, and are *inserted,* for the most part, into the posterior portion of the proximal surface of the olecranon; a band of fibers is, however, continued downward, on the lateral side, over the Anconeus, to blend with the deep fascia of the forearm.

The long head of the Triceps brachii descends between the Teres minor and Teres major, dividing the triangular space between these two muscles and the humerus into two smaller spaces, one triangular, the other quadrangular (Fig. 6–37). The triangular space contains the scapular circumflex vessels; it is bounded by the Teres minor superiorly, the Teres major inferiorly, and the scapular head of the Triceps laterally. The quadrangular space transmits the posterior humeral circumflex vessels and the axillary nerve; it is bounded by the Teres minor and capsule of the shoulder joint superiorly, the Teres major inferiorly, the long head of the Triceps brachii medially, and the humerus laterally.

Action.—Extends the forearm. The long head extends and adducts the arm.

Nerves.—Branches of the radial nerve, containing fibers from the seventh and eighth cervical nerves.

Variations.—A fourth head from the inner part of the humerus; a slip between Triceps and Latissimus dorsi corresponding to the *Dorso-epitrochlearis.*

The **Subanconeus** (*m. articularis cubiti*) is the name given to a few fibers which spring from the deep surface of the lower part of the Triceps brachii, and are inserted into posterior ligament and synovial membrane of the elbow joint.

Group Actions.—Flexion at the elbow is brought about by the Brachialis, a pure flexor, by the Biceps which also has strong supinating action, and by the Brachioradialis, also a pure flexor. In spite of its old name, Supinator longus, the Brachioradialis does not assist in voluntary supination or pronation; it may have incidental action in restoring the forearm to the middle position from the extreme of either one. The Pronator teres contracts, but its action may be synergetic to counteract the supination of the Biceps. Other forearm muscles may be used in very strong flexion or in cases of paralysis, provided the proper position of pronation or supination is first obtained. Extension of the elbow is performed by the Triceps and Anconeus.

V. THE MUSCLES AND FASCIAE OF THE FOREARM

Antebrachial fascia (*fascia antebrachii; deep fascia of the forearm*) (Fig. 6–43.— The portion of the investing fascia which covers the proximal forearm is a strong aponeurotic sheet, closely adherent to the underlying muscles. Proximally, it is continuous with the brachial fascia and is attached to the epicondyles of the humerus and the olecranon. Distally, it is attached to the distal portions of the radius and ulna and is continued into the fascia of the hand. It is attached to the dorsal border of the ulna through most of its length, closing off the flexor and extensor compartments of the forearm. In the proximal two-thirds of the forearm, the underlying muscles utilize the deep surface of the investing fascia for attachment of their fibers, and the area for attachment is increased further by the strong intermuscular septa which extend deeply toward the bones, between the adjacent muscles. The volar fascia is visibly thickened by collagenous bundles, derived from the tendon of the biceps, which fan out medially to form the aponeurosis of the Biceps brachii muscle. The dorsal portion is thickened, even more than the volar, by bundles from the Triceps tendon. Near the distal ends of the radius and ulna, the fascia is abruptly thickened by the addition of prominent, annular, collagenous bundles which form the flexor and extensor retinacula (volar and dorsal carpal ligaments). In the distal third of the forearm, the muscles and tendons are separated from the overlying fascia and from each other by fascial clefts. The antebrachial fascia is pierced in several places by vessels and nerves, the largest aperture being for the branch of the median cubital vein which communicates with the deep veins in the antecubital fossa. The radius, ulna, and interosseous membrane form a septum dividing the forearm into dorsal or extensor and volar or flexor compartments.

In the **dorsal compartment,** a fascial cleft separates the superficial from the deep muscles, especially near the wrist, but it is closed medially and laterally, not communicating with other clefts. In the **palmar compartment,** the fascial cleft is more extensive and may communicate with the fascial clefts of the palm. It is especially evident between the surface of the Pronator quadratus and the overlying muscles and tendons, and is here called the **palmar interosseous cleft.** It continues proximally between the Flexores profundus and superficialis and may follow along the ulnar vessels and nerve to the antecubital fossa. At the wrist it is in contact with the proximal end of the flexor tendon sheaths (radial and ulnar bursae) and may continue distally under these sheaths into the middle palmar cleft.

The antebrachial or forearm muscles may be divided into a **palmar** and a **dorsal group.**

1. The Palmar Antebrachial Muscles

These muscles are divided for convenience of description into two groups, superficial and deep.

The Superficial Group (Fig. 6–40)

> Pronator teres
> Flexor carpi radialis
> Palmaris longus
> Flexor carpi ulnaris
> Flexor digitorum superficialis

The muscles of this group take origin from the medial epicondyle of the humerus by a common tendon; they receive additional fibers from the deep fascia of the

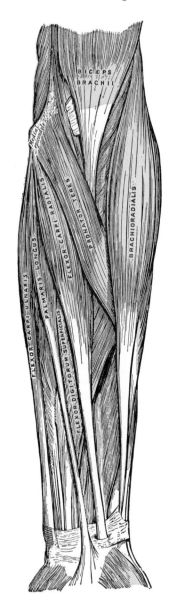

Fig. 6–40.—Palmar aspect of left forearm. Superficial muscles.

forearm near the elbow, and from the septa which pass from this fascia between the individual muscles.

The **Pronator teres** has two heads of origin —humeral and ulnar. The **humeral head,** the larger and more superficial, *arises* immediately proximal to the medial epicondyle, and from the tendon common to the origin of the other muscles; also from the intermuscular septum between it and the Flexor carpi radialis and from the antebrachial fascia. The **ulnar head** is a thin fasciculus, which *arises* from the medial side of the coronoid process of the ulna, and joins the preceding at an acute angle. The median nerve enters the forearm between the two heads of the muscle, and is separated from the ulnar artery by the ulnar head. The muscle passes obliquely across the forearm, and ends in a flat tendon, which is *inserted* into a rough impression at the middle of the lateral surface of the body of the radius. The lateral border of the muscle forms the medial boundary of a triangular hollow, the antecubital fossa, situated in the bend of the elbow joint and containing the brachial artery, median nerve, and tendon of the Biceps brachii.

Action.—Pronates the hand.

Nerve.—A branch of the median nerve, containing fibers from the sixth and seventh cervical nerves.

Variations.—Absence of ulnar head; additional slips from the medial intermuscular septum, from the Biceps and from the Brachialis anterior occasionally occur.

The **Flexor carpi radialis** lies on the medial or ulnar side of the preceding muscle. It *arises* from the medial epicondyle by the common tendon; from the fascia of the forearm; and from the intermuscular septa between it and the Pronator teres laterally, the Palmaris longus medially, and the Flexor digitorum superficialis beneath. Slender and aponeurotic in structure at its commencement, it increases in size, and ends in a tendon which forms rather more than the distal half of its length. This tendon passes through a canal in the lateral part of the flexor retinaculum and runs through a groove on the trapezium bone; the groove is converted into a canal by fibrous tissue, and lined by a synovial sheath. The tendon

is *inserted* into the base of the second metacarpal bone, and sends a slip to the base of the third metacarpal bone. The radial artery, in the distal part of the forearm, lies between the tendon of this muscle and the Brachioradialis.

Action.—Flexes the hand and helps to abduct (radial flex) it.

Nerve.—A branch of the median nerve, containing fibers from the sixth and seventh cervical nerves.

Variations.—Slips from the tendon of the Biceps, the bicipital fascia, the coronoid, and the radius have been found. Its insertion often varies and may be mostly into the annular ligament, the trapezium, or the fourth metacarpal as well as the second or third. The muscle may be absent.

The **Palmaris longus** is a slender, fusiform muscle, lying on the medial side of the preceding. It *arises* from the medial epicondyle of the humerus by the common tendon, from the intermuscular septa between it and the adjacent muscles, and from the antebrachial fascia. It ends in a slender, flattened tendon, which passes over the upper part of the flexor retinaculum, and is *inserted* into the central part of the flexor retinaculum and into the palmar aponeurosis, frequently sending a tendinous slip to the short muscles of the thumb.

Action.—Flexes the hand.

Nerve.—A branch of the median nerve, containing fibers from the sixth and seventh cervical nerves.

Variations.—One of the most variable muscles in the body. This muscle is often absent (about 10 per cent), and is subject to many variations; it may be tendinous proximally and muscular distally; or it may be muscular in the center with a tendon above and below; or it may present two muscular bundles with a central tendon; or finally it may consist solely of a tendinous band. The muscle may be double. Slips of origin from the coronoid process or from the radius have been seen. Partial or complete insertion into the fascia of the forearm, into the tendon of the Flexor carpi ulnaris and pisiform bone, into the scaphoid, and into the muscles of the little finger have been observed.

The **Flexor carpi ulnaris** lies along the ulnar side of the forearm. It *arises* by two heads, humeral and ulnar, connected by a tendinous arch, beneath which the ulnar nerve and posterior ulnar recurrent artery pass. The **humeral head** *arises* from the medial epicondyle of the humerus by the common tendon; the **ulnar head** *arises* from the medial margin of the olecranon and from the proximal two-thirds of the dorsal border of the ulna by an aponeurosis, common to it and the Extensor carpi ulnaris and Flexor digitorum profundus; and from the intermuscular septum between it and the Flexor digitorum superficialis. The fibers end in a tendon which occupies the anterior part of the lower half of the muscle and is *inserted* into the pisiform bone, and is prolonged from this to the hamate and fifth metacarpal bones by the pisohamate and pisometacarpal ligaments; it is also attached by a few fibers to the flexor retinaculum. The ulnar vessels and nerve lie on the lateral side of the tendon of this muscle, in the distal two-thirds of the forearm.

Action.—Flexes and adducts (ulnar flexes) the hand.

Nerve.—A branch of the ulnar nerve, containing fibers from the eighth cervical and first thoracic nerves.

Variations.—Slips of origin from the coronoid. The *Epitrochleo-anconeus,* a small muscle often present runs from the back of the inner condyle to the olecranon, over the ulnar nerve.

The **Flexor digitorum superficialis** (*flexor digitorum sublimis*) (Fig. 6–40) is placed beneath the previous muscle; it is the largest of the muscles of the superficial group, and arises by three heads—humeral, ulnar, and radial. The **humeral head** *arises* from the medial epicondyle of the humerus by the common tendon, from the ulnar collateral ligament of the elbow joint, and from the intermuscular septa between it and the preceding muscles. The **ulnar head** *arises* from the medial side of the coronoid process, proximal to the ulnar origin of the Pronator teres (see Fig. 4–140, page 210). The **radial head** *arises* from the oblique line of the radius, extending from the radial tuberosity to the insertion of the Pronator teres. The muscle speedily separates into two planes of muscular fibers, superficial and deep: the superficial plane divides into two parts which end in tendons for the middle and ring fingers; the deep plane gives off a

muscular slip to join the portion of the superficial plane which is associated with the tendon of the ring finger, and then divides into two parts, which end in tendons for the index and little fingers. As the four tendons thus formed pass beneath the flexor retinaculum into the palm of the hand, they are arranged in pairs, the superficial pair going to the middle and ring fingers, the deep pair to the index and little fingers. The tendons diverge from one another in the palm and form dorsal relations to the superficial palmar arch and digital branches of the median and ulnar nerves. Opposite the bases of the first phalanges each tendon divides into two slips to allow the passage of the corresponding tendon of the Flexor digitorum profundus; the two slips then reunite and form a grooved channel for the reception of the accompanying tendon of the Flexor digitorum profundus. Finally the tendon divides and is inserted into the sides of the second phalanx about its middle.

Action.—Flexes the second phalanx of each finger; by continued action, flexes the first phalanx and hand.

Nerves.—Branches of the median nerve, containing fibers from the seventh and eighth cervical and first thoracic nerves.

Variations.—The radial head, or little finger portion may be absent. Accessory slips may occur from ulnar tuberosity to the index and middle finger portions; from the inner head to the Flexor profundus; from the ulnar or annular ligament to the little finger.

The Deep Group (Fig. 6–41)

Flexor digitorum profundus
Flexor pollicis longus
Pronator quadratus

The **Flexor digitorum profundus** is situated on the ulnar side of the forearm, immediately beneath the superficial Flexors. It *arises* from the proximal three-fourths of the volar and medial surfaces of the body of the ulna, embracing the insertion of the Brachialis proximally, and extending distally to within a short distance of the Pronator quadratus. It also arises from a depression on the medial side of the coronoid process; from the upper three-fourths of the dorsal border of the ulna, by an aponeurosis in common with the Flexor and Extensor carpi ulnaris; and from the ulnar half of the interosseous membrane. The muscle ends in four tendons which run under the transverse carpal ligament dorsal to the tendons of the Flexor digitorum superficialis. Opposite the first phalanges the tendons pass through the openings in the tendons of the Flexor digitorum superficialis, and are finally *inserted* into the bases of the last phalanges. The portion of the muscle for the index finger is usually distinct throughout, but the tendons for the middle, ring, and little fingers are connected together by areolar tissue and tendinous slips, as far as the palm of the hand.

Distal to the metacarpophalangeal joints, the tendons of the Flexores digitorum superficialis and profundus lie in strong ligamentous tunnels, the digital **fibrous tendon sheaths** (page 481 and Fig. 6–46). Each tunnel is lined by the **synovial tendon sheath**, a lubricated layer which is reflected on the contained tendons. Within each digital sheath, the tendons of the superficialis and profoundus are connected to each other and to the phalanges by tendinous bands called **vincula tendinum** (Fig. 6–42). There are two types of vincula: (*a*) the **vincula brevia**, which are two in number in each finger, are fan-shaped expansions near the termination of the tendons, one connecting the superficialis tendon to the palmar surface of the proximal interphalangeal joint and the head of the first phalanx, and the other connecting the profundus tendon to the palmar surface of the second interphalangeal joint and the head of the second phalanx; (*b*) the **vincula longa** are slender, independent bands which are found in two positions: one pair of them in each finger connects the deep surface of the profundus tendon to the subjacent superficialis tendon after the former has passed through the split in the latter; another pair, or a single band, connects the superficialis tendon to the proximal end of the first phalanx.

Action.—Flexes the terminal phalanx of each finger; by continued action flexes the other phalanges and to some extent the hand.

Nerves.—A branch of the palmar interosseous nerve from the median and a branch of the ulnar, containing fibers from the eighth cervical and first thoracic nerves.

Variations.—The index finger portion may

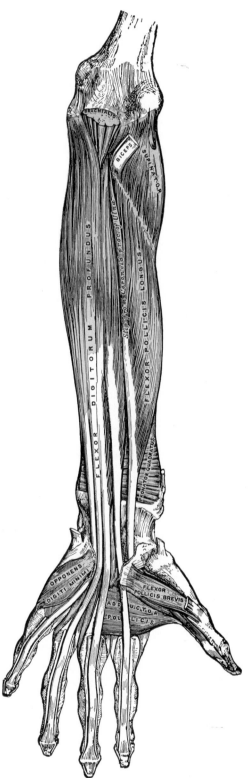

Fig. 6–41.—Palmar aspect of the left forearm. Deep muscles.

arise partly from the upper part of the radius. Slips from the inner head of the Flexor superficialis, medial epicondyle, or the coronoid are found. Connection with the Flexor pollicis longus occurs.

Four small muscles, the Lumbricales, are connected with the tendons of the Flexor profundus in the palm. They will be described with the muscles of the hand.

The **Flexor pollicis longus** is situated on the radial side of the forearm, lying in the same plane as the preceding. It *arises* from the grooved volar surface of the body of the radius, extending from immediately below the tuberosity and oblique line to within a short distance of the Pronator quadratus. It *arises* also from the adjacent part of the interosseous membrane, and generally by a fleshy slip from the medial border of the coronoid process, or from the medial epicondyle of the humerus. The fibers end in a flattened tendon, which passes beneath

the flexor retinaculum, is then lodged between the lateral head of the Flexor pollicis brevis and the oblique part of the Adductor pollicis, and, entering an ossceo-aponeurotic canal similar to those for the Flexor tendons of the fingers, is *inserted* into the base of the distal phalanx of the thumb. The palmar interosseous nerve and vessels pass downward on the palmar aspect of the interosseous membrane between the Flexor pollicis longus and Flexor digitorum profundus.

Action.—Flexes the second phalanx of the thumb; by continued action, flexes the first phalanx, and flexes and adducts the metacarpal.

Nerve.—A branch of the palmar interosseous nerve from the median, containing fibers from the eighth cervical and first thoracic nerves.

Variations.—Slips may connect with Flexor superficialis, or Profundus, or Pronator teres. An additional tendon to the index finger is sometimes found.

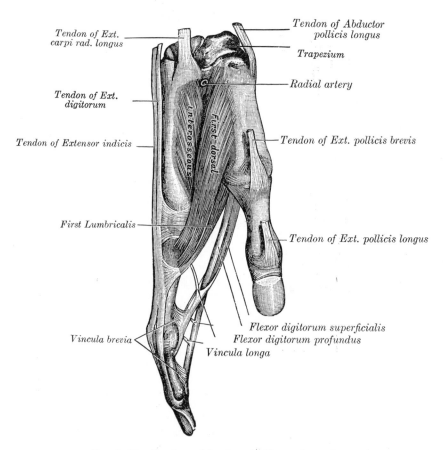

FIG. 6–42.—Tendons of forefinger and vincula tendinum.

The **Pronator quadratus** is a small, flat, quadrilateral muscle, extending across the palmar aspect of the distal parts of the radius and ulna. It *arises* from the pronator ridge on the distal part of the palmar surface of the body of the ulna; from the medial part of the palmar surface of the distal fourth of the ulna; and from a strong aponeurosis which covers the medial third of the muscle. The fibers pass lateralward and slightly distalward, to be inserted into the distal fourth of the lateral border and the palmar surface of the body of the radius. The deeper fibers of the muscle are inserted into the triangular area proximal to the ulnar notch of the radius—an attachment comparable with the origin of the Supinator from the triangular area dorsal to the radial notch of the ulna.

Action.—Pronates the hand.

Nerve.—A branch of the palmar interosseous nerve from the median, containing fibers from the eighth cervical and first thoracic nerves.

Variations.—Rarely absent; split into two or three layers; increased attachment proximally or distally.

2. The Dorsal Antebrachial Muscles

These muscles are divided for convenience of description into two groups, superficial and deep.

The Superficial Group (Fig. 6–44)

Brachioradialis
Extensor carpi radialis longus
Extensor carpi radialis brevis
Extensor digitorum
Extensor digiti minimi
Extensor carpi ulnaris
Anconeus

The **Brachioradialis** (*Supinator longus*) is the most superficial muscle on the radial side of the forearm. It *arises* from the proximal two-thirds of the lateral supracondylar ridge of the humerus, and from the lateral intermuscular septum, being limited proximally by the groove for the radial nerve. Interposed between it and the Brachialis are the radial nerve and the anastomosis between the radial collateral branch of the profunda artery and the radial recurrent. The fibers end proximal to the middle of

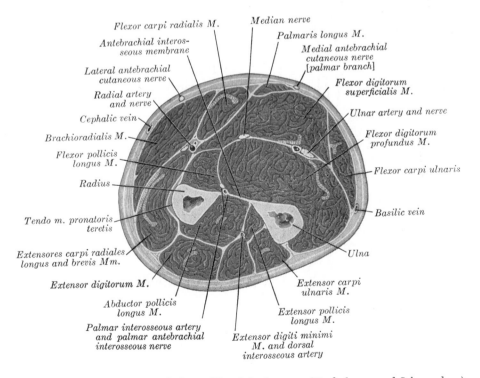

Fig. 6–43.—Cross section through the middle of the forearm. (Eycleshymer and Schoemaker.)

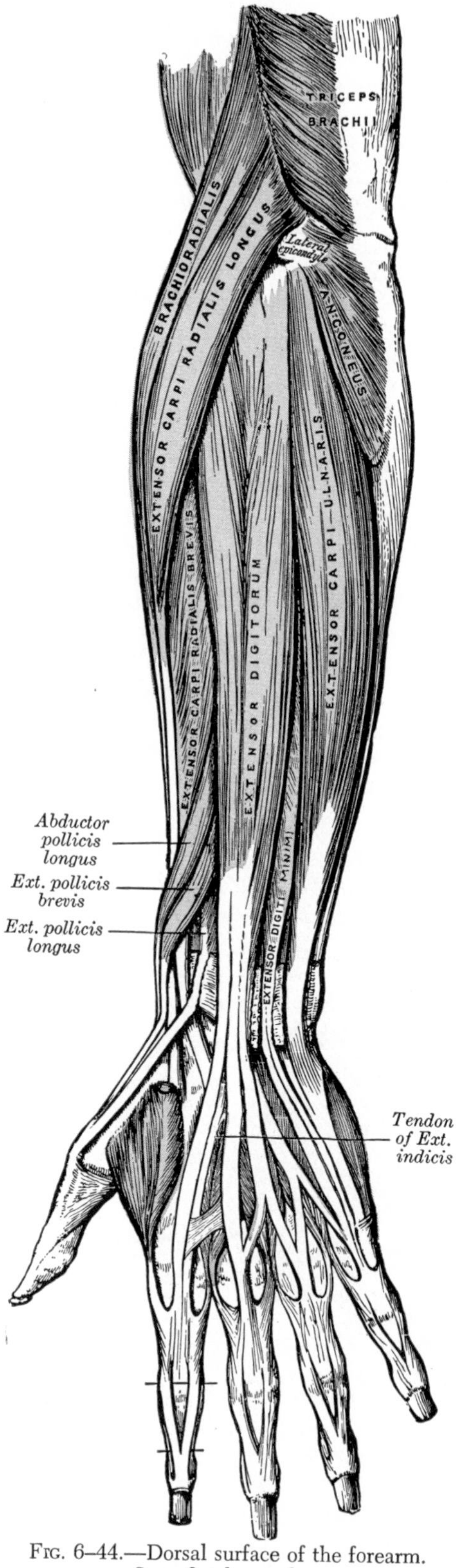

Fig. 6-44.—Dorsal surface of the forearm.
Superficial muscles.

the forearm in a flat tendon, which is *inserted* into the lateral side of the base of the styloid process of the radius. The tendon is crossed near its insertion by the tendons of the Abductor pollicis longus and Extensor pollicis brevis; on its ulnar side is the radial artery.

Action.—Flexes the forearm.

Nerve.—A branch of the radial nerve, containing fibers from the fifth and sixth cervical nerves.

Variations.—Fusion with the Brachialis; tendon of insertion may be divided into two or three slips; insertion partial or complete into the middle of the radius, fasciculi to the tendon of the Biceps, the tuberosity or oblique line of the radius; slips to the Extensor carpi radialis longus or Abductor pollicis longus; absence; rarely doubled.

The **Extensor carpi radialis longus** (*Extensor carpi radialis longior*) is placed partly deep to the Brachioradialis. It *arises* from the distal third of the lateral supracondylar ridge of the humerus, from the lateral intermuscular septum, and by a few fibers from the common tendon of origin of the Extensor muscles of the forearm. The fibers end at the upper third of the forearm in a flat tendon, which runs along the lateral border of the radius, beneath the Abductor pollicis longus and Extensor pollicis brevis; it then passes beneath the extensor retinaculum, where it lies in a groove on the back of the radius common to it and the Extensor carpi radialis brevis, immediately behind the styloid process. It is *inserted* into the dorsal surface of the base of the second metacarpal bone, on its radial side (Fig. 6-42).

Action.—Extends and abducts (radial flexes) the hand.

Nerve.—A branch of the radial nerve, containing fibers from the sixth and seventh cervical nerves.

The **Extensor carpi radialis brevis** (*Extensor carpi radialis brevior*) is shorter and thicker than the preceding muscle, beneath which it is placed. It *arises* from the lateral epicondyle of the humerus, by a tendon common to it and the three following muscles; from the radial collateral ligament of the elbow joint; from a strong aponeurosis which covers its surface; and from the intermuscular septa between it and the adjacent muscles. The fibers end about the middle of the forearm in a flat tendon which is closely connected with that of the preceding muscle, and accompanies it to the wrist; it passes beneath the Abductor pollicis longus and Extensor pollicis brevis, then beneath the extensor retinaculum, and is *inserted* into the dorsal surface of the base of the third metacarpal bone on its radial side. Under the extensor retinaculum the tendon lies on the dorsum of the radius in a shallow groove, to the ulnar side of that which lodges the tendon of the Extensor carpi radialis longus, and separated from it by a faint ridge.

The tendons of the two preceding muscles pass through the same compartment of the dorsal carpal ligament in a single synovial sheath.

Action.—Extends and may abduct the hand.

Nerve.—A branch of the radial nerve, containing fibers from the sixth and seventh cervical nerves.

Variations.—Either muscle may split into two or three tendons of insertion to the second and third or even the fourth metacarpal. The two muscles may unite into a single belly with two tendons. Cross slips between the two muscles may occur. The *Extensor carpi radialis intermedius* rarely arises as a distinct muscle from the humerus, but is not uncommon as an accessory slip from one or both muscles to the second or third or both metacarpals. The *Extensor carpi radialis accessorius* is occasionally found arising from the humerus with or below the Extensor carpi radialis longus and inserted into the first metacarpal, the Abductor pollicis brevis, the First dorsal interosseus, or elsewhere.

The **Extensor digitorum** (*Extensor digitorum communis*) (Fig. 6–44) *arises* from the lateral epicondyle of the humerus, by the common tendon; from the intermuscular septa between it and the adjacent muscles, and from the antebrachial fascia. It divides distally into four tendons, which pass, together with that of the Extensor indicis, through a separate compartment of the extensor retinaculum, within a synovial sheath. The tendons then diverge on the back of the hand, and are *inserted* into the second and third phalanges of the fingers

in the following manner. Opposite the metacarpophalangeal articulation each tendon is bound by fasciculi to the collateral ligaments and serves as the dorsal ligament of this joint; after having crossed the joint, it spreads out into a broad aponeurosis which covers the dorsal surface of the first phalanx and is reinforced, in this situation, by the tendons of the Interossei and Lumbricales. Opposite the first interphalangeal joint this aponeurosis divides into three slips; an intermediate and two collateral: the former is inserted into the base of the second phalanx; and the two collateral, which are continued onward along the sides of the second phalanx, unite by their contiguous margins and are *inserted* into the dorsal surface of the distal phalanx. As the tendons cross the interphalangeal joints, they furnish them with dorsal ligaments. The tendon to the index finger is accompanied by the Extensor indicis, which lies on its ulnar side. On the back of the hand, the tendons to the middle, ring, and little fingers are connected by obliquely placed bands (*connexus intertendinei*). Occasionally the first tendon is connected to the second by a thin transverse band.

Action.—Extends the phalanges, and by continued action, extends the wrist. For more details, especially of extension of proximal phalanges, see Group actions on page 472.

Nerve.—A branch of the deep radial nerve, containing fibers from the sixth, seventh and eighth cervical nerves.

Variations.—An increase or decrease in the number of tendons is common; an additional slip to the thumb is sometimes present.

The **Extensor digiti minimi** (*Extensor digiti quinti proprius*) is a slender muscle placed on the medial side of the Extensor digitorum, with which it is generally connected. It *arises* from the common Extensor tendon by a thin tendinous slip, from the intermuscular septa between it and the adjacent muscles. Its tendon runs through a compartment of the extensor retinaculum dorsal to the distal radioulnar joint, then divides into two as it crosses the hand, and finally joins the expansion of the Extensor digitorum tendon on the dorsum of the first phalanx of the little finger.

Action.—Extends the little finger.

Nerve.—A branch of the deep radial nerve, containing fibers from the sixth, seventh, and eighth cervical nerves.

Variations.—An additional fibrous slip from the lateral epicondyle; the tendon of insertion may not divide or may send a slip to the ring finger. Absence of muscle rare; fusion of the belly with the Extensor digitorum is not uncommon.

The **Extensor carpi ulnaris** lies on the ulnar side of the forearm. It *arises* from the lateral epicondyle of the humerus, by the common tendon, by an aponeurosis from the dorsal border of the ulna in common with the Flexor carpi ulnaris and the Flexor digitorum profundus, and from the deep fascia of the forearm. It ends in a tendon which runs in a groove between the head and the styloid process of the ulna, passing through a separate compartment of the extensor retinaculum, and is *inserted* into the prominent tubercle on the ulnar side of the base of the fifth metacarpal bone.

Action.—Extends and adducts the hand.

Nerve.—A branch of the deep radial nerve, containing fibers from the sixth, seventh, and eighth cervical nerves.

Variations.—Doubling; reduction to tendinous band; insertion partially into fourth metacarpal. In many cases (52 per cent) a slip is continued from the insertion of the tendon anteriorly over the Opponens digiti minimi, to the fascia covering that muscle, the metacarpal bone, the capsule of the metacarpophalangeal articulation, or the first phalanx of the little finger. This slip may be replaced by a muscular fasciculus arising from or near the pisiform.

The **Anconeus** is a small triangular muscle which is placed on the dorsum of the elbow joint, and appears to be a continuation of the Triceps brachii. It *arises* by a separate tendon from the dorsal part of the lateral epicondyle of the humerus; its fibers diverge and are *inserted* into the side of the olecranon, and proximal fourth of the dorsal surface of the body of the ulna.

Action.—Extends the forearm.

Nerve.—A branch of radial nerve, containing fibers from the seventh and eighth cervical nerves.

The Deep Group (Fig. 6–46)

Supinator
Abductor pollicis longus
Extensor pollicis brevis
Extensor pollicis longus
Extensor indicis

The **Supinator** (*Supinator brevis*) (Fig. 6–45) is a broad muscle, curved around the proximal third of the radius. It consists of two planes of fibers, between which the deep branch of the radial nerve lies. The two planes *arise* in common—the superficial one by tendinous and the deeper by muscular fibers—from the lateral epicondyle of the humerus; from the radial collateral ligament of the elbow joint, and the annular ligament; from the ridge on the ulna, which runs obliquely distalward from the dorsal end of the radial notch; from the triangular depression distal to the notch; and from a tendinous expansion which covers the sur-

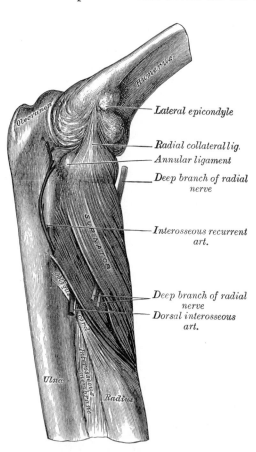

Lateral epicondyle

Radial collateral lig.
Annular ligament

Deep branch of radial nerve

Interosseous recurrent art.

Deep branch of radial nerve
Dorsal interosseous art.

Fig. 6–45.—The Supinator.

face of the muscle. The superficial fibers surround the proximal part of the radius, and are inserted into the lateral edge of the radial tuberosity and the oblique line of the radius, as far distally as the insertion of the Pronator teres. The proximal fibers of the deeper plane form a sling-like fasciculus, which encircles the neck of the radius proximal to the tuberosity and is attached to the back part of its medial surface; the greater part of this portion of the muscle is inserted into the dorsal and lateral surfaces of the body of the radius, midway between the oblique line and the head of the bone.

Action.—Supinates the hand.

Nerve.—Fibers from the deep branch of the radial nerve, containing fibers from the sixth cervical nerve.

The **Abductor pollicis longus** (*Extensor ossis metacarpi pollicis*) lies immediately distal to the Supinator and is sometimes united with it. It *arises* from the lateral part of the dorsal surface of the body of the ulna distal to the insertion of the Anconeus, from the interosseous membrane, and from the middle third of the dorsal surface of the body of the radius. Passing obliquely distalward and lateralward, it ends in a tendon which runs through a groove on the lateral side of the lower end of the radius, accompanied by the tendon of the Extensor pollicis brevis, and is *inserted* into the radial side of the base of the first metacarpal bone (Fig. 6–42). It usually gives off two slips near its insertion: one to the trapezium and the other to blend with the origin of the Abductor pollicis brevis.

Action.—Abducts the thumb and, by continued action, the wrist.

Nerve.—A branch of the deep radial nerve, containing fibers from the sixth and seventh cervical nerves.

Variations.—More or less doubling of muscle and tendon with insertion of the extra tendon into the first metacarpal, the trapezium, or into the Abductor pollicis brevis or Opponens pollicis.

The **Extensor pollicis brevis** (*Extensor primi internodii pollicis*) (Fig. 6–46) lies on the medial side of, and is closely connected with, the Abductor pollicis longus.

It *arises* from the dorsal surface of the body of the radius distal to that muscle, and from the interosseous membrane. Its direction is similar to that of the Abductor pollicis longus, its tendon passing through the same

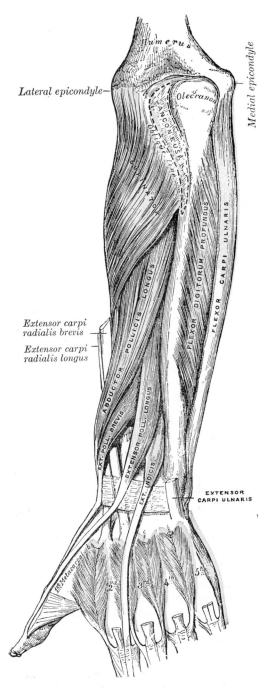

FIG. 6–46.—Dorsal surface of the forearm. Deep muscles.

groove on the lateral side of the distal end of the radius, to be *inserted* into the base of the first phalanx of the thumb (Fig. 6–42).

Action.—Extends the first phalanx of the thumb and, by continued action, abducts the hand.

Nerve.—A branch of the deep radial nerve, containing fibers from the sixth and seventh cervical nerves.

Variations.—Absence; fusion of tendon with that of the Extensor pollicis longus.

The **Extensor pollicis longus** (*Extensor secundi internodii pollicis*) is much larger than the preceding muscle, the origin of which it partly covers. It *arises* from the lateral part of the middle third of the dorsal surface of the body of the ulna distal to the origin of the Abductor pollicis longus, and from the interosseous membrane. It ends in a tendon which passes through a separate compartment in the extensor retinaculum, lying in a narrow, oblique groove on the dorsum of the lower end of the radius. It then crosses obliquely the tendons of the Extensores carpi radiales longus and brevis, and is separated from the Extensor brevis pollicis by a triangular interval (the anatomical snuff-box), in which the radial artery is found, and is finally *inserted* into the base of the last phalanx of the thumb. The radial artery is crossed by the tendons of the Abductor pollicis longus and of the Extensores pollicis longus and brevis.

Action.—Extends the second phalanx of the thumb and, by continued action, abducts the hand.

Nerve.—A branch of the deep radial nerve, containing fibers from the sixth, seventh, and eighth cervical nerves.

The **Extensor indicis** (*Extensor indicis proprius*) is a narrow, elongated muscle, placed medial to, and parallel with, the preceding. It *arises* from the dorsal surface of the body of the ulna below the origin of the Extensor pollicis longus, and from the interosseous membrane. Its tendon passes under the extensor retinaculum in the same compartment as that which transmits the tendons of the Extensor digitorum, and opposite the head of the second metacarpal bone, joins the ulnar side of the tendon of the Extensor digitorum which belongs to the index finger (Fig. 6–46).

Action.—Extends and to some extent adducts the index finger.

Nerve.—A branch of the deep radial nerve, containing fibers from the sixth, seventh, and eighth cervical nerves.

Variations.—Doubling; the ulnar part may pass beneath the dorsal carpal ligament with the Extensor digitorum; a slip from the tendon may pass to the middle finger.

Group Actions.—Flexion at the wrist is brought about by the Flexor carpi ulnaris, Flexor carpi radialis, and Palmaris longus. The Flexores digitorum superficialis and profundus are not used in voluntary flexion at the wrist unless they are prevented from flexing the fingers. The extensors of the wrist are the Extensores carpi radialis longus and brevis, and the Extensor carpi ulnaris. The Extensor digitorum may act if the fingers are prevented from extending. Abduction (radial flexion) at the wrist is performed by the Flexor carpi radialis, Extensores carpi radialis longus and brevis, and Abductor pollicis longus. Adduction (ulnar flexion) at the wrist is performed by the Flexor carpi ulnaris and Extensor carpi ulnaris. Pronation of the hand is brought about by the Pronator teres and Pronator quadratus. Supination is performed by the Biceps and the Supinator. Group actions of muscles inserting on the digits are presented after the description of the muscles of the hand.

VI.   THE MUSCLES AND FASCIAE OF THE HAND

The muscles of the hand are subdivided into three groups: (1) those of the thumb, which occupy the radial side and produce the **thenar eminence;** (2) those of the little finger, which occupy the ulnar side and give rise to the **hypothenar eminence;** (3) those in the middle of the palm and between the metacarpal bones.

The **subcutaneous fascia** (*tela subcutanea*) of the palmar surface of the forearm changes its character abruptly at the distal crease of the wrist from a delicate movable tissue into the tough cushion which covers the palm and palmar surface of the digits. The latter contains a considerable amount of fat, but cannot be separated readily into superficial and deep layers. The adipose tissue is permeated by strong fibrous bands and septa which break it up into small granular lobules and bind it securely to the deep fascia. The dermis is very compact, and not only protects the underlying structures, but also offers resistance to the progress of infectious processes seeking to point toward the surface; at the same time, the vertical direction of the fibrous bands tends to guide the spread of the infection into deeper layers. The subcutaneous fascia is adherent to the deep fascia over the entire palm, but the union is especially strong at the skin creases of the wrist, the major creases of the palm, and the creases of the digits. At the medial and lateral borders of the hand and digits, the fascia changes its character rather abruptly as it becomes continuous with the corresponding layer of the dorsum.

The subcutaneous fascia of the dorsum of the hand and digits is delicate and movable, like that of the forearm with which it is continuous. Its two layers can be identified; the superficial one is thin but may contain a small amount of fat; the deep one is a definite fibrous sheet and supports the superficial veins and cutaneous nerves. It is separated from the deep fascia by a distinct fascial cleft, the dorsal subcutaneous cleft (described below), which imparts the characteristic movability to the skin of the back of the hand.

Deep fascia of the wrist.—The antebrachial fascia at the wrist is thickened into an annular band or cuff which holds the tendons of the forearm muscles close against the wrist. For convenience in description, it is divided into two parts, the flexor retinaculum and the extensor retinaculum.

The **Flexor Retinaculum** (*retinaculum flexorum*).—Two important bands, formerly distinguished as the volar or palmar carpal ligament and the transverse carpal ligament, have been lumped together under the term flexor retinaculum in the revised Paris Nomenclature. Their identity is preserved in this description because of their clinical significance. The additional band, the transverse carpal ligament, is distal to the palmar carpal ligament and lies at a deeper level. It is not a fascial derivative, coming rather from the tendons and ligaments of the carpus, but it will be described here because of its close association with the fascia.

The **Palmar Carpal Ligament** (*Volar carpal ligament*) (Fig. 6–53), not to be confused with the transverse carpal ligament described below, is the distal portion of the investing antebrachial fascia which is abruptly thickened at the wrist by the addition of strong transverse collagenous bundles. It is attached medially and laterally to the styloid processes of the ulna and radius and under it lie the tendons of the flexor muscles. Its distal border is difficult to determine because it merges with the transverse carpal ligament, except where the ulnar artery emerges from under the palmar carpal to lie superficial to the transverse carpal ligament.

The relations of the flexor tendons to each other and to the nerves and blood vessels in the wrist are of great surgical interest because they are frequently severed in industrial injuries and must be sutured back into place. These relations are quite constant unless actual muscular variations or anomalies occur, and are shown in Figure 6–50.

Transverse Carpal Ligament (*retinaculum flexorum; anterior annular ligament*) (Figs. 6–47, 6–53).—The transverse carpal ligament is a thick fibrous band which arches over the deep groove on the palmar surface of the carpal bones, forming a tunnel through which the long flexor tendons and the median nerve pass. It is attached, medially, to the pisiform and the hamulus of the hamate, and laterally, to the tuberosity of the scaphoid, and the medial part of the palmar surface and ridge of the trapezium. Its proximal border is partly merged with the distal border of the palmar carpal ligament, but the latter belongs to a definitely different and more superficial stratum, and is separated from it by the ulnar artery and nerve. It is attached to the palmar aponeurosis, which lies superficial to it, and contributes oblique crossed fibers to the deep surface of the aponeurosis. It is attached to the trapezium in two parts, one on either side of the groove in which the tendon of the Flexor carpi radialis lies. The Flexor carpi ulnaris, at its insertion, contributes tendinous fibers to the ligament and the short muscles of the thumb and little finger arise from it to a large extent.

The **Synovial Sheaths of the Flexor Tendons at the Wrist** (Fig. 6–47).—As the tendons pass under the transverse carpal ligament, they are enclosed in two specialized synovial sacs; the larger one, for all the tendons of the Flexores digitorum superficialis and profundus, is called the **ulnar bursa**; the smaller one, for the Flexor pollicis longus, is called the **radial bursa.** They extend proximally into the forearm for about 2.5 cm beyond the transverse carpal ligament. The radial bursa extends distally to the terminal phalanx of the thumb where

the Flexor pollicis longus inserts. The ulnar bursa continues distally beyond the middle of the palm as the digital sheath for the little finger, but it is greatly reduced in diameter at the middle of the metacarpal bones by the formation of terminal diverticula about the tendons of the second, third, and fourth digits. The tendons to the second, third, and fourth digits, therefore, are without synovial sheaths for a short distance

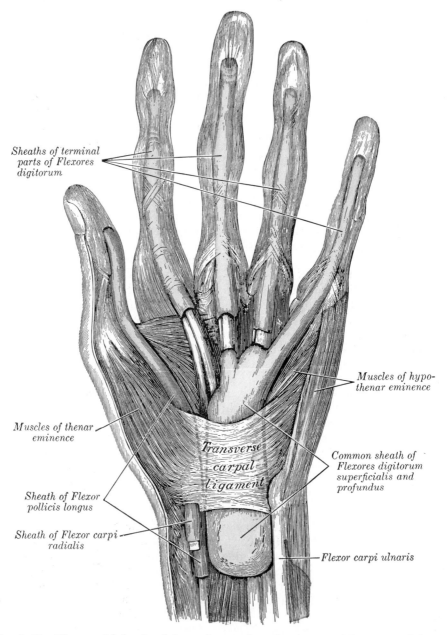

Sheaths of terminal parts of Flexores digitorum

Muscles of hypothenar eminence

Muscles of thenar eminence

Transverse carpal ligament

Common sheath of Flexores digitorum superficialis and profundus

Sheath of Flexor pollicis longus

Sheath of Flexor carpi radialis

Flexor carpi ulnaris

FIG. 6–47.—The synovial sheaths of the tendons on the palmar aspect of the wrist and digits.

in the middle of the palm, but they have independent digital sheaths beginning proximally over the heads of their metacarpal bones and continuing distally to the terminal phalanges, where the profundus inserts.

Extensor Retinaculum (*retinaculum extensorum; dorsal carpal ligament; posterior annular ligament*) (Figs. 6–46, 6–48).—The extensor retinaculum, under which the

extensor tendons lie, is the distal portion of the investing antebrachial fascia which is thickened abruptly by the addition of transverse collagenous bundles. The latter take a somewhat oblique course, extending distalward as they cross from the radial to the ulnar side. The ligament is attached, medially, to the styloid process of the ulna, and to the triangular and pisiform bones,

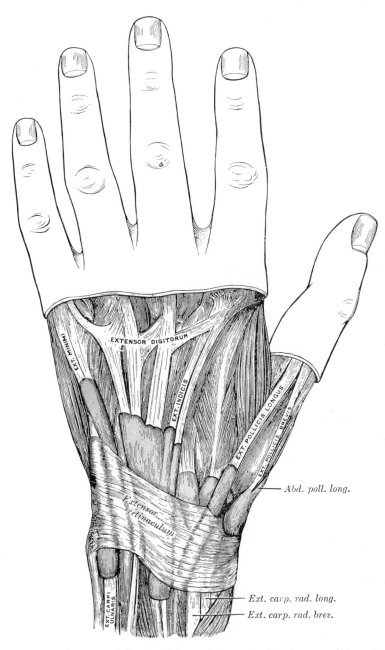

Fig. 6–48.—The synovial sheaths of the tendons on the dorsal aspect of the wrist.

and laterally, to the lateral margin of the radius. Between these medial and lateral borders, it is attached to the ridges on the dorsal surface of the radius.

The **Synovial Sheaths of the Extensor Tendons at the Wrist** (Fig. 6–48).—Between the extensor retinaculum and the carpal bones, six tunnels are formed for the passage of tendons, each tunnel having a separate synovial sheath. One is found in each of the following positions (Fig. 6–50): (1) on the lateral (radial) side of the styloid process of the radius, for the tendons of the Abductor pollicis longus and Extensor pollicis brevis; (2) dorsal to the styloid process, for the tendons of the Extensores carpi radialis longus and brevis; (3) about the middle of the dorsal surface of the radius, for the tendon of the Extensor pollicis longus; (4) more medially, for the tendons of the Extensor digitorum and Extensor indicis; (5) opposite the interval between the radius and ulna, for the Extensor digiti minimi; (6) between the head and styloid process of the ulna, for the tendon of the Extensor carpi ulnaris. The sheaths lining these tunnels all begin proximal to the dorsal carpal ligament; those for the tendons of the Abductor pollicis longus, Extensor pollicis brevis, Extensores carpi radialis, and Extensor carpi ulnaris stop immediately proximal to the bases of the metacarpal bones, while the sheaths for the Extensor digitorum, Extensor indicis, and Extensor digiti minimi are prolonged to the junction of the proximal and intermediate thirds of the metacarpus.

Deep fascia of the palm.—The investing layer of deep fascia in the palm is continuous with the antebrachial fascia which is represented, in the wrist, by the palmar carpal ligament. It is continuous also with the fascia of the dorsum at the borders of the hand, attaching to the fifth metacarpal bone medially, and the first and second metacarpal bones laterally, as it passes over them. The thenar fascia, over the muscular eminence at the radial side of the hand, and the hypothenar fascia, over the eminence of the ulnar side, are similar in texture to the antebrachial fascia, but that in the central part of the palm is greatly strengthened into what is called the palmar aponeurosis.

Palmar Aponeurosis (Fig. 6–49).—The palmar aponeurosis is made up of two components: (a) a thick superficial stratum of longitudinal bundles which are the direct continuation of the tendon of the Palmaris longus, and (b) a thinner deep stratum of transverse fibers which is continuous with the palmar carpal ligament. The two strata are intimately fused and partly interwoven. The deeper portion is securely attached to the transverse carpal ligament and the latter may contribute obliquely running fibers to the aponeurosis. The longitudinal bundles of the superficial stratum form a uniform layer in the proximal part of the palm, but distally, they fan out and are segregated into divergent bands which extend toward the bases of the digits, covering the long flexor tendons. The four bands to the fingers are heavier and more constant than the one to the thumb. Each of these bands has a double termination, the superficial part attaching to the skin, and the deeper part ending in the fibrous flexor tendon sheath. The most superficial fibers attach to the skin at the distal crease of the palm; other superficial fibers terminate at the crease at the base of the digit. The deeper portion of each band contributes to the fibrous tendon sheath in two ways: some of the fibers continue distally into the digit, assisting in the formation of the digital sheath; the greater number of fibers, however, form two arching ligamentous bands which penetrate deeply toward the metacarpal bone on each side of the tendon. They attach to the bone and send fibers to the transverse metacarpal ligament, thus completing the formation of the tunnel which lies on the head of the metacarpal bone. In the central part of the palm, as the longitudinal bands diverge and separate from each other, the intervals between them are occupied by transverse fibers which represent the distal thickening of the deeper stratum of the aponeurosis. These transverse fibers, making up what is called the **superficial transverse metacarpal ligament**, extend as far distally as the heads of the metacarpal bones, but from here to the webs between the digits, the intervals are not covered by aponeurosis. In this distal, uncovered portion, therefore, the digital vessels and nerves, and the tendons of the Lumbricales are more readily

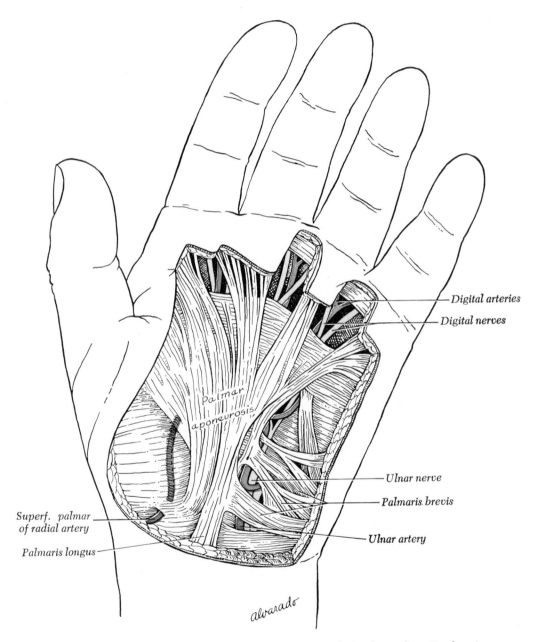

Fig. 6–49.—Superficial dissection of the palm of the hand. (Redrawn from Töndury.)

accessible to the surgeon. The intervals are closed distally by other transverse fibers which occupy and support the webs between the digits and which are variously named the **fasciculi transversi**, the **superficial transverse ligament of the fingers**, the **interdigital ligament**, and the **ligamentum natatorium**. They attach to the digital sheaths at the bases of the proximal phalanges and merge into the fibrous septa of the sides of the fingers known as the **cutaneous ligaments of the phalanges**. The digital vessels and nerves enter the digits deep to the fasciculi transversi and lie against the cutaneous ligaments of the phalanges in their course toward the ends of the fingers. As mentioned above, the band of longitudinal fibers from the palmar aponeurosis which extends toward the thumb is not as robust as the other four. Some of its fibers attach to the longitudinal crease of the palm, many of them fuse with the fascia of the thenar eminence, and a comparatively small number assist in the formation of the sheath for the Flexor pollicis longus tendon. When the Palmaris longus is absent, a condition which occurs in 13 per cent of the hands, the attachment of the palmar aponeurosis to the transverse carpal ligament is strengthened to compensate for the loss of continuity with the Palmaris tendon. The Palmaris brevis is a small but constant muscle which lies superficial to the hypothenar fascia; it has its origin at the ulnar border of the palmar aponeurosis and inserts into the skin of the ulnar border of the palm. The palmar aponeurosis is fused, at its radial border, with the fascial membrane of the thenar eminence, and from this line of union, a membrane is continued deeply into the palm and is attached to the first metacarpal bone, forming the **thenar septum**. Similarly, the aponeurosis is fused with the hypothenar fascia, and from this union a septum is continued deeply to the fifth metacarpal, forming the **hypothenar septum**. These two septa divide the palm into three compartments: a thenar, a hypothenar, and a central compartment (see below).

Digital Tendon Sheaths.—The tendons of the Flexores digitorum superficialis and profundus are held in position along the digits by strong fibrous tunnels (Fig. 6–53).

The tunnels or canals are formed by the palmar surfaces of the phalanges and by strong collagenous bands which arch over the tendons and are attached to the margins of the phalanges on either side. Opposite the middle of the proximal and second phalanges, the bands (*digital vaginal ligaments*) are very strong and their fibers are transverse. Opposite the joints they are much thinner, and consist of annular and cruciate ligamentous fibers. At their proximal ends, the digital sheaths merge with the deeper parts of the palmar aponeurosis. Within each of the five fibrous digital sheaths, there is a synovial tendon sheath; that for the thumb is continuous with the radial bursa; that for the little finger with the ulnar bursa; those for the other three fingers are closed proximally at the metacarpophalangeal joints (Fig. 6–47).

Deep fascia of the dorsum of the hand. —The investing layer of deep fascia of the dorsum of the hand is directly continuous with the antebrachial fascia which is thickened at the wrist by the addition of annular collagenous bundles into the extensor retinaculum. It is continuous, at the ulnar side of the hand, with the hypothenar fascia, after being attached to the dorsum of the fifth metacarpal bone. At the radial border of the hand, it is continuous with the thenar fascia after being attached to the first metacarpal bone. The investing fascia forms the superficial boundary of a flat compartment which contains the tendons of the extensors of the digits.

The fascia of the dorsum of the thumb corresponds to that of the hand proper whose investing and subaponeurotic fasciae fuse into a single layer at its radial border. This single layer, after being attached to the dorsum of the second metacarpal bone, continues laterally over the first Interosseus dorsalis, in the web between the thumb and the index finger, and is attached to the ulnar border of the first metacarpal bone. It separates again into two layers which form a compartment for the extensor tendons of the thumb, and then, reunited into a single sheet, attaches to the radial border of the first metacarpal, where it becomes continuous with the thenar fascia.

Fascial compartments of the hand.—The **thenar compartment** (not to be confused with the thenar fascial cleft or space) occupies the thenar eminence of the palm and contains the short muscles of the thumb with the exception of the Adductores. The boundary is formed by the investing layer of deep fascia and its inward continuation, the thenar septum, which lies between the Adductores and the Flexor brevis. The compartment is closed by the attachment of this fascia proximally, to the carpal bones and transverse carpal ligament; distally, to the first phalanx at the insertion of the enclosed muscles; dorsally, along the subcutaneous border of the first metacarpal bone, and ventrally, by the attachment of the thenar septum along the Adductores. In addition to the Abductor brevis, the Flexor brevis and the Opponens, the compartment contains the first metacarpal bone, the superficial palmar branch of the radial artery, and a portion of the tendon of the Flexor pollicis longus enclosed in the radial bursa. Within the compartment, the muscles are enclosed in their individual fascial sheaths, and for the most part, are separated from each other by fascial clefts, but these clefts do not communicate with each other nor with the major fascial clefts of the palm.

The **hypothenar compartment** occupies the hypothenar eminence and contains the short muscles of the little finger. It is enclosed by the hypothenar investing fascia

and the hypothenar septum which lies between the Flexor digiti minimi and the third Interosseus palmaris. The fascia forms a closed compartment by attaching along the fifth metacarpal bone on both the dorsal and palmar aspects of the muscles which it surrounds. Within the compartment, the individual muscles are enclosed in fascial sheaths of their own and are separated from each other by fascial clefts, but these clefts do not communicate with each other or with the major fascial cleft of the palm.

The **central compartment** is bounded, medially and laterally, by the thenar and hypothenar fascial septa; superficially, by the palmar aponeurosis; and deeply, by a fascial membrane which covers the deep surface of the long flexor tendon mass. It contains the Flexores digitorum superficialis and profundus tendons, the Lumbricales, the superficial palmar arch, the palmar branch of the medial nerve, and the superficial branch of the ulnar nerve. The compartment is narrow proximally, but widens distally as the tendons diverge toward their fingers. It is closed proximally, as far as fascial attachments are concerned, but the tendon sheaths within it extend back into the forearm, and its tissues merge distally with those of the webs and digits.

The **Interosseus-Adductor Compartment.** —The compartments of the palm are sepa-

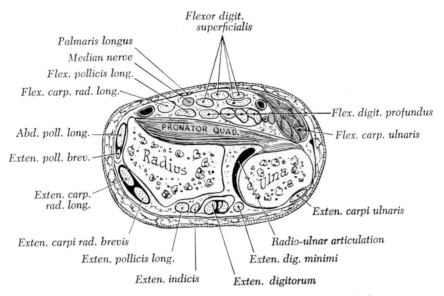

Flexor digit. superficialis

Palmaris longus
Median nerve
Flex. pollicis long.
Flex. carp. rad. long.

Flex. digit. profundus

Abd. poll. long.
PRONATOR QUAD.
Flex. carp. ulnaris

Exten. poll. brev.
Radius
Ulna

Exten. carp. rad. long.

Exten. carpi ulnaris

Exten. carpi rad. brevis
Radio-ulnar articulation
Exten. pollicis long.
Exten. dig. minimi
Exten. indicis
Exten. digitorum

Fig. 6–50.—Transverse section across distal ends of radius and ulna.

rated from the dorsal portion of the hand by a septum which is made up principally of the Interossei and the Adductores pollicis, and which may be called accordingly, the interosseus-adductor compartment. It is enclosed by two fascial membranes which are continuous with each other around its medial and lateral borders. The membrane on the dorsal surface, called the **dorsal interosseous fascia,** covers and is adherent to the dorsal surfaces of the second to fifth metacarpal bones and the intervening dorsal Interossei. The palmar surface is covered by the **palmar interosseous fascia** in the ulnar half, and by the fascia of the Adductores in the radial half. The compartment is closed by the attachment of these membranes at the origins of the muscles proximally, and at their insertions distally. It contains the second, third, fourth, and fifth metacarpal bones, all the Interossei, the Adductores transversus and obliquus, the deep palmar arch, and the deep branch of the ulnar nerve. The Adductores and the Interossei have been placed in the same rather than in separate compartments because the entire muscle mass lies deep to the palmar compartments (Fig. 6–50) and is separated from them by the major fascial cleft of the palm. A fascial cleft may separate the two Adductores from each other and from the first dorsal Interosseus, and these clefts may communicate with the thenar cleft or with the tissue spaces of the web between the thumb and index finger.

The **dorsal tendon compartment** is enclosed, superficially, by the investing fascia, and deeply, by a membrane of fascia covering the deep surface of the extensor tendon mass called the **dorsal subaponeurotic fascia.** The compartment is closed at the sides of the hand by the fusion of the two membranes into a single sheet where both are attached to the dorsum of the second and fifth metacarpal bones. It is closed distally by the fusion of the two layers at the webs between the fingers and their attachment to the joint capsules and tendinous expansions of the digits. It is closed proximally by the fusion with the tendon sheaths which pass under the dorsal carpal ligament. The compartment is separated from the subcutaneous fascia by the subcutaneous fascial cleft, and from the dorsal interosseous fascia by the subaponeurotic fascial cleft.

Fascial Clefts (*fascial spaces*) **of the Palm** (Fig. 6–51).—The term fascial cleft is preferable to fascial space because the latter quite frequently leads to ambiguity and the former agrees with the description in other parts of this book (see pages 373–376). Fascial clefts are planes of cleavage between fascial membranes and should not be confused with fascial compartments which are enclosures formed by fascial membranes and contain muscles, bones, or other structures. Fascial clefts may occur inside of compartments or may lie between compartments.

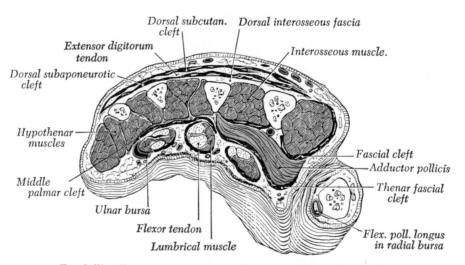

FIG. 6–51.—Transverse section across the wrist and digits. (Kanavel.)

The major fascial cleft of the palm lies between the fascia covering the deep surface of the long flexor tendon mass and the fascia covering the Adductores and the Interossei of the medial portion of the hand; that is, it lies between the central palmar compartment and the interosseus adductor compartment. Delicate membranous septa of variable number and extent attach to the metacarpal bones distally, and extend toward the wrist, producing more or less complete subdivisions of the cleft. The septum which is attached to the middle metacarpal bone is more constant and better developed than the rest and is commonly described as subdividing the cleft into two parts, the middle palmar cleft and the thenar cleft (Kanavel '42).

The **middle palmar cleft** (*middle palmar space*) is triangular in shape and separates the deep surface of the long flexor tendons from the Interossei in the central part of the palm. It lies between the palmar interosseous fascia and the fascia covering the deep surface of the long flexor tendon compartment. The cleft may be closed at the wrist by adhesion between the fascial membranes at the transverse carpal ligament, or it may be continued proximally, deep to the ulnar bursa, and communicate with the anterior interosseous cleft of the forearm. It is closed, medially, by the attachment of the hypothenar septum to the fifth metacarpal bone, and laterally, from the thenar cleft by a transparent fibrous membrane which is attached along the middle metacarpal bone. This membrane, instead of attaching to the flexor tendon mass directly over the middle metacarpal, takes an oblique course toward the region of the second metacarpal, causing the middle palmar cleft to extend into the radial portion of the palm so that it overlaps the thenar cleft to some extent. The membrane between the middle palmar and the thenar cleft may be incomplete proximally, allowing the two clefts to communicate with each other. Closely associated with the middle palmar cleft, acting as diverticula, are the clefts surrounding the second, third, and fourth Lumbricales, called the **lumbrical canals.**

The **thenar cleft** (*thenar space*) (Fig. 6–51) overlies the palmar surface of the Adductores pollicis. It is bounded, medially, by the membrane which is attached to the middle metacarpal bone and which separates it from the middle palmar cleft; laterally, by the thenar septum and the first metacarpal bone; proximally, by the transverse carpal ligament, and distally, by the extent of the Adductor transversus. The thenar cleft is commonly continuous with the cleft between the two Adductores and between the latter and the first dorsal Interosseus. It may communicate with the middle palmar cleft, proximally, and it usually has a diverticulum extending along the first Lumbricalis, the lumbrical canal.

The **Lumbrical canals** are the tubular fascial clefts which separate the Lumbricales from the denser connective tissue which surrounds them. These clefts are closely associated with, or act as diverticula for, the major fascial clefts of the palm in the following way: the cleft of the first Lumbrical is with the thenar cleft, that of the second, third, and fourth with the middle palmar cleft.

The **dorsal subcutaneous cleft** (*dorsal subcutaneous fascial space*) (Fig. 6–51) separates the subcutaneous fascia from the deep fascia. It extends distally out into the fingers and proximally into the forearm. It is closed at the borders of the hand by the attachment of the subcutaneous fascia to the deep fascia of the palm, and it has no communication with the dorsal subaponeurotic cleft.

The **dorsal subaponeurotic fascial cleft** (*dorsal subaponeurotic fascial space*) (Fig. 6–51) separates the dorsal subaponeurotic fascia from the dorsal interosseus fascia. It is closed at the sides of the hand by fusion of the two membranes near their attachment to the second and the fifth metacarpal bones. It is closed distally by the fusion of the two layers and their union with the joint capsules and extensor expansions of the digits. Proximally, it is obliterated by the attachment of the tendon sheaths to the bones and ligaments of the wrist. This cleft does not communicate with the dorsal subcutaneous cleft, nor with the palmar clefts, and it does not pass from the hand into the forearm.

Variations.—The flexor tendons to the index finger and the first Lumbricalis may be enclosed in a separate compartment. In this case, there are two membranes attached to the middle metacarpal bone. One has the usual position, extending obliquely toward the radial side of the hand and overlying the thenar cleft; the other extends directly toward the deep surface of the flexor tendon mass and attaches along the flexors to the middle finger. The index compartment is bounded by these two membranes and the palmar aponeurosis. The portion of the middle palmar cleft which usually lies superficial to the thenar cleft is within the index compartment, and the lumbrical canal for this finger leads into this cleft rather than the thenar cleft.

Surgical Considerations.—An understanding of the relationship of the fascial clefts to each other and to the tendon sheaths may best be reached by a brief review of the probable spread of an infective process, independent of the participation of the blood vessels or lymphatics.

A subcutaneous abscess on the dorsum of the hand or in the webs could be expected to point at the surface locally because of the softness of the tissues; in the palm, it might reach the surface or spread to the webs, but would not be likely to penetrate the palmar aponeurosis. An abscess of the index finger, after penetrating to the deeper tissues, might progress proximally until it arrived in the lumbrical canal, and through this path, reach the thenar cleft. From the other three fingers, it might reach the middle palmar cleft by a similar path. It might progress from either of these deep palmar clefts to the other, or, if the swelling and edema would permit, it might reach the cleft of the forearm. An infection involving a digital flexor tendon sheath, if it occurred in the little finger or the thumb, would quickly follow the sheath into the ulnar or radial bursae, and after a period of time, the latter could rupture into the volar fascial cleft of the forearm. If the infection were in the index finger, the tendon sheath might be expected to rupture into the thenar cleft; if in the middle or fourth finger, it would rupture into the middle palmar cleft. An infection of the dorsal subcutaneous cleft could be expected to reach the surface locally. An infection in the dorsal subaponeurotic cleft would be expected to spread throughout the entire cleft, and then eventually rupture into the webs or at the sides of the hand.

1. **The Thenar Muscles** (Figs. 6–52, 6–53)

Abductor pollicis brevis
Opponens pollicis
Flexor pollicis brevis

Adductor pollicis $\left\{ \begin{array}{l} \text{Caput obliquum} \\ \text{Caput transversum} \end{array} \right.$

The **Abductor pollicis brevis** (*Abductor pollicis*) is a thin, flat muscle, placed most superficially in the thenar region. It *arises* from the transverse carpal ligament, the tuberosity of the scaphoid, and the ridge of the trapezium, frequently by two distinct slips. Running lateralward and distalward, it is *inserted* by a thin, flat tendon into the radial side of the base of the first phalanx of the thumb and the capsule of the metacarpophalangeal articulation.

Action.—Abducts the thumb, that is, draws it away in a plane at right angles to that of the palm of the hand.

Nerve.—A branch of the median nerve, containing fibers from the sixth and seventh cervical nerves.

The **Opponens pollicis** is a small, triangular muscle placed beneath the preceding. It *arises* from the ridge on the trapezium and from the flexor retinaculum, passes distalward and lateralward, and is *inserted* into the whole length of the metacarpal bone of the thumb on its radial side.

Action.—Abducts, flexes, and rotates the metacarpal of the thumb, bringing the thumb out in front of the palm to face the fingers.

Nerve.—A branch of the median nerve, containing fibers from the sixth and seventh cervical nerves.

The **Flexor pollicis brevis** consists of two portions, lateral and medial. The **lateral** and more **superficial portion** *arises* from the distal border of the flexor retinaculum and the distal part of the ridge on the trapezium bone; it passes along the radial side of the tendon of the Flexor pollicis longus, and, becoming tendinous, is *inserted* into the radial side of the base of the proximal phalanx of the thumb; in its tendon of insertion there is a sesamoid bone. The **medial** and **deeper portion** of the muscle is very small, and *arises* from the ulnar side of the first metacarpal bone between the Adductor pollicis (obliquus) and the lateral head of the first Interosseous dorsalis, and is *inserted* into the ulnar side of the base of the first phalanx with the Adductor pollicis (obliquus). The medial part of the Flexor brevis pollicis is sometimes described as the **first Interosseous palmaris**.

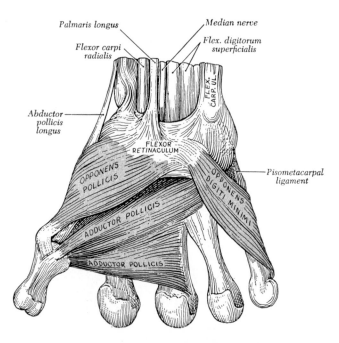

FIG. 6–52.—The muscles of the thumb.

Action.—Flexes and adducts the thumb.

Nerve.—The lateral portion, a branch of the median nerve, containing fibers from the sixth and seventh cervical nerves; the medial portion, a nerve from the deep branch of the ulnar nerve, containing fibers from the eighth cervical and first thoracic nerves.

The **Adductor pollicis, caput obliquus** (*Adductor obliquus pollicis*) arises by several slips from the capitate bone, the bases of the second and third metacarpals, the intercarpal ligaments, and the sheath of the tendon of the Flexor carpi radialis. From this origin the greater number of fibers pass obliquely distalward and converge to a tendon which, uniting with the tendons of the medial portion of the Flexor pollicis brevis and the transverse part of the Adductor, is *inserted* into the ulnar side of the base of the proximal phalanx of the thumb, a sesamoid bone being present in the tendon. A considerable fasciculus, however, passes more obliquely beneath the tendon of the Flexor pollicis longus to join the lateral

portion of the Flexor brevis and the Abductor pollicis brevis.

The **Adductor pollicis, caput transversus** (*Adductor transversus pollicis*) (Fig. 6–52) is the most deeply seated of this group of muscles. It is of a triangular form arising by a broad base from the distal two-thirds of the palmar surface of the third metacarpal bone; the fibers converge, to be *inserted* with the medial part of the Flexor pollicis brevis and Adductor pollicis (obliquus) into the ulnar side of the base of the first phalanx of the thumb.

Action.—Adducts, that is, brings the thumb toward the palm.

Nerve.—A branch of the deep palmar branch of the ulnar, containing fibers from the eighth cervical and first thoracic nerves.

Variations.—The Abductor pollicis brevis is often divided into an outer and an inner part; accessory slips from the tendon of the Abductor pollicis longus or Palmaris longus, more rarely from the Extensor carpi radialis longus, from the styloid process or Opponens pollicis or from the skin over the thenar eminence. The deep head of the Flexor pollicis brevis may be absent or enlarged. The two adductors vary in their relative extent and in the closeness of their connection. The Adductor obliquus may receive a slip from the transverse metacarpal ligament.

2. The Hypothenar Muscles (Figs. 6–52, 53)

Palmaris brevis
Abductor digiti minimi
Flexor digiti minimi brevis
Opponens digiti minimi

The **Palmaris brevis** is a thin, quadrilateral muscle, placed beneath the integument of the ulnar side of the hand. It *arises* by tendinous fasciculi from the transverse carpal ligament and palmar aponeurosis; the fleshy fibers are *inserted* into the skin on the ulnar border of the palm of the hand.

Action.—Draws the skin at the ulnar side of the palm toward the middle of the palm, increasing the height of the hypothenar emi-

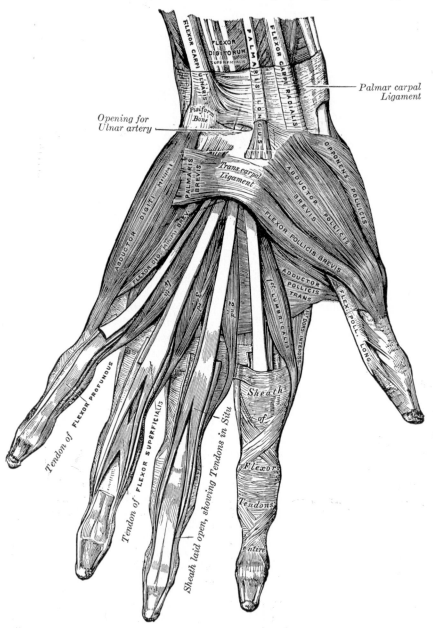

Fig. 6–53.—The muscles of the left hand. Palmar surface.

nence, as in clenching the fist. Holds the hypothenar subcutaneous pad in place, as in catching a ball.

Nerve.—A branch of the ulnar nerve, containing fibers from the eighth cervical nerve.

The **Abductor digiti minimi** (*Abductor digiti quinti*) is situated on the ulnar border of the palm of the hand. It *arises* from the pisiform bone and from the tendon of the Flexor carpi ulnaris, and ends in a flat tendon, which divides into two slips; one is *inserted* into the ulnar side of the base of the first phalanx of the little finger; the other into the ulnar border of the aponeurosis of the Extensor digiti minimi.

Action.—Abducts the little finger and flexes its proximal phalanx.

Nerve.—A branch of the ulnar nerve, containing fibers from the eighth cervical and first thoracic nerves.

The **Flexor digiti minimi brevis** (*Flexor digiti quinti brevis*) lies on the same plane as the preceding muscle, on its radial side. It *arises* from the convex surface of the hamulus of the hamate bone, and the palmar surface of the flexor retinaculum, and is *inserted* into the ulnar side of the base of the first phalanx of the little finger. It is separated from the Abductor, at its origin, by the deep branches of the ulnar artery and nerve. This muscle is sometimes wanting; the Abductor is then, usually, of large size.

Action.—Flexes the little finger.

Nerve.—A branch of the ulnar nerve, containing fibers from the eighth cervical and first thoracic nerves.

The **Opponens digiti minimi** (*Opponens digiti quinti*) (Fig. 6–52) is of a triangular form, and placed immediately beneath the preceding muscles. It *arises* from the convexity of the hamulus of the hamate bone and contiguous portion of the flexor retinaculum; it is inserted into the whole length of the metacarpal bone of the little finger, along its ulnar margin.

Action.—Abducts, flexes, and rotates the fifth metacarpal in bringing the little finger out to face the thumb.

Nerve.—A branch of the ulnar nerve, con-

taining fibers from the eighth cervical and first thoracic nerves.

Variations.—The Palmaris brevis varies greatly in size. The Abductor digiti minimi may be divided into two or three slips or united with the Flexor digiti minimi brevis. Accessory head from the tendon of the Flexor carpi ulnaris, the flexor retinaculum, the fascia of the forearm or the tendon of the Palmaris longus. A portion of the muscle may insert into the metacarpal, or separate slips, the *Pisimetacarpus, Pisiuncinatus* or the *Pisiannularis* muscle may exist.

3. The Intermediate Muscles

Lumbricales
Interossei

The **Lumbricales** (Fig. 6–53) are four small fleshy fasciculi, associated with the tendons of the Flexor digitorum profundus. The first and second *arise* from the radial sides and palmar surfaces of the tendons of the index and middle fingers respectively; the third, from the contiguous sides of the tendons of the middle and ring fingers; and the fourth, from the contiguous sides of the tendons of the ring and little fingers. Each passes to the radial side of the corresponding finger, and opposite the metacarpophalangeal articulation is *inserted* into the tendinous expansion of the Extensor digitorum covering the dorsal aspect of the finger.

Action.—Flex the metacarpophalangeal joints and extend the two distal phalanges.

Nerves.—The first and second Lumbricales by branches of the third and fourth digital branches of the median nerve, containing fibers from the sixth and seventh cervical nerves. The third and fourth by branches of the deep palmar branch of the ulnar, containing fibers from the eighth cervical nerve. The third Lumbricalis may receive twigs from both nerves or all its fibers from the median nerve.

Variations.—The Lumbricales vary in number from two to five or six and there is considerable variation in insertions.

The **Interossei** (Figs. 6–54, 6–55) are so named from occupying the intervals between the metacarpal bones, and are divided into two sets, a dorsal and a palmar.

The **Interossei dorsales** (*Dorsal interossei*) are *four* in number, and occupy the intervals between the metacarpal bones. They are bipenniform muscles, each *arising* by two heads from the adjacent sides of the metacarpal bones, but more extensively from the metacarpal bone of the finger into which the muscle is inserted They are *inserted* into the bases of the proximal phalanges and into the aponeuroses of the tendons of the Extensor digitorum. Between the double origin of each of these muscles is a narrow triangular interval; through the first of these the radial artery passes; through each of the other three a perforating branch from the deep palmar arch is transmitted.

The **first** or **Abductor indicis** is larger than the others. It is flat, triangular in form, and *arises* by two heads, separated by a fibrous arch for the passage of the radial artery from the dorsum to the palm of the hand. The lateral head *arises* from the proximal half of the ulnar border of the first metacarpal bone; the medial head, from almost the entire length of the radial border of the second metacarpal bone; the tendon is *inserted* into the radial side of the index finger. The **second** and **third** are inserted into the middle finger, the former into its radial, the latter into its ulnar side. The **fourth** is *inserted* into the ulnar side of the ring finger.

Action.—The dorsal interossei abduct the fingers from an imaginary line drawn through the axis of the middle finger and flex the metacarpophalangeal joint and extend the two distal phalanges.

Nerves.—Branches from the deep palmar branch of the ulnar, containing fibers from the eighth cervical and first thoracic nerves.

The **Interossei palmares** (*Volar interossei*), three in number, are smaller than the Interossei dorsales, and placed upon the palmar surfaces of the metacarpal bones, rather than between them. Each *arises* from the entire length of the metacarpal bone of one finger, and is *inserted* into the side of the base of the first phalanx and aponeurotic expansion of the Extensor communis tendon to the same finger.

The **first** *arises* from the ulnar side of the second metacarpal bone, and is *inserted* into the same side of the first phalanx of the index finger. The **second** *arises* from the radial side of the fourth metacarpal bone, and is *inserted* into the same side of the

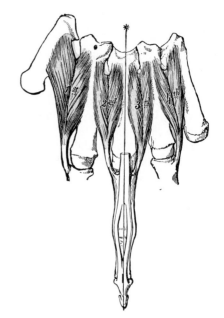

Fig. 6–54.—The Interossei dorsales of left hand.

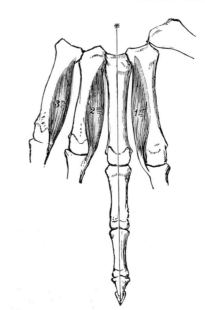

Fig. 6–55.—The Interossei palmares of left hand.

ring finger. The **third** *arises* from the radial side of the fifth metacarpal bone, and is *inserted* into the same side of the little finger. From this account it may be seen that each finger is provided with two Interossei, with the exception of the little finger, in which the Abductor takes the place of one of the pair.

As already mentioned (page 482), the medial head of the Flexor pollicis brevis is sometimes described as the **Interosseus palmaris primus.**

Action.—Adduct the fingers toward an imaginary line through the axis of the middle finger, flex the metacarpophalangeal joint, and extend the two distal phalanges.

Nerves.—Branches from the deep palmar branch of the ulnar, containing fibers from the eighth cervical and first thoracic nerves.

Group Actions.—Flexion of the fingers in grasping an object is performed by the Flexores digitorum superficialis and profundus. The wrist extensors contract synergetically to prevent flexion of the wrist. The characteristic action of the profundus, to flex the terminal phalanx, can be carried out independently as a freak by some individuals. The action of both muscles on the two terminal phalanges can be performed independently of the proximal phalanx by calling into play the synergetic action of the Extensor digitorum on the proximal phalanx. Flexion of the proximal phalanx at the same time as extension of the two distal joints is performed by the Interossei dorsales and palmares, and the Lumbricales.

Extension of the proximal phalanges is performed by the Extensor digitorum, Extensor indicis, and Extensor digiti minimi. The wrist flexors contract synergetically to prevent extension at the wrist. The long extensors have a weak action on the two terminal joints, and must be assisted in this action by the Lumbricales and the Interossei.

Abduction of the fingers is performed by the Interossei dorsales and the Abductor digiti minimi, considering the axis of the middle finger as the center of the hand. Full abduction can be carried out only if the fingers are extended, because of restrictions in the joints. Adduction of the fingers is performed by the Interossei palmares, and can be performed with the fingers either flexed or extended.

The thumb is so placed that its plane of flexion and extension is at right angles to that of the fingers. Flexion and extension of the thumb, therefore, are in the same plane as abduction and adduction of the fingers; abduction and adduction of the thumb are in the same plane as flexion and extension of the fingers. Flexion of the distal phalanx is performed by the Flexor pollicis longus; of the proximal phalanx alone by the Flexor pollicis brevis. Extension of the distal phalanx is performed by the Extensor pollicis longus, the proximal phalanx by the Extensor pollicis brevis. Abduction of the thumb is by the Abductores pollicis longus and brevis; adduction by the Adductor pollicis. It is seldom that extension or abduction are performed independently of each other, most normal activity being a mixture of the two movements. When the thumb is used in grasping, first, as a preliminary, it is abducted by the Abductores, and rotated by the Opponens so that its palmar surface faces the palm of the hand, and then finally, the actual grasping is performed largely by the Flexor pollicis longus.

THE MUSCLES AND FASCIAE OF THE LOWER LIMB

The muscles of the lower limb are subdivided into groups corresponding with the different regions of the limb.

> I. Muscles of the Iliac Region
> II. Muscles of the Thigh
> III. Muscles of the Leg
> IV. Muscles of the Foot

I. THE MUSCLES AND FASCIAE OF THE ILIAC REGION

> Psoas major
> Psoas minor
> Iliacus

The fascia covering the intraäbdominal surface of the Iliacus and Psoas is part of the endoabdominal or internal investing layer of deep fascia, but, as it follows these muscles under the inguinal ligament out into the thigh, it becomes continuous with the fascia lata which is a portion of the external investing fascia. In this way, it forms an important direct continuity between the internal and the external investing layers of deep fascia. The endoabdominal portion of this fascia is covered internally by subserous fascia.

Iliac Fascia (*fascia iliaca*).—The endoabdominal portion of the iliac fascia is attached at its cranial limit to the entire length of the inner lip of the crest of the ilium, along with the muscle. Cranial to that, it is continuous with the definitive transversalis fascia except near the vertebral column where it is continuous with the quadratus lumborum fascia. It is continuous, medially, with the psoas fascia, and, after attaching to the arcuate line of the ilium, it is continued down into the lesser pelvis as the obturator internus fascia. At the inguinal ligament, it meets and fuses with the transversalis fascia of the lateral portion of the anterior abdominal wall to form a single sheet which follows the surface of the Iliacus under the ligament out into the thigh. It is securely attached to the ligament, as it passes under it, and at this point the origins of the Obliquus internus and Transversus abdominis are attached to the iliac fascia as well as to the inguinal ligament. In the thigh, it becomes continuous with the part of the fascia lata over the Sartorius, laterally, and with the iliopectineal fascia, medially.

Psoas fascia.—The cranial extremity of the psoas fascia is intimately blended with the medial lumbocostal arch of the Diaphragm, which stretches from the bodies to the transverse processes of the first or second lumbar vertebrae. It is attached, medially, by a series of arched processes to the intervertebral disks and prominent margins of the vertebrae, and to the upper part of the sacrum. The intervals left between these arched processes and the constricted bodies of the vertebrae transmit the lumbar arteries and veins, and the filaments of the sympathetic trunk. It is continuous laterally with the quadratus lumborum and iliac fasciae. At the inguinal ligament and in the thigh, the psoas and iliac fasciae combine to make up part of a complex called the iliopectineal fascia.

Iliopectineal fascia (*fascia iliopectinea*).—The iliopectineal fascia is made up of three interconnected portions: (*a*) the fascia covering the femoral portions of the iliacus and Psoas; (*b*) the fascia over the proximal portion of the Pectineus, and (*c*) a thickened band which dips down between the Psoas and the femoral vessels as they pass under the inguinal ligament. The fascia as a whole is continuous, under the inguinal ligament, with the endoabdominal iliac, psoas, and transversalis fasciae. Parts (*a*) and (*b*) together form a sheet which is the floor of the femoral (Scarpa's) triangle. At the junction of the iliopsoas and pectineal portions of this fascia, there is a firm attach-

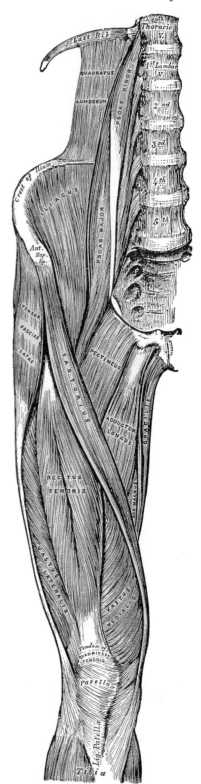

Fig. 6–56.—Muscles of the iliac and anterior femoral regions.

ment to the iliopectineal eminence of the ilium and to the pubocapsular ligament of the hip joint. From the attachment to the iliopectineal eminence, the fascia passes outward as a thickened band (*c*, page 488) between the Psoas and the femoral vessels toward the inguinal ligament to which it becomes attached (Fig. 8–57). This band divides the interval beneath the inguinal ligament, that is, between the ligament and the pelvic bone, into two parts known as the lacuna musculorum and the lacuna vasorum. The **lacuna musculorum** contains the Iliacus and Psoas, and the femoral nerve. The **lacuna vasorum** contains the femoral artery and vein, and the femoral canal (page 653).

The **Iliopsoas** is frequently regarded as a single muscle but it is here divided into its two parts, the Psoas major and the Iliacus, for convenience of description.

The **Psoas major** (*Psoas magnus*) (Fig. 6–56) is a long fusiform muscle placed on the side of the lumbar region of the vertebral column and brim of the lesser pelvis. It *arises* (1) from the ventral surfaces of the bases and caudal borders of the transverse processes of all the lumbar vertebrae; (2) from the sides of the bodies and the corresponding intervertebral disks of the last thoracic and all the lumbar vertebrae by five slips, each of which is attached to the adjacent cranial and caudal margins of two vertebrae, and to the intervertebral disk; (3) from a series of tendinous arches which extend across the constricted parts of the bodies of the lumbar vertebrae between the previous slips; the lumbar arteries and veins, and filaments from the sympathetic trunk pass beneath these tendinous arches. The muscle proceeds downward across the brim of the lesser pelvis, and diminishing gradually in size, passes beneath the inguinal ligament and ventral to the capsule of the hip joint and ends in a tendon which also receives nearly the whole of the fibers of the Iliacus and is *inserted* into the lesser trochanter of the femur. A large bursa, which may communicate with the cavity of the hip joint, separates the tendon from the pubis and the capsule of the joint.

Action.—Flexes the thigh; flexes the lumbar vertebral column and bends it laterally.

Nerves.—Branches of the lumbar plexus, containing fibers from the second and third lumbar nerves.

The **Psoas minor** (*Psoas parvus*) is a long slender muscle, placed ventral to the Psoas major. It *arises* from the sides of the bodies of the twelfth thoracic and first lumbar vertebrae and from the fibrocartilage between them. It ends in a long flat tendon which is *inserted* into the pectineal line and iliopectineal eminence, and, by its lateral border, into the iliac fascia. This muscle is often absent.

Action.—Flexes the pelvis and lumbar vertebral column.

Nerve.—A branch of the first lumbar nerve.

The **Iliacus** is a flat, triangular muscle, which fills the iliac fossa. It *arises* from the superior two-thirds of this fossa, and from the inner lip of the iliac crest; from the anterior sacroiliac and the iliolumbar ligaments, and base of the sacrum dorsally and it reaches as far as the anterior superior and anterior inferior iliac spines, and the notch between them ventrally. The fibers converge to be *inserted* into the lateral side of the tendon of the Psoas major, some of them being prolonged on to the body of the femur for about 2.5 cm distal and anterior to the lesser trochanter.

Action.—Flexes the thigh.

The action of the Iliopsoas on the hip joint is primarily flexion. Electromyography has shown it to be an important postural muscle. On this basis it may be slightly active in either medial or lateral rotation but is not used for either one voluntarily. It should be noticed that rotation of the thigh does not produce rotation of the femur about its long axis; this could be brought about only by displacing the head from the acetabulum. If Galen's experiment is modernized by using a rubber instead of a leather band attached at the origin and insertion, the only mechanically possible action is shown to be medial rotation because the pull is anterior to the hip joint.

Nerves.—Branches of the femoral nerve, containing fibers from the second and third lumbar nerves.

Variations.—The *Iliacus minor* or *Iliocapsularis*, a small detached part of the Iliacus, is frequently present. It arises from the anterior inferior spine of the ilium and is inserted into the lower part of the intertrochanteric line of the femur or into the iliofemoral ligament.

II. THE MUSCLES AND FASCIAE OF THE THIGH

The **subcutaneous fascia** (*tela subcutanea*) forms a prominent layer over the entire thigh. It usually contains a considerable amount of fat, but it varies in thickness in different regions. It is continuous with the subcutaneous fascia of the abdomen, the leg, and, over the gluteal region, with the back. It may be separated into a superficial fatty layer and a deep membranous layer between which are found the superficial vessels and nerves, the superficial inguinal lymph nodes, and the great saphenous vein. In well-nourished individuals, the adipose tissue of the superficial layer is usually divided into two or three subsidiary layers by fibrous membranes which are associated with the emergence of the superficial nerves. The deep or fibrous layer is adherent to the fascia lata a little below the inguinal ligament and along the proximal medial portion of the thigh. It is attached to the margin of the saphenous opening (fossa ovalis), and fills the opening itself with an irregular layer of spongy tissue, called the **fascia cribrosa** (Fig. 9–34) because it is pierced by numerous openings for the passage of the saphenous vein and other blood and lymphatic vessels.

A large subcutaneous bursa is found in the subcutaneous fascia over the patella.

Fascia lata (Fig. 6–57).—The external investing fascia of the thigh is named fascia lata from its broad extent. Its thick, lateral portion is commonly taken to be typical of its texture, but it is thin in some areas where it has not been reinforced by fibrous contributions from the tendons. Proximally, it is continuous with the external abdominal and thoracolumbar fasciae after being attached to the pelvic bone and inguinal ligament; distally, it is continuous with the fascia of the leg. The *medial portion* lies over the Adductor group of muscles and is thin, gray, and not aponeurotic. It is attached to the

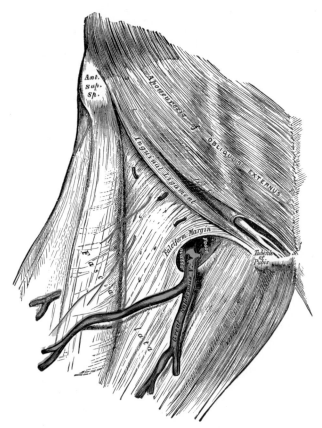

Fɪɢ. 6–57.—The hiatus saphenus (fossa ovalis).

ischial tuberosity and ischiopubic ramus, and beyond this is continuous with the external perineal fascia. At the knee it is thick and aponeurotic, having been strengthened by fibers from the tendon of the Sartorius. The *anterior portion* is attached to the pubic tubercle, inguinal ligament, and anterior superior iliac spine. Just distal to the lateral half of the inguinal ligament, it is a single sheet formed by the fusion of three abdominal fasciae. Of these, the most superficial is the external fascia over the aponeurosis of the Obliquus externus (fascia innominata) which passes superficial to the inguinal ligament; the middle one is the transversalis fascia of the anterior abdominal wall which passes under the inguinal ligament; the internal one also passes under the inguinal ligament and is the continuation of the iliac fascia. Distal to the medial half of the inguinal ligament, the middle and internal layers just mentioned fail to join the fascia lata at the ligament and continue into the thigh as the anterior and posterior portions of the femoral sheath.

The fascia lata in this medial region over the femoral vessels is thickened and laminated and has an opening through it, called the hiatus saphenus (fossa ovalis) (see below), for the passage of the great saphenous vein. The outer lamina is attached to the pubic tubercle along with the inguinal ligament; it has a free falciform margin which crosses the proximal end of the great saphenous vein and spirals distally around the vein to join the deep lamina medial to the vein. The two laminae are separated by a pad of adipose tissue. The anterior portion of the fascia lata is thicker than the medial, but is truly aponeurotic only near the knee where it is reinforced by fibers from the tendons of the Vasti. It is separated from the underlying Sartorius and Quadriceps by a fascial cleft, except near the knee.

The *lateral portion* is a thick strong aponeurosis, containing the tendinous fibers of insertion of the Gluteus maximus and the Tensor fasciae latae. It is attached proximally to the crest of the ilium and the dorsum of the sacrum. Between the iliac crest and the superior border of the Gluteus maximus, it is thickened by vertical tendinous bundles and is known as the **gluteal aponeurosis** (Fig. 6–61) which is used by

the Gluteus medius for part of its origin. At the border of the maximus it splits to enclose the muscle; the external layer of this **gluteal fascia** is thin, is closely bound to the subcutaneous fascia and the muscle, and sends septa down between large bundles of the muscle. In the region over the great trochanter, the muscular fasciculi end in a broad tendon which is imbedded in the fascia lata and is called the **iliotibial band** (*tractus iliotibialis*). Below the anterior part of the iliac crest, the fascia splits to enclose the Tensor fasciae latae which is inserted into the iliotibial band distal to the maximus. The iliotibial band is separated from the underlying Vastus lateralis by a distinct fascial cleft. It is inserted into the tibia and is blended with fibrous expansions from the Vastus lateralis and Biceps femoris. The *posterior portion* of the fascia lata is formed proximally by the union of the two layers of fascia enclosing the Gluteus maximus at its inferior border. It covers the hamstring muscles and the popliteal fossa.

Two strong intermuscular septa (Fig. 6–58) connect the deep surface of the fascia lata with the linea aspera of the femur. The **lateral intermuscular septum** is the stronger; it separates the Vastus lateralis from the Biceps femoris and is used by both muscles for the origin of their fibers. It extends from the insertion of the Gluteus maximus to the lateral condyle. The **medial intermuscular septum** lies between the Vastus medialis and the Adductores and Pectineus. Its superficial portion near the fascia lata splits to enclose the Sartorius and contributes to the formation of the adductor canal (Hunter's canal) about the femoral vessels.

The **hiatus saphenus** (*fossa ovalis*) (Fig. 6–57) is an oval aperture in the fascia lata in the proximal part of the thigh, a little below the medial end of the inguinal ligament. The great saphenous vein passes through it just before it joins the femoral vein. The fascia lata in this part of the thigh is thickened by lamination into two leaves separated by fat. The superficial leaf is attached to the inguinal ligament and pubic tubercle. It ends abruptly in a free border, the **falciform margin of the fossa**, which forms a spiral of one turn beginning at the pubic tubercle, coursing at first in a lateral direction superficial to the vein, then down along the vein, and back medially deep to the vein. Medial to the vein, the superficial leaf merges with the deep leaf. The proximal and lateral part of the falciform margin is called the **cornu superius**, the medial and distal part the **cornu inferius**. The deep leaf is formed by the pectineal, iliopectineal, and iliac fasciae. The hiatus is filled in and covered over by a thickened pad derived from the deep layer of subcutaneous fascia, called the **fascia cribrosa**.

1. The Anterior Femoral Muscles
(Fig. 6–56)

Sartorius

Quadriceps
femoris
$\left\{\begin{array}{l}\text{Rectus femoris}\\\text{Vastus lateralis}\\\text{Vastus medialis}\\\text{Vastus intermedius}\end{array}\right.$

Articularis genus

The **Sartorius**, the longest muscle in the body, is narrow and ribbon-like; it *arises* by tendinous fibers from the anterior supe-

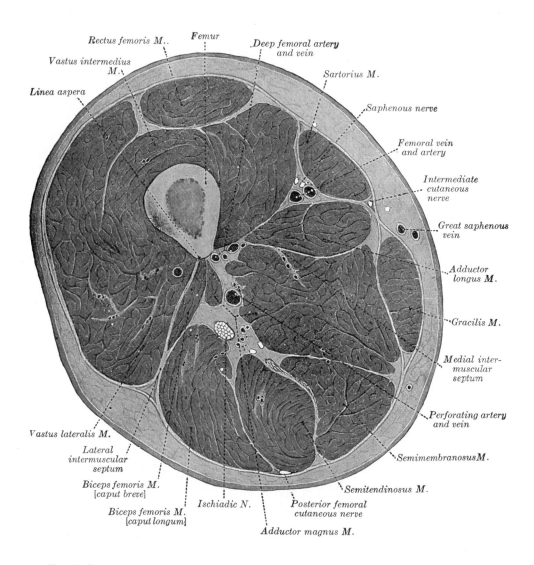

FIG. 6–58.—Cross section through the middle of the thigh. (Eycleshymer and Shoemaker.)

rior iliac spine and the upper half of the notch distal to it. It passes obliquely across the proximal and anterior part of the thigh, from the lateral to the medial side of the limb, then descends vertically, as far as the medial side of the knee, passing posterior to the medial condyle of the femur. It ends in a tendon which curves obliquely anteriorly and expands into a broad aponeurosis, which is *inserted,* anterior to the Gracilis and Semitendinosus, into the proximal part of the medial surface of the body of the tibia, nearly as far forward as the anterior crest. The proximal part of the aponeurosis is curved backward over the proximal edge of the tendon of the Gracilis so as to be inserted posterior to it. An offset from its proximal margin blends with the capsule of the knee joint, and another from its distal border, with the fascia on the medial side of the leg.

Action.—Flexes the thigh and rotates it laterally. Flexes the leg and, after it is flexed, rotates it slightly medially.

Nerves.—Branches, usually two in number, from the femoral nerve containing fibers from the second and third lumbar nerves. The first branch is distributed to the proximal portion

of the muscle and arises in common with the intermediate anterior cutaneous nerve; the second branch is distributed to the distal portion.

Variations.—Slips of origin from the outer end of the inguinal ligament, the notch of the ilium, the iliopectineal line or the pubis occur. The muscle may be split into two parts, and one part may be inserted into the fascia lata, the femur, the ligament of the patella or the tendon of the Semitendinosus. The tendon of insertion may end in the fascia lata, the capsule of the knee joint, or the fascia of the leg. The muscle may be absent.

The **Quadriceps femoris** (*Quadriceps extensor*) includes the four remaining muscles on the anterior thigh. It is the great extensor muscle of the leg, forming a large fleshy mass which covers the front and sides of the femur. It is subdivided into separate portions, which have received distinctive names. One occupying the middle of the thigh, and connected above with the ilium, is called from its straight course the **Rectus femoris.** The other three lie in immediate connection with the body of the femur, which they cover from the trochanters to the condyles. The portion on the lateral side

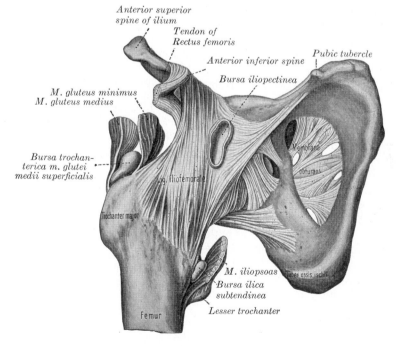

FIG. 6–59.—Muscular insertions and bursae around the hip joint (Rauber-Kopsch, *Lehrbuch u. Atlas d. Anatomie d. Menschen,* 19th Edition, Vol. I, courtesy of Georg Thieme Verlag, Stuttgart).

of the femur is termed the **Vastus lateralis;** that covering the medial side, the **Vastus medialis;** and that between, the **Vastus intermedius** (Fig. 6–56).

The **Rectus femoris** is situated in the middle of the anterior thigh; it is fusiform in shape, and its superficial fibers are arranged in a bipenniform manner, the deep fibers running straight down to the deep aponeurosis. It *arises* by two tendons: one, the anterior or straight, from the anterior inferior iliac spine; the other, the posterior or reflected, from a groove above the brim of the acetabulum. The two unite at an acute angle, and spread into an aponeurosis which is prolonged downward on the anterior surface of the muscle, and from this the muscular fibers arise. The muscle ends in a broad and thick aponeurosis which occupies the distal two-thirds of its posterior surface, and gradually becoming narrowed into a flattened tendon, is *inserted* into the base of the patella.

The **Vastus lateralis** (*Vastus externus*) is the largest part of the Quadriceps femoris. It *arises* by a broad aponeurosis, which is attached to the proximal part of the intertrochanteric line, to the anterior and inferior borders of the greater trochanter, to the lateral lip of the gluteal tuberosity, and to the proximal half of the lateral lip of the linea aspera; this aponeurosis covers the proximal three-fourths of the muscle, and from its deep surface many fibers take origin. A few additional fibers arise from the tendon of the Gluteus maximus, and from the lateral intermuscular septum between the Vastus lateralis and short head of the Biceps femoris. The fibers form a large fleshy mass, which is attached to a strong aponeurosis, placed on the deep surface of the distal part of the muscle; this aponeurosis becomes contracted and thickened into a flat tendon inserted into the lateral border of the patella, blending with the Quadriceps femoris tendon, and giving an expansion to the capsule of the knee joint.

The Vastus medialis and Vastus intermedius appear to be inseparably united, but when the Rectus femoris has been reflected a narrow interval will be observed extending proximally from the medial border of the patella between the two muscles, and the separation may be continued as far as the distal part of the intertrochanteric line, where, however, the two muscles are frequently continuous.

The **Vastus medialis** (*Vastus internus*) *arises* from the distal half of the intertrochanteric line, the medial lip of the linea aspera, the proximal part of the medial supracondylar line, the tendons of the Adductor longus and the Adductor magnus and the medial intermuscular septum. Its fibers are directed distally and anteriorly, and are chiefly attached to an aponeurosis which lies on the deep surface of the muscle and is *inserted* into the medial border of the patella and the Quadriceps femoris tendon, an expansion being sent to the capsule of the knee joint.

The **Vastus intermedius** (*Crureus*) *arises* from the front and lateral surfaces of the body of the femur in its proximal two-thirds and from the distal part of the lateral intermuscular septum. Its fibers end in a superficial aponeurosis which forms the deep part of the Quadriceps femoris tendon.

The **tendons** of the different portions of the Quadriceps unite at the distal part of the thigh, so as to form a single strong tendon which is inserted into the base of the patella, some few fibers passing over it to blend with the ligamentum patellae. More properly, the patella may be regarded as a sesamoid bone, developed in the tendon of the Quadriceps; and the ligamentum patellae, which is continued from the apex of the patella to the tuberosity of the tibia, as the proper tendon of insertion of the muscle, the medial and lateral patellar retinacula (see p. 348) being expansions from its borders. A bursa, which usually communicates with the cavity of the knee joint, is situated between the femur and the portion of the Quadriceps tendon proximal to the patella; another is interposed between the tendon and the proximal part of the anterior tibia; and a third, the **prepatellar bursa,** is placed superficial to the patella itself.

Action.—The entire Quadriceps extends the leg. The Rectus femoris also flexes the thigh.

Nerves.—Branches of the femoral nerve containing fibers from the second, third, and fourth lumbar nerves.

The **Articularis genus** (*Subcrureus*) is a small muscle, usually distinct from the Vastus intermedius, but occasionally blended with it; it *arises* from the anterior surface of the distal part of the body of the femur, and is inserted into the proximal part of the synovial membrane of the knee joint. It sometimes consists of several separate muscular bundles.

Action.—Draws the articular capsule proximally.

Nerve.—A branch of the nerve to the Vastus intermedius.

2. The Medial Femoral Muscles

Gracilis
Pectineus
Adductor longus
Adductor brevis
Adductor magnus

The **Gracilis** (Fig. 6–56) is the most superficial muscle on the medial side of the thigh. It is thin and flattened, broad proximally, narrow and tapering distally. It *arises* by a thin aponeurosis from the anterior margins of the inferior half of the symphysis pubis and the superior half of the pubic arch. The fibers run vertically downward, and end in a rounded tendon which passes posterior to the medial condyle of the femur, curves around the medial condyle of the tibia, where it becomes flattened, and is *inserted* into the upper part of the medial surface of the body of the tibia, distal to the condyle. A few of the fibers of the distal part of the tendon are prolonged into the deep fascia of the leg. At its insertion the tendon is situated immediately above that of the Semitendinosus, and its upper edge is overlapped by the tendon of the Sartorius, with which it is in part blended. It is separated from the tibial collateral ligament of the knee joint by a bursa common to it and the tendon of the Semitendinosus.

Action.—Adducts the thigh. Flexes the leg, and after it is flexed, assists in its medial rotation.

Nerve.—A branch of the anterior division of the obturator nerve containing fibers from the third and fourth lumbar nerves.

The **Pectineus** (Fig. 6–56) is a flat, quadrangular muscle, situated at the anterior part of the proximal and medial aspect of the thigh. It *arises* from the pectineal line, and to a slight extent from the surface of bone anterior to it, between the iliopectineal eminence and tubercle of the pubis, and

from the fascia covering the anterior surface of the muscle; the fibers pass downward, backward, and lateralward, to be inserted into a rough line leading from the lesser trochanter to the linea aspera.

Action.—Flexes and adducts the thigh, and rotates it medially.

Nerve.—Usually a branch of the femoral nerve containing fibers from the second, third, and fourth lumbar nerves. When an accessory obturator is present, one of its branches is distributed to the Pectineus. It may receive a branch from the obturator nerve.

Variations.—The Pectineus may consist of two incompletely separated strata; the lateral or dorsal stratum is supplied by a branch of the femoral nerve or the accessory obturator if present; the medial or ventral stratum when present is supplied by the obturator nerve. The muscle may be attached to or inserted into the capsule of the hip joint.

The **Adductor longus** (Fig. 6–60), the most superficial of the three Adductores, is a triangular muscle lying in the same plane as the Pectineus. It *arises* by a flat, narrow tendon, from the anterior pubis, at the angle of junction of the crest with the symphysis; and soon expands into a broad fleshy belly. This passes downward, backward, and lateralward, and is *inserted*, by an aponeurosis, into the linea aspera, between the Vastus medialis and the Adductor magnus, with both of which it is usually blended.

Action.—Adducts, flexes, and tends to rotate the thigh medially.

Nerve.—A branch of the anterior division of the obturator nerve containing fibers from the third and fourth lumbar nerves.

Variations.—The Adductor longus may be double, may extend to the knee, or be more or less united with the Pectineus.

The **Adductor brevis** (Fig. 6–60) is situated immediately posterior to the two preceding muscles. It is somewhat triangular in form, and *arises* by a narrow origin from the outer surface of the inferior ramus of the pubis, between the Gracilis and Obturator externus. Its fibers, passing backward, lateralward, and downward, are *inserted* by an aponeurosis into the line leading from the lesser trochanter to the linea aspera and

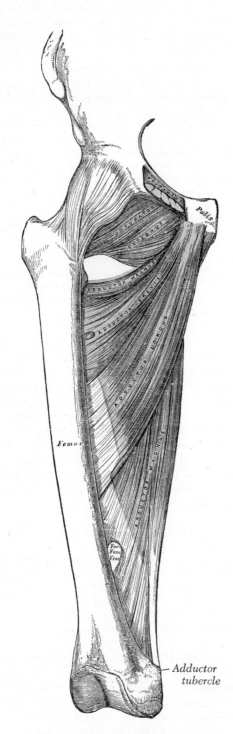

Femur

— Adductor
tubercle

Fig. 6–60.—Deep muscles of the medial
femoral region.

into the proximal part of the linea aspera,
immediately posterior to the Pectineus and
proximal part of the Adductor longus.

Action.—Adducts, flexes, and tends to rotate
the thigh medially.

Nerve.—A branch of the obturator nerve,
usually from its anterior division, containing
fibers from the third and fourth lumbar nerves.

Variations.—The Adductor brevis may be
divided into two or three parts, or it may be
united with the Adductor magnus.

The **Adductor magnus** (Fig. 6–60) is a
large triangular muscle, situated on the
medial side of the thigh. It *arises* from a
small part of the inferior ramus of the pubis,
from the inferior ramus of the ischium, and
from the outer margin of the inferior part
of the tuberosity of the ischium. Those fibers
which arise from the ramus of the pubis are
short, horizontal in direction, and are
inserted into the rough line leading from
the greater trochanter to the linea aspera,
medial to the Gluteus maximus; those from
the ramus of the ischium are directed distalward and lateralward with different degrees
of obliquity, to be *inserted,* by means of a
broad aponeurosis, into the linea aspera
and the proximal part of its medial prolongation. The medial portion of the muscle,
composed principally of the fibers arising
from the tuberosity of the ischium, forms
a thick fleshy mass consisting of coarse
bundles which descend almost vertically,
and end about the distal third of the thigh
in a rounded tendon which is inserted into
the adductor tubercle on the medial condyle
of the femur, and is connected by a fibrous
expansion to the line leading proximally
from the tubercle to the linea aspera. At
the *insertion* of the muscle, there is a series
of osseoaponeurotic openings, formed by
tendinous arches attached to the bone. The
upper four openings are small, and give
passage to the perforating branches of the
profunda femoris artery. The most distal
is of large size, the *hiatus tendineus,* and
transmits the femoral vessels to the popliteal
fossa.

Action.—The entire muscle adducts the thigh
powerfully. The upper portion rotates the thigh
medially and flexes it; the lower portion extends
it powerfully and rotates it laterally.

Nerves.—Branches of the posterior division of the obturator nerve containing fibers from the third and fourth lumbar nerves, and in addition a branch from the sciatic nerve.

Variations.—The Adductor magnus is composed of three superimposed portions. The superior is frequently distinct but the middle and inferior are usually fused. The ischiocondylar or inferior portion is derived from the flexor or hamstring muscles of lower forms and is the portion supplied by the sciatic nerve. The magnus may be fused with the Quadratus femoris, or with either the Adductor longus or brevis.

The **Adductor minimus** is the name given to the proximal portion of the Adductor magnus when it forms a distinct muscle.

3. The Muscles of the Gluteal Region
(Figs. 6–61, 6–62)

Gluteus maximus
Gluteus medius
Gluteus minimus
Tensor fasciae latae
Piriformis
Obturator internus
Gemellus superior
Gemellus inferior
Quadratus femoris
Obturator externus

The **Gluteus maximus,** the most superficial muscle in the gluteal region, is a broad and thick fleshy mass of a quadrilateral shape, and forms the prominence of the buttocks. Its large size is one of the most characteristic features of the muscular system in man, connected as it is with the power he has of maintaining the trunk in the erect posture. The muscle is remarkably coarse in structure, being made up of fasciculi lying parallel with one another and collected together into large bundles separated by fibrous septa. It *arises* from the posterior gluteal line of the ilium, and the rough portion of bone including the crest, immediately superior and dorsal to it; from the posterior surface of the lower part of the sacrum and the side of the coccyx; from the aponeurosis of the Sacrospinalis, the sacrotuberous ligament, and the fascia (gluteal aponeurosis) covering the Gluteus medius. The fibers are directed obliquely distalward and lateralward; those forming the proximal and larger portion of the muscle, together with the superficial fibers of the distal portion, end in a thick tendinous lamina, which passes across the greater trochanter, and is *inserted* into the iliotibial band of the fascia lata; the deeper fibers of the distal portion of the muscle are inserted into the gluteal tuberosity between the Vastus lateralis and Adductor magnus.

Action.—Extends and laterally rotates the thigh. Through the iliotibial band, it braces the knee when the latter is fully extended.

Nerve.—The inferior gluteal nerve, containing fibers from the fifth lumbar and first and second sacral nerves.

Bursae.—Three bursae are usually found in relation with the deep surface of this muscle. One of these, of large size, and generally multilocular, separates it from the greater trochanter; a second, often wanting, is situated on the tuberosity of the ischium; a third is found between the tendon of the muscle and that of the Vastus lateralis.

The **Gluteus medius** (Fig. 6–62) is a broad, thick, radiating muscle, situated on the outer surface of the pelvis. Its posterior third is covered by the Gluteus maximus, its anterior two-thirds by the gluteal aponeurosis, which separates it from the superficial fascia and integument. It *arises* from the outer surface of the ilium between the iliac crest and posterior gluteal line dorsally, and the anterior gluteal line ventrally; it also *arises* from the gluteal aponeurosis covering its outer surface. The fibers converge to a strong flattened tendon which is *inserted* into the oblique ridge on the lateral surface of the greater trochanter. A bursa separates the tendon of the muscle from the surface of the trochanter over which it glides.

Action.—Abducts the thigh and rotates it medially. The anterior portion flexes and rotates medially; the posterior portion extends and rotates laterally.

Nerve.—Branches of the superior gluteal nerve, containing fibers from the fourth and fifth lumbar and first sacral nerves.

Variations.—The posterior border may be more or less closely united to the Piriformis, or some of the fibers end on its tendon.

The **Gluteus minimus** (Fig. 6–62), the smallest of the three Glutei, is placed imme-

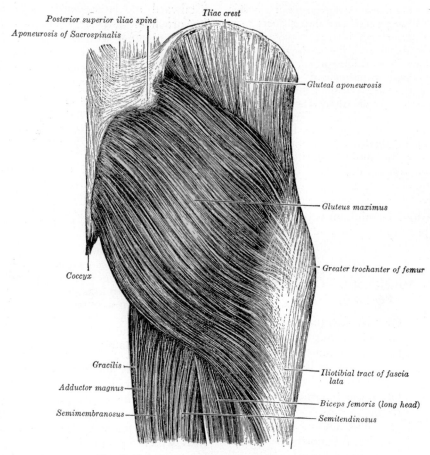

Posterior superior iliac spine

Aponeurosis of Sacrospinalis

Iliac crest

Gluteal aponeurosis

Gluteus maximus

Greater trochanter of femur

Coccyx

Gracilis

Adductor magnus

Semimembranosus

Iliotibial tract of fascia lata

Biceps femoris (long head)

Semitendinosus

Fig. 6–61.—The right Gluteus maximus muscle.

diately deep to the preceding. It is fan-shaped, *arising* from the outer surface of the ilium, between the anterior and inferior gluteal lines, and from the margin of the greater sciatic notch. The fibers converge to the deep surface of a radiated aponeurosis, and this ends in a tendon which is inserted into an impression on the anterior border of the greater trochanter, and gives an expansion to the capsule of the hip joint. A bursa is interposed between the tendon and the greater trochanter. Between the Gluteus medius and Gluteus minimus are the deep branches of the superior gluteal vessels and the superior gluteal nerve. The deep surface of the Gluteus minimus is in relation with the reflected tendon of the Rectus femoris and the capsule of the hip joint.

Action.—Rotates the thigh medially, abducts it, and to some extent, flexes it.

Nerve.—Branch of the superior gluteal nerve, containing fibers from the fourth and fifth lumbar and first sacral nerves.

Variations.—The muscle may be divided into an anterior and a posterior part, or it may send slips to the Piriformis, the Gemellus superior or the outer part of the origin of the Vastus lateralis.

The **Tensor fasciae latae** (*Tensor fasciae femoris*) (Fig. 6–56) *arises* from the anterior part of the outer lip of the iliac crest; from the outer surface of the anterior superior iliac spine, and part of the outer border of the notch below it, between the Gluteus medius and Sartorius; and from the deep surface of the fascia lata. It is *inserted* between the two layers of the iliotibial band of the fascia lata about the junction of the middle and proximal thirds of the thigh.

Action.—Flexes the thigh and, to some extent, rotates it medially.

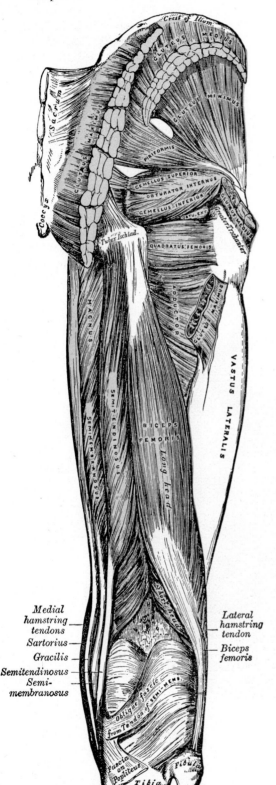

Medial
hamstring
tendons
Sartorius—
Gracilis—
Semitendinosus—
Semi-
membranosus

Lateral
hamstring
tendon
— Biceps
femoris

Fig. 6–62.—Muscles of the gluteal and posterior
femoral regions.

Nerve.—A branch of the superior gluteal nerve to the Gluteus minimus, containing fibers from the fourth and fifth lumbar and first sacral nerves.

The **Piriformis** is a flat muscle, pyramidal in shape, lying almost parallel with the posterior margin of the Gluteus medius. It is situated partly within the pelvis against its posterior wall, and partly at the back of the hip joint. It *arises* from the anterior sacrum by three fleshy digitations, attached to the portions of bone between the first, second, third, and fourth anterior sacral foramina, and to the grooves leading from the foramina: a few fibers also arise from the margin of the greater sciatic foramen, and from the anterior surface of the sacrotuberous ligament. The muscle passes out of the pelvis through the greater sciatic foramen, the upper part of which it fills, and is *inserted* by a rounded tendon into the superior border of the greater trochanter posterior to, but often partly blended with, the common tendon of the Obturator internus and Gemelli (Fig. 6–62).

Action.—Rotates the thigh laterally, abducts and, to some extent, extends it.

Nerve.—One or two branches from the second sacral or the first and second sacral nerves.

Variations.—It is frequently pierced by the common peroneal nerve and thus divided more or less into two parts. It may be united with the Gluteus medius, or send fibers to the Gluteus minimus or receive fibers from the Gemellus superior. It may have only one or two sacral attachments or be inserted into the capsule of the hip joint. It may be absent.

The **Obturator Membrane** (Fig. 5–27, page 315) is a thin fibrous sheet which almost completely closes the obturator foramen. Its fibers are arranged in interlacing bundles mainly transverse in direction; the superior bundle is attached to the obturator tubercles and completes the obturator canal for the passage of the obturator vessels and nerve. The membrane is attached to the sharp margin of the obturator foramen except at its inferior lateral angle, where it is fixed to the pelvic surface of the inferior ramus of the ischium, *i.e.*, within the margin. The two obturator muscles arise partly from the opposite surfaces of this membrane.

The **Obturator internus** (Fig. 6–63) occupies a large part of the interior surface of the anterior and lateral wall of the lesser pelvis, where it surrounds the greater part of the obturator foramen. It *arises* from the internal surface of the superior and inferior rami of the pubis, from the ramus of the ischium, and the part of the lesser pelvic wall formed by the ilium and ischium reaching from the border of the auricular surface of the ilium and the superior part of the greater sciatic notch to the obturator foramen (Fig. 5–27). It also arises from the pelvic surface of the obturator membrane except in the posterior part, from the tendinous arch which completes the canal for the passage of the obturator vessels and nerve, and to a slight extent from the obturator fascia which covers the muscle. The fibers converge rapidly toward the lesser sciatic foramen, and end in four or five tendinous bands on the deep surface of the muscle; these bands make a right angle bend around the grooved surface between the spine and tuberosity of the ischium. This bony surface is covered by smooth cartilage, which serves as a pulley and is separated from the tendon by a bursa, and presents one or more ridges corresponding with the furrows between the tendinous bands. These bands leave the pelvis through the lesser sciatic foramen and unite into a single flattened tendon, which passes horizontally across the capsule of the hip joint,

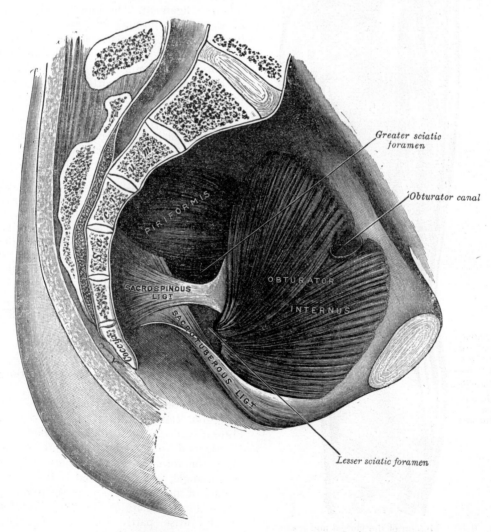

Fig. 6–63.—The left Obturator internus. Pelvic aspect.

and, after receiving the attachments of the Gemelli, is *inserted* into the anterior part of the medial surface of the greater trochanter proximal to the trochanteric fossa. A bursa, narrow and elongated in form, is usually found between the tendon and the capsule of the hip joint; it occasionally communicates with the bursa between the tendon and the ischium.

Action.—Rotates the thigh laterally; extends and abducts when the thigh is flexed.

Nerves.—A special nerve from the sacral plexus, containing fibers from the lumbosacral trunk (fifth lumbar), and first and second sacral nerves.

The **Gemelli** are two small muscles which lie parallel with the tendon of the Obturator internus after it emerges from the lesser sciatic foramen (Fig. 6–62).

The **Gemellus superior**, the smaller of the two, *arises* from the outer surface of the spine of the ischium, blends with the upper part of the tendon of the Obturator internus, and is *inserted* with it into the medial surface of the greater trochanter.

The **Gemellus inferior** *arises* from the tuberosity of the ischium, immediately inferior to the groove for the Obturator internus tendon. It blends with the lower part of the tendon of the Obturator internus, and is *inserted* with it into the medial surface of the greater trochanter.

Action.—Rotate the thigh laterally.

Nerves.—A branch of the nerve to the Obturator internus supplies the superior; a branch of the nerve to the Quadratus femoris supplies the inferior.

Variations.—The Gemelli vary in size; the superior is smaller and is more frequently absent. The inferior is more frequently bound intimately to the Obturator internus. The superior may be fused with the Piriformis or Gluteus minimus, the inferior with the Quadratus femoris.

The **Quadratus femoris** (Fig. 6–62) is a flat, quadrilateral muscle between the Gemellus inferior and the proximal margin of the Adductor magnus; it is separated from the latter by the terminal branches of the medial femoral circumflex vessels. It *arises* from the proximal part of the external border of the tuberosity of the

ischium, and is *inserted* into the proximal part of the linea quadrata—that is, the line which extends vertically distalward from the intertrochanteric crest. A bursa is often found between this muscle and the lesser trochanter.

Action.—Rotates the thigh laterally.

Nerve.—A special branch from the sacral plexus, containing fibers from the lumbosacral trunk (fourth and fifth lumbar) and first sacral nerves.

Variations.—Absence of the Quadratus femoris has been reported in 1 or 2 per cent, but this may be apparent only because it is fused either with the Gemellus inferior or the Adductor magnus. It may be double at its insertion, the posterior part having the usual attachment to the femur, the anterior part attaching to the intertrochanteric crest.

The **Obturator externus** (Fig. 6–64) is a flat, triangular muscle, which covers the outer surface of the anterior wall of the pelvis. It *arises* from the margin of bone immediately around the medial side of the obturator foramen, viz., from the rami of the pubis, and the ramus of the ischium; it also arises from the medial two-thirds of the outer surface of the obturator membrane, and from the tendinous arch which completes the canal for the passage of the obturator vessels and nerves. The fibers springing from the pubic arch extend on to the inner surface of the bone, where they obtain a narrow origin between the margin of the foramen and the attachment of the obturator membrane. The fibers converge as they pass lateralward, and end in a tendon which runs across the posterior surface of the neck of the femur and distal part of the capsule of the hip joint and is *inserted* into the trochanteric fossa of the femur. The obturator vessels lie between the muscle and the obturator membrane; the anterior branch of the obturator nerve reaches the thigh by passing superior to the muscle, and the posterior branch by piercing it.

Action.—Rotates the thigh laterally.

Nerve.—A branch of the obturator nerve, containing fibers from the third and fourth lumbar nerves.

Group Action of Muscles About the Hip Joint.—Extension of the thigh is performed by

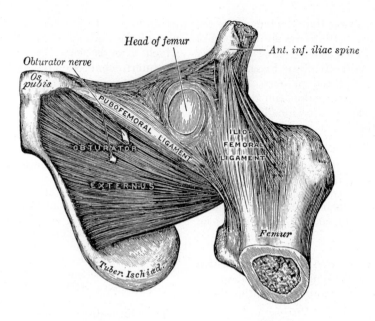

FIG. 6–64.—The Obturator externus.

the Gluteus maximus and Adductor magnus; the former is most powerful in a position of lateral rotation, the latter in medial rotation. The Gluteus medius acts synergetically to neutralize the adduction of the magnus and the lateral rotation of the maximus. The hamstring muscles extend the thigh, but are used for this action only if it accompanies flexion of the leg.

Flexion of the thigh is performed by the Iliopsoas, Tensor fasciae latae, Pectineus, and Sartorius. The action is initiated by the Tensor, Pectineus, and Sartorius; stronger and final action is by the Iliopsoas; the Adductor longus assists. Positions of medial or lateral rotation have only slight effect. The Rectus femoris flexes, but only if the muscle is acting as an extensor at the knee.

Abduction of the thigh is performed by the Gluteus medius and minimus, the latter especially in a position of medial rotation. Adduction is performed by the Adductores magnus, longus, brevis, and the Gracilis.

Lateral rotation of the thigh is performed by the Obturator externus and internus, Gemelli Piriformis, and Quadratus femoris. Other muscles, such as the Gluteus maximus, have incidental power for lateral rotation but are not used for pure rotatory action. Medial rotation is performed by the Gluteus minimus and anterior Gluteus medius. The Tensor fasciae latae, the Adductores and Iliopsoas have incidental power for this action.

4. The Posterior Femoral Muscles (Hamstring Muscles) (Fig. 6–62)

The **Biceps femoris** (*Biceps*) is situated on the posterior and lateral aspect of the thigh. It has two heads of origin; one, the **long head,** *arises* from the distal and inner impression on the posterior part of the tuberosity of the ischium, by a tendon common to it and the Semitendinosus, and from the inferior part of the sacrotuberous ligament; the other, the **short head,** *arises* from the lateral lip of the linea aspera, between the Adductor magnus and Vastus lateralis, extending up almost as high as the insertion of the Gluteus maximus; from the lateral prolongation of the linea aspera to within 5 cm of the lateral condyle; and from the lateral intermuscular septum. The fibers of the long head form a fusiform belly, which passes obliquely distalward and lateralward across the sciatic nerve to end in an aponeurosis which covers the posterior surface of the muscle, and receives the fibers of the short head; this aponeurosis becomes gradually contracted into a tendon which is *inserted* into the lateral side of the head of the fibula, and by a small slip into the lateral condyle of the tibia. At its insertion the tendon divides into two portions which embrace the fibular collateral ligament of the knee joint. From the posterior border of the tendon a thin expansion is given off to the fascia of the leg. The tendon of insertion of this muscle forms the lateral hamstring; the common peroneal nerve descends along its medial border.

Action.—Flexes the leg, and after it is flexed, rotates it laterally. The long head extends the thigh and tends to rotate it laterally.

Nerves.—The long head is supplied by branches, usually two, from the tibial portion of the sciatic nerve, containing fibers from the first three sacral nerves. The nerve to the short head comes from the peroneal portion and contains fibers from the fifth lumbar and first two sacral nerves.

Variations.—The short head may be absent; additional heads may arise from the ischial tuberosity, the linea aspera, the medial supracondylar ridge of the femur or from various other parts. A slip may pass to the Gastrocnemius.

The **Semitendinosus**, remarkable for the great length of its tendon of insertion, is situated at the posterior and medial aspect of the thigh. It *arises* from the distal and medial impression on the tuberosity of the ischium, by a tendon common to it and the long head of the Biceps femoris; it also arises from an aponeurosis which connects the adjacent surfaces of the two muscles to the extent of about 7.5 cm from their origin. The muscle is fusiform and ends a little distal to the middle of the thigh in a long round tendon which lies along the medial side of the popliteal fossa; it then curves around the medial condyle of the tibia and passes over the tibial collateral ligament of the knee joint, from which it is separated by a bursa, and is *inserted* into the proximal part of the medial surface of the body of the tibia, nearly as far anteriorly as its anterior crest. At its insertion it gives off from its distal border a prolongation to the deep fascia of the leg and lies posterior to the tendon of the Sartorius, and distal to that of the Gracilis, to which it is united. A tendinous intersection is usually observed about the middle of the muscle.

Action.—Flexes the leg, and, after it is flexed, rotates it medially; extends the thigh.

Nerves.—Usually two branches of the tibial

portion of the sciatic, containing fibers from the fifth lumbar and first two sacral nerves.

The **Semimembranosus**, so called from its membranous tendon of origin, is situated at the back and medial side of the thigh. It *arises* by a thick tendon from the proximal and outer impression on the tuberosity of the ischium, above and lateral to the Biceps femoris and Semitendinosus. The tendon of origin expands into an aponeurosis which covers the proximal part of the anterior surface of the muscle; from this aponeurosis muscular fibers arise, and converge to another aponeurosis which covers the distal part of the posterior surface of the muscle and becomes the tendon of insertion. It is *inserted* mainly into the horizontal groove on the posterior medial aspect of the medial condyle of the tibia. The tendon of insertion gives off certain fibrous expansions: one, of considerable size, passes upward and lateralward to be *inserted* into the back part of the lateral condyle of the femur, forming part of the oblique popliteal ligament of the knee joint; a second is continued distalward to the fascia which covers the Popliteus muscle, while a few fibers join the tibial collateral ligament of the joint and the fascia of the leg. The muscle overlaps the proximal part of the popliteal vessels.

Action.—Flexes the leg, and, after it is flexed, tends to rotate it medially. Extends the thigh.

Nerve.—Several branches from the tibial portion of the sciatic nerve, containing fibers from the fifth lumbar and first two sacral nerves.

Variations.—It may be reduced or absent, or double, arising mainly from the sacrotuberous ligament and giving a slip to the femur or Adductor magnus.

The tendons of insertion of the two preceding muscles form the medial hamstrings.

Group Actions at the Knee.—Extension of the leg is performed by the Quadriceps femoris; *i.e.*, the Rectus femoris, Vastus lateralis, Vastus medialis, and Vastus intermedius. Flexion is performed by the hamstring muscles; *i.e.*, the Biceps femoris, Semitendinosus, and Semimembranosus, and by the Popliteus. The Sartorius and Gracilis act in full flexion and against resistance. The Gastrocnemius has flexing action but is used more as a protective agent to prevent hyperextension. Lateral rotation of the flexed knee is performed by the Biceps femoris; medial rotation by the Popliteus and, to a lesser extent, by the medial hamstrings.

III. THE MUSCLES AND FASCIAE OF THE LEG

The **deep fascia** (*fascia cruris*) of the leg forms a complete investment to the muscles, and is fused with the periosteum over the subcutaneous surfaces of the bones. It is continuous *proximally* with the fascia lata, and is attached around the knee to the patella, the ligamentum patellae, the tuberosity and condyles of the tibia, and the head of the fibula. *Posteriorly,* it forms the popliteal fascia, covering in the popliteal fossa; here it is strengthened by transverse fibers, and perforated by the small saphenous vein. It receives an expansion from the tendon of the Biceps femoris laterally, and from the tendons of the Sartorius, Gracilis, Semitendinosus, and Semimembranosus medially; *anteriorly,* it blends with the periosteum covering the subcutaneous surface of the tibia, and with that covering the head and malleolus of the fibula; *distally,* it is continuous with the superior extensor retinaculum and flexor retinaculum. It is thick and dense in the proximal and anterior part of the leg, and gives attachment by its deep surface, to the Tibialis anterior and Extensor digitorum longus; but thinner posteriorly, where it covers the Gastrocnemius and Soleus. It gives off from its deep surface, on the lateral side of the leg, two strong intermuscular septa, the **anterior** and **posterior peroneal septa**, which enclose the Peronei longus and brevis, and separate them from the muscles of the anterior and posterior crural regions, and several more slender processes which enclose the individual muscles in each region. A broad transverse intermuscular septum, called the **deep transverse fascia of the leg,** intervenes between the superficial and deep posterior crural muscles.

The muscles of the leg may be divided into three groups: anterior, posterior, and lateral.

1. The Anterior Crural Muscles (Fig. 6–65)

The **Tibialis anterior** (*Tibialis anticus*) is situated on the lateral side of the tibia; it is thick and fleshy proximally, tendinous distally. It *arises* from the lateral condyle and proximal half or two-thirds of the lateral surface of the body of the tibia; from the adjoining part of the interosseous membrane; from the deep surface of the fascia; and from the intermuscular septum between it and the Extensor digitorum longus. The fibers run vertically distalward, and end in a tendon which is apparent on the anterior surface of the muscle at the lower third of the leg. After passing through the most medial compartments of the extensor retinacula, it is *inserted* into the medial and plantar surface of the first cuneiform bone, and the base of the first metatarsal bone. This muscle overlaps the anterior tibial vessels and deep peroneal nerve in the proximal part of the leg.

Action.—Dorsally flexes and supinates (adducts and inverts) the foot.

Nerve.—Branch of the deep peroneal nerve, containing fibers from the fourth and fifth lumbar and first sacral nerves.

Variations.—A deep portion of the muscle is rarely inserted into the talus, or a tendinous slip may pass to the head of the first metatarsal bone or the base of the first phalanx of the great toe. The *Tibiofascialis anterior* is a small muscle from the distal part of the tibia to the transverse or cruciate crural ligaments or deep fascia.

The **Extensor hallucis longus** (*Extensor proprius hallucis*) is a thin muscle, situated between the Tibialis anterior and the Extensor digitorum longus. It *arises* from the anterior surface of the fibula for about the middle two-fourths of its extent, medial to the origin of the Extensor digitorum longus; it also arises from the interosseous membrane to a similar extent. The anterior tibial vessels and deep peroneal nerve lie between it and the Tibialis anterior. The fibers pass distalward, and end in a tendon which occupies the anterior border of the muscle, passes through a distinct compartment in the inferior extensor retinaculum, crosses from the lateral to the medial side of the anterior tibial vessels near the bend of the ankle, and is *inserted* into the base of the distal phalanx of the great toe. Opposite the metatarsophalangeal articulation, the tendon gives off a thin prolongation on either side, to cover the surface of the joint. An expan-

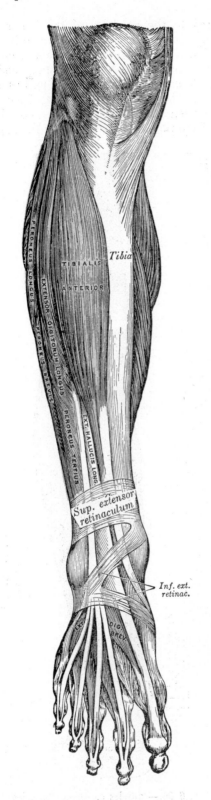

FIG. 6–65.—Muscles of the anterior leg.

sion from the medial side of the tendon is usually inserted into the base of the proximal phalanx.

Action.—Extends proximal phalanx of great toe. Dorsally flexes and supinates the foot.

Nerve.—Branch of the deep peroneal nerve, containing fibers from the fourth and fifth lumbar and first sacral nerves.

Variations.—Occasionally united at its origin with the Extensor digitorum longus. *Extensor ossis metatarsi hallucis,* a small muscle, sometimes found as a slip from the Extensor hallucis longus, or from the Tibialis anterior, or from the Extensor digitorum longus, or as a distinct muscle; it traverses the same compartment of the superior retinaculum with the Extensor hallucis longus.

The **Extensor digitorum longus** is a penniform muscle situated at the lateral part of the anterior leg. It *arises* from the lateral condyle of the tibia; from the proximal three-fourths of the anterior surface of the body of the fibula; from the proximal part of the interosseous membrane; from the deep surface of the fascia; and from the intermuscular septa between it and the Tibialis anterior on the medial, and the Peronei on the lateral side. Between it and the Tibialis anterior are the proximal portions of the anterior tibial vessels and deep peroneal nerve. The tendon passes under the extensor retinacula in company with the Peroneus tertius, and divides into four slips, which run forward on the dorsum of the foot, and are *inserted* into the second and third phalanges of the four lesser toes. The tendons to the second, third, and fourth toes are each joined, opposite the metatarsophalangeal articulation, on the lateral side by a tendon of the Extensor digitorum brevis. The tendons are inserted in the following manner: each receives a fibrous expansion from the Interossei and Lumbricales, and then spreads out into a broad aponeurosis which covers the dorsal surface of the first phalanx: this aponeurosis, at the articulation of the proximal with the middle phalanx, divides into three slips—an intermediate, which is inserted into the base of the second phalanx; and two collateral slips, which, after uniting on the dorsal surface of the second phalanx, are continued onward, to be inserted into the base of the third phalanx.

Action.—Extends the proximal phalanges of the four small toes. Dorsally flexes and pronates the foot.

Nerve.—Branches of the deep peroneal nerve, containing fibers from the fourth and fifth lumbar and first sacral nerves.

Variations.—This muscle varies considerably in the modes of origin and the arrangement of its various tendons. The tendons to the second and fifth toes may be found doubled, or extra slips are given off from one or more tendons to their corresponding metatarsal bones, or to the short extensor, or to one of the interosseous muscles. A slip to the great toe from the innermost tendon has been found.

The **Peroneus tertius** is a part of the Extensor digitorum longus, and might be described as its fifth tendon. The fibers belonging to this tendon *arise* from the distal third or more of the anterior surface of the fibula; from the distal part of the interosseous membrane; and from an intermuscular septum between it and the Peroneus brevis. The tendon, after passing under the superior and inferior extensor retinacula in the same canal as the Extensor digitorum longus, is *inserted* into the dorsal surface of the base of the metatarsal bone of the little toe (Fig. 6–68).

Action.—Dorsally flexes and pronates the foot.

Nerve.—Branch of the deep peroneal nerve, containing fibers from the fourth and fifth lumbar and first sacral nerves.

2. The Posterior Crural Muscles

The muscles of the posterior leg are subdivided into two groups—superficial and deep. Those of the superficial group constitute a powerful muscular mass, forming the calf of the leg. Their large size is one of the most characteristic features of the muscular apparatus in man, and bears a direct relation to his erect posture and his mode of locomotion.

The Superficial Group (Fig. 6–66)

    Gastrocnemius
    Soleus
    Plantaris

The **Gastrocnemius** is the most superficial muscle, and forms the greater part of the calf. It *arises* by two heads which are connected to the condyles of the femur by strong, flat tendons. The **medial** and **larger head** takes its origin from a depression at the proximal and posterior part of the medial condyle and from the adjacent part of the femur. The **lateral head** *arises* from an impression on the side of the lateral condyle and from the posterior surface of the femur

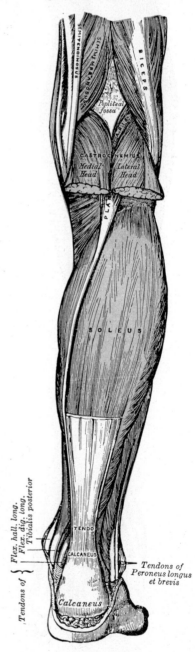

FIG. 6–66.—Muscles of the posterior leg. Superficial layer.

immediately proximal to the lateral part of the condyle. Both heads also *arise* from the subjacent part of the capsule of the knee. Each tendon spreads out into an aponeurosis which covers the posterior surface of that portion of the muscle to which it belongs. From the anterior surfaces of these tendinous expansions, muscular fibers are given off, those of the medial head being thicker and extending more distally than those of the lateral. The fibers unite at an angle in the middle line of the muscle in a tendinous raphe which expands into a broad aponeurosis on the anterior surface of the muscle, and into this the remaining fibers are inserted. The aponeurosis, gradually contracting, unites with the tendon of the Soleus, and forms with it the tendo calcaneus.

Action.—Plantar flexes the foot (points the toe); flexes the leg; tends to supinate the foot.

Nerves.—Branches of the tibial nerve, containing fibers from the first and second sacral nerves.

Variations.—Absence of the outer head of the entire muscle. Extra slips from the popliteal surface of the femur.

The **Soleus** is a broad flat muscle situated immediately deep to the Gastrocnemius. It *arises* by tendinous fibers from the posterior surface of the head of the fibula, and from the proximal third of the posterior surface of the body of the bone; from the popliteal line, and the middle third of the medial border of the tibia; some fibers also arise from a tendinous arch placed between the tibial and fibular origins of the muscle, under which the popliteal vessels and tibial nerve run. The fibers end in an aponeurosis which covers the posterior surface of the muscle, and, gradually becoming thicker and narrower, joins the *insertion* of the Gastrocnemius, and forms with it the tendo calcaneus.

Action.—Plantar flexes the foot.

Nerve.—Branch of the tibialis, containing fibers from the first and second sacral nerves.

Variations.—Accessory head to its lower and inner part usually ending in the tendo calcaneus, or the calcaneus, or the flexor retinaculum.

The Gastrocnemius and Soleus together form a muscular mass which is occasionally described as the **Triceps surae:** its tendon of insertion is the tendo calcaneus.

The **tendo calcaneus** (*tendon of Achilles*), the common tendon of the Gastrocnemius and Soleus, is the thickest and strongest in the body. It is about 15 cm long, and begins near the middle of the leg, but receives fleshy fibers on its anterior surface, almost to its distal end. Gradually becoming contracted distally, it is inserted into the middle part of the posterior surface of the calcaneus, a bursa being interposed between the tendon and the upper part of this surface. The tendon spreads out somewhat at its distal end, so that its narrowest part is about 4 cm above its insertion. It is covered by the fascia and the integument, and is separated from the deep muscles and vessels by a considerable interval filled up with areolar and adipose tissue. Along its lateral side, but superficial to it, is the small saphenous vein.

The **Plantaris** is a small muscle placed between the Gastrocnemius and Soleus. It *arises* from the distal part of the lateral prolongation of the linea aspera, and from the oblique popliteal ligament of the knee joint. It forms a small fusiform belly, from 7 to 10 cm long, ending in a long slender tendon which crosses obliquely between the two muscles of the calf, and runs along the medial border of the tendo calcaneus, to be *inserted* with it into the posterior part of the calcaneus. This muscle is sometimes double, and at other times wanting. Occasionally, its tendon is lost in the retinaculum, or in the fascia of the leg.

Action.—Plantar flexes the foot; flexes the leg.

Nerve.—Branch of the tibial nerve, containing fibers from the fourth and fifth lumbar and first sacral nerves.

The Deep Group (Fig. 6–67)

> Popliteus
> Flexor hallucis longus
> Flexor digitorum longus
> Tibialis posterior

Deep Transverse Fascia.—The deep transverse fascia of the leg is a transversely placed intermuscular septum between the superficial and deep muscles of the posterior leg. At the sides it is connected to the margins of the tibia and fibula. Proximally, where it covers the Popliteus, it is thick and dense, and receives an expansion from the tendon of the Semimembranosus; it is thinner in the middle of the leg; but distally, where it covers the tendons passing behind the malleoli, it is thickened and continuous with the flexor retinaculum (laciniate ligament).

The **Popliteus** is a thin, flat, triangular muscle, which forms the distal part of the floor of the popliteal fossa. It *arises* by a strong tendon about 2.5 cm long, from a depression at the anterior part of the groove on the lateral condyle of the femur, and to a small extent from the oblique popliteal ligament of the knee joint; and is *inserted* into the medial two-thirds of the triangular surface proximal to the popliteal line on the posterior surface of the body of the tibia, and into the tendinous expansion covering the surface of the muscle.

Action.—Flexes the leg and rotates it medially.

Nerve.—Branch of the tibial nerve, containing fibers from the fourth and fifth lumbar and first sacral nerves.

Variations.—Additional head from the sesamoid bone in the outer head of the Gastrocnemius. *Popliteus minor*, rare, origin from femur on the inner side of the Plantaris, insertion into the posterior ligament of the knee joint. *Peroneotibialis*, 14 per cent, origin inner side of the head of the fibula, insertion into the upper end of the oblique line of the tibia, it lies beneath the Popliteus.

The **Flexor hallucis longus** is situated on the fibular side of the leg. It *arises* from the inferior two-thirds of the posterior surface of the body of the fibula, except the most distal 2.5 cm, from the distal part of the interosseous membrane, and from an intermuscular septum between it and the Peronei, laterally, and from the fascia covering the Tibialis posterior, medially. The fibers pass obliquely backward, and end in a tendon which occupies nearly the whole length of the posterior surface of the muscle. This tendon lies in a groove which crosses the posterior surface of the distal end of the tibia, the posterior surface of the talus, and the under surface of the sustentaculum tali of the calcaneus; in the sole of the foot it

runs forward between the two heads of the Flexor hallucis brevis, and is *inserted* into the base of the terminal phalanx of the great toe. The grooves on the talus and calcaneus which contain the tendon of the muscle,

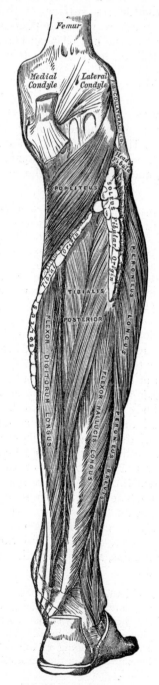

Fig. 6–67.—Muscles of the posterior of the leg. Deep layer.

are converted by tendinous fibers into distinct canals, lined by a synovial sheath. As the tendon passes distally in the sole of the foot, it is situated deep to, and crosses from the lateral to the medial side of the tendon of the Flexor digitorum longus, to which it is connected by a strong tendinous slip.

Action.—Flexes second phalanx of great toe. Plantar flexes and supinates the foot.

Nerve.—Branch of the tibial nerve, containing fibers from the fifth lumbar and first and second sacral nerves.

Variations.—Usually a slip runs to the Flexor digitorum and frequently an additional slip runs from the Flexor digitorum to the Flexor hallucis. *Peroneocalcaneus internus*, rare, origin below or outside the Flexor hallucis from the back of the fibula, passes under the sustentaculum tali with the Flexor hallucis and is inserted into the calcaneus.

The **Flexor digitorum longus** is situated on the tibial side of the leg. At its origin it is thin and pointed, but it gradually increases in size as it descends. It *arises* from the posterior surface of the body of the tibia, from immediately distal to the popliteal line to within 7 or 8 cm of its distal extremity, medial to the tibial origin of the Tibialis posterior; it also arises from the fascia covering the Tibialis posterior. The fibers end in a tendon which runs nearly the whole length of the posterior surface of the muscle. This tendon passes behind the medial malleolus in a groove common to it and the Tibialis posterior, but separated from the latter by a fibrous septum, each tendon being contained in a special compartment lined by a separate synovial sheath. It passes obliquely distalward and lateralward, superficial to the deltoid ligament of the ankle joint, into the sole of the foot (Fig. 6-70), where it crosses superficial to the tendon of the Flexor hallucis longus, and receives from it a strong tendinous slip. It then expands and is joined by the Quadratus plantae, and finally divides into four tendons, which are *inserted* into the bases of the distal phalanges of the second, third, fourth, and fifth toes, each tendon passing through an opening in the corresponding tendon of the Flexor digitorum brevis opposite the base of the proximal phalanx.

Action.—Flexes the terminal phalanges of the four small toes. Plantar flexes and supinates the foot.

Nerve.—Branch of the tibial nerve, containing fibers from the fifth lumbar and first sacral nerves.

Variations.—*Flexor accessorius longus digitorum,* not infrequent, origin from fibula, or tibia, or the deep fascia and ending in a tendon which, after passing beneath the retinaculum joins the tendon of the long flexor or the Quadratus plantae.

The **Tibialis posterior** (*Tibialis posticus*) lies between the two preceding muscles, and is the most deeply seated of the muscles on the back of the leg. It begins proximally by two pointed processes, separated by an angular interval through which the anterior tibial vessels pass to the anterior leg. It *arises* from the whole of the posterior surface of the interosseous membrane, excepting its lowest part; from the lateral portion of the posterior surface of the body of the tibia, between the commencement of the popliteal line proximally and the junction of the middle and lower thirds of the body distally; and from the proximal two-thirds of the medial surface of the fibula; some fibers also arise from the deep transverse fascia, and from the intermuscular septa separating it from the adjacent muscles. In the distal fourth of the leg its tendon passes anterior to that of the Flexor digitorum longus and lies with it in a groove behind the medial malleolus, but enclosed in a separate sheath; it next passes under the flexor retinaculum and over the deltoid ligament into the foot, and then beneath the plantar calcaneonavicular ligament. The tendon contains a sesamoid fibrocartilage, as it runs over the plantar calcaneonavicular ligament. It is *inserted* into the tuberosity of the navicular bone, and gives off fibrous expansions, one of which passes backward to the sustentaculum tali of the calcaneus, others distalward and lateralward to the three cuneiforms, the cuboid, and the bases of the second, third, and fourth metatarsal bones.

Action.—Supinates (adducts and inverts), and plantar flexes the foot.

Nerve.—Branch of the tibial nerve, containing fibers from the fifth lumbar and first sacral nerves.

3. The Lateral Crural Muscles (Fig. 6–68)

Peroneus longus
Peroneus brevis

The **Peroneus longus** is situated at the proximal part of the lateral side of the leg, and is the more superficial of the two muscles. It *arises* from the head and proximal two-thirds of the lateral surface of the body of the fibula, from the deep surface of the fascia, and from the intermuscular septa between it and the muscles on the anterior and posterior leg; occasionally also by a few fibers from the lateral condyle of the tibia. Between its attachments to the head and to the body of the fibula there is a gap through which the common peroneal nerve passes to the anterior leg. It ends in a long tendon, which runs behind the lateral malleolus, in a groove common to it and the tendon of the Peroneus brevis, posterior to which it lies; the groove is converted into a canal by the superior peroneal retinaculum, and the tendons in it are contained in a common synovial sheath. The tendon then extends obliquely forward

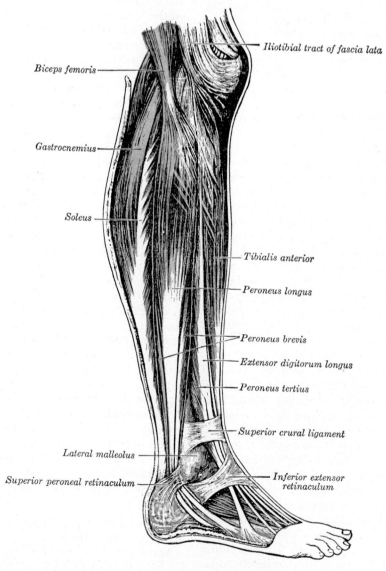

Biceps femoris

Gastrocnemius

Soleus

Iliotibial tract of fascia lata

Tibialis anterior

Peroneus longus

Peroneus brevis

Extensor digitorum longus

Peroneus tertius

Superior crural ligament

Lateral malleolus

Superior peroneal retinaculum

Inferior extensor retinaculum

FIG. 6–68.—The right lateral crural muscles.

across the lateral side of the calcaneus distal to the trochlear process, and the tendon of the Peroneus brevis, and under cover of the inferior peroneal retinaculum. It crosses the lateral side of the cuboid, and then runs on the plantar surface of that bone in a groove which is converted into a canal by the long plantar ligament; the tendon then crosses the sole of the foot obliquely, and is inserted into the lateral side of the base of the first metatarsal bone and the lateral side of the medial cuneiform. Occasionally it sends a slip to the base of the second metatarsal bone. The tendon changes its direction at two points: first, behind the lateral malleolus; secondly, on the cuboid bone; in both of these situations the tendon is thickened, and, in the latter, a sesamoid fibrocartilage (sometimes a bone), is usually developed in its substance.

Action.—Pronates (abducts and everts) and plantar flexes the foot.

Nerve.—Branch of the superficial peroneal nerve, containing fibers from the fourth and fifth lumbar and first sacral nerves.

The **Peroneus brevis** lies under cover of the Peroneus longus, and is a shorter and smaller muscle. It *arises* from the distal two-thirds of the lateral surface of the body of the fibula; medial to the Peroneus longus; and from the intermuscular septa separating it from the adjacent muscles on the anterior and posterior leg. The fibers pass vertically distalward, and end in a tendon which runs behind the lateral malleolus along with but anterior to that of the preceding muscle, the two tendons being enclosed in the same compartment, and lubricated by a common synovial sheath. It then runs distalward on the lateral side of the calcaneus, proximal to the trochlear process and the tendon of the Peroneus longus, and is *inserted* into the tuberosity at the base of the fifth metatarsal bone, on its lateral side.

On the lateral surface of the calcaneus the tendons of the Peronei longus and brevis occupy separate osseoaponeurotic canals formed by the calcaneus and the peroneal retinacula; each tendon is enveloped by a distalward prolongation of the common synovial sheath.

Action.—Pronates (everts and abducts) and plantar flexes the foot.

Nerve.—Branch of the superficial peroneal nerve, containing fibers from the fourth and fifth lumbar and first sacral nerves.

Variations.—Fusion of the two peronei is rare. A slip from the Peroneus longus to the base of the third, fourth or fifth metatarsal bone, or to the Adductor hallucis is occasionally seen.

Peroneus accessorius, origin from the fibula between the longus and brevis, joins the tendon of the longus in the sole of the foot.

Peroneus quinti digiti, rare, origin lower fourth of the fibula under the brevis, insertion into the Extensor aponeurosis of the little toe. More common as a slip of the tendon of the Peroneus brevis.

Peroneus quartus, 13 per cent (Gruber), origin back of fibula between the brevis and the Flexor hallucis, insertion into the peroneal spine of the calcaneum (*peroneocalcaneus externum*), or less frequently into the tuberosity of the cuboid (*peroneocuboideus*).

Group Actions at the Ankle.—Plantar flexion of the foot is brought about by the Gastrocnemius and Soleus, the Plantaris, the Peronei longus and brevis, and the Tibialis posterior. The Peronei and Tibialis posterior act alone if there is no resistance to be overcome. The Flexores digitorum and hallucis come into action to meet great resistance. The Gastrocnemius supinating and the Peronei pronating are synergetic.

Dorsal flexion of the foot is performed by the Tibialis anterior, Extensores digitorum and hallucis longi, and Peroneus tertius. The Peroneus brevis acts synergetically along with the Extensor digitorum and Peroneus tertius to neutralize the inversion of the Tibialis anterior and Extensor hallucis.

Supination of the Foot (combined adduction and inversion).—The Tibiales anterior and posterior are the principal supinators. The posterior adducts more powerfully; the anterior inverts more powerfully. The Gastrocnemius tends to supinate and acts synergetically against the dorsal flexion of the Tibialis anterior.

Pronation of the foot (combined abduction and eversion) is performed by the Peronei. The Peroneus brevis abducts more strongly and acts slightly before the longus; the longus everts more strongly. The Peroneus tertius acts when it is present. The Extensor digitorum longus acts as an emergency muscle.

The Fascia Around the Ankle

Fibrous bands, imbedded in or fused with the external investing layer of deep fascia, bind down the tendons situated anterior and posterior to the ankle in their passage to the foot. The identity of the fibrous bands and the direction of their component fibers is more distinct in some bodies than in others. These bands occupy positions proximal and distal to the ankle and are divisible into three groups: the retinacula for the extensor muscles, superior and inferior, the retinaculum for the flexor muscles, and the retinacula for the peroneal muscles, superior and inferior.

The **Superior Extensor Retinaculum** (*retinaculum mm. extensores superius; transverse crural ligament*) (Figs. 6–65, 6–69) binds down the tendons of the Extensor digitorum longus, Extensor hallucis longus, Peroneus tertius, and Tibialis anterior as they descend anterior to the tibia and fibula; under it are found also the anterior tibial vessels and deep peroneal nerve. It is attached laterally to the distal end of the fibula, and medially to the tibia.

The **Inferior Extensor Retinaculum** (*retinaculum mm. extensores inferius; cruciate crural ligament*) (Figs. 6–65, 6–69, 6–70) is a Y-shaped band placed anterior to the ankle joint, the stem of the Y being attached laterally to the proximal surface of the calcaneus, anterior to the depression for the interosseous talocalcanean ligament; it is directed medialward as a double layer, one lamina passing anterior, the other posterior to the tendons of the Peroneus tertius and Extensor digitorum longus. At the medial border of the latter tendon these two layers join together, forming a compartment in which the tendons are enclosed. From the medial extremity of this sheath the two limbs of the Y diverge: one is directed proximalward and medialward, to be attached to the tibial malleolus, passing over the Extensor hallucis longus and the vessels and nerves, but enclosing the Tibialis anterior by a splitting of its fibers. The other limb extends distalward and medialward, to be attached to the border of the plantar aponeurosis, and passes over the tendons of the Extensor hallucis longus and Tibialis anterior and also the vessels and nerves.

The **Flexor Retinaculum** (*retinaculum mm. flexorum; laciniate ligament*) (Fig. 6–70) is a strong fibrous band, extending from the tibial malleolus proximally to the margin of the calcaneus distally, converting a series of bony grooves in this situation into canals for the passage of the tendons of the Flexor muscles and the posterior tibial vessels and tibial nerve into the sole of the foot. It is continuous by its proximal border with the deep fascia of the leg, and by its distal border with the plantar aponeurosis and the fibers of origin of the Abductor hallucis muscle. Enumerated from the medial side, the four canals which it forms transmit the tendon of the Tibialis posterior; the tendon of the Flexor digitorum longus; the posterior tibial vessels and tibial nerve, which run through a broad space beneath the ligament; and lastly, in a canal formed partly by the talus, the tendon of the Flexor hallucis longus.

The **Superior Peroneal Retinaculum** (*retinaculum mm. peroneorum superius*) (Fig. 6–68) binds down the tendons of the Peronei longus and brevis as they run across the lateral side of the ankle. They are attached proximally to the lateral malleolus and distally to the lateral surface of the calcaneus.

The **Inferior Peroneal Retinaculum** (*retinaculum mm. peroneorum inferius*) (Fig. 6–69) is continuous with the fibers of the inferior extensor retinaculum anteriorly and is attached posteriorly to the lateral surface of the calcaneus. Some of the fibers are fixed to the peroneal trochlea, forming a septum between the tendon of the Peroneus longus and the brevis.

The **Synovial Sheaths of the Tendons Around the Ankle** (Figs. 6–69, 6–70).—All the tendons crossing the ankle joint are enclosed for part of their length in synovial sheaths which have an almost uniform length of about 8 cm each. On the anterior aspect of the ankle the sheath for the Tibialis anterior extends from the proximal margin of the superior extensor retinaculum to the interval between the diverging limbs of the inferior retinaculum; those for the Extensor digitorum longus and Extensor hallucis longus reach proximally to just above the level of the tips of the malleoli, the former being the higher. The sheath of the Extensor hallucis longus is prolonged on to the base of the first metatarsal bone, while that of the Extensor digitorum longus reaches only to the level of the base of the fifth metatar-

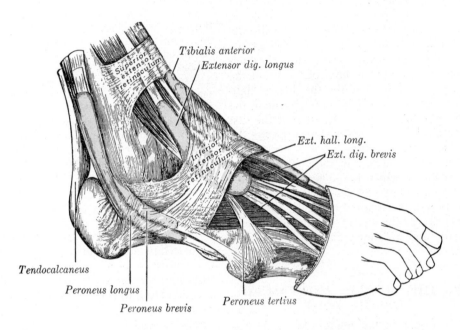

FIG. 6–69.—The synovial sheaths of the tendons around the ankle. Lateral aspect.

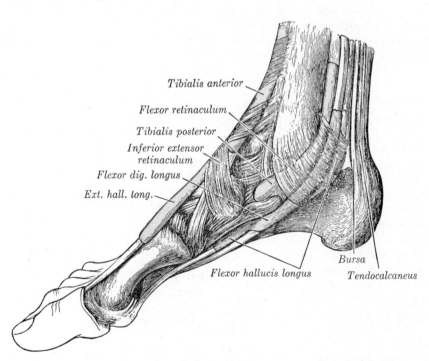

FIG. 6–70.—The synovial sheaths of the tendons around the ankle. Medial aspect.

sal. On the *medial side* of the ankle (Fig. 6–69) the sheath for the Tibialis posterior extends most proximally—to about 4 cm above the tip of the malleolus—while distally it just stops short of the tuberosity of the navicular. The sheath for the Flexor hallucis longus reaches to the level of the tip of the malleolus, while that for the Flexor digitorum longus is slightly more proximal; the former is continued to the base of the first metatarsal, but the latter stops opposite the medial cuneiform bone.

On the *lateral side* of the ankle (Fig. 6–68) a sheath which is single for the greater part of its extent encloses the Peronei longus and brevis. It extends proximally for about 4 cm above the tip of the malleolus and distally and anteriorly for about the same distance.

IV.　THE MUSCLES AND FASCIAE OF THE FOOT

The **fascia on the dorsum** of the foot is a thin membranous layer, continuous above with the superior and inferior extensor retinacula; on either side it blends with the plantar aponeurosis; anteriorly it forms a sheath for the tendons on the dorsum of the foot.

1. The Dorsal Muscle of the Foot

The **Extensor digitorum brevis** (Fig. 6–69) is a broad, thin muscle, which *arises* from the distal and lateral surfaces of the calcaneus, distal to the groove for the Peroneus brevis; from the lateral talocalcaneal ligament; and from the common limb of the cruciate crural ligament. It passes obliquely across the dorsum of the foot, and ends in four tendons. The most medial, which is the largest, is *inserted* into the dorsal surface of the base of the proximal phalanx of the great toe, crossing the dorsalis pedis artery; it is frequently described as a separate muscle—the **Extensor hallucis brevis.** The other three are *inserted* into the lateral sides of the tendons of the Extensor digitorum longus of the second, third, and fourth toes.

Action.—Extends the proximal phalanges of the great and the adjacent three small toes.

Nerve.—Branch of the deep peroneal nerve,

containing fibers from fifth lumbar and first sacral nerves.

Variations.—Accessory slips of origin from the talus and navicular, or from the external cuneiform and third metatarsal bones to the second slip of the muscle, and one from the cuboid to the third slip have been observed. The tendons vary in number and position; they may be reduced to two, or one of them may be doubled, or an additional slip may pass to the little toe. A supernumerary slip ending on one of the metatarsophalangeal articulations, or joining a dorsal interosseous muscle is not uncommon. Deep slips between this muscle and the Dorsal interossei occur.

Plantar Aponeurosis

The **plantar aponeurosis** (*aponeurosis plantaris; plantar fascia*) is of great strength, and consists of pearly white glistening fibers, disposed, for the most part, longitudinally: it is divided into central, lateral, and medial portions.

The **central portion,** the thickest, is narrow proximally and *attached* to the medial process of the tuberosity of the calcaneus, proximal to the origin of the Flexor digitorum brevis. It becomes broader and thinner distally, and divides near the heads of the metatarsal bones into five processes, one for each of the toes. Each of these processes divides opposite the metatarsophalangeal articulation into two strata, superficial and deep. The superficial stratum is *inserted* into the skin of the transverse sulcus which separates the toes from the sole. The deeper stratum divides into two slips which embrace the side of the Flexor tendons of the toes, and blend with the sheaths of the tendons and with the transverse metatarsal ligament, thus forming a series of arches through which the tendons of the short and long Flexors pass to the toes. The intervals left between the five processes allow the digital vessels and nerves and the tendons of the Lumbricales to become superficial. At the point of division of the aponeurosis, numerous transverse fasciculi are superadded; these serve to increase the strength of the aponeurosis at this part by binding the processes together, and connecting them with the integument. The central portion of the plantar aponeurosis is continuous with the lateral and medial portions and sends deeper into the foot, at the lines of junction, two strong vertical intermuscular septa, broader distally than proximally, which separate the intermediate from the lateral and medial plantar groups of muscles; from these again are derived thinner transverse septa which separate the various layers of muscles in this region. The deep surface of this aponeurosis gives origin proximally to the Flexor digitorum brevis.

The lateral and medial portions of the plantar aponeurosis are thinner than the central piece, and cover the sides of the sole of the foot.

The **lateral portion** covers the under surface of the Abductor digiti minimi; it is thin distally and thick proximally, where it forms a strong band between the lateral process of the tuberosity of the calcaneus and the base of the fifth metatarsal bone; it is continuous medially with the central portion of the plantar aponeurosis, and laterally with the dorsal fascia.

The **medial portion** is thin, and covers the under surface of the Abductor hallucis; it is *attached* behind to the flexor retinaculum, and is continuous around the side of the foot with the dorsal fascia, and laterally with the central portion of the plantar aponeurosis.

2. The Plantar Muscles of the Foot

The muscles in the plantar region of the foot may be divided into three groups, in a similar manner to those in the hand. Those of the medial plantar region are connected with the great toe, and correspond with those of the thumb; those of the lateral plantar region are connected with the little toe, and correspond with those of the little finger; and those of the intermediate plantar region are connected with the tendons intervening between the two former groups. But in order to facilitate the description of these muscles, it is more convenient to divide them into four layers, in the order in which they are successively exposed in dissection.

The First Layer (Fig. 6–71)

Abductor hallucis
Flexor digitorum brevis
Abductor digiti minimi

The **Abductor hallucis** lies along the medial border of the foot and covers the origins of the plantar vessels and nerves. It *arises* from the medial process of the tuberosity of the calcaneus, from the flexor retinaculum, from the plantar aponeurosis, and from the intermuscular septum between it and the Flexor digitorum brevis. The fibers end in a tendon which is *inserted,* together with the medial tendon of the Flexor hallucis brevis, into the tibial side of the base of the first phalanx of the great toe.

Action.—Abducts the great toe.

Nerve.—Branch of the medial plantar nerve, containing fibers from the fourth and fifth lumbar nerves.

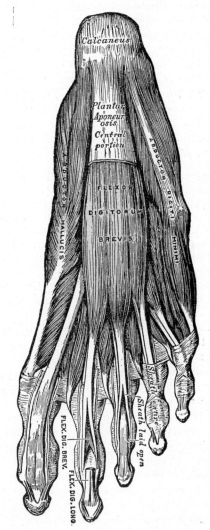

Fig. 6–71.—Muscles of the sole of the foot. First layer.

The **Flexor digitorum brevis** lies in the middle of the sole of the foot, immediately deep to the central part of the plantar aponeurosis, with which it is firmly united. Its deep surface is separated from the lateral plantar vessels and nerves by a thin layer of fascia. It *arises* by a narrow tendon, from the medial process of the tuberosity of the calcaneus, from the central part of the plantar aponeurosis, and from the intermuscular septa between it and the adjacent muscles. It passes distalward, and divides into four tendons, one for each of the four lesser toes. Opposite the bases of the first phalanges, each tendon divides into two slips, to allow the passage of the corresponding tendon of the Flexor digitorum longus; the two portions of the tendon then unite and form a grooved channel for the reception of the accompanying long Flexor tendon. Finally, it divides a second time, and is *inserted* into the sides of the second phalanx about its middle. The mode of division of the tendons of the Flexor digitorum brevis, and of their insertion into the phalanges, is analogous to that of the tendons of the Flexor digitorum superficialis in the hand.

Action.—Flexes the second phalanges of the four small toes.

Nerve.—Branch of the medial plantar nerve, containing fibers from the fourth and fifth lumbar nerves.

Variations.—Slip to the little toe frequently wanting, 23 per cent; or it may be replaced by a small fusiform muscle arising from the long flexor tendon or from the Quadratus plantae.

Fibrous Sheaths of the Flexor Tendons.— The terminal portions of the tendons of the long and short Flexor muscles are contained in osseoaponeurotic canals similar in their arrangement to those in the fingers. These canals are formed dorsally by the phalanges and on the plantar aspect fibrous bands arch across the tendons, and are attached on either side to the margins of the phalanges. Opposite the bodies of the proximal and middle phalanges the fibrous bands are strong, and the fibers are transverse; but opposite the joints they are much thinner and the fibers are directed obliquely. Each canal contains a synovial sheath which is reflected on the contained tendons.

The **Abductor digiti minimi** (*Abductor digiti quinti*) lies along the lateral border of the foot, and is in relation by its medial margin with the lateral plantar vessels and nerves. It *arises,* by a broad origin, from the lateral process of the tuberosity of the calcaneus, from the plantar surface of the calcaneus between the two processes of the tuberosity, from the distal part of the medial process, from the plantar aponeurosis, and from the intermuscular septum between it and the Flexor digitorum brevis. Its tendon, after gliding over a smooth facet on the plantar surface of the base of the fifth metatarsal bone, is *inserted,* with the Flexor digiti minimi brevis, into the fibular side of the base of the first phalanx of the little toe.

Action.—Abducts the small toe.

Nerve.—Branch of the lateral plantar nerve, containing fibers from the first and second sacral nerves.

Variations.—Slips of origin from the tuberosity at the base of the fifth metatarsal. *Abductor ossis metatarsi quinti,* origin external tubercle of the calcaneus, insertion into tuberosity of the fifth metatarsal bone in common with or beneath the outer margin of the plantar fascia.

The Second Layer (Fig. 6–72)

Quadratus plantae
Lumbricales

The **Quadratus plantae** (*Flexor accessorius*) is separated from the muscles of the first layer by the lateral plantar vessels and nerve. It *arises* by two heads, which are separated from each other by the long plantar ligament: the **medial** or **larger head** is muscular, and is attached to the medial concave surface of the calcaneus, below the groove which lodges the tendon of the Flexor hallucis longus; the **lateral head,** flat and tendinous, *arises* from the lateral border of the inferior surface of the calcaneus, distal to the lateral process of its tuberosity, and from the long plantar ligament. The two portions join at an acute angle, and end in a flattened band which is *inserted* into the lateral margin and dorsal and plantar surfaces of the tendon of the Flexor digitorum longus, forming a kind of

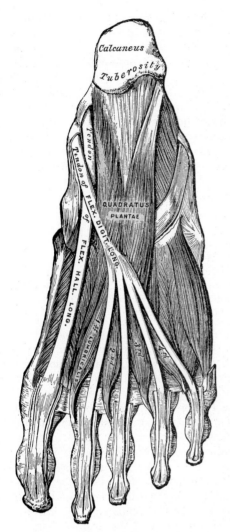

FIG. 6–72.—Muscles of the sole of the foot. Second layer.

The **Lumbricales** are four small muscles, accessory to the tendons of the Flexor digitorum longus, numbered from the medial side of the foot. They *arise* from these tendons, as far back as their angles of division, each springing from two tendons, except the first. The muscles end in tendons which pass distalward on the medial sides of the four lesser toes, and are *inserted* into the expansions of the tendons of the Extensor digitorum longus on the dorsal surfaces of the first phalanges.

Action.—Flex the proximal phalanges and extend the two distal phalanges of the four small toes.

Nerves.—The first Lumbricalis by a branch of the medial plantar nerve, containing fibers from the fourth and fifth lumbar nerves; the other three Lumbricales by branches of the lateral plantar nerve containing fibers from the first and second sacral nerves.

Variations.—Absence of one or more; doubling of the third or fourth. Insertion partly or wholly into the first phalanges.

The Third Layer (Fig. 6–73)

Flexor hallucis brevis
Adductor hallucis
Flexor digiti minimi brevis

The **Flexor hallucis brevis** *arises* by a pointed tendinous process from the medial part of the plantar surface of the cuboid bone, from the contiguous portion of the lateral cuneiform, and from the prolongation of the tendon of the Tibialis posterior which is attached to that bone. It divides distally into two portions, which are inserted into the medial and lateral sides of the base of the first phalanx of the great toe, a sesamoid bone being present in each tendon at its insertion. The **medial portion is** blended with the Abductor hallucis previous to its insertion; the **lateral portion** with the Adductor hallucis; the tendon of the Flexor hallucis longus lies in a groove between the two portions. The lateral portion is sometimes described as the **first Interosseous plantaris.**

Action.—Flexes the proximal phalanx of the great toe.

Nerve.—Branch of the medial plantar nerve, containing fibers from the fourth and fifth lumbar and first sacral nerves.

groove, in which the tendon is lodged. It usually sends slips to those tendons of the Flexor digitorum longus which pass to the second, third, and fourth toes.

Action.—Flexes the terminal phalanges of the four small toes.

Nerve.—Branch of the lateral plantar nerve, containing fibers from the first and second sacral nerves.

Variations.—Lateral head often wanting; entire muscle absent. Variation in the number of digital tendons to which fibers can be traced. Most frequent offsets are sent to the second, third and fourth toes; in many cases to the fifth as well; occasionally to two toes only.

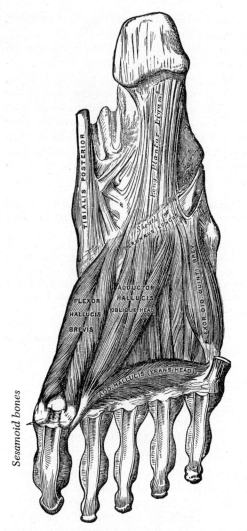

Sesamoid bones

FIG. 6–73.—Muscles of the sole of the foot. Third layer.

Variations.—Origin subject to considerable variation; it often receives fibers from the calcaneus or long plantar ligament. Attachment to the cuboid sometimes wanting. Slip to first phalanx of the second toe.

The **Adductor hallucis** (*Adductor obliquus hallucis*) *arises* by two heads—oblique and transverse. The **oblique head** is a large, thick, fleshy mass, crossing the foot obliquely and occupying the hollow space superficial to the first, second, third, and fourth metatarsal bones. It *arises* from the bases of the second, third, and fourth metatarsal bones, and from the sheath of the tendon of the Peroneus longus, and is *inserted*, together

with the lateral portion of the Flexor hallucis brevis, into the lateral side of the base of the proximal phalanx of the great toe. The **transverse head** (*Transversus pedis*) is a narrow, flat fasciculus which *arises* from the plantar metatarsophalangeal ligaments of the third, fourth, and fifth toes (sometimes only from the third and fourth), and from the transverse ligament of the metatarsus. It is *inserted* into the lateral side of the base of the first phalanx of the great toe, its fibers blending with the tendon of insertion of the oblique head.

Action.—Adducts the great toe.

Nerve.—Branch of the lateral plantar nerve, containing fibers from the first and second sacral nerves.

Variations.—Slips to the base of the first phalanx of the second toe. *Opponens hallucis,* occasional slips from the adductor to the metatarsal bone of the great toe.

The Abductor, Flexor brevis, and Adductor of the great toe, like the similar muscles of the thumb, give off, at their insertions, fibrous expansions to blend with the tendons of the Extensor hallucis longus.

The **Flexor digiti minimi brevis** (*Flexor digiti quinti brevis*) lies superficial to the metatarsal bone of the little toe, and resembles one of the Interossei. It *arises* from the base of the fifth metatarsal bone, and from the sheath of the Peroneus longus; its tendon is *inserted* into the lateral side of the base of the first phalanx of the fifth toe. Occasionally a few of the deeper fibers are inserted into the lateral part of the distal half of the fifth metatarsal bone; these are described by some as a distinct muscle, the **Opponens digiti minimi.**

Action.—Flexes the proximal phalanx of the small toe.

Nerve.—Branch of the lateral plantar nerve, containing fibers from the first and second sacral nerves.

The Fourth Layer
Interossei

The **Interossei** in the foot are similar to those in the hand, with this exception, that they are grouped around the middle line of the *second* digit, instead of that of the *third*. They are seven in number, and consist of two groups, dorsal and plantar.

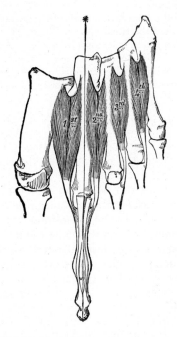

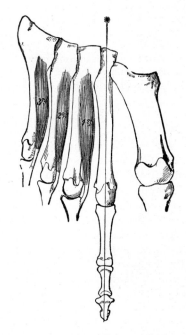

Fig. 6–74.—The Interossei dorsales. Left foot.

Fig. 6–75.—The Interossei plantares. Left foot.

The **Interossei dorsales** (*Dorsal interossei*) (Fig. 6–74), situated between the metatarsal bones, are four bipenniform muscles, each *arising* by two heads from the adjacent sides of the metatarsal bones between which it is placed; their tendons are *inserted* into the bases of the first phalanges, and into the aponeurosis of the tendons of the Extensor digitorum longus. The first is *inserted* into the medial side of the second toe; the other three are *inserted* into the lateral sides of the second, third, and fourth toes. In the angular interval left between the heads of each of the three lateral muscles, one of the perforating arteries passes to the dorsum of the foot; through the space between the heads of the first muscle the deep plantar branch of the dorsalis pedis artery enters the sole of the foot.

Action.—Abduct the toes from the longitudinal axis of the second toe. Flex the proximal and extend the distal phalanges.

Nerves.—Twigs from the deep branch of the lateral plantar nerve, containing fibers from the first and second sacral nerves.

The **Interossei plantares** (*Plantar interossei*) (Fig. 6–75), *three* in number, lie beneath rather than between the metatarsal bones, and each is connected with but one metatarsal bone. They *arise* from the bases and medial sides of the bodies of the third, fourth, and fifth metatarsal bones, and are *inserted* into the medial sides of the bases of the first phalanges of the same toes, and into the aponeuroses of the tendons of the Extensor digitorum longus.

Action.—Adduct the toes toward the axis of the second toe. Flex the proximal and extend the distal phalanges.

Nerve.—Branches of the lateral plantar nerve, containing fibers from the first and second sacral nerves.

Group Actions of Foot Muscles.—Flexion of the distal phalanges of the four small toes is performed by the Flexor digitorum longus and the Quadratus plantae. The latter is really a part of the long flexor correcting the direction of its pull. Flexion of the second phalanges is by the Flexor digitorum brevis. Flexion of the proximal phalanges is by the Interossei and Lumbricales, with added strength given to the small toe by its Flexor brevis and Abductor. Flexion of the second phalanx of the great toe is performed by the Flexor hallucis longus; of the proximal phalanx by the Flexor hallucis brevis, Abductor hallucis, and Adductor hallucis.

Extension of the distal phalanges of all toes

is performed by the Interossei, Lumbricales, Abductor digiti minimi, and Abductor hallucis. Extension of all toes is performed by the Extensor digitorum longus, Extensor hallucis longus, and Extensor digitorum brevis.

Abduction and adduction of the toes is toward the longitudinal axis of the second digit, rather than the third as in the hand. The abductors are the Interossei dorsales, Abductor hallucis, and Abductor digiti minimi. The adductors are the Interossei plantares and Adductor hallucis.

STRUCTURE OF MUSCLES (Figs. 6–76, 6–77)

The **skeletal** or **voluntary** muscles are called **striated muscles** because they consist of long thread-like fibers which appear under the microscope to be crossed by regularly placed, parallel, transverse bands or cross striations. The smallest independent units of the tissue are the muscle fibers. They are just within the limit of visibility with the naked eye, measuring from 0.01 to 0.1 mm in diameter, and from 1 mm to 12 cm in length. They are cylindrical unless crowded against each other, and have either blunt or tapering ends. They do not divide or anastomose, but may occasionally be split for a short distance near their terminations. The maximum length of fiber, except in very long muscles, is approximately 10 cm, and if the muscular fasciculi are of this length or shorter, the majority of fibers extend from one tendon to the other. If the fasciculi are longer, however, the fibers have one termination at a tendon and the other within the muscle, or in the case of a very long muscle like the Sartorius, both terminations may be within the muscle.

Each **muscle fiber** is a multinucleated cell. It has an outer membrane called the **sarcolemma** within which lie the nuclei and the contractile, cross-striated substance. The **nuclei** are flat oval discs, and usually are closely applied to the inner surface of the sarcolemma in adult human fibers. They are irregularly placed throughout the length of the fiber, and vary in number according to the size of the fiber, several hundred occurring in a larger one. About each nucleus there is a small amount of **cytoplasm** which may contain mitochondria and a Golgi complex.

The **cross-striated substance** is composed of a cytoplasmic matrix, the **sarcoplasm**, within which the long filamentous **myo-**

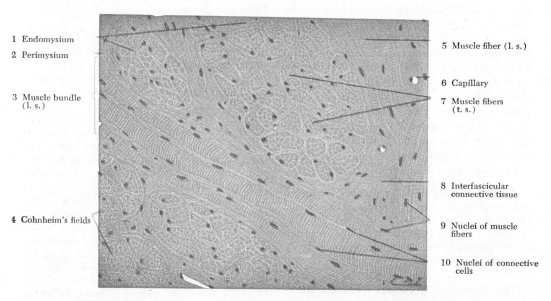

1 Endomysium

2 Perimysium

3 Muscle bundle (l. s.)

4 Cohnheim's fields

5 Muscle fiber (l. s.)

6 Capillary

7 Muscle fibers (t. s.)

8 Interfascicular connective tissue

9 Nuclei of muscle fibers

10 Nuclei of connective cells

Fig. 6–76.—Striated muscle. Muscles of the tongue. 320 ×. Stain: hematoxylin-eosin. (di Fiore, *Atlas of Human Histology,* Lea & Febiger.)

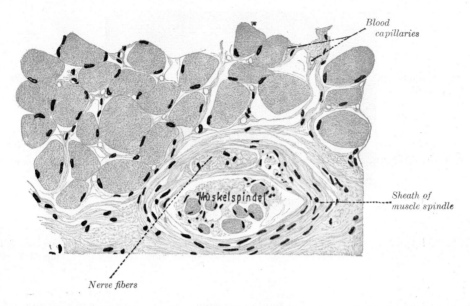

Fig. 6–77.—Cross section of part of a lumbrical muscle from a human hand showing muscle fibers, endomysium, blood capillaries, and a muscle spindle.   Approximately 400×.   (Rauber-Kopsch, *Lehrbuch u. Atlas d. Anatomie d. Menschen,* 19th Ed. Vol. I, courtesy of Georg Thieme Verlag.)

fibrillae are embedded and more or less evenly distributed across the fiber. The fibrillae impart to the fiber a fine longitudinal marking or striation which is less regular and distinct than the cross striation. They are usually 0.002 mm or less in diameter in human muscle. The cross-striated appearance is due to the fact that the myofibrillae are made up of alternating birefringent and monorefringent segments which are evenly spaced at intervals of approximately 0.003 mm, unless the fiber is distorted by contraction. During contraction the cross stripes become narrower or crowded together, and when the fiber is stretched they become wider or farther apart. The cross striations have been given various names and designations, and have been used, along with other lines, bands, disks, and membranes, as the basis for many diverse theories of muscular contraction.

The sarcoplasm contains numerous granular bodies called *sarcosomes,* as well as tiny *droplets of fat,* scattered among the myofibrillae. The fat droplets may be abundant in well-nourished individuals. Glycogen may appear to be concentrated in the sarcosomes in fixed and stained material, but in the living fiber it is diffusely distributed. The relative amount of sarcoplasm and myofibrillae varies in individual fibers. Those with a greater concentration of fibrillae are pale in appearance in the fresh condition; those with more sarcoplasm are darker. These **light and dark fibers** are intermingled in most human muscles, and are not so sharply demarcated as in animals or birds with light and dark flesh.

Most muscles are connected with bones, cartilages, or ligaments, and usually through the intervention of tendons or aponeuroses. The latter may be long, or the muscle may seem to connect directly with the bone or cartilage. Muscles may also attach to the skin (facial muscles), a mucous membrane (tongue), a fibrous plate (ocular muscles), or form circular bands (sphincter muscles).

The **attachment of the ends of the muscle fibers** is the same in principle whether the fasciculi end in a tendon or are connected directly with bones or cartilage. Each muscle fiber has a distinct termination which is rounded, conical, or truncated, and is covered by sarcolemma. The reticular fibrillae of the endomysium become thickened and are closely adherent to the end of the fiber as they pass over it, and they continue on beyond it into the tendon, where they

Fig. 6–78.—Attachment of a muscle fiber from the Flexor digitorum superficialis of a monkey. Reticular connective tissue fibrils are blackened with Masson's silver stain. Photograph. Magnified 900× (Goss).

become the actual substance of the tendon (Fig. 6–78). In muscles which appear to attach directly to bones, cartilages, or other structures, the reticular endomysium becomes continuous with the periosteum, or other fibrous layer in the same manner as with a tendon. When the muscle fibers terminate within instead of at the end of a fasciculus, as in long muscles, the ends may be either blunt or long and tapering, and they are secured in place by the merging of their terminal reticular endomysium with the endomysium of neighboring fibers.

The muscle, considered as a contractile organ, has a parenchyma composed of muscle fibers, and an intrinsic **connective tissue stroma.** Each fiber is surrounded by a very delicate, close-meshed network of reticular connective tissue fibrillae. This net, together with a few collagenous and elastic fibers which bind contiguous muscle fibers together, is known as the **endomysium.** Groups of a dozen or more fibers are brought together into bundles or **fasciculi** and enclosed by a thin lamella of collagenous and elastic fibers known as the perimysium. The **perimysium** also includes all the connective tissue which binds the fas-

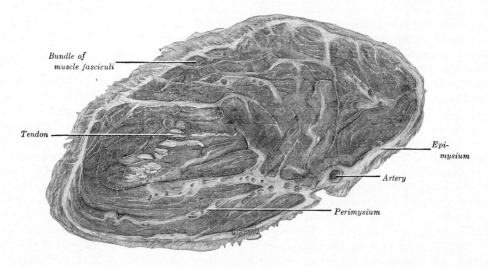

Fig. 6–79.—Cross section of Sartorius from dissected cadaver. 4×.

ciculi into larger groups and forms fibrous septa throughout the muscle. The concentration of connective tissue which envelops the entire muscle is known as the **epimysium** (Fig. 6–79). It should not be confused with the definitive fascial membranes described on page 374. It may be well developed, or quite delicate, as it is where the muscle glides freely under a strong fascial sheet, or it may lose its identity and become fused with a fascial membrane if the latter is used by the muscle fibers for their attachment.

Vessels and Nerves of Striated Muscle.— The **capillaries** of striped muscle are very abundant, and form a sort of rectangular network, the branches of which run longitudinally in the endomysium between the muscular fibers, and are joined at short intervals by transverse anastomosing branches. The larger vascular channels, arteries and veins, are found only in the perimysium between the muscular fasciculi. **Nerves** are profusely distributed to striped muscle. For mode of termination see section on Neurology. The existence of **lymphatic vessels** in striped muscle has not been ascertained, though they have been found in tendons and in the sheaths of the muscles.

REFERENCES

MUSCLE

BAST, T. H. and G. BERG. 1926. The prevention and treatment of "Charley Horse." A study of the changes taking place under this type of athletic injury. Athletic J., April, pg. 15.

GOSS, C. M. 1963. On the anatomy of muscles for beginners by Galen of Pergamon. Anat. Rec., 145, 477-502.

MARKEE, J. E., J. T. LOGUE, M. WILLIAMS, W. B. STANTON, R. N. WRENN, and L. B. WALKER. 1955. Two-joint muscles of the thigh. J. Bone Jt. Surg., 37A, 125-142.

RASCH, P. J. and R. K. BURKE. 1963. *Kinesiology and Applied Anatomy*. 2nd Edition. 503 pages, illustrated. Lea & Febiger, Philadelphia.

TANNER, J. M. 1964. *The Physique of the Olympic Athlete*. 126 pages, illustrated. George Allen and Unwin, Ltd., London.

HISTOLOGY AND EMBRYOLOGY

BINTLIFF, S. and B. E. WALKER. 1960. Radioautographic study of skeletal muscle regeneration. Amer. J. Anat., 106, 233-245.

FAWCETT, D. W. 1960. The sarcoplasmic reticulum of skeletal and cardiac muscle. Circulation, 24, 336-348.

GASSER, R. F. 1967. The development of the facial muscles in man. Amer. J. Anat., 120, 357-376.

GOSS, C. M. 1944. The attachment of skeletal muscle fibers. Amer. J. Anat., 74, 259-290.

MACKAY, B., T. J. HARROP, and A. R. MUIR, 1969. The fine structure of the muscle tendon junction in the rat. Acta Anat., 73, 588-604.

McKENZIE, J. 1962. The development of the sternomastoid and trapezius muscles. Carneg. Instn., Contr. Embryol., 37, 123-129.

SMITH, R. D. 1950. Studies on rigor mortis. I. Observations on the microscopic and submicroscopic structure. Anat. Rec., 108, 185-206.

WELLS, L. J. 1954. Development of the human diaphragm and pleural sacs. Carneg. Instn., Contr. Embryol., 35, 107-134.

MYOLOGY AND VARIATIONS

ANSON, B. J., L. E. BEATON, AND C. B. McVAY. 1938. The pyramidalis muscle. Anat. Rec., 72, 405-411.

ASHLEY, G. T. 1952. The manner of insertion of the pectoralis major muscle in man. Anat. Rec., 113, 301-307.

BABA, M. A. 1954. The accessory tendon of the abductor pollicis longus muscle. Anat. Rec., 119, 541-548.

BEATON, L. E. and B. J. ANSON. 1937. The relation of the sciatic nerve and of its subdivisions to the piriformis muscle. Anat. Rec., 70, 1-6.

BEATON, L. E. and B. J. ANSON. 1942. Variations in the origin of the m. trapezius. Anat. Rec., 83, 41-46.

DAY, M. H. and J. R. NAPIER. 1961. The two heads of flexor pollicis brevis. J. Anat. (Lond.), 95, 123-130.

FROIMSON, A. and K. S. ALFRED. 1961. Sesamoid bone of the subscapularis tendon. J. Bone Jt. Surg., 43-A, 881-884.

GEORGE, R. 1953. Co-incidence of palmaris longus and plantaris muscles. Anat. Rec., 116, 521-524.

GRAY, D. J. 1945. Some anomalous hamstring muscles. Anat. Rec., 91, 33-38.

GREIG, H. W., B. J. ANSON and J. M. BUDINGER. 1952. Variations in the form and attachments of the biceps brachii muscle. Quart. Bull., Northwestern Univ. M. School, 26, 241-244.

LANDSMEER, J. M. F. 1949. The anatomy of the dorsal aponeurosis of the human finger and its functional significance. Anat. Rec., 104, 31-44.

LEWIS, O. J. 1962. The comparative morphology of M. flexor accessorius and the associated long flexor tendons. J. Anat. (Lond.), 96, 321-333.

LOVEJOY, J. F., Jr., and T. P. HARDEN. 1971. Popliteus muscle in man. Anat. Rec., 169, 727-730.

MANTER, J. T. 1945. Variations of the interosseous muscles of the human foot. Anat. Rec., 93, 117-124.

MARTIN, B. F. 1964. Observations of the muscles and tendons of the medial aspect of the sole of the foot. J. Anat. (Lond.), 98, 437-453.

MARTIN, B. F. 1968. The origins of the hamstring muscles. J. Anat. (Lond.), *102*, 345-352.

MEHTA, H. J. and W. U. GARDNER. 1961. A study of lumbrical muscles in the human hand. Amer. J. Anat., *109*, 227-238.

REIMANN, A. F., E. H. DASELER, B. J. ANSON, and L. E. BEATON. 1944. The palmaris longus muscle and tendon. A study of 1600 extremities. Anat. Rec., *89*, 495-505.

SEIB, G. A. 1938. The m. pectoralis minor in American whites and American negroes. Amer. J. Phys. Anthrop., *23*, 389-419.

STEIN, A. H., JR. 1951. Variations of the tendons of insertion of the abductor pollicis longus and the extensor pollicis brevis. Anat. Rec., *110*, 49-55.

STRAUS, W. L., JR. 1942. The homologies of the forearm flexors: Urodeles, lizards, mammals. Amer. J. Anat., *70*, 281-316.

WRIGHT, R. R., W. GREIG, and B. J. ANSON. 1946. Accessory tendinous (peroneal) origin of the first dorsal interosseous muscle. A study of 125 specimens of lower extremity. Quart. Bull., Northwestern M. School, *20*, 339-341.

MUSCLE ACTION AND ELECTROMYOGRAPHY

BASMAJIAN, J. V. 1962. *Muscles Alive, Their Functions Revealed by Electromyography.* xi + 267 pages, illustrated. Williams & Wilkins Company, Baltimore.

BASMAJIAN, J. V. and J. W. BENTZON. 1954. An electromyographic study of certain muscles of the leg and foot in the standing position. Surg. Gynec. Obstet., *98*, 662.

BASMAJIAN, J. V. and C. R. DUTTA. 1961. Electromyography of the pharyngeal constrictors and levator palati in man. Anat. Rec., *139*, 561-563.

BASMAJIAN, J. and G. STECKO. 1963. The role of muscles in arch support of the foot. J. Bone Jt. Surg., *45-A*, 1184-1190.

BEARN, J. G. 1961. An electromyographic study of the trapezius, deltoid, pectoralis major, biceps and triceps muscles, during static loading of the upper limb. Anat. Rec., *140*, 103-107.

BEARN, J. G. 1963. The history of the ideas on the functions of the biceps brachii muscle as a supinator. Medical History, *7*, 32-42.

BOIVIN, G., G. E. WADSWORTH, J. M. LANDSMEER, and C. LONG, II. 1969. Electromyographic kinesiology of the hand: Muscles driving the index finger. Arch. Phys. Med. Habil., *50*, 17-26.

CAMPBELL, E. J. M. 1955. The role of scalene and sternomastoid muscles in breathing in normal subjects. J. Anat. (Lond.), *89*, 378-386.

DAVIS, P. R. and J. D. G. TROUP. 1966. Human thoracic diameters at rest and during activity. J. Anat. (Lond.), *100*, 397-410.

DE SOUSA, O. M., J. LACAZ DE MORAES, and L. DE MORAES VIEIRA. 1961. Electromyographic study of the brachioradialis muscle. Anat. Rec., *139*, 125-131.

DONISCH, E. W., and J. V. BASMAJIAN. 1972. Electromyography of deep back muscles in man. Amer. J. Anat., *133*, 25-36.

HRYCYSHYN, A. W., and J. V. BASMAJIAN. 1972.

Electromyography of the oral stage of swallowing in man. Amer. J. Anat., *133*, 333-340.

JONES, D. S., R. J. BEARGIE, and J. E. PAULY. 1953. An electromyographic study of some muscles of costal respiration in man. Anat. Rec., *117*, 17-24.

LEHR, R. P., P. L. BLANTON, and N. L. BIGGS. 1971. An electromyographic study of the mylohyoid muscle. Anat. Rec., *169*, 651-659.

LORD, F. P. 1937. Movements of the jaw and how they are effected. Internat. J. Orthodontia, *23*, 557-571.

MARMOR, L., C. D. BECHTOL, and C. B. HALL. 1961. Pectoralis major muscle. Function of sternal portion and mechanism of rupture of normal muscle. J. Bone Jt. Surg., *43-A*, 81-87.

MATHESON, A. B., D. C. SINCLAIR, and W. G. SKENE. 1970. The range and power of ulnar and radial deviation of the fingers. J. Anat. (Lond.), *107*, 439-458.

MORRIS, J. M., G. BENNER, and D. B. LUCAS. 1962. An electromyographic study of the intrinsic muscles of the back in man. J. Anat. (Lond.), *96*, 509-520.

MOYERS, R. E. 1950. An electromyographic analysis of certain muscles involved in temporomandibular movement. Amer. J. Orthodont., *36*, 481.

STEENDIJK, R. 1948. On the rotating function of the iliopsoas muscle. Acta Neerl. Morph. Norm. et Pathol., *6*, 175.

SULLIVAN, W. E., O. A. MORTENSEN, M. MILES, and L. S. GREENE. 1950. Electromyographic studies of M. biceps brachii during normal voluntary movement at the elbow. Anat. Rec., *107*, 243-252.

SUNDERLAND, S. 1945. The actions of the extensor digitorum communis, interosseous and lumbrical muscles. Amer. J. Anat., *77*, 189-217.

TRAVILL, A. and J. V. BASMAJIAN. 1961. Electromyography of the supinators of the forearm. Anat. Rec., *139*, 557-560.

TRAVILL, A. A. 1962. Electromyographic study of the extensor apparatus of the forearm. Anat. Rec., *144*, 373-376.

TENDON SHEATHS AND TENDONS

CUMMINS, E. J., B. J. ANSON, B. W. CARR, and R. R. WRIGHT. 1946. The structure of the calcaneal tendon (of Achilles) in relation to orthopedic surgery. Surg. Gynec. Obstet., *83*, 107-116.

GRODINSKY, M. 1930. A study of the tendon sheaths of the foot and their relation to infection. Surg. Gynec. Obstet., *51*, 460-468.

KANAVEL, A. B. 1939. *Infections of the Hand.* 503 pages. 7th edition. Lea & Febiger, Philadelphia.

KAPLAN, E. B. 1945. Surgical anatomy of the flexor tendons of the wrist. J. Bone Jt. Surg., *27*, 368-372.

LEWIS, O. J. 1964. The tibialis posterior tendon in the primate foot. J. Anat. (Lond.), *98*, 209-218.

(MUSCLE) EXPERIMENTAL

COMER, R. D. 1956. An experimental study of the "laws" of muscle and tendon growth. Anat. Rec., *125*, 665-682.

GAY, A. J., JR. and T. E. HUNT. 1954. Reuniting of skeletal muscle fibers after transection. Anat. Rec., *120*, 853-872.

POGOGEFF, I. A. and M. R. MURRAY. 1946. Form and behavior of adult mammalian skeletal muscle in vitro. Anat. Rec., *95*, 321-335.

SZEPSENWOL, J. 1946. A comparison of growth, differentiation, activity and action currents of heart and skeletal muscle in tissue culture. Anat. Rec., *95*, 125-146.

BLOOD VESSELS AND NERVES OF MUSCLES

HARRISON, V. F., and O. A. MORTENSEN. 1962. Identification and voluntary control of single motor unit activity in the tibialis anterior muscle. Anat. Rec., *144*, 109-116.

HOLLINSHEAD, W. H. and J. E. MARKEE. 1946. The multiple innervation of limb muscles in man. J. Bone Jt. Surg., *28*, 721-731.

MARKEE, J. E., W. B. STANTON, and R. N. WRENN. 1952. The intramuscular distribution of the nerves to the muscles of the inferior extremity. Anat. Rec., *112*, 457.

MORRISON, A. B. 1954. The levatores costarum and their nerve supply. J. Anat. (Lond.), *88*, 19-24.

ROBBINS, H. 1963. Anatomical study of the median nerve in the carpal tunnel and etiologies of the carpal-tunnel syndrome. J. Bone Jt. Surg., *45-A*, 953-965.

SUNDERLAND, S. 1945. The innervation of the flexor digitorum profundus and lumbrical muscles. Anat. Rec., *93*, 317-321.

SUNDERLAND, S. 1946. The innervation of the first dorsal interosseous muscle of the hand. Anat. Rec., *95*, 7-10.

TELFORD, I. R. 1941. Loss of nerve endings in degenerated skeletal muscles of young vitamin E deficient rats. Anat. Rec., *81*, 171-182.

FASCIA

HEAD, NECK, AND THORAX

COLLER, F. A. and L. YGLESIAS. 1937. The relation of the spread of infection to fascial planes in the neck and thorax. Surgery, *1*, 323-333.

GAUGHRAN, G. R. L. 1964. Suprapleural membrane and suprapleural bands. Anat. Rec., *148*, 553-559.

GRODINSKY, M. and E. A. HOLYOKE. 1938. The fasciae and fascial spaces of the head, neck, and adjacent regions. Amer. J. Anat., *63*, 367-408.

HENKE, JAK. WILHELM. 1872. Untersuchung der Ausbreitung des Bindegewebes mittelst Kunstlicher. Infiltration. Beiträge zur Anatomie des Menschen mit Beziehung auf Bewegung. Hefte 1. C. F. Winter'schen Verlags duch handlung, Leipzig.

LEE, F. C. 1941. Description of a fascia situated between the serratus anterior muscle and the thorax. Anat. Rec., *81*, 35-41.

LEE, F. C. 1944. Note on a fascia underneath the pectoralis major muscle. Anat. Rec., *90*, 45-49.

ABDOMEN, WALL, AND INGUINAL REGION

ANSON, B. J., E. H. MORGAN, and C. B. MCVAY. 1960. Surgical anatomy of the inguinal region based upon a study of 500 body halves. Surg. Gynec. Obstet., *111*, 707-725.

CHANDLER, S. B. 1950. Studies on the inguinal region. III. The inguinal canal. Anat. Rec., *107*, 93-102.

CHOUKE, K. S. 1935. The constitution of the sheath of the rectus abdominis muscle. Anat. Rec., *61*, 341-349.

CONGDON, E. D., J. N. EDSON, and S. YANITELLI. 1946. Gross structure of the subcutaneous layer of the anterior and lateral truck in the male. Amer. J. Anat., *79*, 399-429.

COOPER, G. W. 1952. Fascial variants of the trigonum lumbale (Petití). Anat. Rec., *114*, 1-8.

DOYLE, J. F. 1971. The superficial inguinal arch. A re-assessment of what has been called the inguinal ligament. J. Anat. (Lond.), *108*, 297-304.

MCVAY, C. B. and B. J. ANSON. 1940. Composition of the rectus sheath. Anat. Rec., *77*, 213-225.

TOBIN, C. E. 1944. The renal fascia and its relation to the transversalis fascia. Anat. Rec., *89*, 295-311.

TOBIN C. E. and J. A. BENJAMIN. 1949. Anatomic and clinical re-evaluation of Camper's, Scarpa's and Colles' fasciae. Surg. Gynec. Obstet., *88*, 545-559.

TOBIN, C. E., J. A. BENJAMIN, and J. C. WELLS. 1946. Continuity of the fasciae lining the abdomen, pelvis, and spermatic cord. Surg. Gynec. Obstet., *83*, 575-596.

PELVIS AND PERINEUM

DAVIES, J. W. 1934. The pelvic outlet—its practical application. Surg. Gynec. Obstet., *58*, 70-78.

MCVAY, C. B. and B. J. ANSON. 1940. Aponeurotic and fascial continuities in the abdomen, pelvis and thigh. Anat. Rec., *76*, 213-231.

ROBERTS, W. H., J. HABENICHT, and G. KRISHINGNER. 1964. The pelvic and perineal fasciae and their neural and vascular relationships. Anat. Rec., *149*, 707-720.

ROBERTS, W. H., and W. H. TAYLOR. 1970. The presacral component of the visceral pelvic fascia and its relation to the pelvic splanchnic innervation of the bladder. Anat. Rec., *166*, 207-212.

TOBIN, C. E. and J. A. BENJAMIN. 1944. Anatomical study and clinical consideration of the fasciae limiting urinary extravasation from the penile urethra. Surg. Gynec. Obstet., *79*, 195-204.

TOBIN, C. E. and J. A. BENJAMIN. 1945. Anatomical and surgical restudy of Denonvilliers' fascia. Surg. Gynec. Obstet., *80*, 373-388.

WASHBURN, S. L. 1957. Ischial callosities as sleeping adaptations. Amer. J. Phys. Anthrop., *15*, 269-276.

WESSON, M. B. 1953. What are Buck's and Colles' fasciae? J. Urol., *70*, 503-511.

UHLENHUTH, E. 1953. *Problems in the Anatomy of the Pelvis.* xiv + 206 pages, illustrated. J. B. Lippincott Co., Philadelphia.

FASCIA OF UPPER LIMB

CONGDON, E. D. and H. S. FISH. 1953. The chief insertion of the bicipital aponeurosis is on the ulna. A study of collagenous bundle patterns of antebrachial fascia and bicipital aponeurosis. Anat. Rec., *116*, 395-408.

DORLING, G. C. 1944. Fascial sling of the scapula and clavicle for dropped shoulder and winged scapula. Brit. J. Surg., *32*, 311-315.

GRAYSON, J. 1941. The cutaneous ligaments of the digits. J. Anat. (Lond.), *75*, 164-165.

GRODINSKY, M. and E. A. HOLYOKE. 1941. The fasciae and fascial spaces of the palm. Anat. Rec., *79*, 435-451.

JONES, F. W. 1942. *The Principles of Anatomy As Seen in the Hand.* 2nd Edition. x + 417 pages. Williams & Wilkins, Publishers, Baltimore.

KAPLAN, E. B. 2nd edition. 1953. *Functional and Surgical Anatomy of the Hand.* xiv + 337 pages, 182 figures. J. B. Lippincott Co., Philadelphia.

LARSEN, R. D. and J. L. POSCH. 1958. Dupuytren's contracture. J. Bone Jt. Surg., *40-A*, 773-792.

FASCIA OF LOWER LIMB

ANSON, B. J. and C. B. McVAY. 1938. The fossa ovalis, and related blood vessels. Anat. Rec., *72*, 399-404.

BELLOCQ, P. and P. MEYER. 1957. Contribution a l'étude de l'aponévrose dorsale du pied (Fascia dorsalis pedis, P.N.A.). Acta Anat., *30*, 67-80.

GRODINSKY, M. 1929. A study of the fascial spaces of the foot and their bearing on infections. Surg. Gynec. Obstet., *49*, 737-751.

KAMEL, R. and F. B. SAKLA. 1961. Anatomical compartments of the sole of the human foot. Anat. Rec., *140*, 57-60.

KAPLAN, E. B. 1958. The iliotibial tract. Clinical and morphological significance. J. Bone Jt. Surg., *40-A*, 817-832.

7 | *The Heart*

THE CARDIOVASCULAR SYSTEM AS A WHOLE

uiding the circulating blood through all parts of the body is a system of vessels composed of the heart, the arteries, the veins, and the capillaries. The **heart,** by its rhythmic contraction, propels the blood through the vessels of all parts of the body. The vessels conducting blood away from the heart are the **arteries.** They ramify greatly, becoming progressively smaller, and end in minute vessels, the **arterioles.** From these vessels the blood is able to carry out its nutrient and absorbing functions by passing through a network of microscopic channels, called **capillaries,** whose walls are very thin and therefore allow the blood to exchange substances with the tissues. From the capillaries, the blood is collected into **venules;** then through **veins** of progressively greater diameter it reaches the heart again. This passage of the blood through the heart and blood vessels is termed the **circulation of the blood.**

The powerful pumping portions of the heart are the two **ventricles,** right and left, separated by a muscular septum. The thick-walled ventricles are made to function efficiently by being quickly and forcibly filled with blood by the contraction of the **atria,** also a right and left to correspond with the ventricles. As the blood is returned to the heart through the veins of the body, it enters the **right atrium** and is forced into the **right ventricle.** From here it is pumped into the capillaries of the lungs through the pulmonary arteries and there is refreshed by giving up carbon dioxide and absorbing oxygen. It is returned by the pulmonary vein to the **left atrium** which forces it into the **left ventricle.** The left ventricle propels

the blood through the aorta and systemic arteries, through the capillaries, and back to the heart again through the veins. The circulation through the **right side of the heart** and the lungs is known as the **lesser** or **pulmonary circulation.** The circulation through the **left heart** and **systemic arteries** and **veins** is known as the **systemic circulation.**

Although the circuit away from the heart and back again characteristically involves only one set of capillaries, an exception is found in the vessels of the abdominal organs. The blood supplied to the spleen, pancreas, stomach, and intestines by the systemic arteries is collected into a large vein, the **portal vein, which enters the liver** and ramifies within it. As the blood passes through the capillary-like **sinusoids,** it exchanges nutrient materials with the liver cells, and is then collected into the hepatic veins which empty into the large systemic vena cava inferior just before it opens into the right atrium. This is known as the **portal circulation.**

The description of the cardiovascular system is divided into three parts for convenience in description, but all parts are, of course, completely dependent upon each other functionally. The three sections deal with the heart, the arteries, and the veins. The capillaries are of more or less uniform structure throughout the body and belong to the subject of histology.

A time-honored division of the subject of Anatomy is that of the Vascular System. Its study is called **angiology** (*angiologia*). It includes two groups of vessels, the blood vascular system and the lymphatic system. Although the lymphatic system drains into the veins, making the two systems interdependent, there are sufficient differences morphologically and functionally to make separate treatment advisable. The lymphatic system is described in the chapter following the cardiovascular system.

18

Development of the Heart

The first **primordium** of the **vascular system** can be recognized very soon after the first appearance of the **coelom** or body cavity within the embryo. The latter occurs on about the twentieth day, at a time when the mesoderm has been formed by the primitive streak but before it has been organized into the primitive segments or somites (See Chapter 1). This first coelom is a U-shaped cavity encircling the cephalic end of the neural folds (Fig. 7–5). It is the **primordium of the pericardial cavity.** Its outer layer, the **somatic mesoderm**, will become the parietal pericardium; its inner layer, the **splanchnic mesoderm**, will develop into the myocardium and epicar-

dium. Scattered cells of mesodermal origin, called **mesenchyme** (*mesenchyma*), migrate between the compact laminae of the splanchnic mesoderm and the entoderm. This mesenchyme proliferates rapidly and forms into strands and sheets which are the **primordium of the endocardium.**

The earliest stages of development of the human heart are very imperfectly known because of the scarcity of human embryos of this age. It is necessary, therefore, to fill in the gaps with information obtained from lower forms. The chick embryo has been a favorite object for these pioneering studies because of its ready accessibility, but the available human embryos have shown that their development follows the pattern of other mammals rather than that of birds. Thus the ventricle of the

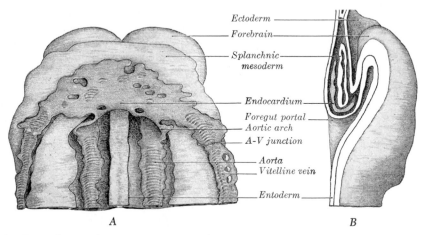

Fig. 7–1.—Rat embryo with 3 somites. *A*, Ventral view with entoderm removed. Lumen of heart and vessels is incomplete. *B*, Median sagittal section of same embryo. Drawings of wax reconstruction made at 400× magnification.

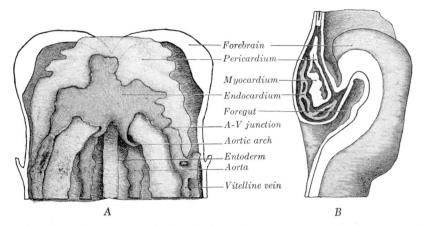

Fig. 7–2.—Rat embryo with 4 somites. *A*, Ventral view with entoderm removed. Compare endocardium with that in Figure 7–1. *B*, Median sagittal section of same embryo.

early mammalian heart is at first more saccular than tubular, and although it absorbs the two lateral tubes, it is differentiated primarily from tissue already present in the median region, not formed by the coalescence of two independent lateral parts, as in the chick. The following account of this early period is based on numerous rat embryos (Goss '52), both living and preserved, which have been used to elucidate the unknown period between an early human embryo containing the earliest stage of the pericardial cavity (Fig. 7–5) and another human embryo with a rather well-developed bent tubular heart (Fig. 7–6) (Davis '27). It is convenient to use the number of the somites as an index of the stage of development because (*a*) they can be counted accurately, (*b*) they retain a relatively stable relationship to the rest

of the embryo, and (*c*) the passage of time can be estimated since each new somite between the third and tenth requires three or four hours for its appearance.

By the time the embryo has acquired its **first three somites** or primitive body segments and before the neural folds have closed into the neural tube, the **endocardial mesenchyme** has spread out in a thin sheet across the midline in the cranial portion of the embryo (Fig. 7–1). It also extends down on each side of the still shallow foregut invagination and a lumen has begun to form in these lateral extensions. The splanchnic mesoderm becomes thickened where it is in contact with the endocardium and is thus

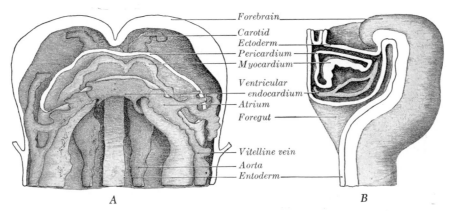

FIG. 7–3.—Rat embryo with 5 somites. *A*, Ventral view with entoderm removed. Ventral part of pericardium cut away to show myocardium. Dotted line indicates extent of endocardium inside ventricle. *B*, Median sagittal section of same embryo. Compare with Figures 7–1 and 7–2.

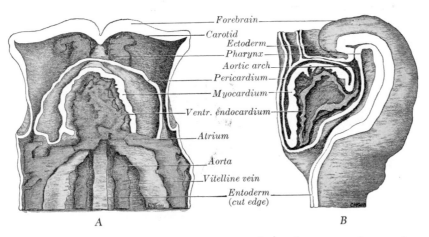

FIG. 7–4.—Rat embryo with 6 somites. *A*, Ventral view with entoderm removed. Ventral part of pericardium and myocardium also removed to show endocardium. *B*, Median sagittal section of same embryo (Goss '52).

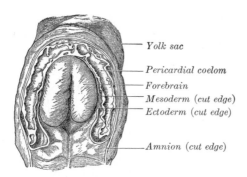

Yolk sac
Pericardial coelom
Forebrain
Mesoderm (cut edge)
Ectoderm (cut edge)

Amnion (cut edge)

FIG. 7–5.—Human embryo with first somite forming, 1.5 mm in length. Dorsal view with amnion removed and ectoderm and mesoderm cut away to show pericardial cavity. Carnegie collection No. 5080. (Redrawn from Davis '27, courtesy Carnegie Institution.)

identifiable as the **primordium of the myocardium.** The myocardial layer is also deeply grooved along the lateral endocardial tubes. Thus, the two lateral endocardial tubes with their partial cloak of myocardium constitute the **primitive lateral hearts** where the first contractions occur. The site of the future atrioventricular junction is marked by a constriction of the lumen of the lateral tube slightly caudal to the level of the foregut portal.

The **median sheet of mesenchyme** is quickly differentiated into endocardium by having the lumen extend into it from the lateral tubes. At the same time, the sheet appears to pull in its outlying parts in much the same manner as an amoeba withdraws its pseudopodia. The result is a sac rather

than a tube, which, with its adjacent myocardial mesoderm, constitutes the **primitive ventricle.** As the endocardial sac expands, it sinks deeply into a pocket in the mesoderm. The pericardial cavity enlarges to accommodate this growth and in a **four somite embryo** the entire cardiac complex bulges ventrally at the foregut portal (Fig. 7–2).

Up to this time the myocardium has been entirely dorsal to the endocardium. The rapid expansion of the pericardial cavity in a ventral direction allows the myocardium to protrude ventrally over the endocardial sac, until, in a **five somite embryo,** the heart lies in a plane at right angles to the axis of the embryo with the arterial end more dorsally placed and the venous end more ventrally placed (Fig. 7–3).

By a continued rapid growth, the ventricular myocardium extends over the entire ventral surface of the endocardium (Fig. 7–4). In a **six somite embryo** it encloses the endocardium except for a narrow interval along the midline dorsally. This change allows the heart to assume a more tubular shape with cranial and caudal extremities. It never quite reaches the stage of a straight tube, as in lower forms, however. The opening of the aortic sac is always dorsal rather than cranial to the heart and the whole tube maintains a curvature with ventral convexity. The result is the marked ventral protrusion of the cardiac complex which is characteristic of early mammalian embryos. When the embryo has acquired **seven**

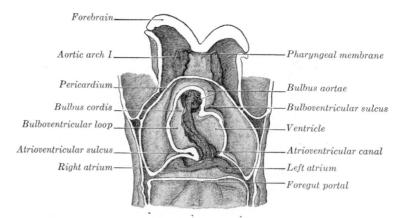

Forebrain
Aortic arch I
Pericardium
Bulbus cordis
Bulboventricular loop
Atrioventricular sulcus
Right atrium

Pharyngeal membrane
Bulbus aortae
Bulboventricular sulcus
Ventricle
Atrioventricular canal
Left atrium
Foregut portal

FIG. 7–6.—Human embryo with 8 somites, 2 mm in length. Ventral view of plaster model of heart and pericardial region. Pericardial wall and myocardium removed to expose endocardium. (Redrawn from Davis '27. Courtesy of Carnegie Institution.)

somites, the myocardium completely surrounds the endocardium. The middle portion of the tubular ventricle becomes free and only the ends are attached, the arterial end by its continuity with the aortic sac and first aortic arches and the venous end by the atria and vitelline veins.

The length of the tubular heart increases much more rapidly than the longitudinal extent of the pericardium. The tube therefore continues to bend, this time in a loop with convexity to the right (**bulboventricular loop**) (Fig. 7–6). The resulting groove on the left side is called the **bulboventricular sulcus** and establishes a subdivision of the tube into the **bulbus cordis** or **conus** and the **primitive ventricle**. The ventricular portion increases rapidly in all diameters,

protruding ventrally and causing another bend at the atrioventricular junction. The opening of the atrium becomes narrowed into the **atrioventricular canal**, entering the ventricle dorsally and from the left. The pericardial cavity expands out over the two lateral portions of the atrium which are then brought closer together and a constriction is formed between the veins and the atrium (Fig. 7–7).

The proximal portions of the veins become part of a common chamber and are called the right and left **horns of the sinus venosus**. The umbilical and common cardinal veins (from the embryo proper) have joined the vitelline veins just before they enter the sinus horns. The continued bending of the bulboventricular loop folds

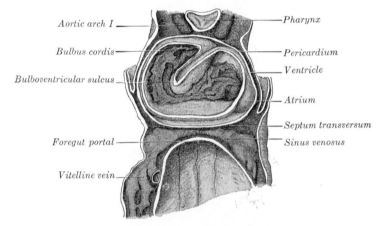

FIG. 7–7.—Human embryo with 11 to 12 somites, 3.09 mm in length. Ventral view of plaster model of heart and pericardial region. Pericardial wall and myocardium removed to expose endocardium. (Redrawn from Davis '27. Courtesy of Carnegie Institution.)

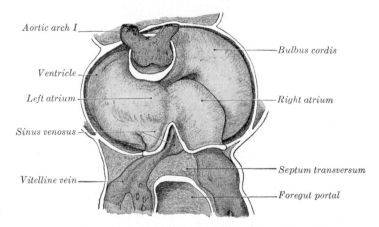

FIG. 7–8.—Human embryo with 20 somites, 3.01 mm in length. Dorsal view of plaster model of heart. (Redrawn from Davis '27. Courtesy of Carnegie Institution.)

the conus against the atrium on the dorsal and cranial aspect of the heart. The conus has remained a relatively narrow tube and as it now presses against the median region of the atrium, the lateral portions bulge out on each side. The resulting two expansions of the atrium are the **primitive right** and

both sides and is thus the **precursor of the inferior vena cava.** The resultant enlargement of the right horn of the sinus venosus shifts the sinoatrial opening to the right. The right anterior cardinal vein also becomes dominant and shifts the opening somewhat cranialward (Fig. 9–7).

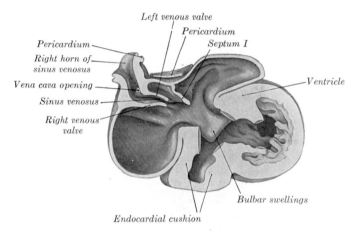

Fig. 7–9.—Model of heart of 6.5-mm human embryo, interior of lower half seen from above. (From J. Tandler, 1912, courtesy of J. B. Lippincott Company.)

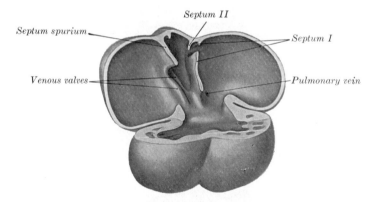

Fig. 7–10.—Model of heart of 9-mm human embryo, ventral view of interior of atria. (From J. Tandler, 1912, courtesy of J. B. Lippincott Company.)

left atria (Fig. 7–13). They represent the first division of the heart into its permanent right and left sides.

The tissue between the pericardial cavity and the foregut portal becomes thickened into a mass called the **septum transversum** (Fig. 7–8). As the liver cords grow into the septum there is a reorganization of the veins. The right vein gradually takes over the drainage of blood from the vitelline, umbilical, and posterior cardinal circulations of

In the narrow part of the atrium between the two expansions to the right and left of the conus, a fold of endocardial tissue grows toward the atrioventricular opening. This is known as **septum I** or the **septum primum.** The shifting of the sinuatrial opening to the right, already mentioned, places this opening to the right of the newly forming septum (Fig. 7–9). As the septum primum progresses toward the ventricles, two swellings appear in the wall of the atrioventricular canal.

They soon join across the opening, making a division into a right and left atrioventricular opening. This is the **septum intermedium.** The narrowing open space between the advancing septum primum and the septum intermedium is the **ostium I** or **ostium primum** (Fig. 7–10). Before the septum primum completely closes this opening between the two atria, which would shut off all flow of blood between them, the

After septum I has joined with the septum intermedium its remaining part becomes thinned out and acts as a flap over the foramen ovale below the crescentic edge of septum II (**valve of the foramen ovale**). After birth, when the foramen ovale is no longer functional, this flap becomes adherent to the limbus and seals the opening. During this development the **pulmonary veins** grow out from the left atrium but

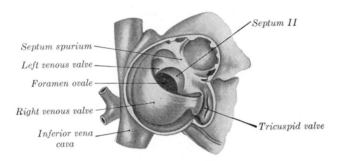

FIG. 7–11.—Interior of right atrium of reconstruction of heart of 25-mm human embryo. (R. H. Licata, Am. J. Anat., *94*, 80, 1954, courtesy of Wistar Press.)

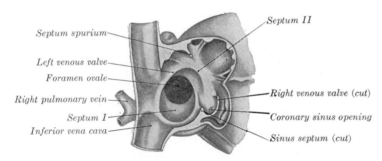

FIG. 7–12.—Interior of right atrium of same heart as in Figure 7–11 with right venous valve removed. (R. H. Licata, Am. J. Anat., *94*, 80, 1954, courtesy of Wistar Press.)

substance of the septum near its attachment to the cranial part of the atrial wall is thinned out and breaks through. This forms **ostium II** or the **ostium secundum** (Fig. 7–10) (Tandler '12).

In the right atrium, a new partition now grows down from the cranial and ventral part of the atrial wall toward the venous opening so as to cover the new opening in septum I. This is **septum II** or **septum secundum** (Fig. 7–11). Its growth ceases before it is complete, leaving an opening called the **foramen ovale**, and its crescentic border persists as the **limbus fossae ovalis**, seen in the adult right atrium (Fig. 7–24).

they are of insignificant size because of the non-functioning condition of the lungs. The left heart, therefore, receives the bulk of its blood through the foramen ovale rather than from these veins.

While the interatrial septa are forming, the slit-like opening of the sinus venosus into the right atrium is guarded by two valves, the **right** and **left venous valves** (*valvulae venosae; sinus folds*) (Fig. 7–9). Above the opening, the two venous valves unite into a single fold, the **septum spurium,** which is a prominent feature at this stage but is of no significance for the eventual partitioning of the heart.

As the atrium increases in size it absorbs the sinus venosus into its walls. The upper part of the sinus then becomes the opening for the **superior vena cava.** The inferior part becomes divided by the further growth of the right venous valve into two openings, the **inferior vena cava** and the **coronary sinus.** The left venous valve disappears except for the lower part which is absorbed into the septum secundum. The upper part of the right venous valve and the septum spurium almost disappear but are retained in the adult heart as the crista terminalis. The lower part of the right valve persists and is divided into two parts by the growth of a transverse fold, the **sinus septum** (Fig. 7–12). The right upper part becomes the **valve of the inferior vena cava.** The lower left part becomes the **valve of the coronary sinus.** Of the horns of the sinus venosus, the right becomes much more prominent since it furnishes the inferior and superior venae cavae. The left horn remains small,

receiving only the left common cardinal vein. The left anterior cardinal dwindles into the **oblique vein** (Fig. 8–16) and the **vestigial fold of Marshall.** The left common cardinal also loses its connection with the body wall and becomes the **coronary sinus,** draining blood only from the substance of the heart.

The primitive ventricle becomes divided by the growth of the **septum inferius** or **interventricular septum.** It grows up from the most prominent part of the ventricular wall, its position on the surface of the heart being indicated by a furrow. Its dorsal part grows more rapidly than the ventral and fuses with the dorsal part of the septum intermedium. The opening between the two ventricles remains for some time but is ultimately closed by the fusion of the interventricular and aortic septa (Fig. 7–13).

The conus, as mentioned above, was at first separated from the ventricle by a deep fold, the bulboventricular sulcus. This fold

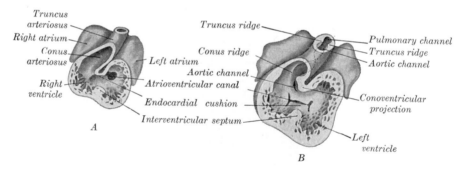

Fig. 7–13.—The conus and ventricle of the hearts of human embryos opened to show stages in the partitioning into pulmonary and aortic outlets. Semischematic drawings. A, 4- to 5-mm embryo; B, 8.8-mm embryo. (T. C. Kramer, Amer. J. Anat., 71, 359, 1942, courtesy of Wistar Press.)

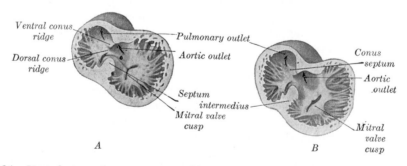

Fig. 7–14.—Ventral views of reconstructions of hearts of human embryos opened to show partitioning of ventricles and formation of interventricular septum. A, 13 mm, Carnegie Embryological Collection No. 841; B, 14.5 mm, modified from Tandler ('12). (From T. C. Kramer, Amer. J. Anat., 71, 361, 1942, courtesy of Wistar Press.)

gradually recedes until the conus and ventricle open freely into each other. This has the effect of allowing the **aortic sac** or the **truncus aortae** to open directly into the ventricle and it comes to lie in line with the path of the growing interventricular septum (Fig. 7–14). The portion of the ventricular wall which had been the conus remains to the right of the interventricular septum and becomes a part of the adult right ventricle (Kramer '42).

the endocardial cushions develop special swellings which become the **semilunar valves.** Two of the valve cusps in each vessel are formed from tissue of the ridges which partition the truncus and they retain this association of position throughout the subsequent rotation and twisting of these vessels (Kerr and Goss '56). The third cusp in each vessel develops independently from dorsal and ventral intercalated swellings (Figs. 7–15, 7–28) (Kramer '42).

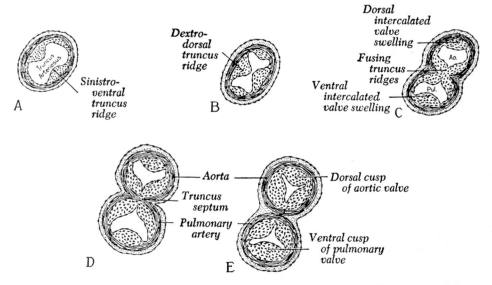

Fɪɢ. 7–15.—Schematic diagrams of the partitioning of the truncus arteriosus and the origin of the aortic and pulmonary semilunar valves. (T. C. Kramer, Amer. J. Anat., *71*, 354, 1942, courtesy of Wistar Press.)

The **truncus arteriosus** or **primitive ascending aorta** has at this time developed both its fourth and its pulmonary pairs of arches. The partitioning of the truncus into the **aorta** and **pulmonary trunk** begins by the growth of two **endothelial cushions** in the region of the fourth pair of arches. The growth of these cushions toward the conus makes a pair of spiral ridges (Fig. 7–14). As the tops of these ridges grow together, partitioning off the truncus, there is a separation of the aortic from the pulmonary trunk. The right ventricle becomes connected with the pulmonary arches which go to the lungs. The left ventricle becomes connected with the fourth arch which forms the aorta.

At the junction of the conus and truncus,

Beginning of Contraction.—The first contractions of the heart have not been observed in human embryos but it can be assumed that they follow the same sequence as in other mammals. In a rat embryo with three somites (Fig. 7–2), two lateral heart tubes are recognizable. On the ventricular side of a slight constriction which marks the future atrioventricular junction, a small group of cells in the splanchnic mesoderm begins contracting with a regular rhythm. In a rat embryo, the left lateral heart initiates the contraction at a rate of 20 to 30 per minute (Goss '38). A few hours later, the right heart commences with a slower rate and a regular rhythm independent of the left side. As the contraction involves the median portion of the ventricle, the rate increases gradually, the left side acting as pacemaker. By the time the rate has reached 90 per minute, a small part of the atrium adjacent to the A–V junction has begun contraction and acts as the pacemaker, but with a distinct interval between the atrial and ventricular contraction (Goss '42b).

Beginning of Circulation.—The primitive heart continues to contract while it is enlarging and differentiating after the three somite stage, but is ineffective for circulation until the embryo has eight or nine somites (Fig. 7–6), a period of fifteen to twenty hours. At this time the atrium is well established, the ventricle is an effective muscular pump, the cells in the blood islands of the yolk sac have acquired hemoglobin, and a continuous lumen has been established through the aortae, umbilical and omphalomesenteric arteries, yolk sac and placental capillaries, and back through the corresponding veins to the heart (Goss '42a).

The development of the arteries and veins is described at the beginning of the corresponding chapters.

PECULIARITIES IN THE VASCULAR SYSTEM OF THE FETUS

The chief peculiarities of the fetal heart are the direct communication between the atria through the foramen ovale, and the large size of the valve of the inferior vena cava. Among other peculiarities the following may be noted: (1) In early fetal life the heart lies immediately below the mandibular arch and is relatively large in size. As development proceeds it is gradually drawn within the thorax, but at first it lies in the middle line; toward the end of pregnancy it gradually becomes oblique in direction. (2) For a time the atrial portion exceeds the ventricular in size, and the walls of the ventricles are of equal thickness; toward the end of fetal life the ventricular portion becomes the larger and the wall of the left ventricle exceeds that of the right in thickness. (3) Its size is large as compared with that of the rest of the body, the proportion at the second month being 1 to 50, and at birth, 1 to 120, while in the adult the average is about 1 to 160.

The **foramen ovale**, situated at the lower part of the atrial septum, forms a free communication between the atria until the end of fetal life. A septum (*septum secundum*) grows down from the upper wall of the atrium to the right of the primary septum in which the foramen ovale is situated; shortly after birth it fuses with the primary septum and the foramen ovale is obliterated.

The **valve of the inferior vena cava** serves to direct the blood from that vessel through the foramen ovale into the left atrium.

The peculiarities in the arterial system of the fetus are the communication between the pulmonary artery and the aorta by means of the ductus arteriosus, and the continuation of the internal iliac (hypogastric) arteries as the umbilical arteries to the placenta.

The **ductus arteriosus** is a short tube, about 1.25 cm in length at birth, and 4.4 mm in diameter. In the early condition it forms the continuation of the pulmonary artery, and opens into the aorta, just beyond the origin of the left subclavian artery; and so conducts the greater amount of the blood from the right ventricle into the aorta. When the branches of the pulmonary artery have become larger relatively to the ductus arteriosus, the latter is chiefly connected to the left pulmonary artery.

The **internal iliac** (*hypogastric*) **arteries** run along the sides of the bladder and thence cranialward on the inside of the anterior abdominal wall to the umbilicus; here they pass out of the abdomen and are continued as the **umbilical arteries** in the umbilical cord to the placenta. They convey the fetal blood to the placenta.

The peculiarities in the venous system of the fetus are the communications established between the placenta and the liver and portal vein, through the umbilical vein; and between the umbilical vein and the inferior vena cava through the ductus venosus.

Fetal Circulation

The fetal blood is returned from the placenta to the fetus by the umbilical vein (Fig. 7–16). This vein enters the abdomen

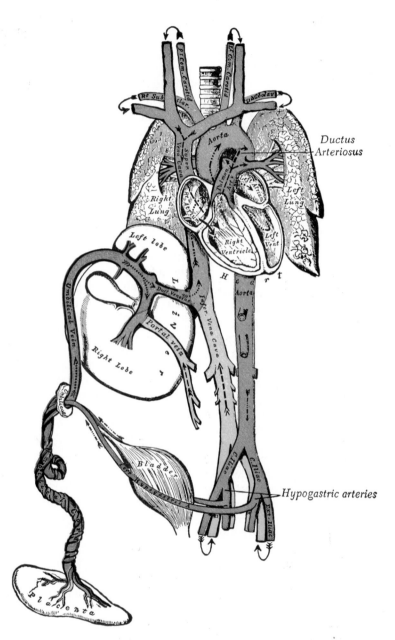

FIG. 7–16.—Plan of the fetal circulation. In this plan the figured arrows represent the kind of blood, as well as the direction which it takes in the vessels. Thus—arterial blood is figured >------->, venous blood, > — — — >; mixed (arterial and venous) blood, > — --- —>.

at the umbilicus, and passes cranialward along the free margin of the falciform ligament of the liver to the caudal surface of that organ, where it gives off two or three branches, one of large size to the left lobe, and others to the lobus quadratus and lobus caudatus. At the **porta hepatis** (*transverse fissure of the liver*) it divides into two branches: of these, the larger is joined by the portal vein, and enters the right lobe; the smaller is continued cranialward, under the name of the **ductus venosus**, and joins the inferior vena cava. The blood, therefore, which traverses the umbilical vein passes to the inferior vena cava in three different ways. A considerable quantity circulates through the liver with the portal venous blood before entering the inferior vena cava by the hepatic veins; some enters the liver directly, and is carried to the inferior vena cava by the hepatic veins; the remainder passes directly into the inferior vena cava through the ductus venosus.

In the inferior vena cava, the blood carried by the ductus venosus and hepatic veins becomes mixed with that returning from the lower extremities and abdominal wall. It enters the right atrium, and, guided by the valve of the inferior vena cava, passes through the foramen ovale into the left atrium, where it mixes with a small quantity of blood returned from the lungs by the pulmonary veins. From the left atrium it passes into the left ventricle; and from the left ventricle into the aorta, by means of which it is distributed almost entirely to the head and upper extremities, a small quantity being probably carried into the descending aorta. From the head and upper limbs the blood is returned by the superior vena cava to the right atrium, where it mixes with a small portion of the blood from the inferior vena cava. From the right atrium it passes into the right ventricle, and thence into the pulmonary artery. The lungs of the fetus being inactive, only a small quantity of the blood of the pulmonary artery is distributed to them by the right and left pulmonary arteries, and returned by the pulmonary veins to the left atrium: the greater part passes through the ductus arteriosus into the aorta, where it mixes with a small quantity of the blood

transmitted by the left ventricle into the aorta. Through this vessel it is in part distributed to the lower limbs and the viscera of the abdomen and pelvis, but the greater amount is conveyed by the umbilical arteries to the placenta.

From the preceding account of the circulation of the blood in the fetus the following facts will be evident: (1) The placenta serves the purposes of nutrition and excretion, receiving the impure blood from the fetus, and returning it purified and charged with additional nutritive material. (2) Nearly all the blood of the umbilical vein traverses the liver before entering the inferior vena cava; hence the large size of the liver, especially at an early period of fetal life. (3) The right atrium is the point of meeting of a double current, the blood in the inferior vena cava being guided by the valve of this vessel into the left atrium, while that in the superior vena cava descends into the right ventricle. At an early period of fetal life it is highly probable that the two streams are quite distinct; for the inferior vena cava opens almost directly into the left atrium, and the valve of the inferior vena cava would exclude the current from the right ventricle. At a later period, as the separation between the two atria becomes more distinct, it seems probable that some mixture of the two streams must take place. (4) The pure blood carried from the placenta to the fetus by the umbilical vein, mixed with the blood from the portal vein and inferior vena cava, passes almost directly to the arch of the aorta, and is distributed by the branches of that vessel to the head and upper limbs. (5) The blood contained in the descending aorta, chiefly derived from that which has already circulated through the head and limbs, together with a small quantity from the left ventricle, is distributed to the abdomen and lower limbs.

Changes in the Vascular System at Birth.—At birth, when respiration is established, an increased amount of blood from the pulmonary artery passes through the lungs, and the placental circulation is cut off. The foramen ovale gradually decreases in size during the first month, but a small opening usually persists until the last third of the first year and often later; the valvular fold above mentioned adheres to the margin of the foramen for the greater part of its circumference, but a slit-like opening is left between the two atria above, and this sometimes persists.

The **ductus arteriosus** begins to contract immediately after respiration is established, and its lumen slowly becomes obliterated; it ultimately degenerates into an impervious cord, the **ligamentum arteriosum,** which connects the left pulmonary artery to the arch of the aorta (Figs. 7–22, 7–23).

Of the **internal iliac (hypogastric) arteries,** the parts extending from the sides of the bladder to the umbilicus become obliterated between the second and fifth days after birth, and persist as fibrous cords

covered with peritoneum, the **medial umbilical ligaments,** which form ridges on the interior surface of the anterior abdominal wall (Fig. 16–54).

The **umbilical vein** and **ductus venosus** are obliterated between the first and fifth days after birth; the former becomes the ligamentum teres, the latter the ligamentum venosum, of the liver. The hepatic half of the ductus venosus may remain open, receive tributaries from the liver and thus function as a hepatic vein in the adult.

THE HEART

The **heart** (*cor*) is a hollow muscular organ shaped like a blunt cone about the size of the fist of the same individual. It rests on the diaphragm between the lower

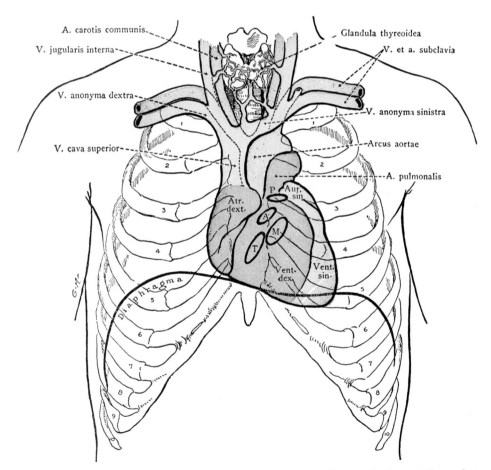

FIG. 7–17.—The heart and cardiac valves projected on the anterior chest wall, showing their relation to the ribs, sternum, and diaphragm. (Eycleshymer and Jones.)

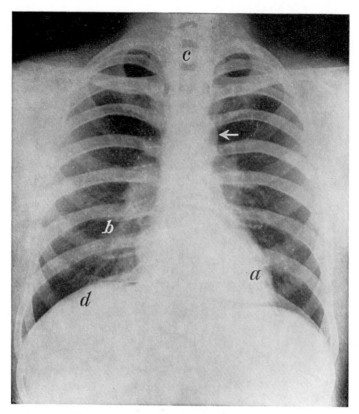

Fig. 7–18.—Anterior view, thorax, heart, and diaphragm. *a,* Apex of heart; *b,* bronchi; *c,* trachea; *d,* diaphragm. The arrow points to the arch of the aorta. (Department of Radiology, University of Pennsylvania.)

part of the two lungs. It is enclosed in a special membrane, the pericardium, and occupies a topographical compartment of the thorax known as the **middle mediastinum** (Fig. 15–28). Its position relative to the chest wall is shown diagrammatically in Figure 7–17 and the position and extent of the cardiac shadow in an x-ray plate are shown in Figure 7-18. It is covered ventrally by the sternum and adjoining parts of the third to sixth costal cartilages. The apex of the cone points caudalward, ventralward, and to the left, about two-thirds of the whole organ being to the left of the median plane.

Size.—The heart, in the adult, measures about 12 cm in length, 8 to 9 cm in breadth at the broadest part, and 6 cm in thickness. Its weight, in the male, varies from 280 to 340 grams; in the female, from 230 to 280 grams. The heart frequently continues to increase in weight and size up to an advanced period of life; this increase may be pathological.

The Heart Wall.—The wall of the heart is composed of three layers, an outer epicardium, a middle myocardium, and an inner endocardium. The surface layer of the **epicardium** is the **serous membrane** or **visceral pericardium**. This is a single sheet of squamous mesothelial cells resting on a lamina propria of delicate connective tissue. Between the serous coat and the myocardium is a layer of heavier fibroelastic connective tissue. The latter is interspersed with **adipose tissue** which fills in the crevices and sulci to give the heart a smooth, rounded contour. The larger blood vessels and the nerves are contained in this layer also. The dark reddish color of the myocardium is visible through the epicardium except where fat has accumulated. The amount of fat varies greatly; it is seldom absent except in emaciated individuals and

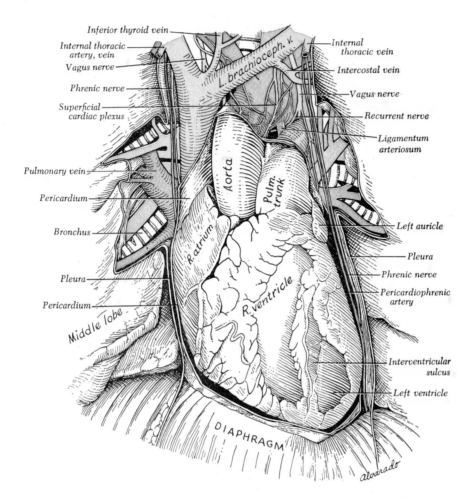

Fig. 7–19.—The sternocostal surface of the heart and great vessels after removal of the pericardium, with the structures at the roots of the lungs exposed.

may almost completely obscure the myocardium in the very obese.

The **myocardium** is composed of layers and bundles of cardiac muscle with a minimum of other tissue except for the blood vessels. It is described in detail in a later section.

The **endocardium** is the interior lining of the heart. Its surface layer is composed of squamous endothelial cells and is continuous with the endothelial lining of the blood vessels. The connective tissue is quite thin

and transparent over the muscular walls of the ventricles but is thickened in the atria and at the attachments of the valves. It contains small blood vessels, parts of the specialized conduction system, and a few bundles of smooth muscle.

Although the heart is freely movable and unattached to the surrounding organs, it is maintained in its proper position in the thorax by continuity with the great blood vessels and by an enclosing membranous sac, the pericardium (Fig. 7–19).

THE PERICARDIUM

The pericardium includes two quite different components, the serous membrane and the fibrous sac.

The **Serous Pericardium** lines the inside of the fibrous sac and covers the outside of the heart. Its characteristics are such that it provides these two structures with smooth and glistening surfaces, completely free and movable although in contact with each other. The serous layer of the heart itself is called the visceral pericardium or the epicardium. It covers the atria and ventricles and extends beyond them out on the great vessels for 2 or 3 cm (see Fig. 7–19). The serous layer lining the fibrous sac is called the parietal pericardium. The two parts of the serous layer are continuous with each other and the point at which the visceral layer ends and folds back on itself to become the parietal layer is called the **reflection of the pericardium.**

The **Pericardial Cavity.**—In a healthy state, the two serous membranes are closely apposed to each other, separated only by enough serous or watery fluid to make their surfaces slippery. This allows the heart to move easily during its contraction in systole and its relaxation in diastole. Since the two surfaces are not attached, there is a potential space between them called the pericardial cavity. After injury or due to disease, fluid may exude into the cavity causing a wide separation between the heart and the outer pericardium.

The extension of the epicardium on the great vessels is in the form of two tubular prolongations. One encloses the aorta and pulmonary trunk and is called the **arterial mesocardium.** The other encloses the venae cavae and the pulmonary veins and is called the **venous mesocardium.** The reflection of the venous mesocardium forms a U-shaped cul-de-sac in the dorsal wall of the pericardial cavity known as the **oblique pericardial sinus.** Between the arterial mesocardium and the venous mesocardium is a pericardium-lined passage named the **transverse pericardial sinus** (Fig. 7–20).

The **Fibrous Pericardium** forms a flask-like sac the neck of which is closed by its attachment to the great vessels just beyond the reflection of the serous pericardium (Fig. 7–19). It is a tough membrane, much thicker than the parietal pleura. Its outer surface is adherent in varying degrees to all the structures surrounding it. It is attached ventrally to the manubrium of the sternum by a fibrous condensation, the **superior pericardiosternal ligament,** and to the Xiphoid process by the **inferior pericardiosternal ligament.** The fibrous tissue intervening between the sac and the vertebral column is the **pericardiovertebral ligament.** The sac is securely attached to the central tendon and muscular part of the left side of the dome of the diaphragm. A thickening of this fibrous attachment in the area of the inferior vena cava has been called the **pericardiophrenic ligament.**

The **ligament of the left vena cava** (*vestigial fold of Marshall*) is a triangular fold covered with serous pericardium stretching from the left pulmonary artery to the atrial wall or the subjacent pulmonary vein. It is formed by the remnant of the proximal part of the left superior vena cava (left anterior cardinal vein and left duct of Cuvier) which becomes obliterated during fetal life. If well developed, it may remain as a fibrous band stretching from the highest left intercostal vein to a small vein draining into the coronary sinus, the **oblique vein of the left atrium** (*oblique vein of Marshall*) (Fig. 8–16).

The lateral parts of the outer surface of the pericardial sac, *i.e.*, the mediastinal surfaces, are apposed to the mediastinal parietal pleura. The two membranes are adherent but not fused and the phrenic nerve with its accompanying blood vessels is held between them as it traverses the thorax. Because of the rounded contour of the heart, the pleural cavity thus partly encircles the pericardial sac extending ventrally between it and the chest wall except for a small triangular area on the left side (Figs. 3–13, 3–17, 15–28). This area corresponds to the caudal portion of the body of the sternum and the medial ends of the left fourth and fifth costal cartilages. In percussion of the chest for physical diagnosis the triangle is called the area of absolute dullness because no lung is present to give it resonance. It is also important clinically as the area through which a needle can be introduced into the pericardial

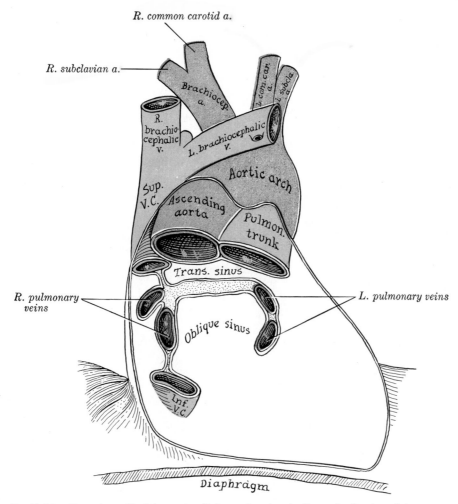

Fig. 7–20.—Posterior wall of the pericardial sac, showing the lines of reflection of the serous pericardium on the great vessels.

cavity for removal of excess fluid, without traversing the pleural cavity or lungs. The dorsal surface of the sac is in relation to the bronchi, esophagus, and descending thoracic aorta. The caudal surface is attached to the dome of the diaphragm. The cranial part of the ventral surface may be in contact with the thymus in a child.

COMPONENT PARTS OF THE HEART

The heart consists of four chambers: two larger ventricles with thick muscular walls making up the bulk of the organ and two smaller atria with thin muscular walls. The septum which separates the ventricles also extends between the atria, subdividing the whole heart into what are called **left** (Latin-sinister) and **right** (Latin-dexter) **halves** or **sides** of the heart. As they lie in the body, however, the right side is mostly ventral or anterior and the left side largely dorsal or posterior.

On the surface of the heart, the **coronary sulcus** (*sulcus coronarius*) (Figs. 7–21, 7–22) encircles the heart between the ventricles, which are toward the apex, and the atria, which are at the base. The sulcus is occupied by the arterial and venous vessels supplying the heart. Its ventral part is not visible because it is covered over by the conus arteriosus. The line of separation

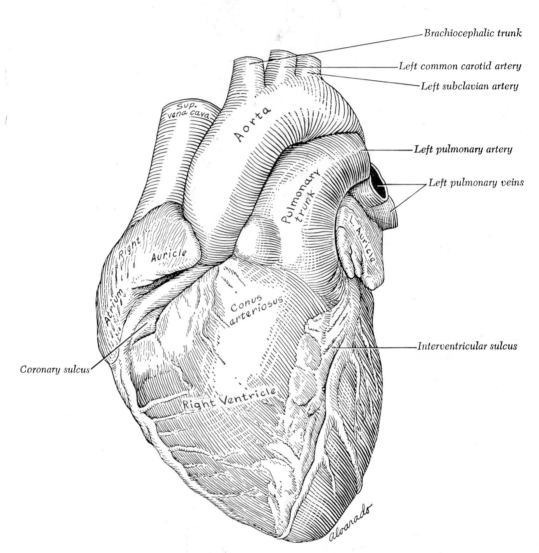

Brachiocephalic trunk

Left common carotid artery

Left subclavian artery

Left pulmonary artery

Left pulmonary veins

Interventricular sulcus

Coronary sulcus

Sup.
vena cava

Aorta

Pulmonary
trunk

Right
Auricle

L. Auricla

Atrium

Conus
arteriosus

Right Ventricle

alvarado

Fɪɢ. 7–21.—Heart. Sternocostal surface, ventral aspect. (Rauber-Kopsch, redrawn.)

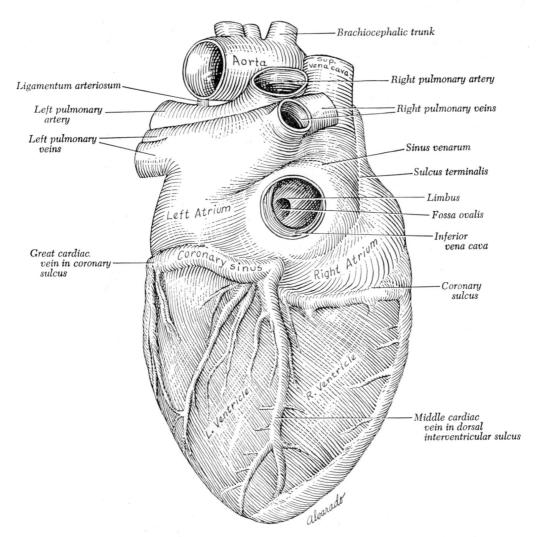

FIG. 7–22.—Heart. Diaphragmatic surface, inferior aspect. (Rauber-Kopsch, redrawn.)

between the two ventricles is marked by the **anterior interventricular sulcus** (*sulcus interventricularis anterior; ventral longitudinal sulcus*) on the sternocostal surface and by the **posterior interventricular sulcus** (*sulcus interventricularis posterior; dorsal longitudinal sulcus*) on the diaphragmatic surface, the two grooves becoming continuous near the apex in the **incisura apicis cordis.**

The **Apex of the Heart** (*apex cordis*) points caudalward, ventralward, and to the left. Although its position changes continually during life the tip remains close to the following point under usual circumstances: deep to the left fifth intercostal space, 8 or 9 cm from the midsternal line, or about 4 cm caudal and 2 cm medial to the left nipple. It is overlapped by an extension of the pleura and lungs as well as by the structures of the anterior thoracic wall.

The **Base of the Heart** (*basis cordis*) (Fig. 7–23) is more difficult to visualize than the apex but represents the base of the blunt cone which is the heart. Since

it is opposite the apex, it faces to the right, cranialward, and dorsalward. It involves mainly the left atrium, part of the right atrium, and the proximal parts of the great vessels. Its superior boundary is at the bifurcation of the pulmonary trunk, its inferior at the coronary sulcus, its right side the sulcus terminalis and its left the oblique vein of the left atrium. The descending thoracic aorta, the esophagus, and the thoracic duct intervene between it and the bodies of the fifth to eighth thoracic vertebrae.

The **Sternocostal Surface** (*facies sternocostalis*) (Fig. 7–21) is occupied by the right atrium and especially its auricula at the right, and the right ventricle with a small part of the left ventricle at the left. It is crossed obliquely caudalward by the coronary sulcus and by the anterior interventricular sulcus.

The **Diaphragmatic Surface** (*facies diaphragmatica*) (Fig. 7–22) is somewhat flattened from its contact with the diaphragm in a preserved cadaver. It involves

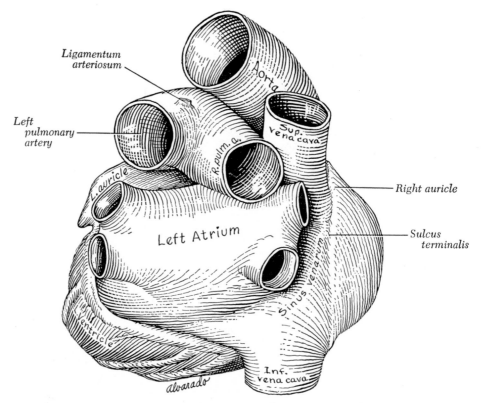

Fig. 7–23.—Base of the heart, dorsal or posterior aspect. (From Rauber-Kopsch, redrawn.)

the two ventricles, the left slightly more than the right, and is crossed obliquely by the posterior interventricular sulcus. It is marked off from the base of the heart by the coronary sulcus.

The **Right Margin of the Heart** (*margo dexter*), longer than the left, describes an arc from the superior vena cava to the apex. Its cranial part is along the right atrium, its caudal part along the right ventricle. The ventricular part is almost horizontal, running along the line of attachment of the diaphragm to the anterior chest wall. It tends to be thin and sharp and is called the **acute margin**. The atrial portion is almost vertical and is situated deep to the third to fifth costal cartilages 1.25 cm to the right of the margin of the sternum.

The **Left Margin** (*obtuse margin*) is formed mainly by the left ventricle and to a small extent by the left atrium. It extends obliquely caudalward to the apex from a point 2.5 cm from the sternal margin in the second left interspace, describing a curve with convexity to the left.

Right Atrium (*atrium dextrum; right auricle*) (Fig. 7–24).—The right atrium appears somewhat larger and its wall, 2 mm in thickness, somewhat thinner than the left. It has a capacity of about 57 ml. It consists

of two parts: (*a*) a principal cavity or the sinus venarum cavarum and (*b*) a hollow appendage, the auricula.

(*a*) The **Sinus Venarum** (*sinus venarum cavarum*) is the part of the cavity between the two venae cavae and the atrioventricular opening. Its wall merges with the two cavae and the interior surface is quite smooth except for certain rudimentary structures described below.

(*b*) The **Right Auricula** (*auricula dextra; right auricular appendage*), shaped something like a dog's ear, is a blind pouch extending cranialward between the superior vena cava and the right ventricle. Its junction with the sinus venarum is marked on the outside by a groove, the **sulcus terminalis** (Fig. 7–22), which corresponds to a ridge on the inside, the **crista terminalis**. The internal surface of the auricula has the muscular bundles raised into distinctive parallel ridges resembling the teeth of a comb and named, therefore, the **musculi pectinati**.

The interior of the right atrium presents the following openings and parts for examination (Fig. 7–24).

The **superior vena cava** (*v.c. cranialis* in Figs. 7–23 and 7–24) opens into the cranial and posterior part of the sinus

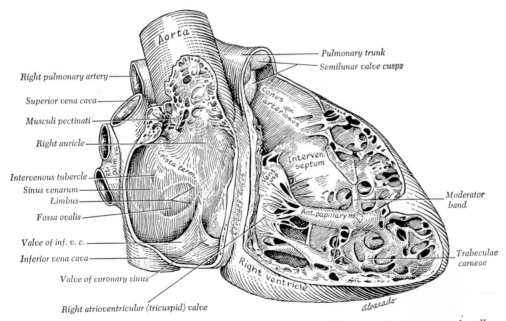

FIG. 7–24.—Interior of right atrium and right ventricle of heart after removal of sternocostal wall. (From Rauber-Kopsch, redrawn.)

venarum. Its orifice faces caudalward and ventralward so that the blood entering the atrium through it is directed toward the atrioventricular opening. It returns the blood from the cranial half of the body. There is no valve at its opening.

The **inferior vena cava** (*v.c. caudalis*) opens into the most caudal part of the sinus venarum near the interatrial septum. It returns blood from the caudal half of the body. Its orifice is larger than that of the superior vena cava and faces cranialward and dorsalward, toward the fossa ovalis. A rudimentary valve extends cranialward on the septum from the opening.

The **valve of the inferior vena cava** (*valvula venae cavae inferioris; Eustachian valve*) (Fig. 7–24) is a single crescentic fold attached along the ventral and left margin of the orifice of the inferior vena cava. Its concave free margin ends in two cornua of which the left is continuous with the ventral margin of the limbus fossae ovalis, and the right spreads out on the atrial wall. The valve is composed of a fold of the membranous lining of the atrium containing a few muscular fibers. In the fetus the valve is more prominent and tends to direct the blood from the inferior cava through the then patent foramen ovale into the left atrium. In the adult it is usually rudimentary and has little if any functional significance. It may be thin and fenestrated, very small, or even entirely obliterated.

The **coronary sinus** (*sinus coronarius*) (Fig. 7–24) opens into the right atrium between the inferior vena cava and the atrioventricular opening. It returns the blood from the substance of the heart itself and has a valve at its orifice. The **valve of the coronary sinus** (*valvula sinus coronarii; Thebesian valve*) is a single semicircular fold of the lining membrane of the atrium attached to the right and inferior lips of the orifice of the coronary sinus. It may be cribriform or double and is seldom effective for more than partial closure of the orifice during contraction of the atrium.

The **foramina venarum minimarum** are the openings of small veins (*venae cordis minimae; Thebesian veins*) which empty their blood directly into the cavity of the atrium. A few larger ones can usually be seen in the septal wall. They were used by the ancients to explain passage of blood from one side of the heart to the other before Harvey (1610 A.D.) discovered the circulation of the blood.

The right atrioventricular opening is the large oval aperture of communication between the atrium and the ventricle. It is described with the right ventricle.

The **interatrial septum** forms the dorsal wall of the right atrium. It contains rudimentary structures which were of significance in the fetus. The **fossa ovalis** (Fig. 7–24) is an oval depression in the septal wall and corresponds to the foramen ovale of the fetal heart. It lies within a triangle established by the openings of the two venae cavae and the coronary sinus. The **limbus fossae ovalis** is the prominent oval margin of the foramen which persists in the adult. It is distinct cranially and at the sides but is deficient caudally. Frequently the cranial part of the limbus does not fuse with the left leaf of the septum, leaving a slit-like opening in the septum through which a probe may be passed into the left atrium. This is spoken of as probe patency of the foramen and is found in 20 to 25 per cent of all hearts. Larger openings are not uncommon also but these defects of the interatrial septum are seldom of functional significance.

The **intervenous tubercle** (*tubercle of Lower*) is a small raised area in the septal wall of the right atrium between the fossa ovalis and the orifice of the superior vena cava. It is more distinct in the hearts of certain quadrupeds than in man and was supposed by Richard Lower (1631-1691) to direct the blood from the superior cava into the atrioventricular opening. It is now of historical interest only.

Right Ventricle (*ventriculus dexter*) (Fig. 7–24).—The right ventricle occupies a large part of the ventral or sternocostal surface of the heart. Its right boundary is the coronary sulcus, its left the anterior longitudinal sulcus. Superiorly, the part of it named the **conus arteriosus** joins the pulmonary trunk. Inferiorly its wall forms the acute margin of the heart and extends around to the diaphragmatic surface for some distance. Its wall is about one-third as thick as that of the left ventricle; it is thickest near the base of the heart and gradually becomes thinner toward the apex. Its capacity is the same as that of the left ventricle, about 85 ml (Latimer '53).

The interior of the right ventricle presents the following openings and parts for examination (Fig. 7–25):

The **right atrioventricular orifice** or **ostium** is an oval aperture about 4 cm in diameter. It is surrounded by a strong fibrous ring and is guarded by the tricuspid valve.

The **right atrioventricular** or **tricuspid valve** (Fig. 7–26) surrounds the orifice with a thin apron which projects into the ventricle in three leaflets or cusps. The **ventral** or **infundibular cusp** (*anterior cusp*) is attached to the ventral wall in the region of the conus arteriosus (*infundibulum*). The **dorsal** or **marginal cusp** (*posterior cusp*) is attached to the part of the ventricle which curves around from the sternocostal to the diaphragmatic surface forming the acute margin of the heart. The **medial** or **septal cusp** (*cuspus septalis*) is attached to the septal wall of the ventricle. The cusps are of unequal size, and small intercalated leaflets may occur between them. The anterior is the largest, the posterior the smallest. The leaflets are composed of strong fibrous tissue, thick in the central part, thin

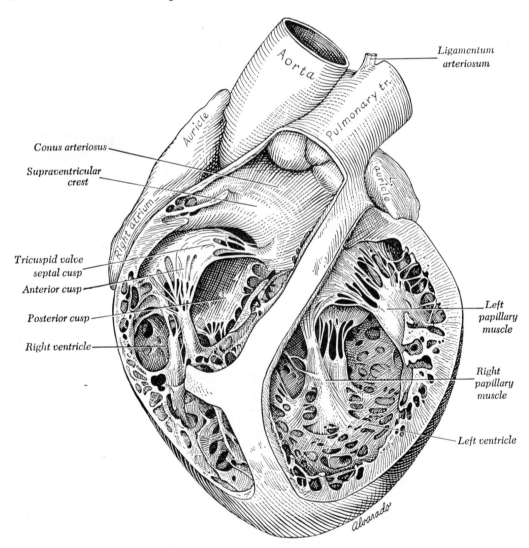

Fig. 7–25.—Interior of right and left ventricles of heart to show atrioventricular valves. (From Rauber-Kopsch, redrawn.)

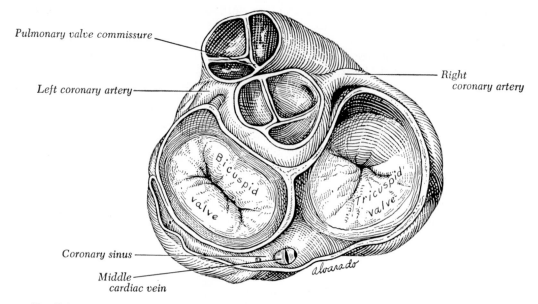

Pulmonary valve commissure

Right coronary artery

Left coronary artery

Coronary sinus

Middle cardiac vein

Fig. 7–26.—Openings and valves of the heart in closed position. Viewed from above after removal of atria and arterial trunks. (From Rauber-Kopsch, redrawn.)

and translucent near the margin. They are roughly triangular in shape, the bases attached to the fibrous ring and the apices projecting into the ventricular cavity. The atrial surface of the leaflets is smoothly covered by the atrial endocardial membrane but the ventricular surface is irregular and the free border presents a ragged edge for the attachment of the chordae tendineae. In order to make the valve as a whole competent to withstand the back pressure and prevent regurgitation of blood into the atrium during ventricular systole, the cusps are held in place by the chordae tendineae and papillary muscles.

The **chordae tendineae** are delicate but strong fibrous cords. They are attached to the apices, margins, and ventricular surfaces of the valve cusps and are anchored in the muscular wall. They number about twenty and are of different length and thickness. The majority are attached to projections of the trabeculae called papillary muscles.

The **trabeculae carneae** (*columna carneae*) are irregular bundles and bands of muscle which project from the inner surface of the ventricle except in the conus arteriosus. They are of three types, some are ridges along the wall, others extend across the lumen for short distances and are covered

on all sides by endocardium, and still others form the special structures, papillary muscles, described below. The surface of the conus is quite smooth and its limit is marked by a ridge of muscular tissue, the **crista supraventricularis** (Fig. 7–25), which extends across the dorsal wall toward the pulmonary trunk from the attachment of the ventral cusp at the atrioventricular ring. The **moderator band** (*trabecula septomarginalis*) is a stout bundle of muscle in the central or apical part of the right ventricle. It crosses the lumen from the base of the anterior papillary muscle to the septum opposite. It is of variable size and constancy in the human heart but is more prominent in some larger mammals where it is supposed to resist over distention of the ventricle. When present it usually contains a branch of the atrioventricular conduction bundle (Truex and Copenhaver '47).

The **papillary muscles** are rounded or conical projections of muscle to whose apices the chordae tendineae are attached. Although their size and number are somewhat variable, the two principal ones are named anterior and posterior. The main **anterior papillary muscle** (*m. pap. ventralis*, Fig. 7–24), the larger, protrudes partly from the ventral and partly from the septal wall.

Its chordae tendineae are attached to the anterior and posterior cusps. Frequently it provides part of the moderator band. The **posterior papillary muscle** is in several parts arising from the posterior wall and is attached to the chordae tendineae of the posterior and septal cusps. A small anterior papillary muscle arising near the septal end of the crista supraventricularis, called the papillary muscle of the conus, is attached to the anterior and septal cusps. Other small papillary muscles are attached to individual chordae tendineae.

The **orifice or ostium of the pulmonary trunk** is a circular opening at the summit of the conus arteriosus. It is close to the ventricular septum, cranial and to the left of the atrioventricular opening. The **pulmonary valve** consists of three **semilunar cusps** formed by duplicatures of the endocardial lining and reinforced by fibrous tissue (Fig. 7–27). They are attached by a curved margin with convexity toward the ventricle, the free border toward the vessel. Behind each **cusp** (*velum*) is a pocket called the **sinus** and the point at which the attachments of two adjacent cusps come together is called a **commissure**. Each cusp has a thickened **nodule** (*nodulus; corpus Arantii*) at the center of the free margin. A thickened band curves in an arc with convexity toward the vessel wall from the nodule to the commissure of attachment. The part of the valve between these arcs and the free margin is thin or even fenestrated and from its crescentic shape is called a **lunula**. The line of contact between the cusps, when the valve is closed, is not at the free margin but by the nodule and lunulae.

The **Left Atrium** (*atrium sinistrum; left auricle*) (Fig. 7–31) is rather smaller than the right and its wall is thicker, measuring about 3 mm. It forms a large part of the surface at the base and dorsal part of the heart. Its separation from the right atrium is not identifiable on the dorsal surface except in a distended heart but the aorta and pulmonary trunk lie between them on the ventral surface. It consists of two parts, (*a*) a principal cavity and (*b*) an auricula.

(*a*) The **principal cavity** contains the openings of the four pulmonary veins, two on each side. They are not guarded by valves. Frequently the two left veins have a common opening. The left atrioventricular opening is rather smaller than the right, and is guarded by the mitral valve. The surface of the principal cavity is smooth. The interatrial septum contains a depression which is bounded below by a crescentic ridge. This is the edge of the **valve of the foramen ovale** (*valvula foraminis ovalis*) the remnant of the septum primum which was fused over the opening of the foramen ovale at birth.

(*b*) The **left auricula** (*auricula sinistra; left auricular appendage*) is somewhat constricted at its junction with the principal cavity. It is longer, narrower, and more curved than the right, and its margins are more deeply indented. It curves ventralward around the base of the pulmonary trunk, only its tip being visible on the sternocostal aspect of the heart (Fig. 7–21). It lies over the beginning of the left coronary artery. Its interior is marked by the muscular ridges of musculi pectinati.

Left Ventricle (*ventriculus sinister*) (Fig. 7–25).—The left ventricle occupies a small part of the sternocostal and about half of the diaphragmatic surface of the heart. Its tip forms the apex of the heart. The left ventricle is longer and more conical in shape than the right and its walls are about three times as thick. In a cross section its cavity is oval or circular in outline. The interior presents two openings, the atrioventricular guarded by the mitral valve, and the aortic guarded by the aortic valve.

The **left atrioventricular opening** (*mitral orifice*) is somewhat smaller than the right. It is encircled by a dense fibrous ring. The **bicuspid** or **mitral valve** (*valvula bicuspidalis [mitralis]*) (Fig. 7–26) surrounds the opening and extends down into the left ventricle in two valve cusps of unequal size. The larger is placed ventrally and to the right, adjacent to the aortic opening, and is called the ventral, anterior, or aortic cusp. The smaller is the dorsal or posterior cusp. Two smaller cusps are usually found at the angles of junction between the main cusps. The chordae tendineae are thicker, stronger, and less numerous than in the right ventricle (Fig. 7–25).

The **trabeculae carneae** are similar to those of the right ventricle but are more numerous and more densely packed, especially at the apex and on the dorsal wall. There are two **papillary muscles,** one attached to the ventral and one to the dorsal wall. They are of large size and end in rounded extremities to which the chordae tendineae are attached. The chordae from each papillary muscle go to both cusps of the valve.

The **aortic opening** is a circular aperture ventral and to the right of the atrioventricular orifice. It is guarded by the **aortic semilunar valves.** The portion of the ventricle immediately adjacent to the aortic orifice is termed the aortic vestibule and has fibrous instead of muscular walls.

The **aortic valve** (Fig. 7–27) is composed of three semilunar cusps, similar to those of the pulmonary valve but larger, thicker, and stronger. The lunulae are more distinct and the noduli or corpora Arantii are thicker and more prominent. Between the cusps and the aortic wall there are dilated pockets called the **aortic sinuses** (*sinuses of Valsalva*). From two of these sinuses the coronary arteries take origin.

The naming of the different cusps and sinuses of the aortic valve presents a problem. As the heart is situated in the body, one of the valves is anterior and the other two right and left posterior. When the heart is removed from the body, as it is at a necropsy, the heart may be held so that the septum is in the middle and the two ventricles exactly right and left. In this position, the aorta has a posterior and right and left anterior cusps (see Fig. 7–26). Another nomenclature which is independent of the position of the heart is based on the origin of the coronary arteries. On this basis there are right and left coronary cusps and a non-coronary cusp. Recently another terminology has been suggested which has the same advantage as the use of the coronary arteries but in addition it names the cusps of the pulmonary valve and is also useful even when there are anomalies of the coronary arteries. In this nomenclature advantage is taken of the embryological development of the aortic septum by division of the truncus arteriosus into the aorta and pulmonary trunk. Two adjacent cusps of both arteries are formed from the aortic septum, causing the commissure between the cusps of the two vessels to be almost exactly opposite. Thus the aortic and pulmonary valves can both be identified as a **right** and **left adjacent** and an **opposite cusps** (Fig. 7–28). (Kerr and Goss '56.)

The **interventricular septum** has an

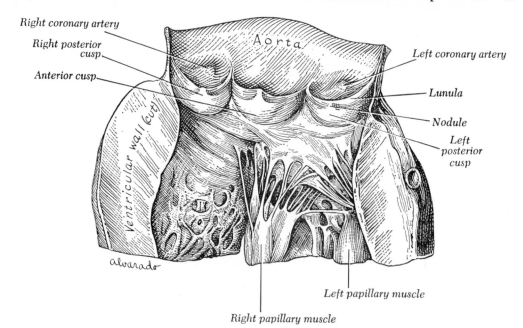

FIG. 7–27.—Aortic opening spread out to show semilunar valve and part of bicuspid valve. (From Rauber-Kopsch, redrawn.)

AORTA

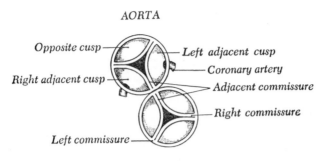

Opposite cusp — Left adjacent cusp

Right adjacent cusp — Coronary artery

Adjacent commissure

Right commissure

Left commissure

PULMONARY TRUNK

FIG. 7–28.—Names of aortic and pulmonary semilunar valve cusps with reference to the adjacent commissures. The examiner holds the specimen so that he looks from the lumen toward the adjacent commissure in naming the valves of either artery. (Kerr and Goss '56.)

oblique position. It has a curvature with convexity to the right, thus completing the oval of the thick left ventricle and encroaching on the cavity of the right ventricle. Its margins correspond to the ventral and dorsal interventricular sulci. The greater part of it is thick and muscular and is called the **muscular interventricular** septum. The cranial part adjoining the atrial septum is thinner and fibrous and is called the **membranous interventricular septum** (Fig. 7–30). This is the last part of the septum to close in embryological development and is the usual site of the defect in a condition known as *patent interventricular septum.*

The fibrous rings which encircle the atrioventricular and arterial orifices merge with the septum membranaceum and compose a resistant core sometimes called the **skeleton of the heart** (Fig. 7–29). In the heart of a large animal such as the ox, it may contain cartilage and bone. Between the right margin of the **left atrioventricular ring**, and **aortic ring** and the **right atrioventricular ring**, the dense tissue forms a triangular mass, the **right fibrous trigone** (*trigonum fibrosum dextrum*). This mass is also the basal thickening of the membranous septum. A fibrous band from the right trigonum and right atrioventricular ring to the posterior side of the conus and anterior aspect of the aortic ring is called the **tendon of the conus.** A smaller trigone, the **trigonum fibrosum sinistrum** lies between the aortic and left atrioventricular rings. The atrioventricular rings and the trigones separate the muscular walls of the atria from those of the ventricles. The bundles of muscular fibers of both chambers, however, use the fibrous tissue for the attachment of their origins and insertions.

The **muscular structure of the heart** consists of bands of fibers which present an exceedingly intricate interlacement. They

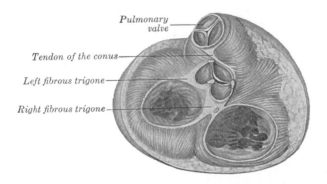

Pulmonary valve

Tendon of the conus —

Left fibrous trigone —

Right fibrous trigone —

FIG. 7–29.—The fibrous skeleton of the heart. The atria have been removed and the heart is viewed toward the ventricles. (Redrawn from Tandler '13, Courtesy of Gustav Fischer Verlag.)

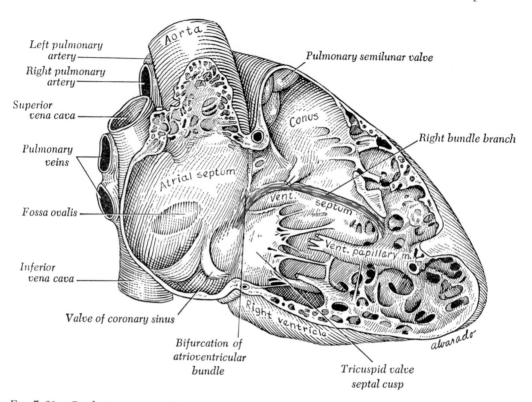

Left pulmonary
artery

Right pulmonary
artery

Superior
vena cava

Pulmonary
veins

Fossa ovalis

Inferior
vena cava

Valve of coronary sinus

Bifurcation of
atrioventricular
bundle

Aorta

Pulmonary semilunar valve

Conus

Right bundle branch

Atrial septum

Vent. septum

Vent. papillary m.

Right ventricle

alvarado

Tricuspid valve
septal cusp

Fig. 7–30.—Conduction system of heart. Interior of right side of heart exposed, atrioventricular bundle
(bundle of His) and right bundle branch in red. (From Rauber-Kopsch, redrawn.)

comprise (*a*) the fibers of the atria, (*b*) the fibers of the ventricles, and (*c*) the atrioventricular bundle of His.

The principal **muscle bundles of the atria** radiate from one central area which surrounds the orifice of the superior vena cava and is for the most part buried in the anterior part of the atrial septum; in front and to the right of the orifice of the vena cava it comes to the external surface. The portion that appears in the groove between the vena cava and the right atrium has been designated the **sinuatrial node;** it is the seat of impulse formation for the atria in the normally beating heart. The portion that is buried in the atrial septum has been named the **septal raphe.** It provides an apparent mechanical support for many of the larger muscle bundles of both atria. The fibers of the sinuatrial node resemble those of the atrioventricular node. With the exception of the interatrial bundle, which connects the anterior surfaces of the two atria, the various muscle bundles are confined to their respec-

tive atria. These bundles radiate from either side of the septal raphe which lies in front of the oval fossa, into the walls of the atria, the sinous venosus and the superior vena cava. There are about fifteen muscle bundles which make up the walls of the two atria. They merge into one another more or less.

The **muscle bundles of the ventricles** probably all arise from the tendinous structures at the base, converge in spiral courses toward the apex for varying distances and then turn spirally upward to be inserted on the opposite side of these same tendinous structures. The superficial fibers pass to the vortex at the apex of the left ventricle before they turn upward while the deep ones turn upward at varying distances without reaching the apex.

The **superficial bulbospiral bundle** arises from the conus, left side of the aortic septum, aortic ring and left atrioventricular ring, passes apicalward and somewhat toward the right to the posterior horn of the vortex of the left ventricle. At their origin the fibers form a broad thin sheet that becomes thick and narrow at the apex where the bundle twists on itself and continues upward in a spiral manner on the inner surface of the left ventricle, spreading out into a thin sheet that is inserted on the opposite side of the tendinous structures from which it arose. These fibers make nearly a double circle around the heart somewhat like a figure 8 that is open at the top. As the fibers pass toward the apex they lie superficial to the deep bulbospiral bundle and as they pass upward from the apex they partly blend and partly pass on the inner side of it in directions nearly at right angles to their superficial fibers.

The **deep bulbospiral bundle** arises immediately beneath the superficial bundle from the left side of the left ostia. The fibers pass downward to the right and enter the septum through the posterior longitudinal sulcus. They then encircle the left ventricle without reaching the apex after turning upon themselves on the apical side of the ring and blend with the fibers of the superficial bundle as they pass spirally upward to be inserted on the opposite sides of the fibrous rings of the left side. These fibers likewise seem to form an open figure 8 with both loops of about the same size.

The **deep sinuspiral bundle** is more especially concerned with the right ventricle although its fibers communicate freely with the papillary muscles of both ventricles. Its fibers arise from the posterior part of the left ostium and pass diagonally into the deeper layer of the wall of the right ventricle where they turn upward to the conus and membranous septum. Some probably pass through the right vortex.

The **interventricular bundles** are represented in part by the longitudinal bundle of the right ventricle which passes through the septum and must be cut in order to unroll the heart, and by the interpapillary bands.

Conduction System of the Heart.—The parts of the conduction system are the sinuatrial node, the atrioventricular node, the atrioventricular bundle, and the terminal conducting fibers or Purkinje fibers. Although they differ from each other somewhat, all are composed of modified cardiac muscle and have the power of spontaneous

rhythmicity and conduction more highly developed than the rest of the heart. Both the ventricles and the atria have an innate power of spontaneous contractility within their muscular tissue which is independent of any nervous influence. The conduction system, however, initiates and superimposes a rhythm with a rapid rate which it transmits to all parts of the heart and one which can be regulated by the nervous system.

The **sinuatrial node** (*S–A node; sinus node; node of Keith and Flack*) is a small knot of modified heart muscle situated in the crista terminalis at the junction of the superior vena cava and right atrium. It receives its name from the fact that this area is developed from the margin of the sinus venosus in the embryo. It is not visible in gross dissections but can be recognized on microscopic examination by certain histological peculiarities. The contraction of the heart is initiated by this node and it is therefore called the "pacemaker" of the heart. A special bundle for the conduction of the impulse from this node to the ventricle has not been identified morphologically. The fibers of the node merge with the atrial musculature which alone appears to be responsible for conducting the impulse to the atrioventricular node.

The **atrioventricular node** (*A–V node; node of Tawara*) (Fig. 7–30) lies near the orifice of the coronary sinus in the septal wall of the right atrium. It is of small size and not being encapsulated by connective tissue can seldom be recognized except by its connection with the A–V bundle.

The **atrioventricular** or **A–V bundle** (*bundle of His*) (Fig. 7–30) begins at the A–V node and follows along the membranous septum toward the left atrioventricular opening a distance of 1 or 2 cm. At about the middle of the septum, it splits into right and left branches which straddle the summit of the muscular part of the ventricular septum. The **right bundle branch** continues under the endocardium toward the apex spreading to all parts of the right ventricle. It breaks up into small bundles of what are called the terminal conducting fibers or Purkinje fibers which become continuous with the muscle of the right ventricle. A large bundle may pass along the moderator band, if this is present, to reach the opposite

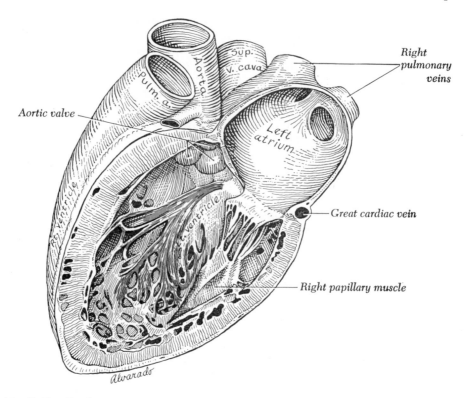

FIG. 7–31.—Conduction system of heart. Interior of left side exposed, left bundle branch in red. (From Rauber-Kopsch, redrawn.)

wall of the ventricle. The **left bundle branch** (Fig. 7–31) penetrates the fibrous septum and comes to lie just under the endocardium of the left ventricle. It fans out on the septal wall more quickly than the right bundle and breaks up into bundles of the terminal conducting fibers of Purkinje which are distributed throughout the left ventricle.

The two bundle branches are surrounded by more or less distinct connective tissue sheaths and may be visible, therefore, in gross examination of the heart. They are more easily seen in a fresh specimen than in a preserved one, and the left branch frequently shows more clearly than the right. The bundle and branches may be demonstrated by injecting India ink into the connective tissue sheath of a fresh specimen, a procedure which is particularly effective with a sheep or ox heart.

The **terminal conducting fibers** or **Purkinje fibers** are different histologically from the nodal fibers but they merge with the nodal tissue on the one hand and with the regular heart muscle on the other. They can only be identified with the aid of a microscope. They run at first in small bundles under the endocardium, then penetrate and ramify as individual fibers throughout the ventricular musculature.

Histology of Cardiac Muscle (Fig. 7–32).— The contractile tissue of the myocardium is composed of fibers with the characteristic cross striations of muscular tissue. The **cross-striated fibrillae** making up the contractile substance resemble those of skeletal muscle very closely but the architecture of the tissue as a whole shows certain differences. The fibers are about one-third as large in diameter as those of skeletal muscle, they are richer in **sarcoplasm,** the nuclei are centrally placed instead of being at the periphery, and the cell membrane is not thickened into a distinctly identifiable sarcolemma. The fibers are composed of a linear series of cellular units whose boundaries are indistinct in most histological preparations. The fibers also branch frequently and are interconnected into a network which is continuous except where the bundles and laminae are attached at their origins and insertions into the fibrous trigone of the heart. Due to the spiral

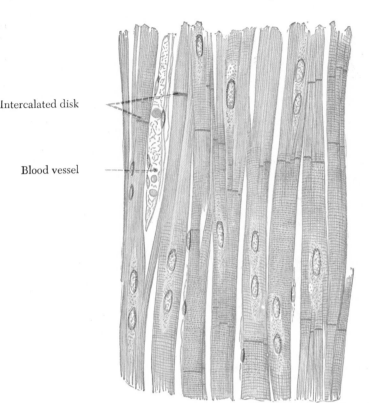

Intercalated disk

Blood vessel

FIG. 7–32.—Longitudinal section of human cardiac muscle fibers, approximately 400 × magnification. (From Rauber-Kopsch, *Lehrbuch u. Atlas d. Anatomie d. Menschen,* 19th Edition, Vol. I, courtesy of Georg Thieme Verlag, Stuttgart, 1955.)

course of these bundles, as described above, any section through the wall of the ventricle will contain many fibers cut obliquely as well as transversely and longitudinally. In the individual fibers, the cross-striated fibrillae are more concentrated at the periphery. The central core contains the nuclei and may contain a concentration of sarcoplasm which is rich in sarcosomes (Fig. 7–33).

Intercalated Disks (Fig. 7–32).—A characteristic feature of freshly preserved cardiac muscle is the presence of intercalated disks. They appear in longitudinally cut fibers as deeply staining lines which cross the fibers in the intervals between the nuclei. They may be seen in hematoxylin and eosin preparations but their visibility is enhanced by special stains. Recent studies with the electron microscope indicate that they represent cell boundaries. They may cross an entire fiber as a straight line or have a step-like configuration due to the refractive qualities of the cross-striated substance.

The connective tissue in cardiac muscle is more sparse than that of skeletal muscle. The fibers are enmeshed in a delicate reticular fibrillar net containing a few elastic fibers. Collagenous fibers are found between the muscular bundles and laminae and accompanying blood vessels. The fibers are attached to the fibrous trigones at the atrioventricular junction in much the same way as skeletal muscle fibers to tendons.

Conduction System.—The sinuatrial node contains slender fusiform cells largely filled with sarcoplasm but containing a few striated fibrillae. They are irregularly grouped together and at the periphery of the node merge with the atrial musculature. The atrioventricular node also contains slender fibers which are branched and irregularly arranged. They merge with the atrial musculature on the one hand and continue into the atrioventricular bundle on the other. In the main bundle and its first two branches the fibers remain slender but in the smaller branches they take on the characteristics of Purkinje fibers. The latter are larger in diameter than ordinary cardiac muscle; they contain relatively few peripherally placed myofibrillae; they have abundant sarcoplasm, and the centrally placed nuclei are larger and more vesicular than those of the usual cardiac

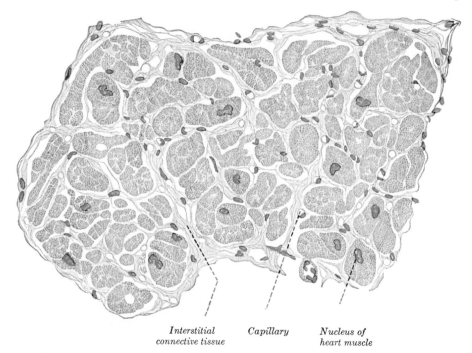

Interstitial   *Capillary*   *Nucleus of*
connective tissue   *heart muscle*

FIG. 7–33.—Cross section of human cardiac muscle. (From Rauber-Kopsch, *Lehrbuch u. Atlas d. Anatomie d. Menschen,* 19th Edition, Vol. I, courtesy of Georg Thieme Verlag, Stuttgart, 1955.)

muscle. As the ramifications of the bundles spread into the ventricular wall the individual Purkinje fibers become continuous with fibers of the main cardiac musculature.

Vessels and Nerves.—The **arteries** supplying the heart are the right and left coronary arteries, described in the next chapter. The **veins** are mostly tributaries of the coronary sinus, described in the chapter on veins.

The **lymphatics** consist of deep and superficial plexuses which form right and left trunks and end in the tracheobronchial nodes (see Lymphatic System).

The **nerves** are cardiac branches of the vagus nerve and sympathetic trunks. These nerves are combined into the cardiac plexus and its ramifications, the coronary plexuses, which accompany the coronary arteries. The sympathetic fibers are postganglionics from the cervical and upper thoracic ganglia. The vagus fibers are preganglionics whose ganglion cells form clusters in the connective tissue of the epicardium of the atria and the interatrial septum (see Nervous System).

Congenital Malformations of the Heart.—Abnormalities may result from the suppression or overgrowth in varying degrees of the individual parts of the heart. The more prevalent abnormalities, however, are the result of partial failure or abnormality in the partitioning

of the heart. A defect in the atrial septum which allows the passage of a probe through the foramen ovale is quite common (20%) but is of no functional significance. Larger defects of the septum are less common and may involve the atrial or ventricular septa or both. Abnormality in the growth of the septum and the endocardial cushions responsible for separating the truncus into aorta and pulmonary trunk results in what are called transposition complexes. Transposition refers to a shifting of the arterial trunks from their proper origins. It is designated according to the degree of displacement as (*a*) overriding, (*b*) partial transposition and (*c*) complete transposition. In overriding, one of the arterial trunks straddles the ventricular septum in which there is a defect. In partial transposition, both arterial trunks arise from the same ventricle. In complete transposition the aorta arises from the right ventricle and the pulmonary trunk from the left ventricle. It is not uncommon for the abnormalities to occur in groups or complexes. One of the better known complexes called the tetralogy of Fallot has a combination of the following four abnormalities: (1) overriding of the aorta, (2) a defect in the ventricular septum, (3) pulmonary stenosis, and (4) hypertrophied right ventricle. The Eisenmenger complex is similar to the tetralogy, having (1) an overriding aorta, (2) septal defect, (3) hypoplasia of the aorta, and (4) right ventricular hypertrophy. These and many other congenital malformations of the heart interfere with the pumping of the proper amount of blood into the pulmonary circulation. In consequence, the blood sent into the systemic arteries is poorly oxygenated, the predominance

of venous blood causes cyanosis, and the individual is popularly known as a "blue baby." A brief outline of cardiac anomalies with many photographs and diagrams is given by Lev ('53), and they are considered in detail in Gould ('53).

REFERENCES

EMBRYOLOGY AND CHANGES AT BIRTH

DAVIS, C. L. 1927. Development of the human heart from its first appearance to the stage found in embryos of twenty paired somites. Carneg. Instn., Contr. Embryol., 19, 245-284, + 8 plates.

EVERETT, N. B. and R. J. JOHNSON. 1951. A physiological and anatomical study of the closure of the ductus arteriosus in the dog. Anat. Rec., 110, 103-111.

GOSS, C. M. 1952. Development of the median coordinated ventricle from the lateral hearts in rat embryos with three to six somites. Anat. Rec., 112, 761-796.

KRAMER, T. C. 1942. The partitioning of the truncus and conus and the formation of the membranous portion of the interventricular septum in the human heart. Amer. J. Anat., 71, 343-370.

LICATA, R. H. 1954. The human embryonic heart in the ninth week. Amer. J. Anat., 94, 73-126.

SHANER, R. F. 1963. Abnormal pulmonary and aortic semilunar valves in embryos. Anat. Rec., 147, 5-13.

SMITH, R. B. 1970. The development of the intrinsic innervation of the human heart between 10 and 70 mm stages. J. Anat. (Lond.), 107, 271-279.

TANDLER, J. 1912. The development of the heart. In: *Manual of Human Embryology* by Keibel and Mall. J. B. Lippincott Company, Philadelphia. 2, 534-570.

VERNALL, D. G. 1962. The human embryonic heart in the seventh week. Amer. J. Anat., 111, 17-24.

DE VRIES, P. A. and J. B. DE C. SAUNDERS. 1962. Development of the ventricles and spiral outflow tract in the human heart. Carneg. Instn., Contr. Embryol., 37, 89-114, +9 plates.

EXPERIMENTAL

COPENHAVER, W. M. 1945. Heteroplastic transplantation of the sinus venosus between two species of Amblystoma. J. exp. Zool., 100, 203-216.

FALES, D. E. 1946. A study of double hearts produced experimentally in embryos of Amblystoma punctatum. J. exp. Zool., 101, 281-298.

FOX, M. H. and C. M. GOSS. 1958. Experimentally produced malformations of the heart and great vessels in rat fetuses. Transposition complexes and aortic arch abnormalities. Amer. J. Anat., 102, 65-92.

GOSS, C. M. 1935. Double hearts produced experimentally in rat embryos. J. exp. Zool., 72, 33-49.

HISTOLOGY AND HISTOGENESIS

BACON, R. L. 1948. Changes with age in the reticular fibers of the myocardium of the mouse. Amer. J. Anat., 82, 469-496.

COPENHAVER, W. M. and R. C. TRUEX. 1952. Histology of the atrial portion of the cardiac conduction system in man and other mammals. Anat. Rec., 114, 601-625.

GOSS, C. M. 1931. "Slow-motion" cinematographs of the contraction of single cardiac muscle cells. Proc. Soc. exp. Biol. (N.Y.), 29, 292-293.

GOSS, C. M. 1933. Further observations on the differentiation of cardiac muscle in tissue cultures. Arch. f. exp. Zellforsch., 14, 175-201.

LEAK, L. V. and J. F. BURKE. 1964. The ultrastructure of human embryonic myocardium. Anat. Rec., 149, 623-650.

MUIR, A. R. 1957. An electron microscope study of the embryology of the intercalated disc in the heart of the rabbit. J. Biophys. Biochem. Cytol., 3, 193-202.

MUIR, A. R. 1965. Further observations of the cellular structure of cardiac muscle. J. Anat. (Lond.), 99, 27-46.

SINCLAIR, J. G. 1957. Synchronous mitosis on a cardiac infarct. Tex. Rep. Biol. Med., 15, 347-352.

TRUEX, R. C. and W. M. COPENHAVER. 1947. Histology of the moderator band in man and other mammals with special reference to the conduction system. Amer. J. Anat., 80, 173-202.

GROSS ANATOMY AND ANOMALIES

COBB, W. M. 1944. Apical pericardial adhesion resembling the reptilian gubernaculum cordis. Anat. Rec., 89, 87-91.

GOULD, S. E. 1953. *Pathology of the Heart.* xiii + 1023 pages. Charles C Thomas, Publishers, Springfield, Illinois.

KERR, A., JR. and C. M. GOSS. 1956. Retention of embryonic relationship of aortic and pulmonary valve cusps and a suggested nomenclature. Anat. Rec., 125, 777-782.

LATIMER, H. B. 1953. The weight and thickness of the ventricular walls in the human heart. Anat. Rec., 117, 713-724.

LEV, M. 1953. *Autopsy Diagnosis of Congenitally Malformed Hearts.* xiv + 194 pages. Charles C Thomas, Publishers, Springfield, Illinois.

MAYER, F. E., A. S. NADAS, and P. A. ONGLEY. 1957. Ebstein's anomaly: Presentation of ten cases. Circulation, 16, 1057-1069.

REEMTSMA, K. and W. M. COPENHAVER. 1958. Anatomic studies of the cardiac conduction system in congenital malformations of the heart. Circulation, *17*, 271-276.

SHANER, R. F. 1949. Malformation of the atrioventricular endocardial cushions of the embryo pig and its relation to defects of the conus and truncus arteriosus. Amer. J. Anat., *84*, 431-455.

TANDLER, J. 1913. Anatomie des Herzens. In *von Bardelebens Handbuch der Anatomie des Menschen.* viii + 292 pages. Gustav Fischer, Jena.

TROYER, J. R. 1961. A multiple anomaly of the human heart and great veins. Anat. Rec., *139*, 509-513.

CONDUCTION SYSTEM

BAERG, R. D. and D. L. BASSETT. 1963. Permanent gross demonstration of the conduction tissue in the dog heart with palladium iodide. Anat. Rec., *146*, 313-317.

ERICKSON, E. E. and M. LEV. 1952. Ageing changes in the human atrioventricular node, bundle, and bundle branches. J. Geront., 7, 1-12.

JAMES, T. N. 1961. Anatomy of the human sinus node. Anat. Rec., *141*, 109-139.

JAMES, T. N. 1961. Morphology of the human atrioventricular node, with remarks pertinent to its electrophysiology. Amer. Heart J., *62*, 756-771.

STOTLER, W. A. and R. A. McMAHON. 1947. The innervation and structure of the conductive system of the human heart. J. Comp. Neurol., 87, 57-84.

TITUS, J. L., G. W. DAUGHERTY, and J. EDWARDS. 1963. Anatomy of the normal human atrioventricular conduction system. Amer. J. Anat., *113*, 407-415.

WHITE, P. D. 1957. The evolution of our knowledge about the heart and its diseases since 1628. Circulation, *15*, 915-923.

WIDRAN, J. and M. LEV. 1951. The dissection of the atrioventricular node, bundle and bundle branches in the human heart. Circulation, *4*, 863-867.

BLOOD AND NERVE SUPPLY

ELLISON, J. P., and T. H. WILLIAMS. 1969. Sympathetic nerve pathways to the human heart, and their variations. Amer. J. Anat., *124*, 149-162.

GROSS, L. and M. A. KUGEL. 1933. The arterial blood vascular distribution to the left and right ventricles of the human heart. Amer. Heart J., *9*, 165-177.

JAMES, T. N. and G. E. BURCH. 1958. The atrial coronary arteries in man. Circulation, *17*, 90-98.

JAMES, T. N. and G. E. BURCH. 1958. Blood supply of the human interventricular septum. Circulation, *17*, 391-396.

MIKHAIL, Y. 1970. Intrinsic nerve supply of the ventricles of the heart. Acta Anat., *76*, 289-298.

RANDALL, W. C., J. A. ARMOUR, D. C. RANDALL, and A. A. SMITH. 1971. Functional anatomy of the cardiac nerves in the baboon. Anat. Rec., *170*, 183-198.

TRUEX, R. C. and A. W. ANGULO. 1952. Comparative study of the arterial and venous systems of the ventricular myocardium with special reference to the coronary sinus. Anat. Rec., *113*, 467-491.

WOOLARD, H. H. 1926. The innervation of the heart. J. Anat. (Lond.), *60*, 345-373.

For additional references on innervation of the heart see Autonomic Nervous System.

SURGERY OF THE HEART AND GREAT VESSELS

BLALOCK, A. 1947. The technique of creation of artificial ductus arteriosus in the treatment of pulmonic stenosis. J. Thorac. Surg., *16*, 244-257.

GROSS, R. E. 1947. Complete division for the patent ductus arteriosus. J. Thorac. Surg., *16*, 314-327.

JAMES, T. N. and G. E. BURCH. 1958. Topography of the human coronary arteries in relation to cardiac surgery. J. Thorac. Surg., *36*, 656-664.

8 | *The Arteries*

aving arrived at the heart, the blood must then be propelled to all parts of the body through the arteries. Two large arteries leave the heart. The pulmonary artery distributes the blood to the lungs constituting the pulmonary system. The aorta, with its branches constituting the systemic arteries, distributes blood to the rest of the body. The tree-like ramifications of the systemic arteries find their way into all the tissues except the nails, epidermis, cornea, mucous membranes, and most cartilage.

The branchings of the arteries follow different patterns. A short trunk may subdivide into several branches at the same point, for example in the celiac or thyrocervical trunk. Several branches may be given off in succession and the main trunk continue, as in the arteries to the limbs. The division may be dichotomous as at the bifurcation of the abdominal aorta into the common iliac arteries.

Although the branches of an artery are smaller than their trunk, the combined cross-sectional area of the two resulting arteries is greater than that of the parent trunk before the division. The combined cross-sectional area of all the arterial branches, therefore, greatly exceeds that of the aorta.

Throughout the body generally the larger arterial branches pursue a fairly straight course, but in certain situations they are tortuous. The facial artery and the arteries of the lips, for example, are extremely tortuous, being accommodated to the movements of the parts. The uterine arteries also are coiled and tortuous, accommodating themselves to the increase of size which the uterus undergoes during pregnancy. The larger arteries usually are deeply placed, occupying the more protected situations such as along the flexor surface of the limbs where they are less exposed to injury.

Anastomosis of Arteries.—The branches of the arteries in many parts of the body open into branches of other arteries of a similar size in what are called **anastomoses,** instead of terminating only in the capillaries. Anastomosis may take place between larger arteries forming arches such as those in the palm of the hand, the arcades of the intestines, or the circle of Willis at the base of the brain (Figs. 8–30, 8–37, 8–46). More frequently the anastomosis is between smaller arteries, one millimeter or less in diameter. The latter are quite numerous about the joints and between the terminal branches of the arterial trunks supplying adjacent areas of the body. They form a definite pattern of connections between neighboring arteries but in any one body all the different anastomoses which are described will not be equally well developed nor will they be easily demonstrable by dissection.

Collateral Circulation.—It is frequently necessary for a surgeon to tie off or ligate an artery in order to prevent excessive bleeding. The anastomoses with branches of adjacent arteries may be more numerous in one part of the artery than in another part and the surgeon will endeavor to place his ligature where advantage can be taken of the most numerous or effective anastomoses. When the supply of blood to the distal part of an artery which has been occluded is effected through anastomosis of its branches, the resulting pathways constitute the collateral circulation. This may take place through anastomoses with branches of the proximal part of the same artery or with branches of a large neighboring artery. In some areas the collateral circulation is very free, as it would be if either the radial or ulnar artery were ligated. In other places the collateral circulation might be carried by very small arteries only. If the part of the body involved can be put at rest, a trickle of blood through small anastomosing channels will keep the tissues alive until the anastomoses expand. The arteries retain their power of growth throughout life and a small anastomosing

artery may enlarge into a main trunk, completely restoring the circulation to the part, if time is allowed for this growth to take place.

With the arteries in certain parts of the body, anastomoses are very limited or lacking. These are called end arteries. When such an artery is blocked by thrombosis or embolism, the tissues in that area are left without blood supply and the resulting condition is called an **infarct**.

DEVELOPMENT OF THE ARTERIES

The primordium of the aorta appears in the very early embryo at about the same time as the heart (Fig. 7–1). Two strands of cells arch dorsally from the endocardial mesenchyme, pass on each side of the foregut invagination, and turn caudalward along the neural groove. The strands beside the foregut are the primordia of the first aortic arches and their continuations are the dorsal aortae. The latter acquire isolated stretches of lumen at the same time as the lateral hearts, *i.e.,* in embryos with three somites. The arches become patent sometime later, at seven or eight somites, just before circulation begins. The umbilical arteries also probably arise from mesenchyme, independently, but after these main channels have become connected and the circulation is established, no further growth appears to be by local differentiation from the mesenchyme.

The circulation becomes functional when the first blood cells are washed out of the blood islands of the yolk sac at about the stage of nine somites (Goss '42). At this time the heart is contracting vigorously; capillaries have connected the aortae with the blood islands which, in turn, drain into the vitelline veins, and the blood cells have elaborated hemoglobin. The umbilical arteries and veins also have become connected and oxygenation of the blood takes place in the primitive placenta. From this time forward, the circulatory system develops by (*a*) budding from existing trunks, (*b*) formation of new capillary networks, and (*c*) selection of parts of the network as arteries and veins. As the bulk of an embryonic part increases in size the capil-

laries are lengthened. This soon reaches a maximum and the hemodynamic forces select certain channels for arterioles and arteries or others for venules and veins. Later the pattern of the arteries and veins changes continually according to the growth requirements of the embryo. Not only do some vessels grow larger but well-established vessels may be superseded by others more favorably placed and in consequence either regress or disappear.

The **first aortic arch** functions until the neural tube is closed and the pharynx begins to differentiate into pouches (Figs. 8–1 and 8–2). As the second pharyngeal pouch forms, a sprout from the aortic sac joins the dorsal aorta passing between the first and second pouches. The first aortic arch diminishes as this second arch develops and the third, fourth, and sixth (pulmonary) arches are formed in a similar manner and disappear or become modified into various adult structures. The fifth arch is never more than a questionable rudiment in the human embryo.

By the time the embryo is 4 mm in length, the first arch has about disappeared, the second arch has formed, reached its maximum development and diminished in size, and the third arch is well developed (Figs. 8–3 and 8–4). Sprouts may be present for the fourth and pulmonary arches. In a 5-mm embryo the third and fourth arches have reached a maximum and the dorsal and ventral sprouts of the pulmonary arches are nearly joined (Figs. 8–6 and 8–7). The pulmonary arches are usually complete in a 6-mm embryo. The right one soon begins to regress and has disappeared in a 12 to

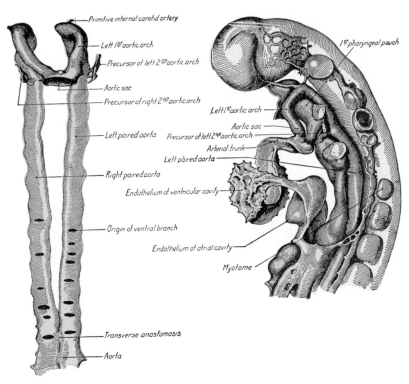

FIG. 8–1

FIG. 8–2

FIGS. 8–1 and 8–2.—Ventral and lateral views of the cranial portion of the arterial system of a 3-mm human embryo. The first aortic arch is at its maximum development and the dorsal and ventral outgrowths, which are to aid in the formation of the second arch, are just appearing. (Congdon, 1922.)

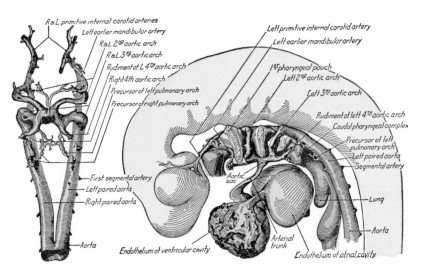

FIG. 8–3

FIG. 8–4

FIGS. 8–3 and 8–4.—Ventral and lateral views of an embryo 4 mm in length, in which the first arch has gone, the second is much reduced, and the third well developed. Dorsal and ventral outgrowths for the fourth and probably the pulmonary arch (fifth) are present. (Congdon, 1922.)

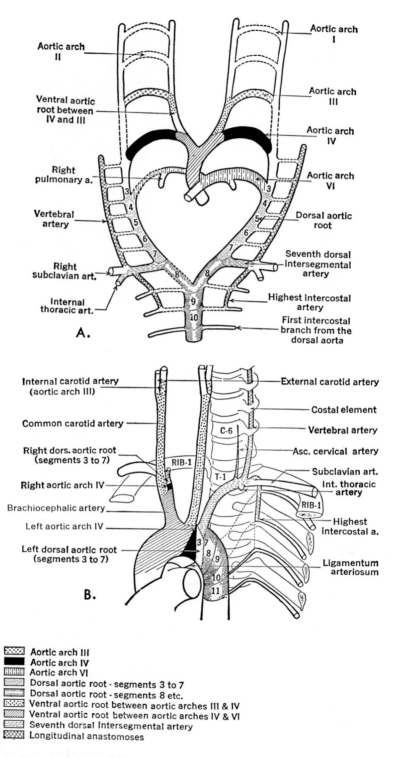

Aortic arch II

Aortic arch I

Ventral aortic root between IV and III

Aortic arch III

Aortic arch IV

Right pulmonary a.

Aortic arch VI

Vertebral artery

Dorsal aortic root

Seventh dorsal intersegmental artery

Right subclavian art.

Highest intercostal artery

Internal thoracic art.

First intercostal branch from the dorsal aorta

A.

Internal carotid artery (aortic arch III)

External carotid artery

Costal element

Common carotid artery

Vertebral artery

Right dors. aortic root (segments 3 to 7)

Asc. cervical artery

Subclavian art.

Right aortic arch IV

Int. thoracic artery

Brachiocephalic artery

Left aortic arch IV

Highest intercostal a.

Left dorsal aortic root (segments 3 to 7)

Ligamentum arteriosum

B.

☒ Aortic arch III
■ Aortic arch IV
▥ Aortic arch VI
▤ Dorsal aortic root - segments 3 to 7
▦ Dorsal aortic root - segments 8 etc.
▨ Ventral aortic root between aortic arches III & IV
▧ Ventral aortic root between aortic arches IV & VI
▤ Seventh dorsal Intersegmental artery
▨ Longitudinal anastomoses

FIG. 8–5.—*A*, Schematic diagram indicating the various components of the embryonic aortic arch complex in the human embryo. Those components which do not normally persist in the adult are indicated by broken outlines. *B*, Adult aorta as seen from the left ventral aspect. (From the original of Figs. 1 and 3, Alexander Barry, Anat. Rec. *111*, 222, 1951.)

13-mm embryo. The third arches also have become incomplete at this time (Figs. 8–8, 8–9, and 8–10).

Aortic Arches.—Although the aortic arches do not persist as such, remnants of them remain as important parts of the arterial system (Fig. 8–5). The **first arch,** even at the very earliest stages, has an extension into the region of the forebrain, the primordium of the **internal carotid artery** (Fig. 8–4). The arch disappears and the primitive carotid becomes a cranial continuation of the dorsal aorta. The **second arch** also disappears early but its dorsal end gives rise to the *stapedial artery* which atrophies in man but persists in some mammals. It passes through the ring of the stapes and divides into supraorbital, infraorbital, and mandibular branches which follow the three divisions of the trigeminal nerve. The infraorbital and mandibular arise from a

common stem, the terminal part of which anastomoses with the external carotid. On the obliteration of the stapedial artery this anastomosis enlarges and forms the **maxillary artery,** and the branches of the stapedial artery are now branches of this vessel. The common stem of the infraorbital and mandibular branches passes between the two roots of the auriculotemporal nerve and becomes the middle meningeal artery; the original supraorbital branch of the stapedial is represented by the orbital twigs of the middle meningeal. The **third arch** provides the **common carotid artery,** and is therefore named the **carotid arch.** Its proximal segment connected with the aortic sac persists as the common carotid and gives rise to the external carotid (Fig. 8–10). The arch itself persists, keeping its connection with the cranial extension of the dorsal aorta but losing its connection with the

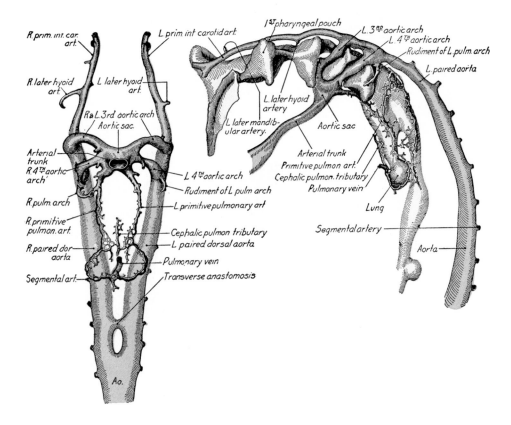

FIG. 8–6

FIG. 8–7

Figs. 8–6 and 8–7.—Ventral and lateral views of a 5-mm embryo. The third and fourth arches are in a condition of maximum development; the dorsal and ventral sprouts for the pulmonary arches have nearly met. The primitive pulmonary arches are already of considerable length. (Congdon, 1922.)

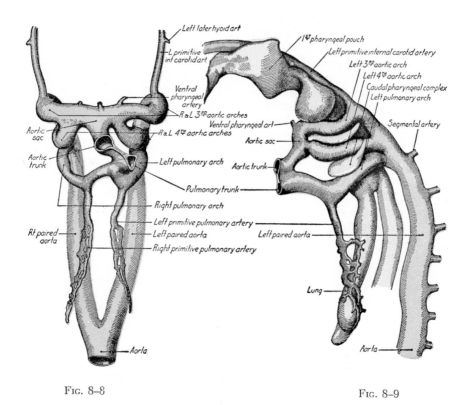

FIG. 8–8

FIG. 8–9

FIGS. 8–8 and 8–9.—Ventral and lateral views of an 11-mm embryo. The pulmonary arches are complete and the right is already regressing. The third arch is bent cranially at its dorsal end. (Congdon, 1922.)

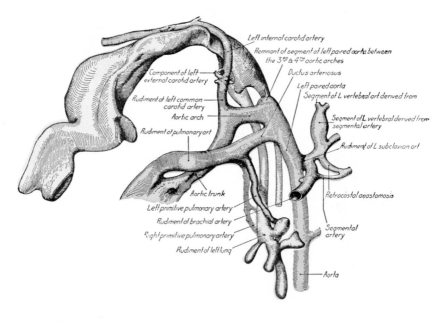

FIG. 8–10.—Lateral view of a 14-mm embryo. The last indications of the aortic arch system are just disappearing. (Congdon, 1922.)

caudal extension toward the fourth arch. The left **fourth arch** persists and provides the basis for the adult **aortic arch** between the aortic sac and the ductus arteriosus. The right fourth arch persists as the proximal part of the *right subclavian artery*. The right **pulmonary arch** disappears. The left pulmonary arch remains continuous with the pulmonary part of the truncus arteriosus as the **pulmonary trunk**. It gives rise to the pulmonary arteries and its distal segment persists as the **ductus arteriosus** until after birth at which time it contracts and gradually becomes fibrosed into the ligamentum arteriosum (Fig. 7–22).

Dorsal Aortae.—The two dorsal aortae remain separate for a short time, but come together at about the 3-mm stage to form a single trunk caudal to the eighth or ninth somites. The segment of the aorta between the third and fourth arches disappears on both sides. The left dorsal aorta becomes the descending aorta continuing caudally from the left fourth arch. The right dorsal aorta retains its continuity with the left fourth arch and becomes the *right subclavian artery* as far as the seventh intersegmental artery. It disappears between this point and the original junction with the left dorsal aorta. The **aortic isthmus is** the part of the aorta between the origin of the left subclavian and the attachment of the ductus arteriosus.

The changes in the location of the heart during development produce certain changes in the aortic arches. The heart originally lies ventral to the most cranial part of the pharynx. It later recedes into the thorax, drawing the aortic arches with it. On the right side the fourth arch recedes to the root of the neck; on the left it is withdrawn into the thorax. The recurrent laryngeal nerves of the vagus originally pass caudal to the pulmonary arches. When the heart recedes the nerves are pulled down by these arches. On the right side the pulmonary arch disappears allowing the nerve to slip up to the next arch, *i.e.*, the fourth, and it thus loops around the adult right subclavian artery. On the left side the pulmonary arch becomes the ductus arteriosus and the left recurrent nerve in the adult loops around the ligamentum arteriosum

and that part of the aorta to which the latter is attached.

The Subclavian and Vertebral Arteries.— Segmental arteries arise from the primitive dorsal aortae and anastomose between successive segments (Fig. 8–5). The seventh segmental artery is of special interest, since it forms the lower end of the vertebral artery and, when the forelimb bud appears, sends a branch to it (the subclavian artery). From the seventh segmental arteries the entire left subclavian and the greater part of the right subclavian are formed. The second pair of segmental arteries accompany the hypoglossal nerves to the brain and are named the *hypoglossal arteries*. Each sends forward a branch which forms the cerebral part of the **vertebral artery** and anastomoses with the posterior branch of the internal carotid. The two vertebrals unite on the ventral surface of the hindbrain to form the basilar artery. Later the hypoglossal artery atrophies and the vertebral is connected with the first segmental artery. The cervical part of the vertebral is developed from a longitudinal anastomosis between the first seven segmental arteries, so that the seventh of these ultimately becomes the source of the artery. As a result of the growth of the upper limb, the subclavian artery increases greatly in size and the vertebral then appears to spring from it.

Recent observations show that several segmental arteries contribute branches to the upper limb bud and form in it a free capillary anastomosis. Of these branches, only one, that derived from the **seventh segmental artery**, persists to form the **subclavian artery**. The subclavian artery is prolonged into the limb as the latter grows and becomes its arterial stem. Although this **axis artery** (Fig. 8–11) is a continuous vessel its parts are named topographically subclavian, axillary, and brachial as far as the forearm. The direct continuation of this stem in the forearm is the palmar interosseous artery. A branch which accompanies the median nerve soon increases in size and forms the main vessel (median artery) of the forearm, while the palmar interosseous diminishes. Later the radial and ulnar arteries are developed as branches of the brachial part of the stem and coincidentally

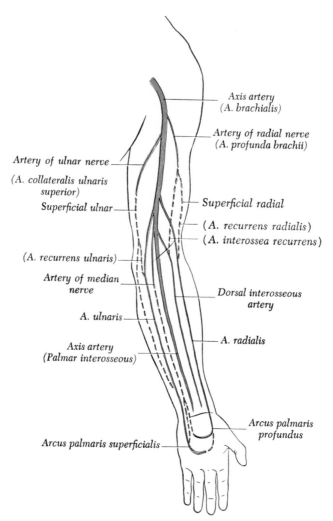

Axis artery
(A. brachialis)

Artery of radial nerve
(A. profunda brachii)

Artery of ulnar nerve

(A. collateralis ulnaris
superior)

Superficial ulnar

Superficial radial

(A. recurrens radialis)
(A. interossea recurrens)

(A. recurrens ulnaris)

Artery of median
nerve

Dorsal interosseous
artery

A. ulnaris

A. radialis

Axis artery
(Palmar interosseous)

Arcus palmaris
profundus

Arcus palmaris superficialis

Fig. 8–11.—Diagram to illustrate the embryonic pattern and subsequent development of the arteries to the upper limb. Names in parentheses are the adult structures. Adapted from De Vries '02.

with their enlargement the median artery recedes; occasionally it persists as a vessel of some considerable size and then accompanies the median nerve into the palm of the hand.

Descending Aorta.—The segmental arteries caudal to the seventh grow out into the body wall and retain their segmental character to become the intercostal and lumbar arteries. In the early embryo, paired ventral branches of the aorta grow out on the yolk sac as the omphalomesenteric arteries. At its caudal end the ventral branches accompanying the allantois become the umbilical arteries. As the gut develops, a number of ventral branches grow into it.

At first they are paired but with the formation of a mesentery the pairs fuse into single stems. These ventral branches are irregularly spaced and by a process of shifting along a rich anastomosis the three main trunks, the celiac, the superior, and the inferior mesenteric, are finally selected. Lateral branches of the aorta grow into the mesonephros and the long course of the testicular and ovarian arteries is the result of the caudal migration of the gonads after their arteries had become established. Later similar lateral branches grow into the kidneys and suprarenal glands.

According to Senior (Fig. 8–12) the

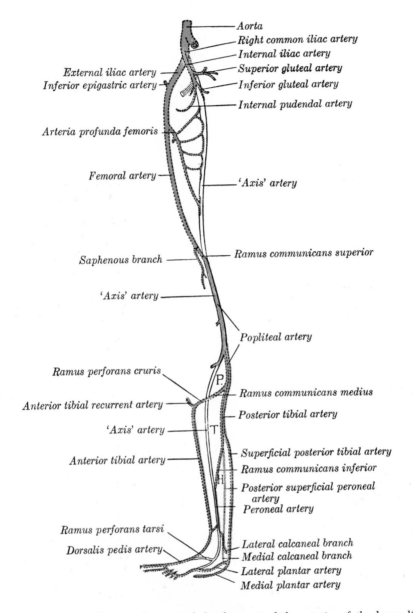

FIG. 8–12.—A diagram to illustrate the general development of the arteries of the lower limb. The letter *P* indicates the position of the Popliteus; *T*, that of the Tibialis posterior; *H*, that of the Flexor hallucis longus. (H. D. Senior, 1919.)

primary arterial trunk or "axis" **artery of the embryonic lower limb** arises from the dorsal root of the umbilical artery, and courses along the posterior surface of the thigh, knee and leg. The femoral artery springs from the external iliac and forms a new channel along the ventral side of the thigh to its communication with the axis artery above the knee. As this channel increases in size, that part of the axis artery proximal to the communication disappears, except its upper end which persists as the inferior gluteal artery. Two other segments of the axial artery persist; one forms the proximal part of the popliteal artery, and the other forms a part of the peroneal artery.

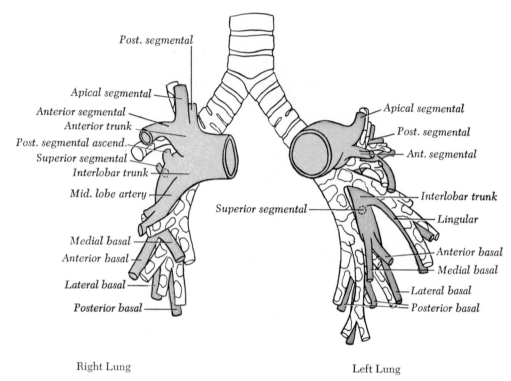

Right Lung   Left Lung

Fig. 8–13.—Diagram showing relation of the branches of the pulmonary artery to the bronchi. (Redrawn after E. A. Boyden, 1955.)

THE PULMONARY TRUNK

The **pulmonary trunk** (*truncus pulmonis; pulmonary artery*) (Figs. 7–19, 8–14) conveys the blood which has given up oxygen, *i.e.*, venous blood, from the heart to the lungs. It is a short, wide vessel, about 5 cm in length and 3 cm in diameter, arising from the conus of the right ventricle. It ascends obliquely, angling dorsally and passes at first ventral and then to the left of the ascending aorta. Near the under surface of the aortic arch at about the level of the fibrocartilage between the fifth and sixth thoracic vertebrae it divides into right and left branches of nearly equal size.

Relations.—This entire vessel is contained within the pericardium. It is enclosed with the ascending aorta in a single tube of the visceral layer of the serous pericardium, which is continued upward upon them from the base of the heart. The fibrous layer of the pericardium is gradually lost upon the external coats of the two branches of the artery. *Ventrally*, the pulmonary trunk is separated from the anterior end of the second left intercostal space

by the pleura and left lung, in addition to the pericardium; at first it is ventral to the ascending aorta, and higher up lies ventral to the left atrium on a plane dorsal to the ascending aorta. On *either side* of its origin is the auricula of the corresponding atrium and a coronary artery, the left coronary artery passing, in the first part of its course, behind the vessel. The superficial part of the cardiac plexus lies above its bifurcation, between it and the arch of the aorta.

The **right pulmonary artery** (*a. pulmonis dextra*) (Fig. 8–13) is longer and slightly larger than the left. As it leaves the bifurcation it curves around the ascending aorta, proceeds horizontally to the right, dorsal to the aorta and superior vena cava and ventral to the right bronchus. At the root of the lung it divides into an anterior trunk and a somewhat larger interlobar trunk. The anterior trunk is the artery to the superior lobe of the lung and crosses ventral to the superior lobe bronchus which in consequence has been designated in Aeby's (1878) nomenclature

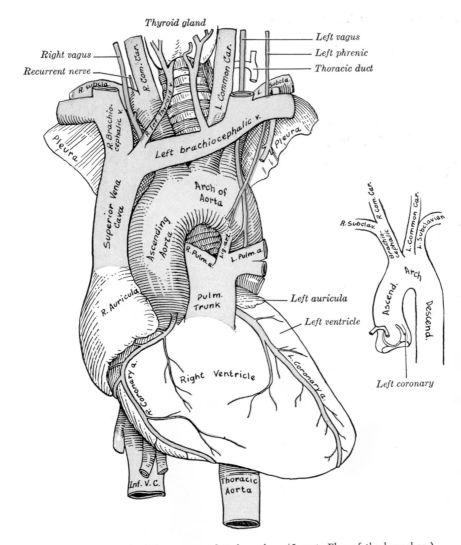

Thyroid gland

Right vagus
Recurrent nerve

Left vagus
Left phrenic
Thoracic duct

R. Subcla.
R. Com. Car.
R. Int. Thyroid
L. Common Car.
L. Subcla.

R. Brachio-cephalic v.

Left brachiocephalic v.

Pleura   Pleura

Superior Vena Cava

Arch of Aorta

Ascending Aorta

R. Pulm. a.   Lig. art.   L. Pulm. a.

Pulm. Trunk

Left auricula
Left ventricle

R. Auricula

R. Coronary a.

Right Ventricle

L. Coronary a.

R. Subclav.   R. Com. Car.   L. Common Car.
Brachio-cephalic   L. Subclavian

Ascend.   Arch   Descend.

Left coronary

Inf. V. C.   Thoracic Aorta

Fig. 8–14.—The arch of the aorta, and its branches. (Insert: Plan of the branches.)

the **eparterial bronchus.** The interlobar trunk furnishes the middle and inferior lobe arteries.

The **anterior trunk** divides into a superior and an inferior division. The **superior division** supplies an **apical** bronchopulmonary segmental (*ramus apicalis*) and a **posterior segmental** artery (*ramus posterior*). The inferior division becomes almost exclusively the **anterior segmental** artery (*ramus anterior*). The posterior artery from the superior division arches over the bronchus to reach its segment and is called a recurrent branch. In addition to its supply from the anterior trunk, the posterior segment also receives an **ascending branch** (*ramus ascendens posterior*) from the interlobar trunk in most instances (see Fig. 15–34).

The **interlobar trunk** lies at the bottom of the fissure between the superior and middle lobes. It represents the continuation of the pulmonary artery as far as the origin of the middle lobe artery. It supplies one or two ascending branches to the posterior segment of the superior lobe.

The **middle lobe** artery is pictured (Fig. 8–13) as a single artery dividing into a medial and a lateral segmental artery but according to Boyden ('55) it is more common for the two segmental arteries to arise separately. The **lateral segmental artery** (*ramus lateralis*) divides into a posterior and an anterior branch. The **medial segmental artery** (*ramus medialis*) divides into a superior and an inferior branch (see Fig. 15–35).

The first branch of the continuation of the interlobar artery into the inferior lobe is the **superior segmental artery** (*ramus superior*). It divides into a lateral branch and a combined medial and superior branch.

The termination of the right pulmonary artery is the **basal segmental artery** (*pars basalis*) which supplies the four basal segments. It divides into two parts, one provides the **medial basal** (*ramus basalis medialis*) and the **anterior basal** (*ramus basalis anterior*) segmental arteries. The other provides the **lateral basal** (*ramus basalis lateralis*) and the **posterior basal** (*ramus basalis posterior*) segmental arteries. The individual segmental arteries of all four segments are quite variable, both in their origin and in the number and size of accessory branches (see Fig. 15–37).

The **left pulmonary artery** (*arteria pulmonis sinistra*) (Fig. 8–13) is shorter and somewhat smaller than the right. In the fetus, however, it is larger and more important because it provides the ductus arteriosus. In the adult the ductus has regressed into a short fibrous cord, the **ligamentum arteriosum**, connecting the left pulmonary artery with the arch of the aorta just distal to the reflection of the pericardium (Figs. 7–19, 8–16). Although it has many similarities with the right, it differs in both general and specific details of branching. It tends to have more separate branches than the right and it does not have a distinct superior and inferior trunk.

Superiorly it is attached to the arch of the aorta, by the ligamentum arteriosum, on the left of which is the left recurrent nerve and on the right the superficial part of the cardiac autonomic plexus. *Inferiorly* it is joined to the superior left pulmonary vein by the **ligament of the left vena cava.** As it enters the root of the lung it arches over the left bronchus and passes dorsal to the bronchi, continuing past the interlobar fissure as the interlobar portion.

The left pulmonary artery has no superior trunk, as on the right, and there is, consequently, no eparterial bronchus. The apicoposterior bronchopulmonary segment of the left upper lobe is supplied by separate apical and posterior arteries. The **apical segmental artery** (*ramus apicalis*) arises from the anterior surface of the left pulmonary artery as the latter arches over the bronchus and from there it runs along the bronchus. It may be represented by two or more branches, one of which may combine with branches to the posterior or anterior segments. The **posterior segmental artery** (*ramus posterior*) arises distal to the apical and continues along its bronchus. The **anterior segmental artery** (*ramus anterior descendens*) may be single or multiple, arising from the superior part of the arched portion of the main artery. Its separate branches supply the anterior and posterior parts of the segment (see Fig. 15–38).

The remainder of the left pulmonary artery resembles an inferior trunk but is named the **interlobar portion.** From its posterior surface the **posterior segmental artery** (*ramus posterior*) arises either as a single branch or as two branches, one of which may supply a branch to the posterior part of the anterior segment (*ramus anterior ascendens*) (see Fig. 15–38).

The **lingular artery** (*ramus lingularis*) arises from the anterior or medial part of the interlobar artery, usually as a single branch. Its branches, the **superior** and **inferior lingular arteries** (*rami lingulares superior et inferior*), however, may be supplied by separate arteries.

The **superior segmental artery** (*ramus superior*) of the left inferior lobe has an unexpected origin from the interlobar artery between the origin of the apical posterior artery of the superior lobe and that of the lingular artery. It divides into three branches which may combine with or supply branches to other inferior lobe segments.

The remainder of the interlobar artery continues as the **basal segmental artery** (*pars basalis*). The **medial basal** and **anterior basal** segmental arteries (*rami basales medialis et anterior*) usually arise from a common trunk. The **lateral basal segment** (*ramus basalis lateralis*) arises more often as a separate branch than from a common trunk with the **posterior basal artery** (*ramus basalis posterior*) or it may be combined as a whole or in part with branches of the superior segmental arteries.

THE AORTA

The aorta is the main trunk of the systemic arteries. At its commencement from the aortic opening of the left ventricle it is about 3 cm in diameter. It ascends toward the neck for a short distance then bends to the left and dorsalward over the root of the left lung. It descends within the thorax on the left side of the vertebral column and passes through the aortic hiatus of the diaphragm into the abdominal cavity. Opposite the caudal border of the fourth lumbar vertebra, considerably diminished in size (about 1.75 cm in diameter) it bifurcates into the two common iliac arteries. The parts of the aorta are the **ascending aorta,** the **arch of the aorta,** and the **thoracic** and **abdominal portions** of the descending aorta.

The Ascending Aorta

The **ascending aorta** (*aorta ascendens*) (Fig. 8–15) is about 5 cm in length. It is covered by the visceral pericardium which encloses it in a common sheath with the pulmonary trunk. It commences at the semilunar valve, on a level with the caudal border of the third costal cartilage dorsal to the left half of the sternum. It curves obliquely to the right, in the direction of the heart's axis, as high as the cranial border of the second right costal cartilage, lying about 6 cm deep to the dorsal surface of

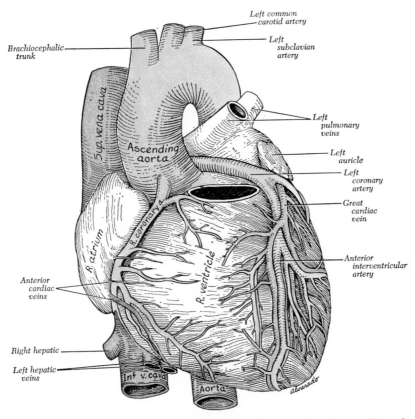

FIG. 8–15.—The coronary blood vessels of the sternocostal aspect of the heart. The pulmonary trunk and pulmonary valve have been removed. (Redrawn from Rauber-Kopsch.)

the sternum. At its origin, opposite the segments of the aortic valve, are three small dilatations called the **aortic sinuses.** At the continuation of the ascending aorta into the aortic arch the caliber of the vessel is increased by a bulging of its right wall, causing the transverse section to present a somewhat oval figure.

Relations.—The ascending aorta is covered at its commencement by the trunk of the pulmonary artery and the right auricula, and more cranially, is separated from the sternum by the pericardium, the right pleura, the ventral margin of the right lung, some loose areolar tissue, and the remains of the thymus; dorsally, it rests upon the left atrium and right pulmonary artery. On the *right side,* it is in relation with the superior vena cava and right atrium, the former lying partly dorsal to it; on the *left side,* with the pulmonary trunk.

Branches of the ascending aorta are the 1) right and 2) left coronary arteries.

1. The **Right Coronary Artery** (*a. coronaria dextra*) (Figs. 8–15, 8–16) arises in the right adjacent (Fig. 7–28) or right posterior aortic sinus (of Valsalva). It passes to the right and toward the apex between the conus arteriosus and right auricula into the coronary sulcus. It follows the sulcus first to the right around the right margin of the heart and then to the left on the diaphragmatic surface, ending in two or three branches beyond the interventricular sulcus.

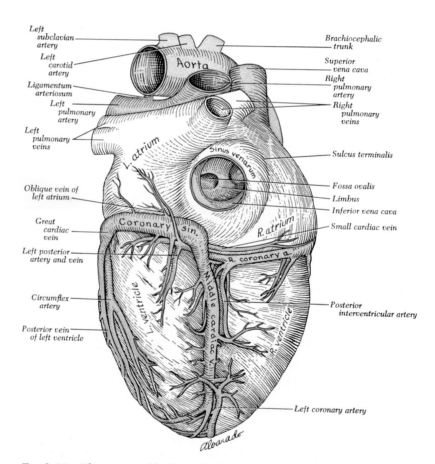

Fig. 8–16.—The coronary blood vessels of the diaphragmatic aspect of the heart. (Redrawn from Rauber-Kopsch.)

a) The **right** or **dorsal interventricular artery** (*ramus interventricularis posterior*) runs down the dorsal sulcus two thirds of the way to the apex supplying branches to both ventricles.

b) The large **marginal branch** arises at the right margin and follows the acute margin, terminating near the apex on the dorsal surface of the right ventricle and supplying both ventral and dorsal surfaces of the right ventricle. It supplies small branches to the right atrium, one of which passes between the right atrium and superior vena cava to supply the *sinuatrial node.* (Gross '21.)

2. The **Left Coronary Artery** (*a. coronaria sinistra*) (Figs. 8–15, 8–16) arises in the left adjacent (Fig. 7–28) or left posterior aortic sinus (of Valsalva) and after a short course under cover of the left auricula bifurcates into the anterior descending branch and the circumflex branch.

a) The **ventral** or **left interventricular artery** (*ramus interventricularis anterior*) passes to the left between the pulmonary trunk and the left auricula to the ventral interventricular sulcus which it follows to the apex, supplying branches to both ventricles.

b) The **circumflex branch** (*r. circumflexus*) follows the left part of the coronary sulcus running first to the left and then to the right reaching nearly as far as the posterior interventricular sulcus. It supplies branches to the left atrium and ventricle.

Variations.—The left coronary artery or its descending branch may arise from the left pulmonary artery, supplying the left ventricle. This anomaly can be diagnosed clinically and usually leads to death in infancy. Rarely both coronaries arise from the pulmonary artery.

Single coronary arteries have been described. The left coronary may be a branch of the right or both arteries may arise in the same aortic sinus. There may be three coronary arteries, the accessory supplying part of the branches of either artery.

Three **patterns of distribution** of the coronary arteries have been described (Schlesinger '40). In half the hearts, the right coronary predominates, in a third, they are equally balanced and in the rest, the left coronary predominates. The sinuatrial node is supplied by a branch of the right coronary in 70 per cent, by the left in 25 per cent, by both in 7 per cent. The *atrioventricular node* is supplied by the right in 92 per cent. The *right bundle branch* generally is supplied by the anterior descending branch of the left coronary; the *left bundle branch* by septal branches of the left and small vessels from the right coronary artery (Gregg '50).

The Arch of the Aorta

The **arch of the aorta** (*arcus aortae; transverse aorta*) (Fig. 8–14) begins at the level of the cranial border of the second sternocostal articulation of the right side, and runs at first cranially, and to the left ventral to the trachea; it is then directed dorsally on the left side of the trachea and finally passes caudalward on the left side of the body of the fourth thoracic vertebra, at the caudal border of which it becomes continuous with the descending aorta. It thus forms two curvatures: one with its convexity cranialward, the other with its convexity ventralward and to the left. Its cranial border is usually about 2.5 cm caudal to the superior border of the manubrium sterni.

Relations.—The arch of the aorta is covered *ventrally* by the pleurae and anterior margins of the lungs, and by the remains of the thymus. As the vessel runs dorsally its *left* side is in contact with the left lung and pleura. On the left side of this part of the arch are four nerves; in order these are, the left phrenic, the caudal of the superior cardiac branches of the left vagus, the superior cardiac branch of the left sympathetic, and the trunk of the left vagus. As the last nerve crosses the arch it gives off its recurrent branch, which hooks around the vessel and then passes cranialward on its right side. The highest left intercostal vein runs obliquely on the left side of the arch, between the phrenic and vagus nerves. On the *right* are the deep part of the cardiac plexus, the right recurrent nerve, the esophagus, and the thoracic duct; the trachea lies dorsal and to the right of the vessel. *Cranially* are the brachiocephalic, left common carotid, and left subclavian arteries, which arise from the convexity of the arch and are crossed close to their origins by the left brachiocephalic vein. *Caudally*, are the bifurcation of the pulmonary artery, the left bronchus, the ligamentum arteriosum, the superficial part of the cardiac plexus, and the left recurrent nerve. As already stated, the ligamentum arteriosum connects the commencement of the left pulmonary artery to the aortic arch.

Between the origin of the left subclavian artery and the attachment of the ductus arteriosus the lumen of the fetal aorta is considerably narrowed, forming what is termed the **aortic isthmus,** while immediately beyond the ductus arteriosus the vessel presents a fusiform dilation which His has named the **aortic spindle**—the point of junction of the two parts being marked in the concavity of the arch by an indentation or angle. These conditions persist, to some extent, in the adult, where His found that the average diameter of the spindle exceeded that of the isthmus by 3 mm.

Variations.—The height to which the aorta rises in the thorax is usually about 2.5 cm below the cranial border of the sternum; but it may ascend nearly to the top of the bone. Occasionally it is found 4 cm, more rarely from 5 to 8 cm caudal to this point. Sometimes the aorta arches over the root of the right lung (right aortic arch) instead of over that of the left, and passes along the right side of the vertebral column, a condition which is found in birds. In such cases all the thoracic and abdominal viscera are transposed. Less frequently the aorta, after arching over the root of the right lung, is directed to its usual position on the left side of the vertebral column; this peculiarity is not accompanied by transposition of the viscera. The aorta occasionally divides, as in some quadrupeds, into an ascending and a descending trunk, the former of which is directed vertically cranialward, and subdivides into three branches, to supply the head and upper limbs. Sometimes the aorta subdivides near its origin into two branches, which soon reunite. In one of these specimens the esophagus and trachea were found to pass through the interval between the two branches; this is the normal condition of the vessel in the Reptilia.

Branches (Figs. 7–20, 8–14).—Three branches are given off from the arch of the aorta: the **brachiocephalic,** the **left common carotid,** and the **left subclavian** (83 to 94 per cent).

Variations.—The branches, instead of arising from the highest part of the arch, may spring from the commencement of the arch or upper part of the ascending aorta; or the distance between them at their origins may be increased or diminished, the most frequent change in this respect being the approximation of the left carotid toward the brachiocephalic artery.

The *number* of the primary branches may be reduced to one, or more commonly two; the left carotid arising from the brachiocephalic artery; or (more rarely) the carotid and subclavian arteries of the left side arising from a left brachiocephalic artery. But the number may be increased to four, from the right carotid and subclavian arteries arising directly from the aorta, the brachiocephalic being absent. In most of these the right subclavian has been found to arise from the left end of the arch; it is the second or third branch given off, instead of the first, in other specimens. Another common form in which there are four primary branches is that in which the left vertebral artery arises from the arch of the aorta between the left carotid and subclavian arteries. Lastly, the number of trunks from the arch may be increased to five or six; in these instances, the external and internal carotids arise separately from the arch, the common carotid being absent on one or both sides. In some few instances six branches have been found, and this condition is associated with the origin of both vertebral arteries from the arch.

When the aorta arches over to the right side, the three branches have a reverse arrangement; the brachiocephalic artery is a left one, and the right carotid and subclavian arise separately. In other instances, where the aorta takes its usual course, the two carotids may be joined in a common trunk, and the subclavians arise separately from the arch, the right subclavian generally arising from the left end.

In some instances other arteries spring from the arch of the aorta. Of these the most common are the bronchial, one or both, and the thyroidea ima; but the internal thoracic and the inferior thyroid have been seen to arise from this vessel.

The Brachiocephalic Trunk

The **brachiocephalic trunk** (*truncus brachiocephalicus; a. Anonyma; innominate artery*) (Figs. 8–14) is the largest branch of the arch of the aorta, and is from 4 to 5 cm in length. It *arises,* on a level with the cranial border of the second right costal cartilage, from the commencement of the arch of the aorta, on a plane ventral to the origin of the left carotid; it ascends cranially, dorsally, and obliquely to the right to the level of the cranial border of the right sternoclavicular articulation, where it divides into the right common carotid and right subclavian arteries.

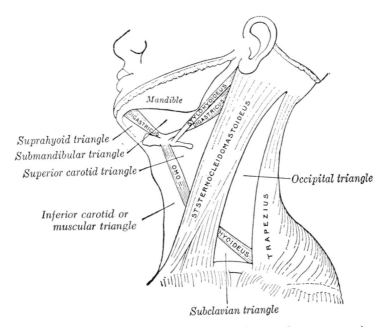

Mandible

Suprahyoid triangle
Submandibular triangle
Superior carotid triangle

Occipital triangle

Inferior carotid or muscular triangle

Subclavian triangle

Fig. 8–17.—The triangles of the neck. The Sternocleidomastoideus separates the anterior from the posterior triangle.

Relations.—*Ventrally*, it is separated from the manubrium sterni by the Sternohyoideus and Sternothyroideus, the remains of the thymus, the left brachiocephalic and right inferior thyroid veins which cross its root, and sometimes the superior cardiac branches of the right vagus. *Dorsally*, it crosses the trachea obliquely. On the *right side* are the right brachiocephalic vein, the superior vena cava, the right phrenic nerve, and the pleura; and on the *left side*, the remains of the thymus, the origin of the left common carotid artery, the inferior thyroid veins, and the trachea.

Branches.—The brachiocephalic trunk usually gives off no branches; but occasionally a small branch, the **thyroidea ima,** arises from it. Sometimes it gives off a **thymic or bronchial branch.**

The **thyroidea ima** (*a. thyroidea ima*) ascends ventral to the trachea to the inferior part of the thyroid gland, which it supplies. It varies greatly in size, and appears to compensate for deficiency or absence of one of the other thyroid vessels. It occasionally arises from the aorta, the right common carotid, the subclavian or the internal thoracic.

Variations.—The brachiocephalic trunk sometimes divides above the level of the sternoclavicular joint; less frequently below it. It may be absent, the right subclavian and the right common carotid then arising directly from the aorta. When the aortic arch is on the right side, the brachiocephalic is directed to the left side of the neck.

THE ARTERIES OF THE HEAD AND NECK

The principal arteries of supply to the head and neck are the two **common carotids;** each divides into two branches, viz., (1) the **external carotid,** supplying the exterior of the head, the face, and the greater part of the neck; (2) the **internal carotid,** supplying to a great extent the parts within the cranial and orbital cavities. The verte-

bral arteries, to be described later, assist in supplying the brain.

The Common Carotid Artery

The **common carotid** arteries (*a. carotis communis*) differ in length and in their mode of origin. The *right* begins at the bifurcation of the brachiocephalic trunk dorsal to the sternoclavicular joint and is confined to the neck. The *left* springs from the highest part of the arch of the aorta to the left of, and on a plane dorsal to, the brachiocephalic trunk, and therefore consists of a thoracic and a cervical portion.

The **thoracic portion of the left common carotid artery** ascends from the arch of the aorta through the superior mediastinum to the level of the left sternoclavicular joint, where it becomes the cervical portion (Fig. 8–30).

Relations.—*Ventrally*, it is separated from the manubrium sterni by the Sternohyoideus and Sternothyroideus, the anterior portions of the left pleura and lung, the left brachiocephalic vein, and the remains of the thymus; *dorsally*, it lies on the trachea, esophagus, left recurrent nerve, and thoracic duct. To its *right side* are the brachiocephalic artery, the trachea, the inferior thyroid veins, and the remains of the thymus; to its *left side* are the left vagus and phrenic nerves, left pleura, and lung. The left subclavian artery is dorsal and slightly lateral to it.

The **cervical portions** of the common carotids resemble each other so closely that one description will apply to both (Fig. 8–18). Each vessel passes obliquely from the sternoclavicular articulation, to the level of the cranial border of the thyroid cartilage, where it divides into the external and internal carotid arteries.

At the inferior part of the neck the two common carotid arteries are separated by a narrow interval which contains the trachea; but at the superior part, the thyroid gland, the larynx and pharynx project ventralward between the two vessels. The common carotid artery is contained in a sheath which is derived from the deep cervical fascia and encloses also the internal jugular vein and vagus nerve, the vein lying lateral to the artery, and the nerve between the artery and vein on a plane dorsal to both. Each of these three structures has a separate compartment within the sheath.

Relations.—At the caudal part of the neck, the common carotid artery is very deep, being *covered* by the integument, subcutaneous fascia, Platysma, and deep cervical fascia, the Sternocleidomastoideus, Sternohyoideus, Sternothyroideus, and Omohyoideus; in the cranial part of its course it is more superficial, being covered merely by the integument, the superficial fascia, Platysma, deep cervical fascia, and medial margin of the Sternocleidomastoideus. When the latter muscle is drawn dorsalward, the artery is seen in a triangular space, the **carotid triangle**, bounded dorsally by the Sternocleidomastoideus, cranially by the Stylohyoideus and posterior belly of the Digastricus, and caudally by the superior belly of the Omohyoideus. This part of the artery is crossed obliquely, from its medial to its lateral side, by the sternocleidomastoid branch of the superior thyroid artery; it is also crossed by the superior and middle thyroid veins which end in the internal jugular. Usually superficial to its sheath, but occasionally contained within it, are the descending hypoglossal and descending cervical nerves and their connecting loop, the ansa cervicalis. The superior thyroid vein crosses the artery near its termination, and the middle thyroid vein a little below the level of the cricoid cartilage; the anterior jugular vein crosses the artery just above the clavicle, but is separated from it by the Sternohyoideus and Sternothyroideus. *Dorsally*, the artery is separated from the transverse processes of the cervical vertebrae by the Longus colli and Longus capitis, the sympathetic trunk being interposed between it and the muscles. The inferior thyroid artery crosses dorsal to the inferior part of the vessel. *Medially*, it is in relation with the esophagus, trachea, and thyroid gland (which overlaps it), the larynx and pharynx, the inferior thyroid artery and recurrent nerve being interposed. *Lateral* to the artery are the internal jugular vein and vagus nerve.

At the inferior part of the neck, the right recurrent nerve crosses obliquely dorsal to the artery; the right internal jugular vein diverges from the artery, but the left approaches and often overlaps it.

The **carotid body** lies deep to the bifurcation of the common carotid artery or somewhat between the two branches. It is a small oval body, 2 to 5 mm in diameter, with a characteristic structure composed of epithelioid cells, abundant nerve fibers, and a delicate fibrous capsule (Figs. 11–119, 11–120). It is part of the visceral afferent system of the body, containing chemoreceptor endings which probably respond to changes in the oxygen content of the blood. It is probably supplied by branches of the vagus nerve.

The **Carotid Sinus.**—The carotid sinus, a slight dilatation of the terminal portion of the common carotid artery and of the internal carotid artery at its origin from the common carotid, or a dilatation 1 cm in length of the internal carotid only, is an important organ for the regulation of systemic blood pressure. Special nervous end organs in its modified wall respond to increase and to decrease in blood pressure and through a reflex arc, probably via the carotid branch of the glossopharyngeal nerve, convey stimuli to the medulla which result in increasing or decreasing the rate of the heart beat.

Variations.—The *right common carotid* may arise above the level of the cranial border of the sternoclavicular articulation; this variation occurs in about 12 per cent; the artery may arise as a separate branch from the arch of the aorta, or in conjunction with the left carotid. The *left common carotid* varies in its origin more than the right. In the majority of abnormal specimens it arises with the brachiocephalic trunk, the point of division occurring higher than usual, the artery dividing opposite or even cranial to the hyoid bone; more rarely, it occurs more caudally, opposite the middle of the larynx, or the lower border of the cricoid cartilage. Very rarely, the common carotid ascends in the neck without any subdivision, either the external or the internal carotid being wanting; and in a few instances the common carotid has been found to be absent, the external and internal carotids arising directly from the arch of the aorta.

The common carotid usually gives off no branch previous to its bifurcation, but it occasionally gives origin to the superior thyroid or its laryngeal branch, the ascending pharyngeal, the inferior thyroid, or, more rarely, the vertebral artery.

Collateral Circulation.—After ligature of the common carotid, the collateral circulation can be perfectly established, by the free communication which exists between the carotid arteries of opposite sides, both without and within the cranium, and by enlargement of the branches of the subclavian artery on the side corresponding to that on which the vessel has been tied. The chief communications outside the skull take place between the superior and inferior thyroid arteries, and the profunda cervicis and ramus descendens of the occipital; the vertebral supplies blood to the branches of the internal carotid within the cranium.

The External Carotid Artery

The **external carotid artery** (*a. carotis externa*) (Fig. 8–18) begins opposite the superior border of the thyroid cartilage and passing cranialward, curves somewhat anteriorly and then inclines dorsally to the space behind the neck of the mandible, where it divides into the superficial temporal and maxillary arteries. It rapidly diminishes in size in its course up the neck, owing to the number and large size of its branches. In the child, it is somewhat smaller than the internal carotid, but in the adult, the two vessels are of nearly equal size. At its origin, this artery is more superficial, and placed nearer the midline than the internal carotid, and is contained within the carotid triangle.

Relations.—The external carotid artery is *covered* by the skin, subcutaneous fascia, Platysma, deep fascia, and anterior margin of the Sternocleidomastoideus; it is crossed by the hypoglossal nerve, by the lingual, ranine, common facial, and superior thyroid veins; and by the Digastricus and Stylohyoideus; more cranially it penetrates into the substance of the parotid gland, where it lies deep to the facial nerve and the junction of the temporal and maxillary veins. *Medial* to it are the hyoid bone, the wall of the pharynx, the superior laryngeal nerve, and a portion of the parotid gland. *Lateral* to it, in the lower part of its course, is the internal carotid artery. *Posterior* to it, near its origin, is the superior laryngeal nerve; and higher up, it is separated from the internal carotid by the Styloglossus and Stylopharyngeus, the glossopharyngeal nerve, the pharyngeal branch of the vagus, and part of the parotid gland.

Branches.—The branches of the external carotid artery arise in the following order:

1. Superior Thyroid
2. Ascending Pharyngeal
3. Lingual
4. Facial
5. Occipital
6. Posterior Auricular
7. Superficial Temporal
8. Maxillary

1. The **Superior Thyroid Artery** (*a. thyroidea superior*) (Fig. 8–18) *arises* from the external carotid artery just caudal to the level of the greater cornu of the hyoid bone and ends in the thyroid gland.

From its origin deep to the anterior border of the Sternocleidomastoideus it runs cranialward and anteriorly for a short

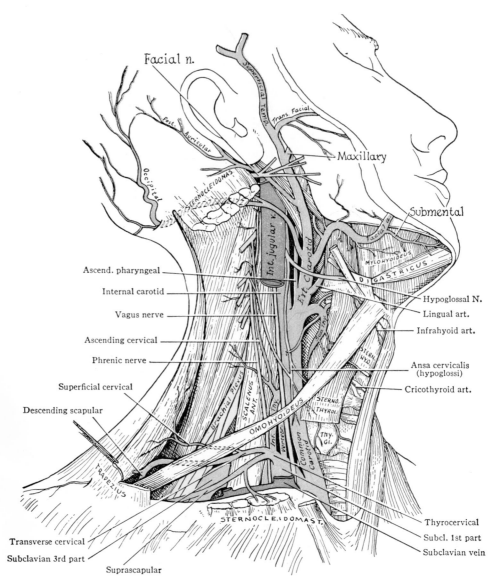

FIG. 8–18.—Superficial dissection of the right side of the neck, showing the carotid and subclavian arteries.

distance in the carotid triangle, where it is covered by the skin, Platysma, and fascia; it then arches caudalward deep to the Omohyoideus, Sternohyoideus, and Sternothyroideus. To its medial side are the Constrictor pharyngis inferior and the external branch of the superior laryngeal nerve.

Variations.—In 16 per cent of cadavers, the superior thyroid arises from the common carotid.

Branches.—It distributes twigs to the adjacent muscles, and usually two main branches to the thyroid gland; one, the larger, supplies principally the ventral surface and on the isthmus of the gland anastomoses with the corresponding artery of the opposite side; a second branch descends on the dorsal surface of the gland and anastomoses with the inferior thyroid artery.

In addition to the glandular branches and

small muscular twigs, the branches of the superior thyroid are:

a) Infrahyoid
b) Sternocleidomastoid
c) Superior laryngeal
d) Cricothyroid

a) The **infrahyoid branch** (*ramus infra-hyoideus; hyoid branch*) is small and runs along the lower border of the hyoid bone deep to the Thyrohyoideus and anastomoses with the vessel of the opposite side.

b) The **sternocleidomastoid branch** (*ramus sternocleidomastoideus; sternomas-toid branch*) runs caudally and laterally across the carotid sheath, and supplies the Sternocleidomastoideus and neighboring muscles and integument; it frequently arises as a separate branch from the external carotid.

c) The **superior laryngeal artery** (*a. laryngea superior*), larger than either of the preceding, accompanies the internal laryngeal branch of the superior laryngeal nerve, deep to the Thyrohyoideus; it pierces the thyrohoid membrane and supplies the muscles, mucous membrane, and glands of the larynx, anastomosing with the branch from the opposite side. It sometimes (13 per cent) arises separately from the external carotid.

d) The **cricothyroid branch** (*ramus crico-thyroideus*) is small and runs transversely across the cricothyroid membrane, communicating with the artery of the opposite side.

2. The **Ascending Pharyngeal Artery** (*a. pharyngea ascendens*) (Fig. 8–22), the smallest branch of the external carotid, is a long, slender vessel, deeply seated in the neck, dorsal to the other branches of the external carotid and to the Stylopharyngeus. It *arises* from the posterior part of the external carotid, near the commencement of that vessel, and ascends vertically between the internal carotid and the side of the pharynx, anterior to the Longus capitis, to the inferior surface of the base of the skull. In 14 per cent of cadavers it arises from the occipital artery.

Branches.—Its branches are:

a) Pharyngeal
b) Palatine
c) Prevertebral
d) Inferior tympanic
e) Posterior meningeal

a) The **pharyngeal branches** (*rami pharyngei*) are three or four in number. These descend to supply the Constrictores pharyngis medius ramifying in its substance and in the mucous membrane lining it. A branch supplies the Stylopharyngeus.

b) The **palatine branch** varies in size, and may take the place of the ascending palatine branch of the facial artery, when that vessel is small. It passes inward upon the Constrictor pharyngis superior, sends ramifications to the soft palate and tonsil and supplies a branch to the auditory tube.

c) The **prevertebral branches** are numerous small vessels, which supply the Longi capitis and colli, the sympathetic trunk, the hypoglossal and vagus nerves, and the lymph nodes; they anastomose with the ascending cervical artery.

d) The **inferior tympanic artery** (*a. tympanica inferior*) is a small branch which passes through a minute foramen in the petrous portion of the temporal bone, in company with the tympanic branch of the glossopharyngeal nerve, to supply the medial wall of the tympanic cavity and anastomose with the other tympanic arteries.

e) The **meningeal branches** are several small vessels, which supply the dura mater. One, the **posterior meningeal**, enters the cranium through the jugular foramen; a second passes through the foramen lacerum; and occasionally a third through the canal for the hypoglossal nerve.

3. The **Lingual Artery** (*a. lingualis*) (Fig. 8–22) *arises* from the external carotid opposite the tip of the greater cornu of the hyoid bone, and between the superior thyroid and facial arteries; it first runs obliquely medialward cranial to the greater cornu of the hyoid bone, then curves caudally and ventrally, forming a loop which is crossed by the hypoglossal nerve. It passes deep to the Digastricus and Stylohyoideus, runs horizontally medial to the Hyoglossus, and finally, ascending almost perpendicularly to the tongue, turns ventrally on its inferior surface as far as the tip, under the name of the **profunda linguae.**

Relations.—Its first, or oblique, portion is superficial, and is contained within the carotid triangle; it rests upon the Constrictor pharyngis medius, and is covered by the Platysma and the fascia of the neck. Its second, or curved, portion also lies upon the Constrictor pharyngis medius, being covered at first by the tendon of the Digastricus and by the Stylohyoideus, and afterward by the Hyoglossus. Its third, or horizontal, portion lies between the Hyoglossus and Genioglossus. The fourth, or terminal part, under the name of the **profunda linguae** (*ranine artery*) runs along the under surface of the tongue to its tip; here it is superficial, being covered only by the mucous membrane; deep to it is the Longitudinalis inferior, and on the medial side the Genioglossus. The hypoglossal nerve crosses the first part of the lingual artery, but is separated from the second part by the Hyoglossus.

Branches.—The branches of the lingual artery are:

  a) Suprahyoid
  b) Dorsal lingual
  c) Sublingual
  d) Deep lingual (Ranine)

a) The **suprahyoid branch** (*ramus suprahyoideus; hyoid branch*) runs along the cranial border of the hyoid bone, supplying the muscles attached to it and anastomosing with its fellow of the opposite side.

b) The **dorsal lingual arteries** (*a. dorsales linguae; rami dorsales linguae*) consist usually of two or three small branches which *arise* under cover of the Hyoglossus; they ascend to the posterior part of the dorsum of the tongue, and supply the mucous membrane in this situation, the glossopalatine arch, the tonsil, soft palate, and epiglottis, anastomosing with the vessels of the opposite side.

c) The **sublingual artery** (*a. sublingualis*) *arises* at the anterior margin of the Hyoglossus and runs anteriorly between the Genioglossus and Mylohyoideus to the sublingual gland. It supplies the gland and gives branches to the Mylohyoideus and neighboring muscles, and to the mucous membrane of the mouth and gums. One branch runs medial to the alveolar process of the mandible in the substance of the gum to anastomose with a similar artery from the other side; another pierces the Mylohyoideus and anastomoses with the submental branch of the facial artery.

d) The **deep lingual artery** (*a. profunda linguae; ranine artery*) is the terminal portion of the lingual artery; it pursues a tortuous course, running along the under surface of the tongue, between the Longitudinalis inferior, and the mucous membrane; it lies on the lateral side of the Genioglossus, accompanied by the lingual nerve. It anastomoses with the artery of the opposite side at the tip of the tongue.

4. The **Facial Artery** (*a. facialis; external maxillary artery*) (Fig. 8–19), *arises* in the carotid triangle a little superior to the lingual artery and, sheltered by the ramus of the mandible, passes obliquely deep to the Digastricus and Stylohyoideus, arches anteriorly to enter a groove on the posterior surface of the submandibular gland where it joins the facial vein. It then becomes quite superficial and winds around the inferior border of the mandible at the anterior edge of the Masseter, to enter the face. It crosses the cheek lateral to the angle of the mouth, runs alongside the nose, and ends at the medial commissure of the eye, under the name of the angular artery. This vessel, both in the neck and on the face, is remarkably tortuous: in the neck, to accommodate itself to the movements of the pharynx in deglutition, and in the face, to the movements of the mandible, lips, and cheeks.

Relations.—In the neck, its origin is covered by the integument, Platysma, and fascia; it then passes deep to the Digastricus and Stylohyoideus muscles, part of the submandibular gland, and frequently the hypoglossal nerve. It lies upon the Constrictores pharyngis medius and superior, the latter of which separates it, at the summit of its arch, from the tonsil. On the face, where it passes over the body of the mandible, it is superficial, lying immediately beneath the Platysma. In its course over the face, it is covered by the integument, the fat of the cheek, and, near the angle of the mouth, by the Platysma, Risorius, and Zygomaticus major. It rests on the Buccinator and Levator anguli oris, and passes either over or under the Levator labii superioris. The anterior facial vein lies lateral to the artery, and takes a more direct course across the face. The branches of the facial nerve cross the artery.

Branches.—The branches of the facial artery may be divided into two sets: those given off in the neck (*cervical*), and those on the face (*facial*).

Cervical Branches	*Facial Branches*
a) Ascending palatine	*e*) Inferior labial
b) Tonsillar	*f*) Superior labial
c) Glandular	*g*) Lateral nasal
d) Submental	*h*) Angular

i) Muscular

a) The **ascending palatine artery** (*a. palatina ascendens*) (Fig. 8–22) *arises* close to the origin of the facial artery and passes up between the Styloglossus and Stylopharyngeus to the side of the pharynx, along which it is continued between the Constrictor pharyngis superior and the Pterygoideus medialis to near the base of the skull. It divides near the Levator veli palatini into two branches: one follows the course of this muscle, and, winding over the upper border of the Constrictor pharyngis superior, supplies the soft palate and the palatine glands, anastomosing with its fellow of the opposite side and with the descending palatine branch of the maxillary artery; the other pierces the Constrictor pharyngis superior and supplies the palatine tonsil and auditory tube, anastomosing with the tonsillar and ascending pharyngeal arteries.

b) The **tonsillar branch** (*ramus tonsillaris*) (Fig. 8–22) ascends between the Pterygoideus medialis and Styloglossus, and then along the side of the pharynx, perforating the Constrictor pharyngis superior, to ramify in the substance of the palatine tonsil and root of the tongue.

c) The **glandular branches** (*rami glandulares; submaxillary branches*) consist of three or four large vessels, which supply the submandibular gland, some being prolonged to the neighboring muscles, lymph nodes, and integument.

d) The **submental artery** (*a. submentalis*),

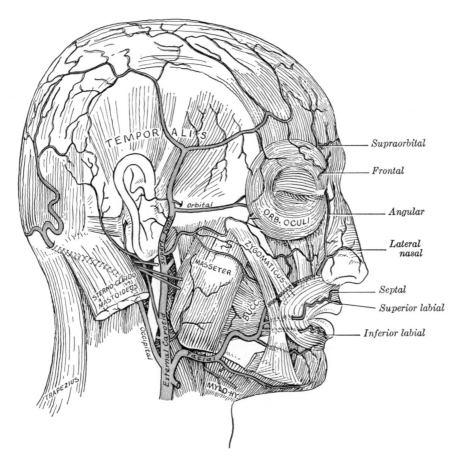

Fig. 8–19.—The arteries of the face and scalp. The muscular tissue of the lips has been cut away, in order to show the course of the labial arteries.

the largest of the cervical branches is given off from the facial artery just as that vessel quits the submandibular gland. It runs anteriorly upon the Mylohyoideus, just inferior to the body of the mandible, and deep to the Digastricus. It supplies the surrounding muscles, and anastomoses with the sublingual artery and with the mylohyoid branch of the inferior alveolar; at the symphysis menti it turns upward over the border of the mandible and divides into a superficial and a deep branch. The superficial branch passes between the integument and Levator labii inferioris, and anastomoses with the inferior labial artery; the deep branch runs between the muscle and the bone, supplies the lip, and anastomoses with the inferior labial and mental arteries.

e) The **inferior labial artery** (a. *labialis inferior*) *arises* near the angle of the mouth. It passes anteriorly beneath the Depressor anguli oris and, penetrating the Orbicularis oris, runs a tortuous course along the edge of the lower lip between this muscle and the mucous membrane. It supplies the labial glands, the mucous membrane, and the muscles of the lower lip; and anastomoses with the artery of the opposite side, and with the mental branch of the inferior alveolar artery.

f) The **superior labial artery** (a. *labialis superior*), larger and more tortuous than the inferior, follows a similar course along the edge of the upper lip, lying between the mucous membrane and the Orbicularis oris, anastomosing with the artery of the opposite side. It supplies the upper lip, and gives off two or three vessels which ascend to the nose; a **septal branch** ramifies on the nasal septum as far as the point of the nose, and an **alar branch** supplies the ala of the nose.

g) The **lateral nasal** branch is derived from the facial as that vessel ascends alongside of the nose. It supplies the ala and dorsum of the nose, anastomosing with its fellow, with the septal and alar branches, with the dorsal nasal branch of the ophthalmic, and with the infraorbital branch of the maxillary.

h) The **angular artery** (a. *angularis*) is the terminal part of the facial. It ascends to the medial angle of the orbit, imbedded in the fibers of the Levator labii superioris alaeque nasi, accompanied by the angular vein. Its branches anastomose with the infraorbital and after supplying the lacrimal sac and Orbicularis oculi, it ends by anastomosing with the dorsal nasal branch of the ophthalmic artery.

i) The **muscular branches** in the neck are distributed to the Pterygoideus medialis and Stylohyoideus, and on the face to the Masseter and Buccinator as well as the muscles of expression.

The **anastomoses of the facial artery** are very numerous, not only with the vessel of the opposite side, but, *in the neck*, with the sublingual branch of the lingual, with the ascending pharyngeal, and by its ascending palatine and tonsillar branches with the palatine branch of the maxillary; *on the face*, with the mental branch of the inferior alveolar as it emerges from the mental foramen, with the transverse facial branch of the superficial temporal, with the infraorbital branch of the maxillary, and with the dorsal nasal branch of the ophthalmic.

Variations.—The facial artery frequently arises in common with the lingual from a truncus linguofacialis. It varies in size and in the extent to which it supplies the face; it occasionally ends as the submental, and not infrequently extends only as high as the angle of the mouth or nose. The deficiency is then compensated for by enlargement of one of the neighboring arteries.

5. The **Occipital Artery** (a. *occipitalis*) (Fig. 8–19) *arises* from the posterior part of the external carotid, opposite the facial, near the lower margin of the posterior belly of the Digastricus, and ends in the posterior part of the scalp.

Course and Relations.—At its origin, it is covered by the posterior belly of the Digastricus and the Stylohyoideus, and the hypoglossal nerve winds around it; it crosses the internal carotid artery, the internal jugular vein, and the vagus and accessory nerves. It next ascends to the interval between the transverse process of the atlas and the mastoid process of the temporal bone, and passes horizontally posterior, grooving the surface of the latter bone, being covered by the Sternocleidomastoideus, Splenius capitis, Longissimus capitis, and Digastricus, and resting upon the Rectus capitis lateralis, the Obliquus superior, and Semispinalis capitis. It then changes its course and runs vertically upward, pierces the fascia connecting the cranial attachment of the Trapezius with the Sternocleidomastoideus, and ascends in a tortuous course in the superficial fascia of the scalp, where it divides into numerous branches, which reach as high as the vertex of the skull and anastomose with the posterior auricular and superficial temporal arteries. Its terminal portion is accompanied by the greater occipital nerve.

Branches.—The branches of the occipital artery are:

    a) Muscular
    b) Sternocleidomastoid
    c) Auricular
    d) Meningeal
    e) Descending
    f) Terminal

a) The **muscular branches** (*rami musculares*) supply the Digastricus, Stylohyoideus, Splenius, and Longissimus capitis.

b) The **sternocleidomastoid artery** (*a. sternocleidomastoidea; sternomastoid artery*) generally *arises* from the occipital close to its commencement, but sometimes springs directly from the external carotid. It passes caudalward and dorsalward over the hypoglossal nerve, and enters the substance of the muscle, in company with the accessory nerve.

c) The **auricular branch** (*ramus auricularis*) supplies the back of the concha and frequently gives off a branch, which enters the skull through the mastoid foramen and supplies the dura mater, the diploë, and the mastoid cells. This latter branch sometimes *arises* directly from the occipital artery, and is then known as the **mastoid branch.**

d) The **meningeal branch** (*ramus meningeus; dural branch*) ascends with the internal jugular vein, and enters the skull through the jugular foramen and condyloid canal, to supply the dura mater in the posterior fossa.

e) The **descending branch** (*ramus descendens; arteria princeps cervicis*) (Fig. 8–22), the largest branch of the occipital, running caudalward in the posterior neck, divides into a superficial and deep portion. The superficial portion is deep to the Splenius, giving off branches which pierce that

muscle to supply the Trapezius and anastomose with the ascending branch of the transverse cervical. The deep portion runs down between the Semispinales capitis and cervicis and anastomoses with the vertebral and with the deep cervical, a branch of the costocervical trunk. The anastomosis between these vessels assists in establishing the collateral circulation after ligature of the common carotid or subclavian artery.

f) The **terminal branches** of the occipital artery are distributed to the back of the head: they are very tortuous, and lie between the integument and Occipitalis, anastomosing with the artery of the opposite side and with the posterior auricular and temporal arteries, and supplying the Occipitalis, the integument, and pericranium. One of the terminal branches may give off a meningeal twig which passes through the parietal foramen.

6. The **Posterior Auricular Artery** (*a. auricularis posterior*) (Fig. 8–18) is small and *arises* from the external carotid, near the Digastricus and Stylohyoideus, opposite the apex of the styloid process. It ascends, under cover of the parotid gland on the styloid process of the temporal bone, to the groove between the cartilage of the ear and the mastoid process, where it divides into its auricular and occipital branches.

Branches.—Besides several small branches to the Digastricus, Stylohyoideus, and Sternocleidomastoideus, and to the parotid gland, this vessel gives off three branches:

    a) Stylomastoid
    b) Auricular
    c) Occipital

a) The **stylomastoid artery** (*a. stylomastoidea*) enters the stylomastoid foramen and supplies the tympanic cavity, the tympanic antrum and mastoid cells, and the semicircular canals. In the young subject a branch from this vessel, the **posterior tympanic artery,** with the anterior tympanic artery from the maxillary, forms a vascular circle which surrounds the tympanic membrane and sends delicate vessels to ramify on that membrane. It anastomoses with the superficial petrosal branch of the middle meningeal artery by a twig which enters the hiatus canalis facialis.

b) The **auricular branch** (*ramus auricu-*

laris) ascends behind the ear, deep to the Auricularis posterior, and is distributed to the back of the auricula, upon which it ramifies minutely, some branches curving around the margin of the cartilage, others perforating it, to supply the anterior surface. It anastomoses with the parietal and anterior auricular branches of the superficial temporal.

c) The **occipital branch** (*ramus occipitalis*) passes posteriorly, over the Sternocleidomastoideus, to the scalp above and behind the ear. It supplies the Occipitalis and the scalp in this situation and anastomoses with the occipital artery.

7. The **Superficial Temporal Artery** (*a. temporalis superficialis*) (Fig. 8–18), the smaller of the two terminal branches of the external carotid, appears, from its direction, to be the continuation of that vessel. It begins in the substance of the parotid gland, posterior to the neck of the mandible, crosses over the posterior root of the zygomatic process of the temporal bone, and about 5 cm above this process divides into two branches, a frontal and a parietal.

Relations.—As it crosses the zygomatic processes, it is covered by the Auricularis anterior muscle, and by a dense fascia; it is crossed by the temporal and zygomatic branches of the facial nerve and one or two veins, and is accompanied by the auriculotemporal nerve, which lies immediately posterior to it. Just above the zygomatic process and in front of the auricle, the superficial temporal artery is quite superficial, being covered only by skin and fascia, and can easily be felt pulsating. This artery is often used for determining the pulse, particularly by anesthetists.

Branches.—Besides some twigs to the parotid gland, to the temporomandibular joint, and to the Masseter muscle, its branches are:

> *a*) Transverse facial
> *b*) Middle temporal
> *c*) Zygomaticoörbital
> *d*) Anterior auricular
> *e*) Frontal
> *f*) Parietal

a) The **transverse facial artery** (*a. transversa faciei*) (Fig. 8–18) is given off from the superficial temporal before that vessel quits the parotid gland. Running anteriorly through the substance of the gland, it passes transversely across the side of the face, between the parotid duct and the inferior border of the zygomatic arch, divides into numerous branches, which supply the parotid gland and duct, the Masseter, and the integument and anastomose with the facial, masseteric, buccal, and infraorbital arteries. This vessel rests on the Masseter, and is accompanied by one or two branches of the facial nerve.

b) The **middle temporal artery** (*a. temporalis media*) arises immediately above the zygomatic arch, and, perforating the temporal fascia, lies against the squama of the temporal bone where it gives branches to the Temporalis, anastomosing with the deep temporal branches of the maxillary.

c) The **zygomaticoörbital artery** (*a. zygomatico-orbitalis*) runs along the superior border of the zygomatic arch, between the two layers of the temporal fascia, to the lateral angle of the orbit. This branch,

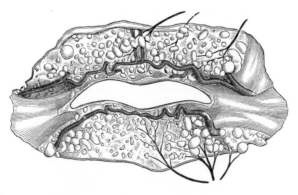

FIG. 8–20.—The labial arteries, the glands of the lips and the nerves of the right side seen from the posterior surface after removal of the mucous membrane. (Poirier and Charpy.)

which may arise from the middle temporal artery, supplies the Orbicularis oculi, and anastomoses with the lacrimal and palpebral branches of the ophthalmic artery (Fig. 8–24).

d) The **anterior auricular branches** (*rami auriculares anteriores*) are distributed to the anterior portion of the auricula, the lobule, and part of the external meatus, anastomosing with the posterior auricular.

e) The **frontal branch** (*ramus frontalis; anterior temporal*) runs tortuously toward the forehead, supplying the muscles, integument, and pericranium in this region, and anastomosing with the supraorbital and frontal arteries.

f) The **parietal branch** (*ramus parietalis; posterior temporal*) larger than the frontal, curves upward and backward on the side of the head, lying superficial to the temporal fascia, and anastomosing with its fellow of the opposite side, and with the posterior auricular and occipital arteries.

8. The **Maxillary Artery** (*a. maxillaris; internal maxillary artery*) (Figs. 8–18, 8–21, not labelled), the larger of the two terminal branches of the external carotid, *arises* deep to the neck of the mandible, and is at first imbedded in the substance of the parotid gland. It passes anteriorly between the ramus of the mandible and the sphenomandibular ligament, and then runs, either superficial or deep to the Pterygoideus lateralis, to the pterygopalatine fossa. It supplies the deep structures of the face, and may be divided into **mandibular, pterygoid,** and **pterygopalatine portions.**

The **first** or **mandibular portion,** when it is to become **superficial** to the muscle, runs anteriorly in a horizontal course, lying parallel with and a little inferior to the auriculotemporal nerve, and crosses the inferior alveolar nerve while running along the inferior border of the Pterygoideus lateralis.

The **second** or **pterygoid portion** arches laterally around the inferior border of the Pterygoideus lateralis and takes an oblique course anteriorly and superiorly under cover of the ramus of the mandible and insertion of the Temporalis, lying on the superficial (frequently on the deep) surface of the muscle. It then penetrates this muscle by passing between its two heads and enters the fossa.

The **third** or **pterygopalatine portion** is a short segment lying in the pterygopalatine fossa as it gives off several important

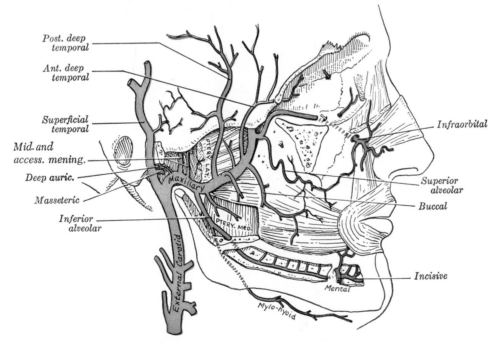

Labels on figure:
Post. deep temporal
Ant. deep temporal
Superficial temporal
Mid. and access. mening.
Deep auric.
Masseteric
Inferior alveolar
External Carotid
Maxillary
PTERY. LAT.
PTERY. MED.
Mylo-hyoid
Mental
Infraorbital
Superior alveolar
Buccal
Incisive

Fig. 8–21.—Plan of branches of the maxillary artery.

branches and makes its way toward the sphenopalatine foramen, passing beside the pterygopalatine ganglion and terminating as the sphenopalatine artery.

The **first** or **mandibular portion,** when it is to become **deep** to the Pterygoid muscle, continues medially as well as anteriorly and somewhat superiorly close to the bony wall of the infratemporal fossa.

The **second** or **pterygoid portion,** when deep or medial to the muscle, lies close to the lateral pterygoid plate which there forms the bottom of the infratemporal fossa. It passes over the inferior alveolar and lingual nerves as they emerge from the foramen ovale and reaches the pterygopalatine fossa by passing between the two heads of the muscle. Occasionally the artery passes deep to the emerging nerves.

The **third portion** is similar in both superficial and deep positions of the artery.

Branches.—The origin of the branches is

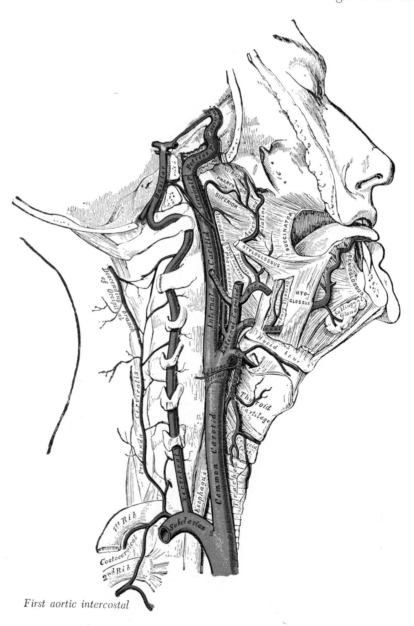

Fig. 8–22.—The internal carotid and vertebral arteries. Right side.

different in the two positions of the artery and will be described separately. Those of the **superficial position** are as follows:

Branches of the First or Mandibular Portion.—

- *a*) Deep auricular
- *b*) Anterior tympanic
- *c*) Inferior alveolar
- *d*) Middle meningeal
- *e*) Accessory meningeal

a) The **deep auricular artery** (*a. auricularis profunda*), a small artery, ascends in the substance of the parotid gland, deep to the temporomandibular articulation, pierces the cartilaginous or bony wall of the external acoustic meatus, and supplies its cuticular lining and the outer surface of the tympanic membrane. It gives a branch to the temporomandibular joint.

b) The **anterior tympanic artery** (*a. tympanica anterior; tympanic artery*) often *arises* in common with the preceding and is also of small size. It passes behind the temporomandibular articulation, enters the tympanic cavity through the petrotympanic fissure, and ramifies upon the tympanic membrane, forming a vascular circle around the membrane with the stylomastoid branch of the posterior auricular, and anastomosing with the artery of the pterygoid canal and with the caroticotympanic branch from the internal carotid.

c) The **inferior alveolar artery** (*a. alveolaris inferior; inferior dental artery*) descends with the inferior alveolar nerve to the mandibular foramen on the medial surface of the ramus of the mandible. It enters the mandibular canal and runs through it to the first premolar tooth where it divides into the mental and incisor branches.

i) The **mylohyoid artery** (*ramus mylohyoideus*) *arises* from the inferior alveolar artery just before the latter enters the mandibular foramen. It runs in the mylohyoid groove on the medial surface of the ramus along with the mylohyoid nerve to the inferior surface of the mylohyoideus which it supplies.

ii) The **incisor branch,** or termination of the artery, continues in the mandibular canal to the midline anteriorly where it gives branches to the incisor teeth and anastomoses with its fellow of the opposite side.

iii) The **mental branch** emerges with the nerve from the mental foramen, supplies the chin, and anastomoses with the submental and inferior labial arteries.

iv) **Dental branches** corresponding to the number of roots of the teeth enter the minute apertures at the extremities of the roots, and supply the pulp of the teeth.

v) A small **lingual branch** may arise from the inferior alveolar near its origin and descend with the lingual nerve to supply the mucous membrane of the mouth.

d) The **middle meningeal artery** (*a. meningea media*) is the largest and most constant of the arteries which supply the dura mater. It ascends between the sphenomandibular ligament and the Pterygoideus lateralis, and between the two roots of the auriculotemporal nerve to the foramen spinosum of the sphenoid bone, through which it enters the cranium; it then, in a groove on the great wing of the sphenoid bone, divides into anterior and posterior branches (Fig. 12–1).

i) The **anterior branch,** the larger, crosses the great wing of the sphenoid, reaches the groove, or canal, in the sphenoidal angle of the parietal bone, and then divides into branches which spread out between the dura mater and internal surface of the cranium, some passing upward as far as the vertex, and others backward to the occipital region (see Fig. 4–64).

ii) The **posterior branch** curves posteriorly on the squama of the temporal bone, and, reaching the parietal some distance anterior to its mastoid angle, divides into branches which supply the posterior part of the dura mater and cranium. The branches of the middle meningeal artery are distributed partly to the dura mater, but chiefly to the bones; they anastomose with the arteries of the opposite side, and with the anterior and posterior meningeal arteries.

e) The **accessory meningeal branch** (*ramus meningeus accessorius; small meningeal*), frequently *arising* from a common trunk with the preceding, enters the skull through the foramen ovale, and supplies the trigeminal ganglion and dura mater.

The middle meningeal, on entering the cranium, gives off the following branches: (1) Numerous small vessels supply the trigeminal ganglion and the dura mater nearby. (2) A **superficial petrosal** branch enters the hiatus of the facial canal, supplies the facial nerve, and anastomoses with the stylomastoid branch of the posterior auricular artery. (3) A **superior tympanic artery** runs in the canal for the Tensor tympani, and supplies this muscle and the lining membrane of the canal. (4) **Orbital branches** pass through the superior orbital fissure or through separate canals in the great wing of the sphenoid, to anastomose with the lacrimal or other branches of the ophthalmic artery. (5) **Temporal branches** pass through foramina in the great wing of the sphenoid, and anastomose in the temporal fossa with the deep temporal arteries.

Branches of the Second or Pterygoid Portion of the Maxillary in its Superficial Position (Fig. 8–21).

　　f) Deep temporal
　　g) Pterygoid
　　h) Masseteric
　　i) Buccal

f) The **deep temporal branches,** two in number, **anterior** and **posterior,** ascend between the Temporalis and the pericranium; they supply the muscle, and anastomose with the middle temporal artery; the anterior communicates with the lacrimal artery by means of small branches which perforate the zygomatic bone and great wing of the sphenoid.

g) The **pterygoid branches** (*rami pterygoidei*), irregular in number and origin, supply the Pterygoidei.

h) The **masseteric artery** (*a. masseterica*) is small and passes lateralward through the mandibular notch to the deep surface of the Masseter. It supplies the muscle, and anastomoses with the masseteric branches of the facial and with the transverse facial artery.

i) The **buccal artery** (*a. buccis; buccinator artery*), a small branch, runs obliquely forward between the Pterygoideus medialis and the insertion of the Temporalis, to the external surface of the Buccinator, to which it is distributed, anastomosing with branches of the facial and with the infraorbital.

Branches of the First and Second Portions Deep to the Pterygoideus Lateralis.

The deep auricular and anterior tympanic branches are unchanged. The masseteric and posterior deep temporal come from a common trunk with the inferior alveolar from the first portion. The middle and accessory meningeals arise from the second portion close to the foramen spinosum. The pterygoid, buccal, and anterior deep temporals also arise from the second portion, either separately or from a common trunk, deep to the muscle.

Branches of the Third or Pterygopalatine Portion (Fig. 8–21).

　　j) Posterior superior alveolar
　　k) Infraorbital
　　l) Greater palatine
　　m) Artery of the pterygoid canal
　　n) Pharyngeal
　　o) Sphenopalatine

j) The **posterior superior alveolar artery** (*a. alveolaris superior posterior; alveolar* or *posterior dental artery*) is given off from the maxillary, frequently in conjunction with the infraorbital, just as the trunk of the vessel is passing into the pterygopalatine fossa. Descending upon the tuberosity of the maxilla, it divides into numerous branches, some of which enter the alveolar canals, to supply the molar and premolar teeth and the lining of the maxillary sinus, while others continue anteriorly on the alveolar process to supply the gums.

k) The **infraorbital artery** (*a. infraorbitalis*) appears, from its direction, to be the continuation of the trunk of the internal maxillary, but often *arises* in conjunction with the posterior superior alveolar. It runs along the infraorbital groove and canal with the infraorbital nerve, and emerges on the face through the infraorbital foramen.

i) **Orbital branches** *arise* in the infraorbital canal which assist in the supply of the lacrimal gland and the Rectus and Obliquus inferior muscles.

ii) The **anterior superior alveolar branches** descend through the anterior alveolar canals to supply the incisor and canine teeth, and the mucous membrane of the maxillary sinus.

iii) The **facial branches,** after emerging from the infraorbital foramen, run upward to the medial angle of the *orbit,* supplying the lacrimal sac and anastomosing with the angular artery. Branches also run toward the *nose* and anastomose with the dorsal nasal branch of the ophthalmic. Other branches descend between the two Levatores labii superiores and anastomose with the *facial,* transverse facial, and buccal arteries.

l) The **greater palatine artery** (NA *a. palatina descendens*) (Fig. 12–12) *arises* in the pterygopalatine fossa and descends with the greater palatine branch of the pterygopalatine nerve in the pterygopalatine canal. After emerging from the greater palatine foramen it runs anteriorly in a groove on the medial side of the alveolar border of the hard palate to the incisive canal. Branches are distributed to the gums, palatine glands, and mucous membrane of the roof of the mouth; a terminal branch anastomoses with the nasopalatine branch of the sphenopalatine artery in the **incisive canal.**

i) The **lesser palatine arteries** *arise* in the pterygopalatine canal and descend through the lesser palatine foramina to supply the soft palate and palatine tonsil, and anastomose with the ascending palatine branch of the facial artery.

m) The **artery of the pterygoid canal** *arises* in the pterygopalatine fossa and enters the pterygoid (Vidian) canal with the corresponding nerve. It is distributed to the upper part of the pharynx, auditory tube, and sphenoidal sinus, sending into the tympanic cavity a small branch which anastomoses with the other tympanic arteries.

n) The **pharyngeal branch,** very small, runs posteriorly in the pharyngeal canal with the pharyngeal branch of the pterygopalatine nerve and is distributed to the auditory tube, the upper part of the pharynx, and the sphenoidal sinus.

o) The **sphenopalatine artery** (*a. sphenopalatina; nasopalatine artery*) enters the nasal cavity through the sphenopalatine foramen at the posterior end of the superior meatus close to the pterygopalatine ganglion and accompanied by the pterygopalatine nerves. Shortly after traversing the foramen it divides into a lateral and a septal branch.

i) The **posterior lateral** nasal branches run anteriorly, spreading over the conchae and meatuses, and assisting in the supply of the frontal, maxillary, ethmoidal, and sphenoidal sinuses.

ii) The **posterior septal branches** arch over the roof of the nasal cavity on the inferior surface of the sphenoid bone and run anteriorly and inferiorly over the nasal septum. One branch, the nasopalatine, runs in a groove on the vomer with the nasopalatine nerve to the incisive canal where it anastomoses with the greater palatine artery.

The Internal Carotid Artery

The **internal carotid artery** (*a. carotis interna*) (Fig. 8–22) begins at the bifurcation of the common carotid opposite the cranial border of the thyroid cartilage, and runs perpendicularly cranialward, ventral to the transverse processes of the first three cervical vertebrae, to reach the inferior surface of the petrous portion of the temporal bone. It next enters the carotid canal which curves abruptly and runs horizontally medialward and anteriorly to the foramen lacerum at the end of the petrous portion. As the artery leaves the canal it curves upward, crossing the cranial part of the foramen lacerum to enter the middle cranial fossa between the lingula and petrosal process of the sphenoid bone. During the first part of its course within the cranial cavity it is suspended between the layers of the dura mater which form the cavernous sinus (Fig. 12–13). It ascends toward the posterior clinoid process, and makes a double curvature like the letter "s" before perforating the part of the dura mater which forms the roof of the cavernous sinus on the medial side of the anterior clinoid process. The artery then passes between the optic and oculomotor nerves, and approaches the anterior perforated substance at the medial end of the lateral cerebral fissure where it gives off its terminal branches.

Relations.—In considering the relations, this vessel may be divided into four portions: **cervical, petrous, cavernous,** and **cerebral.**

The **cervical portion** of the internal carotid is comparatively superficial at its commencement, where it is contained in the carotid triangle, and lies posterior and lateral to the external carotid, overlapped by the Sterno-cleidomastoideus, and covered by the deep fascia, Platysma, and integument: it then passes deep to the parotid gland, being crossed by the hypoglossal nerve, the Digastricus and Stylohyoideus, and the occipital and posterior auricular arteries. More cranially, it is separated from the external carotid by the Styloglossus and Stylopharyngeus, the tip of the styloid process and the stylohyoid ligament, the glosso-pharyngeal nerve and the pharyngeal branch of the vagus. It is in relation, *posteriorly*, with the Longus capitis, the superior cervical ganglion of the sympathetic trunk with its ascending branch to the carotid plexus, and the superior laryngeal nerve; *laterally*, with the internal jugular vein and vagus nerve, the nerve lying on a plane posterior to the artery; *medially*, with the pharynx, superior laryngeal nerve, and ascending pharyngeal artery. At the base of the skull the glossopharyngeal, vagus, accessory, and hypoglossal nerves lie between the artery and the internal jugular vein.

Petrous Portion.—When the internal carotid artery enters the canal in the petrous portion of the temporal bone, it first ascends a short distance, then curves forward and medialward, and again ascends as it leaves the canal to enter the cavity of the skull between the lingula and petrosal process of the sphenoid. The artery lies at first anterior to the cochlea and tympanic cavity; from the latter cavity it is separated by a thin, bony lamella, which is cribriform in the young subject, and often partly absorbed in old age. More anteriorly it is separated from the trigeminal ganglion by a thin plate of bone, which forms the floor of the fossa for the ganglion and the roof of the horizontal portion of the canal. Frequently this bony plate is more or less deficient, and then the ganglion is separated from the artery by fibrous membrane. The artery is separated from the bony wall of the carotid canal by a prolongation of dura mater. It is surrounded by a number of small veins and by filaments of the carotid plexus of nerves, derived from the ascending branch of the superior cervical ganglion of the sympa-thetic trunk.

Cavernous Portion.—In this part of its course, the artery is situated between the layers of the dura mater forming the cavernous sinus, but covered by the lining membrane of the sinus. It at first ascends toward the posterior clinoid process, then passes forward by the side of the body of the sphenoid bone, and again curves upward on the medial side of the anterior clinoid process, and perforates the dura mater

forming the roof of the sinus. This portion of the artery is surrounded by filaments of the sympathetic nerve, and on its lateral side is the abducent nerve.

Cerebral Portion.—Having perforated the dura mater on the medial side of the anterior clinoid process, the internal carotid passes between the optic and oculomotor nerves toward the anterior perforated substance at the medial extremity of the lateral cerebral fissure, where it divides into its terminal or cerebral branches.

Variations.—The length of the internal carotid varies according to the length of the neck, and also according to the point of bifurcation of the common carotid. It arises sometimes from the arch of the aorta. The course of the artery, instead of being straight, may be very tortuous. A few instances are recorded in which this vessel was altogether absent; in one of these the common carotid passed up the neck and gave off the usual branches of the external carotid; the cranial portion of the internal caro-tid was replaced by two branches of the internal maxillary, which entered the skull through the foramen rotundum and foramen ovale, and joined to form a single vessel.

Branches.—The cervical portion of the internal carotid gives off no branches. Those from the other portions are:

From the Petrous Portion
   1. Caroticotympanic
   2. Artery of the Pterygoid Canal
   3. Cavernous
   4. Hypophyseal

From the Cavernous Portion
   5. Ganglionic
   6. Anterior Meningeal
   7. Ophthalmic
   8. Anterior Cerebral
   9. Middle Cerebral

From the Cerebral Portion
   10. Posterior Communicating
   11. Anterior Choroidal

1. The **caroticotympanic branch** (*ramus caroticotympanicus; tympanic branch*) is small; it enters the tympanic cavity through a minute foramen in the carotid canal, and anastomoses with the anterior tympanic branch of the maxillary, and with the stylo-mastoid artery.

2. The **artery of the pterygoid canal** (*a.*

canalis pterygoidei [*Vidii*]; *Vidian artery*) is a small, inconstant branch which passes into the pterygoid canal and anastomoses with a branch of the maxillary artery.

3. The **cavernous branches** are numerous small vessels which supply the hypophysis, the trigeminal ganglion, and the walls of the cavernous and inferior petrosal sinuses. Some of them anastomose with branches of the middle meningeal.

4. The **hypophyseal branches** are one or two minute vessels supplying the hypophysis.

5. The **ganglionic branches** are small vessels to the trigeminal ganglion.

6. The **anterior meningeal branch** (*a. meningea anterior*) is a small branch which passes over the small wing of the sphenoid to supply the dura mater of the anterior cranial fossa; it anastomoses with the meningeal branch from the posterior ethmoidal artery.

7. The **Ophthalmic Artery** (*a. ophthalmica*) (Fig. 8–23) *arises* from the internal carotid, just as that vessel is emerging from the cavernous sinus on the medial side of the anterior clinoid process, and enters the orbital cavity through the optic canal, inferior and lateral to the optic nerve. It then passes over the nerve to reach the medial wall of the orbit, and thence horizontally along the inferior border of the Obliquus superior, and divides it into two terminal branches, the **supratrochlear** and **dorsal nasal**. As the artery crosses the optic nerve it is accompanied by the nasociliary nerve, and is separated from the frontal nerve by the Rectus superior and Levator palpebrae superioris. The artery runs below, rather than above, the optic nerve in 15 per cent of cases.

Branches.—The branches of the ophthalmic artery may be divided into an **orbital group**, distributed to the orbit and surrounding parts; and an **ocular group**, to the muscles and bulb of the eye.

Orbital Group

a) Lacrimal
b) Supraorbital
c) Posterior ethmoidal
d) Anterior ethmoidal
e) Medial palpebral
f) Supratrochlear
g) Dorsal nasal

Ocular Group

h) Central artery of the retina
i) Short posterior ciliary
j) Long posterior ciliary
k) Anterior ciliary
l) Muscular

a) The **lacrimal artery** (*a. lacrimalis*) *arises* close to the optic foramen, not infrequently before it enters the orbit, and is one of the largest branches derived from the ophthalmic. It accompanies the lacrimal nerve along the superior border of the Rectus lateralis, and supplies the lacrimal gland. Its terminal branches, escaping from the gland, are distributed to the eyelids and conjunctiva: of those supplying the eyelids, two are of considerable size and are named the **lateral palpebral arteries;** they run medialward in the upper and lower lids respectively and anastomose with the medial palpebral arteries, forming an arterial circle in this situation. The lacrimal artery gives off one or two **zygomatic branches,** one of which passes through the zygomaticotemporal foramen, to reach the temporal fossa, and anastomoses with the deep temporal arteries; another appears on the cheek through the zygomatico-facial foramen, and anastomoses with the transverse facial. A **recurrent branch** passes backward through the lateral part of the superior orbital fissure to the dura mater, and anastomoses with a branch of the middle meningeal artery. The lacrimal artery is sometimes derived from one of the anterior branches of the middle meningeal artery.

b) The **supraorbital artery** (*a. supraorbitalis*) springs from the ophthalmic as that vessel is crossing over the optic nerve. It passes superiorly on the medial borders of the Rectus superior and Levator palpebrae, and meeting the supraorbital nerve, accompanies it between the periosteum and Levator palpebrae to the supraorbital foramen; passing through this it divides into a superficial and a deep branch, which supply the integument, the muscles, and the pericranium of the forehead, anastomosing with the supratrochlear, the frontal branch of the superficial temporal, and the artery of the opposite side. This artery in the orbit supplies the Rectus superior and the Levator palpebrae, and sends a **branch**

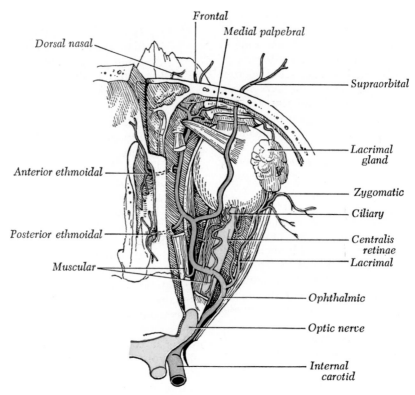

Fig. 8–23.—The ophthalmic artery and its branches.

across the pulley of the Obliquus superior, to supply the parts at the medial palpebral commissure. At the supraorbital foramen it frequently transmits a branch to the diploë.

c) The **posterior ethmoidal artery,** smaller than the anterior, passes through the posterior ethmoidal canal, supplies the posterior ethmoidal air cells, and, entering the cranium, gives off a meningeal branch to the dura mater, and nasal branches which descend into the nasal cavity through apertures in the cribriform plate, anastomosing with branches of the sphenopalatine.

d) The **anterior ethmoidal artery** accompanies the nasociliary nerve through the anterior ethmoidal canal, supplies the anterior and middle ethmoidal cells and frontal sinus, and, entering the cranium, gives off a meningeal branch to the dura mater. The nasal branches descend into the nasal cavity through the slit by the side of the crista galli, and, running along the groove on the inner surface of the nasal

bone, supply branches to the lateral wall and septum of the nose, and a terminal branch which appears on the dorsum of the nose between the nasal bone and the lateral cartilage.

e) The **medial palpebral arteries** (*aa. palpebrales mediales; internal palpebral arteries*), two in number, **superior** and **inferior,** *arise* from the ophthalmic, opposite the pulley of the Obliquus superior; they leave the orbit to encircle the eyelids near their free margins, forming a superior and an inferior arch, between the Orbicularis oculi and the tarsi. The **superior palpebral** anastomoses, at the lateral angle of the orbit, with the zygomaticoörbital branch of the temporal artery and with the superior of the two lateral palpebral branches from the lacrimal artery (Fig. 8–24); the **inferior palpebral** anastomoses, at the lateral angle of the orbit, with the inferior of the two lateral palpebral branches from the lacrimal and with the transverse facial artery, and, at the medial part of the lid, with a branch

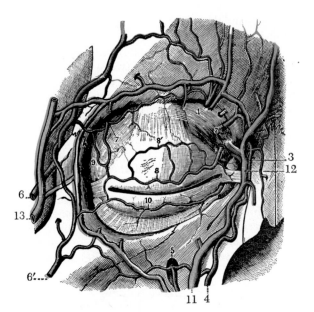

Fig. 8–24.—Blood vessels of the eyelids, anterior view. 1, supraorbital artery and vein; 3, angular artery, the terminal branch of 2, the facial artery; 4, nasal artery; 5, infraorbital artery; 6, anterior branch of the superficial temporal artery; 6', zygomatic branch of the transverse facial artery; 7, lacrimal artery; 8, superior palpebral artery with 8', its external arch; 9, anastomoses of the superior palpebral with the superficial temporal and lacrimal; 10, inferior palpebral artery; 11, facial vein; 12, angular vein; 13, branch of the superficial temporal vein. (Testut.)

from the angular artery. From this last anastomosis a branch passes to the nasolacrimal duct, ramifying in its mucous membrane as far as the inferior meatus of the nasal cavity.

f) The **supratrochlear artery** (*a. frontalis*), one of the terminal branches of the ophthalmic, leaves the orbit at its medial angle with the supratrochlear nerve, and, ascending on the forehead, supplies the integument, muscles, and pericranium, anastomosing with the supraorbital artery, and with the artery of the opposite side.

g) The **dorsal nasal artery** (*a. dorsalis nasi; nasal artery*), the other terminal branch of the ophthalmic, emerges from the orbit above the medial palpebral ligament, and, after giving a twig to the superior part of the lacrimal sac, divides into two branches, one of which crosses the root of the nose, and anastomoses with the angular artery, the other runs along the dorsum of the nose, supplies its outer surface; and anastomoses with the artery of the opposite side, and with the lateral nasal branch of the facial.

h) The **central artery of the retina** (*a. centralis retinae*) is the first and one of the smallest branches of the ophthalmic artery. It runs for a short distance within the dural sheath of the optic nerve, but about 1.25 cm behind the eyeball it pierces the nerve obliquely, and runs forward in the center of its substance to the retina. Its mode of distribution will be described with the anatomy of the eye. It may be a branch of the lacrimal artery (Fig. 8–23). There is an arteria cilioretinalis in 13 per cent of cases (Adachi).

The **ciliary arteries** (*aa. ciliares*) are divisible into three groups, the long and short posterior, and the anterior.

i) The **short posterior ciliary arteries**, from six to twelve in number, arise from the ophthalmic, or its branches, and surrounding the optic nerve, run to the posterior part of the eyeball, pierce the sclera around the entrance of the nerve, and supply the choroid and ciliary processes.

j) The **long posterior ciliary arteries**, two in number, pierce the posterior part of the sclera at some little distance from the optic nerve, and run along either side of the eyeball, between the sclera and choroid, to the ciliary muscle, where they divide into two branches; these form around the circumference of the iris, the **circulus arteriosus major** which sends numerous converging branches in the substance of the iris to its pupillary margin, where they form a second arterial circle, the **circulus arteriosus minor.**

k) The **anterior ciliary arteries** are derived from the muscular branches; they run to the front of the eyeball in company with the tendons of the Recti, form a vascular zone beneath the conjunctiva, and then pierce the sclera a short distance from the cornea and end in the circulus arteriosus major.

l) The **muscular branches** (*rami musculares*), two in number, **superior and inferior,** frequently spring from a common trunk. The **superior,** often wanting, supplies the

Levator palpebrae superioris, Rectus superior, and Obliquus superior. The **inferior,** more commonly present, passes forward between the optic nerve and Rectus inferior, and is distributed to the Recti lateralis, medialis, and inferior, and the Obliquus inferior. This vessel gives off most of the anterior ciliary arteries. Additional muscular branches are given off from the lacrimal and supraorbital arteries, or from the trunk of the ophthalmic.

8. The **Anterior Cerebral Artery** (*a. cerebri anterior*) (Figs. 8–25, 8–27, 8–29) *arises* from the internal carotid, at the medial extremity of the lateral cerebral sulcus. It passes forward and medialward across the anterior perforated substance,

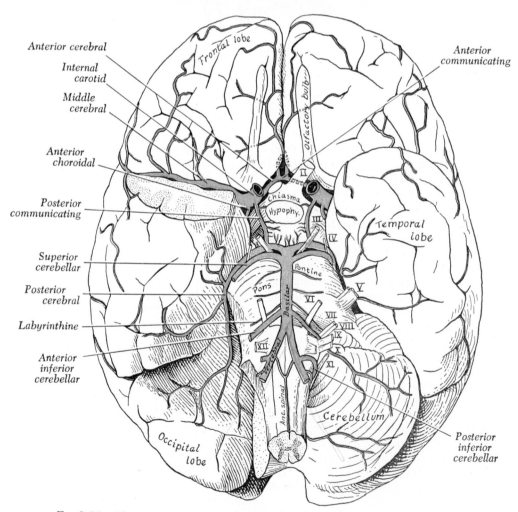

Fig. 8–25.—The arteries of the base of the brain. The temporal pole of the cerebrum and a portion of the cerebellar hemisphere have been removed on the right side.

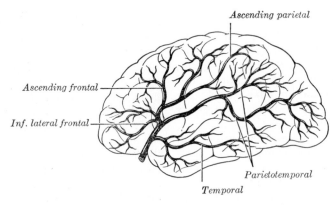

Fig. 8–26.—Branches of the middle cerebral artery to the lateral surface of the cerebral hemisphere. (Modified after Foix.)

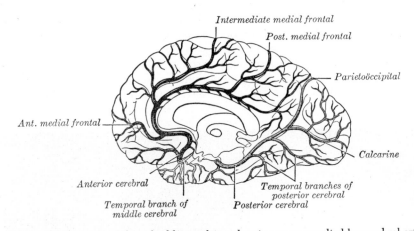

Fig. 8–27.—Medial surface of cerebral hemisphere, showing areas supplied by cerebral arteries.

above the optic nerve, to the commencement of the longitudinal fissure. Here it comes into close relationship with the opposite artery, to which it is connected by a short trunk, the anterior communicating artery. From this point the two vessels run side by side in the longitudinal fissure, curve around the genu of the corpus callosum, and, turning backward, continue along the upper surface of the corpus callosum to its posterior part, where they end by anastomosing with the posterior cerebral arteries.

Branches.—In its first part the anterior cerebral artery gives off twigs which pierce the anterior perforated substance and the lamina terminalis, and supply the rostrum of the corpus callosum and the septum pellucidum. A larger branch, phylogeneti-cally one of the oldest of the cerebral arteries, takes a recurrent course laterally over the anterior perforated substance as the **medial striate artery.** This medial striate artery supplies the lower anterior portion of the basal nuclei, i.e., the lower part of the head of the caudate nucleus, the lower part of the frontal pole of the putamen, the frontal pole of the globus pallidus, and the anterior limb of the internal capsule up to the dorsal limit of the globus pallidus. The **inferior** or **orbital branches** of the anterior cerebral artery are distributed to the orbital surface of the frontal lobe, where they supply the olfactory lobe, gyrus rectus, and internal orbital gyrus. The **anterior** or **prefrontal branches** supply a part of the superior frontal gyrus, and send twigs over the edge of the hemisphere to the superior and middle frontal gyri and upper part of the anterior central gyrus. The **middle branches** supply the corpus callosum, the cingulate gyrus, the medial surface of the superior frontal gyrus, and the upper part of the anterior central gyrus. The **posterior branches** supply the precuneus and adjacent lateral surface of the hemisphere.

The **anterior communicating artery** (*a. communicans anterior*) connects the two anterior cerebral arteries across the commencement of the longitudinal fissure. Sometimes this vessel is wanting, the two arteries joining to form a single trunk, which afterward divides; or it may be wholly, or partially, divided into two. Its length averages about 4 mm, but varies greatly. It gives off some anteromedial branches.

9. The **Middle Cerebral Artery** (*a. cerebri media*) (Figs. 8–26, 8–29), the largest branch of the internal carotid, runs at first lateralward in the lateral cerebral sulcus (*Sylvian fissure*) and then backward and

upward on the surface of the insula, where it divides into a number of branches which are distributed to the lateral surface of the cerebral hemisphere.

Branches. *a*) **Rami centrales.**—At its commencement the middle cerebral artery gives off the *lateral striate arteries* which supply the basal nuclei, *i.e.*, the whole of the putamen except the lower anterior pole, the upper part of the head and the whole of the body of the caudate nucleus, the lateral part of the globus pallidus, and the capsula interna above the level of the globus pallidus. The thalamus is nowhere supplied by branches of the middle cerebral artery.

b) **Rami corticales.**—Branches supplying cortical areas may be designated as follows: An *inferior lateral frontal branch* supplies the inferior frontal gyrus (Broca's convolution) and the lateral part of the orbital surface of the frontal lobe. An *ascending frontal branch* supplies the anterior central gyrus. An *ascending parietal branch* is distributed to the posterior central gyrus

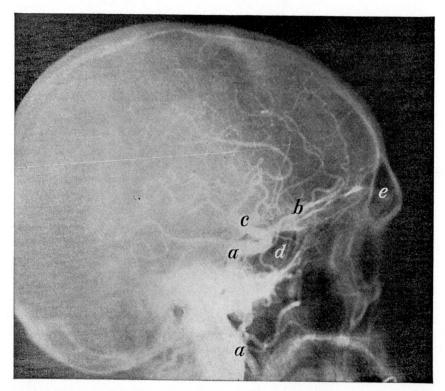

Fɪɢ. 8–28.—Internal carotid artery. *a*, After injection of diotrast into the common carotid artery; *b*, anterior cerebral artery; *c*, middle cerebral; *d*, sphenoidal sinus; *e*, frontal sinus. (Department of Radiology, University of Pennsylvania.)

and the lower part of the superior parietal lobule. A *parietotemporal branch* supplies the supramarginal and angular gyri, and the posterior parts of the superior and middle temporal gyri. *Temporal* branches, two or three in number, are distributed to the lateral surface of the temporal lobe.

10. The **Posterior Communicating Artery** (*a. communicans posterior*) (Fig. 8–29) runs backward from the internal carotid, and anastomoses with the posterior cerebral, a branch of the basilar. It varies in size, being sometimes small, and occasionally so large that the posterior cerebral may be considered as arising from the internal carotid rather than from the basilar. It is frequently larger on one side than on the other. **Branches** of the posterior communicating artery enter the base of the brain between the infundibulum and the optic tract, and supply the genu and about the anterior one-third of the posterior limb of the internal capsule. There are also branches to the anterior one-third of the

thalamus (exclusive of the anterior nucleus) and to the walls of the third ventricle.

11. The **Anterior Choroidal Artery** (*a. chorioidea; choroid artery*) is, next to the middle cerebral artery, the most important source of supply to the internal capsule. The artery *arises* from the internal carotid near the origin of the posterior communicating artery, and takes its course along the optic tract, and around the cerebral peduncle as far as the lateral geniculate body, where its main branches turn to enter the choroid plexus of the inferior horn of the lateral ventricle. In its course it gives branches to the optic tract, the cerebral peduncle, and the base of the brain. These **branches** terminate in the lateral geniculate body, the tail of the caudate nucleus, and the posterior two-thirds of the posterior limb of the internal capsule as far as the dorsal limit of the globus pallidus. The infralenticular and retrolenticular portions of the internal capsule are also vascularized by branches of the anterior choroidal artery.

Plan of Arteries of the Brain

Since the mode of distribution of the vessels of the brain has an important bearing upon a considerable number of the pathological lesions which may occur in this part of the nervous system, it is important to consider a little more in detail the manner in which the vessels are distributed.

The cerebral arteries are derived from the internal carotid and vertebral, which at the base of the brain form a remarkable anastomosis known as the **arterial circle of Willis**. It is formed anteriorly by the anterior cerebral arteries, branches of the internal carotid which are connected together by the anterior communicating; posteriorly by the two posterior cerebral arteries, branches of the basilar which are connected on either side with the internal carotid by the posterior communicating (Figs. 8–25, 8–29). The parts of the brain included within this arterial circle are the lamina terminalis, the optic chiasma, the infundibulum, the tuber cinereum, the corpora mammillaria, and the posterior perforated substance.

The three trunks which together supply each cerebral hemisphere *arise* from the arterial circle of Willis. From its *anterior*

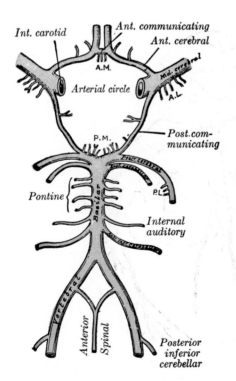

Fig. 8–29.—Diagram of the arterial circulation at the base of the brain. *A.L.* Anterolateral. *A.M.* Anteromedial. *P.L.* Posterolateral. *P.M.* Posteromedial ganglionic branches.

part proceed the two anterior cerebrals, from its *anterolateral* parts the middle cerebrals, and from its *posterior* part the posterior cerebrals. Each of these principal arteries gives origin to the numerous vessels which supply the brain substance. They contribute to a continuous complex network of capillaries, which is of different density in various parts of the central nervous system, probably in conjunction with the varying requirements of the different structures. Thus the gray matter of the brain has a much denser capillary bed than the white matter. The arteries on the surface of the brain anastomose freely, but within the central nervous system arterial anastomoses are rare. There is no evidence for the existence of arteriovenous anastomoses either in the pia or within the central

nervous system. While true anatomical end arteries exist in certain mammals, in man end or terminal arteries do not exist in a strict anatomical sense, *i.e.*, there are no arteries which have no connections with neighboring vessels. The lack of sufficiently large and numerous anastomoses, however, and the high vulnerability of the nervous tissue in case of lack of oxygen, when an artery is occluded, greatly reduces the chances that an efficient collateral circulation might take care of the needs of the ischemic area. This ischemic area corresponds in its extent to the area of supply of the occluded vessel. The arteries of the brain may be considered, therefore, as functional end or terminal arteries, though anatomically they are not strictly what Cohnheim designated as terminal arteries.

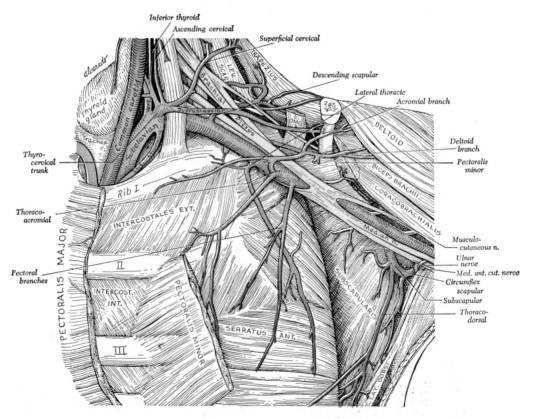

FIG. 8–30.—The subclavian and axillary arteries with their branches. *M. omohyoideus + Pectoral branch. (Redrawn from Rauber-Kopsch.)

THE SUBCLAVIAN ARTERY

The **subclavian artery** (*a. subclavia*) (Fig. 8–30) has a different origin on the two sides. On the right side it *arises* from the brachiocephalic artery deep to the right sternoclavicular articulation; on the left side it springs from the arch of the aorta. The two vessels, therefore, in the first part of their course, differ in length, direction, and relation with neighboring structures.

In order to facilitate the description, each subclavian artery is divided into three parts. The first portion extends from the origin of the vessel to the medial border of the Scalenus anterior; the second lies deep to this muscle; and the third extends from the lateral margin of the muscle to the outer border of the first rib where it becomes the axillary artery. The first portions of the two vessels require separate descriptions; the second and third parts of the two arteries are practically the same.

First Part of the Right Subclavian Artery (Figs. 8–18, 8–22).—The first part of the right subclavian artery *arises* from the brachiocephalic artery, deep to the cranial part of the right sternoclavicular articulation, and passes cranially and laterally to the medial margin of the Scalenus anterior. It arches a little above the clavicle, the extent to which it does so varying in different individuals.

Relations.—It is covered, *ventrally*, by the integument, subcutaneous fascia, Platysma, deep fascia, the clavicular origin of the Sternocleidomastoideus, the Sternohyoideus, and Sternothyroideus, and another layer of the deep fascia. It is crossed by the internal jugular and vertebral veins, by the vagus nerve and the cardiac branches of the vagus and sympathetic, and by the subclavian loop (*ansa subclavia*) of the sympathetic trunk which forms a ring around the vessel. The anterior jugular vein is directed laterally ventral to the artery, but is separated from it by the Sternohyoideus and Sternothyroideus. *Caudal and dorsal* to the artery is the pleura, which separates it from the apex of the lung; *dorsally* is the sympathetic trunk, the Longus colli and the first thoracic vertebra. The right recurrent nerve winds around the proximal part of the vessel.

First Part of the Left Subclavian Artery (Fig. 8–30).—The first part of the left subclavian artery *arises* from the arch of the aorta dorsal to the left common carotid at the level of the fourth thoracic vertebra; it ascends in the superior mediastinum to the root of the neck and then arches laterally to the medial border of the Scalenus anterior.

Relations.—It is in relation, *ventrally*, with the vagus, cardiac, and phrenic nerves, which lie parallel with it, the left common carotid artery, left internal jugular and vertebral veins, and the commencement of the left brachiocephalic vein, and is covered by the Sternothyroideus, Sternohyoideus, and Sternocleidomastoideus; *dorsally*, it is in relation with the esophagus, thoracic duct, left recurrent nerve, inferior cervical ganglion of the sympathetic trunk, and Longus colli; more cranially, however, the esophagus and thoracic duct lie to its right side, the latter ultimately arching over the vessel to join the angle of union between the subclavian and internal jugular veins. *Medial* to it are the esophagus, trachea, thoracic duct, and left recurrent nerve; *lateral* to it, the left pleura and lung.

Second and Third Parts of the Subclavian Artery (Fig. 8–30).—The second **portion of** the subclavian artery lies dorsal to the Scalenus anterior; it is very short, and forms the most cranial part of the arch described by the vessel.

Relations.—It is covered *ventrally*, by the skin, subcutaneous fascia, Platysma, deep cervical fascia, Sternocleidomastoideus, and Scalenus anterior. On the right side of the neck the phrenic nerve is separated from the second part of the artery by the Scalenus anterior, while on the left side it crosses the first part of the artery close to the medial edge of the muscle. *Dorsal* to the vessel are the pleura and the Scalenus medius; *cranially*, the brachial plexus of nerves; *caudally*, the pleura. The subclavian vein lies ventral to the artery, separated from it by the Scalenus anterior.

The **third portion** of the subclavian artery continues its distal course laterally from the lateral margin of the Scalenus anterior to the outer border of the first rib, where it becomes the axillary artery This is the most superficial portion of the vessel, and it crosses the subclavian triangle (Fig. 8–17).

Relations.—It is covered *ventrally* by the skin, the subcutaneous fascia, the Platysma, the supraclavicular nerves, and the deep cervical fascia. The external jugular vein crosses its medial part and receives the transverse scapular, transverse cervical, and anterior jugular veins, which frequently form a plexus ventral to the artery. The nerve to the Subclavius and accessory phrenic cross between the artery and veins. The terminal part of the artery lies dorsal to the clavicle and the Subclavius and is crossed by the transverse scapular vessels. The subclavian vein is ventral to the artery. *Dorsally*, it lies on the inferior trunk of the brachial plexus, which intervenes between it and the Scalenus medius. The upper trunks of the brachial plexus and the Omohyoideus are cranial and lateral to it. Inferiorly, it rests on the upper surface of the first rib.

Variations.—The subclavian arteries vary in their origin, their course, and the height to which they rise in the neck.

The origin of the right subclavian from the brachiocephalic takes place, in some cases, cranial to the sternoclavicular articulation, and occasionally, but less frequently, caudal to that joint. The artery may arise as a separate trunk from the arch of the aorta, and in such cases it may be either the first, second, third, or even the last branch derived from that vessel; in the majority, however, it is the first or last, rarely the second or third. When it is the first branch, it occupies the ordinary position of the brachiocephalic artery; when the second or third, it gains its usual position by passing dorsal to the right carotid; and when the last branch, it arises from the left extremity of the arch, and passes obliquely toward the right side, usually dorsal to the trachea, esophagus, and right carotid, sometimes between the esophagus and trachea, to the cranial border of the first rib, whence it follows its ordinary course. In very rare instances, this vessel arises from the thoracic aorta, as low down as the fourth thoracic vertebra. Occasionally, it perforates the Scalenus anterior; more rarely it passes ventral to that muscle. Sometimes the subclavian vein passes with the artery dorsal to the Scalenus anterior. The artery may ascend as high as 4 cm above the clavicle, or any intermediate point between this and the cranial border of the bone, the right subclavian usually ascending higher than the left.

The left subclavian is occasionally joined at its origin with the left carotid.

The left subclavian artery is more deeply placed than the right in the first part of its course, and, as a rule, does not reach quite as high a level in the neck. The posterior border of the Sternocleidomastoideus corresponds pretty closely to the lateral border of the Scalenus anterior, so that the third portion of the artery, the part most accessible for operation, lies immediately lateral to the posterior border of the Sternocleidomastoideus.

Collateral Circulation.—After ligature of the third part of the subclavian artery, the collateral circulation is established mainly by anastomoses between the suprascapular, the descending ramus of the transverse cervical artery, and the subscapular artery, also by an anastomosis between branches of the internal thoracic and lateral thoracic and subscapular arteries.

Branches.—The branches of the subclavian artery are:

1. Vertebral
1A. Basilar
2. Thyrocervical
3. Internal thoracic
4. Costocervical
5. Descending scapular

The first four branches generally *arise* from the first portion on the left side, but the costocervical trunk more frequently springs from the second portion on the right side. On both sides the first three branches *arise* close together at the medial border of the Scalenus anterior, in the majority of cases leaving a free interval of from 1.25 to 2.5 cm between the commencement of the artery and the origin of the nearest branch. The transverse cervical is much less constant than the other branches, and when present is the only branch of the third portion of the artery.

1. The **Vertebral Artery** (*a. vertebralis*) (Fig. 8–22) is the first and usually the largest branch of the subclavian, arising quite deeply in the neck from the cranial and dorsal surface of the vessel. It angles dorsalward to the transverse process of the sixth cervical vertebra and runs cranialward through the foramina in the transverse processes of the other cervical vertebrae to the inferior surface of the skull. After penetrating the foramen in the atlas it bends abruptly medialward around the lateral surface of the superior articular process and then lies in a groove on the cranial surface of the posterior arch of the atlas. It passes under an arch in the posterior atlantoöccipital membrane (Fig. 5–13, arrow), makes an abrupt bend to enter the cranial cavity through the foramen magnum, and after a short course joins the other vertebral to form the basilar artery.

Relations.—The vertebral artery may be divided into four parts: The **first part** between the Longus colli and the Scalenus anterior, has ventral to it the internal jugular and vertebral veins, and it is crossed by the inferior thyroid artery; the left vertebral is crossed by the thoracic duct also. Dorsal to it are the transverse process of the seventh cervical vertebra, the sympathetic trunk and inferior cervical ganglion. The **second part** runs through the foramina in the transverse processes of the first six cervical vertebrae, and is surrounded by branches from the inferior cervical sympathetic ganglion and by a plexus of veins which unite to form the vertebral vein at the caudal part of the neck. It is situated ventral to the trunks of the cervical nerves, and pursues an almost vertical course as far as the transverse process of the axis, above which it runs lateralward to the foramen in the transverse process of the atlas. The **third part** issues from the latter foramen on the medial side of the Rectus capitis lateralis, and curves around the superior articular process of the atlas, the anterior ramus of the first cervical nerve being on its medial side; it then lies in the groove on the cranial surface of the posterior arch of the atlas, and enters the vertebral canal by passing under the arch in the posterior atlantoöccipital membrane. This part of the artery is covered by the Semispinalis capitis and is contained in the **suboccipital triangle** (Fig. 6–15)—a triangular space bounded by the Rectus capitis posterior major, the Obliquus superior, and the Obliquus inferior. The first cervical or suboccipital nerve lies between the artery and the posterior arch of the atlas. The **fourth part** pierces the dura mater and inclines medially, ventral to the medulla oblongata; it is placed between the hypoglossal nerve and the anterior root of the first cervical nerve and beneath the first digitation of the ligamentum denticulatum. At the inferior border of the pons it unites with the vessel of the opposite side to form the basilar artery.

Branches.—The branches of the vertebral artery may be divided into two sets: those given off in the neck, and those within the cranium.

Cervical Branches

a) Spinal
b) Muscular

Cranial Branches

c) Meningeal
d) Posterior spinal

e) Anterior spinal
f) Posterior inferior cerebellar
g) Medullary

a) **Spinal branches** (*rami spinales*) enter the vertebral canal through the intervertebral foramina, and each divides into two branches. Of these, one passes along the roots of the nerves to supply the spinal cord and its membranes, anastomosing with the other arteries of the spinal cord; the other divides into an ascending and a descending branch, which unite with similar branches from the cranial and caudal arteries, so that two lateral anastomotic chains are formed on the dorsal surfaces of the bodies of the vertebrae, near the attachment of the pedicles. From these anastomotic chains branches are supplied to the periosteum and the bodies of the vertebrae, and others form communications with similar branches from the opposite side; from these communications small twigs arise which join similar branches cranially and caudally, to form a central anastomotic chain on the dorsal surface of the bodies of the vertebrae.

b) **Muscular branches** (Fig. 8–22) are given off to the deep muscles of the neck, where the vertebral artery curves around the articular process of the atlas. They anastomose with the occipital, and with the ascending and deep cervical arteries.

c) The **meningeal branch** (*ramus meningeus; posterior meningeal branch*) springs from the vertebral at the level of the foramen magnum, ramifies between the bone and dura mater in the cerebellar fossa, and supplies the falx cerebelli. It is frequently represented by one or two small branches.

d) The **posterior spinal artery** (*a. spinalis posterior; dorsal spinal artery*) arises from the vertebral, at the side of the medulla oblongata; passing dorsally, it descends on this structure, lying ventral to the dorsal roots of the spinal nerves, and reinforced by a succession of spinal branches it is continued to the end of the spinal cord, and the cauda equina. Branches from the posterior spinal arteries form a free anastomosis around the dorsal roots of the spinal nerves, and communicate, by means of very tortuous transverse branches, with the vessels of the opposite side. Close to its origin each gives off an ascending branch, which ends at the side of the fourth ventricle.

e) The **anterior spinal artery** (*a. spinalis anterior; ventral spinal artery*) (Fig. 8–29) is a small branch, which *arises* near the termination of the vertebral, and, descending ventral to the medulla oblongata, unites with its fellow of the opposite side at the level of the foramen magnum. The single trunk thus formed descends on the ventrum of the spinal cord, and is reinforced by a succession of small branches which enter the vertebral canal through the intervertebral foramina; these branches are derived from the vertebral and the ascending cervical of the inferior thyroid in the neck; from the intercostals in the thorax; and from the lumbar, iliolumbar, and lateral sacral arteries in the abdomen and pelvis. They unite, by means of ascending and descending branches, to form a single ventral median artery, which extends as far as the lower part of the spinal cord, and is continued as a slender twig on the filum terminale. This vessel is placed in the pia mater along the ventral median fissure; it supplies that membrane and the substance of the spinal cord, and sends off branches at its caudal part to be distributed to the cauda equina.

f) The **posterior inferior cerebellar artery** (*a. cerebelli inferior posterior*) (Fig. 8–25), the largest branch of the vertebral, winds posteriorly around the superior part of the medulla oblongata, passing between the origins of the vagus and accessory nerves, over the inferior peduncle to the inferior surface of the cerebellum, where it divides into two branches. The **medial branch** is continued posteriorly to the notch between the two hemispheres of the cerebellum; while the **lateral** supplies the inferior surface of the cerebellum, as far as its lateral border, where it anastomoses with the anterior inferior cerebellar and the superior cerebellar branches of the basilar artery. Branches from this artery supply the choroid plexus of the fourth ventricle.

g) The **medullary arteries** (*bulbar arteries*) are several minute vessels which spring from the vertebral and its branches and are distributed to the medulla oblongata.

1A. The **Basilar Artery** (*a. basilaris*) (Fig. 8–25), so named from its position at the base of the skull, is a single trunk formed by the junction of the two vertebral arteries: it extends from the inferior to the superior border of the pons, lying in its median groove, under cover of the arachnoid. It ends by dividing into the two posterior cerebral arteries.

Its **branches,** on either side, are the following:

a) Pontine
b) Labyrinthine
c) Anterior inferior cerebellar
d) Superior cerebellar
e) Posterior cerebral

a) The **pontine branches** (*rami ad pontem; transverse branches*) are a number of small vessels which come off at right angles from either side of the basilar artery and supply the pons and adjacent parts of the brain.

b) The **labyrinthine artery** (*a. labyrinthi; internal auditory artery*), a long slender branch, *arises* from near the middle of the artery; it accompanies the acoustic nerve through the internal acoustic meatus, and is distributed to the internal ear. It often *arises* from the anterior inferior cerebellar artery.

c) The **anterior inferior cerebellar artery** (*a. cerebelli inferior anterior*) passes posteriorly to be distributed to the anterior part of the inferior surface of the cerebellum, anastomosing with the posterior inferior cerebellar branch of the vertebral.

d) The **superior cerebellar artery** (*a. cerebelli superior*) *arises* near the termination of the basilar. It passes lateralward, immediately inferior to the oculomotor nerve, which separates it from the posterior cerebral artery and winds around the cerebral peduncle, close to the trochlear nerve. Arriving at the superior surface of the cerebellum, it divides into branches which ramify in the pia mater and anastomose with those of the inferior cerebellar arteries. Several branches go to the pineal body, the anterior medullary velum, and the tela choroidea of the third ventricle.

e) The **posterior cerebral artery** (*a. cerebri posterior*) (Figs. 8–25, 8–27, 8–29) is larger than the preceding, from which it is separated near its origin by the oculomotor nerve. Passing lateralward, parallel to the superior cerebellar artery, and receiving the posterior communicating from the internal carotid, it winds around the cerebral peduncle, and reaches the tentorial surface of the occipital lobe of the cerebrum, where it breaks up into branches for the supply of the temporal and occipital lobes. It may be a branch of the internal carotid.

The **branches** of the posterior cerebral are central, choroidal, and cortical.

i) The medial posterior **central branches** are a group of small arteries which *arise* at the commencement of the artery. They enter the posterior perforated substance and supply the medial surface of the thalamus and the walls of the third ventricle. The lateral posterior central *arise* after the posterior cerebral has circled the cerebral peduncle and supply the posterior portion of the thalamus.

ii) The posterior **choroidal branches** run anteriorly beneath the splenium of the corpus callosum and supply the tela choroidea and choroid plexus of the third ventricle.

iii) The **cortical branches** are (*1*) the *temporal,* anterior and posterior, which supply the uncus, fusiform gyrus, and inferior temporal gyrus, (*2*) the *occipital* or calcarine which supply the cuneus, lingual gyrus, and posterior surface of the occipital lobe, and (*3*) the *parietoöccipital* which go to the cuneus and precuneus.

2. The **Thyrocervical Trunk** (*truncus thyrocervicalis; thyroid axis*) (Figs. 8–18, 8–30) is short and thick; it *arises* from the first portion of the subclavian artery, close to the medial border of the Scalenus anterior, and divides almost immediately into the following three branches:

> *a*) Inferior thyroid
> *b*) Suprascapular
> *c*) Transverse cervical

a) The **inferior thyroid artery** (*a. thyroidea inferior*) runs cranially, ventral to the vertebral artery and Longus colli; then loops medialward dorsal to the carotid sheath and its contents, and the sympathetic trunk with the middle cervical ganglion resting upon the vessel. Reaching the caudal border of the thyroid gland it divides into two branches, which supply the caudal parts of the gland, and anastomose with the superior thyroid, and with the corresponding artery of the opposite side. The recur-

rent nerve passes cranialward generally dorsal, but occasionally ventral to the artery.

The **branches** of the inferior thyroid are:

> i) Inferior laryngeal
> ii) Tracheal
> iii) Esophageal
> iv) Ascending cervical
> v) Muscular

i) The **inferior laryngeal artery** (*a. laryngea inferior*) ascends upon the trachea to the dorsal part of the larynx under cover of the Constrictor pharyngis inferior, in company with the recurrent nerve, and supplies the muscles and mucous membrane of this part, anastomosing with the branch from the opposite side, and with the superior laryngeal branch of the superior thyroid artery.

ii) The **tracheal branches** (*rami tracheales*) are distributed upon the trachea, and anastomose caudally with the bronchial arteries.

iii) The **esophageal branches** (*rami oesophagei*) supply the esophagus, and anastomose with the esophageal branches of the aorta.

iv) The **ascending cervical artery** (*a. cervicalis ascendens*) *arises* from the inferior thyroid as that vessel passes dorsal to the carotid sheath, on the anterior tubercles of the transverse processes of the cervical vertebrae in the interval between the Scalenus anterior and Longus capitis. It supplies the muscles of the neck with twigs which anastomose with branches of the vertebral, sending one or two **spinal branches** into the vertebral canal through the intervertebral foramina to be distributed to the bodies of the vertebrae, in the same manner as the spinal branches from the vertebral. It anastomoses with the ascending pharyngeal and occipital arteries.

v) The **muscular branches** supply the infrahyoid muscles, the Longus colli, Scalenus anterior, and Constrictor pharyngis inferior, to the spinal cord and its membranes, and inferior.

b) The **suprascapular artery** (*a. suprascapularis; transverse scapular artery*) passes laterally across the Scalenus anterior and phrenic nerve, being covered by the Sternocleidomastoideus. It crosses the subclavian artery and the brachial plexus, and runs deep to and parallel with the clavicle and Subclavius, and deep to the inferior belly of the Omohyoideus, to the superior border of the scapula. It passes over the superior transverse ligament of the scapula by which it is separated from the suprascapular nerve, to enter the supraspinous fossa. In this situation it lies close to the bone, and ramifies between it and the Supraspinatus, to which it supplies branches. It then winds around the neck of the scapula through the great scapular notch under cover of the inferior transverse ligament, and reaches the infraspinous fossa, where it anastomoses with the scapular circumflex and the descending scapular arteries. Beside distributing branches to the Sternocleidomastoideus, Subclavius, and neighboring muscles, it gives off a **suprasternal branch,** which crosses over the sternal end of the clavicle to the skin of the chest, and an **acromial branch,** which pierces the Trapezius and supplies the skin over the acromion, anastomosing with the thoracoacromial artery. As the artery passes over the superior transverse ligament of the scapula, it sends a branch into the subscapular fossa, where it ramifies beneath the Subscapularis, and anastomoses with the subscapular artery and descending scapular arteries. It also sends **articular branches** to the acromioclavicular and shoulder joint, and a **nutrient artery** to the **clavicle.**

, c) The **transverse cervical artery** (*a. transversa colli*) occurs in two forms with about equal frequency. Its two main parts may have separate origins. When the transverse cervical *arises* as a common source (Fig. 8–18), it passes laterally across the posterior cervical triangle in a position somewhat cranial and dorsal to the suprascapular artery. Medially it crosses superficial to the Scalenus anterior and phrenic nerve and lies deep to the Sternocleidomastoideus. Laterally it crosses the trunks of the brachial plexus covered only by the Platysma, investing layer of deep fascia, and the inferior belly of the Omohyoideus. As it approaches the anterior margin of the Trapezius it divides into a superficial branch and a deep branch.

i) The **superficial branch** (*ramus superficialis; ascending branch; superficial cervi-*

cal) (Figs. 8–18, 8–30) is the principal blood supply of the Trapezius, and lying against its deep surface it divides into an ascending and a descending branch. The former ascends along the anterior border of the Trapezius, distributing branches to it and to neighboring muscles and anastomosing with the superficial part of the descending branch of the occipital artery. The descending branch accompanies the accessory nerve on the deep surface of the Trapezius, supplying the muscle.

ii) The **deep branch** (*ramus profundus; descending branch*) (Fig. 8–31) gives branches to the Levator scapulae and neighboring deep cervical muscles. Passing deep to the Levator it reaches the superior angle of the scapula and passes down along the vertebral border to the inferior angle lying deep to the Rhomboidei and accompanying the dorsoscapular nerve. It supplies muscular branches to the Rhomboidei, Serratus posterior superior, Subscapularis, Supra- and Infraspinatus, and other neighboring muscles. These branches anastomose with the suprascapular, subscapular and circumflex scapular arteries.

iii) The **superficial cervical artery** (*a. cervicalis superficialis*) (Fig. 8–30) corresponds to the superficial branch of the transverse cervical in (i) above, but *arises* from the thyrocervical trunk without sharing its origin with the deep branch. It passes laterally across the posterior triangle of the neck, at first deep to the Sternocleidomastoideus and superficial to the Scalenus anterior and phrenic nerve, then cranialward across the trunks of the brachial plexus to the anterior border of the Trapezius. Here it divides into an ascending branch and a descending branch. The ascending branch supplies the upper part of the Trapezius and neighboring muscles and anastomoses with the superficial descending branch of the occipital artery. The descending branch lies against the deep surface of the Trapezius, supplying it with branches and accompanying the accessory nerve.

iv) The **descending scapular artery** (*a. scapularis descendens; dorsal scapular artery*) (Figs. 8–30, 8–31) is also called in many books the transverse cervical artery. Its distribution corresponds to that of the

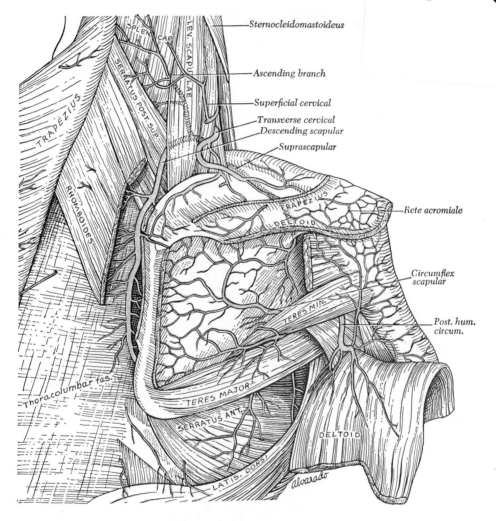

Sternocleidomastoideus

Ascending branch

Superficial cervical

Transverse cervical
Descending scapular

Suprascapular

Rete acromiale

Circumflex
scapular

Post. hum.
circum.

SPLEN. CAP.
LEV. SCAPULAE
SERRATUS POST. SUP.
TRAPEZIUS
RHOMBOIDES
TRAPEZIUS
DELTOID
TERES MIN.
Thoracolumbar fas.
TERES MAJOR
SERRATUS ANT.
DELTOID
LATIS. DORSI
alvarado

Fig. 8–31.—The arteries of the dorsal aspect of the shoulder and neck (Redrawn after Tiedmann.)

deep branch of the transverse cervical in (ii) above. It has an independent *origin* from the third part of the subclavian artery instead of from the thyrocervical trunk. It passes cranialward for a short distance, then loops around the brachial plexus, frequently passing between the anterior and posterior divisions of the upper trunk. It courses caudalward toward the scapula and at the border of the Levator scapulae divides into an ascending and a descending branch. The ascending branch (*ramus ascendens*) supplies the Levator scapulae and neighboring deep cervical muscles. The descending branch (*posterior scapular artery; ramus descendens*), usually double, passes deep to the Levator scapulae to the superior angle of the scapula, and then descends

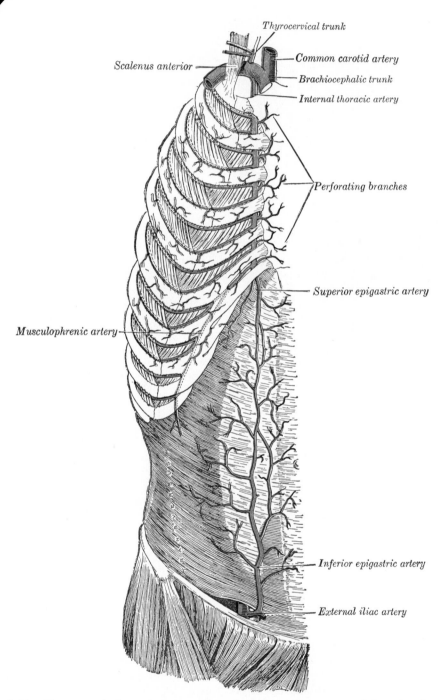

FIG. 8–32.—The internal thoracic artery and its branches. The intercostal branches lie at first between the pleura and the internal intercostal muscles, then between the internal and external intercostal muscles. (The figure does not indicate these relationships.)

ventral to the Rhomboidei along the vertebral border of the bone as far as the inferior angle. The medial division is accompanied by the dorsal scapular nerve. The lateral division lies on the costal surface of the Serratus anterior muscle and is usually larger than the medial division. The descending branch supplies the Rhomboidei, Latissimus dorsi, and Trapezius, and anastomoses with the suprascapular and subscapular arteries, and with the posterior branches of some of the intercostal arteries (Huelke '58).

3. The **Internal Thoracic Artery** (*a. thoracica interna; internal mammary artery*) (Fig. 8–32) *arises* from the first portion of the subclavian opposite the thyrocervical trunk. It descends dorsal to the cartilages of the upper six ribs at a distance of about 1.25 cm from the margin of the sternum, and at the level of the sixth intercostal space divides into the **musculophrenic** and **superior epigastric arteries.**

Relations.—In its descent toward the sternum it is deep to the sternal end of the clavicle, the subclavian and internal jugular veins, and the first costal cartilage. As it enters the thorax it passes close to the lateral side of the brachiocephalic vein, and the phrenic nerve crosses from its lateral to its medial side. Caudal to the first costal cartilage it descends almost vertically to its point of bifurcation. It is covered ventrally by the cartilages of the first six ribs and the intervening Intercostales interni and anterior intercostal membranes, and is crossed by the terminal portions of the first six intercostal nerves. It rests on the pleura, as far as the third costal cartilage; caudal to this level, upon the Transversus thoracis. It is accompanied by a pair of veins that unite to form a single vessel which runs medial to the artery and ends in the corresponding brachiocephalic vein.

Branches.—The branches of the internal thoracic are:

    a) Pericardiophrenic
    b) Mediastinal
    c) Thymic
    d) Sternal
    e) Anterior intercostal
    f) Perforating
    g) Musculophrenic
    h) Superior epigastric

a) The **pericardiophrenic artery** (*a. pericardiacophrenica; a. comes nervi phrenici*) is a long slender branch, which accompanies the phrenic nerve, between the pleura and pericardium, to the diaphragm, to which it is distributed; it anastomoses with the musculophrenic and inferior phrenic arteries (Fig. 7–19).

b) The **mediastinal branches** (*rami mediastinales; anterior mediastinal arteries*) are small vessels, distributed to the areolar tissue and lymph nodes in the anterior mediastinum and supply the cranial part of the ventral surface of the pericardium; the caudal part receives branches from the musculophrenic artery.

c) The **thymic branches** supply the remains of the thymus.

d) The **sternal branches** (*rami sternales*) are distributed to the Transversus thoracis, and to the posterior surface of the sternum.

The mediastinal, pericardial, and sternal branches, together with some twigs from the pericardiacophrenic, anastomose with branches from the intercostal and bronchial arteries, and form a **subpleural mediastinal plexus.**

e) The **anterior intercostal arteries** (*rami intercostales anteriores; intercostal arteries*) (Fig. 8-32) supply the first five or six intercostal spaces; the rest of the spaces are supplied by the musculophrenic. In each space, a small artery passes laterally lying near the caudal margin of the more cranial rib, supplying the muscles and finally anastomosing with the aortic intercostals. At first the artery lies between the pleura and the Internal intercostal, then between the two muscles. A branch may run along the caudal part of the intercostal space and some branches perforate the Externus to supply the Pectorals and the mammary gland.

f) The **perforating branches** (*rami perforantes*) pierce the ventral chest wall in the first five or six intercostal spaces close to the sternum. They penetrate the Internus, external membrane and Pectoralis major and bending laterally as they become more superficial supply the Pectoralis and integument. The arteries of the second, third, and fourth spaces give **mammary branches** to the mamma in the female and are greatly enlarged during lactation.

g) The **musculophrenic artery** (*a. musculophrenica*) is directed obliquely caudalward and lateralward, dorsal to the cartilages of the false ribs; it perforates the diaphragm at the eighth or ninth costal cartilage, and ends, considerably reduced in size, opposite the last intercostal space. It gives off intercostal branches to the seventh, eighth, and ninth intercostal spaces; these diminish in size as the spaces decrease in length, and are distributed in a manner precisely similar to the intercostals from the internal thoracic. The musculophrenic also gives branches to the caudal part of the pericardium, and others which run dorsalward to the diaphragm, and caudalward to the abdominal muscles.

h) The **superior epigastric artery** (*a. epigastrica superior*) continues in the original direction of the internal thoracic; it descends through the interval between the costal and sternal attachments of the diaphragm, and enters the sheath of the Rectus abdominis, at first lying dorsal to the muscle, and then perforating and supplying it, and anastomosing with the inferior epigastric artery from the external iliac. Branches perforate the anterior wall of the sheath of the Rectus, and supply the other muscles of the abdomen and the integument, and a small branch passes ventral to the xiphoid process and anastomoses with the artery of the opposite side. It also gives some twigs to the diaphragm, while from the artery of the right side small branches extend into the falciform ligament of the liver and anastomose with the hepatic artery.

4. The **Costocervical Trunk** (*truncus costocervicalis; superior intercostal artery*) (Fig. 8–22), *arises* from the cranial and dorsal part of the subclavian artery, deep to the Scalenus anterior on the right side, and medial to that muscle on the left side. Passing dorsalward, it gives off the profunda cervicalis and continues as the highest intercostal artery.

a) The **highest intercostal artery** descends under the pleura ventral to the necks of the first and second ribs, and anastomoses with the first aortic intercostal. As it crosses the neck of the first rib it lies medial to the anterior division of the first thoracic nerve, and lateral to the first thoracic ganglion of the sympathetic trunk. In the first intercostal space, it gives off a branch which is distributed in a manner similar to the distribution of the aortic intercostals. The branch for the second intercostal space usually joins with one from the highest aortic intercostal artery. This branch is not constant, but is more commonly found on the right side; when absent, its place is supplied by an intercostal branch from the aorta. Each intercostal gives off a dorsal branch which goes to the dorsal vertebral muscles, and sends a small spinal branch through the corresponding intervertebral foramen to the spinal cord and its membranes.

The **deep cervical artery** (*a. cervicalis profunda*) *arises*, in most cases, from the costocervical trunk, and is analogous to the dorsal branch of an aortic intercostal artery: occasionally it is a separate branch from the subclavian artery. Passing dorsalward, above the eighth cervical nerve and between the transverse process of the seventh cervical vertebra and the neck of the first rib, it runs up the dorsum of the neck, between the Semispinales capitis and cervicis, as high as the axis vertebra, supplying these and adjacent muscles, and anastomosing with the deep division of the descending branch of the occipital, and with branches of the vertebral. It gives off a spinal twig which enters the canal through the intervertebral foramen between the seventh cervical and first thoracic vertebrae.

5. The **Descending Scapular Artery** (*a. scapularis descendens*), the only branch of the third part of the subclavian occurs in half the bodies (see page 606).

THE ARTERIES OF THE UPPER LIMB

The axis artery (page 567) which supplies the upper limb continues from its commencement to the elbow as a single trunk, but different portions of it have received different names according to the regions through which they pass. That part of the vessel which extends from its origin to the outer border of the first rib is termed the subclavian; beyond this point to the distal border of the axilla it is named the **axillary;** and from there to the bend of the elbow it is termed the **brachial,** which ends by dividing into two branches, the **radial** and **ulnar.**

The Axilla

The axilla occupies a pyramidal space between the upper lateral part of the chest and the medial side of the arm (see p. 452).

Boundaries.—The *apex,* which is directed cranialward toward the root of the neck, corresponds to the interval between the outer border of the first rib, the superior border of the scapula, and the dorsal surface of the clavicle, and through it the axillary vessels and accompanying nerves pass. The *base* is broad at the chest but narrow and pointed at the arm; it is formed by the integument and a thick layer of fascia, the **axillary fascia,** extending between the caudal border of the Pectoralis major ventrally, and the caudal border of the Latissimus dorsi dorsally. The *ventral wall* is formed by the Pectorales major and minor, the former covering the whole of this wall, the latter only its central part. The space between the cranial border of the Pectoralis minor and the clavicle is occupied by the coracoclavicular fascia. The *dorsal wall,* which extends somewhat more caudally than the ventral, is formed by the Subscapularis cranially, the Teres major and Latissimus dorsi caudally. On the *medial side* are the first four ribs with their corresponding Intercostals, and part of the Serratus anterior. On the *lateral side,* where the ventral and dorsal walls converge, the space is narrow, and bounded by the humerus, the Coracobrachialis, and the Biceps brachii.

Contents.—It contains the axillary vessels, and the brachial plexus of nerves, with their branches, some branches of the intercostal nerves, and a large number of lymph nodes, together with a quantity of fat and loose areolar tissue. The axillary artery and vein, with the brachial plexus of nerves, extend obliquely along the lateral boundary of the axilla, from its apex to its base, and are placed much nearer to the ventral than to the dorsal wall, the vein lying to the thoracic side of the artery and partially concealing it. At the ventral part of the axilla, in contact with the Pectorals, are the thoracic

branches of the axillary artery, and along the caudal margin of the Pectoralis minor the lateral thoracic artery extends to the side of the chest. Dorsally, in contact with the lower margin of the Subscapularis, are the subscapular vessels and nerves; winding around the lateral border of this muscle are the scapular circumflex vessels; and, close to the neck of the humerus, the posterior humeral circumflex vessels and the axillary nerve curve dorsally to the shoulder. Along the medial or thoracic side no vessel of any importance exists, the cranial part of the space being crossed merely by a few small branches from the highest thoracic artery. There are some important nerves, however, in this situation, viz., the long thoracic nerve, descending on the surface of the Serratus anterior, to which it is distributed; and the intercostobrachial nerve, perforating the ventral part of this wall, and passing across the axilla to the medial side of the arm.

The position and arrangement of the lymph nodes are described in the section on Lymphatics.

The Axillary Artery

The **axillary artery** (*a. axillaris*) (Fig. 8–33), the continuation of the subclavian, commences at the outer border of the first rib, and ends at the distal border of the tendon of the Teres major, where it takes the name of brachial. Its direction varies with the position of the limb; thus the vessel is nearly straight when the arm is directed at right angles with the trunk, but concave cranially when the arm is elevated, and convex when the arm lies by the side. At its origin the artery is very deeply situated, but near its termination is superficial, being covered only by the skin and fascia. To facilitate the description of the vessel it is divided into three portions; the first part lies proximal, the second deep, and the third distal to the Pectoralis minor. Its branches are subject to great variation.

Relations.—The **first portion** of the axillary artery is covered *ventrally* by the clavicular portion of the Pectoralis major and the clavipectoral fascia, and is crossed by the lateral anterior thoracic nerve, and the thoracoacromial and cephalic veins; *dorsal* to it are the first intercostal space, the corresponding Intercostalis externus, the first and second digitations of the Serratus anterior, and the long thoracic and medial anterior thoracic nerves, and the medial cord of the brachial plexus; on its *lateral side* is the brachial plexus, from which it is separated by a little areolar tissue; on its *medial,* or thoracic side, is the axillary vein which overlaps the artery. It is enclosed, together with the axillary vein and the brachial plexus, in a fibrous sheath —the **axillary sheath**—continuous above with the deep cervical fascia.

The **second portion** of the axillary artery is covered, *ventrally,* by the Pectorales major and minor; *dorsal* to it are the posterior cord of the brachial plexus, and some areolar tissue which intervenes between it and the Subscapularis; on the *medial side* is the axillary vein, separated from the artery by the medial cord of the brachial plexus and the medial pectoral nerve; on the *lateral side* is the lateral cord of the brachial plexus. The brachial plexus thus surrounds the artery on three sides, and separates it from direct contact with the vein and adjacent muscles.

The **third portion** of the axillary artery extends from the lower border of the Pectoralis minor to the distal border of the tendon of the Teres major. *Ventrally,* it is covered by the Pectoralis major proximally, but only by the integument and fascia more distally; *dorsally,* it is in relation with the Subscapularis, and the tendons of the Latissimus dorsi and Teres major; on its *lateral side* is the Coracobrachialis, and on its *medial* or thoracic side, the axillary vein. The nerves of the brachial plexus bear the following relations to this part of the artery: on the *lateral side* are the lateral head and the trunk of the median, and the musculocutaneous for a short distance; on the *medial side* the ulnar (between the vein and artery) and medial brachial cutaneous (to the medial side of the vein); *ventrally* are the medial head of the median and the medial antebrachial cutaneous, and *dorsally,* the radial and axillary, the latter only as far as the distal border of the Subscapularis.

Collateral Circulation after Ligature of the Axillary Artery.—If the artery be tied proximal to the origin of the thoracoacromial, the collateral circulation will be carried on by the same branches as after the ligature of the third part of the subclavian; if at a more distal point, between the thoracoacromial and the subscapular, the latter vessel, by its free anastomosis with the suprascapular and transverse cervical branches of the subclavian, will become the chief agent in carrying on the circulation; the lateral thoracic, if it be distal to the ligature, will materially contribute by its anastomoses

with the intercostal and internal thoracic arteries. If the point included in the ligature is distal to the origin of the subscapular artery, it will probably also be distal to the origins of the two humeral circumflex arteries. The chief agents in restoring the circulation will then be the subscapular and the two humeral circumflex arteries anastomosing with the a. profunda brachii.

Branches.—The branches of the axillary are:

From first part
  1. Highest thoracic

From second part
  2. Thoracoacromial
  3. Lateral thoracic

From third part
  4. Subscapular
  5. Posterior humeral circumflex
  6. Anterior humeral circumflex

1. The **Highest Thoracic Artery** (*a. thoracica suprema*) is a small vessel which may *arise* from the thoracoacromial or may be absent. Running medially along the cranial border of the Pectoralis minor, it passes between it and the Pectoralis major to the side of the chest. It supplies branches to these muscles, and to the parietes of the thorax, and anastomoses with the internal thoracic and intercostal arteries.

2. The **Thoracoacromial Artery** (*a. thoracoacromialis; acromiothoracic artery; thoracic axis*) is a short trunk, which *arises* ventrally from the axillary artery, its origin being generally overlapped by the cranial edge of the Pectoralis minor. At the border of this muscle, it pierces the clavipectoral fascia and divides into four branches— pectoral, acromial, clavicular, and deltoid.

(*a*) The **pectoral branch** descends between the two Pectorales, and is distributed to them and to the mamma, anastomosing with the intercostal branches of the internal and lateral thoracics.

(*b*) The **acromial branch** runs laterally superficial to the coracoid process and deep to the Deltoideus, to which it gives branches; it then pierces that muscle and ends on the acromion in an arterial network formed by branches from the suprascapular,

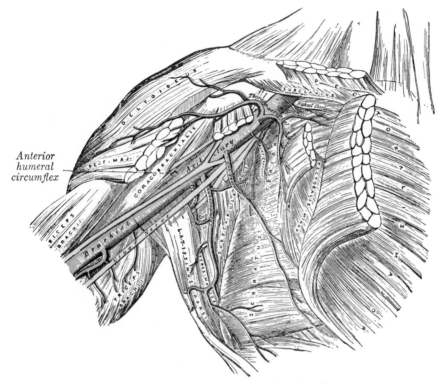

Anterior
humeral
circumflex

FIG. 8–33.—The axillary artery and its branches.

thoracoacromial, and posterior humeral circumflex arteries (Fig. 8–30).

(c) The **clavicular branch** runs medially to the sternoclavicular joint, supplying this articulation, and the Subclavius.

(d) The **deltoid** (*humeral*) **branch**, often arising with the acromial, crosses superficial to the Pectoralis minor and passes in the same groove as the cephalic vein, between the Pectoralis major and Deltoideus, and gives branches to both muscles.

3. The **Lateral Thoracic Artery** (*a. thoracalis lateralis; long thoracic artery; external mammary artery*) (Fig. 8–33) follows the lower border of the Pectoralis minor to the side of the chest, supplying the Serratus anterior and the Pectoralis, and sending branches across the axilla to the axillary lymph nodes and Subscapularis; it anastomoses with the internal thoracic, subscapular, and intercostal arteries, and with the pectoral branch of the thoracoacromial. In the female it supplies an **external mammary branch** (*ramus mammarius lateralis*) which turns around the free edge of the Pectoralis major and supplies the mamma. The lateral thoracic artery is often (60 per cent) a branch either of the thoracoacromial artery or of the subscapular artery. It comes directly from the axillary only in about 30 per cent of cases.

4. The **Subscapular Artery** (*a. subscapularis*) (Fig. 8–30), the largest branch of the axillary artery, *arises* at the distal border of the Subscapularis and after a short course, about 4 cm, divides into (a) the circumflex scapular and (b) thoracodorsal arteries.

(a) The **scapular circumflex artery** (*a. circumflexa scapulae*) (Fig. 8–34) is generally larger than the thoracodorsal artery. It curves around the lateral border of the scapula, traversing the triangular space between the Teres major, the Subscapularis, and the long head of the Triceps. It enters the infraspinous fossa between the Teres minor and the scapula, remaining close to the bone. It supplies branches to the Infraspinatus and anastomoses with the suprascapular and descending scapular arteries. In the triangular space it gives a branch to the Subscapularis and a considerable branch continues along the lateral border of the scapula between the Teres major and minor, supplying these muscles and anastomosing with the descending scapular at the inferior angle of the scapula. It also supplies branches to the long head of the Triceps and the Deltoideus.

(*b*) The **thoracodorsal artery** (*a. thoracodorsalis*) (Fig. 8–30) is the continuation of the subscapular artery through the posterior portion of the axilla along the anterior border of the Latissimus dorsi in company with the thoracodorsal nerve. It gives branches to the Subscapularis, is the principal supply of the Latissimus, and anastomoses with the circumflex scapular and descending scapular arteries. One or two sizable branches cross the axilla to supply the Serratus anterior and intercostal muscles, anastomosing with intercostal, lateral thoracic, and thoracoacromial branches. When the lateral thoracic artery is small or lacking, this branch may take its place.

5. The **Posterior Humeral Circumflex Artery** (*a. circumflexa humeri posterior; posterior circumflex artery*) (Fig. 8–34) *arises* from the axillary artery at the distal border of the Subscapularis, and runs dorsally with the axillary nerve through the quadrangular space bounded by the Subscapularis and Teres minor proximally, the Teres major distally, the long head of the

Triceps brachii medially, and the surgical neck of the humerus laterally. It winds around the neck of the humerus and is distributed to the Deltoideus and shoulder joint, anastomosing with the anterior humeral circumflex and profunda brachii.

6. The **Anterior Humeral Circumflex Artery** (*a. circumflexa humeri anterior; anterior circumflex artery*) (Fig. 8–33) is considerably smaller than the posterior opposite which it *arises* from the lateral side of the axillary artery. It may arise in common with the posterior humeral circumflex, or be represented by three or four very small branches. It runs horizontally, beneath the Coracobrachialis and short head of the Biceps brachii, ventral to the neck of the humerus. On reaching the intertubercular sulcus, it gives off a branch which ascends in the sulcus to supply the head of the humerus and the shoulder joint. The trunk of the vessel is then continued onward deep to the long head of the Biceps brachii and the Deltoideus, and anastomoses with the posterior humeral circumflex artery.

The Brachial Artery

The **Brachial artery** (*a. Brachialis*) (Fig. 8–35) commences at the distal margin of the tendon of the Teres major, and,

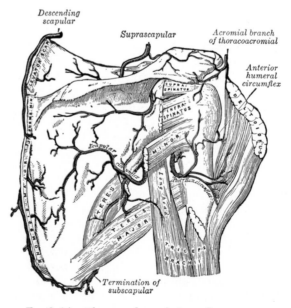

FIG. 8–34.—The scapular and circumflex arteries.

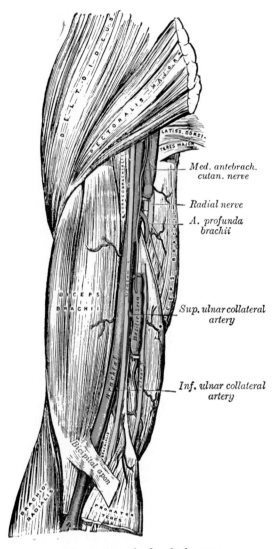

FIG. 8–35.—The brachial artery.

passing down the arm, ends about 1 cm distal to the bend of the elbow, where it divides into the **radial** and **ulnar arteries.** At first the brachial artery lies medial to the humerus, but as it runs down the arm it gradually becomes anterior to the bone, and at the bend of the elbow it lies midway between its two epicondyles.

Relations.—The artery is superficial throughout its entire extent, being covered, *anteriorly,* by the integument and the subcutaneous and deep fasciae; the bicipital fascia (*lacterus fibrosus*) lies anterior to it opposite the elbow and separates it from the vena mediana cubiti; the median nerve crosses from its lateral to its

medial side opposite the insertion of the Coracobrachialis. *Posteriorly,* it is separated from the long head of the Triceps brachii by the radial nerve and a. profunda brachii. It then lies upon the medial head of the Triceps brachii, next upon the insertion of the Coracobrachialis, and lastly on the Brachialis. *Laterally,* it is in relation proximally with the median nerve and the Coracobrachialis, distally with the Biceps brachii, the two muscles overlapping the artery to a considerable extent. *Medially,* its proximal half is in relation with the medial antebrachial cutaneous and ulnar nerves, its distal half with the median nerve. The basilic vein lies on its medial side, but is separated from it in the distal part of the arm by the deep fascia. The artery is accompanied by two venae comitantes, which lie in close contact with it, and are connected together at intervals by short transverse branches.

The Antecubital Fossa.—At the bend of the elbow the brachial artery sinks deeply into a triangular interval, the **antecubital fossa.** The base of the triangle is directed proximally, and is represented by a line connecting the two epicondyles of the humerus; the sides are formed by the medial edge of the Brachioradialis and the lateral margin of the Pronator teres; the floor is formed by the Brachialis and Supinator. This fossa contains the brachial artery, with its accompanying veins; the radial and ulnar arteries; the median nerve; and the tendon of the Biceps brachii. The brachial artery occupies the middle of the fossa, and divides opposite the neck of the radius into the radial and ulnar arteries; it is covered, anteriorly, by the integument, the subcutaneous fascia, and the vena mediana cubiti, the last being separated from the artery by the bicipital fascia (*lacertus fibrosus*). Posteriorly it is the Brachialis which separates it from the elbow joint. The median nerve lies close to the medial side of the artery, proximally, but is separated from it distally by the ulnar head of the Pronator teres. The tendon of the Biceps brachii lies to the lateral side of the artery.

The radial nerve lies normally just outside the fossa between the Supinator and the Brachioradialis, but may be exposed through the fossa by drawing the Brachioradialis laterally.

Variations of the Brachial Artery.—The brachial artery, accompanied by the median nerve, may leave the medial border of the Biceps brachii, and descend toward the medial epicondyle of the humerus; in such cases it usually passes behind the *supracondylar process* of the humerus, from which a fibrous arch is thrown over the artery; it then runs beneath or through the substance of the Pronator teres, to the bend of the elbow. This variation bears considerable analogy with the normal condition of the artery in some of the carnivora; it has been referred to in the description of the humerus (p. 208).

A frequent variation is the **superficial brachial artery,** which may continue into the forearm to form a superficial antebrachial artery. The superficial brachial may rejoin the brachial distally. Frequently the brachial artery divides at a higher level than usual, and the vessels concerned in this high division are three, viz., radial, ulnar, and interosseous. Most frequently the radial is given off high up, the other limb of the bifurcation consisting of the ulnar and interosseous; in some instances the ulnar arises above the ordinary level, and the radial and interosseous form the other limb of the division; occasionally the interosseous arises high up (see Fig. 8–11).

Sometimes, long slender vessels, *vasa aberrantia,* connect the brachial or the axillary artery with one of the arteries of the forearm, or branches from them. These vessels usually join the radial.

Collateral Circulation.—After the application of a ligature to the brachial artery in the upper third of the arm, the circulation is carried on by branches from the humeral circumflex and subscapular arteries anastomosing with ascending branches from the profunda brachii. If the artery be tied *below* the origin of the profunda brachii and superior ulnar collateral, the circulation is maintained by the branches of these two arteries anastomosing with the inferior ulnar collateral, the radial and ulnar recurrents, and the dorsal interosseous.

Branches.—The branches of the brachial artery are (Fig. 8–35):

1. Deep brachial
2. Nutrient of humerus
3. Superior ulnar collateral
4. Inferior ulnar collateral
5. Muscular

1. The **Deep Brachial Artery** (*a. profunda brachii; superior profunda artery*), the largest branch, *arises* from the medial and posterior part of the brachial, just distal to the distal border of the Teres major. It passes deeply into the arm between the long and lateral heads of the Triceps brachii and then accompanies the radial nerve in the spiral groove between the lateral and medial heads of the Triceps, on the posterior aspect of the humerus, and terminates by dividing into the radial collateral and the middle collateral branches. It has five branches:

a) The **ascending branch** (*ramus deltoideus*) runs proximally between the long and lateral heads of the Triceps to anastomose with the posterior humeral circumflex artery.

b) The **radial collateral artery** (*a. collateralis radialis*) frequently described as the terminal portion of the profunda, continues with the radial nerve into the forearm. It lies deep to the lateral head of the Triceps until it reaches the lateral supracondylar ridge where it pierces the lateral intermuscular septum, descends between the Brachioradialis and the Brachialis to the palmar aspect of the lateral epicondyle, and ends by anastomosing with the radial recurrent artery. Just before it pierces the intermuscular septum, it gives off a branch which descends to the posterior aspect of the lateral epicondyle and there joins the anastomosis about the olecranon.

c) The **middle collateral artery** (*a. collateralis media*) enters the substance of the long and medial heads of the Triceps and descends along the posterior aspect of the humerus to the elbow where it anastomoses with the interosseous recurrent and joins the anastomosis about the olecranon.

d) **Muscular branches** supply the Deltoideus and the other muscles between which the artery runs.

e) A **nutrient artery** (*a. nutricia humeri*) to the humerus may enter a canal posterior to the deltoid tuberosity.

2. The **Nutrient Artery** (*a. nutricia humeri*) of the body of the humerus *arises* about the middle of the arm and enters the nutrient canal near the insertion of the Coracobrachialis.

3. The **Superior Ulnar Collateral Artery** (*a. collateralis ulnaris superior; inferior profunda artery*), long and slender, *arises* from the brachial a little distal to the middle of the arm; it frequently springs from the

proximal part of the a. profunda brachii. It pierces the medial intermuscular septum, and descends on the surface of the medial head of the Triceps brachii to the space between the medial epicondyle and olecranon, accompanied by the ulnar nerve, and ends deep to the Flexor carpi ulnaris by anastomosing with the posterior ulnar recurrent, and inferior ulnar collateral. It sometimes sends a branch anterior to the medial epicondyle, to anastomose with the anterior ulnar recurrent.

4. The **Inferior Ulnar Collateral Artery** (*a. collateralis ulnaris inferior; anastomatica magna artery*) *arises* about 5 cm proximal to the elbow. It passes medialward upon the Brachialis, and piercing the medial intermuscular septum, winds around the dorsum of the humerus between the Triceps brachii and the bone, forming, by its junction with the profunda brachii, an arch proximal to the olecranon fossa. As the vessel lies on the Brachialis, it gives off branches which ascend to join the superior ulnar collateral: others descend anterior to the medial epicondyle, to anastomose with

the anterior ulnar recurrent. Posterior to the medial epicondyle a branch anastomoses with the superior ulnar collateral and posterior ulnar recurrent arteries.

5. The **Muscular Branches** (*rami musculares*), three or four in number, are distributed to the Coracobrachialis, Biceps brachii, and Brachialis.

The Anastomosis Around the Elbow Joint (*rete articulare cubiti*) (Fig. 8–36).—The vessels engaged in this anastomosis may be conveniently divided into those anterior to and those posterior to the medial and lateral epicondyles of the humerus. The branches anastomosing anterior to the medial epicondyle are: the anterior branch of the inferior ulnar collateral, the anterior ulnar recurrent, and the anterior branch of the superior ulnar collateral. Those posterior to the medial epicondyle are: the inferior ulnar collateral, the posterior ulnar recurrent, and the posterior branch of the superior ulnar collateral. The branches anastomosing anterior to the lateral epicondyle are: the radial recurrent and the radial collateral branch of the profunda brachii. Those posterior to

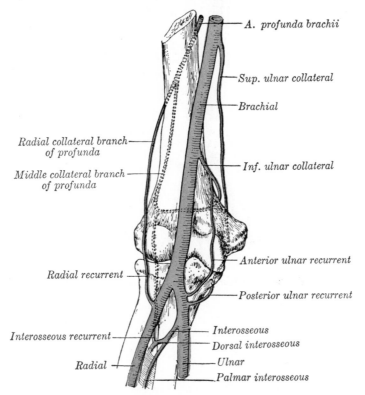

Fig. 8–36.—Diagram of the anastomosis around the elbow joint.

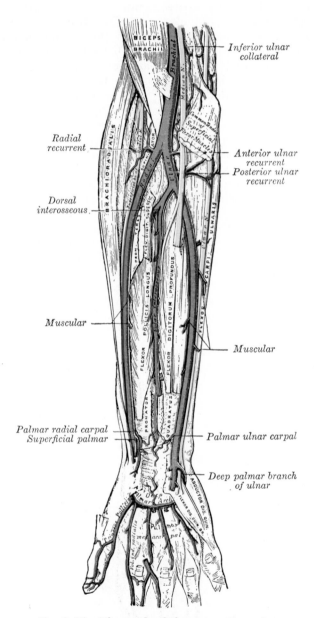

Fig. 8–37.—Ulnar and radial arteries. Deep view.

the lateral epicondyle (perhaps better described as situated between the lateral epicondyle and the olecranon) are: the inferior ulnar collateral, the interosseous recurrent, and the middle collateral branch of the profunda brachii. There is also an arch of anastomosis proximal to the olecranon, formed by the interosseous recurrent joining with the inferior ulnar collateral and posterior ulnar recurrent (Fig. 8–36).

The Radial Artery

The **radial artery** (*a. Radialis*) (Fig. 8–37) appears, from its direction, to be a continuation of the brachial, but it is smaller in caliber than the ulnar. It commences at the bifurcation of the brachial, just distal to the bend of the elbow, and passes along the radial side of the forearm to the wrist. It then winds dorsally, around the lateral side of the carpus, deep to the tendons of the Abductor pollicis longus and Extensores pollicis longus and brevis to the proximal end of the space between the metacarpal bones of the thumb and index finger. Finally it passes between the two heads of the first Interosseous dorsalis into the palm of the hand, to form the deep palmar arch.

Relations.—(*a*) *In the forearm* the artery extends from the neck of the radius to the forepart of the styloid process, being placed to the medial side of the body of the bone proximally, and in front of it distally. Its proximal part is overlapped by the fleshy belly of the Brachioradialis; the rest of the artery is superficial, being covered by the integument and the subcutaneous and deep fasciae. In its course, it lies upon the Biceps Brachii tendon, the Supinator, the Pronator teres, the radial origin of the Flexor digitorum superficialis, the Flexor pollicis longus, the Pronator quadratus, and the distal end of the radius. In the proximal third of its course it lies between the Brachioradialis and the Pronator teres; in the distal third of its course it lies between the Brachioradialis and Flexor carpi radialis. The superficial branch of the radial nerve is close to the lateral side of the artery in the middle third of its course, and some filaments of the lateral antebrachial cutaneous nerve run along the distal part of the artery as it winds around the wrist. The vessel is accompanied by a pair of venae comitantes throughout its whole course. As the artery lies against the distal part of the radius, it is very superficial and is the

place which is usually chosen for *taking the pulse*.

(*b*) *At the wrist* the artery reaches the back of the carpus by passing between the radial collateral ligament of the wrist and the tendons of the Abductor pollicis longus and Extensor pollicis brevis. It then descends on the scaphoid and trapezium, and before disappearing between the heads of the first Interosseus dorsalis is crossed by the tendon of the Extensor pollicis longus. In the interval between the two Extensores pollicis (the anatomical "snuff box") it is crossed by the digital rami of the superficial branch of the radial nerve which go to the thumb and index finger.

(*c*) *In the hand,* it passes from the proximal end of the first interosseous space, between the heads of the first Interosseus dorsalis, transversely across the palm between the Adductor pollicis obliquus and Adductor pollicis transversus, but sometimes piercing the latter muscle, to the base of the metacarpal bone of the little finger, where it anastomoses with deep palmar branch from the ulnar artery, completing the **deep palmar arch** (Fig. 12–49).

Variations.—The origin of the radial artery is, in nearly one case in eight, higher than usual. In the forearm it deviates less frequently from its normal position than the ulnar. It has been found lying on the deep fascia instead of beneath it. It has also been observed on the surface of the Brachioradialis, instead of under its medial border; and in turning around the wrist, it has been seen lying on, instead of beneath, the Extensor tendons of the thumb. A large *median* artery may replace the radial in the formation of the palmar arches.

Branches.—The branches of the radial artery may be divided into three groups, corresponding with the three regions in which the vessel is situated.

In the Forearm
- *a*) Radial recurrent
- *b*) Muscular
- *c*) Palmar carpal
- *d*) Superficial palmar

At the Wrist
- *e*) Dorsal carpal
- *f*) First dorsal metacarpal

In the Hand
- *g*) Princeps pollicis
- *h*) Radialis indicis
- *i*) Deep palmar arch
- *j*) Palmar metacarpal
- *k*) Perforating
- *l*) Recurrent

a) The **radial recurrent artery** (*a. recurrens radialis*) *arises* immediately distal to the elbow. It ascends between the branches of the radial nerve, lying on the Supinator and then between the Brachioradialis and Brachialis, supplying these muscles and the elbow joint, and anastomosing with the radial collateral branch of the profunda brachii.

b) The **muscular branches** (*rami musculares*) are distributed to the muscles on the radial side of the forearm.

c) The **palmar carpal branch** (*ramus carpeus palmaris; volar radial carpal artery*) is a small vessel which *arises* near the distal border of the Pronator quadratus and, running across the palmar aspect of the carpus, anastomoses with the palmar carpal branch of the ulnar artery. This anastomosis is joined by a branch from the palmar interosseous proximally, and by recurrent branches from the deep palmar arch distally, thus forming a **palmar carpal network** which supplies the articulations of the wrist.

d) The **superficial palmar branch** (*ramus palmaris superficialis; superficialis volar artery*) *arises* from the radial artery where this vessel is about to wind around the lateral side of the wrist. It passes through, occasionally over, the muscles of the ball of the thumb, which it supplies, and anastomoses with the terminal portion of the ulnar artery, completing the superficial palmar arch. This vessel varies considerably in size: usually it is very small, and ends in the muscles of the thumb; sometimes it is as large as the continuation of the radial.

e) The **dorsal carpal branch** (*ramus carpeus dorsalis; posterior radial carpal artery*) (Fig. 8–39) is a small vessel which *arises* deep to the Extensor tendons of the thumb; crossing the carpus transversely toward the medial border of the hand, it anastomoses with the dorsal carpal branch of the ulnar and with the palmar and dorsal interosseous arteries to form a **dorsal carpal network**. From this network are given off three slender **dorsal metacarpal arteries,** which run distally on the second, third and fourth Interossei dorsales and bifurcate into the dorsal digital branches for the supply of the adjacent sides of the middle, ring, and little fingers respectively, communicat-

ing with the proper palmar digital branches of the superficial palmar arch. Near their origins they anastomose with the deep palmar arch by the **proximal perforating arteries,** and near their points of bifurcation with the common palmar digital vessels of the superficial palmar arch by the **distal perforating arteries.**

f) The **first dorsal metacarpal** *arises* just before the radial artery passes between the two heads of the first Interosseous dorsalis and divides almost immediately into two branches which supply the adjacent sides of the thumb and index finger; the radial side of the thumb receives a branch directly from the radial artery.

g) The **arteria princeps pollicis** (Fig. 8–37) *arises* from the radial just as it turns medialward to the deep part of the palm; it descends between the first Interosseous dorsalis and Adductor pollicis obliquus, along the ulnar side of the metacarpal bone of the thumb to the base of the first phalanx, where it lies beneath the tendon of the Flexor pollicis longus, and divides into two branches. These make their appearance between the medial and lateral insertions of the Adductor pollicis obliquus and run along the sides of the thumb, forming on the palmar surface of the last phalanx an arch from which branches are distributed to the integument and subcutaneous tissue of the thumb.

h) The **arteria radialis indicis** (Fig. 8–37) *arises* close to the preceding, descends between the first Interosseus dorsalis and Adductor pollicis transversus, and runs along the radial side of the index finger to its extremity, where it anastomoses with the proper digital artery, supplying the ulnar side of the finger. At the distal border of the Adductor pollicis transversus this vessel anastomoses with the princeps pollicis, and gives a communicating branch to the superficial palmar arch. The a. princeps pollicis and a. palmaris indicis radialis may spring from a common trunk termed the **first palmar metacarpal artery.**

i) The **deep palmar arch** (*arcus palmaris profundus*) (Fig. 8–37) is the terminal part of the radial artery which anastomoses with the deep palmar branch of the ulnar. It lies upon the carpal extremities of the metacarpal bones and on the Interossei, being

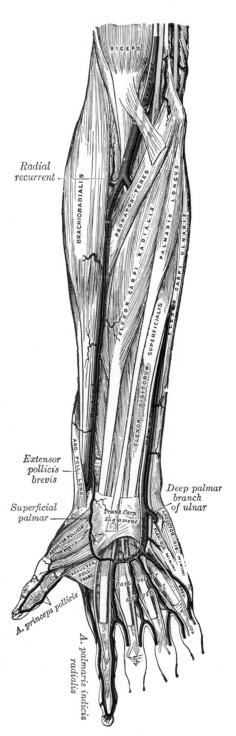

Fig. 8–38.—The radial and ulnar arteries.

covered by the Adductor pollicis obliquus, the Flexor tendons of the fingers, and the Lumbricales. The deep palmar arch lies deep to the ulnar nerve in 63 per cent of cases; superficial, that is to say, palmar to the ulnar nerve, in 34 per cent of cases. It is occasionally (2.5 per cent) double and encircles the ulnar nerve (Fig. 12–49).

j) The **palmar metacarpal arteries** (*aa. metacarpeae palmares; palmar interosseous arteries*), three or four in number, *arise* from the convexity of the deep palmar arch. They run distally upon the Interossei, and anastomose at the clefts of the fingers with the common digital branches of the superficial palmar arch.

k) The **perforating branches** (*rami perforantes*), three in number, pass dorsally from the deep palmar arch, through the second, third, and fourth interosseous spaces and between the heads of the corresponding Interossei dorsales, to anastomose with the dorsal metacarpal arteries.

l) The **recurrent branches** *arise* from the concavity of the deep palmar arch. They ascend on the palmar aspect of the wrist, supply the intercarpal articulations, and end in the palmar carpal network.

The Ulnar Artery

The **ulnar artery** (*a. ulnaris*) (Fig. 8–38), the larger of the two terminal branches of the brachial, *begins* a little distal to the bend of the elbow, and, passing obliquely distalward, reaches the ulnar side of the forearm at a point about midway between the elbow and the wrist. It then runs along the ulnar border to the wrist, crosses the flexor retinaculum on the radial side of the pisiform bone, and immediately beyond this bone becomes the superficial palmar arch.

Relations.—(*a*) *In the forearm.*—In its *proximal half*, it is deeply seated, being covered by the Pronator teres, Flexor carpi radialis, Palmaris longus, and Flexor digitorum superficialis; it lies upon the Brachialis and Flexor digitorum profundus. The median nerve is in relation with the medial side of the artery for about 2.5 cm and then crosses the vessel, being separated from it by the ulnar head of the Pronator teres. In the *distal half* of the forearm it lies upon the Flexor digitorum profundus, being covered by the integument and the subcutaneous and deep fasciae, and placed between the Flexor carpi ulnaris and Flexor digitorum superficialis. It is accompanied by two venae comitantes, and is overlapped in its middle third by the Flexor carpi ulnaris; the ulnar nerve lies on the medial side of the distal two-thirds of the artery, and the palmar cutaneous branch of the ulnar nerve descends on the distal part of the vessel to the palm of the hand.

(*b*) *At the wrist* (Fig. 8–38) the ulnar artery is covered by the integument and the palmar carpal ligament, and lies upon the flexor retinaculum. On its medial side is the pisiform bone, and, somewhat dorsal to the artery, the ulnar nerve.

Variations.—The ulnar artery varies in its origin in the proportion of about one in thirteen cases; it may arise about 5 to 7 cm below the elbow, but frequently higher. Variations in the position of this vessel are more common than in the radial. When its origin is normal, the course of the vessel is rarely changed. When it arises high up, it is almost invariably superficial to the Flexor muscles in the forearm, lying commonly beneath the fascia, more rarely between the fascia and integument. In a few cases, its position was subcutaneous in the upper part of the forearm, and subaponeurotic in the lower part.

Branches.—The branches of the ulnar artery may be arranged in the following groups:

In the Forearm
- *a*) Anterior ulnar recurrent
- *b*) Posterior ulnar recurrent
- *c*) Common interosseous
- *d*) Muscular

At the Wrist
- *e*) Palmar carpal
- *f*) Dorsal carpal

In the Hand
- *g*) Deep palmar
- *h*) Superficial palmar arch
- *i*) Common palmar digital

a) The **anterior ulnar recurrent artery** (*a. recurrens ulnaris, ramus anterior*) (Fig. 8–37) *arises* immediately distal to the elbow joint, runs proximally between the Brachialis and Pronator teres, supplies twigs to those muscles, and, anterior to the medial epicondyle, anastomoses with the superior and inferior ulnar collateral arteries (Fig. 8–36).

b) The **posterior ulnar recurrent artery** (*a. recurrens ulnaris, ramus posterior*) (Fig. 8–37) is much large, and *arises* somewhat more distally than the preceding. It passes dorsally and medialward on the Flexor

digitorum profundus, dorsal to the Flexor digitorum superficialis, and ascends dorsal to the medial epicondyle of the humerus. Between this process and the olecranon, it lies beneath the Flexor carpi ulnaris, and ascending between the heads of that muscle, in relation with the ulnar nerve, it supplies the neighboring muscles and the elbow joint, and anastomoses with the superior and inferior ulnar collateral and the interosseous recurrent arteries (Fig. 8–36).

c) The **common interosseous artery** (*a. interossea communis*) (Fig. 8–37), about 1 cm in length, *arises* immediately distal to the tuberosity of the radius, and, passing dorsally to the proximal border of the interosseous membrane, divides into two branches, the palmar and dorsal interosseous arteries.

i) The **palmar interosseous artery** (*a. interossea palmaris; volar interosseous artery*) (Fig. 8–37), passes down the forearm on the palmar surface of the interosseous membrane. It is accompanied by the palmar interosseous branch of the median nerve, and is overlapped by the contiguous margins of the Flexor digitorum profundus and Flexor pollicis longus, giving off **muscular branches**, and the **nutrient arteries** of the radius and ulna. At the proximal border of the Pronator quadratus it pierces the interosseous membrane and reaches the dorsum of the forearm, where it anastomoses with the dorsal interosseous artery (Fig. 8–39). It then descends, in company with the terminal portion of the dorsal interosseous nerve, to the dorsum of the wrist to join the dorsal carpal network. The palmar interosseous artery gives off a slender branch, the **arteria mediana**, which accompanies the median nerve, and gives offsets to its substance; this artery is sometimes much enlarged, and runs with the nerve into the palm of the hand. Before it pierces the interosseous membrane the palmar interosseous sends a branch distally dorsal to the Pronator quadratus to join the palmar carpal network.

ii) The **dorsal interosseous artery** (*a. interossea dorsalis; posterior interosseous artery*) (Fig. 8–39) passes dorsally either over or between the oblique cord and the proximal border of the interosseous membrane. It appears between the contiguous

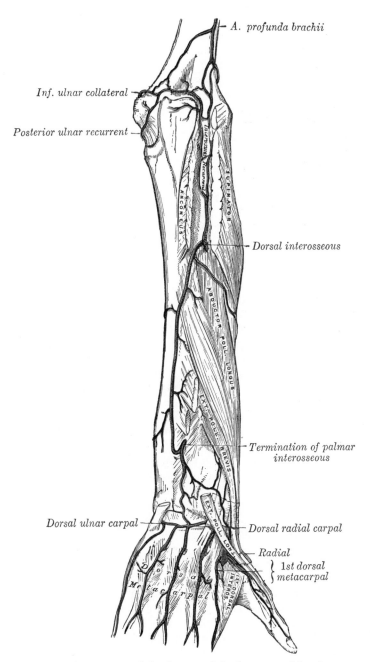

Fɪɢ. 8–39.—Arteries of the dorsum of the forearm and hand.

borders of the Supinator and the Abductor pollicis longus, and runs down the back of the forearm between the superficial and deep layers of muscles, to both of which it distributes branches. Upon the Abductor pollicis longus and the Extensor pollicis brevis, it is accompanied by the dorsal interosseous nerve. At the distal part of the forearm it anastomoses with the termination of the palmar interosseous artery, and with the dorsal carpal network. It gives off, near its origin, the **interosseous recurrent artery**, which ascends to the interval between the lateral epicondyle and olecranon, on or through the fibers of the Supinator, but deep to the Anconeus, and anastomoses with the middle collateral branch of the profunda brachii, the posterior ulnar recurrent and the inferior ulnar collateral.

d) The **muscular branches** (*rami musculares*) are distributed to the muscles along the ulnar side of the forearm.

e) The **palmar carpal branch** (*ramus carpeus palmaris; anterior ulnar carpal artery*) is a small vessel which crosses the palmar aspect of the carpus deep to the tendons of the Flexor digitorum profundus, and anastomoses with the corresponding branch of the radial artery.

f) The **dorsal carpal branch** (*ramus carpeus dorsalis; posterior ulnar carpal artery*) arises immediately proximal to the pisiform bone, and winds dorsally beneath the tendon of the Flexor carpi ulnaris, it passes across the dorsal surface of the carpus beneath the Extensor tendons, to anastomose with a corresponding branch of the radial artery. Immediately after its origin, it gives off a small branch which runs along the ulnar side of the fifth metacarpal bone, and supplies the ulnar side of the dorsal surface of the little finger.

g) The **deep palmar branch** (*ramus palmaris profundus; profunda branch*) (Fig. 8–38) passes between the Abductor digiti minimi and Flexor digiti minimi brevis and

through the origin of the Opponens digiti minimi; it anastomoses with the radial artery, and completes the deep palmar arch (Fig. 8–37).

h) The **superficial palmar arch** (*arcus palmaris superficialis; superficial volar arch*) (Fig. 8–38) is formed by the ulnar artery, and is usually completed by the superficial palmar branch of the radial. It may be completed by a branch from the a. palmaris indicis radialis or the a. princeps pollicis, or the median artery. The arch curves across the palm, its convexity distally (Fig. 12–48).

Relations.—The superficial palmar arch is covered by the skin, the Palmaris brevis, and the palmar aponeurosis. It lies upon the flexor retinaculum, the Flexor digiti minimi brevis and Opponens digiti minimi, the tendons of the Flexor digitorum superficialis, the Lumbricales, and the divisions of the median and ulnar nerves.

i) Three **common palmar digital arteries** (*aa. digitales palmares communes; volar digital arteries*) (Fig. 8–38) *arise* from the convexity of the superficial arch and proceed distally on the second, third, and fourth Lumbricales. Each receives the corresponding palmar metacarpal artery and then divides into a pair of **proper palmar digital arteries** (*aa. digitales palmar propriae; collateral digital arteries*) which run along the contiguous sides of the index, middle, ring, and little fingers dorsal to the corresponding digital nerves; they anastomose freely in the subcutaneous tissue of the finger tips and by smaller branches near the interphalangeal joints. Each gives off a couple of dorsal branches which anastomose with the dorsal digital arteries, and supply the soft parts on the dorsum of the middle and ungual phalanges, including the matrix of the finger nail. The proper palmar digital artery for the ulnar side of the little finger springs from the ulnar artery under cover of the Palmaris brevis.

THE ARTERIES OF THE TRUNK

The Descending Aorta

The **descending aorta** is divided into two portions, the **thoracic** and **abdominal**, in correspondence with the two great cavities of the trunk in which it is situated.

The Thoracic Aorta

The **thoracic aorta** (*aorta thoracica*) (Fig. 8–40) is contained in the posterior mediastinum. It begins at the caudal border of the fourth thoracic vertebra where it is

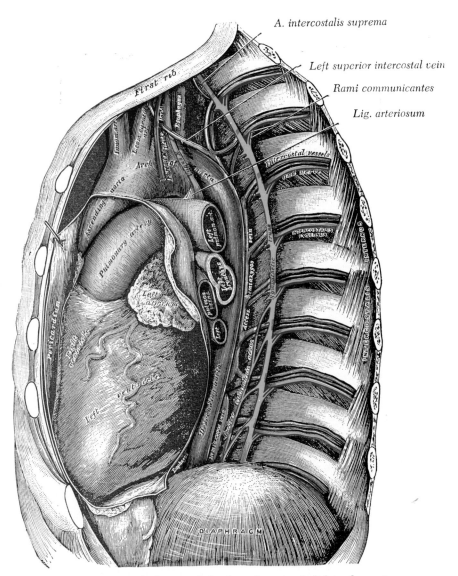

A. intercostalis suprema

Left superior intercostal vein

Rami communicantes

Lig. arteriosum

Fɪɢ. 8–40.—The heart and the thoracic aorta. Left lateral aspect.

continuous with the aortic arch, and ends ventral to the caudal border of the twelfth at the aortic hiatus in the diaphragm. At its commencement, it is situated on the left of the vertebral column; it approaches the median line as it descends; and, at its termination, lies directly ventral to the column. The vessel describes a curve which is concave ventrally, and as the branches given off from it are small, its diminution in size is inconsiderable.

Relations.—It is in relation, *ventrally*, with the root of the left lung, the pericardium, the esophagus, and the diaphragm; *dorsally*, with the vertebral column and the hemiazygos veins; on the *right side*, with the azygos vein and thoracic duct; on the *left side*, with the left pleura and lung. The esophagus, with its accompanying plexus of nerves, lies on the right side of the aorta cranially; but at the caudal part of the thorax it is placed ventral to the aorta, and, close to the diaphragm, is situated on its left side.

Branches of the Thoracic Aorta.—

Visceral

1. Pericardial
2. Bronchial
3. Esophageal
4. Mediastinal

Parietal

5. Posterior Intercostal
6. Subcostal
7. Superior phrenic

1. The **Pericardial Branches** (*rami pericardiaci*) consist of a few small vessels which are distributed to the dorsal surface of the pericardium.

2. The **Bronchial Arteries** (*aa. bronchiales*) (Fig. 12–24) vary in number, size, and origin. There is as a rule only one **right bronchial artery** which *arises* from the first aortic intercostal, or from the cranial left bronchial artery. The **left bronchial arteries** are usually two in number, and *arise* from the thoracic aorta. The cranial left bronchial arises opposite the fifth thoracic vertebra, the caudal just below the level of the left bronchus. Each vessel runs on the dorsal part of its bronchus, dividing and subdividing along the bronchial tubes, supplying them, the areolar tissue of the lungs, the bronchial lymph nodes, and the esophagus.

3. The **Esophageal Arteries** (*rami esophagei*) (Fig. 12–24), four or five in number, *arise* from the ventral aspect of the aorta, and pass obliquely caudalward to the esophagus, forming a chain of anastomoses along that tube, anastomosing with the esophageal branches of the inferior thyroid arteries, and with ascending branches from the left inferior phrenic and left gastric arteries.

4. The **Mediastinal Branches** (*rami mediastinales*) are numerous small vessels which supply the lymph glands and loose areolar tissue in the posterior mediastinum.

5. **Posterior Intercostal Arteries** (*aa. intercostales posteriores; aortic intercostals*). —There are usually nine pairs of aortic intercostal arteries. They *arise* from the dorsal aspect of the aorta, and are distributed to the caudal nine intercostal spaces, the first two spaces being supplied by the highest intercostal artery, a branch of the costocervical trunk of the subclavian. The **right aortic intercostals** are longer than the left, on account of the position of the aorta on the left side of the vertebral column; they pass across the vertebrae dorsal to the esophagus, thoracic duct, and vena azygos, and are covered by the right lung and pleura. The **left aortic intercostals** run dorsally on the sides of the vertebrae and are covered by the left lung and pleura; the cranial two vessels are crossed by the highest left intercostal vein, the caudal vessels by the hemiazygos veins. The further course of the intercostal arteries is practically the same on both sides. Opposite the heads of the ribs the sympathetic trunk passes ventral to them, and the splanchnic nerves also descend ventral to the caudal arteries. Each artery then divides into a ventral and dorsal rami.

a) The **ventral ramus** crosses the corresponding intercostal space obliquely toward the angle of the more cranial rib, and thence is continued ventrally in the costal groove. It is placed at first between the pleura and the posterior intercostal membrane, then it pierces this membrane, and lies between it and the Intercostalis externus as far as the angle of the rib; from this onward it runs between the Intercostalis externus and Intercostalis internus, and anastomoses ventrally with the intercostal branch of the internal thoracic or the musculophrenic.

Each artery is accompanied by a vein and a nerve, the former being cranial and the latter caudal to the artery, except in the cranial spaces, where the nerve is at first cranial to the artery. The first aortic intercostal artery anastomoses with the intercostal branch of the costocervical trunk, and may form the chief supply of the second intercostal space. The most caudal two intercostal arteries are continued anteriorly from the intercostal spaces into the abdominal wall, and anastomose with the subcostal, superior epigastric, and lumbar arteries.

Branches.—The ventral rami give off the following branches:

    i) Collateral intercostal
    ii) Muscular
    iii) Lateral cutaneous
    iv) Mammary

i) The **collateral intercostal branch** comes off from the intercostal artery near the angle of the rib, and descends to the cranial border of the rib below, along which it courses to anastomose with the intercostal branch of the internal thoracic.

ii) **Muscular branches** are given to the Intercostales and Pectorales and to the Serratus anterior; they anastomose with the highest and lateral thoracic branches of the axillary artery.

iii) The **lateral cutaneous branches** accompany the lateral cutaneous branches of the thoracic nerves.

iv) **Mammary branches** are given off by the vessels in the third, fourth, and fifth spaces. They supply the mamma, and increase considerably in size during the period of lactation.

b) The **dorsal ramus** runs dorsally through a space which is bounded by the necks of the ribs, medially by the body of a vertebra, and laterally by an anterior costotransverse ligament. It gives off a **spinal branch** which enters the vertebral canal through the intervertebral foramen and is distributed to the spinal cord and its membranes and the vertebrae. It then courses over the transverse process with the dorsal division of the thoracic nerve, supplies branches to the muscles of the back and cutaneous branches which accompany the corresponding cutaneous branches of the dorsal division of the nerve.

6. The **Subcostal Arteries,** so named because they lie caudal to the last ribs, constitute the most caudal pair of branches derived from the thoracic aorta, and are in series with the intercostal arteries. Each passes along the caudal border of the twelfth rib dorsal to the kidney and ventral to the Quadratus lumborum muscle, and is accompanied by the twelfth thoracic nerve. It then pierces the posterior aponeurosis of the Transversus abdominis, and, passing between this muscle and the Obliquus internus, anastomoses with the superior epigastric, caudal intercostal, and lumbar arteries. Each subcostal artery gives off a dorsal branch which has a distribution similar to that of the dorsal ramus of an intercostal artery.

7. The **Superior Phrenic Branches** are small and *arise* from the caudal part of the thoracic aorta; they are distributed to the dorsal part of the cranial surface of the diaphragm, and anastomose with the musculophrenic and pericardiacophrenic arteries.

Variations of the aorta are described on page 576.

A small **aberrant artery** is sometimes found *arising* from the right side of the thoracic aorta near the origin of the right bronchial. It passes upward and to the right behind the trachea and the esophagus, and may anastomose with the highest right intercostal artery. It represents the remains of the right dorsal aorta, and in a small proportion of cases is enlarged to form the first part of the right subclavian artery.

The Abdominal Aorta

The **abdominal aorta** (*aorta abdominalis*) (Fig. 8–41) *begins* at the aortic hiatus of the diaphragm, ventral to the caudal border of the body of the last thoracic vertebra, and, descending ventral to the vertebral column, ends on the body of the fourth lumbar vertebra, commonly a little to the left of the midline, by dividing into the two common iliac arteries. It diminishes rapidly in size, in consequence of the many large branches which it gives off. As it lies upon the bodies of the vertebrae, the curve which

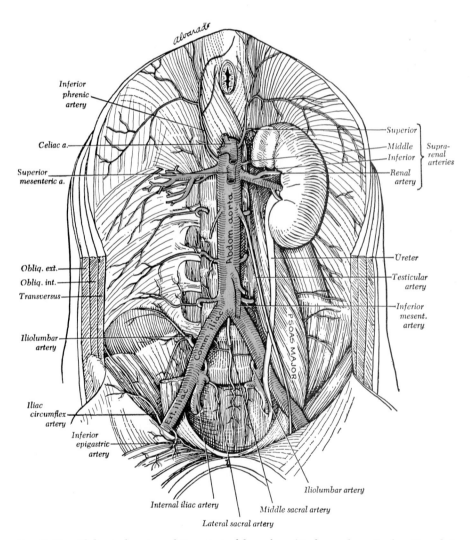

Fig. 8–41.—Abdominal aorta and its principal branches. (Redrawn from Rauber-Kopsch.)

it describes is convex ventrally, the summit of the convexity corresponding to the third lumbar vertebra.

Relations.—The abdominal aorta is covered, *ventrally*, by the lesser omentum and stomach, dorsal to which are the branches of the celiac artery and the celiac plexus; caudal to these, by the lienal vein, the pancreas, the left renal vein, the inferior part of the duodenum, the mesentery and aortic plexus. *Dorsally*, it is separated from the lumbar vertebrae and intervertebral disks by the anterior longitudinal ligament and left lumbar veins. On its *right side* are the azygos vein, cisterna chyli, thoracic duct, and the right crus of the diaphragm —the last separating it from the proximal part of the inferior vena cava and right celiac ganglion; the inferior vena cava is in contact with the aorta caudally. On the *left side* are the left crus of the diaphragm, the left celiac ganglion, the ascending part of the duodenum, and some coils of the small intestine.

Collateral Circulation.—The collateral circulation would be carried on by the anastomoses between the internal thoracic and the inferior epigastric; by the free communication between the superior and inferior mesenterics, if the ligature were placed between these vessels; or by the anastomosis between the inferior mesenteric and the internal pudendal, when (as is more common) the point of ligature is below the origin of the inferior mesenteric; and possibly by the anastomoses of the lumbar arteries with the branches of the internal iliac.

Branches.—The branches of the abdominal aorta may be divided into three sets: visceral, parietal, and terminal.

	A. Celiac
	B. Superior mesenteric
	C. Inferior mesenteric
Visceral Branches	D. Middle suprarenals
	E. Renals
	F. Testicular
	G. Ovarian

	H. Inferior phrenics
Parietal Branches	I. Lumbars
	J. Middle sacral

Terminal Branches   K. Common iliacs

Of the visceral branches, the celiac artery and the superior and inferior mesenteric arteries are unpaired, while the suprarenals, renals, testicular, and ovarian are paired.

Of the parietal branches the phrenics and lumbars are paired; the middle sacral is unpaired. The terminal branches are paired.

A. The **Celiac Artery** (*truncus celiacus; celiac axis*) (Figs. 8–42, 8–43), a thick trunk, 7 to 20 mm in diameter, *arises* from the aorta just caudal to the aortic hiatus of the diaphragm. It is covered by the peritoneum of the dorsal wall of the lesser sac (omental bursa). On its right side are the right celiac ganglion and the cardiac end of the stomach. A short distance caudal to it, commonly 1.25 cm, the aorta gives origin to the superior mesenteric artery. After a short course it usually divides into three large branches: 1. the Left Gastric, 2. the Common Hepatic, and 3. the Splenic.

Variations.—The most constant feature of this artery is the variability of its branching and the routes by which blood reaches the organs which it principally supplies, according to Michels ('53) whose descriptions and statistics are followed in this account. The stomach, liver, pancreas, duodenum, and spleen are the names of these organs and they can conveniently be used to designate the important variations.

A *gastrohepatosplenic trunk* has the classical branches: a left gastric, a common hepatic, and a splenic artery. This is the usual variety (89 per cent). It is complete, that is, it gives origin to all its expected branches in 64.5 per cent and in the remainder there are supernumerary or accessory branches.

A *hepatosplenic trunk* (3.5 per cent) has hepatic and splenic arteries but the left gastric arises independently from the aorta or from the hepatic or splenic arteries.

A *gastrosplenic trunk* (5.5 per cent) has the left gastric and splenic arteries but the hepatic arises from the aorta or from the superior mesenteric artery.

A *hepatogastric trunk* (1.5 per cent) has the left gastric and hepatic arteries but the splenic arises from the aorta or from the superior mesenteric artery.

In rare instances the celiac trunk and superior mesenteric arteries are combined.

1. The **Left Gastric Artery** (*arteria gastrica sinistra*) is most commonly the first branch of the celiac trunk, arising near its middle part. It is much larger (4 to 5 mm in diameter) than the right gastric artery. Frequently (25 per cent) it *arises* from the termination of the celiac trunk together with the hepatic and splenic arteries,

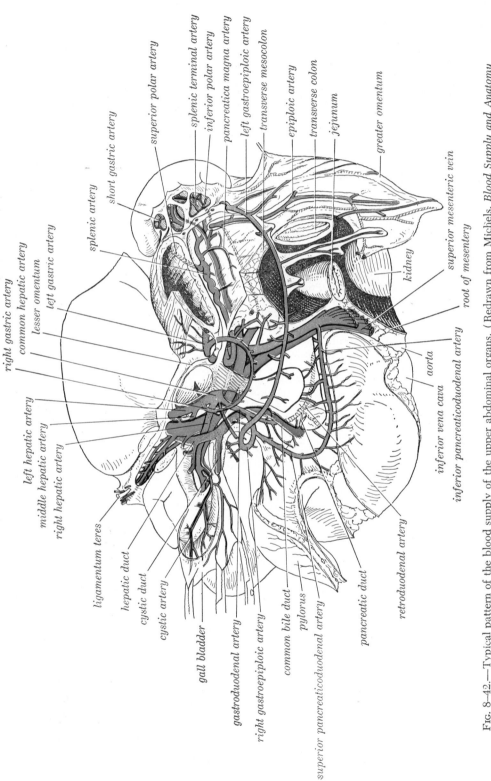

Fig. 8–42.—Typical pattern of the blood supply of the upper abdominal organs. (Redrawn from Michels, *Blood Supply and Anatomy of the Upper Abdominal Organs*, 1955.)

forming a tripod. It is covered by that part of the parietal peritoneum of the dorsal abdominal wall which forms the dorsal layer of the lesser sac (omental bursa). It runs ventrally, cranially and to the left in a gentle curve, raising a crescentic ridge of peritoneum, the left gastropancreatic fold, and reaches the ventral wall of the lesser sac near the cardiac end of the stomach, accompanied by the coronary vein. Here it reverses its direction and lying between the two leaves of the lesser omentum bifurcates into two main branches.

a) The **ventral branch** is distributed to the anterior surface of the stomach in two or three subdivisions.

b) The **dorsal branch** runs along the lesser curvature of the stomach, giving branches to the posterior surface of this organ and usually anastomosing with the right gastric artery.

c) A **cardioesophageal branch** arises before the bifurcation and is distributed to the cardia and esophagus in one to three subdivisions.

Variations.—The left gastric artery was a branch of the aorta in 2.5 per cent, of a gastrosplenic trunk in 4.5 per cent, and a hepatogastric trunk in 1.5 per cent. An accessory left gastric arose from the splenic in 6 per cent, the left hepatic in 3 per cent and from the celiac in 2 per cent (200 bodies, Michels '53).

An **aberrant left hepatic artery** is given off just proximal to the bend and bifurcation in 23 per cent, half of which (11.5 per cent) replace the left hepatic entirely and half are accessory. It may be 3 to 5 mm in diameter. It runs cranially about 6 cm to near the esophagus, lying between the two layers of the lesser omentum; it crosses the caudate lobe and enters the liver substance at the fissure for the ligamentum venosum. Because of this frequent origin of the left hepatic, the left gastric was named the gastrohepatic artery by Haller (1756).

The inferior phrenics may arise from the left gastric.

2. The **Common Hepatic Artery** (*arteria hepatica communis*) (Fig. 8–42) arising from the celiac trunk is slightly smaller (7 to 8 mm) than the splenic artery. Its first part is horizontal, running from left to right along the cranial border of the head of the pancreas to the pylorus or first part of the duodenum. Dorsal to the duodenum it gives off the gastroduodenal and turns cranially as the vertical portion. It is covered by the peritoneum of the dorsal wall of the lesser sac (omental bursa) which is raised thereby into the right hepatopancreatic fold. It then passes through the hepatoduodenal ligament (lesser omentum) to the porta hepatis. As it lies between the two layers of the lesser omentum, it is ventral to the epiploic foramen having the common bile duct to its right and the portal vein dorsal to it. At a variable distance from the liver it gives rise to three terminal branches— right, left, and middle hepatic arteries.

Variations.—Of 200 bodies, 41.5 per cent had one or more aberrant hepatic arteries, 50.5 per cent had none; several bodies had 2 and a few 3 aberrants. The common hepatic arose from the aorta in 3, from the superior mesenteric in 5, and the left gastric in 1 (Michels '53).

Branches.—(*a*) Gastroduodenal, (*b*) Right gastric, (*c*) Right hepatic, (*d*) Left hepatic, (*e*) Middle hepatic.

(*a*) The **gastroduodenal artery** (*arteria gastroduodenalis*) (Fig. 8–42) arises from the common hepatic trunk, usually about halfway between the celiac origin and its division into the hepatic branches. At first it is to the left of the common bile duct, then crosses either dorsal or ventral to it and descends dorsal to the first part of the duodenum to the caudal border of the pylorus where it divides into the right gastroepiploic and superior pancreatico-duodenal arteries.

Variations.—The gastroduodenal artery arose from the left hepatic artery (11 per cent), the right hepatic (7 per cent), the middle hepatic (1 per cent), a replaced hepatic trunk (3.5 per cent) or from the celiac or superior mesenteric arteries.

Branches.—i) Retroduodenal, ii) Right gastroepiploic, iii) Superior pancreatico-duodenal.

i) The **retroduodenal artery** is the first branch of the gastroduodenal. It *arises* above the duodenum, runs dorsal to it along the left side of the common bile duct, crosses ventral to the supraduodenal portion of the duct and forms the U-shaped dorsal pancreaticoduodenal arcade. The arcade supplies branches to all four parts of the duodenum and the head of the pancreas, and anastomoses with a dorsal branch of the inferior pancreaticoduodenal from the superior mesenteric artery. The retroduodenal supplies branches to the bile duct, the first part of the duodenum and the head of the pancreas.

Variations.—The retroduodenal artery arose from the common hepatic in 4 per cent, right hepatic in 2 per cent, aberrant right hepatic 3 per cent, and was absent in 1 per cent.

The **supraduodenal artery** was identified in all but 6 per cent as a branch to the upper part of the first portion of the duodenum; it arose from the retroduodenal in 50 per cent, from the gastroduodenal in 25 per cent and the remainder from proper hepatic branches.

ii) The **right gastroepiploic artery** (*arteria gastroepiploica dextra*) (Fig. 8–43) is the larger of the two terminal branches of the gastroduodenal and is much larger than the left gastroepiploic artery. It passes from right to left at a variable distance from the greater curvature of the stomach. It lies between the two layers of the gastrocolic ligament or the ventral two layers of the greater omentum when these are not adherent to the colon. It gives off a large ascending **pyloric branch** near its origin, and at its termination usually anastomoses with the left gastroepiploic artery. It supplies a number of ascending branches to the stomach and descending branches to the greater omentum. The **long ventral epiploic branches** extend around the free border of the omentum and in the dorsal layer of the omentum may anastomose with other gastric, or with pancreatic arteries.

iii) The **superior pancreaticoduodenal artery** (*arteria pancreaticoduodenalis superior*) (Fig. 8–42) *arises* from the gastroduodenal artery as the latter passes dorsal

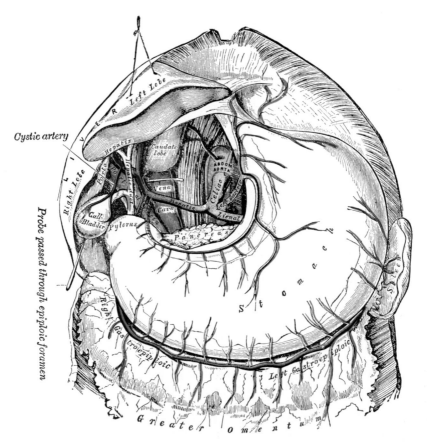

Fig. 8–43.—The celiac artery and its branches; the liver has been raised, and the lesser omentum and anterior layer of the greater omentum removed.

to the first part of the duodenum. It makes a loop on the ventral surface of the pancreas, runs along the groove between the pancreas and the descending portion of the duodenum, sinks into the substance of the pancreas, and dorsal to the head of the pancreas anastomoses with the inferior pancreaticoduodenal artery, a branch of the superior mesenteric artery. The loop supplies branches to the ventral surface of all three parts of the duodenum. The **ventral pancreaticoduodenal arcade** formed by the anastomosis of the superior and inferior arteries supplies numerous branches to the pancreas and duodenum.

(*b*) The **right gastric artery** (*arteria gastrica dextra*) (Fig. 8–42) is much smaller and less constant than the left gastric. It *arises* from the common hepatic between the gastroduodenal and proper hepatic arteries and passes between the layers of the lesser omentum to the pylorus, which it supplies with branches. It runs to the left along the lesser curvature of the stomach supplying an extensive area with anterior and posterior branches and anastomoses with the dorsal branch of the left gastric artery.

Variations.—The right gastric arose from the common hepatic in 40 per cent, the left hepatic in 40.5 per cent, the right hepatic in 5.5 per cent, the middle hepatic in 5 per cent, the gastroduodenal in 8 per cent.

(*c*) The **right hepatic artery** (*arteria hepatica propria, ramus dexter*) (Fig. 8–42) usually lies ventral to the portal vein and crosses dorsal to the hepatic duct to enter the cystic triangle. It gives off the cystic artery and divides into two main branches before entering the right lobe of the liver (Fig. 16–94).

Variations.—Of 200 bodies, 52 had aberrant right hepatic arteries, 36 replacing the usual artery and 16 accessory to it. The most frequent source was the superior mesenteric artery (Michels '53).

The **cystic artery** (*arteria cystica*) (Fig. 8–43) *arises* from the right hepatic artery to the right of the hepatic duct in the triangle bounded by the cystic duct, hepatic duct, and the liver (**cystic triangle of Calot**). At the gallbladder it divides

into a superficial branch which supplies the free surface and a deep branch which supplies the embedded surface of the gallbladder. The branches anastomose and supply twigs to the adjacent liver substance.

Variations.—The cystic artery had a single origin in 75 per cent, double in 25 per cent. In 18 per cent the cystic artery arose to the left of the hepatic duct, not in the cystic triangle, and had to cross the duct to reach the gallbladder. In 5 per cent its origin was from an artery other than the right hepatic. In most of the double cystic arteries, both superficial and deep arose in the cystic triangle; the superficial was more variable than the deep, originating from the right hepatic to the left of the duct or from another artery.

(*d*) The **left hepatic artery** (*arteria hepatica propria, ramus sinistra*) (Fig. 8–42) usually divides into two branches, an upper and a lower, supplying twigs to the capsule of the liver and the caudate lobe before entering the substance of the left lobe (Fig. 16–94).

Variations.—Of 200 bodies, 54 had aberrant left hepatic arteries, 31 replacing the usual artery and 23 accessory. The majority of these arose from the left gastric artery.

(*e*) The **middle hepatic artery** enters the fossa for the round ligament and accompanies the middle hepatic duct to the quadrate lobe. It is the principal supply for this lobe and sends twigs to the round ligament and sometimes to the left lobe. It *arises* from the right hepatic in 45 per cent, left 45 per cent and from other sources in 10 per cent (Fig. 16–94).

3. The **Splenic** or **Lienal Artery** (*a. lienalis*) (Fig. 8–44), the largest branch of the celiac, passes horizontally to the left along the pancreas to reach the spleen. It varies from 8 to 32 cm in length and is usually tortuous. The *first part*, before it reaches the pancreas, is a short arc, swinging to the right and caudally, then across the aorta to the cranial border of the pancreas. The *second part* lies in a groove on the dorsal cranial surface of the body of the pancreas. It is the most tortuous part, being undulating and having one or more loops or coils, but seldom is it far from the pancreas. The *third part* leaves the cranial border to cross the ventral surface of the pancreas obliquely

and in most subjects divides into its superior and inferior terminal arteries. The *fourth part* is between the tail of the pancreas and the hilum of the spleen, and, as just mentioned, in the majority of individuals is represented by terminal arteries (Michels '42) (Fig. 8–44).

The splenic artery is covered by the peritoneum of the dorsal wall of the omental bursa (lesser sac) and lies, therefore, dorsal to the stomach. Its fourth part crosses ventral to the cranial part of the left kidney and enters the hilum by passing in the substance of the phrenicolienal or lienorenal ligaments.

Branches.—Its branches are:

  a) Pancreatic
  b) Left gastroepiploic
  c) Short gastric
  d) Splenic

a) **Pancreatic branches.**—In addition to numerous twiglike branches to the pancreas, the splenic artery gives rise to three larger branches: i) dorsal pancreatic, ii) pancreatica magna, and iii) caudae pancreatis (Fig. 8–42).

i) The **dorsal pancreatic artery** typically *arises* from the first part of the splenic, but it commonly has other origins. It gives a number of twigs to the neck and body of the pancreas. It has two small right branches, one curving ventrally along the head of the pancreas supplying it and anastomosing with the gastroduodenal or its branches; the other supplies the uncinate process by plexiform branches and anastomoses with the inferior pancreaticoduodenal artery. A branch often runs caudally to below the pancreas and anastomoses with superior mesenteric branches. The main

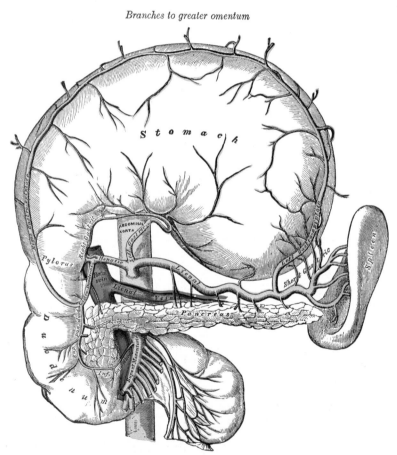

Branches to greater omentum

Fig. 8–44.—The celiac artery and its branches; the stomach has been raised and the peritoneum removed.

left branch is of considerable size and becomes the transverse pancreatic artery.

The **transverse pancreatic artery** takes a course to the left for about two-thirds of the length of the dorsocaudal surface of the pancreas and enters the substance of the gland to anastomose with other pancreatic branches from the splenic artery. It runs parallel with the pancreatic duct for some distance supplying it and neighboring gland with branches. It has long and short branches which descend in the dorsal layer of the great omentum as the posterior epiploic arteries which anastomose with the anterior epiploics of the gastroepiploic arcade.

Variations.—The dorsal pancreatic arose from the splenic artery in 39 per cent, the celiac in 22 per cent, the hepatics in 19 per cent, the superior mesenteric in 14 per cent, and the gastroduodenal in 2 per cent in 200 bodies (Michels '53). The **descending branch** of the dorsal pancreatic may be quite large and supply the middle colic artery.

ii) The **arteria pancreatica magna** *arises* from the second part of the splenic artery. It is the largest pancreatic branch (2 to 4 mm diameter) and branching to right and left anastomoses with other pancreatic arteries.

iii) The **arteria caudae pancreatis** *arises* from the third part of the splenic artery or one of its terminal branches. It supplies branches to the tail of the pancreas, anastomoses with the magna and dorsal pancreatics and supplies an accessory spleen when one is present.

b) The **left gastroepiploic artery** (*a. gastroepiploica sinistra*), the largest branch, *comes* from the third part of the splenic artery or from its inferior terminal artery. It reaches the stomach through the pancreaticolienal and gastrolienal ligaments and courses along the greater curvature of the stomach from left to right within the anterior part of the great omentum. Its branches are distributed to both surfaces of the stomach and to the greater omentum.

c) The **short gastric arteries** (*aa. gastricae breves; vasa brevia*) consist of from five to seven small branches, which *arise* from the end of the lienal artery, and from its terminal divisions. They pass from left to right,

between the layers of the gastrolienal ligament, and are distributed to the greater curvature of the stomach, anastomosing with branches of the left gastric and left gastroepiploic arteries.

d) **Splenic Branches.**—The splenic artery divides about 3.5 cm from the spleen into a superior and an inferior terminal branch. These **terminal** branches may divide into a number or a few branches before entering the spleen, and there may be an intermediate terminal branch. The branching of the inferior terminal is more complicated than the superior, and it may give origin to the left gastroepiploic and inferior polar arteries. Arteries to the poles of the spleen are of frequent occurrence. The **superior polar** usually has its origin from the main splenic artery and may come from the celiac axis. **Inferior polar** arteries are more frequent than superior and arise from the left gastroepiploic most frequently, but may come from the splenic or inferior terminal branches. Cross anastomoses between the larger branches at the hilum are common (Michels '42).

B. The **Superior Mesenteric Artery** (*a. mesenterica superior*) (Fig. 8–45) is a large vessel which supplies the whole length of the small intestine, except the superior part of the duodenum; it also supplies the cecum and the ascending part of the colon and about one-half of the transverse part of the colon. It *arises* from the ventral surface of the aorta, about 1.25 cm caudal to the celiac artery, at the level of the first lumbar vertebra, and is crossed at its origin by the lienal vein and the neck of the pancreas. It passes ventral to the processus uncinatus of the head of the pancreas and inferior part of the duodenum, and descends between the layers of the mesentery to the right iliac fossa, where, considerably diminished in size, it anastomoses with one of its own branches, viz., the ileocolic. In its course it crosses ventral to the inferior vena cava, the right ureter and Psoas major, and forms an arch, the convexity of which is directed to the left side. It is accompanied by the superior mesenteric vein which lies to its right side, and it is surrounded by the superior mesenteric plexus of nerves. Occasionally it arises from the aorta by a common trunk with the coeliac axis.

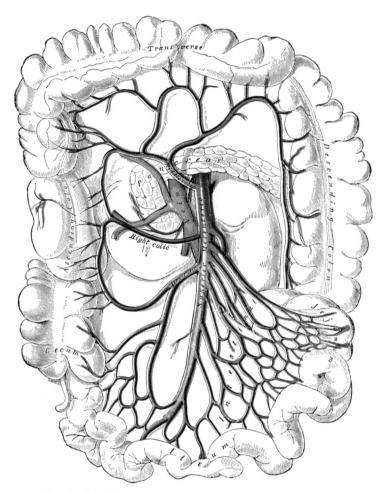

Fig. 8–45.—The superior mesenteric artery and its branches.

Branches.—Its branches are:

1. Inferior pancreaticoduodenal
2. Intestinal
3. Ileocolic
4. Right colic
5. Middle colic

1. The **Inferior Pancreaticoduodenal Artery** (Fig. 8–42) (*a. pancreaticoduodenalis inferior*) *arises* from the superior mesenteric or from its first intestinal branch, opposite the cranial border of the inferior part of the duodenum. It courses to the right between the head of the pancreas and duodenum, and then ascends to anastomose with the superior pancreaticoduodenal artery. It distributes branches to the head of the pancreas and to the descending and inferior parts of the duodenum.

2. The **Intestinal Arteries** (*aa. ejunales et ilei; vasa intestini tenuis*) *arise* from the convex side of the superior mesenteric artery, are usually from twelve to fifteen in number, and are distributed to the jejunum and ileum. They run nearly parallel with one another between the layers of the mesentery, each vessel dividing into two branches which unite with adjacent branches, forming a series of arches, the convexities of which are directed toward the intestine (Fig. 8–46). From this first set of arches branches arise, which unite with similar branches from above and below, thus forming a second series of arches; from the lower branches of the artery, a third, a fourth, or even a fifth series of arches may be formed, diminishing in size the nearer they approach the intes-

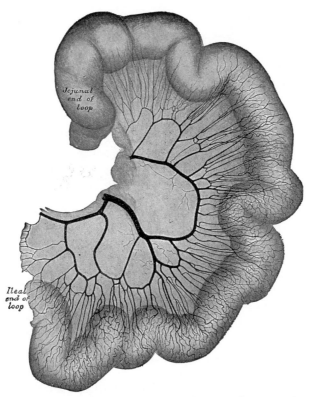

Fig. 8–46.—Loop of small intestine showing distribution of intestinal arteries. (From a preparation by Mr. Hamilton Drummond.) The vessels were injected while the gut was *in situ;* the gut was then removed, and an *x*-ray photograph taken.

tine. In the short, upper part of the mesentery only one set of arches exists, but as the depth of the mesentery increases, second, third, fourth, or even fifth groups are developed. From the terminal arches numerous small straight vessels arise which encircle the intestine, upon which they are distributed, ramifying between its coats. From the intestinal arteries small branches are given off to the lymph nodes and other structures between the layers of the mesentery.

3. The **Ileocolic Artery** (*a. ileocolica*) is the most caudal branch *arising* from the concavity of the superior mesenteric artery. It passes caudally and to the right toward the right iliac fossa, where it divides into a superior and an inferior branch; the inferior anastomoses with the end of the superior mesenteric artery, the superior with the right colic artery.

The inferior branch of the ileocolic runs toward the superior border of the ileocolic junction and supplies the following branches

(Fig. 8–45): (*a*) **colic**, which pass upward on the ascending colon; (*b*) **anterior** and **posterior cecal**, which are distributed to the cecum; (*c*) an **appendicular artery**, which descends dorsal to the termination of the ileum and enters the mesenteriole of the vermiform appendix; it runs near the free margin of this mesenteriole and ends in branches which supply the appendix (Fig. 8–47); and (*d*) **ileal**, which run cranially and to the left on the caudal part of the ileum, and anastomose with the termination of the superior mesenteric.

4. The **Right Colic Artery** (*a. colica dextra*) *arises* from about the middle of the concavity of the superior mesenteric artery, or from a stem common to it and the ileocolic. It passes to the right covered by the peritoneum, and ventral to the right internal spermatic or ovarian vessels, the right ureter and the Psoas major, toward the middle of the ascending colon; sometimes the vessel lies at a higher level, and crosses the descending part of the duodenum and the

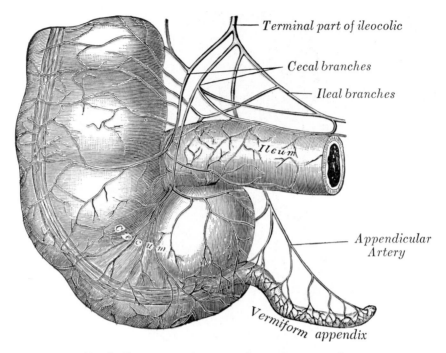

Terminal part of ileocolic

Cecal branches

Ileal branches

Ileum

Appendicular
Artery

Cecum

Vermiform appendix

FIG. 8–47.—Arteries of cecum and vermiform appendix.

caudal end of the right kidney. At the colon it divides into a descending branch, which anastomoses with the ileocolic, and an ascending branch, which anastomoses with the middle colic. These branches form arches, from the convexity of which vessels are distributed to the ascending colon.

5. The **Middle Colic Artery** (*a. colica media*) *arises* from the superior mesenteric just caudal to the pancreas and, passing ventrally between the layers of the transverse mesocolon, divides into two branches, right and left; the former anastomoses with the right colic; the latter with the left colic, a branch of the inferior mesenteric. The arches thus formed are placed about two fingers' breadth from the transverse colon, to which they distribute branches.

C. The **Inferior Mesenteric Artery** (*a. mesenterica inferior*) (Fig. 8–48) supplies the left half of the transverse part of the colon, the whole of the descending and iliac parts of the colon, the sigmoid colon, and the greater part of the rectum. It is smaller than the superior mesenteric, and *arises* from the aorta, about 3 or 4 cm above its division into the common iliacs and close to the caudal border of the third part of the duodenum, at the level of the middle of

the third lumbar vertebra. It passes caudally covered by the peritoneum, lying at first ventral to and then on the left side of the aorta. It crosses the left common iliac artery and is continued into the lesser pelvis under the name of the **superior rectal artery,** which descends between the two layers of the sigmoid mesocolon and ends on the upper part of the rectum.

Branches.—Its branches are:

1. Left colic
2. Sigmoid
3. Superior rectal

1. The **Left Colic Artery** (*a. colica sinistra*) runs to the left covered by the peritoneum and ventral to the Psoas major, and after a short, but variable, course divides into an ascending and a descending branch; the stem of the artery or its branches cross the left ureter and left testicular vessels. The ascending branch crosses ventral to the left kidney and ends, between the two layers of the transverse mesocolon, by anastomosing with the middle colic artery; the descending branch anastomoses with the highest sigmoid artery. From the arches formed by these anastomoses, branches are distributed to the descending

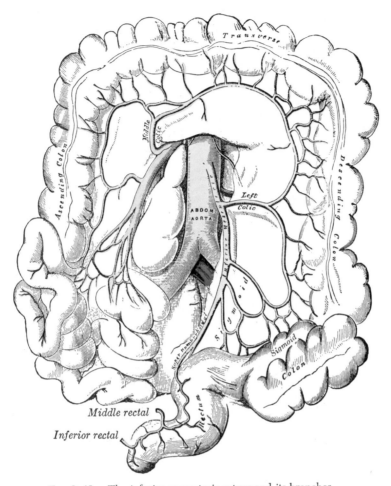

Fig. 8–48.—The inferior mesenteric artery and its branches.

colon and the left part of the transverse colon.

2. The **Sigmoid Arteries** (*aa. sigmoideae*) (Fig. 8–48), two or three in number, run obliquely caudalward and to the left behind the peritoneum and ventral to the Psoas major, ureter, and testicular vessels. Their branches supply the caudal part of the descending colon, the iliac colon, and the sigmoid or pelvic colon; anastomosing cranially with the left colic, and caudally with the superior rectal artery.

3. The **Superior Rectal Artery** (*a. rectalis superior; superior hemorrhoidal artery*) (Fig. 8–48), the continuation of the inferior mesenteric, descends into the pelvis between the layers of the mesentery of the sigmoid colon, crossing, in its course, the left common iliac vessels. It divides, oppo-

site the third sacral vertebra, into two branches, which descend one on either side of the rectum, and about 10 or 12 cm. from the anus break up into several small branches. These pierce the muscular coat of the bowel and run caudally, as straight vessels, placed at regular intervals from each other in the wall of the gut between its muscular and mucous coats, to the level of the Sphincter ani internus; here they form a series of loops around the lower end of the rectum, and anastomose with the middle rectal branches of the internal iliac, and with the inferior rectal branches of the internal pudendal.

D. The **Middle Suprarenal Arteries** (*aa. suprarenales mediae; middle capsular arteries; suprarenal arteries*) are two small vessels which *arise,* one from either side of the aorta, opposite the superior mesenteric artery. They pass lateralward and slightly upward, over the crura of the diaphragm, to the suprarenal glands, where they anastomose with suprarenal branches of the inferior phrenic and renal arteries. In the fetus these arteries are of large size.

E. The **Renal Arteries** (*aa. renales*) (Fig. 8–41) are two large trunks, which *arise* from the two sides of the aorta, immediately caudal to the superior mesenteric artery, at the level of the disk between the first and second lumbar vertebrae. Each is directed across the crus of the diaphragm, so as to form nearly a right angle with the aorta. The right is longer than the left, on account of the position of the aorta; it passes dorsal to the inferior vena cava, the right renal vein, the head of the pancreas, and the descending part of the duodenum. The left is somewhat more cranial than the right; it lies dorsal to the left renal vein, the body of the pancreas and the lienal vein, and is crossed by the inferior mesenteric vein. Before reaching the hilum of the kidney, each artery divides into four or five branches; the greater number of these lie between the renal vein and ureter, the vein being ventral, the ureter dorsal, with one or more branches usually dorsal to the ureter. Each vessel gives off some small **inferior suprarenal branches** to the suprarenal gland, the ureter, and the surrounding cellular tissue and muscles.

Variations.—One or two **accessory renal arteries** are frequently found (23 per cent), more especially on the left side. They usually arise from the aorta, and may come off above or below the main artery, the former being the more common position. Instead of entering the kidney at the hilum, they usually pierce the upper or lower part of the organ.

F. The **Testicular Arteries** (*aa. testiculares; internal spermatic arteries*) (Fig. 8–41) are two long slender vessels which *arise* from the ventral surface of the aorta a little caudal to the renal arteries. Each passes obliquely caudalward and lateralward under the peritoneum, resting on the Psoas major, the right testicular lying ventral to the inferior vena cava and dorsal to the middle colic and ileocolic arteries and the terminal part of the ileum, the left dorsal

to the left colic and sigmoid arteries and the iliac colon. Each crosses obliquely over the ureter and the distal part of the external iliac artery to reach the deep inguinal ring, through which it passes, and accompanies the other constituents of the spermatic cord along the inguinal canal to the scrotum, where it becomes tortuous, and divides into several branches. Two or three of these accompany the ductus, and supply the epididymis, anastomosing with the artery of the ductus deferens; others pierce the back part of the tunica albuginea, and supply the substance of the testis. The testicular artery supplies one or two small branches to the ureter, and in the inguinal canal gives one or two twigs to the Cremaster.

G. The **Ovarian Arteries** (*aa. ovaricae*) (Fig. 9–33) are the arteries in the female corresponding to the testicular in the male. They are shorter than the testicular, and do not pass out of the abdominal cavity. The origin and course of the first part of each artery are the same as those of the testicular, but on arriving at the superior opening of the lesser pelvis, the ovarian artery passes medially between the two layers of the suspensory ligament of the ovary and of the broad ligament of the uterus, to be distributed to the ovary. Small branches are given to the ureter and the uterine tube, and one passes on to the side of the uterus and anastomoses with the uterine artery. Other offsets are continued on the round ligament of the uterus, through the inguinal canal, to the integument of the labium majus and groin.

At an early period of fetal life, when the testes or ovaries lie by the side of the vertebral column, caudal to the kidneys, the testicular or ovarian arteries are short; but with the descent of these organs into the scrotum or lesser pelvis, the arteries are gradually lengthened.

H. The **Phrenic Arteries** (*aa. phrenicae, inferior phrenic arteries*) (Fig. 8–41) are two small vessels, which supply the diaphragm but present much variety in their origin. They may *arise* separately from the ventral surface of the aorta, immediately proximal to the celiac artery or from one of the renal arteries, or by a common trunk, which may spring either from the aorta or from the celiac artery. They diverge from each other across the crura of the diaphragm, and then run obliquely lateralward upon its under surface. The left phrenic passes dorsal to the esophagus, and runs ventrally on the left side of the esophageal hiatus. The right phrenic passes dorsal to the inferior vena cava, and along the right side of the foramen which transmits that vein. Near the dorsal part of the central tendon each vessel divides into a medial and a lateral branch. The **medial branch** curves ventrally, and anastomoses with its fellow of the opposite side, and with the musculophrenic and pericardiacophrenic arteries. The **lateral branch** passes toward the side of the thorax, and anastomoses with the caudal intercostal arteries, and with the musculophrenic. The lateral branch of the right phrenic gives off a few vessels to the inferior vena cava; and the left one, some branches to the esophagus. Each vessel gives off **superior suprarenal branches** to the suprarenal gland of its own side. The spleen and the liver also receive a few twigs from the left and right vessels respectively.

I. The **Lumbar Arteries** (*aa. lumbales*) are in series with the intercostals. They are usually four in number on either side, and *arise* from the dorsum of the aorta, opposite the bodies of the cranial four lumbar vertebrae. A fifth pair, small in size, is occasionally present arising from the middle sacral artery. The lumbar arteries run lateralward on the bodies of the lumbar vertebrae, dorsal to the sympathetic trunk, to the intervals between the adjacent transverse processes, and are then continued into the abdominal wall. The arteries of the right side pass dorsal to the inferior vena cava, and the upper two on each side run dorsal to the corresponding crus of the diaphragm. The arteries of both sides pass beneath the tendinous arches which give origin to the Psoas major, and are then continued dorsal to this muscle and the lumbar plexus. They now cross the Quadratus lumborum, the upper three arteries running dorsal to, the last usually ventral to, the muscle. At the lateral border of the Quadratus lumborum they pierce the posterior aponeurosis of the Transversus abdominis and are carried ventrally between this muscle and the Obliquus internus. They

anastomose with the caudal intercostal, the subcostal, the iliolumbar, the deep iliac circumflex, and the inferior epigastric arteries.

Branches.—In the interval between the adjacent transverse processes each lumbar artery gives off a **posterior ramus** which is continued dorsally between the transverse processes and is distributed to the muscles and skin of the back; it furnishes a **spinal branch** which enters the vertebral canal and is distributed in a manner similar to the spinal branches of the posterior rami of the intercostal arteries (page 627). **Muscular branches** are supplied from each lumbar artery and from its posterior ramus to the neighboring muscles.

J. The **Middle Sacral Artery** (*a. sacralis mediana*) (Fig. 8–50) is a small vessel, which *arises* from the dorsal aspect of the aorta, a little proximal to its bifurcation. It descends in the midline ventral to the fourth and fifth lumbar vertebrae, the sacrum and coccyx, and from it, minute branches are said to pass to the posterior surface of the rectum. On the last lumbar vertebra it anastomoses with the lumbar branch of the iliolumbar artery; ventral to the sacrum it anastomoses with the lateral sacral arteries, and sends offsets into the anterior sacral foramina. It is crossed by the left common iliac vein, and is accompanied by a pair of venae comitantes; these unite to form a single vessel, which opens into the left common iliac vein.

K. The **Common Iliac Arteries** (*aa. iliacae communes*) (Figs. 8–41, 8–50).—The abdominal aorta divides on the left side of the body of the fourth lumbar vertebra into the two common iliac arteries, each about 5 cm in length. They diverge laterally from the end of the aorta, pass caudalward and divide opposite the intervertebral fibrocartilage between the last lumbar vertebra and the sacrum, into two branches, the **external** and **internal iliac arteries;** the former supplies the lower limb; the latter, the viscera and parietes of the pelvis.

The **right common iliac artery** (Fig. 8–49) is usually somewhat longer than the left, and passes more obliquely across the body of the last lumbar vertebra. Ventral to it are the peritoneum, the small intestines, branches of the sympathetic nerves, and, at its point of division, the ureter. Dorsally, it is separated from the bodies of the fourth and fifth lumbar vertebrae, and the intervening disk, by the terminations of the two common iliac veins and the commencement of the inferior vena cava. *Laterally,* it is in relation with the inferior vena cava, the right common iliac vein, and with the Psoas major. *Medial* to it is the left common iliac vein.

The **left common iliac artery** is in relation, *ventrally,* with the peritoneum, the small intestines, branches of the sympathetic nerves, the superior rectal artery, and is crossed at its point of bifurcation by the ureter. It rests on the bodies of the fourth and fifth lumbar vertebrae, and the intervening disk. The left common iliac vein lies partly *medial* to, and partly *dorsal* to the artery; *laterally,* the artery is in relation with the Psoas major.

Branches.—The common iliac arteries give off small branches to the peritoneum, Psoas major, ureters, and the surrounding areolar tissue, and occasionally give origin to the iliolumbar, or accessory renal arteries.

Variations.—The *point of origin* varies according to the bifurcation of the aorta. In three-fourths of a large number of cases, the aorta bifurcated either upon the fourth lumbar vertebra, or upon the disk between it and the fifth; the bifurcation being, in one case out of nine, below, and in one out of eleven, above this point. In about 80 per cent the aorta bifurcated within 1.25 cm above or below the level of the crest of the ilium; more frequently below than above.

The *point of division* varies greatly. In two-thirds of a large number of cases it was between the last lumbar vertebra and the upper border of the sacrum, being above that point in one case out of eight, and below it in one case out of six. The left common iliac artery divides more distally oftener than the right.

The *relative lengths,* also, of the two common iliac arteries vary. The right common iliac was the longer in sixty-three cases; the left in fifty-two; while they were equal in fifty-three. The length of the arteries varied, in five-sevenths of the cases examined, from 3.5 to 7.5 cm; in about half of the remaining cases the artery was longer, and in the other half, shorter, the minimum length being less than 1.25 cm, the maximum, 11 cm. In rare instances, the right common iliac has been found wanting, the

external and internal iliac arising directly from the aorta.

Collateral Circulation.—The principal agents in carrying on the collateral circulation after the application of a ligature to the common iliac are: the anastomoses of the rectal branches of the internal iliac with the superior rectal from the inferior mesenteric; of the uterine, ovarian, and vesical arteries of the opposite sides; of the lateral sacral with the middle sacral artery; of the inferior epigastric with the internal thoracic, inferior intercostal, and lumbar arteries; of the deep iliac circumflex with the lumbar arteries; of the iliolumbar with the last lumbar artery; of the obturator artery, by means of its pubic branch, with the vessel of the opposite side and with the inferior epigastric.

The Internal Iliac or Hypogastric Artery

The **internal iliac** or **hypogastric artery** (*a. iliaca interna*) (Figs. 8–49, 8–50) *begins* at the bifurcation of the common iliac, opposite the lumbosacral articulation, and follows a short, curved course of about 4 cm toward the greater sciatic foramen. It is the most variable artery in the body in terms of the pattern of its branches, but in slightly more than half the bodies it divides into a ventral and a dorsal division. It supplies the walls of the pelvis, the pelvic viscera, the buttock, the genital organs, and part of the medial thigh.

In the fetus, the internal iliac artery is twice as large as the external iliac, and is the direct continuation of the common iliac. It ascends along the side of the bladder, and runs cranialward on the inside of the anterior wall of the abdomen to reach the umbilicus, converging toward its fellow of the opposite side. Having passed through the umbilical opening, the two arteries, here termed **umbilical,** enter the umbilical cord, where they are coiled around the umbilical vein, and ultimately ramify in the placenta.

After birth, when the placental circulation ceases, only the pelvic portion of the umbilical artery remains patent, becoming in the adult the internal iliac and first part of the superior vesical artery. The remainder of the vessel is converted into a solid fibrous cord, the **lateral umbilical ligament** (*obliterated hypogastric artery*) which extends from the pelvis to the umbilicus.

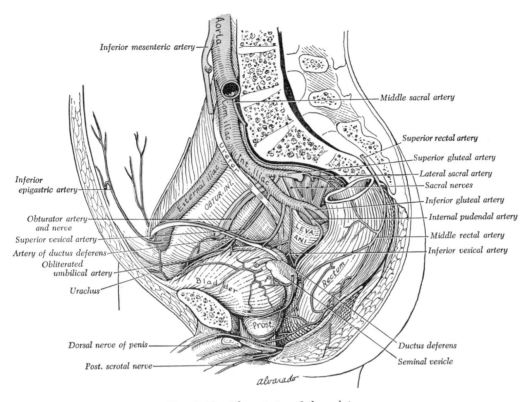

FIG. 8–49.—The arteries of the pelvis.

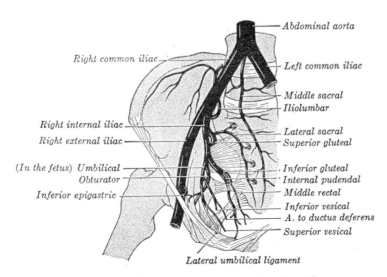

FIG. 8–50.—The internal iliac artery and its branches.

Collateral Circulation.—The circulation after ligature of the internal iliac artery is carried on by the anastomoses of the uterine and ovarian arteries; of the vesical arteries of the two sides; of the rectal branches of the internal iliac with those from the inferior mesenteric; of the obturator artery, by means of its pubic branch, with the vessel of the opposite side, and with the inferior epigastric and medial femoral circumflex; of the circumflex and perforating branches of the profunda femoris with the inferior gluteal; of the superior gluteal with the posterior branches of the lateral sacral arteries; of the iliolumbar with the last lumbar; of the lateral sacral with the middle sacral; and of the iliac circumflex with the iliolumbar and superior gluteal.

Branches.—The branches of the internal iliac are subject to great variation in their origin (see below). The division into a ventral and dorsal trunk is seldom complete and any two branches may arise from a common trunk. The distribution of the individual branches is more constant, however, and may be grouped as follows: visceral, ventral parietal, and dorsal parietal.

Visceral Branches
  1. Umbilical
  2. Inferior vesical
  3. Middle rectal
  4. Uterine

Ventral Parietal
  5. Obturator
  6. Internal pudendal

Dorsal Parietal
  7. Iliolumbar
  8. Lateral sacral
  9. Superior gluteal
  10. Inferior gluteal

Variations.—The patterns of variation in the origin of branches were recorded in 130 consecutive dissections (Ashley and Anson '41). In the commonest pattern (58 per cent) the inferior gluteal and internal pudendal were derived from a common stem with the umbilical and superior gluteal arising above them by separate stems. In 17 per cent the inferior and superior gluteals arose from one stem, the umbilical and pudendal from another stem. In 10 per cent the pudendal, inferior gluteal, and umbilical arose by a common stem, the superior gluteal separately. In 8 per cent the umbilical and pudendal arose separately above

a common stem for the gluteals; the pudendal appeared to be the continuation of the hypogastric, a condition described by some authors as the primitive type found in lower animals. Five other patterns are described but their appearance was less frequent than the above. The obturator was probably the most variable branch and a few of its variations are described with that artery.

1. The **Umbilical Artery** (*a. umbilicalis*) in the adult is the remains of the intrapelvic and intraabdominal portions of the fetal umbilical artery. It retains its lumen for a short distance beyond the internal iliac artery, giving off the superior vesical artery. Beyond this it is a fibrous cord which runs cranialward beside the bladder to the umbilicus and is named the *lateral umbilical* or *lateral vesicoumbilical ligament* (Fig. 8–49).

a) The **superior vesical artery** (*a. vesicalis superior*) supplies a number of small branches to the cranial portion of the bladder. A small branch, the **artery to the ductus deferens**, accompanies the ductus in its course to the testis where it anastomoses with the testicular artery. One or more (*b*) **middle vesical arteries** are distributed to the fundus of the bladder and the seminal vesicles. Small branches supply the **ureter.**

2. The **Inferior Vesical Artery** (*a. vesicalis inferior*) frequently *arises* in common with the middle rectal, and is distributed to the fundus of the bladder, the prostate, and the vesiculae seminales. The branches to the prostate communicate with the corresponding vessels of the opposite side.

3. The **Middle Rectal Artery** (*a. rectalis media; middle hemorrhoidal artery*) may *arise* with the preceding vessel. It is distributed to the rectum, anastomosing with the inferior vesical and with the superior and inferior rectal arteries. It gives offsets to the vesiculae seminales and prostate.

4. The **Uterine Artery** (*a. uterina*) (Figs. 8–51, 9–33) *arises* from the medial surface of the internal iliac, continues medialward on the Levator ani toward the cervix uteri and about 2 cm from the cervix it crosses the ureter, to which it supplies a small branch. Reaching the side of the uterus it ascends in a tortuous manner between the two layers of the broad ligament to the junction of the uterine tube and uterus. It then runs

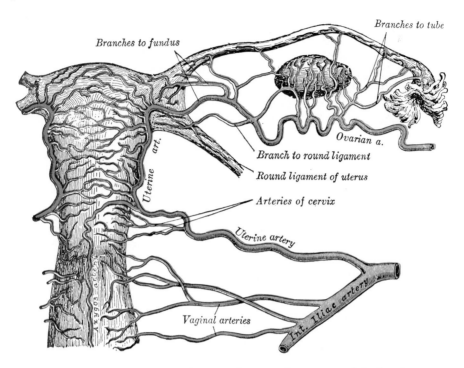

Fig. 8–51.—The arteries of the internal organs of generation of the female, seen from behind. (After Hyrtl.)

lateralward toward the hilum of the ovary, and ends by anastomosing with the ovarian artery. It supplies branches to the cervix uteri, one of which descends either on the ventral or the dorsal surface of the vagina, anastomoses with branches of the vaginal arteries, and forms with them a median longitudinal vessel, the **azygos artery of the vagina.** There are usually two azygos arteries, one of which runs down the ventral and the other the dorsal surface of the vagina. The uterine supplies numerous branches to the body of the uterus, and from its terminal portion twigs are distributed to the uterine tube and the round ligament of the uterus before anastomosing with the ovarian artery.

a) The **vaginal artery** (*a. vaginalis*) *arises* from the uterine just before it reaches the cervix and descends upon the vagina, supplying its mucous membrane. It sends branches to the bulb of the vestibule, the fundus of the bladder, and the contiguous part of the rectum. It assists in forming the azygos arteries of the vagina, and is frequently represented by two or three

branches. It may be an independent branch of the internal pudendal or have a common origin with the inferior vesical.

5. The **Obturator Artery** (*a. obturatoria*) (Fig. 8–49) *arises* from the ventral or medial surface of the internal iliac artery and runs ventralward on the lateral wall of the pelvis to the obturator canal. It leaves the pelvic cavity through the canal and in the thigh divides into anterior and posterior branches. Within the pelvic cavity the obturator is medial to the obturator fascia covering the Obturator internus. It is crossed medially by the ureter and ductus deferens, is accompanied by the obturator nerve, and is covered by the peritoneum.

Branches.—*Inside the pelvis* the obturator artery gives off:

a) **Iliac branches** to the iliac fossa, which supply the bone and the Iliacus, and anastomose with the iliolumbar artery.

b) A **vesical branch,** which runs backward to supply the bladder.

c) A **pubic branch,** which is given off from the vessel just before it leaves the pelvic cavity. The pubic branch ascends upon the

inside of the pubis, communicating with the opposite corresponding vessel, and with the inferior epigastric artery.

Outside the pelvis, the obturator artery divides at the margin of the obturator foramen, into an anterior and a posterior branch which encircle the foramen under cover of the Obturator externus.

d) The **anterior branch** runs ventralward on the outer surface of the obturator membrane curving caudalward along the anterior margin of the foramen. It distributes branches to the Obturator externus, Pectineus, Adductores, and Gracilis, anastomosing with the posterior branch and with the medial femoral circumflex artery.

e) The **posterior branch** follows the posterior margin of the foramen and divides into two branches. One runs anteriorly on the inferior ramus of the ischium where it anastomoses with the anterior branch of the obturator. The other branch gives twigs to the muscles attached to the ischial tuberosity and anastomoses with the inferior gluteal. The **acetabular** or articular branch enters the hip joint through the acetabular notch, ramifies in the fat at the bottom of the acetabulum and sends a twig along the ligamentum capitis to the head of the femur.

Variations.—The obturator artery arose from the internal iliac or one of its branches in 60 to 70 per cent of 320 bodies (Pick, Anson and Ashley '42). It arose from the main trunk in 23 per cent, from the anterior division in 20 per cent, from the posterior division in 3 per cent, from the superior gluteal in 11 per cent, from the inferior gluteal in 9 per cent, and from the internal pudendal or external iliac in 2 or 3 per cent. The origin from the inferior epigastric occurred in 27 per cent (Fig. 8–52), and in this position might curve along the lacunar ligament and be endangered during an operation for strangulated femoral hernia.

6. The **Internal Pudendal Artery** (*a. pudenda interna; internal pudic artery*) supplies the perineum and external organs of generation, and although the course of the artery is the same in the two sexes, the vessel is smaller in the female than in the male, and the distribution of its branches somewhat different (Fig. 8–50).

The **internal pudendal artery in the male** passes caudally toward the inferior border of the greater sciatic foramen, and emerges from the pelvis between the Piriformis and Coccygeus; it then crosses the ischial spine, and reenters the pelvis through the lesser sciatic foramen. The artery crosses the Obturator internus as the latter forms the lateral wall of the ischiorectal fossa and gradually approaches the margin of the inferior ramus of the ischium. It passes ventrally between the two layers of the fascia of the urogenital diaphragm, then runs ventrally along the medial margin of the inferior ramus of the pubis, and about 1.25 cm dorsal to the pubic arcuate ligament it divides into the **dorsal and deep arteries of the penis,** but it may pierce the superficial fascia of the urogenital diaphragm before doing so.

Relations.—Within the pelvis, it lies ventral to the Piriformis muscle, the sacral plexus of nerves, and the inferior gluteal artery. Crossing the ischial spine, it is covered by the Gluteus maximus and overlapped by the sacrotuberous ligament. Here the pudendal nerve lies to the medial side and the nerve to the Obturator internus to the lateral side. In the perineum it lies on the lateral wall of the ischiorectal fossa in a canal (*Alcock's canal*) formed by the splitting of the obturator fascia. It is accompanied by a pair of venae comitantes and the pudendal nerve.

Variations.—The internal pudendal artery is sometimes smaller than usual, or fails to give off one or two of its usual branches; in such cases the deficiency is supplied by branches derived from an additional vessel, the **accessory pudendal,** which generally arises from the internal pudendal artery before its exit from the greater sciatic foramen. It passes forward along the lower part of the bladder and across the side of the prostate to the root of the penis, where it perforates the urogenital diaphragm, and gives off the branches usually derived from the internal pudendal artery. The deficiency most frequently met with is that in which the internal pudendal ends as the artery of the

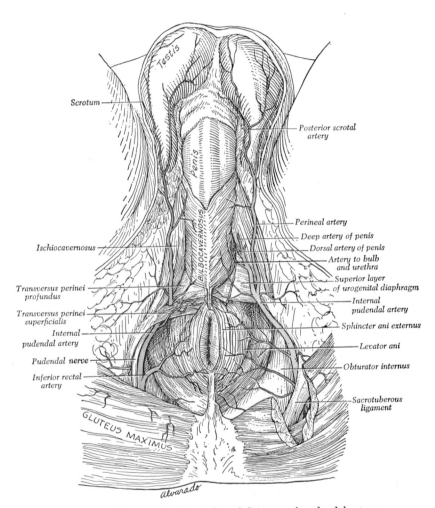

Fig. 8–53.—The superficial branches of the internal pudendal artery.

urethral bulb, the dorsal and deep arteries of the penis being derived from the accessory pudendal. The internal pudendal artery may also end as the perineal, the artery of the urethral bulb being derived, with the other two branches, from the accessory vessel. Occasionally the accessory pudendal artery is derived from one of the other branches of the internal iliac artery, most frequently the inferior vesical or the obturator.

Branches.—The branches of the internal pudendal artery (Fig. 8–53) are:

    a) Muscular
    b) Inferior rectal
    c) Perineal
    d) Artery of the bulb
    e) Urethral
    f) Deep artery of the penis
    g) Dorsal artery of the penis

a) The **muscular branches** consist of two sets: one given off in the pelvis; the other as the vessel crosses the ischial spine. The former consists of several small offsets which supply the Levator ani, the Obturator internus, the Piriformis, and the Coccygeus. The branches given off outside the pelvis are distributed to the adjacent parts of the Gluteus maximus and external rotator muscles. They anastomose with branches of the inferior gluteal artery.

b) The **inferior rectal artery** (*a. rectalis inferior; inferior hemorrhoidal artery*) arises from the internal pudendal as it passes above the ischial tuberosity. Piercing its fascial sheath (*Alcock's canal*), it divides into two or three branches which cross the ischiorectal fossa, and are distributed to the muscles and integument of the anal region, and send offshoots around the lower edge of the Gluteus maximus to the skin of the buttock. They anastomose with the corresponding vessels of the opposite side, with the superior and middle rectal, and with the perineal artery.

c) The **perineal artery** (*a. perinei; superficial perineal artery*) arises from the internal pudendal, ventral to the preceding branches, crosses either over or under the Transversus perinei superficialis, and continues parallel with the pubic arch in the interspace between the Bulbocavernosus and Ischiocavernosus. It supplies both of these muscles and finally divides into several posterior scrotal branches which are distributed to the skin and dartos tunic of the scrotum. As it crosses the Transversus perinei superficialis it gives off the transverse perineal artery which runs transversely on the cutaneous surface of the muscle, and anastomoses with the corresponding vessel of the opposite side and with the inferior rectal arteries. It supplies the Transversus perinei superficialis and the structures between the anus and the bulb of the penis.

d) The **artery of the bulb** (*a. bulbi penis*) is a short vessel of large caliber which *arises* from the internal pudendal between the two layers of fascia of the urogenital diaphragm; it passes medialward, pierces the inferior fascia of the urogenital diaphragm, and gives off branches which ramify in the bulb of the penis and in the posterior part of the corpus spongiosum penis. It gives off a small branch to the bulbourethral gland.

e) The **urethral artery** (*a. urethralis*) *arises* a short distance anterior to the artery of the urethral bulb. It runs medially, pierces the inferior fascia of the urogenital diaphragm, and enters the corpus cavernosum penis, in which it is continued forward to the glans penis.

f) The **deep artery of the penis** (*a. profunda penis; artery to the corpus cavernosum*) (Fig. 8–53), one of the terminal branches of the internal pudendal, *arises* from that vessel while it is situated between the two fasciae of the urogenital diaphragm; it pierces the superficial layer, and, entering the crus penis obliquely, continues distally in the center of the corpus cavernosum penis, to which its branches are distributed.

g) The **dorsal artery of the penis** (*a. dorsalis penis*), the other terminal branch of the internal pudendal, *arises* in the deep perineal compartment just lateral to the Bulbocavernosus and pierces the superficial fascia of the urogenital diaphragm between the crus penis and the pubic arch. It passes between the two layers of the suspensory ligament and reaching the dorsum of the penis it runs distally to the glans, sending branches to the glans and prepuce. On the penis it lies between the dorsal nerve and deep dorsal vein. It supplies the integument and fibrous sheath of the corpus cavernosum and anastomoses with the deep artery.

The **internal pudendal artery in the female** is smaller than in the male. Its origin and course are similar, and there is considerable analogy in the distribution of its branches. The perineal artery supplies the labia pudendi; the artery of the bulb supplies the bulbus vestibuli and the erectile tissue of the vagina; the deep artery of the clitoris supplies the corpus cavernosum clitoridis; and the dorsal artery of the clitoris supplies the dorsum of that organ, and ends in the glans and prepuce of the clitoris.

7. The **Iliolumbar Artery** (*a. iliolumbalis*), (Fig. 8–50) a posterior branch of the internal iliac, turns upward behind the obturator nerve and the external iliac vessels, to the medial border of the Psoas major dorsal to which it divides into a lumbar and an iliac branch.

a) The **lumbar branch** (*ramus lumbalis*) supplies the Psoas major and Quadratus lumborum, anastomoses with the last lumbar artery, and sends a small **spinal branch** through the intervertebral foramen between the last lumbar vertebra and the sacrum, into the vertebral canal, to supply the cauda equina.

b) The **iliac branch** (*ramus iliacus*) descends to supply the Iliacus; some offsets, running between the muscle and the bone, anastomose with the iliac branches of the obturator; one of these enters an oblique canal to supply the bone, while others run along the crest of the ilium, distributing branches to the gluteal and abdominal muscles, and anastomosing in their course with the superior gluteal, iliac circumflex, and lateral femoral circumflex arteries.

8. The **Lateral Sacral Arteries** (*aa. sacrales laterales*) (Fig. 8–50) *arise* from the posterior part of the internal iliac; there are usually two, a **superior** and an **inferior**.

a) The **superior** passes medialward, and, after anastomosing with branches from the middle sacral, enters the first or second anterior sacral foramen, supplies branches to the contents of the sacral canal, and, escaping by the corresponding posterior sacral foramen, is distributed to the skin and muscles on the dorsum of the sacrum, anastomosing with the superior gluteal.

b) The **inferior** runs obliquely across the Piriformis and the sacral nerves to the medial side of the anterior sacral foramina, descends on the sacrum, and anastomoses over the coccyx with the middle sacral and the opposite lateral sacral artery. In its course it gives off branches, which enter the anterior sacral foramina; these, after supplying the contents of the sacral canal, emerge by the posterior sacral foramina, and are distributed to the muscles and skin on the dorsal surface of the sacrum, anastomosing with the gluteal arteries.

9. The **Superior Gluteal Artery** (*a. glutea superior; gluteal artery*) (Fig. 8–54) is the largest branch of the internal iliac, and appears to be the posterior continuation of the vessel. It is a short artery which runs dorsalward between the lumbosacral trunk and the first sacral nerve, and, passing out of the pelvis above the superior border of the Piriformis, immediately divides into a **superficial** and a **deep branch**. Within the pelvis it gives off a few branches to the Iliacus, Piriformis, and Obturator internus, and before quitting that cavity, a nutrient artery which enters the ilium.

a) The **superficial branch** enters the deep surface of the Gluteus maximus, and divides into numerous branches, some of which supply the muscle and anastomose with the inferior gluteal, while others perforate its tendinous origin, and supply the integument covering the dorsal surface of the sacrum, anastomosing with the posterior branches of the lateral sacral arteries.

b) The **deep branch** lies deep to the Gluteus medius and almost immediately subdivides into two. Of these, the **superior division**, continuing the original course of the vessel, passes along the superior border of the Gluteus minimus to the anterior superior spine of the ilium, anastomosing with the deep iliac circumflex artery and the ascending branch of the lateral femoral circumflex artery. The **inferior division** crosses the Gluteus minimus obliquely to the greater trochanter, distributing branches to the Glutei and anastomoses with the lateral femoral circumflex artery. Some branches pierce the Gluteus minimus and supply the hip joint.

10. The **Inferior Gluteal Artery** (*a. glutea inferior; sciatic artery*) (Fig. 8–54), is dis-

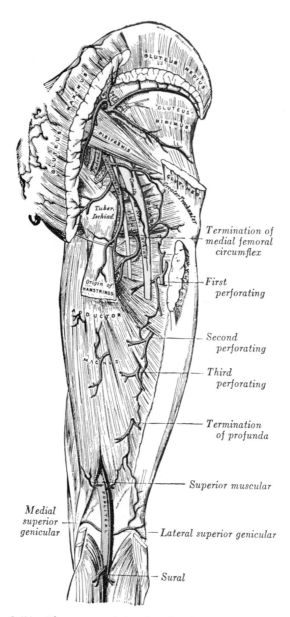

Fig. 8–54.—The arteries of the gluteal and posterior femoral regions.

tributed chiefly to the buttocks and back of the thigh. It passes posteriorly between the first and second sacral nerves, or between the second and third sacral nerves, and then descends between the piriformis and coccygeus muscles through the lower part of the sciatic foramen to the gluteal region. It then descends in the interval between the greater trochanter of the femur and tuberosity of the ischium, accompanied by the sciatic and posterior femoral cutaneous nerves and covered by the Gluteus maximus, is continued down the posterior thigh, supplying the skin, and anastomosing with branches of the perforating arteries.

Inside the pelvis it distributes branches to the Piriformis, Coccygeus, and Levator ani; some branches which supply the fat around the rectum, and occasionally take the place of the middle rectal artery; and vesical branches, to the fundus of the bladder, vesiculae seminales, and prostate. *Outside the pelvis* it gives off the following branches.

    a) Muscular
    b) Coccygeal
    c) Comitans nervi ischiadici
    d) Anastomotic
    e) Articular
    f) Cutaneous

a) The **muscular branches** supply the Gluteus maximus, anastomosing with the superior gluteal artery in the substance of the muscle, the lateral rotators, anastomosing with the internal pudendal artery; and the muscles attached to the tuberosity of the ischium, anastomosing with the posterior branch of the obturator and the medial femoral circumflex arteries.

b) The **coccygeal branches** run medialward, pierce the sacrotuberous ligament, and supply the Gluteus maximus, the integument, and other structures on the back of the coccyx.

c) The **arteria comitans nervi ischiadici** is a long, slender vessel, which accompanies the sciatic nerve for a short distance; it then penetrates it, and runs in its substance to the distal part of the thigh.

d) The **anastomotic** is directed distally across the external rotators, and assists in forming the so-called **crucial anastomosis** by joining with the first perforating and

medial and lateral femoral circumflex arteries.

e) The **articular branch**, generally derived from the anastomotic, is distributed to the capsule of the hip joint.

f) The **cutaneous branches** are distributed to the skin of the buttock and posterior thigh.

The External Iliac Artery

The **external iliac artery** (*a. iliaca externa*) (Fig. 8–49) is larger than the internal iliac, and passes obliquely distally and laterally along the medial border of the Psoas major, from the bifurcation of the common iliac to a point beneath the inguinal ligament midway between the anterior superior spine of the ilium and the symphysis pubis, where it enters the thigh and becomes the femoral artery.

Relations.—*Anteriorly and medially,* the artery is in relation with the peritoneum, subperitoneal areolar tissue, the termination of the ileum and frequently the vermiform appendix on the right side, and, on the left, the sigmoid colon. In the female it is crossed at its origin by the ovarian vessels, and occasionally by the ureter. The testicular vessels lie for some distance upon it near its termination, and it is crossed in this situation by the genital branch of the genitofemoral nerve and the deep iliac circumflex vein; the ductus deferens in the male, and the round ligament of the uterus in the female, curve across its medial side. *Dorsally,* it is in relation with the medial border of the Psoas major, from which it is separated by the iliac fascia. At the proximal part of its course, the external iliac vein lies partly dorsal to it, but more distally lies entirely to its medial side. *Laterally,* it rests against the Psoas major, from which it is separated by the iliac fascia. Numerous lymphatic vessels and lymph nodes lie on the ventral and on the medial side of the vessel.

Collateral Circulation.—The principal anastomoses in carrying on the collateral circulation, after the application of a ligature to the external iliac, are: the iliolumbar with the iliac circumflex; the superior gluteal with the lateral femoral circumflex; the obturator with the medial femoral circumflex; the inferior gluteal with the first perforating and circumflex branches of the profunda artery; and the internal pudendal with the external pudendal. When the obturator arises from the inferior epigastric, it is supplied with blood by branches, from either the internal iliac, the lateral sacral, or the internal pudendal. The inferior epigastric receives its supply from the internal thoracic and lower intercostal arteries, and from the internal iliac by the anastomoses of its branches with the obturator.

Branches.—Besides several small branches to the Psoas major and the neighboring lymph nodes, the external iliac gives off two branches of considerable size:

1. Inferior epigastric
2. Deep iliac circumflex

1. The **Inferior Epigastric Artery** (*a. epigastrica inferior; deep epigastric artery*) (Fig. 8–57) *arises* from the external iliac, immediately superior to the inguinal ligament. It curves ventrally in the subperitoneal tissue, and then ascends obliquely along the medial margin of the deep inguinal ring; continuing its course cranialward it pierces the transversalis fascia, and passing ventral to the linea semicircularis, ascends between the Rectus abdominis and the posterior lamella of its sheath. It finally divides into numerous branches, which anastomose, above the umbilicus, with the superior epigastric branch of the internal thoracic and with the lower intercostal arteries (Fig. 8–32). As the inferior epigastric artery passes obliquely along the margins of the deep inguinal ring, it lies dorsal to the commencement of the spermatic cord. The ductus deferens, as it leaves the spermatic cord in the male, or the round ligament of the uterus in the female, winds around the lateral and dorsal aspects of the artery.

Branches.—The branches of the vessel are: the **cremasteric artery** (*external spermatic artery*), which accompanies the spermatic cord, and supplies the Cremaster and other coverings of the cord, anastomosing with the testicular artery (in the female it is very small and accompanies the round ligament); a **pubic branch** which runs along the inguinal ligament, and then

descends along the medial margin of the femoral ring to the internal surface of the pubis, and there anastomoses with the pubic branch of the obturator artery; **muscular branches**, some of which are distributed to the abdominal muscles and peritoneum, anastomosing with the iliac circumflex and lumbar arteries; branches which perforate the tendon of the Obliquus externus, and supply the integument, anastomosing with branches of the superficial epigastric.

Variations.—The origin of the inferior epigastric may take place from any part of the external iliac between the inguinal ligament and a point 6 cm above it; or it may arise below this ligament, from the femoral. It frequently springs from the external iliac, by a common trunk with the obturator. Sometimes it arises from the obturator, the latter vessel being furnished by the internal iliac, or it may be formed of two branches, one derived from the external iliac, the other from the internal iliac (see Fig. 8–52).

2. The **Deep Iliac Circumflex Artery** (*a. circumflexa ilium profunda*) *arises* from the lateral aspect of the external iliac nearly opposite the inferior epigastric artery. It ascends obliquely lateralward deep to the inguinal ligament, contained in a fibrous sheath formed by the junction of the transversalis fascia and iliac fascia, to the anterior superior iliac spine, where it anastomoses with the ascending branch of the lateral femoral circumflex artery. It then pierces the transversalis fascia and passes along the inner lip of the crest of the ilium to about its middle, where it perforates the Transversus, and runs dorsally between that muscle and the Obliquus internus, to anastomose with the iliolumbar and superior gluteal arteries. Opposite the anterior superior spine of the ilium it gives off a large branch, which ascends between the Obliquus internus and Transversus muscles, supplying them, and anastomosing with the lumbar and inferior epigastric arteries (Fig. 8–57, not labelled).

THE ARTERIES OF THE LOWER LIMB

The artery which supplies the greater part of the lower limb is the direct continuation of the external iliac. It runs as a

single trunk from the inguinal ligament to the distal border of the Popliteus, where it divides into two branches, the **anterior** and

posterior tibial. The proximal part of the main trunk is named the **femoral,** the distal part the **popliteal.**

The Femoral Artery

The **femoral artery** (*a. femoralis*) (Figs. 8–59, 8–60) begins immediately distal to the inguinal ligament, midway between the anterior superior spine of the ilium and the symphysis pubis, and passes distally along the anterior and medial side of the thigh. It ends at the junction of the middle with the lower third of the thigh, where it passes through an opening in the Adductor magnus to become the popliteal artery. The vessel, at the proximal part of the thigh, lies anterior to the hip joint; in the distal part of its course it lies to the medial side of the body of the femur, and between these two parts, where it crosses the angle between the head and body, the vessel is some distance from the bone. The first 4 cm of the vessel is

enclosed, together with the femoral vein, in a fibrous sheath—the **femoral sheath.** In the upper third of the thigh the femoral artery is contained in the **femoral triangle** (*Scarpa's triangle*), and in the middle third of the thigh, in the **adductor canal** (*Hunter's canal*).

The **femoral sheath** (*crural sheath*) (Figs. 8–55, 8–56) is formed by a prolongation under the inguinal ligament of the fasciae which line the abdomen, the transversalis fascia being continued down anterior to the femoral vessels and the iliac fascia posterior to them. The sheath assumes the form of a short funnel, the wide end of which is directed superiorly, while the inferior, narrow end fuses with the adventitia of the vessels, about 4 cm distal to the inguinal ligament. It is strengthened anteriorly by a band termed the **deep crural arch** (page 429). The lateral wall of the sheath is perforated by the femoral branch of the genitofemoral nerve; the medial wall is

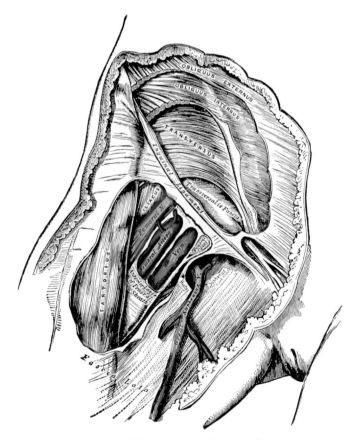

Fig. 8–55.—Femoral sheath laid open to show its three compartments.

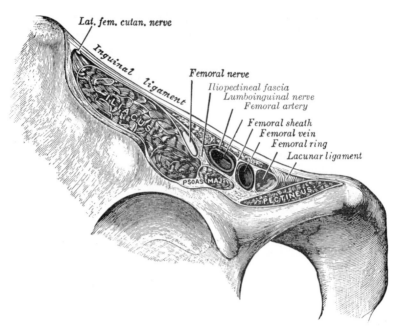

Fig. 8–56.—Structures passing between the inguinal ligament and pelvic bone.

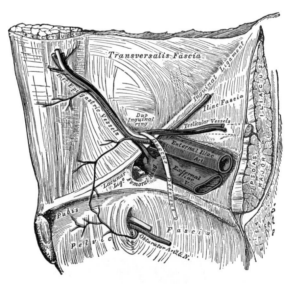

Fig. 8–57.—The relations of the femoral and abdominal inguinal rings, seen from within the abdomen. Right side.

pierced by the great saphenous vein and by some lymphatic vessels. The sheath is divided by two vertical partitions which stretch between its anterior and posterior walls. The lateral compartment contains the femoral artery, the intermediate the femoral vein, whereas the medial and smallest compartment is named the **femoral canal**, and contains some lymphatic vessels and a lymph node (Cloquet's or Rosenmüller's) embedded in a small amount of areolar tissue. The femoral canal is conical and measures about 1.25 cm in length. Its base, directed superiorly and named the **femoral ring**, is oval in form, its long diameter being directed transversely and measuring about

1.25 cm. The femoral ring (Figs. 8–55, 8–56) is bounded *anteriorly* by the inguinal ligament, *posteriorly* by the Pectineus covered by the pectineal fascia, *medially* by the crescentic base of the lacunar ligament, and *laterally* by the fibrous septum on the medial side of the femoral vein. The spermatic cord in the male and the round ligament of the uterus in the female lie immediately superior to the anterior margin of the ring, and the inferior epi-gastric vessels are close to its superior and lateral angle. The femoral ring is closed by a somewhat condensed portion of the sub-peritoneal fatty tissue, named the **septum femorale** (*crural septum*), the abdominal surface of which supports a small lymph node and is covered by the parietal layer of the peritoneum. The septum femorale is pierced by numerous lymphatic vessels passing from the deep inguinal to the external iliac lymph nodes. The parietal

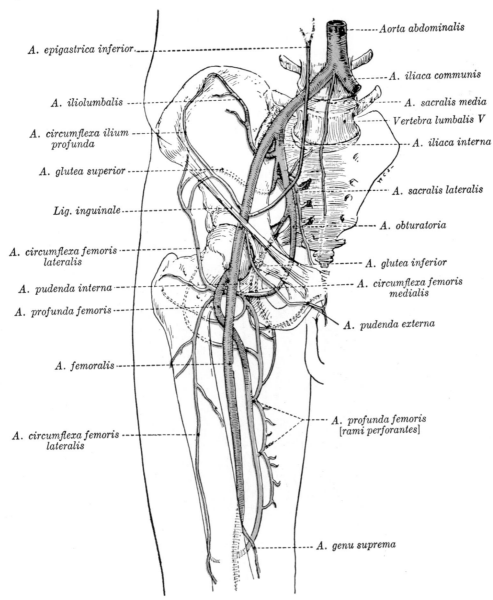

A. epigastrica inferior

A. iliolumbalis

A. circumflexa ilium profunda

A. glutea superior

Lig. inguinale

A. circumflexa femoris lateralis

A. pudenda interna

A. profunda femoris

A. femoralis

A. circumflexa femoris lateralis

Aorta abdominalis

A. iliaca communis

A. sacralis media

Vertebra lumbalis V

A. iliaca interna

A. sacralis lateralis

A. obturatoria

A. glutea inferior

A. circumflexa femoris medialis

A. pudenda externa

A. profunda femoris [rami perforantes]

A. genu suprema

Fig. 8–58.—Collateral circulation about the hip and the upper part of the right thigh. (Eycleshymer and Jones.)

peritoneum covering it presents a slight depression named the **femoral fossa**. The femoral nerve is not enclosed by the femoral sheath.

The **femoral triangle** (*trigonum femorale; Scarpa's triangle*) (Fig. 8–59) corresponds to the depression seen immediately distal to the fold of the groin. Its apex is directed distally, and the sides are formed laterally by the Sartorius, medially by the Adductor longus, and proximally by the inguinal ligament. The floor of the space is formed from its lateral to its medial side by the Iliacus, Psoas major, Pectineus, and in some cases a small part of the Adductor brevis; it is divided into two nearly equal parts by the femoral vessels, which extend from near the middle of its base to its apex: the artery giving off its superficial and deep branches, the vein receiving the deep femoral and great saphenous tributaries. On

the lateral side of the femoral artery is the femoral nerve with its branches. Besides the vessels and nerves, this space contains some fat and lymphatics.

The **adductor canal** (*canalis adductorius; Hunter's canal*) is a fascial tunnel in the middle third of the thigh, extending from the apex of the femoral triangle to the opening in the Adductor magnus. It is bounded anteriorly and laterally by the Vastus medialis; posteriorly by the Adductores longus and magnus. Its superficial covering is a strong fascia which lies deep to the Sartorius and which extends from the Vastus medialis across the femoral vessels to the Adductores longus and magnus. The canal contains the femoral artery and vein, and the saphenous nerve.

Relations of the Femoral Artery.—In the *femoral triangle* (Fig. 8–59) the artery is superficial. *Anterior* to it are the skin and subcu-

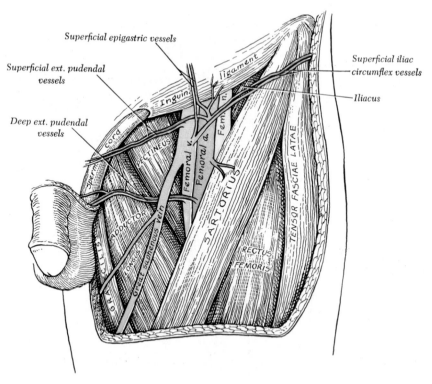

Fig. 8–59.—The left femoral triangle.

taneous fascia, the superficial subinguinal lymph nodes, the superficial iliac circumflex vein, the superficial layer of the fascia lata and the anterior part of the femoral sheath. The femoral branch of the genitofemoral nerve courses for a short distance within the lateral compartment of the femoral sheath, and lies at first anterior and then lateral to the artery. Near the apex of the femoral triangle the medial branch of the anterior femoral cutaneous nerve crosses the artery from its lateral to its medial side.

Posterior to the artery are the posterior part of the femoral sheath, the pectineal fascia, the medial part of the tendon of the Psoas major, the Pectineus and the Adductor longus. The artery is separated from the capsule of the hip joint by the tendon of the Psoas major, from the Pectineus by the femoral vein and profunda vessels, and from the Adductor longus by the femoral vein. The nerve to the Pectineus passes medialward behind the artery. On the *lateral* side of the artery, but separated from it by some fibers of the Psoas major, is the femoral nerve. The femoral vein is on the *medial* side of the proximal part of the artery, but is posterior to the vessel in the distal part of the femoral triangle.

In the *adductor canal* (Fig. 8–60) the artery is more deeply situated, being covered by the integument, the superficial and deep fasciae, the Sartorius and the fibrous roof of the canal; the saphenous nerve crosses from its lateral to its medial side. Posterior to the artery are the Adductores longus and magnus; anterior and lateral to it is the Vastus medialis. The femoral vein lies posterior to the proximal part, and lateral to the distal part of the artery.

Variations.—A few cases have been recorded in which the femoral artery was absent, its place being supplied by the inferior gluteal artery which accompanied the sciatic nerve to the popliteal fossa. The external iliac in these cases was small, and terminated in the profunda femoris. The femoral vein is occasionally placed along the medial side of the artery throughout the entire extent of the femoral triangle; or it may be split so that a large vein is placed on either side of the artery for a greater or lesser distance.

Collateral Circulation.—After ligature of the femoral artery, the main channels for carrying on the circulation are the anastomoses between —(1) the superior and inferior gluteal branches of the internal iliac with the medial and lateral femoral circumflex and first perforating branches of the profunda femoris; (2) the obturator branch of the internal iliac with the medial femoral circumflex of the profunda; (3)

the internal pudendal of the internal iliac with the superficial and deep external pudendal of the femoral; (4) the deep iliac circumflex of the external iliac with the lateral femoral circumflex of the profunda and the superficial iliac circumflex of the femoral, and (5) the inferior gluteal of the internal iliac with the perforating branches of the profunda (Fig. 8–58).

Branches.—The branches of the femoral artery are:

1. Superficial epigastric
2. Superficial iliac circumflex
3. Superficial external pudendal
4. Deep external pudendal
5. Muscular
6. Profunda femoris
7. Descending genicular artery

1. The **Superficial Epigastric Artery** (*a. epigastrica superficialis*) arises from the femoral artery about 1 cm below the inguinal ligament, and, passing through the femoral sheath and the fascia cribrosa, turns superficial to the inguinal ligament, and ascends between the two layers of the subcutaneous fascia of the abdominal wall nearly to the umbilicus. It distributes branches to the superficial subinguinal lymph nodes; the subcutaneous fascia, and the integument; it anastomoses with branches of the inferior epigastric, and with its fellow opposite (Fig. 8–59).

2. The **Superficial Iliac Circumflex Artery** (*a. circumflexa ilium superficialis*), smallest of the cutaneous branches, arises close to the preceding, and piercing the fascia lata, runs lateralward, parallel with the inguinal ligament, to the crest of the ilium; it divides into branches which supply the integument of the groin, and the superficial subinguinal lymph nodes, anastomosing with the deep iliac circumflex, superior gluteal and lateral femoral circumflex arteries.

3. The **Superficial External Pudendal Artery** (*a. pudenda externa superficialis; superficial external pudic artery*) arises from the medial side of the femoral artery, close to the preceding vessels, and, after piercing the femoral sheath and fascia cribrosa, courses medialward, across the spermatic cord (or round ligament in the female), to be distributed to the integument on the lower part of the abdomen, the **penis**

and scrotum in the male, and the labium majus in the female, anastomosing with branches of the internal pudendal.

4. The **Deep External Pudendal Artery** (*a. pudenda externa profunda; deep external pudic artery*) (Fig. 8–59), more deeply seated than the preceding, passes medialward across the Pectineus and the Adductor longus muscles; it is covered by the fascia lata, which it pierces at the medial side of the thigh, and is distributed, in the male, to the integument of the scrotum and perineum, in the female to the labium majus; its branches anastomose with the scrotal (or labial) branches of the perineal artery.

5. **Muscular Branches** (*rami musculares*) are supplied by the femoral artery to the Sartorius, Vastus medialis, and Adductores.

6. The **Profunda Femoris Artery** (*a. profunda femoris; deep femoral artery*) (Fig. 8–60) is a large vessel *arising* from the lateral and posterior part of the femoral artery, from 2 to 5 cm below the inguinal ligament. At first it lies lateral to the femoral artery; it then runs posterior to it and the femoral vein to the medial side of the femur, and, passing distally posterior to the Adductor longus, ends at the distal third of the thigh in a small branch, which pierces the Adductor magnus, and is distributed on the posterior thigh to the hamstring muscles. The terminal part of the profunda is sometimes named the **fourth perforating artery.**

Relations.—*Posterior* to it are the Iliacus, Pectineus, Adductor brevis, and Adductor magnus. *Anteriorly* it is separated from the femoral artery by the femoral and profunda veins and by the Adductor longus. *Laterally*, the origin of the Vastus medialis intervenes between it and the femur.

Variations.—This vessel sometimes arises from the medial side, and, more rarely, from the back of the femoral artery; but a more important peculiarity, from a surgical point of view, is that relating to the height at which the vessel arises. In three-fourths of a large number of specimens it arose from 2.25 to 5 cm below the inguinal ligament; in a few specimens the distance was less than 2.25 cm; more rarely, opposite the ligament; and in one specimen above the inguinal ligament, from the external iliac. Occasionally the distance between the origin of the vessel and the inguinal ligament exceeds 5 cm.

Branches.—The profunda gives off the following branches (Fig. 8–58):

  a) Medial femoral circumflex
  b) Lateral femoral circumflex
  c) Perforating
  d) Muscular

a) The **medial femoral circumflex artery** (*a. circumflexa femoris medialis; internal circumflex artery*) *arises* from the medial aspect of the profunda, and winds around the medial side of the femur, passing first between the Pectineus and Psoas major, and then between the Obturator externus and the Adductor brevis. At the proximal border of the Adductor brevis it gives off an **ascending branch** which goes to the Adductores, the Gracilis, and Obturator externus, and anastomoses with the obturator artery; the **transverse branch** descends beneath the Adductor brevis, to supply it and the Adductor magnus. The continuation of the vessel passes posteriorly and divides into superficial, deep, and acetabular branches.

i) The **superficial branch** appears between the Quadratus femoris and proximal border of the Adductor magnus, and anastomoses with the inferior gluteal, lateral femoral circumflex, and first perforating arteries (*cruciate anastomosis*).

ii) The **deep branch** runs obliquely cranialward upon the tendon of the Obturator externus and in front of the Quadratus femoris toward the trochanteric fossa, where it anastomoses with twigs from the gluteal arteries.

iii) The **acetabular branch** *arises* opposite the acetabular notch and enters the hip joint beneath the transverse ligament in company with an articular branch from the obturator artery; it supplies the fat in the bottom of the acetabulum, and continues along the round ligament to the head of the femur.

b) The **lateral femoral circumflex artery** (*a. circumflexa femoris lateralis; external circumflex artery*) *arises* from the lateral side of the profunda, passes horizontally between the divisions of the femoral nerve, and posterior to the Sartorius and Rectus femoris, and divides into ascending, transverse, and descending branches.

i) The **ascending branch** passes superiorly deep to the Tensor fasciae latae, to the lat-

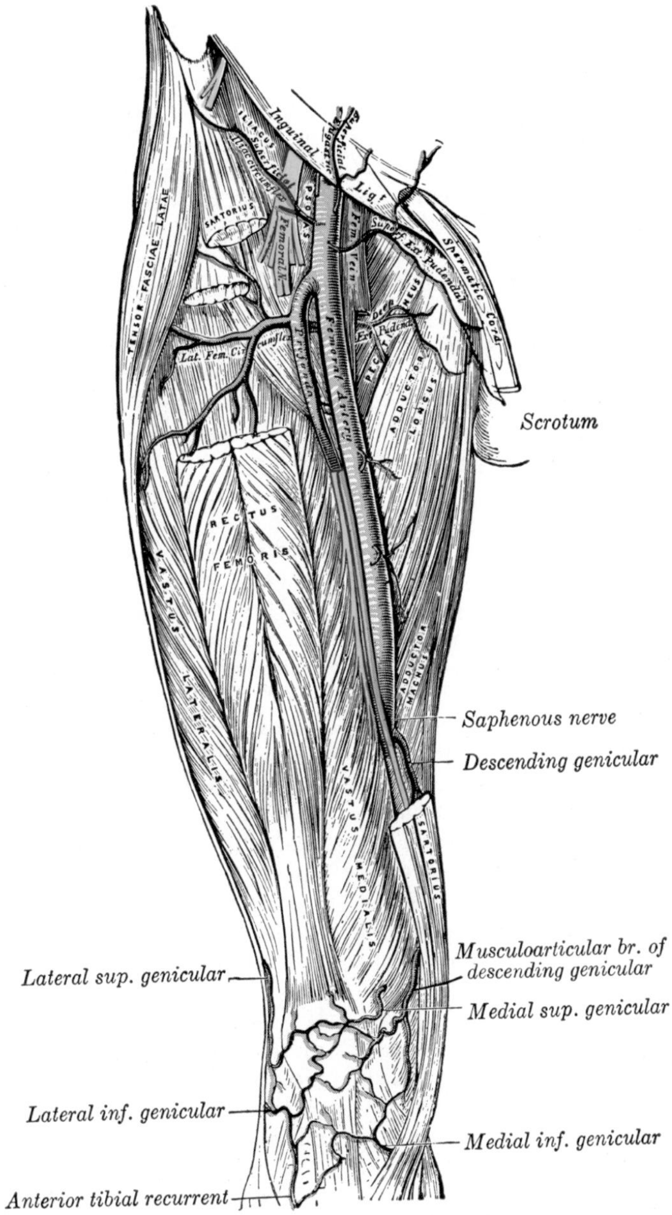

FIG. 8-60.—The femoral artery.

eral aspect of the hip, and anastomoses with the terminal branches of the superior gluteal and deep iliac circumflex arteries.

ii) The **descending branch** runs distally deep to the Rectus femoris, upon the Vastus lateralis, to which it gives offsets; one long branch descends in the muscle as far as the knee, and anastomoses with the superior lateral genicular branch of the popliteal artery. It is accompanied by the branch of the femoral nerve to the Vastus lateralis.

iii) The **transverse branch**, the smallest branch if present, but often absent, passes lateralward over the Vastus intermedius, pierces the Vastus lateralis, and winds around the femur, just distal to the greater trochanter, anastomosing on the back of the thigh with the medial femoral circumflex inferior gluteal and first perforating arteries.

Variations.—The medial circumflex artery arises independently from the femoral in from 19 to 26.5 per cent, according to various authors. The lateral circumflex artery arises independently from the femoral as frequently as 18 per cent of the time. The two circumflex arteries may arise by a common trunk from the profunda femoris.

c) The **perforating arteries** (Fig. 8–54), usually three in number, are so named because they perforate the Adductor magnus to reach the back of the thigh. They pass posteriorly close to the linea aspera of the femur under cover of small tendinous arches in the muscle. The first is given off proximal to the Adductor brevis, the second anterior to that muscle, and the third immediately distal to it.

i) The **first perforating artery** (a perforans prima) passes posteriorly between the Pectineus and Adductor brevis (sometimes perforating the latter); it pierces the Adductor magnus close to the linea aspera. It gives branches to the Adductores brevis and magnus, Biceps femoris, and Gluteus maximus, and anastomoses with the inferior gluteal, medial and lateral femoral circumflex and second perforating arteries (Fig. 8–58).

ii) The **second perforating artery** (a. perforans secunda), larger than the first, pierces the tendons of the Adductores brevis and magnus, and divides into ascending and descending branches, which supply the posterior femoral muscles, anastomosing with the first and third perforating. The second artery frequently *arises* in common with the first. The **nutrient artery** of the femur is usually given off from the second perforating artery; when two nutrient arteries exist, they usually spring from the first and third perforating vessels.

iii) The **third perforating artery** (a. perforans tertia) pierces the Adductor magnus, and divides into branches which supply the posterior femoral muscles; anastomosing proximally with the higher perforating arteries, and distally with the terminal branches of the profunda and the muscular branches of the popliteal. The termination of the profunda artery, already described, is sometimes termed the **fourth perforating artery.**

d) Numerous **muscular branches** *arise* from the profunda; some of these end in the Adductores, others pierce the Adductor magnus, give branches to the hamstrings, and anastomose with the medial femoral circumflex artery and with the superior muscular branches of the popliteal.

7. The **Descending Genicular Artery** (a. genus descendens; highest genicular artery; anastomotica magna artery) (Fig. 8–60) *arises* from the femoral just before it passes through the opening in the tendon of the Adductor magnus, and immediately divides into a saphenous and an articular branch.

a) The **saphenous branch** pierces the aponeurotic covering of the adductor canal, and accompanies the saphenous nerve to the medial side of the knee. It passes between the Sartorius and Gracilis, and, piercing the fascia lata, is distributed to the integument of the proximal and medial part of the leg, anastomosing with the medial inferior genicular artery.

b) The **articular branch** descends in the substance of the Vastus medialis and anterior to the tendon of the Adductor magnus, to the medial side of the knee, where it anastomoses with the medial superior genicular artery and anterior recurrent tibial artery. A branch from this vessel crosses proximal to the patellar surface of the femur, forming an anastomotic arch with the lateral superior genicular artery, and supplying branches to the knee joint.

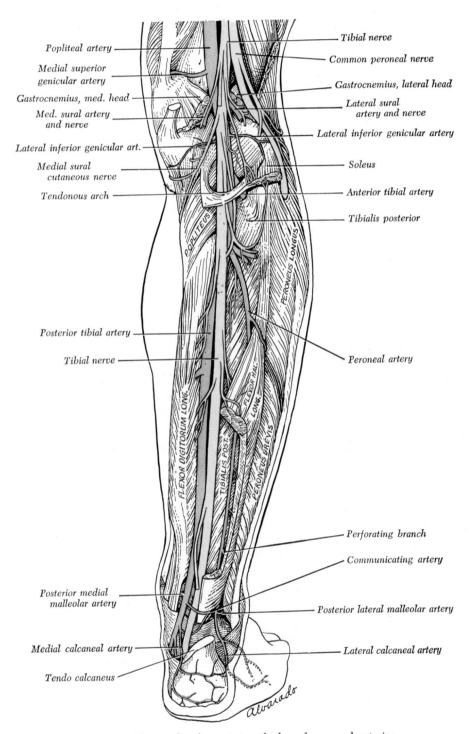

FIG. 8–61.—The popliteal, posterior tibial, and peroneal arteries.

The Popliteal Artery

The **popliteal artery** (*a. poplitea*) (Fig. 8–61) is the continuation of the femoral, and courses through the popliteal fossa. It extends from the opening in the Adductor magnus, at the junction of the middle and distal thirds of the thigh, to the intercondylar fossa of the femur, and continues distally to the distal border of the Popliteus, where it divides into **anterior** and **posterior tibial arteries.**

The **popliteal fossa** (Fig. 8–61) is a lozenge-shaped space, posterior to the knee joint. Laterally it is bounded by the Biceps femoris proximally, and by the Plantaris and the lateral head of the Gastrocnemius distally; medially it is limited by the Semitendinosus and Semimembranosus, and by the medial head of the Gastrocnemius. The floor is formed by the popliteal surface of the femur, the oblique popliteal ligament of the knee joint, the proximal end of the tibia, and the fascia covering the Popliteus; the fossa is covered by the fascia lata.

Contents.—The popliteal fossa contains the popliteal vessels, the tibial and the common peroneal nerves, the termination of the small saphenous vein, the distal part of the posterior femoral cutaneous nerve, the articular branch from the obturator nerve, a few small lymph nodes, and a considerable amount of fat. The tibial nerve descends through the middle of the fossa, lying under the deep fascia and crossing the vessels posteriorly from the lateral to the medial side. The common peroneal nerve descends on the lateral side of the upper part of the fossa, close to the tendon of the Biceps femoris. On the floor of the fossa are the popliteal vessels, the vein being superficial to the artery and united to it by dense areolar tissue; the vein is a thick-walled vessel, at first lateral to the artery, and then crossing it posteriorly to gain its medial side below; sometimes it is double, the artery lying between the two veins, which are usually connected by short transverse branches. The articular branch from the obturator nerve descends upon the artery to the knee joint. The popliteal lymph nodes, six or seven in number, are imbedded in the fat; one lies beneath the popliteal fascia near the termination of the external saphenous vein, another between the popliteal

artery and the back of the knee joint, while others are placed alongside the popliteal vessel. Arising from the artery, and passing off from it at right angles, are its genicular branches.

Relations.—Anterior to the artery are the popliteal surface of the femur (which is separated from the vessel by some fat), the back of the knee joint, and the fascia covering the Popliteus. Posteriorly, it is overlapped by the Semimembranosus proximally, and is covered by the Gastrocnemius and Plantaris distally. In the middle part of its course the artery is separated from the integument and fascia by a quantity of fat, and is crossed from the lateral to the medial side by the tibial nerve and the popliteal vein, the vein being between the nerve and the artery and closely adherent to the latter. On its *lateral side*, proximally, are the Biceps femoris, the tibial nerve, the popliteal vein, and the lateral condyle of the femur; distally, the Plantaris and the lateral head of the Gastrocnemius. On its *medial* side, proximally, are the Semimembranosus and the medial condyle of the femur; distally, the tibial nerve, the popliteal vein, and the medial head of the Gastrocnemius. The relations of the popliteal lymph nodes to the artery are described above.

Variations in Point of Division.—Occasionally the popliteal artery divides into its terminal branches opposite the knee joint. The anterior tibial under these circumstances usually passes anterior to the Popliteus.

Unusual Branches.—The artery sometimes divides into the anterior tibial and peroneal, the posterior tibial being wanting, or very small. Occasionally it divides into three branches, the anterior and posterior tibial, and peroneal.

Branches.—The branches of the popliteal artery are:

1. Superior muscular
2. Sural
3. Cutaneous
4. Medial superior genicular
5. Lateral superior genicular
6. Middle genicular
7. Medial inferior genicular
8. Lateral inferior genicular

1. The two or three **Superior Muscular Branches** *arise* from the proximal part of the artery, and are distributed to the distal parts of the Adductor magnus and hamstring muscles, anastomosing with the endings of the profunda femoris.

2. The **Sural Arteries** (*aa. surales; inferior*

muscular arteries) are two large branches, **medial** and **lateral**, which are distributed to the Gastrocnemius, Soleus, and Plantaris. They *arise* from the popliteal artery opposite the knee joint.

3. The **Cutaneous Branches** *arise* either from the popliteal artery or from some of its branches; they descend between the two heads of the Gastrocnemius, and, piercing the deep fascia, are distributed to the skin of the back of the leg. One branch usually accompanies the small saphenous vein.

The **superior genicular arteries** (*aa. genus superiores; superior articular arteries*) (Figs. 8–61, 8–62), two in number, *arise* one on either side of the popliteal, and wind around the femur immediately proximal to its condyles to the front of the knee joint.

4. The **Medial Superior Genicular** runs anterior to the Semimembranosus and Semitendinosus, proximal to the medial head of the Gastrocnemius, and passes deep to the tendon of the Adductor magnus. It divides into two branches, one of which supplies the Vastus medialis, anastomosing with the descending genicular and medial inferior genicular arteries; the other ramifies close to the surface of the femur, supplying it and the knee joint, and anastomosing with the lateral superior genicular artery. The medial superior genicular artery is frequently of small size, a condition which is associated with an increase in the size of the descending genicular.

5. The **Lateral Superior Genicular** passes proximal to the lateral condyle of the femur, deep to the tendon of the Biceps femoris, and divides into a superficial and a deep branch; the superficial branch supplies the Vastus lateralis, and anastomoses with the descending branch of the lateral femoral circumflex and the lateral inferior genicular arteries; the deep branch supplies the lower part of the femur and knee joint, and forms an anastomotic arch across the anterior aspect of the bone with the descending genicular and the medial inferior genicular arteries.

6. The **Middle Genicular Artery** (*a. genus media; azygos articular artery*) is a small

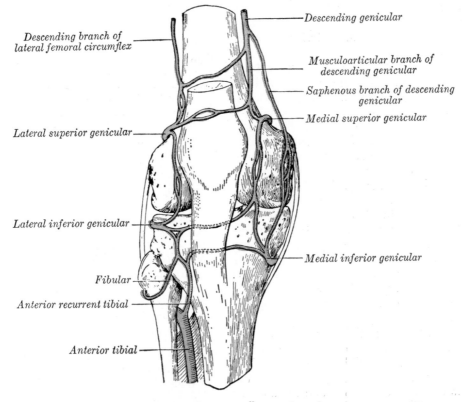

FIG. 8–62.—Circumpatellar anastomosis.

branch, *arising* opposite the back of the knee joint. It pierces the oblique popliteal ligament, and supplies the ligaments and synovial membrane in the interior of the articulation.

The **inferior genicular arteries** (*aa. genus inferiores; inferior articular arteries*) (Figs. 8–61, 8–62), two in number, *arise* from the popliteal under cover of the Gastrocnemius.

7. The **Medial Inferior Genicular** first descends along the proximal margin of the Popliteus, to which it gives branches; it then passes below the medial condyle of the tibia, deep to the tibial collateral ligament, at the anterior border of which it ascends to the anterior and medial side of the joint, to supply the proximal end of the tibia and the articulation of the knee, anastomosing with the lateral inferior and medial superior genicular arteries.

8. The **Lateral Inferior Genicular** runs lateralward above the head of the fibula to the front of the knee joint, passing deep to the lateral head of the Gastrocnemius, the fibular collateral ligament, and the tendon of the Biceps femoris. It ends by dividing into branches which anastomose with the medial inferior and lateral superior genicular arteries, and with the anterior recurrent tibial artery.

The **Anastomosis Around the Knee Joint** (Fig. 8–62).—Around the patella and on the contiguous ends of the femur and tibia, is an intricate network of vessels forming a superficial and a deep plexus. The **superficial plexus** is situated in the subcutaneous fascia around the patella, and forms three well-defined arches: one, above the proximal border of the patella, in the loose connective tissue over the Quadriceps femoris; the other two, distal to the level of the patella, are situated in the fat posterior to the ligamentum patellae. The **deep plexus**, which forms a close network of vessels, lies on the distal end of the femur and proximal end of the tibia around their articular surfaces, and sends numerous offsets into the interior of the joint. The arteries which form this plexus are the two medial and the two lateral genicular branches of the popliteal, the descending genicular, the descending branch of the lateral femoral circumflex, and the anterior recurrent tibial.

The Anterior Tibial Artery

The **anterior tibial artery** (*a. tibialis anterior*) (Fig. 8–63) commences at the bifurcation of the popliteal, at the distal border of the Popliteus, passes anteriorly between the two heads of the Tibialis posterior, and through the aperture above the proximal border of the interosseous membrane, to the deep part of the front of the leg: it here lies close to the medial side of the neck of the fibula. It then descends on the anterior surface of the interosseous membrane, gradually approaching the tibia; at the distal part of the leg it lies on this bone, and then anterior to the ankle joint, where it is more superficial, and becomes the **dorsalis pedis.**

Relations.—In the proximal two-thirds of its extent, the anterior tibial artery rests upon the interosseous membrane; in the distal third, upon the front of the tibia, and the anterior ligament of the ankle joint. In the proximal third of its course, it lies between the Tibialis anterior and Extensor digitorum longus; in the middle third between the Tibialis anterior and Extensor hallucis longus. At the ankle it is crossed from the lateral to the medial side by the tendon of the Extensor hallucis longus, lying between it and the first tendon of the Extensor digitorum longus. It is covered in the proximal two-thirds by the muscles which lie on either side of it, and by the deep fascia; in the distal third, by the integument and fascia, and the extensor retinaculum.

The anterior tibial artery is accompanied by a pair of venae comitantes which lie on either side of the artery; the deep peroneal nerve, coursing around the lateral side of the neck of the fibula, comes into relation with the lateral side of the artery shortly after it has reached the front of the leg; about the middle of the leg the nerve is anterior to the artery; at the lower part it is generally again on the lateral side.

Variations.—This vessel may be diminished in size, may be deficient to a greater or less extent or may be entirely wanting, its place being supplied by perforating branches from the posterior tibial, or by the perforating branch of the peroneal artery. The artery occasionally deviates toward the fibular side of the leg, regaining its usual position at the anterior part of the ankle. In rare instances the vessel has been found to approach the surface in the middle of the leg, being covered merely by the integument and fascia distal to that point.

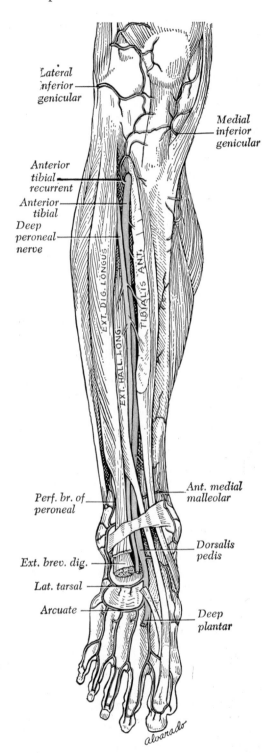

Fig. 8–63.—Anterior tibial and dorsalis pedis arteries.

Branches.—The branches of the anterior tibial artery are:

1. Posterior tibial recurrent
2. Fibular
3. Anterior tibial recurrent
4. Muscular
5. Anterior medial malleolar
6. Anterior lateral malleolar

1. The **Posterior Tibial Recurrent Artery** (*a. recurrens tibialis posterior*) an inconstant branch, is given off from the anterior tibial before that vessel passes through the interosseous space. It ascends anterior to the Popliteus, which it supplies, and anastomoses with the inferior genicular branches of the popliteal artery, giving an offset to the tibiofibular joint.

2. The **Fibular Artery** is sometimes derived from the anterior tibial, sometimes from the posterior tibial. It passes lateralward, around the neck of the fibula, through the Soleus, which it supplies, and ends in the substance of the Peroneus longus.

3. The **Anterior Tibial Recurrent Artery** (*a. recurrens tibialis anterior*) arises from the anterior tibial, as soon as that vessel has passed through the interosseous space; it ascends in the Tibialis anterior, ramifies on the front and sides of the knee joint, and assists in the formation of the patellar plexus by anastomosing with the genicular branches of the popliteal, and with the descending genicular artery.

4. The **Muscular Branches** (*rami musculares*) are numerous; they are distributed to the muscles which lie on either side of the vessel, some piercing the deep fascia to supply the integument, others passing through the interosseous membrane, and anastomosing with branches of the posterior tibial and peroneal arteries.

5. The **Anterior Medial Malleolar Artery** (*a. malleolaris anterior medialis; internal malleolar artery*) arises about 5 cm proximal to the ankle joint, and passes posterior to the tendons of the Extensor hallucis longus and Tibialis anterior, to the medial side of the ankle, upon which it ramifies, anastomosing with branches of the posterior tibial and medial plantar arteries and with the medial calcaneal from the posterior tibial.

6. The **Anterior Lateral Malleolar Artery** (*a. malleolaris anterior lateralis; external*

malleolar artery) passes deep to the tendons of the Extensor digitorum longus and Peroneus tertius and supplies the lateral side of the ankle, anastomosing with the perforating branch of the peroneal artery and with ascending twigs from the lateral tarsal artery.

The arteries around the ankle joint anastomose freely with one another and form networks below the corresponding malleoli. The **medial malleolar network** (*rete maleolare mediale*) is formed by the anterior medial malleolar branch of the anterior tibial, the medial tarsal branches of the dorsalis pedis, the posterior medial malleolar and medial calcaneal branches of the posterior tibial and branches from the medial plantar artery. The **lateral malleolar network** (*rete maleolare laterale*) is formed by the anterior lateral malleolar branch of the anterior tibial, the lateral tarsal branch of the dorsalis pedis, the perforating and the lateral calcaneal branches of the peroneal, and twigs from the lateral plantar artery. The anterior medial and anterior lateral malleolar arteries often branch from the dorsalis pedis.

The Dorsalis Pedis Artery

The **arteria dorsalis pedis** (*a. dorsalis pedis*) (Fig. 8–63), the continuation of the anterior tibial, passes anteriorly from the ankle joint along the tibial side of the dorsum of the foot to the proximal part of the first intermetatarsal space, where it divides into two branches, the **first dorsal metatarsal** and the **deep plantar**.

Relations.—This vessel rests upon the front of the articular capsule of the ankle joint, the talus, navicular, and intermediate cuneiform bones, and the ligaments connecting them, being covered by the integument, fascia and retinaculum, and crossed near its termination by the first tendon of the Extensor digitorum brevis. On its *tibial side* is the tendon of the Extensor hallucis longus; on its *fibular side*, the first tendon of the Extensor digitorum longus, and the termination of the deep peroneal nerve. It is accompanied by two veins.

Variations.—The dorsal artery of the foot may be larger than usual, to compensate for a deficient plantar artery; or its terminal branches to the toes may be absent, the toes then being supplied by the medial plantar; or ₁

its place may be taken by a large perforating branch of the peroneal artery, in 3 per cent of bodies. It frequently curves lateralward, lying lateral to the line between the middle of the ankle and the first interosseous space. In 12 per cent of bodies the dorsalis pedis is so small that it may be spoken of as absent. The arcuate artery is a vessel of significant size in only 50 per cent of bodies (Huber '41).

Branches.—The branches of the arteria dorsalis pedis are:

1. Lateral tarsal
2. Medial tarsal
3. Arcuate
4. First dorsal metatarsal
5. Deep plantar

1. The **Lateral Tarsal Artery** (*a. tarsea lateralis; tarsal artery*) *arises* from the dorsalis pedis as that vessel crosses the navicular bone; it passes laterally in an arched direction, lying upon the tarsal bones, and covered by the Extensor digitorum brevis; it supplies this muscle and the articulations of the tarsus, and anastomoses with branches of the arcuate, anterior lateral malleolar and lateral plantar arteries, and with the perforating branch of the peroneal artery.

2. The **Medial Tarsal Arteries** (*aa. tarseae mediales*) are two or three small branches which ramify on the medial border of the foot and join the medial malleolar network.

3. The **Arcuate Artery** (*a. arcuata; metatarsal artery*) *arises* a little anterior to the lateral tarsal artery; it passes lateralward, over the bases of the metatarsal bones, deep to the tendons of the Extensor digitorum brevis, its direction being influenced by its point of origin; and it anastomoses with the lateral tarsal and lateral plantar arteries. This vessel gives off the **second, third,** and **fourth dorsal metatarsal arteries,** which run distally upon the corresponding Interossei dorsales; in the clefts between the toes, each divides into two **dorsal digital branches** for the adjoining toes. At the proximal parts of the interosseous spaces these vessels receive the posterior perforating branches from the plantar arch, and at the distal parts of the spaces they are joined by the anterior perforating branches from the plantar metatarsal arteries. The fourth dorsal meta-

tarsal artery gives off a branch which supplies the lateral side of the fifth toe.

4. The **First Dorsal Metatarsal Artery** (*a. dorsalis hallucis*) runs distally on the first Interosseous dorsalis, and at the cleft between the first and second toes divides into two branches, one of which passes deep to the tendon of the Extensor hallucis longus, and is distributed to the medial border of the great toe; the other bifurcates to supply the adjoining sides of the great and second toes.

5. The **Deep Plantar Artery** (*ramus plantaris profundus; communicating artery*) penetrates into the sole of the foot, between the two heads of the first Interosseous dorsalis, and unites with the termination of the lateral plantar artery, to complete the plantar arch. It sends a branch along the medial side of the great toe, and continues distally along the first interosseous space as the **first plantar metatarsal artery**, which bifurcates for the supply of the adjacent sides of the first and second toes.

The Posterior Tibial Artery

The **posterior tibial artery** (*arteria tibialis posterior*) (Fig. 8–61) begins at the distal border of the Popliteus, opposite the interval between the tibia and fibula. It descends obliquely and as it approaches the tibial side of the leg, lying posterior to the tibia in the lower part of its course, it is situated midway between the medial malleolus and the medial process of the calcaneal tuberosity. Under cover of the origin of the Abductor hallucis it divides into the **medial** and **lateral plantar arteries**.

Relations.—The posterior tibial artery lies successively upon the Tibialis posterior, the Flexor digitorum longus, the tibia, and the posterior part of the ankle joint. It is covered by the deep transverse fascia of the leg, which separates it proximally from the Gastrocnemius and Soleus; at its termination it is covered by the Abductor hallucis. In the distal third of the leg, where it is more superficial, it is covered only by the integument and fascia, and runs parallel with the medial border of the tendo calcaneus. It is accompanied by two veins, and by the tibial nerve, which lies at first to the medial side of the artery, but soon crosses it posteriorly, and in the greater part of its course is on its lateral side.

Behind the medial malleolus, the tendons, blood vessels, and nerve are arranged, under cover of the retinaculum mm. flexorum, in the following order from the medial to the lateral side: the tendons of the Tibialis posterior and Flexor digitorum longus, lying in the same groove, posterior to the malleolus, the former being the more medial. Next is the posterior tibial artery with a vein on either side of it; and lateral to the vessels is the tibial nerve; about 1.25 cm nearer the heel is the tendon of the Flexor hallucis longus.

Variations.—The posterior tibial is not infrequently smaller than usual, or absent, its place being supplied by a large peroneal artery, which either joins the small posterior tibial artery, or continues alone to the sole of the foot.

Branches.—The branches of the posterior tibial artery are:

1. Peroneal
2. Nutrient (tibial)
3. Muscular
4. Posterior medial malleolar
5. Communicating
6. Medial calcaneal
7. Medial plantar
8. Lateral plantar

1. The **Peroneal Artery** (*a. peronea*) (Fig. 8–61) is deeply seated on the fibular side of the back of the leg. It *arises* from the posterior tibial, about 2.5 cm distal to the lower border of the Popliteus, passes obliquely toward the fibula, and then descends along the medial side of that bone, contained in a fibrous canal between the Tibialis posterior and the Flexor hallucis longus, or in the substance of the latter muscle. It then runs posterior to the tibiofibular syndesmosis and divides into lateral calcaneal branches which ramify on the lateral and posterior surfaces of the calcaneus.

It is covered, in the proximal part of its course, by the Soleus and deep transverse fascia of the leg; distally, by the Flexor hallucis longus.

Variations.—The peroneal artery may arise 7 or 8 cm distal to the Popliteus, or from the posterior tibial high up, or even from the popliteal. Its size is more frequently increased than diminished; and then it either reinforces the posterior tibial by its junction with it, or altogether takes the place of the posterior tibial in the distal part of the leg and foot, the latter vessel existing only as a short muscular branch.

In those rare instances where the peroneal artery is smaller than usual, a branch from the posterior tibial supplies its place; and a branch from the anterior tibial compensates for the diminished perforating artery. In one specimen the peroneal artery was entirely wanting. The perforating branch is sometimes enlarged (3 per cent of cases), and takes the place of the dorsalis pedis artery. In 50 per cent of bodies, it anastomoses with a lateral branch of the anterior tibial, proximal to the anterior lateral malleolar.

Branches.—The branches of the peroneal artery are:

 a) Muscular
 b) Nutrient (fibular)
 c) Perforating
 d) Communicating
 e) Posterior lateral malleolar
 f) Lateral calcaneal

a) The **muscular branches** of the peroneal artery go to the Soleus, Tibialis posterior, Flexor hallucis longus, and Peronei.

b) The **nutrient artery** (*a. nutricia fibulae*) is directed distally into the fibula.

c) The **perforating branch** (*ramus perforans; anterior peroneal artery*) pierces the interosseous membrane, about 5 cm proximal to the lateral malleolus, to reach the anterior part of the leg, where it anastomoses with the anterior lateral malleolar; it then passes down anterior to the tibiofibular syndesmosis, gives branches to the tarsus, and anastomoses with the lateral tarsal.

d) The **communicating branch** (*ramus communicans*) is given off from the peroneal about 2.5 cm from its distal end, and joins the communicating branch of the posterior tibial.

e) The **posterior lateral malleolar branches** (*rami malleolares laterales*) are small branches which wind around the lateral malleolus and join the lateral malleolar network.

f) The **lateral calcaneal** (*rami calcanei*) are the terminal branches of the peroneal artery; they pass to the lateral side of the heel, and communicate with the lateral malleolar and, on the back of the heel, with the medial calcaneal arteries.

2. The **Nutrient Artery** (*a. nutricia tibiae*) of the tibia *arises* from the posterior tibial near its origin and after supplying a few muscular branches, enters the nutrient canal of the bone, which it traverses obliquely distalward. This is the largest nutrient artery of bone in the body.

3. The **Muscular Branches** of the posterior tibial are distributed to the Soleus and deep muscles along the back of the leg.

4. The **Posterior Medial Malleolar Artery** (*a. malleolaris posterior medialis; internal malleolar artery*) is a small branch which winds around the tibial malleolus and ends in the medial malleolar network.

5. The **Communicating Branch** (*ramus communicans*) runs transversely across the posterior aspect of the tibia, about 5 cm proximal to its distal end, deep to the Flexor hallucis longus, and joins the communicating branch of the peroneal.

6. The **Medial Calcaneal** (*rami calcanei mediales; internal calcaneal*) are several large arteries which *arise* from the posterior tibial just before its division; they pierce the flexor retinaculum and are distributed to the fat and integument posterior to the tendo calcaneus and about the heel, and to the muscles on the tibial side of the sole, anastomosing with the peroneal and medial malleolar and, on the back of the heel, with the lateral calcaneal arteries.

7. The **Medial Plantar Artery** (*a. plantaris medialis; internal plantar artery*) (Figs. 8–64, 8–65), the smaller of the two terminal branches of the posterior tibial, passes anteriorly along the medial side of the foot. It is at first situated deep to the Abductor hallucis, and then between it and the Flexor digitorum brevis, both of which it supplies. At the base of the first metatarsal bone, where it is much diminished in size, it passes along the medial border of the first toe, anastomosing with the first dorsal metatarsal artery. Small superficial digital branches accompany the digital branches of the medial plantar nerve and join the plantar metatarsal arteries of the first three spaces.

8. The **Lateral Plantar Artery** (*a. plantaris lateralis; external plantar artery*), much larger than the medial, passes first obliquely lateralward to the base of the fifth metatarsal bone, then curves medialward to the interval between the bases of the first and second metatarsal bones, forming the

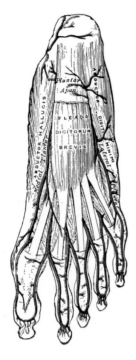

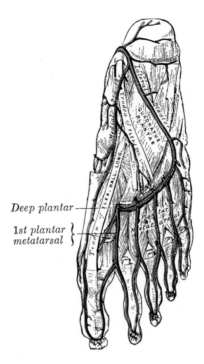

Deep plantar —

1st plantar
metatarsal

FIG. 8–64.—The plantar arteries. Superficial view.   FIG. 8–65.—The plantar arteries. Deep view.

plantar arch. As this artery passes lateralward, it is first placed between the calcaneus and Abductor hallucis, and then between the Flexor digitorum brevis and Quadratus plantae. As it runs distally along the base of the little toe it lies more superficially between the Flexor digitorum brevis and Abductor digiti minimi, covered by the plantar aponeurosis and integument. The remaining portion of the vessel is deeply situated; it extends from the base of the fifth metatarsal bone to the proximal part of the first interosseous space, where it unites with the deep plantar branch of the dorsalis pedis artery, thus completing the **plantar arch.**

Branches.—The plantar arch, besides distributing numerous branches to the muscles, integument, and fasciae in the sole, gives off the following branches:

a) Perforating
b) Plantar metatarsal arteries

a) The **perforating branches** (*rami perforantes*) are *three* in number; they penetrate through the proximal parts of the second, third, and fourth interosseous spaces, between the heads of the Interossei dorsales, and anastomose with the dorsal metatarsal arteries.

b) The **plantar metatarsal arteries** (*aa. metatarseae plantares; digital branches*) are *four* in number, and run distally between the metatarsal bones and in contact with the Interossei. Each divides into a pair of plantar digital arteries which supply the adjacent sides of the toes. Near their points of division each gives off an **anterior perforating branch** to join the corresponding dorsal metatarsal artery. The **first plantar metatarsal artery** (*arteria princeps hallucis*) springs from the junction between the lateral plantar and deep plantar arteries and sends a digital branch to the medial side of the great toe. The digital branch for the lateral side of the little toe *arises* from the lateral plantar artery near the base of the fifth metatarsal bone.

Histology.—The histology of arteries, veins, and blood is described at the end of Chapter 9 on Veins.

REFERENCES

ARTERIES

CLARA, M. 1956. *Die arterio-venosen Anastomosen, Anatomie, Biologie, Pathologie.* 2nd edition. vii + 315 pages, illustrated. Springer-Verlag, Wien.

DeBAKEY, M. E., E. S. CRAWFORD, O. CREECH, JR., and D. A. COOLEY. 1957. Arterial homografts for peripheral arteriosclerotic occlusive disease. Circulation, *15*, 21-30.

GOSS, C. M. 1961. On anatomy of veins and arteries by Galen of Pergamon. Anat. Rec., *141*, 355-366.

JONES, R. J. 1963. *Evolution of the Atherosclerotic Plaque.* xiii + 360 pages, illustrated. The University of Chicago Press, Chicago.

McDONALD, D. A. 1960. *Blood Flow in Arteries.* xi + 328 pages, illustrated. The Williams & Wilkins Company, Baltimore.

ORBINSON, J. L. and D. E. SMITH. 1963. *The Peripheral Blood Vessels.* xii + 357 pages, illustrated. The Williams & Wilkins Company, Baltimore.

POYNTER, C. M. 1923. Congenital anomalies of the arteries and veins of the human body with bibliography. Nebraska Univ. Studies, 1923.

QUIRING, D. P. 1949. *Collateral Circulation (Anatomical Aspects).* 142 pages. Lea & Febiger, Philadelphia.

SODICOFF, M. and R. T. BINHAMMER. 1964. The time of origin of the parabiotic anastomosis. Anat. Rec., *148*, 625-629.

EMBRYOLOGY

ALLAN, F. D. 1961. An histological study of the nerves associated with the ductus arteriosus. Anat. Rec., *139*, 531-537.

AUËR, J. 1948. The development of the human pulmonary vein and its major variations. Anat. Rec., *101*, 581-594.

CONGDON, E. D. 1922. Transformation of the aortic arch system during the development of the human embryo. Carneg. Instn., Contr. Embryol., *14*, 47-110.

DE VRIESE, B. 1902. Recherches sur l'évolution des vaisseaux sanguins des membres chez l'homme. Arch. de Biol., *18*, 665-730.

GOLDSMITH, J. B. and H. W. BUTLER. 1937. The development of the cardiac-coronary circulatory system. Amer. J. Anat., *60*, 185-202.

MOFFAT, D. B. 1961. The development of the posterior cerebral artery. J. Anat. (Lond.), *95*, 485-494.

MORRIS, E. D. and D. B. MOFFAT. 1956. Abnormal origin of the basilar artery from the cervical part of the internal carotid and its embryological significance. Anat. Rec., *125*, 701-712.

NOBACK, G. J. and I. REHMAN. 1941. The ductus arteriosus in the human fetus and newborn infant. Anat. Rec., *81*, 505-527.

PADGET, D. H. 1954. Designation of the embryonic intersegmental arteries in reference to the vertebral artery and subclavian stem. Anat. Rec., *119*, 349-356.

SENIOR, H. D. 1919. The development of the arteries of the human lower extremity. Amer. J. Anat., *25*, 55-95.

CORONARY ARTERIES

CHANDER, S. and I. JIT. 1957. Single coronary artery. J. anat. Soc. India, 6, 116-118.

CHASE, R. E. and C. F. DE GARIS. 1939. Arteriae coronariae (cordis) in the higher primates. Amer. J. phys. Anthrop., *24*, 427-448.

CHINN, J. and M. A. CHINN. 1961. Report of an accessory coronary artery arising from the pulmonary artery. Anat. Rec., *139*, 23-28.

GREGG, D. E. 1950. *Coronary Circulation in Health and Disease.* 227 pages. Lea & Febiger, Philadelphia.

GROSS, L. and M. A. KUGEL. 1933. The arterial blood vascular distribution to the left and right ventricles of the human heart. Amer. Heart J., *9*, 165-177.

HALPERN, M. H. and M. M. MAY. 1958. Phylogenetic study of the extracardiac arteries to the heart. Amer. J. Anat., *102*, 469-480.

JAMES, T. N. 1961. *Anatomy of the Coronary Arteries.* xi + 2111 pages, illustrated. Paul B. Hoeber, Inc., New York.

JAMES, T. N. and G. E. BURCH. 1958. The atrial coronary arteries in man. Circulation, *17*, 90-98.

SMOL'YANNIKOV, A. V. and T. A. NADDACHINA. 1964. Anomalies of coronary arteries of the heart. Fed. Proc. Translation supplement, *23*, T679-T685.

TRUEX, R. C., F. G. NOLAN, R. C. TRUEX, JR., H. P. SCHNEIDER, and H. I. PERLMUTTER. 1961. Anatomy and pathology of the whale heart with special reference to the coronary circulation. Anat. Rec., *141*, 325-353.

AORTA AND AORTIC ARCHES

BARRY, A. 1951. The aortic arch derivatives in the human adult. Anat. Rec., *111*, 221-238.

BLALOCK, A. 1946. Operative closure of patent ductus arteriosus. Surg. Gynec. Obstet., *82*, 113-114.

DE GARIS, C. F. 1941. The aortic arch in primates. Amer. J. phys. Anthrop., *28*, 41-74.

FOX, M. H. and C. M. GOSS. 1958. Experimentally produced malformations of the heart and great vessels in rat fetuses. Transposition complexes and aortic arch abnormalities. Amer. J. Anat., *102*, 65-92.

GROSS, R. E. and P. F. WARE. 1946. The surgical significance of aortic arch anomalies. Surg. Gynec. Obstet., *83*, 435-448.

McDONALD, J. J. and B. J. ANSON. 1940. Variations in the origin of arteries derived from the aortic arch, in American whites and negroes. Amer. J. phys. Anthrop., *27*, 91-108.

POYNTER, C. W. M. 1916. Arterial anomalies pertaining to the aortic arches and the branches arising from them. Nebraska Univ. Studies, *16*, 229-345.

SHANER, R. F. 1956. The persisting right sixth aortic arch of mammals, with a note on fetal coarctation. Anat. Rec., *125*, 171-184.

SINCLAIR, J. G. and N. D. SCHOFIELD. 1944. Anomalies of the cardio-pulmonary circuit compensated without a ductus arteriosus. Anat. Rec., *90*, 209-216.

WOODBURNE, R. T. 1951. A case of right aortic arch and associated venous anomalies. Anat. Rec., *111*, 617-628.

GROSS ANATOMY AND ANOMALIES OF
HEAD AND NECK

ADAMS, W. E. 1955. The carotid sinus complex, "parathyroid" III and thymo-parathyroid bodies, with special reference to the Australian opossum, Trichosurus vulpecula. Amer. J. Anat., *97*, 1-58.

ADAMS, W. E. 1957. On the possible homologies of the occipital artery in mammals, with some remarks on the phylogeny and certain anomalies of the subclavian and carotid arteries. Acta Anat. (Basel), *29*, 90-113.

ALLAN, F. D. 1952. An accessory or superficial inferior thyroid artery in a full term infant. Anat. Rec., *112*, 539-542.

BALDWIN, B. A. 1964. The anatomy of the arterial supply to the cranial regions of the sheep and ox. Amer. J. Anat., *115*, 101-117.

BAUMEL, J. J. and D. Y. BEARD. 1961. The accessory meningeal artery of man. J. Anat. (Lond.), *95*, 386-402.

BLAIR, C. B., JR., K. NANDY, and G. H. BOURNE. 1962. Vascular anomalies of the face and neck. Anat. Rec., *144*, 251-257.

DE LA TORRE, E. and M. G. NETSKY. 1960. Study of persistent primitive maxillary artery in human fetus: Some homologies of cranial arteries in man and dog. Amer. J. Anat., *106*, 185-195.

EVANS, T. H. 1956. Carotid canal anomaly: Other instances of absent internal carotid artery. Med. Times, *84*, 1069-1072.

KIRCHNER, J., E. YANAGISAWA, and E. S. CRELIN. 1961. Surgical anatomy of the ethmoidal arteries. Arch. Otolaryng., *74*, 382-386.

LEE, I. N. 1955. Anomalous relationship of the inferior thyroid artery. Anat. Rec., *122*, 499-506.

McCULLOUGH, A. W. 1962. Some anomalies of the cerebral arterial circle (of Willis) and related vessels. Anat. Rec., *142*, 537-543.

REED, A. F. 1943. The relations of the inferior laryngeal nerve to the inferior thyroid artery. Anat. Rec., *85*, 17-24.

SCHARRER, E. 1940. Arteries and veins in the mammalian brain. Anat. Rec., *78*, 173-196.

SEYDEL, H. G. 1964. The diameters of the cerebral arteries of the human fetus. Anat. Rec., *150*, 79-88.

SHANKLIN, W. M. and N. A. AZZAM. 1963. A study of valves in the arteries of the rodent brain. Anat. Rec., *147*, 407-413.

ARTERIES OF THORAX

ALEXANDER, W. F. 1946. The course and incidence of the lateral costal branch of the internal mammary artery. Anat. Rec., *94*, 446.

CHASE, R. E. and C. F. DE GARIS. 1939. Arteriae coronariae (cordis) in the higher primates. Amer. J. phys. Anthrop., *24*, 427-448.

KROPP, B. N. 1951. The lateral costal branch of the internal mammary artery. J. thorac. Surg., *21*, 421-425.

O'RAHILLY, R., H. DEBSON, and T. S. KING. 1950. Subclavian origin of bronchial arteries. Anat. Rec., *108*, 227-238.

SCIACCA, A. and M. CONDORELLI, 1960. *Involution of the Ductus Arteriosus.* 52 pages, illustrated. S. Karger, Basel.

SHAPIRO, A. L. and G. L. ROBILLARD. 1950. The esophageal arteries. Their configurational anatomy and variations in relation to surgery. Ann. Surg., *131*, 171-185.

TOBIN, C. E. 1960. Some observations concerning the pulmonic vasa vasorum. Surg. Gynec. Obstet., *111*, 297-303.

ARTERIES OF THE UPPER LIMB

COLEMAN, S. S. and B. J. ANSON. 1961. Arterial patterns in the hand based upon a study of 650 specimens. Surg. Gynec. Obstet., *113*, 409-424.

DASELER, E. H. and B. J. ANSON. 1959. Surgical anatomy of the subclavian artery and its branches. Surg. Gynec. Obstet., *108*, 149-174.

EDWARDS, E. A. 1960. Organization of the small arteries of the hand and digits. Amer. J. Surg., *99*, 837-846.

HUELKE, D. F. 1959. Variation in the origins of the branches of the axillary artery. Anat. Rec., *135*, 33-41.

HUELKE, D. F. 1962. The dorsal scapular artery— A proposed term for the artery to the rhomboid muscles. Anat. Rec., *142*, 57-61.

MARKEE, J. E., J. WRAY, J. NORK, and F. McFALLS. 1961. A quantitative study of the vascular beds of the hand. J. Bone Jt. Surg., *43-A*, 1187-1196.

McCORMACK, L. J., E. W. CAULDWELL and B. J. ANSON. 1953. Brachial and antebrachial arterial patterns. A study of 750 extremities. Surg. Gynec. Obstet., *96*, 43-54.

MULLER, ERIK. 1903. Beitrage zur Morphologie des Gefasssystems. I. Die Arm Arterien des Menschen. Anat. Heften., *22*, 379-575.

O'RAHILLY, R., H. DEBSON, and T. S. KING. 1950. Subclavian origin of bronchial arteries. Anat. Rec., *108*, 227-238.

PICK, J. 1958. The innervation of the arteries in the upper limb of man. Anat. Rec., *130*, 103-124.

ROHLICH, K. 1934. Uber die Arteria transversa colli des Menschen. Anat. Anz., *79*, 37.

SLAGER, R. F. and K. P. KLASSEN. 1958. Anomalous right subclavian artery arising distal to a coarctation of the aorta. Ann. Surg., *147*, 93-97.

WEATHERSBY, H. T. 1956. Unusual variation of the ulnar artery. Anat. Rec., *124*, 245-248.

ARTERIES OF THE ABDOMEN

BASMAJIAN, J. V. 1954. The marginal anastomoses of the arteries to the large intestine. Surg. Gynec. Obstet., *99*, 614-616.

BAYLIN, G. J. 1939. Collateral circulation following an obstruction of the abdominal aorta. Anat. Rec. 75, 405-408.

BENTON, R. S. and W. B. COTTER. 1963. A hitherto undocumented variation of the inferior mesenteric artery in man. Anat. Rec., 145, 171-173.

CAULDWELL, E. W. and B. J. ANSON. 1943. The visceral branches of the abdominal aorta: Topographical relationships. Amer. J. Anat., 73, 27-58.

CLAUSEN, H. J. 1955. An unusual variation in origin of the hepatic and splenic arteries. Anat. Rec., 123, 335-340.

CRELIN, E. S., JR. 1948. An unusual anomalous blood vessel connecting the renal and internal spermatic arteries. Anat. Rec., 102, 205-212.

EDWARDS, L. F. 1941. The retroduodenal artery. Anat. Rec., 81, 351-355.

HEALEY, J., P. SCHROY, and R. SORENSEN. 1953. The intrahepatic distribution of the hepatic artery in man. J. Internat. Coll. Surg., 20, 133.

LAUFMAN, H., R. E. BERGGREN, T. FINLEY, and B. J. ANSON. 1960. Anatomical studies of the lumbar arteries: with reference to the safety of translumbar aortography. Ann. Surg., 152, 621-634.

MICHALS, N. A. 1953. Variational anatomy of the hepatic, cystic, and retroduodenal arteries. A.M.A. Arch Surg., 66, 20-32.

MICHELS, N. A. 1953. Collateral arterial pathways to the liver after ligation of the hepatic artery and removal of the celiac axis. Cancer, 6, 708-724.

MICHELS, N. A. 1955. *Blood Supply and Anatomy of the Upper Abdominal Organs, with a Descriptive Atlas.* xiv + 581 pages, illustrated. J. B. Lippincott Co., Philadelphia.

MICHELS, N. A. 1962. The anatomic variations of the arterial pancreaticoduodenal arcades: Their import in regional resection involving the gallbladder, bile ducts, liver, pancreas, and parts of the small and large intestines. J. Int. Coll. Surg., 37, 13-40.

MICHELS, N. A., P. SIDDHARTH, P. L. KORNBLITH, and W. W. PARKE. 1965. The variant blood supply to the descending colon, rectosigmoid and rectum based on 400 dissections. Its importance in regional resections: a reivew of medical literature. Dis. Colon, Rectum, 8, 251-278.

MILLOY, F. J., B. J. ANSON, and DAVID K. McAFFEE. 1960. The rectus abdominis muscle and the epigastric arteries. Surg. Gynec. Obstet., 110, 293-302.

TOBIN, C. E. 1952. The branchial arteries aud their connections with other vessels in the human lung. Surg. Gynec. Obstet., 95, 741-750.

WOODBURNE, R. T. and L. L. OLSEN. 1951. The arteries of the pancreas. Anat. Rec., 111, 255-270.

ARTERIES OF LOWER LIMB

HUBER, J. F. 1941. The arterial network supplying the dorsum of the foot. Anat. Rec., 80, 373-391.

KEEN, J. A. 1961. A study of the arterial variations in the limbs, with special reference to symmetry of vascular patterns. Amer. J. Anat., 108, 245-261.

MANSFIELD, A. O. and J. M. HOWARD. 1964. Absence of both common iliac arteries. A case report. Anat. Rec., 150, 363-381.

PICK, J. W., B. J. ANSON, and F. L. ASHLEY. 1942. The origin of the obturator artery. Amer. J. Anat., 70, 317-343.

SANDERS, R. J. 1963. Relationships of the common femoral artery. Anat. Rec., 145, 169-170.

VANN, H. M. 1943. A note on the formation of the plantar arterial arch of the human foot. Anat. Rec., 85, 269-275.

WEATHERSBY, H. T. 1955. The artery of the index finger. Anat. Rec., 122, 57-64.

WILLIAMS, C. D., C. H. and L. R. MARTIN. 1934. Origin of the deep and circumflex femoral group of arteries. Anat. Rec., 60, 189.

9 | *The Veins*

eins convey the blood from the capillaries to the heart. There are two distinct systems, the **pulmonary** and the **systemic**, corresponding to the two circulatory cycles which the blood must make, away from and back to the heart.

The **Pulmonary Veins** (*venae pulmonales*) belong to the lesser or pulmonary circulation and return the oxygenated blood from the capillaries of the lungs to the left atrium of the heart.

The **Systemic Veins** (*venae systemates*) belong to the greater or systemic circulation and convey the blood from the remainder of the body to the right atrium of the heart. There are two subdivisions of the systemic veins, one composed of the coronary and caval tributaries, the other comprising the portal vein and its tributaries.

The **Portal Vein** (*vena portae*) is interposed between certain abdominal viscera and the liver. Its special feature is that it both begins and ends in capillaries, *i.e.*, it receives the blood from the capillaries of the digestive tube, spleen, and pancreas and delivers it to the capillary-like sinusoids of the liver whence it makes its way to the general systemic system through the hepatic tributaries of the inferior vena cava.

DEVELOPMENT OF THE VEINS

The earliest stages of development of the veins are described more completely in the chapter on embryology and the chapter on the heart.

The **Parietal Veins.**—The first veins to appear in the body of the embryo are two short transverse veins, the **common cardinal veins** (*ducts of Cuvier*) which open into the right and left horns of the sinus venosus. Each of these veins receives a vein from the cranial and caudal portions of the embryo (Fig. 9–1). The cranial tributaries, called the **anterior cardinal veins** (*precardinal veins*), return blood from the head and soon become the primitive jugular veins (Fig. 9–2). The caudal tributaries, called the **posterior cardinal veins** (*postcardinal veins*), return blood from the parietes of the trunk, from the mesonephroi (Wolffian bodies), and from the lower limb buds. This primitive double (right and left) plan of veins is converted into the single caval systems by the formation of a series of transverse connecting veins.

Inferior Vena Cava.—The blood from the lower limbs is collected by the right and left iliac veins which, in the earlier stages of development, open into the corresponding right and left posterior cardinal veins. Later, a transverse connection, the **left common iliac vein,** is developed between the caudal parts of the two posterior cardinals (Fig. 9–4) and through this the blood is carried into the right posterior cardinal vein. The portion of the left posterior cardinal, caudal to the left renal vein atrophies and disappears up to the point of entrance of the left testicular vein. The portion cranial to the left renal vein persists as the **hemiazygos, accessory hemiazygos,** and caudalward portion of the **highest left intercostal vein.** The right posterior cardinal vein forms a large venous trunk, lying along the dorsal abdominal wall, which forms the caudal part of the inferior vena cava up to the level of the renal veins. Cranial to the level of the renal veins the right posterior cardinal vein remains close to the vertebral

(673)

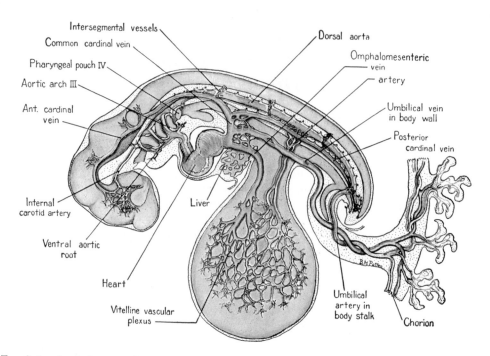

Fig. 9–1.—Semischematic diagram to show basic vascular plan of human embryo at the end of first month. For the sake of simplicity the paired vessels are shown only on side toward observer. (Patten, *Human Embryology*, 1953, courtesy of The Blakiston Div., McGraw-Hill Book Co.)

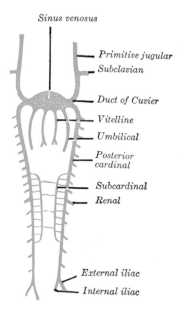

Fig. 9–2.—Scheme of arrangement of parietal veins.

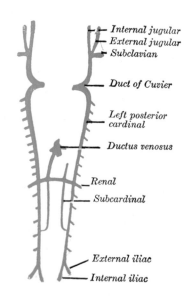

Fig. 9–3.—Scheme showing early stages of development of the inferior vena cava.

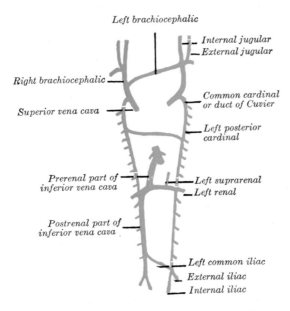

FIG. 9–4.—Diagram showing development of main cross branches between jugulars and between cardinals.

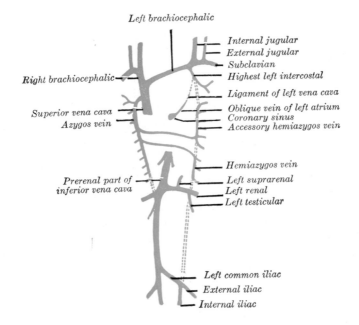

FIG. 9–5.—Diagram showing completion of development of the parietal veins.

column and persists as the **azygos vein.** The hemiazygos vein becomes connected with the azygos by the formation of a transverse branch ventral to the vertebral column (Figs. 9–4, 9–5).

The further development of the inferior vena cava is associated with the formation of two additional veins, the **subcardinal veins** (Figs. 9–2, 9–3) which are parallel with and ventral to the posterior cardinal veins. They arise as longitudinal anastomosing channels which link up the tributaries from the mesentery to the posterior cardinal veins. The two subcardinals for a time are connected by several cross branches lying ventral to the aorta, but only one of these transverse channels persists. At the same level a cross connection is established between the posterior cardinals and subcardinals of the two sides (Fig. 9–3). The right subcardinal caudal to this cross connection disappears, whereas the part cranial to it, the prerenal portion, forms a connection with the ductus venosus at the point of opening of the hepatic veins. Another pair of longitudinal veins is formed at a later period, dorsal to the posterior cardinal veins. These are the **supracardinals** which gradually take over the drainage of blood from the region caudal to the renal veins. The prerenal part of the subcardinal communicates with the supracardinal and enlarges rapidly, thus forming a single trunk, the **inferior vena cava** (Fig. 9–4) which consists, therefore, of the proximal part of the ductus venosus, the prerenal part of the right subcardinal vein, the postrenal part of the supracardinal vein, and the cross branch between these two veins.

The left subcardinal vein disappears except for the part immediately cranial to the renal vein which is retained as the **left suprarenal vein.** The testicular or ovarian vein opens into the postrenal part of the corresponding posterior cardinal vein. This portion of the right posterior cardinal, as already explained, forms the caudal part of the inferior vena cava, so that the **right testicular vein** opens directly into the inferior cava. The postrenal segment of the left posterior cardinal disappears, with the exception of the portion between the testicular and renal vein, so that the **left testicular** vein drains into the left renal vein (Fig. 9–5).

Superior Vena Cava.—The anterior cardinal veins receiving the blood from the head and brain through the primitive jugular veins and from the limb buds through the subclavian veins increase rapidly with the growth of the embryo. As described above, the posterior cardinal veins lose much of their importance and become the azygos veins, leaving anterior cardinals as the chief source of blood for the common cardinals. The latter gradually assume an almost vertical course when the heart is brought down into the thorax. The common cardinals are originally of the same diameter and are frequently termed the **right** and **left superior venae cavae.** Later a transverse connecting vein, the **left brachiocephalic** or **innominate vein,** conveys the blood from the left side into the right superior vena cava (Figs. 9–4, 9–5). The right anterior cardinal between the azygos vein and the union with the left brachiocephalic becomes the upper part of the **adult superior vena cava.** The lower part of the superior cava, that is, the part between the azygos vein and the heart, is formed by the right common cardinal vein. The left anterior cardinal vein caudal to the transverse left brachiocephalic and the common cardinal of that side regress, the brachiocephalic part becoming the **vestigial fold of Marshall** and the common cardinal the **oblique vein of the left atrium** (*oblique vein of Marshall*) (Fig. 9–5). Both right and left superior venae cavae are present in some animals, and occasionally are found in adult human beings. The oblique vein of the left atrium drains across the dorsal surface of the left atrium to open into the **coronary sinus** which represents the persistent left horn of the sinus venosus.

The Visceral Veins.—The visceral veins are the two **vitelline** or **omphalomesenteric veins** bringing the blood from the yolk sac, and the two **umbilical veins** returning the blood from the placenta; these four veins open close together into the sinus venosus.

The **Vitelline Veins** run cranialward at first ventral to, and subsequently on either side of, the intestinal canal. They unite on the ventral aspect of the canal, and beyond

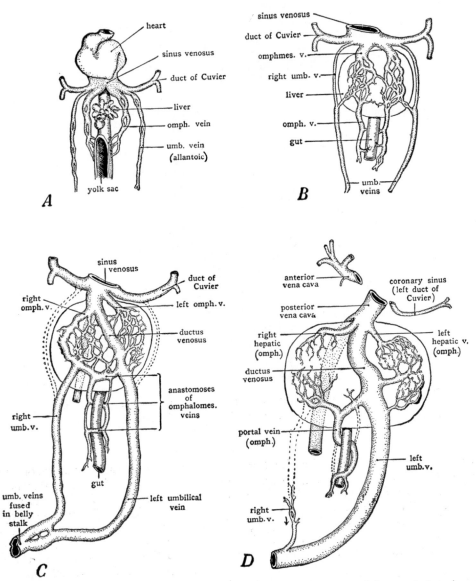

FIG. 9–6.—Diagrams showing development of portal circulation from omphalomesenteric veins, and changes by which blood returning from placenta by way of umbilical veins is rerouted through liver. A, Based on conditions in pig embryos of 3 to 4 mm—applicable to human embryos of fourth week. B, Based on pig embryos of about 6 mm—applicable to human embryos of fifth week. C, Based on pig embryos of 8 to 9 mm—applicable to human embryos early in sixth week. D, Based on pig embryos of 20 mm and above—applicable to human embryos of seven weeks and older. (Patten, *Human Embryology,* 2nd Ed. 1953, courtesy of The Blakiston Div., McGraw-Hill Book Co.)

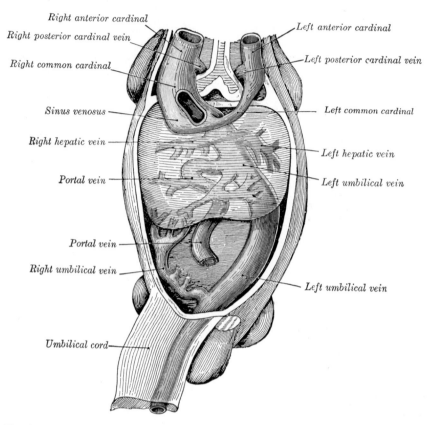

Right anterior cardinal

Right posterior cardinal vein

Right common cardinal

Sinus venosus

Right hepatic vein

Portal vein

Portal vein

Right umbilical vein

Umbilical cord

Left anterior cardinal

Left posterior cardinal vein

Left common cardinal

Left hepatic vein

Left umbilical vein

Left umbilical vein

FIG. 9–7.—Human embryo with heart and anterior body wall removed to show the sinus venosus and its tributaries. (After His.)

this are connected to each other by two anastomotic branches, one on the dorsal, and the other on the ventral aspect of the duodenal portion of the intestine, which is thus encircled by two venous rings (Fig. 9–6); into the middle or dorsal anastomosis the superior mesenteric vein opens. The portions of the veins above the upper ring become interrupted by the developing liver and are broken up by it into a plexus of small capillary-like vessels termed **sinusoids** (Minot). The branches conveying the blood to this plexus are named the **venae advehentes,** and become the branches of the portal vein; while the vessels draining the plexus into the sinus venosus are termed the **venae revehentes,** and form the future hepatic veins (Figs. 9–6, 9–7). Ultimately the left vena revehens no longer communicates directly with the sinus venosus, but opens into the right vena revehens. The

persistent part of the upper venous ring, above the opening of the superior mesenteric vein, forms the trunk of the portal vein.

The two **Umbilical Veins** fuse early to form a single trunk in the body stalk, but remain separate within the embryo and pass forward to the sinus venosus in the side walls of the body. Like the vitelline veins, their direct connection with the sinus venosus becomes interrupted by the developing liver, and thus at this stage all the blood from the yolk sac and placenta passes through the substance of the liver before it reaches the heart. The right umbilical and right vitelline veins shrivel and disappear; the left umbilical, on the other hand, becomes enlarged and opens into the upper venous ring of the vitelline veins. With the atrophy of the yolk sac the left vitelline vein also undergoes atrophy and disappears. Finally a direct branch is established between this ring and the right hepatic vein; this branch is named the **ductus venosus**, and, enlarging rapidly, it forms a wide channel through which most of the blood, returned from the placenta, is carried directly to the heart without passing through the liver (Fig. 7–16). A small proportion of the blood from the placenta, however, is conveyed from the left umbilical vein to the liver through the left vena advehens. The left umbilical vein and the ductus venosus undergo atrophy and obliteration after birth, and form respectively the **ligamentum teres** and **ligamentum venosum** of the liver.

THE PULMONARY VEINS

The **pulmonary veins** (*venae pulmonales*) (Fig. 9–8), two from each lung, return the oxygenated blood to the left atrium of the heart. The capillary networks in the walls of the air sacs join together to form one vessel for each lobule. These vessels unite successively to form a single trunk for each lobe, three for the right and two for the left lung. The vein from the middle lobe of the right lung usually unites with that from the superior lobe so that ultimately two trunks are formed from each lung. They perforate the fibrous layer of the pericardium and open separately into the cranial part of the left atrium. The three veins of the right lung may remain separate and not infrequently the two left veins end by a common opening.

At the roots of the lungs, the superior pulmonary veins lie caudal and ventral to the pulmonary arteries. The inferior veins are situated in the most caudal part of the hilum on a plane dorsal to the superior veins (Figs. 15–30, 15–31). They both perforate the fibrous layer of the pericardium after a very short course and open separately into the cranial part of the left atrium (Fig. 9–9). Within the pericardial sac, only the ventral surfaces are invested by the serous membrane. The right pulmonary veins pass dorsal to the right atrium and superior vena cava, the left ventral to the descending thoracic aorta.

Within the lung the ramifications of the veins do not accompany the bronchi and arteries in the central regions of the seg-

ments. Instead they have two usual positions, superficial near the pleura or between the bronchopulmonary segments and deeper or intrasegmental. Their branchings are quite variable in both the larger and the smaller tributaries. The bronchopulmonary segments are described in Chapter 15 (Boyden '55).

The **right superior pulmonary vein** (*v. pulmonis superior dextra*) has three tributaries for the superior lobe and one for the middle lobe.

The **apical segmental vein** (*ramus apicalis*) lies near the mediastinal pleura. It has a tributary draining the apex and a tributary separating the apical from the anterior segments.

The **posterior segmental vein** (*r. posterior*) is the largest vein of the superior lobe and has an interlobar and a central division. The central division has an apical, a posterior, and an intersegmental tributary which is a prominent vein of the posterior surface. The interlobar division lies near the pleura between the superior and middle lobes.

The **anterior segmental vein** (*r. anterior*) has two mediastinal tributaries, superior and inferior which straddle the anterior ramus of the anterior bronchus of the right superior lobe.

The **right middle lobe vein** (*v. lobi medii*) is usually a single tributary of the right superior pulmonary vein but its two segmental veins may remain separate. The lateral segmental vein splits into a tributary for the lateral segment and a tributary separating the segments. The medial segmental vein lies near the mediastinal surface and has two tributaries, one subpleural and one intrasegmental.

The **right inferior pulmonary vein** (*v. pulmonis inferior dextra*) occupies the most inferior part of the hilum and has two main tributaries, a superior segmental and a basal vein.

The **superior segmental vein** (*r. superior*) begins near the medial side of the middle lobe bronchus and is formed by two or three tributaries with intrasegmental and intersegmental tributaries.

The **common basal vein** (*r. basalis communis*) is a large short trunk draining the four basal segments. It has two principal tributaries, a superior basal and an inferior basal. The **superior basal** tributary (*r. basalis superior*) lies deep to the surface of the lobe and drains both the anterior basal and lateral basal segments by intrasegmental and intersegmental tributaries. The **inferior basal** tributary (*r. basalis inferior*) drains posterior basal and lateral basal segments. The medial basal segment is drained by several tributaries opening into both the superior basal and inferior basal veins.

The **left superior pulmonary vein** (*v. pulmonis superior sinistra*) has two divisions, superior and lingular, which correspond roughly to the right superior and middle lobe veins. The superior division receives the apicoposterior and anterior segmental veins, the lingular division receives superior and inferior segmental veins.

The **apicoposterior segmental vein** (*r. apicoposterior*) has two large tributaries, apical and posterior. The apical vein receives both intrasegmental from the apicoposterior segment and intersegmental tributaries from both segments.

The **posterior segmental vein** (*r. posterior*)

also receives intrasegmental and intersegmental tributaries.

The **lingular division vein** (*r. lingularis*) drains the superior lingular and inferior lingular segments. The superior has posterior and anterior intersegmental tributaries. The tributaries of the inferior lingular vein are superior and inferior. The latter may drain into the inferior pulmonary vein in some instances.

The **left inferior pulmonary vein** (*v. pulmonis inferior sinistra*) receives a smaller superior segmental vein and a larger common basal segmental vein. The **superior segmental vein** (*r. superior*) lies on the medial side of the superior lobe bronchus and has three tributaries, a central intrasegmental and two intersegmental tributaries separating the segment from its neighbors.

The **common basal vein** (*r. basalis communis*) receives the **superior basal vein** (*r. basalis superior*) which largely drains the anterior basal segment by intrasegmental and intersegmental tributaries and the **inferior basal vein** (*r. basalis inferior*) which

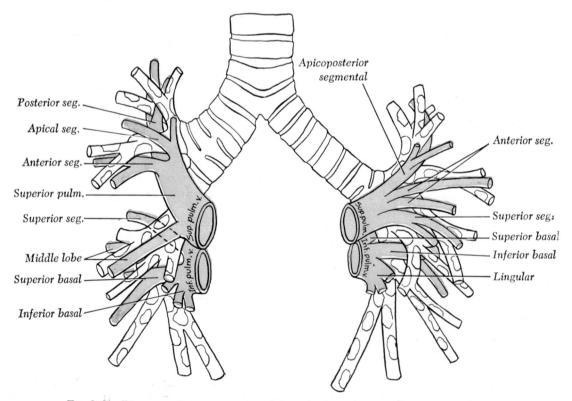

FIG. 9–8.—Diagrammatic representation of the pulmonary veins in relation to the bronchi. (Redrawn after *Nomina Anatomica*, 1968, Excerpta Medica Foundation.)

drains the posterior basal and lateral basal segments by intrasegmental and intersegmental tributaries. The veins of the medial basal segment are medial and lateral tributaries one or both of which may enter the common basal vein. The medial basal has two tributaries, one intrasegmental and one intersegmental separating the medial basal from the posterior basal segment. The anterior basal vein has an intrasegmental and an intersegmental tributary which usually lies deep to the pulmonary ligament and separates the medial basal from the lateral basal segments. The lateral basal vein is primarily an intersegmental vessel in the depth of the lobe between the lateral basal and posterior basal bronchi but also has an intrasegmental tributary. The posterior basal vein is intrasegmental, lying in the crotch between the branches of the posterior basal segmental bronchus and draining their regions by two tributaries.

THE SYSTEMIC VEINS

The **veins** commence by minute plexuses which receive the blood from the capillaries. The tributaries arising from these plexuses unite together into trunks, and these, in their passage toward the heart, constantly increase in size as they receive tributaries, or join other veins. The veins are larger in caliber and altogether more numerous than the arteries; hence, the entire capacity of the venous system is much greater than that of the arterial; the capacity of the pulmonary veins, however, only slightly exceeds that of the pulmonary arteries. The veins when full are cylindrical like the arteries, but their walls are thin and they collapse when the vessels are empty. In the cadaver they may be interrupted at intervals by nodular enlargements which indicate the existence of valves in their interior. They anastomose very freely with one another, especially in certain regions of the body; and these communications exist between the larger trunks as well as between the smaller branches. Thus, between the venous sinuses of the cranium, and between the veins of the neck, where obstruction would be attended with imminent danger to the cerebral venous system, large and frequent anastomoses are found. The same free communication exists between the veins

throughout the whole extent of the vertebral canal, and between the veins composing the various venous plexuses in the abdomen and pelvis.

The systemic venous channels are of three types according to location, superficial and deep veins, and venous sinuses.

The **Superficial Veins** (*cutaneous veins*) are found between the layers of the subcutaneous fascia immediately beneath the skin; they return the blood from deeper structures by perforating the deep fascia.

The **Deep Veins** accompany the arteries, and are usually enclosed in the same sheaths with those vessels. The larger arteries—such as the axillary, subclavian, popliteal, and femoral—usually have only one accompanying vein. With the smaller arteries—as the radial, ulnar, brachial, tibial, peroneal—they exist generally in pairs, one lying on each side of the vessel, and are called **venae comites** or **venae comitantes.** In certain organs of the body, however, the deep veins do not accompany the arteries; for instance, the veins in the skull and vertebral canal, the hepatic veins in the liver, and the larger veins returning blood from the bones.

Venous Sinuses, found in the interior of the skull, consist of canals formed by separation of the two layers of the dura mater; their outer coat consists of fibrous tissue, their inner of an endothelial layer continuous with the lining of the veins. The coronary sinus is the terminal part of the veins of the heart. The sinus venosus is part of the embryonic heart.

The systemic veins may be arranged into three groups: (1) The coronary system or veins of the heart. (2) The superior vena cava and its tributaries, or the veins of the upper limbs, head, neck, and thorax, and (3) The inferior vena cava and its tributaries, or the veins of the lower limbs, abdomen, and pelvis.

THE VEINS OF THE HEART

The **veins of the heart** (*venae cordis*), with the exception of certain small vessels opening directly into the chambers, are tributaries of the coronary sinus.

The **Coronary Sinus** (*sinus coronarius*) (Figs. 8–16, 9–9) is a wide venous channel about 2.25 cm in length situated in the posterior part of the coronary sulcus, and covered by muscular fibers from the left atrium. It ends in the right atrium between the opening of the inferior vena cava and the atrioventricular aperture, its orifice being guarded by a single incompetent semilunar valve, the **valve of the coronary sinus** (*valve of Thebesius*).

Tributaries.—Its tributaries are: (1) the great, (2) small, and (3) middle cardiac veins, (4) the posterior vein of the left ventricle, and (5) the oblique vein of the left atrium, all of which, except the last, are provided with valves at their orifices.

1. The **Great Cardiac Vein** (*v. cordis magna; left coronary vein*) (Fig. 8–15) begins at the apex of the heart and ascends along the anterior interventricular sulcus to the base of the ventricles. It then curves to the left in the coronary sulcus, and reaching the back of the heart, opens into the left extremity of the coronary sinus. It receives tributaries from the left atrium and from both ventricles: one, the **left marginal vein,** is of considerable size, and ascends along the left margin of the heart.

2. The **Small Cardiac Vein** (*v. cordis parva; right coronary vein*) (Fig. 8–16) runs in the coronary sulcus between the right atrium and ventricle, and opens into the right extremity of the coronary sinus. It receives blood from the back of the right atrium and ventricle; the **right marginal vein** ascends along the right margin of the heart and joins it in the coronary sulcus, or opens directly into the right atrium.

3. The **Middle Cardiac Vein** (*v. cordis media*) (Fig. 8–16) commences at the apex of the heart, ascends in the posterior interventricular sulcus, receives tributaries from both ventricles, and ends in the coronary sinus near its right extremity.

4. The **Posterior Vein of the Left Ventricle** (*v. posterior ventriculi sinistri*) (Fig. 8–16) runs on the diaphragmatic surface of the left ventricle, accompanying the circumflex branch of the left coronary artery, to the coronary sinus; it may end in the great cardiac vein.

5. The **Oblique Vein of the Left Atrium** (*v. obliqua atrii sinistri, oblique vein of Marshall*) (Fig. 9–9) is a small vessel which

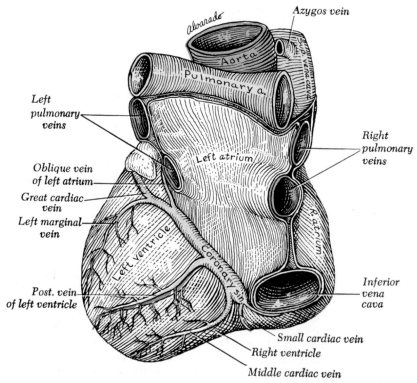

Fɪɢ. 9–9.—Base and diaphragmatic surface of heart.

descends obliquely on the back of the left atrium and ends in the coronary sinus near its left extremity; it is continuous cranialward with the **ligament of the left vena cava** (*lig. venae cavae sinistrae; vestigial fold of Marshall*), and the two structures form the remnant of the left common cardinal vein.

The following cardiac veins do not end in the coronary sinus:

6. The **Anterior Cardiac Veins** (Fig. 8–15), comprising three or four small vessels which collect blood from the ventral aspect of the right ventricle, open into the right atrium. The right marginal vein frequently opens into the right atrium, and is therefore sometimes regarded as belonging to this group.

7. The **Smallest Cardiac Veins** (*venae cordis minimae; veins of Thebesius*) consist of a number of minute veins which arise in the muscular wall of the heart; the majority open into the atria, but a few end in the ventricles.

THE SUPERIOR VENA CAVA

The **superior vena cava** (*vena cava superior*) (Figs. 7–19, 8–14, 9–27, 12–78) drains the blood from the cranial half of the body. In caliber, it is the second largest vein in the body, about 2 cm in diameter, but it is only some 7 cm in length. It is formed by the junction of the two brachiocephalic veins at the level of the first intercostal space close behind the sternum on the right side. From this junction it transports the blood caudalward, lying deep to the second rib cartilage and intercostal space at the right border of the sternum, and empties into the cranial part of the right atrium at the level of the third right costal cartilage. It makes a slight curve with the convexity to the right. The part of the vessel nearer

to the heart, amounting to about half its length, is within the pericardial sac and is covered by the serous pericardium (Fig. 7–19).

Relations.—Ventral to it are the anterior margins of the right lung and pleura with the pericardium intervening; these separate it from the first and second intercostal spaces and from the second and third right costal cartilages; dorsal to it are the root of the right lung and the right vagus nerve. On its right side are the phrenic nerve and right pleura; on its left side, the commencement of the brachiocephalic artery and the ascending aorta, the latter overlapping it. Just before it pierces the pericardium, it receives the azygos vein and several small veins from the pericardium and other contents of the mediastinum. The superior vena cava has no valves.

Variations.—Sometimes the brachiocephalic veins open separately into the right atrium; in such instances the right vein takes the ordinary course of the superior vena cava; the left vein— left superior vena cava, as it is then termed —which may communicate by a small branch with the right one, passes ventral to the root of the left lung, and, turning to the dorsum of the heart, ends in the right atrium. This occasional condition in the adult is due to the persistence of the early fetal condition, and is the normal state of things in birds and some mammals.

Tributaries.—In addition to its two large tributaries of origin, it has draining into it one large and one or two small veins which will be described with the veins of the thorax.

In describing the tributaries of the vena cava and other large veins, their function, that of transporting the blood toward the heart, will be emphasized by describing them on a regional basis. Thus, the brachiocephalic veins drain the blood (1) from the head and neck and (2) from the upper limb.

VEINS OF THE HEAD AND NECK

The **veins of the head and neck** may be divided into the following groups:

Veins of the Face

A. Superficial Veins of the Face
B. Deep Veins of the Face

Veins of the Cranium

C. Veins of the Brain
D. Sinuses of the Dura Mater
E. Diploic Veins
F. Emissary Veins

A. Superficial Veins of the Face

1. Facial
2. Superficial temporal
3. Posterior auricular
4. Occipital
5. Retromandibular

These veins are tributaries of the internal and external jugular veins.

1. The **Facial Vein** (*v. facialis, anterior facial vein*) (Figs. 9–10, 9–11, 9–19) drains the blood from the superficial structures of the face. Its first part, named the angular vein, begins at the union of the frontal and supraorbital veins. It accompanies the facial artery but has a less tortuous course. From the angular, it passes deep to the Zygomaticus major and minor, follows the border of the Masseter to the inferior border of the body of the mandible, and curves around the bone into the neck. In the neck it has a communication with the anterior jugular vein and with the external jugular before emptying into the internal jugular. In former terminology the communication with the external jugular was named the *posterior facial* and the segment emptying into the internal jugular was named the *common facial vein*. It has **anastomoses** with the cavernous sinus, (*a*) by way of the angular, supraorbital, and superior ophthalmic veins (Fig. 9–15) and (*b*) through its tributary the deep facial vein, pterygoid plexus, and inferior ophthalmic vein.

The facial vein has no valves. In cases of acne about the nose and mouth, the blood in the vein may become thrombosed. The infected thrombus can progress against the flow of blood, finally reaching the cavernous sinus through its anastomoses, causing meningitis.

Tributaries. (a) The **angular vein** (*vena angularis*), formed by the junction of the frontal and supraorbital veins, at the root of the nose receives the infraorbital, superior and inferior palpebral, and external nasal veins (Fig. 8–24).

i. The **frontal vein** (*v. frontalis*) begins in a plexus on the forehead and scalp which communicates with the frontal tributaries of the superficial temporal vein. It lies near the vein of the opposite side in its course to the root of the nose where the two veins communicate by a transverse communication before it joins the supraorbital. Occasionally the two veins join on the upper forehead to form a prominent single vein which bifurcates at the root of the nose to join the two angular veins.

ii. The **supraorbital vein** (*v. supraorbitalis*) also begins by anastomosing with a frontal tributary of the superficial temporal. It lies superficial to the Frontalis, which it drains, and joins the frontal at the medial angle of the orbit. As it passes over the supraorbital notch in the frontal bone it forms an anastomosis with the superior ophthalmic vein and receives the frontal diploic vein through a foramen at the bottom of the notch.

(*b*) The **deep facial vein** (*v. facialis profunda*) (Figs. 9–10, 9–11, 9–18) empties into the facial at the area where the Zygomaticus crosses the border of the Masseter. It begins as a large anastomosis with the pterygoid plexus (Fig. 9–11) and receives small tributaries from the Buccinator, Zygomaticus, Masseter, and other neighboring structures.

Other tributaries of the facial vein in its course across the face are external nasal (*vv. nasales externae*), superior labial (*v. labialis superior*), inferior labial (*v. labialis inferior*), parotid (*vv. parotideae*), and external palatine (*v. palatina externa*). In the neck it receives a submental (*v. submentalis*) and submandibular vein (*v. submandibularis*).

2. The **Superficial Temporal Vein** (*v. temporalis superficialis*) (Fig. 9–10) begins in the plexus on the vertex and side of the head where it anastomoses with the vein of the other side and the frontal, supraorbital, posterior auricular, and occipital veins. Frontal and parietal tributaries unite near the ear to form the trunk of the vein which crosses the posterior root of the zygomatic arch and then enters the substance of the parotid gland to unite with the maxillary vein to form the retromandibular vein.

Tributaries.—It is joined with the **middle temporal vein** (*v. temporalis media*) (Fig. 9–10) and receives the **transverse facial** (*v. facialis transversa*) and **anterior auricular** (*v. auricularis anterior*) veins.

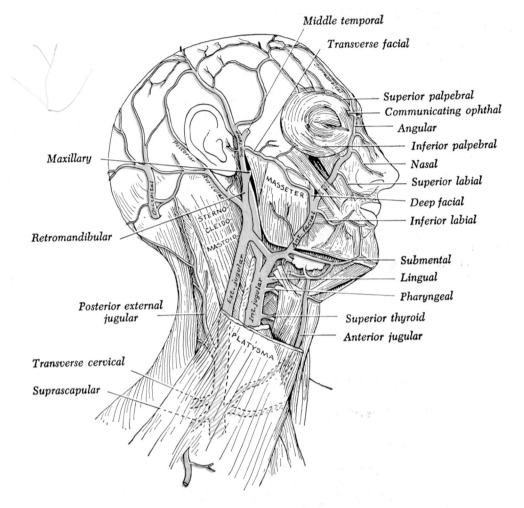

FIG. 9–10.—Superficial veins of the head and neck.

3. The **Posterior Auricular Vein** (*v. auricularis posterior*) (Fig. 9–10) begins on the side of the head in anastomoses with the occipital and superficial temporal veins. It passes posterior to the auricula and joins the posterior division of the retromandibular vein. It receives tributaries from the back of the ear and the stylomastoid vein (*v. stylomastoidea*).

4. The **Occipital Vein** (*v. occipitalis*) (Fig. 9–10) begins in a plexus on the posterior part of the vertex where it anastomoses with posterior auricular and superficial temporal tributaries. As a single vessel it lies beside the occipital artery and greater occipital nerve and with them pierces the cranial attachment of the Trapezius. It enters the

suboccipital triangle and, forming a plexus, joins the deep cervical and vertebral veins. The parietal emissary vein connects it with the superior sagittal sinus, the mastoid emissary vein connects it with the transverse sinus, and through the occipital diploic vein it is connected with the confluens sinuum. Occasionally it follows the occipital artery and empties into the internal jugular or it may join the posterior auricular and drain into the external jugular vein.

5. The **Retromandibular vein** (*v. retromandibularis*) (Fig. 9–10) is formed by the junction of the superficial temporal and maxillary veins. It runs through the substance of the parotid gland superficial to the external carotid artery but deep to the facial

nerve, between the ramus of the mandible and the Sternocleidomastoideus. It has a large communication with the facial vein which was formerly called the posterior facial vein; the common trunk from these two veins was called the common facial vein. It receives small tributaries from the parotid gland and Masseter and joins the posterior auricular vein to form the beginning of the external jugular vein.

B. Deep Veins of the Face

    1. Maxillary
    2. Pterygoid plexus

These veins are tributaries of the internal and external jugular veins.

1. The **Maxillary Vein** (*v. maxillaris*) (Fig. 9–10) is a short trunk which accompanies the first part of the maxillary artery passing between the condyle of the mandible and the sphenomandibular ligament. It is formed by a confluence of the veins in the pterygoid plexus and ends by joining the superficial temporal to form the retromandibular vein.

2. The **Pterygoid Plexus** (*plexus pterygoideus*) (Fig. 9–11) is an extensive network of veins situated between the Temporalis and Pterygoideus lateralis, between the two Pterygoidei, and extending out between surrounding structures in the infratemporal fossa. It receives tributaries corresponding with branches of the maxillary artery, namely, the inferior alveolar vein (*v. alveolaris inferior*), the middle meningeal vein (*v. meningea media*), deep temporal veins (*vv. temporales profundae*), masseteric vein (*v. masseterica*), buccal vein (*v. buccalis*),

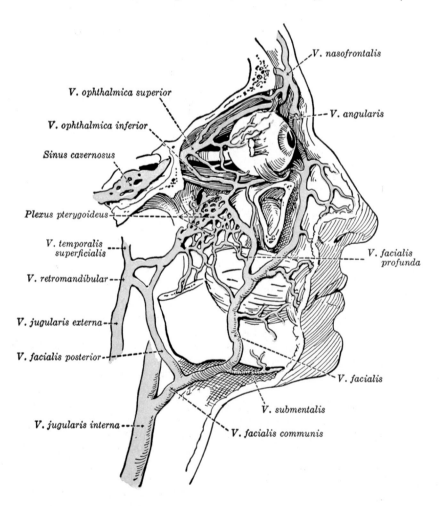

Fig. 9–11.—Principal veins of face and orbit. (Eycleshymer and Jones.)

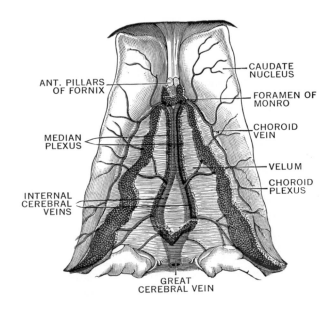

ANT. PILLARS
OF FORNIX

CAUDATE
NUCLEUS

FORAMEN OF
MONRO

CHOROID
VEIN

MEDIAN
PLEXUS

VELUM

CHOROID
PLEXUS

INTERNAL
CEREBRAL
VEINS

GREAT
CEREBRAL VEIN

FIG. 9–12.—The great cerebral and internal cerebral veins (of Galen).
(Poirier and Charpy.)

posterior superior alveolar veins (*vv. alve-olares posteriores superiores*), pharyngeal veins (*vv. pharyngeae*), descending palatine vein (*v. palatina descendens*), infraorbital vein (*v. infraorbitalis*), vein of the pterygoid canal (*v. canalis pterygoideae*), and sphenopalatine vein (*v. sphenopalatina*). The sphenopalatine vein, accompanying the artery, drains the veins of the nasal septum and the extensive plexuses of the erectile tissue in the mucous membrane of the nasal conchae. Its tributaries anastomose with the inferior ophthalmic and ethmoidal veins and may communicate, through the cribriform plate and foramen cecum of the ethmoid bone, with the superior sagittal sinus.

The pterygoid plexus communicates with the cavernous sinus through the foramen of Vesalius, through the rete foraminae ovalis, emissary veins of the foramen lacerum, and inferior ophthalmic vein. It communicates with the facial vein through the deep facial and angular veins.

VEINS OF THE CRANIUM

C. Veins of the Brain
D. Sinuses of the Dura Mater
E. Diploic Veins
F. Emissary Veins

C. The Veins of the Brain

The **veins of the brain** are tributaries of the internal jugular. They possess no valves, and their walls, owing to the absence of muscular tissue, are extremely thin. They pierce the arachnoid membrane and the inner or meningeal layer of the dura mater, and open into the cranial venous sinuses. They may be divided into two sets, **cerebral** and **cerebellar**.

The **Cerebral Veins** (*vv. cerebri*) are divisible into (1) external and (2) internal groups according to whether they drain the outer surfaces and empty into the venous sinuses or drain the inner parts of the hemispheres and empty into the great vein of Galen.

1. The **External Veins** are *a*) the superior, *b*) middle, and *c*) inferior cerebral veins.

a) The **superior cerebral veins** (*vv. cerebri superiores*), eight to twelve in number, drain the superior, lateral, and medial surfaces of the hemispheres, most of them being lodged in the sulci but some run across the gyri. They open into the superior sagittal sinus; the anterior vein runs nearly at right angles to the sinus; the posterior and larger veins are directed more

23

rostralward and open into the sinus in a direction opposed to the current of the blood contained within it.

b) The **middle cerebral vein** (*v. cerebri media superficialis; superficial Sylvian vein*) begins on the lateral surface of the hemisphere, and, running along the lateral cerebral sulcus, ends in the cavernous or the sphenoparietal sinus. It is connected (i) with the superior sagittal sinus by the **vena anastomotica superior** (*great anastomotic vein of Trolard*), which opens into one of the superior cerebral veins; and (ii) with the transverse sinus by the **vena anastomotica inferior** (*posterior anastomotic vein of Labbé*), which courses over the temporal lobe.

c) The **inferior cerebral veins** (*vv. cerebri inferiores*), of small size, drain the inferior surfaces of the hemispheres. Those on the orbital surface of the frontal lobe join the superior cerebral veins, and through these open into the superior sagittal sinus; those of the temporal lobe anastomose with the middle cerebral and basal veins, and join the cavernous, sphenoparietal, and superior petrosal sinuses.

2. The **Internal Cerebral Veins** are: *a*) the great cerebral vein, *b*) the internal cerebral veins, *c*) the thalamostriate vein, *d*) the choroid vein, and *e*) the basal vein.

a) The **great cerebral vein** (*v. cerebri magna; great vein of Galen*) (Figs. 9–12, 9–14), formed by the union of the two internal cerebral veins, is a short median trunk which curves posteriorly around the splenium of the corpus callosum and ends by joining the inferior sagittal sinus at the anterior extremity of the straight sinus. (See Goss '61 for Galen's description.)

b) The **internal cerebral veins** (*vv. cerebri internae; veins of Galen; deep cerebral veins*) (Fig. 9–12) drain the deep parts of the hemisphere and are two in number; each is formed near the interventricular foramen by the union of the **thalamostriate** and **choroid veins**. They run backward parallel with one another, between the layers of the tela chorioidea of the third ventricle, and beneath the splenium of the corpus callosum, where they unite to form a short trunk, the **great cerebral vein;** just before their union each receives the corresponding basal vein.

c) The **thalamostriate vein** (*v. thalamostriata; vena corporis striati*) commences in the groove between the corpus striatum and thalamus, receives numerous veins from both of these parts, and unites behind the crus fornicis with the choroid vein, to form one of the internal cerebral veins.

d) The **choroid vein** runs along the whole length of the choroid plexus, and receives veins from the hippocampus, the fornix, and the corpus callosum.

e) The **basal vein** is formed at the anterior perforated substance by the union of (i) a small **anterior cerebral vein** which accompanies the anterior cerebral artery, (ii) the **deep middle cerebral vein** (*deep Sylvian vein*), which receives tributaries from the insula and neighboring gyri, and runs in the lower part of the lateral cerebral sulcus, and (iii) the **inferior striate veins,** which leave the corpus striatum through the anterior perforated substance. The basal vein passes posteriorly around the cerebral peduncle, and ends in the internal cerebral vein; it receives tributaries from the interpeduncular fossa, the inferior horn of the lateral ventricle, the hippocampal gyrus, and the midbrain.

The **Cerebellar Veins** are placed on the surface of the cerebellum, and are disposed in two sets, superior and inferior. The **superior cerebellar veins** (*vv. cerebelli superiores*) pass either rostralward and medialward, across the superior vermis, to end in the straight sinus and the internal cerebral veins, or lateralward to the transverse and superior petrosal sinuses. The **inferior cerebellar veins** (*vv. cerebelli inferiores*), of large size, end in the transverse, superior petrosal, and occipital sinuses.

D. The Sinuses of the Dura Mater

The **sinuses of the dura mater** (*sinus durae matris*) are venous channels which drain the blood from the brain into the internal jugular vein; they are devoid of valves, and are situated between the two layers of the dura mater but are lined by endothelium continuous with that of the veins into which they drain. They may be divided into two groups: (1) posterior superior, and (2) an anterior inferior.

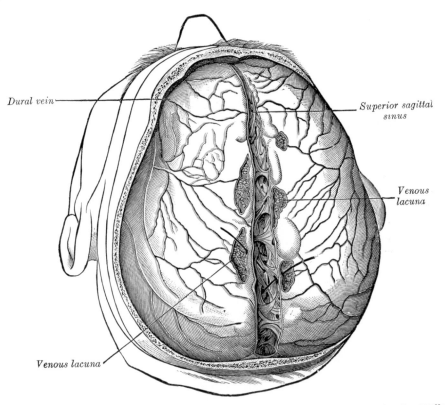

Dural vein

Superior sagittal sinus

Venous lacuna

Venous lacuna

F<small>IG</small>. 9–13.—Superior sagittal sinus laid open after removal of the skull cap. The chordae Willisii are clearly seen. The venous lacunae are also well shown; from two of them probes are passed into the superior sagittal sinus. (Poirier and Charpy.)

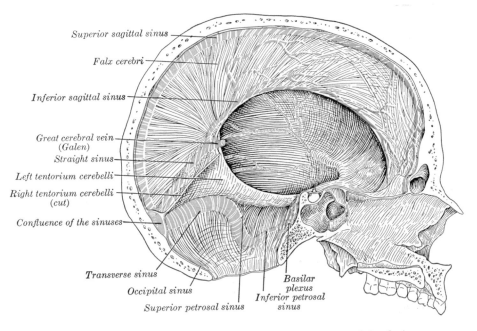

Superior sagittal sinus

Falx cerebri

Inferior sagittal sinus

Great cerebral vein (Galen)

Straight sinus

Left tentorium cerebelli

Right tentorium cerebelli (cut)

Confluence of the sinuses

Transverse sinus

Occipital sinus

Superior petrosal sinus

Basilar plexus

Inferior petrosal sinus

F<small>IG</small>. 9–14.—Sagittal section of the skull, showing the sinuses of the dura.

1. The **posterior superior group** comprises the

a) Superior sagittal
b) Inferior sagittal
c) Straight
d) Transverse
e) Occipital
f) Confluence of the sinuses

a) The **superior sagittal sinus** (*sinus sagittalis superior; superior longitudinal sinus*) (Figs. 9–13, 9–14) occupies the attached or convex margin of the falx cerebri. Commencing at the foramen cecum, through which it receives a vein from the nasal cavity, it runs posteriorly, grooving the inner surface of the frontal bone, the adjacent margins of the two parietals, and the superior division of the cruciate eminence of the occipital; near the internal occipital protuberance it deviates to one side (usually the right), and is continued as the corresponding transverse sinus. It is triangular in section, gradually increasing in size as it passes posteriorly. Its inner surface presents the openings of the superior cerebral veins. Their orifices are concealed by fibrous folds and numerous fibrous bands (*Chordae Willisii*) extending transversely across the inferior angle of the sinus. Small openings in the wall communicate with irregularly shaped venous spaces, **lacunae laterales** (*venous lacunae*), in the dura mater near the sinus. They include a small frontal, a large parietal, and an occipital, intermediate in size between the other two. Most of the cerebral veins from the outer surface of the hemisphere open into these lacunae, and numerous **arachnoid granulations** (*Pacchionian bodies*) project into them. The superior sagittal sinus receives the superior cerebral veins, veins from the diploë, and, near the posterior extremity of the sagittal suture, the anastomosing emissary veins from the pericranium which pass through the parietal foramina, and veins from the dura mater.

Numerous anastomoses exist between this sinus and the veins of the nose, scalp, and diploë.

b) The **inferior sagittal sinus** (*sinus sagittalis inferior; inferior longitudinal sinus*) (Fig. 9–14) is contained in the posterior half or two-thirds of the free margin of the falx cerebri. It is cylindrical, increases in size as it passes posteriorly, and ends in the straight sinus. It receives several veins from the falx cerebri, and occasionally a few from the medial surfaces of the hemispheres.

c) The **straight sinus** (*sinus rectus; tentorial sinus*) (Fig. 9–14) is situated at the line of junction of the falx cerebri with the tentorium cerebelli. It is triangular in section, increases in size as it proceeds posteriorly from the end of the inferior sagittal sinus to the transverse sinus of the side opposite to that in which the superior sagittal sinus terminates. Beside the inferior sagittal sinus, it receives the great cerebral vein (*great vein of Galen*) and the superior cerebellar veins. A few transverse fibrous bands cross its interior.

d) The **transverse sinuses** (*sinus transversi; lateral sinuses*) (Figs. 9–15, 12–1) are of large size and begin at the internal occipital protuberance; one, generally the right, being the direct continuation of the superior sagittal sinus, the other of the straight sinus. Each transverse sinus passes horizontally lateralward and rostralward, describing a slight curve with superior convexity, to the base of the petrous portion of the temporal bone, and lies, in this part of its course, in the attached margin of the tentorium cerebelli. When it leaves the tentorium it becomes the **sigmoid sinus** which curves inferiorly and medialward to reach the jugular foramen, where it ends in the internal jugular vein. In its course it rests upon the squama of the occipital, the mastoid angle of the parietal, the mastoid part of the temporal, and just before its termination, the jugular process of the occipital; the portion which occupies the groove on the mastoid part of the temporal is the **sigmoid sinus**. The transverse sinuses are frequently of unequal size, that formed by the superior sagittal sinus being the larger. On transverse section the horizontal portion exhibits a prismatic, the curved portion a semicylindrical form. They receive the blood from the superior petrosal sinuses at the base of the petrous portion of the temporal bone; they anastomose with the veins of the pericranium by means of the mastoid and condyloid emissary veins; and they receive some of the inferior cerebral and inferior cerebellar veins, and some veins

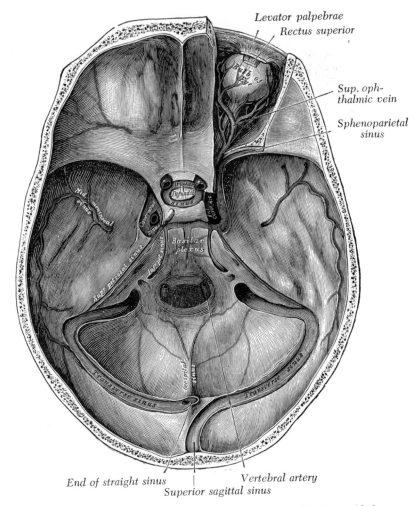

Levator palpebrae
Rectus superior

Sup. oph-
thalmic vein

Sphenoparietal
sinus

End of straight sinus   Vertebral artery
Superior sagittal sinus

FIG. 9–15.—The sinuses at the base of the skull. See also Figure 12–1.

from the diploë. The **petrosquamous sinus,** when present, runs along the junction of the squama and petrous portion of the temporal, and opens into the transverse sinus.

e) The **occipital sinus** (*sinus occipitalis*) (Fig. 9–15) is the smallest of the cranial sinuses. It is situated in the attached margin of the falx cerebelli, and is generally single, but occasionally there are two. It commences around the margin of the foramen magnum by several small venous channels, one of which joins the terminal part of the transverse sinus; it communicates with the posterior internal vertebral venous plexuses and ends in the confluence of the sinuses.

f) The **confluence of the sinuses** (*confluens sinuum; torcular Herophili*) (Fig. 9–14) is the dilated junction of three tributary sinuses, the superior sagittal, the straight, and the occipital with the two large transverse sinuses. The direction of currents within the confluence is such that the right transverse sinus usually receives the greater part of its blood from the superior sagittal sinus and the left from the straight sinus.

2. The **anterior inferior group of sinuses** includes the:

a) Cavernous
b) Intercavernous
c) Superior petrosal
d) Inferior petrosal
e) Basilar plexus

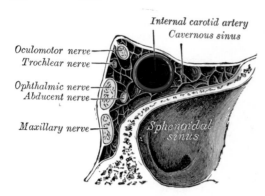

Internal carotid artery
Cavernous sinus

Oculomotor nerve
Trochlear nerve

Ophthalmic nerve
Abducent nerve

Maxillary nerve

Sphenoidal sinus

Fig. 9–16.—Oblique section through the cavernous sinus.

a) The **cavernous sinus** (*sinus cavernosus*) (Figs. 9–15, 9–16) is so named because of the resemblance of its trabeculated structure to that of the corpus cavernosum penis. It is irregular in shape, filling in the space between the body of the sphenoid bone and the dura forming the medial boundary of the middle cranial fossa, from the superior orbital fissure to the apex of the petrous portion of the temporal bone. In separate fibrous sheaths derived from the lateral wall of the sinus, are the following nerves, beginning with the most superior: the oculomotor, trochlear, ophthalmic and maxillary divisions of the trigeminal. Suspended by fibrous trabeculae between the lateral wall of the sinus and the sphenoid bone are the abducent nerve and the internal carotid artery with its surrounding cavernous plexus of the sympathetic. The blood flows around these structures but is separated from them by fibrous sheaths and endothelium.

The cavernous sinus receives the superior and inferior ophthalmic veins, some of the cerebral veins, and the sphenoparietal sinus. It has anastomoses with the transverse sinus through the superior petrosal sinus, with the pterygoid plexus through the foramen of Vesalius, foramen ovale, and foramen lacerum, with the angular and facial vein through the superior ophthalmic vein, and the two sinuses anastomose with each other through the intercavernous sinuses. The cavernous sinus drains into the inferior petrosal sinus and thence into the internal jugular vein.

i) The **superior ophthalmic vein** (*v. oph-*thalmica superior*) (Fig. 9–11) begins at the inner angle of the orbit in a vein named the nasofrontal which anastomoses anteriorly with the angular vein; it passes posteriorly through the superior part of the orbit, and receives tributaries corresponding to the branches of the ophthalmic artery. Forming a short single trunk, it passes between the two heads of the Rectus lateralis and through the medial part of the superior orbital fissure, and ends in the cavernous sinus.

ii) The **inferior ophthalmic vein** (*v. ophthalmica inferior*) begins in a venous network at the anterior part of the floor and medial wall of the orbit; it receives some veins from the Rectus inferior, Obliquus inferior, lacrimal sac and eyelids, runs posteriorly in the inferior part of the orbit and divides into two branches. One of these passes through the inferior orbital fissure and joins the pterygoid venous plexus, while the other enters the cranium through the superior orbital fissure and ends in the cavernous sinus, either by a separate opening, or more frequently in common with the superior ophthalmic vein.

b) The **intercavernous sinuses** (*sinus intercavernosi*), anterior and posterior, connect the two cavernous sinuses across the midline. The **anterior** passes anterior to the hypophysis cerebri, the **posterior** posterior to it, and they form with the cavernous sinuses a venous circle (**circular sinus**) around the hypophysis. The anterior one is usually the larger of the two; either one may be absent.

c) The **superior petrosal sinus** (*sinus petrosus superior*) (Fig. 9– 15), small and narrow, connects the cavernous with the transverse sinus. It runs lateralward and backward, from the posterior end of the cavernous sinus over the trigeminal nerve, and lies in the attached margin of the tentorium cerebelli and in the superior petrosal sulcus of the temporal bone; it joins the transverse sinus where the latter curves into the sigmoid sinus on the inner surface of the mastoid part of the temporal. It receives cerebellar and inferior cerebral veins, and veins from the tympanic cavity.

d) The **inferior petrosal sinus** (*sinus petrosus inferior*) (Fig. 9–15) is situated in the inferior petrosal sulcus formed by the junction of the petrous part of the temporal with the basilar part of the occipital. It begins in the posterior inferior part of the cavernous sinus, and, passing through the anterior part of the jugular foramen, ends in the superior bulb of the internal jugular vein. The inferior petrosal sinus receives the internal auditory veins and also veins from the medulla oblongata, pons, and inferior surface of the cerebellum.

The exact relation of the parts to one another in the jugular foramen is as follows: the inferior petrosal sinus lies medially and anteriorly with the meningeal branch of the ascending pharyngeal artery; the transverse sinus is situated at the lateral and posterior part of the foramen with a meningeal branch of the occipital artery, and between the two sinuses are the glosso-pharyngeal, vagus, and accessory nerves. These three sets of structures are divided from each other by two processes of fibrous tissue. The junction of the inferior petrosal sinus with the internal jugular vein takes place on the lateral aspect of the nerves.

e) The **basilar plexus** (*plexus basilaris; transverse or basilar sinus*) (Fig. 9–15) consists of several interlacing venous channels between the layers of the dura mater over the basilar part of the occipital bone, and serves to connect the two inferior petrosal sinuses. It communicates with the anterior vertebral venous plexus.

E. The Diploic Veins

The **diploic veins** (*venae diploicae*) (Fig. 9–17) occupy channels in the diploë of the cranial bones. They are large and exhibit at irregular intervals pouch-like dilatations; their walls are thin, and formed of endothelium resting upon a layer of elastic tissue.

So long as the cranial bones are separable from one another, these veins are confined to particular bones; but when the sutures are obliterated, they unite with each other, and increase in size. They communicate with the meningeal veins and the sinuses of the dura mater, and with the veins of the pericranium. They consist of (1) the **frontal,** which anastomoses with the supraorbital vein and the superior sagittal sinus; (2) the **anterior temporal,** which, confined chiefly to the frontal bone, anastomoses with the sphenoparietal sinus and one of the deep temporal veins through an aperture in the great wing of the sphenoid; (3) the **posterior temporal,** which is in the parietal bone, and ends in the transverse sinus, through an aperture at the mastoid angle of the parietal bone or through the mastoid foramen; and

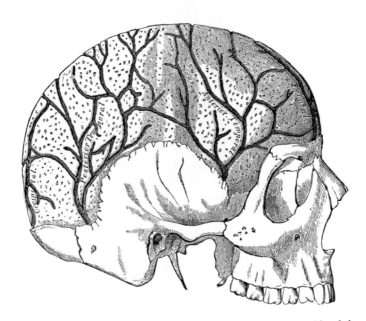

Fig. 9–17.—Veins of the diploë displayed by the removal of the outer table of the skull.

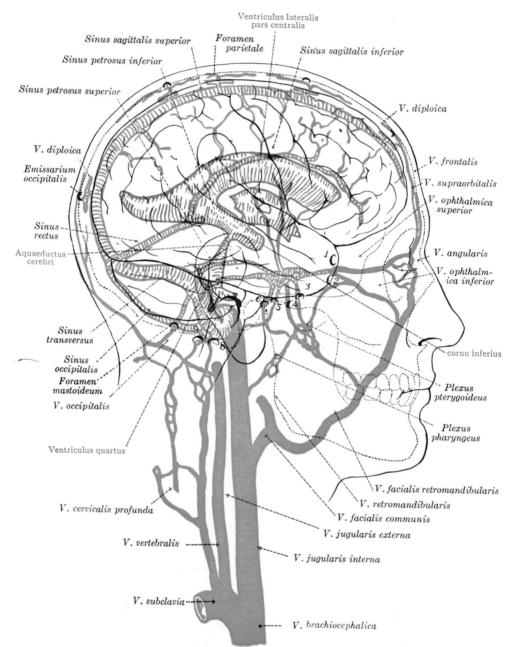

Fig. 9–18.—The veins of the head and neck projected upon the skull and brain in place, showing the emissary veins. The cerebral ventricles are in red. The veins in their extracranial portions are shown in deep blue; the intracranial portions in light blue. The numbers indicate the foramina through which these veins traverse the cranium. These are: *1*. Fissura orbitalis superior. *2*. Fissura orbitalis inferior. *3*. Foramen ovale. *4*. Foramen spinosum. *5*. Foramen lacerum. *6*. Canalis caroticus. *7*. Foramen jugulare. *8*. Canalis hypoglossi. *9*. Canalis condyloideus. The black inverted crescents indicate openings through which emissary veins pass. (Eycleshymer and Jones.)

(4) the **occipital,** the largest of the four, which is confined to the occipital bone, and opens either externally into the occipital vein, or internally into the transverse sinus or into the confluence of the sinuses (*torcular Herophili*).

F. The Emissary Veins

The **emissary veins** (*venae emissariae*) (Fig. 9–18) pass through various foramina and openings in the cranial wall and establish anastomoses between the sinuses of the dura inside the skull and the veins on the exterior of the skull. Some are always present, others only occasionally. The principal emissary veins are the following: (1) A *mastoid emissary vein,* usually present, runs through the mastoid foramen and may traverse a diploic vein for a short distance, connecting the transverse sinus with the posterior auricular or with the occipital vein. (2) A *parietal emissary vein* passes through the parietal foramen and usually through the diploic veins in order to connect the superior sagittal sinus with the veins of the scalp. (3) A network of minute veins (*rete canalis hypoglossi*) traverses the hypoglossal canal and joins the transverse sinus with the vertebral vein and deep veins of the neck. (4) An inconstant *condyloid emissary vein* passes through the condyloid canal and connects the transverse sinus with the deep veins of the neck. (5) A network of veins (*rete foraminis ovalis*) connects the cavernous sinus with the pterygoid plexus through the foramen ovale. (6) Two or three small veins run through the *foramen lacerum* and connect the cavernous sinus with the pterygoid plexus. (7) The emissary vein of the *foramen of Vesalius* connects the same parts. (8) An *internal carotid plexus* of veins traverses the carotid canal between the cavernous sinus and the internal jugular vein. (9) A vein is transmitted through the *foramen cecum* and connects the superior sagittal sinus with the veins of the nasal cavity. Other anastomoses between the sinuses of the dura and the external veins not classed as emissary veins are the connection between the cavernous sinus and the facial vein through the superior ophthalmic and angular veins, and the connection

between the inferior petrosal sinus and the vertebral veins through the basilar plexus.

THE VEINS OF THE NECK

The **veins of the neck** (Figs. 9–10, 9–19) return the blood from the head and face as well as from the cervical structures (see outline page 687).

> G. External Jugular
> H. Internal Jugular
> I. Vertebral

The internal jugular is a direct tributary of the brachiocephalic; the external jugular and vertebral are tributaries of the subclavian.

G. The External Jugular Vein

The **external jugular vein** (*v. jugularis externa*) (Figs. 9–10, 9–19, 9–27) receives the greater part of the blood from the exterior of the cranium and the deep parts of the face, being formed by the junction of the retromandibular with the posterior auricular vein. It commences in the substance of the parotid gland, on a level with the angle of the mandible, and runs perpendicularly down the neck, in the direction of a line drawn from the angle of the mandible to the middle of the clavicle at the posterior border of the Sternocleidomastoideus. In its course it crosses the Sternocleidomastoideus obliquely, and in the subclavian triangle perforates the deep fascia to end in the subclavian vein, lateral or ventral to the Scalenus anterior. It is separated from the Sternocleidomastoideus by the investing layer of the deep cervical fascia, and is covered by the Platysma, the subcutaneous fascia, and the integument; it crosses the anterior cutaneous cervical nerve, and its cranial half runs parallel with the great auricular nerve. The external jugular vein varies in size, bearing an inverse proportion to the other veins of the neck; it is occasionally double. It is provided with two pairs of valves, the inferior pair being placed at its entrance into the subclavian vein, the superior in most cases about 4 cm above the clavicle. The portion of vein between the two sets of valves is often dilated, and is termed the **sinus.** These valves do not prevent the regurgitation of the blood, or the passage of injection distally.

Tributaries.—The external jugular vein receives the occipital, the posterior external jugular, and, near its termination, the transverse cervical, suprascapular, and anterior jugular veins; in the substance of the parotid, a large branch of communication from the internal jugular joins it.

a) The **posterior external jugular vein** (*v. jugularis posterior*) begins in the occipital region and returns the blood from the skin and superficial muscles in the cranial and dorsal part of the neck, lying between the Splenius and Trapezius. It runs down the dorsal part of the neck, and opens into the middle third of the external jugular vein.

b) The **anterior jugular vein** (*v. jugularis anterior*) begins near the hyoid bone by the confluence of several superficial veins from the submandibular region. It descends between the median line and the anterior border of the Sternocleidomastoideus, and, at the inferior part of the neck, passes deep to that muscle to open into the termination of the external jugular, or, in some instances, into the subclavian vein (Figs. 9–10, 9–19). It varies considerably in size, bearing usually an inverse proportion to the external jugular; most frequently there are two anterior jugulars, a right and left, but sometimes only one. Its tributaries are some laryngeal veins, and occasionally a small thyroid vein. Just above the sternum the two anterior jugular veins communicate by a transverse trunk, the **jugular venous arch**, which receives tributaries from the inferior thyroid veins; each also communicates with the internal jugular. There are no valves in this vein.

c) The **transverse cervical veins** (Fig. 9–19) drain the blood from the trapezius and

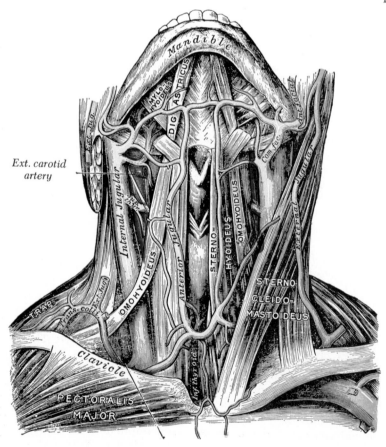

Ext. carotid
artery

Subclavian vein

Fig. 9–19.—The veins of the neck, anterior aspect. (Spalteholz.)

other superficial and deep structures of the surrounding area, corresponding to and accompanying the artery of the same name. They drain into the external jugular just before the latter joins the subclavian vein, or they may open into the subclavian independently.

d) The **suprascapular vein** (*transverse scapular vein*) accompanies the artery of the same name and opens into the external jugular near its termination or into the subclavian vein.

H. The Internal Jugular Vein

The **internal jugular vein** (*v. jugularis interna*) collects the blood from the brain, the face, and the neck. It is directly continuous with the transverse sinus in the posterior compartment of the jugular foramen, at the base of the skull. At its origin it is somewhat dilated, and this dilation is called the **jugular bulb**. It runs down the side of the neck in a vertical direction, lying at first lateral to the internal carotid artery, and then lateral to the common carotid, and at the root of the neck unites with the subclavian vein to form the brachiocephalic vein; a little superior to its termination is a second dilatation, the inferior bulb. At first, it lies upon the Rectus capitis lateralis, posterior to the internal carotid artery and the nerves passing through the jugular foramen; more inferiorly, the vein and artery lie upon the same plane, the glossopharyn-geal and hypoglossal nerves passing anteriorly between them; the vagus descends between the vein and the artery in the same sheath, and the accessory runs obliquely dorsalward, superficial or deep to the vein. At the root of the neck the right internal jugular vein is placed at a little distance from the common carotid artery, and crosses the first part of the subclavian artery, whereas the left internal jugular vein usually overlaps the common carotid artery. The left vein is generally smaller than the right, and each contains a pair of valves which are placed about 2.5 cm above the termination of the vessel.

Tributaries.—This vein receives in its course the inferior petrosal sinus, the facial, lingual, pharyngeal, superior and middle thyroid, and sometimes the occipital veins. The thoracic duct on the left side and the right lymphatic duct on the right side open into the angle of union of the internal jugular and subclavian veins.

a) The **inferior petrosal sinus** (*sinus petrosus inferior*) leaves the skull through the anterior part of the jugular foramen, and joins the superior bulb of the internal jugular vein (Fig. 9–15).

b) The **lingual vein** (*v. lingualis*) (Fig. 9–20) opens into the internal jugular near the hyoid bone, receiving blood from the tongue through two or three tributaries. The **dorsal lingual veins** (*vv. dorsalis linguae*) begin at the dorsum of the tongue and pass posteriorly as the venae comitantes of the

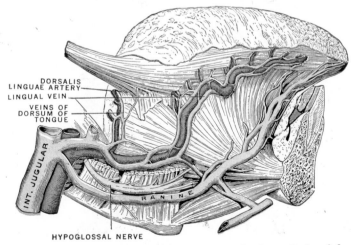

DORSALIS LINGUAE ARTERY
LINGUAL VEIN
VEINS OF DORSUM OF TONGUE
INT. JUGULAR
RANINE
HYPOGLOSSAL NERVE

Fig. 9–20.—Veins of the tongue. The hypoglossal nerve has been displaced downward in this preparation. (Testut after Hirschfeld.)

arteries of the same name. The **deep lingual vein** (*v. profunda linguae*) begins near the tip of the tongue with an anastomosis with its fellow of the opposite side and continues as the vena comitans of the artery of the same name. The **vena comitans** of the **hypoglossal nerve** (*v. comitans n. hypoglossi; ranine vein*) begins near the mucous membrane of the inferior surface of the tip of the tongue and accompanies the hypoglossal nerve, retaining its superficial position at first, then passing along the lateral surface of the Hypoglossus; it frequently is a large vein, and it commonly empties into the facial vein or even into the internal jugular without joining the lingual. A small tributary, the **sublingual vein,** anastomoses with the submental vein.

The **pharyngeal veins** (*vv. pharyngeae*) begin in the **pharyngeal plexus** on the outer surface of the pharynx, and, after receiving some posterior meningeal veins and the vein of the pterygoid canal, end in the internal jugular. They occasionally open into the facial, lingual, or superior thyroid vein.

The **superior thyroid vein** (*v. thyreoidea superioris*) begins in the substance and on the surface of the thyroid gland, by tributaries corresponding with the branches of the superior thyroid artery, and ends in the cranial part of the internal jugular vein. It receives the superior laryngeal and cricothyroid veins.

The **middle thyroid vein** (Fig. 9–21) collects the blood from the inferior part of the thyroid gland, and after being joined by some veins from the larynx and trachea, ends in the caudal part of the internal jugular vein.

I. The Vertebral Vein

The **vertebral vein** (*v. vertebralis*) (Fig. 9–22) does not emerge through the foramen magnum with the vertebral artery, but is formed in the suboccipital triangle, from numerous small tributaries which spring from the internal vertebral venous plexuses and issue from the vertebral canal superior to the posterior arch of the atlas. They unite with small veins from the deep

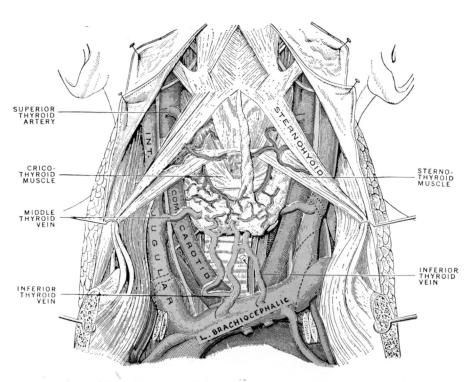

FIG. 9–21.—The middle and inferior thyroid veins. (Poirier and Charpy.)

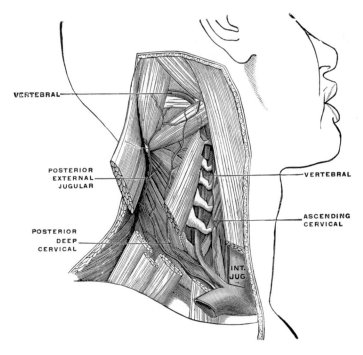

FIG. 9–22.—The vertebral vein. (Poirier and Charpy.)

muscles at the cranial part of the dorsum of the neck, and form a vessel which enters the foramen in the transverse process of the atlas. Forming a dense plexus around the vertebral artery, they descend in the canal formed by the foramina transversaria of the cervical vertebrae. This plexus ends in a single trunk which emerges from the foramen transversarium of the sixth or seventh cervical vertebra, and opens at the root of the neck into the dorsal part of the brachiocephalic vein near its origin, its mouth being guarded by a pair of valves. On the right side, it crosses the first part of the subclavian artery.

Tributaries.—The vertebral vein communicates with the transverse sinus by a vein which passes through the condyloid canal, when that canal exists. It receives tributaries from the occipital vein and from the prevertebral muscles, from the internal and external vertebral venous plexuses, from the anterior vertebral and the deep cervical veins; close to its termination it is sometimes joined by the first intercostal vein.

a) The **anterior vertebral vein** (*ascending cervical vein*) commences in a plexus around the transverse processes of the cranial cervical vertebrae, descends in company with the ascending cervical artery between the Scalenus anterior and Longus capitis muscles, and opens into the terminal part of the vertebral vein.

b) The **accessory vertebral vein,** when present, arises from the venous plexus on the vertebral artery and passes through the foramen transversarium of the seventh cervical vertebra before running ventralward to join the brachiocephalic vein.

c) The **deep cervical vein** (*v. cervicalis profunda; posterior vertebral or posterior deep cervical vein*) accompanies its artery between the Semispinales capitis and cervicis. It begins in the suboccipital region by communicating branches from the occipital vein and by small veins from the deep muscles at the back of the neck. It receives tributaries from the plexuses around the spinous processes of the cervical vertebrae, and terminates in the inferior part of the vertebral vein.

THE VEINS OF THE UPPER LIMB

The **veins of the upper limb** are composed of two sets, superficial and deep, which are distinguished by their topographic position but which anastomose with each other at frequent intervals, thus forming several parallel channels of drainage from any one area. The superficial veins are spread immediately beneath the integument, between the superficial and deep strata of the subcutaneous fascia. The deep veins accompany the arteries, are the venae comitantes, and are usually given the same names as their arteries. Both sets are provided with valves, more numerous in the deep than in the superficial, but the walls of the superficial are thicker because they are more exposed to mechanical pressure and trauma.

The Superficial Veins of the Upper Limb

Two large veins, the cephalic and the basilic, receive the blood from the superficial veins of the upper limb, conveying it to the axillary vein and thence into the subclavian and brachiocephalic veins. The cephalic vein occupies the radial side of the forearm and lateral side of the arm; the basilic vein occupies the ulnar and medial sides.

The **superficial veins of the upper limb** are:

      1. Cephalic
      2. Basilic
      3. Median antebrachial
      4. Dorsal digital

1. The **Cephalic Vein** (*v. cephalica*) (Fig. 9–23) begins in the radial part of the dorsal venous network of the hand and winds proximalward around the radial border of the forearm, receiving tributaries from both palmar and dorsal surfaces. Just distal to the antecubital fossa it has a wide anastomosis with the median cubital vein. It continues proximally along the lateral side of the fossa in the groove between the Brachioradialis and the Biceps brachii. It crosses superficial to the musculocutaneous nerve and ascends in the groove along the lateral border of the Biceps brachii. In the proximal third of the arm it passes between the Pectoralis major and Deltoideus, where it is accom-

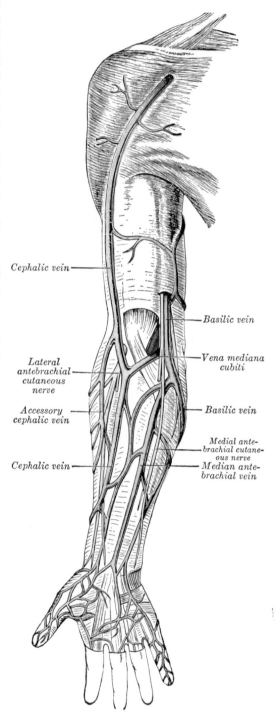

Cephalic vein

Lateral antebrachial cutaneous nerve

Accessory cephalic vein

Cephalic vein

Basilic vein

Vena mediana cubiti

Basilic vein

Medial antebrachial cutaneous nerve

Median antebrachial vein

FIG. 9–23.—The superficial veins of the upper limb.

panied by the deltoid branch of the thoracoacromial artery. In the triangular interval between the origins of these two muscles from the clavicle (Mohrenheim's triangle), it runs deep to the Major, and just proximal to the Pectoralis minor, it pierces the clavipectoral fascia and, crossing the axillary artery, ends in the axillary vein just caudal to the clavicle. Sometimes it communicates with the external jugular vein by a branch which ascends ventral to the clavicle.

The **accessory cephalic vein** (*v. cephalica accessoria*) arises either from a small tributary plexus on the dorsum of the forearm or from the ulnar side of the dorsal venous network. It remains on the radial side of the cephalic and joins it at the elbow. The accessory cephalic may spring from the cephalic proximal to the wrist and join it again more proximally. A large oblique anastomosis frequently connects the basilic and cephalic veins on the dorsum of the forearm.

2. The **Basilic Vein** (*v. basilica*) (Fig. 9–23) begins in the ulnar part of the dorsal venous network. It runs proximally on the posterior surface of the ulnar side of the forearm and inclines toward the anterior surface distal to the elbow, where it is joined by the vena mediana cubiti. It ascends obliquely in the groove between the Biceps brachii and Pronator teres and crosses the brachial artery, from which it is separated by the bicipital aponeurosis; filaments of the medial antebrachial cutaneous nerve pass both anterior and dorsal to this portion of the vein. It then runs proximally along the medial border of the Biceps brachii, perforates the deep fascia a little distal to the middle of the arm, and, ascending on the medial side of the brachial artery to the distal border of the Teres major joins the brachial to form the axillary vein.

3. The **Median Antebrachial Vein** (*v. mediana antebrachii*) drains the venous plexus on the palmar surface of the hand. It ascends on the ulnar side of the anterior forearm and ends in the **median cubital vein** (*vena mediana cubiti*); not uncommonly it divides into two vessels, one of which joins the basilic, the other the cephalic, distal to the elbow, and there is a wide anastomosis with the deep veins.

There is great variation in the superficial veins of the forearm. There is a reciprocal relationship between the cephalic and the basilic veins; either one may predominate or even be lacking. The median antebrachial vein may be absent as a definite vessel. The median cubital vein may split into a distinct Y; one arm of the Y draining into the cephalic, the other into the basilic. In this case one branch is called the **median cephalic vein,** the other the **median basilic vein.**

Clinical Considerations.—The veins of the proximal forearm are usually the ones employed in venipuncture. One of the veins of the median cubital complex regularly has a large vessel anastomosing with the deep veins (Fig. 9–25). This anastomosis holds the superficial vein in place and will prevent it from slipping away from the point of the needle.

4. The **Dorsal Digital Venous Network** (*rete venosum dorsale manus*).—The dorsal digital veins pass along the sides of the fingers and are joined to one another by oblique communicating branches. Those from the adjacent sides of the fingers unite to form three dorsal metacarpal veins (Fig. 9–24), which end in a dorsal venous network on the back of the hand. The radial part of the network is joined by the dorsal digital vein from the radial side of the index finger and by the dorsal digital veins of the thumb, and is prolonged proximally as the cephalic vein. The ulnar part of the network receives the dorsal digital vein of the ulnar side of the little finger and is continued proximally as the basilic vein. An anastomosing vein frequently makes an additional connection with either the cephalic or basilic vein in the middle of the forearm.

The **palmar digital veins** (Fig. 9–23) on each finger are connected to the dorsal digital veins by oblique intercapitular veins. They drain into a venous plexus which is situated over the thenar and hypothenar eminences and across the palmar surface of the wrist.

The Deep Veins of the Upper Limb

The **deep veins** follow the course of the arteries, accompanying them as their venae comitantes. They are generally arranged in

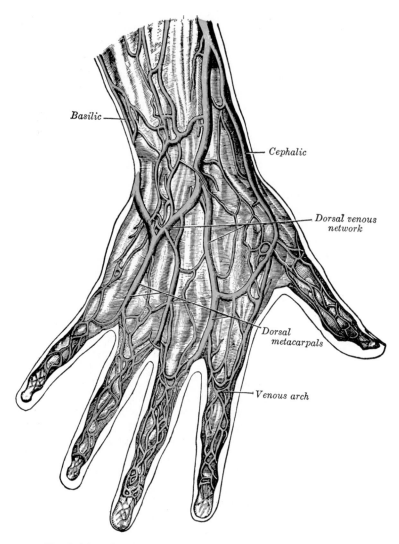

Basilic

Cephalic

Dorsal venous network

Dorsal metacarpals

Venous arch

FIG. 9–24.—The veins on the dorsum of the hand. (Bourgery.)

pairs, situated on either side of the corresponding artery, and are connected at intervals by short transverse anastomoses.

Deep Veins of the Hand.—Each of the superficial and deep palmar arterial arches is accompanied by a pair of venae comitantes which constitute the **superficial** and **deep palmar venous arches,** and receive the veins corresponding to the branches of the arterial arches; thus the common palmar digital veins, formed by the union of the proper palmar digital veins, open into the superficial, and the palmar metacarpal veins into the deep palmar venous arches. The dorsal **metacarpal veins** receive perforating branches from the palmar metacarpal veins and end in the radial veins and the superficial veins on the dorsum of the wrist.

The **Deep Veins of the Forearm** (Fig. 9–25) are the venae comitantes of the radial and ulnar arteries and constitute respectively the proximal continuations of the deep and superficial palmar venous arches; they unite in the bend of the elbow to form the brachial veins. The radial veins are smaller than the ulnar and receive the dorsal metacarpal veins. The ulnar veins receive tributaries from the deep palmar venous arches and communicate with the superficial veins at the wrist; near the elbow they receive the palmar and dorsal interosseous veins and send a large communicating branch (*profunda vein*) to the vena mediana cubiti.

The **Brachial Veins** (*vv. brachiales*) are placed on either side of the brachial artery, receiving tributaries corresponding with the branches given off from that vessel; near the distal margin of the Subscapularis, they join the axillary vein; the medial one frequently is absorbed by the basilic vein and in the rest of its proximal course the latter appears to be one of the brachial comitantes.

These deep veins have numerous anastomoses, not only with each other, but also with the superficial veins.

The **Axillary Vein** (*v. axillaris*) (Fig. 9–26) begins at the junction of the basilic and brachial veins near the distal border of the Teres major, and ends at the outer border of the first rib, by becoming the subclavian vein. In addition to the tributaries which correspond to the branches of the axillary

artery, it receives the cephalic vein near its termination and may receive an additional deeper brachial comitans near its beginning. It lies on the medial side of the artery, which it partly overlaps; between the two vessels are the medial cord of the brachial plexus, the median, the ulnar, and the pectoral nerves. It is provided with a pair of valves opposite the distal border of the Subscapularis; valves are also found at the ends of the cephalic and subscapular veins.

The **Subclavian Vein** (*v. subclavia*), the continuation of the axillary, extends from the lateral border of the first rib to the sternal end of the clavicle, where it unites with the internal jugular to form the brachiocephalic vein. It is in relation, *ventrally*, with the clavicle and Subclavius; *dorsally* and *cranially* with the subclavian artery, from which it is separated medially by the Scalenus anterior and the phrenic nerve. *Inferiorly*, it rests in a depression on the first rib and upon the pleura. It is usually provided with a pair of valves,

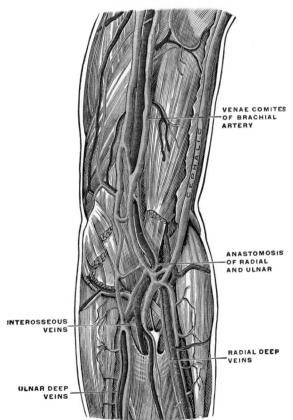

FIG. 9–25.—The deep veins at the elbow. (Bourgery.)

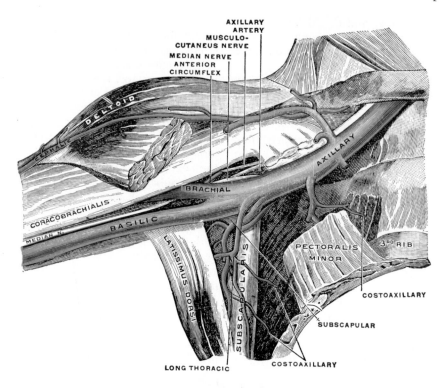

FIG. 9–26.—The veins of the right axilla, ventral view. (Spalteholz.)

which are situated about 2.5 cm from its termination.

The subclavian vein occasionally rises in the neck to a level with the third part of the subclavian artery, and occasionally passes with this vessel dorsal to the Scalenus anterior.

Tributaries.—The subclavian vein receives the external jugular vein, sometimes the anterior jugular vein, and occasionally a small branch from the cephalic which ascends ventral to the clavicle. At its angle of junction with the internal jugular, the left subclavian vein receives the thoracic duct, and the right subclavian vein the right lymphatic duct (Figs. 10–6, 10–8).

THE VEINS OF THE THORAX

The special vein of the thorax which drains the blood from the thoracic wall and intercostal spaces is the azygos vein. It is the first tributary of the superior vena cava, joining the latter as it passes ventral to the root of the lung. The other principal tributaries of the superior vena cava are the brachiocephalic veins, two large collecting trunks which receive the blood from the head, neck, and upper limbs.

The superior vena cava is described on page 683, veins of the heart on page 682, and the pulmonary veins on page 679.

The **veins of the thorax** are:

    1. Azygos
    2. Right brachiocephalic
    3. Left brachiocephalic
    4. Internal thoracic
    5. Inferior thyroid
    6. Highest intercostal
    7. Veins of vertebral column

    1. The **Azygos Vein** (*v. azygos; vena azygos major*) (Fig. 9–27) begins opposite the first or second lumbar vertebra, by a

tributary, the ascending lumbar vein (page 710), sometimes by a branch from the right renal vein, or from the inferior vena cava. It enters the thorax through the aortic hiatus in the diaphragm, and passes along the right side of the vertebral column to the fourth thoracic vertebra, where it arches ventralward over the root of the right lung, and ends in the superior vena cava, just before that vessel pierces the pericardium. In the aortic hiatus, it lies on the right side of the aorta with the thoracic duct; in the thorax it lies upon the intercostal arteries, to the right of the aorta and thoracic duct, and is partly covered by pleura.

Tributaries.—It receives the right subcostal and intercostal veins, the first three or four of these latter opening by a common stem, the right superior intercostal vein. It receives the hemiazygos veins, several esophageal, mediastinal, and pericardial veins, and, near its termination, the right bronchial vein. A few imperfect valves are found in the azygos vein; but the valves of its tributaries are complete.

a) The **posterior intercostal veins** (*vv. intercostales posteriores*) are so named to distinguish them from the anterior intercostal veins which are tributaries of the internal thoracic veins. They accompany the intercostal arteries, one in each intercostal space. They lie along the cranial border of the space in a groove on the caudal border of the rib, in a position cranial to the artery. Each vein receives a **dorsal tributary** from the muscles and cutaneous area of the back and a **spinal tributary** joining the vertebral plexuses of the vertebral column and spinal cord. The posterior intercostal veins anastomose with the anterior veins of the internal thoracic, and with the lateral thoracic or thoracoepigastric veins.

The posterior intercostal veins of the right side, except the first two or three, open into the azygos vein. The veins of the right first and second, occasionally also the third, intercostal spaces join to form the **right highest intercostal** which usually empties into the azygos but may accompany the superior intercostal and drain into the right brachiocephalic. The left highest intercostal vein receives blood from the first two or three left intercostal spaces and empties

into the left brachiocephalic vein. The other left posterior intercostal veins usually join the hemiazygos or accessory hemiazygos, but it is not uncommon for several of them, especially in the midthorax, to cross the vertebral column and open into the azygos individually.

The **subcostal veins** correspond to the intercostal veins but lie in the abdominal wall along the caudal border of the twelfth rib. The one on the right side is a tributary of the azygos, the one on the left of the hemiazygos vein.

b) The **hemiazygos vein** (*v. hemiazygos; vena azygos minor inferior*) begins in the left ascending lumbar vein. It enters the thorax, through the left crus of the diaphragm, and, ascending on the left side of the vertebral column, as high as the ninth thoracic vertebra, passes horizontally across the vertebral column, dorsal to the aorta, esophagus, and thoracic duct, to end in the azygos vein. It receives the caudal four or five intercostal veins, the subcostal vein of the left side, and some esophageal and mediastinal veins.

c) The **accessory hemiazygos vein** (*v. hemiazygos accessoria; vena azygos minor superior*) descends on the left side of the vertebral column, and either crosses the body of the eighth thoracic vertebra to join the azygos vein or ends in the hemiazygos. It varies inversely in size with the highest left intercostal vein and when it is small, or altogether wanting, the left highest intercostal vein may extend as low as the fifth or sixth intercostal space.

In obstruction of the vena cava, the azygos and hemiazygos veins are an important means by which the venous circulation is carried on, connecting as they do the superior and inferior venae cavae, and communicating with the common iliac veins by the ascending lumbar veins and with many of the tributaries of the inferior vena cava.

d) The **bronchial veins** (*vv. bronchiales*) return the blood from the larger bronchi, and from the structures at the roots of the lungs; that of the right side opens into the azygos vein, near its termination; that of the left side, into the highest left intercostal or the accessory hemiazygos vein. A considerable quantity of the blood which is carried

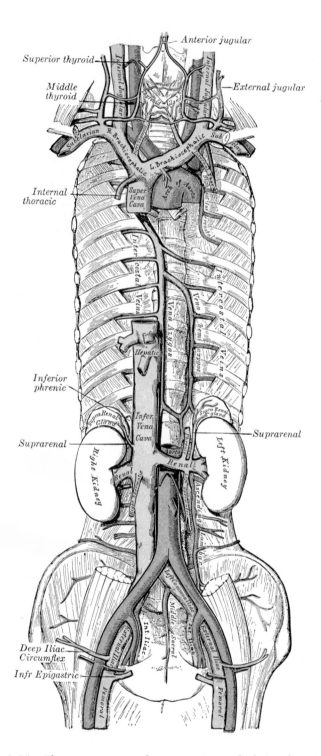

FIG. 9–27.—The venae cavae and azygos veins, with their tributaries.

to the lungs through the bronchial arteries is returned to the left side of the heart through the pulmonary veins.

The **Brachiocephalic Veins** (*venae brachiocephalicae; innominate veins*) (Fig. 9–27), are two large trunks, right and left, on either side of the root of the neck. They are formed by the junction of the internal jugular and subclavian veins of the corresponding side and they terminate by uniting to form the superior vena cava. Because the vena cava is situated toward the right, the left brachiocephalic is longer than the right.

2. The **Right Brachiocephalic Vein** (*v. brachiocephalica dextra; right innominate vein*) is a short vessel, about 2.5 cm in length, which begins behind the sternal end of the clavicle, and, passing almost vertically caudalward, joins with the left brachiocephalic vein just below the cartilage of the first rib, close to the right border of the sternum, to form the superior vena cava. It lies ventral and to the right of the brachiocephalic artery; on its right side are the phrenic nerve and the pleura, which are interposed between it and the apex of the lung. This vein, at its commencement, receives the right vertebral vein; and more caudally, the right internal thoracic and right inferior thyroid veins, and sometimes the vein from the first intercostal space.

3. The **Left Brachiocephalic Vein** (*v. brachiocephalica sinistra; left innominate vein*), about 6 cm in length, from deep to the sternal end of the clavicle runs obliquely distalward and to the right, dorsal to the upper half of the manubrium sterni, to the sternal end of the first right costal cartilage, where it unites with the right brachiocephalic vein to form the superior vena cava. It is separated from the manubrium sterni by the Sternohyoideus and Sternothyroideus, the thymus or its remains, and some loose areolar tissue. Dorsal to it are the three large arteries, brachiocephalic, left common carotid, and left subclavian, arising from the aortic arch, together with the vagus and phrenic nerves. The left brachiocephalic vein may occupy a higher level, crossing the jugular notch and lying directly in front of the trachea.

Tributaries.—Both veins have as tributaries the vertebral, the internal thoracic, and the inferior thyroid. The left also has the left highest intercostal veins, and occasionally some thymic and pericardiac veins. The right vein may receive the vein from the first intercostal space.

4. The **Internal Thoracic Vein** (*internal mammary vein*) is the vena comitans of the internal thoracic artery, and receives tributaries corresponding to the branches of the artery. It then unites to form a single trunk, which runs up on the medial side of the artery and ends in the corresponding brachiocephalic vein. The **superior phrenic vein,** *i.e.,* the vein accompanying the pericardiacophrenic artery, usually opens into the internal thoracic vein.

5. The **Inferior Thyroid Veins** (*vv. thyroideae inferiores*) (Fig. 9–21), two, frequently three or four, in number, arise in the venous plexus on the thyroid gland, communicating with the middle and superior thyroid veins. They form a plexus ventral to the trachea, deep to the Sternothyroidei. From this plexus, a left vein descends and joins the left brachiocephalic trunk, and a right vein passes obliquely downward and to the right across the brachiocephalic artery to open into the right brachiocephalic vein, just at its junction with the superior vena cava; sometimes the right and left veins open by a common trunk in the latter situation. These veins receive esophageal, tracheal, and inferior laryngeal veins, and are provided with valves at their terminations into the brachiocephalic veins.

6. The **Highest Intercostal Veins** (*v. intercostalis suprema; superior intercostal veins*) (right and left) drain the blood from the upper two or three intercostal spaces. The right vein (*v. intercostalis suprema dextra*) passes downward and opens into the vena azgos; the left vein (*v. intercostalis suprema sinistra*) runs across the arch of the aorta and the origins of the left subclavian and left common carotid arteries and opens into the left brachiocephalic vein. It usually receives the left bronchial vein, and sometimes the left superior phrenic vein. It has a prominent anastomosis with the accessory hemiazygos vein and drains all the upper left spaces when the latter is missing (Fig. 10–6).

7. The **Veins of the Vertebral Column** (Figs. 9–28, 9–29).—The veins which drain the blood from the vertebral column, the

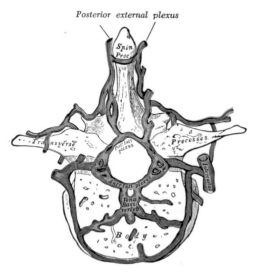

FIG. 9–28.—Transverse section of a thoracic vertebra, showing the vertebral venous plexuses.

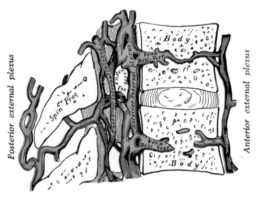

FIG. 9–29.—Median sagittal section of two thoracic vertebrae, showing the vertebral venous plexuses.

neighboring muscles, and the meninges of the spinal cord form intricate plexuses extending along the entire length of the column; these plexuses may be divided into two groups, external and internal, according to their positions inside or outside the vertebral canal. The plexuses of the two groups anastomose freely with each other and end in the intervertebral veins.

The plexuses and veins of the vertebral system are:

    a) External plexus
    b) Internal plexus
    c) Basivertebral veins
    d) Intervertebral veins
    e) Spinal cord veins

a) The **external vertebral venous plexuses** (*plexus venosi vertebrales externi; extraspinal veins*), best marked in the cervical region, consist of anterior and posterior plexuses which anastomose freely with each other. The **anterior external plexuses** lie ventral to the bodies of the vertebrae, communicate with the basivertebral and intervertebral veins, and receive tributaries from the bodies of vertebrae. The **posterior external plexuses** are placed partly on the dorsal surfaces of the vertebral arches and their processes, and partly between the deep dorsal muscles. They are best developed in the cervical region, and there anastomose with the vertebral, occipital, and deep cervical veins.

b) The **internal vertebral venous plexuses** (*plexus venosi vertebrales interni; intraspinal veins*) lie within the vertebral canal between the dura mater and the vertebrae, and receive tributaries from the bones and from the spinal cord. They form a closer network than the external plexuses, and, running mainly in a vertical direction, form four longitudinal veins, two anterior and two posterior; they therefore may be divided into anterior and posterior groups. The **anterior internal plexuses** consist of large veins which lie on the posterior surfaces of the vertebral bodies and intervertebral disks on either side of the posterior longitudinal ligament; under cover of this ligament they are connected by transverse branches into which the basivertebral veins open. The **posterior internal plexuses** are placed, on either side of the middle line ventral to the vertebral arches and ligamenta flava, and anastomose by veins passing through these ligaments with the posterior external plexuses. The anterior and posterior plexuses communicate freely with each other by a series of **venous rings** (*retia venosa vertebrarum*), one opposite each vertebra. Around the foramen magnum they form an intricate network which opens into the vertebral veins and is connected above with the occipital sinus, the basilar plexus, the condyloid emissary vein, and the rete canalis hypoglossi.

c) The **basivertebral veins** (*vv. basivertebrales*) emerge from the foramina on the dorsal surfaces of the vertebral bodies. They are contained in large, tortuous channels in the substance of the bones, similar in every respect to those found in the diploë of the cranial bones. They communicate through small openings on the ventral and lateral sides of the bodies of the vertebrae with the anterior external vertebral plexuses, and converge dorsally into the principal canal, sometimes double toward its posterior part, and open by valved orifices into the transverse branches which unite the anterior internal vertebral plexuses. They become greatly enlarged in advanced age.

d) The **intervertebral veins** (*vv. intervertebrales*) accompany the spinal nerves through the intervertebral foramina; they receive the veins from the spinal cord, drain the internal and external vertebral plexuses and end in the vertebral, intercostal, lumbar, and lateral sacral veins, their orifices being provided with valves.

e) The **veins of the spinal cord** (*vv. spinales; veins of the medulla spinalis*) are situated in the pia mater and form a minute, tortuous, venous plexus. They emerge chiefly from the median fissures of the cord and are largest in the lumbar region. In this plexus there are (1) two median longitudinal veins, one anterior to the anterior fissure, and the other posterior to the posterior sulcus of the cord, and (2) four lateral longitudinal veins which run dorsal to the nerve roots. They end in the intervertebral veins. Near the base of the skull they unite, and form two or three small trunks, which communicate with the vertebral veins, and then end in the inferior cerebellar veins, or in the inferior petrosal sinuses.

Practical Considerations.—According to Batson (1940) the veins of the vertebral column constitute a system paralleling the main caval system. He reached this conclusion as a result of x-ray studies of human cadavers and living animals. A thin solution of radiopaque material which he injected into the dorsal vein of the penis in a cadaver found its way readily into the veins of the entire vertebral column, the skull, and the interior of the cranium. The material drained from the dorsal vein of the penis into the prostatic plexus and then followed communications with the veins of the sacrum, ilium, lumbar vertebrae, upper femur, and the venae vasorum of the large femoral blood vessels, without traversing the main caval tributaries. Similarly, material injected into a small breast vein found its way into the veins of the clavicle, the intercostal veins, the head of the humerus, cervical vertebrae, and dural sinuses without following the caval paths. Thorium dioxide injected into the dorsal vein of the penis of an anaesthetized monkey drained into the caval system when the animal was undisturbed, but if its abdomen was put under pressure with a binder, simulating the increased intra-abdominal pressure of coughing or straining, the material drained into the veins of the vertebrae. Batson believes that the spread of metastases from tumors and abscesses, in many cases such as the metastases to the pelvic bones from the prostate, can be explained only through the channels of the vertebral venous system and its extensive communication with the caval system. When the pressure within the thorax and abdomen is increased by coughing or straining, the blood may flow along the vertebral system rather than the caval and it may even be forced into the vertebral veins from the viscera.

THE VEINS OF THE ABDOMEN AND PELVIS

The **Inferior Vena Cava** (*v. cava inferior*) (Fig. 9–27), returns the blood from the parts below the diaphragm to the heart. It is formed by the junction of the two common iliac veins, at the right side of the body of the fifth lumbar vertebra. It ascends along the vertebral column, on the right side of the aorta, and, having reached the liver, is continued in a groove on its dorsal surface. It then perforates the diaphragm between the median and right portions of its central tendon; it subsequently inclines ventralward and medialward for about 2.5 cm,

and, piercing the fibrous pericardium, it receives a covering of serous pericardium and opens into the right atrium. Ventral to its atrial orifice is a semilunar valve, the **valve of the inferior vena cava** (*Eustachian valve*) which is rudimentary in the adult, but is of large size and exercises an important function in the fetus.

Relations.—The *abdominal portion* of the inferior vena cava is in relation *ventrally* with the right common iliac artery, the mesentery, the right internal spermatic artery, the inferior part of the duodenum, the pancreas, the common bile duct, the portal vein, and the posterior surface of the liver; the last partly overlaps and occasionally completely surrounds it; *dorsally*, with the vertebral column, the right Psoas major, the right crus of the diaphragm, the right inferior phrenic, suprarenal, renal and lumbar arteries, right sympathetic trunk and right celiac ganglion, and the medial part of the right suprarenal gland; on the *right side*, with the right kidney and ureter; on the *left side*, with the aorta, right crus of the diaphragm, and the caudate lobe of the liver.

The *thoracic portion* is situated partly inside and partly outside the pericardial sac. The *extrapericardial part* is separated from the right pleura and lung by a fibrous band, named the **right phrenicopericardiac ligament.** This ligament, often feebly marked, is attached to the margin of the caval opening in the diaphragm and to the pericardium at the root of the right lung. The *intrapericardiac part* is very short, and is covered by the serous layer of the pericardium.

Variations.—This vessel is sometimes placed on the left side of the aorta, as high as the left renal vein, and, after receiving this vein, crosses over to its usual position on the right side; or it may be placed altogether on the left side of the aorta, and in such a case the abdominal and thoracic viscera, together with the great vessels, are all transposed. Most of the variations are due to the persistence of the left lumbar supracardinal vein of the embryo. Occasionally the inferior vena cava joins the azygos vein, which is then of large size. In such cases, the superior vena cava receives all the blood from the body before transmitting it to the right atrium, except the blood from the hepatic veins, which passes directly into the right atrium.

Collateral Circulation.—The inferior vena cava below the renal arteries is sometimes ligated for thrombosis with favorable results. The channels and anastomoses carrying the collateral circulation are as follows: (1) the vertebral veins, (2) anastomoses between the lumbar veins and the ascending lumbar of the azygos system, (3) anastomoses between the superior and inferior rectal veins, (4) the thoracoepigastric vein, which connects the superficial inferior epigastric with the lateral thoracic vein, (5) several anastomoses with the portal system of veins.

Tributaries.—The inferior vena cava receives blood from the two common iliacs and the following veins:

1. Lumbar
2. Testicular
3. Ovarian
4. Renal
5. Suprarenal
6. Inferior phrenic
7. Hepatic

1. The **Lumbar Veins** (*vv. lumbales*), *four* in number on each side, collect the blood by dorsal tributaries from the muscles and integument of the loins, and by abdominal tributaries from the walls of the abdomen, where they communicate with the epigastric veins. At the vertebral column, they receive veins from the vertebral plexuses, and then pass ventrally, around the sides of the bodies of the vertebrae, dorsal to the Psoas major, and end in the dorsal part of the inferior cava. The left lumbar veins are longer than the right, and pass dorsal to the aorta. The lumbar veins are connected by a longitudinal vein which passes ventral to the transverse processes of the lumbar vertebrae, and is called the **ascending lumbar vein;** it is most frequently the origin of the corresponding azygos or hemiazygos vein, and provides anastomoses between the common iliac, iliolumbar, and azygos or hemiazygos veins of its own side of the body (Fig. 9–30).

2. The **Testicular Veins** (*spermatic veins*) (Fig. 9–31).—A number of small veins emerge from the back of the testis, receive tributaries from the epididymis, and unite to form a convoluted plexus called the **pampiniform plexus,** which constitutes the greater mass of the spermatic cord; the numerous vessels composing this plexus ascend along the cord, anterior to the ductus deferens. Distal to the superficial inguinal ring they unite to form three or four veins, which pass along the inguinal canal, and, entering the abdomen through the deep inguinal ring, coalesce to form two veins, which ascend on the Psoas major, under cover of the peritoneum, lying on either side of the testicular artery and unite to form a single vein. The **right testicular vein** opens into the inferior vena cava at an acute angle; the **left testicular vein** into the left renal vein, at a right angle. The testicular veins are provided with valves. The left testicular vein passes dorsal to the iliac colon, and is thus exposed to pressure from the contents of that part of the bowel.

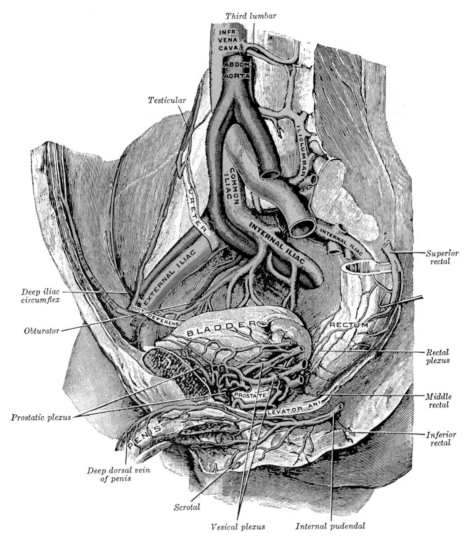

Fig. 9–30.—The veins of the right half of the male pelvis. (Spalteholz.)

3. The **Ovarian Veins** (*vv. ovaricae*) (Fig. 9–33) correspond with the testicular in the male; they form a plexus in the broad ligament near the ovary and uterine tube, and communicate with the uterine plexus. They end in the same way as the testicular veins in the male. Valves are occasionally found in these veins. Like the uterine veins, they become much enlarged during pregnancy.

4. The **Renal Veins** (*vv. renales*) (Fig. 9–27) are of large size, and pass ventral to the renal arteries. The left is longer than the right, and passes ventral to the aorta,

just caudal to the origin of the superior mesenteric artery. It receives the left testicular and left inferior phrenic veins, and generally, the left suprarenal vein. The left opens into the inferior vena cava at a slightly higher level than the right.

5. The **Suprarenal Veins** (*vv. suprarenales*) are two in number: the right ends in the inferior vena cava; the left, in the left renal or left inferior phrenic vein.

6. The **Inferior Phrenic Veins** (*vv. phrenicae inferiores*) follow the course of the inferior phrenic arteries; the right ends in the inferior vena cava; the left is often

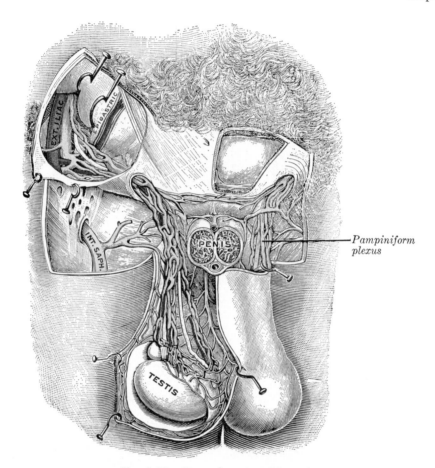

Pampiniform plexus

Fig. 9–31.—Testicular veins. (Testut.)

represented by two branches, one of which ends in the left renal or suprarenal vein, while the other passes ventral to the esophageal hiatus in the diaphragm and opens into the inferior vena cava.

7. The **Hepatic Veins** (*vv. hepaticae*) commence in the substance of the liver from the central or intralobular and sublobular veins which unite to form larger and larger veins, arranged in two groups. The **upper group** usually consists of three large veins, which converge toward the posterior surface of the liver, and open into the inferior vena cava, while that vessel is situated in the groove on the dorsal part of the liver. The veins of the **lower group** vary in number, and are of small size; they come from the right and caudate lobes. The tributaries of the hepatic veins run singly, and are in direct contact with the hepatic tissue. They are destitute of values (Fig. 16–95).

THE VEINS OF THE PELVIS AND PERINEUM

The **Common Iliac Veins** (*vv. iliacae communes*) (Fig. 9–30) are the distal continuation of the inferior vena cava at its bifurcation. They are formed by the union of the internal and external iliac veins, ventral to the sacroiliac articulation. Passing obliquely cranialward toward the right side, they end upon the fifth lumbar vertebra by uniting with each other at an acute angle to form the inferior vena cava. The **right common iliac** is shorter than the left, nearly vertical in direction, and ascends dorsal and then lateral to its corresponding artery. The **left common iliac**, longer and more oblique in its course, is at first situated on the medial side of the corresponding artery, and then dorsal to the right common iliac artery. Each common iliac receives the iliolumbar, and sometimes the lateral sacral veins. The left receives, in addition, the middle sacral vein. No valves are found in these veins.

The **middle sacral veins** (*vv. sacrales mediales*) accompany the corresponding artery along the hollow of the sacrum, and join to form a single vein, which ends in the left common iliac vein; sometimes in the angle of junction of the two iliac veins.

Variations.—The left common iliac vein, instead of joining with the right in its usual position, occasionally ascends on the left side of the aorta as high as the kidney, where, after receiving the left renal vein, it crosses over the aorta, and then joins with the right vein to form the vena cava. In these bodies, the two common iliacs are connected by a small communicating branch at the spot where they are usually united. This variation represents a persisting left lumbar supracardinal (Fig. 9–5).

Internal Iliac Vein

The **internal iliac vein** (*v. iliaca interna; hypogastric vein*) (Figs. 9–27, 9–30) begins near the upper part of the greater sciatic foramen, passes cranialward dorsal and slightly medial to the corresponding artery and, at the brim of the pelvis, joins with the external iliac to form the common iliac vein.

Tributaries.—With the exception of the fetal umbilical vein which passes cranialward and dorsalward from the umbilicus to the liver, and the iliolumbar vein which usually joins the common iliac vein, the tributaries of the internal iliac vein correspond with the branches of the internal iliac artery. It receives the following:

1. Superior gluteal
2. Inferior gluteal
3. Internal pudendal
4. Obturator
5. Lateral sacral
6. Middle rectal
7. Dorsal veins of penis
8. Vesical
9. Uterine
10. Vaginal

1. The **Superior Gluteal Veins** (*vv. glutaeae superiores; gluteal veins*) are venae

comitantes of the superior gluteal artery; they receive tributaries from the buttock corresponding with the branches of the artery, and enter the pelvis through the greater sciatic foramen, above the Piriformis, and frequently unite before ending in the internal iliac vein.

2. The **Inferior Gluteal Veins** (*vv. glutaeae inferiores; sciatic veins*), or venae comitantes of the inferior gluteal artery, begin on the proximal part of the posterior thigh, where they anastomose with the medial femoral circumflex and first perforating veins. They enter the pelvis through the lower part of the greater sciatic foramen and join to form a single stem which opens into the distal part of the internal iliac vein.

3. The **Internal Pudendal Veins** (*internal pudic veins* (Fig. 9–30) are the venae comitantes of the internal pudendal artery. They begin in the deep veins of the penis which issue from the corpus cavernosum penis, accompany the internal pudendal artery, and unite to form a single vessel, which ends in the internal iliac vein. They receive the veins from the urethral bulb, and the perineal and inferior rectal veins. The deep dorsal vein of the penis communicates with the internal pudendal veins, but ends mainly in the pudendal plexus.

4. The **Obturator Vein** (*v. obturatoria*) begins in the proximal portion of the adductor region of the thigh and enters the pelvis through the upper part of the obturator foramen. It runs dorsalward and cranialward on the lateral wall of the pelvis below the obturator artery, and then passes between the ureter and the internal iliac artery, to end in the internal iliac vein.

5. The **Lateral Sacral Veins** (*vv. sacrales laterales*) accompany the lateral sacral arteries on the anterior surface of the sacrum and end in the internal iliac vein.

6. The **Middle Rectal Vein** (*middle hemorrhoidal vein*) takes origin in the rectal plexus and receives tributaries from the bladder, prostate, and seminal vesicle; it runs lateralward on the pelvic surface of the Levator ani to end in the internal iliac vein.

The **rectal plexus** (*hemorrhoidal plexus*) (Fig. 9–30) surrounds the rectum, and communicates ventrally with the vesical plexus in the male, and the uterovaginal plexus in

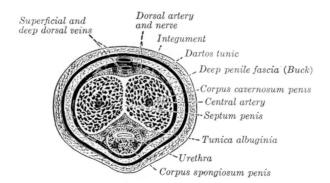

Fig. 9–32.—The penis in transverse section showing the blood vessels.

the female. It consists of two parts, an **internal** in the submucosa, and an **external** outside of the muscular coat. The internal plexus presents a series of dilated pouches which are arranged in a circle around the tube, immediately above the anal orifice, and are connected by transverse branches.

The lower part of the external plexus is drained by the **inferior rectal** veins into the internal pudendal vein; the middle part by the **middle rectal** vein which joins the internal iliac vein; and the upper part by the **superior rectal** vein which forms the commencement of the inferior mesenteric vein, a tributary of the portal vein. A free communication between the portal and systemic venous systems is established through the rectal plexus.

The veins of the rectal plexus are contained in very loose, connective tissue, so that they get less support from surrounding structures than most other veins, and are less capable of resisting increased blood pressure.

7. The **Dorsal Veins of the Penis** (*vv. dorsales penis*) (Figs. 9–30, 9–32) are two in number, a superficial and a deep. The **superficial vein** drains the prepuce and skin of the penis, and, running proximally in the subcutaneous tissue, inclines to the right or left, and opens into the corresponding superficial external pudendal vein, a tributary of the great saphenous vein. The **deep vein** lies beneath the deep fascia of the penis; it receives the blood from the glans penis and corpora cavernosa penis and courses dorsalward in the midline between the dorsal arteries; near the root of the penis

it passes between the two parts of the suspensory ligament and then through an aperture between the arcuate pubic ligament and the transverse ligament of the pelvis, and divides into two branches, which enter the pudendal plexus. The deep vein also anastomoses distal to the symphysis pubis with the internal pudendal vein.

8. The **Vesical Veins** (*vv. vesicales*) envelop the caudal part of the bladder and base of the prostate. They form the **vesical plexus** and anastomose with the pudendal and prostatic plexuses. They drain by means of several veins into the internal iliac veins.

The **pudendal plexus** (*plexus pudendalis; vesicoprostatis plexus*) (Fig. 9–30) lies deep to the arcuate pubic ligament and the symphysis pubis, and ventral to the bladder and prostate. Its chief tributary is the deep dorsal vein of the penis, but it also receives tributaries from the bladder and prostate. It anastomoses with the vesical plexus and with the internal pudendal vein and drains into the vesical and internal iliac veins. The **prostatic veins** form a well-marked **prostatic plexus** which lies partly in the fascial sheath of the prostate and partly between the sheath and the prostatic capsule. It anastomoses with the pudendal and vesical plexuses and with tributaries of the vertebral veins.

9. The **Uterine Veins** (*vv. uterinae*) (Fig. 9–33) lie along the sides and superior angles of the uterus between the two layers of the broad ligament. They form the **uterine plexuses** and anastomose with the ovarian

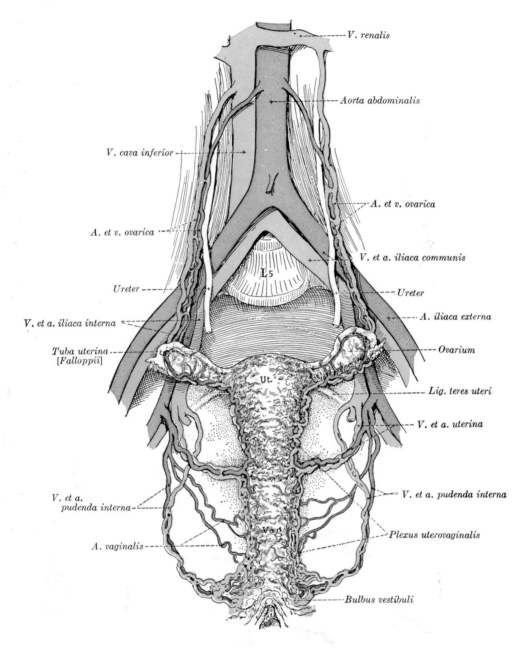

Fig. 9–33.—The blood vessels of the female pelvis showing chief source of blood supply of the uterus and vagina. (Eycleshymer and Jones.)

and vaginal plexuses. They drain through a pair of uterine veins on either side: these arise from the inferior part of the plexuses, opposite the external orifice of the uterus, and open into the corresponding internal iliac vein.

10. The **Vaginal Veins** (*vv. vaginales*) are placed at the sides of the vagina; they form the **vaginal plexuses** which anastomose with the uterine, vesical, and rectal plexuses, and are drained by a vein on either side, into the internal iliac veins.

THE VEINS OF THE LOWER LIMB

The **External Iliac Vein** (*v. iliaca externa*) receives the blood from the lower limb and the inferior part of the anterior abdominal wall. It is the cranialward continuation of the femoral vein, and begins under the inguinal ligament, and, passing cranialward along the brim of the lesser pelvis, ends opposite the sacroiliac articulation, by uniting with the internal iliac vein to form the common iliac vein. On the right side, it lies at first medial to the artery: but, as it passes cranialward, gradually inclines dorsal to it. On the left side, it lies altogether on the medial side of the artery. It frequently contains one, sometimes two valves.

Tributaries.—The external iliac vein receives the (1) inferior epigastric, (2) deep iliac circumflex, and (3) pubic veins (Fig. 9–30).

1. The **Inferior Epigastric Vein** (*v. epigastrica inferior; deep epigastric vein*) is formed by the union of the venae comitantes of the inferior epigastric artery, which anastomose cranially with the superior epigastric vein; it joins the external iliac about 1.25 cm proximal to the inguinal ligament.

2. The **Deep Iliac Circumflex Vein** (*v. circumflexa ilii profunda*) is formed by the union of the venae comitantes of the deep iliac circumflex artery, and joins the external iliac vein about 2 cm above the inguinal ligament.

3. The **Pubic Vein** communicates with the obturator vein in the obturator foramen, and ascends on the dorsum of the pubis to the external iliac vein.

The veins of the lower limb are subdivided, like those of the upper, into two sets, **superficial** and **deep**; the superficial veins are placed beneath the integument between the two layers of subcutaneous

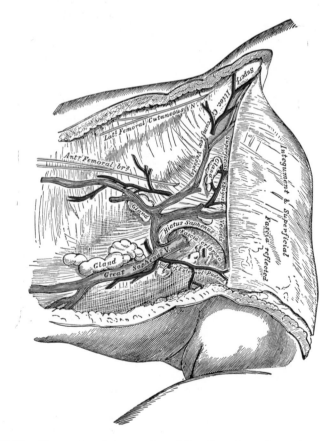

Fig. 9–34.—The great saphenous vein and its tributaries at the saphenous hiatus.

fascia; the deep veins accompany the arteries. Both sets of veins are provided with valves, which are more numerous in the deep than in the superficial set. Valves are also more numerous in the veins of the lower than in those of the upper limb.

The Superficial Veins of the Lower Limb

The **superficial veins** of the lower limb are the **great** and **small saphenous veins** and their tributaries.

The **Great Saphenous Vein** (*v. saphena magna; internal or long saphenous vein*) (Figs. 9–34, 9–35), the longest vein in the body, begins in the medial marginal vein of the dorsum of the foot and ends in the femoral vein about 3 cm below the inguinal ligament. It ascends anterior to the tibial malleolus and along the medial side of the leg in relation with the saphenous nerve. It runs proximally posterior to the medial condyles of the tibia and femur and along the medial side of the thigh and, passing through the *hiatus saphenus (fossa ovalis)*, ends in the femoral vein.

Tributaries.—At the ankle it receives tributaries from the sole of the foot through the medial marginal vein; in the leg it anastomoses freely with the small saphenous vein, communicates with the anterior and posterior tibial veins and receives many cutaneous veins; in the thigh it anastomoses with the femoral vein and receives numerous tributaries; those from the medial and posterior parts of the thigh frequently unite to form a large **accessory saphenous vein** which joins the main vein at a variable level. Near the saphenous hiatus it is joined by the superficial epigastric, superficial iliac circumflex, and superficial external pudendal veins.

The **thoracoepigastric vein** is the name given to the anastomosis between the superficial epigastric vein and the lateral thoracic vein. It runs along the anterior or lateral aspect of the trunk, covered only by the skin and subcutaneous fascia.

Clinical Considerations.—The thoracoepigastric vein provides an important collateral channel in obstruction of the inferior vena cava or portal vein. When this collateral is well developed, the vein becomes greatly dilated and tortuous, forming what is known as a *caput medusae*.

The valves in the great saphenous vein vary from ten to twenty in number; they are more numerous in the leg than in the thigh.

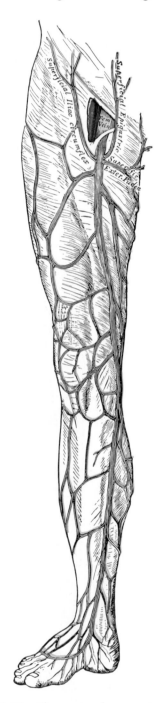

FIG. 9–35.—The great saphenous vein and its tributaries.

The **small saphenous vein** (*v. saphena parva; external or short saphenous vein*) (Fig. 9–36) begins posterior to the lateral malleolus as a continuation of the lateral marginal vein; it first ascends along the lateral margin of the tendo calcanei and then crosses it to reach the middle of the back of the leg. Running directly proximalward, it perforates the deep fascia in the distal part of the popliteal fossa, and ends in the popliteal vein between the heads of the Gastrocnemius. It anastomoses with the deep veins on the dorsum of the foot, and receives numerous large tributaries from the back of the leg. Before it pierces the deep fascia, it joins a vessel which runs proximally and anteriorly to anastomose with the great saphenous vein. The small saphenous vein possesses from nine to twelve valves, one of which is always found near its termination in the popliteal vein. In the distal third of the leg the small saphenous vein is in close relation with the sural nerve, in the proximal two-thirds with the medial sural cutaneous nerve.

On the **dorsum of the foot** the **dorsal digital veins** receive, in the clefts between the toes, the **intercapitular veins** from the plantar cutaneous venous arch and join to form short **common digital veins** which unite across the distal ends of the metatarsal bones in a **dorsal venous arch.** Proximal to this arch is an irregular venous network which receives tributaries from the deep veins and is joined at the sides of the foot by a **medial** and a **lateral marginal vein,** formed mainly by the union of branches from the superficial parts of the sole of the foot.

On the **sole of the foot** the superficial veins form a **plantar cutaneous venous arch** which extends across the roots of the toes and opens at the sides of the foot into the medial and lateral marginal veins. Proximal to this arch is a **plantar cutaneous venous network** which is especially dense in the fat beneath the heel; this network communicates with the cutaneous venous arch and with the deep veins, but is chiefly drained into the medial and lateral marginal veins.

The Deep Veins of the Lower Limb

The **deep veins** of the lower limb accompany the arteries and their branches; they possess numerous valves.

The **Plantar Digital Veins** (*vv. digitales plantares*) *arise* from plexuses on the plantar surfaces of the digits, and, after sending **intercapitular veins** to join the dorsal digital veins, unite to form four **metatarsal veins;** these run proximally in the metatarsal spaces, anastomose, by means of perforating veins, with the veins on the dorsum of the foot, and unite to form the **deep plantar venous arch** which lies alongside the plantar arterial arch. From the deep plantar venous arch the **medial** and **lateral plantar veins** run proximally close to the corresponding arteries and, after anastomosing with the great and small saphenous veins, posterior to the medial malleolus, unite to form the posterior tibial veins.

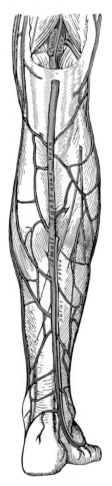

Fig. 9–36.—The small saphenous vein.

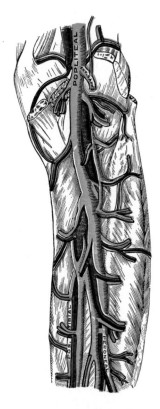

The **Posterior Tibial Veins** (*vv. tibiales posteriores*) accompany the posterior tibial artery, and are joined by the **peroneal veins.**

The **Anterior Tibial Veins** (*vv. tibiales anteriores*) are the upward continuation of the venae comitantes of the dorsalis pedis artery. They leave the anterior leg by passing between the tibia and fibula, over the interosseous membrane, and unite with the posterior tibial, to form the **popliteal vein.**

The **Popliteal Vein** (*v. poplitea*) (Fig. 9–37) is formed by the junction of the anterior and posterior tibial veins at the distal border of the Popliteus; it ascends through the popliteal fossa to the aperture in the Adductor magnus, where it becomes the femoral vein. In the distal part of its course it is placed medial to the artery; between the heads of the Gastrocnemius it is superficial to that vessel; but proximal to the knee joint, it is close to its lateral side. It receives tributaries corresponding to the branches of the popliteal artery, and it also receives the small saphenous vein. The valves in the popliteal vein are usually four in number.

The **Femoral Vein** (*v. femoralis*) accompanies the femoral artery through the proximal two-thirds of the thigh. In the distal part of its course it lies lateral to the artery; proximally, it is deep to it; and at the inguinal ligament, it lies on its medial side, and on the same plane. It receives numerous muscular tributaries, and about 4 cm below the inguinal ligament is joined by the v. profunda femoris; near its termination it is joined by the great saphenous vein. The valves in the femoral vein are three in number.

The **deep femoral vein** (*v. profunda femoris*) receives tributaries corresponding to the perforating branches of the profunda artery, and through these establishes anastomoses with the popliteal vein distally and the inferior gluteal vein proximally. It also receives the medial and lateral femoral circumflex veins.

THE PORTAL SYSTEM OF VEINS

The **portal system** (Fig. 9–38) includes all the veins which drain the blood from the abdominal part of the digestive tube and from the spleen, pancreas, and gallbladder. From these viscera the blood is conveyed to the liver by the **portal vein.** In the liver this vein ramifies like an artery and ends in capillary-like vessels termed **sinusoids**, from which the blood is conveyed to the inferior vena cava by the **hepatic veins.** From this it will be seen that the blood of the portal system passes through two sets of minute vessels, viz., (*a*) the capillaries of the digestive tube, spleen, pancreas, and gallbladder; and (*b*) the sinusoids of the liver. In the adult the portal vein and its tributaries are destitute of valves; in the fetus and for a short time after birth valves can be demonstrated in the tributaries of the portal vein; as a rule they soon atrophy and disappear, but in some subjects they persist in a degenerate form.

The **portal vein** (*vena portae*) is about 8 cm in length, and is formed at the level of the second lumbar vertebra by the junction of the superior mesenteric and splenic veins, the union of these veins taking place ventral to the inferior vena cava and dorsal to the neck of the pancreas. It passes dorsal to the superior part of the duodenum and then ascends in the right border of the lesser omentum to the right portion of the porta hepatis, where it divides into a right and a left branch, which accompany the corresponding branches of the hepatic artery into the substance of the liver. In the lesser omentum it is placed dorsal to and between

the common bile duct and the hepatic artery, the former lying to the right of the latter. It is surrounded by the hepatic plexus of nerves, and is accompanied by numerous lymphatic vessels and some lymph nodes. The **right branch** of the portal vein enters the right lobe of the liver, but before doing so generally receives the cystic vein. The **left branch**, longer but of smaller caliber than the right, crosses the left sagittal fossa, gives branches to the caudate lobe, and then enters the left lobe of the liver. As it crosses the left sagittal fossa it is joined ventrally by a fibrous cord, the **ligamentum teres** (*obliterated umbilical*

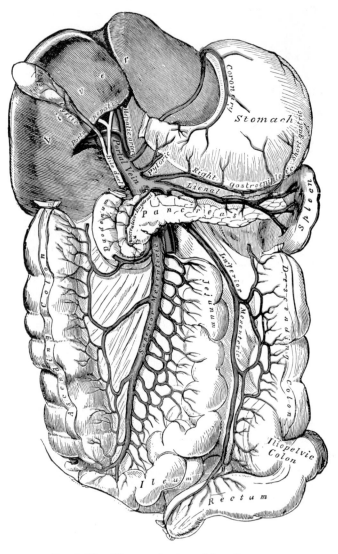

FIG. 9–38.—The portal vein and its tributaries.

vein), and is united to the inferior vena cava by a second fibrous cord, the **ligamentum venosum** (*obliterated ductus venosus*) (Fig. 16–93).

Tributaries.—The tributaries of the portal vein are:

    1. Lienal (splenic)
    2. Superior mesenteric
    3. Coronary
    4. Pyloric
    5. Cystic
    6. Parumbilical

1. The **Lienal Vein** (*v. lienalis; splenic vein*) commences by five or six large tributaries which return the blood from the spleen. These unite to form a single vessel, which passes from left to right, grooving the superior and dorsal part of the pancreas, caudal to the lienal artery, and ends dorsal to the neck of the pancreas by uniting at a right angle with the superior mesenteric to form the portal vein. The lienal vein is of large size, but is not tortuous like the artery.

Tributaries.—The lienal vein receives the short gastric veins, the left gastroepiploic vein, the pancreatic veins, and the inferior mesenteric veins.

a) The **short gastric veins** (*vv. gastricae breves*), four or five in number, drain the fundus and left part of the greater curvature of the stomach, and pass between the two layers of the gastrolienal ligament to end in the lienal vein or in one of its large tributaries.

b) The **left gastroepiploic vein** (*v. gastroepiploica sinistra*) receives tributaries from both ventral and dorsal surfaces of the stomach and from the greater omentum; it runs from right to left along the greater curvature of the stomach and ends in the lienal vein.

c) The **pancreatic veins** (*vv. pancreaticae*) consist of several small vessels which drain the body and tail of the pancreas, and open into the trunk of the lienal vein.

d) The **inferior mesenteric vein** (*v. mesenterica inferior*) returns blood from the rectum, and the sigmoid and descending parts of the colon. It begins in the rectum as the **superior rectal vein,** which has its origin in the rectal plexus, and through this

plexus communicates with the middle and inferior rectal veins. The superior rectal vein leaves the lesser pelvis and crosses the left common iliac vessels with the superior rectal artery, and is continued cranialward as the inferior mesenteric vein. This vein lies to the left of its artery, and ascends under cover of the peritoneum and ventral to the left Psoas major; it then passes dorsal to the body of the pancreas and opens into the lienal vein; sometimes (10 per cent) it ends in the angle of union of the lienal and superior mesenteric veins, or drains into the superior mesenteric vein.

Tributaries.—The inferior mesenteric vein receives the **sigmoid veins** from the sigmoid colon and iliac colon, and the **left colic vein** from the descending colon and left colic flexure.

2. The **Superior Mesenteric Vein** (*v. mesenterica superior*) returns the blood from the small intestine, from the cecum, and from the ascending and transverse portions of the colon. It begins in the right iliac fossa by the union of the veins which drain the terminal part of the ileum, the cecum, and vermiform process, and ascends between the two layers of the mesentery on the right side of the superior mesenteric artery. In its cranialward course it passes ventral to the right ureter, the inferior vena cava, the inferior part of the duodenum, and the lower portion of the head of the pancreas. Dorsal to the neck of the pancreas it unites with the lienal vein to form the portal vein.

Tributaries.—Besides the tributaries which correspond with the branches of the superior mesenteric artery, viz., the **intestinal, ileocolic, right colic,** and **middle colic veins,** the superior mesenteric vein is joined by the right gastroepiploic and pancreaticoduodenal veins.

a) The **right gastroepiploic vein** (*v. gastroepiploica dextra*) receives tributaries from the greater omentum and from parts of both ventral and dorsal surfaces of the stomach; anastomosing with the left gastroepiploic it runs from left to right along the greater curvature of the stomach between the two layers of the greater omentum.

b) The **pancreaticoduodenal veins** (*vv. pancreaticoduodenales*) accompany their corresponding arteries; the inferior of the two frequently joins the right gastroepiploic vein.

3. The **Coronary Vein** (*v. coronaria ventriculi; gastric vein*) receives tributaries from both surfaces of the stomach; it runs from right to left along the lesser curvature of the stomach, between the two layers of the lesser omentum, to the esophageal opening of the stomach, where it receives some esophageal veins. It then turns caudalward and passes from left to right dorsal to the peritoneum of the lesser sac (omental bursa) and ends in the portal vein.

4. The **Pyloric Vein** is of small size, and runs from left to right along the pyloric portion of the lesser curvature of the stomach, between the two layers of the lesser omentum, to end in the portal vein.

5. The **Cystic Veins** (*vv. cysticae*) are several small vessels which drain the blood from the gallbladder, and, accompanying the cystic duct, usually end in the right branch of the portal vein.

6. **Parumbilical Veins** (*vv. parumbilicales*).—In the course of the ligamentum teres of the liver and of the middle umbilical ligament, small veins (*parumbilical*) are found which establish an anastomosis between the veins of the anterior abdominal wall and the portal, internal and common iliac veins. The best marked of these small veins is one which commences at the umbilicus and runs dorsalward and cranialward in, or on the surface of, the ligamentum teres between the layers of the falciform ligament to end in the left portal vein.

Collateral venous circulation to relieve portal obstruction in the liver is effected mainly: (*a*) by anastomoses between the gastric veins and the esophageal veins which empty into the azygos system, and (*b*) by anastomoses between the inferior mesenteric veins and the rectal veins that empty into the internal iliac veins. Other possible collaterals which are not of much practical importance are (*c*) the accessory portal system of Sappey, branches of which pass in the round and falciform ligaments to unite with the superior and inferior epigastric, and internal thoracic veins, and through the diaphragmatic veins with the azygos; a single large vein, shown to be a parumbilical vein, may pass from the hilum of the liver by the round ligament to the umbilicus, producing there a group of prominent varicose veins known as the *caput medusae;* and (*d*) the veins of Retzius, which connect the intestinal veins with the inferior vena cava and its retroperitoneal branches.

HISTOLOGY OF THE BLOOD VESSELS

Arteries.—The arteries are composed of three coats: (1) an internal coat, the tunica intima; (2) a middle coat, the tunica media, and (3) an external coat, the tunica adventitia (Fig. 9–39). The composition and relative thickness of these coats is different in arteries of different size, a medium-sized artery, 3 to 4 mm in diameter, such as the distal portion of the radial artery has the characteristic features well represented.

(1) The **inner coat** (*tunica intima*) (Figs. 9–40, 9–41) is composed of: (a) a pavement membrane of **endothelium**, continuous with the endothelium of capillaries on the one hand and the endocardium of the heart on the other. The endothelium is a single layer of simple squamous or plate-like cells, polygonal, oval, or fusiform in shape, with rounded oval or flattened nuclei. The outlines of the cells can be brought out by treatment with silver nitrate. (b) A **subendothelial layer** of delicate connective tissue intervenes between the endothelium, and (c) the **internal elastic membrane.** The latter consists of a network of elastic fibers arranged more or less longitudinally, leaving elongated apertures or perforations which give it a fenestrated appearance and account for its name, the *fenestrated membrane* (of Henle). This membrane forms the chief thickness of the intima. It is thrown into folds when the artery is empty and in a microscopic cross section appears as a wavy line, glassy and almost unstained with hematoxylin and eosin but stained heavily with special stains for elastic fibers (Fig. 9–39).

(2) The **middle coat** (*tunica media*) (Figs. 9–39, 9–40) makes up the bulk of the wall of an artery. It is composed of lamellae of smooth muscle cells and elastic tissue which are disposed circularly around the vessel. The thickness of this coat as well as its composition varies with the size of the vessel.

(3) The **external coat** (*tunica adventitia*) (Figs. 9–39, 9–40) consists of areolar connective tissue with a fine feltwork of collagenous and elastic fibers. The elastic tissue is more abundant adjacent to the tunica media and is sometimes called the **tunica elastica externa.**

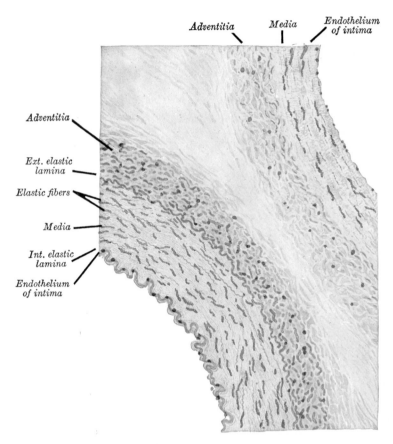

FIG. 9–39.—Cross section of parts of the walls of adjacent medium-sized artery and vein. Artery, lower left; vein, upper right.

The tunica adventitia contains the arteries and veins supplying the vessel walls, the vasa vasorum, in larger arteries and it also contains lymphatic vessels and the nerve fibers on their way to the cells of the smooth muscle coat.

Although the arteries of different sizes make gradual transitions into each other certain characteristic features should be mentioned. **Small arteries** have the different coats greatly reduced in thickness. The inner coat is composed primarily of the endothelium, the inner elastic membrane being scarcely identifiable. The middle coat contains only a single layer of smooth muscle cells (Fig. 9–41) in the smallest arterioles but it is relatively thick in small arteries (Fig. 9–40) and the elastic fibers are very sparse or non-existent. The outer coat is also very thin, being composed of fine collagenous and reticular fibrils.

Medium-sized arteries have already been described as the typical arteries and include most of the named arteries which distribute the blood to the different organs and parts of the

body. The muscular coat is especially well developed and by its innervation through the vasomotor nerves from the sympathetic system it controls the flow of blood into a particular area. In the somewhat smaller arteries the tunica media is almost entirely muscular, but in larger ones more and more of the muscle is replaced by elastic tissue. The thickness of the adventitia is variable; in arteries from protected areas such as those in the abdominal or cranial cavities it is relatively thin but it is much thickened in exposed areas such as the limbs. The bundles of nerve fibers are numerous in the adventitia of many arteries and on the arteries supplying the abdominal organs frequently form an incomplete coat.

Large arteries have a thickened intima and in older people this layer is likely to contain plaques of cholesterol or calcium salts and other pathological changes. The internal elastic layer is made up of a number of strata rather than a single membrane. The middle coat contains relatively large amounts of elastic tissue inter-

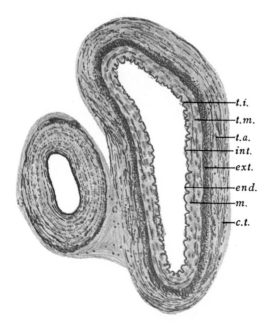

t.i.

t.m.

t.a.

int.

ext.

end.

m.

c.t.

Fig. 9–40.—Cross section of a small artery and accompanying vein from a 22-year-old executed man. The artery is labeled. *c.t.*, connective tissue; *end.*, endothelium; *ext.*, external elastic membrane; *int.*, internal elastic membrane; *m.*, smooth muscle cells; *t.a.*, tunica adventitia; *t.i.*, tunica intima; *t.m.*, tunica media. × 60. (Redrawn from Sobotta.)

spersed between muscular lamellae and held in place by areolar connective tissue. The adventitia is quite thick and contains well-developed vasa vasorum, bundles of nerves and frequent lymphatic vessels. The **aorta** (Fig. 9–42) has these features most highly developed but they are also visible in the brachio-cephalic, carotid, subclavian, axillary, vertebral, and iliac arteries. The pulmonary trunk and arteries resemble the aorta but all coats are thinner.

The Capillaries.—The smaller arterial branches (excepting those of the cavernous structure of the sexual organs, of the splenic pulp, and of the placenta) terminate in networks of vessels which pervade nearly every tissue of the body. These vessels, from their

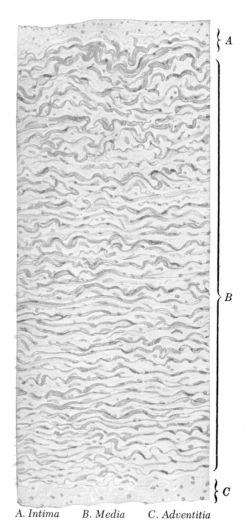

A. Intima   B. Media   C. Adventitia

Fig. 9–42.—A portion of a transverse section of the wall of the aorta. *A*, Intima; *B*, Media; *C*, Adventitia.

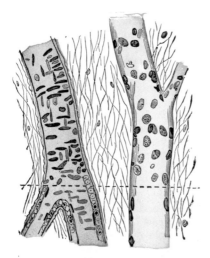

Fig. 9–41.—Small artery and vein, pia mater of sheep. × 250. Surface view above the interrupted line; longitudinal section below. Artery in red; vein in blue.

minute size, are termed capillaries. They are interposed between the smallest branches of the arteries and the commencing veins, constituting a network, the branches of which maintain the same diameter throughout; the meshes of the network are more uniform in shape and size than those formed by the anastomoses of the small arteries and veins.

The *diameters* of the capillaries vary in the different tissues of the body, the usual size being about 8 μ. The smallest are those of the brain and the mucous membrane of the intestines; and the largest those of the skin and the marrow of bone, where they are stated to be as large as 20 μ in diameter. The *form* of the capillary net varies in the different tissue, the meshes being generally rounded or elongated.

The *rounded form of mesh* is most common, and prevails where there is a dense network, as in the lungs, in most glands and mucous membranes and in the cutis; the meshes are not of an absolutely circular outline, but more or less angular, sometimes nearly quadrangular, or polygonal, or more often irregular.

Elongated meshes are observed in the muscles and nerves, the meshes resembling parallelograms in form, the long axis of the mesh running parallel with the long axis of the nerve or muscle. Sometimes the capillaries have a *looped arrangement;* a single vessel projecting from the common network and returning after forming one or more loops, as in the papillae of the tongue and skin.

The number of the capillaries and the size of the meshes determine the degree of vascularity of a part. The closest network and the smallest interspaces are found in the lungs and in the choroid coat of the eye. In these situations the interspaces are smaller than the capillary vessels themselves. In the intertubular plexus of the kidney, in the conjunctiva, and in the cutis, the interspaces are from three to four times as large as the capillaries which form them; and in the brain from eight to ten times as large as the capillaries in the long diameters of the meshes, and from four to six times as large in their transverse diameters. In the adventitia of arteries the width of the meshes is ten times that of the capillary vessels. As a general rule, the more active the function of the organ, the closer is its capillary net and the larger its supply of blood; the meshes of the network are very narrow in all growing parts, in the glands, and in the mucous membranes, wider in bones and ligaments which are comparatively inactive; blood vessels are nearly absent in tendons, in which very little organic change occurs after their formation. In the liver the capillaries take a more or less radial

course toward the intralobular vein, and their walls may be incomplete, so that the blood comes into direct contact with the liver cells. These vessels in the liver are not true capillaries but "sinusoids;" they are developed by the growth of columns of liver cells into the blood spaces of the embryonic organ, have an irregular lumen and have very little, if any, connective tissue covering.

Structure.—The wall of a capillary consists of a fine transparent endothelial layer, composed of cells joined edge to edge by an interstitial cement substance, and continuous with the endothelial cells which line the arteries and veins. When stained with nitrate of silver the edges which bound the endothelial cells are brought into view (Fig. 9-43). These cells are of large size and of an irregular polygonal or lanceolate shape, each containing an oval nucleus which may be displayed by carmine or hematoxylin.

In many situations a delicate sheath or envelope of branched nucleated connective tissue cells is found around the simple capillary tube, particularly in the larger ones; and in other places, especially in the glands, the capillaries are invested with reticular connective tissue.

Veins.—The veins have the same three coats as the arteries, **tunica intima, tunica media,** and **tunica adventitia.** The thickness and composition of the three coats are different, however, and at various points the veins contain valves (Figs. 9-39, 9-44).

In the **smallest veins,** three coats are scarcely distinguishable (Fig. 9-41). The endothelium is supported by delicate reticular and elastic

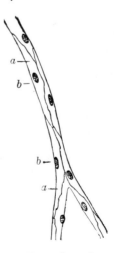

Fig. 9-43.—Capillaries from the mesentery of a guinea pig, after treatment with solution of nitrate of silver. *a*, Cells. *b*, Their nuclei.

fibers which merge with the connective tissue of the surrounding area. In slightly larger veins (0.4 mm diameter) there is a middle coat composed of connective tissue and circular smooth muscle cells and an adventitia composed of areolar connective tissue.

In **medium-sized veins** the tunica intima contains endothelial cells with nuclei more oval than those in the arteries. It is supported by delicate connective tissue and a network of elastic fibers takes the place of the fenestrated membrane of the arteries. The valves are part of the intima. The tunica media is composed of a thick layer of collagenous and elastic fibers intermingled with a variable number of circular smooth muscle cells. The elastic and muscular elements are much less numerous in veins than in arteries. The tunica adventitia consists of areolar connective tissue with longitudinal elastic fibers. In the **largest veins** the outer coat is from two to five times thicker than the middle coat, and contains a large number of longitudinal muscular fibers. These are most distinct in the inferior vena cava, especially at the termination of this vein in the heart, in the trunks of the hepatic veins, in all the large trunks of the portal vein, and in the external iliac, renal, and azygos veins. In the renal and portal veins they extend through the whole thickness of the outer coat, but in the other veins mentioned a layer of connective and elastic tissue is found external to the muscular fibers. All the large veins which open into the heart are covered for a short distance with a layer of striped muscular tissue continued on to them from the heart. Muscular

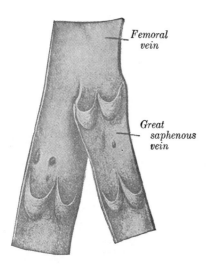

Femoral vein

Great saphenous vein

FIG. 9–44.—The proximal portions of the femoral and great saphenous veins laid open to show the valves. About two-thirds natural size.

tissue is wanting: (1) in the veins of the maternal part of the placenta; (2) in the venous sinuses of the dura mater and the veins of the pia mater of the brain and spinal cord; (3) in the veins of the retina; (4) in the veins of the cancellous tissue of bones; (5) in the venous spaces of the corpora cavernosa. The veins of the above-mentioned parts consist of an internal endothelial lining supported on one or more layers of areolar tissue.

Valves.—Most veins are provided with valves which serve to prevent the reflux of the blood. Each valve is formed by a reduplication of the inner coat, strengthened by connective tissue and elastic fibers, and is covered on both surfaces with endothelium, the arrangement of which differs on the two surfaces. On the surface of the valve next to the wall of the vein the cells are arranged transversely; while on the other surface, over which the current of blood flows, the cells are arranged longitudinally in the direction of the current. Most commonly two such valves are found placed opposite each other, more especially in the smaller veins or in the larger trunks at the point where they are joined by smaller branches; occasionally there are three and sometimes only one. The valves are semilunar. They are attached by their convex edges to the wall of the vein; the concave margins are free, directed in the course of the venous current, and lie in close apposition with the wall of the vein as long as the current of blood takes its natural course; if, however, any regurgitation takes place, the valves become distended, their opposed edges are brought into contact, and the current is interrupted. The wall of the vein on the cardiac side of the point of attachment of each valve is expanded into a pouch or sinus, which gives to the vessel, when injected or distended with blood, a knotted appearance.

The valves are very numerous in the veins of the limbs, especially of the lower limbs, these vessels having to conduct the blood against the force of gravity. They are absent in the very small veins, *i.e.*, those less than 2 mm in diameter, also in the venae cavae, hepatic, renal, uterine, and ovarian veins. A few valves are found in each testicular vein, and one also at its point of junction with the renal vein or inferior vena cava respectively. The cerebral and spinal veins, the veins of the cancellated tissue of bone, the pulmonary veins, and the umbilical vein and its branches are also destitute of valves. A few valves are occasionally found in the azygos and intercostal veins. Rudimentary valves are found in the tributaries of the portal venous system.

The veins, like the arteries, are supplied with nutrient vessels, **vasa vasorum.** Nerves also are distributed to them in the same manner as to the arteries, but in much less abundance.

BLOOD

Blood consists of a fluid medium called the **plasma** in which are suspended minute structures called the **formed elements** of the blood. The formed elements include (1) the red blood corpuscles (RBC), (2) the white blood cells (WBC), and (3) the platelets. Very minute droplets of fat called chylomicra are also present, especially after the ingestion of fatty foods.

The **blood plasma** after the formed elements are removed by centrifugation is a clear, somewhat viscous, slightly yellowish fluid rich in dissolved proteins. One of these, fibrinogen, is precipitated out as very minute threads called fibrin when the blood clots. After the blood has clotted and the fibrin as well as the formed elements has been removed, the remaining fluid resembles plasma but is called serum.

The **red blood corpuscles** or **erythrocytes** are biconcave disks with an average diameter of 7.7 micra (μ) and a thickness of 1.9 μ. Variations in size of more than 1 or 2 μ are rare in normal blood, but in anemia they may be enlarged into macrocytes (megalocytes), or reduced into microcytes. Individual erythrocytes have a pale yellowish color when viewed in a thin film of fresh blood under the high power of a microscope, but when superimposed into several layers they take on their characteristic reddish hue. They have no nucleus and appear homogeneous when viewed with either the light or the electron microscopes. They are composed principally of an iron-containing protein responsible for the color called hemoglobin and a lipoid constituent which appears to be in higher concentration in the surface membrane. The semipermeable properties of this membrane allow the erythrocyte to imbibe fluid by osmosis from a hypotonic medium, with the result that it enlarges and becomes spherical in shape. In the hypertonic medium it shrivels into a spine-covered sphere, a process called **crenation.** The membrane can be torn with a microneedle and the hemoglobin released into the surrounding medium. The membrane also ruptures when erythrocytes are suspended in distilled water, leaving a faintly visible remnant called a *ghost cell.*

In a thin, dried film of blood or **blood smear,** the erythrocytes are colored an orange pink by the eosin of Wright's stain (eosin and methylene blue). In a thin film of fresh blood stained supravitally with neutral red or brillant cresyl blue, a few erythrocytes contain dark-staining inclusions. Such cells are called **reticulocytes.** There may be 0.1 per cent in normal blood but their number is increased in conditions in which new erythrocytes are being formed rapidly in the bone marrow.

The number of erythrocytes per cubic millimeter of blood is 5,500,000 to 5,000,000 in men and 5,000,000 to 4,500,000 in women. A range of 4 to 6 million is considered normal and 8 million may be found in individuals living at high altitudes. There may be less than one million in severe anemia.

White Blood Cells, White Corpuscles, or **Leukocytes** are of different sizes, the majority being about 10 μ in diameter. They are much less numerous than erythrocytes, in a healthy individual from 7,000 to 12,000 per cubic millimeter of blood. The leukocytes are true cells, having a nucleus and cytoplasm. Several different types are recognized and are named according to (1) the condition of the nucleus as polymorphonuclear or mononuclear, (2) the presence or absence of granules in the cytoplasm as granulocytic and agranulocytic and (3) whether the granules stain with acid or basic dyes as neutrophilic, acidophilic, or basophilic. The **polymorphonuclear leukocytes** are also **granulocytes** and are given the specific names (a) **polymorphonuclear neutrophils,** (b) **eosinophils** (*acidophils*), and (c) **basophils.** The **mononuclear cells** are **agranular** and of two types based on the size, the shape of the nucleus, and the quality of the cytoplasm. The **lymphocytes** are smaller, have round nuclei, and clear cytoplasm. The **large mononuclears** or **monocytes** have crescentic nuclei and a very finely granular cytoplasm.

The leukocytes are actively amoeboid cells. Living cells can be examined in a thin film of blood under high magnification with the microscope and may be recognized by their appearance and activity. If the stage of the microscope is warmed to body temperature, the polymorphonuclear cells immediately become motile. They send out pseudopods of clear protoplasm and the granular cytoplasm and the nuclei stream after them. The neutrophils are **phagocytic,** especially for bacteria, both in the body and *in vitro.* In a dilute solution of neutral red, the granules segregate the dye and become red or brown in color. The lymphocytes in a fresh preparation are at first rounded and immobile, but after a few hours take up active locomotion. The nucleus is carried along in the front of the cell rather than behind as in the granulocytes. In a thin film of blood **supravitally stained** with a dilute solution of neutral red and janus green, shadowy, bluish-green, rod-shaped mitochondria may be seen in the hollow of the bean-shaped nucleus, and one or two tiny round neutral red vacuoles in the lymphocytes. The monocytes send out filamentous or wavy pseudopods, but undergo very little locomotion. They are actively **phagocytic** and in a supravital stain of neutral red and janus green they develop a number of relatively large red vacuoles. A few small round mitochondria occupy a deep hollow in one side of the nucleus.

Blood is usually examined in a thin dry film or **blood smear,** stained with eosin and methylene blue (a Romanovsky stain such as Wright's stain). The polymorphonuclear **neutrophils** are about 10 μ in diameter, have a dark nucleus with one to five lobes and a cytoplasm filled with small lavender or purplish granules. They are the most numerous leukocytes (60 per cent) and in an infection large numbers massed together are called pus. The polymorphonuclear **eosinophils** usually have a nucleus with two lobes and the cytoplasm is filled with large granules stained bright red with the eosin. They are not numerous (2 per cent) in normal blood but are increased in parasitic and allergic diseases. The polymorphonuclear **basophils** are the least numerous of leukocytes (0.5 per cent or less) and have a three or four lobed nucleus and large dark purple granules in the cytoplasm. The **lymphocytes** range from 6 to 9 μ in diameter and are in moderate numbers (20 per cent). The nucleus is round and stains

deeply. The cytoplasm is a transparent light blue and varies in amount according to the size of the cell. A few dark granules may be present. The **monocytes** are the largest leukocytes, 10 to 15 μ in diameter, but are not very numerous (4 per cent). The nucleus is bean- or sausage-shaped and the abundant cytoplasm has an appearance like bluish ground glass.

Blood platelets (*thromboplastids, thrombocytes*) are small (2 to 4 μ) masses of protoplasm with central granular cytoplasm but no nucleus. They may be colorless, round or stellate, and, in a fresh film, form the centers of radiating threads of fibrin. In eosin and methylene blue they stain much like the cytoplasm of neutrophils. Their number varies greatly from 200,000 to 800,000.

The average number of formed elements in 1 cubic millimeter of blood:

Erythrocytes	4,500,000 to 5,500,000
Leukocytes	6,000 to 10,000
Platelets	200,000 to 800,000

A **differential count** of white blood cells gives the following averages:

Polymorphonuclear neutrophils	50 to 75%
Polymorphonuclear eosinophils	2 to 4%
Polymorphonuclear basophils	0.5%
Lymphocytes	20 to 40%
Monocytes	3 to 8%

REFERENCES

EMBRYOLOGY

AÜER, J. 1948. The development of the human pulmonary vein and its major variations. Anat. Rec., *101,* 581-594.

McCLURE, C. F. W. and G. S. HUNTINGTON. 1929. The mammalian vena cava posterior. Amer. anat. Mem., *15,* 5-149.

PADGET, D. H. 1956. The cranial venous system in man in reference to development, adult configuration, and relation to the arteries. Amer. J. Anat., *98,* 307-356.

SHANER, R. F. 1961. The development of the bronchial veins, with special reference to anomalies of the pulmonary veins. Anat. Rec., *140,* 159-165.

WILLIAMS, A. F. 1953. The formation of the popliteal vein. Surg. Gynec. Obstet., *97,* 769-772.

VEINS

BATSON, O. V. 1940. The function of the vertebral veins and their rôle in the spread of metastases. Ann. Surg., *112,* 138-149.

BATSON, O. V. 1942. The rôle of the vertebral veins in metastatic processes. Ann. intern. Med., *16,* 38-45.

BOL'SHAKOV, O. P. 1964. Macroscopic and microscopic structural features of cavernous sinus. Fed. Proc., *23,* T308-T311.

BROWNING, H. C. 1953. The confluence of dural venous sinuses. Amer. J. Anat., *93,* 307-330.

CHARLES, C. M., T. L. FINLEY, R. D. BAIRD, and J. S. COPE. 1930. On the termination of the circumflex veins of the thigh. Anat. Rec., *46,* 125-132.

DOUGLASS, B. E., A. H. BAGGENSTOSS, and W. H. HOLLINSHEAD. 1950. The anatomy of the portal vein and its tributaries. Surg. Gynec. Obstet., *91,* 562-576.

EDWARDS, E. A. and J. D. ROBUCK, JR. 1947. Applied anatomy of the femoral vein and its tributaries. Surg. Gynec. Obstet., *85,* 547-557.

GIBSON, J. B. 1959. The hepatic veins in man and their sphincter mechanisms. J. Anat. (Lond.), *93,* 368-379.

GOSS, C. M. 1962. On anatomy of veins and arteries by Galen of Pergamon. Anat. Rec., *141,* 355-366.

HOLLINSHEAD, W. H. and J. A. McFARLANE. 1953. The collateral venous drainage from the kidney following occlusion of the renal vein in the dog. Surg. Gynec. Obstet., *97,* 213-219.

KEEN, J. A. 1941. The collateral venous circulation in a case of thrombosis of the inferior vena cava, and its embryological interpretation. Brit. J. Surg., 29, 105-114.

MASSOPUST, L. C. and W. D. GARDNER. 1950. Infrared photographic studies of the superficial thoracic veins in the female. Surg. Gynec. Obstet., 91, 717-727.

MULLARKY, R. E. 1965. The Anatomy of Varicose Veins. xvi + 89 pages, illustrated. Charles C Thomas, Springfield, Illinois.

NANDY, K. and C. B. BLAIR, JR. 1965. Double superior venae cavae with completely paired azygos veins. Anat. Rec., 151, 1-9.

SHAH, A. C. and SRIVASTAVA. 1966. Fascial canal for the small saphenous vein. J. Anat. (Lond), 100, 411-413.

VARIATIONS AND ANOMALIES

BECKER, F. F. 1962. A singular left sided inferior vena cava. Anat. Rec., 143, 117-120.

BORUCHOW, I. B. and J. JOHNSON. 1972. Obstructions of the vena cava. Review. Surg. Gyn. Obst., 134, 115-121.

BRANTIGAN, O. C. 1947. Anomalies of the pulmonary veins; their surgical significance. Surg. Gynec. Obstet., 84, 653-658.

VAN CLEAVE, C. D. 1931. A multiple anomaly of the great veins and interatrial septum in a human heart. Anat. Rec., 50, 45-51.

CONN, L. C., J. CALDER, J. W. MacGREGOR, and R. F. SHANER. 1942. Report of a case in which all pulmonary veins from both lungs drain into the superior vena cava. Anat. Rec., 83, 335-340.

DiDIO, L. J. A. 1961. The termination of the vena gastrica sinistra in 220 cadavers. Anat. Rec., 141, 141-144.

FRIEDMAN, S. M. 1945. Report of two unusual venous abnormalities (left postrenal inferior vena cava; postaortic left innominate vein). Anat. Rec., 92, 71-76.

REIS, R. H. and G. ESENTHER. 1959. Variations in the pattern of renal vessels and their relation to the type of posterior vena cava in man. Amer. J. Anat., 104, 293-318.

VALVES IN VEINS

BASMAJIAN, J. V. 1952. The distribution of valves in the femoral, external iliac, and common iliac veins and their relationship to varicose veins. Surg. Gynec. Obstet., 95, 537-542.

EDWARDS, E. A. 1936. The orientation of venous valves in relation to body surfaces. Anat. Rec., 64, 369-385.

EDWARDS, E. A. and J. E. EDWARDS. 1943. The venous valves in thromboangiitis obliterans. Arch. Path., 35, 242-252.

KAMPMEIER, O. F. and C. BIRCH. 1927. The origin and development of the venous valves, with particular reference to the saphenous district. Amer. J. Anat., 38, 451-499.

POWELL, T. and R. B. LYNN. 1951. The valves of the external iliac, femoral, and upper third of the popliteal veins. Surg. Gynec. Obstet., 92, 453-455.

HISTOLOGY OF BLOOD VESSELS

CHAMBERS, R. and B. W. ZWEIFACH. 1944. Topography and function of the mesenteric capillary circulation. Amer. J. Anat., 75, 173-205.

CHAMBERS, R. and B. W. ZWEIFACH. 1947. Intercellular cement and capillary permeability. Physiol. Rev., 27, 436-463.

CLARK, E. R. and E. L. CLARK. 1943. Caliber changers in minute blood-vessels observed in the living mammal. Amer. J. Anat., 73, 215-250.

HIBBS, R. G., G. E. BURCH, and J. H. PHILLIPS. 1958. The fine structure of the small blood vessels of normal human dermis and subcutis. Amer. Heart J., 56, 662-670.

ZWEIFACH, B. W. 1959. The microcirculation of the blood. Sci. Amer., 200, 54-60.

BLOOD AND HEMATOPOIESIS

ACKERMAN, A. G. 1962. Histochemical observations on the oxidative enzyme activity and hemoglobin synthesis in the developing erythrocytic cells of the embryonic liver. Anat. Rec., 144, 239-244.

BLOOM, W. and G. W. BARTELMEZ. 1940. Hematopoiesis in young human embryos. Amer. J. Anat., 67, 21-54.

CRAFTS, R. C. 1946. Effects of hypophysectomy, castration, and testosterone propionate on hematopoiesis in the adult male rat. Endocrinology, 39, 401-413.

HAYES, M. A. and B. L. BAKER. 1952. The effect of prolonged administration of adrenocortical extract on bone marrow cytology. Univ. Mich. med. Bull., 18, 109-114.

HUNT, T. E., E. A. HUNT, and D. CURRY. 1961. Basophil leucocytes and mast cells in the rabbit before and after intravenous injection of compound 48/80. Anat. Rec., 140, 123-133.

OSOGOE, B. and K. OMURA. 1950. Transplantation of hematopoietic tissue into the circulating blood. II. Injection of bone marrow into normal rabbits, with special reference to the histogenesis of extramedullary foci of hematopoiesis. Anat. Rec., 108, 663-686.

POLLAK, O. J. 1951. Grouping, Typing, and Banking of Blood. xiv + 163 pages. Charles C Thomas, Springfield, Illinois.

WINTROBE, M. M. 1966. Clinical Hematology, 1287 pages, 6th edition. Lea & Febiger, Philadelphia.

10 | *The Lymphatic System*

ymph is a clear, watery fluid contained in a system of vessels almost as pervasive as that for the blood. The peripheral parts of the system do not communicate with the blood vessels, but the lymph is finally emptied into the blood stream by special communications at the junction of the jugular and subclavian veins at both sides of the neck. The endothelium of the veins at these points is continuous into the lymphatic vessels.

The study of the lymphatic system in the dissecting room is very unsatisfactory because the thinness of the walls of the vessels and their small size make them indistinguishable from the surrounding connective tissues. Most of the information concerning the anatomy of the system has been obtained by laborious, special investigations with the injection of colored mass into the very minute vessels. Injection into the larger vessels is ineffective because of the presence of numerous valves.

The **lymphatic system** consists of (1) an extensive capillary network which collects lymph in the various organs and tissues; (2) an elaborate system of collecting vessels which carry the lymph from the lymphatic capillaries to the blood stream, opening into the great veins at the root of the neck; (3) a number of firm rounded bodies called lymph nodes which are placed like filters in the paths of the collecting vessels; (4) certain lymphatic organs which resemble the lymph nodes, that is, tonsils and solitary or aggregated lymphatic nodules; (5) the spleen, and (6) the thymus. Another element which might be added is the lymphoid, or, as it is often called, adenoid tissue. It is recognizable only with the aid of a microscope and consists of reticular or areolar connective tissue which contains an accumulation of lymphocytes but lacks the organized lymphatic nodules. The spleen is generally recognized as a lymphatic organ because it contains lymphatic nodules. The thymus, on the other hand, has been described previously with the ductless glands, but since no evidence of glandular function has been found, and since it is made up principally of lymphoid tissue, it is here classed as a lymphatic organ. The lymphatic capillaries and collecting vessels are lined throughout by a continuous layer of endothelial cells, forming thus a closed system. The lymphatic vessels of the small intestine receive the special designation of **lacteals** or **chyliferous vessels**; they differ in no respect from the lymphatic vessels generally except that during the process of digestion they contain a milk-white fluid, the **chyle.**

Development

The earliest lymphatic endothelium probably arises from venous endothelium as sprouts from several regions of the primitive veins. The first sprouts come from the primitive right and left internal jugular veins at their junctions with the subclavians. They grow, branch, and anastomose to form the **jugular lymph sacs** (Fig. 10–1). From these sacs lymphatic capillary plexuses spread to the neck, head, arms, and thorax. The more direct channels of the plexuses enlarge and form the lymphatic vessels. The larger vessels acquire smooth muscular coats with nerve connections, and exhibit contractility. Each jugular sac retains at least one connection with its jugular vein. From the left one the upper part of the thoracic duct develops. At a slightly older stage (8th week) lymphatic sprouts from the primitive vena cava and mesonephric veins form the unpaired

(731)

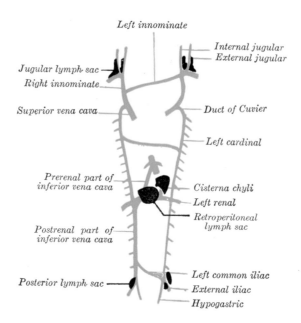

FIG. 10–1.—Scheme showing relative positions of primary lymph sacs based on the
description given by Florence Sabin ('16).

retroperitoneal lymph sac. From it, lymphatics spread to the abdominal viscera and diaphragm. The sac establishes connections with the cisterna chyli and loses its connections with the veins.

At about the same time another series of sprouts from the primitive veins of the Wolffian bodies forms the **cisterna chyli.** It gives rise to the cisterna chyli and to part of the thoracic duct. It joins that part of the duct which develops from the left jugular sac, and loses all its connections with the veins. At a slightly later stage (8th week) endothelial sprouts are given off from the primitive iliac veins at their junctions with the posterior cardinal veins. They form the paired **posterior lymph sacs.** From these sacs lymphatics spread to the abdominal wall, pelvic region, and legs. They join the cisterna chyli and lose all connections with the veins.

According to this view, all lymphatic endothelium is derived from venous endothelium which it resembles in many respects. An improbable opposing view derives lymphatic endothelium from mesenchymal cells which are supposed to flatten out into endothelium about small isolated accumulations of fluid in the neighborhood of the

large primitive veins. These little sacs are supposed to join to form the first lymphatic capillaries and lymph sacs and to acquire secondary connections with the veins. It might be pointed out that this supposed origin of lymphatic endothelium is analogous to the formation of the mesothelium of joint cavities and that true lymphatic endothelium resembles venous endothelium, not mesothelium. Endothelium forms tubes which acquire muscle walls. Mesothelial sacs do neither.

The **primary lymph nodes** begin to develop during the third month in capillary lymphatic plexuses formed out of the large lymph sacs. Secondary lymph nodes develop later, and even after birth, in peripherally located capillary lymphatic plexuses. Lymphocytes are already present in the blood stream and tissues long before lymph nodes begin to appear. The earliest stages are obscure. It seems probable that lymphocytes lodge in special regions of the capillary plexuses and multiply to form larger and larger masses of lymphocytes which become the cortical lymphoid nodules and medullary cords. Parallel with this multiplication of lymphocytes, the surrounding mesenchyme forms a connective tissue capsule from which trabeculae carrying blood vessels grow into the lymphoid tissue. In the larger trabeculae, collagenous fibers are laid down and continued as reticular fibrils into the smaller trabeculae. The mesenchymal cells of the larger trabeculae become fibroblasts, those accompanying the reticular fibrils are known as reticular cells. Just what happens to the lymphatic endothelium of the capillary plexus within the developing node is obscure. Some authors believe it forms the cortical and medullary sinuses. Others believe that the endothelial lining becomes incomplete and that the reticular cells are washed by the lymph. With this idea goes the theory that some of the reticular cells round up and become the first lymphocytes of the developing node and that they retain this potency throughout life.

The Lymphatic Vessels

Lymphatic Capillary Plexuses.—The networks which collect lymph from the intercellular fluid may be said to constitute the beginning of the lymphatic system, since from these plexuses arise the lymphatic vessels which conduct the lymph centrally through one or more lymph nodes to the thoracic duct or to the right lymphatic duct.

The number, the size, and the richness of the capillary plexuses differ in different regions and organs. Where abundant, they are usually arranged in two or more anastomosing layers. Most capillaries are without valves (Fig. 10–2).

Lymphatic capillaries are especially abundant in the **dermis** of the skin. They form a continuous network over the entire surface of the body with the exception of the cornea. The dermis has a superficial plexus, without valves, connected by many anastomoses with a somewhat wider, coarser deep plexus with a few valves. The former sends blind ends into the papillae. The plexuses are especially rich over the palmar surface of the hands and fingers, the plantar surface of the feet and toes, the conjunctiva, the scrotum, the vulva, and around the orifices where the skin becomes continuous with the mucous membranes (Fig. 10–3).

Lymphatic capillary plexuses are abundant in the **mucous membranes** of the respiratory and digestive systems. They form a continuous network from the nares and lips to the anus. In most places, there is a subepithelial plexus in the mucosa which anastomoses freely with a coarser plexus in the submucosa. Blind ends extend between the tubular glands of the stomach and into the villi of the intestine. The latter are the

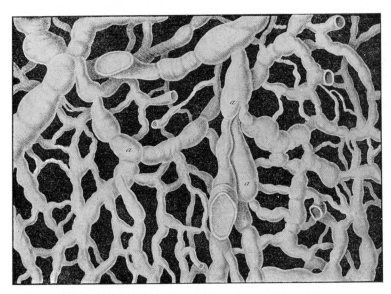

Fig. 10–2.—Lymph capillaries from the human scrotum, showing also transition from capillaries to the collecting vessels *a, a.* × 20 dia. (Teichmann.)

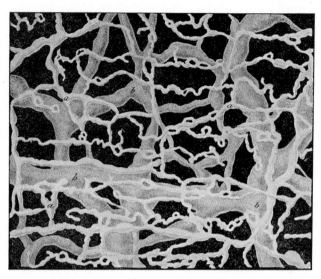

Fig. 10–3.—Lymph capillaries of the dermis from the inner border of the sole of the human foot. *a, a,* outer layer; *b, b,* inner layer. × 30 dia. (Teichmann.)

lacteals. They have a smooth muscle coat and are contractile. Those portions of the alimentary canal covered by peritoneum have a subserous capillary plexus beneath the mesothelium which anastomoses with the submucosal set.

The **lungs** have a rich subserous plexus which connects with the deep plexuses within the lungs. The latter accompany the bronchi and bronchioles. Its capillaries do not extend to the alveoli.

The **salivary glands, pancreas** and **liver** possess deep lymphatic plexuses. They are perilobular and do not extend between the epithelial cells. The gallbladder, cystic, hepatic, and common bile ducts have rich plexuses in the mucosa. The deep lymphatics of the liver deliver a copious supply of lymph. The liver and gallbladder have rich subserous plexuses.

The **kidney** has a rich network in the capsule and a deep plexus between the tubules of the parenchyma. The renal pelvis and the ureters have rich networks in the mucosa and muscular layer which are continuous with similar plexuses of the urinary bladder. The male urethra has a dense plexus in the mucosa. The capillaries are especially abundant around the navicular fossa.

The **testis**, epididymis, ductus deferens, seminal vesicles and prostate have super-ficial capillary plexuses. Both the testis and prostate have deep interstitial plexuses.

The **ovary** has a rich capillary plexus in the parenchyma. Capillaries are absent in the tunica albuginea. The uterine tubes and uterus have mucous and muscular plexuses as well as serous and subserous ones. The vagina has a rich fine-meshed plexus in the mucosa and a coarser one in the muscular layer.

Lymphatic capillary plexuses are abundant beneath the mesothelial lining of the pleural, peritoneal, and pericardial cavities and the joint capsules. The dense connective tissues of tendons, ligaments, and periosteum are richly supplied with plexuses. The **heart** has a rich subepicardial plexus and a subendocardial one. The myocardium has a uniform capillary plexus which anastomoses with both the subendocardial and subepicardial ones. There are no collecting trunks in the myocardium.

Very fine capillary plexuses have been described about the fibers of **skeletal muscle**. Their occurrence in bone and bone marrow is not settled.

No lymphatic capillaries have been found in the central nervous system, the meninges, the eyeball and orbital fat, the cornea, the internal ear, cartilage, epidermis, and the spleen.

Collecting Vessels.—Lymphatic vessels of the first order arise in the lymphatic capillary plexuses which they drain. They all enter lymph nodes as afferent vessels. From these nodes, efferent vessels usually pass to one or to a series of lymph nodes before they join the thoracic duct or the right lymphatic duct. Lymphatic vessels usually accompany blood vessels and anastomose frequently. The larger collecting vessels often extend over long distances without change of caliber. The lymphatic vessels are exceedingly delicate, and their coats are so transparent that the fluid they contain is readily seen through them. They are interrupted at intervals by constrictions, which correspond to the situations of valves in their interior and give them a knotted or beaded appearance.

The lymphatic vessels are arranged into a **superficial** and a **deep set.** On the surface of the body the **superficial** lymphatic vessels are placed immediately beneath the integument, accompanying the superficial veins; they join the deep lymphatic vessels in certain situations by perforating the deep fascia. In the interior of the body they lie in the submucous areolar tissue, throughout the whole length of the digestive, respiratory, and genitourinary tracts; and in the subserous tissue of the thoracic and abdominal walls. Plexiform networks of minute lymphatic vessels are found interspersed among the proper elements and blood vessels of the several tissues; the vessels composing the network, as well as the meshes between them, are much larger than those of the capillary plexus. From these networks small vessels emerge, which pass, either to a neighboring node, or to join some larger lymphatic trunk. The **deep** lymphatic vessels, fewer in number but larger than the superficial, accompany the deep blood vessels. The lymphatic vessels of any part or organ exceed the veins in number, but in size they are much smaller. Their anastomoses also, especially those of the large trunks, are more frequent, and are effected by vessels equal in diameter to those which they connect, the continuous trunks retaining the same diameter.

Structure of Lymphatic Vessels.—The larger lymphatic vessels are composed of three coats.

The *internal* coat is thin, transparent, slightly elastic, and consists of a layer of elongated endothelial cells with wavy margins by which the contiguous cells are dovetailed into one another; the cells are supported on an elastic membrane. The *middle* coat is composed of smooth muscular and fine elastic fibers, disposed in a transverse direction. The *external* coat consists of connective tissue, intermixed with smooth muscular fibers longitudinally or obliquely disposed; it forms a protective covering to the other coats, and serves to connect the vessel with the neighboring structures. In the smaller vessels there are no muscular or elastic fibers, and the wall consists only of a connective tissue coat, lined by endothelium. The thoracic duct has a more complex structure than the other lymphatic vessels; it presents a distinct subendothelial layer of branched cells, similar to that found in the arteries; in the middle coat there is, in addition to the muscular and elastic fibers, a layer of connective tissue with its fibers arranged longitudinally. The lymphatic vessels are supplied by **nutrient vessels,** which are distributed to their outer and middle coats, and many unmyelinated nerves in the form of a fine plexus of fibrils surround them.

The **valves** of the lymphatic vessels are formed of thin layers of fibrous tissue covered on both surfaces by endothelium which presents the same arrangement as on the valves of veins (page 726). In form the valves are semilunar; they are attached by their convex edges to the wall of the vessel, the concave edges being free and directed along the course of the contained current. Usually two such valves, of equal size, are found opposite each other, but occasionally exceptions occur, especially at or near the anastomoses of lymphatic vessels, and one valve may be of small size and the other increased in proportion.

In the lymphatic vessels the valves are placed at much shorter intervals than in the veins. They are most numerous near the lymph nodes, and are found more frequently in the lymphatic vessels of the neck and upper extremity than in those of the lower extremity. The wall of the lymphatic vessel immediately above the point of attachment of each segment of a valve is expanded into a pouch or sinus which gives to these vessels, when distended, the knotted or beaded appearance already referred to. Valves are wanting in the vessels composing the plexiform network in which the lymphatic vessels originate.

The **lymph** is propelled by contractions of the vessel walls in the larger vessels, but by the motion of surrounding structures in smaller

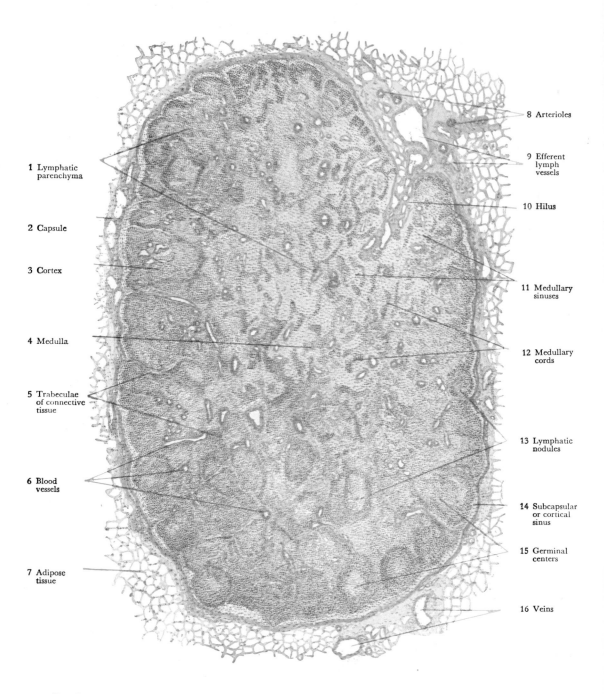

1 Lymphatic
  parenchyma

2 Capsule

3 Cortex

4 Medulla

5 Trabeculae
  of connective
  tissue

6 Blood
  vessels

7 Adipose
  tissue

8 Arterioles

9 Efferent
  lymph
  vessels

10 Hilus

11 Medullary
   sinuses

12 Medullary
   cords

13 Lymphatic
   nodules

14 Subcapsular
   or cortical
   sinus

15 Germinal
   centers

16 Veins

FIG. 10–4.—Panoramic view of section through a lymph node. Stained with hematoxylin and eosin. × 32.
(di Fiore, *An Atlas of Human Histology,* Lea & Febiger.)

ones. The valves prevent the backward flow. The segments between the valves contract twelve to eighteen times per minute in the mesenteric lymphatic vessels of the rat.

Lymph Nodes (*nodi lymphatici; lymph glands*).—The lymph nodes are small oval or bean-shaped bodies, situated in the course of lymphatic vessels so that the lymph filters through them on its way to the blood. Each generally presents on one side a slight depression—the hilum—through which the blood vessels enter and leave and efferent lymphatic vessels emerge. Afferent lymphatic vessels enter the organ at various places on the periphery. When a fresh lymph node is cut open, it displays a lighter **cortical** part and a darker **medullary** part. The cortical part is deficient at the hilum.

Structure of Lymph Nodes (Fig. 10–4).—A lymph node consists of enormous numbers of **lymphocytes** densely packed into masses that are partially subdivided into a series of cortical nodules and medullary cords by anastomosing connective tissue trabeculae bordered by lymph sinuses. The **trabeculae** extend into the node from the connective tissue **capsule** and the hilum. An extensive network of **reticular fibers** extends from the trabeculae to all parts of the node. Continuous, narrow endothelial-lined spaces just beneath the capsule and bordering the trabeculae are known as the subcapsular, cortical, and medullary **sinuses** (Fig. 10–5). The afferent vessels open into the subcapsular, and the efferent vessels arise from the medullary sinuses. The lymph flow is retarded by the relatively enormous extent of the sinuses and the great numbers of reticular fibers which cross them in all directions. Numerous reticular cells and macrophages cling to the reticular fibers. Some authors believe that the sinuses

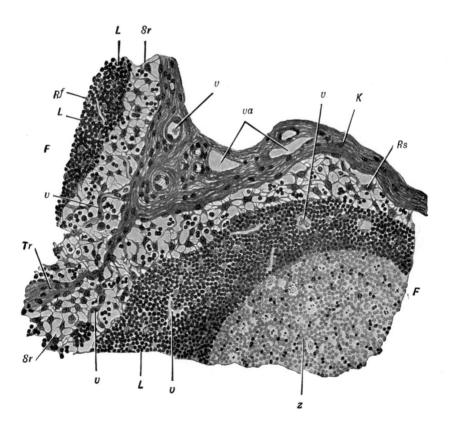

Fig. 10–5.—Portion of cortex of a lymph node of a dog. *K*, capsule; *Tr*, trabecula; *F*, periphery of follicle with a germinal center, z; *Sr*, subcapsular sinuses; *v*, blood vessels; *va*, afferent lymphatic vessels; *L*, small lymphocytes; *Rf*, reticular cells in the follicles and *Rs* in the sinus. Hematoxylin-eosin-azure stain. × 187. (Bloom and Fawcett, *Textbook of Histology*, courtesy of W. B. Saunders Company.)

are lined by reticular cells and not by lymphatic endothelium.

The capsule consists of dense bundles of collagenous fibers, networks of elastic fibers, and a few smooth muscle cells. It is thickened at the hilum. The collagenous fibers are continued into the trabeculae and are continuous with the reticular fibers of the sinuses and nodules.

Lymphocytes increase by division and enter the lymph. The nodules or follicles in the cortical portion of the gland frequently show central areas where mitotic nuclei indicate multiplication of the lymphocytes. These areas are termed **germinal centers.** The cells composing them have more abundant protoplasm than the peripheral cells and consist of large and medium-sized lymphocytes.

Blood vessels enter the hilum and their branches traverse the trabeculae and medullary cords to the cortex. A rich capillary plexus extends throughout the pulp.

Nerves, probably vasomotor, enter the hilum and accompany the blood vessels.

Hemolymph or **hemal nodes** are found in certain animals, notably ruminants, but probably only occur as pathological structures in man. They are quite small and resemble lymph nodes in structure, but are likely to be without afferent and efferent vessels and are engorged with blood.

Lymph.—Lymph, found only in the closed lymphatic vessels, is a transparent, colorless, or slightly yellow watery fluid of specific gravity about 1.015; it closely resembles the blood plasma, but is more dilute. When it is examined under the microscope, leukocytes of the lymphocyte class are found floating in the transparent fluid; they are always increased in number after the passage of the lymph through lymphoid tissue, as in lymph nodes. Lymph should be distinguished from "tissue fluid" which is found outside the lymphatic vessels in the tissue spaces.

THE THORACIC DUCT

The **thoracic duct** (*ductus thoracicus*) (Figs. 12–24, 12–25, 10–6, 10–18) conveys the lymph from the greater part of the body into the blood. It is the common trunk of all the lymphatic vessels except those on the right side of the head, neck, and thorax, the right upper limb, the right lung, right side of the heart, and the diaphragmatic surface of the liver. In the adult it varies in length from 38 to 45 cm and extends from the second lumbar vertebra to the root of the neck. It begins in the abdomen by a dilatation, the **cisterna chyli,** which is situated on the ventral surface of the body of the second lumbar vertebra, to the right side of and dorsal to the aorta, by the side of the right crus of the diaphragm. It enters the thorax through the aortic hiatus of the diaphragm, and ascends through the posterior mediastinum between the aorta and azygos vein. Dorsal to it in this region are the vertebral column, the right intercostal arteries, and the hemiazygos veins as they cross to open into the azygos vein; ventral to it are the diaphragm, esophagus, and pericardium, the last being separated from it by a recess of the right pleural cavity. Opposite the fifth thoracic vertebra, it inclines toward the left

side, enters the superior mediastinum, and ascends dorsal to the aortic arch and the thoracic part of the left subclavian artery and between the left side of the esophagus and the left pleura, to the superior orifice of the thorax. Passing into the root of the neck it forms an arch which rises about 3 or 4 cm above the clavicle and crosses ventral to the subclavian artery, the vertebral artery and vein, and the thyrocervical trunk or its branches. It also passes ventral to the phrenic nerve and the medial border of the Scalenus anterior, and dorsal to the left common carotid artery, vagus nerve, and internal jugular vein. It ends by opening into the angle of junction of the left subclavian vein with the left internal jugular vein.

The thoracic duct is about 3 to 5 mm in diameter at its commencement, but it diminishes considerably in caliber in the middle of the thorax, and is again dilated just before its termination. It is generally flexuous, and constricted at intervals so as to present a varicose appearance. Not infrequently it divides in the middle of its course into two vessels of unequal size which soon reunite, or into several branches which form a plexiform interlacement. It occa-

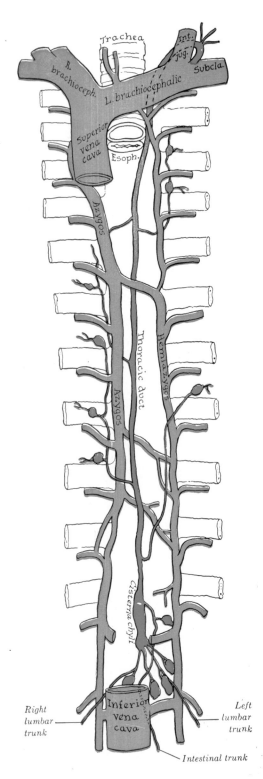

FIG. 10–6.—The thoracic duct.

sionally divides at its upper part into two branches, right and left, the left ending in the usual manner, while the right opens into the right subclavian vein, in connection with the right lymphatic duct. The thoracic duct has several valves; at its termination it is provided with a pair, the free borders of which are turned toward the vein, so as to prevent the passage of venous blood into the duct.

The **cisterna chyli** (*receptaculum chyli*) (Fig. 10–7) is the dilated origin of the thoracic duct which receives the two lumbar lymphatic trunks, right and left, and the intestinal lymphatic trunk. The **lumbar trunks** are formed by the union of the efferent vessels from the lateral aortic lymph nodes. They receive the lymph from the lower limbs, from the walls and viscera of the pelvis, from the kidneys and suprarenal nodes and the deep lymphatics of the greater part of the abdominal wall. The **intestinal trunk** receives the lymph from the stomach and intestine, from the pancreas and spleen, and from the visceral surface of the liver.

Tributaries.—Opening into the thoracic duct near the cisterna, on either side, is a descending trunk from the posterior intercostal lymph nodes of the caudal six or seven intercostal spaces. In the thorax the duct is joined, on either side, by a trunk which drains the upper lumbar lymph nodes and pierces the crus of the diaphragm. It also receives the efferents from the posterior mediastinal lymph nodes and from the posterior intercostal lymph nodes of the upper six left spaces. In the neck it is joined by the **left jugular** and **left subclavian trunks,** and sometimes by the **left bronchomediastinal trunk;** the last-named, however, usually opens independently into the junction of the left subclavian and internal jugular veins.

The **right lymphatic duct** (*ductus lymphaticus dexter*) (Fig. 10–8), about 1.25 cm in length, courses along the medial border of the Scalenus anterior at the root of the neck and ends in the right subclavian vein, at its angle of junction with the right internal jugular vein. Its orifice is guarded by two semilunar valves, which prevent the passage of venous blood into the duct.

Tributaries.—The right lymphatic duct receives the lymph from the right side of the head and neck through the **right jugular trunk**; from the right upper extremity through the **right subclavian trunk**; from the right side of the thorax, right lung, right side of the heart, and part of the convex surface of the liver, through the **right bronchomediastinal trunk.** These three collecting trunks frequently open separately in the angle of union of the two veins.

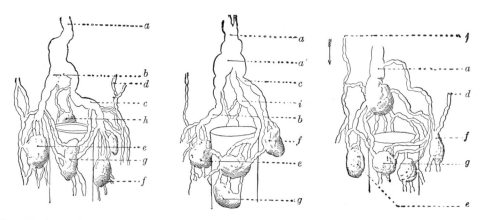

Fig. 10–7.—Modes of origin of thoracic duct. (Poirier and Charpy.) *a,* Thoracic duct. *a',* Cisterna chyli. *b, c,* Efferent trunks from lateral aortic node. *d,* An efferent vessel which pierces the left crus of the diaphragm. *e, f,* Lateral aortic nodes. *h,* Retroaortic nodes. *i,* Intestinal trunk. *j,* Descending branch from intercostal lymphatics.

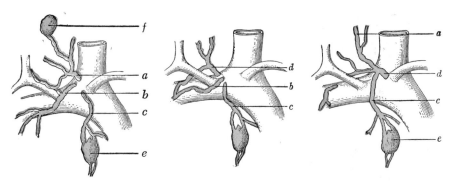

Fig. 10–8.—Terminal collecting trunks of right side. *a,* Jugular trunk. *b,* Subclavian trunk. *c,* Bronchomediastinal trunk. *d,* Right lymphatic trunk. *e,* Node of internal mammary chain. *f,* Node of deep cervical chain. (Poirier and Charpy.)

THE LYMPHATICS OF THE HEAD, FACE, AND NECK

The **Lymphatic Vessels of the Scalp** (Fig. 10–9) are divisible into (*a*) those of the frontal region, which terminate in the anterior auricular and parotid nodes; (*b*) those of the temporoparietal region, which end in the parotid and posterior auricular nodes; and (*c*) those of the occipital region, which terminate partly in the occipital nodes and partly in a trunk which runs down along the posterior border of the Sternocleidomastoideus to end in the inferior deep cervical nodes.

The **Lymphatic Vessels of the Auricula and External Acoustic Meatus** are also divisible into three groups: (*a*) an anterior, from the lateral surface of the auricula and

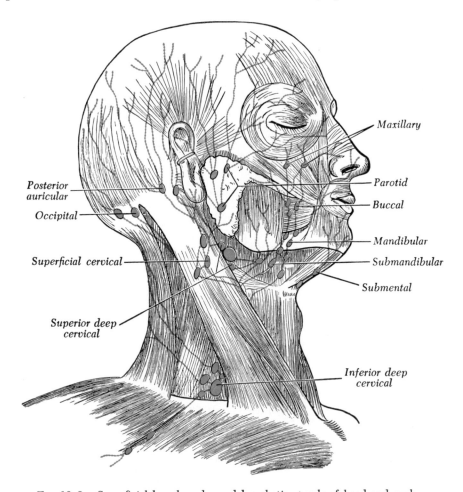

Fig. 10–9.—Superficial lymph nodes and lymphatic vessels of head and neck.

anterior wall of the meatus to the anterior auricular nodes; (b) a posterior, from the margin of the auricula, the upper part of its cranial surface, the internal surface and posterior wall of the meatus to the posterior auricular and superior deep cervical nodes; (c) an inferior, from the floor of the meatus and from the lobule of the auricula to the superficial and superior deep cervical nodes.

The **Lymphatic Vessels of the Face** (Fig. 10–10) are more numerous than those of the scalp. Those from the eyelids and conjunctiva terminate partly in the submandibular but mainly in the parotid nodes. The vessels from the posterior part of the cheek also pass to the parotid nodes, while those from the anterior portion of the cheek, the side of the nose, the upper lip, and the lateral portions of the lower lip

end in the submandibular nodes. The deeper vessels from the temporal and infratemporal fossae pass to the deep facial and superior deep cervical nodes. The deeper vessels of the cheek and lips end, like the superficial, in the submandibular nodes. Both superficial and deep vessels of the central part of the lower lip run to the submental nodes.

Lymphatic Vessels of the Nasal Cavities. —Those from the anterior parts of the nasal cavities communicate with the vessels of the integument of the nose and end in the submandibular nodes; those from the posterior two-thirds of the nasal cavities and from the accessory air sinuses pass partly to the retropharyngeal and partly to the superior deep cervical nodes.

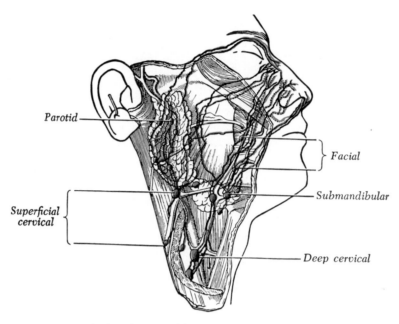

Parotid

Facial

Submandibular

Superficial cervical

Deep cervical

Fig. 10–10.—The lymphatics and lymph nodes of the face. (After Küttner.)

Lymphatic Vessels of the Mouth.—The vessels of the gums pass to the submandibular nodes; those of the hard palate are continuous in front with those of the upper gum, but pass backward to pierce the Constrictor pharyngis superior and end in the superior deep cervical and subparotid nodes; those of the soft palate pass backward and lateralward and end partly in the retropharyngeal and subparotid, and partly in the superior deep cervical nodes. The vessels of the anterior part of the floor of the mouth pass either directly to the inferior nodes of the superior deep cervical group, or indirectly through the submental nodes; from the rest of the floor of the mouth the vessels pass to the submandibular and superior deep cervical nodes.

The **lymphatic vessels of the palatine tonsil,** usually three to five in number, pierce the buccopharyngeal fascia and Constrictor pharyngis superior and pass between the Stylohyoideus and internal jugular vein to the most superior deep cervical node (Fig. 10–11). They end in a node which lies at the side of the posterior belly of the Digastricus, on the internal jugular vein and is commonly called the **tonsillar node.** Occa-

sionally one or two additional vessels run to small nodes on the lateral side of the vein under cover of the Sternocleidomastoideus.

The **lymphatic vessels of the tongue** (Figs. 10–11, 10–12) are drained chiefly into the deep cervical nodes lying between the posterior belly of the Digastricus and the superior belly of the Omohyoideus; one node situated at the bifurcation of the common carotid artery is so intimately associated with these vessels that it is known as the **principal node of the tongue.** The lymphatic vessels of the tongue may be divided into four groups: (1) apical, from the tip of the tongue to the suprahyoid nodes and principal node of the tongue; (2) lateral, from the margin of the tongue—some of these pierce the Mylohyoideus to end in the submandibular nodes, others pass down on the Hyoglossus to the superior deep cervical nodes; (3) basal, from the region of the vallate papillae to the superior deep cervical nodes; and (4) median, a few of which perforate the Mylohyoideus to reach the submandibular nodes, while the majority turn around the posterior border of the muscle to enter the superior deep cervical nodes.

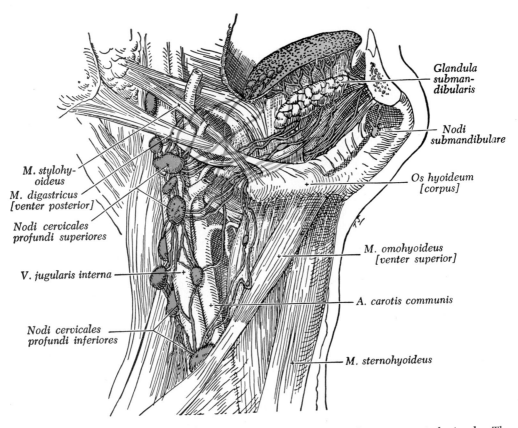

FIG. 10–11.—The deep cervical lymphatic nodes and vessels of the right upper cervical triangle. The lymphatic drainage of the tongue is shown. (Eycleshymer and Jones.)

Glandula subman- dibularis

Nodi submandibulare

M. stylohy- oideus

M. digastricus [venter posterior]

Nodi cervicales profundi superiores

V. jugularis interna

Nodi cervicales profundi inferiores

Os hyoideum [corpus]

M. omohyoideus [venter superior]

A. carotis communis

M. sternohyoideus

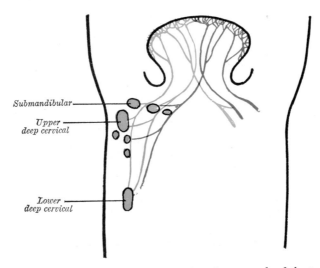

Submandibular

Upper deep cervical

Lower deep cervical

FIG. 10–12.—A diagram to show the course of the central lymphatic vessels of the tongue to the lymph nodes on both sides of the neck. (Jamieson and Dobson.)

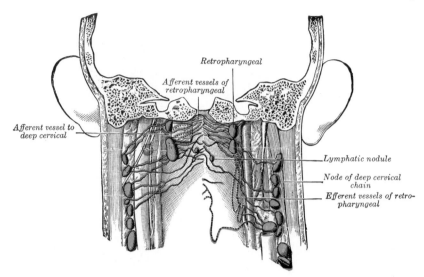

FIG. 10–13.—Lymphatics and lymph nodes of the pharynx. (Poirier and Charpy.)

The Lymph Nodes of the Head

The lymph nodes of the head (Fig. 10–9) are arranged in the following groups:

Occipital
Posterior Auricular
Anterior Auricular
Parotid
Facial
Deep Facial
Lingual
Retropharyngeal

The **Occipital Nodes** (*nodi lymphatici occipitales*), one to three in number, are placed on the back of the head close to the margin of the Trapezius and resting on the insertion of the Semispinalis capitis. Their afferent vessels drain the occipital region of the scalp, while their efferents pass to the superior deep cervical nodes.

The **Posterior Auricular Nodes** (*nodi lymphatici retroauriculares; mastoid nodes*), usually two in number, are situated on the mastoid insertion of the Sternocleidomastoideus, deep to the Auricularis posterior. Their afferent vessels drain the posterior part of the temporoparietal region, the upper part of the cranial surface of the auricula or pinna, and the posterior part of the external acoustic meatus; their efferents pass to the superior deep cervical nodes.

The **Anterior Auricular Nodes** (*nodi lymphatici parotidei superficiales; superficial parotid or preauricular nodes*), from one to three in number, lie immediately anterior to the tragus. Their afferents drain the lateral surface of the auricula and the skin of the adjacent part of the temporal region; their efferents pass to the superior deep cervical nodes.

The **Parotid Nodes** (*nodi lymphatici parotidei*) (Fig. 10–10) form two groups in relation with the parotid salivary gland, a group imbedded in the substance of the gland, and a group of subparotid nodes lying on the lateral wall of the pharynx. Occasionally small nodes are found in the subcutaneous tissue over the parotid gland. Their afferent vessels drain the root of the nose, the eyelids, the frontotemporal region, the external acoustic meatus and the tympanic cavity, possibly also the posterior parts of the palate and the floor of the nasal cavity. The efferents of these nodes pass to the superior deep cervical nodes. The afferents of the subparotid nodes drain the nasal part of the pharynx and the posterior parts of the nasal cavities; their efferents pass to the superior deep cervical nodes.

The **Facial Nodes** (Fig. 10–9) comprise three groups: (*a*) **infraorbital** or **maxillary**, scattered over the infraorbital region from the groove between the nose and cheek to the zygomatic arch; (*b*) **buccal,** one or more placed on the Buccinator opposite the angle of the mouth; (*c*) **mandibular,** on the outer surface of the mandible, in front of the Masseter and in contact with the facial artery and vein. Their afferent vessels drain the eyelids, the conjunctiva, and the skin and mucous membrane of the nose and cheek; their efferents pass to the submandibular nodes.

The **Deep Facial Nodes** (*nodi lymphatici buccales; internal maxillary glands*) (Fig. 10–10) are placed deep to the ramus of the mandible, on the outer surface of the Pterygoideus lateralis, in relation to the maxillary artery. Their afferent vessels drain the temporal and infratemporal fossae and the nasal part of the pharynx; their efferents pass to the superior deep cervical nodes.

The **Lingual Nodes** (*nodi lymphatici linguales*) are two or three small nodules lying on the Hyoglossus and under the Genioglossus. They form merely substations in the course of the lymphatic vessels of the tongue.

The **Retropharyngeal Nodes** (Fig. 10–13), from one to three in number, lie in the buccopharyngeal fascia, behind the upper part of the pharynx and anterior to the arch of the atlas, being separated, however, from the latter by the Longus capitis. Their afferents drain the nasal cavities, the nasal part of the pharynx, and the auditory tubes; their efferents pass to the superior deep cervical nodes.

The Lymphatic Vessels of the Neck

The **lymphatic vessels of the skin and muscles of the neck** (Fig. 10–14) pass to the deep cervical nodes. From the upper part of the *pharynx* (Fig. 10–13) the lymphatic vessels pass to the retropharyngeal, from the lower part to the deep cervical nodes. From the *larynx* two sets of vessels arise, an upper and a lower. The vessels of the upper set pierce the hyothyroid membrane and join the superior deep cervical nodes. Of the lower set, some pierce the conus elasticus and join the pretracheal and prelaryngeal nodes; others run between the cricoid and first tracheal ring and enter the inferior deep cervical nodes. The lymphatic vessels of the *thyroid gland* consist of two

sets, an upper, which accompanies the superior thyroid artery and enters the superior deep cervical nodes, and a lower, which runs partly to the pretracheal nodes and partly to the small paratracheal nodes which accompany the recurrent nerves. These latter nodes receive also the lymphatic vessels from the cervical portion of the trachea.

The Lymph Nodes of the Neck

The lymph nodes of the neck include the following groups:

> Submandibular
> Submental
> Superficial Cervical
> Anterior Cervical
> Deep Cervical

The **Submandibular Nodes** (*nodi lymphatici submandibulares*) (Fig. 10–14), three to six in number, are placed beneath the body of the mandible in the submandibular triangle, and rest on the superficial surface of the submandibular salivary gland. One node, the **middle node of Stahr,** which lies on the external maxillary artery as it turns over the mandible, is the most constant of the series; small lymph nodes are sometimes found on the deep surface of the submandibular salivary gland. The afferents of the submandibular nodes drain the medial palpebral commissure, the cheek, the side of the nose, the upper lip, the lateral part of the lower lip, the gums, and the anterior part of the margin of the tongue; efferent vessels from the facial and submental nodes also enter the submandibular nodes. Their efferent vessels pass to the superior deep cervical nodes.

The **Submental** or **Suprahyoid Nodes** (*nodi lymphatici submentales*) are situated between the anterior bellies of the Digastrici. Their afferents drain the central portions of the lower lip and floor of the mouth and the apex of the tongue; their efferents pass partly to the submandibular nodes and partly to a node of the deep cervical group situated on the internal jugular vein at the level of the cricoid cartilage.

The **Superficial Cervical Nodes** (*nodi lymphatici cervicales superficiales*) lie in close relationship with the external jugular vein

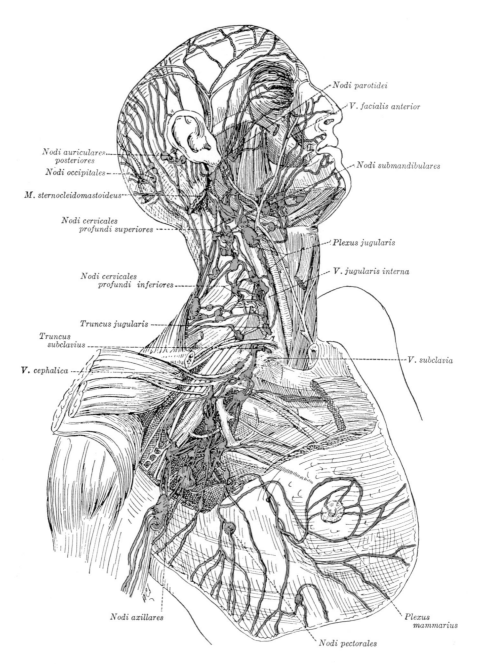

Fig. 10–14.—The deep lymphatic nodes and vessels of the right side of the head and neck, and of the mammary and axillary regions. (Eycleshymer and Jones.)

as it emerges from the parotid gland, and, therefore, superficial to the Sternocleidomastoideus. Their afferents drain the lower parts of the auricula and parotid region, while their efferents pass around the anterior margin of the Sternocleidomastoideus to join the superior deep cervical nodes.

The **Anterior Cervical Nodes** form an irregular and inconstant group ventral to the larynx and trachea. They may be divided into (*a*) a **superficial set**, placed on the anterior jugular vein; (*b*) a **deeper set**, which is further subdivided into prelaryngeal, on the middle cricothyroid ligament, and pretracheal, ventral to the trachea. This deeper set drains the lower part of the larynx, the thyroid gland, and the cranial part of the trachea; its efferents pass to the most inferior of the superior deep cervical nodes.

The **Deep Cervical Nodes** (*nodi lymphatici cervicales profundi*) (Figs. 10–9, 10–11) are numerous and of large size; they form a chain along the carotid sheath lying by the side of the pharynx, esophagus, and trachea, and extending from the base of the skull to the root of the neck. They are usually described in two groups:

a) The **superior deep cervical nodes** lie under the Sternocleidomastoideus in close relation with the accessory nerve and the internal jugular vein, some of the nodes lying anterior to and others posterior to the vessel. The superior deep cervical nodes drain the occipital portion of the scalp, the auricula, the back of the neck, a considerable part of the tongue, the larynx, thyroid

gland, trachea, nasal part of the pharynx, nasal cavities, palate, and esophagus. They receive also the efferent vessels from all the other nodes of the head and neck, except those from the inferior deep cervical nodes. The most superior of these deep cervical nodes (Fig. 10–11) lies just below the superior belly of the Digastricus in a rather superficial position which allows it to be palpated with ease in the living patient. It is called the **subdigastric or tonsillar node** because it becomes swollen when the tonsil or pharynx is inflamed and is therefore valuable in physical diagnosis.

b) The **inferior deep cervical nodes** extend beyond the posterior margin of the Sternocleidomastoideus into the supraclavicular triangle, where they are closely related to the brachial plexus and subclavian vein. The inferior deep cervical nodes drain the back of the scalp and neck, the superficial pectoral region, part of the arm, and occasionally, part of the superior surface of the liver. In addition, they receive vessels from the superior deep cervical nodes. The efferents of the superior deep cervical nodes pass partly to the inferior deep cervical nodes and partly to a trunk which unites with the efferent vessel of the inferior deep cervical nodes and forms the **jugular trunk**. On the right side, this trunk ends in the junction of the internal jugular and subclavian veins; on the left side it joins the thoracic duct. A few minute **paratracheal nodes** are situated alongside the recurrent nerves on the lateral aspects of the trachea and esophagus.

THE LYMPHATICS OF THE UPPER LIMB

The Lymphatic Vessels of the Upper Limb

The lymphatic vessels of the upper limb are divided into two sets, superficial and deep.

The **superficial lymphatic vessels** commence (Fig. 10–15) in the lymphatic plexus which everywhere pervades the skin; in the hand the meshes of the plexus are much finer in the palm and on the flexor aspect of the digits than elsewhere. The **digital plexuses** are drained by a pair of vessels which run on the sides of each digit, and

incline backward to reach the dorsum of the hand. From the dense plexus of the palm, vessels pass in different directions, toward the wrist, distally to join the digital vessels, medialward to join the vessels on the ulnar border of the hand, and lateralward to those on the thumb. Several vessels from the central part of the plexus unite to form a trunk which passes around the metacarpal bone of the index finger to join the vessels on the back of that digit and on the back of the thumb. Running proximally from around

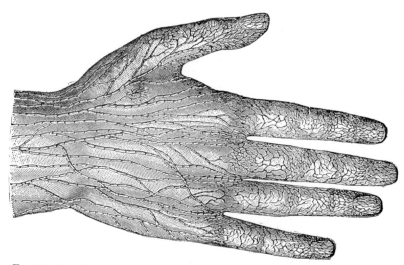

FIG. 10–15.—Lymphatic vessels of the dorsal surface of the hand. (Sappey.)

the wrist, the lymphatic vessels are collected into radial, median, and ulnar groups, which accompany respectively the cephalic, median, and basilic veins in the forearm (Fig. 10–16). A few of the ulnar lymphatics end in the *supratrochlear nodes,* but the majority pass directly to the lateral group of *axillary nodes.* Some of the radial vessels are collected into a trunk which ascends with the cephalic vein to the deltoideopectoral nodes; the efferents from this group pass either to the subclavicular axillary nodes or to the inferior cervical nodes.

The **deep lymphatic vessels** accompany the deep blood vessels. In the forearm, they consist of four sets, corresponding with the radial, ulnar, palmar, and dorsal interosseous arteries; they communicate at intervals with the superficial lymphatics, and some of them end in the nodes which are occasionally found beside the arteries. In their course proximally, a few end in the nodes which lie upon the brachial artery; but most of them pass to the lateral group of axillary nodes.

The Lymph Nodes of the Upper Limb

The lymph nodes of the upper limb are divided into two sets, **superficial** and **deep.**

The **superficial lymph nodes** are few and of small size. One or two **supratrochlear nodes** (Fig. 10–16) are placed above the medial epicondyle of the humerus, medial

to the basilic vein. Their afferents drain the middle, ring, and little fingers, the medial portion of the hand, and the superficial area over the ulnar side of the forearm; these vessels are, however, in free communication with the other lymphatic vessels of the forearm. Their efferents accompany the basilic vein and join the deeper vessels. One or two **deltoideopectoral nodes** are found beside the cephalic vein, between the Pectoralis major and Deltoideus, immediately below the clavicle. They are situated in the course of the external collecting trunks of the arm.

The **deep lymph nodes** are chiefly grouped in the axilla, although a few may be found in the forearm, in the course of the radial, ulnar, and interosseous vessels, and in the arm along the medial side of the brachial artery.

The **Axillary Nodes** (*nodi lymphatici axillares*) (Figs. 10–14, 10–17) are of large size, vary from twenty to thirty in number, and may be arranged in the following groups:

a) Lateral
b) Anterior or pectoral
c) Posterior or subscapular
d) Central or intermediate
e) Medial or subclavicular

a) A **lateral group** of from four to six glands lies in relation to the medial and posterior aspects of the axillary vein; the afferents of these nodes drain the whole arm with the exception of that portion whose vessels accompany the cephalic vein. The efferent vessels pass partly to the central and subclavicular groups of axillary nodes and partly to the inferior deep cervical nodes.

b) An **anterior** or **pectoral group** consists of four or five nodes along the lateral border of the Pectoralis minor, in relation with the lateral thoracic artery. Their afferents drain the skin and muscles of the anterior and lateral thoracic walls, and the central and lateral parts of the mamma; their efferents pass partly to the central and partly to the subclavicular groups of axillary nodes.

c) A **posterior** or **subscapular group** of six or seven nodes is placed along the lower margin of the posterior wall of the axilla in the course of the subscapular artery. The afferents of this group drain the skin and muscles of the dorsal part of the neck and thoracic wall; their efferents pass to the central group of axillary nodes.

d) A **central** or **intermediate group** of three or four large nodes is imbedded in the adipose tissue near the base of the axilla. Its afferents are the efferent vessels of all the preceding groups of axillary nodes; its efferents pass to the subclavicular group.

e) A **medial** or **subclavicular group** of six to twelve nodes is situated partly posterior

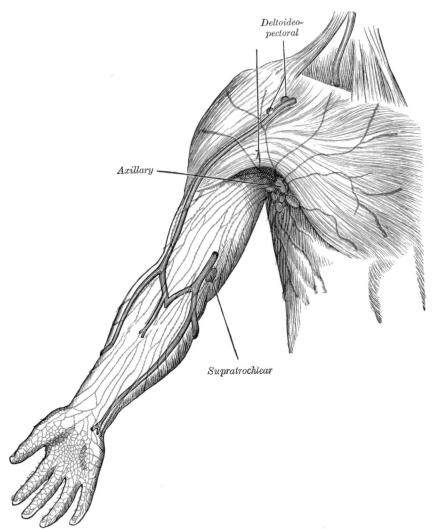

Fig. 10–16.—The superficial lymph nodes and lymphatic vessels of the upper limb.

to the cranial portion of the Pectoralis minor and partly cranial to the border of this muscle. Its only direct territorial afferents are those which accompany the cephalic vein and one which drains the peripheral part of the mamma, but it receives the efferents of all the other axillary nodes. The efferent vessels of the subclavicular group unite to form the **subclavian trunk**, which opens either directly into the junction of the internal jugular and subclavian veins or into the jugular lymphatic trunk; on the left side it may end in the thoracic duct. A few efferents from the subclavicular nodes usually pass to the inferior deep cervical nodes.

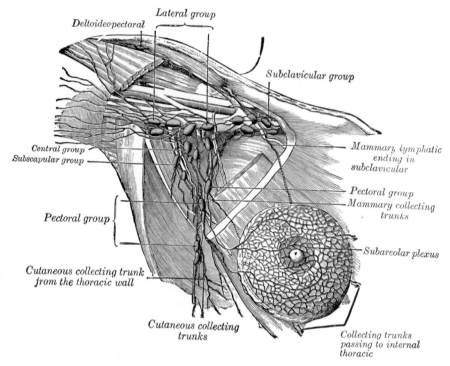

FIG. 10–17.—Lymphatics of the mamma, and the axillary nodes (semidiagrammatic). (Poirier and Charpy.)

THE LYMPHATICS OF THE TRUNK

The Parietal Lymphatic Vessels of the Thorax

The **superficial lymphatic vessels of the thoracic wall** ramify beneath the skin and converge to the axillary nodes. Those over the Trapezius and Latissimus dorsi run ventralward and unite to form about ten or twelve trunks which end in the subscapular group. Those over the pectoral region, including the vessels from the skin covering the peripheral part of the mamma, run dorsalward, and those over the Serratus anterior cranialward, to the pectoral group. Others near the lateral margin of the sternum pass inward between the rib cartilages and end in the sternal nodes, while the vessels of opposite sides anastomose across the ventral surfaces of the sternum. A few vessels from the cranial part of the pectoral region ascend over the clavicle to the supraclavicular group of cervical nodes.

The **Lymphatic Vessels of the Mamma** (Fig. 10–17) originate in a plexus in the interlobular spaces and on the walls of the galactophorous ducts. Those from the central part of the gland pass to an intricate plexus situated beneath the areola, a plexus which receives also the lymphatics from the skin over the central part of the gland and those from the areola and nipple. Its efferents are collected into two trunks which pass to the *pectoral group of axillary nodes*. The vessels which drain the medial part of the mamma pierce the thoracic wall and end in the *sternal nodes*. A vessel occasionally emerges from the cranial part of the mamma and, piercing the Pectoralis major, terminates in the *subclavicular nodes*.

The **deep lymphatic vessels of the thoracic wall** (Fig. 10–18) are of three types:

Lymphatics of the muscles
Intercostal lymphatics
Lymphatics of the diaphragm

The **lymphatics of the muscles** which lie on the ribs: most of these end in the axillary nodes, but some from the Pectoralis major pass to the sternal nodes.

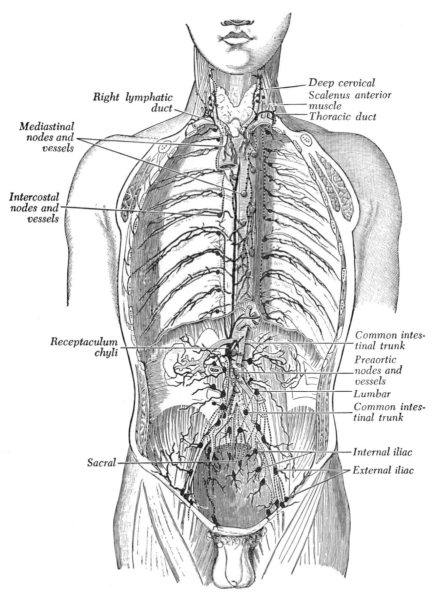

FIG. 10–18.—Deep lymph nodes and vessels of the thorax and abdomen (diagrammatic). Afferent vessels are represented by continuous lines, and efferent and internodular vessels by dotted lines. (Cunningham.)

25

The **intercostal lymphatics** drain the Intercostales and parietal pleura. Those draining the Intercostales externi run dorsalward and, after receiving the vessels which accompany the posterior branches of the intercostal arteries, end in the *intercostal nodes*. Those of the Intercostales interni and parietal pleura consist of a single trunk in each space. These trunks run ventralward in the subpleural tissue and the upper six open separately into the *sternal nodes* or into the vessels which unite them; those of the lower spaces unite to form a single trunk which terminates in the most caudal of the *sternal nodes*.

The **lymphatic vessels of the diaphragm** form two plexuses, one on its thoracic and another on its abdominal surface. These plexuses anastomose freely with each other, and are best marked on the parts covered respectively by the pleura and peritoneum. That on the **thoracic surface** communicates with the lymphatics of the costal and mediastinal parts of the pleura, and its efferents consist of three groups: (*a*) anterior, passing to the nodes which lie near the junction of the seventh rib with its cartilage; (*b*) middle, to the nodes on the esophagus and to those around the termination of the inferior vena cava; and (*c*) posterior, to the nodes which surround the aorta at the point where this vessel leaves the thoracic cavity.

The **abdominal surface of the diaphragm** has a plexus of fine vessels which anastomose with the lymphatics of the liver and, at the periphery of the diaphragm, with those of the subperitoneal tissue. The efferents from the right half of this plexus terminate partly in a group of nodes on the trunk of the corresponding inferior phrenic artery, while others end in the right *lateral aortic nodes*. Those from the left half of the plexus pass to the pre- and lateral aortic nodes and to the nodes on the terminal portion of the esophagus.

The Lymph Nodes of the Thorax

The **lymph nodes of the thorax** may be divided into parietal and visceral—the former being situated in the thoracic wall, the latter in relation to the viscera.

The **parietal lymph nodes** include:

> Sternal nodes
> Intercostal nodes
> Diaphragmatic nodes

The **Sternal Nodes** (*nodi lymphatici sternales; internal mammary nodes*) are placed at the anterior ends of the intercostal spaces, by the side of the internal thoracic artery. They derive afferents from the mamma, from the deeper structures of the anterior abdominal wall above the level of the umbilicus, from the diaphragmatic surface of the liver through a small group of nodes which lie dorsal to the xiphoid process, and from the deeper parts of the ventral portion of the thoracic wall. Their efferents usually unite to form a single trunk on either side; this may open directly into the junction of the internal jugular and subclavian veins, or that of the right side may join the right subclavian trunk, and that of the left the thoracic duct.

The **Intercostal Nodes** (*nodi lymphatici intercostales*) occupy the dorsal parts of the intercostal spaces, in relation to the intercostal vessels. They receive the deep lymphatics from the posterolateral aspect of the chest; some of these vessels are interrupted by small lateral intercostal nodes. The efferents of the nodes in the caudal four or five spaces unite to form a trunk which descends and opens either into a cisterna chyli or into the commencement of the thoracic duct. The efferents of the nodes in the upper spaces of the left side end in the thoracic duct; those of the corresponding right spaces, in the right lymphatic duct.

The **Diaphragmatic Nodes** lie on the thoracic aspect of the diaphragm, and consist of three sets, anterior, middle, and posterior.

The **anterior set** comprises (*a*) two or three small nodes dorsal to the base of the xiphoid process which receive afferents from the convex surface of the liver, and (*b*) one or two nodes on either side near the junction of the seventh rib with its cartilage, which receive lymphatic vessels from the ventral part of the diaphragm. The efferent vessels of the anterior set pass to the *sternal nodes*.

The **middle set** consists of two or three nodes on either side close to where the phrenic nerves enter the diaphragm. On the right side some of the nodes of this group lie within the fibrous sac of the pericardium, ventral to the termination of the inferior vena cava. The afferents of this set are derived from the middle part of the diaphragm, those on the right side also receiving afferents from the convex surface of the liver. Their efferents pass to the posterior mediastinal nodes.

The **posterior set** consists of a few nodes situated on the back of the crura of the diaphragm, and connected on the one hand with the lumbar nodes and on the other with the posterior mediastinal nodes.

The Visceral Lymphatic Vessels in the Thorax

The lymphatic vessels of the thoracic viscera comprise:

Lymphatics of heart and pericardium
Lymphatics of lungs and pleura
Lymphatics of thymus
Lymphatics of esophagus

The **Lymphatic Vessels of the Heart** consist of two plexuses: (*a*) **deep,** immediately under cover of the endocardium; and (*b*) **superficial,** subjacent to the visceral pericardium. The deep plexus opens into the superficial, the efferents of which form right and left collecting trunks. The **left trunks,** two or three in number, ascend in the anterior interventricular sulcus, receiving, in their course, vessels from both ventricles. On reaching the coronary sulcus they are joined by a large trunk from the diaphragmatic surface of the heart, and then unite to form a single vessel which ascends between the pulmonary artery and the left atrium and ends in one of the *tracheobronchial nodes.* The **right trunk** receives its afferents from the right atrium and from the right border and diaphragmatic surface of the right ventricle. It ascends in the posterior interventricular sulcus and then runs forward in the coronary sulcus, and passes dorsal to the pulmonary artery, to end in one of the *tracheobronchial nodes* (Fig. 10–19).

The **Lymphatic Vessels of the Lungs** originate in two plexuses, a superficial and a deep. The **superficial plexus** is placed beneath the pulmonary pleura. The **deep plexus** accompanies the branches of the pulmonary vessels and the ramifications of the bronchi. In the case of the larger bronchi the deep plexus consists of two networks—one, submucous, beneath the mucous membrane, and another, peribronchial, outside the walls of the bronchi. In the smaller bronchi there is but a single plexus, which extends as far as the bronchioles, but fails to reach the alveoli, in the walls of which there are no lymphatic vessels. The superficial efferents turn around the borders of the lungs and the margins of their fissures, and converge to end in some nodes situated at the hilum; the deep efferents are conducted to the hilum along the pulmonary vessels and bronchi, and end in the *tracheobronchial nodes.* Little or no anastomosis occurs between the superficial and deep lymphatics of the lungs, except in the region of the hilum.

The **Lymphatic Vessels of the Pleura** consist of two sets—one in the visceral and another in the parietal part of the membrane; those of the visceral pleura drain into the superficial efferents of the lung. The **lymphatics** of the **parietal pleura** have three modes of ending: (*a*) those of the costal portion join the lymphatics of the Intercostales interni and so reach the *sternal nodes;* (*b*) those of the diaphragmatic part are drained by the efferents of the diaphragm; (*c*) those of the mediastinal portion terminate in the *posterior mediastinal nodes.*

The **Lymphatic Vessels of the Thymus** end in the anterior mediastinal, tracheobronchial, and sternal nodes.

The **Lymphatic Vessels of the Esophagus** form a plexus around that tube, and the collecting vessels from the plexus drain into the *posterior mediastinal nodes.*

The Lymph Nodes of the Thoracic Viscera

The **visceral lymph nodes** consist of three groups:

Anterior mediastinal
Posterior mediastinal
Tracheobronchial

The **Anterior Mediastinal Nodes** (*nodi lymphatici mediastinales anteriores*) are

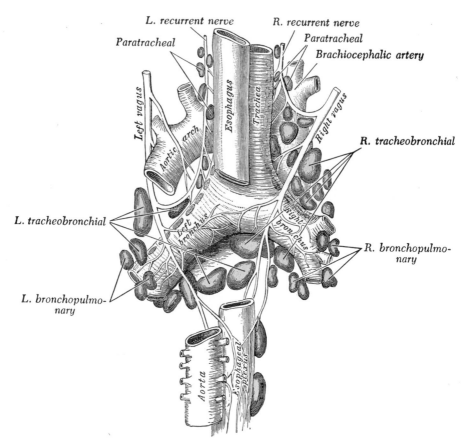

Fig. 10–19.—The tracheobronchial lymph nodes. (From a figure designed by M. Hallé.)

placed in the ventral part of the superior mediastinum, ventral to the aortic arch and in relation to the brachiocephalic veins and the large arterial trunks which arise from the aortic arch. They receive afferents from the thymus and pericardium, and from the sternal nodes; their efferents unite with those of the tracheobronchial nodes, to form the **right** and **left bronchomediastinal trunks.**

The **Posterior Mediastinal Nodes** (*nodi lymphatici mediastinales posteriores*) lie dorsal to the pericardium in relation to the esophagus and descending thoracic aorta. Their afferents are derived from the esophagus, the dorsal part of the pericardium, the diaphragm, and the convex surface of the liver. Their efferents mostly end in the *thoracic duct,* but some join the tracheobronchial nodes.

The **Tracheobronchial Nodes** (Fig. 10–19) form four main groups: (*a*) **tracheal,** on either side of the trachea; (*b*) **bronchial,** in the angles between the lower part of the trachea and bronchi and in the angle between the two bronchi; (*c*) **broncho-pulmonary,** in the hilum of each lung; and (*d*) **pulmonary,** in the lung substance, on the larger branches of the bronchi. The afferents of the tracheobronchial nodes drain the lungs and bronchi, the thoracic part of the trachea and the heart; some of the efferents of the posterior mediastinal nodes also end in this group. Their efferent vessels ascend upon the trachea and unite with efferents of the sternal and anterior mediastinal nodes to form the **right** and **left bronchomediastinal trunks.** The right bronchomediastinal trunk may join the right lymphatic duct, and the left the thoracic duct, but more frequently they open independently of these ducts into the junction of the internal jugular and subclavian veins of their own side.

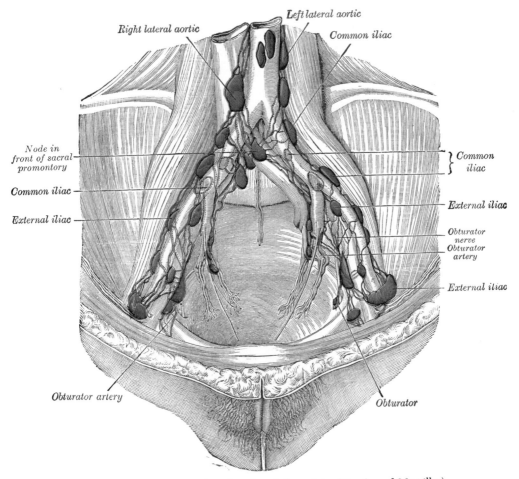

FIG. 10–20.—The parietal lymph nodes of the pelvis. (Cunéo and Marcille.)

In all town dwellers there are continually being swept into the bronchi and alveoli large quantities of inhaled dust and black carbonaceous pigment. Particles which are not picked up by dust cells, caught in the mucus, and discharged, are taken up by phagocytes and transported in the lymphatic vessels to the lymph nodes. At first the nodes are moderately enlarged, firm, inky black, and gritty on section; later they enlarge still further, often becoming fibrous from the irritation set up by the minute foreign bodies with which they are crammed, and may break down into a soft slimy mass or may calcify.

The Parietal Lymphatic Vessels of the Abdomen and Pelvis

The lymphatic vessels of the walls of the abdomen and pelvis may be divided into two sets, superficial and deep.

The **superficial vessels** follow the course of the superficial blood vessels and converge to the superficial inguinal nodes; those derived from the integument of the ventral abdomen below the umbilicus follow the course of the superficial epigastric vessels, and those from the sides of the lumbar part of the abdominal wall pass along the crest of the ilium, with the superficial iliac circumflex vessels. The superficial lymphatic vessels of the gluteal region turn horizontally around the buttock, and join the *superficial inguinal* and *subinguinal nodes.*

The **deep vessels** run alongside the principal blood vessels. Those of the parietes of the pelvis, which accompany the superior and inferior gluteal, and obturator vessels, follow the course of the internal iliac artery, and ultimately join the *lateral aortic nodes.*

The Parietal Lymph Nodes of the Abdomen and Pelvis

The lymph nodes of the abdomen and pelvis may be divided, from their situations, into (*a*) **parietal**, lying behind the peritoneum and in close association with the larger blood vessels; and (*b*) **visceral**, which are found in relation to the visceral arteries.

The **parietal nodes** (Figs. 10–18, 10–20, 10–21) include the following groups:

External Iliac   Iliac Circumflex
Common Iliac   Internal Iliac
Epigastric   Sacral

$$\text{Lumbar} \left\{ \begin{array}{l} \text{Lateral Aortic} \\ \text{Preaortic} \\ \text{Retroaortic} \end{array} \right.$$

The **External Iliac Nodes**, from eight to ten in number, lie along the external iliac vessels. They are arranged in three groups, one on the lateral, another on the medial, and a third on the anterior aspect of the vessels; the third group is sometimes absent. Their principal afferents are derived from the inguinal and subinguinal nodes, the deep lymphatics of the abdominal wall below the umbilicus and of the adductor region of the thigh, and the lymphatics from the glans penis vel clitoridis, the membranous urethra, the prostate, the fundus of the bladder, the cervix uteri, and upper part of the vagina.

The **Common Iliac Nodes**, four to six in number, are grouped at the sides and dorsal to the common iliac artery, one or two being placed distal to the bifurcation of the aorta, ventral to the fifth lumbar vertebra. They drain chiefly the internal and external iliac nodes, and their efferents pass to the lateral aortic nodes.

The **Epigastric Nodes** (*nodi lymphatici epigastrici*), three or four in number, are placed alongside the caudal portion of the inferior epigastric vessels.

The **Iliac Circumflex Nodes**, two to four in number, are situated along the course of the deep iliac circumflex vessels; they are sometimes absent.

The **Internal Iliac Nodes** (*nodi lymphatici iliaci interni; hypogastric nodes*) (Fig. 10–21) surround the internal iliac vessels, and receive the lymphatics corresponding to the distribution of the branches of the internal iliac artery, *i.e.*, they receive lymphatics from all the pelvic viscera, from the deeper parts of the perineum, including the membranous and cavernous portions of the urethra, and from the buttock and dorsum of the thigh. An **obturator node** is sometimes seen in the upper part of the obturator foramen. The efferents of the internal iliac group end in the common iliac nodes.

The **Sacral Nodes** are placed in the concavity of the sacrum, in relation to the middle and lateral sacral arteries; they receive lymphatics from the rectum and posterior wall of the pelvis.

The **Lumbar Nodes** (*nodi lymphatici lumbales*) (Fig. 10–18) are very numerous, and consist of right and left lateral aortic, preaortic, and retroaortic groups.

The **right lateral aortic nodes** are situated partly ventral to the inferior vena cava, near the termination of the renal vein, and partly dorsal to it on the origin of the Psoas major, and on the right crus of the diaphragm. The **left lateral aortic nodes** form a chain on the left side of the abdominal aorta ventral to the origin of the Psoas major and left crus of the diaphragm. The nodes on either side receive (*a*) the efferents of the common iliac nodes, (*b*) the lymphatics from the testis in the male and from the ovary, uterine tube, and body of the uterus in the female; (*c*) the lymphatics from the kidney and suprarenal node; and (*d*) the lymphatics draining the lateral abdominal muscles and accompanying the lumbar veins. Most of the efferent vessels of the lateral aortic nodes converge to form the **right and left lumbar trunks** which join the cisterna chyli, but some enter the pre- and retroaortic nodes, and others pierce the crura of the diaphragm to join the caudal end of the thoracic duct. The **preaortic nodes** lie ventral to the aorta, and may be divided into **celiac, superior mesenteric,** and **inferior mesenteric** groups, arranged around the origins of the corresponding arteries. They receive a few vessels from the lateral aortic nodes, but their principal afferents are derived from the viscera supplied by the three arteries with which they are associated. Some of their efferents pass to the retroaortic nodes, but the majority unite to form the **intestinal trunk,** which

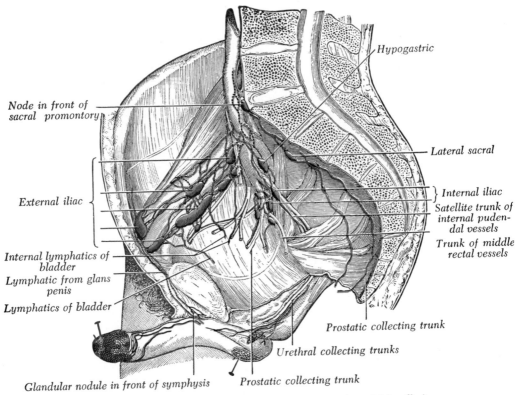

Fig. 10–21.—Iliopelvic nodes (lateral view). (Cunéo and Marcille.)

enters the cisterna chyli. The **retroaortic nodes** are placed caudal to the cisterna chyli, on the bodies of the third and fourth lumbar vertebrae. They receive lymphatic trunks from the lateral and preaortic node, and their efferents end in the cisterna chyli.

The Lymphatic Vessels of the Abdominal and Pelvic Viscera

The lymphatic vessels of the abdominal and pelvic viscera consist of those of the:

    Stomach
    Duodenum
    Omentum
    Jejunum and ileum
    Cecum and appendix
    Colon
    Anus and rectum
    Liver
    Gallbladder
    Pancreas

The lymphatic vessels of the subdiaphragmatic portion of the digestive tube are situated partly in the mucous membrane and partly in the seromuscular coats, but as the former system drains into the latter, the two may be considered as one.

The **Lymphatic Vessels of the Stomach** (Figs. 10–22, 10–23) are continuous at the cardiac orifice with those of the esophagus, and at the pylorus with those of the duodenum. They mainly follow the blood vessels, and may be arranged in four sets. (*a*) Those of the first set accompany the branches of the left gastric artery, receive tributaries from a large area on either surface of the stomach, and terminate in the *superior gastric nodes.* (*b*) Those of the second set drain the fundus and body of the stomach on the left of a line drawn vertically from the esophagus; they accompany, more or less closely, the short gastric and left gastroepiploic arteries, and end in the *pancreaticolienal nodes.* (*c*) The vessels of the third set drain the right portion of the greater curvature as far as the pyloric portion, and end in the *inferior gastric nodes,* the efferents of which pass to the

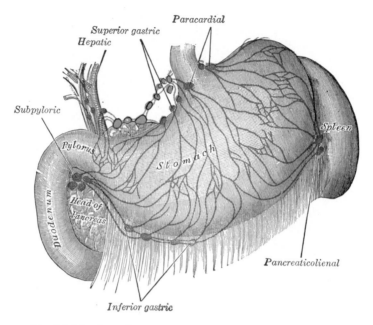

Fig. 10–22.—Lymphatics of stomach. (Jamieson and Dobson.)

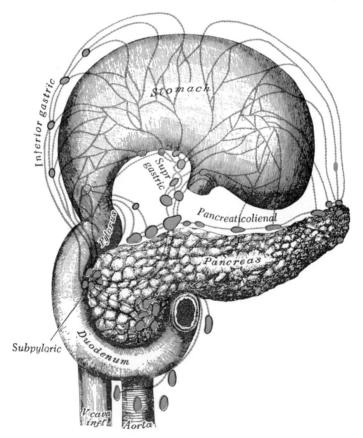

Fig. 10–23.—Lymphatics of stomach. The stomach has been turned upward.
(Jamieson and Dobson.)

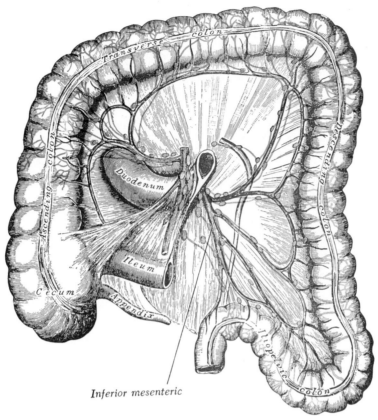

FIG. 10–24.—Lymphatics of colon. (Jamieson and Dobson.)

subpyloric group. (*d*) Those of the fourth set drain the pyloric portion and pass to the *hepatic* and *subpyloric nodes,* and to the *superior gastric nodes.*

The **Lymphatic Vessels of the Duodenum** (Fig. 10–23) consist of an anterior and a posterior set, which open into a series of small *pancreaticoduodenal nodes* on the anterior and posterior aspects of the groove between the head of the pancreas and the duodenum. The efferents of these nodes run in two directions, cranialward to the *hepatic nodes* and caudalward to the *preaortic nodes* around the origin of the superior mesenteric artery.

The **Lymphatic Vessels of the Omentum** accompany the arteries and veins and are richly supplied with valves.

The **Lymphatic Vessels of the Jejunum and Ileum** are termed **lacteals,** from the milk-white fluid they contain during intestinal digestion. They run between the layers of the mesentery and enter the *mesenteric*

nodes, the efferents of which end in the *preaortic nodes.*

The **Lymphatic Vessels of the Vermiform Appendix and Cecum** (Figs. 10–25, 10–26) are numerous, since in the wall of the appendix there is a large amount of adenoid tissue. From the body and tail of the vermiform appendix eight to fifteen vessels ascend between the layers of the mesenteriole, one or two being interrupted in the node which lies between the peritoneal layers. They unite to form three or four vessels, which end partly in the lower and partly in the upper nodes of the *ileocolic chain.* The vessels from the root of the appendix and from the cecum consist of an anterior and a posterior group. The anterior vessels pass ventral to the cecum, and end in the *anterior ileocolic nodes* and in the nodes of the ileocolic chain; the posterior vessels ascend over the dorsum of the cecum and terminate in the *posterior ileocolic nodes* and in the lower nodes of the ileocolic chain.

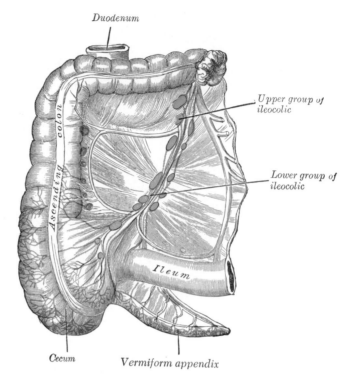

FIG. 10–25.—The lymphatics of cecum and vermiform appendix from the front.
(Jamieson and Dobson.)

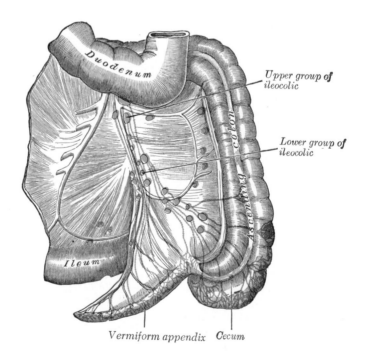

FIG. 10–26.—The lymphatics of cecum and vermiform appendix from behind.
(Jamieson and Dobson.)

Lymphatic Vessels of the Colon (Fig. 10–24).—The lymphatic vessels of the ascending and transverse parts of the colon finally end in the mesenteric nodes, after traversing the *right colic* and *mesocolic nodes*. Those of the descending and iliac sigmoid parts of the colon are interrupted by the small nodes on the branches of the left colic and sigmoid arteries, and ultimately end in the *preaortic nodes* around the origin of the inferior mesenteric artery.

Lymphatic Vessels of the Anus, Anal Canal, and Rectum.—The lymphatics from the **anus** pass ventralward and end with those of the integument of the perineum and scrotum in the *superficial inguinal nodes;* those from the **anal canal** accompany the middle and inferior rectal arteries, and end in the internal iliac nodes; while the vessels from the **rectum** traverse the pararectal nodes and pass to those in the sigmoid mesocolon; the efferents of the latter terminate in the *preaortic nodes* around the origin of the inferior mesenteric artery.

The **Lymphatic Vessels of the Liver** are divisible into two sets, superficial and deep. The former arise in the subperitoneal areolar tissue over the entire surface of the organ, and may be grouped into (*a*) those on the convex surface, (*b*) those on the visceral surface.

(*a*) On the **convex surface:** The vessels from the dorsal part of this surface reach their terminal nodes by three different routes; the vessels of the middle set, five or six in number, pass through the vena caval foramen in the diaphragm and end in one or two *inferior caval nodes* which are situated around the terminal part of the vein; a few vessels from the left side pass dorsalward toward the esophageal hiatus, and terminate in the paracardial group of *superior gastric nodes;* the vessels from the right side, one or two in number, run on the abdominal surface of the diaphragm, and, after crossing its right crus, end in the *preaortic nodes* which surround the origin of the celiac artery. From the portions of the right and left lobes adjacent to the falciform ligament, the lymphatic vessels converge to form two trunks, one of which accompanies the inferior vena cava through the diaphragm, and ends in the nodes around the terminal part of this vessel; the

other runs caudalward and ventralward, and, turning around the inferior sharp margin of the liver, accompanies the upper part of the ligamentum teres, and ends in the superior *hepatic nodes*. From the anterior surface a few additional vessels turn around the inferior sharp margin to reach the *superior hepatic nodes.*

(*b*) On the **visceral surface:** The vessels from this surface mostly converge to the porta hepatis, and accompany the deep lymphatics, emerging from the porta to the *hepatic nodes;* one or two from the posterior parts of the right and caudate lobes accompany the inferior vena cava through the diaphragm, and end in the *inferior caval nodes.*

The deep lymphatics converge to ascending and descending trunks. The ascending trunks accompany the hepatic veins and pass through the diaphragm to end in the *inferior caval nodes* around the terminal part of the vein. The descending trunks emerge from the porta hepatis, and end in the *hepatic nodes.*

The **Lymphatic Vessels of the Gallbladder** pass to the *hepatic nodes* in the porta hepatis; those of the **common bile duct** to the *hepatic nodes* alongside the duct and to the upper *pancreaticoduodenal nodes.*

The **Lymphatic Vessels of the Pancreas** follow the course of its blood vessels. Most of them enter the *pancreaticolienal nodes,* but some end in the *pancreaticoduodenal nodes,* and others in the preaortic nodes near the origin of the superior mesenteric artery (Fig. 10–23).

The **lymphatic vessels of the spleen** and **suprarenal glands.**

The **Lymphatic Vessels of the Spleen,** both superficial and deep, pass to the *pancreaticolienal nodes* (Fig. 10–22).

The **Lymphatic Vessels of the Suprarenal Glands** usually accompany the suprarenal veins, and end in the *lateral aortic nodes;* occasionally some of them pierce the crura of the diaphragm and end in the nodes of the posterior mediastinum.

The **lymphatic vessels of the urinary organs** (Fig. 10–27).

The **Lymphatic Vessels of the Kidney** form three plexuses: one in the substance of the kidney, a second beneath its fibrous capsule, and a third in the perinephric fat;

the second and third communicate freely with each other. The vessels from the plexus in the kidney substance converge to form four or five trunks which issue at the hilum. Here they are joined by vessels from the plexus under the capsule, and, following the course of the renal vein, end in the *lateral aortic nodes.* The perinephric plexus is drained directly into the more cranial of the lateral aortic nodes.

The **Lymphatic Vessels of the Ureter** run in different directions. Those from its proximal portion end partly in the efferent

ventral and another from the dorsal surface of the bladder. The vessels from the ventral surface pass to the *external iliac nodes,* but in their course minute nodes are situated. These minute nodes are arranged in two groups, an **anterior vesical,** ventral to the bladder, and a **lateral vesical,** in relation to the lateral umbilical ligament. The vessels from the dorsal surface pass to the *internal, external,* and *common iliac nodes;* those draining the superior part of this surface traverse the *lateral vesical nodes.*

The **Lymphatic Vessels of the Prostate**

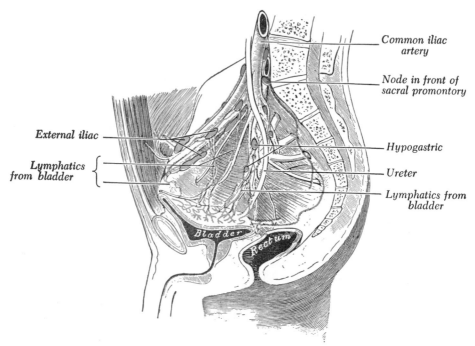

External iliac

Lymphatics from bladder

Common iliac artery

Node in front of sacral promontory

Hypogastric

Ureter

Lymphatics from bladder

Bladder

Rectum

FIG. 10–27.—Lymphatics of the bladder. (Cunéo and Marcille.)

vessels of the kidney and partly in the *lateral aortic nodes;* those from the portion immediately above the brim of the lesser pelvis are drained into the *common iliac nodes;* whereas the vessels from the intrapelvic portion of the tube either join the efferents from the bladder, or end in the *internal iliac nodes.*

The **Lymphatic Vessels of the Bladder** (Fig. 10–27) originate in two plexuses, an intra- and an extramuscular, it being generally admitted that the mucous membrane is devoid of lymphatics. The efferent vessels are arranged in two groups, one from the

(Fig. 10–28) terminate chiefly in the *internal iliac* and *sacral nodes,* but one trunk from the posterior surface ends in the *external iliac nodes,* and another from the anterior surface joins the vessels which drain the membranous part of the urethra.

Lymphatic Vessels of the Urethra.—The lymphatics of the cavernous portion of the urethra accompany those of the glans penis, and terminate with them in the *deep subinguinal* and *external iliac nodes.* Those of the membranous and prostatic portions, and those of the whole urethra in the female, pass to the *internal iliac nodes.*

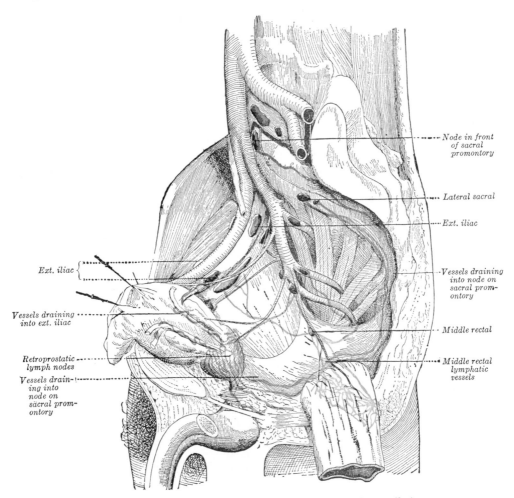

Node in front
of sacral
promontory

Lateral sacral

Ext. iliac

Vessels draining
into node on
sacral prom-
ontory

Middle rectal

Middle rectal
lymphatic
vessels

Ext. iliac {

Vessels draining
into ext. iliac

Retroprostatic
lymph nodes

Vessels drain-
ing into
node on
sacral prom-
ontory

FIG. 10–28.—Lymphatics of the prostate. (Cunéo and Marcille.)

The **lymphatic vessels of the reproductive organs.**

The **Lymphatic Vessels of the Testes** consist of two sets, superficial and deep, the former commencing on the surface of the tunica vaginalis, the latter in the epididymis and body of the testis. They form from four to eight collecting trunks which ascend with the testicular veins in the spermatic cord and along the ventral surface of the Psoas major to the level where the testicular vessels cross the ureter and end in the *lateral* and *preaortic groups* of *lumbar nodes.*

The **Lymphatic Vessels of the Ductus Deferens** pass to the *external iliac nodes;* those of the **vesiculae seminales** partly to the *internal* and partly to the *external iliac nodes.*

The **Lymphatic Vessels of the Ovary** are similar to those of the testis, and ascend with the ovarian artery to the *lateral* and *preaortic nodes.*

The **Lymphatic Vessels of the Uterine Tube** pass partly with those of the ovary and partly with those of the uterus.

The **Lymphatic Vessels of the Uterus** (Fig. 10–29) consist of two sets, superficial and deep, the former being placed beneath the peritoneum, the latter in the substance of the organ. The lymphatics of the **cervix uteri** run in three directions: transversely to the *external iliac nodes*, posterolaterally to the *internal iliac nodes*, and posteriorly to the *common iliac nodes.* The majority of the vessels of the **body** and **fundus** of the uterus pass lateralward in the broad ligaments, and are continued up with the ovarian vessels to the lateral and *preaortic nodes;* a few, however, run to the external iliac nodes, and one or two to the superficial inguinal nodes. In the unimpregnated uterus the lymphatic vessels are very small, but during gestation they are greatly enlarged.

The **Lymphatic Vessels of the Vagina** are carried in three directions: those of the upper part of the vagina to the *external iliac nodes,* those of the middle part to the *internal iliac nodes,* and those of the middle and lower part to the *common iliac nodes.* On the course of the vessels from the middle and lower parts small nodes are situated. Some lymphatic vessels from the lower part of the vagina join those of the vulva and pass to the *superficial inguinal nodes.* The lymphatics of the vagina anastomose with those of the cervix uteri, vulva, and rectum, but not with those of the bladder.

Lymphatic Vessels of the Perineum and External Genitals.—The lymphatic vessels of the perineum, of the integument of the penis, and of the scrotum (or vulva) follow the course of the external pudendal vessels, and end in the *superficial inguinal* and *subinguinal nodes.* Those of the glans penis vel clitoridis terminate partly in the deep subinguinal nodes and partly in the *external iliac nodes* (Fig. 10–21).

The Lymph Nodes of the Abdominal and Pelvic Viscera

The **visceral nodes** are associated with the branches of the celiac, superior and inferior mesenteric arteries. Those related to the branches of the celiac artery form three sets:

> Gastric
> Hepatic
> Pancreaticolienal

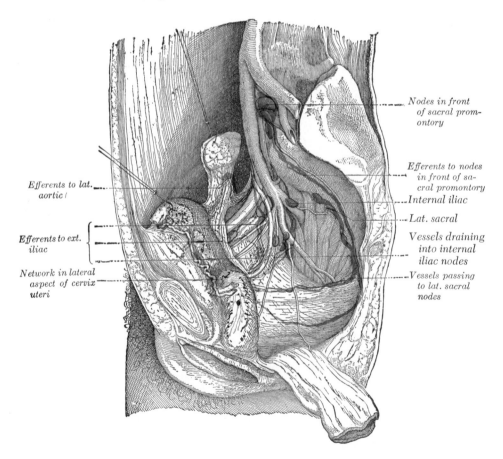

FIG. 10–29.—Lymphatics of the uterus. (Cunéo and Marcille.)

The **Gastric Nodes** (Figs. 10–22), 10–23) consist of two sets, **superior** and **inferior**.

The **Superior Gastric Nodes** (*nodi lymphatici gastrici sinistri*) accompany the left gastric artery and are divisible into three groups: (*a*) **upper,** on the stem of the artery; (*b*) **lower,** accompanying the descending branches of the artery along the cardiac half of the lesser curvature of the stomach, between the two layers of the lesser omentum; and (*c*) **paracardial** outlying members of the gastric nodes, disposed in a manner comparable to a chain of beads around the neck of the stomach. They receive their afferents from the stomach; their efferents pass to the *celiac group of preaortic nodes.*

The **Inferior Gastric Nodes** (*nodi lymphatici gastrici dextri; right gastroepiploic glands*), four to seven in number, lie between the two layers of the greater omentum along the pyloric half of the greater curvature of the stomach.

The **Hepatic Nodes** (*nodi lymphatici hepaticae*) (Fig. 10–22), consist of the following groups: (*a*) **hepatic,** on the stem of the hepatic artery, and extending along the common bile duct, between the two layers of the lesser omentum, as far as the porta hepatis; the **cystic node,** a member of this group, is placed near the neck of the gallbladder; (*b*) **subpyloric,** four or five in number, in close relation to the bifurcation of the gastroduodenal artery, in the angle between the superior and descending parts of the duodenum; an outlying member of this group is sometimes found above the duodenum on the right gastric (pyloric) artery. The nodes of the hepatic chain receive afferents from the stomach, duodenum, liver, gallbladder, and pancreas; their efferents join the *celiac group of preaortic nodes.*

The **Pancreaticolienal Nodes** (*nodi lymphatici pancreaticolienales; splenic glands*) (Fig. 10–23) accompany the lienal (splenic) artery, and are situated in relation to the posterior surface and upper border of the pancreas; one or two members of this group are found in the gastrolienal ligament. Their afferents are derived from the stomach, spleen, and pancreas, their efferents join the *celiac group of preaortic nodes.*

The **superior mesenteric nodes** may be divided into three principal groups:

> Mesenteric
> Ileocolic
> Mesocolic

The **Mesenteric Nodes** (*nodi lymphatici mesenterici*) lie between the layers of the mesentery. They vary from one hundred to one hundred and fifty in number, and may be grouped into three sets: one lying close to the wall of the small intestine, among the terminal twigs of the superior mesenteric artery; a second, in relation to the loops and primary branches of the vessels; and a third along the trunk of the artery.

The **Ileocolic Nodes** (Figs. 10–25, 10–26), from ten to twenty in number, form a chain around the ileocolic artery, but show a tendency to subdivision into two groups, one near the duodenum and another on the lower part of the trunk of the artery. Where the vessel divides into its terminal branches the chain is broken up into several groups: (*a*) **ileal,** in relation to the ileal branch of the artery; (*b*) **anterior ileocolic,** usually of three nodes, in the ileocolic fold, near the wall of the cecum; (*c*) **posterior ileocolic,** mostly placed in the angle between the ileum and the colon, but partly lying dorsal to the cecum at its junction with the ascending colon; (*d*) a single node, between the layers of the mesenteriole of the vermiform appendix; (*e*) **right colic,** along the medial side of the ascending colon.

The **Mesocolic Nodes** (*nodi lymphatici mesocolici*) are numerous, and lie between the layers of the transverse mesocolon, in close relation to the transverse colon; they are best developed in the neighborhood of the right and left colic flexures. One or two small nodes are occasionally seen along the trunk of the right colic artery and others are found in relation to the trunk and branches of the middle colic artery.

The superior mesenteric nodes receive afferents from the jejunum, ileum, cecum, vermiform appendix, and the ascending and transverse parts of the colon; their efferents pass to the *preaortic nodes.*

The **Inferior Mesenteric Nodes** (Fig. 10–24) consist of: (*a*) small nodes on the branches of the left colic and sigmoid arteries; (*b*) a group in the sigmoid mesocolon, around the superior rectal artery; and (*c*) a **pararectal** group in contact with the muscular coat of the rectum. They drain the descending, iliac and sigmoid parts of the colon and the upper part of the rectum, their efferents pass to the *preaortic nodes*.

THE LYMPHATICS OF THE LOWER LIMB

The Lymphatic Vessels of the Lower Limb

The lymphatic vessels of the lower limb consist of two sets, superficial and deep, and in their distribution correspond closely with the veins.

The **superficial lymphatic vessels** lie in the subcutaneous fascia, and are divisible into two groups: a medial, which follows the course of the great saphenous vein, and a lateral, which accompanies the small saphenous vein. The vessels of the **medial group** (Fig. 10–31) are larger and more numerous than those of the lateral group, and commence on the tibial side and dorsum of the foot; they ascend both anterior and posterior to the medial malleolus, run up the leg with the great saphenous vein, pass with it posterior to the medial condyle of the femur, and accompany it to the groin, where they end in the *subinguinal group of superficial nodes*. The vessels of the **lateral group** *arise* from the fibular side of the foot; some ascend anterior to the leg, and, just

below the knee, cross the tibia to join the lymphatics on the medial side of the thigh, others pass posterior to the lateral malleolus and, accompanying the small saphenous vein, enter the *popliteal nodes*.

The **deep lymphatic vessels** are few in number, and accompany the deep blood vessels. In the leg, they consist of three sets, the anterior tibial, posterior tibial, and peroneal, which accompany the corresponding blood vessels, two or three with each artery; they enter the *popliteal lymph nodes*.

The deep lymphatic vessels of the gluteal and ischial regions follow the course of the corresponding blood vessels. Those accompanying the superior gluteal vessels end in a node which lies on the intrapelvic portion of the superior gluteal artery near the upper border of the greater sciatic foramen. Those following the inferior gluteal vessels traverse one or two small nodes which lie caudal to the Piriformis muscle, and end in the *internal iliac nodes*.

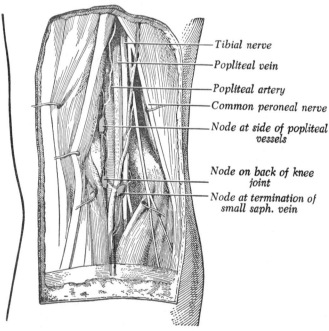

— Tibial nerve
— Popliteal vein
— Popliteal artery
— Common peroneal nerve
— Node at side of popliteal vessels
— Node on back of knee joint
— Node at termination of small saph. vein

Fɪɢ. 10–30.—Lymph nodes of popliteal fossa. (Poirier and Charpy.)

The Lymph Nodes of the Lower Limb

The lymph nodes of the lower limb consist of the following:

Anterior tibial node
Popliteal nodes
Inguinal nodes

The **Anterior Tibial Node** (*nodus lymphaticus tibialis anterior*) is small and inconstant. It lies on the interosseous membrane in relation to the proximal part of the anterior tibial vessels, and constitutes a substation in the course of the anterior tibial lymphatic trunks.

The **Popliteal Nodes** (*nodi lymphatici poplitei*) (Fig. 10–30), small in size and some six or seven in number, are imbedded in the fat contained in the popliteal fossa. One lies immediately beneath the popliteal fascia, near the terminal part of the small saphenous vein, and drains the region from which this vein derives its tributaries. Another is placed between the popliteal artery and the posterior surface of the knee joint; it receives the lymphatic vessels from the knee joint together with those which accompany the genicular arteries. The others lie at the sides of the popliteal vessels, and receive as afferents the trunks which accompany the anterior and posterior tibial vessels. The efferents of the popliteal nodes pass almost entirely alongside the femoral vessels to the deep inguinal nodes, but a few may accompany the great saphenous vein, and end in the nodes of the superficial subinguinal group.

The **Inguinal Nodes** (*nodi lymphatici inguinales*) (Fig. 10–31), from twelve to twenty in number, are situated at the upper part of the femoral triangle. They may be divided into two groups by a horizontal line at the level of the termination of the great saphenous vein; those lying proximal to this line are termed the **superficial inguinal nodes,** and those distal to it the **subinguinal nodes,** the latter group consisting of a *superficial* and a *deep* set.

The **Superficial Inguinal Nodes** form a chain immediately distal to the inguinal ligament. They receive as afferents lymphatic vessels from the integument of the penis, scrotum, perineum, buttock, and abdominal wall below the level of the umbilicus.

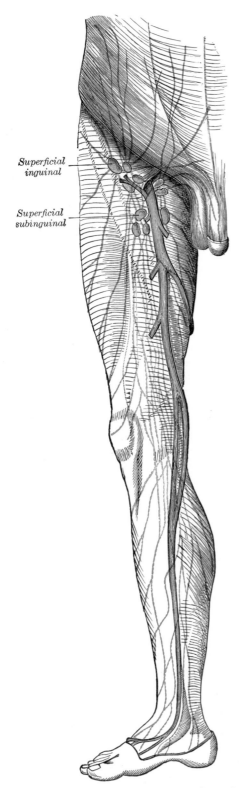

Superficial inguinal

Superficial subinguinal

FIG. 10–31.—The superficial lymph nodes and lymphatic vessels of the lower limb.

The **Superficial Subinguinal Nodes** (*nodi lymphatici subinguinales superficiales*) are placed on either side of the proximal part of the great saphenous vein; their afferents consist chiefly of the superficial lymphatic vessels of the lower limb; but they also receive some of the vessels which drain the integument of the penis, scrotum, perineum, and buttock.

The **Deep Subinguinal Nodes** (*nodi lymphatici subinguinales profundae*) vary from one to three in number, and are placed deep to the fascia lata, on the medial side of the femoral vein. When three are present, the most distal is situated just below the junction of the great saphenous and femoral veins, the middle in the femoral canal, and the most proximal in the lateral part of the femoral ring. The middle one is the most inconstant of the three, but the highest, the **node of Cloquet or Rosenmüller**, is also frequently absent. They receive as afferents the deep lymphatic trunks which accompany the femoral vessels, the lymphatics from the glans penis vel clitoridis, and also some of the efferents from the superficial subinguinal nodes.

THE SPLEEN

The **spleen** (*lien*) is situated principally in the left hypochondriac region (Figs. 3–17, 3–19, 3–20), but its cranial extremity extends into the epigastric region (Fig. 3–25). It lies between the fundus of the stomach and the diaphragm. It is soft, of very friable consistency, highly vascular, and of a dark purplish color. During fetal life and shortly after birth it gives rise to new red blood corpuscles but the evidence that this function is retained in adult life is not satisfactory. It is supposed to be an organ for the destruction of red blood corpuscles and the preparation of new hemoglobin from the iron thus set free.

The size and weight of the spleen are liable to very extreme variations at different periods of life, in different individuals, and in the same individual under different conditions. *In the adult* it is usually about 12 cm in length, 7 cm in breadth, and 3 or 4 cm in thickness. The spleen increases in weight from 17 grams or less during the first year to 170 grams at twenty years, and then slowly decreases to 122 grams at seventy-six to eighty years. Male spleens weigh more than female ones and spleens from whites weigh more than those from Negroes. The variation in weight of adult spleens is from 100 to 250 grams, and in extreme instances, 50 to 400 grams. The size of the spleen is increased during and after digestion, and varies according to the state of nutrition of the body, being large in well fed, and small

in starved animals. In malarial fever it becomes much enlarged, weighing occasionally as much as 9 kilos.*

Development.—The spleen appears about the fifth week as a localized thickening of the mesoderm in the dorsal mesogastrium near the tail of the pancreas. With the change in position of the stomach the spleen is carried to the left, and comes to lie dorsal to the stomach and in contact with the left kidney. The part of the dorsal mesogastrium which intervened between the spleen and the greater curvature of the stomach forms the gastrosplenic ligament.

Relations in Adult.—The **diaphragmatic surface** (*facies diaphragmatica; external or phrenic surface*) is convex, smooth, and is directed cranialward, dorsalward, and to the left, except at its cranial end, where it is directed slightly medialward. It is in relation with the caudal surface of the diaphragm, which separates it from the ninth, tenth, and eleventh ribs of the left side, and the intervening caudal border of the left lung and pleura.

The **visceral surface** (Fig. 10–32) is divided by a ridge into a **gastric**, a **renal**, and a **colic portion**.

The **gastric surface** (*facies gastrica; anterior surface*), which is directed ventralward,

*These data are taken from the monograph entitled *The Structure and Use of the Spleen* by Henry Gray which was published in 1854 and which won the triennial "Astley Cooper Prize" of £300 in 1853.

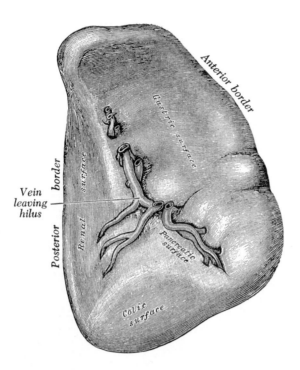

Fig. 10–32.—The visceral surface of the spleen.

cranialward, and medialward, is broad and concave, and is in contact with the dorsal wall of the stomach, and caudal to this with the tail of the pancreas. It presents near its medial border a long fissure, or more frequently a series of depressions termed the **hilum**. This is pierced by several irregular apertures, for the entrance and exit of vessels and nerves.

The **renal surface** (*facies renalis*) is directed medialward and caudalward. It is somewhat flattened, is considerably narrower than the gastric surface, and is in relation with the cranial part of the ventral surface of the left kidney and occasionally with the left suprarenal gland.

The **colic surface** (*facies colica*) is flat, triangular in shape, and rests upon the left flexure of the colon and the phrenicocolic ligament, and is generally in contact with the tail of the pancreas.

The **anterior border** (*margo anterior*) is free, sharp, and thin, and is often notched, especially caudally; it separates the diaphragmatic from the gastric surface. The **posterior border** (*margo posterior*), more

rounded and blunt than the anterior, separates the renal from the diaphragmatic surface; it corresponds to the lower border of the eleventh rib and lies between the diaphragm and left kidney. The intermediate margin is the ridge which separates renal and gastric surfaces. The **inferior border** (*internal border*) separates the diaphragmatic from the colic surface.

The **posterior extremity** (*extremitas posterior; superior extremity*) is directed toward the vertebral column, where it lies on a level with the eleventh thoracic vertebra. The **anterior extremity** (*extremitas anterior; inferior extremity*) corresponds with the colic surface.

The visceral **peritoneum** is closely adherent to the fibrous capsule of the spleen except at the hilum. At the hilum the peritoneum is reflected over both surfaces of the phrenicolienal and lienorenal ligaments stretching dorsalward, and over the gastrolienal ligament stretching ventralward to the greater curvature of the stomach. These ligaments are composed of a thin membrane of connective tissue conveying the lienal artery and its branches, the short gastric and left gastroepiploic arteries. Since these ligaments form the left boundary of the omental bursa (lesser sac), their outer surfaces are covered by peritoneum of the greater sac and their inner surfaces by peritoneum of the omental bursa. The spleen itself is covered by greater sac peritoneum only, unless the gastrolienal and phrenico-

the lines of reflection of the phrenicolienal and gastrolienal ligaments.

The **fibroelastic coat** (*tunica albuginea*) invests the organ, and at the hilum is prolonged inward upon the vessels in the form of sheaths. From these sheaths, as well as from the inner surface of the fibroelastic coat, numerous small fibrous bands, **trabeculae** (Fig. 10–33), are given off in all directions; these constitute the framework of the spleen. The spleen therefore

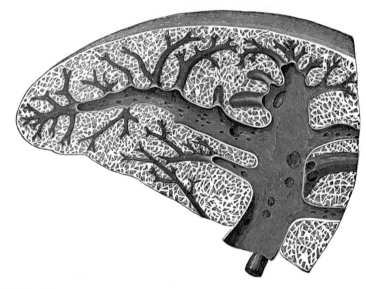

Fig. 10–33.—Transverse section of the spleen, showing the trabecular tissue and the splenic vein and its tributaries. (From Henry Gray, 1854.)

lienal ligaments fail to meet at the hilum, in which condition a small area within the hilum may be covered by peritoneum from the omental bursa.

Frequently in the neighborhood of the spleen, and especially in the gastrolienal ligament and greater omentum, small nodules of splenic tissue may be found, either isolated or connected to the spleen by thin bands of splenic tissue. They are known as **accessory spleens** (*lien accessorius; supernumerary spleen*). They vary in size from that of a pea to that of a plum.

Structure.—The spleen is invested by two coats: an **external serous** and an **internal fibroelastic** coat.

The **external or serous coat** (*tunica serosa*) is derived from the peritoneum; it is thin, smooth, and in the human subject intimately adherent to the fibroelastic coat. It invests the entire organ, except at the hilum and along

consists of a number of small subdivisions formed by the trabeculae and the contained **splenic pulp.**

The fibroelastic coat, the sheaths of the vessels, and the trabeculae are composed of white and yellow elastic fibrous tissues. It is owing to the presence of the elastic tissue that the spleen possesses a considerable amount of ability to undergo the great variations in size that it presents under certain circumstances. In addition to these constituents of this tunic, there is found in man a small amount of smooth muscle and in some mammalia, *e.g.*, dog, pig, and cat, a large amount, so that the trabeculae appear to consist chiefly of muscular tissue.

The **splenic pulp** (*pulpa lienis*) is a soft mass of a dark reddish-brown color, resembling grumous blood; it consists of a fine reticulum of fibers, continuous with those of the trabeculae, to which are applied flat, branching cells. The meshes of the reticulum are filled with blood, and many lymphocytes, polymorphonuclear

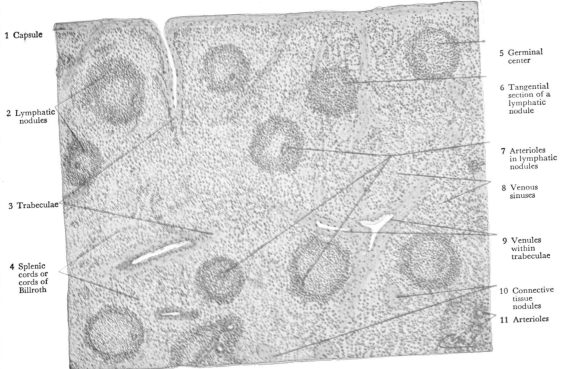

1 Capsule

5 Germinal center

2 Lymphatic nodules

6 Tangential section of a lymphatic nodule

7 Arterioles in lymphatic nodules

8 Venous sinuses

3 Trabeculae

9 Venules within trabeculae

4 Splenic cords or cords of Billroth

10 Connective tissue nodules

11 Arterioles

FIG. 10–34.—A section through a portion of the human spleen, stained with hematoxylin and eosin. × 50. (di Fiore, *Atlas of Human Histology*, Lea & Febiger.)

neutrophiles, and monocytes or macrophages. The latter are often large and contain ingested and partly digested red blood cells and pigment. In the spleens of young individuals, giant cells may also be found, each containing numerous nuclei or one compound nucleus. Nucleated red blood corpuscles have also been found in the spleens of young animals.

Blood Vessels of the Spleen.—The **splenic artery** is remarkable for its large size in proportion to the size of the organ, and also for its tortuous course. It divides into six or more branches, which enter the hilum and ramify throughout its substance (Fig. 10–34), receiving sheath of fibrous tissue from the trabeculae. Similar sheaths also invest the nerves and veins. Each of the larger branches of the artery supplies chiefly that region of the organ in which the branch ramifies, having no anastomosis with the majority of the other branches (Fig. 8–43).

Each branch runs in the transverse axis of the organ, from within outward, diminishing in size during its transit, and giving off smaller branches, some of which pass to the anterior, others to the posterior part. These ultimately leave the trabecular sheaths, and terminate in

the proper substance of the spleen in small tufts or **pencils of minute arterioles** (*penicilli*).

Some authors claim that the arterioles are connected with venules by capillaries and that they have a complete endothelial lining. Most authors, however, maintain that the endothelial connections are incomplete. In this open type of circulation the blood moves slowly through the spleen pulp and bathes the reticular network of the **sinuses.** The flow through the pulp is controlled by rhythmic contractions and relaxation of individual arterioles and groups of them.

The altered coat of the arterioles, consisting of lymphoid tissue, presents here and there thickenings of a spheroidal shape, the **lymphatic nodules** (*folliculi lymphatici lienales; Malpighian bodies of the spleen*). These bodies vary in size from about 0.25 to 1 mm in diameter. They are merely local expansions or hyperplasiae of the lymphoid tissue, of which the external coat of the smaller arteries of the spleen is formed. They are most frequently found surrounding the arteriole, which thus seems to tunnel them, but occasionally they grow from one side of the vessel only, and present the appearance of a sessile **bud grow-**

ing from the arterial wall. These bodies are visible to the naked eye on the surface of a fresh section of the organ, appearing as minute dots of a semiopaque whitish color in the dark substance of the pulp. In minute structure they resemble the lymphoid tissue of lymph nodes, consisting of a delicate reticulum, in the meshes of which lie ordinary lymphocytes (Fig. 10–34). The reticulum is made up of extremely fine fibrils, and is comparatively open in the center of the corpuscle, becoming more dense at its periphery. The cells which it encloses are possessed of ameboid movement. When treated with carmine they become deeply stained, and can be easily distinguished from those of the pulp.

The smaller **veins** unite to form larger ones; these do not accompany the arteries, but soon enter the trabecular sheaths of the capsule, and by their junction form six or more branches, which emerge from the hilum, and, uniting, constitute the lienal vein, the largest radicle of the portal vein.

The **veins** are remarkable for their numerous anastomoses, whereas the arteries seldom anastomose.

The **lymphatics** are described on page 761.

The **nerves** are derived from the celiac plexus and are chiefly unmyelinated. They are distributed to the blood vessels and to the smooth muscle of the capsule and trabeculae.

THE THYMUS

The thymus in an infant (Fig. 10–35) is a prominent organ occupying the ventral superior mediastinum, but the thymus in an adult of advanced years may be scarcely recognizable because of atrophic changes. During its growth period, it has much of the appearance and texture of a gland, and previously it has been classified as one of the ductless glands, but a definite glandular function has not been established. It is included here among the lymphatic organs because it resembles them structurally in being composed largely of lymphocytes and because its only known function is that of producing lymphocytes.

Development.—The thymus appears in the form of two flask-shaped entodermal diverticula, which arise, one on either side, from the third branchial pouch (Fig. 18–1), and extend lateralward and dorsalward into the surrounding mesoderm ventral to the ventral aortae. Here they meet and become joined to each other by connective tissue, but there is never any fusion of the thymus tissue proper. The pharyngeal opening of each diverticulum is soon obliterated, but the neck of the flask persists for some time as a cellular cord. By further proliferation of the cells, buds are formed which become elongated and greatly branched, lose their epithelial character, and become the cellular reticulum which forms the core of the lobules. The lobules remain separated from each other by loose connective tissue and

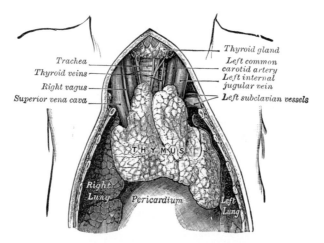

Fig. 10–35.—The thymus of a full-term fetus; exposed *in situ*.

the numerous small lymphocytes which make their appearance probably wander in from the surrounding mesenchyme. As the lobules increase in size, there is a differentiation into a peripheral zone or **cortex** in which the small lymphocytes are concentrated, and a central **medulla** in which the cellular reticulum predominates. The thymus attains a weight of 12 to 14 grams before birth, but it does not reach its greatest relative size until the age of two years. It continues to grow until puberty, at which time it reaches its greatest absolute size, weighing about 35 grams.

length, 4 cm in width, and 6 mm in thickness. In animals, it is the neck sweetbread of the butcher shop.

Structure.—The two lobes are composed of numerous lobules, varying from 0.5 to 2 mm in diameter, separated from each other by delicate connective tissue, and visible with the unaided eye. With low magnification, two zones of tissue can be seen within each lobule, an outer **cortex** and an inner **medulla**. The cortical part is made up almost entirely of small lymphocytes, held in place by reticular tissue composed of reticular (argyrophilic) fibers with relatively few reticular cells. The medul-

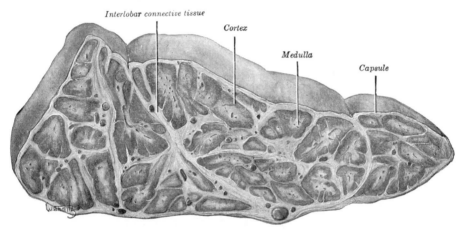

Fig. 10–36.—Section of thymus from full-term fetus, lightly stained with hematoxylin. × 5.

Anatomy.—The thymus consists of two lateral lobes held in close contact by connective tissue which also encloses the whole organ in a distinct capsule. It is situated partly in the thorax and partly in the neck, extending from the fourth costal cartilage to the caudal border of the thyroid gland. In the neck it lies ventral and lateral to the trachea, deep to the origins of the Sternohyoidei and Sternothyroidei. In the thorax it occupies the anterior portion of the superior mediastinum (Fig. 15–28, page 1144); superficial to it is the sternum, and deep to it are the great vessels and the cranial part of the fibrous pericardium. The two lobes generally differ in size and shape, the right frequently overlapping the left. It is of a pinkish gray color, soft and lobulated, measuring approximately 5 cm in

lary portion contains much fewer lymphocytes, its reticulum is more cellular, and it contains the characteristic **thymic** or **Hassall's corpuscles.** These corpuscles are from 0.03 to 0.1 mm in diameter; they have a central core of granular cells surrounded by concentric lamellae of epithelioid cells. Although the lobules appear to be distinct from one another when viewed in a cross section, it has been found by careful study that the medullary tissue forms a continuous system of central stalks and branching cords.

Involution.—After puberty the thymus undergoes involution. Usually this is a gradual process called age involution, but it may be superseded by a rapid accidental involution due to starvation or acute disease. The small lymphocytes of the cortex disappear and the reticular tissue becomes compressed. The disappearing thymic tissue is likely to be replaced by adipose tissue, but the connective tissue

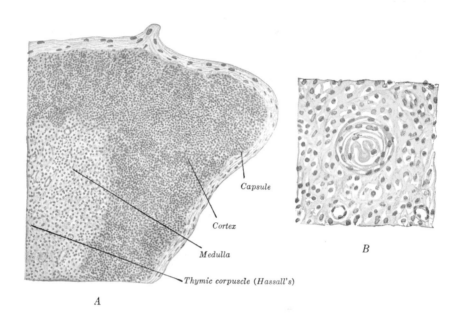

FIG. 10–37.—*A*, A section through a portion of the thymus of a kitten. × 120. *B*, A concentric corpuscle of Hassall. × 350. Stained with hematoxylin and eosin.

capsule may persist, retaining its original shape and approximate size so that, in an older individual what appears to be a yellowish-colored thymus, when sectioned, will reveal only small islands of thymic tissue surrounded by fat.

Function.—No function of the thymus is known except that of producing small lymphocytes, and possibly plasma cells. The association of the growth of the organ with the growth of the individual and its involution after sexual maturity have made many investigators attempt to discover some endocrine function, but definite confirmation of claims along these lines is lacking except that its involution and regeneration may be influenced by known endocrine factors.

Vessels and Nerves.—The **arteries** supplying the thymus are derived from the internal thoracic, and from the superior and inferior thyroids. The **veins** end in the left brachiocephalic vein, and in the thyroid veins. The **lymphatics** end in the anterior mediastinal, tracheobronchial and sternal nodes. The **nerves** are exceedingly minute; they are derived from the vagi and sympathetic. Branches from the descendens hypoglossi and phrenic reach the investing capsule, but do not penetrate into the substance of the gland.

REFERENCES

EMBRYOLOGY

HOLYOKE, E. A. 1936. The role of the primitive mesothelium in the development of the mammalian spleen. Anat. Rec., *65*, 333-349.

HUNTINGTON, G. S. 1911. *The Anatomy and Development of the Systemic Lymphatic Vessels in the Domestic Cat.* American Anatomical Memoir No. 1, Wistar Institute of Anatomy and Biology, Philadelphia, 175 pages.

KAMPMEIER, O. F. 1928. The genetic history of the valves in the lymphatic system of man. Amer. J. Anat., *40*, 413-457.

LEWIS, F. T. 1909. On the cervical veins and lymphatics in four human embryos. With an interpretation of anomalies of the subclavian and jugular veins in the adult. Amer. J. Anat., *9*, 33-42.

SABIN, F. R. 1916. The origin and development of the lymphatic system. Johns Hopk. Hosp. Rep., *17*, 347-440.

LYMPHATIC SYSTEM AND LYMPHOCYTES

RÉNYI-VÁMOS, F. 1960. *Das Innere Lymphgefasssystem der Organe.* 447 pages, illustrated. Publishing House of the Hungarian Academy of Sciences.

RUSZNYÁK, I., M. FÖLD, and G. SZABO. 1960. *Lymphatics and Lymph Circulation, Physiology and Pathology.* 853 pages, illustrated. Pergamon Press, Ltd., New York.

SHREWSBURY, M. M., JR. and W. O. REINHARDT. 1955. Relationships of adrenals, gonads, and thyroid to thymus and lymph nodes, and to blood and thoracic duct leukocytes. Blood, *10*, 633-645.

LYMPHATIC VESSELS

ALLEN, L. 1943. The lymphatics of the parietal tunica vaginalis propria of man. Anat. Rec., *85*, 427-433.

BLAIR, J. B., E. A. HOLYOKE, and R. R. BEST. 1950. A note on the lymphatics of the middle and lower rectum and anus. Anat. Rec., *108*, 635-644.

BUTLER, H. and K. BALANKURA. 1952. Preaortic thoracic duct and azygos veins. Anat. Rec., *113*, 409-419.

DIXON, F. W. and N. L. HOERR. 1944. Lymphatic drainage of paranasal sinuses. Laryngoscope, *54*, 165-175.

FOHMANN, V. 1833. Mémoire sur le vaisseaux lymphatiques de la peau, des membranes muqueuses, sereuses, du tissu nerveux et musculaire. J. Desoer, Liége.

GUSEV, A. M. 1964. Lymph vessels of human conjunctiva. Fed. Proc., *23*, T1099-T1102.

PAPP, M., P. ROHLICH, I. RUSZNYÄK, and I. TÖRÖ. 1964. Central chyliferous vessel of intestinal villus. Fed. Proc., *23*, T155-T158.

PATEK, P. R. 1939. The morphology of the lymphatics of the mammalian heart. Amer. J. Anat., *64*, 203-249.

PIERCE, E. C. 1944. Renal lymphatics. Anat. Rec., *90*, 315-335.

REIFFENSTUHL, G. 1964. *The Lymphatics of the Female Genital Organs.* Trans. by Leslie D. EKVALL. ix + 165 pages, illustrated. J. B. Lippincott Company, Philadelphia.

ROUVIERE, H. *Anatomy of the Lymphatic System.* Trans. by M. J. Tobias. ix + 318 pages. Edwards Brothers, Ann Arbor.

SPIRIN, B. A. 1964. Internal lymphatic system of penis. Fed. Proc., *23*, T159-T167.

VAN PERNIS, P. A. 1949. Variations of the thoracic duct. Surg., *26*, 806-809.

WALLACE, S., L. JACKSON, B. SCHAFFER, J. GOULD, R. R. GREENING, A. WEISS, and S. KRAMER. 1961. Lymphangiograms: their diagnostic and therapeutic potential. Radiology, *76*, 179-199.

YOFFEY, J. M. and F. C. COURTICE. 1956. *Lymphatics, Lymph and Lymphoid Tissue.* vii + 510 pages. Harvard University Press, Cambridge.

LYMPH NODES

BAKER, B. L., D. J. INGLE, and C. R. LI. 1951. The histology of the lymphoid organs of rats treated with adrenocorticotropin. Amer. J. Anat., *88*, 313-349.

FURUTA, W. J. 1948. The histologic structure of the lymph node capsule at the hilum. Anat. Rec., *102*, 213-223.

PINTO, S. 1946. Technica de la injection de linfaticos en el animal vivo. Medicine, Madrid, *14*, 453.

REINHARDT, W. O. 1946. Growth of lymph nodes, thymus and spleen, and output of thoracic duct lymphocytes in the normal rat. Anat. Rec., *94*, 197-211.

SPLEEN

GALINDO, B. and J. A. FREEMAN. 1963. Fine structure of splenic pulp. Anat. Rec., *147*, 25-29.

GRAY, H. 1854. *On the Structure and Use of the Spleen.* xix + 380 pages, illustrated. J. W. Parker & Son, London.

KRUMBHAAR, E. B. and S. W. LIPPINCOTT. 1939. Postmortem weight of "normal" human spleen at different ages. Am. J. Med. Sci., *197*, 344-358.

MACKENZIE, D. W., JR., A. O. WHIPPLE, and M. P. WINTERSTEINER. 1942. Studies on the microscopic anatomy and physiology of living transilluminated mammalian spleens. Amer. J. Anat., *68*, 397-456.

MICHELS, N. A. 1942. The variational anatomy of the spleen and splenic artery. Amer. J. Anat., *70*, 21-72.

PECK, H. M. and N. L. HOERR. 1951. The intermediary circulation in the red pulp of the mouse spleen. Anat. Rec., *109*, 447-477.

SNOOK, T. 1950. A comparative study of the vascular arrangements in mammalian spleens. Amer. J. Anat., *87*, 31-77.

WILLIAMS, R. G. 1961. Studies of the vasculature in living autografts of spleen. Anat. Rec., *140*, 109-121.

THYMUS

CSABA GY., I. TÖRÖ, E. KAPA. 1960. Provoked tissue reaction of the thymus in tissue culture. Acta Morph. Acad. Sci. Hungaricae., *9*, 197-202 + 4 colored plates.

GRÉGOIRE. C. 1943. Regeneration of the involuted thymus after adrenalectomy. J. Morph., *72*, 239-261.

NORRIS, E. H. 1938. The morphogenesis and histogenesis of the thymus gland in man. Carneg. Instn. Contr. Embryol., *27*, 191-207 + 7 plates.

SMITH, C. and H. T. PARKHURST. 1949. Studies on the thymus of the mammal. II. A comparison of the staining properties of Hassall's corpuscles and of thick skin of the guinea pig. Anat. Rec., *103*, 649-673.

TÖRÖ, I. 1961. The cytology of the thymus gland. Folia biol. (Krakow), *7*, 145-149.

11 | *The Central Nervous System*

THE NERVOUS SYSTEM AS A WHOLE

The nervous system is an extensive and complicated organization of structures by which the internal reactions of the individual are correlated and integrated and by which his adjustments with his environment are controlled. It is separated arbitrarily for convenience in description, into two large divisions, the central nervous system and the peripheral nervous system.

The central nervous system, the subject of this chapter, is composed of the brain and spinal cord.

The peripheral nervous system is composed of the nerves, ganglia, and end organs which connect the central nervous system with all the other parts of the body. It is the subject of the next two chapters; the first deals with the peripheral nervous system proper and the second with the specialized sense organs.

DEVELOPMENT OF THE NERVOUS SYSTEM

The earliest stages of development of the nervous system are included in Chapter One and that account may be used as a background for the following description.

The entire nervous system is of ectodermal origin. Its first rudiment is the **neural plate** with its folds and **neural groove** which extend along the axis of the embryo (Fig. 2–12). The rostral part of the plate is broad and forms the brain. The narrow caudal part forms the spinal cord. The **neural folds** curl dorsally until their prominent edges meet and fuse in the midline to form the **neural tube** (Fig. 2–14). The cavity of the tube is retained in the ventricles of the brain and the central canal of the spinal cord. Certain of the malformations of the nervous system can be traced to this very early stage of development. If the neural groove fails to close or, having closed, fails to be separated completely from the overlying ectoderm, the development of the supporting bones is also abnormal and the condition known as spina bifida results (Duncan '57).

After formation of the neural tube, further development occurs by mitotic division in and near the layer of cells lining the central cavity, the medullary epithelium. All of the elements in the central nervous system except the microglia, meninges, and blood vessels are derived from this source. The primitive neural tube has three layers (Fig. 11–2): 1) The **ependymal layer** lines the lumen and becomes the ependyma of the adult nervous system. The processes of its cells extend out toward the periphery, acting as supporting elements. 2) The **mantle layer** is the middle bulky cellular portion of the tube and contains (*a*) the **germinal cells** whose proliferation provides the neuroblasts which develop into the nerve cells of the central gray substance and (*b*) **spongioblasts** which develop into the supporting elements or neuroglia (Fig. 11–1). 3) The **marginal layer** is not cellular and becomes the outer white substance when the fibers grow out from the mantle layer and accumulate at the periphery.

Development of the Spinal Cord.—As the lateral walls of the neural tube increase in thickness, the germinal cells remain near the ependyma. They are round or oval cells and many of the nuclei can be seen in stages of mitotic division in slide preparations. The daughter cells, **neuroblasts**, migrate out into the mantle layer and begin their differentiation into young

(777)

nerve cells. The lumen of the tube becomes altered by the formation of a longitudinal groove, the **sulcus limitans** (Fig. 11–3A), along the middle portion of each side. The sulcus marks the subdivision of the wall of the tube into a dorsal or alar lamina and a ventral or basal lamina. These laminae represent an early functional differentiation, the **alar lamina** or **alar plate** becoming the sensory portion of the gray substance and

the **basal lamina** or **basal plate** becoming the motor portion in later development (Figs. 11–2, 11–5).

The basal lamina becomes thickened, and in a cross section the mantle layer appears as an expanded zone between the marginal and ependymal layers. This thickening is the rudiment of the **ventral gray column** or **ventral horn.** Its neuroblasts send their axons out through the marginal layer, form-

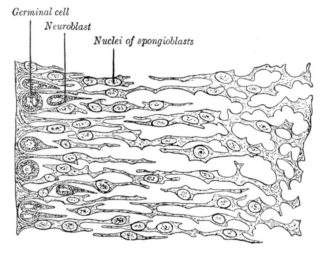

Fig. 11–1.—Transverse section of the spinal cord of a human embryo at the beginning of the fourth week. The left edge of the figure corresponds to the lining of the central canal. (His.)

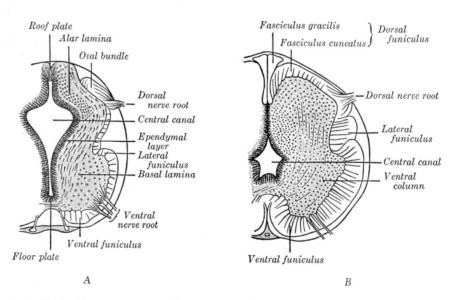

Fig. 11–2.—Transverse sections through the spinal cords of human embryos. (His.) A, aged about four and a half weeks, B, aged about three months.

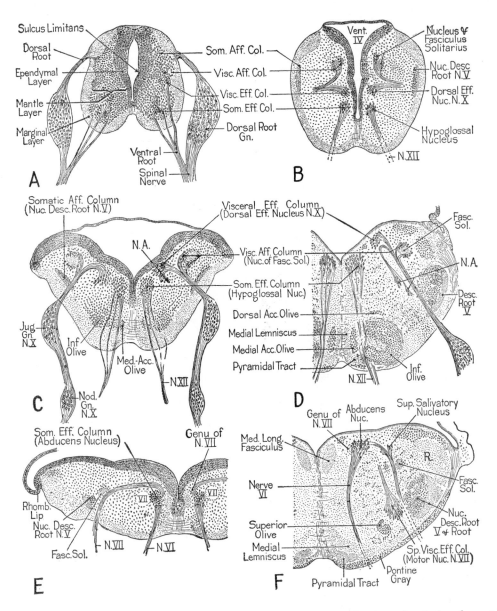

FIG. 11–3.—Cross sections of cord and medulla comparing manner of origin of spinal and cranial nerves. *A*, Thoracic cord of 14.8-mm embryo to show origin of typical spinal nerve. *B*, Lower part of medulla for comparison with cord structure. *C*, Medulla of 15-mm embryo at level of upper rootlets of nerves X and XII. *D*, Medulla of 73-mm embryo at level of upper rootlets of nerves X and XII. Compare with *C*. *E*, Medulla of 15-mm embryo at level of pons showing the roots of nerves VI and VII. *F*, Medulla of 73-mm embryo at level of pons showing the roots of nerves VI and VII. Compare with *E*.

Abbreviations: Acc., accessory; Aff., afferent; Col., column; Desc., descending; Eff, efferent; Fasc. Sol., fasciculus solitarius; Gn., ganglion; Inf., inferior; Jug., jugular; N.A., nucleus ambiguus; note that in the younger stage (*C*) it is not separated from the general afferent column as happens later (*D*). The nucleus ambiguus, since it supplies skeletal muscle of branchiomeric origin rather than smooth muscle, is classified as a *special* visceral efferent nucleus; Nuc., nucleus; R., restiform body; Rhomb., rhombic; Sp., special; Visc., visceral. (Courtesy of B. M. Patten, *Human Embryology*, 2nd Ed. Copyright, 1953. Blakiston Div., McGraw-Hill Book Co.)

ing the ventral roots of the spinal nerves. The alar lamina also thickens gradually, becoming the **dorsal gray column** or **dorsal horn.** Later, the axons of many of its cells grow cranialward to assist in forming the dorsal and lateral funiculi; a few cross to the opposite side, forming the ventral white commissure.

At about the end of the fourth week nerve fibers begin to appear in the marginal layer. The first to develop are the short intersegmental fibers from neuroblasts in the mantle layer. Fibers from the cells in the spinal ganglia grow into the marginal layer and by the sixth week have formed a well-defined oval bundle in the peripheral part of the alar lamina. This bundle gradually

of the nerves and the bulk of the peripheral field its nerve supplies. Experimental evidence indicates that segmentation in the nervous system is determined to a large extent by that of the mesodermal somites (Ditwiler '36, Chapter 12). In later development the long projection tracts growing into the marginal layer obliterate the segmental appearance.

Up to the third month of fetal life the spinal cord occupies the entire length of the vertebral canal. After this, the canal lengthens more rapidly than the cord. By the sixth month the caudal end of the cord reaches only as far as the upper part of the sacrum; at birth it is on a level with the third lumbar vertebra, and in the adult it

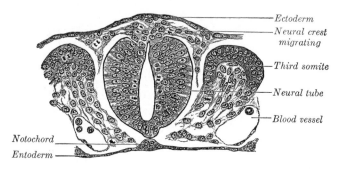

Fig. 11–4.—Neural crest formation. Transverse section through third somite of 9 somite rat embryo.

increases in size and spreads toward the midline forming the rudiment of the dorsal funiculus. Long intersegmental fibers appear about the third month and corticospinal fibers about the fifth month. The nerve fibers acquire their myelin sheaths at different times somewhat later, the dorsal and ventral roots about the fifth month, the corticospinal after the ninth month. The expanding growth of the ventral horn and ventral funiculus of the two sides causes these parts to bulge out beyond the floor plate, leaving a deep groove, the ventral median fissure.

Segmentation.—The thickening of the wall of the neural tube is not uniform. At the level of the spinal nerves it is greater than in between, producing a beaded appearance and establishing the **neuromeres** or segments. The relative thickness of the segments varies according to the size

lies at the disk between the first and second lumbar vertebrae.

The **Spinal Nerves.**—The outgrowth of axons from neuroblasts in the basal lamina to form the ventral roots has been mentioned. The dorsal roots are formed by the ingrowth of fibers from the spinal ganglion cells.

The **Spinal Ganglia.**—Before the neural groove is closed into the neural tube of the early embryo, the prominent margin of the neural fold contains a ridge of special cells known as the **ganglionic ridge** or **neural crest** (see Chapter 1). When the neural folds meet in the midline, the two ridges fuse along the line of closure (Fig. 11–4). The cells of the crest proliferate, separate from the neural tube, and migrate out from the narrow space between the ectoderm and neural tube. They gather into groups of cells corresponding to each somite, and become the primitive spinal ganglia. The sensory components of cranial nerves V, VII, IX, and X are developed from the neural crest in the head region (Yntema '44).

The cells of the ganglia are of two kinds, like those of the mantle layer of the cord, neuroblasts and spongioblasts. The **neuroblasts** send out processes both centrally and peripherally. The central ones form the dorsal roots, the peripheral ones join the ventral roots in forming the spinal nerves. Eventually the two processes come from a single stem of the cell which in consequence is called a unipolar cell.

The **spongioblasts** of the ganglion develop into **sheath cells** which not only form capsules around the nerve cell bodies in the ganglion but also migrate out along the nerves to form the **neurilemma** of peripheral nerves. In the peripheral growth of the spinal nerves the axons grow out first as naked protoplasmic processes. The spongioblasts of the ganglion migrate out on these nerve fibers, proliferate by mitosis, and provide *neurilemma* or *Schwann sheath cells* for the peripheral nerves. Later **myelin** develops by the spiral infolding of the surface membrane of the sheath cells of the larger fibers (Geren '54, Harrison '24).

Development of the Brain.—The cephalic portion of the neural tube, even before it is completely closed, enlarges and begins to change its shape (Fig. 2–15). The unequal growth of the different parts in size and thickness, together with the formation of certain flexures, very early establishes three recognizable regions within the primitive brain: the forebrain, midbrain, and hindbrain (Figs. 2–20, 11–6). The first of the flexures to appear is the **ventral cephalic flexure** in the region of the midbrain (Fig. 11–7). In it, the forebrain makes a U-shaped bend ventrally over the cranial end of the notochord and foregut causing the midbrain to protrude dorsally as the most prominent part of the brain. The second bend, the **cervical flexure,** appears at the junction of the hindbrain and spinal cord, where these two parts form a right angle with each other (Fig. 11–7). This flexure gradually diminishes as the body posture changes and the head becomes erect and it disappears some time after the

fifth week. The third bend is named the **pontine flexure** because it occurs in the region of the future pons (Fig. 11–9).

The sulcus limitans which has been described as dividing the lateral wall of the spinal cord into alar and basal laminae is clearly distinguishable extending through the hindbrain (Fig. 11–5) and midbrain but its termination in the forebrain is not established.

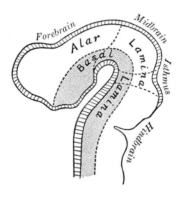

Fig. 11–5.—Diagram to illustrate the alar and basal laminae of brain. The dotted line represents the sulcus limitans. (His.)

The **Hindbrain or Rhombencephalon.**—At the time of appearance of the cephalic flexure, the hindbrain is as long as the other two parts combined. Just caudal to the midbrain, its cavity is constricted into the *isthmus rhombencephali* (Fig. 11–5). The rest of the hindbrain is divided into a cranial portion, the **metencephalon,** and a caudal portion, the **myelencephalon.** The metencephalon develops into the cerebellum and pons, the myelencephalon into the medulla oblongata (Fig. 11–8).

The **medulla oblongata** is at first similar to the spinal cord, with roof plate, floor plate, and side walls of alar and basal laminae. The roof plate becomes greatly stretched out at an early stage (Fig. 11–3) but the floor plate holds the side walls together so that they flatten out into a plate-like structure. The sulcus limitans persists but now separates a laterally placed alar lamina from a more medially placed basal lamina. The functional pattern remains similar to the spinal cord with a sensory alar

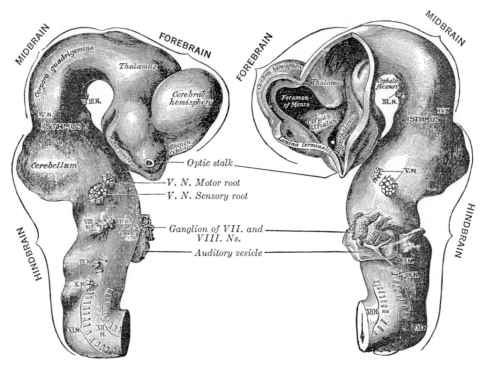

FIG. 11–6.—Exterior of brain of human embryo of four and a half weeks. (From model by His.)

Fig. 11–7.—Brain of human embryo of four and a half weeks showing interior of forebrain. (From model by His.)

plate, a motor basal plate and with autonomic nuclei along the sulcus limitans.

Sensory fibers from the neural crest ganglia of the glossopharyngeal and vagus nerves form a bundle opposite the sulcus limitans. This is the **tractus solitarius** and corresponds to the oval bundle of the primitive spinal cord (Fig. 11–2). At first it is applied to the outer surface of the alar lamina, but it becomes buried by the overgrowth of neighboring parts. At about five weeks, the part of the alar lamina next to the floor plate bends over laterally, forming the **rhombic lip** (Figs. 11–3E, 11–8). The rhombic lip folds down over the main part of the alar lamina, fuses with it, and buries the tractus solitarius and spinal root of the trigeminal nerve. Later, the nodulus and flocculus of the cerebellum are developed from the rhombic lip (Fig. 11–11).

Although the basal plate corresponds to the ventral horn of the spinal cord in giving rise to motor fibers, the cells become arranged in groups or nuclei instead of continuous columns. In addition, neuroblasts migrate from the alar plate and rhombic lip into the basal lamina and become aggregated into the **olivary nuclei** (Fig. 11–3,D). Many of the fibers from these cells cross the midline of the floor plate, constituting the rudiment of the *raphe of the medulla*. The accumulation of these cells and fibers in the ventral part of the basal plate pushes the motor nuclei deeply into the interior and in the adult they are found close to the interior lumen. The change is still further accentuated by the development of the pyramids at about the fourth month and by the fiber connections of the cerebellum. On the floor of the fourth ventricle, the rhomboid fossa, a series of six temporary transverse grooves appear. The most cephalic or first and second lie over the trigeminal nucleus, the third over the facial, the fourth over the abducent, the fifth over the glossopharyngeal, and the sixth over the nucleus of the vagus nerve.

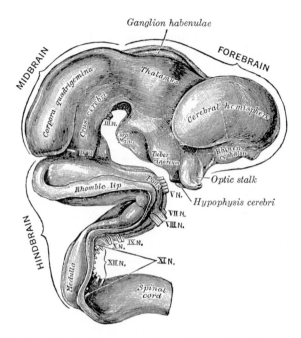

FIG. 11–8.—Exterior of brain of human embryo of five weeks. (From model by His.)

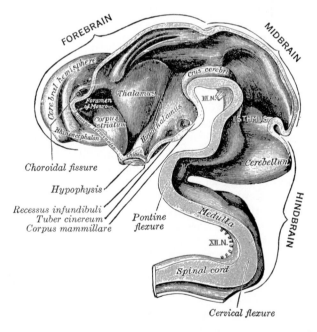

FIG. 11–9.—Interior of brain of human embryo of five weeks. (From model by His.)

The **Cerebellum.**—The alar laminae of the cephalic portion of the hindbrain become thickened into plates which fuse in the midline (Fig. 11–10). The resulting thick lamina roofs over the cephalic part of the hindbrain vesicle as the rudiment of the cerebellum, the outer surface of which is at first smooth. **Fissures** appear first in the vermis and floccular regions during the third month. The fissures on the hemisphere appear during the fifth month. The order of

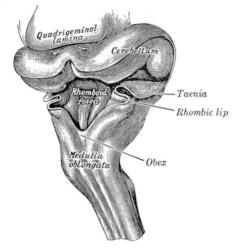

Fɪɢ. 11–10.—Hindbrain of a human embryo of three months—viewed from behind and partly from left side. (From model by His.)

appearance of the primitive fissures does not correspond to their relative importance in the adult. The best marked of the early fissures are as follows: (*a*) the *fissura posterolateralis*, first to develop, is formed by the folding of the rhombic lip and separates the flocculonodular lobe from the corpus cerebelli; (*b*) the *fissura prima* is between the culmen and declive; (*c*) the *fissura secunda* is between the future pyramid and uvula. A transverse furrow, the *incisura fastigii*, crosses the ventricular surface, producing a recess like a tent in the roof of the fourth ventricle (Fig. 11–14).

The **Midbrain** or **Mesencephalon.**—The midbrain exists for a time as a thin-walled vesicle of some size (Figs. 11–5, 11–9). Its cavity, relatively reduced in size, becomes the adult cerebral aqueduct. The basal laminae thicken into the **cerebral peduncles**

which enlarge rapidly after the fourth month. Neuroblasts near the aqueduct very early establish the oculomotor and trochlear nuclei. The alar laminae invade the roof plate which then becomes thickened into the **colliculi** (*corpora quadrigemina*). Some of the cells form the mesencephalic root of the trigeminal nerve.

The **Forebrain** or **Prosencephalon.**—The early forebrain has the same parts as similar stages of the spinal cord and medulla, namely, thick lateral walls connected by thin roof and floor plates (Fig. 11–5). The division of the rostral end into alar and basal laminae, however, is not clear. Some authors consider the hypothalamus a derivative of the basal lamina, others maintain that the entire forebrain is derived from the alar lamina.

At a very early period, before the closure of the cranial part of the neural tube, the **optic vesicles** appear as diverticula on each side of the forebrain (Fig. 2–16). They communicate with the forebrain vesicle by wide openings at first. Later the proximal part of the vesicle is narrowed into the **optic stalk** (Figs. 13–7, 13–9) and the peripheral part is expanded into the **optic cup.** The cavity of the vesicle later disappears; the stalk is invaded by nerve fibers and becomes the optic nerve and tract; the cup becomes the retina (Fig. 13–11).

After closure of the neural tube, the lateral walls of the forebrain grow much faster than the median portion resulting in a large pouch, the **cerebral hemisphere,** on each side (Fig. 11–6). The cavity of the pouch is the rudiment of the lateral ventricle and its communication with the forebrain vesicle is the future interventricular foramen (of Monro) (Fig. 11–7). The rostral part of the forebrain, including the hemispheres, is the **telencephalon** or **endbrain.** The more caudal part is called the **diencephalon** or **between brain.** The cavity of the median part of the forebrain vesicle is the third ventricle and the median portion of the rostral wall of the vesicle is a thin plate, the lamina terminalis.

The **Diencephalon.**—The diencephalon gives rise to the thalamus, metathalamus, epithalamus, and hypothalamus. A groove in the lateral wall of this part of the forebrain vesicle, the **sulcus hypothalamicus** (*cf.*

Fig. 11–9) is considered by many authorities to be the cephalic continuation of the sulcus limitans. Since the cephalic end of this groove bifurcates, the exact termination of the sulcus limitans is a matter of controversy. The **thalamus** arises as a thickening of the cephalic two-thirds of the alar plate. For some time it is visible as a prominence on the external surface of the brain (Fig. 11–6), but it later becomes buried by the overgrowth of the hemispheres. The thalami of the two sides protrude medially into the ventricular cavity and in many brains eventually join across the midline in a small commissure-like junction of gray substance, the **adhesio interthalamica** (*massa intermedia*). The **metathalamus** comprises the medial and lateral geniculate bodies which appear as slight prominences on the outer surface of the alar lamina. The **epithalamus** includes the epiphysis or pineal body, the posterior commissure, and the trigonum habenulae. The pineal body arises as an evagination of the roof plate between the thalamus and the colliculi (Fig. 11–11). The trigonum habenulae develops in the roof plate just rostral to the pineal and the

posterior commissure develops just caudal to it.

The **hypothalamus** (Fig. 11–7) is developed from the ventral part of the basal plate and the floor plate of the diencephalon. It comprises the tuber cinereum (Fig. 11–9) with the attached neurohypophysis, the optic vesicles and chiasma, and the mammillary bodies.

The roof plate of the diencephalon rostral to the pineal body remains thin and epithelial in character. It later combines with the pia mater over it to form the **choroid plexus** of the third ventricle.

Telencephalon.—The **Cerebral Hemispheres** increase rapidly in size and ultimately overlap the structures of the mid- and hindbrains. Each hemisphere has three fundamental parts, the rhinencephalon, the corpus striatum, and the neopallium.

Histogenesis.—The wall of the hemisphere remains typical of the primitive neural tube for some time, with three layers, a thick ependymal and mantle zone and a thin marginal zone. During the third month neuroblasts migrate peripherally from the ependymal and mantle zones. They collect

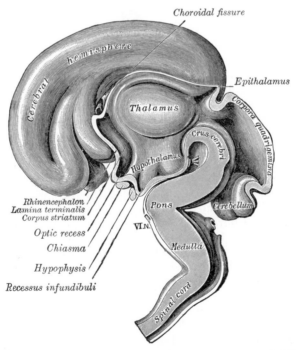

Fig. 11–11.—Median sagittal section of brain of human embryo of three months. (From model by His.)

in the deeper part of the marginal zone thus forming the outer layer of gray substance of the **cortex** or **pallium**. The white substance of the hemisphere is formed by the growth of fibers from the central nuclei of the corpus striatum into the outer part of the mantle zone. Later fibers from the newly formed outer cortex grow into the area also. The fibers in the white substance of the hemisphere begin to acquire myelin sheaths at the time of birth and the process continues until puberty (Ward, '54).

the lumen is obliterated and the remaining stalk becomes the primitive olfactory bulb and tract. The proximal part of the olfactory lobe is connected with the parolfactory area adjacent to the lamina terminalis on the medial surface of the hemisphere by the gyrus olfactorius medialis.

The caudal part of the primitive olfactory ridge develops into the anterior perforated substance and piriform lobe (Fig. 11–12). At the beginning of the fourth month the piriform lobe appears as a curved elevation

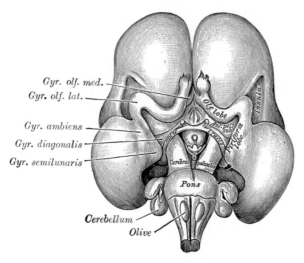

Fig. 11–12.—Inferior surface of brain of embryo at beginning of fourth month. (From Kollmann.)

The **rhinencephalon** or nose brain is phylogenetically the oldest part of the telencephalon and the part of the cerebral cortex included in it is called the **archipallium**. It forms almost the entire hemisphere in fishes, amphibians, and reptiles, but is poorly developed in man. In the embryo it first appears as a longitudinal ridge with a corresponding internal furrow at the rostral extremity of the hemisphere close to the lamina terminalis (Figs. 11–6, 11–9). It is separated from the lateral surface of the hemisphere by the external rhinal fissure but is continuous caudally with the future temporal lobe. The rostral part of the ridge is the **primitive olfactory lobe**, the caudal part the **piriform lobe** (Fig. 11–12). From the rostral end, a hollow stalk grows forward, retaining for a time its connection with the ventricle. During the third month

which continues caudally into the temporal lobe. The connection between the olfactory lobe and piriform lobe is the **lateral olfactory gyrus**. It makes a sharp bend in the position of the threshold of the future insula (*limen insulae*) with which it later forms a connection. The caudal part of the piriform lobe is composed of the gyrus ambiens and gyrus semilunaris which are absorbed into the uncus of the adult gyrus hippocampus. The development of the fornix is described with that of the commissures.

The **corpus striatum** appears as a thickening in the ventral wall of the hemisphere between the optic stalk and the interventricular foramen, close against the thalamus (Figs. 11–7, 11–9). By the second month it is seen as a swelling in the floor of the lateral ventricle extending back to the occipital pole of the primitive hemisphere.

When this portion of the hemisphere is expanded into the temporal lobe, a part of the corpus striatum is carried into the roof of the inferior horn where it becomes the tail of the caudate nucleus. During the fourth and fifth months, an invasion by fibers of the internal capsule incompletely divides the corpus striatum into the **caudate** and **lentiform nuclei.**

which is immediately continuous with the roof plate of the diencephalon lies over the primitive interventricular foramen and extends caudally. This border remains thin and of an ependymal or epithelial character. It is invaginated into the medial wall of the lateral ventricle to form the **choroid fissure** (Figs. 11–9, 11–14). Mesodermal tissue from the outer surface of the hemisphere,

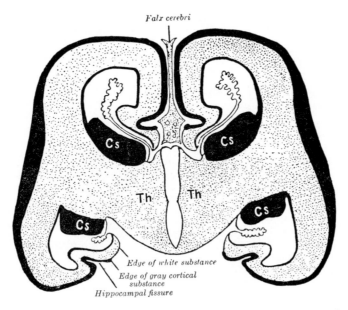

Fig. 11–13.—Diagrammatic coronal section of brain to show relations of neopallium. (After His.) *Cs*, Corpus striatum. *Th*, Thalamus.

The **neopallium** or non-olfactory part of the cerebral cortex takes up the greater part of the hemisphere. Its cavity, the **primitive lateral ventricle,** expands in all directions, but more especially dorsally and caudally. By the third month the hemispheres cover the diencephalon, by the sixth month they overlap the midbrain and by the eighth they reach the hindbrain. The median lamina uniting the two hemispheres does not share in their expansion and remains as the thin roof of the third ventricle. Thus the hemispheres become separated by a deep cleft, the forerunner of the longitudinal fissure. The cavity of the ventricle is gradually drawn out into three prolongations which represent the future anterior, inferior, and posterior horns.

The part of the wall along the medial border of the primitive lateral ventricle

the rudiment of the pia mater, spreads into the fissure between the two layers of the ependyma to form the rudiment of the **tela choroidea.** The blood vessels accompanying the mesoderm form the choroid plexus which almost completely fills the cavity of the ventricle for several months. The tela choroidea lying over the roof of the diencephalon also invaginates the thin epithelial covering to form the choroid plexus of the third ventricle.

The medial wall of the hemisphere along a line parallel with the choroid fissure becomes pushed or folded into the cavity of the lateral ventricle. The prominence inside the ventricle is the rudiment of the **hippocampus** and the groove outside is the **hippocampal fissure** (Fig. 11–13). The outer gray substance of the cortex ends along the prominence between the hippocampal fissure and the choroid fissure leaving the white substance exposed along a thin edge which is continuous with the epithelium of the choroid plexus (Fig. 11–13). With the growth of the temporal lobe rostrally the hippocampal fissure and parts associated with it extend from the interventricular foramen to the end of the inferior horn of the ventricle. The thickened edge of gray substance becomes the **gyrus dentatus**, the **fasciola cinerea**, and the **supra-** and **subcallosal gyri**. The free edge of the white substance forms the **fimbria hippocampi** and the body and crus of the **fornix**.

fibers from the hippocampus pass into the ventral part of the lamina terminalis and, arching over the thalamus to the corpora mammillaria, develop into the **fornix**. The anterior portion between the corpus callosum and fornix is not invaded by commissural fibers, and becomes the **septum pellucidum**.

Sulci and Gyri.—The outer surface of the cerebral hemisphere is at first smooth. Later it exhibits a number of convolutions, or gyri, separated from each other by sulci, most of which make their appearance during the sixth or seventh months

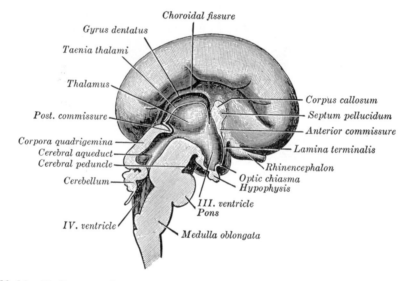

Fɪɢ. 11–14.—Median sagittal section of brain of human embryo of four months. (Marchand.)

The **Commissures.**—The development of the posterior commissure has already been described (page 785). The corpus callosum, the fornix, and the anterior commissure arise from the lamina terminalis (Fig. 11–14). At about the fourth month a small thickening appears in this lamina, immediately in front of the interventricular foramen. The lower part of this thickening is soon constricted off, and fibers grow into it to form the **anterior commissure**. The upper part continues to grow caudally with the hemispheres, and is invaded by two sets of fibers. Transverse fibers, extending between the hemispheres, pass into its dorsal part, which is now differentiated as the **corpus callosum**. Longitudinal

of fetal life. The term complete sulcus is applied to such grooves as involve the entire thickness of the cerebral wall, and thus produce corresponding eminences in the ventricular cavity. The other sulci affect only the superficial part of the wall, and therefore leave no impressions in the ventricle. The complete sulci are the choroidal and hippocampal already described, the calcarine, which produces the swelling in the ventricle known as the calcar avis, and the collateral sulcus with its corresponding eminence in the ventricular cavity. The central sulcus (*fissure of Rolando*) is developed in two parts, the intraparietal sulcus in four parts, and the cingulate sulcus in two or three parts. The **lateral cerebral** or **Sylvian sulcus** differs from all the other sulci in its mode of development. It appears about the third month as a depression, the **Sylvian fossa,** on the lateral surface of the hemisphere (Fig. 11–15). The floor of this fossa becomes the **insula** and its adherence to the subjacent corpus striatum prevents this part of the cortex from expanding at

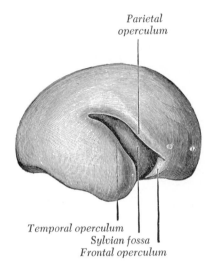

Parietal operculum

Temporal operculum
Sylvian fossa
Frontal operculum

FIG. 11–15.—Outer surface of cerebral hemisphere of human embryo of about five months.

the same rate as the portions which surround it. The neighboring parts of the hemisphere therefore gradually grow over and cover in the insula, constituting the temporal, parietal, frontal, and orbital **opercula** of the adult brain. The frontal and orbital opercula are the last to form, but by the end of the first year after birth the insula is completely submerged. The sulci separating the opposed margins of the opercula constitute the adult lateral cerebral sulcus.

The **Autonomic Nervous System.**—The cells in the central gray of the spinal cord occupying a lateral position between the afferent groups of the dorsal horn and the somatic efferent groups of the ventral horn send their axons out through the ventral roots (Fig. 11–16,A). These axons leave the spinal nerves in the thoracic and first two lumbar segments to enter the sympathetic ganglia. They are the preganglionic fibers and together with a few afferent fibers from the spinal ganglia become the white rami communicantes. In the medulla and

pons, the nerve cells in the lateral part of the basal plate send axons out with the facial, glossopharyngeal, and vagus nerves. These fibers meet the various peripheral ganglia associated with them and become the parasympathetic preganglionic fibers for smooth muscle, heart, and glands. In the sacral region, the lateral horn cells grow out into the pelvic nerves and become the preganglionics of the sacral portion of the parasympathetic system.

Autonomic Ganglia.—When the cells of the *neural crest* (Fig. 11–4) gather into the primordia of the spinal ganglia along the primitive neural tube, a certain number of cells continue to migrate ventrally (Yntema and Hammond '54). In the region of the spinal cord these migrating cells collect into groups along the lateral aspect of the aorta and become the ganglia of the sympathetic trunk. Others migrate farther toward the alimentary tube and become the celiac and associated ganglia. The cells from the cranial neural crest migrate in a similar fashion and become the parasympathetic ganglia. As mentioned above, axons from the central nervous system reach these sympathetic and parasympathetic ganglia to become the preganglionic fibers. The cells in the ganglia grow out to their various specific organs and become the postganglionic fibers.

The **Cranial Nerves.**—With the exception of the olfactory, optic, and acoustic nerves, which are considered elsewhere (see Chapter 13), the cranial nerves are developed in a manner similar to the spinal nerves. The sensory or afferent nerves are derived from the cells of the ganglion rudiments of the neural crest. The central processes of these cells grow into the brain and form the roots of the nerves, while the peripheral processes extend outward and constitute their fibers of distribution (Fig. 11–2). It has been seen, in considering the development of the medulla oblongata (page 781), that the tractus solitarius (Fig. 11–3), derived from the fibers which grow inward from the ganglion rudiments of the glossopharyngeal and vagus nerves, is the homologue of the oval bundle in the cord, which had its origin in the dorsal nerve roots. The motor or efferent nerves arise as outgrowths of the neuroblasts situated in the basal laminae

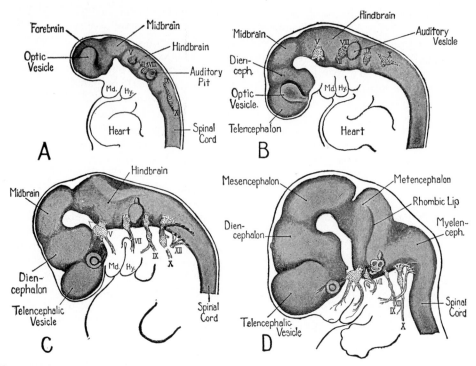

Fig. 11–16.—Four stages in early development of brain and cranial nerves of young human embryos. (Adapted from various sources, primarily figures by Streeter and reconstructions in the Carnegie Collection.)

A, At 20 somites—based on the Davis embryo—probable fertilization age of three and a half weeks.
B, At 4 mm, fertilization age of about four weeks.
C, At 8 mm, fertilization age a little over five weeks.
D, At 17 mm, fertilization age about seven weeks.

The cranial nerves are indicated by the appropriate Roman numerals:

V, Trigeminal; VII, facial; VIII, acoustic; IX, glossopharyngeal; X, vagus; XI, accessory; XII, hypoglossal. Abbreviations: Hy., hyoid arch; Md., mandibular arch. (Courtesy of B. M. Patten, *Human Embryology*, 2nd Ed. Copyright, 1953. Blakiston Div., McGraw-Hill Book Co.)

of the mid- and hindbrain. They are grouped into two sets, according to whether they spring from the medial or the lateral parts of the basal lamina. To the former set belong the oculomotor, trochlear, abdu-cent, and hypoglossal nerves; to the latter, the accessory and the motor fibers of the trigeminal, facial, glossopharyngeal, and vagus nerves (Figs. 11–3, 11–16) (Humphrey '52, Brown '56).

The Parts of the Adult Central Nervous System

The central nervous system is a single continuous structure which is divided for convenience in description into two parts: the brain or encephalon and the spinal cord or medulla spinalis. Both parts are composed of two substances or types of tissue: the gray substance or gray matter containing primarily nerve cells and their closely related processes, and the white substance or white matter composed of bundles or masses of nerve fibers, predominantly myeli-nated. The nerve cells or neurons are the control centers. By means of various processes, they form connections with each other called synapses. Their processes reach more remote nerve cells within the central nervous system as the fibers of the white substance and they reach the structures of the body with which they are associated functionally by means of the peripheral nervous system.

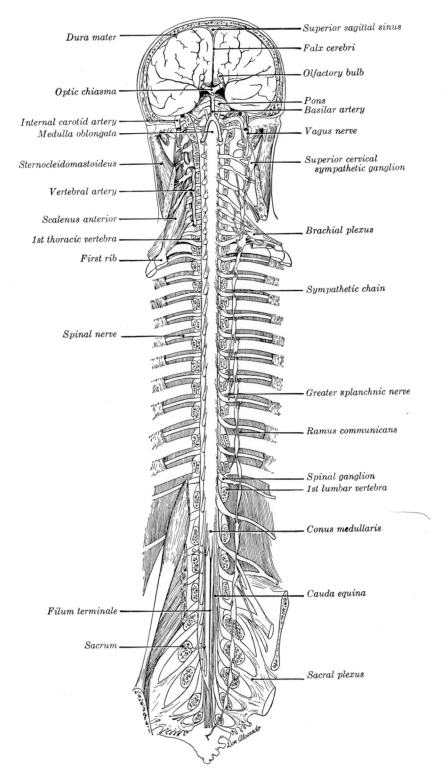

FIG. 11–17.—Dissection to show the ventral or anterior aspect of the brain and spinal cord in situ. (Redrawn from Hirschfeld and Leveille.)

THE SPINAL CORD

The **spinal cord** (*medulla spinalis*) (Fig. 11–17) is an elongated, nearly cylindrical structure approximately 1 cm in diameter, with an average length of 42 to 45 cm and weight of 30 gm. It lies within the vertebral canal, its cranial end continuous with the medulla oblongata of the brain at the upper border of the atlas. Its caudal extremity, at the lower border of the first or upper border of the second lumbar vertebra, tapers to a point, forming the **conus medullaris**. The

closely invested by the dura, is adherent to it, and extends beyond the apex of the sheath to the dorsal part of the first coccygeal vertebra to which it is attached. The filum consists mainly of fibrous tissue continuous with the pia mater. A few nerve fibers adherent to the outer surface probably represent rudimentary second and third coccygeal nerves. The central canal of the spinal cord continues down into the filum for 5 or 6 cm.

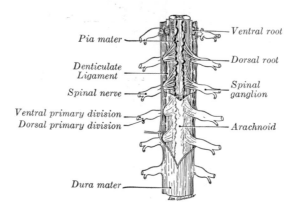

Fig. 11–18.—Several segments of the thoracic spinal cord showing the meninges and spinal roots. Dorsal aspect.

position of the cord varies somewhat with the movements of the vertebral column, being drawn upward slightly when the column is flexed. At birth the lower end lies at the third lumbar vertebra but it gradually recedes during childhood until it reaches the adult position.

The spinal cord is enclosed by the three protective membranes called **meninges**, the dura mater, the arachnoid, and the pia mater, structures which are described at the end of this chapter (Fig. 11–18).

The **filum terminale** is a delicate filament which continues down the vertebral canal from the apex of the conus medullaris to the first segment of the coccyx (Fig. 11–17). It is approximately 20 cm in length. Its first 15 cm, contained within the tubular sheath of the dura mater and surrounded by the nerves of the cauda equina, is called the **filum terminale internum**. The lower part, the **filum terminale externum**, is

External Configuration of the Cord

Enlargements.—The spinal cord is slightly flattened dorsoventrally and its diameter is increased by enlargements in two areas, cervical and lumbar (Fig. 11–19). The **cervical enlargement** (*intumescentia cervicalis*) extends from the third cervical to the second thoracic vertebra, its maximum circumference (about 38 mm) being on a level with the sixth pair of cervical nerves. It corresponds to the origin of the large nerves which supply the upper limbs. The **lumbar enlargement** (*intumescentia lumbalis*) begins at the level of the ninth thoracic vertebra and reaches its maximum circumference (about 33 mm) opposite the last thoracic vertebra, after which it tapers rapidly into the conus medullaris. It corresponds to the origin of nerves which supply the lower limbs.

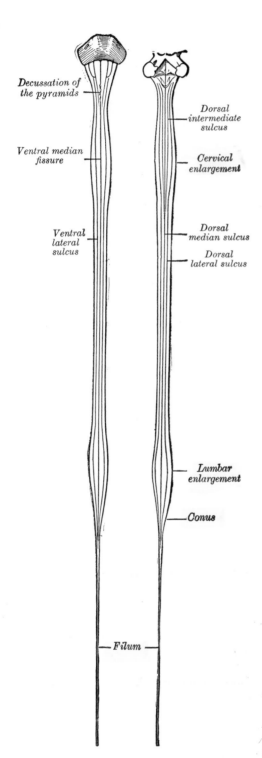

Decussation of
the pyramids

Dorsal
intermediate
sulcus

Ventral median
fissure

Cervical
enlargement

Ventral
lateral
sulcus

Dorsal
median sulcus

Dorsal
lateral sulcus

Lumbar
enlargement

Conus

Filum

FIG. 11–19.—Diagrams of the spinal cord.

Fissures and Sulci (Fig. 11–19).—A ventral or anterior median fissure and a dorsal or posterior median sulcus divide the spinal cord longitudinally into symmetrical right and left halves (Fig. 11–20).

The **Ventral Median Fissure** (*fissura mediana anterior*) (Fig. 11–19) has an average depth of about 3 mm, increasing caudally. It contains a double fold of pia mater and its floor is formed by transverse white substance, the ventral white commissure (Fig. 11–20).

The **Dorsal Median Sulcus** (*sulcus medianus posterior*) (Fig. 11–19) is a shallow groove marking the position of the **dorsal median septum** (Fig. 11–20), a thin sheet of neuroglial tissue which penetrates more than half way into the substance of the cord, effectively separating the dorsal portion into right and left halves.

The **Dorsal Lateral Sulcus** (*sulcus lateralis posterior*) (Figs. 11–19, 11–20) is a longitudinal furrow corresponding to the position of the attachments of the dorsal roots of the spinal nerves. In the upper thoracic and cervical regions of the cord a groove, the **dorsal intermediate sulcus** (*sulcus intermedius posterior*), lies between the dorsal median and dorsal lateral sulci. It marks the separation between the fasciculi gracilis and cuneatus. The **ventral lateral sulcus** is an indistinct shallow groove where the ventral roots of the spinal nerves are attached.

Segments of the Cord.—Thirty-one pairs of spinal nerves originate from the spinal cord, each with a ventral or anterior root and a dorsal or posterior root. The pairs are grouped as follows: 8 cervical, 12 thoracic, 5 lumbar, 5 sacral, and 1 coccygeal. The cord itself also is divided for convenience of description into cervical, thoracic, lumbar, and sacral regions to correspond with the nerve roots. It is customary also to speak of the part of the cord giving rise to each pair of nerves as a **spinal segment** or **neuromere** although the distinguishing marks, except for the roots, are not retained after embryonic life. The segments vary in their extent along the cord, in the cervical region averaging 13 mm, in the midthoracic region 26 mm, whereas they diminish rapidly in the lumbar and sacral region from 15 to 4 mm. Because the cord is much

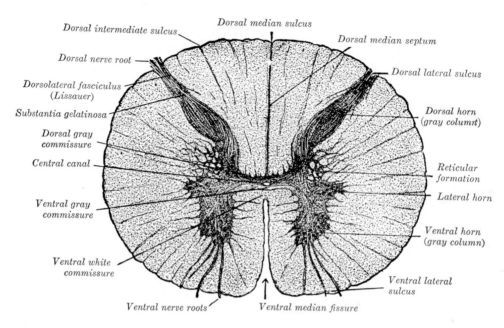

Dorsal intermediate sulcus

Dorsal median sulcus

Dorsal nerve root

Dorsal median septum

Dorsolateral fasciculus (Lissauer)

Dorsal lateral sulcus

Substantia gelatinosa

Dorsal horn (gray column)

Dorsal gray commissure

Central canal

Reticular formation

Lateral horn

Ventral gray commissure

Ventral horn (gray column)

Ventral white commissure

Ventral lateral sulcus

Ventral nerve roots

Ventral median fissure

Fig. 11–20.—Transverse section of the spinal cord in the midthoracic region.

shorter than the vertebral column, the more caudally placed nerve roots take a progressively more oblique direction, so that the lumbar and sacral nerves pass almost vertically downward within the dura for some distance before reaching their foramina of exit. The resulting collection of rootlets and nerves beyond the termination of the cord is called the **cauda equina** (Fig. 11–17).

Spinal Nerve Roots.—Each spinal nerve has two roots: (*a*) a dorsal or posterior root, also called the sensory root because its fibers bring impulses to the cord and (*b*) a ventral or anterior root, also called the motor root because its fibers carry impulses from the cord out to muscles and other structures (page 946).

The **Dorsal Roots** (*radix dorsalis; sensory root*) (Fig. 11–21) are attached in linear series along the dorsal lateral sulcus of the cord by six or eight **rootlets** (*fila radicularia*). They contain the central processes of spinal ganglion cells whose peripheral processes extend out to the sensory endings in the skin, muscles, tendons, viscera, *etc.* The rootlets are divided into two groups as they enter the cord, (*a*) a *medial bundle* containing mainly large myelinated fibers and (*b*)

a *lateral bundle* containing fine myelinated and unmyelinated fibers.

The **Ventral Roots** (*radix ventralis; anterior root*) (Fig. 11–21) are attached to the spinal cord along the ventral lateral sulcus, in two or three irregular rows of rootlets spread over a strip about 2 or 3 mm in width. They contain the axons of the cells in the ventral and lateral cell columns of the central gray matter of the cord.

Internal Structure of the Spinal Cord

The spinal cord is composed of two principal parts extending the length of the cord, an inner core of gray substance where nerve cells predominate and an outer layer of white substance where myelinated nerve fibers predominate.

Gray Substance (*substantia grisea centralis; gray matter*) (Fig. 11–20).—The central gray core is arranged in two large lateral masses connected across the midline by a narrow but continuous strip, the **gray commissure.** In cross section this gives a configuration similar to the letter H. Each lateral portion splays outward like a crescent, or better, like an inverted comma

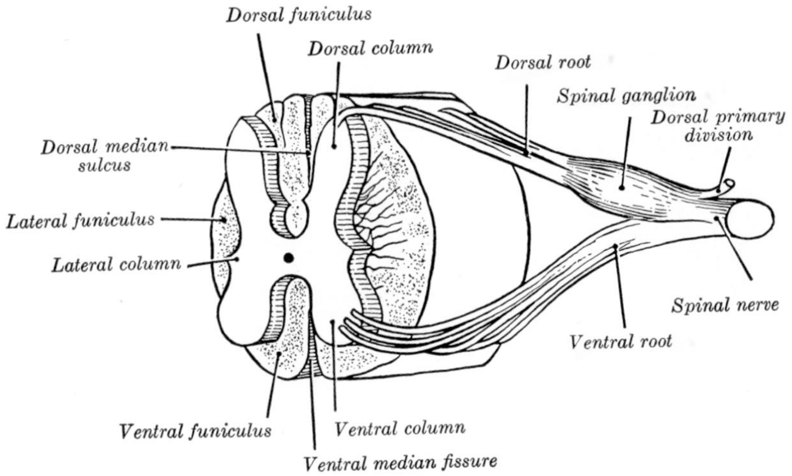

Fig. 11–21.—Diagram of an upper thoracic segment of the spinal cord with spinal roots and ganglion.

(Fig. 11–20). The commissure or crossbar of the H is closer to the ventral than the dorsal surface of the cord and contains the central canal. The part of the lateral mass extending dorsalward from the commissure is called the **dorsal** or **posterior column** (*columna posterior*) (Fig. 11–21); that extending ventralward, the **ventral** or **anterior column** (*columna anterior*). Since they are usually studied in cross sections of the cord, they are commonly referred to as **dorsal** and **ventral horns** (*cornua anterius et posterius*) (see below).

The quantity of gray matter varies markedly at different levels of the cord and its configuration is more or less characteristic at each level (Fig. 11–22). In the thoracic region it is small, surrounded by a relatively large amount of white matter. It is increased in the cervical and lumbar enlargements, and its proportion to the white substance is greatest in the conus medullaris. In the cervical region the dorsal column is narrow, the ventral broad; in the thoracic region both columns are attenuated and the lateral column becomes evident; in the lumbar enlargement both ventral and dorsal columns are expanded, and in the conus medullaris they are merged into an oval outline with a broad commissure.

The centrally placed gray substance of the spinal cord consists of nerve cells with their dendrites and a dense feltwork of nerve fibers, the **neuropil** (Fig. 11–106), supported and held together by neuroglia (Fig. 11–107). The nerve fibers are the axons not only of the local nerve cells, but also of cells in other parts of the nervous system which synapse with these cells. Relatively few of the fibers are myelinated because they lose their sheaths at some distance before they terminate. In cross sections of the cord prepared with the Nissl stain (LaVelle '56) which colors the cell bodies, it can be seen that cells of recognizable types tend to be arranged in groups or nuclei which have a functional significance (Fig. 11–23). The individual groups or columns may extend through all or part of the length of the cord.

The **Dorsal Horn** (*cornu posterius; posterior horn; posterior column*) in a cross section (Fig. 11–20) projects farther and is more slender than the ventral horn. It reaches almost to the surface in the dorsal lateral sulcus, from which it is separated by a thin layer of white substance, the dorsolateral fasciculus. The dorsal extremity is also capped by a crescentic mass of translucent gelatinous tissue containing many small nerve cells, named the substantia gelatinosa (of Rolando). The cells of the dorsal horn belong to the sensory side of a neuronal or reflex arc and are concerned with receiving and relaying impulses from the dorsal root fibers of spinal nerves.

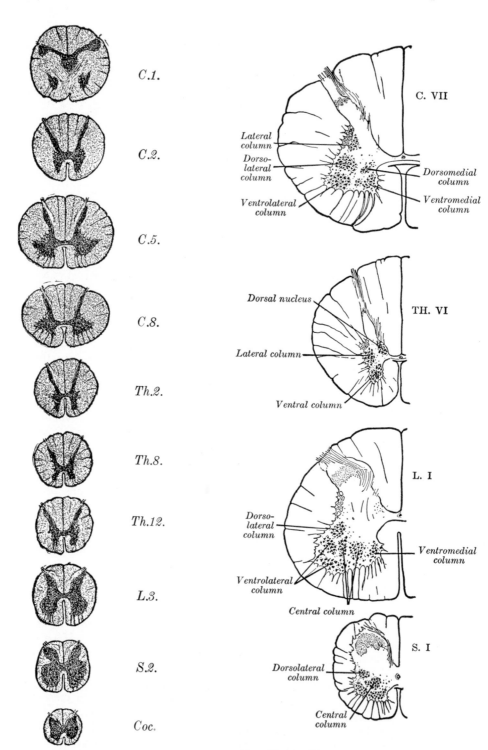

C.1.

C.2.

C.5.

C.8.

Th.2.

Th.8.

Th.12.

L.3.

S.2.

Coc.

C. VII

Lateral column

Dorso-lateral column

Dorsomedial column

Ventrolateral column

Ventromedial column

Dorsal nucleus

TH. VI

Lateral column

Ventral column

L. I

Dorso-lateral column

Ventromedial column

Ventrolateral column

Central column

S. I

Dorsolateral column

Central column

FIG. 11–22.—Transverse sections of the spinal cord at its different levels.

FIG. 11–23.—Transverse sections of the spinal cord at different levels to show the arrangement of the principal cell columns.

Nuclei or **Columns of Cells of the Dorsal Horn** (Fig. 11–23).—(1) The **substantia gelatinosa** (*of Rolando*) extends the entire length of the cord and into the medulla oblongata where it becomes the spinal nucleus of the trigeminal nerve. It contains numerous small cells whose axons end in adjacent columns, both gray and white. The fibers entering it are from the lateral bundle of the dorsal rootlets by way of the dorsolateral tract or tract of Lissauer (Fig. 11–25). (2) The **dorsal nucleus** or **Clarke's column** (Fig. 11–23) extends from the last cervical or first thoracic to the second or third lumbar segments, gradually decreasing in size both above and below a maximum at the twelfth thoracic. It is represented, however, in other regions by aggregates of cells, a cervical nucleus at the third cervical and a sacral at the middle or lower sacral region of the cord. It is a well-defined oval area in the medial part of the base of the dorsal horn, containing medium and large, oval or piriform cells. It receives fibers from the dorsal funiculus and the axons of its cells course laterally and turn cranialward in the dorsal spinocerebellar tract. (3) The **nucleus proprius** extends throughout the length of the cord in the body and basal regions of the dorsal horn. The cells are of medium and small size with occasional large solitary cells.

The **Ventral Horn** (*cornu anterius; anterior horn; anterior column*) (Figs. 11–20, 11–21) is broader and does not reach as near the surface as the dorsal horn. It is expanded in the cervical and lumbar enlargements to correspond with the innervation of the limbs (Fig. 11–22).

Columns of Cells of the Ventral Horn.— The nerve cells of the ventral horn are large polygonal cells with long branching dendrites. The axons of these cells leave the cord through the ventral roots and traverse the peripheral nerves to the various muscles, giving off collaterals near their termination to a large number of motor nerve end plates (125 to 400). The cells are arranged in columnar groups which extend longitudinally through varying numbers of segments (Fig. 11–23). In general the trunk musculature is represented medially and the limb musculature laterally. This is emphasized by the number of large columns placed

laterally at the levels of the cervical and lumbosacral enlargements of the cord (Fig. 11–23). The nuclear groups have been mapped out with greater refinement (Elliott '42) but functional correlation with the muscles and muscle groups has been established in only a few instances.

The **phrenic nucleus** appears to be a well-defined group of cells at an intermediate position in the most ventral part of the ventral horn. The cells are arranged in globular clusters, like a string of beads, extending from the lowest part of the third cervical through the upper part of the sixth cervical segment of the cord. There is no central decussation but fibers do cross from the left to the right nerve at the level of the pericardium (Warwick and Mitchell '56).

Lateral Horn (*cornu laterale; lateral or intermediolateral gray column*) (Fig. 11–23).—In the thoracic and upper lumbar segments, the portion of the gray mass lateral to the commissure protrudes into the white substance to form the intermediate lateral gray column or lateral horn.

Its cells are fusiform or stellate and of medium size. Their axons pass out through the ventral roots and thence through the white rami communicantes into the sympathetic trunk (page 1010). They are the preganglionic fibers of the **sympathetic** or **thoracolumbar outflow** of the autonomic nervous system.

The column is again differentiated in the third and fourth sacral segments where the axons of its cells enter the pelvic nerves as the sacral portion of the **parasympathetic** or **craniosacral outflow** of the autonomic system. In the upper cervical region of the cord the column is also differentiated and in the medulla it is represented by the dorsal nucleus of the vagus and other autonomic nuclei extending rostrally parallel with the sulcus limitans. The axons of certain other cells of the lateral column pass into the ventral and lateral funiculi.

Reticular Formation (*formatio reticularis*) (Fig. 11–20).—Between the lateral and dorsal horns the gray substance sends irregular slender projections into the white substance giving an appearance of a network in cross section and hence it is called the reticular formation. It is best developed in the cervical region and is continuous with the reticular formation of the medulla. It consists of medium and small cells scattered between strands of unmyelinated fiber which insinuate themselves into the white matter at the base of the dorsal horn.

The **Central Canal** (*canalis centralis*) (Fig. 11–20) runs throughout the entire length of the spinal cord, marking a division of the narrow strip of gray matter (*substantia intermedia centralis*) into a **ventral** and a **dorsal gray commissure.** The former is thin and in contact with the ventral white

White Substance.—The white substance of the spinal cord consists of myelinated and unmyelinated fibers in great numbers embedded in a spongework of neuroglia. The majority of fibers pursue a longitudinal course, but many cross from one side to the other in the ventral white commissure. Many of the longitudinal fibers are grouped into bundles with a common function, origin, and destination. These are called **fiber tracts.** The fibers of neighboring tracts tend to intermingle, so that they can seldom be identified without special methods.

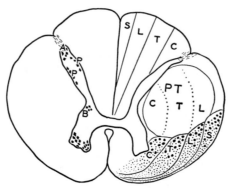

Fig. 11–24.—Semi-schematic drawing to show on the left side the cell groups of the posterior horn, and on the right side the arrangement of the spinothalamic and other tracts in the lower cervical region. The heavy dots represent fibers concerned with temperature; the medium-sized dots, fibers mediating pain, and the fine dots, fibers carrying touch and pressure impulses. Note the overlapping and topical arrangement of the fibers. *A,* apical group of large ganglion cells of the posterior horn; *B,* basal group of large ganglion cells of the posterior horn; *C,* fibers from the cervical segment of the spinal cord; *L,* fibers from the lumbar segment of the spinal cord; *P,* pericornual groups of large ganglion cells of the posterior horn; *PT,* tractus pyramidalis; *S,* fibers from the sacral segment of the spinal cord, and *T,* fibers from the thoracic segment of the spinal cord. (Walker, courtesy of Arch. Neur. and Psych.)

commissure. The dorsal gray commissure reaches to the dorsal median septum. The central canal continues cranialward into the medulla oblongata where it opens out into the fourth ventricle. Caudally, it reaches for a short distance into the filum terminale and in the conus medullaris there is a fusiform dilatation 8 to 10 mm in length and triangular in cross section, the **terminal ventricle** (*ventriculus terminalis*). The canal is filled with spinal fluid and is lined by the ependyma composed of ciliated, columnar epithelium. Just outside of the ependyma is the **substantia gelatinosa centralis,** consisting mainly of neuroglia and the processes of the ependymal cells, but also containing a few nerve cells and fibers.

Lamination.—During the development of the brain and spinal cord, fibers are usually added to the various tracts in an orderly manner. The new ingrowing fibers are placed beside the older ones, building up the tract toward the side from which they have approached it, by a process known as lamination. A clear example is found in the nuclei gracilis and cuneatus. The longest fibers beginning at the caudal end of the cord are placed against the dorsal median sulcus. Fibers are added laterally until the most cephalic part of the body is represented in the most lateral part of the posterior funiculus (Fig. 11–24).

The cord is also subdivided into groups of tracts more or less closely associated functionally. These are the **fasciculi.** The entire white matter is divided into the three funiculi, dorsal, ventral, and lateral (Fig. 11–21).

Dorsal Funiculus (*funiculus posterior*) (Fig. 11–21).—The dorsal funiculus contains two large ascending fasciculi, (1) the fasciculus gracilis, (2) the fasciculus cuneatus; (3) a small descending fasciculus, the comma fasciculus, and (4) an intersegmental fasciculus, the dorsal proper fasciculus.

(1) The **fasciculus gracilis** (*tract of Goll*) (Fig. 11–25) lies next to the dorsal median septum and extends throughout the cord. It increases in size as the myelinated fibers from the medial strand of the dorsal roots from the lumbar and lower thoracic segments are added to the sacral. The fibers from the lowest segments retain their medial position and by a process of lamination, the fibers from successively higher segments are placed more laterally (Fig. 11–24). Some fibers may give collaterals to the nucleus dorsalis, nucleus proprius, and ventral horn and finally end in the nucleus gracilis of the medulla. Others may terminate in these nuclei.

(2) The **fasciculus cuneatus** (*tract of Burdach*) (Fig. 11–25) is similar to the fasciculus gracilis in position, function, and origin and together they might, therefore, be considered as one fasciculus. They are,

however, separated by the dorsal intermediate septum (Fig. 11–20) cranial to the sixth thoracic segment. Fibers from the upper thoracic and cervical dorsal roots enter the fasciculus cuneatus and by the same process of lamination as in the gracilis, the fibers of the higher segments are placed more laterally. Those that reach the medulla end in the nucleus cuneatus. Both gracile and cuneate fasciculi mediate discriminatory sensation; proprioception from muscles, tendons, and joints; tactile from the skin, and the vibratory sense.

(3) The **fasciculus interfascicularis** or **comma fasciculus** (*of Schultze*) (Fig. 11–25) is found in the cervical and upper thoracic region squeezed in between the fasciculus gracilis and cuneatus. In the lower thoracic region the fibers occupy a peripherally median position in the **septomarginal tract**; in the lumbar region, a deeper median position in the **oval area of Flechsig**; and in the sacral region in the **triangle of Phillipe-Gombault**. These tracts carry descending fibers; some are from the dorsal roots and some are intersegmental from cells in the dorsal horn.

(4) The **dorsal proper fasciculus** (*fasciculus proprius dorsalis; posterior ground bundle*) (Fig. 11–25) lies close to the

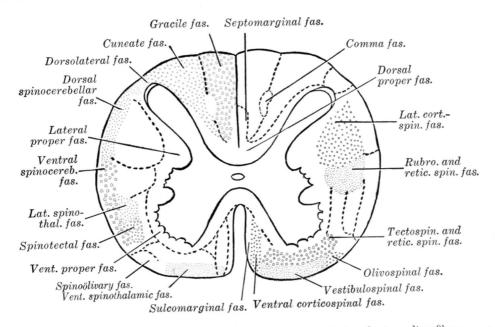

FIG. 11–25.—Diagram of the principal fasciculi of the spinal cord. Ascending fibers in blue, descending in red.

central gray substance. It receives fibers from cells in the dorsal horn which bifurcate into ascending and descending branches. They are intersegmental, running for varying distances and sending off collaterals and terminals to the gray matter.

Lateral Funiculus.—The lateral funiculus contains the following: A, ascending tracts or fasciculi: (1) the dorsal spinocerebellar tract, (2) ventral spinocerebellar tract, (3) lateral spinothalamic tract, (4) spinotectal tract, (5) dorsolateral tract (of Lissauer), and (6) lateral proper fasciculus; B, descending fasciculi: (7) the lateral corticospinal tract (pyramidal tract), (8) rubrospinal tract, (9) the olivospinal tract, and (10) the lateral reticulospinal tract.

(1) The **dorsal spinocerebellar tract** (*tract of Flechsig; tractus spinocerebellaris posterior*) is a flattened peripheral band extending from the tip of the dorsal horn about half way ventrally (Fig. 11–25). It begins at about the second or third lumbar nerve, increasing in size as it ascends to the inferior cerebellar peduncle and vermis of the cerebellum. Its fibers are of large size, derived from the dorsal nucleus (Clarke's column) mainly of the same side.

(2) The **ventral spinocerebellar tract** (*tract of Gowers; tractus spinocerebellaris anterior*) forms a band which skirts the periphery from the dorsal spinocerebellar almost to the ventral lateral sulcus (Fig. 11–25). It begins at about the level of the third lumbar nerves, and can be followed through the medulla and pons to the superior cerebellar peduncle and vermis. Its fibers come from cells in the dorsal horn and intermediate gray matter of the same or opposite sides. They are proprioceptive fibers whose impulses do not reach the level of consciousness.

(3) The **lateral spinothalamic tract** (Fig. 11–25) lies deep to the ventral spinocerebellar tract. It conveys impulses for pain and temperature. Fibers entering the cord through the lateral strand of the dorsal root cross the dorsolateral tract (of Lissauer) and end about cells in the substantia gelatinosa within one or two segments. Axons from these cells cross in the ventral white commissure to this tract and ascend to the thalamus. Within the tract, the fibers for temperature are placed more dorsally than those for pain. The tract increases as it ascends, fibers being added by lamination to the deeper part from higher segments (Fig. 11–24).

(4) The **spinotectal tract** (Fig. 11–25) is ventral to the lateral spinothalamic tract and contains axons from cells in the dorsal horn of the opposite side. They come to lie in the medial portion of the lateral lemniscus and end in the superior colliculus of the midbrain.

(5) The **dorsolateral tract** (*tract of Lissauer*) (Fig. 11–25) lies at the tip of the dorsal horn, ventral and lateral to the incoming dorsal roots. It contains fine myelinated and unmyelinated fibers, some from the lateral strand of the dorsal roots and some from the cells in the dorsal horn. They divide into short ascending and descending branches which end mostly in the substantia gelatinosa (Earle '52).

(6) The **lateral proper fasciculus** (*fasciculus proprius lateralis; lateral ground bundle*) (Fig. 11–25) consists chiefly of intersegmental fibers which arise from cells in the gray substance, and, after a longer or shorter course, reenter the gray substance and ramify in it. These fibers make up a moderately thick layer next to the gray substance. In addition, certain pathways, either long tracts or neuron chains, from autonomic centers in the brain stem to autonomic centers in the cord run in this area, near to the lateral horn.

(7) The **lateral corticospinal tract** (*crossed pyramidal tract*) (Figs. 11–25, 11–31) extends throughout the cord and is the principal pathway for voluntary movement. The fibers are axons of the pyramidal cells in the cortex which traverse the brain in the pyramidal tract. In the pyramidal decussation of the medulla 70 per cent of the fibers cross to the opposite side and become the lateral corticospinal tract. (The 30 per cent uncrossed are the ventral corticospinal tract; see below.) The fibers give off many collaterals and synapse either with the motor cells in the ventral horn or with other cells in the base of the dorsal horn. The fibers are arranged by lamination, the cervical being the more medially placed (Fig. 11–24) (Lassek '54).

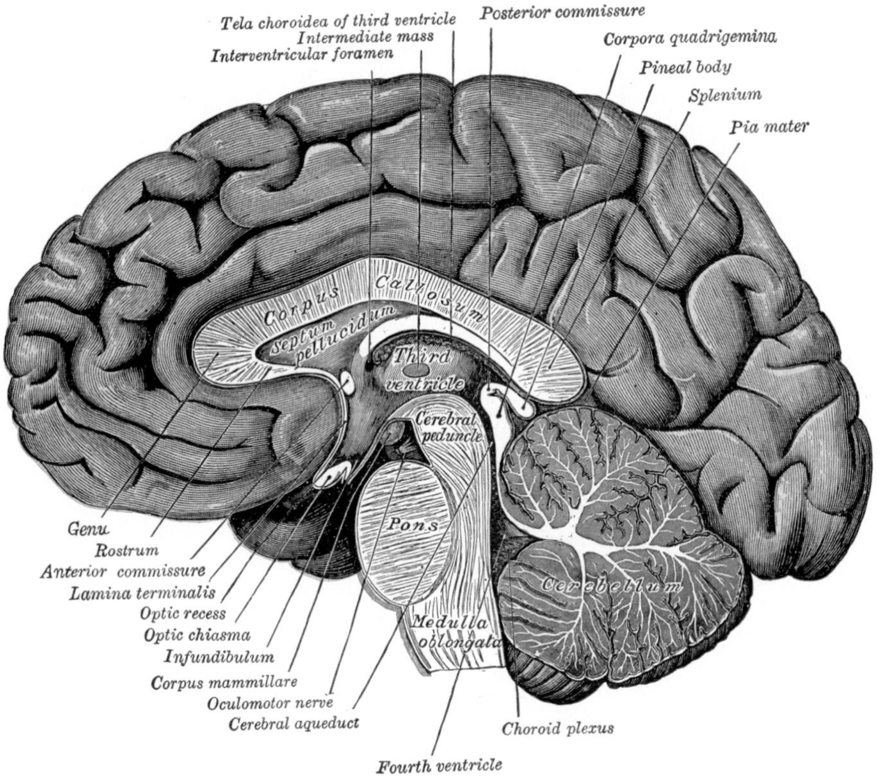
Tela choroidea of third ventricle
Intermediate mass
Interventricular foramen
Posterior commissure
Corpora quadrigemina
Pineal body
Splenium
Pia mater
Corpus Callosum
Septum pellucidum
Third ventricle
Cerebral peduncle
Pons
Cerebellum
Medulla oblongata
Genu
Rostrum
Anterior commissure
Lamina terminalis
Optic recess
Optic chiasma
Infundibulum
Corpus mammillare
Oculomotor nerve
Cerebral aqueduct
Choroid plexus
Fourth ventricle

FIG. 11–26.—Median sagittal section of brain. The relations of the pia mater are indicated by the red color.

(8) The **rubrospinal tract** (Fig. 11–25) is less conspicuous in man than in most mammals. Its fibers arise in the red nucleus of the midbrain, cross to the opposite side, and probably terminate in relation to the motor horn cells either directly or through intercalated neurons.

(9) The **olivospinal tract** (Fig. 11–25) arises in the vicinity of the inferior olivary nucleus in the medulla oblongata, and is seen only in the cervical region of the spinal cord. It lies at the periphery close to the most lateral of the ventral spinal nerve rootlets.

(10) The **lateral reticulospinal tract** (Fig. 11–37) is made up of fibers not clearly segregated from other fibers in the ventral part of the lateral funiculus near the rubrospinal tract. The predominant number of fibers arise from the cells of the medullary reticular formation. They enter the medial part of the ventral horn and probably terminate in relation to internuncial neurons (Truex and Carpenter '64).

Ventral Funiculus.—The ventral funiculus contains the following tracts or fasciculi: A, ascending tracts: (1) ventral spinothalamic tract, (2) ventral proper fasciculus; B, descending tracts: (3) ventral corticospinal tract, (4) vestibulospinal tract, (5) ventral reticulospinal tract, (6) tectospinal tract, and (7) sulcomarginal fasciculus.

(1) The **ventral spinothalamic tract** (*tractus spinothalamicus anterior*) (Fig. 11–25) occupies a peripheral position near the ventral median fissure more or less intermingled with the vestibulospinal tract. It receives fibers originating in the dorsal gray matter of the opposite side which have crossed in the ventral white commissure. The fibers turn cranialward, reaching the thalamus directly or by relays of other cells.

They convey tactile impulses of touch and pressure with considerable overlapping of segments and bilateral representation.

(2) The **ventral proper fasciculus** (*fasciculus proprius ventralis; anterior ground bundle*) (Fig. 11–25) includes a rather wide band of white substance next to the gray substance of the ventral horn. It consists of (*a*) longitudinal intersegmental fibers which arise from cells in the gray substance, especially the medial group of the ventral horn (Fig. 11–23) and after a longer or shorter course reenter the gray substance; (*b*) commissural fibers which cross in the ventral white commissure to the gray of the opposite side; (*c*) fibers of autonomic neuron chains from the brain stem.

(3) The **ventral corticospinal tract** (*direct pyramidal tract*) (Fig. 11–25) lies close to the ventral median fissure and is present only in the cranial part of the spinal cord, gradually increasing from a midthoracic level. It consists of fibers arising from pyramidal cells of the cortex of the same side.

(4) The **vestibulospinal tract** (Fig. 11–25) is placed peripherally in much the same area as the ventral spinothalamic tract. Its fibers arise from cells in the lateral vestibular nucleus (Reiters) and some of them can be traced as far as the sacral region. Their terminals and collaterals end among the motor cells of the ventral horn and are concerned with the maintenance of tonus and equilibrium.

(5) The **ventral reticulospinal tract** (Fig. 11–37) arises from cells in the reticular formation of the pons and higher levels of the brain stem. It occupies a position not clearly segregated in the medial part of the ventral funiculus and its fibers terminate in relation to ventral horn cells in a manner similar to the lateral reticulospinal tract.

(6) The **tectospinal tract** (Fig. 11–25) is situated partly in the ventral and partly in the lateral funiculus. It is mainly derived from cells in the superior and inferior colliculi of the opposite side which cross the median raphe in the fountain decussation of Meynert and form the ventral longitudinal bundle of the reticular formation in the brain stem. Its terminals and collaterals end directly or indirectly among the anterior motor horn cells. It is primarily concerned with visual reflexes.

(7) The **sulcomarginal fasciculus** (Fig. 11–25) occupies a narrow marginal band along the border of the ventral median fissure. It is the caudal continuation as far as the upper thoracic cord of the medial longitudinal fasciculus of the medulla. It takes part in the vestibular and visual reflexes, coordinating movements of the head and eyes.

THE BRAIN STEM AND CEREBELLUM

The brain stem is the part of the brain connecting the spinal cord with the forebrain and cerebrum. It is composed of the medulla oblongata, pons, and mesencephalon. The cerebellum is attached to the medulla and pons with the fourth ventricle intervening.

THE MEDULLA OBLONGATA

The **medulla oblongata** or **bulb**, as it is commonly called in combined names such as the corticobulbar fibers, is continuous caudally with the spinal cord and rostrally with the pons (Fig. 11–26). Its dorsal surface fits into the fossa between the hemispheres of the cerebellum. Its approximate measurements are 3 cm long, 2 cm wide, and 1.25 cm thick. The central canal continues upward from the spinal cord through its caudal half, but in the cranial half the medulla is split open at the dorsal median sulcus, expanding the canal into the fourth ventricle (Fig. 11–35).

The **ventral median fissure** (*fissura mediana anterior*) (Fig. 11–35) continues along the medulla to a small pocket at the caudal border of the pons called the **foramen cecum**. It is crossed just rostral to the junction with the spinal cord by a series of oblique bundles which form the pyramidal decussation. Above the decussation delicate **ventral external arcuate fibers** emerge from the fissure and run on the surface to the inferior cerebellar peduncle.

The **dorsal median sulcus** (*sulcus medianus posterior*) (Fig. 11–35) in the caudal half of the medulla gradually becomes shallower and finally ends where the central canal opens out into the fourth ventricle. The floor of the ventricle, known as the rhomboid fossa, also is marked by a longitudinal groove, the median sulcus.

The **ventral lateral sulcus** (*sulcus lateralis anterior*) and the **dorsal lateral sulcus** (*sulcus lateralis posterior*) correspond to the same sulci in the spinal cord. The fibers

of the hypoglossal nerve represent the upward continuation of the ventral spinal nerve roots and emerge in linear series with them from the ventral lateral sulcus. The accessory, vagus, and glossopharyngeal nerves correspond to the dorsal roots and attach to the dorsal lateral sulcus. Although the three districts on each lateral half of the medulla appear to be continuous with the funiculi of the spinal cord, they do not correspond exactly because of changes which take place within the medulla.

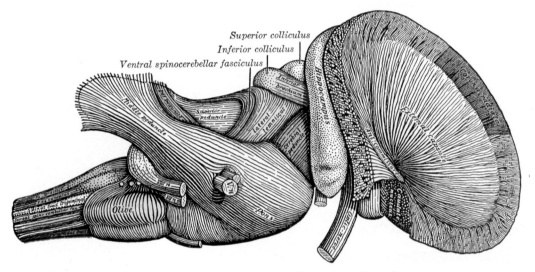

FIG. 11–27.—Superficial dissection of brain stem. Lateral view.

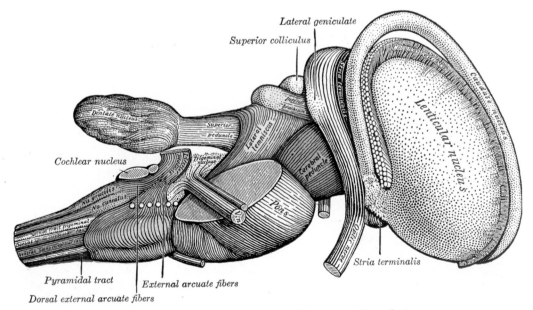

FIG. 11–28.—Partial dissection of brain stem. Lateral view.

The **pyramid** (*pyramis medullae oblongatae*) (Fig. 11–34) is a rounded prominence which lies between the ventral median fissure and the ventral lateral sulcus and appears to be continuous with the ventral funiculus of the spinal cord. It contains mainly motor fibers from the cerebral cortex, corticobulbar and corticospinal fibers. About two-thirds of these fibers cross the midline in bundles, forming the **pyramidal decussation** (*decussatio pyramidum*) (Fig. 11–34) and become the lateral corticospinal tract of the spinal cord.

They obliterate the ventral median fissure thus marking the point of transition from the spinal cord to medulla. The uncrossed one-third of the fibers enter the ventral corticospinal tract.

The **olive** (*oliva*) (Figs. 11–30, 11–37) is a prominent oval mass about 1.25 cm long in what corresponds to an upward continuation of the lateral funiculus of the spinal cord. It lies between the rootlets of the spinal accessory, vagus, and glossopharyngeal nerves, in the dorsal lateral sulcus on the one hand, and those of the

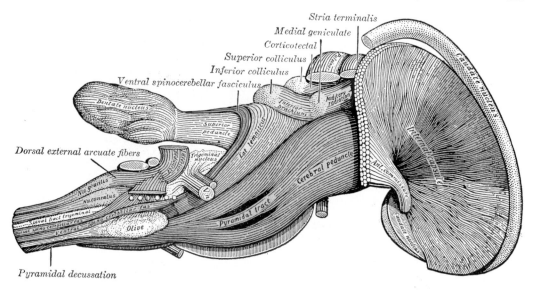

Fig. 11–29.—Deeper dissection of brain stem. Lateral view.

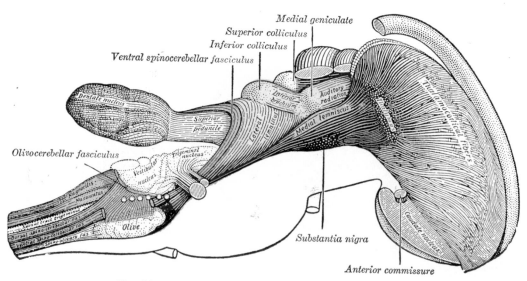

Fig. 11–30.—Deep dissection of brain stem. Lateral view.

hypoglossal nerve in the ventral lateral sulcus on the other. In the depression between the upper end of the olive and the pons are the roots of the statoacoustic and facial nerves (Fig. 11–34).

The dorsal district of the medulla differs in its rostral and caudal halves. The lower half appears to be an upward continuation of the dorsal funiculus of the spinal cord, and lies between the dorsal median sulcus and the dorsal lateral sulcus. The upper end of the fasciculus gracilis forms an elongated swelling, named the **clava** (*tuberculum nuclei gracilis*) (Fig. 11–35) containing the nucleus gracilis. That of the fasciculus cuneatus forms the **cuneate tubercle** (*tuberculum nuclei cuneati*), containing the nucleus cuneatus. Just lateral to the fasciculus cuneatus is an elevation produced by the substantia gelatinosa (of Rolando) which ends about 1.25 cm below the pons in the **tuber cinereum** (*tubercle of Rolando*).

The rostral half of the dorsal district of the medulla is composed of two prominent ridges which diverge to help form the lateral boundaries of the **rhomboid fossa** (Fig. 11–35). They contain the **inferior cerebellar peduncles** or **restiform bodies** (Figs. 11–27, 11–28) which connect the spinal cord and medulla with the cerebellum.

Internal Structure of Medulla Oblongata

The medulla changes rapidly as it makes the transition from the spinal cord to the brain stem. The decussation of the pyramids produces an alteration in the position of the tracts from lateral to ventral and separates part of the ventral horn from the rest of the gray matter. The internal arcuate fibers from the nuclei gracilis and cuneatus sweep ventrally around the central gray matter. They separate it from the dorsal gray horn, and decussate between the

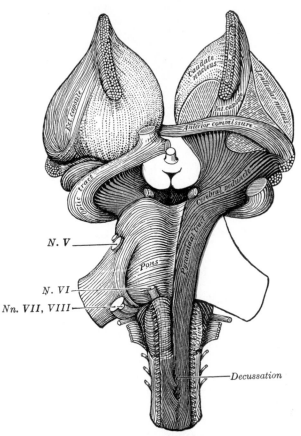

Fig. 11–31.—Superficial dissection of brain stem. Ventral view.

central gray and the pyramidal tracts. New structures are added, the large inferior olives and the nuclei of certain cranial nerves. The central canal opens out into the fourth ventricle and the central gray becomes spread out in the floor of the rhomboid fossa. As the internal arcuate and other fibers traverse the gray matter, arcuate fibers (Figs. 11–32, 11–36). Others continue as the **dorsal external arcuate fibers** (Figs. 11–28, 11–33) into the inferior cerebellar peduncle.

The **lateral** or **accessory cuneate nucleus** occupies a position superior and superficial to the nucleus cuneatus. It begins at the caudal end of the olive, enlarging as the

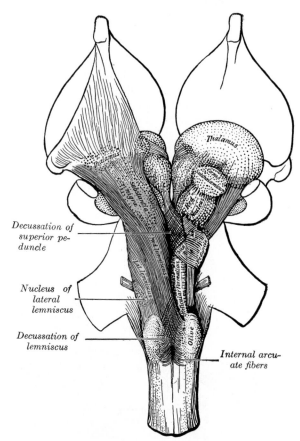

Decussation of superior peduncle

Nucleus of lateral lemniscus

Decussation of lemniscus

Internal arcuate fibers

FIG. 11–32.—Deep dissection of brain stem. Ventral view.

they disperse it into smaller masses, forming the reticular substance.

The **gray substance of the medulla oblongata.**—(1) The **nucleus gracilis** and **nucleus cuneatus** are buried deep to the fibers of the corresponding fasciculi in the clava and cuneate tubercles (Figs. 11–35, 11–36). They extend from the caudal end of the medulla to the level of the hypoglossal nerve. They are composed of small and medium-sized cells about which the fibers of the fasciculi terminate. Most of the axons of these cells pass ventrally into the **internal**

nucleus cuneatus decreases and extends into the medial margin of the inferior cerebellar peduncle. It contains large round or polygonal cells whose axons pass through the peduncle into the cerebellum.

(2) The **inferior olivary nucleus** (*nucleus olivaris*) is a characteristically folded gray lamina with an open part or **hilum** facing medially (Fig. 11–37). The axons of its cells pass out in the medullary white substance of the hilum in a fascicle called the **peduncle of the olive.** These are the olivocerebellar fibers. They cross the raphe in

the midline, become part of the system of internal arcuate fibers, pass partly through and partly around the olivary nucleus of the other side and enter the inferior cerebellar peduncle. The fibers are smaller than the internal arcuate fibers of the medial lemniscus and probably are accompanied by uncrossed olivocerebellar fibers. They are distributed to the cerebellar cortex in orderly sequence and there is a close functional association between the olive and the cerebellum. Fibers passing from cerebellum to olive have been described. The fibers of the **thalamoölivary tract** (*central tract of the tegmentum*) which may or may not originate in the thalamus surround the

olive in a dense capsule and terminate in relation to the cells of its gray nucleus. Many collaterals from the reticular formation and pyramid enter the olive, and fibers from the spinal cord enter by uncertain paths. The **medial** and **dorsal accessory olivary nuclei** are flattened plates of gray substance placed near the main inferior olivary nucleus (Fig. 11–37).

As the gray substance of the rostral part of the medulla becomes spread out by the expansion of the central canal into the fourth ventricle, the motor nuclei take up their positions medially and the sensory nuclei laterally in the floor of the rhomboid fossa.

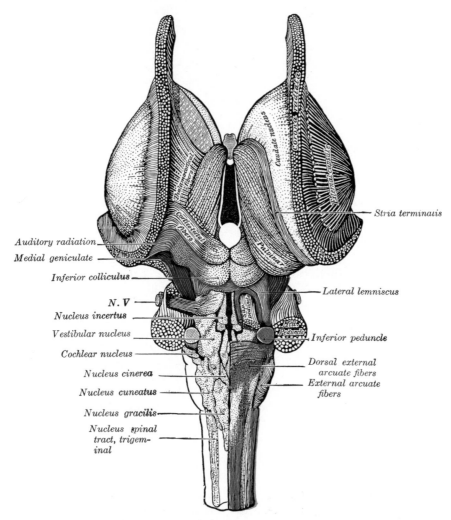

Fig. 11–33.—Dissection of brain stem. Dorsal view. The nuclear masses of the medulla are taken from model by Weed, Carnegie Publication, No. 19.

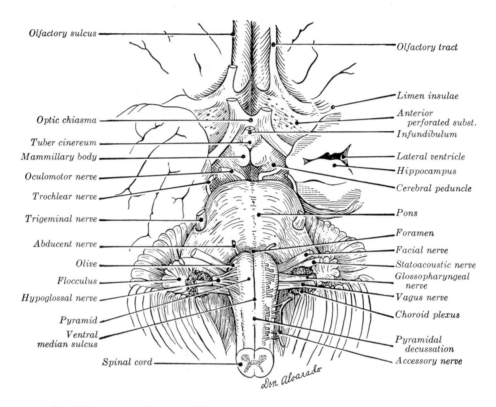

Olfactory sulcus

Olfactory tract

Limen insulae

Anterior perforated subst.

Optic chiasma

Infundibulum

Tuber cinereum

Mammillary body

Lateral ventricle

Oculomotor nerve

Hippocampus

Trochlear nerve

Cerebral peduncle

Trigeminal nerve

Pons

Abducent nerve

Foramen

Facial nerve

Olive

Statoacoustic nerve

Flocculus

Glossopharyngeal nerve

Hypoglossal nerve

Vagus nerve

Pyramid

Choroid plexus

Ventral median sulcus

Pyramidal decussation

Spinal cord

Accessory nerve

Don Alvarado

Fig. 11–34.—Ventral aspect of the brain stem, showing position of cranial nerves.

(3) The **hypoglossal nucleus** (*nucleus n. hypoglossi*) (Fig. 11–36) occupies the central gray column ventral to the central canal in the closed part of the medulla and extends up into the open part in the floor of the rhomboid fossa. Here it produces an elevation, the **trigonum hypoglossi,** in the medial eminence of the **calamus scriptorius** (Fig. 11–50).

(4) The **nucleus ambiguus** (Fig. 11–36) is the rostral continuation of the dorsal lateral group of cells in the ventral horn of the spinal cord. It is the motor nucleus of the special visceral motor fibers of the glossopharyngeal, vagus, and accessory nerves which supply the striated muscles of the pharynx and larynx.

(5) The **dorsal motor nucleus of the vagus** (*nucleus dorsalis n. vagi; nucleus alae cinereae*) (Fig. 11–36) occupies the **ala cinerea** of the rhomboid fossa (Fig. 11–50). It is a long column of cells lying lateral to the hypoglossal nucleus and extending from the decussation of the medial lemniscus

through the medulla as far rostrally as the upper limit of the olive (Figs. 11–37, 11–41). It is the general visceral efferent nucleus of the vagus nerve supplying parasympathetic fibers to the smooth muscle and glands of the thorax and abdomen and inhibitory fibers to the heart.

(6) The **nucleus of the spinal tract of the trigeminal nerve** (*nucleus tractus spinalis n. trigemini*) is a direct continuation of the substantia gelatinosa (of Rolando) in the spinal cord. It lies between the nucleus cuneatus and the reticular substance (Fig. 11–36) and is surrounded by the fibers of its tract, the spinal tract of the trigeminal nerve.

(7) **Nucleus of the tractus solitarius** is a slender nucleus extending the entire length of the medulla, lying lateral to the dorsal nucleus of the vagus (Fig. 11–36). The caudal ends of the nuclei of the two sides are united dorsal to the central canal in the closed part of the medulla. Its cells are scattered around and among the fibers of its

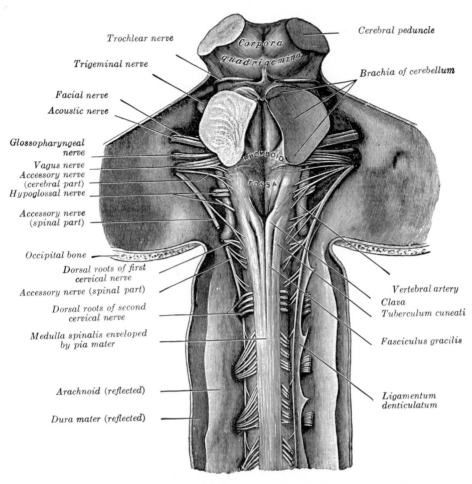

Trochlear nerve

Trigeminal nerve

Facial nerve

Acoustic nerve

Glossopharyngeal
nerve

Vagus nerve

Accessory nerve
(cerebral part)

Hypoglossal nerve

Accessory nerve
(spinal part)

Occipital bone

Dorsal roots of first
cervical nerve

Accessory nerve (spinal part)

Dorsal roots of second
cervical nerve

Medulla spinalis enveloped
by pia mater

Arachnoid (reflected)

Dura mater (reflected)

Corpora
quadrigemina

RHOMBOID

FOSSA

Cerebral peduncle

Brachia of cerebellum

Vertebral artery

Clava

Tuberculum cuneati

Fasciculus gracilis

Ligamentum
denticulatum

FIG. 11–35.—The upper part of the medulla spinalis, and the hind- and midbrains.
Dorsal aspect.

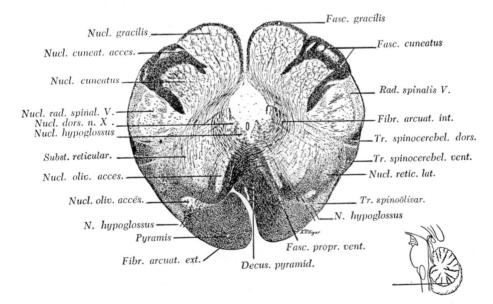

Nucl. gracilis

Nucl. cuneat. acces.

Nucl. cuneatus

Nucl. rad. spinal. V.

Nucl. dors. n. X.

Nucl. hypoglossus

Subst. reticular.

Nucl. oliv. acces.

Nucl. oliv. accës.

N. hypoglossus

Pyramis

Fibr. arcuat. ext.

Decus. pyramid.

Fasc. gracilis

Fasc. cuneatus

Rad. spinalis V.

Fibr. arcuat. int.

Tr. spinocerebel. dors.

Tr. spinocerebel. vent.

Nucl. retic. lat.

Tr. spinoölivar.

N. hypoglossus

Fasc. propr. vent.

FIG. 11–36.—Transverse section of medulla oblongata at the decussation of the pyramids. Nuclear groups
represented diagrammatically in color. (From Villiger-Addison, courtesy of J. B. Lippincott Co.)

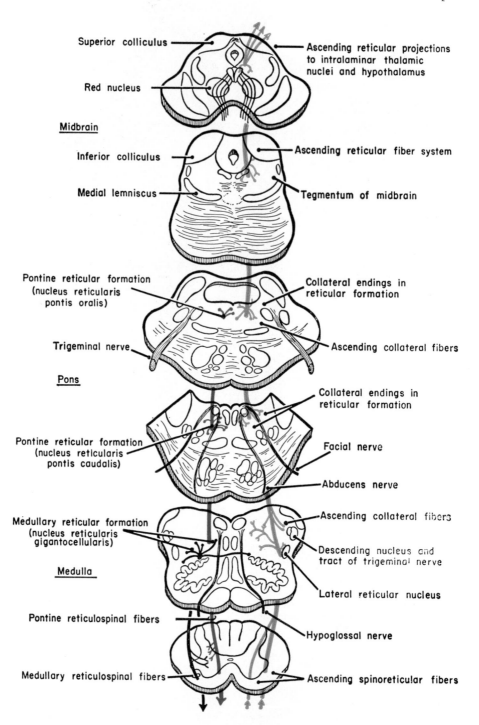

Fig. 11–37.—Schematic diagram of ascending and descending reticular fibers systems. Ascending spinoreticular and collateral reticular projections are shown on the right in *blue*. Pontine reticulospinal fibers of the medial reticulospinal tract are on the left in *red*. Medullary reticulospinal fibers of the lateral reticulospinal tract are in *black*. (Truex and Carpenter, 1964, *Strong and Elwyn's Human Neuroanatomy*, courtesy of The Williams & Wilkins Co.)

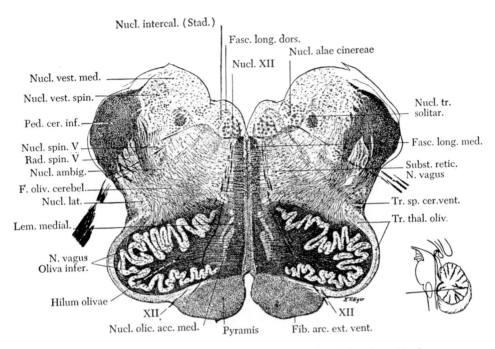

Fig. 11–38.—Transverse section of medulla oblongata at the middle of the olive. Nuclear groups represented diagrammatically in color. (From Villiger-Addison, courtesy of J. B. Lippincott Co.)

tract. It is especially concerned with taste in the facial, glossopharyngeal and vagus nerves and includes the sensory nucleus of these two nerves.

(8) The **reticular formation** (*substantia reticularis*) (Figs. 11–32, 11–37) of the medulla oblongata is a continuation of that of the spinal cord. It is composed of small groups of cells scattered among bundles of fibers. In the portion medial to the emerging fibers of the hypoglossal nerve, fibers predominate, and it is called the *white reticular substance.* This contains the medial lemniscus, tectospinal tract, and medial longitudinal fasciculus. The part lateral to the hypoglossal fibers and dorsal to the olive in which nuclei predominate is called the *gray reticular substance.* Its nuclei include the **lateral reticular nucleus** and the **nucleus magnocellularis** as well as nuclei of cranial nerves. It contains the internal arcuate fibers, the rubrospinal, thalamoölivary, ventral and dorsal spinocerebellar, spinothalamic, and spinotectal tracts.

(9) The **nuclei arcuati** are groups of cells serially continuous with the nuclei pontis

which lie superficial to the pyramid among the ventral external arcuate fibers.

(10) The **nucleus intercalatus** (*of Staderini*) (Fig. 11–38) occupies a position near the midline at the caudal end of the rhomboid fossa. It lies between the dorsal motor nucleus of the vagus and the hypoglossal nucleus. Some of its limits are not clearly defined, making it appear to be continuous with the medial vestibular nucleus and nucleus prepositus rostrally. It receives fibers from the vestibular nucleus, the glossopharyngeal and vagus nerves, and from the dorsal longitudinal bundle (of Schütz), close to which it lies. Its axons enter the reticular formation, and the hypoglossal and vagus nuclei. Connections with the dorsal bundle and the nucleus salivatorius probably give it a visceral function.

(11) The **nucleus prepositus** occupies about the same position as the nucleus intercalatus and may have the same function. It extends between the hypoglossal and abducens nuclei.

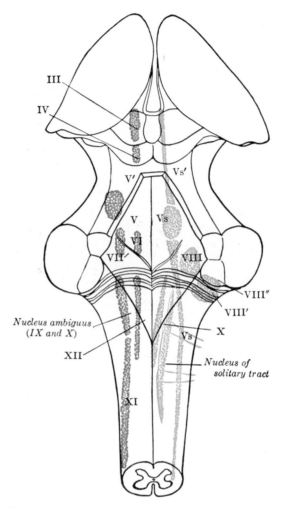

FIG. 11–39.—The cranial nerve nuclei schematically represented; dorsal view. Motor nuclei in red, sensory in blue. (The olfactory and optic centers are not represented.)

The **White Substance of the Medulla Oblongata.**—(1) The **internal arcuate fibers** (*fibrae arcuatae internae*) (Fig. 11–36) are mainly from two sources, (*a*) the nuclei gracilis and cuneatus, and (*b*) the inferior olive. The gracile and cuneate fibers are larger than the olivary fibers. They curve around the central gray toward the raphe and, as they cross the midline, form the **decussation of the medial lemniscus** (*decussatio lemniscorum*).

(2) The **medial lemniscus** (Figs. 11–32, 11–42) occupies the angle between the pyramid and the inferior olive, beside the median raphe. The fibers running parallel are arranged so that those from the nucleus cuneatus (upper limb and trunk) are in the ventral part of the bundle; those from the nucleus gracilis (leg and lower trunk) in the dorsal part. As the band reaches the upper end of the olive in its cephalic course, it swings laterally and spreads out into a thin ribbon dorsal to the pontine nuclei.

(3) The **ventral spinothalamic tract** (Fig. 11–25) is deflected dorsally by the pyramidal decussation and continues rostrad just dorsal to the medial lemniscus. Where the latter swings laterally in front of the olive, the ventral spinothalamic tract also turns laterally and accompanies the lateral spinothalamic tract to the thalamus.

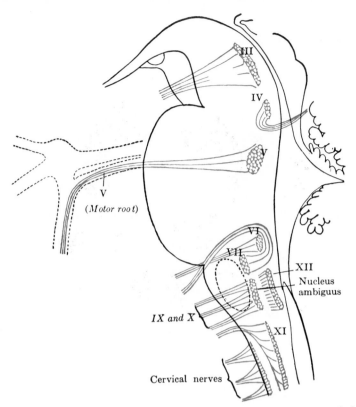

Fig. 11–40.—Nuclei of origin of cranial motor nerves schematically represented; lateral view.

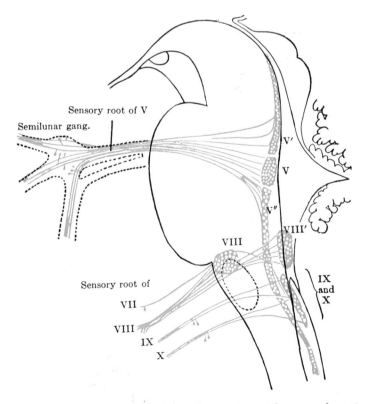

Fig. 11–41.—Primary terminal nuclei of the afferent (sensory) cranial nerves schematically represented; lateral view. The olfactory and optic centers are not represented.

(813)

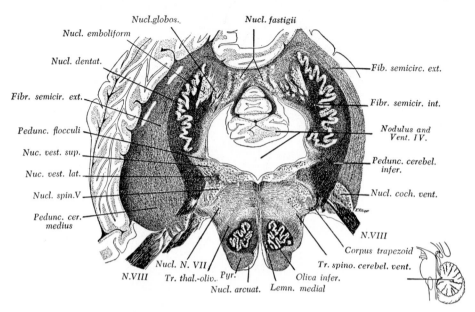

Nucl. emboliform
Nucl. dentat.
Fibr. semicir. ext.
Pedunc. flocculi
Nuc. vest. sup.
Nuc. vest. lat.
Nucl. spin.V
Pedunc. cer. medius
Nucl.globos.
Nucl. fastigii
Fib. semicirc. ext.
Fibr. semicir. int.
Nodulus and Vent. IV.
Pedunc. cerebel. infer.
Nucl. coch. vent.
N.VIII
Corpus trapezoid
Tr. spino. cerebel. vent.
Oliva infer.
Lemn. medial
Nucl. arcuat.
Pyr.
Tr. thal.-oliv.
N.VIII
Nucl. N. VII

FIG. 11–42.—Transverse section through medulla oblongata and cerebellar peduncles at the level of the eighth cranial nerve. Nuclear groups represented diagrammatically in color. (From Villiger-Addison, courtesy of J. B. Lippincott Co.)

(4) The **dorsal spinocerebellar tract** (Fig. 11–30) leaves the lateral district of the medulla to enter the inferior cerebellar peduncle of the same side on its way to the vermis. The ventral spinocerebellar tract (Fig. 11–29) retains its lateral position in the medulla until it passes deep to the external arcuate fibers (Fig. 11–28). It lies between the olive and the roots of the vagus and glossopharyngeal nerves.

(5) The two **pyramids** (Fig. 11–38) contain the motor fibers from the precentral motor cortex, the corticobulbar fibers to the nuclei of cranial nerves, and the corticospinal to the ventral horns of the cord. At the caudal limit of the medulla, about two-thirds cross in the pyramidal decussation (Fig. 11–31) and continue in the cord as the lateral corticospinal tract. The fibers remaining on the same side, those more laterally placed, continue as the ventral corticospinal tract.

(6) The **tectospinal tract** lies close to the midline in the upper part of the medulla just dorsal to the medial lemniscus. It becomes more laterally placed as it courses caudad through the region of the decussations and finally reaches its ventral position in the spinal cord.

(7) The **medial longitudinal fasciculus** (*fasciculus longitudinalis medialis*) (Fig. 11–38) occupies a position close to the midline between the tectospinal tract and dorsal gray substance. It is a continuation upward of the sulcomarginal fasciculus of the spinal cord.

(8) The **inferior cerebellar peduncle** or **restiform body** (*pedunculus cerebellaris inferior*) (Figs. 11–27, 11–38) contains the following fibers: (*a*) the dorsal spinocerebellar tract, (*b*) the olivocerebellar fibers, (*c*) fibers from the ventral arcuate nuclei, (*d*) fibers from the lateral reticular nucleus, (*e*) dorsal external arcuate fibers, (*f*) fibers from the vestibular nuclei, and (*g*) probable cerebellobulbar, cerebelloölivary, and cerebelloreticular fibers.

THE PONS

The **pons** (*pons Varolii*) produces a prominent swelling with well-defined borders on the ventral surface of the brain stem, between the medulla oblongata and the cerebral peduncles of the midbrain (Fig. 11–26). The dorsal surface, where it forms the cephalic part of the rhomboid fossa

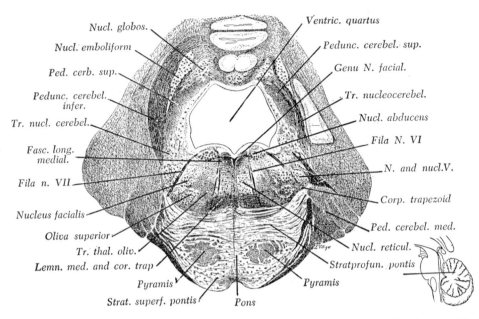

FIG. 11–43.—Transverse section through pons and cerebellum at the level of the facial nerve. Nuclear groups represented diagrammatically in color. (From Villiger-Addison, courtesy of J. B. Lippincott Co.)

(Fig. 11–50), is hidden by the cerebellum. The ventral prominence consists of transverse strands arched across the midline like a bridge and gathered at each side into a compact bundle or arm, the middle cerebellar peduncle. The midline of the pons has a shallow depression, the sulcus basilaris, in which the basilar artery lies, and the entire structure rests upon the clivus of the sphenoid bone. At its caudal border are the origins of the abducent, facial, and statoacoustic nerves. At its cephalic border, the origin of the trigeminal nerve marks the boundary between the pons and the middle cerebellar peduncle (Fig. 11–27).

Structure of the Pons.—The pons is composed of two parts which differ in appearance and structure (Fig. 11–43). The ventral or basilar portion consists of masses of transverse fibers, separated here and there by longitudinal bundles and small nuclei. The dorsal portion, called the tegmentum, is an upward continuation of the reticular formation of the medulla, with similar tracts and nuclei.

Nuclei of the Tegmentum of the Pons.—(1) The **nucleus of the abducens nerve** (*nucleus n. abducentis; N. VI*) forms part of the dorsal gray substance in the medial

eminence of the floor of the fourth ventricle, deep to the colliculus facialis (Fig. 11–50). The axons of its cells penetrate the pons close to the median raphe and emerge on the ventral surface at the caudal border of the pons (Fig. 11–34).

(2) The **nucleus of the facial nerve** (*nucleus n. facialis; N. VII*) is situated deep in the reticular formation lateral to the abducens nucleus (Fig. 11–43). Its fibers take a tortuous course, first toward the floor of the fourth ventricle. Next, they loop around the abducens nucleus forming the internal genu and causing the elevation of the facial colliculus (Fig. 11–50). Finally they pass laterally and ventrally, penetrating the substance of the pons, and emerge at its caudal border between the olive and inferior cerebellar peduncle (Fig. 11–34).

(3) The **motor nucleus of the trigeminal nerve** (*nucleus motorius n. trigemini; N. V*) is situated in the rostral part of the pons, close to its dorsal surface along the lateral margin of the fourth ventricle (Fig. 11–43).

(4) The **sensory nuclei of the trigeminal nerve** represent a cephalic continuation of the sensory column of the spinal cord known as the substantia gelatinosa (of Rolando). The fibers entering the pons from the

trigeminal (semilunar, Gasserian) ganglion divide into ascending and descending branches. The descending branches go to the caudal nucleus, the **nucleus of the spinal tract of the trigeminal nerve** (Fig. 11–38) described as part of the medulla (page 808). Its axons cross the midline and run upward to the thalamus just medial to the lateral spinothalamic tract. The ascending branches enter the **main sensory nucleus of the trigeminal nerve** (*nucleus sensorius superior n. trigemini; N. V*) which lies lateral to the motor nucleus close to the superior cerebellar peduncle (Fig. 11–43). Its axons cross to the opposite side to join the medial margin of the medial lemniscus on their way to the thalamus. These two nuclei correspond to the posterior gray columns of the cord. The **mesencephalic root** (*tractus mesencephalicus*) of the trigeminal nerve consists of axons from cells in the central gray around the aqueduct and is included in the description of the midbrain.

(5) The **nucleus of the cochlear division** of the eighth nerve (*nucleus n. statoacustici*) in the caudal part of the pons is divided into two parts, a **dorsal** and a **ventral** **nucleus** (*nuclei cochleares, ventralis et dorsalis*) (Fig. 11–42). They are continuous with each other and lie on the dorsal and lateral aspect of the inferior cerebellar peduncle, producing an eminence, the acoustic tubercle (Fig. 11–50).

(6) The **superior olive** consists of two or three small masses of gray substance in the tegmentum of the pons rostral to the inferior olive and just dorsolateral to the trapezoid body (Fig. 11–43). It receives terminals and collaterals from the cochlear nuclei of the same and opposite sides by way of the trapezoid body. From its fusiform cells axons pass dorsally as the peduncle of the superior olive, and cross in the midline ventral to the abducens nucleus. The termination of these fibers is not well established; some enter the medial longitudinal fasciculus and others join the lateral lemniscus on both sides (Stotler '53).

(7) The **nuclei of the vestibular division** (*nuclei vestibulares*) of the eighth nerve occupy a large area in the lateral part of

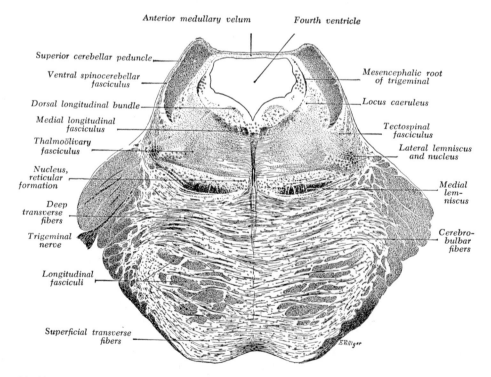

Fig. 11–44.—Transverse section through rostral part of pons. Nuclear groups represented diagrammatically in color. (From Villiger-Addison, courtesy of J. B. Lippincott Co.)

the floor of the fourth ventricle (Fig. 11–42). (*a*) The **medial nucleus** (*nucleus medialis; chief, dorsal, or triangular nucleus of Schwalbe*) is the largest and occupies most of the **area acoustica** (*area vestibularis*) of the rhomboid fossa (Fig. 11–50). (*b*) The **lateral vestibular nucleus** (*nucleus lateralis; nucleus of Deiters*) lies close to the inferior cerebellar peduncle at the entrance of the vestibular nerve. (*c*) The **superior vestibular nucleus** (*nucleus superior; nucleus of Bechterew*) is just caudal to the motor nucleus of the trigeminal nerve. It occupies the wall of the ventricle and is almost continuous with the central nuclei of the cerebellum. (*d*) The **inferior nucleus** (*nucleus inferior; nucleus of the descending tract*) is medial to the inferior cerebellar peduncle and dorsal to the trigeminal spinal tract extending caudally almost to the nucleus cuneatus.

The **Ventral or Basilar Part of the Pons** (*pars basilaris pontis*).—The transverse fibers are divided into two groups, superficial and deep, by the scattered bundles of longitudinal fasciculi (Fig. 11–44). The **superficial transverse fibers** (*fibrae pontis superficialis*) form a thick layer on the ventral surface of the pons (Figs. 11–43, 11–44). They are gathered into a massive bundle at each side, the brachium pontis or middle cerebellar peduncle. The **deep transverse fibers** (*fibrae pontis profundae*) partly interweave with the longitudinal fasciculi and partly lie dorsal to them. They also enter the middle cerebellar peduncle at each side.

The **longitudinal fasciculi** enter the pons from the cerebral peduncles of the midbrain and split up into smaller bundles, separated by transverse fibers and small pontine nuclei (Fig. 11–29). They are composed of two types of fibers. (*a*) The corticospinal fibers pass through the pons in bundles which are gathered together again at its caudal limit to form the pyramid of the medulla. They give collaterals to the pontine nuclei. (*b*) The corticopontine fibers end in the pontine nuclei of the same side and are relayed by the transverse fibers through the middle cerebellar peduncle to the cerebellum of the opposite side.

The **pontine nuclei** (*nuclei pontis*) (Fig. 11–43) are small groups of cells scattered between the transverse fibers. They are relay stations in the corticopontocerebellar path.

The **tracts traversing the tegmental part of the pons** are chiefly the upward continuation of those of the reticular formation of the medulla: (1) the medial lemniscus, (2) the spinothalamic tracts, (3) the ventral spinocerebellar tract, (4) the thalamo-olivary fasciculus, (5) the medial longitudinal fasciculus (Fig. 11–44). In addition there are the aberrant pyramidal fibers and the trapezoid body.

Aberrant pyramidal fibers leave the medial margins of the cerebral peduncles in the midbrain, turn dorsally, leaving the main bundles of fibers from the cortex, and run caudalward just medial to the medial lemniscus. Most of them cross, but some remain ipsilateral, and terminate in the motor nuclei of the cranial nerves.

The **trapezoid body** (*corpus trapezoideum*) is formed by a concentration of transverse fibers which show prominently in sections of the caudal part of the pons (Fig. 11–43). They occupy the ventral portion of the tegmentum where they mingle with the fibers of the medial lemniscus, crossing them at right angles. They arise primarily in the ventral and dorsal cochlear nuclei, cross the midline and turn rostrally at the superior olive of the opposite side to form the bulk of the lateral lemniscus. This is part of the central acoustic pathway.

THE CEREBELLUM

The cerebellum rests on the floor of the posterior cranial fossa (Fig. 4–64) and occupies the interval between the brain stem and the occipital lobes of the cerebrum (Fig. 11–26). The tentorium cerebelli separates it from the occipital lobes, and the fourth ventricle intervenes between it and the brain stem. It is attached to the latter, however, by three peduncles (Fig. 11–47): (1) the superior peduncle or brachium conjunctivum connecting it with the midbrain, (2) the middle peduncle or brachium pontis connecting with the pons, and (3) the inferior peduncle or restiform body connecting with the medulla. Its average weight is between 140 and 150 gm; the ratio of its

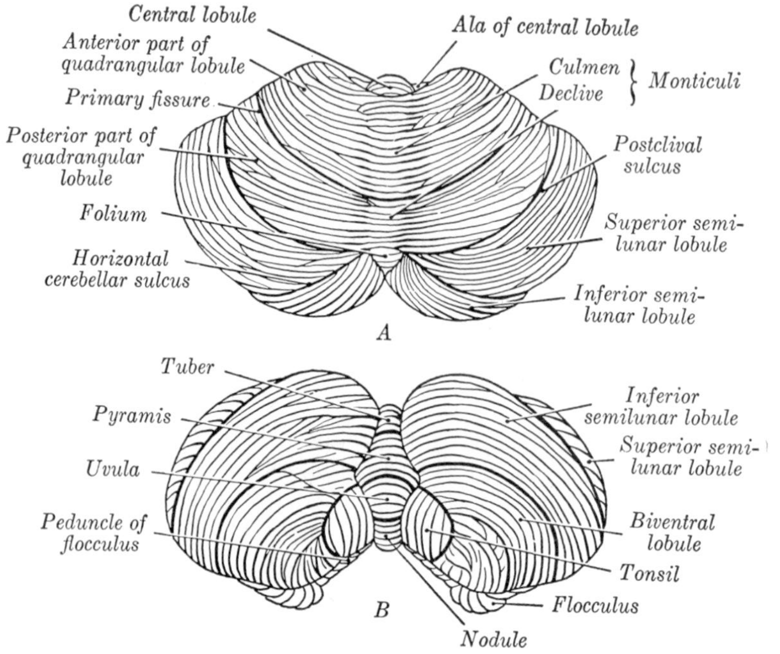

Fig. 11–45.—*A*, Dorsal surface of the cerebellum; *B*, Ventral surface of the cerebellum so oriented with relation to *A* as to indicate the continuity of folium and tuber (after Sobotta-McMurrich).

size compared with the cerebrum is about 1 to 8 in the adult and 1 to 20 in the infant.

The cerebellum is composed of a narrow median portion, the **vermis** (Fig. 11–45), and two **hemispheres** which protrude laterally and posteriorly. On the superior surface, which is rather flat, the vermis forms a slight ridge, the **monticulus.** Posteriorly, a deep cleft between the hemispheres, called the posterior cerebellar notch, is occupied by the falx cerebelli. More anteriorly over the medulla, the space between the hemispheres is called the **vallecula cerebelli** (Buchanan '37).

The surface of the cerebellum, the **cortex,** is thrown up into numerous parallel ridges called **folia cerebelli** which are separated by **fissures** (*fissurae cerebelli*). The latter cut deeply into the substance and ramify like the leaves of a plant into what is called the **arbor vitae** (Fig. 11–50). The fissures divide the cerebellum into lobes and lobules.

The superior part of the **vermis** beginning at the anterior medullary velum is composed of the lingula, the central lobule (*lobus centralis*), the culmen and declive of the *monticulus,* and the folium. The inferior vermis comprises the tuber, pyramis, uvula, and nodulus (Fig. 11–45).

Correspondingly, the **hemispheres** contain the vinculum as a lateral continuation of the lingula, separated by the precentral fissure from the ala of the central lobule. The anterior and posterior semilunar lobules, continuous with the culmen and declive, are separated from the ala and central lobule by the preculminate fissure and from each other by the anterior superior fissure or **fissura prima.** The superior and inferior semilunar lobules, continuous with the folium and tuber, are separated from the declive and posterior semilunar lobule by the posterior superior fissure and from each other by the horizontal sulcus.

Based upon comparative anatomy and embryology a more useful and functionally correlated division of the cerebellum (Larsell '47) is as follows (Fig. 11–48): The small flocculonodular lobe is separated from the corpus cerebelli or the remainder of the cerebellum by the **posterolateral** and **uvulo-**

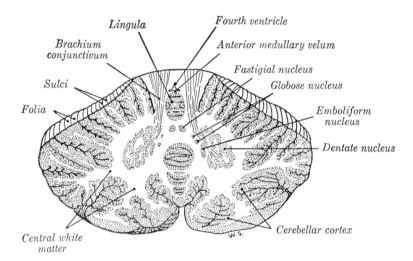

Fig. 11–46.—Horizontal section of the cerebellum showing the arrangement of the cortical gray matter and the locations of the central nuclei within the white matter (after Sobotta-McMurrich).

nodular fissures. The corpus cerebelli is divided by the fissura prima into anterior and posterior lobes. The lobules of the anterior lobe are the lingula, central lobule, and culmen with their lateral connections. The posterior lobe contains the lobulus simplex or declive and posterior semilunar lobule; the ansiform lobule comprises the superior and inferior semilunar lobules, called crus I and crus II, and the folium and tuber of the vermis.

Internal Structure of the Cerebellum

The cerebellum consists of an outer layer of gray substance, the cortex, and an inner core of white substance, in which certain nuclear groups are imbedded (Fig. 11–46).

White Substance of the Cerebellum.—The **central white substance** (*corpus medullare*) extends out in the **lobules** and **laminae,** forming plate-like cores (*laminae albae*) for the **folia.** It consists of the fibrae propriae and the projection fibers which pass through the superior, middle, and inferior peduncles.

The **inferior cerebellar peduncle** or **restiform body** (*pedunculus cerebellaris inferior*) (Fig. 11–47) on each side forms the lateral wall of the fourth ventricle in its caudal half. It connects the cerebellum with the medulla oblongata and contains largely afferent fibers. (1) The dorsal spinocerebellar tract is a proprioceptive path from the nucleus dorsalis of the same side of the

spinal cord which ends mainly in the superior vermis. (2) The dorsal external arcuate fibers are also proprioceptive from the nuclei gracilis and cuneatus of the same side. (3) The olivocerebellar tract from the inferior olivary nucleus crosses for the most part and goes to the cortex of the hemisphere, the vermis, and the central nuclei. (4) The ventral external arcuate fibers, crossed and uncrossed, come from the arcuate and reticular nuclei of the medulla. (5) The vestibulocerebellar tract, a proprioceptive path, is derived partly from direct fibers from the vestibular nuclei. They occupy the medial segment of the peduncle, along with fibers from the sensory nuclei of the cranial nerves, and end in the cortex of the vermis and in the central nuclei. (6) Also there are fibers from the central nuclei of the cerebellum to the vestibular nuclei and nuclei of the reticular formation.

The **middle cerebellar peduncle** or **brachium pontis** (*pedunculus cerebellaris medius*) (Fig. 11–47) of each side contains the transverse fibers of the pons which originate in the pontine nuclei of the opposite and the same side. They are distributed to all parts of the cerebellar cortex, completing the extensive corticopontocerebellar pathway which is particularly important in muscular synergy.

The **superior cerebellar peduncle** or **brachium conjunctivum** (*pedunculus cerebellaris superior*) (Fig. 11–47) forms the lateral wall on each side of the cephalic portion of the fourth ventricle. The bulk of its fibers are efferent, mainly from the nucleus dentatus (Fig. 11–42), which penetrate deeply into the tegmentum of the mesencephalon and cross in the **decussation** of the superior peduncle. They are part of the cerebellorubrospinal pathway. The fibers of the ventral spinocerebellar tract (Gower) enter the cerebellum by arching over the main part of the superior peduncle. They end in the medial portion of the anterior lobe. The uncinate bundle (of Russell) of the fastigiobulbar tract also hooks around the superior peduncle just ventral to the

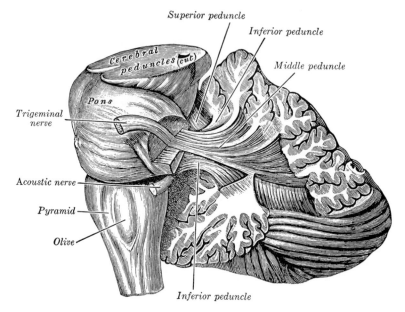

Fɪɢ. 11–47.—Dissection showing the projection fibers of the cerebellum. (After E. B. Jamieson.)

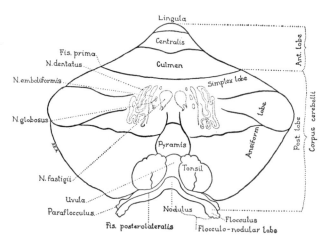

Fɪɢ. 11–48.—Diagrammatic representation of the lobes of the cerebellum redrawn after Larsell with the cerebellar nuclei as illustrated by Jakob projected in dotted lines.

spinocerebellar tract. Its fibers arise in the nucleus fastigii, cross, and reach the vestibular and reticular nuclei.

The **Fibrae Propriae.**—(1) Commissural fibers cross the midline at the rostral and caudal parts of the vermis, connecting the two sides. (2) Association and arcuate fibers connect parts of the same side of the cerebellum.

Cerebellar Cortex (*cortex cerebelli*).— The gray substance of the cortex forms a characteristic covering for the white substance or medullary core of the primary and secondary laminae of the folia.

Microscopic Appearance of Cortex.—The cortex consists of two layers, an external molecular layer and an internal granule cell layer, separated by an incomplete stratum of Purkinje cells (Fig. 11–49).

The cells of the external gray or **molecular layer** (*stratum moleculare*) are arranged in two strata: An outer layer of small cells with branched axons and an inner layer containing the basket cells. The axons of the basket cells run for some distance parallel with the surface of the folium, giving off collaterals which pass vertically toward the Purkinje cells. They ramify about the bodies of the latter in basket-like networks.

The **Purkinje cells** have large flask-shaped cell bodies which lie in a single stratum at the separation of the molecular and nuclear layers. From the neck of the flask one or more large dendrites extend out into the

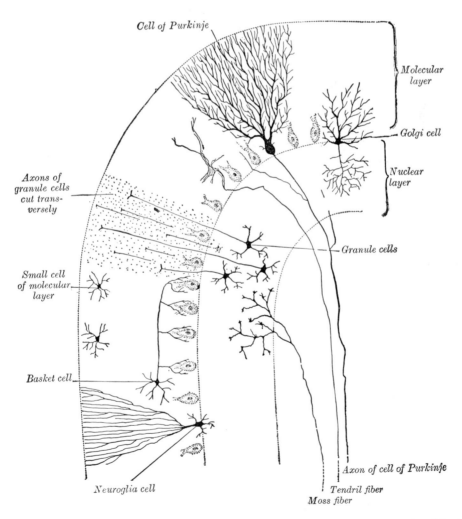

FIG. 11–49.—Transverse section of a cerebellar folium. (Diagrammatic, after Cajal and Kölliker.)

molecular layer forming extensive and characteristic arborizations (Fig. 11–49). Both cell and arborization are flattened, narrow in the direction of the long axis of the folium but spread out widely in a plane perpendicular to this. The axon of the Purkinje cell leaves the bottom of the flask and runs toward the center of the folium. It passes through the nuclear layer, becomes myelinated, and traverses the white substance to the central nuclei of the cerebellum. It gives off fine collaterals as it passes through the granular layer, some of which run back into the molecular layer.

The **granular layer** (*statum granulosum; nuclear layer*) (Fig. 11–48) contains numerous small nerve cells of a reddish-brown color called granule cells. Most of the cells are spherical and are provided with short spider-like dendrites. Their axons pass peripherally into the molecular layer, bifurcate, and run at right angles to the axis of the folium. In the outer part of the granular layer there are larger Golgi type II cells, whose dendrites ramify in the molecular layer and axons in the nuclear layer.

The great majority of afferent fibers to the cerebellum terminate in one of two ways. (1) Fibers end in the nuclear layer by dividing into numerous branches or moss-like appendages around the cells. They were called **moss fibers** by Ramon y Cajal, and appear to come from the inferior peduncle. (2) The clinging or **tendril fibers** can be traced into the molecular layer where they cling to the branches of the Purkinje cells. They were said to come mainly from the middle peduncle.

The **central gray substance** of the cerebellum is divided into four nuclei on each side, a large nucleus dentatus and three small nuclei, emboliformis, globosus, and fastigii (Fig. 11–42) (Chambers and Sprague '55).

The **nucleus dentatus** (Fig. 11–50) is situated near the center of the central mass of white substance in each hemisphere. It consists of a folded gray lamina, very similar to the inferior olive, with a hilum facing medially through which most of the fibers emerge on their way to the superior peduncle.

The **nucleus emboliformis** lies just medial to the dentatus, partly covering its hilum.

The **nucleus globosus** consist of several small, irregular groups of cells between the emboliformis and fastigii. Some authors regard the emboliformis and globosus as parts of one nucleus, the **nucleus interpositus**. The **nucleus fastigii** is situated close to the midline at the rostral end of the superior vermis and immediately over the roof of the fourth ventricle, from which it is separated by a thin layer of white substance.

The Fourth Ventricle

The **fourth ventricle** (*ventriculus quartus*) (Fig. 11–50) is a flattened diamond-shaped cavity containing spinal fluid. It is the remains of the cavity of the embryonic neural tube in the hindbrain, that is, the region of the cerebellum, pons, and upper medulla. The narrow rostral angle opens into the cerebral aqueduct, and the narrow caudal angle into the central canal of the closed portion of the medulla. Its roof is composed of structures associated with the attachment of the cerebellum; its floor is the rhomboid fossa of the brain stem; its lateral walls are narrowed to mere boundaries.

The rostral portion of the **roof** or **dorsal wall** (*tegmen ventriculi quarti*) is formed by the superior cerebellar peduncles and the anterior medullary velum (Figs. 11–50, 11–51). Caudally it is formed by the posterior medullary velum, tela choroidea and choroid plexus, the taenia of the fourth ventricle, the lingula, and the obex. The superior peduncles are separated as they emerge from the central white substance of the cerebellum but converge as they approach the inferior colliculi. Between them is the **anterior medullary velum**, a thin white lamina which is continuous with the vermis and over which the lingula is prolonged. The **frenulum veli**, from between the inferior colliculi, joins the anterior velum where the trochlear nerves emerge.

The **posterior medullary velum** (*velum medullare posterius*) (Fig. 11–51) is a thin sheet composed of ependyma on the inside and pia mater on the outside. It stretches from the cerebellar attachment of the anterior medullary velum at the middle of the

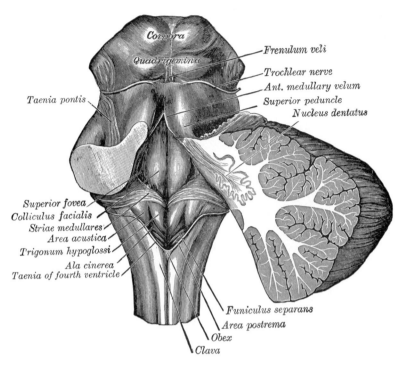

FIG. 11–50.—The floor of the fourth ventricle or rhomboid fossa after the structures of the roof have been cut or pulled aside.

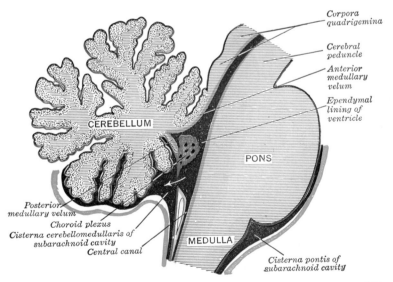

FIG. 11–51.—Scheme of roof of fourth ventricle. The arrow is in the foramen of Magendie.

ventricle down to the clava, where it is called the **taenia ventriculi quarti**, becoming gradually narrower and ending caudally in the **obex**. Each lateral angle of the diamond-shaped cavity of the fourth ventricle is prolonged into the **lateral recess** which lies over the striae medullares and under the choroid plexus.

The **choroid plexus** (*plexus chorioideus ventriculi quarti*) (Fig. 11–51) is an elongated tuft of tortuous blood vessels chiefly of capillary size, protruding into the fourth ventricle from its roof. The vessels belong to the layer of pia mater, the vascular coat of the brain, but are covered everywhere with the epithelial ependyma lining the ventricle. The plexus on each side is shaped like an inverted letter L with a median part close to the midline and a lateral part extending out into and somewhat beyond the lateral recess (Fig. 11–34).

Foramina or **Openings in the Roof of the Fourth Ventricle.**—There are three openings in the roof of the fourth ventricle through which the spinal fluid produced by the choroid plexuses can escape into the subarachnoid space. A **median aperture** (*apertura mediana ventriculi quarti; foramen of Magendie*) is just in front of the obex. Two **lateral apertures** (*aperturae laterales ventriculi quarti; foramina of Luschka*) are located at the extremities of the lateral recesses.

The **rhomboid fossa** (*fossa rhomboidea*) (Fig. 11–50) occupies the floor or ventral wall of the fourth ventricle. It is formed by the dorsal surface of the pons and of the cephalic or open half of the medulla (Fig. 11–35). It is divided into a triangular cephalic portion with its apex at the cerebral aqueduct, a triangular caudal portion with its apex at the central canal of the medulla, and an intermediate portion prolonged outward into the lateral recesses. The apical part of the caudal portion is called the **calamus scriptorius** from its resemblance to the point of a quill pen.

The fossa is divided into symmetrical halves by a **median sulcus**. On each side of this sulcus is the **medial eminence** (*eminentia medialis*) which is bounded laterally by the **sulcus limitans** (Fig. 11–50). In the cephalic portion, the medial eminence occupies most of the fossa and contains an

oval swelling, the **colliculus facialis** (Fig. 11–50), which overlies the root of the facial nerve as it encircles the nucleus of the abducens nerve. In the caudal portion, the medial eminence contains the **trigonum hypoglossi** (Fig. 11–50), overlying the nucleus of the hypoglossal nerve and the nucleus intercalatus.

Between the colliculus facialis and the middle cerebellar peduncle the sulcus limitans widens into a flattened depression, the **superior fovea** (Fig. 11–50). Similarly, at the level of the trigonum hypoglossi is a depression, the **inferior fovea**. The rostral portion of the sulcus is called the **locus coeruleus** because of a bluish color imparted to it by an underlying patch of deeply pigmented nerve cells termed the *substantia ferruginea*. Lateral to the sulcus and foveae, in the intermediate portion of the pons, is a rounded elevation named the **area acustica** (*area vestibularis*) (Fig. 11–50) which extends out into the lateral recess as the **tuberculum acusticum**. The **striae medullares** (Fig. 11–50) are raised bundles which cross the area acustica and medial eminence to disappear in the median sulcus. Between the trigonum hypoglossi and the caudal part of the area acustica is a dark triangular area, the **ala cinerea**, or **triangle of the vagus nerve** (*trigonum n. vagi*). The caudal part of the ala cinerea is crossed by a narrow translucent ridge, the *funiculus separans*, and between the latter and the clava is a small tongue-shaped region, the **area postrema** (Fig. 11–50) (Brizzee '54).

THE MIDBRAIN OR MESENCEPHALON

The mesencephalon or midbrain is a short constricted segment of the brain stem connecting the pons and cerebellum with the forebrain. Its ventral portion contains the cerebral peduncles; its dorsal portion, called the **tectum**, contains the corpora quadrigemina (Fig. 11–35). The cerebral aqueduct, connecting the fourth and third ventricles, is surrounded by the central gray stratum of the midbrain.

The two prominent ridges formed by the **cerebral peduncles** (Fig. 11–34) diverge as they approach the cerebral hemispheres, leaving a triangular depressed area, the interpeduncular fossa. The floor of this fossa is called the posterior perforated substance due to the numerous apertures for small blood vessels which penetrate it. At the angle between the medial surface of the peduncle and the perforated substance is a longitudinal furrow, the oculomotor sulcus, from which the oculomotor nerve emerges.

The corpora quadrigemina (tectum of midbrain) are four rounded prominences, the right and left superior, and right and left inferior colliculi (Figs. 11–50, 11–58), separated by a cruciate sulcus. At the upper end of the longitudinal part of this sulcus is a slight depression for the pineal body. At the lower end a white band, the frenulum veli, continues down into the anterior medullary velum. On each side of this band the trochlear nerve emerges and swings around the outside of the peduncles. The superior colliculi are larger than the inferior and oval in shape; the inferior are more prominent and hemispherical.

The superior brachium (brachium colliculi superius) (Fig. 11–58) extends lateralward from the superior colliculus. It passes under the pulvinar of the thalamus and above the medial geniculate body to reach the lateral geniculate body and the optic tract. Its fibers connect the occipital lobe or visual cortex with the superior colliculus. The inferior brachium (brachium colliculi inferius) passes in a rostral and ventral direction to the medial geniculate body (Fig. 11–58). Its fibers connect the lateral lemniscus, medial geniculate body, and temporal lobe or auditory area of the cerebral cortex.

Internal Structure of the Mesencephalon

The cerebral peduncle is separated by the substantia nigra into a ventral part, the crusta (crus cerebri) or basis pedunculi, and a more dorsal part, the tegmentum (Fig. 11–54). The crusta is made up principally of longitudinal bundles of efferent fibers from the cells of the cerebral cortex, the corticospinal and corticobulbar tracts. The fibers from the motor cortex occupy the middle three-fifths and continue mainly as the pyramids in the medulla oblongata. The frontopontine fibers occupy the medial fifth, and the parietotemporopontine fibers

the lateral fifth. In addition the fibers of the aberrant pyramidal tracts run in the most medial, lateral, and deepest layers of the crusta.

The substantia nigra (Fig. 11–54) is a layer of gray substance containing numerous deeply pigmented nerve cells. It extends from the rostral part of the pons to the subthalamus. The zona compacta is the layer of gray substance next to the tegmentum. The zona reticulata is adjacent to the crusta and sends scattered prolongations between the bundles of its fibers.

The Tegmentum of the Midbrain is continuous with that of the pons and contains most of the same gray columns and fiber tracts as well as additional ones. It is also continued without sharp demarcation into the subthalamus of the diencephalon.

Nuclei of the Midbrain.—(1) The nucleus of the mesencephalic root of the trigeminal nerve (nucleus tr. mesencephalicus n. trigemini) (Fig. 11–54) forms a scattered strand in the lateral part of the central gray which surrounds the aqueduct. It also extends down along the lateral angle of the fourth ventricle to the main trigeminal nucleus. The axons of its cells gather at the lateral border of the nucleus and reach the motor root of the nerve, probably serving as its proprioceptive fibers.

(2) The nucleus of the trochlear nerve (nucleus n. trochlearis) (Fig. 11–53) is at the level of the inferior colliculus. It is a compact group of cells in the ventral part of the central gray close to the midline and to the medial longitudinal fasciculus. The trochlear fibers arch dorsally around the central gray, decussate in the anterior medullary velum, and emerge at the dorsal surface of the brain stem just caudal to the inferior colliculus (Fig. 11–34).

(3) The nucleus of the oculomotor nerve (nucleus n. oculomotorius) (Fig. 11–54) appears in a cross section in the interval between the V-shaped section of the medial longitudinal fasciculi of the two sides, immediately rostral to the trochlear nucleus. It extends as far as the rostral limit of the superior colliculus. Its cells are somewhat separated by fibers of the medial longitudinal fasciculus and one group of cells bridges the midline in what is called the nucleus of Perlia (see page 869). Its fibers run ven-

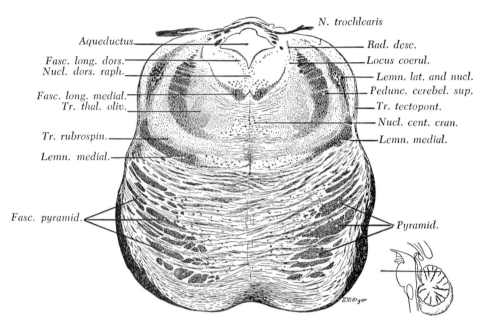

Fig. 11–52.—Transverse section through transition between pons and mesencephalon. Nuclear groups represented diagrammatically in color. (From Villiger-Addison, courtesy of J. B. Lippincott Co.)

trally through the region of the red nucleus and emerge medial to the basis pedunculi (Fig. 11–62).

(4) The **central gray** (*substantia grisea centralis*) surrounding the aqueduct is continuous with the periventricular system of the hypothalamus. The nerve cells are small and arranged in a number of groups. The *nucleus of Darkschewitsch* in the ventrolateral margin at the transition between the midbrain and hypothalamus is better defined and contains larger cells. It is closely associated with the medial longitudinal fasciculus topographically and probably functionally.

(5) The **reticular formation** is continuous with the reticular nucleus of the thalamus, the zona incerta, and the lateral region of the hypothalamus. Central, lateral, and dorsolateral nuclei have been defined. The *interstitial nucleus* (of Cajal) is located at the dorsomedial border of the capsule of the red nucleus, somewhat encroaching on the lateral border of the medial longitudinal fasciculus. It was early described as part of the nucleus of Darkschewitsch, but it is separated from the latter by a thin lamina of fibers. It is frequently known as the **nucleus of the medial longitudinal fasciculus** (Ingram and Ranson '35).

The reticular formation has come into prominence during the last decade because it has been shown to be an important link in the inhibitory, stimulatory, and facilitatory control of motor and autonomic responses (Magoun and Rhines '46, Magoun '52 and '54). The cells of the reticular formation from the thalamus to the medulla are connected by both ascending and descending fibers (see Fig. 11–37). For further details consult chapter by French in Section 1 Volume 2 of *Handbook of Physiology*, 1964.

(6) The **red nucleus** (*nucleus ruber*) (Fig. 11–54) is a prominent feature of the midbrain, occupying a large part of the tegmentum. It is an egg-shaped mass extending from the caudal limit of the superior colliculus into the subthalamic region. It is circular in cross section and contains mainly small and medium-sized cells which contain the reddish pigment responsible for the name given to the nucleus. It also contains less numerous large cells. It receives most of the fibers of the superior cerebellar peduncle and in addition some fibers from the frontal cortex. The smaller cells send their axons into the tegmentum to form the rubroreticular tract. The axons of the large cells take a caudal

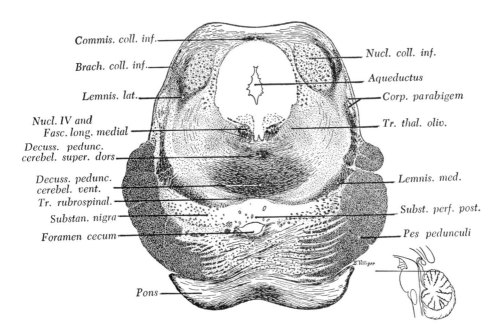

Commis. coll. inf.

Brach. coll. inf.

Lemnis. lat.

Nucl. IV and Fasc. long. medial

Decuss. pedunc. cerebel. super. dors

Decuss. pedunc. cerebel. vent.

Tr. rubrospinal.

Substan. nigra

Foramen cecum

Nucl. coll. inf.

Aqueductus

Corp. parabigem

Tr. thal. oliv.

Lemnis. med.

Subst. perf. post.

Pes pedunculi

Pons

F‍ɪɢ. 11–53.—Transverse section through mesencephalon at inferior colliculus. Nuclear groups represented diagrammatically in color. (From Villiger-Addison, courtesy of J. B. Lippincott Co.)

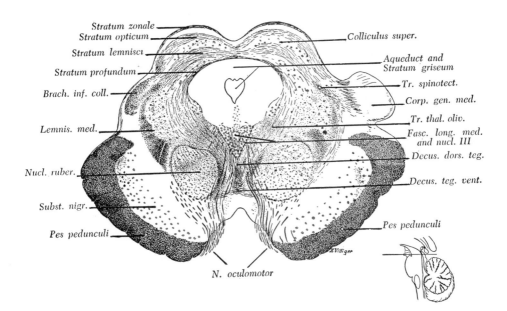

Stratum zonale

Stratum opticum

Stratum lemnisci

Stratum profundum

Brach. inf. coll.

Lemnis. med.

Nucl. ruber.

Subst. nigr.

Pes pedunculi

Colliculus super.

Aqueduct and Stratum griseum

Tr. spinotect.

Corp. gen. med.

Tr. thal. oliv.

Fasc. long. med. and nucl. III

Decus. dors. teg.

Decus. teg. vent.

Pes pedunculi

N. oculomotor

F‍ɪɢ. 11–54.—Transverse section through mesencephalon at superior colliculus. Nuclear groups represented diagrammatically in color. (From Villiger-Addison, courtesy of J. B. Lippincott Co.)

course and become the rubrospinal tract. It also sends fibers to the lateral ventral nucleus of the thalamus (Carpenter and Pines '57, Stern '38).

(7) The **interpeduncular ganglion** (*nucleus interpeduncularis*) is a group of cells of uncertain function in the posterior perforated substance of the interpeduncular fossa.

The **Tectum of the Midbrain** (*lamina tecti*) is composed of four rounded prominences, the **corpora quadrigemina** or the superior and inferior colliculi of the two sides (Fig. 11–50).

(8) The **superior colliculus** (*colliculus superior*) contains four or five strata (Fig. 11–54). (*a*) The outermost layer, the **stratum zonale,** is composed of fine white fibers, the majority of which are received from the optic tract and visual cortex through the superior brachium. (*b*) The second layer, the **stratum griseum** (*stratum cinereum*), is of gray substance consisting mostly of small multipolar nerve cells about which most of the fibers of the two adjacent strata terminate. (*c*) Deeper is the **stratum opticum** containing many fibers and scattered large multipolar cells. The fibers are from the optic tract, lateral geniculate body, and cortex. The majority send their axons peripherally into the stratum griseum. (*d*) The **stratum lemnisci** contains interspersed large multipolar cells and fibers mainly from the stratum opticum. (*e*) The **stratum profundum,** sometimes included in the stratum lemnisci, is a thin layer adjacent to the central gray and is composed of fibers from the spinotectal tract.

The superior colliculi of the two sides are connected by the commissure of the superior colliculus and the posterior commissure. They give rise to the tectobulbar and tectospinal tracts and are connected with the reticular formation, substantia nigra, zona incerta, and possibly the pontine nuclei. They are primary visual centers especially for coordination of eye movements (page 870).

(9) The **inferior colliculus** (*colliculus inferior*) (Fig. 11–53) consists of a compact nucleus of medium and small multipolar cells surrounded by a capsule of white fibers derived from the lateral lemniscus. Most of these fibers end in the nucleus of

the same side but some cross in the commissure of the inferior colliculus. The axons of the cells in the nucleus pass through the inferior brachium to the lateral geniculate body. Fibers also cross in the commissure to the opposite side. Scattered large cells medial to the main nucleus send axons through the stratum profundum to the tectobulbar and tectospinal tracts. There are also connections between the inferior colliculus and the substantia nigra, the nuclei of the reticular formation, and the pontine nuclei. It is a center for correlation and control of reflexes in response to sound.

The **Pretectal Region** is a transition zone between the superior colliculus and thalamus. It contains nuclei which receive fibers from the optic tract and sends axons to the Edinger-Westphal nucleus of the oculomotor nerve.

The **Fiber Tracts of the Tegmentum of the Midbrain.**—(1) The **medial longitudinal fasciculus** (*fasciculus longitudinalis medialis*) (Fig. 11–54) lies close to the midline just ventral to and in close association with the oculomotor and trochlear nuclei. It occupies the same relative position in the pons and medulla oblongata and continues into the spinal cord as the sulcomarginal fasciculus. The rostral origin of this tract is in the interstitial nucleus of Cajal in the reticular formation at the junction of the midbrain and hypothalamus. It receives a large number of its fibers from the vestibular nuclei and contributes fibers to the oculomotor, trochlear, abducens, and accessory nerve nuclei, and to the motor horn cells in the cervical spinal cord. It is the tract for correlation of eye, head, and neck movements, especially in response to sensations from the semicircular canals of the ear.

(2) The **medial lemniscus** (*lemniscus medialis*) together with the **trigeminothalamic lemniscus** forms a large bundle in the ventral and lateral part of the tegmentum in the more caudal part of the mesencephalon. In the cephalic position it is displaced dorsally and laterally close to the medial geniculate body and its inferior brachium.

(3) The **lateral lemniscus** (*lemniscus lateralis*) in the caudal part of the midbrain lies close to the surface in the tegmentum, dorsal to the medial lemniscus. Its fibers run dorsally as well as rostrally and sur-

round the nucleus of the inferior colliculus. Most of them end there but some continue in the inferior quadrigeminal brachium into the medial geniculate body.

(4) The **thalamoölivary tract** is in the reticular substance ventral and lateral to the central gray.

(5) The **rubrospinal tract** arises in the caudal part of the red nucleus and promptly decussates in the **ventral tegmental decussation** (Fig. 11–54). It passes laterally to a position just medial to the medial lemniscus where it turns caudally and continues into the lateral funiculus of the spinal cord.

(6) The **tectospinal** and **tectobulbar tracts** arise in the superior colliculus. The fibers sweep around the central gray in the stratum profundum of the colliculus and cross in the **dorsal tegmental decussation** (*fountain decussation of Meynert*) (Fig. 11–54). Turning caudally they lie ventral to the medial longitudinal fasciculus and enter the ventral funiculus of the spinal cord.

(7) The **decussation of the superior cerebellar peduncle** (*decussatio pedunculorum cerebellarium superiorum*) occupies the tegmentum ventral to the central gray at the level of the inferior colliculus. After crossing, most of the fibers ascend to the red nucleus and neighboring reticular substance. A few in the medial capsule of the red nucleus continue on to the nucleus ventralis lateralis of the thalamus.

THE FOREBRAIN OR PROSENCEPHALON

The forebrain or prosencephalon consists of: (1) the diencephalon or between brain and (2) the telencephalon or endbrain.

The Diencephalon

The diencephalon connects the midbrain with the cerebral hemispheres, and corresponds to the structures which bound the third ventricle. It comprises (1) the thalamus, (2) the metathalamus, (3) the epithalamus, (4) the subthalamus, and (5) the hypothalamus (Hess '57, Kuhlenbeck '54, Sawyer *et al.* '54).

The **Thalamus** (*dorsal thalamus*) (Figs. 11–55, 11–56) is a large ovoid mass about 4 cm long forming most of the side wall of the third ventricle and extending caudally for some distance beyond it. Its medial and superior surfaces are exposed in the ventricles and its inferior and lateral surfaces buried against other structures. The anterior or rostral extremities of the parts on the two sides lie close to the midline where they are narrow and form the posterior boundary of the interventricular foramina. The posterior extremities are thicker and diverge from each other extending out over the superior colliculi and having the pineal body between them. The medially prominent rounded posterior portion is the **pulvinar** (Fig. 11–55). The superior surface is free, covered only by the pia mater of the transverse cerebral fissure forming the tela choroidea (Fig. 11–26). In a groove between the thalamus and caudate nucleus, the sulcus terminalis, are the stria and vena terminalis (Fig. 11–26). A shallow furrow in the position of the overlying fornix divides the dorsal surface into a medial portion, covered by the tela, and a lateral portion which forms part of the floor of the lateral ventricle (Fig. 11–85). The superior surface is separated from the medial surface by the taeniae thalami (Fig. 11–56), the torn edges of the ependymal roof of the third ventricle, which meet in the stalk of the pineal body. The protruding medial surfaces of the thalamus are usually joined across the narrow intervening third ventricle by a bridge known as the **massa intermedia** (*adhaesio interthalamicus; middle commissure*) (Figs. 11–55, 11–57). The inferior surface of the thalamus rests upon and is continuous with the upward prolongation of the tegmentum of the midbrain. The lateral surface is separated from the corpus striatum by the occipital part of the internal capsule (Fig. 11–57).

Structure.—The thalamus consists chiefly of gray substance, but its upper surface is covered by a layer of white substance, named the **stratum zonale**, and just under its lateral surface by a similar layer termed the **external medullary lamina**. Its gray substance is subdivided into three parts—

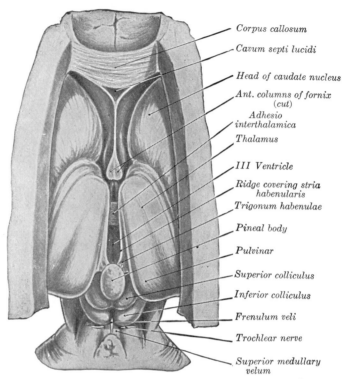

FIG. 11–55.—The thalami, exposed from above. The trunk and splenium of the corpus callosum, most of the septum lucidum, the body of the fornix, the tela choroidea with its contained plexuses and the epithelial roof of the third ventricle have all been removed.

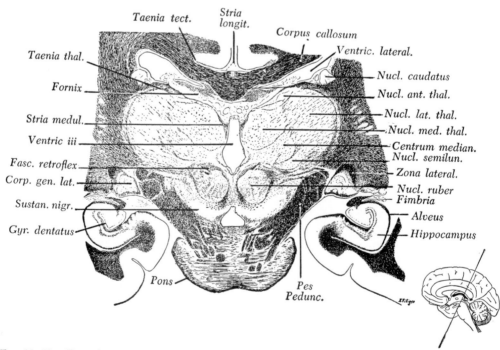

FIG. 11–56.—Frontal section through basal part of cerebral hemisphere and midbrain. Weigert myelin stain. (From Villiger-Addison, courtesy of J. B. Lippincott Co.)

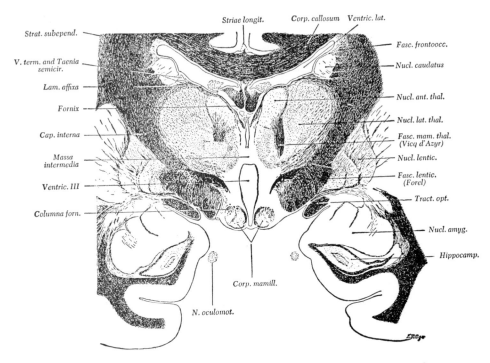

Striae longit.   *Corp. callosum*   *Ventric. lat.*

Strat. subepend.

V. term. and Taenia semicir.

Lam. affixa

Fornix

Cap. interna

Massa intermedia

Ventric. III

Columna forn.

Fasc. frontoocc.

Nucl. caudatus

Nucl. ant. thal.

Nucl. lat. thal.

Fasc. mam. thal. (Vicq d'Azyr)

Nucl. lentic.

Fasc. lentic. (Forel)

Tract. opt.

Nucl. amyg.

Hippocamp.

Corp. mamill.

N. oculomot.

FIG. 11–57.—Frontal section through thalamus and corpus striatum. Weigert myelin stain. (From Villiger-Addison, courtesy of J. B. Lippincott Co.)

anterior, medial, and lateral—by a white layer, the **internal medullary lamina.** The nuclei of which the thalamus is composed may be divided into the following groups: (1) Midline; (2) Anterior; (3) Medial; (4) Lateral; and (5) Posterior (Dekaban '53, Toncray and Krieg '46, Walker '38).

1. **Midline Nuclei.**—The thin periventricular gray substance under the ependyma of the third ventricle is continuous with the central gray around the aqueduct and in a number of brains bridges the cavity in the massa intermedia. It contains groups of cells and is connected with the hypothalamus and cortex serving a visceral function.

2. **Anterior Nuclei.**—There are three anterior nuclei, dorsal, ventral, and medial. All three receive connections with the mammillary body through the mammillothalamic tract (bundle of Vicq d'Azyr) and its fibers are projected to the gyrus cinguli. It is an olfactory center.

3. **Medial Nuclei.**—There are several nuclei in the medial group, but the function of all is not known. The medial dorsal nucleus has two main parts, ventromedial

and dorsolateral. The former is connected with the periventricular system and the hypothalamus; the latter has projections to area 9 (Brodmann) of the frontal cortex. The centrum medianum is a well-defined structure near the caudal end of the thalamus. Its connections are obscure and probably represent associations between other parts of the thalamus.

4. **Lateral nuclei** form a large mass rostral to the pulvinar between the external and internal medullary laminae. The anterior ventral nucleus receives fibers from the globus pallidus through the ansa lenticularis and fasciculus thalamicus and sends others to the corpus striatum. The lateral ventral nucleus receives the dentatorubrothalamic tract and reticulothalamic fibers. Its efferent fibers go to the motor cortex, areas 4 and 6 (Brodmann).

The posterior ventral nucleus has two important subdivisions. (*a*) The **nucleus ventralis posteromedialis** (*nucleus semilunaris, n. arcuatus*) is a crescentic mass ventral to the centrum medianum. It receives the secondary trigeminal tract and its axons go to the somesthetic area of the postcentral gyrus for the face, areas 3, 1, 2. (*b*) The **nucleus ventralis posterolateralis** is a large area, the terminus of the spinothalamic tract and medial lemniscus, and projects to the sensory cortex, areas 3, 1, 2 of Brodmann (Fig. 11–69).

The nucleus lateralis dorsalis and nucleus lateralis posterior are connected with other parts of the thalamus and the parietal lobe.

The **nucleus reticularis** is a thin plate between the external medullary lamina and the internal capsule. It is continuous with the reticular formation of the mesencephalon through the zona incerta. Its connections are similar to neighboring nuclei.

5. **Posterior Nuclei.**—The major nucleus in this group is the **pulvinar**. It may be divided into lateral, medial, and inferior parts. It is projected to the parietal lobe, particularly to regions near the somatic sensory, auditory, and visual projection areas. It receives fibers from the same cortical areas and from the neighboring thalamic nuclei. It is an association area.

The **Metathalamus** is made up of the geniculate bodies, a medial and a lateral (Fig. 11–58) on each side. The **medial geniculate body** (*corpus geniculatum mediale*) is a small oval tubercle between the pulvinar, colliculi, and cerebral peduncle. It is connected with the inferior colliculus by the inferior brachium (Fig. 11–58) and is a relay station between the latter and the auditory cortex.

The **lateral geniculate body** (*corpus geniculatum laterale*) is an oval elevation of the lateral aspect of the posterior end of the thalamus. It is connected with the superior colliculus by the superior brachium. Most of the fibers of the optic tract end here (Fig. 11–58) and are relayed to the visual cortex surrounding the calcarine fissure.

The **Epithalamus** comprises the trigonum habenulae, the pineal body, and the posterior commissure (Fig. 11–60).

The **trigonum habenulae** is a triangular area at the posterior end of the taenia thalami close to the pineal body (Fig. 11–58). It contains the habenular nucleus and receives fibers from the stria medullaris which runs with the taenia thalami over the surface of the thalamus. The **habenular commissure** (Fig. 11–60), connecting the habenulae of the two sides, lies just in front on the pineal body.

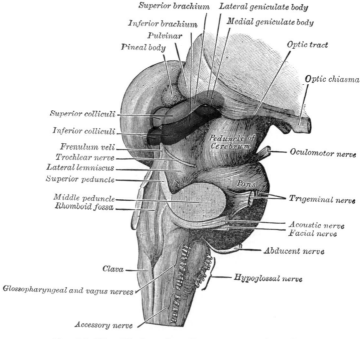

Fig. 11–58.—Hind- and midbrains; posterolateral view.

The **pineal body** (*corpus pineale*) (Figs. 11–55, 11–58) is a gland-like structure about 8 mm long, shaped like a pine cone, lying in a pocket between the superior colliculi, the pulvinar, and the splenium of the corpus callosum. It is attached by a stalk which is composed of two laminae, a superior and inferior which are separated by the pineal recess of the third ventricle. The inferior lamina is continuous with the posterior commissure, the superior with the habenular commissure (Gardner '53).

of fibers known as the fields H, H_1, and H_2 of Forel.

The **nucleus subthalamicus** (*corpus luysii*) is an oval, cylindrical mass lying on the dorsal surface of the transition from the internal capsule to cerebral peduncle. It receives fibers from the globus pallidus and peduncle as a part of the descending pathway from the corpus striatum.

The **zona incerta** is continuous with and similar to the reticular formation of the midbrain and the reticular nucleus of the

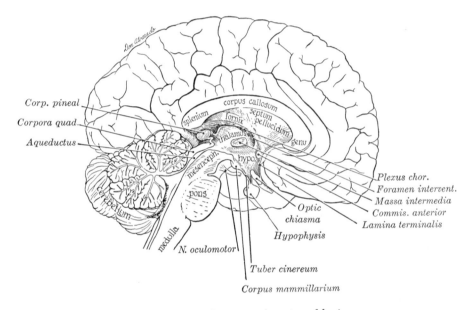

Corp. pineal

Corpora quad.

Aqueductus

corpus callosum

splenium   fornix   Septum pellucidum

thalamus   genu

mesenceph.   hypo.

Plexus chor.

Foramen intervent.

Massa intermedia

Commis. anterior

Lamina terminalis

pons

cerebellum

medulla

N. oculomotor

Optic chiasma

Hypophysis

Tuber cinereum

Corpus mammillarium

Fig. 11–59.—Median sagittal section of brain.

The **posterior commissure** (Fig. 11–60) is a rounded band of white fibers crossing the midline at the junction of the aqueduct with the third ventricle just anterior and superior to the superior colliculi. Some of its fibers connect these colliculi but the function of the rest is uncertain.

The **Subthalamus** (*tegmentum diencephali; ventral thalamus*) (Figs. 11–57, 11–60) is the transition zone between the thalamus (dorsal thalamus) and the tegmentum mesencephali. It is squeezed in between the cerebral peduncle and the mammillary area with the hypothalamus medial and rostral to it. The red nucleus and the substantia nigra are prolonged upward into it. It contains the nucleus subthalamicus and masses

thalamus. It has numerous connections, especially with the tectum, red nucleus, and tegmentum of the midbrain. It is a correlation center for optic and vestibular impulses relayed to the globus pallidus (Papez '42, '51).

Rostral to the red nucleus is a dense mass of longitudinally running fibers known as the **tegmental field** (*field H*) **of Forel**. The medial fibers are composed of dentato-, rubro-, and reticulothalamic connections. They continue rostrally as the **thalamic fasciculus**, field H_1 of Forel, which enters the medial part of the external medullary lamina and ends in the nucleus ventralis lateralis of the thalamus. The lateral fibers of the tegmental field come from field H_2

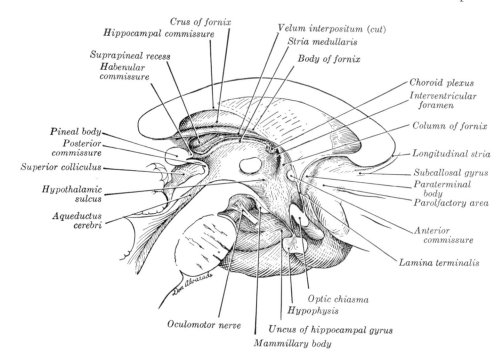

Crus of fornix
Hippocampal commissure
Velum interpositum (cut)
Stria medullaris
Suprapineal recess
Habenular commissure
Body of fornix
Choroid plexus
Interventricular foramen
Pineal body
Posterior commissure
Column of fornix
Superior colliculus
Longitudinal stria
Hypothalamic sulcus
Subcallosal gyrus
Paraterminal body
Parolfactory area
Aqueductus cerebri
Anterior commissure
Lamina terminalis
Optic chiasma
Hypophysis
Oculomotor nerve
Uncus of hippocampal gyrus
Mammillary body

FIG. 11–60.—Median sagittal section of brain stem and region of third ventricle, part of Figure 11–59 enlarged.

which is rostral and lateral. It is composed of fibers from the globus pallidus in the **ansa lenticularis** and **fasciculus lenticularis** which bend around the medial edge of the internal capsule on the way to the red nucleus and reticular substance.

The **Hypothalamus** (Figs. 11–59, 11–61) forms the ventral and rostral part of the wall of the third ventricle where it is marked off from the thalamus by the hypothalamic sulcus. It also includes the externally visible parts, (1) the corpora mammillaria, (2) tuber cinereum, (3) infundibulum, (4) hypophysis, and (5) optic chiasma.

The **mammillary bodies** (*corpora mammillaria*) are two round masses about 5 mm in diameter placed close to the midline near the cerebral peduncles. The **tuber cinereum** is the lamina of gray substance situated between the mammillary bodies and the optic chiasma (Figs. 11–34, 11–62). Laterally it is bounded by the optic tracts and cerebral peduncles. From its undersurface a hollow conical process, the **infundibulum** (Figs. 11–26, 11–34), projects downward and forward and ends in the posterior lobe

of the hypophysis cerebri or pituitary body (Figs. 11–60, 11–62). At the floor of the ventricle the infundibulum produces a raised area, the *median eminence*, and it narrows into the stem of attachment to the hypophysis. The median eminence, the stem, and the infundibular process or neural lobe make up an entity of similar tissue known as the **neurohypophysis** (see Hypophysis in last chapter). The **optic chiasma**, named from its resemblance to the Greek letter X, is situated at the junction of the floor and anterior wall of the third ventricle (Fig. 11–62). From the chiasma the optic nerves extend rostrally beside the hypophysis to enter the orbit. The optic tracts pass backward from the chiasma across the cerebral peduncles, between the anterior perforated substance and the tuber cinereum (Fig. 11–62), to the lateral geniculate bodies (Fig. 11–58).

Structure of the Hypothalamus.—The **periventricular area** (*nucleus paraventricularis*) connects that of the thalamus with the central gray around the aqueduct. It contains a few groups of small cells and fine

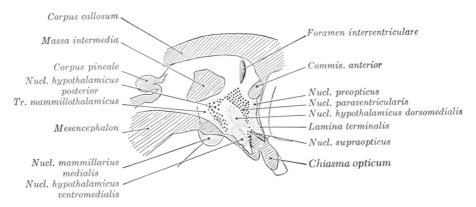

FIG. 11–61.—The nuclei of the hypothalamus in relation to the lateral wall of the third ventricle. (From Le Gros Clark '36, courtesy of Journal of Anatomy.)

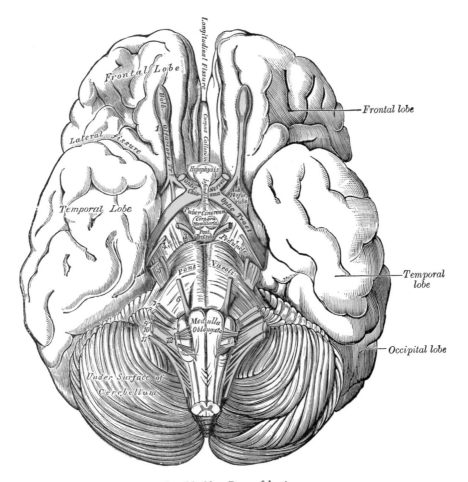

FIG. 11–62.—Base of brain.

fibers which pass to the nuclei of the pons and medulla through the dorsal longitudinal fasciculus (of Schütz) (Ingram '49).

The **supraoptic nucleus** consists of deeply staining moderately large cells and a rich capillary bed of particular pattern. Afferents come from the paraventricular nucleus and efferents constitute the supraopticohypophyseal tract.

Of the nuclei in the infundibular region, the **ventromedial hypothalamic nucleus** receives connections from the medial forebrain bundle, from the stria terminalis, from the periventricular system, from the fornix and from the globus pallidus. Its efferents enter the periventricular system. Closely related to this nucleus is the **dorsomedial nucleus** and the dorsal hypothalamic area.

The **posterior hypothalamic area** is a moderately extensive zone receiving fibers from the medial forebrain bundle, the fornix, the periventricular system and probably the tegmentum. It is thought to be the main origin of descending efferent hypothalamic systems which run rather diffusely scattered through the lateral parts of the tegmentum of the pons and medulla. They are largely short neuron chains and connect with the lateral and anterolateral columns of the cord.

The **mammillary region** contains the mammillary complex and scattered cells in the premammillary and supramammillary areas. The **medial mammillary nucleus** is an ovoid mass of cells producing the protuberance of the mammillary body. The **lateral mammillary nucleus** is smaller and fits into the angle between the medial nucleus and the base of the brain. The **nucleus intercalatus** is a separate group of slightly different cells in the dorsal part of the nucleus. Both medial and lateral nuclei receive fibers from the fornices of both sides, the fibers decussating just above the mammillary bodies. They also receive fibers from the medial forebrain bundle, from the thalamus and from the inferior mammillary peduncle from the tegmentum of the midbrain. The chief efferents from the mammillary nuclei form the principal mammillary peduncle which runs dorsorostrally to the thalamus as the **mammillothalamic tract** (bundle of Vicq d'Azyr). A short distance above the mammillary body descend-

ing collaterals are given off to form the **mammillotegmental tract.**

The **lateral hypothalamic area** extends from the lateral preoptic area to the tegmentum of the midbrain. It contains the **medial forebrain bundle,** scattered groups of cells, and two or three lateral tuberal nuclei (*nuclei tuberales*) (Crosby and Woodburne '51).

The **Preoptic Area** is derived from the telencephalon rostral to the hypothalamus, but it is closely associated with the latter functionally. The medial preoptic area is related to the middle region of the hypothalamus and rostrally to the nuclei associated with the stria terminalis, anterior commissure, and parolfactory areas. The lateral preoptic area contains the medial forebrain bundle and is distinguishable from the lateral hypothalamic area only by its position. The connections of both areas are with olfactory areas, the hypothalamus, and the anterior and medial thalamic nuclei.

The Third Ventricle

The **third ventricle** (*ventriculus tertius*) (Fig. 11–55) is a narrow median cleft between the thalami of the two hemispheres. It is lined by ependyma and filled with spinal fluid. The rostral end has an opening on each side into the lateral ventricles, the **interventricular foramen** (of Monro), and at the posterior end it is continuous with the fourth ventricle through the cerebral aqueduct (of Sylvius).

The lateral wall, its most extensive surface, is divided into an upper and a lower part by the **hypothalamic sulcus** (Fig. 11–60). The upper part contains the columns of the fornix and the thalamus, the lower the subthalamus and hypothalamus. In most brains the thalami of the two sides come into contact at their most prominent medial eminence. The resulting continuity of thalamic substance bridges the third ventricle as the **adhesio interthalamica** (*massa intermedia*) which appears like a commissure in a medial sagittal section (Fig. 11–61) but does not function as one.

The rostral boundary is formed by the lamina terminalis and anterior commissure. The part of the roof just above the anterior commissure is the inferior edge of the septum pellucidum. Posterior to this, the roof is above the interventricular foramen and continues back along the fornix to the splenium of the corpus callosum. It is composed of the **tela choroidea** or **velum interpositum** and the **choroid plexus.** The posterior boundary is formed by the habenular commissure, the pineal stalk, and the posterior commissure. Below the posterior commissure is the opening of the cerebral aqueduct. The floor of the ventricle is composed of the tegmentum of the cerebral peduncles and the structures of the hypothalamus.

Certain recesses make the outline of the ventricle irregular. Between the lamina terminalis and the optic chiasma is the optic recess. Into the stalk of the hypophysis is the infundibular recess and into the stalk of the pineal is the pineal recess. Above the habenular commissure is the suprapineal recess and above the anterior commissure between the columns of the fornix is a slight recess called the vulva cerebri or triangular recess.

THE TELENCEPHALON

Cerebrum.—The cerebrum is the large rounded structure occupying most of the cranial cavity. A deep median sagittal groove, the longitudinal cerebral fissure, divides it into right and left cerebral hemispheres. At the bottom of this fissure, the hemispheres are connected with each other by the great central white commissure, the corpus callosum (Figs. 11–59, 11–62). The internal structures of the hemispheres merge with those of the diencephalon and further continuity with the brain stem is established through the cerebral peduncles (Fig. 11–60).

The **Cerebral Hemispheres** are composed of (1) the extensive outer gray substance or cerebral cortex, (2) the underlying white substance, the centrum semiovale, (3) internally located masses of central gray substance separated by intervening white laminae, collectively known as the basal ganglia and corpus striatum, and (4) certain centrally and medially placed structures which are collectively known as the rhinencephalon.

CEREBRAL CORTEX

Each hemisphere presents a **convex** superior and lateral **surface** (*facies superolateralis cerebri*), a flat **medial surface** (*facies medialis cerebri*), and an irregular **inferior** or **basal surface** (*facies inferior cerebri*). A distinct medial or **superior margin** separates the lateral and medial surfaces; a more rounded **lateral margin** separates the lateral and inferior surfaces. The anterior end is the **frontal pole,** the posterior end the **occipital pole,** and the anterior end of the laterally protruding temporal lobe is the **temporal pole.** The surface of the hemisphere is marked by numerous irregular grooves, the fissures and sulci, with intervening rounded eminences, the convolutions or gyri. At first glance they appear quite irregular, but they can, with study, be fitted into a basic plan. The two hemispheres of the same brain differ from each other and they in turn from every other brain.

The **Sulcus Lateralis** (*lateral cerebral fissure; fissure of Sylvius*) (Fig. 11–63) is the well-marked groove on the lateral surface separating the larger superior masses of the frontal and parietal regions from the temporal region of the hemisphere. It is a complete fissure, isolating much of the temporal lobe and in the depths of its posterior portion is hidden the insula. At the tip of the temporal pole, it is widened into the lateral fossa. A short anterior part of the sulcus is called its **stem,** the long backward extension (about 7 cm) is called the **posterior ramus.** At the junction of the stem and posterior ramus two short branches (about 2 cm) form a V-shaped extension into the frontal lobe, the **anterior horizontal** (*ramus anterior*) and **anterior ascending** (*ramus ascendens*) **rami** (Fig. 11–63).

The **Central Sulcus** (*fissure of Rolando; sulcus centralis*) (Figs. 11–63, 11–65) begins on the medial surface of the hemisphere at about the middle of the superior margin and runs downward and forward on the lateral surface until it almost meets the lateral cerebral sulcus about 2.5 cm posterior to the anterior ascending ramus. It has two sinuous curves, a superior genu with convexity posteriorly, and an inferior genu with convexity anteriorly (Fig. 11–63).

The **Calcarine Sulcus** (*sulcus calcarinus;*

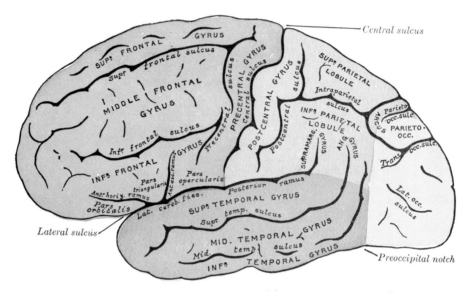

FIG. 11–63.—Lateral surface of left cerebral hemisphere, viewed from the side.

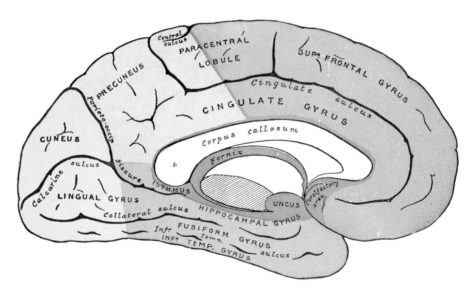

FIG. 11–64.—Medial surface of left cerebral hemisphere. Lobes in color.

calcarine fissure) (Fig. 11–64) is on the medial surface at the posterior part of the hemisphere. It begins anteriorly just below the splenium of the corpus callosum and runs with an upward arch toward the occipital pole. It is divided into an anterior and a posterior part by its junction with the parietoöccipital sulcus. The anterior part is called a complete fissure, its depth causing an elevation in the wall of the posterior horn of the lateral ventricle, the calcar avis (Figs. 11–87, 11–90).

The **Parietoöccipital Sulcus** (*sulcus parietooccipitalis; parietoöccipital fissure*) (Fig. 11–64), on the medial surface of the occipital region, appears to be an upward continuation of the anterior part of the calcarine sulcus. It crosses the superior margin of the hemisphere about 5 cm from the occipital pole and has a short (1.25 cm) lateral segment. It usually covers a submerged gyrus.

The **Cingulate Sulcus** (*sulcus cinguli*) (Fig. 11–64), on the medial surface of the hemisphere, follows the curve of the superior margin from the rostrum of the corpus callosum to a point just posterior to the central fissure. Here it turns upward and just crosses the superior margin.

The **Collateral Sulcus** (*sulcus collateralis; collateral fissure*) (Fig. 11–64) is seen on the inferior or tentorial surface of the hemisphere. Its course is parallel with the inferior lateral margin, from near the occipital pole almost to the temporal pole.

The **Sulcus Circularis** (Fig. 11–67) is buried in the depths of the lateral cerebral fissure and surrounds the insula under the frontal, parietal, and temporal opercula.

Lobes of the Hemispheres (Fig. 11–63).— The cerebral cortex is divided into the following lobes: frontal, occipital, temporal, parietal, and central (insular) lobes.

The **Frontal Lobe** (*lobus frontalis*) is the largest, occupying part of the lateral, medial, and inferior surfaces. It extends posteriorly to the central sulcus and inferiorly to the lateral fissure.

On the lateral surface are three sulci (Fig. 11–63); the **precentral sulcus** and the **superior** and **inferior frontal** sulci running from the precentral toward the frontal pole. There are four gyri: precentral and superior, middle, and inferior frontal gyri.

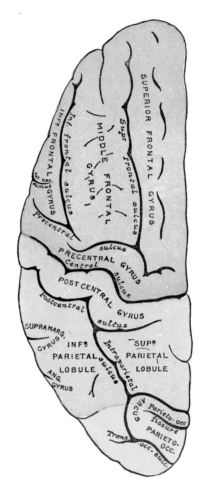

Fig. 11–65.—The lateral surface of left cerebral hemisphere, viewed from above.

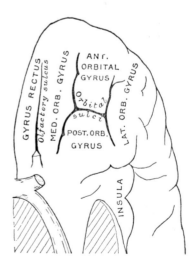

Fig. 11–66.—Orbital surface of left frontal lobe.

The **precentral gyrus** lies between the **central** and **precentral sulci,** extending from the lateral sulcus to the superior margin of the hemisphere. The **superior frontal gyrus** lies superior to the **superior frontal sulcus** and spreads over the superior margin of the hemisphere to its medial surface. It is usually subdivided by a longitudinal interrupted sulcus, the **paramedian sulcus.** The **middle frontal gyrus** lies between the superior and inferior frontal sulci. It also is subdivided by a longitudinal sulcus, the middle frontal sulcus, which bifurcates widely at its anterior extremity.

The **inferior frontal gyrus** lies below the inferior frontal sulcus and spreads around the lateral margin onto the orbital part of the inferior surface of the hemisphere. It is subdivided by the rami of the lateral sulcus (Fig. 11–63) into an anterior, a triangular, and a basal or opercular part. The left inferior frontal gyrus is usually more highly convoluted than the right and is referred to as **Broca's speech area.**

The inferior or orbital surface is divided by an H-shaped **orbital sulcus** into four gyri; medial, anterior, lateral, and posterior **orbital gyri** (Fig. 11–66). The olfactory sulcus, a groove which lodges the olfactory tract, separates the **gyrus rectus** from the medial orbital gyrus.

The medial surface of the frontal lobe is limited posteriorly by a line drawn downward and anteriorly to the corpus callosum from the point where the central sulcus cuts the superior margin. It includes the superior frontal (marginal) gyrus and part of the **cingulate gyrus.** The posterior part of the superior frontal gyrus continuing somewhat into the parietal lobe may be marked off as the **paracentral lobule.**

The **Occipital Lobe** (*lobus occipitalis*) (Fig. 11–63) occupies a comparatively small pyramidal portion at the occipital pole, with medial, lateral, and inferior surfaces. The medial surface (Fig. 11–64) is bounded anteriorly by the parietoöccipital sulcus and a line extending it to a shallow indentation of the lateral margin named the **preoccipital notch.** The posterior calcarine sulcus divides the medial surface of the lobe into a wedge-shaped area, the **cuneus,** and the **lingual gyrus.**

The anterior boundary of the lateral surface is a line from the lateral part of the parietoöccipital sulcus to the preoccipital notch (Fig. 11–63). A lateral occipital sulcus divides the lateral surface into a **superior** and **inferior occipital gyrus.** Part of the arcus parietoöccipitalis lies between the parietoöccipital sulcus and the transverse occipital sulcus.

The inferior or tentorial surface of the occipital lobe (Fig. 11–64) is limited by an imaginary transverse line across the preoccipital notch. The posterior segment of the collateral sulcus separates the adjacent parts of the lingual and **fusiform gyri.**

The **Temporal Lobe** (*lobus temporalis*) (Fig. 11–63) is separated from the frontal lobe by the lateral sulcus, from the occipital lobe by the line from the preoccipital notch to the parietoöccipital sulcus and from the parietal lobe by a line from the posterior extremity of the lateral sulcus toward the occipital pole.

The lateral surface of the lobe (Fig. 11–63) is subdivided into three parallel gyri, superior, middle, and inferior temporal gyri, by two sulci, superior (parallel sulcus) and middle temporal sulci.

The superior surface is hidden and extends deeply into the hemisphere as the inferior limit of the lateral sulcus. It overlaps the insula as the **temporal operculum** (Fig. 11–67). Three or four gyri run from the posterior end of the lateral sulcus obliquely outward and forward, they are the **transverse temporal gyri,** one of which is the cortical center for hearing.

The inferior surface of the temporal lobe (Fig. 11–64) is continuous with the tentorial surface of the occipital lobe, the separation being marked by the preoccipital notch. It is transversed by the inferior temporal sulcus and gyrus, the **fusiform gyrus,** and the collateral sulcus. On the medial side of the collateral sulcus are the lingual gyrus posteriorly and the hippocampal gyrus anteriorly.

The **hippocampal gyrus** (*gyrus parahippocampalis*) lies between the collateral sulcus and the hippocampal sulcus. Posteriorly it is continuous above with the cingulate gyrus through the isthmus (Fig. 11–64) and below with the lingual gyrus.

Its anterior extremity is curved into a hook, the **uncus** (Fig. 11–64), which is part of the rhinencephalon and is separated from the temporal pole by the incisura temporalis.

The **Hippocampal Sulcus** extends from the splenium of the corpus callosum to the inside of the hook of the uncus. It is a complete sulcus and gives rise to the prominence of the hippocampus in the inferior horn of the lateral ventricle.

the superior lobule, the **arcus parietoöccipitalis** encircles the end of the parietoöccipital sulcus. In the inferior parietal lobule, the **supramarginal gyrus** arches over the upturned end of the lateral sulcus (Fig. 11–63). The **angular gyrus** arches over the upturned end of the superior temporal sulcus.

On the medial surface, the parietooccipital sulcus separates the parietal from

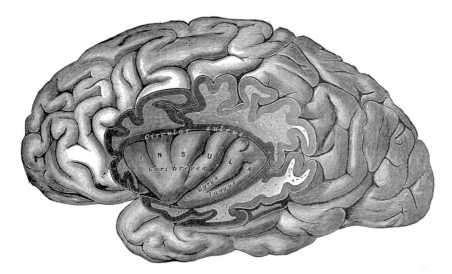

Fig. 11–67.—The insula of the left side, exposed by removing the opercula.

The **Parietal Lobe** (*lobus parietalis*) (Fig. 11–63) occupies parts of the lateral and medial surfaces of the hemisphere. On the lateral surface the central sulcus separates it from the frontal lobe, the parietooccipital sulcus from the occipital lobe and an imaginary line from the posterior ramus of the lateral sulcus toward the occipital pole separates it from the temporal lobe. The **postcentral sulcus** runs parallel with the central sulcus, leaving between them the **postcentral gyrus** (*gyrus centralis posterior*). The sulcus may be in two parts, an upper and a lower. From about the middle of the postcentral sulcus the **intraparietal sulcus** extends posteriorly toward the occipital pole and joins the transverse occipital sulcus. The part of the lobe posterior to the postcentral sulcus is divided by the horizontal intraparietal sulcus into a **superior** and an **inferior parietal lobule**. In

the occipital lobe. The anterior boundary is an imaginary line from the end of the central sulcus toward the middle of the corpus callosum. The upturned end of the sulcus cinguli divides the arc into a **precuneus** or **quadrate lobule** and a **paracentral lobule**.

The **Insula** (*central lobe*) lies hidden in the depths of the lateral sulcus and can be seen only if the lips of the sulcus are bent back or cut away (Fig. 11–67). These lips are parts of three lobes, frontal, parietal, and temporal. They are separated by the rami of the lateral sulcus, and are named the frontal, parietal, and temporal opercula. The frontal operculum may also be described in two parts, an orbital and a frontal operculum, separated by the anterior horizontal ramus of the lateral sulcus. The insula is encircled and separated from the opercula by a deep circular sulcus. When the opercula are cut away, the insula appears as a triangular area with an apex, the **limen insulae,** directed toward the anterior perforated substance. It is crossed by the gyrus longus and gyri breves (Frontera '56).

Structure of the Cerebral Cortex

It is customary to separate the cortex or pallium into the olfactory cortex, called the archipallium because it is phylogenetically old, and the non-olfactory younger cortex called the neopallium. The archipallium is described as part of the rhinencephalon. In addition to the distinction based upon olfaction and topographic position, it is possible to make a distinction on the basis of structure. The neopallium is called the isocortex because it all has a similar basic pattern in the distribution of its cellular and fibrous layers. The olfactory cortex is called the allocortex because its parts have quite different patterns.

Layers of the Isocortex (Fig. 11–68).— The isocortex is divided into six zones or layers:

Layer I. The outermost layer is the **plexiform layer** of Cajal (*lamina zonalis; molecular layer*). It is a narrow zone of white substance composed largely of myelinated fibers running tangentially. It contains the terminations of the apical dendrites of the pyramidal cells of deeper layers, horizontal cells, and granule cells with short axons.

Layer II. The **external granular layer** (*lamina granularis externa*) contains the shafts of pyramidal cells, and large numbers of granule cells (Golgi type II cells) with short axons and small pyramidal cells.

Layer III. The **layer of pyramidal cells** (*lamina pyramidalis*) has two zones, an outer with medium-sized cells and an inner with larger pyramidal cells.

Layer IV. The **internal granular layer** (*lamina granularis interna*) contains a large number of small stellate cells with short axons and scattered small pyramidal cells. It is the site of the fibers of the outer band of Baillarger and contains the synapses with pyramidal cells of layer III.

Layer V. The **ganglionic layer** (*lamina ganglionaris*) includes the inner layer of large pyramidal cells and the horizontal fibers of the inner band of Baillarger. This layer contains the giant cells of Betz in the precentral cortex.

Layer VI. The **polymorphic layer** (*lamina multiformis*) contains cells of irregular shape whose axons enter the subjacent white substance of the hemisphere.

Functional Localization

Special Types of Cerebral Cortex.—There is much evidence that individual areas of the human cerebral cortex do not function independently without the influence of other areas or of even the entire hemisphere. Nevertheless, study of the cellular components (cytoarchitectonics), or the arrangement of fibers (myeloarchitectonics) of different areas, partial extirpation at operation, disease, electrical stimulation under local anaesthesia, and animal experimentation have made it possible to give a functional identification to certain areas. Brodmann ('09) mapped forty-seven different areas as the result of a cytoarchitectonic survey with the Nissl stain for cells (Fig. 11–69). The Vogts ('09) described over two hundred areas on the basis of differences in the pattern made by the myelinated fibers. Von Economo ('29) has mapped out the cortex into areas containing five fundamental types of architecture. These maps are useful for topographic purposes but are only partly of value for functional identification (Von Bonin '50, Brody '55).

The **precentral** cortex or **motor area** includes the precentral gyrus and the posterior part of the frontal gyri. It has received particular attention because its stimulation with electrodes causes contraction of the voluntary muscles. It corresponds to areas 4 and 6 of Brodmann (Fig. 11–69) and is characterized histologically by the absence of the granular layer, and the presence of large pyramidal cells in layer V.

The motor area (area 4, Fig. 11–69) contains the **giant pyramidal cells of Betz** in the ganglionic layer. It was thought formerly that these cells were responsible for the fibers of the pyramidal tract and that they controlled voluntary motion, but other pyramidal cells contribute extensively to the tract and normal voluntary activity

I. *Plexiform layer*

II. *External granular layer*

III. *Pyramidal cell layer*

IV. *Internal granular layer*

V. *Ganglionic layer*

VI. *Polymorphic layer*

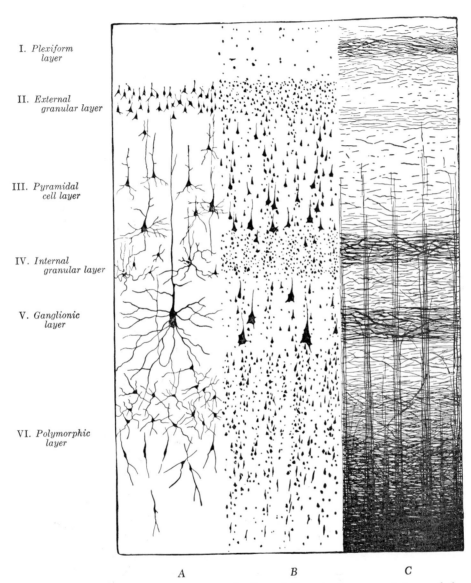

A   B   C

Fig. 11–68.—A diagram showing the layers of cells and fibers in the gray substance of the cortex of the human cerebral hemisphere, according to the histological methods of Golgi, Nissl and Weigert. *A*, Stained by the method of Golgi; *B*, by that of Nissl; *C*, by that of Weigert. (After Brodmann: from Luciani's *Physiology*, Macmillan & Co., Ltd.)

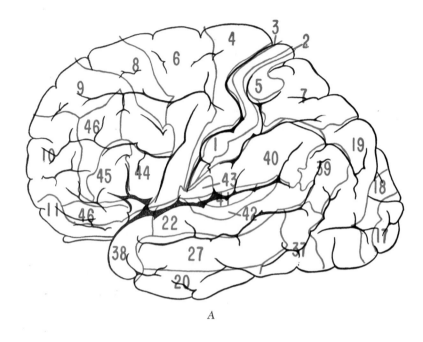

A

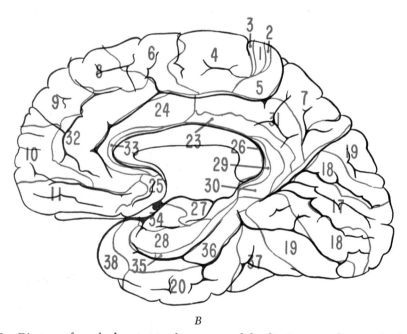

B

Fig. 11–69.—Diagram of cerebral cortex to show areas of localization according to Brodmann '09. A, Lateral aspect of hemisphere; B, Medial.

requires associations with other parts of the cortex. It is an important motor area, nevertheless, because its removal causes paralysis of voluntary muscles, especially of the opposite side of the body. The different parts of the body are represented in sequence along the central sulcus. The lower limb area is at the superior margin of the hemisphere, the hand is half way around on the convex surface, and the face

jugate movements of the eyes or of the eyes and head. The premotor area gives rise to some of the fibers making up the group heterogeneously called extrapyramidal fibers.

Suppressor zones have been described in monkeys and other animals (McCulloch '44) in areas 6, 8, 2, 19, and 24. When electrical stimulation is applied to one of them, electrical evidence of cortical activity

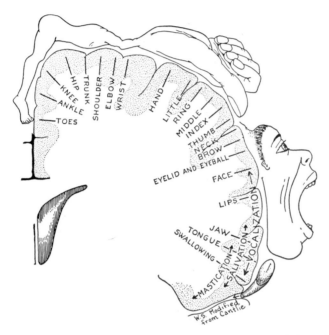

Fig. 11–70.—Motor homunculus illustrating motor representation in area 4 (anterior central gyrus). (After Penfield and Rasmussen, *Cerebral Cortex of Man*, The Macmillan Co.)

and mouth are at the lateral sulcus. The size of the areas varies, those parts of the body with delicate complicated movements such as the hand being greater (Fig. 11–70).

The **premotor** or **precentral area** (area 6, Fig. 11–69) is like the motor area (area 4) except that it lacks the giant cells of Betz. The two areas stand in intimate relationship to each other functionally. The evidence indicates that motor responses elicited by stimulation of this area are produced by transmission through area 4 and involve larger groups of muscles in more complicated acts. In front of area 6 is area 8, electrical stimulation of which evokes con-

is suppressed in that area and gradually spreads throughout the cortex, taking as long as a half hour. Stimulation of area 8 may cause cessation of muscular contractions and suppress further motor response while the stimulus is applied. The connections involved are imperfectly understood.

The **frontal area** extending rostralward from the precentral area differs from the latter histologically in having the granular layer restored. The layer of large pyramidal cells and the polymorphous layer are reduced. This area contains extensive associations with other parts of the cortex and with the thalamus. The surgical operation of *lobotomy*, which isolates the area from the rest of the brain, especially the thalamus, has been used in the treatment of severe psychoses with generally favorable results (Freeman and Watts '48). Experimental evidence has also shown that stimulation of this area causes autonomic effects on the circulation, respiration, pupillary reaction, and other visceral activity (Beach *et al.* '55, Krieg '54).

layer, the breadth of the outer line of Baillarger and the large pyramidal cells. In the caudal part the large pyramidal cells are reduced in size and number and the granular layer is wider but not so dense.

The postcentral area receives exteroceptive and proprioceptive afferent fibers from the spinal cord and brain stem relayed through the thalamus. The different parts of

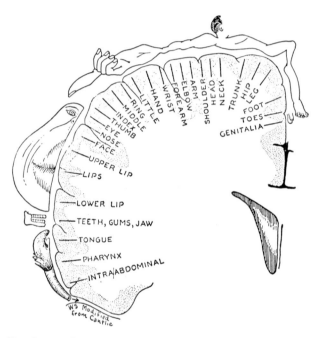

FIG. 11–71.—Sensory homunculus showing representation in the sensory cortex (after Penfield and Rasmussen, *Cerebral Cortex of Man*, The Macmillan Co.).

A **speech area** (*Broca's area*) is located in the frontal operculum (area 44, Fig. 11–69). It is better developed in the left hemisphere of right-handed persons and its destruction results in the inability to make articulate speech although the vocal organs are intact. This is called motor aphasia and the area is considered in consequence to be the motor speech area.

The **postcentral area** (areas 3, 1, 2. Fig. 11–69) is the sensory or conscious somesthetic area. It occupies all but the lowest part of the postcentral gyrus on the convex surface of the hemisphere and extends into the adjoining paracentral lobule of the medial surface. The rostral part, area 3, is characterized by the density of the granular

the body are represented in a sequence similar to that of the precentral area. The area for the lower part of the body is at the superior margin of the hemisphere and the face and mouth at the lateral sulcus. The more sensitive parts of the body, such as the face and mouth, have a larger area (Fig. 11–71).

The **visual sensory** or **striate area** (area 17, Fig. 11–69) occupies the walls of the posterior part of the calcarine sulcus, extending into the cuneus above, the lingual gyrus below, and around the occipital pole to the lateral surface of the hemisphere. It has three distinguishing histological features: (1) The outer band of Baillarger is broad, prominent, and visible to the naked eye. It is named the stria of Gennari and gives the name striate area to this part of the cortex. (2) The outer layer of large pyramidal cells has large stellate cells instead of pyramidal cells. (3) The inner layer of large pyramidal cells contains the solitary cells of Meynert. It is the cortical center for vision, receiving fibers from the lateral geniculate body. Color, size, form, motion, illumination, and transparency are all recognized and determined in this area.

Owing to the decussation of part of the fibers of the optic nerves, the visual cortex of one hemisphere receives impressions from the temporal part of the retina of the same side and the nasal part of the opposite side. The upper lip of the calcarine sulcus is associated with the upper quadrants of the retina, the lower lip with the lower quadrants. The macula is represented in the occipital pole and clinical evidence indicates that it is bilateral (Fulton '49).

The **visual psychic** or **parastriate area** (areas 18 and 19) surrounds the visual sensory area and is histologically similar to it. This area is responsible for elaboration of visual impressions and association of them with past experience for recognition and identification. It also relates eye movements to visual impressions.

The **auditory sensory area** (areas 41 and 42, Fig. 11–69) occupies the transverse temporal gyri and part of the superior temporal gyrus. It is characterized histologically by having an unusually large number of giant cells in the outer layer of large pyramidal cells, and by having great numbers of fibers. It receives fibers from the medial geniculate body through the auditory radiation of the sublentiform limb of the internal capsule. Auditory impressions reach this area as sounds which can be differentiated for loudness, quality, and pitch.

The **auditory psychic area** (area 22) occupies the superior temporal gyrus and surrounds the auditory sensory area and has a smaller number of giant cells. In this area auditory impressions are interpreted with respect to their probable source and associated with past experience.

The **parietal area** (areas 5, 7, 39 and 40) is situated between the visual, auditory, and somatic areas. It is characterized by the

absence of large pyramidal cells and by the breadth of the inner band of Baillarger. It correlates and blends impressions from the surrounding sensory areas.

WHITE SUBSTANCE OF THE CEREBRAL HEMISPHERE

The outer layer of the hemisphere, the gray matter or cortex, has been described. The internal portion, except for the central gray nuclei of the basal ganglia, is composed of white matter and is called the **centrum semiovale.**

The fibers of the centrum are divided into three classes: (1) projection fibers which connect the cortex with the more caudal parts of the brain and the spinal cord; (2) commissural fibers which connect the two hemispheres, and (3) association fibers which connect the different parts of the cortex.

(1) **Projection fibers** are both afferent and efferent fibers which pass through the corona radiata and internal capsule (Fig. 11–72).

(2) The **Commissural fibers** pass through (*a*) the corpus callosum, (*b*) the anterior commissure and the hippocampal commissure, commissura fornicis described as part of the rhinencephalon, and (*c*) the posterior commissure, described with the diencephalon.

(*a*) The **Corpus Callosum** (Figs. 11–55, 11–56, 11–74) is the great central white commissure connecting the two hemispheres. Its size and shape can be seen on a median sagittal section (Fig. 11–59). At the anterior end is the **genu** where it is bent back on itself around the anterior end of the septum pellucidum. The inferior limb of the genu tapers rather rapidly as the **rostrum** which is continuous with the lamina terminalis at the anterior commissure. The greater part of the corpus callosum extending posteriorly from the genu is called the **body** or **trunk** (*truncus corporis callosi*). The posterior end is enlarged into the thick, convex, free edge, the **splenium,** which overlaps the tela choroidea of the third ventricle and overhangs the epiphysis and midbrain.

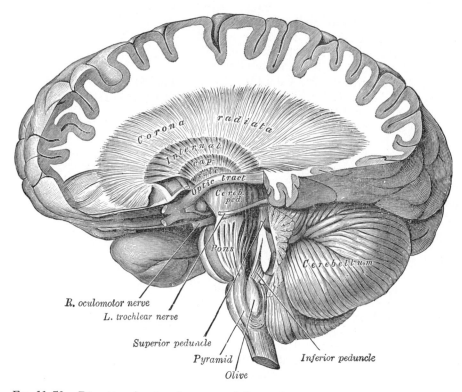

Fig. 11–72.—Dissection showing the course of the cerebrospinal fibers. (E. B. Jamieson.)

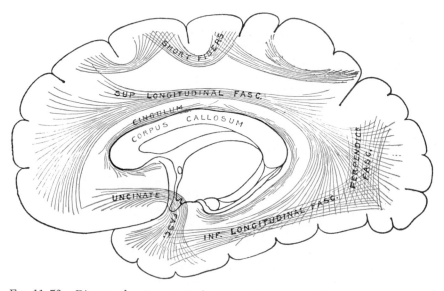

Fig. 11–73.—Diagram showing principal systems of association fibers in the cerebrum.

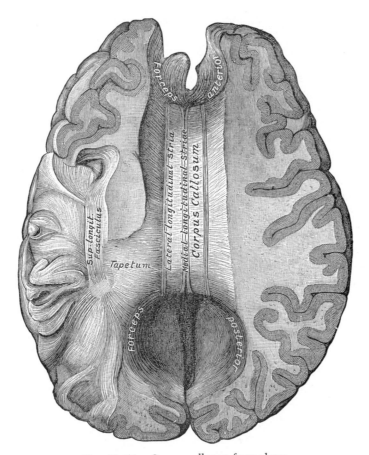

FIG. 11–74.—Corpus callosum from above.

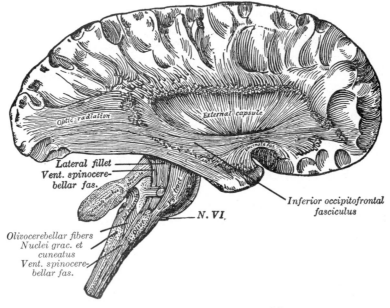

FIG. 11–75.—Deep dissection of cortex and brain stem.

The superior surface of the corpus callosum, at the bottom of the longitudinal cerebral fissure, is about 2.5 cm wide (Fig. 11–74). It is overlapped on both sides by the gyrus cinguli from which it is separated by the slit-like callosal sulcus (Fig. 11–64). It is covered by a thin layer of gray matter, the **indusium griseum** (*supracallosal gyrus*) and the white strands of the medial and lateral longitudinal striae (page 861). The inferior surface in the midline is attached to the septum pellucidum anteriorly and to the body of the fornix posteriorly. The lateral portions of the inferior surface form the roof of the lateral ventricles.

The fibers of the corpus callosum mingle with the projection fibers of the corona radiata as they radiate out to the cortex. The fibers to the frontal lobe from the genu of the corpus callosum constitute the **forceps anterior** (*forceps minor*) (Fig. 11–73) and those from the splenium to the occipital lobe the **forceps posterior** (*forceps major*). The fibers from the remaining body of the corpus callosum are called the **tapetum**.

(3) The **Association fibers** are of two kinds, short association fibers connecting adjacent gyri and long association fibers connecting more distant parts of the hemisphere.

The short association fibers lie immediately beneath the gray matter of the cortex (Fig. 11–74).

The long association fibers form more or less distinct bundles and are given the following descriptive names:

(*a*) The **uncinate fasciculus** curves around the lateral cerebral sulcus (fissure) of Sylvius and unites the frontal lobe with the anterior end of the temporal lobe.

(*b*) The **cingulum** is a fasciculus contained within the cingulate gyrus (Fig. 11–64). From the anterior perforated substance rostrally it follows the curve around the corpus callosum and ends in the hippocampal gyrus.

(*c*) The **superior longitudinal fasciculus** lies above the lentiform nucleus and the insula and extends from the frontal lobe anteriorly to the occipital lobe posteriorly and the temporal lobe inferiorly.

(*d*) The **inferior longitudinal fasciculus** connects the temporal and occipital lobes,

running along the lateral walls of the inferior and posterior horns of the lateral ventricle.

(*e*) The **superior occipitofrontal fasciculus** passes between the lateral aspect of the caudate nucleus and the corona radiata connecting the frontal with the occipital and temporal lobes.

(*f*) The **inferior occipitofrontal fasciculus** connecting the frontal lobe with the occipital lobe passes between the lentiform nucleus and the uncinate fasciculus at the base of the external capsule (Fig. 11–75).

(*g*) The **perpendicular fasciculus** runs vertically through the rostral part of the occipital lobe, connecting the inferior parietal lobule with the fusiform gyrus (Figs. 11–64, 11–73).

(*h*) The **fornix** and other parts of the rhinencephalon contain association fibers (page 862).

BASAL GANGLIA AND CORPUS STRIATUM

The **basal ganglia** represent the central gray matter of the telencephalon. They lie between the thalamus of the diencephalon, and the centrum semiovale or white matter of the hemisphere. They include the caudate nucleus, the lentiform nucleus, the claustrum, and the amygdaloid body.

The **corpus striatum** includes the caudate and lentiform nuclei, and a broad band of white fibers separating them, the internal capsule. Its name is derived from the striped appearance produced by narrow strands of gray matter crossing the white matter between the two nuclear masses (Fig. 11–78).

The **Caudate Nucleus** (*nucleus caudatus; caudatum*) (Fig. 11–77) is a prominent feature of the lateral ventricle (Figs. 11–79, 11–87) and forms an arch following its curvature. The anterior portion or head is enlarged and occupies most of the lateral wall of the anterior horn of the ventricle. Its narrow posterior portion, the tail, is separated from the thalamus by the stria terminalis and terminal vein (Fig. 11–87). Its rostral extremity is continuous with the putamen (Figs. 11–77, 11–79) and its caudal extremity ends in the amygdaloid complex (Fig. 11–77).

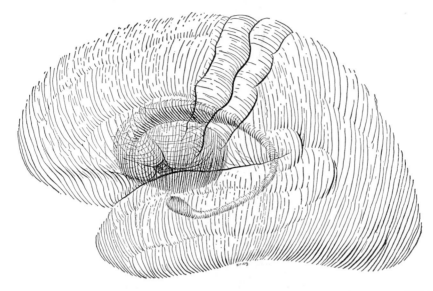

FIG. 11–76.—Phantom of the corpus striatum within the cerebral hemisphere. (Courtesy of W. J. Krieg's *Functional Neuroanatomy,* 2nd Ed. 1953. Blakiston Div., McGraw-Hill Book Co.)

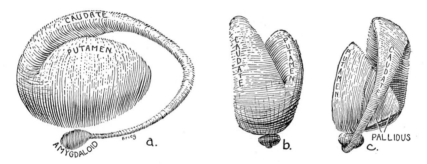

FIG. 11–77.—The corpus striatum of the left side with the internal capsule omitted; (*a*) lateral aspect; (*b*) rostral aspect; (*c*) caudal aspect. (Courtesy of W. J. Krieg's *Functional Neuroanatomy,* 2nd Ed. 1953; Blakiston Div., McGraw-Hill Book Co.)

The **Lentiform Nucleus** (*nucleus lentiformis; lenticular nucleus*) (Fig. 11–81) is a wedge-shaped mass of gray substance lying lateral to the thalamus and caudate nucleus (Fig. 11–78). It is divided by a layer of white matter, the lateral medullary lamina, into two parts: the putamen, which is larger and more lateral, and the globus pallidus.

The **Putamen** (Figs. 11–77, 11–80) is separated from the caudate nucleus by the anterior limb of the internal capsule except at its rostral extremity where the two nuclei are joined. Its lateral surface conforms roughly to the insula but it is separated from the latter by the external capsule and claustrum (Fig. 11–78).

The **Globus Pallidus** (Fig. 11–80) is separated from the putamen by the medial lamina and from the thalamus by the posterior limb of the internal capsule. It is divided into two parts, a medial and a lateral, by the medullary lamina.

The **Claustrum** (Fig. 11–78) is a thin layer of gray matter between the putamen and the insula. It is separated from the putamen by a lamina of white matter, the **external capsule** (*capsula externa*). It is

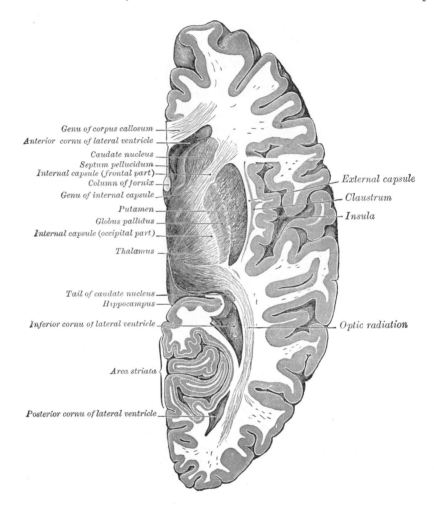

Genu of corpus callosum
Anterior cornu of lateral ventricle
Caudate nucleus
Septum pellucidum
Internal capsule (frontal part)
Column of fornix
Genu of internal capsule
Putamen
Globus pallidus
Internal capsule (occipital part)
Thalamus
Tail of caudate nucleus
Hippocampus
Inferior cornu of lateral ventricle
Area striata
Posterior cornu of lateral ventricle
External capsule
Claustrum
Insula
Optic radiation

FIG. 11–78.—Horizontal section of right cerebral hemisphere.

regarded as a detached portion of the insula. Its lateral surface is somewhat irregular, following the gyri and sulci of the insula (Fig. 11–78), and is separated from the outer gray cortex of the insula by a layer of white fibers, the **capsula extrema**, which corresponds to the cortical band of Baillarger.

The **Amygdaloid Body** (*corpus amygdaloideum; amygdaloid nucleus; amygdala*) (Fig. 11–56) is an ovoid mass of gray matter about the size and shape of an almond in the roof of the rostral end of the lateral ventricle. It is closely associated with the cortex covering it, the uncus of the gyrus hippocampi. The tail of the caudate nucleus ends in it and it is in contact with a part of the putamen and the claustrum. It is divided by delicate white laminae into several groups of nuclei.

The **Internal Capsule** (*capsula interna*) (Figs. 11–72, 11–81) is a thick lamina of white matter situated between the thalamus, caudate, and lentiform nuclei. Between the caudate nucleus and thalamus it is bent almost into a right angle. The angle is called the **genu** (*genu capsulae internae*). The portion rostral to the genu, the **anterior limb** (*crus anterior*) is between the caudate nucleus and the lentiform nucleus. The portion caudal to the genu, the **posterior limb** (*crus posterius*), is between the thalamus and the lentiform nucleus. The part of the posterior limb which curves around the caudal end of the lentiform nucleus is called the **retrolenticular part** (*pars retro-*

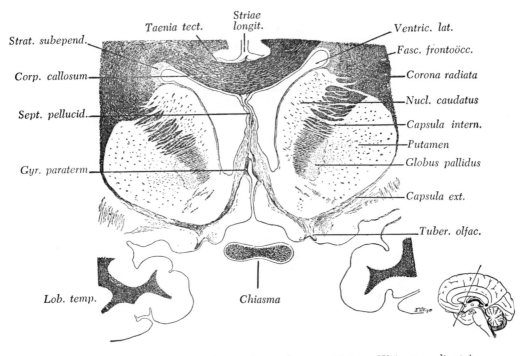

FIG. 11–79.—Frontal section through rostral part of corpus striatum. Weigert myelin stain.
(From Villiger-Addison, courtesy of J. B. Lippincott Co.)

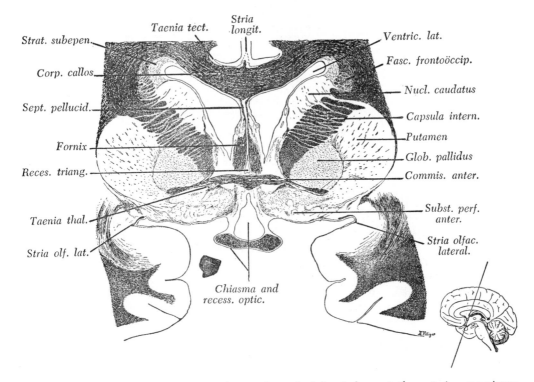

FIG. 11–80.—Frontal section through basal part of cerebral hemisphere at the anterior commissure.
Weigert myelin stain. (From Villiger-Addison, courtesy of J. B. Lippincott Co.)

lentiformis); and the part which lies ventral to the lentiform nucleus is the *pars sublentiformis.*

The **anterior limb** of the internal capsule contains: thalamocortical fibers from the lateral nucleus of the thalamus to the frontal lobe; corticothalamic fibers from the frontal lobe to the thalamus; frontopontine fibers from the frontal lobe to the nuclei of the pons; probably collaterals from the above tracts to the caudate and putamen; and fibers from the caudate to the putamen. The **genu** contains in addition to corticothalamic

optic radiation (geniculocalcarine tract) from the lateral geniculate body to the visuosensory area of the occipital lobe; fibers from the cortex to the superior colliculus; corticothalamic fibers from the temporal and occipital lobes to the lateral nucleus of the thalamus.

The fibers of the internal capsule radiate widely as they pass to and from the various parts of the cerebral cortex, forming the corona radiata (Fig. 11–72) and intermingling with the fibers of the corpus callosum.

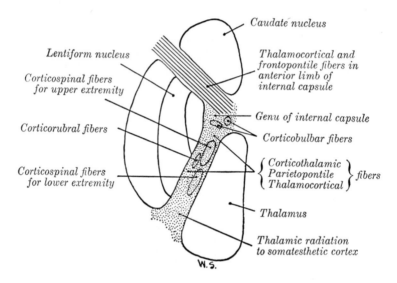

Fig. 11–81.—Diagram showing the locations of the various functional groups of fibers in the internal capsule as seen in a horizontal section. (Everett, *Functional Neuroanatomy,* 1965. Lea & Febiger.)

and thalamocortical fibers, corticobulbar fibers to the motor nuclei of the cranial nerves. The adjoining region of the **posterior limb** contains in addition to corticothalamic and thalamocortical fibers, corticospinal and corticorubral fibers. The corticospinal fibers to the motor nuclei of the muscles of the arm are nearer to the genu than those to the leg. The **retrolenticular part** contains thalamocortical fibers, sensory fibers from the lateral nucleus of the thalamus to the postcentral gyrus. The **sublenticular part** contains: the temporopontine fibers from the cortex of the temporal lobe to the pontine nuclei; the auditory radiation from the medial geniculate body to the audiosensory area of the transverse temporal gyrus; the

The **External Capsule** (*capsula externa*) (Fig. 11–78) is a lamina of white fibers between the putamen and the claustrum. It contains association and projection fibers associated with neighboring and adjacent tracts and nuclei similar to the parts of the internal capsule.

The Lateral Ventricle

The right and left lateral ventricles (*ventriculus lateralis*) (Fig. 11–82) are the cavities inside the cerebral hemispheres. Although they are the most cephalic representatives of the ventricular system, they are not called first and second ventricles, as would be expected to account for the

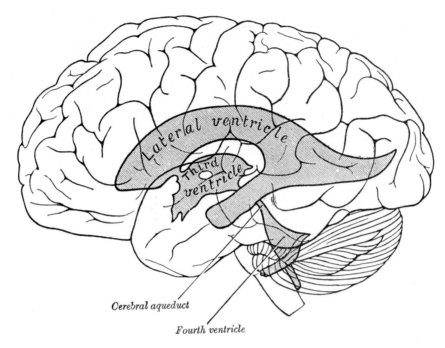

Cerebral aqueduct

Fourth ventricle

Fig. 11–82.—Scheme showing relations of the ventricles to the surface of the brain.

numerical designation of the third ventricle. The lateral ventricle is medially placed within the hemisphere, separated from its fellow of the opposite side by a thin vertical partition, the septum pellucidum (Fig. 11–56). It is lined by ependyma, contains spinal fluid, and communicates with the third ventricle through the interventricular foramen of Monro. It is quite irregular in shape, consisting of a central part or body, and three prolongations, the anterior, posterior, and inferior horns (Figs. 11–85, 11–86).

The **central part** or **body** (*pars centralis*) extends from the interventricular foramen to the splenium of the corpus callosum. Its roof is formed by the undersurface of the corpus callosum, its medial wall by the septum pellucidum, and its floor by the caudate nucleus, thalamus, stria and vena terminalis, the choroid plexus, and the lateral part of the fornix (Fig. 11–87).

The **anterior horn** (*cornu anterius*) (Fig. 11–87) is the part of the ventricle which extends into the frontal lobe beyond the interventricular foramen. It curves around the rostral end of the caudate nucleus and is confined by the genu and rostrum of the corpus callosum.

The **posterior horn** (*cornu posterius*) (Fig. 11–87) is the somewhat narrowed extension of the lateral ventricle into the occipital lobe. Its roof is formed by the fibers of the corpus callosum passing to the temporal and occipital cortex. On its medial wall a longitudinal eminence, the **calcar avis,** is the result of the deep infolding of the calcarine sulcus. The forceps posterior of the corpus callosum causes another projection, the **bulb of the posterior horn,** above the calcar avis, and the collateral sulcus produces a prominence lateral to the calcar avis, the **collateral trigone.**

The **inferior horn** (*cornu inferius*) (Fig. 11–91) is the extension of the lateral ventricle into the temporal lobe, reaching rostrally to within 2.5 cm of the temporal pole. It curves around the caudal end of the thalamus and follows the arch of the tail of the caudate nucleus. Its roof is formed chiefly by the inferior surface of the tapetum, but the tail of the caudate nucleus and the **stria terminalis** also extend forward in the roof to the rostral extremity of the horn. The floor is formed by the hippocampus, the **fimbria hippocampi,** the collateral eminence and the choroid plexus. When

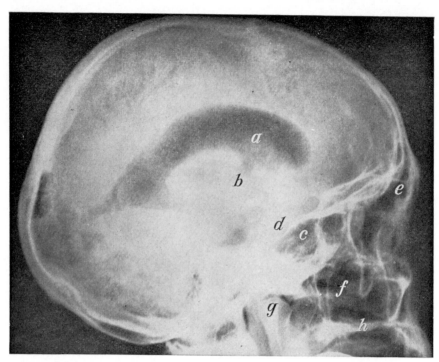

Fig. 11–83.—Adult head. *a*, Lateral ventricle injected with air; *b*, third ventricle, interventricular foramen mid-way between *a* and *b*; *c*, sphenoidal sinus; *d*, sella turcica; *e*, frontal sinus; *f*, maxillary sinus; *g*, condyle of mandible; *h*, hard palate. (Department of Radiology, University of Pennsylvania.)

the choroid plexus is torn away, a cleft-like opening is left along the medial wall of the horn. This is the inferior part of the choroid fissure.

The **interventricular foramen** (*foramen interventriculare; foramen of Monro*) (Fig. 11–60) is the opening between the lateral and third ventricle. It is situated between the columns of the fornix and the anterior end of the thalamus.

The **septum pellucidum** (Fig. 11–59) is a narrow partition between the two lateral ventricles. It is attached to the undersurface of the body of the corpus callosum, to the concave surface of the genu and rostrum of the corpus callosum, and to the body of the fornix. It is composed of two thin sheets, the *laminae septi pellucidi*, with a narrow cleft between, the *cavum septi pellucidi*. The cleft is not connected with the ventricles, but the lateral surfaces of the septum form part of the wall of the lateral ventricles and are therefore covered with ependyma.

The **choroid plexus** (*plexus choroideus ventriculi lateralis*) (Fig. 11–87) extends

from the interventricular foramen, where it is continuous with the plexus of the third ventricle, through the body and to the rostral end of the inferior horn. It does not extend into the anterior and posterior horns. The part in the body of the ventricle projects into the lateral wall from under the lateral edge of the fornix as an extension of the tela choroidea of the third ventricle through a cleft called the choroid fissure (Fig. 11–89). The part in the inferior horn lies in the concavity of the hippocampus and overlaps the fibria hippocampi from which it is reflected over to the roof of the horn (Fig. 11–87). The plexus consists of minute villous processes or tufts of blood vessels brought into the tela choroidea by the pia mater (Fig. 11–89). The tufts are covered everywhere by a layer of epithelial cells derived from the ependyma. The arteries of the plexus are: (*a*) the anterior choroidal, a branch of the internal carotid which enters the plexus at the end of the inferior horn, and (*b*) the posterior choroidal, one or two small branches of the posterior cerebral which pass under the

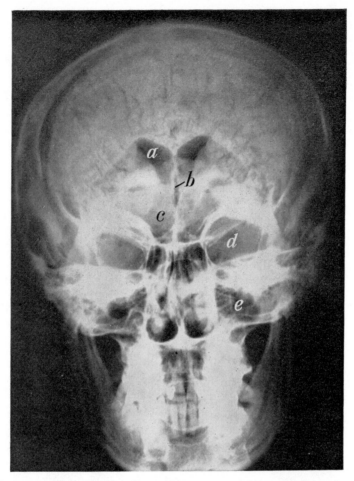

Fig. 11–84.—Brain ventricles injected with air. Note extension of air into sulci. *a*, Lateral ventricle; *b*, third ventricle; *c*, frontal sinus; *d*, orbit; *e*, maxillary sinus. (Department of Radiology, University of Pennsylvania.)

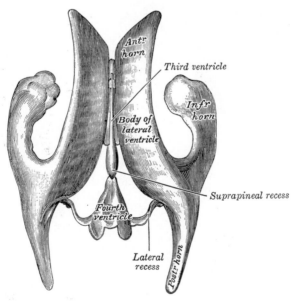

Fig. 11–85.—Drawing of a cast of the ventricular cavities, viewed from above. (Retzius.)

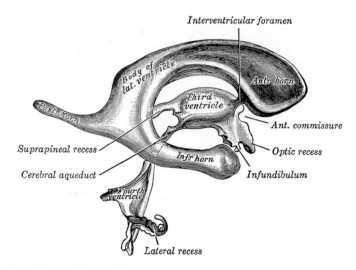

FIG. 11–86.—Drawing of a cast of the ventricular cavities, viewed from the side. (Retzius.)

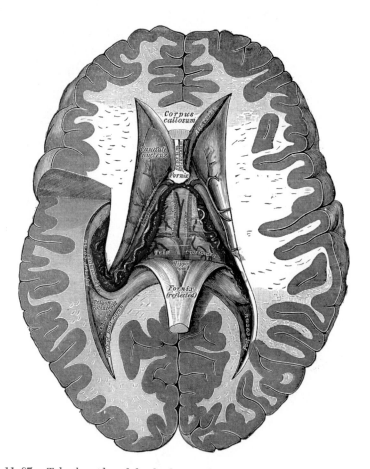

FIG. 11–87.—Tela choroidea of the third ventricle, and the choroid plexus of the left lateral ventricle, exposed from above. (See Figs. 11–11, 11–13.)

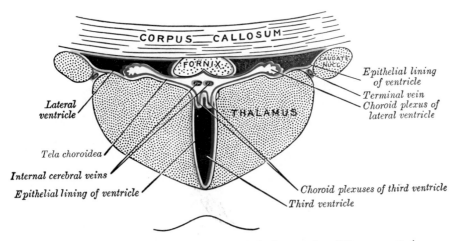

FIG. 11–88.—Coronal section of lateral and third ventricles. (Diagrammatic.)

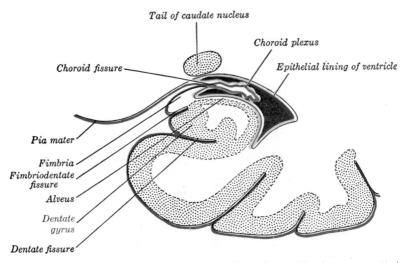

FIG. 11–89.—Coronal section of inferior horn of lateral ventricle. (Diagrammatic.)

splenium. The veins of the choroid plexus unite into a tortuous vein which drains rostrally to the interventricular foramen where it joins the vena terminalis to form the internal cerebral vein (of Galen).

The **choroid fissure** is the cleft-like space which remains when the choroid plexus is torn away and the epithelial lining of the ventricle is severed. Its extent is the same as the plexus, from the interventricular foramen to the tip of the inferior horn. Near the foramen, the fissure is between the lateral edge of the fornix and the upper surface of the thalamus. At the beginning of the inferior horn it is between the com-

mencement of the fimbria hippocampi and the caudal end of the thalamus. In the rest of the inferior horn it is between the fimbria in the floor and the stria terminalis in the roof.

THE RHINENCEPHALON

The **Rhinencephalon** or smell brain is composed of a rather heterogeneous complex of structures concerned with the reception and conduction of olfactory impulses. In lower vertebrates, most of the pallium or cortex is taken up by the rhinencephalon and because of its great phylogenetic age

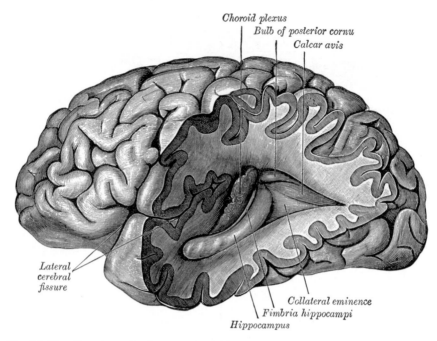

FIG. 11–90.—Posterior and inferior cornua of left lateral ventricle exposed from the side.

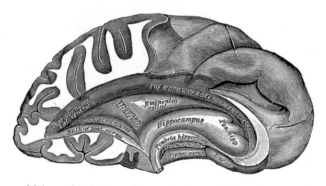

FIG. 11–91.—Temporal lobe with inferior and posterior horns of lateral ventricle viewed from above.

is called the archipallium. In man, the exuberant growth of the cortex not involved in olfaction, called the neopallium, has submerged the archipallium and relegated it to more or less hidden areas of the medial and inferior surfaces of the hemispheres. The rhinencephalon includes: (1) the olfactory bulb, (2) the olfactory tract and striae, (3) the anterior perforated substance, (4) the piriform area, (5) the hippocampal formation, (6) the paraterminal and parolfactory areas, and (7) the fornix.

(1) The **Olfactory Bulb** (*bulbus olfactorius*) (Figs. 11–62, 13–1) rests on the cribriform plate of the ethmoid bone through the foramina of which it receives the fila olfactoria of the olfactory nerve (page 1038). The bulb is connected with the hemisphere by the olfactory tract. The olfactory tract and bulb are really outgrowths of the brain, similar to the optic nerve, rather than peripheral nerves. In some lower animals and in the human embryo they contain a cavity linked with the ventricle but in the human adult the cavity is obliterated, its place being taken by neuroglia.

(2) The **Olfactory Tract** (*tractus olfac-*

torius) (Fig. 11–34) joins the inferior surface of the frontal lobe just rostral to the anterior perforated substance, forming a triangular expansion, the **olfactory trigone.** Its fibers then diverge into three strands, the medial, intermediate, and lateral olfactory striae. The **medial olfactory stria** is augmented by a small amount of gray matter and becomes the **medial olfactory gyrus.** It reaches the medial surface of the hemisphere, turns upward, and merges with the paraterminal and parolfactory area just rostral to the lamina terminalis

rarely in man it contains a small oval elevation, the **olfactory tubercle** (*tuberculum olfactorium*). In this area parts of the basal ganglia come close to the surface.

(4) The **Piriform Area** (*pyriform lobe*) includes the lateral olfactory gyrus, the limen insulae, the uncus, and the adjacent part of the hippocampal gyrus (Fig. 11–35). In the brain of a four-month fetus, the piriform lobe and its association with other parts of the rhinencephalon show more clearly before the neopallium overshadows them (Fig. 11–12).

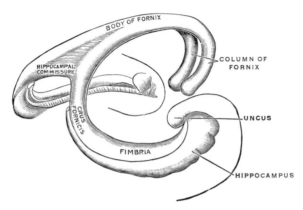

FIG. 11–92.—Diagram of the fornix. (Spitzka.)

(Fig. 11–34). The **lateral olfactory stria** with its accompanying gray matter, the **lateral olfactory gyrus,** bends laterally toward the exposed approach to the insula, the **limen insulae.** Here it bends back sharply and continues medially into the piriform area and uncus. An **intermediate olfactory stria** between the medial and lateral striae is occasionally present, blending directly with the anterior perforated substance.

(3) The **Anterior Perforated Substance** (Fig. 11–34) is a flattened, depressed area on the basal surface of the hemisphere just rostral to the optic tract. It is named from the numerous minute holes left in its surface by the withdrawal of the many small penetrating blood vessels when the pia mater is removed. Its rostral boundary is the olfactory trigone and striae. The posterior boundary is the optic tract, parallel with which is a ridge on its surface, the **diagonal band of Broca.** In certain animals but

(5) The **Hippocampal Formation** comprises (*a*) the paraterminal gyrus, (*b*) the indusium griseum, the longitudinal striae of the corpus callosum and the diagonal band of Broca, (*c*) the hippocampus, and (*d*) the gyrus dentatus.

(*a*) The **paraterminal gyrus** (*subcallosal gyrus*) (Fig. 11–60) is the thin sheet of gray matter which covers the undersurface of the rostrum of the corpus callosum. It is continuous inferiorly with the medial olfactory gyrus. Superiorly it follows around the genu of the corpus callosum and becomes the supracallosal gyrus.

(*b*) The **indusium griseum** (*supracallosal gyrus*) is a thin sheet of gray matter which covers the superior surface of the corpus callosum. It is continuous laterally under the callosal sulcus with the gyrus cinguli. Posteriorly it continues into the **gyrus fasciolaris** (*fasciola cinerea*) at the splenium of the corpus callosum (Fig. 11–93). The medial and lateral longitudinal striae represent the white matter of the vestigial supracallosal gyrus. They are two ridges on the superior surface of the corpus callosum (Fig. 11–73). Anteriorly they sweep around the genu, enter the paraterminal gyrus, and emerge as a single band which courses laterally at the posterior part of the anterior perforated substance as the **diagonal band** (*of Broca*).

(Fig. 11–93) is a narrow crenated strip of cortex between the fimbria hippocampi and the hippocampal gyrus. It lies in the depths of the hippocampal sulcus which separates it from the hippocampal gyrus. The fimbrio-dentate sulcus separates it from the fimbria. It is continued posteriorly, under the splenium of the corpus callosum, as the delicate gyrus fasciolaris (*fasciola cinerea*) which

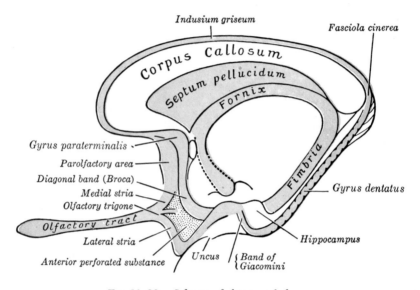

FIG. 11–93.—Scheme of rhinencephalon.

(*c*) The **hippocampus** (*Ammon's horn; cornu ammonis*), as its picturesque names imply, is a peculiarly shaped protuberance in the floor of the inferior horn of the lateral ventricle (Figs. 11–87, 11–91, 11–92). Its rostral extremity, the **pes hippocampi**, is enlarged and exhibits three or four rounded projections, the digitations of the hippocampus. Its substance is continuous with the hippocampal gyrus, the hippocampus having pushed inward at the hippocampal sulcus and curled over on itself (Fig. 11–96). Its internal structure resembles a very specialized cerebral cortex, but its ventricular surface is covered by a layer of white fibers, the **alveus**. The fibers of the alveus converge toward the medial border of the hippocampus at the choroid fissure to form the **fimbria of the hippocampus** (*fimbria hippocampi*).

(*d*) The **gyrus dentatus** (*dentate fascia*)

in turn is continuous with the indusium griseum. Anteriorly it enters the notch of the uncus and bends medially as the **band of Giacomini**.

(6) The **Paraterminal Body** (*gyrus paraterminalis; precommissural area; septal area*) (Fig. 11–60) is immediately anterior to the lamina terminalis. It includes the parolfactory area and is continuous with the paraterminal gyrus and the medial olfactory gyrus. From the subcallosal area, the diagonal band (*of Broca*) swings ventrally and laterally across the anterior perforated substance just rostral to the optic tract and ends in the amygdaloid complex.

(7) The **Fornix** occupies a position close to the median plane and follows part of the undersurface of the arch of the corpus callosum over the thalamus (Figs. 11–92, 11–93). It is composed of two stout bands, separated at both ends but joined together

at the middle. Its parts are the crura, the hippocampal commissure, the body, and the columns. The **crus fornicis** begins posteriorly as a continuation of the fimbria of the hippocampus. It arches upward over the thalamus, closely applied to the inferior surface of the corpus callosum and inclines toward the midline. The two crura are connected by a lamina of transverse fibers variously called the **commissura fornicis** (*hippocampal commissure*) the **psalterium,** or the **lyra.** At the highest part of their arch the crura are fused into the **body of the fornix.** The body of the fornix lies above the tela choroidea and ependymal roof of the third ventricle (Fig. 11–87) and is attached to the lower borders of the septum pellucidum (Fig. 11–93) and the under-surface of the corpus callosum. Anteriorly, above the interventricular foramina, the body divides again into the **columns** or **anterior pillars of the fornix.** They bend downward, forming the anterior boundary of the interventricular foramina, become buried behind the wall of the third ventricle and end in the mammillary bodies.

The **anterior commissure** (Fig. 11–60) is a bundle of white fibers which crosses the midline just rostral to the columns in the anterior wall of the third ventricle at the junction of the lamina terminalis and rostrum of the corpus callosum. Its constituent fiber bundles are twisted like a rope and extend laterally ventral to the anterior limb of the internal capsule and between the anterior perforated substance and the ventral part of the lentiform nucleus (Fig. 11–80). Most of the fibers belong to the rhinencephalon, connecting the piriform areas and amygdalae. Fibers have been traced to the external capsule and temporal neopallium (Fox *et al.*, '48). A second part of the commissure connecting the olfactory lobes of the two sides is prominent in some animals (Brodal '48) but is much reduced in man.

Structure of the Rhinencephalon

The **Olfactory Bulb.**—A section through the olfactory bulb shows it to be quite definitely stratified (Fig. 11–94). (1) The *outer layer,* concentrated on the inferior surface adjacent to the cribriform plate, contains unmyelinated fibers, the central processes from the olfactory neuroepithelial cells of the nasal mucosa (page 1038). (2) The *glomerular layer* contains numerous spheroidal structures called glomeruli which are formed by the interlacement of the ends of the unmyelinated olfactory fibers with the dendrites of the mitral and tufted cells. (3) The *molecular layer* contains a dense stratum of large **mitral cells** whose dendrites enter the glomeruli and whose axons penetrated to a deeper layer and enter the olfactory tract. Mingled with

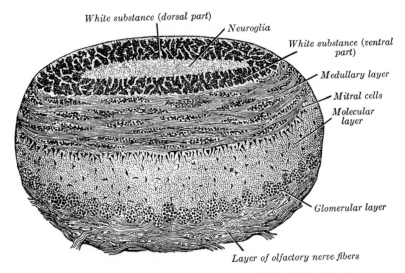

White substance (dorsal part)

Neuroglia

White substance (ventral part)

Medullary layer

Mitral cells

Molecular layer

Glomerular layer

Layer of olfactory nerve fibers

FIG. 11–94.—Coronal section of olfactory bulb. (Schwalbe.)

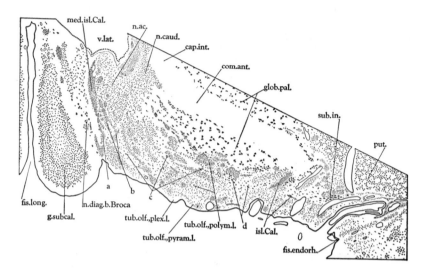

Fɪɢ. 11–95.—Drawing showing the mid portion of the human tuberculum olfactorium. *Cap. int.,* capsula interna; *com. ant.,* commissura anterior; *d,* large neurons of polymorph layer; *fis. endorh.,* fissura endorhinalis; *fis. long.,* fissura longitudinalis; *g. subcal.,* gyrus subcallosus; *isl. Cal.,* island of Calleja; *glob. pal.,* globus pallidus; *med. isl. Cal.,* medial island of Calleja; *n. ac.,* nucleus accumbens; *n. caud.,* nucleus caudatus; *n. diag. b. Broca,* nucleus of the diagonal band of Broca; *put.,* putamen; *sub. in.,* substantia innominata; *tub. olf., plex. l.,* tuberculum olfactorium plexiform layer; *tub. olf. polym. l.,* tuberculum olfactorium polymorph layer; *tub olf., pyram. l.,* tuberculum olfactorium pyramidal layer; *v. lat.,* ventriculus lateralis. (Crosby and Humphrey, courtesy of Jour. Comp. Neur.)

the granule cells of the molecular layer are **tufted cells** whose dendrites synapse with the unmyelinated olfactory fibers in tufts and glomeruli. Their axons are smaller than mitral cell axons. They cross to the olfactory bulb of the opposite side in the anterior commissure and form synapses with the mitral cells. (4) An *inner granule cell layer* or medullary layer contains mitral and tufted cell axons and granule cells. (5) The *layer of white substance* contains the myelinated axons of the mitral cells on their way to the olfactory tract. (6) An *inner core of neuroglia* is a remnant of the lining of the cavity which was present in the embryo.

The **anterior perforated substance** and **olfactory tubercle** show three moderately well-defined layers (Fig. 11–95): (1) an external molecular layer, (2) a layer of pyramidal cells, and (3) an irregular layer of polymorphic cells. In the third layer two specialized groups of cells are found. Several densely packed groups of small well-stained cells are named the islands of Calleja. A band of large cells just under the ventral surface of the globus pallidus is the substantia innominata of Reichert. The

nucleus accumbens forms the floor of the caudal part of the anterior horn of the lateral ventricle, between the head of the caudate nucleus and the anterior perforated substance.

The **piriform area** shows a definitely stratified arrangement, which differs in certain respects from the cortex of the neopallium. The molecular layer is unusually broad and contains a large number of tangential fibers. The adjoining layer contains two varieties of cells, each arranged in clumps or cell nests. The larger cells average 28 μ in diameter, are stellate in form, and poor in Nissl bodies. The smaller cells, although pyramidal in form, are smaller than the cells of the layer of small pyramidal cells in the neopallium. The third layer is deep, and the cells which it contains are chiefly pyramidal in shape, and their apices point obliquely to the surface. These cells are especially rich in basal dendrites. The fourth layer is much narrower and contains remarkably few cells, some resembling those of the third layer, while others are small and stellate in shape. The fifth layer is broader and contains cells which resemble the pyramidal elements of the third layer. A sixth, deeper, layer of fusiform cells is also present. This area receives the olfactory neurons of the second order and gives rise to those of the third order, which proceed to the hippocampal formation.

The **hippocampus** is more primitive in its structure than the piriform area and consists essentially of three layers (Fig. 11–96). It represents a portion of the cortex which has been rolled into the inferior horn of the lateral ventricle and its superficial cortical layer lies in relation to the hippocampal fissure and the dentate gyrus. (1)

The superficial or *molecular layer* is unusually broad and is densely packed with tangential fibers. It is usually described as consisting of a superficial part, the *stratum moleculare*, and a deep part, the *stratum lacunosum*. (2) This is succeeded by a broad layer of large pyramidal cells, which give off long apical dendrites into the molecular layer. Their basal axons run centrally through the succeeding polymorphous layer and pass into the subjacent white substance, which here constitutes the alveus. The numerous apical dendrites which are crowded together in the super-

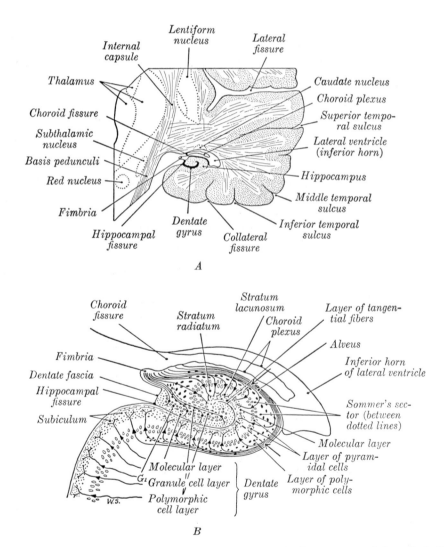

A

B

Fig. 11–96.—*A*, Frontal section through the temporal lobe and adjacent areas of the right half of the brain to show the structure and relations of the hippocampus. *B*, Diagrammatic representation of the histologic structure of the hippocampus. (Everett, *Functional Neuroanatomy*, Lea & Febiger.)

ficial part of this pyramidal layer have given rise to its subdivision into a *stratum radiatum* (or dendritic part) and a *stratum lucidum* (or cellular part). (3) The third layer contains *polymorphous cells* some of which are cells of Martinotti. Here, as elsewhere, they send their axons into the molecular layer. Others are aberrant pyramidal cells, and still others send their axons into the pyramidal layer where they end by arborizing around the pyramidal cells. The white fibers of the alveus cover the polymorphous layer and separate it from the ependyma on the free ventricular surface of the hippocampus (Everett '71, Kaada *et al.* '53).

The **dentate gyrus** also consists of three layers, a molecular layer, a granular layer, and a polymorphous layer. (1) The molecular stratum is well developed, and receives the dendrites of the cells of the second layer. (2) These are, for the most part, small granule cells but a number of large pyramidal cells are found among them. The axons of these cells traverse the third, or polymorphous layer, and then enter the adjoining molecular layer of the hippocampus, through which they pass to reach the pyramidal layer where they terminate by arborizing around the large pyramidal cells. These axons are characterized by small varicosities as they run in the pyramidal layer. (3) The third, or polymorphous, layer contains many Golgi Type II cells and many cells which send their axons through the adjoining layers of the hippocampus to reach the alveus.

The neurons of the hippocampus appear to be particularly vulnerable in infections, and in degenerative and toxic processes. This may be associated with its peculiar blood supply (Nilges '44) particularly in the area marked Sommer's sector in Figure 11–96.

CRANIAL NERVE NUCLEI AND THEIR CONNECTIONS

I. The **Olfactory Nerves** (*first cranial nerve; N. I*) or sensory nerves of smell are short bundles of fibers from the nasal mucous membrane which pass through the cribriform plate of the ethmoid bone to reach the olfactory bulb. The structure resembling a nerve on the basal aspect of the frontal lobe (Fig. 11–62) is the olfactory tract. The long processes of the neuroepithelial cells in the olfactory nerves described in the chapter on Sense Organs (page 1038) form synapses (1) with the mitral cells in the glomeruli of the olfactory bulb (Fig. 13–1) and (2) with the tufted cells of the bulb. The axons of the tufted cells cross to the bulb of the opposite side in the anterior commissure, reinforcing the sensory impulses.

The axons of the mitral cells traverse the olfactory tract and are distributed to different parts of the rhinencephalon through three strands, the lateral, intermediate, and medial olfactory striae (Fig. 11–60). The final destinations of the impulses are mainly the hippocampal cortex, the fornix, and the mammillary bodies and thence either association paths through the thalamus to the cortex or motor and reflex paths to the brain stem. A great many of the collaterals and terminal fibers which interconnect all parts of the rhinencephalon are not understood, but the following basic pathways are quite well established (see Rhinencephalon for figure references).

Lateral Olfactory Stria.—The greater number of olfactory bulb mitral cell axons in the lateral olfactory stria traverse the piriform area and enter the uncus and rostral part of the hippocampal gyrus. Fibers arising here pass to the hippocampus and dentate gyrus. Axons of the pyramidal cells of the hippocampus, either by way of the alveus or more directly as perforant fibers, reach the fimbria of the hippocampus of the opposite side through the commissura fornicis (*psalterium; hippocampal commissure*). Most of the fibers continue through the columns of the fornix to the mammillary bodies (see below). Others of the mitral cell axons in the lateral stria end in the olfactory trigone and anterior perforated substance and are relayed to the amygdaloid nucleus.

The **mammillary bodies** contain the terminations of the fibers of the columns of the fornix. The two principal tracts from these bodies are (*a*) the mammillotegmental tract (of Gudden) to the nuclei of the tegmentum of the pons and medulla oblongata; and (*b*) the mammillothalamic tract (bundle of Vicq d'Azyr) which goes to the anterior nucleus of the thalamus. From the thalamus associations are made through the cingulum with the cingulate gyrus and thence with many parts of the hemisphere.

Amygdaloid Nuclei.—The amygdaloid nuclei are interconnected with the anterior perforated substance, the piriform area, the parolfactory area, the corpus striatum, and the cortex of the temporal lobe. The amygdaloid bodies of the two sides are connected with each other through the anterior commissure (Jiminez-Castellanos '49).

The **diagonal band of Broca** (*diagonal gyrus of the rhinencephalon*) forms a slight ridge across the caudal part of the anterior perforated substance, just rostral to and parallel with the optic tract. It extends from the parolfactory area on the medial surface of the hemisphere to the amygdaloid nuclei and serves to connect these two centers.

The **stria terminalis** arises in the amygdaloid body and is visible in the wall of the lateral ventricle as a strand in the groove between the thalamus and the tail of the caudate nucleus. Separate components pass to the parolfactory area, to the habenular nuclei, and to the hypothalamus of the same and opposite sides.

Intermediate Olfactory Stria.—The axons of olfactory bulb mitral cells in the intermediate olfactory stria are relayed in the anterior perforated substance to the nuclei of the parolfactory area and subcallosal gyrus. Fibers from here course through the stria medullaris thalami to the habenular nuclei. Some fibers cross to the opposite side in the habenular commissure. Most fibers pass through the habenulopeduncular tract (*fasciculus retroflexus of Meynert*) and through the midbrain to the posterior perforated substance and end in the interpeduncular nucleus and dorsal tegmental nucleus.

Medial Olfactory Stria.—The axons of olfactory bulb mitral cells in the medial olfactory stria enter the subcallosal area and paraterminal gyrus. Fibers from the

nuclei in this area pass through the indusium griseum and longitudinal striae to the hippocampus and make other connections with olfactory centers through paths similar to those followed by fibers from the other olfactory striae.

The **subcallosal area,** in addition to the above connections, is the source of a prominent tract called the medial forebrain bundle. Fibers from this bundle pass to the tuber cinereum, the brain stem and the mammillary bodies.

The **paraterminal gyrus** receives fibers from most of the same sources as the subcallosal area and by its continuity with the indusium griseum sends fibers through the medial and lateral longitudinal striae caudalward around the corpus callosum. At the splenium the two striae come together as the gyrus fasciolaris. The fibers enter the dentate gyrus and thence follow the paths within the hippocampal complex.

II. The **Optic Nerve** (*nervus opticus; second cranial nerve; N. II*) (page 907) consists chiefly of coarse myelinated fibers which arise in the ganglionic layer of the retina (page 1055). The majority are third neurons in the visual pathway. In the optic disc, the fibers from the macula lutea make up the temporal half, but in the nerve they soon take a central position having the fibers from the peripheral retinal quadrants placed around them. At the optic chiasma the fibers from the medial half of the retina cross to the optic tract of the opposite side; the fibers from the temporal half of the retina stay on the same side. Thus the fibers from the right half of both retinas pass to the right hemisphere and those from the left half to the left hemisphere. Approximately 60 per cent of the fibers cross.

Primary Visual Centers.—Most of the fibers of the optic tract terminate in (1) the lateral geniculate body; other fibers continue through the superior brachium to (2) the superior colliculus and to (3) the pretectal region.

(1) The **lateral geniculate body** receives the visual sensory fibers in an orderly manner (Fig. 11–97). The axons of medium-sized pigmented nerve cells leave its dorsal rostral surface, run rostrally and then laterally through the retrolenticular portion of the internal capsule as the genicu-

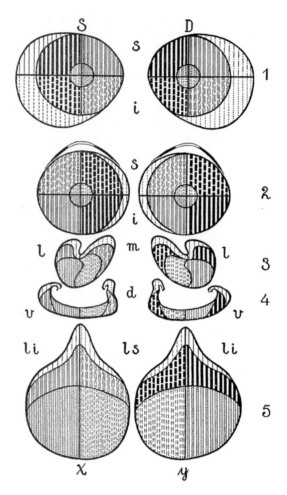

Fig. 11–97.—Diagrammatic illustrations of the projection of the various quadrants of the visual fields. Left (S) and right (D) sides of the visual fields and of the afferent visual apparatus.

1. Both fields of vision with upper (s) and lower (i), nasal and temporal halves; the smaller inner circles represent the "central" or macular portions (their relative side is somewhat exaggerated); the large circles represent the peri- or extramacular portions of the binocular visual fields; the outermost lightly shaded sickle-shaped zones represent the monocular portions of the visual fields.

2. Left and right retinae with their upper (s) and lower (i), nasal and temporal halves; smaller and larger circles and the monocular portions as above.

3. Schematic cross section through the left and right geniculate bodies; their internal margins (m) close to the thalamus; (l) their external margins; their concave contours in the figure facing upward represent their ventral margins.

4. Cross sections through the left and right visual radiation (external sagittal strata of the parieto-occipital lobes); their dorsal horizontal branches (d), their ventral horizontal branches (v) with perpendicular or vertical branches (in the figure horizontal) connecting both horizontal branches.

5. Left and right visual projection cortex, the area striata of Elliot Smith, field 17 of Brodmann, each subdivided into an upper (ls) and a lower half (li) corresponding with the upper and lower lips of the calcarine sulcus. The dividing lines, vertical in the figure, and terminating at the letters x and y, correspond in their upper parts to the bottom of the calcarine sulcus and to the horizontal meridians of both visual fields dividing the upper from the lower extramacular quadrants; in their lower parts these lines correspond to horizontal meridians dividing the upper from lower macular quadrants. The points where these lines reach the posterior limits of both striate areas, x and y in the figure, correspond to both points of fixation in the visual fields. The vertical lines or meridians dividing the left from the right homonymous halves of the macular portion of the visual fields correspond to the posterior (lower in the figure) circumference of the striate areas close to the lettes x and y. (Polyak, University of California Press, 1932.)

localcarine fasciculus. They pass caudally and medially to terminate in the visual cortex or area striata (area 17, Fig. 11–69) in the immediate neighborhood of the calcarine sulcus. Some of the fibers make a detour over the inferior horn of the ventricle before turning back to the occipital lobe. The representation of visual fields is illustrated in Figure 11–97. The cortex of the two sides is connected by commissural fibers in the optic radiation and splenium of the corpus callosum. Association fibers connect this area with other regions of the cortex and some fibers go back to the geniculate body. Clinical evidence indicates that the foveae are represented bilaterally in the cortex.

(2) The **superior colliculus** receives fibers from the optic tract through the superior brachium as the visual afferent arm for reflex control of the ocular muscles. There is a point-to-point relationship between the retina and the colliculus similar to that with the geniculate body. It is the origin of the tectobulbar and tectospinal tracts and its other connections are described with the oculomotor nerve.

(3) In the **pretectal region** the fibers form synapses with cells whose axons pass ventrally around the rostral end of the central gray of the midbrain to the oculomotor nucleus. They terminate in the nucleus of Edinger-Westphal as the afferent arm of reflex pupillary constriction in response to light.

III. The **Oculomotor Nerve** (*n. oculomotorius; N. III; third cranial nerve*) (page 909) contains somatic motor fibers for innervation of the Levator palpebrae and all the extraocular muscles except the Obliquus superior and the Rectus lateralis. It also contains autonomic fibers for the Ciliaris muscle and the Sphincter pupillae and proprioceptive fibers from the above-mentioned extrinsic muscles.

The **oculomotor nucleus** (*nucleus n. oculomotorius*) (Fig. 11–54) lies in the gray substance ventral to the aqueduct. It is from 6 to 10 mm in length, the rostral portion extending under the floor of the third ventricle for a short distance and the caudal end reaching to the level of the trochlear nucleus. It is intimately related to the medial longitudinal fasciculus which

lies against its ventrolateral aspect and many of its cells lie among the fibers of the posterior longitudinal fasciculus (of Schütz) as well. The fibers from the nucleus are collected into bundles which pass across the medial longitudinal fasciculus, the tegmentum, the red nucleus, and the medial margin of the substantia nigra in a series of curves, and finally emerge from the oculomotor sulcus medial to the cerebral peduncle (Fig. 11–60).

The oculomotor nucleus is in two parts, a large-celled somatic nucleus and a smaller-celled autonomic nucleus. The **somatic nucleus** in turn is composed of more or less definite groups of neurons for the individual muscles which can be mapped topographically (Fig. 11–98). The nucleus representing the Rectus medialis is the most ventral, the Rectus inferior most dorsal, and the Obliquus inferior intermediate between them. The Rectus superior is medial to the others in the caudal two-thirds only of the somatic nucleus. The Levator palpebrae is represented by the caudal central nucleus. The fibers from the nuclei for the Rectus inferior, Obliquus inferior, and Rectus medialis are uncrossed, going to the ipsilateral eye. Those from the Rectus superior nucleus are crossed and those from the Levator nucleus are bilateral.

The **autonomic nucleus of the oculomotor nerve** (Fig. 11–98) is rostral and dorsal to the somatic nucleus. It is composed not only of the lateral portion known as the *nucleus of Edinger-Westphal* but also of a median portion containing a similar small motor type of cell. The fibers from the nucleus are uncrossed, traverse the inferior division of the oculomotor nerve and enter the ciliary ganglion as its preganglionic fibers. Experiments with monkeys indicate that 96 per cent of these preganglionic fibers are for the innervation of the Ciliaris muscle and only 3 or 4 per cent for the Sphincter pupillae (Warwick '56).

Experiments with monkeys indicate that a central *nucleus of Perlia* is inconstant in occurrence and when present is composed of neurons supplying the Rectus superior. It cannot, therefore, be a **center** for control of **convergence**, as is usually stated (Warwick '56). It is probable that this delicate adaptive mechanism is controlled by a correlating center, perhaps in the reticular formation like that for conjugate movements.

The **proprioceptive fibers in the oculomotor nerve** are probably from the mesencephalic nucleus of the trigeminal nerve which occupies a position adjacent to the motor nuclei of all the extraocular muscles.

The fibers from the cerebral cortex controlling eye movements arise chiefly from area 8 of Brodmann (Fig. 11–69) for voluntary movement, and from the upper part of area 19, for fixation of gaze in

nuclei of the reticular formation; (3) interconnections between the nuclei of the eye muscles and head- and neck-turning nuclei; (4) from proprioceptive systems in the cervical cord. Connections from the tectum of the mesencephalon are described under the heading of the superior colliculus.

The **superior colliculus** is a center for coordination of eye movements. It receives (1) fibers from the optic tract, (2) sensory fibers from the spinal cord in the spinotectal tract, (3) fibers from the central sensory path of the trigeminal nerve, (4) central

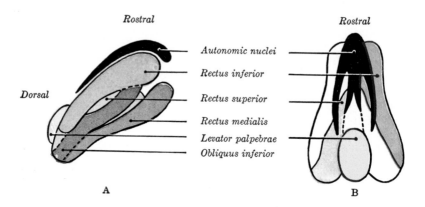

Rostral   *Rostral*

Dorsal

— Autonomic nuclei

— Rectus inferior

— Rectus superior

— Rectus medialis
— Levator palpebrae
— Obliquus inferior

A   **B**

Fig. 11–98.—Diagrams showing the representation of the right extraocular muscles in the oculomotor nucleus of a monkey. A, Right lateral aspect; B, Dorsal aspect. (Redrawn from Warwick, R., 1953, J. comp. Neurol.)

response to a visual stimulus. Other fibers appear to come from parts of the parietal and temporal lobes. They leave the peduncles in the rostral part of the mesencephalon. It is doubtful whether many of them pass directly to the oculomotor nuclei. The majority apparently terminate in nuclei of the reticular substance where **conjugate movements** of the eyes are integrated. A center for conjugate lateral deviation of gaze lies close to the abducens nucleus, and one for conjugate upward deviation of gaze apparently lies dorsolateral to the oculomotor nucleus. From such centers connections pass to the eye muscle nuclei both directly and through the medial longitudinal fasciculus. The last-mentioned fasciculus also carries the following types of connections: (1) from the vestibular nuclei, crossed and uncrossed; (2) from various

auditory fibers from the lateral lemniscus for reflex movements of the eyes in response to sound, (5) fibers from the visual cortex through the corticotectal part of the optic radiations, (6) fibers from the stria medullaris thalami of the opposite side which pass through the habenular commissure for primary and cortical olfactory associations.

The descending **efferent fibers of the superior colliculus** arise from large cells in the stratum opticum and stratum lemnisci. They pass around the ventral aspect of the central gray substance where most of them cross in the dorsal tegmental decussation (fountain decussation of Meynert) and then turn caudalward to form the tectobulbar and tectospinal tracts. From the tectobulbar tract, terminals and collaterals are given off to the oculomotor, trochlear, and abducens nuclei and to the motor nucleus of the facial nerve. The tectospinal fibers end by terminals and collaterals either directly or indirectly among the ventral motor horn cells of the spinal cord, especially in the cervical segments, to coordinate head and neck movements with eye movements. Many collaterals from both tracts go to the red nucleus. Probably no fibers pass from the superior colliculus to the visual sensory cortex.

The **pupillary reflexes** are of two kinds: (1) in response to light and (2) associated with accommodation. The response to light is through the pretectal nuclei of the hypothalamus and the autonomic nucleus of the oculomotor. The contraction of the ciliary muscle and pupillary sphincter which accompanies the convergence for accommodation is probably initiated by proprioceptive impulses from the ocular muscles, and correlated in nuclei of the reticular formation.

IV. The **Trochlear Nerve** (*n. trochlearis; fourth cranial nerve; N. IV*) (page 911) contains somatic motor fibers for the Obliquus superior muscle of the eye. The trochlear nucleus (Fig. 11–53) is a small oval mass in the central gray of the cerebral aqueduct at the level of the inferior colliculus. The axons from the nucleus start caudalward in the tegmentum but turn abruptly dorsalward before reaching the pons. They pass into the anterior medullary velum, cross to the opposite side, and emerge from the velum immediately caudal to the inferior colliculus.

The nucleus receives terminals from the medial longitudinal fasciculus and other sources similar to the oculomotor nucleus and it also contains similar proprioceptive connections.

V. The **Trigeminal Nerve** (*n. trigeminus; fifth cranial nerve; N. V*) (page 912) contains somatic sensory, special visceral efferent, and proprioceptive fibers. They pass from their nuclei through the lateral part of the tegmentum of the pons (Fig. 11–29) then between the fibers of the middle cerebellar peduncle (Fig. 11–44) and emerge at the middle of the line of transition from the pons to the peduncle (Fig. 11–62).

The terminal sensory nucleus has an enlarged rostral end, the main sensory nucleus, and a long slender caudal portion, the nucleus of the spinal tract of the trigeminal nerve which becomes continuous with the substantia gelatinosa (of Rolando) of the spinal cord.

The **main sensory nucleus** is lateral to the motor nucleus and ventral to the superior cerebellar peduncle (Figs. 11–43, 11–42). It is primarily for discriminative sense. It receives the short ascending branches of the fibers from the cells of the trigeminal (semilunar) ganglion. The axons of its large and medium-sized cells cross to the opposite side and form two tracts, a ventral and a dorsal. The fibers of the ventral tract, sometimes called the **trigeminal lemniscus,** are myelinated and collect in a band along the dorsal medial margin of the medial lemniscus. The dorsal tract passes through the reticular substances near the central gray around the aqueduct. Both tracts terminate in the nucleus ventralis posteromedialis of the thalamus and are relayed to the postcentral cerebral cortex.

The descending branches of the sensory root fibers course through the pons and medulla in the **spinal tract of the trigeminal nerve** (Figs. 11–38, 11–39). They end by collaterals and terminals in the nucleus of the spinal tract (Figs. 11–40, 11–42) as far down as the second cervical segment. It is primarily the sensory nucleus for pain and temperature. The second neurons from the nucleus cross to the opposite side in the reticular substance, join the medial margin of the lateral spinothalamic tract, and terminate in the nucleus ventralis posteromedialis of the thalamus. In addition to these projection tracts there are numerous connections from the sensory trigeminal nuclei to the motor nuclei of the medulla and pons. The lateral third of the reticular formation is particularly related to the trigeminal system and contains many secondary and tertiary centers. There are numerous commissural fibers of this system in the rostral part of the pons. The somatic sensory fibers of the vagus, the glossopharyngeal, and the facial nerves probably end in the nucleus of the descending tract of the trigeminal and their cortical impulses are probably carried up in the central sensory path of the trigeminal.

The **mesencephalic root** (*descending*

root of the trigeminal) arises from unipolar cells arranged in scattered groups in a strand at the lateral edge of the central gray matter surrounding the upper end of the fourth ventricle and the cerebral aqueduct (*nucleus mesencephalicus n. trigemini*). The cells develop from the alar lamina and are large, round, unipolar, and without dendrites like cells of the sensory type. The axons give off collaterals to the motor nucleus and pass between the motor and main sensory nuclei to enter the motor root and pass into the mandibular branch of the nerve. They are probably proprioceptive for the muscles of mastication. It is thought that the proprioceptive fibers for the extrinsic ocular muscles also arise in this nucleus.

The **motor nucleus** (Fig. 11–44) is situated in the rostral part of the pons near the lateral angle of the fourth ventricle. It is serially homologous with the facial nucleus, and the nucleus ambiguus. The axons arise from large pigmented multipolar cells. The voluntary motor control for the muscles of mastication is mediated through aberrant pyramidal tract fibers, more from the opposite than the same side. The reflex control is provided by collaterals and terminals (1) from the sensory nucleus of the trigeminal of the same and a few from the opposite side by way of the trigeminothalamic tract; (2) from the mesencephalic root of the trigeminal; and (3) from nuclei in the reticular formation. Many of these connections have interposed association neurons.

VI. The **Abducent Nerve** (*n. abducens; sixth cranial nerve; N. VI*) (page 924) contains somatic motor fibers for the Rectus lateralis muscle of the eye. The nucleus of this nerve is serially homologous with those of the oculomotor and hypoglossal nerves, and the ventral column of the spinal cord. It is situated close to the floor of the fourth ventricle deep to the facial colliculus (Figs. 11–40, 11–41, 11–44).

It receives voluntary impulses from the aberrant pyramidal tract and has reflex connections with the medial longitudinal fasciculus as well as other connections similar to those of the oculomotor and trochlear nerves. The nerve probably contains proprioceptive fibers from the mesencephalic nucleus of the trigeminal.

VII. The **Facial Nerve** (*n. facialis; seventh cranial nerve; N. VII*) (page 924) contains somatic and visceral afferent, special visceral afferent (taste) and general and special visceral efferent fibers. It is in two parts, the motor root and the **nervous intermedius.** The intermedius is more lateral and lies between the motor root and the eighth nerve. It includes the sensory and parasympathetic components and is sometimes called the glossopalatine nerve (page 926).

The **somatic sensory** fibers are few in number and join the auricular branch of the vagus to supply the external acoustic meatus. They arise from cells in the geniculate ganglion and their central termination is probably with the spinal tract of the trigeminal nerve.

The **visceral afferent** fibers are probably the fibers which reach the mucous membrane of the pharynx, nose, and palate through the greater petrosal nerve (Foley et al. '46). They arise from cells in the geniculate ganglion and in the medulla join the tractus solitarius.

The **taste fibers** arising from cells in the geniculate ganglion join the tractus solitarius and terminate in the rostral portion of the nucleus of the tractus solitarius. There is some evidence that fibers from the nucleus join the medial lemniscus and reach the thalamus (Allen '23). There are connections with the salivatory nucleus, the nucleus ambiguus, the hypoglossal nucleus, the dorsal nucleus of the vagus, and with the spinal cord through the reticulospinal tract (Allen '27).

The **special visceral efferent** fibers supply the striated muscles of facial expression. The **motor nucleus of the facial nerve,** deep in the reticular formation of the pons, is serially homologous with the nucleus ambiguus and the lateral part of the ventral horn of the spinal cord. Axons from its cells leave the dorsal surface of the nucleus and in the first part of their course continue dorsally and medially until they reach the rhomboid fossa. They turn sharply and run rostrally, dorsal to the medial longitudinal fasciculus, along the medial side of the abducent nucleus. They are gathered into a compact bundle and turning abruptly to the side arch over the nucleus nervi abducentis dorsally, producing the elevation in the rhomboid fossa called the facial colliculus. The twist around the abducent nucleus is called the internal genu or genu of the root of the facial nerve. The second part of its course takes the root ventralward, lateralward, and caudalward to the superficial emergence of the nerve from a recess between the olive and inferior cerebellar peduncle at the caudal border of the pons.

Within the nucleus, the large multipolar motor-type cells are arranged in small groups representing the various facial muscles (Szentágothai '48). The cells which innervate the lower part of the face receive connections from the aberrant pyramidal system which are entirely crossed. Those to the upper part of the face are both crossed and uncrossed. Reflex connections are formed with the nucleus of the spinal tract of the trigeminal, with the cochlear nuclei, and with numerous other nuclei and mechanisms in the reticular formation of the medulla and pons, some of them concerned with emotional expression.

General visceral efferent or **autonomic fibers** arise from the superior salivatory nucleus, usually described as a group of small cells dorsomedial to the facial nucleus. These preganglionic parasympathetic fibers join the nervus intermedius. They are distributed to the submandibular ganglion by way of the chorda tympani for innervation of the submandibular and sublingual glands, and to the pterygopalatine ganglion by way of the greater petrosal nerve for innervation of the lacrimal gland and glands in the nasal and palatine mucosa.

VIII. The **Acoustic Nerve** (*n. vestibulocochlearis; statoacustic nerve; otic nerve; eighth cranial nerve; N. VIII*) (page 931) consists of two nerves, the cochlear nerve or the nerve of hearing, and the vestibular nerve or the sensory nerve of equilibration.

The **cochlear nerve** (Fig. 11–99) is composed of fibers from the spiral ganglion which bifurcate as they enter the cochlear nuclei. The short ascending branches end in the ventral cochlear nucleus; the longer descending branches end in the dorsal

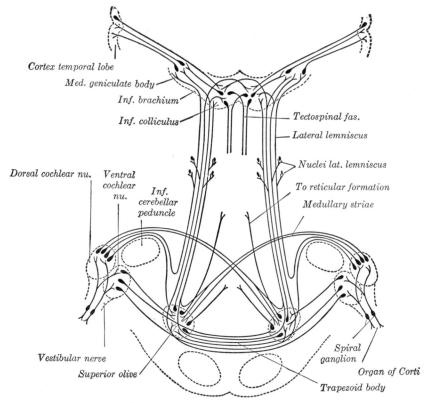

Cortex temporal lobe
Med. geniculate body
Inf. brachium
Inf. colliculus
Tectospinal fas.
Lateral lemniscus
Nuclei lat. lemniscus
Dorsal cochlear nu.
Ventral cochlear nu.
Inf. cerebellar peduncle
To reticular formation
Medullary striae
Vestibular nerve
Superior olive
Spiral ganglion
Organ of Corti
Trapezoid body

Fig. 11–99.—Connections of cochlear nerve.

cochlear nucleus. The terminations of the fibers are very orderly, different parts of the cochlea being represented in different parts of the cochlear nuclei. Each branch of the entering fibers gives off many collaterals, usually about 50. In most cases one, and not more than three, of these collaterals ends in a giant synapse which makes contact with as much as two-thirds of the area of the cell body of the second neuron.

The **dorsal cochlear nucleus** forms a projection on the dorsal and lateral aspect of the inferior cerebellar peduncle, the acoustic tubercle. Axons of the second order from its large fusiform cells pass over the peduncle and cross the floor of the fourth ventricle as the striae medullares acousticae. These fibers run under the gray substance in the floor of the ventricle, through the upper margin of the reticular substance. Many of them cross in the median raphe, sink into the reticular formation and join the trapezoid body or the lateral lemniscus or terminate in the superior olivary nucleus on the side opposite their origin. They pass upward in the lateral lemniscus and end by terminals and collaterals either in the nuclei of the lateral lemniscus, in the inferior colliculus, or in the medial geniculate body. Some of the fibers in the striae medullares do not cross the midline but dip into the reticular formation and end in the superior olivary nucleus of the same side or join the lateral lemniscus and pass upward in it to end by terminals and collaterals in the nuclei of the lateral lemniscus, in the inferior colliculus, and in the medial geniculate body of the same side.

The **ventral cochlear nucleus** is continuous with the dorsal cochlear nucleus. Axons of the second order from its cells pass horizontally in the trapezoid body; here some of them end in the superior olivary nucleus of the same side; others cross the midline and end in the superior olivary nucleus of the opposite side or pass by these nuclei, giving off collaterals to them, and join the lateral lemniscus. They are distributed with fibers from the dorsal cochlear nucleus to the nuclei of the lateral lemniscus and to the inferior colliculus by collaterals and terminals. Other fibers from the ventral cochlear nucleus pass dorsal to the inferior peduncle and then dip into the substance of the pons

to join the trapezoid body or the superior olivary nucleus of the same side.

Fibers from the **trapezoid body** and **superior olive** ascend in the lateral lemniscus. Many fibers are relayed in the nuclei of the lateral lemniscus on their way to the inferior colliculus and medial geniculate body. Others are relayed in the inferior colliculus and pass through the inferior brachium to the medial geniculate body. Fibers from large cells in the inferior colliculus pass through the deep white layer into the tegmentum of the same and opposite sides and descend to lower motor centers, perhaps with the tectospinal fasciculus. A large proportion of the axons from the medial geniculate body pass rostrally beneath the optic tract to join the corona radiata and auditory radiation to the cortex of the superior temporal gyrus.

The **vestibular nerve** (page 932) arises from the bipolar cells of the vestibular ganglion. The central fibers enter the medulla between the restiform body and the spinal tract of the trigeminal nerve (Fig. 11–42) and bifurcate into ascending and descending branches. The ascending branches go to the medial, lateral, and superior nuclei and a few through the inferior peduncle of the cerebellum to the nucleus fastigii and the vermis. The fibers from the semicircular canals go to the superior and rostral part of the medial nucleus. Those from the maculae of the saccule and utricle go to the lateral and caudal part of the medial nucleus. The descending branches constitute the descending or spinal root of the vestibular nerve and terminate in the nucleus of the same name.

The connections of the **medial nucleus** are widespread through the medulla and pons to various nuclei of the reticular formation, to the motor nuclei of the cranial nerves, and to autonomic centers and nuclei. It contributes many fibers, both ascending and descending, to the medial longitudinal fasciculus of both sides (Buchanan '37).

The **superior nucleus** is particularly associated with the vermis, the flocculonodular lobe, and the central nuclei of the cerebellum. It contributes ascending fibers to the medial longitudinal fasciculus of the same side, and descending fibers to the vestibulospinal tract of the same side.

The **lateral nucleus** contains large polygonal cells whose axons at first run medially into the reticular formation and then turn caudalward. They shift to a more ventral position as they descend and in the spinal cord become the direct vestibulospinal tract. This is apparently the chief antigravity mechanism of the central nervous system. Collaterals from the axons ascend through the restiform body to terminate in the vermis. It also sends ascending and descending fibers to the medial longitudinal fasciculus of both sides.

The **nucleus of the spinal root of the vestibular nerve** contains large, medium, and small cells. It has widespread connections through the medulla and pons, particularly with nuclei of the reticular formation. It contributes descending fibers to the medial longitudinal fasciculus of both sides. Certain of the descending fibers form the crossed vestibulospinal tract which runs in the sulcomarginal fasciculus of the spinal cord.

IX. The **Glossopharyngeal Nerve** (*nervus glossopharyngeus, ninth cranial nerve; N. IX*) (page 932) contains general and special visceral afferent, and general and special visceral efferent fibers. It is similar to the vagus and is connected with the medulla by rootlets arranged in series with those of the vagus. There may be some somatic afferents similar to the auricular branch of the vagus.

The **special visceral afferent fibers** for taste arise from cells in the superior and inferior ganglia and after entering the medulla pass through the tractus solitarius to the nucleus solitarius. The nucleus has connections described with the facial nerve.

The **general visceral afferent fibers** also join the tractus solitarius.

The **special visceral efferent fibers** innervate the Stylopharyngeus muscle. They arise in the cephalic end of the nucleus ambiguus and have connections similar to those described with the vagus nerve.

The **general visceral efferent** or **parasympathetic preganglionic fibers** arise in the nucleus salivatorius inferior and have connections similar to those described with the facial nerve. They supply secretomotor fibers to the parotid gland by relay in the otic ganglion.

X. The **Vagus Nerve** (*nervus vagus; tenth cranial nerve; N. X*) (page 936) contains somatic and visceral afferent, general and special visceral efferent, and special visceral afferent (taste) fibers. The afferent fibers have their cells of origin in the jugular and nodose ganglia. They enter the medulla accompanied by the motor fibers through eight or ten rootlets in the groove between the olive and the inferior cerebellar peduncle (Fig. 11–38) (Anderson and Berry '56).

The **somatic sensory fibers** supply part of the external acoustic meatus and a small area on the back of the ear. They probably join the spinal tract of the trigeminal nerve and have connections with the thalamus and sensory area of the cortex. Their descending fibers probably establish relations with motor nuclei of the spinal cord and medulla.

The **visceral afferent fibers** join the tractus solitarius and end in its nucleus. Their terminals make associations with centers in the reticular formation, especially those concerned with respiration, vasomotor control, cardiac activity, etc., and are relayed to motor nuclei in the medulla and spinal cord. They probably have connections with the dorsal motor nucleus of the vagus.

Taste fibers in the vagus nerve from a few taste buds on the epiglottis and larynx follow the pathways described for taste in the facial and glossopharyngeal nerves.

The **special visceral efferent fibers** of the vagus innervate the striated muscles of the pharynx and larynx. They arise in the nucleus ambiguus and after a short dorsal course turn back ventrally and laterally to emerge with the sensory fibers by several rootlets. The nucleus receives fibers either directly or indirectly through collaterals or terminals from the aberrant pyramidal tract of the opposite side. Reflex collaterals are also derived from the central tracts of the trigeminal, glossopharyngeal, vagus, and spinal nerves.

The autonomic fibers arise from cells in the **dorsal nucleus** (*Nucleus of the ala cinerea*). They are **preganglionic parasympathetics** for the smooth muscle and glands of the esophagus, stomach, small intestine, upper colon, gallbladder, pancreas, and lungs and inhibitory fibers to the heart. The nucleus receives terminals of visceral afferent fibers and from many other sources not exactly known. There are connections with the hypothalamus, probably through the dorsal longitudinal fasciculus of Schütz, as well as with other nuclei and tracts in the reticular formation.

XI. The **Accessory Nerve** (*nervus accessorius; eleventh cranial nerve; N. XI*) (page 944) has two parts, the spinal part and the cranial or bulbar part. The cranial part is accessory to the vagus nerve, has an origin like the vagus, joins the vagus immediately after leaving the cranium and its fibers are distributed as branches of the vagus.

The **cranial part** contains special and general visceral efferent fibers which arise in the nucleus ambiguus. This nucleus appears to be an upward continuation of the lateral cell groups of the ventral horn of the spinal cord. The fibers leaving the nucleus pass ventral to the spinal tract of the trigeminal and emerge in a series of rootlets from the posterior lateral sulcus, in series with the roots of the vagus. Through the nucleus ambiguus the cranial part makes connections with opposite aberrant pyramidal tract and with the terminal sensory nuclei of the cranial nerves.

The **visceral efferent fibers** are few in number and arise in the dorsal nucleus of the vagus. They join the vagus at the jugular foramen and are distributed as part of the vagus nerve.

The **spinal part** arises from lateral cell groups in the ventral horn of the first five or six cervical segments of the spinal cord. It is somatic motor and is described on page 944.

The nucleus of origin of the spinal part receives either directly or indirectly terminals and collaterals from the pyramidal and aberrant pyramidal tracts. It also receives fibers from the medial longitudinal fasciculus for coordination of head and eye movements and probably from the rubrospinal and vestibulospinal tracts. It is connected indirectly with spinal somatic sensory nerves by association fibers of the fasciculi proprii.

XII. The **Hypoglossal Nerve** (*nervus hypoglossus; twelfth cranial nerve; N. XII*) (page 945) contains somatic efferent fibers for innervation of the muscles of the tongue. Although Langworthy ('24) places

the proprioceptive fibers to the tongue in the hypoglossal nerve, Pearson ('45) does not confirm this opinion.

The **hypoglossal nucleus** lies near the central canal in the caudal closed part of the medulla and under the trigonum hypoglossi of the floor of the rhomboid fossa. The cells of both nuclei send long dendritic processes across the midline to the opposite nucleus. The axons of the large multipolar cells pass ventrally through the reticular formation medial to the inferior olive and emerge as a number of rootlets from the ventral lateral sulcus in series with the motor rootlets of the spinal cord.

The hypoglossal nucleus receives, either directly or indirectly, numerous collaterals and terminals from the contralateral aberrant pyramidal tract and a few from the ipsilateral tract for voluntary control from the cerebral cortex. Many reflex collaterals are received from the secondary sensory paths of the trigeminal, facial, glossopharyngeal, and vagus nerves. Collaterals from the medial longitudinal fasciculus and the tectobulbar tract are said to enter the nucleus.

SPINAL NERVES—COMPOSITION AND CONNECTIONS

The spinal nerves consist of fibers conveying somatic and visceral afferent, and somatic and visceral efferent impulses.

The afferent fibers, both somatic and visceral, are processes of the spinal ganglion cells (see page 947) which enter the spinal cord through the dorsal root. The somatic afferents include those commonly called sensory for pain, temperature, and touch and deep sensation from muscles and joints. The visceral afferent fibers probably mediate reflex mechanisms which do not reach consciousness.

The somatic efferent fibers are motor for the voluntary or skeletal muscles of the body and are processes of anterior horn cells which leave the spinal cord through the ventral roots. Visceral efferent fibers, processes of lateral horn cells, also leave by the ventral roots and make up the preganglionic fibers of the sympathetic system and the sacral part of the parasympathetic system. They carry motor impulses to involuntary muscles of blood vessels and other viscera and secretory impulses to glands.

The **dorsal root fibers** entering the spinal cord promptly divide into ascending and descending branches. They make up the bulk of the posterior funiculus and give off a number of collaterals within the cord. Many of the ascending branches reach the brain, forming the tracts described below. The remainder of the ascending branches, together with the descending branches and the collaterals, form synapses with cells in the various parts of the gray matter of the cord itself. The axons of these cells may pass out into the lateral and ventral funiculi and turn upward to reach the brain. A large number complete their connections with the motor fibers within the cord, however, establishing a large number and variety of spinal cord reflexes (Fig. 11–100).

Somatic Afferent Paths.—The fibers in the dorsal roots are not segregated according to functional types until they enter the spinal cord. A partial separation is provided, however, by the formation of the two strands within the rootlets, a medial and a lateral strand. The medial strand

contains mainly large myelinated fibers which carry touch, proprioception and vibratory sense. The lateral strand contains small fibers mostly not myelinated, which carry pain and temperature sensibility. Within the cord there is a further separation of fibers into functional pathways.

The sense of touch is carried by two different paths: (1) tactile discrimination, that is, (*a*) the ability to distinguish two points close together, (*b*) the exact location of a stimulus, and (*c*) stereognosis or recognition of the shape of an object; (2) touch and pressure of a diffuse nature.

Tactile discrimination is carried by large myelinated fibers through the medial strand of the dorsal rootlet into the dorsal funiculus. They ascend in the cuneate and gracile fasciculi to the corresponding nuclei. The secondary axons cross the midline in the internal arcuate fibers and pass through the medial lemniscus to the posterolateral ventral nucleus of the thalamus. They are relayed from here through the internal capsule and corona radiata to the somatic sensory area of the cortex in the postcentral gyrus (area 3, 1, 2, Fig. 11–69).

Diffuse touch and pressure fibers traverse the medial strand of the dorsal rootlets and

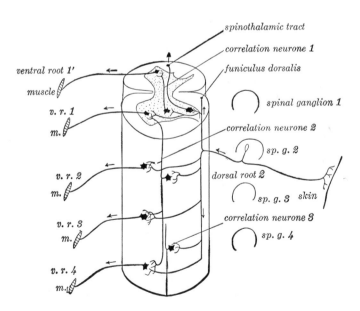

Fig. 11–100.—Diagram of the spinal cord reflex apparatus. Some of the connections of a single afferent neuron from the skin (*d.r.2*) are indicated; *d.r.2*, dorsal root from second spinal ganglion; *m*, muscles; *sp.g. 1* to *sp.g. 4*, spinal ganglia; *v.r. 1'* to *v.r. 4*, ventral roots. (After Herrick.)

enter the dorsal funiculus. As they run in the dorsal funiculus they give off collaterals to the gray substance of the successive levels through which they pass. In this way they reach the gray substance in a number of segments above their entrance. The collaterals synapse with neurons in the dorsal gray horn whose axons cross the midline in the ventral white commissure and form the ventral spinothalamic tract of the opposite side. These fibers join the medial lemniscus in the medulla and accompany the tactile fibers to the thalamus and cortex.

Pain and temperature fibers in the dorsal root are small unmyelinated and myelinated fibers which enter the cord through the lateral strand of the rootlet and end in the substantia gelatinosa of Rolando within one or two segments. The axons of the second neurons cross in the ventral white commissure and ascend in the lateral spinothalamic tract to the posterolateral ventral nucleus of the thalamus. An accessory interrupted pathway of intercalated neurons probably runs parallel with this tract. From the thalamus fibers pass through the caudal limb of the internal capsule to the somesthetic area of the postcentral gyrus of the cortex. Unmyelinated fibers of the posterior roots which turn into the dorsolateral tract (of Lissauer) ascend or descend for short distances and end in the substantia gelatinosa. They are part of the mechanism for reflexes associated with pain within the cord.

Proprioceptive fibers have two important destinations, the cerebrum and the cerebellum. The proprioceptive fibers are large myelinated fibers entering the cord through the medial strand of the dorsal rootlets. Those to the cerebrum ascend in the fasciculus gracilis and cuneatus to the nuclei of these names in the medulla oblongata. Axons from these nuclei cross to the opposite side in the internal arcuate fibers and their decussation. They then turn rostralward in the medial lemniscus and end in the posterolateral ventral nucleus of the thalamus. The third order neurons pass through the internal capsule and corona radiata to the somatic sensory area of the postcentral gyrus of the cortex.

The proprioceptive fibers to the cerebellum travel by three different pathways. (1)

Fibers traverse the gracile and cuneate fasciculi of the posterior funiculus along with the cerebral fibers and end in the lateral cuneate nucleus. Axons from cells in this nucleus pass along the dorsal external arcuate fibers through the inferior cerebellar peduncle of the same side to the cerebellar cortex. (2) Other proprioceptive fibers in the fasciculus cuneatus give off terminals or collaterals to the nucleus dorsalis of the spinal cord. Axons from the nucleus dorsalis pass to the dorsal spinocerebellar tract (of Flechsig) of the same side. At the level of the olive the fibers curve under the external arcuate fibers to the inferior cerebellar peduncle. They give off collaterals to the dentate nucleus and finally terminate in the cortex of the dorsal and rostral portion of the vermis on the same side. (3) Still other terminals and collaterals from fibers in the fasciculus cuneatus synapse with cells in the dorsal horn gray substance in or near the dorsal nucleus of the other side. They cross in the white and gray commissures and pass with fibers from the same side through the lateral funiculus to the ventral spinocerebellar tract (of Gowers). This tract ascends until it passes under the external arcuate fibers dorsal to the olive, and then joins the lateral edge of the lateral lemniscus. At the level of the motor nucleus of the trigeminal nerve it crosses over the superior cerebellar peduncle and turns abruptly caudad. It follows along the medial border of the peduncle, enters cerebellum, and ends in the vermis of the same and opposite sides.

The **visceral afferent fibers** have their cell bodies in the spinal ganglia. The peripheral branches probably traverse the rami communicantes from the sympathetic nerves of the viscera. The central fibers divide into ascending and descending branches in the spinal cord. Their terminations and connections are not known. Some fibers accompanying the peripheral sympathetics are pain fibers which enter the lateral spinothalamic tract with the other pain fibers.

Somatic Efferent Fibers to the Spinal Cord.—The large multipolar cells in the ventral horn of the spinal cord receive impulses from a number of higher centers. They are the cells whose axons terminate in the motor end plates of the skeletal muscle fibers and are called the **lower motor neurons** or the final common pathway for motor response. These same neurons receive fibers from other parts of the gray substance of the cord to complete the various spinal cord reflexes.

The **Pyramidal Tract.**—Voluntary control of muscles is conveyed by axons of the pyramidal cells in the precentral (area 4, Fig. 11–69) and adjacent cortex. As they pass through the corona radiata they are gathered together and occupy the rostral two-thirds of the occipital part of the internal capsule. They continue through the cerebral peduncle as the corticospinal fibers, occupying the middle three-fifths of the crusta. They split up into several bundles as they pass through the pons but come together again as the pyramids of the medulla oblongata. In the caudal part of the medulla, two-thirds to three-fourths of the fibers, those nearest the median fissure, cross to the other side in bundles as the decussation of the pyramids. They continue down the spinal cord as the lateral corticospinal or crossed pyramidal tract. The remaining fibers, more laterally placed, do not decussate but continue down the cord as the ventral corticospinal or direct (uncrossed) pyramidal tract. The lateral corticospinal fibers terminate either directly or through intercalated neurons in synapses with the ventral horn cells. The fibers of the ventral corticospinal tract cross to the opposite side in the ventral white commissure before they form connections with the ventral horn cells. A few fibers of both tracts remain uncrossed and terminate on the same side of the cord. The axons of the ventral horn cells pass out through the ventral roots to their specific muscles.

The **Rubrospinal Tract.**—The coordination and reflex control of motion from the cerebellum are conveyed by the cerebellorubral fibers in the superior cerebellar peduncle. From the red nucleus, fibers cross in the ventral tegmental decussation of Forel and as the rubrospinal tract pass through the pons, medulla, and lateral funiculus of the spinal cord. The rubrospinal fibers end either directly or indirectly through terminals and collaterals about the motor cells of the ventral horn of the side opposite the red nucleus of origin. A few are said to remain on the same side. The afferent arms of the cerebellar connections are described on page 819.

The **Tectospinal Tract.**—Reflexes of coordination with vision and probably with hearing are conveyed by the tectospinal tract. The axons of large cells in the stratum opticum and stratum lemnisci of the superior colliculus cross the median raphe of the midbrain in the fountain decussation of Meynert and descend in the tegmentum. Some of the fibers are said not to cross. The tectospinal tract continues through the reticular formation of the pons and medulla, becoming more or less intermingled with the medial longitudinal fasciculus, and in the spinal cord splits into a larger medial tectospinal tract and a smaller lateral tectospinal tract which runs with the rubrospinal tract. The fibers end either directly or indirectly by terminals and collaterals in relation with the motor cells of the ventral horn.

The **Direct Vestibulospinal Tract.**—The coordination and reflex control of equilibrium are conveyed through the vestibulospinal tract. Axons of cells in the lateral vestibular nucleus (magnocellular of Deiters) descend in the ventral funiculus of the cord and terminate directly or indirectly in relation to the ventral horn cells. These fibers are uncrossed and extend as far as the sacral region of the cord. They tend to mingle with the fibers of the spinothalamic, tectospinal tracts, and the fasciculi proprii.

Meninges of the Brain and Spinal Cord

The brain and spinal cord are enclosed by three membranes, (1) an outer tough protective membrane, the dura mater, (2) an inner, more delicate, fibrous membrane, the pia mater, which carries the blood vessels to the brain and cord, and (3) an intermediate spiderweb-like structure which, together with the spinal fluid, fills in the space between the other two.

The Dura Mater

The dura mater, also known as the **pachymeninx**, is composed of dense, fibrous connective tissue with collagenous bundles arranged in interlacing layers. It is a dual structure with an inner meningeal layer and an outer periosteal layer. Although these layers are continuous at the foramen magnum their disposition over the brain and spinal cord is different and will be described separately.

Cranial Dura Mater (*dura mater encephali*) (Fig. 9–13ff.). The two layers of the dura over the brain are tightly fused together except in certain places where they are separated to provide space for the venous sinuses (page 688) and where the inner layer forms fibrous septa between parts of the brain. The outer surface is closely applied to the inner surface of the cranial bones. It sends many fine fibrous and vascular projections into the bony substance which give it a hairy appearance when the bones are stripped away. The attachment is more secure over the sutures and at the base of the skull. The internal surface of the dura is smooth, unattached, and covered with a layer of mesothelium lining the subdural space. The inner layer, leaving the outer layer at the position of certain venous sinuses, sends partitions inward between larger divisions of the brain named (*a*) the falx cerebri, (*b*) the tentorium cerebelli, and (*c*) the falx cerebelli. In addition, it forms smaller projections and partitions, the diaphragma sellae and the pocket for the trigeminal ganglion (cavum Meckelii).

The **falx cerebri** (Fig. 9–14) is a strong membrane extending down into the longitudinal fissure between the two cerebral hemispheres. It is attached to the skull bones along the midline of the inner surface of the cranial vault from the christa galli to the internal occipital protuberance where it becomes continuous with the tentorium cerebelli (Fig. 9–14). At this attachment it is separated from the outer layer of dura, leaving space for the superior sagittal sinus (Figs. 9–13, 11–103). The inner free margin of the falx contains the inferior sagittal sinus (Fig. 9–14).

The **tentorium cerebelli** (Figs. 9–14, 12–1) is a transverse shelf of dura mater separating the cerebellum from the occipital part of the cerebral hemispheres. It is attached laterally and posteriorly to the transverse sinuses. Anteriorly it is attached along the

superior border of the petrous portion of the temporal bone and to the posterior clinoid process of the sphenoid bone, leaving a narrow space for the superior petrosal sinus (Figs. 9–14, 9–15). It slopes upward toward the midline where it is continuous with the falx cerebri and there forms the straight sinus (Fig. 9–13). Its free border extends from the junction with the free border of the falx cerebri to the anterior clinoid process, curving laterally to leave a large oval opening, the incisura tentorii, for passage of the cerebral peduncles.

The **falx cerebelli** is a small triangular process of dura which is attached to the lower division of the vertical crest on the inner surface of the occipital bone with its free border projecting into the posterior cerebellar notch between the two cerebellar hemispheres.

The **diaphragma sellae** (Fig. 9–14) connects the clinoid attachments of the two sides of the tentorium cerebelli. It forms a roof over the hypophysis lying in the sella turcica. A circular opening in the center, which allows passage of the infundibulum, is surrounded by the circular or intercavenous sinus (Fig. 9–15).

The **Spinal Dura Mater** (*dura mater spinalis*) (Fig. 11–101) forms a loose sheath around the spinal cord and corresponds to the inner or meningeal layer of the cranial dura. The outer or periosteal layer is interrupted at the foramen magnum and is represented below this point by the periosteum of the vertebrae which lines the vertebral canal. A considerable interval, the epidural space, intervenes between the spinal dura and the vertebral canal. It contains a quantity of loose areolar tissue and a plexus of veins; the veins correspond in position with the cranial dural sinuses. The spinal dura is attached to the circumference of the foramen magnum, to the second and third cervical vertebrae, and by fibrous slips to the posterior longitudinal ligament, especially near the caudal end of the vertebral canal. The tubular sheath of the dura is much larger than is necessary for its contents, most of the interval between it and the cord being occupied by subarachnoid space (Fig. 11–103). The cavity of the tube ends at the level of the second sacral

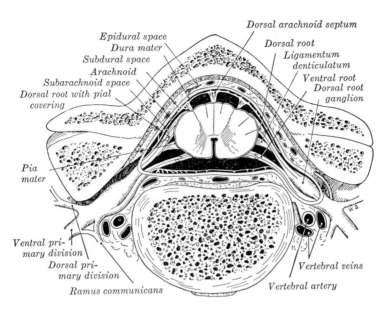

Dorsal arachnoid septum

Epidural space
Dura mater
Subdural space
Arachnoid
Subarachnoid space
Dorsal root with pial
covering

Dorsal root
Ligamentum
denticulatum
Ventral root
Dorsal root
ganglion

Pia
mater

Ventral pri-
mary division
Dorsal pri-
mary division
Ramus communicans

Vertebral veins
Vertebral artery

Fig. 11–101.—Cross section of spinal cord in the spinal canal showing its meningeal coverings and the manner of exit of the spinal nerves. (After Rauber in Everett, *Functional Neuroanatomy*, 6th Edition, 1971, Lea & Febiger.)

vertebra at which point the dura closely invests the filum terminale and is attached with the latter to the back of the coccyx, blending with the periosteum. The caudal part of the sheath, from the conus medullaris at the second lumbar to the second sacral vertebra, is occupied by the cauda equina. As the spinal nerves on each side pass through the intervertebral foramina they are covered by prolongations of the dura. These sheaths are short in the cephalic part of the vertebral column, but gradually become longer more caudally.

The Pia Mater

The pia mater is a delicate connective tissue membrane, closely applied to the brain and spinal cord, and carrying the rich network of blood vessels which supply the nervous tissue. It is attached to the nervous tissue especially where the minute vessels penetrate the pia-glial membrane. The outer surface is covered by ramifications of the arachnoid villi. The perineurium of the nerves leaving the brain and cord is reinforced by fibers from the pia mater which blend with the dura at the exit of the nerves.

The two delicate membranes, the pia

mater and the arachnoid, taken together form the **leptomeninges.**

The **Cranial Pia Mater** (*pia mater encephali*) (Fig. 11–26) invests the entire surface of the brain, dipping into the fissures and sulci of the cerebral and cerebellar hemispheres. It extends into the transverse cerebral fissure where it forms the tela choroidea of the third ventricle and combines with the ependyma to form the choroid plexuses of the third and lateral ventricles. It also passes over the roof of the fourth ventricle and forms its tela choroidea and choroid plexus.

The **Spinal Pia Mater** (*pia mater spinalis*) (Figs. 11–18, 11–101) is thicker, firmer, and less vascular than the cranial pia mater. It consists of two layers, the outer or additional one being composed of longitudinally arranged collagenous fibers. The inner layer is intimately adherent to the entire surface of the spinal cord and sends a septum into the anterior median fissure. The fibers of the outer layer are concentrated along this fissure into a stout glistening band, the **linea splendens.** Another concentration of fibers along each side forms the denticulate ligament. At the caudal end of the spinal cord, the pia mater is prolonged into the filum terminale which blends with the dura mater at the second sacral vertebra and continues caudally to the coccyx where it fuses with the periosteum. It secures the caudal end of the spinal cord, is called the central ligament of the spinal cord, and assists in maintaining the cord in position during movements of the body.

The **denticulate ligament** (*ligamentum denticulatum*) (Fig. 11–18) is a fibrous band of pia mater extending the entire length of the spinal cord on each side between the dorsal and ventral spinal nerve roots. Its lateral border has a festooned appearance due to its attachment to the dura at regular intervals. There are twenty-one of these points of attachment, the most cranial one at the foramen magnum between the vertebral artery and the hypoglossal rootlets and the most caudal at the conus medullaris.

The Arachnoid

The arachnoid (Fig. 11–18) is a delicate, avascular membrane lying between the dura mater and pia mater. It is separated from the dura by the subdural space and the pia by the subarachnoid space containing the cerebrospinal fluid.

The **Cranial Arachnoid** (*arachnoidea encephali*) (Fig. 11–102) is closely applied to the inner surface of the dura over the brain but is separated from it by a thin film of fluid in the subdural space. Its surface adjacent to the dura is covered with a layer of mesothelium. It does not dip into the sulci or fissures except to follow the falx and tentorium. The inner surface of the arachnoid membrane is connected with pia by delicate fibrous threads, the arachnoid trabeculae, which traverse the subarachnoid space intervening between these two membranes.

The **Spinal Arachnoid** (*arachnoidea spinalis*) (Fig. 11–101) is a tubular membrane, loosely investing the spinal cord. It is continuous with the cranial arachnoid and caudally encloses the cauda equina.

The **Subarachnoid Space** (*cavum subarachnoidea*) (Fig. 11–102) is narrow on the surface of the hemispheres. Over the summit of the gyri the pia and arachnoid are in close contact, and together are called the piaärachnoid membrane. The arachnoid bridges over the sulci, leaving a wider space. At the base of the brain the arachnoid is somewhat thicker and in places leaves wide intervals of space, the subarachnoid cisternae, which are named from their positions.

The **cisterna cerebellomedullaris** (*cisterna magna*) (Fig. 11–102) results from the arachnoid bridging over the interval between the projecting caudal part of the

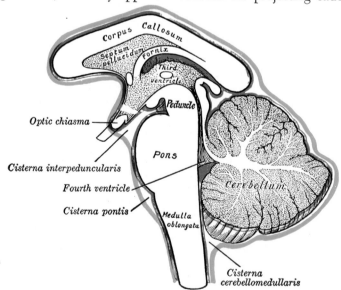

Fig. 11–102.—Diagram showing the positions of the three principal subarachnoid cisternae.

cerebellar hemispheres and the medulla oblongata, and is continuous with the spinal subarachnoid space. The **cisterna pontis** over the ventral aspect of the pons contains the basilar artery. The **cisterna interpeduncularis** (*cisterna basalis*) is a wide cavity between the two temporal lobes including the interpeduncular fossa. It contains the circle of Willis. It extends rostrally over the optic chiasma, forming the *cisterna chiasmatis*. The **cisterna fossae cerebri lateralis** occupies the area over the stem of the lateral sulcus and contains the lateral cerebral artery. The **cisterna venae magnae cerebri** occupies the interval between the splenium of the corpus callosum and the superior surface of the cerebellum. It reaches in between the layers of the tela choroidea of the third ventricle and contains the great cerebral vein of Galen.

The **ventricular system** of the brain (page 854) opens out into the subarachnoid space through the fourth ventricle by three holes. The foramen of Majendie (Fig. 11–51) is in the caudal part of the roof of the fourth ventricle in the midline. The two foramina of Luschka are at the ends of the lateral recesses of the fourth ventricle (Fig. 11–50), between the flocculus of the cerebellum and the glossopharyngeal nerve.

The **spinal part of the subarachnoid space** is a wide interval, occupying much of the space within the dural sheath. It is incompletely divided by the longitudinal subarachnoid septum which connects the arachnoid and pia along the posterior median sulcus of the cord. It is further subdivided by the denticulate ligament of the pia mater.

The **arachnoid granulations** (*Pacchionian bodies; arachnoid villi; glandulae Pacchioni*) (Fig. 11–102) are berry-like tufts of arachnoid which protrude into the superior sagittal sinus or venous lacunae associated with it. They do not occur during infancy and are rare before the third year but usually are found after the seventh year. They increase in number and size as age advances. They push against the dura and eventually cause absorption of bone, leaving depressions in the inner table of the calvaria. The wall of a villus is composed only of arachnoid and endothelium of the venous sinus; the cerebrospinal fluid in the subarachnoid space passes through this thin membrane and is taken up by the blood stream.

The Cerebrospinal Fluid

The cerebrospinal fluid fills the ventricles of the brain and occupies the subarachnoid space (Fig. 11–103). It is a clear watery fluid, similar to tissue fluid and lymph in composition but exhibiting certain quantitative differences. Normally it is elaborated by the choroid plexus of the ventricles of the brain. There is some evidence that it may have other sources and the volume of its production can be altered by changes in blood pressure relationships.

The fluid drains from the lateral ventricles through the interventricular foramina of Monro into the third ventricle. This fluid, combined with that produced by the choroid plexus of the third ventricle, passes through the cerebral aqueduct of Sylvius into the fourth ventricle. Fluid escapes from the fourth ventricle through the openings in its roof, the median foramen of Magendie and the two lateral foramina of Luschka. The fluid in the central canal of the spinal cord apparently is produced by the ependymal lining and drains into the fourth ventricle. An obstruction of the foramina between the ventricles or of the exit from the fourth ventricle causes an accumulation of fluid in the ventricles with the resultant condition known as **hydrocephalus** (Sweet and Locksley '53).

From the foramina of the fourth ventricle the fluid enters the subarachnoid space and its various cysternae. The fluid is absorbed by the blood stream through the arachnoid villi (pacchionian bodies) which protrude into the superior sagittal sinus (Fig. 9–13). Small amounts of fluid may escape through the perineural spaces of the cranial and spinal nerves and reach the lymphatic capillaries.

Lumbar Puncture.—The spinal fluid is usually removed for diagnostic purposes from the subarachnoid space surrounding the cauda equina of the spinal cord. A needle inserted in the midline between the spines of the third and fourth lumbar vertebrae will enter the subarachnoid space,

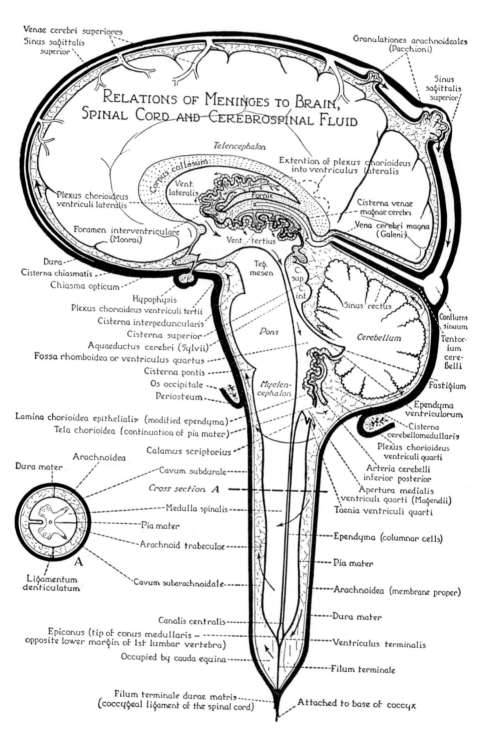

FIG. 11-103.—The relations of the meninges to the brain, spinal cord, and cerebrospinal fluid. (From Rasmussen, *The Principal Nervous Pathways*, courtesy of the Macmillan Company.)

avoiding the spinal cord and insinuating itself among the nerves of the cauda equina (Fig. 11–17). **Cistern puncture** is performed by inserting a needle between the atlas and the occipital bone, entering the cisterna cerebellomedullaris or cisterna magna (Fig. 11–102).

Spinal anaesthesia is administered by introducing the anaesthetic into the subarachnoid space by lumbar puncture.

HISTOLOGY OF THE NERVOUS SYSTEM

The basic elements of the nervous system are the nerve cells or neurons with their processes and certain special supporting and protective structures.

The Neuron (Fig. 11–104).—The nerve cell has a nucleus and cell body or **perikaryon,** with the usual organelles and inclusion bodies. Characteristically the nucleus is large and vesicular with a prominent nucleolus in which the sex element may be clearly visible (Barr, Bertram and Lindsay '50). Adult nerve cells are never seen in mitosis. The cytoplasm contains clear-cut mitochondria, and a Golgi apparatus can be demonstrated with proper technique. The specialized structures of nerve cells include neurofibrillae and Nissl granules. The delicate **neurofibrillae** can be demonstrated with special stains and with the electron microscope but are not clearly visible in living mammalian cells. A chromophilic substance is usually visible in clumps known as Nissl granules or **tigroid bodies** when the tissue is stained with certain basic dyes (Fig. 11–104). Pigment granules of various shades of yellow, brown, black, and red are found in some nerve cells, especially in particular nuclear masses such as the substantia nigra and the red nucleus.

Nissl granules have become important structures in neurological research. Many nerve cells of a particular functional type have a characteristic size, shape, and distribution of the Nissl bodies, and this may be used as corroborative evidence in identifying the function of a nuclear group. When the axon of a nerve cell is cut, the number of Nissl bodies diminishes by a process known as chromatolysis. This is used extensively in following nerve tracts after injury or experimental interruption.

Processes of neurons are of two kinds, **dendrites** (*dendrons*) and axons. There may be one or several dendrites. They are branched prolongations of the cytoplasm and contain all the structures found in the perikaryon, *i.e.*, Nissl bodies, mitochondria, fibrillae, pigment, etc. They vary greatly in size, shape and number with different types of neurons. They divide and subdivide into smaller and smaller branches. Many short processes called **gemmules** project from them and produce a ragged appearance. They are naked, without myelin or other sheath. Some are more or less characteristic such as the long apical dendrites of pyramidal cells in the cerebral cortex, or the Purkinje cells of the cerebellum (Figs. 11–49, 11–68).

Axons or Axis Cylinders (Fig. 11–104A).— Typically, a nerve cell has only one axon, but this may give off a number of branches or collaterals during its course. Unlike the dendrite, the axon has a uniform small diameter and smooth surface except near its termination. It contains neurofibrillae and mitochondria but not Nissl bodies nor granules. At its junction with the cell body there is commonly a low projection, also devoid of Nissl bodies, called the **axon hillock.** The unipolar neurons of the spinal ganglia are different from most others in having no dendrites and a single axon which divides into a central and a peripheral process. The cells in the cochlear and vestibular ganglia are bipolar, with two axons and no dendrites (Hartmann '54).

Neuron Doctrine and Conduction of Impulses.—The neuron theory was given its final corroboration by Harrison ('07) when he watched the axons grow out of tadpole neuroblasts in tissue cultures. A corollary of the doctrine is that the nervous impulse travels in a particular direction with respect to the processes of the nerve cell. The impulse is received by the dendrites and passed out to the next cell in the chain or to the end organ by the axon. The contact between two nerve cells is the **synapse.**

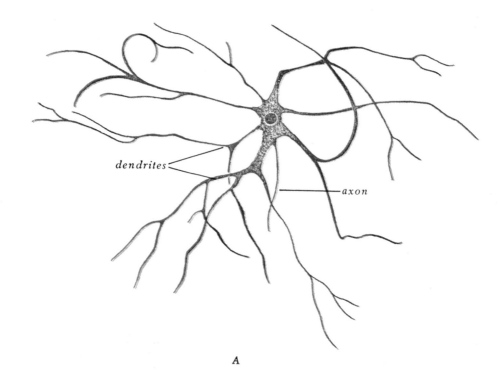

A

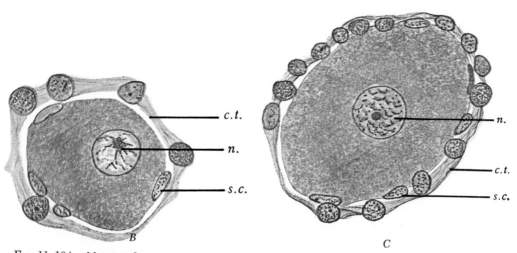

B

C

Fig. 11–104.—Motor and sensory neurons. *A,* Motor neuron from the ventral horn of a human spinal cord. Macerated preparation. × 200. *B* and *C,* Sensory ganglion cells from a section through the spinal ganglion of a 22-year-old man. *c.t.,* connective tissue capsule; *n.,* nucleus and nucleolus of neuron; *s.c.,* nucleus of sheath (*Schwann*) cell. × 500. (Redrawn from Sobotta.)

The axon breaks up into a number of **telodendria** which terminate in special knobs or endings against the dendrites or perikaryon of the next nerve cell in the path of the impulse. Exceptions are found in the unipolar and bipolar sensory cells, in which both receiving and transmitting processes are called axons.

Nerve Fibers (Fig. 11–107).—A nerve fiber is the axon of a nerve cell and may be short or very long, depending on its location and function. Nerve fibers usually have protective sheaths except at their origin from the perikaryon or near their termination. Longer naked fibers are found in the neuropil of the gray substance of the central nervous system (Fig. 11–108). Fibers with sheaths are of two types, myelinated (medullated) and unmyelinated (non-medullated). The myelinated fibers in the central nervous system make up the white substance and are embedded in a matrix of the neuroglia. The fibers of the peripheral nerves, both myelinated and unmyelinated, have cellular sheaths called neurilemma.

Myelin Sheath (Figs. 11–105, 11–106).—Myelin is a lipoidal substance which gives a whitish appearance to fresh nerves because of its high refractivity. Recent studies with the electron microscope have shown that the myelin occurs in multiple thin lamellae (about 100 Å) wrapped around the axon (Robertson '54). In fresh or living nerve, the myelin sheath appears structureless except for certain irregular artifacts. The myelin sheaths in peripheral nerves have interruptions at more or less regular intervals 1 or 2 mm apart known as **nodes of Ranvier.** In the central nervous system nodes occur rarely and only at a

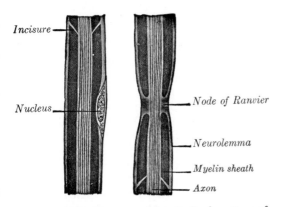

Incisure

Nucleus

Node of Ranvier

Neurolemma

Myelin sheath

Axon

Fig. 11–105.—Diagram of longitudinal sections of myelinated nerve fibers. Osmic acid.

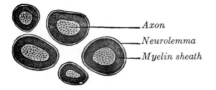

Axon

Neurolemma

Myelin sheath

Fig. 11–106.—Transverse sections of myelinated nerve fibers. Osmic acid.

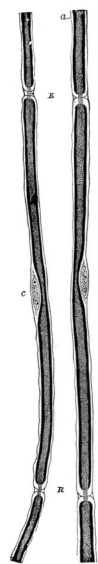

Fig. 11–107.—Diagram of myelinated nerve fibers stained with osmic acid. × 425. (Schafer.) R, Nodes of Ranvier. a, Neurilemma. c, Nucleus.

point of bifurcation of a fiber. The myelin sheaths generally vary in diameter from 2 to 10 micra, but large and smaller ones do occur (Bodian '51).

Myelin colors poorly in histological preparations of the commonly used cytoplasmic and nuclear stains. It blackens readily with osmic acid, however, and can be stained quite selectively by special methods such as the Weigert stain. This coloration of the myelin sheath is used extensively in the study of the fiber tracts of the central nervous system and makes possible such pictures as Figure 11–39.

Neurilemma.—The neurilemma or **sheath of Schwann** is composed of a series of very attenuated nucleated cells. On myelinated nerves, there is a neurilemma cell for each internodal segment between nodes of Ranvier. The nucleus is oval and flattened and lies in a slight depression in the myelin. The neurilemma cells of unmyelinated fibers are difficult to distinguish from the fibers except by the presence of their nuclei unless the

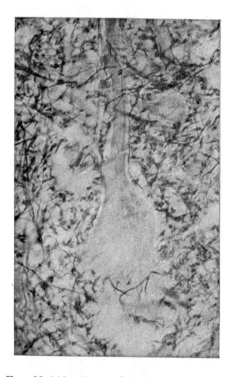

Fig. 11–108.—Neuropil from ventral horn of spinal cord. The large motor cell with nucleus is very faintly stained; the nerve fibers and terminal buttons are blackened with silver. Photograph by Charles O'Connor. Magnified 15000 ×.

nerve fibers themselves have been brought out by some special method such as a silver stain.

Since nerve fibers are protoplasmic threads of minute size, 1 or 2 micra in diameter, they are very difficult to distinguish from surrounding tissue in ordinary histological preparations. It is possible, however, to blacken them with silver and it is by this means that their morphological details have been studied. Another staining method is intravital methylene blue but both this and silver techniques are notoriously capricious.

Gray Substance of the Central Nervous System.—The gray substance of the brain and spinal cord is composed of nerve cells, naked nerve fibers, dendrites, and the supporting tissue neuroglia, all of which taken together is called the **neuropil.** In the usual histological preparations stained with hematoxylin and eosin, the tissue has a rather homogeneous appearance with a preponderance of eosin-stained cytoplasm and scattered nuclei. The nuclei of the nerve cells can be distinguished by their greater size, vesicular appearance, and prominent nucleoli. The neuroglia are described below. Blood vessels appear as in other tissues except that their walls tend to be thinner, and in sections including the surface of the brain, the pia mater and arachnoid can be seen. As mentioned above, the nerve cell bodies can be emphasized by the use of Nissl stain, the nerve fibers by silver methods, and the neuroglia by still other special methods.

White Substance.—The white substance is composed of fibers. The myelinated fibers predominate in bulk but may be outnumbered by unmyelinated fibers which are less prominent because of their small size. The fibers are supported by the neuroglia and blood vessels and at the surface the leptomeninges offer additional support. The myelin sheaths are easily distinguished by their circular shape, their size, and their refractivity although they stain but lightly in hematoxylin and eosin preparations. The only nuclei visible are those of the neuroglia and vascular tissue.

Neuroglia.—The supporting tissue of the central nervous system includes three cellular elements, two of which, astrocytes and oligodendrocytes, are of ectodermal origin derived from the spongioblasts of the embryonic neural tube (page 777). The third, microgliocyte, is of mesodermal origin, apparently being a specialized connective tissue histiocyte (Glees '55).

(1) **Astrocytes** or **macroglia** are moderately large cells with numerous processes which radiate out from the cell body. In most instances one or more processes have terminal expansions which attach to blood vessels or the pia mater. The marginal astrocytes together with the pia form the pia-glial membrane, which invests the brain and spinal cord and accompanies penetrating blood vessels as a cuff to considerable

depths. There are two types of **astrocytes,** **fibrous** and **protoplasmic** (Fig. 11–109). The former are found chiefly in the white matter and, with appropriate fixation and staining, appear to contain fibers which run through the protoplasm of the cell body and processes. The **protoplasmic astrocytes** occur chiefly in the gray matter and their processes branch profusely. The nuclei of the astrocytes are moderately large, oval in shape, and contain scattered chromatin granules, but no nucleolus. The astrocytes provide the repair mechanism and replace lost tissue by forming glial scars.

(2) **Oligodendrocytes** or **oligodendroglia** are somewhat smaller than the astrocytes

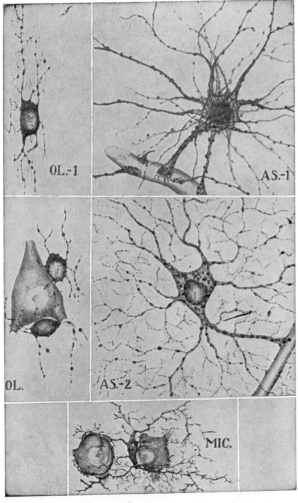

Fig. 11–109.—Interstitial cells of the central nervous system. *AS.–1,* Fibrous astrocyte with perivascular feet on vessel; *AS.–2,* protoplasmic astrocytes; *MIC.,* microglia; *OL.,* oligodendroglia. (Penfield and Cone, in Cowdry's *Special Cytology,* courtesy of Paul B. Hoeber, Inc.)

and have fewer processes. They are found either in close association with smaller blood vessels, or as satellite cells closely applied to large nerve cells, or in rows between bundes of fibers in the white matter (Fig. 11–109). In the latter situation, their processes clasp the nerve fibers. It is thought that the oligodendroglia play a metabolic role in the formation and preservation of the myelin sheaths of the nerve fibers in the central nervous system. The nuclei of these cells are round to oval, usually smaller in size but richer in chromatin than the nuclei of the astrocytes. They have no nucleoli.

(3) **Microgliocytes** or **microglia** are found diffusely through both the gray and white matter. Normally they are small cells, with two or more finely branching, feathery processes (Fig. 11–109). The nucleus is small and varies in shape from round to oblong or angular. It stains deeply but contains no nucleolus. The microglia are the scavengers of the nervous system and become actively ameboid and phagocytic in case of injury and death of the other elements.

Peripheral Nerves.—The peripheral nerves are of two types depending on whether their fibers are predominantly myelinated or whether they are almost exclusively unmyelinated. The former are the spinal nerves and their branches and, with certain exceptions, the cranial nerves.

The latter are the nerves of the autonomic system.

Myelinated nerves (*medullated nerves*) are recognizable by the characteristic structure of the myelin sheaths. In cross sections they can be recognized by their large circular outline with the axon in the center (Fig. 11–106). In paraffin sections either the myelin sheath or the axon may be somewhat shrunken and displaced. In longitudinal sections the nodes of Ranvier are a distinguishing feature. Usually a nerve shortens or retracts during fixation with the result that the fibers have a wavy course and many of them are cut across or with varying degrees of obliquity.

Connective tissue sheath.—In addition to the special sheaths on individual fibers, peripheral nerves have well-organized connective tissue sheaths (Fig. 11–110). Around and between the individual fibers is a sheath of reticular connective tissue fibrils called the **endoneurium.** The nerve fibers are collected into cylindrical bundles with an outer wrapping of the collagenous and elastic fibers of areolar tissue. This is called the **perineurium.** Large nerves have these secondary bundles bound together with an outside sheath called the **epineurium.** The connective tissue contains the usual cells, fibroblasts, macrophages, etc., and capillaries, arteries, veins, and lymphatic vessels appropriate to the size of the nerve.

Unmyelinated nerves (*non-medullated*)

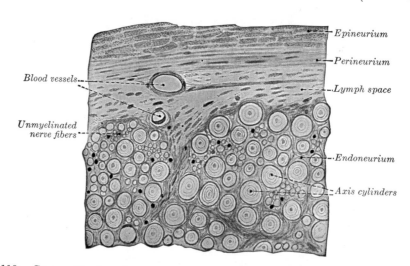

FIG. 11–110.—Cross section from human sciatic nerve. (From Rauber-Kopsch, *Lehrbuch u. Atlas d. Anatomie d. Menschen*, 19th Edition, Vol. II, courtesy of Georg Thieme Verlag, Stuttgart, 1955.)

have the same basic connective tissue elements as myelinated nerves except that the nerves are usually much smaller. They make up the trunks and branches of the autonomic nervous system. In cross sections there is little of characteristic structure except the connective tissue sheath. In longitudinal section there is a considerable resemblance to non-striated muscle when a comparison is made in the usual histological slides. The unmyelinated fibers (Remak fibers) can be stained with silver and appear abundant in many so-called myelinated cranial and spinal nerves as well as in the unmyelinated nerves of the sympathetic and parasympathetic systems.

Ganglia.—The nerve cells occurring outside the central nervous system are chiefly collected into groups in ganglia although many individual cells and very small groups are found, especially in association with the alimentary organs. Ganglia are of two types, (1) sensory ganglia on the dorsal roots of spinal nerves and on the sensory roots of the trigeminal, facial, glossopharyngeal, and vagus nerves, and (2) the autonomic ganglia of the sympathetic and parasympathetic systems. The special sensory ganglia of the statoacoustic nerve are described in the chapter on the sense organs.

Sensory Ganglia (Fig. 11–109).—In a sensory ganglion, the cluster of ganglion cells and traversing fibers is invested by a smooth, firm, closely adhering, membranous envelope of dense areolar tissue continuous with the perineurium of the nerves. Numerous septa penetrate the ganglia and carry blood vessels to supply it. Each nerve cell has a nucleated sheath which is continuous with the neurilemma of its axon. The nerve cells in the ganglia of the spinal nerves are piriform in shape, and have a single process, the axon, which divides a short distance from the cell, while still in the ganglion, in a T-shaped manner. One branch runs in the posterior root and enters the spinal cord: the larger peripheral branch joins the spinal nerve and runs uninterruptedly to its end organ in the skin or muscle. The larger unipolar neurons have coiled or split axons near the cell body, but the axon straightens into a single myelinated fiber before it divides. The smaller and more numerous unipolar cells may or may not have the proximal part of the axon coiled. They have unmyelinated axons which divide in a T-like manner into a fine central and coarse peripheral branch. The central branches go to the spinal cord and the peripheral ones to the skin, a few to muscles (Hess '55, Adamstone *et al.* '53).

Autonomic Ganglia contain the cell bodies of multipolar postganglionic neurons with one axon and several dendrites (Fig. 11–112). Myelinated (preganglionic) and unmyelinated (postganglionic) fibers end,

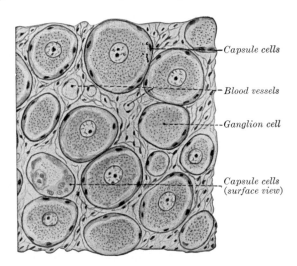

Fig. 11–111.—Section of the sixth cervical spinal ganglion of man. (From Rauber-Kopsch *Lehrbuch u. Atlas d. Anatomie d. Menschen,* 19th Edition, Vol. II, courtesy of Georg Thieme Verlag, Stuttgart, 1955.)

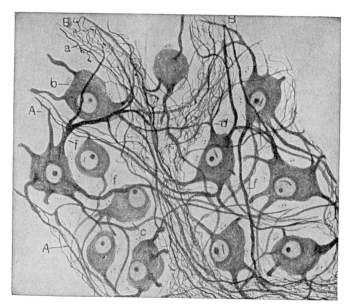

FIG. 11–112.—A portion of the stellate ganglion of a one and one-half-year-old child. *B*, Stout protoplasmic tract of articulation; *f*, very long primordial dendrite; *d*, very short dendrite; *c*, monopolar sympathetic cell; *a*, preganglionic fibers; *A*, tracts passing through the ganglion. Several of the neurons show intranuclear rods. (deCastro in Penfield's *Cytology and Cellular Pathology of the Nervous System*, courtesy of Paul B. Hoeber, Inc.)

traverse, or arise in the ganglion. Sheath cells form capsules similar to those in the sensory ganglia around the bodies of the ganglion cells. In addition, the sheath cells extend along the dendrites forming what are known as protoplasmic bands or cords, in which little structure is visible. Inside these protoplasmic cords, the preganglionic fibers, having lost their myelin sheaths, end in an extraordinarily dense synaptic relationship with the dendrites of the neurons in the ganglion (Fig. 11–112).

Peripheral Termination of Nerves

Nerve fibers may terminate peripherally as simple naked fibers or, more frequently, they are enclosed in more or less elaborate end organs. For convenience they are divided into the two large functional groups of sensory or receptor and motor or effector endings.

Sensory Nerve Terminations or **Receptor Organs.**—It is customary to divide the sensory organs according to function also, into those of general and special sensibility. The special senses of vision, hearing, equilibration, taste, and smell are dealt with in the chapter on sense organs. The receptors of the general sensations of heat, cold, pain, touch, and proprioception are widely distributed throughout the body.

Free nerve endings occur chiefly in the epidermis and in the epithelium covering certain mucous membranes; they are seen in the stratified squamous epithelium of the cornea, in the root sheaths and papillae of the hairs, and around the bodies of the sudoriferous glands. As the nerve fiber approaches its termination, the myelin sheath disappears, leaving only the axon surrounded by the neurilemma. At its termination the fiber loses its neurilemma, and consists of an axon which breaks up into its constituent fine varicose fibrillae. These often present regular varicosities, anastomose with each other, and end in small knobs or disks between the epithelial cells. They are probably the endings for sensations of pain and temperature.

The special sensory end organs exhibit great variety in size and shape, but have the common feature of a capsule enveloping the terminal nerve fibrillae. Included in this group are the end bulbs of Krause, the corpuscles of Pacini, of Golgi and Mazzoni,

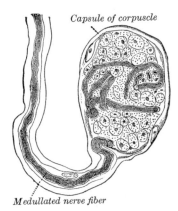

Capsule of corpuscle

Medullated nerve fiber

FIG. 11–113.—End bulb of Krause. (Klein.)

of Wagner and Meissner, and the neuro-tendinous and neuromuscular spindles.

The **end bulbs of Krause** (Fig. 11–113) are minute cylindrical or oval bodies, consisting of a capsule formed by the expansion of the connective tissue sheath of a medullated fiber. They contain a soft semifluid core in which the axon terminates either in a bulbous extremity or in a coiled plexiform mass. End bulbs are found in the conjunctiva of the eye (where they are spheroidal in shape in man, but cylindrical in most other animals), in the mucous membrane of the lips and tongue, and in the epineurium of nerve trunks. They are also found in the penis and the clitoris, and have received the name of genital corpuscles; in these situations they have a mulberry-like appearance, being constricted by connective tissue septa into from two to six knob-like masses. In the synovial membranes of certain joints, e.g., those of the fingers, rounded or oval end bulbs occur, and are designated articular end bulbs.

The **Pacinian corpuscles** (Fig. 11–114) are found in the subcutaneous, submucous, and subserous connective tissue of many parts of the body. They are especially numerous in the palm of the hand, sole of the foot, the genital organs, about joints, and in the mesentery about the pancreas. Each corpuscle is attached to the end of a single nerve fiber and when dissected from fresh tissue is visible to the naked eye as a white bulb 2 to 4 mm in diameter. Its structure is very characteristic, consisting of a number of concentric lamellae, in cross sec-

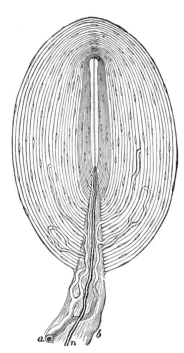

FIG. 11–114.—Pacinian corpuscle, with its system of capsules and central cavity. *a*, Arterial twig, ending in capillaries, which form loops in some of the intercapsular spaces, and one penetrates to the central capsule. *b*, The fibrous tissue of the stalk. *n*, Nerve tube advancing to the central capsule, there losing its white matter, and stretching along the axis to the opposite end, where it ends by a tuberculated enlargement.

tion reminiscent of an onion. In its elongated central core the nerve fiber loses, first its myelin sheath and then its neurolemma among specialized central layers.

The **corpuscles of Golgi** and **Mazzoni** are found in the subcutaneous tissue of the pulp of the fingers. They differ from Pacinian corpuscles in that their capsules are thinner, their contained cores thicker, and in the latter the axons ramify more extensively and end in flat expansions.

The **tactile corpuscles** of Meissner (Fig. 11–115) occur in the papillae of the corium of the hand and foot, the front of the forearm, the skin of the lips, the mucous membrane of the tip of the tongue, the palpebral conjunctiva, and the skin of the mammary papilla. They are small oval bodies with a connective tissue capsule and tiny plates stacked one above the other. The nerve fiber penetrates the capsule,

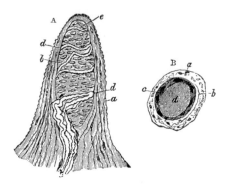

FIG. 11–115.—Papilla of the hand, treated with acetic acid. × 350. A, Side view of a papilla of the hand. *a,* Cortical layer. *b,* Tactile corpuscle of Meissner. *c,* Small nerve of the papilla, with neurilemma. *d,* Its two nervous fibers running with spiral coils around the tactile corpuscle. *e,* Apparent termination of one of these fibers. B, A tactile papilla seen from above so as to show its transverse section. *a,* Cortical layer. *b,* Nerve fiber. *c,* Outer layer of the tactile body, with nuclei. *d,* Clear interior substance.

spirals through the interior, and ends in globular enlargements.

Corpuscles of Ruffini.—Ruffini described a special variety of nerve endings in the subcutaneous tissue of the human finger (Fig. 11–116). They are principally situated at the junction of the corium with the subcutaneous tissue. They are oval in shape, and consist of strong connective tissue sheaths, inside which the nerve fibers divide into numerous branches, which show varicosities and end in small free knobs.

The **neurotendinous spindles** (*organs of Golgi*) are chiefly found near the junctions of tendons and muscles. Each is enclosed in a capsule which contains a number of enlarged tendon fasciculi (intrafusal fasciculi). One or more nerve fibers perforate the side of the capsule and lose their medullary sheaths; the axons subdivide and end between the tendon fibers in irregular disks or varicosities (Fig. 11–117).

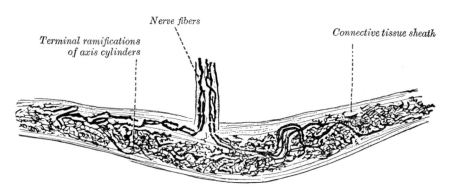

FIG. 11–116.—Nerve ending of Ruffini. (After A. Ruffini.)

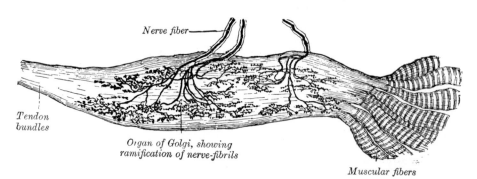

FIG. 11–117.—Organ of Golgi (neurotendinous spindle) from the human tendo calcaneus. (After Ciaccio.)

The **neuromuscular spindles** are present in the majority of voluntary muscles. They consist of small bundles of rather delicate muscular fibers (intrafusal fibers), invested by a capsule within which the sensory nerve fibers terminate. These neuromuscular spindles vary in length from 0.8 to 5 mm, and have a fusiform appearance. One to three or four large myelinated nerve fibers enter the fibrous capsule, divide several times, and, losing their myelin sheaths, ultimately end in naked axons encircling the intrafusal fibers by flattened expansions, or irregular ovoid or rounded disks (Fig. 11–118). Neuromuscular spindles have not yet been demonstrated in the tongue muscles, and only a few exist in the ocular muscles.

General Visceral Receptors.—Myelinated afferent fibers and probably unmyelinated afferent fibers traverse the sympathetic nerves and rami communicantes to the spinal ganglia. They convey impulses from receptors in the abdominal and thoracic viscera. Similar afferents traverse the vagus and glossopharyngeal nerves to their sensory ganglia and branches of the pelvic nerves contain visceral afferents from pelvic viscera. Afferent fibers from the lungs and heart especially are found in the vagus.

Special Visceral Receptors for reflex control of respiration and the circulation are supplied by branches of the vagus and glossopharyngeal nerves. The **carotid body** (*glomus caroticum*) is an oval mass about 5 mm in length situated at the bifurcation of the common carotid artery. It has a fibrous capsule and contains cords and clumps of epithelioid cells among which numerous nerve fibers branch into special endings (Figs. 11–119, 11–120). The afferent fibers join the vagus through the intercarotid plexus and pharyngeal branches. They are chemoreceptors, sensitive to the concentration of carbon dioxide in the blood, and assist in reflex control of respiration. The **aortic bodies** (*glomera aortica*) are similar to the carotid bodies. They have been found in mammals and probably occur in man. The right aortic body is situated at the junction of the right subclavian and right common carotid arteries. The left body is situated in the angle between the left subclavian artery and the aorta. The afferent fibers join the cardiac branches of the vagus.

The **carotid sinus** is a slight enlargement of the beginning of the internal carotid or the immediately adjacent part of the common carotid artery. Its walls contain special sinuses and elaborate networks and terminations of nerves. The afferent fibers reach the glossopharyngeal nerve through the carotid sinus nerve (nerve of Hering). The receptors are sensitive to changes in blood pressure and enter into reflex control of the circulation.

Motor nerves are supplied to both striated and smooth muscular fibers. In the **smooth** or **involuntary muscles** the nerves are derived from the autonomic system and are composed mainly of unmyelinated fibers. Near their terminations they divide into numerous branches, which communicate and form intimate plexuses. There are,

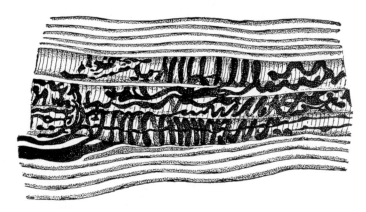

FIG. 11–118.—Middle third of a terminal plaque in the muscle spindle of an adult cat. (After Ruffini.)

however, no fusions of axons or their branches. There is considerable variation in the relationship of these peripheral plexuses to the muscles and glands which the fibers innervate and also in the number and distribution of the ganglion cells which may be associated with them. From these plexuses minute branches are given off which divide and break up into the ultimate fibrillae of which the nerves are composed. These fibrillae course between the involuntary muscle cells and gland cells and terminate on the surfaces of the cells.

The **motor nerves** to the **striated** or **voluntary muscles** are derived from the cranial and spinal nerves, and are composed mainly of myelinated fibers. The nerve enters the sheath of the muscle and breaks up into fibers or bundles of fibers which gradually divide until, as a rule, a single nerve fiber reaches a single muscular fiber. As the nerve terminates in a special expansion, called a **motor end plate** (Fig. 11–121), it loses its myelin sheath, ramifying like roots of a tree. The neurilemma merges with the sarcolemma of the muscle, and the axon makes an elaborate synaptic contact with the muscular fiber (Robertson '56, Cole '57).

Degenerative Changes.—Various types of degenerative change occur in neurons due to disease, trauma, senility, etc. Shrinkage and dissolution of cell bodies with consequent degeneration of the axons is a normal phenomenon which starts about the twentieth year and becomes increasingly pronounced with age. Symptoms resulting from such loss of neurons are rarely apparent until after the fifth decade. Three types of degeneration of neurons have been of value for study of the anatomy of the nervous system.

(1) If the cell body is destroyed the axon degenerates, or if an axon is cut that part severed from the cell body degenerates. This is known as **Wallerian degeneration.** Changes occur first at the cut end and progress peripherally. In fishes and amphibians this distal progression is slow and may take days to weeks, but in mammals it is rapid and requires only hours to days, so that at times it appears as though all parts of the severed axon were degenerating simultaneously. There is at first swelling with loss of internal structure. The axon becomes

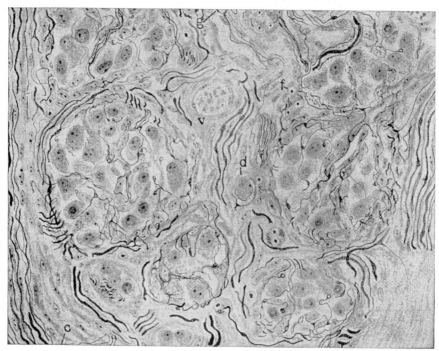

Fig. 11–119.—Human carotid body showing nerve endings and fibers among epithelioid cells. (De Castro, *Travaux du Laboratorie de Recherches Bibliogiques de Université de Madrid.*)

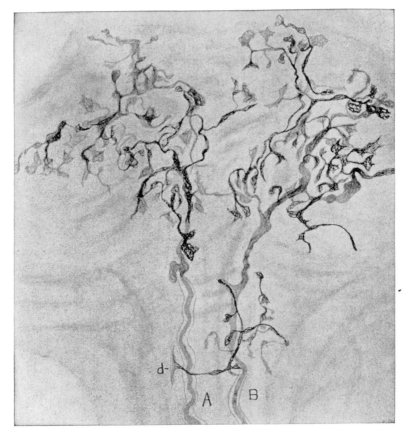

Fig. 11–120.—Receptor, Type II, of human carotid sinus. *A, B,* afferent medullated fibers to glosso-pharyngeal nerve. (De Castro, *Travaux du Laboratorie de Recherches Biologiques de Université de Madrid.*)

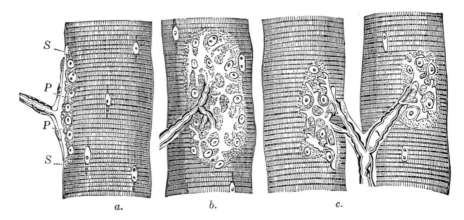

Fig. 11–121.—Muscular fibers of *Lacerta viridis* with the terminations of nerves. *a,* Seen in profile. *P, P,* The nerve end plates. *S, S,* The base of the plate, consisting of a granular mass with nuclei. *b,* The same as seen in looking at a perfectly fresh fiber, the nervous ends being probably still excitable. (The forms of the variously divided plate can hardly be represented in a woodcut by sufficiently delicate and pale contours to reproduce correctly what is seen in nature.) *c,* The same as seen two hours after death from poisoning by curare.

tortuous, then fragmented and soon disappears. The myelin breaks down into globules of fatty material which can be stained by the Marchi method. This change is at its maximum between two and three weeks. The degenerating myelin is gradually removed by the phagocytic cells. Fat globules, free and intracellular, can be found, for as long as three months, in the neighborhood of the degenerated fibers. In the central nervous system, no true regeneration occurs and the lost fibers are replaced by glia, chiefly astrocytes. The scar so formed does not stain by the Weigert method of staining normal myelin sheaths, and when the degenerated fibers form a compact bundle, such as the crossed corticospinal tract, their course can be followed by this method. Since these methods depend on changes in the myelin they have to be supplemented by silver stains on normal material in order to determine the exact termination of the axons.

Attempts to follow degenerating axons by silver and other stains, as well as attempts to identify degenerating synapses after lesions of the axons, have not proved satisfactory.

In peripheral nerves degeneration of the axon and myelin proceeds as described above. The sensory end organs, motor end plates, and striped muscles degenerate more slowly. Smooth muscles and glands undergo certain functional changes. The sheath of Schwann cells at first multiply markedly, and later become indistinguishable from connective tissue cells except by their arrangement in rows. Regeneration normally occurs in a manner closely resembling the original development of the peripheral nerves, though there are liable to be abnormalities in motor and in discriminatory distribution, which persist.

(2) The second type of degeneration important in anatomical investigations is known as **retrograde degeneration.** This term is applied to the changes which occur in the cell body and in the proximal portion of an axon which has been transected. There are wide variations in the extent to which the degeneration may proceed. Many cells show no demonstrable change. In others a condition known as **chromatolysis** occurs, which consists in fragmentation and loss of the Nissl bodies and in nuclear changes. The nucleus becomes shrunken, irregular, and eccentrically placed, surrounded by a clear zone and the remaining Nissl substance appears in a finely divided, powdery form around the periphery of the cell. Chromatolysis is usually well developed in six to ten days after the lesion, although early changes may be noted in twenty-four hours. In certain cells the condition may remain unchanged for years; other cells rapidly recover; a few types of cells, usually belonging to highly specialized systems, proceed to complete degeneration and disappear. In the latter conditions the proximal portions of the axons also degenerate. In peripheral nerves retrograde degeneration of the axons usually extends only one or two segments proximal to the injury.

(3) The third type of degeneration is known as **transynaptic.** It occurs only in the cells of certain systems and is probably best illustrated in the case of the visual system. When the optic tract is cut the next neurons in the pathway to the cortex, namely the cells of the lateral geniculate body, undergo chromatolysis which proceeds to disintegration and disappearance.

Retrograde and transynaptic degeneration have been widely used for determining the origins of fibers, both of peripheral nerves and of tracts in the central nervous system, and for studies on the interrelations of various parts of the forebrain. The methods obviously have their limitations, but with due care give reliable results where they are applicable.

REFERENCES

GENERAL

VON BONIN, G. 1963. *The Evolution of the Human Brain.* xiv + 92 pages, 7 figures. The University of Chicago Press, Chicago.

BURR, H. S. 1958. Design in the nervous system. Anat. Rec., *131,* 405-415.

CHUSID, J. G. and J. J. MACDONALD. 1964. *Correlative Neuroanatomy and Functional Neurology.* iv + 409 pages, illustrated. Lange Medical Publications, Los Altos, California.

EVERETT, N. B. 1971. *Functional Neuroanatomy* Including an Atlas of the Brain Stem. 420 pages, 304 illustrations. Lea & Febiger, Philadelphia.

KOPSCH, F. 1955. *Rauber-Kopsch Lehrbuch und Atlas der Anatomie des Menschen.* 19th Edition. 2 Volumes. Georg Thieme Verlag, Stuttgart.

KRIEG, W. J. S. 1953. *Functional Neuroanatomy.* xviii + 659 pages. The Blakiston Company, New York.

MAGOUN, H. W. 1959-60. *Neurophysiology. Section 1, Handbook of Physiology,* edited by J. Field. 3 Volumes. xiii + 779 pages; v + 781-1439; v + 1441-2013 pages, illustrated. The Williams and Wilkins Company, Baltimore.

PEELE, T. 1961. *The Neuroanatomic Basis for Clinical Neurology.* 2nd Edition. vii + 662 pages, illustrated. McGraw-Hill Book Company, Inc., New York.

RASMUSSEN, A. T. 1954. *The Principal Nervous Pathways.* 4th Edition. ix + 73 pages. The Macmillan Company, New York.

ROBERTS, M. and J. HANAWY. 1970. *Atlas of the Human Brain Stem in Section.* xv + 95 pages, 40 illustrations. Lea & Febiger, Philadelphia.

TRUEX, R. C. and M. B. CARPENTER. 1969. *Human Neuroanatomy.* 6th edition, xiv + 673 pages, illustrated. The Williams & Wilkins Co., Baltimore.

ZEMAN, W. and J. R. M. INNES. 1963. *Craigie's Neuroanatomy of the Rat.* ix + 230 pages, illustrated. Academic Press, New York.

EMBRYOLOGY

BROWN, J. W. 1956. The development of the nucleus of the spinal tract of V in human fetuses of 14 to 21 weeks of menstrual age. J. comp. Neurol., *106,* 393-423.

BUEKER, E. D. 1948. Implantation of tumors in the hind limb field of the embryonic chick and the developmental response of the lumbosacral nervous system. Anat. Rec., *102,* 369-389.

BURR, H. S. 1932. An electro-dynamic theory of development suggested by studies of proliferation rates in the brain of Amblystoma. J. comp. Neurol., *56,* 347-371.

CONEL, J. L. 1942. The origin of the neural crest. J. comp. Neurol., *76,* 191-215.

HARRISON, R. G. 1924. Neuroblast versus sheath cell in the development of peripheral nerves. J. comp. Neurol., *37,* 123-205.

HOGG, I. D. 1944. The development of the nucleus dorsalis (Clarke's column). J. comp. Neurol., *81,* 69-95.

HUMPHREY, T. 1952. The spinal tract of the trigeminal nerve in human embryos between 7½ and 8½ weeks of menstrual age and its relation to early fetal behavior. J. comp. Neurol., *97,* 143-209.

HUMPHREY, T. 1963. The development of the anterior olfactory nucleus of human fetuses. In *Progress in Brain Research.* Vol. 3. Edited by W. Bargmann and J. P. Schade. Pages 170-190.

KIMMEL, D. L. 1941. Development of the afferent components of the facial, glossopharyngeal and vagus nerves in the rabbit embryo. J. comp. Neurol., *74,* 447-471.

PATTEN, B. M. 1953. *Human Embryology.* 2nd Edition. xvii + 798 pages. The Blakiston Company, New York.

STRONG, L. H. 1961. The first appearance of vessels within the spinal cord of the mammal: Their developing patterns as far as partial formation of the dorsal septum. Acta Anat., *44,* 80-108.

YNTEMA, C. L. and W. S. HAMMOND. 1954. The origin of intrinsic ganglia of trunk viscera from vagal neural crest in the chick embryo. J. comp. Neurol., *101,* 515-541.

SPINAL CORD

CHAKRAVORTY, B. G. 1971. Arterial supply of the cervical spinal cord (with special reference to the radicular arteries). Anat. Rec., *170,* 311-329.

EARLE, K. M. 1952. The tract of Lissauer and its possible relation to the pain pathway. J. comp. Neurol., *96,* 93-111.

ELLIOTT, H. C. 1942-43. Studies on the motor cells of the spinal cord. I. Distribution in the normal human cord. II. Distribution in the normal human fetal cord. Amer. J. Anat., *70,* 95-117 and 72, 29-38.

GILLILAN, L. A. 1958. The arterial blood supply of the human spinal cord. J. comp. Neurol., *110,* 75-103.

HUMPHREY, T. 1950. Intramedullary sensory ganglion cells in the roof plate area of the embryonic human spinal cord. J. comp. Neurol., *92,* 333-399.

KITAI, S. T. and F. MORIN. 1962. Microelectrode study of dorsal spinocerebellar tract. Amer. J. Physiol., *203,* 799-802.

LASSEK, A. M. 1954. *The Pyramidal Tract: Its Status in Medicine.* v + 166 pages, Charles C Thomas, Publisher, Springfield.

MEHLER, W. R. 1962. The anatomy of the so-called "pain tract" in man and an analysis of the course and distribution of the fasciculus anterolateralis. Basic Res. in Paraplegia, 26-55.

WARWICK, R. and G. A. G. MITCHELL. 1956. The phrenic nucleus of the macaque. J. comp. Neurol., *105,* 553-585.

BRAIN STEM AND CEREBELLUM

ANDERSON, F. D. and C. M. BERRY, 1956. An oscillographic study of the central pathways of the vagus nerve in the cat. J. comp Neurol., *106,* 163-181.

BROWN, J. O. 1943. The nuclear pattern of the non-tectal portions of the midbrain and isthmus in the dog and cat. J. comp. Neurol., *78,* 365-405.

BUCHANAN, A. R. 1937. The course of the secondary vestibular fibers in the cat. J. comp. Neurol., *67,* 183-204.

CARPENTER, M. B., G. M. BRITTIN, and J. PINES. 1958. Isolated lesions of the fastigial nuclei in the cat. J. comp. Neurol., *109,* 65-89.

CARPENTER, M. B. and J. PINES. 1957. The rubrobulbar tract: Anatomical relationships, course, and terminations in the rhesus monkey. Anat. Rec., *128,* 171-185.

CHAMBERS, W. W. and J. M. SPRAGUE. 1955. Functional localization in the cerebellum. I. Organization in longitudinal cortico-nuclear zones and their contribution to the control of posture, both extrapyramidal and pyramidal. J. comp. Neurol., 103, 105-129.

FEREMUTSCH, K. 1965. *Mesencephalon*. Primatologia, Vol. 2, Teil 2, Lieferung 5. v + 174 pages, 52 figures. S. Karger, Basel.

INGRAM, W. R. and S. W. RANSON. 1935. The nucleus of Darkschewitsch and nucleus interstitialis in the brain of man. J. nerv. ment. Dis., 81, 125-137.

LARSELL, O. 1947. The development of the cerebellum in man in relation to its comparative anatomy. J. comp. Neurol., 87, 85-129.

MACLEAN, P. D., R. H. DENNISTON, and S. DUA. 1963. Further studies on cerebral representation of penile erection: Caudal thalamus, midbrain, and pons. J. Neurophysiol., 26, 273-293.

MACLEAN, P. D., S. DUA, and R. H. DENNISTON. 1963. Cerebral localization for scratching and seminal discharge. Arch. Neurol., 9, 485-497.

PEARSON, A. A. 1939. The hypoglossal nerve in human embryos. J. comp. Neurol., 71, 21-39.

RASMUSSEN, A. T. and W. T. PEYTON. 1948. The course and termination of the medial lemniscus in man. J. comp. Neurol., 88, 411-424.

RASMUSSEN, G. L. 1946. The olivary peduncle and other fiber projections of the superior olivary complex. J. comp. Neurol., 84, 141-219.

RUSSELL, G. V. 1957. The brainstem reticular formation. Tex. Rep. Biol. Med., 15, 332-337.

SCHÜTZ, H. 1891. Anatomische Untersuchungen über den Faserverlauf in zentralen Höhlengrau und de Nervenfaserschund in denselben bei der progressiven Paralyse der Irren. Arch. Psychiat. Nervenkr., 22, 527-587.

STERN, K. 1938. Note on the nucleus ruber magnocellularis and its efferent pathway in man. Brain, 61, 284-289.

STOTLER, W. A. 1953. An experimental study of the cells and connections of the superior olivary complex of the cat. J. comp. Neurol., 98, 401-431.

SZENTÁGOTHAI, J. 1948. The representation of facial and scalp muscles in the facial nucleus. J. comp. Neurol., 88, 207-220.

VILLIGER, E. 1925. *Brain and Spinal Cord*. 3rd American Edition, W. H. F. Addison, Editor. x + 335 pages. J. B. Lippincott Company, Philadelphia.

DIENCEPHALON

BLEIER, R. *The Hypothalamus of the Cat. A Cytoarchitectonic Atlas with Horsley-Clarke Co-ordinates.* x + 109 pages, 30 plates. The Johns Hopkins Press, Baltimore.

BRIZZEE, K. R. 1954. A comparison of cell structure in the area postrema, supraoptic crest, and intercolumnar tubercle with notes on the neurohypophysis and pineal body in the cat. J. comp. Neurol., 100, 699-715.

CROSBY, E. C. and R. T. WOODBURNE. 1951. The mammalian midbrain and isthmus regions. Part II. The fiber connections. C. The hypothalamo-tegmental pathways. J. comp. Neurol., 94, 1-32.

DEKEBAN, A. 1953. Human thalamus: An anatomical, developmental and pathological study. I. Division of the adult human thalamus into nuclei by use of the cyto-myelo-architectonic method. J. comp. Neurol., 99, 639-683.

FOX, C. A. 1943. The stria terminalis, longitudinal association bundle and precommissural fornix fibers in the cat. J. comp. Neurol., 79, 277-295.

GARDNER, J. H. 1953. Innervation of pineal gland in hooded rat. J. comp. Neurol., 99, 319-329.

HUNT, W. E. and J. L. O'LEARY. 1952. Form of thalamic response evoked by peripheral nerve stimulation. J. comp. Neurol., 97, 491-514.

INGRAM, W. R. 1940. Nuclear organization and chief connections of the primate hypothalamus. Ass. Res. nerv. Dis. Proc., 20, 195-244.

KUHLENBECK, H. 1954. *The Human Diencephalon —A Summary of Development, Structure, Function and Pathology.* 230 pages. S. Karger, Basel.

KUHLENBECK, H. 1961. Evolution of the diencephalon in vertebrates. Progr. in Med., 39, 119-120.

KUHLENBECK, H. 1962. Morphologic significance of the so-called peripeduncular nucleus in the mammalian brain. Anat. Rec., 142, 314.

MAGOUN, H. W. and M. RANSON. 1942. The supraoptic decussations in the cat and monkey. J. comp. Neurol., 76, 435-459.

NAUTA, W. 1960. Limbic system and hypothalamus: Anatomical aspects. Physiol. Rev., 40, 102-104.

SAWYER, C. H., J. W. EVERETT, and J. D. GREEN. 1954. The rabbit diencephalon in stereotaxic coordinates. J. comp. Neurol., 101, 801-824.

TONCRAY, J. E. and W. J. S. KRIEG. 1946. The nuclei of the human thalamus: a comparative approach. J. comp. Neurol., 85, 421-459.

TELENCEPHALON

BARNARD, J. W. and C. N. WOOLSEY. 1956. A study of localization in the corticospinal tracts of monkey and rat. J. comp. Neurol., 105, 25-50.

BEACH, F. A., A. ZITRIN, and J. JAYNES. 1955. Neural mediation of mating in male cats. II. Contributions of the frontal cortex. J. exp. Zool., 130, 381-401.

BILLENSTIEN, D. C. 1953. The vascularity of the motor cortex of the dog. Anat. Rec., 117, 129-144.

BRODMAN, K. 1909. *Vergleichende Lokalisationslehre der Grosshirnrinde in ihren Prinzipien dargestellt auf Grund des Zellenbaues.* x + 324 pages. Reprinted 1925 by J. A. Barth, Leipzig.

BRODY, H. 1955. Organization of the cerebral cortex. III. A study of aging in the human cerebral cortex. J. comp. Neurol., 102, 511-556.

CAREY, J. H. 1957. Certain anatomical and functional interrelations between the tegmentum of the midbrain and the basal ganglia. J. comp. Neurol., 108, 57-90.

FRONTERA, J. G. 1956. Some results obtained by electrical stimulation of the cortex of the island of Reil in the brain of the monkey (Macaca mulatta). J. comp. Neurol., 105, 365-394.

HARMAN, P. J. and C. M. BERRY. 1956. Neuroanatomical distribution of action potentials evoked by photic stimuli in cat fore- and midbrain. J. comp. Neurol., 105, 395-416.

KRIEG, W. J. S. 1954. Connections of the Frontal Cortex of the Monkey. xi. + 299 pages. Charles C Thomas, Publisher, Springfield.

LAUER, E. W. 1945. The nuclear pattern and fiber connections of certain basal telencephalic centers in the Macaque. J. comp. Neurol., 82, 215-254.

McCULLOCH, W. S. 1949. Cortico-cortical connections. In Bucy's The Precentral Motor Cortex, 2nd Edition. xiv + 615 pages. University of Illinois Press, Urbana.

NARNARINE-SINGH, D., G. GEDDES, and J. B. HYDE. 1972. Sizes and numbers of arteries and veins in normal human neopallium. J. Anat. (Lond.), 111, 171-179.

NAUTA, W. 1962. Neural associations of the amygdaloid complex in the monkey. Brain, 85, 505-520.

PEELE, T. L. 1942. Cytoarchitecture cf individual parietal areas in the monkey (Macaca mulatta) and the distribution of the efferent fibers. J. comp. Neurol., 77, 693-737.

PENFIELD, W. and T. RASMUSSEN. 1950. The Cerebral Cortex of Man; A Clinical Study of Localization of Function. xv + 248 pages. The Macmillan Company, New York.

RASMUSSEN, A. T. 1943. The extent of recurrent geniculo-calcarine fibers (loop of Archambault and Meyer) as demonstrated by gross brain dissection. Anat. Rec., 85, 277-284.

TOMASCH, J. 1954. Size, distribution, and number of fibers in the human corpus callosum. Anat. Rec., 119, 119-135.

RHINENCEPHALON

ALLEN, W. F. 1948. Fiber degeneration in Ammon's horn resulting from extirpations of the piriform and other cortical areas and from transection of the horn at various levels. J. comp. Neurol., 88, 425-438.

ALLISON, A. C. 1953. The structure of the olfactory bulb and its relationship to the olfactory pathways in the rabbit and the rat. J. comp. Neurol., 98, 309-353.

HARRISON, J. M. and M. LYON. 1957. The role of the septal nuclei and components of the fornix in the behavior of the rat. J. comp. Neurol., 108, 121-137.

JIMINEZ-CASTELLANOS, J. 1949. The amygdaloid complex in monkey studied by reconstructional methods. J. comp. Neurol., 91, 507-526.

MACCHI, G. 1951. The ontogenetic development of the olfactory telencephalon in man. J. comp. Neurol., 95, 245-305.

NILGES, R. G. 1944. The arteries of the mammalian cornu ammonis. J. comp. Neurol., 80, 177-190.

RAMON Y CAJAL, S. 1955. Studies on the Cerebral Cortex (Limbic Structure). Translated from Spanish by Lisbeth M. Kraft. xi + 179 pages. The Year Book Publishers, New York.

RIOCH, D. McK. and C. BRENNER. 1938. Experiments on the corpus striatum and rhinencephalon. J. comp. Neurol., 68, 491-507.

HISTOLOGY AND HISTOCHEMISTRY

ADAMSTONE, F. B. and A. B. TAYLOR. 1953. Structure and physical nature of the cytoplasm of living spinal ganglion cells of the adult rat. J. morph., 92, 513-529.

BAJUSZ, E. and G. JASMIN. 1964. Major Problems in Neuroendocrinology, an International Symposium. viii + 471 pages, illustrated. The Williams & Wilkins Company, Baltimore.

BARR, M. L., L. F. BERTRAM, and H. A. LINDSAY. 1950. The morphology of the nerve cell nucleus, according to sex. Anat. Rec., 107, 283-297.

BODIAN, D. 1951. A note on nodes of Ranvier in the central nervous system. J. comp. Neurol., 94, 475-483.

BUEKER, E. D. 1943. Intracentral and peripheral factors in the differentiation of motor neurons in transplanted lumbo-sacral spinal cords of chick embryos. J. exp. Zool., 93, 99-129.

CRESWELL, G. F., D. J. REIS, and P. D. MACLEAN. 1964. Aldehyde-fuchsin positive material in brain of squirrel monkey (Saimiri sciureus). Amer. J. Anat., 115, 543-557.

CSILLIK, B. 1965. Functional Structure of the Post-Synaptic Membrane in the Myoneural Junction. 154 pages, 65 figures. Akadémiai Kiadó, Budapest.

D'ANGELO, C., M. ISSIDORIDES, and W. M. SHANKLIN. 1956. A comparative study of the staining reactions of granules in the human neuron. J. comp. Neurol., 106, 487-505.

DONAHUE, S. 1964. A relationship between fine structure and function of blood vessels in the central nervous system of rabbit fetuses. Amer. J. Anat., 115, 17-26.

DUNCAN, D. 1957. Electron microscope study of the embryonic neural tube and notochord. Tex. Rep. Biol. Med., 15, 367-377.

FORD, D. H. and S. KANTOUNIS. 1957. The localization of neurosecretory structures and pathways in the male albino rabbit. J. comp. Neurol., 108, 91-107.

GEREN, B. B. 1954. The formation from the Schwann cell surface of myelin in the peripheral nerves of chick embryos. Exp. Cell Res., 7, 558-562.

HARTMANN, J. F. 1954. Electron microscopy of motor nerve cells following section of axones. Anat. Rec., 118, 19-33.

HOERR, N. L. 1936. Cytological studies by the Altmann-Gersh freezing-drying method. III. The preexistence of neurofibrillae and their disposition in the nerve fiber. IV. The structure of the myelin sheath of nerve fibers. Anat. Rec., 66, 81-95.

LA VELLE, A. 1956. Nucleolar and Nissl substance development in nerve cells. J. comp. Neurol., 104, 175-205.

PENFIELD, W. 1932. *Cytology and Cellular Pathology of the Nervous System.* 3 Volumes. P. B. Hoeber, Inc., New York.

ROBERTSON, J. D. 1956. The ultrastructure of a reptilian myoneural junction. J. biophys. biochem. Cytol., *2*, 381-394.

SCHMITT, F. O. 1957. The fibrous protein of the nerve axon. J. cell. comp. Physiol., *49*, 165-174.

NEUROGLIA, MENINGES AND SPINAL FLUID

DI CHIRO, G. 1961. *An Atlas of Detailed Normal Pneumoencephalographic Anatomy.* xi. + 328 pages, 283 figures. Charles C Thomas, Publisher, Springfield.

FLYGER, G. and U. B. E. HJELMQUIST. 1957. Normal variations in the caliber of the human cerebral aqueduct. Anat. Rec., *127*, 151-162.

GLEES, P. 1955. *Neuroglia: Morphology and Function.* xii + 111 pages. Charles C Thomas, Publisher, Springfield.

HARVEY, S. C. and H. S. BURR. 1926. The development of the meninges. Arch. Neurol. Psychiat. (Chicago), *15*, 545-565.

KIMMEL, D. L. 1961. Innervation of spinal dura mater and dura mater of the posterior cranial fossa. Neurology (Minneap.), *11*, 800-809.

MILLEN, J. W. and D. H. M. WOOLLAM. 1962. *The Anatomy of the Cerebrospinal Fluid.* viii + 151 pages, 82 figures. Oxford University Press, New York.

NAKAI, J. 1963. *Morphology of Neuroglia.* ix + 198 pages, illustrated. Igaku Shoin Ltd., Tokyo, Japan.

SCHULTZ, R. L., E. A. MAYNARD and D. C. PEASE. 1957. Electron microscopy of neurons and neuroglia of cerebral cortex and corpus callosum. Amer. J. Anat., *100*, 369-407.

STRONG, L. H. 1956. Early development of the ependyma and vascular pattern of the fourth ventricular choroid plexus in the rabbit. Amer. J. Anat., *99*, 249-290.

SWEET, W. H. and H. B. LOCKSLEY. 1953. Formation, flow and reabsorption of cerebrospinal fluid in man. Proc. Soc. exp. Biol. (N.Y.), *84*, 397-402.

REGENERATION

BUEKER, E. D. and C. E. MAYERS. 1951. The maturity of peripheral nerves at the time of injury as a factor in nerve regeneration. Anat. Rec., *109*, 723-743.

CHAN-NAO, L. and D. SCOTT, JR. 1958. Regeneration in the dorsal spino-cerebellar tract of the cat. J. comp. Neurol., *109*, 153-167.

CLARK, E. R. and E. L. CLARK. 1947. Microscopic studies on the regeneration of medullated nerves in the living mammal. Amer. J. Anat., *81*, 233-268.

MOYER, E. K., D. L. KIMMEL and L. W. WINBORNE. 1953. Regeneration of sensory spinal nerve roots in young and in senile rats. J. comp. Neurol., *98*, 283-307.

TURNER, R. S. 1943. Chromatolysis and recovery of efferent neurons. J. comp. Neurol., *79*, 73-78.

PERIPHERAL NERVE ENDINGS

ADAMS, W. E. 1958. *The Comparative Morphology of the Carotid Body and Carotid Sinus.* xviii + 272 pages. Charles C Thomas, Springfield.

CAUNA, N. 1958. Structure of digital touch corpuscles. Acta Anat., *32*, 1-23.

CAUNA, N. 1959. The mode of termination of the sensory nerves and its significance. J. comp. Neurol., *113*, 169-209.

CAUNA, N. 1960. The distribution of cholinesterase in the cutaneous receptor organs, especially touch corpuscles of the human finger. J. Histochem. Cytochem., *8*, 367-375.

CAUNA, N. and G. MANNAN. 1959. Development and postnatal changes of digital pacinian corpuscles (Corpuscula lamellosa) in the human hand. J. Anat. (Lond.), *93*, 271-286.

CAUNA, N. and L. L. ROSS. 1959. Observations on the fine structure of Meissner's corpuscles. J. Anat. (Lond.), *93*, 574.

GARNER, C. M. and D. DUNCAN. 1958. Observations on the fine structure of the carotid body. Anat. Rec., *130*, 691-709.

HERING, H. E. 1927. Die Karotissinusreflexe auf Herz und Gefasse vom normalphysiologischen, pathologisch-physiologischen und klinischen Standpunkt. Th. Steinkopff, Dresden.

HOFFMAN, H. and J. H. W. BIRRELL. 1958. The carotid body in normal and anoxic states; an electron microscopic study. Acta Anat., *32*, 297-311.

HOLLINSHEAD, W. H. 1941. Chemoreceptors in the abdomen. J. comp. Neurol., *74*, 269-285.

KOUSDA, T. 1951. Study on the pressoreceptor area of the aortic arch. Nippon J. Angio-Cardiol., *15*, 4-10.

MERRILLEES, N. C. R., S. SUNDERLAND, and W. HAYHOW. 1950. Neuromuscular spindles in the extraocular muscles in man. Anat. Rec., *108*, 23-30.

NEFF, W. H. 1965. *Contributions to Sensory Physiology.* Vol. 1, x + 274 pages, illustrated. Academic Press, Inc., Publishers, New York.

PALLIE, W., K. NISHI, and C. OURA. 1970. The Pacinian corpuscle, its vascular supply and inner core. Acta Anat., *77*, 508-520.

PICK, J. 1959. The discovery of the carotid body. J. Hist. Med., *14*, 61-73.

REGER, J. F. 1958. The fine structure of neuromuscular synapses of gastrocnemii from mouse and frog. Anat. Rec., *130*, 7-23.

ROSS, L. L. 1957. A cytological and histochemical study of the carotid body of the cat. Anat. Rec., *129*, 433-455.

SETO, H. 1963. *Studies on the Sensory Innervation (Human Sensibility).* 2nd Edition. x + 523 pages, + 7, + 537 figures. Igaku Shoin, Ltd., Tokyo and Charles C Thomas, Publisher, Springfield.

TAKASHI, M. 1957. On the development of the complex pattern of Pacinian corpuscle distributed in the human retroperitoneum. Anat. Rec., *128*, 665-678.

12 | *The Peripheral Nervous System*

ervous impulses are conveyed to and from the nervous system by the ramifications of the **Peripheral Nervous System.** It is composed of nerve fibers, ganglia, and end organs. Afferent or sensory fibers carry toward the Central Nervous System the impulses arising from stimulation of sensory end organs. Efferent or motor fibers carry impulses from the Central Nervous System to the muscles and other responsive organs. The somatic fibers, both afferent and efferent, are associated with the general body, typified by the bones, skeletal muscles, and skin. The visceral fibers are also both afferent and efferent, and are associated with the internal organs, vessels, and mucous membranes.

The Peripheral Nervous System consists of (*a*) the cranial nerves, (*b*) the spinal nerves, and (*c*) the sympathetic nervous system. These three morphological subdivisions are not independent functionally, but combine and communicate with each other to supply both the somatic and visceral parts with both afferent and efferent fibers. For example, the efferent impulses for control of the viscera are carried partly by the sympathetic system and partly by portions of certain cranial and sacral nerves. The latter are grouped under the name parasympathetic or craniosacral system, and both sympathetic and parasympathetic systems together make up the visceral motor or autonomic nervous system, described in later pages.

THE CRANIAL NERVES

The **cranial nerves** (*nervi craniales; cerebral nerves*) are attached to the base of the brain (Fig. 11–62) and make their passage from the cranial cavity through various openings or foramina in the skull (Fig. 12–1). There are twelve pairs, and beginning with the most anterior, they are designated by Roman numerals and named as follows:

palatine from the facial, and the **Equilibratory** from the acoustic, making fifteen pairs in all (Hardesty '33).

In addition to the commonly accepted cranial nerves, there is a small nerve named the Nervus Terminalis, classified morphologically by some comparative anatomists as the first cranial nerve.

I. Olfactory	V. Trigeminal	IX. Glossopharyngeal
II. Optic	VI. Abducent	X. Vagus
III. Oculomotor	VII. Facial	XI. Accessory
IV. Trochlear	VIII. Acoustic	XII. Hypoglossal

Since certain of these nerves, particularly V, VII, and VIII, contain two or more distinct functional components, their parts have been given the status of independent nerves by some authorities. In this classification, the **Masticatory nerve** would be separated from the trigeminal, the **Glosso-**

The **Nervus Terminalis** (*terminal nerve*) originates from the cerebral hemisphere in the region of the olfactory trigone, courses anteriorly along the medial surface of the olfactory tract and bulb to the lateral surface of the crista galli, and passes through the anterior part of the cribriform plate of

Outline of the Cranial Nerves

Nerves	Components	Function	Central Connection	Cell Bodies	Peripheral Distribution
I. Olfactory	Afferent / Special visceral	Smell	Olfactory bulb and tract	Olfactory epithelial cells	Olfactory nerves
II. Optic	Afferent / Special somatic	Vision	Optic nerve and tract	Ganglion cells of retina	Rods and cones of retina
III. Oculomotor	Efferent / Somatic	Ocular movement	Nucleus III	Nucleus III	Branches to Levator palpebrae, Rectus superior, medius, inferior, Obliquus inferior
	Efferent / General visceral	Contraction of pupil and accommodation	Nucleus of Edinger-Westphal	Nucleus of Edinger-Westphal	Ciliary ganglion; Ciliaris and Sphincter pupillae
	Afferent / Proprioceptive	Muscular sensibility	Nucleus mesencephalicus V	Nucleus mesencephalicus V	Sensory endings in ocular muscles
IV. Trochlear	Efferent / Somatic	Ocular movement	Nucleus IV	Nucleus IV	Branches to Obliquus superior
	Afferent / Proprioceptive	Muscular sensibility	Nucleus mesencephalicus V	Nucleus mesencephalicus V	Sensory endings in Obliquus superior
V. Trigeminal	Afferent / General somatic	General sensibility	Trigeminal sensory nucleus	Trigeminal ganglion (Gasserian)	Sensory branches of ophthalmic maxillary and mandibular nerves to skin and mucous membranes of face and head
	Efferent / Special visceral	Mastication	Motor V nucleus	Motor V nucleus	Branches to Temporalis, Masseter, Pterygoidei, Mylohyoideus, Digastricus, Tensores tympani and palatini
	Afferent / Proprioceptive	Muscular sensibility	Nucleus mesencephalicus V	Nucleus mesencephalicus V	Sensory endings in muscles of mastication
VI. Abducent	Efferent / Somatic	Ocular movement	Nucleus VI	Nucleus VI	Branches to Rectus lateralis
	Afferent / Proprioceptive	Muscular sensibility	Nucleus mesencephalicus V	Nucleus mesencephalicus V	Sensory endings in Rectus lateralis
VII. Facial	Efferent / Special visceral	Facial expression	Motor VII nucleus	Motor VII nucleus	Branches to facial muscles, Stapedius, Stylohyoideus, Digastricus
	Efferent / General visceral	Glandular secretion	Nucleus salivatorius	Nucleus salivatorius	Greater petrosal nerve, pterygopalatine ganglion, with branches of maxillary V to glands of nasal mucosa. Chorda tympani, lingual nerve, submandibular ganglion, submandibular and sublingual glands
	Afferent / Special visceral	Taste	Nucleus tractus solitarius	Geniculate ganglion	Chorda tympani, lingual nerve, taste buds, anterior tongue
	Afferent / General visceral	Visceral sensibility	Nucleus tractus solitarius	Geniculate ganglion	Great petrosal, chorda tympani and branches

Nerve		Function	Nucleus	Cells of origin	Distribution
VIII. Acoustic	Afferent General somatic	Cutaneous sensibility	Nucleus spinal tract of V	Geniculate ganglion	With auricular branch of vagus to external ear and mastoid region
	Afferent Special somatic	Hearing	Cochlear nuclei	Spiral ganglion	Organ of Corti in cochlea
	Afferent Proprioceptive	Sense of equilibrium	Vestibular nuclei	Vestibular ganglion	Semicircular canals, saccule, and utricle
IX. Glosso-pharyngeal	Afferent Special visceral	Taste	Nucleus tractus solitarius	Inferior ganglion IX	Lingual branches, taste buds, posterior tongue
	Afferent General visceral	Visceral sensibility	Nucleus tractus solitarius	Inferior ganglion IX	Tympanic nerve to middle ear, branches to pharynx and tongue, carotid sinus nerve
	Efferent General visceral	Glandular secretion	Nucleus salivatorius	Nucleus salivatorius	Tympanic, lesser petrosal nerves, otic ganglion, with auriculotemporal V to parotid gland
	Efferent Special visceral	Swallowing	Nucleus ambiguus	Nucleus ambiguus	Branch to Stylopharyngeus
X. Vagus	Efferent General visceral	Involuntary muscle and gland control	Dorsal motor nucleus X	Dorsal motor nucleus X	Cardiac nerves and plexus; ganglia on heart. Pulmonary plexus; ganglia, respiratory tract. Esophageal, gastric, celiac plexuses; myenteric and submucous plexuses, muscles and glands of digestive tract down to transverse colon
	Efferent Special visceral	Swallowing and phonation	Nucleus ambiguus	Nucleus ambiguus	Pharyngeal branches, superior and inferior laryngeal nerves
	Afferent General visceral	Visceral sensibility	Nucleus tractus solitarius	Inferior ganglion X	Fibers in all cervical, thoracic, and abdominal branches; carotid and aortic bodies
	Afferent Special visceral	Taste	Nucleus tractus solitarius	Inferior ganglion X	Branches to region of epiglottis and taste buds
	Afferent General somatic	Cutaneous sensibility	Nucleus spinal tract V	Superior ganglion X	Auricular branch to external ear and meatus
XI. Accessory	Efferent Special visceral	Swallowing and phonation	Nucleus ambiguus	Nucleus ambiguus	Bulbar portion, communication with vagus, in vagus branches to muscles of pharynx and larynx
	Efferent Special somatic	Movements of shoulder and head	Lateral column of upper cervical spinal cord	Lateral column of upper cervical spinal cord	Spinal portion, branches to Sternocleidomastoideus and Trapezius
XII. Hypoglossal	Efferent General somatic	Movements of tongue	Nucleus XII	Nucleus XII	Branches to extrinsic and intrinsic muscles of tongue

the ethmoid bone. It is a compact bundle beside the olfactory tract, a close plexus beside the bulb, and a loose plexus on the crista galli, where it is embedded in the dura mater some distance above the cribriform bone. Within the cranium, filaments join the olfactory bundles and *vomeronasal nerves,* and apparently pass to the mucous membrane of the septum along with them. The majority of filaments form a single strand which passes through the cribriform plate anterior to the vomeronasal nerves, and is distributed to the membrane near the anterior superior border of the nasal septum. In the nasal cavity, the nerve communicates with the medial nasal branch of the anterior ethmoidal branch of the ophthalmic division of the trigeminal nerve.

In the embryo, a loose mass of fibers and cells in the portion of the nerve medial to the rostral end of the olfactory bulb is called the **ganglion terminale.** The nerve

fibers are unmyelinated and ganglion cells are scattered in groups along the peripheral course of the nerve. The cells in the ganglion and along the nerve have been described by different authors as unipolar, bipolar, and multipolar (Pearson '41).

The central connections of the nervus terminalis end in the septal nuclei, the olfactory lobe, the posterior precommissural region, and the anterior portion of the supraoptic region of the brain. Those in the first three are sensory, those in the supraoptic region may be preganglionic autonomic fibers associated with the serous glands (of Bowman) or with blood vessels in the olfactory membrane (Larsell '50).

I. THE OLFACTORY NERVE

The **olfactory nerve** (*nervus olfactorius; first nerve*) (Fig. 12–2), or nerve of smell,

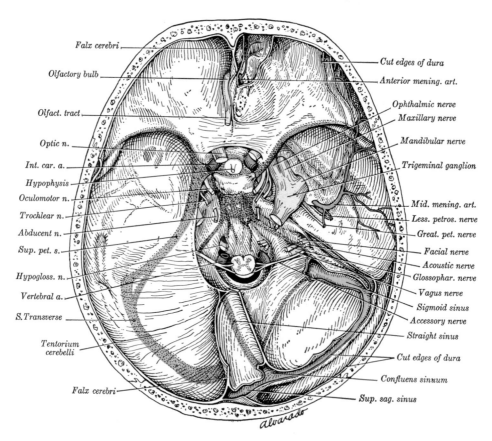

Fig. 12–1.—Interior of base of skull, showing dura mater, dural sinuses and exit of cranial nerves. (Redrawn from Töndury.)

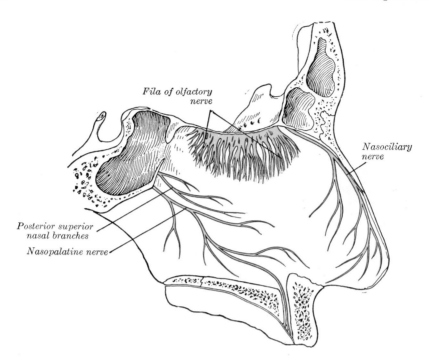

*Fila of olfactory
nerve*

*Nasociliary
nerve*

*Posterior superior
nasal branches*

Nasopalatine nerve

Fig. 12–2.—The nerves of the right side of the septum of the nose.

in its properly restricted sense is represented on both sides of the nasal cavity by a number of bundles of nerve fibers which are the central processes of the neuroepithelial cells in the olfactory mucous membrane of the nose. These cells are described in more detail as part of the organ of smell in the chapter on Sense Organs. The nerve fibers make a plexiform network in the mucous membrane of the superior nasal concha and the part of the septum opposite, and then are gathered into approximately twenty bundles which pass through the foramina of the cribriform plate of the ethmoid bone in two groups. A lateral group comes from the concha, a medial group from the septum. The fibers are unmyelinated, and the bundles have connective tissue sheaths derived from the tissues of the dura, arachnoid, and pia mater. After passing through the foramina, the fibers end in the glomeruli of the olfactory bulb, an oval mass measuring approximately 3 mm × 15 mm, which rests against the intracranial surface of the cribriform plate of the ethmoid (Fig. 12–1).

Although the olfactory bulb and its connection with the brain, the olfactory tract, have a gross appearance like that of a nerve, they are more accurately classed as parts of the brain. They are components of the rhinencephalon, the portion of the brain associated with the sense of smell, which is described in the chapter on the Central Nervous System, pages 859-863.

The **vomeronasal nerve,** which is an important part of the olfactory system in macrosmatic animals, is present in the human fetus but disappears before birth. The nerve fibers originate in the olfactory epithelial cells of the vomeronasal organ (organ of Jacobson), a rudiment of which may persist in man; they pass upward in the submucous tissue of the nasal septum and through the cribriform plate of the ethmoid bone to end in the accessory olfactory bulb.

II. THE OPTIC NERVE

The **optic nerve** (*n. opticus; second nerve*) (Fig. 12–3), or nerve of sight, consists mainly of the axons or central processes of the cells in the ganglionic layer of the retina. Within the bulb of the eye,

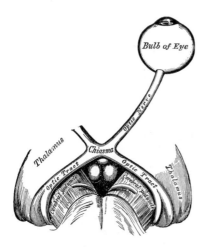

FIG. 12–3.—The left optic nerve and the optic tracts.

these axons lie in the stratum opticum or layer of nerve fibers of the retina (see next chapter). They converge toward the optic papilla, or disk, which is 3 mm medial to the posterior pole of the bulb, and are gathered there into small bundles which pierce the choroid and sclerotic coats by means of the many small foramina of the lamina cribrosa sclerae, to become the optic nerve. The nerve, as it courses posteriorly toward the brain, traverses the central region of the orbit, passes through the optic canal (foramen), and then, approaching the nerve of the other side, joins it to form the optic chiasma. From the chiasma, the fibers are continued in the optic tracts, which diverge from each other to reach the base of the brain near the cerebral peduncle. The central connections of the fibers are described in the chapter on the Central Nervous System, page 867.

The optic nerve has four portions, contained in: (a) the bulb, (b) the orbit, (c) the optic canal, and (d) the cranial cavity. The **intraocular portion** is very short, about 1 mm, and the nerve fibers within it are unmyelinated until they pass through the lamina cribrosa, when they become myelinated and supported by neuroglia.

The **orbital portion**, 3 to 4 mm in diameter, is from 20 to 30 mm long, and has a slightly sinuous course which allows greater length for unrestricted movement of the eyeball. It is invested by sheaths derived from the dura,

arachnoid, and pia mater, all three of which fuse and become continuous with the sclera at the lamina cribrosa; the dura extends as far back as the cranial cavity, the arachnoid somewhat farther, and the pia all the way to the chiasma. The pia closely ensheathes the nerve, and sends numerous septa into its substance, carrying the blood supply. Between the sheaths are subarachnoid and subdural spaces similar to and continuous with those of the cranial cavity, the subarachnoid ending in a cul-de-sac at the lamina cribrosa. As it traverses the orbit, the optic nerve is surrounded by the posterior part of the fascia bulbi (Tenon's capsule), the orbital fat, and, in its anterior two-thirds, by the ciliary nerves and arteries. Toward the posterior part of the orbit, it is crossed obliquely by the nasociliary nerve, the ophthalmic artery, the superior ophthalmic vein, and the superior division of the oculomotor nerve. Farther toward the roof of the orbit are the Rectus superior and Levator palpebrae superioris muscles, and the trochlear and frontal nerves. Inferior to it are the inferior division of the oculomotor nerve and the Rectus inferior; medial to it the Rectus medialis; and lateral to it the abducent nerve and Rectus lateralis, and, in the posterior part of the orbit, the ciliary ganglion and the ophthalmic artery (Figs. 8–23, 12–6). As it passes into the optic canal, accompanied by the ophthalmic artery, it is surrounded by the anulus tendineus communis (of Zinn), which serves as the origin of the ocular muscles. A short distance behind the bulb, the optic nerve is pierced by the central artery of the retina and its accompanying vein. These vessels enter the bulb through an opening in the lamina cribrosa and supply the retina.

Within the optic canal, the ophthalmic artery lies inferior to the nerve, just after it has branched off of the internal carotid. Medially, separated by a thin plate of bone, is the sphenoidal air sinus. If the pneumatization of the bone is extensive, the nerve may be almost completely surrounded by the sphenoidal sinus or the posterior ethmoidal cells. Superior to the nerve, the three sheaths, dura, arachnoid, and pia, are fused to each other, to the nerve, and to the periosteum of the bone, fixing the nerve and preventing it from being forced back and forth in the foramen. The subarachnoid and subdural spaces are present only below the nerve.

The **intracranial portion** of the nerve (Fig. 12–1) rests on the anterior part of the cavernous sinus, and on the diaphragma sellae which overlies the hypophysis. The part of the brain above the nerve is the anterior perforated substance. The internal carotid artery approaches it laterally, and then is directly inferior to it at the origin of the ophthalmic artery. The anterior cerebral artery crosses superior to it.

The **Optic Chiasma** (*chiasma opticum*) (Fig. 12–3), as the name indicates, resembles the Greek letter "X," from the convergence of the optic nerves in front and the divergence of the optic tracts behind. It rests upon the tuberculum sellae of the sphenoid bone and on the diaphragma sellae of the dura. It is continuous superiorly with the lamina terminalis, and posteriorly with the tuber cinereum and the infundibulum of the hypophysis. Lateral to it, on each side, are the anterior perforated substance and the internal carotid artery.

The **Optic Tract** (*tractus opticus*) (Fig. 11–62), leaving the chiasma, diverges from its fellow of the opposite side until it reaches the cerebral peduncle, across whose undersurface it winds obliquely. It is adjacent to the tuber cinereum and peduncle, thus having contact with the third ventricle. As it terminates, a shallow groove divides it into a medial and lateral root. The lateral root contains fibers of visual function and ends in the lateral geniculate body. The medial root probably is auditory rather than visual, carrying commissural fibers connecting the two medial geniculate bodies (commissure of Gudden and probably other non-visual commissures, those of Meynert and Darkschewitsch).

The fibers lying medially in the optic nerve cross in the chiasma and continue in the optic tract of the other side. The lateral fibers are uncrossed, and continue to the brain in the optic tract of the same side. The fibers from the two sides become intermingled in the tract (Mayer '40).

The optic nerve corresponds to a tract of fibers within the brain rather than to the other cranial nerves because of its embryological development and its structure. It is developed from a diverticulum of the lateral aspect of the forebrain; its fibers probably are third in the chain of neurons from the receptors to the brain; they are supported by neuroglia instead of having neurilemmal sheaths, and the nerve has three sheaths prolonged from the corresponding meninges of the brain (see Sense Organ chapter).

III. THE OCULOMOTOR NERVE

The **oculomotor nerve** (*n. oculomotorius; third nerve*) (Figs. 12–6, 12–7) supplies the Levator palpebrae superioris, the extrinsic muscles of the eye except the Obliquus superior and Rectus lateralis, and the intrinsic muscles with the exception of the Dilatator pupillae. Its superficial origin (Fig. 11–62) is from the midbrain at the oculomotor sulcus on the medial side of the cerebral peduncle. Its deep origin and central connections within the brain are described with the Central Nervous System, page 969. The oculomotor nerve is traditionally considered to be a motor nerve, containing special somatic efferent fibers for the ocular muscles and parasympathetic fibers for the ciliary ganglion, but experimental evidence indicates that it contains proprioceptive fibers from cells in the brain stem near the motor cells (Corbin and Oliver '42).

As it emerges from the brain into the posterior cranial fossa, the nerve is invested with pia mater, and is bathed in the cerebrospinal fluid of the cisterna interpeduncularis (basalis). It passes between the superior cerebellar and posterior cerebral arteries near the termination of the basilar artery. Lateral to the posterior clinoid process, it is covered with arachnoid, and between the anterior and posterior clinoid processes it pierces the dura (Fig. 12–1) by passing between the free and attached borders of the tentorium cerebelli. It runs anteriorly, embedded in the lateral wall of the cavernous sinus (Fig. 12–13), superior to the other orbital nerves (see page 924), and enters the superior orbital fissure between the two heads of the Rectus lateralis (Fig. 12–4) in two divisions, a superior and an inferior. Within the fissure, the trochlear, lacrimal, and frontal nerves lie superior to it, the abducent nerve inferior and lateral to it, and the nasociliary nerve passes between its two divisions.

Two **communications** join the oculomotor as it runs along the wall of the cavernous sinus, (*a*) one from the **cavernous plexus** of the sympathetic, carrying postganglionic fibers from the superior cervical ganglion, and (*b*) one from the **ophthalmic** division of the trigeminal nerve.

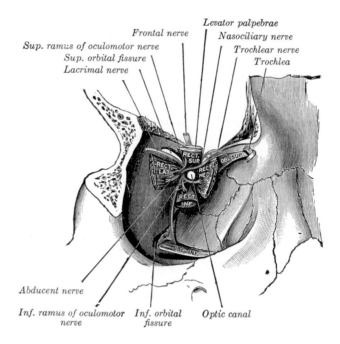

Frontal nerve

Sup. ramus of oculomotor nerve
Sup. orbital fissure
Lacrimal nerve

Levator palpebrae
Nasociliary nerve
Trochlear nerve
Trochlea

Abducent nerve

Inf. ramus of oculomotor
nerve

Inf. orbital
fissure

Optic canal

FIG. 12–4.—Dissection showing origins of right ocular muscles, and nerves entering by
the superior orbital fissure.

The **superior division** (Fig. 12–6), smaller than the inferior, passes medialward across the optic nerve, and sends branches to: (*a*) the **Rectus superior,** and (*b*) the **Levator palpebrae superioris.** The branch to the Levator contains sympathetic postganglionic fibers from the cavernous plexus, which continue anteriorly to reach the non-striated muscle attached to the superior tarsus.

The **inferior division** has four branches: (*a*) to the **Rectus medialis,** passing inferior to the optic nerve, (*b*) to the **Rectus inferior,** and (*c*) to the **Obliquus inferior.** The latter runs forward between the inferior and lateral Recti, and gives off a short, rather thick branch, (*d*) the **root of the ciliary ganglion.**

The **Ciliary Ganglion** (*g. ciliare; ophthalmic or lenticular ganglion*) (Figs. 12–5, 12–6) is a very small (1 or 2 mm in diameter) parasympathetic ganglion whose preganglionic fibers come from the oculomotor nerve and whose postganglionic fibers carry motor impulses to the Ciliaris and Sphincter pupillae muscles. It is situated about 1 cm from the posterior boundary of the orbit, close to the lateral surface of

the optic nerve and between it and the Rectus lateralis and the ophthalmic artery.

The **parasympathetic motor root** (*radix oculomotorii; short root*) of the ganglion is a rather short, thick nerve, which may be double. It comes from the branch of the inferior division of the oculomotor nerve which supplies the Obliquus inferior and is connected with the posterior inferior angle of the ganglion. It contains preganglionic parasympathetic fibers from the Edinger-Westphal group of cells of the third nerve nucleus, which form synapses in the ganglion with cells whose postganglionic fibers pass through the short ciliary nerves into the bulb.

Two communications were formerly called roots of the ganglion: (*a*) a **communication** with the **nasociliary nerve** (*radix longa or long root; sensory root*) joins the posterior superior angle of the ganglion, and contains sensory fibers which traverse the short ciliary nerves on their way from the cornea, iris, and ciliary body, without forming synapses in the ganglion; (*b*) a **communication** with the **sympathetic** (*sympathetic root*) is a slender filament from

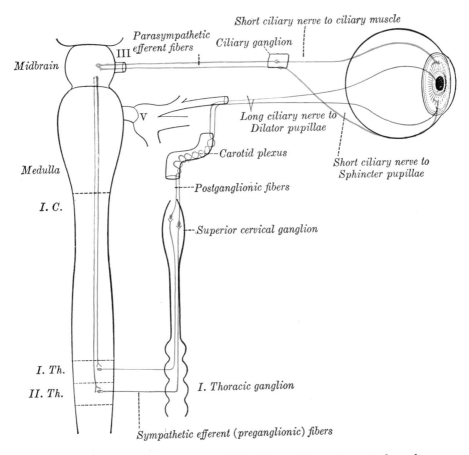

Fig. 12–5.—Autonomic connections of the ciliary and superior cervical ganglia. Parasympathetic blue, sympathetic red.

the cavernous plexus, and frequently it is blended with the communication with the nasociliary. It contains postganglionic fibers from the superior cervical ganglion which pass through the ciliary ganglion without forming synapses and reach the Dilatator pupillae muscle and the blood vessels of the bulb through the short ciliary nerves.

The branches of the ganglion are the **short ciliary nerves,** from six to ten delicate filaments, which leave the anterior part of the ganglion in two bundles, superior and inferior. They run anteriorly with the ciliary arteries in a wavy course, one set above and the other below the optic nerve, and are accompanied by the long ciliary nerves from the nasociliary. They pierce the sclera at the posterior part of the bulb, pass anteriorly in delicate grooves on the inner surface of the sclera, and are distributed to the ciliary

body, iris, and cornea. The parasympathetic postganglionic fibers supply the Ciliaris and Sphincter pupillae muscles. The short ciliary nerves also carry sympathetic post-ganglionic fibers to the Dilatator pupillae muscle, and sensory fibers from the cornea, iris, and ciliary body.

IV. THE TROCHLEAR NERVE

The **trochlear nerve** (*n. trochlearis; fourth nerve*) (Fig. 12–7) is the smallest of the cranial nerves and supplies the Obliquus superior oculi. Its superficial origin (Fig. 11–58) is from the surface of the anterior medullary velum at the side of the frenulum veli, immediately posterior to the inferior colliculus. Its deep origin and central connections are described with the Central Nervous System on page 871.

The nerve is directed laterally across the superior cerebellar peduncle, and winds around the cerebral peduncle (Fig. 11–62) near the pons. It runs anteriorly along the free border of the tentorium cerebelli for 1 or 2 cm (Fig. 12–1), pierces the dura just posterior to the posterior clinoid process, and, without changing the direction of its course, takes its place in the lateral wall of the cavernous sinus (Fig. 12–13) between the oculomotor nerve and the ophthalmic division of the trigeminal to which it becomes firmly attached by connective tissue. It crosses the oculomotor nerve in a slightly upward course, enters the orbit through the superior orbital fissure, and becomes superior to the other nerves. In the orbit it passes medialward, above the origin of the Levator palpebrae superioris (Fig. 12–4), and finally enters the orbital surface of the Obliquus superior.

In the lateral wall of the cavernous sinus the trochlear nerve forms **communications** with the cavernous plexus of the sympathetic and with the ophthalmic division of the trigeminal.

Variations.—Occasionally sensory branches appear to come from the trochlear nerve, but they are probably aberrant fibers which have come from the ophthalmic nerve.

V. THE TRIGEMINAL NERVE

The **trigeminal nerve** (*n. trigeminus; fifth or trifacial nerve*) is the largest of the cranial nerves, and is the great cutaneous sensory nerve of the face, the sensory nerve to the mucous membranes and other internal structures of the head, and the motor nerve of the muscles of mastication.

The Roots.—The trigeminal nerve traditionally has two roots, a larger sensory root and a smaller motor root (Fig. 12–6). Recent investigations have revealed a third or intermediate root.

The **sensory root** (*radix sensoria, portio major*) is composed of a large number of fine bundles in close association with each other which enter the pons through the lateral part of its ventral surface (Fig. 11–58). The nerve fibers in the sensory root are the central processes of the ganglion cells in the trigeminal (semilunar) ganglion.

The **motor root** (*radix motoria, portio minor*) emerges in several bundles from the surface of the pons approximately 2 to 5 mm medial and anterior (rostral) to the sensory root. The motor root includes, in addition to the motor fibers, proprioceptive sensory fibers from mesencephalic central root of the nerve (Corbin and Harrison '40).

The **intermediate root,** composed of one or more bundles, has an origin from the surface of the pons quite distinct from that of the other two roots (Jannetta and Rand '66). The bundles arise between the motor and sensory roots, separated from them by 1 to 5 mm (Vidic and Stephanos '69). The internal fiber connections, function, and surgical importance of this root are not as yet explained.

The three roots pass anteriorly in the posterior cranial fossa, under the shadow of the tentorium where the latter is attached to the petrous portion of the temporal bone, to reach the trigeminal ganglion. The deep origin and central connections of these roots are described with the Central Nervous System on page 871.

The **Trigeminal Ganglion** (*ganglion semilunare; Gasserian ganglion*) (Fig. 12–8) lies in a pocket of dura mater (*cavum Meckelii*) which occupies the trigeminal impression near the apex of the petrous portion of the temporal bone (Burr and Robinson '25). The ganglionic mass is flat and semilunar in shape, measuring approximately 1 cm × 2 cm (Fig. 12–8), with the central processes of the cells leaving the concavity and the peripheral fibers the convexity of the crescent. The ganglion is lateral to the posterior part of the cavernous sinus and the internal carotid artery at the foramen lacerum, and the central fibers pass inferior to the superior petrosal sinus. The motor root, being medial to the sensory, passes beneath the ganglion, that is, between it and the petrous bone, and leaves the skull through the foramen ovale with the mandibular nerve. The greater petrosal nerve also passes between the ganglion and the bone. The ganglion receives filaments from the carotid plexus of the sympathetic, and it gives off minute branches to the tentorium cerebelli, and to the dura mater of the middle cranial fossa.

The peripheral fibers from the ganglion are collected into three large divisions: 1.

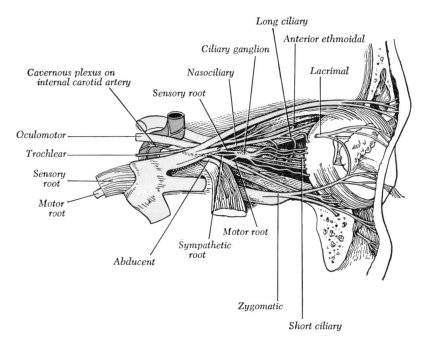

FIG. 12–6.—Nerves of the orbit. Side view.

the Ophthalmic, 2. the Maxillary, and 3. the Mandibular Nerves. The ophthalmic and maxillary remain sensory, but the mandibular becomes mixed, being joined by the motor root just outside the skull.

Four small ganglia are associated with these nerves and have previously been described with them: the ciliary ganglion with the ophthalmic, the pterygopalatine with the maxillary, and the otic and submandibular with the mandibular. These ganglia are not a part of the trigeminal complex functionally, however, and the trigeminal fibers which communicate with them pass through without synapses. The ganglia are parasympathetic and are therefore described with the nerves which supply their motor roots with preganglionic fibers, viz., the ciliary ganglion with the oculomotor nerve (page 911), the pterygopalatine and submandibular ganglia with the facial nerve (page 927), and the otic ganglion with the glossopharyngeal nerve (page 934).

1. The Ophthalmic Nerve

The **Ophthalmic Nerve** (*n. ophthalmicus*) (Figs. 12–6, 12–7), or first division of the trigeminal, leaves the anterior superior part

of the trigeminal ganglion, and enters the orbit through the superior orbital fissure. It is a flattened band about 2.5 cm long and

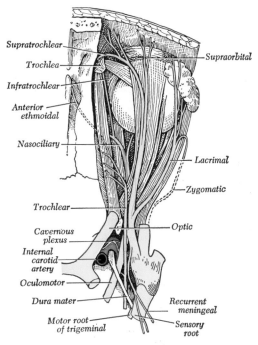

FIG. 12–7.—Nerves of the orbit. Seen from above.

lies in the lateral wall of the cavernous sinus, inferior to the oculomotor and trochlear nerves (Fig. 12–13). It is a sensory nerve supplying the bulb of the eye, conjunctiva, lacrimal gland, part of the mucous membrane of the nose and paranasal sinuses, and the skin of the forehead, eyelids, and nose.

The ophthalmic nerve is joined by **communicating filaments** from the **cavernous plexus** of the **sympathetic,** and communicates with the **oculomotor, trochlear,** and **abducent** nerves. It gives off a recurrent filament to the dura mater and just before it passes through the superior orbital fissure it divides into three branches: frontal, lacrimal, and nasociliary.

The **branches and communications** of the ophthalmic nerve are:

1. The **tentorial branch** (*r. tentorii*) (Fig. 12–7, *recurrent filament*) arises near the ganglion, passes across and is adherent to the trochlear nerve, and runs between the layers of the tentorium to which it is distributed.

2. The **Lacrimal Nerve** (*n. lacrimalis*) (Fig. 12–7) is the smallest of the three branches of the ophthalmic. It passes anteriorly in a separate tube of dura mater, and enters the orbit through the narrowest part of the superior orbital fissure. In the orbit it runs along the superior border of the Rectus lateralis close to the periorbita, and enters the lacrimal gland with the lacrimal artery, giving off filaments to supply the gland and adjacent conjunctiva. Finally it pierces the orbital septum, and ends in the skin of the upper eyelid, joining with filaments of the facial nerve.

In the orbit, through a **communication with the zygomatic branch of the maxillary nerve** (Figs. 12–6, 12–8), it receives postganglionic parasympathetics which are the secretomotor fibers for the lacrimal gland. These fibers pass from their cells of origin in the pterygopalatine ganglion, through the pterygopalatine nerves to the maxillary nerve, then along the zygomatic and zygomaticotemporal nerves, finally traversing the communication mentioned above, and being distributed with the branches of the lacrimal nerve to the gland. The preganglionic fibers reach the ganglion from the facial nerve by way of the greater petrosal and Vidian nerves.

Variations.—The lacrimal nerve is occasionally absent, and its place is then taken by the zygomaticotemporal branch of the maxillary. Sometimes the latter is absent and a communication of the lacrimal is substituted for it.

3. The **Frontal Nerve** (*n. frontalis*) (Fig. 12–7) is the largest branch of the ophthalmic, and may be regarded, both from its size and direction, as the continuation of the nerve. It enters the orbit through the superior orbital fissure, continues rostrally between the Levator palpebrae superioris and the periorbita, and, at a variable distance approximately halfway to the supraorbital margin, it divides into a large supraorbital and a small supratrochlear branch.

a) The **supratrochlear nerve** (*n. supratrochlearis*) (Fig. 12–7) bends medially to pass superior to the pulley of the Obliquus superior and gives off a filament which communicates with the infratrochlear branch of the nasociliary. It pierces the orbital fascia, sends filaments to the conjunctiva and skin of the medial part of the upper lid, passes deep to the Corrugator and Frontalis, and divides into branches which pierce the muscles to supply the skin of the lower and mesial part of the forehead.

b) The **supraorbital nerve** (*n. supraorbitalis*) (Fig. 12–7), the continuation of the frontal nerve, leaves the orbit through the supraorbital notch or foramen. It gives filaments to the upper lid and continues upon the forehead, dividing into medial and lateral branches beneath the Frontalis. The **medial branch,** sometimes called the **frontal branch** (*ramus frontalis*), is smaller; it pierces the muscle and supplies the scalp as far as the parietal bone. The larger **lateral branch** pierces the galea aponeurotica and supplies the scalp nearly as far back as the lambdoidal suture (Fig. 12–18). The supraorbital nerve may divide before leaving the orbit, in which case the lateral branch occupies the supraorbital notch or foramen and the medial or frontal branch may have a notch of its own.

c) The **branch** to the **frontal sinus.**—In the supraorbital notch, a small filament pierces the bone to supply the mucous **membrane** of the frontal sinus.

4. The **Nasociliary Nerve** (*n. nasociliaris; nasal nerve*) (Fig. 12–7) is intermediate in size between the frontal and lacrimal, and is more deeply placed in the orbit. It enters the orbit between the two heads of the Rectus lateralis, and between the superior and inferior divisions of the oculomotor nerve. It passes across the optic nerve and runs obliquely inferior to the Rectus superior and Obliquus superior, to the medial wall of the orbital cavity. Here it passes through the anterior ethmoidal foramen as the **anterior ethmoidal nerve** (*n. ethmoideus anterior*), and enters the cranial cavity just superior to the cribriform plate of the ethmoid bone. It runs along a shallow groove on the lateral margin of the plate, and, penetrating the bone through a slit at the side of the crista galli, it enters the nasal cavity. It supplies branches to the mucous membrane of the nasal cavity (Figs. 12–2, 12–12) and finally it emerges between the inferior border of the nasal bone and the lateral nasal cartilage as the external nasal branch.

The **branches** of the nasociliary nerve are as follows:

a) The **communication** with the **ciliary ganglion** (*ramus communicans g. ciliare; long or sensory root of the ganglion; radix longa ganglii ciliaris*) (Fig. 12–6) usually arises from the nasociliary between the two heads of the Rectus lateralis. It runs anteriorly on the lateral side of the optic nerve, and enters the posterior superior angle of the ciliary ganglion. It contains sensory fibers which pass through the ganglion without synapses and continue on into the bulb by way of the short ciliary nerves. It is sometimes joined by a filament from the cavernous plexus of the sympathetic, or from the superior ramus of the oculomotor nerve. The ciliary ganglion is described with the oculomotor nerve (page 910).

b) The **long ciliary nerves** (*nn. ciliares longi*), two or three in number, are given off from the nasociliary as it crosses the optic nerve. They accompany the short ciliary nerves from the ciliary ganglion, pierce the posterior part of the sclera, and, running anteriorly between it and the choroid, are distributed to the iris and cornea. In addition to afferent fibers, the long ciliary nerves probably contain sympa-

thetic fibers from the superior cervical ganglion to the Dilatator pupillae muscle, which passes through the communication between the cavernous plexus and the ophthalmic nerve.

c) The **infratrochlear nerve** (*n. infratrochlearis*) (Fig. 12–7) is given off from the nasociliary just before it enters the anterior ethmoidal foramen. It runs anteriorly along the superior border of the Rectus medialis, and is joined near the pulley of the Obliquus superior by a filament from the supratrochlear nerve. It then passes to the medial angle of the eye, and supplies the skin of the eyelids and side of the nose, the conjunctiva, lacrimal sac, and caruncula lacrimalis.

d) The **ethmoidal branches** supply the mucous membrane of the sinuses. The **posterior ethmoidal nerve** (*n. ethmoidalis posterior*) leaves the orbit through the posterior ethmoidal foramen and supplies the posterior ethmoidal and the sphenoidal sinuses. The **anterior ethmoidal branches** (*rr. ethmoidales anterior*) are filaments which are given off as the nerve passes through the anterior ethmoidal foramen, and supply the anterior ethmoidal and frontal sinuses.

e) The **internal nasal branches** supply the mucous membrane of the anterior part of the septum (*rr. mediales*) and lateral wall (*rr. laterales*) of the nasal cavity (Figs. 12–2, 12–12).

f) The **external nasal branch** emerges between the nasal bone and the lateral nasal cartilage, passes deep to the Nasalis muscle, and supplies the skin of the ala and apex of the nose (Fig. 12–8).

2. The Maxillary Nerve

The **Maxillary Nerve** (*n. maxillaris; superior maxillary nerve*) (Fig. 12–8), or second division of the trigeminal, arises from the middle of the trigeminal ganglion, is intermediate between the other two divisions in size and position, and, like the ophthalmic, is entirely sensory. It supplies the skin of the middle portion of the face, lower eyelid, side of the nose, and upper lip (Fig. 12–10); the mucous membrane of the nasopharynx, maxillary sinus, soft palate, tonsil and roof of the mouth, the upper gums and

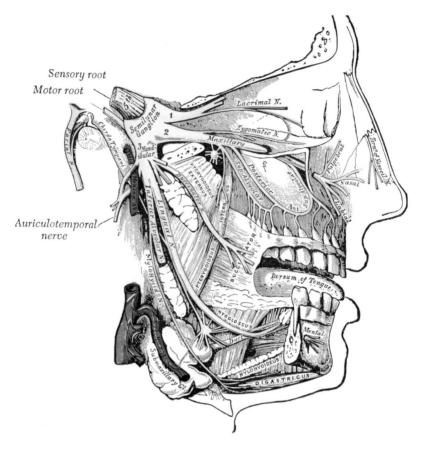

Fig. 12–8.—Distribution of the maxillary and mandibular nerves, and the submandibular ganglion.

teeth. It passes horizontally rostralward, at first in the inferior part of the lateral wall of the cavernous sinus and then beneath the dura, to the foramen rotundum, through which it leaves the cranial cavity. From the foramen rotundum, it crosses the pterygopalatine fossa, inclines lateralward in a groove on the posterior surface of the maxilla, and enters the orbit through the inferior orbital fissure. In the posterior part of the orbit it becomes the **infraorbital nerve**, lies in the infraorbital groove, and, continuing rostralward, dips into the infraorbital canal. It emerges into the face through the infraorbital foramen, where it is deep to the Levator labii superioris, and divides into branches for the skin of the face, nose, lower eyelid, and upper lip.

The branches of the maxillary nerve may be divided into four groups, those given off: (A) in the cranium, (B) in the pterygopala-

tine fossa, (C) in the infraorbital canal, and (D) in the face.

A. **Branches** in the cranium:

1. The **Middle Meningeal Nerve** (*n. meningeus medius; meningeal or dural branch*) is given off from the maxillary nerve directly after its origin from the trigeminal (semilunar) ganglion; it accompanies the middle meningeal artery and supplies the dura mater.

B. **Branches** in the pterygopalatine fossa:

2. The **Zygomatic Nerve** (*n. zygomaticus; temporomalar nerve; orbital nerve*) (Fig. 12–6) arises in the pterygopalatine fossa, enters the orbit by the inferior orbital fissure, and divides into two branches, zygomaticotemporal and zygomaticofacial.

a) The **zygomaticotemporal branch** (*r. zygomaticotemporalis; temporal branch*) runs along the lateral wall of the orbit in a groove in the zygomatic bone, and, passing through a small foramen or through the sphenozygomatic suture, enters the temporal fossa. It runs upward between the bone and substance of the Temporalis muscle, pierces the temporal fascia about 2.5 cm above the zygomatic arch, is distributed to the skin of the side of the forehead, and communicates with the facial nerve and with the auriculotemporal branch of the mandibular nerve. As it pierces the temporal fascia, it gives off a slender twig, which runs between the two layers of the fascia to the lateral side of the orbit (Fig. 12–18).

Before it leaves the orbit, it sends a **communication to the lacrimal nerve** through which the postganglionic parasympathetic fibers from the pterygopalatine ganglion reach the lacrimal gland (Fig. 12–8).

b) The **zygomaticofacial branch** (r. zygomaticofacialis; malar branch) passes along the inferior lateral angle of the orbit, through the zygomatic bone by way of the zygomaticoörbital and zygomaticofacial foramina, emerges upon the face, and, perforating the Orbicularis oculi, supplies the skin on the prominence of the cheek. It joins with the facial nerve and with the inferior palpebral branches of the infraorbital (Fig. 12–18).

Variations.—The two branches are variable in size, a deficiency in one being made up by the other or by the lacrimal or infraorbital nerves.

3. **The Pterygopalatine Nerves** (nn. pterygopalatini; sphenopalatine nerves) (Figs. 12–8, 12–9) are two short trunks which unite at the pterygopalatine ganglion, and then are redistributed into a number of branches. Formerly these trunks were called the sensory roots of the ganglion, and their peripheral branches were listed as the branches of distribution of the ganglion. Since the great majority of the fibers in the trunks are trigeminal somatic afferents which merely pass beside or through the ganglion without synapses, the branches are listed here as belonging to the maxillary nerve rather than the ganglion. The ganglion is described with the facial nerve.

The pterygopalatine nerves serve also as important functional communications between the ganglion and the maxillary nerve. Postganglionic parasympathetic secretomotor fibers from the ganglion pass through them and back along the main maxillary nerve to the zygomatic nerve, through which they are routed to the lacrimal nerve and lacrimal gland. Other fibers from the ganglion accompany the branches of distribution of the maxillary nerve to the glands of the nasal cavity and palate.

The **branches of distribution** from the pterygopalatine nerves are divisible into

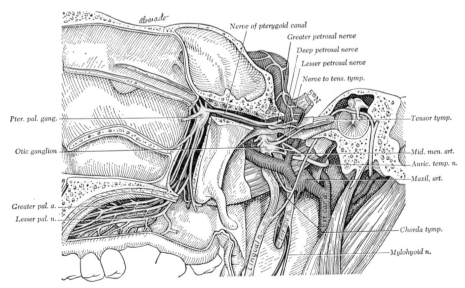

FIG. 12–9.—The pterygopalatine ganglion and nasal and palatine nerves.

four groups, viz., (*a*) orbital, (*b*) palatine, (*c*) posterior superior nasal, and (*d*) pharyngeal.

a) The **orbital branches** (*rr. orbitales; ascending branches*) are two or three delicate filaments which enter the orbit by the inferior orbital fissure, and supply the periosteum. Filaments pass through foramina in the frontoëthmoidal suture to supply the mucous membrane of the posterior ethmoidal and sphenoidal sinuses.

b) The **greater palatine nerve** (*n. palatinus anterior; anterior palatine nerve; descending branch*) (Figs. 12–9, 12–12) passes through the pterygopalatine canal, emerges upon the hard palate through the greater palatine foramen, and divides into several branches, the longest of which passes anteriorly in a groove in the hard palate nearly as far as the incisor teeth. It supplies the gums and mucous membrane of the hard palate and adjacent parts of the soft palate, and communicates with the terminal filaments of the nasopalatine nerve.

i) **Posterior inferior nasal branches** leave the nerve while it is in the canal, enter the nasal cavity through openings in the palatine bone, and ramify over the inferior nasal concha and middle and inferior meatuses.

ii) The **lesser palatine nerves** (*nn. palatini medius et posterior*) (Fig. 12–9) emerge through the lesser palatine foramina and distribute branches to the soft palate, uvula, and tonsil. They join with the tonsillar branches of the glossopharyngeal nerve to form a plexus around the tonsil (*circulus tonsillaris*). Many of the somatic afferent fibers contained in the lesser palatine nerves belong to the facial nerve, have their cells in the geniculate ganglion, and traverse the greater petrosal and nerve of the pterygoid canal (*Vidian*) (page 927).

c) The **posterior superior nasal branches** (*rr. nasales posteriores superiores*) enter the posterior part of the nasal cavity by the sphenopalatine foramen and supply the mucous membrane covering the superior and middle conchae, the lining of the posterior ethmoid sinuses, and the posterior part of the septum. One branch, longer and larger than the others, and named the **naso-palatine nerve,** passes across the roof of the nasal cavity inferior to the ostium of the sphenoidal sinus to reach the septum. It runs obliquely forward and downward, lying between the mucous membrane and periosteum of the septum, to the incisive canal (Fig. 12–2). It passes through the

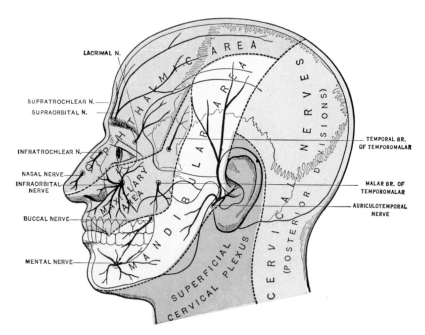

Fig. 12–10.—Sensory areas of the head, showing the general distribution of the three divisions of the fifth nerve. (Modified from Testut.)

canal and communicates with the corresponding nerve of the opposite side and with the greater palatine nerve.

d) The **pharyngeal branch** (*r. pharyngeus; pterygopalatine nerve*) (Fig. 12–9) leaves the posterior part of the pterygopalatine ganglion. It passes through the pharyngeal canal with the pharyngeal branch of the maxillary artery, and is distributed to the mucous membrane of the nasal part of the pharynx posterior to the auditory tube.

4. The **Posterior Superior Alveolar Branches** (*rr. alveolares superiores posteriores; posterior superior dental branches*) (Fig. 12–8) arise from the trunk of the nerve just before it enters the infraorbital groove; there are generally two, but sometimes they arise by a single trunk. They cross the tuberosity of the maxilla and give off several twigs to the gums and neighboring parts of the mucous membrane of the cheek. They then enter the posterior alveolar canals on the infratemporal surface of the maxilla, and, passing anteriorly in the substance of the bone, communicate with the middle superior alveolar nerve, and give off branches to the lining membrane of the maxillary sinus and three twigs to each molar tooth; these twigs enter the foramina at the apices of the roots of the teeth.

C. Branches in the infraorbital canal:

5. The **Middle Superior Alveolar Branch** (*r. alveolaris superior medius; middle superior dental branch*) is given off from the nerve in the posterior part of the infraorbital canal, and runs downward and forward in a canal in the lateral wall of the maxillary sinus to supply the two premolar teeth. It forms a superior dental plexus with the anterior and posterior superior alveolar branches.

6. The **Anterior Superior Alveolar Branch** (*r. alveolaris superior anterior; anterior superior dental branch*) (Fig. 12–8), of considerable size, is given off from the nerve just before its exit from the infraorbital foramen; it courses in a canal in the anterior wall of the maxillary sinus, and divides into branches which supply the incisor and canine teeth. It communicates with the middle superior alveolar branch, and gives off a nasal branch, which passes through a minute canal in the lateral wall of the inferior meatus, and supplies the mucous

membrane of the anterior part of the inferior meatus and the floor of the nasal cavity, communicating with the nasal branches from the pterygopalatine nerves.

The **infraorbital nerve** emerges through the infraorbital foramen and supplies the following branches:

D. **Branches in the face:**

7. The **Inferior Palpebral Branches** (*rr. palpebrales inferiores; palpebral branches*) (Fig. 12–8) pass superiorly, deep to the Orbicularis oculi, and supply the skin and conjunctiva of the lower eyelid, joining at the lateral angle of the orbit with the facial and zygomaticofacial nerves.

8. The **External Nasal Branches** (*rr. nasales externi*) (Fig. 12–8) supply the skin of the side of the nose and of the septum mobile nasi, and join with the terminal twigs of the nasociliary nerve.

9. The **Superior Labial Branches** (*rr. labiales superiores; labial branches*) (Fig. 12–8), the largest and most numerous, pass deep to the Levator labii superioris, and are distributed to the skin of the upper lip, to the mucous membrane of the mouth, and to the labial glands. They communicate immediately inferior to the orbit with filaments from the facial nerve, forming with them the **infraorbital plexus.**

3. The Mandibular Nerve

The **Mandibular Nerve** (*n. mandibularis; inferior maxillary nerve*) (Figs. 12–8, 12–11, 12–12), or third and largest division of the trigeminal, is a mixed nerve and has two roots: a *large sensory root* arising from the inferior angle of the trigeminal ganglion, and a *small motor root* (the entire motor root of the trigeminal). The sensory fibers supply the skin of the temporal region, auricula, external meatus, cheek, lower lip, and lower part of the face; the mucous membrane of the cheek, tongue, and mastoid air cells; the lower teeth and gums; the mandible and temporomandibular joint; and part of the dura mater and skull. The motor fibers supply the muscles of mastication (Masseter, Temporalis, Pterygoidei), the Mylohyoideus and anterior belly of the Digastricus, and the Tensores tympani and veli palatini. The two roots leave the middle cranial fossa through the foramen ovale, the motor part medial to the sensory, and unite just outside the skull. The main trunk thus formed is very short, 2 or 3 mm, and divides into a smaller anterior and a larger posterior division. The otic ganglion lies close against the medial surface of the nerve, just outside of the foramen ovale where the two roots fuse, and it surrounds the origin of the medial pterygoid nerve (Fig. 12–20).

A **communication** to the **otic ganglion** from the internal pterygoid nerve was formerly called a root of the ganglion, but the fibers pass through without synapses.

Variations.—The two divisions are separated by a fibrous band, the pterygospinous ligament, which may become ossified and the anterior division then passes through a separate opening in the bone, the pterygospinous foramen (foramen of Civanini).

A. The branches of the main trunk of the nerve are as follows:

1. The **Ramus Meningeus** (*nervus spinosus, recurrent branch*) enters the skull through the foramen spinosum with the middle meningeal artery. It divides into two branches which accompany the anterior and posterior divisions of the artery and supply the dura mater. The anterior branch communicates with the meningeal branch of the maxillary nerve; the posterior sends filaments to the mucous membrane of the mastoid air cells.

2. The **Medial Pterygoid Nerve** (*n. pterygoideus medialis*) (Fig. 12–20) is a slender branch which penetrates the otic ganglion, and, after a short course, enters the deep surface of the muscle. It has two small branches which have a close association with the otic ganglion, and have been described as branches of the ganglion, but their fibers pass through the ganglion without interruption, as follows:

a) The **nerve** to the **Tensor veli palatini** (Fig. 12–9) enters the muscle near its origin.

b) The **nerve** to the **Tensor tympani** lies close to and nearly parallel with the lesser superficial petrosal nerve, and penetrates the cartilage of the auditory tube to supply the muscle.

B. The anterior division of the mandibular nerve (Fig. 12–8) receives a small contribution of sensory fibers and all of the motor fibers from the motor root except those in the medial pterygoid and mylohyoid nerves. Its branches supply the muscles of mastication and the skin and mucous membrane of the cheek as follows:

3. The **Masseteric Nerve** (*n. massetericus*) (Fig. 12–11) passes lateralward, above the Pterygoideus lateralis, to the mandibular notch, which it crosses with the masseteric artery to enter the Masseter near its origin from the zygomatic arch. It gives a filament to the temporomandibular joint.

4. The **Deep Temporal Nerves** (*nn. temporales profundi*) (Fig. 12–11) are usually two, anterior and posterior, but a third or intermediate may be present. The anterior deep temporal frequently is given off from the buccal nerve; it emerges with the latter between the two heads of the Pterygoideus lateralis and turns superiorly into the anterior portion of the Temporalis. The posterior deep temporal, and the intermediate if present, pass over the superior border of the Pterygoideus lateralis close to the bone of the temporal fossa, and enter the deep surface of the muscle. The posterior sometimes arises in common with the masseteric nerve.

5. The **Lateral Pterygoid Nerve** (*n. pterygoideus lateralis*) enters the deep surface of the muscle. It frequently arises in conjunction with the buccal nerve.

6. The **Buccal Nerve** (*n. buccalis; buccinator nerve; long buccal nerve*) (Fig. 12–11) passes between the two heads of the Pterygoideus lateralis to reach its superficial surface, follows or penetrates the inferior part of the Temporalis, and emerges from under the anterior border of the Masseter. It ramifies on the surface of the Buccinator, forming a plexus of communications with the buccal branches of the facial nerve, supplies the skin of the cheek over this muscle, and sends penetrating branches to supply the mucous membrane of the mouth and part of the gums in the same area.

Variations.—The buccal nerve may supply a branch to the Pterygoideus lateralis as it passes through that muscle, and frequently it gives off the anterior deep temporal nerve. It may arise from the trigeminal ganglion, passing through its own foramen; it may be a branch of the inferior alveolar nerve; or it may be replaced by a branch of the maxillary nerve.

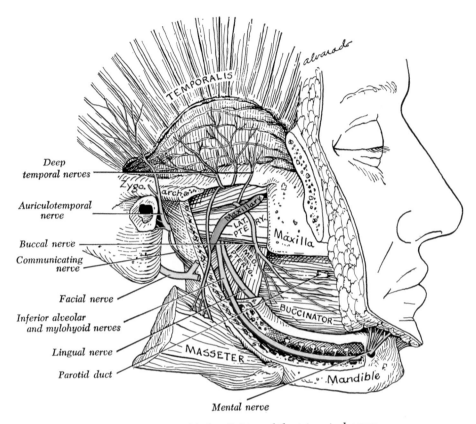

Deep
temporal nerves

Auriculotemporal
nerve

Buccal nerve

Communicating
nerve

Facial nerve

Inferior alveolar
and mylohyoid nerves

Lingual nerve

Parotid duct

Mental nerve

FIG. 12–11.—Mandibular division of the trigeminal nerve.

C. The posterior division of the mandibular nerve is mainly sensory, but it has a small motor component. Its **branches** are as follows:

7. The **Auriculotemporal Nerve** (*n. auriculotemporalis*) (Fig. 12–8) generally arises by two roots which join after encircling the middle meningeal artery close to the foramen spinosum. It runs posteriorly, deep to the Pterygoideus lateralis, along the medial side of the neck of the mandible, and turns superiorly with the superficial temporal artery, between the auricula and the condyle of the mandible, under cover of the parotid gland. Escaping from beneath the gland, it passes over the root of the zygomatic arch, and divides into superficial temporal branches (Fig. 12–18). The branches and communications of the auriculotemporal nerve are as follows:

a) The **communications** with the **facial nerve** (Fig. 12–11), usually two of considerable size, pass anteriorly from behind the neck of the mandible, and join the facial nerve in the substance of the parotid gland at the posterior border of the Masseter. They carry sensory fibers which accompany the zygomatic, buccal, and mandibular branches of the facial nerve and supply the skin of these areas.

b) **Communications** with the **otic ganglion** (Fig. 12–20) join the roots of the auriculotemporal nerve close to their origin. They carry postganglionic parasympathetic fibers whose preganglionics come from the glossopharyngeal nerve and supply the parotid gland with secretomotor fibers.

c) The **anterior auricular branches** (*nn. auriculares anteriores*) (Fig. 12–11), usually two, supply the skin of the anterior superior part of the auricula, principally the helix and tragus.

d) The **branches** to the **external acoustic meatus** (*nn. meatus acoustici*) (Fig. 12–11), two in number, enter the meatus between its bony and cartilaginous portions and

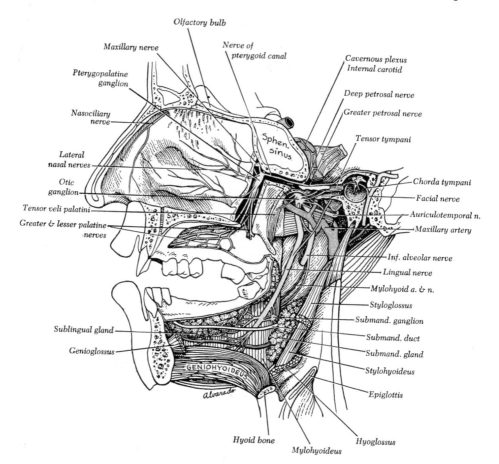

FIG. 12–12.—Deep dissection of the region of the face viewed from its medial aspect, showing the pterygo-palatine, otic, and submandibular ganglia and associated structures. (Redrawn from Töndury.)

supply the skin lining it; the upper one sends a filament to the tympanic membrane.

e) The **articular branches** consist of one or two twigs which enter the posterior part of the temporomandibular joint.

f) The **parotid branches** (*rami parotidei*) supply the parotid gland, carrying the parasympathetic postganglionic fibers transmitted by the communication between the auriculotemporal nerve and the otic ganglion.

g) The **superficial temporal branches** (*rr. temporales superficiales*) accompany the superficial temporal artery to the vertex of the skull; they supply the skin of the temporal region and communicate with the facial and zygomaticotemporal nerves.

8. The **Lingual Nerve** (*n. lingualis*) (Figs. 12–8, 12–12, 12–36) is at first deep to the Pterygoideus lateralis, running parallel with the inferior alveolar nerve, lying medial and anterior to it, and frequently joined to it by a branch which may cross the maxillary artery. The chorda tympani nerve joins it here also. The lingual nerve runs between the Pterygoideus medialis and the mandible, and then crosses obliquely over the Constrictor pharyngis superior and the Styloglossus to reach the side of the tongue. It passes between the Hyoglossus and deep part of the submandibular gland, and finally, crossing the lateral side of the submandibular duct runs along the undersurface of the tongue to its tip, lying immedi-

ately beneath the mucous membrane (Fig. 12–12). Its communications and branches are as follows:

a) The **chorda tympani** (Figs. 12–8, 12–12), a branch of the facial nerve, joins the lingual posteriorly at an acute angle, 1 or 2 cm from the foramen ovale. It carries special sensory fibers for taste and parasympathetic preganglionic fibers for the submandibular ganglion.

b) The **communications** with the **submandibular ganglion** are usually two or more short nerves by which the ganglion seems to be suspended (Fig. 12–12). The proximal nerves carry the preganglionic parasympathetic fibers communicated to the lingual by the chorda tympani. The distal communication contains postganglionic fibers for distribution to the sublingual gland.

c) **Communications** with the **hypoglossal nerve** (Fig. 12–36) form a plexus at the anterior margin of the Hyoglossus.

d) The **branches of distribution** supply the mucous membrane of the anterior two-thirds of the tongue, the adjacent mouth and gums, and the sublingual gland. The taste buds of the anterior two-thirds of the tongue are supplied by the fibers communicated through the chorda tympani.

9. The **Inferior Alveolar Nerve** (*n. alveolaris inferior; inferior dental nerve*) (Fig. 12–11) accompanies the inferior alveolar artery, at first deep to the Pterygoideus lateralis, and then between the sphenomandibular ligament and the ramus of the mandible, to the mandibular foramen. It enters the mandibular canal through the foramen, and passes anteriorly within the bone as far as the mental foramen, where it divides into two terminal branches. The branches of the nerve are as follows:

a) The **mylohyoid nerve** (*n. mylohyoideus*) (Fig. 12–8) leaves the inferior alveolar nerve just before it enters the mandibular foramen, and continues inferiorly and anteriorly in a groove on the deep surface of the ramus of the mandible to reach the Mylohyoideus. It supplies this muscle and crosses its superficial surface to reach the anterior belly of the Digastricus (Fig. 12–8), which it supplies also.

b) The **dental branches** form a plexus within the bone, and supply the molar and premolar teeth, filaments entering the pulp

canal of each root through the apical foramen and supplying the pulp of the tooth.

c) The **incisive branch** is one of the terminal branches. It continues anteriorly within the bone, after the mental nerve separates from it, and forms a plexus which supplies the canine and incisor teeth.

d) The **mental nerve** (*n. mentalis*) (Fig. 12–11), the other terminal branch, emerges from the bone at the mental foramen, and divides beneath the Depressor anguli oris' muscle into three branches; one is distributed to the skin of the chin, the other two to the skin and mucous membrane of the lower lip. These branches communicate freely with branches of the facial nerve.

The otic and submandibular ganglia, although closely associated with the mandibular nerve, are not connected functionally, and are described, therefore, with the nerves which supply them with preganglionic fibers; the otic ganglion with the glossopharyngeal nerve and the submandibular with the facial nerve.

Variations.—The inferior alveolar and lingual nerves may form a single trunk, or they may have communications of variable size; the chorda tympani may appear to join the inferior alveolar and a later communication carry the fibers to the lingual. The inferior alveolar nerve is occasionally perforated by the maxillary artery. It may have accessory roots from other branches of the mandibular, or have a separate root from the trigeminal ganglion. The mylohyoid nerve may communicate with the lingual and it has been described as sending filaments to the Depressor anguli oris, Platysma, submandibular gland, or integument below the chin.

Trigeminal Nerve Pain (*tic douloureux*). —The trigeminal nerve is more frequently the seat of severe neuritic or neuralgic pain than any other nerve in the body. The pain of a localized infection or irritation may be confined to that area, but quite commonly this is not the case. Involvement of an internal branch is likely to set up severe distress in a related cutaneous area by referred pain. As a general rule the diffusion of pain over the branches of the nerve is confined to one of the main divisions, although in severe cases it may radiate over the other main divisions.

The commonest example of this condition is the neuralgia which is often associated with dental caries—here, although the tooth itself may not appear to be painful, the most distressing referred pains may be experienced, and those be at once relieved by treatment of the affected tooth. With the ophthalmic nerve, severe supraorbital pain is commonly associated with acute glaucoma or with frontal or ethmoidal sinusitis. Malignant growths or empyema of the maxillary sinus, and diseased conditions in the nasal cavity, as well as dental caries, may cause neuralgia of the second division. Pain in the mandibular division is likely to be in the ear or other distribution of the auriculotemporal nerve although the actual disease may involve one of the lower teeth or the tongue.

When a focus of infection or irritation cannot be found, as is all too frequently the case, various measures may be taken to interrupt the pain fibers. Local injection of alcohol into the painful nerve may give temporary relief, and injection of a main division close to the ganglion has been performed with a certain measure of success. The main divisions have been incised surgically, and the entire ganglion removed in intractable cases. In the latter operation bleeding is likely to be dangerous and the nerves to the muscles of mastication may be paralyzed. In more recent operations the motor root is spared by cutting the sensory root inside the cranium before it reaches the ganglion.

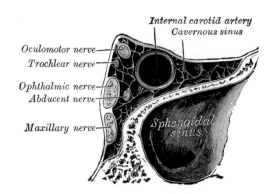

FIG. 12–13.—Oblique section through the right cavernous sinus.

VI. THE ABDUCENT NERVE

The **abducent nerve** (*n. abducens; sixth nerve*) (Fig. 12–6) supplies the Rectus lateralis oculi. Its superficial origin (Fig. 11–62) is in the furrow between the inferior border of the pons and the superior end of the pyramid of the medulla oblongata. The deep origin and central connections are described on page 872. It pierces the dura mater on the dorsum sellae of the sphenoid bone, runs through a notch below the posterior clinoid process and traverses the cavernous sinus lateral to the internal carotid artery (see below). It enters the orbit through the superior orbital fissure, superior to the ophthalmic vein, from which it is separated by a lamina of dura mater. After passing between the two heads of the Rectus lateralis, it enters its ocular surface.

Communication with the **sympathetic** system is by several filaments from the carotid and cavernous plexuses, and with the **trigeminal** by a filament from the ophthalmic nerve.

Relation of Orbital Nerves to the Cavernous Sinus

Embedded in the lateral wall of the cavernous sinus, in order beginning with the most superior, are the oculomotor, trochlear, ophthalmic, and maxillary nerves (Fig. 12–13). The abducent nerve is suspended by connective tissue trabeculae within the sinus, lateral to the internal carotid artery and medial to the ophthalmic nerve. The maxillary nerve is related to the posterior portion of the sinus, and soon diverges from the other nerves. As the nerves approach the superior orbital fissure, the oculomotor and ophthalmic divide into branches, and the abducent nerve approaches the others, so that their relative positions are considerably changed.

VII. THE FACIAL NERVE

The **facial nerve** (*n. facialis; seventh nerve*) has two roots of unequal size (Fig. 12–14). The larger is the **motor root;** the smaller, lying between the motor root and the acoustic nerve, is called the **nervus intermedius** (*pars intermedia; nerve of Wrisberg*), and contains special sensory fibers for taste and parasympathetic fibers. The superficial origin of both roots is at the inferior border of the pons (Fig. 11–62) in the recess between the olive and the root being medial, the acoustic nerve lateral, inferior cerebellar peduncle, the motor and the intermedius between. The deep origin and central connections within the medulla are described on page 872.

The facial is the motor nerve to the

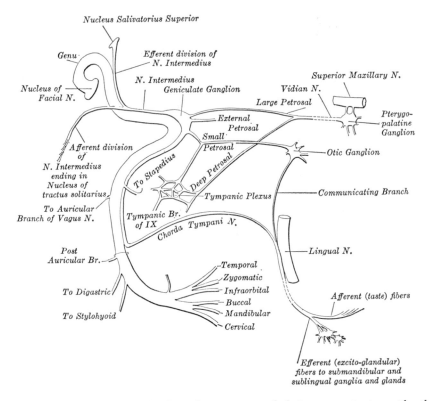

Nucleus Salivatorius Superior

Genu

*Efferent division of
N. Intermedius*

*Nucleus of
Facial N.*

N. Intermedius

*N. Intermedius
Geniculate Ganglion*

Superior Maxillary N.

Vidian N.

Large Petrosal

*External
Petrosal*

*Small
Petrosal*

*Pterygo-
palatine
Ganglion*

*Afferent division
of
N. Intermedius
ending in
Nucleus of
tractus solitarius*

To Stapedius

Deep Petrosal

Otic Ganglion

*To Auricular
Branch of Vagus N.*

Tympanic Plexus

Communicating Branch

*Tympanic Br.
of IX*

Chorda Tympani N.

*Post
Auricular Br.*

Lingual N.

To Digastric

To Stylohyoid

Temporal
Zygomatic
Infraorbital
Buccal
Mandibular
Cervical

Afferent (taste) fibers

*Efferent (excito-glandular)
fibers to submandibular and
sublingual ganglia and glands*

Fig. 12–14.—Plan of the facial and intermediate nerves and their communication with other nerves.

muscles of facial expression, to those in the scalp and external ear, to the Buccinator, Platysma, Stapedius, Stylohyoideus, and posterior belly of the Digastricus. The sensory part supplies the anterior two-thirds of the tongue with taste, and parts of the external acoustic meatus, soft palate, and adjacent pharynx with general sensation. The parasympathetic part supplies secreto-motor fibers for the submandibular, sublingual, lacrimal, nasal, and palatine glands.

From their superficial attachment, the two roots pass lateralward with the acoustic nerve into the internal acoustic meatus (Fig. 12–1). At the fundus of the meatus (Fig. 4–90), the facial nerve separates from the acoustic and enters the substance of the petrous portion of the temporal bone, through which it runs a serpentine course in its own canal, the **facial canal** (*aqueductus Fallopii*). At first it continues lateralward in the region between the cochlea and semicircular canals (Fig 13–58); near the tympanic cavity it makes an abrupt bend posteriorly, runs in the medial

wall of the cavity just above the oval window (Fig. 13–50), covered by a thin plate of bone which causes a slight prominence in the wall (Fig. 13–51), and then dips down beside the mastoid air cells to reach the stylomastoid foramen. At the point mentioned, where it changes its course abruptly, there is an exaggeration of the bend into a U-shaped structure named the **geniculum;** at this point also, the two roots become fused and the nerve is swollen by the presence of the geniculate ganglion (Fig. 12–15). As it emerges from the stylomastoid foramen, the nerve runs anteriorly in the substance of the parotid gland, crosses the external carotid artery, and divides at the posterior border of the ramus of the mandible into two primary branches, a superior, the temporofacial, and an inferior, the cervicofacial, from which numerous offsets, in a plexiform arrangement, **parotid plexus,** are distributed over the head, face, and upper part of the neck, supplying the superficial muscles in these regions.

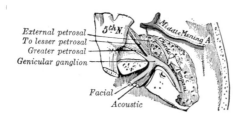

External petrosal
To lesser petrosal
Greater petrosal
Genicular ganglion
5ᵗʰN.
Middle Mening. A.
Facial
Acoustic

FIG. 12–15.—The geniculate ganglion and facial nerve in the temporal bone.

The **Geniculate Ganglion** (*g. geniculi*) (Fig. 12–15) is a small fusiform swelling of the geniculum, where the facial nerve bends abruptly backward at the hiatus of the facial canal. It is the sensory ganglion of the facial nerve. The central processes of its unipolar ganglion cells reach the brain stem through the **nervus intermedius;** the majority of the peripheral processes pass to the taste buds of the anterior two-thirds of the tongue through the chorda tympani and lingual nerves; a considerable number of peripheral processes pass through the greater superficial petrosal and lesser palatine nerves to the soft palate; and a smaller number join the auricular branch of the vagus to supply the skin of the external acoustic meatus and mastoid process (Foley et al. '46).

The **glossopalatine nerve** is the name given by some authorities (Hardesty '33) to that portion of the facial nerve contributed by the **nervus intermedius.** It comprises, therefore, the sensory part, including the geniculate ganglion, the chorda tympani, and greater petrosal nerve; and the parasympathetic part, including the submandibular and pterygopalatine ganglia and their branches. Although this separation is suggested by the similarity between this complex and the glossopharyngeal nerve, it has not been generally adopted.

Communications of the Facial Nerve.—
A. In the internal acoustic meatus, communications with the **acoustic nerve** probably contain fibers which leave the facial, run with the acoustic for a short distance, and return to the facial nerve.
B. At the geniculate ganglion, it communicates (*a*) with the **otic ganglion** by filaments which join the lesser petrosal nerve; (*b*) with the **sympathetic fibers** on

the middle meningeal artery (*external superficial petrosal nerve*).
C. In the facial canal, just before it leaves the stylomastoid foramen, it communicates with the **auricular branch of the vagus** as the latter runs across it in the substance of the bone.
D. After its exit from the Stylomastoid foramen, it communicates with (*a*) the **glossopharyngeal** nerve, (*b*) the **vagus** nerve, and (*c*) the **great auricular** nerve from the cervical plexus. (*d*) The **auriculotemporal** nerve from the mandibular division of the trigeminal usually sends two **communications** of considerable size which pass anteriorly from behind the neck of mandible to join the facial nerve in the substance of the parotid gland (Fig. 12–11). They carry sensory fibers which accompany the terminal zygomatic, buccal, and mandibular branches (see below) to the skin of these areas.
E. Peripheral branches communicate: behind the ear, with the **lesser occipital** nerve; on the face, with the **trigeminal branches;** and in the neck, with the **cervical cutaneous** nerve.
A. The branches from the geniculate ganglion of the facial nerve:
1. The **Greater Petrosal Nerve** (*n. petrosus major; greater superficial petrosal nerve*) arises from the geniculate ganglion (Fig. 12–15), and, after a short course in the bone, emerges through the hiatus of the facial canal (Fig. 4–89) into the middle cranial fossa. It runs rostralward beneath the dura mater and the trigeminal ganglion in a sulcus on the anterior surface of the petrous portion of the temporal bone, passes superior to the cartilage of the auditory tube which fills the foramen lacerum, crosses the lateral side of the internal carotid artery, and unites with the deep petrosal nerve to form the nerve of the pterygoid canal (Vidian nerve) (Figs. 12–9, 12–12). The greater petrosal nerve is a mixed nerve, containing sensory and parasympathetic fibers. The parasympathetic fibers, from the nervus intermedius, become the motor root of the pterygopalatine ganglion. The bulk of the nerve consists of sensory fibers which are the peripheral processes of cells in the geniculate ganglion and are distributed to the soft palate through the lesser

palatine nerves, with a few filaments to the auditory tube.

The **nerve of the pterygoid canal** (*n. canalis pterygoidei [Vidii]; Vidian nerve*) (Fig. 12–9), formed by the union of the greater petrosal and deep petrosal nerves at the foramen lacerum, enters the posterior opening of the pterygoid canal with the corresponding artery, and is joined by a small *ascending sphenoidal branch* from the otic ganglion. The bony wall of the canal commonly causes a ridge in the floor of the sphenoidal sinus, and while the nerve is in the canal it gives off one or two filaments for the mucous membrane of the sinus. From the anterior opening of the canal, the nerve crosses the pterygopalatine fossa, and enters the pterygopalatine ganglion.

The **Pterygopalatine Ganglion** (*g. pterygopalatinum; sphenopalatine ganglion; Meckel's ganglion*) (Figs. 12–9, 12–12) is deeply placed in the pterygopalatine fossa, just inferior to the maxillary nerve as the latter crosses the fossa close to the pterygopalatine foramen. It is triangular or heart-shaped, about 5 mm in length, is embedded in the fibrous tissue between the neighboring bones, and is closely attached to the pterygopalatine branches of the maxillary division of the trigeminal nerve. It is a parasympathetic ganglion relaying chiefly secretomotor impulses from the facial (Fig. 12–16).

The **parasympathetic root** (*visceral efferent*) of the pterygopalatine ganglion is the greater petrosal nerve, and its continuation, the nerve of the pterygoid canal. The fibers

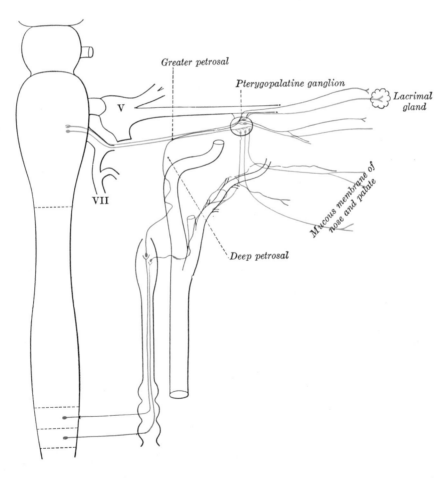

Fig. 12–16.—Autonomic connections of the pterygopalatine and superior cervical ganglia. Parasympathetic blue, sympathetic red.

are preganglionic parasympathetic fibers which leave the brain stem in the nervus intermedius.

Communications of the Pterygopalatine Ganglion:

a) Two short trunks from the **maxillary nerve,** the pterygopalatine nerves, are commonly called the sensory root of the ganglion although they have no such functional connection with it. They contain mainly sensory fibers from the trigeminal ganglion which pass through or beside the pterygopalatine ganglion without synapses, and continue on their way to the mucous membrane of the nasal cavity and palate. These trunks are important communications for the ganglion, however, since they are traversed by the postganglionic fibers of distribution on their way to the maxillary nerve, whence they reach the lacrimal gland and the small glands of the nasal cavity and palate.

b) The **deep petrosal nerve** is commonly called the sympathetic root of the pterygopalatine ganglion, although it is merely a communication between the ganglion and the sympathetic system. It contains postganglionic fibers from the superior cervical sympathetic ganglion, by way of the carotid plexus, which pass through the pterygopalatine ganglion without synapses and accompany the branches of the pterygopalatine nerves (trigeminal branches) to their destination in the mucous membrane of the nasal cavity and palate.

The **branches of distribution** from the **pterygopalatine ganglion,** containing the postganglionic fibers of the cells within the ganglion, are not independent nerves for the most part, but find their way to their destinations by accompanying other nerves, mainly the branches of the maxillary nerve.

(*a*) The **fibers** for the **lacrimal gland** pass back to the main trunk of the maxillary nerve through the pterygopalatine nerves. They leave the maxillary through the zygomatic and zygomaticotemporal nerves, pass through a communication between the latter and the lacrimal nerve in the orbit, and are then distributed to the gland (Fig. 12–6).

(*b*) Fibers for the **small glands** of the mucous membrane of the **nasal cavity, pharynx,** and **palate** join the greater and lesser palatine nerves, the posterior superior

nasal branches and the pharyngeal branch, and are distributed with them.

B. Branches of the facial nerve within the facial canal:

2. The **nerve to the Stapedius muscle** (*n. stapedius*) arises from the facial nerve as it passes downward in the posterior wall of the tympanum, and reaches the muscle through a minute opening in the base of the pyramid (see Middle Ear).

3. The **Chorda Tympani Nerve** (*n. chorda tympani*) (Figs. 12–8, 12–12) arises from the part of the facial nerve which runs vertically downward in the posterior wall of the tympanum just before it reaches the stylomastoid foramen. Entering its own canal in the bone about 6 mm above the stylomastoid foramen, the chorda passes back cranialward almost parallel with the facial nerve but diverges toward the lateral wall of the tympanum. It emerges through an aperture in the posterior wall of the tympanum (*iter chordae posterius*) between the base of the pyramid and the attachment of the tympanic membrane. It runs horizontally along the lateral wall of the tympanum covered by the thin mucous membrane, and, lying against the tympanic membrane, crosses the attached manubrium of the malleus (Fig. 13–49). It leaves the tympanic cavity near the anterior border of the membrane through the *iter chordae anterius,* traverses a canal in the petrotympanic fissure (*canal of Huguier*), and emerges from the skull on the medial surface of the spine of the sphenoid bone. After crossing the spine, usually in a groove, and being joined by a small communication from the otic ganglion, it unites with the lingual nerve at an acute angle between the Pterygoideus medialis and lateralis.

The bulk of the fibers of the chorda tympani are special visceral afferents for taste which are distributed with the branches of the lingual nerve to the anterior two-thirds of the tongue. It also contains preganglionic parasympathetic fibers (secretomotor) from the nervus intermedius, which terminate in synapses with cells in the submandibular ganglion (Foley '45).

The **Submandibular Ganglion** (*g. submandibulare; submaxillary ganglion*) (Figs. 12–8, 12–12) is a small mass, 2 to 5 mm in diameter, situated above the deep portion

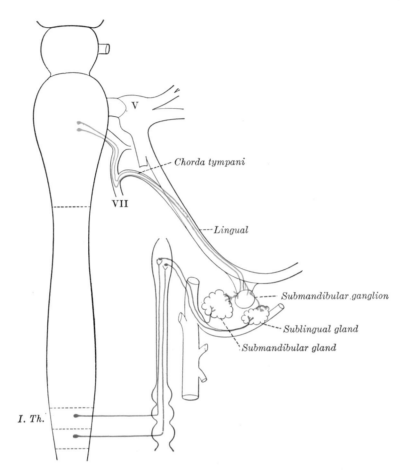

FIG. 12–17.—Autonomic connections of the submandibular and superior cervical ganglia. Parasympathetic blue, sympathetic red.

of the submandibular gland, on the Hyoglossus, near the posterior border of the Mylohyoideus, and suspended from the lower border of the lingual nerve by two filaments approximately 5 mm in length. The proximal filament is the **parasympathetic root** which conveys fibers originating in the nervus intermedius and communicated to the lingual by the chorda tympani. These are preganglionic visceral efferent fibers (secretomotor) whose postganglionic fibers innervate the submandibular, sublingual, lingual, and neighboring small salivary glands (Fig. 12–17).

The **branches of distribution** are (a) five or six filaments distributed to the **submandibular gland** and its duct, (b) to the **small glands** about the floor of the mouth,

and (c) the distal filament attaching the ganglion to the lingual nerve which communicates the fibers distributed to the **sublingual** and small lingual **glands** with the terminal branches of the lingual nerve. Small groups of ganglion cells are constantly found in the stroma of the submandibular gland, usually near the larger branches of the duct, and are considered to be functionally a part of the submandibular ganglion.

A **communication** with the **sympathetic** bundles on the facial artery have been called the *sympathetic root of the ganglion,* but the fibers are postganglionic and have no synapses in the ganglion.

Visceral afferent fibers passing through the root to the lingual and thence to the chorda tympani have been called the *sensory root,* but they have no synapses in the ganglion and their cell bodies are in the geniculate ganglion.

C. Branches of the facial nerve in the face and neck:

4. The **Posterior Auricular Nerve** (*n. auricularis posterior*) (Fig. 12–18) arises close to the stylomastoid foramen, and runs cranialward anterior to the mastoid process; here it is joined by a filament from the auricular branch of the vagus, and communicates with the posterior branch of the great auricular, and with the lesser occipital. Between the external acoustic meatus and mastoid process it divides into auricular and occipital branches. The auricular branch supplies the Auricularis posterior and the intrinsic muscles on the cranial surface of the auricula. The occipital

branch, the larger, passes backward along the superior nuchal line of the occipital bone, and supplies the Occipitalis.

5. The **Digastric Branch** (*r. digastricus*) arises close to the stylomastoid foramen, and divides into several filaments which supply the posterior belly of the Digastricus.

6. The **Stylohyoid Branch** (*r. stylohyoideus*) frequently arises in conjunction with the digastric branch; it is long and slender, and enters the Stylohyoideus about its middle.

Parotid plexus.—The terminal portion of the facial nerve, within the substance of the parotid gland, divides into a temporofacial and a cervicofacial division which in turn, either within the gland or after leaving it, break up into a plexus and supply the

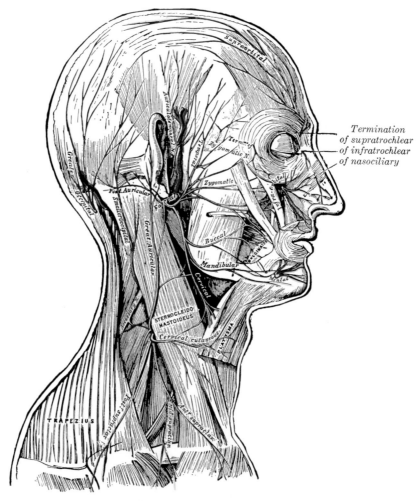

Termination
of supratrochlear
of infratrochlear
of nasociliary

FIG. 12–18.—The nerves of the scalp, face, and side of neck.

facial muscles in the different regions as the following nerves:

7. The **Temporal Branches** (*rr. temporales*) (Fig. 12–18) cross the zygomatic arch to the temporal region, supplying the Auriculares anterior and superior. They communicate with the zygomaticotemporal branch of the maxillary, and with the auriculotemporal branch of the mandibular division of the trigeminal nerve. The more anterior branches supply the Frontalis, the Orbicularis oculi, and the Corrugator, and join the supraorbital and lacrimal branches of the ophthalmic nerve.

8. The **Zygomatic Branches** (*rr. zygomatici; malar branches*) run across the face in the region of the zygomatic arch to the lateral angle of the orbit, where they supply the Orbicularis oculi, and communicate with filaments from the lacrimal nerve of the ophthalmic division and the zygomaticofacial branch of the maxillary division of the trigeminal nerve. The lower zygomatic branches commonly join the deep buccal branches and assist in forming the infraorbital plexus (Anson '50).

9. The **Buccal Branches** (*rr. buccales; infraorbital branches*), of larger size than the rest, pass horizontally rostralward to be distributed inferior to the orbit and around the mouth. The superficial branches run beneath the skin and superficial to the muscles of the face, which they supply: some are distributed to the Procerus, communicating at the medial angle of the orbit with the infratrochlear and nasociliary branches of the ophthalmic nerve. The deep branches, commonly reinforced by zygomatic branches, pass deep to the Zygomaticus and the Levator labii superioris, supplying them and forming an infraorbital plexus with the infraorbital branch of the maxillary division of the trigeminal nerve (Fig. 12–18). These branches also supply the small muscles of the nose. The more inferior deep branches supply the Buccinator and Orbicularis oris, and communicate with filaments of the buccal branch of the mandibular division of the trigeminal nerve.

10. The **Mandibular Branch** (*r. marginalis mandibulae*) passes rostralward deep to the Platysma and Depressor anguli oris, supplying the muscles of the lower lip and chin, and communicating with the mental branch of the inferior alveolar nerve.

11. The **Cervical Branch** (*r. colli*) runs rostralward deep to the Platysma which it supplies, and forms a series of arches across the side of the neck over the suprahyoid region. One branch joins the cervical cutaneous nerve from the cervical plexus.

VIII. **THE ACOUSTIC NERVE**[*]

The **acoustc nerve** (*n.* NA *vestibulocochlearis; n. statoacusticus; n. octavius; eighth nerve*) (Fig. 11–62) consists of two distinct sets of fibers, the **Cochlear and Vestibular Nerves**, which differ in their peripheral endings, central connections, functions, and time of myelination. These two portions of the acoustic nerve are joined into a common trunk which enters the internal acoustic meatus with the facial nerve (Fig. 12–15). Centrally, the acoustic nerve divides into a lateral (cochlear) root and a medial (vestibular) root. As it passes distally in the internal auditory meatus, it divides into the various branches which are distributed to the receptor areas in the membranous labyrinth (see Sense Organ chapter). Both divisions of this nerve are sensory and the fibers arise from bipolar ganglion cells.

Cochlear Nerve.—The cochlear nerve or root, the nerve of hearing, arises from bipolar cells in the *spiral ganglion of the cochlea*, situated near the inner edge of the osseous spiral lamina. The peripheral fibers pass to the organ of Corti. The central fibers pass through the modiolus and then through the foramina of the tractus spiralis foraminosus or through the foramen centrale into the lateral or outer end of the internal acoustic meatus. The nerve passes along the internal acoustic meatus with the vestibular nerve and across the subarachnoid space, just above the flocculus, almost directly medialward toward the inferior peduncle to terminate in the cochlear nuclei.

The cochlear nerve is placed lateral to the vestibular root. Its fibers end in two nuclei: one, the central cochlear nucleus, lies immediately in front of the inferior

[*] This BNA term will be used until the NA becomes stabilized.

cerebellar peduncle; the other, the dorsal cochlear nucleus, tuberculum acusticum, somewhat lateral to it (see Central Nervous System).

Vestibular Nerve.—The vestibular nerve or root, the nerve of equilibration, arises from bipolar cells in the *vestibular ganglion* (*ganglion of Scarpa*), which is situated in the superior part of the lateral end of the internal acoustic meatus. The peripheral fibers divide into three branches: the superior branch passes through the foramina in the area vestibularis superior and ends in the utricle and in the ampullae of the superior and lateral semicircular ducts; the fibers of the inferior branch traverse the foramina in the area vestibularis inferior and end in the saccule; the posterior branch runs through the foramen singulare and supplies the ampulla of the posterior semicircular duct.

The fibers of the vestibular nerve enter the medulla oblongata, pass between the inferior cerebellar peduncle and the spinal tract of the trigeminal, and bifurcate into ascending and descending branches. The descending branches form the spinal root of the vestibular nerve and terminate in the associated nucleus. The ascending branches pass to the medial, lateral, and superior vestibular nuclei, and to the nucleus fastigii and the vermis (see Central Nervous System).

IX. THE GLOSSOPHARYNGEAL NERVE

The **glossopharyngeal nerve** (*n. glossopharyngeus; ninth nerve*) (Fig. 12–22), as its name implies, is distributed to the tongue and pharynx. It is a mixed nerve, its sensory fibers being both visceral and somatic, and its motor fibers both general and special visceral efferents. The somatic afferent fibers supply the mucous membrane of the pharynx, fauces, palatine tonsil, and posterior part of the tongue; special visceral afferents supply the taste buds of the posterior part of the tongue; general visceral afferents supply the blood pressure receptor of the carotid sinus. The special visceral efferent fibers supply the Stylopharyngeus; the general visceral efferents are mainly

secretomotor for the parotid and small glands in the mucous membrane of the posterior part of the tongue and neighboring pharynx. The superficial origin is by three or four rootlets in series with those of the vagus nerve, attached to the superior part of the medulla oblongata in the groove between the olive and the inferior peduncle (Figs. 11–35, 11–62). The deep origin and central connections are described on page 875.

From its superficial origin the nerve passes lateralward across the flocculus to the jugular foramen, through which it passes, lateral and anterior to the vagus and accessory nerves (Fig. 12–22), in a separate sheath of dura mater, and lying in a groove on the lower border of the petrous portion of the temporal bone. After its exit from the skull, it runs anteriorly between the internal jugular vein and internal carotid artery, superficial to the latter vessel and posterior to the styloid process and its muscles. It follows the posterior border of the Stylopharyngeus for 2 or 3 cm, then curves across its superficial surface to the posterior border of the Hyoglossus, and penetrates more deeply to be distributed to the palatine tonsil, the mucous membrane of the fauces and base of the tongue, and the glands of that region. The portion of the nerve which lies in the jugular foramen has two enlargements, the superior and inferior ganglia.

The **Superior Ganglion** (*g. superius; jugular ganglion*) is situated in the upper part of the groove in which the nerve is lodged during its passage through the jugular foramen. It is very small, may be absent, and is usually regarded as a detached portion of the inferior ganglion.

The **Inferior Ganglion** (*g. petrosum; petrous ganglion*) is situated in a depression in the lower border of the petrous portion of the temporal bone. These ganglia contain the cell bodies for the sensory fibers of the nerve (Fig. 12–21).

Communications.—(1) The communications with the **vagus nerve** are two filaments, one joining the auricular branch, the other the superior ganglion. (2) The superior cervical **sympathetic** ganglion communicates with the inferior ganglion. (3) The communication with the **facial nerve** is between the trunk of the glossopharyngeal below the inferior ganglion and the facial nerve after its exit from the stylomastoid foramen; it perforates the posterior belly of the Digastricus.

The **branches of the glossopharyngeal nerve** are as follows:

1. The **Tympanic Nerve** (*n. tympanicus; nerve of Jacobson*) (Fig. 12–19) supplies parasympathetic fibers to the parotid gland, through the otic ganglion, and sensory fibers to the mucous membrane of the middle ear. It arises from the inferior ganglion, enters a small canal through an opening in the bony ridge which separates the carotid canal from the jugular fossa on the inferior surface of the petrous portion of the temporal bone. After a short cranialward course in the bone, it enters the tym-

panic cavity by an aperture in its floor near the medial wall. It continues upward in a groove on the surface of the promontory (Fig. 13–50), helps to form the tympanic plexus, reenters a canaliculus at the level of the processus cochleariformis, passes internal to the semicanal for the Tensor tympani, and continues as the lesser petrosal nerve (Rosen '50).

The **Tympanic Plexus** lies in grooves on the surface of the promontory (Fig. 13–50), and is formed by the junction of the tympanic and caroticotympanic nerves. The **caroticotympanic nerves, superior** and **inferior,** are communications from the carotid plexus of the sympathetic which enter the tympanic cavity by perforating the wall of the carotid canal. The plexus communicates with the greater petrosal nerve by a

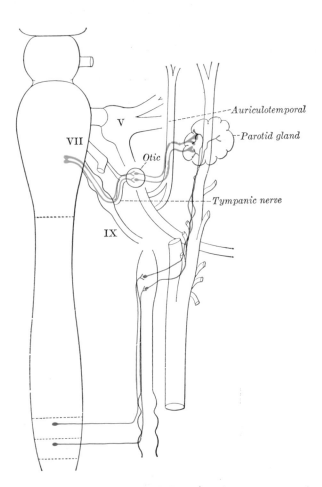

FIG. 12–19.—Autonomic connections of the otic and superior cervical ganglia. Parasympathetic blue, sympathetic red.

filament which passes through an opening on the labyrinthic wall, in front of the fenestra vestibuli.

a) **Sensory branches** are distributed through the plexus to the mucous membrane of the fenestra ovalis, fenestra rotunda, tympanic membrane, auditory tube, and mastoid air cells.

b) The **lesser petrosal nerve** (*n. petrosus superficialis minor*) (Figs. 12–9, 12–20) is the terminal branch or continuation of the tympanic nerve beyond the plexus. After

mandibular division of the trigeminal, immediately outside of the foramen ovale, and it has the origin of the medial pterygoid nerve embedded in it. It is lateral to the cartilaginous portion of the auditory tube, anterior to the middle meningeal artery, and posterior to the origin of the Tensor veli palatini.

The **root** of the **otic ganglion,** which is **parasympathetic,** is the lesser petrosal nerve. It contains preganglionic fibers from the nucleus salivatorius inferior in the

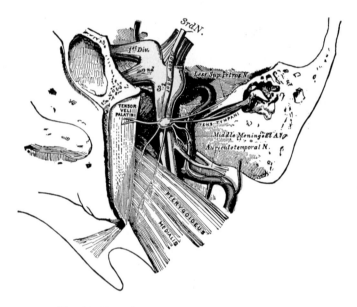

FIG. 12–20.—The otic ganglion and its branches.

penetrating the bone medial to the Tensor tympani, it emerges into the cranial cavity on the superior surface of the petrous portion of the temporal bone, immediately lateral to the hiatus of the facial canal. It leaves the cranial cavity again through the fissure between the petrous portion and the great wing of the sphenoid, or through a small opening in the latter bone, and terminates in the otic ganglion as its visceral motor or parasympathetic root. In the canal it is joined by a filament from the geniculate ganglion of the facial nerve.

The **Otic Ganglion** (*g. oticum*) (Figs. 12–12, 12–20) is a flattened, oval, or stellate ganglion, 2 to 4 mm in diameter, closely approximated to the medial surface of the

medulla oblongata, principally through the glossopharyngeal but probably partly through the facial nerve.

Communications of the Otic Ganglion.— (*a*) A communication with the **sympathetic** network on the middle meningeal artery has been called the sympathetic root of the ganglion, but these fibers are already post-ganglionic and pass through the ganglion without synapses. (*b*) A communication with the **medial pterygoid nerve** has been described as a motor root, and the continuation of the fibers to the Tensor veli palatini and Tensor tympani as branches of the ganglion, but they are trigeminal fibers which pass through the ganglion (see page 920). (c) A communication with the

mandibular nerve has been called a sensory root, but the fibers have no functional connection with the ganglion. (*d*) A slender filament, the **sphenoidal branch**, connects with the nerve of the pterygoid canal, and a small branch communicates with the **chorda tympani.**

Branches of Distribution of the Otic Ganglion.—The postganglionic fibers arising in the otic ganglion pass mainly through a communication with the auriculotemporal nerve and are distributed with its **branches** to the **parotid gland.** Other filaments probably accompany other nerves to reach small glands in the mouth and pharynx.

2. The **Carotid Sinus Nerve** (*carotid nerve; nerve of Hering*) (Fig. 12–21) arises from the main trunk of the glossopharyngeal nerve just beyond its emergence from the jugular foramen, and communicates with the nodose ganglion or the pharyngeal branch of the vagus near its origin. Its continuation or principal branch runs down the anterior surface of the internal carotid artery to the carotid bifurcation and terminates in the wall of the dilated portion of the artery at this point, called the carotid sinus, supplying it with afferent fibers for its blood pressure receptors. It has a rather constant branch which joins the intercarotid plexus, formed principally by vagus and sympathetic branches, or communicates with these nerves independently, and reaches the carotid body. Glossopharyngeal fibers may traverse the plexus and its branches to the carotid body on their way to the carotid sinus (Sheehan *et al.* '41). Its functional association with the carotid body is questionable.

3. **Pharyngeal Branches** (*rr. pharyngei*) are three or four filaments which join

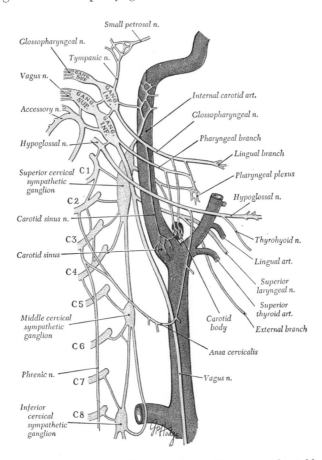

FIG. 12–21.—Carotid sinus nerve, carotid body, internal carotid artery, and neighboring cranial, spinal, and sympathetic nerve connections. Diagrammatic.

pharyngeal branches of the vagus and sympathetic opposite the Constrictor pharyngis medius, to form the **pharyngeal plexus** (Fig. 12–21). Branches from the plexus penetrate the muscular coat of the pharynx and supply its muscles and mucous membrane; the exact contribution of the glossopharyngeal is uncertain.

4. The **Branch** to the **Stylopharyngeus** (*r. stylopharyngeus*) is its only muscular branch.

5. The **Tonsillar Branches** (*rr. tonsillares*) supply the palatine tonsil, forming around it a network from which filaments are distributed to the soft palate and fauces, where they communicate with the lesser palatine nerves.

6. The **Lingual Branches** (*rr. linguales*) are two in number; one supplies the vallate papillae with afferent fibers for taste, and general afferents to the mucous membrane at the base of the tongue; the other supplies the mucous membrane and glands of the posterior part of the tongue, and communicates with the lingual nerve.

Neuralgic pain or "tic douloureux" of the glossopharyngeal nerve occurs in the ear, throat, base of the tongue, rim of the palate, and the lower lateral and posterior part of the pharynx. The most common trigger zone is the tonsillar fossa (Pastore and Meredith '49).

X. THE VAGUS NERVE

The **vagus nerve** (*n. vagus; tenth nerve; pneumogastric nerve*) (Figs. 12–22, 12–37), named from its wandering course, is the longest of the cranial nerves, and has the most extensive distribution, passing through the neck and thorax into the abdomen. It has both somatic and visceral afferent fibers, and general and special visceral efferent fibers. The somatic sensory fibers supply the skin of the posterior surface of the external ear and the external acoustic meatus; the visceral afferent fibers supply the mucous membrane of the pharynx, larynx, bronchi, lungs, heart, esophagus, stomach, intestines, and kidney. General visceral efferent fibers (parasympathetic) are distributed to the heart, and supply the non-striated muscle and glands of the esophagus, stomach, trachea, bronchi, biliary

tract, and most of the intestine. Special visceral efferent fibers supply the voluntary muscles of the larynx, pharynx, and palate (except the Tensor), but most of the latter fibers originate in the cranial part of the accessory nerve.

The superficial origin of the vagus is composed of eight or ten rootlets attached to the medulla oblongata in the groove between the olive and the inferior peduncle, inferior to those of the glossopharyngeal and superior to those of the accessory nerve (Fig. 11–62). The deep origin and central connections are described on page 875. The rootlets unite into a flat cord which passes beneath the flocculus of the cerebellum to the jugular foramen. The nerve leaves the cranial cavity through this opening, accompanied by the accessory nerve and contained in the same dural sheath, but separated by a septum from the glossopharyngeal nerve, which lies anteriorly. This portion of the vagus presents two enlargements, the superior and inferior ganglia which are the **Sensory Ganglia** of the nerve.

The **Superior Ganglion** (*g. superius; g. of the root; jugular ganglion*) (Fig. 12–21) is a spherical swelling, about 4 mm in diameter, of the vagus nerve as it lies in the jugular foramen. The central processes of its unipolar (sensory) ganglion cells enter the medulla, usually in three or four large independent rootlets, slightly dorsal to the motor rootlets. Most of the peripheral processes of the ganglion cells enter the auricular branch of the vagus, but a few probably are distributed with the pharyngeal branches.

The **Inferior Ganglion** (*g. inferius; g. of the trunk; nodose ganglion*) (Fig. 12–23) forms a fusiform swelling about 2.5 cm long on the vagus nerve after its exit from the jugular foramen and about 1 cm distal to the superior ganglion. The central processes of its unipolar (sensory) cells pass through the superior ganglion without traversing the region occupied by cells, and frequently accompany motor fibers in the rootlets for a short distance but enter the medulla slightly dorsal to them, in line with the superior rootlets. Some of the peripheral processes of the ganglion cells make up the internal ramus of the superior laryngeal nerve and the rest are distributed

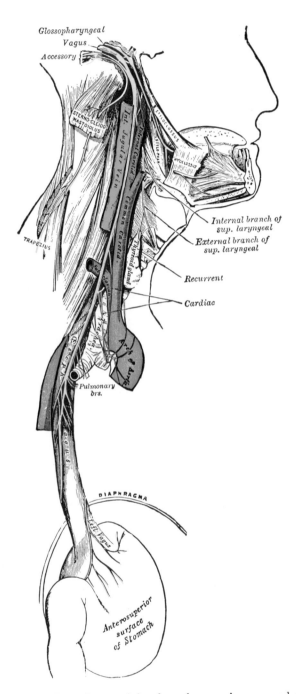

Fig. 12–22.—Course and distribution of the glossopharyngeal, vagus, and accessory nerves.

with other branches of the vagus to the larynx, trachea, bronchi, esophagus, and other thoracic and abdominal viscera.

Communications.—A. At the superior ganglion, several delicate filaments communicate with (1) the cranial portion of the **accessory** nerve; (2) with the inferior ganglion of the **glossopharyngeal**; (3) with the **facial** by means of the auricular branch; and (4) with the superior cervical ganglion of the **sympathetic** (*jugular nerve*).

B. The cranial part of the **accessory** nerve joins the vagus just proximal to the inferior ganglion, and is the source of the greater part of the fibers in the motor branches of the vagus to the pharynx and larynx.

C. At the inferior ganglion, communication is with (1) the **hypoglossal** nerve, (2) the superior cervical **sympathetic** ganglion, and (3) the loop between the **first** and **second cervical** spinal nerves.

The resulting **vagus nerve trunk**, after it has been joined by the cranial part of the accessory nerve just distal to the inferior ganglion, passes vertically down the neck within the carotid sheath, deep to and between the internal jugular vein and the internal and common carotid arteries. Beyond the root of the neck the course of the nerve differs on the two sides of the body.

The **right vagus** crosses the first part of the subclavian artery, lying superficial to it and between it and the brachiocephalic vein (Fig. 12–24), and continues along the side of the trachea to the dorsal aspect of the root of the lung, where it spreads out in the posterior pulmonary plexus (Fig. 12–25). Below this plexus, it splits into cords which enter into the formation of the plexus on the dorsal aspect of the esophagus. After sending communications to the left vagus, these cords unite with each other and with communications from the left vagus to form a single trunk, the posterior vagus nerve, before passing through the esophageal hiatus in the diaphragm (Fig. 12–25). Below the diaphragm, the posterior vagus continues along the lesser curvature of the stomach on its posterior surface for a short distance, and divides into a celiac and several gastric branches.

The **left vagus** enters the thorax between the left carotid and subclavian arteries, deep to the left brachiocephalic vein (Fig. 12–37). It crosses the left side of the arch of the aorta (Fig. 12–24), angling dorsally in its caudalward course, passes between the aorta and the left pulmonary artery just distal to the ligamentum arteriosum, and reaches the dorsal aspect of the root of the lung where it flattens out into the posterior pulmonary plexus (Fig. 12–25). It reaches the esophagus as a variable number of strands which follow down the ventral aspect of the esophagus, send communications to the right vagus, and usually unite with each other and with substantial communications from the right vagus to form a single trunk, the anterior vagus nerve, before passing through the diaphragm. Below the diaphragm, the anterior vagus, on the anterior aspect of the stomach, divides into an hepatic and several gastric branches (Jackson '49).

A. Branches in the jugular fossa:

1. The **Meningeal Branch** (*r. meningeus; dural branch*) is a recurrent filament which arises at the superior ganglion, and is distributed to the dura mater in the posterior cranial fossa.

2. The **Auricular Branch** (*r. auricularis; nerve of Arnold*) arises from the superior ganglion and soon communicates by a filament with the inferior ganglion of the glossopharyngeal. It passes posterior to the internal jugular vein, and when it reaches the lateral wall of the jugular fossa, enters the mastoid canaliculus, which crosses the facial canal in the bone about 4 mm superior to the stylomastoid foramen, and communicates with the facial nerve. It is a somatic afferent nerve and reaches the surface by passing through the tympanomastoid fissure. It divides into two branches: (*a*) one joins the posterior auricular nerve, and (*b*) the other is distributed to the skin of the back of the auricula and to the posterior part of the external acoustic meatus.

B. Branches of the vagus nerve in the neck:

3. The **Pharyngeal Branches** (*rr. pharyngei*), usually two, arise at the upper part of the inferior ganglion, and contain sensory fibers from the ganglion and motor fibers from the communication with the accessory

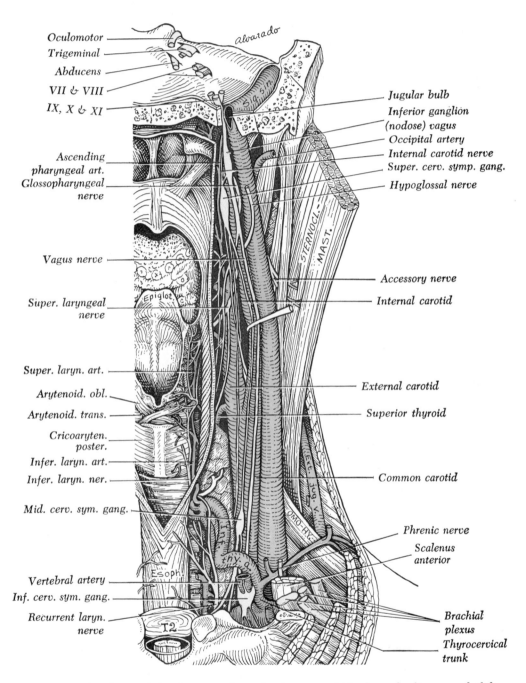

Oculomotor
Trigeminal
Abducens
VII & VIII
IX, X & XI

Ascending
pharyngeal art.
Glossopharyngeal
nerve

Vagus nerve

Super. laryngeal
nerve

Super. laryn. art.
Arytenoid. obl.
Arytenoid. trans.
Cricoaryten.
poster.
Infer. laryn. art.
Infer. laryn. ner.
Mid. cerv. sym. gang.

Vertebral artery
Inf. cerv. sym. gang.
Recurrent laryn.
nerve

alvarado

Jugular bulb
Inferior ganglion
(nodose) vagus
Occipital artery
Internal carotid nerve
Super. cerv. symp. gang.
Hypoglossal nerve

Accessory nerve
Internal carotid

External carotid
Superior thyroid

Common carotid

Phrenic nerve
Scalenus
anterior

Brachial
plexus
Thyrocervical
trunk

Fig. 12–23.—Dorsal view of the pharynx and associated nerves and blood vessels after removal of the cervical vertebrae and part of the occipital bone. (Redrawn from Töndury.)

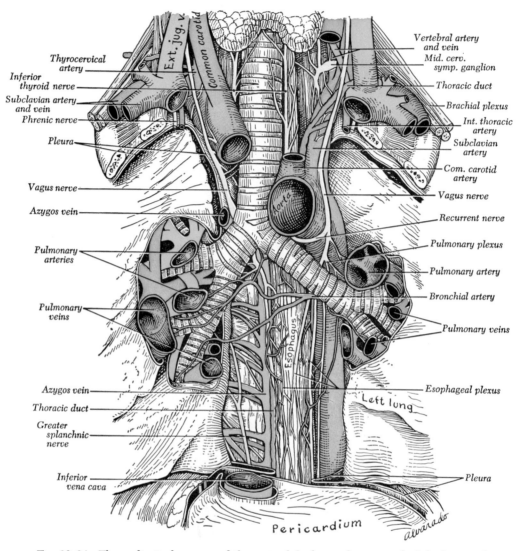

Fig. 12–24.—The mediastinal organs and the roots of the lungs after removal of the heart and pericardium. (Redrawn from Töndury)

nerve. They pass across the internal carotid artery to the superior border of the Constrictor pharyngis medius where they divide into several bundles which join branches of the glossopharyngeal, sympathetic, and external branch of the superior laryngeal to form the **pharyngeal plexus** (Fig. 12–21).

a) Through the plexus, **branches** are distributed to the muscles and mucous membrane of the **pharynx,** and the muscles of the **soft palate,** except the Tensor veli palatini.

b) The **nerves** to the **carotid body** are filaments from the pharyngeal and possibly **from** the superior laryngeal branches which join with similar filaments from the glossopharyngeal nerve and the superior cervical sympathetic ganglion to form the intercarotid plexus between the internal and external carotid arteries at the bifurcation. The vagus fibers are visceral afferents which terminate in the carotid body, a chemoreceptor sensitive to changes in oxygen tension of the blood, located at the carotid bifurcation (Sheehan *et al.* '41).

4. The **Superior Laryngeal Nerve** (*n. laryngeus superior*) (Fig. 12–22) arises near the inferior end of the inferior ganglion, passes caudalward and medialward deep to the internal carotid artery and

along the pharynx toward the superior cornu of the thyroid cartilage. It has a communication with the superior cervical sympathetic ganglion and may contribute to the intercarotid plexus. It terminates by dividing into a smaller external and a larger internal branch.

a) The **external branch** (*r. externus*) (Fig. 12–22) continues caudalward beside the larynx, deep to the Sternothyroideus, and supplies motor fibers to the Cricothyroideus muscle and part of the Constrictor pharyngis inferior. It contributes fibers to the pharyngeal plexus and communicates with the superior sympathetic cardiac nerve.

b) The **internal branch** (*r. internus*) swings anteriorly to reach the thyrohyoid membrane which it pierces with the superior laryngeal artery. It supplies sensory fibers to the mucous membrane and parasympathetic secretomotor fibers to the associated glands through branches to the epiglottis, base of the tongue, aryepiglottic fold, and the larynx as far caudalward as the vocal folds. A filament passes caudalward beneath the mucous membrane on the inner surface of the thyroid cartilage and joins the recurrent nerve (Fig. 12–23).

5. The **Superior Cardiac Branches** (*rr. cardiaci superiores; cervical cardiac branches*), two or three in number, arise from the vagus at the superior and inferior parts of the neck. The superior branches are small, and communicate with the cardiac branches of the sympathetic. They can be traced to the deep part of the cardiac plexus. The inferior branch arises at the root of the neck just cranial to the first rib. On the right side it passes ventral or lateral to the brachiocephalic artery and joins the deep part of the cardiac plexus. On the left side it passes across the left side of the arch of the aorta, and joins the superficial part of the cardiac plexus.

6. The **Recurrent Nerve** (*n. recurrens; inferior or recurrent laryngeal nerve*) (Fig. 12–25), as its name implies, arises far caudally and runs back cranialward in the neck to its destination, the muscles of the larynx. The origin and early part of its course are different on the two sides. On the right side, it arises in the root of the neck, as the vagus crosses superficial to the first part of the subclavian artery. It loops under the arch

of this vessel and passes dorsal to it to the side of the trachea and esophagus (Fig. 12–74). On the left side, the recurrent nerve arises in the cranial part of the thorax, as the vagus crosses the left side of the arch of the aorta (Fig. 12–75). Just distal to the ligamentum arteriosum, it loops under the arch and passes around it to the side of the trachea. The further course on the two sides is similar; it passes deep to the common carotid artery, and along the groove between the trachea and esophagus, medial to the overhanging deep surface of the thyroid lobe. Here it comes into close relationship with the terminal portion of the inferior thyroid artery. It runs under the caudal border of the Constrictor pharyngis inferior, enters the larynx through the cricothyroid membrane deep to the articulation of the inferior cornu of the thyroid with the cricoid cartilage, and is distributed to all the muscles of the larynx except the Cricothyroideus. Its branches are as follows:

a) **Cardiac branches** are given off as the nerve loops around the subclavian artery or the aorta, and are described below as the inferior cardiac branches of the vagus.

b) **Tracheal** and **esophageal branches**, more numerous on the left than on the right, are distributed to the mucous membranes and muscular coats (Fig. 12–23).

c) **Pharyngeal branches** are filaments to the Constrictor pharyngis inferior.

d) Sensory and secretomotor filaments, which reach the recurrent through the communication with the internal branch of the superior laryngeal, supply the mucous membrane of the larynx below the vocal folds.

e) The **inferior laryngeal nerves** are the terminal branches which supply motor fibers to all the intrinsic muscles of the larynx except the Cricothyroideus.

Variations.—When the right subclavian artery arises from the descending aorta, the recurrent nerve arises in the neck and passes directly to the larynx.

C. Branches of the Vagus Nerve in the Thorax:

7. The **Inferior Cardiac Branches** (*rr. cardiaci inferiores; thoracic cardiac branches*) arise on the right side from the

trunk of the vagus as it lies by the side of the trachea and from the recurrent nerve, and on the left side from the recurrent only. They end in the deep part of the cardiac plexus (Fig. 7–19).

a) The *visceral efferent fibers to the heart* in all the cardiac branches are pregangli-

onic. After passage through the cardiac and coronary plexuses (see page 1036) these fibers form synapses with groups of ganglion cells in the heart wall, and the postganglionic fibers terminate about the conduction system and musculature of the heart.

b) The *visceral afferent fibers* from cells

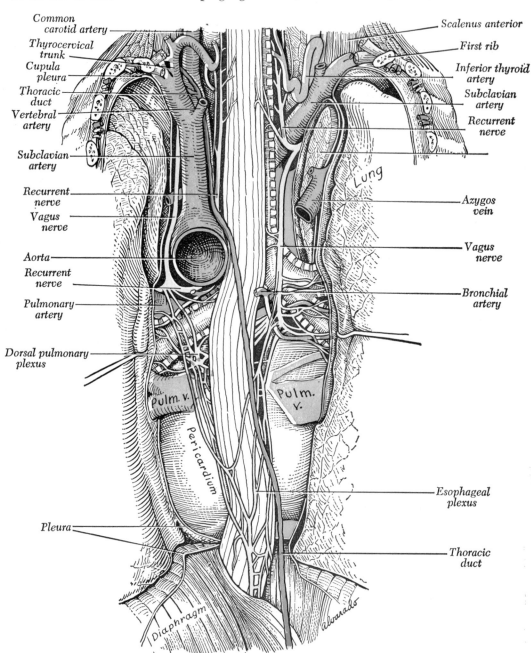

Fig. 12–25.—Dorsal view of the mediastinal structures and roots of the lungs, after removal of the vertebral column, ribs, and thoracic aorta. (Redrawn from Töndury.)

in the inferior ganglion traverse the cardiac plexus and cardiac nerves, supplying the **heart** and **great vessels.**

c) The *visceral afferent fibers* supplying the **aortic bodies** (*glomera aortica*) are carried mainly by the cardiac branches of the right vagus, and those supplying the **supracardial bodies** (*aortic paraganglia*) mainly by those of the left vagus. These bodies are chemoreceptors similar to the carotid body; the cell bodies of the afferent fibers are in the inferior ganglion (Hollinshead '39, '40).

The Depressor Nerve or Nerve of Cyon.—In some animals, the afferent fibers of the vagus from the heart and great vessels are largely contained in a separate nerve whose stimulation causes depression of the activity of the heart. In man, these fibers are probably contained in the inferior cardiac branches (Mitchell '53).

8. The **Anterior Bronchial Branches** (*rr. bronchiales anteriores; anterior or ventral pulmonary branches*) are two or three small nerves on the ventral surface of the root of the lung which join with filaments from the sympathetic to form the **anterior pulmonary plexus.** From this plexus filaments follow the ramifications of the bronchi and pulmonary vessels, or communicate with the cardiac or posterior pulmonary plexuses.

9. The **Posterior Bronchial Branches** (*rr. bronchiales posteriores; posterior or dorsal pulmonary branches*) (Fig. 12–25) are numerous offshoots from the main trunk of the vagus as it passes dorsal to the root of the lung. The vagus itself in this region is flattened and spread out so that it combines with the bronchial branches and with the sympathetic communications to form what is called the **posterior pulmonary plexus.** The plexus has communications with the cardiac, aortic, and esophageal plexuses and its branches follow the ramifications of the bronchi and pulmonary vessels.

a) The *visceral efferent fibers* form synapses with small groups of ganglion cells in the walls of the bronchi, and the postganglionic fibers terminate in the non-striated muscle and glands of the bronchi (Larsell '51).

b) Afferent fibers supply the lungs and bronchi.

10. The **Esophageal Branches** (*rr. esoph-* *agei*) (Fig. 12–25) consist of superior filaments from the recurrent nerve and inferior branches from the trunk of the vagus and the esophageal plexus. They contain both visceral efferent (parasympathetic preganglionic) and visceral afferent fibers.

The esophageal plexus (*pl. esophageus anterior et posterior*).—The fibers of the vagus caudal to the root of the lung split into several bundles (Fig. 12–25), usually two to four larger bundles and a variable number of smaller parallel and communicating strands on each side, which spread out on the esophagus and become partly embedded in its adventitial coat. Filaments are given off for the innervation of the esophagus and there are communications with the splanchnic nerves and sympathetic trunk, forming as a whole the esophageal plexus. The bundles from the left vagus (Fig. 12–24) gradually swing around to the ventral surface of the esophagus, those from the right (Fig. 12–25) to the dorsal surface. Just cranial to the diaphragm, the larger bundles from the left vagus usually combine with one or two strands from the right on the ventral surface of the esophagus to form a single trunk, and this newly constituted vagus, which is not strictly the equivalent of the left vagus, passes through the esophageal hiatus of the diaphragm as the **anterior vagus** (*truncus vagalis anterior*). A similar combination on the dorsal surface of the esophagus, mostly right vagus but with a communication from the left, passes through the esophageal hiatus as the **posterior vagus** (*truncus vagalis posterior*) (Jackson '49).

D. Branches of the Anterior and Posterior Vagi in the Abdomen (Figs. 12–71, 12–72):

The **Branches** in the abdomen contain both *visceral efferent* (parasympathetic preganglionic, and *visceral afferent* fibers. The cells of the *postganglionic* fibers of the stomach and intestines are in the myenteric plexus (of Auerbach) and the submucous plexus (of Meissner); those of the glands are either in small local groups of ganglion cells or possibly in the celiac plexus.

11. **Gastric Branches** (*rr. gastrici*).—Usually four to six branches are given off by both anterior and posterior vagi, at the cardiac end of the stomach. They fan out

over their respective surfaces of the fundus and body, and penetrate the wall to be distributed to the myenteric and submucous plexuses. On both anterior and posterior surfaces, one branch, longer than the others, follows along the lesser curvature and has been called the *principal nerve of the lesser curvature;* it is distributed to the pyloric vestibule rather than the pylorus itself.

12. The **Hepatic Branches** (*rr. hepatici*) from the anterior are larger than those from the posterior vagus. They cross from the stomach to the liver in the lesser omentum, and continue along the fissure for the ductus venosus to the porta hepatis where they give off right and left branches to the liver. The large hepatic branch of the anterior vagus contributes to the plexus on the hepatic artery and has the following branches:

a) **Branches to the gallbladder** and bile ducts come from the hepatic branches or the plexus on the artery.

b) A **pancreatic branch** runs dorsalward to its destination.

c) A **branch** along the right gastric artery is distributed to the **pylorus** and the first part of the **duodenum.**

d) A **branch** accompanies the gastroduodenal artery and right gastroepiploic artery and is distributed to the **duodenum** and **stomach.**

13. **Celiac Branches.**—A large terminal division of the posterior vagus follows the left gastric artery or runs along the crus of the diaphragm to the celiac plexus. The terminal branches cannot be followed once the nerve has entered the plexus, but the vagus fibers, through the secondary plexuses, reach the duodenum, pancreas, kidney, spleen, small intestine, and large intestine as far as the splenic flexure. Before the nerve enters the plexus it may give a branch to the superior mesenteric artery or to the aortic plexus (see Celiac Plexus).

XI. THE ACCESSORY NERVE

The **accessory nerve** (*n. accessorius; eleventh nerve; spinal accessory nerve*) (Fig. 12–22) is a motor nerve consisting of two parts, a cranial and a spinal part.

A. The **Cranial part** (*r. internus; accessory portion*) arises by four or five delicate **cranial rootlets** (*radices craniales*) from the side of the medulla oblongata, inferior to and in series with those of the vagus (Fig. 11–35). From its deep origin, described on page 876, and from its destination, it might well be considered a part of the vagus. It runs lateralward to the jugular foramen, where it interchanges fibers with the spinal part or becomes united with it for a short distance, and has one or two filaments of communication with the superior ganglion of the vagus. It then passes through the jugular foramen, separates from the spinal part and joins the vagus as the **ramus internus,** just proximal to the inferior ganglion. Its fibers are distributed through the pharyngeal branch of the vagus to the Musculus uvulae, Levator veli palatini, and the Constrictores pharyngis and through the superior and inferior laryngeal branches of the vagus to the muscles of the larynx and the esophagus.

B. The **Spinal part** (*r. externus; spinal portion*) originates from motor cells in the lateral part of the ventral column of gray substance of the first five cervical segments of the spinal cord. The fibers pass through the lateral funiculus, emerge on the surface as the **spinal rootlets** (*radices spinales*) (Fig. 11–35,) and join each other seriatim as they follow up the cord between the ligamentum denticulatum and the dorsal rootlets of the spinal nerves. The nerve passes through the foramen magnum into the cranial cavity, crosses the occipital bone to the jugular notch, and penetrates the dura mater over the jugular bulb as the **ramus externus** (Fig. 12–1). It passes through the jugular foramen lying in the same sheath of dura as the vagus, but separated from it by a fold of the arachnoid. In the jugular foramen, it interchanges fibers with the cranial part or joins it for a short distance and separates from it again. At its exit from the foramen, it turns dorsalward, lying ventral to the internal jugular vein in two-thirds and dorsal to it in one-third of the bodies. It passes posterior to the Stylohyoideus and Digastricus to the cranial part of the Sternocleidomastoideus, which it pierces, and then courses obliquely caudalward across the posterior triangle of the neck to the ventral border of the Trapezius (Fig. 12–22). In the posterior triangle it is covered only by the outer investing layer of deep fascia, the subcutaneous fascia and the skin (Fig. 12–35). It com-municates with the **second, third,** and **fourth cervical nerves,** and assuming a plexiform arrangement continues on the deep surface of the Trapezius almost to its caudal border (Fig. 12–22). Experimental observations with monkeys (Corbin and Harrison '38) indicate that the communications with the cervical nerves carry proprioceptive sensory fibers from cells in the dorsal root ganglia of the spinal nerves. The branches of the accessory nerve are as follows:

1. **Sternocleidomastoideus branches** are given off as the nerve penetrates this muscle.

2. **Trapezius branches** are supplied from the part of the nerve lying deep to the muscle.

Variations.—The lower limit of the origin of the spinal part may vary from C 3 to C 7. It may pass beneath the Sternocleidomastoideus without piercing it, and in one instance it ended in that muscle, the Trapezius being supplied by the third and fourth cervical nerves.

XII. THE HYPOGLOSSAL NERVE

The **hypoglossal nerve** (*n. hypoglossus; twelfth nerve*) (Figs. 12–21, 12–36) is the motor nerve of the tongue. Its superficial origin from the medulla oblongata is by a series of rootlets in the ventrolateral sulcus between the pyramid and the olive (Fig. 11–62). The deep origin is described on page 876.

The rootlets are collected into two bundles which perforate the dura mater separately, opposite the hypoglossal canal in the occipital bone, and unite after their passage through it; in some instances the canal is divided by a small bony spicule. As the nerve emerges from the skull, it is deeply placed beneath the internal carotid artery and internal jugular vein, and is closely bound to the vagus nerve. It runs caudalward and forward between the vein and artery, becomes superficial to them near the angle of the mandible, loops around the

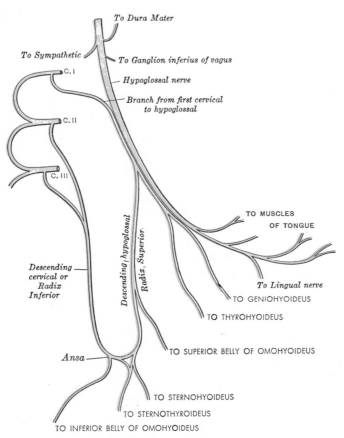

FIG. 12–26.—Plan of hypoglossal and first three cervical nerves.

occipital artery, and passes anteriorly across the external carotid and lingual arteries caudal to the tendon of the Digastricus (Fig. 12–36). It curves slightly cranialward above the hyoid bone, and passes deep to the tendon of the Digastricus and the Stylohyoideus, between the Mylohyoideus and the Hyoglossus, and continues anteriorly among the fibers of the Genioglossus as far as the tip of the tongue, distributing branches to the intrinsic muscles.

Communications.—(1) The communications with the **vagus** take place close to the skull, numerous filaments passing between the hypoglossal and the inferior ganglion of the vagus through the mass of connective tissue which binds the two nerves together. (2) As the nerve winds around the occipital artery, it communicates by a filament with the **pharyngeal plexus**. (3) The communication with the **sympathetic** takes place opposite the atlas by branches of the superior cervical ganglion. (4) The communications with the **lingual** take place near the anterior border of the Hyoglossus by numerous filaments which lie upon the muscle. (5) The communication which takes place opposite the atlas, between the hypoglossal and the loop connecting the anterior primary divisions of the **first** and **second cervical nerves**, is especially significant because it contains the motor fibers for the nerves to the supra- and infrahyoid muscles. This communication probably also contains sensory fibers from the cranialmost cervical dorsal root ganglia (Pearson '39).

Branches.—1. **Meningeal branches** are minute filaments which are given off in the hypoglossal canal and pass back to the dura mater of the posterior cranial fossa. They probably contain sensory fibers communicated to the hypoglossal from the loop between the first and second cervical nerves.

2. The **Descending Hypoglossal** (*radix superior ansae cervicalis; r. descendens; descendens hypoglossi*) (Figs. 12–26, 12–36) is a long slender branch which leaves the hypoglossal as it loops around the occipi-

tal artery. It runs along the superficial surface of the carotid sheath to the middle of the neck, gives a branch to the superior belly of the Omohyoideus, and becomes the medial arm of a loop,[*] the **ansa hypoglossi** (see page 958). The lateral arm of the loop is the descending cervical nerve (communicantes cervicales) (*radix inferior*) from the second and third cervicals. The branches from the loop supply (1) the inferior belly of the Omohyoideus, (2) the Sternohyoideus, and (3) the Sternothyroideus. The fibers in the descending hypoglossal originate in the first cervical spinal segment, not in the hypoglossal nucleus, and pass through the communication between the first cervical nerve and the hypoglossal described above.

3. The **Thyrohyoid Branch** and the **Geniohyoid Branch** are also made up of fibers from the first cervical nerve. They leave the hypoglossal near the posterior border of the Hyoglossus. The thyrohyoid branch runs obliquely across the greater cornu of the hyoid bone to reach the muscle which it supplies.

4. The **Muscular Branches** containing true hypoglossal fibers are distributed to the Styloglossus, Hyoglossus, Genioglossus, and intrinsic muscles of the tongue. At the under surface of the tongue, numerous slender branches pass cranialward into the substance of the organ to supply its intrinsic muscles. The branches of the hypoglossal to the tongue probably contain proprioceptive sensory fibers with cells of origin in the first (if present) and second cervical dorsal root ganglia.

[*]The Paris Nomenclature has changed the name of this loop to ansa cervicalis. The names ansa hypoglossi, descending hypoglossal, and descending cervical are retained here because (1) there are a number of cervical loops and (2) its greatest importance is to the surgeon who would identify it by its connection with the hypoglossal nerve rather than the cervical nerves. Knowledge of its cervical origin has been obtained largely through clinical observations and animal experimentation, not from dissection.

THE SPINAL NERVES

The **spinal nerves** (*nervi spinales*) arise from the spinal cord within the spinal canal and pass out through the intervertebral foramina. The thirty-one pairs are grouped as follows: 8 Cervical; 12 Thoracic; 5 Lumbar; 5 Sacral; 1 Coccygeal. The first cervical leaves the vertebral canal between the occipital bone and the atlas and is therefore called the suboccipital nerve; the eighth leaves between the seventh cervical and first thoracic vertebrae.

Roots of the Spinal Nerves (Fig. 11–21). —Each spinal nerve is attached to the spinal cord by two roots, a ventral or motor root, and a dorsal or sensory root.

The **Ventral Root** (*radix anterior; anterior root; motor root*) emerges from the ventral surface of the spinal cord as a number of **rootlets** or filaments (*fila radicularia*) which usually combine to form two bundles near the intervertebral foramen.

The **Dorsal Root** (*radix posterior; posterior root; sensory root*) is larger than the ventral root because of the greater size and number of its rootlets; these are attached along the ventral lateral furrow of the spinal cord and unite to form two bundles which enter the spinal ganglion.

The dorsal and ventral roots unite immediately beyond the spinal ganglion to form the spinal nerve which then emerges through the intervertebral foramen. Both nerve roots receive a covering from the pia mater, and are loosely invested by the arachnoid, the latter being prolonged as far as the points where the roots pierce the dura mater. The two roots pierce the dura separately, each receiving from this membrane a sheath which becomes continuous with the connective tissue of the epineurium after the roots join to form the spinal nerve.

The **Spinal Ganglion** (*ganglion spinale, dorsal root ganglion*) (Fig. 11–21) is a collection of nerve cells on the dorsal root of the spinal nerve. It is oval in shape and proportional in size to the dorsal root on which it is situated; it is bifid medially where it is joined by the two bundles of rootlets. The ganglion is usually placed in the intervertebral foramen, immediately outside the dura mater, but there are exceptions to this rule; the ganglia of the first and second cervical nerves lie on the vertebral arches of the axis and atlas respectively, those of the sacral nerves are inside the vertebral canal, and that of the coccygeal nerve is within the sheath of the dura mater.

Size and Direction.—In the cervical region, the roots of the first four nerves are small, the last four large and the dorsal roots are three

times as large as the ventral, a large than in any other region; their indi ments are also larger than in the ven The first cervical is an exception, its do being smaller than its ventral. The roots first and second cervical nerves are short run nearly horizontally to their exits from vertebral canal. The roots of the third to th eighth nerves run obliquely caudalward, the obliquity and length successively increasing, but the distance between the attachment to the spinal cord and the exit from the canal never exceeds the height of one vertebra.

In the thoracic region the roots, with the exception of the first, are small and the dorsal is only slightly larger than the ventral. They increase successively in length, and in the more caudal thoracic region descend in contact with the spinal cord for a distance equal to the height of at least two vertebrae before they emerge from the vertebral canal.

In the lumbar and superior sacral regions are found the largest roots, with the most numerous individual filaments. The roots of the coccygeal nerve are the smallest. The roots of the lumbar, sacral, and coccygeal nerves run vertically caudalward, and, since the spinal cord ends near the lower border of the first lumbar vertebra, the roots of the successive segments are increasingly long. The name cauda equina is given to the resulting collection of nerve roots below the termination of the spinal cord. The largest nerve roots, and consequently the largest spinal nerves, are attached to the cervical and lumbar swellings of the spinal cord and are the nerves largely distributed to the upper and lower limbs.

The **Gray Rami Communicantes** (*postganglionic rami*) contain the postganglionic fibers from the adjacent sympathetic chain ganglia. These fibers are visceral efferents running in the nerves and their branches toward the periphery where they supply the smooth muscle in the blood vessel walls, the Arrectores pilorum muscles, and the sweat glands. Since there are not so many sympathetic ganglia as there are spinal nerves, some ganglia supply rami to more than one nerve. Although variations are common, a simple plan may be given as follows: The first four cervical nerves receive their rami from the superior cervical ganglion, the fifth and sixth from the middle, and the seventh and eighth cervical nerves from the inferior cervical ganglion. The first ten thoracic nerves receive rami from

which the preganglionic fibers from the spinal cord reach the sympathetic chain and are thus the roots of the sympathetic ganglia. They arise from the twelve thoracic and first two lumbar nerves only and usually join the sympathetic chain at or near a ganglion. They leave the ventral primary division of the spinal nerve soon after it has emerged from the intervertebral foramen.

A small **Meningeal Branch** is given off from each spinal nerve immediately after it emerges from the intervertebral foramen. This branch reenters the vertebral canal through the foramen, supplies afferent fibers to the vertebrae and their ligaments, and carries sympathetic postganglionic fibers to the blood vessels of the spinal cord and its membranes.

the eleventh ... m a single ... and sacral ... variable ... ond only

... ves two ... ami communi- ... nerve just distal to ... n the dorsal and ventral ... the thoracic and upper lumbar ... are regularly medial to the white ıami communicantes which are the branches of the spinal nerves carrying preganglionic fibers to the sympathetic chain. The ventral and dorsal primary divisions (described below) may each receive a ramus, but if the rami join the ventral division only, some of the fibers course back centrally until they can reach the dorsal division (Dass '52).

The **White Ramus Communicans** (*ramus communicans alba; preganglionic ramus*) is the branch of the spinal nerve through

Primary Divisions.—The spinal nerve splits into its two primary divisions, ventral and dorsal, almost as soon as the two roots join and both divisions receive fibers from both roots.

DORSAL PRIMARY DIVISIONS OF THE SPINAL NERVES

The **dorsal primary divisions** (*rami dorsales*) are smaller, as a rule, than the ventral divisions. As they arise from the spinal nerve, they are directed dorsalward, and,

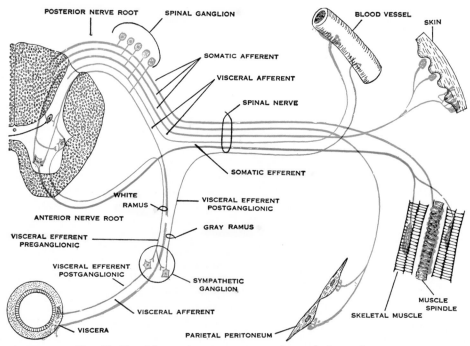

Fig. 12–27.—Scheme showing structure of a typical spinal nerve.

with the exception of those of the first cervical, the fourth and fifth sacral, and the coccygeal, divide into **medial** and **lateral branches** for the supply of the muscles and skin of the dorsal part of the neck and trunk (Fig. 12–52).

The Cervical Nerves

In the **first cervical** or **suboccipital nerve** the dorsal primary division is larger than the ventral. It emerges from the spinal canal superior to the posterior arch of the atlas and inferior to the vertebral artery to enter the suboccipital triangle (Fig. 6–15). It supplies the muscles which bound this triangle, viz., the Rectus capitis posterior major, and the Obliquus superior and inferior, and it gives branches to the Rectus capitis posterior minor and the Semispinalis capitis. A filament from the branch to the Obliquus inferior joins the dorsal division of the second cervical nerve.

Variations.—The first nerve occasionally has a cutaneous branch which accompanies the occipital artery to the scalp and communicates with the greater and lesser occipital nerves.

The dorsal division of the **second cervical nerve** is much larger than the ventral and is the greatest of the cervical dorsal divisions. It emerges between the posterior arch of the atlas and the lamina of the axis, caudal to the Obliquus inferior, which it supplies. It communicates with the first cervical and then divides into a large medial branch and a small lateral branch.

a) The **greater occipital nerve** (*n. occipitalis major*) (Figs. 12–28, 12–35) is the name given to the medial branch because of its size and distribution. It crosses obliquely between the Obliquus inferior and the Semispinalis capitis, pierces the latter and the Trapezius near their attachments to the occipital bone, and becomes subcutaneous (Fig. 12–28). It communicates with the third cervical, runs upward on the back of the head with the occipital artery, and divides into branches which supply the scalp over the vertex and top of the head, communicating with the lesser occipital nerve. It gives muscular branches to the Semispinalis capitis.

The **lesser occipital nerve** (*n. occipitalis minor*) is described with the ventral primary divisions on page 956.

b) The **lateral branch** (*r. lateralis; external branch*) supplies branches to the Splenius and Semispinalis capitis, and often com-

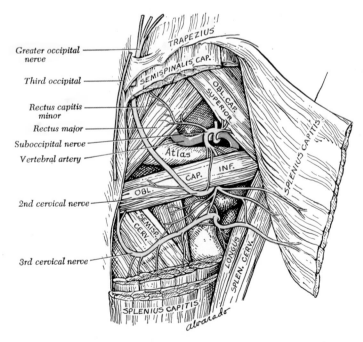

Greater occipital nerve

Third occipital

Rectus capitis minor

Rectus major

Suboccipital nerve

Vertebral artery

2nd cervical nerve

3rd cervical nerve

FIG. 12–28.—Dorsal primary divisions of the upper three cervical nerves.

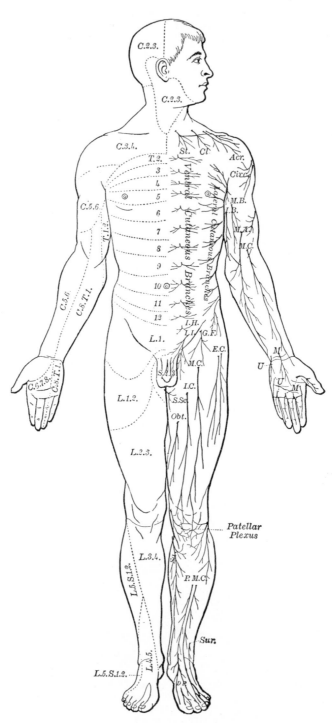

Fig. 12–29.—Distribution of cutaneous nerves. Ventral aspect.

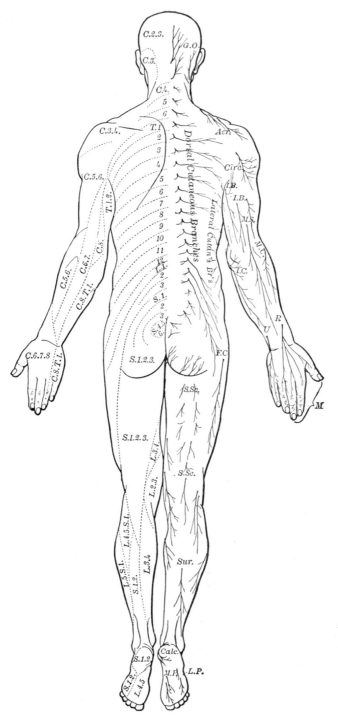

Fig. 12–30.—Distribution of cutaneous nerves. Dorsal aspect.

Figs. 12–29, 12–30 (and others like them) which show areas of skin supplied by each of the dorsal root ganglia are diagrammatic. The work of Sherrington has demonstrated that the "sensory root field" of a particular dorsal root ganglion overlaps that of the zones or "dermatomes" supplied by the ganglion above and below (Fig. 12–32). In fact fibers carrying different modalities, *i.e.*, pain and touch, vary in the amount of this overlap. This interesting subject is discussed by Fulton, J. F. (*Physiology of the Nervous System,* Oxford University Press, New York, 1938, see pages 34–38).

municates with the lateral branch of the third nerve.

The dorsal division of the **third cervical nerve** is intermediate in size between the second and fourth. Its medial branch runs between the Semispinalis capitis and cervicis, pierces the Splenius and Trapezius, and gives branches to the skin.

The **third occipital nerve** (*n. occipitalis tertius; least occipital nerve*) is the cutaneous part of the third nerve. It pierces the Trapezius medial to the greater occipital nerve, with which it communicates, and is distributed to the skin of the lower part of the back of the head (Fig. 12–28). The lateral branch communicates with that of the second cervical, supplies the same muscles, and gives a branch to the Longissimus capitis.

The dorsal primary divisions of the **fourth to eighth cervical nerves** divide into medial and lateral branches. The medial branches of the fourth and fifth run between the Semispinalis capitis and cervicis which they supply, and near the spinous processes of the vertebrae pierce the Splenius and Trapezius to end in the skin (Fig. 12–31). Those of the lower three nerves are small and end in the Semispinalis cervicis and capitis, Multifidus, and Interspinales. The lateral branches of the lower five nerves supply the Splenius, Iliocostalis cervicis, and Longissimus capitis and cervicis.

The dorsal divisions of the first, and the medial branches of the dorsal divisions of the second and third cervical nerves are sometimes joined by communicating loops to form a **posterior cervical plexus** (Cruveilhier). The greater and lesser occipital nerves vary reciprocally with each other; the greater may communicate with the great auricular or posterior auricular nerves, and a branch to the auricula has been observed. The cutaneous branch of the fifth nerve may be lacking and the lower cervical nerves occasionally have cutaneous twigs.

The Thoracic Nerves

The **dorsal primary divisions** of all the thoracic nerves have medial and lateral branches, but the cutaneous branches in the more cranial are different from those in the more caudal thorax.

a) The **medial branch** (*r. medialis; internal branch*) (Fig. 12–52) from the dorsal divisions of the **upper six thoracic nerves** passes between the Semispinalis and Multifidus, supplying them, pierces the Rhomboidei and Trapezius, and, approaching the surface close to the spinous process of the vertebra (Fig. 12–31), extends out laterally to the skin over the back. The medial branches of the lower six nerves end in the Transversospinales and Longissimus muscles, usually without cutaneous branches.

b) The **lateral branches** (*r. lateralis; external branch*) run through or deep to the Longissimus to the interval between it and the Iliocostalis and supply these muscles. They gradually increase in size from the first to the twelfth; the cranial six end in the muscles, but the more **caudal six** have **cutaneous branches** which pierce the Serratus posterior inferior and the Latissimus dorsi along the line of junction between the fleshy and aponeurotic portions of the latter muscle (Fig. 12–31).

The cutaneous portions of both medial and lateral branches have a caudalward course which becomes more pronounced caudally, that of the twelfth nerve reaching down to the skin of the buttocks. The cutaneous part of the first thoracic may be lacking. Both medial and lateral branches of some nerves may have cutaneous fibers, especially those of the sixth, seventh, and eighth (Fig. 12–32).

The Lumbar Nerves

a) The **medial branches** of the dorsal primary divisions of the lumbar nerves run close to the articular processes of the vertebrae and end in the Multifidus.

b) The **lateral branches** supply the Sacrospinalis. The upper three give off cutaneous nerves which pierce the aponeurosis of the Latissimus dorsi at the lateral border of the Sacrospinalis and cross the posterior part of the iliac crest to be distributed, as the **superior clunial nerves** (*nn. clunium superiores*), to the skin of the buttocks as far as the greater trochanter (Fig. 12–31).

The Sacral and Coccygeal Nerves

The dorsal divisions of the sacral nerves are small, and diminish in size distally; they emerge, except the last, through the posterior sacral foramina under cover of the Multifidus.

a) The **medial branches** of the first three are small, and end in the Multifidus.

b) The **lateral branches** of the first three join with one another and with the last lumbar and fourth sacral to form loops on the dorsal surface of the Sacrum (Fig. 12–33). From these loops branches run to the dorsal surface of the sacrotuberous ligament and form a second series of loops under the Gluteus maximus. From this second series two or three cutaneous branches pierce the Gluteus maximus along a line from the posterior superior iliac spine to the tip of

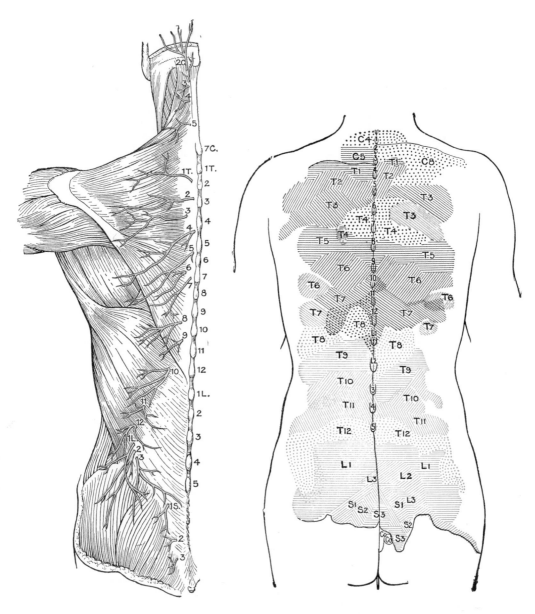

Fig. 12–31.—Diagram of the distribution of the cutaneous branches of the dorsal divisions of the spinal nerves.

Fig. 12–32.—Areas of distribution of the cutaneous branches of the dorsal divisions of the spinal nerves. The areas of the medial branches are in black, those of the lateral in red. (H. M. Johnston.)

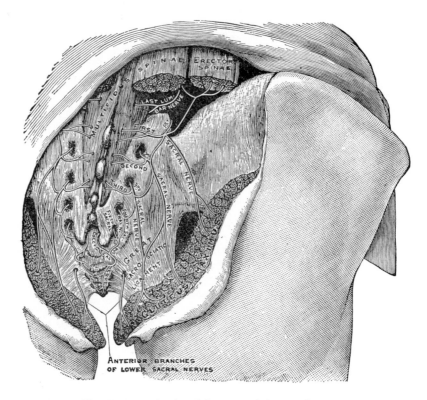

FIG. 12–33.—The dorsal divisions of the sacral nerves.

the coccyx, and supply the skin over the medial part of the buttocks.

The dorsal divisions of the last two sacral and the coccygeal nerves do not divide into medial and lateral branches, but unite with each other on the back of the sacrum, to form loops which then supply the skin over the coccyx.

VENTRAL PRIMARY DIVISIONS OF THE SPINAL NERVES

The **ventral primary divisions** (*rr. ventrales*) of the spinal nerves supply the ventral and lateral parts of the trunk and all parts of the limbs. They are for the most part larger than the dorsal divisions. In the thoracic region they remain independent of one another, but in the cervical, lumbar, and sacral regions they unite near their origins to form plexuses.

The Cervical Nerves (nn. cervicales)

The **ventral division** of the **first cervical** or **suboccipital nerve** issues from the vertebral canal cranial to the posterior arch of the atlas and runs rostralward around the lateral aspect of the superior articular process, medial to the vertebral artery. In most instances it is medial and anterior to the Rectus capitis lateralis, but occasionally it pierces the muscle.

The **ventral divisions** of the other **cervical nerves** pass outward between the anterior and posterior Intertransversarii, lying on the grooved cranial surfaces of the transverse processes of the vertebrae. The first four cervical nerves form the cervical plexus; the last four, together with the first thoracic, form the brachial plexus. They all receive gray rami communicantes from the sympathetic chain.

The Cervical Plexus

The **cervical plexus** (*plexus cervicalis*) (Fig. 12–34) is formed by the ventral pri-

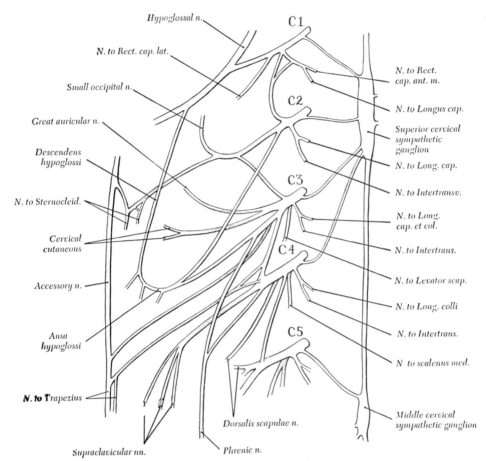

Fig. 12–34.—Plan of the cervical plexus.

mary divisions of the first four cervical nerves; each nerve, except the first, divides into a superior and an inferior branch, and the branches unite to form three loops. The sympathetic rami may join the nerves or the loops. The plexus is situated opposite the cranial four cervical vertebrae, ventro-lateral to the Levator scapulae and Scalenus medius and deep to the Sternocleidomas-toideus.

The cervical plexus has communications with certain cranial nerves, and muscular and cutaneous branches which may be arranged in tabular form as follows, the numbers indicating the segmental compo-nents:

A. COMMUNICATIONS OF THE CERVICAL PLEXUS

1. with Vagus nerve	C 1, 2
2. with Hypoglossal nerve	C 1, 2
3. with Accessory nerve	C 2, 3, 4

B. SUPERFICIAL OR CUTANEOUS BRANCHES

1. Lesser occipital	C 2
2. Great auricular	C 2, 3
3. Anterior cutaneous°	C 2, 3
4. Supraclavicular	C 3, 4

C. DEEP OR MUSCULAR BRANCHES

5. Rectus capitis anterior	C 1, 2
and lateralis	C 1
6. Longus capitis	C 1, 2, 3
and colli	C 2, 3, 4
7. Geniohyoideus	C 1, (2)
Thyrohyoideus	C 1, (2)
Omohyoideus (superior)	C 1, (2)
8. Sternohyoideus	C 2, 3
Sternothyroideus	C 2, 3
Omohyoideus (inferior)	C 2, 3
9. Phrenic	C 3, 4, 5
10. Sternocleidomastoideus	C 2, 3
11. Trapezius	C 3, 4
12. Levator scapulae	C 3, 4
13. Scalenus medius	C 3, 4

°This name conforms with names in more caudal segments.

A. Communications of the Cervical Plexus:

1. The **communication** with the **vagus nerve** is between the loop connecting the first and second nerves and the inferior ganglion.

2. The **communication** with the **hypoglossal nerve** (Figs. 12–26, 12–34) is a short bundle which leaves the loop between the first and second nerves. The great bulk of its fibers is from the first nerve and runs distally with the hypoglossal for 3 or 4 cm and leaves it again as the descending hypoglossal (*ramus superior*). It contains motor and proprioceptive fibers for certain of the hyoid muscles and will be described below as part of the ansa hypoglossi (*ansa cervicalis*).

3. The **communications** with the **accessory nerve** (Fig. 12–34) leave the cervical plexus at several points: *a*) one leaves the loop between the second and third nerves, frequently appearing to come from the lesser occipital nerve (C 2), and joins the fibers of the accessory which supply the Sternocleidomastoideus; (*b*) a bundle leaves the third nerve, sometimes in association with the great auricular nerve, and joins the accessory fibers to the Trapezius; (*c*) one or two bundles leave the fourth nerve and join the accessory directly or enter into a network on the deep surface of the Trapezius. These communications contain proprioceptive sensory fibers (Corbin and Harrison '39).

B. Superficial or Cutaneous Branches of the Cervical Plexus:

1. The **Lesser Occipital Nerve** (*n. occipitalis minor*) (Fig. 12–35) arises from the second cervical nerve or the loop between the second and third, and ascends along the Sternocleidomastoideus, curving around its posterior border. Near the insertion of the muscle on the cranium, it perforates the deep fascia, and is continued upward along the side of the head behind the ear, supplying the skin and communicating with the greater occipital and the great auricular nerves, and the posterior auricular branch of the facial. It has an auricular branch which supplies the upper and back part of the auricula, communicating with the mastoid branch of the great auricular.

Variations.—The lesser occipital nerve varies reciprocally with the greater occipital; it is frequently duplicated or may be wanting.

2. The **Great Auricular Nerve** (*n. auricularis magnus*) (Fig. 12–35), larger than the preceding, arises from the second and third nerves, winds around the posterior border of the Sternocleidomastoideus, and, after perforating the deep fascia, ascends on the surface of that muscle but deep to the Platysma, and divides into an anterior and a posterior branch.

a) The **anterior branch** (*r. anterior; facial branch*) is distributed to the skin of the face over the parotid gland. It communicates in the substance of the gland with the facial nerve.

b) The **posterior branch** (*r. posterior; mastoid branch*) supplies the skin over the mastoid process and back of the auricula except its upper part; a filament pierces the auricula to reach its lateral surface, where it is distributed to the lobule and lower part of the concha. It communicates with the smaller occipital, the auricular branch of the vagus, and the posterior auricular branch of the facial nerve.

3. The **Anterior Cutaneous Nerve** (*n. transversus colli; n. cutaneus colli*) (Fig. 12–35) arises from the second and third cervical nerves and bends around the posterior border of the Sternocleidomastoideus about at its middle. Crossing the surface of the muscle obliquely as it runs horizontally ventralward, it perforates the deep fascia, passing deep to the external jugular vein, and the Platysma, and divides into ascending and descending branches.

a) The **ascending branches** (*rr. superiores*) pass cranialward to the submandibular region, pierce the Platysma, and are distributed to the cranial, ventral, and lateral parts of the neck. One filament accompanies the external jugular vein toward the angle of the mandible and communicates with the cervical branch of the facial nerve under cover of the Platysma.

b) The **descending branches** (*rr. inferiores*) pierce the Platysma and are distributed to the skin of the ventral and lateral parts of the neck as far down as the sternum.

4. The **Supraclavicular Nerves** (*nn. supraclaviculares; descending branches*) (Fig.

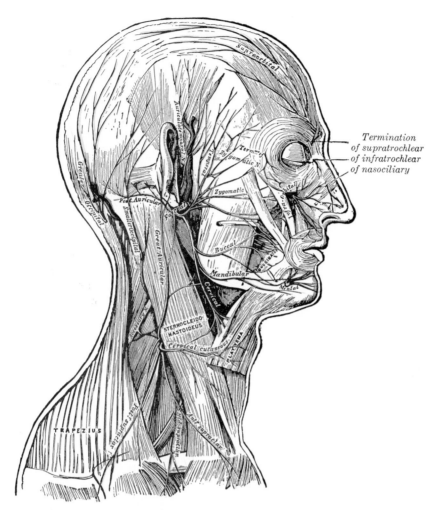

Termination
of supratrochlear
of infratrochlear
of nasociliary

Fɪɢ. 12–35.—The nerves of the scalp, face, and side of neck.

12–35) arise from the third and fourth, mainly the fourth, cervical nerves. They emerge from under the posterior border of the Sternocleidomastoideus and cross the posterior triangle of the neck under cover of the investing layer of deep fascia. Near the clavicle they perforate the fascia and Platysma in three bundles or groups—anterior, middle, and posterior.

a) The **medial supraclavicular nerves** (*anterior supraclavicular nerves; suprasternal nerves*) cross the external jugular vein, the clavicular head of the Sternocleidomastoideus, and the clavicle to supply the skin of the medial infraclavicular region as far as the midline. They furnish one or two filaments to the sternoclavicular joint.

b) The **intermediate supraclavicular nerves** (*middle supraclavicular nerves*) cross the clavicle and supply the skin over the Pectoralis major and Deltoideus, communicating with the cutaneous branches of the more cranial intercostal nerves.

c) The **lateral supraclavicular nerves** (*posterior supraclavicular nerves; supraacromial nerves*) pass obliquely across the outer surface of the Trapezius and the acromion, and supply the skin of the cranial and dorsal parts of the shoulder.

Variations.—One of the middle supraclavicular nerves may perforate the clavicle.

C. Deep or Muscular Branches of the Cervical Plexus:

5. **Branches** to the **Rectus capitis anterior** and **Rectus capitis lateralis** come from the loop between the first and second nerves.

6. **Branches** to the **Longus capitis** and **Longus colli** are given off separately; for the capitis from the first, second, and third; for the colli from the second, third, and fourth nerves.

7. **Branches** to the **hyoid musculature** (Fig. 12–36) are described above (page 945) as branches of the hypoglossal nerve, but the fibers do not come from the hypoglossal nucleus in the brain; they originate instead from the cervical segments of the spinal cord. Fibers communicated to the hypoglossal from the first cervical, or loop between the first and second, leave the hypoglossal again (1) appearing to be individual branches to the **Geniohyoideus**, **Thyrohyoideus** or superior belly of the **Omohyoideus** or (2) as the **radix superior**

(*descendens hypoglossi*) to join the **radix inferior** (*descendens cervicalis*) from the second and third cervical nerves to form the **ansa cervicalis** (*ansa hypoglossi*).

a) The **nerve** to the **Geniohyoideus** leaves the hypoglossal nerve distally and has a course similar to that of the Thyrohyoid nerve.

b) The **nerve** to the **Thyrohyoideus** (Fig. 12–36) leaves the hypoglossal nerve near the posterior border of the Hypoglossus and runs obliquely across the greater cornu of the hyoid bone and enters the muscle as a slender filament. It leaves the hypoglossal distal to the origin of the radix superior of the ansa.

c) The **fibers** for the **superior belly** of the **Omohyoideus** leave the superior root before it forms the ansa and reach the muscle as a slender filament.

8. The **Ansa Cervicalis** (*ansa hypoglossi*,

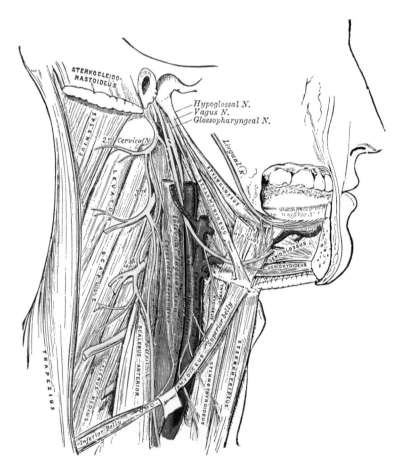

Fig. 12–36.—Hypoglossal nerve, cervical plexus, and their branches.

page 945) is a loop of slender nerves which may be somewhat plexiform, ventral and lateral to the common carotid artery and internal jugular vein. It is formed by fibers from the first cervical spinal segment which leave the hypoglossal as the **radix superior** (*descendens hypoglossi*, page 945) and the **radix inferior** (*descendens cervicalis*, page 946) from the second and third cervical nerves. It usually lies superficial to the carotid sheath at about the level of the cricoid cartilage.

a) The **nerves** to the **Sternohyoideus** and **Sternothyroideus** leave the convexity of the loop and run down the superficial surface of the carotid artery to enter their muscles at the root of the neck. They may be separate filaments or be combined into a single nerve for a variable distance.

b) The **nerve** to the **inferior belly** of the **Omohyoideus** leaves the convexity of the loop and runs lateralward across the neck deep to the intermediate tendon of the muscle to reach the inferior belly.

Variations.—The ansa is quite variable in position; it may occur at any level. Frequently it is high near the bifurcation of the carotid, in which circumstance it may be within the carotid sheath. The descendens hypoglossi may appear to arise wholly or in part from the vagus. A branch to the Sternocleidomastoideus and filaments entering the thorax to join the vagus or sympathetic have been described.

9. The **Phrenic Nerve** (*n. phrenicus; internal respiratory nerve of Bell*) (Fig. 12–37) is generally known as the motor nerve to the Diaphragm, but it contains about half as many sensory as motor fibers, and it should not be forgotten that the lower thoracic nerves also contribute to the innervation of the Diaphragm. The phrenic nerve originates chiefly from the fourth cervical nerve but is augmented by fibers from the third and fifth nerves. It lies on the ventral surface of the Scalenus anterior, gradually crossing from its lateral to its medial border (Fig. 12–36). Under cover of the Sternocleidomastoideus, it is crossed by the inferior belly of the Omohyoideus and the transverse cervical and suprascapular vessels. It continues with the Scalenus anterior between the subclavian vein and artery, and, as it enters the thorax, it crosses

the origin of the internal thoracic artery and is joined by the pericardiophrenic branch of this artery. It passes caudalward over the cupula of the pleura and ventral to the root of the lung, then along the lateral aspect of the pericardium, between it and the mediastinal pleura, until it reaches the diaphragm, where it divides into its terminal branches. At the root of the neck it is joined by a communication from the sympathetic trunk.

The right nerve is more deeply placed, is shorter, and runs more vertically caudalward than the left. In the upper part of the thorax it is lateral to the right brachiocephalic vein and the superior vena cava (Figs. 12–37, 12–78).

The left nerve is longer than the right because of the inclination of the heart toward the left and because of the more caudal position of the diaphragm on this side. At the root of the neck it is crossed by the thoracic duct, and in the superior mediastinum it lies between the left common carotid and subclavian arteries, and is lateral to the vagus as it crosses the left side of the arch of the aorta (Figs. 12–37, 12–79).

a) The **pleural branches** of the phrenic are very fine filaments supplied to the costal and mediastinal pleura over the apex of the lung.

b) The **pericardial branches** are delicate filaments to the upper part of the pericardium.

c) The **terminal branches** pass through the diaphragm separately, and, diverging from each other, are distributed on the abdominal surface, supplying the Diaphragma muscle and sensory fibers to the peritoneum. On the right side, a branch near the inferior vena cava communicates with the phrenic plexus which accompanies the inferior phrenic artery from the celiac plexus, and where they join there is usually a small ganglion, the **phrenic ganglion.** On the left, there is a communication with the phrenic plexus also, but without a ganglion.

The **accessory phrenic nerve** is described on page 963.

Variations.—The phrenic nerve may receive fibers from the descendens cervicalis, or from the second or sixth nerves. At the root of the neck or in the thorax it may be joined by an

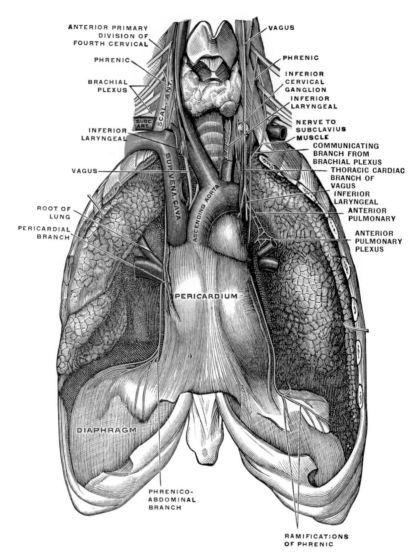

FIG. 12–37.—The phrenic nerve and its relations with the vagus nerve.

accessory phrenic from the fifth nerve or from the nerve to the Subclavius. It may arise from the nerve to the Subclavius or give a branch to that muscle. It may pass ventral to the subclavian vein or perforate it.

10. The **branches** to the **Sternocleidomastoideus** may be independent or partly be communications with the accessory nerve (described above). They are proprioceptive sensory rather than motor nerves and are derived from the second and third nerves (Corbin and Harrison '38).

11. The **branches** to the **Trapezius** are like those to the Sternocleidomastoideus and are derived from the third and fourth nerves.

12. **Muscular branches** to the **Levator Scapulae** are supplied by the third and fourth nerves.

13. **Branches** to the **Scalenus medius** are from the third and fourth nerves.

THE BRACHIAL PLEXUS

The **brachial plexus** (*plexus brachialis*) (Figs. 12–38, 12–39), as its name implies, supplies the nerves to the upper limb. It is formed by the ventral primary divisions of the fifth to eighth cervical and the first thoracic nerves. A communicating loop from the fourth to the fifth cervical, and one from the second to the first thoracic also usually contribute to the plexus. It lies in the lateral part of the neck in the clavicular region, extending from the Scalenus anterior to the axilla.

Components.—The brachial plexus is composed of roots, trunks, divisions, cords, and terminal nerves. The **roots** of the brachial plexus are provided by the anterior primary divisions of the last four cervical and the first thoracic nerves. The trunks are formed from these roots and are named according to their position relative to each other. The **superior trunk** is formed by the union of the fifth and sixth cervical nerves as they emerge between the Scalenus medius and anterior. The **middle trunk** is formed by the seventh cervical alone. The **inferior trunk** is formed by the union of the eighth cervical and first thoracic nerves. The trunks, after a short course, split into

anterior and **posterior divisions.** The anterior and posterior divisions of the superior and middle trunks are about equal in size, but the posterior division of the inferior trunk is much smaller than the anterior because it receives a very small or no contribution from the first thoracic nerve. The **cords,** formed from these divisions, are named according to their relation to the axillary artery: **lateral, medial,** and **posterior.** The anterior divisions of the superior and middle trunks are united into the lateral cord. The anterior division of the inferior trunk becomes the medial cord. The posterior divisions of all three trunks are united into the posterior cord. The cords in turn break up into the nerves which are the terminal branches.

Sympathetic Contributions to the Brachial Plexus.—The ventral primary divisions of the spinal nerves which enter into the brachial plexus obtain their sympathetic fibers in the form of the gray rami communicantes from the sympathetic chain. The fifth

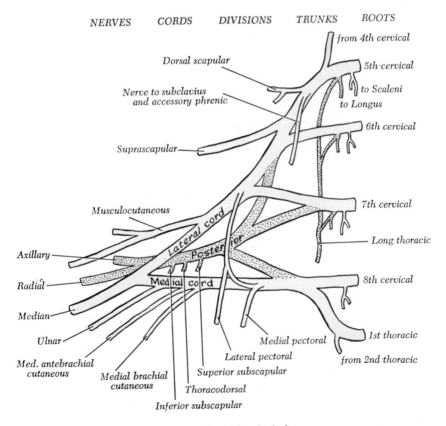

Fig. 12–38.—Plan of brachial plexus.

and sixth nerves receive fibers from the middle cervical ganglion; the sixth, seventh, and eighth nerves from the inferior cervical or stellate ganglion; the first and second thoracic nerves from the stellate or the first and second thoracic ganglia.

Relations.—In the neck, the brachial plexus lies in the posterior triangle, being covered by the skin, Platysma, and deep fascia; it is crossed by the supraclavicular nerves, the inferior belly of the Omohyoideus, the external jugular vein, and the transverse cervical artery. The roots emerge between the Scaleni anterior and medius, cranial to the third part of the subclavian artery, while the trunk formed by the union of the eighth cervical and first thoracic is placed dorsal to the artery; the plexus next passes dorsal to the clavicle, the Subclavius, and the transverse scapular vessels, and lies upon the first digitation of the Serratus anterior, and the Subscapularis. In the axilla it is placed lateral to the first portion of the axillary artery; it surrounds the second part of the artery, one cord lying medial to it, one lateral to it, and one dorsal to it; in the axilla it gives off its terminal branches to the upper limb.

Variations of the brachial plexus are of several types: (*a*) variations in the contributions of the spinal nerves to the roots of the plexus, (*b*) variations in the formation of the trunks, divisions, or cords, (*c*) variations in the origin or combination of the branches, (*d*) variations in the relation to the artery.

(*a*) The fourth cervical nerve contributes to two-thirds of the plexuses, and T 2 contributes to more than one-third. When the contribution from C 4 is large and that from T 2 lacking, the plexus appears to have a more cranial position and has been termed prefixed. Similarly, when the contribution from T 2 is large and from C 4 lacking, the plexus appears to have a more caudal position and has been termed post-fixed. It is doubtful whether this shifting of position is more common than one in which the plexus is spread out to include both C 4 and T 2, or contracted to exclude both (Kerr '18). Variations in the contribution to the plexus may be correlated with the position of the limb bud at the time the nerves first grow into it in the embryo (Miller and Detwiler '36), and many variations are similar to the usual conditions found in the different primates (Miller '34).

(*b*) The trunks vary little in their formation from the cervical roots, but the superior and inferior trunks especially may appear to be absent because the nerves split into dorsal and ventral divisions before they combine into trunks. The cords in these instances are formed from the divisions of the nerves but the sources of the fibers can be readily traced and made to correspond with the usual pattern. Many of these instances may be the result of too vigorous a removal of the connective tissue sheaths of the nerves in dissection. The medial cord may receive a contribution from the middle trunk and the lateral cord may receive fibers from C 8 or the lower trunk.

(*c*) The median nerve may have small heads either medial or lateral, in addition to the usual two. It appears to receive fibers from all segments entering the plexus in most instances. Many peculiarities involve combined origins of the median and musculocutaneous with separation into definitive nerves or branches farther down the arm; thus the musculocutaneous gives a branch to the median in the arm in a fourth of the bodies, but a branch from the median to the ulnar is much less frequent. The musculocutaneous frequently receives fibers from C 4 in addition to the usual C 5 and C 6, and appears to receive fibers from C 7 in only two-thirds. The nerve to the Coracobrachialis is a branch of the lateral cord or some part of the plexus (exclusive of C 8 and T 1) other than the musculocutaneous in almost half of the bodies. The ulnar nerve may have a lateral head from the lateral cord, the lateral head of the median, or from C 7; in two-thirds of the plexuses the ulnar may receive fibers from C 7 or possibly more cranial segments. The radial and axillary nerves may be formed from the trunks and divisions without the presence of a true posterior cord. The radial nerve appears to receive fibers from all segments contributing to the plexus in most instances, but participation of C 4 and T 1 is probably incidental because C 4 may not enter the plexus and in a few instances T 1 can be definitely excluded. The axillary nerve probably receives no fibers from C 8 and T 1 and the contribution from C 7 is undetermined (Kerr '18).

(*d*) Among the variations in the relationship between the axillary artery and the brachial plexus, the most common is an artery superficial to the median nerve; also, the median nerve may be split by a branch of the artery. An aberrant axillary artery, *i.e.*, one not derived from the seventh segmental, has a different relation to the roots of the plexus depending on whether it was derived from a segmental artery cranial or caudal to the seventh. The cords of the plexus may be split by arterial branches, and communicating loops of nerves may be formed around the artery or its branches (Miller '39).

A. BRANCHES FROM THE CERVICAL NERVES

1. To the phrenic nerve   C 5
2. To Longus colli and the scaleni   C 5, 6, 7, 8
3. Accessory phrenic   C 5

B. BRANCHES FROM THE ROOTS

4. Dorsal scapular   C 5
5. Long thoracic   C 5, 6, 7

C. BRANCHES FROM THE TRUNKS

6. Nerve to the Subclavius   C 5, 6
7. Suprascapular   C 5, 6

D. BRANCHES FROM THE CORDS

8. Pectoral   C 5, 6, 7, 8, T 1
9. Subscapular   C 5, 6
10. Thoracodorsal   C, 6, 7, 8
11. Axillary   C 5, 6
12. Medial brachial cutaneous   C 8, T 1
13. Medial antebrachial cutaneous   C 8, T 1

E. TERMINAL NERVES

14. Musculocutaneous   C 5, 6, 7
15. Median   C 6, 7, 8, T 1
16. Ulnar   C 8, T 1
17. Radial   C 5, 6, 7, 8, (T 1)

The ventral primary divisions of the fifth to eighth cervical nerves give branches before they enter into the plexus, and the brachial plexus may be divided into the branches which arise from the roots, from the trunks, or from the cords, and into the terminal nerves. From their topographical relation to the clavicle, the branches may be divided into supra- and infraclavicular branches.

A. **Branches** of the anterior primary divisions of the last four cervical and first thoracic nerves **before** they enter the brachial plexus:

1. The fifth cervical may contribute to the **phrenic nerve** at its origin.

2. **Muscular branches** from each of the four lower cervical nerves are supplied to the Longus colli and the **Scalenus anterior, medius,** and **posterior.**

3. The **Accessory Phrenic Nerve** (Fig. 12–37) is an inconstant branch which may come from the nerve to the Subclavius or from the fifth nerve. It passes ventral to the subclavian vein and joins the phrenic nerve at the root of the neck or in the thorax, forming a loop around the vein.

Surgical Considerations.—The operation of resection of the phrenic nerve for immobilization of the diaphragm may be only partially successful if the accessory is not resected also. In avulsion of the phrenic nerve, the subclavian vein is in danger of being torn by the loop between the accessory and the phrenic nerves.

Supraclavicular Branches

B. **Branches** from the **roots** (spinal nerves):

4. The **Dorsal Scapular Nerve** (*n. dorsalis scapulae; nerve to the Rhomboidei; posterior scapular nerve*) (Fig. 12–38) arises from the fifth cervical nerve near the intervertebral foramen, frequently in common with a root of the long thoracic nerve. It pierces the Scalenus medius and runs dorsally as well as caudalward on the deep surface of the Levator scapulae to the vertebral border of the scapula. It supplies the Rhomboideus major and minor and, along with the third and fourth nerves, gives a branch to the Levator.

5. The **Long Thoracic Nerve** (*n. thoracicus longus; external respiratory nerve of Bell; posterior thoracic nerve*) (Fig. 12–39) is the nerve to the Serratus anterior. It arises by three roots: from the fifth, sixth, and seventh cervical nerves; those from the fifth and sixth join just after they pierce the Scalenus medius; the seventh joins them at the level of the first rib. It runs caudalward dorsal to the brachial plexus and axillary vessels, and continues along the lateral surfaces of the Serratus anterior, under cover of the Subscapularis. It sends branches to all the digitations of the Serratus; the fibers from the fifth nerve supply the upper part, the sixth the middle, and the seventh the lower part of the muscle.

Variations.—The root from the fifth nerve may remain independent; the root from the seventh may be lacking.

C. Branches from the Trunks:

6. The **Nerve to the Subclavius** (*n. subclavius*) is a small branch which arises from the superior trunk, although its fibers are mainly from the fifth nerve, and passes ventral to the distal part of the plexus, the subclavian artery, and the subclavian vein to reach the Subclavius muscle (Fig. 12–37).

Variations.—The accessory phrenic may arise from the subclavius nerve or the phrenic nerve may supply the branch to the Subclavius.

7. The **Suprascapular Nerve** (*n. suprascapularis*) (Fig. 12–40) arises from the superior trunk and takes a more or less direct course across the posterior triangle to the scapular notch, passing dorsal to the inferior belly of the Omohyoideus and the anterior border of the Trapezius. It passes through the notch, under the superior transverse ligament, runs deep to the Supraspinatus, and around the lateral border of the spine of the scapula into the infraspinatous fossa. In the supraspinatous fossa it gives two branches, one to the Supraspinatus and the other an articular filament to the shoulder joint. In the infraspinatous fossa it gives two branches to the Infraspinatus, and filaments to the shoulder joint and scapula.

Infraclavicular Branches

The infraclavicular branches arise from the three cords of the brachial plexus, but it should be emphasized that a particular branch of any cord need not contain fibers from all the cervical nerves contributing to that cord. For example, the axillary nerve, from the posterior cord, contains fibers from C 5 and 6 only, not from C 5, 6, 7, 8, and T 1. Likewise, a branch from one of the larger terminal nerves may not contain fibers from all the cervical segments contributing to that nerve; the branch of the radial nerve to the Supinator, for example, contains only fibers from C 6.

D. **Branches from the cords:**

8. The **Pectoral Nerves** (*nn. pectorales; thoracales anteriores*) (Fig. 12–46) are two nerves, one lateral and one medial to the axillary artery, which arise at the level of the clavicle and supply the pectoral muscles.

a) The **lateral (superior) pectoral nerve** (*n. pectoralis lateralis*) is so named because it is lateral to the artery and arises from the lateral cord of the brachial plexus or from the anterior divisions of the superior and middle trunks just before they unite into the cord. It passes superficial to the first part of the axillary artery and vein, sends

a communicating branch to the inferior pectoral branch, and then pierces the clavipectoral fascia to reach the deep surface of the clavicular and cranial sternocostal portions of the Pectoralis major.

b) The **medial (inferior) pectoral nerve** (*n. pectoralis medialis*) is so named because its origin is from the medial cord of the brachial plexus, medial to the artery. Its origin is more lateral in position with respect to the midline of the body than that of the superior branch. It passes between the axillary artery and vein, gives a branch which joins the communication from the superior branch to form a plexiform loop around the artery, and enters the deep surface of the Pectoralis minor. It supplies this muscle and two or three of its branches continue through the muscle to supply the more caudal part of the Pectoralis major. The most distal branch may pass around the border of the minor. The loop gives off branches which supply both muscles.

9. The **Subscapular Nerves** (*nn. subscapulares*) (Fig. 12–39), usually two in number, arise from the posterior cord of the brachial plexus, deep in the axilla.

a) The **superior subscapular** (*short subscapular*), the smaller of the two, enters the superior part of the Subscapularis and is frequently double.

b) The **inferior subscapular** supplies the distal part of the Subscapularis and ends in the Teres major.

Variations.—The nerve to the Teres major may be a separate branch of the posterior cord, or, more rarely, of the axillary nerve.

10. The **Thoracodorsal Nerve** (*n. thoracodorsalis; middle or long subscapular nerve*) (Fig. 12–39) is a branch of the posterior cord of the brachial plexus, usually arising between the two subscapular nerves. It follows the course of the subscapular and thoracodorsal arteries along the posterior wall of the axilla, under cover of the ventral border of the Latissimus dorsi, and terminates in branches which supply this muscle.

11. The **Axillary Nerve** (*n. axillaris; circumflex nerve*) (Fig. 12–40) is the last branch of the posterior cord of the brachial plexus before the latter becomes the radial nerve. It passes over the insertion of the

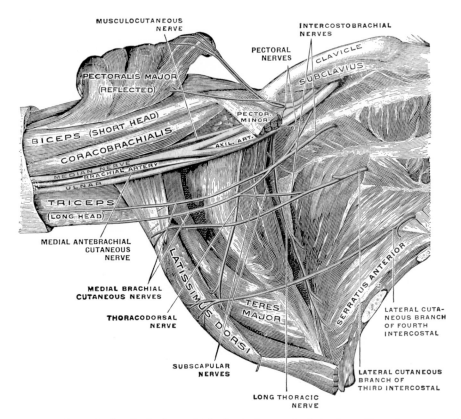

Fig. 12–39.—The right brachial plexus (infraclavicular portion) in the axillary fossa; viewed from below and in front. The Pectoralis major and minor muscles have been in large part removed; their attachments have been reflected. (Spalteholz.)

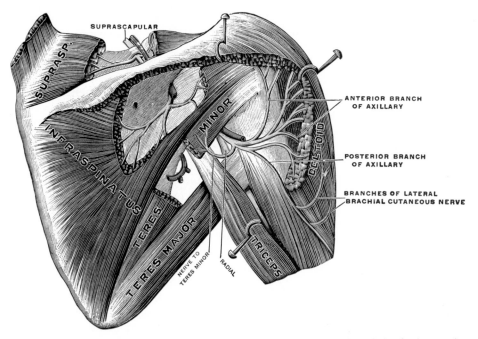

Fig. 12–40.—Suprascapular and axillary nerves of right side, seen from behind. (Testut.)

Subscapularis, dorsal to the axillary artery, crosses the Teres minor, and leaves the axilla, accompanied by the posterior humeral circumflex artery, by passing through the quadrilateral space bounded by the surgical neck of the humerus, the Teres major, Teres minor, and the long head of the Triceps. It divides into two branches.

a) The **posterior branch** (*lower branch*) supplies the Teres minor and the posterior

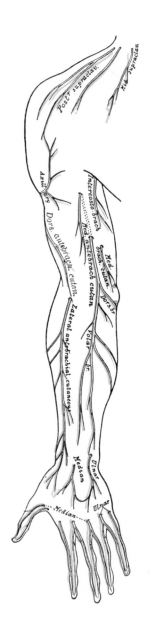

Fig. 12–41.—Cutaneous nerves of right upper limb.
Anterior view.

part of the Deltoideus, and then pierces the deep fascia at the posterior border of the Deltoideus as the **lateral brachial cutaneous nerve,** to supply the skin over the distal two-thirds of the posterior part of this muscle and over the adjacent long head of the Triceps brachii.

b) The **anterior branch** (*upper branch*) winds around the surgical neck of the humerus with the posterior humeral circumflex vessels, under cover of the Deltoideus, as far as its anterior border. It supplies this muscle and sends a few small cutaneous filaments to the skin covering its distal part.

c) **Articular filaments** leave the nerve near its origin and in the quadrilateral space, and supply the anterior inferior part of the capsule of the shoulder joint.

12. The **Medial Brachial Cutaneous Nerve** (*n. cutaneus brachii medialis; nerve of Wrisberg*), a small nerve, arises from the medial cord of the brachial plexus and is distributed to the medial side of the arm. It passes through the axilla, at first lying dorsal, then medial to the axillary vein and brachial artery. It pierces the deep fascia in the middle of the arm and is distributed to the skin of the arm as far as the medial epicondyle and olecranon. A part of it forms a loop with the intercostobrachial nerve in the axilla, and there is a reciprocal relationship in size between these two nerves. It also communicates with the ulnar branch of the medial antebrachial cutaneous nerve or it may be a branch of the latter nerve.

13. The **Medial Antebrachial Cutaneous Nerve** (*n. cutaneus antebrachii medialis*) (Figs. 12–41 to 12–44, 12–46) arises from the medial cord of the brachial plexus, medial to the axillary artery. Near the axilla, it gives off a filament which pierces the fascia and supplies the skin over the Biceps nearly as far as the elbow. The nerve runs down the ulnar side of the arm medial to the brachial artery, pierces the deep fascia with the basilic vein about the middle of the arm, and divides into an anterior and an ulnar branch.

a) The **ulnar branch** (*r. ulnaris; posterior branch*) passes obliquely distalward on the medial side of the basilic vein, anterior to the medial epicondyle of the humerus to the dorsum of the forearm, and continues

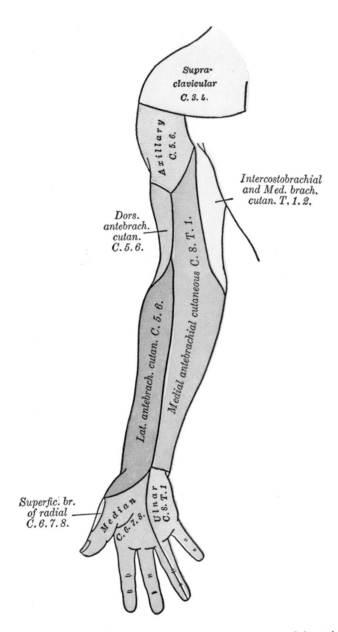

Fig. 12–42.—Diagram of segmental distribution of the cutaneous nerves of the right upper limb. Anterior view.

on its ulnar side as far as the wrist, supplying the skin. It communicates with the medial brachial cutaneous, the dorsal antebrachial cutaneous, and the dorsal branch of the ulnar nerve.

b) The **anterior branch** (*r. anterior*) is larger and passes, usually superficial but occasionally deep, to the median basilic vein. It continues on the anterior part of the ulnar side of the forearm, distributing filaments to the skin as far as the wrist, and communicating with the palmar cutaneous branch of the ulnar nerve.

E. The **terminal branches** of the **brachial plexus** are the musculocutaneous, median, ulnar, and radial nerves.

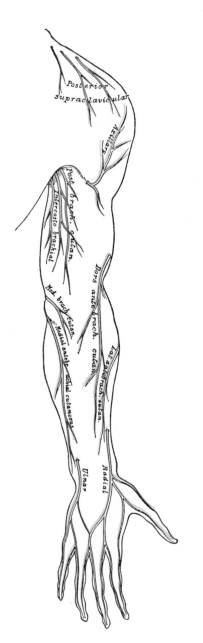

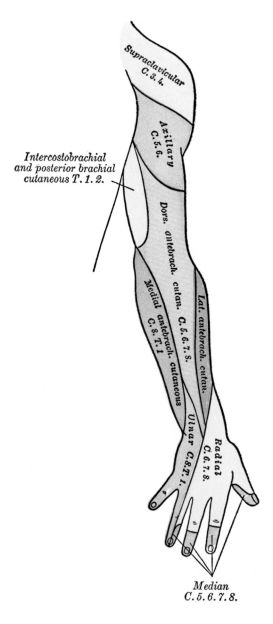

Fig. 12–43.—Cutaneous nerves of right upper limb. Posterior view.

Fig. 12–44.—Diagram of segmental distribution of the cutaneous nerves of the right upper limb. Posterior view.

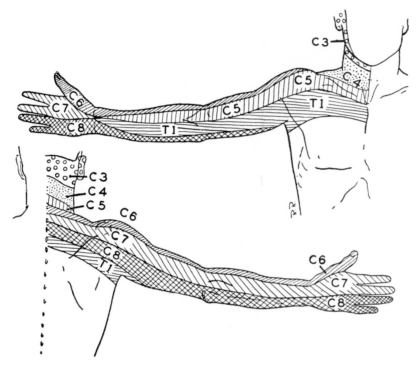

FIG. 12–45.—Dermatome chart of the upper limb of man outlined by the pattern of hyposensitivity from loss of function of a single nerve root. (From the original of Fig. 7, Keegan and Garrett, courtesy of Anatomical Record, *102*, 415.)

The Musculocutaneous Nerve

14. The **musculocutaneous nerve** (*n. musculocutaneus*) (Fig. 12–46) is formed by the splitting of the lateral cord of the brachial plexus at the inferior border of the Pectoralis minor into two branches, the other branch being the lateral root of the median nerve. It pierces the Coracobrachialis, and, lying between the Brachialis and the Biceps brachii, crosses to the lateral side of the arm. A short distance above the elbow it pierces the deep fascia lateral to the tendon of the Biceps and continues into the forearm as the lateral antebrachial cutaneous nerve.

a) The **branch** to the **Coracobrachialis** leaves the nerve close to its origin.

b) **Muscular branches** are supplied to the **Biceps** and the greater part of the **Brachialis**.

c) An **articular filament** given off from the nerve to the Brachialis supplies the elbow joint.

d) A **filament** to the **humerus** enters the nutrient foramen with the artery.

e) The **lateral antebrachial cutaneous nerve** (*n. cutaneus antebrachii lateralis*) (Figs. 12–41 to 12–44) passes deep to the cephalic vein and divides opposite the elbow joint into an anterior and a dorsal branch.

i) The **anterior branch** (*volar branch*) follows along the radial border of the forearm to the wrist, and supplies the skin over the radial half of its anterior surface. At the wrist it is superficial to the radial artery, and some of its filaments pierce the deep fascia to follow the vessel to the dorsal surface of the carpus. It terminates in cutaneous filaments at the thenar eminence after communicating with the superficial branch of the radial and the palmar cutaneous branch of the median nerve.

ii) The **dorsal branch** (*posterior branch*) passes distally along the dorsal part of the radial surface of the forearm, supplying the skin almost to the wrist, and communicating with the dorsal antebrachial cutaneous nerve and the superficial branch of the radial.

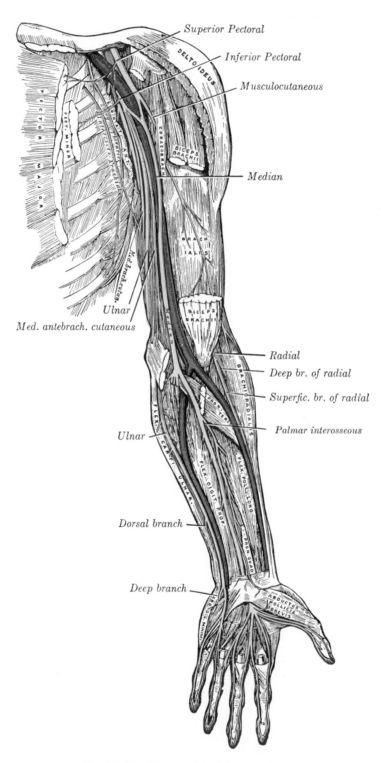

Fɪɢ. 12–46.—Nerves of the left upper limb.

Variations.—The musculocutaneous and median nerves present frequent irregularities in their origins from the lateral cord of the plexus. The branch to the Coracobrachialis may be a separate nerve. In this condition the musculocutaneous may continue with the median for a variable distance before it passes under the Biceps. Some of the fibers of the median may run for some distance in the musculocutaneous before they join their proper trunk; less frequently the reverse is the case and fibers of the musculocutaneous run with the median. It may give a branch to the Pronator teres or it may supply the dorsum of the thumb in the absence of the superficial branch of the radial.

The Median Nerve

15. The **median nerve** (*n. medianus*) (Fig. 12–46) is the nerve to the radial side of the flexor portion of the forearm and hand. It takes its origin from the brachial plexus by two large roots, one from the lateral and one from the medial cord. The roots at first lie on each side of the third part of the axillary artery, then embracing it, they unite on its ventral surface to form the trunk of the nerve. In its course down the arm it accompanies the brachial artery, to which it is at first lateral, but it gradually crosses the ventral surface of the artery in the middle or distal part of the arm and lies medial to it at the bend of the elbow, where it is deep to the bicipital fascia and superficial to the Brachialis. In the forearm it passes between the two heads of the Pronator teres, being separated from the ulnar artery by the deep head. It continues distally between the Flexores digitorum superficialis and profundus almost to the flexor retinaculum where it becomes more superficial and lies between the tendons of the Flexor digitorum superficialis and the Flexor carpi radialis. It is deep to the tendon of the Palmaris longus and slightly ulnaward of it, and it is the most superficial of the structures which pass through the tunnel under the flexor retinaculum. In the palm of the hand it is covered only by the skin and the palmar aponeurosis and rests on the tendons of the flexor muscles (Fig. 12–47). Immediately after emerging from under the retinaculum it becomes enlarged and flattened and splits into muscular and digital branches.

Branches.—The median nerve has no branches above the elbow joint unless, as occasionally happens, the nerve to the Pronator teres arises there.

a) **Articular branches** to the elbow joint are one or two twigs given off as the nerve passes the joint.

b) **Muscular branches** (*rr. musculares*) leave the nerve near the elbow and supply all the superficial muscles of the anterior part of the forearm except the Flexor carpi ulnaris, *i.e.*, the **Pronator teres, Flexor carpi radialis, Palmaris longus** and **Flexor digitorum superficialis.**

c) The **anterior interosseous nerve** (*n. interosseus anterior*) (Fig. 12–46) accompanies the anterior interosseous artery along the anterior surface of the interosseous membrane in the interval between the Flexor pollicis longus and the Flexor digitorum profundus, ending in the Pronator quadratus and the wrist joint. It supplies all the deep anterior muscles of the forearm except the ulnar half of the profundus, *i.e.*, the radial half of the **Flexor digitorum profundus,** the **Flexor pollicis longus,** and the **Pronator quadratus.**

d) The **palmar branch of the median** (*r. cutaneus palmaris n. mediani*) (Fig. 12–42) pierces the antebrachial fascia distal to the wrist and divides into a medial and lateral branch. The medial branch supplies the skin of the palm and communicates with the palmar cutaneous branch of the ulnar nerve. The lateral branch supplies the skin over the thenar eminence and communicates with the lateral antebrachial cutaneous nerve.

e) The **muscular branch** in the hand (Fig. 12–48) is a short stout nerve which leaves the radial side of the median nerve, sometimes in company with the first common palmar digital nerve, just after the former passes under the flexor retinaculum. It supplies the muscles of the thenar eminence with the exception of the deep head of the short flexor, *i.e.*, the **Abductor pollicis brevis,** the **Opponens pollicis** and the superficial head of the **Flexor pollicis brevis.**

f) The **first common palmar digital nerve** (Figs. 12–48, 12–50) divides into three proper palmar digital nerves (*digital collaterals*), two of which supply the sides of the thumb, the third gives a **twig to the first Lumbricalis** and continues as the proper palmar digital for the radial side of the index finger.

g) The **second common palmar digital nerve** gives a twig to the **second Lumbricalis** and, continuing to the web between the index and middle fingers, splits into proper digital nerves for the adjacent sides of these fingers.

h) The **third common palmar digital nerve** occasionally gives a twig to the third Lumbricalis, in which case it has a double innervation; it communicates with a branch of the ulnar nerve, and continues to the web between the middle and ring fingers where it splits into proper digital nerves for the adjacent sides of these digits.

i) The **proper digital nerves** (*digital collaterals*) (Figs. 12–47, 12–48, 12–50) supply the skin of the palmar surface and the dorsal surface over the terminal phalanx of their digits. At the end of the digit each nerve terminates in two branches, one of which ramifies in the skin of the ball, the other in the pulp under the nail. They communicate with the dorsal digital branches of the superficial radial, and in the fingers they are superficial to the corresponding arteries.

Variations.—The relation of the median nerve to the two heads of the Pronator teres varies from that described in 16 per cent; it may pass deep to the humeral head in the absence of an ulnar head, or deep to the ulnar head, or split the humeral head (Jamieson and Anson '52). There is overlapping of territory in the innervation of the Flexor digitorum profundus by the median and ulnar nerves in 50 per cent; it is twice as common for the median to encroach on the ulnar. The portion of the Profundus attached to the index finger is the only one constantly supplied by one nerve, the median. In the majority of cases the Profundus and Lumbricalis of a particular digit are innervated by the same nerve. Encroachment of the median on the ulnar is less common for the Lumbricales than the Profundus. The median nerve may supply the first dorsal Interosseus (Sunderland '45, '46).

The Ulnar Nerve

16. The **ulnar nerve** (*n. ulnaris*) (Fig. 12–46) occupies a superficial position along the medial side of the arm and is the nerve to

the muscles and skin of the ulnar side of the forearm and hand. It is the terminal continuation of the medial cord of the brachial plexus, after the medial head of the median has separated from it. It is medial, at first to the axillary, and then to the brachial artery as far as the middle of the arm, and is parallel with and not far distant from the median and medial antebrachial cutaneous nerves. In the middle of the arm it angles dorsally, pierces the medial intermuscular septum, and follows along the medial head of the Triceps to the groove between the olecranon and the medial epicondyle of the humerus. In this position it is covered only by the skin and fascia and can readily be palpated as the "funny bone" of the elbow. It is accompanied in its course through the distal half of the arm by the superior ulnar collateral artery and the ulnar collateral branch of the radial nerve. It enters the forearm between the two heads of the Flexor carpi ulnaris and continues between this muscle and the Flexor digitorum profundus half way down the forearm. In the proximal part of the forearm it is separated from the ulnar artery by a considerable distance, but in the distal half it lies close to its ulnar side, radialward from the Flexor carpi ulnaris, and covered only by the skin and fascia. Proximal to the wrist it gives off a large dorsal branch and continues into the hand where it has muscular and digital branches.

Branches.—The ulnar nerve usually has no branches proximal to the elbow. Distal to the elbow its branches are as follows:

a) The **articular branches** to the elbow joint are several small filaments which leave the nerve as it lies in the groove between the olecranon and the medial epicondyle of the humerus.

b) **Muscular branches** (*rr. musculares*), two in number, arise near the elbow, and supply the **Flexor carpi ulnaris** and the ulnar half of the **Flexor digitorum profundus.**

c) The **palmar cutaneous branch** of the ulnar (*r. cutaneus palmaris*) arises near the middle of the forearm and accompanies the ulnar artery into the hand. It gives filaments to the artery, perforates the flexor retinaculum, and ends in the skin of the palm, communicating with the palmar branch of the median nerve.

d) The **dorsal branch** (*r. dorsalis manus*) (Figs. 12–43, 12–46) arises in the distal half of the forearm, and reaches the dorsum of the wrist by passing between the Flexor carpi ulnaris and the ulna. It pierces the deep fascia and divides into two **dorsal digital nerves** and a metacarpal communi-cating branch. The more medial digital nerve supplies the ulnar side of the little finger; the other digital branch, the adjacent sides of the little and ring fingers. The metacarpal branch supplies the skin of that area and continues toward the web between the ring and middle fingers where it joins a similar branch of the superficial radial to supply the adjacent sides of these two

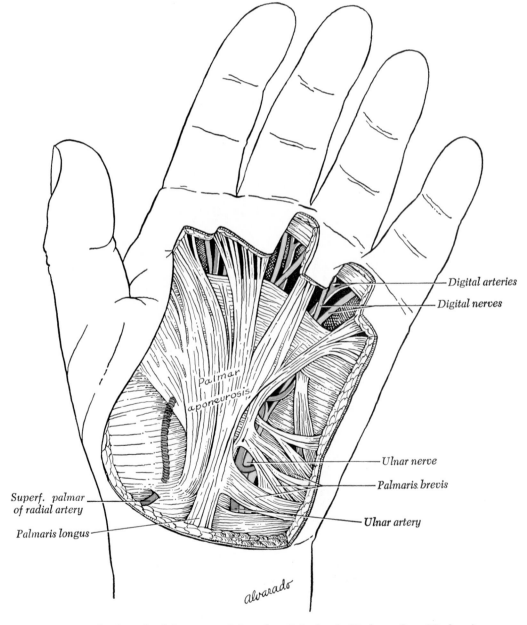

FIG. 12–47.—Superficial dissection of the palm of the hand. (Redrawn from Töndury.)

fingers. On the little finger the dorsal digital branches extend only as far as the base of the terminal phalanx, and on the ring finger as far as the base of the second phalanx; the more distal parts of these digits are supplied by the dorsal branches of the proper palmar digital nerves from the ulnar.

e) The **palmar branch** (*r. palmaris*) or terminal portion of the ulnar nerve crosses the ulnar border of the wrist in company with the ulnar artery, superficial to the flexor retinaculum and under cover of the Palmaris brevis, and divides into a superficial and deep branch.

i) The **superficial branch** (*r. superficialis*) (Figs. 12–48, 12–50) supplies the **Palmaris brevis** and the skin of the hypothenar eminence, and divides into **digital branches**.

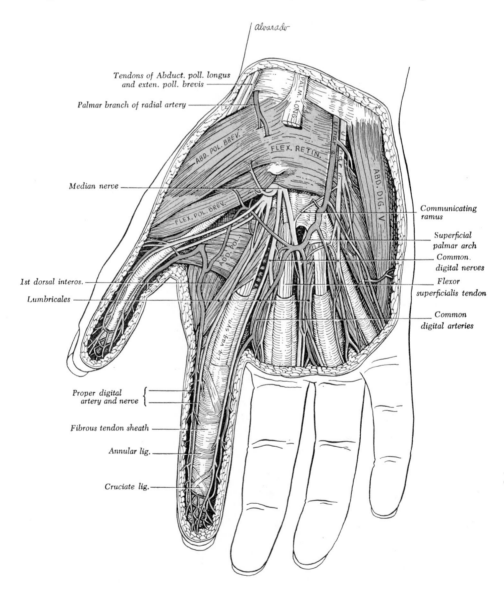

Fig. 12–48.—Dissection of the palm of the hand, showing the superficial palmar arch, median and ulnar nerves, and synovial sheaths. (Redrawn from Töndury.)

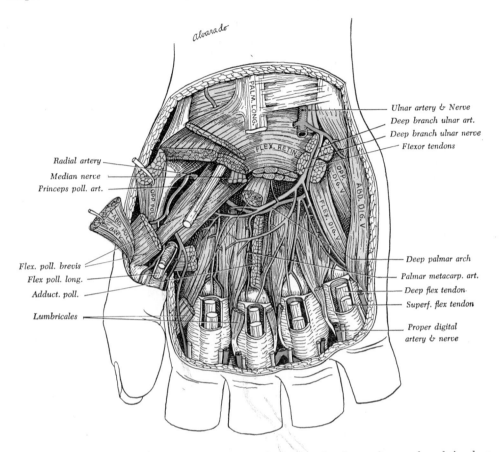

FIG. 12–49.—Deep dissection of the palm of the hand, showing the deep palmar arch and the deep palmar branch of the ulnar nerve. (Redrawn from Töndury.)

A proper palmar digital branch goes to the ulnar side of the little finger; a common palmar digital divides into proper digital branches for the adjacent sides of the **little and ring fingers** and communicates with the branches of the median nerve. The proper digital branches are distributed to the fingers in the same manner as those of the median.

ii) The **deep branch** (*r. profundus*) (Figs. 12–48, 12–50) passes between the Abductor digiti minimi and Flexor digiti minimi brevis accompanied by the deep branch of the ulnar artery; it then pierces the Opponens digiti minimi and follows the course of the deep palmar arch across the Interossei, deep to the midpalmar and thenar fascial clefts. Near its origin it gives branches to the **three small muscles of the little finger,** and as it crosses the hand it supplies the **third and fourth Lumbricales** and **all the Interossei,** both palmar and dorsal. It ends by supplying the **Adductores pollicis** and the deep head of the **Flexor pollicis brevis.** It also sends articular filaments to the wrist joint.

Variations.—The ulnar nerve may pass in front of the medial epicondyle. It frequently has a communication with the median nerve in the forearm, rarely with the medial antebrachial cutaneous, median, or musculocutaneous in the arm. It may send muscular branches to the medial head of the Triceps, the Flexor digitorum superficialis, the first and second Lumbricales and the superficial head of the Flexor pollicis brevis. It may have deficiencies on the dorsum of the hand which are supplied by the radial or it may encroach upon the area usually supplied by that nerve (Sunderland and Hughes '46). See variations of the median nerve and brachial plexus.

The Radial Nerve

17. The **radial nerve** (*n. radialis; musculo-spiral nerve*) (Fig. 12–51), the largest branch of the brachial plexus, is the continuation of the posterior cord, and supplies the extensor muscles of the arm and forearm, as well as the skin covering them. It crosses the tendon of the Latissimus dorsi, deep to the axillary artery, and, after passing the inferior border of the Teres major, it winds around the medial side of the humerus and enters the substance of Triceps between the medial and long heads. It take a spiral course down the arm close to the humerus in the groove which separates the origins of the medial and lateral heads of the Triceps, accompanied by the arteria profunda brachii. Having reached the lateral side of the arm, it pierces the lateral intermuscular septum and runs between the Brachialis and Brachioradialis anterior to the lateral epicondyle, where it divides into superficial and deep branches (Fig. 12–46).

A. **Branches of the radial nerve** in the arm are both muscular and cutaneous:

a) The **medial muscular branches** arise

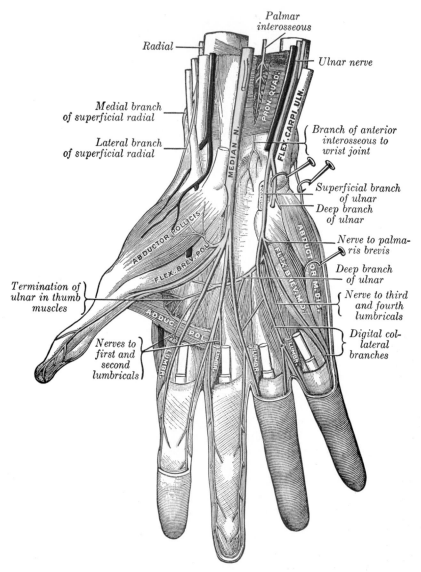

FIG. 12–50.—Deep palmar nerves. (Testut.)

in the axilla and supply the **medial** and **long heads** of the **Triceps.** That to the medial head is a long filament which accompanies the ulnar nerve and superior ulnar collateral artery as far as the lower third of the arm and is therefore named the **ulnar collateral nerve.**

b) The **posterior brachial cutaneous nerve** (*n. cutaneus brachii posterior; internal cutaneous branch of the musculospiral*) arises in the axilla to the medial side of the arm, supplying the skin on the dorsal surface nearly as far as the olecranon. In its course it crosses dorsal to, and communicates with, the intercostobrachial nerve.

c) The **posterior muscular branches** arise from the nerve as it lies in the spiral groove of the humerus, and supply the **medial** and **lateral heads** of the **Triceps** and the **Anconeus.** The nerve to the latter muscle

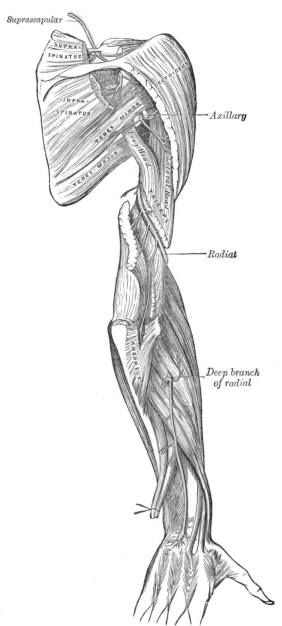

Fig. 12–51.—The suprascapular, axillary, and radial nerves.

is a long, slender filament which lies buried in the substance of the medial head of the Triceps.

d) The **posterior antebrachial cutaneous nerve** (*n. cutaneus antebrachii posterior; external cutaneous branch of the musculospiral*) perforates the lateral head of the Triceps at its attachment to the humerus and divides into proximal and distal branches.

i) The **proximal** and smaller **branch** lies close to the cephalic vein and supplies the skin of the dorsal part of the distal half of the arm.

ii) The **distal branch** pierces the deep fascia below the insertion of the Deltoideus and continues along the lateral side of the arm and elbow, and dorsal side of the forearm to the wrist. It supplies the skin in its course and near its termination communicates with the dorsal branch of the lateral antebrachial cutaneous nerve.

e) The **lateral muscular branches** supply the **Brachioradialis, Extensor carpi radialis longus,** and the lateral part of the **Brachialis.**

f) **Articular branches** to the elbow come from the radial, between the Brachialis and Brachioradialis, from the ulnar collateral nerve, and from the nerve to the Anconeus.

B. The **branches** of the **radial nerve** in the **forearm** are:

g) The **Superficial Branch of the Radial Nerve** (*r. superf. n. radialis; radial nerve; superficial radial*) (Fig. 12–46) runs along the lateral border of the forearm under cover of the Brachioradialis. In the proximal third of the forearm it gradually approaches the radial artery; in the middle third it lies just radialward to the artery, and in the lower third it quits the artery and angles dorsally under the tendon of the Brachioradialis toward the dorsum of the wrist where it pierces the deep fascia and divides into two branches.

i) The **lateral branch** is smaller, supplies the skin of the radial side and ball of the thumb, and communicates with the palmar branch of the lateral antebrachial cutaneous nerve.

ii) The **medial branch** communicates, above the wrist, with the dorsal branch of the lateral antebrachial cutaneous nerve, and, on the dorsum of the hand, with the dorsal branch of the ulnar nerve. It then

divides into four **dorsal digital nerves** (Fig. 12–43) which are distributed as follows: (*1*) supplies the ulnar side of the thumb, (*2*) the radial side of the index finger, (*3*) the adjacent sides of the index and middle fingers, and (*4*) communicates with the dorsal branch of the ulnar nerve, and supplies the adjacent sides of the middle and ring fingers.

h) The **Deep Branch of the Radial Nerve** (*r. profundus; deep radial; posterior interosseous nerve*) (Fig. 12–51) winds to the dorsum of the forearm around the radial side of the radius between the planes of fibers of the Supinator (Fig. 6–45, page 470), and continues between the superficial and deep layers of muscles to the middle of the forearm. Considerably diminished in size and named the **dorsal interosseous nerve,** it lies on the dorsal surface of the interosseous membrane, and under cover of the Extensor pollicis longus continues to the dorsum of the carpus where it ends in a gangliform enlargement.

i) **Muscular Branches.**—Those to the Extensor carpi radialis brevis and the Supinator are given off before the nerve turns dorsally. After passing through the Supinator, branches are given to the Extensor digitorum, Extensor digiti minimi, Extensor carpi ulnaris, the two Extensores and Abductor longus pollicis, and the Extensor indicis.

ii) **Articular filaments** from the terminal enlargement are distributed to the ligaments and articulations of the carpus and metacarpus.

Variations.—The radial nerve may pass through the quadrilateral space with the axillary nerve, and when the profunda brachii supplies the nutrient artery of the humerus it may be accompanied by a filament from the radial. The nerve to the Brachialis is inconstant. The deep branch may pass superficial to the entire Supinator. There is great variation in the distribution and overlapping of the radial and ulnar nerves on the back of the hand; the primitive arrangement appears to be that the radial supplies three and a half, the ulnar one and a half digits (Sunderland '45, '46).

Thoracic Nerves (nn. Thoracales)

The **ventral primary divisions** (*rami ventrales; anterior divisions*) (Fig. 12–52) of

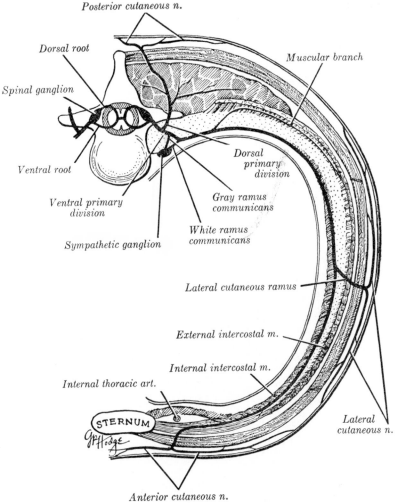

Posterior cutaneous n.

Dorsal root

Muscular branch

Spinal ganglion

Dorsal primary division

Ventral root

Gray ramus communicans

Ventral primary division

White ramus communicans

Sympathetic ganglion

Lateral cutaneous ramus

External intercostal m.

Internal intercostal m.

Internal thoracic art.

STERNUM

Lateral cutaneous n.

Anterior cutaneous n.

Fig. 12–52.—Plan of a typical intercostal nerve.

the thoracic nerves number twelve on each side. The first eleven, situated between the ribs, are termed **intercostals;** the twelfth lies below the last rib and is called the **subcostal** (*n. subcostalis* BR). The intercostal nerves are distributed chiefly to the parietes of the thorax and abdomen, and differ from the other spinal nerves, in that each pursues an independent course, *i.e.,* they do not enter into a plexus. The first two contribute to the upper limb as well as to the thorax; the next four are limited in their distribution to the thorax; the lower five supply the parietes of the thorax and the abdomen. The twelfth thoracic is distributed to the abdominal wall and the skin of the buttock.

Rami Communicantes.—Each thoracic nerve contributes preganglionic sympathetic fibers to the sympathetic chain through a **white ramus communicans,** and receives postganglionic fibers from the chain ganglia through **gray rami communicantes.** Both rami are attached to the spinal nerves near their exit from the intervertebral foramina, the gray rami being more medial than the white. For more details see the section on the Sympathetic System.

The First Thoracic Nerve.—The ventral primary division of the first thoracic nerve divides immediately into two parts: (*a*) the larger part leaves the thorax ventral to the neck of the first rib, and becomes one of the roots of the brachial plexus (described above); (*b*) the smaller part becomes the **first intercostal nerve.** The first intercostal runs along the first intercostal space to the sternum, perforates the muscles and deep fascia, and ends as the first anterior cutaneous nerve of the thorax. It has no lateral cutaneous branch, as a rule, but it may have a communication with the intercostobra-chial branch of the second nerve. A communication between the first and second nerves inside the thorax is of frequent occurrence, and contains postganglionic sympathetic fibers from the second or even the third thoracic sympathetic ganglion (Kirgis and Kuntz '42).

The Superior Thoracic Nerves (Figs.

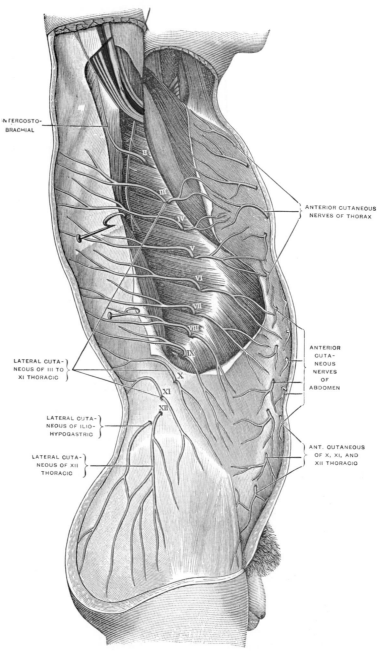

Fɪɢ. 12–53.—Cutaneous distribution of thoracic nerves. (Testut.)

12–53, 12–54, 12–78).—The ventral primary divisions of the second to the sixth thoracic nerves and the intercostal portion of the first thoracic are known as the **thoracic intercostal nerves** (*nn. intercostales*). They pass ventralward in the intercostal spaces with the intercostal vessels; from the vertebral column to the angles of the ribs, they lie between the pleura and the posterior intercostal membrane; from the angle to the middle of the ribs, they pass between the Intercostales interni and externi. They then enter the substance of the Intercostales interni where they remain concealed until they reach the costal cartilages, where they again emerge on the inner surface of the muscle and lie between it and the pleura. The nerves are inferior to the vessels and with them are at first close to the rib above, but as they proceed ventrally may approach

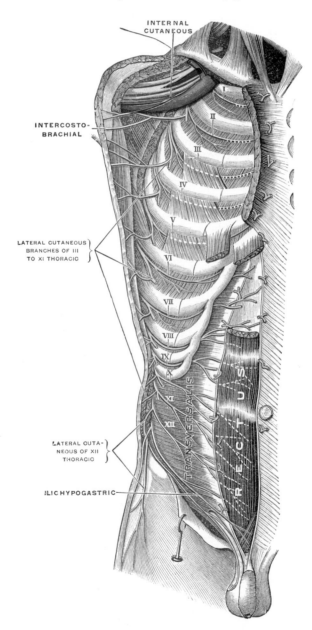

Fig. 12–54.—Intercostal nerves, the superficial muscles having been removed. (Testut.)

the middle of the intercostal space. Near the sternum they pass ventral to the Transversus thoracis and the internal thoracic vessels, and near the sternum pierce the Intercostalis internus, the anterior intercostal membrane, the Pectoralis major and pectoral fascia, terminating as the **anterior cutaneous nerves** of the thorax. They have short medial and longer lateral branches which supply the skin and mamma.

Muscular branches supply the Intercostales interni and externi, the Subcostales, the Levatores costarum, the Serratus posterior superior, and the Transversus thoracis. At the ventral part of the thorax some of these branches cross the costal cartilages from one intercostal space to another.

Cutaneous Branches.—The **lateral cutaneous nerves** (*rr. cutanei laterales*) (Figs. 12–53, 12–54) arise from the intercostal nerves about midway between the vertebral column and the sternum, pierce the Intercostales externi and Serratus anterior, and divide into anterior and posterior branches. The anterior branches supply the skin of the lateral and ventral part of the chest and mammae; those of the fifth and sixth nerves also supply the cranial digitations of the Obliquus externus abdominis. The posterior branches supply the skin over the Latissimus dorsi and scapular region.

The **intercostobrachial nerve** (Figs. 12–39, 12–41, 12–46) arises from the second intercostal nerve as if it were a lateral cutaneous nerve, but it fails to divide into an anterior and a posterior branch. After piercing the Intercostales and the Serratus anterior, it crosses the axilla, embedded in the adipose tissue, to reach the medial side of the arm. It forms a loop of communication with the medial brachial cutaneous nerve of the brachial plexus and assists it in supplying the skin of the medial and posterior part of the arm. It also communicates with the posterior brachial cutaneous branch of the radial nerve.

An intercostobrachial branch frequently arises from the third intercostal nerve, supplying the axilla and medial side of the arm.

The Lower Thoracic Nerves (Figs. 12–53, 12–54).—The ventral primary divisions of the seventh to eleventh thoracic nerves are continued beyond the intercostal spaces into the anterior abdominal wall and are named the **thoracoabdominal intercostal nerves.** They have the same arrangement as the upper nerves as far as the anterior ends of the intercostal spaces, where they pass dorsal to the costal cartilages. They run ventralward between the Obliquus internus and Transversus abdominis, pierce the sheath of the Rectus abdominis, supply this muscle, and terminate as the **anterior cutaneous branches.**

1. **Muscular branches** supply the Intercostales, the Obliqui and Transversus abdominis, and the last three the Serratus posterior inferior.

2. The **lateral cutaneous branches** (Figs. 12–53, 12–54) arise midway along the nerves, pierce the Intercostales externi and the Obliquus externus in line with the lateral cutaneous branches of the upper intercostals, and divide into anterior and posterior branches. The anterior branches give branches to the digitations of the Obliquus externus, and extend caudalward and ventralward nearly as far as the margin of the Rectus abdominis, supplying the skin. The posterior branches pass dorsally to supply the skin over the Latissimus dorsi.

3. The **anterior cutaneous branches** penetrate the anterior layer of the sheath of the Rectus abdominis, and divide into medial and lateral branches which supply the skin of the anterior part of the abdominal wall.

The anterior primary division of the **twelfth thoracic** (Figs. 12–53, 12–54, 12–57) or **subcostal nerve** is larger than those above it. It runs along the inferior border of the twelfth rib, often communicates with the first lumbar nerve, and passes under the lateral lumbocostal arch. It crosses the ventral surface of the Quadratus lumborum, penetrates the Transversus abdominis, and continues ventralward to be distributed in the same manner as the lower intercostal nerves. It communicates with the iliohypogastric nerve of the lumbar plexus and gives a branch to the Pyramidalis. The lateral cutaneous branch is large and does not divide into anterior and posterior branches. It perforates the Obliqui, passes downward over the crest of the ilium ventral to a similar branch of the iliohypogastric nerve, and is distributed to the skin of the anterior part of the gluteal region, some filaments extending as far down as the greater trochanter.

The Lumbar Nerves (nn. lumbales)

The **ventral primary divisions** (*rr. anteriores*) are increasingly large as they are placed more caudally in the lumbar region. Their course is lateralward and caudalward, either under cover of the Psoas major or between its fasciculi. The first three nerves and the larger part of the fourth are connected together by communicating loops, and they are frequently joined by a communication from the twelfth thoracic, forming the lumbar plexus. The smaller part of the fourth joins the fifth nerve to form the lumbosacral trunk which enters into the

formation of the sacral plexus. The fourth nerve, because it is divided between the two plexuses, has been named the **nervus furcalis.**

Rami Communicantes.—Only the first two lumbar nerves contribute preganglionic sympathetic fibers to the sympathetic chain through **white rami communicantes.** All the lumbar nerves receive postganglionic fibers from the sympathetic chain through **gray rami communicantes.** For more details see the section on the Sympathetic Nervous System.

Muscular branches are supplied to the Psoas major and the Quadratus lumborum from the ventral primary divisions of the lumbar nerves before they enter the lumbar plexus.

THE LUMBOSACRAL PLEXUS

The **Lumbosacral Plexus** (*p. lumbosacralis*) is the name given to the combination of all the ventral primary divisions of the lumbar, sacral, and coccygeal nerves.

The lumbar and sacral plexuses supply the lower limb, but in addition the sacral nerves supply the perineum through the pudendal plexus, and the coccygeal region through

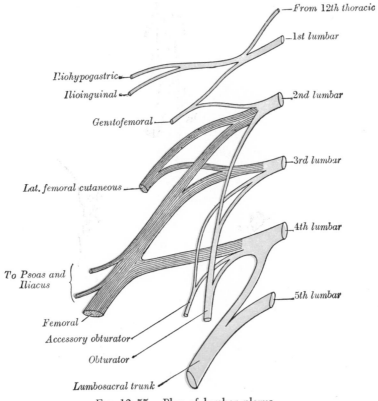

—*From 12th thoracic*

—*1st lumbar*

Iliohypogastric

Ilioinguinal

—*2nd lumbar*

Genitofemoral

—*3rd lumbar*

Lat. femoral cutaneous

—*4th lumbar*

To Psoas and Iliacus

—*5th lumbar*

Femoral

Accessory obturator

Obturator

Lumbosacral trunk

FIG. 12–55.—Plan of lumbar plexus.

the coccygeal plexus. For convenience in description, these plexuses will be considered separately.

The Lumbar Plexus

The **lumbar plexus** (*plexus lumbalis*) (Fig. 12–55) is formed by the ventral primary divisions of the first three and the greater part of the fourth lumbar nerves, with a communication from the twelfth thoracic usually joining the first lumbar nerve. It is situated on the inside of the posterior abdominal wall, either dorsal to the Psoas major or among its fasciculi, and ventral to the transverse processes of the lumbar vertebrae. The lumbar plexus is not an intricate interlacement like the bra-

chial plexus, but its branches usually arise from two or three nerves, so that the resulting junctions between adjacent nerves have the appearance of loops.

The manner in which the plexus is formed is the following (Figs. 12–55, 12–56): the first lumbar nerve, usually supplemented by a communication from the twelfth thoracic, splits into a cranial and a caudal branch; the cranial forms the iliohypogastric and ilioinguinal; the caudal and smaller branch unites with a branch from the second lumbar to form the genitofemoral nerve. The remainder of the second, and the third and fourth nerves each split into a small ventral and a large dorsal portion; the ventral portions unite into one nerve, the obturator. The dorsal portions of the

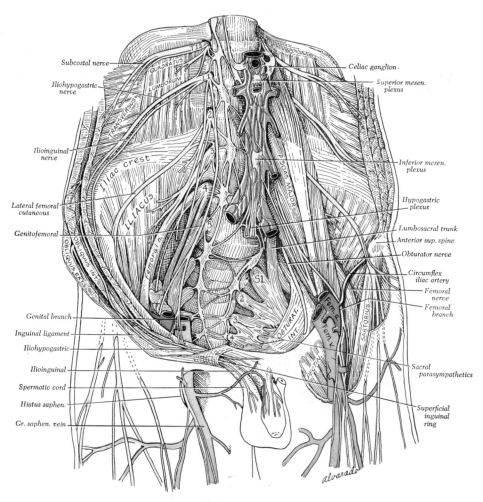

FIG. 12–56.—The lumbar plexus and its branches.

second and third nerves each divide again into unequal branches; the two smaller branches unite to form the lateral femoral cutaneous nerve; the two larger branches join the dorsal portion of the fourth nerve to form the femoral nerve. The accessory obturator, present in about one out of five individuals, comes from the third and fourth. A considerable part of the fourth nerve joins the fifth lumbar in the lumbo-sacral trunk.

The **branches** of the **lumbar plexus** are as follows:

1. Iliohypogastric	L 1, (T 12)
2. Ilioinguinal	L 1
3. Genitofemoral	L 1, 2
4. Lateral femoral cutaneous	L 2, 3
5. Obturator	L 2, 3, 4
6. Accessory obturator	L 3, 4
7. Femoral	L 2, 3, 4

These branches may be divided into two groups according to their distribution. The first three supply the caudal part of the parietes of the abdominal wall; the last four supply the anterior thigh and medial part of the leg.

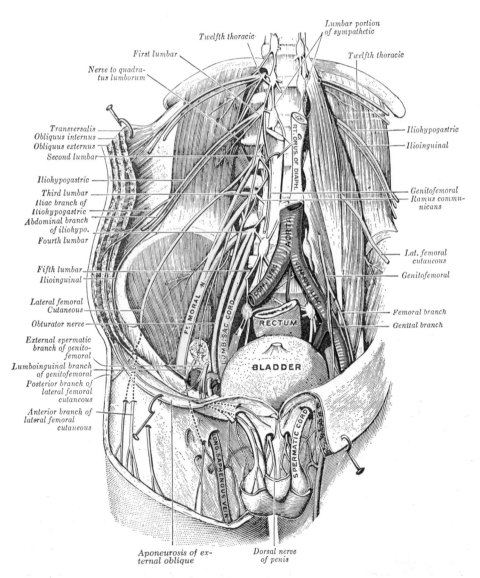

FIG. 12–57.—Deep and superficial dissection of the lumbar plexus. (Testut.)

1. **The Iliohypogastric Nerve** (*n. iliohypogastricus*) (Figs. 12–56, 12–58) arises from the first lumbar and from the communication with the twelfth thoracic when this is present. It emerges from the upper part of the lateral border of the Psoas major, crosses the Quadratus lumborum to the crest of the ilium, and penetrates the posterior part of the Transversus abdominis near the crest of the ilium. Between the Transversus and the Obliquus internus it divides into a lateral and an anterior cutaneous branch.

a) The **lateral cutaneous branch** (*r. cutaneus lateralis; iliac branch*) pierces the Obliqui internus and externus immediately above the iliac crest, and is distributed to the skin of the gluteal region posterior to the lateral cutaneous branch of the twelfth thoracic (Figs. 12–57, 12–64), these two nerves being inversely proportional in size.

b) The **anterior cutaneous branch** (*r. cutaneus anterior; hypogastric branch*) (Figs. 12–54, 12–58) continues its course between the Obliquus internus and Transversus, pierces the former, and becomes subcutaneous by passing through a perforation in the aponeurosis of the Obliquus externus about 2 cm above the superficial inguinal ring. It is distributed to the skin of the hypogastric region (Fig. 12–27).

c) **Muscular branches** are supplied to the Obliquus internus and Transversus.

2. The **Ilioinguinal Nerve** (*n. ilioinguinalis*) (Figs. 12–56, 12–57) arises from the lateral border of the Psoas major just caudal to the iliohypogastric and follows a similar course obliquely across the fibers of the Quadratus lumborum to the crest of the ilium. It penetrates the Transversus near the anterior part of the crest, communicates with the iliohypogastric nerve, and pierces the Obliquus internus, distributing filaments to it. It then accompanies the spermatic cord through the superficial inguinal ring, and is distributed to the skin over the proximal and medial part of the thigh, the root of the penis and the scrotum in the male (Fig. 12–57), and the mons pubis and labium majus in the female.

Muscular twigs are supplied to the Obliquus internus and Transversus.

The optimum point for blocking the iliohypogastric and ilioinguinal nerves with local anes-

thetic is 4 to 6 cm posterior to the anterior superior spine of the ilium, along the lateral aspect of the external lip of the crest, where the nerves perforate the Transversus abdominis (Jamieson *et al.* '52).

3. **The Genitofemoral Nerve** (*n. genitofemoralis; genitocrural nerve*) (Figs. 12–56, 12–57, 12–59) arises from the first and second lumbar nerves, passes caudalward through the substance of the Psoas major until it emerges on its ventral surface opposite the third or fourth lumbar vertebra, where it is covered by transversalis fascia and peritoneum. On the surface of the muscle, or occasionally within its substance, it divides into a genital and a femoral branch.

a) The **genital branch** (*r. genitalis; external spermatic nerve*) continues along the Psoas major to the inguinal ligament, where it either pierces the transversalis and internal spermatic fascia or passes through the internal inguinal ring to reach the spermatic cord. It lies against the dorsal aspect of the cord, supplies the Cremaster, and is distributed to the skin of the scrotum and adjacent thigh. In the female it accompanies the round ligament of the uterus.

b) The **femoral branch** (*r. femoralis; lumboinguinal nerve*) lies on the Psoas major, lateral to the genital branch, and passes under the inguinal ligament with the external iliac artery. It enters the femoral sheath, superficial and lateral to the artery, then pierces the sheath and fascia lata to supply the skin of the proximal part of the anterior surface of the thigh. It gives a filament to the femoral artery and communicates with the anterior cutaneous branches of the femoral nerve (Figs. 12–58, 12–59).

Variations.—The iliohypogastric or ilioinguinal nerves may arise from a common trunk or the ilioinguinal may join the iliohypogastric at the iliac crest, the latter nerve then supplying the missing branches. The ilioinguinal may be lacking and the genital branch supply its branches, or the genital branch may be absent and the ilioinguinal substitute for it. The femoral branch and lateral femoral cutaneous nerves, or the anterior cutaneous branches of the femoral may substitute for each other to a greater or lesser extent.

4. The **Lateral Femoral Cutaneous Nerve** (*n. cutaneus femoralis lateralis; external cutaneous nerve*) (Figs. 12–55, 12–58) arises from the dorsal portions of the ventral primary divisions of the second and third lumbar nerves. It emerges from the lateral border of the Psoas major about its middle, and runs across the Iliacus obliquely toward the anterior superior iliac spine. It then passes under the inguinal ligament and over the Sartorius into the subcutaneous tissues of the thigh, dividing into an anterior and a posterior branch (Fig. 12–57).

a) The **anterior branch** becomes superficial about 10 cm distal to the inguinal ligament, and is distributed to the skin of the lateral and anterior parts of the thigh as far as the knee (Fig. 12–58). The terminal filaments communicate with the anterior cutaneous branches of the femoral nerve and the infrapatellar branches of the saphenous nerve, forming the **patellar plexus.**

b) The **posterior branch** pierces the fascia lata and subdivides into filaments which pass posteriorly across the lateral and posterior surfaces of the thigh, supplying the skin from the level of the greater trochanter to the middle of the thigh (Fig. 12–64).

5. The **Obturator Nerve** (*n. obturatorius*) (Figs. 12–55, 12–56), the motor nerve to the Adductores, arises by three roots from the ventral portions of the second, third, and fourth lumbar nerves; that from the third is the largest, and that from the second is often very small. It emerges from the medial border of Psoas major near the brim of the pelvis, under cover of the common iliac vessels, passes lateral to the hypogastric vessels and the ureter, and runs along the lateral wall of the lesser pelvis, to enter the upper part of the obturator foramen with the obturator vessels. As it enters the thigh it divides into anterior and posterior branches which are separated by some of the fibers of the Obturator externus and the Adductor brevis.

a) The **anterior or superficial branch** (*r. anterior*) (Fig. 12–60) communicates with the accessory obturator nerve, when this is present, and passes over the superior border of the Obturator externus, deep to the Pectineus and Adductor longus and superficial to the Adductor brevis. At the distal border of the Adductor longus it communicates with the anterior cutaneous and saphenous branches of the femoral nerve, forming the **subsartorial plexus,** and terminates in filaments accompanying the femoral artery. Its branches are as follows:

i) An **articular branch** to the hip joint is given off near the obturator foramen.

ii) **Muscular branches** are supplied to the Adductor longus, Gracilis, and usually to the Adductor brevis; in rare instances it gives a branch to the Pectineus.

iii) A **cutaneous branch** is occasionally found as a continuation of the communication with the anterior cutaneous and saphenous branches. It emerges from beneath the inferior border of the Adductor longus, continues along the posterior margin of the Sartorius to the medial side of the knee, where it pierces the deep fascia, communicates with the saphenous nerve, and is distributed to the skin of the medial side of the proximal half of the leg.

b) The **posterior branch** (*r. posterior*) pierces the anterior part of the Obturator externus, passes posterior to the Adductor brevis and anterior to the Adductor magnus, and divides into muscular and articular branches.

i) **Muscular branches** are given (*a*) to the Obturator externus as it passes between its fibers, (*b*) to the Adductor magnus, and (*c*) to the Adductor brevis when it is not supplied by the anterior branch.

ii) The **articular branch** for the knee joint either perforates the Adductor magnus or passes under the arch through which the femoral artery passes and enters the popliteal fossa. It accompanies the popliteal artery, supplying it with filaments, and reaches the back of the knee joint where it perforates the oblique popliteal ligament and is distributed to the synovial membrane.

6. The **Accessory Obturator Nerve** (*n. obturatorius accessorius*) (Fig. 12–56), present in about 29 per cent, is of small size. It arises from the ventral part of the third and fourth lumbar nerves, follows along the medial border of the Psoas major, and crosses over the superior ramus of the pubis instead of going through the obturator foramen with the obturator nerve. It passes deep to the Pectineus, supplying it with branches, communicates with the anterior

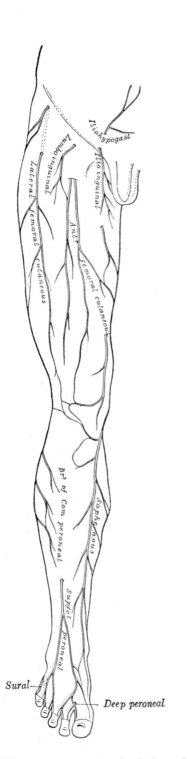

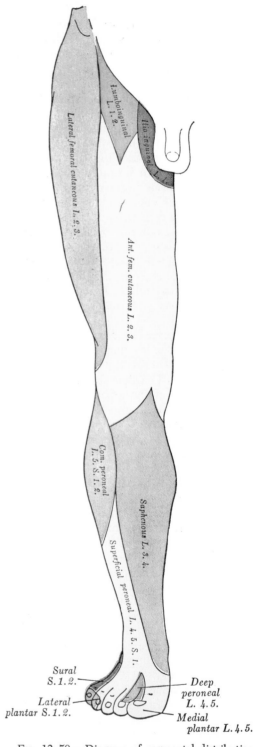

Fig. 12–58.—Cutaneous nerves of right lower limb. Anterior view.

Fig. 12–59.—Diagram of segmental distribution of the cutaneous nerves of the right lower limb. Anterior view.

branch of the obturator nerve, and sends a branch to the hip joint.

Variations.—When the obturator nerve has a cutaneous branch, the medial cutaneous branch of the femoral is correspondingly small. When the accessory obturator is lacking, the obturator supplies two nerves to the hip joint; the accessory may be very small and be lost in the capsule of the hip joint.

7. The **Femoral Nerve** (*n. femoralis; anterior crural nerve*) (Figs. 12–55 to 12–60), the largest branch of the lumbar plexus and the principal nerve of the anterior part of the thigh, arises from the dorsal portions of the ventral primary divisions of the second, third, and fourth lumbar nerves. It emerges through the fibers of the Psoas major at the distal part of its lateral border, and passes down between it and the Iliacus, being covered by the iliac portion of the transversalis fascia. It passes under the inguinal ligament, lateral to the femoral artery, and breaks up into branches soon after it enters the thigh.

a) **Muscular branches** to the Iliacus are given off within the abdomen.

b) The **anterior cutaneous branches** (*rr. cutanei anteriores*) are two large nerves, the intermediate and medial cutaneous nerves (Fig. 12–58).

i) The **intermediate cutaneous nerve** (*ramus cutaneus anterior; middle cutaneous nerve*) pierces the fascia lata (and generally the Sartorius) about 7.5 cm distal to the inguinal ligament, and divides into two branches which descend in immediate proximity along the anterior thigh, to supply the skin as far distally as the anterior knee. Here they communicate with the medial cutaneous nerve and the infrapatellar branch of the saphenous, to form the **patellar plexus.** In the proximal part of the thigh the lateral branch of the intermediate cutaneous communicates with the femoral branch of the genitofemoral nerve.

ii) The **medial cutaneous nerve** (*ramus cutaneus anterior; internal cutaneous nerve*) passes obliquely across the proximal part of the sheath of the femoral artery, and divides anterior to, or at the medial side of that vessel, into two branches, an anterior and a posterior. Before dividing, it gives

off a few filaments which pierce the fascia lata to supply the integument of the medial side of the thigh, accompanying the great saphenous vein. One of these filaments passes through the saphenous opening; a second becomes subcutaneous about the middle of the thigh; a third pierces the fascia at its lower third. The **anterior branch** runs distalward on the Sartorius, perforates the fascia lata at the distal third of the thigh, and divides into two branches: one supplies the integument as far distally as the medial side of the knee; the other crosses to the lateral side of the patella, communicating in its course with the infrapatellar branch of the saphenous nerve. The posterior branch descends along the medial border of the Sartorius muscle to the knee, where it pierces the fascia lata, communicates with the saphenous nerve, and gives off several cutaneous branches. It then passes down to supply the integument of the medial side of the leg. Beneath the fascia lata, at the lower border of the Adductor longus, it joins to form a plexiform network (*subsartorial plexus*) with branches of the saphenous and obturator nerves.

Variations.—When the communicating branch from the obturator nerve is large and continued to the integument of the leg, the posterior branch of the medial cutaneous is small, and terminates in the plexus, occasionally giving off a few cutaneous filaments.

c) The **nerve to the Pectineus** arises immediately below the inguinal ligament, and passes deep to the femoral sheath to enter the anterior surface of the muscle; it is often double.

d) The **nerve to the Sartorius** arises in common with the intermediate cutaneous nerve and enters the deep surface of the proximal part of the muscle.

e) The **saphenous nerve** (*n. saphenus*) (Figs. 12–58, 12–60, 12–64) is the largest and longest branch of the femoral nerve, and supplies the skin of the medial side of the leg. It passes deep to the Sartorius in company with the femoral artery, and lies anterior to the artery, crossing from its lateral to its medial side, within the fascial covering of the adductor canal (Fig. 8–61). At the tendinous arch in the Adduc-

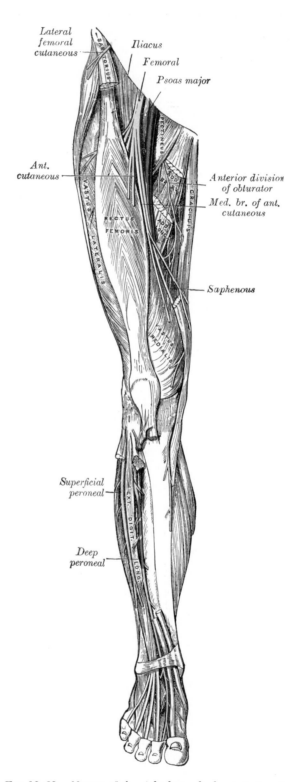

FIG. 12–60.—Nerves of the right lower limb. Anterior view.

tor magnus, it quits the artery, penetrates the fascial covering of the adductor canal, continues along the medial side of the knee deep to the Sartorius, and pierces the fascia lata between the tendons of the Sartorius and Gracilis to become subcutaneous. It then accompanies the great saphenous vein along the tibial side of the leg, and at the medial border of the tibia in the distal third of the leg, it divides into two terminal branches.

The **branches** of the saphenous nerve are:

i) A branch joins the medial cutaneous and obturator nerves in the middle of the thigh to form the **subsartorial plexus.**

ii) A large **infrapatellar branch,** given off at the medial side of the knee, pierces the Sartorius and fascia lata, and is distributed to the skin in front of the patella. Above the knee it communicates with the anterior cutaneous branches of the femoral nerve;

below the knee with other branches of the saphenous; and, on the lateral side of the joint, with branches of the lateral femoral cutaneous nerve to form the **patellar plexus** (*plexus patellae*).

iii) **Branches below the knee** are distributed to the skin of the anterior and medial side of the leg, communicating with the medial cutaneous nerve and the cutaneous branch of the obturator if present. One branch continues along the margin of the tibia, and ends at the ankle. The other terminal branch passes anterior to the ankle, and is distributed to the medial side of the foot, as far as the ball of the great toe, communicating with the medial branch of the superficial peroneal nerve.

f) **Branches to the Quadriceps Femoris** (Fig. 12–60).—The branch to the Rectus femoris enters the proximal part of the deep surface of the muscle. The branch to the

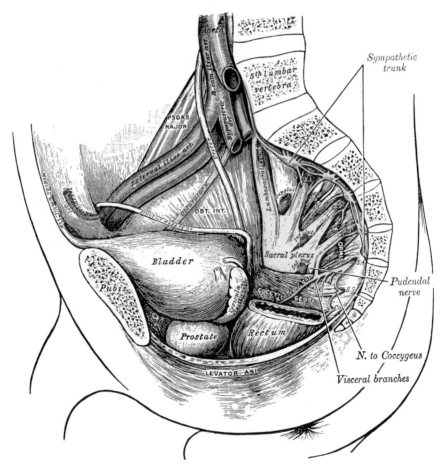

FIG. 12–61.—Dissection of side wall of pelvis showing sacral and pudendal plexuses. (Testut.)

Vastus lateralis, of large size, accompanies the descending branch of the lateral femoral circumflex artery to the distal part of the muscle. The branch to the Vastus medialis runs parallel with the saphenous nerve, lateral to the femoral vessels and outside of the adductor canal, and enters the muscle about at its middle. The branches to the Vastus intermedius, two or three in number, enter the anterior surface of the muscle about the middle of the thigh; a filament from one of these descends through the muscle to the Articularis genu and the knee joint.

  g) The **articular branch** to the **hip joint** is derived from the nerve to the Rectus femoris.

  h) **Articular branches** to the **knee joint** are three in number. (1) A long slender filament derived from the nerve to the Vastus lateralis penetrates the capsule of the joint on its anterior aspect. (2) A filament derived from the nerve to the Vastus medialis can usually be traced downward on the surface of this muscle; it pierces the muscle and accompanies the articular branch of the descending genicular artery to the medial side of the articular capsule, which it penetrates, to supply the synovial membrane. (3) The branch to the Vastus intermedius which supplies the Articularis genu also is distributed to the knee joint.

The Sacral and Coccygeal Nerves
(nn. sacrales et coccygeus)

  The **ventral primary divisions** (*rr. anteriores*) of the sacral and coccygeal nerves

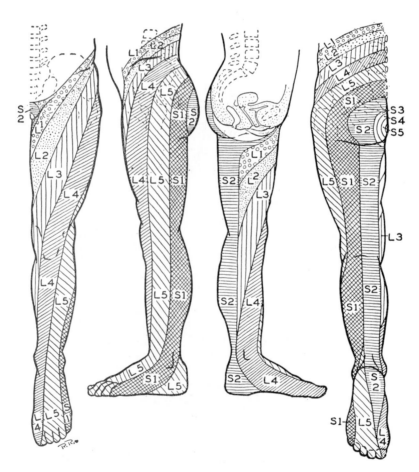

Fɪɢ. 12–62.—Dermatome chart of the lower limb of man outlined by the pattern of hyposensitivity from loss of function of a single nerve root. (From the original of Figure 9; Keegan and Garrett, courtesy of Anatomical Record, *102*, 417.)

enter into the formation of the sacral and pudendal plexuses. Those from the first four sacral nerves enter the pelvis through the anterior sacral foramina, the fifth between the sacrum and coccyx, and the coccygeals below the first piece of the coccyx. The first and second sacrals are large; the third, fourth, and fifth diminish progressively in size.

The Sacral Plexus

The sacral plexus (*plexus sacralis*) (Figs. 12–61, 12–63, 12–70) is formed by the lumbosacral trunk from the fourth and fifth lumbar, and by the first, second, and third sacral nerves. They converge toward the caudal part of the greater sciatic foramen, and unite into a large flattened band, most of which is continued into the thigh as the

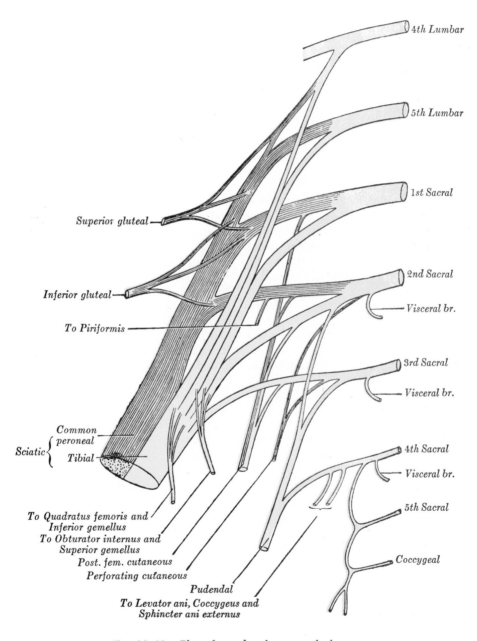

FIG. 12–63.—Plan of sacral and coccygeal plexuses.

sciatic nerve. The plexus lies against the posterior and lateral wall of the pelvis, between the Piriformis and the internal iliac vessels which are embedded in the pelvic subserous fascia. The nerves entering the plexus, with the exception of the third sacral, split into ventral and dorsal portions, and the branches arising from them are as follows:

	Ventral Portions	Dorsal Portions
1. Nerve to Quadratus femoris and Gemellus inferior	L 4, 5, S 1	
2. Nerve to Obturator internus and Gemellus superior	L 5, S 1, 2	
3. Nerve to Piriformis		S (1), 2
4. Superior gluteal		L 4, 5, S 1
5. Inferior gluteal		L 5, S 1, 2
6. Posterior femoral cutaneous	S 2, 3	S 1, 2
7. Perforating cutaneous		S 2, 3
8. Sciatic — Tibial	L 4, 5, S 1, 2, 3	
— Common peroneal		L 4, 5, S 1, 2
9. (Pudendal)	S 2, 3, 4	

1. The **Nerve to the Quadratus femoris and Gemellus inferior,** from the ventral portions of L 4, 5, and S 1, leaves the pelvis through the greater sciatic foramen, distal to the Piriformis, and ventral to the sciatic nerve. It runs ventral or deep to the tendon of the Obturator internus and Gemelli, and enters the deep surface of the Quadratus and Gemellus inferior.

An **articular branch** is given to the hip joint.

2. The **Nerve to the Obturator internus and Gemellus superior** (Fig. 12–66), from the ventral portions of L 5, and S 1, 2, leaves the pelvis through the greater sciatic foramen distal to the Piriformis. It gives off the branch to the Gemellus superior, then crosses the spine of the ischium, reenters the pelvis through the lesser sciatic foramen, and enters the pelvic surface of the Obturator internus.

3. The **Nerve to the Piriformis,** from the portions division of S 2, or S 1 and S 2, enters the ventral surface of the muscle; it may be double.

4. The **Superior Gluteal Nerve** (*n. gluteus superior*) (Figs. 12–66, 12–70), from the dorsal portions of L 4, 5, and S 1, leaves the pelvis through the greater sciatic foramen proximal to the Piriformis, and, accompanying the superior gluteal vessels and their branches, divides into a superior and an inferior branch.

a) The **superior branch** is distributed to the Gluteus minimus.

b) The **inferior branch** crosses the Gluteus minimus, gives filaments to the Gluteus medium and minimus, and ends in a branch to the Tensor fasciae latae.

5. The **Inferior Gluteal Nerve** (*n. gluteus inferior*), from the dorsal portions of L 5, and S 1, 2, leaves the pelvis through the greater sciatic foramen distal to the Piriformis, and enters the deep surface of the Gluteus maximus.

6. The **Posterior Femoral Cutaneous Nerve** (*n. cutaneus femoralis posterior; small sciatic nerve*) (Figs. 12–64, 12–66) is distributed to the skin of the perineum and posterior surface of the thigh and leg. It arises from the dorsal portions of S 1 and 2, and from the ventral divisions of S 2 and 3, and leaves the pelvis through the greater sciatic foramen distal to the Piriformis. It accompanies the inferior gluteal artery to the inferior border of the Gluteus maximus, and runs down the posterior thigh, superficial to the long head of the Biceps femoris and deep to the fascia lata to the back of the knee. It pierces the deep fascia and accompanies the small saphenous vein to the middle of the back of the leg, its terminal twigs communicating with the sural nerve.

a) The **gluteal branches** (*nn. clunium inferiores*), three or four in number, turn upward around the lower border of the Gluteus maximus, and supply the skin covering the lower and lateral part of that muscle.

b) The **perineal branches** (*rr. perineales*)

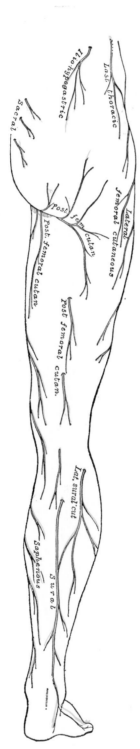

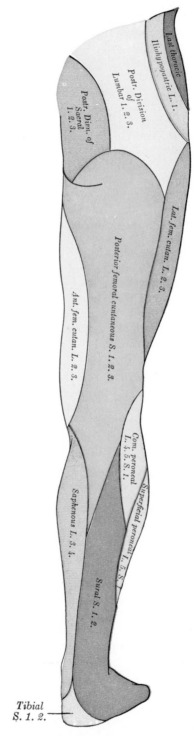

Fig. 12–64.—Cutaneous nerves of right lower limb. Posterior view.

Fig. 12–65.—Diagram of the segmental distribution of the cutaneous nerves of the right lower limb. Posterior view.

(Fig. 12–66) arise at the lower border of the Gluteus maximus, run medially over the origin of the hamstrings toward the groove between the thigh and perineum, and pierce the deep fascia to supply the skin of the external genitalia and adjacent proximal medial surface of the thigh. One long branch, the inferior pudendal (long scrotal nerve) (Fig. 12–70), runs ventrally in the fascia of the perineum to the skin of the scrotum and base of the penis in the male, and of the labium majus in the female, communicating with the posterior scrotal and inferior rectal branches of the pudendal nerve.

c) The **femoral branches** consist of numerous filaments from both sides of the nerve which are distributed to the skin of the posterior and medial sides of the thigh and the popliteal fossa (Fig. 12–64).

d) The **sural branches** are usually two terminal twigs which supply the skin of the posterior leg to a varying extent, and communicate with the sural nerve.

Variations.—When the tibial and peroneal nerves arise separately, the posterior femoral cutaneous also arises from the sacral plexus in two parts. The ventral portion accompanies the tibial nerve and gives off the perineal and medial femoral branches; the dorsal portion passes through the Piriformis with the peroneal and supplies the gluteal and lateral femoral branches. The inferior pudendal branch may pierce the sacrotuberous ligament. The sural branches may be lacking or may extend down as far as the ankle.

7. The **Perforating Cutaneous Nerve** (*n. clunium inferior medialis*) arises from the posterior surface of the second and third sacral nerves. It pierces the caudal part of the sacrotuberous ligament, winds around the inferior border of the Gluteus maximus, and is distributed to the skin over the medial and caudal parts of that muscle.

Variations.—The perforating cutaneous nerve is lacking in one third of the bodies; its place may be taken by a branch of the posterior femoral cutaneous nerve, or by a branch from S 3 and 4, or S 4 and 5. It may pierce the Gluteus maximus as well as the ligament, or instead of piercing the ligament it may accompany the pudendal nerve or go between the

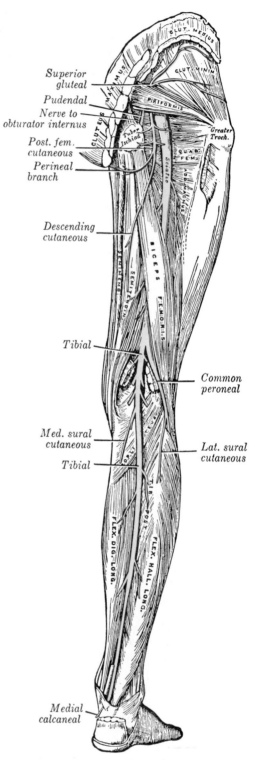

Fig. 12–66.—Nerves of the right lower limb.*
Posterior view.

*N.B.—In this diagram the medial and lateral sural cutaneous are not in their normal position. They have been displaced by the removal of the superficial muscles.

muscle and the ligament. It may arise in common with the pudendal nerve.

The Sciatic Nerve

8. The **sciatic nerve** (*n. ischiadicus; great sciatic nerve*) (Fig. 12–66) is the largest nerve in the body. It supplies the skin of the foot and most of the leg, the muscles of the posterior thigh, all the muscles of the leg and foot, and contributes filaments to all the joints of the lower limb. It is the continuation of the main part of the sacral plexus, arising from L 4, 5, and S 1, 2, 3. It passes out of the pelvis through the greater sciatic foramen, and extends from the inferior border of the Piriformis to the distal third of the thigh, where it splits into two large terminal divisions, the tibial and common peroneal nerves. In the proximal part of its course it rests upon the posterior surface of the ischium, between the ischial tuberosity and the greater trochanter of the femur, and crosses the Obturator internus, Gemelli, and Quadratus femoris. It is accompanied by the posterior femoral cutaneous nerve and the inferior gluteal artery, and is covered by the Gluteus maximus. More distally it lies upon the Adductor magnus, and is crossed obliquely by the long head of the Biceps femoris.

The tibial and common peroneal nerves represent two divisions within the sciatic nerve which are manifest at the origin of the nerve and preserve their identity throughout its length although combined into one large nerve by a common connective tissue sheath. The tibial division takes its origin in the sacral plexus from the ventral divisions of L 4, 5 and S 1, 2, 3; the peroneal division originates from the dorsal divisions of L 4, 5 and S 1, 2. In some bodies these two divisions remain separate throughout their course, no true sciatic nerve being formed.

A. The **branches** of the **sciatic nerve** before it splits into the tibial and common peroneal nerves are as follows:

a) **Articular branches** (*rr. articulares*) arise from the proximal part of the nerve and supply the hip joint, perforating the posterior part of the capsule. They may arise from the sacral plexus.

b) The **muscular branches** (*rr. muscu-lares*) to the hamstrings, viz., the long head of the Biceps femoris, the Semitendinosus, and the Semimembranosus, and to the Adductor magnus come from the tibial division; the branch to the short head of the Biceps comes from the common peroneal.

B. The **Tibial Nerve** (*n. tibialis; internal popliteal nerve*) (Fig. 12–66), the larger of the two terminal divisions of the sciatic nerve, is composed of fibers from the ventral portions of L 4, 5, and S 1, 2, 3. It continues in the same direction as the sciatic nerve, at first deep to the long head of the Biceps, then through the middle of the popliteal fossa covered by adipose tissue and fascia. After crossing the Popliteus, it passes between the heads of the Gastrocnemius and under the Soleus. It remains deep to these muscles down to the medial margin of the tendo calcaneus, along which it runs to the flexor retinaculum, and there divides into the medial and lateral plantar nerves. At its origin it is some distance lateral to the popliteal artery and vein, but as it continues in a straight course it crosses superficial to the vessels in the popliteal fossa, and, lying medial to them, passes with them under the tendinous arch formed by the Soleus. It accompanies the posterior tibial artery, at first being medial but soon crossing the artery and lying lateral to it down to the ankle (Fig. 8–61).

The **branches** of the tibial nerve are as follows:

a) **Articular branches** (*rr. articulares*) supply the knee and ankle joints.

i) Three branches, accompanying the superior and inferior medial genicular, and the middle genicular arteries, pierce the ligaments and supply the **knee joint**. The superior branch is inconstant.

ii) Just proximal to the bifurcation at the flexor retinaculum, an articular branch is given off to the **ankle joint**.

b) **Muscular branches** (*rr. musculares*) are supplied to the muscles of the posterior leg. Branches arising as the nerve lies between the two heads of the Gastrocnemius are distributed to (*a*) both heads of the Gastrocnemius, (*b*) Plantaris, (*c*) Soleus, and (*d*) Popliteus; the latter turns around the distal border and is distributed to the deep surface of the muscle. Arising more distally, either separately or by a common trunk are branches to (*e*) the Soleus, (*f*) Tibialis posterior, (*g*) Flexor digitorum longus, and (*h*) Flexor hallucis longus; the branch to the last muscle accompanies the peroneal artery; that to the Soleus enters the deep surface of the muscle.

c) The **medial sural cutaneous nerve** (*n. cutaneus surae medialis; n. communicans tibialis*) remains superficial in the groove between the two heads of the Gastrocnemius, accompanied by the small saphenous vein, and, at about the middle of the back of the leg, pierces the deep fascia and is joined by the communicating ramus of the lateral sural cutaneous branch of the peroneal nerve, to form the sural nerve.

i) The **sural nerve** (*n. suralis; short saphenous nerve*) (Fig. 12–64) is formed by the union of the medial sural cutaneous nerve and the communicating ramus of the lateral sural cutaneous nerve (peroneal anastomotic) (Fig. 12–64). Lying with the small saphenous vein near the lateral margin of the tendo calcaneus, it continues distally to the interval between the lateral malleolus and the calcaneus, supplying branches to the skin of the back of the leg, and communicating with the posterior femoral cutaneous nerve. The nerve turns anteriorly below the lateral malleolus, and is continued as the **lateral dorsal cutaneous nerve** along the lateral side of the foot and little toe (Fig. 12–58), communicating on the dorsum of the foot with the intermediate dorsal cutaneous nerve, a branch of the superficial peroneal.

d) The **medial calcaneal branches** (*rr. calcanei mediales; internal calcaneal branches*) (Fig. 12–66) perforate the flexor retinaculum, and are distributed to the skin of the heel and medial side of the sole of the foot.

e) The **medial plantar nerve** (*n. plantaris medialis; internal plantar nerve*) (Fig. 12–67), the larger of the two terminal branches of the tibial nerve, accompanies the medial plantar artery. From its origin under the flexor retinaculum it passes deep to the Abductor hallucis, and, appearing between this muscle and the Flexor digitorum brevis, gives off the proper plantar digital nerve to the great toe, and finally divides opposite the bases of the metatarsal bones into three common digital nerves.

The **branches** of the medial plantar nerve are as follows:

i) The **plantar cutaneous branches** pierce the plantar aponeurosis between the Abduc-

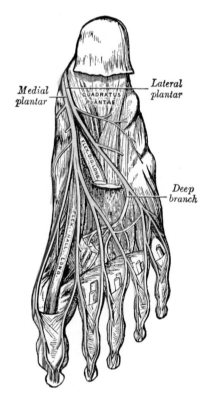

Fig. 12–67.—The plantar nerves.

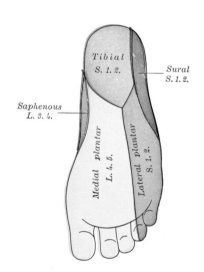

Fig. 12–68.—Diagram of the segmental distribution of the cutaneous nerves of the sole of the foot.

tor hallucis and the Flexor digitorum brevis and are distributed to the skin of the sole of the foot (Fig. 12–68).

ii) **Muscular branches** for the Abductor hallucis and Flexor digitorum brevis arise from the trunk of the nerve and enter the deep surfaces of the muscles. A branch for the Flexor hallucis brevis springs from the proper digital nerve to the medial side of the great toe. A branch for the first Lumbricalis comes from the first common digital nerve.

iii) The **articular branches** supply the joints of the tarsus and metatarsus.

iv) The **proper digital nerve** of the great toe pierces the plantar aponeurosis posterior to the tarsometatarsal joint, sends a branch to the Flexor hallucis brevis, and is distributed to the skin of the medial side of the great toe (Fig. 12–67).

v) The **three common digital nerves** (*nn. digitales plantares communes*) pass between the divisions of the plantar aponeurosis, and each splits into two **proper digital nerves** (*nn. plantares digitales propriae*). Those of the first common digital supply the adjacent sides of the great and second toes; those of the second, the adjacent sides of the second and third toes; those of the third, the adjacent sides of the third and fourth toes. The first common digital gives a twig to the first Lumbricalis. The third communicates with the lateral plantar nerve. Each proper digital nerve gives off cutaneous and articular filaments along the digit, finally terminating in the ball of the toe, and opposite the distal phalanx sends a dorsal branch which is distributed to the structures around the nail.

f) The **lateral plantar nerve** (*n. plantaris lateralis; external plantar nerve*) (Figs. 12–67, 12–68) supplies the skin of the fifth and lateral half of the fourth toes, as well as most of the deep muscles of the foot, its distribution being similar to that of the ulnar nerve in the hand. It passes distally with the lateral plantar artery to the lateral side of the foot, lying between the Flexor digitorum brevis and Quadratus plantae, and in the interval between the former muscle and the Abductor digiti minimi, divides into a superficial and a deep branch.

The **branches** of the lateral plantar nerve are as follows:

i) **Muscular branches** to the Quadratus plantae and the Abductor digiti minimi are given off before the division into superficial and deep branches.

ii) The **superficial branch** (*r. superficialis*) splits into a common digital nerve and a nerve which supplies the proper digital for the lateral side of the little toe, and muscular branches to the Flexor digiti minimi brevis and the two Interossei of the fourth intermetatarsal space. The common digital nerve has a communication with the third common digital branch of the medial plantar nerve and divides into two proper digital nerves which are distributed to the adjacent sides of the fourth and fifth toes.

iii) The **deep branch** (*r. profundus; muscular branch*) accompanies the lateral plantar artery on the deep surface of the tendons of the Flexores longi and the Adductor hallucis, and supplies all the Interossei except those in the fourth intermetatarsal space, the second, third, and fourth Lumbricales, and the Adductor hallucis.

C. The **Common Peroneal Nerve** (*n. peronaeus communis; external popliteal nerve; peroneal nerve*) (Fig. 12–66), the smaller of the two terminal divisions of the sciatic nerve, is composed of fibers from the dorsal portions of the fourth and fifth lumbar, and first and second sacral nerves. It runs obliquely along the lateral side of the popliteal fossa, close to the medial border of the Biceps femoris and between that muscle and the lateral head of the Gastrocnemius, to the head of the fibula. It winds around the neck of the fibula and passes deep to the Peroneus longus, where it divides into the superficial and deep peroneal nerves.

The **branches** of the common peroneal nerve are as follows:

a) The **articular branches** (*rr. articulares*) to the knee are three in number: (*a*) one accompanies the superior lateral genicular artery to the knee, occasionally arising from the trunk of the sciatic; (*b*) one accompanies the inferior lateral genicular artery to the joint; and (*c*) the third (recurrent) articular nerve is given off at the point of division of the common peroneal nerve and accompanies the anterior tibial recurrent artery (Fig. 8–64) through the substance of the Tibialis anterior to the anterior part of the knee (Fig. 12–60, not labeled).

b) The **lateral sural cutaneous nerve** (*n. cutaneus surae lateralis; lateral cutaneous branch*) is distributed to the skin of the posterior and lateral surfaces of the leg (Fig. 12–65) and has an important communicating ramus:

The **communicating ramus** (*n. communicans fibularis; peroneal anastomotic nerve*) (Figs. 12–64, 12–66) arises near the head of the fibula, crosses superficial to the lateral head of the Gastrocnemius, and in the middle of the leg joins the medial sural cutaneous nerve to form the sural nerve.

c) The **deep peroneal nerve** (*n. peroneus profundus; anterior tibial nerve*) (Figs. 12–60, 12–69), arising from the bifurcation of the common peroneal nerve between the fibula and the Peroneus longus, continues deep to the Extensor digitorum longus to the anterior surface of the interosseous membrane. It meets the anterior tibial artery in the proximal third of the leg (Fig. 8–63) and continues with it distally, passes under the extensor retinaculum, and terminates at the ankle in a medial and a lateral branch.

i) **Muscular branches** are supplied in the leg to (*1*) the Tibialis anterior, (*2*) Extensor digitorum longus, (*3*) Peroneus tertius, and (*4*) Extensor hallucis longus.

ii) An **articular branch** is supplied to the ankle joint.

iii) The **lateral terminal branch** (*external or tarsal branch*) passes across the tarsus deep to the Extensor digitorum brevis, and, having become enlarged like the dorsal interosseous nerve at the wrist, supplies the Extensor digitorum brevis. (*1*) Three minute interosseous branches are given off from the enlargement; they supply the tarsal joints and the metatarsophalangeal joints of the second, third, and fourth toes. (*2*) A muscular filament is sent to the second Interosseus dorsalis from the first of these interosseous branches.

iv) The **medial terminal branch** (*internal branch*) accompanies the dorsalis pedis artery along the dorsum of the foot, and, at the first interosseous space, divides into two dorsal digital nerves (Fig. 12–70) which supply adjacent sides of the great and second toes, communicating with the medial dorsal cutaneous branch of the

superficial peroneal nerve. An interosseous branch, given off before it divides, enters the first space, supplying the metatarsophalangeal joint of the great toe and sending a filament to the first Interosseus dorsalis.

d) The **superficial peroneal nerve** (*n. peroneus superficialis; musculocutaneous nerve*) (Figs. 12–69, 12–70) passes distally between the Peronei and the Extensor digitorum longus, pierces the deep fascia in the lower third of the leg, and divides into a medial and intermediate dorsal cutaneous nerve.

i) **Muscular branches** are given off in its course between the muscles to the Peroneus longus and brevis.

ii) **Cutaneous filaments** are supplied to the skin of the lower part of the leg.

iii) The **medial dorsal cutaneous nerve** (*n. cutaneus dorsalis medialis; internal dorsal cutaneous branch*) (Fig. 12–69) passes in front of the ankle joint and divides into two dorsal digital branches. The medial one supplies the medial side of the great toe and communicates with the deep peroneal nerve. The lateral one supplies the adjacent sides of the second and third toes. It also supplies the skin of the medial side of the foot and ankle, and communicates with the saphenous nerve.

iv) The **intermediate dorsal cutaneous nerve** (*n. cutaneus dorsalis intermedius; external dorsal cutaneous branch*) (Fig. 12–69) passes along the lateral part of the dorsum of the foot, supplying the skin of the lateral side of the foot and ankle and communicating with the sural nerve. It terminates by dividing into two dorsal digital branches, one of which supplies the adjacent sides of the third and fourth toes, the other the adjacent sides of the fourth and little toes. Frequently some of the lateral branches of the superficial peroneal nerve are absent, and their places are then taken by branches of the sural nerve.

The Pudendal Plexus

The **pudendal plexus** (*plexus pudendus*) (Figs. 12–63, 12–70) is formed from the anterior branches of the second and third and all of the fourth sacral nerves. It is sometimes considered to be a part of the

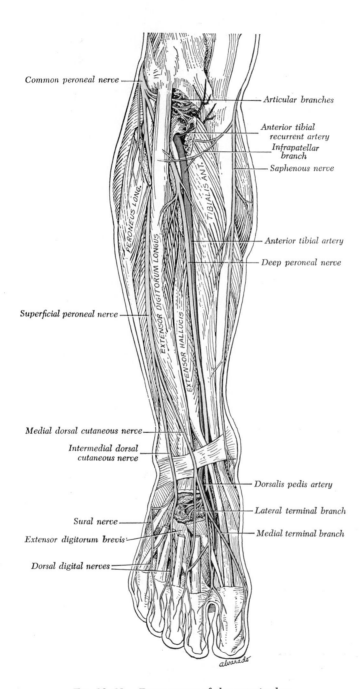

F I G. 12–69.—Deep nerves of the anterior leg.

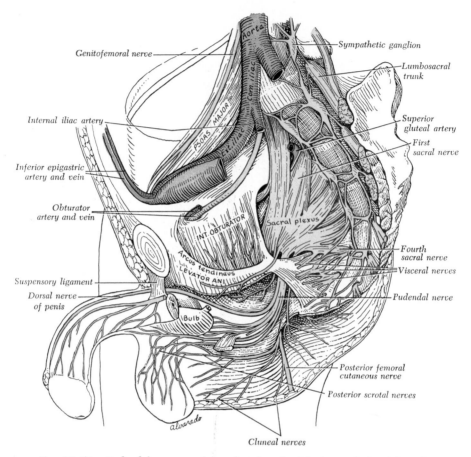

Genitofemoral nerve

Sympathetic ganglion

Lumbosacral trunk

Internal iliac artery

Superior gluteal artery

First sacral nerve

Inferior epigastric artery and vein

Obturator artery and vein

INT. OBTURATOR

Sacral plexus

Arcus tendineus LEVATOR ANI

Fourth sacral nerve

Visceral nerves

Suspensory ligament

Dorsal nerve of penis

Bulb

Pudendal nerve

Posterior femoral cutaneous nerve

Posterior scrotal nerves

Cluneal nerves

alvarado

FIG. 12–70.—Pudendal nerve, and sacral and pudendal plexus of the right side.

sacral plexus. It lies in the posterior hollow of the pelvis, on the ventral surface of the Piriformis.

The **branches** of the pudendal plexus are as follows:

1. The **Visceral Branches** (*rr. viscerales*) arise from the second, third, and fourth sacral nerves, and contain parasympathetic visceral efferent preganglionic fibers which form synapses with the cells in the small scattered ganglia located in or near the walls of the pelvic viscera, and visceral afferent fibers from the pelvic organs. These nerves join branches of the hypogastric plexus of the sympathetic and the sympathetic chain to form the pelvic plexus, which lies in the deeper portion of the pelvic subserous fascia (Ashley and Anson '46).

The various elements of the pelvic plexus cannot be followed by dissection, but the destination of the sacral parasympathetic fibers has been traced by clinical observation and animal experimentation as follows:

a) The **branches** to the **bladder, prostate, and seminal vesicles** approach these organs from their posterior and lateral sides, and terminate in groups of ganglion cells in the pelvic plexus or in the walls of the organs. The postganglionic fibers are efferents to the muscles and glands except the sphincters, to which they are inhibitory.

b) The **branches** to the **uterus** reach the ganglia in the uterine plexus. The postganglionic fibers are inhibitory except during pregnancy when their function is said to be reversed.

c) The **branches** to the **external genitalia** leave the pelvic plexus to join the pudendal nerve, and are distributed through its branches to the corpora cavernosa, causing active dilatation of the cavernous blood sinuses.

d) The **branches** to the **alimentary tract** consist of (i) filaments which go directly to the rectum from the pelvic plexus, and (ii) fibers which pass through the hypogastric plexus, the hypogastric nerves, and the inferior mesenteric plexus to reach the descending and sigmoid colon. These are preganglionic fibers which have synapses with the cells in the myenteric and submucous plexuses and are efferent to this part of the intestine except for the sphincter ani internus, to which they are inhibitory.

2. The **Muscular Branches** (Fig. 12–61), derived mainly from the fourth (sometimes from the third and fifth) sacral, enter the pelvic surfaces of the Levator ani and Coccygeus; the nerve to the Sphincter ani externus (perineal branch) reaches the ischiorectal fossa by piercing the Coccygeus or by passing between it and the Levator.

3. The **Pudendal Nerve** (*n. pudendus; internal pudic nerve*) (Figs. 8–53, 12–70) arises from the second, third, and fourth sacral nerves, and, passing between the Piriformis and Coccygeus, leaves the pelvis through the distal part of the greater sciatic foramen. It then crosses the spine of the ischium, and reenters the pelvis through the lesser sciatic foramen. It accompanies the internal pudendal vessels along the lateral wall of the ischiorectal fossa in a tunnel formed by a splitting of the obturator fascia (*Alcock's canal*), and, as it approaches the urogenital diaphragm, splits into two terminal branches.

The **branches** of the pudendal nerve are as follows:

a) The **inferior rectal nerve** (*inferior hemorrhoidal nerve*) arises from the pudendal before its terminal division or occasionally directly from the sacral plexus, and coursing medially across the ischiorectal fossa with the inferior rectal vessels, breaks up into branches which are distributed to the Sphincter ani externus, and the integument around the anus. Branches of this nerve communicate with the perineal branch of the posterior femoral cutaneous and with the posterior scrotal nerves.

b) The **perineal nerve** (*n. perinei*) (Fig. 12–70), the larger and more superficial of the two terminal branches of the pudendal, accompanies the perineal artery and at the urogenital diaphragm divides into superficial and deep branches.

i) The **superficial branches** of the perineal nerve are two in number, the medial and lateral posterior scrotal (or labial) nerves (*nn. scrotales [labiales] posteriores*). They pierce the fascia of the urogenital diaphragm, and run ventrally along the lateral part of the urogenital triangle in company with the posterior scrotal branches of the perineal artery (Fig. 8–53). They are distributed to the skin of the scrotum in the male, or of the labium majus in the female, and the lateral branch communicates with the perineal branch of the posterior femoral cutaneous nerve.

ii) The **deep branch** is mainly muscular, supplying branches to the Transversus perinei superficialis, Bulbocavernosus, Ischiocavernosus, Transversus perinei profundus and Sphincter urethrae membranacea. The nerve to the bulb is given off from the nerve to the Bulbocavernosus, pierces this muscle, and supplies the corpus cavernosum urethrae and the mucous membrane of the urethra.

c) The **dorsal nerve** of the **penis** (*n. dorsalis penis*) (Fig. 12–70), the deeper terminal branch of the pudendal nerve, accompanies the internal pudendal artery along the ramus of the ischium, and then runs ventrally along the margin of the inferior ramus of the pubis, lying between the superficial and deep layers of fascia of the urogenital diaphragm. Piercing the superficial layer, it gives a branch to the corpus cavernosum penis, and continuing forward in company with the dorsal artery of the penis between the layers of the suspensory ligament, it runs along the dorsum of the penis, and is distributed to the skin of that organ, ending in the glans penis.

The **dorsal nerve** of the **clitoris** (*n. dorsalis clitoridis*) is smaller and has a distribution corresponding to that of the penis in the male.

The **Coccygeal Plexus** (*plexus coccygeus*) (Fig. 12–63) is formed by the coccygeal nerve with communications from the fourth and fifth sacral nerves. From this delicate plexus, a few fine filaments, the Anococcygeal Nerves, pierce the sacrotuberous ligament and supply the skin in the region of the coccyx.

THE VISCERAL NERVOUS SYSTEM

The **Visceral Nervous System** or visceral portion of the peripheral nervous system (*vegetative nervous system; involuntary nervous system; major sympathetic system; plexiform nervous system*) comprises the whole complex of fibers, nerves, ganglia, and plexuses by means of which impulses are conveyed from the central nervous system to the viscera and from the viscera to the central nervous system. It has the usual two groups of fibers necessary for reflex connections: (*a*) afferent fibers, receiving stimuli and carrying impulses toward the central nervous system, and (*b*) efferent fibers, carrying impulses from the appropriate centers to the active effector organs, which, in this instance, are the non-striated muscle, cardiac muscle, and glands of the body.

Attention is called to a new use of the term sympathetic system in this edition. In former editions it was used for the visceral nervous system and included both afferent and efferent, autonomic and parasympathetic components. Although the central nervous system is involved in all nervous or reflex control of the viscera, with the possible exception of the enteric plexus, it is more convenient for the purposes of description to separate the peripheral from the central portions. It must be emphasized, however, that the separation is artificial and should not be carried over into physiological considerations. In the present edition it is used in the restricted meaning of the thoracolumbar division of the visceral efferent system. The terms sympathetic afferent and autonomic afferent will be replaced by visceral afferent.

THE VISCERAL AFFERENT FIBERS

The **visceral afferent fibers** cannot be separated into a morphologically independent system because, like the somatic sensory fibers, they have their cell bodies situated in the sensory ganglia of the cerebrospinal nerves. The distinction between somatic and visceral afferent fibers is one of peripheral distribution rather than one of fundamental anatomical and physiological significance. The visceral afferents, however, commonly have modalities of sensation which are different from those of somatic afferents, and most of them are either vaguely localized or have no representation in consciousness. The visceral efferent fibers make reflex connections with both somatic and visceral afferents, and somatic efferents may have reflex connections with visceral afferents. The number and extent of the visceral afferents are not clearly established, and the peripheral processes reach the ganglia by various routes. Many traverse the branches and plexuses of the autonomic system, most of them accompany blood vessels for at least a part of their course, and a certain number run in the cerebrospinal nerves.

Referred Pain.—Although many, perhaps most, of the physiological impulses carried by visceral afferent fibers fail to reach consciousness, pathological conditions or excessive stimulation may bring into action those which carry pain. The central nervous system has a poorly developed power of localizing the source of such pain, and by some mechanism not clearly understood, the pain may be referred to the region supplied by the somatic afferent fibers whose central connections are the same as those of the visceral afferents. For example, the visceral afferents from the heart enter the upper thoracic nerves, and impulses traversing them may cause painful sensations in the axilla, down the ulnar surface of the arm, and in the precordial region. The study of clinical cases of referred pain has been very useful in tracing the path of afferent fibers from the various viscera, and a knowledge of these paths may be of great assistance to the diagnostician in locating a pathological process.

The visceral afferent fibers are summarized below; for more complete descriptions of the individual nerves mentioned, consult the accounts given elsewhere.

The Head.—Visceral afferents from endings on the peripheral blood vessels of the face and scalp probably accompany the branches of the external carotid artery to the superior cervical ganglion, and through communicating rami to the spinal nerves. Afferents on the blood vessels of the brain and meninges may accompany the branches of the internal carotid and vertebral arteries, passing through the upper cervical spinal nerves or possibly the ninth and tenth cranial nerves (Christensen *et al.* '52).

The Nose and Nasal Cavity.—There is some evidence that a few visceral afferent fibers from the nose are brought to the brain by the nervus terminalis (Larsell '50). Others traverse the branches of the spheno-palatine and palatine nerves, to reach the facial nerve through the Vidian and greater superficial petrosal nerves.

The Mouth and Pharynx.—The visceral afferents from the mouth, pharynx, and salivary glands pass through the pharyngeal plexus to the glossopharyngeal, vagus, and facial nerves.

The Neck.—Visceral afferents from the larynx, trachea, esophagus, and thyroid gland are carried by the vagus or reach the sympathetic trunk through the pharyngeal plexus and pass through rami communicantes to the cervical or upper thoracic nerves.

The carotid sinus nerve carries the visceral afferents from the pressoreceptor endings in the carotid sinus through the glossopharyngeal nerve. The chemoreceptor afferents from the carotid body reach the vagus nerve through branches of the pharyngeal or superior laryngeal nerves.

The Thorax.—The visceral afferents from the thoracic wall and parietal pleura join the intercostal nerves, after following the arteries for variable distances, and thus enter the spinal ganglia. Those from the parietal pericardium join either the phrenic nerve or the intercostal nerves.

Visceral afferents from the heart and the origins of the great vessels enter the cardiac plexus and either join the branches of the vagus or reach the upper thoracic spinal nerves and their ganglia by way of sympathetic branches, the sympathetic trunk, and rami communicantes.

The aortic bodies (glomera aortica) are chemoreceptors similar to the carotid bodies; their afferent fibers run in the right vagus and have cell bodies in the inferior ganglion (Hollinshead '39). The supracardial bodies (aortic paraganglia) are also chemoreceptors with their afferent fibers in the left vagus and cell bodies in the inferior ganglion (Hollinshead '40).

The Depressor Nerve.—In some animals, the afferent fibers of the vagus from the heart and great vessels are contained in a separate nerve whose stimulation depresses the activity of the heart. In man, these fibers are probably contained in the cardiac branches of the recurrent nerves (Mitchell '53).

Visceral afferents from the lungs, bronchi, and pulmonary pleura, through the pulmonary plexuses, reach either the vagus nerve, or, through the sympathetic branches and rami communicantes, the spinal nerves and their ganglia.

The Abdomen.—Visceral afferent fibers from the abdominal wall and parietal peritoneum probably accompany the arteries in part of their course and finally, through the spinal nerves, reach the spinal ganglia. The myelinated fibers from the Pacinian corpuscles in the mesentery and about the pancreas at its base run in the thoracic splanchnic nerves, then through the sympathetic trunk, and finally over the white rami communicantes to the spinal nerves and ganglia (Sheehan '33).

Visceral afferent fibers from the stomach, small intestine, cecum, appendix, ascending and transverse colon, liver, gallbladder, bile ducts, pancreas, and suprarenals traverse the celiac plexus and its secondary plexuses and branches, mainly accompanying the arteries, pass through the splanchnic nerves, the sympathetic trunk, and rami communicantes to reach the spinal nerves and ganglia. Some of these afferents may enter the vagus nerve (Mitchell '53).

The visceral afferents from the kidney, ureter, testis and ductus deferens, ovary and uterine tube traverse the renal and celiac plexuses or parts of their secondary plexuses, pass through the lower thoracic and upper lumbar splanchnic nerves to the sympathetic trunk, and thence through white rami communicantes to the spinal nerves and ganglia (Christensen et al. '51).

The Pelvis.—The visceral afferent fibers from the descending colon, sigmoid, and rectum traverse the pelvic plexus, hypogastric nerves and plexus, the inferior mesenteric plexus, celiac plexus, and lumbar splanchnic nerves on their way to the sympathetic trunk, white rami communicantes, and spinal nerves and ganglia. Others from the rectum pass through the pelvic plexus into the visceral branches of the second, third, and fourth sacral nerves and their ganglia.

Visceral afferents from the **bladder, prostate, seminal vesicles,** and **urethra** pass through the pelvic plexus and through the hypogastric nerves and plexuses, splanchnic nerves, and sympathetic rami, into the lumbar ganglia or through the visceral branches of the sacral nerves into the sacral ganglia.

Visceral afferents from the **uterus** traverse the pelvic plexus, hypogastric nerves and plexus, lumbar splanchnic nerves, sympathetic trunk, rami communicantes, and lumbar spinal nerves and their ganglia.

Visceral afferent fibers from the **external genitalia** pass through either the pelvic plexus or the pudendal nerve and reach the sacral nerves and their ganglia.

The Upper Extremity.—Visceral afferent fibers accompany the peripheral blood vessels for some distance, but may join the larger branches of the brachial plexus, and reach the dorsal root ganglia through the spinal nerves, or they may follow the paths of the sympathetic fibers and reach the dorsal ganglia, especially the first two or three thoracic, through the white rami communicantes.

The Lower Extremity.—Visceral afferents accompany the peripheral vessels and the femoral artery to the aortic plexus, then through the lumbar splanchnic nerves to the rami communicantes, spinal nerves and ganglia. Others may join the tibial or peroneal nerves and traverse the sacral and lumbar sympathetic trunk and rami communicantes to reach the lumbar nerves and their ganglia (Kuntz '51).

THE AUTONOMIC OR VISCERAL EFFERENT SYSTEM

The **Visceral Efferent** portions of the peripheral nervous system are combined into a morphological and physiological entity called the **Autonomic Nervous System** (*Systema Nervosum Autonomicum*). The fundamental morphological difference between the visceral and somatic motor systems is that two neurons are required to transmit an impulse from the central nervous system to the active effector organ in the viscera, whereas only a single neuron is required to carry an impulse from the central nervous system to a skeletal muscle fiber. As the name autonomic implies, this system has a certain amount of independence because, in most individuals, it is not under direct voluntary command. It is controlled by neurons within the central nervous system, nevertheless, and is connected with the latter at various levels. The enteric plexus is the only portion of the visceral system which seems to carry out reflex responses without involving the central nervous system (Kuntz '53).

The **Autonomic Nervous System** (*systema nervosum autonomicum*) is composed of two divisions or systems which differ from each other morphologically and which are for the most part antagonistic to each other physiologically. The morphological differences have to do (*a*) with the manner in which the two systems are connected with the central nervous system and (*b*) with the location of their ganglia. The sympathetic or thoracolumbar system is connected with the central nervous system through the thoracic and upper lumbar segments of the spinal cord, and its ganglia tend to be placed near the spinal column rather than near the viscera innervated. The parasympathetic or craniosacral system is connected with the central nervous system through certain of the cranial nerves and through the middle three sacral segments of the spinal cord, and its ganglia tend to be placed peripherally near the organs innervated.

The sympathetic and parasympathetic systems both innervate many of the same organs, and in this double innervation the two systems are usually antagonistic to each other physiologically. No consistent rule can be given for the effect of each, but in general the sympathetic system mobilizes the energy for sudden activity such as that in rage or flight; for example, the pupils dilate, the heart beats faster, the peripheral blood vessels constrict and the blood pressure rises. The parasympathetic system aims more toward restoring the reserves; for example, the pupils contract, the heart beats more slowly, and the alimentary tract and its glands become active.

The two systems frequently travel together, especially in the thorax, abdomen,

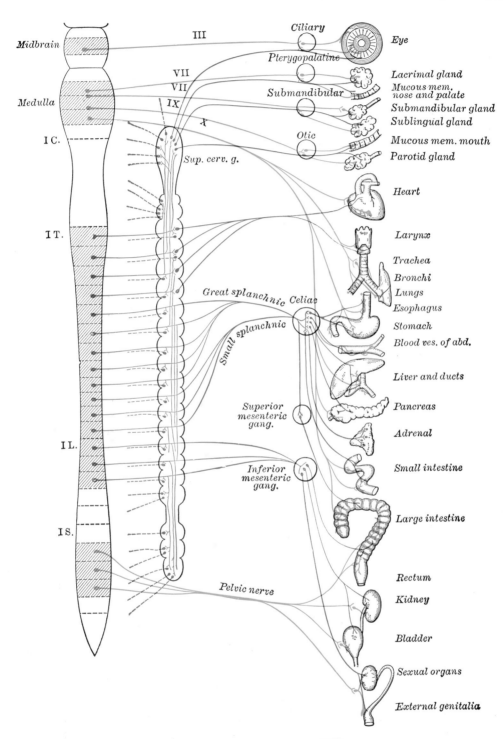

Fɪɢ. 12–71.—Diagram of efferent autonomic nervous system. *Blue,* cranial and sacral outflow, parasympathetic. *Red,* thoracolumbar outflow, sympathetic. ----------, postganglionic fibers to spinal and cranial nerves to supply vasomotors to head, trunk and limbs, motor fibers to smooth muscles of skin and fibers to sweat glands. (Modified after Meyer and Gottlieb.) This is only a diagram and does not accurately portray all the details of distribution.

and pelvis, with the result that extensive plexuses are formed which contain the fibers of both. The arrangement of the bundles within these plexuses is very complicated and the identity of individual fibers cannot be determined with certainty. For the purposes of description, therefore, a third subdivision of the autonomic system is recognized—the great autonomic plexuses.

PARASYMPATHETIC SYSTEM

The parasympathetic system is the **craniosacral portion** of the **autonomic nervous system** and contains visceral efferent fibers which originate in certain cranial nerves and in the sacral portion of the spinal cord.

A. The Cranial Portion of the Parasympathetic System

The **cranial outflow** includes fibers in the oculomotor, facial, glossopharyngeal, and vagus nerves. These nerves have been described in previous pages and the details will be repeated here only as far as they apply to the visceral efferent fibers.

The **Oculomotor Nerve** contains efferent fibers for the non-striated muscle making up the Ciliaris and Sphincter pupillae muscles of the eyeball (for diagram, see Fig. 12–5). The preganglionic fibers arise from cells in the Edinger-Westphal nucleus located in the anterior part of the oculomotor nucleus in the tegmentum of the midbrain. They run in the inferior division of the oculomotor nerve to the ciliary ganglion (Fig. 12–6) and there form synapses with the ganglion cells. The postganglionic fibers proceed in the short ciliary nerves to the eyeball, penetrate the sclera, and reach the muscles named above.

The **Facial Nerve** contains efferent fibers for the lacrimal gland, the submandibular and sublingual glands, and the many small glands in the mucous membrane of the nasal cavity, palate, and tongue (for diagrams, see Figs. 12–16, 12–17). The preganglionic fibers arise from cells in the superior salivatory nucleus in the reticular formation, dorsomedial to the facial nucleus in the pons, and leave the brain in the nervus intermedius.

1. **Pterygopalatine Ganglion** (Fig. 12–12).—Certain of the preganglionic fibers branch from the facial nerve at the geniculum via the greater petrosal nerve, course through the pterygoid canal, and terminate by forming synapses with the cells in the pterygopalatine ganglion. Some of the postganglionic fibers reach the **lacrimal gland** via the maxillary, zygomatic, and lacrimal nerve route; others accompany the branches of the maxillary nerve to the **glands in the mucous membrane of the nasal cavity and nasopharynx**; and still others accompany the palatine nerves to the **glands of the soft palate, tonsils, uvula, roof of the mouth, and upper lip.**

2. **Submandibular Ganglion** (Fig. 12–12).—Other preganglionic fibers leave the facial nerve in the chorda tympani and with it join the lingual nerve to reach the submandibular ganglion. They form synapses in the ganglion or with groups of ganglion cells in the substance of the gland. The postganglionic fibers form the secretomotor supply to the **submandibular** and the **sublingual glands.**

3. Filaments **Communicating** with the **Otic Ganglion** may contribute preganglionic fibers which join those from the glossopharyngeal nerve to supply the parotid gland.

The **Glossopharyngeal Nerve** contains efferent fibers for the parotid gland and small glands in the mucous membrane of the tongue and floor of the mouth (for diagram see Fig. 12–21). The preganglionic fibers arise in the inferior salivatory nucleus in the medulla oblongata, traverse the tympanic and lesser petrosal nerves, and form synapses in the **otic ganglion** (Fig. 12–12). Most of the postganglionic fibers join the auriculotemporal nerve, and are distributed with its branches to the **parotid gland,** providing its secretomotor fibers. Other postganglionic fibers are said to supply the glands of the mucous membrane of the tongue and floor of the mouth.

The **Vagus Nerve** contains efferent fibers for the non-striated muscle and glands of the bronchial tree, of the alimentary tract as far as the transverse colon, of the gallbladder and bile ducts, the pancreas, and inhibitory fibers for the heart. The preganglionic fibers arise from cells in the dorsal motor nucleus of the vagus in the medulla oblongata, and run in the vagus nerve and its branches to ganglia situated in or near the organs innervated.

1. The **Heart** (Figs. 7–19, 12–77).—The preganglionic fibers for the heart reach the cardiac plexus by way of the superior and inferior cardiac nerves of the vagus, and are distributed by branches of the plexus to the ganglion cells in the heart wall. The ganglion cells form numerous clusters in the connective tissue of the epicardium on the surface of the atrium, on the auricular appendages, and in the interatrial septum. The postganglionic fibers terminate in relation to the specialized muscular elements in the sinoatrial and atrioventricular nodes, and atrioventricular bundle and its branches as far as the Purkinje fibers (Stotler and McMahon '47).

2. The **Lungs.**—The preganglionic fibers for the bronchi leave the main trunk of the vagus nerve in the thorax as the anterior and posterior bronchial branches. They traverse the anterior and posterior pulmonary plexuses (Figs. 12–24, 12–25), and terminate in the clusters of ganglion cells scattered along the ramifications of the bronchial tree. The postganglionic fibers are distributed to the bronchial musculature and bronchial glands (Larsell '51).

3. The **Alimentary Tract** receives, through various branches, preganglionic fibers which end in the **myenteric plexus** of Auerbach and the **submucous plexus** of Meissner, forming synapses with the ganglion cells scattered in groups throughout these plexuses. The postganglionic fibers are the efferents for the muscular walls and the secreting cells of the tunica mucosa.

a) The preganglionic fibers reach the upper part of the **esophagus** through the recurrent nerve, and the lower part, below the hilum of the lung, through branches from the esophageal plexus (Fig. 12–76).

b) The **stomach** receives an average of four branches from the anterior vagus and six from the posterior vagus. The pylorus and duodenum receive fibers from the hepatic branch of the anterior vagus.

c) The **small intestine**, cecum, appendix vermiformis, ascending and transverse colon receive fibers from the posterior vagus which join the celiac plexus (Fig. 12–80)

and accompany the branches of the superior mesenteric artery (Jackson '49).

d) The **gallbladder** and bile ducts receive preganglionic fibers through the vagus branches in the celiac plexus which traverse the gastrohepatic ligament and terminate in the small clusters of ganglion cells in the wall of the gallbladder and in the region adjacent to the bile ducts. The postganglionic fibers are the efferents for the muscular walls and mucous membrane.

e) The **pancreas** receives fibers through the hepatic branches of the anterior and posterior vagi, and through the branches of the celiac plexus which accompany the arteries supplying this organ.

B. The Sacral Portion of the Parasympathetic System

The cells which give rise to the sacral outflow are in the second, third, and fourth sacral segments of the spinal cord, and pass out with the corresponding sacral nerves. They leave the sacral nerves in the visceral branches and join the pelvic plexus (Fig. 12–81) in the deeper portions of the pelvic subserous fascia. Branches from this plexus contain preganglionic fibers for the scattered ganglia in or near the walls of the various pelvic viscera (Ashley and Anson '46).

a) The **branches** to the **bladder, prostate** and **seminal vesicles** supply efferent fibers to these organs except for the fibers to the sphincter which are inhibitory.

b) The **branches** to the **uterus** and **vagina** reach ganglia in the uterovaginal plexus; the postganglionic fibers being inhibitory except during pregnancy when their function is said to be reversed.

c) The **branches** of the **pelvic plexus** which join the pudendal nerve and are distributed to the **external genitalia** cause active dilatation of the cavernous blood sinuses of the erectile corpora.

The visceral branches of the sacral nerves in laboratory animals are concentrated in a single trunk called the "**nervus erigens**" because of this function, but the term is not directly applicable to the conditions in man because there is no single nerve such as that used for experimentation in the animals.

d) **Branches** from the **pelvic plexus**

containing preganglionic fibers join the **hypogastric nerve,** and through it and the inferior mesenteric plexus are distributed to the **descending** and **sigmoid colon** and **rectum.** They are efferent to this part of the intestine except for the sphincter ani internus, for which they are inhibitory. Small branches from the pelvic plexus go directly to the rectum.

SYMPATHETIC SYSTEM

The **sympathetic system** (*systema nervosum sympathicum*) (Fig. 12–72) receives its fibers of connection with the central nervous system through the **thoracolumbar outflow** of visceral efferent fibers. These fibers are the axons of cells in the lateral column of gray matter in the thoracic and upper lumbar segments of the spinal cord. They leave the cord through the ventral roots of the spinal nerves and traverse short communications to reach the sympathetic trunk, where they may terminate in the chain ganglia of the trunk itself, or they may continue into the collateral ganglia of the prevertebral plexuses. They are the preganglionic fibers, and are mostly of the small myelinated variety (3 μ or less in diameter). The postganglionic fibers are the axons of the cells in these chain and collateral ganglia and are generally unmyelinated. They are distributed to the heart, nonstriated muscle, and glands all over the body, which they reach by way of communications with the cerebrospinal nerves, by way of various plexuses, and by their own visceral branches of distribution.

Variability is a prominent characteristic of the sympathetic system, and, although a description which will correspond with even the majority of individuals is impossible, certain general principles of organization can be recognized. These are incorporated in the account which follows and the general description is supplemented and made specific by giving a few of the common variations. Many of the details concerning the paths taken by the fibers are still unknown, and we must rely heavily upon information obtained from animal experimentation although it may not have been confirmed by clinical observations with human patients.

The **Sympathetic Trunk** consists of a series of ganglia called the central or chain ganglia (Fig. 12–76), connected by intervening cords and extending along the lateral aspect of the vertebral column from the base of the skull to the coccyx. The cranial end of the trunk proper is formed by the superior cervical ganglion (Fig. 12–72), but there is a direct continuation into the head, the internal carotid nerve. The caudal ends of the two trunks converge at the coccyx and may merge into a single ganglion, the ganglion impar. Cross connections between the two cords in the sacral region are frequent, but rarely occur above the fifth lumbar.

The trunk contains, in addition to the ganglia, the preganglionic fibers which are small (1 to 3 μ in diameter) and myelinated, the postganglionic fibers which are mostly unmyelinated, and a smaller number of afferent fibers which are both myelinated (medium, 5 μ, and large, 10 μ) and unmyelinated. All types of fibers may run up or down in the trunk for the distance between two or many segments. The cords intervening between the ganglia are usually single except in the lower cervical region, but doubling is frequent in any part of the trunk although it very rarely extends farther than between two adjacent ganglia. Small collections of ganglion cells occur outside of the major ganglia, and may be microscopic in size, or grossly visible. They are called intermediary ganglia, and are found in the roots or branches of the trunk and even in the spinal nerves close to these communicating rami (Kuntz and Alexander '50).

Ganglia.—The **central** or **chain ganglia** of the sympathetic trunk are rounded, fusiform or irregular in shape, with diameters usually ranging from 1 to 10 mm, but neighboring ganglia may fuse into larger masses and the superior cervical is always larger. They contain multipolar neurons whose processes are the postganglionic fibers.

The **roots** of the **ganglia** of the sympathetic trunk are commonly called the **white rami communicantes** (*rami communicantes albi*) because of the whitish color imparted to them by the preponderance of myelinated fibers which they contain. These myelinated fibers are mainly the small (1 to 3 μ in diameter) preganglionic axons of the

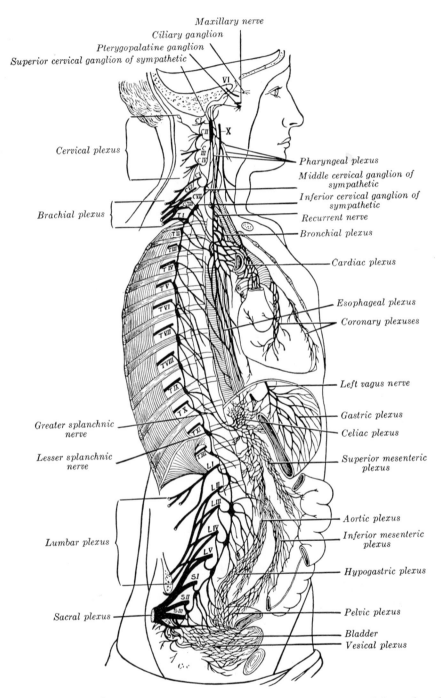

Fig. 12–72.—The right sympathetic chain and its connections with the thoracic, abdominal, and pelvic plexuses. (After Schwalbe.)

thoracolumbar outflow whose cell bodies are in the lateral column of gray matter in the spinal cord. They emerge from the spinal cord with the somatic motor fibers in the ventral roots of the spinal nerves of the thoracic, and the first and second lumbar segments. Many of the preganglionic fibers in each root fail to make synaptic connections in the ganglia at the level of entrance; some travel cranialward to the cervical ganglia, some caudalward to the lumbar and sacral ganglia, and many in the lower thoracic and upper lumbar levels pass out of the trunk and reach the celiac and related collateral ganglia through the splanchnic nerves. One preganglionic fiber may give collaterals to several of the chain ganglia and may terminate about as many as 15 to 20 ganglionic neurons.

The roots of the ganglia or white rami communicantes are to be distinguished from the branches to the spinal nerves or gray rami communicantes with which they are closely related. The white rami leave the anterior primary divisions of the spinal nerves close outside of the intervertebral foramina but are regularly more distal than the gray rami. They contain a number of medium and large myelinated fibers, probably afferents, and some unmyelinated fibers in addition to the small myelinated preganglionic fibers. In the lower thoracic and upper lumbar regions the white rami take an oblique course from the nerve of one segment to the ganglion of the segment below and have been called, therefore, the oblique rami, as opposed to the transverse gray rami (Botar '32). They are more likely to be attached to the intervening cords than the gray rami. The white rami, varying from 0.5 to 2 cm, are longer in the lower thoracic and lumbar region where they are also usually double or triple.

The sympathetic trunk has branches of distribution, containing the postganglionic fibers originating in the ganglia, and branches of communication, containing preganglionic fibers which pass through the trunk, without synapses, on their way to the collateral ganglia in the abdomen.

A. The **branches of distribution** are of several types: (1) branches to the spinal nerves, (2) branches to the cranial nerves, (3) branches accompanying arteries, (4) separate branches to individual organs, and (5) branches to the great autonomic plexuses.

1. The **Branches** to the **Spinal Nerves** are commonly called the **gray rami communi-**

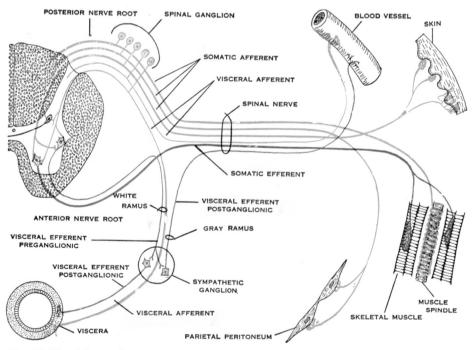

Fig. 12–73.—Scheme showing structure of a typical spinal nerve. Afferent, *blue;* efferent, *red.*

cantes (Fig. 12–73) because the preponderance of unmyelinated fibers gives them a more grayish cast than the roots or white rami, which are adjacent to them in the thoracic area. These branches contribute postganglionic fibers, most of which accompany the cutaneous branches of the spinal nerves to supply the Arrectores pilorum muscles, the sweat glands, and the vasoconstrictor fibers of the peripheral blood vessels.

Gray rami join all the spinal nerves, while the white rami arise only from the thoracic and first two lumbar segments. The gray ramus usually leaves the trunk at a ganglion near the same level as the spinal nerve, and in the thoracic and lumbar regions, where there are both gray and white rami, it is regularly proximal to the white.

2. **Branches** to the **Cranial Nerves** may go directly to the nerves or they may pass through plexuses on blood vessels. Direct communications are encountered to the superior and inferior ganglia of the vagus, the inferior ganglion of the glossopharyngeal, and the hypoglossal nerves.

3. **Branches Accompanying the Arteries** are too numerous to list here, but will be given with the detailed description of the different parts of the system in later pages. Prominent examples are the nerves on the internal and external carotid arteries and their branches.

4. **Branches to Individual Organs** may take an independent course but they commonly pass through plexuses, for example, the cardiac branches, or accompany blood vessels for some distance, for example, those to the bulb of the eye.

5. **Branches to the Cardiac, Pulmonary,** and **Pelvic Plexuses** probably contain postganglionic fibers, but most of the branches from the trunk to the abdominal plexuses probably contain preganglionic fibers.

B. The principal **branches of communication** containing preganglionic fibers are the **splanchnic nerves.** They branch from the ganglia or the trunk of the thoracic and lumbar regions and supply the fibers which are the roots to the celiac, aorticorenal, and mesenteric ganglia. The postganglionic fibers from these ganglia supply the various abdominal and pelvic organs.

Divisions of the Sympathetic System.— The sympathetic nervous system is divided into portions according to topographical position as follows: (1) cephalic, (2) cervical, (3) thoracic, (4) abdominal or lumbar, and (5) pelvic. These parts are not independent and are chosen merely for convenience in description. In addition to these portions, concerned primarily with the sympathetic trunk and ganglia, there are the autonomic plexuses which are described separately.

The Cephalic Portion of the Sympathetic System

The **cephalic portion** (*pars cephalica s. sympathici*) of the sympathetic system contains no part of the ganglionated cord, but is formed largely by a direct cephalic prolongation from the superior cervical ganglion, named the internal carotid nerve. In addition, there are nerves accompanying the vertebral artery and the various branches of the external carotid artery which supply fibers to many structures in the head, such as the pilomotor muscles, sweat glands, and peripheral arteries of the face, and the salivary and other glands.

The **Internal Carotid Nerve** (Fig. 12–23), arising from the cephalic end of the superior cervical ganglion, accompanies the internal carotid artery, and, entering the carotid canal in the petrous portion of the temporal bone, divides into two branches which lie against the medial and lateral aspects of the artery. The lateral branch, the larger of the two, distributes filaments to the carotid artery and forms the internal carotid plexus. The medial branch also distributes filaments to the artery and, following the artery to the cavernous sinus, forms the cavernous plexus.

A. The **Internal Carotid Plexus** (*pl. caroticus internus; carotid plexus*) (Fig. 12–21) is the continuation of the lateral branch of the internal carotid nerve and surrounds the lateral aspect of the artery, occasionally containing a small carotid ganglion. In addition to filaments to the artery, it has the following branches:

1. A **Communication** with the **Trigeminal Nerve** joins the latter at the trigeminal ganglion.

2. A **Communication** joins the **Abducent Nerve** as it lies near the lateral aspect of the internal carotid artery.

3. The **Deep Petrosal Nerve** (*n. petrosus profundus*) (Fig. 12–9) leaves the plexus at the lateral side of the artery, passes through the cartilage of the auditory tube which fills the foramen lacerum, and joins the greater petrosal nerve to form the nerve of the pterygoid canal (Vidian nerve). In the pterygopalatine fossa, the Vidian nerve joins the pterygopalatine ganglion and its contribution through the deep petrosal nerve has been called the sympathetic root of the ganglion, but the sympathetic fibers, already postganglionic, pass through or beside the ganglion without synapses, and are distributed to the glands and blood vessels of the pharynx, nasal cavity, and palate by accompanying the branches of the maxillary nerve (for diagram, see Fig. 12–16).

4. The **Caroticotympanic Nerves** (*nn. caroticotympanici superior et inferior*) are two or three filaments which pass through foramina in the bony wall of the carotid canal and join the tympanic plexus on the promontory of the middle ear (Fig. 13–48).

B. The **Cavernous Plexus** (*pl. cavernosus*) (Fig. 12–12) is the continuation of the medial branch of the internal carotid nerve, and lies inferior and medial to the part of the internal carotid artery enclosed by the cavernous sinus. It communicates with the adjacent cranial nerves and continues along the artery to its terminal branches.

1. The **Communication** with the **Oculomotor Nerve** enters the orbit and joins the nerve at its point of division.

2. A **Communication** joins the **Trochlear Nerve** as it lies in the lateral wall of the cavernous sinus.

3. The **Communication** with the **Ophthalmic** division of the **trigeminal** joins the latter nerve at its inferior surface.

4. **Fibers** to the **Dilator pupillae muscle** of the iris traverse the communication with the ophthalmic nerve, accompany the nasociliary nerve and the long ciliary nerves to the posterior part of the bulb where they penetrate the sclera and run forward to the iris (for diagram, see Fig. 12–4).

5. The **Communication** with the **Ciliary Ganglion** leaves the anterior part of the cavernous plexus and enters the orbit through the superior orbital fissure. It may join the nasociliary nerve and reach the ganglion through the latter's branch to the ganglion, or it may take an independent course to the ganglion.

6. **Filaments** to the **Hypophysis** accompany its blood vessels.

7. The **Terminal Filaments** from the internal carotid and cavernous plexuses continue along the anterior and middle cerebral arteries, and the ophthalmic artery. Fibers on the cerebral arteries may be traced to the pia mater; those on the ophthalmic artery accompany all its branches in the orbit. The filaments on the anterior communicating artery may connect the sympathetic nerves of the right and left sides.

C. The **External Carotid Nerves** (*nn. carotici externi*), which are branches of the superior cervical ganglion, send filaments out along all the branches of the external carotid artery. The filaments on the facial artery join the **submandibular ganglion;** they have formerly been called the sympathetic root of the ganglion, but they pass through it without forming synapses and supply the submandibular and probably the sublingual glands (for diagram, see Fig. 12–17). The network of filaments on the middle meningeal artery gives off the **small deep petrosal nerve,** which has been called the sympathetic root of the **otic ganglion,** but its fibers pass through the ganglion without synapses, some accompanying the auriculotemporal nerve to the parotid gland (for diagram, see Fig. 12–19), others forming the **external superficial petrosal nerve** which is a communication with the geniculate ganglion. Filaments on the facial, superficial temporal, and other arteries which are distributed to the skin supply the Arrectores pilorum muscles, and sweat glands, as well as the muscles constricting the arteries themselves.

The Cervical Portion of the Sympathetic System

The **cervical portion** (*pars cervicalis s. sympathici*) of the sympathetic trunk consists of three ganglia, superior, middle, and inferior, connected by intervening cords

(Fig. 12–74). It is ventral to the transverse processes of the vertebrae, close to the carotid artery, being embedded in the fascia of the carotid sheath itself, or in the connective tissue between the sheath and the Longus colli and capitis. It receives no roots or white rami communicantes from the cervical spinal nerves; its preganglionic fibers enter the trunk through the white rami from the upper five thoracic spinal nerves, mainly the second and third, and travel cranialward in the trunk to the three ganglia. The trunk also contains postganglionic fibers from various sources, and visceral afferent fibers with their cell bodies in the dorsal root ganglia.

The **Superior Cervical Ganglion** (*g. cervicale superius*) (Fig. 12–22), much larger than the other cervical ganglia and usually the largest of all the trunk ganglia, is approximately 28 mm long and 8 mm wide, fusiform in shape, frequently broad and flat, and occasionally constricted into two or more parts. It is embedded in the connective tissue between the carotid sheath and the prevertebral fascia over the Longus capitis, at the level of the second cervical vertebra. It is the cephalic end of the sympathetic chain, and is connected with the middle ganglion caudally by a rather long interganglionic cord. It is believed to be formed by the coalescence of sympathetic primordia from the cranial four cervical segments of the body.

Its **branches** are as follows:

1. The **Internal Carotid Nerve** (described above, page 1013) leaves the cephalic pole of the ganglion, and serves as a direct continuation of the sympathetic trunk into the head.

2. **Communications** with the **Cranial Nerves** are delicate filaments which join (*a*) the inferior ganglion of the glossopharyngeal nerve, (*b*) the superior and inferior ganglia of the vagus, and (*c*) the hypoglossal nerve. The jugular nerve is a filament which passes cranialward to the base of the skull and divides to join the inferior ganglion of the glossopharyngeal and the superior ganglion of the vagus nerve.

3. **Branches** to the **cranial two to four cervical Spinal Nerves** are the **gray rami communicantes** of these nerves. They course lateralward and dorsalward, and have been called the lateral or external branches of the ganglion. The branches to any one nerve are variable and may be multiple or absent.

a) The **branches** to the **first** and **second nerves** are constantly present. Since the spinal nerves in the neck are connected by loops, as parts of the cervical plexus, the branch to the first nerve may join the loop between the first and second nerves, and the branch to the second may join the loop between the second and third nerves. There may be two to four branches to each of these nerves.

b) The **third cervical nerve** receives a branch (ramus) from the ganglion in the majority of individuals, but in many instances it comes from the trunk below the superior ganglion. A branch for the third nerve often forms a loop with a lower branch which then supplies roots to both the third and fourth nerves. Branches may join the third nerve itself or the loop between it and the fourth nerve.

c) The **fourth nerve** receives a branch from the ganglion only occasionally. Its rami frequently arise in common with those of either the third, fifth, or even the sixth nerves, and may come from the trunk or from nerves accompanying the vertebral artery.

4. **Pharyngeal Branches** (*rr. laryngopharyngei*), commonly four to six, leave the medial aspect of the ganglion, and, in their course toward the pharynx, communicate with pharyngeal branches of the glossopharyngeal and vagus nerves opposite the Constrictor pharyngis medius to form the **pharyngeal plexus.** Some of the filaments form a plexus on the lateral wall of the pharynx, others travel in the substance of the prevertebral fascia to the dorsum of the pharynx and form a posterior pharyngeal plexus. Filaments communicate with the superior laryngeal nerve (Braeucker '23).

5. The **Nerves** to the **External Carotid Artery** usually are two relatively large bundles which form a network about the artery and continue as secondary branches to the common carotid artery and to the branches of the external carotid. The latter are described as part of the cephalic sympathetic system (page 1014).

6. The **Intercarotid Plexus** receives one or two branches, either from the ganglion or from the external carotid nerves. They communicate with filaments from the

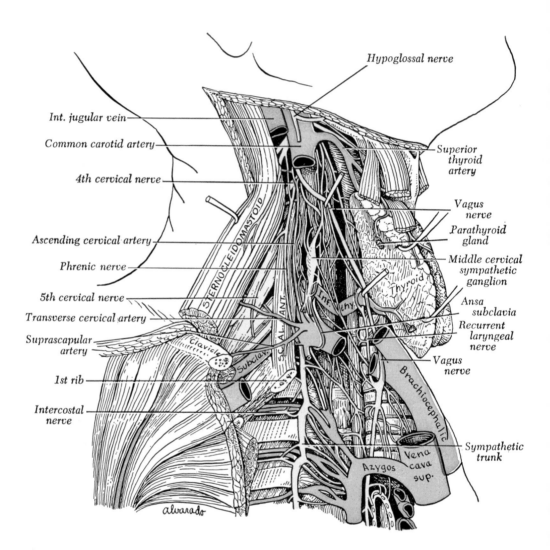

Fig. 12–74.—Cervical portion of the sympathetic nervous system on the right side, with the common carotid artery and internal jugular vein removed and with the vagus nerve and thyroid gland drawn aside. (Redrawn from Töndury.)

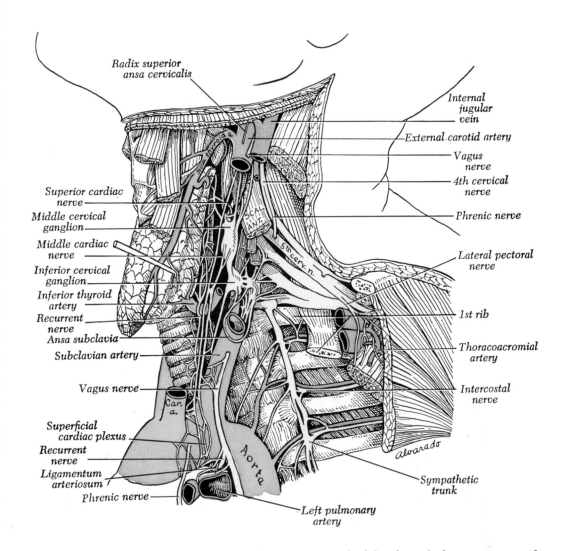

Fig. 12–75.—Cervical portion of the sympathetic system on the left side, with the common carotid artery, internal jugular vein, vagus nerve, and subclavian artery partly removed, and the thyroid gland drawn forward. (Redrawn from Töndury.)

pharyngeal branch of the vagus and the carotid branch of the glossopharyngeal in the region of the carotid bifurcation, and are distributed in the plexus to the carotid sinus and the carotid body, probably serving a vasomotor function.

7. The **Superior Cardiac Nerve** (*n. cardicus superior*) (Figs. 12–74, 12–75) arises by two or three filaments from the ganglion, and occasionally also by a filament from the trunk between the superior and middle ganglia. It runs down the neck in the connective tissue of the posterior layers of the carotid sheath superficial to the Longus colli, and crosses ventral or dorsal to the inferior thyroid artery and recurrent nerve. The course of the nerves on the two sides then differs. The right nerve, at the root of the neck, passes either ventral or dorsal to the subclavian artery, and along the innominate artery to the deep part of the cardiac plexus. The left nerve passes ventral to the common carotid and across the left side of the arch of the aorta, to reach the superficial part of the cardiac plexus. The superior cardiac nerves may communicate with the middle and inferior cardiac sympathetic nerves, with the cardiac branches of the vagus, the external branch of the superior laryngeal nerve, the recurrent nerve, the thyroid branch of the middle ganglion, the nerves on the inferior thyroid artery, and the tracheal and anterior pulmonary plexuses.

The **Middle Cervical Ganglion** (*g. cervicale medium*) (Fig. 12–75), the smallest of the three cervical ganglia, is quite variable in size, form, and position, and may be either absent or double. It probably represents a fusion of the two sympathetic primordia corresponding to the fifth and sixth cervical nerves. When single, the ganglion may have a high position, at the level of the transverse process of the sixth cervical vertebra (the **carotid** or *Chassaignac's tubercle*), or a low position nearer the level of the seventh cervical vertebra (Axford '28; Jamieson *et al.* '52). In the high position it lies on the Longus colli above the cranial bend of the inferior thyroid artery. In the low position it lies in close association with the ventral or ventromedial aspect of the vertebral artery, 1 to 3 cm from the latter's origin. The middle cervical

ganglion has no white ramus communicans; its preganglionic fibers probably leave the spinal cord through the white rami of the second and third thoracic nerves and reach the ganglion through the intervening sympathetic trunk.

Variations.—The middle cervical ganglion was absent in 5, single in 10 and double in 10 of 25 dissections (Pick and Sheehan '46). It was present in 53 and double in two of 100 body halves (Jamieson *et al.* '52), with 64 per cent in the high position. A small ganglion in the low position may have no branches. Several small thickenings may occur along the trunk between the superior and inferior ganglia. The ganglion may be split, surrounding the inferior thyroid artery (Axford '28).

The **intermediate cervical sympathetic ganglion** (of Jonnesco '23) corresponds to a middle ganglion in the low position described above. According to Saccomanno ('43) it is of more constant appearance than a middle ganglion in the high position. A ganglion in the low position has been called the **thyroid ganglion** because of its close relationship to the inferior thyroid artery (Jamieson *et al.* '52). The **vertebral ganglion** is a name given to a ganglion in the low position, on the deep part of the loop of the ansa subclavia, included as part of the stellate ganglionic configuration (Pick and Sheehan '46; Woollard and Norrish '33).

The **branches** of the middle cervical ganglion are as follows:

1. The **Branches** (gray rami communicantes) are constantly supplied to the **fifth** and **sixth** nerves. A ganglion in the high position may send branches to the seventh nerve.

a) The **gray rami** of the **fifth cervical nerve** number from one to three. (1) The most constant arises from the middle ganglion or the trunk just above it, runs upward and laterally across the Scalenus anterior, winds around the carotid tubercle and along a groove in the transverse process of the fifth cervical vertebra to the fifth nerve. It may pierce the Scalenus anterior, it may divide, one branch going to the sixth nerve, or it may be prolonged to give a branch to the fourth or even the third nerve (Axford '28). (2) A ramus present in the majority of individuals leaves the trunk just above the carotid tubercle, pierces the Longus colli either medial or lateral to the vertebral artery and receives a communication from the nerve of the vertebral artery. (3) An inconstant ramus is a branch of that of the sixth nerve which accompanies the vertebral artery.

b) The **rami** of the **sixth cervical nerve** may be from two to four in number, one from the middle and one from the inferior ganglion being constant. (1) A short ramus from the middle ganglion runs cranialward through the Longus colli just above the carotid tubercle to join the sixth nerve as it lies in its groove, medial to the vertebral artery. (2) A long fine branch from the middle ganglion or the trunk just above it crosses the vertebral artery and Scalenus anterior, and joins the nerve lateral to the carotid tubercle, sometimes continuing on to the fifth nerve.

2. The **Middle Cardiac Nerve** (*n. cardiacus medius; great cardiac nerve*) (Fig. 12–75), the largest of the three cardiac nerves in the neck, arises from the middle cervical ganglion, from the trunk between the middle and inferior ganglia, or both. As it descends dorsal to the common carotid artery, it communicates with the superior cardiac nerve and the inferior laryngeal nerve. On the right side, at the root of the neck, it goes either deep or superficial to the subclavian artery, continues along the trachea, communicates with the recurrent nerve, and joins the right side of the deep part of the cardiac plexus. The left nerve enters the thorax between the common carotid and subclavian arteries, and joins the left side of the deep part of the cardiac plexus.

3. **Thyroid Nerves.**—Branches from the middle cervical ganglion form a plexus on the inferior thyroid artery, supply the thyroid gland, join the plexus on the common carotid artery, and may communicate with the inferior laryngeal and external branch of the superior laryngeal nerves (Braeucker '22).

The trunk between the middle and inferior cervical ganglia is constantly double, the two strands enclosing the subclavian artery. The superficial strand is usually much longer than the deep, forms a loop about the artery supplying it with branches, and is called the **ansa subclavia** (of Vieussens) (Fig. 12–75). Since it is a rather constant feature, it may be used to identify and distinguish the middle and inferior ganglia or the components representing them. Occasionally the loop is formed about

the vertebral artery instead of the subclavian, or there may be individual loops about both.

The **Inferior Cervical Ganglion** (*g. cervical inferius*) (Fig. 12–75) is situated between the base of the transverse process of the seventh cervical vertebra and the neck of the first rib, on the medial side of the costocervical artery. In most instances it is incompletely separated from or fused with the first thoracic ganglion, but it will be described as it appears when discrete, and the fused ganglion, called the stellate, also will be described below. It is larger than the middle ganglion, is irregular in shape, and probably represents the fusion of sympathetic primordia corresponding to the seventh and eighth cervical nerves. It has no white ramus, but receives its preganglionic fibers from the thoracic part of the trunk through its connection with the first thoracic ganglion.

Variations.—The inferior cervical ganglion was independent in 5 out of 25 (Pick and Sheehan '46) and in 18 out of 100 cases (Jamieson *et al.* '52). A white ramus has been described joining the eighth cervical nerve and the sympathetic (Pearson '52).

The **branches** of the inferior cervical ganglion are as follows:

1. **Branches** (**gray rami communicantes**) are constantly supplied to the **sixth, seventh, and eighth cervical nerves.**

a) The **sixth cervical nerve** commonly receives branches from the inferior cervical ganglion as well as from the middle cervical ganglion. (1) A constant, rather thick ramus from the deep part of the inferior cervical ganglion runs cranialward along the medial aspect of the vertebral artery, ventral to the vertebral veins and lateral to the Longus colli, and enters the foramen in the transverse process of the sixth cervical vertebra with the vertebral artery. It communicates with the plexus on the vertebral artery and supplies rami for the sixth and seventh, sometimes the fifth or even more cephalic nerves. (2) An inconstant ramus from the inferior ganglion is similar to the last, but pierces the Scalenus anterior instead of passing through the foramen (Axford '28).

b) The **seventh cervical nerve** receives from two to five branches (gray rami) from the inferior cervical ganglion. (1) A constant, well-defined branch 15 to 25 mm long crosses ventral to the eighth cervical nerve, either deep to, superficial to, or piercing the Scalenus anterior. It may be composed of two or three parallel filaments. (2) A constant branch accompanies the vertebral artery and is shared by the sixth nerve. (3) A frequent branch lies close to the vertebral vein, crosses the eighth cervical nerve, to which it may give filaments, and enters the foramen in the seventh cervical vertebra with the vein (Axford '28).

c) The **eighth cervical nerve** receives from two to five rather short branches (averaging 10 mm in length) from the inferior cervical ganglion. (1) A constant, well-defined, thick branch runs cranialward and lateralward, often across the neck of the first rib, and joins the eighth cervical nerve deep to the Scalenus anterior. It is dorsal to the first part of the subclavian artery and is intimately related to the superior intercostal artery. It may be represented by two to four parallel filaments. (2) A constant short thick branch from the upper pole of the inferior ganglion runs vertically cranialward, medial and dorsal to the vertebral artery, a few millimeters lateral to the Longus colli. It passes ventral to the transverse process of the first thoracic vertebra and medial to the first costocentral articulation, and joins the eighth cervical nerve as it emerges from its foramen. (3) A frequent ramus accompanies the vertebral vein with a similar branch to the seventh nerve.

2. The **Inferior Cardiac Nerve** (*n. cardiacus inferior*) (Fig. 12–75) arises from either the inferior cervical ganglion, the first thoracic ganglion, the stellate ganglion, or the ansa subclavia. It passes deep to the subclavian artery and along the anterior surface of the trachea to the deep cardiac plexus. It communicates with the middle cardiac nerve and the recurrent laryngeal nerve, and supplies twigs to various cervical structures (Saccomanno '43).

3. **Vertebral Nerve.**—The branches which accompany the vertebral artery through the vertebral foramina are of considerable size. They join similar branches from the first thoracic ganglion to form the vertebral nerve which continues into the cranial cavity on the basilar, posterior cerebral, and cerebellar arteries. Communications between the vertebral nerve and the cervical spinal nerves frequently serve as rami of these nerves (Christensen *et al.* '52).

The **Junction of the Cervical with the Thoracic portion** of the sympathetic trunk requires special consideration, first, because

the lowest cervical and highest thoracic ganglia are usually fused, and second, because the trunk makes an abrupt change of direction at this point. The cervical portion of the trunk lies upon the ventral aspect of the transverse processes, but is also in a plane ventral to the vertebral bodies on account of the latter's small size and the presence of the Longus colli muscle. The trunk drops back dorsalward as it enters the thorax, winding around the transverse process of the seventh cervical vertebra to reach the neck of the first rib.

The **Stellate Ganglion** (*cervicothoracic ganglion*) (Fig. 12–75) is the name given to the ganglionic mass which results when, as is usually the case, the inferior cervical and first thoracic ganglia are fused. It is quite variable in size and form, occasionally including the middle cervical or the second thoracic ganglia, and is located between the eighth cervical and first thoracic nerves. Its branches and communications are modifications of those which would be found if the component ganglia remained separate. The white ramus communicans, therefore, comes from the first thoracic nerve, or from the second also if the mass includes the second ganglion.

The **branches** of the **stellate ganglion** which supply the gray rami to the spinal nerves include: (*a*) a frequent branch to the sixth nerve, (*b*) a constant branch to the seventh, and (*c*) constant double or multiple branches to the eighth cervical, (*d*) first thoracic, and (*e*) second thoracic nerves. The branch to the vertebral artery is a large one leaving the superior border of the ganglion, and forming the major portion of the vertebral nerve. Other branches are similar to those described for the independent ganglia (Kirgis and Kuntz '42).

Variations.—In 25 bodies, the inferior cervical ganglion was independent in 5, fused with the first thoracic in 17, and with both the first and second thoracic in 3 (Pick and Sheehan '46). A stellate ganglion was present in 82 of 100 (Jamieson *et al.* '52). When the second thoracic is fused with the first, the inferior cervical is more likely to be independent.

Surgical Considerations.—The branches (gray rami) from the stellate ganglion to the eighth cervical and first thoracic nerves carry the bulk of the sympathetic fibers to the upper limb

(Kirgis and Kuntz '42). Other branches also carry fibers, however, and Woollard and Norrish ('33) recommend that the middle cervical and second thoracic ganglia be included in the "stellate complex," and that the description of the independent ganglia be abandoned. The sympathetic rami of the brachial plexus are described in more detail on page 961.

Comparative Anatomical Considerations.— The stellate ganglion in cats includes the inferior cervical and first three thoracic ganglia (Saccomanno '43) and in rhesus monkeys the inferior cervical and first two thoracic ganglia (Sheehan and Pick '43). The function of the stellate ganglion in control of the heart, as it is revealed by animal experimentation with dogs, cats, and monkeys, does not agree entirely with that revealed by clinical observations. A partial explanation of this is provided by the fact that more thoracic ganglia are included in the stellate complex and a much greater bulk of the accelerator nerve arises from the ganglion in these animals than in man.

The Thoracic Portion of the Sympathetic System

The **thoracic portion** of the **sympathetic trunk** (*pars thoracalis s. sympathici*) (Figs. 12–76, 12–77) contains a series of ganglia which correspond approximately to the thoracic spinal nerves, but the coalescence of adjacent ganglia commonly reduces the number to fewer than twelve. The ganglia are oval, fusiform, triangular, or irregular in shape; they lie against the necks of the ribs in the upper thorax; gradually they become more ventrally placed in the lower thorax, and finally they lie at the sides of the bodies of the lowest thoracic vertebrae. The trunk is covered by the costal portion of the parietal pleura, and the interganglionic cords are between the pleura and the intercostal vessels which they cross.

The roots of the ganglia are the white rami communicantes which are supplied by each spinal nerve to the corresponding ganglion or the trunk nearby. They contain predominantly small myelinated fibers (1 to 3 μ in diameter) which leave the spinal cord through the ventral roots of the spinal nerves, and whose cell bodies are in the lateral column of gray matter in the spinal cord. A large number of the preganglionic fibers fail to make synapses in their ganglia of entrance. From the upper five roots,

these fibers take a cranial direction in the trunk and, for the most part, terminate in the cervical ganglia. Many of the fibers in the caudal six or seven rami traverse the trunk for a variable distance and then emerge into the splanchnic nerves which are the roots of the celiac and related ganglia.

The **first thoracic ganglion,** when independent, is larger than the rest, is elongated or crescentic in shape, and, because of the change in direction of the trunk as it passes from the neck into the thorax, the ganglion is elongated dorsoventrally. It lies at the medial end of the first intercostal space, or ventral to the neck of the first rib, medial to the costocervical arterial trunk. It is usually combined with the inferior cervical ganglion into a stellate ganglion (see above) or it may coalesce with the second thoracic ganglion. The second to the tenth ganglia lie opposite the intervertebral disk or the cranial border of the next more caudal vertebra, slightly lower than the corresponding spinal nerve. In the majority of individuals, the last thoracic ganglion, lying on the body of the twelfth vertebra, is larger and by its connection to both the eleventh and twelfth nerves takes the place of these two ganglia.

Variations.—The first thoracic ganglion was independent of the stellate in 5, the second thoracic in 22 out of 25 instances. Fusion occurred between the 3rd and 4th three times, 4th and 5th five times, 5th and 6th once, 6th and 7th once, 7th and 8th four times, 8th and 9th twice, 9th and 10th twice (Pick and Sheehan '46). Small accessory ganglia occur at or near the junction of the communicating rami with the thoracic nerves, especially the upper four. Those at the white rami may provide sympathetic pathways to spinal nerves without traversing the sympathetic trunk (Ehrlich and Alexander '51).

The **branches** of the **thoracic** trunk are of three varieties: gray rami communicantes, visceral branches, and the splanchnic nerves.

1. The **Branches** which form the **Gray Rami Communicantes** are supplied to each spinal nerve. Usually two or three short branches are sent to each corresponding spinal nerve, and occasionally a slender branch reaches the next more caudal gan-

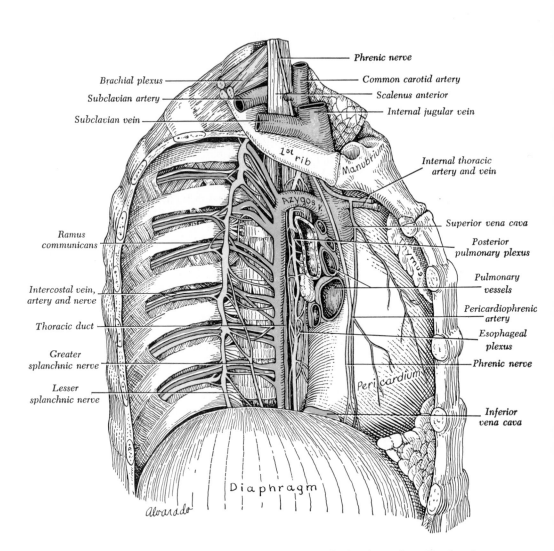

FIG. 12–76.—The mediastinum from the right side. (Redrawn from Töndury.)

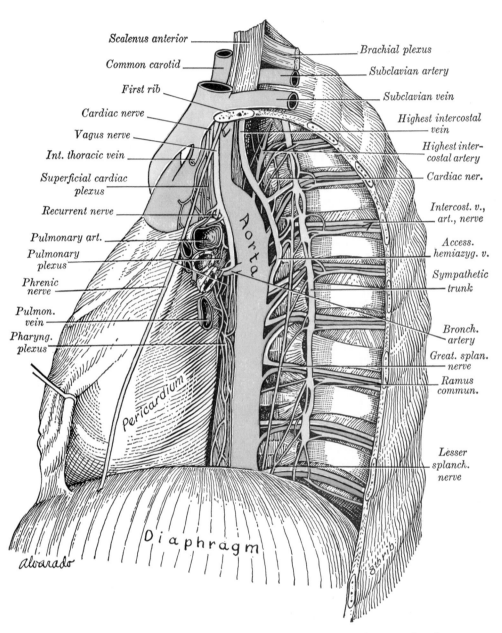

Scalenus anterior

Common carotid

First rib

Cardiac nerve

Vagus nerve

Int. thoracic vein

Superficial cardiac plexus

Recurrent nerve

Pulmonary art.

Pulmonary plexus

Phrenic nerve

Pulmon. vein

Pharyng. plexus

Brachial plexus

Subclavian artery

Subclavian vein

Highest intercostal vein

Highest intercostal artery

Cardiac ner.

Intercost. v., art., nerve

Access. hemiazyg. v.

Sympathetic trunk

Bronch. artery

Great. splan. nerve

Ramus commun.

Lesser splanch. nerve

Aorta

Pericardium

Diaphragm

alvarado

8th rib

Fig. 12–77.—The mediastinum from the left side. (Redrawn from Töndury.)

glion. When there are two or more branches to a single nerve, one branch may go to the anterior and one to the posterior primary division of the nerve. When the branches go only to the anterior primary division, the fibers turn back within the nerve to reach the posterior division (Dass '52). The gray rami are regularly proximal and transverse, the white rami distal and oblique.

2. **Visceral branches** are supplied to the cardiac, pulmonary, esophageal and aortic plexuses.

a) **Branches** to the **cardiac plexus** from the first five thoracic ganglia are variable in number both in different individuals and between the two sides in the same individual, and may be fifteen or twenty in some cases. The larger ones usually come from ganglia, smaller ones from intervening cords. They course medially close to the intercostal artery and vein, usually between them and in the same connective tissue sheath, supplying filaments to them and communicating with the esophageal and pulmonary plexuses. From the right side they approach the deep part of the plexus, between the esophagus and the lateral aspect of the aorta. From the left side, they pass dorsal to the aorta and approach the deep part of the plexus from the right side (Kuntz and Morehouse '30). Branches from the third, fourth, and fifth ganglia are more abundant than the first two and branches may come from the sixth and seventh, but the latter usually enter the aortic network. The cross-sectional area of the thoracic cardiac nerves, as they enter the plexus, is twice as large as that of the cervical cardiac nerves (Saccomanno '43).

b) Delicate **esophageal branches** from several of the thoracic ganglia, both upper and lower, may follow the intercostal vessels to the esophagus and join the plexus formed by the vagus, or filaments may be supplied by the cardiac and aortic plexuses, or by the splanchnic nerves.

c) The **posterior pulmonary plexus** receives, from the second, third, and fourth ganglia, twigs which follow the intercostal arteries to the hilum of the lung.

d) **Branches** to the **aortic network** come from the last five or six thoracic ganglia, from the cardiac plexus, and from the splanchnic nerves. Branches from these bundles accompany the branches of the artery, and probably supplement the splanchnic nerves.

3. The **Splanchnic Nerves** arise from the caudal six or seven thoracic and first lumbar ganglia. They are not, strictly speaking, true branches of the ganglia, since they contain but a small number of postganglionic fibers. They are composed principally of myelinated fibers, and accordingly have a whitish color and firm consistency similar to those of the somatic nerves. The small myelinated fibers, 1 to 3 μ in diameter, which predominate are the preganglionic fibers which pass through the chain ganglia without synapses, to become the roots of the celiac and related ganglia. There are appreciable numbers of large myelinated fibers also which are probably visceral afferents with their cell bodies in the dorsal root ganglia of the spinal nerves. Many of the fibers of all types probably come from spinal cord segments higher than the ganglia from which the branches arise.

a) The **Greater Splanchnic Nerve** (*n. splanchnicus major*) (Figs. 12–76, 12–77) is formed by contributions from the fifth (or sixth) to the ninth (or tenth) thoracic ganglia, which leave the ganglia in a medial direction and angle across the vertebral bodies obliquely in their caudalward course. They are combined into a single nerve which pierces the crus of the diaphragm and, after making an abrupt bend or loop ventralward, ends in the celiac ganglion by entering the lateral border of its principal mass (Fig. 12–78). A small splanchnic ganglion occurs commonly in the nerve at the level of the eleventh or twelfth thoracic vertebra; it is considered to be part of the celiac ganglion formed by cells which failed to migrate as far as the large ganglion during embryonic development. Preganglionic fibers to the suprarenal glands are conveyed by the splanchnic nerves and pass through the celiac plexus without synapses in the ganglion.

b) The **Lesser Splanchnic Nerve** (*n. splanchnicus minor*) (Fig. 12–77) is formed by branches of the ninth and tenth thoracic ganglia, or from the cord between them. It pierces the crus of the diaphragm with the greater splanchnic nerve and ends in the aorticorenal ganglion.

c) The **Lowest Splanchnic Nerve** (*n. splanchnicus imus; least splanchnic nerve*), when present, is a branch of the last thoracic ganglion or of the lesser splanchnic nerve. It passes through the diaphragm with the sympathetic trunk and ends in the renal plexus.

Variations.—The uppermost branch to the splanchnics in 25 dissections was the 4th, once; 5th, twice; 6th, eleven times; 7th, seven times; 8th, four times (Pick and Sheehan '46). Filaments from the upper thoracic and stellate ganglia, from the cardiac nerves, or from the branches to the pulmonary and aortic plexuses sometimes continue down to join the celiac plexus and have been considered to be a fourth splanchnic nerve. Lumbar splanchnic nerves are described below.

The Abdominal Portion of the Sympathetic System

The **abdominal portion** of the **sympathetic trunk** (*pars abdominalis s. sympathici; lumbar portion of the ganglionated cord*) (Fig. 12–79) is situated ventral to the bodies of the lumbar vertebrae, along the medial margin of the Psoas major. The cord connecting the last thoracic and first lumbar ganglia bends ventrally as it passes under the medial lumbocostal arch of the diaphragm, bringing the trunk rather abruptly into its ventral relationship with the lumbar vertebrae. The left trunk is partly concealed by the aorta, the right by the inferior vena cava.

The **lumbar ganglia** have no fixed pattern. The number varies from two to six, with four or five occurring in three-fourths of the trunks, but massive fusions are frequent and two examples with four ganglia may bear no resemblance to each other. Although the five individual lumbar ganglia should not be expected in any particular instance, each one occurs with sufficient frequency to make an anatomical description possible. The numbering of the ganglia is based upon the spinal nerves with which they are connected as well as upon the relationship to the vertebrae (Pick and Sheehan '46).

The **roots of the lumbar ganglia** (*white rami communicantes*) are found only as far as the second lumbar spinal nerve, the caudal limit of the thoracolumbar outflow.

The preganglionic fibers for the rest of the lumbar, and the sacral and coccygeal ganglia run caudally in the trunk, mainly from these first two lumbar roots. One or two roots (white rami) are supplied to each of the first three lumbar ganglia (or their representatives in fused ganglia) by the spinal nerve one segment above; the roots take an oblique course caudalward while the branches to the spinal nerves (gray rami) take a transverse course (Botar '32). Thus the twelfth thoracic nerve sends roots to the first lumbar ganglion, the first lumbar nerve to the second ganglion, and the second nerve to the third ganglion.

The ganglia of the lumbar trunk, when independently represented, lie on the bodies of the corresponding vertebrae or the intervertebral disks caudally. The first ganglion is close to or partly concealed by the medial lumbocostal arch. The ganglion on the second lumbar vertebra is the most constant, largest, and most easily palpated and identified by the surgeon. The fifth ganglion is relatively inaccessible to the surgeon because of the common iliac vessels (Cowley and Yeager '49).

Variations.—In 25 bodies, the first lumbar ganglion (identified by its rami) was independent in 13, fused with other ganglia in 10, and separated into two parts in 2; the second ganglion was missing in 2, independent in 12, fused in 7, and split in 4; the third ganglion was independent in 2, fused in 17, split in 4, and connected only with L 3 nerve in 3; the fourth ganglion was independent in one, fused in 12, split in 12, and of these 11 connected with L 4 only; the fifth ganglion was independent in 4, fused in 3, split in 18, and of these 15 connected with L 5 only (Pick and Sheehan '46).

The **branches** of the **lumbar trunk** may be divided into three groups: (1) the branches which are the gray rami communicantes, (2) the lumbar splanchnic nerves, and (3) the visceral branches through the celiac plexus.

1. The **branches** which are the **gray rami communicantes** are supplied to each of the lumbar nerves. They take a transverse path, in contrast to the oblique path of the white rami, and they are more proximal than the white rami in the segments where both are present. They are longer than those in the thoracic region because the lumbar trunk

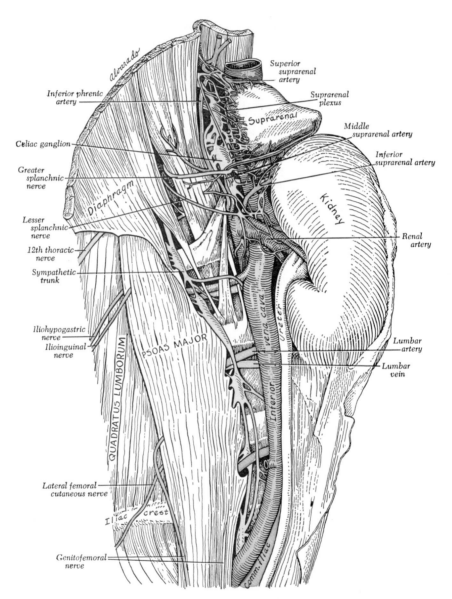

Fig. 12–78.—The lumbar portion of the right sympathetic trunk, the celiac ganglion, splanchnic nerves, suprarenal gland, and kidney. (Redrawn from Töndury.)

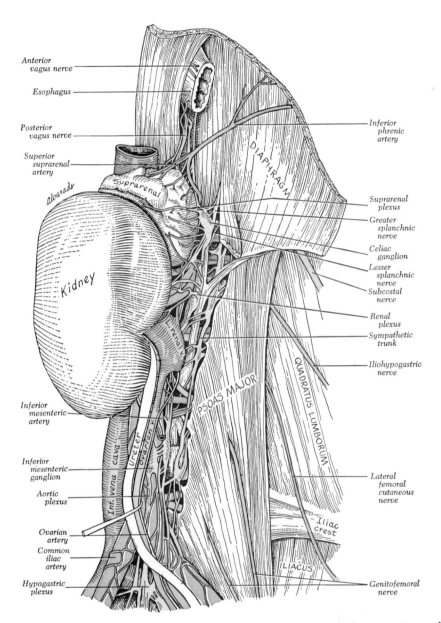

Fig. 12–79.—The lumbar portion of the left sympathetic trunk, celiac and mesenteric ganglia, splanchnic nerves, hypogastric plexus, suprarenal gland, and kidney. (Redrawn from Töndury.)

is more ventrally placed, at some distance from the spinal nerves, and they commonly accompany the lumbar arteries under the fibrous arches of the Psoas, frequently splitting, doubling, and rejoining (Kuntz and Alexander '50).

2. The **lumbar splanchnic nerves** are two to four relatively short branches of the lumbar trunk at the level of the first, second, and third lumbar vertebrae, and are, therefore, caudal to the last root of the lesser splanchnic nerve. They pass either medialward, caudalward, and ventralward to join the aortic network, or, in the case of the most caudal lumbar splanchnic, caudalward and ventralward around the aorta to the inferior mesenteric ganglion or the hypogastric nerve; on the right side they pass between the aorta and the inferior vena cava (Trumble '34). Frequently they contain small groups of ganglion cells which are believed to be displaced from the sympathetic trunk (Harris '43).

3. The **Celiac Ganglion** (*g. celiaca; semilunar ganglion*) (Fig. 12–80) comprises two masses of ganglionic tissue approximately 2 cm in diameter, superficially resembling lymph nodes, lying ventral and lateral to the abdominal aorta on each side, at the level of the first lumbar vertebra. The ganglia are irregular in shape, are usually partly dispersed into several small ganglionic masses, and are connected with each other across the midline by a dense network of bundles, especially caudal to the celiac artery. The ganglia lie on the ventral surface of the crura of the diaphragm, close to the medial border of the suprarenal glands. The right ganglion is covered by the inferior vena cava; the left is covered by the peritoneum of the lesser sac in close relation to the pancreas.

The **aorticorenal** and **superior mesenteric ganglia** are masses which are more or less completely detached from the caudal portion of the celiac ganglion but represent portions of the larger celiac ganglionic complex. The aorticorenal ganglion lies at the origin of the renal artery, the superior mesenteric ganglion at the origin of the corresponding artery.

The **roots** of these ganglia are the **splanchnic nerves** (see above). Each greater splanchnic nerve enters the dorsal and lateral border of the main celiac ganglion (Fig. 12–80); the lesser splanchnic nerve enters the aorticorenal ganglion (Fig. 12–78) and the lowest splanchnic joins the renal plexus. They contain preganglionic fibers from the last six or seven thoracic spinal cord segments which pass through the sympathetic trunk without synapses.

The **postganglionic fibers** arising in the celiac ganglia form an extensive plexus (the celiac plexus) of nerve bundles and filaments which branch off into a number of subsidiary plexuses, in general following the branches of the abdominal aorta. The nerves from the celiac plexus pass down the aorta in the form of a network which is penetrated by the inferior mesenteric artery. The more cranial lumbar splanchnic nerves make a stout contribution to this net and join the caudal lumbar splanchnic nerves to form thick ganglionated nerve bundles on each side of the midline. These bundles converge and meet at the bifurcation of the aorta, with a free decussation of fibers, and then continue as the **right** and **left hypogastric nerves**. Small ganglionic nodes are present, especially if the network is compressed (Trumble '34).

The **Inferior Mesenteric Ganglion** is more difficult to define in man than in many animals, but a considerable amount of ganglionic tissue is almost invariably present at the origin of the inferior mesenteric artery. The roots of the ganglion are provided by nerves from the celiac plexus, the celiac roots, and by the lumbar splanchnic nerves.

The inferior mesenteric ganglion in cats (Harris '43) is composed of three distinct masses arranged in a triangle about the origin of the inferior mesenteric artery. Ganglia are found in dogs, guinea pigs, rabbits, and monkeys also (Trumble '34).

The **branches** of the **inferior mesenteric ganglion** are (*a*) nerves which accompany the inferior mesenteric artery and its **branches** to supply the **colon,** and (*b*) fibers which join each hypogastric nerve and continue from the bifurcation to join the pelvic plexus. The hypogastric nerve crosses the medial side of the ureter and contributes to the ureteric network of nerves. It contains mainly fine unmyelinated fibers but has many medium myelinated fibers (4 to

6μ) and a few large ones, probably afferent
The hypogastric nerves fan out into an
extensive network just under the parietal
peritoneum in the subserous fascia. They
supply the rectal, vesical, prostatic, ureteric,
and ductus deferens nerves (Ashley and
Anson '46).

The Pelvic Portion of the Sympathetic System

The **pelvic portion** of the **sympathetic
trunk** (*pars pelvina s. symphathici*) (Fig.

12–81) lies against the ventral surface of
the sacrum, medial to the sacral foramina.
It is the direct continuation of the lumbar
trunk and contains four or five ganglia,
smaller than those in other parts of the
chain. Fusion of adjacent ganglia is quite
common and cords connecting the trunks
of the two sides across the midline are of
regular occurrence. There are no white
rami communicantes in the sacral region;
small myelinated preganglionic fibers from
the second, third, and fourth sacral nerves
enter the pelvic plexus but they belong to

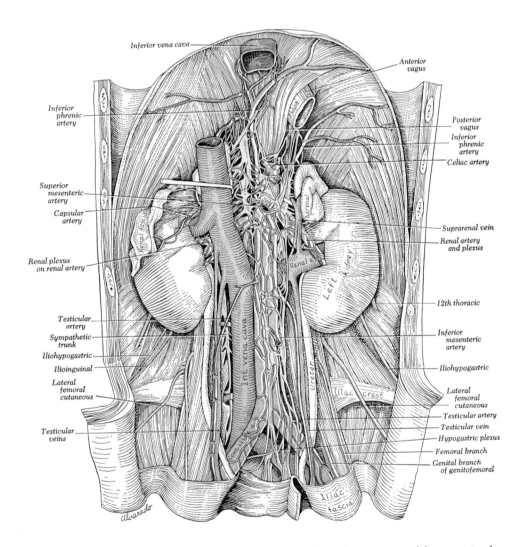

Fig. 12–80.—View of the posterior abdominal wall, showing the celiac, aortic, and hypogastric plexuses of autonomic nerves. (Redrawn from Töndury.)

the parasympathetic or craniosacral outflow rather than the sympathetic. The coccygeal ganglion is the most caudal ganglion of the sympathetic trunk; it is commonly a single ganglion, the **ganglion impar,** representing a fusion of the ganglia of the two sides, and usually lies in the midline but may be at one side.

The **branches** of the **sacral** and **coccygeal ganglia** which are the **gray rami communicantes** of the **sacral spinal nerves** are supplied to each of the sacral and the coccygeal nerves. In the majority of instances, each ganglion, or its representative in a fused ganglion, supplies rami to two adjacent spinal nerves (Pick and Sheehan '46).

Visceral branches in variable numbers join the hypogastric and pelvic plexuses, and are supplied through them to the pelvic viscera and blood vessels (Trumble '34).

THE GREAT AUTONOMIC PLEXUSES

The two subdivisions of the autonomic nervous system, the sympathetic and parasympathetic, are combined into extensive plexuses in the thorax, abdomen, and pelvis, named respectively the cardiac plexus, the celiac plexus, and the pelvic plexus. Experimental and clinical observations have made it possible to trace the sympathetic and parasympathetic components to some extent, but on the morphological evidence of dissections, it is almost impossible to distinguish the ultimate paths of the fibers belonging to the two systems. These plexuses also contain visceral afferent fibers, described in earlier pages.

The Cardiac Plexus

The **cardiac plexus** (*plexus cardiacus*) (Fig. 7–19) is situated at the base of the heart, close to the arch of the aorta, and is traditionally subdivided into a superficial and a deep part for topographical reasons although the functional associations do not justify this division. The sympathetic contribution is largely postganglionic, the parasympathetic largely preganglionic with scattered groups of ganglion cells.

The **superficial part** of the **cardiac plexus** (Fig. 12–77) lies in the arch of the aorta somewhat on the left side between it and the bifurcation of the pulmonary artery. It is formed by the superior cervical cardiac branch of the left sympathetic and the lower of the two superior cardiac branches of the left vagus. A small ganglion, the cardiac ganglion (of Wrisberg), is occasionally found in this plexus at the right side of the ligamentum arteriosum; it is probably a parasympathetic ganglion receiving preganglionic fibers from the vagus.

The **branches** of the **superficial cardiac plexus** are as follows: (*a*) to the **deep cardiac plexus,** (*b*) to the **anterior coronary plexus,** and (*c*) to the **left anterior pulmonary plexus.**

The **deep part** of the **cardiac plexus** is situated deep to the arch of the aorta, between it and the bifurcation of the trachea and cranial to the pulmonary artery. It is much more extensive than the superficial part and receives all the cardiac branches of both vagi and sympathetic trunks except the two mentioned above which enter the superficial part (left superior sympathetic and left lower superior of the vagus); it also receives the lower cardiac branches of the vagus and the recurrent nerve, visceral rami from the upper cranial thoracic sympathetic ganglia, and commmunications from the superficial part of the cardiac plexus.

The **branches** of the **deep part** of the **cardiac plexus** may be divided into right and left halves.

The right half of the deep cardiac plexus gives branches which follow the right pulmonary artery; those ventral to the artery are more numerous and, after contributing a few filaments to the anterior pulmonary plexus, continue into the anterior coronary plexus; those dorsal to the artery distribute a few filaments to the right atrium, and are then continued onward to form part of the posterior coronary plexus.

The left half of the deep cardiac plexus communicates with the superficial plexus, gives filaments to the left atrium and to the anterior pulmonary plexus, and is then continued into the greater part of the posterior coronary plexus.

1. The **Posterior Coronary Plexus** (*pl. coronarius posterior; left coronary plexus*) is larger than the anterior, and accompanies

the left coronary artery. It is formed chiefly by filaments from the left half, and by a few from the right half of the deep plexus. It is distributed to the left atrium and ventricle.

2. The **Anterior Coronary Plexus** (*pl. coronarius anterior; right coronary plexus*) is formed partly from the superficial and partly from the deep parts of the cardiac plexus. It accompanies the right coronary artery, and is distributed to the right atrium and ventricle (Stotler and McMahon '47).

The Celiac Plexus

The **celiac plexus** (*plexus coeliacus; solar plexus*) (Figs. 12–78, 12–80) is situated at the level of the upper part of the first lumbar vertebra, and contains two large ganglionic masses and a dense network of fibers surrounding the roots of the celiac and superior mesenteric arteries. The denser part of the plexus lies between the suprarenal glands, on the ventral surface of the crura of the diaphragm and abdominal aorta, and dorsal to the stomach and the omental bursa, but it has extensive prolongations caudalward on the aorta and out along its branches. The celiac ganglia are described with the abdominal portion of the sympathetic system (page 1028). The preganglionic parasympathetic fibers reach the plexus through the anterior (left) and posterior (right) vagi on the stomach. The preganglionic sympathetic fibers reach the celiac, aorticorenal, and superior mesenteric ganglia through the greater and lesser splanchnic nerves. These nerves also supply preganglionic fibers to the cells in the medulla of the suprarenal glands, which correspond developmentally to postganglionic neurons.

The **secondary plexuses** and prolongations from the celiac plexus are as follows:

1. The **Phrenic Plexus** (*pl. phrenicus*) (Fig. 12–80) accompanies the inferior phrenic artery to the diaphragm. It arises from the superior part of the celiac plexus, and is larger on the right side than on the left. It communicates with the phrenic nerve; at the point of junction with the right phrenic, near the vena caval foramen in the diaphragm, there may be a small ganglion, the **phrenic ganglion**. The phrenic plexus

gives filaments to the inferior vena cava, inferior phrenic arteries, and to the suprarenal and hepatic plexuses.

2. The **Hepatic Plexus** (*pl. hepaticus*) accompanies the hepatic artery, ramifying upon its branches, and upon those of the portal vein in the substance of the liver. A considerable network accompanies the gastroduodenal artery and is continued as the inferior gastric plexus on the right gastroepiploic artery along the greater curvature of the stomach, communicating with the splenic plexus. Extensions from the hepatic plexus supply the pancreas and gallbladder, and there is a communication with the phrenic plexus.

3. The **Splenic Plexus** (*pl. lienalis*), containing mainly fibers from the left celiac ganglion and the anterior (left) vagus, accompanies the splenic artery to the spleen and in its course sends filaments along its branches, especially to the pancreas.

4. The **Superior Gastric Plexus** (*pl. gastricus superior; gastric or coronary plexus*) accompanies the left gastric artery along the lesser curvature of the stomach and communicates with the anterior (left) vagus.

5. The **Suprarenal Plexus** (*pl. suprarenalis*) (Fig. 12–78) is composed principally of short, rather stout, branches from the celiac ganglion, with some contributions from the greater splanchnic nerve and the phrenic plexus. The fibers it contains are predominantly preganglionic sympathetics which pass through the celiac ganglion without synapses and are distributed to the medulla of the suprarenal gland where the cells are homologous with postganglionic neurons. Postganglionic fibers to the blood vessels are present also (Swinyard '37).

6. The **Renal Plexus** (*pl. renalis*) (Fig. 12–78) is formed by filaments from the celiac plexus, the aorticorenal ganglion, the aortic plexus, and the smallest splanchnic nerve. It accompanies the renal artery into the kidney, giving some filaments to the spermatic plexus and to the inferior vena cava on the right side (Christensen *et al.* '51).

7. The **Spermatic Plexus** (*pl. spermaticus*) receives filaments from the renal and aortic plexuses, and accompanies the testicular

artery to the testis. In the female the ovarian plexus (*pl. arteriae ovaricae*) arises from the renal plexus, accompanies the ovarian artery, and is distributed to the ovary, uterine tubes, and the fundus of the uterus.

8. The **Superior Mesenteric Plexus** (*pl. mesentericus superior*) is essentially the lower part of the celiac plexus; it may appear more or less detached from the rest of the plexus and frequently it contains a separate ganglionic mass, the superior mesenteric ganglion (*g. mesentericum superius*). Its vagus fibers come principally from the posterior vagus. It surrounds and accompanies the superior mesenteric artery being distributed with the latter's pancreatic, intestinal, ileocolic, right colic, and middle colic branches, which supply the corresponding organs.

9. The **Abdominal Aortic Plexus** (*pl. aorticus abdominalis; aortic plexus*) (Fig. 12–80) is formed from both right and left celiac plexuses and ganglia, and from the lumbar splanchnic nerves. It lies upon the ventral and lateral surfaces of the aorta between the origins of the superior and inferior mesenteric arteries. From this plexus arise the spermatic, inferior mesenteric, external iliac, and hypogastric plexuses, and filaments to the inferior vena cava.

10. The **Inferior Mesenteric Plexus** (*pl. mesentericus inferior*) surrounds the origin of the inferior mesenteric artery and contains a ganglion (Fig. 12–80), the inferior mesenteric ganglion, or thickened bundles which contain ganglion cells. It is derived from the aortic plexus, through whose celiac and lumbar splanchnic contributions it receives preganglionic as well as postganglionic fibers. It surrounds the inferior mesenteric artery, and divides into a number of subsidiary plexuses which accompany its branches, the left colic, sigmoid, and superior rectal, to the corresponding organs (Harris '43).

11. The **Superior Hypogastric Plexus** (*pl. hypogastricus; hypogastric plexus; presacral nerve*) (Fig. 12–80) is the caudalward continuation of the aortic and inferior mesenteric plexuses. It extends from the level of the fourth lumbar to the first sacral vertebrae and lies in the subserous fascia just under the peritoneum. It is at first ventral to the aorta, then between the common iliac arteries, crosses the left common iliac vein and enters the pelvis to lie against the middle sacral vessels and the vertebrae. At the first sacral vertebra it divides into two parts, the hypogastric nerves.

The **hypogastric nerve** may be a single rather large nerve or several bundles forming a parallel network. It lies medial and dorsal to the common and internal iliac arteries, crosses the branches of the latter, and enters the inferior hypogastric plexus.

12. The **Inferior Hypogastric Plexus** is a fan-like expansion from the hypogastric nerves at the proximal part of the rectum and bladder in the subserous fascia just above the sacrogenital fold. It receives filaments from the sacral portion of the sympathetic chain and from the deeper parts of the pelvic plexus.

The Pelvic Plexus

The **pelvic plexus** (Fig. 12–81) of the autonomic system is formed by the hypogastric plexus, by rami from the sacral portion of the sympathetic chain, and by the visceral branches of the second, third, and fourth sacral nerves. Through its secondary plexuses it is distributed to all the pelvic viscera.

1. The **middle rectal plexus** (*pl. rectalis medius; middle hemorrhoidal plexus*) is contained in the tissue of the sacrogenital fold and is therefore the most superficial part of the pelvic plexus, with relation to the peritoneum. It is usually independent of the middle rectal artery and, except for its terminal filaments to the caudal sigmoid colon and rectum, is several centimeters from the bowel itself. From its superficial position, it appears to be the continuation of the inferior hypogastric plexus, but the latter has a number of other continuations and the rectal plexus supplies the lower bowel with parasympathetic fibers through contributions from the visceral branches of the sacral nerves. It communicates with the superior rectal branches from the inferior mesenteric plexus (Ashley and Anson '46).

2. The **vesical plexus** (*pl. vesicalis*) arises from the anterior part of the pelvic plexus, its fibers derived from the superficial or hypogastric network and the deeper

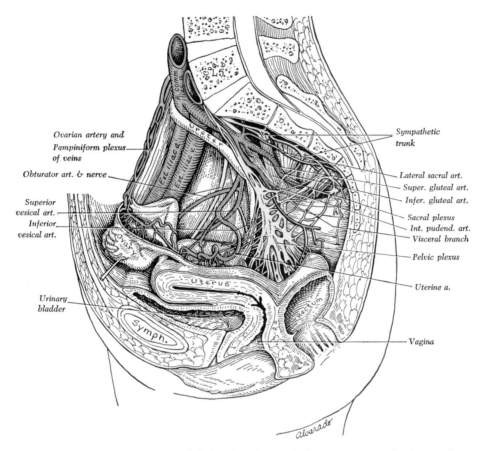

Ovarian artery and
Pampiniform plexus
of veins

Obturator art. & nerve

Superior
vesical art.
Inferior
vesical art.

Urinary
bladder

Sympathetic
trunk

Lateral sacral art.
Super. gluteal art.
Infer. gluteal art.

Sacral plexus
Int. pudend. art.
Visceral branch

Pelvic plexus

Uterine a.

Vagina

FIG. 12–81.—Sagittal section of an adult female pelvis, with the peritoneum and subserous fascia partially dissected away to show the pelvic plexus of autonomic nerves. (Redrawn from Töndury.)

bundles from the sacral nerves. The filaments are divisible into periureteric, prostatic, seminal vesicle, and lateral vesical groups.

3. The **prostatic plexus** (*pl. prostaticus*), from the deeper ventral part of the pelvic plexus, is composed of larger nerves which are distributed to the prostate, seminal vesicles, and corpora cavernosa. The nerves supplying the corpora consist of two sets, the greater and lesser cavernous nerves, which arise from the ventral part of the prostatic plexus, join with branches of the pudendal nerve, and pass beneath the pubic arch. Filaments at the base of the gland supply the prostatic and membranous urethra, the ejaculatory ducts, and the bulbourethral glands.

a) The **greater cavernous nerve** (*n. cavernosus penis major*) passes distalward along the dorsum of the penis, joins the dorsal

nerve of the penis, and is distributed to the corpus cavernosum penis.

b) The **lesser cavernous nerves** (*nn. cavernosi penis minores*) perforate the fibrous covering of the penis near its root, and are distributed to the corpus spongiosum penis and the penile urethra.

4. The **vaginal plexus** arises from the caudal part of the pelvic plexus, and is distributed to the walls of the vagina, and to the erectile tissue of the vestibule and clitoris.

5. The **uterine plexus** arises from the caudal portion of the pelvic plexus, and approaches the uterus from its caudal and lateral aspect in the base of the broad ligament, in the same region as the uterine artery. It is distributed to its musculature, supplies filaments to the uterine tube, and communicates with the ovarian plexus (Curtis *et al.* '42).

REFERENCES

CRANIAL NERVES

III. CORBIN, K. B. and R. K. OLIVER. 1942. The origin of fibers to the grape-like endings in the insertion third of the extra-ocular muscles. J. comp. Neurol., 77, 171-186.

V. AUGUSTINE, J. R., B. VIDIĆ, and P. A. YOUNG. 1971. An intermediate root of the trigeminal nerve in the dog. Anat. Rec., 169, 697-703.

V. BURR, H. S. and G. B. ROBINSON. 1925. An anatomical study of the Gasserian ganglion, with particular reference to the nature and extent of Meckel's cave. Anat. Rec., 29, 269-282.

V. CARTER, R. B. and E. N. KEEN. 1971. The intramandibular course of the inferior alveolar nerve. J. Anat. (Lond.), 108, 433-440.

V. CORBIN, K. B. and F. HARRISON. 1940. Function of mesencephalic root of fifth cranial nerve. J. Neurophysiol., 3, 423-435.

V. FOLEY, J. O., H. R. PEPPER, and W. H. KESSLER. 1946. The ratio of nerve fibers to nerve cells in the geniculate ganglion. J. comp. Neurol., 85, 141-148.

V. HENDERSON, W. R. 1965. The anatomy of the Gasserian ganglion and the distribution of pain in relation to injection and operation for trigeminal neuralgia. Ann. R. Coll. Surg., 37, 346-373.

V. ROBERTS, W. H. and N. B. JORGENSEN. 1962. A note on the distribution of the superior alveolar nerves in relation to the primary teeth. Anat. Rec., 141, 81-84.

V. VIDIĆ, B. and J. STEFANATOS. 1969. The roots of the trigeminal nerve and their fiber components. Anat. Rec., 163, 330.

VII. BRUESCH, S. R. 1944. The distribution of myelinated afferent fibers in the branches of the cat's facial nerve. J. comp. Neurol., 81, 169-191.

VII. FOLEY, J. O. 1945. The sensory and motor axons of the chorda tympani. Proc. Soc. exp. Biol. Med., 60, 262-267.

VII. GOMEZ, H. 1961. The innervation of lingual salivary glands. Anat. Rec., 139, 69-76.

VII. LAURENSON, R. D. 1964. A rapid exposure of the facial nerve. Anat. Rec., 150, 317-318.

VII. McCORMACK, L. J., E. W. CAULDWELL, and B. J. ANSON. 1945. The surgical anatomy of the facial nerve with special reference to the parotid gland. Surg. Gynec. Obstet., 80, 620-630.

VIII. GACEK, R. R. and G. L. RASMUSSEN. 1961. Fiber analysis of the statoacoustic nerve of guinea pig, cat, and monkey. Anat. Rec., 139, 455-463.

IX. GUTH, L. 1957. The effects of glossopharyngeal nerve transection on the circumvallate papilla of the rat. Anat. Rec., 128, 715-731.

IX. PASTORE, P. N. and J. M. MEREDITH. 1949. Glossopharyngeal neuralgia. Arch. Otolaryng., 50, 789-794.

IX. ROSEN, S. 1950. The tympanic plexus. Arch. Otolaryng., 52, 15-18.

IX. SHEEHAN, D., J. H. MULHOLLAND, and B. SHAFIROFF. 1941. Surgical anatomy of the carotid sinus nerve. Anat. Rec., 80, 431-442.

IX. WILLIS, A. G. and J. D. TANGE. 1959. Studies on the innervation of the carotid sinus of man. Amer. J. Anat., 104, 87-113.

X. ARMSTRONG, W. G. and J. W. HINTON. 1951. Multiple divisions of the recurrent laryngeal nerve: an anatomic study. Arch. Surg., 62, 532-539.

X. HOFFMAN, H. H. and H. N. SCHNITZLEIN. 1961. The numbers of nerve fibers in the vagus nerve of man. Anat. Rec., 139, 429-435.

X. HOLLINSHEAD, W. H. 1939. The origin of the nerve fibers to the glomus aorticum of the cat. J. comp. Neurol., 71, 417-426.

X. HOLLINSHEAD, W. H. 1940. The innervation of the supracardial bodies in the cat. J. comp. Neurol., 73, 37-48.

X. JACKSON, R. G. 1949. Anatomy of the vagus nerves in the region of the lower esophagus and the stomach. Anat. Rec., 103, 1-18.

X. MOHR, P. D. 1969. The blood supply of the vagus nerve. Acta Anat., 73, 19-26.

X. RÉTHI, A. 1951. Histological analysis of the experimentally degenerated vagus nerve. Acta Morphol., 1, 221-230.

X. SCHNITZLEIN, H. N., L. C. ROWE, and H. H. HOFFMAN. 1958. The myelinated component of the vagus nerves in man. Anat. Rec., 131, 649-667.

X. SUNDERLAND, S. and W. E. SWANEY. 1952. The intraneural topography of the recurrent laryngeal nerve in man. Anat. Rec., 114, 411-426.

XI. CORBIN, K. B. and F. HARRISON. 1938. Proprioceptive components of cranial nerves. The spinal accessory nerve. J. comp. Neurol., 69, 315-328.

HARDESTY, I. 1933. The nervous system. In *Morris' Human Anatomy.* 9th edition. Pages 1000-1038. P. Blakiston's Sons and Company, Philadelphia.

LARSELL, O. 1950. The nervus terminalis. Ann. Otol. (St. Louis), 59, 414-438.

PEARSON, A. A. 1941. The development of the nervus terminalis in man. J. comp. Neurol., 75, 39-66.

SPINAL NERVES

BARRY, A. 1956. A quantitative study of the prenatal changes in angulation of the spinal nerves. Anat. Rec., 126, 97-110.

CORBIN, K. B. and F. HARRISON. 1939. The sensory innervation of the spinal accessory and tongue musculature in the rhesus monkey. Brain, 62, 191-197.

HINSEY, J. C. 1933. The functional components of the dorsal roots of spinal nerves. Quart. Rev. Biol., 8, 457-464.

Hogg, I. D. 1941. Sensory nerves and associated structures in the skin of human fetuses of 8 to 14 weeks of menstrual age correlated with functional capability. J. comp. Neurol., 75, 371-410.

Keegan, J. J. and F. D. Garrett. 1948. The segmental distribution of the cutaneous nerves in the limbs of man. Anat. Rec., 102, 409-437.

Kozlov, V. I. 1969. Formation and structure of the spinal nerve. Acta Anat., 73, 321-350.

Moyer, E. K. and B. F. Kaliszewski. 1958. The number of nerve fibers in motor spinal nerve roots of young, mature and aged cats. Anat. Rec., 131, 681-699.

Moyer, E. K. and D. L. Kimmel. 1948. The repair of severed motor and sensory spinal nerve roots by the arterial sleeve method of anastomosis. J. comp. Neurol., 88, 285-317.

Pearson, A. A., R. W. Sauter, and J. J. Bass. 1963. Cutaneous branches of the dorsal primary rami of the cervical nerves. Amer. J. Anat., 112, 169-180.

Roofe, P. G. 1940. Innervation of the annulus fibrosus and posterior longitudinal ligament. Arch. Neurol. Psychiat., 44, 100-103.

Stilwell, D. L., Jr. 1957. Regional variations in the innervation of deep fasciae and aponeuroses. Anat. Rec., 127, 635-653.

Sunderland, S. and K. C. Bradley. 1949. The cross-sectional area of peripheral nerve trunks devoted to nerve fibers. Brain, 72, 428-449.

Brachial Plexus

Blunt, M. J. 1959. The vascular anatomy of the median nerve in the forearm and hand. J. Anat. (Lond.), 93, 15-22.

Fullerton, P. M. 1963. The effect of ischaemia on nerve conduction in the carpal tunnel syndrome. J. Neurol. Neurosurg. Psychiat., 26, 385-397.

Gitlin, G. 1957. Concerning the gangliform enlargement ('Pseudoganglion') on the nerve to the teres minor muscle. J. Anat. (Lond.), 91, 466-470.

Harris, W. 1939. The Morphology of the Brachial Plexus. xviii + 117 pages. Oxford University Press, London.

Jamieson, R. W. and B. J. Anson. 1952. The relation of the median nerve to the heads of origin of the pronator teres muscle. Quart. Bull. Northw. Univ. Med. Sch., 26, 34-35.

Kasai, T. 1963. About the N. cutaneus brachii lateralis inferior. Amer. J. Anat., 112, 305-309.

Kerr, A. T. 1918. The brachial plexus of nerves in man, the variations in its formation and branches. Amer. J. Anat., 23, 285-395.

Miller, M. R., H. J. Ralston, III, and M. Kasahara. 1958. The pattern of cutaneous innervation of the human hand. Amer. J. Anat., 102, 183-217.

Miller, R. A. and S. R. Detwiler. 1936. Comparative studies upon the origin and development of the brachial plexus. Anat. Rec., 65, 273-292.

Stilwell, D. L., Jr. 1957. The innervation of deep structures of the hand. Amer. J. Anat., 101, 75-99.

Sunderland, S. 1945. The innervation of the flexor digitorum profundus and lumbrical muscles. Anat. Rec., 93, 317-321.

Sunderland, S. 1946. The innervation of the first dorsal interosseous muscle of the hand. Anat. Rec., 95, 7-10.

Sunderland, S. and G. M. Bedbrook. 1949. The relative sympathetic contribution to individual roots of the brachial plexus in man. Brain, 72, 297-301.

Lumbosacral Plexus

Beaton, L. E. and B. J. Anson. 1938. The sciatic nerve and the piriformis muscle: their interrelation a possible cause of coccygodynia. J. Bone Jt. Surg., 20, 686-688.

Day, M. H. 1964. The blood supply of the lumbar and sacral plexuses in the human foetus. J. Anat. (Lond.), 98, 105-116.

Harrison, V. F. and O. A. Mortensen. 1962. Identification and voluntary control of single motor unit activity in the tibialis anterior muscle. Anat. Rec., 144, 109-116.

Huelke, D. F. 1957. A study of the formation of the sural nerve in adult man. Amer. J. phys. Anthrop., 15, 137-147.

Jamieson, R W., L. L. Swigart, and B. J. Anson. 1952. Points of parietal perforation of the ilioinguinal and iliohypogastric nerves in relation to optimal sites for local anaesthesia. Quart. Bull. Northw. Univ. Med. Sch., 26, 22-26.

Miller, M. R. and M. Kasahara. 1959. The pattern of cutaneous innervation of the human foot. Amer. J. Anat., 105, 233-255.

Stillwell, D. L., Jr. 1957. The innervation of deep structures of the foot. Amer. J. Anat., 101, 59-73.

Autonomic System

Ashley, F. L. and B. J. Anson. 1946. The pelvic autonomic nerves in the male. Surg. Gynec. Obstet., 82, 598-608.

Braeucker, W. 1923. Die Nerven der Schilddrüse und der Epithelkörperchen. Anat. Anz., 56, 225-249.

Curtis, A. H., B. J. Anson, F. L. Ashley, and T. Jones. 1942. The anatomy of the pelvic autonomic nerves in relation to gynecology. Surg. Gynec. Obstet., 75, 743-750.

Harris, A. J. 1943. An experimental analysis of the inferior mesenteric plexus. J. comp. Neurol., 79, 1-17.

Kuntz, A. 1953. The Autonomic Nervous System. 4th edition. 605 pages. Lea & Febiger, Philadelphia.

Kuz'mina, S. V. 1964. Structural organization of inferior mesenteric ganglion. Fed. Proc., 23, T706-T710.

Mitchell, G. A. G. 1951. The intrinsic renal nerves. Acta Anat., 13, 1-15.

Mitchell, G. A. G. 1953. Anatomy of the Autonomic Nervous System. xvi + 356 pages. Williams & Wilkins Company, Baltimore.

MIZERES, N. J. 1963. The cardiac plexus in man. Amer. J. Anat., *112*, 141-151.

SOUTHAM, J. A. 1959. The inferior mesenteric ganglion. J. Anat. (Lond.), *93*, 304-308.

SULKIN, N. M. and A. KUNTZ. 1950. A histochemical study of the autonomic ganglia of the cat following prolonged preganglionic stimulation. Anat. Rec., *108*, 255-277.

SWINYARD, C. A. 1937. The innervation of the suprarenal glands. Anat. Rec., *68*, 417-429.

SYMPATHETIC NERVES

AXFORD, M. 1928. Some observations on the cervical sympathetic ganglia. J. Anat. (Lond.), *62*, 301-318.

BECKER, R. F. and J. A. GRUNT. 1957. The cervical sympathetic ganglia. Anat. Rec., *127*, 1-14.

BOTAR, J. 1932. La chaine sympathique latero-vertebrale lombaire, ses ganglions et ses rameaux communicants chez le nouveau-ne. Ann. anat. Path., *9*, 449-455.

CHRISTENSEN, K., E. LEWIS, and A. KUNTZ. 1951. Innervation of the renal blood vessels in the cat. J. comp. Neurol., *95*, 373-385.

CHRISTENSEN, K., E. H. POLLEY, and E. LEWIS. 1952. The nerves along the vertebral artery and innervation of the blood vessels of the hindbrain of the cat. J. comp. Neurol., *96*, 71-91.

DASS, R. 1952. Sympathetic components of the dorsal primary divisions of human spinal nerves. Anat. Rec., *113*, 493-501.

EBBESSON, S. O. E. 1963. A quantitative study of human superior cervical sympathetic ganglia. Anat. Rec., *146*, 353-356.

EDWARDS, E. A. 1951. Operative anatomy of the lumbar sympathetic chain. Angiology, *2*, 184-198.

EDWARDS, L. F. and R. C. BAKER. 1940. Variations in the formation of the splanchnic nerves in man. Anat. Rec., *77*, 335-342.

FOLEY, J. O. and H. N. SCHNITZLEIN. 1957. The contribution of individual thoracic spinal nerves to the upper cervical sympathetic trunk. J. comp. Neurol., *108*, 109-120.

JAMIESON, R. W., D. B. SMITH, and B. J. ANSON. 1952. The cervical sympathetic ganglia. Quart. Bull. Northw. Univ. Med. Sch., *26*, 219-227.

JONNESCO, T. 1923. *Le Sympathique Cervicothoracique.* 91 pages. Masson, Paris.

KISNER, W. H. and H. MAHORNER. 1952. An evaluation of lumbar sympathectomy. Amer. Surg., *18*, 30-35.

KUNTZ, A. and W. F. ALEXANDER. 1950. Surgical implications of lower thoracic and lumbar independent sympathetic pathways. Arch. Surg., *61*, 1007-1018.

KUNTZ, A. and A. MOREHOUSE. 1930. Thoracic sympathetic cardiac nerves in man. Their relation to cervical sympathetic ganglionectomy. Arch. Surg., *20*, 607-613.

MUSTALISH, A. C. and J. PICK. 1964. On the innervation of the blood vessels in the human foot. Anat. Rec., *149*, 587-590.

PEARSON, A. A. 1952. The connections of the sympathetic trunk in the cervical and upper thoracic levels in the human fetus. Anat. Rec., *106*, 231.

PICK, J. and D. SHEEHAN. 1946. Sympathetic rami in man. J. Anat. (Lond.), *80*, 12-20.

REED, A. F. 1951. The origins of the splanchnic nerves. Anat. Rec., *109*, 341.

SACCOMANNO, G. 1943. The components of the upper thoracic sympathetic nerves. J. comp. Neurol., *79*, 355-378.

SHASHIRINA, M. I. 1964. Structural organization of superior cervical sympathetic ganglion of the cat. Fed. Proc., *23*, T711-T714.

TSOURAS, S. 1949. Anatomical observations on the lumbar chain ganglia and the sympathetic system. Greek Medical Progress, *3*, 3.

WRETE, M. 1959. The anatomy of the sympathetic trunks in man. J. Anat. (Lond.), *93*, 448-459.

13 | *The Organs of the Senses*

ndings of the sensory nerves of the peripheral nervous system are found in end organs which may be divided into two large groups: (1) the organs of the special senses of smell, taste, sight, and hearing, and (2) the sensory endings of general sensation of heat, cold, touch, pain, pressure, etc.

THE SPECIAL SENSES

THE ORGAN OF SMELL (ORGANUM OLFACTUS)

The sensory endings for the sense of smell are located in the nose, and for this reason the entire nose has, by long tradition, been described in this chapter. In the present edition, however, all but the olfactory area will be described in the chapter on the Respiratory System because the nose and nasal cavity are of more importance as air passages than for the sense of smell and they are customarily referred to by clinicians as the upper respiratory tract.

The **olfactory region** is located in the most superior part of both nasal fossae, and occupies the mucous membrane covering the superior nasal concha and the septum opposite (Fig. 13–1), and is confined, therefore, to an area of the fossa of which the walls are formed by the ethmoid bone.

The olfactory sensory endings are the least specialized of the special senses. They are modified epithelial cells liberally scattered among the columnar epithelium of the

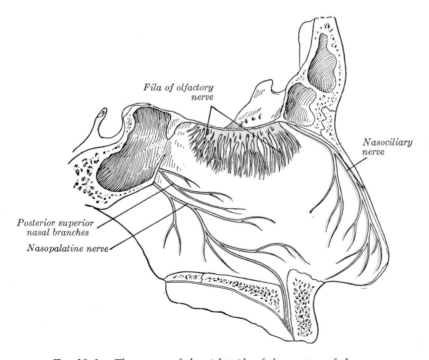

Fila of olfactory nerve

Nasociliary nerve

Posterior superior nasal branches

Nasopalatine nerve

Fig. 13–1.—The nerves of the right side of the septum of the nose.

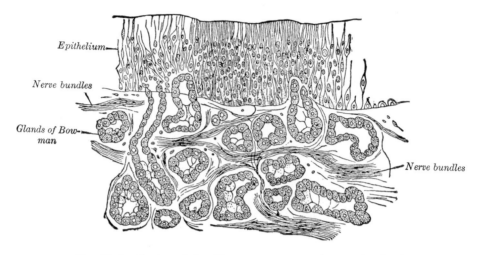

Epithelium

Nerve bundles

Glands of Bow-man

Nerve bundles

FIG. 13–2.—Section of the olfactory mucous membrane. (Cadiat.)

mucous membrane. The sensory cells are known as **olfactory cells,** the other epithelial cells as **supporting cells,** and although the epithelium appears pseudostratified, the supporting cells are not ciliated and goblet cells are lacking (Fig. 13–2).

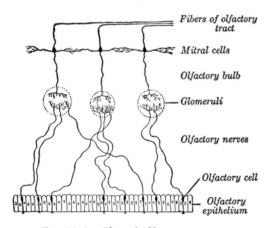

Fibers of olfactory tract

Mitral cells

Olfactory bulb

Glomeruli

Olfactory nerves

Olfactory cell

Olfactory epithelium

FIG. 13–3.—Plan of olfactory neurons.

The **olfactory cells** are bipolar in form, with a small amount of cytoplasm surrounding a large spherical nucleus. The slender peripheral or superficial process extends to the surface of the epithelial membrane and sends out beyond the surface a tuft of very fine processes known as **olfactory hairs.** The central or deep process finds its way through the basement membrane and in the under-

lying connective tissue joins neighboring processes to form the bundles of unmyelinated fibers of the **olfactory nerves** (Fig. 13–3). Mingled with the nerve bundles in the subepithelial tissue are numerous branched tubular glands of a serous secreting type (glands of Bowman), which keep the membrane protected with moisture.

The bundles of nerve fibers form a plexus in the submucosa and are finally collected into about twenty nerves which pass through the openings in the cribriform plate of the ethmoid bone as the **fila olfactoria.** The nerve fibers end by forming synapses with processes of the mitral cells in the glomeruli of the olfactory bulb (Fig. 13–3, also see page 863).

THE ORGAN OF TASTE
(ORGANUM GUSTUS)

The peripheral taste organs (*gustatory organs*) are the **taste buds** (*gustatory caliculi*) distributed over the tongue and occasionally on adjacent parts. They are spherical or ovoid nests of cells embedded in the stratified squamous epithelium (Fig. 13–5), and are present in large numbers on the sides of the vallate papillae (*papillae vallatae*) and to a lesser extent on the opposed walls of the fossae around them (Fig. 13–4). They are also found over the sides and back of the tongue, especially on

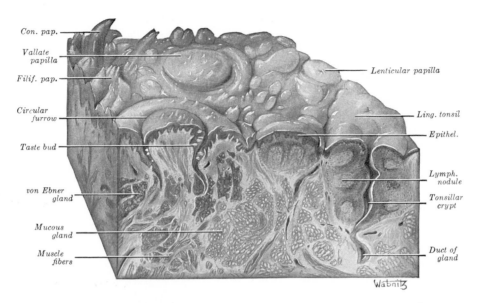

Con. pap.
Vallate papilla
Filif. pap.
Circular furrow
Taste bud
von Ebner gland
Mucous gland
Muscle fibers

Lenticular papilla
Ling. tonsil
Epithel.
Lymph. nodule
Tonsillar crypt
Duct of gland

Watritz

Fig. 13–4.— Section of posterior part of dorsum of tongue showing circumvallate papillae and lingual tonsil. (Redrawn from Braus.)

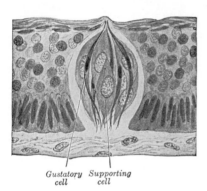

Gustatory   Supporting
cell   cell

Fig. 13–5.—Section through a taste bud. Semidiagrammatic. × 450.

the fungiform papillae (Fig. 16–36). They are plentiful over the fimbriae linguae, and are occasionally present on the oral surface of the soft palate and on the posterior surface of the epiglottis.

Each **taste bud** (*caliculus gustatorius*) occupies an ovoid pocket which extends through the thickness of the epithelium. It has two openings, one at the surface and the other at the basement membrane. The cells within the bud are of two kinds, **gustatory cells** and **supporting cells** (Fig. 13–5). The supporting cells are elongated, extending between the basement membrane and the surface; they form an outer shell for the taste bud, arranged like the staves of a wooden cask, and are also scattered through the bud between the gustatory cells. The latter occupy the interior portion of the bud; they are spindle-shaped with a large round central nucleus. The peripheral end of each gustatory cell protrudes through the opening at the surface, the **gustatory pore** (*porus gustatorius*), a delicate hair-like process, the **gustatory hair.** The central end of the gustatory cell does not end in an axon, as in the case of the olfactory cell, but remains within the taste bud where it has intimate contact with many fine terminations of nerves which pass into the taste bud through an opening in the basement membrane. The nerves are myelinated until they reach the taste buds but lose their sheaths as they enter the bud.

Nerves of Taste.—The posterior third of the tongue, including the taste buds on the vallate papillae, is supplied by the glosso-pharyngeal nerve. The fibers for taste to the anterior two-thirds leave the brain in the nervus intermedius, are continued in the chorda tympani, and reach the tongue in the branches of the lingual nerve. It is believed that taste buds on the epiglottis are supplied by the vagus nerve.

THE ORGAN OF SIGHT: THE EYE

The **bulb of the eye** (*bulbus oculi; eyeball*), or **organ of sight** (*organum visus*), is contained in a bony cavity, the orbit, composed of parts of the frontal, maxillary, zygomatic, sphenoid, ethmoid, lacrimal, and palatine bones. It is embedded in the orbital fat, but is separated from it by a thin membranous sac, the fascia bulbi. Associated with it are certain accessory structures, viz., the muscles, fasciae, eyebrows, eyelids, conjunctiva, and lacrimal apparatus.

Development.—The retina and optic nerve come from the forebrain; the lens from the overlying ectoderm; and the accessory structures from the mesenchyma. The eyes begin to develop as a pair of diverticula from the lateral aspects of the forebrain. These diverticula make their appearance before the closure of the anterior end of the neural tube; after the closure of the tube they are known as the **optic vesicles**. They project toward the sides of the head, and the peripheral part of each expands to form a hollow bulb, while the proximal part remains narrow and constitutes the **optic stalk**. When the peripheral part of the optic vesicle comes in contact with the overlying ectoderm, the latter thickens, invaginates, becomes severed from the ectoderm, and

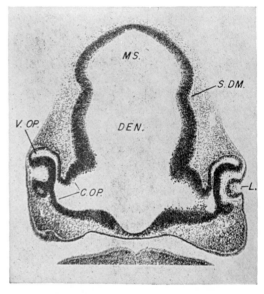

Fig. 13–7.—Brain of 6.7-mm human embryo. *C.OP.*, conus opticus; *DEN.*, diencephalon; *L.*, lens; *MS.*, mesencephalon; *S.DM.*, sulcus di-mesencephalicus; *V.OP.*, optic vesicle. (Carnegie Embryo No. 6502. Bartelmez and Dekaban, 1962.)

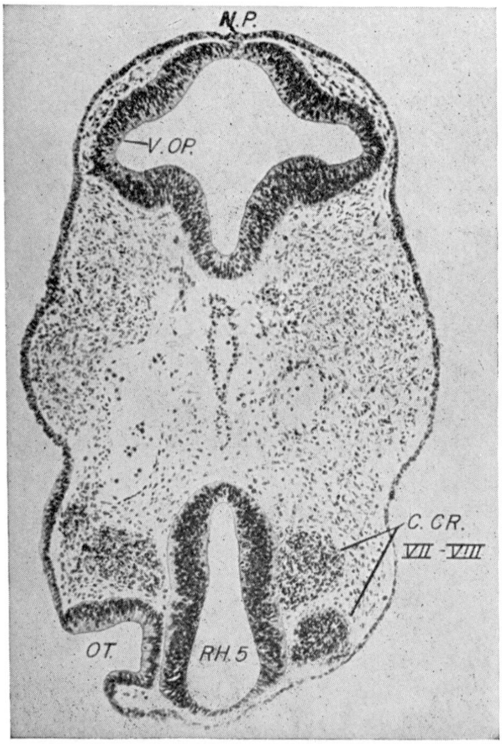

Fig. 13–6.—Brain of 3.9-mm human embryo, 22-23 somites. *C.CR.*, neural crest of nerves VII and VIII; *N.P.*, neuropore; *OT.*, otic vesicle; *RH*, rhombencephalon; *V.OP.*, optic vesicle. (Carnegie embryo No. 8943. Bartelmez and Dekeban, Contribution to Embryology, Vol. 37, 1962.)

forms the **lens vesicle**. At the same time the optic vesicle partially encircles the lens vesicle, forming the **optic cup**. Its two layers are continuous with each other at the cup margin, which ultimately overlaps the front of the lens except for the future aperture of the pupil (Figs. 13–6 to 13–9). The process of invagination also causes an infolding of the posterior surface of the vesicle and the optic stalk which produces the **choroidal fissure** (Fig. 13–10). Mesenchyma and the retinal blood vessels (hyaloid artery) grow into the fissure. The fissure closes during the seventh week and the two-layered optic cup and optic stalk become complete. Sometimes the choroidal fissure persists, and when this occurs the choroid and iris in the region of the fissure remain undeveloped, giving rise to the condition known as *coloboma* of the choroid or iris.

The **retina** is developed from the optic cup. The outer stratum of the cup persists as a single layer of cells which assume a columnar shape, acquire pigment, and form the pigmented layer of the retina; the pig-

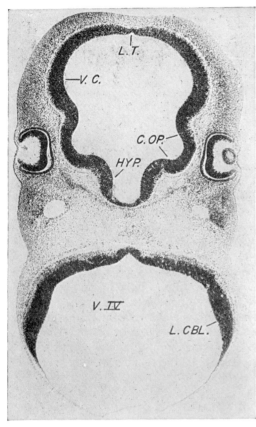

Fig. 13–8.—Brain of 7.5-mm human embryo. C.OP., conus opticus; HYP., hypophysis; L. CBL., lamina cerebellaris; L.T., lamina terminalis; V.C., cerebral vesicle; V.IV., fourth ventricle. (Carnegie Embryo No. 6506. Bartelmez and Dekaban, 1962.)

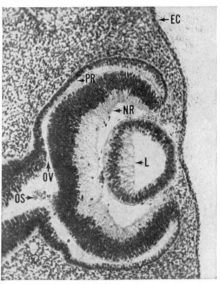

Fig. 13–9.—Optic cup and lens vesicle, 8-mm human embryo. EC, ectoderm; L, lens; NR, neural part of retina; OS, optic stalk; OV, optic vesicle; PR, pigment layer of retina. (Courtesy of Doctor Aeleta Barber.)

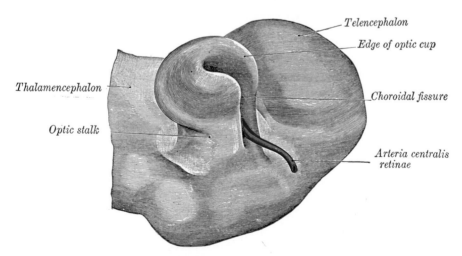

Fig. 13–10.—Optic cup and choroidal fissure seen from below, from a human embryo of about four weeks. (Kollmann.)

ment first appears in the cells near the edge of the cup. The cells of the inner stratum proliferate and form a layer of considerable thickness from which the nervous elements and the sustentacular fibers of the retina are developed. In that portion of the cup which overlaps the lens the inner stratum is not differentiated into nervous elements, but forms a layer of columnar cells which is applied to the pigmented layer, and these two strata form the **pars ciliaris** and **pars iridica retinae.**

The cells of the inner or retinal layer of the optic cup become differentiated into spongioblasts and germinal cells, and the latter by their subdivisions give rise to neuroblasts. From the spongioblasts the sustentacular fibers of Müller, the outer and inner limiting membranes, together with the groundwork of the molecular layers of the retina are formed. The neuroblasts become arranged to form the ganglionic and nuclear layers. The layer of rods and cones is first developed in the central part of the optic cup, and from there gradually extends toward the cup margin. All the layers of the retina are completed by the eighth month of fetal life.

The optic stalk is converted into the **optic nerve** by the obliteration of its cavity and the growth of nerve fibers into it. Most of these fibers are centripetal, and grow backward into the optic stalk from the nerve cells of the retina, but a few extend in the opposite direction and are derived from nerve cells in the brain. The fibers of the optic nerve receive their myelin sheaths about the tenth week after birth. The **optic chiasma** is formed by the meeting and partial decussation of the fibers of the two optic nerves. Behind the chiasma the fibers grow backward as the optic tracts to the thalami and midbrain.

The **crystalline lens** is developed from the lens vesicle, which recedes within the margin of the cup, and becomes separated by mesoderm from the overlying ectoderm. The cells forming the posterior wall of the vesicle lengthen and are converted into the lens fibers, which grow forward and fill up the cavity of the vesicle (Fig. 13–11). The cells forming the anterior wall retain their cellular character, and form the epithelium on the anterior surface of the adult lens.

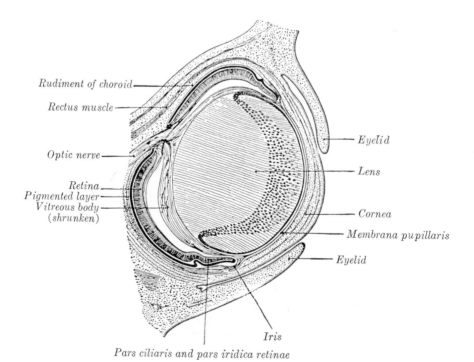

Rudiment of choroid

Rectus muscle

Optic nerve

Retina
Pigmented layer
Vitreous body
(shrunken)

Eyelid

Lens

Cornea

Membrana pupillaris

Eyelid

Iris

Pars ciliaris and pars iridica retinae

FIG. 13–11.—Horizontal section through the eye of an eighteen-day-old rabbit embryo. × 30. (Kölliker.)

The Hyaloid Artery.—A capillary net continuous with the primitive choroid net enters the choroid fissure. As the fissure closes, connections with the choroid net are all cut off except at the edge of the cup. The vessel enclosed in the optic stalk is the hyaloid artery. Its branches surround the deep surface of the lens and drain into the choroid net at the margin of the cup. As the vitreous body increases, the hyaloid supplies branches to it. By the second month the lens is invested by a vascular mesodermal capsule, the **capsula vasculosa lentis;** the blood vessels supplying the posterior part of this capsule are derived from the hyaloid artery; those for the anterior part from the anterior ciliary arteries; the portion of the capsule which covers the front of the lens is named the **pupillary membrane.** By the sixth month all the vessels of the capsule are atrophied except the hyaloid artery, which disappears during the ninth month; the position of this artery is indicated in the adult by the hyaloid canal, which reaches from the optic disk to the posterior surface of the lens. With the loss of its blood vessels the capsula vasculosa lentis disappears, but sometimes the pupillary membrane persists at birth, giving rise to the condition termed *congenital atresia of the pupil.*

The Central Artery of the Retina.—By the fourth month branches of the hyaloid artery and veins which have developed during the third month begin to spread out in the retina and reach the ora serrata by the eighth month. After atrophy of the vitreous part of the hyaloid vessels, the proximal part in the optic nerve and retina becomes the central artery of the retina.

The **choroid** is analogous to the leptomeninges of the brain and spinal cord. It develops from mesenchyma between the sclera (dura) and the optic cup (an extension of the brain wall). The mesenchyma is invaded by capillaries from the ciliary vessels. They form a rich plexus over the outer surface of the optic cup and connect with the hyaloid capillary plexus in the pupillary region. The mesenchyma forms a loose network of fibroblasts, pigmented cells, and collagenous fibrils, partially separated by mesothelial-lined, fluid-containing spaces.

The **Canal of Schlemm** (*sinus venosus sclerae*) and the circular vessels of the iris develop from the anterior extension of the plexus of the choroid. The former is analogous to a venous sinus of the dura mater; the anterior chamber to the subarachnoid spaces. The drainage of aqueous humor via the pectinate villi into the canal of Schlemm is analogous to the drainage of the subarachnoid fluid via the arachnoid granulations into the dural sinuses.

The **vitreous body** (Fig. 13–12) develops between the lens and the optic cup as the two structures become separated. Some authors believe that at the beginning, the retina, and perhaps the lens also, play roles in the formation of the vitreous body; others believe that the retina plays the sole role; and still others, that both retina and invading mesenchyma are involved. Fixed and stained preparations show throughout the vitreous body a delicate network of fibrils continuous with the long processes of stellate mesenchymal cells and with the retina. Later the fibrils are limited to the ciliary region where they are supposed to form the **zonula ciliaris.**

The **sclera** (Fig. 13–12) is derived from the mesenchyma surrounding the optic cup. A dense layer of collagenous fibers, continuous with the sheath of the optic nerve, is formed by the fibroblasts. The sclera is analogous to the dura mater.

Most of the **cornea** (Fig. 13–12) is derived from mesenchyma which invades the region between the lens and ectoderm. The overlying ectoderm becomes the corneal epithelium. The endothelial (mesothelial) layer comes from mesenchymal cells which line the corneal side of the cleft (anterior chamber) which develops between cornea and pupillary membrane. The factors responsible for the transparency of the cornea are unknown.

The **anterior chamber** of the eye appears as a cleft in the mesoderm separating the lens from the overlying ectoderm. The layer of mesoderm anterior to the cleft forms the substantia propria of the cornea, that posterior to the cleft the stroma of the iris and the pupillary membrane.

Membrana Pupillaris.—In the fetus, the pupil is closed by a delicate vascular membrane, the **membrana pupillaris,** which

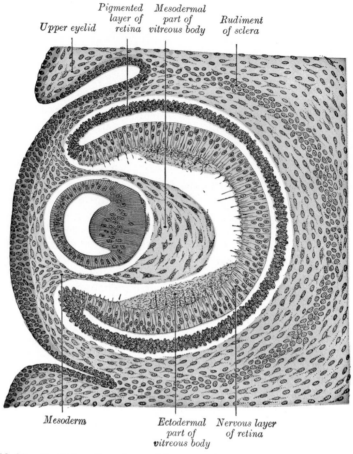

Upper eyelid — *Pigmented layer of retina* — *Mesodermal part of vitreous body* — *Rudiment of sclera*

Mesoderm — *Ectodermal part of vitreous body* — *Nervous layer of retina*

Fig. 13–12.—Sagittal section of eye of human embryo of six weeks. (Kollmann.)

divides the space in which the iris is suspended into two distinct chambers. The vessels of this membrane are partly derived from those of the margin of the iris and partly from those of the capsule of the lens; they have a looped arrangement, and converge toward each other without anastomosing. About the sixth month the membrane begins to disappear by absorption from the center toward the circumference, and at birth only a few fragments are present; in exceptional cases it persists.

The fibers of the **ciliary muscle** are derived from the mesoderm, but those of the Sphincter and Dilatator pupillae are of ectodermal origin, being developed from the cells of the pupillary part of the optic cup.

The **eyelids** are formed as small cutaneous folds (Figs. 13–11, 13–12), which, about the middle of the third month, come together and unite in front of the cornea. They remain united until about the end of the sixth month.

The **lacrimal sac** and **nasolacrimal duct** result from a thickening of the ectoderm in the groove, **nasoöptic furrow**, between the lateral nasal and maxillary processes. This thickening forms a solid cord of cells which sinks into the mesoderm; during the third month the central cells of the cord break down, and a lumen, the nasolacrimal duct, is established. The lacrimal ducts arise as buds from the upper part of cord of cells and secondarily establish openings (*puncta lacrimalia*) on the margins of the lids. The **epithelium** of the cornea and conjunctiva, and that which lines the ducts and alveoli of the lacrimal gland, is of ectodermal origin, as are also the **eyelashes** and the lining cells of the glands which open on the lid margins.

THE BULB OF THE EYE

The bulb of the eye is composed of segments of two spheres of different sizes. The anterior segment, the cornea, is from a small sphere, is transparent, and forms about one-sixth of the bulb. The posterior segment is from a larger sphere, is opaque, and forms about five-sixths of the bulb. The term **anterior pole** is applied to the central point of the anterior curvature of the bulb, and that of **posterior pole** to the central point of its posterior curvature; a line joining the two poles forms the **optic axis.** The axes of the two bulbs are nearly parallel, and therefore do not correspond to the axes of the orbits, which diverge lateralward. The optic nerves follow the direction of the axes of the orbits, and are therefore not parallel; each enters its eyeball 3 mm to the nasal side and a little below the level of the posterior pole. The bulb measures rather more in its transverse and anteroposterior diameters than in its vertical diameter, the former amounting to about 24 mm, the latter to about 23.5 mm; in the female all three diameters are rather less than in the male; its

anteroposterior diameter at birth is about 17.5 mm, and at puberty from 20 to 21 mm.

The Tunics of the Eye (Fig. 13–13)

The bulb is composed of three tunics: (1) An outer fibrous tunic, consisting of the **sclera** posteriorly and the **cornea** anteriorly; (2) an intermediate vascular, pigmented tunic, comprising the **choroid, ciliary body,** and **iris;** and (3) an internal nervous tunic, the **retina.**

1. **The Fibrous Tunic** (*tunica fibrosa bulbi*) (Fig. 13–13).—The sclera is opaque, and constitutes the posterior five-sixths of the tunic; the cornea is transparent, and forms the anterior sixth.

The **Sclera** has received its name from its density and hardness; it is a tough, inelastic membrane, which maintains the size and form of the bulb. Its thicker posterior part measures 1 mm. Its **external surface** is white, and is quite smooth, except at the points were the Recti and Obliqui are

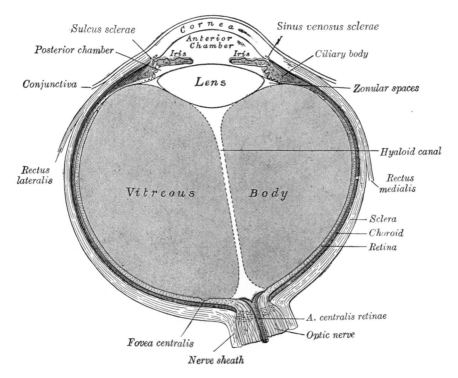

Fig. 13–13.—Horizontal section of the eyeball.

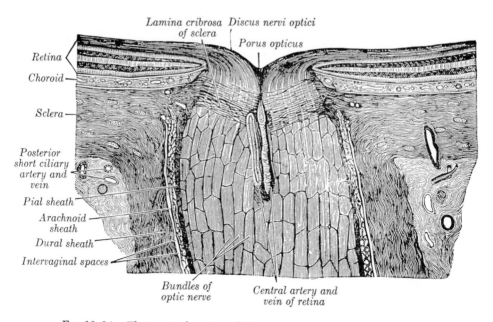

Fig. 13–14.—The terminal portion of the optic nerve as it traverses the lamina cribrosa of the sclera.

inserted into it. Its anterior part is covered by the conjunctival membrane; the rest of the surface is separated from the fascia bulbi (capsule of Tenon) by the loose connective tissue of a fascial cleft (see page 375) which permits rotation of the eyeball within the fascia bulbi. Its **inner surface** is brown in color and marked by grooves, in which the ciliary nerves and vessels are lodged; it is loosely attached to the pigmented lamina suprachoroidea of the choroid. Posteriorly it is pierced by the optic nerve, and is continuous through the fibrous sheath of this nerve with the dura mater. The optic nerve passes through the **lamina cribrosa sclerae** (Fig. 13–14) in which the minute orifices transmit the nervous filaments, and the fibrous septa dividing them from one another are continuous with those which separate the bundles of nerve fibers. One of these openings, larger than the rest and occupying the center of the lamina, transmits the central artery and vein of the retina. Around the entrance of the optic nerve are numerous small apertures for the transmission of the ciliary vessels and nerves, and about midway between the cribrosa and the sclerocorneal junction are four or five large apertures for the transmission of veins

(**venae vorticosae**). The sclera is directly continuous with the cornea anteriorly, the line of union being termed the **sclerocorneal junction.**

Near the junction the inner surface of the sclera projects into a circular ridge, the **scleral spur.** To it is attached the ciliary muscle and the iris. Anterior to this ridge a circular depression, the scleral sulcus, is crossed by trabecular tissue which separates the angle of the anterior chamber from the **sinus venosus sclerae** (canal of Schlemm). The spaces of the trabecular tissue (spaces of Fontana) connect on one side with the anterior chamber of the eye and on the other with the pectinate villi. The aqueous humor filters through the walls of the villi into the sinus venosus sclerae.

The **Cornea,** the transparent part of the external tunic, is almost circular in outline, occasionally a little broader in the transverse than in the vertical direction. It is convex anteriorly and projects like a dome beyond the sclera. Its degree of curvature varies in different individuals, and in the same individual at different periods of life, being more pronounced in youth than in advanced life. The cornea is dense and of uniform thickness throughout; its poste-

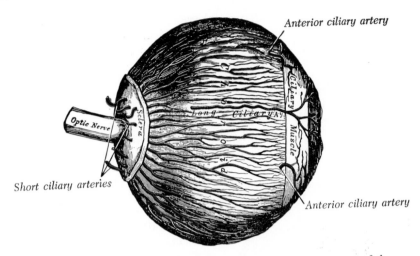

FIG. 13–15.—The arteries of the choroid and iris. The greater part of the
sclera has been removed. (Enlarged.)

rior circumference is perfectly circular in
outline, and exceeds the anterior slightly
in diameter. Immediately anterior to the
sclerocorneal junction the cornea bulges
inward as a thickened rim.

2. The **Vascular Tunic** (*tunica vasculosa bulbi; uvea*) (Figs. 13–15, 13–16, 13–
17) of the eye is composed of the choroid,
the ciliary body, and the iris.

The choroid invests the posterior five-
sixths of the bulb, and extends as far
anteriorly as the ora serrata of the retina.
The ciliary body connects the choroid to
the circumference of the iris. The iris is a
circular diaphragm behind the cornea, and
presents near its center a rounded aperture,
the **pupil**.

The **Choroid** (*choroidea*) is a thin, highly
vascular membrane, of a dark brown or
chocolate color, investing the posterior
five-sixths of the bulb; it is pierced posteriorly by the optic nerve, and in this
situation is firmly adherent to the sclera.
Its outer surface is loosely connected by
the lamina suprachoroidea with the sclera;
its inner surface is attached to the pigmented layer of the retina.

The **Ciliary Body** (*corpus ciliare*) extends
from the ora serrata of the retina to the
outer edge of the iris and the sclerocorneal
junction. It consists of the thickened vascular tunic of the eye and the ciliary muscle.
Its inner surface is covered by the thin, pigmented **ciliary part of the retina**. The

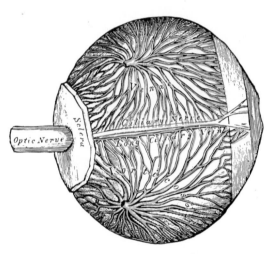

FIG. 13–16.—The veins of the choroid. (Enlarged.)

suspensory ligament of the lens is attached
to the ciliary body. The ciliary body comprises two zones, the orbiculus ciliaris and
the ciliary processes (Fig. 13–18).

The **orbiculus ciliaris** is 4 mm wide and
extends from the ora serrata to the ciliary
processes. Its thickness increases as it
approaches the ciliary processes owing to
the increase in thickness of the ciliary muscle. The choroid layer is thicker here than
over the optical part of the retina. Its inner
surface presents numerous small, radially
arranged ridges, the ciliary processes.

The **ciliary processes** (*processus ciliares*) are formed by the inward folding of the various layers of the choroid, *i.e.*, the choroid proper and the lamina basalis, and are received between corresponding foldings of the suspensory ligament of the lens. They are arranged radially on the posterior surface of the iris, forming a sort of frill around the margin of the lens (Fig. 13–18). They vary from sixty to eighty in number; the larger ones are about 2.5 mm in length, and the smaller, consisting of about one-third of the entire number, are situated in spaces between them, but without regular arrangement. They are attached by their periphery to three or four of the ridges of the orbiculus ciliaris, and are continuous with the layers of the choroid. Their other extremities are free and rounded, and are directed toward the circumference of the lens. Anteriorly, they are continuous with the periphery of the iris. Their posterior surfaces are connected with the suspensory ligament of the lens.

The **Ciliaris muscle** (*m. ciliaris; ciliary muscle*) consists of unstriped fibers; it forms a grayish, semitransparent, circular band, about 3 mm broad, on the outer surface of the anterior part of the choroid. It is thickest anteriorly, and consists of two sets of fibers, **meridional** and **circular.** The meridional fibers, much the more numerous, arise from the posterior margin of the scleral spur; they run posteriorly, and are attached to the ciliary processes and orbiculus ciliaris. The circular fibers are internal to the meridional ones, and in a meridional section appear as a triangular zone behind the filtration angle and close to the circumference of the iris. They are well developed in hypermetropic, but are rudimentary or absent in myopic eyes. The Ciliaris muscle is the chief agent in accommodation, *i.e.*, in adjusting the eye to the vision of near

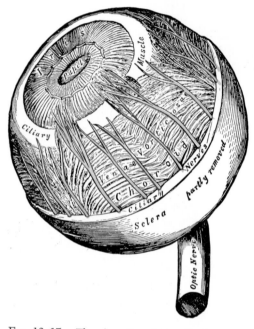

Fig. 13–17.—The choroid and iris. (Enlarged.)

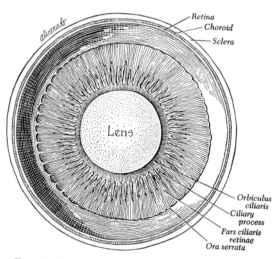

Fig. 13–18.—Interior of anterior half of bulb of eye.

objects. When it contracts it draws the ciliary processes centripetally, relaxes the suspensory ligament of the lens, and thus allows the lens to become more convex.

The **Iris.**—The iris has received its name from its various colors in different individuals. It is a circular, contractile disk suspended in the aqueous humor between the cornea and lens, and perforated a little to the nasal side of its center by a circular aperture, the **pupil.** By its periphery it is continuous with the ciliary body, and is also connected with the posterior elastic lamina of the cornea by means of the pectinate ligament; its flattened surfaces are anterior and posterior, the anterior toward the cornea, the posterior toward the ciliary processes and lens. The iris divides the space between the lens and the cornea into an anterior and a posterior chamber. The **anterior chamber** of the eye is bounded anteriorly by the posterior surface of the cornea; posteriorly by the iris and the central part of the lens. The **posterior chamber** is a narrow chink posterior to the peripheral part of the iris, and anterior to the suspensory ligament of the lens and the ciliary processes (Fig. 13–13). In the adult the two chambers communicate through the pupil, but in the fetus, up to the seventh month, they are separated by the *membrana pupillaris* (page 1043).

3. The **Internal Tunic** (*tunica interna bulbi*) is a continuous membrane, but has a posterior nervous part, the retina proper, and an anterior non-nervous part, the **pars ciliaris** and **pars iridica retinae.**

The **Retina** is a delicate nervous membrane, upon which the images of external objects are received. Its outer surface is in contact with the choroid; its inner with the vitreous body. Posteriorly, it is continuous with the optic nerve; it gradually diminishes in thickness anteriorly, and extends nearly as far as the ciliary body, where it appears to end in a jagged margin, the **ora serrata.** Here the nervous tissues of the retina end, but a thin prolongation of the membrane extends anteriorly over the back of the ciliary processes and iris, forming the **pars ciliaris retinae** and **pars iridica retinae** already referred to. This forward prolonga-

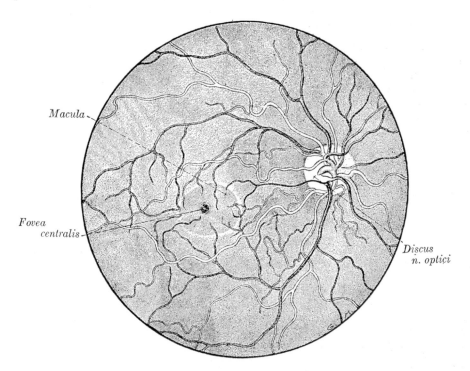

Fig. 13–19.—Interior of posterior half of right eye as viewed through an ophthalmoscope. The distribution of central vessels of the retina is shown (veins darker than arteries) and their relation to the optic disk. The area of most acute vision, the macula, is shown. (Eycleshymer and Jones.)

tion consists of the pigmentary layer of the retina together with a stratum of columnar epithelium. The retina is soft, semitransparent, and of a purple tint in the fresh state, owing to the presence of a coloring material named **rhodopsin** or **visual purple**, but it soon becomes clouded, opaque, and bleached when exposed to sunlight. Exactly in the center of the posterior part of the retina, corresponding to the optical axis of the eye, and at a point in which the sense of vision is most perfect, is an oval yellowish area, the **macula**; in the macula is a central depression, the **fovea centralis** (Fig. 13–18). At the fovea centralis the retina is exceedingly thin, and the dark color of the choroid is distinctly seen through it. About 3 mm to the nasal side of the macula is the **optic disk** (*discus nervi optici*) which marks the exit of the optic nerve. A depression in the center of the disk, the porus opticus, marks the point of entrance of the arteria centralis retinae. The disk is the only part of the surface of the retina which is insensitive to light, and it is termed the **blind spot** (Fig. 13–19).

Structure of the Tunics of the Bulb

Structure of the Sclera.—The sclera is formed of bundles and laminae of collagenous fibers intermixed with fine elastic fibers, fibroblasts and other connective tissue cells. Its *vessels* are not numerous, the capillaries being of small size, uniting at long and wide intervals. Its *nerves* are derived from the ciliary nerves.

Structure of the Cornea (Fig. 13–20).—The cornea consists of five layers: (1) the **corneal epithelium,** continuous with that of the conjunctiva; (2) the **anterior lamina;** (3) the **substantia propria;** (4) the **posterior lamina;** and (5) the **endothelium** (mesothelium) of the anterior chamber.

(1) The **corneal epithelium** (*epithelium anterius corneae; anterior layer*) covers the front of the cornea and consists of several layers of cells. The cells of the deepest layer are columnar; then follow two or three layers of polyhedral cells, the majority of which are prickle cells similar to those found in the stratum mucosum of the cuticle. Lastly, there are three or four layers of squamous cells, with flattened nuclei.

(2) The **anterior lamina** (*lamina limitans anterior; anterior limiting layer; Bowman's membrane*) consists of extremely closely interwoven fibrils, similar to those found in the substantia propria, but containing no corneal corpuscles. It may be regarded as a condensed part of the substantia propria.

(3) The **substantia propria corneae** is fibrous, tough, unyielding, and perfectly transparent. It is composed of about sixty flattened superimposed lamellae made up of bundles of modified connective tissue, the fibers of which are directly continuous with those of the sclera. The fibers of each lamella are for the most part parallel with one another, but at right angles to those of adjacent lamellae, from one lamella to the next. The lamellae are held together by an interstitial cement substance, except for the **corneal spaces** which contain the **corneal corpuscles,** modified fibroblasts resembling the space in form, but not entirely filling them.

(4) The **posterior elastic lamina** (*lamina limitans posterior; membrane of Descemet; membrane of Demours*) covers the posterior surface of the substantia propria, and is an elastic, transparent homogeneous membrane, of extreme thinness, which is not rendered opaque by either water, alcohol, or acids. When stripped from the substantia propria it curls up, or rolls upon itself with the attached surface innermost.

At the margin of the cornea the posterior elastic lamina breaks up into fibers which form the trabecular tissue already described. Some of the fibers of this trabecular tissue are continued into the substance of the iris, forming the **pectinate ligament of the iris** (*Lig. pectinatum anguli iridocornealis*), while others are connected with the forepart of the sclera and choroid.

(5) The **endothelium** of the anterior chamber (*endothelium camerae anterioris; posterior layer; corneal endothelium*) is a mesothelial, rather than an endothelial layer. It covers the posterior surface of the elastic lamina, is reflected on to the front of the iris, and also lines the spaces of the angle of the iris; it consists of a single stratum of polygonal, flattened, nucleated cells.

Vessels and Nerves.—The cornea is a non-vascular structure; the capillary vessels ending in loops at its circumference are derived from the anterior ciliary arteries. Lymphatic vessels have not yet been demonstrated in it, but are represented by the channels in which the bundles of nerves run; these channels are lined by an endothelium. The **nerves** are numerous and are derived from the ciliary nerves. Around the periphery of the cornea they form an *annular plexus*, from which fibers enter the substantia propria. They lose their myelin sheaths and ramify throughout its substance in a delicate network, and their terminal filaments

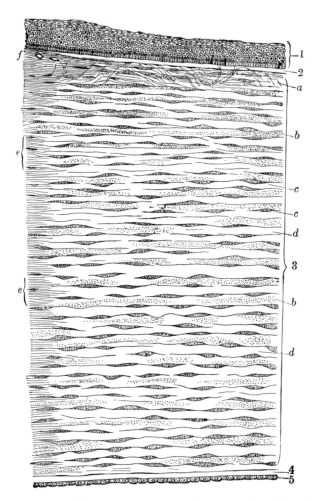

Fig. 13–20.—Vertical section of human cornea from near the margin. (Waldeyer.) Magnified. 1. Epithelium. 2. Anterior lamina. 3. Substantia propria. 4. Posterior elastic lamina. 5. Endothelium of the anterior chamber. *a*, Oblique fibers in the anterior layer of the substantia propria. *b*, Lamellae, the fibers of which are cut across, producing a dotted appearance. *c*, Corneal corpuscles appearing fusiform in section. *d*, Lamellae, the fibers of which are cut longitudinally. *e*, Transition to the sclera, with more distinct fibrillation, and surmounted by a thicker epithelium. *f*, Small blood vessels cut across near the margin of the cornea.

form a firm and closer plexus on the surface of the cornea proper, beneath the epithelium. This is termed the *subepithelial plexus,* and from it fibrils are given off which ramify between the epithelial cells, forming an *intra-epithelial plexus.*

Structure of the Choroid.—The choroid consists mainly of a dense capillary plexus, and of small arteries and veins carrying blood to and returning it from this plexus. On its external surface is a thin membrane, the **lamina suprachoroidea,** composed of delicate non-vascular lamellae—each lamella consisting of a network of fine elastic fibers among which are branched pigment cells. The potential spaces between the lamellae are lined by mesothelium, and open freely into the **perichoroidal space** (*spatium perichoroideale*).

Internal to this lamina is the **choroid proper,** consisting of two layers: an outer, composed of small arteries and veins, with pigment cells interspersed between them; and an inner, consisting of a capillary plexus.

The **outer layer** (*lamina vasculosa*) consists, in part, of the larger branches of the short ciliary arteries which run between the veins before they bend inward to end in the capillaries, but is formed principally of veins, named, from their arrangement, the **venae vorticosae** (Figs. 13–16, 13–17). They converge to four or five equidistant trunks, which pierce the sclera about midway between the sclerocorneal junction and the entrance of the optic nerve. Interspersed between the vessels are dark star-staped pigment cells, the processes of which, communicating with those of neighboring cells, form a delicate network or stroma, which loses its pigmentary character toward the inner surface of the choroid.

The **inner layer** (*lamina choriocapillaris*) consists of an exceedingly fine capillary plexus, formed by the short ciliary vessels; the network is closer and finer in the posterior than in the anterior part of the choroid. About 1.25 cm behind the cornea its meshes become larger, and are continuous with those of the ciliary processes. These two laminae are connected by a **stratum intermedium** consisting of fine elastic fibers. On the inner surface of the lamina choriocapillaris is a very thin, structureless, or faintly fibrous membrane, called the **lamina basalis;** it is closely connected with the stroma of the choroid, and separates it from the pigmentary layer of the retina.

One of the functions of the choroid is to provide nutrition for the retina, and to convey vessels and nerves to the ciliary body and iris.

Tapetum.—This name is applied to the outer and posterior part of the choroid, which in many animals presents an iridescent appearance.

Structure of the Ciliary Processes.—The ciliary processes (Fig. 13–18) are similar in structure to the choroid, but the vessels are larger, and have chiefly a longitudinal direction. Their posterior surfaces are covered by a bilaminar layer of black pigment cells, which is continued anteriorly from the retina, and is named the **pars ciliaris retinae.** In the stroma of the ciliary processes there are also stellate pigment cells, but these are not so numerous as in the choroid itself.

Structure of the Iris.—The iris is composed of the following structures:

1. A layer of flattened mesothelial cells (*endothelium camerae anterius*) is placed on a delicate hyaline basement membrane on the anterior surface (*facies anterior*). This layer is continuous with the mesothelium covering the posterior elastic lamina of the cornea, and in individuals with dark-colored irides the cells contain pigment granules.

2. The **stroma** (*stroma iridis*) of the iris consists of fibers and cells. The former are made up of delicate collagenous bundles; a few fibers at the circumference of the iris have a circular direction, but the majority radiate toward the pupil, forming by their interlacement, delicate meshes, in which the vessels and nerves are contained. Interspersed between the bundles of connective tissue are numerous branched cells with fine processes. In dark eyes many of them contain pigment granules, but in blue eyes and the eyes of albinos they are unpigmented.

3. The **muscular fibers** are involuntary, and consist of circular and radiating fibers. The **circular fibers** form the **Sphincter pupillae;** they are arranged in a narrow band about 1 mm in width which surrounds the margin of the pupil toward the posterior surface of the iris; those near the free margin are closely aggregated; those near the periphery of the band are somewhat separated and form incomplete circles. The **radiating fibers** form the **Dilatator pupillae;** they converge from the circumference toward the center, and blend with the circular fibers near the margin of the pupil.

4. The posterior surface of the iris is of a deep purple tint, being covered by two layers of pigmented columnar epithelium, continuous at the periphery of the iris with the pars ciliaris retinae. This pigmented epithelium is named the **pars iridica retinae.**

The color of the iris is produced by the reflection of light from dark pigment cells underlying a translucent tissue, and is therefore determined by the amount of the pigment and its distribution throughout the texture of the iris. The number and the situation of the pigment cells differ in different irides. In the albino, pigment is absent; in the various shades of blue eyes the pigment cells are confined to the posterior surface of the iris, whereas in gray, brown, and black eyes pigment is found also in the cells of the stroma and in those of the endothelium on the front of the iris.

ciliary arteries, having reached the attached margin of the iris, divides into an upper and lower branch; these anastomose with corresponding branches from the opposite side and thus encircle the iris; into this vascular circle, **circulus arteriosus major,** the anterior ciliary arteries pour their blood, and from it vessels converge to the free margin of the iris, and there communicate and form a second circle, **circulus arteriosus minor** (Fig. 13–21).

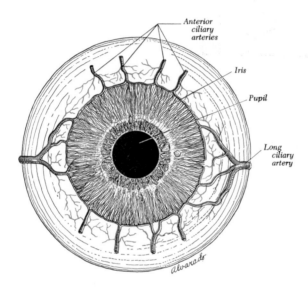

FIG. 13–21.—Iris, anterior view.

The iris may be absent, either in part or altogether as a congenital condition, and in some instances the pupillary membrane may remain persistent, although it is rarely complete. The iris may be the seat of a malformation, termed *coloboma,* which consists in a deficiency or cleft, due in a great number of cases to an arrest in development. In these cases the cleft is found at the lower aspect, extending directly downward from the pupil, and the gap frequently extends through the choroid to the porus opticus. In some rarer cases the gap is found in other parts of the iris, and is not then associated with any deficiency of the choroid.

Vessels and Nerves.—The **arteries of the iris** are derived from the long and anterior ciliary arteries, and from the vessels of the ciliary processes (see page 595). Each of the two long

The **nerves of the choroid and iris** (Fig. 13–16) are the long and short ciliary; the former being branches of the nasociliary nerve, the latter of the ciliary ganglion. They pierce the sclera around the entrance of the optic nerve, run forward in the perichoroidal space, and supply the blood vessels of the choroid. After reaching the iris they form a plexus around its attached margin; from this are derived unmyelinated fibers which end in the Sphincter and Dilatator pupillae. Other fibers from the plexus end in a network on the anterior surface of the iris. The fibers derived through the motor root of the ciliary ganglion from the oculomotor nerve supply the Sphincter, while those derived from the sympathetic supply the Dilatator.

Structure of the Retina (Figs. 13–22, 13–23).—The retina consists of an outer pigmented layer and an inner nervous stratum or retina proper.

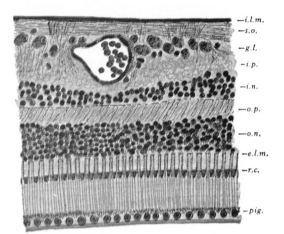

FIG. 13–22.—Section of human retina. *e.l.m.*, external limiting membrane; *g.l.*, ganglionic layer; *i.l.m.*, internal limiting membrane; *i.n.*, inner nuclear layer; *i.p.*, inner plexiform layer; *o.n.*, outer nuclear layer; *o.p.*, outer plexiform layer; *pig.*, pigment cell layer; *r.c.*, layer of rods and cones; *s.o.*, stratum opticum. × 500. (Redrawn from Sobotta.)

Retina Proper.—The nervous structures of the retina proper are supported by a series of non-nervous or sustentacular fibers. In histological sections made perpendicularly to the surface of the retina, ten layers are identified and named from within outward as follows:

1. Internal limiting membrane
2. Stratum opticum, layer of nerve fibers
3. Ganglion cell layer
4. Inner plexiform layer
5. Inner nuclear layer
6. Outer plexiform layer
7. Outer nuclear layer
8. External limiting membrane
9. Layer of rods and cones
10. Pigmented layer

1. The **internal limiting membrane** is a thin cribriform layer formed by the sustentacular fibers (page 1056).

2. The **stratum opticum** or **layer of nerve fibers** is formed by the expansion of the fibers of the optic nerve; it is thickest near the **porus opticus,** gradually diminishing toward the **ora serrata.** As the nerve fibers pass through the lamina cribrosa sclerae they lose their myelin sheaths and are continued onward through the choroid and retina as simple axons. When they reach the internal surface of the retina they

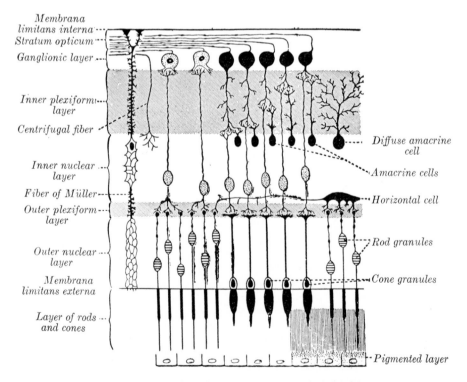

FIG. 13–23.—Plan of retinal neurons. (After Cajal.)

radiate from their point of entrance over this surface grouped in bundles, and in many places arranged in plexuses. Most of the fibers are centripetal, and are the direct continuations of the axons of the cells of the ganglionic layer, but a few of them are centrifugal and ramify in the inner plexiform and inner nuclear layers, where they end in enlarged extremities.

3. The **ganglion cell layer** consists of a single layer of large ganglion cells, except in the macula, where there are several strata. The cells are somewhat flask-shaped; the rounded internal surface of each resting on the stratum opticum sends an axon into the fiber layer. From the opposite end numerous dendrites extend into the inner plexiform layer, where they branch and form flattened arborizations at different levels. The ganglion cells vary much in size, and the dendrites of the smaller ones as a rule arborize in the inner plexiform layer as soon as they enter it, while those of the larger cells ramify close to the inner nuclear layer.

4. The **inner plexiform layer** is made up of a dense reticulum of minute fibrils formed by the interlacement of the dendrites of the ganglion cells with those of the cells of the inner nuclear layer; within this reticulum a few branched spongioblasts are embedded.

5. The **inner nuclear layer** is made up of a number of closely packed cells, of which there are three varieties, viz., bipolar cells, horizontal cells, and amacrine cells.

The **bipolar cells,** by far the most numerous, are round or oval in shape, and each is prolonged into an inner and an outer process. They are divisible into rod bipolars and cone bipolars. The inner processes of the **rod bipolars** run through the inner plexiform layer and arborize around the bodies of the cells of the ganglionic layer; their outer processes end in the outer plexiform layer in tufts of fibrils around the button-like ends of the inner processes of the rod granules. The inner processes of the **cone bipolars** ramify in the inner plexiform layer in contact with the dendrites of the ganglionic cells.

The **horizontal cells** lie in the outer part of the inner nuclear layer and possess somewhat flattened cell bodies. Their dendrites divide into numerous branches in the outer plexiform layer, while their axons run horizontally for some distance and finally ramify in the same layer.

The **amacrine cells** are placed in the inner part of the inner nuclear layer, and are so named because they have not yet been shown to possess axons. Their dendrites undergo

extensive ramification in the inner plexiform layer.

The cell bodies and nuclei of the sustentacular fibers are in this layer.

6. The **outer plexiform layer** is much thinner than the inner; but, like it, consists of a dense network of minute fibrils derived from the processes of the horizontal cells of the preceding layer, and the outer processes of the rod and cone bipolar cells, which ramify in it, forming arborizations around the enlarged ends of the rod fibers and with the branched foot plates of the cone fibers.

7. The **outer nuclear layer,** like the inner nuclear layer, contains several strata of oval cell bodies; they are of two kinds: rod and cone nuclei, so named on account of their being respectively connected with the rods and cones of the next layer. The **rod nuclei** are much more numerous, and are placed at different levels throughout the layer. Prolonged from either extremity of each cell is a fine process. The outer process is continuous with a single rod of the layer of rods and cones; the inner ends in the outer plexiform layer in an enlarged extremity, and is embedded in the tuft into which the outer processes of the rod bipolar cells break up. In its course it presents numerous varicosities. The **cone nuclei,** fewer in number than the rod nuclei, are placed close to the membrana limitans externa, through which they are continuous with the cones of the layer of rods and cones. They contain a pyriform nucleus which almost completely fills the cell. From the inner extremity, a thick process passes into the outer plexiform layer, and there expands into a pyramidal enlargement or foot plate from which are given off numerous fine fibrils that come in contact with the outer processes of the cone bipolars.

8. The **external limiting membrane** is formed by the sustentacular cells (see page 1056).

9. **The layer of rods and cones** (*Jacob's membrane*).—The elements composing this layer are of two kinds, **rods** and **cones,** the former being much more numerous than the latter except in the macula (Figs. 13–22, 13–23).

The **rods** are cylindrical 40 to 60 μ in length, of 2 μ thickness, and are arranged perpendicular to the surface. Each rod consists of two segments, an outer and inner, of about equal length. The outer segment is more slender, strongly retractive, is marked by transverse striae, and tends to break up into a number of thin disks superimposed on one another; it also exhibits faint longitudinal markings. The inner part of the inner segment is indistinctly granular; its outer part presents a longitudinal striation representing the ellipsoid which is composed of fine, bright, highly refracting fibrils. The visual purple or **rhodopsin,** found in the outer segments, gives a purple or reddish color in a dark-adapted eye, and is the pigment for detection of low light intensity.

The **cones** are conical or flask-shaped, their broad ends resting upon the membrana limitans externa, the narrow, pointed extremity being turned to the choroid. Like the rods, each is made up of two segments, outer and inner; the outer segment is a short conical process, which, like the outer segment of the rod, exhibits transverse striae. The inner segment resembles the inner segment of the rods in structure, but with a larger, thicker ellipsoid, and bulging deep granular part. The pigment in the outer segment of the cones is **iodopsin** which is for color vision.

Supporting Framework of the Retina.—The nervous layers of the retina are connected by a supporting framework, formed by the **sustentacular fibers of Müller;** these fibers pass through all the nervous layers, except that of the rods and cones. Each begins on the inner surface of the retina by an expanded, often forked, base, which sometimes contains a spheroidal body staining deeply with hematoxylin, the edges of the bases of adjoining fibers being united to form the **membrana limitans interna.** As the fibers pass through the nerve fiber and ganglionic layers they give off a few lateral branches; in the inner nuclear layer they give off numerous lateral processes for the support of the bipolar cells, while in the outer nuclear layer they form a network around the rod and cone fibrils, and unite to form the **membrana limitans externa** at the bases of the rods and cones. At the level of the inner nuclear layer each sustentacular fiber contains a clear oval nucleus.

Macula (*macula lutea*) **and Fovea Centralis.** —In the macula the nerve fibers are wanting as a continuous layer, the ganglionic layer consists of several strata of cells, and the rods are lacking. The cones are longer and narrower than in other parts, and in the outer nuclear layer there are only cone nuclei, the processes of which are very long and arranged in curved lines. In the fovea centralis the only parts present are (1) the cones; (2) the outer nuclear layer, the cone fibers of which are almost horizontal in direction; (3) an exceedingly thin inner plexiform layer. The pigmented layer is thicker and its pigment more pronounced than elsewhere. The color of the macula seems to imbue all the layers except that of the cones; it is of a rich yellow, deepest toward the center of the macula.

10. The **pigmented layer** consists of a single stratum of cells. When viewed from the outer surface these cells are smooth and hexagonal in shape; when seen in section each cell consists of an outer non-pigmented part containing a large oval nucleus and an inner pigmented portion which extends as a series of straight thread-like processes between the rods, this being especially the case when the eye is exposed to light. In the eyes of albinos the cells of this layer are destitute of pigment.

At the **ora serrata** (Fig. 13–18) the nervous layers of the retina end abruptly, and the retina is continued onward as a single layer of columnar cells covered by the pigmented layer. This double layer is known as the **pars ciliaris retinae,** and can be traced forward from the ciliary processes on to the back of the iris, where it is termed the **pars iridica retinae.**

The **arteria centralis retinae** (Fig. 13–19) and its accompanying vein pierce the optic nerve, and enter the bulb of the eye through the porus opticus. The artery immediately bifurcates into an upper and a lower branch, and each of these again divides into a medial or nasal and a lateral or temporal branch, which at first run between the hyaloid membrane and the nervous layer; but they soon enter the latter, and pass anteriorly, dividing dichotomously. From these branches a minute capillary plexus is given off, which does not extend beyond the inner nuclear layer. The macula receives two small branches (superior and inferior macular arteries) from the temporal branches and small twigs directly from the central artery; these do not, however, reach as far as the fovea centralis, which has no blood vessels. The branches of the arteria centralis retinae do not anastomose with each other—in other words they are terminal arteries. In the fetus, a small vessel, the *arteria hyaloidea,* passes forward as a continuation of the arteria centralis retinae through the vitreous humor to the posterior surface of the capsule of the lens.

The Refracting Media

The refracting media are five, viz.:

The Cornea (see page 1046)
Aqueous humor
Vitreous body
Zonula ciliaris
Crystalline lens

The **Aqueous Humor** (*humor aqueus*) fills the anterior and posterior chambers (*camerae anterior et posterior*). It is small in quantity, has an alkaline reaction, and consists mainly of water. The aqueous humor is secreted by the ciliary process. The fluid passes through the posterior chamber and the pupil into the anterior chamber. From the angle of the anterior chamber it passes into the spaces of Fontana to the pectinate villi through which it is filtered into the venous canal of Schlemm.

The **Vitreous Body** (*corpus vitreum*) fills the concavity of the pars optica retinae to which it is firmly adherent, especially at the ora serrata. It is transparent, semigelatinous, and is hollowed anteriorly for the lens. Some indications of the hyaloid canal between the optic nerve and lens may persist.

No blood vessels penetrate the vitreous body, so that its nutrition must be carried on by vessels of the retina and ciliary processes, situated upon its exterior.

The **Zonula Ciliaris** (*zonule of Zinn, suspensory ligament of the lens*) consists of a series of straight fibrils which radiate from the ciliary body to the lens. It is attached to the capsule of the lens a short distance anterior to its equator. Scattered delicate fibers are also attached to the region of the equator itself. This ligament retains the lens in position, and is relaxed by the contraction of the meridional fibers of the Ciliaris muscle, so that the lens is allowed to become more convex. Posterior to the suspensory ligament there is a sacculated canal, the **spatia zonularia** (*canal of Petit*), which encircles the equator of the

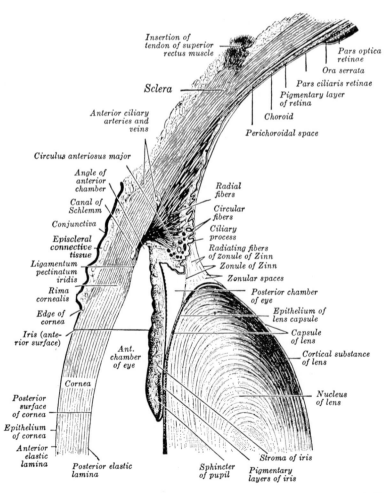

Fig. 13–24.—The upper half of a sagittal section through the front of the eyeball.

lens; it can be easily inflated through a fine blowpipe inserted under the suspensory ligament.

The **Crystalline Lens** (*lens*), enclosed in its capsule, is situated immediately between the iris and the vitreous body, encircled by the ciliary processes, which slightly overlap its margin.

The **capsule of the lens** (*capsula lentis*) is a transparent, structureless membrane which closely surrounds the lens, and is thicker in front than behind. It is brittle but highly elastic, and when ruptured the edges roll up with the outer surface innermost. It rests in the hyaloid fossa in the anterior part of the vitreous body posteriorly, and is in contact with the free border of the iris anteriorly, but recedes from it at the circumference, thus forming the posterior chamber of the eye. It is retained in its position chiefly by the suspensory ligament of the lens, already described.

The **lens** is a transparent, biconvex body, the convexity of its anterior being less than that of its posterior surface. The central points of these surfaces are termed respectively the **anterior** and **posterior poles**; a line connecting the poles constitutes the **axis** of the lens, while the marginal circumference is termed the **equator**.

Structure.—The lens is made up of soft cortical substance and a firm, central part, the **nucleus** (Fig. 13–25). Faint lines (*radii lentis*) radiate from the poles to the equator. In the adult there may be six or more of these lines, but in the fetus there are only three and they diverge from each other at angles of 120° (Fig. 13–26); on the anterior surface one line ascends vertically and the other two diverge downward; on the posterior surface one ray descends vertically and the other two diverge upward. They correspond with the free edges of an equal number of septa composed of an amorphous substance, which dip into the substance of the lens, and mark the interruptions of a series of concentrically arranged laminae (Fig. 13–25). Each lamina is built up of a number of hexagonal, ribbon-like lens fibers, the edges of which are more or less serrated— the serrations fitting between those of neighboring fibers, while the ends of the fibers come into apposition at the septa. The fibers run in a curved manner from the septa on the anterior surface to those on the posterior surface. No fibers pass from pole to pole; they are arranged

FIG. 13–25.—The crystalline lens, hardened and divided. (Enlarged.)

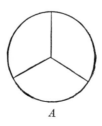
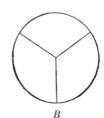

FIG. 13–26.—Diagram to show the direction and arrangement of the radiating lines on the front and back of the fetal lens. *A*, From the front. *B*, From the back.

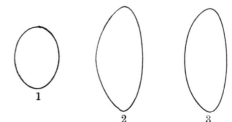

FIG. 13–27.—Profile views of the lens at different periods of life. 1. In the fetus. 2. In adult life. 3. In old age.

in such a way that those which begin near the pole on one surface of the lens end near the peripheral extremity of the plane on the other, and *vice versa*. The fibers of the outer layers of the lens are nucleated, and together form a nuclear layer, most distinct toward the equator. The anterior surface of the lens is covered by a layer of transparent, columnar, nucleated epithelium. At the equator the cells become elongated, and their gradual transition into lens fibers can be traced (Fig. 13–28).

In the fetus, the lens is nearly spherical, and has a slightly reddish tint; it is soft and breaks down readily on the slightest pressure. A small branch from the arteria centralis retinae runs forward, as already mentioned, through the vitreous body to the posterior part of the capsule of the lens, where its branches radiate and form a plexiform network, which covers the posterior surface of the capsule, and they are continuous around the margin of the capsule with the vessels of the pupillary membrane, and with those of the iris. **In the adult,** the lens is colorless, transparent, firm in texture, and devoid

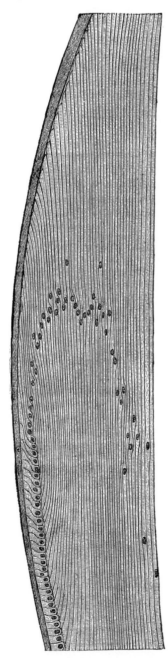

Fig. 13–28.—Section through the margin of the lens, showing the transition of the epithelium into the lens fibers. (Babuchin.)

of vessels. **In old age** it becomes flattened on both surfaces, slightly opaque, of an amber tint, and increased in density (Fig. 13–27).

Vessels and Nerves of the Bulb (Fig. 13–15).—The **arteries** of the eye are the long, short, and anterior ciliary arteries, and the arteria centralis retinae. They have already been described (see page 595).

The **ciliary veins** are seen on the outer surface of the choroid, and are named, from their arrangement, the *venae vorticosae* (Fig. 13–16); they converge to four or five equidistant trunks which pierce the sclera midway between the sclerocorneal junction and the porus opticus. Another set of veins accompanies the anterior ciliary arteries. All these veins open into the ophthalmic veins.

The **ciliary nerves** are derived from the nasociliary nerve and from the ciliary ganglion.

THE ACCESSORY ORGANS OF THE EYE

The **accessory organs of the eye** (*organa oculi accessoria*) include the **ocular muscles,** the **fasciae,** the **eyebrows,** the **eyelids,** the **conjunctiva,** and the **lacrimal apparatus.**

The **Ocular Muscles** (*musculi oculi*).— The ocular muscles are:

> Levator palpebrae superioris
> Rectus superior
> Rectus inferior
> Rectus medialis
> Rectus lateralis
> Obliquus superior
> Obliquus inferior

The **Levator palpebrae superioris** (Fig. 13–29) is a thin, flat muscle *arising* from the inferior surface of the small wing of the sphenoid, where it is separated from the optic foramen by the origin of the Rectus superior. At its origin it is narrow and tendinous, but becomes broad and fleshy and ends anteriorly in a wide aponeurosis which splits into three lamellae. The superficial lamella blends with the superior part of the orbital septum, and is prolonged over the superior tarsus to the palpebral part of the Orbicularis oculi, and to the deep surface of the skin of the upper eyelid. The middle lamella, largely made up of non-striated muscular fibers, is inserted into the superior margin of the superior tarsus, while the deepest lamella blends with an expansion from the sheath of the Rectus superior and with it is attached to the superior fornix of the conjunctiva.

The four **Recti** (Fig. 13–30) *arise* from a fibrous ring (*anulus tendineus communis*) which surrounds the superior, medial, and inferior margins of the optic foramen and encircles the optic nerve (Fig. 13–31). The ring is completed by a tendinous bridge prolonged over the inferior and medial part

the nasociliary nerve, the abducent nerve, and the ophthalmic vein. Although these muscles present a common origin and are inserted in a similar manner into the sclera, there are certain differences between their length and breadth. The Rectus medialis is the broadest, the Rectus lateralis the longest, and the Rectus superior the thinnest and narrowest.

The **Obliquus superior oculi** (*superior oblique*) is a fusiform muscle, placed at the

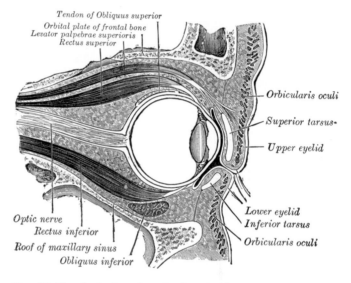

FIG. 13–29.—Sagittal section of right orbital cavity.

of the superior orbital fissure and attached to a tubercle on the margin of the great wing of the sphenoid, bounding the fissure. Two specialized parts of this fibrous ring may be made out: a lower, the *ligament* or *tendon of Zinn*, which gives origin to the **Rectus inferior**, part of the **Rectus medialis,** and the lower head of origin of the **Rectus lateralis**; and an upper, which gives origin to the **Rectus superior,** the rest of the Rectus medialis, and the upper head of the Rectus lateralis. This upper band is sometimes termed the *superior tendon of Lockwood.* Each muscle passes anteriorly in the position implied by its name, to be *inserted* by a tendinous expansion into the sclera, about 6 mm from the margin of the cornea. Between the two heads of the Rectus lateralis is a narrow interval, through which pass the two divisions of the oculomotor nerve,

superior and medial side of the orbit. It *arises* immediately above the margin of the optic foramen, superior and medial to the origin of the Rectus superior, and, passing anteriorly, ends in a rounded tendon which plays in a fibrocartilaginous ring or pulley attached to the trochlear fovea of the frontal bone. The contiguous surfaces of the tendon and ring are lined by a delicate synovial sheath, and enclosed in a thin fibrous investment. The tendon bends at more than a right angle around the pulley, passes beneath the Rectus superior to the lateral part of the bulb of the eye, and is *inserted* into the sclera, posterior to the equator of the eyeball, the insertion of the muscle lying between the Rectus superior and Rectus lateralis.

The **Obliquus inferior oculi** (*inferior oblique*) is a thin, narrow muscle, placed

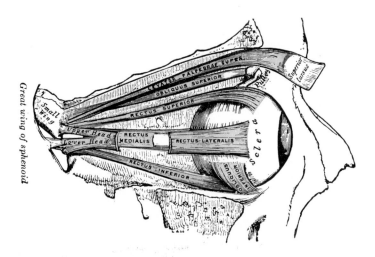

FIG. 13–30.—Muscles of the right orbit.

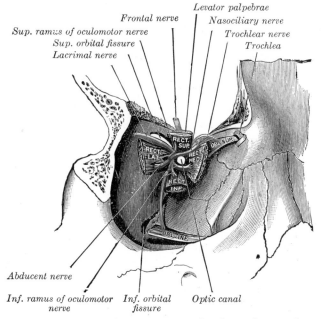

FIG. 13–31.—Dissection showing origins of right ocular muscles, and nerves entering by the superior orbital fissure.

near the anterior margin of the floor of the orbit. It *arises* from the orbital surface of the maxilla, lateral to the lacrimal groove. Passing lateralward, at first between the Rectus inferior and the floor of the orbit, and then between the bulb and the Rectus lateralis, it is *inserted* into the lateral part of the sclera between the Rectus superior and Rectus lateralis, near to, but somewhat posterior to, the insertion of the Obliquus superior.

medialis, on the other hand, produces a purely horizontal movement. If any two neighboring Recti of one eye act together they carry the globe of the eye in the diagonal of these directions, viz., upward and medialward, upward and lateralward, downward and medialward, or downward and lateralward. Sometimes the corresponding Recti of the two eyes act in unison, and at other times the opposite Recti act together. Thus, in turning the eyes to the right, the Rectus lateralis of the right eye will act in unison with the Rectus medialis of the

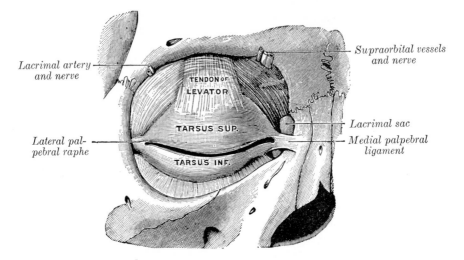

Lacrimal artery
and nerve

Lateral pal-
pebral raphe

TENDON OF LEVATOR

TARSUS SUP.

TARSUS INF.

Supraorbital vessels
and nerve

Lacrimal sac

Medial palpebral
ligament

Fig. 13–32.—The tarsi and their ligaments. Right eye; anterior view.

Nerves.—The Levator palpebrae superioris, Obliquus inferior, and the Recti superior, inferior, and medialis are supplied by the oculomotor nerve; the Obliquus superior, by the trochlear nerve; the Rectus lateralis, by the abducent nerve (Fig. 13–31).

Actions.—The Levator palpebrae *raises* the upper eyelid, and is the direct antagonist of the Orbicularis oculi. The four Recti are attached to the bulb of the eye in such a manner that, acting singly, they will turn its corneal surface in a direction corresponding to their names. The movement produced by the Rectus superior or Rectus inferior is not quite a simple one, because the axis of the orbit, and hence their direction of pull, diverges from the median plane, causing their pull to be accompanied by a certain deviation medialward, with a slight amount of rotation. These latter movements are corrected by the Obliqui, the Obliquus inferior correcting the medial deviation caused by the Rectus superior and the Obliquus superior that caused by the Rectus inferior. The contraction of the Rectus lateralis or Rectus

left eye; but if both eyes are directed to an object in the middle line at a short distance, the two Recti mediales will act in unison. The movement of circumduction, as in looking around a room, is performed by the successive actions of the four Recti. The Obliqui rotate the eyeball on its anteroposterior axis, the superior directing the cornea downward and lateralward, and the inferior directing it upward and lateralward; these movements are required for the correct viewing of an object when the head is moved laterally, as from shoulder to shoulder, in order that the picture may fall in all respects on the same part of the retina of either eye.

A layer of non-striped muscle, the **Orbitalis muscle** (of H. Müller), may be seen bridging across the inferior orbital fissure.

The **Fascia Bulbi** (*vaginae bulbi; capsule of Tenon*) is a thin membrane which envelops the eyeball from the optic nerve to the ciliary region, separating it from the orbital fat and forming a socket in which it plays.

Its inner surface is smooth, and is separated from the outer surface of the sclera by the **periscleral fascial cleft**. This space between the fascia and sclera (*spatium intervaginale*) is continuous with the subdural and subarachnoid cavities, and is traversed by delicate bands of connective tissue which extend between the fascia and the sclera. The fascia is perforated by the ciliary vessels and nerves, and fuses with the sheath of the optic nerve and with the sclera around the lamina cribrosa. Anteriorly it blends

and lateralis are strong, especially that from the latter muscle, and are attached to the lacrimal and zygomatic bones respectively. As they probably check the actions of these two Recti they have been named the **medial** and **lateral check ligaments**. A thickening of the lower part of the fascia bulbi, named the **suspensory ligament of the eye,** is slung like a hammock below the eyeball, being expanded in the center, and narrow at the ends which are attached to the zygomatic and lacrimal bones respectively.

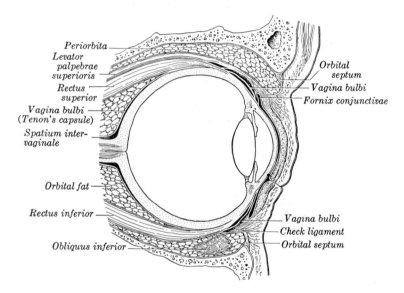

Fig. 13–33.—Fascia bulbi (capsule of Tenon). Sagittal section, semidiagrammatic. (Redrawn from *Surgical Anatomy of the Human Body* by J. B. Deaver, 1926. P. Blakiston's Son & Co.)

with the ocular conjunctiva, and with it is attached to the ciliary region of the bulb. It is prolonged over the tendons of the individual ocular muscles as a tubular sheath (*fascia musculares*). The sheath of the tendon of the Obliquus superior is carried as far as the fibrous pulley of that muscle; that on the Obliquus inferior reaches as far as the floor of the orbit, to which it gives off a slip. The sheaths on the Recti are gradually lost in the perimysium, but they give off important expansions. The expansion from the Rectus superior blends with the tendon of the Levator palpebrae; that of the Rectus inferior is attached to the inferior tarsus. The expansions from the sheaths of the Recti medialis

The **Periorbita** is the name given to the periosteum of the orbit. It is loosely connected to the bones and can be readily separated from them. It is continuous with the dura mater by processes which pass through the optic foramen and superior orbital fissure, and with the sheath of the optic nerve. At the margin of the orbit, a process from it assists in forming the **orbital septum.** It has two other processes, one to enclose the lacrimal gland, the other to hold the pulley of the Obliquus superior in position.

The **Orbital Septum** (*septum orbitale; palpebral ligament*) (Fig. 13–33) is a membranous sheet continuous with the periorbita where it is attached to the edge of the orbit.

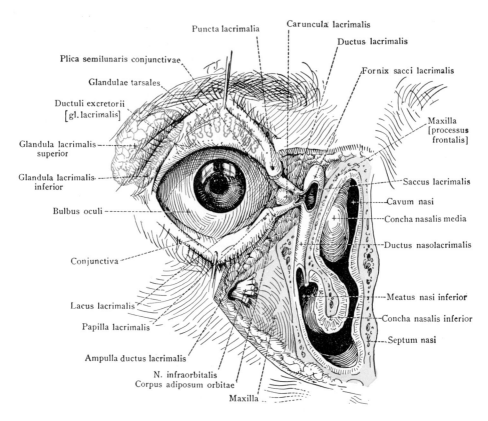

Fig. 13–34.—Topography of the lacrimal apparatus. The lacrimal and tarsal glands are shown in blue. The nose and a portion of the face have been cut away. (Eycleshymer and Jones.)

In the upper eyelid it blends by its peripheral circumference with the tendon of the Levator palpebrae superioris and the superior tarsus, in the lower eyelid with the inferior tarsus. Medially it is thin, and, becoming separated from the medial palpebral ligament, is fixed to the lacrimal bone immediately behind the lacrimal sac. The septum is perforated by the vessels and nerves which pass from the orbital cavity to the face and scalp. The eyelids are richly supplied with blood.

The **Eyebrows** (*supercilia*) are two arched eminences which surmount the superior margins of the orbits. The eyebrows consist of thickened integument, connected beneath with the Orbicularis oculi, Corrugator, and Frontalis muscles, and support numerous short, thick hairs.

The **Eyelids** (*palpebrae*) are two thin, movable covers, placed in front of the eye, protecting it from injury by their closure.

The upper eyelid is larger, and more movable than the lower, and is furnished with an elevator muscle, the Levator palpebrae superioris. When the eyelids are open, an elliptical space, the **palpebral fissure** (*rima palpebrarum*), is left between their margins, the angles of which correspond to the junctions of the upper and lower eyelids, and are called the **palpebral commissures or canthi.**

The **lateral palpebral commissure** (*external* or *lateral canthus*) is more acute than the medial, and the eyelids here lie in close contact with the bulb of the eye. The **medial palpebral commissure** (*internal* or *medial canthus*) is prolonged for a short distance toward the nose, and the two eyelids are separated by a triangular space, the **lacus lacrimalis** (Fig. 13–34). At the basal angles of the lacus lacrimalis, on the margin of each eyelid, is a small conical elevation, the **lacrimal papilla,** the apex of which is pierced by a small orifice, the **punctum lacrimale,** the commencement of the lacrimal duct.

The **Eyelashes** (*cilia*) are attached to the free edges of the eyelids; they are short, thick, curved hairs, arranged in a double or triple row: those of the upper eyelid, more numerous and longer than those of the lower, curve upward; those of the lower eyelid curve downward, so that they do not

The **Conjunctiva** is the mucous membrane of the eye. It lines the inner surfaces of the eyelids or palpebrae, and is reflected over the anterior part of the sclera.

The **palpebral portion** (*tunica conjunctiva palpebrarum*) is thick, opaque, highly vascular, and covered with numerous papillae, its deeper part presenting a considerable amount of lymphoid tissue. At the margins of the lids it becomes continuous

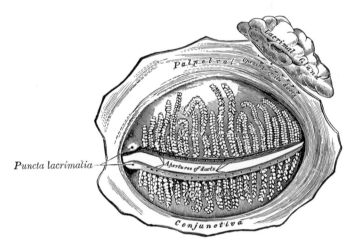

Fig. 13–35.—The tarsal glands, etc., seen from the inner surface of the eyelids.

interlace when the lids are closed. Near the attachment of the eyelashes are the openings of a number of glands, the **ciliary glands** arranged in several rows close to the free margin of the lid; they are regarded as enlarged and modified sudoriferous glands.

The **Tarsal Glands** (*glandulae tarsales; Meibomian glands*) (Figs. 13–35, 13–36) are situated upon the inner surfaces of the eyelids, between the tarsi and conjunctiva, and may be distinctly seen through the latter when the eyelids are everted, presenting an appearance similar to parallel strings of pearls. There are about thirty in the upper eyelid, and somewhat fewer in the lower. They are embedded in grooves in the inner surfaces of the tarsi, and correspond in length with the breadth of these plates; they are, consequently, longer in the upper than in the lower eyelid. Their ducts open on the free margins of the lids by minute apertures.

with the lining membrane of the ducts of the tarsal glands, and, through the lacrimal ducts, with the lining membrane of the lacrimal sac and nasolacrimal duct. At the lateral angle of the upper eyelid the ducts of the lacrimal gland open on its free surface; and at the medial angle it forms a semilunar fold, the **plica semilunaris** (Fig. 13–36). The line of reflection of the conjunctiva from the upper eyelid on to the bulb of the eye is named the **superior fornix,** and that from the lower lid the **inferior fornix.**

The **bulbar portion** (*tunica conjunctiva bulbi*) of the conjunctiva is loosely connected over the sclera; it is thin, transparent, destitute of papillae, and only slightly vascular. At the circumference of the *cornea,* the conjunctiva becomes the epithelium of the cornea, already described (see page 1050). *Lymphatics* arise in the conjunctiva in a delicate zone around the cornea, and run to the scleral conjunctiva. Lymphoid follicles are found in the conjunctiva,

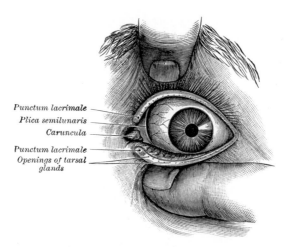

Punctum lacrimale
Plica semilunaris
Caruncula
Punctum lacrimale
Openings of tarsal glands

Fɪɢ. 13–36.—Anterior view of left eye with eyelids separated to show medial canthus.

chiefly situated near the medial palpebral commissure. They were first described by Brush, in his description of Peyer's patches of the small intestine, as "identical structures existing in the under eyelid of the ox."

In and near the fornices, but more plentiful in the upper than in the lower eyelid, a number of convoluted tubular glands open on the surface of the conjunctiva.

The **Caruncula Lacrimalis** is a small, reddish protuberance, situated at the medial palpebral commissure, and filling up the **lacus lacrimalis.** It consists of a small island of skin containing sebaceous and sudoriferous glands, and is the source of the whitish secretion which constantly collects in this region. A few slender hairs are attached to its surface. Lateral to the caruncula is a slight semilunar fold of conjunctiva, the concavity of which is directed toward the cornea; it is called the **plica semilunaris.** Smooth musclar fibers have been found in this fold, and in some of the domesticated animals it contains a thin plate of cartilage.

The **nerves** in the conjunctiva are numerous and form rich plexuses. According to Krause they terminate in a peculiar form of tactile corpuscle, which he terms "terminal bulbs."

The **Lacrimal Apparatus** (*apparatus lacrimalis*) (Fig. 13–34) consists of (*a*) the **lacrimal gland,** which secretes the tears, and its excretory ducts, which convey the fluid to the surface of the eye; (*b*) the **lacrimal ducts,** the **lacrimal sac,** and the

nasolacrimal duct, by which the fluid is conveyed into the cavity of the nose.

The **Lacrimal Gland** (*glandula lacrimalis*) (Figs. 13–34, 13–35) is superior and lateral to the bulb and is lodged in the lacrimal fossa, on the medial side of the zygomatic process of the frontal bone. It is of an oval form, about the size and shape of an almond, and consists of two portions. The **orbital part** (*pars orbitalis; superior lacrimal gland*) is connected to the periosteum of the orbit by a few fibrous bands, and rests upon the tendons of the Recti superioris and lateralis, which separate it from the bulb of the eye. The **palpebral part** (*pars palpebralis; inferior lacrimal gland*) is separated from the orbital part by a fibrous septum, and projects into the back part of the upper eyelid, where its deep surface is related to the conjunctiva. The ducts of the glands, from six to twelve in number, run obliquely beneath the conjunctiva for a short distance, and open along the upper and lateral half of the superior conjunctival fornix.

The **Lacrimal Ducts** (*ductus lacrimales; lacrimal canals*).—The lacrimal ducts, one in each eyelid, commence at minute orifices, termed **puncta lacrimalia,** on the summits of the **papillae lacrimales,** seen on the margins of the lids at the lateral extremity of the lacus lacrimalis. The **superior duct,** the smaller and shorter of the two, at first ascends, and then bends at an acute angle, and passes medialward and downward to the lacrimal sac. The **inferior duct** at first descends, and then runs almost horizontally to the lacrimal sac. At the angles they are dilated into **ampullae**; their walls are dense in structure and their mucous lining is covered by stratified squamous epithelium, placed on a basement membrane. Outside the latter is a layer of striped muscle, continuous with the lacrimal part of the Orbicularis oculi; at the base of each lacrimal papilla the muscular fibers are circularly arranged and form a kind of sphincter.

The **Lacrimal Sac** (*saccus lacrimalis*).—The lacrimal sac is the upper dilated end of the nasolacrimal duct, and is lodged in a deep groove formed by the lacrimal bone and frontal process of the maxilla. It is oval in form and measures from 12 to 15 mm in length; its upper end is closed and rounded; its lower is continued into the nasolacrimal duct. Its superficial surface is covered by a fibrous expansion derived from the medial palpebral ligament, and its deep surface is crossed by the lacrimal part of the Orbicularis oculi (page 381), which is attached to the crest on the lacrimal bone.

The **Nasolacrimal Duct** (*ductus nasolacrimalis; nasal duct*).—The nasolacrimal duct is a membranous canal, about 18 mm in length, which extends from the lower part of the lacrimal sac to the inferior meatus of the nose, where it ends by a somewhat expanded orifice, provided with an imperfect valve, the **plica lacrimalis** (*Hasneri*), formed by a fold of the mucous membrane. It is contained in an osseous canal, formed by the maxilla, the lacrimal bone, and the inferior nasal concha; it is narrower in the middle than at either end, and is directed downward, backward, and a little lateralward. The mucous lining of the lacrimal sac and nasolacrimal duct is covered with columnar epithelium, which in places is ciliated.

Structure of the Eyelids (Fig. 13–37).—The eyelids are composed of the following structures from outer to inner surface: integument, areolar tissue, fibers of the Orbicularis oculi, tarsus, orbital septum, tarsal glands, and conjunctiva. The upper eyelid has, in addition, the aponeurosis of the Levator palpebrae superioris (Fig. 13–32).

The **integument** is extremely thin, and con-

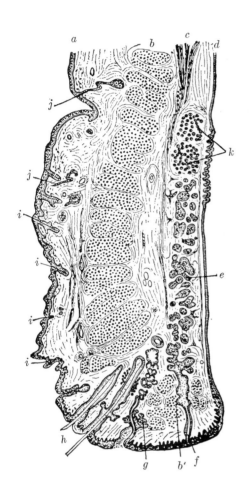

FIG. 13–37.—Sagittal section through the upper eyelid. (After Waldeyer.) *a*, Skin. *b*, Orbicularis oculi. *b'*, Marginal fasciculus of Orbicularis (ciliary bundle). *c*, Levator palpebrae. *d*, Conjunctiva. *e*, Tarsus. *f*, Tarsal gland. *g*, Ciliary gland. *h*, Eyelashes. *i*, Small hairs of skin. *j*, Sweat glands. *k*, Posterior tarsal glands.

tinuous at the margins of the eyelids with the conjunctiva.

The **subcutaneous areolar tissue** is very lax and delicate, and seldom contains any fat.

The **palpebral fibers of the Orbicularis oculi** are thin, pale in color, and possess an involuntary as well as a voluntary action.

The **tarsi** (*tarsal plates*) (Fig. 13–32) are two thin, elongated plates of dense connective tissue about 2.5 cm in length; one is placed in each eyelid, and contributes to its form and support. The **superior tarsus** (*tarsus superior; superior tarsal plate*), the larger, is of a semilunar form, about 10 mm in breadth at the center, and gradually narrowing toward its extremities. To the anterior surface of this plate the aponeurosis of the Levator palpebrae superioris is attached. The **inferior tarsus** (*tarsus inferior; inferior tarsal plate*), the smaller, is thin, elliptical in form, and has a vertical diameter of about 5 mm. The free or ciliary margins of these plates are thick and straight. The attached or orbital margins are connected to the circumference of the orbit by the orbital septum. The lateral angles are

attached to the zygomatic bone by the lateral palpebral raphe. The medial angles of the two plates end at the lacus lacrimalis, and are attached to the frontal process of the maxilla by the medial palpebral ligament (page 382).

Structure of the Tarsal Glands.—The tarsal glands are modified sebaceous glands, each consisting of a single straight tube or follicle, with numerous small lateral diverticula. The tubes are supported by a basement membrane, and are lined at their mouths by stratified epithelium; the deeper parts of the tubes and the lateral offshoots are lined by a layer of polyhedral cells.

Structure of the Lacrimal Gland.—In structure and general appearance the lacrimal resembles the serous salivary glands.

Structure of the Lacrimal Sac.—The lacrimal sac consists of a fibrous elastic coat, lined internally by mucous membrane: the latter is continuous, through the lacrimal ducts, with the conjunctiva, and through the nasolacrimal duct with the mucous membrane of the nasal cavity.

THE ORGAN OF HEARING AND EQUILIBRATION: THE EAR

The **ear** (*organum vestibulocochleare; organum oticum; organum stato-acousticum*), or **organ of hearing and equilibration,** is divisible into three parts: the **external ear,** the **middle ear** or **tympanic cavity,** and the **internal ear** or **labyrinth.**

The **Development of the Ear.**—The first rudiment of the internal ear appears shortly after that of the eye, in the form of a patch of thickened ectoderm, the **auditory plate,** over the region of the hindbrain. The auditory plate becomes depressed and converted into the **auditory pit** (Fig. 13–38). The mouth of the pit is then closed, and the **auditory vesicle** is formed (Fig. 13–39); from this the epithelial lining of the membranous labyrinth is derived. From the vesicle certain diverticula are given off which form the various parts of the membranous labyrinth (Fig. 13–40). One from the middle part forms the ductus and saccus endolymphaticus, another from the anterior end gradually elongates, and, forming a tube coiled on itself, becomes the cochlear duct, the vestibular extremity of which is subsequently constricted to form the canalis reuniens. Three others appear as disk-like

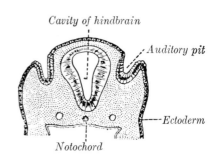

FIG. 13–38.—Section through the head of a human embryo, about twelve days old, in the region of the hindbrain. (Kollmann.)

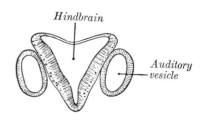

FIG. 13–39.—Section through hindbrain and auditory vesicles of an embryo more advanced than that of Figure 13–38. (After His.)

evaginations on the surface of the vesicle; the central parts of the walls of the disks coalesce and disappear, while the peripheral portions persist to form the semicircular ducts; of these the superior is the first and the lateral the last to be completed. The central part of the vesicle represents the membranous vestibule, and is subdivided by a constriction into a smaller ventral part, the saccule, and a larger dorsal and posterior part, the utricle. This subdivision is effected by a fold which extends deeply into the proximal part of the ductus endolymphaticus, with the result that the utricle and saccule ultimately communicate with each other by means of a Y-shaped canal. The saccule opens into the cochlear duct, through the canalis reuniens, and the semicircular ducts communicate with the utricle.

The mesodermal tissue surrounding the various parts of the epithelial labyrinth is converted into a cartilaginous capsule, and this is finally ossified to form the bony labyrinth. Between the cartilaginous capsule and the epithelial structures is a stratum of mesodermal tissue which is differentiated into three layers, viz., an outer, forming the periosteal lining of the bony labyrinth; an inner, in direct contact with the epithelial structures; and an intermediate, consisting of gelatinous tissue: by the absorption of this latter tissue the perilymphatic spaces are developed. The modiolus and osseous spiral lamina of the cochlea are not preformed in cartilage but are ossified directly from connective tissue.

The **middle ear** and **auditory tube** are developed from the first pharyngeal pouch. The entodermal lining of the dorsal end of this pouch is in contact with the ectoderm of the corresponding pharyngeal groove; by the extension of the mesoderm between these two layers the tympanic membrane is formed. During the sixth or seventh month the tympanic antrum appears as an upward and backward expansion of the tympanic cavity. With regard to the exact mode of development of the ossicles of the middle ear there is some difference of opinion. The view generally held is that the **malleus** is developed from the proximal end of the mandibular (Meckel's) cartilage (Fig. 2–29), the **incus** in the proximal end of the mandibular arch, and that the **stapes**

is formed from the proximal end of the hyoid arch. The malleus, with the exception of its anterior process, is ossified from a single center which appears near the neck of the bone; the anterior process is ossified separately in membrane and joins the main part of the bone about the sixth month of fetal life. The incus is ossified from one center which appears in the upper part of its long crus and ultimately extends into its lenticular process. The stapes first appears as a ring (*anulus stapedius*) encircling a small vessel, the stapedial artery, which subsequently undergoes atrophy; it is ossified from a single center which appears in its base.

The **external ear** and **external acoustic meatus** are developed from the first branchial groove. The lower part of this groove extends inward as a funnel-shaped tube (primary meatus) from which the cartilaginous portion and a small part of the roof of the osseous portion of the meatus are developed. From the lower part of the funnel-shaped tube an epithelial lamina extends downward and inward along the inferior wall of the primitive tympanic cavity; by the splitting of this lamina the inner part of the meatus (secondary meatus) is produced, while the inner portion of lamina forms the cutaneous stratum of the tympanic membrane.

The **auricula** or **pinna** is developed by the gradual differentiation of tubercles which appear around the margin of the first branchial groove (see Chapter 2 and Fig. 2–39).

The **acoustic nerve** rudiment appears about the end of the third week as a group of ganglion cells closely applied to the cranial edge of the auditory vesicle. Whether these cells are derived from the ectoderm adjoining the auditory vesicle, or have migrated from the wall of the neural tube, is as yet uncertain. The ganglion gradually splits into two parts, the **vestibular ganglion** and the **spiral ganglion.** The peripheral branches of the vestibular ganglion pass in two divisions, the pars superior giving rami to the superior ampulla of the superior semicircular duct, to the lateral ampulla and to the utricle; and the pars inferior giving rami to the saccule and the posterior ampulla. The proximal fibers

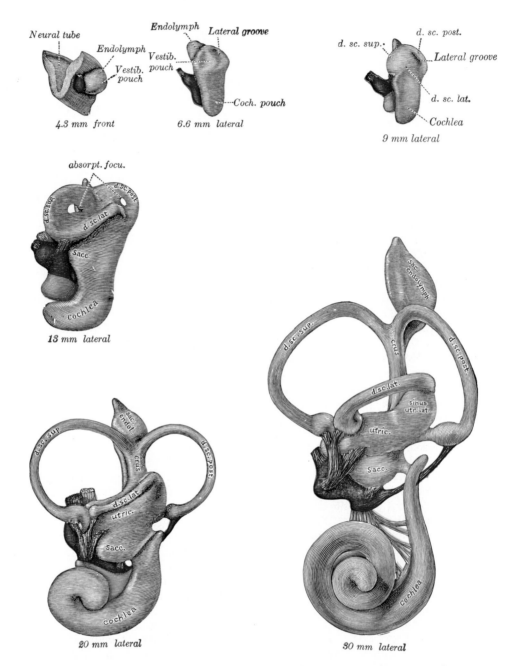

FIG. 13–40.—Lateral views of membranous labyrinth and acoustic complex. × 25 dia. (Streeter.) *absorpt. focu,* area of wall where absorption is complete; *amp.,* ampulla membranacea; *crus,* crus commune; *d. sc. lat.,* ductus semicircularis lateralis; *d. sc. post.,* ductus semicircularis posterior; *d. sc. sup.,* ductus semicircular superior; *coch. or cochlea,* ductus cochlearis; *duct. endolymph,* ductus endolymphaticus; *d. reuniens,* ductus reuniens Henseni; *endol. or endolymphs,* appendix endolymphaticus; *rec. utr.,* recessurs utriculi; *sacc.,* sacculus; *sac endol.,* saccus endolymphaticus; *sinu utr. lat.,* sinus utriculi lateralis; *utric.,* utriculus; *vestib. p.,* vestibular pouch.

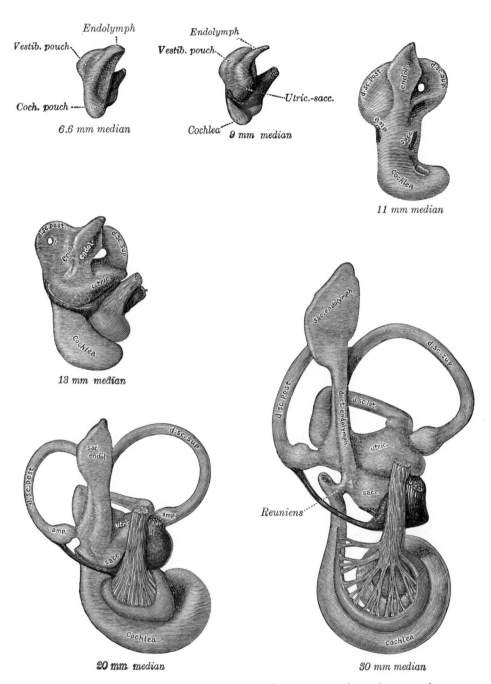

Fig. 13–41.—Median views of membranous labyrinth and acoustic complex in human embryos. × 25 dia. (Streeter.)

of the vestibular ganglion form the vestibular nerve; the proximal fibers of the spiral ganglion form the cochlear nerve.

THE EXTERNAL EAR

The **external ear** consists of the expanded portion named the **auricula** or **pinna**, and the **external acoustic meatus.** The former projects from the side of the head and serves to collect the vibrations of the air by which sound is produced; the latter leads inward from the bottom of the auricula and conducts the vibrations to the tympanic cavity.

The **Auricula** or **Pinna** (Fig. 13–42) is of an ovoid form, with its larger end directed upward. Its lateral surface is irregularly concave, directed slightly anteriorly, and presents numerous eminences and depressions to which names have been assigned. The prominent rim of the auricula is called the **helix;** where the helix turns inferiorly, a small tubercle, the **auricular tubercle of Darwin,** is frequently seen; this tubercle is very evident about the sixth month of fetal life when the whole auricula has a close resemblance to that of some of the adult monkeys. Another curved prominence, parallel with and anterior to the helix, is called the **antihelix;** this divides above into two crura, between which is a triangular depression, the **fossa triangularis.** The narrow curved depression between the helix and the antihelix is called the **scapha;** the antihelix describes a curve around a deep, capacious cavity, the **concha,** which is par-

tially divided into two parts by the **crus** or commencement of the helix; the superior part is termed the **cymba conchae,** the inferior part the **cavum conchae.** Anterior to the concha, and projecting posteriorly over the meatus, is a small pointed eminence, the **tragus,** so called from its being generally covered on its undersurface with a tuft of hair, resembling a goat's beard. Opposite the tragus, and separated from it by the **intertragic notch,** is a small tubercle, the **antitragus.** Below this is the **lobule,** composed of areolar and adipose tissues, and wanting the firmness and elasticity of the rest of the auricula.

The cranial surface of the auricula presents elevations which correspond to the depressions on its lateral surface and after which they are named, *e.g.,* **eminentia conchae, eminentia triangularis,** etc.

Structure.—The auricula is composed of a thin plate of yellow elastic cartilage, covered with integument, and connected to the surrounding parts by ligaments and muscles; and to the commencement of the external acoustic meatus by fibrous tissue.

The **skin** is thin, closely adherent to the cartilage, and covered with fine hairs furnished with sebaceous glands, which are most numerous in the concha and scaphoid fossa. On the tragus and antitragus the hairs are strong and numerous. The skin of the auricula is continuous with that lining the external acoustic meatus.

The **cartilage of the auricula** (*cartilago auriculae; cartilage of the pinna*) (Figs. 13–42, 13–43) consists of a single piece; it gives

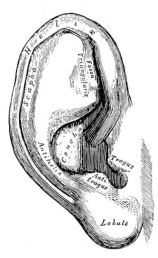

Fig. 13–42.—The auricula. Lateral surface.

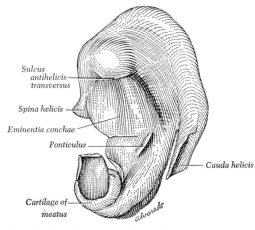

Fig. 13–43.—Cranial surface of cartilage of right auricula.

form to this part of the ear, and upon its surface are found the eminences and depressions above described. It is absent from the lobule; it is deficient, also, between the tragus and beginning of the helix, the gap being filled up by dense fibrous tissue. At the front part of the auricula, where the helix bends upward, is a small projection of cartilage, called the *spina helicis,* while in the lower part of the helix the cartilage is prolonged downward as a tail-like process, the **cauda helicis;** this is separated from the antihelix by a fissure, the **fissura anti-tragohelicina.** The cranial aspect of the cartilage exhibits a transverse furrow, the **sulcus antihelicis transversus,** which corresponds with the inferior crus of the antihelix and separates the eminentia conchae from the eminentia triangularis. The eminentia conchae is crossed by a vertical ridge (*ponticulus*), which gives attachment to the Auricularis posterior muscle. In the cartilage of the auricula are two fissures, one behind the crus helicis and another in the tragus.

The **ligaments of the auricula** (*ligamenta auricularia* [*Valsalva*]; *ligaments of the pinna*) consist of two sets: (1) **extrinsic,** connecting it to the side of the head; (2) **intrinsic,** connecting various parts of its cartilage together.

(1) The **extrinsic ligaments** are two in number, anterior and posterior. The *anterior ligament* extends from the tragus and spina helicis to the root of the zygomatic process of the temporal bone. The *posterior ligament* passes from the posterior surface of the concha to the outer surface of the mastoid process.

(2) The chief **intrinsic ligaments** are: (*a*) a strong fibrous band, stretching from the tragus to the commencement of the helix, completing the meatus in front, and partly encircling the boundary of the concha; and (*b*) a band between the antihelix and the cauda helicis. Other less important bands are found on the cranial surface of the pinna.

The **muscles of the auricula** (Fig. 13–44) consist of two sets: (1) the **extrinsic,** which connect it with the skull and scalp and move the auricula as a whole; and (2) the **intrinsic,** which extend from one part of the auricle to another.

(1) The **extrinsic muscles** are the Auriculares anterior, superior, and posterior, described with the facial muscles on page 380.

(2) The **intrinsic muscles** are:

Helicis major
Helicis minor
Tragicus
Antitragicus
Transversus auriculae
Obliquus auriculae

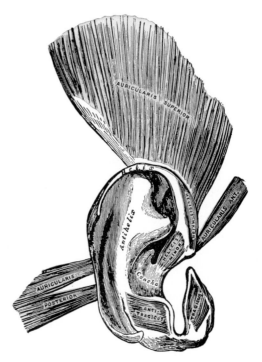

Fig. 13–44.—The muscles of the auricula.

The *Helicis major* is a narrow vertical band situated upon the anterior margin of the helix. It *arises* from the spina helicis, and is *inserted* into the anterior border of the helix, just where it is about to curve backward.

The *Helicis minor* is an oblique fasciculus, covering the crus helicis.

The *Tragicus* is a short, flattened vertical band on the lateral surface of the tragus.

The *Antitragicus arises* from the outer part of the antitragus, and is inserted into the cauda helicis and antihelix.

The *Transversus auriculae* is placed on the cranial surface of the pinna. It consists of scattered fibers, partly tendinous and partly muscular, extending from the eminentia conchae to the prominence corresponding with the scapha.

The *Obliquus auriculae,* also on the cranial surface, consists of a few fibers extending from the upper and back part of the concha to the convexity immediately above it.

Nerves.—The Auriculares anterior and superior and the intrinsic muscles on the lateral surface are supplied by the temporal branch of the facial nerve, the Auricularis posterior and the intrinsic muscles on the cranial surface by the posterior auricular branch of the same nerve.

The **arteries of the auricula** are the posterior

auricular from the external carotid, the anterior auricular from the superficial temporal, and a branch from the occipital artery.

The **veins** accompany the corresponding arteries.

The **sensory nerves** are: the great auricular, from the cervical plexus; the auricular branch of the vagus; the auriculotemporal branch of the mandibular nerve; and the lesser occipital from the cervical plexus.

The **External Acoustic Meatus** (*meatus acusticus externus; external auditory canal or meatus*) extends from the bottom of the concha to the tympanic membrane (Figs. 13–45, 13–46). It is about 4 cm in length if measured from the tragus; from the bottom of the concha its length is about 2.5 cm. It forms an S-shaped curve, and is directed at first inward, forward, and slightly upward (*pars externa*); it then passes inward and backward (*pars media*), and lastly is carried inward, forward, and slightly downward (*pars interna*). It is an oval cylindrical canal, the greatest diameter being directed downward and backward at the external orifice, but nearly horizontally at the inner end. It presents two constrictions, one near the inner end of the cartilaginous portion, and another, the **isthmus,** in the osseous portion, about 2 cm from the bottom of the concha. The tympanic membrane, which closes the inner end of the meatus, is obliquely directed; in consequence of this the floor and anterior wall of the meatus are longer than the roof and posterior wall.

The external acoustic meatus is formed partly by cartilage and membrane, and partly by bone, and is lined by skin.

The **cartilaginous portion** (*meatus acusticus externus cartilagineus*) is about 8 mm in length; it is continuous with the cartilage of the auricula, and firmly attached to the circumference of the auditory process of the temporal bone. The cartilage is deficient at the cranial and posterior part of the meatus, its place being supplied by fibrous membrane; two or three deep fissures are present in the anterior part of the cartilage.

The osseous portion (*meatus acusticus externus osseus*) is about 16 mm in length, and is narrower than the cartilaginous portion. It is directed inward and a little forward, forming in its course a slight curve the convexity of which is upward and backward. Its inner end is smaller than the outer, and sloped, the anterior wall projecting beyond the posterior for about 4 mm; it is marked, except at its cranial part, by a narrow groove, the **tympanic sulcus,** in which the circumference of the tympanic membrane is attached. Its outer end is dilated and rough in the greater part of its circumference, for the attachment of the cartilage of the auricula. The anterior and inferior parts of the osseous portion are formed by a curved plate of bone, the tympanic part of the temporal, which, in the fetus, exists as a separate ring (**anulus tympanicus**), incomplete at its cranial part (page 174).

The **skin** lining the meatus is very thin; adheres closely to the cartilaginous and osseous portions of the tube, and covers the outer surface of the tympanic membrane. After maceration, the thin pouch of epidermis, when withdrawn, preserves the form of the meatus. In the thick subcutaneous tissue of the cartilaginous part of the meatus are numerous ceruminous glands which secrete the ear-wax; their structure resembles that of the sudoriferous glands.

Relations of the Meatus.—Anterior to the osseous part is the condyle of the mandible, but the latter is frequently separated from the cartilaginous part by a portion of the parotid gland. The movements of the jaw influence to some extent the lumen of this latter portion. Posterior to the osseous part are the mastoid air cells, separated from the meatus by a thin layer of bone.

The **arteries** supplying the meatus are branches from the posterior auricular, maxillary, and temporal.

The **nerves** are chiefly derived from the auriculotemporal branch of the mandibular nerve and the auricular branch of the vagus.

THE MIDDLE EAR OR TYMPANUM

The **middle ear** or **tympanic cavity** (*cavum tympani; drum; tympanum*) is an irregular, laterally compressed space within the temporal bone. It is filled with air, which is conveyed to it from the nasal part of the pharynx through the auditory tube. The cavity contains a chain of tiny movable

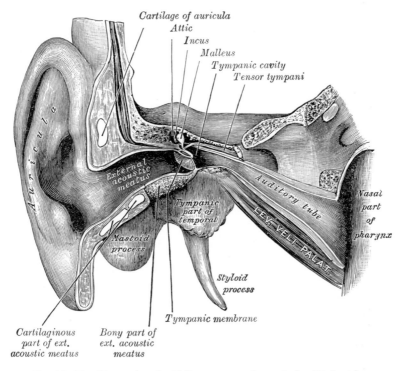

Cartilage of auricula
Attic
Incus
Malleus
Tympanic cavity
Tensor tympani

Auricula

External acoustic meatus

Auditory tube

Nasal part of pharynx

Tympanic part of temporal

LEV. VELI PALAT.

Mastoid process

Styloid process

Tympanic membrane

Cartilaginous part of ext. acoustic meatus

Bony part of ext. acoustic meatus

FIG. 13–45.—External and middle ear, opened anteriorly. Right side.

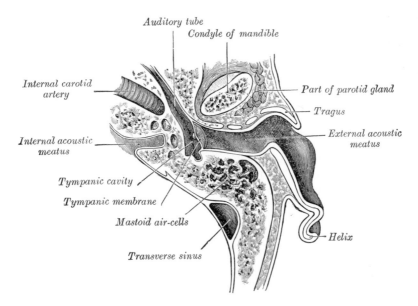

Auditory tube
Condyle of mandible

Internal carotid artery

Part of parotid gland

Tragus

Internal acoustic meatus

External acoustic meatus

Tympanic cavity

Tympanic membrane

Mastoid air-cells

Helix

Transverse sinus

FIG. 13–46.—Horizontal section through left ear; upper half of section.

bones which bridge from its lateral to its medial wall, and serve to convey the vibrations communicated to the tympanic membrane across the cavity to the internal ear.

The tympanic cavity consists of two parts: the **tympanic cavity proper,** opposite the tympanic membrane, and the **attic** or **epitympanic recess,** cranial to the level of the membrane; the latter contains the upper half of the malleus and the greater part of the incus. Including the attic, the vertical and anteroposterior diameters of the cavity are each about 15 mm. The transverse diameter measures about 6 mm superiorly and 4 mm inferiorly; opposite the center of the tympanic membrane it is only about 2 mm. The tympanic cavity is bounded laterally by the tympanic membrane; medially, by the lateral wall of the internal ear; it communicates, posteriorly, with the tympanic antrum and through it with the mastoid air cells, and anteriorly with the auditory tube (Fig. 13–45).

The **Tegmental Wall** or **Roof** (*paries tegmentalis*) is formed by a thin plate of bone, the **tegmen tympani,** which separates the tympanic from the cranial cavity. It is situated on the anterior surface of the petrous portion of the temporal bone close to its angle of junction with the squama temporalis; it is prolonged posteriorly so as to roof in the tympanic antrum, and anteriorly to cover in the semicanal for the Tensor tympani muscle. Its lateral edge corresponds with the remains of the petrosquamous suture.

The **Jugular Wall** or **Floor** (*paries jugularis*) is narrow, and consists of a thin plate of bone (**fundus tympani**) which separates the tympanic cavity from the jugular fossa. It presents, near the labyrinthic wall, a small aperture for the passage of the tympanic branch of the glossopharyngeal nerve.

The **Membranous** or **Lateral Wall** (*paries membranaceus; outer wall*) is formed mainly by the tympanic membrane, partly by the ring of bone into which this membrane is inserted. The ring of bone is incomplete at its upper part, forming a notch (*of Rivinus*), close to which are three small apertures: the **iter chordae posterius,** the **petrotympanic fissure,** and the **iter chordae anterius.**

The **iter chordae posterius** (*apertura tympanica canaliculi chordae*) is situated in the angle of junction between the mastoid and membranous wall of the tympanic cavity immediately posterior to the tympanic membrane and on a level with the superior end of the manubrium of the malleus; it leads into a minute canal in the mastoid bone, which descends almost parallel with the facial nerve, and meets the facial canal at an acute angle near the stylomastoid foramen. Through it the chorda tympani nerve enters the tympanic cavity (Fig. 13–49).

The **petrotympanic fissure** (*fissura petrotympanica; Glaserian fissure*) (Fig. 13–49) opens just superior and anterior to the ring of bone into which the tympanic membrane is inserted; in this situation it is a mere slit about 2 mm in length. It lodges the anterior process and anterior ligament of the malleus, and gives passage to the anterior tympanic branch of the maxillary artery.

The **iter chordae anterius** (*canal of Huguier*) is placed at the medial end of the petrotympanic fissure; through it the chorda tympani nerve leaves the tympanic cavity.

The **Tympanic Membrane** (*membrana tympani*) (Figs. 13–47, 13–49) separates the tympanic cavity from the bottom of the external acoustic meatus. It is a thin, semitransparent membrane, nearly oval in form, directed very obliquely inferiorly and medially so as to form an angle of about 55 degrees with the floor of the meatus. Its longest diameter, almost vertical, measures from 9 to 10 mm; its shortest diameter measures from 8 to 9 mm. The greater part of its circumference is thickened, and forms a **fibrocartilaginous ring** which is fixed in the **tympanic sulcus** at the inner end of the meatus. This sulcus is deficient superiorly at the **notch of Rivinus,** and from the ends of this notch two bands, the **anterior** and **posterior malleolar folds,** are prolonged to the lateral process of the malleus. The small, somewhat triangular part of the membrane situated superior to these folds is lax and thin, and is named the **pars flaccida** as opposed to the rest of the membrane, called the **pars tensa.** The manubrium of the malleus is firmly attached to the medial surface of the membrane as far as its center, which it draws inward toward the tympanic cavity; the lateral surface of the membrane is thus

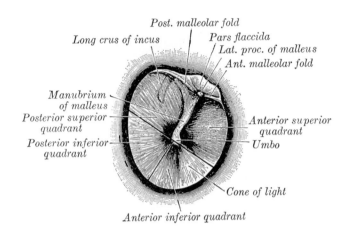

Post. malleolar fold

Long crus of incus

Pars flaccida

Lat. proc. of malleus

Ant. malleolar fold

Manubrium of malleus

Posterior superior quadrant

Posterior inferior quadrant

Anterior superior quadrant

Umbo

Cone of light

Anterior inferior quadrant

Fig. 13–47.—Right tympanic membrane as seen through a speculum.

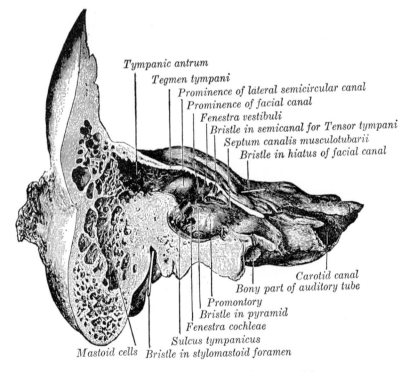

Tympanic antrum

Tegmen tympani

Prominence of lateral semicircular canal

Prominence of facial canal

Fenestra vestibuli

Bristle in semicanal for Tensor tympani

Septum canalis musculotubarii

Bristle in hiatus of facial canal

Carotid canal

Bony part of auditory tube

Promontory

Bristle in pyramid

Fenestra cochleae

Sulcus tympanicus

Mastoid cells   *Bristle in stylomastoid foramen*

Fig. 13–48.—Coronal section of right temporal bone.

Fɪɢ. 13–49.—The right membrana tympani with the malleus and the chorda tympani, viewed from within. (Spalteholz.)

concave, and the most depressed part of this concavity is named the **umbo.**

Structure.—The tympanic membrane is composed of three strata: a **lateral** (*cutaneous*), an **intermediate** (*fibrous*), and a **medial** (*mucous*). The **cutaneous stratum** is derived from the integument lining the meatus. The **fibrous stratum** consists of two layers: a **radiate stratum,** the fibers of which diverge from the manubrium of the malleus, and a **circular stratum,** the fibers of which are plentiful around the circumference but sparse and scattered near the center of the membrane. Branched or dendritic fibers are also present, especially in the posterior half of the membrane. The **mucous stratum** is the internal surface of the membrane derived from the lining of the cavity.

Vessels and Nerves.—The **arteries** of the tympanic membrane are derived from the deep auricular branch of the maxillary, which ramifies beneath the cutaneous stratum; and from the stylomastoid branch of the posterior auric-

ular, and tympanic branch of the maxillary, which are distributed on the mucous surface. The superficial **veins** open into the external jugular; those on the deep surface drain partly into the transverse sinus and veins of the dura mater, and partly into a plexus on the auditory tube. The membrane receives its chief **nerve supply** from the auriculotemporal branch of the mandibular; the auricular branch of the vagus, and the tympanic branch of the glossopharyngeal also supply it.

The **Labyrinthic** or **Medial Wall** (*paries labyrinthicus; inner wall*) (Fig. 13–48) is vertical in direction, and presents for examination the **fenestrae vestibuli** and **cochleae,** the **promontory,** and the **prominence of the facial canal.**

The **fenestra vestibuli** (*fenestra ovalis; oval window*) is a reniform opening leading from the tympanic cavity into the vestibule of the internal ear; its long diameter is horizontal, and its convex border is upward. In the intact body it is occupied by the base of the stapes, the circumference of which is secured by the annular ligament to the margin of the foramen (Fig. 13–51).

The **fenestra cochleae** (*fenestra rotunda; round window*) is situated inferior and a little posterior to the fenestra vestibuli, from which it is separated by a rounded elevation, the **promontory**. It is placed at the bottom of a funnel-shaped depression and, in the macerated bone, leads into the cochlea of the internal ear; in the intact body it is closed by a membrane, the **secondary tympanic membrane**, which is concave toward the cochlea. This membrane consists of three layers: an external, or mucous, derived from the mucous lining of the tympanic cavity; an internal, from the lining membrane of the cochlea; and an intermediate, or fibrous layer.

The **promontory** (*promontorium*) is a rounded prominence, formed by the projection outward of the first turn of the cochlea; it is placed between the fenestrae, and its surface is furrowed by small grooves for the nerves of the tympanic plexus. A minute spicule of bone frequently connects the promontory to the pyramidal eminence.

The **prominence of the facial canal** (*prominentia canalis facialis; prominence of aqueduct of Fallopius*) indicates the position of the bony canal in which the facial nerve is contained; this canal traverses the labyrinthic wall of the tympanic cavity superior to the fenestra vestibuli, and posterior to that opening curves nearly

vertically inferiorly along the mastoid wall (Fig. 13–50).

The **Mastoid** or **Posterior Wall** (*paries mastoidus*) presents for examination the **entrance to the tympanic antrum,** the **pyramidal eminence,** and the **fossa incudis.**

The **entrance to the antrum** is a large irregular aperture, which leads posteriorly from the epitympanic recess into a considerable air space, named the **tympanic** or **mastoid antrum** (see page 170). The antrum leads posteriorly and inferiorly into the **mastoid air cells,** which vary considerably in number, size, and form; the antrum and mastoid air cells are lined by mucous membrane, continuous with that lining the tympanic cavity. On the medial wall of the entrance to the antrum is a rounded eminence, situated superior and posterior to the prominence of the facial canal; it corresponds with the position of the ampullated ends of the superior and lateral semicircular canals.

The **pyramidal eminence** (*eminentia pyramidalis; pyramid*) is situated immediately posterior to the fenestra vestibuli, and anterior to the vertical portion of the facial canal; it is hollow, and contains the Stapedius muscle; its summit projects anteriorly toward the fenestra vestibuli, and is pierced by a small aperture which transmits the tendon of the muscle. The cavity in the pyramidal eminence is prolonged posteriorly toward the facial canal, and communicates with it by a minute aperture which transmits a branch of the facial nerve to the Stapedius muscle.

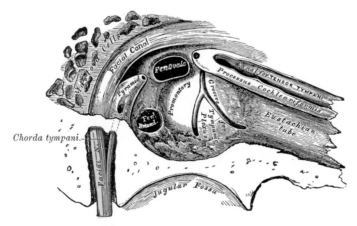

FIG. 13–50.—View of the medial wall of the tympanum (enlarged). (See Fig. 13–48.)

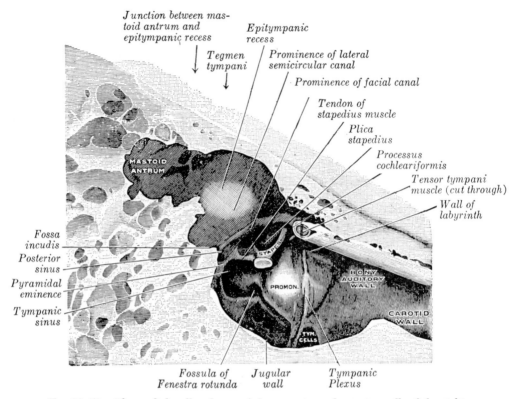

Junction between mas-
toid antrum and
epitympanic recess

Epitympanic
recess

Tegmen
tympani

Prominence of lateral
semicircular canal

Prominence of facial canal

Tendon of
stapedius muscle

Plica
stapedius

Processus
cochleariformis

Tensor tympani
muscle (cut through)

Wall of
labyrinth

Fossa
incudis

Posterior
sinus

Pyramidal
eminence

Tympanic
sinus

Fossula of   Jugular   Tympanic
Fenestra rotunda   wall   Plexus

FIG. 13–51.—The medial wall and part of the posterior and anterior walls of the right
tympanic cavity, lateral view. (Spalteholz.) (See Fig. 13–48.)

The **fossa incudis** is a small depression in the inferior and posterior part of the epitympanic recess; it lodges the short crus of the incus.

The **Carotid** or **Anterior Wall** (*paries caroticus*) corresponds with the carotid canal, from which it is separated by a thin plate of bone perforated by the tympanic branch of the internal carotid artery, and by the caroticotympanic nerve which connects the sympathetic plexus on the internal carotid artery with the tympanic plexus on the promontory. At the superior part of the anterior wall are the orifice of the semicanal for the Tensor tympani muscle and the tympanic orifice of the auditory tube, separated from each other by a thin horizontal plate of bone, the **septum canalis musculotubarii.** These canals run from the tympanic cavity medialward to the retiring angle between the squama and the petrous portion of the temporal bone.

The **semicanal for the Tensor tympani** (*semicanalis m. tensoris tympani*) is the superior and the smaller of the two; it is cylindrical and lies beneath the tegmen tympani. It extends on to the labyrinthic wall of the tympanic cavity and ends immediately above the fenestra vestibuli. The **septum canalis musculotubarii** (*processus cochleariformis*) passes posteriorly, forming its lateral wall and floor; it expands above the anterior end of the fenestra vestibuli and terminates there by curving lateralward so as to form a pulley over which the tendon of the muscle passes.

The **Auditory Tube** (*tuba auditiva; Eustachian tube*) is the channel through which the tympanic cavity communicates with the nasal part of the pharynx. Its length is about 36 mm, and in its medialward course, forms an angle of about 45 degrees with the sagittal plane and one of from 30 to 40 degrees with the horizontal plane. Its walls are partly of bone and partly of cartilage and fibrous tissue (Fig. 13–45).

The **osseous portion** (*pars ossea tubae auditivae*) is about 12 mm in length. It begins in the carotid wall of the tympanic cavity inferior to the septum canalis musculotubarii, and, gradually narrowing, ends at the angle of junction of the squama and the petrous portion of the temporal bone, its extremity presenting a jagged margin which serves for the attachment of the cartilaginous portion.

The **cartilaginous portion** (*pars cartilaginea tubae and auditivae*), about 24 mm in length, is contained in a triangular plate of elastic fibrocartilage, the apex of which is attached to the margin of the medial end of the osseous portion of the tube, while its base lies directly under the mucous membrane of the nasal part of the pharynx, where it forms an elevation, the **torus tubarius** or cushion, posterior to the pharyngeal orifice of the tube. The superior edge of the cartilage is bent over laterally forming a groove or furrow open inferiorly and laterally. The part of the canal lying in this furrow is completed by fibrous membrane. The cartilage lies in a groove between the petrous part of the temporal and the great wing of the sphenoid, which ends opposite the middle of the medial pterygoid plate. The cartilaginous and bony portions of the tube are not in the same plane, the former inclining inferiorly a little more than the latter. The diameter of the tube is not uniform throughout, being greatest at the pharyngeal orifice, least at the junction of the bony and cartilaginous portions, and again increased toward the tympanic opening; the narrowest part of the tube is termed the **isthmus**. The position and relations of the **pharyngeal orifice** are described with the nasal part of the pharynx. The mucous membrane of the tube is continuous with that of the nasal part of the pharynx, and with that of the tympanic cavity. It is covered with ciliated epithelium and is thin in the osseous portion, while in the cartilaginous portion it contains many mucous glands and near the pharyngeal orifice a considerable amount of adenoid tissue, the **tubal tonsil**. The tube is opened during deglutition by the Salpingopharyngeus and Dilatator tubae. The latter arises from the hook of the cartilage and from the membranous part of the tube, and blends below with the Tensor veli palatini.

The Auditory Ossicles (Ossicula Auditus)

The tympanic cavity contains a chain of three movable ossicles, the **malleus, incus,** and **stapes**. The first is attached to the tympanic membrane, the last to the circumference of the fenestra vestibuli, the incus being placed between and connected to both by delicate articulations (Fig. 13–55).

The **Malleus** (Fig. 13–52), so named from its fancied resemblance to a hammer, consists of a **head, neck,** and three processes, viz., the **manubrium,** the **anterior** and **lateral processes**.

The **head** (*caput mallei*) is the large superior extremity of the bone; it is oval in shape, and articulates posteriorly with the incus, being free in the rest of its extent. The facet for articulation with the incus is constricted near the middle, and consists of a superior larger and inferior smaller part, which are bent nearly at a right angle to each other. Opposite the constriction the inferior margin of the facet projects in the form of a process, the **cog-tooth** or **spur of the malleus**.

The **neck** (*collum mallei*) is the narrow contracted part joining the head with a prominence to which the two processes are attached.

The **manubrium mallei** (*handle*) is connected by its lateral margin with the tympanic membrane. It is directed slightly posteriorly as well as inferiorly; it decreases in size toward its free end, which is slightly curved and flattened transversely. On its medial side, near the head end, is a slight projection, into which the tendon of the Tensor tympani is inserted.

The **anterior process** (*processus anterior; processus gracilis*) is a delicate spicule directed anteriorly toward the petrotympanic fissure, to which it is connected by ligamentous fibers. In the fetus this is the longest process of the malleus, and is in direct continuity with the cartilage of Meckel.

The **lateral process** (*processus lateralis; processus brevis*) is a slight conical projection, which springs from the root of the manubrium; it is directed laterally, and is attached to the superior part of the tympanic membrane and, by means of the

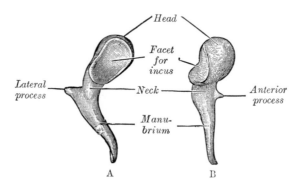

FIG. 13–52.—Left malleus. A, Posterior. B, Medial view.

anterior and posterior malleolar folds, to the extremities of the notch of Rivinus.

The **Incus** (Fig. 13–53) has received its name from its resemblance to an anvil, but it is more like a premolar tooth, with two roots, which differ in length, and are widely separated from each other. It consists of a **body** and **two crura**.

The **body** (*corpus incudis*) is somewhat cubical but compressed transversely. On its anterior surface is a deeply concavo-convex facet, which articulates with the head of the malleus.

The two crura diverge from each other nearly at right angles.

The **short crus** (*crus breve; short process*), somewhat conical in shape, projects almost horizontally backward, and is attached to the **fossa incudis**, in the posterior part of the epitympanic recess.

The **long crus** (*crus longum; long process*) descends nearly vertically parallel to the manubrium of the malleus, and, bending medialward, ends in a rounded projection, the **lenticular process**, which is tipped with cartilage, and articulates with the head of the stapes.

The **Stapes** (Fig. 13–54), so called from its resemblance to a stirrup, consists of a **head, neck, two crura**, and a **base**.

The **head** (*caput stapedis*) presents a depression, which is covered by cartilage, and articulates with the lenticular process of the incus.

The **neck**, the constricted part of the bone succeeding the head, has the insertion of the tendon of the Stapedius muscle attached to it.

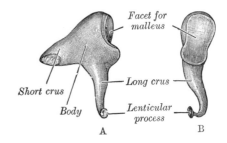

FIG. 13–53.—Left incus. A, Medial. B, Anterior view.

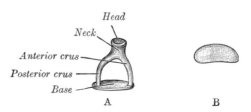

FIG. 13–54.—A, Left stapes. B, Base of stapes, medial surface.

The **two crura** (*crus anterius and crus posterius*) diverge from the neck and are connected at their ends by a flattened oval plate, the **base** (*basis stapedis*), which forms the foot-plate of the stirrup and is fixed to the margin of the fenestra vestibuli by a ring of ligamentous fibers. Of the two crura the anterior is shorter and less curved than the posterior.

Articulations of the Auditory Ossicles (*articulationes ossiculorum auditus*).—The incudomalleolar joint is a saddle-shaped diarthrosis; it is surrounded by an articular capsule, and the joint cavity is incompletely divided into two by a wedge-shaped articular disk or meniscus. The incudostapedial joint is an enarthrosis, surrounded by an articular capsule; some observers have described an articular disk or meniscus in this joint; others regard the joint as a syndesmosis.

Ligaments of the Ossicles (*ligamenta ossiculorum auditus*).—The ossicles are connected with the walls of the tympanic cavity by ligaments: three for the malleus, and one each for the incus and stapes.

The **anterior ligament of the malleus** (*lig. mallei anterius*) is attached by one end to the neck of the malleus, just above the anterior process, and by the other to the anterior wall of the tympanic cavity, close to the petrotympanic fissure, some of its fibers being prolonged through the fissure to reach the spina sphenoidalis.

The **superior ligament of the malleus** (*lig. mallei superius*) is a delicate, round bundle which descends from the roof of the epitympanic recess to the head of the malleus.

The **lateral ligament of the malleus** (*lig. mallei laterale; external ligament of the malleus*) is a triangular band passing from the posterior part of the notch of Rivinus to the head of the malleus. Helmholtz

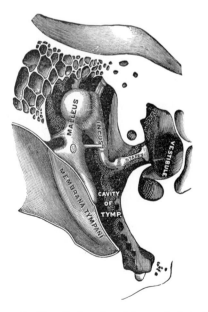

FIG. 13–55.—Chain of ossicles and their ligaments, anterior view of a vertical transverse section of the tympanum. (Testut.)

described the anterior ligament and the posterior part of the lateral ligament as forming together the **axis ligament** around which the malleus rotates.

The **posterior ligament of the incus** (*lig. incudis posterius*) is a short, thick band connecting the end of the short crus of the incus to the fossa incudis.

A **superior ligament of the incus** (*lig. incudis superius*) has been described, but it is little more than a fold of mucous membrane.

The vestibular surface and the circumference of the base of the stapes are covered with hyaline cartilage; that encircling the base is attached to the margin of the fenestra vestibuli by a fibrous ring, the **annular ligament of the base of the stapes** (*lig. annulare stapedis*).

The **Muscles of the Tympanic Cavity** (*musculi ossiculorum auditus*) are the Tensor tympani and Stapedius (Figs. 13–45, 13–51).

The **Tensor tympani** is contained in the bony canal above the osseous portion of the auditory tube, from which it is separated by the septum canalis musculotubarii. It *arises* from the cartilaginous portion of the auditory tube and the adjoining part of the great wing of the sphenoid, as well as from the osseous canal in which it is contained. Passing posteriorly through the canal, it ends in a slender tendon which enters the tympanic cavity, makes a sharp bend around the extremity of the septum, the **processus cochleariformis**, and is inserted into the manubrium of the malleus, near its root. It is supplied by a branch of the mandibular nerve which passes through the otic ganglion.

The **Stapedius** *arises* from the wall of the conical cavity inside of the pyramid; its tendon emerges from the orifice at the apex of the eminence, and is *inserted* into the posterior surface of the neck of the stapes. It is supplied by a branch of the facial nerve.

Actions.—The Tensor tympani draws the tympanic membrane medialward, and thus increases its tension. The Stapedius pulls the head of the stapes posteriorly tilting the base and possibly increasing the tension of the fluid within the internal ear. Both muscles have a snubbing action, reducing the oscillations of the ossicles, protecting the inner ear from injury during a loud noise such as that of a riveting machine.

The **Mucous Membrane of the Tympanic Cavity** is continuous with that of the pharynx, through the auditory tube. It invests the auditory ossicles, and the muscles and nerves contained in the tympanic cavity; forms the medial layer of the tympanic membrane, and the lateral layer of the secondary tympanic membrane, and is reflected into the tympanic antrum and mastoid cells, which it lines throughout. It forms several folds, which extend from the walls of the tympanic cavity to the ossicles; of these, one descends from the roof of the cavity to the head of the malleus and superior margin of the body of the incus, a second invests the Stapedius muscle; other folds invest the chorda tympani nerve and the Tensor tympani muscle. These folds separate off pouch-like cavities, and give the interior of the tympanum a somewhat honey-combed appearance. One of these pouches, the **pouch of Prussak**, is well marked and lies between the neck of the malleus and the membrana flaccida. Two other recesses may be mentioned: they are formed by the mucous membrane which envelops the chorda tympani nerve and are named the **anterior** and **posterior recesses of Troltsch**. In the tympanic cavity this membrane is pale, thin, slightly vascular, and covered for the most part with columnar ciliated epithelium, but over the pyramidal eminence, ossicles, and tympanic membrane it possesses a flattened non-ciliated epithelium. In the tympanic antrum and mastoid cells its epithelium is also non-ciliated. In the osseous portion of the auditory tube the membrane is thin; but in the cartilaginous portion it is very thick, highly vascular, and provided with numerous mucous glands; the epithelium which lines the tube is columnar and ciliated.

Vessels and Nerves.—The **arteries** are six in number. Two of them are larger than the others, viz., the tympanic branch of the maxillary, which supplies the tympanic membrane; and the stylomastoid branch of the posterior auricular, which supplies the back part of the tympanic cavity and mastoid cells. The smaller arteries are—the petrosal branch of the middle meningeal, which enters through the hiatus of the facial canal; a branch from the ascending pharyngeal, and another from the artery of the pterygoid canal, which accompany the auditory tube; and the tympanic branch from the internal carotid, given off in the carotid canal and perforating the thin anterior wall of the tympanic cavity. The **veins** terminate in the pterygoid plexus and the superior petrosal sinus. The **nerves** constitute the tympanic

plexus, which ramifies upon the surface of the promontory and the chorda tympani which merely passes through the cavity. The plexus is formed by (1) the tympanic branch of the glossopharyngeal; (2) the caroticotympanic nerves; (3) the smaller petrosal nerve; and (4) a branch which joins the greater petrosal.

(1) The **tympanic branch of the glossopharyngeal** (*Jacobson's nerve*) enters the tympanic cavity by an aperture in its floor close to the labyrinthic wall, and divides into branches which ramify on the promontory and enter into the formation of the tympanic plexus.

(2) The **superior and inferior caroticotympanic nerves** from the carotid plexus of the sympathetic pass through the wall of the carotid canal, and join the plexus. The branch to the greater petrosal passes through an opening on the labyrinthic wall, anterior to the fenestra vestibuli.

(3) The **smaller petrosal nerve** is a continuation of the tympanic branch of the glossopharyngeal nerve beyond the tympanic plexus. It penetrates the bone near the geniculate ganglion of the facial nerve with which it communicates by a filament, continues through the bone, and enters the middle cranial fossa through a small aperture situated lateral to the hiatus of the facial canal on the anterior surface of the petrous portion of the temporal bone. It leaves the middle fossa through a fissure or foramen of its own or through the foramen ovale, and ends in the otic ganglion, constituting its root and supplying it with preganglionic parasympathetic fibers.

(4) The **branches of distribution** of the tympanic plexus are supplied to the mucous membrane of the tympanic cavity; a branch passes to the fenestra vestibuli, another to the fenestra cochleae, and a third to the auditory tube. The smaller petrosal may be looked upon as the continuation of the tympanic branch of the glossopharyngeal through the plexus to the otic ganglion.

In addition to the tympanic plexus there are the nerves supplying the muscles. The Tensor tympani is supplied by a branch from the mandibular which passes through the otic ganglion, and the Stapedius by a branch from the facial.

The **chorda tympani nerve** crosses the tympanic cavity (Fig. 13–49). It leaves the facial canal, about 6 mm before the facial nerve emerges from the stylomastoid foramen. It runs back upward in a canal of its own, almost parallel with the facial canal and enters the tympanic cavity through the iter chordae posterius, where it becomes invested with mucous membrane. It traverses the tympanic cavity across the medial surface of the tympanic membrane and over the upper part of the manubrium of the malleus to the carotid wall, where it makes exit through the iter chordae anterius (*canal of Huguier*).

THE INTERNAL EAR OR LABYRINTH

The **internal ear** (*auris interna*) is the essential part of the organ of hearing, receiving the ultimate distribution of the acoustic nerve. It is called the **labyrinth,** from the complexity of its shape, and consists of two parts: the **osseous labyrinth,** a series of cavities within the petrous part of the temporal bone, and the **membranous labyrinth,** a series of communicating membranous sacs and ducts, contained within the bony cavities.

The **Osseous Labyrinth** (*labyrinthus osseus*) (Figs. 13–56, 13–57) consists of three parts: the **vestibule, semicircular canals,** and **cochlea.** These are cavities hollowed out of the substance of the bone, and lined by periosteum; they contain a clear fluid, the **perilymph,** in which the membranous labyrinth is suspended.

The **Vestibule** (*vestibulum*) is the central part of the osseous labyrinth, and is situated medial to the tympanic cavity, posterior to the cochlea, and anterior to the semicircular canals. It is somewhat ovoid in shape, but flattened transversely; it measures about 5 mm sagittally, the same vertically, and about 3 mm across. In its *lateral* or *tympanic wall* is the **fenestra vestibuli,** closed, in the intact body, by the base of the stapes

and annular ligament. On its *medial wall* is a small circular depression, the **recessus sphericus,** which is perforated at its anterior and inferior part by several minute holes (**macula cribrosa media**) for the passage of filaments of the acoustic nerve to the saccule; and posterior to this depression is an oblique ridge, the **crista vestibuli,** the anterior end of which is named the **pyramid of the vestibule.** This ridge bifurcates to enclose a small depression, the **fossa cochlearis,** which is perforated by a number of holes for the passage of filaments of the acoustic nerve which supply the vestibular end of the ductus cochlearis. At the posterior part of the medial wall is the orifice of the **aquaeductus vestibuli,** which extends to the posterior surface of the petrous portion of the temporal bone. It transmits a small vein, and contains a tubular prolongation of the membranous labyrinth, the **ductus endolymphaticus,** which ends in a cul-de-sac between the layers of the dura mater within the cranial cavity. On the superior *wall* or *roof* is a transversely oval depression, the **recessus ellipticus,** separated from the recessus sphericus by the crista vestibuli already mentioned. The pyramid and adjoining part of the recessus ellipticus are perforated by a number of holes (**macula cribrosa superior**). The apertures in the pyramid transmit the nerves to the utricle; those in the recessus ellipticus the nerves to the ampullae of the superior and lateral semicircular ducts. Posteriorly are the five orifices of the semicircular canals. Anteriorly is an elliptical opening, which communi-

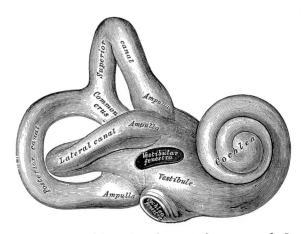

Fig. 13–56.—Right osseous labyrinth with spongy bone removed. Lateral view.

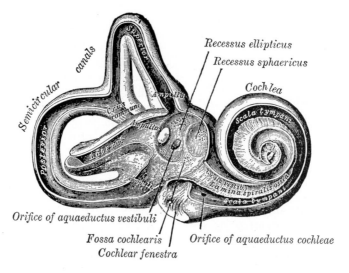

FIG. 13–57.—Interior of right osseous labyrinth.

cates with the scala vestibuli of the cochlea.

The **Bony Semicircular Canals** (*canales semicirculares ossei*) are three in number, **superior, posterior,** and **lateral,** and are situated superior and posterior to the vestibule. They are unequal in length, compressed from side to side, and each describes the greater part of a circle. Each measures about 0.8 mm in diameter, and presents a dilatation at one end, called the **ampulla,** which measures more than twice the diameter of the tube. They open into the vestibule by five orifices, one of the apertures being common to two of the canals.

The **superior semicircular canal** (*canalis semicircularis anterior*), 15 to 20 mm in length, is vertical in direction, and is placed transversely to the long axis of the petrous portion of the temporal bone, on the anterior surface of which its arch forms a round projection. It describes about two-thirds of a circle. Its lateral extremity is ampullated, and opens into the superior part of the vestibule; the opposite end joins with the superior part of the posterior canal to form the **crus commune,** which opens into the superior and medial part of the vestibule.

The **posterior semicircular canal** (*canalis semicircularis posterior*), also vertical, is directed posteriorly, nearly parallel to the posterior surface of the petrous bone; it is the longest of the three, measuring from 18 to 22 mm; its inferior or ampullated end

opens into the inferior and posterior part of the vestibule, its superior into the crus commune already mentioned.

The **lateral** or **horizontal canal** (*canalis semicircularis lateralis; external semicircular canal*) is the shortest of the three. It measures from 12 to 15 mm, and its arch is directed horizontally lateralward; thus each semicircular canal stands at right angles to the other two. Its ampullated end corresponds to the superior and lateral angle of the vestibule, just above the fenestra vestibuli, where it opens close to the ampullated end of the superior canal; its opposite end opens at the superior and posterior part of the vestibule. The lateral canal of one ear is very nearly in the same plane as that of the other, while the superior canal of one ear is nearly parallel to the posterior canal of the other.

The **Cochlea** (Figs. 13–56, 13–57).—The cochlea bears some resemblance to a common snail shell; it forms the anterior part of the labyrinth, is conical in form, and placed almost horizontally in front of the vestibule; its **apex** (*cupula*) is directed anteriorly and laterally, with a slight inclination inferiorly, toward the labyrinthic wall of the tympanic cavity; its **base** corresponds with the bottom of the internal acoustic meatus, and is perforated by numerous apertures for the passage of the cochlear division of the acoustic nerve. It

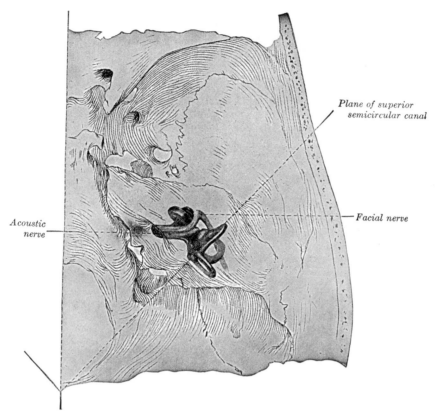

*Plane of superior
semicircular canal*

Facial nerve

*Acoustic
nerve*

Fig. 13–58.—Position of the right bony labyrinth of the ear in the skull, viewed from above. The temporal bone is considered transparent and the labyrinth drawn in from a corrosion preparation. (Spalteholz.)

measures about 5 mm from base to apex, and its breadth across the base is about 9 mm. It consists of a conical-shaped central axis, the **modiolus**; of a canal, the inner wall of which is formed by the central axis, wound spirally around it for two turns and three quarters, from the base to the apex; and of a delicate lamina, the **osseous spiral lamina**, which projects from the modiolus, and, following the windings of the canal, partially subdivides it into two. In the intact body a membrane, the **basilar membrane**, stretches from the free border of this lamina to the outer wall of the bony cochlea and completely separates the canal into two passages, which, however, communicate with each other at the apex of the modiolus by a small opening, named the **helicotrema**.

The **modiolus** is the conical central axis or pillar of the cochlea. Its base is broad, and appears at the bottom of the internal acoustic meatus, where it corresponds with the area cochleae; it is perforated by numerous orifices, which transmit filaments of the cochlear division of the acoustic nerve; the nerves for the first turn and a half pass through the foramina of the tractus spiralis foraminosus; those for the apical turn, through the foramen centrale. The canals of the tractus spiralis foraminosus pass through the modiolus and successively bend outward to reach the attached margin of the lamina spiralis ossea. Here they become enlarged, and by their apposition form the **spiral canal of the modiolus**, which follows the course of the attached margin of the osseous spiral lamina and lodges the **spiral ganglion** (*ganglion of Corti*). The foramen centrale is continued into a canal which runs through the middle of the modiolus to its apex. The modiolus diminishes rapidly in size in the second and succeeding coil.

The bony canal of the cochlea takes two

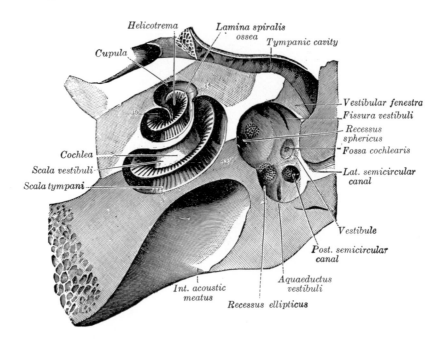

Helicotrema   Lamina spiralis
                    ossea   Tympanic cavity
Cupula

Vestibular fenestra
Fissura vestibuli
Recessus sphericus
Fossa cochlearis

Cochlea
Scala vestibuli
Scala tympani

Lat. semicircular canal

Vestibule

Post. semicircular canal

Int. acoustic meatus   Aquaeductus vestibuli
Recessus ellipticus

FIG. 13–59.—The cochlea and vestibule, viewed from above. All the hard parts which form the roof of the internal ear have been removed with the saw.

and three-quarter turns around the modiolus. It is about 30 mm in length, and diminishes gradually in diameter from the base to the summit, where it terminates in the **cupula,** which forms the apex of the cochlea. The beginning of this canal is about 3 mm in diameter; it diverges from the modiolus toward the tympanic cavity and vestibule, and presents three openings. One, the **fenestra cochleae,** communicates with the tympanic cavity—in the intact body this aperture is closed by the **secondary tympanic membrane;** another, of an elliptical form, opens into the vestibule. The third is the aperture of the aquaeductus cochleae, leading to a minute funnel-shaped canal, which opens on the inferior surface of the petrous part of the temporal bone and transmits a small vein, and also forms a communication between the subarachnoid cavity and the scala tympani.

The **osseous spiral lamina** (*lamina spiralis ossea*) is a bony shelf or ledge which projects from the modiolus into the interior of the canal, and, like the canal, takes two and three-quarter turns around the modiolus. It reaches about half way toward the outer

wall of the tube, and partially divides its cavity into two passages or scalae, of which the superior is named the **scala vestibuli,** while the inferior is termed the **scala tympani.** Near the summit of the cochlea the lamina ends in a hook-shaped process, the **hamulus laminae spiralis;** this assists in forming the boundary of a small opening, the **helicotrema,** through which the two scalae communicate with each other. From the spiral canal of the modiolus numerous canals pass outward through the osseous spiral lamina as far as its free edge. In the inferior part of the first turn a second bony lamina, the **secondary spiral lamina,** projects inward from the outer wall of the bony tube; it does not, however, reach the primary osseous spiral lamina, so that if viewed from the vestibule a narrow fissure, the **vestibule fissure,** is seen between them.

The *osseous labyrinth* is lined by an exceedingly thin fibrous membrane; its attached surface is rough and fibrous, and closely adherent to the bone; its free surface is smooth and pale, covered with a layer of epithelium, and secretes a thin, limpid fluid, the **perilymph.** A delicate tubular process of this membrane is prolonged along the aqueduct of the cochlea to the inner surface of the dura mater.

The Membranous Labyrinth

The **Membranous Labyrinth** (*labyrinthus membranaceus*) (Figs. 13–60, 13–61, 13–62) is lodged within the bony cavities just described, and has the same general form as these; it is, however, considerably smaller, and is partly separated from the bony walls by a quantity of fluid, the **perilymph.** In certain places it is fixed to the walls of the cavity. The membranous labyrinth contains fluid, the **endolymph,** and on its walls the ramifications of the acoustic nerve are distributed.

Within the osseous vestibule the membranous labyrinth does not quite preserve the form of the bony cavity, but consists of two membranous sacs, the **utricle,** and the **saccule.**

The **Utricle** (*utriculus*), the larger of the two, is of an oblong form, compressed transversely, and occupies the superior and posterior part of the vestibule, lying in contact with the recessus ellipticus and the part inferior to it. That portion which is lodged in the recess forms a sort of pouch or cul-de-sac, the floor and anterior wall of which are thickened, and form the **macula acustica utriculi,** which receives the utricular filaments of the acoustic nerve. The cavity of the utricle communicates behind with the semicircular ducts by five orifices.

From its anterior wall is given off the **ductus utriculosaccularis,** which opens into the ductus endolymphaticus.

The **Saccule** (*sacculus*) is the smaller of the two vestibular sacs; it is globular in form, and lies in the recessus sphericus near the opening of the scala vestibuli of the cochlea. Its anterior part exhibits an oval thickening, the **macula acustica sacculi,** to which are distributed the saccular filaments of the acoustic nerve. Its cavity does not directly communicate with that of the utricle. From the posterior wall a canal, the **ductus endolymphaticus,** is given off; this duct is joined by the ductus utriculosaccularis, and then passes along the aquaeductus vestibuli and ends in a blind pouch (*saccus endolymphaticus*) on the posterior surface of the petrous portion of the temporal bone, where it is in contact with the dura mater. From the inferior part of the saccule a short tube, the **ductus reuniens** (*canalis reuniens of Hensen*) passes inferiorly and opens into the ductus cochlearis near its vestibular extremity (Fig. 13–60).

The **Semicircular Ducts** (*ductus semicirculares; membranous semicircular canals*) (Figs. 13–61, 13–62) are about one-fourth of the diameter of the osseous canals, but in number, shape, and general form they are precisely similar, and each presents at one end an ampulla. They open by five orifices into the utricle, one opening being common to the medial end of the superior and the superior end of the posterior duct. In the ampullae the wall is thickened, and

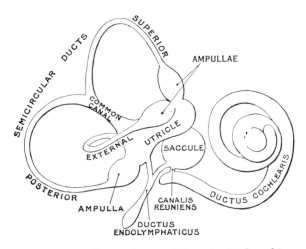

Fig. 13–60.—The membranous labyrinth. (Enlarged.)

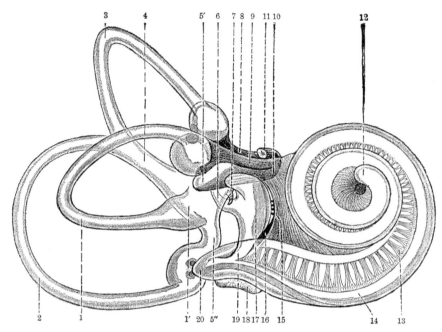

FIG. 13–61.—Right human membranous labyrinth, removed from its bony enclosure and viewed from the anterolateral aspect. (G. Retzius.)

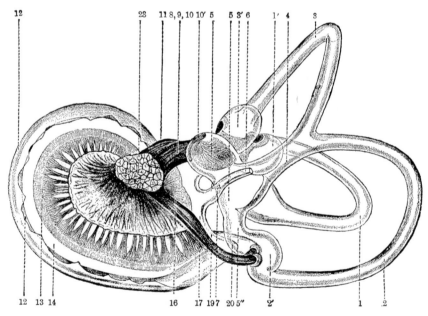

FIG. 13–62.—The same from the posteromedial aspect. 1. Lateral semicircular canal; 1′, its ampulla; 2. Posterior canal; 2′, its ampulla. 3. Superior canal; 3′, its ampulla. 4. Conjoined limb of superior and posterior canals (*sinus utriculi superior*). 5. Utricle. 5′. Recessus utriculi. 5″. Sinus utriculi posterior. 6. Ductus endolymphaticus. 7. Canalis utriculosaccularis. 8. Nerve to ampulla of superior canal. 9. Nerve to ampulla of lateral canal. 10. Nerve to recessus utriculi (in Fig. 13–61, the three branches appear conjoined). 10′. Ending of nerve in recessus utriculi. 11. Facial nerve. 12. Lagena cochleae. 13. Nerve of cochlea within spiral lamina. 14. Basilar membrane. 15. Nerve fibers to macula of saccule. 16. Nerve to ampulla of posterior canal. 17. Saccule. 18. Secondary membrane of tympanum. 19. Canalis reuniens. 20. Vestibular end of ductus cochlearis. 23. Section of the facial and acoustic nerves within internal acoustic meatus. The separation between them is not apparent in the section. (G. Retzius.)

projects into the cavity as a fiddle-shaped, transversely placed elevation, the **crista ampularis,** in which the nerves end.

The utricle, saccule, and semicircular ducts are held in position by numerous fibrous bands which stretch across the space between them and the bony walls.

Structure (Fig. 13–63).–The walls of the utricle, saccule, and semicircular ducts consist of three layers. The *outer layer* is a loose and flocculent structure, apparently composed of ordinary fibrous tissue containing blood vessels and some pigment cells. The *middle layer,* thicker and more transparent, forms a homogeneous membrana propria, and presents on its internal surface, especially in the semicircular ducts, numerous papilliform projections, which, on the addition of acetic acid, exhibit an appearance of longitudinal fibrillation. The *inner layer* is formed of polygonal nucleated epithelial cells. In the maculae of the utricle and saccule, and in the cristae ampulares of the semicircular ducts, the middle coat is thickened and the epithelium is columnar, and consists of **supporting cells** and **hair cells.** The former are fusiform, and their deep ends are attached to the membrana propria, while their free extremities are united to form a thin cuticle. The **neuroepithelium** or hair cells are flask-shaped, and their deep, rounded ends do not reach the membrana propria, but lie between the supporting cells. The deep part of each contains a large nucleus, while its more superficial part is granular and pigmented. The free end is surmounted by a long, tapering, hair-like filament, which projects into the cavity. The filaments of the acoustic nerve enter these parts, and, having pierced the outer and middle layers, they lose their myelin sheaths, and their axons ramify between the hair cells.

Two small rounded bodies termed **otoconia** (*statoconia*) each consisting of a mass of minute crystalline grains of carbonate of lime, held together in a mesh of gelatinous tissue, are suspended in the endolymph in contact with the free ends of the hairs projecting from the maculae.

The **Ductus Cochlearis** (*membranous cochlea; scala media*).–The ductus cochlearis consists of a spirally arranged tube enclosed in the bony canal of the cochlea and lying along its outer wall.

As already stated, the osseous spiral lamina extends only part of the distance between the modiolus and the outer wall of the cochlea, while the **basilar membrane** (*membrana spiralis*) stretches from its free edge to the outer wall of the cochlea, and completes the roof of the scala tympani. A second and more delicate membrane, the

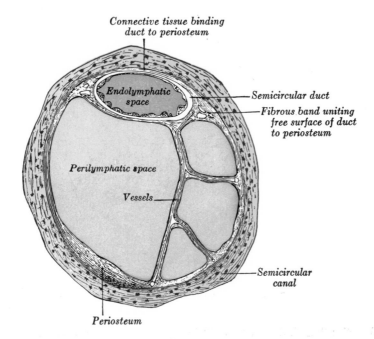

Connective tissue binding duct to periosteum

Endolymphatic space

Semicircular duct

Fibrous band uniting free surface of duct to periosteum

Perilymphatic space

Vessels

Semicircular canal

Periosteum

FIG. 13–63.–Transverse section of a human semicircular canal and duct. (After Rüdinger.)

vestibular membrane (*Reissner's membrane*), extends from the thickened periosteum covering the osseous spiral lamina to the outer wall of the cochlea, where it is attached at some little distance above the outer edge of the basilar membrane. A canal is thus shut off between the scala tympani and the scala vestibuli; this is the **ductus cochlearis** or **scala media** (Fig. 13–64). It is triangular on transverse section, its roof being formed by the vestibular membrane, its outer wall by the periosteum lining the bony canal, and its floor by the membrana spiralis and the outer part of the lamina spiralis ossea. Its extremities are closed; the upper is termed the **lagena** and is attached to the cupula at the upper part of the helicotrema; the inferior is lodged

in the recessus cochlearis of the vestibule. Near the inferior end the ductus cochlearis is brought into continuity with the saccule by a narrow, short canal, the **ductus reuniens** (Fig. 13–60).

The **Spiral Organ of Corti** rests upon the membrana basilaris. The vestibular membrane is thin and homogeneous, and is covered on its superior and inferior surfaces by a layer of epithelium. The periosteum, forming the outer wall of the ductus cochlearis, is greatly thickened and altered in character, and is called the **spiral ligament**. It projects inward below as a triangular prominence, the **basilar crest**, which gives attachment to the outer edge of the basilar membrane; immediately above the crest is a concavity, the **sulcus spiralis externus**. The

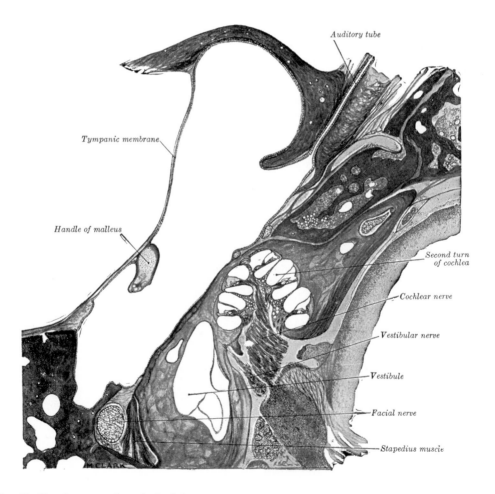

Auditory tube

Tympanic membrane

Handle of malleus

Second turn of cochlea

Cochlear nerve

Vestibular nerve

Vestibule

Facial nerve

Stapedius muscle

Fig. 13–64.—A section through the left temporal bone. (Drawn from a section prepared at the Fehrens Institute, kindly lent by Prof. J. Kirk.)

upper portion of the spiral ligament contains numerous capillary loops and small blood vessels, and is termed the **stria vascularis.**

The osseous spiral lamina consists of two plates of bone, and between these are the canals for the transmission of the filaments of the acoustic nerve. On the upper plate of that part of the lamina which is outside the vestibular membrane, the periosteum is thickened to form the **limbus laminae spiralis** (Fig. 13–65); this ends externally in a concavity, the **sulcus spiralis internus,** which represents, on section, the form of the letter C; the upper part, formed by the overhanging extremity of the limbus, is named the **vestibular lip;** the lower part, prolonged and tapering, is called the **tym-**

panic lip, and is perforated by numerous foramina for the passage of the cochlear nerves. The upper surface of the vestibular lip is intersected at right angles by a number of furrows, between which are numerous elevations; these present the appearance of teeth along the free surface and margin of the lip, and have been named **dentes acoustici.** The limbus is covered by a layer of what appears to be squamous epithelium, but the deeper parts of the cells with their contained nuclei occupy the intervals between the elevations and between the auditory teeth. This layer of epithelium is continuous on the one hand with that lining the sulcus spiralis internus, and on the other with that covering the under surface of the vestibular membrane.

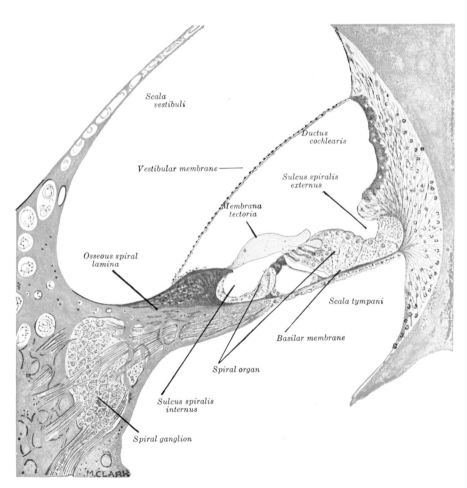

FIG. 13–65.—A section through the second turn of the cochlea indicated in the previous figure. (Mallory's stain.)

Basilar Membrane.—The basilar membrane stretches from the tympanic lip of the osseous spiral lamina to the basilar crest and consists of two parts, an inner and an outer. The inner is thin, and is named the **zona arcuata:** it supports the spiral organ of Corti. The outer is thicker and striated, and is termed the **zona pectinata.** The under surface of the membrane is covered by a layer of vascular connective tissue; one of the vessels in this tissue is somewhat larger than the rest, and is named the **vas spirale.**

Rods of Corti.—Each of these consists of a base or foot-plate and elongated part or body, and an upper end or head; the body of each rod is finely striated, but in the head there is an oval non-striated portion which stains deeply with carmine. Occupying the angles between the rods and the basilar membrane are nucleated cells which partly envelop the rods and extend on to the floor of Corti's tunnel; these may be looked upon as the undifferentiated parts of the cells from which the rods have been formed.

The **inner rods** number nearly 6000, and their bases rest on the basilar membrane close

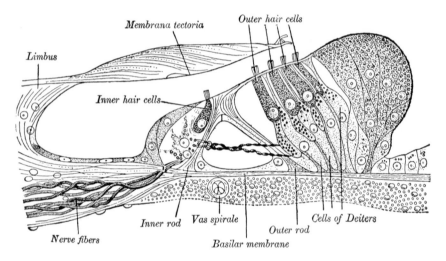

Fig. 13–66.—Section through the spiral organ of Corti. Magnified. (G. Retzius.)

The **Spiral Organ** (*organum spirale*) or **Organ of Corti** (Fig. 13–66) is composed of a series of epithelial structures placed upon the inner part of the basilar membrane. The average length is 31.5 mm. The more central of these structures are two rows of rod-like bodies, the **inner and outer rods** or **pillars of Corti.** The bases of the rods are supported on the basilar membrane, those of the inner row at some distance from those of the outer; the two rows incline toward each other and, coming into contact superiorly, enclose between them and the basilar membrane a triangular tunnel, the **tunnel of Corti.** On the inner side of the inner rods is a single row of hair cells, and on the outer side of the outer rods three or four rows of similar cells, together with certain supporting cells termed the cells of Deiters and Hensen. The free ends of the outer hair cells occupy a series of apertures in a net-like membrane, the **reticular membrane,** and the entire organ is covered by the tectorial membrane.

to the tympanic lip of the sulcus spiralis internus. The shaft or body of each is sinuously curved and forms an angle of about 60 degrees with the basilar membrane. The head resembles the proximal end of the ulna and presents a deep concavity which accommodates a convexity on the head of the outer rod. The headplate, or portion overhanging the concavity, overlaps the headplate of the outer rod.

The **outer rods,** nearly 4000 in number, are longer and more obliquely set than the inner, forming with the basilar membrane an angle of about 40 degrees. Their heads are convex internally; they fit into the concavities on the heads of the inner rods and are continued outward as thin flattened plates, termed **phalangeal processes,** which unite with the phalangeal processes of Deiters' cells to form the reticular membrane.

Hair Cells.—The hair cells are short columnar cells; their free ends are on a level with the heads of Corti's rods, and each is surmounted by about twenty hair-like processes arranged in the form of a crescent with its concavity directed inward. The deep ends of the cells reach about half way along Corti's rods, and each contains a large nucleus; in contact with the deep ends of the hair cells are the terminal filaments of the cochlear division of the acoustic nerve. The *inner* hair cells, about 3500 in number, are arranged in a single row on the medial side of the inner rods, and their diameters being greater than those of the rods it follows that each hair cell is supported by more than one rod. The free ends of the inner hair cells are encircled by a cuticular membrane which is fixed to the heads of the inner rods. Adjoining the inner hair cells are one or two rows of columnar supporting cells, which, in turn, are continuous with the cubical cells lining the sulcus spiralis internus. The *outer* hair cells number about 12,000, and are nearly twice as long as the inner. In the basal coil of the cochlea they are arranged in three regular rows; in the apical coil, in four somewhat irregular rows. The receptors for high tones are located in the basal turn of the cochlea.

Between the rows of the outer hair cells are rows of supporting cells, called the **cells of Deiters;** their expanded bases are planted on the basilar membrane, while the opposite end of each presents a clubbed extremity or **phalangeal process.** Immediately to the outer side of Deiters' cells are five or six rows of columnar cells, the **supporting cells of Hensen.** Their bases are narrow, while their upper parts are expanded and form a rounded elevation on the floor of the ductus cochlearis. The columnar cells lying outside Hensen's cells are termed the **cells of Claudius.** A space exists between the outer rods of Corti and the adjacent hair cells; this is called the **space of Nuel.**

The **reticular lamina** (Fig. 13–66) is a delicate framework perforated by rounded holes which are occupied by the free ends of the outer hair cells. It extends from the heads of the outer rods of Corti to the external row of the outer hair cells, and is formed by several rows of "minute fiddle-shaped cuticular structures," called **phalanges,** between which are circular apertures containing the free ends of the hair cells. The innermost row of phalanges consists of the phalangeal processes of the outer rods of Corti; the outer rows are formed by the modified free ends of Deiters' cells.

Covering the sulcus spiralis internus and the spiral organ of Corti is the **tectorial membrane,** which is attached to the limbus laminae spiralis close to the inner edge of the vestibular mem-

brane. Its inner part is thin and overlies the auditory teeth of Huschke; its outer part is thick, and along its lower surface, opposite the inner hair cells, is a clear band, named **Hensen's stripe,** due to the intercrossing of its fibers. The lateral margin of the membrane is much thinner. It consists of fine colorless fibers embedded in a transparent matrix (the matrix may be a variety of soft keratin), of a soft collagenous, semisolid character with marked adhesiveness. The general transverse direction of the fibers inclines from the radius of the cochlea toward the apex.

The **acoustic nerve** (*n. vestibulocochlearis; n. statoacusticus; nerve of hearing*) divides near the bottom of the internal acoustic meatus into an anterior or cochlear and a posterior or vestibular branch.

The **vestibular nerve** (*n. vestibularis*) supplies the utricle, the saccule, and the ampullae of the semicircular ducts. On the trunk of the nerve, within the internal acoustic meatus, is a ganglion, the **vestibular ganglion** (*ganglion of Scarpa*); the fibers of the nerve arise from the cells of this ganglion. On the distal side of the ganglion the nerve splits into a superior, an inferior, and a posterior branch. The filaments of the *superior branch* are transmitted through the foramina in the area vestibularis superior, and end in the macula of the utricle and in the ampullae of the superior and lateral semicircular ducts; those of the *inferior branch* traverse the foramina in the area vestibularis inferior, and end in the macula of the saccule. The *posterior branch* runs through the foramen singulare at the posteroinferior part of the bottom of the meatus and divides into filaments for the supply of the ampulla of the posterior semicircular duct.

The **cochlear nerve** (*n. cochlearis*) divides into numerous filaments at the base of the modiolus; those for the basal and middle coils pass through the foramina in the tractus spiralis foraminosus; those for the apical coil through the canalis centralis, and the nerves bend outward to pass between the lamellae of the osseous spiral lamina. Occupying the spiral canal of the modiolus is the **spiral ganglion of the cochlea** (*ganglion of Corti*) (Fig. 13–65), consisting of bipolar nerve cells, which constitute the cells of origin of this nerve. Reaching the outer edge of the osseous spiral lamina, the fibers of the nerve pass through the foramina in the tympanic lip; some end by arborizing around the bases of the inner hair cells, while others pass between Corti's rods and across the tunnel, to end in a similar manner in relation to the outer hair cells. The cochlear nerve gives off a vestibular branch to supply the vestibular end of the ductus cochlearis; the filaments of this branch pass through the foramina in the fossa cochlearis.

Vessels.—The **arteries of the labyrinth** are the internal auditory, from the basilar, and the stylomastoid, from the posterior auricular. The internal auditory artery divides at the bottom of the internal acoustic meatus into two branches: cochlear and vestibular. The cochlear branch subdivides into twelve or fourteen twigs, which traverse the canals in the modi-

olus, and are distributed, in the form of a capillary network, in the lamina spiralis and basilar membrane. The vestibular branches are distributed to the utricle, saccule, and semicircular ducts.

The **veins** of the vestibule and semicircular canals accompany the arteries, and, receiving those of the cochlea at the base of the modiolus, unite to form the internal auditory veins which end in the posterior part of the superior petrosal sinus or in the transverse sinus.

REFERENCES

Organs of the Senses

MATHEWS, L. and M. KNIGHT. 1963. *The Senses of Animals.* 240 pages, illustrated. Philosophical Library, New York.

NEFF, W. H. 1965. *Contributions to Sensory Physiology.* Vol. 1, x + 274 pages, illustrated. Academic Press, Inc., New York.

Taste and Smell

BRADLEY, R. M. and I. B. STERN. 1967. The development of the human taste bud during the fetal period. J. Anat. (Lond.), *101,* 743-752.

FISHMAN, I. Y. 1957. Single fiber gustatory impulses in rat and hamster. J. cell. comp. Physiol., *49,* 319-334.

GUTH, L. 1958. Taste buds on the cat's circumvallate papilla after reinnervation by glossopharyngeal, vagus, and hypoglossal nerves. Anat. Rec., *130,* 25-37.

LALONDE, E. R. and J. A. EGLITIS. 1961. Number and distribution of taste buds on the epiglottis, pharynx, larynx, soft palate and uvula in a human newborn. Anat. Rec., *140,* 91-95.

Eye

ADELMANN, H. B. 1936. The problem of cyclopia. Quart. Rev. Biol., *11,* 284-304.

BARBER, A. 1955. *Embryology of the Human Eye.* 236 pages, 193 figures. C. V. Mosby, St. Louis, Missouri.

BARTELMEZ, G. W. and A. S. DEKABAN. 1962. The early development of the human brain. Carneg. Instn., Contr. Embryol., *37,* 13-32 + 30 plates.

BEATIE, J. C. and D. L. STILLWELL, JR. Innervation of the eye. Anat. Rec., *141,* 45-61.

BODEMER, C. W. 1958. The origin and development of the extrinsic ocular muscles in the trout (Salmo trutta). J. Morph., *102,* 119-155.

CHARNWOOD, L. 1950. *An Essay on Binocular Vision.* vii + 117 pages, illustrated. Hafner Publishing Company, New York.

COOPER, S. and P. M. DANIEL. 1949. Muscle spindles in human extrinsic eye muscles. Brain, *72,* 1-24 + 4 plates.

COULOMBRE, A. J. and E. S. CRELIN. 1958. The role of the developing eye in the morphogenesis of the avian skull. Amer. J. phys. Anthrop., *16,* 25-37.

DETWILER, S. R. 1955. The eye and its structural adaptations. Proc. Amer. Philos. Soc., *99,* 224-238.

GÉNIS-GÁLVEZ, J. M. 1957. Innervation of the ciliary muscle. Anat. Rec., *127,* 219-230.

GILBERT, P. W. 1957. The origin and development of the human extrinsic ocular muscles. Carneg. Instn. Contr. Embryol., *36,* 59-78 + 14 plates.

GILLILAN, L. A. 1961. The collateral circulation of the human orbit. Arch. Ophthal., *65,* 684-694.

HOGAN, M. J., J. A. ALVARADO, and J. E. WEDDELL. 1971. Histology of the Human Eye. xiii + 687 pages, illustrated. W. B. Saunders Company, Philadelphia.

JAKUS, M. A. 1964. *Ocular Fine Structure, Selected Electron Micrographs.* ix + 204 pages, 91 plates. Little, Brown and Company, Boston.

MANN, I. C. 1957. *Developmental Abnormalities of the Eye.* xi + 419 pages. J. B. Lippincott Company, Philadelphia.

O'RAHILLY, R. 1961. A simple demonstration of the features of the bony orbit. Anat. Rec., *141,* 315-316.

POLYAK, S. L. 1941. *The Retina.* Univ. of Chicago Press, Chicago.

PRINCE, J. H. 1956. *Comparative Anatomy of the Eye.* iv + 418 pages. Charles C Thomas, Publisher, Springfield.

REYER, R. W. 1954. Regeneration of the lens in the amphibian eye. Quart. Rev. Biol., *29,* 1-46.

STONE, L. S. 1960. Regeneration of the lens, iris, and neural retina in a vertebrate eye. Yale J. Biol. Med., *32,* 464-473.

WOOLF, D. 1956. A comparative cytological study of the ciliary muscle. Anat. Rec., *124,* 145-164.

WOLFF, E. 1954. *The Anatomy of the Eye and Orbit.* 4th Edition. H. K. Lewis & Company, Ltd., London.

Ear

ALTMANN, F. 1950. Normal development of the ear and its mechanics. Arch. Otolaryng. (Chicago), *52,* 725-766.

ALTMANN, F. 1951. Malformations of the Eustachian tube, the middle ear, and its appendages. Arch. Otolaryng. (Chicago), *54,* 241-266.

AMJAD, A. H., A. A. SHEER, and J. ROSENTHAL. 1969. Human internal auditory canal. Arch. Otolaryngology, *89*, 709-714.

ANSON, B. J. and T. H. BEST. 1955. The ear and the temporal bone. Development and adult structure. Otolaryngology, *1*, 1-111.

ANSON, B. J. and T. H. BAST. 1956. Development and adult anatomy of the auditory ossicles in relation to the operation for mobilization of the stapes in otosclerotic deafness. Laryngoscope, *66*, 785-795.

ANSON, B. J., J. S. HANSON, and S. F. RICHANY. 1960. Early embryology of the auditory ossicles and associated structures in relation to certain anomalies observed clinically. Ann. Otol., *69*, 427-447.

BLEVINS, C. E. 1963. Innervation of the tensor tympani muscle of the cat. Amer. J. Anat., *113*, 287-301.

BLEVINS, C. E. 1964. Studies on the innervation of the stapedius muscle of the cat. Anat. Rec., *149*, 157-171.

ERULKAR, S. D., M. L. SHELANSKI, B. L. WHITSEL, and P. OGLE. 1964. Studies of muscle fibers of the tensor tympani of the cat. Anat. Rec., *149*, 279-297.

GRAVES, G. O. and L. F. EDWARDS. 1944. The eustachian tube. A review of its descriptive, microscopic, topographic and clinical anatomy. Arch. Otolaryng. (Chicago), *39*, 359-397.

GUSSEN, R. 1968. Articular and internal remodeling in the human otic capsule. Am. J. Anat., *122*, 397-418.

McLELLAN, M. S. and C. H. WEBB. 1961. Ear studies in the newborn infant. J. Pediat., *58*, 523-527.

SCHUKNECHT, H. K. 1950. A clinical study of auditory damage following blows to the head. Ann. Otolaryng. (St. Louis), *59*, 331-358.

SPECTOR, B. 1944. Storage of trypan blue in the internal ear of the rat. Anat. Rec., *88*, 83-89.

SUNDERLAND, S. 1945. The arterial relations of the internal auditory meatus. Brain, *68*, 23-27.

YNTEMA, C. L. 1933. Experiments on the determination of the ear ectoderm in the embryo of Amblystoma punctatum. J. exp. Zool., *65*, 317-357.

14 | *The Integument*

overing the surface of the body and sheltering it from injurious influences in the environment is the **Skin** or **Integument**. It protects the deeper tissues from injury, from drying, and from invasion by foreign organisms; it contains the peripheral endings of many of the sensory nerves; it plays an important part in the regulation of the body temperature, and has also limited excretory and absorbing powers. It consists principally of a layer of dense connective tissue, named the **dermis** (*corium, cutis vera*) and an external covering of epithelium, termed the **epidermis** or **cuticle**. On the surface of the former layer are **sensitive** and **vascular papillae**; within or beneath it are certain organs with special functions: namely, the **sudoriferous** and **sebaceous glands**, and the **hair follicles**.

Development.—The epidermis and its appendages, consisting of the hairs, nails, sebaceous and sweat glands, are developed from the ectoderm, whereas the corium or true skin is of mesodermal origin. About the fifth week the epidermis consists of two layers of cells, the deeper one corresponding to the rete mucosum. The subcutaneous fat appears about the fourth month, and the papillae of the true skin about the sixth. A considerable desquamation of epidermis takes place during fetal life, and this desquamated epidermis, mixed with sebaceous secretion, constitutes the **vernix caseosa**, with which the skin is smeared during the last three months of fetal life. The nails are formed at the third month, and begin to project from the epidermis about the sixth. The hairs appear between the third and fourth months in the form of solid downgrowths of the deeper layer of the epidermis, the growing extremities of which become inverted by papillary projections from the corium. The central cells of the solid downgrowths undergo alteration to form the hair, while the peripheral cells are retained to form the lining cells of the hair follicle. About the fifth month the fetal hairs (**lanugo**) appear first on the head and then on the other parts; they drop off after birth, and give place to the permanent hairs. The cellular structures of the sudoriferous and sebaceous glands are formed from the ectoderm, whereas the connective tissue and blood vessels are derived from the mesoderm. All the sweat glands are fully formed at birth; they begin to develop as early as the fourth month.

Structure.—The **epidermis, cuticle, or scarf skin** is non-vascular, consists of stratified epithelium, and is accurately molded over the papillary layer of the dermis. It varies in thickness in different parts. In some situations, as in the palms of the hands and soles of the feet, it is thick, hard, and horny in texture. The more superficial layers of cells, called the **horny layer** (*stratum corneum*), may be separated by maceration from a deeper stratum, which is called the **stratum mucosum,** and which consists of several layers of differently shaped cells. The free surface of the epidermis is marked by a network of linear furrows of variable size, dividing the surface into a number of polygonal or lozenge-shaped areas. Some of these furrows are large, as are those opposite the flexures of the joints, and correspond to folds in the dermis. In other situations, as upon the back of the hand, they are exceedingly fine, and intersect one another at various angles. Upon the palmar surfaces of the hands and fingers, and upon the soles of the feet, the epidermal ridges are very distinct, and are disposed in curves; they depend upon the large size and peculiar arrangements of the papillae upon which the epidermis is placed. These ridges increase friction between contact surfaces to prevent slipping in walking or prehension. The direction of the ridges is at right angles to the force that tends to produce slipping or to the resultant of such forces when these forces vary in direction. In each individual the lines on the tips of the fingers and thumbs form distinct patterns unlike those of any other person. A method of determining the identity of a person is based on this fact by making impressions or **fingerprints** of these lines. The deep surface of the epidermis is accurately molded upon the papillary layer of the dermis, the papillae being covered by a basement membrane. When the epidermis is lifted off by maceration and inverted it presents on its under surface a number of pits or depressions corresponding to the papillae, and ridges corresponding to the intervals between them. Fine tubular prolongations are continued from this layer into the ducts of the sudoriferous and sebaceous glands.

The **stratified squamous epithelium** of the epidermis is composed of several layers named according to various categories such as shape of cells, texture, composition, and position. Beginning with the deepest they are named, (*a*) stratum basale, (*b*) stratum spinosum, (*c*) stratum granulosum, (*d*) stratum lucidum, and (*e*) stratum corneum.

(*a*) The **stratum basale,** the deepest layer, is composed of columnar or cylindrical cells, giving it an alternate name **stratum cylindricum.** The ends of the cells in contact with the basement membrane have toothlike protoplasmic projections which fit into sockets in the membrane and appear to anchor the cells to the underlying dermis. The cells of this layer undergo division by mitosis, supplying new cells to make up for the continual loss of surface layers from abrasion. This layer has been appropriately named the **stratum germinativum.**

(*b*) The **stratum spinosum** is composed of several layers of polygonal cells, the number depending upon the area of the body from which the skin is taken. As a result of the slight shrinkage caused by technical procedures, these cells in ordinary histological preparations appear to have *cytoplasmic bridges* connecting them with their neighbors. When they are pulled apart, they appear to have minute *spines* on their surface, giving them the name *prickle cells.* Electron microscopic studies have shown that each cell is adherent to its neighbors at particular points called desmosomes, and that these points are drawn out into the spines by shrinkage. The electron studies have also shown that the cytoplasm of these cells contains minute fibrils, some of which are oriented toward

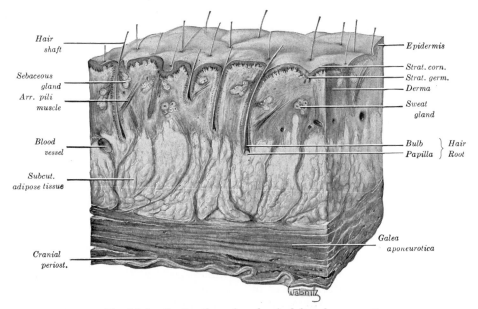

F<small>IG</small>. 14–1.—Section through scalp of adult cadaver. × **5.**

the desmosomes. In light microscopic preparations these fibrils may be visible and, since they are associated with the desmosomes, give the impression of fibrils running between cells and are called **tonofibrils.**

(c) The **stratum granulosum** is composed of two or three rows of flattened cells, lying parallel with the surface. They contain numerous quite large granules which stain deeply with hematoxylin. They are composed of **keratohyalin,** a substance which apparently is transformed into keratin in more superficial layers.

(d) The **stratum lucidum** appears to be a homogeneous translucent band, much thinner than the strata on either side of it. The cells contain droplets of **eleidin** and their nuclei and cell boundaries are not visible.

(e) The **stratum corneum** is composed of squamous places or scales fused together to make the outer horny layer. These plates are the remains of the cells and contain a fibrous protein **keratin.** The most superficial layer sloughs off or "desquamates." The thickness of this layer is correlated with the trauma to which an area is subjected, being very thick on the palms and soles, but thin over protected areas.

The black **color of the skin** in the Negro, and the tawny color among some of the white races, are due to the presence of pigment in the cells of the epidermis. This pigment is more especially distinct in the cells of the stratum basale, and is similar to that found in the cells of the pigmentary layer of the retina. As the cells approach the surface and desiccate, the color becomes partially lost; the disappearance of the pigment from the superficial layers of the epidermis is, however, difficult to explain.

The pigment (**melanin**) consists of dark brown or black granules of very small size, closely packed together within the cells, but not involving the nucleus.

The main purpose served by the epidermis is that of protection. As the surface is worn away new cells are supplied and thus the true skin, the vessels and nerves which it contains are defended from damage.

The **Dermis, Corium, Cutis Vera** or **True Skin** is tough, flexible, and elastic. It varies in thickness in different parts of the body. Thus it is very thick in the palms of the

hands and soles of the feet; thicker on the dorsal aspect of the body than on the ventral, and on the lateral than on the medial sides of the limbs. In the eyelids, scrotum, and penis it is exceedingly thin and delicate.

It consists of felted connective tissue, with a varying amount of elastic fibers and numerous blood vessels, lymphatics, and nerves. The connective tissue is arranged in two layers: a **deeper** or **reticular,** and a **superficial** or **papillary** (Fig. 14-2).

Smooth muscle cells are found in the dermis and the subcutaneous layers of the scrotum, penis, labia majora, and nipples. In the mammary papilla, the smooth muscle cells are disposed in circular bands and radiating bundles, arranged in superimposed laminae. In parts of the skin where there are hairs, discrete bundles of smooth muscle called the Arrectores pilorum are attached in the superficial layers of the corium and near the base of each hair follicle.

The **papillary layer** (*stratum papillare; superficial layer; corpus papillare of the corium*) consists of numerous small, highly sensitive, and vascular eminences, the **papillae,** which rise perpendicularly from its surface. The papillae are minute conical eminences, having rounded or blunted extremities, occasionally divided into two or more parts, and are received into corresponding pits on the under surface of the cuticle. On the general surface of the body, more especially in parts endowed with slight sensibility, they are few in number, and exceedingly minute; but in some situations, as upon the palmar surfaces of the hands and fingers, and upon the plantar surfaces of the feet and toes, they are long, of large size, closely aggregated together, and arranged in parallel curved lines, forming the elevated ridges seen on the free surface of the epidermis. Each ridge contains two rows of papillae, between which the ducts of the sudoriferous glands pass outward to open on the summit of the ridge. Each papilla consists of very small and closely interlacing bundles of finely fibrillated tissue, with a few elastic fibers; within this tissue is a capillary loop, and in some papillae, especially in the palms of the hand and the fingers, there are tactile corpuscles (Fig. 14-3).

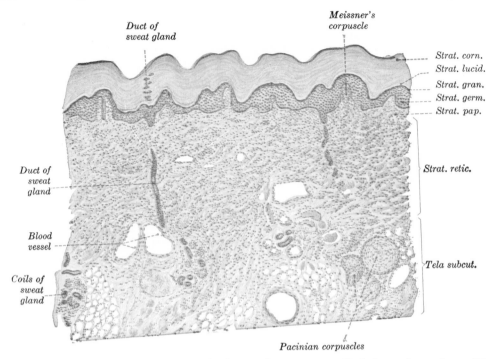

FIG. 14–2.—Section through the skin of the human foot cut perpendicularly to the surface. (From Rauber-Kopsch, *Lehrbuch u. Atlas d. Anatomie d. Anatomie d. Menschen*, 19th Edition, Vol. II, courtesy Georg Thieme Verlag, Stuttgart, 1955.)

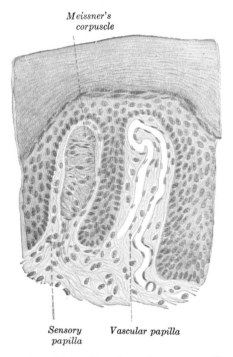

FIG. 14–3.—Vascular and sensory papillae in the skin of the human foot. (From Rauber-Kopsch, *Lehrbuch u. Atlas d. Anatomie d. Menschen*, 19th Edition, Vol. II, courtesy Georg Thieme Verlag, Stuttgart, 1955.)

The **reticular layer** (*stratum reticulare; deep layer*) consists of fibroelastic connective tissue, composed chiefly of collagenous bundles, but containing yellow elastic fibers in varying number in different parts of the body. The cells it contains are principally fibroblasts and histiocytes, but other types may be found. Near the papillary layer the collagenous bundles are small and compactly arranged; in the deeper layers they are larger and coarser and between their meshes are sweat glands, sebaceous glands, hair shafts or follicles, and small collections of fat cells. The deep surface of the reticular layer merges with the adipose tissue of the subcutaneous fascia (*tela subcutanea*).

Cleavage Lines of the Skin (*Langer's lines*).—When a penetrating wound is made with a sharp conical instrument it does not leave a round hole in the skin, as might be expected, but a slit such as would be expected from a flat blade. Maps of the directions of these slits from puncture wounds over all parts of the body have been made from dissecting room material (Figs. 14–4, 14–5). These maps indicate that there are definite lines of **tension** or **cleavage lines** within the skin which are characteristic for

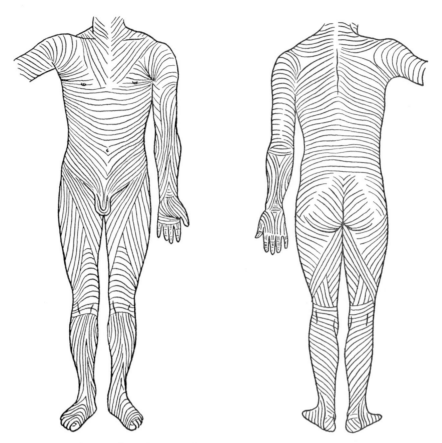

FIG. 14–4.—Cleavage lines (Langer's lines) of the skin. Trunk and extremities. (Eller.)

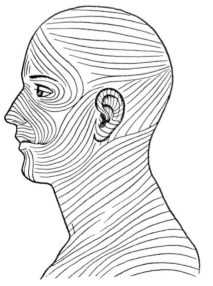

FIG. 14–5.—Cleavage lines (Langer's lines) of the skin. Head and neck. (Eller.)

each part of the body. In microscopic sections cut parallel with these lines, most of the collagenous bundles of the reticular layer are cut longitudinally, while in sections cut across the lines, the bundles are in cross section. The cleavage lines correspond closely with the crease lines on the surface of the skin in most parts of the body. The pattern of the cleavage lines, according to Cox (1941) varies with body configuration, but is constant for individuals of similar build, regardless of age. There are limited areas of the body in which the orientation of the bundles is irregular and confused. The cleavage lines are of particular interest to the surgeon because an incision parallel to the lines heals with a fine linear scar while an incision across the lines may set up irregular tensions which result in an unsightly scar.

The **arteries** supplying the skin form a network in the subcutaneous tissue, branches of which supply the sudoriferous glands, the hair follicles, and the fat. Other branches unite in a plexus immediately beneath the dermis; from this plexus, fine capillary vessels pass into the papillae, forming, in the smaller ones, a single capillary loop, but in the larger, a more or less convoluted vessel. The **lymphatic vessels** of the skin form two networks, superficial and deep, which communicate with each other and with those of the subcutaneous tissue by oblique branches.

The **nerves** of the skin terminate partly in the epidermis and partly in the corium; their different modes of ending are described on pages 893 to 897.

THE APPENDAGES OF THE SKIN

The appendages of the skin are the **nails,** the **hairs,** and the **sudoriferous** and **sebaceous glands** with their ducts.

The **Nails** (*ungues*) (Fig. 14–6) are flattened, elastic structures of a horny texture, placed upon the dorsal surfaces of the terminal phalanges of the fingers and toes.

Each nail is convex on its outer surface, concave within, and is implanted by a portion, called the **root,** into a groove in the skin; the exposed portion is called the **body,** and the distal extremity the **free edge.** The nail is firmly adherent to the corium, being accurately molded upon its surface; the part beneath the body and root of the nail is called the **nail matrix,** because from it the nail is produced. Under the greater part of the body of the nail, the matrix is thick, and raised into a series of longitudinal ridges which are very vascular, and the color is seen through the transparent tissue. Near the root of the nail, the papillae are smaller, less vascular, and have no regular arrangement, and here the tissue of the nail is not firmly adherent to the connective tissue stratum but only in contact with it; hence this portion is of a whiter color, and is called the **lunula** on account of its shape.

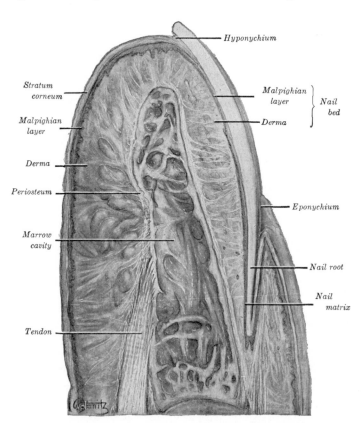

Fig. 14–6.—Median longitudinal section through finger of adult cadaver. The marrow has been removed to show bony trabeculae. × 5.

The cuticle as it passes forward on the dorsal surface of the finger or toe is attached to the surface of the nail a little in advance of its root; at the extremity of the finger it is connected with the undersurface of the nail; both epidermic structures are thus directly continuous with each other. The superficial, horny part of the nail consists of a greatly thickened stratum lucidum, the stratum corneum forming merely the thin cuticular fold (**eponychium**) which overlaps the lunula; the deeper part consists of the stratum mucosum. The cells in contact with the papillae of the matrix are columnar in form and arranged perpendicularly to the surface; those which succeed them are of a rounded or polygonal form, the more superficial ones becoming broad, thin, and flattened, and so closely packed as to make the limits of the cells very indistinct. The nails grow in length by the proliferation of the cells of the stratum germinativum at the root of the nail, and in thickness from that part of the stratum germinativum which underlies the lunula.

Hairs (*pili*) are found on nearly every part of the surface of the body, but are absent from the palms of the hands, the soles of the feet, the dorsal surfaces of the terminal phalanges, the glans penis, the inner surface of the prepuce, and the inner surfaces of the labia. They vary much in length, thickness, and color in different parts of the body and in different races of mankind. In some parts, as in the skin of the eyelids, they are so short as not to project beyond the follicles containing them; in others, as upon the scalp, they are of considerable lengths and in some other parts, such as the eyelashes, the hairs of the pubic region, and the whiskers and beard, they are remarkable for their thickness. Straight hairs are stronger than curly hairs, and present on transverse section a cylindrical or oval outline; curly hairs, on the other hand, are flattened. A hair consists of a **root**, the part implanted in the skin; and a **shaft** or **scapus**, the portion projecting from the surface.

The **root of the hair** (*radix pili*) ends in an enlargement, the **hair bulb**, which is whiter in color and softer in texture than

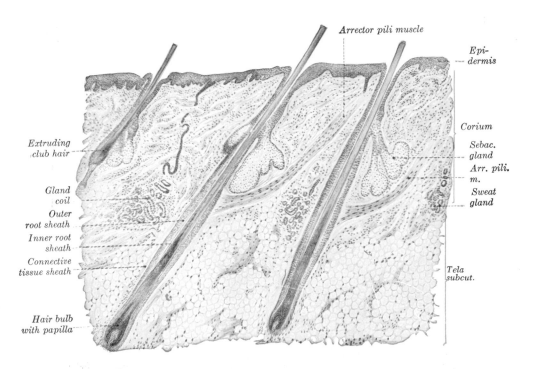

Arrector pili muscle

*Epi-
dermis*

*Extruding
club hair*

Corium

*Sebac.
gland*

*Arr. pili.
m.*

*Gland
coil*

*Outer
root sheath*

*Inner root
sheath*

*Connective
tissue sheath*

*Sweat
gland*

*Tela
subcut.*

*Hair bulb
with papilla*

Fɪɢ. 14–7.—Section of human scalp with hairs cut longitudinally. (From Rauber-Kopsch, *Lehrbuch u. Atlas d. Anatomie d. Menschen,* 19th Edition, Vol. II, courtesy Georg Thieme Verlag, Stuttgart, 1955.)

the shaft, and is lodged in a follicular involution of the epidermis called the **hair follicle** (Fig. 14–7). When the hair is of considerable length the follicle extends into the subcutaneous cellular tissue. The hair follicle commences on the surface of the skin with a funnel-shaped opening, and passes inward in an oblique or curved direction—the latter in curly hairs—to become dilated at its deep extremity, where it corresponds with the hair bulb. Opening into the follicle near its free extremity, are the ducts of one or more sebaceous glands. At the bottom of each hair follicle is a small conical, vascular eminence or **papilla,** similar in every respect to those found upon the surface of the skin; it is continuous with the dermic layer of the follicle, and is supplied with nerve fibrils. The hair follicle consists of two coats—an **outer** or **dermic,** and an **inner** or **epidermic.**

The **outer** or **dermic coat** is formed mainly of fibrous tissue; it is continuous with the dermis, is highly vascular, and is supplied by numerous minute nervous filaments. It consists of three layers (Fig. 14–8). The most internal is a hyaline basement membrane, which is well marked in the larger hair follicles, but is not very distinct in the follicles of minute hairs; it is limited to the

deeper part of the follicle. Outside this is a compact layer of fibers and spindle-shaped cells arranged circularly around the follicle; this layer extends from the bottom of the follicle as high as the entrance of the ducts of the sebaceous glands. Externally a thick layer of connective tissue is arranged in longitudinal bundles, forming a more open texture and corresponding to the reticular part of the dermis; in this are contained the blood vessels and nerves.

The **inner** or **epidermic coat** is closely adherent to the root of the hair, and consists of two strata named respectively the **outer** and **inner root sheaths;** the former of these corresponds with the stratum mucosum of the epidermis, and resembles it in the rounded form and soft character of its cells; at the bottom of the hair follicle these cells become continuous with those of the root of the hair. The inner root sheath consists of (1) a delicate cuticle next to the hair, composed of a single layer of imbricated scales with atrophied nuclei; (2) one or two layers of horny, flattened, nucleated cells, known as **Huxley's layer;** and (3) a single layer of cubical cells with clear flattened nuclei, called **Henle's layer.**

The **hair bulb** is molded over the papilla and composed of polyhedral epithelial cells,

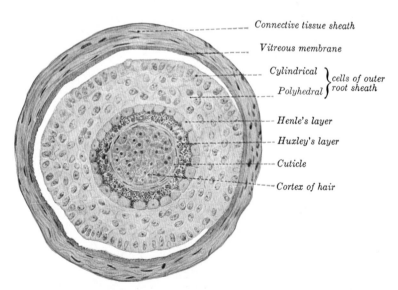

Connective tissue sheath

Vitreous membrane

Cylindrical } *cells of outer*
Polyhedral } *root sheath*

Henle's layer

Huxley's layer

Cuticle

Cortex of hair

Fig. 14–8.—Cross section through hair and root sheath. The red granules in Huxley's layer are keratohyalin. The space between hair and sheath is an artifact. (From Rauber-Kopsch, *Lehrbuch u. Atlas d. Anatomie d. Menschen,* 19th Edition, Vol. II, courtesy Georg Thieme Verlag, Stuttgart, 1955.)

which as they pass upward into the root of the hair become elongated and spindle-shaped, except some in the center which remain polyhedral. Some of these latter cells contain pigment granules which give rise to the color of the hair. It occasionally happens that these pigment granules completely fill the cells in the center of the bulb; this gives rise to the dark tract of pigment often found, of greater or lesser length, in the axis of the hair.

The **shaft of the hair** (*scapus pili*) consists, from within outward, of three parts, the medulla, the cortex, and the cuticle. The **medulla** is usually wanting in the fine hairs covering the surface of the body, and commonly in those of the head. It is more opaque and deeper colored than the cortex when viewed by transmitted light; but when viewed by reflected light it is white. It is composed of rows of polyhedral cells, containing granules of eleidin and frequently air spaces. The **cortex** constitutes the chief part of the shaft; its cells are elongated and united to form flattened fusiform fibers which contain pigment granules in dark hair, and air in white hair. The **cuticle** consists of a single layer of flat scales which overlap one another from deep to superficial.

Connected with the hair follicles are minute bundles of involuntary muscular fibers, termed the **Arrectores pilorum.** They *arise* from the superficial layer of the dermis and are inserted into the hair follicle, deep to the entrance of the duct of the sebaceous gland. They are placed on the side toward which the hair slopes, and by their action diminish the obliquity of the follicle and make the hair stand erect (Fig. 14–7). The sebaceous gland is situated in the angle which the Arrector muscle forms with the superficial portion of the hair follicle, and contraction of the muscle thus tends to squeeze the sebaceous secretion from the duct of the gland.

The **Sebaceous Glands** (*glandulae sebaceae*) are small, sacculated, glandular organs, lodged in the substance of the dermis. They are found in most parts of the skin, but are especially abundant in the scalp and face; they are also very numerous around the apertures of the anus,

nose, mouth, and external ear, but are wanting in the palms of the hands and soles of the feet. Each gland consists of a single duct, more or less capacious, which emerges from a cluster of oval or flask-shaped alveoli which vary from two to five in number, but in some instances there may be as many as twenty. Each alveolus is composed of a transparent basement membrane, enclosing a number of epithelial cells. The outer or marginal cells are small and polyhedral, and are continuous with the cells lining the duct. The remainder of the alveolus is filled with larger cells, containing lipid, except in the center, where the cells have become disintegrated, leaving a cavity filled with their débris and a mass of fatty matter, which constitutes the **sebum cutaneum.** The ducts open most frequently into the hair follicles, but occasionally upon the general surface, as in the labia minora and the free margin of the lips. On the nose and face the glands are of large size, distinctly lobulated, and often become much enlarged from accumulation of pent-up secretion. The tarsal glands of the eyelids are elongated sebaceous glands with numerous lateral diverticula.

The **Sudoriferous** or **Sweat Glands** (*glandulae sudoriferae*) (Fig. 14–9) are found in almost every part of the skin. Each consists of a single tube, the deep part of which is irregularly coiled into an oval or spherical ball, named the **body** of the gland, while the superficial part, or **duct,** traverses the dermis and cuticle and opens on the surface of the skin by a funnel-shaped aperture. In the superficial layers of the dermis the duct is straight, but in the deeper layers it is convoluted or even twisted; where the epidermis is thick, as in the palms of the hands and soles of the feet, the part of the duct which passes through it is spirally coiled. The size of the glands varies. They are especially large in those regions where the amount of perspiration is great, as in the axillae, where they form a thin, mammillated layer of a reddish color, which corresponds exactly to the situation of the hair in this region; they are large also in the groin. Their number varies. They are very plentiful on the palms of the hands, and on the

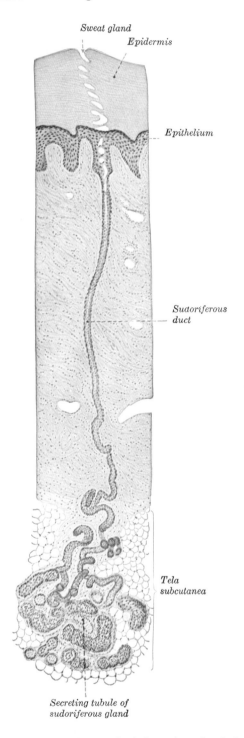

Sweat gland

Epidermis

Epithelium

Sudoriferous duct

Tela subcutanea

Secreting tubule of sudoriferous gland

FIG. 14–9.—Sweat gland from the sole of the human foot. (From Rauber-Kopsch, *Lehrbuch u. Atlas d. Anatomie d. Menschen,* 19th Edition, Vol. II, courtesy Georg Thieme Verlag, Stuttgart, 1955.)

soles of the feet, where the orifices of the ducts are exceedingly regular, and open on the curved ridges; they are least numerous in the neck and back. On the palm there are about 370 per square centimeter; on the back of the hand about 200; forehead 175; breast, abdomen and forearm 155, and on the leg and back from 60 to 80 per square centimeter. Krause estimates the total number at about 2,000,000. The average number of sweat glands per square centimeter of skin area in various races as shown by the fingers is as follows (Clark and Llamon '17):

American (white)	558.2
American (Negro)	597.2
Filipino	653.6
Moro	684.4
Negrito (adult)	709.2
Hindu	738.2
Negrito (youth)	950.0

They are absent in the deeper portion of the external auditory meatus, the prepuce and the glans penis. The tube, both in the body of the gland and in the duct, consists of two layers—an outer of fine **areolar tissue,** and an inner of **epithelium.** The outer layer is thin and is continuous with the superficial stratum of the dermis. In the body of the gland the epithelium consists of a single layer of cubical cells, between the deep ends of which the basement membrane is a layer of longitudinally or obliquely arranged non-striped muscular fibers. The ducts are destitute of muscular fibers and are composed of a basement membrane lined by two or three layers of polyhedral cells; the lumen of the duct is coated by a thin cuticle. When the epidermis is carefully removed from the surface of the dermis, the ducts may be pulled out from the dermis in the form of short, thread-like processes. The ceruminous glands of the external acoustic meatus, the ciliary glands at the margins of the eyelids, the circumanal glands and probably the mammary glands are modified sudoriferous glands. The average quantity of sweat secreted in twenty-four hours varies from 700 to 900 grams.

REFERENCES

SKIN

BREATHNACH, A. S. 1964. Observations on cytoplasmic organelles in Langerhans cells of human epidermis. J. Anat. (Lond.), *98*, 265-270.

KATZBERG, A. A. 1958. The area of the dermo-epidermal junction in human skin. Anat. Rec., *131*, 717-726.

MATOLTSY, A. G. and S. J. SINESI. 1957. A study of the mechanism of keratinization of human epidermal cells. Anat. Rec., *128*, 55-68.

ODLAND, G. F. 1950. The morphology of the attachment between the dermis and the epidermis. Anat. Rec., *108*, 399-413.

SARKANY, I. and G. A. CARON. 1965. Microtopography of the human skin. J. Anat. (Lond.), *99*, 350-364.

SELBY, C. C. 1955. An electron microscope study of the epidermis of mammalian skin in thin sections. I. Derma-epidermal junction and basal cell layer. J. biophys. biochem. Cytol., *1*, 429-444.

TAUSSIG, J. and G. D. WILLIAMS. 1940. Skin color and skin cancer. Arch. Path., *30*, 721-730.

HAIR, NAILS, AND GLANDS

BAKER, B. L. 1951. The relationship of the adrenal, thyroid, and pituitary glands to the growth of hair. Ann. N. Y. Acad. Sci., *53*, 690-707.

BIRBECK, M. S. C., E. H. MERCER, and N. A. BARNICOT. 1956. The structure and formation of pigment granules in human hair. Exp. Cell Res., *10*, 505-514.

BIRBECK, M. S. C. and E. H. MERCER. 1957. The electron microscopy of the human hair follicle. I. Introduction and the hair cortex. II. The hair cuticle. III. The inner root sheath and trichohyaline. J. biophys. biochem. Cytol., *3*, 203-230.

BUTCHER, E. O. 1946. Hair growth and sebaceous glands in skin transplanted under the skin and into the peritoneal cavity in the rat. Anat. Rec., *96*, 101-109.

BUTCHER, E. O. 1951. Development of the pilary system and the replacement of hair in mammals. Ann. N. Y. Acad. Sci., *53*, 508-516.

CLARK, E. and R. H. LLAMON. 1917. Observations on the sweat glands of tropical and northern races. Anat. Rec., *12*, 139-147.

DIXON, A. D. 1961. The innervation of hair follicles in the mammalian lip. Anat. Rec., *140*, 147-158.

DUGGINS, O. H. 1954. Age changes in head hair from birth to maturity. IV. Refractive indices and birefringence of the cuticle of hair of children. Am. J. phys. Anthrop., *12*, 89-114.

HAMILTON, J. B. 1942. Male hormone stimulation is prerequisite and an incitant in common baldness. Amer. J. Anat., *71*, 451-480.

HAMILTON, J. B. 1951. Patterned loss of hair in man: types and incidence. Ann. N. Y. Acad. Sci., *53*, 708-728.

JOHNSON, P. L. and G. BEVELANDER. 1946. Glycogen and phosphatase in the developing hair. Anat. Rec., *95*, 193-199.

SULZBERGER, M. B., F. HERRMANN, R. KELLER, and B. V. PISHA. 1950. Studies of sweating: III. Experimental factors in influencing the function of the sweat ducts. J. invest. Derm., *14*, 91-112.

SUNDERLAND, S. and L. J. RAY. 1952. The effect of denervation on nail growth. J. Neurol. Neurosurg. Psychiat., *15*, 50-53.

THOMAS, P. K. and D. G. FERRIMAN. 1957. Variation in facial and pubic hair growth in white women. Amer. J. phys. Anthrop., *15*, 171-180.

TROTTER, M. 1938. A review of the classifications of hairs. Amer. J. phys. Anthrop., *24*, 105-126.

TROTTER, M. 1939. Classifications of hair color. Amer. J. phys. Anthrop., *25*, 237-260.

TROTTER, M. and O. H. DUGGINS. 1948. Age changes in head hair from birth to maturity. I. Index and size of hair of children. Amer. J. phys. Anthrop., 489-506.

TROTTER, M. and O. H. DUGGINS. 1950. Age changes in head hair from birth to maturity. III. Cuticular scale counts of hair of children. Amer. J. phys. Anthrop., n.s., *8*, 467-484.

15 | *The Respiratory System*

espiration, or exchange of gaseous substance between the air and the blood stream, is brought about in the Respiratory System (*apparatus respiratorius; respiratory apparatus*): the nose, nasal passages, nasopharynx, larynx, trachea, bronchi and lungs. The pleura, pleural cavities and the topography of other structures in the thorax are also described in this chapter. The muscular actions associated with respiration are discussed with the description of the diaphragm in the chapter on Muscles.

Development of the Respiratory System. —The development of the nose is described in the chapter on Embryology.

The primordium of the principal respiratory organs appears as a median longitudinal groove in the ventral wall of the pharynx in an embryo of the third week (3 to 4 mm). The groove deepens and its lips fuse to form a tube, the **laryngotracheal tube** (Fig. 15–1), the cranial end of which opens into the pharynx by a slit-like aperture formed by the persistent ventral part of the groove. The tube is lined by entoderm from which the epithelial lining of the respiratory tract is developed. The cranial part of the tube becomes the larynx, its next succeeding part the trachea, and from its caudal end two lateral outgrowths arise, the right and left **lung buds**, from which the bronchi and lungs are developed.

The first rudiment of the larynx consists of two **arytenoid swellings**, which appear on either side of the cranial end of the laryngotracheal groove, and are continuous ventrally with a transverse ridge (**furcula of His**) which lies between the ventral ends of the third branchial arches and from which the epiglottis is subsequently developed (Figs. 16–4, 16–5). After the separation of

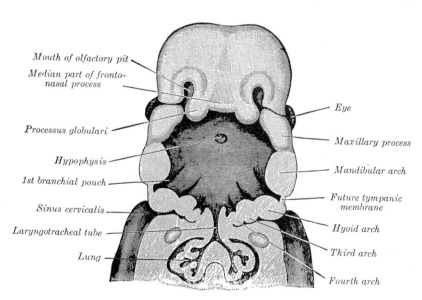

Mouth of olfactory pit

Median part of fronto-nasal process

Processus globulari

Hypophysis

1st branchial pouch

Sinus cervicalis

Laryngotracheal tube

Lung

Eye

Maxillary process

Mandibular arch

Future tympanic membrane

Hyoid arch

Third arch

Fourth arch

FIG. 15–1.—The head and neck of a thirty-two-day-old human embryo seen from the ventral surface. The floor of the mouth and pharynx has been removed. (His.)

the trachea from the esophagus the arytenoid swellings come into contact with each
other and with the back of the epiglottis,
and the entrance to the larynx assumes the
form of a T-shaped cleft. The margins of
the cleft adhere to each other and the laryngeal entrance is for a time occluded.

The mesodermal wall of the tube becomes
condensed to form the cartilages of the
larynx and trachea. The arytenoid swellings
are differentiated into the arytenoid and
corniculate cartilages, and the folds joining
them to the epiglottis form the aryepiglottic
folds in which the cuneiform cartilages are
developed as derivatives of the epiglottis.
The thyroid cartilage appears as two lateral
plates, each chondrified from two centers
and united in the ventral midline by membrane in which an additional center of
chondrification develops. The cricoid cartilage arises from two cartilaginous centers
which soon unite ventrally and gradually
grow around to fuse on the dorsal aspect of
the tube.

The right and left lung buds grow out
dorsal to the common cardinal veins, and are
at first symmetrical, but their ends soon
become lobulated, three lobules appearing
on the right, and two on the left; these
subdivisions are the early indications of
the corresponding lobes of the lungs (Figs.
15–2, 15–3). The buds undergo further subdivision and ramification, and ultimately end
in minute expanded extremities—the infundibula of the lung. After the sixth month the
air sacs begin to make their appearance on
the infundibula in the form of minute
pouches.

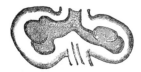

Fig. 15–2.—Lung buds from a human embryo
of about four weeks, showing commencing lobu-
lations. (His.)

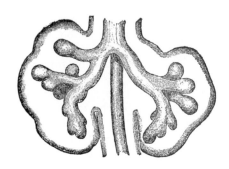

Fig. 15–3.—Lungs of a human embryo more
advanced in development. (His.)

During the course of their development
the lungs migrate in a caudal direction, so
that by the time of birth the bifurcation of
the trachea is opposite the fourth thoracic
vertebra. As the lungs grow they project
into that part of the celom which will ultimately form the pleural cavities, and the
superficial layer of the mesoderm enveloping the lung rudiment expands on the
growing lung and is converted into the
pulmonary pleura.

The pulmonary arteries are derived from
the sixth aortic arches (Figs. 8–3, 8–8).

THE UPPER RESPIRATORY TRACT

THE NOSE

The **external nose** (*nasus externus*) is
shaped like a pyramid, its free angle termed
the **apex**. The two elliptical orifices, the
nares, are separated from each other by a
median septum, the **columna**. The interior
surface of the **anterior nares or nostrils** is
provided with a number of stiff hairs, or
vibrissae, which arrest the passage of
foreign substances carried with the current
of air intended for respiration. The lateral
surfaces of the nose form, by their union in

the middle line, the **dorsum nasi**, the direction of which varies considerably in different
individuals; the upper part of the dorsum
is supported by the nasal bones, and is
named the **bridge**. The lateral surface ends
below in rounded eminences, the **alae nasi**.

Structure.—The framework of the external
nose is composed of bones and cartilages; it
is covered by the integument, and is lined by
mucous membrane.

The **bony framework** occupies the upper
part of the organ; it consists of the nasal bones,
and the frontal processes of the maxillae.

The **cartilaginous framework** (*cartilagines nasi*) consists of five large pieces, viz., the **cartilage of the septum**, the **two lateral** and the **two greater alar cartilages**, and several smaller pieces, the **lesser alar cartilages** (Figs. 15–4, 15–5, 15–6, 15–7). The various cartilages are connected to each other and to the bones by a tough fibrous membrane.

The **cartilage of the septum** (*cartilago septi nasi*) is thicker at its margins than at its center, and separates the nasal cavities. Its anterior margin is connected with the nasal bones, and is continuous with the anterior margins of the lateral cartilages; inferiorly, it is connected to the medial crura of the greater alar cartilages by fibrous tissue. Its posterior margin is connected with the perpendicular plate of the ethmoid; its inferior margin with the vomer and the palatine processes of the maxillae.

It may be prolonged posteriorly (especially in children) as a narrow process, the **sphenoidal process**, for some distance between the vomer and perpendicular plate of the ethmoid. The septal cartilage does not reach as far as the lowest part of the nasal septum. This is formed by the medial crura of the greater alar cartilages.

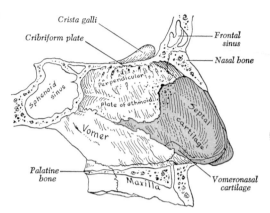

FIG. 15–6.—Bones and cartilages of septum of nose. Right side.

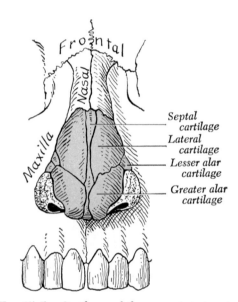

FIG. 15–7.—Cartilages of the nose. Anterior view.

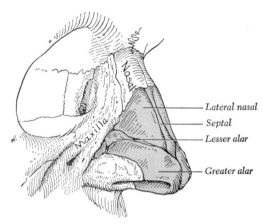

FIG. 15–4.—Cartilages of the nose. Side view.

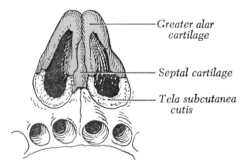

FIG. 15–5.—Cartilages of the nose. Inferior view.

The **lateral cartilage** (*cartilago nasi lateralis; upper lateral cartilage*) is situated below the inferior margin of the nasal bone, and is flattened, and triangular in shape. Its anterior margin is thicker than the posterior, and is continuous above with the cartilage of the septum, but separated from it below by a narrow fissure; its superior margin is attached to the nasal bone and the frontal process of the maxilla; its inferior margin is connected by fibrous tissue with the greater alar cartilage.

The **greater alar cartilage** (*cartilago alaris major; lower lateral cartilage*) is a thin, flexible plate situated immediately below the preceding, and bent upon itself in such a manner as to form the medial and lateral walls of the naris of its own side. The portion which forms the **medial wall** (*crus mediale*) is loosely connected with the corresponding portion of the opposite cartilage, the two forming, together with the thickened integument and subjacent tissue, the **septum mobile nasi.** The part which forms the **lateral wall** (*crus laterale*) is curved to correspond with the ala of the nose; it is oval, flattened, and narrow posteriorly where it is connected with the frontal process of the maxilla by a tough fibrous membrane, in which are found three or four small cartilaginous plates, the **lesser alar cartilages** (*cartilagines alares minores; sesamoid cartilages*). Superiorly, it is connected by fibrous tissue to the lateral cartilage and anterior part of the cartilage of the septum; inferiorly, it falls short of the margin of the naris, the ala being completed by fatty and fibrous tissue covered by skin. Anteriorly, the greater alar cartilages are separated by a notch which corresponds with the apex of the nose.

The **muscles** acting on the external nose are described in Chapter 6.

The **integument** of the dorsum and sides of the nose is thin, and loosely connected with the subjacent parts; but over the tip and alae it is thicker and more firmly adherent, and is furnished with a large number of sebaceous

follicles, the orifices of which are usually quite distinct.

The **arteries** of the external nose are the alar and septal branches of the facial, which supply the alae and septum; the dorsum and sides are supplied from the dorsal nasal branch of the ophthalmic and the infraorbital branch of the maxillary. The **veins** end in the anterior facial and ophthalmic veins.

The **nerves** for the muscles of the nose are derived from the facial; the skin receives branches from the infratrochlear and nasociliary branches of the ophthalmic, and from the infraorbital of the maxillary.

The Nasal Cavity

The **nasal cavity** (*cavum nasi*) is divided by the median septum into two symmetrical and approximately equal chambers, the **nasal fossae.** They have their external openings through the nostrils or nares, and they open into the nasopharynx through the choanae. The nares are oval apertures measuring about 1.5 cm anteroposteriorly and 1 cm transversely. The choanae are

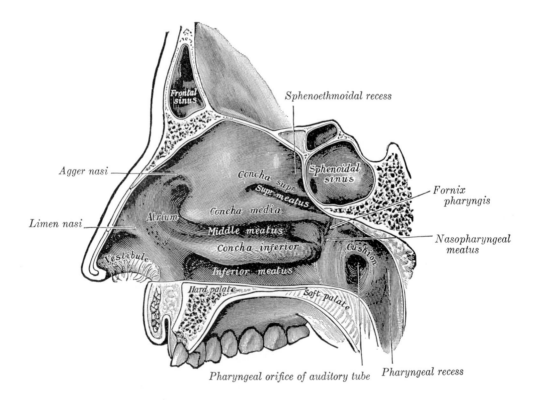

Fig. 15–8.—Lateral wall of nasal cavity.

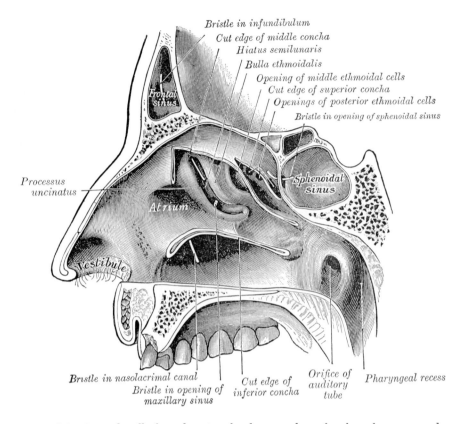

Bristle in infundibulum
Cut edge of middle concha
Hiatus semilunaris
Bulla ethmoidalis
Opening of middle ethmoidal cells
Cut edge of superior concha
Openings of posterior ethmoidal cells
Bristle in opening of sphenoidal sinus

Frontal sinus

Sphenoidal sinus

Processus uncinatus

Atrium

Vestibule

Bristle in nasolacrimal canal
Bristle in opening of maxillary sinus
Cut edge of inferior concha
Orifice of auditory tube
Pharyngeal recess

FIG. 15–9.—Lateral wall of nasal cavity; the three nasal conchae have been removed.

two oval openings measuring approximately 2.5 cm vertically and 1.5 cm transversely.

The bony boundaries of the nasal cavity are described in the chapter on Osteology, page 146.

Inside the aperture of the nostril is a slight dilatation, the **vestibule,** bounded laterally by the ala and lateral crus of the greater alar cartilage, and medially by the medial crus of the same cartilage. It is lined by skin containing hairs and sebaceous glands, and extends as a small recess toward the apex of the nose. Each nasal fossa is divided into two parts: an **olfactory region,** consisting of the superior nasal concha and the opposed part of the septum, and a **respiratory region,** which comprises the rest of the cavity.

Lateral Wall (Figs, 15–8, 15–9). On the lateral wall are the **superior, middle,** and **inferior nasal conchae,** overhanging the corresponding nasal passages or **meatuses.** Above the superior concha is a narrow

recess, the **sphenoethmoidal recess,** into which the sphenoidal sinus opens. The **superior meatus** is a short oblique passage extending about half way along the superior border of the middle concha; the posterior ethmoidal cells open into the anterior of this meatus. The **middle meatus** is continued anteriorly into a shallow depression, situated above the vestibule and named the **atrium** of the middle meatus. The lateral wall of this meatus is fully displayed only by raising or removing the middle concha. On it is a rounded elevation, the **bulla ethmoidalis,** and anterior and inferior to this is a curved cleft, the **hiatus semilunaris.**

The **bulla ethmoidalis** (Fig. 15–9) is caused by the bulging of the middle ethmoidal cells which open on or immediately above it, and the size of the bulla varies with that of its contained cells.

The **hiatus semilunaris** is the narrow curved opening into a deep pocket, the

infundibulum. It is bounded inferiorly by the sharp concave margin of the **uncinate process** of the ethmoid bone, and superiorly by the bulla ethmoidalis. The anterior ethmoidal cells open into the anterior part of the infundibulum. In slightly over 50 per cent of subjects the infundibulum is directly continuous with the **frontonasal duct** or passage leading from the frontal air sinus. When the anterior end of the uncinate process fuses with the anterior part of the bulla, this continuity is interrupted and the frontonasal duct then opens directly into the anterior end of the middle meatus.

In the bottom of the infundibulum, below the bulla ethmoidalis, and partly hidden by the inferior end of the uncinate process, is the **ostium maxillare,** or opening from the maxillary sinus. This opening is placed near the roof of the sinus. An **accessory** opening from the sinus is frequently present below the posterior end of the middle nasal concha. The **inferior meatus** is inferior and lateral to the inferior nasal concha; the nasolacrimal duct opens into this meatus under cover of the anterior part of the concha.

Medial Wall (Fig. 15–6).—The medial wall or septum is frequently more or less deflected from the median plane, thus lessening the size of one nasal fossa and increasing that of the other; ridges or spurs of bone growing into one or the other fossa from the septum are also sometimes present. Immediately over the incisive canal at the inferior edge of the cartilage of the septum is a depression, the **nasopalatine recess.** In the septum close to this recess a minute orifice may be discerned; it leads posteriorly into a blind pouch, the rudimentary **vomeronasal organ of Jacobson,** which is supported by a strip of cartilage, the **vomeronasal cartilage.** This organ is well developed in many of the lower animals, where it apparently plays a part in the sense of smell, since it is supplied by twigs of the olfactory nerve and lined by epithelium similar to that in the olfactory region of the nose.

The **roof** of the nasal cavity is narrow from side to side, except at its posterior part, and may be divided into sphenoidal, ethmoidal, and frontonasal parts, after the bones which form it.

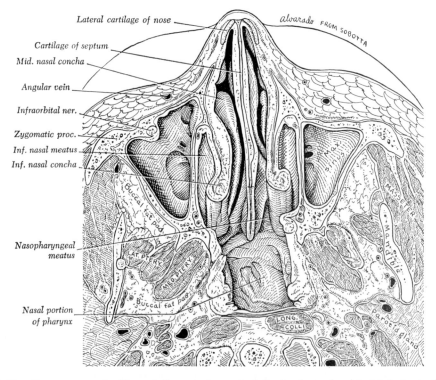

Fig. 15–10.—Transverse section through the anterior part of the head at a level just inferior to the apex of the dens (odontoid process). Inferior view.

The **floor** is concave from side to side; its anterior three-fourths are formed by the palatine process of the maxilla, its posterior fourth by the horizontal process of the palatine bone.

The **Mucous Membrane** (*membrana mucosa nasi*).—The nasal mucous membrane lines the nasal cavity, and is intimately adherent to the periosteum or perichondrium. It is continuous with the skin through the nares, and with the mucous membrane of the nasal part of the pharynx through the choanae. From the nasal fossa its continuity with the conjunctiva may be traced through the nasolacrimal and lacrimal ducts; and with the frontal, ethmoidal, sphenoidal, and maxillary sinuses, through the several openings in the meatuses. Continuity with the middle ear is established with the nasopharynx through the auditory tube. The mucous membrane is thickest, and most vascular, over the nasal conchae. It is also thick over the septum, but it is very thin in the meatuses, on the floor of the nasal cavity, and in the various sinuses.

Owing to the thickness of the greater part of this membrane, the nasal cavities are much narrower, and the middle and inferior nasal conchae appear larger and more prominent than in the skeleton; also the various apertures communicating with the meatuses are considerably narrowed.

Structure of the Mucous Membrane.—The mucous membrane covering the *respiratory portion* of the nasal cavity has a pseudostratified, ciliated, columnar epithelium, liberally interspersed with goblet cells. Beneath the basement membrane of the epithelium, the areolar connective tissue is infiltrated with lymphocytes which may be so numerous that they form a diffuse lymphoid tissue. The lamina propria is composed of a layer of glands toward the surface and a layer of blood vessels next to the periosteum. The glands may be large or small and contain either mucous or serous alveoli. They have individual openings on the surface, and their secretion forms a protective layer over the membrane which serves to warm and moisten the inspired air. In the deeper layers of the lamina propria, especially over the conchae, the dilated veins and blood spaces form a rich plexus which bears a superficial resemblance to erectile tissue; they easily become engorged as the result of irritation or inflammation, causing the membrane to swell,

encroaching upon the lumen of the meatuses and even occluding the ostia of the sinuses.

The mucous membrane of the paranasal sinuses is columnar ciliated, similar to that of the nasal fossae, but is much thinner and for the most part is lacking in glands in the lamina propria.

The **olfactory portion** of the mucous membrane of the nasal cavity is described at the beginning of Chapter 13.

Vessels and Nerves.—The **arteries** of the nasal cavities are the anterior and posterior ethmoidal branches of the ophthalmic, which supply the ethmoidal cells, frontal sinuses, and roof of the nose; the sphenopalatine branch of the maxillary, which supplies the mucous membrane covering the conchae, the meatuses and septum; the septal branch of the superior labial of the facial; the infraorbital and alveolar branches of the maxillary, which supply the lining membrane of the maxillary sinus; and the pharyngeal branch of the same artery is distributed to the sphenoidal sinus. The ramifications of these vessels form a close plexiform network, beneath and in the substance of the mucous membrane.

Some **veins** open into the sphenopalatine vein; others join the anterior facial vein; some accompany the ethmoidal arteries, and end in the ophthalmic veins; and, lastly, a few communicate with the veins on the orbital surface of the frontal lobe of the brain, through the foramina in the cribriform plate of the ethmoid bone. When the foramen cecum is patent it transmits a vein to the superior sagittal sinus.

The **lymphatics** are described in Chapter 10.

The **nerves** of ordinary sensation are: the nasociliary branch of the ophthalmic, filaments from the anterior alveolar branch of the maxillary, the nerve of the pterygoid canal, the nasopalatine, the anterior palatine, and nasal branches of the pterygopalatine.

The nasociliary branch of the ophthalmic distributes filaments to the anterior part of the septum and lateral wall of the nasal cavity. Filaments from the anterior alveolar nerve supply the inferior meatus and inferior concha. The nerve of the pterygoid canal supplies the upper and back part of the septum and superior concha, and the upper nasal branches from the pterygopalatine have a similar distribution. The nasopalatine nerve supplies the middle of the septum. The anterior palatine nerve supplies the lower nasal branches to the middle and inferior conchae.

The **olfactory,** the special nerves of the sense of smell, are a number of fine filaments distributed to the mucous membrane of the olfactory region. Their fibers arise from the bipolar

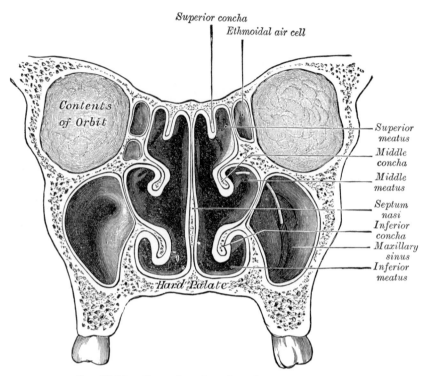

Superior concha
Ethmoidal air cell

Contents of Orbit

Superior meatus
Middle concha
Middle meatus
Septum nasi
Inferior concha
Maxillary sinus
Inferior meatus

Hard Palate

Fig. 15–11.—Coronal section of nasal cavities.

olfactory cells and are destitute of medullary sheaths. They unite into fasciculi which form a plexus beneath the mucous membrane and then ascend in grooves or canals in the ethmoid bone. They pass into the skull through the foramina in the cribriform plate of the ethmoid and enter the olfactory bulb, in which they ramify and form synapses with the dendrites of the mitral cells (see Organs of the Senses).

The Accessory Sinuses of the Nose

The **accessory sinuses** or **air cells of the nose** (*sinus paranasales*) (Figs. 15–9, 15–10, 15–11, 15–12) are the **frontal, ethmoidal, sphenoidal,** and **maxillary;** they vary in size and form in different individuals, and are lined by ciliated mucous membrane directly continuous with that of the nasal cavity.

The **Frontal Sinuses** (*sinus frontales*), situated behind the superciliary arches, are rarely symmetrical, and the septum between them frequently deviates to one side or the other of the middle line. Their average measurements are as follows: height, 3 cm; breadth, 2.5 cm; depth, 2.5 cm. A large frontal sinus may extend out over most of

the orbit. Each opens into the anterior part of the corresponding middle meatus of the nose through the frontonasal duct which enters the anterior part of the middle meatus. Absent at birth, they are generally fairly well developed between the seventh and eighth years, but reach their full size only after puberty.

The **Ethmoidal Air Cells** (*cellulae ethmoidales*) consist of numerous small thin-walled cavities occupying the ethmoidal labyrinth and completed by the frontal, maxilla, lacrimal, sphenoidal, and palatine bones. They lie between the upper parts of the nasal cavities and the orbits, and are separated from these cavities by thin bony laminae. On either side they are arranged in three groups, **anterior, middle,** and **posterior.** The anterior and middle groups open into the middle meatus of the nose, the former by way of the infundibulum, the latter on or above the bulla ethmoidalis. The posterior cells open into the superior meatus under cover of the superior nasal concha; sometimes one or more open into the sphenoidal sinus. The ethmoidal cells begin to develop after fetal life.

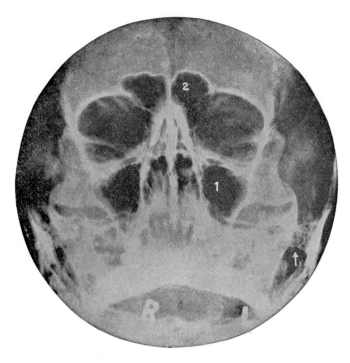

Fig. 15–12.—Adult skull. Frontal view. *1*, Maxillary sinus; *2*, frontal sinus. The arrow is directed towards the mastoid air cells

The **Sphenoidal Sinuses** (*sinus sphenoidales*) (Fig. 15–9), contained within the body of the sphenoid, vary in size and shape, and owing to the lateral displacement of the intervening septum they are rarely symmetrical. The following are their average measurements: vertical height, 2.2 cm; transverse breadth, 2 cm; depth 2.2 cm. When exceptionally large they may extend into the roots of the pterygoid processes or great wings, and may invade the basilar part of the occipital bone. Each sinus communicates with the sphenoethmoidal recess by means of an ostium in the upper part of its anterior wall. They are present as minute cavities at birth, but their main development takes place after puberty.

The **Maxillary Sinus** (*sinus maxillaris; antrum of Highmore*), the largest of the accessory sinuses of the nose, is a pyramidal cavity in the body of the maxilla. Its base is formed by the lateral wall of the nasal cavity, and its apex extends into the zygomatic process. Its roof or orbital wall is frequently ridged by the bony wall of the infraorbital canal; its floor is formed by the alveolar process and is usually 1 to 10 mm below the level of the floor of the nose. Projecting into the floor are several conical elevations corresponding with the roots of the first and second molar teeth, and in some cases the floor is perforated by one or more of these roots. The size of the sinus varies in different skulls, and even on the two sides of the same skull. The adult capacity varies from 9.5 cc to 20 cc, average about 14.75 cc. The following measurements are those of an average-sized sinus: vertical height opposite the first molar tooth, 3.75 cm; transverse breadth, 2.5 cm; anteroposterior depth, 3 cm. In the superior part of its medial wall is an ostium through which it communicates with the lower part of the infundibulum; a second or accessory orifice is frequently present in the middle meatus posterior to the first. The maxillary sinus appears as a shallow groove on the medial surface of the bone about the fourth month of fetal life, but does not reach its full size until after the second dentition.

THE LARYNX

The **larynx** or **organ of voice** is the part of the air passage which connects the pharynx with the trachea. It produces a considerable projection in the middle line of the neck called the Adam's apple. It forms the inferior part of the anterior wall of the pharynx, and is covered by the mucous lining of that cavity. Its vertical extent corresponds to the fourth, fifth, and sixth cervical vertebrae, but it is placed somewhat higher in the female and also during childhood; on either side of it lie the great vessels of the neck. The average measurements of the adult larynx are as follows:

	In males mm	In females mm
Length	44	36
Transverse diameter . . .	43	41
Anteroposterior diameter . .	36	26
Circumference	136	112

Until puberty the larynx of the male differs little in size from that of the female. In the female its increase after puberty is only slight; in the male it undergoes considerable increase; all the cartilages are enlarged and the thyroid cartilage becomes prominent in the middle line of the neck, while the length of the rima glottidis is nearly doubled.

The larynx is broad above, where it presents the form of a triangular box flattened dorsally and at the sides, and bounded ventrally by a prominent vertical ridge. Caudally, it is narrow and cylindrical. It is composed of cartilages, which are connected together by ligaments and moved by numerous muscles. It is lined by mucous membrane continuous with that of the pharynx and the trachea.

The **Cartilages of the Larynx** (*cartilagines laryngis*) (Fig. 15–13) are nine in number, three single and three paired, as follows:

> Thyroid
> Cricoid
> Two Arytenoid
> Two Corniculate
> Two Cuneiform
> Epiglottis

The **Thyroid Cartilage** (*cartilago thyroidea*) (Fig. 15–13) is the largest cartilage of

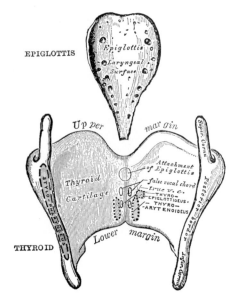

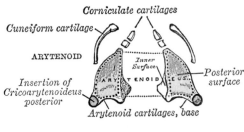

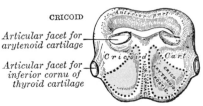

Fig. 15–13.—The cartilages of the larynx. Posterior view.

the larynx. It consists of two laminae, the anterior borders of which are fused with each other at an acute angle in the middle line of the neck, and form a subcutaneous projection named the **laryngeal prominence** (Adam's apple). This prominence is most distinct at its cranial part, and is larger in the male than in the female. Immediately above it the laminae are separated by a V-shaped notch, the **superior thyroid notch**. The laminae are irregularly quadrilateral in shape, and their posterior angles are prolonged into processes termed the **superior** and **inferior cornua.**

On the *outer surface* of each lamina an **oblique line** runs caudalward and ventralward from the superior thyroid tubercle situated near the root of the superior cornu, to the inferior thyroid tubercle on the caudal border. This line gives attachment to the Sternothyroideus, Thyrohyoideus, and Constrictor pharyngis inferior. The *inner surface* is smooth; cranially and dorsally it is slightly concave and covered by mucous membrane. Ventrally, in the angle formed by the junction of the laminae, are attached the stem of the epiglottis, the ventricular and vocal ligaments, the Thyroarytenoidei, Thyroepiglottici and Vocales muscles, and the thyroepiglottic ligament.

The *cranial border* is concave dorsally and convex ventrally; it gives attachment to the corresponding half of the thyrohyoid membrane. The *caudal border* is concave dorsally, and nearly straight ventrally, the two parts being separated by the inferior thyroid tubercle. A small part of it in and near the midline is connected to the cricoid cartilage by the middle cricothyroid ligament. The *dorsal border,* thick and rounded, receives the insertions of the Stylopharyngeus and Pharyngopalatinus. It ends cranially, in the superior cornu, and caudally, in the inferior. The **superior cornu** is long and narrow, directed cranialward, and ends in a conical extremity, which gives attachment to the lateral thyrohyoid ligament. The **inferior cornu** is short and thick; it is directed caudalward, with a slight inclination, and presents, on the medial side of its tip, a small oval articular facet for articulation with the side of the cricoid cartilage.

During infancy the laminae of the thyroid cartilage are joined to each other by a narrow, lozenge-shaped strip, named the **intrathyroid cartilage.** This strip extends from the cranial to the caudal border of the cartilage in the midline, and is distinguished from the laminae by being more transparent and more flexible.

The **Cricoid Cartilage** (*cartilago cricoidea*) (Fig. 15–13) is smaller, but thicker and stronger than the thyroid, and forms the caudal and dorsal parts of the wall of the larynx. It consists of two parts: a **poste-** rior **quadrate lamina,** and a narrow **anterior arch,** one-fourth or one-fifth of the depth of the lamina.

The **lamina** (*lamina cartilaginis cricoideae; posterior portion*) is deep and broad, and measures about 2 or 3 cm craniocaudally; on its dorsal surface, in the middle line, is a vertical ridge to the caudal part of which are attached the longitudinal fibers of the esophagus; and on either side of this a broad depression for the Cricoarytenoideus posterior.

The **arch** (*arcus cartilaginis cricoideae; anterior portion*) is narrow, convex, and measures vertically from 5 to 7 mm; it affords attachment externally, ventrally, and at the sides to the Cricothyroidei, and dorsally, to part of the Constrictor pharyngis inferior.

On either side, at the junction of the lamina with the arch, is a small round articular surface, for articulation with the inferior cornu of the thyroid cartilage.

The *caudal border* of the cricoid cartilage is horizontal, and connected to the highest ring of the trachea by the cricotracheal ligament. The *cranial border* runs obliquely cranialward and dorsalward, owing to the great depth of the lamina. It gives attachment, ventrally, to the middle cricothyroid ligament; at the side, to the conus elasticus and the Cricoarytenoidei laterales; dorsally, it presents, in the middle, a shallow notch, and on either side of this is a smooth, oval, convex surface for articulation with the base of an arytenoid cartilage. The inner surface of the cricoid cartilage is smooth, and lined by mucous membrane.

The **Arytenoid Cartilages** (*cartilagines arytenoideae*) (Fig. 15–13) are two in number, situated at the cranial border of the lamina of the cricoid cartilage, at the dorsum of the larynx. Each is pyramidal in form, and has three surfaces, a base, and an apex.

The *dorsal surface* is triangular, smooth, concave, and gives attachment to the Arytenoidei obliquus and transversus. The *ventrolateral surface* is somewhat convex and rough. On it, near the apex of the cartilage, is a rounded elevation (**colliculus**) from which a ridge (**crista arcuata**) curves at first dorsalward and then caudalward and ventralward to the vocal process. The caudal part of this crest intervenes between two depressions or **foveae,** an upper, triangular, and a lower oblong in shape; the latter gives attachment to the Vocalis muscle. The *medial surface* is narrow, smooth, and flattened, covered by mucous membrane, and forms the lateral boundary of the intercartilaginous part of the rima glottidis.

The **base** of each cartilage is broad, and on it is a concave smooth surface, for articulation with the cricoid cartilage. Its lateral angle is short, rounded, and prominent; it projects dorsalward, and is termed the **muscular process;** it receives the insertion of the Cricoarytenoideus posterior dorsally, and the Cricoarytenoideus lateralis ventrally. Its anterior angle, also prominent, but more pointed, projects horizontally ventralward; it gives attachment to the vocal ligament, and is called the **vocal process** (Fig. 15–16).

The **apex** of each cartilage is pointed, curved dorsalward and medialward, and surmounted by a small conical, cartilaginous nodule, the **corniculate cartilage.**

The **Corniculate Cartilages** (*cartilagines corniculatae; cartilages of Santorini*) are two small conical nodules consisting of yellow elastic cartilage, which articulate with the summits of the arytenoid cartilages and serve to prolong them dorsalward and medialward. They are situated in the posterior parts of the aryepiglottic folds of mucous membrane, and are sometimes fused with the arytenoid cartilages.

The **Cuneiform Cartilages** (*cartilagines cuneiformes; cartilages of Wrisberg*) are two small, elongated pieces of yellow elastic cartilage, placed one on either side, in the aryepiglottic fold, where they give rise to small whitish elevations on the surface of the mucous membrane, just ventral to the arytenoid cartilages.

The **Epiglottis** (*cartilago epiglottica*) (Fig. 15–13) is a thin lamella of yellow elastic cartilage, shaped like a leaf, which projects obliquely upward behind the root of the tongue, and ventral to the entrance to the larynx. The free extremity is broad and rounded; the attached part or stem is long, narrow, and connected by the **thyro-**

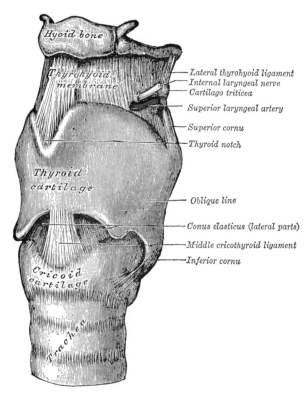

Fɪɢ. 15–14.—The ligaments of the larynx. Anterior and slightly lateral view.

epiglottic ligament to the angle formed by the two laminae of the thyroid cartilage, a short distance caudal to the superior thyroid notch. The caudal part of its anterior surface is connected to the cranial border of the body of the hyoid bone by an elastic ligamentous band, the **hyoepiglottic ligament.**

The *anterior* or *lingual surface* is curved ventralward, and covered on its cranial, free part by mucous membrane which is reflected on to the sides and root of the tongue, forming a median and two lateral **glossoepiglottic folds;** the lateral folds are partly attached to the wall of the pharynx. The depressions between the epiglottis and the root of the tongue, on either side of the median fold, are named the **valleculae** (Fig. 15–18). The caudal part of the anterior surface lies dorsal to the hyoid bone, the hyothyroid membrane, and upper part of the thyroid cartilage, but is separated from these structures by a mass of fatty tissue.

The *posterior* or *laryngeal surface* is smooth, concave from side to side, concavo-convex vertically; its caudal part projects dorsalward as an elevation, the **tubercle** or **cushion.** The surface of the cartilage is indented by a number of small pits, in which mucous glands are lodged. To its sides the aryepiglottic folds are attached.

Structure.—The corniculate and cuneiform cartilages, the epiglottis, and the apices of the arytenoids at first consist of hyaline cartilage, but later elastic fibers are deposited in the matrix, converting them into yellow elastic cartilage, which shows little tendency to calcification. The thyroid, cricoid, and the greater part of the arytenoids consist of hyaline cartilage, and become more or less ossified as age advances. Ossification commences about the twenty-fifth year in the thyroid cartilage, and somewhat later in the cricoid and arytenoids; by the sixty-fifth year these cartilages may be completely converted into bone.

Ligaments.—The ligaments of the larynx (Figs. 15–14, 15–15) are **extrinsic,** *i.e.,* those

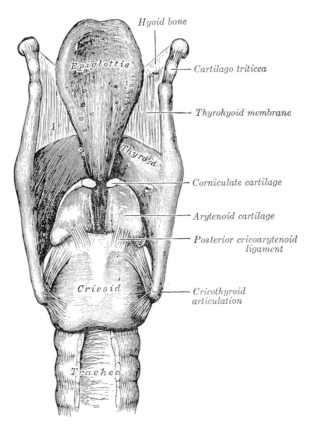

FIG. 15–15.—Ligaments of the larynx. Posterior view.

connecting the thyroid cartilage and epiglottis with the hyoid bone, and the cricoid cartilage with the trachea; and **intrinsic,** those which connect the several cartilages of the larynx to each other.

Extrinsic Ligaments.—The ligaments connecting the thyroid cartilage with the hyoid bone are the thyrohyoid membrane, and a middle and two lateral thyrohyoid ligaments.

The **Thyrohyoid Membrane** (*membrana thyrohyoidea; hyothyroid membrane*) is a broad, fibroelastic layer, attached to the cranial border of the thyroid cartilage and to the front of its superior cornu, and to the cranial margin of the dorsal surface of the body and greater cornua of the hyoid bone, thus passing dorsal to the posterior surface of the body of the hyoid, and being separated from it by a bursa which facilitates the upward movement of the larynx during deglutition. Its middle thicker part is termed the **middle thyrohyoid ligament** (*ligamentum thyrohyoideum medium; middle hyothyroid ligament*); its lateral thinner portions are pierced by the superior laryngeal vessels and the internal branch of the superior laryngeal nerve. Its anterior surface is in relation with the Thyrohyoideus, Sternohyoideus, and Omohyoideus, and with the body of the hyoid bone.

The **Thyrohyoid Ligament** (*ligamentum thyrohyoideum; lateral hyothyroid ligament*) is a round elastic cord, which forms the posterior border of the thyrohyoid membrane and passes between the tip of the superior cornu of the thyroid cartilage and the extremity of the greater cornu of the hyoid bone. A small cartilaginous nodule (*cartilago triticea*), sometimes bony, is frequently found in it.

The epiglottis is connected with the hyoid bone by an elastic band, the **hyoepiglottic ligament** (*ligamentum hyoepiglotticum*), which extends from the anterior surface of the epiglottis to the upper border of the body of the hyoid bone. The glossoepiglottic folds of mucous membrane (page 1123) may also be considered as extrinsic ligaments of the epiglottis.

The **Cricotracheal Ligament** (*ligamentum cricotracheale*) connects the cricoid cartilage with the first ring of the trachea. It resembles the fibrous membrane which connects the cartilaginous rings of the trachea to each other.

Intrinsic Ligaments.—Beneath the mucous membrane of the larynx is a broad sheet of fibrous tissue containing many elastic fibers, and termed the **elastic membrane of the larynx.** It is subdivided on either side by the interval between the ventricular and vocal ligaments; the cranial portion extends between the arytenoid cartilage and the epiglottis and is often poorly defined; the caudal part is a well-marked membrane forming, with its fellow of the opposite side, the conus elasticus which connects the thyroid, cricoid, and arytenoid cartilages to one another.

The **Conus Elasticus** (*cricothyroid membrane*) is composed mainly of yellow elastic tissue. It consists of an anterior and two lateral portions. The **anterior part** or **middle cricothyroid ligament** (*ligamentum cricothyroideum medium; central part of cricothyroid membrane*) is thick and strong, and connects together the ventral parts of the contiguous margins of the thyroid and cricoid cartilages. It is overlapped on either side by the Cricothyroidei, but between these is subcutaneous; it is crossed horizontally by a small anastomotic arterial arch formed by the junction of the two cricothyroid arteries, branches of which pierce it. The **lateral portions** are thinner and lie close under the mucous membrane of the larynx; they extend from the superior border of the cricoid cartilage to the inferior margin of the vocal ligaments, with which they are continuous. These ligaments may therefore be regarded as the free borders of the lateral portions of the conus elasticus, and extend from the vocal processes of the arytenoid cartilages to the angle of the thyroid cartilage about midway between its cranial and caudal borders.

An **articular capsule,** strengthened posteriorly by a well-marked fibrous band, encloses the articulation of the inferior cornu of the thyroid with the cricoid cartilage on either side.

Each arytenoid cartilage is connected to the cricoid by a capsule and a posterior cricoarytenoid ligament. The **capsule** (*capsula articularis cricoarytenoidea*) is thin and loose, and is attached to the margins of the articular surfaces. The **posterior cricoary-**

tenoid ligament (*ligamentum cricoarytenoideum posterius*) extends from the cricoid to the medial and dorsal part of the base of the arytenoid.

The **thyroepiglottic ligament** (*ligamentum thyroepiglotticum*) is a long, slender, elastic cord which connects the stem of the epiglottis with the angle of the thyroid cartilage, immediately beneath the superior thyroid notch, cranial to the attachment of the ventricular ligaments.

Movements.—The articulation between the inferior cornu of the thyroid cartilage and the cricoid cartilage on either side is a diarthrodial one, and permits of rotatory and gliding movements. The rotatory movement is one in which the cricoid cartilage rotates upon the inferior cornua of the thyroid cartilage around an axis passing transversely through both joints. The gliding movement consists in a limited shifting of the cricoid on the thyroid in different directions.

The articulation between the arytenoid cartilages and the cricoid is also a diarthrodial one, and permits of two varieties of movement: one is a rotation of the arytenoid on a vertical axis, whereby the vocal process is moved lateralward or medialward, and the rima glottidis increased or diminished; the other is a gliding movement, and allows the arytenoid cartilages to approach or recede from each other; from the direction and slope of the articular surfaces lateral gliding is accompanied by a forward and downward movement. The two movements of gliding and rotation are associated, the medial gliding being connected with the medialward rotation, and the lateral gliding with lateralward rotation. The posterior cricoarytenoid ligaments limit the forward movement of the arytenoid cartilages on the cricoid.

Interior of the Larynx (Figs. 15–16, 15–17).—The **cavity of the larynx** (*cavum laryngis*) extends from the laryngeal entrance to the caudal border of the cricoid cartilage where it is continuous with that of the trachea. It is divided into two parts by the projection of the vocal folds, between which is a narrow triangular fissure or opening, the **rima glottidis.** The portion of the cavity of the larynx above the vocal folds is called the **vestibule;** it is wide and triangular in shape, its base or anterior wall presenting, however, about its center the backward projection of the tubercle of the epiglottis. It contains the ventricular folds,

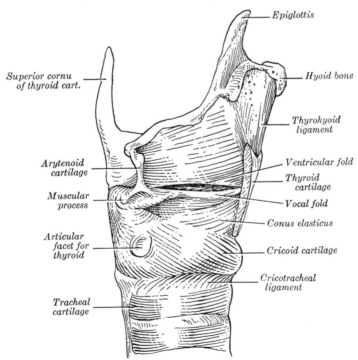

FIG. 15–16.—A dissection to show the right half of the conus elasticus. The right lamina of the thyroid cartilage and the subjacent muscles have been removed.

and between these and the vocal folds are the **ventricles of the larynx.** The portion caudal to the vocal folds is at first of an elliptical form, but widens out, assumes a circular form, and is continuous with the tube of the trachea.

The **entrance of the larynx** (Fig. 15–18) is bounded, ventrally, by the epiglottis; dorsally, by the apices of the arytenoid cartilages, the corniculate cartilages, and the interarytenoid notch; and on either side, by a fold of mucous membrane, enclosing ligamentous and muscular fibers, stretched between the side of the epiglottis and the apex of the arytenoid cartilage; this is the **aryepiglottic fold,** on the posterior part of the margin of which the cuneiform cartilage forms a more or less distinct whitish prominence, the **cuneiform tubercle.**

The **Ventricular Folds** (*plicae ventriculares; superior or false vocal cords*) are two thick folds of mucous membrane, each enclosing a narrow band of fibrous tissue, the **ventricular ligament** which is attached ventrally to the angle of the thyroid cartilage immediately below the attachment of the epiglottis, and dorsally to the ventrolateral surface of the arytenoid cartilage, a short distance above the vocal process. The caudal border of this ligament, enclosed in mucous membrane, forms a free crescentic margin, which constitutes the upper boundary of the ventricle of the larynx.

The **Vocal Folds** (*plicae vocales; inferior or true vocal cords*) enclose two strong bands, named the **vocal ligaments** or vocal cords (*ligamenta vocales; inferior thyroarytenoid*). Each ligament consists of a band of yellow elastic tissue, attached ventrally to the angle of the thyroid cartilage, and dorsally to the vocal process of the arytenoid. Its caudal border is continuous with the thin lateral part of the conus elasticus. Its cranial border forms the boundary of the ventricle of the larynx. Laterally, the Vocalis muscle lies parallel with it. The ligament is covered medially by mucous membrane which is extremely thin and closely adherent to its surface.

The **Ventricle of the Larynx** (*ventriculus laryngis* [*Morgagnii*]; *laryngeal sinus*) is a fusiform fossa, bounded by the free crescentic edge of the ventricular fold, by the straight margin of the vocal fold, and by

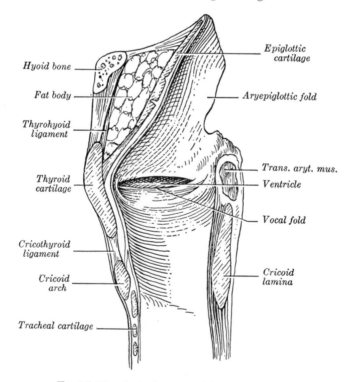

Hyoid bone

Fat body

Thyrohyoid ligament

Thyroid cartilage

Cricothyroid ligament

Cricoid arch

Tracheal cartilage

Epiglottic cartilage

Aryepiglottic fold

Trans. aryt. mus.

Ventricle

Vocal fold

Cricoid lamina

FIG. 15–17.—Sagittal section of larynx, right half.

the mucous membrane covering the corresponding Thyroarytenoideus. The anterior part of the ventricle leads by a narrow opening into a cecal pouch of mucous membrane of variable size called the **appendix.**

The **appendix of the laryngeal ventricle** (*appendix ventriculi laryngis; laryngeal saccule*) is a membranous sac, placed between the ventricular fold and the inner surface of the thyroid cartilage, occasionally extending as far as its cranial border or even higher. On the surface of its mucous membrane are the openings of sixty or seventy mucous glands, which are lodged in the submucous areolar tissue. This sac is enclosed in a fibrous capsule, continuous caudally with the ventricular ligament. Its medial surface is covered by a few delicate muscular fasciculi, which *arise* from the apex of the arytenoid cartilage and become lost in the aryepiglottic fold of mucous membrane; laterally it is separated from the

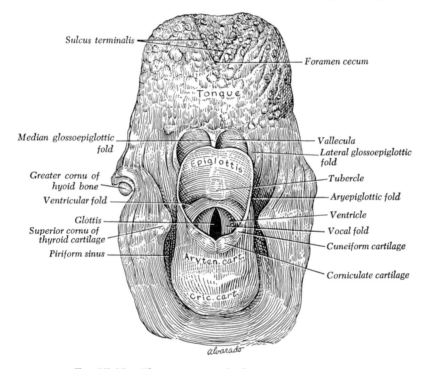

Fig. 15–18.—The entrance to the larynx, posterior view.

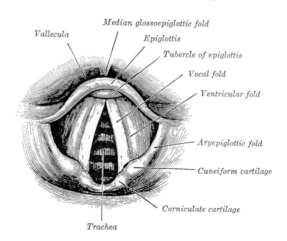

Fig. 15–19.—Laryngoscopic view of interior of larynx.

thyroid cartilage by the Thyroepiglotticus. These muscles compress the sac, and express the secretion it contains upon the vocal folds to lubricate their surfaces.

The **Rima Glottidis** (Fig. 15–19) is the elongated fissure or opening between the vocal folds ventrally, and the bases and vocal processes of the arytenoid cartilages dorsally. It is therefore subdivided into a larger anterior intramembranous part (*glottis vocalis*), which measures about three-fifths of the length of the entire aperture, and a posterior intercartilaginous part (*glottis respiratoria*). Posteriorly it is limited by the mucous membrane passing between the arytenoid cartilages. The rima glottidis is the narrowest part of the cavity of the larynx, and its level corresponds with the bases of the arytenoid cartilages. Its length, in the male, is about 23 mm; in the female from 17 to 18 mm. The width and shape of the rima glottidis vary with the movements of the vocal cords and arytenoid cartilages during respiration and phonation. In the condition of rest, *i.e.*, when these structures are uninfluenced by muscular action, as in quiet respiration, the intermembranous part is triangular, with its apex in front and its base behind—the latter being represented by a line, about 8 mm long, connecting the anterior ends of the vocal processes, while the medial surfaces of the arytenoids are parallel to each other, and hence the intercartilaginous part is rectangular. During extreme adduction of the vocal folds, as in the emission of a high note, the intermembranous part is reduced to a linear slit by the apposition of the vocal folds, while the intercartilaginous part is triangular, its apex corresponding to the anterior ends of the vocal processes of the arytenoids, which are approximated by the medial rotation of the cartilages. Conversely in extreme abduction of the vocal folds, as in forced inspiration, the arytenoids and their vocal processes are rotated lateralward, and the intercartilaginous part is triangular in shape but with its apex directed backward. In this condition the entire glottis is somewhat lozenge-shaped, the widest part of the aperture corresponding with the attachments of the vocal folds to the vocal processes.

Muscles.—The muscles of the larynx are

extrinsic, passing between the larynx and parts around, described in Chapter 6; and *intrinsic*, confined entirely to the larynx.

The intrinsic muscles are:

> Cricothyroideus
> Cricoarytenoideus posterior
> Cricoarytenoideus lateralis
> Arytenoideus
> Thyroarytenoideus

The **Cricothyroideus** (*Cricothyroid*) (Fig. 15–20), triangular in form, *arises* from the front and lateral part of the cricoid cartilage; its fibers are arranged in two groups. The caudal fibers constitute a **pars obliqua** and slant backward to the anterior border of the inferior cornu; the anterior fibers, forming a **pars recta,** run cranialward, backward, and lateralward to the posterior part of the lower border of the lamina of the thyroid cartilage. The medial borders of the muscles on the two sides are separated by a triangular interval, occupied by the middle cricothyroid ligament.

The **Cricoarytenoideus posterior** (*posterior cricoarytenoid*) (Fig. 15–20) *arises* from the broad depression on the corresponding half of the posterior surface of the lamina of the cricoid cartilage; its fibers run cranialward and lateralward, and converge to be *inserted* into the back of the muscular

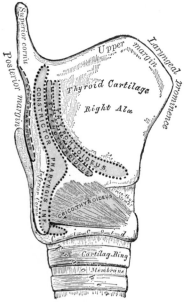

FIG. 15–20.—Side view of the larynx, showing muscular attachments.

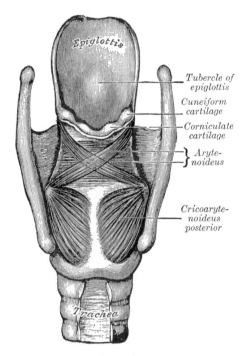

FIG. 15–21.—Muscles of larynx. Posterior view.

Aryepiglotticus. The **Arytenoideus transversus** crosses transversely between the two cartilages.

The **Thyroarytenoideus** (*Thyroarytenoid*) (Figs. 15–22, 15–23) is a broad, thin, muscle

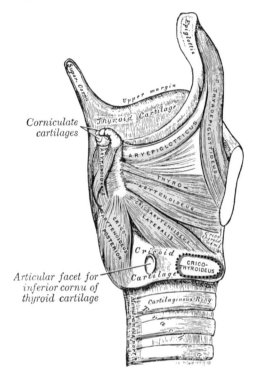

FIG. 15–22.—Muscles of larynx. Side view. Right lamina of thyroid cartilage removed.

process of the arytenoid cartilage. The uppermost fibers are nearly horizontal, the middle oblique, and the lowest almost vertical.

The **Cricoarytenoideus lateralis** (*lateral cricoarytenoid*) (Fig. 15–22) is smaller than the preceding, and of an oblong form. It *arises* from the cranial border of the arch of the cricoid cartilage, and, passing obliquely upward and backward, is *inserted* into the front of the muscular process of the arytenoid cartilage.

The **Arytenoideus** (Fig. 15–21) is a single muscle, filling up the posterior concave surfaces of the arytenoid cartilages. It *arises* from the posterior surface and lateral border of one arytenoid cartilage, and is inserted into the corresponding parts of the opposite cartilage. It consists of oblique and transverse parts. The **Arytenoideus obliquus,** the more superficial, forms two fasciculi, which pass from the base of one cartilage to the apex of the opposite one, and therefore cross each other like the limbs of the letter X; a few fibers are continued around the lateral margin of the cartilage, and are prolonged into the aryepiglottic fold; they are sometimes described as a separate muscle, the

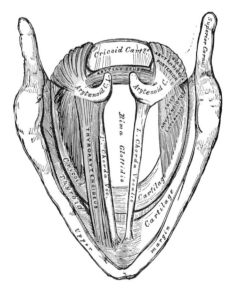

FIG. 15–23.—Muscles of the larynx, seen from above. (Enlarged.)

which lies parallel with and lateral to the vocal fold, and supports the wall of the ventricle and its appendix. It *arises* from the caudal half of the angle of the thyroid cartilage, and from the middle cricothyroid ligament. Its fibers pass dorsalward and lateralward, to be *inserted* into the base and anterior surface of the arytenoid cartilage. The more medial fibers of the muscle can be differentiated as a band which is inserted into the vocal process of the arytenoid cartilage, and into the adjacent portion of its anterior surface; it is termed the **Vocalis,** and lies parallel with the vocal ligament, to which it is adherent.

A considerable number of the fibers of the Thyroarytenoideus are prolonged into the aryepiglottic fold, where some of them become lost, while others are continued to the margin of the epiglottis. They have received a distinctive name, **Thyroepiglotticus,** and are sometimes described as a separate muscle. A few fibers extend along the wall of the ventricle from the lateral wall of the arytenoid cartilage to the side of the epiglottis and constitute the **Ventricularis** muscle.

Actions.—The actions of the muscles of the larynx may be conveniently divided into two groups: 1. Those which open and close the glottis. 2. Those which regulate the degree of tension of the vocal folds.

The *Cricoarytenoidei posteriores* separate the vocal folds, and, consequently, **open** the glottis by rotating the arytenoid cartilages outward around a vertical axis passing through the cricoarytenoid joints, so that their vocal processes and the vocal folds attached to them become widely separated.

The *Cricoarytenoidei laterales* **close** the glottis by rotating the arytenoid cartilages inward, so as to approximate their vocal processes.

The *Arytenoideus* approximates the arytenoid cartilages, and thus **closes** the opening of the glottis, especially at its back part.

The *Cricothyroidei* produce **tension** and elongation of the vocal folds by drawing up the arch of the cricoid cartilage and tilting back the upper border of its lamina; the distance between the vocal processes and the angle of the thyroid is thus increased, and the folds are consequently elongated.

The *Thyroarytenoidei,* consisting of two parts having different attachments and different directions, are rather complicated in their action. Their main use is to draw the arytenoid cartilages forward toward the thyroid, and thus shorten and relax the vocal folds. Their lateral portions rotate the arytenoid cartilage inward, and thus narrow the rima glottidis by bringing the two vocal folds together. Certain minute fibers of the vocalis division, inserting obliquely upon the vocal ligament and designated as the **aryvocalis muscle,** are considered by Strong ('35) to be chiefly responsible for the control of pitch, through their ability to regulate the length of the vibrating part of the vocal folds.

The manner in which the entrance of the larynx is closed during deglutition is referred to on page 1194.

Mucous Membrane.—The mucous membrane of the larynx is continuous above with that lining the mouth and pharynx, and is prolonged through the trachea and bronchi into the lungs. It lines the posterior surface and the upper part of the anterior surface of the epiglottis, to which it is closely adherent, and forms the aryepiglottic folds which bound the entrance of the larynx. It lines the whole of the cavity of the larynx; forms, by its reduplication, the chief part of the ventricular fold, and, from the ventricle, is continued into the ventricular appendix. It is then reflected over the vocal ligament, where it is thin, and very intimately adherent, covers the inner surface of the conus elasticus and cricoid cartilage, and is ultimately continuous with the lining membrane of the trachea. The anterior surface and the upper half of the posterior surface of the epiglottis, the upper part of the aryepiglottic folds and the vocal folds are covered by stratified squamous epithelium; all the rest of the laryngeal mucous membrane is covered by columnar ciliated cells, but patches of stratified squamous epithelium are found in the mucous membrane above the glottis.

Glands.—The mucous membrane of the larynx is furnished with numerous mucus-secreting glands, the orifices of which are found in nearly every part; they are very plentiful upon the epiglottis, being lodged in little pits in its substance; they are also found in large numbers along the margin of the aryepiglottic fold, in front of the arytenoid cartilages, where they are termed the **arytenoid glands.** They exist also in large numbers in the ventricular appendages. None are found on the free edges of the vocal folds.

Vessels.—The chief **arteries** of the larynx are the laryngeal branches derived from the superior and inferior thyroid. The **veins** accompany the arteries; those accompanying the superior laryngeal artery join the superior thyroid vein which opens into the internal jugular vein;

while those accompanying the inferior laryngeal artery join the inferior thyroid vein which opens into the brachiocephalic vein. The **lymphatic vessels** consist of two sets, superior and inferior. The former accompany the superior laryngeal artery and pierce the thyrohyoid membrane, to end in the nodes situated near the bifurcation of the common carotid artery. Of the latter, some pass through the middle cricothyroid ligament and open into nodes lying in front of that ligament or in front of the upper part of the trachea, while others pass to the deep cervical nodes and to the nodes accompanying the inferior thyroid artery.

Nerves.—The **nerves** are derived from the vagus nerve by way of the internal and external branches of the superior laryngeal nerve and from the recurrent nerve. The internal laryngeal branch is sensory. It enters the larynx by piercing the posterior part of the thyrohyoid membrane above the superior laryngeal vessels, and divides into a branch which is distributed to both surfaces of the epiglottis, a second to the aryepiglottic fold, and a third, the largest, which supplies the mucous membrane over the back of the larynx and communicates with the recurrent nerve. The external laryngeal branch supplies the Cricothyroideus. The recurrent nerve passes cranialward beneath the caudal border of the Constrictor pharyngis inferior immediately dorsal to the cricothyroid joint. It supplies all the muscles of the larynx except the Cricothyroideus. The sensory branches of the laryngeal nerves form subepithelial plexuses, from which fibers end between the cells covering the mucous membrane.

Over the posterior surface of the epiglottis, in the aryepiglottic folds, and less regularly in some other parts, taste buds similar to those in the tongue are found.

Fibers from the sympathetic supply the blood vessels and glands.

THE TRACHEA AND BRONCHI

The **trachea** or **windpipe** (Fig. 15–24) is a cartilaginous and membranous tube, extending from the larynx, on a level with the sixth cervical vertebra, to the cranial border of the fifth thoracic vertebra where it divides into the two bronchi. The trachea is nearly but not quite cylindrical, being flattened dorsally; it measures about 11 cm in length; its diameter, from side to side, is from 2 to 2.5 cm, being always greater in the male than in the female. In the child the trachea is smaller, more deeply placed, and more movable than in the adult.

Relations.—The *ventral surface* of the trachea is covered, **in the neck,** by the isthmus of the thyroid gland, the inferior thyroid veins, the arteria thyroidea ima (when that vessel exists), the Sternothyroideus and Sternohyoideus muscles, the cervical fascia, and, more superficially, by the anastomosing branches between the anterior jugular veins; **in the thorax,** it is covered by the manubrium sterni, the remains of the thymus, the left brachiocephalic vein, the aortic arch, the brachiocephalic and left common carotid arteries, and the deep cardiac plexus. *Dorsally* it is in contact with the esophagus. *Laterally,* **in the neck,** it is in relation with the common carotid arteries, the right and left lobes of the thyroid gland, the inferior thyroid arteries, and the recurrent nerves; **in the thorax,** it lies in the superior mediastinum, and is in relation on the right side with the pleura and right vagus, and near the root of the neck with the brachiocephalic artery; on its left side are the left recurrent nerve, the aortic arch, and the left common carotid and subclavian arteries.

The **operation for tracheotomy,** when performed as an emergency, is best carried out by piercing the neck 1 cm below the cricoid cartilage and opening the trachea between the second and third cartilaginous rings. The thyroid and cricoid cartilages are avoided because of subsequent scarring of the larynx. The isthmus of the thyroid gland is usually caudal to the first two tracheal cartilages and no blood vessels of appreciable size are present. Caudal to the thyroid gland the trachea is situated much deeper than above it because of Burns's space, and large veins are likely to be present.

The **Right Bronchus** (*bronchus dexter*) is wider, shorter, and less abrupt in its divergence from the trachea than the left. It is about 2.5 cm long and enters the right lung nearly opposite the fifth thoracic vertebra. The azygos vein arches over it and the pulmonary artery lies at first inferior and then ventral to it. It gives rise to three subsidiary bronchi, one to each of the lobes. The superior lobe bronchus comes off above the pulmonary artery and was called by Aeby, therefore, the **eparterial bronchus.** The bronchi to the middle and inferior

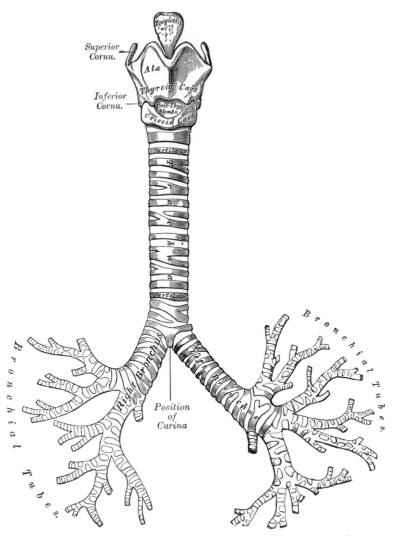

Fig. 15–24.—Front view of cartilages of larynx, trachea, and bronchi.

lobes separate below the pulmonary artery and are accordingly hyparterial in position.

The **right superior lobe bronchus** divides into three branches named, according to the bronchopulmonary segments which they enter, the bronchus for the apical segment, for the posterior segment, and for the anterior segment. The **right middle lobe bronchus** divides into two branches, the bronchus for the lateral and for the medial segments. The **right inferior lobe bronchus** first gives off the bronchus to the superior segment, and then divides into four bronchi for the basal segments, the medial basal, anterior basal, lateral basal, and posterior basal segments.

The **Left Bronchus** (*bronchus sinister*) is smaller in caliber but about twice as long as the right (5 cm). It passes under the aortic arch, and crosses ventral to the esophagus, thoracic duct, and descending aorta. It is superior to the pulmonary artery at first, then dorsal, and finally passes inferior to the artery before it divides into the bronchi for the superior and inferior lobes. Both lobar bronchi, therefore, are hyparterial in position.

The **left superior lobe bronchus** divides into two branches, one of which is distributed to a portion of the left lung corresponding to the right superior lobe, the other to a portion corresponding to the right middle lobe. These two branches are called the superior division and inferior division bronchi to differentiate them from segmental bronchi. The *superior division bronchus* of the left superior lobe divides into branches for the apical posterior segment and the anterior segment. The *inferior division bronchus* divides into bronchi for the superior and inferior segments. The **left inferior lobe bronchus** gives off first the bronchus for the superior segment and then divides into three branches for basal segments, the anterior medial basal, lateral basal, and posterior basal segments.

The picture seen through a bronchoscope may be reproduced if a section is made across the trachea and a bird's-eye view taken of its interior (Fig. 15–25). At the bottom of the trachea, the septum which separates the two bronchi is visible as a

spur (in bronchoscopic terminology) and is named in this case the carina. The **carina** is placed to the left of the middle line and the right bronchus appears as a more direct continuation of the trachea than the left. Because of this asymmetry and because the right bronchus is larger in diameter than the left, foreign bodies which enter the trachea have a tendency to drop into the right bronchus rather than into the left.

Structure (Fig. 15–26).—The trachea and extrapulmonary bronchi are composed of imperfect rings of hyaline cartilage, fibrous tissue, muscular fibers, mucous membrane, and glands.

The **Cartilages** of the trachea vary from sixteen to twenty in number: each forms an imperfect ring, which occupies the anterior two-thirds or so of the circumference of the

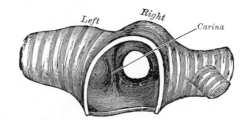

Fig. 15–25.—Bifurcation of the trachea, viewed from above, the interior showing the carina as it would be seen through a bronchoscope.

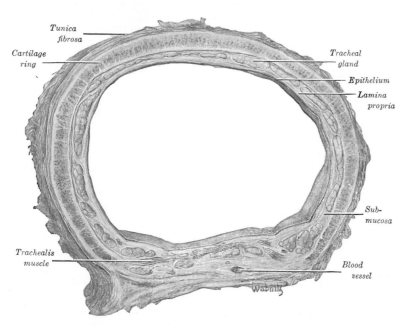

Fig. 15–26.—Cross section of trachea. Adult cadaver. × 5.

trachea, being deficient behind, where the tube is completed by fibrous tissue and non-striated muscular fibers. The cartilages are placed horizontally above each other, separated by narrow intervals. They measure about 4 mm in depth and 1 mm in thickness. Their outer surfaces are flattened in a vertical direction, but the internal are convex, the cartilages being thicker in the middle than at the margins. Two or more of the cartilages often unite, partially or completely, and they are sometimes bifurcated at their extremities. They are highly elastic, but may become calcified in advanced life. In the right bronchus the cartilages vary in number from six to eight; in the left, from nine to twelve. They are shorter and narrower than those of the trachea, but have the same shape and arrangement. The special tracheal cartilages are the first and the last (Fig. 15–24).

The **first cartilage** is broader than the rest, and often divided at one end; it is connected by the cricotracheal ligament with the caudal border of the cricoid cartilage, with which, or with the succeeding cartilage, it is sometimes blended.

The **last cartilage** is thick and broad in the middle, because its lower border is prolonged into a triangular hook-shaped process, which curves downward and backward between the origins of the two bronchi. It ends on each side in an imperfect ring, which encloses the commencement of each bronchus. The cartilage above the last is somewhat broader than the others at its center.

The **Fibrous Membrane.**—The cartilages are enclosed in an elastic fibrous membrane, which consists of two layers; one, the thicker, passing over the outer surface of the ring, the other over the inner surface; at the cranial and caudal margins of the cartilages the two layers blend together to form a single membrane, which connects the rings with one another. In the space between the ends of the rings dorsally, the membrane forms a single layer also.

The **Muscular Tissue** consists of two layers of non-striated muscle, longitudinal and transverse, between the ends of the cartilages. The **longitudinal fibers** are external, and consist of a few scattered bundles. The **transverse fibers** (Trachealis muscle) are arranged internally in branching and anastomosing bands which extend more or less transversely (Fig. 15–26).

The **Mucous Membrane** is continuous above with that of the larynx, and below with that of the bronchi. It consists of areolar and lymphoid tissue, and presents a well-marked basement membrane, supporting a stratified epithelium, the surface layer of which is columnar and ciliated, while the deeper layers are composed

of oval or rounded cells. Beneath the basement membrane there is a distinct layer of longitudinal elastic fibers with a small amount of intervening areolar tissue. The submucous layer is composed of a loose meshwork of connective tissue, containing large blood vessels, nerves, and mucous glands; the ducts of the latter pierce the overlying layers and open on the surface (Fig. 15–27).

Vessels and Nerves.—The trachea is supplied with blood by the inferior thyroid **arteries.** The **veins** end in the thyroid venous plexus. The **nerves** are derived from the vagus and the recurrent nerves, and from the sympathetic; they are distributed to the Trachealis muscles and between the epithelial cells.

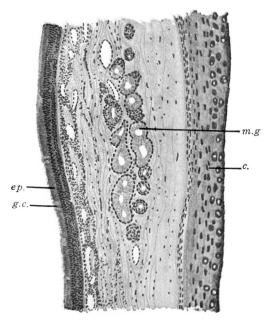

Fig. 15–27.—Section through wall of human trachea. *c*, cartilage; *e.p.*, pseudostratified ciliated epithelium; *g.c.*, goblet cell; *m.g.*, mucous gland. × 8. (Sobotta.)

THE PLEURA

Each lung is invested by an exceedingly delicate serous membrane, the **pleura,** which is arranged in the form of a closed invaginated sac. A portion of the serous membrane covers the surface of the lung and dips into the fissures between its lobes; it is called the **pulmonary pleura** (*pleura pulmonis*). The rest of the membrane lines the inner surface of the chest wall, covers the diaphragm, and is reflected over the structures occupying the middle of the thorax; this portion is termed the **parietal pleura** (*pleura parietalis*). The two layers are continuous with each other around and below the root of the lung; in health they are in actual contact with each other, but the potential space between them is known as the **pleural cavity** (*cavum pleurae*). When the lung collapses or when air or fluid collects between the two layers, the cavity becomes apparent. The right and left pleural sacs are entirely separate from each other; between them are all the thoracic viscera except the lungs, and they touch each other only for a short distance ventrally, opposite the second and third pieces of the sternum; the interval in the middle of the thorax between the two sacs is termed the mediastinum.

Different portions of the parietal pleura have received special names which indicate their position: thus, that portion which lines the inner surfaces of the ribs and Intercos-

tals is the **costal pleura;** that clothing the convex surface of the diaphragm is the **diaphragmatic pleura;** that which rises into the neck, over the summit of the lung, is the **cupula of the pleura** (*cervical pleura*); and that which is applied to the other thoracic viscera is the **mediastinal pleura.**

Relations of the Parietal Pleura (Figs. 15–28, 15–29).—Commencing at the sternum, the pleura passes lateralward, lining the inner surfaces of the costal cartilages, ribs, and Intercostals. At the dorsal part of the thorax it passes over the sympathetic trunk and its branches, and is reflected upon the sides of the bodies of the vertebrae, where it is separated by a narrow interval, the posterior mediastinum, from the opposite pleura. From the vertebral column the pleura passes to the side of the pericardium, which it covers to a slight extent before it covers the dorsal part of the root of the lung, from the caudal border of which a triangular sheet descends vertically toward the

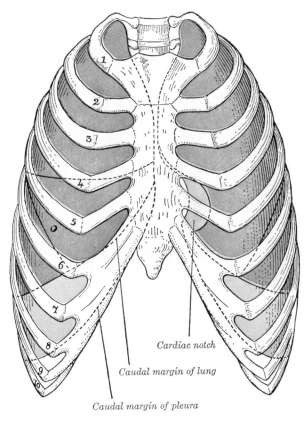

Cardiac notch

Caudal margin of lung

Caudal margin of pleura

Fig. 15–28.—Front view of thorax, showing the relations of the pleurae and lungs to the chest wall. Phrenicocostal and costomediastinal sinus in blue; lungs in purple.

diaphragm. This sheet is the posterior layer of a fold, known as the **pulmonary ligament**. From the dorsal portion of the lung root, the pleura may be traced over the costal surface of the lung, the apex and base, and also over the sides of the fissures between the lobes, on to its mediastinal surface and the ventral part of its root. It is continued from the caudal margin of the root as the anterior layer of the pulmonary ligament, and from this it is reflected on to the pericardium (**pericardial pleura**), and from it to the dorsal surface of the sternum. Cranial to the level of the root of the lung, however, the mediastinal pleura passes uninterruptedly from the vertebral column to the sternum over the structures in the superior mediastinum. It covers the cranial surface of the diaphragm and extends, ventrally, as low as the costal cartilage of the seventh rib; at the side of the chest, to the caudal border of the tenth rib on the left side and to the cranial border of the same rib on the right side; and dorsally, it reaches as low as the twelfth rib, and sometimes even to the transverse process of the first lumbar vertebra. The **cupula** projects through the superior opening of the thorax into the neck, extending from 2.5 to 5 cm above the sternal end of the first rib; this portion of the sac is strengthened by a dome-like expansion of fascia (**Sibson's fascia**), attached ventrally to the inner border of the first rib, and dorsally to the anterior border of the transverse process of the seventh cervical vertebra. This is covered and strengthened by a few spreading muscular fibers derived from the Scaleni (see Figs. 15–30, 15–31), called the Scalenus minimus.

In the ventral part of the chest, the two pleural sacs behind the manubrium are separated by an angular interval; the reflection being represented by a line drawn from the sternoclavicular articulation to the midpoint of the junction of the manubrium with the body of the sternum. From this point the two pleurae descend in close con-

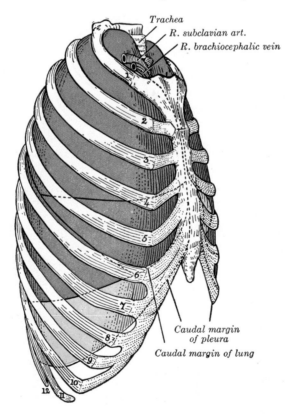

Trachea
R. subclavian art.
R. brachiocephalic vein

Caudal margin of pleura
Caudal margin of lung

FIG. 15–29.—Lateral view of thorax, showing the relations of the pleurae and lungs to the chest wall. Phrenicocostal and costomediastinal sinus in blue; lungs in purple.

tact to the level of the fourth costal cartilages, and the line of reflection on the right side is continued caudalward in nearly a straight line to the xiphoid process, and then turns lateralward, while on the left side the line of reflection diverges lateralward and is continued caudalward, close to the left border of the sternum, as far as the sixth costal cartilage. The caudal limit of the pleura is on a considerably lower level than the corresponding limit of the lung, but does not extend to the attachment of the diaphragm, so that caudal to the line of reflection from the chest wall on to the diaphragm the pleura is in direct contact with the rib cartilages and the Intercostales interni. Moreover, in ordinary inspiration the thin inferior margin of the lung does not extend as far as the line of the pleural reflection, with the result that the costal and diaphragmatic pleurae are here in contact, the intervening narrow slit being termed the **phrenicocostal sinus** (*recessus costodiaphragmaticus*). A similar condition exists behind the sternum and rib cartilages, where the anterior thin margin of the lung falls short of the line of pleural reflection, and where the slit-like cavity between the two layers of pleura forms what is called the **costomediastinal sinus** (*recessus costomediastinalis*).

The line along which the right pleura is reflected from the chest wall to the diaphragm starts ventrally, immediately below the seventh sternocostal joint, and runs caudalward and dorsalward behind the seventh costal cartilage so as to cross the tenth rib in the midaxillary line, from which it is prolonged to the level of the spinous process of the twelfth thoracic vertebra. The reflection of the left pleura follows at first the ascending part of the sixth costal cartilage, and in the rest of its course is slightly lower than that of the right side.

The free surface of the pleura is smooth, glistening, and moistened by a serous fluid; its attached surface is intimately adherent to the lung, and to the pulmonary vessels as they emerge from the pericardium; it is also adherent to the cranial surface of the diaphragm: throughout the rest of its extent it is easily separable from the adjacent parts.

The right pleural sac is shorter, wider, and reaches higher in the neck than the left.

Pulmonary Ligament (*ligamentum pulmonale; ligamentum latum pulmonis*).—The root of the lung is covered by pleura, except at its caudal border where the investing layers come into contact. Here they form a sort of mesenteric fold, the pulmonary ligament, which extends between the mediastinal surface of the lung and the pericardium. Just above the diaphragm the ligament ends in a free falciform border. It serves to retain the lower part of the lung in position.

Structure of Pleura.—Like other serous membranes, the pleura is composed of a single layer of flattened mesothelial cells resting upon a delicate connective tissue membrane, beneath which lies a stroma of collagenous tissue containing several prominent networks of yellow elastic fibers. Blood vessels, lymphatics, and nerves are distributed in the substance of the pleura.

Vessels and Nerves.—The **arteries of the pleura** are derived from the intercostal, internal thoracic, musculophrenic, thymic, pericardiac, pulmonary, and bronchial vessels. The **veins** correspond to the arteries. The **lymphatics** are described on page 753. The **nerves** of the parietal pleura are derived from the phrenic, intercostal, vagus and sympathetic nerves; those of the pulmonary pleura from the vagus and sympathetic through the pulmonary plexuses at the hilum of the lung. Kölliker states that nerves accompany the ramifications of the bronchial arteries in the pulmonary pleura.

THE MEDIASTINUM

The **mediastinum** is interposed as a partition in the median portion of the thorax, separating the parietal pleural sacs of the two lungs (Fig. 15–31). It extends from the sternum ventrally to the vertebral column dorsally and comprises all the thoracic viscera, except the lungs and pleurae, embedded in a thickening and expansion of the subserous fascia of the thorax. It is divided arbitrarily, for the purposes of description, into cranial and caudal parts by a plane which extends from the sternal angle to the caudal border of the fourth thoracic vertebra. The cranial part is

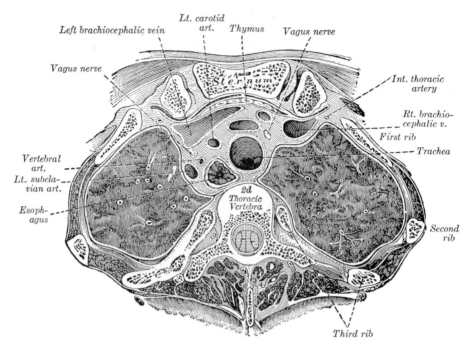

FIG. 15–30.—Transverse section through the upper margin of the second thoracic vertebra, showing the superior mediastinum. (Braune.)

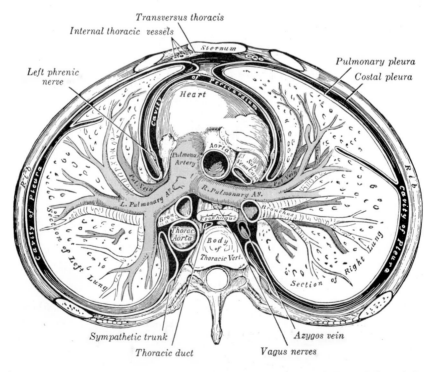

FIG. 15–31.—A transverse section of the thorax showing the contents of the middle and the posterior mediastinum. The pleural and pericardial cavities are exaggerated since normally there is no space **between** parietal and visceral pleurae and between pericardium and heart.

named the **superior mediastinum;** the caudal part is again subdivided into three parts: the **anterior mediastinum,** ventral to the pericardium; the **middle mediastinum,** containing the pericardium; and the **posterior mediastinum,** dorsal to the pericardium.

The **superior mediastinum** (Fig. 15–30) is bounded by the superior aperture of the thorax; by the plane of the superior limit of the pericardium; by the manubrium; by the upper four thoracic vertebrae; and laterally, by the mediastinal pleurae of the two lungs. It contains the origins of the Sternohyoidei and Sternothyroidei and the lower ends of the Longi colli; the aortic arch; the brachiocephalic artery and the thoracic portions of the left common carotid and the left subclavian arteries; the brachiocephalic veins and the upper half of the superior vena cava; the left highest intercostal vein; the vagus, cardiac, phrenic, and left recurrent nerves; the trachea, esophagus, and thoracic duct; the remains of the thymus, and some lymph nodes.

The **anterior mediastinum** (Fig. 15–31) is bounded ventrally by the body of the sternum and, because of the position of the heart, the left Transversus thoracis muscle and parts of the fourth, fifth, sixth, and seventh costal cartilages. It is bounded dorsally by the parietal pericardium and extends caudalward as far as the diaphragm. Besides a few lymph nodes and vessels, it contains only a thin layer of subserous fascia

which is separated from the endothoracic or deep fascia superiorly by a fascial cleft, but there is a firm attachment in its caudal part which forms the pericardiosternal ligament.

The **middle mediastinum** (Fig. 15–31) is the broadest part of the interpleural septum. It contains the heart enclosed in the pericardium, the ascending aorta, the lower half of the superior vena cava with the azygos vein opening into it, the pulmonary trunk dividing into its two branches, the right and left pulmonary veins, and the phrenic nerves.

The **posterior mediastinum** (Fig. 15–31) is an irregularly shaped mass running parallel with the vertebral column, and, because of the slope of the diaphragm, extends caudally beyond the pericardium. It is bounded ventrally, by the pericardium and, more caudally, by the diaphragm; dorsally, by the vertebral column from the lower border of the fourth to the twelfth thoracic vertebra; and on either side, by the mediastinal pleurae. It contains the thoracic part of the descending aorta, the azygos and hemiazygos veins, the vagus and splanchnic nerves, the bifurcation of the trachea and the two bronchi, the esophagus, the thoracic duct, and many large lymph nodes.

The bifurcation of the trachea, the two bronchi, and the roots of the two lungs are included in the middle mediastinum by some authorities.

THE LUNGS (PULMONES)

The **lungs** are the essential organs of respiration; they are placed on either side within the thorax, and separated from each other by the heart and other contents of the mediastinum (Fig. 12–76). The substance of the lung is of a light, porous, spongy texture; it floats in water, and crepitates when handled, owing to the presence of air in the alveoli. It is also highly elastic, hence the retracted state of these organs when they are removed from the closed cavity of the thorax. The surface is smooth, shining, and marked out into numerous polyhedral areas, indicating the lobules of the organ: each of these areas is crossed by numerous lighter lines.

At birth the lungs are pinkish-white in color; in adult life the color is a slaty gray, mottled in patches, and as age advances, this mottling assumes a black color. The coloring matter consists of granules of a carbon deposited in the areolar tissue near the surface of the organ. It increases in quantity as age advances, and is more abundant in males than in females. As a rule, the posterior border of the lung is darker than the anterior.

The right lung usually weighs about 625 gm, the left 567 gm, but much variation is met according to the amount of blood or serous fluid they may contain. The lungs are heavier in the male than in the female, their proportion to the body being 1 to 37 in the former and 1 to 43 in the latter. The vital capacity, the quantity of air that can be exhaled by the deepest expiration after making the deepest inspiration, varies greatly with the individual; an average for an adult man is 3700 cc. The total volume of the fully expanded lungs is about 6500 cc; this includes both tissues and contained air. The tidal air, the amount of air breathed in or out during quiet respiration, is about 500 cc for the adult man. Various calculations indicate that the total epithelial area of the respiratory and non-respiratory surfaces during ordinary deep inspiration of the adult is not greater than 70 square meters.

Each lung is conical in shape, and presents for examination an **apex**, a **base**, three **borders**, and two **surfaces.**

The **apex** (*apex pulmonis*) is rounded,

and extends into the root of the neck, reaching from 2.5 to 5 cm above the level of the sternal end of the first rib. A sulcus produced by the subclavian artery as it curves ventral to the pleura runs cranialward and lateralward immediately below the apex.

The **base** (*basis pulmonis*) is broad, concave, and rests upon the convex surface of the diaphragm, which separates the right lung from the right lobe of the liver, and the left lung from the left lobe of the liver, the stomach, and the spleen. Since the diaphragm extends higher on the right than on the left side, the concavity on the base of the right lung is deeper than that on the left. Laterally and dorsally, the base is bounded by a thin, sharp margin which projects for some distance into the phrenicocostal sinus of the pleura, between the lower ribs and the costal attachment of the diaphragm. The base of the lung descends

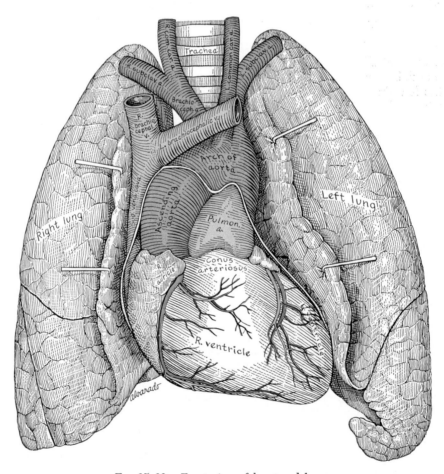

Fig. 15–32.—Front view of heart and lungs.

during inspiration and ascends during expiration.

Surfaces.—The **costal surface** (*facies costalis; external or thoracic surface*) is smooth, convex, of considerable extent, and corresponds to the form of the cavity of the chest, being deeper dorsally than ventrally. It is in contact with the costal pleura, and presents, in specimens which have been hardened *in situ*, slight grooves corresponding with the overlying ribs.

The **Mediastinal surface** (*facies mediastinalis; inner surface*) is divided into a mediastinal and a vertebral portion. The **mediastinal portion** is in contact with the mediastinal pleura. It presents a deep concavity, the **cardiac impression**, which accommodates the pericardium. Dorsal and cranial to this concavity is a slight depression named the **hilum**, through which the structures which form the root of the lung enter and leave.

On the **right lung** (Fig. 15–33), immediately cranial to the hilum, is an arched furrow which accommodates the azygos

vein. Running cranialward, and then arching lateralward some little distance below the apex, is a wide groove for the superior vena cava and right brachiocephalic vein; dorsal to the hilum and the attachment of the pulmonary ligament is a vertical groove for the esophagus; this groove becomes less distinct caudally, owing to the inclination of the lower part of the esophagus to the left of the midline. Ventral and to the right of the lower part of the esophageal groove is a deep concavity for the extrapericardiac portion of the thoracic part of the inferior vena cava.

On the **left lung** (Fig. 15–34), immediately cranial to the hilum, is a well-marked curved furrow produced by the aortic arch, and running cranialward from this toward the apex is a groove accommodating the left subclavian artery; a slight impression ventral to the latter and close to the margin of the lung lodges the left innominate vein. Dorsal to the hilum and pulmonary ligament is a vertical furrow produced by the descending aorta, and ventral to this, near

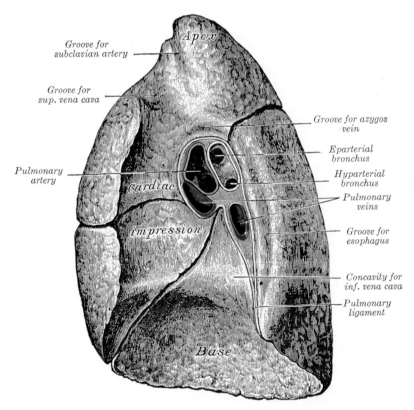

Fig. 15–33.—Mediastinal surface of right lung.

the base of the lung, the lower part of the esophagus causes a shallow impression.

Borders.—The **inferior border** (*margo inferior*) is thin and sharp where it separates the base from the costal surface and extends into the phrenicocostal sinus; medially, where it divides the base from the mediastinal surface, it is blunt and rounded.

The **anterior border** (*margo anterior*) is thin and sharp, and overlaps the ventral surface of the heart and pericardium. The anterior border of the right lung is almost vertical and projects into the costomediastinal sinus; that of the left presents an angular notch, the **cardiac notch,** in which the pericardium comes into contact with the sternum. Opposite this notch the anterior margin of the left lung is situated some little distance lateral to the line of reflection of the corresponding part of the pleura.

Fissures and Lobes of the Lungs.—The **right lung** is divided into three lobes, superior, middle, and inferior, by two interlobar fissures.

The **oblique fissure** (*fissura obliqua*) separates the inferior from the middle and

superior lobes, and corresponds with the oblique fissure in the left lung. Its direction is more vertical, however, and it cuts the lower border about 7.5 cm dorsal to its extremity.

The **horizontal fissure** separates the superior from the middle lobe. It begins in the previous fissure near the dorsal border of the lung, and, running horizontally, cuts the ventral border on a level with the sternal end of the fourth costal cartilage; on the mediastinal surface it may be traced dorsal to the hilum. The middle lobe, the smallest lobe of the right lung, is wedge-shaped, and includes the lower part of the ventral border and the ventral part of the base of the lung.

The right lung is shorter by 2.5 cm than the left because of the diaphragm rising higher on the right side and is broader because of the inclination of the heart to the left side; its total capacity is greater, however, and it weighs more than the left lung.

The **left lung** is divided into two lobes, a superior and an inferior, by the **oblique fissure** (*fissura obliqua*), which extends from the costal to the mediastinal surface of the

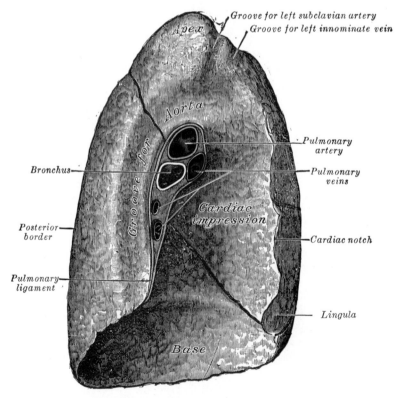

FIG. 15–34.—Mediastinal surface of left lung.

lung both above and below the hilum. As seen on the surface, this fissure begins on the mediastinal surface of the lung at the hilum, and runs dorsalward and cranialward to the posterior border, which it crosses at a point about 6 cm below the apex. It then extends caudalward and ventralward over the costal surface, to reach the lower border near its extremity, and its further course can be followed across the mediastinal surface as far as the hilum. The superior lobe includes the apex, the anterior border, a considerable part of the costal surface, and the greater part of the mediastinal surface of the lung. The inferior lobe is larger than the superior and comprises almost the whole of the base, a large portion of the costal surface, and the greater part of the vertebral part of the medial surface.

The **Root of the Lung** (*radix pulmonis*) (Figs. 15–33, 15–34) is formed by the bronchus, pulmonary artery, pulmonary veins, bronchial arteries and veins, pulmonary plexuses of nerves, lymphatic vessels, and bronchial lymph nodes. These structures are all embedded in mediastinal connective tissue and the entire mass is encircled by a reflection of the pleura. It corresponds to the hilum, which is near the center of the mediastinal surface of the lung, dorsal to the cardiac impression and closer to the posterior than the anterior border. The root of the right lung lies dorsal to the superior vena cava and the right atrium, and the azygos vein arches over it (Fig. 12–76). The root of the left lung is ventral to the descending aorta and inferior to the aortic arch (Fig. 12–77). The phrenic nerve, the pericardiophrenic artery and vein, and the anterior pulmonary plexus of nerves are ventral, while the vagus nerve and its posterior pulmonary plexus are dorsal to the root of the lung on both sides. Caudal to the root of each lung, the reflection of the pleura from mediastinum to lung is prolonged downward toward the diaphragm as the pulmonary ligament (page 1137).

The chief structures of the roots of both lungs have a similar relation to each other in a dorsoventral direction, but there is a difference in their superior and inferior relations on the two sides. Thus the pulmonary veins are ventral, the bronchi dorsal, and the pulmonary arteries between, on both

sides. On the right side, the superior lobe bronchus is superior, the pulmonary artery is slightly lower, next are the bronchi to the middle and inferior lobes, and most inferior is the pulmonary vein (Fig. 15–33). On the left side, the pulmonary artery is superior, the pulmonary veins inferior, and the bronchus between (Fig. 15–34).

Further Subdivision of the Lung.—The importance of certain smaller units of structure of the lungs, called bronchopulmonary segments, has been emphasized by the thoracic surgeon, bronchoscopist, and radiologist. In order to interpret these units correctly, one should give particular attention to the concept that the lung is fundamentally the aggregate of all the branchings of the bronchus. According to this concept, a bronchopulmonary segment would be that portion of the lung to which any particular bronchus is distributed and the term might conceivably be applied to the lobule, supplied by its lobular bronchus. In actual practice, however, it is customary to restrict the term bronchopulmonary segment to the portion of the lung supplied by the direct branches of the lobar bronchi (and of the division bronchi in the case of the left superior lobe). These segments are as definite as the lobes and their relative extent and position can be demonstrated by introducing different colored gelatin into their bronchi. It is possible, in many cases, to follow the delicate connective tissue between them and dissect them apart. Another demonstration of their fundamental nature is given by the fact that the majority of extra fissures follow the planes of separation between the segments.

The Bronchopulmonary Segments (Fig. 15–35).—The bronchopulmonary segments are named according to their positions in the lobes, and the bronchus to each is named after its segment. The right superior lobe has three segments, an apical, a posterior, and an anterior segment. The right middle lobe has a lateral and a medial segment. The right inferior lobe has a superior and four basal segments, the medial basal, anterior basal, lateral basal, and posterior basal segments. The left superior lobe is first separated into two divisions, the superior division corresponding to the right superior lobe and an inferior division, corresponding

Right Lung		**Left Lung**	
LOBES	SEGMENTS	LOBES	SEGMENTS
Superior	Apical Posterior Anterior	Superior	Superior Division { Apicalposterior / Anterior Inferior (Lingular) Division { Superior / Inferior
Middle	Lateral Medial		
Inferior	Superior Medial Basal Anterior Basal Lateral Basal Posterior Basal	Inferior	Superior Anteromedial Basal Lateral Basal Posterior Basal

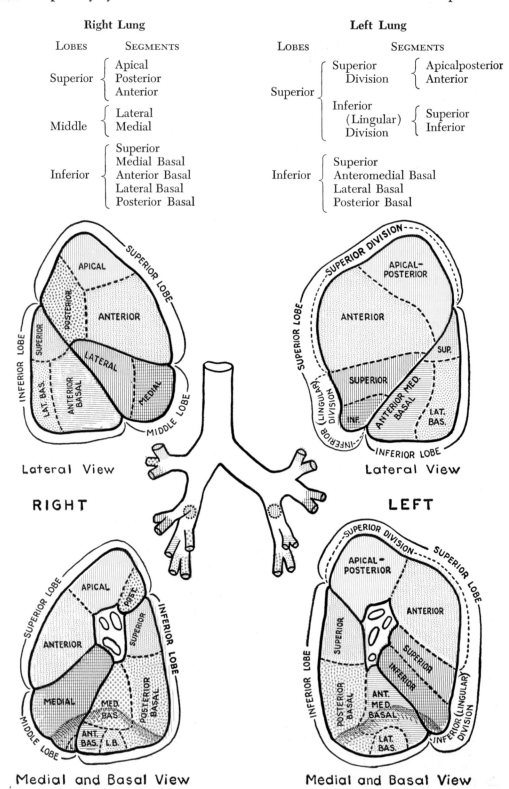

FIG. 15–35.—The bronchopulmonary segments. The segmental branches of the bronchi are shown in corresponding colors. (After J. F. Huber, 1947.)

to the right middle lobe. The superior division of the left superior lobe has an apical-posterior and an anterior segment. The inferior division of the left superior lobe has a superior and an inferior segment. The left inferior lobe has a superior and three basal segments, the anteromedial basal, lateral basal, and posterior basal segments.

Variations.—The branching of the lobar bronchi to form segmental bronchi is reasonably constant, according to Huber (1947). The superior lobe bronchus is somewhat more constant than the inferior lobe bronchus. In approximately 95 per cent of the specimens, it is possible to identify three segmental bronchi coming from the right superior lobe bronchus, although in some of these cases, two segmental bronchi may seem to arise from a very short common stem. The size of the lung segment supplied by a particular bronchus may be larger or smaller than expected, even when the branching appears at first glance to follow the usual pattern, because it may have exchanged a smaller branch bronchus with an adjacent segmental bronchus. In the case of the lower lobe bronchus, the superior and the medial basal segmental bronchi are

about as constant as the branches in the superior lobe, but there is more variation in the anterior basal, lateral basal, and posterior basal segments. In somewhat less than 50 per cent of the specimens, branches come from the posterior aspect of the inferior lobe bronchus between the superior segmental and the basal segmental bronchi, or from the stem below the anterior basal segmental bronchus.

The branchings of the bronchi are designated according to the parts of the lungs which they ventilate. The two main bronchi correspond to the right and left lungs. The right bronchus is subdivided into three lobar bronchi for the superior, middle, and inferior lobes; the left bronchus is subdivided into two, for the superior and inferior lobes. Each lobar bronchus is subdivided into branchings for the bronchopulmonary segments depicted in Figure 15-36.

For the purpose of relating the smaller branchings of the bronchi to the ramifications of the pulmonary arteries and veins, they have been named and numbered by Boyden ('55) as follows (corresponding structures of right and left lung are placed opposite each other) (see Fig. 15-36):

THE SEGMENTAL BRONCHI

Right Superior Lobe

B^1—Apical bronchus
  B^1a—apical ramus
  B^1b—anterior ramus
B^2—Anterior bronchus
  B^2a—posterior ramus
    B^2a1—superior subramus
    B^2a2—inferior subramus
  B^2b—anterior ramus
B^3—Posterior bronchus
  B^3a—apical ramus
  B^3b—posterior ramus

Right Middle Lobe

B^4—Lateral bronchus
  B^4a—posterior ramus
  B^4b—anterior ramus
B^5—Medial bronchus
  B^5a—superior ramus
  B^5b—inferior ramus

Left Superior Lobe
Superior Division

B^{1+3}—Apical posterior bronchus
  B^1a, B^3a—apical rami
  B^1b, B^3b—anterior and posterior rami
B^2—Anterior bronchus
  B^2a—posterior ramus

  B^2b—anterior ramus

Left Superior Lobe
Inferior (Lingular) Division

B^4—Superior lingular bronchus
  B^4a—posterior ramus
  B^4b—anterior ramus
B^5—Inferior lingular bronchus
  B^5a—superior ramus
  B^5b—inferior ramus

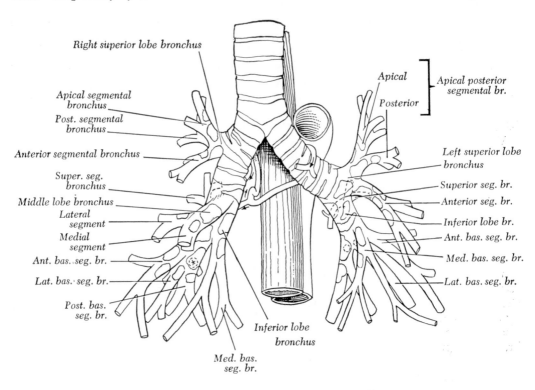

FIG. 15–36.—Ventral view of bronchial tree illustrating prevailing mode of branching. (Redrawn from E. A. Boyden, 1955, *Segmental Anatomy of the Lungs,* courtesy of Blakiston Division, McGraw-Hill Book Co., Inc.)

Right Inferior Lobe	*Left Inferior Lobe*
B^6—Superior bronchus	B^6—Superior bronchus
B^6a—medial ramus	B^6a—medial ramus
B^6a1—paravertebral subramus	B^6a1—paravertebral subramus
B^6a2—posterior subramus	B^6a2—posterior subramus
B^6b—superior ramus	B^6b—superior ramus
B^6c—lateral ramus	B^6c—lateral ramus
B^*—Subsuperior bronchus	B^*—Subsuperior bronchus
$BX^*(10)$—accessory	$BX^*(9)$, $BX^*(10)$—accessory
subsuperior bronchus	subsuperior bronchus
B^7—Medial basal bronchus	B^7—Medial basal bronchus
B^7a—anterior ramus	B^7a—lateroanterior ramus
B^7b—medial ramus	B^7b—medioanterior ramus
B^8—Anterior basal bronchus	B^8—Anterior basal bronchus
B^8a—lateral ramus	B^8a—lateral ramus
B^8b—basal ramus	B^8b—basal ramus
B^9—Lateral basal bronchus	B^9—Lateral basal bronchus
B^9a—lateral ramus	$BX^*(9)$—accessory subsuperior ramus
B^9b—basal ramus	B^9b—basal ramus $= B^9$
B^{10}—Posterior basal bronchus	B^{10}—Posterior basal bronchus
$BV^*(10)$—accessory	$BX^*(10)$—accessory
subsuperior ramus	subsuperior ramus
B^{10}a—laterobasal ramus	B^{10}a—laterobasal ramus
B^{10}b—mediobasal ramus	B^{10}b—mediobasal ramus

Blood Vessels of the Lungs (Figs. 15–36 to 15–41).—The vessels which serve the special function of the lungs, that is, of aerating the blood, are the pulmonary arteries and veins and their connecting capillaries. The blood supply for the purposes of nutrition to the tissues of the lungs is provided by the bronchial arteries.

The **Right** and **Left Pulmonary Arteries,** formed by the bifurcation of the pulmonary trunk, convey the deoxygenated blood to the right and left lungs. They divide into branches which accompany the bronchi, coursing chiefly along their dorsal surface. The bronchopulmonary segments are supplied by main intrasegmental branches of the pulmonary arteries, which are single, for the most part, but which may arise as common trunks for adjacent segments (Boyden, 1945). The artery for one segment is likely to supply small branches to the neighboring segments. Distal to the alveolar duct, branches are distributed to each atrium, from which arise smaller radicles terminating in a dense capillary network in the walls of the alveoli.

THE SEGMENTAL ARTERIES

Right Superior Lobe

A^1—Apical segmental artery
  A^1a—apical ramus
  A^1b—anterior ramus
A^2—Anterior segmental artery
  A^2a—posterior ramus
    A^2a1—superior subramus
    A^2a2—inferior subramus
  A^2b—anterior ramus
A^3—Posterior segmental artery
  A^3a—apical ramus
  A^3b—posterior ramus

Middle Lobe

A^4—Lateral segmental artery

  A^4a—posterior ramus
  A^4b—anterior ramus
A^5—Medial segmental artery

  A^5a—superior ramus
  A^5b—inferior ramus

Right Inferior Lobe

A^6—Superior segmental artery
  A^6a—medial ramus
  A^6b—superior ramus
  A^6c—lateral ramus
A^7—Medial basal artery
  A^7a—anterior ramus
  A^7b—medial ramus
A^8—Anterior basal artery
  A^8a—lateral ramus
  A^8b—basal ramus
A^9—Lateral basal artery
  A^9a—lateral ramus
  A^9b—basal ramus
A^{10}—Posterior basal artery
  $A^{10}a$—laterobasal ramus
  $A^{10}b$—mediobasal ramus

Left Superior Lobe
Superior Division

A^{1+3}—Apical posterior segmental artery
  A^1a—apical ramus
  A^1b—anterior ramus
  A^3a—apical ramus
  A^3b—posterior ramus
A^2—Anterior segmental artery
  A^2a—posterior ramus
  A^2b—anterior ramus

Left Superior Lobe
Inferior or Lingular Division

A^4—Superior lingular artery
  (*a. lingularis superior*)
  A^4a—pars posterior
  A^4b—pars anterior
A^5—Inferior lingular artery
  (*a. lingularis inferior*)
  A^5a—pars superior
  A^5b—pars inferior

Left Inferior Lobe

A^6—Superior segmental artery
  A^6a—medial ramus
  A^6b—superior ramus
  A^6c—lateral ramus
A^7—Medial basal artery
  A^7a—anterior ramus
  A^7b—medial ramus
A^8—Anterior basal artery
  A^8a—lateral ramus
  A^8b—basal ramus
A^9—Lateral basal artery
  A^9a—lateral ramus
  A^9b—basal ramus
A^{10}—Posterior basal artery
  $A^{10}a$—laterobasal ramus
  $A^{10}b$—mediobasal ramus

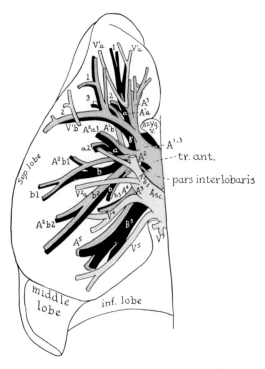

FIG. 15–37.—Dissection of mediastinal surface of right superior lobe showing the prevailing pattern. The conventional colors are followed: red for arteries, blue for veins, and black for bronchi. (Redrawn from E. A. Boyden, 1955, *Segmental Anatomy of the Lungs,* courtesy of Blakiston Division, McGraw-Hill Book Co., Inc.)

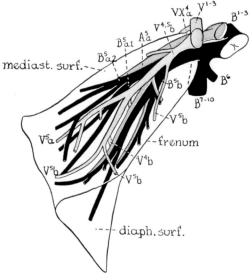

FIG. 15–38.—Dissection of mediastinal surface of middle lobe. (Redrawn from E. A. Boyden, 1955, *Segmental Anatomy of the Lungs,* courtesy of Blakiston Division, McGraw-Hill Book Co., Inc.)

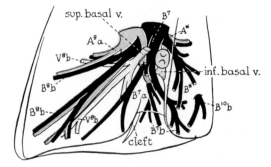

FIG. 15–39.—Dissection of mediastinal surface of right inferior lobe; the mode of branching of the inferior pulmonary vein is atypical. (Redrawn from E. A. Boyden, 1955, *Segmental Anatomy of the Lungs,* courtesy of Blakiston Division, McGraw-Hill Book Co., Inc.)

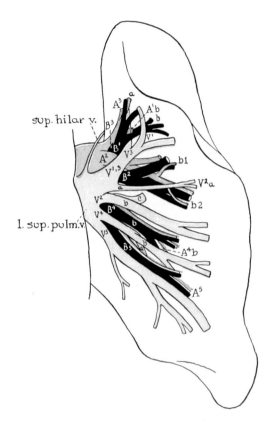

FIG. 15–40.—Dissection of mediastinal surface of left superior lobe. (Redrawn from E. A. Boyden, 1955, *Segmental Anatomy of the Lungs*, courtesy of Blakiston Division, McGraw-Hill Book Co., Inc.)

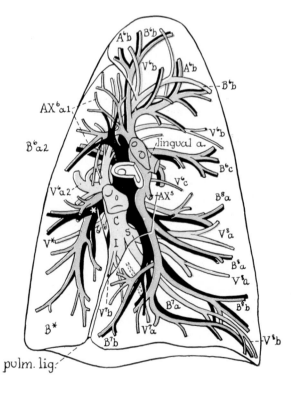

FIG. 15–41.—Dissection of mediastinal surface of fairly typical left inferior lobe. (Redrawn from E. A. Boyden, 1955, *Segmental Anatomy of the Lungs*, courtesy of Blakiston Division, McGraw-Hill Book Co., Inc.)

THE SEGMENTAL VEINS

Right Superior Lobe

V^1—Apical segmental vein
  V^1a—apical ramus
  V^1b—anterior ramus
V^3—Posterior segmental vein
  V^3a—apical ramus
  V^3b—posterior ramus
  V^3c—posterior intersegmental ramus
  V^3d—anterior ramus
V^2—Anterior segmental vein
  V^2a—superior ramus
  V^2b—Inferior ramus

Left Superior Lobe
Superior Division

V^{1-3}—Apical posterior segmental vein
  V^1—apical vein
    V^1a—apical ramus
    V^1b—anterior ramus
  V^3—posterior vein
    V^3a—apical ramus
    V^3b—posterior ramus
    V^3c—posterior intersegmental ramus

V^2—Anterior segmental vein
  V^2a—superior ramus
  V^2b—inferior ramus
  V^2c—posterior ramus

Middle Lobe

V^4—Lateral segmental vein
  V^4a—posterior ramus
  V^4b—anterior ramus
V^5—Medial segmental vein
  V^5a—superior ramus
  V^5b—inferior ramus

Inferior Division

V^4—Superior lingular segmental vein
  V^4a—posterior ramus
  V^4b—anterior ramus
V^5—Inferior lingular segmental vein
  V^5a—superior ramus
  V^5b—inferior ramus

Right Inferior Lobe

V^6—Superior segmental vein
  V^6a—medial ramus
  V^6b—superior ramus
  V^6c—lateral ramus
V^7—Medial basal segmental vein
  V^7a—anterior ramus
  V^7b—medial ramus
V^8—Anterior basal segmental vein
  V^8a—lateral ramus
  V^8b—basal ramus
V^9—Lateral basal segmental vein
  V^9a—lateral ramus
  V^9b—basal ramus
V^{10}—Posterior basal segmental vein
  V^{10}a—laterobasal ramus
  V^{10}b—mediobasal ramus

Left Inferior Lobe

V^6—Superior segmental vein
  V^6a—medial ramus
  V^6b—superior ramus
  V^6c—lateral ramus
V^7—Medial basal segmental vein
  V^7a—lateroanterior ramus
  V^7b—medioanterior ramus
V^8—Anterior basal segmental vein
  V^8a—lateral ramus
  V^8b—basal ramus
V^9—Lateral basal segmental vein
  V^9a—lateral ramus
  V^9b—basal ramus
V^{10}—Posterior basal segmental vein
  V^{10}a—laterobasal ramus
  V^{10}b—mediobasal ramus

The alveoli are lined by a continuous layer of pulmonary alveolar epithelium. The nuclei of the epithelial cells protrude into the air spaces, and the perinuclear cytoplasm attenuates quite abruptly into thin sheets of cytoplasm. The cytoplasmic sheets, averaging about 0.2 μ in thickness, rest on a basement membrane and face on the alveolar air spaces. The alveolar walls contain blood capillaries, collagenous, reticular, and elastic connective tissue fibers. The barrier between the capillary blood and alveolar air includes two thin layers of cytoplasm, alveolar epithelium and capillary endothelium with adherent basement membranes for each. Tissue space between these two membranes may be potential or real, but they are not adherent and variable amounts of interstitial elements may separate them.

The fetal lung resembles a gland in that the alveoli have a small lumen and are lined by cubical epithelium. After the first respiration the alveoli become distended, and the epithelium takes on the characteristics of the adult.

The **bronchial arteries** supply blood for the nutrition of the lung; the right lung usually receives a single artery and the left lung two. They are derived from the ventral side of the upper part of the thoracic aorta or from the upper aortic intercostal arteries. Some are distributed to the bronchial glands and to the walls of the bronchi and pulmonary vessels; those supplying the bronchi extending as far as the respiratory bronchioles, where they form capillary plexuses which unite with similar plexuses formed by the pulmonary artery, both of which give rise to small venous trunks forming one of the sources of the pulmonary vein. Others are distributed in the interlobular areolar tissue, and end partly in the deep, partly in the superficial, bronchial veins. Lastly, some ramify upon the surface of the lung, beneath the pleura, where they form a capillary network.

The **bronchial vein** is formed at the root of the lung, receiving superficial and deep veins from a limited area about the hilum, the larger part of the blood supplied by the bronchial arteries being returned by the pulmonary veins. It ends on the right side in the azygos vein, and on the left side in the highest intercostal or in the accessory hemiazygos vein.

The **lymphatics** are described on page 753.

Nerves.—The lungs are supplied from the anterior and posterior pulmonary plexuses, formed chiefly by branches from the sympathetic and vagus. The filaments from these plexuses accompany the bronchial tubes, supplying efferent fibers to the bronchial muscle and afferent fibers to the bronchial mucous membrane and probably to the alveoli of the lung. Small ganglia are found upon these nerves.

REFERENCES

Nose and Paranasal Sinuses

ALI, M. Y. 1965. Histology of the human nasopharyngeal mucosa. J. Anat. (Lond.), 99, 657-672.

BLANTON, P. L. and N. L. BIGGS. 1969. Eighteen hundred years of controversy: The paranasal sinuses. Am. J. Anat., 124, 135-147.

CAUNA, N., K. H. HINDERER, and R. T. WENTGES. 1969. Sensory receptor organs of the human nasal respiratory mucosa. Am. J. Anat., 124, 187-209.

FABRICANT, N. D. and G. CONKLIN. 1965. *The Dangerous Cold—Its Cures and Complications.* ix + 179 pages, illustrated. The Macmillan Company, New York.

JACOBS, M. H. 1947. Anatomic study of the maxillary sinus from the standpoint of the oral surgeon. J. oral Surg., 5, 282-291.

REVSKOI, Y. K. 1965. Variations in frontal sinus structure and their significance in selection of pilots. Fed. Proc., 24, T948-T950.

ROSE, J. M., C. M. POMERAT, and B. DANES. 1949. Tissue culture studies of ciliated nasal mucosa in man. Anat. Rec., 104, 409-419.

ROSEN, M. D. and B. G. SARNAT. 1954. A comparison of the volume of the left and right maxillary sinuses in dogs. Anat. Rec., 120, 65-71.

SCHAEFFER, J. P. 1920. *The Nose, Paranasal Sinuses, Nasolacrimal Passageways, and Olfactory Organ in Man.* xxii + 370 pages. Blakiston Company, Philadelphia.

SELLERS, L. M. 1949. The frontal sinus—A problem in diagnosis and treatment. The Mississippi Doctor, 27, 317-320.

WARBRICK, J. G. 1960. The early development of the nasal cavity and upper lip in the human embryo. J. Anat. (Lond.), 94, 351-362.

Larynx, Trachea, and Bronchi

BLANDING, J. D., JR., R. W. OGILVIE, C. L. HOFFMAN, and W. H. KNISELY. 1964. The gross morphology of the arterial supply to the trachea, primary bronchi, and esophagus of the rabbit. Anat. Rec., 148, 611-614.

FISHER, A. W. F. 1964. The intrinsic innervation of the trachea. J. Anat. (Lond.), 98, 117-124.

GRAY, F. W. and C. M. WISE. 1959. *The Bases of Speech.* 3rd Edition. xiii + 562 pages, 77 figures. Harper and Brothers, Publishers, New York.

KEENE, M. F. L. 1961. Muscle spindles in human laryngeal muscles. J. Anat. (Lond.), 95, 25-29.

KING, B. T. and R. L. GREGG. 1948. An anatomical reason for the various behaviors of paralyzed vocal cords. Ann. Otol., (St. Louis), 57, 925-944.

LATARJE, M. 1954. La vascularisation sanguinea des bronches. Les bronches, 4, 145.

MILLER, R. A. 1941. The laryngeal sacs of an infant and an adult gorilla. Amer. J. Anat., 69, 1-17.

SMITH, E. I. 1957. The early development of the trachea and esophagus in relation to atresia of the esophagus and tracheoesophageal fistula. Carneg. Instn., Contr. Embryol., 36, 41-58 + 4 plates.

STRONG, L. H. 1935. The mechanism of laryngeal pitch. Anat. Rec., 63, 13-28.

TURNER, R. S. 1962. A note of geometry of the tracheal bifurcation. Anat. Rec., 143, 189-194.

MILLER, W. S. 1937. *The Lung.* xiv + 209 pages. Charles C Thomas, Springfield.

NEGUS, V. 1965. *The Biology of Respiration.* xi + 228 pages, 154 figures. The Williams & Wilkins Company, Baltimore.

THOMAS, L. B. and E. A. BOYDEN. 1952. Agenesis of the right lung. Surgery, 31, 429-435.

TOBIN, C. E. 1952. Methods of preparing and studying human lungs expanded and dried with compressed air. Anat. Rec., 114, 453-465.

WELLS, L. J. 1954. Development of the human diaphragm and pleural sacs. Carneg. Instn. Contr. Embryol., 35, 107-134.

WOODBURNE, R. T. 1947. The costomediastinal border of the left pleura in the procordial area. Anat. Rec., 97, 197-210.

LUNGS AND PLEURA

ALTMAN, P. L., J. F. GIBSON, JR., and C. C. WANG. 1958. *Handbook of Respiration.* xv + 403 pages. D. S. Dittmer and R. M. Greebe, Editors. W. B. Saunders Company, Philadelphia.

BOYDEN, E. A. 1967. Notes on the development of the lung in infancy and early childhood. Am. J. Anat., 121, 749-762.

DRINKER, C. K. 1954. *The Clinical Physiology of the Lungs.* ix + 84 pages. Charles C Thomas, Springfield.

ENGEL, S. 1962. *Lung Structure.* x + 300 pages, 424 figures. Charles C Thomas, Springfield.

FENN, W. O. and H. RAHN, 1965. *Handbook of Physiology—Respiration.* Sec. 3, Vol. 2, viii + pgs. 927-1696, illustrated. American Physiological Society and Williams & Wilkins Company, Baltimore.

HARBORD, R. P. and R. WOOLNER. 1959. *Symposium on Pulmonary Ventilation.* 109 pages, 28 figures. Williams & Wilkins Company, Baltimore.

VON HAYEK, H. 1960. *The Human Lung,* translated by Vernon E. Krahl. xii + 372 pages, 276 figures. Hafner Publishing Company, New York.

HEUCK, F. 1959. *Die Streifenatelektasen der Lunge,* viii + 108 pages, 64 figures. Georg Thieme Verlag, Stuttgart.

HUGHES, G. M. 1963. *Comparative Physiology of Vertebrate Respiration.* xiii + 145 pages, 39 figures. Harvard University Press, Cambridge, Mass.

KRAHL, V. E. 1955. Current concept of the finer structure of the lung. Arch. intern. Med., 96, 342-356.

LACHMAN, E. 1942. A comparison of the posterior boundaries of lungs and pleura as demonstrated on the cadaver and on the roentgenogram of the living. Anat. Rec., 83, 521-542.

LACHMAN, E. 1946. The dynamic concept of thoracic topography: A critical review of present day teaching of visceral anatomy. Amer. J. Roentgenol., 56, 419-440.

McLAUGHLIN, R. F., W. S. TYLER, and R. O. CANADA. 1961. A study of the subgross pulmonary anatomy in various mammals. Amer. J. Anat., 108, 149-165.

BRONCHOPULMONARY SEGMENTS

BOYDEN, E. A. 1953. A critique of the international nomenclature of bronchopulmonary segments. Dis. Chest, 23, 266-269.

BOYDEN, E. A. 1955. *Segmental Anatomy of the Lungs.* xviii + 276 pages, illustrated. The Blakiston Division, McGraw-Hill Book Company, Inc.

JACKSON, C. L. and J. F. HUBER. 1943. Correlated applied anatomy of the bronchial tree and lungs with a system of nomenclature. Dis. Chest, 9, 319-326 + 1 plate.

TOBIN, C. E. and M. O. ZARIQUIEY. 1950. Bronchopulmonary segments and blood supply of the human lung. Med. Radiogr. Photogr., 26, 38-45.

BLOOD VESSELS, NERVES, AND LYMPHATICS

ALEXANDER, H. L. 1933. The autonomic control of the heart, lungs, and bronchi. Ann. intern Med., 6, 1033-1043.

FINDLAY, C. W., JR. and H. C. MAIER. 1951. Anomalies of the pulmonary vessels and their surgical significance. Surg., 29, 604-641.

KOHN, K. and M. RICHTER. 1958. *Die Lungenarterienbahn bei angeborenen Herzfehlern.* viii + 112 pages. Georg Thieme Verlag, Stuttgart.

LARSELL, O. 1922. The ganglia, plexuses, and nerve-terminations of the mammalian lung and pleura pulmonalis. J. comp. Neurol., 35, 97-132.

OLIVEROS, L. G. 1959. *Veins of the Lungs.* 280 pages, 44 figures + plates. Universidad de Salamanca, Salamanca, Spain.

SIMER, P. H. 1952. Drainage of pleural lymphatics. Anat. Rec., 113, 269-283.

SPENCER, H. and D. LOEF. 1964. The innervation of the human lung. J. Anat. (Lond.), 98, 599-609.

TOBIN, C. E. 1952. The bronchial arteries and their connections with other vessels in the human lung. Surg. Gynec. Obstet., 95, 741-750.

TOBIN, C. E. 1954. Lymphatics of the pulmonary alveoli. Anat. Rec., 120, 625-635.

16 | *The Digestive System*

he **Digestive System** (*apparatus digestorius; organs of digestion*), providing the apparatus for the digestion of food, consists of the digestive tube and certain accessory organs.

The **digestive tube** (*alimentary canal*) is a musculomembranous tube, about 9 meters long, extending from the mouth to the anus, and lined throughout its entire extent by mucous membrane. It has received different names in the various parts of its course: at its commencement is the **mouth**, where provision is made for the mechanical division of the food (*mastication*), and for its admixture with a fluid secreted by the salivary glands (*insalivation*); beyond this are the organs of deglutition, the **pharynx** and the **esophagus**, which convey the food into the **stomach**, in which it is stored for a time and in which also the first stages of the digestive process take place; the stomach is followed by the **small intestine**, which is divided for purposes of description into three parts, the **duodenum**, the **jejunum**, and **ileum**. In the small intestine the process of digestion is completed and the resulting products are absorbed into the blood and lacteal vessels. Finally the small intestine ends in the **large intestine**, which is made up of **cecum, colon, rectum**, and **anal canal**, the last terminating on the surface of the body at the **anus**.

The accessory organs are the **teeth**, for purposes of mastication; the three pairs of **salivary glands**—the **parotid, submandibular**, and **sublingual**—the secretion from which mixes with the food in the mouth and converts it into a bolus and acts chemically on one of its constituents; the **liver** and **pancreas**, two large glands in the abdomen, the secretions of which, in addition to that of numerous minute glands in the walls of the alimentary canal, assist in the process of digestion.

The Development of the Digestive Tube

The primitive digestive tube consists of two parts, viz.: (1) the **foregut**, within the cephalic flexure, and dorsal to the heart; and (2) the **hindgut**, within the caudal flexure (Fig. 16–1). Between these is the wide opening of the yolk sac, which is gradually narrowed and reduced to a small foramen leading into the vitelline duct. At first the foregut and hindgut end blindly. The anterior end of the foregut is separated from the stomodeum by the buccopharyngeal membrane (Fig. 16–1); the hindgut ends in the cloaca, which is closed by the cloacal membrane.

The **Mouth.**—The mouth is developed partly from the stomodeum, and partly from the floor of the anterior portion of the foregut. By the growth of the head end of the embryo, and the formation of the cephalic flexure, the pericardial area and the buccopharyngeal membrane come to lie on the ventral surface of the embryo. With the further expansion of the brain, and the forward bulging of the pericardium, the buccopharyngeal membrane is depressed between these two prominences. This depression constitutes the **stomodeum** (Fig. 16–1). It is lined by ectoderm, and is separated from the anterior end of the foregut by the buccopharyngeal membrane. This membrane is devoid of mesoderm, being formed by the apposition of the stomodal ectoderm with the foregut entoderm; at the end of the third week it disappears, and thus a communication is established between mouth and the future pharynx. No trace of the membrane is found in the adult, and the communication just mentioned must not be confused with the permanent isthmus

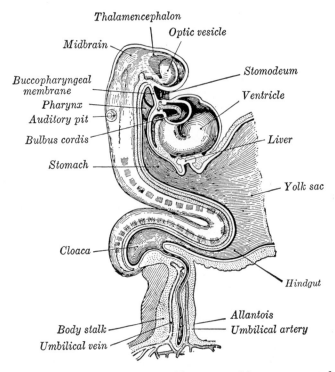

Fig. 16–1.—Human embryo about fifteen days old. Brain and heart represented from right side. Digestive tube and yolk sac in median section. (After His.)

faucium. The lips, teeth, and gums are formed from the walls of the stomodeum, but the tongue is developed in the floor of the pharynx.

The visceral arches extend in a ventral direction between the stomodeum and the pericardium; and with the completion of the mandibular arch and the formation of the maxillary processes, the mouth assumes the appearance of a pentagonal orifice. The orifice is bounded cranially by the fronto-nasal process, caudally by the mandibular arch, and laterally by the maxillary proc-esses (Fig. 16–2). With the inward growth and fusion of the palatine processes (Figs. 2–37, 2–38), the stomodeum is divided into an upper nasal and a lower buccal part. Along the free margins of the processes bounding the mouth cavity a shallow groove appears; this is termed the **primary labial groove,** and from the bottom of it a down-growth of ectoderm takes place into the underlying mesoderm. The central cells of the ectodermal downgrowth degenerate and **a secondary labial groove** is formed; by the

deepening of this, the lips and cheeks are separated from the alveolar processes of the maxillae and mandible.

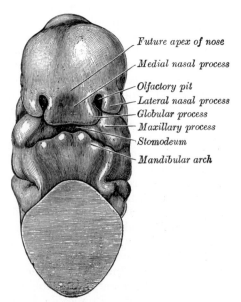

Fig. 16–2.—Head end of human embryo of about thirty to thirty-one days. (From model by Peters.)

The **Salivary Glands.**—The salivary glands arise as buds from the epithelial lining of the mouth; the parotid appears during the fourth week in the angle between the maxillary process and the mandibular arch; the submandibular appears in the sixth week, and the sublingual during the ninth week in the hollow between the tongue and the mandibular arch.

The **Tongue** (Figs. 16–3 to 16–6) is developed in the floor of the pharynx, and consists of an anterior or buccal and a posterior or pharyngeal part which are separated in the adult by the V-shaped sulcus terminalis. During the third week there appears, immediately dorsal to the ventral ends of the two halves of the mandibular arch, a rounded swelling named the **tuberculum impar,** which may help to form the buccal part of the tongue or may be purely a transitory structure. From the ventral ends of the fourth arch there arises a second and larger elevation, in the center of which is a median groove or furrow. This elevation was named by His the **furcula,** and is at first separated from the tuberculum impar by a depression, but later by a ridge, the **copula,** formed by the forward growth and fusion of the ventral ends of the second and third arches. The posterior or pharyngeal part of the tongue is developed from the copula, which

extends forward in the form of a V, so as to embrace between its two limbs the buccal part of the tongue. At the apex of the V a pit-like invagination occurs, to form the thyroid gland, and this depression is represented in the adult by the **foramen cecum** of the tongue. In the adult the union of the anterior and posterior parts of the tongue

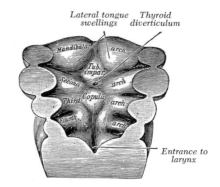

Fɪɢ. 16–4.—Floor of pharynx of human embryo about twenty-six days old. (From model by Peters.)

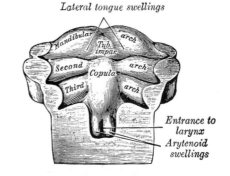

Fɪɢ. 16–5.—Floor of pharynx of human embryo of about the end of the fourth week. (From model by Peters.)

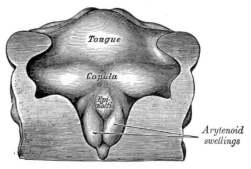

Fɪɢ. 16–6.—Floor of pharynx of human embryo about thirty days old. (From model by Peters.)

Fɪɢ. 16–3.—Same embryo as shown in Figure 16–2, with front wall of pharynx removed.

is marked by the V-shaped **sulcus termi-nalis,** the apex of which is at the foramen cecum, while the two limbs run lateralward and forward, parallel to, but a little behind, the vallate papillae.

The **palatine tonsils** are developed from the dorsal angles of the second branchial pouches. The entoderm which lines these pouches grows in the form of a number of solid buds into the surrounding mesoderm.

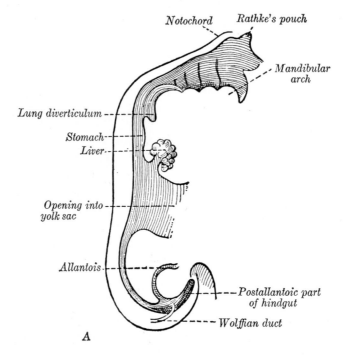

Notochord   *Rathke's pouch*

Mandibular arch

Lung diverticulum

Stomach
Liver

Opening into yolk sac

Allantois

Postallantoic part of hindgut

Wolffian duct

A

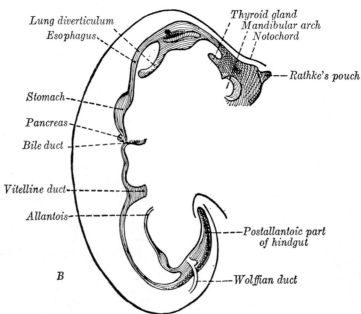

Lung diverticulum
Esophagus

Thyroid gland
Mandibular arch
Notochord

Rathke's pouch

Stomach

Pancreas

Bile duct

Vitelline duct

Allantois

Postallantoic part of hindgut

Wolffian duct

B

Fig. 16–7.—Sketches in profile of two stages in the development of the human digestive tube. (His.) A × 30. B × 20.

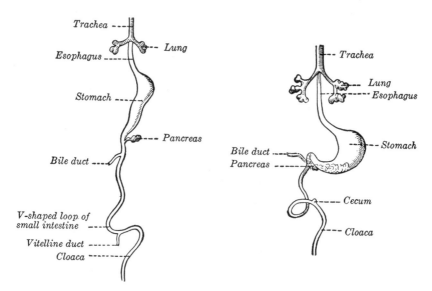

FIG. 16–8.—Ventral view of two successive stages in the development of the digestive tube. (His.)

These buds become hollowed out by the degeneration and casting off of their central cells, and by this means the tonsillar crypts are formed. Lymphoid cells accumulate around the crypts, and become grouped to form the lymphoid follicles; the latter, however, are not well defined until after birth.

The Further Development of the Digestive Tube.—The cranial part of the foregut becomes dilated to form the pharynx (Fig. 16–1), in relation to which the branchial arches are developed (see page 31); the succeeding part remains tubular, and with the descent of the stomach is elongated to form the esophagus. About the fourth week a fusiform dilatation, the future stomach, makes its appearance, and beyond this the gut opens freely into the yolk sac (Fig. 16–7, A and B). The opening is at first wide, but is gradually narrowed into a tubular stalk, the **yolk stalk** or **vitelline duct.** Between the stomach and the mouth of the yolk sac the liver diverticulum appears. From the stomach to the rectum the alimentary canal is attached to the notochord by a band of mesoderm, from which the common mesentery of the gut is subsequently developed.

The stomach has an additional attachment, that to the ventral abdominal wall as far as the umbilicus by the septum transversum. The cranial portion of the septum takes part in the formation of the diaphragm, whereas the caudal portion, into which the liver grows, forms the **ventral mesogastrium** (Fig. 16–10). As the stomach undergoes further dilatation, its two curvatures become recognizable (Figs. 16–7, B and 16–8), the greater toward the vertebral column and the lesser toward the anterior wall of the abdomen, with its two surfaces looking to the right and left respectively. Caudal to the stomach the gut undergoes great elongation, and forms a V-shaped loop which projects ventralward; from the bend or angle of the loop the vitelline duct passes into the umbilicus (Fig. 16–9). For a time a considerable part of the loop extends beyond the abdominal cavity into the umbilical cord, but by the end of the third month it is withdrawn within the cavity. With the lengthening of the tube, the mesoderm which attaches it to the future vertebral column and carries the blood vessels for the supply of the gut, is thinned and drawn out to form the **posterior common mesentery.** The portion of this mesentery attached to the greater curvature of the stomach is named the **dorsal mesogastrium,** and the part which suspends the colon is termed the **mesocolon** (Fig. 16–10).

About the sixth week a diverticulum of the gut appears just caudal to the opening of the vitelline duct, and indicates the future

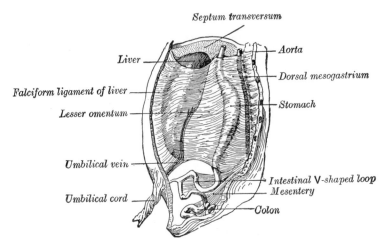

FIG. 16–9.—The primitive mesentery of a six-week-old human embryo, half schematic. (Kollmann.)

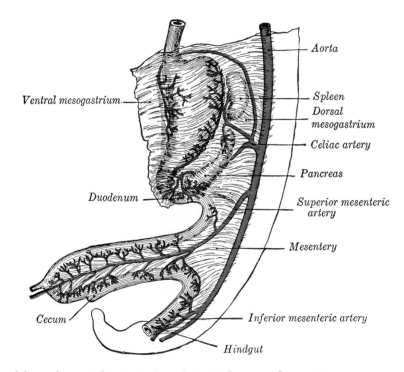

FIG. 16–10.—Abdominal part of digestive tube and its attachment to the primitive or common mesentery.
Human embryo of six weeks. (After Toldt.)

cecum and vermiform appendix. The part of the loop on the distal side of the cecal diverticulum increases in diameter and forms the future ascending and transverse portions of the large intestine. Until the fifth month the cecal diverticulum has a uniform caliber, but from this time onward its distal part remains rudimentary and

forms the vermiform appendix, while its proximal part expands to form the cecum. Changes also take place in the shape and position of the stomach. Its dorsal part or greater curvature, to which the dorsal mesogastrium is attached, grows much more rapidly than its ventral part or lesser curvature to which the ventral mesogastrium is fixed. Also, the greater curvature swings toward the left, so that the right surface of the stomach becomes directed dorsalward and the left surface ventralward (Fig. 16–11), a change in position which explains why the left vagus nerve is found on the ventral, and the right vagus on the dorsal surface of the stomach. The dorsal mesogastrium being attached to the greater curvature must necessarily follow its movements, and hence it becomes greatly elongated and drawn lateralward and ventralward from the vertebral column, and, as in the case of the stomach, the right surfaces of both the dorsal and ventral mesogastria are now directed dorsalward, and the left ventralward.

In this way a pouch, the **lesser sac** (*bursa omentalis*), is formed dorsal to the stomach. This increases in size as the digestive tube undergoes further development, the entrance to the pouch becoming the future **foramen epiploicum** or **foramen of Winslow.** The duodenum is developed from that part of the tube which immediately succeeds the stomach; it undergoes little elongation, being more or less fixed in position by the liver and pancreas, which arise as diverticula from it. The duodenum is first suspended by a mesentery, and projects ventralward

in the form of a loop. The loop and its mesentery are subsequently displaced by the transverse colon, so that the right surface of the duodenal mesentery is directed dorsalward, and, adhering to the parietal peritoneum, is lost. The remainder of the digestive tube becomes greatly elongated, and as a consequence the tube is coiled on itself, and this elongation demands a corresponding increase in the width of the intestinal attachment of the mesentery, which becomes folded.

At this stage the small and large intestines are attached to the vertebral column by a continuous common mesentery, the coils of the small intestine falling to the right of the middle line, the large intestine to the left side* (Fig. 16–11).

The gut is now rotated counterclockwise upon itself, so that the large intestine is carried ventral to the small intestine, and the cecum is placed immediately caudal to the liver. About the sixth month the cecum

*Sometimes this condition persists throughout life, and it is then found that the duodenum does not cross from the right to the left side of the vertebral column, but lies entirely on the right side of the median plane, where it is continued into the jejunum; the arteries to the small intestine (*aa. intestinales*) also arise from the right instead of the left side of the superior mesenteric artery.

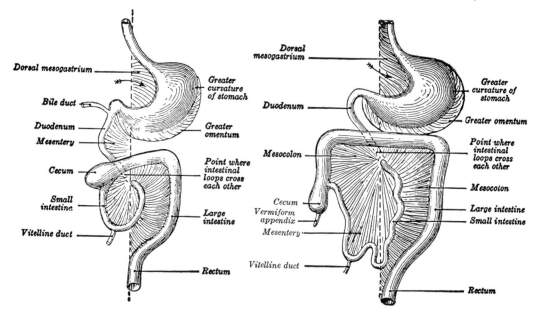

Fig. 16–11.—Diagrams to illustrate two stages in the development of the digestive tube and its mesentery. The arrow indicates the entrance to the bursa omentalis. The ventral mesogastrium has been eliminated.

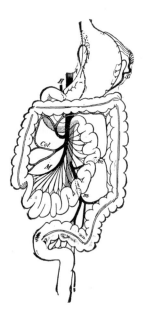

Fig. 16–12.—Final disposition of the intestines and their vascular relations. (Jonnesco.) *A,* Aorta. *H,* Hepatic artery. *M, Col.,* Branches of superior mesenteric artery. *m, m',* Branches of inferior mesenteric artery. *S,* Splenic artery.

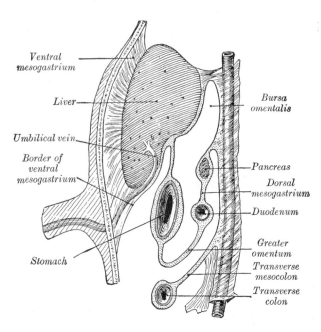

Ventral *mesogastrium*

Liver

Umbilical vein

Border of ventral mesogastrium

Stomach

Bursa omentalis

Pancreas

Dorsal mesogastrium

Duodenum

Greater omentum

Transverse mesocolon

Transverse colon

Fig. 16–13.—Schematic figure of the bursa omentalis, etc. Human embryo of eight weeks. (Kollmann.)

descends into the right iliac fossa, and the large intestine forms an arch consisting of the ascending, transverse, and descending portions of the colon—the transverse portion crossing ventral to the duodenum and lying just caudal to the greater curvature of the stomach, the coils of the small intestine being disposed within this arch (Fig. 16–12). Sometimes the caudalward progress of the cecum is arrested, so that in the adult it may be found lying immediately caudal to the liver instead of in the right iliac region.

Further changes take place in the lesser sac (bursa omentalis) and in the common mesentery, which give rise to the peritoneal relations seen in the adult. The bursa omentalis, which at first reaches only as far as the greater curvature of the stomach, grows caudalward to form the greater omentum, this extension lying ventral to the transverse colon and the coils of the small intestine (Fig. 16–13). Before the pleuroperitoneal opening is closed, the bursa omentalis

sends a diverticulum cranialward on either side of the esophagus; the left diverticulum soon disappears, but the right is constricted off and persists in most adults as a small sac lying within the thorax on the right side of the caudal end of the esophagus.

The ventral layer of the transverse mesocolon is at first distinct from the dorsal layer of the greater omentum, but ultimately the two blend, and hence the greater omentum appears as if attached to the transverse colon (Fig. 16–14). The mesenteries of the ascending and descending parts of the colon disappear in the majority of cases, while that of the small intestine assumes the oblique attachment characteristic of its adult condition.

The lesser omentum is formed, as indicated above, by the mesoderm or **ventral mesogastrium** which attaches the stomach and duodenum to the anterior abdominal wall. The subsequent growth of the liver separates this leaf into two parts, viz., the lesser omentum between the stomach and

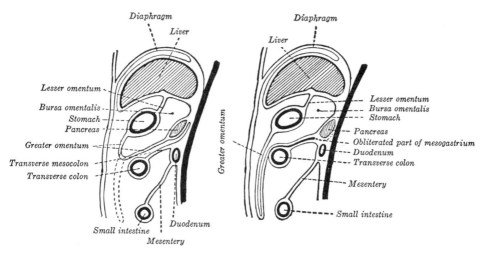

Fig. 16–14.—Diagrams to illustrate the development of the greater omentum and transverse mesocolon.

liver, and the falciform and coronary ligaments between the liver and the abdominal wall and diaphragm (Fig. 16–13).

The **Rectum and Anal Canal.**—The hindgut is at first prolonged caudalward into the body stalk as the tube of the allantois; but, with the growth and flexure of the caudal end of the embryo, the body stalk, with its contained allantoic tube, is carried cranialward to the ventral aspect of the body, and consequently a bend is formed at the junction of the hindgut and allantois. This bend becomes dilated into a pouch, which constitutes the **entodermal cloaca**; the hindgut opens into its dorsal part and the allantois extends out ventrally from its ventral part. At a later stage the mesonephric (Wolffian) and paramesonephric (Müllerian) ducts open into its ventral portion. The cloaca is, for a time, shut off from the exterior by the **cloacal membrane**, formed by the apposition of the ectoderm and entoderm, and reaching, at first, as far cranialward as the future umbilicus. The mesoderm subsequently progresses caudalward to form the lower part of the abdominal wall and symphysis pubis. By the growth of the surrounding tissues the cloacal membrane comes to lie at the bottom of a depression, which is lined by ectoderm and named the **ectodermal cloaca** (Fig. 16–15).

The entodermal cloaca is divided into a dorsal and a ventral part by means of a partition, the **urorectal septum** (Fig. 16–

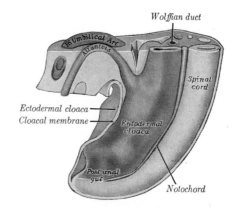

Fig. 16–15.—Tail end of human embryo from fifteen to eighteen days old. (From model by Keibel.)

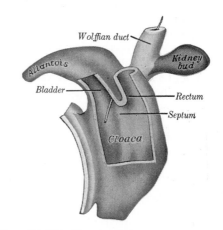

Fig. 16–16.—Cloaca of human embryo from twenty-five to twenty-seven days old. (From model by Keibel.)

16), which grows caudalward from the ridge separating the allantoic from the cloacal opening of the intestine and ultimately fuses with the cloacal membrane and divides it into an anal and a urogenital part. The dorsal part of the cloaca forms the rectum, and the ventral part forms the urogenital sinus and bladder. For a time a communication named the **cloacal duct** exists between the two parts of the cloaca below the urorectal septum; this duct occasionally persists as a passage between the rectum and urethra. The anal canal is formed by an invagination of the ectoderm behind the urorectal septum. This invagination is termed the **proctodeum,** and it meets with the entoderm of the hindgut and with it forms the **anal membrane.** By the absorption of this membrane the anal canal becomes continuous with the rectum (Fig. 17–12). A small part of the hindgut projects caudalward beyond the anal membrane; it is named the **post anal gut** (Fig. 16–15), and usually becomes obliterated.

THE DIGESTIVE TUBE

THE MOUTH

The **cavity of the mouth** (*cavum oris; oral or buccal cavity*) (Fig. 16–17) is placed at the commencement of the digestive tube; it is a nearly oval shaped cavity which consists of two parts: an outer, smaller portion, the **vestibule,** and an inner, larger part, the **mouth cavity proper.**

The **Vestibule** (*vestibulum oris*) is a slit-like space, bounded externally by the lips and cheeks, internally by the gums and teeth. It communicates with the surface of the body by the **rima** or **orifice of the mouth.** Superiorly and inferiorly, it is limited by the reflection of the mucous membrane from the lips and cheeks to the gum covering the upper and lower alveolar arch respectively. It receives the secretion from the parotid salivary glands, and communicates, when the jaws are closed, with the mouth cavity proper by an aperture on either side behind the molar teeth, and by narrow clefts between opposing teeth.

The **Mouth Cavity Proper** (*cavium oris proprium*) (Fig. 16–17) is bounded laterally and ventrally by the alveolar arches with their contained teeth; dorsally, it communicates with the pharynx by a constricted aperture termed the **isthmus faucium.** It is roofed in by the hard and soft palates, while the greater part of the floor is formed by the tongue, the remainder by the reflection of the mucous membrane from the sides and under surface of the tongue to the gum lining the inner aspect of the mandible. It receives the secretion from the submandibular and sublingual salivary glands.

The **Lips** (*labia oris*) (Fig. 16–17), the two fleshy folds which surround the rima or orifice of the mouth, are covered externally by integument and internally by mucous membrane, between which are found the Orbicularis oris muscle, the labial vessels, some nerves, areolar tissue, and fat, and numerous small labial glands. The inner surface of each lip is connected in the middle line to the corresponding gum by a median fold of mucous membrane, the **frenulum.**

The **Cheeks** (*buccae*) form the sides of the face, and are continuous anteriorly with the lips. They are composed externally of integument; internally of mucous membrane; and between the two of a muscular stratum, besides a large quantity of fat (*corpus adiposum buccae*) areolar tissue, vessels, nerves, and buccal glands.

The **labial glands** (*glandulae labiales*) are situated between the mucous membrane and the Orbicularis oris, around the orifice of the mouth. They are globular in form and about the size of small peas; their ducts open by minute orifices upon the mucous membrane. In structure they resemble the salivary glands.

The **buccal glands** are placed between the mucous membrane and Buccinator muscle: they are similar in structure to the labial glands, but smaller. About five, of a larger size than the rest, are placed between the Masseter and Buccinator muscles around the distal extremity of the parotid duct; their ducts open in the mouth opposite the last molar tooth. They are called **molar glands.**

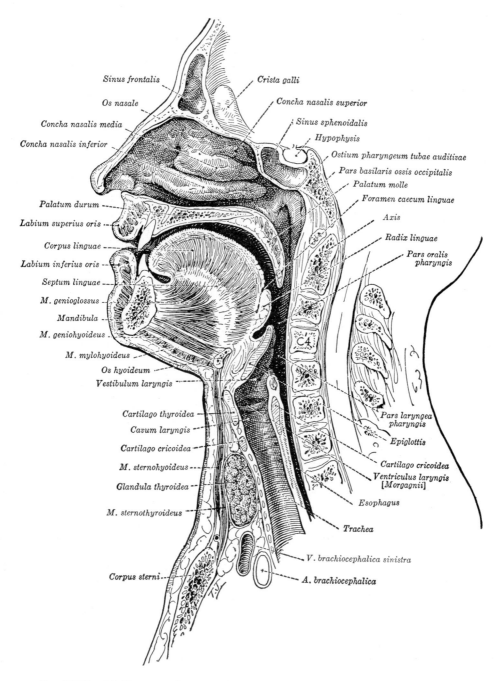

Fig. 16–17.—Median sagittal section of head and neck showing nasal, pharyngeal, and laryngeal cavities. (Eycleshymer and Jones.)

Structure.—The **mucous membrane** lining the cheek is reflected above and below upon the gums, and is continuous behind with the lining membrane of the soft palate. Opposite the second molar tooth of the maxilla is a papilla, on the summit of which is the aperture of the parotid duct (*papilla parotidea*). The principal muscle of the cheek is the Buccinator; but other muscles enter into its formation, viz., the Zygomaticus, Risorius, and Platysma.

The **Gums** (*gingivae*) are composed of dense fibrous tissue, closely connected to the periosteum of the alveolar processes, and surrounding the necks of the teeth. They are covered by smooth and vascular mucous membrane, which is remarkable for its limited sensibility. Around the necks of the teeth this membrane presents numerous fine papillae, and is reflected into the alveoli, where it is continuous with the periosteal membrane lining these cavities.

The **Palate** (*palatum*) forms the roof of the mouth; it consists of two portions, the **hard palate** anteriorly, the **soft palate** posteriorly.

The **Hard Palate** (*palatum durum*) (Fig. 16–17) forms the roof of the mouth and separates the oral and nasal cavities. It is bounded anteriorly and at the sides by the alveolar arches and gums; posteriorly, it is continuous with the soft palate. Its bony support (*palatum osseum*), formed by the palatine process of the maxilla and the horizontal part of the palatine bone (Fig. 16–19), is covered by a dense structure, formed by the periosteum and mucous membrane of the mouth, which are intimately adherent. Along the middle line is a linear raphe, which ends anteriorly in a small papilla (*papilla incisiva*) corresponding with the incisive canal. On either side and anterior to the raphe the mucous membrane is thick, pale in color, and corrugated; posteriorly, it is thin, smooth, and of a deeper color; it is covered with stratified squamous epithelium, and furnished with numerous palatal glands, which lie between the mucous membrane and the surface of the bone.

The **Soft Palate** (*palatum molle*) (Fig. 16–17) is suspended from the posterior border of the hard palate. It consists of a fold of mucous membrane enclosing muscular fibers, an aponeurosis, vessels, nerves, lymphoid tissue, and mucous glands. These

are described on page 1189. When elevated, as in swallowing and in sucking, it completely separates the nasal cavity and nasopharynx from the posterior part of the oral cavity and the oral portion of the pharynx (Fig. 16–17). When occupying its usual position, *i.e.*, relaxed and pendent, its anterior surface is concave, continuous with the roof of the mouth, and marked by a median raphe. Its posterior surface is convex, and continuous with the mucous membrane covering the floor of the nasal cavities. It is attached to the posterior margin of the hard palate, and its sides are blended with the pharynx. Its posterior border is free and hangs like a curtain between the mouth and pharynx, termed the **palatine velum** (*velum palatinum*).

Hanging from the middle of its posterior border is a small, conical, pendulous process, the **palatine uvula.** Arching lateralward from the base of the uvula on either side are two curved folds of mucous membrane, containing muscular fibers, called the **arches** or **pillars of the fauces** (Fig. 16–42).

THE TEETH

The **Teeth** (*dentes*) (Figs. 16–18 to 16–24). Two sets of teeth make their appearance at different periods of life. Those of the first set appear in infancy, and are called the **deciduous** or **milk teeth.** Those of the second set appear in childhood, continue until old age, and are named **permanent teeth.**

The **deciduous teeth** are twenty in number: four incisors, two canines, and four molars, in each jaw.

The **permanent teeth** are thirty-two in number: four incisors, two canines, four premolars, and six molars, in each jaw.

The **dental formulae** may be represented as follows:

General Characteristics.—Each tooth consists of three portions: the **crown,** projecting above the gum; the **root,** embedded in the alveolus; and the **neck,** the constricted portion between the crown and root (Fig. 16–24).

The **roots** of the teeth are firmly implanted in depressions within the alveoli; these depressions are lined with periosteum which invests the tooth as far as the neck.

Deciduous Teeth

		mol.	can.	in.	in.	can.	mol.	
Upper jaw	. . . . .	2	1	2	2	1	2	} Total 20
Lower jaw	. . . . .	2	1	2	2	1	2	

Permanent Teeth

		mol.	pr. mol.	can.	in.	in.	can.	pr. mol.	mol.	
Upper jaw	. . . . .	3	2	1	2	2	1	2	3	} Total 32
Lower jaw	. . . . .	3	2	1	2	2	1	2	3	

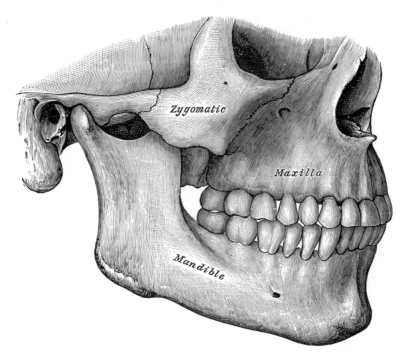

FIG. 16–18.—Side view of the teeth and jaws.

At the margins of the alveoli, the periosteum is continuous with the fibrous structure of the gums.

In consequence of the curve of the dental arch, terms such as anterior and posterior, as applied to the teeth, are misleading and confusing. Special terms are therefore used to indicate the different surfaces of a tooth: the surface directed toward the lips or cheek is known as the **labial** or **buccal surface;** that directed toward the tongue is described as the **lingual surface;** those surfaces which touch neighboring teeth are termed **surfaces of contact.** In the case of the incisor and canine teeth the surfaces of contact are medial and lateral; in the premolar and molar teeth they are anterior and posterior.

The superior dental arch is larger than the inferior, so that in the normal condition the teeth in the maxillae slightly overlap those of the mandible both anteriorly and at the sides. Since the upper central incisors are wider than the lower, the other teeth in the upper arch are thrown somewhat posteriorly, and the two sets do not quite

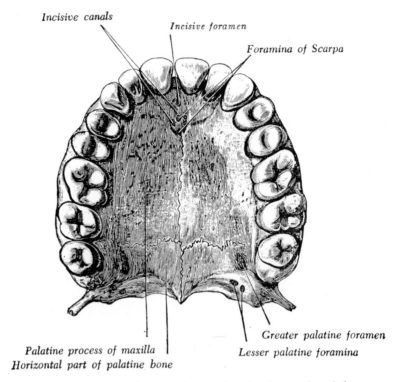

Incisive canals

Incisive foramen

Foramina of Scarpa

Greater palatine foramen

Lesser palatine foramina

Palatine process of maxilla
Horizontal part of palatine bone

FIG. 16–19.—Permanent teeth of upper dental arch, seen from below.

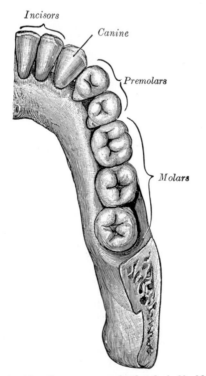

Incisors

Canine

Premolars

Molars

FIG. 16–20.—Permanent teeth of right half of lower dental arch, seen from above.

correspond to each other when the mouth is closed: thus the upper canine tooth rests partly on the lower canine and partly on the first premolar, and the cusps of the upper molar teeth lie behind the corresponding cusps of the lower molar teeth. The two series, however, end at nearly the same point posteriorly; this is mainly because the molars in the upper arch are the smaller.

The Permanent Teeth (*dentes permanentes*) (Figs. 16–20, 16–21).—The **Incisors** (*dentes incisivi; incisive or cutting teeth*) are so named from their presenting a sharp cutting edge, adapted for biting the food. They are eight in number, and form the four front teeth in each dental arch.

The **crown** is directed vertically, and is chisel-shaped, being bevelled at the expense of its lingual surface, so as to present a sharp horizontal cutting edge, which, before being subjected to attrition, presents three small prominent points separated by two slight notches. It is convex, smooth, and highly polished on its labial surface; concave on its lingual surface, where, in the teeth of the upper arch, it is frequently marked by an inverted V-shaped eminence, situated near the gum. This is known as the **basal ridge** or **cingulum.** The **neck** is constricted. The **root** is long, single, conical, transversely flattened, thicker anteriorly, and slightly grooved on either side in the longitudinal direction.

The **upper incisors** are larger and stronger than the lower, and are directed obliquely downward and forward. The central ones are larger than the lateral, and their roots are more rounded.

jects beyond the level of the other teeth. The **root** is single, but longer and thicker than that of the incisors, conical in form, compressed laterally, and marked by a slight groove on each side.

The **upper canine teeth** (popularly called *eye teeth*) are larger and longer than the lower, and usually present a distinct basal ridge.

The **lower canine teeth** (popularly called *stomach teeth*) are placed nearer the middle line than the upper, so that their summits

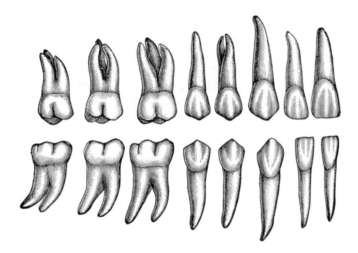

Fig. 16–21.—Permanent teeth. Right side. (Burchard.)

The **lower incisors** are smaller than the upper; the central ones are smaller than the lateral, and are the smallest of all the incisors. They are placed vertically and are somewhat bevelled anteriorly, where they have been worn down by contact with the overlapping edge of the upper teeth. The cingulum is absent.

The **Canine Teeth** (*dentes canini*) are four in number, two in the upper, and two in the lower arch, one being placed lateral to each lateral incisor. They are larger and stronger than the incisors, and their roots sink deeply into the bones, and cause well-marked prominences upon the surface of the alveolar arch.

The **crown** is large and conical, very convex on its labial surface, a little hollowed and uneven on its lingual surface, and tapering to a blunted point or cusp, which pro-

correspond to the intervals between the upper canines and the lateral incisors.

The **Premolars** or **Bicuspid teeth** (*dentes premolares*) are eight in number, four in each arch. They are situated lateral to and posterior to the canine teeth, and are smaller and shorter than the canine teeth.

The **crown** is compressed anteroposteriorly, and surmounted by two pyramidal eminences or cusps, a labial and a lingual, separated by a groove; hence their name **bicuspid.** Of the two cusps the labial is the larger and more prominent. The **neck** is oval. The **root** is generally single, compressed, and presents anteriorly and posteriorly a deep groove, which indicates a tendency in the root to become double. The apex is generally bifid.

The **upper premolars** are larger, and present a greater tendency to the division

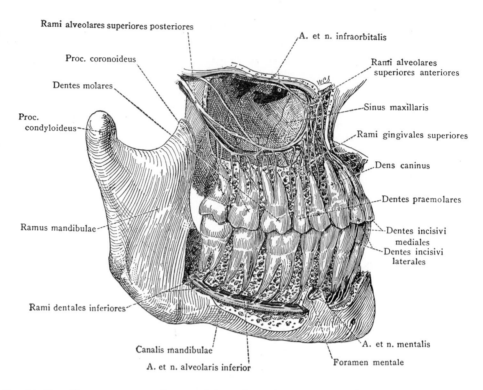

Rami alveolares superiores posteriores

Proc. coronoideus

Dentes molares

Proc.
condyloideus

Ramus mandibulae

Rami dentales inferiores

Canalis mandibulae

A. et n. alveolaris inferior

A. et n. infraorbitalis

Rami alveolares
superiores anteriores

Sinus maxillaris

Rami gingivales superiores

Dens caninus

Dentes praemolares

Dentes incisivi
mediales
Dentes incisivi
laterales

A. et n. mentalis

Foramen mentale

FIG. 16–22.—The permanent teeth, viewed from the right. The external layer of bone has been partly removed and the maxillary sinus has been opened to show the blood and nerve supply to the teeth. (Eycleshymer and Jones.)

of their roots than the lower; this is especially the case in the first upper premolar.

The **Molar Teeth** (*dentes molares*) are the largest of the permanent set, and their broad crowns are adapted for grinding and crushing the food. They are twelve in number; six in each arch, three being placed posterior to the second premolars.

The **crown** of each is nearly cubical in form, convex on its buccal and lingual surfaces, flattened on its surfaces of contact; it is surmounted by four or five tubercles, or cusps, separated from each other by a cruciate depression; hence the molars are sometimes termed **multicuspids**. The **neck** is distinct, large, and rounded.

The **Upper Molars.**—As a rule the first is the largest, and the third the smallest of the upper molars. The crown of the first has usually four tubercles; that of the second, three or four; that of the third, three. Each upper molar has three roots, and of these two are buccal and nearly parallel to each other; the third is lingual and diverges

from the others. The roots of the third molar (*dens serotinus* or *wisdom tooth*) are more or less fused together.

The **Lower Molars.**—The lower molars are larger than the upper. On the crown of the first there are usually five tubercles; on those of the second and third, four or five. Each lower molar has two roots, an anterior, nearly vertical, and a posterior, directed obliquely backward; both roots are grooved longitudinally, indicating a tendency to division. The two roots of the third molar (*dens serotinus* or *wisdom tooth*) are more or less united.

The **Deciduous Teeth** (*dentes decidui; temporary or milk teeth*) (Fig. 16–23).— The deciduous are smaller than, but, generally speaking, resemble in form, the teeth which bear the same names in the permanent set. The posterior of the two molars is the largest of all the deciduous teeth, and is succeeded by the first molar. The first upper molar has only three cusps—two labial, one lingual; the second upper molar has four cusps. The first lower molar has four cusps; the second lower molar has five. The roots of the deciduous molars are smaller and more divergent than those of the permanent molars, but in other respects bear a strong resemblance to them.

Structure of the Teeth.—In a vertical section of a tooth (Figs. 16–24, 16–25), the **pulp**

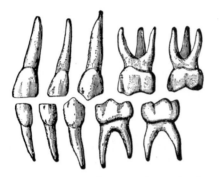

FIG. 16–23.—Deciduous teeth. Left side.

cavity is seen in the interior of the crown and the center of each root; it opens by a minute orifice at the extremity of the latter. It contains the **dental pulp,** a loose connective tissue richly supplied with vessels and nerves, which enter the cavity through the small aperture, the *foramen apicis dentis,* at the point of each root. Some of the cells of the pulp are arranged as a layer on the wall of the pulp cavity; they are named the **odontoblasts,** and during the development of the tooth, are columnar in shape, but later on, after the dentin is fully formed, they become flattened and resemble osteoblasts. Each has two fine processes, the outer one passing into a dental canaliculus, the inner being continuous with the processes of the connective tissue cells of the pulp matrix.

The solid portion of the tooth consists of (1) the **ivory** or **dentin,** which forms the bulk of the tooth; (2) the **enamel,** which covers the exposed part of the crown; and (3) a thin layer of bone, the **cement** or **crusta petrosa,** which is disposed on the surface of the root.

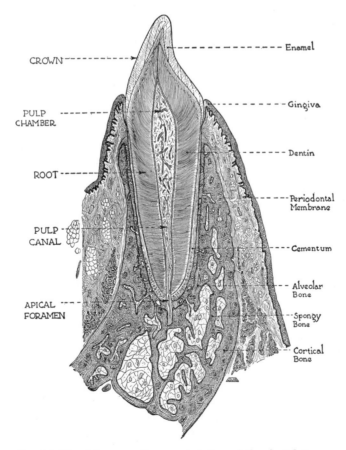

FIG. 16–24.—Diagrammatic representation of the dental tissue. (Schour, *Noyes' Oral Histology and Embryology.*)

The **dentin** (*dentinum; substantia eburnea; ivory*) (Fig. 16–24) forms the principal mass of a tooth. It is a modification of osseous tissue, from which it differs, however, in structure. Microscopically it consists of a number of minute wavy and branching tubes, the **dental canaliculi,** embedded in a dense homogeneous substance, the **matrix.**

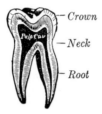

FIG. 16–25.—Vertical section of a molar tooth.

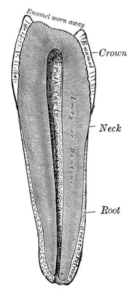

FIG. 16–26.—Vertical section of a premolar tooth. (Magnified.)

The **dental canaliculi** (*dentinal tubules*) (Fig. 16–27) are placed parallel with one another, and open at their inner ends into the pulp cavity. In their course to the periphery they present two or three curves, and are twisted on themselves in a spiral direction. These canaliculi vary in direction: thus in a tooth of the mandible they are vertical in the upper portion of the crown, becoming oblique and then horizontal in the neck and upper part of the root, while toward the lower part of the root they are inclined downward. In their

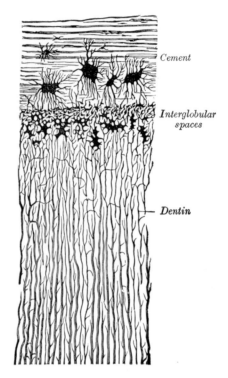

FIG. 16–27.—Transverse section of a portion of the root of a canine tooth. × 300.

course they divide and subdivide dichotomously, and, especially in the root, give off minute branches, which join together in loops in the matrix, or end blindly. Near the periphery of the dentin, the finer ramifications of the canaliculi terminate imperceptibly by free ends. The dental canaliculi have definite walls, consisting of an elastic homogeneous membrane, the **dentinal sheath** (*of Neumann*), which resists the action of acids; they contain slender cylindrical prolongations of the odontoblasts, the **dentinal fibers** (*Tomes' fibers*).

The **matrix** (*intertubular dentin*) is translucent, and contains the chief part of the inorganic matter of the dentin. In it are a number of fine fibrils, which are continuous with the fibrils of the dental pulp. After the organic matter has been removed by steeping a tooth in weak acid, the remaining organic matter may be torn into laminae which run parallel with the pulp cavity, across the direction of the tubes. A section of dry dentin often displays a series of somewhat parallel lines—the **incremental lines** (*of Salter*). These lines are composed of imperfectly calcified dentin arranged in layers. In consequence of the imperfection in the calcifying process, little irregular cavities are left, termed **interglobular spaces** (*spatia interglobu-*

laria) (Fig. 16–17). Normally a series of these spaces is found toward the outer surface of the dentin, where they form a layer which is sometimes known as the **granular layer**. They have received their name from the fact that they are surrounded by minute nodules or globules of dentin. Other curved lines may be seen parallel to the surface. These are the *lines of Schreger,* and are due to the optical effect of simultaneous curvature of the dentinal fibers.

The **enamel** (*substantia adamantina*) is the hardest and most compact part of the tooth, and forms a thin crust over the exposed part of the crown, as far as the commencement of the root. It is thickest on the grinding surface of the crown, until worn away by attrition, and becomes thinner toward the neck. It consists of minute hexagonal rods or columns termed **enamel fibers** or **enamel prisms** (*prismata adamantina*). They lie parallel with one another, resting by one extremity upon the dentin, which presents a number of minute depressions for their reception, and forming the free surface of the crown by the other extremity. The columns are directed vertically on the summit of the crown, horizontally at the sides; they are about 4 μ in diameter, and pursue a more or less wavy course. Each column is a six-sided prism and presents numerous dark transverse shadings; these shadings are probably due to the manner in which the columns are developed in successive stages, producing shallow constrictions, as will be subsequently explained. Another series of lines, having a brown appearance, the **parallel striae** or **colored lines** (*of Retzius*), is seen on section. According to Ebner, they are produced by air in the interprismatic spaces; others believe that they are the result of true pigmentation. Numerous minute interstices intervene between the enamel fibers near their dentinal ends, a provision calculated to allow the permeation of fluids from the dental canaliculi into the substance of the enamel.

The **crusta petrosa** or **cement** (*substantia ossea*) is disposed as a thin layer on the roots of the teeth, from the termination of the enamel to the apex of each root, where it is usually very thick. In structure and chemical composition it resembles bone. It contains, sparingly, the lacunae and canaliculi which characterize true bone; the lacunae placed near the surface receive the canaliculi radiating from the side of the lacunae toward the periodontal membrane; and those more deeply placed join with the adjacent dental canaliculi. In the thicker portions of the crusta petrosa, the lamellae and Haversian canals peculiar to bone are also found.

As age advances, the cement increases in thickness, and gives rise to those bony growths or exostoses so common in the teeth of the aged; the pulp cavity also becomes partially filled up by a hard substance, intermediate in structure between dentin and bone (*osteodentin*, Owen; *secondary dentin*, Tomes). It appears to be formed by a slow conversion of the dental pulp, which shrinks, or even disappears.

Development of the Teeth

In describing the development of the teeth, the mode of formation of the deciduous teeth must first be considered, and then that of the permanent series (Figs. 16–28 to 16–31).

Development of the Deciduous Teeth.—The development of the deciduous teeth begins

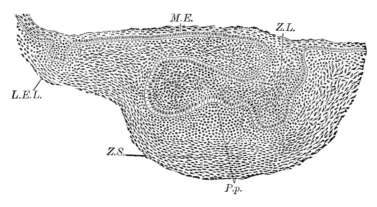

FIG. 16–28.—Sagittal section through the first lower deciduous molar of a human embryo 30 mm long. (Röse.) × 100. *L.E.L.*, Labiodental lamina, here separated from the dental lamina. *Z.L.*, Placed over the shallow dental furrow, points to the dental lamina, which is spread out below to form the enamel germ of the future tooth. *P.p.*, Bicuspid papilla, capped by the enamel germ. *Z.S.*, Condensed tissue forming dental sac. *M.E.*, Mouth epithelium.

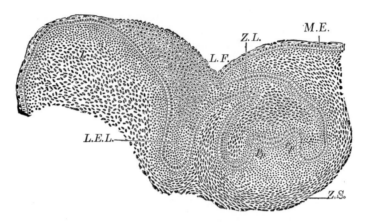

Fig. 16–29.—Similar section through the canine tooth of an embryo 40 mm long. (Röse.) × 100. *L.F.*, Labiodental furrow. The other lettering as in Fig. 16–28.

about the sixth week of fetal life as a thickening of the epithelium along the line of the future jaw, the thickening being due to a rapid multiplication of the more deeply situated epithelial cells. As the cells multiply they extend into the subjacent mesoderm, and thus form a ridge or strand of cells embedded in mesoderm.

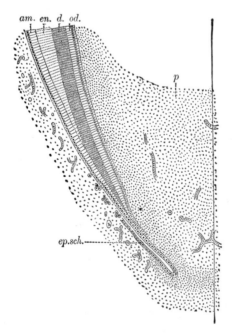

Fig. 16–30.—Longitudinal section of the lower part of a growing tooth, showing the extension of the layer of ameloblasts beyond the crown to mark off the limit of formation of the dentin of the root. (Röse.) *am.*, Ameloblasts, continuous below with *ep. sch.*, the epithelial sheath of Hertwig. *d.*, Dentin. *en.*, Enamel. *od.*, Odontoblasts. *p.*, Pulp.

About the seventh week a longitudinal splitting or cleavage of this strand of cells takes place, and it becomes divided into two strands; the separation begins anteriorly and extends laterally, the process occupying four or five weeks. Of the two strands thus formed, the **labial** forms the **labiodental lamina**; while the other, the **lingual**, is the ridge of cells in connection with which the teeth, both deciduous and permanent, are developed. Hence it is known as the **dental lamina** or **common dental germ**. It forms a flat band of cells, which grows into the substance of the embryonic jaw, at first horizontally inward, and then, as the teeth develop, vertically, *i.e.*, upward in the upper jaw, and downward in the lower jaw. While still maintaining a horizontal direction it has two edges—an *attached edge*, continuous with the epithelium lining the mouth, and a *free edge*, projecting inward, and embedded in the mesodermal tissue of the embryonic jaw. Along its line of attachment to the buccal epithelium is a shallow groove, the **dental furrow**.

About the ninth week the dental lamina begins to develop enlargements along its free border. These are ten in number in each jaw, and each corresponds to a future deciduous tooth. They consist of masses of epithelial cells; and the cells of the deeper part—that is, the part farthest from the margin of the jaw—increase rapidly and spread out in all directions. Each mass thus comes to assume a club shape, connected with the general epithelial lining of the mouth by a narrow neck, embraced by mesoderm. They are now known as **special dental germs** (Fig. 16–28). After a time the lower expanded portion inclines outward, so as to form an angle with the superficial constricted portion, which is sometimes known as the neck of the special dental germ. About the tenth week the mesodermal tissue deep to these special dental germs becomes differentiated into papillae; these grow upward, and come in contact with the epithelial cells of the special dental germs, which become folded over them like a hood or cap. There is, then, at this stage a papilla (or papillae) which has already begun to assume somewhat the shape of the crown of the future tooth, and from which the dentin and pulp of the tooth are formed, surmounted by a dome or cap of epithelial cells from which the enamel is derived.

nent teeth—*i.e.*, the ten anterior ones in each jaw. Here the same process goes on as has been described in connection with those of the deciduous teeth: that is, they recede into the substance of the gum behind the germs of the deciduous teeth. As they recede they become club-shaped, form expansions at their distal extremities, and finally meet papillae, which have been formed in the mesoderm, just in the same manner as was the case in the deciduous teeth. The apex of each papilla indents the dental germ, which encloses it, and, forming a cap for it, becomes converted into the enamel,

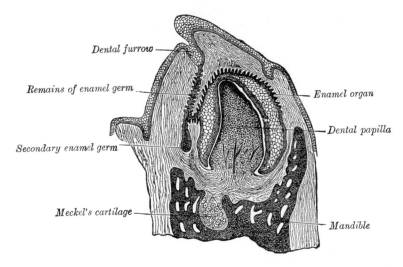

FIG. 16–31.—Vertical section of the mandible of an early human fetus. × 25.

In the meantime, while these changes have been going on, the dental lamina has been extending posteriorly behind the special dental germ corresponding to the second deciduous molar tooth, and at about the seventeenth week it presents an enlargement, the *special dental germ*, for the first permanent molar, soon followed by the formation of a papilla in the mesodermal tissue for the same tooth. This is followed, about the sixth month after birth, by a further extension posteriorly of the dental lamina, with the formation of another enlargement and its corresponding papilla for the second molar. And finally the process is repeated for the third molar, its papilla appearing about the fifth year of life.

After the formation of the special dental germs, the dental lamina undergoes atrophic changes and becomes cribriform, except on the lingual and lateral aspects of each of the special germs of the temporary teeth, where it undergoes a local thickening forming the special dental germ of each of the successional perma-

while the papilla forms the dentin and pulp of the permanent tooth.

The **special dental germs** consist at first of rounded or polyhedral epithelial cells: after the formation of the papillae, these cells undergo a differentiation into three layers. Those which are in immediate contact with the papilla become elongated, and form a layer of well-marked columnar epithelium coating the papilla. They are the cells which form the enamel fibers, and are therefore termed **enamel cells** or **ameloblasts** (Fig. 16–30). The cells of the outer layer of the special dental germ, which are in contact with the inner surface of the dental sac, presently to be described, are much shorter, cubical in form, and are named the **external enamel epithelium**. All the intermediate round cells of the dental germ between these two layers undergo a peculiar change. They become stellate in shape and develop processes, which unite to form a network into which fluid is secreted; this has the appearance of a jelly, and to it the name of enamel pulp

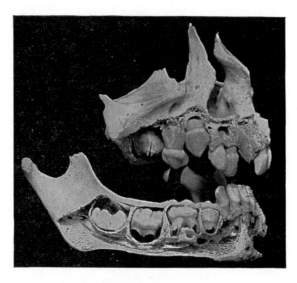

Fig. 16–32.—Maxilla and mandible at about one year. (Noyes.)

is given. This transformed special dental germ is now known under the name of **enamel organ** (Fig. 16–31).

While these changes are going on, a sac is formed around each enamel organ from the surrounding mesodermal tissue. This is known as the **dental sac,** and is a vascular membrane of connective tissue. It grows up from below, and thus encloses the whole tooth germ; as it grows it causes the neck of the enamel organ to atrophy and disappear, so that all communication between the enamel organ and the superficial epithelium is cut off. At this stage there are vascular papillae surmounted by caps of epithelial cells, the whole being surrounded by membranous sacs.

Formation of the Enamel (Fig. 16–30).—The enamel is formed exclusively from the enamel cells or ameloblasts of the special dental germ, either by direct calcification of the columnar cells, which become elongated into the hexagonal rods of the enamel; or, as is more generally believed, as a secretion from the ameloblasts, within which calcareous matter is subsequently deposited.

The process begins at the apex of each cusp, at the ends of the enamel cells in contact with the dental papilla. Here a fine globular deposit takes place, being apparently shed from the end of the ameloblasts. It is known by the name of the **enamel droplet,** and resembles keratin in its resistance to the action of mineral acids. This droplet then becomes fibrous and calcifies and forms the first layer of the enamel; a second droplet now appears and calcifies, and

so on; successive droplets of keratin-like material are shed from the ameloblasts and form successive layers of enamel, the ameloblasts gradually receding as each layer is produced, until at the termination of the process they have almost disappeared.

The intermediate cells of the enamel pulp atrophy and disappear, so that the newly formed calcified material and the external enamel epithelium come into apposition. This latter layer, however, soon disappears on the emergence of the tooth beyond the gum. After its disappearance the crown of the tooth is still covered by a distinct membrane, which persists for some time. This is known as the **cuticula dentis,** or **Nasmyth's membrane,** and is believed to be the last-formed layer of enamel derived from the ameloblasts, which has not become calcified. It forms a horny layer, which may be separated from the subjacent calcified mass by the action of strong acids. It is marked by the hexagonal impressions of the enamel prisms, and, when stained by nitrate of silver, shows the characteristic appearance of epithelium.

Formation of the Dentin (Fig. 16–30).— While these changes are taking place in the epithelium to form the enamel, contemporaneous changes occurring in the differentiated mesoderm of the dental papillae result in the formation of the dentin. As before stated, the first germs of the dentin are the papillae, corresponding in number to the teeth, formed from the soft mesodermal tissue which bounds the depressions containing the special enamel germs.

The papillae grow upward into the enamel germs and become covered by them, both being enclosed in a vascular connective tissue, the **dental sac,** in the manner above described. Each papilla then constitutes the formative pulp from which the dentin and permanent pulp are developed; it consists of rounded cells and is very vascular, and soon begins to assume the shape of the future tooth. The next step is the appearance of the **odontoblasts,** which have a relation to the development of the teeth similar to that of the osteoblasts to the formation of bone. They are formed from the cells of the periphery of the papilla—that is to say, from the cells in immediate contact with the ameloblasts of the special dental germ. These cells become elongated, one end of the elongated cell resting against the epithelium of the special dental germs, the other being tapered and often branched.

By the direct transformation of the peripheral ends of these cells, or by a secretion from them, a layer of uncalcified matrix (**prodentin**) is formed which caps the cusp or cusps, if there are more than one, of the papillae. This matrix becomes fibrillated, and in it islets of calcification make their appearance, and coalescing give rise to a continuous layer of calcified material which covers each cusp and constitutes the first layer of dentin. The odontoblasts, having thus formed the first layer, retire toward the center of the papilla, and, as they do so, produce successive layers of dentin from their peripheral extremities—that is to say, they form the dentinal matrix in which calcification subsequently takes place. As they thus recede from the periphery of the papilla, they leave behind them filamentous processes of cell protoplasm, provided with finer side processes; these are surrounded by calcified material, and thus form the dental canaliculi, and, by their side branches, the anastomosing canaliculi: the processes of protoplasm contained within them constitute the **dentinal fibers** (*Tomes' fibers*).

In this way the entire thickness of the dentin is developed, each canaliculus being completed throughout its whole length by a single odontoblast. The central part of the papilla does not undergo calcification, but persists as the pulp of the tooth. In this process of formation of dentin it has been shown that an uncalcified matrix is first developed, and that in this matrix islets of calcification appear which subsequently blend together to form a cap to each cusp: in like manner, successive layers are produced which ultimately become blended with each other. In certain places this blending is not complete, portions of the matrix remaining uncalcified

between the successive layers; this gives rise to little spaces, which are the interglobular spaces alluded to above.

Formation of the Cement.—The root of the tooth begins to be formed shortly before the crown emerges through the gum, but is not completed until some time afterward. It is produced by a downgrowth of the epithelium of the dental germ, which extends almost as far as the situation of the apex of the future root, and determines the form of this portion of the tooth. This fold of epithelium is known as the **epithelial sheath,** and on its papillary surface odontoblasts appear, which in turn form dentin, so that the dentin formation is identical in the crown and root of the tooth. After the dentin of the root has been developed, the vascular tissues of the dental sac begin to break through the epithelial sheath, and spread over the surface of the root as a layer of bone-forming material. In this osteoblasts make their appearance, and the process of ossification goes on in identically the same manner as in the ordinary intramembranous ossification of bone. In this way the cement is formed, and consists of ordinary bone containing canaliculi and lacunae.

Formation of the Alveoli.—About the fourteenth week of embryonic life the dental lamina becomes enclosed in a trough or groove of mesodermal tissue, which at first is common to all the dental germs, but subsequently becomes divided by bony septa into loculi, each loculus containing the special dental germ of a deciduous tooth and its corresponding permanent tooth. After birth each cavity becomes subdivided, so as to form separate loculi (the future alveoli) for the deciduous tooth and its corresponding permanent tooth. Although at one time the whole of the growing tooth is contained in the cavity of the alveolus, the latter never completely encloses it, since there is always an aperture over the top of the crown filled by soft tissue, by which the dental sac is connected with the surface of the gum, and which in the permanent teeth is called the **gubernaculum dentis.**

Development of the Permanent Teeth.—The permanent teeth develop in two sets: (1) those which replace the deciduous teeth, and which, like them, are ten in number in each jaw: these are the **successional permanent teeth;** and (2) those which have no deciduous predecessors, but are added to the temporary dental series. These are three in number on either side in each jaw, and are termed **superadded permanent teeth.** They are the three molars of the permanent set, the molars of the deciduous set being replaced by the premolars of the permanent set. The development of the successional permanent teeth—the ten anterior ones in either jaw—has already been indicated. During their development the permanent teeth, enclosed in their sacs, come to be placed on the lingual side of the deciduous teeth and more distant from the margin of the future gum, and, as already stated, are separated from them by bony partitions. As the crown of the permanent tooth grows, absorption of these bony partitions and of the root of the deciduous tooth takes place, and finally nothing but the crown of the deciduous tooth remains. This is shed or removed, and the permanent tooth takes its place.

The superadded permanent teeth are developed in the manner already described, by extensions of the posterior part of the dental lamina in each jaw.

Eruption of the Teeth.—When the calcification of the different tissues of the tooth is sufficiently advanced to enable it to bear the pressure to which it will be afterward subjected, eruption takes place, the tooth making its way through the gum. The gum is absorbed by the pressure of the crown of the tooth against it, which is itself pressed toward the surface by the increasing size of the root. At the same time the septa between the dental sacs ossify, and constitute the alveoli; these firmly embrace the necks of the teeth, and afford them a solid basis of support.

The eruption of the deciduous teeth commences about the seventh month after birth, and is completed about the end of the second year, the teeth of the lower jaw preceding those of the upper.

The following are the most usual times of eruption:

Lower central incisors	6 to 9 months
Upper incisors	8 to 10 months
Lower lateral incisors and first molars	15 to 21 months
Canines	16 to 20 months
Second molars	20 to 24 months

There are, however, considerable variations in these times; thus, according to Holt:

At the age of	1 year	a child should have	6 teeth
"	1½ years	" "	12 "
"	2 "	" "	16 "
"	2½ "	" "	20 "

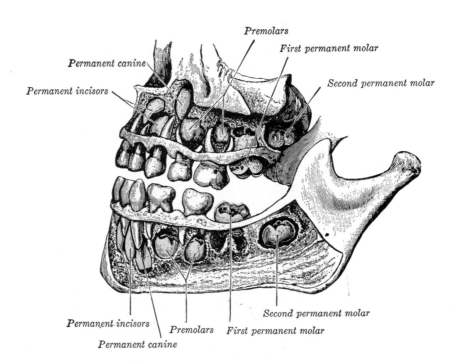

Fig. 16–33.—The teeth of a child aged about seven years. The permanent teeth are colored *blue*.

Calcification of the permanent teeth proceeds in the following order in the lower jaw (in the upper jaw it takes place a little later). The first molar, soon after birth; the central and lateral incisors, and the canine, about six months after birth; the premolars, at the second year, or a little later; the second molar, about the end of the second year; the third molar, about the twelfth year.

The eruption of the permanent teeth takes place at the following periods, the teeth of the lower jaw preceding those of the upper by short intervals:

First molars	6th year
Two central incisors	7th year
Two lateral incisors	8th year

First premolars	9th year
Second premolars	10th year
Canines	11th to 12th year
Second molars . . .	12th to 13th year
Third molars . . .	17th to 25th year

Toward the sixth year, before the shedding of the deciduous teeth begins, there are twenty-four teeth in each jaw, viz., the ten deciduous teeth and the crowns of all the permanent teeth except the third molars (Fig. 16–33). The third molars (*wisdom teeth; dentes serotini*) are irregular in their eruption and may be badly oriented or buried in bone to such an extent that they must be removed surgically. Not infrequently one or all four may fail entirely to develop.

THE TONGUE

The **Tongue** (*lingua*) is the principal organ of the sense of taste, an important organ of speech, and it assists in the mastication and deglutition of the food. It is situated in the floor of the mouth, within the curve of the body of the mandible.

The **Root** (*radix linguae; base*) is its posterior part by which it is connected with the hyoid bone by the Hyoglossi and Genioglossi muscles and the hyoglossal membrane; with the epiglottis by three folds (*glossoepiglottic*) of mucous membrane; with the soft palate by the glossopalatine arches; and with the pharynx by the Constrictores pharyngis superiores and the mucous membrane.

The **Apex** (*apex linguae; tip*) is the somewhat attenuated anterior end which rests against the lingual surfaces of the lower incisor teeth.

The **Inferior Surface** (*facies inferior linguae; undersurface*) (Fig. 16–39) is connected with the mandible by the mucous membrane which is reflected over the floor of the mouth to the lingual surface of the gum. The mucous membrane between the floor and the tongue in the middle line is elevated into a distinct vertical fold, the **frenulum linguae.** On either side lateral to the frenulum is a slight fold of the mucous membrane, the **plica fimbriata,** the free edge of which occasionally exhibits a series of fringe-like processes.

The apex of the tongue, part of the inferior surface, the sides, and dorsum are free.

The **Dorsum of the Tongue** (*dorsum linguae*) (Fig. 16–34) is convex and marked by a **median sulcus,** which divides it into symmetrical halves; this sulcus ends posteriorly, about 2.5 cm from the root of the organ, in a depression, the **foramen cecum,** from which a shallow groove, the **sulcus terminalis,** runs lateralward and forward on either side to the margin of the tongue. The part of the dorsum of the tongue anterior to this groove, forming about two-thirds of its surface, looks upward, and is rough and covered with papillae; the posterior third looks posteriorly, and is smoother, and contains numerous mucous glands and lymph follicles (**lingual tonsil**). The foramen cecum is the remains of the cranial part of the **thyroglossal duct** or diverticulum from which the thyroid gland is developed; the pyramidal lobe of the thyroid gland indicates the position of the lower part of the duct.

Papillae of the Tongue

The **papillae of the tongue** (Fig. 16–35) are projections of the corium, thickly distributed over the anterior two-thirds of its dorsum, giving to this surface its characteristic roughness. The varieties of papillae are the **papillae vallate, papillae fungiformes, papillae filiformes,** and **papillae simplices.**

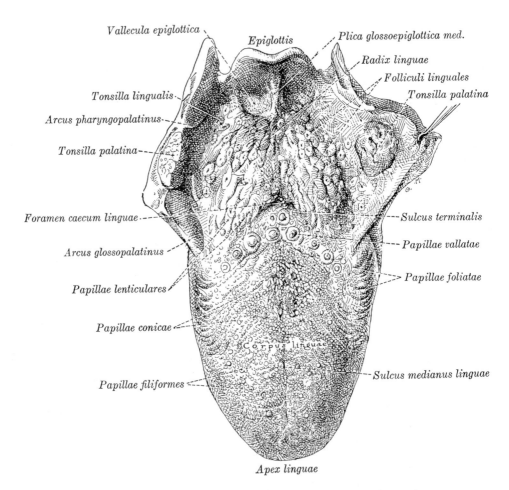

Vallecula epiglottica

Epiglottis

Plica glossoepiglottica med.

Radix linguae

Folliculi linguales

Tonsilla lingualis

Tonsilla palatina

Arcus pharyngopalatinus

Tonsilla palatina

Foramen caecum linguae

Sulcus terminalis

Papillae vallatae

Arcus glossopalatinus

Papillae foliatae

Papillae lenticulares

Papillae conicae

Corpus linguae

Sulcus medianus linguae

Papillae filiformes

Apex linguae

Fig. 16–34.—The dorsum of the tongue. (Eycleshymer and Jones.)

The **papillae vallatae** (*circumvallate papillae*) (Fig. 16–35) are of large size, and vary from eight to twelve in number. They are situated on the dorsum of the tongue immediately anterior to the foramen cecum and sulcus terminalis, forming a row on either side, like the limbs of the letter V. Each papilla consists of a projection of mucous membrane from 1 to 2 mm wide, attached to the bottom of a circular depression of the mucous membrane, the margin of which is elevated to form a wall (*vallum*); between this and the papilla is a circular sulcus or furrow. The papilla is shaped like a truncated cone, the smaller end being attached to the tongue; the broader part is free, projecting a little above the surface of the tongue and being studded with nu-

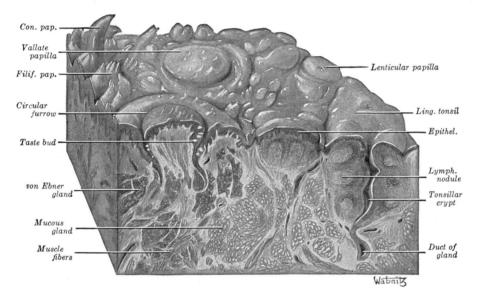

Fig. 16–35.—Section of posterior part of dorsum of tongue showing circumvallate papillae and lingual tonsil. (Redrawn from Braus.)

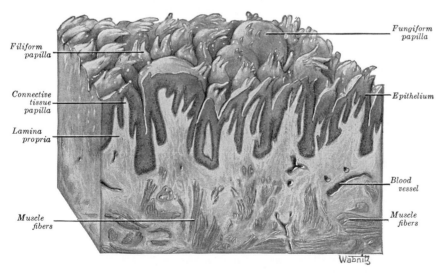

Fig. 16–36.—Section of anterior part of dorsum of tongue showing fungiform and filiform papillae. (Redrawn from Braus.)

merous small secondary papillae and covered by stratified squamous epithelium. The taste buds are especially numerous on the walls of the papilla within the circular furrow.

The **papillae fungiformes** (*fungiform papillae*) (Fig. 16–36), more numerous than the preceding, are found chiefly at the sides and apex, but are scattered irregularly and sparingly over the dorsum. They are easily recognized, among the other papillae, by their large size, rounded eminences, and deep red color. They are narrow at their attachment to the tongue, but broad and rounded at their free extremities, and covered with secondary papillae.

The **papillae filiformes** (*filiform or conical papillae*) (Fig. 16–36) cover the anterior two-thirds of the dorsum. They are very minute, filiform in shape, and arranged in lines parallel with the two rows of the papillae vallatae, except at the apex of the organ, where their direction is transverse. Projecting from their apices are numerous filamentous processes, or secondary papillae; these are of a whitish tint owing to the thickness and density of the epithelium of which they are composed, which has here undergone a peculiar modification, the cells having become cornified and elongated into dense, imbricated, brushlike processes. They contain also a number of elastic fibers, which render them firmer and more elastic than the papillae of mucous membrane generally. The larger and longer papillae of this group are sometimes termed **papillae conicae.**

The **papillae simplices** are similar to those of the skin, and cover the entire mucous membrane of the tongue, as well as the larger papillae. They consist of closely set microscopic projections into the layer of epithelium, each containing a capillary loop.

Muscles of the Tongue

The **tongue** is divided into lateral halves by a median fibrous septum which extends throughout its entire length and is fixed inferiorly to the hyoid bone. In either half there are two sets of muscles, extrinsic and intrinsic; the former have their origins outside the tongue, the latter are contained entirely within it.

The **extrinsic muscles** (Fig. 16–37) are:

1. Genioglossus
2. Hyoglossus
3. Chondroglossus
4. Styloglossus
5. Palatoglossus

1. The **Genioglossus** (*Geniohyoglossus*) is a flat fan-shaped muscle close to and parallel with the median plane. It *arises* by a short tendon from the superior mental spine on the inner surface of the symphysis menti, immediately above the Geniohyoideus. The inferior fibers extend downward, to be attached by a thin aponeurosis to the upper part of the body of the hyoid bone, a few passing between the Hyoglossus and Chondroglossus to blend with the Constrictores pharyngis; the middle fibers pass posteriorly, and the superior ones upward, to enter the whole length of the under surface of the tongue, from the root to the apex. The muscles of the two sides are separated at their insertions by the median fibrous septum of the tongue; anteriorly, they are more or less blended, the fasciculi decussating in the median plane.

2. The **Hyoglossus** *arises* from the side of the body and from the whole length of the greater cornu of the hyoid bone, and passes almost vertically upward to enter the side of the tongue, between the Styloglossus and Longitudinalis inferior. The fibers arising from the body of the hyoid bone overlap those from the greater cornu.

3. The **Chondroglossus** is sometimes described as a part of the Hyoglossus, but is separated from it by fibers of the Genioglossus, which pass to the side of the pharynx. It is about 2 cm long, and *arises* from the medial side and base of the lesser cornu and contiguous portion of the body of the hyoid bone, and passes directly upward to blend with the intrinsic muscular fibers of the tongue, between the Hyoglossus and Genioglossus.

A small slip of muscular fibers is occasionally found, arising from the cartilago triticea in the lateral thyrohyoid ligament and entering the tongue with the most posterior fibers of the Hyoglossus.

4. The **Styloglossus**, the shortest and smallest of the three styloid muscles, *arises*

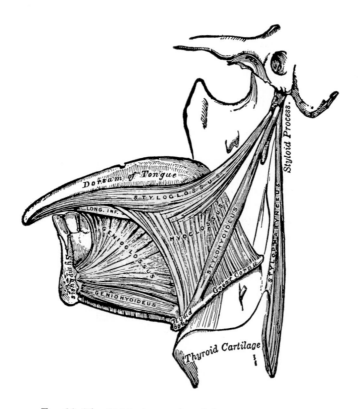

FIG. 16–37.—Extrinsic muscles of the tongue. Left side.

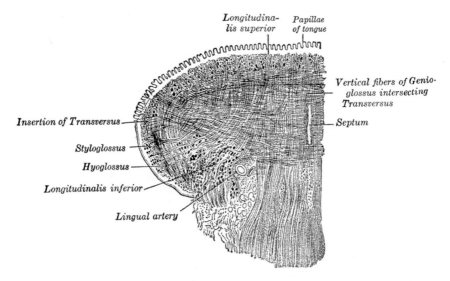

FIG. 16–38.—Coronal section of tongue, showing intrinsic muscles. (Altered from Krause.)

from the anterior and lateral surfaces of the styloid process, and from the stylomandibular ligament. Curving down anteriorly between the internal and external carotid arteries, it divides upon the side of the tongue into two portions: one, longitudinal, enters the side of the tongue near its dorsal surface, blending with the fibers of the Longitudinalis inferior anterior to the Hyoglossus; the other, oblique, overlaps the Hyoglossus and decussates with its fibers.

5. The **Palatoglossus** (*glossopalatinus*), although one of the muscles of the tongue, is more closely associated with the soft palate both in situation and function; it has consequently been described with the muscles of that structure (page 1194).

The **intrinsic muscles** (Fig. 16–38) are:

1. Longitudinalis superior
2. Longitudinalis inferior
3. Transversus
4. Verticalis

1. The **Longitudinalis linguae superior** (*Superior lingualis*) is a thin stratum of oblique and longitudinal fibers immediately underlying the mucous membrane on the dorsum of the tongue. It *arises* from the submucous fibrous layer close to the epiglottis and from the median fibrous septum, and runs anteriorly to the edges of the tongue.

2. The **Longitudinalis linguae inferior** (*Inferior lingualis*) is a narrow band situated on the inferior surface of the tongue between the Genioglossus and Hyoglossus. It extends from the root to the apex of the tongue, some of its fibers being connected with the body of the hyoid bone, others blending with the fibers of the Styloglossus.

3. The **Transversus linguae** (*Transverse lingualis*) consists of fibers which *arise* from the median fibrous septum and pass lateralward to be *inserted* into the submucous fibrous tissue at the sides of the tongue.

4. The **Verticalis linguae** (*Vertical lingualis*) is found only at the borders of the anterior part of the tongue, its fibers extending from the upper to the under surface.

Nerves.—The muscles of the tongue described above are supplied by the hypoglossal nerve.

Actions.—The movements of the tongue, although numerous and complicated, may be understood by carefully considering the direction of the fibers of its muscles. The Genioglossi, by means of their posterior fibers, draw the root of the tongue anteriorly, and protrude the apex from the mouth. The anterior fibers draw the tongue back into the mouth. The two parts acting in their entirety draw the tongue downward, so as to make its superior surface concave from side to side, forming a channel along which fluids may pass toward the pharynx, as in sucking. The Hyoglossi depress the tongue, and draw down its sides. The Styloglossi draw the tongue upward and backward. The Glossopalatini draw the root of the tongue upward.

The intrinsic muscles are mainly concerned in altering the shape of the tongue, whereby it becomes shortened, narrowed, or curved in different directions; thus, the Longitudinalis superior and inferior tend to shorten the tongue, but the former, in addition, turns the tip and sides upward so as to render the dorsum concave, while the latter pulls the tip downward and renders the dorsum convex. The Transversus narrows and elongates the tongue, and the Verticalis flattens and broadens it. The complex arrangement of the muscular fibers of the tongue, and the various directions in which they run, give to this organ the power of assuming the forms necessary for the enunciation of the different consonantal sounds.

Structure of the Tongue.—The tongue is invested by mucous membrane and a submucous fibrous layer.

The **mucous membrane** (*tunica mucosa linguae*) covering the inferior surface of the organ is thin, smooth, and identical in structure with that lining the rest of the oral cavity. The mucous membrane of the dorsum of the tongue posterior to the foramen cecum and sulcus terminalis is thick and freely movable over the subjacent parts. It contains a large number of lymphoid follicles, which together constitute the **lingual tonsil** (Fig. 16–35). Each follicle forms a rounded eminence, the center of which is perforated by a minute orifice leading into a funnel-shaped cavity or recess; around this recess are grouped numerous oval or rounded nodules of lymphoid tissue, each enveloped by a capsule derived from the submucosa, and opening into the bottom of the recesses are the ducts of mucous glands. The mucous membrane on the anterior part of the dorsum of the tongue is thin, intimately adherent to the muscular tissue, and presents numerous minute surface eminences, the **papillae** of the tongue. It consists of a layer of connective tissue, the

corium, covered with epithelium (Figs. 16–35, 16–36).

The epithelium is stratified squamous, similar to that of the skin: and each papilla has a separate investment from root to summit. The deepest cells may sometimes be detached as a separate layer, corresponding to the rete mucosum, but they never contain pigment.

The **corium** consists of a dense feltwork of fibrous connective tissue, with numerous elastic fibers, firmly connected with the fibrous tissue forming the septa between the muscular bundles of the tongue. It contains the ramifications of the numerous vessels and nerves from which the papillae are supplied, large plexuses of lymphatic vessels, and the glands of the tongue.

Structure of the Papillae.—The papillae apparently resemble in structure those of the cutis, consisting of cone-shaped projections of connective tissue, covered with a thick layer of stratified squamous epithelium, and containing one or more capillary loops among which nerves are distributed in great abundance. If the epithelium be removed, it will be found that they are not simple elevations like the papillae of the skin, for the surface of each is studded with minute conical processes which form secondary papillae. In the papillae vallatae, the nerves are numerous and of large size; in the papillae fungiformes they are also numerous, and end in a plexiform network, from which brush-like branches proceed; in the papillae filiformes, their mode of termination is similar.

Glands of the Tongue.—The tongue is provided with mucous and serous glands.

The **mucous glands** are similar in structure to the labial and buccal glands. They are found especially at the back part behind the vallate papillae, but are also present at the apex and marginal parts. In this connection the anterior lingual glands require special notice. They are situated on the under surface of the apex of the tongue (Fig. 16–39), one on either side of the frenulum, where they are covered by a fasciculus of muscular fibers derived from the Styloglossus and Longitudinalis inferior. They are from 12 to 25 mm long, and about 8 mm broad, and each opens by three or four ducts on the under surface of the apex.

The **serous glands** (*v. Ebner's glands*) occur only at the back of the tongue in the neighborhood of the taste-buds, their ducts opening for the most part into the fossae of the vallate papillae. These glands are racemose, the duct of each branching into several minute ducts, which end in alveoli lined by a single layer of more or less columnar epithelium. Their secretion is of a watery nature, and probably assists in the distribution of the substance to be tasted over the taste area.

The **septum** consists of a vertical layer of fibrous tissue, extending throughout the entire length of the median plane of the tongue, though not quite reaching the dorsum. It is thicker behind than in front, and occasionally contains a small fibrocartilage, about 6 mm in length. It is well displayed by making a vertical section across the organ.

The **hyoglossal membrane** is a strong fibrous lamina, which connects the under surface of the root of the tongue to the body of the hyoid bone. This membrane receives some of the fibers of the Genioglossi, anteriorly.

Taste-buds, the end-organs of the gustatory sense, are scattered over the mucous membrane of the mouth and tongue at irregular intervals. They occur especially in the sides of the vallate papillae. They are described under the Organs of the Senses.

Vessels and Nerves.—The main **artery** of the tongue is the lingual branch of the external carotid, but the facial and ascending pharyngeal also give branches to it. The **veins** open into the internal jugular.

The **lymphatics of the tongue** have been described on page 742.

The **sensory nerves of the tongue** are: (1) the lingual branch of the mandibular, which is distributed to the papillae at the anterior part and sides of the tongue, and forms the nerve of ordinary sensibility for its anterior two-thirds; (2) the chorda tympani branch of the facial, which runs in the sheath of the lingual, and is generally regarded as the nerve of taste for the anterior two-thirds; this nerve is a continuation of the sensory root of the facial (*nervus intermedius*); (3) the lingual branch of the glossopharyngeal, which is distributed to the mucous membrane at the base and sides of the tongue, and to the papillae vallatae, and which supplies both gustatory filaments and fibers of general sensation to this region; (4) the superior laryngeal, which sends some fine branches to the root near the epiglottis.

The Salivary Glands

Three pairs of large salivary glands pour their secretion into the mouth; they are the **parotid, submandibular,** and **sublingual** (Fig. 16–40).

The **Parotid Gland** (*glandula parotis*), the largest of the three, varies in weight from 14 to 28 gm. It lies upon the side of the face, immediately below and in front

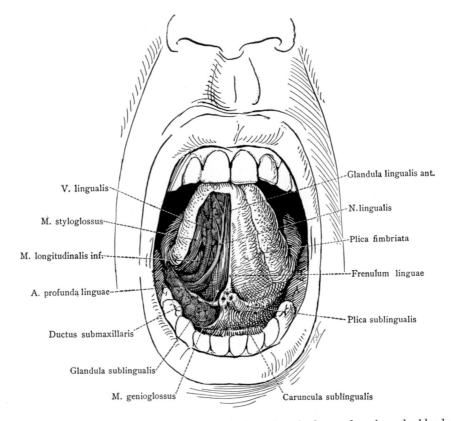

V. lingualis

M. styloglossus

M. longitudinalis inf.

A. profunda linguae

Ductus submaxillaris

Glandula sublingualis

M. genioglossus

Glandula lingualis ant.

N. lingualis

Plica fimbriata

Frenulum linguae

Plica sublingualis

Caruncula sublingualis

Fig. 16–39.—The inferior surface of the tongue, with the right side dissected to show the blood vessels, nerve and salivary gland. (Eycleshymer and Jones.)

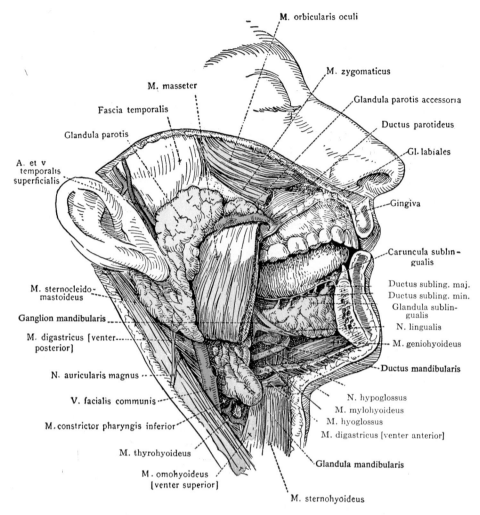

FIG. 16–40.—The salivary glands in a dissection of the right side of the face.
(Eycleshymer and Jones.)

of the external ear. The main portion of the gland is superficial, somewhat flattened and quadrilateral in form, and is placed between the ramus of the mandible, the mastoid process and Sternocleidomastoideus. Superiorly, it is broad and reaches nearly to the zygomatic arch; inferiorly, it tapers somewhat to about the level of a line joining the tip of the mastoid process to the angle of the mandible. The remainder of the gland is irregularly wedge-shaped, and extends deeply inward toward the pharyngeal wall.

The gland is enclosed within a capsule continuous with the deep cervical fascia; the layer covering the superficial surface is dense and closely adherent to the gland;

a portion of the fascia, attached to the styloid process and the angle of the mandible, is thickened to form the *stylomandibular ligament* which intervenes between the parotid and submandibular glands.

The **anterior surface** of the gland is molded on the posterior border of the ramus of the mandible, clothed by the Pterygoideus medialis and Masseter. The inner lip of the groove dips, for a short distance, between the two Pterygoid muscles, while the outer lip extends for some distance over the superficial surface of the Masseter; a small portion of this lip immediately below the zygomatic arch is usually detached, and is named the **accessory part** (*glandula parotis accessoria*) of the gland.

The **posterior surface** is grooved longitudinally and abuts against the external acoustic meatus, the mastoid process, and the anterior border of the Sternocleidomastoideus.

The **superficial surface**, slightly lobulated, is covered by the integument, the superficial fascia containing the facial branches of the great auricular nerve and some small lymph glands, and the fascia which forms the capsule of the gland.

The **deep surface** extends inward by means of two processes, one of which lies on the Digastricus, styloid process, and the styloid group of muscles, and projects under the mastoid process and Sternocleidomastoideus; the other is situated anterior to the styloid process, and sometimes passes into the posterior part of the mandibular fossa behind the temporomandibular joint. The deep surface is in contact with the internal and external carotid arteries, the internal jugular vein, and the vagus and glossopharyngeal nerves.

The gland is separated from the pharyngeal wall by some loose connective tissue.

Structures within the Gland.—The *external carotid artery* lies at first on the deep surface, and then in the substance of the gland. The artery gives off its *posterior auricular* branch which emerges from the gland posteriorly; it then divides into its terminal branches, the *maxillary* and *superficial temporal;* the former runs anteriorly deep to the neck of the mandible; the latter runs upward across the zygomatic arch and gives off its *transverse facial* branch which emerges from the anterior part of the gland. Superficial to the arteries are the *superficial temporal* and *maxillary veins,* uniting to form the *retromandibular vein.* A large anastomosing vein emerges from the gland and unites with the facial to form the *common facial* vein; the remainder of the vein unites in the gland with the posterior auricular to form the *external jugular* vein. On a still more superficial plane is the **facial nerve,** the branches of which emerge from the borders of the gland. Branches of the *great auricular nerve* pierce the gland to join the facial, while the *auriculotemporal nerve* issues from the upper part of the gland.

The **parotid duct** (*ductus parotideus; Stensen's duct*) is about 7 cm long. It begins at the anterior part of the gland, crosses the Masseter, and at the anterior border of this muscle turns medially, nearly at a right angle, passes through the corpus adiposum of the cheek and pierces the Buccinator. It runs for a short distance obliquely forward between the Buccinator and mucous membrane of the mouth, and opens upon the oral surface of the cheek by a small orifice, opposite the second upper molar tooth. While crossing the Masseter, it receives the duct of the accessory gland; in this position it lies between the branches of the facial nerve; the accessory part of the gland and the transverse facial artery are superior to it.

Structure.—The parotid duct has a wall of considerable thickness; its canal is about the size of a crow quill (3 to 4 mm), but at its orifice on the oral surface of the cheek its lumen is greatly reduced in size. It consists of a thick external fibrous coat which contains contractile fibers, and of an internal or mucous coat lined with low columnar epithelium.

Vessels and Nerves.—The **arteries** supplying the parotid gland are derived from the external carotid, and from the branches given off by that vessel in or near its substance. The **veins** empty themselves in the external jugular, through some of its tributaries. The **lymphatics** end in the superficial and deep cervical lymph nodes, passing in their course through two or three nodes, placed on the surface and in the substance of the parotid. The **nerves** are derived from the plexus of the *sympathetic* on the external carotid artery, and from the auriculotemporal nerve. The fibers from the latter nerve are cranial *parasympathetics* derived from the glossopharyngeal, and possibly from the facial, through the otic ganglion. The sympathetic fibers are regarded as chiefly vasoconstrictors, the parasympathetic fibers as secretory.

The **Submandibular Gland** (*glandula mandibularis; submaxillary gland*) (Fig. 16–40) is rounded in form and about the size of a walnut. A considerable part of it is situated in the submandibular triangle, reaching anteriorly to the anterior belly of the Digastricus and posteriorly to the stylomandibular ligament, which intervenes between it and the parotid gland. It extends superiorly under the inferior border of the body of the mandible; inferiorly, it usually overlaps the intermediate tendon of the

Digastricus and the insertion of the Stylohyoideus, and from its deep surface a tongue-like *deep process* extends anteriorly above the Mylohyoideus muscle.

Its **superficial surface** consists of an upper and a lower part. The **upper part** is directed superficially, and lies partly against the submandibular depression on the inner surface of the body of the mandible, and partly on the Pterygoideus medialis. The **lower part** is covered by the skin, superficial fascia, Platysma, and deep cervical fascia; it is crossed by the facial vein and by filaments of the facial nerve; in contact with it, near the mandible, are the submandibular lymph nodes.

The **deep surface** is in relation with the Mylohyoideus, Hyoglossus, Styloglossus, Stylohyoideus, and posterior belly of the Digastricus; in contact with it are the mylohyoid nerve and the mylohyoid and submental vessels. The facial artery is embedded in a groove in the posterior border of the gland.

The **deep process** of the gland extends anteriorly between the Mylohyoideus laterally, and the Hyoglossus and Styloglossus medially; above it are the lingual nerve and submandibular ganglion; below it the hypoglossal nerve and its accompanying vein.

The **submandibular duct** (*ductus submandibularis; Wharton's duct; submaxillary duct*) is about 5 cm long, and its wall is much thinner than that of the parotid duct. It begins by numerous branches from the deep surface of the gland, and runs anteriorly between the Mylohyoideus and the Hyoglossus and Genioglossus, then between the sublingual gland and the Genioglossus, and opens by a narrow orifice on the summit of a small papilla, the **caruncula sublingualis,** at the side of the frenulum linguae (Fig. 16–39). On the Hyoglossus it lies between the lingual and hypoglossal nerves, but at the anterior border of the muscle it is crossed laterally by the lingual nerve; the terminal branches of the lingual nerve ascend on its medial side.

Vessels and Nerves.—The **arteries** supplying the submandibular gland are branches of the facial and lingual. Its **veins** follow the course of the arteries. The **secretomotor nerves** are from cranial parasympathetic fibers of the facial, which pass via the chorda tympani and submandibular ganglion; the **sympathetic** (*vas-*

omotor) fibers come from the superior cervical ganglion by way of plexuses on the external carotid and facial arteries.

The **Sublingual Gland** (*glandula sublingualis*) (Fig. 16–40) is the smallest of the three glands. It is situated beneath the mucous membrane of the floor of the mouth, at the side of the frenulum linguae, in contact with the sublingual depression on the inner surface of the mandible, close to the symphysis. It is narrow, flattened, shaped somewhat like an almond, and weighs nearly 2 gm. It is in relation, *superiorly,* with the mucous membrane; *inferiorly,* with the Mylohyoideus; *posteriorly,* with the deep part of the submandibular gland; *laterally,* with the mandible; and *medially,* with the Genioglossus, from which it is separated by the lingual nerve and the submandibular duct. Its excretory ducts are from eight to twenty in number. One or more join to form the **larger sublingual duct** (*duct of Bartholin*), which opens into the submandibular duct. Of the **small sublingual ducts** (*ducts of Rivinus*), some join the submandibular duct; others open separately into the mouth, on the elevated crest of mucous membrane (*plica sublingualis*), caused by the projection of the gland, on either side of the frenulum linguae.

Vessels and Nerves.—The sublingual gland is supplied with blood from the sublingual and submental arteries. Its nerves are derived in a manner similar to those of the submandibular gland.

Structure of the Salivary Glands.—The salivary glands are compound racemose glands consisting of numerous lobes, which are made up of smaller lobules, connected together by dense areolar tissue, vessels, and ducts. Each lobule consists of the ramifications of a single duct, the branches ending in dilated ends or alveoli on which the capillaries are distributed. The alveoli are enclosed by a basement membrane, which is continuous with the membrana propria of the duct and consists of a network of branched and flattened nucleated cells.

The alveoli of the salivary glands are of two kinds, which differ in the appearance of their secreting cells, in their size, and in the nature of their secretion. (1) The mucous variety secretes a viscid fluid, which contains mucin; (2) the serous variety secretes a thinner and more watery fluid. The *sublingual gland* con-

sists of *mucous,* the *parotid* of *serous alveoli.*
The *submandibular* contains *both mucous* and
serous alveoli, the latter, however, preponder-
ating.

The cells in the **mucous alveoli** are cuboidal
in shape. In the fresh condition they contain
large granules of mucinogen. In hardened
preparations a delicate protoplasmic network
is seen, and the cells are clear and transparent.
The nucleus is usually situated near the basement membrane, and is flattened.

In some alveoli are seen peculiar crescentic
bodies, lying between the cells and the membrana propria. They are termed the **crescents
of Gianuzzi,** or the **demilunes of Heidenhain**
(Fig. 16–41), and are composed of polyhedral
granular cells. Fine canaliculi pass between
the mucus-secreting cells to reach the demilunes.

In the **serous alveoli** the cells almost completely fill the cavity, so that there is hardly
any lumen perceptible; they contain secretory
granules embedded in a closely reticulated protoplasm (Fig. 16–41)). The cells are more
cubical than those of the mucous type; the
nucleus of each is spherical and placed near
the center of the cell, and the granules are
smaller.

Both mucous and serous cells vary in appearance according to whether the gland is in a
resting condition or has been recently active.
In the former case the cells are large and
contain many secretory granules; in the latter
case they are shrunken and contain few
granules, chiefly collected at the inner ends of
the cells.

The **ducts** are lined at their origins by epithelium which differs little from the pavement
form. As the ducts enlarge, the epithelial cells
change to the columnar type, and the part of
the cell next the basement membrane is finely
striated.

The lobules of the salivary glands are richly
supplied with blood vessels which form a dense
network in the interalveolar spaces. Fine
plexuses of nerves are also found in the interlobular tissue. The nerve fibrils pierce the
basement membrane of the alveoli, and end
in branched varicose filaments between the
secreting cells. In the hilum of the submandibular gland there is a collection of nerve cells
termed **Langley's ganglion.**

Accessory Glands.—Besides the salivary
glands proper, numerous other glands are found
in the mouth. Many of these glands are found
at the posterior part of the dorsum of the tongue
behind the vallate papillae, and also along its
margins as far forward as the apex. Others lie
around and in the palatine tonsil between its

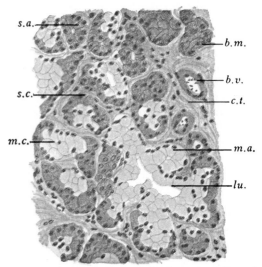

Fɪɢ. 16–41.—Section of the human sublingual
gland. *b.m.,* basement membrane; *b.v.,* blood vessel;
c.t., connective tissue stroma; *lu.,* lumen; *m.a.,* mucous alveolus; *m.c.,* mucus-secreting cells; *s.a.,*
serous alveolus; *s.c.,* serous crescent (crescent of
Gianuzzi or demilune of Heidenhain). × 250.
(Sobotta.)

crypts, and large numbers are present in the
soft palate, the lips, and cheeks. These glands
are of the same structure as the larger salivary
glands, and are of the mucous or mixed type.

THE FAUCES

The aperture by which the mouth communicates with the pharynx is called the
Fauces. The **Isthmus Faucium** (Fig. 16–42)
is bounded, superiorly, by the soft palate;
inferiorly, by the dorsum of the tongue; and
on either side, by the palatoglossal arch.

The **palatoglossal arch** (*arcus glossopalatinus; anterior pillar of fauces*) curves
downward from the soft palate near the
uvula to the side of the base of the tongue,
and is formed by the projection of the
Palatoglossus with its covering mucous
membrane.

The **palatopharyngeal arch** (*arcus pharyngopalatinus; posterior pillar of fauces*)
is blended with the free border of the soft
palate near the uvula and curves downward to the side of the pharynx; it is formed
by the projection of the Palatopharyngeus,
covered by mucous membrane. The two
arches diverge inferiorly to form a triangu-

lar interval, in which the palatine tonsil is lodged.

The **Palatine Tonsil** (*tonsilla palatina; tonsil*) (Figs. 16–42, 16–43) is a prominent mass of lymphatic tissue situated at the side of the fauces between the palatoglossal and palatopharyngeal arches. Each tonsil consists of an aggregation of lymphatic nodules underlying the mucous membrane between the arches. The lymphatic tissue does not completely fill the interval between the two arches, however; a small depression, the **supratonsillar fossa** occupies the upper part of the interval. The tonsil extends for a variable distance deep to the palatoglossal arch. The covering fold of mucous membrane reaches across the supratonsillar

fossa, between the two arches, sometimes termed the **plica semilunaris;** the remainder of the fold is called the **plica triangularis.** Between the plica triangularis and the surface of the tonsil is a space known as the **tonsillar sinus,** which may be obliterated by its walls becoming adherent. A large portion of the tonsil is below the level of the surrounding mucous membrane, *i.e.*, is embedded, while the remainder projects as the visible tonsil (Fig. 16–43, *A*). In the child the tonsils are relatively (and frequently absolutely) larger than in the adult, and about one-third of the tonsil is embedded. After puberty the embedded portion diminishes considerably in size and the tonsil assumes a disk-like form. The shape and

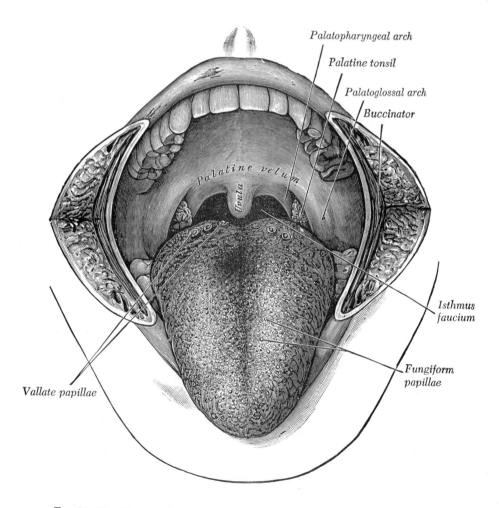

Fig. 16–42.—The mouth cavity. The cheeks have been slit transversely and the tongue pulled forward.

size of the tonsil vary considerably in different individuals.

The **medial surface** of the tonsil is free except anteriorly, where it is covered by the plica triangularis; it presents from twelve to fifteen orifices (*fossulae tonsillares*) leading into **crypts** (*cryptae tonsillares*) which may branch and extend deeply into the tonsillar substance.

The **lateral** or **deep surface** is adherent to a fibrous capsule which is continued into the plica triangularis. It is separated from the inner surface of the Constrictor pharyngis superior usually by some loose connective tissue; this muscle intervenes between the tonsil and the facial artery with its tonsillar and ascending palatine branches. The

internal carotid artery lies posterior and lateral to the tonsil at a distance of 20 to 25 mm.

The palatine tonsils form part of a circular band of lymphatic tissue (Waldeyer's ring) which guards the opening into the digestive and respiratory tubes (Fig. 16–43). The anterior part of the ring is formed by the submucous lymphatic collections (**lingual tonsil**) on the posterior part of the tongue; the lateral portions consist of the palatine tonsils and tubal tonsils at the openings of the auditory tubes, while the ring is completed posteriorly by the pharyngeal tonsil on the posterior wall of the pharynx. In the intervals between these main masses are smaller collections of lymphoid tissue.

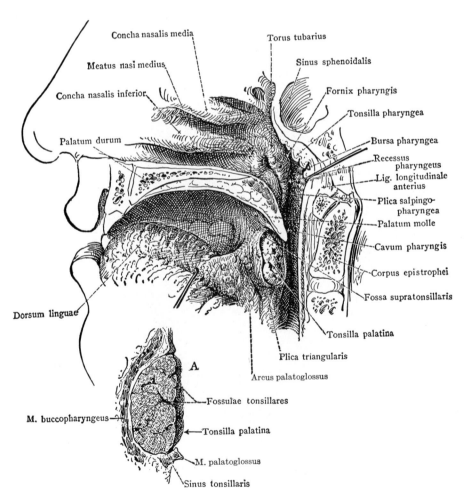

Fɪɢ. 16–43.—The oral and nasal pharynx in median sagittal section, showing palatine and pharyngeal tonsil. *A*, Detail of palatine tonsil in frontal section. (Eycleshymer and Jones.)

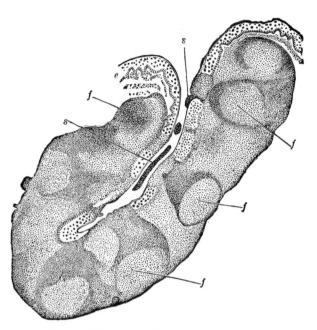

Fig. 16–44.—Section through one of the crypts of the tonsil. (Stöhr.) Magnified. *e*, Stratified epithelium of general surface, continued into crypt. *f*, *f*, Nodules of lymphoid tissue—opposite each nodule numbers of lymph cells are passing into or through the epithelium. *s*, *s*, Cells which have thus escaped to mix with the saliva as salivary corpuscles.

Structure (Fig. 16–44).—Stratified squamous epithelium, like that of the palate and oral pharynx, covers the free surface of the tonsil and extends down into its substance to form the lining of the crypts. Each **crypt** is surrounded by a layer of lymphatic tissue containing numerous scattered lymphatic nodules whose germinal centers are especially prominent in children and young adults. A thin connective tissue capsule, derived from the submucosa of the pharynx, encloses the whole tonsil and sends delicate septa in between the lymphatic tissue layers surrounding the crypts. The epithelium of the crypts is so invaded by leukocytes in many places that it is scarcely distinguishable from the lymphatic tissue. Polymorphonuclear leukocytes from the blood as well as lymphocytes penetrate the epithelium, and when they are found as free swimming cells in the saliva, they are known as *salivary corpuscles*. Small mucous glands occur in the submucosa about the tonsil but their ducts, as a rule, do not open into the crypts.

Vessels and Nerves.—The **arteries** supplying the tonsil are the dorsalis linguae from the lingual, the ascending palatine and tonsillar from the facial, the ascending pharyngeal from the external carotid, the descending palatine branch of the maxillary, and a twig from the small meningeal.

The **veins** forming the tonsillar plexus, on the lateral side of the tonsil, drain into the pterygoid plexus or the facial vein.

The **lymphatic vessels**, beginning in the dense network of capillaries surrounding the lymphatic tissue, penetrate the pharyngeal wall (page 742) and pass to the deep cervical nodes. The largest of these nodes, lying beside the posterior belly of the Digastricus, is especially associated with the tonsil and is easily palpated when the latter is inflamed.

The **nerves** are derived from the middle and posterior palatine branches of the maxillary, and from the glossopharyngeal.

Palatine Aponeurosis.—Attached to the posterior border of the hard palate is a thin, firm fibrous lamella which supports the muscles and gives strength to the soft palate. Laterally it is continuous with the pharyngeal aponeurosis.

Muscles of the Palate.—The muscles of the palate (Fig. 16–45) are:

1. Levator veli palatini
2. Tensor veli palatini
3. Musculus uvulae
4. Palatoglossus
5. Palatopharyngeus

1. The **Levator veli palatini** (*Levator palati*) is situated lateral to the choanae and deep to the torus tubarius. It *arises* from the inferior surface of the apex of the petrous part of the temporal bone and from the medial lamina of the cartilage of the auditory tube. After passing above the superior concave margin of the Constrictor pharyngis superior it spreads out in the palatine velum, its fibers extending obliquely downward and medialward to the middle line, where they blend with those of the opposite side.

2. The **Tensor veli palatini** (*Tensor palati*) is a thin, ribbon-like muscle placed lateral and anterior to the Levator. It *arises* by a flat lamella from the scaphoid fossa at the base of the medial pterygoid plate, from the spine of the sphenoid and from the lateral wall of the cartilage of the auditory tube. Descending vertically between the medial pterygoid plate and the Pterygoideus medialis it ends in a tendon which winds around the pterygoid hamulus, being retained in this situation by some of the fibers of origin of the Pterygoideus medialis. Between the tendon and the hamulus is a small bursa. The tendon then passes medialward and is *inserted* into the palatine aponeurosis and into the surface behind the transverse ridge on the horizontal part of the palatine bone.

3. The **Musculus uvulae** (*Azygos uvulae*) *arises* from the posterior nasal spine of the palatine bones and from the palatine aponeurosis; it descends to be inserted into the uvula.

4. The **Palatoglossus** (*glossopalatinus*) is a small fleshy fasciculus, narrower in the middle than at either end, forming, with the mucous membrane covering its surface, the palatoglossal arch. It *arises* from the anterior surface of the soft palate, where it is continuous with the muscle of the opposite side, and passing down, anterior to the palatine tonsil, is *inserted* into the side of the tongue, some of its fibers spreading over the dorsum, and others passing deeply to intermingle with the Transversus linguae.

5. The **Palatopharyngeus** (*pharyngopalatinus*) is a long, fleshy fasciculus narrower in the middle than at either end, forming, with the mucous membrane covering its surface, the palatopharyngeal arch. It is separated from the Palatoglossus by a triangular interval, in which the palatine tonsil is lodged. It *arises* from the soft palate, where it is divided into two fasciculi by the Levator veli palatini and Musculus uvulae. The **posterior fasciculus** lies in contact with the mucous membrane, and joins with that of the opposite muscle in the middle line; the **anterior fasciculus,** the thicker, lies in the soft palate between the Levator and Tensor, and joins in the middle line the corresponding part of the opposite muscle. Passing down posterior to the palatine tonsil, the Palatopharyngeus joins the Stylopharyngeus, and is inserted with that muscle into the posterior border of the thyroid cartilage, some of its fibers being lost on the side of the pharynx and others passing across the middle line posteriorly, to decussate with the muscle of the opposite side.

Nerves.—The Tensor veli palatini is supplied by a branch of the fifth cranial nerve; the remaining muscles of this group are in all probability supplied by the bulbar portion of the accessory nerve through the pharyngeal plexus.

Deglutition.—During the *first stage* of deglutition, the bolus of food is driven back into the fauces by the pressure of the tongue against the hard palate, the base of the tongue being, at the same time, retracted, and the larynx raised with the pharynx. During the *second stage* the entrance to the larynx is closed by the drawing forward of the arytenoid cartilages toward the cushion of the epiglottis—a movement produced by the contraction of the thyroarytenoidei, the Arytenoidei, and the Arytenoepiglottidei.

After leaving the tongue the bolus passes on to the posterior or laryngeal surface of the epiglottis, and glides along this for a certain distance; then the Palatoglossi, the constrictors of the fauces, contract behind it; the palatine velum is slightly raised by the Levator veli palatini, and made tense by the Tensor veli palatini; and the Palatopharyngei, by their contraction, pull the pharynx upward over the bolus, and come nearly together, the uvula filling up the slight interval between them. By these means the food is prevented from passing into the nasal part of the pharynx; at the same time, the Palatopharyngei form an inclined plane, directed obliquely downward and backward along the undersurface of which the bolus descends into the lower part of the pharynx. The Salpingopharyngei raise the upper and lateral parts of the pharynx—*i.e.,* those parts

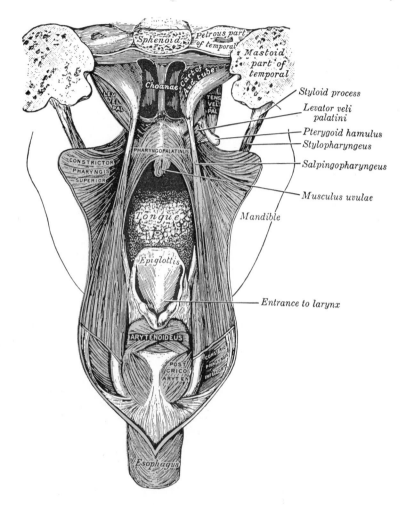

Fig. 16–45.—Dissection of the muscles of the palate, posterior aspect.

which are above the points where the Stylopharyngei are attached to the pharynx.

Mucous Membrane.—The mucous membrane of the soft palate is thin, and covered with stratified squamous epithelium on both surfaces, except near the pharyngeal ostium of the auditory tube, where it is columnar and ciliated. According to Klein, the mucous membrane on the nasal surface of the soft palate in the fetus is covered throughout by columnar ciliated epithelium, which subsequently becomes squamous except at its free margin. Beneath the mucous membrane on the oral surface of the soft palate is a considerable amount of lymphoid tissue. The palatine glands form a continuous layer on its posterior surface and around the uvula.

Vessels and Nerves.—The **arteries** supplying the palate are the descending palatine branch of the maxillary, the ascending palatine branch of the facial, and the palatine branch of the ascending pharyngeal. The **veins** end chiefly in the pterygoid and tonsillar plexuses. The **lymphatic vessels** pass to the deep cervical nodes. The **sensory nerves** are derived from the palatine, nasopalatine, and glossopharyngeal nerves.

THE PHARYNX

The **pharynx** is that part of the digestive tube which is placed posterior or dorsal to the nasal cavities, mouth, and larynx. It is a musculomembranous tube, extending from the inferior surface of the skull to the level of the cricoid cartilage ventrally, and that of the sixth cervical vertebra dorsally.

The cavity of the pharynx (Fig. 16–43) is about 12.5 cm long, and broader in the transverse than in the anteroposterior diameter. Its greatest breadth is immediately below the base of the skull, where it projects on either side, behind the pharyngeal ostium of the auditory tube, as the **pharyngeal recess** (*fossa of Rosenmüller*); its narrowest point is at its termination in the esophagus. It is limited, *superiorly,* by the body of the sphenoid and basilar part of the occipital bone; *inferiorly,* it is continuous with the esophagus; *posteriorly,* it is separated by a fascial cleft from the cervical portion of the vertebral column, and the prevertebral fascia covering the Longus colli and Longus capitis muscles; *anteriorly,* it is incomplete,

and is attached in succession to the medial pterygoid plate, pterygomandibular raphe, mandible, tongue, hyoid bone, and thyroid and cricoid cartilages; *laterally,* it is connected to the styloid processes and their muscles, and is in contact with the common and internal carotid arteries, the internal jugular veins, the glossopharyngeal, vagus, and hypoglossal nerves, and the sympathetic trunks, and with small parts of the medial Pterygoidei. Seven cavities communicate with it, viz., the two nasal cavities, the two tympanic cavities, the mouth, the larynx, and the esophagus. The cavity of the pharynx may be subdivided into three parts: **nasal, oral,** and **laryngeal** (Fig. 16–17).

The **Nasal Part of the Pharynx** (*pars nasalis pharyngis; nasopharynx*) (Fig. 16–46) lies posterior to the nose and above the

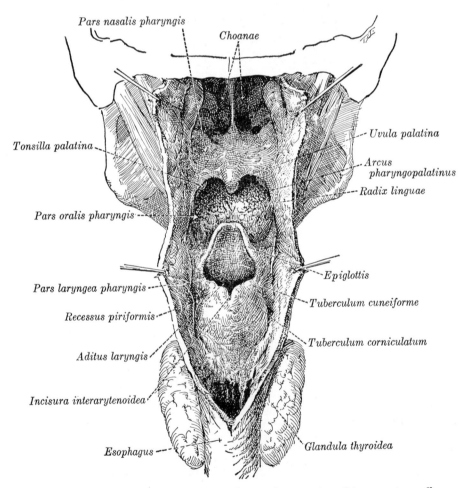

FIG. 16–46.—The pharynx viewed through a median incision of its posterior wall. (Eycleshymer and Jones.)

level of the soft palate: it differs from the oral and laryngeal parts of the pharynx in that its cavity always remains patent. It communicates through the choanae with the nasal cavities anteriorly. On its lateral wall is the **pharyngeal ostium of the auditory tube,** and bounded posteriorly by the **torus,** a prominence caused by the protrusion of the medial end of the cartilage of the tube under the mucous membrane. A vertical fold of mucous membrane, the **salpingo-pharyngeal fold,** stretching down from the torus contains the Salpingopharyngeus muscle. A second and smaller fold, the **salpingo-palatine fold,** stretches from the upper part of the torus to the palate. Behind the ostium of the auditory tube is a deep recess, the **pharyngeal recess** (*fossa of Rosenmüller*) (Fig. 16–43). On the posterior wall is a prominence produced by lymphatic tissue, which is known as the **pharyngeal tonsil;** during childhood it is likely to be hypertrophied into a considerable mass when it is called **adenoids** (Fig. 16–43). Above the pharyngeal tonsil, in the middle line, an irregular flask-shaped depression of the mucous membrane sometimes extends up as far as the basilar process of the occipital bone; it is known as the **pharyngeal bursa.**

The **Oral Part of the Pharynx** (*pars oralis pharyngis*) reaches from the soft palate to the level of the hyoid bone. It opens into the mouth anteriorly, through the isthmus faucium. In its lateral wall, between the two palatine arches, is the **faucial or palatine tonsil.**

The **Laryngeal Part of the Pharynx** (*pars laryngea pharyngis*) reaches from the hyoid bone to the lower border of the cricoid cartilage, where it is continuous with the esophagus. Anteriorly, the entrance of the larynx is formed by the epiglottis, laterally its boundaries are the aryepiglottic folds. On either side of the laryngeal orifice is a recess, termed the **sinus piriformis,** which is bounded medially by the aryepiglottic fold, laterally by the thyroid cartilage and thyrohyoid membrane.

Muscles of the Pharynx.—The muscles of the pharynx (Fig. 16–47) are:

1. Constrictor inferior
2. Constrictor medius
3. Constrictor superior
4. Stylopharyngeus
5. Salpingopharyngeus
6. Palatopharyngeus

1. The **Constrictor pharyngis inferior** (*Inferior constrictor*) (Figs. 16–47, 16–48), the thickest of the three constrictors, *arises* from the sides of the cricoid cartilage in the interval between the Cricothyroideus anteriorly, and the articular facet for the inferior cornu of the thyroid cartilage posteriorly. It *arises* also from the oblique line on the side of the lamina of the thyroid cartilage, from the surface dorsal to this nearly as far as the posterior border and from the inferior cornu. From these origins the fibers spread dorsalward and medialward to be *inserted* with the muscle of the opposite side into the fibrous raphe in the posterior median line of the pharynx. The inferior fibers are horizontal and continuous with the circular fibers of the esophagus; the rest ascend, increasing in obliquity, and overlap the Constrictor medius.

2. The **Constrictor pharyngis medius** (*Middle constrictor*) (Figs. 16–47, 16–48) is a fan-shaped muscle, smaller than the preceding. It *arises* from the whole length of the superior border of the greater cornu of the hyoid bone, from the lesser cornu, and from the stylohyoid ligament. The fibers diverge from their origin: the lower ones descend beneath the Constrictor inferior; the middle fibers pass transversely; and the upper fibers ascend and overlap the Constrictor superior. It is *inserted* into the posterior median fibrous raphe, blending in the middle line with the muscle of the opposite side.

3. The **Constrictor pharyngis superior** (*Superior constrictor*) (Figs. 16–47, 16–48) is a quadrilateral muscle, thinner and paler than the other two. It *arises* from the inferior third of the posterior margin of the medial pterygoid plate and its hamulus, from the pterygomandibular raphe, from the alveolar process of the mandible above the posterior end of the mylohyoid line, and by a few fibers from the side of the tongue. The fibers curve posteriorly to be inserted into the median raphe, being also prolonged by means of an aponeurosis to the pharyngeal spine on the basilar part of the occipital bone. The superior fibers curve below the

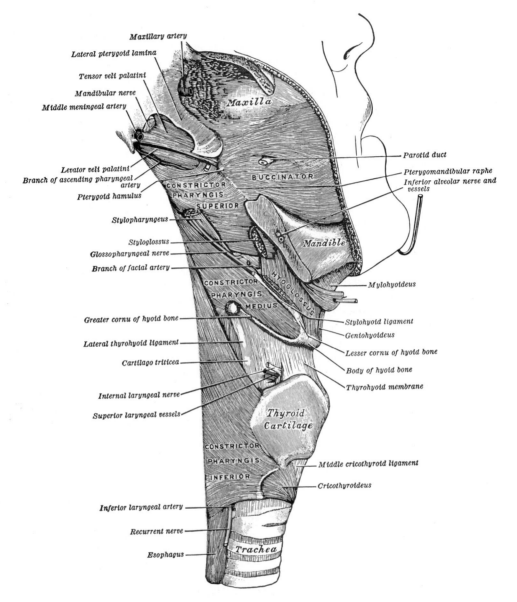

Fig. 16–47.—The Buccinator and muscles of the pharynx.

Levator veli palatini and the auditory tube. The interval between the upper border of the muscle and the base of the skull is closed by the pharyngeal aponeurosis, and is known as the **sinus of Morgagni.**

4. The **Stylopharyngeus** (Fig. 16–37), a long, slender muscle, *arises* from the medial side of the base of the styloid process, passes downward along the side of the pharynx between the Constrictores superior and

medius, and spreads out beneath the mucous membrane. Some of its fibers are lost in the Constrictor muscles, while others, joining with the Palatopharyngeus, are inserted into the posterior border of the thyroid cartilage. The glossopharyngeal nerve runs on the lateral side of this muscle, and crosses over it to reach the tongue.

5. The **Salpingopharyngeus** (Fig. 16–45) *arises* from the inferior part of the auditory

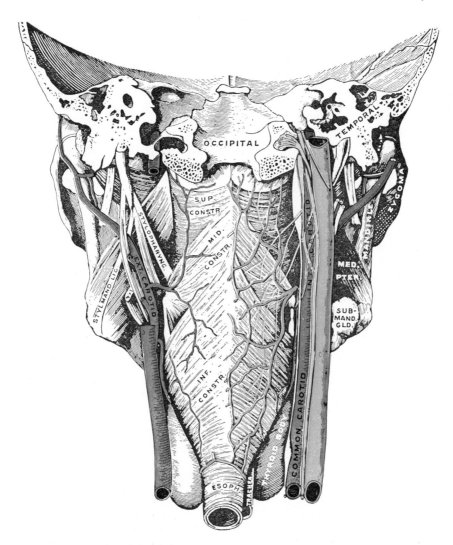

Fig. 16–48.—Muscles of the pharynx, together with the associated vessels and nerves, posterior aspect. (Modified after Testut.)

tube near its orifice; it passes downward and blends with the posterior fasciculus of the Palatopharyngeus.

6. The **Palatopharyngeus** is described above.

Nerves.—The Constrictores and Salpingopharyngeus are supplied by branches from the pharyngeal plexus, the Constrictor inferior by additional branches from the external laryngeal and recurrent nerves, and the Stylopharyngeus by the glossopharyngeal nerve.

Actions.—When deglutition is about to be performed, the pharynx is drawn upward and dilated in different directions, to receive the food propelled into it from the mouth. The Stylopharyngei, which are much farther removed from one another at their origin than at their insertion, draw the sides of the pharynx upward and lateralward, and so increase its transverse diameter; its breadth in the anteroposterior direction is increased by the larynx and tongue being carried forward in their ascent. As soon as the bolus of food is received in the pharynx, the elevator muscles relax, the pharynx descends, and the Constrictores contract upon the bolus, and convey it downward into the esophagus.

Structure.—The pharynx is composed of three coats: **mucous, fibrous,** and **muscular.**

The **pharyngeal aponeurosis,** or **fibrous coat,** is situated between the mucous and muscular layers. It is thick superiorly where the muscular fibers are wanting, and is firmly connected to the basilar portion of the occipital and the petrous portions of the temporal bones. As it descends it diminishes in thickness, and is gradually blended with the fascia. A strong fibrous band attached to the pharyngeal spine on the inferior surface of the basilar portion of the occipital bone forms the median raphe which serves as the insertion for all three of the Constrictores pharyngis.

The **mucous coat** is continuous with that lining the nasal cavities, the mouth, the auditory tubes, and the larynx. In the nasal part of the pharynx it is covered by columnar ciliated epithelium; in the oral and laryngeal portions the epithelium is stratified squamous. Beneath the mucous membrane are found racemose mucous glands; they are especially numerous at the cranial part of the pharynx around the orifices of the auditory tubes.

THE ESOPHAGUS

The **esophagus** or **gullet** (Fig. 12–25) is a muscular canal, about 23 to 25 cm long, extending from the pharynx to the stomach. It begins in the neck at the inferior border of the cricoid cartilage, opposite the sixth cervical vertebra, descends along the ventral aspect of the vertebral column, passes through the superior and posterior mediastina and the diaphragm, and, entering the abdomen, ends at the cardiac orifice of the stomach, opposite the eleventh thoracic vertebra. The general direction of the esophagus is vertical; but it presents two slight curves in its course. At its commencement it is placed in the middle line; but it inclines to the left side as far as the root of the neck, gradually passes to the middle line again at the level of the fifth thoracic vertebra, and finally deviates to the left as it passes ventralward to the esophageal hiatus in the diaphragm. The esophagus also presents anteroposterior flexures corresponding to the curvatures of the cervical and thoracic portions of the vertebral column. It is the narrowest part of the digestive tube, and is most contracted at its commencement, and at the point where it passes through the diaphragm.

Relations.—The **cervical portion** of the esophagus is in relation, *ventrally,* with the trachea; and at the lower part of the neck, where it projects to the left side, with the thyroid gland; dorsally, it rests upon the vertebral column and Longus colli muscles; *on either side* it is in relation with the common carotid artery (especially the left, as it inclines to that side), and parts of the lobes of the thyroid gland; the recurrent nerves ascend between it and the trachea; to its left side is the thoracic duct.

The **thoracic portion** of the esophagus is at first situated in the superior mediastinum between the trachea and the vertebral column, a little to the left of the median line. It then passes dorsal and to the right of the aortic arch, and descends in the posterior mediastinum along the right side of the descending aorta, then runs ventral and a little to the left of the aorta, and enters the abdomen through the diaphragm at the level of the tenth thoracic vertebra. Just before it perforates the diaphragm it presents a distinct dilatation. It is in relation, *ventrally,* with the trachea, the left bronchus, the pericardium, and the diaphragm; *dorsally,* it rests upon the vertebral column, the Longus colli muscles, the right aortic intercostal arteries, the thoracic duct, and the hemiazygos veins; and near the diaphragm, upon the ventral surface of the aorta. On its *left* side, in the superior mediastinum, are the terminal part of the aortic arch, the left subclavian artery, the thoracic duct, and left pleura, while running upward in the angle between it and the trachea is the left recurrent nerve; caudally, it is in relation with the descending thoracic aorta. On its *right* side are the right pleura, and the azygos vein which it overlaps. Below the roots of the lungs the vagi descend in close contact with it, the right nerve passing down dorsal, and the left nerve ventral to it, the two nerves forming a plexus around the tube.

In the caudal part of the posterior mediastinum the thoracic duct lies to the right side of the esophagus; more cranially, it is placed dorsal to it, and, crossing about the level of the fourth thoracic vertebra, is continued upward on its left side.

The **abdominal portion** of the esophagus lies in the esophageal groove on the posterior surface of the left lobe of the liver. It measures about 1.25 cm in length, and only its ventral and left aspects are covered by peritoneum. It

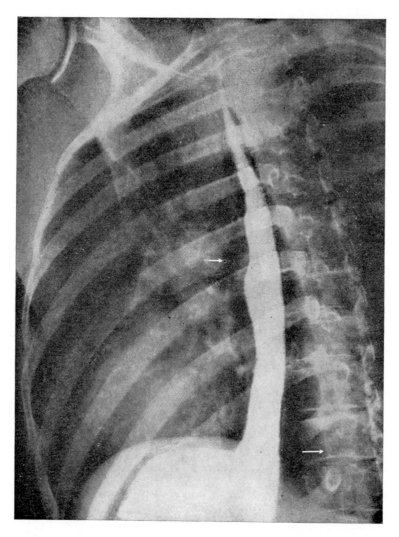

FIG. 16–49.—Esophagus during the passage of a barium meal. Note that in the upper part of the esophagus longitudinal folds in the mucous membrane can be identified. The upper arrow points to the shadow of the right bronchus; the lower arrow indicates the tenth thoracic vertebra. Note that the lower part of the esophagus inclines forward away from the vertebral column.

is somewhat conical with its base applied to the upper orifice of the stomach, and is known as the **antrum cardiacum.**

Structure (Fig. 16–50).—The esophagus has four coats: an **external** or **fibrous,** a **muscular,** a **submucous** or **areolar,** and an **internal** or **mucous coat.**

The **muscular coat** (*tunica muscularis*) is composed of two planes of considerable thickness: an external of longitudinal and an internal of circular fibers.

The *longitudinal fibers* are arranged, at the commencement of the tube, in three fasciculi: one ventral, which is attached to the vertical

ridge on the posterior surface of the lamina of the cricoid cartilage by the *tendo cricoësophageus;* and one at either side, which is continuous with the muscular fibers of the pharynx: as they descend they blend together, and form a uniform layer, which covers the outer surface of the tube.

Accessory slips of muscular fibers pass between the esophagus and the left pleura, where the latter covers the thoracic aorta, or the root of the left bronchus, or the back of the pericardium.

The *circular fibers* are continuous above with the Constrictor pharyngis inferior; their direc-

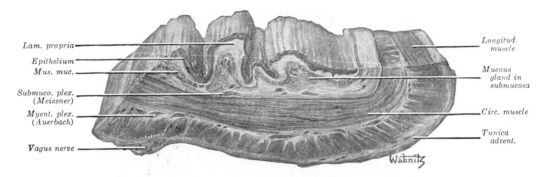

Lam. propria
Epithelium
Mus. muc.
Submuco. plex.
(Meissner)
Myent. plex.
(Auerbach)
Vagus nerve

Longitud.
muscle
Mucous
gland in
submucosa
Circ. muscle
Tunica
advent.

Fig. 16–50.—Cross section through lower part of esophagus of adult cadaver. × 5.

tion is transverse at the cranial and caudal parts of the tube, but oblique in the intermediate part.

The muscular fibers in the cranial part of the esophagus are of a red color, and consist chiefly of striated muscle; the intermediate part is mixed and the lower part, with rare exceptions, contains only smooth muscle.

The **areolar** or **submucous coat** (*tela submucosa*) loosely connects the mucous and muscular coats. It contains blood vessels, nerves, and mucous glands. The **esophageal glands** (*glandulae esophageae*) are small compound racemose glands of the mucous type: they are lodged in the submucous tissue, and each opens upon the surface by a long excretory duct.

The **mucous coat** (*tunica mucosa*) is thick, of a reddish color cranially, and pale caudally. It is disposed in longitudinal folds, which disappear on distention of the tube. Its surface is covered throughout with a thick layer of stratified squamous epithelium. Beneath the mucous membrane, between it and the areolar coat, is a layer of longitudinally arranged non-striped muscular fibers. This is the **muscularis mucosae.** At the commencement of the esophagus it is absent, or represented by only a few scattered bundles; more caudally it forms a considerable stratum.

Vessels and Nerves.—The **arteries** supplying the esophagus are derived from the inferior thyroid branch of the thyrocervical trunk, from the descending thoracic aorta, from the bronchial arteries, from the left gastric branch of the celiac artery, and from the left inferior phrenic of the abdominal aorta. They have for the most part a longitudinal direction.

The **veins** end in the inferior thyroid, azygos, hemiazygos, and gastric veins, thereby forming an important anastomosis between the portal and systemic venous systems.

The **nerves** are derived from the recurrent vagus, supplying the striated musculature of the organ, and from the vagus and sympathetic trunks which supply fibers to the smooth musculature; these cranial parasympathetic and sympathetic fibers form plexuses between the two layers of the muscular coat, and in the submucosa, as in the stomach and intestines.

THE ABDOMEN

The **abdomen** contains the largest cavity in the body. The cranial boundary is formed by the diaphragm which extends over it as a dome, so that the cavity extends high into the bony thorax, reaching on the right side, in the mammary line, to the cranial border of the fifth rib; on the left side it falls below this level by about 2.5 cm. The caudal boundary is formed principally by the Levator ani and Coccygeus or diaphragm of the pelvis. In order to facilitate description, it is artificially divided into two parts: a cranial and larger part, the abdomen proper; and a caudal and smaller part, the pelvis, the limit between them being marked by the superior aperture of the lesser pelvis.

The **abdomen proper** differs from the other great cavities of the body in being bounded for the most part by muscles and fasciae, so that it can vary in capacity and shape according to the condition of the viscera which it contains; but, in addition to this, the abdomen varies in form and extent with age and sex. In the adult male, with moderate distention of the viscera, it is oval in shape, but at the same time flattened dorsoventrally. In the adult female, with a fully developed pelvis, it is ovoid with the narrower pole cranialward and in young children it is also ovoid but with the narrower pole caudalward.

Boundaries.—It is bounded *ventrally* and at the sides by the abdominal muscles and the Iliacus; *dorsally* by the vertebral column and the Psoas and Quadratus lumborum muscles; *cranially* by the diaphragm; *caudally* by the plane of the superior aperture of the lesser pelvis. The muscles forming the boundaries of the cavity are lined upon their inner surfaces by transversalis fascia.

The abdomen contains the greater part of the digestive tube, the liver and pancreas, the spleen, the kidneys, and the suprarenal glands. Most of these structures, as well as the wall of the cavity in which they are contained, are more or less covered by an extensive and complicated serous membrane, the peritoneum.

The **apertures in the walls of the abdomen,** for the transmission of structures to or from it, are *cranially,* the **vena caval opening,** the **aortic hiatus,** and the **esophageal hiatus.** *Caudally,* there are two apertures on either side: one for the passage of the **femoral vessels** and lumboinguinal nerve, and the other for the transmission of the **spermatic cord** in the male, and the round ligament of the uterus in the female. In the fetus, the **umbilicus** transmits the umbilical vessels, the allantois, and the vitelline duct through the ventral wall.

Regions.—A simple but effective division of the abdomen, and one which is commonly used by clinicians, is established by two lines intersecting at right angles; the median sagittal line and a transverse line through the umbilicus divide the abdomen into **four quadrants,** an upper and lower right, and an upper and lower left. The more accurate and elaborate division described and depicted in the third chapter (page 73) will be used for reference in many descriptions of abdominal structures.

The following description is based upon the inspection of a cadaver. It must be kept continuously in mind that the abdominal organs are extremely movable during life and that the arbitrary placement of this account is adopted in order that a basic knowledge may be obtained for future adaptation to the conditions which will be seen in patients. It will be convenient to refer frequently to the second chapter, Figures 3–17, 3–19, 3–20, and 3–23, in order to relate the organs to the skeleton.

Disposition of the Abdominal Viscera.— When the anterior abdominal wall of the cadaver is removed, only a few of the undisturbed organs are plainly visible (Fig. 16–51). Cranially and to the right is the **liver,** with the falciform ligament marking its ventral surface into the right and left lobes. The narrow left lobe extends across the median line, partly concealing the lesser curvature of the stomach. The **stomach** occupies the left cranial region more or less completely depending on the degree of its distention. The tip of the **gallbladder** may protrude from the caudal border of the right lobe of the liver, and between this and the stomach is the pylorus and first part of the **duodenum.** Draped caudalward from the greater curvature of the stomach and the first part of the duodenum, a filmy apron, the **greater omentum** (or epiploon), extends caudalward, concealing most of the other intestines. Caudally and on the right, parts of the descending and sigmoid **colon** are usually visible and a variable number of coils of the small intestine appear in the caudal part of the cavity. Just above the pubis, the urinary **bladder** is visible if partly distended.

The **large** and **small intestines** are brought into view if the omentum is lifted and turned upward over the stomach and liver (Fig. 16–52). The large intestine forms an arch with the ascending colon on the right, the descending and sigmoid on the left, and the transverse colon, with the omentum attached, forming the top of the arch. The small intestine may be lifted and drawn to the left to display its attachment to the mesentery and the attachment of the colon to the transverse mesocolon. The blind end of the ascending colon, the **cecum,** lies in the hollow of the right ilium, and near the ileocecal junction the vermiform **appendix** may or may not be visible.

Further inspection requires removal of some of the organs and structures, namely, the small intestine except for the duodenum

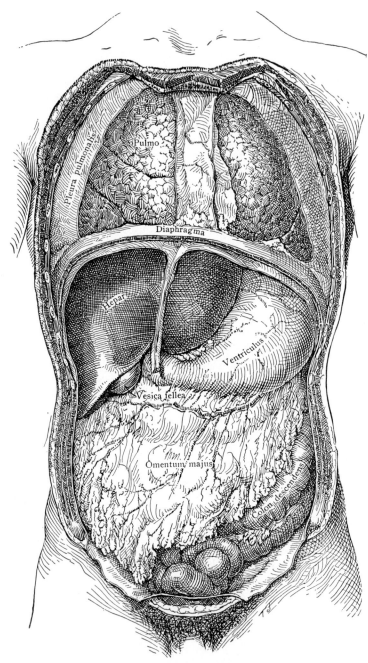

F<small>IG</small>. 16-51.—Ventral view of the thoracic and abdominal viscera in position after removal of the anterior thoracic and abdominal walls. (Eycleshymer and Jones.)

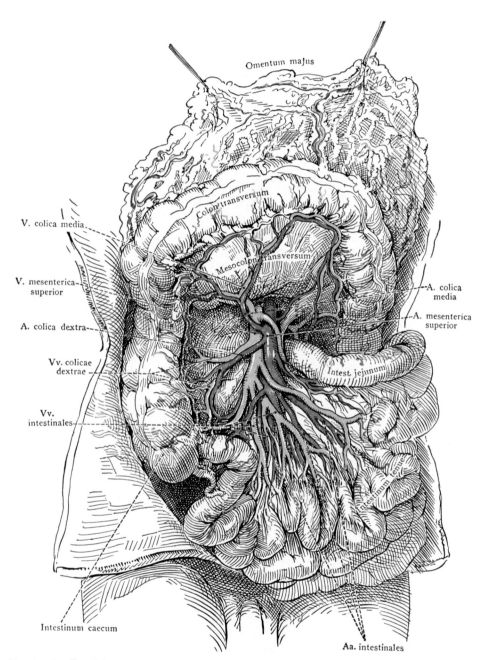

Fig. 16–52.—Small and large intestines with their mesenteries and blood vessels viewed after the greater omentum has been drawn upward over the chest. (Eycleshymer and Jones.)

and terminal ileum, the greater omentum, transverse colon with its mesocolon, and the left lobe of the liver (Fig. 16–53). It is then possible to see the terminal part of the esophagus as it passes through the diaphragm and joins the cardiac end of the stomach. The filmy **lesser omentum** stretches between the lesser curvature of the stomach and the liver and duodenum. The head of the **pancreas** lies in the loop formed by the other three parts of the duodenum. The **spleen** can be brought into view between the greater curvature of the stomach and the diaphragm if the stomach is pulled forcibly

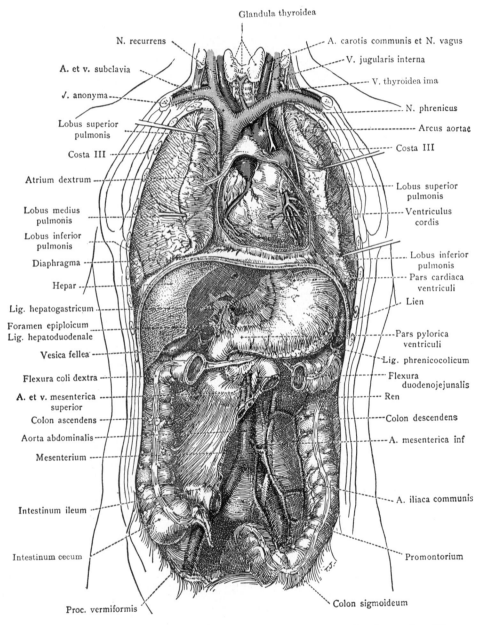

Glandula thyroidea

N. recurrens

A. et v. subclavia

J. anonyma

Lobus superior pulmonis

Costa III

Atrium dextrum

Lobus medius pulmonis

Lobus inferior pulmonis

Diaphragma

Hepar

Lig. hepatogastricum

Foramen epiploicum
Lig. hepatoduodenale

Vesica fellea

Flexura coli dextra

A. et v. mesenterica superior

Colon ascendens

Aorta abdominalis

Mesenterium

Intestinum ileum

Intestinum cecum

Proc. vermiformis

A. carotis communis et N. vagus

V. jugularis interna

V. thyroidea ima

N. phrenicus

Arcus aortae

Costa III

Lobus superior pulmonis

Ventriculus cordis

Lobus inferior pulmonis

Pars cardiaca ventriculi

Lien

Pars pylorica ventriculi

Lig. phrenicocolicum

Flexura duodenojejunalis

Ren

Colon descendens

A. mesenterica inf

A. iliaca communis

Promontorium

Colon sigmoideum

FIG. 16–53.—Ventral view of the thoracic and abdominal viscera partially dissected. The anterior pleurae and pericardium have been removed; the structures at the root of the neck dissected. The left lobe of the liver, the greater omentum, transverse colon, jejunum and ileum have been removed. (Eycleshymer and Jones.)

toward the liver. Since it is placed rather far dorsally it may not be seen but it can easily be palpated. If the subject is not obese, a smooth rounded prominence just caudal to the right and left flexures of the colon will represent the caudal poles of the two **kidneys.** The sigmoid colon is completely visible and its junction with the rec-

tum may be detected. In the female, if the sigmoid colon is drawn to the left and caudalward, the **uterus** lying in the middle line and the uterine tubes on each side may be brought into view. In the middle line extending caudally, the prominence caused by the abdominal **aorta** and its bifurcation into the iliac arteries should be discernible.

THE PERITONEUM

The **peritoneum** (*tunica serosa*) is the most extensive serous membrane in the body, and consists, in the male, of a closed sac, a part of which is applied against the abdominal parietes, while the remainder is reflected over the contained viscera. In the female the peritoneum is not a closed sac, since the free ends of the uterine tubes open directly into the peritoneal cavity. The part which lines the abdominal wall is named the **parietal peritoneum;** that which is reflected over the contained viscera constitutes the **visceral peritoneum.** The *free surface* of the membrane is a smooth layer of flattened mesothelium, lubricated by a small quantity of serous fluid, which allows the viscera to glide freely against the wall of the cavity or upon each other with the least possible friction. The *attached* surface is connected to the viscera and inner surface of the parietes by means of areolar tissue, termed the **subserous fascia.** The parietal portion is separated by a fascial cleft from the transversalis fascia lining of the abdomen and pelvis, but is more closely adherent to the under surface of the diaphragm, and also in the midline of the anterior wall.

The space between the parietal and visceral layers of the peritoneum is named the **peritoneal cavity;** but under normal conditions this cavity is merely a potential one, since the parietal and visceral layers are in contact. The peritoneal cavity is divided by a narrow constriction into a greater sac and a lesser sac. The **greater sac,** commonly referred to simply as the **peritoneal cavity,** is related to the great majority of the abdominal structures. The **lesser sac** is named the **omental bursa** and is related only to the dorsal surface of the stomach and closely surrounding structures. The constriction between the two sacs is between

the liver and duodenum and is named the **epiploic foramen** or *foramen of Winslow* (Figs. 16–53, 16–56).

Certain of the abdominal viscera are completely surrounded by peritoneum and are suspended from the wall by a thin sheet of peritoneal-covered connective tissue carrying the blood vessels. These sheets are given the general name of mesenteries. Other viscera are more closely attached to the abdominal wall and are only partly covered by peritoneum rather than suspended; these viscera are said to be **retroperitoneal.** No term corresponding to retroperitoneal has been adopted for the suspended viscera; intraperitoneal is unsatisfactory because this would imply that they are inside the peritoneal cavity. The term **mesentery,** when used specifically, refers only to the peritoneal suspension of the small intestine; "entery" referring to the intestine and "mes-" to the peritoneum. Thus the names of the suspending folds of other organs use only the "meso," for example, the mesocolon, mesoappendix, mesovarium.

The complicated disposition and variations of the peritoneum in the adult body can be understood only through frequent reference to the changes in position and attachment of the viscera during embryonic and fetal development described in the first part of this chapter.

The **Parietal Peritoneum.**—The peritoneum covering the internal surface of the ventral abdominal wall is smooth and relatively uninterrupted by folds and attachments. Certain vestiges of embryonic structures, however, have remained. Between the pubic bone and the umbilicus, a fibrous band in the midline, the middle umbilical ligament, is the remains of the urachus, and

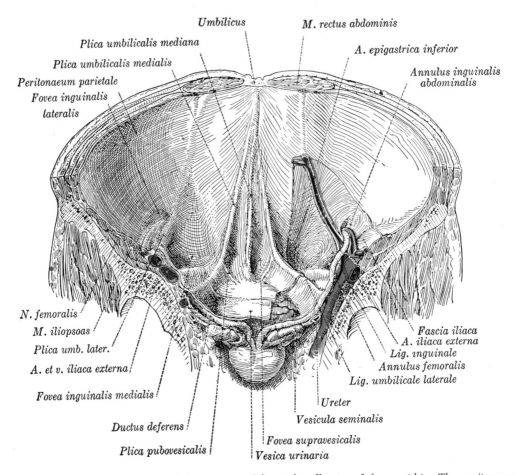

Fig. 16–54.—The lower portion of the anterior abdominal wall, viewed from within. The peritoneum has been partially removed from the right side. (Eycleshymer and Jones.)

three or four centimeters lateral to it are the lateral umbilical ligaments, fibrous remnants of the obliterated umbilical arteries. The fold of peritoneum over the urachal remnant is the **middle umbilical fold** (*plica umb. mediana*) (Fig. 16–54); the fold over the arterial remnants is the **medial umbilical fold** (*plica umb. medialis*). Three or four centimeters lateral to the latter fold is the **lateral umbilical fold** (*plica umb. lateralis*) which is produced by a slight protrusion of the inferior epigastric artery and the interfoveolar ligament. The contracted bladder may have transverse folds extending laterally from its fundus, the **transverse vesical folds** (*plicae ves. laterales*).

Extending cranialward from the umbilicus is the **falciform ligament** of the liver. It is a thin fibrous sheet with a fibrous cord,

the ligamentum teres, or remnant of the umbilical vein, in its free crescentic border. Its attachment to the wall is narrow from the umbilicus to the cranial surface of the liver where its two leaves separate and the reflections extend laterally to become the anterior leaf of **the coronary ligament of the liver** (*lig. coronarium hepatis*).

The **anterior leaf** of the coronary ligament is the anterior boundary of an irregularly oval or triangular area of liver surface approximately 15 cm in diameter, which is without peritoneal covering and is called, therefore, the **bare area** of the liver. The posterior boundary of this area is the **posterior leaf** of the coronary ligament where the peritoneum is reflected from the liver to the diaphragmatic portion of the dorsal abdominal wall. Between the anterior and posterior coronary leaves, the reflection extends laterally into a crescentic fold on each side, the **right** and **left triangular ligaments of** the liver (*lig. triangulare dextrum et sinistrum*) (Fig. 16–55).

On the *left side*, beginning at the falciform and left triangular ligaments on the *anterior abdominal wall,* the peritoneum is uninterrupted by attachments as it passes around the lateral wall and over part of the posterior wall, until it reaches the **gastrophrenic** and **phrenicolienal ligaments.** The latter represent the adult position of the fetal dorsal mesogastrium. The phrenicolienal ligament angles laterally and its more caudal part is reflected from the cranial pole of the left kidney and then meets the splenic or left flexure of the colon. The peritoneum at the reflection of the flexure is prolonged laterally for several centimeters into a narrow fold, the **phrenicocolic ligament,** representing the left lateral extent of the greater omentum.

The peritoneum of the *anterior* and *left lateral abdominal wall* from this ligament down into the pelvis and into the rectovesical fossa is uninterrupted by attachments. It is reflected over the descending colon without mesentery or mesocolon, then the **sigmoid mesocolon** swings laterally into a reflection over the rectum.

The peritoneum of the *right side* of the *anterior* and *lateral abdominal wall* extending caudally into the iliac region is uninterrupted by attachments. Along the right side of the posterior abdominal wall, there is a reflection to the ascending colon and cecum. Between the right triangular ligament and the attachment of the right or hepatic flexure of the colon an unattached peritoneal surface 3 or 4 cm wide extends medialward and cranialward between the anterior leaf of the coronary ligament of the liver cranially, and the **transverse mesocolon** and reflection over the superior part of the duodenum caudally. Medial to the duodenum, the peritoneum of the posterior wall covers the inferior vena cava and this structure marks the position of the **epiploic foramen** which leads toward the left into the **lesser sac** or **omental bursa.**

The reflections of peritoneum from the posterior wall which form the *boundaries of the lesser sac,* beginning at the vena cava,

are as follows: The coronary ligament of the liver, along the caudate lobe to the diaphragmatic end of the lesser omentum, the esophagus and the gastrophrenic ligament, the phrenicolienal ligament, containing the tail of the pancreas, the transverse mesocolon, and the first part of the duodenum (Fig. 16–57).

The reflection of peritoneum to the ascending colon extends caudalward along the *right dorsolateral wall* to the cecum in the right iliac fossa. After rounding the blind end of the cecum the peritoneum from the dorsal wall is reflected to the mesentery of the ileum and jejunum. The attachment of the mesentery extends medialward as well as cranialward until it reaches the junction of the duodenum with the jejunum at the caudal leaf of the transverse mesocolon. A triangular area of dorsal wall to the right of the mesentery is thus completely surrounded by peritoneal reflections, viz., the ascending colon laterally, the transverse mesocolon cranially and the mesentery medially. The peritoneum in this triangular area covers part of the right kidney and duodenum.

The peritoneum of the dorsal wall between the mesentery, transverse mesocolon, and descending and sigmoid colons extends into the right side of the pelvis between the rectum and bladder. Here it follows closely the surfaces of the pelvic viscera and the inequalities of the pelvic walls, and presents important differences in the two sexes.

(*a*) **In the male** (Fig. 16–55) it encircles the sigmoid colon, from which it is reflected to the posterior wall of the pelvis as a fold, the **sigmoid mesocolon.** It then leaves the sides, and finally the ventral surface of the rectum, and is continued on to the cranial ends of the seminal vesicles and the bladder; on either side of the rectum it forms the **pararectal fossa,** which varies in size with the distention of the rectum. Ventral to the rectum the peritoneum forms the **rectovesical excavation** which is limited laterally by peritoneal folds extending from the sides of the bladder to the rectum and sacrum. These folds are known from their position as the **rectovesical** or **sacrogenital folds.** The peritoneum of the ventral pelvic wall covers the superior surface of the bladder,

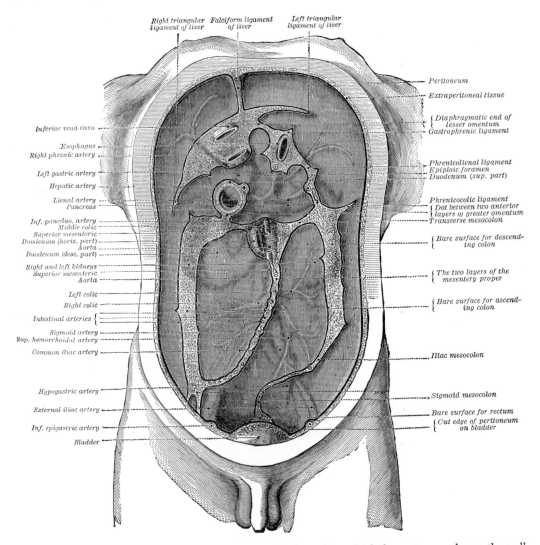

Right triangular ligament of liver — Falciform ligament of liver — Left triangular ligament of liver

Peritoneum

Extraperitoneal tissue

{ Diaphragmatic end of lesser omentum
Gastrophrenic ligament

Inferior vena cava

Phrenicolienal ligament
Epiploic foramen
Duodenum (sup. part)

Esophagus
Right phrenic artery

Left gastric artery

Hepatic artery

Phrenicocolic ligament
{ Dot between two anterior layers of greater omentum
Transverse mesocolon

Lienal artery
Pancreas

Inf. pancduo. artery
Middle colic
Superior mesenteric
Duodenum (horiz. part)
Aorta
Duodenum (desc. part)

{ Bare surface for descending colon

Right and left kidneys
Superior mesenteric
Aorta

{ The two layers of the mesentery proper

Left colic
Right colic

{ Bare surface for ascending colon

Intestinal arteries {

Sigmoid artery
Sup. hemorrhoidal artery

Common iliac artery

Iliac mesocolon

Hypogastric artery

Sigmoid mesocolon

External iliac artery

Bare surface for rectum
{ Cut edge of peritoneum on bladder

Inf. epigastric artery

Bladder

FIG. 16–55.—Diagram devised by Delépine to show the lines along which the peritoneum leaves the wall of the abdomen to invest the viscera.

and on either side of this viscus forms a depression, termed the **paravesical fossa,** which is limited laterally by the fold of peritoneum covering the ductus deferens. The size of this fossa is dependent on the state of distention of the bladder; when the bladder is empty, a variable fold of peritoneum, the **plica vesicalis transversa,** divides the fossa into two portions. Between the paravesical and pararectal fossae the only elevations of the peritoneum are those produced by the ureters and the internal iliac vessels (Fig. 17–15).

(b) **In the female,** pararectal and paravesical fossae similar to those in the male are present: the lateral limit of the paravesical fossa is the peritoneum investing the round ligament of the uterus. The rectovesical excavation is, however, divided by the uterus and vagina into a small anterior vesicouterine and a large, deep, posterior rectouterine excavation. The sacrogenital folds form the margins of the latter, and are continued on to the dorsum of the uterus to form a transverse fold, the **torus uterinus.** The **broad ligaments** extend from the sides of the uterus to the lateral walls of the pelvis; they contain in their free margins the uterine tubes, and suspended from their posterior layers the ovaries. On the lateral pelvic wall dorsal to the attachment of the broad ligament, in the angle between the elevations produced by the diverging internal iliac and external iliac vessels is a slight fossa, the **ovarian fossa,** in which the ovary normally lies (Fig. 17–59).

The **Omental Bursa or Lesser Peritoneal Sac** (*bursa omentalis*) is so named because a part of its wall is formed by the two omenta (to be described below). On the dorsal abdominal wall the peritoneum of the greater sac is continuous with that of the lesser sac as it crosses the ventral surface of the inferior vena cava a short distance to the right of the midline (Fig. 16–57). This passage of continuity between the two sacs is the **epiploic foramen** or *foramen of Winslow.*

The **epiploic foramen** (Fig. 16–56) is a peritoneal-covered passage which usually will admit two fingers. It can easily be located by the probing finger in the abdomi-

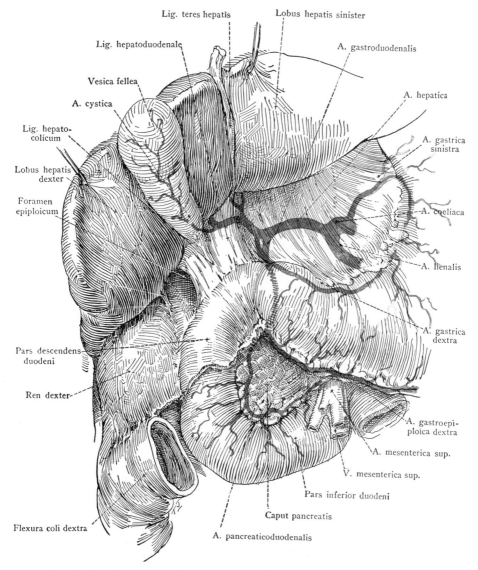

FIG. 16–56.—The epiploic foramen (of Winslow) and neighboring structures. (Eycleshymer and Jones.)

nal cavity by pushing its way between the inferior vena cava and the free edge of the hepatoduodenal ligament in the region between the neck of the gallbladder and the first part of the duodenum (Fig. 16–56). It is bounded *ventrally* by the free border of the lesser omentum, with the common bile duct, hepatic artery, and portal vein between its two layers; *dorsally* by the peritoneum covering the inferior vena cava; *cranially* by the peritoneum on the caudate process of the liver, and *caudally* by the peritoneum covering the commencement of the duodenum and the hepatic artery before it passes between the two layers of the lesser omentum.

The **boundaries** of the **omental bursa** will now be evident. It is bounded *ventrally* by the caudate lobe of the liver, the lesser omentum, the stomach, and the greater omentum. *Dorsally*, it is limited by the greater omentum, the transverse colon, the transverse mesocolon, the ventral surface of the pancreas, the left suprarenal gland, and the cranial end of the left kidney. To the right of the esophageal opening of the stomach it is formed by that part of the diaphragm which supports the caudate lobe of the liver. *Laterally*, the bursa extends from the epiploic foramen to the hilum of the spleen, where it is limited by the phrenicolienal and gastrolienal ligaments.

The omental bursa, therefore, consists of a series of pouches or **recesses** to which the following terms are applied: (1) the **vestibule**, a narrow channel continued from the epiploic foramen, over the head of the pancreas to the gastropancreatic fold; this fold extends from the omental tuberosity of the pancreas to the right side of the fundus of the stomach, and contains the left gastric artery and coronary vein; (2) the **superior omental recess**, between the caudate lobe of the liver and the diaphragm; (3) the **lienal recess**, between the spleen and the stomach; (4) the **inferior omental recess**, which comprises the remainder of the bursa (Fig. 16–57).

In the fetus the bursa reaches as low as the free margin of the greater omentum, but in the adult its vertical extent is usually more limited owing to adhesions between the layers of the omentum. During a considerable part of fetal life the transverse colon is suspended from the posterior abdominal wall by a mesentery of its own, the greater omentum passing ventral to the colon (Fig. 16–13). This condition occasionally persists throughout life, but as a rule adhesion occurs between the mesentery of the transverse colon and the posterior layer of the greater omentum, with the result that the colon appears to receive its peritoneal covering by the splitting of the two posterior layers of the latter fold (Fig. 16–14). In the adult the omental bursa intervenes between the stomach and the structures on which that viscus lies, and performs therefore the functions of a serous bursa for the stomach.

The **visceral peritoneum** covers all or part of each abdominal organ and forms the various mesenteries, omenta, and ligaments associated with them.

Mesenteries, omenta, and **peritoneal ligaments.**—Numerous peritoneal folds extend between the various organs or connect them to the parietes; they serve to hold the viscera in position, and, at the same time, enclose the vessels and nerves proceeding to them.

The **mesenteries** are: the mesentery proper, the transverse mesocolon, and the sigmoid mesocolon. In addition to these there are sometimes present an ascending and a descending mesocolon.

The **mesentery proper** (*mesenterium*) is the broad, fan-shaped fold of peritoneum which connects the convolutions of the jejunum and ileum with the dorsal wall of the abdomen. Its *root*—the part connected with the structures ventral to the vertebral column—is about 15 cm long extending obliquely from the duodenojejunal flexure at the left side of the second lumbar vertebra to the right sacroiliac articulation (Fig. 16–55). Its *intestinal border* is about 6 meters long; and here the two layers separate to enclose the intestine, and form its peritoneal coat. Its cranial end is narrow but it widens rapidly to about 20 cm, and is thrown into numerous plaits or folds. It suspends the small intestine, and contains between its layers the intestinal branches of the superior mesenteric artery, with their accompanying tributaries of the portal vein and plexuses of nerves, the lacteal vessels, and mesenteric lymph nodes.

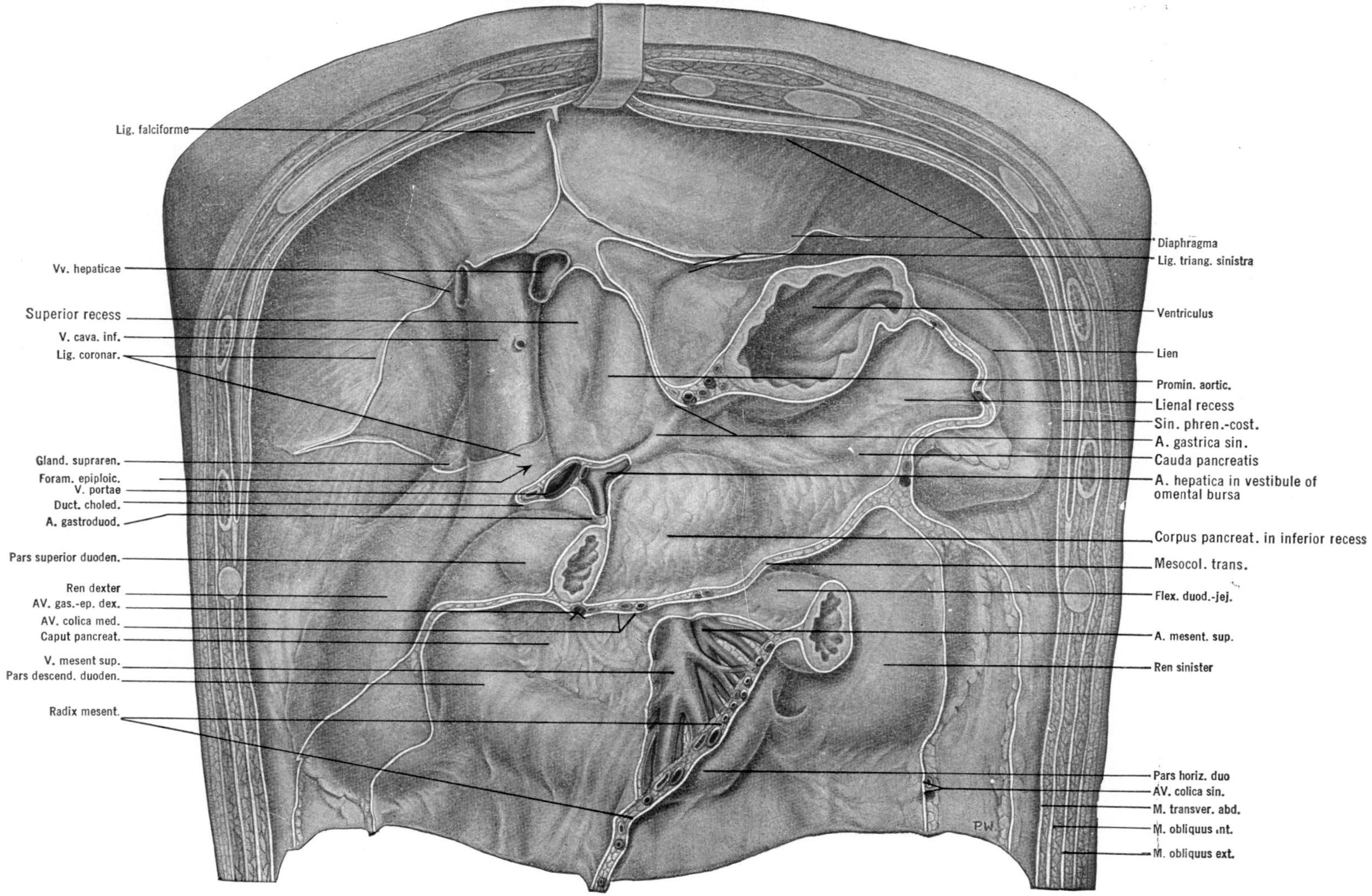
Lig. falciforme
Vv. hepaticae
Superior recess
V. cava. inf.
Lig. coronar.
Gland. supraren.
Foram. epiploic.
V. portae
Duct. choled.
A. gastroduod.
Pars superior duoden.
Ren dexter
AV. gas.-ep. dex.
AV. colica med.
Caput pancreat.
V. mesent sup.
Pars descend. duoden.
Radix mesent.
Diaphragma
Lig. triang. sinistra
Ventriculus
Lien
Promin. aortic.
Lienal recess
Sin. phren.-cost.
A. gastrica sin.
Cauda pancreatis
A. hepatica in vestibule of omental bursa
Corpus pancreat. in inferior recess
Mesocol. trans.
Flex. duod.-jej.
A. mesent. sup.
Ren sinister
Pars horiz. duo
AV. colica sin.
M. transver. abd.
M. obliquus int.
M. obliquus ext.
P.W.

FIG. 16–57.—View of the posterior wall of the lesser sac of the peritoneum (omental bursa), showing the attachment of the ligaments of the liver, the hepatoduodenal ligament, the root of the mesentery and transverse mesocolon, and reflection of peritoneum from the stomach. The aorta, pancreas, kidneys, spleen, and duodenum are covered by peritoneum. (Töndury, *Angewandte und topographische Anatomie*, courtesy of Georg Thieme Verlag.)

The **transverse mesocolon** (*mesocolon transversum*) is a broad fold, which connects the transverse colon to the dorsal wall of the abdomen. It is continuous with the greater omentum along the ventral surface of the transverse colon. Its two peritoneal layers diverge along the anterior border of the pancreas. It contains between its layers the vessels which supply the transverse colon.

The **sigmoid mesocolon** (*mesocolon sigmoideum*) is the fold of peritoneum which retains the sigmoid colon in connection with the pelvic wall. Its line of attachment forms a U-shaped curve, the apex of the curve being placed about the point of division of the left common iliac artery. It is continuous with the *iliac mesocolon* on the left and ends in the median plane at the level of the third sacral vertebra over the rectum. The sigmoid and superior rectal vessels run between the two layers of this fold.

In most cases the peritoneum covers only the ventral surface and sides of the ascending and descending parts of the colon. Sometimes, however, these are surrounded by the serous membrane and attached to the posterior abdominal wall by an ascending and a descending mesocolon respectively. A fold of peritoneum, the **phrenicocolic ligament,** is continued from the left colic flexure to the diaphragm opposite the tenth and eleventh ribs; it forms a pocket which supports the spleen, and therefore has received the name of *sustentaculum lienis*.

The **appendices epiploicae** are small pouches of the peritoneum filled with fat and situated along the colon and upper part of the rectum. They are chiefly appended to the transverse and sigmoid parts of the colon.

There are two omenta, the lesser and the greater.

The **lesser omentum** (*omentum minus; small omentum; gastrohepatic omentum*) extends to the liver from the lesser curvature of the stomach and the commencement of the duodenum. It is continuous with the two layers of peritoneum which cover respectively the ventral and dorsal surfaces of the stomach and first part of the duodenum. The two peritoneal layers leave the stomach and duodenum as a thin membrane and ascend to the porta of the liver, to the left of the porta the omentum is attached to the bottom of the fossa for the ductus venosus, along which it extends to the diaphragm, where the two layers separate to embrace the end of the esophagus. At the right, the omentum ends in a free margin which constitutes the ventral boundary of the epiploic foramen. The portion of the lesser omentum between the liver and stomach is termed the **hepatogastric ligament;** that between the liver and duodenum is the **hepatoduodenal ligament** (Fig. 16–56). Between the two layers of the hepatoduodenal ligament, close to the right free margin, are the hepatic artery, the common bile duct, the portal vein, lymphatics, and the hepatic plexus of nerves—all these structures being enclosed in a fibrous capsule (*Glisson's capsule*). The right and left gastric vessels run between the layers of the lesser omentum where they are attached to the stomach.

The **greater omentum** (*omentum majus; gastrocolic omentum*) (Fig. 16–51) is a filmy apron draped over the transverse colon and coils of the small intestine. It is attached along the greater curvature of the stomach and the first part of the duodenum; its left border is continuous with the gastrolienal ligament. If it is lifted and turned back cranialward over the stomach and liver, one sees that it is adherent to the transverse colon along the latter's whole length across the abdomen (Fig. 16–52). It may be a single membrane, covered with peritoneum on both surfaces, from the colon to its free caudal border, but a variable amount of it near the colon often is composed of two peritoneal-covered membranes with a pocket in between. If it contains this pocket, only the dorsal of the two membranes is attached to the colon and the pocket opens into the omental bursa. This condition is best understood by a reference to the conditions in the fetus depicted in the diagrams of Figure 16–14. The membrane forming the omentum is thin, transparent, and fenestrated, except where there are blood vessels and accumulations of fat. The membrane between the stomach and the colon is the **gastrocolic ligament.** The right and left gastroepiploic blood vessels run between its two peritoneal layers near their attachment to the greater curvature of the stomach, and must be avoided when the ligament is cut to gain access to the omental bursa.

The greater omentum is a remarkable structure. It is extremely movable in the living individual, and seems to have the ability to spread itself in areas where its presence is useful for the bodily economy. For example, in a patient with a ruptured appendix, the omentum may be found covering the area, walling off the infection into an abscess and preventing generalized peritonitis. Occasionally it finds its way into the sac of a hernia, where it may form

Peritoneal Recesses or **Fossae** (*retroperitoneal fossae*).—In certain parts of the abdominal cavity there are recesses of peritoneum forming culs-de-sacs or pouches, which are of surgical interest in connection with the possibility of the occurrence of *intraabdominal* or *retroperitoneal herniae* (Mayo *et al.* '41). The largest of these is the omental bursa (already described), but several others, of smaller size, require mention, and may be divided into three groups, viz.: duodenal, cecal, and intersigmoid.

1. **Duodenal Fossae** (Figs. 16–58, 16–59). —Three are fairly constant, viz.: (*a*) The

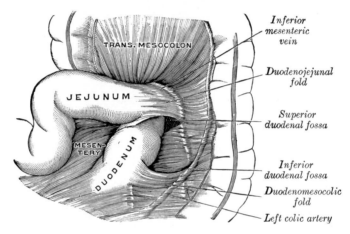

FIG. 16–58.—Superior and inferior duodenal fossae. (Poirier and Charpy.)

adhesions, close up the sac, and produce a "spontaneous cure" of the hernia. In a cadaver it may be free and spread over the intestines or may be bunched up near the colon or into a corner. Adhesions, except along the stomach and colon, are abnormal and the result of inflammatory processes.

Ligaments.—The term ligament has two meanings. When applied to structures related to a joint, it is a strong fibrous cord or sheet. When applied to a serous membrane, it is merely a layer of serous membrane and has little or no tensile strength. The **ligaments** of the **peritoneum** are the parts of the membrane extending between two structures and usually are named from these two structures, the gastrolienal ligament, for example. The ligaments are described in detail along with the organs to which they are related.

inferior duodenal fossa, present in about 75 per cent of bodies, is situated opposite the third lumbar vertebra on the left side of the ascending portion of the duodenum. It is bounded by a thin sharp fold of peritoneum and the tip of the index finger introduced into the fossa passes for some little distance caudalward behind the ascending portion of the duodenum. (*b*) The **superior duodenal fossa**, present in about 50 per cent of bodies, often coexists with the inferior. It lies on the left of the ascending portions of the duodenum, at the level of the second lumbar vertebra. It extends cranialward behind the sickle-shaped duodenojejunal fold, and has a depth of about 2 cm. (*c*) The **duodenojejunal fossa** exists in about 20 per cent of bodies. It lies between the right and left duodenomesocolic folds, extending cranial-

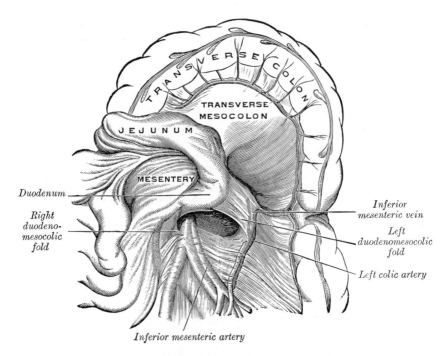

FIG. 16–59.—Duodenojejunal fossa. (Poirier and Charpy.)

ward behind the duodenojejunal junction toward the pancreas between the aorta on the right and kidney on the left. It has a depth of from 2 to 3 cm; its orifice is nearly circular and will admit the tip of the finger. (*d*) The **paraduodenal fossa**, rarely found, lies a short distance to the left of the ascending portion of the duodenum behind a peritoneal fold which contains the inferior mesenteric vein. When this fossa is very large, containing most of the small intestine, its presence has been explained by an abnormal rotation of the gut in embryonic development (Mayo *et al.* '41). A similar condition is usual in squirrel monkeys. (*e*) The **retroduodenal fossa**, only occasionally present, lies dorsal to the horizontal and ascending parts of the duodenum and ventral to the aorta.

2. **Cecal Fossae** (*pericecal folds or fossae*).—There are three principal pouches or recesses in the neighborhood of the cecum (Figs. 16–60 to 16–62): (*a*) The **superior ileocecal fossa** is formed by a fold of peritoneum, arching over the branch of the ileocolic artery which supplies the ileocolic junction. The fossa is a narrow chink situated between the mesentery of the small

intestine, the ileum, and the small portion of the cecum behind. (*b*) The **inferior ileocecal fossa** is situated behind the angle of junction of the ileum and cecum. It is formed by the ileocecal fold of peritoneum (*bloodless fold of Treves*), the upper border of which is fixed to the ileum, opposite its mesenteric attachment, while the lower border, passing over the ileocecal junction, joins the **mesenteriole** of the **vermiform appendix** (or *process*), and sometimes the appendix itself. Between this fold and the mesenteriole of the appendix is the **inferior ileocecal fossa**. It is bounded cranially by the dorsal surface of the ileum and the mesentery; ventrally and caudally by the ileocecal fold, and dorsally by the upper part of the mesenteriole of the vermiform appendix. (*c*) The **cecal fossa** is situated immediately behind the cecum, which has to be raised to bring it into view. It varies much in size and extent. In some cases it is sufficiently large to admit the index finger, and extends upward behind the ascending colon in the direction of the kidney; in others it is merely a shallow depression. It is bounded on the right by the **cecal fold**,

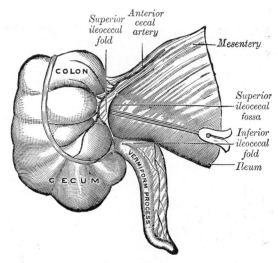

FIG. 16–60.—Superior ileocecal fossa. (Poirier and Charpy.)

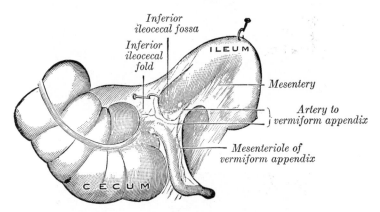

FIG. 16–61.—Inferior ileocecal fossa. The cecum and ascending colon have been drawn lateralward and downward, the ileum upward and backward, and the vermiform appendix downward. (Poirier and Charpy.)

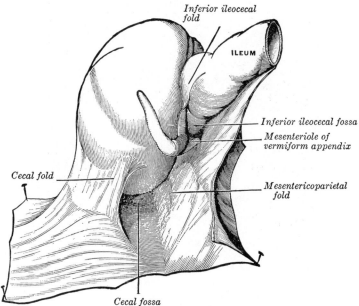

FIG. 16–62.—The cecal fossa. The ileum and cecum are drawn dorsalward and cranialward. (Souligoux.)

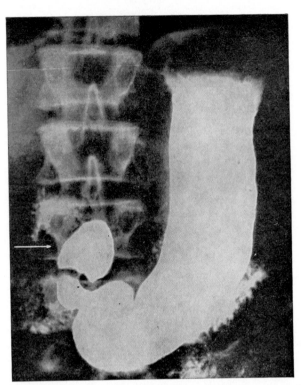

Fig. 16–63.—Normal stomach after a barium meal. The tone of the muscular wall is good and supports the weight of the column in the body of the organ. The arrow points to the duodenal cap, below which a gap in the barium indicates the position of the pylorus.

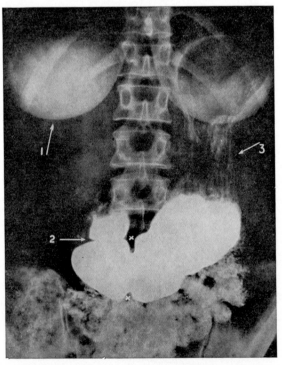

Fig. 16–64.—Atonic stomach after a barium meal. Note that this stomach contains the same amount of barium as the stomach in Fig. 16–63. Arrow 1 points to the shadow of the right breast; arrow 2, to the pylorus; arrow 3, to the upper part of the body of the stomach, where longitudinal folds can be seen in the mucous membrane. × × marks a wave of peristalsis.

which is attached by one edge to the abdominal wall from the caudal border of the kidney to the iliac fossa and by the other to the posterolateral aspect of the colon. In some instances additional fossae, the **retrocecal fossae**, are present.

3. The **Intersigmoid Fossa** (*recessus intersigmoideus*) (Fig. 16–84) is constant in the fetus and during infancy, but disappears in a certain percentage of individuals as age advances. When the sigmoid colon is drawn cranialward, the left surface of the sigmoid mesocolon is exposed, and on it will be seen a funnel-shaped recess of the peritoneum lying on the external iliac vessels in the interspace between the Psoas and Iliacus muscles. This is the *intersigmoid fossa*, which lies dorsal to the sigmoid mesocolon, and ventral to the parietal peritoneum. The fossa varies in size; in some instances it is a mere dimple, whereas in others it will admit the whole of the index finger.

THE STOMACH

The **stomach** (*ventriculus; gaster*) is situated in the right upper quadrant of the abdomen, partly covered by the ribs. It lies in a recess in the epigastric and left hypochondriac regions bounded by the anterior abdominal wall and the diaphragm, between the liver and the spleen.

The **shape and position** of the stomach are so greatly modified by changes within itself and in the surrounding viscera that no one form can be described as typical. The chief modifications are determined by (1) the amount of the stomach contents, (2) the stage which the digestive process has reached, (3) the degree of development of the gastric musculature, and (4) the condition of the adjacent intestines. It is possible, however, by comparing a series of stomachs to determine certain points more or less common to all (Figs. 16–65, 16–66).

Openings.—The opening from the esophagus into the stomach is known as the **cardiac orifice** (*ostium cardiacum*), so named from

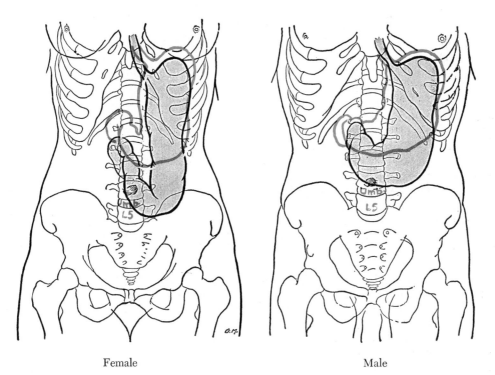

Female   Male

Fɪɢ. 16–65.—Average position of the stomach based on x-ray studies. Standing position in black; reclining position in red. (Eycleshymer and Jones.)

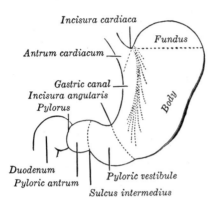

Fig. 16–66.—Diagram showing the subdivisions of the human stomach. (F. T. Lewis.)

its close relationship with the part of the diaphragm upon which the heart rests. The short abdominal portion of the esophagus is situated to the left of the middle line at the level of the tenth thoracic vertebra. It is curved sharply to the left and dilated into the **antrum cardiacum.** The right margin of the esophagus is continuous with the lesser curvature of the stomach; the left margin joins the greater curvature at an acute angle, forming the **incisura cardiaca.**

The opening of the stomach into the duodenum is the **pyloric orifice** (*ostium pyloricum*) which lies to the right of the middle line at the level of the cranial border of the first lumbar vertebra. Its position is usually indicated on the surface of the stomach by a circular groove, the duodenopyloric constriction.

Curvatures.—The **lesser curvature** (*curvatura ventriculi minor*) (Fig. 16–67) forms the right or concave border of the stomach. Nearer its pyloric than its cardiac end is a well-marked notch, the **incisura angularis** (Fig. 16–66), which varies somewhat in position with the state of distention; it serves to separate the stomach into a right and a left portion. The hepatogastric ligament (*lesser omentum*) which contains the left gastric artery and the right gastric branch of the hepatic artery is attached to the lesser curvature.

The **greater curvature** (*curvatura ventriculi major*) is directed to the left and ventralward, and is four or five times as long as the lesser curvature. Starting from the cardiac orifice at the incisura cardiaca,

it arches cranialward, and to the left; the highest point of the convexity is on a level with the sixth left costal cartilage. From this level it curves more gradually caudalward, with a slight convexity to the left as low as the cartilage of the ninth rib; it then continues to the right, to end at the pylorus. Directly opposite the incisura angularis of the lesser curvature the greater curvature presents a dilatation, which is the **pyloric vestibule;** this dilatation is limited on the right by a slight groove, the **sulcus intermedius,** which is about 2.5 cm from the pyloric ostium. The portion between the sulcus intermedius and the pyloric ostium is termed the **pyloric antrum.** Attached along the greater curvature are certain mesenteric membranes derived from the embryonic dorsal mesogastrium; from the fundus, the gastrophrenic ligament extends to the diaphragm; from the cranial part of the body, the gastrolienal ligament runs to the spleen; and the greater omentum is suspended from the remainder of the curvature.

Surfaces.—When the stomach is in a contracted condition, the two surfaces between the curvatures face somewhat cranialward and caudalward, but when it is partly distended they face ventralward and dorsalward (Fig. 16–53).

The **ventral surface** (*paries anterior; anterosuperior surface*) (Fig. 16–53) is covered with peritoneum of the greater sac. At the left and cranially it is against the diaphragm which separates it from the base of the left lung, the heart, and the seventh, eighth, and ninth ribs and the corresponding intercostal spaces. The right portion is in relation with the left and quadrate lobes of the liver and the anterior abdominal wall. When the stomach is empty, the transverse colon may lie on the caudal part of this surface.

The **dorsal surface** (*paries posterior; posteroinferior surface*) (Fig. 16–57) is covered with peritoneum of the lesser sac or omental bursa. It is in relation with the diaphragm, the spleen, the left suprarenal gland, the cranial part of the left kidney, the ventral surface of the pancreas, the left colic flexure, and the transverse mesocolon. These structures form a shallow **stomach bed** in which the organ rests (Fig. 16–57).

The transverse mesocolon separates the stomach from the duodenojejunal junction and the rest of the small intestine. When the stomach is distended, especially if the patient is in the upright posture, the pyloric half of the stomach slides caudalward over the mesocolon and colon and may reach into the pelvis (Figs. 16–63, 16–64).

Component Parts of the Stomach (Fig. 16–66).—A plane passing through the incisura angularis on the lesser curvature divides the stomach into a left portion or **body** and a right or **pyloric portion.** The cranial portion of the body is the **fundus** which is marked off from the remainder of the body by a plane passing horizontally through the cardiac orifice. To the right of a plane through the sulcus intermedius at right angles to the long axis of this portion is the pyloric antrum (Fig. 16–66), ending in the thickened muscular ring, the **pyloric sphincter** (*m. sphincter pylori*).

If the stomach be examined during the process of digestion, it will be found divided by a muscular constriction into a large dilated left portion and a narrow contracted tubular right portion. The constriction is in the body of the stomach, and does not follow any of the anatomical landmarks; indeed, it shifts gradually toward the left as digestion progresses, *i.e.*, more of the body is gradually absorbed into the tubular part.

Position of the Stomach.—The position of the stomach varies with the posture, with the amount of the stomach contents and with the condition of the intestines on which it rests. According to Moody, radiographs of the normal erect living body show the ordinary range of variation of the most caudal part of the greater curvature to be from 7.3 cm above to 13.5 cm below the interiliac line in males and from 6.5 cm above to 13.7 cm below the line in females. It is below the interiliac line in 74.4 per cent of males and in 87 per cent of females. With the body horizontal the most caudal part of the greater curvature is in males from 16.5 cm above to 7.3 cm below the interiliac line and in females 15.5 cm above to 8.4 cm below the line. The most common position in the erect male (26 per cent) is 2.6 to 5 cm below and in the horizontal male (22.4 per cent) 2.5 to 5 cm above the interiliac line. In the erect female the most common position (22.4 per cent) is 5 to 7.5 cm below and in the horizontal female (24 per cent) 2.5 to 5 cm above the interiliac line (Figs. 3–29 to 3–33).

The position of the pylorus in the erect living body of the male varies from 14.5 cm above to 8 cm below and in the female from 15 cm above to 2.5 cm below the interiliac line. The range of position in regard to the sagittal axis of the erect body varies in males from 8.8 cm to the right to 2 cm to the left of the axis. In 84 per cent it is to the right of the axis. In females the position ranges from 6 cm to the right to 2.6 cm to the left of the sagittal axis. In 18.5 per cent it is to the right. The most common position in both males and females is from 2.5 to 5 cm to the right.

Interior of the Stomach.—When examined after death, the stomach is usually fixed at some temporary stage of the digestive process. A common form is that shown in Figure 16–67. If the viscus be laid open by a section through

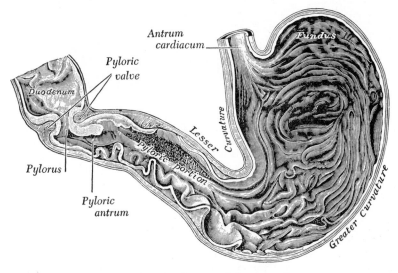

Fig. 16–67.—Interior of the stomach.

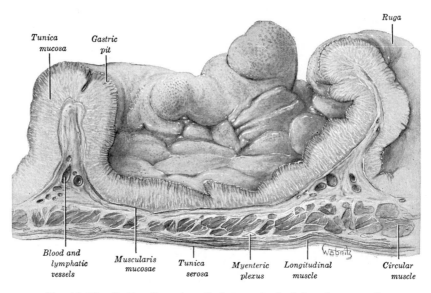

FIG. 16–68.—Section through wall of stomach of adult cadaver. × 5.

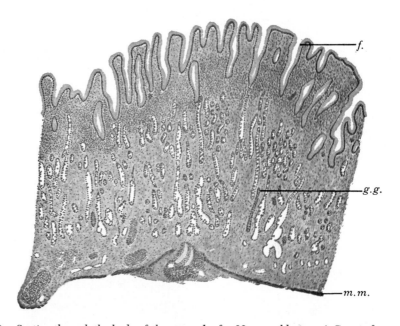

FIG. 16–69.—Section through the body of the stomach of a 22-year-old man. *f*, Gastric fovea; *g.g.*, gastric gland; *m.m.*, muscularis mucosae. × 50. (Sobotta.)

the plane of its two curvatures, it is seen to consist of two segments: (*a*) a large globular portion on the left and (*b*) a narrow tubular part on the right. These correspond to the clinical subdivisions of fundus and pyloric portions already described, and are separated by a constriction which indents the body and greater curvature, but does not involve the lesser curvature. To the left of the cardiac orifice is the incisura cardiaca: the projection of this notch into the cavity of the stomach increases as the organ distends, and has been supposed to act as a valve preventing regurgitation into the esophagus. In the pyloric portion are seen: (*a*) the elevation corresponding to the incisura angularis, and (*b*) the circular projection from the duodenopyloric constriction which forms the pyloric valve; the separation of the pyloric antrum from the rest of the pyloric part is scarcely indicated.

Structure.—The wall of the stomach has four coats: **mucous, submucous, muscular,** and **serous.**

The **mucous membrane** (*tunica mucosa*) (Fig. 16–68) lining the stomach has a soft, velvety appearance and pinkish color in the fresh state, and is thrown into thick folds, known as **rugae,** which tend to have a longitudinal direction along the lesser curvature and at the pylorus but elsewhere somewhat resemble a honeycomb. The folds involve both the mucosa and submucosa but are transient and movable, and are gradually obliterated as the stomach is distended.

The surface of the membrane does not appear smooth when examined with a lens because

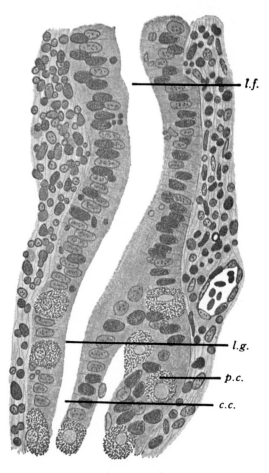

Fig. 16–70.—Higher magnification of transition from fovea to gastric gland shown in Figure 16–69. *c.c.,* Chief cell; *l.f.,* lumen of fovea; *l.g.,* lumen of gland; *p.c.,* parietal cell. × 5000. (Sobotta.)

closely scattered everywhere are the openings of the **gastric pits** or **foveolae.** On the cut surface of the stomach wall, the mucous membrane appears quite thick (5 mm) because tubular gastric glands extend down into it from the foveolae (Fig. 16–69).

The **lining epithelium** which covers the surface and extends down into the foveolae is composed of tall columnar cells of characteristic and rather uniform appearance called **theca cells.** They secrete mucus, and within the part of the cell toward the surface the precursor of the mucus can be seen in a shallow pocket or theca, much smaller and more difficult to identify than the pocket of a goblet cell. They do not have a cuticular border like that seen in the cells of the intestine, and there are no goblet cells.

The **gastric glands** are simple tubular glands which open in groups of two or three into the bottoms of the gastric pits. Three kinds may be distinguished: (*a*) Fundic Glands, (*b*) Cardiac Glands, and (*c*) Pyloric Glands, but the last two, being local modifications, will be described with the appropriate parts. The **fundic glands** are frequently called simply gastric glands because they are the most characteristic of the stomach and occur throughout the fundus and body. The epithelial cells are of two types, chief cells and parietal cells. The **chief cells** are again subdivided into neck chief cells and body chief cells. The **neck chief cells** provide a transition between the lining epithelium of the foveolae and the secreting part of the glands. They are cuboidal or columnar in shape, have a basophilic cytoplasm and, since they appear to be the source of new epithelial cells, they frequently contain mitotic nuclei. The **body chief cells** extend down to the bottom of the glands and resemble the neck chief cells but are slightly more basophilic and have a prominent basal striation. They contain many small secretion granules which can be made clearly visible with special stains (Bowie '40), and are the cells which secrete pepsin. The **parietal cells** are the most characteristic cells in the gastric glands. Their name originated from the fact that they are pushed back against the basement membrane. They do not form a continuous layer but are scattered all along the walls of the glands, separated by several chief cells and usually overlapped by parts of the neighboring body chief cells which intervene between them and the internal surface of the gland. They are four or five times as large as chief cells, have a granular, intensely acidophilic cytoplasm and are the source of the hydrochloric acid in the gastric juice (Fig. 16–70). The tissue surrounding and interven-

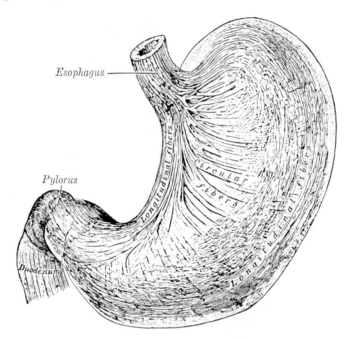

FIG. 16–71.—The longitudinal and circular muscular fibers of the stomach, ventral and superior aspect. (Spalteholz.)

ing between the gastric pits and the glands is composed of areolar connective tissue, and blood and lymphatic capillaries with scattered lymphocytes or occasionally even lymphatic nodules. The boundary between the mucosa and submucosa is marked by a thin sheet of smooth muscle cells, the **muscularis mucosae,** made up of an inner circular and outer longitudinal layer.

The **tela submucosa** is composed of areolar connective tissue, and blood and lymphatic vessels, and extends up into the rugae.

The **muscular coat** (*tunica muscularis*) of the stomach has the two layers of smooth muscle fibers characteristic of the digestive tube, inner circular and outer longitudinal, but in addition has a layer of oblique fibers. The inner circular layer is well represented over the entire organ (Fig. 16–68). The outer longitudinal layer is not so uniform as the circular and is more concentrated along the lesser curvature and the greater curvature (Fig. 16–71). The oblique fibers are internal to the circular fibers, chiefly at the cardiac end of the stomach, and spread over the ventral and dorsal surfaces (Figs. 16–71, 16–72).

The **serous coat** (*tunica serosa*) is composed of a small amount of areolar tissue connecting the mesothelial layer of the peritoneum to the muscular coat. It contains some of the larger blood vessels and lymphatics. A narrow strip along the lesser and greater curvatures, where the two omenta are attached, is not covered by peritoneum.

The **cardiac portion** of the stomach has certain peculiarities. The stratified squamous epithelium of the esophagus is abruptly replaced by the columnar epithelium of the stomach. The **cardiac glands** are longer, more twisted, and contain no parietal cells; they resemble somewhat the esophageal glands and secrete mucus. Aggregations of lymph nodes or lymphatic nodules are not uncommon. The muscular layers are continuous with those of the esophagus.

The **pylorus** is distinctly marked by the thickening of the circular layer into the **pyloric sphincter** which acts as a valve to close the lumen. The lining epithelium makes an abrupt transition from the gastric type with its theca cells to the intestinal epithelium with striated cuticular border and interspaced goblet cells. The **pyloric glands** are devoid of parietal cells, are longer, more tortuous, and have the appearance of mucus-secreting cells. At the transition from stomach to duodenum, the gastric glands can be distinguished from the Brunner glands of the duodenum because the former lie in the tunica mucosa, that is, inside the muscularis mucosae, whereas the latter are in the submucosa.

Vessels and Nerves.—The arteries supplying the stomach are: the left gastric, the right gastric and right gastroepiploic branches of the

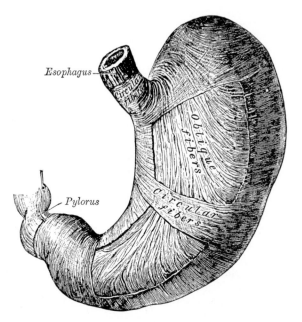

FIG. 16–72.—The oblique muscular fibers of the stomach, ventral and superior aspect. (Spalteholz.)

hepatic, and the left gastroepiploic and short gastric branches of the lienal. They supply the muscular coat, ramify in the submucous coat, and are finally distributed to the mucous membrane. The arrangement of the vessels in the mucous membrane is somewhat peculiar. The arteries break up at the base of the gastric tubules into a plexus of fine capillaries which run upward between the tubules, anastomosing with each other, and ending in a plexus of larger capillaries, which surround the mouths of the tubes, and also form hexagonal meshes around the ducts.

From these the **veins** arise, and pursue a straight course downward, between the tubules, to the submucous tissue; they end either in the lienal and superior mesenteric veins, or directly in the portal vein.

The **lymphatics** are numerous: they consist of a superficial and a deep set, and pass to the lymph nodes found along the two curvatures of the organ (page 757).

The **nerves** are the terminal branches of the right and left vagi, the former usually being distributed upon the dorsal, and the latter upon the ventral part of the organ; numerous sympathetic fibers arising chiefly from the various subdivisions of the celiac plexus accompany the different blood vessels to the organ. According to Michell ('40) small sympathetic filaments may also arise directly from the phrenic and splanchnic trunks. Nerve plexuses are found in the submucous coat and between the layers of the muscular coat as in the intestine. From these plexuses fibrils are distributed to the muscular tissue and the mucous membrane.

THE SMALL INTESTINE

The **small intestine** (*intestinum tenue*), extending from the pylorus to the ileocecal junction, is about 7 meters long, and gradually diminishes in diameter from its commencement to its termination. It is contained in the central and caudal part of the abdominal cavity, and is surrounded cranially and at the sides by the large intestine; a portion of it extends below the superior aperture of the pelvis and lies ventral to the rectum. It is in relation, ventrally, with the greater omentum and abdominal parietes, and is connected to the vertebral column by a reduplication of peritoneum, the **mesentery.** The small intestine is divisible into three portions: the **duodenum,** the **jejunum,** and the **ileum.**

The **Duodenum** (Fig. 16–57) has received

its name from being about equal in length to the breadth of twelve fingers (25 cm). It is the shortest, the widest, and the most fixed part of the small intestine, and has no mesentery, being only partially covered by peritoneum. It has almost a circular course so that its termination is not far removed from its starting point. For purposes of description the duodenum may be divided into four portions: **superior, descending, horizontal,** and **ascending.**

The **superior portion** (*pars superior; first portion*) is about 5 cm long, beginning at the pylorus and extending as far as the neck of the gallbladder. It is the most movable of the four portions. It is almost completely covered by peritoneum, but a small part of its posterior surface near the neck of the gallbladder and the inferior vena cava is without covering; the cranial border of its first half has the hepatoduodenal ligament attached to it; the caudal border, the greater omentum. It is in such close relation with the gallbladder that it is usually found to be stained by bile after death, especially on its ventral surface. It is in relation cranialward with the quadrate lobe of the liver and the gallbladder; dorsally with the gastroduodenal artery, the common bile duct, and the portal vein; and caudally with the head and neck of the pancreas.

The **descending portion** (*pars descendens; second portion*) is from 7 to 10 cm long, and extends from the neck of the gallbladder, on a level with the first lumbar vertebra, along the right side of the vertebral column to the cranial border of the body of the fourth lumbar vertebra. It is crossed in its middle third by the transverse

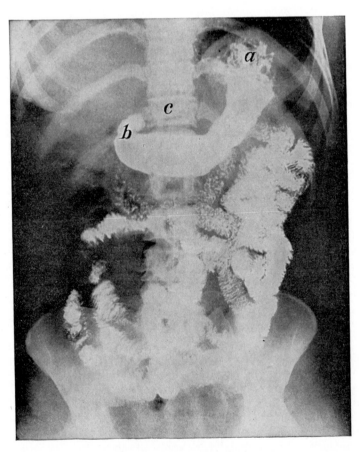

Fig. 16–73.—Stomach and small intestines after a barium meal. *a*, Barium has settled out of the fundus of the stomach; *b*, pylorus; *c*, thoracic vertebra XII. (Department of Radiology, University of Pennsylvania.)

colon, the posterior surface of which is not covered by peritoneum and is connected to the duodenum by a small quantity of connective tissue. The supra- and infra-colic portions are covered ventrally by peritoneum, the infracolic part by the right leaf of the mesentery, but dorsally it is not covered by peritoneum. The descending portion is in relation, ventrally, with the gallbladder, the duodenal impression on the right lobe of the liver and the transverse colon; dorsally, it has a variable relation to the right kidney in the neighborhood of the hilum, and is connected to it by loose areolar tissue. The renal vessels, the inferior vena cava, and the Psoas are also dorsal to it. At its *medial* side are the head of the pancreas and the common bile duct; to its *lateral* side is the right colic flexure. The common bile duct and the pancreatic duct together perforate the medial side of this portion of the intestine (Fig. 16–108), some 7 to 10 cm from the pylorus; the accessory pancreatic duct sometimes pierces it about 2 cm proximal to this.

The **horizontal portion** (*pars horizontalis; third or preaortic or transverse portion*) is from 5 to 7.5 cm long. It begins at the right side of the fourth lumbar vertebra and passes from right to left, with a slight inclination cranialward, ventral to the great vessels and crura of the diaphragm, and joins the ascending portion ventral to the abdominal aorta. It is crossed by the superior mesenteric vessels and the mesentery. Its ventral surface is covered by peritoneum, except near the middle line, where it is crossed by the superior mesenteric vessels. Its posterior surface is not covered by peritoneum, except toward its left extremity, where the posterior layer of the mesentery may sometimes be found covering it to a variable extent. This surface rests upon the right crus of the diaphragm, the inferior vena cava, and the aorta. The cranial surface is in relation with the head of the pancreas.

The **ascending portion** (*pars ascendens; fourth portion*) of the duodenum is about 2.5 cm long. It ascends on the left side of the aorta, as far as the level of the cranial border of the second lumbar vertebra, where it turns abruptly ventralward to become

the jejunum, forming the **duodenojejunal flexure.** It lies ventral to the left Psoas major and left renal vessels, and is covered ventrally, and partly at the sides, by peritoneum continuous with the left portion of the mesentery.

The superior part of the duodenum, as stated above, is somewhat movable, but the rest is practically fixed, and is bound down to neighboring viscera and the posterior abdominal wall by the peritoneum. In addition to this, the duodenojejunal flexure is held in place by a fibrous and muscular band, the **Musculus suspensorius duodeni** or **ligament of Treitz.** This structure commences in the connective tissue around the celiac artery and right crus of the diaphragm, and passes caudalward to be inserted into the superior border of the duodenojejunal curve and a part of the ascending duodenum. It possesses, according to Treitz, smooth muscular fibers mixed with the fibrous tissue of which it is principally made up (Haley and Peden '43). It is of little importance as a muscle, but acts as a suspensory ligament.

Vessels and Nerves.—The **arteries** supplying the duodenum are the right gastric and superior pancreaticoduodenal branches of the hepatic, and the inferior pancreaticoduodenal branch of the superior mesenteric.

The **veins** end in the lienal and superior mesenteric.

The **nerves** are derived from the celiac plexus.

Jejunum and Ileum (Fig. 16–52).—The remainder of the small intestine is named **jejunum** and **ileum,** the former term being given to the proximal two-fifths and the latter to the distal three-fifths. There is no morphological line of distinction between the two, and the division is arbitrary; the character of the intestine gradually undergoes a change so that a portion of the bowel taken from the first part of the jejunum would present characteristic and marked differences from the last part of the ileum.

The **Jejunum** (*intestinum jejunum*) is wider, its diameter being about 4 cm, and its wall is thicker, more vascular, and of a deeper color than the ileum. The circular folds (*plicae circulares; valvulae conniventes*) of its mucous membrane are large and thickly set, and its villi are larger than in the ileum. The aggregated lymph nodules are usually absent in the proximal part of the jejunum, and in the distal part are less frequently found than in the ileum, and are smaller and tend to assume a circular form. By grasping the jejunum between the finger and thumb one can feel the circular folds through the walls but this is not true of the lower part of the ileum; it is possible in this way to distinguish the upper from the lower part of the small intestine.

The **Ileum** (*intestinum ileum*) is narrower, its diameter being 3.75 cm, and its coats thinner and less vascular than those of the jejunum. It possesses but few circular folds, and they are small and disappear entirely toward its caudal end, but **aggregated lymph nodules** (*Peyer's patches*) are larger and more numerous. The jejunum for the most part occupies the umbilical and left iliac regions, whereas the ileum occupies chiefly the umbilical, hypogastric, right iliac, and pelvic regions. The terminal part of the ileum usually lies in the pelvis, from which it ascends over the right Psoas and right iliac vessels; it ends in the right iliac fossa by opening into the medial side of the commencement of the large intestine.

The jejunum and ileum are attached to the posterior abdominal wall by an extensive fold of peritoneum, the **mesentery,** which allows the freest motion, so that each coil can accommodate itself to changes in form and position. The mesentery is fan-shaped: its posterior border or root, about 15 cm long, is attached to the posterior abdominal wall from the left side of the body of the second lumbar vertebra to the right sacroiliac articulation, crossing successively the horizontal part of the duodenum, the aorta, the inferior vena cava, the ureter, the right Psoas muscle (Fig. 16–53). Its breadth between its vertebral and intestinal borders averages about 20 cm, and is greater in the middle than at its ends. Between the two layers of which it is composed are contained blood vessels, nerves, lacteals, and lymph glands, together with a variable amount of fat.

Variations.—Meckel's Diverticulum (*diverticulum ilei*) is the name given to a pouch which projects from the lower part of the ileum in about 2 per cent of subjects. Its average

position is about 1 meter proximal to the ileocecal valve, and its average length about 5 cm. Its caliber is generally similar to that of the ileum, and its blind extremity may be free or may be connected with the abdominal wall or with some other portion of the intestine by a fibrous band. It represents the remains of the proximal part of the vitelline duct, the duct of communication between the yolk sac and the primitive digestive tube in early fetal life.

Structure.—The internal surface has two types of irregularities or projections which are characteristic of the small intestine. They are the large **circular folds** and the minute **villi** (Fig. 16–74).

The **circular folds** (*plicae circulares; valvulae conniventes; valves of Kerkring*) (Fig. 16–74) are valvelike folds which project into the lumen from 3 to 10 mm. The majority extend transversely around the inside of the cylinder of the intestine for about one-half to two-thirds of its circumference, but others complete the circle or form a spiral extending more than once around, even making two or three turns. The folds are also of different height, the high ones tending to alternate with low ones. The size and frequency of the folds are different in the three parts of the small intestine. The folds are formed by both the tela submucosa and tunica mucosa. The core of submucosal connective tissue is quite firm, making these folds permanent structures which are not obliterated by distention as are the transient rugae of the stomach.

The **villi** (*villi intestinales*) (Fig. 16–74) are tiny fingerlike projections, of a size just at the borderline of visibility with the naked eye, crowded together over the entire mucous surface and giving it a velvety appearance. They are quite irregular in size and shape, are larger in some parts of the intestine than in others and become considerably flattened out by distention of the intestine. The villi are entirely made up of tissue belonging to the tunica mucosa.

The wall of the small intestine is composed of four coats: **mucous, submucous, muscular,** and **serous.**

The **mucous membrane** (*tunica mucosa*) is composed of the villi, the intestinal glands, a connective tissue framework, and a muscularis mucosae. The surface epithelium covering the villi is of a simple columnar type in which the majority of cells have a characteristic striated free border. Recent observations with the electron microscope have demonstrated that the striated appearance is due to innumerable closely set projections of cytoplasm which are too small (diameter 0.08 μ) to be distinguished with the best light microscopes (Granger and Baker '50). Scattered liberally among the striated epithelial cells are numerous mucus-secreting goblet cells.

Structure of the Villi.—Each villus has a core of delicate areolar and reticular connective tissue which provides a basement membrane for the epithelium and supports the rich network of capillary blood vessels and the usually single lymphatic vessel. The lymphatic capillary begins blindly near the tip of the villus, occupies a more or less central position, and opens into the lymphatic vessels in the submucosa. This central lymphatic capillary is

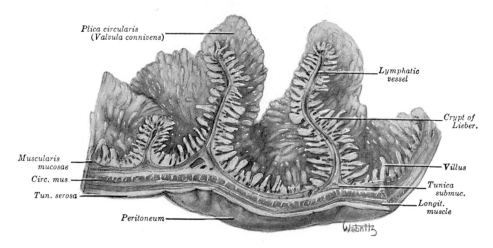

Fig. 16–74.—Section through wall of small intestine (jejunum) of adult cadaver. × 5.

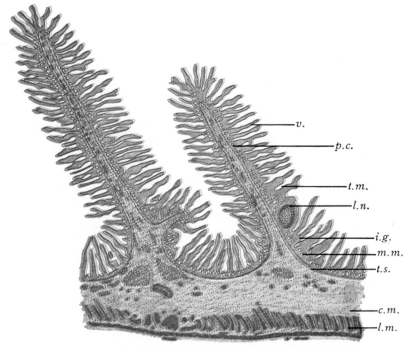

Fig. 16–75.—A longitudinal section through the jejunum of a 24-year-old man, including section through two plicae circulares. *c.m.*, circular muscle; *i.g.*, intestinal gland; *l.m.*, longitudinal muscle; *l.n.*, lymphatic nodule; *m.m.*, muscularis mucosae; *p.c.*, plica circularis; *t.m.*, tunica mucosa; *v.*, villus. × 15. (Sobotta.)

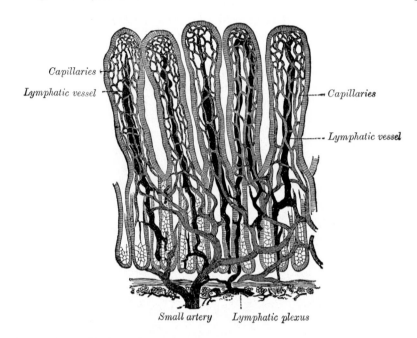

Fig. 16–76.—Villi of small intestine, showing blood vessels and lymphatic vessels. (Cadiat.)

called a **lacteal,** the name having been given because it is filled with a white milky fluid, known as chyle, during the digestion of a meal rich in fat. Scattered single strands of smooth muscle run parallel with the lacteal and appear to be extensions of the muscularis mucosae into the villus.

The **intestinal glands** (*glandulae intestinales; crypts of Lieberkühn*) are simple tubular glands which open into the depressions between the villi and form a rather uniform layer of glandular tissue between the bases of the villi and the muscularis mucosae. The striated surface epithelium and goblet cells extend quite far down into the crypts, but at the bottom or fundus is a group of glandular secreting cells known as the **cells of Paneth.** These cells contain large secretion granules which stain a bright red with eosin, and it is probable that they secrete the digestive enzymes of the small intestine. Mitotic divisions are frequently observed in the cells of the wall of the crypt and it is believed that proliferation of these cells makes up for the loss of surface cells from the natural attrition (Hunt '51).

The **muscularis mucosae** is a thin sheet of non-striated muscle cells, composed of inner circular and outer longitudinal layers at the boundary between the mucosa and submucosa.

The **submucous coat** (*tela submucosa*) is composed of fibroelastic and areolar connec-

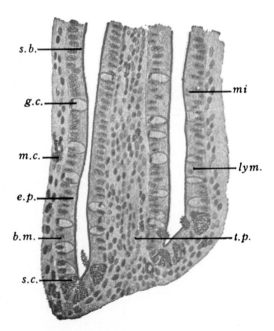

Fig. 16–77.—Intestinal glands (crypts of Lieberkühn) from human duodenum. *b.m.,* basement membrane; *e.p.,* epithelium; *g.c.,* goblet cell; *lym.,* lymphocyte; *m.c.,* mast cell; *mi.,* mitosis; *s.b.,* striated border; *s.c.,* glandular secreting cell (Paneth cell); *t.p.,* tunica propria. × 300. (Sobotta.)

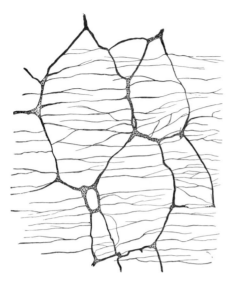

FIG. 16–78.—The myenteric plexus from the rabbit. × 50.

FIG. 16–79.—The plexus of the submucosa from the rabbit. × 50.

tive tissue. It is a strong layer, forming the core of the circular folds. It contains the blood vessels and lymphatics which supply the mucous membrane and, near the muscularis, the **submucous nerve plexus of Meissner** (Fig. 16–79). Small collections of lymphocytes and solitary lymphatic nodules may occur in any part of the small intestine, and groups of nodules, known as Peyer's patches, occur in the ileum. The lymphatic nodules usually occupy the submucosa, infiltrate the muscularis mucosae, and extend out to the free surface, often appearing to obliterate some of the villi (Fig. 16–81).

The **muscular coat** (*tunica muscularis*) is composed of the two layers, usual in the alimentary tube, outer longitudinal and inner circular. Between these two layers is a net of nervous tissue containing non-myelinated nerve fibers and ganglion cells (Fig. 16–78) known as the **myenteric** or **Auerbach's plexus.** The muscularis is somewhat thicker in the proximal than in the distal part of the intestine.

The **serous coat** (*tunica serosa*) is composed of the peritoneum and the areolar connective tissue connecting it to the muscular coat. The small intestine is covered by peritoneum except along the narrow strip or border attached to the mesentery, and the parts of the duodenum which are retroperitoneal.

Special Features.—Duodenum.—The circular folds are not found in the first 2.5 to 5 cm beyond the pylorus, but in the descending part, distal to the openings of the bile and pancreatic ducts, they are especially large and numerous.

The villi are also especially large and numerous in the duodenum. The common bile duct and the pancreatic duct pierce the left side of the descending portion about 7 to 10 cm from the pylorus (Fig. 12–108). At the opening of the ducts there is a thickening of their muscle coats commonly called the **sphincter of Oddi** (*sphincter ampullae hepatopancreaticae*) which usually causes a protrusion, the **papilla of Vater,** at the proximal end of a longitudinal fold (*plica longitudinalis duodeni*) (see comment on page 1227). There is a slight depression and absence of the circular folds around the papilla (Fig. 16–80).

The **duodenal glands** (*glandulae duodenales; Brunner's glands*) differ from the other intestinal glands in occupying the submucosa. The usual intestinal glands or crypts are found in the duodenum and the ducts from the duodenal glands, after penetrating the muscularis mucosae, open into the bottoms of occasional crypts. The duodenal glands are compound tubulo-alveolar glands; they resemble the pyloric glands of the stomach in appearance but are larger, and, as mentioned, they lie in the submucosa. The cells stain lightly in histological preparations, but they give some of the reactions of mucus. The duodenal glands are largest and most numerous near the pylorus, forming there a rather thick complete layer, but they diminish in the horizontal portion and disappear near the duodenojejunal junction.

Jejunum.—The circular folds and villi are almost as large and numerous in the proximal part of the jejunum as in the duodenum but they gradually decrease in size and number toward the ileum.

Ileum.—The circular folds and villi are smaller and less numerous in the ileum than in the jejunum, and toward the terminal part the folds may be widely scattered or even lacking.

Aggregated lymphatic nodules or **Peyer's patches** (*noduli lymphatici aggregati; Peyer's*

*glands; agminated follicles; tonsillae intesti-
nales*) are groups of lymphatic nodules spread
out as a single layer in the mucous membrane
of the wall of the ileum opposite the mesenteric
attachment. The patches are circular or oval,
approximately 1 cm wide and may extend
along the intestine for 3 to 5 cm. They are
largest and most frequent in the distal ileum
but are occasionally seen even in the jejunum.
They can be recognized in gross specimens at
autopsy as thickened whitish patches where
the circular folds are absent and the villi are
very sparse or lacking. In the dissecting room,
where the subjects are usually of advanced
age, the patches are difficult to identify because

of the atrophy of the lymphatic tissue which
takes place in older individuals.

Vessels and Nerves.—The jejunum and ileum
are supplied by the **superior mesenteric artery,**
the intestinal branches of which, having
reached the attached border of the bowel, run
between the serous and muscular coats, with
frequent inosculations, to the free border where
they also anastomose with other branches
running around the opposite surface of the gut.
From these vessels numerous branches are
given off which pierce the muscular coat, supplying it and forming an intricate plexus in the
submucous tissue. From this plexus minute
vessels pass to the glands and villi of the

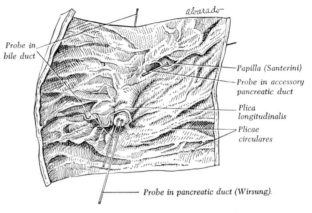

Fɪɢ. 16–80.—Mucous membrane of the descending portion of the duodenum, showing the
plica longitudinalis and the duodenal papilla (of Vater). (Redrawn from Spalteholz.)

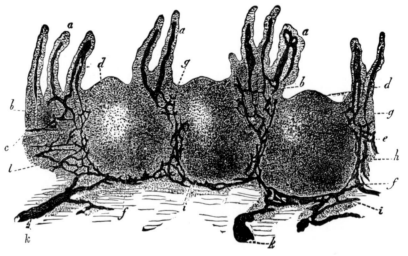

Fɪɢ. 16–81.—Vertical section of a human aggregated lymphatic nodule, injected through its lymphatic
canals. *a*, Villi with their chyle passages. *b*, Intestinal glands. *c*, Muscularis mucosae. *d*, Cupola or
apex of solitary nodule. *e*, Mesial zone of nodule. *f*, Base of nodule. *g*, Points of exit of the lacteals from
the villi, and entrance into the true mucous membrane. *h*, Retiform arrangement of the lymphatics in the
mesial zone. *i*, Course of the latter at the base of the nodule. *k*, Confluence of the lymphatics opening into
the vessels of the submucous tissue. *l*, Follicular tissue of the latter.

mucous membrane. The **veins** have a course and arrangement similar to that of the arteries.

The **lymphatics** of the small intestine (lacteals) are arranged in two sets, those of the mucous membrane and those of the muscular coat. The lymphatics of the villi commence in these structures in the manner described above. They form an intricate plexus in the mucous and submucous tissue, being joined by the lymphatics from the lymph spaces at the bases of the solitary nodules, and from this pass to larger vessels at the mesenteric border of the gut. The lymphatics of the muscular coat are situated to a great extent between the two layers of muscular fibers, where they form a close plexus; throughout their course they communicate freely with the lymphatics from the mucous membrane, and empty themselves in the same manner as these into the origins of the lacteal vessels at the attached border of the gut.

The **nerves** of the small intestines are derived from the plexuses of autonomic nerves around the superior mesenteric artery, representing cranial parasympathetic fibers of the vagus and postganglionic sympathetic fibers from the celiac plexus. From this source they run to the **myenteric plexus** (*Auerbach's plexus*) (Fig. 16–78) of nerves and ganglia situated between the circular and longitudinal muscular fibers from which the nervous branches are distributed to the muscular coats of the intestine. From this a secondary plexus, the **plexus of the submucosa** (*Meissner's plexus*) (Fig. 16–79) is derived, and is formed by branches which have perforated the circular muscular fibers. This plexus lies in the submucous coat of the intestine; it also contains ganglia from which nerve fibers pass to the muscularis mucosae and to the mucous membrane. The nerve bundles of the submucous plexus are finer than those of the myenteric plexus.

THE LARGE INTESTINE

The **large intestine** (*intestinum crassum*) (Figs. 16–53, 16–54) extends from the ileum to the anus. It is about 1.5 meters long, being one-fifth of the whole extent of the intestinal canal. Its caliber, when not unduly distended with feces, is largest at its com-

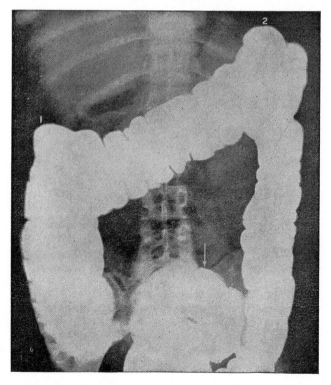

FIG. 16–82.—Large intestine after a barium enema. *1*, Right colic flexure; *2*, left colic flexure. The arrow points to the pelvic colon. Note the sacculations of the gut, and the different levels of the two flexures.

mencement at the cecum and gradually diminishes as far as the rectum, where there is a dilatation of considerable size just above the anal canal. It differs from the small intestine in its sacculated form, in possessing certain appendages to its external coat, the **appendices epiploicae,** and in its longitudinal muscular fibers being arranged in three **longitudinal bands** or **taeniae.** The large intestine describes an arch around the convolutions of the small intestine. It commences in the right iliac region, ascends through the right lumbar and hypochondriac regions to the caudal surface of the liver, and here takes a bend to the left forming the **right colic flexure.** It passes transversely across the epigastric and umbilical regions to the left hypochondriac region, where it bends caudalward, forming the **left colic flexure,** and descends through the left lumbar and iliac regions toward the pelvis. In the left iliac region it forms the

sigmoid flexure, and from this it is continued along the posterior wall of the pelvis to the anus. The large intestine is divided into the **cecum, appendix, colon, rectum,** and **anal canal.**

The **Cecum** (*intestinum caecum*) (Fig. 16–84), the commencement of the large intestine, is the large blind pouch extending caudalward beyond the ileocecal valve. Its size is variously estimated by different authors, but on an average it is 6.25 cm in length and 7.5 cm in breadth. It is situated in the right iliac fossa, cranial to the lateral half of the inguinal ligament: it rests on the Iliacus, and usually lies in contact with the ventral abdominal wall, but the greater omentum and, if the cecum be empty, some coils of small intestine may lie ventral to it. The common position of the cecum in the erect living body is not in the right iliac fossa but in the cavity of the true pelvis. As a rule, it is entirely enveloped by peri-

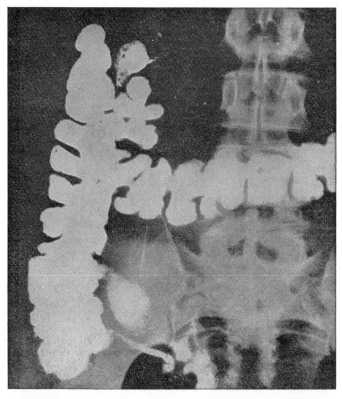

Fig. 16–83.—Part of the large intestine after a barium meal. Note the vermiform appendix, which passes from the medial side of the cecum medially and slightly downward into the true pelvis. At a slightly higher level the terminal part of the ileum can be recognized. The first part of the transverse colon runs downward in front of, and slightly medial to, the ascending colon, before it turns to the left.

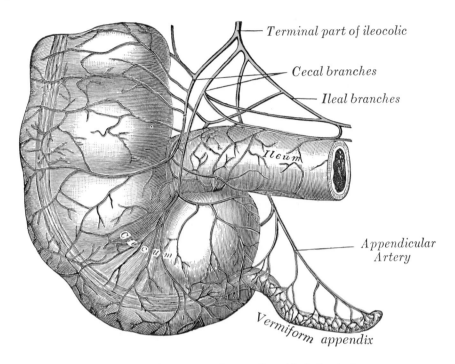

— Terminal part of ileocolic

— Cecal branches

— Ileal branches

Appendicular Artery

FIG. 16–84.—The cecum and appendix, with their arteries.

toneum, but in a certain number of cases (5 per cent), part of the posterior surface is connected to the iliac fascia by connective tissue. The cecum lies quite free in the abdominal cavity and enjoys a considerable amount of movement, so that it may become herniated into the right inguinal canal, and has occasionally been found in an inguinal hernia on the left side.

The cecum varies in shape, but, according to Treves, in man it may be classified under one of four types. In the *first* of his four types (about 2 per cent) the cecum is conical and the appendix rises from its apex. The three longitudinal bands start from the appendix and are equidistant from each other. In the *second* type, the conical cecum has become quadrate by the growing out of a saccule on either side of the anterior longitudinal band. These saccules are of equal size, and the appendix arises from between them, instead of from the apex of a cone (about 3 per cent). The *third* type is the normal type in man. Here the two saccules, which in the second type are uniform, have grown at unequal rates: the right with greater rapidity than the left. In conse-

quence of this an apparently new apex has been formed by the caudalward growth of the right saccule, and the original apex, with the appendix attached, is pushed over to the left toward the ileocecal junction. The three longitudinal bands still start from the base of the vermiform appendix, but they are no longer equidistant from each other, because the right saccule has grown between the anterior and posterior bands, pushing them over to the left. This type occurs in about 90 per cent. The *fourth* type is merely an exaggerated condition of the third; the right saccule is still larger, and at the same time the left saccule has become atrophied, so that the original apex of the cecum, with the vermiform appendix, is close to the ileocecal junction, and the anterior band courses medialward to the same situation (about 4 per cent).

The **Appendix Vermiformis** (*processus vermiformis*) (Fig. 16–85) is a long, worm-shaped tube, 5 to 10 mm in diameter which starts from what was originally the apex of the cecum, and may project in one of several directions: cranialward behind the cecum; to the left behind the ileum and

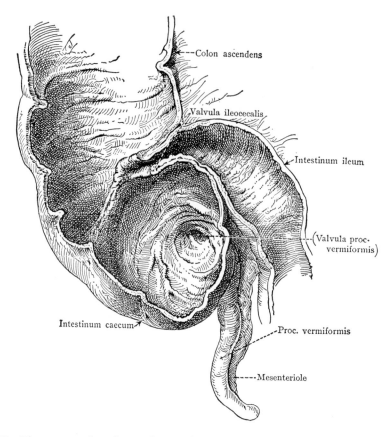

FIG. 16–85.—The cecum, colic valve, and appendix vermiformis, with anterior wall of terminal ileum and cecum removed. (Eycleshymer and Jones.)

mesentery; or caudalward into the lesser pelvis. It varies from 2 to 20 cm in length, its average being about 8 cm. It is suspended by a peritoneal mesenteriole derived from the left leaf of the mesentery which is more or less triangular in shape, and as a rule extends along the entire length of the tube. Between its two layers and close to its free margin lies the appendicular artery (Fig. 16–84). The lumen of the vermiform appendix is small, extends throughout the whole length of the tube, and communicates with the cecum by an orifice which is distal to the ileocecal opening. It is sometimes guarded by a semilunar valve formed by a fold of mucous membrane, but this is by no means constant.

The **Ileocecal Valve** (*valva ileocecalis*) (Fig. 16–85).—The ileum ends by opening into the medial part of the large intestine, at the point of junction of the cecum with the colon. The opening is guarded by a valve, consisting of two segments or lips, which project into the lumen of the large intestine. If the intestine has been inflated and dried, the lips are of a semilunar shape, one toward the colon, the other toward the cecum. At the ends of the aperture the two segments of the valve coalesce, and are continued as narrow membranous ridges around the canal for a short distance, forming the **frenula of the valve.** In the fresh condition, or in specimens which have been hardened *in situ*, the lips project as thick cushion-like folds into the lumen of the large gut; the opening between them may present the appearance of a slit or may be somewhat oval in shape.

The **Colon** is divided into four parts: the **ascending, transverse, descending,** and **sigmoid.**

The **Ascending Colon** (*colon ascendens*)

passes cranialward from its commencement at the cecum, to the caudal surface of the right lobe of the liver where it is lodged in a shallow depression, the **colic impression,** to the right of the gallbladder; here it bends abruptly to the left, forming the **right colic** (*hepatic*) **flexure** (*flexura coli dextra*) (Fig. 16–53). It does not have a mesentery but is retained in contact with the posterior wall of the abdomen by the peritoneum which covers its ventral surface and sides, its dorsal surface being connected by loose areolar tissue with the Iliacus, Quadratus lumborum, aponeurotic origin of Transversus abdominis, and with the ventral and lateral part of the right kidney. It is in relation, ventrally, with the convolutions of the ileum and the abdominal parietes.

The **Transverse Colon** (*colon transversum*) the longest and most movable part of the colon, passes from the right colic flexure in the right hypochondriac region across the abdomen into the left hypochondriac region, where it curves sharply on itself beneath the caudal end of the spleen, forming the **left colic** (*splenic*) **flexure** (*flexura coli sinistra*). In the erect posture in the majority of males it is festooned caudalward from 7.5 to 10 cm below the interiliac line and in the majority of females from 10 to 12.5 cm below the line. It is completely invested by peritoneum, except for the **transverse mesocolon** which is connected to the inferior border of the pancreas by a large duplicature of that membrane. It is in relation, by its cranial surface, with the liver and gallbladder, the greater curvature of the stomach, and the spleen; by its caudal surface, with the small intestine; by its ventral surface, with the anterior layers of the greater omentum and the abdominal parietes; its dorsal surface is in relation from right to left with the descending portion of the duodenum, the head of the pancreas, and some of the convolutions of the jejunum and ileum.

The **left colic** or **splenic flexure** (Fig. 16–53) is situated at the junction of the transverse and descending parts of the colon, and is in relation with the caudal end of the spleen and the tail of the pancreas; the flexure is so acute that the end of the transverse colon usually lies in contact with the front of the descending colon. It lies at a

more cranial level than, and on a plane posterior to, the right colic flexure, and is attached to the diaphragm, opposite the tenth and eleventh ribs, by a peritoneal fold, named the **phrenicocolic ligament,** which assists in supporting the spleen (see page 1209).

The **Descending Colon** (*colon descendens*) passes caudalward through the left hypochondriac and lumbar regions along the lateral border of the left kidney. At the caudal end of the kidney it turns medialward toward the lateral border of the Psoas, and then descends, in the angle between Psoas and Quadratus lumborum, to the crest of the ilium, where it becomes the iliac colon. The peritoneum covers its ventral surface and sides, while its dorsal surface is connected by areolar tissue with the caudal and lateral part of the left kidney, the aponeurotic origin of the Transversus abdominis, and the Quadratus lumborum (Fig. 16–55). It is smaller in caliber and more deeply placed than the ascending colon, and is more covered with peritoneum on its posterior surface than is the ascending colon. The part of the descending colon in the left iliac fossa may be called the **iliac colon** (Fig. 16–86). It is about 12 to 15 cm long. It begins at the level of the iliac crest, and ends in the sigmoid colon at the superior aperture of the lesser pelvis. It curves medialward ventral to the Iliopsoas and is covered by peritoneum on its sides and anterior surface only.

The **Sigmoid Colon** (*colon sigmoideum; pelvic colon; sigmoid flexure*) (Fig. 16–86) forms a loop which averages about 40 cm in length, and normally lies within the pelvis, but on account of its freedom of movement may be displaced into the abdominal cavity. It begins at the superior aperture of the lesser pelvis, where it is continuous with the iliac colon, and forms one or two loops before it reaches the level of the third piece of the sacrum, where it bends downward and ends in the rectum. It is completely surrounded by peritoneum, and is attached to the pelvic wall by an extensive mesentery, the **sigmoid mesocolon,** which gives it a considerable range of movement in its central portion. *Dorsal to the sigmoid colon are the external iliac vessels, the left Piriformis, and left sacral*

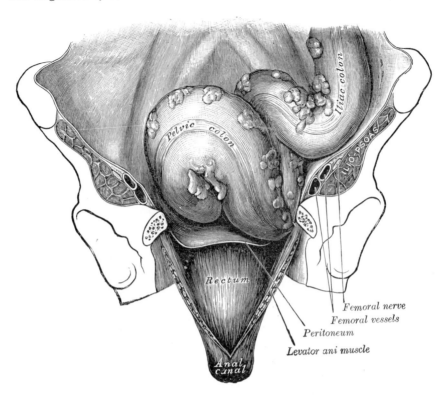

FIG. 16–86.—Iliac colon, sigmoid or pelvic colon, and rectum seen from the front, after removal of pubic bones and bladder.

plexus of nerves; *ventrally*, it is separated from the bladder in the male, and the uterus in the female, by some coils of the small intestine.

The **Rectum** (*intestinum rectum*) (Fig. 16–87) is continuous with the sigmoid colon at the level of the third sacral vertebra; it passes caudalward, lying in the sacrococcygeal curve, and ends in the anal canal. It presents two dorsoventral curves: a cranial, with its convexity dorsalward, and a caudal, with its convexity ventralward. Two lateral curves are also described, one to the right, opposite the junction of the third and fourth sacral vertebrae, and the other to the left, opposite the left sacrococcygeal articulation; they are, however, of little importance. The rectum is about 12 cm long, and at its commencement its caliber is similar to that of the sigmoid colon, but near its termination it is dilated to form the **rectal ampulla**. The rectum has no sacculations comparable to the

haustrae of the colon, but when the lower part is contracted, its mucous membrane is thrown into a number of longitudinal folds. There are certain permanent transverse folds, of a semilunar shape, **plicae transversales recti** (*Houston's valves*) (Fig. 16–88). They are usually three in number, but sometimes a fourth is found, and occasionally only two are present. One is situated near the commencement of the rectum, on the right side; a second, about 3 cm beyond the first, extends inward from the left side of the tube; a third, the largest and most constant, projects dorsalward, opposite the fundus of the urinary bladder. When a fourth is present, it is situated nearly 2.5 cm above the anus on the left and posterior wall of the tube. These folds are about 12 mm in width, and contain some of the circular fibers of the gut. In the empty state of the intestine they overlap each other, as Houston remarks, so effectually as to require considerable maneuvering to conduct a bougie or the finger along the canal. Their

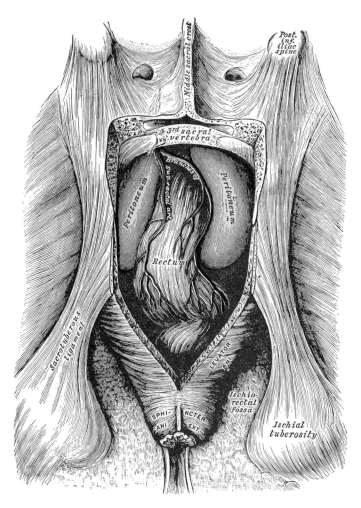

FIG. 16–87.—The posterior aspect of the rectum exposed by removing the lower part of the sacrum and the coccyx.

use seems to be, "to support the weight of fecal matter, and prevent its urging toward the anus, where its presence always excites a sensation demanding its discharge."

The peritoneum covers the ventral and lateral surface of proximal two-thirds of the rectum, but distally, the ventral only; from the latter it is reflected on to the seminal vesicles in the male and the posterior vaginal wall in the female.

The level at which the peritoneum is reflected from the rectum to the viscus ventral to it is of considerable surgical importance in connection with the removal of the distal part of the rectum. In the male, the distance of the rectovesical excavation

from the anus is about 7.5 cm, *i.e.*, the height to which an ordinary index finger can reach. In the female, the rectouterine excavation is about 5.5 cm from the anal orifice. The rectum is surrounded by a dense fascia loosely attached to the rectal wall by areolar tissue, that is, there is a fascial cleft which allows distention.

Relations of the Rectum.—The proximal part of the rectum is in relation, *dorsally*, with the superior rectal vessels, the left Piriformis, and left sacral plexus of nerves, which separate it from the pelvic surfaces of the sacral vertebrae; in its distal part it lies directly on the sacrum, coccyx, and Levatores ani, a dense fascia alone

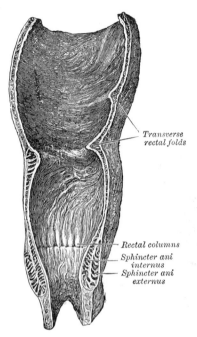

*Transverse
rectal folds*

Rectal columns
*Sphincter ani
internus*
*Sphincter ani
externus*

FIG. 16–88.—Coronal section of rectum
and anal canal.

intervening; *ventrally,* it is separated, in the
male, from the fundus of the bladder; in the
female, from the intestinal surface of the uterus
and its appendages, by some convolutions of
the small intestine, and frequently by the sigmoid colon; it is in relation in the male with
the triangular portion of the fundus of the
bladder, the vesiculae seminales, and ductus
deferentes, and more anteriorly with the posterior surface of the prostate; in the female,
with the posterior wall of the vagina.

The **Anal Canal** (*pars analis recti*) (Figs.
16–87, 16–91, 16–92), or terminal portion of
the large intestine, begins at the level of
the apex of the prostate, and ends at the
anus. It measures from 2.5 to 4 cm in length
and forms an angle with the lower part of
the rectum. It has no peritoneal covering,
but is invested by the Sphincter ani internus,
supported by the Levatores ani, and surrounded at its termination by the Sphincter
ani externus. Dorsal to it is a mass of muscular and fibrous tissue, the **anococcygeal
body**; ventral to it, in the male, but separated by the perineal center, are the
membranous portion of the urethra and
bulb of the penis, and the fascia of the
urogenital diaphragm; and in the female

it is separated from the lower end of the
vagina by a mass of muscular and fibrous
tissue, named the **perineal body.**

The proximal half of the anal canal presents a number of vertical folds produced
by an infolding of the mucous membrane
around a plexus of veins known as the
rectal columns [*Morgagni*] (Fig. 16–88).
They are separated from each other by
furrows (**rectal sinuses**), which end distally in small valve-like folds, termed **anal
valves**, which join together the distal ends
of the rectal columns. When these columns
are swollen and inflamed they are known
as hemorrhoids.

Structure of the Large Intestine.—The wall
of the cecum and colon has certain folds and
irregularities which are characteristic and
which show up most prominently on the
internal surface, but, unlike those in the small
intestine, these folds include all four layers and
may be seen on the external surface. The longitudinal bands of the muscular coat, which will
be described below, cause a puckering of the
wall, so that between them it is bulged out into
sacculations called **haustrae**. The wall between
the haustrae is thrown into folds which have
a crescentic form on the interior of the colon
and are called the **semilunar folds** (*plicae semilunares*) in contrast with the plicae circulares
of the small intestine.

The large intestine has four coats: **mucous,
submucous, muscular** and **serous.**

The **mucous coat** (*tunica mucosa*) is smooth,
that is, devoid of villi, and covers the inner
surface of haustrae and semilunar folds in a
coat of uniform thickness. The surface is covered with simple columnar epithelium containing large numbers of goblet cells. The glands
of the large intestine are simple, straight, tubular glands containing the same type of epithelium as the surface; they are packed quite
closely together and they open on the surface
in tiny round holes which can readily be seen
with a hand lens (Fig. 16–89). There is a
delicate **muscularis mucosae** composed of inner
circular and outer longitudinal fibers. Collections of lymphocytes and solitary lymphatic
nodules are of frequent occurrence, especially
near the colic valve and in the rectum.

The **submucous coat** (*tela submucosa*) is a
rather uniform layer of areolar tissue containing blood and lymphatic vessels and connecting
the mucosa with the muscularis.

The **muscular coat** (*tunica muscularis*) is
composed of the usual inner circular and outer
longitudinal layers of non-striated muscle.

The circular fibers form a thin layer over the cecum and colon; it is somewhat thickened in the semilunar folds between the haustrae, uniformly thickened in the rectum, and, in the anal canal, constitutes the strong circular non-striated muscle, the **Sphincter ani internus.**

The longitudinal muscle fibers are concentrated into three flat longitudinal bands about equally spaced and about 12 mm in width. They are easily seen in a gross specimen (Fig. 16–53) and are called **taeniae coli.** They have specific positions in relation to the position of the colon itself: (1) the posterior taenia is placed

It is incomplete on the ascending and descending colons, where they are attached to the posterior abdominal wall, and on the rectum.

Special Features. Ileocecal Valve.—Each lip of the valve is formed by a reduplication of the mucous membrane and of the circular muscular fibers of the intestine, the longitudinal fibers and peritoneum being continued uninterruptedly from the small to the large intestine.

The surfaces of the valve directed toward the ileum are covered with villi, and present the characteristic structure of the mucous membrane of the small intestine, while those turned

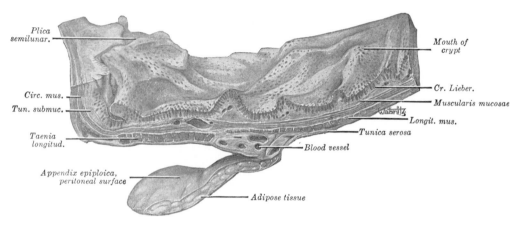

Plica semilunar.

Mouth of crypt

Circ. mus.

Tun. submuc.

Cr. Lieber.

Muscularis mucosae

Longit. mus.

Tunica serosa

Taenia longitud.

Blood vessel

Appendix epiploica, peritoneal surface

Adipose tissue

Fig. 16–89.—Section through colon of adult cadaver. × 5.

along the attached border, (2) the anterior taenia is the one easily visible on the exposed surface of the ascending and descending colon but is covered by the attachment of the greater omentum on the transverse colon, and (3) the lateral taenia is found on the medial side of the ascending and descending colon and on the dorsal aspect of the transverse colon. The anterior taenia is a useful guide for locating the position of the appendix vermiformis because the latter is a direct extension from it (Fig. 16–84).

The taeniae are shorter than the other coats of colon and cecum, causing the intervening wall to bulge into the sacculations known as **haustrae** which are typical of this part of the intestine (Fig. 16–52) and are responsible for the deep depressions in its outline in an x-ray (Fig. 16–83). Between adjacent haustrae are the crescentic folds or plicae semilunares which encroach on the lumen of the intestine.

The **serous coat** (*tunica serosa*), derived from the peritoneum, is complete over the cecum, appendix, transverse colon, and sigmoid colon except for their mesenteric attachments.

toward the large intestine are destitute of villi, and marked with the orifices of the numerous tubular glands peculiar to the mucous membrane of the large intestine. These differences in structure continue as far as the free margins of the valve. It is generally maintained that this valve prevents reflux from the cecum into the ileum, but in all probability it acts as a sphincter around the end of the ileum and prevents the contents of the ileum from passing too quickly into the cecum.

The **appendices epiploicae** are characteristic features of the large intestine and may be used for its identification. They are small, rounded irregular masses of fat, averaging 0.5 to 1.0 cm in diameter, almost completely covered by peritoneum and suspended from the surface of the colon and cecum by slender stalks (Fig. 16–89). They are usually attached along the taeniae, and are most numerous on the transverse colon.

Appendix or Vermiform Process.—The appendix has the same four coats as the colon. The epithelial lining and the glands are similar, but the glands are much fewer and the mucosa and submucosa are much thickened and almost entirely occupied by lymphatic nodules and lymphocytes (Fig. 16–90). The longitudinal fibers of the tunica muscularis are evenly distributed, not arranged into taeniae, as in the colon, and the circular muscle is more prominent than the longitudinal.

Rectum.—The mucous membrane in the rectum is thicker and more vascular than that in the colon and is more loosely attached to the muscularis, as in the esophagus. The longi-

anal sphincter, and diverging from this to the walls of the pelvis is the Levator ani. Just beneath the integument at the anal orifice is a more delicate striated muscle, the Corrugator cutis ani; its fibers are closely associated with the tributaries of the inferior rectal veins draining the plexuses in the rectal columns and spasm of its fibers may seriously retard the venous drainage.

Vessels and Nerves (Fig. 16–92).—The **arteries** supplying the colon are derived from

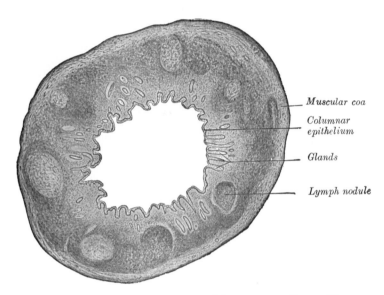

Muscular coa
Columnar epithelium
Glands
Lymph nodule

FIG. 16–90.—Transverse section of human appendix. × 20.

tudinal fibers of the tunica muscularis are spread out into a layer which completely surrounds the rectum, but is thicker on the anterior and posterior walls.

Anal Canal.—The mucous membrane in the anal canal is thick and vascular. Beneath the longitudinal folds or **rectal columns** of Morgagni are dilated veins, often knotted and tortuous, where the tributaries of the superior and inferior rectal veins anastomose. The epithelium changes abruptly at about 1.5 to 2 cm above the anal opening; the transition is marked by a white line, below which the epithelium is stratified squamous continuous with the skin. In the region of this line there are the openings of the **anal glands**, very much enlarged modified skin glands (Fig. 16–91).

The proper circular muscle layer, continuous with that of the rectum, is greatly thickened in the anal canal, forming the non-striated **internal anal sphincter** (*Sphincter ani internus*). In the tissue surrounding the anus is a circular ring of striated muscle, the external

the colic and sigmoid branches of the mesenteric arteries. They give off large branches which ramify between and supply the muscular coats, and, after dividing into small vessels in the submucous tissue, pass to the mucous membrane. The rectum is supplied by the superior rectal branch of the inferior mesenteric, and the anal canal by the middle rectal from the internal iliac and the inferior rectal from the internal pudendal artery. The superior rectal, the continuation of the inferior mesenteric, divides into two branches, which run down either side of the rectum to within about 12.5 cm of the anus; they here split up into about six branches, which pierce the muscular coat and descend between it and the mucous membrane in the longitudinal direction, parallel with each other as far as the Sphincter ani internus, where they anastomose with the other rectal arteries and form a series of loops around the anus. The **veins** of the rectum commence in a plexus of vessels which surrounds the anal canal. In the vessels forming this plexus are

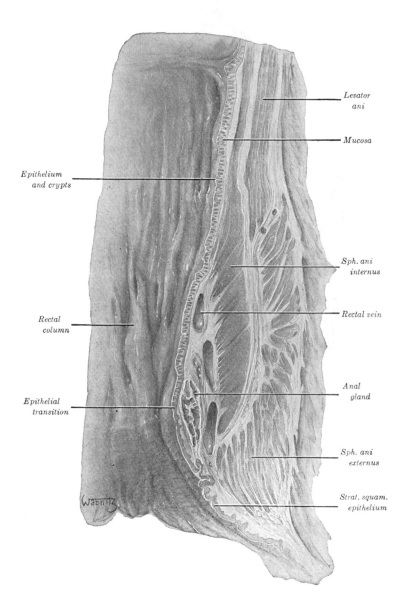

Levator
ani

Mucosa

Epithelium
and crypts

Sph. ani
internus

Rectal
column

Rectal vein

Anal
gland

Epithelial
transition

Sph. ani
externus

Strat. squam.
epithelium

FIG. 16–91.—Section of the rectum and anus of an adult cadaver.
Lightly stained with hematoxylin. × 5.

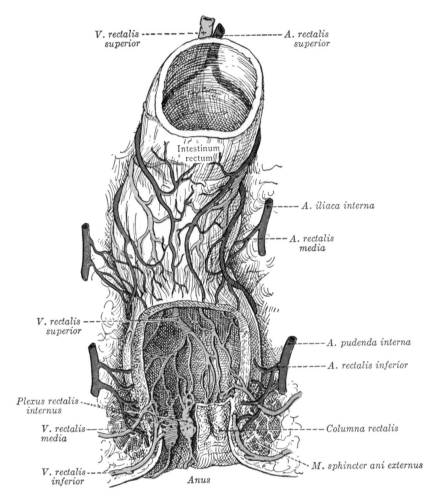

V. rectalis ---------------------+------------------- A. rectalis
superior   superior

Intestinum
rectum

------ A. iliaca interna

------ A. rectalis
media

V. rectalis ---
superior

------- A. pudenda interna

------- A. rectalis inferior

Plexus rectalis ---
internus

V. rectalis---   ------ Columna rectalis
media

------ M. sphincter ani externus

V. rectalis ---
inferior   Anus

Fig. 16–92.—The blood supply of the rectum. A portion of the anterior wall has been cut away to show the rectal columns and the internal hemorrhoidal plexus. (Eycleshymer and Jones.)

smaller saccular dilatations just within the margin of the anus; from the plexus about six vessels of considerable size are given off. These ascend between the muscular and mucous coats for about 12.5 cm, running parallel to each other; they then pierce the muscular coat (Fig. 16–92), and unite to form a single trunk, the superior rectal vein. This arrangement is termed the **rectal** or **hemorrhoidal plexus**; it communicates with the tributaries of the middle and inferior rectal veins at its commencement, thus establishing a communication between the systemic and portal circulations. The **lymphatics** of the large intestine are described on page 761. The **nerves** to that region of the colon supplied by the superior mesenteric artery are derived in the same manner as those for the small intestine; those to the more distal portions of the colon, and to the rectum, are derived from sympathetic and sacral parasympathetic fibers through the inferior mesenteric and hypogastric plexuses (see pages 1028, 1032). They are distributed in a similar way to those found in the small intestine.

THE LIVER

The **liver** (*hepar*), the largest gland in the body, is situated in the cranial and right parts of the abdominal cavity, occupying almost the whole of the right hypochondrium, the greater part of the epigastrium, and not uncommonly extending into the

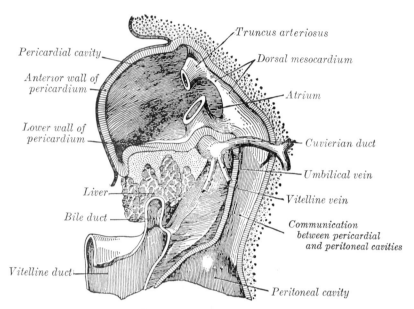

FIG. 16-93.—Liver with the septum transversum. Human embryo 3 mm long.
(After model and figure by His.)

left hypochondrium as far as the mammary line.

Development.—The liver arises in the form of a diverticulum or hollow outgrowth from the ventral surface of that portion of the primitive gut which afterward becomes the descending part of the duodenum (Fig. 16-93). This diverticulum is lined by entoderm, and grows cranialward and ventralward into the septum transversum, a mass of mesoderm between the vitelline duct and the pericardial cavity. Two solid buds of cells which extend into the tissues represent the right and the left lobes of the liver. The solid buds of cells grow into columns or cylinders, termed the **hepatic cylinders,** which branch and anastomose to form a close meshwork. This network proceeds to invade the vitelline and umbilical veins, and eventually breaks up these vessels into a series of capillary-like vessels which ramify in the meshes of the cellular network and ultimately form the **sinusoids** (Minot) of the liver (see page 1253). The continued growth and ramification of the hepatic cylinders gradually produce the mass of the liver. The original diverticulum from the duodenum becomes the common bile duct, and from this the cystic duct and gall-bladder arise as a solid outgrowth which later acquires a lumen. The opening of the common duct is at first in the ventral wall of the duodenum; later, owing to the rotation of the gut, the opening is carried to the left and then dorsalward to the position it occupies in the adult.

As the liver undergoes enlargement, both it and the ventral mesogastrium of the foregut are gradually differentiated from the septum transversum, and from the caudal surface of the latter the liver projects caudalward into the abdominal cavity. By the growth of the liver the ventral mesogastrium is divided into two parts, of which the ventral forms the falciform and coronary ligaments, and the dorsal the lesser omentum. About the third month the liver almost fills the abdominal cavity, and its left lobe is nearly as large as its right. From this period the relative development of the liver is less active, more especially that of the left lobe, which actually undergoes some degeneration and becomes smaller than the right; but up to the end of fetal life the liver remains relatively larger than in the adult.

The **adult liver,** in the male, weighs from 1.4 to 1.6 kilograms, in the female from 1.2

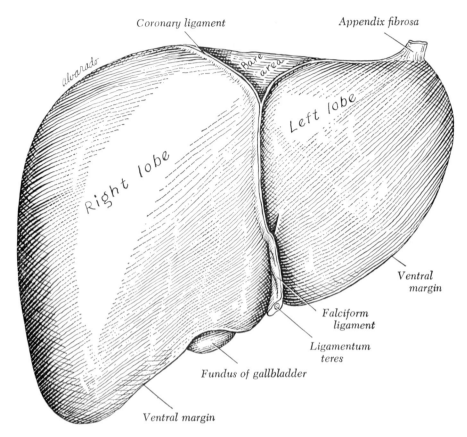

FIG. 16–94.—Ventral or anterior portion of diaphragmatic surface of the liver.
(Redrawn from Rauber-Kopsch.)

to 1.4 kilograms. It is relatively much larger in the fetus than in the adult, constituting, in the former, about one-eighteenth, and in the latter about one thirty-sixth of the entire body weight. Its greatest transverse measurement is from 20 to 22.5 cm. Vertically, near its lateral or right surface, it measures about 15 to 17.5 cm; its greatest dorsoventral diameter from 10 to 12.5 cm is on a level with the cranial end of the right kidney. Opposite the vertebral column this measurement is reduced to about 7.5 cm. Its consistency is that of a soft solid; it is friable, easily lacerated and highly vascular; its color is a dark reddish brown, and its specific gravity is 1.05.

It is irregularly hemispherical in shape with an extensive, relatively smooth, convex diaphragmatic surface and a more irregular concave visceral surface. The diaphragmatic surface has four parts: ven-

tral, superior, dorsal, and right portions. The human liver has four lobes: a large right lobe, a smaller left lobe, and much smaller caudate and quadrate lobes.

The **Diaphragmatic Surface** (*facies diaphragmatica*) has the following portions: ventral, superior, dorsal, and right.

The **ventral** or **anterior portion** (*pars anterior*) (Fig. 16–94) is separated by the diaphragm from the sixth to tenth ribs and their costal cartilages on the right side and from the seventh and eighth cartilages on the left. In the median region it lies dorsal to the xyphoid process and that part of the muscular anterior abdominal wall between the diverging costal margins. It is completely covered by peritoneum except along the line of attachment of the falciform ligament.

The **superior portion** (*pars superior*) (Fig. 16–95) is separated by the dome of

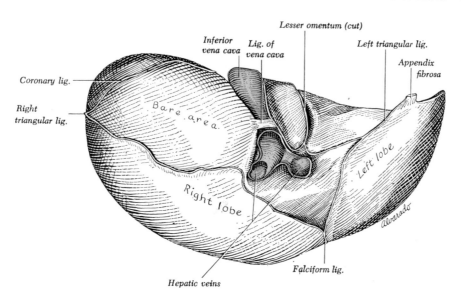

Fig. 16–95.—Superior portion of diaphragmatic surface of liver. (Redrawn from Rauber-Kopsch.)

the diaphragm from the pleura and lungs on the right and the pericardium and heart on the left. The area near the heart is marked by a shallow concavity, the **cardiac fossa** (*impressio cardiaca*). The surface is mostly covered by peritoneum but along its dorsal part it is attached to the diaphragm by the superior reflection of the coronary ligament which separates the part covered with peritoneum from the so-called bare area.

The **dorsal** or **posterior portion** (*pars posterior*) (Fig. 16–95) is broad and rounded on the right but narrow on the left. The central part presents a deep concavity which is molded to fit against the vertebral column and crura of the diaphragm. Close to the right of this concavity the **inferior vena cava** lies almost buried in its **fossa** (*sulcus venae cavae*). Two or three centimeters to the left of the vena cava is the narrow **fossa for the ductus venosus** (*fissura lig. venosi*). The caudate lobe lies between these two fossae. To the right of the vena cava and partly on the visceral surface is a small triangular depressed area, the **suprarenal impression** for the right suprarenal gland. To the left of the fossa for the ductus venosus is the **esophageal groove** for the antrum cardiacum of the esophagus.

A large part of the dorsal portion of the diaphragmatic surface is not covered by peritoneum. It is attached to the diaphragm by loose connective tissue. The uncovered area, frequently called the **bare area** (*area nuda*), is bounded by the superior and inferior reflections of the coronary ligament.

The **right portion** (*pars dextra*) merges with the other three parts of the diaphragmatic surface and continues down to the right margin which separates it from the visceral surface.

The **Visceral Surface** is concave, facing dorsalward, caudalward, and to the left. It contains several fossae and impressions for neighboring viscera. A prominent marking of the left central part is the **porta hepatis,** a fissure for the passage of the blood vessels and bile duct. The visceral surface is covered by peritoneum except where the gallbladder is attached to it and at the porta. The right lobe, lying to the right of the gallbladder, has three impressions. Farthest to the right is the **colic impression**, a flattened or shallow area for the right colic flexure, more dorsally a larger and deeper hollow is the **renal impression** for the right kidney, and the **duodenal impression** is a narrow and poorly marked area lying along the neck of the gallbladder. Between the gallbladder and **fossa for the umbilical vein** is the quadrate lobe. It is in relation with the pyloric end of the stomach, the superior

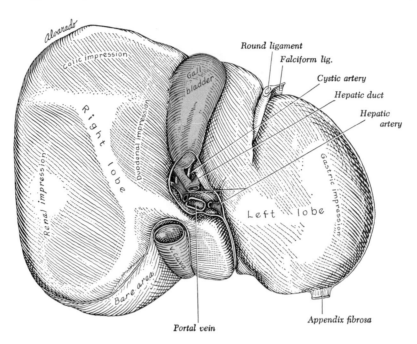

Fᴵɢ. 16–96.—Visceral surface of the liver. (Redrawn from Rauber-Kopsch.)

portion of the duodenum, and the transverse colon. The left lobe, lying to the left of the umbilical vein fossa, has two prominent markings. A large hollow extending out to the margin is the **gastric impression** for the ventral surface of the stomach. Toward the right it merges into a rounded eminence, the **tuber omentale,** which fits into the lesser curvature of the stomach and lies over the ventral surface of the lesser omentum. Just ventral to the inferior vena cava is a narrow strip of liver tissue, the **caudate process,** which connects the right inferior angle of the caudate lobe to the right lobe. Its peritoneal covering forms the ventral boundary of the epiploic foramen.

The **Inferior Border** (*margo inferior*) is thin and sharp, and marked opposite the attachment of the falciform ligament by a deep notch, the **umbilical notch** (*incisura lig. teretis*), and opposite the cartilage of the ninth rib by a second notch for the fundus of the gallbladder. In adult males this border generally corresponds with the lower margin of the thorax in the right mammary line; but in women and children it usually projects below the ribs. In the erect position it often extends below the interiliac line.

The **left extremity of the liver** is thin and flattened from above downward.

Fissures and Fossae.—The **left sagittal fossa** (*longitudinal fissure*) is a deep groove in the visceral surface which extends from the notch on the inferior margin of the liver to the cranial border of the organ. It is not named in the NA, but is worthy of mention because it separates the right and left lobes. The porta joins it, at right angles, and divides it into two parts. The ventral part is the **fissure for the ligamentum teres** (*fissura lig. teretis*) which lodges the umbilical vein in the fetus, and its remains (the ligamentum teres) in the adult; it lies between the quadrate lobe and the left lobe of the liver, and is often partially bridged over by a prolongation of the hepatic substance, the **pons hepatis.** The dorsal part, or **fossa for the ductus venosus,** lies between the left lobe and the caudate lobe; it lodges in the fetus, the ductus venosus, and in the adult a slender fibrous cord, the **ligamentum venosum,** the obliterated remains of that vessel.

The **porta** or **transverse fissure** (*porta hepatis*) (Fig. 16–97) is a short but deep fissure, about 5 cm long, extending transversely across the visceral surface of the left portion of the right lobe, nearer its

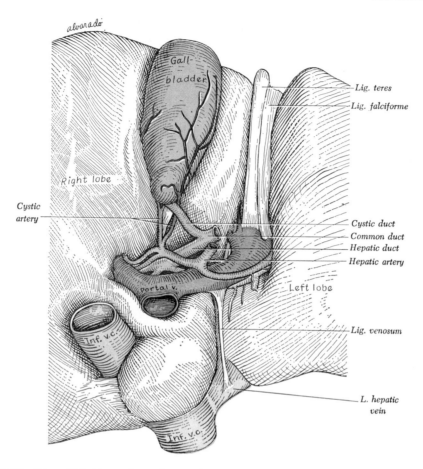

FIG. 16–97.—The gallbladder and portal area of the visceral surface of the liver with blood vessels and ducts exposed. (Redrawn from Rauber-Kopsch.)

dorsal surface than its ventral border. It joins nearly at right angles with the left sagittal fossa, and separates the quadrate lobe ventrally from the caudate lobe and process dorsally. It transmits the portal vein, the hepatic artery and nerves, and the hepatic duct and lymphatics. The hepatic duct lies ventral and to the right, the hepatic artery to the left, and the portal vein dorsal and between the duct and artery.

The **fossa for the gallbladder** (*fossa vesicae felleae*) is a shallow, oblong fossa, placed on the visceral surface of the right lobe, parallel with the left sagittal fossa. It extends from the inferior free margin of the liver, which is notched by it, to the right extremity of the porta.

The **fossa for the inferior vena cava** (*sulcus venae cavae*) is a short deep depres-

sion, occasionally a complete canal in consequence of the substance of the liver surrounding the vena cava. It lies on the posterior surface between the caudate lobe and the bare area of the liver, and is separated from the porta by the caudate process. The orifices of the hepatic veins perforate the floor of this fossa to enter the inferior vena cava.

Lobes.—The **right lobe** (*lobus hepatis dexter*) is six times as large as the left. It occupies the right hypochondrium, and is separated from the left lobe on its diaphragmatic surface by the falciform ligament, and by the left sagittal fossa on its visceral surface. It is of a somewhat quadrilateral form, its visceral and posterior surfaces being marked by three fossae: the porta and the fossae for the gallbladder and inferior vena cava, which separate its left part into two smaller lobes, the **quadrate** and **caudate** lobes. The impressions on the right lobe have already been described.

The **quadrate lobe** (*lobus quadratus*) is situated on the visceral surface of the right lobe, bounded ventrally by the inferior margin of the liver; dorsally by the porta; on the right, by the fossa for the gallbladder; and on the left, by the fossa for the umbilical vein. It is oblong in shape, its dorsoventral diameter being greater than its transverse.

The **caudate lobe** (*lobus caudatus; Spigelian lobe*) is situated upon the dorsal surface of the right lobe of the liver, opposite the tenth and eleventh thoracic vertebrae. It is bounded, inferiorly, by the porta; on the right, by the fossa for the inferior vena cava; and, on the left, by the fossa for the ductus venosus. It is nearly vertical in position, and is somewhat concave in the transverse direction. The **caudate process** is a small elevation of the hepatic substance extending obliquely lateralward, from the lower extremity of the caudate lobe to the visceral surface of the right lobe. It is situated dorsal to the porta, and separates the fossa for the gallbladder from the commencement of the fossa for the inferior vena cava.

The **left lobe** (*lobus hepatis sinister*) is smaller and more flattened than the right. It is situated in the epigastric and left hypochondriac regions. Its cranial surface is slightly convex and is molded to the diaphragm; its caudal surface presents the gastric impression and omental tuberosity already referred to.

Internal Lobes and Segments.—The liver substance is divisible internally on the basis of function and surgical importance into lobes and segments which do not coincide completely with the divisions established by external markings. These subdivisions must take into account the two great functions of the liver: (1) that of a gland with the bile passages serving as its ducts and (2) that of a vascular and storage organ. As a vascular organ it is necessary to devise units or segments including both inflow and outflow of blood, *i.e.*, the branchings of both portal and hepatic veins. This is complicated (1) by the fact that the portal veins have peripheral and the hepatic veins central

positions in the lobules and greater subdivisions and (2) by the fact that the portal vein has a bilateral type of distribution, whereas the hepatic veins have three main stems. It has been customary, however, to describe the internal lobes and segments on the basis of the portal venous radicles along with the bile ducts and to superimpose upon this a description based upon the hepatic venous radicles.

The Hepatic Triad.—The three structures of the triad, bile duct, portal vein, and hepatic artery, gather in the hepatoduodenal ligament ventral to the epiploic foramen (of Winslow) at the porta hepatis with the hepatic duct placed ventrally and to the right, the hepatic artery to the left, and the portal vein dorsally between the artery and duct. After their primary branching into right and left at the porta they continue in a similar relationship throughout the organ. They are contained within a fibrous sheath with prolongations to enclose their smallest branchings which is called the **perivascular fibrous capsule** (*capsula fibrosa perivascularis; Glisson's capsule*) (Fig. 16–102).

The primary branching of the portal vein and hepatic artery at the porta establishes a right internal lobe and a left internal lobe. The **right internal lobe** (*lobus hepatis dexter*) comprises most of the right lobe of external marking described above. It is subdivided into a **posterior segment** (*segmentum posterius*) more superior and somewhat larger than an **anterior segment** (*segmentum anterius*). Both segments are divided into superior and inferior portions.

The **left internal lobe** (*lobus hepatis sinister*) includes the left, the caudate, and the quadrate lobes of external marking. The division of the left lobe into a **medial segment** (*segmentum mediale*) and a **lateral segment** (*segmentum laterale*) is marked on the surface by the attachment of the falciform ligament, the ligamentum teres, and the ligamentum venosum. The two segments have superior and inferior portions.

Vascular segments.—The three principal hepatic veins, right, middle, and left, provide the basis for a selection of parts of the internal segments described above into **true vascular segments,** a right, a left, and two middle segments. A variable number of

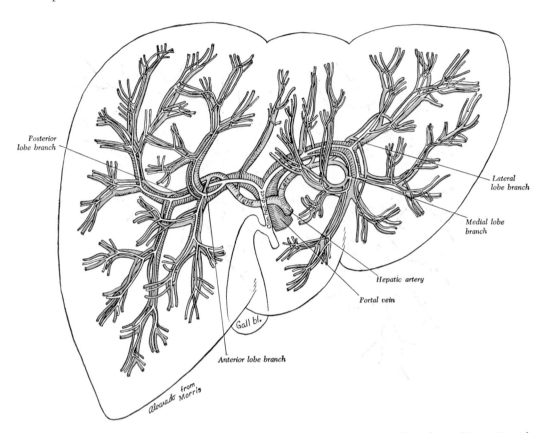

Posterior
lobe branch

Lateral
lobe branch

Medial lobe
branch

Hepatic artery

Portal vein

Gall bl.

Anterior lobe branch

Alvarado from Morris

Fig. 16–98.—Intrahepatic distribution of the hepatic artery, portal vein, and biliary ducts. (From *Morris'
Human Anatomy*, 12th edition, 1966. Courtesy of Blakiston Division of McGraw-Hill Book Company.)

minor hepatic veins assist in the drainage of these vascular segments (Fig. 16–99).

The **right dorsal** or **dorsocaudal vascular segment** corresponds with the dorsal segment of the internal right lobe and is drained by the right hepatic vein.

The **left lateral vascular segment** corresponds with the lateral segment of the left internal lobe and is drained by the left hepatic vein.

The **ventral middle vascular segment** contains both the anterior segment of the right lobe and the medial segment of the internal left lobe. It is drained by the middle hepatic vein.

The **dorsal middle vascular segment** contains the caudate lobe and caudate process, parts of the left internal lobe. It is drained usually by two minor hepatic veins, the cranial caudate hepatic vein and the caudal caudate hepatic vein.

Ligaments.—The liver is connected to the under surface of the diaphragm and to the ventral wall of the abdomen by five ligaments; four of these—the **falciform,** the **coronary,** and the two **lateral**—are peritoneal folds; the fifth, the **round ligament,** is a fibrous cord, the obliterated umbilical vein. The liver is also attached to the lesser curvature of the stomach by the hepatogastric and to the duodenum by the hepatoduodenal ligament (see page 1214).

The **falciform ligament** (*ligamentum falciforme hepatis*) is situated in a parasagittal plane, but lies obliquely so that one surface faces ventralward and is in contact with the peritoneum dorsal to the right Rectus and the diaphragm, while the other is directed dorsalward and is in contact with the left lobe of the liver. It is attached by its left margin to the abdominal surface of the diaphragm, and the dorsal surface of the sheath of the right Rectus as far caudalward as the umbilicus; by its right margin

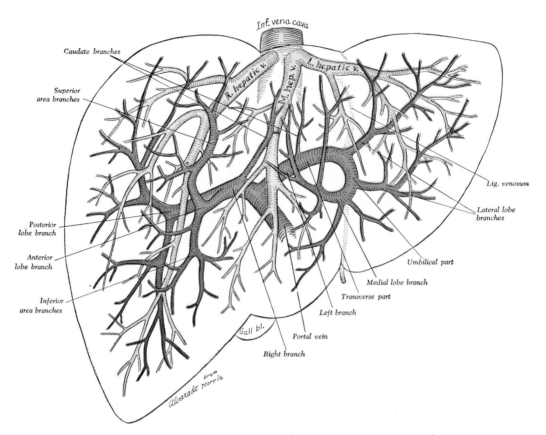

Fig. 16–99.—Intrahepatic distribution of the hepatic and portal veins. (From *Morris' Human Anatomy*, 12th edition, 1966. Courtesy of Blakiston Division of McGraw-Hill Book Company.)

it extends from the notch on the inferior margin of the liver, as far dorsalward as the bare area. It is composed of two layers of peritoneum closely united together. Its base or free edge contains between its layers the round ligament and the parumbilical veins.

The **coronary ligament** (*ligamentum coronarium hepatis*) consists of an anterior and a posterior layer. The *anterior layer* is formed by the reflection of the peritoneum from the cranial margin of the bare area of the liver to the under surface of the diaphragm, and is continuous with the right layer of the falciform ligament. The *posterior layer* is reflected from the caudal margin of the bare area on to the right kidney and suprarenal gland, and is termed the **hepatorenal ligament**.

The **triangular ligaments** (*lateral ligaments*) are two in number, right and left.

The **right triangular ligament** (*ligamentum triangulare dextrum*) is situated at the right extremity of the bare area, and is a small fold which passes to the diaphragm,

being formed by the apposition of the anterior and posterior layers of the coronary ligament.

The **left triangular ligament** (*ligamentum triangulare sinistrum*) is a fold of some considerable size, which connects the posterior part of the upper surface of the left lobe to the diaphragm; its anterior layer is continuous with the left layer of the falciform ligament. It terminates on the left in a strong fibrous band, the **appendix fibrosa hepatis.**

The **round ligament** (*ligamentum teres hepatis*) is a fibrous cord resulting from the obliteration of the umbilical vein. It ascends from the umbilicus, in the free margin of the falciform ligament, to the umbilical notch of the liver, from which it may be traced in its proper fossa on the inferior surface of the liver to the porta, where it becomes continuous with the *ligamentum venosum.*

Fixation of the Liver.—Several factors contribute to the maintenance of the liver in place. The attachments of the liver to the diaphragm by the coronary and triangular ligaments and the intervening connective tissue of the bare area; the intimate connection of the inferior vena cava by the connective tissue and hepatic veins supports the posterior part of the liver. The lax falciform ligament certainly gives no support though it probably limits lateral displacement. During descent of the diaphragm with deep breathing, the liver rolls ventralward, shifting the inferior border caudalward so that it can be palpated.

Vessels and Nerves.—The vessels connected with the liver are: the **hepatic artery,** the **portal vein,** and the **hepatic veins.**

The **hepatic artery** and **portal vein,** accompanied by numerous nerves, ascend to the porta, between the layers of the lesser omentum. The *bile duct* and the lymphatic vessels descend from the porta between the layers of the same omentum. The relative positions of the three structures are as follows: the bile duct lies to the right, the hepatic artery to the left, and the portal vein dorsal to and between the other two (Fig. 16–108). They are enveloped in a loose areolar tissue, the **fibrous capsule of Glisson** (*capsula fibrosa perivascularis*) which accompanies the vessels in their course through the portal canals in the interior of the organ (Fig. 16–102).

The **hepatic veins** (Fig. 16–99) convey the blood from the liver, and are described on page 712. They have very little cellular investment, and what there is binds their parietes closely to the walls of the canals through which they run, so that, on section of the organ, they remain widely open and are solitary, and may be easily distinguished from the branches of the portal vein which are more or less collapsed, and always accompanied by an artery and duct.

The **lymphatic vessels** of the liver are described under the lymphatic system.

The **nerves** of the liver are derived from the right and left vagus and the celiac plexus of the sympathetic. The fibers form plexuses along the hepatic artery and portal vein, enter the porta, and accompany the vessels and ducts to the interlobular spaces. The hepatic vessels are said to receive only sympathetic fibers, while both sympathetic and parasympathetic fibers are distributed to the walls of the bile ducts and gallbladder, where they form plexuses similar to the enteric plexuses of the intestinal wall.

Structure of the Liver.—The substance of the liver or parenchyma is composed of lobules, held together by an extremely fine areolar tissue, in which ramify the portal vein, hepatic

artery, hepatic veins, lymphatics, and nerves, the whole being invested by a serous and a fibrous coat.

The **serous coat** (*tunica serosa*) is derived from the peritoneum, and invests the greater part of the surface of the organ. It is intimately adherent to the fibrous coat.

The **fibrous coat** (*areolar coat*) lies beneath the serous investment, and covers the entire surface of the organ. It is difficult of demonstration, except where the serous coat is deficient. At the porta it is continuous with the fibrous capsule of Glisson, and on the surface of the organ with the areolar tissue separating the lobules.

The **lobules** (*lobuli hepatis*) form the principal mass of the parenchyma. Their outlines, about 2 mm in diameter, give a mottled appearance to the surface of the organ. They are roughly hexagonal in shape, with their columns of cells clustered around an intralobular vein, the smallest radicle of the hepatic vein. The adjacent faces of these neighboring hexagonal (or more irregularly polygonal) lobules are fitted together with a minimum of delicate connective tissue. In the pig, the individual lobules have complete connective tissue capsules and the hexagonal shape is more evident than in the human liver.

The **portal canal** is the name given to the channel through the parenchyma by which the smallest radicles of the portal vein, hepatic artery, and bile duct are distributed. These three are bound together by delicate connective tissue, the **capsula fibrosa perivascularis** or **Glisson's capsule.** They are situated between the lobules and in a microscopic section cut perpendicularly through an intralobular vein, there will be three portal canals at the periphery of the lobule, about equally distant from each other at three angles of the hexagon. Polygonal shapes such as the pentagon in Figure 16–100 are very common, however.

Microscopic Appearance (Fig. 16–100).— Each lobule consists of a mass of cells, **hepatic cells,** arranged in irregular radiating columns and plates between which are the blood channels (*sinusoids*). Between the cells are also the minute bile capillaries. Therefore, in the lobule there are all the essentials of a secreting gland; that is to say: (1) **cells,** by which the secretion is formed; (2) **blood vessels,** in close relation with the cells, containing the blood from which the secretion is derived; (3) **ducts,** by which the secretion, when formed, is carried away.

1. The **hepatic cells** are polyhedral in form. They vary in size from 12 to 25 μ in diameter. They contain one or sometimes two distinct nuclei. The nucleus exhibits an intranuclear

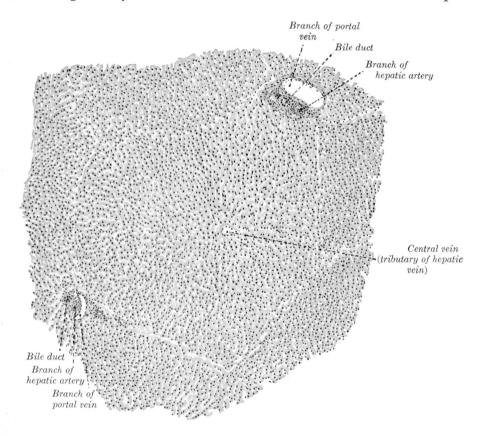

Fig. 16–100.—Human liver lobule surrounded by parts of six neighboring lobules. (From Rauber-Kopsch, *Lehrbuch u. Atlas d. Anatomie d. Menschen,* 19th Edition, Vol. II, courtesy Georg Thieme Verlag, Stuttgart, 1955.)

network and one or two refractile nucleoli. The cells usually contain granules, some of which are protoplasmic, while others consist of glycogen, fat, or an iron compound. In the lower vertebrates, *e.g.,* frog, the cells are arranged in tubes with the bile duct forming the lumen and blood vessels externally.

2. The **Blood Vessels.**—The blood in the capillary plexuses around the liver cells is brought to the liver principally by the portal vein, but also to a certain extent by the hepatic artery.

The **hepatic artery,** entering the liver at the porta with the portal vein and hepatic duct, ramifies with these vessels through the portal canals. It gives off **vaginal branches,** which ramify in the fibrous capsule of Glisson, and appear to be destined chiefly for the nutrition of the coats of the vessels and ducts. It also gives off **capsular branches,** which reach the surface of the organ, ending in its fibrous coat in stellate plexuses. Finally, it gives off **interlobular branches,** which form a plexus outside

each lobule, to supply the walls of the interlobular veins and the accompanying bile ducts. From these plexuses capillaries join directly with the sinusoids of the liver lobule at its periphery.

The **portal vein** also enters at the porta, and runs through the portal canals (Fig. 16–99), enclosed in Glisson's capsule, dividing in its course into branches, which finally break up in the interlobular spaces, into the **interlobular plexuses,** which give off portal venules which divide into small branches and twigs as they pass to the surfaces of the lobules to join the hepatic sinusoids (Fig. 16–101).

Hepatic sinusoids are large, richly anastomosing, modified capillary channels lying between the cords of liver cells. They traverse the liver lobule from its periphery to the intralobular or central vein. At the periphery of the lobule they connect with interlobular branches of the portal vein and hepatic artery. Thus all the blood which enters the liver passes through the sinusoids to the central veins. The sinusoids are

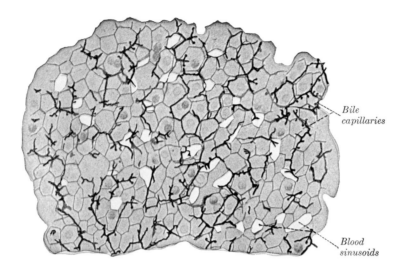

FIG. 16–101.—Bile capillaries in the liver of a rabbit demonstrated by silver chromate, counterstained with alum carmine. (From Rauber-Kopsch, *Lehrbuch u. Atlas d. Anatomie d. Menschen,* 19th Edition, Vol. II, courtesy Georg Thieme Verlag, Stuttgart, 1955.)

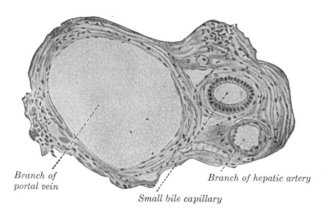

FIG. 16–102.—Cross section of a portal canal with Glisson's capsule of interlobular connective tissue surrounding branches of the portal vein, hepatic artery, and hepatic duct. (From Rauber-Kopsch, *Lehrbuch u. Atlas d. Anatomie d. Menschen,* 19th Edition, Vol. II, courtesy Georg Thieme Verlag, Stuttgart, 1955.)

lined by modified endothelium and contain many macrophages (v. Kupffer cells) attached to their walls.

Hepatic Veins.—At the center of the lobule, the sinusoids empty into one vein, of considerable size, which runs down the center of the lobule from apex to base, and is called the **intralobular** or **central vein.** At the base of the lobule this vein opens directly into the **sublobular vein,** with which the lobule is connected. The sublobular veins unite to form larger and larger trunks, and end at last in the hepatic veins; these converge to form three large trunks

which open into the inferior vena cava while that vessel is situated in its fossa on the posterior surface of the liver.

3. The **bile ducts** commence by little passages in the liver cells which communicate with canaliculi termed **intercellular biliary passages** (*bile capillaries*). These passages are merely little channels or spaces left between the contiguous surfaces of two cells, or in the angle where three or more liver cells meet (Fig. 16–101), and they are always separated from the blood capillaries by at least half the width of a liver cell. The channels thus formed radiate

to the circumference of the lobule, and open into the interlobular bile ducts which run in Glisson's capsule, accompanying the portal vein and hepatic artery (Fig. 16–102). These join with other ducts to form two main trunks, which leave the liver at the porta, and by their union form the **hepatic duct.**

Structure of the Ducts.—The walls of the biliary ducts consist of a connective tissue coat, in which are muscle cells, arranged both circularly and longitudinally, and an epithelial layer, consisting of short columnar cells resting on a distinct basement membrane.

Excretory Apparatus of the Liver.—The excretory apparatus of the liver consists of (1) the **hepatic duct,** formed by the junction of the two main ducts, which pass out of the liver at the porta; (2) the **gallbladder,** which serves as a reservoir for the bile; (3) the **cystic duct,** or the duct of the gallbladder; and (4) the **common bile duct,** formed by the junction of the hepatic and cystic ducts.

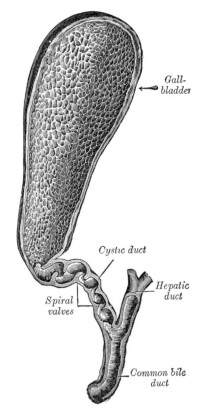

Fig. 16–103.—The gallbladder and bile ducts laid open. (Spalteholz.)

1. The **Hepatic Duct** (*ductus hepaticus*) (Fig. 16–98).—Two main trunks of nearly equal size, one from the right, the other from the left lobe, unite to form the hepatic duct, which passes to the right for about 4 cm, between the layers of the lesser omentum, where it is joined at an acute angle by the cystic duct, to form the common bile duct. The hepatic and part of the common duct are accompanied by the hepatic artery and portal vein.

2. The **Gallbladder** (*vesica fellea*) (Fig. 16–103) is a conical or pear-shaped musculomembranous sac, lodged in a fossa on the visceral surface of the right lobe of the liver and extending from near the right extremity of the porta to the inferior border of the organ. It is from 7 to 10 cm in length, 2.5 cm in breadth at its widest part, and holds from 30 to 35 cc. It is divided into a fundus, body, and neck. The **fundus,** or broad extremity, is directed caudalward, and projects beyond the inferior border of the liver; the **body** and **neck** are directed cranialward and dorsalward to the left. The surface of the gallbladder is attached to the liver by connective tissue and vessels. The caudal surface is covered by peritoneum, which is reflected on to it from the surface of the liver. Occasionally the whole of the organ is invested by the serous membrane, and is then connected to the liver by a kind of mesentery.

Relations.—The **body** is in relation with the commencement of the transverse colon; and farther dorsally usually with the descending portion of the duodenum, but sometimes with the superior portion of the duodenum or pyloric end of the stomach. The **fundus** is in relation, ventrally, with the abdominal parietes, immediately below the ninth costal cartilage; dorsally with the transverse colon. The **neck** is narrow, and curves upon itself like the letter S; at its point of connection with the cystic duct it presents a well-marked constriction.

Structure (Fig. 16–104).—The gallbladder consists of three coats: **serous, fibromuscular,** and **mucous.**

The **external** or **serous coat** (*tunica serosa vesicae felleae*) is derived from the peritoneum; it completely invests the fundus, but covers the body and neck only on their caudal surfaces.

The **fibromuscular coat** (*tunica muscularis vesicae felleae*), a thin but strong layer forming the framework of the sac, consists of dense

Labels in figure: Gall-bladder; Cystic duct; Hepatic duct; Spiral valves; Common bile duct

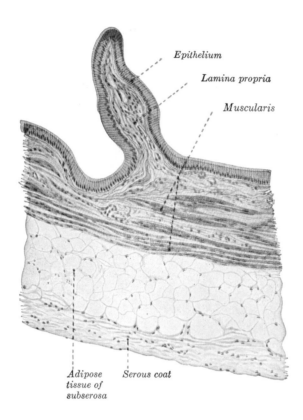

Epithelium

Lamina propria

Muscularis

Adipose tissue of subserosa *Serous coat*

FIG. 16–104.—Cross section through the wall of the human gallbladder. (From Rauber-Kopsch, *Lehrbuch u. Atlas d. Anatomie d. Menschen,* 19th Edition, Vol. II, courtesy Georg Thieme Verlag, Stuttgart, 1955.)

fibrous tissue, which interlaces in all directions, and is mixed with smooth muscular fibers, disposed chiefly in a longitudinal direction, a few running transversely.

The **internal** or **mucous coat** (*tunica mucosa vesicae felleae*) is loosely connected with the fibrous layer. It is generally of a yellowish-brown color, and is elevated into minute rugae. Opposite the neck of the gallbladder the mucous membrane projects inward in the form of oblique ridges or folds, forming a sort of spiral valve.

The mucous membrane is continuous through the hepatic duct with the mucous membrane lining the ducts of the liver, and through the common bile duct with the mucous membrane of the duodenum. It is covered with very high columnar epithelium, and secretes mucin; in some animals it secretes a nucleoprotein instead of mucin.

Vessels and Nerves.—The **arteries** to the gallbladder are derived from the cystic artery; its **veins** drain into liver capillaries and into the portal vein. The **lymphatics** are described

under the lymphatic system. The **nerves** are described on page 1031.

3. The **Cystic Duct** (*ductus cysticus*).—The cystic duct, about 4 cm long, runs dorsalward, caudalward, and to the left from the neck of the gallbladder, and joins the hepatic duct to form the common bile duct. The mucous membrane lining its interior is thrown into a series of crescentic folds, from five to twelve in number, similar to those found in the neck of the gallbladder. They project into the duct in regular succession, and are directed obliquely around the tube, presenting much the appearance of a continuous spiral valve. They constitute the **spiral valve** (*Heister*), which is found only in primates and represents a device to prevent distention or collapse of the cystic duct with changing pressures in the gallbladder or common duct, associated with the assumption of an erect posture. When

the duct is distended, the spaces between the folds are dilated, so as to give to its exterior a twisted appearance.

4. The **Common Bile Duct** (*ductus chole-dochus*) (Fig. 16–108).—The common bile duct is formed by the junction of the cystic and hepatic ducts; it is about 7.5 cm long, and of the diameter of a goose-quill, or soda fountain straw.

It descends along the right border of the lesser omentum dorsal to the superior portion of the duodenum, ventral to the portal vein, and to the right of the hepatic artery; it then runs in a groove near the right border of the posterior surface of the head of the pancreas; here it is situated ventral to the inferior vena cava, and is occasionally completely embedded in the pancreatic substance. At its termination it is closely associated with the terminal portion of the pancreatic duct as it passes obliquely through the muscular and mucous coats of the duodenum. The walls of the terminal portions of both ducts are thickened by the presence of a sphincter muscle, the sphincter of Oddi, which usually causes a protru-sion into the lumen of the duodenum, the **duodenal papilla** or papilla of Vater. A common orifice for the two ducts is present in about 60 per cent and they have separate openings in about 40 per cent. The ducts are narrowed rather than widened as they traverse the papilla and the length of common channel shared by bile and pancreatic ducts is less than one half the papilla in 75 per cent. The term "ampulla of Vater" should therefore be discarded in favor of the "papilla of Vater" (Sterling '55).

Structure.—The coats of the large biliary ducts are an **external** or **fibrous,** and an **internal** or **mucous.** The **fibrous coat** is composed of strong fibroareolar tissue, with a certain amount of muscular tissue, arranged, for the most part, in a circular manner around the duct. The **mucous coat** is continuous with the lining membrane of the hepatic ducts and gallbladder, and also with that of the duodenum; and, like the mucous membrane of these structures, its epithelium is of the columnar variety. It is provided with numerous mucous glands, which are lobulated and open by minute orifices scattered irregularly in the larger ducts.

THE PANCREAS

The **pancreas** (Fig. 16–57) is situated transversely across the posterior wall of the abdomen, in the epigastric and left hypochondriac regions. Its length varies from 12.5 to 15 cm; its weight in the female is 84.88 ± 14.95 grams and in the male 90.41 ± 16.08 grams (Schaefer '26). It is the structure known as the belly sweetbread in the butcher shop.

Development (Figs. 16–105, 16–106).—The pancreas is developed in two parts, a dorsal and a ventral. The former arises as a diverticulum from the dorsal aspect of the duodenum a short distance above the hepatic diverticulum, and growing cranialward and dorsalward into the dorsal mesogastrium, forms a part of the head and uncinate process and the whole of the body and tail of the pancreas. The ventral part appears in the form of a diverticulum from the primi-

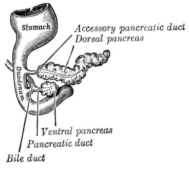

Fig. 16–105.—Pancreas of a human embryo of five weeks. (Kollmann.)

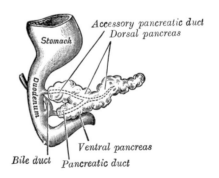

Fig. 16–106.—Pancreas of a human embryo at end of sixth week. (Kollmann.)

tive bile duct and forms the remainder of the head and uncinate process of the pancreas. The duct of the dorsal part (**accessory pancreatic duct**) therefore opens independently into the duodenum, while that of the ventral part (**pancreatic duct**) opens with the common bile duct. About the sixth week the two parts of the pancreas meet and fuse and a communication is established between their ducts. After this has occurred the terminal part of the accessory duct, *i.e.*, the part between the duodenum and the point of meeting of the two ducts, undergoes little or no enlargement, while the pancreatic duct increases in size and forms the main duct of the gland. The opening of the accessory duct into the duodenum is sometimes obliterated, and even when it remains patent it is probable that the whole of the pancreatic secretion is conveyed through the pancreatic duct.

At first the pancreas is between the two layers of the dorsal mesogastrium, which give to it a complete peritoneal investment, and its surfaces look to the right and left. With the change in the position of the stomach the dorsal mesogastrium is drawn

caudalward and to the left, and the right side of the pancreas is directed dorsalward and the left ventralward. The right surface becomes applied to the posterior abdominal wall, and the peritoneum which covered it undergoes absorption; thus, in the adult, the gland appears to lie behind the peritoneal cavity.

The **Adult Pancreas** is a compound racemose gland, analogous in its structures to the salivary glands, though softer and less compactly arranged than those organs. It is long and irregularly prismatic in shape; its right extremity, being broad, is called the **head,** and is connected to the main portion of the organ, or **body,** by a slight constriction, the **neck,** while its left extremity gradually tapers to form the **tail.**

Relations.—The **Head** (*caput pancreatis*) is lodged within the curve of the duodenum. Its cranial border is overlapped by the superior part of the duodenum and its caudal overlaps the horizontal part; its right and left borders overlap and insinuate themselves around the descending and ascending parts of the duodenum respectively. The angle of junction of the caudal and left

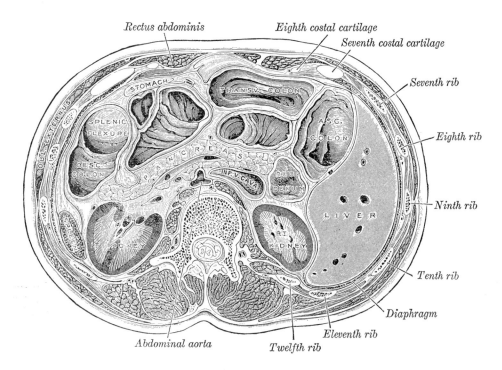

Fig. 16–107.—Transverse section through the middle of the first lumbar vertebra, showing the relations of the pancreas. (Braune.)

lateral borders forms a prolongation, termed the **uncinate process.** In the groove between the duodenum and the right lateral and caudal borders are the anastomosing superior and inferior pancreaticoduodenal arteries (Fig. 8–42); the common bile duct descends to its termination in the descending part of the duodenum close to the right border.

Anterior Surface.—The greater part of the right half of this surface is in contact with the transverse mesocolon, only areolar tissue intervening. From its cranial part the **neck** springs, its right limit being marked by a groove for the gastroduodenal artery. The caudal part of the right half, below the transverse colon, is covered by peritoneum continuous with the inferior layer of the transverse mesocolon, and is in contact with the coils of the small intestine. The superior mesenteric artery passes down ventral to the left half across the uncinate process; the superior mesenteric vein runs cranialward on the right side of the artery and, dorsal to the neck, joins with the lienal vein to form the portal vein (Fig. 9–39).

The **posterior surface** is in relation with the inferior vena cava, the common bile duct, the renal veins, the right crus of the diaphragm, and the aorta.

The **Neck** is about 2.5 cm long, and is directed at first ventralward, and then to the left to join the body. Its ventral surface supports the pylorus; its dorsal surface is in relation with the commencement of the portal vein; on the right it is grooved by the gastroduodenal artery.

The **Body** (*corpus pancreatis*) is prismatic in shape, and has three surfaces: **anterior, posterior,** and **inferior,** and three borders: **superior, anterior,** and **inferior.**

The **anterior surface** (*facies anterior*) is somewhat concave, and is covered by the dorsal surface of the stomach which rests upon it, the two organs being separated by the omental bursa. Where it joins the neck there is a well-marked prominence, the **tuber omentale,** which abuts against the posterior surface of the lesser omentum.

The **posterior surface** (*facies posterior*) is devoid of peritoneum, and is in contact with the aorta, the lienal vein, the left kidney and its vessels, the left suprarenal

gland, the origin of the superior mesenteric artery, and the crura of the diaphragm.

The **inferior surface** (*facies inferior*) is narrow on the right but broader on the left, and is covered by peritoneum; it is in contact with the duodenojejunal flexure and some coils of the jejunum; its left extremity rests on the left colic flexure.

The **superior border** (*margo superior*) is blunt and flat to the right, narrow and sharp to the left, near the tail. It commences on the right in the omental tuberosity, and is in relation with the celiac artery, from which the hepatic artery courses to the right just above the gland, while the lienal artery runs toward the left in a groove along this border.

The **anterior border** (*margo anterior*) separates the anterior from the inferior surface, and along this border the two layers of the transverse mesocolon diverge from each other, one passing over the anterior surface, the other over the inferior surface.

The **inferior border** (*margo inferior*) separates the posterior from the inferior surface; the superior mesenteric vessels emerge under its right extremity.

The **Tail** (*cauda pancreatis*) is the narrow part extending to the left as far as the caudal part of the gastric surface of the spleen. It lies in the phrenicolienal ligament, and is in contact with the left colic flexure.

The **Pancreatic Duct** (*ductus pancreaticus; duct of Wirsung*) extends transversely from left to right through the substance of the pancreas (Fig. 16–108). It commences by the junction of the small ducts of the lobules situated in the tail of the pancreas, and, running from left to right through the body, it receives the ducts of the various lobules composing the gland. Considerably augmented in size, it reaches the neck, and turning caudalward, dorsalward, and to the right, it comes into relation with the common bile duct, which lies to its right side; leaving the head of the gland, it passes very obliquely through the muscular and mucous coats of the duodenum, and ends by an orifice common to it and the common bile duct upon the summit of the duodenal papilla, situated at the medial side of the descending portion of the duodenum, 7.5 to 10 cm below the pylorus. In 40 per cent of the bodies the pancreatic duct and the

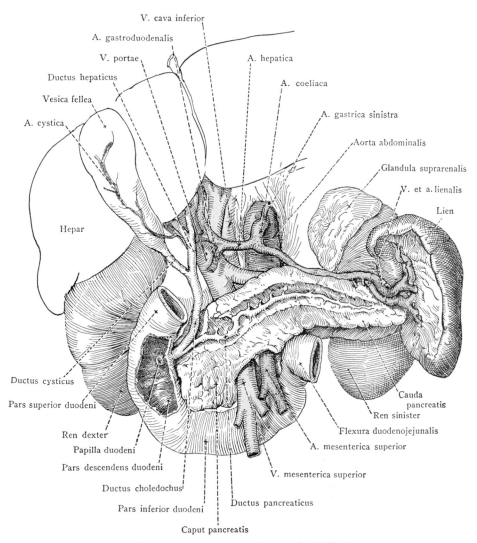

V. cava inferior
A. gastroduodenalis
V. portae
Ductus hepaticus
Vesica fellea
A. cystica
Hepar
A. hepatica
A. coeliaca
A. gastrica sinistra
Aorta abdominalis
Glandula suprarenalis
V. et a. lienalis
Lien
Ductus cysticus
Pars superior duodeni
Ren dexter
Papilla duodeni
Pars descendens duodeni
Ductus choledochus
Pars inferior duodeni
Caput pancreatis
Cauda pancreatis
Ren sinister
Flexura duodenojejunalis
A. mesenterica superior
V. mesenterica superior
Ductus pancreaticus

FIG. 16–108.—The pancreas, pancreatic duct, and neighboring structures.
(Eycleshymer and Jones.)

common bile duct open separately into the duodenum. Frequently there is an additional duct, which is given off from the pancreatic duct in the neck of the pancreas and opens into the duodenum about 2.5 cm above the duodenal papilla. It receives the ducts from the lower part of the head, and is known as the **accessory pancreatic duct** (*duct of Santorini*) (see Development).

Function.—The pancreas is a gland of both external (exocrine) and internal (endocrine) secretion. The greater part of its bulk is formed by the exocrine gland. Its secretion, the **pancreatic juice,** is conveyed by the pancreatic

duct to the duodenum where its several enzymes aid in the digestion of proteins, carbohydrates, and fats. The endocrine gland is formed by small clumps of cells known as islands of Langerhans scattered throughout the pancreas. The secretion of these cells, called **insulin,** is taken up by the blood stream and is an important factor in the control of sugar metabolism in the body.

Structure.—The **exocrine pancreas** is a compound tubulo-acinar or racemose gland, resembling the parotid gland in microscopic structure. It has a connective tissue covering but no distinct capsule. It is made up of lobes and lobules which are identifiable but indistinctly marked with connective tissue in the human pancreas. The main **pancreatic duct** extends throughout the length of the gland receiving smaller ducts from the lobes and lobules along its course (Fig. 16–108). Although two ducts are formed in the embryo, they are usually combined into a single system in the adult (see page 1259). The main duct is lined with cylindrical epithelium containing occasional goblet cells; the smaller intermediate and intercalated ducts have lower cuboidal cells. The **secreting acini** are almost filled by the protruding apices of the glandular cells which are of the serous or zymogenic type. The basal portion of these cells, resting on a basement membrane, contains the nucleus and a *basophilic substance* which has a striated appearance and is identified as the cytoplasmic

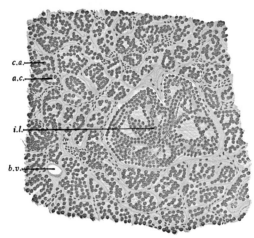

FIG. 16–109.—Section of portion of lobule from human pancreas. *ac.*, acinus of exocrine secreting gland; *b.v.*, blood vessel; *c.a.*, centroacinar cells; *i.l.*, island of endocrine-secreting cells (island of Langerhans). × 200. (Redrawn from Sobotta.)

reticulum by the electron microscope. The apical portion of these cells contains *zymogen granules*, easily seen in the usual eosin-stained preparations because of their red color. During digestion the granules become reduced in number. A characteristic feature of the pancreas is the presence of **centroacinar cells**. They represent a continuation of the terminal duct into the secreting acini. They are smaller than the secreting cells and have a clear cytoplasm (Munger '58, Oram '55).

The **Island of Langerhans** can be identified in the usual histological preparations as a cluster of less deeply stained cells. They are scattered throughout the gland but usually appear near the center of a lobule. The cells are arranged in cords or plates one or two cells thick between which there are abundant capillaries. By the use of special stains it has been shown that there are three types of cells: **alpha cells, beta cells** and **delta cells.** The beta cells are most abundant, and are the source of insulin (Ferner '52, Lacy '57).

Vessels.—The **arteries** of the pancreas are mainly branches of the splenic (page 634) which form several arcades with the pancreatic branches of the gastroduodenal (page 632) and superior mesenteric arteries (Michels '55). The **veins** are tributaries of the splenic and superior mesenteric portions of the portal vein. The **lymphatics** drain into the regional nodes associated with the major arteries to the gland.

Nerves.—The autonomic innervation is both parasympathetic and sympathetic through the splenic subdivision of the celiac plexus. The **sympathetic nerves** are postganglionic; the **parasympathetic nerves** are preganglionics from the vagus which synapse with ganglion cells scattered throughout the gland. **Sensory nerves** to the pancreas are mainly conveyed by way of the splanchnic nerves (Alvarado '55).

REFERENCES

EMBRYOLOGY AND GENERAL

HUNTINGTON, G. S. 1903. *The Anatomy of the Human Peritoneum and Abdominal Cavity.* vii + 292 pages. Lea Brothers and Company, Philadelphia.

JIT, I. 1952. The development and the structure of the suspensory muscle of the duodenum. Anat. Rec., *113*, 395-407.

KRAUS, B. S. and R. E. JORDAN. 1965. *The Human Dentition Before Birth.* 218 pages, 128 figures. Lea & Febiger, Philadelphia.

MICHELS, N. A. 1955. *Blood Supply and Anatomy of the Upper Abdominal Organs, with a Descriptive Atlas.* xiv + 581 pages. J. B. Lippincott Company, Philadelphia.

PROVENZA, V. 1964. *Oral Histology Inheritance and Development.* xiv + 548 pages, illustrated. J. B. Lippincott Company, Philadelphia.

SCHWEGLER, R. A., JR. and E. A. BOYDEN. 1937. The development of the pars intestinalis of the common bile duct in the human fetus with special reference to the origin of the ampulla of Vater and the sphincter of Oddi. I. The involution of the ampulla. Anat. Rec., *67*, 441-467. II. The early development of the musculus proprius. Anat. Rec., 68, 17-41. III. The composition of the musculus proprius. Anat. Rec., *68*, 193-219.

SMITH, E. I. 1957. The early development of the trachea and esophagus in relation to atresia of the esophagus and tracheoesophageal fistula. Carneg. Instn., Contr. Embryol., *36*, 41-58.

MOUTH, PHARYNX, AND ESOPHAGUS

BENNETT, G. A. and R. C. HUTCHINSON. 1946. Experimental studies on the movements of the mammalian tongue. II. The protrusion mechanism of the tongue (dog). Anat. Rec., *94*, 57-83.

BUTLER, H. 1951. The veins of the oesophagus. Thorax, *6*, 276-296.

GARN, S. M., K. KOSKI, and A. B. LEWIS. 1957. Problems in determining the tooth eruption sequence in fossil and modern man. Amer. J. Phys. Anthrop., *15*, 313-331.

LERCHE, W. 1950. *The Esophagus and Pharynx in Action: A Study of Structure in Relation to Function.* xii + 222 pages. Charles C Thomas, Springfield.

MUNGER, B. L. 1964. Histochemical studies on seromucous and mucous secreting cells of human salivary glands. Amer. J. Anat., *115*, 411-429.

SAUER, M. E. 1951. The cricoesophageal tendon. A recommendation for its inclusion in official anatomical nomenclature. Anat. Rec., *109*, 691-697.

SHACKLEFORD, J. M. and C. E. KLAPPER. 1962. A sexual dimorphism of hamster submaxillary mucin. Anat. Rec., *142*, 495-503.

SHAPIRO, H. H. 1954. *Maxillofacial Anatomy with Practical Applications.* xiv + 392 pages. J. B. Lippincott Company, Philadelphia.

SPRAGUE, J. M. 1944. The innervation of the pharynx in the rhesus monkey, and the formation of the pharyngeal plexus in primates. Anat. Rec., *90*, 197-208.

STRONG, L. H. 1956. Muscle fibers of the tongue functional in consonant production. Anat. Rec., *126*, 61-79.

SWIGART, L. L., R. G. SIEKERT, W. C. HAMBLEY, and B. J. ANSON. 1950. The esophageal arteries— an anatomic study of 150 specimens. Surg. Gynec. Obstet., *90*, 234-243.

TANDLER, B. 1962. Ultrastructure of the human submaxillary gland. I. Architecture and histological relationships of the secretory cells. Amer. J. Anat., *111*, 287-307.

TELFORD, I. R. 1946. Pigment studies on the incisor teeth of vitamin E deficient rats of the Long-Evans strain. Proc. Soc. exp. Biol. Med., *63*, 89-91.

STOMACH

BAKER, B. L. and G. D. ABRAMS. 1954. Effect of hypophysectomy on the cytology of the fundic glands of the stomach and on the secretion of pepsin. Amer. J. Physiol., *177*, 409-412.

BOWIE, D. J. 1940. The distribution of the chief or pepsin-forming cells in the gastric mucosa of the cat. Anat. Rec., *78*, 9-17.

CAREY, J. M. and W. H. HOLLINSHEAD. 1955. An anatomic study of the esophageal hiatus. Surg. Gynec. Obstet., *100*, 196-200.

HUNT, T. E. 1957. Mitotic activity in the gastric mucosa of the rat after fasting and refeeding. Anat. Rec., *127*, 539-550.

HUNT, T. E. and E. A. HUNT. 1962. Radioautographic study of proliferation in the stomach of the rat using thymidine-H³ and compound 48/80. Anat. Rec., *142*, 505-517.

LACHMAN, E. 1957. Roentgenologic manifestations of emotional disturbances in the stomach. Amer. J. Roentgenol., *77*, 162-166.

MITCHELL, G. A. G. 1940. A macroscopic study of the nerve supply of the stomach. J. Anat. (Lond.), *75*, 50-63.

MOODY, R. O., R. G. VAN NUYS, and C. H. KIDDER. 1929. The form and position of the empty stomach in healthy young adults as shown in roentgenograms. Anat. Rec., *43*, 359-379.

WEBER, J. 1958. The basophilic substance of the gastric chief cells and its relation to the process of secretion. Acta Anat., *33*, Suppl. 31, 1-79.

SMALL AND LARGE INTESTINE

BROWN, J. O. and R. J. ECHENBERG. 1964. Mucosal reduplications associated with the ampullary portion of the major duodenal papilla in humans. Anat. Rec., *150*, 293-301.

BUIRGE, R. E. 1943. Gross variations in the ileocecal valve. A study of the factors underlying incompetency. Anat. Rec., *86*, 373-385.

FRIEDMAN, S. M. 1946. The position and mobility of the duodenum in the living subject. Amer. J. Anat., *79*, 147-165.

HABER, J. J. 1947. Meckel's diverticulum. Amer. J. Surg., *73*, 468-485.

HALEY, J. C. and J. K. PEDEN. 1943. The suspensory muscle of the duodenum. Amer. J. Surg., *59*, 546-550.

JACOBSON, L. F. and R. J. NOER. 1952. The vascular pattern of the intestinal villi in various laboratory animals and man. Anat. Rec., *114*, 85-101.

LOW, A. 1907. A note on the crura of the diaphragm and the muscle of Treitz. J. Anat. Physiol., *42*, 93-96.

NOER, R. J. 1943. The blood vessels of the jejunum and ileum: A comparative study of man and certain laboratory animals. Amer. J. Anat., *73*, 293-334.

SHAH, M. A. and M. SHAH. 1946. The arterial supply of the vermiform appendix. Anat. Rec., *95*, 457-460.

STEPHENS, D. F. 1963. *Congenital Malformations of the Rectum, Anus and Genito-Urinary Tracts.* xvi + 371 pages, 227 figures. The Williams and Wilkins Company, Baltimore.

WAKELEY, C. P. G. and R. J. GLADSTONE. 1928. The relative frequency of the various positions of the appendix vermiformis as ascertained by an analysis of 5000 cases. J. Anat. (Lond.), *63*, 157-158.

WOTTON, R. 1963. Lipid absorption. In *International Review of Cytology*. Edited by G. H. Bourne. *15*, 399-416. Academic Press, Inc. New York.

PERITONEUM AND HERNIA

BARON, M. A. 1941. Structure of the intestinal peritoneum in man. Amer. J. Anat., *69*, 439-496.

GARDNER, J. H., E. A. HOLYOKE, and R. P. GIOVACCHINI. 1957. Cleavage lines of the visceral and parietal peritoneum. Anat. Rec., *127*, 241-256.

MADDEN, J. L. 1956. Anatomic and technical considerations in the treatment of esophageal hiatal hernia. Surg. Gynec. Obstet., *102*, 187-194.

MAYO, C. W., L. K. STALKER, and J. M. MILLER. 1941. Intra-abdominal hernia. Review of 39 cases in which treatment was surgical. Ann. Surg., *114*, 875-885.

LIVER, BILE DUCTS, AND GALLBLADDER

ALEXANDER, W. F. 1940. The innervation of the biliary system. J. comp. Neurol., *72*, 357-370.

DAVID, H. 1954. Submicroscopic Ortho- and Patho-Morphology of the Liver. 2 volumes. Text Volume, 412 pages. *Atlas Volume of Electron-photomicrographs*, 361 illustrations. Pergamon Press, New York.

DI DIO, L. J. A. and E. A. BOYDEN. 1962. The choledochoduodenal junction in the horse—a study of the musculature around the ends of the bile and pancreatic ducts in a species without a gall bladder. Anat. Rec., *143*, 61-69.

ELIAS, H. and D. PETTY. 1952. Gross anatomy of the blood vessels and ducts within the human liver. Amer. J. Anat., *90*, 59-111.

GOOR, D. A. and P. A. EBERT. 1972. Anomalies of the biliary tree: Report of a repair of accessory bile duct and review of literature. Arch. Surg., *104*, 302-309.

HAMMOND, W. S. 1939. On the origin of the cells lining the liver sinusoids in the cat and the rat. Amer. J. Anat., *65*, 199-227.

HARD, W. L. and R. K. HAWKINS. 1950. The role of the bile capillaries in the secretion of phosphatase by the rabbit liver. Anat. Rec., *106*, 395-411.

HEALY, J. E., JR. and P. C. SCHROY. 1953. Anatomy of the biliary ducts within the human liver; analysis of the prevailing pattern of branchings and the major variations of the biliary ducts. Arch. Surg., *66*, 599-616.

HJORTSJÖ, C. H. 1951. The topography of the intrahepatic duct system. Acta Anat., *11*, 599-615.

JOHNSON, F. E. and E. A. BOYDEN. 1952. The effect of double vagotomy on the motor activity of the human gall bladder. Surgery, *32*, 591-601.

MICHELS, N. A. 1953. Variational anatomy of the hepatic, cystic and retroduodenal arteries. Arch. Surg., *66*, 20-32.

POPPER, H. P. and F. SCHAFFNER. 1957. *Liver: Structure and Function.* xv + 777 pages. The Blakiston Division, McGraw-Hill Book Company, New York.

RAPPAPORT, A. M. 1958. The structural and functional unit in the human liver (liver acinus). Anat. Rec., *130*, 673-689.

SEVERN, C. B. 1972. A morphological study of the development of the human liver. II. Establishment of liver parenchyma, extrahepatic ducts, and associated venous channels. Amer. J. Anat., *133*, 85-107.

STERLING, J. A. 1954. The common channel for bile and pancreatic ducts. Surg. Gynec. Obstet., *98*, 420-424.

SUTHERLAND, S. D. 1966. The intrinsic innervation of the gall bladder in Macaca rhesus and Cavia porcellus. J. Anat. (Lond.), *100*, 261-268.

VANDAMME, J. P., J. BONTE, and G. VAN DER SCHUEREN. 1969. A revaluation of hepatic and cystic arteries. The importance of the aberrant hepatic branches. Acta Anat., *73*, 192-209.

WILLIAMS, W. L. 1948. Vital staining of damaged liver cells. I. Reactions to acid azo dyes following acute chemical injury. Anat. Rec., *101*, 133-147.

PANCREAS

BENCOSME, S. A. 1955. The histogenesis and cytology of the pancreatic islets in the rabbit. Amer. J. Anat., *96*, 103-151.

CONKLIN, J. L. 1962. Cytogenesis of the human fetal pancreas. Amer. J. Anat., *111*, 181-193.

DAWSON, W. and J. LANGMAN. 1961. An anatomical radiological study of the pancreatic duct pattern in man. Anat. Rec., *139*, 59-68.

FERNER, H. 1952. *Das Inselsystem des Pankreas.* xi + 186 pages. Georg Thieme Verlag, Stuttgart.

GOLDSTEIN, M. B. and E. A. DAVIS, JR. 1968. The three dimensional architecture of the islets of Langerhans. Acta Anat., *71*, 161-171.

HARD, W. L. 1944. The origin and differentiation of the alpha and beta cells in the pancreatic islets of the rat. Amer. J. Anat., *75*, 369-403.

KUNTZ, A. and C. A. RICHINS. 1949. Effects of direct and reflex nerve stimulation of the exocrine secretory activity of pancreas. J. Neurophysiol., *12*, 29-35.

LACY, P. E. 1957. Electron microscopic identification of different cell types in the islets of Langerhans of the guinea pig, rat, rabbit and dog. Anat. Rec., *128*, 255-267.

LATTA, J. S. and H. T. HARVEY. 1942. Changes in the islets of Langerhans of the albino rat induced by insulin administration. Anat. Rec., *82*, 281-295.

MUNGER, B. L. 1958. A phase and electron microscopic study of cellular differentiation in pancreatic acinar cells of the mouse. Amer. J. Anat., *103*, 1-33.

RICHINS, C. A. 1945. The innervation of the pancreas. J. Comp. Neurol., *83*, 223-236.

SINGH, I. 1963. The terminal part of the accessory pancreatic duct and its musculature in the Rhesus monkey. J. Anat. (Lond.), *97*, 107-110 + 1 plate.

WINBORN, W. B. 1963. Light and electron microscopy of the islets of Langerhans of the Saimiri monkey pancreas. Anat. Rec., *147*, 65-93.

The Urogenital System

rogenital or Genitourinary is the name given to the system or apparatus consisting of (*a*) the **urinary organs** for the formation and discharge of the urine, and (*b*) the **genital organs,** which are concerned with the process of reproduction.

Development of the Urinary Organs

The urogenital glands and ducts are developed from the intermediate cell mass which is situated between the primitive segments and the lateral plates of mesoderm. The permanent organs of the adult are preceded by structures which, with the exception of the ducts, disappear almost entirely before the end of fetal life. These paired structures are: the **pronephros,** the **mesonephros,** and the **mesonephric** and **paramesonephric ducts.** The pronephros disappears very early. The structural elements of the mesonephros almost entirely degenerate, but in their place the genital gland develops. The mesonephric duct remains as the duct of the male genital gland, the paramesonephric as that of the female. The final kidney is a new organ, the **metanephros.**

The **Pronephros and Mesonephric** (*Wolffian*) **Duct** (Fig. 17–1).—In the lateral part of the intermediate cell mass, immediately under the ectoderm, in the region from the fifth cervical to the third thoracic segments, a series of short evaginations from each segment grows dorsolaterally and caudally, fusing successively caudalward to form the **pronephric duct** (Fig. 17–1). This continues to grow caudally until it opens into the ventral part of the cloaca; beyond the pronephros it is termed the **mesonephric** or **Wolffian duct.**

The original evaginations form a series of transverse tubules, each of which communicates by means of a funnel-shaped ciliated opening, the **nephrostome,** with the coelomic cavity, and in the course of each duct a **glomerulus** also is developed. A secondary glomerulus is formed ventral to each of these, and the complete group constitutes the **pronephros.** The pronephros undergoes rapid atrophy and disappears in 4-mm embryos except for the pronephric ducts which persist as the excretory ducts of the succeeding kidneys, the mesonephroi.

The **Mesonephros** (Fig. 17–1).—On the medial side of the mesonephric duct, from the sixth cervical to the third lumbar segments, a series of tubules, the **mesonephric tubules,** is developed; at a later stage in development they increase in number by outgrowths from the original tubules. These tubules first appear as solid masses of cells which later develop a lumen; one end grows toward and finally opens into the mesonephric duct, the other dilates and is invaginated by a tuft of capillary blood vessels to form a glomerulus. The tubules collectively constitute the **mesonephros** or **Wolffian body.** By the fifth or sixth week this body produces an elongated spindle-shaped eminence, termed the **urogenital fold,** which projects into the coelomic cavity at the side of the dorsal mesentery, reaching from the septum transversum cranially to the fifth lumbar segment caudally. The reproductive glands develop in the urogenital folds. The mesonephric bodies are the permanent kidneys in fishes and amphibians, but in reptiles, birds, and mammals, they atrophy and for the most part disappear synchronously with the development of the permanent kidneys. The atrophy begins during the sixth or seventh week and rapidly proceeds, so that by the beginning of the fourth month only the ducts and a few of the tubules remain.

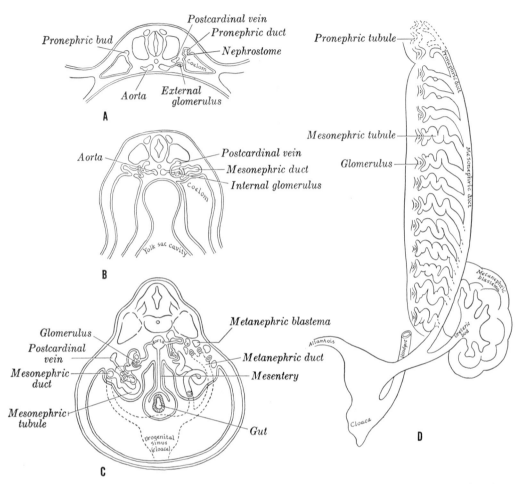

FIG. 17–1.—Pronephros, mesonephros and metanephros. *A,* Diagrammatic cross section of embryo through somites 7 to 14. Earlier stages of development on left. *B,* Typical cross section through somites containing mesonephros. Earlier stage of development on left. *C,* Typical cross section through caudal portion of the mesonephros and including part of the metanephros. Excretory ducts and cloaca outlined in dash line. *D,* Ventrolateral view of embryonic excretory system (schematic). (Redrawn from *Essentials of Embryology* by F. D. Allan, Oxford University Press, 1960.)

The Metanephros and the Permanent Kidney.—The rudiments of the permanent kidneys make their appearance about the end of the first or the beginning of the second month. Each kidney has a two-fold origin, part arising from the metanephros, and part as a diverticulum from the caudal end of the mesonephric duct, close to where the latter opens into the cloaca (Figs. 17–9, 17–10).

The metanephros arises in the intermediate cell mass, caudal to the mesonephros, which it resembles in structure. The diverticulum from the mesonephric duct grows dorsalward and cranialward along the dorsal abdominal wall, where its blind extremity expands and subsequently divides into several buds which form the rudiments of the pelvis and calyces of the kidney; by continued growth and subdivision it gives rise to the collecting tubules of the kidney. The caudal portion of the diverticulum becomes the ureter. The secretory tubules are developed from the metanephros, which is molded over the growing end of the diverticulum from the mesonephric duct.

The tubules of the metanephros, unlike those of the pronephros and mesonephros, do not open into the mesonephric duct. One end expands to form a glomerulus, while

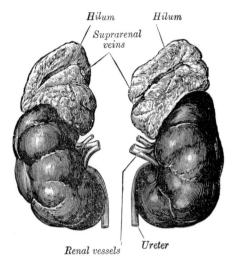

FIG. 17–2.—The kidneys and suprarenal glands of a newborn child. Anterior aspect.

after birth, and traces of it may be found even in the adult. The kidney of the ox and many other animals remains lobulated throughout life.

The **Urinary Bladder.**—The bladder is formed partly from the endodermal cloaca and partly from the ends of the mesonephric ducts; the allantois takes no share in its formation. After the rectum has separated from the dorsal part of the cloaca, the ventral part becomes subdivided into three portions: (1) an anterior **vesicourethral portion,** continuous with the allantois—into this portion the mesonephric ducts open; (2) an intermediate narrow channel, the **pelvic portion;** and (3) a posterior **phallic portion,** closed externally by the urogenital membrane. The second and third parts together constitute the **urogenital sinus** (Figs. 17–

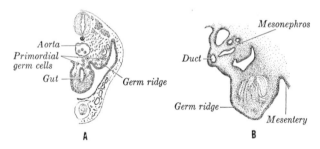

FIG. 17–3.—*A,* Diagrammatic cross section to show sites and migration of primordial germ cells from mesentery to gonad. *B,* Undifferentiated gonad. (Redrawn from *Essentials of Embryology* by F. D. Allan, Oxford University Press, 1960.)

the rest of the tubule rapidly elongates to form the convoluted and straight tubules, the loops of Henle, and the connecting tubules; these last establish communication with the collecting tubules derived from the ultimate ramifications of the ureteric diverticulum. The mesoderm around the tubules becomes condensed to form the connective tissue of the kidney. The ureter opens at first into the caudal end of the mesonephric duct; after the sixth week it is separated from the mesonephric duct, and opens independently into the part of the cloaca which ultimately becomes the bladder (Figs. 17–9, 17–10).

The secretory tubules of the kidney become arranged into pyramidal masses or lobules (Fig. 17–2). The lobulated condition of the kidneys exists for some time

11, 17–12). The vesicourethral portion absorbs the ends of the mesonephric ducts and the associated ends of the renal diverticula, and these give rise to the trigone of the bladder and part of the prostatic urethra. The remainder of the vesicourethral portion forms the body of the bladder and part of the prostatic urethra; its apex is prolonged to the umbilicus as a narrow canal (the **urachus**), which later is obliterated and becomes the middle umbilical ligament.

Development of the Generative Organs

Genital Glands.—The first appearance of the genital gland is essentially the same in the two sexes, and consists in a thickening of the epithelial layer which lines the peritoneal cavity on the medial side of the urogenital fold. The thick plate of epithelium pushes the mesoderm before it and forms a distinct projection, the **genital ridge** (Fig. 17–3). From it the testis in the male and the ovary in the female are developed. At first the mesonephros and genital ridge are suspended by a common mesentery, but as the embryo grows the genital ridge gradually becomes pinched off from the mesonephros, with which it is at first continuous, though it still remains connected to the remnant of this body by a fold of peritoneum, the **mesorchium** or **mesovarium**. About the seventh week the distinction of sex in the genital ridge begins to be perceptible.

latory duct, while the seminal vesicle arises during the third month as a lateral diverticulum from its caudal end. A large part of the cranial portion of the mesonephros atrophies and disappears, but a few tubules may persist as the appendix of the epididymis, vestigial structures which end blindly. From the remainder of the cranial tubules the efferent ducts of the testis form.

The caudal tubules are represented by the ductuli aberrantes, and by the paradidymis.

Descent of the Testes.—The testes, at an early period of fetal life, are placed at the dorsal part of the abdominal cavity, covered by the peritoneum, and each is attached by a peritoneal fold, the **mesorchium**, to the mesonephros.

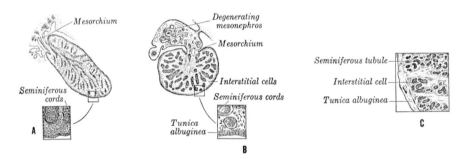

Fig. 17–4.—Differentiation of testis; *A, B,* and *C* represent progressively later stages. (Redrawn from *Essentials of Embryology* by F. D. Allan, Oxford University Press, 1960.)

The **Testis** (Fig. 17–4).—At first the testis is a collection of cells derived from the coelomic epithelium, with a surface covering and a central mass. As sexual differentiation begins, a series of cords appears in the central mass, surrounded by a concentration of cells which is converted later into the tunica albuginea. The tunica separates the cords from the surface epithelium, excluding it from any part in forming the parenchyma of the testis. The more peripheral parts of the cords develop into the seminiferous tubules; the central parts run together toward the future mediastinum testis, forming a network which becomes the rete testis.

The **Male Mesonephric Duct** (*ductus mesonephricus; Wolffian duct*) (Fig. 17–5). —In the male, the mesonephric duct (see page 1265) persists, and forms the epididymis, the ductus deferens and the ejacu-

From the ventral part of the mesonephros a fold of peritoneum termed the **inguinal fold** grows ventralward to meet and fuse with a peritoneal fold, the **inguinal crest,** which grows dorsalward from the anterolateral abdominal wall. The testis thus acquires an indirect connection with the anterior abdominal wall, and at the same time a portion of the peritoneal cavity lateral to these fused folds is marked off as the future saccus vaginalis.

In the inguinal crest a peculiar structure, the **gubernaculum testis,** makes its appearance. This is at first a slender band, extending from that part of the skin of the groin which afterward forms the scrotum through the inguinal canal to the body and epididymis of the testis. As development advances, the peritoneum covering the gubernaculum forms two folds, one above the testis and the other below it. The one cranial to the

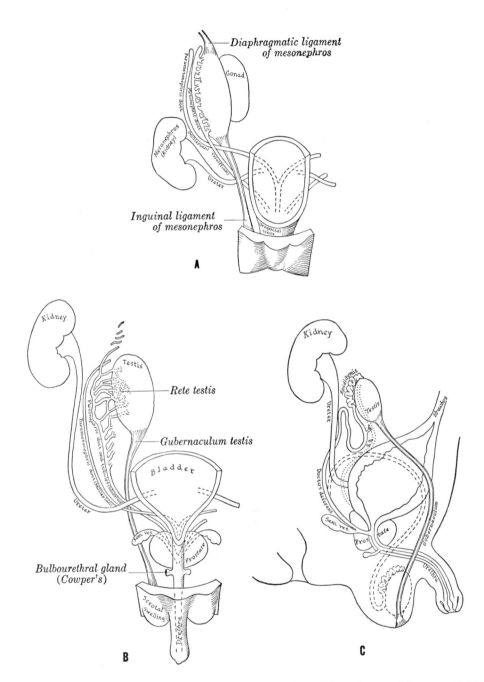

Fig. 17–5.—Diagrammatic representation of the differentiation of the male genital ducts. *A,* Undifferentiated stage. *B,* Early differentiation. *C,* The condition just prior to birth when testis is undescended. Dash lines indicate position of testis after descent and dotted lines the position of the obliterated paramesonephric duct. Vestigial structures retained from the latter are labeled. (Redrawn from *Essentials of Embryology* by F. D. Allan, Oxford University Press, 1960.)

testis is the **plica vascularis**, and contains ultimately the testicular vessels; the caudal one, the **plica gubernatrix**, contains the lower part of the gubernaculum, which has now grown into a thick cord. It ends caudally at the abdominal inguinal ring beside a tube of peritoneum, the saccus or processus vaginalis, which protrudes itself down the inguinal canal. By the fifth month the cranial part has disappeared, but the caudal part now consists of a central core of unstriped muscle fiber, and an outer layer of striped elements, connected with the abdominal wall. The main portion of the gubernaculum is attached to the skin at the

cranialward displacement of the testis. In addition, the gubernaculum may actually shorten, in which case it might exert traction on the testis and tend to displace it toward the scrotum. By the end of the eighth month the testis has reached the scrotum, preceded by the processus vaginalis, which communicates by its upper extremity with the peritoneal cavity. Just before birth the upper part of the saccus vaginalis normally becomes closed, and this obliteration extends gradually downward to within a short distance of the testis. The peritoneum surrounding the testis is then entirely cut off from the general peritoneal

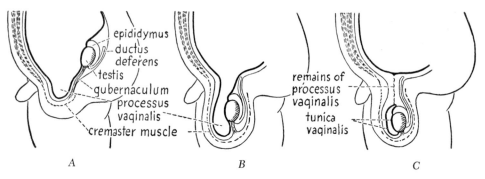

FIG. 17–6.—Diagram illustrating descent of the testis: *A*, Before descent. The processus vaginalis is present before descent begins, the testis lying behind the peritoneum. *B*, Descent nearly complete but processus vaginalis not obliterated. *C*, Processus vaginalis obliterated except for the terminal portion, which persists as the tunica vaginalis of the adult.

point where the scrotum develops and as the pouch forms, most of the caudal end of the gubernaculum is carried with it; other bands extend to the medial side of the thigh and to the perineum.

The tube of peritoneum constituting the **saccus** or **processus vaginalis** projects itself downward into the inguinal canal, and emerges at the superficial inguinal ring, pushing before it a part of the Obliquus internus and the fascia of the Obliquus externus, which form respectively the Cremaster muscle and the external spermatic fascia. It forms a gradually elongating sack, which eventually reaches the bottom of the scrotum, and behind this sack the testis descends. Since the growth of the gubernaculum is not commensurate with the growth of the body of the fetus, the latter increasing more rapidly in relative length, it has been assumed that this prevents

cavity and becomes the **tunica vaginalis** (Fig. 17–6).

Occasionally the testis fails to descend or the descent is incomplete, a condition known as *cryptorchidism*. In such cases administration of an extract of the pituitary gland may cause the testis to occupy its normal position. This would seem to indicate that the descent may to some extent be under hormonal control.

The **Prostate.**—The prostate arises between the third and fourth months as a series of solid diverticula from the epithelium lining the urogenital sinus and vesicourethral part of the cloaca. These buds arise in five distinct groups, grow rapidly in length, and soon acquire a lumen. Eventually the prostatic urethra and ejaculatory ducts are embedded in a five-lobed gland the parts of which are called the median, anterior, posterior and lateral lobes.

The lateral lobes are the largest. There are no distinct dividing lines between the parts of the formed gland, and the divisions are important only because of their individual peculiarities in disease processes. **Skene's ducts** in the female urethra are regarded as the homologues of the prostatic glands.

The **bulbourethral glands of Cowper** in the male and **greater vestibular glands of Bartholin** in the female also arise as diverticula from the epithelial lining of the urogenital sinus.

The **Ovary** (Fig. 17-7).—The ovary, formed from the genital ridge, is at first a collection of cells derived from the coelomic epithelium; later the mass is differentiated into a central part of medulla covered by a surface layer, the **germinal epithelium.** Between the cells of the germinal epithelium a number of larger cells, the **primitive ova,**

the primitive germinal epithelium to be carried into the gonad with the sex cords and later develop either into ova in the female or sperm cells in the male. Some authors deny their existence while others claim they all degenerate and take no part in formation of the adult sex cells.

The **Female Mesonephros** and **Mesonephric Duct** (see page 1265) atrophy. The remains of the mesonephric tubules may be divided into three groups. One group of tubules from the cranial portion of the mesonephros persists and produces one or more **vesicular appendices** in the fringes of the uterine tube. The middle and largest group, together with a segment of the mesonephric duct, persist as the **epoöphoron** (*organ of Rosenmüller*). Persistent portions of the mesonephric duct are known as **Gartner's ducts** (Fig. 17-55).

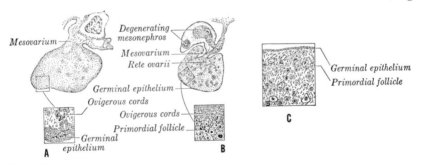

FIG. 17-7.—Differentiation of ovary; *A, B,* and *C* represent progressively later stages. (Redrawn from *Essentials of Embryology* by F. D. Allan, Oxford University Press, 1960.)

are found. These are carried into the subjacent stroma by bud-like ingrowths (**genital cords**) of the germinal epithelium. The surface epithelium ultimately forms the permanent epithelial covering of this organ; it soon loses its connection with the central mass, and a tunica albuginea develops between them. The ova are chiefly derived from the cells of the central mass; these are separated from one another by the growth of connective tissue in an irregular manner; each ovum acquires a covering of connective tissue (follicle) cells, and in this way the rudiments of the ovarian follicles are formed.

Primordial germ cells (Fig. 17-3) are, according to some authors, set aside at a very early age from the somatic cells. They are first recognized in the yolk sac. Later they migrate through the mesentery into

These may exist as part of the epoöphoron or as isolated segments as far as the hymen. The third and most caudal group of remaining mesonephric ducts constitutes the **paroöphoron,** which usually disappears completely before the adult stage. Any one of these vestigial tubules which persists in the adult as a stalked vesicle is called an **hydatid.**

The **Paramesonephric Ducts** (*ductus paramesonephrici; Müllerian ducts*) (Fig. 17-8).—Shortly after the formation of the mesonephric ducts this second pair of ducts is developed. Each arises on the lateral aspect of the corresponding mesonephric duct as a tubular invagination of the cells lining the coelom (Fig. 17-8). The orifice of the invagination remains patent, and undergoes enlargement and modification to form the abdominal ostium of the uterine tube. The ducts pass caudalward lateral to the mesonephric ducts, but toward the caudal end of the embryo they cross to the medial side of these ducts, and thus come to lie side by side between and caudal to the latter—the four ducts forming what is termed the **genital cord.** The parameso- nephric ducts end in an epithelial elevation, the **paramesonephric eminence,** on the ventral part of the cloaca between the orifices of the mesonephric ducts; at a later date they open into the cloaca in this situation.

In the male the paramesonephric (*Müllerian*) ducts atrophy, but traces of their

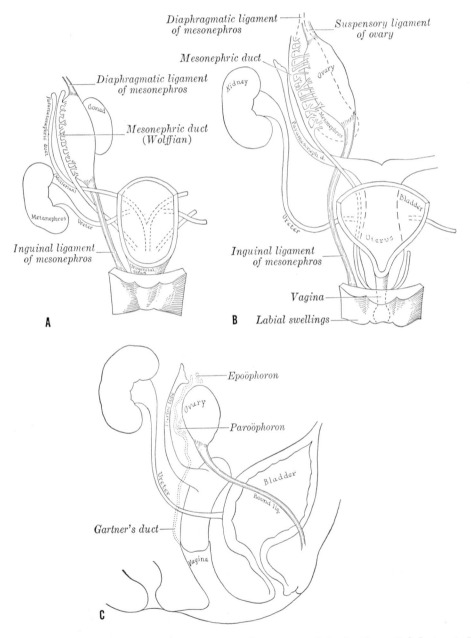

Fig. 17–8.—Diagrammatic representation of the differentiation of the female genital ducts. *A,* Undifferentiated stage. *B,* Early differentiation. *C,* The condition existing at birth. In the latter, the dotted lines indicate the position of the mesonephric duct and tubules. Vestigial structures derived from these are labeled. (Redrawn from *Essentials of Embryology* by F. D. Allan, Oxford University Press, 1960.)

cranial ends are represented by the **appendices testis** (*hydatids of Morgagni*), while their caudal portions fuse to form the utriculus in the floor of the prostatic portion of the urethra (Fig. 17–45).

In the female the paramesonephric (*Müllerian*) ducts persist and undergo further development. The portions which lie in the genital ridge fuse to form the uterus and vagina; the parts cranial to this ridge remain separate, and each forms the corresponding uterine tube—the abdominal ostium of which is developed from the cranial extremity of the original tubular invagination from the coelom. The fusion of the paramesonephric ducts begins in the third month, and the septum formed by their fused medial walls disappears. Entodermal epithelium of the urogenital sinus invades the region where the vagina forms, replacing the paramesonephric epithelium almost entirely, and for a time the vagina is represented by a solid rod of epithelial cells, but in fetuses of five months the lumen reappears. About the fifth month an annular constriction marks the position of the neck of the uterus, and after the sixth month the walls of the uterus begin to thicken. A ring-like outgrowth of epithelium occurs at the caudal end of the uterus and marks the future vaginal fornices. The hymen arises at the site of the paramesonephric eminence. It represents the separation between vagina and urogenital sinus.

Descent of the Ovaries (Fig. 17–8).—In the female there is also a gubernaculum, which effects a considerable change in the position of the ovary, though not so extensive a change as in that of the testis. The gubernaculum in the female lies in contact

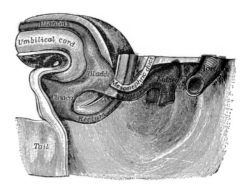

Fig. 17–10.—Tail end of human embryo thirty-two to thirty-three days old. (From model by Keibel.)

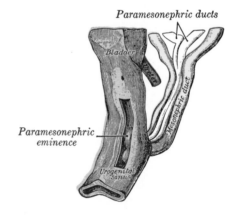

Fig. 17–11.—Urogenital sinus of female human embryo eight and a half to nine weeks old. (From model by Keibel.)

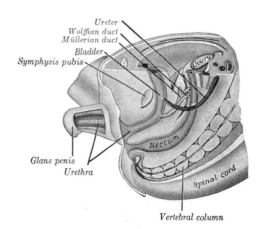

Fig. 17–12.—Tail end of human embryo, from eight and a half to nine weeks old. (From model by Keibel.)

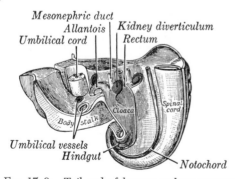

Fig. 17–9.—Tail end of human embryo twenty-five to twenty-nine days old. (From model by Keibel.)

with the fundus of the uterus and acquires adhesions to this organ; thus the ovary is prevented from descending below this level. The part of the gubernaculum between the ovary and the uterus becomes ultimately the proper ligament of the ovary, while the part between the uterus and the labium majus forms the round ligament of the uterus. A pouch of peritoneum analogous to the saccus vaginalis in the male accompanies it along the inguinal canal; it is called the **canal of Nuck.**

In rare cases the gubernaculum may fail to develop adhesions to the uterus, and then the ovary descends through the inguinal canal into the labium majus, and under these circumstances its position resembles that of the testis.

The **External Organs of Generation** (Fig. 17–13).—As already stated (page 1163), the cloacal membrane, composed of ectoderm and endoderm, originally reaches from the umbilicus to the tail. The growing mesoderm extends to the midventral line for some distance caudal to the umbilicus, and forms the lower part of the abdominal wall; it ends caudally in a prominent swelling, the **cloacal tubercle.** Dorsal to this tubercle

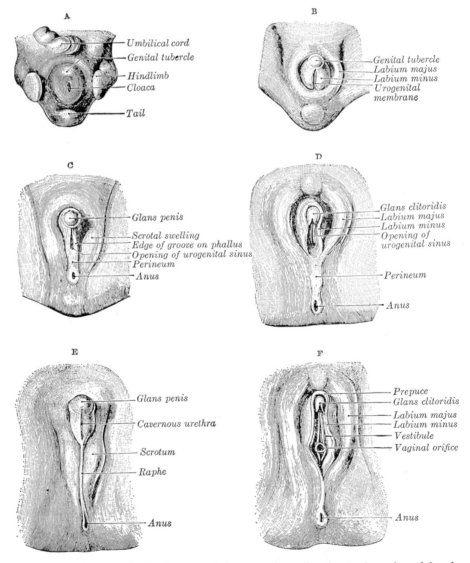

Fig. 17–13.—Stages in the development of the external sexual organs in the male and female.
(Drawn from the Ecker-Ziegler models.)

the urogenital part of the cloacal membrane separates the ingrowing sheets of mesoderm.

The first rudiment of the penis (or clitoris) is a structure termed the **phallus;** it is derived from the phallic portion of the cloaca which has extended on the end and sides of the under surface of the cloacal tubercle. The terminal part of the phallus representing the future glans becomes solid; the remainder, which is hollow, is converted into a longitudinal groove by the absorption of the urogenital membrane.

In the female a deep groove forms around the phallus and separates it from the rest of the cloacal tubercle, which is now termed the **genital tubercle.** The tissue at the sides of the genital tubercle grows caudalward as the **genital swellings,** which ultimately form the labia majora; the tubercle itself becomes the mons pubis. The labia minora arise by the continued growth of the lips of the groove on the under surface of the phallus; the remainder of the phallus forms the clitoris.

In the male the early changes are similar, but the pelvic portion of the cloaca undergoes much greater development, pushing before it the phallic portion. The genital swellings extend around between the pelvic portion and the anus, and form a scrotal area; during the changes associated with the descent of the testes this area is drawn out to form the scrotal sacs. The penis is developed from the phallus. As in the female, the urogenital membrane undergoes absorption, forming a channel on the under surface of the phallus; this channel extends only as far forward as the corona glandis.

The **corpora cavernosa** of the penis (or clitoris) and of the urethra arise from the mesodermal tissue in the phallus; they are at first dense structures, but later vascular spaces appear in them, and they gradually become cavernous.

The **prepuce** in both sexes is formed by the growth of a solid plate of ectoderm into the superficial part of the phallus; on coronal section this plate presents the shape of a horseshoe. By the breaking down of its more centrally situated cells the plate is split into two lamellae, and a cutaneous fold, the prepuce, is liberated and forms a hood over the glans. Adherent prepuce is not a secondary adhesion, but a hindered central desquamation.

The **Urethra.**—As already described, in both sexes the phallic portion of the cloaca extends on to the under surface of the cloacal tubercle as far forward as the apex. At the apex the walls of the phallic portion come together and fuse, the lumen is obliterated, and a solid plate, the **urethral plate,** is formed. The remainder of the phallic portion is for a time tubular, and then, by the absorption of the urogenital membrane, it establishes a communication with the exterior; this opening is the **primitive urogenital ostium,** and it extends forward to the corona glandis.

In the female this condition is largely retained; the portion of the groove on the clitoris broadens out while the body of the clitoris enlarges, and thus the adult urethral opening is situated behind the base of the clitoris.

In the male, by the greater growth of the pelvic portion of the cloaca a longer urethra is formed, and the primitive ostium is carried distalward with the phallus, but it still ends at the corona glandis. Later it closes from proximally to distally. Meanwhile the urethral plate of the glans breaks down centrally to form a median groove continuous with the primitive ostium. This groove also closes proximally, so that the external urethral opening is shifted distalward to the end of the glans (Fig. 17–12).

THE URINARY ORGANS

The urinary organs comprise the **kidneys,** which produce the urine, the **ureters,** or ducts, which convey urine to the **urinary bladder,** where it is for a time retained; and the **urethra,** through which it is discharged from the body.

The Kidneys

The **kidneys** (*renes*) are situated in the dorsal part of the abdomen, one on either side of the vertebral column, covered by the peritoneum, and surrounded by a mass

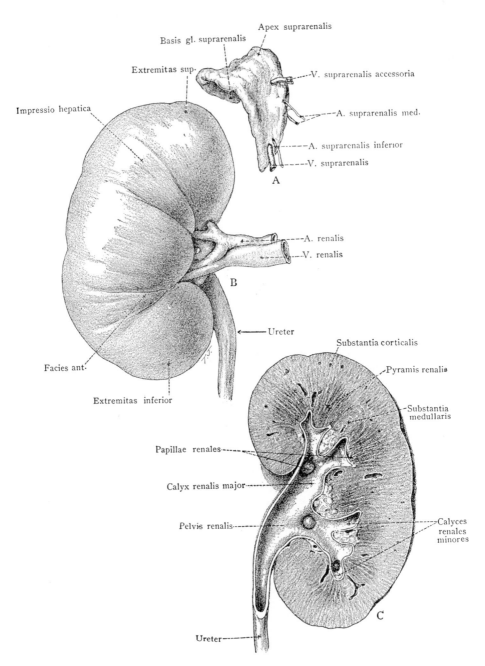

FIG. 17–14.—The right kidney and suprarenal gland. *A*, Suprarenal gland. *B*, Kidney, surface view. *C*, Kidney, longitudinal section showing pelvis. (Eycleshymer and Jones.)

of fat and loose areolar tissue. Their cranial extremities are on a level with the cranial border of the twelfth thoracic vertebra, their caudal extremities on a level with the third lumbar. The right kidney is usually slightly more caudal than the left, probably due to the presence of the liver. The long axis of each kidney is parallel with the vertebral column. Each kidney is about 11.25 cm in length, 5 to 7.5 cm in breadth, and rather more than 2.5 cm in thickness. The left is somewhat longer, and narrower, than the right. The weight of the kidney in the adult male varies from 125 to 170 gm, in the adult female from 115 to 155 gm. The combined weight of the two kidneys in proportion to that of the body is about 1 to 240. The kidneys in the newborn are about three times as large in proportion to the body weight as in the adult.

The kidney has a characteristic form similar to a bean, and presents for examination two surfaces, two borders, and a cranial and caudal extremity (Fig. 17–14).

Surfaces.—The **ventral surface** (*facies anterior*) (Fig. 17–15) of each kidney is convex, and looks ventralward and slightly lateralward. Its relations to adjacent viscera differ so completely on the two sides that separate descriptions are necessary.

Ventral surface of right kidney (Fig. 17–17).—A narrow portion at the cranial extremity is in relation with the right suprarenal gland. A large area just caudal to this and involving about three-fourths of the surface lies in the renal impression on the visceral surface of the liver, and a narrow but somewhat variable area near the medial border is in contact with the descending part of the duodenum. The caudal part of the ventral

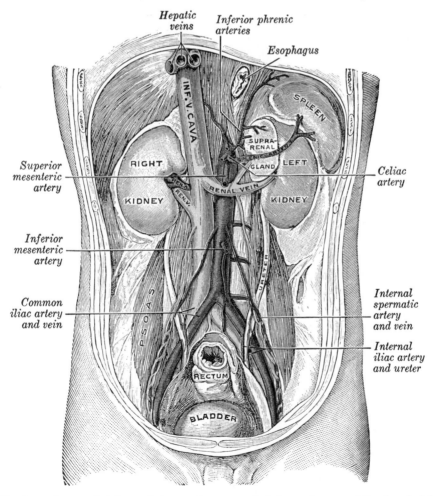

Fig. 17–15.—Ventral view of abdominal viscera after removal of the peritoneum of the dorsal abdominal wall, showing kidneys, suprarenal glands, and great vessels. (Corning.)

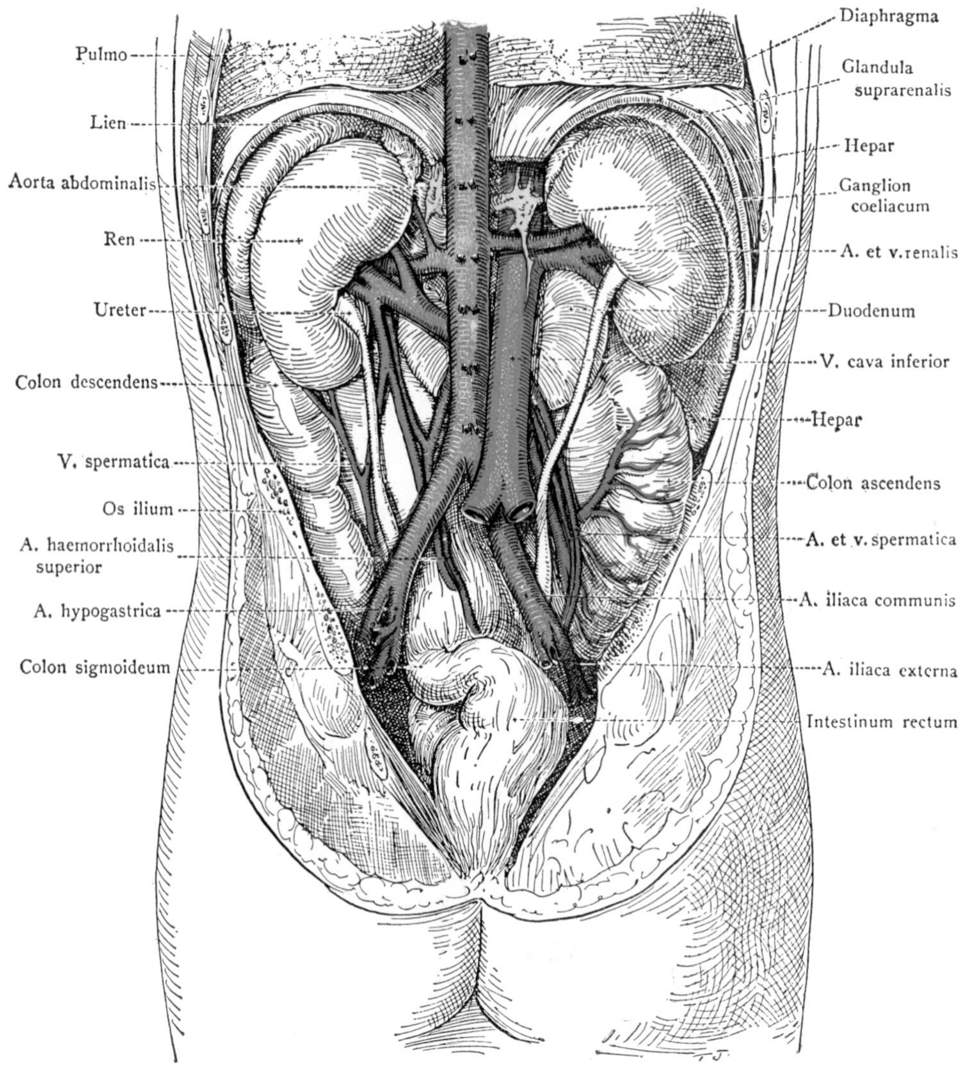

FIG. 17–16.—Dissection of abdominal viscera, dorsal view showing relation of the kidneys.
(After Corning's *Topographischen Anatomie* in Eycleshymer and Jones.)

surface is in contact laterally with the right colic flexure, and medially, as a rule, with the small intestine. The areas in relation with the liver and small intestine are covered by peritoneum; the suprarenal, duodenal, and colic areas are devoid of peritoneum.

Ventral surface of left kidney (*facies anterior*) (Fig. 17–17).—A small area along the cranial part of the medial border is in relation with the left suprarenal gland, and close to the lateral border is a long strip in contact with the renal impression on the spleen. A somewhat quadrilateral field, about the middle of the ventral surface, marks the site of contact with the body of the pancreas, on the deep surface of which are the lienal vessels. Above this is a small triangular portion, between the suprarenal and splenic areas, in contact with the posterior surface of the stomach. Caudal to the pancreatic area the lateral part is in relation with the left colic flexure, the medial with the small intestine. The areas in contact with the stomach and spleen are covered by the peritoneum of the omental bursa, while that in relation to the small intestine is covered by the peritoneum of the greater sac; dorsal to the latter are some branches of the left colic vessels. The suprarenal, pancreatic, and colic areas are devoid of peritoneum.

The **dorsal surface** (*facies posterior*) (Figs. 17–16, 17–18) of each kidney is directed dorsalward and medialward. It is embedded in areolar and fatty tissue and entirely devoid of peritoneal covering. It lies upon the diaphragm, the medial and lateral lumbocostal arches, the Psoas major, the Quadratus lumborum, and the tendon of the Transversus abdominis, the subcostal, and one or two of the upper lumbar arteries, and the last thoracic, iliohypogastric, and ilioinguinal nerves. The cranial extremity

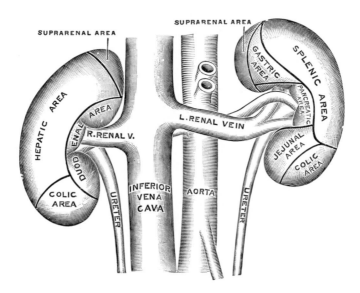

Fɪɢ. 17–17.—The ventral surfaces of the kidneys, showing the areas of contact of neighboring viscera.

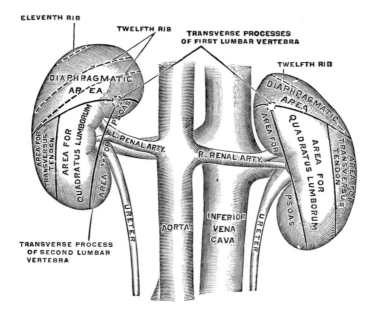

Fɪɢ. 17–18.—The dorsal surfaces of the kidneys, showing areas of relation to the parietes.

of the right kidney rests upon the twelfth rib, the left usually on the eleventh and twelfth. The diaphragm separates the kidney from the pleura, which dips down to form the phrenicocostal sinus, but frequently the muscular fibers of the diaphragm are defective or absent over a triangular area immediately above the lateral lumbocostal arch, and when this is the case the perinephric areolar tissue is in contact with the diaphragmatic pleura.

Borders.—The **lateral border** (*margo lateralis; external border*) is convex, and is directed toward the posterolateral wall of the abdomen. On the left side it is in contact, at its cranial part, with the spleen.

The **medial border** (*margo medialis; internal border*) is concave in the center and convex toward either extremity; it is directed ventralward and a little caudalward. Its central part presents a deep longitudinal fissure, bounded by prominent overhanging ventral and dorsal lips. This fissure is named the **hilum,** and transmits the vessels, nerves, and ureter. Above the hilum the medial border is in relation with the suprarenal gland; below the hilum, with the ureter.

The relative position of the main structures in the hilum is as follows: the vein is ventral, the artery in the middle, and the ureter dorsal and directed caudalward. Frequently, however, branches of both artery and vein are placed dorsal to the ureter.

Extremities.—The **cranial extremity** (*extremitas superior*) is thick and rounded, and is nearer the median line than the caudal; it is surmounted by the suprarenal gland, which covers also a small portion of the ventral surface.

The **caudal extremity** (*extremitas inferior*) is smaller and thinner than the superior and farther from the median line. It extends to within 5 cm of the iliac crest.

Renal Fascia (*fascia renalis*) (Figs. 17–19 to 17–22).—The kidney and its vessels are embedded in a mass of fatty tissue, termed the adipose capsule or *perirenal fat,* which is thickest at the margins of the kidney and is prolonged through the hilum into the renal sinus. The kidney and the adipose capsule together are enclosed in a specialized lamination of the subserous

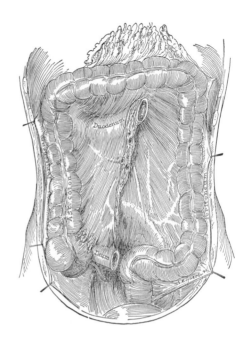

Fig. 17–19.—Visceral and parietal peritoneum associated with large intestine. Small intestine and mesentery removed. (Tobin, 1944.)

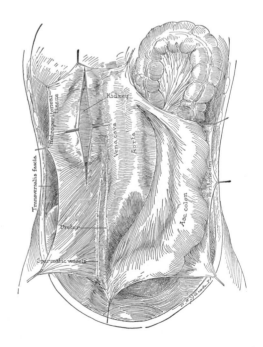

Fig. 17–20.—Peritoneum associated with ascending colon dissected free and displaced to expose the deeper stratum of subserous fascia associated with kidney and great vessels. (Tobin, 1944.)

fascia called the renal fascia. It occupies a position between the internal investing layer of deep fascia (transversalis, endoabdominal fascia) and the stratum of subserous fascia associated with the intestine and its blood vessels (Fig. 17–20). In forming the renal fascia, the subserous fascia of the lateral abdominal wall splits into two fibrous lamellae near the lateral border of the kidney (Fig. 17–22). Both lamellae extend medially, the ventral one over the ventral surface of the kidney, the dorsal one over the dorsal surface. The anterior lamella continues over the renal vessels and aorta to join the similar membrane of the other side. The posterior lamella also continues across the middle line but lies deep to the aorta, and is there more adherent to the underlying deep fascia than in the region of the kidney. The renal fascia is connected to the fibrous tunic of the kidney by numerous trabeculae which traverse the adipose capsule, and are strongest near the lower end of the organ. Dorsal to the fascia renalis is a considerable quantity of fat, which constitutes the *paranephric body (pararenal fat)*.

Fixation of the Kidney.—The kidneys are not rigidly fixed to the abdominal wall and, since they are in contact with the diaphragm, move with it during respiration. They are held in position by the renal fascia described above and by the large renal arteries and veins. That the adipose capsule and the paranephric fat body play an important part in holding the kidney in posi-

tion is indicated by the occurrence of a condition called movable kidney in emaciated individuals (see page 1287).

General Structure of the Kidney.—The kidney is invested by a fibrous tunic or capsule which forms a firm, smooth covering to the organ. The tunic can be easily stripped off, but in doing so numerous fine

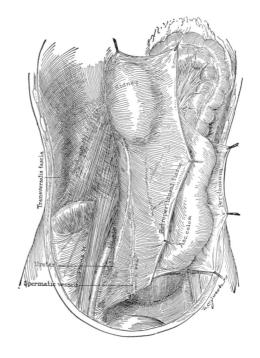

FIG. 17–21.—Deeper stratum of subserous fascia dissected free and displaced to expose the transversalis fascia. (Tobin, 1944.)

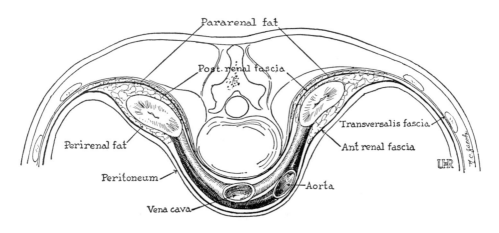

FIG. 17–22.—Transverse section, showing relations of renal fascia. (Tobin, 1944).

processes of connective tissue and small blood vessels are torn through. When the capsule is stripped off, the surface of the kidney is found to be smooth and of a deep red color. In infants, fissures extending for some depth may be seen on the surface of the organ, a remnant of the lobular construction of the gland. If a vertical section of the kidney be made from its convex to its concave border, it will be seen that the hilum expands into a central cavity, the **renal sinus;** this contains the cranial part of the renal pelvis and the calyces, surrounded by some fat in which are embedded the branches of the renal vessels and nerves. The renal sinus is lined by a prolongation of the fibrous tunic, which is continuous with the covering of the pelvis of the kidney around the lips of the hilum. The **minor renal calyces,** from four to thirteen in number, are cup-shaped tubes, each of which embraces usually one but occasionally two or more of the renal papillae; they unite to form two or three short tubes, the major calyces, and these in turn join to

form a funnel-shaped sac, the **renal pelvis.** Spirally arranged muscles surround the calyces which may have a milking action on these tubes, thereby aiding the flow of urine into the renal pelvis. As the pelvis leaves the renal sinus it diminishes rapidly in caliber and merges insensibly into the ureter, the excretory duct of the kidney.

The kidney is composed of an internal **medullary** and an external **cortical substance** (Fig. 17–23).

The **medullary substance** (*substantia medullaris*) consists of a series of striated conical masses, termed the **renal pyramids.** They vary from eight to eighteen in number and have their bases directed toward the circumference of the kidney, while their apices converge toward the renal sinus, where they form prominent papillae projecting into the lumen of the minor calyces.

The **cortical substance** (*substantia corticalis*) is reddish brown in color and soft and granular in consistency. It lies immediately beneath the fibrous tunic, arches over the bases of the pyramids, and dips in

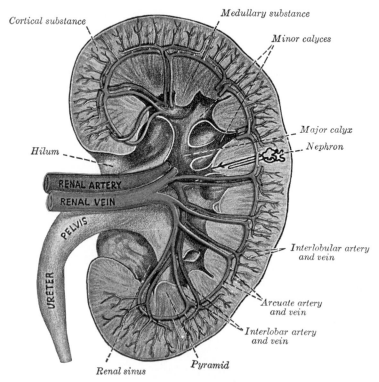

Fig. 17–23.—Diagram of a vertical section through the kidney. Nephron and blood vessels greatly enlarged.

between adjacent pyramids toward the renal sinus. The parts dipping in between the pyramids are named the **renal columns** (Bertini), while the portions which connect the renal columns to each other and intervene between the bases of the pyramids and the fibrous tunic are called the **cortical arches**. If the cortex be examined with a lens, it will be seen to consist of a series of lighter colored, conical areas, termed the radiate part, and a darker colored intervening substance, which from the complexity of its structure is named the convoluted part. The rays gradually taper toward the circumference of the kidney, and consist of a series of outward prolongations from the base of each renal pyramid.

Minute Anatomy.—The **renal tubules** (Fig. 17–23), of which the kidney is for the most part made up, commence in the cortical substance, and after pursuing a very circuitous course through the cortical and medullary substances, finally end at the apices of the renal pyramids by open mouths, so that the fluid which they contain is emptied, through the calyces, into the pelvis of the kidney. If the surface of one of the papillae be examined with a lens, it will be seen to be studded over with minute openings, the orifices of the renal tubules, from sixteen to twenty in number, and if pressure be made on a fresh kidney, urine will be seen to exude from these orifices. The tubules commence in the cortex and renal columns as the **renal corpuscles** or **Malpighian bodies**. They are small rounded masses, deep red in color, varying in size, but averaging about 0.2 mm in diameter. Each of these bodies is composed of two parts: a central **glomerulus** of vessels, and a double-walled membranous envelope, the **glomerular capsule** (*capsule of Bowman*), which is the invaginated pouch-like commencement of a renal tubule.

The **glomerulus** is a tuft of non-anastomosing capillaries, among which there is a scanty amount of connective tissue. This capillary tuft is derived from an arteriole, the *afferent vessel*, which enters the capsule, generally at a point opposite to that at which the capsule joins the tubule (Fig. 17–24). Upon entering the capsule the afferent arteriole divides into from 2 to 10 primary branches, which in turn subdivide into about 50 capillary loops which generally do not anastomose. These loops are from 300 to 500 μ in length. The capillaries join to form the *efferent arteriole*, which leaves Bowman's capsule adjacent to the afferent ves-

sel, the latter generally being the larger of the two. The total surface area of the capillaries of all glomeruli is about 1 square meter.

For a variable distance before the afferent arteriole enters the glomerulus the muscle cells of the media adjacent to the distal convolution of its own nephron are modified, appearing as relatively large afibrillar cells. This structure is said to present evidence of glandular activity. At times it appears partially to invest the artery in a nest-like group of cells embedded in a delicate fibrillar network. It is variously known as the **juxtaglomerular apparatus** (Goormaghtigh), the **Polkissen** (Zimmerman) and the **periarterial pad** (Edwards). Its exact function has not been determined; neither is it known whether or not all afferent arteries contain it, but it is common in man. The efferent arteriole has only circular, smooth muscle fibers in its wall, which may be a means of regulating glomerular blood pressure.

The **glomerular** or Bowman's capsule, which surrounds the glomerulus, consists of a double-walled sac. The outer wall (parietal layer) is continuous with the inner wall (visceral layer) at the points of entrance and exit of the afferent and efferent vessels respectively. The cavity between the two layers is continuous with the lumen of the proximal convoluted tubule. The parietal layer is smooth. The visceral layer covers the glomerulus and dips in between the capillary loops, almost completely surrounding each one. Both layers of the capsule consist of flattened epithelial cells which have a basement membrane. This covers the outer surface of the parietal layer and is continuous with that of the tubule cells. The basement membrane of the visceral layer is in contact with the glomerular capillaries. Microdissection experiments of living glomeruli show them to lie in a gelatinous matrix. The walls of the capillaries, the overlying cells of the visceral layer of the capsule and the gelatinous matrix are now thought to constitute a filter mechanism through which non-protein constituents of the blood plasma can enter the tubule.

A **renal tubule**, beginning with the capsule of Bowman as it surrounds the glomerulus and ending where the tubule joins the excretory duct or collecting tubule, constitutes a **nephron** —the structural and functional unit of the kidney. There are about 1,250,000 of these units in each kidney.

A tubule presents during its course, many changes in shape and direction, and is contained partly in the cortical and partly in the medullary substance. At its junction with the glomerular capsule it exhibits a somewhat constricted portion, which is termed the **neck.**

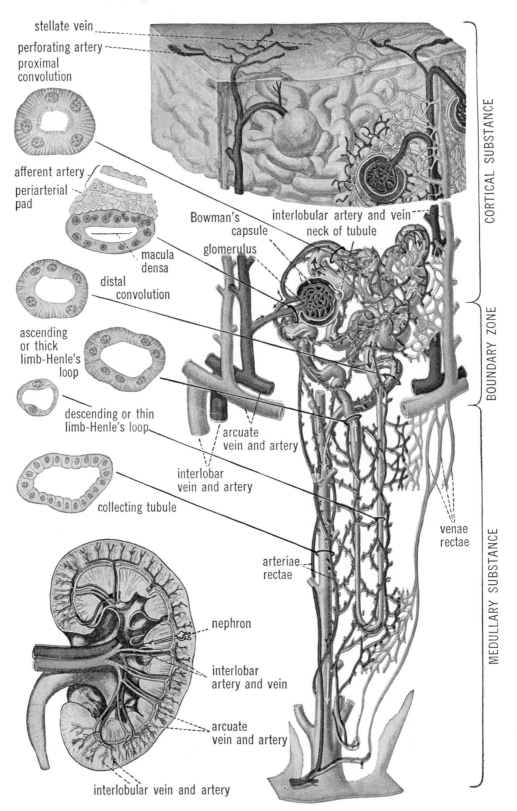

stellate vein

perforating artery

proximal
convolution

afferent artery

periarterial
pad

macula
densa

distal
convolution

ascending
or thick
limb-Henle's
loop

descending or thin
limb-Henle's loop

collecting tubule

Bowman's
capsule

glomerulus

interlobular artery and vein
neck of tubule

CORTICAL SUBSTANCE

BOUNDARY ZONE

MEDULLARY SUBSTANCE

arcuate
vein and artery

interlobar
vein and artery

venae
rectae

arteriae
rectae

nephron

interlobar
artery and vein

arcuate
vein and artery

interlobular vein and artery

FIG. 17–24.—Diagram of a portion of kidney lobule illustrating a nephron, typical histological sections of the various divisions of a nephron, and the disposition of the renal vessels. The section of the collecting tubule is reproduced at a lower magnification than the divisions of the nephron. (Courtesy of R. G. Williams.)

Beyond this the tubule becomes convoluted, and pursues a considerable course in the cortical substance constituting the **proximal convoluted tube.** The convolutions disappear as the tube approaches the medullary substance in a more or less spiral manner. Throughout this portion of its course the renal tubule is contained entirely in the cortical substance, and presents a fairly uniform caliber. It now enters the medullary substance, suddenly becomes much smaller, quite straight, and dips down for a variable depth into the pyramid, constituting the thin or **descending limb of Henle's loop.** Bending on itself, it forms what is termed the **loop of Henle** and ascending, it becomes suddenly enlarged, forming the thick or **ascending limb of Henle's loop,** which enters the cortical substance where it again becomes dilated, and tortuous. It is now called the **distal convoluted tubule.** The terminal part of the ascending limb of Henle's loop crosses in contact with or sometimes lies parallel to the afferent arteriole of its own glomerulus. The turns of the distal convoluted tubule into which it merges lie among the coils of the proximal portion of the nephron and terminate in a narrow part which enters a collecting tubule.

The **straight** or **collecting tubes** commence in the radiate part of the cortex, where they receive the curved ends of the distal convoluted tubules. They unite at short intervals with one another, the resulting tubes presenting a considerable increase in caliber, so that a series of comparatively large tubes passes from the bases of the rays into the renal pyramids. In the medulla the tubes of each pyramid converge to join a central tube (*duct of Bellini*) which finally opens on the summit of one of the papillae; the contents of the tube are therefore discharged into one of the minor calyces.

Structure of the Renal Tubules.—The various parts of the nephron present quite different cellular appearances and these appearances vary depending upon the functional state of the cells. The **proximal convoluted tubule** is about 14 mm long and 59 μ in diameter. It is composed of one layer of large cuboidal cells with central spherical nuclei. The cells dovetail laterally with one another and the lateral cell limits are rarely seen. The distal ends of the cells bulge into the lumen and are covered by a brush border. The cytoplasm is abundant and coarsely granular. Parallel striations in the cytoplasm perpendicular to the basement membrane are due to mitochondria.

The transition between the epithelium of the proximal convoluted tubule and the thin segment is abrupt. The **thin segment** may be absent in nephrons beginning near the surface

of the kidney. It is composed of squamous cells with pale-staining cytoplasm and flattened nuclei. The epithelial change from the descending limb to the **ascending** or **thick limb** is also quite abrupt. The cells become cuboidal and deeper-staining, with perpendicular striations in the basal parts but without distinct cell boundaries or brush borders.

At about the junction of the ascending limb with the distal convoluted tubule the nephron comes into contact with the juxtaglomerular cells of its own afferent arteriole. Here the epithelium of the tubule is greatly modified, having high cells and crowded nuclei. This constitutes the **macula densa** or epithelial plaque, the function of which is unknown.

Transition to the tortuous **distal convoluted tubule** is gradual. This segment of the nephron is about 5 mm long and 35 μ in diameter. The cells are lower and the lumen larger than in the proximal tubule. They do not have a brush border and the boundaries are fairly distinct. The distal convoluted tubule merges into a short connecting segment which joins the collecting or excretory tubule.

The collecting tubules have a typical epithelium which is quite different from that in the various portions of the nephron. In the smallest tubes the cells are cuboidal and distinctly outlined with round nuclei and clear cytoplasm. As the tubules become larger the cells are higher, finally becoming tall columnar in the ducts of Bellini. The columnar epithelium becomes continuous with the cells covering the surface of the papillae.

The length of the nephron varies from 30 to 38 mm, while the length of the collecting tubules is estimated at from 20 to 22 mm.

The Renal Blood Vessels (Fig. 17–24).—The kidney is plentifully supplied with blood by the renal artery, a large branch of the abdominal aorta. Before entering the kidney substance the number and disposition of the branches of the renal artery exhibit great variation. In most cases the renal artery divides into two primary branches, a larger anterior and a smaller posterior. The anterior branch supplies exclusively the anterior or ventral half of the organ and the posterior supplies the posterior or dorsal part. Therefore there is a line (Bröedel's line) in the long axis of the lateral border of the kidney which passes between the two main arterial divisions and in which there are no large vessels, a feature which is utilized to minimize hemorrhage when nephrotomy is done. The primary branches subdivide and diverge until they come to lie on the anterior and posterior aspects respectively of the calyces. Further subdivisions occur which enter the kidney substance and run between the pyramids. These are known as **interlobar arteries.** When these vessels reach the corticomedullary zone they make more or less well-defined arches over the bases of the pyramids and are then called **arcuate arteries.** These vessels give off a series of branches called **interlobular arteries.**

The interlobular arteries and the terminal parts of the arcuate vessels run vertically and nearly parallel towards the cortex and periphery of the kidney. The interlobular arteries may terminate: (1) as an afferent glomerular artery to one or more glomeruli; (2) in a capillary plexus around the convoluted tubules in the cortices without relation to the glomerulus, therefore a nutrient artery, and (3) as a perforating capsular vessel. Divisions classified under (2) must be regarded as exceptional.

The most important and numerous branches of the interlobular arteries are the **afferent glomerular vessels.** These break up into capillary loops, the glomerulus, within Bowman's capsule. The loops unite to form the efferent arteriole. The **efferent glomerular vessel** forms a plexus about the convoluted tubule and part of Henle's loop and sends one or more branches toward the pelvis, the **arteria recta,** which supplies the collecting tubules and loops of Henle.

The arteriae rectae are derived chiefly from the efferent arterioles of glomeruli located in the boundary zone. They are frequently known as **arteriae rectae spuriae** to distinguish them from a few straight vessels arising directly from the arcuate or interlobular arteries without relation to glomeruli and hence called **arteriae rectae verae.** It is likely that the so-called arteriae rectae verae were once derived from an efferent glomerular vessel but that the glomerulus atrophied, thus giving them the appearance of true nutrient branches from the arcuate or interlobular arteries. No arteriae rectae are derived from the vessels of the cortical zone. It has not been determined whether or not there are anastomoses between the capillaries of adjacent nephrons. However, there is free anastomosis between the branches of the arteriae rectae. The arteriae rectae or straight arteries surround the limbs of Henle and pass down between the straight collecting tubules, where they form terminal plexuses around the tubes.

This plexus drains into the **venae rectae** which in turn carry the blood to the interlobular veins, thence into the arcuate veins, then into the interlobar veins and finally into the renal veins, which discharge into the inferior vena cava. It will be noted that the chief renal arteries have counterparts in the venous sys-

tem. All veins from the dorsal half of the kidney cross to the ventral half between the minor calyces to join the ventral collecting veins before leaving the kidney. Whatever arterial anastomoses there are within the kidney occur beyond the glomeruli, the renal arteries and their branches therefore are terminal as far as the arteriae rectae. An exception to this condition is found in cases in which arteriovenous anastomoses have been described. These connections have been found in three locations: between the arteries and veins of the sinus renalis, between the subcapsular vessels and between the interlobular arteries and veins. In the latter position particularly arterial blood could reach the tubules of the nephron by retrograde flow through the veins. The constancy with which this type of anastomosis occurs or its rôle in the circulation of the kidney has not been accurately determined. Anastomosis between veins is very rich. The perforating capsular vessels, the terminations of interlobular arteries, form connections with non-renal vessels in the fat which surrounds the kidney. They frequently drain into the subcapsular or **stellate veins,** which in turn drain into the interlobular veins.

The **lymphatics** of the kidney are described in Chapter 10.

Nerves of the Kidney.—The nerves of the kidney, although small, are about fifteen in number. They have small ganglia developed upon them, and are derived from the renal plexus, which is formed by branches from the celiac plexus, the lower and outer part of the celiac ganglion and aortic plexus, and from the lesser and lowest splanchnic nerves. They communicate with the testicular plexus, a circumstance which may explain the occurrence of pain in the testis in affections of the kidney. They accompany the renal artery and its branches, and are distributed to the blood vessels and to the cells of the urinary tubules.

Connective Tissue (*intertubular stroma*).— Although the tubules and vessels are closely packed, a small amount of connective tissue, continuous with the fibrous tunic, binds them firmly together and supports the blood vessels, lymphatics, and nerves.

Variations.—Malformations of the kidney are not uncommon. There may be an entire absence of one kidney, but, according to Morris, the number of these cases is "excessively small": or there may be congenital atrophy of one kidney, when the kidney is very small, but usually healthy in structure. These cases are of great importance, and must be duly taken into account when nephrectomy is contemplated. A more common malformation is where

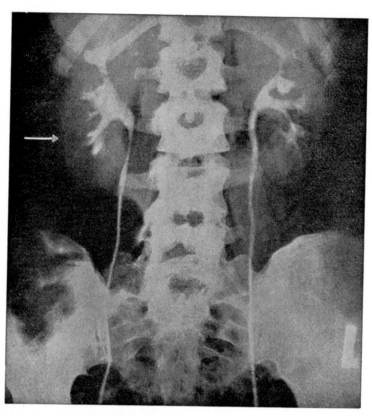

FIG. 17–25.—Ureters, pelves and minor calyces after intravenous injection of uroselectan. Note cupping of minor calyces; the relation of the ureter to the transverse processes of the lumbar vertebrae, and the Psoas major. The arrow points to the shadow of the right kidney. Anterior view.

the two kidneys are fused together. They may be joined together only at their lower ends by means of a thick mass of renal tissue, so as to form a horseshoe-shaped body, or they may be completely united, forming a disk-like kidney, from which two ureters descend into the bladder. These fused kidneys are generally situated in the middle line of the abdomen, but may be displaced as well. In some mammals, *e.g.*, ox and bear, the kidney consists of a number of distinct lobules; this lobulated condition is characteristic of the kidney of the human fetus, and traces of it may persist in the adult. Sometimes the pelvis is duplicated, while a double ureter is not very uncommon. In some rare instances a third kidney may be present.

One or both kidneys may be misplaced as a congenital condition, and remain fixed in this abnormal position. They are then very often misshapen. They may be situated higher, though this is very uncommon, or lower than normal or removed farther from the vertebral column than usual; or they may be displaced into the iliac fossa, over the sacro-iliac joint,

on to the promontory of the sacrum, or into the pelvis between the rectum and bladder or by the side of the uterus. In these latter cases they may give rise to very serious trouble. The kidney may also be displaced as a congenital condition, but may not be fixed; it is then known as a *floating kidney*. It is believed to be due to the fact that the kidney is completely enveloped by peritoneum which then passes backward to the vertebral column as a double layer, forming a mesonephron which permits movement. The kidney may also be misplaced as an acquired condition; in these cases the kidney is mobile in the tissues by which it is surrounded, moving with the capsule in the perinephric tissues. This condition is known as *movable kidney*, and is more common in the female than in the male. It occurs in badly nourished people, or in those who have become emaciated from any cause. It must not be confounded with the *floating kidney*, which is a congenital condition due to the development of a mesonephron. The two conditions cannot, however, be distinguished until the abdomen is opened or the kidney explored from the loin. Accessory renal arteries entering one or both poles of the kidney instead of at the hilum are fairly common.

The Ureters

The **ureters** (Figs. 17–15, 17–16) are the two tubes which convey the urine from the kidneys to the urinary bladder. The renal portion of each tube commences within the sinus of the corresponding kidney as a number of short cup-shaped tubes, termed **calyces**, which encircle the renal papillae. Since a single calyx may enclose more than one papilla the calyces are generally fewer in number than the pyramids—the former varying from seven to thirteen, the latter from eight to eighteen. The calyces join to form two or three short tubes, and these unite to form a funnel-shaped dilatation named the **renal pelvis.** The pelvis is situated partly inside and partly outside the renal sinus where it becomes continuous with the ureter, usually on a level with the spinous process of the first lumbar vertebra.

The **Ureter Proper** varies in length from 28 to 34 cm, the right being about 1 cm shorter than the left. It is a thick-walled narrow tube, not of uniform caliber, varying from 1 mm to 1 cm in diameter. It runs caudalward and medialward on the Psoas major muscle and, entering the pelvic cavity, finally opens into the fundus of the bladder.

The **abdominal part** (*pars abdominalis*) lies behind the peritoneum on the medial part of the Psoas major embedded in the subserous fascia, and is crossed obliquely by the testicular vessels. It enters the pelvic cavity by crossing either the termination of the common, or the commencement of the external, iliac vessels (Fig. 17–15).

At its origin the *right* ureter is usually covered by the descending part of the duodenum, and in its course lies to the right of the inferior vena cava, and is crossed by the right colic and ileocolic vessels, while near the superior aperture of the pelvis it passes dorsal to the caudal part of the mesentery and the terminal part of the ileum. The *left* ureter is crossed by the left colic vessels, and near the superior aperture of the pelvis passes behind the sigmoid colon and its mesentery.

The **pelvic part** (*pars pelvina*) runs at first caudalward on the lateral wall of the pelvic cavity, along the anterior border of the greater sciatic notch and under cover of the peritoneum. It lies ventral to the internal iliac artery medial to the obturator nerve and the obturator, inferior vesical, and middle rectal arteries. Opposite the lower part of the greater sciatic foramen it inclines medialward, and reaches the lateral angle of the bladder, where it is situated ventral to the upper end of the seminal vesicle; here the ductus deferens crosses to its medial side, and the vesical veins surround it. Finally, the ureters run obliquely for about 2 cm through the wall of the bladder and open by slit-like apertures into the cavity of the viscus at the lateral angles of the trigone. When the bladder is distended the openings of the ureters are about 5 cm apart, but when it is empty and contracted the distance between them is diminished by one-half. Owing to their oblique course through the coats of the bladder, the upper and lower walls of the terminal portions of the ureters become closely applied to each other when the viscus is distended, and, acting as valves, prevent regurgitation of urine from the bladder. There are three points in the course of the ureter where it normally undergoes constriction: (1) at the ureteropelvic junction, average diameter 2 mm; (2) at the place where it crosses the iliac vessels, 4 mm, and (3) where it joins the bladder, 1 to 5 mm. Between these points the abdominal ureter averages 10 mm in diameter and the pelvic ureter 5 mm.

In the **female**, the ureter forms, as it lies in relation to the wall of the pelvis, the posterior boundary of a shallow depression named the **ovarian fossa**, in which the ovary is situated. It then runs medialward and ventralward on the lateral aspect of the cervix uteri and upper part of the vagina to reach the fundus of the bladder. In this part of its course it is accompanied for about 2.5 cm by the uterine artery, which then crosses over the ureter and ascends between the two layers of the broad ligament. The ureter is about 2 cm distant from the side of the cervix of the uterus.

Structure (Fig. 17–26).—The ureter is composed of three coats: **fibrous, muscular,** and **mucous coats.**

The **fibrous coat** (*tunica adventitia*) is continuous at one end with the fibrous tunic of the kidney on the floor of the sinus, while at the other it is lost in the fibrous structure of the bladder.

In the renal pelvis the **muscular coat** (*tunica muscularis*) consists of two layers, longitudinal and circular: the longitudinal fibers become lost upon the sides of the papillae at the extremities of the calyces; the circular fibers may be traced surrounding the medullary substance in the same situation. In the ureter proper the muscular fibers are very distinct, and are arranged in three layers: an external longitudinal, a middle circular, and an internal, less distinct than the other two, but having a general longitudinal direction. According to Kölliker this internal layer is found only in the neighborhood of the bladder.

The **mucous coat** (*tunica mucosa*) is smooth, presenting only a few longitudinal folds which become effaced by distention. It is continuous with the mucous membrane of the bladder below, while it is prolonged over the papillae of the kidney above. Its epithelium is of a transitional character, and resembles that found in the bladder (see Fig. 17–33). It consists of several layers of cells, of which the innermost— that is to say, the cells in contact with the urine —are somewhat flattened, with concavities on their deep surfaces into which the rounded ends of the cells of the second layer fit. These, the intermediate cells, more or less resemble columnar epithelium, and are pear-shaped, with rounded internal extremities which fit into the concavities of the cells of the first layer, and narrow external extremities which are wedged in between the cells of the third layer. The external or third layer consists of conical or oval cells varying in number in different parts, and presenting processes which extend down into the basement membrane. Beneath the epithelium, and separating it from the muscular coats, is a dense layer of fibrous tissue containing many elastic fibers.

Vessels and Nerves.—The **arteries** supplying the ureter are branches from the renal, testicular, internal iliac, and inferior vesical.

The **nerves** are derived from the inferior mesenteric, testicular, and pelvic plexuses. The lower one-third of the ureter contains nerve cells which are probably incorporated in vagus efferent chains. The afferent supply of the ureter is contained in the eleventh and twelfth thoracic and first lumbar nerves. The vagus supply to the ureter probably also has afferent components.

Variations.—The upper portion of the ureter is sometimes double; more rarely it is double the greater part of its extent, or even completely so. In such cases there are two openings into the bladder. Asymmetry in these variations is common.

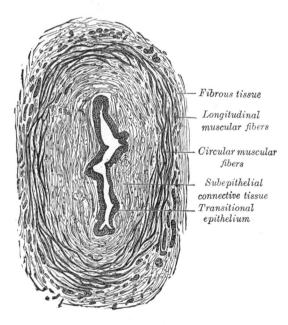

Fibrous tissue

Longitudinal
muscular fibers

Circular muscular
fibers

Subepithelial
connective tissue

Transitional
epithelium

Fig. 17–26.—Transverse section of ureter.

The Urinary Bladder

The **urinary bladder** (*vesica urinaria*) (Fig. 17–27) is a musculomembranous sac which acts as a reservoir for the urine; and as its size, position, and relations vary according to the amount of fluid it contains, it is necessary to study it as it appears (*a*) when *empty*, and (*b*) when *distended*. In both conditions the position of the bladder varies with the condition of the rectum, being pushed upward and forward when the rectum is distended.

The Empty Bladder.—When hardened *in situ*, the empty bladder has the form of a flattened tetrahedron, with its vertex tilted ventralward. It presents a fundus, a vertex, a superior and an inferior surface. The **fundus** (Fig. 17–44) is triangular in shape, and is directed caudalward and dorsalward toward the rectum, from which it is separated by the rectovesical fascia, the vesiculae seminales, and the terminal portions of the ductus deferentes. The **vertex** is directed ventralward toward the upper part of the symphysis pubis, and from it the middle umbilical ligament is continued cranialward on the anterior abdominal wall to the umbilicus. The peritoneum covering the ligament is the middle umbilical fold.

The **superior surface** is triangular, bounded on either side by a lateral border which separates it from the inferior surface, and by a posterior border, represented by a line joining the two ureters, which intervenes between it and the fundus. The lateral borders extend from the ureters to the vertex, and from them the peritoneum is carried to the walls of the pelvis. On either side of the bladder the peritoneum shows a depression, named the **paravesical fossa** (Fig. 16–54). The superior surface is covered by peritoneum, and is in relation with the sigmoid colon and some of the coils of the small intestine. When the bladder is empty and firmly contracted, this surface is convex and the lateral and posterior borders

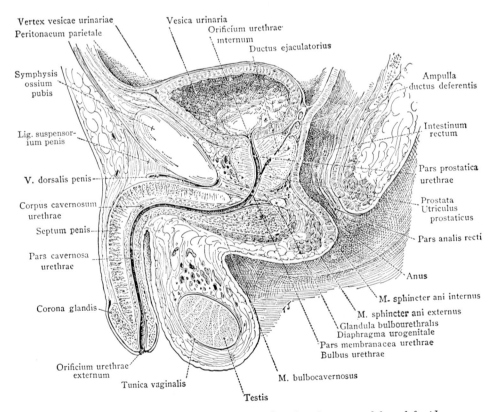

FIG. 17–27.—Median sagittal section through male pelvis, viewed from left side. (Eycleshymer and Jones.)

are rounded; whereas if the bladder be relaxed it is concave, and the interior of the viscus, as seen in a median sagittal section, presents the appearance of a V-shaped slit with a shorter posterior and a longer anterior limb—the apex of the V corresponding with the internal orifice of the urethra.

The **inferior surface** is uncovered by peritoneum. It may be divided into a posterior or prostatic area and two lateral surfaces. The prostatic area rests upon and is in direct continuity with the base of the prostate; and from it the urethra emerges. The lateral portion of the inferior surface is directed caudalward and lateralward and is separated from the symphysis pubis by the prevesical fascial cleft (*cavum Retzii*).

When the bladder is empty it is placed entirely within the pelvis, below the level of the obliterated hypogastric arteries, and below the level of those portions of the ductus deferentes which are in contact with the lateral wall of the pelvis; after they cross the ureters the ductus deferentes come into contact with the fundus of the bladder. As the viscus fills, its fundus, being more or less fixed, is only slightly depressed; while its superior surface gradually rises into the abdominal cavity, carrying with it its peritoneal covering, and at the same time rounding off the posterior and lateral borders.

The Distended Bladder.—When the bladder is moderately full, it contains about 0.5 liter and assumes an oval form; the long

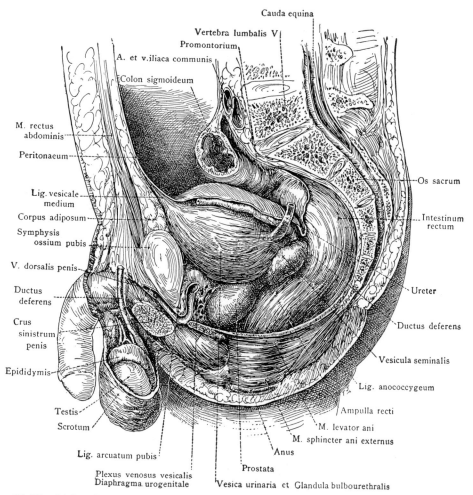

Fig. 17–28.—Male pelvic organs and perineum seen from left side after removal of left pelvic wall. Bladder and rectum moderately distended. (Eycleshymer and Jones.)

diameter of the oval measures about 12 cm and is directed cranialward and ventralward. In this condition it presents a posterosuperior, and anteroinferior, and two lateral surfaces, a fundus and a summit.

The **posterosuperior surface** is covered by peritoneum: dorsally, it is separated from the rectum by the rectovesical excavation, while its anterior part is in contact with the coils of the small intestine.

The **anteroinferior surface** is devoid of peritoneum, and rests against the pubic bones, above which it is in contact with the

back of the anterior abdominal wall. The caudal parts of the lateral surfaces are destitute of peritoneum, and are in contact with the lateral walls of the pelvis. The line of peritoneal reflection from the lateral surface is raised to the level of the obliterated hypogastric artery.

The **fundus** undergoes little alteration in position, being only slightly lowered. It exhibits, however, a narrow triangular area, which is separated from the rectum merely by the rectovesical fascia. This area is bounded below by the prostate, above by

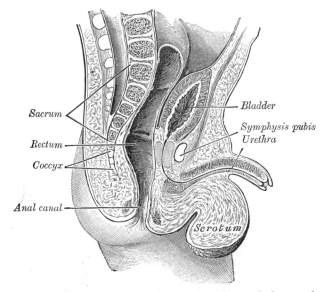

FIG. 17–29.—Sagittal section through the pelvis of a newly born male child.

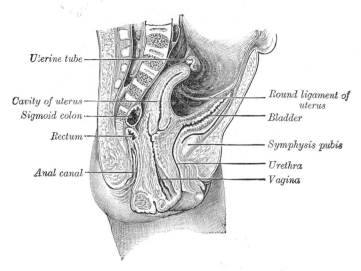

FIG. 17–30.—Sagittal section through the pelvis of a newly born female child.

the rectovesical fold of peritoneum, and laterally by the ductus deferentes. The ductus deferentes frequently come in contact with each other above the prostate, and under such circumstances the lower part of the triangular area is obliterated. The line of reflection of the peritoneum from the rectum to the bladder appears to undergo little or no change when the latter is distended; it is situated about 10 cm from the anus.

The **summit** is directed cranialward and ventralward above the point of attachment of the middle umbilical ligament, and hence the peritoneum which follows the ligament, forms a pouch of varying depth between the summit of the bladder and the anterior abdominal wall.

The Bladder in the Child (Figs. 17–29, 17–30).—In the newborn child the internal urethral orifice is at the level of the upper border of the symphysis pubis; the bladder therefore lies relatively at a much higher level in the infant than in the adult. Its anterior surface "is in contact with about the lower two-thirds of that part of the abdomi-

nal wall which lies between the symphysis pubis and the umbilicus" (Symington). Its fundus is clothed with peritoneum as far as the level of the internal orifice of the urethra. Although the bladder of the infant is usually described as an abdominal organ, Symington has pointed out that only about one-half of it lies above the plane of the superior aperture of the pelvis. Disse maintains that the internal urethral orifice sinks rapidly during the first three years, and then more slowly until the ninth year, after which it remains stationary until puberty, when it again slowly descends and reaches its adult position.

The Female Bladder (Fig. 17–31).—In the female, the bladder is in relation dorsally with the uterus and the upper part of the vagina. It is separated from the anterior surface of the body of the uterus by the vesicouterine excavation, but below the level of this excavation it is connected to the front of the cervix uteri and the upper part of the anterior wall of the vagina by areolar tissue. When the bladder is empty the uterus rests upon its superior surface.

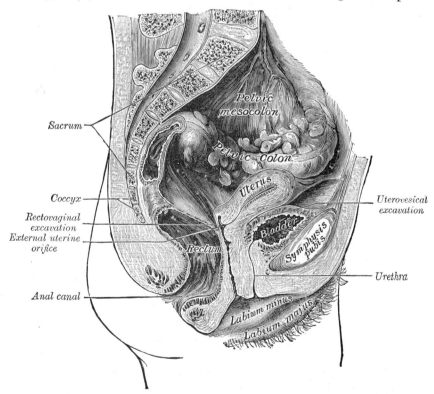

FIG. 17–31.—Median sagittal section of female pelvis.

The female bladder is said by some to be more capacious than that of the male, but probably the opposite is the case.

Ligaments.—The bladder is held in position by ligamentous attachments at its inferior portion or base, that is, near the exit of the urethra, and at the vertex. The remainder of the wall, enclosed in subserous fascia, is free to move during the expansion and contraction of filling and emptying.

The base of the bladder is attached to the internal investing layer of deep fascia on the pubic bone by strong fibrous bands which may contain muscle fibers, the Pubo-vesicales. In the male, because the prostate is firmly bound to the bladder in this region, these attachments are between the prostate and pubic bone rather than directly to the bladder, and are named the **medial** and **lateral puboprostatic ligaments** (see page 439). In the female the attachments are directly between bladder and pubis and are therefore called the **pubovesical ligaments.**

The base of the bladder is secured posteriorly to the side of the rectum and the sacrum by condensations of the subserous fascia underlying the sacrogenital folds. They are called the **rectovesical ligaments,** or, since they may contain smooth muscle bundles, the **Rectovesicales** muscles.

The **middle umbilical ligament** is a fibrous or fibromuscular cord, the remains of the urachus (page 1267) which extends from the vertex of the bladder to the umbilicus. It is broad at its attachment to the bladder and becomes narrow as it nears the umbilicus.

In addition to these fibrous or true ligaments, there are a series of folds, where the peritoneum is reflected from the bladder to the abdominal wall, called **false ligaments of the bladder.** Anteriorly there are three folds: the **middle umbilical fold** on the middle umbilical ligament, and two **lateral umbilical folds** on the obliterated umbilical arteries. The reflections of the peritoneum on to the side wall of the pelvis form the **lateral false ligaments,** while the sacrogenital folds constitute **posterior false ligaments.**

Interior of the Bladder (Fig. 17–32).— The mucous membrane lining the greater part of the bladder is loosely attached to

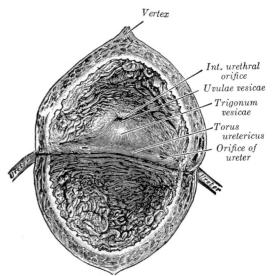

Vertex

Int. urethral orifice
Uvulae vesicae
Trigonum vesicae
Torus uretericus
Orifice of ureter

Fig. 17–32.—The interior of bladder.

the muscular coat, and appears wrinkled or folded when the bladder is contracted: in the distended condition of the bladder the folds are effaced. Over a small triangular area, termed the **trigonum vesicae,** immediately above and behind the internal orifice of the urethra, the mucous membrane is firmly bound to the muscular coat, and is always smooth. The anterior angle of the trigonum vesicae is formed by the internal orifice of the urethra: its posterolateral angles by the orifices of the ureters. Stretching dorsal to the latter openings is a slightly curved ridge, the **torus uretericus,** forming the base of the trigone and produced by an underlying bundle of non-striped muscular fibers. The lateral parts of this ridge extend beyond the openings of the ureters, and are named the **plicae uretericae;** they are produced by the terminal portions of the ureters as they traverse the bladder wall obliquely. When the bladder is illuminated the torus uretericus appears as a pale band and forms an important guide during the operation of introducing a catheter into the ureter.

The **orifices of the ureters** are placed at the posterolateral angles of the trigonum vesicae, and are usually slit-like in form. In the contracted bladder they are about 2.5 cm apart and about the same distance from the internal urethral orifice; in the

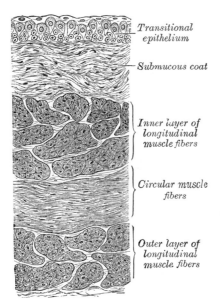

FIG. 17–33.—Vertical section of bladder wall.

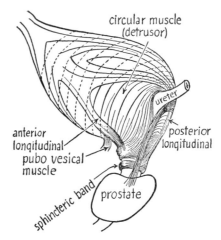

FIG. 17–34.—Diagram of the muscle of the
bladder. (After McCrea.)

distended viscus these measurements may
be increased to about 5 cm.

The **internal urethral orifice** is placed at
the apex of the trigonum vesicae, in the
most dependent part of the bladder, and is
usually somewhat crescentic in form; the
mucous membrane immediately behind it
presents a slight elevation, the **uvula
vesicae**, caused by the middle lobe of the
prostate.

Structure (Fig. 17–33).—The bladder is
composed of the four coats: **serous, muscular,
submucous,** and **mucous coats.**

The **serous coat** (*tunica serosa*) is a partial
one and is derived from the peritoneum. It
invests the superior surface and the upper parts
of the lateral surfaces, and is reflected from
these on to the abdominal and pelvic walls.

The **muscular coat** (*tunica muscularis*) consists of three layers of smooth muscular fibers:
an external layer, composed of fibers having
for the most part a longitudinal arrangement;
a middle layer, in which the fibers are arranged,
more or less, in a circular manner; and an
internal layer, in which the fibers have a general
longitudinal arrangement (Fig. 17–34).

The *fibers of the external layer* arise from
the posterior surface of the body of the pubis
in both sexes (*musculi pubovesicales*), and in
the male from the adjacent part of the prostate
and its capsule. They pass, in a more or less
longitudinal manner, up the inferior surface
of the bladder, over its vertex, and then descend

along its fundus to become attached to the
prostate in the male, and to the front of the
vagina in the female. At the sides of the bladder
the fibers are arranged obliquely and intersect
one another. This layer has been named the
Detrusor urinae muscle.

The *fibers of the middle circular layer* are
very thinly and irregularly scattered on the
body of the organ, and, although to some extent
placed transversely to the long axis of the
bladder, are for the most part arranged
obliquely. Toward the lower part of the
bladder, around the internal urethral orifice,
they are disposed in a thick circular layer, forming the **Sphincter vesicae,** which is continuous
with the muscular fibers of the prostate.

The *internal longitudinal layer* is thin, and
its fasciculi have a reticular arrangement
but with a tendency to assume for the most
part a longitudinal direction. Two bands of
oblique fibers, originating behind the orifices
of the ureters, converge to the back part of the
prostate, and are inserted by means of a fibrous
process, into the middle lobe of that organ.
They are the **muscles of the ureters,** described
by Sir C. Bell, who supposed that during the
contraction of the bladder they serve to retain
the oblique direction of the ureters, and so
prevent the reflux of the urine into them.

The **submucous coat** (*tela submucosa*) consists of a layer of areolar tissue, connecting
together the muscular and mucous coats, and
intimately united to the latter.

The **mucous coat** (*tunica mucosa*) is thin,
smooth, and of a pale rose color. It is continuous above through the ureters with the lining
membrane of the renal tubules, and below with
that of the urethra. The loose texture of the
submucous layer allows the mucous coat to

be thrown into folds or *rugae* when the bladder is empty. Over the trigonum vesicae the mucous membrane is closely attached to the muscular coat, and is not thrown into folds, but is smooth and flat. The epithelium covering it is of the transitional variety, consisting of a superficial layer of polyhedral flattened cells, each with one, two, or three nuclei; beneath these is a stratum of large club-shaped cells, with their narrow extremities directed downward and wedged in between smaller spindle-shaped cells, containing oval nuclei (Fig. 17–33). The epithelium varies accordingly as the bladder is distended or contracted. In the former condition the superficial cells are flattened and those of the other layers are shortened; in the latter they present the appearance described above. There are no true glands in the mucous membrane of the bladder, though certain mucous follicles which exist, especially near the neck of the bladder, have been regarded as such.

Vessels and Nerves.—The **arteries** supplying the bladder are the superior, middle and inferior vesical, derived from the internal iliac artery. The obturator and inferior gluteal arteries also supply small visceral branches to the bladder, and in the female additional branches are derived from the uterine and vaginal arteries.

The **veins** form a complicated plexus on the inferior surface, and fundus near the prostate, and end in the internal iliac veins.

The **lymphatics** are described in Chapter 10.

The **nerves** of the bladder are (1) fine medullated fibers from the third and fourth sacral nerves, and (2) non-medullated fibers from the hypogastric plexus. They are connected with ganglia in the outer and submucous coats and are finally distributed, all as non-medullated fibers, to the muscular layer and epithelial lining of the viscus.

Variations.—A defect of development, in which the bladder is implicated, is known under the name of *extroversion of the bladder*. In this condition the lower part of the abdominal wall and the anterior wall of the bladder are wanting, so that the fundus of the bladder presents on the abdominal surface, and is pushed forward by the pressure of the viscera within the abdomen, forming a red vascular tumor on which the openings of the ureters are visible. The penis, except the glans, is rudimentary and is cleft on its dorsal surface, exposing the floor of the urethra, a condition known as *epispadias*. The pelvic bones are also arrested in development.

The Male Urethra

The **male urethra** (*urethra virilis*) (Fig. 17–35) extends from the internal urethral orifice in the urinary bladder to the external urethral orifice at the end of the penis. It presents a double curve in the ordinary relaxed state of the penis (Fig. 17–27). Its length varies from 17.5 to 20 cm, and it is divided into three portions, the **prostatic, membranous,** and **cavernous,** the structure and relations of which are essentially different. Except during the passage of the urine or semen, the greater part of the urethral canal is a mere transverse cleft or slit, with its surfaces in contact; at the external orifice the slit is vertical, in the membranous portion irregular or stellate,

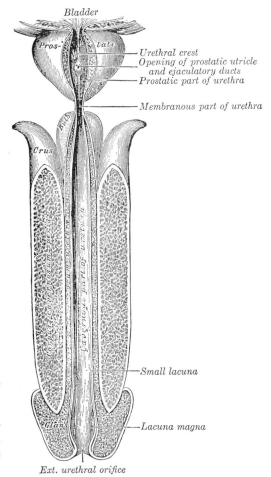

Fig. 17–35.—The male urethra laid open on its anterior (upper) surface.

and in the prostatic portion somewhat arched.

The **prostatic portion** (*pars prostatica*), the widest and most dilatable part of the canal, is about 3 cm long. It runs almost vertically through the prostate from its base to its apex, lying nearer its anterior than its posterior surface; the form of the canal is spindle-shaped, being wider in the middle than at either extremity, and narrowest below, where it joins the membranous portion. A transverse section of the canal as it lies in the prostate is horseshoe-shaped, with the convexity directed ventralward.

Upon the posterior wall or floor is a narrow longitudinal ridge, the **urethral crest**, formed by an elevation of the mucous membrane and its subjacent tissue. It is from 15 to 17 mm in length, and about 3 mm in height. On either side of the crest is a slightly depressed fossa, the **prostatic sinus**, the floor of which is perforated by numerous apertures, the **orifices of the prostatic ducts** from the lateral lobes of the prostate; the ducts of the middle lobe open behind the crest. At the distal part of the urethral crest, below its summit, is a median elevation, the **colliculus seminalis or verumontanum** upon or within the margins of which are the orifices of the prostatic utricle and the slit-like openings of the ejaculatory ducts.

The **prostatic utricle** (*sinus pocularis*) forms a cul-de-sac about 6 mm long, which runs upward and backward in the substance of the prostate behind the middle lobe. Its walls are composed of fibrous tissue, muscular fibers, and mucous membrane, and numerous small glands open on its inner surface. It was called by Weber the **uterus masculinus**, from its being developed from the united lower ends of the atrophied paramesonephric or Müllerian ducts, and therefore is homologous with the uterus and vagina in the female.

The **membranous portion** (*pars membranacea*) is the shortest, least dilatable, and, with the exception of the external orifice, the narrowest part of the canal. It extends between the apex of the prostate and the bulb of the penis, perforating the urogenital diaphragm about 2.5 cm from the pubic symphysis. The dorsal part of the urethral bulb lies in apposition with the superficial layer of the urogenital diaphragm, but its upper portion diverges somewhat from this fascia: the anterior wall of the membranous urethra is thus prolonged for a short distance ventral to the urogenital diaphragm; it measures about 2 cm in length, while the posterior wall which is between the two fasciae of the diaphragm is only 1.25 cm long.

The membranous portion of the urethra is completely surrounded by the fibers of the Sphincter urethrae. Ventral to it the deep dorsal vein of the penis enters the pelvis between the transverse ligament of the pelvis and the arcuate pubic ligament; on either side near its termination are the bulbourethral glands.

The **cavernous portion** (*pars cavernosa; penile or spongy portion*) is the longest part of the urethra, and is contained in the corpus cavernosum. It is about 15 cm long, and extends from the termination of the membranous portion to the external urethral orifice. Commencing at the superficial layer of the urogenital diaphragm it passes ventralward and upward to the front of the symphysis pubis, and then, in the flaccid condition of the penis, it bends downward. It is narrow, and of uniform size in the body of the penis, measuring about 6 mm in diameter; it is dilated proximally, within the bulb, and again anteriorly within the glans penis, where it forms the **fossa navicularis urethrae**.

The **external urethral orifice** (*orificium urethrae externum; meatus urinarius*) is the most contracted part of the urethra; it is a vertical slit, about 6 mm long, bounded on either side by two small labia.

The lining membrane of the urethra, especially on the floor of the cavernous portion, presents the orifices of numerous mucous glands and follicles situated in the submucous tissue, and named the **urethral glands** (*Littré*). Besides these there are a number of small pit-like recesses, or **lacunae** of varying sizes. Their orifices are directed distalward, so that they may easily intercept the point of a catheter in its passage along the canal. One of these lacunae, larger than the rest, is situated on the upper surface of the fossa navicularis; it is called the **lacuna magna.** The bulbourethral glands open into the cavernous portion about 2.5 cm distal to the inferior fascia of the urogenital diaphragm.

Structure.—The urethra is composed of mucous membrane, supported by a submucous tissue which connects it with the various structures through which it passes.

The **mucous coat** is continuous with the mucous membrane of the bladder, ureters, and kidneys; externally, with the integument covering the glans penis, and is prolonged into the ducts of the glands which open into the urethra, viz., the bulbourethral glands and the prostate; and into the ductus deferentes and vesiculae seminales, through the ejaculatory ducts. In the cavernous and membranous portions the mucous membrane is arranged in longitudinal folds when the tube is empty. Small papillae are found upon it, near the external urethral orifice; its epithelial lining is of the columnar variety except near the external orifice, where it is squamous and stratified.

The **submucosa** has a characteristic structure. It is composed of a thick stroma of connective tissue very rich in elastic fibers. These fibers connect freely with the spongy tissue of the penis which prevents ready removal of the mucosa in this region. However, in the membranous and prostatic portions, which change but little during erection, the urethra is quite free and may on dissection be stripped readily.

Vessels and Nerves.—The **urethral artery,** a branch of the internal pudendal artery in the perineum, supplies the membranous and penile urethra. The **veins** of the urethra and corpus spongiosum drain into the deep vein of the penis and the pudendal plexus. The **lymphatics** are described on page 762. The **nerves** come from the pudendal nerve.

Congenital Defects.—The defect most frequently met is one in which there is a cleft of the floor of the urethra owing to an arrest of union in the middle line. This is known as *hypospadias,* and the cleft may vary in extent. The simplest and most common form is where the deficiency is confined to the glans penis. The urethra ends at the point where the extremity of the prepuce joins the body of the penis, in a small valve-like opening. The prepuce is also cleft on its undersurface and forms a sort of hood over the glans. There is a depression on the glans in the position of the normal meatus. This condition produces no disability and requires no treatment. In more severe cases the cavernous portion of the urethra is cleft throughout its entire length, and the opening of the urethra is at the point of junction of the penis and scrotum. The under surface of the penis in the middle line presents a furrow lined by a moist mucous membrane, on either side of which is often more or less dense fibrous tissue stretching from the glans to the opening of the urethra, which prevents complete erection from taking place. Great discomfort is induced during micturition, and sexual intercourse is impossible. The condition may be remedied by a series of plastic operations. The worst form of this condition is where the urethra is deficient as far back as the perineum, and the scrotum is cleft. The penis is small and bound down between the two halves of the scrotum, so as to resemble an hypertrophied clitoris. The testes are often undescended. The condition of parts, therefore, very much resembles the external organs of generation of the female, and many children, the victims of this malformation, have been brought up as girls. The halves of the scrotum, deficient of testes, resemble the labia, the cleft between them looks like the orifice of the vagina, and the diminutive penis is taken for an enlarged clitoris.

A much more uncommon form of malformation is where there is an apparent deficiency of the upper wall of the urethra; this is named *epispadias.* The deficiency may vary in extent; when it is complete the condition is associated with extroversion of the bladder. In less extensive cases, where there is no extroversion, there is an infundibuliform opening into the bladder. The penis is usually dwarfed and turned upward, so that the glans lies over the opening. Congenital stricture is also occasionally met with, and in such cases multiple strictures may be present throughout the whole length of the cavernous portion.

The Female Urethra

The **female urethra** (*urethra muliebris*) (Fig. 17–31) is a narrow membranous canal, about 4 cm long, extending from the bladder to the external orifice in the vestibule. It is placed dorsal to the symphysis pubis, embedded in the anterior wall of the vagina, and its direction is obliquely downward and forward; it is slightly curved with the concavity directed forward. Its diameter when undilated is about 6 mm. It perforates the fasciae of the urogenital diaphragm, and its external orifice is situated directly ventral to the vaginal opening and about 2.5 cm dorsal to the glans clitoridis. The lining membrane is thrown into longitudinal folds, one of which, placed along the floor of the canal, is termed the **urethral crest.** Many small urethral glands open into the urethra. The largest of these are the paraurethral glands (*of Skene*) the ducts of which open just within the urethral orifice.

Structure.—The female urethra consists of three coats: **muscular, erectile,** and **mucous.**

The **muscular coat** is continuous with that of the bladder; it extends the whole length of the tube, and consists of circular fibers. In addition to this, between the superior and inferior fasciae of the urogenital diaphragm, the female urethra is surrounded by the Sphincter urethrae, as in the male.

A **thin layer of spongy erectile tissue,** containing a plexus of large veins, intermixed with bundles of unstriped muscular fibers, lies immediately beneath the mucous coat.

The **mucous coat** is continuous externally with that of the vulva, and internally with that of the bladder. It is lined by stratified squamous epithelium, which becomes transitional near the bladder. Its external orifice is surrounded by a few mucous follicles.

Vessels and Nerves.—The **arteries** are derived from the inferior vesical and internal pudendal arteries. The **veins** drain into the vesical and vaginal veins. The **lymphatics** are described on page 762. The **nerves** are from the pelvic plexus and pudendal nerves.

THE MALE GENITAL ORGANS

The **male genitals** (*organa genitalia virilia*) include the **testes,** the **ductus deferentes,** the **vesiculae seminales,** the **ejaculatory ducts,** and the **penis,** together with the following accessory structures, viz., the **prostate** and the **bulbourethral glands.**

The Testes and Their Coverings
(Figs. 17–36, 17–37)

The **testes** are two parenchymatous organs which produce the semen; they are suspended in the scrotum by the spermatic cords. At an early period of fetal life the testes are contained in the abdominal cavity, behind the peritoneum. Before birth they descend to the inguinal canal, along which they pass and, emerging at the superficial inguinal ring, descend into the scrotum, becoming invested in their course by coverings derived from the serous, muscular, and fibrous layers of the abdominal parietes, as well as by the scrotum.

The **coverings of the testes** are:

> Skin   ⎫
> Dartos tunic ⎬ Scrotum
>   ⎭
> External spermatic fascia
> Cremasteric layer
> Internal spermatic fascia
> Tunica vaginalis

The **Scrotum** is a cutaneous pouch containing the testes and parts of the spermatic cords. It is divided on its surface into two

lateral portions by a ridge or **raphe,** which is continued ventralward to the under surface of the penis, and dorsalward, along the middle line of the perineum to the anus. The external aspect of the scrotum varies under different circumstances: thus, under the influence of warmth, and in old and debilitated persons, it becomes elongated and flaccid; but, under the influence of cold, and in the young and robust, it is short, corrugated, and closely applied to the testes. Of the two lateral portions the left hangs lower than the right, to correspond with the greater length of the left spermatic cord.

The **scrotum** consists of two layers, the **skin** or **integument** and the **dartos tunic.**

The **integument** is very thin, of a brownish color, and generally thrown into folds or rugae. It is provided with sebaceous follicles, the secretion of which has a characteristic odor, and is beset with thinly scattered, crisp kinky hairs, the roots of which are visible through the skin.

The **dartos tunic** (*tunica dartos*) contains a thin layer of non-striped muscular fibers, continuous, around the base of the scrotum, with the two layers of the superficial fascia of the groin and the perineum; it sends inward a septum, which divides the scrotal pouch into two cavities for the testes, and extends between the raphe and the under surface of the penis, as far as its root.

The dartos tunic is closely united to the skin externally, but is separated from the subjacent parts by a distinct fascial cleft,

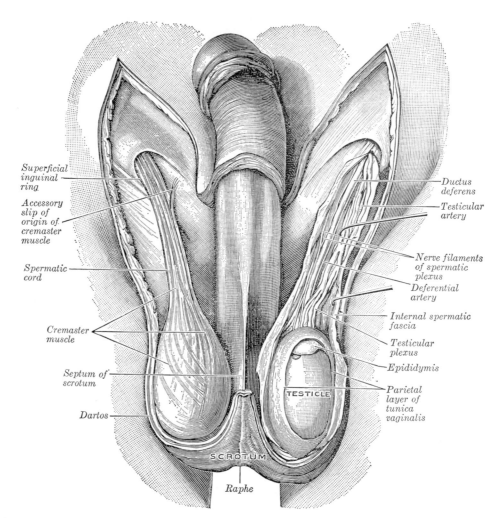

Superficial inguinal ring

Accessory slip of origin of cremaster muscle

Spermatic cord

Cremaster muscle

Septum of scrotum

Dartos

Ductus deferens

Testicular artery

Nerve filaments of spermatic plexus

Deferential artery

Internal spermatic fascia

Testicular plexus

Epididymis

Parietal layer of tunica vaginalis

TESTICLE

SCROTUM

Raphe

FIG. 17–36.—The scrotum. The penis has been turned upward and the anterior wall of the scrotum has been removed. On the right side, the spermatic cord, the internal spermatic fascia, and the Cremaster muscles are displayed; on the left side the internal spermatic fascia has been divided by a longitudinal incision passing along the front of the cord and the testicle, and a portion of the parietal layer of the tunica vaginalis has been removed to display the testicle, and a portion of the head of the epididymis covered by the visceral layer of the tunica vaginalis. (Toldt.)

upon which it glides with the greatest facility. It contains no fat and is highly vascular.

The **external spermatic fascia** (*intercrural or intercolumnar fascia*) is a thin membrane prolonged distalward over the cord and testis. It is continuous, at the superficial inguinal ring, with the deep fascia covering the aponeurosis of the Obliquus externus abdominis (*fascia innominata of Gallaudet*) and is therefore part of the external investing fascia of the body. It is separated from the enclosing dartos by a fascial cleft.

The **cremasteric layer** consists of the scattered bundles of the Cremaster connected into a continuous membrane by the cremasteric fascia. It forms the *middle spermatic layer* and corresponds to the Obliquus internus abdominis and its fasciae (page 423).

The **internal spermatic fascia** (*infundibuliform fascia; tunica vaginalis communis*) is a thin membrane, often difficult to separate from the preceding, but more easily separated from the cord and testis which it encloses. It is continuous at the deep

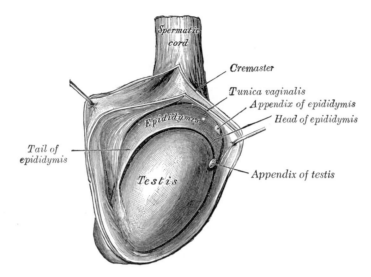

FIG. 17–37.—The right testis, exposed by laying open the tunica vaginalis.

inguinal ring with the transversalis fascia (page 429).

The **tunica vaginalis** is described with the testes.

Vessels and Nerves.—The **arteries** supplying the coverings of the testes are: the superficial and deep external pudendal branches of the femoral, the superficial perineal branch of the internal pudendal, and the cremasteric branch from the inferior epigastric. The **veins** follow the course of the corresponding arteries. The **lymphatics** end in the inguinal lymph nodes. The **nerves** are the ilioinguinal and genitofemoral branches of the lumbar plexus, the two superficial perineal branches of the pudendal nerve, and the pudendal branch of the posterior femoral cutaneous nerve.

The **inguinal canal** (*canalis inguinalis*) is described on page 430.

The **spermatic cord** (*funiculus spermaticus*) (Fig. 17–38) extends from the deep inguinal ring, where the structures of which it is composed converge, to the testis. In the abdominal wall the cord passes obliquely along the inguinal canal, lying at first inferior to the Obliquus internus, and superior to the fascia transversalis, but nearer the pubis, it rests upon the inguinal and lacunar ligaments, having the aponeurosis of the Obliquus externus ventral to it, and the inguinal falx dorsal to it. It then escapes at the ring, and descends nearly

vertically into the scrotum. The left cord is rather longer than the right, consequently the left testis hangs somewhat lower than the right.

Structure of the Spermatic Cord.—The spermatic cord is composed of arteries, veins, lymphatics, nerves, and the excretory duct of the testis. These structures are connected together by the **innermost spermatic fascia** which is continuous with the subserous fascia of the abdomen at the deep inguinal ring, and are invested by the layers brought down by the testis in its descent.

The **arteries of the cord** are: the testicular and external spermatics, and the artery to the ductus deferens.

The *testicular artery,* a branch of the abdominal aorta, escapes from the abdomen at the deep inguinal ring, and accompanies the other constituents of the spermatic cord along the inguinal canal and through the superficial inguinal ring into the scrotum. It then descends to the testis, and, becoming tortuous, divides into several branches, two or three of which accompany the ductus deferens and supply the epididymis, anastomosing with the artery of the ductus deferens: the others supply the substance of the testis.

The *external spermatic artery* is a branch of the inferior epigastric artery. It accompanies the spermatic cord and supplies the coverings of the cord, anastomosing with the testicular artery.

The *artery of the ductus deferens,* a branch of the superior vesical, is a long, slender vessel,

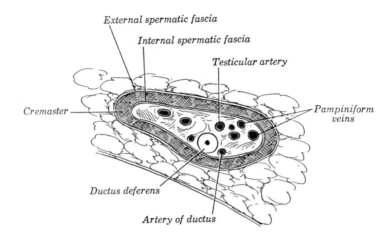

External spermatic fascia

Internal spermatic fascia

Testicular artery

Cremaster

Pampiniform veins

Ductus deferens

Artery of ductus

FIG. 17–38.—Section across the spermatic cord as it lies in the scrotum canal.

which accompanies the ductus deferens, ramifying upon its coats, and anastomosing with the testicular artery near the testis.

The **testicular veins** (Fig. 9–32) emerge from the back of the testis, and receive tributaries from the epididymis: they unite and form a convoluted plexus, the **plexus pampiniformis,** which forms the chief mass of the cord; the vessels composing this plexus are very numerous, and ascend along the cord in front of the ductus deferens; below the superficial inguinal ring they unite to form three or four veins, which pass along the inguinal canal, and, entering the abdomen through the abdominal inguinal ring, coalesce to form two veins. These again unite to form a single vein, which opens on the right side into the inferior vena cava, at an acute angle, and on the left side into the left renal vein, at a right angle.

The **lymphatic vessels** are described on page 763.

The **nerves** are the spermatic plexus from the sympathetic, joined by filaments from the pelvic plexus which accompany the artery of the ductus deferens.

The scrotum forms an admirable covering for the protection of the testes. These bodies, lying suspended and loose in the cavity of the scrotum and surrounded by serous membrane, are capable of great mobility, and can therefore easily slip about within the scrotum and thus avoid injuries from blows or squeezes. The skin of the scrotum is very elastic and capable of great distention, and on account of the looseness and amount of subcutaneous tissue, the scrotum becomes greatly enlarged in cases of edema, to which this part is especially liable as a result of its dependent position.

The **Testes** are of an oval form (Fig. 17–37), compressed laterally and are suspended in the scrotum by the spermatic cords. They average 4 to 5 cm in length, 2.5 cm in breadth, and 3 cm in the anteroposterior diameter. The weight of one gland varies from 10.5 to 14 gm. It has an oblique position in the scrotum; the cranial extremity is directed ventralward and a little lateralward; the caudal, dorsalward and a little medialward; the anterior convex border looks ventralward and caudalward, the posterior or straight border, to which the cord is attached, dorsalward and cranialward.

The anterior border and lateral surfaces, as well as both extremities of the organ, are convex, free, smooth, and invested by the visceral layer of the tunica vaginalis. The posterior border, to which the cord is attached, receives only a partial investment from that membrane. Lying upon the lateral edge of this posterior border is a long, narrow, flattened body, named the **epididymis.**

The **epididymis** consists of a central portion or **body;** an upper enlarged extremity, the **head** (*globus major*); and a lower pointed extremity, the **tail** (*globus minor*), which is continuous with the ductus deferens, the **duct of the testis.** The head is intimately connected with the upper end of the testis by means of the efferent ductules of the gland; the tail is connected with the lower end by cellular tissue, and a reflection of the tunica vaginalis. The lateral surface, head and tail of the epididymis are free and covered by the serous membrane; the body is also completely invested by it, except along its posterior border, while between the body and the testis is a pouch, named the **sinus of the epididymis** (*digital fossa*). The epididymis is connected to the back of the testis by a fold of the serous membrane.

Appendages of the Testis and Epididymis.—On the proximal extremity of the testis, just beneath the head of the epididymis, is a minute oval, sessile body, the **appendix of the testis** (*hydatid of Morgagni*); it is the remnant of the upper end of the Müllerian duct. On the head of the epididymis is a second small stalked appendage (sometimes duplicated); it is named the **appendix of the epididymis** (*pedunculated hydatid*), and is usually regarded as a detached efferent duct.

The testis is invested by three tunics: the **tunica vaginalis, tunica albuginea,** and **tunica vasculosa.**

The **Tunica Vaginalis** (*tunica vaginalis propria testis*) is the serous covering of the testis. It is a pouch of serous membrane, derived from the saccus vaginalis of the peritoneum, which in the fetus preceded the descent of the testis from the abdomen into the scrotum. After its descent, that portion of the pouch which extends from the deep inguinal ring to near the upper part of the gland becomes obliterated; the lower portion remains as a closed sac, which invests the testis, and may be described as consisting of a **visceral** and a **parietal** lamina (Fig. 17–37).

The **visceral lamina** (*lamina visceralis*) covers the greater part of the testis and epididymis, connecting the latter to the testis by means of a distinct fold. From the posterior border of the gland it is reflected on to the internal surface of the scrotal coverings.

The **parietal lamina** (*lamina parietalis*) is more extensive than the visceral, extending cranialward for some distance ventral and medial to the cord, and reaching below the testis. The inner surface of the tunica vaginalis is smooth, and covered by a layer of mesothelial cells. The interval between the visceral and parietal laminae constitutes the cavity of the tunica vaginalis.

The obliterated portion of the saccus vaginalis may generally be seen as a fibrous thread lying in the loose areolar tissue around the spermatic cord; sometimes this may be traced as a distinct band from the upper end of the inguinal canal, where it is connected with the peritoneum, down to the tunica vaginalis; sometimes it gradually becomes lost on the spermatic cord. Occasionally no trace of it can be detected. In some cases it happens that the pouch of peritoneum does not become obliterated, but the sac of the peritoneum communicates with the tunica vaginalis. This may give rise to one of the varieties of oblique inguinal hernia (page 1307). In other cases the pouch may contract, but not become entirely obliterated; it then forms a minute canal leading from the peritoneum to the tunica vaginalis.

The **Tunica Albuginea** is the fibrous covering of the testis. It is a dense membrane, of a bluish-white color, composed of bundles of white fibrous tissue which interlace in every direction. It is covered by the tunica vaginalis, except at the points of attachment of the epididymis to the testis, and along its posterior border, where the testicular vessels enter the gland. It is applied to the tunica vasculosa over the glandular substance of the testis, and, at its posterior border, is reflected into the interior of the gland, forming an incomplete vertical septum, called the **mediastinum testis** (*corpus Highmori*).

The **mediastinum testis** extends from the upper to near the lower extremity of the gland, and is wider above than below. From its front and sides numerous imperfect septa (*trabeculae*) are given off, which radiate toward the surface of the organ, and are attached to the tunica albuginea. They divide the interior of the organ into a number of incomplete lobules which are somewhat cone-shaped, being broad at their bases at the surface of the gland, and becoming narrower as they converge to the mediastinum. The mediastinum supports the vessels and ducts of the testis in their passage to and from the substance of the gland.

The **Tunica Vasculosa** is the vascular layer of the testis, consisting of a plexus of blood vessels, held together by delicate areolar tissue. It clothes the inner surface of the tunica albuginea and the different septa in the interior of the gland and therefore forms an internal investment to all the spaces of which the gland is composed.

Structure.—The glandular structure of the testis consists of numerous lobules. Their number, in a single testis, is estimated by Berres at 250, and by Krause at 400. They differ in size according to their position, those in the middle of the gland being larger and longer. The lobules (Fig. 17–39) are conical in shape, the base being directed toward the circumference of the organ, the apex toward the mediastinum. Each lobule is contained in one of the intervals between the fibrous septa which extend between the mediastinum testis and the tunica albuginea, and consists of from one to three, or more, minute convoluted tubes, the **tubuli seminiferi.** The tubules may be separately unraveled by careful dissection under water, and may be seen to commence either by free cecal ends or by anastomotic loops (Fig. 17–40). They are supported by loose connective tissue which contains here and there groups of "interstitial cells" containing yellow pigment granules. The total number of tubules is estimated by Lauth at 840, and the average length of each is 70 to 80 cm. Their diameter varies from 0.12 to 0.3 mm. The tubules are pale in color in early life, but in old age they acquire a deep yellow tinge from containing much fatty matter. Each tubule consists of a basement layer formed of laminated connective tissue containing numerous elastic fibers with flattened cells between the layers and covered externally by a layer of flattened epithelioid cells.

Within the basement membrane are epithelial cells arranged in several irregular layers, which are not always clearly separated, but which may be arranged in three different groups. Among these cells may be seen the **spermatozoa** in different stages of development.

(1) Lining the basement membrane and forming the outer zone is a layer of cubical cells, with small nuclei; some of these enlarge to become **spermatogonia.** The nuclei of some of the spermatogonia may be in process of mitotic division (*karyokinesis*, page 15), and, in consequence of this, daughter cells are formed, which constitute the second zone (Fig. 17–41).

(2) Within this layer, a number of larger polyhedral cells with clear nuclei are arranged

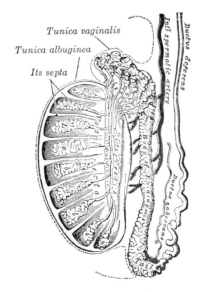

FIG. 17–39.—Vertical section of the testis, to show the arrangement of the ducts.

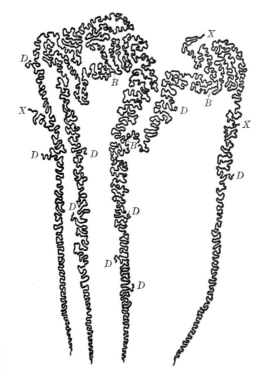

FIG. 17–40.—Four seminiferous tubules from human testis showing anastomosing loops. *B,* branching or fork; *D,* diverticulum; *X,* broken end. (Johnson, Anat. Rec., 1934; courtesy of Wistar Institute.)

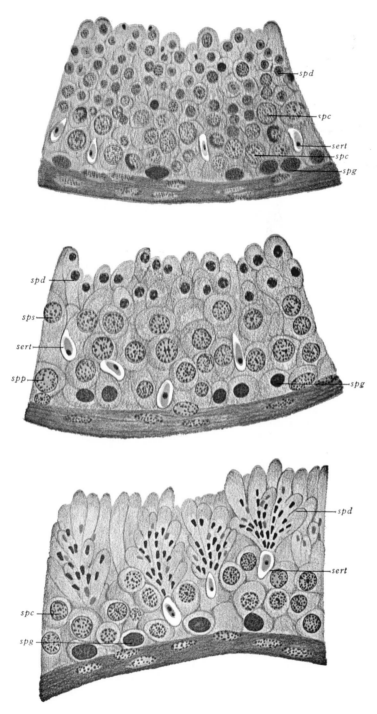

Fig. 17–41.—Sections from seminiferous tubules from an adult human male, showing three stages of development. *1.* Spermatogonia and spermatids predominate. *2.* Spermatocytes predominate. *3.* Maturing spermatozoa or spermatoblast predominate. *sert.*, supporting cells (*Sertoli cells*); *spc,* spermatocytes; *spd,* spermatids; *spg,* spermatogonia; *spp,* primary spermatocytes; *sps,* secondary spermatocytes. × 400. (Redrawn from Sobotta.)

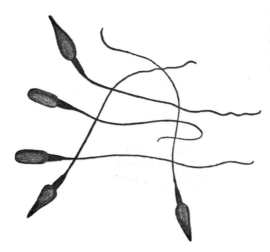

Fig. 17–42.—Mature spermatozoa from an adult human male. × 1000. (Redrawn from Sobotta.)

in two or three layers; these are the **intermediate cells** or **spermatocytes**. Most of these cells are in a condition of karyokinetic division, and the cells which result from this division form those of the next layer, the **spermatids**.

(3) The third layer of cells consists of the spermatids, and each of these, without further subdivision, becomes a **spermatozoön**. The spermatids are small polyhedral cells, the nucleus of each of which contains half the usual number of chromosomes. In addition to these three layers, other cells, termed the **supporting cells** (*cells of Sertoli*), are seen. They are elongated and project inward from the basement membrane toward the lumen of the tube. As development of the spermatozoa proceeds the latter group themselves around the inner extremities of the supporting cells. The nuclear portion of the spermatid, which is partly embedded in the supporting cell, is differentiated to form the head of the spermatozöon, while part of the cell protoplasm forms the middle piece and the tail is produced by an outgrowth from the double centriole of the cell. Ultimately the heads are liberated and the spermatozoa are set free (Fig. 17–42). The structure of the spermatozoön is described on page 15.

In the apices of the lobules, the tubules become less convoluted, assume a nearly straight course, and unite together to form from twenty to thirty larger ducts, of about 0.5 mm in diameter, and these, from their straight course, are called **tubuli recti** (Fig. 17–39).

The **tubuli recti** enter the fibrous tissue of the mediastinum, and pass upward and backward, forming, in their ascent, a close network of anastomosing tubes which are merely channels in the fibrous stroma, lined by flattened epithelium, and having no proper walls; this constitutes the **rete testis**. At the upper end of the mediastinum, the vessels of the rete testis terminate in from twelve to fifteen or twenty ducts, the **ductuli efferentes;** they perforate the tunica albuginea, and carry the seminal fluid from the testis to the epididymis. Their course is at first straight; they then become enlarged, and exceedingly convoluted, and form a series of conical masses, the **coni vasculosi**, which together constitute the head of the epididymis. Each cone consists of a single convoluted duct, from 15 to 20 cm in length, the diameter of which gradually decreases from the testis to the epididymis. Opposite the bases of the cones the efferent vessels open at narrow intervals into a single duct, which constitutes, by its complex convolutions, the body and tail of the epididymis. When the convolutions of this tube are unraveled, it measures upward of 6 meters in length; it increases in diameter and thickness as it approaches the ductus deferens. The convolutions are held together by fine areolar tissue, and by bands of fibrous tissue.

The tubuli recti have very thin walls; like the channels of the rete testis they are lined by a single layer of flattened epithelium. The ductuli efferentes and the tube of the epididymis have walls of considerable thickness, on account of the presence in them of muscular tissue, which is principally arranged in a circular manner. These tubes are lined by columnar ciliated epithelium.

Vessels and **Nerves.**—The **arteries** to the testis are described on page 640, the **veins** on page 710, the **lymphatics** on page 763, and the **nerves** on page 1031.

Variations.—The testis, developed in the lumbar region, may be arrested or delayed in its transit to the scrotum (*cryptorchism*). It may be retained in the abdomen; or it may be arrested at the deep inguinal ring, or in the inguinal canal; or it may just pass out of the superficial inguinal ring without finding its way to the bottom of the scrotum. When retained in the abdomen it gives rise to no symptoms other than the absence of the testis from the scrotum, but when it is retained in the inguinal canal it is subjected to pressure and may become inflamed and painful. The retained testis is probably functionally useless, so that a man in whom both testes are retained (*anorchism*) is sterile, though he may not be impotent. The absence of one testis is termed *monorchism*. When a testis is retained in the inguinal canal it is often complicated with a congenital hernia, the funicular process of the peritoneum not being obliterated.

In addition to the cases above described, where there is some arrest in the descent of the testis, this organ may descend through the inguinal canal, but may miss the scrotum and assume some abnormal position. The most common form is where the testis, emerging at the subcutaneous inguinal ring, slips down between the scrotum and thigh and comes to rest in the perineum. This is known as *perineal ectopia testis*. With each variety of abnormality in the position of the testis, it is very common to find concurrently a congenital hernia, or, if a hernia be not actually present, the funicular process is usually patent, and almost invariably so if the testis is in the inguinal canal.

The testis, finally reaching the scrotum, may occupy an abnormal position in it. It may be inverted, so that its posterior or attached border is directed forward and the tunica vaginalis is situated behind.

Fluid collections of a serous character are very frequently found in the scrotum. To these the term *hydrocele* is applied. The most common form is the ordinary *vaginal hydrocele*, in which the fluid is contained in the sac of the tunica vaginalis, which is separated, in its normal condition, from the peritoneal cavity by the whole extent of the inguinal canal. In another form, the *congenital hydrocele*, the fluid is in the sac of the tunica vaginalis, but this cavity communicates with the general peritoneal cavity, its tubular process remaining pervious. A third variety, known as an *infantile hydrocele*, occurs in those cases where the tubular process becomes obliterated only at its upper part, at or near the deep inguinal ring. It resembles the vaginal hydrocele except in shape, the collection of fluid extending up the cord into the inguinal canal. Fourthly, the funicular process may become obliterated both at the abdominal inguinal ring and above the epididymis, leaving a central unobliterated por-

tion, which may become distended with fluid, giving rise to a condition known as the *encysted hydrocele of the cord.*

Congenital Hernia.—Some varieties of oblique inguinal hernia (Fig. 17–43) depend upon congenital defects in the saccus vaginalis, the pouch of peritoneum which precedes the descent of the testis.

Normally this pouch is closed before birth, closure commencing at two points, viz., at the abdominal inguinal ring and at the top of the epididymis, and gradually extending until the whole of the intervening portion is converted into a fibrous cord. From failure in the completion of this process, variations in the relation of the hernial protrusion to the testis and tunica vaginalis are produced; these constitute distinct varieties of inguinal hernia, viz., the hernia of the funicular process and the complete congenital variety. Mitchell states that of 40 stillborn infants examined, 7 showed complete obliteration of the saccus vaginalis on both sides, 12 showed sacs completely patent on both sides, and 21 exhibited all intermediate degrees of obliteration.

Where the saccus vaginalis remains patent throughout, the cavity of the tunica vaginalis communicates directly with that of the peritoneum. The intestine descends along this pouch into the cavity of the tunica vaginalis which constitutes the sac of the hernia, and the gut lies in contact with the testis. Though this form of hernia is termed *complete congenital*, the term does not imply that the hernia existed at birth, but merely that a condition is present which may allow the descent of the hernia at any moment. As a matter of fact, congenital herniae frequently do not appear until adult life. Where the processus vaginalis is occluded at the lower point only, *i.e.*, just above the testis, the intestine descends into the pouch of peritoneum as far as the testis, but is prevented from entering the sac of the tunica vaginalis by the septum which has formed

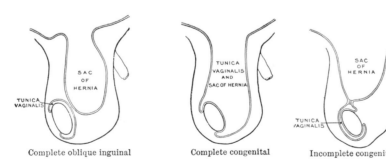

Complete oblique inguinal   Complete congenital   Incomplete congenital

Fig. 17–43.—Varieties of oblique inguinal hernia. See also Figure 17–6.

between it and the pouch. This is known as *hernia into the funicular process* or *incomplete congenital hernia;* it differs from the former in that instead of enveloping the testis it lies above it.

The Ductus Deferens

The **ductus deferens** (*vas deferens; seminal duct*), the excretory duct of the testis, is the continuation of the canal of the epididymis. Commencing at the lower part of the tail of the epididymis it is at first very tortuous, but gradually becoming less twisted it ascends along the posterior border of the testis and medial side of the epididymis, and traverses the inguinal canal to the abdominal inguinal ring as a constituent of the spermatic cord (Figs. 17–38, 16–54). Here it separates from the other structures of the cord, curves around the lateral side of the inferior epigastric artery, and ascends for about 2.5 cm ventral to the external iliac artery (Fig. 8–49). It is next directed dorsalward and slightly caudalward, and, crossing the external iliac vessels obliquely, enters the pelvic cavity, where it lies between the peritoneal membrane and the lateral wall of the pelvis, and descends on the medial side of the obliterated umbilical artery and the obturator nerve and vessels. It then crosses ventral to the ureter, and, reaching the medial side of this tube, bends to form an acute angle, and runs medialward and slightly ventralward between the fundus of the bladder and the upper end of the seminal vesicle (Fig. 17–44). Reaching the medial side of the seminal vesicle, it is directed caudalward and medialward in contact with it, gradually approaching the opposite ductus. Here it lies between the fundus of the bladder and the rectum, where it is enclosed, together with the seminal vesicle, in a sheath derived from the rectovesical portion of the subserous fascia. Lastly, it is directed downward to the base of the prostate, where it becomes greatly narrowed, and is joined at an acute angle by the duct of the seminal vesicle to form the ejaculatory duct, which traverses the prostate behind its middle lobe and opens into the prostatic portion of the urethra, close to the orifice of the prostatic utricle.

The ductus deferens presents a hard and cord-like sensation to the palpating fingers; its walls are dense, and its canal is extremely small. At the fundus of the bladder it becomes enlarged and tortuous, and this portion is termed the **ampulla.** A small triangular area of the fundus of the bladder, between the ductus deferentes laterally and the peritoneum of the bottom of the rectovesical excavation, is in contact with the rectum.

Ductuli Aberrantes.—A long narrow tube, the **ductulus aberrans inferior** (*vas aberrans of Haller*), is occasionally found connected with the lower part of the canal of the epididymis, or with the commencement of the ductus deferens. Its length varies from 3.5 to 35 cm, and it may become dilated toward its extremity; more commonly it retains the same diameter throughout. Its structure is similar to that of the ductus deferens. Occasionally it is found unconnected with the epididymis. A second tube, the **ductulus aberrans superior,** occurs in the head of the epididymis; it is connected with the rete testis.

Paradidymis (*organ of Giraldés*).—This term is applied to a small collection of convoluted tubules, situated in front of the lower part of the cord above the head of the epididymis. These tubes are lined with columnar ciliated epithelium, and probably represent the remains of a part of the Wolffian body.

Structure.—The ductus deferens consists of three coats: (1) an **external** or **areolar coat;** (2) a **muscular coat** which in the greater part of the tube consists of two layers of unstriped muscular fiber: an outer, longitudinal in direction, and an inner, circular; but in addition to these, at the commencement of the ductus, there is a third layer, consisting of longitudinal fibers, placed internal to the circular stratum, between it and the mucous membrane; (3) an **internal** or **mucous coat,** which is pale, and arranged in longitudinal folds. The mucous coat is lined by columnar epithelium which is non-ciliated throughout the greater part of the tube; a variable portion of the testicular end of the tube is lined by two strata of columnar cells and the cells of the superficial layer are ciliated.

Vessels and Nerves.—The **artery of the ductus deferens,** a branch of the superior vesical artery, is a long slender vessel which accompanies the ductus, ramifying upon its coats and anastomosing with the testicular artery near the testis. Its ampulla is supplied from the middle and inferior vesical and middle rectal arteries. The veins drain into the pampiniform plexus, the vesical veins, and prostatic plexus. The **lymphatics** are described on page 763. The **nerves** are from the pelvic plexus accompanying the artery.

The Seminal Vesicles

The **seminal vesicles** (*vesiculae seminales*) (Fig. 17–44) are two lobulated membranous pouches, placed between the fundus of the bladder and the rectum, which secrete a fluid to be added to the secretion of the testes. Carefully executed necropsies have shown that in man the seminal vesicles do not store sperm, as the name implies. They are usually about 7.5 cm long, but vary in size, not only in different individuals, but also in the same individual on the two sides. The **ventral surface** is in contact with the fundus of the bladder, extending from near the termination of the ureter to the base of the prostate. The **dorsal surface** rests upon the rectum, from which it is separated by the rectovesical fascia. The **upper extremities** of the two vesicles diverge from each other, and are in relation with the ductus deferentes and the terminations of the ureters, and are partly covered by peritoneum. The **lower extremities** are pointed,

and converge toward the base of the prostate. The ampullae of the ductus deferentes lie along their medial margin.

Each vesicle consists of a single tube, coiled upon itself, and giving off several irregular blind diverticula; the separate coils, as well as the diverticula, are bound together by fibrous tissue. When uncoiled, the tube is about the diameter of a quill (5 mm) and varies in length from 10 to 15 cm; it ends superiorly in a cul-de-sac; its inferior extremity becomes constricted into a narrow straight duct, which joins with the corresponding ductus deferens to form the ejaculatory duct.

Structure.—The vesiculae seminales are composed of three coats: an **external** or **areolar coat;** a **middle** or **muscular coat,** thinner than in the ductus deferens and arranged in two layers, an outer longitudinal and inner circular; an **internal** or **mucous coat,** which is pale, of a whitish brown color, and presents a delicate reticular structure. The epithelium is columnar, and in the diverticula goblet cells are present, the secretion of which increases the bulk of the seminal fluid.

Vessels and Nerves.—The **arteries** supplying the vesiculae seminales are derived from the middle and inferior vesical and middle rectal. The **veins** and **lymphatics** accompany the

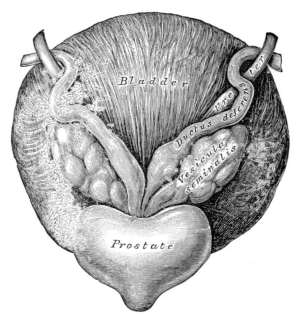

Fig. 17–44.—Fundus of the bladder with the vesiculae seminales.

arteries. The **nerves** are derived from the pelvic plexuses.

The Ejaculatory Duct

The **ejaculatory duct** (*ductus ejaculatorius*) (Figs. 17–27, 17–45), one on either side of the middle line, is formed by the

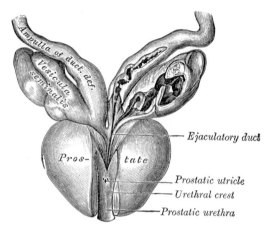

FIG. 17–45.—Vesiculae seminales and ampullae of ductus deferentes, seen from the front. The anterior walls of the left ampulla, left seminal vesicle, and prostatic urethra have been cut away.

union of the duct from the vesicula seminalis with the ductus deferens, and is about 2 cm long. It commences at the base of the prostate, and runs ventralward and caudalward between its middle and lateral lobes, and along the sides of the prostatic utricle, to end in a slit-like orifice close to or just within the margins of the utricle. The ducts diminish in size, and also converge, toward their terminations.

Structure.—The coats of the ejaculatory ducts are extremely thin. They are: an **outer fibrous layer,** which is almost entirely lost after the entrance of the ducts into the prostate; a **layer of muscular fibers** consisting of a thin outer circular, and an inner longitudinal, layer; and **mucous membrane.**

The Penis

The **penis** is attached to the front and sides of the pubic arch. In the flaccid condition it is cylindrical in shape, but when erect assumes the form of a triangular prism with rounded angles, one side of the prism

forming the dorsum. It is composed of three cylindrical masses of cavernous tissue bound together by fibrous tissue and covered with skin. Two of the masses are lateral, and are known as the **corpora cavernosa penis;** the third is median, and is termed the **corpus spongiosum penis** which contains the greater part of the urethra (Figs. 17–47, 17–48).

The integument covering the penis is remarkable for its thinness, its dark color, its looseness of connection with the deeper parts of the organ, and its absence of adipose tissue. At the root of the penis it is continuous with that over the pubes, scrotum, and perineum. At the neck it leaves the surface and becomes folded upon itself to form the **prepuce** or foreskin. The internal layer of the prepuce is directly continuous, along the line of the neck, with the integument over the glans. Immediately behind the external urethral orifice it forms a small secondary reduplication, attached along the bottom of a depressed median raphe, which extends from the meatus to the neck; this fold is termed the **frenulum** of the prepuce. The integument covering the glans is continuous with the urethral mucous membrane at the orifice; it is devoid of hairs, but projecting from its free surface are a number of small, sensitive papillae. Scattered glands are present on the neck of the penis and inner layer of the prepuce, the **preputial glands** (Tyson). They secrete a sebaceous material of very peculiar odor, which readily undergoes decomposition; when mixed with discarded epithelial cells it is called **smegma.** The prepuce covers a variable amount of the glans, and is separated from it by a potential space—the **preputial space**—which presents two shallow fossae, one on either side of the frenulum.

Fascia.—The **subcutaneous fascia of the penis** is directly continuous with that of the scrotum and, like it, contains a **dartos** tunic with its layer of scattered smooth muscle cells. It is not divisible into a superficial and deep layer and contains no adipose tissue. A fascial cleft between the superficial and deep fasciae gives the skin great movability.

The **deep fascia of the penis** (*Buck's fascia*) (Fig. 17–46) forms a tubular investment for the shaft of the penis as far distally

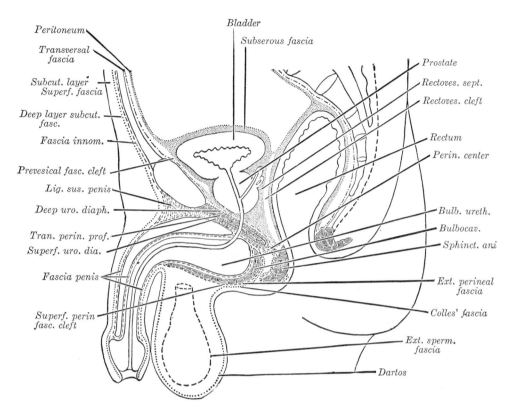

Peritoneum

Transversal
fascia

Subcut. layer
Superf. fascia

Deep layer subcut.
fasc.

Fascia innom.

Prevesical fasc. cleft

Lig. sus. penis

Deep uro. diaph.

Tran. perin. prof.

Superf. uro. dia.

Fascia penis

Superf. perin
fasc. cleft

Bladder

Subserous fascia

Prostate

Rectoves. sept.

Rectoves. cleft

Rectum

Perin. center

Bulb. ureth.

Bulbocav.

Sphinct. ani

Ext. perineal
fascia

Colles' fascia

Ext. sperm.
fascia

Dartos

Fig. 17–46.—Fasciae of pelvis and perineum in median sagittal section. Diagram.

as the corona glandis. Proximally, it invests the crura and bulb and is firmly attached with them to the ischiopubic rami and superficial layer of the urogenital diaphragm. At the anterior or distal extremities of the Bulbocavernosus and Ischiocavernosi, it splits into a superficial and deep lamina; the superficial lamina covers the superficial surface of these muscles as the external perineal fascia of the perineum (page 439); the deep lamina is the continuation of the proper deep fascia of the penis (Buck's fascia). A septum of fascia extends inward between the corpora cavernosa and corpus spongiosum penis providing separate tubular investments for these columns of erectile tissue.

Clinical Considerations.—The deep fascia of the penis (Buck's fascia) encloses the organ in a strong capsule. An abscess, a hematoma, or an extravasation of urine from rupture of the penile urethra would be confined to the penis by this envelope. A rupture of the urethra in its membranous portion, however, would allow urine to enter the fascial cleft

which is between the deep and superficial fasciae and which is continuous with the cleft in the scrotum, under Colles' fascia, and under Scarpa's fascia (see pages 436, 441).

Corpora Cavernosa Penis (Fig. 17–47).— The distal three-fourths of these two cylindrical masses of erectile tissue are intimately bound together and make up the greater part of the shaft of the penis. At the pubic symphysis, however, their posterior portions diverge from each other as two gradually tapering structures called the **crura.**

The corpora retain a uniform diameter in the shaft and terminate anteriorly in a bluntly rounded extremity approximately 1 cm from the end of the penis, being embedded in a cap formed by the glans penis.

The corpora cavernosa penis are surrounded by a strong fibrous envelope consisting of superficial and deep fibers. The superficial fibers are longitudinal in direction, and form a single tube which encloses both corpora; the deep fibers are arranged circularly around each corpus, and form by

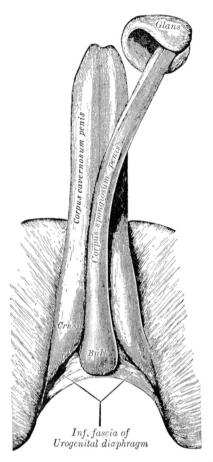

FIG. 17–47.—The constituent cavernous cylinders of the penis. The glans and anterior part of the corpus spongiosum are detached from the corpora cavernosa penis and turned to one side.

nosum penis. The crus is firmly bound to the ramus of the ischium and pubis and is enclosed by the fibers of the Ischiocavernosus muscle.

The **Corpus Spongiosum Penis** (*corpus cavernosum urethrae*) (Fig. 17–47) is the part of the penis which contains the penile urethra. Its middle portion, in the shaft of the penis, is a uniform cylinder somewhat smaller than a corpus cavernosum penis. At each end it is markedly expanded, the distal extremity forming the glans penis, the proximal the bulb of the penis. Between the expansions, it lies in the groove on the under surface of the corpora cavernosa penis.

The **glans penis** is the anterior end of the corpus spongiosum expanded into an obtuse cone very similar to the cap of a mushroom. It is molded over and securely attached to the blunt extremity of the corpora cavernosa penis, and extends farther over their dorsal than their ventral surfaces. Its periphery is larger in diameter than the shaft, projecting in a rounded border, the **corona glandis**. Proximal to the corona is a constriction forming the **retroglandular sulcus** and the **neck of the penis**. At the summit of the glans is the slit-like external orifice of the urethra.

The **bulb** (*bulbus penis*) is the conical enlargement of the proximal 4 or 5 cm of the corpus spongiosum. It is just superficial to the urogenital diaphragm, the superficial layer of which is blended with its fibrous capsule and is called the ligament of the bulb. It is enclosed by the fibers of the Bulbocavernosus.

The urethra enters the corpus spongiosum 1 or 2 cm from the posterior extremity of the bulb by piercing the dorsal surface, *i.e.*, the surface which is blended with the urogenital diaphragm. The posterior, most expanded portion of the bulb, accordingly, projects backward toward the anus beyond the entrance of the urethra.

Ligaments.—(See page 417.) The **ligamentum fundiforme penis** is an extensive thickening of the deep layer of subcutaneous fascia (Scarpa's) of the anterior abdominal wall just above the pubis where it is firmly attached to the rectus sheath. The fibrous bands extend down to the dorsum and sides of the root of the penis. The **suspensory ligament of the penis**, shorter than the

their junction in the median plane the **septum of the penis**. This is thick and complete behind, but is imperfect in front, where it consists of a series of vertical bands arranged like the teeth of a comb; it is therefore named the **septum pectiniforme**. A shallow groove which marks their junction on the upper surface lodges the deep dorsal vein of the penis, while a deeper and wider groove between them on the under surface contains the corpus spongiosum penis.

Each **crus penis**, the tapering posterior portion of a corpus cavernosum penis, terminates just ventral to the tuberosity of the ischium in a bluntly pointed process and, as it meets its fellow, it presents a slight enlargement, the bulb of the corpus caver-

above, is a strong fibrous triangle, derived from the external investing deep fascia, which attaches the dorsum of the root of the penis to the inferior end of the linea alba, the symphysis pubis, and the arcuate pubic ligament. Serving as ligaments also are the attachments of the crura to the ischiopubic rami and of the bulb to the urogenital diaphragm described above.

Muscles.—The voluntary muscles of the penis are the Bulbocavernosus, Ischiocavernosus, and the Transversus perinei superficialis and are described on page 441.

Structure of the Penis.—From the internal surface of the fibrous envelope, tunica albuginea, of the corpora cavernosa penis, as well as from the sides of the septum, numerous bands or cords are given off, which cross the interior of these corpora cavernosa in all directions, subdividing them into a number of separate compartments, and giving the entire structure a spongy appearance (Fig. 17–51). These bands and cords are called **trabeculae,** and consist of white fibrous tissue, elastic fibers,

and plain muscular fibers. In them are contained numerous arteries and nerves.

The component fibers which form the trabeculae are larger and stronger around the circumference than at the centers of the corpora cavernosa; they are also thicker proximally than distally. The interspaces or cavernous spaces (blood sinuses), on the contrary, are larger at the center than at the circumference, their long diameters being directed transversely. They contain blood, and are lined by a layer of flattened cells similar to the endothelial lining of veins.

The fibrous envelope of the corpus spongiosum is thinner, whiter in color, and more elastic than that of the corpora cavernosa penis. The trabeculae are more delicate, nearly uniform in size, and the meshes between them smaller than in the corpora cavernosa penis: their long diameters, for the most part, corresponding with that of the penis. The external envelope or outer coat of the corpus spongiosum is formed partly of unstriped muscular fibers, and a layer of the same tissue immediately surrounds the canal of the urethra. The corpus spongiosum with its expanded end, the glans

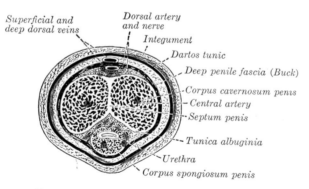

Superficial and deep dorsal veins
Dorsal artery and nerve
Integument
Dartos tunic
Deep penile fascia (Buck)
Corpus cavernosum penis
Central artery
Septum penis
Tunica albuginia
Urethra
Corpus spongiosum penis

Fig. 17–48.—Transverse section of the penis.

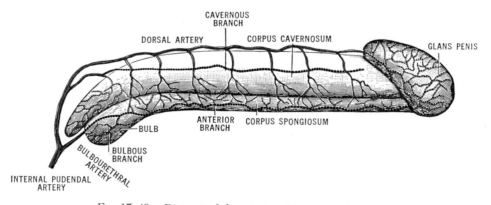

CAVERNOUS BRANCH
DORSAL ARTERY
CORPUS CAVERNOSUM
GLANS PENIS
ANTERIOR BRANCH
CORPUS SPONGIOSUM
BULB
BULBOUS BRANCH
BULBOURETHRAL ARTERY
INTERNAL PUDENDAL ARTERY

Fig. 17–49.—Diagram of the arteries of the penis. (Testut.)

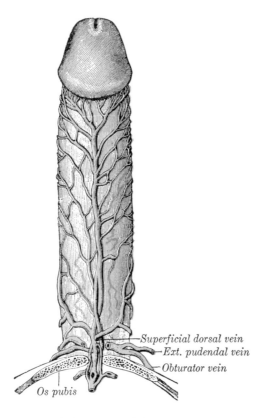

-Superficial dorsal vein
-Ext. pudendal vein
-Obturator vein

Os pubis

Fig. 17–50.—Veins of the penis. (Testut.)

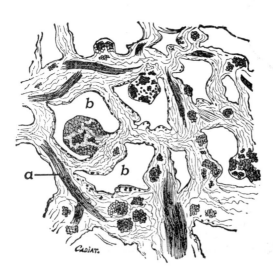

Fig. 17–51.—Section of corpus cavernosum penis in a non-distended condition. (Cadiat.) *a,* Trabeculae of connective tissue, with many elastic fibers and bundles of plain muscular tissue; *b,* Blood sinuses.

penis, may be readily dissected away from the corpora cavernosa penis.

Under cerebral or spinal stimuli the supply of arterial blood to the blood sinuses or interspaces is increased. This increase in size of sinuses in turn produces compression of the deep veins of the penis due to the elasticity of Buck's fascia. Erection is therefore due to a mechanical engorgement of the blood sinuses.

Vessels and Nerves.—Most of the blood to the penis is supplied by the internal pudendal artery, a branch of the internal iliac artery. The **arteries** (Fig. 17–49) supplying the cavernous spaces are the deep arteries of the penis and branches from the dorsal arteries of the penis, which perforate the fibrous capsule, along the upper surface, especially near the distal end of the organ. On entering the cavernous structure the arteries divide into branches, which are supported and enclosed by the trabeculae. Some of these arteries end in a capillary network, the branches of which open directly into the cavernous spaces; others assume a tendril-like appearance, and form convoluted and somewhat dilated vessels, which were named by Müller **helicine arteries.** They open into the spaces, and from them are also given off small capillary branches to supply the trabecular structure. They are bound down in the spaces by fine fibrous processes, and are most abundant in the proximal part of the corpora cavernosa (Fig. 17–48).

The blood from the cavernous spaces is returned by a series of **veins** (Fig. 17–50), some of which emerge from the base of the glans penis and converge on the dorsum of the organ to form the deep dorsal vein; others pass out on the upper surface of the corpora cavernosa and join the same vein; some emerge from the under surface of the corpora cavernosa penis and, receiving branches from the corpus spongiosum, wind around the sides of the penis to end in the deep dorsal vein; but the greater number pass out at the root of the penis and join the prostatic plexus. Batson has demonstrated that the deep dorsal vein of the penis has connection with the vertebral veins, hence it is possible for metastases from cancerous involvement of the pelvic viscera or external genitalia to make their way to the vertebrae or even to the skull and brain without going through the heart and lungs. Pyogenic organisms may be transported by the same route. (See page 709.)

The **lymphatic vessels of the penis** are described on page 764.

The **nerves** are derived from the pudendal nerve and the pelvic plexuses. On the glans and bulb some filaments of the cutaneous nerves have Pacinian bodies connected with them, and, according to Krause, many of them end in peculiar endbulbs.

The Prostate

The **prostate** (*prostata; prostate gland*) (Figs. 17–27, 17–28, 17–52) is a firm, partly glandular and partly muscular body, which is placed immediately below the internal urethral orifice and around the commencement of the urethra. It is situated in the pelvic cavity, below the caudal part of the symphysis pubis, cranial to the deep layer of the urogenital diaphragm, and ventral to the rectum, through which it may be distinctly felt, especially when enlarged. It is about the size and shape of a large chestnut (measurements below), and presents for examination a **base**, an **apex**, a **posterior**, an **anterior**, and two **lateral surfaces.**

The **base** (*basis prostatae*) is applied to the caudal surface of the bladder. The greater part of this surface is directly continuous with the bladder wall; the urethra penetrates it nearer its ventral than its dorsal border.

The **apex** (*apex prostatae*) is directed caudalward, and is in contact with the deep layer of the urogenital diaphragm.

Surfaces.—The **posterior surface** (*facies posterior; dorsal surface; rectal surface*) is flattened from side to side and slightly convex craniocaudally; it is separated from the rectum by its sheath and the important rectovesicle (*Denonvilliers'*) fascia, which corresponds in origin and fate to the processus vaginalis in the inguinal region. It is about 4 cm distant from the anus. Near its cranial border there is a depression through which the two ejaculatory ducts enter the prostate. This depression serves to divide the posterior surface into a lower larger and an upper smaller part. The cranial smaller part constitutes the **middle lobe** of the prostate and intervenes between the ejaculatory ducts and the urethra; it varies greatly in size, and in some cases is destitute of glandular tissue. The caudal larger portion sometimes presents a shallow median furrow, which imperfectly separates it into a **right** and a **left lateral lobe**: these form the main mass of the gland and are directly continuous with each other dorsal to the urethra. Cranial to the urethra they are connected by a band which is named the **isthmus**: this consists of the same tissues as the capsule and is devoid of glandular substance.

The **anterior surface** (*facies anterior; ventral surface; pubic surface*) measures about 2.5 cm craniocaudally but is narrow and convex from side to side. It is placed

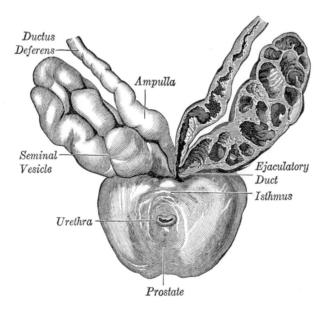

Fig. 17–52.—Prostate with seminal vesicles and seminal ducts, viewed ventrally and cranially. (Spalteholz.)

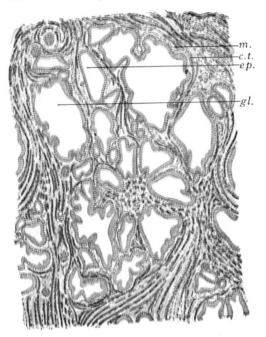

FIG. 17–53.—Section through prostate of a 22-year-old man. *c.t.*, Connective tissue stroma; *ep.*, glandular epithelium; *gl.*, lumen of gland; *m.*, smooth muscle fiber in stroma. × 50. (Sobotta.)

and the ejaculatory ducts (Fig. 17–43). The urethra usually lies along the junction of its anterior with its middle third. The ejaculatory ducts pass obliquely through the posterior part of the prostate, and open into the prostatic portion of the urethra.

Structure.—The prostate is immediately enveloped by a thin but firm fibrous capsule distinct from that derived from the subserous fascia, and separated from it by a plexus of veins. This capsule is firmly adherent to the prostate and is structurally continuous with the stroma of the gland, being composed of the same tissues: non-striped muscle and fibrous tissue. The substance of the prostate is of a pale reddish-gray color, of great density, and not easily torn. It consists of glandular substance and muscular tissue (Fig. 17–53).

The **muscular tissue,** according to Kölliker, constitutes the proper stroma of the prostate; the connective tissue being very scanty, and simply forming, between the muscular fibers, thin trabeculae in which the vessels and nerves of the gland ramify. The muscular tissue is arranged as follows: immediately beneath the fibrous capsule is a dense layer, which forms an investing sheath for the gland: secondly, around the urethra, as it lies in the prostate, is another dense layer of circular fibers, continuous above with the internal layer of the muscular coat of the bladder, and blending below with the fibers surrounding the membranous portion of the urethra. Between these two layers strong bands of muscular tissue, which decussate freely, form meshes in which the glandular structure of the organ is embedded. In that part of the gland which is situated in front of the urethra the muscular tissue is especially dense, and here there is little or no gland tissue; while in that part which is behind the urethra the muscular tissue presents a wide-meshed structure, which is densest at the base of the gland—that is, near the bladder—becoming looser and more sponge-like toward the apex of the organ.

The **glandular substance** is composed of numerous follicular pouches, the lining of which frequently shows papillary elevations. The follicles open into elongated canals, which join to form from twelve to twenty small excretory ducts. They are connected together by areolar tissue, supported by prolongations from the fibrous capsule and muscular stroma, and enclosed in a delicate capillary plexus. The epithelium which lines the canals and the terminal vesicles is of the columnar variety. The prostatic ducts open into the floor of the prostatic portion of the urethra, and are lined by two layers of epithelium, the inner layer con-

about 2 cm dorsal to the pubic symphysis, from which it is separated by a plexus of veins and a quantity of loose fat. It is connected to the pubic bone on either side by the puboprostatic ligaments. The urethra emerges from this surface a little cranial and ventral to the apex of the gland.

The **lateral surfaces** are prominent, and are covered by the ventral portions of the Levatores ani, which are, however, separated from the gland by a plexus of veins.

The prostate measures about 4 cm transversely at the base, 2 cm in its anteroposterior diameter, and 3 cm in its vertical diameter. Its weight is about 20 gm. It is held in its position by the puboprostatic ligaments; by the deep layer of the urogenital diaphragm, which invests the prostate and the commencement of the membranous portion of the urethra, and by the anterior portions of the Levatores ani, which pass dorsalward from the pubis and embrace the sides of the prostate. These portions of the Levatores ani, from the support they afford to the prostate, are named the **Levatores prostatae.**

The prostate is perforated by the urethra

sisting of columnar and the outer of small cubical cells. Small colloid masses, known as **amyloid bodies** (*corpora amylacea*), are often found in the gland tubes.

Vessels and Nerves.—The **arteries** supplying the prostate are derived from the internal pudendal, inferior vesical, and middle rectal. Its **veins** form the prostatic plexus which surrounds the sides and base of the gland. The plexus receives the dorsal vein of the penis and empties into the internal iliac veins. It has anastomotic connections with the vertebral system of veins through which metastases of carcinoma may reach the bones or even the brain. The **lymphatics** are described on page 762.

The **nerves** are derived from the pelvic plexus.

Bulbourethral Glands

The **bulbourethral glands** (*glandulae bulbourethrales; Cowper's glands*) (Fig. 17–27) are two small, rounded, and some-

what lobulated bodies, of a yellow color, about the size of peas, placed dorsal and lateral to the membranous portion of the urethra, between the two layers of the fascia of the urogenital diaphragm. They lie close to the bulb, and are enclosed by the transverse fibers of the Sphincter urethrae.

The excretory duct of each gland, nearly 2.5 cm long, passes obliquely forward beneath the mucous membrane, and opens by a minute orifice on the floor of the cavernous portion of the urethra about 2.5 cm in front of the urogenital diaphragm.

Structure.—Each gland is made up of several lobules, held together by a fibrous investment. Each lobule consists of a number of acini, lined by columnar epithelial cells, opening into one duct, which joins with the ducts of other lobules outside the gland to form the single excretory duct.

THE FEMALE GENITAL ORGANS

The **female genital organs** (*organa genitalia femininae*) consist of an internal and an external group. The **internal organs** are situated within the pelvis, and consist of the **ovaries,** the **uterine tubes,** the **uterus,** and the **vagina.** The **external organs** are superficial to the urogenital diaphragm and below the pubic arch. They comprise the **mons pubis,** the **labia majora et minora pudendi,** the **clitoris,** the **bulbus vestibuli,** and the **greater vestibular glands.**

The Ovaries

The **ovaries** (*ovaria*) are homologous with the testes in the male. They are two nodular bodies, situated one on either side of the uterus in relation to the lateral wall of the pelvis, and attached to the broad ligament of the uterus, dorsal and caudal to the uterine tubes (Fig. 17–54). The ovaries are of a grayish-pink color, and present either a smooth or a puckered uneven surface. They are each about 4 cm in length, 2 cm in width, and about 8 mm in thickness, and weigh from 2 to 3.5 gm. Each ovary presents a lateral and a medial surface, a cranial or tubal and a caudal or uterine extremity, and a ventral or meso-

varian and a dorsal free border. It lies in a shallow depression, named the **ovarian fossa,** on the lateral wall of the pelvis; this fossa is bounded by the external iliac vessels, the obliterated umbilical artery, and the ureter. The exact position of the ovary has been the subject of considerable difference of opinion, and the description given here applies to the ovary of the nulliparous woman. The ovary becomes displaced during the first pregnancy, and probably never again returns to its original position. In the erect posture the long axis of the ovary is vertical.

The *tubal extremity* is near the external iliac vein. Attached to it are the ovarian fimbriae of the uterine tube and a fold of peritoneum, the **suspensory ligament of the ovary,** which is directed cranialward over the iliac vessels and contains the ovarian vessels. The *uterine end* is directed caudalward toward the pelvic floor; it is usually narrower than the tubal, and is attached to the lateral angle of the uterus, immediately behind the uterine tube, by a rounded cord termed the **ligament of the ovary** (*ligamentum ovarii proprium*), which lies within the broad ligament and contains some nonstriped muscular fibers.

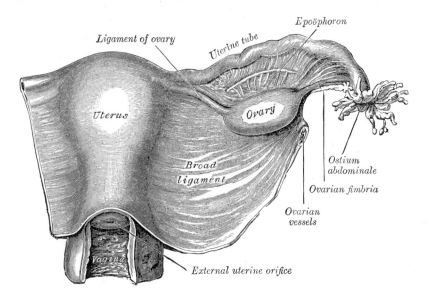

Fig. 17–54.—Uterus and right broad ligament, seen dorsally. The broad ligament has been spread out and the ovary drawn caudalward.

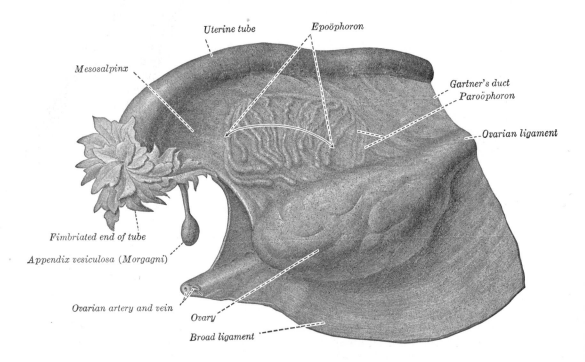

Fig. 17–55.—Broad ligaments of adult, showing remains of the Wolffian body and duct. (Modified from Farre.)

The *lateral surface* is in contact with the parietal peritoneum, which lines the ovarian fossa; the *medial surface* is to a large extent covered by the fimbriated extremity of the uterine tube. The *mesovarian border* is straight and is directed toward the obliterated umbilical artery, and is attached to the dorsal surface of the broad ligament by a short fold named the **mesovarium.** Between the two layers of this fold the blood vessels and nerves pass to the hilum of the ovary. The *free border* is convex, and is directed toward the ureter. The uterine tube arches over the ovary, running along its mesovarian border, then curving over its tubal pole, and finally passing along its free border and medial surface.

The **Epoöphoron** (*parovarium; organ of Rosenmüller*) (Fig. 17–55) lies in the mesosalpinx between the ovary and the uterine tube, and consists of a few short tubules (**ductuli transversi**) which converge toward the ovary while their opposite ends open into a rudimentary duct, the **ductus longitudinalis epoöphori** (*duct of Gartner*).

The **Paroöphoron** consists of a few scattered rudimentary tubules, best seen in the child, situated in the broad ligament between the epoöphoron and the uterus. The ductuli transversi of the epoöphoron and the tubules of the paroöphoron are remnants of the tubules of the Wolffian body or mesonephros; the ductus longitudinalis epoöphori is a persistent portion of the mesonephric or Wolffian duct. In the fetus the ovaries are situated, like the testes, in the lumbar region, near the kidneys, but they gradually descend into the pelvis (page 1271).

Structure (Fig. 17–56).—The surface of the ovary is covered by a layer of columnar cells which constitutes the **germinal epithelium** (*of Waldeyer*). This epithelium which is in linear continuity with the peritoneum gives to the ovary a dull gray color as compared with the shining smoothness of the peritoneum; and

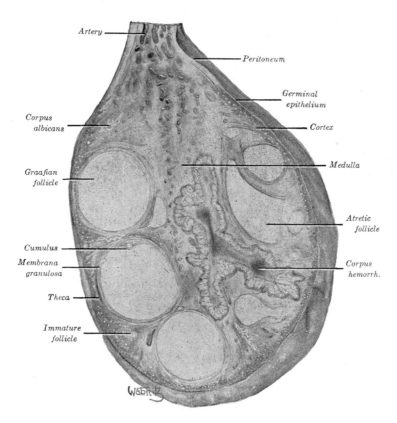

Fig. 17–56.—Section through ovary of twenty-three-year-old woman. Lightly stained with hematoxylin. × 5.

the transition between the squamous epithelium of the peritoneum and the columnar cells which cover the ovary is marked by a line near the attached border of the ovary. The ovary consists of a number of vesicular ovarian follicles embedded in the meshes of a stroma or framework.

The **stroma** is a peculiar soft tissue consisting for the most part of spindle-shaped cells with a small amount of ordinary connective tissue abundantly supplied with blood vessels. At the surface of the organ this tissue is much condensed, and forms a layer (**tunica albuginea**) composed of short connective tissue fibers, with fusiform cells between them.

Ovarian Follicles (Fig. 17–56).—The stroma immediately beneath the tunica albuginea containing a large number of minute vesicles, about 35 μ in diameter, constitutes the **cortex** of the ovary. These vesicles are the **primary follicles** (*folliculi ovarici primarii*). They contain an ovum about 20 μ in diameter surrounded by a single layer of follicular cells. When these follicles begin to mature, the ovum swells and the follicular cells become larger and proliferate by mitosis. When the follicular cells form many layers, a cavity forms at one side filled with fluid. This is the beginning of **vesicular** or **ripening follicles** (*folliculi ovarici vesiculosi*). These recede from the surface toward a highly vascular stroma in the center of the organ, termed the **medullary substance** (*zona vasculosa of Waldeyer*). This stroma forms the tissue of the hilum by which the ovary is attached, and through which the blood vessels enter; it does not contain follicles.

When the follicle is mature, it forms a clear vesicle 10 to 12 mm in diameter and is known as a **Graafian follicle**. The follicular cells are arranged in a layer 3 or 4 cells thick around the large accumulation of follicular fluid. At one side of the follicle they form a mass of cells protruding into the cavity, constituting the **cumulus oophorus** (*discus proligerus*). The ovum, now about 100 μ in diameter, is in the center of the cumulus, with the layer of follicular cells immediately surrounding ranged in a regular row known as the **corona radiata** (Fig. 2–10).

Theca folliculi (Fig. 17–57).—During the stages of growth of the follicle, the cells of the stroma around it become flattened into a sheath or theca. When the follicle is mature the theca develops an outer and inner stratum, the **theca interna** and **theca externa**.

Ovulation.—The mature follicle presses against the tunica albuginea, thinning it and bulging out on the surface of the ovary. Under the influence of hormones of the hypophysis,

at the proper time the follicle ruptures, allowing the fluid to escape into the abdominal cavity and the ovum to be swept into the fimbriated opening of the uterine tube. The cavity of the follicle collapses but contains some fluid tinged with blood which gives it the name **corpus hemorrhagicum** (Fig. 17–56). The remaining folliculur cells become greatly enlarged and take on a yellow color, forming the **corpus luteum**. The cells of the theca interna also enlarge forming *theca lutein cells* in contrast to the *follicular lutein cells*. If the ovum is fertilized, the corpus luteum continues to grow, attaining a size of 30 mm by the end of nine months. This is a **corpus luteum graviditatis.**

When the ovum is not fertilized, the follicle forms a much smaller **corpus luteum menstruationis**. After the corpus degenerates, the connective tissue forms a folded whitish scar known as a **corpus albicans** (Fig. 17–56).

The periodic changes in the ovary and uterus during the menstrual cycle are outlined on page 1329.

Atresia of Follicles.—Many of the follicles do not reach full maturity and rupture; they undergo instead a degenerative process known as atresis. In smaller follicles, the ovum disintegrates and is absorbed with the degenerating follicular cells. With large follicles, after the follicular cells degenerate, the theca cells form a layer around the follicle and are absorbed into the stroma.

Vessels and Nerves.—The **arteries** of the ovaries and uterine tubes are the ovarian from the aorta. Each anastomoses freely in the mesosalpinx, with the uterine artery, giving

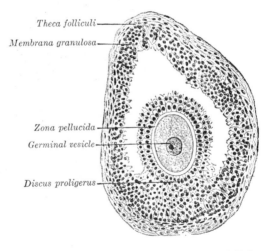

FIG. 17–57.—Section of a vesicular ovarian follicle. × 50.

some branches to the uterine tube, and others which traverse the mesovarium and enter the hilum of the ovary. The **veins** emerge from the hilum in the form of a plexus, the **pampiniform plexus;** the ovarian vein is formed from this plexus, and leaves the pelvis in company with the artery. The **lymphatics** are described on page 763. The **nerves** are derived from the hypogastric or pelvic plexus, and from the ovarian plexus, the uterine tube receiving a branch from one of the uterine nerves.

The Uterine Tube

The **uterine tubes** (*tuba uterina; Fallopian tube; oviduct*) (Figs. 17–58, 17–59, 17–61) convey the ova from the ovaries to the cavity of the uterus. They are bilateral, extending from the superior lateral angle of the uterus to the side of the pelvis. Each one is suspended by a mesenteric peritoneal fold, called the **mesosalpinx,** which comprises the upper free margin and adjacent movable portion of the broad ligament. Each tube is about 10 cm long, and consists of three portions: (1) the **isthmus,** or medial constricted third; (2) the **ampulla,** or intermediate dilated portion, which curves over the ovary; and (3) the **infundibulum** with its **abdominal ostium,** surrounded by **fimbriae,** one of which, the

ovarian fimbria, is attached to the ovary. The uterine tube is directed lateralward as far as the uterine pole of the ovary, and then ascends along the mesovarian border of the ovary to the tubal pole, over which it arches; finally it turns downward and ends in relation to the free border and medial surface of the ovary. The uterine opening is minute, and will admit only a fine bristle; the abdominal opening is somewhat larger. In connection with the fimbriae of the uterine tube, or with the broad ligament close to them, there are frequently one or more small pedunculated vesicles. These are termed the **appendices vesiculosae** (*hydatids of Morgagni*).

Structure.—The uterine tube consists of three coats: **serous, muscular,** and **mucous.** The **external** or **serous coat** is peritoneal. The **middle** or **muscular coat** consists of an external longitudinal and an internal circular layer of non-striped muscular fibers continuous with those of the uterus. The **internal** or **mucous coat** is continuous with the mucous lining of the uterus, and, at the abdominal ostium of the tube, with the peritoneum. It is thrown into longitudinal folds, **plicae tubariae,** which, in the ampulla, are much more extensive than in the isthmus. The lining epithelium is columnar and ciliated. This form of epithelium

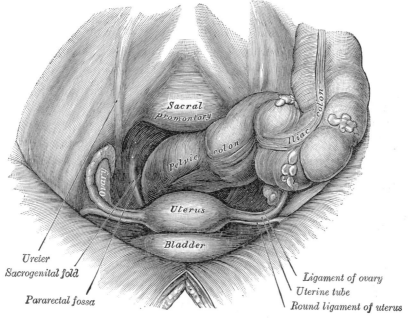

Fig. 17–58.—Female pelvis and its contents, seen from above and in front.

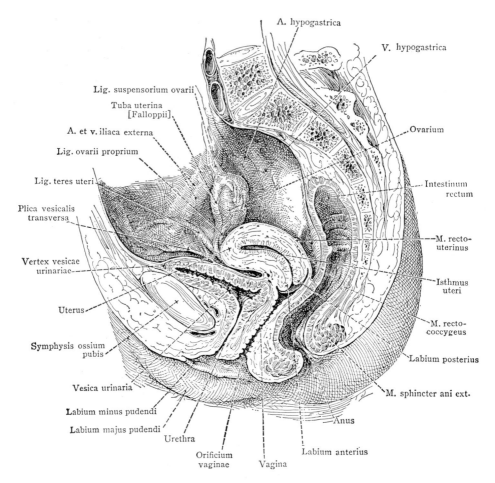

A. hypogastrica

V. hypogastrica

Lig. suspensorium ovarii

Tuba uterina [Falloppii]

A. et v. iliaca externa

Lig. ovarii proprium

Lig. teres uteri

Plica vesicalis transversa

Vertex vesicae urinariae

Uterus

Symphysis ossium pubis

Vesica urinaria

Labium minus pudendi

Labium majus pudendi

Urethra

Orificium vaginae

Vagina

Ovarium

Intestinum rectum

M. recto-uterinus

Isthmus uteri

M. recto-coccygeus

Labium posterius

M. sphincter ani ext.

Anus

Labium anterius

FIG. 17–59.—Median sagittal section of female pelvis. The bladder is empty, the uterus and vagina slightly dilated. (Eycleshymer and Jones.)

is also found on the inner surface of the fimbriae, while on the outer or serous surfaces of these processes the epithelium merges into the mesothelium of the peritoneum.

Fertilization of the ovum is believed (page 17) to occur in the tube, and the fertilized ovum is then normally passed on into the uterus; the ovum, however, may adhere to and undergo development in the uterine tube, giving rise to the commonest variety of *ectopic gestation.* In such cases the amnion and chorion are formed, but a true decidua is never present, and the gestation usually ends by extrusion of the ovum through the abdominal ostium, although it is not uncommon for the tube to rupture into the peritoneal cavity, this being accompanied by severe hemorrhage, and needing surgical interference.

Vessels and Nerves.—The **arteries** supplying the uterine tube are branches of the uterine and ovarian arteries. One branch from the uter-

ine, much stronger than the others, is of special surgical importance (Sampson's artery). It arises just before the uterine anastomoses with the ovarian and continues along the tube to its fibriated extremity giving off branches along its course. The **veins** drain into the uterine plexus. The **lymphatics** drain partly with the uterine and partly with the ovarian vessels.

The **nerves** are derived from the pelvic plexus of sympathetic and parasympathetic sacral nerves.

The Uterus

The **uterus** (*womb*) (Figs. 17–54, 17–58, 17–59) is a thick-walled, hollow, muscular organ situated in the pelvis between the bladder and the rectum. The uterine tubes open into its upper abdominal part and at its perineal end its cavity communicates

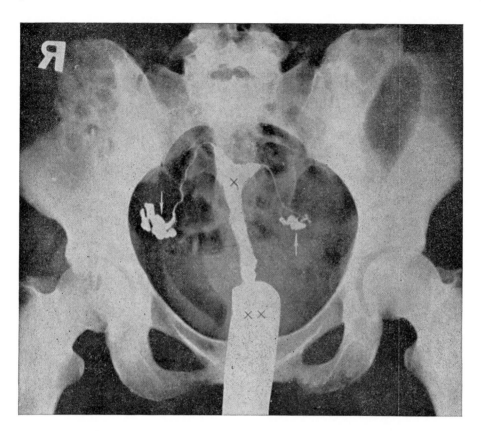

Fig. 17–60.—Genital tract in the female, after an injection of barium sulfate into the uterus. ✕, Body of uterus. Note the two cornua leading to the uterine tubes. ✕ ✕, Speculum in vagina. The arrows indicate the infundibula of the uterine tubes. Some of the barium has passed through the pelvic opening of the tube into the general peritoneal cavity.

with that of the vagina. It changes remarkably in size, structure, and position during pregnancy, and, although it returns again almost to its former condition, the following description is based on the uterus of an adult virgin.

The uterus is pear-shaped, somewhat flattened dorsoventrally, and elongated with its long axis parallel with the median plane but curved to correspond with the axis of the pelvis. It measures about 7.5 cm in length, 5 cm in breadth at its cranial part and nearly 2.75 cm in thickness; it weighs 30 to 40 gm.

The uterus is divided structurally and functionally into two parts, the body and the cervix. On the surface, about midway between the two ends, is a slight constriction known as the **isthmus.** Corresponding to this on the inside is a narrowing of the cavity, formerly known as the **internal** orifice of the uterus. The part cranial to the isthmus is the body, caudal to it the cervix.

The **Body** (*corpus uteri*) (Fig. 17–61).— The part of the body which extends toward the abdomen as a free, rounded extremity above the entrance of the uterine tubes is called the **fundus.** The body gradually narrows from the fundus to the isthmus.

The ventral or **vesical surface** (*facies vesicalis*) faces toward the bladder, and the peritoneum covering its surface is reflected, at the junction with the cervix, into the vesicouterine excavation and thence to the bladder.

The dorsal or **intestinal surface** (*facies intestinalis*) is more convex and the peritoneum covering its surface is more extensive than that of the vesical surface. It extends over the cervix and on over the cranial portion of the vagina. The peri-

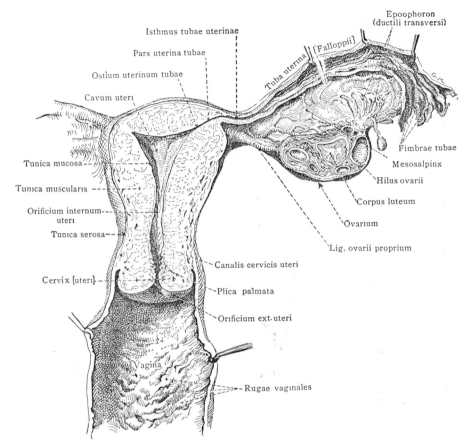

FIG. 17–61.—Section through the vagina, uterus, uterine tube and ovary.
(Eycleshymer and Jones.)

toneal excavation between it and the rectum is the rectouterine fossa and is usually occupied by coils of small intestine.

The **fundus** is convex in all directions and its free surface, covered by peritoneum, is usually in contact with coils of small intestine or a distended sigmoid colon.

The **lateral margins** (*margo lateralis*) are slightly convex. At the upper end of each the uterine tube joins the uterine wall. Ventral and caudal to this point the round ligament of the uterus is fixed, while dorsal to it is the attachment of the ligament of the ovary. These three structures lie within a fold of peritoneum which is reflected from the margin of the uterus to the wall of the pelvis, and is named the **broad ligament.**

The **Cervix** (*cervix uteri; neck*) is the portion of the uterus between the isthmus and the vagina. It is about 2 cm in length, and the end toward the perineum protrudes

into the cavity of the vagina as a free extremity. The attachment of the vagina around its periphery divides it into a vaginal and a supravaginal portion.

The **supravaginal portion** (*portio supravaginalis cervicis*) is separated ventrally from the bladder by fibrous tissue (**parametrium**), which extends also on to its sides and lateralward between the layers of the broad ligaments. The uterine arteries reach the margins of the cervix in this fibrous tissue, where they cross over the ureters. The ureters, on either side, run ventralward in the parametrium about 2 cm from the cervix. Dorsally, the supravaginal cervix is covered by peritoneum, which is prolonged caudalward on the posterior vaginal wall before it is reflected over to the rectum.

The **vaginal portion** (*portio vaginalis*) of the cervix projects into the cavity of the vagina, the free end resting against the dorsal wall. On its rounded extremity is a round or oval depression which marks the **uterine orifice** (*ostium uteri; external os*) leading into the cavity of the cervix. The orifice is bounded by two lips, the **ventral lip** (*labium anterius*) and the **dorsal lip** (*labium posterius*). Between the dorsal lip and the vaginal wall is the **posterior fornix**, the most cranial part of the vaginal cavity, and between the ventral lip and the vaginal wall is the **anterior fornix** (Fig. 17–59).

The **cavity of the uterus** (*cavum uteri*) is compressed dorsoventrally into a mere slit. The orifices of the uterine tubes and the internal surface of the fundus form the base of a triangle whose apex is the isthmus, leading into the canal of the cervix. The total length of the uterine cavity from the external orifice to the fundus is about 6.25 cm.

The **Canal of the Cervix** (*canalis cervicis uteri*) (Fig. 17–61) is somewhat flattened and wider in the middle than at either extremity. It communicates through the isthmus with the cavity of the body, and through the external orifice with the vaginal cavity. The wall of the canal presents an anterior and a posterior longitudinal ridge, from each of which proceed a number of small oblique columns, the **palmate folds** (*plicae palmatae*), giving the appearance of branches from the stem of a tree; to this arrangement the name **arbor vitae uterina** is applied. The folds on the two walls are not exactly opposed, but fit between one another so as to close the cervical canal.

Ligaments.—The principal ligaments of the uterus are the **broad ligaments**, the **round ligaments**, the **uterosacral ligaments**, and the **cardinal ligaments** (*ligamenta transversalia colli*). In addition to these, there are certain peritoneal folds which are called ligaments: the **vesicouterine, rectouterine**, and **sacrogenital ligaments**.

The **Broad Ligaments** (*ligamentum latum uteri*) (Figs. 17–54, 17–55) are two fibrous sheets, covered on both surfaces with peritoneum, which extend from each side of the uterus to the lateral wall and bottom of the pelvis. The uterus and these lateral extensions together form a septum across the cavity of the pelvis, dividing it into a **vesicouterine** and a **rectouterine fossa** (Figs.

17–58, 17–59, not labeled). The broad ligament is thicker at its inferior pelvic attachment than toward its free border. Between the two peritoneal sheets or *leaves of the ligament* are:

1. Parametrium
2. Uterine artery
3. Uterine tube
4. Round ligament
5. Epoöphoron and Paroöphoron
6. (Ovary)
7. (Ureter)

1. The **parametrium** is the extension of the subserous connective tissue of the uterus laterally into the broad ligament. The name has been applied also to the whole broad ligament below the attachment of the ovary, but according to the NA the latter is the **mesometrium**. It contains scattered smooth muscle bundles, is anchored to the lateral pelvic wall, and is continuous with the cardinal ligament of the intrapelvic fascia.

2. The **uterine artery** of each side enters the base of the broad ligament at the latter's attachment to the lateral wall of the pelvis and traverses the pelvis between the two leaves close to their reflection to the pelvic floor. It crosses the ureter just before it reaches the cervical portion of the uterus and in the parametrium it follows the lateral border of the uterus up to the isthmus of the uterine tube. It follows the tube laterally and anastomoses with the ovarian artery. The **ovarian artery** crosses the external iliac vessels in a vertical direction, enters the most superior lateral portion of the broad ligament enclosed in a fibrous cord, the **suspensory ligament of the ovary** (*ligamentum suspensorium ovarii; infundibulopelvic ligament*), and follows the attached border of the ovary to the anastomoses with the uterine artery.

3. The superior free border of the broad ligament is occupied by the **uterine tube** except at its lateral extremity where it forms the suspensory ligament of the ovary which also attaches the infundibulum of the tube to the lateral wall of the pelvis. The free portion of the broad ligament extending down as far as the attachment of the ovary is called the **mesosalpinx**. It is less

fixed than the rest of the ligament and affords a movable mesenteric support for the tube.

4. The **round ligaments** are described below with the other ligaments of the uterus.

5. The **epoöphoron** and **paroöphoron** lie in the mesosalpinx.

6. The **ovary** is secured to the posterior surface of the broad ligament by a mesenteric attachment, the **mesovarium,** derived from the posterior leaf. It does not lie, therefore, between the leaves of the broad ligament. Folds of the posterior leaf at each end of the ovary cover the **ligamentum suspensorium ovarii** and the **ligamentum ovarii proprium.**

7. The **ureter** crosses the attached inferior border of the broad ligament obliquely, as it courses along the pelvic floor toward the base of the bladder. It comes to within 1 or 2 cm of the uterus at this point and lies close to the uterine artery, between the latter and the pelvic diaphragm.

The **Round Ligament** (*ligamentum teres uteri*) is a flattened band attached to the superior part of the lateral border of the uterus just caudal and ventral to the isthmus of the uterine tube. It traverses the pelvis between the leaves of the broad ligament but causes a prominent protrusion of the anterior leaf only. It reaches the pelvic wall lateral to the lateral vesicoumbilical fold, ascends over the external iliac vessels and inguinal ligament, and penetrates the abdominal wall through the deep inguinal ring. It passes through the inguinal canal and its constituent fibers spread out to help form the substance of the labia majora. In the fetus, the peritoneum is prolonged in the form of a tubular process for a short distance into the inguinal canal beside the ligament. This process is called the **canal of Nuck** (Fig. 17–62). It is generally obliterated in the adult, but sometimes remains pervious even in advanced life. It is analogous to the saccus vaginalis, which precedes the descent of the testis.

The **Cardinal Ligament** (*ligamentum transversum colli* [Mackenrodt]) is a fibrous sheet of the subserous fascia embedded in the adipose tissue on each side of the lower cervix uteri and vagina. To form it, the fasciae over the ventral and dorsal walls of

the vagina and cervix come together at the lateral border of these organs, and the resulting sheet extends across the pelvic floor as a deeper continuation of the broad ligament. As the sheet reaches the lateral portion of the pelvic diaphragm, it forms ventral and dorsal extensions which are attached to the internal investing layer of deep fascia (supra-anal fascia) on the inner surface of the Levator ani, Coccygeus and Piriformis. This attachment is commonly visible as a white line 2 or 3 cm below the arcus tendineus of the Levator ani, and is called the **arcus tendineus of the pelvic fascia** (see page 435). The ventral extension is continuous with the tissue supporting the bladder. The dorsal extension blends with the uterosacral ligaments. The vaginal arteries cross the pelvis in close association with this ligament, giving it additional substance and support, and bundles of smooth muscle may be embedded in it (Fig. 9–33).

The **Uterosacral Ligament** is a prominent fibrous band of subserous fascia which takes a curved course along the lateral wall of the pelvis from the cervix uteri to the sacrum. It is a posterior continuation of the tissue which forms the cardinal ligament. It is attached to the deep fascia and periosteum of the sacrum and contains a bundle of smooth muscle named the **Rectouterinus.** The ligaments on the two sides project out from the wall as crescentic shelves which narrow the diameter of the cavity ventral to the lower rectum and mark it off as the *cul-de-sac of Douglas.*

The **vesicouterine fold** or **anterior ligament** is the reflection of peritoneum from the anterior surface of the uterus, at the junction of the cervix and body, to the posterior surface of the bladder.

The **rectovaginal fold** or **posterior ligament** is the peritoneum reflected from the wall of the posterior fornix of the vagina on to the anterior surface of the rectum.

The **sacrogenital** or **rectouterine folds** (*plica rectouterina* [Douglas]) are two crescentic folds which cover the uterosacral ligaments.

The **rectouterine excavation** (*excavatio rectouterina; cavum Douglasi; pouch or cul-de-sac of Douglas*) is a deep pouch formed by the most inferior or caudal portion of the parietal peritoneum. Its ventral boundary is the supravaginal cervix and posterior fornix of the vagina; dorsal, the rectum, and lateral, the sacrogenital folds covering the uterosacral ligaments.

Support of the Uterus.—The principal support of the uterus is the pelvic diaphragm, especially the Levator ani and its investing layers of fascia, and unless it is intact the other structures are unable to carry out their supporting function. The uterus is held in its proper position within the pelvis by its attachment to the vagina and by the cardinal, broad, and uterosacral ligaments. The blood vessels reinforce these ligaments. The round ligaments and the peritoneal folds are of relatively slight importance as mechanical supports. The padding of adipose tissue about the ligaments and organs, in well-nourished individuals, is an important element of support also. There is a great variation in the size and development of the supporting structures in different individuals, and they may be thickened or strengthened in response to physiological and pathological changes.

Position of the Uterus.—The form, size, and situation of the uterus vary at different periods of life and under different circumstances.

In the fetus and infant the uterus is contained in the abdominal cavity, projecting beyond the superior aperture of the pelvis (Fig. 17–62). The cervix is considerably larger than the body.

At puberty the uterus is piriform in shape, and weighs from 14 to 17 gm. It has descended into the pelvis, the fundus being just below the level of the superior aperture of this cavity. The palmate folds are distinct, and extend to the upper part of the cavity of the organ.

The position of the uterus *in the adult* is liable to considerable variation. With the bladder and rectum empty the body of the uterus is nearly horizontal when the individual is standing. The fundus is about 2 cm behind the symphysis pubis and slightly cranial to it. The uterus and vagina are at an angle of about 90° with each other. The external os is half way between the spines of the ischia. As the bladder fills, the uterus is bent back toward the sacrum.

During menstruation the organ is enlarged, more vascular, and its surfaces rounder; the external orifice is rounded, its labia swollen, and the lining membrane of the body thickened, softer, and of a darker color.

During pregnancy the uterus becomes enormously enlarged, and in the eighth month reaches the epigastric region. The increase in size is partly due to growth of preexisting muscle, and partly to development of new fibers.

After parturition the uterus nearly regains its usual size, weighing about 42 gm, but its cavity is larger than in the virgin state; its vessels are tortuous, and its muscular layers are more defined; the external orifice is more marked, and its edges present one or more fissures.

In old age the uterus becomes atrophied, and paler and denser in texture; a more distinct constriction separates the body and cervix. The isthmus is frequently, and the ostium occasionally, obliterated, while the lips almost entirely disappear.

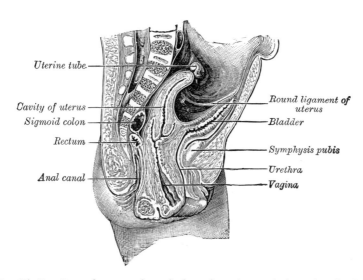

FIG. 17–62.—Sagittal section through the pelvis of a newly born female child.

Structure.—The uterus is composed of three coats: an **external** or **serous**, a **middle** or **muscular**, and an **internal** or **mucous**.

The **serous coat** (*tunica serosa; perimetrium*) is derived from the peritoneum; it invests the fundus and the whole of the intestinal surface of the uterus, but covers the vesical surface only as far as the junction of the body and cervix. In the lower fourth of the intestinal surface the peritoneum, though covering the uterus, is not closely connected with it, being separated from it by a layer of loose cellular tissue and some large veins.

The **muscular coat** (*tunica muscularis, myometrium*) forms the chief bulk of the substance of the uterus. In the virgin it is dense, firm, and of a grayish color, and cuts almost like cartilage. It is thick opposite the middle of the body and fundus, and thin at the orifices of the uterine tubes. It consists of bundles of unstriped muscular fibers, disposed in a thick, felt-like structure, intermixed with areolar tissue, blood vessels, lymphatic vessels, and nerves. Muscle fibers are continued on to the uterine tube, the round ligament, and the ligament of the ovary, some passing at each side into the broad ligament, and others running backward from the cervix into the uterosacral ligaments. During pregnancy the muscular tissue becomes more prominently developed, the fibers being greatly enlarged.

The **mucous membrane** (*tunica mucosa, endometrium*) (Fig. 17–63) is smooth, and closely adherent to the subjacent muscular

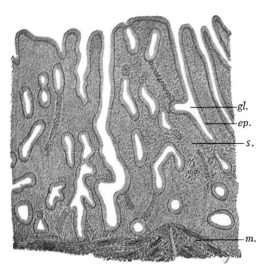

Fig. 17–63.—Vertical section of mucous membrane of human uterus. *ep.*, epithelium; *gl.*, lumen of gland; *m.*, muscularis; *s.*, connective tissue stroma. (Sobotta.)

tissue. It is continuous through the fimbriated extremity of the uterine tubes, with the peritoneum, and, through the uterine ostium, with the lining of the vagina.

In the *body of the uterus* the mucous membrane is smooth, soft, of a pale red color, lined by a single layer of high columnar ciliated epithelium, and presents, when viewed with a lens, the orifices of numerous tubular follicles, arranged perpendicularly to the surface. The structure of the corium differs from that of ordinary mucous membranes, and consists of an embryonic nucleated and highly cellular form of connective tissue in which run numerous large lymphatics. In it are the tube-like **uterine glands.**

In the *cervix* the mucous membrane is sharply differentiated from that of the uterine cavity. It is thrown into numerous oblique ridges, which diverge from an anterior and posterior longitudinal raphe. In the upper two-thirds of the canal, the mucous membrane is provided with numerous deep glandular follicles which secrete a clear viscid alkaline mucus; and, in addition, extending through the whole length of the canal is a variable number of little cysts, which have become occluded and distended with retained secretion. They are called the **ovula Nabothi.** The mucous membrane covering the lower half of the cervical canal presents numerous papillae. The epithelium of the upper two-thirds is cylindrical and ciliated, but below this it loses its cilia, and gradually changes to stratified squamous epithelium close to the ostium uteri. On the vaginal surface of the cervix the epithelium is stratified squamous similar to that lining the vagina.

Vessels and Nerves.—The **arteries** of the uterus are the uterine, from the internal iliac; and the ovarian, from the abdominal aorta (Fig. 8–51). They are remarkable for their tortuous course in the substance of the organ, and for their frequent anastomoses. The termination of the ovarian artery meets that of the uterine artery, and forms an anastomotic trunk from which branches are given off to supply the uterus, their disposition being circular. The **veins** are of large size, and correspond with the arteries. They end in the uterine plexuses. In the impregnated uterus the arteries carry the blood to, and the veins convey it away from, the intervillous space of the placenta (see page 46). The **lymphatics** are described on page 768. The **nerves** are derived from the hypogastric and ovarian plexuses, and from the third and fourth sacral nerves. Afferent fibers from the uterus enter the spinal cord solely through the eleventh and twelfth thoracic nerves.

The **Menstrual Cycle.**—After puberty and throughout the childbearing period the endometrium undergoes periodic changes called the **menses.** The manifestation of this cycle is bleeding from the uterus, and the beginning of the bleeding is used as the point of time from which the cycle is measured. The common length of a cycle is twenty-eight days, but this may vary markedly from individual to individual and between cycles of the same individual.

The changes in the uterus are closely correlated with cyclic changes in the ovary, and the two organs must be considered together. Although the changes are a continuous process, the cycle is usually divided into four phases, viz: (*a*) proliferative, (*b*) secretory or progravid, (*c*) premenstrual, and (*d*) menstrual.

a) The **proliferative phase.**—The portion of the endometrium adjacent to the myometrium, known as the *basalis*, remains after the menstrual flow allowing the glands and lining epithelium to be restored to an inactive condition such as that shown in Figure 17–63. After the cessation of the menstrual flow, a hormone from the adenohypophysis, the **follicle-stimulating hormone,** causes the growth of one of the ovarian follicles (page 1319). As the follicle enlarges it produces the hormone **estrogen** which stimulates the proliferation of the endometrium, greatly enlarging the uterine glands.

b) The **secretory** or **progravid phase.**— When the ovarian follicle has ripened into a mature Graafian follicle it ruptures, frequently at the middle of the cycle, liberating the ovum. The ruptured follicle, under the influence of the pituitary **luteinizing hormone,** develops into a corpus luteum whose proliferating cells produce the hormone **progesterone.** The progesterone stimulates the enlarged endometrial glands to secrete and prepare the uterus for the reception of the fertilized ovum.

c) The **premenstrual phase.**—If the ovum is fertilized, it continues to grow into the corpus luteum of pregnancy, but if the ovum is not fertilized, the corpus luteum begins to degenerate and it no longer produces its hormones. The endometrium responds by changes in the blood supply,

deterioration of the tissues, and fragmentation of the glands and epithelium.

d) The **menstrual phase.**—The portion of the endometrium containing the enlarged glands and abundant capillaries is known as the *functionalis* in distinction from the bottoms of the glands and tissue adjacent to the myometrium, known as the *basalis*. During the menstrual discharge, the functionalis is sloughed away and the ruptured vessels produce the *bleeding* of the menstrual flow. The basalis remains to establish the new endometrium for the next succeeding cycle.

The Vagina

The **vagina** (Figs. 17–61, 17–64) extends from the vestibule to the uterus, and is situated dorsal to the bladder and ventral to the rectum; its axis forms an angle of over 90° with that of the uterus. Its walls are ordinarily in contact, and the usual shape of its lower part on transverse section is that of an H, the transverse limb being slightly curved forward or backward, while the lateral limbs are somewhat convex toward the median line; its middle part has the appearance of a transverse slit. Its length is 6 to 7.5 cm along its ventral wall, and 9 cm along its dorsal wall. It is constricted at its commencement, dilated in the middle, and narrowed near its uterine extremity; it surrounds the vaginal portion of the cervix uteri, its attachment extending higher up on the dorsal than on the ventral wall of the uterus. The recess dorsal to the cervix is the **posterior fornix;** the smaller recesses, ventral and lateral, are called the **anterior** and **lateral fornices.**

Relations.—The **ventral surface** of the vagina is in relation with the fundus of the bladder, and with the urethra. Its **dorsal surface** is separated from the rectum by the rectouterine excavation in its upper fourth, and by the rectovaginal fascia in its middle two-fourths; the lower fourth is separated from the anal canal by the perineal body. As the terminal portions of the ureters pass ventralward and medialward to reach the fundus of the bladder, they run close to the lateral fornices of the vagina, and as they enter the bladder are slightly in front of the ventral fornix.

Structure.—The vagina consists of an **internal mucous lining** and a **muscular coat** separated by a layer of erectile tissue.

The **mucous membrane** (*tunica mucosa*) is continuous above with that lining the uterus. Its inner surface presents two longitudinal ridges, one on its anterior and one on its posterior wall. These ridges are called the **columns of the vagina** and from them numerous transverse ridges or rugae extend outward on either side. These rugae are divided by furrows of variable depth, giving to the mucous membrane the appearance of being studded over with conical projections or papillae; they are most numerous near the orifice of the vagina, especially before parturition. The epithelium covering the mucous membrane is of the stratified squamous variety. The submucous tissue is very loose, and contains a plexus of large veins, together with smooth muscular fibers derived from the muscular coat. It contains a number of mucous crypts, but no true glands.

The **muscular coat** (*tunica muscularis*) consists of two layers: an external longitudinal, which is by far the stronger, and an internal circular layer. The longitudinal fibers are continuous with the superficial muscular fibers of the uterus. The strongest fasciculi are those attached to the rectovesical fascia on either side. The two layers are not distinctly separable from each other, but are connected by oblique decussating fasciculi, which pass from the one layer to the other. In addition to this, the vagina at its lower end is surrounded by the erectile tissue of the bulb of the vestibule and a band of striped muscular fibers, the **Bulbocavernosus** (see page 443).

External to the muscular coat is a layer of connective tissue, containing a large plexus of blood vessels.

Vessels and Nerves.—The **vaginal artery** arises from the uterine or from the adjacent internal iliac artery, supplies the mucous membrane, and anastomoses with the uterine, inferior vesicle, and middle rectal arteries. The branch of the uterine to the cervix descends on the dorsal or ventral wall forming the azygos artery of the vagina which anastomoses with the vaginal (Fig. 8–51). The **veins** run along both sides forming plexuses with the uterine, vesical, and rectal veins and ending in a vein which opens into the internal iliac vein. The **lymphatics** are described on page 764.

The **nerves** are derived from the plexus and from the pudendal nerve.

The External Genital Organs

The **external genital organs** of the female (*partes genitales externae muliebres*) (Fig. 17–64) are: (1) the **mons pubis**, (2) the **labia majora**, (3) the **labia minora**, (4) the **clitoris**, (5) the **vestibule of the vagina**, (6) the **bulb of the vestibule**, and (7) the **greater vestibular glands**. The term **pudendum** or **vulva**, as generally applied, includes all these parts.

1. The **Mons Pubis** (*commissura laborum anterior; mons Veneris*), the rounded eminence in front of the pubic symphysis, is formed by a collection of fatty tissue beneath the integument. It becomes covered with hair at the time of puberty.

2. The **Labia Majora** (*labia majora pudendi*) are two prominent longitudinal cutaneous folds which extend caudalward and dorsalward from the mons pubis and form the lateral boundaries of a fissure or cleft, the **pudendal cleft** (*rima pudendi*), into which the vagina and urethra open. Each labium has two surfaces, an outer, pigmented and covered with strong, crisp hairs; and an inner, smooth and beset with large sebaceous follicles. Between the two there is a considerable quantity of areolar tissue, fat, and a tissue resembling the dartos tunic of the scrotum, besides vessels, nerves, and glands. The labia are thicker in front, where they form by their meeting the **anterior labial commissure**. Posteriorly they are not really joined, but appear to become lost in the neighboring integument, ending close to, and nearly parallel with, each other. Together with the connecting skin between them, they form the **posterior labial commissure** or posterior boundary of the pudendum. The labia majora correspond to the scrotum in the male.

3. The **Labia Minora** (*labia minora pudendi; nymphae*) are two small folds, situated between the labia majora, and extending from the clitoris obliquely backward for about 4 cm on either side of the orifice of the vagina, between which and the labia majora they end; in the virgin the posterior ends of the labia minora are usually joined across the middle line by a fold of skin, named the **frenulum of the labia** or **fourchette**. Anteriorly, each labium minus divides into two portions: the upper division passes above the clitoris to meet its fellow of the opposite side, forming a fold which overhangs the glans clitoridis, and is named the **preputium clitoridis**; the lower

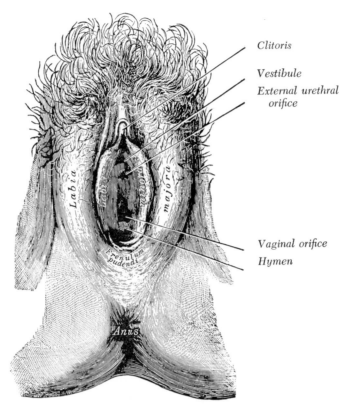

Clitoris

Vestibule

*External urethral
    orifice*

Vaginal orifice

Hymen

Fɪɢ. 17–64.—External genital organs of female. The labia minora have been drawn apart.

division passes beneath the clitoris and becomes united to its undersurface, forming, with the corresponding structure of the opposite side, the **frenulum of the clitoris.** On the opposed surfaces of the labia minora are numerous sebaceous follicles.

4. The **Clitoris** is an erectile structure, homologous with the penis. It is situated beneath the anterior labial commissure, partially hidden between the anterior ends of the labia minora. The **free extremity** (*glans clitoridis*) is a small rounded tubercle. It consists of two corpora cavernosa, composed of erectile tissue enclosed in a dense layer of fibrous membrane; each corpus is connected to the rami of the pubis and ischium by a crus similar to that of the penis. The clitoris is provided, like the penis, with a suspensory ligament and with two small muscles, the Ischiocavernosi, which are inserted into the crura of the clitoris.

5. The **Vestibule** (*vestibulum vaginae*).— The cleft between the labia minora and behind the glans clitoridis is named the

vestibule of the vagina: in it are seen the urethral and vaginal orifices and the openings of the ducts of the greater vestibular glands.

The **external urethral orifice** (*orificium urethrae externum; urinary meatus*) is placed about 2.5 cm dorsal to the glans clitoridis and immediately ventral to that of the vagina; it usually assumes the form of a short, sagittal cleft with slightly raised margins.

The **vaginal orifice** is a median slit caudal and dorsal to the opening of the urethra; its size varies inversely with that of the **hymen.**

The **hymen** is a thin fold of mucous membrane situated at the orifice of the vagina; the inner edges of the fold are normally in contact with each other, and the vaginal orifice appears as a cleft between them. The hymen varies much in shape. When stretched, its commonest form is that of a ring, generally broadest posteriorly; sometimes it is represented by a semilunar fold, with its concave margin turned toward the pubes. Occasionally it is cribriform, or its free margin forms a membranous fringe. It may be entirely absent, or may form a complete septum across the lower end of the vagina; the latter condition is known as an **imperforate hymen.** It may persist after copulation, so that its presence cannot be considered a sign of virginity. When the hymen has been ruptured, small rounded elevations known as the **carunculae hymenales** are found as its remains. Between the hymen and the frenulum of the labia is a shallow depression, named the **navicular fossa.**

6. The **Bulb of the Vestibule** (*bulbus vestibuli; vaginal bulb*) is the homologue of the bulb and adjoining part of the corpus spongiosum penis of the male, and consists of two elongated masses of erectile tissue, placed one on either side of the vaginal orifice and united to each other in front by a narrow median band termed the **pars intermedia.** Each lateral mass measures a little over 2.5 cm in length. Their posterior ends are expanded and are in contact with the greater vestibular glands; their anterior ends are tapered and joined to each other by the pars intermedia; their deep surfaces are in contact with the superficial layer of the urogenital diaphragm; superficially they are covered by the Bulbocavernosus.

7. The **Greater Vestibular Glands** (*glandulae vestibularis majores; Bartholin's glands*) are the homologues of the bulbourethral glands in the male. They consist of two small, roundish bodies situated one on either side of the vaginal orifice in contact with the posterior end of each lateral mass of the bulb of the vestibule. Each gland opens by means of a duct, about 2 cm long, immediately lateral to the hymen, in the groove between it and the labium minus.

Vessels and Nerves.—The **arteries** are derived from the internal pudendal artery through the perineal, artery of the bulb, and deep and dorsal arteries of the clitoris. The **veins** drain into the pudendal plexus which empties into the vaginal and inferior vesical veins. The **lymphatics** follow the external pudendal vessels and drain into the inguinal or external iliac nodes. The **nerves** are derived from the pudendal nerve and the pelvic plexus.

Mammary Gland

The mammary gland (*mamma; breast*) is an accessory of the reproductive system in function, since it secretes milk for nourishment of the infant, but structurally and developmentally it is closely related to the integument. It reaches its typical exquisite development in women during the early childbearing period but is present only in a rudimentary form in infants, children, and men.

In the adult nullipara, each mamma forms a discoidal, hemispherical, or conical eminence on the anterior chest wall, extending from the second to the sixth or seventh rib, and from the lateral border of the sternum into the axilla. It protrudes 3 to 5 cm from the chest wall, and its craniocaudal diameter, approximately 10 to 12 cm, is somewhat less than its transverse diameter. Its average weight is 150 to 200 gm, increasing to 400 or 500 gm during lactation. The left mamma is generally slightly larger than the right.

The **glandular tissue** forms fifteen or twenty lobes arranged radially about the nipple, each lobe having its own individual excretory duct. The glandular tissue does not occupy the entire eminence called the breast; a variable but considerable amount of adipose tissue fills out the stroma between and around the lobes. The central portion is predominantly glandular, the peripheral predominantly fat. The connective tissue stroma in many places is concentrated into fibrous bands which course vertically through the substance of the breast, attaching the deep layer of the subcutaneous fascia to the dermis of the skin. These bands are known as **suspensory ligaments of the breast** or **Cooper's ligaments.** The entire breast is contained within the subcutaneous fascia. The deep surface is separated from the underlying external investing layer of deep fascia by a fascial cleft which allows considerable mobility. The deep surface of the breast is concave, molded over the ventral chest wall mostly in contact with the pectoral fascia, but laterally with the axillary and serratus anterior fascia, and inferiorly it may reach the Obliquus externus and Rectus abdominis.

The **Mammary Papilla** (*papilla mam-*

mae) or **Nipple** projects as a small cylindrical or conical body, a little below the center of each breast at about the level of the fourth intercostal space. It is perforated at the tip by fifteen or twenty minute openings, the apertures of the lactiferous ducts. The characteristic skin of the nipple, pigmented, wrinkled, and roughened by papillae, extends outward on the surface of the breast for 1 or 2 cm to form the **areola.** The color of the nipple and areola in nulliparae varies from rosy pink to brown, depending on the complexion of the individual. During the second month and progressing through pregnancy, the skin becomes darker and the areola becomes larger. Following lactation the pigmentation diminishes but is never entirely lost and may be used to differentiate nulliparous from parous individuals.

The **areola** (*areola mammae*) is made rough by the presence of numerous large sebaceous glands which produce small elevations of its surface. These **areolar glands** (*glands of Montgomery*) secrete a lipoid material which lubricates and protects the nipple during nursing. The subcutaneous tissue of the areola contains circular and radiating smooth muscle bundles which cause the nipple to become erect in response to stimulation.

Development.—The primordium of the mamma is first recognizable during the sixth week of intrauterine life as a bandlike thickening of the ectoderm of the anterolateral body wall. It extends from the axilla to the inguinal region and is called the **milk line.** The thickening of the ectoderm pushes into the subcutaneous mesoderm as it enlarges. After the eighth week, only the portion of the ridge destined to become mamma is identifiable. During the remainder of fetal life the epithelial cells proliferate, gradually forming buds and cords of cells projecting into the subcutaneous tissue, and by birth, in both sexes, little more than the main ducts have formed. The glands remain in this infantile condition in the male.

Adolescent Hypertrophy.—In the female there is but slight change from the infantile condition until the approach of puberty. At this time the mamma enlarges due to an increase in glandular tissue, particularly the ducts, and to a deposit of adipose tissue.

The mammary papilla and areola enlarge, increase slightly in pigmentation, acquire smooth muscle and become sensitive. After the onset of the menses, with each period there is a change in the mamma. In the premenstrual phase there is a vascular engorgement, increase in the glands, and enlargement of their lumen. During the postmenstrual phase the gland regresses and then remains in an inactive stage until the next premenstrual phase.

Hypertrophy of Pregnancy.—Visible enlargement of the breast begins after the second month of pregnancy and is accompanied by increased pigmentation and enlargement of the papilla, areola, and areolar glands. The duct system develops first, reaching its completion during the first six months; the acini and secreting portion follow during the last three months. The adipose tissue is almost completely replaced by parenchyma.

The secretion from the mammary gland during the first two or three days after parturition is thin and yellowish and is called **colostrum.** The secretion of true milk begins on the third or fourth day and continues through the nursing period.

Involution after Lactation.—At the termination of nursing, the gland gradually regresses by loss of the glandular tissue; the ducts and acini return to their former size and number, and the interstices are filled with adipose tissue. There is a slight decrease in size of the breast as a whole and it tends to become more flabby and pendulous than the nulliparous breast. The pigmentation of the nipple and areola decrease but do not entirely disappear.

Menopausal Involution.—At the end of the childbearing period, the mammae regress and the glandular tissue reverts toward the infantile condition. The adipose tissue disappears more slowly, especially in obese individuals, but eventually a senile atrophy occurs which leaves the mamma a shriveled pendulous fold of skin.

Hormonal Relationships.—The hypertrophy of puberty and the cyclic engorgements accompanying menstruation are responses to variations in the concentration of the ovarian sex hormones from the follicles and corpora lutea. The hypertrophy of pregnancy is in response to increase in

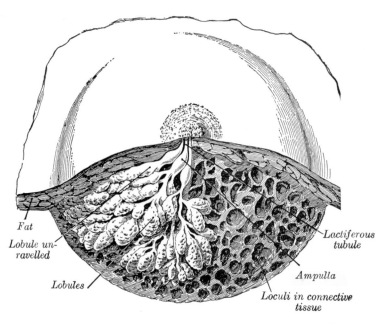

Fat

*Lobule un-
ravelled*

Lobules

*Lactiferous
tubule*

Ampulla

*Loculi in connective
tissue*

Fɪɢ. 17–65.—Dissection of the lower half of the mamma during the period of lactation. (Luschka.)

the corpus luteum hormone. The presence of the ovary is necessary only during the first part of pregnancy; later, the hormones are supplied by the placenta. Lactation is a response to the lactogenic hormone of the hypophysis, but is influenced by the nervous system through the stimulus of suckling. Suppression of the ovarian hormones after the menopause results in the involution. Quite frequently the mammary glands in the newborn of both sexes secrete a fluid called "witches' milk," under the influence of the hormones passed through the placenta from the maternal circulation.

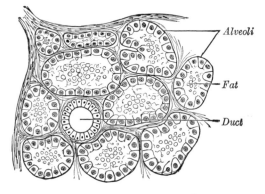

Alveoli

Fat

Duct

Fɪɢ. 17–66.—Section of portion of mamma.

Structure (Figs. 17–65, 17–66).—The mamma consists of gland tissue, of fibrous tissue connecting its lobes, and of fatty tissue in the intervals between the lobes. The gland tissue, when freed from fibrous tissue and fat, is of a pale reddish color, firm in texture, flattened and thicker in the center than at the circumference. The subcutaneous surface of the mamma presents numerous irregular processes which project toward the skin and are joined to it by bands of connective tissue. It consists of numerous lobes, and these are composed of lobules, connected together by areolar tissue, blood vessels, and ducts. The smallest lobules consist of a cluster of rounded alveoli, which open into the smallest branches of the lactiferous ducts; these ducts unite to form larger ducts, and these

end in a single canal, corresponding with one of the chief subdivisions of the gland. The number of excretory ducts varies from fifteen to twenty; they are termed the **tubuli lactiferi.** They converge toward the areola, beneath which they form dilatations or **ampullae,** which serve as reservoirs for the milk, and, at the base of the papillae, become contracted, and pursue a straight course to its summit, perforating it by separate orifices considerably narrower than the ducts themselves. The ducts are composed of areolar tissue containing longitudinal and transverse elastic fibers; muscular fibers are entirely absent; they are lined by columnar epithelium resting on a basement membrane. The epithelium of the mamma differs according to the state of activity of the organ. In the gland of a woman who is not pregnant or suckling, the alveoli are very small and solid, being filled with a mass of granular polyhedral cells. During pregnancy the alveoli enlarge, and the cells undergo rapid multiplication. At the commencement of lactation, the cells in the center of the alveolus undergo fatty degeneration, and are eliminated in the first milk as **colostrum corpuscles.** The peripheral cells of the alveolus remain, and form a single layer of granular, short columnar cells, with spherical nuclei, lining the basement membrane. The cells during the state of activity of the gland are capable of forming, in their interior, oil globules, which are then ejected into the lumen of the alveolus, and constitute the milk globules. When the acini are distended by the accumulation of the

secretion the lining epithelium becomes flattened.

The **fibrous tissue** invests the entire surface of the mamma. Bands of fibrous tissue traverse the gland and connect the overlying skin to the underlying deep layer of subcutaneous fascia. These constitute the ligaments of Cooper.

The **fatty tissue** covers the surface of the gland, and occupies the interval between its lobes. It usually exists in considerable abundance, and determines the form and size of the areola and papilla (Fig. 17–67).

Vessels and Nerves.—The **arteries** supplying the mammae are derived from the thoracic branches of the axillary, the intercostals, and the internal thoracic. The **veins** describe an anastomotic circle around the base of the papilla, called by Haller the **circulus venosus.**

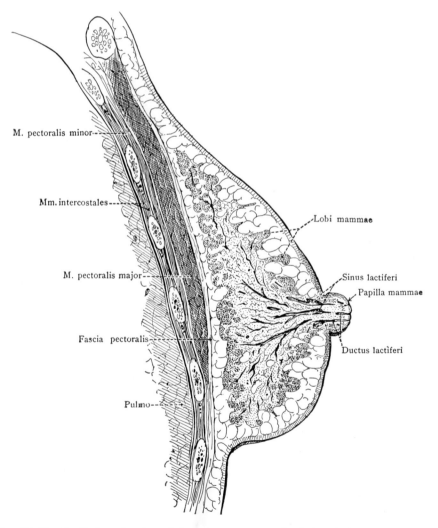

M. pectoralis minor

Mm. intercostales

M. pectoralis major

Fascia pectoralis

Pulmo

Lobi mammae

Sinus lactiferi

Papilla mammae

Ductus lactiferi

Fig. 17–67.—Sagittal section through lactating mammary gland. (Eycleshymer and Jones.)

From this, large branches transmit the blood to the circumference of the gland, and end in the axillary and internal thoracic veins. The **lymphatics** are described in Chapter 10. The **nerves** are derived from the anterior and lateral cutaneous branches of the fourth, fifth, and sixth thoracic nerves.

Variations.—Independent of the physiological variations, conditions of underdeveloped breasts (*hypomastia*), hypertrophy, and inequality on the two sides are quite common. Variations in position, cranially or caudally, are not infrequent, as might be expected from the method of embryological development. Absence of the breast, *amastia*, is very rare; an increase in the number of mammae, **polymastia**, is not as rare. The supernumerary mammae occur somewhere along the milk line in most instances, but other locations have been reported. Not infrequently they secrete milk during normal periods of lactation. When only the nipples of the super-

numerary mammae are present the condition is **polythelia**; the extra nipples occur along the milk line in the majority of cases and are found in males, though less frequently than in females.

Racial variations may be due to genetic influences which cause the discoidal, hemispherical, pear-shaped, or conical forms to predominate. On the other hand, they may be due to intentional practices such as suppressing development by tight brassiéres at adolescence or tremendously elongating them by manipulation to make nursing convenient for an infant strapped on the back.

Gynecomastia is a condition in which the mammae of a male are enlarged. In pseudo-gynecomastia the increase is due to adipose tissue. In the true gynecomastia, to some extent the epithelial tissue but more especially the firmer connective tissue elements are involved. It probably has a background of endocrine dysfunction but this is not clearly understood.

REFERENCES

KIDNEY

ANSON, B. J., J. W. PICK, and E. W. CAULDWELL. 1942. The anatomy of commoner renal anomalies: Ectopic and horseshoe kidneys. J. Urol., *47*, 112-132.

CAMERON, G. and R. CHAMBERS. 1938. Direct evidence of function in kidney of an early human fetus. Amer. J. Physiol., *123*, 482-485.

CARLSON, B. M. and C. F. MORGAN. 1963. Effects of boiling and of lyophilization upon the ability of frog kidney implants to form accessory limbs in urodeles. The Physiologist, *6*, 154.

DAVIES, J. 1954. Cytological evidence of protein absorption in fetal and adult mammalian kidneys. Amer. J. Anat., *94*, 45-71.

GROTH, C. G. 1972. Landmarks in clinical renal transplantation. Surg. Gynec. Obstet., *134*, 323-328.

MOFFAT, D. B. and J. FOURMAN. 1964. Ectopic glomeruli in the human and animal kidney. Anat. Rec., *149*, 1-11.

MOODY, R. O. and R. G. VAN NUYS. 1940. The position and mobility of the kidneys in healthy young men and women. Anat. Rec., *76*, 111-133.

SMITH, H. W. 1956. *Principles of Renal Physiology.* x + 237 pages. Oxford University Press, New York.

SULKIN, N. M. 1949. Cytologic studies of the remaining kidney following unilateral nephrectomy in the rat. Anat. Rec., *105*, 95-111.

THOENES, W. 1964. Mikromorphologie des Nephron nach temporarer Ischamie. vii + 95 pages, 42 figures. Georg Thieme Verlag, Stuttgart.

ZWEMER, R. L. and R. M. WOTTON. 1944. Fat excretion in the guinea pig kidney. Anat. Rec., *90*, 107-114.

URETER, BLADDER, AND URETHRA

GOSLING, J. A. 1970. The musculature of the upper urinary tract. Acta Anat., *75*, 408-422.

HEIMBURGER, R. F. 1949. The sacral innervation of the human bladder. Arch. Neurol. Psychiat., *62*, 686-687.

HUNTER, DE W. T. JR. 1954. A new concept of urinary bladder musculature. J. Urol., *71*, 695-704.

LANGWORTHY, O. R. and E. L. MURPHY. 1939. Nerve endings in the urinary bladder. J. comp. Neurol., *71*, 487-509.

MEYER, R. 1946. Normal and abnormal development of the ureter in the human embryo—A mechanical consideration. Anat. Rec., *96*, 355-371.

ROBSON, M. C. and E. B. RUTH. 1962. Bilocular bladder: An anatomical study of a case: with a consideration of urinary tract anomalies. Anat. Rec., *142*, 63-71.

SIMON, H. E. and N. A. BRANDEBERRY. 1946. Anomalies of the urachus: Persistent fetal bladder. J. Urol., *55*, 401-408.

TROTTER, M. and J. C. FINERTY. 1948. An anomalous urinary bladder. Anat. Rec., *100*, 259-269.

TYLER, D. E. 1962. Stratified squamous epithelium in the vesical trigone and urethra: Findings correlated with the menstrual cycle and age. Amer. J. Anat., *111*, 319-335.

WESSON, M. B. 1950. Anatomy, physiology, embryology, and congenital abnormalities of the bladder. In *Cyclopedia of Medicine, Surgery, and Specialties.* Vol. 2, 211-229. F. A. Davis Company, Philadelphia.

WOODBURNE, R. T. 1961. The sphincter mechanism of the urinary bladder and the urethra. Anat. Rec., *141*, 11-20.

WOODBURNE, R. T. 1965. The ureter, ureterovisceral junction, and vesical trigone. Anat. Rec., *151*, 243-249.

WOODBURNE, R. T. and J. LAPIDES. 1972. The urethral lumen during peristalsis. Amer. J. Anat., *133*, 255-258.

FASCIA

BENJAMIN, J. A. and C. E. TOBIN. 1951. Abnormalities of the kidneys, ureters, and perinephric fascia: Anatomic and clinical study. J. Urol., *65*, 715-731.

TOBIN, C. E. 1944. The renal fascia and its relation to the transversalis fascia. Anat. Rec., *89*, 295-311.

TOBIN, C. E. and J. A. BENJAMIN. 1945. Anatomical and surgical restudy of Denonvilliers' fascia. Surg. Gynec. Obstet., *80*, 373-388.

UHLENHUTH, E., W. M. WOLFE, E. M. SMITH, and E. B. MIDDLETON. 1948. The rectogenital septum. Surg. Gynec. Obstet., *86*, 148-163.

UHLENHUTH, E. 1953. *Problems in the Anatomy of the Pelvis.* xiv + 206 pages, 82 illustrations. J. B. Lippincott Company, Philadelphia.

BLOOD VESSELS AND NERVES

ANSON, B. J., E. W. CAULDWELL, J. W. PICK, and L. E. BEATON. 1947. The blood supply of the kidney, suprarenal gland, and associated structures. Surg. Gynec. Obstet., *84*, 313-320.

BOYER, C. C. 1956. The vascular pattern of the renal glomerulus as revealed by plastic reconstruction from serial sections. Anat. Rec., *125*, 433-441.

FINE, H. and E. N. KEEN. 1966. The arteries of the human kidney. J. Anat. (Lond.), *100*, 881-894.

HARMAN, P. J. and H. DAVIES. 1948. Intrinsic nerves in the mammalian kidney. I. Anatomy in mouse, rat, cat, and macaque. J. comp. Neurol., *89*, 225-243.

LJUNGQVIST, A. and C. LAGERGREN. 1962. Normal intrarenal arterial pattern in adult and ageing human kidney. J. Anat. (Lond.), *96*, 285-300.

MITCHELL, G. A. G. 1950. The nerve supply of the kidneys. Acta Anat., *10*, 1-37.

PICK, J. W. and B. J. ANSON. 1940. The renal vascular pedicle. J. Urol., *44*, 411-434.

REPRODUCTIVE SYSTEM
GENERAL

ASDELL, S. A. 1964. *Patterns of Mammalian Reproduction.* xiii + 670 pages. Cornell University Press, New York.

BLANDAU, R. J. and D. L. ODOR. 1949. The total number of spermatozoa reaching various segments of the reproductive tract in the female albino rat at intervals after insemination. Anat. Rec., *103*, 93-109.

BURROWS, H. 1949. *Biological Actions of Sex Hormones.* 2nd edition. xiv + 616 pages. Cambridge University Press, London.

HOOKER, C. W. 1946. Reproduction. Ann. Rev. Physiol., *8*, 467-498.

MARKEE, J. E. 1951. Physiology of reproduction. Ann. Rev. Physiol., *13*, 367-396.

MITCHELL, G. A. G. 1938. The innervation of the ovary, uterine tube, testis and epididymis. J. Anat. (Lond.), *72*, 508-517.

REYNOLDS, E. L. and P. GROTE. 1948. Sex differences in the distribution of tissue components in the human leg from birth to maturity. Anat. Rec., *102*, 45-53.

STIEVE, P. H. 1952. *Der Einfluss des Nervensystems auf Bau and Tätigkeit der Geschlechtsorgane des Menschen.* vii + 191 pages. Georg Thieme Verlag, Stuttgart.

EMBRYOLOGY AND ANOMALIES

DEY, F. L. 1943. Genital changes in female guinea pigs resulting from destruction of the median eminence. Anat. Rec., *87*, 85-90.

EVERETT, N. B. 1943. Observational and experimental evidences relating to the origin and differentiation of the definitive germ cells in mice. J. exp. Zool., *92*, 49-91.

GILLMAN, J. 1948. The development of the gonads in man, with a consideration of the role of fetal endocrines and the histogenesis of ovarian tumors. Carneg. Instn., Contr. Embryol., *32*, 83-131 + 6 plates.

HOLYOKE, E. A. 1949. The differentiation of embryonic gonads transplanted to the adult omentum in the albino rat. Anat. Rec., *103*, 675-699.

HOOKER, C. W. and L. C. STRONG. 1944. Hermaphroditism in rodents, with a description of a case in the mouse. Yale J. Biol. Med., *16*, 341-352.

MONROE, C. W. and B. SPECTOR. 1962. The epithelium of the hydatid of morgagni in the human adult female. Anat. Rec., *142*, 189-193.

PATTON, W. H. G. 1948. A case of true hermaphroditism in an adult African. Anat. Rec., *101*, 479-485.

PRICE, D. 1947. An analysis of the factors influencing growth and development of the mammalian reproductive tract. Physiol. Zool., *20*, 213-247.

YOUNG, H. H. 1937. *Genital Abnormalities, Hermaphroditism and Related Adrenal Diseases.* 649 pages. The Williams & Wilkins Company, Baltimore.

MALE REPRODUCTIVE SYSTEM

BASSETT, E. G. 1961. Observations on the retractor clitoridis and retractor penis muscles of mammals, with special reference to the ewe. J. Anat. (Lond.), *95*, 61-77.

BURNS, R. K. 1945. Bisexual differentiation of the sex ducts in opossums as a result of treatment with androgen. J. exp. Zool., *100*, 119-140.

CALABRISI, P. 1956. The nerve supply of the erectile cavernous tissue of the genitalia in the human embryo and fetus. Anat. Rec., *125*, 713-723.

CHANG, K. S. F., F. K. HSU, S. T. CHAN, and Y. B. CHAN. 1960. Scrotal asymmetry and handedness. J. Anat. (Lond.), *94*, 543-548.

CHRISTENSEN, G. C. 1954. Angioarchitecture of the canine penis and the process of erection. Amer. J. Anat., *95*, 227-261.

CLARK, G. and J. A. GAVAN. 1962. Skeletal effects of prepubertal castration in the male chimpanzee. Anat. Rec., *143*, 179-181.

CLERMONT, Y. 1963. The cycle of the seminiferous epithelium in man. Amer. J. Anat., *112*, 35-51.

CONGDON, E. D. and J. M. ESSENBERG. 1955. Subcutaneous attachments of the human penis and scrotum. A study of 55 series of gross sections. Amer. J. Anat., *97*, 331-357.

CRELIN, E. S. and D. K. BLOOD. 1961. The influence of the testes on the shaping of the bony pelvis in mice. Anat. Rec., *140*, 375-379.

DEYSACH, L. J. 1939. The comparative morphology of the erectile tissue of the penis with especial emphasis on the probable mechanism of erection. Amer. J. Anat., *64*, 111-131.

FARRIS, E. J. 1950. *Human Fertility and Problems of the Male.* xvi + 211 pages. Author's Press, Inc., Palisades Park, N. J.

FAWCETT, D. W. and M. H. BURGOS. 1960. Studies on the fine structure of the mammalian testis. II. The human interstitial tissue. Amer. J. Anat., *107*, 245-269.

HAMILTON, J. B. 1942. Male hormone stimulation is prerequisite and an incitant in common baldness. Amer. J. Anat., *71*, 451-480.

HAMILTON, J. B. 1946. A secondary sexual character that develops in men but not in women upon ageing of an organ present in both sexes. Anat. Rec., *94*, 466-467.

HAMILTON, J. B. 1948. The role of testicular secretions as indicated by the effects of castration in man and by studies of pathological conditions and the short lifespan associated with maleness. Recent. Progr. Hormone Res., *3*, 257-322.

KUNTZ, A. and R. E. MORRIS, JR. 1946. Components and distribution of the spermatic nerves and the nerves of the vas deferens. J. comp. Neurol., *85*, 33-44.

MOREHEAD, J. R. and C. F. MORGAN. 1963. Renewal of spermatogenesis following 28-day cryptorchidism in the rat as affected by injections of testosterone propionate. Anat. Rec., *145*, 262.

NARBAITZ, R. 1962. The primordial germ cells in the male human embryo. Carneg. Instn. Contr. Embryol., *37*, 117-119 + 5 plates.

PFEIFFER, C. A. and A. KIRSCHBAUM. 1943. Relation of interstitial cell hyperplasia to secretion of male hormone in the sparrow. Anat. Rec., *85*, 211-227.

RUKSTINAT, G. J. and R. J. HASTERLIK. 1939. Congenital absence of the penis. Arch. Path. (Chicago), *27*, 984-993.

TOBIN, C. E. and J. A. BENJAMIN. 1944. Anatomical study and clinical consideration of the fasciae limiting urinary extravasation from the penile urethra. Surg. Gynec. Obstet., *79*, 195-204.

WELLS, L. J. 1943. Descent of the testis; anatomical and hormonal considerations. Surgery, *14*, 436-472.

WESSON, M. B. 1945. The value of Buck's and Colles' fasciae. J. Urol., *53*, 365-372.

WILLIAMS, R. G. 1949. Some responses of living blood vessels and connective tissue to testicular grafts in rabbits. Anat. Rec., *104*, 147-161.

OVARIES AND OVULATION

BASSETT, D. L. 1943. The changes in the vascular pattern of the ovary of the albino rat during the estrous cycle. Amer. J. Anat., *73*, 251-291.

DUKE, K. L. 1947. The fibrous connective tissue of the rabbit ovary from sex differentiation to maturity. Anat. Rec., *98*, 507-525.

EVERETT, J. W. 1945. The microscopically demonstrable lipids of cyclic corpora lutea in the rat. Amer. J. Anat., *77*, 293-323.

EVERETT, J. W. 1956. Functional corpora lutea maintained for months by autografts of rat hypophyses. Endocrinology, *58*, 786-796.

EVERETT, N. B. 1942. The origin of ova in the adult opossum. Anat. Rec., *82*, 77-91.

FARRIS, E. J. 1946. The time of ovulation in the monkey. Anat. Rec., *95*, 337-345.

GREEN, J. A. 1957. Some effects of advancing age on the histology and reactivity of the mouse ovary. Anat. Rec., *129*, 333-347.

GREENWALD, G. S. 1964. Ovarian follicular development in the pregnant hamster. Anat. Rec., *148*, 605-609.

HILL, R. T., E. ALLEN, and T. C. KRAMER. 1935. Cinemicrographic studies of rabbit ovulation. Anat. Rec., *63*, 239-245.

KENT, H. A., JR. 1962. Polyovular follicles and multinucleate ova in the ovaries of young hamsters. Anat. Rec., *143*, 345-349.

KREHBIEL, R. and J. C. PLAGGE. 1962. Distribution of ova in the rat uterus. Anat. Rec., *143*, 239-241.

LANGMAN, L. and H. S. BURR. 1942. Electrometric timing of human ovulation. Amer. J. Obstet. Gynec., *44*, 223-230.

MARVIN, H. N. 1947. Diestrus and the formation of corpora lutea in rats with persistent estrus, treated with desoxycorticosterone acetate. Anat. Rec., *98*, 383-391.

MOSSMAN, H. W., M. J. KOERING and D. FERRY, JR. 1964. Cyclic changes of interstitial gland tissue of the human ovary. Amer. J. Anat., *115*, 235-255.

PANKRATZ, D. S. 1938. Some observations on the graafian follicles in an adult human ovary. Anat. Rec., *71*, 211-219.

PINCUS, G. 1939. The comparative behavior of mammalian eggs *in vivo* and *in vitro*. IV. The development of fertilized and artificially activated rabbit eggs. J. exp. Zool., *82*, 85-129.

WATZKA, M. 1957. Weibliche Genitalorgan: Das Ovarium. In *Handbuch der mikroskopischen Anatomie des Menschen.* (von Möllendorff). vol. 7 pt. 3, iv + 178 pages. Springer-Verlag, Berlin.

UTERUS AND MENSTRUATION

ALDEN, R. H. 1942. The oviduct and egg transport in the albino rat. Anat. Rec., *84*, 137-169.

ALDEN, R. H. 1945. Implantation of the rat egg. I. Experimental alteration of uterine polarity. J. exp. Zool., *100*, 229-235.

BO, W. J. 1956. The relationship between vitamin A deficiency and estrogen in producing uterine metaplasia in the rat. Anat. Rec., *124*, 619-627.

FORD, D. H. 1956. A study of the changes in vaginal alkaline phosphatase activity during the estrous cycle in adult and in young "first-estrous" rats. Anat. Rec., *125*, 261-277.

GARDNER, W. U. 1955. Localization of strain differences in vaginal sensitivity to estrogens. Anat. Rec., *121*, 297-298.

GOSS, C. M. 1962. Galen's On the Anatomy of the Uterus. Anat. Rec., *144*, 77-83.

JARCHO, J. 1946. Malformations of the uterus; review of the subject, including embryology, comparative anatomy, diagnosis and report of cases. Amer. J. Surg., *71*, 106-166.

MARKEE, J. E. 1940. Menstruation in intraocular endometrial transplants in the rhesus monkey. Carneg. Instn., Contr. Embryol., *28*, 221-308 + 7 plates.

PAPANICOLAOU, G. N., H. F. TRAUT, and A. A. MARCHETTI. 1948. *The Epithelia of Woman's Reproductive Organs.* vi + 53 pages. The Commonwealth Fund, New York.

REYNOLDS, S. R. M. 1965. *Physiology of the Uterus.* xxvi + 619 pages, illustrated. Hafner Publishing Company, New York.

SONG, JOSEPH. 1964. *The Human Uterus, Morphogenesis and Embryological Basis for Cancer.* xi + 196 pages, 150 figures. Charles C Thomas, Publisher, Springfield.

FEMALE PELVIS AND FASCIA

CURTIS, A. H., B. J. ANSON, F. L. ASHLEY, and T. JONES. 1942. The blood vessels of the female pelvis in relation to gynecological surgery. Surg. Gynec. Obstet., *75*, 421-423.

GREULICH, W. W. and H. THOMS. 1944. The growth and development of the pelvis of individual girls before, during, and after puberty. Yale J. Biol. Med., *17*, 91-98.

MAHRAN, M. and A. M. SALEH. 1964. The microscopic anatomy of the hymen. Anat. Rec., *149*, 313-318.

POWER, R. M. H. 1944. The exact anatomy and development of the ligaments attached to the cervix uteri. Surg. Gynec. Obstet., *79*, 390-396.

THOMS, H. and W. W. GREULICH. 1940. A comparative study of male and female pelves. Amer. J. Obstet. Gynec., *39*, 56-62.

UHLENHUTH, E. 1953. *Problems in the Anatomy of the Pelvis.* xiv + 206 pages. J. B. Lippincott Company, Philadelphia.

MAMMAE

AGATE, F. J., JR. 1952. The growth and secretory activity of the mammary glands of the pregnant rhesus monkey (Macaca mulatta) following hypophysectomy. Amer. J. Anat., *90*, 257-283.

CHOLNOKY, T. DE. 1939. Supernumerary breast. Arch. Surg., *39*, 926-941.

GIACOMETTI, L. and W. MONTAGNA. 1962. The nipple and areola of the human female breast. Anat. Rec., *144*, 191-197.

KARSNER, H. T. 1946. Gynecomastia. Amer. J. Path., *22*, 235-315.

MALINIAC, J. W. 1943. Arterial blood supply of the breast. Arch. Surg., *47*, 329-343.

ROMANO, S. A. and E. M. McFETRIDGE. 1938. The limitations and dangers of mammography by contrast mediums. J.A.M.A., *110*, 1905-1910.

THOREK, M. 1942. *Plastic Surgery of the Breast and Abdominal Wall.* xiii + 446 pages. Charles C Thomas, Springfield.

TRAURIG, H. H. and C. F. MORGAN. 1962. Autoradiographic investigation of mouse mammary gland growth by the incorporation of tritiated thymidine into its epithelial components. Anat. Rec., *142*, 286-287.

WILLIAMS, W. L. 1942. Normal and experimental mammary involution in the mouse as related to the inception and cessation of lactation. Amer. J. Anat., *71*, 1-41.

WILLIAMS, W. L. 1945. The effects of lactogenic hormone on post parturient unsuckled mammary glands of the mouse. Anat. Rec., *93*, 171-183.

18 | *The Endocrine Glands*

Endocrine glands or duct-less glands (*glandulae sine ductibus*) are grouped together because of their common characteristic of not having ducts and of discharging their specific secretions, called hormones, into the blood stream.

The **hormones** are chemical messengers, and although they are carried by the blood stream to all parts of the body, only certain organs or types of cells are able to respond to their stimulation. The specific organ which does respond to a particular hormone is called its **target organ.** There are a number of hormones besides the ones secreted by the endocrine glands described in this chapter. For example, the islands of Langerhans in the pancreas secrete insulin; the gonads, androgens and estrogens; and the mucous membrane of the gastrointestinal tract, secretin. These structures are described in other chapters.

The endocrine glands included in this chapter are the thyroid and parathyroid glands, the hypophysis or pituitary gland, the suprarenal gland, and the paraganglia. The pineal gland is included because it is ductless although its endocrine activity is not established. The thymus was at one time classed with the ductless glands but is now included in the lymphatic system.

THE THYROID GLAND

Development.—The thyroid gland is developed from a median diverticulum of the ventral wall of the pharynx which appears about the fourth week on the summit of the tuberculum impar, but later is found in the furrow immediately caudal to the tuberculum (Fig. 16–3). It grows distalward and caudalward as a tubular duct, which bifurcates and subsequently subdivides into a series of cellular cords, from which the isthmus and lateral lobes of the thyroid gland are developed. The connection of the median diverticulum with the pharynx is termed the **thyroglossal duct;** its continuity is subsequently interrupted and it undergoes degeneration, its cranial end being represented by the foramen cecum of the tongue, and its caudal by the pyramidal lobe of the thyroid gland.

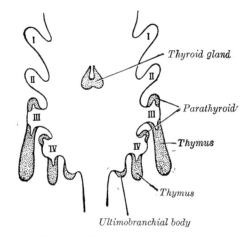

FIG. 18–1.—Scheme showing development of brachial epithelial bodies. (Modified from Kohn.) *I, II, III, IV.* Branchial pouches.

The **Thyroid Gland** (*glandula thyroidea*) (Fig. 18–2) is a highly vascular organ, situated at the front of the neck. It consists of right and left lobes connected across the middle line by a narrow portion, the isthmus. Its weight is somewhat variable, but is usually about 30 grams. It is slightly heavier in women and it becomes enlarged during pregnancy.

The **lobes** (*lobuli gl. thyroideae*) are conical in shape, the apex of each being directed cranialward and lateralward as far as the

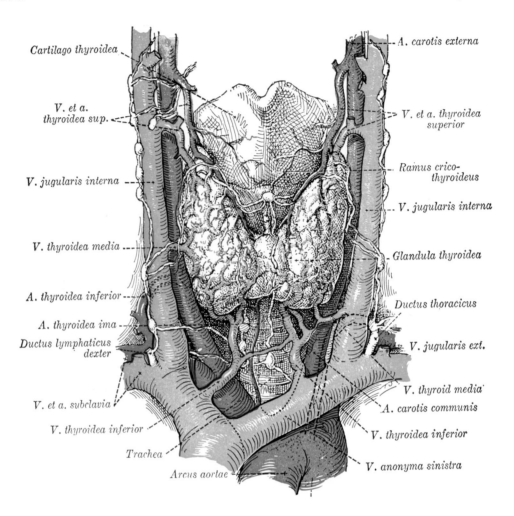

Cartilago thyroidea

A. carotis externa

V. et a. thyroidea sup.

V. et a. thyroidea superior

Ramus crico-thyroideus

V. jugularis interna

V. jugularis interna

V. thyroidea media

Glandula thyroidea

A. thyroidea inferior

A. thyroidea ima

Ductus thoracicus

Ductus lymphaticus dexter

V. jugularis ext.

V. thyroid media

A. carotis communis

V. et a. subclavia

V. thyroidea inferior

V. thyroidea inferior

Trachea

V. anonyma sinistra

Arcus aortae

Fig. 18–2.—Blood supply of the thyroid gland, viewed from in front. The thoracic duct and principal lymphatics are also shown. (Ecyleshymer and Jones.)

junction of the middle with the caudal third of the thyroid cartilage; the base looks caudalward, and is on a level with the fifth or sixth tracheal ring. Each lobe is about 5 cm long; its greatest width is about 3 cm, and its thickness about 2 cm. The **lateral or superficial surface** is convex, and covered by the skin, the subcutaneous and deep fasciae, the Sternocleidomastoideus, the superior belly of the Omohyoideus, the Sternohyoideus and Sternothyroideus, and beneath the last muscle by the visceral layer of the deep fascia, which forms a capsule for the gland. The **deep** or **medial surface** is molded over the underlying structures, viz, the trachea, the Constrictor pharyngis

inferior and posterior part of the Crico-thyroideus, the esophagus (particularly on the left side of the neck), the superior and inferior thyroid arteries, and the recurrent nerves. The **ventral border** is thin, and inclines obliquely from above downward toward the middle line of the neck, while the **dorsal border** is thick and overlaps the common carotid artery, and, as a rule, the parathyroids.

The **isthmus** (*isthmus gl. thyroideae*) connects the lower thirds of the lobes together; it measures about 1.25 cm in breadth, and the same in depth, and usually covers the second and third rings of the trachea. Its situation and size present many variations. In the middle line of the neck it is covered by the skin and fascia, and close to the middle line, on either side, by the Sternothyroideus. Across its cranial border runs an anastomotic branch uniting the two superior thyroid arteries; at its caudal border are the inferior thyroid veins. Sometimes the isthmus is altogether wanting.

A third lobe, of conical shape, called the **pyramidal lobe,** frequently arises from the cranial part of the isthmus, or from the adjacent portion of either lobe, but most commonly the left, and ascends as far as the hyoid bone. It is occasionally quite detached, or may be divided into two or more parts.

A fibrous or muscular band is sometimes found attached to the body of the hyoid bone, and to the isthmus of the gland, or its pyramidal lobe. When muscular, it is termed the **Levator glandulae thyroideae.**

Small detached portions of thyroid tissue are sometimes found in the vicinity of the lateral lobes or above the isthmus; they are called **accessory thyroid glands** (*glandulae thyroideae accessoriae*).

Structure.—The thyroid gland is invested by a thin capsule of connective tissue, which projects into its substance and imperfectly divides it into masses of irregular form and size. When the fresh organ is cut open, its interior appears to be of a brownish-red color, and made up of a number of closed vesicles containing colloid and separated from each other by intermediate connective tissue (Fig. 18–3).

The thyroid follicles (acini, vesicles) of the adult are closed sacs of inconstant shape and size. They are normally of microscopic dimensions, although macroscopically visible in colloid goiter. The follicles located peripherally may be larger than those centrally placed. The cuboidal epithelium of the normal thyroid rests directly on the delicate connective tissue surrounding the follicle. No basement membrane can be seen. The capillaries and lymphatics are thus in close contact with the secretory epithelium of the gland. The follicle normally is filled with colloid which contains the active principle of the gland, thyroxin. The thyroid epithelium may secrete this hormone directly into the colloid-filled lumen of the follicle where it is stored, or the hormone may be secreted directly into the capillaries. The stored colloid may be absorbed and liberated into the capillaries.

The hormone, thyroxin, requires iodine for its elaboration and is normally obtained in adequate amounts in the diet. Proper functioning of the thyroid gland and normal histological structure are dependent upon the adequacy of available iodine. The thyroid itself is activated or regulated by another hormone, the thyrotrophic hormone of the anterior pituitary gland. Removal of the thyroid results in a marked reduction of the oxidative processes of the body. This lowered metabolic rate is characteristic of hypothyroidism. In infancy and childhood the thyroid gland is essential to normal growth of the body.

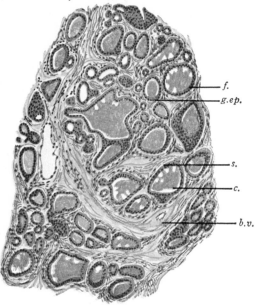

Fig. 18–3—Section of thyroid gland of a 21-year-old man. *b.v.*, blood vessel; *c.*, colloid; *f.*, follicle; *g.ep.*, glandular epithelium; *s.* connective tissue stroma. × 120. (Sobotta.)

Vessels and Nerves (Fig. 18–4).—The **arteries** supplying the thyroid gland are the superior and inferior thyroids and sometimes an additional branch (thyroidea ima) from the brachiocephalic artery or the arch of the aorta, which ascends upon the front of the trachea. The arteries are remarkable for their large size and frequent anastomoses. The **veins** (Fig. 18–2) form a plexus on the surface of the gland and on the front of the trachea; from this plexus the superior, middle, and inferior thyroid veins arise; the superior and middle end in the internal jugular, the inferior in the brachiocephalic vein. The capillary blood vessels form a dense plexus in the connective tissue around

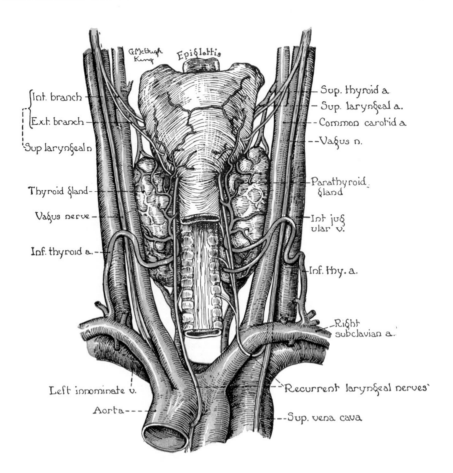

G.McHugh King

Epiglottis

Int. branch
Ext. branch
Sup laryngeal n

Thyroid gland
Vagus nerve

Inf. thyroid a.

Sup. thyroid a.
Sup. laryngeal a.
Common carotid a
Vagus n.

Parathyroid gland

Int. jugular v.

Inf. thy. a.

Right subclavian a.

Left innominate v.

Aorta

Recurrent laryngeal nerves

Sup. vena cava

Fig. 18–4.—Posterior view of larynx, trachea, thyroid, and parathyroids. Relations of thyroid arteries and laryngeal nerves are shown. (Nordland [1930]; Surgery, Gynecology and Obstetrics.)

the vesicles, between the epithelium of the vesicles and the endothelium of the lymphatics, which surround a greater or smaller part of the circumference of the vesicle. The **lymphatic vessels** run in the interlobular connective tissue, not uncommonly surrounding the arteries which they accompany, and communicate with a network in the capsule of the gland; they may contain colloid material. They end in the thoracic and right lymphatic trunks. The **nerves** are derived from the middle and inferior cervical ganglia of the sympathetic.

THE PARATHYROID GLANDS

Development.—The parathyroid bodies are developed as outgrowths from the third and fourth branchial pouches (Fig. 18–1).

Anatomy.—The **parathyroid glands** (Fig. 18–4) are small brownish-red bodies, situated as a rule between the dorsal borders of the lateral lobes of the thyroid gland and its capsule. They differ from it in structure, being composed of masses of cells arranged in a more or less columnar fashion with numerous intervening capillaries. They measure on an average about 6 mm in length, and from 3 to 4 mm in breadth, and usually present the appearance of flattened oval disks. They are divided according to their situation, into **superior** and **inferior.** The superior, usually two in number, are the more constant in position, and are situated, one on either side, at the level of the caudal border of the cricoid cartilage, beside the junction of the pharynx and esophagus. The inferior, also usually two in number, may be applied to the caudal edge of the lateral lobes, or placed at some little distance caudal to the thyroid gland, or found in relation to one of the inferior thyroid veins.

In man, they number four as a rule; fewer than four were found in less than 1 per cent of over a thousand persons (Pepere), but more than four (five or six) in over 33 per cent of 122 bodies examined by Civalleri. In addition, numerous minute islands of parathyroid tissue may be found scattered in the connective tissue and fat of the neck around the parathyroid glands proper, and quite distinct from them.

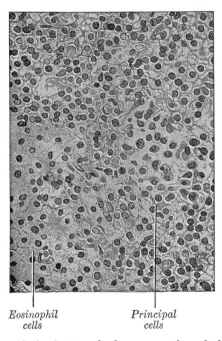

Eosinophil   *Principal*
cells   *cells*

Fig. 18–5.—Section of a human parathyroid gland to show principal and eosinophil cells. × 250.

Structure.—Microscopically the parathyroids consist of intercommunicating columns of cells supported by connective tissue containing a rich supply of blood capillaries. Most of the cells are clear, but some, larger in size, contain oxyphil granules. Vesicles containing colloid have been described as occurring in the parathyroid.

The parathyroids secrete a hormone, parathyrin or parathormone, necessary for calcium metabolism. The tetany which follows parathyroidectomy can be relieved by feeding or injecting calcium salts or parathyroid extracts.

HYPOPHYSIS CEREBRI OR PITUITARY GLAND

Development.—The hypophysis has a dual origin which corresponds to the two distinct parts of the adult gland. At an early embryonic stage, about the fourth week, when the neural tube and the primitive digestive tube are still in close proximity, outgrowths or diverticula from both tubes come in contact with each other to form the primordium of the hypophysis. The part from the neural tube is in the diencephalic floor, the region of the hypothalamus (Figs. 11–9, 11–11), and is called the **infundibular process.** The part from the alimentary tube is from the portion of the future pharynx developed from ectoderm, the stomodaeum, and is known as **Rathke's pouch** (Fig. 16–7). The infundibular process retains a connection with the hypo-

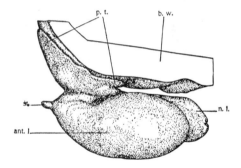

Fig. 18–6.—Model of hypophysis and adjacent brain wall from a thirty-day embryo, viewed from the left side. × 25. *b, w.*, brain wall; *p. t.*, pars tuberalis; *st.*, stalk; *ant. l.*, anterior lobe; *n. l.*, neural lobe. (Atwell, Amer. J. Anat.; courtesy of Wistar Institute.)

thalamus as the stalk, preserves some resemblance to neural tissue, and becomes the *neural lobe.* The cells of Rathke's pouch become adherent to the infundibular process and grow partly around it, forming the *pars tuberalis* (Fig. 18–6). The remainder of the pouch forms a double layered cup and the connection with the pharynx gradually disappears. The rostral portion of the cup thickens greatly and becomes the glandular anterior lobe or *pars distalis.* The other layer of the cup remains adherent to the neural lobe, thickens but little, and develops into the *pars intermedia.* The cavity of the original diverticulum eventually disappears except for the vestigial lumen,

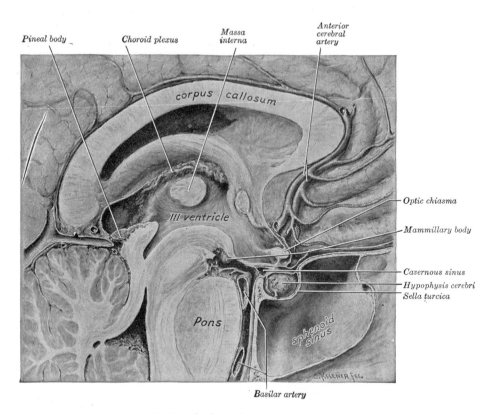

FIG. 18–7.—The hypophysis cerebri in position.

in the form of one or more narrow vesicles of variable length in the pars intermedia.

The **Hypophysis Cerebri** or **Pituitary Gland** (*glandula pituitaria*) is couched in the sella turcica or hypophyseal fossa of the body of the sphenoid bone. It is attached to the hypothalamus by the **stalk** or **infundibulum.** After the brain has been removed from the cranial cavity, the cut end of the stalk is visible protruding through a hole in the diaphragma sellae (Fig. 12–1). The latter is a shelf of dura mater stretching between the clinoid processes and covering over the sella turcica. The two principal portions of the gland, the anterior and posterior lobes, are not easily distinguished in the whole gland but are visible in a median sagittal section because of a difference in color.

The hypophysis is a small gland, 1.2 to 1.5 cm in its greatest diameter which is from side to side; its rostrocaudal diameter is approximately 1 cm, and its thickness is

0.5 cm. Its weight in an adult male is from 0.5 to 0.6 gm but varies according to the stature rather than the weight of an individual. It is larger in women, and since there is a slight increase during pregnancy, in a multipara it may weigh as much as 1 gm.

Relations.—Many important structures lie in close proximity to the hypophysis (Figs. 12–1, 18–7). The internal carotid artery emerges from the dural covering of the cavernous sinus immediately lateral to it. The intercavernous and circular sinuses are enclosed in the diaphragma sellae above it. The optic nerves, chiasma, and optic tracts lie between it and the bulk of the brain. In skulls with large sphenoid sinuses, the hypophysis may be separated from this air space only by a thin plate of bone and the dural lining of the sella. The practical importance of these relations is emphasized in patients with tumors of the hypophysis. They may compress the carotid arteries or the optic nerves causing changes in the retina known as choked disks which are visible with the opthalmoscope.

The **hypophysis** is divided into two parts on the basis of embryologic development, adult morphology, and function. The naming of these main parts and of their subdivisions results in some overlapping which is best shown in a table.

acid dyes, named **acidophiles,** and those that take up basic dyes, the **basophiles.** The basophiles can again be subdivided into beta and delta types by their coloration with certain dyes. There is no definite pattern of distribution of the two chromophiles, either in different parts of the gland or within individual cords. The chromophobes are smaller than the chromophiles, less discrete, and tend to accumulate in

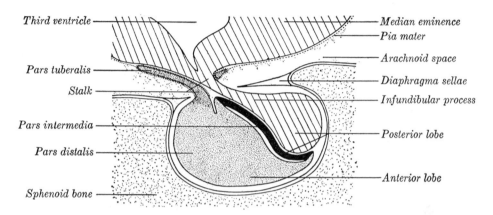

FIG. 18–8.—Diagrammatic median sagittal section of the hypophysis cerebri or pituitary gland. (Atwell, Amer. J. Anat., '18)

Major Divisions and Subdivisions of the Mammalian Hypophysis

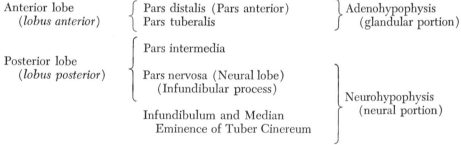

| Anterior lobe (*lobus anterior*) | { Pars distalis (Pars anterior) Pars tuberalis | } Adenohypophysis (glandular portion) |
| Posterior lobe (*lobus posterior*) | { Pars intermedia Pars nervosa (Neural lobe) (Infundibular process) Infundibulum and Median Eminence of Tuber Cinereum | } Neurohypophysis (neural portion) |

The **anterior lobe** (*lobus anterior*) or **adenohypophysis** is larger than the posterior, occupying about three-fourths of the gland. It appears darker in a sagittal section because of its vascularity and composition of glandular tissue.

Structure.—The anterior lobe or **pars distalis** is composed of cords of epithelial cells richly supplied with sinusoidal capillaries. The epithelial cells are of two main types, those that are colored by various stains, the **chromophiles,** and those that remain pale or unstained, the **chromophobes.** The chromophiles in turn are of two kinds, those whose granules take up

the center of cords and clumps. The lack of mitotic division in the hypophysis and certain experimental observations indicate that the chromophiles undergo cycles of secretory activity in which they discharge their specific granules and pass through a resting stage in the form of chromophobes.

The **posterior lobe** (*lobus posterior*) or **neurohypophysis** is pale in color, corresponding to the overlying brain with which it is continuous by means of the infundibulum and median eminence of the tuber cinereum.

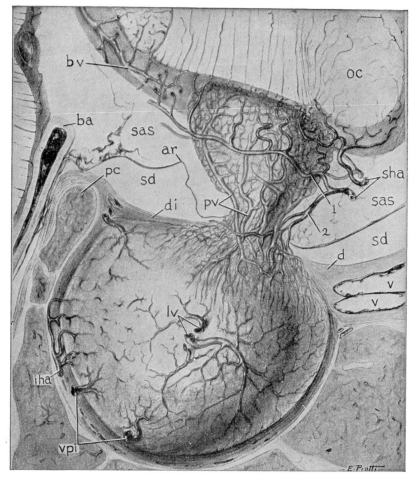

Fɪɢ. 18–9.—Schematic drawing of the hypophysis of the adult rhesus monkey. *ar,* arachnoid membrane; *ba,* basilar artery; *bv,* basilar vein; *d,* dura; *di,* sellar diaphragm; *iha,* inferior hypophyseal artery; *lv,* lateral hypophyseal veins; *oc,* optic chiasma; *pc,* posterior clinoid process; *pv,* portal venules; *sas,* subarachnoid space; *sd,* subdural space; *sha,* superior hypophyseal arteries (1, branches to hypophyseal stalk; 2, branches to anterior lobe); *v,* dural veins; *vpi,* veins of infundibular process. (Wislocki [1938], *Pituitary Gland,* Assn. for Research in Nervous and Mental Diseases, The Williams & Wilkins Co.)

Structure.—The specific cell is the pituicyte which resembles neuroglia in shape due to its long processes. The outlines of the cells and their processes are only distinguishable in special preparations. They contain a variable number of granules and lipid globules. The inclusions blackened by osmic acid correspond with the neurosecretion which is stored by them but produced by the nerve cells of the supraoptic and paraventricular nuclei of the hypothalamus.

The **pars tuberalis** is a layer of cuboidal cells covering the stalk and neighboring area. It is richly supplied with arterial blood.

The **pars intermedia** is composed of a thin layer of epithelial cells between the neural lobe and the pars distalis. It encloses the remnants of the vestigial lumen of Rathke's pouch which appear as vesicles containing a "colloid" resembling that of the thyroid gland in histological preparations but having no relation to it chemically. The lining cells also may be ciliated.

Blood supply.—The **arteries** to the hypophysis are the superior hypophyseal arteries from the internal carotid or posterior communicating arteries and the inferior hypophyseal arteries which are also branches of the internal carotid but traverse the cavernous sinus. The branches of the superior arteries supply the stalk and adjacent parts of the anterior lobe. The branches of the inferior arteries supply the posterior lobe. The blood supply of the pars distalis is mainly through a **portal system** of veins. The blood from the capillaries of the pars tuberalis and adjacent stalk collect into veins which pass along the stalk and break up into the numerous sinusoidal capillaries of the pars distalis (Harris '55).

The **veins** of the hypophysis are the lateral hypophyseal veins which drain into the cavernous and intercavernous sinuses.

Nerves.—The pars distalis has no specific innervation. Fibers from the superior cervical ganglion of the sympathetic system have been traced along the blood vessels but have not been conclusively associated with the glandular cells (Green '51).

The neurohypophysis is supplied by fibers from the supraoptic and paraventricular nuclei of the hypothalamus. Similar osmiophilic neurosecretory granules are found in the cells of these nuclei and in their processes which extend down to the posterior lobe in the **hypothalamico-hypophyseal** system in the infundibulum (Palay '53). Experimental evidence indicates that the hormones extracted from the neurohypophysis originate from these neurosecretions and are merely stored in the gland.

Function.—The hypophysis supplies a number of hormones. Among those of the anterior lobe are a growth hormone, somatotrophin, affecting general body growth; a thyrotrophic hormone acting on the thyroid gland; an adrenocorticotrophic hormone (ACTH) acting on the suprarenal cortex; two gonadotrophic hormones, one stimulating ovarian follicles (FSH) and another stimulating the lutein cells (LH) (see page 1329), and a hormone (prolactin) promoting milk secretion by the mammary gland. Attempts to associate the specific types of cells with the elaboration of these hormones have been partly successful. The growth hormone is produced by the acidophiles. Tumors of the gland composed of acidophiles are found in patients with a tremendous overgrowth of various parts of the body in a condition called acromegaly, and their absence is notable in pituitary dwarfs. Acidophiles also appear to be responsible for prolactin. Experimental evidence points to the basophiles as the source of the follicle-stimulating, thyroid-stimulating, and luteinizing hormones. The source of adrenotrophic hormone is controversial but may be a specific cell developed from chromophobes (Siperstein '63).

From the neurohypophysis two hormones have been extracted: oxytocin (Pituitrin) stimulates the contraction of smooth muscle and is sometimes used in obstetrical practice to make the uterus contract; the antidiuretic principle or vasopressin inhibits diuresis by the kidneys and also raises the blood pressure. As mentioned above, it is doubtful that the neural lobe elaborates these hormones. A lack of the antidiuretic principle in patients produces a condition called diabetes insipidus.

THE SUPRARENAL OR ADRENAL GLAND

The suprarenal gland in man and other mammals is a combination of two distinct glands which remain independent in fishes and other more primitive vertebrates. The two organs in the human gland are fused together but remain distinct and identifiable as the cortex and medulla of the adult gland.

Development.—The **cortex** of the suprarenal gland is first recognizable in an embryo of the sixth week as a groove in the *coelom* at the base of the mesentery near the cranial end of the mesonephros. The cells at the bottom of the groove proliferate rapidly to form a mass in the mesenchyme extending toward the aorta. During the seventh and eighth weeks the cells become arranged into cords with dilated blood spaces between; the connection with the coelomic mesothelium is lost, and a capsule of connective tissue encloses the gland. During the remainder of fetal development the cortical tissue is composed of two zones, an outer zone of more undifferentiated cells and an inner zone of cell cords which appear to be differentiated and active in secretion. After birth the inner zone atrophies and the outer zone is differentiated into the three zones of the adult gland.

The **medulla** of the suprarenal gland is developed from cells of the *neural crest* which migrate ventrally along with the cells which form the sympathetic ganglia. These cells later detach themselves from the ganglia and become small knots of glandular cells scattered along the vertebral column. During the seventh and eighth weeks a large group of the neural cells, migrating along the suprarenal vein, invades the cortex, thus establishing the primordium of the suprarenal medulla. These glandular cells of sympathetic origin contain a substance, probably the precursor of the specific secretion, epinephrine, which is colored brown by chromic acid. This has given them their name of **chromaffin** or **pheochrome cells.** Many of the small masses of chromaffin cells mentioned above persist throughout life and are given the name **paraganglia.**

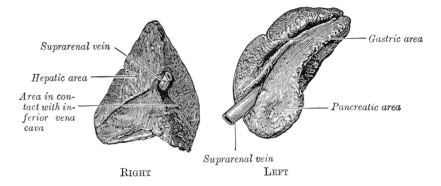

FIG. 18–10.—Suprarenal glands, ventral view.

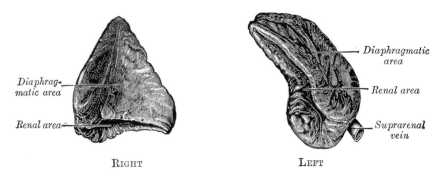

FIG. 18–11.—Suprarenal glands, dorsal view.

The **suprarenal gland** (*glandula supra-renalis; adrenal gland*) (Figs. 12–78, 12–79), as the name suggests, is located at the cranial pole of each of the two kidneys. They are quite different in shape on the two sides (Figs. 18–10, 18–11). The right gland resembles a pyramid, or the cocked hat of colonial days; the left is semilunar in form and tends to be slightly larger. The length and width vary from 3 to 5 cm and the thickness 4 to 6 mm. The average weight is 3.5 to 5 gm.

Relations.—The **right suprarenal** is situated dorsal to the inferior vena cava and right lobe of the liver, and ventral to the diaphragm and cranial end of the right kidney. The *ventral surface* has two areas: a medial, narrow, and non-peritoneal, which lies cranial to the inferior vena cava; and a lateral, somewhat triangular, in contact with the liver. The cranial part of the latter surface is devoid of peritoneum, and is in relation with the bare area of the liver near its caudal and medial angle, while its caudal

portion is covered by peritoneum, reflected onto it from the inferior layer of the coronary ligament; occasionally the duodenum overlaps the inferior portion. A little below the apex, and near the ventral border of the gland, is a short furrow termed the **hilum,** from which the suprarenal vein emerges to join the inferior vena cava. The *posterior surface* is divided into cranial and caudal parts by a curved ridge: the cranial, slightly convex, rests upon the diaphragm; the caudal, concave, is in contact with the cranial end and the adjacent part of the ventral surface of the kidney.

The **left suprarenal** is crescentic in shape, its concavity being adapted to the medial border of the cranial part of the left kidney. Its *ventral surface* has two areas: a cranial one, covered by the peritoneum of the omental bursa, which separates it from the cardiac end of the stomach, and sometimes from the cranial extremity of the spleen; and a caudal one, which is in contact with the pancreas and lienal artery, and is therefore not covered by the peritoneum. On the ventral surface, near its caudal end, is a furrow or **hilum,** from which the suprarenal vein emerges. Its *posterior surface* presents a vertical ridge, which divides it into two areas; the lateral area rests on the kidney, the medial and smaller on the left crus of the diaphragm.

The surface of the suprarenal gland is surrounded by areolar tissue containing much fat, and closely invested by a thin fibrous capsule, which is difficult to remove on account of the numerous fibrous processes and vessels entering the organ through the furrows on its ventral surface and base.

Accessory suprarenals (*glandulae suprarenales accessoriae*) are of frequent occurrence in the connective tissue near the main gland, but may occur near the testis or the ovary where small groups of embryonic cells might have migrated with the mesonephros, especially in certain hermaphroditic conditions. They are usually small round bodies composed of cortical cells. Larger ones may have the cortical zones represented and, more rarely, contain central medullary tissue.

Structure.—The suprarenal gland consists of two portions, an outer **cortex** taking up the greater part of the organ and an inner **medulla** (Fig. 18–12). In the fresh condition the outer cortex is a deep yellow color, the inner part a dark red or brown, and the medulla a pale

pink. At the transition between the cortex and medulla, there is no connective tissue barrier and the cords of the two types of cells are intermingled for a short distance.

The **cortical portion** (*substantia corticalis*) consists of a fine connective-tissue network, in which is embedded the glandular epithelium. The epithelial cells are polyhedral in shape and possess rounded nuclei; many of the cells contain coarse granules, others lipoid globules. Owing to differences in the arrangement of the cells, three distinct zones can be made out: (1) the **zona glomerulosa,** situated beneath the capsule, consists of cells arranged in rounded groups, with here and there indications of an alveolar structure; the cells of this zone are very granular, and stain deeply. (2) The **zona fasciculata,** continuous with the zona glomerulosa, is composed of columns of cells arranged in a radial manner; these cells contain finer granules and in many instances globules of lipoid material. (3) The **zona reticularis,** in contact with the medulla, consists of cylindrical masses of cells irregularly arranged; these cells often contain pigment granules which give this zone a darker appearance than the rest of the cortex (Fig. 18–13).

The **medullary portion** (*substantia medullaris*) is extremely vascular, and consists of large chromaffin cells arranged in a network. The irregular polyhedral cells have a finely granular cytoplasm that is probably concerned with the secretion of epinephrine. In the meshes of the cellular network are large anastomosing venous sinuses (sinusoids) which are in close relationship with the chromaffin or medullary cells. In many places the endothelial lining of

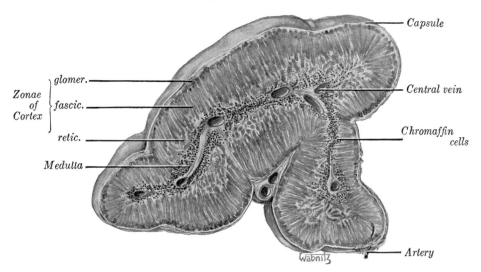

FIG. 18–12.—Section through suprarenal gland of adult cadaver. Lightly stained with hematoxylin. × 5.

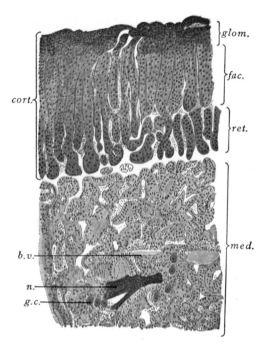

FIG. 18–13.—Section of human suprarenal gland. *b.v.*, central vein; *cort.*, cortex; *fac.*, zona fasciculata; *glom.*, zona glomerulosa; *ret.*, zona reticularis; *med.*, medulla; *g.c.*. ganglion cell; *n.*, nerve fibers. × 100. (Redrawn from Sobotta.)

the blood sinuses is in direct contact with the medullary cells. Some authors consider the endothelium absent in places and here the medullary cells are directly bathed by the blood. This intimate relationship between the chromaffin cells and the blood stream undoubtedly facilitates the discharge of the internal secretion into the blood. There is a loose meshwork of supporting connective tissue containing non-striped muscle fibers. This portion of the gland is richly supplied with non-medullated nerve fibers, and here and there sympathetic ganglia are found.

Vessels and Nerves.—The suprarenals are highly vascular organs. The arteries are numerous and of comparatively large size. These arteries are derived from the aorta and the inferior phrenic and the renal arteries. Three sets of branches penetrate the capsule, one of which breaks up in capillaries which supply the capsule. The second set breaks up into capillaries which supply the cell cords of the cortex and empty into veins in the medulla. The third set of arteries traverses the cortex to supply the medulla only and break up into the sinusoids of the medulla. Venous blood from the capsule

is collected into veins of the capsule. Blood from the two other sets of arteries is gathered into the central vein in the medulla which emerges at the hilum as the suprarenal vein. On the right side this vein opens into the inferior vena cava, on the left into the left renal vein.

Lymphatics accompany the large blood vessels and end in the lumbar nodes.

The **nerves** are exceedingly numerous, and are derived from the celiac and renal plexuses. They enter the lower and medial part of the capsule, traverse the cortex, and end around the cells of the medulla. They have numerous small ganglia in the medullary portion of the gland.

Function.—The adrenal medulla elaborates an internal secretion, epinephrine (Adrenalin), which has definite sympathomimetic actions such as causing a constriction of arterioles, acceleration of the heart rate, contraction of the radial muscle of the iris. In man an injection of the drug epinephrine hydrochloride results in a rise in systolic blood pressure, pulse rate, an increase in the minute volume of the heart and in volume of respiration. These effects are transitory and disappear after one or two hours. A rise in blood sugar and basal metabolic rate follows injection of this drug. This is due to an increased enzymatic breakdown of glycogen in liver and muscle. The action of epinephrine is of brief duration, the hormone being rapidly inactivated in the body (Cori).

The adrenal cortex elaborates one or more hormones essential to maintenance of life. Various fractions serve in the maintenance of physiological "steady states," in regulation of the distribution of water and electrolytes and in many aspects of carbohydrate metabolism and muscular efficiency. Research in this field is very active.

Paraganglia, Glomera, and "Glands" or "Bodies"

The various structures which have been given these names are for the most part small, difficult to demonstrate, and frequently misnamed. The following is an attempt to sort them out and to give their appropriate names and synonyms.

Paraganglia

The **paraganglia** (*chromaffin bodies; pheochrome bodies*) are small groups of chromaffin cells connected with the ganglia of the sympathetic trunk and the ganglia of the celiac, renal, suprarenal, aortic and hypogastric plexuses. They are some-times found in connection with the ganglia of other sympathetic plexuses.

Aortic Bodies or Lumbar Paraganglia

The **aortic glands** or **bodies** (*corpora para-aortica, glands or organs of Zucker-kandl*) are the largest of these groups of chromaffin cells and measure in the newborn about 1 cm in length. They lie one on either side of the aorta in the region of the inferior mesenteric artery. They decrease in size with age and after puberty are only visible with the microscope. Other groups of chro-maffin cells have been found associated with the sympathetic plexuses of the abdomen independently of the ganglia.

Glomus Coccygeum

The **glomus coccygeum** (*coccygeal gland or body; Luschka's gland*) is placed ventral

to, or immediately distal to, the tip of the coccyx. It is about 2.5 mm in diameter and is irregularly oval in shape; several smaller nodules are found around or near the main mass.

It consists of irregular masses of round or polyhedral cells of each mass being grouped around a dilated sinusoidal capillary vessel. Each cell contains a large round or oval nucleus, the protoplasm surrounding which is clear, and is not stained by chromic salts. It is not a chromaffin paraganglion and since its structure is not like that of the chemoreceptors (carotid bodies) its func-tion is unknown (Hollinshead '42).

The **glomus caroticum** or **carotid body** is not a paraganglion but a chemoreceptor and is described on page 895.

The **glomus aorticum** is better named the **cardioaortic body** or glomus aorticum supra-cardiale and is a chemoreceptor similar to the carotid body, page 895.

REFERENCES

ENDOCRINE GLANDS AND ENDOCRINOLOGY

FRANKEL, H. H., R. R. PATEK, and S. BERNICK. 1962. Long term studies of the rat reticuloendo-thelial system and endocrine gland responses to foreign particles. Anat. Rec., *142*, 359-373.

GORBMAN, A. and H. A. BERN. 1962. *A Textbook of Comparative Endocrinology.* xv + 468 pages, illustrated. John Wiley & Sons, Inc., New York.

HALL, P. F. 1959. *The Functions of the Endocrine Glands.* xviii + 290 pages, 77 figures. W. B. Saunders Company, Philadelphia.

MITSKEVICH, M. S. 1957. *Glands of Internal Secre-tion in the Embryonic Development of Birds and Mammals.* x + 304 pages, illustrated. Trans-lated from Russian by Israel Program for Scien-tific Translations for the National Science Foundation.

TALBOT, N. B., E. H. SOBEL, J. W. McARTHUR, and J. D. CRAWFORD. 1952. *Functional Endocrinol-ogy from Birth through Adolescence.* xxx + 638 pages. Harvard University Press, Cambridge.

WILKINS, L. 1950. *The Diagnosis and Treatment of Endocrine Disorders in Childhood and Ado-lescence.* xx + 384 pages. Charles C Thomas, Springfield.

YOUNG, W. C. 1961. *Sex and Internal Secretions.* 3rd Edition. Vol. 1, xxiv + 704 pages, Vol. 2, vi + 707-1609 pages, illustrated. Williams & Wilkins Company, Baltimore.

SUPRARENAL GLAND

ALDEN, R. H. and J. S. DAVIS. 1962. Role of adrenals in uterine lipid metabolism. Anat. Rec., *142*, 53-56.

BOURNE, G. H. 1949. *The Mammalian Adrenal Gland.* vii + 239 pages. Oxford University Press, London.

CROWDER, R. E. 1957. The development of the adrenal gland in man, with special reference to origin and ultimate location of cell types, and evidence in favor of the "cell migration" theory. Carneg. Instn. Contr. Embryol., *36*, 193-210 + 8 plates.

HOUSER, R. G., F. A. HARTMAN, R. A. KNOUFF, and F. W. McCOY. 1962. Adrenals in some panama monkeys. Anat. Rec., *142*, 41-51.

MacFARLAND, W. E. and H. A. DAVENPORT. 1941. Adrenal innervation. J. comp. Neurol., *75*, 219-233.

SWINYARD, C. A. 1943. Growth of the human supra-renal glands. Anat. Rec., *87*, 141-150.

UOTILA, U. U. 1940. The early embryological devel-opment of the fetal and permanent adrenal cortex in man. Anat. Rec., *76*, 183-203.

WIESEL, J. 1902. Beiträge zur Anatomie und Entwicklung der menschlicken Nebenniere. Anat. Hefte, *19*, 481-522.

SUPRARENAL CORTEX

CAHILL, G. F., M. M. MELICOW, and H. DARBY. 1942. Adrenal cortical tumors. Surg. Gynec. Obstet., *74*, 281-305.

DOUGHERTY, T. F. and A. WHITE. 1945. Functional alterations in lymphoid tissue induced by adrenal cortical secretion. Amer. J. Anat., *77*, 81-116.

EVERETT, N. B. 1949. Autoplastic and homoplastic transplants of the rat adrenal cortex and medulla to the kidney. Anat. Rec., *103*, 335-347.

HOAR, R. M. and A. J. SALEM. 1962. The production of congenital malformations in guinea pigs by adrenalectomy. Anat. Rec., *143*, 157-167.

HOERR, N. L. 1936. Histological studies on lipins. II. A cytological analysis of the liposomes in the adrenal cortex of the guinea pig. Anat. Rec., *66*, 317-342.

HUNT, T. E. and E. A. HUNT. 1964. The proliferative activity of the adrenal cortex using a radioautographic technic with Thymidine-H³. Anat. Rec., *149*, 387-395.

JONES, I. C. 1957. *The Adrenal Cortex.* x + 316 pages. Cambridge University Press, New York.

KNOUFF, R. A., J. B. BROWN, and B. M. SCHNEIDER. 1941. Correlated chemical and histological studies of the adrenal lipids. I. The effect of extreme muscular activity on the adrenal lipids of the guinea pig. Anat. Rec., *79*, 17-38.

SALMON, T. N. and R. L. ZWEMER. 1941. A study of the life history of cortico-adrenal gland cells of the rat by means of trypan blue injections. Anat. Rec., *80*, 421-429.

SHERIDAN, M. N. and W. D. BELT. 1964. Fine structure of the guinea pig adrenal cortex. Anat. Rec., *149*, 73-97.

WILLIAMS, R. G. 1947. Studies of adrenal cortex: Regeneration of the transplanted gland and the vital quality of autogenous grafts. Amer. J. Anat., *81*, 199-231.

WOTTON, R. M. and R. L. ZWEMER. 1943. A study of the cytogenesis of cortico-adrenal cells in the cat. Anat. Rec., *86*, 409-416.

ZWEMER, R. L. and R. TRUSZKOWSKI. 1937. The importance of corticoadrenal regulation of potassium metabolism. Endocrinology, *21*, 40-49.

SUPRARENAL MEDULLA AND CHROMAFFIN CELLS

BROWN, W. J., L. BARJAS, and H. LATTA. 1971. The ultrastructure of the human adrenal medulla. Anat. Rec., *169*, 173-183.

CORI, C. F. and A. D. WELCH. 1941. The Adrenal Medulla. In: *Glandular Physiology and Therapy.* Amer. Med. Assn., Chicago, 307-326.

COUPLAND, R. E. 1952. The prenatal development of the abdominal para-aortic bodies in man. J. Anat. (Lond.), *86*, 357-372 + 4 plates.

EULER, U. S. VON. 1951. Hormones of the sympathetic nervous system and the adrenal medulla. Brit. med. J., *1*, 105-108.

FREEMAN, N. E., R. H. SMITHWICK, and J. C. WHITE. 1934. Adrenal secretion in man. Amer. J. Physiol., *107*, 529-534.

HOLLINSHEAD, W. H. 1940. Chromaffin tissue and paraganglia. Quart. Rev. Biol., *15*, 156-171.

YATES, R. D. 1964. A light and electron microscopic study correlating the chromaffin reaction and granule ultrastructure in the adrenal medulla of the Syrian Hamster. Anat. Rec., *149*, 237-249.

SUPRARENAL BLOOD, LYMPH AND NERVE SUPPLY

ANSON, B. J., E. W. CAULDWELL, J. W. PICK, and L. E. BEATON. 1947. The blood supply of the kidney, suprarenal gland, and associated structures. Surg. Gynec. Obstet., *84*, 313-320.

DOLISHNII, N. V. 1965. Plastic properties of the arteries of the adrenals. Fed. Proc., *24*, T941-T944.

GAGNON, R. 1956. The venous drainage of the human adrenal gland. Rev. Canad. Biol., *14*, 350-359.

HARRISON, R. G. and M. J. HOEY. 1960. *The Adrenal Circulation.* vii + 77 pages, 56 figures. Charles C Thomas, Publisher, Springfield.

MERKLIN, R. J. 1962. Arterial supply of the suprarenal gland. Anat. Rec., *144*, 359-371.

MERKLIN, R. J. 1966. Suprarenal gland lymph drainage [man]. Am. J. Anat., *119*, 359-374.

MIKHAIL, Y. and Z. MAHRAN. 1965. Innervation of the cortical and medullary portions of the adrenal gland of the rat during postnatal life. Anat. Rec., *152*, 431-437.

HYPOPHYSIS (PITUITARY GLAND)

APLINGTON, H. W., JR. 1962. Cellular changes in the pituitary of necturus following thyroidectomy. Anat. Rec., *143*, 133-145.

CRAFTS, R. C. and B. S. WALKER. 1947. The effects of hypophysectomy on serum and storage iron in adult female rats. Endocrinology, *41*, 340-346.

ETKIN, W. 1943. The developmental control of pars intermedia by brain. J. exp. Zool., *92*, 31-47.

GREEN, H. T. 1957. The venous drainage of the human hypophysis cerebri. Amer. J. Anat., *100*, 435-469.

GREEN, J. D. 1951. The comparative anatomy of the hypophysis, with special reference to its blood supply and innervation. Amer. J. Anat., *88*, 225-311.

HARRIS, G. W. 1955. *Neural Control of the Pituitary Gland.* ix + 298 pages. The Williams & Wilkins Company, Baltimore.

HEGRE, E. S. 1946. The developmental stage at which the intermediate lobe of the hypophysis becomes determined. J. exp. Zool., *103*, 321-333.

HUNT, T. E. 1949. Mitotic activity in the hypophysis of the rat during pregnancy and lactation. Anat. Rec., *105*, 361-373.

HUNT, T. E. 1951. The effect of hypophyseal extract on mitotic activity of the rat hypophysis. Anat. Rec., *111*, 713-725.

IFFT, J. D. 1953. The effect of superior cervical ganglionectomy on the cell population of the rat adenohypophysis and on the estrous cycle. Anat. Rec., *117*, 395-404.

MARKEE, J. C., C. H. SAWYER, and W. H. HOLLINSHEAD. 1948. Adrenergic control of the release of luteinizing hormone from the hypophysis of the rabbit. Recent Prog. Hormone Res., *2*, 117-131.

MARKEE, J. E., C. H. SAWYER, and W. H. HOLLINSHEAD. 1964. Activation of the anterior hypophysis by electrical stimulation in the rabbit. Endocrinology, *38*, 345-357.

MCCONNELL, E. M. 1953. The arterial blood supply of the human hypophysis cerebri. Anat. Rec., *115*, 175-203.

McGRATH, P. 1971. The volume of the human pharyngeal hypophysis in relation to age and sex. J. Anat. (Lond.), *110*, 275-282.

McNARY, W. F., JR. 1957. Progressive cytological changes in the hypophysis associated with endocrine interaction following exposure to cold. Anat. Rec., *128*, 233-253.

MESSIER, B. 1965. Number and distribution of thyrotropic cells in the mouse pituitary gland. Anat. Rec., *153*, 343-348.

RASMUSSEN, A. T. 1936. The Proportions of the Various Subdivisions of the Normal Adult Human Hypophysis Cerebri, etc. In: Res. Publ. Ass. nerv. Ment. Dis., The Pituitary Gland. Williams & Wilkins, Baltimore, 118-150.

SAWYER, C. H. 1947. Cholinergic stimulation of the release of melanophore hormone by the hypophysis in salamander larvae. J. exp. Zool., *106*, 145-179.

SHANKLIN, W. M. 1951. The histogenesis and histology of an integumentary type of epithelium in the human hypophysis. Anat. Rec., *109*, 217-231.

STANFIELD, J. P. 1960. The blood supply of the human pituitary gland. J. Anat. (Lond.), *94*, 257-273.

SZENTAGOTHAI, J., B. FLERKO, B. HESS, and B. HALASZ. 1962. *Hypothalamic Control of the Anterior Pituitary.* 330 pages, 125 figures. Publishing House of the Hungarian Academy of Sciences, Budapest.

WISLOCKI, G. B. 1938. The Vascular Supply of the Hypophysis Cerebri of the Rhesus Monkey and Man. In: Res. Publ. Ass. nerv. ment. Dis., The Pituitary Gland. Williams & Wilkins, Baltimore, 48-68.

THYROID

Brookhaven Symposia in Biology No. 7. 1955. *The Thyroid.* vii + 271 pages. Brookhaven National Laboratory, Upton, New York.

HUNT, T. E. 1944. Mitotic activity in the thyroid gland of female rats. Anat. Rec., *90*, 133-138.

NORDLAND, M. 1930. The larynx as related to surgery of the thyroid based on an anatomical study. Surg. Gynec. Obstet., *51*, 449-459.

PLAGGE, J. C. 1943. Effects of hypotonic solutions upon the living thyroid gland. Anat. Rec., *87*, 345-353.

RAMSAY, A. J. and G. A. BENNETT. 1943. Studies on the thyroid gland. I. The structure, extent and drainage of the "lymph-sac" of the thyroid gland (Felis domestica). Anat. Rec., *87*, 321-339.

RICHTER, K. M. 1944. Some new observations bearing on the effect of hyperthyroidism on genital structure and function. J. morph., *74*, 375-393.

SHEPARD, T. H., H. J. ANDERSEN, and H. ANDERSEN. 1964. The human fetal thyroid. I. Its weight in relation to body weight, crown-rump length, foot length and estimated gestation age. Anat. Rec., *148*, 123-128.

SHEPARD, T. H., H. ANDERSEN, and H. J. ANDERSEN. 1964. Histochemical studies of the human fetal thyroid during the first half of fetal life. Anat. Rec., *149*, 363-379.

WELLER, G. L., JR. 1933. Development of the thyroid, parathyroid and thymus glands in man. Carneg. Instn. Contr. Embryol., *24*, 93-139 + 4 plates.

WILLIAMS, R. G. 1944. Some properties of living thyroid cells and follicles. Amer. J. Anat., *75*, 95-119.

PARATHYROID GLANDS

BAKER, B. L. 1942. A study of the parathyroid glands of the normal and hypophysectomized monkey (Macaca mulatta). Anat. Rec., *83*, 47-73.

DeROBERTIS, E. 1940. The cytology of the parathyroid gland of rats injected with parathyroid extract. Anat. Rec., *78*, 473-495.

FLEISCHMANN, W. 1951. *Comparative Physiology of the Thyroid and Parathyroid Glands.* v + 78 pages. Charles C Thomas, Springfield.

HEINBACH, W. F., JR. 1933. A study of the number and location of the parathyroid glands in man. Anat. Rec., *57*, 251-261.

NORRIS, E. H. 1937. The parathyroid glands and the lateral thyroid in man: Their morphogenesis, histogenesis, topographic anatomy and prenatal growth. Carneg. Instn. Contr. Embryol., *26*, 247-294 + 6 plates.

NORRIS, E. H. 1946. Anatomical evidence of prenatal function of the human parathyroid glands. Anat. Rec., *96*, 129-141.

RAYBUCK, H. E. 1952. The innervation of the parathyroid glands. Anat. Rec., *112*, 117-123.

PINEAL

CLAUSEN, H. J. and B. MOFSHIN. 1939. The pineal eye of the lizard (Anolis carolinensis). A photoreceptor as revealed by oxygen consumption studies. J. cell. comp. Physiol., *14*, 29-41.

SULLENS, W. E. and M. D. OVERHOLSER. 1941. Pinealectomy in successive generations of rats. Endocrinology, *28*, 835-839.

(Additional references on the Pineal and Hypothalamus are on page 900.)

Bibliographic Index

Edwards, L. F., 672, 1036, 1097
Eglitis, J. A., 1096
Elias, H., 1264
Elliott, H. C., 899
Ellison, J. P., 560
Engel, S., 1154
Enlow, D. H., 283
Erickson, E. E., 560
Erulkar, S. D., 1097
Esenther, G., 729
Essenberg, J. M., 1338
Etkin, W., 1354
Etter, L. E., 284
Euler, U. S. von, 1354
Evans, F. G., 283, 284
Evans, T. H., 671
Everett, J. W., 50, 900, 1338
Everett, N. B., 559, 898, 1337,
    1338, 1353

Fabricant, N. D., 1153
Fales, D. E., 559
Farris, E. J., 1338
Fawcett, D. W., 49, 285, 523,
    1338
Fawcett, E., 282, 283
Felix, M. D., 49
Fenn, W. O., 1154
Feremutsch, K., 900
Ferner, H., 1264
Ferriman, D. G., 1109
Ferry, D., Jr., 1338
Figge, J. H. F., 93
Findlay, C. W., Jr., 1154
Fine, H., 1337
Finerty, J. C., 50, 1336
Finley, T., 672
Finley, T. L., 728
Fish, H. S., 526
Fisher, A. W. F., 1153
Fishman, I. Y., 1096
Fleischmann, W., 1355
Flerko, B., 1355
Flyger, G., 902
Fohmann, V., 775
Föld, M., 774
Foley, J. O., 1034, 1036
Ford, D. H., 901, 1339
Fourman, J., 1336
Fox, C. A., 900
Fox, M. H., 559, 670
Francis, C. C., 282, 283, 368
Frankel, H. H., 1353
Frantz, C. H., 284
Freda, V. J., 50
Freeman, J. A., 49, 775
Freeman, N. E., 1354
Friedman, S. M., 729, 1263
Froimson, A., 523
Frontera, J. G., 901
Frost, H. M., 283
Fullerton, P. M., 1035
Furuta, W. J., 775

Gacek, R. R., 1034
Gagnon, R., 1354

Galindo, B., 775
Gardner, E., 282, 284, 368, 369
Gardner, J. E., 1264
Gardner, J. H., 900
Gardner, W. D., 729
Gardner, W. U., 524, 1339
Garn, S. M., 1263
Garner, C. M., 902
Garrett, F. D., 1035
Gasser, R. F., 523
Gaughran, G. R. L., 525
Gavan, J. A., 1338
Gay, A. J., Jr., 525
Geddes, G., 901
Génis-Gálvez, J. M., 1096
George, R., 523
George, W. C., 51
Geren, B. B., 901
Giacometti, L., 1339
Gibson, J. B., 728
Gibson, J. F., Jr., 1154
Gilbert, P. W., 1096
Gillilan, L. A., 899, 1096
Gillman, J., 1337
Giovacchini, R. P., 1264
Gitlin, G., 51, 1035
Gladston, R. J., 1263
Glees, P., 902
Glick, D., 285
Goerttler, K., 50, 93
Goldenberg, R. R., 285
Goldsmith, J. B., 670
Goldstein, M. B., 1264
Goldstein, M. S., 283
Gomez, H., 1034
Goor, D. A., 1264
Gorbman, A., 1353
Gosling, J. A., 1336
Goss, C. M., 523, 559, 670, 728,
    1339
Gould, J., 775
Gould, S. E., 559
Graves, G. O., 1097
Gray, D. J., 282, 283, 284, 368,
    523
Gray, D. L., 282
Gray, F. W., 1153
Gray, H., 775
Grayson, J., 526
Green, H. T., 1354
Green, J. A., 1338
Green, J. D., 900, 1354
Greene, L. S., 524
Greening, R. R., 775
Greenwald, G. S., 1338
Greep, R. O., 50
Gregg, D. E., 670
Gregg, R. L., 1154
Grègoire, C., 775
Greig, H. W., 523
Greig, W., 524
Greulich, W. W., 282, 1339
Grodinsky, M., 524, 525, 526
Gross, J., 285
Gross, L., 560, 670

Gross, R. E., 560, 670
Grote, P., 1337
Groth, C. G., 1336
Grunt, J. A., 1036
Gusev, A. M., 775
Gussen, R., 1097
Guth, L., 1034, 1096

Habenicht, J., 525
Haber, J. J., 1263
Hackett, G. S., 368
Hadley, L. A., 283
Haines, R. W., 283, 368, 369
Halasz, B., 1355
Haley, J. C., 1263
Hall, C. B., 524
Hall, D., 285
Hall, P. F., 1353
Halls, A. A., 369
Halpern, M. H., 670
Ham, A. W., 50
Hambley, W. C., 1263
Hamerman, D., 368
Hamilton, J. B., 1109, 1338
Hamilton, W. J., 50
Hammond, W. S., 899, 1264
Hamre, C. J., 284
Hamshere, R. M., 369
Hanaway, J., 899
Handforth, J. R., 284
Hanson, J. S., 1097
Harbord, R. P., 1154
Hard, W. L., 1264
Harden, T. P., 523
Hardesty, I., 1034
Harman, P. J., 901, 1336
Harris, A. J., 1035
Harris, G. W., 1354
Harris, W., 1035
Harris, W. H., 283
Harrison, F., 1034
Harrison, J. M., 901
Harrison, R. G., 49, 899, 1354
Harrison, V. F., 525, 1035
Harrop, T. J., 523
Hart, V. L., 369
Hartman, C. G., 49, 50
Hartman, F. A., 1353
Hartmann, J. F., 901
Harvey, H. T., 1264
Harvey, S. C., 902
Hasterlik, R. J., 1338
Havers, C., 283
Hawkins, R. K., 1264
Haxton, H., 284
Haxton, H. A., 369
von Hayek, H., 1154
Hayes, M. A., 729
Hayhow, W., 902
Healey, J., 672
Healy, J. E., Jr., 1264
Hegre, E. S., 1354
Heimburger, R. F., 1336
Heinbach, W. F., Jr., 1355
Henderson, R. S., 284
Henderson, W. R., 1034

Subject Index